FRUIT CROPS

FRUIT CROPS
Diagnosis and Management of Nutrient Constraints

Edited By

A.K. SRIVASTAVA, Ph.D.
Principal Scientist (Soil Science), ICAR-Central Citrus Research Institute, Nagpur, India

CHENGXIAO HU, Ph.D.
Professor & Principal Scientist (Citrus nutrition and fertilization), Institute of Citrus Science and College of Resources & Environment, Huazhong Agricultural University, Wuhan, China

ELSEVIER

Elsevier
Radarweg 29, PO Box 211, 1000 AE Amsterdam, Netherlands
The Boulevard, Langford Lane, Kidlington, Oxford OX5 1GB, United Kingdom
50 Hampshire Street, 5th Floor, Cambridge, MA 02139, United States

Library of Congress Cataloging-in-Publication Data

A catalog record for this book is available from the Library of Congress

British Library Cataloguing-in-Publication Data

A catalogue record for this book is available from the British Library

ISBN: 978-0-12-818732-6

For information on all Elsevier publications
visit our website at https://www.elsevier.com/books-and-journals

Publisher: Charlotte Cockle
Acquisition Editor: Nancy Maragioglio
Editorial Project Manager: Redding Morse
Production Project Manager: Swapna Srinivasan
Cover Designer: Miles Hitchen

Typeset by SPi Global, India

Contents

Contributors

Vojtěch Adam Department of Chemistry and Biochemistry, Faculty of AgriSciences, Mendel University in Brno, Brno, Czech Republic

Riaz Ahmad Department of Horticulture, Bahauddin Zakariya University, Multan, Pakistan

Selena Ahmed The Food and Health Lab, Department of Health and Human Development, Montana State University, Bozeman, MT, United States

Kashif Akram Department of Food Sciences, Cholistan University of Veterinary & Animal Sciences, Bahawalpur, Pakistan

M.S. Alam Horticulture Division, Bangladesh Institute of Nuclear Agriculture, Mymensingh, Bangladesh

Paula Alayón Luaces Fruticultura, Facultad de Ciencias Agrarias, Universidad Nacional del Nordeste, Corrientes, Argentina

Wasayf J. Almalki Department of Chemical and Biological Engineering, The University of Sheffield, Sheffield, United Kingdom

Muhammad Akbar Anjum Department of Horticulture, Bahauddin Zakariya University, Multan, Pakistan

Chrysovalantou Antonopoulou Department of the Agricultural Development, School of Agricultural and Forestry Sciences, Democritus University of Thrace, Orestiada, Greece

Jiří Antošovský Department of Agrochemistry, Soil Science, Microbiology and Plant Nutrition, Faculty of AgriSciences, Mendel University in Brno, Brno, Czech Republic

Margarida Arrobas Centro de Investigação de Montanha (CIMO), Instituto Politécnico de Bragança, Bragança, Portugal

Ignácio Aspiazú State University of Montes Claros, Montes Claros, Brazil

Cuihua Bai College of Natural Resources and Environment, South China Agricultural University, Guangzhou, People's Republic of China

Elena Baldi Department of Agricultural and Food Sciences, Alma Mater Studiorum University of Bologna, Bologna, Italy

Allen V. Barker University of Massachusetts, Amherst, MA, United States

Betina Pereira de Bem Instituto Federal de Santa Catarina (IFSC), Urupema, Brazil

Adalberto Benavides-Mendoza Autonomous Agricultural University Antonio Narro, Department of Horticulture, Saltillo, Mexico

Maja Benkovi University of Zagreb, Faculty of Food Technology and Biotechnology, Zagreb, Croatia

Renato Vasconcelos Botelho Agronomy Department, State University of Midwestern Paraná—Unicentro, Paraná, Brazil

Alberto Fontanella Brighenti Empresa de Pesquisa Agropecuária e Extensão de Rural de Santa Catarina (Epagri), São Joaquim, Brazil

Gustavo Brunetto Universidade Federal de Santa Maria (UFSM), Campus Universitário, Centro de Ciências Rurais, Departamento de Solos, Camobi, Santa Maria, Brazil

David R. Bryla U.S. Department of Agriculture, Agricultural Research Service, Horticultural Crops Research Unit, Corvallis, OR, United States

Hakan Burhan Sen Research Group, Department of Biochemistry, Faculty of Arts and Science, Dumlupınar University, Evliya Çelebi Campus, Kütahya, Turkey

Thomas O. Butler Department of Chemical and Biological Engineering, The University of Sheffield, Sheffield, United Kingdom

Miaomiao Cai College of Resources and Environment/ Micro-element Research Center/Hubei Provincial Engineering Laboratory for New Fertilizers, Huazhong Agricultural University; Key Laboratory of Horticultural Plant Biology (HZAU), MOE, Wuhan, People's Republic of China

Rodolfo Canet Center for the Development of Sustainable Agriculture, Valencian Institute of Agricultural Research, Valencia, Spain

Luciano Cavani Department of Agricultural and Food Sciences, Alma Mater Studiorum University of Bologna, Bologna, Italy

James Chapman School of Medical and Applied Sciences, Central Queensland University, Rockhampton, QLD, Australia

John M. Chater Department of Botany and Plant Sciences, University of California, Riverside, CA, United States

Christos Chatzissavvidis Department of the Agricultural Development, School of Agricultural and Forestry Sciences, Democritus University of Thrace, Orestiada, Greece

Jianjun Chen Department of Environmental Horticulrture and Mid-Florida Research and Education Center, University of Florida, Institute of Food and Agricultural Sciences, Apopka, FL, United States

Li-Song Chen Institute of Plant Nutritional Physiology and Molecular Biology, College of Resources and Environment, Fujian Agriculture and Forestry University, Fuzhou, China

Marlise Nara Ciotta Empresa de Pesquisa Agropecuária e Extensão de Rural de Santa Catarina (Epagri), Lages, Brazil

Jucinei José Comin Universidade Federal de Santa Catarina (UFSC), Departamento de Engenharia Rural, Florianópolis, Brazil

Lessandro De Conti Instituto Federal de Educação, Ciência e Tecnologia Farroupilha—Campus Santo Augusto, Santo Augusto, Brazil

Márcio Cleber de Medeiros Corrêa Federal University of Ceará, Fortaleza, Brazil

Juan Manuel Covarrubias-Ramírez CESAL-INIFAP, Saltillo, México

Daniel Cozzolino School of Science, RMIT University, Melbourne, VIC, Australia

Sjoerd E.A.T.M. van der Zee Soil Physics and Land Management Group, Wageningen University, Wageningen, The Netherlands; School of Chemistry, Monash University, Melbourne, VIC, Australia

José Aridiano Lima de Deus Institute of Technical Assistance and Rural Extension of Paraná (EMATER-PR), Curitiba, Brazil

Sara Di Lonardo Research Institute on Terrestrial Ecosystems—Italian National Research Council (IRET-CNR), Sesto Fiorentino, Italy

Bartolomeo Dichio Department of European and Mediterranean Cultures: Architecture, Environment and Cultural Heritage (DiCEM), University of Basilicata, Matera, Italy

Zhihao Dong College of Resources and Environment/ Micro-element Research Center/Hubei Provincial Engineering Laboratory for New Fertilizers, Huazhong Agricultural University; Key Laboratory of Horticultural Plant Biology (HZAU), MOE, Wuhan, People's Republic of China

Ladislav Ducsay Department of Agrochemistry and Plant Nutrition, The Faculty of Agrobiology and Food Resources, Slovak University of Agriculture in Nitra, Nitra, Slovakia

Madeleine F. Dupont School of Science, RMIT University, Melbourne, VIC, Australia

Aaron Elbourne School of Science, RMIT University, Melbourne, VIC, Australia

Fatima Elmusa Sen Research Group, Department of Biochemistry, Faculty of Arts and Science, Dumlupınar University, Evliya Çelebi Campus, Kütahya, Turkey

Jeanette M. Van Emon EVE Sciences, Henderson, NV, United States

Hassan Etesami Agriculture & Natural resources Campus, Faculty of Agricultural Engineering & Technology, Department of Soil Science, University of Tehran, Tehran, Iran

Róger Fallas-Corrales Soil Physics and Land Management Group, Wageningen University, Wageningen, The Netherlands

Umar Farooq Department of Food Science & Technology, Muhammad Nawaz Shareef University of Agriculture, Multan, Pakistan

Laura Olivia Fuentes-Lara Autonomous Agricultural University Antonio Narro, Department of Animal Nutrition, Saltillo, Mexico

José E. Gaiad Fruticultura, Facultad de Ciencias Agrarias, Universidad Nacional del Nordeste, Corrientes, Argentina

Bin Gao Department of Agricultural and Biological Engineering, University of Florida, Institute of Food and Agricultural Sciences, Gainesville, FL, United States

Maciej Gąstoł Department of Pomology and Apiculture, Agricultural University in Kraków, Kraków, Poland

Melanie D. Gomez Herrera Fruticultura, Facultad de Ciencias Agrarias, Universidad Nacional del Nordeste, Corrientes, Argentina

Fulya Gulbagca Sen Research Group, Department of Biochemistry, Faculty of Arts and Science, Dumlupınar University, Evliya Çelebi Campus, Kütahya, Turkey

Peng Guo Institute of Plant Nutritional Physiology and Molecular Biology, College of Resources and Environment, Fujian Agriculture and Forestry University, Fuzhou; College of Life Sciences, Henan Agricultural University, Zhengzhou, China

Zafar Hayat Department of Animal Sciences, CVAS-University of Veterinary and Animal Sciences, Jhang, Pakistan

Jia-Dong He College of Horticulture and Gardening, Yangtze University, Jingzhou, China

Chengxiao Hu College of Resources and Environment/Microelement Research Center/Hubei Provincial Engineering Laboratory for New Fertilizers, Huazhong Agricultural University; Key Laboratory of Horticultural Plant Biology (HZAU), MOE, Wuhan, People's Republic of China

Antonio Ibacache Instituto de Investigaciones Agropecuarias (INIA), Centro Regional de Investigación Intihuasi, La Serena, Chile

Diego S. Intrigliolo Irrigation Deparment, CEBAS-CSIC, Murcia; CSIC Associated Unit "Riego en la agricultura Mediterránea," Instituto Valenciano de Investigaciones Agrarias, Moncada, Spain

Maria Del Rosario Jacobo-Salcedo CENID RASPA-INIFAP, Gómez Palacio, México

Byoung Ryong Jeong Horticulture Major, Division of Applies Life Science (BK21 Plus Program), Graduate School, Gyeongsang National University, Jinju, Republic of Korea

Wei Jia College of Resources and Environment/Microelement Research Center/Hubei Provincial Engineering Laboratory for New Fertilizers, Huazhong Agricultural University; Key Laboratory of Horticultural Plant Biology (HZAU), MOE, Wuhan, People's Republic of China

Huan-Xin Jiang Institute of Plant Nutritional Physiology and Molecular Biology, College of Resources and Environment, Fujian Agriculture and Forestry University, Fuzhou, China

Antonio Juárez-Maldonado Autonomous Agricultural University Antonio Narro, Department of Botany, Saltillo, Mexico

Tamara Jurina University of Zagreb, Faculty of Food Technology and Biotechnology, Zagreb, Croatia

Davie Kadyampakeni University of Florida, Institute of Food and Agricultural Sciences, Citrus Research and Education Center, Lake Alfred, FL, United States

Evangelos Karagiannis   Laboratory of Pomology, Department of Agriculture, Aristotle University of Thessaloniki, Thessaloniki, Greece

Jasenka Gajdoš Kljusurić   University of Zagreb, Faculty of Food Technology and Biotechnology, Zagreb, Croatia

Jinxue Li   College of Resources and Environment/Microelement Research Center/Hubei Provincial Engineering Laboratory for New Fertilizers, Huazhong Agricultural University; Key Laboratory of Horticultural Plant Biology (HZAU), MOE, Wuhan, People's Republic of China

Wenhuan Liu   National Engineering Research Center for Citrus Technology/Citrus Research Institute, Southwest University-Chinese Academy of Agricultural Sciences, Chongqing, China

Arcângelo Loss   Universidade Federal de Santa Catarina (UFSC), Centro de Ciencias Agrarias, Florianopolis, Brazil

Cledimar Rogério Lourenzi   Universidade Federal de Santa Catarina (UFSC), Departamento de Engenharia Rural, Florianópolis, Brazil

Donglin Luo   College of Natural Resources and Environment, South China Agricultural University, Guangzhou, People's Republic of China

YanYan Ma   National Engineering Research Center for Citrus Technology/Citrus Research Institute, Southwest University-Chinese Academy of Agricultural Sciences, Chongqing, China

Rui Machado   ICAAM—Mediterranean Institute for Agricultural and Environmental Sciences; Crop Sciences Department, School of Science and Technology, University of Évora, Évora, Portugal

Victor Martins Maia   State University of Montes Claros, Montes Claros, Brazil

Belén Martínez-Alcántara   Certification Section, Plant Health Service, Ministry of Agriculture, Environment, Climate Change and Rural Development, Valencia, Spain

Renato de Mello Prado   São Paulo State University—UNESP, Jaboticabal, Brazil

George Wellington Bastos de Melo   Embrapa Uva e Vinho, Bento Gonçalves, Brazil

Donald J. Merhaut   Department of Botany and Plant Sciences, University of California, Riverside, CA, United States

Michail Michailidis   Laboratory of Pomology, Department of Agriculture, Aristotle University of Thessaloniki, Thessaloniki, Greece

Nebojša Milošević   Department of Pomology and Fruit Breeding, Fruit Research Institute, Čačak, Republic of Serbia

Tomo Milošević   Department of Fruit Growing and Viticulture, Faculty of Agronomy, University of Kragujevac, Čačak, Republic of Serbia

Alba N. Mininni   Department of European and Mediterranean Cultures: Architecture, Environment and Cultural Heritage (DiCEM), University of Basilicata, Matera, Italy

Athanassios Molassiotis   Laboratory of Pomology, Department of Agriculture, Aristotle University of Thessaloniki, Thessaloniki, Greece

Isidro Morales   National Polytechnic Institute, CIIDIR-Oaxaca, Oaxaca, Mexico

Babak Motesharezadeh   Department of Soil Science, University of Tehran, Tehran, Iran

Seyed Majid Mousavi   Soil and Water Research Institute, Agricultural Research, Education and Extension Organization (AREEO), Karaj, Iran

Marcelo Marques Lopes Müller   Agronomy Department, State University of Midwestern Paraná—Unicentro, Paraná, Brazil

William Natale   Federal University of Ceará, Fortaleza, Brazil

Erika Nava-Reyna   CENID RASPA-INIFAP, Gómez Palacio, México

Rolf Nestby   Division Food and Society (Horticulture), Norwegian Institute of Bioeconomy (NIBIO), Ås, Norway

Danúbia Aparecida Costa Nobre   State University of Montes Claros, Montes Claros, Brazil

Kenneth Nyombi   Makerere University, College of Agricultural and Environmental Sciences, Kampala, Uganda

Dámaris Leopoldina Ojeda-Barrios   Autonomous University of Chihuahua, Laboratory of Plant Physiology, Chihuahua, Mexico

Fernanda Soares Oliveira   State University of Montes Claros, Montes Claros, Brazil

İbrahim Ortaş   University of Cukurova, Faculty of Agriculture, Department of Soil Science and Plant Nutrition, Adana, Turkey

Gloria Padmaperuma   Department of Chemical and Biological Engineering, The University of Sheffield, Sheffield, United Kingdom

Léon Etienne Parent   Department of Soil and Agri-food Engineering, Université Laval, Québec, QC, Canada

Víctor Manuel Parga-Torres   CESAL-INIFAP, Saltillo, México

Margarita Parra   Irrigation Deparment, CEBAS-CSIC, Murcia, Spain

Betânia Vahl de Paula   Universidade Federal de Santa Maria (UFSM), Campus Universitário, Centro de Ciências Rurais, Departamento de Solos, Camobi, Santa Maria, Brazil

Rodinei Facco Pegoraro   Federal University of Minas Gerais, Belo Horizonte, Brazil

Ana Pérez-Piqueres   Center for the Development of Sustainable Agriculture, Valencian Institute of Agricultural Research, Valencia, Spain

Raffaella Petruccelli   Institute of BioEconomy—Italian National Research Council (IBE-CNR), Sesto Fiorentino, Italy

Aoife Power   School of Medical and Applied Sciences, Central Queensland University, Rockhampton, QLD, Australia

John E. Preece   National Clonal Germplasm Repository, USDA-ARS, University of California, Davis, CA, United States

Fangying Qiu   National Engineering Research Center for Citrus Technology/Citrus Research Institute, Southwest University-Chinese Academy of Agricultural Sciences, Chongqing, China

Ana Quiñones Center for the Development of Sustainable Agriculture, Valencian Institute of Agricultural Research, Valencia, Spain

M.A. Rahim Department of Horticulture, Bangladesh Agriculture University, Mymensingh, Bangladesh

R.A. Ram ICAR-Central Institute for Subtropical Horticulture, Lucknow, India

Hermann Restrepo-Diaz Departamento de Agronomia, Facultad de Ciencias Agrarias, Universidad Nacional de Colombia, Bogotá, Colombia

Jorge B. Retamales Head ISHS Division Vine and Berry Fruits, Viña del Mar, Chile

Felipe Klein Ricachenevsky Universidade Federal de Santa Maria (UFSM), Departamento de Biologia, Santa Maria, Brazil

Patrizia Ricciuti Department of Soil, Plant and Food Sciences (DiSSPA), University of Bari 'Aldo Moro', Bari, Italy

M. Ângelo Rodrigues Centro de Investigação de Montanha (CIMO), Instituto Politécnico de Bragança, Bragança, Portugal

Isabel Rodríguez-Carretero Center for the Development of Sustainable Agriculture, Valencian Institute of Agricultural Research, Valencia, Spain

Danilo Eduardo Rozane São Paulo State University, UNESP, Registro, Brazil

José S. Rubio-Asensio Irrigation Deparment, CEBAS-CSIC, Murcia, Spain

Pavel Ryant Department of Agrochemistry, Soil Science, Microbiology and Plant Nutrition, Faculty of AgriSciences, Mendel University in Brno, Brno, Czech Republic

Alefsi David Sánchez-Reinoso Departamento de Agronomia, Facultad de Ciencias Agrarias, Universidad Nacional de Colombia, Bogotá, Colombia

Alberto Sandoval-Rangel Autonomous Agricultural University Antonio Narro, Department of Horticulture, Saltillo, Mexico

Eva Sapáková Department of Languages, Faculty of Regional Development and International Studies, Mendel University in Brno, Brno, Czech Republic

Djalma Eugênio Schmitt Universidade Federal de Santa Catarina (UFSC), Curitibanos, Brazil

Fatih Sen Sen Research Group, Department of Biochemistry, Faculty of Arts and Science, Dumlupınar University, Evliya Çelebi Campus, Kütahya, Turkey

Ricardo Serralheiro ICAAM—Mediterranean Institute for Agricultural and Environmental Sciences; Agricultural Engineering Department, School of Science and Technology, University of Évora, Évora, Portugal

Afshan Shafi Department of Food Science & Technology, Muhammad Nawaz Shareef University of Agriculture, Multan, Pakistan

Bo Shu College of Horticulture and Gardening, Yangtze University, Jingzhou, China

Faqih A.B. Ahmad Shuhaili Department of Chemical and Biological Engineering, The University of Sheffield, Sheffield, United Kingdom

Petr Škarpa Department of Agrochemistry, Soil Science, Microbiology and Plant Nutrition, Faculty of AgriSciences, Mendel University in Brno, Brno, Czech Republic

Adriano Sofo Department of European and Mediterranean Cultures: Architecture, Environment and Cultural Heritage (DiCEM), University of Basilicata, Matera, Italy

Giovambattista Sorrenti Department of Agricultural and Food Sciences, Alma Mater Studiorum University of Bologna, Bologna, Italy

André Luiz Kulkamp de Souza Empresa de Pesquisa Agropecuária e Extensão Rural de Santa Catarina (Epagri), Videira, Brazil

Matheus Severo de Souza Kulmann Universidade Federal de Santa Maria (UFSM), Campus Universitário, Centro de Ciências Rurais, Departamento de Solos, Camobi, Santa Maria, Brazil

A.K. Srivastava Indian Council of Agricultural Research-Central Citrus Research Institute, Nagpur, India

Lincon Oliveira Stefanello Universidade Federal de Santa Maria (UFSM), Campus Universitário, Centro de Ciências Rurais, Departamento de Solos, Camobi, Santa Maria, Brazil

Alyssa L. Stewart The Food and Health Lab, Department of Health and Human Development, Montana State University, Bozeman, MT, United States

Margie L. Stratton*

Qiling Tan College of Resources and Environment/Microelement Research Center/Hubei Provincial Engineering Laboratory for New Fertilizers, Huazhong Agricultural University, Wuhan, People's Republic of China

Ning Tang Institute of Plant Nutritional Physiology and Molecular Biology, College of Resources and Environment, Fujian Agriculture and Forestry University, Fuzhou; Research Institute for Special Plants, Chongqing University of Arts and Sciences, Chongqing, China

Georgia Tanou Institute of Soil and Water Resources, ELGO-DEMETER, Thessaloniki, Greece

Adriele Tassinari Universidade Federal de Santa Maria (UFSM), Campus Universitário, Centro de Ciências Rurais, Departamento de Solos, Camobi, Santa Maria, Brazil

Tadeu Luis Tiecher Instituto Federal Farroupilha, Campus Alegrete, Alegrete, Brazil

Moreno Toselli Department of Agricultural and Food Sciences, Alma Mater Studiorum University of Bologna, Bologna, Italy

Vi Khanh Truong School of Science, RMIT University, Melbourne, VIC, Australia

Matjaž Turinek University of Maribor, Maribor, Slovenia

Ana Jurinjak Tušek University of Zagreb, Faculty of Food Technology and Biotechnology, Zagreb, Croatia

*Retired

Seetharaman Vaidyanathan Department of Chemical and Biological Engineering, The University of Sheffield, Sheffield, United Kingdom

Davor Valinger University of Zagreb, Faculty of Food Technology and Biotechnology, Zagreb, Croatia

Tripti Vashisth University of Florida, Institute of Food and Agricultural Sciences, Citrus Research and Education Center, Lake Alfred, FL, United States

Nicolás Verdugo-Vásquez Instituto de Investigaciones Agropecuarias (INIA), Centro Regional de Investigación Intihuasi, La Serena, Chile

Zonghua Wang Institute of Oceanography, Minjiang University; Fujian University Key Laboratory for Plant-Microbe Interaction, Fujian Agriculture and Forestry University, Fuzhou, China

Xiangying Wei Institute of Oceanography, Minjiang University; Fujian University Key Laboratory for Plant-Microbe Interaction, Fujian Agriculture and Forestry University, Fuzhou, China; Department of Environmental Horticulrture and Mid-Florida Research and Education Center, University of Florida, Institute of Food and Agricultural Sciences, Apopka, FL, United States

Qiang-Sheng Wu College of Horticulture and Gardening, Yangtze University, Jingzhou, China

Cristos Xiloyannis Department of European and Mediterranean Cultures: Architecture, Environment and Cultural Heritage (DiCEM), University of Basilicata, Matera, Italy

Lin-Tong Yang Institute of Plant Nutritional Physiology and Molecular Biology, College of Resources and Environment, Fujian Agriculture and Forestry University, Fuzhou, China

Lixian Yao College of Natural Resources and Environment, South China Agricultural University, Guangzhou, People's Republic of China

Jovani Zalamena Instituto Federal do Rio Grande do Sul (IFRS)—Campus Restinga, Porto Alegre, Brazil

Ting Zhan College of Resources and Environment/Micro-element Research Center/Hubei Provincial Engineering Laboratory for New Fertilizers, Huazhong Agricultural University; Key Laboratory of Horticultural Plant Biology (HZAU), MOE, Wuhan, People's Republic of China

Yuanyuan Zhao College of Resources and Environment/Micro-element Research Center/Hubei Provincial Engineering Laboratory for New Fertilizers, Huazhong Agricultural University; Key Laboratory of Horticultural Plant Biology (HZAU), MOE, Wuhan, People's Republic of China

Yongqiang Zheng Citrus Research Institute, Southwest University-Chinese Academy of Agricultural Sciences, Chongqing, China

Vasileios Ziogas Institute of Olive Tree, Subtropical Plants and Viticulture, Hellenic Agricultural Organization (H.A.O.)—Demeter, Chania, Greece

Andrés Zurita-Silva Instituto de Investigaciones Agropecuarias (INIA), Centro Regional de Investigación Intihuasi, La Serena, Chile

Editors' biography

Dr. A.K. Srivastava, received his MSc (Ag) and PhD in soil science from Banaras Hindu University in 1984 and 1988, respectively, and is currently the principal scientist (Soil Science) at Central Citrus Research Institute (formerly, National Research Centre for Citrus), Nagpur under the Indian Council of Agricultural Research, New Delhi. He has extensively pursued research work on different aspects of citrus nutrition including nutrient constraints analysis of citrus orchards by developing DRIS-based soil-plant nutrient diagnostics, orchard efficiency modeling, targeted yield-based site-specific nutrient management exploiting spatial variability in soil fertility, citrus rhizosphere-specific microbial consortium and soil carbon loading, INM module, fertigation scheduling, nutrient mapping using geospatial tools, nutrient dynamic studies, transformation of soil microbial biomass nutrients within citrus rhizosphere, and soil fertility map as decision support tool for fertilizer recommendation. He has handled 30 projects (13 as principal investigator and 17 as coprincipal investigator), exclusively on FRUIT NUTRITION.

Awards: He has been credited with a large number of publications including **161 research papers (102 papers in Indian journals and 59 papers in foreign journals), 49 policy review papers (32 in Indian journals and 17 in foreign journals), delivered 102 invited lead/keynote lectures, and 252 abstracts in seminar/symposium/conference.** He is the recipient of numerous awards including S.N. Ranade Award for Excellence in Micronutrient Research, FAI Silver Jubilee Award, International Plant Nutrition Institute-FAI Award, Netaji Subhash Chandra Bose Award for Excellence, National Magnum Foundation Award for Excellence, Dhirubhai Morarji Award, B.L. Jain Award, World Aqua Foundation Award, Best Citrus Scientist Award, etc. In addition, his biography appeared in two world-famous biographical compilations.

Academic fellowships: He is the life member of as many as **32 academic societies** besides being an honorary member of World Association of Soil and Water Conservationists. He is also the author of books like Citrus: Soil and Climate, Citrus Nutrition published by IBDC, Lucknow and editor of book titled Advances in Citrus Nutrition by Springer-Verlag, Netherlands. He has been **inducted as a Fellow of 10 professional academic societies** (Maharashtra Academy of Sciences, National Environmental Science Academy, Environmental Research Academy, United Writers Association, Indian Society of Citriculture, Indian Society of Soil Science, Indian Society of Agricultural Chemists, Confederation of Horticultural Associations of India, Gramin Vikas Society, and National Academy of Biological Sciences). He is a member of Asian Council of Science Editors at Seoul, Korea.

Editorial assignments: He is an editor-in-chief of Research Journal on Earth Sciences and International Journal of Horticultural and Crop Science Research, Current Horticulture; honorary editor, Agricultural Science Digest; executive editor, Agritechnology; regional editor of International Journal of Food, Agriculture, and Environment; and Advances in Horticultural Science and a member of editorial board of prestigious peer-reviewed journals including **Journal of Plant Nutrition, Communications in Soil Science and Plant Analysis (Taylor & Francis, USA), Journal of Agronomy and Crop Sciences (Crop Sci. Soc. Am.), and associate editor, Agronomy Journal (Am. Soc. Agron.) and Scientia Agricola (Brazilian Acad. Agri. Sci.) to name a few** and 20 national journals. He is also a regular paper setter on advance courses on soil fertility and soil chemistry across **nine agricultural universities** in India. He is also a member of the management committee of **three ICAR-based organizations**, viz., ICAR-NRC on Seed Spices, Ajmer, Rajasthan and ICAR-NRC on Pomegranate, Solapur, Maharashtra; ICAR-Indian Institute of Vegetable Research, Varanasi, Uttar Pradesh; and ICAR-National Bureau of Soil Survey and Land Use planning, Nagpur, Maharashtra in addition as a member of Technical Advisory Committee, Central Institute of Horticulture, Nagaland, India. He is a panel member of a committee constituted by Ministry of Agriculture, Government of India, New Delhi for Codex Standards for Organic Farming and Integrated Nutrient Management. He is also an **adjunct faculty** of three agricultural universities and a **visiting professor** at Huazhong Agricultural University, Hubei and Yangtze University, Jingzhou, China.

Foreign assignments: He was invited as a citrus expert by Government of Nepal in 1999 under Indo-Nepal MOU, as a keynote speaker in World Citrus Congress held at Wuhan, China in 2008, and as a resource person for Soil–Plant Interaction Workshop in 2018 at Huazhong Agriculture University (Wuhan) and Yangtze University (Jingzhou), China. He was also invited as a lead speaker in International Symposium by Huazhong Agriculture University China in November 2018 and is invited as a keynote speaker at International Symposium on Mineral of Fruit crops scheduled to be held on June 7–11, 2020 in Jerusalem (Israel).

More information may be obtained at his **URLs**: www.aksrivastavacitrus.com http://livedna.org/91.949, scholar.google.co.in/citations and contacted at **email**: aksrivas2007@gmail.com. **Mob**: 91-9422458020/07709150513.

Dr. Chengxiao Hu has been the director of the International Cooperation and Exchange Division of Huazhong Agricultural University (and the Dean of the International College), and has received his BA, MS, and PhD degree in agronomy from Huazhong Agricultural University in 1988, 1991, and 1999, respectively. He joined Huazhong Agricultural University in July 1991, and was promoted as associate professor and professor, respectively, in 1997 and 2002. His research fields include citrus and some other crop rotation formula fertilization technology system, agricultural product safety production and nutrient management, crop molybdenum nutrition mechanism and application of molybdenum fertilizer technology, urban sludge agricultural pollution control principles and technologies, mechanisms of mid- and micronutrient functions, especially Mo, Se of horticultural plants for edible quality development particularly the sugar/acid ratio, Vitamin C, etc.. He has presided over more than 40 projects, including the National Natural Science Foundation, Scientific and Technological Support Program of the Ministry of Science and Technology, 948 projects of the Ministry of Agriculture, the Ministry of Education, and the New Zealand ASIA2000 fund.

Awards: He has won the second prize of Science and Technology Progress Award by Education Ministry in 2017; the second prize of Hubei Science and Technology Progress Award in 1998, 2016, and 2018; excellent patent project and the Hubei outstanding doctoral dissertation for five times by the Education Department of Hubei Province. Two papers won the first prize of provincial excellent scientific papers and two papers won the second prize. The book *Practical Formula Fertilization Technology* coedited by him won the second prize of the 4th Excellent Science Works Award (Science Book) in Hubei Province in 2002. He is credited with five invention patents and one utility model patent. He has published more than 280 papers in various domestic and foreign journals, and 92 articles were included in SCI. He has edited or participated in the compilation of five books.

Academic fellowships: For more than a decade he has been the director of the New Fertilizer Engineering Laboratory of Hubei Province, the deputy director of the Academic Committee of the Key Laboratory of cultivated land conservation in the middle and lower reaches of the Yangtze River, the deputy director of the Chinese Academy of Plant Nutrition and Fertilizer Education Work Committee, the executive member of the council of the Chinese Citrus Society, the executive director of the Hubei Soil and Fertilizer Society, and the director of the Hubei Fertilizer Application Association. He was selected as the youth expert of Huazhong Agricultural University (1998), young and middle-aged experts with outstanding contributions to Hubei province (2001), the high level talent project of the new century of Hubei province (2003), the "excellent talent support program in the new century" from Education Ministry (2004), Senior Members of the Ministry of Agriculture (2006) and the "fifteenth" Advanced Individuals in the Field of Intelligence (2006) at the national level.

Editorial assignments: Professor Hu has served as a member of the Degree Committee of the People's Government of Hubei Province, a member of the Expert Committee of Textbook Construction of the Textbook Office in the Ministry of Agriculture, a member of China Fertilizer and Soil Conditioner Standardization Technical Committee, and as deputy chairman of the Education Working Committee of the Chinese Society of Plant Nutrition and Fertilizer.

Professor Hu can be reached at his **email**: hucx@mail.hzau.edu.cn and contacted at **Tel**: +86 27 8728 8840.

Preface

Fruit crops have been cultivated for centuries, both commercially and in amateur orchards as a major part of agricultural production. Agriculture is conceived as one of the humanity's crowning achievements and one of the central dynamics in the rise of human civilization. "An apple a day keeps the doctor away" is a popular adage, signifies the importance of fruit crops in the human diet (Chapter 1). Presently, fruit crops have touched 675 million metric tons (114 million metric tons of bananas) of production with most popular fruit varieties such as bananas and apples followed by grapes and oranges, offering a promising alternative to nutritional security options, besides easing the load on otherwise heavy per capita consumption of cereal crops. As we race toward a global population of 10 billion, the business as usual for fruit farming no longer appears as a viable option.

Entangled in multiple stresses, establishing a sustainable production system is the key challenge of present time fruit science. Decline in soil fertility due to nutrient mining is the major constraint limiting the productivity of fruit crops. Consistent reduction in nutrient density of different fruit crops is an indication of the nutrient mining-induced decline in productivity over time (Chapter 2). Fruit crops by the virtue of their perennial nature of woody framework (Nutrients locked therein), extended demand for nutrient supply across physiological growth stages, differential root distribution pattern (root volume distribution), and preferential requirement of some nutrients over others, collectively make them nutritionally more efficient than annual crops (Chapter 3).

Fruit crops by the virtue of being considered most nutrient responsive in nature, often develop certain overlapping morphogenetic symptoms under nutrient-capped scenario (Chapter 4), where ecophysiology of growing the fruit crops play a decisive role (Chapter 5). Perennial fruit trees play an important role in the carbon cycle of terrestrial ecosystems and sequestering atmospheric CO_2. An increase in yield of fruit crops such as apple, grape, banana, pineapple, mango, citrus, etc. in response to elevated CO_2 concentration has been extensively studied. It remains to be investigated, how accurate estimation of orchard C budget vis-à-vis timescale and feedback mechanisms of changes in soil carbon pool and steady-state level under specific fruit crop in order to expand potential of C credits through perennial fruit crops (Chapter 6). Fruit crops are undoubtedly one such group of perennial crops potentially very promising while looking at effective options for neutralizing (atmospheric carbon dioxide offset) the increasing menace of climate change-related issues (Chapter 7).

Plant nutritionists across the globe are on their toes to find ways and means to identify nutrient constraints as early in standing crop season as possible while dealing with fruit crops. Exciting progress has been made over the years, and accordingly, the basis of nutrient management strategy has experienced many paradigm shifts. While doing so, it is being increasingly felt to have some diagnostic tool to identify nutrient constraint as and when it originates by capturing the signals released at the subcellular level. On the other hand, conventionally used diagnostic tools of identifying nutrient constraints such as leaf analysis, soil analysis, juice analysis, and to some extent, metalloenzyme-based biochemical analysis, all have been under continuous use and refinement. And, therefore, the development of nutrient diagnostics is an extremely complex exercise. The issue becomes still quite complex under the soil conditions facing multiple nutrient deficiencies.

Not surprisingly, proximal sensing through spectral signatures of crop canopies in the orchards are more complex and often quite dissimilar from those of single green leaves measured under carefully controlled conditions. Even when leaf spectral properties remain relatively constant throughout the season, canopy spectra change dynamically depending upon variation in soil type, vegetation, and architectural arrangement of plant components. Vegetation indices provide a very simple yet an effective method for extracting the green plant quantity signal from complex canopy spectra. Narrower band indices such as the photochemical reflectance index, water band index, and normalized pigment chlorophyll ratio index are examples of reflectance indices that are correlated with certain physiological plant responses, and hold promise for diagnosing water and nutrient stress.

Nondestructive methods of identification of nutrient constraints, especially spectroscopic methods (Chapter 8), hold a definite edge, capable of sensing nutrient deficit as a biological nutrient sensor (Chapter 9) to track the genesis of nutrient deficiency on a real-time basis. Ironically, micronutrient deficiencies are diagnosed through a specific pattern of chlorosis, e.g., Fe versus Mn or Fe/Mn versus Zn backed up by nutrient concentration, capturing symptomatic pattern of chlorosis via spectral norms (signatures), irrespective of crop species further limit this concept toward more wider application. In the light of these developments, a relatively new concept popularly known as "Nutriomics" has emerged, revealing some lesser-known facts about fruit nutrition as a function of genomics (Chapter 10). By contrast, the better acknowledged methods of nutrient constraints diagnosis exploiting the merits of destructive methods of analysis like leaf analysis is by far the most widely used diagnostic tool (Chapter 11), of which many other developments have taken place (Chapter 12) to add better precision-based interpretation. However, among destructive methods of diagnostics tools, none of them is capable of identifying the nutritional disorders in the current seasons crop, thereby, aiming the outcome of diagnosis supposedly effective in next season crop. Flower analysis, though still in infancy stage, holds a better promise (Chapter 13), since it offers a comparatively longer time from anthesis to fruit maturity to schedule the fertilizer recommendation without compromising with either fruit yield or any of fruit quality parameters.

Growing fruit crops under diverse agro-pedological conditions confronted with multiple limitations is a considerable challenge with respect to the deficiency of calcium (Chapter 14) and boron (Chapter 15) or toxicity of boron (Chapter 16) and aluminum (Chapter 17), which need some strategic reorientation in our nutrient management options. On the other hand, the beneficial nutrients such as selenium (Chapter 18) and silicon (Chapter 19) have of late attracted the researchers, to be part of fruit nutrition program. These two nutrients hold a strong synergism with nutrients such as potassium, iron, manganese, calcium, etc. with varying agronomic implications. However, the full potential of such crunch nutrients could be realized only when floor management is properly looked into using suitable cover crops, serving multiple soil fertility functions (Chapter 20) through an effective rootstock-scion combination to optimize quality production, besides prolonging the orchard longevity (Chapter 21).

Rhizosphere security (soil security) is the call of the day these days, where physicochemical and biological properties of the soil inhibited by the roots are shaped in accordance to crop metabolism. Ecological significance of rhizosphere in terms of genetic, functional, and metabolic responses is another dimension in fruit crops, which need an incisive analysis (Chapter 22). However, it remains to be seen, whether or not and to what extent, such ecosystem service functions of rhizosphere are governed by different soil microbial communities. Arbuscular mycorrhizal fungi (AMF) is one of the most influential soil inhabiting fungi able to establish symbiotic relationship with more than 90% of the plants representing terrestrial ecosystem. How does mycorrhizosphere of fruit crops aid in unraveling the hidden facts about fruit nutrition through elevated synthesis of glomalin-related-soil-protein (Glycoprotein) are chronologically analyzed (Chapter 23) in the backdrop of some striking breakthroughs about the role of AMF in fruit nutrition (Chapter 24) to develop fruit trees with the desired biochemical and physiological preparedness as a result of mycorrhization to resist against a variety of other abiotic and biotic stresses. This is where mycorrhization of fruit trees could develop a better nutrient sink vis-à-vis quality fruit production. Recently, microbial inoculation has assumed a much greater significance, ever since depleting soil organic carbon has assumed an alarming proportion to facilitate soil fertility and plant nutrition act in a coordinated manner. In this pursuit, microbial consortium showed a clear cut superiority over single or dual microbial inoculation (Chapter 25) which has an added advantage of regulating the rhizosphere functional dynamics through biofertigation (Chapter 26) as the newest concept of fertigation, a little known nutrient supply system in fruit crops. Hence, these attempts are likely to provide some plausible answers with regard to top environmental problems, viz., microbial diversity loss, ecosystem collapse, and climate change to later tailor them into organic production system.

Regenerative farming using fruit crops could be quite thrilling and remunerative for long-term production sustainability. Adopting organic way (Chapter 27) and biodynamic method of nutrient management (Chapter 28) using cosmic energy-based calendar operation are likely to throw up new vistas of addressing different soil fertility constraints toward an optimized performance. However, foliar nutrition (Chapter 29) is the way forward approach to produce nutrient-dense fruit crops. But, nutrients aligning through phloem mobility and phloem immobility further pose some uncompromisable limitations, with the results, foliar feeding of nutrients still remains a formidable challenge. Overcoming the ever-increasing frequency of different soil fertility constraints, use of soilless method of cultivation is gradually gaining momentum in perennial fruit crops, called open-field hydroponics (Chapter 30), though not a popular concept but has many challenges to overcome before open-field hydroponics becomes a popular and conventional method of fruit crops growing.

Sustaining soil fertility with respect to fruit crops is another core agenda where biochar (essentially charcoal having carbon residence time in soil extending for >100 years) proved its utility, mostly under tropical environment with acid soils, imparting an additional liming value to biochars. The much value-added biochars have been derived from banana and orange sun-dried peels. Therefore, biochars need to be utilized for expanding carbon sequestration potential of soil, improving soil nutrient balances, especially in alkaline soils, soil-crop health under typical long-term field conditions. Additionally, biochar augers so well in organic production system, need to look afresh (Chapter 31). Ironically, stress and plant nutrition hardly complement each other. Despite quantum of researches dedicated to salinity responses of fruit crops, physiological basis of salinity tolerance is yet little understood (Chapter 32) at molecular level, another core area of research in fruit crops has been addressed so beautifully. Managing salinity stress toward better performance of fruit crops is always tricky issue. Improving the level of plant nutrition plays a combative role in moderating the impact of salinity exposure to fruit crops (Chapter 33) and has been highlighted with the help of some success stories.

Considering the thumping success of trunk nutrition, will it not be more advisable to analyze the xylem sap or phloem tissue for chemical and microbial constituents, since the signal transduction for various nutrients functioning mediate through these tissues only. Such attempts could provide some meaningful clues about the presence or absence of those signals to be later utilized in understanding the underlying principles of nutrient stress-induced warning mechanism. These studies could lay the solid foundation for developing some probe linked to transpiration stream of plant to act as early warning system for identifying deficiencies of various nutrients (Chapter 34). Use of nano-fertilizers (synthesized or modified form of traditional fertilizers), though still not a popular option to conventionally used fertilizers, offer some definite promise toward elevated use efficiency of applied fertilizers (Chapter 35) through proper delivery system utilizing different types of nano-fertilizers. However, issues relating health hazards need thorough studies with regard to nano-fertilizes to be really effective cropwide. The concept of nutrient-use-efficiency applied on the principles of 4R Nutrient Stewardship (Right amount of fertilizers using right source is applied at right time and right place) provides an ultimate framework guide to fertilizer use to any crop, with fruit crops being no exception (Chapter 36). Such attempt is slated toward increased production, profitability, and environmental safety. A further understanding of nutrient-microbe synergy provides a solid foundation in unlocking the productivity potential of fruit crops, besides safeguarding the soil health and possibility of doubling the yield coupled with nutrient-use-efficiency as central theme. With the availability of more technical know-how on combined use of organic manures, prolonged shelf life of microbial bio-fertilizers, and inorganic chemical fertilizers, an understanding on nutrient acquisition and regulating the water relations would help switch orchards to better CO_2 sink (expanding carbon capturing capacity of rhizosphere), so that a more sustainable fruit-based integrated crop production system could be evolved (Chapter 37). A comprehensive comparative study of organic versus inorganic fertilizers will be a booster to add strength to such integrated approaches (Chapter 38), where use of slow-release fertilizers can be stitched quite effectively to match with nutrient demand with critical growth stages (Chapter 39), a prerequisite to another form of nutrient-use-efficiency, known as nutrient utilization efficiency.

Correct diagnosis of nutrient constraints and their management are the two contrasting pillars of any successful fruit nutrition program. One of the most complex issues about fruit nutrition is the time taken to respond to fertilizer application in fruit crops such as citrus, mango, litchi, pomegranate, grapes, guava to name a few. It is because of erroneous diagnosis of nutrient constraints or big canopy size that dictates the nutritional behavior of these fruit crops at different developmental phases during long orchards, life. An assessment on annual nutrient export from orchard, quantum of nutrients locked into the trees skeletal framework and ability to distinguish between nutrient remobilized within tree canopy and externally applied fertilizer sources, senile nature of trees etc., singly or collectively govern the nutrient responsiveness of these fruit crops. An extensive attempt has been made to address the diagnosis and management of nutrient constraints in some of the premier fruit crops such as berries (Chapter 40), stone fruits (Chapter 41), papaya (Chapter 42), mango (Chapter 43), banana (Chapter 44), litchi (Chapter 45), pomegranate (Chapter 46), grapes (Chapter 47), guava (Chapter 48), citrus (Chapter 49), and pineapples (Chapter 50). These fruit-based chapters would go a long way in enriching the literature through state of the art compilation and in-depth analysis to bring out the long pending issues to limelight and offer a long-term solution for those researchers and practitioners involved in fruit nutrition.

We place on record our sincere acknowledgment to all the learned researchers/scientists having contributed their chapters and standing by us for so long. We also wish to thank acquisition editor, Dr. Nancy Maragioglio; Mr. Redding Morse and Ms. Swapna Srinivasan during the course of this book, an exciting and educative journey through fruit nutrition to both of us as editors of such massive effort. We earnestly hope, this book will attract a worldwide readership as a popular source of literature on Fruit Nutrition.

A.K. Srivastava
ICAR-Central Citrus Research Institute, India

Chengxiao Hu
Huazhong Agricultural University, Wuhan, China

CHAPTER

1

Fruits and nutritional security

Umar Farooq[a,*], Afshan Shafi[a], Kashif Akram[b], Zafar Hayat[c]

[a]Department of Food Science & Technology, Muhammad Nawaz Shareef University of Agriculture, Multan, Pakistan
[b]Department of Food Sciences, Cholistan University of Veterinary & Animal Sciences, Bahawalpur, Pakistan
[c]Department of Animal Sciences, CVAS-University of Veterinary and Animal Sciences, Jhang, Pakistan
*Corresponding author. E-mail: ufd302003@gmail.com

1 General

The concept of nutritional security is difficult to define due to its complex, broad, and multidimensional nature. Food availability, affordability, access, safety, and its stability are the basic pillars or dimensions of food security. It also has multidisciplinary nature with the involvement of a variety of stakeholders with national and international status (Candel, 2014; Hendriks, 2015). Food availability refers to the supply of quality food with sufficient quantity, and access is concerned with socioeconomic status of individuals to purchase appropriate foods to meet nutritional requirements. Similarly, the stability in food security is referred to the achievement of a situation where an individual or whole population has access to adequate food all the time (FAO, 2006). These pillars of food security are interconnected, for example, food access is not possible without food availability and food utilization is linked with food access (Hendriks, 2015). When personal needs of sufficient, safe, and wholesome food are fulfilled for healthy and active life all the time, then the person is considered as food secured. As per definitions of food security, only a person should not have access to food; instead, the food must also fulfill the energy and nutritional requirements of the body to prevent the situation of malnutrition. In current scenario, the food security has become a major issue not only for the developing countries but also for the developed countries. Not only the solutions for such a complex problem should consider the environmental and technical perspectives, but also the nations should look at the economic, social, and political aspects to handle the situation of food insecurity (Termeer et al., 2015).

The concerns of food security are not only focused on the prevailing conditions but also related to the future challenges of feeding of rapidly increasing world population (IFPRI, 2015). The research findings have indicated that there is a continuous experience of food insecurity. The first or primary indicator of food insecurity is considered to be the shortage or poverty, which reflects the issues related to food availability and access. To cover such situation, the people

A.K. Srivastava, Chengxiao Hu (eds.)
Fruit Crops: Diagnosis and Management of Nutrient Constraints
https://doi.org/10.1016/B978-0-12-818732-6.00001-0

try to find out the ways to cut food consumption. This leads to the usage of cheaper and energy dense food commodities to fulfill the energy requirements of the body, which results in hidden hunger through malnutrition of specific nutrients especially the micronutrients. The deficiencies of such vital nutrients lead to acute hunger, and such situation is known as acute food insecurity (Hendriks, 2015).

There are a number of identified reasons for food insecurity, and the major focus of the nations is to improve the economic status of the individual so that each person may have the capacity of purchasing. The other major target is to ensure food availability to feed the whole world. To combat such situations of food insecurity, a number of programs are being launched by the government and nongovernment organizations to fight against food and nutrition insecurity (Tanumihardjo et al., 2007). However, these programs have been found to be little fruitful in reducing food insecurity and failed to address the challenges of nutrition insecurity (Lear et al., 2014; Shisana et al., 2014). These programs have been unable to combat both food and nutrition insecurities (Khoury et al., 2014). The basic reason behind the situations is basically the lack of food diversification. The people rely on only limited foods specially the staple foods of their respective regions, which lead to nutritional insecurities. Although different programs of food fortification and supplementation have been launched all over the world, however, these programs are also limited to the fortification of specific targeted nutrients, and ultimately, the consumer fails to get all necessary nutrients required for a healthy life (Popkin et al., 2012). Thus, the consumption of only staple food over a long period leads to a number of health diseases and disorders due to the situation of under nutrition (Smith and Haddad, 2015; Papathakis and Pearson, 2012). Due to the fact, about 800 million people all over the world are considered to be undernourished with two billion people suffering from micronutrient deficiencies. Similarly, due to unbalanced diet, 1.9 billion people are overweight all over the globe, and one out of every three persons is malnourished (IFPRI, 2015).

The poor diets mostly based on staple foods are the common sources of hidden hunger as such diets no doubt provide enough energy for body but fail to provide all essential nutrients like vitamins and minerals. The people suffering from hidden hunger have not enough awareness about the importance of balanced diet, or they may not have enough access to wide range of nutritious foods (animal and plant based) due to any reason. The poverty and high prices could be among the basic factors that tend the consumer to continue longtime intake of staple foods with reduced or even zero intake of nonstaple foods, which result in nutrition insecurity (Bouis et al., 2011). The major reasons of nutritional food insecurity are highlighted in Fig. 1.1. These situations necessitate the intake of nonstaple foods like fruits to meet the needs of nutrient-based food security. Fruits and vegetables have important role in the provision of a healthy diet, and daily intake of such nonstaple foods helps to control and manage a number of human diseases and health disorders. With intake of fruits and vegetables, 2.7 million lives could be saved through the prevention of chronic diseases along with alleviation of nutritional deficiencies related to micronutrients (WHO, 2003).

Different strategies have been adopted to combat the situation of hidden hunger. The strategies may include fortification, biofortification, supplementation, and diversification. The diversification in diet and diet pattern seems to be one of the effective methods to control hidden hunger. This diet diversification has also positive effect

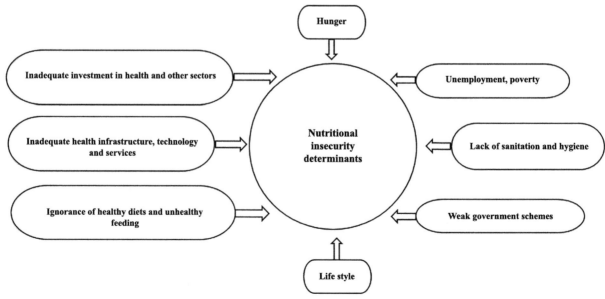

FIG. 1.1   Determinants of nutritional insecurity. *Modified from Aijaz, R., 2017. Preventing hunger and malnutrition in India. ORF Issue Brief 182, pp. 1–12.*

on child nutritional outcomes even when controlling for socioeconomic factors. The dietary diversification no doubt ensures a healthy and balanced diet with combination of macro- and micronutrients through a wide range of food choices including fruits.

2 Fruits

Fruits are the essential part of healthy lifestyle and crucial part of safe and healthy diet. As per recommendations of the WHO, it has been reported that the less consumption of fruits leads toward various metabolic disorders especially cardiovasculars such as 11% heart strokes and 31% ischemic heart disease globally. On the basis of aforementioned facts, it was also predicted that daily consumption of fruits may can protect more than 2.7 million people annually. Therapeutic significant of fruits also proved might be due to the characteristics of low calories, high nutritional contents, dietary fibers, and robust biologically active compound. These characteristics of fruits make them able to cure a number of disorders and improve health status of community (Du et al., 2011).

2.1 Nutritional components of fruits

Fruits are not only the most acceptable for their delightful taste but also possess a number of therapeutic benefits containing frequent nutrients (Buachan et al., 2014; Slavin and Lloyd, 2012). A wide variety of fruits contain significant amount of ascorbic acid or vitamin C like citrus fruits (orange and mandarin). Vitamin C is present in abundant quantity in citrus fruits that only one mandarin can accomplish the recommended daily allowance (RDA) of vitamin C for a normal human being. Similarly, other fruits are rich with some other specific nutrients. Generally, fruits contain 70%–80% moisture, 1.5% proteins, 13%–15% carbohydrates, up to 6% dietary fibers, 501 mg minerals, and up to 90 mg vitamins (Fig. 1.2). But this composition varies with fruit to fruit and variety to variety such as mango fruit that contains nearly 81% moisture, 0.4% fat, 0.6% protein, and 0.8% fibers.

It also contains nearly 17% of carbohydrates. Fruits also contain a perishable amount of minerals like magnesium, potassium, sodium, phosphorus, and sulfur. Similarly, guava comprises 77%–86% moisture content, whereas the remaining nutrients include crude fiber (2.8%–5.5%), protein (0.9%–1.0%), fat (0.1-%0.5%), minerals (0.43%–0.7%), and carbohydrate (9.5%–10%), and it also contains vitamins and minerals. This composition of guava fruit differs significantly with season, maturity stage, production technology, and variety (Mandal et al., 2009; Jiménez-Escrig et al., 2001). It is a good source of dietary fiber, dietary minerals (potassium, manganese, and copper), and vitamins (A, C, and folic acid) (Hassimotto et al., 2005). This fruit is also called "super fruit" as it has considerable amounts of vitamin A and C (Suntornsuk et al., 2002). Except vitamin C, a good quantity of other nutrients such as folate carotenoids and potassium is also present in many fruits. β-Cryptoxanthin and β-carotene are the known precursors of

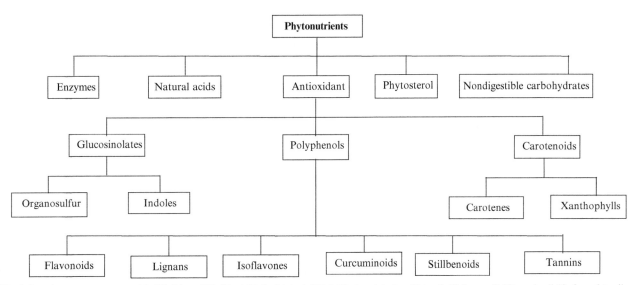

FIG. 1.2   Phytonutrients tree. *Modified from FFL (Food Fit for Living), 2016. Phytonutrients—Nature's Unknown Soldiers. Available from: http://www. foodfitforliving.com/thisweekatfffl/2016/1/9/week-13-phytonutrients-atures-unknown-soldiers. (Accessed 27 January 2019).*

TABLE 1.1 Nutritional composition of various fruits (National Nutrient Database for Standard Reference, 2018).

Nutrient	Apple	Guava	Mango	Citrus	Pomegranate	Pear	Peach	Banana	Watermelon	Apricot
Carbohydrates	17.3 g	14.32 g	25 g	11.8 g	18.70 g	27 g	16.7 g	22.84%	21 g	11.12 g
Proteins	0.3 g	2.55 g	1 g	0.9 g	1.67 g	1 g	1.4 g	1.09%	1 g	1.4 g
Lipids	0.2 g	0.95 g	0.5 g	0.1 g	1.17 g	0%	0.4 g	0.33%	0 g	0.39 g
Vitamins	–	–	–	70 mg	–	15%	22%	–	–	–
Minerals	–	–	–	7%	–	2%	–	484 mg	270 mg	–
Fiber	3.0 g	5.4 g	3 g	2.4 g	4.0 g	5%	3.1 g	2.6%	1 g	–

vitamin A and are classified under the category of carotenoids (Aldeguer et al., 2014). The nutritional composition of various fruits is shown in Table 1.1.

2.1.1 Lipids

A molecule of nutritive fat normally contains a number of fatty acids (having long chains of hydrogen and carbon atoms), attached with glycerol. They are normally found as triglycerides (three fatty acids bonded with one glycerol backbone). In human diet, a minimum of two fatty acids is important. A suitable balance of essential fatty acids—omega-6 and omega-3 fatty acids—looks also important for health, though conclusive experimental demonstration has been elusive. Among these "omega" long-chain polyunsaturated fatty acids are substrates for a class of eicosanoids called as prostaglandins, which play a major role in the human body. Fruits like banana, grapes, custard apples, ber, and cashew nut are good sources of fat (Chadha, 2007).

2.1.2 Water

Water essential to normal body functions as a vehicle for carrying other nutrients. The human body consists of 60% water, which is crucial for the proper physiology of the human body. Fruits are the richest source of water as they contain 70%–80% water contents (Desjardins, 2007).pt?>

2.1.3 Proteins

Proteins are the basis of many animal body structures (e.g., muscles, skin, and hair) and form the enzymes, which catalyze chemical reactions throughout the body. These foods are the building blocks of the body. These are important for body development. Lack of proteins in the body is responsible for stunted growth, increased chances of diseases, and lethargy. Protein molecules consist of amino acids having nitrogen and sometimes sulfur (during burning of protein, a distinctive smell is produced due to these components, such as the keratin in hair). To produce new proteins (protein retention) and exchange damaged proteins (maintenance), amino acids must be required in the human body. In digestive juices, amino acids are soluble in small intestine, where they are absorbed into the blood. They cannot be stored in the body after absorption, so they are either metabolized as required or excreted in the urine. The average adult requires 1 gram of protein per kilogram of body weight per day; children may require two to three times of this amount. Cashew nut, almond, filbert, pecan, pistachio, and walnut are rich in protein. Cashew nut is the richest source of protein among fruits (Kazi et al., 2015).

2.1.4 Carbohydrates

Carbohydrates are among the most important nutritional components of fruit with dominating contents of glucose, fructose, and sucrose. Among fiber constituents, pectin is the major component present in fruits, which makes 65%–70% of the total fiber. Other components of the fiber like cellulose, lignin, hemicellulose, and gums are also part of total fiber. Due to improper blood glucose metabolism, hyperglycemia (high blood glucose) or hypoglycemia (low blood glucose) may appear (Mayes et al., 2011).

2.1.5 Ascorbic acid (vitamin C)

Vitamin C is a category of ascorbic acid and is included in the group of water-soluble vitamins. In the human body, the vitamin C takes part in the formation of collagen, which is the main component of connective tissues. In the consequences, the deficiency of this vitamin causes weakness of tissues. Vitamin C also aids in the iron absorption. By using citrus products, we can control cold and anemia. The antioxidative activity of vitamin C is well recognized, and due to this behavior, it is considered to be responsible in preventing the oxidation of fatty acids, protein, and DNA. As these radicals are responsible for diseases like cancer, cardiovascular diseases (CVD), and cataract formation, vitamin C can play role in the management of these diseases as well. The 10 g daily intake of vitamin C is considered to be effective for prohibition of its deficiencies. High amount of vitamin C may be hazardous as it may produce danger of iron load (Ford and Giles, 2000).

2.1.6 Folic acid

Folic acid is also known as foliate. This water-soluble vitamin is associated with cell metabolism. Folic acid helps in DNA, RNA, and hemoglobin formation and has key role in anemia prevention. By using 400 μg folic acid, the defects of natural tube can be prevented, and 225 mL of fresh orange juice has 75 μg folic acid (NIH, 2018).

2.1.7 Potassium

Potassium controls the acid-base balance of the body. Blood pressure of the body is normally associated with minerals. According to daily requirement allowance, 2000 mg of potassium should be taken on daily basis. By using citrus fruit products and juices, we can improve the intake of potassium. A 235 mg of potassium comes by drinking 225 mL glass of orange juice (Duarte and Paull, 2015).

2.2 Bioactive compounds in fruits

It plays a role against free radicals and prevents the body from free radicals. It protects against the degenerative sicknesses such as melanoma and cataracts. It also plays a role in boosting immune system, iron absorption, and bovine collagen development. Bovine collagen plays a role in bone fortification and ligament and wound healing (Liu et al., 2012).

2.2.1 Phytochemicals

Fruits are known as the amazing natural medicines due to the presence of many bioactive compounds like flavonoids, vitamins, minerals, anthocyanins, and other compounds (Halliwell, 2006). These bioactive compounds prevent fruit from pathogens and are responsible for fruit flavor and fruit color. At the same time, these play an important role in the prevention of many chronic diseases. However, the debatable discussion is whether the extracts from fruits rich of phytochemicals have equal beneficial approach toward human health as achievable through whole food having phytochemical or the mixture of foods. It has been investigated by the researchers that different portions of fruits have different levels of phytochemicals. For example, the apple peel only contributes 0.4% antioxidant activity due to vitamin C as compared with the whole antioxidant activity that clearly indicates that the most of the antioxidant activity is contributed by other compounds like phenolics and flavonoids (Liu, 2003). So, all the phytochemicals play an important role in functional outcomes of the fruits. The classification of phytonutrients presents in fruits and other food commodities is elaborated in Fig. 1.2. Studies also predicted that the two phytochemicals (quercetin and ellagic acid) in strawberries were found responsible for anticancer and antimicrobial activity by blocking the suppress progression, initiation of carcinogenesis, and tumor proliferation (Denny and Buttriss, 2007).

2.2.2 Phenolics

The phytochemicals like phenolic are present in fruits, and due to protective biological functions of the phenolic, the fruits have prime importance in diet therapies and should be included in diet pattern. The phenolic compounds are even present in the by-products of fruits and in the fruit industry wastes. A number of phenolic compounds have been found in the peels of various fruits (Schieber et al., 2001). The fruits like grapes, apples, raspberries, cranberries, and strawberries and also their drinks such as orange and apple juices are good sources of phenolics. These compounds offer very strong facts of antimicrobials uniqueness (Urquiaga and Leighton, 2000).

2.2.3 Phenolic acid

Apple contains phenolic acids that are dihydrochalcones (phloretin glycosides), caffeic acid (chlorogenic acid), and p-coumaric acid (p-coumaryl-quinic acid); both are present with quinic acid in their esterified form (Awad et al., 2000; Thielen et al., 2005).

2.2.4 Flavonoids

Flavonoids are composed of carbon atoms with 15 in number, and these carbon atoms are ordered in a C6-C3-C6 configuration. These are considered as low-molecular-weight compounds. Physically, flavonoids are made up of two rings (aromatic), which are united through three carbon bridges leading to heterocyclic ring structure (Balasundram et al., 2006).

Flavonoids (or bioflavonoid) are placed in the list of secondary metabolites of plant and are abundantly present in the plant cells, which are responsible for photosynthesis. These compounds are also present in different parts of a plant like leaves, stem, fruit, seeds, and flower. They are responsible for the color of the flowers. As per classification of IUPAC, these compounds can be further categorized into flavones (e.g., rutin and quercetin), isoflavonoids, and neoflavonoids (Viuda-Martos et al., 2010).

The two main subtypes of apple polyphenols are flavonoids and phenolic acids. Apple is a good source of these compounds, and the most important flavonoids in apple are epicatechin, catechin, quercetins, and proanthocyanidins, which are their oligomers. These types of phenols have astringency and bitterness in apple (Thielen et al., 2005).

2.2.5 Tannins

Tannins are polyphenolic compounds that are high in molecular weight and soluble in water. Due to the presence of large number of hydroxyl groups, the tannins have the ability to bind with carbohydrates and proteins but up to limited levels. Tannins are further classified in to two main groups: nonhydrolyzable tannins and hydrolyzable tannins. Hydrolyzable tannins are actually gallic acid esters; after consumption, they are converted into gallic acid and then absorb into digestive tract. Condensed tannins are high in molecular weight compound and contain catechin polymers (Hassanpour et al., 2011).

2.2.6 Anthocyanins

Anthocyanins are the most important and prevalent group of flavonoids found in pomegranate arils. These compounds are responsible for the red color of juice and fruit. Anthocyanins are water soluble and are polyhydroxyl and polymethoxyl glucosidic derivatives of 2-phenylbenzopyrylium salt or flavylium salts. They are present in fruits and are titled as delphinidin, malvidin, cyanidin, pelargonidin, petunidin, and peonidin. Due to the number and location of hydroxyl group, each anthocyanin was different from other. The difference in each anthocyanin arises from the position and number of hydroxyl groups, the location, the level of methylation of the hydroxyl groups, number and nature of attached sugars, and aliphatic or aromatic acids bound to the sugars in the molecule (Afaq et al., 2005).

3 Therapeutic benefits of fruits

There is a strong consideration regarding utilization of fruits and positive impact on health status (Table 1.2). Epidemiological research studies have demonstrated the health-promoting activities of fruits such as anticancer properties, asthma-curing ability, cardiovascular protective effect, antioxidative, antihypercholesterolemic, immunity boosting effect, antidiabetic, and digestion-improving characteristics (Saito et al., 2002).

3.1 Anticancer properties

Many fruits have anticancer properties due to the presence of bioactive compounds, which work against cancerous cells. Sun and Liu (2008) strongly recommended that the frequent consumption of apple minimizes the threat of the occurrence of cancer especially breast and colon cancers. Another study was conducted by Liu et al. (2008) to expose the effect of apple extract on cancerous cells in rats. Results of the study revealed that apple extract has the potential to reduce the tumor size in rats. They suggested on the basis of the results of their study that whole apple (peel, pomace, and pulp) has the potential to stop the growth of cancerous cells and to reduce the possibility of various types of cancers like lung cancer, prostate cancer, and colon cancer.

Sun and Liu (2008) performed a research experiment on anticancer effect of bioactive compounds of various parts of apple. They basically separated biologically active compounds such as phenolics and flavonoids from the different parts of apple. Then, these bioactive compounds were evaluated for their potential physiological effects. Results indicated that three phenolic components and six flavonoid compounds were found to be present in apple peel.

TABLE 1.2 Health benefits of some phytonutrients present in fruits (Abuajah et al., 2014; Tanveer et al., 2017; Dolkar et al., 2017).

Phytonutrients	Compound	Sources	Health benefits
FLAVONOIDS			
	Quercetin	Grape fruit, apple, cranberries	Works as antidiabetic by mediation of glucose transporter
	Kaempferol	Strawberries, gooseberries, cranberries, grapefruit, apple	Possesses antidiabetic activity due to involvement in improved insulin secretion
	Myricetin	Grapes, berries, walnuts	Helps in controlling diabetes by improving insulin sensitivity
	Rutin	Citrus fruits, oranges, lemon, berries, limes, grapefruit, peaches, apples	Has free radical scavenging activity, helpful in the prevention of oxidative stress and inhibition of lipid peroxidation
	Hesperidin	Citrus fruits	Effectively increases the concentration of glycogen and glycolysis (hepatic)
	Naringenin	Citrus fruits	Acts as antihyperlipidemic by increasing expression of LDL receptor, reduces the glucose uptake from intestine
	Tangeritin	Citrus	Increases phosphorylation and alters the secretions of leptin, adiponectin, etc.
	Catechin	Grapes, apple juice	Has antidiabetic potential and prevents high blood glucose levels
	Epigallocatechin gallate	Pomegranate juice	Helps enhance the expression of insulin receptor
	Anthocyanins and anthocyanidins	Bilberry, raspberry, strawberry, black currants, peach, plum	Improve the insulin sensitivity, neutralize free radicals and ultimately reduce risk of cancer, work as antiinflammatory
	Flavonones	Citrus fruit	Possess antioxidant and anticancer activity
	Flavonols	Apple	Serve as antioxidant
	Phenolic acids	Berries	Help in the prevention of cancer and have antioxidant activity
	Resveratrol	Dark grapes, raisins, peanut, berries	Has antidiabetic and antihyperlipidemic potentials
	Curcumin	Turmeric root	Has antioxidative, antiinflammatory, and anticlotting potentials
	Ferulic acid	Apple, pears, citrus fruit	Improves eye vision and is effective against heart diseases
	Proanthocyanidines	Cranberries	Help in the management of cardiovascular disease and urinary tract disorders
CAROTENOIDS			
	Lycopene	Grapefruit, guava, papaya, watermelon	Has antioxidative and anticancer activities
	α- and β-carotene	Some fruits	Has ability to neutralize free radicals
	Zeaxanthin	Citrus fruits	Provides protection to eyes against muscular degeneration
DIETARY FIBER			
	Insoluble fiber	Nuts	Has anticancer potentials
	Soluble fiber	Some fruit	Possesses anticancer activities, improves digestive tract and management of cardiovascular diseases

The therapeutic evaluation of bioactive compounds showed significant antiproliferative potential and strong antioxidative property against the lines of cancer cells.

Similarly, citrus fruits contain a perishable amount of vitamin C, which possessed an excellent anticancer properties. Vitamin C or ascorbic acid has marvelous defensive mode of action against cancer or carcinogens (Rafiq et al., 2018).

3.2 Cardiovascular protection

Cardiovascular disorders including heart attack, stroke, and vein blockage are increasing day by day globally. One of the main reasons of these disorders is unhealthy diet pattern (Rye and Barter, 2014). Auclair et al. (2008) recommended the frequent consumption of fruits in daily diet after performing research experiment on cardiovascular diseases. In the reported study, the effect of fruit extract was evaluated against atheroscrotic lesions in mouse model. The results of the experiment revealed that fruit extract has a strong potential to reduce the atherosclerotic lesions (fatty build up) about 38% in atheroscrotic mouse. Results concluded that fruits possessed significant therapeutic characteristics, and these therapeutic properties of fruits might be due to the presence of phenolic contents.

Likewise, Griep et al. (2013b) illustrated the importance of citrus fruit and its products on the basis of medical evidences. They suggested on the basis of their results that citrus fruits reduce the risk of cardiovascular disorders and regulate blood pressure. They also recommended the consumption of citrus fruits in combination with other fruits, which might be more effective.

Chan et al. (2006) reported that apricot phenolic compounds like lycopene, chlorogenic acid, and β-carotene alleviate the oxidation of low-density lipoproteins (LDL), thus improving the antioxidant or defense mechanism of body. This antioxidant mechanism diminishes the danger of emerging atherosclerosis and coronary heart disease.

3.3 Hypercholesterolemia

Hypercholesterolemia is defined as high level of cholesterol in blood stream that results in cardiovascular diseases. Whole fruits have extraordinary nutraceutical value because nutrients are present in different parts of fruits, so it can prevent atherosclerosis more effectively (Chahoud et al., 2004). Total dietary fiber both soluble and insoluble and some other bioactive compounds are also responsible to manage hypercholesterolemia (Gorinstein et al., 2000). Various fruits have the ability to reduce plasma cholesterol. For example, medical evidences proved that the consumption of 100-g persimmon regularly inhibits the formation of plaque/occlusion in blood arteries (Gorinstein et al., 2000).

One of the major reasons of occurrence of the heart diseases is, elevated levels of homocysteine in the blood, which causes toxicity in vascular walls. High consumption of citrus fruit increases the folate level in blood plasma and decreases the rise homocysteine levels. This process ultimately controlled the blood LDL level and adjusted the blood pressure (Griep et al., 2013a). In the same way, apricot contains substantial quantity of fiber both soluble and insoluble forms (Ishaq et al., 2009). Soluble dietary fiber present in apricot lowers LDL cholesterol (Anderson et al., 2009; Aller et al., 2004).

3.4 Brain health

A strong association exists between consumption of fruits and brain health (Vauzour et al., 2017). Van-de-Rest et al. (2015) emphasized on the high consumption of fruits for the better cognition in adults. In vivo and in vitro mechanistic investigations revealed the neuron protective effect of fruits, and fruits also motivate neurogenesis. A randomized controlled medium-term experimental study was conducted by Pujos-Guillot et al. (2013) to assess the metabolomic effect of citrus juice. In this study, 500 mL/day of citrus juice was administrated to 12 human subjects for the time period of 1 month. At the end of the experimental period, the results influenced the positive effect of citrus consumption on metabolomics. In the similar way, recent studies found the positive effect of apple juice on mice having Alzheimer's disease. This study also comprises on 1-month duration. Results indicated a significant improvement in the brain (Chan et al., 2006).

3.5 Immunity

Immunity is actually the balanced condition with satisfactory biological defense system that has the ability to fight against infections, diseases, and disorders and possesses sufficient tolerance to avoid different autoimmune diseases. Fruits contain a number of nutritive and bioactive compounds that are responsible to boost the immune system of human beings (Hosseini et al., 2018). Fruits are rich in antioxidants and soluble fiber that provide a healthy digestive system that results a strong immunity (Aprikian, 2003). Apple contains a good quantity of digestible fibers that possess prebiotic characteristics, and these prebiotics promote the gut health by increasing the civilization of beneficial gut microbiota (Loo, 2004). Fruit's soluble fiber strengthens the immune system of human beings preventing from many diseases (Sherry, 2010).

3.6 Diabetes

Elevated level of plasma glucose is referred as diabetes. Globally, the prevalence of diabetic complications has increased day by day, and diabetes is also becoming the reason of other chronic diseases. So, its prevention is becoming the issue of serious scientific concern. Recent research documented that the management of diabetic complication can be managed by utilization of fruits. It was also observed that consumption of one apple per day reduces the 28% possibility of diabetes (Song et al., 2005). Methodically, it is workable that fruit's polyphenol particularly catechins and phloretin glucoside (dihydrochalcones) may be effective against diabetes and thus declines the onset of diabetes especially type 2 diabetes (Marks et al., 2009).

3.7 Infectious diseases

Phytochemicals of fruits possessed significant potential to protect against infectious agents. Fruit and their extracts have strong identified antimicrobial properties (Nagesh and Shanthamma, 2009). Various clinical studies have indicated that fruit's bioactive compounds have the potential compared with the commercial antimicrobials to kill the microbes by disrupting their cell membranes, depriving the substrate, or inactivating the enzymes. Through this mechanism, these bioactive compounds of fruits block the mechanism of microbes (Rupasinghe, 2003).

Aiswarya et al. (2011) conducted an experimental trial to evaluate the antimicrobial potential of aqueous and alcoholic extracts of apple. In this reported study, they compared the effect on Gram-positive and Gram-negative bacteria through disc diffusion method. They found a significant zone of inhibition of apple extracts against both Gram-positive and Gram-negative bacteria, which was comparable with commercial antibiotic (ciprofloxacin).

Du et al. (2011) evaluated the antimicrobial potential against *Salmonella enterica*, *Escherichia coli*, and *Listeria monocytogenes* through in vitro experiment. Apple skin powder was used in edible film/coating to evaluate the potential of inhibition against *S. enterica*, *E. coli*, and *L. monocytogenes*. The edible film containing apple peel showed the highest zone of inhibition against *L. monocytogenes*. The lowest concentration required to stop this microbe was 1.5%.

Malaviya and Mishra (2011) performed a comprehensive research experiment to assess antimicrobial potential of alcoholic and aqueous extracts of various fruits like apple, pomegranate, guava, and oranges against fungi and bacteria. On comparison of both extracts, results showed that the aqueous extracts indicated greater antimicrobial activity in contrast to alcoholic extracts. They were active against 76.4% of the full amount of Gram-negative bacteria that included 93.75% *E. coli*, 83.33% *Klebsiella* sp., and 73.68% *Pseudomonas* sp.

4 Conclusion

By reviewing the importance of fruits and their nutrients in the maintenance of proper physiology for improved health status of community, it is concluded that consumption of fruits in adequate amount is necessary to maintain a healthy life. Various parts of fruits possessed therapeutic benefits to the human beings because they contain essential nutrients and biologically active components. For the aim to provide a healthy status and a safe health to the community, nutritional security is crucial nowadays. To preserve the nutrients or maintain the nutritional security, a positive and robust role of food scientist is the dire need of the current era.

5 Future perspectives

The role and importance of fruits to insure nutritional food insecurity cannot be neglected at any cost. The fruits can provide a number of necessary nutrients required for a healthy human body. These fruits not only provide the necessary nutrients but also provide many health benefits beyond the nutrients due to the presence of bioactive and functional compounds in different parts of the fruits. Reasonable data and information are available on nutritional composition of different fruits and their role in human health. However, the composition of fruits varies due to production technology, season, soil conditions, and variety. There is a need of scientific work to improve the quality of fruits especially the production of nutrient-enriched varieties of fruits with reduced antinutritional compounds. Moreover, there is need of more awareness among the community regarding the importance of balanced diet and role of fruits for the management of nutritional deficiencies and human diseases.

References

Abuajah, C.I., Ogbonna, A.C., Osuji, C.M., 2014. Functional components and medicinal properties of food: a review. J. Food Sci. Technol. https://doi.org/10.1007/s13197-014-1396-5.

Afaq, F., Saleem, M., Krueger, C.G., Reed, J.D., Mukhtar, H., 2005. Anthocyanin and hydrolyzable tannin-rich pomegranate fruit extract modulates MAPK and NF-kappa B pathways and inhibits skin tumorigenesis in CD-1 mice. Int. J. Cancer 113, 423–433.

Aiswarya, G., Reza, K.H., Radhika, G., Farook, S.M., 2011. Study for antibacterial activity of cashew apple (*Anacardium occidentale*) extracts. Pharm. Lett. 3 (1), 193–200.

Aldeguer, M., López-Andreo, M., Gabaldón, J.A., Puyet, A., 2014. Detection of mandarin in orange juice by single-nucleotide polymorphism qPCR assay. Food Chem. 145, 1086–1091.

Aller, R., de Luis, D.A., Izaola, O., La Calle, F., del Olmo, L., Fernandez, L., Hernandez, J.G., 2004. Effect of soluble fiber intake in lipid and glucose leaves in healthy subjects: a randomized clinical trial. Diabetes Res. Clin. Pract. 65 (1), 7–11.

Anderson, J.W., Baird, P., Davis, R.H., Ferreri, S., Knudtson, M., Koraym, A., Waters, V., Williams, C.L., 2009. Health benefits of dietary fiber. Nutr. Rev. 67 (4), 188–205.

Aprikian, O., 2003. Apple pectin and a polyphenol-rich apple concentrate are more effective together than separately on cecal fermentations and plasma lipids in rats. J. Nutr. 133 (6), 1860–1865.

Auclair, S., Silberberg, M., Gueux, E., Morand, C., Mazur, A., Milenkovic, D., Scalbert, A., 2008. Apple polyphenols and fibers attenuate atherosclerosis in apolipoprotein E-deficient mice. J. Agric. Food Chem. 56 (14), 5558–5563.

Awad, M.A., Dejager, A., Westing, V.L.M., 2000. Flavonoid and chlorogenic acid levels in apple fruit: characterization of variation. Sci. Hortic. 83, 249–263.

Balasundram, N., Sundram, K., Samman, S., 2006. Phenolic compounds in plants and agri-industrial by-products: antioxidant activity, occurrence, and potential uses. Food Chem. 99, 191–203.

Bouis, H.E., Eozenou, P., Rahman, A., 2011. Food prices, household income, and resource allocation: socioeconomic perspectives on their effects on dietary quality and nutritional status. Food Nutr. Bull. 21 (1), S14–S23.

Buachan, P., Chularojmontri, L., Wattanapitayakul, S.K., 2014. Selected activities of *Citrus maxima* Merr. fruits on human endothelial cells: enhancing cell migration and delaying cellular aging. Nutrients 6 (4), 1618–1634.

Candel, J., 2014. Food security governance: a systematic literature review. Food Sec. 6, 585–601.

Chadha, K.L., 2007. Handbook of Horticulture. ICAR, Pusa, New Delhi.

Chahoud, G., Aude, Y., Mehta, J., 2004. Dietary recommendations in the prevention and treatment of coronary heart disease: do we have the ideal diet yet? Am. J. Cardiol. 94, 1260–1267.

Chan, A., Graves, V., Shea, T.B., 2006. Apple juice concentrate maintains acetylcholine levels following dietary compromise. J. Alzheimers Dis. 9, 287–291.

Denny, A., Buttriss, J., 2007. Plant Foods and Health: Focus on Plant Bioactives. European Food Information Resource (EuroFIR) Consortium. Funded under the EU 6th Framework Food Quality and Safety Thematic Priority. Contract FOOD-CT-2005-513944.

Desjardins, Y., 2007. First international symposium on human health effects of fruits and vegetables. Acta Hortic. 744, 470.

Dolkar, D., Bakshi, P., Wali, V.K., Sharma, V., Shah, R.A., 2017. Fruits as nutraceuticals. Ecol. Environ. Conserv. 23, S113–S118.

Du, W.X., Olsen, C.W., Avena-Bustillos, R.J., Friedman, M., McHugh, T.H., 2011. Physical and antibacterial properties of edible films formulated with apple skin polyphenols. J. Food Sci. 76 (2), 149–155.

Duarte, O., Paull, R.E., 2015. Exotic Fruits and Nuts of New World. CAB International, Wallingford.

FAO, 2006. Policy Briefing No. 2. Available from: http://www.fao.org/forestry/13128-0e6f36f27e0091055 bec28ebe830f46b3.pdf.

Ford, E.S., Giles, W.H., 2000. Serum vitamins, carotenoids, and angina pectoris: findings from the National Health and Nutrition Examination Survey III. Ann. Epidemiol. 10, 106–116.

Gorinstein, S., Kulasek, G.W., Bartnikowska, E., Leontowicz, M., Zemser, M., Morawiec, M., Trakhtenberg, S., 2000. The effects of diets, supplemented with either whole persimmon or phenol-free persimmon, on rats fed cholesterol. Food Chem. 70, 303–308.

Griep, L.M., Wolbers, F., de-Wagenaar, B., ter-Braak, P.M., Weksler, B.B., Romero, I.A., Couraud, P.O., Vermes, I., van-der-Meer, A.D., van-den-Berg, A., 2013a. BBB on chip: microfluidic platform to mechanically and biochemically modulate blood-brain barrier function. Biomed. Microdevices 15 (1), 145–150.

Griep, L.M.O., Stamler, J., Chan, Q., Van-Horn, L., Steffen, L.M., Miura, K., Elliott, P., 2013b. Association of raw fruit and fruit juice consumption with blood pressure: the INTERMAP study1-4. Am. J. Clin. Nutr. 97 (5), 1083–1091.

Halliwell, B., 2006. Oxidative stress and neurodegeneration: where are we now? J. Neurochem. 97 (6), 1634–1658.

Hassanpour, S., Sadaghian, M., Maherisis, N., Eshratkhah, B., Chaichisemsari, M., 2011. Effect of condensed tannin on controlling faecal protein excretion in nematode-infected sheep: in vivo. J. Am. Sci. 7 (5), 896–900.

Hassimotto, N.M.A., Genovese, M.I., Lajolo, F.M., 2005. Antioxidant activity of dietary fruits, vegetables, and commercial frozen fruit pulps. J. Agric. Food Chem. 53 (8), 2928–2935.

Hendriks, S., 2015. The food security continuum: a novel tool for understanding food insecurity as a range of experiences. Food Sec. 7, 609–619.

Hosseini, B., Berthon, B.S., Saedisomeolia, A., Starkey, M.R., Collison, A., Wark, P.A.B., Wood, L.G., 2018. Effects of fruit and vegetable consumption on inflammatory biomarkers and immune cell populations: a systematic literature review and meta-analysis. Am. J. Clin. Nutr. 108 (1), 136–155.

IFPRI (International Food Policy Research Institute), 2015. Global nutrition report 2015: Actions and accountability to advance nutrition and sustainable development. Washington, DC. Available from: http://ebrary.ifpri.org/cdm/ref/collection/p15738coll2/id/129443.

Ishaq, S., Rathore, H.A., Majeed, S., Awan, S., Shah, S.Z.A., 2009. Studies on the physico-chemical and organoleptic characteristics of apricot (*Prunus armeniaca* L.) produced in Rawalakot, Azad Jammu and Kashmir during storage. Pak. J. Nutr. 8 (6), 856–860.

Jiménez-Escrig, A., Rincón, M., Pulido, R., Saura-Calixto, F., 2001. Guava fruit (*Psidium guajava* L.) as a new source of antioxidant dietary fiber. J. Agric. Food Chem. 49 (11), 5489–5493.

Kazi, N.A., Yadav, J.P., Agale, M.G., 2015. Nutritional value of fruits. SRJIS, 3 (16), 2937–2943.

Khoury, C.K., Bjorkman, A.D., Dempewolf, H., Ramirez-Villegas, J., Guarino, L., Jarvis, A., Rieseberg, L.H., Struik, P.C., 2014. Increasing homogeneity in global food supplies and the implications for food security. Proc. Natl. Acad. Sci. U.S.A. 111, 4001–4006.

Lear, S.A., Teo, K., Gasevic, D., Zhang, X., Poirier, P.P., Rangarajan, S., Seron, P., Kelishadi, R., Tamil, A.M., Kruger, A., Iqbal, R., Swidan, H., Gómez-Arbeláez, D., Yusuf, R., Chifamba, J., Kutty, V.R., Karsıdag, K., Kumar, R., Li, W., Szuba, A., Avezum, A., Diaz, R., Anand, S.S., Rosengren, A., Yusuf, S., 2014. The association between ownership of common household devices and obesity and diabetes in high, middle and low income countries. Can. Med. Assoc. J. 186, 258–266.

Liu, R.H., 2003. Health benefits of fruits and vegetables are from additive and synergistic combination of phytochemicals. Am. J. Clin. Nutr. 78, 517S–520S.

Liu, J.R., Dong, H.W., Chen, B.Q., Zhao, P., Liu, R.H., 2008. Fresh apples suppress mammary carcinogenesis and proliferative activity and induce apoptosis in mammary tumors of the Sprague-Dawley rat. J. Agric. Food Chem. 57 (1), 297–304.

Liu, Y.Q., Emily, H., Sherry, A.T., 2012. History, global distribution, and nutritional importance of citrus fruits. Compr. Rev. Food Sci. Food Saf. 11, 530–545.

Loo, V.J.A., 2004. Prebiotics promote good health: the basis, the potential, and the emerging evidence. J. Clin. Gastroenterol. 38 (6), 70–75.

Malaviya, A., Mishra, N., 2011. Antimicrobial activity of tropical fruits. Biol. Forum Int. J. 3 (1), 1–4.

Mandal, S., Sarkar, R., Patra, P., Nandan, C.K., Das, D., Bhanja, S.K., Islam, S.S., 2009. Structural studies of a heteropolysaccharide (PS-I) isolated from hot water extract of fruits of *Psidium guajava* (Guava). Carbohydr. Res. 344 (11), 1365–1370.

Marks, S.C., Mullen, W., Borges, G., Crozier, A., 2009. Absorption, metabolism and excretion of cider dihydrochalcones in healthy humans and subjects with an ileostomy. J. Agric. Food Chem. 57, 10–15.

Mayes, S., Massawe, F.J., Alderson, P.G., Roberts, J.A., Azam-Ali, S.N., Hermann, M., 2011. The potential for underutilized crops to improve security of food production. J. Exp. Bot. 29, 1–5.

Nagesh, K.S., Shanthamma, C., 2009. Antibacterial activity of *Curculigo orchioides* rhizome extract on pathogenic bacteria. Afr. J. Microbiol. Res. 3 (1), 005–009.

National Nutrient Database for Standard Reference, 2018. Fruits and Fruit Juices. United States Department of Agriculture. Agric Res Ser 3.9.5.1_2018-12-22.

NIH (National Institutes of Health), 2018. Folate, Retrieved From Folate Fact Sheet for Health Professionals. Available from: https://ods.od.nih.gov/factsheets/Folate-HealthProfessional/. (Accessed 26 November 2018).

Papathakis, P.C., Pearson, K.E., 2012. Food fortification improves the intake of all fortified nutrients, but fails to meet the estimated dietary requirements for vitamins A and B6, riboflavin and zinc, in lactating South African women. Public Health Nutr. 15, 1810–1817.

Popkin, B.M., Adair, L.S., Ng, S.W., 2012. Global nutrition transition and the pandemic of obesity in developing countries. Nutr. Rev. 70, 3–21.

Pujos-Guillot, E., Hubert, J., Martin, J.F., Lyan, B., Quintana, M., Claude, S., Chabanas, B., Rothwell, J.A., Bennetau-Pelissero, C., Scalbert, A., Comte, B., Hercberg, S., Morand, C., Galan, P., Manach, C., 2013. Mass spectrometry-based metabolomics for the discovery of biomarkers of fruit and vegetable intake: citrus fruit as a case study. J. Proteome Res. 12, 1645–1659.

Rafiq, S., Rajkumari-Kaul, S.A., Bashir, S.N., Nazir, F., Nayik, G.A., 2018. Citrus peel as a source of functional ingredient: a review. J. Saudi Soc. Agric. Sci. 17 (4), 351–358.

Rupasinghe, H.P.V., 2003. Using change for success: Fruit-based bioproduct research at the Nova Scotia Agricultural College. Annual Report 2003 of the Nova Scotia Fruit Growers' Association, pp. 66–69.

Rye, K.A., Barter, P.J., 2014. Cardioprotective functions of HDLs. J. Lipid Res. 55 (2), 168–179.

Saito, T., Miyake, M., Toba, M., Okamatsu, H., Shimizu, S., Noda, M., 2002. Inhibition by apple polyphenols of ADP-ribosyltransferase activity of cholera toxin and toxin-induced fluid accumulation in mice. Microbiol. Immunol. 46, 249–255.

Schieber, A., Stintzing, F.C., Carle, R., 2001. Byproducts of plant food processing as a source of functional compounds-recent developments. Trends Food Sci. Technol. 12, 401–413.

Sherry, C.L., 2010. Sickness behavior induced by endotoxin can be mitigated by the dietary soluble fiber, pectin, through up-regulation of IL-4 and Th2 polarization. Brain Behav. Immun. 24 (4), 631–640.

Shisana, O., Labadarios, D., Rehle, T., Simbayi, L., Zuma, K., Dhansay, A., Reddy, P., Parker, W., Hoosain, E., Naidoo, P., Hongoro, C., Mchiza, Z., Steyn, N.P., Dwane, N., Makoae, M., Maluleke, T., Ramlagan, S., Zungu, N., Evans, M.G., Jacobs, L., Faber, M., 2014. The South African National Health and Nutrition Examination Survey (SANHANES-1). HSRC Press, Cape Town.

Slavin, J.L., Lloyd, B., 2012. Health benefits of fruits and vegetables. Adv. Nutr. 3 (4), 506–516.

Smith, L.C., Haddad, L., 2015. Reducing child undernutrition: past drivers and priorities for the post-MDG era. World Dev. 68, 180–204. https://doi.org/10.1016/j.worlddev.2014.11.014.

Song, Y., Manson, J., Buring, J., Sesson, H., Lin, S., 2005. Associations of dietary flavonoids with risk of type 2 diabetes and markers of insulin resistance and systemic inflammation in women: a prospective and cross sectional analysis. J. Am. Coll. Nutr. 24, 376–384.

Sun, J., Liu, R.H., 2008. Apple phytochemical extracts inhibit proliferation of estrogen-dependent and estrogen-independent human breast cancer cells through cell cycle modulation. J. Agric. Food Chem. 56 (24), 11661–11667.

Suntornsuk, L., Gritsanapun, W., Nilkamhank, S., Paochom, A., 2002. Quantitation of vitamin C content in herbal juice using direct titration. J. Pharm. Biomed. Anal. 28 (5), 849–855.

Tanumihardjo, S.A., Anderson, C., Kaufer-Horwitz, M., Bode, L., Emenaker, N.J., Haqq, A.M., Satia, J.A., Silver, H.J., Stadler, D.D., 2007. Poverty, obesity, and malnutrition: an international perspective recognizing the paradox. J. Am. Diet. Assoc. 107, 1966–1972.

Tanveer, A., Akram, K., Farooq, U., Hayat, Z., Shafi, A., 2017. Management of diabetic complications through fruit flavonoids as a natural remedy. Crit. Rev. Food Sci. Nutr. 57 (7), 1411–1422.

Termeer, C., Dewulf, A., Breman, G., Stiller, S., 2015. Governance capabilities for dealing wisely with wicked problems. Adm. Soc. 47 (6), 680–710.

Thielen, C., Will, F., Zacharias, J., Dietrich, H., Jacob, H., 2005. Distribution of dihydrochalcones and flavonols in apple tissue and comparison between fruit and juice. In: Proceedings of European Symposium on Apple Processing 68, pp. 16–18.

Urquiaga, I., Leighton, F., 2000. Plant polyphenol antioxidants and oxidative stress. Biol. Res. 33, 55–64.

Van-de-Rest, O., Berendsen, A.A., Haveman-Nies, A., de-Groot, L.C., 2015. Dietary patterns, cognitive decline, and dementia: a systematic review. Adv. Nutr. 6, 154–168.

Vauzour, D., Camprubi-Robles, M., Miquel-Kergoat, S., Andres-Lacueva, S.B., Barberger-Gateau, P., Bowman, G.L., Caberlotto, L., Clarke, R., Hogervorst, P., Kiliaan, A.J., Lucca, U., Manach, C., Minihane, A.M., Mitchell, E.S., Perneczky, R., Perry, H., Rousse, A.M., Ramirez, M., 2017. Nutrition for the ageing brain: towards evidence for an optimal diet. Ageing Res. Rev. 35, 222–240.

Viuda-Martos, M., Fernandez-Lopez, J., Perez-Alvarez, J.A., 2010. Pomegranate and its many functional components as related to human health: a review. Comp. Rev. Food Sci. Food Saf. 9, 635–654.

WHO, 2003. Diet, nutrition and the prevention of chronic diseases. Report of a Joint FAO/WHO Expert Consultation, Geneva, WHO Technical Report Series, No. 916.

Further reading

Aijaz, R., 2017. Preventing Hunger and Malnutrition in India. ORF Issue Brief, 182, pp. 1–12.

Bommarco, R., Kleijn, D., Potts, S.G., 2013. Ecological intensification: harnessing ecosystem services for food security. Trends Ecol. Evol. 28, 230–238.

FAO, 2014. Panorama of Food Security in Latin America and Caribbean, Santiago de Chile. Available from: http://www.fao.org/3/a-i4230e.pdf.

FAO, 2015. Regional Overview of Food Insecurity Latin America and the Caribbean. The region has reached the international hunger targets. Available from:http://www.fao.org/3/a-i4636e.pdf.

FAOSTAT, 2011. FAO Statistical Databases. FAO, Rome. Available from: http://faostat.fao.org. (Accessed 15 September 2011).

FFL (Food Fit for Living), 2016. Phytonutrients—Nature's Unknown Soldiers. Available from: http://www.foodfitforliving.com/thisweekatfffl/2016/1/9/week-13-phytonutrients-atures-unknown-soldiers. (Accessed 27 January 2019).

He, X., Liu, R.H., 2008. Phytochemicals of apple peels: isolation, structure elucidation, and their antiproliferative and antioxidant activities. J. Agric. Food Chem. 56 (21), 9905–9910.

Meade, B., Rosen, S., 2013. International Food Security Assessment, 2013-2023, GFA-24. US Department of Agriculture, Economic Research Service, United States.

2

Nutrient density of fruit crops as a function of soil fertility

Allen V. Barker[a],*, *Margie L. Stratton*[†]

[a]University of Massachusetts, Amherst, MA, United States
*Corresponding author. E-mail: barker@umass.edu

1 Defining nutrient density and its importance

Nutrient density is a term applied to describe the proportion of nutrients in foods. Commonly, the nutrients are the ones identified as mineral nutrients or chemical elements that are necessary for plant and animal life. Nutrients also include organic components of foods to include vitamins, proteins, fats, carbohydrates, fibers, antioxidants, and other essential constituents for human health. Nutrient density of foods has been used to develop dietary guidelines in the consumption of foods (Arsenault et al., 2012; Drewnowski, 2010, 2017; Kennedy et al., 2008; Mobley et al., 2009). However, the term has been noted as being difficult to define universally (Nicklas et al., 2014). Classifying or ranking of foods is called nutrient profiling (Drewnowski and Fulgoni, 2014). Nutrient profile models are said to be useful for assessing nutrient density if information only on the nutritional quality of the food is available (Scarborough and Raynor, 2014; Scarborough et al., 2007).

Several different models have been developed to provide numerical scores to guide consumers in choosing healthful foods. These rankings are called nutritional quality indexes, nutrient density scores, and by various other classes (Table 2.1) (Nicklas et al., 2014). The US Department of Agriculture maintains food composition data bases that list nutrient composition of foods (United States Department of Agriculture, 2018). This model is a nutrient adequacy score based on nutrients per unit of fresh weight or per serving. The units of measurement in nutrient profiling for consumers often are nutrient concentrations in milligrams per 100 g or serving size of food. Rankings of nutritional adequacy have included, as noted earlier, nutrients per unit weight (nutrient adequacy score), per calorie (nutrient density score), per unit cost (nutrient-to-price ratio), or ratios of recommended components to restricted components of food (nutrient quality index) (Darmon et al., 2005; Drewnowski, 2010; Drewnowski, 2015; Rampersaud, 2007; Scheidt and Daniel, 2004). In applications of cost-to-price ratios, fruits and vegetables, for example, are expensive sources of dietary energy but provide nutrients at a reasonable cost. Drewnowski and Rehm (2013) showed that beans and starchy vegetables,

[†] Retired

TABLE 2.1 Models for assessing nutrient density in foods.

Model	Base	References
Nutrient adequacy score	Nutrients per unit weight or serving	Drewnowski (2005, 2010), Kennedy et al. (2008), USDA (2018)
Nutrient density score	Nutrients per kcal	Anonymous (2018), Darmon et al. (2005), Di Noia (2014), Drewnowski (2017)
Unit cost	Nutrient-to-price ratio	Darmon et al. (2005), Drewnowski (2015), Drewnowski and Rehm (2013)
Bioavailable nutrient density	Nutrient absorbed per kcal	Hallberg (1981)
Nutrient quality index	Positive and negative nutrients; nutrients divided by detrimental constituents	Arsenault et al. (2012), Drewnowski (2017), Drewnowski and Fulgoni (2014), Rampersaud (2007), Scheidt and Daniel (2004)

including white potatoes, were cheaper per 100 kcal (Calories; the capitalized calorie is 1000 calories or 1 kcal) than were dark-green and deep-yellow vegetables. Potatoes and beans had the lowest-cost sources of potassium and fiber among frequently consumed vegetables. Hallberg (1981) stated that an important measure of the nutritive value of a food could be obtained if the amount of a nutrient absorbed is related to the energy content of the meal studied (bioavailable nutrient density) and applied this concept to absorption of iron. A class called powerhouse vegetables was developed to identify foods providing 10% or more daily value per 100 kcal (Calorie) for 17 nutrients (potassium, fiber, protein, calcium, iron,

TABLE 2.2 Powerhouse fruits and vegetables by ranking of nutrient density scores.

Crops	Nutrient density score[a]	Crops	Nutrient density score[a]
Watercress	100.00	Brussels sprouts	32.23
Chinese cabbage	91.99	Scallion	27.35
Chard	89.27	Kohlrabi	25.92
Beet green	87.08	Cauliflower	25.13
Spinach	86.43	Cabbage	24.51
Chicory	73.36	Carrot	22.60
Leaf lettuce	70.73	Tomato	20.37
Parsley	65.59	Lemon	18.72
Romaine lettuce	63.48	Iceberg lettuce	18.28
Collard green	62.49	Strawberry	17.59
Turnip green	62.12	Radish	16.91
Mustard green	61.39	Winter squash (all)	13.89
Endive	60.44	Orange	12.91
Chive	54.80	Lime	12.23
Kale	49.07	Grapefruit (pink and red)	11.64
Dandelion green	46.34	Rutabaga	11.58
Red pepper	41.26	Turnip	11.43
Arugula	37.65	Blackberry	11.39
Broccoli	34.89	Leek	10.69
Pumpkin	33.82	Sweet potato	10.51
		Grapefruit (white)	10.47

[a] *Calculated as the mean of percent daily values (D) (based on a 2000 kcal/day diet) for 17 nutrients (potassium, fiber, protein, calcium, iron, thiamin, riboflavin, niacin, folate, zinc, and vitamins A, B_6, B_{12}, C, D, E, and K) as provided per 100 kcal of food. Scores above 100 were capped at 100 (indicating that the food provides, on average, 100% DV of the qualifying nutrients per 100 kcal).*

From Di Noia, J., 2014. Defining powerhouse fruits and vegetables: a nutrient density approach. Prev. Chronic Dis. 11, 130390. https://doi.org/10.5888/pcd11.130390.

thiamin, riboflavin, niacin, folate, zinc, and vitamins A, B_6, B_{12}, C, D, E, and K) (Di Noia, 2014). Di Noia (2014) provides listings of fruits and vegetables that are rated by a nutrient adequacy score (mean of percent daily values) divided by an energy score (100 kcal of food). A table that lists these scores is presented (Table 2.2). The Aggregate Nutrient Density Index (ANDI) is similar to the nutrient adequacy score and ranks foods according to nutrients per kcal (Anonymous, 2018). This index measures mineral elements, carotenoids, lutein, zeaxanthin, lycopene, fiber, folate, glucosinolates, and antioxidants. The resulting scores are based on kcal. A low-Calorie food with high nutrients scores higher than an energy-rich food high in Calories. The ANDI scores are adjusted so that the highest score is 1000. Kale has a score of 1000, and many fruits and vegetables have scores under 100. A person would starve for energy on a diet based on high-scoring foods in the ANDI in a system based on nutrients per kcal. Many recommendations for selecting nutrient-dense foods are from this list. Nutritional data are usually presented for raw (uncooked) forms of foods.

In scientific literature of agronomists and horticulturists, the units commonly are percent nutrient per dry weight of food or total content in milligrams per kilogram of food. In this chapter, nutrient density will be discussed in units of dry weight or in total content. For mineral nutrients, a nutrient-dense food is one that has a high concentration of an essential element in the dry or fresh weight of food. Mineral nutrients are fundamentally chemical elements that are derived primarily from the soil (Gupta and Gupta, 2014). Plant-based foods generally contain the mineral nutrients established as essential for human nutrition. They provide much of our skeletal structure and are critical for body processes as essential as constituents and cofactors for enzymes. Humans cannot utilize most foods without critical minerals for enzymes responsible for digestion and absorption. The elements are required in the forms present in crops and are classified into macronutrients and micronutrients or trace minerals based on the relative amounts required. This classification is based upon their requirement rather than on their relative importance. All nutrients are required absolutely so that no one nutrient is more important than another.

Fruit quality is mentioned often in publications and is a term with many characteristics but one that includes fruit contents of mineral nutrients and other components of fruit composition and the relationship of mineral nutrients to various chemical and physical properties of fruits (Chater and Garner, 2019; Labate et al., 2018; Soliman et al., 2018; Vashisth et al., 2017; Zhang et al., 2018; Zhou et al., 2018).

In recent years, concern has developed over reported declines in the concentrations of nutrients in fruits and vegetables. This decline is attributed to the dilution effect of increased yields due to promotion of carbohydrate accumulation in the foods derived from plants without an equal increase in the mineral nutrients in the foods (Davis, 2009; White et al., 2009). The changes also could be due to anomalies of measurement or sampling, changes in the food system, changes in the varieties grown, or changes in agricultural practice (Mayer, 1997). Content of nutrients in food is a product of the concentration of the nutrients times the mass of the food produced by a crop. On this basis, large fruits usually have the same content or higher content than smaller fruits. The decline in concentration may be addressed through fertilization and plant breeding or with supplements in the diet. White et al. (2009) noted that the dilution effect could be addressed through applications of fertilizers and that it should be possible to increase mineral concentrations in potato tubers with a combination of genotypes with high mineral concentrations with fertilization to deliver minerals to the diet without compromising yields. Nagar et al. (2018) studied genetic variability and association of fruit yield with 9 mineral nutrients (sodium, potassium, calcium, magnesium, iron, manganese, copper, zinc, and selenium) in 76 genotypes of pumpkin (*Cucurbita moschata* L.) and reported high heritability for accumulation of mineral nutrients and fruit yield and that fruit yield had a positive association with mineral nutrient content. Nutrient depletion of soils lowers soil fertility and limits crop yields and poses a potential threat to global food security and agricultural sustainability (Tan et al., 2005). Tan et al. (2005) reviewed literature on depletion of nutrients caused by high production levels and decline in fertilizer use in recent decades in many developed, less developed, and developing countries and concluded that soil fertility problems associated with nutrient depletion by crop production without fertilization to return nutrients to the soil are worldwide. This depletion of nutrients from soils, called mining of nutrients, could lower the nutrient contents in plant-derived foods (Lyne and Barak, 2000). On the other hand, Marles (2017) reviewed the mineral nutrient compositions of vegetables, fruits, and grains and reported that over time various changes in sources of information, crop varieties, sampling, laboratory analysis, and statistical treatment of data affect values of reported nutrient contents in plant-derived foods and consequently mineral nutrient concentrations in foods are not declining. Marles (2017) noted also that comparisons with matching archived soil samples showed that nutrient content of soils has not declined in locations where intensively cultivated land has been fertilized. He concluded that the benefits of increased yield to supply food for expanding populations outweighed nutrient dilution effects and that people would have adequate diets by eating recommended amounts of fruits and vegetables from fertilized fields.

Consumers likely are not eating the recommended amounts of fruits and vegetables for healthful diets. Dietary guidelines for Americans indicated that nutrients such as folate, magnesium, potassium, dietary fiber, and vitamins A, C, and K are underconsumed in the United States (Simon, 2012). Jani et al. (2009) noted that from the median amounts of the various recipes fed to Indian children, intakes of nutrients, especially calcium and iron, were low. Vegetables and fruits are major sources of several of these organic micronutrients and minerals as well as many other phytonutrients. The American diet is said to be rich in energy but poor in nutrient consumption, and to improve the intake of nutrients, consumers are recommended to replace some foods in their diets with nutrient-rich options, such as fruits and vegetables (Drewnowski, 2005; Drewnowski et al., 2008; Foote et al., 1999; Keatinge et al., 2011; Liu, 2013). However, affordability is an issue in increasing the amounts of fruits and vegetables in diets, and many foods such as milk, potatoes, citrus juices, cereals, and beans have more favorable nutrient-price ratios than many fruits and vegetables, and shoppers with low socioeconomic status purchase calories in inexpensive forms that are high in fat or carbohydrate and are not nutrient rich (Appelhans et al., 2012; Drewnowski, 2009). On the other hand, a report showed that, although fruits and vegetables are an expensive source of dietary energy, they provide nutrients at a reasonable cost (Darmon et al., 2005).

Adequacy of nutrients in foods can be met in part by increasing the production of crops that are particularly rich in nutrients and by biofortification through enhancing the nutrient contents in fruits, vegetables, and grains (Davies et al., 1981; Arsenault et al., 2015; Bouis, 2001; Bouis and Welch, 2010; Graham et al., 2007; Grusak, 2002; Rengela et al., 1999; Wang et al., 2008; Welch and Graham, 2004). Biofortification may occur through selection of cultivars of fruits, vegetables, and grains that are nutrient rich and through fertilization of soils. Moore et al. (2012) noted that diet patterns that combined intakes across food groups led to greater improvements in overall nutritional adequacy among girls in several age groups; however, they noted that more than 90% of the girls failed to consume the recommended amounts of fruit, vegetables, and dairy products. The vast majority of girls of all ages had inadequate intakes of calcium, magnesium, potassium, and vitamins D and E, but they consumed >750 kcal/day from the category of solid fat and added sugars far in excess of the recommended maximum intakes. Oyarzun et al. (2001) explored the nutritional adequacy of food patterns based on the main staples around the world—corn, wheat, rice, and potato. The authors suggested that people have their diets modified by the addition of foods high in vitamins and minerals, based on food table information on composition.

2 Soil fertility and nutrient-dense crops

2.1 Crop nutrition and fruit composition

Nutrients must be present in the soil and capable of entering into solution in water for plants to absorb the nutrients. Fertilization is necessary to maintain nutrients levels in soils at adequate concentrations to nourish a crop and in turn to provide nutrient-rich produce. Concentration of nutrients in leaves is more responsive to variations in nutrient availability in soils than concentrations in fruits; consequently, more studies are conducted on composition of leaves than on concentrations of nutrients in fruits. Measurements of nutrient concentrations in leaves may be an indication of nutrient concentrations in fruits since nutrients may move from leaves to fruits (Dris et al., 1999; Tomala, 1997). Ferguson (1980) measured the contents of calcium, magnesium, potassium, and phosphorus in kiwifruit (*Actinidia chinensis* Planch.) over the fruit growing period. The maximum calcium accumulation in fruits was reached sooner than for the other nutrients. Potassium and phosphorus moved into the fruit over the entire growing season. Some elements such as calcium and boron do not move from leaves to fruits; however, and nutrients in leaves may not relate to concentrations in fruits. Silber et al. (2018) recommended that analysis of fruits for mineral contents should determine fertilizer management for avocado (*Persea americana* Mill.). Nestby et al. (2005) reviewed results that suggest a direct connection between nutrient uptake and fruit quality. This chapter emphasizes concentrations in fruits, but attention is given to leaves occasionally since nutrients from leaves might enter into fruits or were the only source of information concerning soil fertility and nutrient accumulation in plants for a given source.

Lester et al. (2010) reviewed and reported on nutrient contents in fruits. Potassium among many plant nutrients stood out as having the strongest influence on fruit properties that determine marketability, consumer preference, and the concentration of critically important human health–associated phytonutrients. A case study (Lester et al., 2010) with muskmelon (*Cucumis melo* L.) compared commercial potassium fertilizers and foliar fertilization on fruit marketability (maturity, yield, firmness, soluble solids, sugars, and relative sweetness), consumer preference attributes (sugar content, sweetness, and texture), and phytochemical concentrations (potassium, ascorbic acid, and β-carotene concentrations). Lester et al. (2010) noted that glycine-complexed potassium (potassium metalosate) and potassium sulfate combined with a silicone-based surfactant, applied weekly, as a foliar spray, during fruit development had a strong impact on improving marketability attributes, as well as quality attributes of composition. Among several foliar applied potassium salts studied under field conditions, salts with relatively low salt indices appeared to have the greatest impacts on fruit quality if applied during the mid- to late-season fruit development periods.

Aghili et al. (2009) investigated the nutritional and quality indices of greenhouse cucumber (*Cucumis sativa* L.) in relationship with fruit nutrient concentrations. Fruit firmness showed a positive correlation with calcium concentration. Fruit ascorbic acid and citric acid were influenced positively by potassium concentration in fruits. The green color intensity of the fruit skin was correlated positively with magnesium concentration. Potassium was considered as the most important effective element affecting cucumber quality. Also, Amiri and Fallahi (2007) working with grape (*Vitis vinifera* L.) noted that potassium application influenced yield, fruit set, cluster weight, cluster number, and berry size and quality attributes of soluble solid concentrations than other mineral elements (N, Mg, Fe, and Zn). Soliman et al. (2018) reported that the nitrogen and potassium concentrations of figs (*Ficus carica* L.) increased, whereas phosphorus concentration decreased with increasing amount of potassium fertilization.

Foliar fertilization as a supplement to supplying nutrients to soil is a method of application to correct deficiencies and to feed plants with part of their needs. Foliar fertilization might be a method to boost nutrients in fruits. Roosta and Hamidpour (2013) reported that foliar application to tomato (*Lycopersicon esculentum* Mill.) of potassium, magnesium, iron, manganese, zinc, and copper increased their concentrations in the leaves of aquaponic-treated plants and that foliar spray of potassium, iron, manganese, zinc, and copper increased the concentrations of these elements in the fruits of hydroponically grown plants.

Mature pomegranate trees (*Punica granatum* L.) were treated during fruit set with foliar sprays of zinc sulfate, magnesium sulfate, or potassium nitrate and analyzed for mineral nutrient concentrations, antioxidant activity, total phenolics, total soluble solids, and titratable acidity (Chater and Garner, 2018, 2019). Fruits of zinc-treated trees had greater zinc concentrations, but no other significant differences in fruit mineral nutrient concentrations were detected following foliar spraying. Nutrients in foliage were increased by the sprays. Antioxidant activity, total dissolved solids, and fruit size did not vary with treatments. Splitting of fruits was reduced by all sprays at one of the treatment sites. Foliar sprays may be important to enhance nutrient supply to crops during fruit development. Maity et al. (2017) noted that changes in concentrations of nutrients in plants varied with fruit development. The concentration of nitrogen, phosphorus, potassium, sulfur, iron, zinc, and boron in leaves decreased, whereas calcium, magnesium, manganese, and copper concentration increased during fruit growth and development. As fruit mass increased, biomass accumulation in fruit exceeded nutrient accumulation, and concentration of nutrients in fruits, except Mg, decreased until maturity. During fruit enlargement, demand for nitrogen, phosphorus, potassium, iron, copper, and zinc was high, whereas requirement for calcium, magnesium, and sulfur was high during late stages of fruit development.

Solhjoo et al. (2017) investigated foliar spraying of potassium chloride, potassium sulfate, potassium nitrate, calcium chloride, or combinations of the potassium and calcium sprays on fruit quality of apple (*Malus domestica* Borkh.). Spraying with potassium salts, calcium chloride, and their combinations increased fruit weight, sugar, anthocyanin, and potassium concentrations in fruits and fruit firmness. A combined foliar application of calcium chloride and each of the potassium sources was more effective on the improvement of fruit color, firmness, potassium or calcium accumulation, and K/Ca ratio than if potassium or calcium was applied alone.

Caretto et al. (2018) noted with tomato that a small-fruited cultivar had higher concentrations of total soluble solids and vitamin E than larger-fruited cultivars. Increased potassium levels in the nutrient solution increased total soluble solids, vitamin E, and reducing sugar contents and titratable acidity in the fruits. Serio et al. (2007) reported previously that lycopene content of tomato increased with increasing potassium concentration in the nutrient solution and that a higher lycopene concentration occurred in a small-fruited cultivar than in large-fruited cultivars.

Nava et al. (2007) noted that nitrogen and potassium are in higher concentrations than other macronutrients in apple (*M. domestica* Borkh.) fruits and are associated most frequently with changes in fruit quality. Nitrogen fertilization negatively affected fruit color, flesh firmness, and total soluble solids concentration. Fruit color and total soluble solids were increased by potassium fertilization. Buwalda and Meekings (1990) likewise reported that potassium was the major nutrient accumulating in leaves and fruits of Japanese pear (*Pyrus serotina* Rehd.). Most of the potassium accumulated was in the fruit.

Javaria et al. (2012) concluded that increasing potassium concentration in fruits resulted in improved quality parameters of tomato. They showed that total solids, sugars, and titratable acidity increased with increasing amounts of potassium fertilization. Also, lycopene, vitamin C, and total soluble solids increased with increased application of potassium. Significant relationships occurred between fertilizer rates and surface redness, tissue redness, firmness, crispness, mealiness, sweetness, sourness, and flavor. Hunsche et al. (2003) in a 9-year trial with apple noted that increased potassium fertilization of soil raised fruit mass, diameter, titratable acidity, red color, and potassium concentrations but decreased flesh firmness. Fruits with low potassium concentrations had higher internal breakdown than fruits with higher potassium concentrations. Weight losses during storage, ground color, and rot incidence were not affected by potassium fertilization.

Kadir (2004) studied the response of apple trees to frequent applications of calcium chloride. Trees were sprayed one to eight times with foliar applications of calcium chloride. More than six applications of calcium chloride gave an increase in fruit weight, dimensions, and skin redness and a suppression of scald. Applications had no effect on percent of soluble solids. Increase in fruit quality was attributed to a linear increase in calcium concentrations in fruit and leaf tissues. The increases in calcium concentrations compared with untreated trees were accompanied by increases in potassium, magnesium, phosphorus, and nitrogen concentrations. Fallahi and Simons (1996) reported that leaf and fruit nitrogen, potassium, calcium, and manganese concentrations correlated more often with fruit quality parameters than other elements.

The highest yield and largest fruits of apple were obtained with soil and foliar applications of nitrogen (urea) and zinc (zinc sulfate) in combination (Amiri et al., 2008). Foliar application of nitrogen or zinc alone resulted in low-quality fruits with russeting and lowered soluble solids. No difference between fruit mineral nutrient concentration and yield except for nitrogen occurred among several treatments of foliar and ground applied treatments. The authors concluded that foliar applied nutrients might be more efficient than soil applied, but a combination of soil and foliar applications is recommended for apple tree nutrient management.

Tekaya et al. (2014) reported that foliar fertilization of olive (*Olea europaea* L.) with sprays of nitrogen or combinations of nitrogen and boron, magnesium, sulfur, and manganese, applied during the growing season caused a decrease of total phenols and tocopherols concentrations in fruits and gave qualitative changes in the profiles of minerals, phenols, tocopherols, and carbohydrates. This study showed that fertilization caused a significant decrease of total phenol and tocopherol contents and induced qualitative changes in the profiles of minerals, phenols, tocopherols, and carbohydrates. The results showed an association between environmental concentrations of nutrients and chemical composition of olive fruits.

Ye et al. (2014) studied the effects of continuous cropping on absorption of 11 essential elements and fruit yield and also examined the effects of foliar application of urea and KH_2PO_4 and iron, boron, zinc, and manganese on nutrient absorption with continuous cropping. Continuous cropping caused a suppression in fruit yield compared with peppers grown in rotation soil and affected the absorption of phosphorus, potassium, magnesium, iron, boron, zinc, manganese, and copper and transport of these elements to the aerial parts of the plant, although the elemental concentrations in continuously cropped soil were not lower than those in the soil than with rotation soil. The impact of continuous cropping on the accumulation of micronutrients was greater than for macro elements. Pepper fruit yields were increased by foliar application of urea and KH_2PO_4 or foliar application of iron, boron, zinc, and manganese.

Munaretto et al. (2018) reported that foliar application of silicon promoted higher levels of anthocyanins in the fruits of two strawberry (*Fragaria* × *ananassa* Duchesne) cultivars and increased with the applied dose. The application of silicon was also efficient in maintaining fruit firmness of both cultivars. Haghshenas et al. (2018) reported that, in addition to element concentrations, the ratio of nutrient elements in solution has a role in growth, productivity, quality, and nutrient accumulation. They studied the effects of different potassium-calcium ratios (1.6, 1.4, 1.2, 1, 0.85, and 0.6) in hydroponics with coconut fiber perlite medium on the properties of strawberry fruit. The highest fruit pH, electrical conductivity, total soluble solids, titratable acidity ratio, vitamin C and ellagic acid contents, and fruit color resulted from a 1.4 K/Ca ratio. The K/Ca ratio of 1.6 produced the highest content of protein, and the 0.85 ratio was the most effective treatment for fruit firmness. The authors concluded that nutrient solution with K/Ca ratios between 1 and 1.6 were suitable for producing strawberry.

Many investigators have noted that the composition of fruits varies with species and cultivars of crops. The fruit of pumpkin (*Cucurbita maxima* L.) was noted as being good for supplementing human diets with fiber and minerals such as potassium, calcium, phosphorus, magnesium, sulfur, silicon, iron, and zinc (Czech et al., 2018). Mineral contents varied with cultivar of pumpkin. Zhou et al. (2018) investigated the fruit quality of six citrus cultivars at five stages of ripening and growing in the same orchard. The mineral nutrient concentrations of the cultivars during fruit development were 0.94%–1.92% N, 0.11%–0.23% P, 1.03%–1.37% K, 0.31%–1.15% Ca, 0.11%–0.29% Mg, 3.97–72.34 mg Fe/kg, 1.93–10.64 mg Mn/kg, 1.56–10.73 mg Cu/kg, and 0.90–16.80 mg Zn/kg. The results reported that only magnesium was correlated with sugar and organic acid components, negatively and positively, respectively. Vashisth et al. (2017) reported that phenolic, flavonoid, anthocyanin contents and antioxidant capacity decreased as nitrogen fertilization increased with subtropical peach (*Prunus persica* L.). The authors suggested that subtropical peach phytochemical composition can be affected by different cultivars and tree age and can be managed with cultural practices like nitrogen fertilization and harvest time to produce fruits with altered or desired nutritional composition.

Labate et al. (2018) noted that potassium, magnesium, and calcium supplies influence fruit quality factors of color, shape, firmness, hollow fruit, acidity, and uniformity of ripening. The investigators studied 52 diverse, commercially important cultivars of tomato for concentrations of potassium, magnesium, calcium, and sodium in fruits. The cultivars showed differences in cation concentrations that supported a genetic basis for potassium, magnesium, calcium, and sodium concentrations. The work suggested that cultivars with favorable cation nutrient profiles for human health and fruit quality can be developed.

Nachtigall and Dechen (2006) measured the concentrations of nutrient element in apple leaves and fruits. In fruits, the initial nutrient concentrations during fruit development from bloom to harvest initially decreased quickly, underwent slow and continuous decreases, and then remained almost constant until the end of fruit maturation suggesting nutrient dilution during fruit growth. Potassium was the nutrient present in highest quantities in apple fruits and consequently was the most removed from the soil. Storey and Treeby (2000) followed the concentrations of the macronutrients (K, Ca, Mg, Na, P, and S) and micronutrients (Fe, B, Zn, Mn, and Cu) in whole fruit, pulp, rind, and albedo of navel orange fruit The concentrations of most elements in whole fruits and fruit parts decreased during fruit development. In whole fruit, potassium concentrations decreased throughout fruit development, whereas the calcium concentration increased during early stages of fruit growth and then decreased progressively.

Clark and Smith (1990) recorded changes in the concentrations of macronutrients and micronutrients in the fruit of persimmon (*Diospyros kaki* L.f.). The concentration of nutrients in fruit (skin, flesh, and seeds inclusive) declined during growth. Concentrations of calcium, copper, and iron in skin were consistently greater than in flesh, whereas those of phosphorus and sulfur were consistently smaller than corresponding concentrations in flesh throughout the season. In mature fruit, the highest concentrations were in seeds for all nutrients except calcium, which was more concentrated in skin. Ozenc and Ozenc (2014) recommended that iron fertilization be a practice to improve some nut properties and mineral composition of hazelnut (*Corylus avellana* L.). Phosphorus, iron, and boron contents of nut increased significantly with iron fertilization. Total oil, kernel percentage, and kernel weight increased, and the amounts of empty and wrinkle nuts decreased with iron fertilization.

Kolota and Balbierz (2017) reported that intensive nitrogen fertilization of scallop squash (*Cucurbita pepo* L.) caused enhancement of the total nitrogen and nitrate accumulation in fruits but did not change the concentrations of phosphorus, potassium, calcium, or magnesium. Fruits harvested at 3–6 cm diameter size contained higher concentrations of total nitrogen, phosphorus, potassium, magnesium, and nitrate than larger ones with diameters of 6.1–12 cm or >12 cm. At the maximum concentration of nitrates in fruits harvested early (317 mg NO_3^-/kg fw), scallop squash is a minor contributor to dietary nitrate intake by consumers.

Ayaz et al. (2015) evaluated the nutritional content of seven commercial eggplant (*Solanum melongena* L.) fruits in terms of fatty acid, mineral, sugar, organic acid, amino acid, and polyamine contents. Linoleic acid (range, 39.14%– 53.81%, ave. 45%) was the most abundant fatty acid, potassium (range, 1556–3172 mg/kg fw, ave. 2332) was the most abundant mineral nutrient, malic acid was the major organic acid (range, 130–387 mg/g fw, ave. 157), and the major sugar was fructose (range, 1243–1380 mg/100 g fw, ave. 1350.88). The major polyamine was putrescine (11.54 and 25.70 nmol/g fw, ave. 18), and the major amino acid was glutamine (148–299 mg/100 g fw, ave. 219.74). They suggested that these results may contribute to further studies aiming to develop nutrient-rich varieties of eggplant in breeding programs.

In hydroponic production, the marketable yield on minicucumbers increased with nutritional level of variable nutrient solutions from 25% to 100% of full-strength solution (Maboko et al., 2017). Mineral nutrient content (P, Fe, and Mn) was improved by full nutrition. Maboko et al. (2013) on the other hand reported that the application of arbuscular mycorrhizal fungi in hydroponic production did not enhance plant growth, yield, or mineral nutrient concentrations in tomato fruits. Irrigation can have an effect on composition of fruits, and the effects depend on the composition of the irrigation water. Benko et al. (2011) observed that the lowest nitrogen (2.27% dry matter) of tomato was in fruits harvested from plants irrigated with water with low calcium nitrate concentration, whereas higher amounts of nitrogen accumulated if high calcium nitrate concentrations were applied. Cultivars of tomato differed in dry matter, soluble solids, vitamin C, phosphorus, and potassium contents. Covre et al. (2018) reported that irrigation during the fruiting phase of conilon coffee (*Coffea canephora* Pierre ex A. Froehner) increased the accumulation of nutrients in the fruits and leaves. The macronutrients in greatest quantities were nitrogen, potassium, and calcium regardless of the irrigation treatment.

Carbon dioxide (CO_2) concentrations in the air have been reported to affect nutrient concentrations in foods derived from plants. Elevated atmospheric CO_2 enhances the yield of vegetables and could also affect their nutritional quality. Dong et al. (2018) in a review of 57 articles and 1015 observations showed that elevated CO_2 (540 and 1200 μL/L) compared with ambient concentrations (≥200 and ≤450 μL/L) decreased the concentrations of magnesium, iron, and zinc but maintained the concentrations of phosphorus, potassium, sulfur, copper, and manganese of vegetables (fruits, leafy vegetables, stems, and roots). The paper cited investigations that demonstrated that the high CO_2 concentrations decreased all of the mineral concentrations in grain crops. However, the calcium concentration of vegetables was increased under elevated CO_2. Studies cited in the review reported that elevated CO_2 decreased the mineral concentration by a dilution effect from the increased biomass or by restricted transpiration. The review noted that the high CO_2 increased the concentrations of fructose, glucose, total soluble sugar, total antioxidant capacity, total phenols, total flavonoids, ascorbic acid, and calcium in the edible part of vegetables. Pinero et al. (2018) reported that elevations in atmospheric CO_2 (800 vs 400 μL/L) had little effect on mineral accumulation in sweet pepper (*Capsicum annuum* L.) fruits and that calcium spray on foliage had only minor effects on minerals in fruits, whereas a spray of NaCl increased manganese, chloride, and sodium and lowered boron, phosphorus, potassium, and calcium concentrations in the fruits.

Amendments with organic matter to soils may affect nutrient composition of fruits. Ochmian et al. (2009) reported that substrates of cocoa husk or sawdust mixed with soil gave variations in nutrient content of the media and affected nutrient concentration of blueberry (*Vaccinium corymbosum* L.) fruits and leaves. Sewage sludge is rich in organic matter and plant nutrients, but its use as a fertilizer on soil might lead to inorganic and biological contaminants in foods derived from plants. Mota et al. (2018) reported that with fertilization with sewage sludge in pineapple (*Ananas comosus* Merr.) plantations, zinc and copper contents in the soil, and the zinc content in the leaves of the pineapple plants increased. Sewage sludge did not change the contents of copper or zinc in the fruits. Pineapple fruits did not show microbiological contamination above the limits established by Brazilian legislation.

Nogueirol et al. (2018) studied cadmium accumulation in tomato leaves with treatments of 0, 50, and $100\,\mu mol\,Cd/L$ and NO_3^-/NH_4^+ proportions of 100/0, 70/30, and 50/50 in the nutrient solution. The largest accumulation of cadmium occurred in the shoots at the NO_3^-/NH_4^+ ratio of 70/30 and $100\,\mu mol/L$ Cd. The authors noted that if a plant is to be used as a food source as is the case of tomato, the all-nitrate source is the best alternative because it resulted in higher cadmium accumulation in the roots rather than translocation to the shoots and subsequently to the fruit. Likewise, Geilfus (2018) concluded from a review of the literature that if crops are cultivated on soils that are salinized by chloride, nitrate fertilization could be a strategy to suppress uptake of chloride by means of an antagonistic anion-anion uptake competition. He noted that high concentrations of chloride in the soil can increase phytoavailability of the heavy metal cadmium, accumulating in wheat (*Triticum aestivum* L.) grains above dietary intake thresholds.

2.2 Wastewater irrigation and fertilization

The use of nutrient-rich wastewater in the fertilization and irrigation of agricultural crops has been investigated as a way to increase crop productivity and product quality and to limit environmental pollution by recycling the wastewater rather than disposing of it. Also, due to the scarcity of water in many arid areas, considerable interest is given to recycling wastewater streams such as treated urban wastewater for irrigation in the agricultural sector. Concerns that arise from the use of wastewaters in irrigation are contamination of soils and vegetables with metallic elements and contamination of produce with bacteria.

Vivaldi et al. (2017) studied use of municipal wastewaters on production of nectarine (*P. persica* L.). They concluded that plots irrigated with the wastewaters showed positive impacts on soluble solids and acidity of fruits and early ripening, but fruits had lower mineral nutritional values. The effects were attributed to higher soil pH that lower micronutrients availability and to high concentration of nitrogen and potassium. Jorge et al. (2017) tested dairy cattle wastewater made up of washing water, urine, feces, and milk for the production of tomato. The highest tomato production and foliar nutrient accumulation were observed if 400% of the nitrogen amount recommended for tomato was used. Nutrient accumulation in leaves increased as a function of the nitrogen applied. The authors concluded that nitrogen fertilization can be provided with dairy cattle wastewater but that with a complementary mineral fertilization to supply plants with adequate amounts of phosphorus and potassium was needed. Martel-Valles et al. (2017) showed that irrigation with mixtures of fresh water and produced water from the gas industry of northern Mexico suppressed fruit, stem, and leaf fresh weight and suppressed minerals in plant tissues, except for sodium. However, plants irrigated with a mixture of produced and fresh water showed increased translocation of essential minerals from the leaves to the fruit and partially contributed to alleviate the negative effects of produced water on fruit yield. Almuktar and Schotz (2016) irrigated pepper with urban wastewater that was high in orthophosphate, ammonia, and potassium and in contamination levels of total coliforms, *Salmonella* spp., and *Streptococcus* spp. No elemental contamination was detected in the soils due to irrigation with the wastewater. Potassium and zinc accumulation in the yield of produce were elevated. Fruits harvested from plants irrigated by wastewater showed high contamination by total coliforms, especially for fruits that were located close to the contaminated soil surface.

Onal and Topcuoglu (2012) observed in a greenhouse experiment with strawberry (*Fragaria × ananassa* Duchesne) that if the application of sewage sludge increased mineral nutrient contents and heavy metal accumulation also increased. With the fruit, N, P, K, Mg, Fe, Zn, Cu, Pb, and Cd contents were increased, but Ca, Mn, Ni, and Cr concentrations did not change. The quantity of the accumulation of nutrients and heavy metals leaves was more than in the fruits. The authors concluded that as a result of the heavy metal accumulation, it is necessary to be careful about recycling of sewage sludge in agriculture.

Emongor et al. (2008) observed that tomato plants irrigated with secondary sewage effluent had higher fruit nitrogen, phosphorus, magnesium, calcium, iron, and sodium than plants irrigated with tap water. Tomato plants irrigated with secondary sewage effluent produced higher fruit yields and number of fruits per plants than plants irrigated with tap water. The fruits from plants irrigated with sewage water had no fecal coliforms. It was concluded that secondary sewage effluent can be useful for irrigation and fertilization of horticultural crops such as tomatoes as it contains a cheap source of available plant macronutrients and micronutrients.

2.3 Hydroponics

In hydroponic crop production, nutrients are in solution, and the nutrient supply is balanced and concentrated to supply nutrients to provide nutrient-rich foods from plants. Nutrient supply is not confounded by factors that limit nutrient availability in soil, and nutrient accumulation hydroponically grown fruits can be high. Hydroponics is studied also as a tool to investigate effects of nutrients in a medium on their entry into leaves and fruits.

Maboko and Du Plooy (2017) studied effects of concentrations of nutrient solutions in hydroponic production of tomatoes. Cherry tomatoes produced lower marketable yield at 25% or 50% strength nutrient solutions than with 75% or 100% strength solutions; Foliar fertilizer application did not have an effect on tomato fruit yield but improved the total plant dry weight. Fruit content of potassium, phosphorus, calcium, and zinc increased with increased solution strength. In cherry tomatoes, yield and fruit content were equal to 100% nutrition if the 75%-strength solutions were combined with foliar fertilizer application. The authors suggested that reduced-strength nutrient solutions can maintain yield and quality of tomatoes, whereas application of foliar fertilizer alone had a limited effect. The authors suggested that the results of their research may allow for a suppression of nutrient wastage and a cost saving.

Colla et al. (2004) studied nutrient accumulation in plant parts of zucchini squash (*C. pepo* L.) in hydroponics. Dry weight of plants or parts increased with increasing of the total concentrations of cations or anions in the nutrient solutions. The concentrations of each nutrient in plant parts increased as a consequence of increasing the proportion or concentration of the same nutrient in the medium. Nutrient concentrations of vegetative parts of plant increased more than those of fruits. Colla et al. (2003) evaluated the effects of substrates of rockwool, pumice, perlite, and coir on plant growth, yield, fruit quality, and mineral nutrient composition of cucumber (*Cucumis sativus* L.). Plants grown in coir, perlite, or pumice yielded more than those grown in rockwool. Fruit quality in terms of dry matter, soluble solid content, titratable acidity, and pH was not affected by treatments. Concentrations of potassium and sodium were highest in nutrient solution and fruits of coir treatment, whereas perlite and pumice treatments had highest Ca concentration in nutrient solution, leaves, and fruits. Phosphorus and magnesium were highest in nutrient solution and plant organs of the coir treatment.

Rubio et al. (2010) reported that pepper fruit appearance improved with Ca^{2+} addition to the root medium in hydroponics. Fertilization with K^+ increased fruit acidity and decreased maturity index, which could improve fruit storability. Low Ca^{2+} or high K^+ levels reduced root and shoot growth by dry matter measurements. Therefore, an adequate management of fertilization with Ca^{2+} and K^+ could improve the yield and quality of pepper fruits grown in soilless culture. Zahedifar et al. (2012) studied the effects of nitrogen nutrition and salinity on tomato growth and composition in hydroponics. With nitrogen supplied as ammonium chloride and monoammonium phosphate and salinity imparted as sodium chloride and calcium chloride, number of fruits and root length suppressed at high salinity levels. In the plant parts, phosphorus content was highest in fruits and lowest in roots. Phosphorus accumulation in fruit decreased with increased salinity applications with no nitrogen supplied. Nitrogen applications mitigated the detrimental effects of salinity at low levels of salinity but not at high levels. Nitrogen application decreased iron, zinc, copper, and manganese concentration. Application of nitrogen and salinity levels increased the citric acid content of tomato fruits. Vitamin C content of fruits was not influenced by nitrogen or salinity. Lima et al. (2018) studied the growth of sweet pepper (*C. annuum* L.) cultivars in saline hydroponics water (2.0–6.5 dS/m). The increasing salinity level of the nutrient solution suppressed fruit yield and fruit potassium, phosphorus, and sulfur and increased the sodium and chlorine concentrations in the leaves.

2.4 Organic fertilization

Many studies have been made to compare the composition of plants grown under organic or conventional methods of cultivation. Rajput et al. (2017) evaluated the combined effect of organic (farm manure or composted sugar cane waste) and inorganic (NPK) fertilizers on nutrient composition of banana (*Musa acuminata* Colla). Application of full NPK (500 N-250 P_2O_5-500 K_2O kg/ha) compared with N alone (500 kg/ha) increased fruit phosphorus from 0.10% to

0.16%, potassium from 1.07% to 1.85%, and zinc from 1.99 m/kg to 2.26 mg/kg. Other nutrients (Cu, Fe, and Mn) showed no substantial change with the regimes of fertilization. Fertilization with farm manures or composted press-mud (sugar mill solid waste) was compared with no fertilization. In the bananas, nitrogen was 0.88% with no fertilization, 1.19% with farm manure, and 1.32% with compost; P was 0.11% with no fertilization, 0.12% with farm manure, and 0.19% with compost; K was 0.82% with no fertilization, 1.45% with farm manure, and 1.29% with compost; Fe was 13.9 mg/kg with no fertilization, 19.4 mg/kg with farm manure, and 20.6 mg/kg with compost. Generally, the concentrations of these elements were not significantly different with fertilization from the manure or the compost. The concentrations of copper, manganese, and zinc were unaffected by the regime of fertilization.

Aliyu (2000) compared fertilization of pepper with livestock and poultry manure with the use of the manures supplemented with conventional nitrogen fertilizers. Mineral nitrogen supplementation resulted in higher concentrations of nitrogen, phosphorus, and potassium in pepper fruits. Bastos de Melo et al. (2016) fertilized peach trees with 0, 9, 18, 36, 72, or 144 L of organic compost per plant per year for 2 years. The application of compost to the soil increased the yields per plant and per hectare in the two treatments with the highest compost additions, but the organic compost had little effect on the composition of the fruit at harvest or after 30 days of storage. Elmann et al. (2016) studied the effects of organic and conventional fertilizers on the quality of sweet red peppers and noted that the levels of antioxidants, polyphenols, flavonoids, and vitamin C and the general quality of the peppers were not significantly affected by the different fertilization practices or by long storage.

Gastol and Domagala-Swiatkiewicz (2012) noted that organic or conventional husbandry influenced the mineral content of juices of apple, pear, or black currant (*Ribes nigrum* L.) with variation of effects occurring among the species. Organic apple juices had lower content of S, Na, Cu, B, and Ni than conventional fruit juices. Organic farming method gave higher accumulation of Ca, Mg, P, Na, Zn, Cu, B, Cd, and Ni in organic currant juices. With pear, the only difference was with magnesium content for which organic juices had lower amount of this element.

Peuke (2009) revealed no consistent association of soil nutrient composition on the elemental composition of grape leaves or grape juice suggesting that other soil characteristics in addition to the mineral supply influence the elemental composition of grape leaves and grape juice. Beltran-Gonzalez et al. (2008) observed that organic farming of mandarin oranges (*Citrus reticulata* Blanco) resulted in juices with higher contents of minerals and carotenoids and better sensory quality than conventional farming. They noted that mandarin orange juices can be considered as a good source of some nutrients such as potassium and antioxidant compounds.

Hargreaves et al. (2008) assessed the effectiveness of ruminant and municipal solid waste compost and compost teas made from these composts on selected soil, leaf, and fruit properties of raspberry (*Rubus* spp. L.). Foliar tea application was as effective as compost addition in raspberry production, but concentrations of K in leaf and fruits were lower in tea-treated raspberries. The yield, total antioxidant capacity, and vitamin C content of fruits were not affected treatment. Barker et al. (2017a,b) reported that mineral nutrient of cabbage (*Brassica oleracea* var. *capitata* L.) composition did not vary between modern or heirloom cultivars or among fertilization regimes with conventional chemical or organic fertilizers or with compost but varied among cultivars, suggesting that cultivar selection could lead to production of nutrient-rich cabbage. Neither mass of heads nor days to maturation of crops affected nutrient composition.

2.5 Fruiting disorders and nutrient composition

Mineral nutrients influence fruit quality in many ways including expressions of disorders that occur during development or storage of fruits (Table 2.3). Peet (2009) reviewed causes of physiological disorders of tomato fruits. Causes were associated with genetic susceptibility, environmental factors, watering practices, nutrition, and cultural practices such as pruning and training. The disorders were into groups according to causes into nutrient imbalances (blotchy ripening, and graywall); calcium in the fruit (gold fleck or speck and blossom-end rot); temperature extremes (catfacing, boat fruit, rough fruit, puffiness, and sunscald); genetics (green or yellow shoulder), and watering (cracking, russeting, rain check, and shoulder check). Nitrogen, potassium, phosphorous, calcium, and boron often are correlated to apple quality and disorders, but calcium likely has the largest impact on fruit quality (Ferguson and Boyd, 2002). Mineral analyses of leaves are useful to diagnose apple tree deficiencies but often are related poorly to fruit quality. Using analysis of fruit alone in combination with leaf analysis often permits more precise prediction of fruit quality than leaf analysis alone.

Fallahi et al. (2008) reported that increasing of fruit nitrogen of apple was related inversely to fruit yellow or red color and positively associated with fruit respiration and ethylene evolution. Calcium was related imprecisely to bitter pit and fruit firmness. Potassium fertilization with irrigation increased size, yield, acidity, and color of fruits but decreased firmness at harvest. Multiple sprays of soluble calcium often reduced bitter pit and usually increased

TABLE 2.3 Fruiting disorders caused by deficiencies of plant nutrients, element associated with the disorder, and fruits on which the disorder is common.

Disorder	Deficient element	Fruit	References
Bitter pit	Calcium	Apple	Dris and Niskanen (1999), Fallahi et al. (2008), Michalezuk (1999), Torres et al. (2017)
Internal breakdown	Calcium	Apple, mango, and others	Lin et al. (2012), Na et al. (2013), Raymond et al. (1998), Seo et al. (2015)
Water core	Calcium	Apple	Bakshi et al. (2005), Conway et al. (2002), Lord and Damon (1966), Yildiz et al. (2013)
Crease	Calcium	Orange	Storey et al. (2002)
Split fruit	Calcium	Apple	Opara and Tadesse (2000)
Blossom-end rot	Calcium	Tomato, pepper, mango, and others	Bar-Tal et al. (2001), Rached et al. (2018), Xiong et al. (2017)
Blotchy ripening	Potassium	Tomato	Hartz et al. (1999, 2005)
Yellow shoulder	Potassium	Tomato	Zhang et al. (2018)

calcium concentrations in cortical tissue. Early-season calcium-containing sprays often were more effective than later sprays at reducing bitter pit; however, later applications had a greater influence on fruit calcium concentration. The boron concentration of apple fruit was much more strongly affected by early-season boron sprays than was boron in leaves.

Michalezuk (1999) noted that low calcium content is a major cause of poor storage quality of apples. Apples with low calcium contents are susceptible to development of bitter pit, which can be controlled by increasing calcium contents of the fruit. This action was achieved by multiple spraying with calcium chloride in the summer and by summer pruning. Water core of apple is a physiological disorder that occurs with apples on trees. Water core is noted by water-soaked areas of the cortex and causes the tissue to become translucent. It may dissipate during storage. Calcium deficiency has been associated with development of water core (Bakshi et al., 2005; Conway et al., 2002; Lord and Damon, 1966; Yildiz et al., 2013).

Calcium is accepted widely as the main factor responsible for blossom-end rot (BER) in tomato fruit, but reactive oxygen species have been proposed to initiate BER appearance in tomatoes and other fruit-bearing vegetables (Rached et al., 2018). Rached et al. (2018) reported that ascorbate, the major antioxidant in tomato fruits, is generally lower in green fruit with BER than in non-BER fruit. If salt stress was applied, BER-resistant cultivars showed a significant increase in ascorbate, but pre-BER calcium concentrations did not associate with the BER susceptibility of the cultivars. Xiong et al. (2017) noted that blossom-end rot of tomato could not be prevented by culture of the plants in different media of coir, rockwool, or peat-vermiculite media. Tadesse et al. (1999) grew peppers in hydroponics with elevated electrical conductivity imparted by adding potassium chloride or calcium chloride to the basic nutrient solution. High nutrient conductivities resulted in high occurrence of blossom-end rot and were related to suppression of calcium absorption and enhancement of accumulation of potassium and magnesium in the fruit. Adding of calcium promoted the accumulation of calcium by the fruit and restricted the incidence of blossom-end rot.

Bar-Tal et al. (2001) with tomato grown in hydroponics reported that the total high-quality fruit yields increased with increasing nitrate-ammonium ratios but that these properties decreased with ammonium concentrations above 2 mmol/L. The total and high-quality fruit yields decreased sharply as the ammonium concentration in the solution increased above 2 mmol/L. An increase in the ammonium concentration in the solution caused a suppression of calcium concentration in the leaves and fruits and increased incidence of BER.

Zhang et al. (2018) noted that nutrient solutions enhanced with potassium limited the appearance of yellow shoulder disorder of tomato. Gunter (2010) linked yellow shoulder to low potassium supply during fruit development. An investigation indicated that potassium supplied at early flowering limited internal white tissue and increased the redness of fruit. Hartz et al. (1999, 2005) reported that the addition of potassium in irrigation water or to the soil improved color of tomato fruits but that foliar applications were ineffective or inconsistent.

Torres et al. (2017) discussed that fruit mineral analysis at harvest is a predictive method to assess the risk of bitter pit in apple. The authors evaluated multivariate analyses and linear correlations between calcium and bitter pit and suggested that 60 days after full bloom was the correct time to perform early mineral analysis. The calcium concentration in the fruits either at an early stage or at harvest was predictive of bitter pit development. The boron concentration showed a negative correlation with calcium concentration and a positive correlation with incidence of bitter pit. Other nutrients (nitrogen, potassium, and magnesium) showed little effect on the prediction models. The accuracy of the multivariate model was not better than analysis of calcium alone if the occurrence of bitter pit was high. A calcium threshold at 60 days after full bloom equal to or greater than 11.0 mg/100 g fresh weight indicated a low risk of bitter pit (<10% of incidence).

Dris and Niskanen (1999) studied the effects of preharvest sprays with calcium chloride on apple. The sprays increased firmness of fruits and titratable acidity but lowered soluble solids and incidence of physiological storage disorders. Makus and Morris (1998) noted, however, that foliar spray of calcium glutarate, soil incorporated gypsum, supply of calcium nitrate with irrigation, or a combination of these treatments did not affect nutrients except manganese in the fruit. Seo et al. (2015) associated a browning disorder of Asian pear (*Pyrus pyrifolia* Nak.) with nutrient content. They reported that calcium and potassium concentrations were high in the flesh and peel of severely browned fruits whereas magnesium was low in the peel and flesh. The K/Mg ratio was an effective predictor of the browning disorder, and the investigators recommended that the K/Mg ratio should be lower than 10 to avoid the disorder. Na et al. (2013) reported that fruiting shoots of Fuyu persimmon (*D. kaki* Thumb.) without browning had greater calcium and lower potassium concentrations than the shoots with browning fruits. Proximal flesh had a greater calcium concentration than middle and distal end parts in calyx-end browning fruits.

Storey et al. (2002) proposed a relationship between low rind calcium concentration of orange (*Citrus × sinensis* Osbeck) and crease disorder (albedo breakdown). They reported that concentrations of Ca in rind, albedo, and flavedo tissues were correlated negatively with incidence of crease. Pulp Ca concentrations also were negatively correlated, but ratios of K/Ca and Mg/Ca in pulp were correlated positively with the incidence of crease. Lin et al. (2012) showed that mango (*Mangifera indica* L.) fruit with lumpy tissue had the highest level of starch concentration (18.5%) and pitting necrosis and the lowest level of calcium content (0.008%). Raymond et al. (1998) noted differences in composition of young mango fruits and ripe mango fruits, but internal breakdown of the fruits was not linked to calcium deficiency and calyx-end splitting of apple fruits was related to nutrient content of the fruits during their development (Opara and Tadesse, 2000). Split fruit had higher concentrations of magnesium and potassium and lower concentrations of calcium and phosphorus than nonsplit fruits.

2.5.1 Rootstocks

Rootstocks and mycorrhizal associations have been studied as modifications of plant-soil relationships that may affect composition of foods derived from plants. Rootstocks influence the development of scions in several ways affecting traits such as vegetative vigor, stress tolerance, and fruit yield and quality, including mineral composition (Kullaj, 2018). Reighard et al. (2012) reported differences among rootstocks for peach (*P. persica* Batsch) scion leaf nitrogen, phosphorus, potassium, calcium, magnesium, zinc, manganese, sulfur, and sodium and also reported differences among rootstocks for fruit concentrations of phosphorus, manganese, and sulfur, with additional small differences occurring for potassium, magnesium, calcium, and copper. The authors suggested that differences in mineral accumulation among specific rootstock cultivars and species may be indicators of efficiency for absorbing and transporting nutrients to peach scion cultivars.

Kiczorowski et al. (2018) reported that a cultivar of apple grafted on a rootstock that gave the slowest growth of trees gave the highest concentrations of potassium, sodium, magnesium, and copper in apple flesh and peels. Other constituents such as dry matter, crude protein, fiber, easily hydrolysable sugars, and vitamin C also varied with type of rootstock. Nava et al. (2018) reported that the leaf calcium concentration of apple differed among rootstocks; however, no differences occurred among rootstocks in terms of calcium content in the fruits.

Djidonou et al. (2017) with tomato reported an increase in nitrogen, potassium, and calcium in grafted plants and an increase in plant vegetative growth and fruit yields. The enhancement in total nutrient accumulation was related to increased biomass rather than higher nutrient concentration. Accumulation of phosphorus was not influenced by the rootstocks. In general, grafting with the interspecific rootstocks maintained fruit soluble solid content and total titratable acidity, concentrations of vitamin C, carotenoids, and total phenolics at levels comparable with nongrafted plants.

Sharma et al. (2016) noted that grapefruit (*Citrus paradisi* Macf.) gave fruits with different soluble sugar and phenol contents and increased the accumulation of phosphorus, potassium, magnesium, copper, iron, and manganese with variation among rootstocks and cultivars of grapefruit. Dubey and Sharma (2016) noted variation in nutrient accumulation in leaves of lemon (*Citrus limon* Burm.) grafted on various rootstocks. Jaskani et al. (2014) also reported that scion or rootstock significantly affect leaf nutrient composition (N, Na, Cl, and B) of citrus.

Neilsen and Havipson (2014) reported that few problems occurred in achieving adequate leaf nitrogen, boron, and copper regardless of rootstock, whereas leaf zinc, phosphorus, and magnesium, and fruit calcium often did not achieve sufficiency with some rootstocks. Rootstocks were identified with superior and inferior abilities to accumulate individual nutrients, but few were superior for more than a single nutrient.

Motosugi et al. (1995) studied the effects of rootstocks and ammonium or nitrate nutrition on apple fruits. Fruits of trees grown with nitrate nitrogen were larger and had better surface color, higher soluble solids, higher levels of sugars, and higher acidity than those fed with ammonium nitrogen. Fruits from trees MM.106 or M.26 rootstocks treated with ammonium nitrogen had higher nitrogen but lower calcium concentrations than trees treated with nitrate nitrogen. However, with *Malus prunifolia* rootstocks, the form of nitrogen supplied had no influence on the surface color, juice quality, and mineral of fruit.

2.5.2 Mycorrhizal associations and fruit composition

The majority of plants have mycorrhizal fungi associated with them (Entry et al., 2002). Mycorrhizal fungi are agronomically significant because they form relationships with the roots of a host plant in a symbiotic association. The host provides the fungus with energy from soluble carbon sources, and the fungus provides the host with an increased capacity to absorb water and nutrients from the soil. Mycorrhizae can be particularly beneficial to the host plant in nutrient-poor soils. Mycorrhizae are important in extracting phosphorus from soils and providing phosphorus to the host plant (Smith et al., 2011). Mycorrhizae can impart resistance or tolerance to soil-borne diseases (Azcón-Aguilar and Barea, 1996), soil drought (Odunayo et al., 2014), and salt stress (Garg and Chandel, 2011) to plants. Rouphael et al. (2015) discussed the benefits of mycorrhizal fungi on plant health, nutritional value, and quality of horticultural crops and suggested future research on mycorrhizal fungi.

Bethlenfalvay et al. (1994) grew pea (*Pisum sativum* L.) plants in a greenhouse with different regimes of hydroxyapatite fertilization with or without the mycorrhizal fungus *Glomus mosseae* Gerd. and Trappe. The mycorrhizal infections suppressed the yield of peas. Lipid and protein contents of seeds were highly correlated with phosphorus content in all plants. Protein-concentration ratios did not differ in nonmycorrhizal plants. Ratios of mycorrhizal plants were highest at an intermediate P level and coincided with the highest intensity of root colonization and the greatest suppression of seed yield relative to the nonmycorrhizal plants at the same level of phosphorus supply. Different patterns of phosphorus accumulation in seeds and different relationships between seed phosphorus concentrations and phosphorus content and protein and lipid composition occurred in mycorrhizal and nonmycorrhizal plants. Bethlenfalvay et al. (1997) determined the effects of arbuscular mycorrhizal fungi on seed protein, lipid, and phosphorus and yield of soybean (*Glycine max* Merr.) grown under a high-nitrogen regime in pot culture with or without phosphorus fertilization. Phosphorus concentration was highest, and that of lipids was lowest in mycorrhizal plants. Seed yield and nutrient composition did not differ with intensity of root colonization. The seed protein-lipid ratio was highly correlated with seed phosphorus concentration.

Karagiannidis and Hadjisavva-Zinoviadi (1998) reported that inoculation increased improved plant vegetative growth and increased grain yield and seed size compared with uninoculated plants of durum wheat (*Triticum durum* Desf.). Phosphorus concentration in shoot tissue was increased by mycorrhizal association, but potassium, calcium, and magnesium accumulation was similar for mycorrhizal and nonmycorrhizal plants. The concentrations of manganese, zinc, iron, and copper were lower in the mycorrhizal plants than in the nonmycorrhizal plants. The phosphorus and magnesium concentrations of the grain of mycorrhizal plants were increased, but the calcium concentration was suppressed. The concentration of the heavy metals was either decreased (Mn, Fe, Co, Cr, Ni, and Pb) or was unchanged (Zn, Cu) by inoculation.

Karagiannidis et al. (2007) noted that the macronutrients phosphorus, potassium, and boron were higher in leaves of grapevines (*V. vinifera* L.) and that nonmycorrhizal plants had higher concentrations of zinc, manganese, iron, and copper. No differences occurred in soluble solids (degrees Brix) in fruits between among mycorrhizal and nonmycorrhizal plants. Al-Karaki and Hammad (2001) observed that shoot dry matter and fresh fruit yields were higher in tomato inoculated with arbuscular mycorrhizal fungi inoculated than uninoculated plants grown with or without salt stress from sodium chloride. Fruit contents of phosphorus, potassium, zinc copper, and iron were higher in inoculated plants than in uninoculated plants. Fruit sodium concentrations were lower in inoculated than uninoculated plants grown under salt-stressed conditions. The results of this study indicated that inoculated plants had greater tolerance to salt stress.

3 Future research

Adequacy of nutrients in diets for humans can be met in part by increasing the production of crops that are rich in nutrients and with biofortification of foods by enhancing the nutrient contents of fruits, vegetables, and grains.

Research presented in this chapter noted that cultivated varieties (cultivars) of fruits differed among species of plants used for foods. Through an enhanced review of literature on nutrient accumulation among cultivars, current knowledge can be identified. Further research on nutrient accumulation in plants will identify existing cultivars that produce nutrient-rich foods. Attention can be directed toward assessments of currently available cultivars to identify genetically based traits. This task will be large considering the many cultivars of crops that will need to be investigated and that the capacity to conduct the investigations will be limited by time and resources. However, from these investigations and from current knowledge, a list of nutrient-rich foods derived from plants can be made for immediate application. Plant-derived foods can serve as dietary sources of all essential elements required by humans, but nutrient mineral concentrations are low in some plants and in some parts of plants. Corn, wheat, and rice, which are staple food crops, are low in mineral nutrient contents. Much research has addressed ways to increase the mineral content of these foods, and these efforts have included classical breeding. Progress in the future should employ methods of biotechnology to bring about changes in nutrient improvements in the mineral status of plants.

Research should be directed toward enhancing nutrient absorption by plants. Delivery of nutrients from the soil into the plants must be studied. Studies of root morphology are needed as a step to initiate these investigations. Perhaps, selection or breeding of crops with extensive fibrous roots will raise nutrient absorption by plants. Rootstocks have been shown to affect nutrient absorption by crops. New investigations on the effects of rootstocks are needed, especially for fruits from perennial woody crops. These investigations might be extended to herbaceous, annual crops to develop models for nutrient absorption or to improve nutrient accumulation in these crops. Rootstocks and mycorrhizal associations have been studied as modifications of plant-soil relationships that may affect composition of foods derived from plants. Rootstocks influence the development of scions in several ways affecting traits such as vegetative vigor, stress tolerance, and fruit yield and quality, including mineral composition Differences in mineral accumulation among specific rootstock cultivars and species may be indicators of efficiency for absorbing and transporting nutrients. Economics of modifications with rootstocks must be evaluated for each type of plant, however.

The use and study of mycorrhizal fungal symbionts with agronomic and horticultural crops have increased in recent times. The hyphae of the fungi associate with roots, expand the root system, and enhance nutrient uptake by increased nutrient acquisition from soils. However, information of this enhancement on leaf and fruit composition was limited in the preparation of this chapter. Therefore, further research on farming practices including fertilization and pest control combined with inoculation with efficient mycorrhizal strains and selection of host-fungal combinations is needed. Most of the studies reported in the scientific literature were conducted under controlled conditions, and the response of the association may vary significantly in the outside environment and should be investigated.

Nutrient depletion of soils and inadequate fertilization contribute to insufficient accumulations of nutrients in plant parts that are consumed in diets of humans. Fertilizers are needed to maintain soil fertility to ensure that nutrition of crops is adequate to provide nutrient-rich foods. More information is needed on methods of fertilization to produce nutrient-rich foods and to determine if the amounts of fertilization to provide nutrients for optimum or maximum yields are sufficient to provide nutrient-rich produce or if fertilization should be extended to provide luxury consumption of nutrients. Fertilization for luxury consumption may in turn enhance nutrient accumulation in plant-derived foods. The economics of this development must be assessed. Hydroponic production of foods is a good tool to assess the potential for nutrient accumulation in crops since factors of soil properties do not limit accumulation of nutrients in the plants with hydroponic production.

Analysis of leaves usually is a better assessment of soil fertility than analysis of fruits is. The association of leaf analysis with content of nutrients in fruits is in need of further investigation. Potassium and magnesium are mobile in plants and are redistributed effectively within the plant. Calcium concentrations in fruits, seeds, and tubers are generally low because of the lack of distribution of this nutrient in the phloem and limited transpiration by these organs to effect xylem transport. Transport processes for redistribution of nutrients in plants need further investigation to identify and characterize mechanisms facilitating distribution of potassium, magnesium, and especially calcium between plant organs. Assessments of the effects of mineral fertilization and plant breeding strategies on nutrient accumulation are needed to develop genotypes of plants with effective mechanisms for transport of nutrient between leaves and fruits and tubers. Some elements, especially calcium as a nutrient in foods, are not transported from leaves to fruits due to the lack of movement of the elements in the phloem. Many fruits develop disorders of calcium deficiency due to inadequate transport of nutrients in the xylem and phloem and to fruits. Calcium concentrations in developing fruits decline with time due to dilution of the calcium with carbohydrates with fruit growth and due to limited transport of calcium into the fruits. Perhaps, through selection, breeding, and genetic engineering, transport of calcium in the phloem to fruits may be developed and will increase calcium in fruits so that fruits are a good source of mineral nutrition in diets. The interference of calcium oxalate on calcium transport to fruits may be restricted by modifications of plant metabolism to form malate, for example, instead of oxalate, which precipitates calcium. Genetic studies are needed to determine if calcium transport to fruits can be increased to facilitate accumulation of calcium in fruits.

Many strategies are available to improve the quality of plant-derived foods with respect to nutrient contents. Attention should be given to assessment and to employment of these strategies.

References

Aghili, F., Khoshgoftarmanesh, A.H., Afyuni, M., Mobli, M., 2009. Relationships between fruit mineral nutrients concentrations and some fruit quality attributes in greenhouse cucumber. J. Plant Nutr. 32 (12), 1994–2007.

Aliyu, L., 2000. Effect of organic and mineral fertilizers on growth, yield and composition of pepper (*Capsicum annuum* L.). Biol. Agric. Hortic. 18 (1), 29–36.

Al-Karaki, G.N., Hammad, R., 2001. Mycorrhizal influence on fruit yield and mineral content of tomato grown under salt stress. J. Plant Nutr. 24 (8), 1311–1323.

Almuktar, S.A.A.A.N., Schotz, M., 2016. Mineral and biological contamination of soil and *Capsicum annuum* irrigated with recycled domestic wastewater. Agric. Water Manag. 167, 95–109.

Amiri, M.E., Fallahi, E., 2007. Influence of mineral nutrients on growth, yield, berry quality, and petiole mineral nutrient concentrations of table grape. J. Plant Nutr. 30 (3), 463–470.

Amiri, M., Fallahi, E., Golchin, A., 2008. Influence of foliar and ground fertilization on yield, fruit quality, and fruit mineral nutrients in apple. J. Plant Nutr. 31 (3), 515–525.

Anonymous, 2018. Nutrient Rich. Heathy Eating With Joel Fuhrman MD and John Allen Mollenhauer. https://nutrientrich.com/1/aggregate-nutrient-density-index-andi-score.html (Accessed 1 January 1).

Appelhans, B.M., Milliron, B.-J., Woolf, K., Johnson, T.J., Pagoto, S.L., Schneider, K.L., Whited, M.C., Ventrelle, J.C., 2012. Socioeconomic status, energy cost, and nutrient content of supermarket food purchases. Am. J. Prev. Med. 42 (4), 398–402.

Arsenault, J.E., Fulgoni, V.L., Hersey, J.C., Muth, M.K., 2012. A novel approach to selecting and weighting nutrients for nutrient profiling of foods and diets. J. Acad. Nutr. Diet. 112 (120), 1968–1975.

Arsenault, J.E., Hijmans, R.J., Brown, K.H., 2015. Improving nutrition security through agriculture: an analytical framework based on national food balance sheets to estimate nutritional adequacy of food supplies. Food Sec. 7 (3), 693–707.

Ayaz, F.A., Colak, N., Topuz, M., Tarkowski, P., Jaworek, P., Seiler, G., Inceer, H., 2015. Comparison of nutrient content in fruit of commercial cultivars of eggplant (*Solanum melongena* L.). Pol. J. Food Nutr. Sci. 65 (4), 251–259.

Azcón-Aguilar, C., Barea, J.M., 1996. Arbuscular mycorrhizas and biological control of soil-borne plant pathogens—an overview of the mechanisms involved. Mycorrhiza 6 (6), 457–464.

Bakshi, P., Masoodi, F.A., Chauhan, G.S., Shah, T.A., 2005. Role of calcium in post-harvest life of temperate fruits: a review. J. Food Sci. Technol. Mysore 42, 1–8.

Barker, A.V., Eaton, T.E., Meagy, M.J., Jahanzad, E., Bryson, G.M., 2017a. Enrichment of mineral nutrient content of cabbage through selection of cultivars and soil fertility regimes. J. Plant Nutr. 40 (17), 1465–1474.

Barker, A.V., Eaton, T.E., Meagy, M.J., Jahanzad, E., Bryson, G.M., 2017b. Variation of mineral nutrient contents of modern and heirloom cultivars of cabbage in different regimes of soil fertility. J. Plant Nutr. 40 (17), 2432–2439.

Bar-Tal, A., Keinan, M., Aloni, B., Karni, L., Oserovitz, Y., Gantz, S., et al., 2001. Relationship between blossom-end rot and water availability and calcium fertilization in bell pepper production. Acta Horticult. 554, 97–104.

Bastos de Melo, G.W., Sete, P.B., Ambrosine, V.G., Freitas, R.F., Basso, A., Brunetto, G., 2016. Nutritional status, yield and composition of peach fruit subjected to the application of organic compost. Acta Sci. Agron. 38 (1), 103–109.

Beltran-Gonzalez, F., Perez-Lopez, A.J., Lopez-Nicolas, J.M., Carbonell-Barrachina, A.A., 2008. Effects of agricultural practices on instrumental colour, mineral content, carotenoid composition, and sensory quality of mandarin orange juice, cv. Hernandina. J. Sci. Food Agric. 88 (10), 1731–1738.

Benko, B., Borosic, J., Fabek, S., Toth, N., Voca, S., Poljak, M., 2011. Tomato quality and mineral content dependent on cultivar and nutrient solution composition. Acta Hortic. 960, 269–276.

Bethlenfalvay, G.J., Mihara, K.L., Schreiner, R.P., 1994. Mycorrhizae alter protein and lipid contents and yield of pea seeds. Crop Sci. 34 (4), 998–1003.

Bethlenfalvay, G.J., Schreiner, R.P., Mihara, K.L., 1997. Mycorrhizal fungi effects on nutrient composition and yield of soybean seeds. J. Plant Nutr. 20 (4–5), 581–591.

Bouis, H.E., 2001. Plant breeding: a new tool for fighting micronutrient malnutrition. J. Nutr. 132 (3), 491S–494S.

Bouis, H.E., Welch, R.M., 2010. Biofortification—a sustainable agricultural strategy for reducing micronutrient malnutrition in the Global South. Crop Sci. 50 (S-1), S20–S32.

Buwalda, J.G., Meekings, J.S., 1990. Seasonal accumulation of mineral nutrients in leaves and fruit of Japanese pear (*Pyrus serotina* Rehd.). Sci. Hortic. 41 (3), 209–222.

Caretto, S., Parente, A., Serio, F., Santamaria, P., 2018. Influence of potassium and genotype on vitamin e content and reducing sugar of tomato fruits. HortScience 53 (7), 2048–2051.

Chater, J.M., Garner, L.C., 2018. Foliar nutrient applications to 'Wonderful' pomegranate (*Punica granatum* L.). II. Effects on leaf nutrient status and fruit split, yield and size. Sci. Hortic. 242, 207–213.

Chater, J.M., Garner, L.C., 2019. Foliar nutrient applications to 'Wonderful' pomegranate (*Punica granatum* L.). I. Effects on fruit mineral nutrient concentrations and internal quality. Sci. Hortic. 244, 421–427.

Clark, C.J., Smith, G.S., 1990. Seasonal changes in the composition, distribution and accumulation of mineral nutrients in persimmon fruit. Sci. Hortic. 52 (1–2), 99–111.

Colla, G., Saccardo, F., Rea, E., Pierandrei, F., Salerno, A., 2003. Effects of substrates on yield, quality and mineral composition of soilless-grown cucumbers. Acta Hortic. 614, 205–229.

Colla, G., Rouphael, Y., Molle, A.L., Saccardo, F., Graifenberg, A., Giustiniani, L., 2004. Influences of mineral nutrition on growth and elemental composition of hydroponically grown zucchini plants. Acta Hortic. 644, 399–407.

Conway, W.S., Sams, C.E., Hickey, K.D., 2002. Proceedings of the International Symposium on Foliar Nutrition of Perennial Fruit Plants 594. pp. 413–419.

Covre, A.M., Partelli, F.L., Bonomo, R., Tomaz, M.A., Ramalho, C., 2018. Impacts of water availability on macronutrients in fruit and leaves of conilon coffee. Pesq. Agrop. Brasiliera 53 (9), 1025–1037.

Czech, A., Stepniowska, S., Wiacek, D., Sujak, A., Grela, E.R., 2018. The content of selected nutrients and minerals in some cultivars of *Cucurbita maxima*. Br. Food J. 120 (10), 2261–2269.

Darmon, M., Darmon, N., Maillot, M., Derenowski, A., 2005. A nutrient density standard for vegetables and fruits: nutrients per calorie and nutrients per unit cost. J. Am. Diet. Assoc. 105 (12), 1881–1887.

Davies, J.N., Hobson, G.E., McGlasson, W.B., 1981. The constituents of tomato fruit—the influence of environment, nutrition, and genotype. CRC Crit. Rev. Food Sci. Nutr. 15 (3), 205–280.

Davis, D.R., 2009. Declining fruit and vegetable nutrient composition: what is the evidence? HortScience 44 (1), 15–19.

Di Noia, J., 2014. Defining powerhouse fruits and vegetables: a nutrient density approach. Prev. Chronic Dis. 11, 130390. https://doi.org/10.5888/pcd11.130390.

Djidonou, D., Zhao, X., Brecht, J.K., Cordasco, K.M., 2017. Influence of interspecific hybrid rootstocks on tomato growth, nutrient accumulation, yield, and fruit composition under greenhouse conditions. HortTechnology 27 (6), 868–877.

Dong, J.L., Gruda, N., Lam, S.K., Li, X., Duan, Z.Q., 2018. Effects of elevated CO_2 on nutritional quality of vegetables: a review. Front. Plant Sci. 9, 924. https://doi.org/10.3389/fpls.2018.00924.

Drewnowski, A., 2005. Concept of a nutritious food: toward a nutrient density score. Am. J. Clin. Nutr. 82 (4), 721–732.

Drewnowski, A., 2009. Defining nutrient density: development of the nutrient rich food index. J. Am. Coll. Nutr. 28, 421S–426S.

Drewnowski, A., 2010. The nutrient rich foods index helps to identify healthy, affordable foods. Am. J. Clin. Nutr. 91 (4), 1095S–1101S.

Drewnowski, A., 2015. Nutrition economics: how to eat better for less. J. Nutr. Sci. Vitaminol. 61, S69–S71.

Drewnowski, A., 2017. Uses of nutrient profiling to address public health needs: from regulation to reformulation. Proc. Nutr. Soc. 76 (3), 220–229.

Drewnowski, A., Fulgoni, V.L., 2014. Nutrient density: principles and evaluation tools. Am. J. Clin. Nutr. 99 (5), 1223S–1228S.

Drewnowski, A., Rehm, C.C., 2013. Vegetable cost metrics show that potatoes and beans provide most nutrients per penny. PLoS One 8 (5), e63277. https://doi.org/10.1371/journal.pone.0063277.

Drewnowski, A., Fulgoni, V.L., Young, M.K., Pitman, S., 2008. Nutrient-rich foods: applying nutrient navigation systems to improve public health. J. Food Sci. 73 (9), H222–H228.

Dris, R., Niskanen, R., 1999. Calcium chloride sprays decrease physiological disorders following long-term cold storage of apple. Plant Foods Hum. Nutr. 54 (2), 159–171.

Dris, R., Niskanen, R., Fallahi, E., 1999. Relationships between leaf and fruit minerals and fruit quality attributes of apples grown under northern conditions. J. Plant Nutr. 22 (12), 1839–1851.

Dubey, A.K., Sharma, R.M., 2016. Effect of rootstocks on tree growth, yield, quality and leaf mineral composition of lemon (*Citrus limon* (L.) Burm.). Sci. Hortic. 200, 131–136.

Elmann, A., Garra, A., Alkalai-Tuvia, S., Fallik, E., 2016. Influence of organic and mineral-based conventional fertilization practices on nutrient levels, anti-proliferative activities and quality of sweet red peppers following cold storage. Isr. J. Plant Sci. 63 (1), 51–57.

Emongor, V.E., Macheng, B.J., Kefilwe, S., 2008. Effects of secondary sewage effluent on the growth, development, fruit yield and quality of tomatoes (*Lycopersicon lycopersicum* (L.) Karten). Acta Hortic. 944, 29–39.

Entry, J.A., Rygiewicz, P.T., Watrud, L.S., Donnelly, P.K., 2002. Influence of adverse soil conditions on the formation and function of Arbuscular mycorrhizas. Adv. Environ. Res. 7 (1), 123–138.

Fallahi, E., Simons, B.R., 1996. Interrelations among leaf and fruit mineral nutrients and fruit quality in 'Delicious' apples. J. Tree Fruit Prod. 1 (1), 15–25.

Fallahi, E., Fallahi, B., Neilsen, G.H., Neilsen, D., Peryea, F.J., 2008. Effects of mineral nutrition on fruit quality and nutritional disorders in apples. Acta Hortic. 868, 49–59.

Ferguson, I.B., 1980. Movement of mineral nutrients into developing fruits of the kiwifruit (*Actinidia chinensis* Planch). N. Z. J. Agric. Res. 23 (3), 349–353.

Ferguson, I.B., Boyd, L.M., 2002. Inorganic nutrients and fruit quality. In: Knee, M. (Ed.), Fruit Quality and Its Biological Basis. CRC Press, Boca Raton, FL, pp. 17–45.

Foote, J.A., Giuliano, A.R., Harris, R.B., 1999. Older adults need guidance to meet nutritional recommendations. J. Am. Coll. Nutr. 19 (5), 628–640.

Garg, N., Chandel, S., 2011. Effect of mycorrhizal inoculation on growth, nitrogen fixation, and nutrient uptake in *Cicer arietinum* (L.) under salt stress. Turk. J. Agric. For. 35 (2), 205–214.

Gastol, M., Domagala-Swiatkiewicz, I., 2012. Comparative study on mineral content of organic and conventional apple, pear and black currant juices. Acta Sci. Pol. Hortoru. Cultus 11 (3), 3–14.

Geilfus, C.M., 2018. Review on the significance of chlorine for crop yield and quality. Plant Sci. 270, 112–122.

Graham, R.D., Welch, R.M., Saunders, D.A., Ortiz-Monasterio, I., Bouis, H.E., Bonierbale, M., et al., 2007. Nutritious subsistence food systems. Adv. Agron. 92, 1–74.

Grusak, M., 2002. Enhancing mineral content in plant food products. J. Am. Coll. Nutr. 21 (3), 178S–183S.

Gunter, C.C., 2010. Potassium application timing and method for the reduction of yellow shoulder in processing tomato. Acta Horticult. 852, 291–296.

Gupta, U.C., Gupta, S., 2014. Sources and deficiency diseases of mineral nutrients in human health and nutrition: a review. Pedosphere 24 (1), 13–38.

Haghshenas, M., Arshad, M., Nazarideljou, M., 2018. Different K:Ca ratios affected fruit color and quality of strawberry 'Selva' in soilless system. J. Plant Nutr. 41 (2), 243–252.

Hallberg, L., 1981. A new concept applied in the interpretation of food iron-absorption data. Am. J. Nutr. 1, 123–147.

Hargreaves, J., Adl, M.S., Warman, P.R., Rupasinghe, H.P.V., 2008. The effects of organic amendments on mineral element uptake and fruit quality of raspberries. Plant Soil 308 (102), 213–226.

Hartz, T.K., Miyao, G., Mullen, R.J., Cahn, M.D., Valencia, J., Brittan, K.L., 1999. Potassium requirements for maximum yield and fruit quality of processing tomato. J. Am. Soc. Hortic. Sci. 124 (2), 199–204.

Hartz, T.K., Johnstone, P.R., Francis, D.M., Miyao, E.M., 2005. Processing tomato yield and fruit quality improved with potassium fertigation. HortScience 40 (6), 1862–1867.

Hunsche, M., Brackmann, A., Ernani, P.R., 2003. Effect of potassium fertilization on the postharvest quality of 'Fuji' apples. Pesq. Agrop. Brasileira 38 (4), 489–496.

Jani, R., Udipi, S.A., Ghugre, P.S., 2009. Mineral content of complementary foods. Indian J. Pediatr. 76 (1), 37–44.

Jaskani, M.J., Shafqat, W., Tahir, T., Khurshid, T., Ur-Rahman, H., Saqib, M., 2014. Effect of rootstock types on leaf nutrient composition in three commercial citrus scion cultivars of Pakistan under the ASLP Citrus Project. Acta Hortic. 1128, 131–136.

Javaria, S., Khan, M.Q., Bakhsh, I., 2012. Effect of potassium on chemical and sensory attributes of tomato fruit. J. Anim. Plant Sci. 22 (4), 1081–1085.

Jorge, M.F., Pinho, C.F., Nascentes, A.L., Alves, D.G., Almeida, G.V., Silva, J.B.G., Silva, L.D.B., 2017. Tomato fertigation with dairy cattle wastewater. Hortic. Bras. 35 (2), 230–234.

Kadir, S.A., 2004. Fruit quality at harvest of "Jonathan" apple treated with foliarly-applied calcium chloride. J. Plant Nutr. 27 (11), 1991–2006.

Karagiannidis, N., Hadjisavva-Zinoviadi, S., 1998. The mycorrhizal fungus Glomus mosseae enhances growth, yield and chemical composition of a durum wheat variety in 10 different soils. Nutr. Cycl. Agroecosyst. 52 (1), 1–7.

Karagiannidis, N., Nikolaou, N., Ipsilantis, I., Zioziou, E., 2007. Effects of different N fertilizers on the activity of Glomus mosseae and on grapevine nutrition and berry composition. Mycorrhiza 18 (1), 43–50.

Keatinge, J.D.H., Yang, R.Y., Hughes, J.D., Easdown, W.J., Holmer, R., 2011. The importance of vegetables in ensuring both food and nutritional security in attainment of the millennium development goals. Food Sec. 3 (4), 491–501.

Kennedy, E., Racsa, P., Dallal, G., Lichtenstein, A.H., Goldberg, J., Jacques, P., Hyatt, R., 2008. Alternative approaches to the calculation of nutrient density. Nutr. Rev. 66 (12), 703–709.

Kiczorowski, P., Kiczorowska, B., Krawiec, M., Kaplan, M., 2018. Influence of different rootstocks on basic nutrients, selected minerals, and phenolic compounds of apple cv. 'Sampion'. Acta Sci. Pol. Hortoru. Cultus 17 (4), 167–180.

Kolota, E., Balbierz, A., 2017. Effects of nitrogen fertilization on the mineral status of plants and the chemical composition of scallop squash fruits. J. Elem. 22 (4), 1255–1267.

Kullaj, E., 2018. Rootstocks for improved postharvest quality of fruits: recent advances. In: Siddiqui, M.W. (Ed.), Preharvest Modulation of Posthar-vest Fruit and Vegetable Quality. Academic Press, New York, pp. 189–207.

Labate, J.A., Breska, A.P., Robertson, L.D., King, B.A., 2018. Genetic differences in macro-element mineral concentrations among 52 historically important tomato varieties. Plant Genetic Resour. 16 (4), 343–351.

Lester, G.E., Jifon, J.L., Makus, D.J., 2010. Impact of potassium nutrition on postharvest fruit quality: melon (Cucumis melo L) case study. Plant Soil 335 (1–2), 117–131.

Lima, N.D., Silva, E.F.D.E., Menezes, D., Camara, T.R., Willadino, L.G., 2018. Fruit yield and nutritional characteristics of sweet pepper grown under salt stress in hydroponic system. Rev. Caatinga 31 (2), 297–305.

Lin, H.L., Shiesh, C.C., Chen, P.J., 2012. Physiological disorders in relation to compositional changes in mango (Mangifera indica L. 'Chiin Hwang') fruit. Acta Hortic. 984, 357–363.

Liu, R.H., 2013. Health-promoting components of fruits and vegetables in the diet. Adv. Nutr. 4 (3), 384S–392S.

Lord, W.J., Damon, R.A., 1966. Internal breakdown development in water-cored delicious apples during storage. Proc. Am. Soc. Hortic. Sci. 88, 94–97.

Lyne, J.W., Barak, P., 2000. Are Depleted Soils Causing a Reduction in the Mineral Content of Food Crops? https://soils.wisc.edu/facstaff/barak/poster_gallery/minneapolis2000a/ (Accessed 1 January 1).

Maboko, M.M., Du Plooy, C.P., 2017. Response of hydroponically grown cherry and fresh market tomatoes to reduced nutrient concentration and foliar fertilizer application under shadenet conditions. HortScience 52 (4), 57–578.

Maboko, M.M., Bertling, I., Du Plooy, C.P., 2013. Effect of arbuscular mycorrhiza and temperature control on plant growth, yield, and mineral con-tent of tomato plants grown hydroponically. HortScience 48 (12), 1470–1477.

Maboko, M.M., Du Plooy, C.P., Chiloane, S., 2017. Yield and mineral content of hydroponically grown mini-cucumber (Cucumis sativus L) as affected by reduced nutrient concentration and foliar fertilizer application. HortScience 52 (12), 1728–1733.

Maity, A., Babu, K.D., Sarkar, A., Pal, R.K., 2017. Seasonality of nutrients vis-a-vis fruit quality of pomegranate cv. Bhagwa on vertisol. J. Plant Nutr. 40 (9), 1351–1363.

Makus, D.J., Morris, J.R., 1998. Preharvest calcium applications have little effect on mineral distribution in ripe strawberry fruit. HortScience 33 (1), 64–66.

Marles, R.J., 2017. Mineral nutrient composition of vegetables, fruits and grains: the context of reports of apparent historical declines. J. Food Compos. Anal. 56, 93–103.

Martel-Valles, J.F., Benavides-Mendoza, A., Valdez-Aguilar, L.A., 2017. Mineral composition and growth responses of tomato (Solanum lycopersicum L.) plants to irrigation with produced waters from the oil industry. J. Plant Nutr. 40 (12), 1743–1754.

Mayer, A.-M., 1997. Historical changes in the mineral content of fruits and vegetables. Br. Food J. 99 (6), 207–211.

Michalezuk, L., 1999. The effect of summer pruning and calcium foliar spraying of 'Jonagold' and 'Sampion' apple trees on fruit storage quality. Acta Hortic. 485, 287–291.

Mobley, A.R., Kramer, D., Nicholls, J., 2009. Putting the nutrient-rich foods index into practice. J. Am. Coll. Nutr. 28 (4), 427S–435S.

Moore, L.L., Singer, M.R., Qureshi, M.M., Bradlee, M.L., Daniels, S.R., 2012. Food group intake and micronutrient adequacy in adolescent girls. Nutrients 4 (11), 1692–1708.

Mota, M.F.C., Pegoraro, R.F., dos Santos, S.R., Maia, V.M., Sampaio, R.A., Kondo, M.K., 2018. Contamination of soil and pineapple fruits under fertilization with sewage sludge. Rev. Bras. Eng. Agric. Ambient. 22 (5), 320–325.

Motosugi, H., Gao, Y.P., Sugiura, A., 1995. Rootstock effects on fruit-quality of Fuji apples grown with ammonium or nitrate-nitrogen in sand culture. Sci. Hortic. 61 (3–4), 205–214.

Munaretto, L.M., Botelho, R.V., Resende, J.T., Schwarz, K., Sato, A.J., 2018. Productivity and quality of organic strawberries pre-harvest treated with silicon. Hortic. Bras. 36 (1), 40–46.

Na, Y.G., Kim, W.S., Choi, H.S., 2013. Meteorological factors and calcium concentration on browning of a 'Fuyu' persimmon fruit. J. Plant Nutr. 36 (2), 259–274.

Nachtigall, G.R., Dechen, A.R., 2006. Seasonality of nutrients in leaves and fruits of apple trees. Sci. Agric. 63 (5), 493–501.

Nagar, A., Sureja, A.M., Kar, A., Bhardwaj, R., Krishnan, S.G., Das Munshi, A., 2018. Profiling of mineral nutrients and variability study in pumpkin (*Cucurbita moschata*) genotypes. Agric. Res. 7 (2), 225–231.

Nava, G., Dechen, A.R., Nachtigall, G.R., 2007. Nitrogen and potassium fertilization affect apple fruit quality in southern Brazil. Commun. Soil Sci. Plant Anal. 39 (1–2), 96–107.

Nava, G., Ciotta, M.N., Pasa, N.D., Boneti, J.I.D., 2018. Mineral composition of leaves and fruits of apple, 'Fuji' on different rootstocks of Sao Joaquim. SC. Rev. Bras. Frutic. 40, 1–10 doi:10-1590/0100-29452018685.

Neilsen, G., Havipson, C., 2014. 'Honeycrisp' apple leaf and fruit nutrient concentration is affected by rootstock during establishment. J. Am. Pomol. Soc. 68 (4), 178–189.

Nestby, R., Lieten, F., Pivot, D., Raynal Lacroix, C., Tagliavini, M., 2005. Influence of mineral nutrients on strawberry fruit quality and their accumulation in plant organs. Int. J. Fruit Sci. 5 (1), 139–156.

Nicklas, T.A., Drewnowski, A., O'Neil, C.E., 2014. The nutrient density approach to healthy eating: challenges and opportunities. Public Health Nutr. 17 (12), 2626–2636.

Nogueirol, R.C., Monteiro, F.A., de Souza Jr., J.C., Azevedo, R.A., 2018. NO_3^-/NH_4^+ proportions affect cadmium bioaccumulation and tolerance of tomato. Environ. Sci. Pollut. Res. 25 (14), 13916–13928.

Ochmian, I., Grajkowski, J., Mikiciuk, G., Ostrowska, K., Chelpinski, P., 2009. Mineral composition of high blueberry leaves and fruits depending on substrate type used for cultivation. J. Elem. 14 (3), 509–516.

Odunayo, J.O., Odebode, A.C., Babalola, B.J., Afolayan, E.T., Onu, C.P., 2014. Potentials of arbuscular mycorrhiza fungus in tolerating drought in maize (*Zea mays* L.). Am. J. Plant Sci. 5 (6), 779–786.

Onal, M.K., Topcuoglu, B., 2012. The effects of sewage sludge on the plant nutrients and heavy metal contents of strawberry (Fragaria x ananassa Duch) plant. Acta Hortic. 944, 87–91.

Opara, L.U., Tadesse, T., 2000. Fruit growth and mineral element accumulation in Pacific Rose (TM) apple in relation to orchard management factors and calyx-end splitting. J. Plant Nutr. 23 (8), 1079–1093.

Oyarzun, M.T., Uauy, R., Olivares, S., 2001. Food based approaches to improve vitamin and mineral nutritional adequacy. Arch. Latinoam. Nutr. 51 (1), 7–18.

Ozenc, N., Ozenc, D.B., 2014. Effect of iron fertilization on nut traits and nutrient composition of 'Tombul' hazelnut (*Corylus avellana* L.) and its potential value for human nutrition. Acta Agric. Scand. Sect. B Soil Plant Sci. 64 (7), 633–643.

Peet, M.M., 2009. Physiological disorders in tomato fruit development. Acta Hortic. 821, 151–159.

Peuke, A.D., 2009. Nutrient composition of leaves and fruit juice of grapevine as affected by soil and nitrogen fertilization. J. Plant Nutr. Soil Sci. 172 (4), 557–564.

Pinero, M.C., Perez-Jimenez, M., Lopez-Marin, J., del Amor, F.M., 2018. Fruit quality of sweet pepper as affected by foliar ca applications to mitigate the supply of saline water under a climate change scenario. J. Sci. Food Agric. 98 (3), 1071–1078.

Rached, M., Pierre, B., Yves, G., Matsukura, C., Ariizumi, T., Ezura, H., Fukuda, N., 2018. Differences in blossom-end rot resistance in tomato cultivars is associated with total ascorbate rather than calcium concentration in the distal end part of fruits per se. Hortic. J. 87 (3), 372–381.

Rajput, A., Memon, M., Memon, K.S., Tunio, S., Sial, T.A., Khan, M.A., 2017. Nutrient composition of banana fruit as affected by farm manure, composted pressmud and mineral fertilizers. Pak. J. Bot. 49 (1), 101–108.

Rampersaud, G.C., 2007. A comparison of nutrient density scores for 100% fruit juices. J. Food Sci. 72 (4), S261–S266.

Raymond, L., Schaffer, B., Brecht, J.K., Hanlon, E.A., 1998. Internal breakdown, mineral element concentration, and weight of mango fruit. J. Plant Nutr. 21 (5), 871–889.

Reighard, G.L., Bridges, W., Rauh, B., Mayer, N.A., 2012. Prunus rootstocks influence peach leaf and fruit nutrient content. Acta Hortic. 984, 117–124.

Rengela, Z., Batten, G.D., Crowley, D.E., 1999. Agronomic approaches for improving the micronutrient density in edible portions of field crops. Field Crop Res. 60 (1–2), 27–40.

Roosta, H.R., Hamidpour, M., 2013. Aquaponic and hydroponic systems: effect of foliar application of some macro- and micro-nutrients. J. Plant Nutr. 3 (13), 2070–2083.

Rouphael, Y., Franken, P., Schneider, C., Swartz, D., Giovanetti, M., Agnolucci, M., et al., 2015. Arbuscular mycorrhizal fungi act as biostimulants in horticultural crops. Sci. Hortic. 196, 91–108.

Rubio, J.S., Garcia-Sanchez, F., Flores, P., Navarro, J.M., Martinez, V., 2010. Yield and fruit quality of sweet pepper in response to fertilisation with Ca^{2+} and K^+. Span. J. Agric. Res. 8 (1), 170–177.

Scarborough, P., Raynor, M., 2014. When nutrient profiling can (and cannot) be useful. Public Health Nutr. 17 (12), 2637–2640.

Scarborough, P., Raynor, M., Stockley, L., 2007. Developing nutrient profile models: a systematic approach. Public Health Nutr. 10 (4), 330–336.

Scheidt, D.M., Daniel, E., 2004. Composite index for aggregating nutrient density using food labels: ratio of recommended to restricted food components. J. Nutr. Educ. Behav. 36 (1), 35–39.

Seo, H.J., Chen, P.A., Lin, S.Y., Choi, J.H., Kim, W.S., Haung, T.B., Roan, S.F., Chen, I.Z., 2015. The potassium to magnesium ratio enables the prediction of internal browning disorder during cold storage of Asian pears. Korean J. Hortic. Sci. Technol. 33 (4), 535–541.

Serio, F., Leo, J.J., Parente, A., Santamaria, P., 2007. Potassium nutrition increases the lycopene content of tomato fruit. J. Hortic. Sci. Biotechnol. 82 (6), 941–945.

Sharma, R.M., Dubey, A.K., Awasthi, O.P., Kaur, C., 2016. Growth, yield, fruit quality and leaf nutrient status of grapefruit (*Citrus paradisi* Macf.): variation from rootstocks. Sci. Hortic. 210, 41–48.

Silber, A., Naor, A., Cohen, H., Bar-Noy, Y., Hechieli, N., Levi, M., et al., 2018. Avocado fertilization: matching the periodic demand for nutrients. Sci. Hortic. 241, 231–240. https://doi.org/10.1016/j.scienta.2018.06.09.

Simon, P.W., 2012. Progress toward increasing intake of dietary nutrients from vegetables and fruits: the case for a greater role for the horticultural sciences. HortScience 49 (2), 112–115.

Smith, S.E., Jacobsen, I., Grunlund, M., Smith, F.A., 2011. Roles of arbuscular mycorrhizas in phosphorus nutrition: interactions between pathways of phosphorus uptake in arbuscular mycorrhizal roots have important implications for understanding and manipulating plant phosphorus acquisition. Plant Physiol. 156 (3), 1050–1057.

Solhjoo, S., Gharaghani, A., Fallahi, E., 2017. Calcium and potassium foliar sprays affect fruit skin color, quality attributes, and mineral nutrient concentrations of 'Red Delicious' apples. Int. J. Fruit Sci. 17 (4), 358–372.

Soliman, S.S., Alebidia, A.I., Al-Obeed, R.S., Al-Saif, A.M., 2018. Effect of potassium fertilizer on fruit quality and mineral composition of fig (*Ficus carica* l. cv. Brown Turky). Pak. J. Bot. 50 (5), 1753–1758.

Storey, R., Treeby, M.T., 2000. Seasonal changes in nutrient concentrations of navel orange fruit. Sci. Hortic. 84 (1–2), 67–82.

Storey, R., Treeby, M.T., Milne, D.J., 2002. Crease: another Ca deficiency-related fruit disorder? J. Hortic. Sci. Biotechnol. 77 (5), 565–571.

Tadesse, T., Nichols, M.A., Fisher, K.J., 1999. Nutrient conductivity effects on sweet pepper plants grown using a nutrient film technique 2. Blossom-end rot and fruit mineral status. N. Z. J. Crop. Hortic. Sci. 27 (3), 239–247.

Tan, Z.X., Lal, R., Wiebe, K.D., 2005. Global soil nutrient depletion and yield reduction. J. Sustain. Agric. 26 (1), 123–146.

Tekaya, M., Mechri, B., Cheheb, H., Attia, F., Chraief, I., Ayachi, M., Boujneh, D., Hammami, M., 2014. Changes in the profiles of mineral elements, phenols, tocopherols and soluble carbohydrates of olive fruit following foliar nutrient fertilization. LWT-Food Sci. Technol. 59 (2), 1047–1053.

Tomala, K., 1997. Predicting storage ability of 'Cortland' apples. Acta Hortic. 448, 67–73.

Torres, E., Recasens, I., Avila, G., Lordan, J., Jaume, S., 2017. Early stage fruit analysis to detect a high risk of bitter pit in 'Golden Smoothee'. Sci. Hortic. 219, 98–106.

United States Department of Agriculture, 2018. USDA Food Composition Data Bases. Agricultural Research Service. https://ndb.nal.usda.gov/ndb/ (Accessed 1 January 1).

Vashisth, T., Olmstead, M.A., Olmstead, J., Colquhoun, T.A., 2017. Effects of nitrogen fertilization on subtropical peach fruit quality: organic acids, phytochemical content, and total antioxidant capacity. J. Am. Soc. Hortic. Sci. 142 (5), 393–404.

Vivaldi, G.A., Stellacci, A.M., Vitti, C., Rubino, P., Pedrero, F., Camposeo, S., 2017. Nutrient uptake and fruit quality in a nectarine orchard irrigated with treated municipal wastewaters. Desalin. Water Treat. 71, 312–320.

Wang, Z.-H., Li, S.-X., Malhi, S., 2008. Effects of fertilization and other agronomic measures on nutritional quality of crops. J. Sci. Food Agric. 88 (1), 7–23.

Welch, R.M., Graham, R.D., 2004. Breeding for micronutrients in staple food crops from a human nutrition perspective. J. Exp. Bot. 55 (396), 353–354.

White, P.J., Bradshaw, J.E., Dale, M.F.B., Ramsay, G., Hammond, J.P., Broadley, M.R., 2009. Relationships between yield and mineral concentrations in potato tubers. HortScience 44 (1), 6–11.

Xiong, J., Tian, Y.Q., Wang, J.G., Liu, W., Chen, Q., 2017. Comparison of coconut coir, rockwool, and peat cultivations for tomato production: nutrient balance, plant growth and fruit quality. Front. Plant Sci. 8, 1327. https://doi.org/10.3389/fpls.2017.01327.

Ye, X.H., Zhao, Z.L., Zhou, Q., Zhao, H.B., Guo, C.M., Shi, L.L., Guo, J.W., 2014. Malabsorption of mineral nutrients and effects of foliar fertilization on continuously cropped *Capsicum annuum* L. var. *annuum*. Pak. J. Bot. 46 (5), 1781–1788.

Yildiz, H., Erturk, U., Yerlikaya, C., 2013. The effect of calcium treatments on fruit quality and some physiological disorders on Granny Smith (*Malus domestica* Borkh.) apple cultivar. Acta Hortic. 981, 691–697.

Zahedifar, M., Ronaghi, A., Moosavi, A.A., Shirazi, S.S., 2012. Influence of nitrogen and salinity levels on the fruit yield and chemical composition of tomato in a hydroponic culture. J. Plant Nutr. 35 (14), 2211–2221.

Zhang, Y.T., Suzuki, K., Liu, H.C., Nukaya, A., Kiriiwa, Y., 2018. Fruit yellow-shoulder disorder as related to mineral element uptake of tomatoes grown in high temperature. Sci. Hortic. 242, 25–29.

Zhou, Y., He, W.Z., Zheng, W.L., Tan, Q.L., Xie, Z.Z., Zheng, C.S., Hu, C.X., 2018. Fruit sugar and organic acid were significantly related to fruit Mg of six citrus cultivars. Food Chem. 259, 278–285.

Further reading

Casero, T., Benavides, A.L., Puy, J., Recasens, I., 2010. Interrelation between fruit mineral content and pre-harvest calcium treatments on 'Golden Smoothee' apple quality. J. Plant Nutr. 33 (1), 27–37.

Gurel, S., Sahan, Y., Basar, H., 2014. Antioxidative properties of olive fruits (*Olea europaea* L) from 'Gemlik' variety and relationship with soil properties and mineral composition. Oxid. Commun. 37 (4), 985–1004.

3

Nutrient redistribution in fruit crops: Physiological implications

Renato Vasconcelos Botelho*, Marcelo Marques Lopes Müller

Agronomy Department, State University of Midwestern Paraná—Unicentro, Paraná, Brazil
*Corresponding author. E-mail: rbotelho@unicentro.br

1 Introduction

Essential macroelements, such as nitrogen (N), phosphorus (P), potassium (K), calcium (Ca), magnesium (Mg), and sulfur (S), play important roles in plants as constituents of plant tissues or in buffer systems: they activate enzymes and regulate the osmotic pressure and membrane permeability (Marschner, 1995). Insufficient supply, inadequate phenological application, and macronutrient mobility can cause deficiency and stimulate disorders that are usually characteristic of that element (Tromp, 2005). An excess, on the other hand, can stimulate growth, causing an imbalance between nutrients (Marschner, 1995). In this way, it is important to know when certain macroelements are uptake during the different phenological stages and for which organs/tissues they are distributed.

Weinbaum et al. (2001) suggest that sequential excavation along with biomass determinations and nutrient analyses is the only method that can actually indicate seasonal patterns of nutrient uptake. Several studies have been conducted using this principle to determine the seasonal uptake of nutrients and distribution of macroelements in fruit plants, such as apple trees (Mason and Whitfield, 1960; Meynhardt et al., 1967), peach trees (Stassen et al., 1981, 1997), and pear trees (Stassen and North, 2005).

One of the techniques also used to evaluate the nutrient dynamics in agricultural crops is the use of isotopes. Isotopes are atoms of the same chemical element that have different numbers of neutrons in the nucleus but contain the same number of protons. Isotopes can be stable, and because they do not emit radiation, it occurs in nature more abundantly (^{39}K, ^{1}H, ^{12}C, ^{14}N, ^{16}O, and ^{32}S). There are also radioisotopes (^{40}K, ^{17}O, ^{18}O, ^{33}S, ^{34}S, and ^{36}S), which may be naturally occurring in small amounts in nature, or artificial, which are man-made in special apparatus. The latter, because they are generally heavier, are used as tracers in research with compounds containing the element of interest (Boaretto et al., 2004). The use of this technique in the research of plants allows the metabolic monitoring in these living beings. It is possible to observe the effect of microorganisms and also the uptake of nutrients using labeled fertilizers and to measure the amount existing and consumed by the plants (Pereira and Benedito, 2007).

A.K. Srivastava, Chengxiao Hu (eds.)
Fruit Crops: Diagnosis and Management of Nutrient Constraints
https://doi.org/10.1016/B978-0-12-818732-6.00003-4

In nature, stable N-element isotopes, for example, occur in almost constant proportions of ^{14}N (99.634%) and ^{15}N (0.366%) and are determined in the laboratory by mass spectrometry (Maximo et al., 2005). Even though several transformations of N at the same time occur in the soil, the use of the stable isotope ^{15}N allows to evaluate these individual transformations of N (Trivelin, 2002).

The use of stable isotopes demands higher costs; however, although the cost factor represents a disadvantage, this is overcome by the aspects of safety, absence of radioactive waste, and, mainly, reliability in the results in long-term experiments.

The tracer method with the stable isotope ^{15}N, used in studies of N transformation processes in soil, is classified as tracer ^{15}N technique, in which a substrate or source is labeled with ^{15}N, and it monitors in time the movement of the isotope in the system, in a qualitative and/or quantitative way. The expression nitrogen derived from the fertilizer source (Ndfs), obtained by principles of isotope dilution with ^{15}N, allows the identification of the contribution of the source of interest applied to the soil (Reichardt et al., 2009).

The use of the stable ^{15}N isotope technique allows accurate information on the dynamics of nitrogen in the soil-plant system. If we use a ^{15}N-labeled green manure, for example, the percentage and quantity of this nutrient derived from the green manure and its percentage in different parts of the crop can be determined in soil and planted crop in a sequence.

2 Nutritional requirements in fruit plants

Mineral nutrition is one of the most important factors in fruit production, and minerals are responsible for several functions in plants such as energetic processes, enzymatic activation, and osmotic regulation of membranes (Marschner, 1995; Faust, 1989). According to Kangueehi (2008), the nutritional requirement for fruit production is fundamental especially in high-density orchards.

2.1 Macronutrients

Among the macronutrients, nitrogen (N) is the element that plants require in greater quantity in relation to the other minerals. This is a component of the plant cell that is part of molecules of amino acids, proteins, nucleic acids, and other organic compounds and, therefore, plays an important role in the metabolic processes of plants (Souza and Fernandes, 2006). In peach trees, for example, Malavolta (1980) found that the macronutrient extracted in greater quantity was N, followed by K.

Available in soil, N is found in various forms including ammonium, nitrate, amino acids, peptides, and complex insoluble forms. In general, plants differ in their preference for N sources, but mainly, the uptake is in inorganic forms, such as nitrate (NO_3^-) or ammonium (NH_4^+) (Souza and Fernandes, 2006).

In the assimilation of N by the roots, the amino acids are transported to the leaves by transpiration flow via xylem (Marschner, 1995). However, it can also be transported through the plasma membrane of certain cells, in other forms, such as smaller peptides, purine bases, pyrimidines, and their derivatives (Gillissen et al., 2000). N deficiency in plants is characterized by reduced growth and light green coloration on leaves, and normally, symptoms are visible throughout the plant. The leaves have premature senescence. During leaf senescence, N is transported via phloem to perennial organs of the trees before foliar abscission (Castagnoli et al., 1990). In N-deficient plants, fruits are generally small and have early maturation. On the other hand, excess nitrogen predisposes the plant to premature fruit drop and reduces pulp firmness and period of cold storage (Basso et al., 2003). Moreover, N applications in summer should be avoided because it stimulates growth, which causes shading and delays the end of shoot growth, which negatively affects fruit quality (Kangueehi, 2008).

Phosphorus (P) leaf contents in most plants occur in concentrations between 0.1% and 0.4%, much lower than the levels normally found for N and K (Tisdale et al., 1993). The function of P is the constitution of structural macromolecules, and in most of the nucleic acids, it also participates in the transfer of energy (Araújo and Machado, 2006; Marschner, 1995). This element is not very mobile in the soil, because its availability is affected by the pH, the uptake capacity and soil fixation. The P forms uptake by the plants are HPO_4^{2-} and $H_2PO_4^-$, varying the availability according to soil pH (Basso et al., 2003). P deficiency is marked by delayed growth of plants and roots, delayed flowering, inhibition of sprouting of lateral buds, and decrease in number of fruits. The toxicity of P may be accentuated by Zn deficiency (Araújo and Machado, 2006). Hudina and Stampar (2002) report that foliar application of P results in "William's" pears with higher amounts of sugars and organic acids, which improves fruit quality.

In banana plantations, P is among the most exported nutrients in fruit production, accounting for about 56% of the total uptake (Lahav, 1995). Moreira and Fageria (2008) verified that except N, P, and Mn in the fruits, K, S, Fe, and Zn are contained in the pseudostem.

Potassium (K) is the most abundant cation in the cytoplasm of cells and plays a role in the activation of enzymes, protein synthesis, stomatal movement, photosynthesis, and cell extension. It is also the main cation involved in the establishment of cellular turgor and the maintenance of cellular electric neutrality (Taiz and Zeiger, 2009; Marschner, 1995). The K is mobile in the phloem, resulting in a good distribution for the fruit pulp. Studies indicate that fruit trees have higher K uptake per unit root dry mass than nonfruit trees (Kangueehi, 2008). In the soil, the availability of K is influenced by the exchange capacity of cations (ETC), organic matter, and fertilizers. High concentrations of K in soil inhibit Ca absorption, causing "bitter pit" in apples and pears (Basso et al., 2003). In studies with pears cv. Forelle, Stassen and North (2005) found higher amounts of K, Ca, and Mg in the roots when grafted on quince "A," compared with those grafted on the "BP1" pear hybrid. The results of K effects are similar in pears and apples, increasing quality attributes such as fruit size, flesh firmness, soluble solid content, sugars, and organic acids, yield in juice, taste, while the excess of K stimulates a series of physiological disturbances, such as "bitter pit" and "watercore" (Hudina and Stampar, 2002).

Sulfur (S) is incorporated into amino acids, proteins and coenzymes, and is also necessary for the synthesis of other compounds, including coenzyme A and vitamin B (Neilsen and Neilsen, 2003; Taiz and Zeiger, 2009). Sulfur uptake by the roots of the plants occurs almost exclusively in the form of sulfate ion (SO_4^{2-}), but in small amounts, sulfate can be absorbed by the leaves of plants; nevertheless, this element can be toxic at high concentrations (Tisdale et al., 1993). The process of storage and mobilization of this element requires seasonal changes in the regulation of sulfate uptake and its transport by xylem, which is partially independent of the stored sulfur (Herschbach et al., 2000).

Deficiency of S causes delayed growth and chlorosis in plants. Most of the time, S deficiency is mistaken for N deficiency. However, although similar, S deficiency symptoms occur first in the youngest leaves of plants (Tisdale et al., 1993; Marschner, 1995), while N is translocated to the new parts. In apple trees, Kangueehi (2008) found that the requirement of S is similar to the required amount of P. Comparatively with annual plants, in perennial plants, the regulation of sulfate nutrition appears to be much more complex.

Calcium (Ca) is the most important mineral element in postharvest conservation in fruits, especially apples and pears that are stored for long periods. It is also important in other fruit types that need high levels of Ca to ensure slow ripening and longer shelf life (Kangueehi, 2008). This mineral element is uptake by the plants in the form of Ca^{2+} ion and is found in the leaves in concentrations between 0.2% and 1.0%. It has an important function in the structure and permeability of cell membranes in cell division. In the presence of Ca also occurs the regulated uptake of other cations, due to its ionic charge (Taiz and Zeiger, 2009).

Ca is slightly mobile in the plant. Adequate N supply favors Ca uptake. Low pH and high availability of K and Mg induce the lack of Ca in the plant. However, the availability of B helps the translocation of Ca in the plant and for the fruits. The use of gypsum is a recommended practice as a complementary source of calcium to improve the Ca/Mg ratio in unbalanced soils (Basso et al., 2003). Ca deficiency is manifested by retarded growth and necrosis of shoots and apex of the roots. Low concentrations of Ca in the pulp of the fruits can cause physiological disturbances such as "bitter pit," "water core," and "internal breakdown" in pears and apples (Kangueehi, 2008; Marschner, 1995).

Magnesium (Mg) is a required mineral element by fruit plants in quantities lower than Ca^{2+}, and its uptake is reduced by competition with cations such as K^+, NH_4^+, Ca^{2+}, Mn^{2+}, and H^+ (Marschner, 1995). It is the main constituent of the chlorophyll molecule that plays an important role in the enzymatic activation of ribulose diphosphate carboxylase/oxygenase (rubisco) (Taiz and Zeiger, 2009; Malavolta, 1980). The concentration of Mg in the leaves increases during the last 6 weeks before the leaves fall, from the translocation of the permanent parts of the plant, when it is lost in the fall of the leaves. According to Kangueehi (2008), the loss of approximately 44% of the Mg present in the tree is linked to leaf fall.

2.2 Micronutrients

Micronutrients are considered essential elements for the growth of plants and are uptaken in small quantities (milligram per kilogram of dry matter of the plant), and spraying foliar or soil applications can guarantee the need of the plant. Most of the micronutrients are immobile in the phloem, and the symptoms appear mainly in the new leaves (Neilsen and Neilsen, 2003).

Copper (Cu) is located in the chloroplasts of plants where it participates in the photosynthetic reactions. It is also found in enzymes involved with protein and carbohydrate metabolism. Copper is uptake in the Cu^{2+} ionic form and moves slowly in the plant. Cu deficiency causes inhibition of xylem lignification, which can lead to wilting and winding of new leaves, with turgor decrease in the petioles and stalks, and chlorosis (Gil Salaya, 2000; Kangueehi, 2008). Plants rarely have Cu deficiency, and their availability is adequate in most soils. However, Cu deficiency can occur in plants grown on soils with low total Cu or in soils with high organic matter content. Among the micronutrients, Cu deficiency is the most difficult to diagnose due to the interference of other elements, such as P, Fe, Mo, Zn, and S. In citrus and other fruit trees, applications in excess of phosphate fertilizers can cause deficiency of Cu (Dechen and Nachtigall, 2006). Iron (Fe) is associated with chloroplasts where it plays a role in the synthesis of chlorophyll, electron transport chain (cytochrome and ferredoxin), N_2 fixation, and respiration. Its deficiency decreases chlorophyll and photosynthesis, affecting the activity of the enzymes like rubisco and others, in the respiration, and synthesis of proteins reduces the unloading of the cellular tissue in the phloem. In fruits, it strongly reduces its size and quality (Gil Salaya, 2000; Taiz and Zeiger, 2009). Plants exude substances that form complexes with iron for uptake, such as malic acid and others, which mobilize the iron from the soil to the roots. Fe^{3+} should be reduced to Fe^{2+} in the membranes of the absorbent cells (Taiz and Zeiger, 2009).

Manganese (Mn) is involved in several processes in the plant, such as oxidation (photolysis of water in chloroplasts), reduction of nitrate, and activation of several enzymes, including chlorophyll synthesis (Taiz and Zeiger, 2009). It is generally considered in immobile element in the plant; it is preferentially supplied to the young or growing tissue, thus manifesting its deficiency in the older leaves in the form of irregular spots on the margins between the ridges (Kangueehi, 2008).

The amount of zinc (Zn) is small in the plants; however, about 60 enzymes with Zn participation have been identified, and therefore, it plays important roles in the plant. It has great importance in the role of enzyme cofactor in the production of the indoleacetic acid (IAA) (Mengel and Kirkby, 1982). Zn deficiency results in short internodes, and leaves have a smaller size and internerve chlorosis. Sprouting is also deficient along the branches. The availability varies according to soil type, organic matter content, and pH (Basso et al., 2003).

3 Nutrient dynamics in fruit plants

The plants present variable nutrient demand throughout the phenological stages, and the knowledge of this dynamics is important to guarantee an adequate development of the fruit species. In addition, during the cycle of perennial plants, there are several translocations and storage of nutrients, which guarantee the nutritional balance of plant tissues. In this sense, several research works have already been carried out, seeking to understand these processes and provide subsidies for the proper management of orchards.

3.1 Nutrient uptake

Among several factors, the genetic can be determinant in nutrient uptake in fruit plants. According to Stassen and North (2005), "Forelle" pear trees presented higher levels of nitrogen and phosphorus in the leaves and of potassium, calcium, and magnesium in the roots when grafted on "A" quince, compared with those grafted on the pear hybrid "BP1." On the other hand, Singh et al. (2005) observed large variation in nutrient uptake among 13 selections of Asian pear trees grafted on "Kainth" rootstock.

In relation to the seasonality of the demand for nutrients, Terblanche (1972) found three periods of N and P uptake in apple trees: during the growth of the branches, 6–9 weeks before the beginning of the fall of the leaves and during the fall of the leaves. Stassen et al. (1981) reported two periods in which N increased rapidly throughout the plant. The first period started 3 weeks before sprouting and reached a maximum of 3 weeks prior to stall growth, while the second period was 3 weeks before the end of leaf fall until 3 weeks after leaf fall.

In apple trees, Meynhardt et al. (1967) observed a relatively sharp increase in the total P content of the root system from 14 days after dormancy release until the growth of the branches stopped. This was followed by a fainter increase until the beginning of the leaf fall. Mason and Whitfield (1960) found that P uptake began at the equivalent of the end of October in the Southern Hemisphere and peaked in January. Conradie (1981) observed two periods of P uptake in vines: shortly after the of dormancy release until the beginning of berries ripening and again after harvest until the period of leaf fall.

Mason and Whitfield (1960) reported that K, Ca, and Mg were uptake during most of the growth period of the branches. Terblanche (1972) found that Ca and Mg were uptake during the same period as N and P; however, the patterns were different. Ca and Mg were uptake at a higher rate at 9 weeks before the beginning of leaf fall and at a slower rate at the leaf fall when compared with N and P. In addition, these same authors reported two periods of K: the first was during most of the growth stage of the branches, and the second was during the fall of the leaves. In grapevines, Conradie (1981) found only one period of K uptake (shortly after dormancy release to harvest), however, there were two Ca and Mg uptake peaks: the first from dormancy release to berry ripening; another less noticeable during the 6-week period before leaf fall for Ca and during leaf fall period for Mg.

3.2 Nutrient redistribution in the plant

Most of the nutrients exported from the soil are directed to the fruits, so the need to supply the nutrients of the soil is related to the yield of the crops. It should be noted that leaching losses can also occur, especially under conditions of high precipitation or irrigation; in addition, nutrients are also removed by pruning, and a certain amount is fixed in the permanent structure of the tree (Kotzè, 2001; Stassen and North, 2005).

According to Verlindo et al. (2014), in pear trees of cvs. Cascatense and Tenra, the highest amount of K was immobilized in the roots and in the branches and smaller amounts in the trunk. Ca was less accumulated in the branches. The N, P, Mg, and S elements did not present significant differences between the different permanent parts (Table 3.1). The macronutrients exported by the fruits were in the following order $K>N>P=S=Mg=Ca$ and $N>K>P=S=Mg=Ca$, for the cultivars Cascatense and Tenra, respectively.

Nitrogen assimilated by leaves is stored as protein, but cannot be mobilized and translocated from apple leaves before their fall (Neilsen and Neilsen, 2003). Titus and Kang (1982) and Millard (1996) reported that 23%–50% of the nitrogen is redistributed from the leaves before senescence, as soon as the growth of the branches occurs or as soon as the leaves fall, but occurs predominantly from 3 to 4 weeks before the leaves fall. Different authors reported that N, P, and K are redistributed from the leaves to the permanent parts before the end of leaf fall and stored as reserves to promote new growth in the next cycle when the leaf and root activities are not enough developed to support the new growth (Terblanche, 1972; Conradie, 1981; Titus and Kang, 1982; Millard, 1996; Neilsen and Neilsen, 2003).

In a study by Stassen and North (2005), in pear trees cv. Forelle, the highest N contents were found in the leaves. Therefore, the application of N must be during the autumn resulting in its accumulation in permanent parts of the tree as reserves and redistributed at the beginning of the season to the new flow of growth (Stassen et al., 1981). In other crops, as in peach trees, the total N used for new growth during the first 25–30 days from sprouting comes from the remobilization of the permanent parts (Rufat and Dejong, 2001).

According to Leão (2001), nutrient contents and nutritional status of the grapevine can be influenced by climate, soil, vineyard management, cultivars and rootstocks, sanitary status and fertilization (Giovannini, 1999). According to

TABLE 3.1 Yearly macronutrients mobilization (g/ha) by permanent parts of pear trees cv. Cascatense and Tenra, in density planting of 2500 plants/ha (Guarapuava-PR, Brazil, 2011).

Plant parts	Dry weight (t/ha)	N	P	K	Ca	Mg	S
ROOTS							
Cascatense	3674.2	4415.4	410.1	3052.0	3345.4	886.6	485.0
Tenra	3096.5	4184.7[ns]	405.1[ns]	3050.8[ns]	4673.3[a]	1026.2[ns]	430.5[a]
Trunk							
Cascatense	5912.9	5150.3	298.7	1459.9	3250.7	930.6	534.6
Tenra	5731.7	5482.2[ns]	361.4[a]	1674.4[ns]	2965.6[ns]	704.8[a]	520.6[ns]
BRANCHES							
Cascatense	3790.1	4585.7	485.4	3074.3	1598.3	482.2	371.4
Tenra	2902.0	4096.5[ns]	461.8[ns]	2472.0[a]	1914.2[a]	530.2[ns]	316.2[ns]
Total							
Cascatense	13,377.3	14,151.3	1194.2	7586.3	8194.4	2299.3	1391.0
Tenra	11,730.4[ns]	13,763.4[ns]	1228.4[ns]	7197.2[ns]	9553.1[a]	2261.2[ns]	1267.4[a]

[a] Meaningful at 5% probability.
ns, no significance. The values are averages of seven vegetative cycles.

Brunetto et al. (2005), the initial sprouting of the grapevine has an important contribution of the N element, which is responsible for the initial growth, with most of this element coming from the roots. There is also the use of N reserves at other stages of vine development. Using ^{15}N-labeled fertilizer, it was determined that the N accumulated in the fruits was derived first from the N stored in roots and mature shoots (Conradie, 1981). The amount of N remobilized of the roots, trunk, and other permanent structures is dependent on the age of the plants, time of the year, and conditions of growth.

Brunetto et al. (2006) observed that during the vegetative growth of vines cvs. Riesling Itálico and Chardonnay, a part of the N uptake by the roots was transported to the growth points and the rest was used to replenish the reserves of N in the perennial parts. Similarly, Gazolla Neto et al. (2006) reported that the N uptake by the vines can be redistributed and/or stored in the perennial parts of the plant, preferably in the roots, for the next cycle. Thus, it is important that the N supply occurs during the period when the plants show increased dry matter in roots and leaves. Natale and Marchal (2002) showed that N uptake efficiency varied according to mineral fertilizer source, and ammonium sulfate was more efficient than urea at 12 and 20 days after application, but there was no change in weight of dry matter and in the amount of N of the plants of *Citrus mitis* Bl.

In grapevines, absolute amounts of K in the roots are smaller than N, although in the trunk and in the branches, K is generally larger than N. Only a small amount of K is remobilized from the roots, but none of the K on the trunk and branches is remobilized for other organs. The fruits are the largest drains of K after beginning the growth of the berries. Several studies have shown that the remobilization of K from the leaves to the fruits occurs if the canopy of the vine is extremely dense. Other studies have shown that there may be some redistribution of K from the main axis of the branch to the bunches. However, most of the K found in fruits is uptake from the soil (Mullins et al., 2000). In perennial plants, it is usually assumed that a considerable fraction of the mineral content of senescent leaves is translocated to the permanent structures. In vines, the leaves contain a large proportion of the contents of N total, but little is remobilized for the trunk or for the roots. Less than 5 g N/plant was translocated back to the vine, although approximately 30 g N/vine (33 kg/ha) remained in the fallen leaves (Williams et al., 1987).

3.3 Nutrient recycling by green manure

The cultivation of green manure as a soil cover is a practice adopted worldwide, using winter or summer species. Several species are used as green manure and leguminous are noteworthy because they develop symbiotic associations with atmospheric nitrogen-fixing bacteria that are available to the soil (Teodoro et al., 2011). These plants used as soil cover also contribute to the reduction of infestation by invasive plants, besides providing improvements in the chemical, physical, and biological properties of the soil (Muzilli et al., 1992). These species can be used in isolation or with several species together, which is called a green manure consortium. According to Derpsch and Calegari (1992), the use of soil cover crops as long as possible throughout the year is essential to reduce erosion, increase soil water infiltration, and nutrient cycling.

Several studies have demonstrated the importance of green manure as nutrient recyclers in agroecosystems (Wutke et al., 1998; Faria et al., 2007; Teodoro et al., 2011). Faria et al. (2004), in a research carried out with the cultivation of leguminous species in a vineyard of cv. Italia, verified that green manure increased the cation exchange capacity of the soil, the organic matter content, and exchangeable calcium in the 0–10 cm layer. However, the effects of green manuring were not consistent with grape quality and productivity.

Barradas et al. (2001), in an experiment with eight species of green manure, observed that white lupine and common vetch were the species that accumulated the highest total N in the whole plant, with amounts of 251.6 and 228.1 kg/ha, respectively. Black oat accumulated in the aerial part, little more than 100 kg/ha. When the accumulation of N in the roots was evaluated, annual ryegrass was one of the most outstanding species, reaching 37.1 kg/ha. A similar result was obtained by Aita et al. (2004) that found higher N availability in the soil after common vetch cultivation than with black oat, and this result was attributed to the low C/N ratio of legumes.

In orchards, the practice of cultivating green manure between the lines of plants aims to protect the soil from erosion and to improve its fertility. Paulino et al. (2009), working with biological fixation of nitrogen in organic mango and soursop orchards, evaluated that the Gliricidia had a higher capacity of biological fixation of nitrogen than the crotalaria and that these species can add to the system amount of N higher than required by these fruit species. Sunflower seedlings transferred to soursop 22.5% of N and Gliricidia 40%.

Seyr (2011), evaluating the production of dry mass produced and the nutrient content of the green manure cultivated between the lines of banana cv. Nanicão, found that black oats in intercropping cultivation were the ones that produced the most dry mass. Green manures (black oats, hairy vetch, blue lupine, and fodder turnip) were the ones that recycled the most N, P, and K of the soil, and that these coverages provided some of the nutrients needed for banana development, reducing production costs. Wutke et al. (2005) observed that planting between lines, different species of green manure, such as black oats, peas and lupins in autumn/winter, and dwarf muck in summer on the interlines of a vineyard, did not show negative effects on total soluble solids pH and titratable acidity in "Niagara" grapes.

4 The role of silicon in fruit plant nutrition

Silicon (Si) is the second most abundant element in the Earth's crust, losing only to oxygen. It occurs mainly as an inert mineral from the sands, quartz (SiO_2 pure), kaolinite, micas, feldspar, and other silicate clay minerals (Mengel and Kirkby, 1982). In the soil solution, Si is present as monosilicic acid (H_4SiO_4) largely in the undissociated form that is easily uptake by plants (Raven, 1983). Thus, in spite of an acid, the Si has basic behavior, being represented by some authors as $Si(OH_4)$ instead of H_4SiO_4 (Mengel and Kirkby, 1982; Savant et al., 1997).

The main soil factors that can influence Si soil content are as follows: mineralogical and textural composition, nutrient cycling process, soil acidity, and predominance of ions in the solution. Although not an essential element, it is a micronutrient beneficial to several plants, providing greater protection to environmental, biotic and abiotic stresses, such as pest and disease attack and resistance to water stress (Bertalot et al., 2008). Silicon fertilization has been used in several countries and is considered beneficial not only in conventional agriculture but also in organic and biodynamic agriculture, and its uptake can benefit many crops (Korndorfer et al., 2001).

Si uptake from the soil by the plant is done passively, by mass flow, accompanying water uptake, in the form of monosilicic acid (H_4SiO_4) (Yoshida, 1965). During the active process, water is lost through transpiration, and Si is deposited in plant tissues, and no translocation occurs to newer tissues (Miyake and Takahashi, 1983; Ma and Takahashi, 1990). According to Epstein (1994), this element accumulates in the tissues of all plants and represents from 0.1% to 10% of the dry matter of the plants, concentrating on the support tissues of the stem and leaves (Ma et al., 2001).

In rice plants using the nuclear magnetic resonance technique, it was possible to identify that the form of Si translocated by xylem is monosilicic acid whose concentration in the xylem is transiently high (Mitani et al., 2005) and its distribution is not uniform in the different parts of the plant. In addition, many of the silicon found in plants are polymerized as $SiO_2 \cdot nH_2O$ and can be found in the cell wall, lumen of cells, and intercellular spaces (Sangster et al., 2001). However, considerable amounts of Si can be found in organic compounds because they are bind to proteins, lipids, phenols, and polysaccharides (Fauteux et al., 2005).

Generally, the Si is deposited in tissues where the loss of water is greater, considering that this fact is connected indirectly to the passive transport of this element. However, many studies indicate an active transport system for Si in roots of rice plants involving membrane-carrying proteins (Ma et al., 2004). Another work where the passive Si hypothesis is contested shows that in cucumber plants the absorbed Si content is twice as high as the transpiration rate, thus not having much relation between the mechanisms of transpiration and uptake (Liang et al., 2005).

Plants can be classified as accumulators and nonaccumulators of Si, according to their abilities, and are evaluated according to the molar ratio Si/Ca found in the tissues. In ratios above 1.0, the plants are considered accumulators (rice, sugar cane, wheat, sorghum, and grasses in general); between 1.0 and 0.5, are considered intermediate (cucurbitaceous and soybean); and less than 0.5, nonaccumulators (dicotyledons in general) (Miyake and Takahashi, 1983; Ma et al., 2001).

The Si gives several benefits for the plants, including a greater tolerance of the plants to the attack of insects and diseases, reduction in the transpiration, and greater photosynthetic rate by the improvement of the foliar architecture (Rodrigues et al., 2003). Under conditions of water stress, Si can induce an increase in the rate of superoxide dismutase enzyme activity, thus reducing stress (Schmidt et al., 1999). This element can also act in the stomatal movement in response to environmental stimuli by regulating the water potential of the epidermal cells (Agarie et al., 1998). According to Oliveira and Castro (2002), the accumulation of Si in the leaf also causes reduction in the transpiration and causes that the requirement of water by the plants is smaller, due to the formation of a double layer of silica, which diminishes transpiration by diminishing the opening of the stomata limiting the loss of water.

Cucumber plants grown in nutrient solution, with a concentration of 100 mg/L of SiO_2, presented an increase in chlorophyll content, greater leaf mass (dry and fresh), delay in senescence, and greater stiffness of older leaves (Adatia and Besford, 1986). In a trial with grapevines cv. Blue Bangalore, plants treated with 4 mL/L of SiO_2 showed a longer leaf length, greater leaf area, and higher chlorophyll content. The number of bunches per plant and productivity were higher with the dose of 6 mL/L (Bhavya et al., 2011). In the micropropagation of strawberry with three different sources of Si in the concentration 1 g/L, the propagules showed an increase in dry and fresh mass, chlorophyll content, leaf tissue thickness, epicuticular wax deposition, and the formation of silicon deposition in cells, regardless of the source used (Braga et al., 2009).

The use of Si in the fertilization confers greater structural rigidity to the tissues, making it difficult to enter pathogenic fungi hyphae and the attack of phytophagous insects, besides increasing the concentration of phenolic compounds, acting not only as a physical but also as a biochemical barrier (Korndorfer et al., 2002). According to Epstein (1994), Si can provide changes in leaf anatomy that function as mechanical barriers, with thicker epidermal cells. This happens due to its accumulation in the cells of the epidermis. The effect of Si when incorporated into the cell wall is similar to that of lignin, which is a structural component resistant to compression. Thus, the plant acquires a better leaf architecture, which provides greater penetration of sunlight, greater CO_2 uptake, decreasing transpiration, and increasing the photosynthetic rate (Korndorfer et al., 2002; Korndorfer and Datnoff, 2000).

According to Terry and Joyce (2004), Si can induce some mechanisms of resistance in plants, such as the synthesis of phenolic compounds, lignin, suberin, and callose in the cell wall. Some species, such as tomato and vine, are poorly capable of transporting the silicon from the root system to the shoot; however, Bowen et al. (1992) have shown that Si applied via foliar can control diseases in these crops. In a greenhouse experiment with the hybrid grapevine cultivar LN33, whose treatments were two concentrations of K_2HPO_4 solution (K = 7.4 mM and PO_4 = 5.5 mM and K = 22 mM and PO_4 = 16 mM) and a solution of potassium silicate containing 17 mM Si (2.6 g/L) in vine plants inoculated with *Uncinula necator*, a 14% reduction in the number of powdery mildew colonies was observed in plants sprayed with Si, 24 h prior to inoculation. By means of electron microscopy the authors verified a thick layer of Si on the surface of the pulverized leaves, preventing the growth of the hypha of the pathogen. The reduction of vineyard mildew severity was due to the physical barrier found on the leaf.

Ferreira et al. (2013) verified a reduction of up to 85% in the incidence of bacteriosis (*Xanthomonas arboricola* pv. *pruni*) in plum trees cv. Pluma 7, by biweekly foliar applications of SiO_2 in doses between 1 and 8 g/L (Fig. 3.1), possibly due to the deposition of Si in the cell wall, acting as a physical barrier that hinders bacterial infection, besides inducing some mechanisms of resistance in plants, such as the synthesis of phenolic compounds. With papaya plants, foliar applications of organomineral fertilizer and silica clay at doses of 2 mL and 4 mL/L, respectively, a 20% reduction of incidence of black papaya caused by *Asperisporium caricae* (Speg) Maubl. (Pratissoli et al., 2007). Ribeiro Júnior et al. (2006) verified control if wilt-inducing fungus in cacao (*Verticillium dahliae*) with 1.5 mL/L potassium silicate.

5 Association with microorganisms

Little is known about the action of the factors that promote the growth caused by microorganisms of the soil in the plants. According to Harman et al. (2004), fungi are capable of acting as biostimulants of root growth, promoting root development through phytohormones, thus improving nutrient assimilation and increasing resistance to unfavorable biotic factors, besides degrading sources of nutrients that will be important for the development of the vegetable. Because of a reduced amount of absorbent hairs, blueberry roots form symbioses with arbuscular mycorrhizal fungi (AMF), which receive energetic substrates produced by the plant in exchange for increased absorption of nutrients. This association occurs in a large number of plant species in natural habitats (Wilcox, 1996).

In blueberry orchards in southern Brazil, species of the genus *Acaulospora* and *Glomus* were identified as the most commonly found AMF (Farias, 2012). In the same region, Pinotti et al. (2011) found that the most frequent mycelial fungi observed in the soil samples of blueberry, blackberry, and raspberry were *Penicillium* sp., *Aspergillus* sp., *Clonostachys rosea*, and *Trichoderma* sp.

These fungi are also important to release and provide greater utilization of nutrients for associated plants (Hobbie and Horton, 2007). In addition to the higher vegetative growth, plants with *Trichoderma harzianum* showed higher concentrations of micronutrients (Cu, Zn, Mn, and Fe) in the roots (Yedidia et al., 2001). The increase in phosphorus uptake was also verified in coffee plants, in addition to larger total dry matter and foliar area, in those grown on conditioned substrates with *Trichoderma asperellum* (Jesus et al., 2011).

An alternative when there is no natural occurrence of AMF is the inoculation of the soil or the planting substrate of the seedlings. Camargo et al. (2010) verified that the inoculation of biological agents, composed of certain bacteria and fungi, favors a higher dry mass of the micropropagated plant root system of the blueberry cv. Bluebelle. De Silva et al. (2000) verified that the inoculation with *Gliocladium virens* or with *Pseudomonas fluorescens* in sterile soil increased the leafy area of blueberry plants of the Highbush group, whereas with *T. harzianum*, there was reduction. The number of leaves was also significantly lower in *T. harzianum*-treated seedlings.

FIG. 3.1 Area under the curve of disease progress (AUCDP) of bacteriosis caused by *Xanthomonas arboricola* pv. *pruni* in plum trees cv. Pluma 7, treated with different silicon, in the first (A) and second cycle (B) (Ferreira et al., 2013).

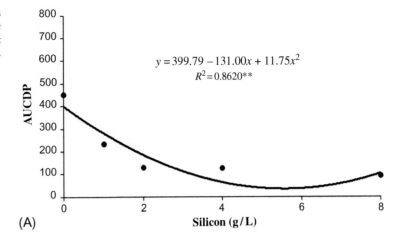

$$y = 399.79 - 131.00x + 11.75x^2$$
$$R^2 = 0.8620^{**}$$

(A)

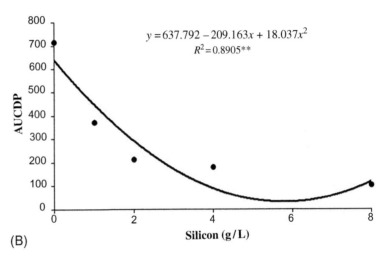

$$y = 637.792 - 209.163x + 18.037x^2$$
$$R^2 = 0.8905^{**}$$

(B)

The fungus *Trichoderma* spp. belonging to the class of the Mitosporic fungi, subclass Hyphomycetes, order Moniliales, family Moniliacea (Samuel and Hadavi, 1996), is a free-living fungus that reproduces asexually, more frequently in tropical and temperate soils, highly interactive in environments such as soil, root, and leaf systems. This genus presents great economic importance for agriculture as it acts as a control agent for diseases of cultivated plants, besides promoting growth and inducing defense response in plants (Lima et al., 2011). The effect of plant development has important implications such as decreasing the growth period and thus the permanence of seedlings in the nursery, increasing plant productivity and production and improving plant vigor at biotic and/or abiotic stresses (Hajieghrari et al., 2010). *Trichoderma* spp. is a microorganism that participates in the decomposition and mineralization of plant residues, contributing to the availability of nutrients to the plants.

Howell and Okon (1987) confirms that the presence of *Trichoderma* spp. in the soil makes the nutrients soluble, allowing greater and faster uptake, with this, soils containing *Trichoderma* spp. have a higher humic content, originating from the lignin that is decomposed by this microorganism. In this way, the increase of the root area of the plant occurs accompanied by the increase of the fresh mass in crops that are treated with *Trichoderma* sp. In pineapple seedlings (*Ananas comosus*) inoculation with *Trichoderma* spp. caused an increase in the weight of the root. Jesus et al. (2011), with the objective of evaluating the biofertilizer of the fungus *T. asperellum* on the development and growth of *Coffea arabica* L. (coffee) seedlings, verified that 5% of the soil biofertilizer provided a 62% increase in total dry biomass of coffee seedlings.

The promotion of plant growth is possibly due to the ability of *Trichoderma* to synthesize indoleacetic acid (AIA). The production of this hormone at low levels was reported in Eucalyptus minicuts treated with the CEN 262 isolate showing an increase of 137%, 145%, and 43% of the aerial part, of the root, and at the height of the plants, respectively, compared with control (Filho et al., 2008).

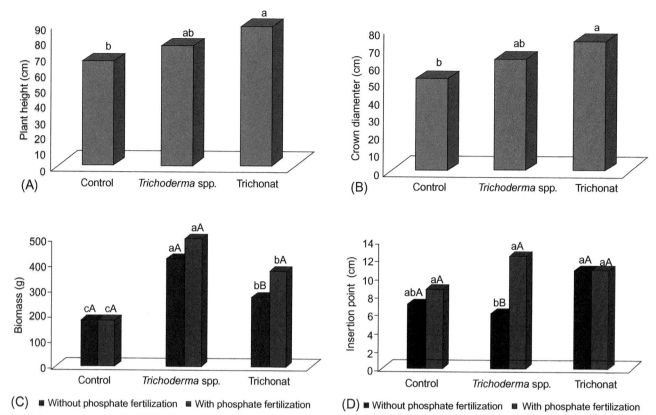

FIG. 3.2 Plant height (A), crown diameter (B), biomass (C), and the insertion point of the first branches (D) of *Eugenia pyriformis* subjected to several *Trichoderma* spp. inocula (Soldan et al., 2018).

The solubilization of the nutrients needed by plants may also be related to the promotion of growth. Altomare et al. (1999) show that the *T. harzianum* showed an ability to solubilize nutrients through compounds such as phosphate rocks, manganese oxide, iron oxide, and metallic zinc. Soldan et al. (2018) verified that fruit trees of *Eugenia pyriformis* inoculated with *Trichoderma* showed higher vegetative growth (Fig. 3.2) and higher values of N, P, K, Ca, Mg, and S.

Hoyos-Carvajal et al. (2009) evaluated the production of metabolites of 101 isolates of *Trichoderma* spp. of Colombia, and 20% of them were able to produce soluble forms of phosphate rock phosphate, 8% of the samples showed ability to produce consistent siderophores to convert iron to soluble forms. In works by Prates et al. (2007), testing the application of the fungus *Trichoderma* spp. in the substrate together with the biweekly spray of the "Pera" orange plants verified a higher concentration of potassium (44.3 g/kg) and sulfur (3.5 g/kg).

Nitrogen-fixing bacteria also develop associations with plant roots, collaborating in the process of nutrient absorption, and are represented by species of the genus *Rhizobium, Azospirillum,* and *Frankia* (Table 3.1). In trees of the species *Handroanthus impetiginosus,* inoculation with the bacterium *Azospirillum brasilense* promoted the in vitro rooting of the seedlings, decreasing the demand for auxins in the medium (Larraburu and Llorente, 2015).

6 Future line of research

Considering the information up to the present time, it is evident that the nutrition of fruit plants is a very complex subject, demanding deep knowledge of the diverse species and cultivars since the requirements are quite distinct, needing an increasing detail for the improvement of the nutritional balance and, consequently, higher fruit yields and quality. Studies that determine the requirements for canopy and rootstock cultivars, taking into account the influence of edaphoclimatic conditions and orchard management, may improve fertilizer adjustments for higher yields. Interactions with soil microbiology are still poorly understood, although their importance is known; therefore, studies in this area should also be advanced.

References

Adatia, M.H., Besford, A.T., 1986. The effects of silicon on cucumber plants grown in recirculating nutrient solution. Ann. Bot. 58 (3), 343–351.

Agarie, S., Hanaoka, N., Ueno, O., 1998. Effects of silicon on transpiration and leaf conductance in rice plants (*Oryza sativa* L.). Plant Prod. Sci. 1 (2), 89–95.

Aita, C., Giacomini, S.J., Hübner, A.P., Chiapinotto, I.C., Fries, M.R., 2004. Consorciação de plantas de cobertura antecedendo o milho em plantio direto. I – dinâmica do nitrogênio no solo. Rev. Bras. Ciênc. Solo 28, 739–749.

Altomare, C., Norvell, W.A., Björkman, T., Harman, G.E., 1999. Solubilization of phosphates and micronutrients by the plant–growth–promoting– and biocontrol fungus *Trichoderma harzianum* Rifai 1295-22. Appl. Environ. Microbiol. 65, 2926–2933.

Araújo, A.P., Machado, C.T.T., 2006. Fósforo. In: Fernandes, M.S. (Ed.), Mineral Nutrition of Plants. Brazilian Soil Science Society, Viçosa, pp. 253–280.

Barradas, C.A.A., Freire, L.R., Almeida, D.L., De-Polli, H., 2001. Winter green manure behaviour in the Rio de Janeiro sierra region. Pesq. Agropec. Bras. 36 (12), 1461–1468.

Basso, C., Freire, C.J.S., Suzuki, A., 2003. Soils, fertilization and nutrition. In: Quezada, A.C., Nakasu, B.H., Herter, F.G. (Eds.), Pear: Production. Embrapa, Brasília, pp. 55–67.

Bertalot, M.J.A., Carvalho-Pupatto, J.G., Rodrigues, E.M., Mendes, R.D., Buso, D., 2008. Alternative Control of Strawberry Plant Diseases. Brasilian Association of Biodynamic Agriculture.

Bhavya, H.K., Nache Gowda, V., Janagath, S., Sreenivas, K.N., Prakash, N.B., 2011. Effect of foliar silicic acid and boron acid in Bangalore blue grapes. In: 5th International Conference on Silicon in Agriculture, September 13-18, Beijing, China.

Boaretto, A.E., Trivelin, P.C.O., Muraoka, T., 2004. Use of isotopes as markers in soil fertility and plant nutrition. In: FERTBIO: 2004, Lajes, Brazil. Brazilian Soil Science Society, pp. 1–75.

Bowen, P., Menzies, J., Ehret, D., 1992. Soluble silicon sprays inhibit powdery mildew development on grape leaves. J. Am. Soc. Hortic. Sci. 117 (6), 906–912.

Braga, F.T., Nunes, C.F., Favero, A.C., Pasqual, M., Carvalho, J.G., Castro, E.M., 2009. Anatomical characteristics of micropropagated strawberry seedlings with different silicon sources. Pesq. Agropec. Bras. 44 (2), 128–132.

Brunetto, G., Kaminski, J., Melo, G.W.B., Gatiboni, L.C., Urquiaga, S., 2005. Absorption and redistribution of nitrogen applied via leaf on young vines. Rev. Bras. Frutic. 27 (1), 110–114.

Brunetto, G., Kaminski, J., Melo, G.W., Brunning, F., Mallmann, F.J.K., 2006. Nitrogen fate in 'Chardonnay' and 'Rhenish Riesling' vines when applied on swelling of the buds. Rev. Bras. Frutic. 28 (3), 497–500.

Camargo, S.S., Pelizza, T.R., Souza, A.L.K., Affonso, L.B., Schuch, M.W., 2010. Agentes biológicos inseridos durante etapa de aclimatização de plantas micro propagadas de mirtileiro bluebelle. In: XII Enpos, II Mostra Científica, Pelotas, Anais. Available in:http://www2.ufpel.edu.br/cic/2010/cd/pdf/CA/CA_00577.pdf.

Castagnoli, S.P., Dejong, T.M., Weibaum, A.S., Johnson, R.S., 1990. Autum foliage applications of ZnSO4 reduced leaf nitrogen remobilization in peach and nectarine. J. Am. Soc. Hortic. Sci. 155 (1), 79–83.

Conradie, W.J., 1981. Seasonal uptake of nutrients by 'Chenin Blanc' in sand culture: II. Phosphorus, potassium, calcium and magnesium. S. Afr. J. Enol. Vitic. 2, 7–13.

Dechen, A.R., Nachtigall, G.R., 2006. Micronutrientes. In: Fernandes, M.S. (Ed.), Mineral Nutrition of Plants. Brazilian Soil Science Society, Viçosa, pp. 327–352.

Derpsch, R., Calegari, A., 1992. 80 p. (IAPAR. Circular, 73). Plants for Winter Green Manuring. Agronomic Institute of Paraná, Londrina .

De Silva, A., Patterson, K., Rothrock, C., Moore, J., 2000. Growth promotion of highbush blueberry by fungal and bacterial inoculants. Hortscience 35 (7), 1228–1230.

Epstein, E., 1994. The anomaly of silicon in plant biology. Proc. Natl. Acad. Sci. U.S.A. 91, 11–17.

Faria, C.M.B., Soares, J.M., Leão, P.C.S., 2004. Green manuring with legumes on vine crop in the middle San Francisco River Region. R. Bras. Ci. Solo 28, 641–648.

Faria, C.M.B., Costa, N.D., Faria, A.F., 2007. Chemical attributes of an Argisol and melon yield through the use of green manures, liming and fertilization. R. Bras. Ci. Solo 31, 299–307.

Farias, D.H., 2012. Diversity of arbuscular mycorrhizal fungi in orchards and seedling growth of micropropagated blueberry. Thesis (Master)Universidade Federal de Pelotas, Pelotas, 74 pp.

Faust, M., 1989. 338 p. Physiology of Temperate Zone Fruit Trees. John Wiley & Sons, New York.

Fauteux, F., Rémus-Borel, W., Menzies, J.G., 2005. Silicon and plant disease resistance against pathogenic fungi. FEMS Microbiol. Lett. 249, 1–6.

Ferreira, S.G.M., Botelho, R.V., Faria, C.M.D.R., Mateus, M.A.F., Zaluski, W.L., 2013. Development and phytosanitary of plums treated with silicon in organic system. Rev. Bras. Frutic. 35, 1059–1065.

Filho, M.R.C., Mello, S.C.M., Santos, R.P., Menêzes, J.E., 2008. Evaluation of Trichoderma isolates in growth production, in vitro indoleacetic acid production and endophytic colonization of eucalyptus seedlings. Res. Dev. Bull. 226, 1–16.

Gazolla Neto, A., Giacobbo, C.L., Pazzin, D., Fachinello, J.C., 2006. Efeito da adubação na qualidade e produtividade de pêssegos cultivar maciel em Pelotas/RS. Available in:http://www.ufpel.edu.br/cic/2006/arquivos/CA_00092.rtf. (Accessed 14 September 2019).

Gil Salaya, G.F., 2000. 590 p. Fruit Production: Temperate and Subtropical Clime Fruits and Wine Grape, second ed. Catholic University of Chile editions, Santiago.

Gillissen, B., Bürkle, L., André, B., Kúhn, C., Rentsch, D., Brandi, D., Frommer, W.B., 2000. A new family of high-affinity transporters for adenine, cytosine, and purine derivatives in arabidopsis. Plant Cell 12 (2), 291–300.

Giovannini, E., 1999. 364 p. Production of Grapes for Wine, Juice and Table. Ed. Renascença, Porto Alegre.

Hajieghrari, B., Tarabi-Giglou, M., Mohammadi, M.R., Davari, M., 2010. Biological potantial of some Iranian Trichoderma isolates in the control of soil borne plant pathogenic fungi. Afr. J. Biotechnol 7 (8), 967–972.

Harman, G.E., Petzoldt, R., Comis, A., Chen, J., 2004. Interactions between *Trichoderma harzianum* strain T22 and maize inbred line Mo17 and effects of these interactions on diseases caused by *Pythium ultimum* and *Colletotrichum graminicola*. Plant Physiol. 94 (2), 146–153.

Herschbach, C., Zalm, E.V.D., Schneider, A., Jouanin, L., Kok, L.J., Rennenberg, H., 2000. Regulation of sulfur nutrition in wild-type and transgenic poplar over-expressing g-glutamylcysteine synthetase in the cytosol as affected by atmospheric H2S. Plant Physiol. 124 (1), 461–472.

Hobbie, E.A., Horton, T.R., 2007. Evidence that saprotrophic fungi mobilise carbon and mycorrhizal fungi mobilize nitrogen during litter decomposition. New Phytologist. 173, 447–449.

Howell, C.R., Okon, Y., 1987. Orillia: Kloepper, J. Recent results of greenhouse and field trials on bacterial-induced plant growth promotion with no obvious symptoms of plant disease. In: First International Workshop on Plant Growth-Promoting Rhizobacteria, 1, pp. 29–33.

Hoyos-Carvajal, L., Orduz, S., Bissett, J., 2009. Growth stimulation in bean (*Phaseolus vulgaris* L.) by Trichoderma. Biol. Control 51, 409–416.

Hudina, M., Stampar, F., 2002. Effect of phosphorus and potassium foliar fertilization on fruit quality of pears. Acta Hortic. 594, 487–493.

Jesus, E.P., Souza, C.H.E., Pomella, A.W.V., Costa, R.L., Seixas, L., Silva, R.B., 2011. Evaluation of the potential of *Trichoderma asperellum* as a substrate conditioner for the production of coffee seedlings. Cerrado Agroc. 2, 7–19.

Kangueehi, G.N., 2008. Nutrient Requirement and Distribution of Intensively Grown 'Brookfield Gala' Apple Trees (Dissertation). University of Stellenbosch, South Africa.

Korndorfer, G.H., Datnoff, L.E., 2000. Role of silicon in the production of sugarcane. In: SECAP 2000, Piracicaba Sugarcane Seminar, 5, Piracicaba, Brazil.

Korndorfer, G.H., Snyder, G.H., Ulloa, M., Powell, G., Datnoff, L.E., 2001. Calibration of soil and plant silicon analysis for rice production. J. Plant Nutr. 7, 1071–1084.

Korndorfer, G.H., Pereira, H.S., Camargo, M.S., 2002. Role of silicon in the production of sugarcane. STAB 21 (2), 6–9.

Kotzė, W.A.G., 2001. Voeding van bladwisselende vrugtebome, bessies, neute en ander gematigde klimaat gewasse in Suid Afrika. LNR Infruitec-Nietvoorbij, Stellenbosch.

Lahav, E., 1995. Banana nutrition. In: Gowen, S. (Ed.), Bananas and Plantains. Chapman & Hall, London, pp. 258–316.

Larraburu, E.E., Llorente, B.E., 2015. *Azospirillum brasilense* enhances in vitro rhizogenesis of *Handroanthus impetiginosus* (pink lapacho) in different culture media. Ann. Forest Sci. 72, 219–229.

Leão, P.C.S., 2001. Table Grape Production—Technical Aspects. Embrapa Semi-Arid (Petrolina, PE). Embrapa Technological Information, Brasília. 128 p.

Liang, Y., Si, J., Romheld, V., 2005. Silicon uptake transport it is Na active process in cucumber. In: Silicon in Agriculture Conference, vol. 3, Uberlândia. Anais... Uberlândia, p. 106.

Lima, R.L.S., Severino, L.S., Gheyi, H.R., Sofiatti, V., Arriel, N.H.C., 2011. Effect of phosphate fertilization on the growth and macronutrient content of jatropha seedlings. Agron. Sci. J. 42 (4), 950–956.

Ma, J., Takahashi, E., 1990. Effect of silicon on the growth and phosphorus uptake of rice. Plant Soil 126, 115–119.

Ma, J.F., Miyaky, Y., Takahashi, E., 2001. Silicon as a benefical elemento for crop plants. In: Datnoff, L.E., Snyder, G.H., Korndorfer, G.H. (Eds.), Silicon in Agriculture. Elsiever Science, The Netherlands, pp. 17–39 (Chapter 2).

Ma, J.F., Mitani, N., Nagao, S., Konishi, S., Tamai, K., 2004. Characterization of the silicone uptake system and molecular mapping of the silicone transporter gene in rice. Plant Physiol. 136 (2), 3284–3289.

Malavolta, E., 1980. Elementos de nutrição mineral de plantas. Agronômica Ceres, Piracicaba, 251 pp.

Marschner, H., 1995. Mineral Nutrition of Higher Plants. Academic Press, San Diego, CA. 889 p.

Mason, A.C., Whitfield, A.B., 1960. Seasonal changes in the uptake and distribution of mineral elements in apple trees. J. Hortic. Sci. 35, 34–55.

Maximo, E., Bendassolli, J.A., Trivelin, P.C.O., Rossete, A.L.R.M., Oliveira, C.R., Prestes, C.V., 2005. Production of double-labeled ammonium sulphate with the stable isotopes ^{15}N and ^{34}S. Quim Nova 28 (2), 211–216.

Mengel, K., Kirkby, E.A., 1982. Principles of Plant Nutrition, third ed. International Potash Institute, Worblaufen-Bern.

Meynhardt, J.T., Heyns, C.F.G., Strydom, D.K., Marais, P.G., 1967. The seasonal uptake of P-labelled phosphate by apple trees grown in sand culture. S. Afr. J. Agric. Sci. 10, 875–881.

Millard, P., 1996. Ecophysiology of the internal cycling of nitrogen for tree growth. J. Plant Nutr. Soil Sci. 159, 1–10.

Mitani, N., Ma, J.F., Iwashita, T., 2005. Characterization of silicone uptake system and isolation of LS11 gene from rice root. In: Silicon in Agriculture Conference, Uberlândia. Anais..., p. 124.

Miyake, Y., Takahashi, E., 1983. Effects of silicone on growth of solution cultured cucumber plants. Soil Sci. Plant Nutr. 29, 71–83.

Moreira, A., Fageria, N.K., 2008. Potential of Brazilian Amazon soils for food and fiber production. Dyn. Soil Dyn. Plant 2, 82–88.

Mullins, M.G., Bouquet, A., Williams, L.E., 2000. Biology of the Grapevine. University Press, Cambridge.

Muzilli, O., Lugão, S.M.B., Fidalski, J., Soares Jr., D., Ribeiro, M.F.S., Fagundes, A.C., 1992. Green Manure for the Improvement of the Soils Occupied With Laundries in the Region of the Caiuá Sandstone. IAPAR, Londrina (Informe da pesquisa, 101).

Natale, W., Marchal, J., 2002. Absorption and redistribution of nitrogen (^{15}N) in *Citrus mitis* Bl. Rev. Bras. Frutic. 24 (1), 183–188.

Neilsen, G.H., Neilsen, D., 2003. Nutritional requirements of apple. In: Ferree, D.C., Warrington, I.J. (Eds.), Apples: Botany, Production and Uses. CAB International, Cambridge, pp. 267–301.

Oliveira, L.A., Castro, N.M., 2002. Occurrence of silica in the leaves of *Curatella americana* L. and Davilla elliptica St. Hil. Rev. Horiz. Cient. (4), 1–16 Avaliable from: www.propp.ufu.br/revistaeletronica/B/OCORRENCIA.pdf.

Paulino, G.M., Alves, B.J.R., Barroso, D.G., Urquiaga, S., Espindola, J.A.A., 2009. Biological fixation and transfer of nitrogen by legumes in organic mango and soursop orchard. Pesq. Agropec. Bras. 44 (12), 1598–1607

Pereira, A.L., Benedito, E., 2007. Stable Isotopes in ecological studies: methods, applications and perspectives. Biosci. J. 13, 16–27.

Pinotti, M.M.Z., Santos, J.C.P., Klauberg Filho, O., Michelluti, D.J., Castro, D.R.L., 2011. Isolation of soil fungi associated to the cultures of blackberry, raspberry and blueberry in Brazilian Southern highland. Rev. Bras. Agroecol. 6 (1), 67–80.

Prates, H.S., Lavres Junior, J., Rossi, M.L., 2007. Mineral composition of citrus seedlings with applications of Trichoderma spp. Agron. Inform. 118, 1–7

Pratissoli, D., Almeida, G.D., Jesus Júnior, W.C., Vicentini, V.B., Holtz, A.M., Cocheto, J.G., 2007. Organomineral fertilizer and silica clay as inductors of resistance to papaya smallpox. Idesia 25 (2), 63–67.

Raven, J.A., 1983. The transport and function of silicone in plants. Biol. Rev. 58 (2), 179–207.

Reichardt, K., Silva, A.L., Fenilli, T.A.B., Timm, L.C., Bruno, I.P., Volpe, C.A., 2009. Relationship between nitrogen fertilization and soil water conditions for a coffee crop in Piracicaba, SP. Coffee Science 4 (1), 41–55.

Ribeiro Júnior, P.M., de Resende, M.L.V., Pereira, R.B., Cavalcanti, F.R., Amaral, D.R., de Pádua, M.A., 2006. Fosfito de potássio na indução de resistência a *Verticillium dahlia* Kleb., em mudas de cacaueiro (*Theobroma cacao* L.). Ciênc. Agrotec. 30 (4), 629–636.

Rodrigues, F.A., Benhamou, N., Datnoff, L.E., Jones, J.B., Bélanger, R.R., 2003. Ultrastructural and cytochemical aspects of silicone-mediated rice blast resistance. Phytopathology 93, 535–546.

Rufat, J., Dejong, T.M., 2001. Estimating seasonal nitrogen dynamics in peach trees in response to nitrogen availability. Tree Physiol. 21, 1133–1140.

Samuel, N., Hadavi, N., 1996. Trichoderma: review of biology and systematics of the genus. J. Gen. Microbiol. 100, 923–935.

Sangster, A.G., Hodson, M.J., Tubb, H.J., 2001. Silicon deposition in higher plants. In: Datnoff, L.E., Snyder, G.H., Korndorfer, G.H. (Eds.), Silicon in Agriculture. Elsiever Science Publishing, Amsterdam, pp. 85–114.

Savant, N.K., Snyder, G.D., Datnoff, L.E., 1997. Silicon in management and sustainable rice production. Adv. Agron. 58, 151–199.

Schmidt, R.E., Zhang, G.Y., Chalmers, D.R., 1999. Response of photosynthesis and superoxide dismutase to sílica applied to creeping bentgrass grown under two fertility levels. J. Plant Nutr. 22, 1763–1773.

Seyr, L., 2011. Soil Management and Bagging of the Bunch in Banana Orchard 'Nanicão'. (Dissertation), State University of Londrina, Londrina.

Singh, T., Sandhu, A.S., Singh, R., Dhillon, W.S., 2005. Vegetative and fruiting behavior of semi-soft pear strains in relation to nutrient status. Acta Horticult. 696, 289–293.

Soldan, A., Watzlawick, L.F., Botelho, R.V., Faria, C.M.D.R., Maia, A.J., 2018. Development of forestry species inoculated with *Trichoderma* spp fertilized with rock phosphate. Floresta Ambiente 25 (4), 1–8.

Souza, S.R., Fernandes, M.S., 2006. Nitrogen. In: Fernandes, M.S. (Ed.), Mineral Nutrition of Plants. Brazilian Soil Science Society, Viçosa, pp. 215–252.

Stassen, P.J.C., North, M.S., 2005. Nutrient distribution and requirement of 'Forelle' pear trees on two rootstocks. Acta Hortic. 671, 493–500.

Stassen, P.J.C., Terblanche, J.H., Strydom, D.K., 1981. The effect of time and rate of nitrogen application on development and composition of peach trees. Agroplantae 13, 56–61.

Stassen, P.J.C., van Vuuren, B.H.P.J., Davie, S.J., 1997. Macro elements in mango trees: uptake and distribution. In: S.A. Mango Grower's Association Yearbook.vol. 17. p. 16.

Taiz, L., Zeiger, E., 2009. Plant Physiology, fourth ed. Porto Alegre, Artmed. 819 p.

Teodoro, R.B., Oliveira, F.L., Silva, D.M.N., Fávero, C., Quaresma, M.A.L., 2011. Aspectos agronômicos de leguminosas para adubação verde no Cerrado do Alto do Jequitinhonha. Rev. Bras. Ciênc. Solo 35, 635–643.

Terblanche, J.H., 1972. Seasonal Uptake and Distribution of Ten Nutrient Elements by Young Apple Trees Grown in Sand Culture. (PhD dissertation) University of Stellenbosch, Stellenbosch.

Terry, L.A., Joyce, D.C., 2004. Elicitors induced disease resistance in postharvest horticultural crops: a brief review. Postharvest Biol. Technol. 32, 113.

Tisdale, S.L., Nelson, W.L., Beaton, J.D., Havlin, J.L., 1993. Soil Fertility and Fertilizers, fifth ed. Macmillan, United States of America.

Titus, J.S., Kang, S., 1982. Nitrogen metabolism, translocation and recycling in apple trees. Hortic. Rev. 4, 204–246.

Trivelin, P.C.O., 2002. ^{15}N isotope dilution technique in studies of nitrogen transformations in the soil: mineralization, nitrification and immobilization gross rates. In: CENA/USP Piracicaba.

Tromp, J., 2005. Mineral nutrition. In: Tromp, J., Webster, A.D., Wertheim, S.J. (Eds.), Fundamentals of Temperate Zone Tree Fruit Production. Backhuys Publishers, Leiden, pp. 55–64.

Verlindo, A., Botelho, R.V., Kawakami, J., Sato, A.J., Calgaro, F., Rombola, A.D., MarchI, T., 2014. Exportation and immobilization of nutrients by pear trees from the cultivars Cascatense and Tenra. Curr. Agric. Sci. Technol. 20, 26–36.

Weinbaum, S.A., Brown, P.H., Rosecrance, R.C., Picchioni, G.A., Niederholzer, F.J.A., Youseffi, F., Muraoka, T.T., 2001. Necessity for whole tree excavations in determining patterns and magnitude of macronutrient uptake by mature deciduous fruit trees. Acta Hortic. 564, 41–49.

Wilcox, H.E., 1996. Mycorrhizae. In: Waisel, Y., Eshel, A., Kafkafi, U. (Eds.), Plants Roots. Marcel Dekker, New York, 1002 pp.

Williams, L.E., Biscay, P.J., Smith, R.J., 1987. Effect of interior canopy defoliation on berry composition and potassium distribution of Thompson Seedless grapevines. Am. J. Enol. Vitic. 38, 287–292.

Wutke, E.B., Fancelli, A.L., Pereira, J.C.V.N.A., Ambrosano, G.M.B., 1998. Irrigated bean yield in rotation with grain crops and green manures. Bragantia. 57(2).

Wutke, E.B., Terra, M.M., Pires, E.J.P., Costa, F., Secco, I.L., Ribeiro, I.J.A., 2005. Influência da cobertura vegetal do solo na qualidade dos frutos de videira Niagara Rosada. Rev. Bras. Frutic. 27 (3), 434–439.

Yedidia, I.S., Kapulnik, Y., Chet, I., 2001. Effect of *Trichoderma harzianum* on microelement concentrations and increased growth of cucumber plants. Plant Soil 235, 235–242.

Yoshida, S., 1965. Chemical aspects of the role of silicone in physiology of the rice plant. Bull. Natl. Inst. Agric. 5, 1–58.

Further reading

Brunetto, G., Ceretta, C.A., Kaminski, J., Melo, G.W.B., Lourenzi, C.R., Furlanetto, V., Moraes, A., 2007. Application of nitrogen in vines in the Gaúcha Campaign: productivity and chemical characteristics of grape must. Cienc. Rur. 37 (2), 389–393.

Buck, G.B., 2006. Foliar Potassium Silicate and the Incidence of Blast on Rice. (Dissertation). Federal University of Uberlândia, Uberlândia.

Giacomini, S.J., Aita, C., Chiapinotto, I.C., Hübner, A.P., Marques, M.G., Cadore, F., 2004. Intercrop of cover crops preceding corn crop in no-till system. II—nitrogen accumulated by corn and grain yield. R. Bras. Ci. Solo 28, 751–762.

Haynes, R.J., Goh, K., 1980. Distribution and budget of nutrients in a commercial applemorchard. Plant Soil 56, 445–457.

Kotz, W.A.G., Villiers, J., 1989. Seasonal uptake and distribution of nutrient elements by kiwifruit vines. 1. Macronutrients. S. Afr. J. Plant Soil 6, 256–264.

Meyer, G., Bigirimana, J., Elad, Y., Hofte, M., 1998. Induced systemic resistance in Trichoderma harzianum T39 biocontrol of Botrytis cinera. Eur. J. Plant Pathol. 4, 279–286.

Santi, A., Amado, T.J.C., Acosta, J.A.A., 2003. Nitrogen fertilization in black oats. I—Influence on dry matter production and nutrient cycling under no-tillage system. Rev. Bras. Cienc. Solo 27, 1075–1083.

Stassen, P.J.C., 1987. Macro-element content and distribution in peach trees. Decid. Fruit Grow. 37, 245–249.

Tagliavini, M., Marangoni, B., 2000. Major nutritional issues in deciduous fruit orchards of Northern Italy. Hortic. Technol. 12, 26–31.

Takahashi, E., Ma, J.F., Miyake, Y., 1990. The possibility of silicon as an essential element for higher plants. Comments Agric. Food Chem. 2 (99), 22.

Trivelin, P.C.O., 2005. Stable isotopes. Mass spectrometry for determination of isotopic ratio of low atomic number elements (IRMS). The tracer technique and isotopic dilution. In: CENA-USP, Piracicaba. Available from: http://www.nutricaodeplantas.agr.br/site/ensino/pos/Palestras_Cena/Trivelin/DiluicaoIsotopica.doc.

Wang, S.Y., Galletta, G.J., 1998. Foliar application and potassium silicate induces metabolic changes in strawberry plants. J. Plant Nutr. 21 (1), 157–167.

4

Plant nutrition and physiological disorders in fruit crops

Yongqiang Zheng[a,*], *YanYan Ma*[b], *Wenhuan Liu*[b], *Fangying Qiu*[b]

[a]Citrus Research Institute, Southwest University-Chinese Academy of Agricultural Sciences, Chongqing, China
[b]National Engineering Research Center for Citrus Technology/Citrus Research Institute, Southwest University-Chinese Academy of Agricultural Sciences, Chongqing, China
*Corresponding author. E-mail: zhengyq@swu.edu.cn

1 Introduction

Apart from insects, pests, and diseases, fruit crops like citrus are very prone to develop different physiological disorders. Some physiological disorders may occur due to one factor either nutritional such as creasing (Bower, 2004) or weather conditions such as sunburn. However, most of the disorders associated with more than one factors like environment and nutrition such as on-tree oleocellosis (Zheng et al., 2016, 2018). In this chapter, we have focused on the classification, criteria of distinguishing, mineral markers, preharvest factors, diagnosis, and management of the important physiological disorders such as alternate bearing, fruit drop, granulation, oleocellosis, cold scald, and stem-end browning, which cause economic losses worldwide (Mishra et al., 2016; Zheng et al., 2011).

A.K. Srivastava, Chengxiao Hu (eds.)
Fruit Crops: Diagnosis and Management of Nutrient Constraints
https://doi.org/10.1016/B978-0-12-818732-6.00004-6

2 Definition of physiological disorders

Physiological disorders are a group of disorders affecting fruit quality and fruit crop caused by nonpathogenic factors (Fig. 4.1) such as environment stresses (temperature, relative humidity, water stress, air pollutants, etc.), nutrient deficiencies or toxicities, chemicals (herbicides, pesticides, etc.), and some genetic factors, which result in abnormal external or internal conditions and abnormal growth pattern of fruits (Wallace, 1934). Therefore, the physiological disorders are distinguished from plant diseases caused by plant pathogens (fungi, bacteria, and viruses) and animal herbivores (insects) (Schutzki and Cregg, 2007). Although the symptoms of physiological disorders may appear disease-like, they can usually be prevented by altering environmental conditions. However, once a plant shows symptoms of a physiological disorder, it is likely that the season's growth or yield will be reduced.

3 Classification of physiological disorders

The productivity and quality of fruits are affected to a greater extent due to physiological disorders and nutrient constraints (Srivastava, 2013b). The physiological disorders of fruit crops occur at both preharvest and postharvest periods according to the occurrence stage (Chikaizumi, 2002; Lafuente and Zacarias, 2006), which can be categorized in various types on the basis of causal factors (Ladaniya, 2008; Mishra et al., 2016; Schutzki and Cregg, 2007).

Among these disorders, nutrient constraints have been recorded in nearly all the geographical regions where commercial citrus plantations have been established. As can be seen from Table 4.1, the most often encountered constraints result from inadequate supplies of macronutrients (typically N, P, or K) (Agustí et al., 2014) and

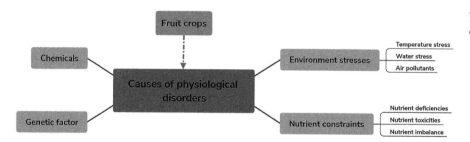

FIG. 4.1   Causes can result in physiological disorders in citrus plantations.

TABLE 4.1   Occurrence of nutrient deficiencies in citrus plantations.

	N	P	K	Ca	Mg	S	Fe	Mn	B	Cu	Zn	Mo
Argentina	●				●		●			●	●	
Australia	●	●						●	●	●	●	
Brazil		●	●	●	●						●	
Chile	●	●				●		●		●		
China		●		●			●	●	●		●	●
Egypt	●	●					●	●				
Iran							●			●	●	
Israel				●	●		●				●	
India	●	●		●	●		●		●		●	
Italy	●		●		●					●		
Japan	●	●	●		●						●	
Spain	●	●	●	●	●		●			●	●	
The United States	●	●	●		●		●	●	●	●	●	●

TABLE 4.2 Occurrence period and relative cause of the important physiological disorders in citrus plantations.

Disorders	Occurrence stage	Temperature	Water stress	Nutrient constraints	Air pollutants	Chemicals	Genetic factor	References
Alternate bearing	Preharvest	●		●			●	Sharma et al. (2019)
Fruit drop	Preharvest	●	●	●	●		●	Goren (1993), Mehouachi et al. (1995), Ogata et al. (2002), Thompson and Taylor (1969)
Fruit splitting	Preharvest	●	●	●			●	Cronje et al. (2013)
Rind breakdown	Preharvest Postharvest	●	●	●			●	Alferez et al. (2001)
Oleocellosis	Preharvest Postharvest	●	●	●			●	Knight et al. (2002), Montero et al. (2012), Zheng et al. (2016), Zheng et al. (2018)
Creasing	Preharvest		●	●			●	Bower (2004), Hussain and Singh (2015)
Granulation	Preharvest Postharvest	●	●					Ritenour et al. (2004)

micronutrients constraints (particularly Fe, Mn, Cu, and Zn), which resulted from the application of fertilizers containing only macronutrients (i.e., N, P, K, Ca, Mg, and S), instances of primary B deficiency (Liu et al., 2013; Wu et al., 2018) and Mo deficiency (Stewart and Leonard, 1951; Tao et al., 2016) in China and the United States have been documented.

Moreover, some important physiological disorders like alternate bearing, fruit drop, splitting, rind breakdown, oleocellosis, creasing, granulation, and chilling injury, which are almost related to environmental factors (temperature and water stress), nutrient constraints, and genetic factor (Table 4.2) (Agustí et al., 2014; de Oliveira and Vitória, 2011).

4 Visual symptoms of nutrient deficiency disorders

The morphological descriptors are commonly used to identify nutritional disorders at citrus orchard level (Srivastava, 2013a; Srivastava and Singh, 2006). Fortunately, the symptoms of nutrient deficiency follow patterns of expression, which depend on the mobility of nutrients in plants (Table 4.3) with typical leaf color illustrations of nutrient constraints that can easily be distinguished (Fig. 4.2), and the degree of deficiency is measured by the severity of symptoms and number of growth terminals affected (Srivastava and Singh, 2009).

TABLE 4.3 Mobility of nutrients in the phloem in plants.

Highly mobile	Variably mobile	Immobile
Nitrogen	Copper	Boron
Phosphorus	Magnesium	Calcium
Potassium	Molybdenum	Iron
	Sulfur	Manganese
	Zinc	

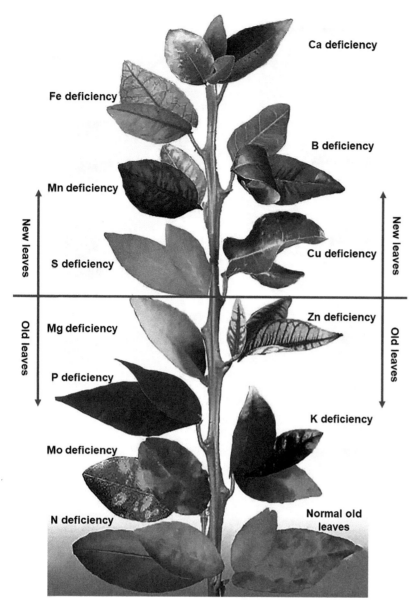

FIG. 4.2 Patterns of symptom organs used to establish nutrient deficiencies diagnosis.

5 Visual symptoms of some important physiological disorders

Sometimes physiological disorders also directly related to malfunctions of fruit development induced by environmental factors (Figs. 4.3 and 4.4). The most extensive description of physiological disorders in citrus is given by Agustí et al. (2002, 2014) and Zheng et al. (2010a) for extensive knowledge.

5.1 Alternate bearing

Alternate bearing is a problematic phenomenon that occurs in fruit crops (Monselise and Goldschmidt, 1982). In citrus, the "on" crop is characterized by a large number of small fruit in one season followed by the "off" crop typically consists of few and large fruit (Agusti et al., 1992; Monselise and Goldschmidt, 1982; Schaffer et al., 1985).

5.2 Fruit drop

In citrus, there are three successive abscission waves affecting flowers, and developing fruitlets can be distinguished, which occur flower and ovary abscission (Fig. 4.3A), June drop (Fig. 4.3B), and preharvest drop

FIG. 4.3 The three successive abscission waves occur during (A) flower and ovary abscission, (B) June drop, and (C) preharvest drop.

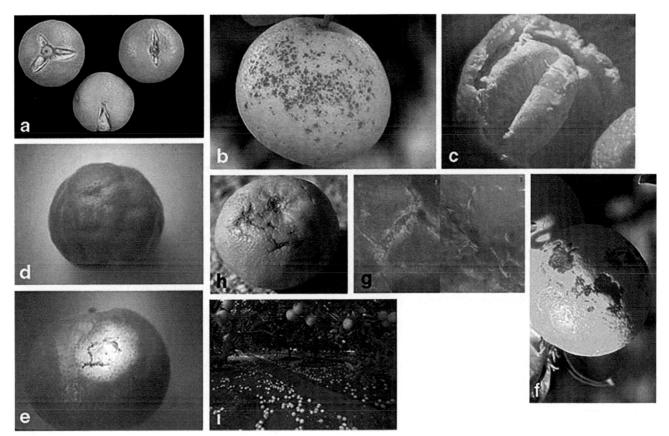

FIG. 4.4 Physiological citrus fruit disorders. (A) Splitting in "Nova" mandarin. (B) Peel pitting in "Marsh" grapefruit. (C) Puffing in Satsuma fruit. (D) Creasing in Clementine mandarin. (E) Albedo breakdown in a clementine fruit showing creasing. (F) Navel rind stain. (G and H) Peel senescence in clementine mandarin. (I) Navel fruit abscission (Agustí et al., 2014).

(Fig. 4.3C), respectively, resulting in extremely low fruit set (<0.5%) even in healthy, well-managed citrus orchards (Iglesias et al., 2006; Rebolledo et al., 2012).

5.3 Fruit splitting or cracking

Citrus fruits, especially acid limes and mandarins, develop a fissure of the peel, usually developing from the stylar end and reaching, or even extending beyond, the equatorial zone (Fig. 4.4A) (Cronje et al., 2013; Mesejo et al., 2016).

5.4 Cold pitting (peel pitting)

Cold pitting or peel pitting is a physiological disorder usually occur during postharvest storage stage, which starts on fruit as discrete areas forming sunken reddish-brown (dark gray in the print version) to black lesions that tend to coalesce producing larger depressions of affected areas (Fig. 4.4B), but in some varieties, such as "Fortune" mandarin and "Marsh" grapefruit, it also can appear during preharvest stage (Cronje et al., 2017).

5.5 Puffing

Puffing is characterized by the disintegration of the deepest cell layers of the albedo tissue that gives rise to separation aerial spaces between peel and pulp, which results in a cracked and low-resistant albedo in mature fruits (Fig. 4.4C). This disorder takes place only in a few mandarin varieties, which are susceptible to puffing, like "Satsuma" mandarin or "Oroval" clementine mandarin (Ibáñez et al., 2014; Martinelli et al., 2015).

5.6 Creasing (albedo breakdown)

Creasing, a peel-related preharvest disorder, affects different cultivars of sweet orange including Washington Navel (Greenberg et al., 2010), Valencia (Jona et al., 1989), and Nova mandarins (Greenberg et al., 2006). The disorder consists of depressions on the flavedo that alternate with healthy areas that turn bulky (Fig. 4.4D and E) due to the separation of adjacent cells rather than cleavage of individual cells (Jones et al., 1967; Storey and Treeby, 1994). However, the disorder sometimes takes place with no damage to the cells so that cells retain their turgor, but many cells are irreparable damaged and lose their turgor and wall collapses.

5.7 Navel rind stain (rind breakdown)

The navel orange varieties like "Navelina," "Washington," and "Lane late" have been proved to be highly sensitive to rind breakdown during ripening under Mediterranean climatic conditions (Zaragoza and Alonso, 1975) and postharvest storage (Alferez et al., 2001; Cronje et al., 2011). The disorder begins at the flavedo-albedo union area, where the cells become dehydrated and flattened, and finally die, which result in reddish-brown sunken areas (dark gray in the print version), especially for dry areas partially covering the exposed portion of the mature fruits (Fig. 4.4F). A similar disorder has also been described in grapefruit, "Fallglo" tangerine (Petracek et al., 1998).

5.8 Oleocellosis

Oleocellosis (or oil spotting) is a physiological disorder that occurs after the rupture of the peel oil gland, causing obviously visible pitting due to the released oil, which is phytotoxic to pericarp cells (Shomer and Erner, 1989), and citrus fruit is highly sensitive to oleocellosis during both the postharvest storage and on-tree ripening, termed as postharvest oleocellosis (PHO) and on-tree oleocellosis (OTO), respectively (Zheng et al., 2016, 2018). In addition, the symptoms were significantly different in different stages. At the stage of fruit expansion, OTO results in light green (dark gray in the print version) and smaller than 0.8 cm due to the low maturity of fruit (Fig. 4.5A), and PHO often appears >1.0 cm in lesions of oleocellosis with yellow spots (white spots in the print version) (Fig. 4.5B) due to various mechanical injuries during harvesting, handling, and marketing. Sometimes, the PHO spots color deepening to brown (dark gray in the print version) with the extension of storage time (Fig. 4.5C).

5.9 Granulation

Granulation (also called crystallization or section drying) is a serious physiological disorders in most of the citrus-growing countries. In citrus, this disorder affects fruit juice sacs, which turn gray and become hard, dry, and enlarged, with little extractable juice (Fig. 4.6) (Ritenour et al., 2004). Sweet orange cultivars (pineapple, Washington Navel, Hamlin, blood red, Mosambi, and Valencia late), mandarin cultivars (Kaula, Nagpur, and Dancy), and citrus hybrid in China (Huangguogan) suffer from granulation more than others like grapefruit, pummelo, lemons, limes, tangors, and tangelos (Singh, 2001; Xiong et al., 2017).

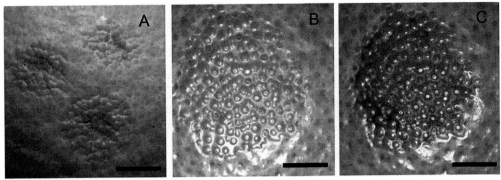

FIG. 4.5 Oleocellosis symptoms in different periods. (A) Fruit coloring stage; (B) fruit harvest time after mechanical injury; (C) postharvest storage period (Zheng et al., 2010a).

FIG. 4.6 Granulation of navel orange.

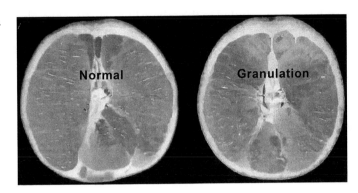

5.10 Predicting disorder incidence

In the last 30 years, the occurrence mechanism of some physiological and nutrient disorders is understood clearly, so that the incidence of these disorders can be reduced by predicting potential incidence according to the significant relationships between causal factors and disorder incidence, or premature expression of the disorder by artificial means.

In view of OTO in citrus, preharvest factors, such as those contributing to the strong pit relationship between diurnal range of rind oil release pressure (ΔRORP) and the incidence of OTO per tree (IOPT) have been proved, and ΔRORP and ROS scavenging capacity could be selected as indicators to assess IOPT and SOPT, respectively (Zheng et al., 2018). Jim Hill (The South Australian Research and Development Institute (SARDI)) has developed inexpensive instruments to measure relevant environmental conditions and the rind turgor pressure in citrus. These instruments were sold as an "oleocellosis prediction kit" (Jim, 2004). However, fruits need to be destroyed for this procedure. Zheng et al. (2010b) have formed the basis for commercial disorder prediction schemes by using VNIR reflectance spectroscopy. These results provide fundamental and practical knowledge for the development of a non-destructive, fast, and accurate technology for classifying fruit oleocellosis based on spectral reflectance (Fig. 4.7).

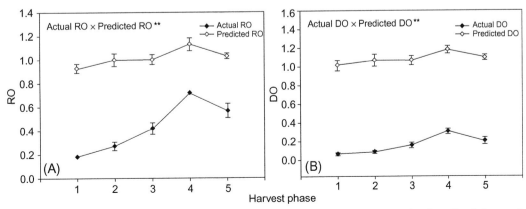

FIG. 4.7 Changes in fruit oleocellosis attributes (actual and predicted rate of oleocellosis (A) and actual and predicted degree of oleocellosis (B)) with different harvest phases. (**) Significant at $P < .01$ (Zheng et al., 2010b).

Similar attempts to use magnetic resonance imaging (MRI) to study and predict fruit splitting in citrus (Zur et al., 2017), and the power of MRI can predict splitting probability as early as 2 months before split fruit.

6 Causes and management of important physiological disorders

There will inevitably be a reduction in the yield and quality once symptoms appear in citrus orchards (Srivastava, 2013a). Therefore, it is necessary to learn about the causes and management approach of important physiological disorders.

6.1 Alternate bearing

6.1.1 Causes

Although the problem has been attributed to the causes like genetic, physiological, environmental, nutritional, and hormonal (Sharma et al., 2019; Stander et al., 2018), the existing studies have not reached a consistent conclusion on the causal relationship between alternate bearing and nutrient constraints and Stander et al. (2018) indicated that changes in leaf macronutrient concentration in leaves can be considered a consequence of, rather than a cause of, differences in fruit load from one season to another. However, it has been proved that cool temperatures around 15–20°C and good light conditions induce greater flowering and fruit setting, respectively, in several tropical and subtropical perennial fruit trees including citrus. As a sequence, the citrus tree produces heavy crop and gets nutritionally exhausted, unable to put forth new flush in one season, which result in heavy fruit loading in "on" year followed "off" year in the next season (Sharma et al., 2019).

6.1.2 Control

The key to the prevention and control alternate bearing is to keep the number of fruit bearing reasonably according to the ratio of leaf to fruit with the aid of suitable pruning, thinning, and proper fertilization.

6.2 Fruit drop

6.2.1 Causes

The causes attributed to fruit drop in citrus are the lack of fertilization, mechanical shock, insects, disease, high temperature, rainfall, and defective irrational practices. The most pronounced stages of fruit drop occur when the fruits are at marble stage. It lasts for a month after full bloom. On the onset of hot summer weather during May-June, the second wave of intense fruit drop occurs, while preharvest drop occurs during ripening period, which lasts from August to January.

6.2.2 Control

The method of control depends upon the causes of the drop and the variety of the fruit. To reduce the preharvest drop, NAA (10 ppm) is sprayed from August till October at monthly interval.

6.3 Fruit splitting or cracking

6.3.1 Causes

This is mainly a physiological disorder and is largely attributed to high atmospheric humidity following heavy rains or heavy irrigation during hot weather. Two types of splitting, namely, radial and transverse, have been noticed. Radial cracking is more common than transverse one. Partial splitting is more prevalent, while splitting down to inner core is rather rare. Often, the cracked surface of the fruit gets infected by disease causing organisms such as *Aspergillus*, *Alternaria*, *Fusarium*, and *Penicillium*, which lead to partial rotting and early fruit dropping from trees.

6.3.2 Control

The disease can be minimized by frequent and light irritations during the dry and hot periods and early picking of fruits soon after maturity.

6.4 Cold pitting (peel pitting)

6.4.1 Causes

The cause of preharvest peel pitting is not well known, although cold and dry winds, low temperature, and relative humidity have been suggested as responsible for pitting. These climatic conditions change the physiological properties of membranes and cuticles and modify the water balance of injured areas.

6.4.2 Control

The application of calcium nitrate just before or at fruit color break has been shown to be effective in controlling preharvest peel pitting of "Fortune" mandarin. There is evidence of a relationship between the reduction of peel pitting and the decrease of water permeability associated with the use of calcium nitrate.

6.5 Puffing

6.5.1 Causes

The cause of puffing has been related to the water exchange regulation through the peel. Accordingly, high values of RH together with high temperatures at fruit color break increase the appearance and intensity of puffing, particularly after a period of drought.

6.5.2 Control

The application of 10 mg/L of GA_3 before fruit color break reduces the occurrence of puffing in Satsuma mandarin. The GA_3 treatment prevents the late growth of the peel and increases the compactness of the albedo. The addition of nitrogen compounds reinforces the effect of GA_3. The main internal fruit characteristics are not modified by such treatments.

6.6 Creasing

6.6.1 Causes

The cause of creasing is not yet clearly understood. Climatic factors, cultural practices, and endogenous factors have been related with this physiological disorder. Several mineral elements have been also related to creasing, with molybdenum (Mo) being of critical importance.

6.6.2 Control

The application of GA_3 (10–20 mg/L) at early stages of fruit development or just prior to fruit color break reduce considerably the incidence of creasing. As for puffiness, the addition of nitrogen compounds reinforces its effect. It had a strong inhibition effect on color development when applied close to color break.

6.7 Navel rind stain (rind breakdown)

The cause of this physiological disorder has been related to nutritional imbalances, drought, and rainy periods in alternation with cold periods. The incidence of navel rind stain varies in intensity from year to year, among orchards and even among varieties, affecting up to 50% of mature fruits in some cases, such as "Navelate" in Spain. Fruit position on the tree has shown as important factor in developing rind breakdown, fruits outside of canopy being most sensitive fruits, and the outside face of fruit being more sensitive than the inside face. In "Navelate" oranges stored at 20°C, transference of fruit from low (45%) to high (95%) RH starts or aggravates the incidence of this disorder.

6.7.1 Control

Nowadays, we have not effective treatments to control it. However, rootstock plays an important influence in the development of the disorder. Carrizo citrange is more susceptible than Cleopatra mandarin, and it, in turn, is more than sour orange. This dependence has been related to rootstock influence on water transpiration capacity, supported by the histological study of fruit peduncle.

6.8 Oleocellosis

6.8.1 Causes

It is caused by rind oil release when oil cells get ruptured during harvesting or during handling from the field to the pack house. Careful harvesting and handling reduce incidence of oleocellosis. Rind oil from ruptured cells discolors the skin making the fruit unmarketable.

6.8.2 Control

Best way to reduce its incidence is to cure the fruit overnight at a temperature of 18–20°C for 12–24 h, before the fruit is moved from the field to the pack house.

6.9 Granulation

6.9.1 Causes

The terms *granulation, crystallization*, and *dry end* are used to describe this trouble. It is much more prevalent in larger sized fruits than in small fruit, in young than in old trees, and in humid than in dry areas. Several factors like luxuriant growth, rootstock and the variety, frequent irrigation, mineral constituents in plant tissue, time of harvest, and exposure to sunlight are found to be associated with this malady. And the plant tissues contain high Ca and Mn and low P and B in the areas with high incidence of granulation. The incidence is relatively high in the fruits of younger plants as compared with those in older plants. The vigorous rootstocks like rough lemon increase the incidence of granulation as compared with less vigorous rootstocks. Late maturity and persistent cold weather throughout the period of maturity have been found to increase the incidence of granulation.

6.9.2 Control

The incidence of granulation could be reduced to 50% by applying two to three sprays of NAA (300 ppm) in the months of August, September, and October. Spraying of GA 15 ppm followed by NAA 300 ppm in October and November also reduce granulation.

7 Future lines of research

In fact, most physiological disorders are linked to plant nutrition, which is the external expression of multiple factors such as physiological activity, dosage, and interaction between elements and external environment of elements in the tree, so it is a systematic project to control the occurrence of these disorders through plant nutrition management. Although the researchers at home and abroad have done some work in the relationship between these disorders and mineral elements, as a system engineering, there are still many deficiencies, such as insufficient research on the mechanism of interaction between elements, and the application research is not universal.

Combined with the existing problems in the current research, our future research focus is (1) to strengthen the research on the mechanism of interaction. So far, the effects of plant nutrition on controlling these disorders are generally judged by external indicators such as disease index, but the research on its mechanism, such as the change of enzyme activity and metabolic process, is less. To promote the application of these disorders' regulations from the perspective of plant nutrition management, it is necessary to clarify the relationship between plant nutrition and disorders at the molecular level. (2) Much researches have been done to study the interaction of elements, but the research on the interaction of elements from the perspective of plant nutrition on disorders is not systematic. Using nutrition coordination of the prevention and cure of the occurrence of these disorders, in addition to fruit crop diversity and complexity of the pathogen, should also fully consider the elements of individual action and interaction, clear the role of each factor, and coordinate the relationship between each element and fruit trees; the guarantee of reasonable application with various nutritional elements can not only provide sufficient nutrients to the tree body and can effectively improve the disorders resistance of fruit tree. (3) At present, most of the research results are limited to the theoretical application, and the case of combining theory with production success is rarely reported. It is feasible to regulate mineral nutrition to improve the disorder resistance of fruit trees, and balanced nutrition, as a supplement of chemical control, agricultural control, and biological control measures, has great development potential and research space and is a new approach worthy of discussion. On the basis of further enriching the content of studying disorders of fruit tree from the perspective of plant nutrition and elevating it to the height of the discipline, the theory should be applied to the production practice of reducing pesticide and fertilizer consumption, reducing environmental pollution and increasing yield.

References

Agusti, M., Almela, V., Pons, J., 1992. Effects of girdling on alternate bearing in citrus. J. Hortic. Sci. 67, 203–210.

Agustí, M., Martinez-Fuentes, A., Mesejo, C., 2002. Citrus fruit quality. Physiological basis and techniques of improvement. Agrociencia 6, 1–16.

Agustí, M., Mesejo, C., Reig, C., Martínez-Fuentes, A., 2014. Citrus production. In: Dixon, G.R., Aldous, D.E. (Eds.), Horticulture: Plants for People and Places. In: Production Horticulture, vol. 1. Springer Netherlands, Dordrecht, pp. 159–195.

Alferez, F., Tadeo, F.R., Zacarías, L., Agustí, M., Juan, M., Almela, V., 2001. Histological and physiological characterization of rind breakdown of 'Navelate' sweet orange. Ann. Bot. 88, 415–422.

Bower, J.P., 2004. The physiological control of citrus creasing. 632 edInternational Society for Horticultural Science (ISHS), Leuven, pp. 111–115.

Chikaizumi, S., 2002. Occurrence of rind disorders of 'Ootani' iyo (Citrus iyo hort. ex Tanaka, var. Ootani) fruit during the pre-harvest and storage periods. J. Jpn. Soc. Hortic. Sci. 71, 13–18.

Cronje, P.J.R., Barry, G.H., Huysamer, M., 2011. Postharvest rind breakdown of 'Nules Clementine' mandarin is influenced by ethylene application, storage temperature and storage duration. Postharvest Biol. Technol. 60, 192–201.

Cronje, P.J., Stander, O.P., Theron, K.I., 2013. Fruit splitting in citrus. Hortic. Rev. 41, e200.

Cronje, P.J.R., Zacarias, L., Alferez, F., 2017. Susceptibility to postharvest peel pitting in Citrus fruits as related to albedo thickness, water loss and phospholipase activity. Postharvest Biol. Technol. 123, 77–82.

de Oliveira, J.G., Vitória, A.P., 2011. Papaya: nutritional and pharmacological characterization, and quality loss due to physiological disorders. An overview. Food Res. Int. 44, 1306–1313.

Goren, R., 1993. Anatomical, physiological, and hormonal aspects of abscission in citrus. Hortic. Rev. 15, 145–182.

Greenberg, J., Kaplan, I., Fainzack, M., Egozi, Y., Giladi, B., 2006. Effects of auxins sprays on yield, fruit size, fruit splitting and the incidence of creasing of 'nova' mandarin. 727 edInternational Society for Horticultural Science (ISHS), Leuven, pp. 249–254.

Greenberg, J., Holtzman, S., Fainzack, M., Egozi, Y., Giladi, B., Oren, Y., Kaplan, I., 2010. Effects of NAA and GA$_3$ sprays on fruit size and the incidence of creasing of 'Washington' navel orange. International Society for Horticultural Science (ISHS), Leuven, pp. 273–279.

Hussain, Z., Singh, Z., 2015. Involvement of polyamines in creasing of sweet orange [Citrus sinensis (L.) Osbeck] fruit. Sci. Hortic. 190, 203–210.

Ibáñez, A.M., Martinelli, F., Reagan, R.L., Uratsu, S.L., Vo, A., Tinoco, M.A., Phu, M.L., Chen, Y., Rocke, D.M., Dandekar, A.M., 2014. Transcriptome and metabolome analysis of Citrus fruit to elucidate puffing disorder. Plant Sci. 217–218, 87–98.

Iglesias, D.J., Tadeo, F.R., Primo-Millo, E., Talon, M., 2006. Carbohydrate and ethylene levels related to fruitlet drop through abscission zone A in citrus. Trees 20, 348–355.

Jim, H., 2004. Oleocellosis Prediction Tools. http://www.australiancitrusgrowers.com.au/PDFs/resources/Oleocellosis Prediction Tools.pdf.

Jona, R., Goren, R., Marmora, M., 1989. Effect of gibberellin on cell-wall components of creasing peel in mature 'Valencia' orange. Sci. Hortic. 39, 105–115.

Jones, W., Embleton, T., Garber, M., Cree, C., 1967. Creasing of orange fruit. Hilgardia 38, 231–244.

Knight, T.G., Klieber, A., Sedgley, M., 2002. Structural basis of the rind disorder oleocellosis in Washington navel orange (Citrus sinensis L. Osbeck). Ann. Bot. 90, 765–773.

Ladaniya, M.S., 2008. Physiological disorders and their management. In: Ladaniya, M.S. (Ed.), Citrus Fruit. Academic Press, San Diego, CA, pp. 451–XXI.

Lafuente, M.T., Zacarias, L., 2006. Postharvest physiological disorders in citrus fruit. Stewart Postharvest Rev. 1, 1–9.

Liu, G.D., Wang, R.D., Liu, L.C., Wu, L.S., Jiang, C.C., 2013. Cellular boron allocation and pectin composition in two citrus rootstock seedlings differing in boron-deficiency response. Plant Soil 370, 555–565.

Martinelli, F., Ibanez, A.M., Reagan, R.L., Davino, S., Dandekar, A.M., 2015. Stress responses in citrus peel: comparative analysis of host responses to Huanglongbing disorders and puffing disorder. Sci. Hortic. 192, 409–420.

Mehouachi, J., Serna, D., Zaragoza, S., Agusti, M., Talon, M., Primo-Millo, E., 1995. Defoliation increases fruit abscission and reduces carbohydrate levels in developing fruits and woody tissues of Citrus unshiu. Plant Sci. 107, 189–197.

Mesejo, C., Reig, C., Martinez-Fuentes, A., Gambetta, G., Gravina, A., Agusti, M., 2016. Tree water status influences fruit splitting in Citrus. Sci. Hortic. 209, 96–104.

Mishra, D.S., Tripathi, A., Nimbolkar, P.K., 2016. Review on physiological disorders of tropical and subtropical fruits: causes and management approach. Int. J. Agric. Environ. Biotechnol. 9, 925–935.

Monselise, S.P., Goldschmidt, E.E., 1982. Alternate bearing in fruit trees. Hortic. Rev. 5, 129–173.

Montero, C.R.S., Schwarz, L.L., dos Santos, L.C., dos Santos, R.P., Bender, R.J., 2012. Oleocellosis incidence in citrus fruit in response to mechanical injuries. Sci. Hortic. 134, 227–231.

Ogata, T., Hirota, T., Shiozaki, S., Horiuchi, S., Kawase, K., Ohashi, M., 2002. Effects of aminoethoxyvinylglycine and high temperatures on fruit set and fruit characteristics of heat-cultured satsuma mandarin. J. Jpn. Soc. Hortic. Sci. 71, 348–354.

Petracek, P.D., Dou, H., Pao, S., 1998. The influence of applied waxes on postharvest physiological behavior and pitting of grapefruit. Postharvest Biol. Technol. 14, 99–106.

Rebolledo, R., García-Luis, A., Guardiola, B., Luis, J., 2012. Effect of 2.4-D exogenous application on the abscission and fruit growth in Sweet orange. var. Salustiana. Agron. Colomb. 30, 34–40.

Ritenour, M.A., Albrigo, L.G., Burns, J.K., Miller, W.M., 2004. Granulation in Florida citrus. Proc. Fla. State Hortic. Soc. 117, 358–361.

Schaffer, A.A., Goldschmidt, E.E., Goren, R., Galili, D., 1985. Fruit-set and carbohydrate status in alternate and nonalternate bearing citrus cultivars. J. Am. Soc. Hortic. Sci. 110, 574–578.

Schutzki, R.E., Cregg, B., 2007. Abiotic Plant Disorders: Symptoms, Signs and Solutions: A Diagnostic Guide to Problem Solving. Michigan State University Extension.

Sharma, N., Singh, S.K., Mahato, A.K., Ravishankar, H., Dubey, A.K., Singh, N.K., 2019. Physiological and molecular basis of alternate bearing in perennial fruit crops. Sci. Hortic. 243, 214–225.

Shomer, I., Erner, Y., 1989. The nature of oleocellosis in citrus-fruits. Bot. Gaz. 150, 281–288.

Singh, R., 2001. 65-year research on citrus granulation. Indian J. Hortic. 58, 112–144.

Srivastava, A.K., 2013a. Nutrient deficiency symptomology in citrus: an effective diagnostic tool or just an aid for post–mortem analysis. Agric. Adv. 2, 177–194.

Srivastava, A.K., 2013b. Recent developments in diagnosis and management of nutrient constraints in acid lime. Sci. J. Agric. (Agric. Adv.) 2, 86–96.

Srivastava, A.K., Singh, S., 2006. Diagnosis of nutrient constraints in citrus orchards of humid tropical India. J. Plant Nutr. 29, 1061–1076.

Srivastava, A.K., Singh, S., 2009. Citrus decline: soil fertility and plant nutrition. J. Plant Nutr. 32, 197–245.

Stander, O.P.J., Barry, G.H., Cronjé, P.J.R., 2018. The significance of macronutrients in alternate bearing 'Nadorcott' mandarin trees. HortScience 53, 1600–1609.

Stewart, I., Leonard, C.D., 1951. Citrus Nutrition Studies. Molybdenum Deficiency. .

Storey, R., Treeby, M.T., 1994. The morphology of epicuticular wax and albedo cells of orange fruit in relation to albedo breakdown. J. Hortic. Sci. 69, 329–338.

Tao, M.M., Ying, H., Sun, Q., Tan, Q.L., Li, C., Hu, C.X., 2016. Soil and plant nutrient status of citrus orchard in Hubei and Hunan Province. Hubei Agric. Sci.. 55, 3289–3293, 3297.

Thompson, C.R., Taylor, O.C., 1969. Effects of air pollutants on growth, leaf drop, fruit drop, and yield of citrus trees. Environ. Sci. Technol. 3, 934–940.

Wallace, T., 1934. II. Some physiological disorders of fruit trees. Ann. Appl. Biol. 21, 322–333.

Wu, X.W., Lu, X.P., Riaz, M., Yan, L., Jiang, C.C., 2018. Boron deficiency and toxicity altered the subcellular structure and cell wall composition architecture in two citrus rootstocks. Sci. Hortic. 238, 147–154.

Xiong, B., Ye, S., Qiu, X., Liao, L., Sun, G.C., Luo, J.Y., Dai, L., Rong, Y., Wang, Z.H., 2017. Exogenous spermidine alleviates fruit granulation in a citrus cultivar (Huangguogan) through the antioxidant pathway. Acta Physiol. Plant.. 39.

Zaragoza, S., Alonso, E., 1975. El manchado de la corteza de los agrios: estudio preliminar en la variedad 'Navelate'. Manchas pre-recolección. INIA Ser. Prot. Veg. 4, 31–32.

Zheng, Y.Q., He, S.L., Yi, S.L., Zhou, Z.Q., Mao, S.S., Zhao, X.Y., Deng, L., 2010a. Characteristics and oleocellosis sensitivity of citrus fruits. Sci. Hortic. 123, 312–317.

Zheng, Y.Q., He, S.L., Yi, S.L., Zhou, Z.Q., Mao, S.S., Zhao, X.Y., Deng, L., 2010b. Predicting oleocellosis sensitivity in citrus using VNIR reflectance spectroscopy. Sci. Hortic. 125, 401–405.

Zheng, Y.Q., Deng, L., He, S.L., Zhou, Z.Q., Yi, S.L., Zhao, X.Y., Wang, L.A., 2011. Rootstocks influence fruit oleocellosis in 'Hamlin' sweet orange (Citrus sinensis L. Osbeck). Sci. Hortic. 128, 108–114.

Zheng, Y., Jia, X.-m., Yang, Q., Liu, Y.-m., Xie, R.-j., Ma, Y.-y., He, S.-l., Deng, L.J.A.P.P., 2016. Role of Ca2+ and calmodulin in on-tree oleocellosis tolerance of Newhall navel orange. Acta Physiol. Plant. 38, 194.

Zheng, Y., Wang, Y., Yang, Q., Liu, Y., Xie, R., He, S., Deng, L., Yi, S., Lv, Q., Ma, Y., 2018. Modulation of tolerance of "Hamlin" sweet orange grown on three rootstocks to on-tree oleocellosis by summer plant water balance supply. Sci. Hortic. 238, 155–162.

Zur, N., Shlizerman, L., Ben-Ari, G., Sadka, A., 2017. Use of magnetic resonance imaging (MRI) to study and predict fruit splitting in citrus. Hortic. J. 86, 151–158.

5

Ecophysiology of fruit crops: A glance at its impact on fruit crop productivity

Hermann Restrepo-Diaz*, Alefsi David Sánchez-Reinoso

Departamento de Agronomia, Facultad de Ciencias Agrarias, Universidad Nacional de Colombia, Bogotá, Colombia

*Corresponding author. E-mail: hrestrepod@unal.edu.co

1 Introduction

Fruit crop production can be understood as an open and complex system in which growth, development, and subsequent yield are conditioned by factors such as the sowing system, the environmental conditions, the type of soil, the crop management, and the interaction between said factors. In this regard, it is very important to study and understand the mechanisms and physiological processes that would influence yield to optimize possible management techniques and try to comprehend the expression or response of different genotypes to environmental variation. Therefore, ecophysiology or environmental physiology of crops is a discipline that studies the interaction between crops and their physical, chemical, and biological environment that, in turn, helps to be the basis of the dynamics and productivity of an agroecosystem.

In this context, ecophysiology studies from the perspective of crop production, in general, have tried to answer a series of questions, such as the following:

- Why do genotypes of the same species present differences in yield?
- What plant mechanisms and/or processes are involved in the acclimation response of a genotype to an environment?
- What environmental factor may be negatively or positively influencing the yield of a crop in a given region?
- What species can be cultivated or introduced in certain production areas?

To answer or explain this series of questions, ecophysiology works have been carried out from the perspectives of agronomic, phenological, morphological, photosynthetic, biochemical, and even genetic analyses. Great progress has been made during the last decades, which has been reflected in advances in fruit crops productivity. However, agriculture is currently facing new problems due to the different global climate change scenarios, which has led to the establishment of new guidelines such as the development of crop models in contrasting environments and comparative studies of genetic variation. Additionally, it is important to continue the

A.K. Srivastava, Chengxiao Hu (eds.)
Fruit Crops: Diagnosis and Management of Nutrient Constraints
https://doi.org/10.1016/B978-0-12-818732-6.00005-8

59

characterization of crop species at the physiological, biochemical, and molecular levels, taking into account the change of environmental conditions and conducting studies over time to characterize the response of agroecosystems to climate change.

As a result, the objective of this chapter is to analyze the role of ecophysiology and show the effect of some environmental factors (solar radiation, temperature, water availability, and wind) on the physiological and agronomic behavior of fruit crops.

2 Light

Light is the main necessary resource for the growth and development of a fruit tree. However, fruit species change their light requirements, since excess or deficiency can be harmful for the crop (Table 5.1). In this regard, the plant's photosynthetic rate is strongly related to the solar radiation intercepted by the plant. Similarly, light influences other processes such as organ formation or appearance, flowering, source-sink relationships, and finally yield.

The length of the day refers to the amount of time a plant is exposed to sunlight. Consequently, the photoperiod is a growth factor that influences the flowering of some fruit crops, which can be divided into three groups: (i) short-day plants that need day lengths shorter than a certain critical maximum (e.g., pineapple and strawberry), (ii) long-day plants that need short darkness periods to flower (e.g., most of the temperate zone fruits), and (iii) neutral day plants that bloom regardless of the length of the day (e.g., guava, papaya, citrus, and banana). Knowledge of the photoperiod has been useful for the production of certain fruit species around the world (Ghosh et al., 2016).

Light also plays an important role in the formation, development, and quality of the fruit. In this regard, it has been reported that a large amount of solar radiation induces changes in the carbohydrate distribution to fruits (Fischer et al., 2012). Similarly, incident radiation during fruit development influences fruit quality. A summary of the effect of light level is presented in Table 5.2.

3 Temperature

Temperature is one of the most limiting factors that influence fruit trees growth and yield in both temperate and tropical regions. In this regard, climate change will cause an increase in temperature from 2°C to 4°C at the end of the 21st century (Lobell and Gourdji, 2012). High temperatures usually promote fruit abscission, causing yield reduction in crops (Stephenson, 1981). Consequently, a good understanding of the acclimation responses of plants to temperature stress conditions (high or low temperatures) is important to specify the behavior of genotypes in their environment and, thus, develop agronomic management strategies and/or phenotypic characterization techniques that help genetic improvement.

To understand the acclimation responses of plants to changes in temperature, it is important to consider that critical temperature levels are also different for cultivated species, between genotypes of a species and phenological stages. A temperature variation can lead to a series of changes in photosynthesis, leaf respiration, plant growth and development, source-sink relationships, among others. In this regard, the combination of excess radiation and high temperatures is often the most limiting factor that affects the growth and final yield of the crop in tropical regions. Additionally, extreme temperatures (high or low) can produce considerable damage such as burning of leaves, branches and stems, leaf senescence and abscission, growth inhibition of stems and roots, fruit damage and discoloration, and low production (Hatfield and Prueger, 2015; Wahid et al., 2007). Photosynthesis and respiration are the growth-related processes that show more continuous changes when temperature increases. On the other hand, the type of response to thermal stress may vary between phenological stages. During the vegetative stage, for example, a high temperature can damage the leaf gas exchange properties. Regarding the reproductive stage, a short period of thermal stress may favor the increase of flower abortion; additionally, damage in the development, viability. and germination of pollen grains and fruit setting has been reported (Hatfield and Prueger, 2015; Wahid et al., 2007).

TABLE 5.1 Photosynthetic characteristics of some fruit species.

Crop	Light saturation point (µmol/m/s)	Maximum photosynthesis (µmol CO$_2$/m/s)	Reference
Citrus sinensis	750–1000	15–22	Baldocchi and Amthor (2001)
Passiflora edulis Sims	230–420 370–1160	9–20[a] 9–16[b]	Perez Martinez and Melgarejo (2015)
Carica papaya	1900	25–30	Baldocchi and Amthor (2001)
Musa spp.	700–2000	5–25	Bréda (2003)
Carica papaya	2000	25–30	Jiménez et al. (2014)
Vitis vinifera	1500	10–15	Tombesi et al. (2019)

[a] *Values taken at the flowering stage.*
[b] *Values taken at the fruiting stage.*

TABLE 5.2 Effect of light during some physicochemical characteristics of fruits exposed to different radiation levels.

Species	Radiation level	Reference
Avocado	• Fruits exposed to greater solar radiation require 1.5 days longer to ripen. • Avocados from the Hass cultivar have shown a higher content of calcium (100%), magnesium (50%), and potassium (60%) when exposed directly to sunlight. • Fruit firmness is lower when the fruit develops with low radiation. • Avocados that develop with low light can present greater damage due to cold during storage.	Moretti et al. (2010)
Cape gooseberry	• Red light reduces the content of total phenolic compounds in fruits. • Blue light increases the content of soluble solids in fruits.	da Silva et al. (2016)
Satsuma mandarin	• Red light increases total carotenoid content in fruits.	Ma et al. (2011)
Lulo	• Long days during flowering accelerate the beginning of harvest, generating a shorter harvest period and a greater number of fruits per plant.	Messinger and Lauerer (2015)
Pineapple guava	• Reduction of degrees Brix under shading. • A higher ripening index in fruits with greater exposure to sunlight.	Martínez-Vega et al. (2008)

Source-sink relationships may also suffer alterations with temperature variations. Low temperatures can affect the transport of sugars through the phloem. In general, cold stress can condition the sink's strength, since enzymes responsible for carbohydrate accumulation (sucrose synthetase, glucokinase, and soluble starch synthetases) and sucrose hydrolysis (invertases) are inhibited, which results in a poor development of the harvestable organ (Lemoine et al., 2013; Tarkowski and Van den Ende, 2015). Also, a thermal stress (due to high or low temperature) alters the source-sink relationships because it can reduce the photoassimilates supply of the source, since it is related to photosynthesis. Under thermal stress conditions, there is a reduction of the photosynthetic rate, alterations of the thylakoid membranes, reduction of the carbon cycle, and generation of reactive oxygen species (Lemoine et al., 2013; Wahid et al., 2007). On the other hand, the temperature during fruit development plays an important role to determine fruit quality. For example, high temperatures cause an increase in sugar content and low levels of tartaric acid in grapes. Additionally, the oil content of "Hass" avocado fruits was higher when they developed at a high temperature, with a higher concentration of palmitic acid (Moretti et al., 2010). It is also important to note that thermal stress, coupled with high solar radiation, can cause physiopathies in fruits such as sunburn, which reduces the product's commercial value (Fig. 5.1). The effect of some physiological disorders on fruits due to the combination of light and temperature is described in Table 5.3.

FIG. 5.1 Sunburn damage in lulo plants (*Solanum quitoense*). Photograph: Hermann Restrepo-Díaz.

TABLE 5.3 Physiopathies due to sunburn in fruit tree crops.

Crop	Physiopathy symptoms
Banana passion fruit	Dark brown spots located in the exposed part, which in turn constitute sources of entry of pathogens to the fruit.
Sweet granadilla	
Passion fruit	
Avocado	Brown spots on the skin and pulp; the fruit's susceptibility to fall increases.
Pineapple	The fruit's pulp presents a translucent appearance; it can also present soft spots.
Mandarin	Asymmetric flattening of the area exposed to the sun; the bark turns yellow and acquires a dry and hard texture.
Banana	Brown spots located in the area exposed to the sun.
Tamarillo	Brown spots on the skin and pulp; the fruit's susceptibility to fall increases.
Lulo	Brown spots on the peel; affected leaves have a stiff appearance.

Based on Chabbal, M.D., Piccoli, A.B., Martínez, G.C., Avanza, M.M., Mazza, S.M., Rodríguez, V.A., 2014. Aplicaciones de caolín para el control del golpe de sol en mandarino'okitsu'. Cultiv. Trop. 35 (1), 50–56; Fischer, G., Casierra-Posada, F., Piedrahíta, W., 2009. Ecofisiología de las especies pasifloráceas cultivadas en Colombia. Cultivo, poscosecha y comercialización de las pasifloráceas en Colombia: maracuyá, granadilla, gulupa y curuba. Revista Colombiana de Ciencias Hortícolas. Bogotá, pp. 45–67; Moretti, C.L., Mattos, L.M., Calbo, A.G., Sargent, S.A., 2010. Climate changes and potential impacts on postharvest quality of fruit and vegetable crops: a review. Food Res. Int. 43 (7), 1824–1832; and Muchui, M.N., Mathooko, F.S., Njoroge, C.K., Kahangi, E.M., Onyango, C.A., Kimani, E.M., 2017. Effect of perforated blue polyethylene bunch covers on selected postharvest quality parameters of tissuecultured bananas (Musa spp.) cv. Williams in Central Kenya. J. Stored Prod. Postharvest Res. 1 (3), 29–41.

4 Water

Water is important to maintain the activity of physiological processes and the transport through the plant membranes. Consequently, knowledge on the severity and duration of a water deficit helps to identify the possible impacts on fruit tree productivity. Water stress is a subject that has been extensively studied, and its effects on growth, phenology, water relations, nutrient content, assimilation, photosynthesis, and respiration have been characterized. Likewise, improving crop tolerance to drought conditions has acquired great importance, since it is a promising tool to face water availability problems caused by climate change and/or climate variability. On the other hand, rainfall intensity and frequency have also suffered alterations in their patterns, which have generated waterlogging problems in certain agricultural areas in tropical regions. For this reason, this part of the chapter will mention some of the effects of abiotic stresses such as drought and waterlogging.

Water stress effects cover a wide range of responses, from molecular levels to plant community. The greatest effect of water deficit is the decrease of the photosynthetic rate, since its drop affects foliar expansion, decouples the

TABLE 5.4 Reduction in crop yield due to the interaction between water stress and the phenological stage.

Crop	Phenological stage	Yield reduction (%)	Reference
Banana	Flower initiation	8–28	Van Asten et al. (2011)
Nectarine	Fruit formation	15–25	Naor et al. (1999)
Peach	Beginning of the reproductive stage	23–37	Berman and DeJong (1996)
Pear	Reproductive	20–40	Naor (2001)

photosynthetic apparatus, advances leaf senescence, reduces rubisco efficiency, and favors the expression of genes associated with the abscisic acid (Al-Absi and Archbold, 2016; Salazar et al., 2015). On the other hand, water deficit produces an increase in root respiration, causing an imbalance in the use of carbon sources (da Silva et al., 2011). Similarly, source-sink relationships of crops can be affected, since the level of carbohydrates in the leaves is altered by water stress. Sucrose and hexose levels increase, while starch levels decrease. These changes in carbohydrate levels can help to provide energy and be osmolytes to maintain the metabolic activity of cells and leaves under low water availability conditions (Lemoine et al., 2013).

A low water content in the soil also produces a low absorption of nutrients and their reduction in the plant tissues because it mainly affects the mass flow and nutrient translocation from the roots to the aerial part, which results in the affectation of the plant's metabolic activities. It has been reported that water deficit is the main cause of losses in banana production in Africa, especially because nutrient intake is negatively affected (Van Asten et al., 2011). On the other hand, it has been found that the use of techniques such as deficit irrigation can improve yield and quality of fruit trees (Yang et al., 2017). With this practice, the plants adapt different physiological strategies such as the increase of water use efficiency and root volume in the soil, and the reduction of the plant's water potential through the production of osmolytes such as proline and soluble sugars such as sucrose, arabinose, and glucose (Galindo et al., 2018). Additionally, the positive effects of a good nutritional status or the addition of nutrients have been reported to help crops to face water deficit periods. In tamarillo, a good potassium nutritional status can help the plant to cope with water deficit conditions while maintaining its water use efficiency (Clavijo-Sánchez et al., 2015). From fruit trees production, the interaction between phenological stages and water stress on plant yield has been studied. In this context, these studies have allowed characterizing what stages are more susceptible to a water deficit, which would negatively affect fruit trees yield and productivity. The effects of the interaction between water deficit and phenological stages in some fruit crops of economic importance are summarized in Table 5.4.

The low oxygen availability due to excess water is another factor to consider in the production of fruit species. Climate change has altered the normal rainfall cycle, flooding arable land and affecting agricultural production. In this sense, prolonged periods of waterlogging stress can cause yield losses between 10% and 40% (Aldana et al., 2014). Among the plant physiological responses to waterlogging periods are the alterations in the leaf gas exchange relationships. The first symptom is a stomatal closure that influences passive water absorption. Likewise, there is a drop in transpiration, which causes leaf wilting and senescence. Stem growth, leaf area, fruit setting and fruit yield may also be affected. Aerenchyma production that can lead to a thickening of the stem is another response that plants have to get oxygen. Finally, studies characterizing the physiological responses of horticultural species cultivated in the Andes have become important during the last decade because climate change has caused heavy rains in dry seasons. Additionally, phenomena such as "La Niña" have brought greater crop waterlogging in agricultural areas of Colombia, especially in flat or riverside areas (Aldana et al., 2014; Baracaldo et al., 2014). In this respect, a summary of the physiological responses of some Solanaceae species to waterlogging is presented in Table 5.5.

TABLE 5.5 Waterlogging effects on the physiological behavior of some crops of economic importance belonging to the Solanaceae family in the Andean region.

Crop	Physiological response	Reference
Cape gooseberry	Waterlogging periods of more than 2 days begin to reduce leaf area. The stem diameter is smaller after 4 days of exposure. Decrease of flower buds, flowers, and fruits with increased stress.	Aldana et al. (2014)
Tamarillo	Lower growth, lower efficient use of nitrogen, reduction of leaf area, greater distribution of assimilates to stems.	Flórez-Velasco et al. (2015)
Lulo	Chlorophyll content decreases. Reductions in stomatal conductance, transpiration, and efficient use of nitrogen. Larger diameter of the stem. Lower leaf succulence.	Betancourt-Osorio et al. (2016)

TABLE 5.6 Acclimatization responses of woody and nonwoody plants to loading caused by wind speed.

Level	Nonwoody plants	Woody plants
Cell	• Increased lignification • Increased structural collenchyma • Increased strengthening sclerenchyma • Increase in developmental rate	• Reduced length and diameter of vessels • Reduced length of wood fibers • Increase in xylem production
Leaves, flowers, and fruits	• Smaller, thicker leaves • More and smaller stomata • Increase in sclerenchyma cells • Decrease in the number of flowers, fruits, and seeds	• Smaller leaves • Thicker cuticle • Decreased seed production
Branches, stems, and roots	• Increased collenchyma • Larger phloem vessels • Wider stems • Thicker stem walls • Reinforcement of stems • Shorter stems • Increased root/shoot ratio	• Larger branch diameter • More flexible stem and branch material • Shorter stems • Increase in secondary growth • Increased root/shoot ratio
Whole plant	• Reorientation	• Reorientation • Increased porosity of tissues

Based on Gardiner, B., Berry, P., Moulia, B., 2016. Wind impacts on plant growth, mechanics and damage. Plant Sci. 245, 94–118.

5 Wind

Wind can be one of the environmental stresses that strongly affect growth, development, and yield of crops. Although there has been extensive research on this subject during the last decades, the response of crops to the effect of wind is still not clear. This may be due to the fact that wind not only causes physical damage but also affects the gaseous exchange properties of leaves and the heat exchange (Onoda and Anten, 2011). Plants exposed to wind exhibit anatomical and morphological changes such as increased leaf thickness, decreased leaf area, and height. A lower leaf growth is possibly due to a lower stomatal conductance. Additionally, continuous gusts of wind can generate a decrease in the height of the plant. Likewise, source-sink relationships can be affected because the distribution of carbohydrates assigned to leaf growth is directed to stems and roots to support the plant, causing a possible impact on crop yield.

On the other hand, wind can cause physical damage by abrasion between the leaf tissues, leaf removal from the plant, and leaf wear by particles suspended in the air. The abrasion between leaf tissues is caused by the friction generated by wind, which produces a disturbance in the epicuticular waxes that eventually increases leaf transpiration causing tissue dehydration. Regarding leaf removal, it is more frequent in plants with leaves of greater area such as corn, potato, and lulo. In the case of banana, high wind speeds cause tearing and subsequent fall of the leaf blade, producing changes in the physiology of the leaf. In addition, a high wind speed reduces the thickness of the boundary layer, which leads to a lower difference in vapor pressure in the leaf, affecting the stomatal conductance and transpiration. On the other hand, in the initial stages of the crop, it is more likely that abrasion may occur due to the particles suspended in the air, which cause defoliation, a fact that may affect crop yield (Cleugh et al., 1998; Turner et al., 2007).

The acclimation responses of crops to wind damage can be divided between nonwoody and woody plants. In nonwoody plants, greater lignification, thickening of the cuticle, and shorter and thicker stems may occur. As for woody plants, there is a reduction in leaf sizes, a larger branch diameter, an increase in root/shoot ratio, and secondary growth (Gardiner et al., 2016). A summary of the acclimation responses of woody and nonwoody plants to wind loading at different levels is presented in Table 5.6.

6 Future line of research

All human activity has directly affected the vegetation cover on the terrestrial surface, causing a land-use change during the last years (Sage and Kubien, 2003). In consequence, a global trend is expanding fruit tree crops towards marginal, degraded, and flooding arable lands where water shortage or waterlogging and unfavorable temperatures would constitute major constraints to crop yield (DaMatta et al., 2007). Also, climate change has caused negative effects on the growth, development, production and quality of fruit crops, which undoubtedly will be difficult to predict and generalize because the physiological processes of plants are multidimensional (Fischer et al., 2016).

The ecophysiology plays an important role in advancing the frontier of knowledge essential to predict crop productivity and acclimatization of fruit crops to marginal lands, biotic stress, abiotic stress, and global climate change. In this sense, several authors suggest that future lines of research in fruit crops should be emphasized in the following areas: (1) basic research to address the mechanistic basis of plant stress tolerance (genetics, biochemistry, and molecular biology) and precision in plant phenotyping (Ainsworth et al., 2016), (2) advance in the characterization of canopy responses of fruit trees to environmental stresses (DaMatta, 2007), (3) mathematical models to predict fruit trees responses to future scenarios such as flooding, droughts, and elevated CO_2 (Fischer et al., 2016), and (4) works focused on strategies combining observations and mechanistic modeling to advance our knowledge on the interaction between the fruit crop system and the environmental change (Damm et al., 2018).

7 Conclusion

Knowledge of plant ecophysiology is an important source of information to understand the effect of the natural changing environmental conditions and their impact on fruit crop productivity, as well as to develop new approaches to improve fruit trees response to new environmental conditions through phenotyping tools, canopy characterization and mathematical models, maintaining commercial agriculture profitable and guaranteeing humankind's supply of goods and services.

References

Ainsworth, E.A., Bernacchi, C.J., Dohleman, F.G., 2016. Focus on ecophysiology. Plant Physiol. 172 (2), 619–621.

Al-Absi, K.M., Archbold, D.D., 2016. Apple tree responses to deficit irrigation combined with periodic applications of particle film or abscisic acid. Horticulturae 2 (4), 1–16.

Aldana, F., García, P.N., Fischer, G., 2014. Effect of waterlogging stress on the growth, development and symptomatology of cape gooseberry (*Physalis peruviana* L.) plants. Rev. Acad. Colomb. Cienc. Exact. Fís. Natur. 38 (149), 393–400.

Baldocchi, D.D., Amthor, J.S., 2001. Canopy photosynthesis: history. In: Terrestrial Global Productivity. Academic Press, New York, NY, pp. 9–31.

Baracaldo, A., Carvajal, R., Romero, A.P., Prieto, A., García, F., Fischer, G., Miranda, D., 2014. El anegamiento afecta el crecimiento y producción de biomasa en tomate chonto (*Solanum lycopersicum* L.), cultivado bajo sombrío. Rev. Colomb. Cienc. Hortic. 8 (1), 92–102.

Berman, M.E., DeJong, T.M., 1996. Water stress and crop load effects on fruit fresh and dry weights in peach (*Prunus persica*). Tree Physiol. 16 (10), 859–864.

Betancourt-Osorio, J., Sanchez-Canro, D., Restrepo-Diaz, H., 2016. Effect of nitrogen nutritional statuses and waterlogging conditions on growth parameters, nitrogen use efficiency and chlorophyll fluorescence in tamarillo seedlings. Not. Bot. Horti Agrobot. Cluj-Napoca 44 (2), 375–381.

Bréda, N.J., 2003. Ground-based measurements of leaf area index: a review of methods, instruments and current controversies. J. Exp. Bot. 54 (392), 2403–2417.

Clavijo-Sánchez, N., Flórez-Velasco, N., Restrepo-Díaz, H., 2015. Potassium nutritional status affects physiological response of tamarillo plants (*Cyphomandra betacea* Cav.) to drought stress. J. Agric. Sci. Technol. 17, 1839–1849.

Cleugh, H.A., Miller, J.M., Böhm, M., 1998. Direct mechanical effects of wind on crops. Agrofor. Syst. 41 (1), 85–112.

DaMatta, F.M., 2007. Ecophysiology of tropical tree crops: an introduction. Braz. J. Plant Physiol. 19 (4), 239–244.

Damm, A., Paul-Limoges, E., Haghighi, E., Simmer, C., Morsdorf, F., Schneider, F.D., van der Tol, C., Migliavacca, M., Rascher, U., 2018. Remote sensing of plant-water relations: an overview and future perspectives. J. Plant Physiol. 227, 3–19.

da Silva, E.C., Nogueira, R.J.M.C., da Silva, M.A., de Albuquerque, M.B., 2011. Drought stress and plant nutrition. Plant Stress 5 (Special Issue 1), 32–41.

da Silva, D.F., Pio, R., Rodrigues-Soares, J.D., de Siqueira-Elias, H.H., Villa, F., Boas, V., de Barros, E.V., 2016. Light spectrum on the quality of fruits of physalis species in subtropical area. Bragantia 75 (3), 371–376.

Fischer, G., Almanza-Merchán, P.J., Ramírez, F., 2012. Source-sink relationships in fruit species. A review. Rev. Colomb. Cienc. Hortic. 6 (2), 238–253.

Fischer, G., Ramírez, F., Casierra-Posada, F., 2016. Ecophysiological aspects of fruit crops in the era of climate change. A review. Agron. Colomb. 34 (2), 190–199.

Flórez-Velasco, N., Balaguera-López, H.E., Restrepo-Díaz, H., 2015. Effects of foliar urea application on lulo (*Solanum quitoense* cv. septentrionale) plants grown under different waterlogging and nitrogen conditions. Sci. Hortic. 186, 154–162.

Galindo, A., Collado-González, J., Griñán, I., Corell, M., Centeno, A., Martín-Palomo, M.J., Girón, I.F., Rodríguez, P., Cruz, Z.N., Memmi, H., Carbonell-Barrachina, A.A., Hernández, F., Torrecillas, A., Moriana, A., Pérez-Lopez, D., 2018. Deficit irrigation and emerging fruit crops as a strategy to save water in Mediterranean semiarid agrosystems. Agric. Water Manag. 202, 311–324.

Gardiner, B., Berry, P., Moulia, B., 2016. Wind impacts on plant growth, mechanics and damage. Plant Sci. 245, 94–118.

Ghosh, A., Dey, K., Das, S., Dutta, P., 2016. Effect of light on flowering of fruit crops. Adv. Life Sci. 5 (7), 2597–2603.

Hatfield, J.L., Prueger, J.H., 2015. Temperature extremes: effect on plant growth and development. Weather Clim. Extrem. 10, 4–10.

Jiménez, V.M., Mora-Newcomer, E., Gutiérrez-Soto, M.V., 2014. Biology of the papaya plant. In: Genetics and Genomics of Papaya. Springer, New York, NY, pp. 17–33.

Lemoine, R., La Camera, S., Atanassova, R., Dédaldéchamp, F., Allario, T., Pourtau, N., Bonnemain, J.L., Laloi, M., Coutos-Thévenot, P., Maurosset, L., Faucher, M., Girousse, C., Lemonnier, P., Parrilla, J., Durand, M., 2013. Source-to-sink transport of sugar and regulation by environmental factors. Front. Plant Sci. 4, 272.

Lobell, D.B., Gourdji, S.M., 2012. The influence of climate change on global crop productivity. Plant Physiol. 160 (4), 1686–1697.

Ma, G., Zhang, L., Kato, M., Yamawaki, K., Kiriiwa, Y., Yahata, M., Ikoma, Y., Matsumoto, H., 2011. Effect of blue and red LED light irradiation on β-cryptoxanthin accumulation in the flavedo of citrus fruits. J. Agric. Food Chem. 60 (1), 197–201.

Martínez-Vega, R.R., Fischer, G., Herrera, A., Chaves, B., Quintero, O.C., 2008. Physico-chemical characteristics of pineapple guava fruits as influenced by canopy position. Rev. Colomb. Cienc. Hortíc. 2 (1), 21–32.

Messinger, J., Lauerer, M., 2015. Solanum quitoense, a new greenhouse crop for Central Europe: flowering and fruiting respond to photoperiod. Sci. Hortic. 183, 23–30.

Moretti, C.L., Mattos, L.M., Calbo, A.G., Sargent, S.A., 2010. Climate changes and potential impacts on postharvest quality of fruit and vegetable crops: a review. Food Res. Int. 43 (7), 1824–1832.

Naor, A., 2001. Irrigation and crop load influence fruit size and water relations in field-grown 'Spadona' pear. J. Am. Soc. Hortic. Sci. 126 (2), 252–255.

Naor, A., Klein, I., Hupert, H., Grinblat, Y., Peres, M., Kaufman, A., 1999. Water stress and crop level interactions in relation to nectarine yield, fruit size distribution, and water potentials. J. Am. Soc. Hortic. Sci. 124 (2), 189–193.

Onoda, Y., Anten, N.P., 2011. Challenges to understand plant responses to wind. Plant Signal. Behav. 6 (7), 1057–1059.

Perez Martinez, L.V., Melgarejo, L.M., 2015. Photosynthetic performance and leaf water potential of gulupa (Passiflora edulis Sims, Passifloraceae) in the reproductive phase in three locations in the Colombian Andes. Acta Biol. Colomb. 20 (1), 183–194.

Sage, R.F., Kubien, D.S., 2003. *Quo vadis* C4? An ecophysiological perspective on global change and the future of C4 plants. Photosynth. Res. 77, 209–225.

Salazar, C., Hernández, C., Pino, M.T., 2015. Plant water stress: associations between ethylene and abscisic acid response. Chil. J. Agric. Res. 75, 71–79.

Stephenson, A.G., 1981. Flower and fruit abortion: proximate causes and ultimate functions. Annu. Rev. Ecol. Syst. 12 (1), 253–279.

Tarkowski, Ł.P., Van den Ende, W., 2015. Cold tolerance triggered by soluble sugars: a multifaceted countermeasure. Front. Plant Sci. 6 (203), 1–7.

Tombesi, S., Cincera, I., Frioni, T., Ughini, V., Gatti, M., Palliotti, A., Poni, S., 2019. Relationship among night temperature, carbohydrate translocation and inhibition of grapevine leaf photosynthesis. Environ. Exp. Bot. 157, 293–298.

Turner, D.W., Fortescue, J.A., Thomas, D.S., 2007. Environmental physiology of the bananas (Musa spp.). Braz. J. Plant Physiol. 19 (4), 463–484.

Van Asten, P.J.A., Fermont, A.M., Taulya, G., 2011. Drought is a major yield loss factor for rainfed East African highland banana. Agric. Water Manag. 98 (4), 541–552.

Wahid, A., Gelani, S., Ashraf, M., Foolad, M.R., 2007. Heat tolerance in plants: an overview. Environ. Exp. Bot. 61 (3), 199–223.

Yang, H., Du, T., Qiu, R., Chen, J., Wang, F., Li, Y., Wang, C., Gao, L., Kang, S., 2017. Improved water use efficiency and fruit quality of greenhouse crops under regulated deficit irrigation in northwest China. Agric. Water Manag. 179, 193–204.

Further reading

Chabbal, M.D., Piccoli, A.B., Martínez, G.C., Avanza, M.M., Mazza, S.M., Rodríguez, V.A., 2014. Aplicaciones de caolín para el control del golpe de sol en mandarino 'okitsu'. Cultiv. Trop. 35 (1), 50–56.

Fischer, G., Casierra-Posada, F., Piedrahíta, W., 2009. Ecofisiología de las especies pasifloráceas cultivadas en Colombia. Cultivo, poscosecha y comercialización de las pasifloráceas en Colombia: maracuyá, granadilla, gulupa y curuba. Revista Colombiana de Ciencias Hortícolas, Bogotá, pp. 45–67.

Muchui, M.N., Mathooko, F.S., Njoroge, C.K., Kahangi, E.M., Onyango, C.A., Kimani, E.M., 2017. Effect of perforated blue polyethylene bunch covers on selected postharvest quality parameters of tissue-cultured bananas (Musa spp.) cv. Williams in Central Kenya. J. Stored Prod. Postharvest Res. 1 (3), 29–41.

6

Estimating carbon fixation in fruit crops

Ana Pérez-Piqueres[a],*, Belén Martínez-Alcántara[b],
Isabel Rodríguez-Carretero[a], Rodolfo Canet[a], Ana Quiñones[a]

[a]Center for the Development of Sustainable Agriculture, Valencian Institute of Agricultural Research, Valencia, Spain
[b]Certification Section, Plant Health Service, Ministry of Agriculture, Environment, Climate Change and Rural Development, Valencia, Spain
*Corresponding author. E-mail: quinones_ana@gva.es

1 Introduction

The latest Intergovernmental Panel on Climate Change Report predicts an unequivocal warming of the climate system (IPCC, 2014). This global warming is due mainly to anthropogenic greenhouse gas emissions. Carbon dioxide (CO_2) is probably the most critical of the greenhouse gases in the atmosphere as it accounts for the largest proportion of "trace gases." Currently, it is responsible for 60% of the improved greenhouse effect (Barros et al., 2014). In this situation, fruit tree orchards can play an essential role in the carbon (C) cycle of terrestrial ecosystems and can contribute to C sequestration (Quiñones et al., 2013) similarly to that described in forestry (Navar, 2009; Djomo et al., 2011; McEwan et al. (2011); Alvarez et al., 2012).

Indeed all plant ecosystems can sequestrate CO_2 from air, but their effectiveness depends on the ability of C transformation into forms with long residence times (Prentice et al., 2001). Unlike annual plants, which can sequestrate more C than forest systems (Watson et al., 2000), but for shorter periods (Jasson et al., 2010), evergreen tree species can bring about a sustainable ecosystem based on their high C fixation potential. This capacity is due to different aspects, such as tree longevity (>30 yr), their year-round activity in maintaining living roots, even in winter, vigorous growth; deeper and more extensive root systems; high leaf area index; and slow leaf turnover (Liguori et al., 2009).

In the ecosystem carbon balance, C stored in soil also contributes to the sequestration capacity of ecosystems (Ryan and Law, 2005). Aboveground litter, belowground litter, and rhizosphere respiration are the most important components of changes in this stored carbon. Therefore, an evaluation of plants' carbon-sinking capacity in natural systems would be useful for the research community worldwide, especially in the Kyoto Protocol context (IGBP Terrestrial Carbon Working Group, 1998). Half of world's land surface is used for agricultural purpose (FAO, 2017), hence the importance of ecological studies on the potential sink capacity of evergreen trees in the EU's natural environment.

A.K. Srivastava, Chengxiao Hu (eds.)
Fruit Crops: Diagnosis and Management of Nutrient Constraints
https://doi.org/10.1016/B978-0-12-818732-6.00006-X

Quantification of the amount of CO_2 removed from the atmosphere through net carbon exchange is known as net primary production (NPP), which is, in CO_2 exchange terms, the amount of carbon (C) retained by plants after assimilation through photosynthesis and autotrophic respiration (Clark et al., 2001) or the total biomass production in a defined area per time unit (Nakicenovic et al., 2007).

Luyssaert et al. (2007) described a hierarchical framework for primary production quantification or C sequestration based on direct measurements of the main NPP components.

$$C \text{ sequestration} = C_{yield} + C_{foliage} + C_{wood} + C_{prunned + litter} + C_{root} + C_{ac}$$

These components are yield (C_{yield}), foliage ($C_{foliage}$), wood (C_{wood}; including branches and trunk), litter organs ($C_{prunned+litter}$), and root (C_{root}; including coarse and fine roots) production. Yield production in forests is included in the reproductive organ compartment, but given its importance in fruit trees, it has been considered an individual component of C fixation.

Apart from these measurable components, C sequestration also includes a variety of additional components and processes that are more difficult to measure and are often ignored. These components are called C_{ac} and include the carbon invested in understory plant growth, as well as the carbon lost through herbivores emitted as volatile organic compounds (VOC) and methane (CH_4) and that exuded from roots or transferred to mycorrhizae. The overall average C_{ac} production and loss were estimated to be 11% by Randerson et al. (2002). In tropical forests, C_{ac} can easily reach 20% of the sum of $C_{foliage}$, C_{wood}, and C_{root} (Clark et al., 2001) but no published data on fruit trees are available.

In forest ecosystems, mean annual primary production can range from $145 \, gC/m^2$ (1.45 Mg/ha/yr) described at a peatland site (Schulze et al., 2010) to $864 \, gC/m^2$ (8.64 Mg/ha/yr) found in tropical humid evergreen forests (Luyssaert et al., 2007). In an extensive work, Djomo et al. (2011) quantified the current annual increase of carbon biomass in a moist tropical forest with similar values (an average of 2.73 Mg/ha/yr).

As indicated earlier, other tree ecosystems could act as carbon sinks by sequestering C in biomass. Therefore, C sequestration estimations are required following the environmental interests established in the Kyoto Protocol. Facini et al. (2007) reported a similar behavior to that observed in forest ecosystems, with C fixation values between 3 and 8 Mg/ha/yr. Tagliavini et al. (2008) in intensive apple orchards, Montanaro et al. (2009) in peach trees, and Wu et al. (2012) in apple orchards measured values that came close to 5 Mg/ha/yr, mainly in the structural component of perennial orchards (twigs, branches, and stems).

This chapter focuses on the importance of citrus in C fixation accounted for by Quiñones et al. (2013) in extensive experiment and carried out in 2013. It also shows the capability of other important fruit crops in C sequestration analyzed by numerous authors (apple, olive, mango, rambutan, santol, etc.).

2 Citrus trees' carbon sequestration capability

Citrus comprises a large group of plants of the genus *Citrus* and other related genera or hybrids of the *Rutaceae* family, represented by oranges, mandarins, lemons, limes, grapefruits, and citrons. Most originate from subtropical and tropical regions of Japan, Southeast Asia, including western areas of India, Bangladesh, Philippines and Indonesia (Webber, 1967). From Asia, citrus plants had spread to North Africa, Australia, and Southeast Europe by the year 1500 and then later to America (Zaragoza, 2007).

Despite these perennial plants' mesophytic characteristics, they are cultivated in many world regions under different edaphic and climatic conditions, and between latitudes 40°N and 40°S (Agustí, 2003), whose production exceeds other fruit groups, such as apples, bananas, grapes, pears, or pineapples. Globally, the main producers are China and Brazil (29.6 and $19.0 \cdot 10^6$ Mg, respectively), followed by the United States ($9.4 \cdot 10^6$ Mg). In the 2013–14 marketing year, the world citrus area covered $8.3 \cdot 10^6$ ha, with a production of $121 \cdot 10^6$ Mg, of which 57% were oranges, 26% were mandarins, and 11% were grapefruits (Sanfeliu, 2016).

In Mediterranean countries, citrus is the second largest EU fruit crop after apple (*Malus domestica* Borkh.) and represents an approximate 20% of world production ($24.2 \cdot 10^6$ Mg) (FAO, 2016). Spain is the leading citrus-producing country in this area, with nearly 60% of EU production, and with citrus orchards covering around $300 \cdot 10^3$ ha, plus a production of $6 \cdot 10^6$ Mg, of which up to 60% and 25% are developed in the Valencian Community (E Spain) and Andalusia (S Spain), respectively.

2.1 Experimental site, culture practice, and experimental design

The experiment was carried out over 12 growing seasons (1998–2009) in three field experimental stands cv. Navelina scion *Citrus sinensis* L. (Osb.) grafted onto Carrizo citrange rootstock *C. sinensis* × *Poncirus trifoliata* L. (Raf.), from 2 to 14 years old (an 8-year-old tree can be considered an adult), which represents over 60% of total orange production. The life span of these trees is around 30–40 years.

The experimental sites were located in Puzol (Valencia, latitude 39° 34′N; longitude 00° 24′W), Vinaroz (Castellón, latitude 40° 32′N; longitude 00° 28′W) and Almonte (Huelva, 37° 27′N; longitude 6° 51′W) in Spain's main producing areas. Soils presented different textures (sandy loam, silty sandy loam, and clay, respectively, according to National Resources Conservation Sources, USDA), with pH ranging from 8.2 to 8.6, and organic matter content of 1.03%, 1.34%, and 1.91%, respectively. Trees were planted by a spacing of 6×4 m (417 tree/ha). The mean annual rainfall was around 460 mm, and concentrated mainly in autumn and spring. The average monthly temperature ranged from 10.6°C in winter to 24.5°C in summer.

During a 12-year growth period, trees received an annual supply of fertilizers from March to October according to the age of trees and the cultural practices used in the area (Quiñones et al., 2011). The amount of water applied to each tree was the equivalent to the total seasonal crop evapotranspiration (ETc) (Doorenbos and Pruitt, 1977). These irrigation water requirements were met by the effective rainfall of the entire year, plus irrigation water from a well. The volume of water applied weekly to each tree was calculated by the following expression:

$$ETc = ETo \times Kc$$

where ETc is crop evapotranspiration, ETo is the reference crop evapotranspiration under standard conditions, and Kc is the crop coefficient. ETo was determined according to Penman-Monteith (Allen et al., 1998). The crop coefficient (Kc) integrates the effect of characteristics that distinguish a typical field crop from the reference grass under standard conditions (excellent agronomic and soil water requirements, having a constant appearance and complete ground cover) and was determined using the information described by Castel and Buj (1992). The volume of water was supplied through eight pressure-compensating emitters per tree, placed every 88 cm in two drip lines, located on both sides of the trunk and producing a wetted area of almost 40% (Keller and Karmeli, 1974).

Groups of trees (at least four uniform trees were used in the selected areas) aged 2, 3, 4, 5, 6, 8, 10, 12, and 14 years were randomly chosen for sampling. The selected trees were healthy and displayed normal growth development of citrus trees (medium size and quality yield).

Each year, pruning practice was undertaken before or during flowering (spring) with emphasis placed on removing weak/dead branches, crossover limbs and water shoots. Aboveground litter and pruning organs were weighed, cut in a mechanical tree chipper, and incorporated uniformly into the soil surface. Pruning obviously reduces the total aboveground biomass and, thus, complicates biomass or C sequestration estimations (Youkhana and Idol, 2011). To avoid weeds from competing with crops for space, water, and nutrients and to facilitate agricultural tasks, traditionally effective herbicides were used in citrus groves. At least three herbicide treatments were applied each year, depending on weed development. In this study, C emissions due to these agricultural practices were not included in the C balance.

2.2 Sample collection and measurements

To evaluate C fixation on biomass, groups of trees were extracted at harvest time (November) for 12 consecutive years in each experimental area. Previously, the yields of all the trees were weighed, and a sample of 60 fruits per replication was collected at random and weighed. The young organs (fruits, leaves, twigs of new shoots, and fibrous roots) and old organs (leaves and twigs of previous years, trunk, and coarse roots) of trees were separated, directly weighed in the field on scales with a 250-kg capacity and sampled to quantify the total dry biomass. Trees were completely excavated to retain most coarse roots. Prior to excavation, 12 soil cores (10 cm in diameter with a depth of 120 cm; 15 cm increases) were taken from each quarter of the wetted area of each tree. Fibrous roots (<2 mm in diameter) were washed to remove soil. The representative samples from each fraction were taken, and they were all washed in a nonionic detergent solution, followed by several rinses in deionized water, weighed, frozen in liquid nitrogen, and freeze dried (TELSTAR LyoAlfa 6, Madrid, Spain). Dry samples were weighed and ground in a water-refrigerated mill (IKA M 20, IKA Labortechnik, Staufen, Germany) before the analysis. In all the trees, litter (petals, calyces, ovaries, young fruits, and old leaves) was collected fortnightly from the beginning of April to mid-June, that is, from the beginning of flowering until the end of fruit setting. Leaf abscission was negligible after this period.

Determination of the total C concentration in the plant samples was performed by an Elemental analyzer (NC 2500 Thermo Finnigan, Bremen, Germany). The carbon contents of the various components and of the whole tree were obtained by multiplying their dry weight by the corresponding mean carbon concentration.

2.3 Direct measurements of carbon sequestration components

For the C sequestration (annual primary production) calculations, we used an adapted equation of Luyssaert et al. (2007) and Navarro et al. (2008) based on the direct measurements of the main components. So the annual C increment rate for each part of the tree in an orchard was counted as a C sink (annual primary production).

$$C \, \text{sequestration} = C_{\text{yield}} + C_{\text{foliage}} + C_{\text{wood}} + C_{\text{prunned + litter}} + C_{\text{coarse root}} + C_{\text{fibrous root}} + C_{\text{ac}}$$
$$= \Delta DM/dt + L + Y$$

where ΔDM is annual dry matter increment, t is the time interval, L takes into account the annual litter organs, and Y is yield.

The biomass exported for the fruit plantation (C_{yield}) was calculated as the carbon incorporated into the biomass yield. C_{yield} was not included in the aboveground sequestration compartment in the equation proposed by Luyssaert et al. (2007) because, in tropical forests, reproductive organs represent a scarce value of net primary production.

The C_{foliage} component is the net annual production of foliage (new and old flush leaves), which was determined by adding the amount of new leaves to the annual increase of old leaves.

$$C_{\text{foliage}} = C_{\text{new flush leaves}} + \Delta C_{\text{old flush leaves}}$$
$$= C_{\text{new flush leaves}} + [(C_{\text{old leaves}}(t + \Delta t) - C_{\text{mean old leaves}}(t))/\Delta t]$$

Wood production (C_{wood}) includes the biomass produced each year by twigs, and the annual C biomass increases in branches and stems. The growth cycle in citrus includes three main flushes: spring, summer, and autumn (Rocuzzo et al., 2012). Thus, new tree organs are formed each year (twigs with new leaves, flowers, and fruits) and preexisting organs display and enhanced size through secondary growth, for example, trunk and branches (Quiñones et al., 2011):

$$C_{\text{wood}} = C_{\text{twigs}} + \Delta C_{\text{branches + trunk}}$$
$$= C_{\text{twigs}} + [(C_{\text{branches}}(t + \Delta t) - C_{\text{mean branches}}(t))/\Delta t] + [(C_{\text{trunk}}(t + \Delta t) - C_{\text{mean trunk}}(t))/\Delta t]$$

The term $C_{\text{pruned+litter}}$ includes the annual primary production in pruning and litter organs (flowers, calyces, little fruits, and old leaves) that can be incorporated into the organic matter pool. Total decomposition would take place within 1 year either through transformation into a stable organic matter (C sequestration compartment) or emission to the atmosphere as CO_2 (the C emission compartment). This study considered a value of 0.25 for the humus transformation coefficient (Leclerc, 1995). Therefore, the C sequestrated in these organs was calculated as

$$C_{\text{pruned + litter}} = \text{Annual Primary Production}_{\text{pruned + litter}} \times 0.25$$

Coarse and fine roots ($C_{\text{coarse root}} + C_{\text{fibrous root}}$) were recorded as separate components and were added to obtain belowground sequestration. The citrus tree root system consists of a relatively shallow well-branched framework of woody laterals and fine fibrous roots (Castle, 1980a). Fibrous roots are usually the most densely concentrated near the soil surface (0–45 cm), while a few roots are found below 90 cm (Castle, 1980a,b; Zhang et al., 1996). To estimate the carbon fixation in these organs, large and small roots (12-3, 3-1, and <1 cm) and fibrous roots (off-white) were separated.

$$C_{\text{roots}} = C_{\text{coarse root}} + C_{\text{fibrous root}} = C_{\text{fibrous root}} + [(C_{\text{coarse roots}}(t + \Delta t) - NPP_{\text{mean coarse roots}}(t))/\Delta t]$$

C fixation was quantified by adding aboveground plus belowground carbon sequestration. In addition to these measurable components, C fixation in biomass also included a variety of additional components and processes that were more difficult to measure (Luyssaert et al., 2007). These components were called C_{ac} and were not quantified in this manuscript, but they were estimated following Clark et al. (2001).

2.4 Statistical analysis

The data for the direct measurement of the C sequestration components are summarized in the figures and tables as the means of three replicates (with four trees per replica) ± standard deviation. ANOVA and means were separated by the LSD test at the 0.05 level to determine the significance of the differences among the variables measured at each harvesting age using the Statistical Analysis System Institute Inc. software (SAS, Cary, North Carolina, the United States). Prior to any statistical analyses, all the data were examined with normal probability plots to determine distribution.

3 Carbon capture of long residence woody, leaf, fruit, and roots

Annual primary production in the Navelina orange tree components (Table 6.1) showed fruit yield to be major sinks of C fixation in trees, with 10.5-kg C/tree/yr (4.4-MgC/ha/yr) accounting for about 40% of the total C sequestration in citrus trees. Bhatnagar et al. (2016) found similar C storage values in fruits (Table 6.2), between 32% and 41% of the total in Nagpur mandarin (*Citrus reticulata* Blanco). Structural organs (twigs, branches, and stems) were another important sink for C partitioning with values of around 6-kg C/tree/yr (2.5-MgC/ha/yr), if we take into account only 25% of the carbon stored in the pruned and abscised organs. Wu et al. (2012) also obtained a potent sink in plants in stems and branch, followed by roots of the total biomass produced in 18-year-old apple trees (Table 6.3). However at this age, trees reach the peak of C sequestration capability, which begins to decline thereafter.

The mean values of the adult trees were 8.1 and 1.2 MgC/ha/yr in the above- and belowground compartments, respectively. Aboveground C fixation doubled the values found by Liguori et al. (2009) in citrus with different plant densities (5 MgC/ha/yr) because the annual C fixation in these trees was underestimated as C storage in the trunk organ was not accounted for in the latter. In our case, trunk primary production exceeded 3 MgC/ha/yr (the fraction included in the wooden compartment). Bhatnagar et al. (2016) counted a similar C sequestration value in 6-year-old Nagpur mandarin trees (217 gC/ha/yr) than that found in trees of a similar age (Table 6.2).

In addition to these measurable components, C storage in the citrus tree biomass also includes a variety of additional components and processes that are more difficult to measure (Luyssaert et al., 2007). These components were called C_{ac} and include the carbon invested in understory plant growth, lost through herbivores, and the carbon emitted as volatile organic compounds and methane and exuded from roots. In this manuscript, understory C sequestration was reduced with numerous herbicide treatments, which prevented weed development. The carbon lost through herbivores was zero as access by animals was prevented. Moreover, volatile compounds and root exudates were not quantified. Clark et al. (2001) considered the C_{ac} component to be 20% of the sum of $C_{foliage}$, C_{wood}, and C_{roots}. In our case, the value of this compartment accounted for about 1 MgC/ha/yr in adult trees (Table 6.1).

Plantation age influences C storage. Fig. 6.1 shows the annual increase in primary production in both new and old organs. This C storage was quantified through C fixation in new organs (abscised reproductive organs, fruits, new flush leaves and twigs, and fibrous roots), plus an annual increase in C content in old organs (abscised old leaves, old leaves, branches, trunk, and coarse root). Carbon sequestration was the result of both the development of new organs and the growth of old organs, which are annual in young trees. However, in adult trees (aged more than 8 years), this increase was due mainly to the development of new flush organs, with values reaching 27-kgC/tree (1125 gC/m^2/yr; 11.3 MgC/ha/yr).

In this study, the carbon concentration in the different compartments ranged between 39.2% in fine roots and 44.6% in stems (data not shown). Many authors have used a mean carbon concentration value from 50% to 46% to estimate C sequestration (Brown et al., 1989; Malhi et al., 1999; Nasi et al., 2009; Djomo et al., 2011) in forest biomass, which could have led to an overestimation of about 5% depending on the value. Liguori et al. (2009) reported in citrus trees a value that came close to that obtained in our analysis by increasing the accuracy of the results.

Our results showed that citrus ecosystems display a similar behavior to that of temperate forests, which sequestrate between 600 and 2500 gC/m^2/yr (Whittaker, 1975) or 100–6600 gC/m^2 in European forests (Valentini et al., 2000). Therefore, citrus orchards behave as an important carbon sink. In adult rubber trees, the annual carbon accumulation values based on the Gompertz model were roughly 6–8 depending on the study area (Wauters et al., 2008).

4 Contribution of different fruit orchards to carbon storage

Other authors have studied potential carbon sequestration in several fruit trees, mainly woody perennials species, for it to be key in the mitigation of climate change by sequestering atmospheric CO_2 (Marín et al., 2016), similarly to that observed in forests (Table 6.4). These authors found, for example, that cocoa ecosystems fixed atmospheric C in the total biomass at a rate between 8.3 and 17.7 Mg. The maximum C storage was obtained in cocoa systems with forest and fruit trees (coffee walnut, cedar, avocado, bay, plum, and papaya). For this C sequestration, our results support the priority of C allocation in vegetative organs, as suggested by Navarro et al. (2008) in other fruit trees (palm trees), which amounted to 16.1-MgC/ha/yr. In another assay (Janiola and Marin, 2016), more vigorous tropical trees gave higher C rates, with santol plantation of 32-year-old trees stored 75-MgC/ha/yr, followed by 15-year-old mango trees (45-Mg C).

TABLE 6.1 Carbon sequestration in Navelina orange trees components (gC/tree/yr) for trees of different age.[a]

Tree component/age of trees (in years)	2	3	4	5	6	8	10	12	14
Fruit yield			27.4 ± 8.1e	285.7 ± 49.8e	872.6 ± 88.3d	3560.9 ± 598.2c	5343.3 ± 737.1b	10,211.5 ± 1009.2a	10,762.9 ± 1063.7a
Foliage	25.6 ± 2.2e[b]	37.8 ± 3.6e	108.7 ± 15.0e	288.7 ± 18.3e	1189.7 ± 166.6d	1750.2 ± 117.1c	2194.0 ± 171.8b	4229.6 ± 293.8a	4889.2 ± 336.2a
Wood	33.8 ± 1.9d	34.2 ± 5.5d	54.5 ± 9.3d	387.4 ± 26.2d	2096.5 ± 170.4c	1687.1 ± 100.4c	2494.5 ± 288.8b	4985.7 ± 1177.9a	3285.5 ± 1034.5a
Pruning/abscised[c]	1.1 ± 0.2f	4.6 ± 0.3f	19.7 ± 2.9f	58.3 ± 10.9f	187.7 ± 29.8e	460.7 ± 127.5d	844.9 ± 117.6c	1019.1 ± 110.5b	1482.4 ± 254.1a
Additional components	12.8 ± 1.3	18.0 ± 2.6	48.6 ± 6.5	161.8 ± 23.2	807.4 ± 36.2	1056.1 ± 52.3	1613.6 ± 230.4	2658.3 ± 158.3	2821.1 ± 145.8
Aboveground[d]	73.3 ± 3.7e	94.6 ± 7.7e	258.9 ± 23.0e	1181.9 ± 46.8e	5153.9 ± 287.6d	8514.9 ± 291.2c	12,490.3 ± 757.2b	23,104.2 ± 1682.4a	23,241.0 ± 1698.8a
Coarse root	19.9 ± 2.3d	11.1 ± 4.6d	120.0 ± 15.5d	268.8 ± 23.8d	1130.4 ± 103.8c	1238.8 ± 114.3b	277.2 ± 183.9d	1656.7 ± 413.7a	1039.9 ± 378.8b
Fine root	10.7 ± 1.7e	27.9 ± 4.5e	48.5 ± 5.6e	186.6 ± 15.2d	521.9 ± 86.1c	983.2 ± 67.2b	1100.9 ± 96.4b	1521.2 ± 326.9a	1523.2 ± 327.4a
Belowground[e]	30.6 ± 3.0e	39.0 ± 7.2e	168.5 ± 19.3e	455.4 ± 31.3e	1652.4 ± 166.9d	2222.0 ± 283.6b	1378.1 ± 196.9c	3177.9 ± 433.7a	2563.1 ± 373.7b
Total gC/tree/yr	103.9 ± 6.1e	133.6 ± 13.4e	427.4 ± 34.5e	1637.3 ± 59.3e	6806.3 ± 288.4d	10,736.9 ± 629.4c	13,868.4 ± 837.8b	26,282.2 ± 1989.4a	25,804.1 ± 1947.6a
Total gC/m²/yr[f]	4.33	5.57	17.81	68.22	283.60	447.37	577.85	1095.09	1075.17
Total MgC/ha/yr	0.04	0.06	0.18	0.68	2.84	4.47	5.78	10.95	10.75

[a] Each value is a mean of three values ± standard error.
[b] ANOVA, significance of differences among the variables measured due to age of trees (in row). Means were separated by Fishers least significant difference (LSD) test at the 0.05 level.
[c] Abscised reproductive organs: Petals, ovaries, and calyces.
[d] Aboveground term includes carbon storage in yield, new and old flush leaves, prunnings and liter organ, twigs, branches and stem.
[e] Belowground term includes carbon sequestration in coarse and fibrous roots.
[f] Carbon fixation per square meter and year in plants grown at spacing of $6 \times 4\,m = xNPP$ (gC/tree/yr)/24 (m²/tree).

TABLE 6.2 Carbon storage in citrus trees components.[a]

Tree component	C sequestration (kg C/tree)
Fruit	2.47[b]
Leaves	0.63
Twigs	0.76
Bark	0.02
Branch	1.44
Stem	0.72
Aboveground[c]	6.04
Roots	1.76
Total	**7.80**
Total g C/m²/yr [b]	**217**

[a] Adapted from Bhatnagar et al. (2016). Carbon Storage = Biomass × % Carbon (0.45% was taken as an average of carbon percent in plants parts).
[b] Carbon fixation per square meter and year in plants grown at spacing of 6 m × 6 m = C sequestration (g C/tree/yr)/36 (m²/tree).
[c] Aboveground term includes carbon storage in yield, new and old flush leaves, prunnings and liter organ, twigs, branches and stem.

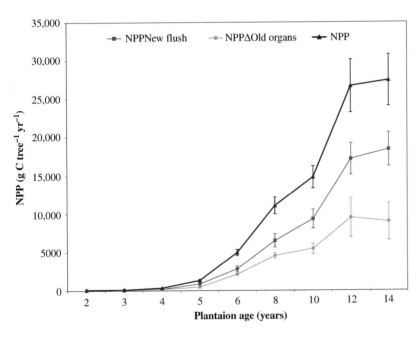

FIG. 6.1 Annual increase in C sequestration in new flush leaves plus annual increase in old organs of Navelina orange trees for trees of different ages. Each value is mean of the three values ± standard error.

TABLE 6.3 Carbon sequestration in apple trees components of 18 years old trees.[a]

Tree component	Biomass (kg/tree)	C sequestration (g C/m²)[b]
Fruit	19.32	592.5
Leaf	5.48	168.1
Branch	81.11	2487.4
Stem	48.44	1485.5
Aboveground	154.35	4733.4
Coarse root	28.03	859.6
Fine root (estimated)	41.70	1278.8
Belowground	69.73	2138.4
Total	224.08	6871.8

[a] Adapted from Wu et al. (2012).
[b] Carbon sequestration per square meter in plants grown at spacing of 3 × 5 m (0.46 g C/g DW was taken in this study).

TABLE 6.4 Carbon fixed by tree ecosystems as net primary production.

Trees	NPP (Mg C/ha/yr)
Agroforestry system and citrus[a]	17.7
Tropical palm plantation[b]	16.1
Tropical forests[c]	12.5
Citrus trees[d]	11.4
Kiwi fruits[e]	8.0
Temperate forests[c]	7.7
Tropical savanna and grasslands[c]	5.4
Mediterranean shrublands[c]	5.0
Apple orchards[f]	5.2–13.3
Temperate grasslands[c]	3.7

[a] *Marín et al. (2016).*
[b] *Navarro et al. (2008).*
[c] *Grace (2004).*
[d] *Quiñones et al. (2013).*
[e] *Facini et al. (2007).*
[f] *Wu et al. (2012).*

The similar rates that we found in citrus were accounted for by Pernice et al. (2006), who obtained about 1750 g C/m^2/yr in 1.14 · 10^6 ha of olive trees. In the same way, Rossi et al. (2007) accounted for 1160 g C/m^2 in kiwi cv Hayard over a 7-month period. Tagliavini et al. (2008) and Liguori et al. (2009) also confirmed that Italian orchards (citrus, wine grape, apple, olive, peach, and orange) represented a significant sink for atmospheric C.

Agricultural practices may lead to an increment in carbon sinks, such as organic fertilization, or the use of mycorrhizal fungi (Shi et al., 2017) or biochar. Indeed, several studies have demonstrated that organic fertilization improves physicochemical soil properties (Melero et al., 2007; Evanylo et al., 2008; Bravo et al., 2012) and stimulated root development (Baldi et al., 2010). Bravo et al. (2012) have shown that compost applications promote root and shoot growth in nectarine trees by increasing C fixation in a study carried out using a ^{13}C isotope-labeled technique.

These results suggest that perennial crops, citrus among them, and forests may act as a remarkable means of C storage in biomass. Therefore, this pool of fixed C could significantly help to reduce CO_2 in the atmosphere. Thus, agriculture could play a key role in mitigating climate change.

5 Future line of research

Nowadays, agricultural practices are being modified to be more sustainable. Researches will focus on analyzing the effect of increasing soil organic matter, highly degraded in recent years, on the carbon sequestration capacity of agricultural ecosystems.

References

Agustí, M., 2003. (in Spanish). Citricultura. Mundi-Prensa, Madrid.

Allen, R.G., Pereira, L.S., Raes, D., Smith, M., 1998. Crop evapotranspiration (guidelines for computing crop water requirements). In: FAO Irrigation and Drainage, paper N° 56. FAO (Food and Agriculture Organization of the United Nations), Rome.

Alvarez, E., Duque, A., Saldarriaga, J., Cabrera, K., de las Salas, G., del Valle, I., Lema, A., Moreno, F., Orrego, S., Rodríguez, L., 2012. Tree aboveground biomass allometries for carbon stocks estimation in the natural forests of Colombia. For. Ecol. Manag. 267, 297–308. https://doi.org/10.1016/j.foreco.2011.12.013.

Baldi, E., Toselli, M., Eissenstat, D.M., Marangoni, B., 2010. Organic fertilization leads to increased peach root production and lifespan. Tree Physiol. 30, 1373–1382.

Barros, V.R., Field, C.B., Dokken, D.J., Mastrandrea, M.D., Mach, K.J., Bilir, T.E., White, L.L. (Eds.), 2014. Climate Change 2014: Impacts, Adaptation, and Vulnerability. Part B: Regional Aspects. Contribution of Working Group II to the Fifth Assessment Report of the Intergovernmental Panel on Climate Change. Cambridge University Press, Cambridge and New York.

Bhatnagar, P., Singh, J., Chauhan, P.S., Sharma, M.K., Meena, C.B., Jain, M.C., 2016. Carbon assimilation potential of Nagpur mandarin (*Citrus reticulata* Blanco.). Int. J. Sci. Environ. Technol. 5 (3), 1402–1409.

Bravo, K., Toselli, M., Baldi, E., Marcolini, G., Sorrenti, G., Quartieri, M., Marangoni, B., 2012. Effect of organic fertilization on carbon assimilation and partitioning in bearing nectarine trees. Sci. Hortic. 137, 100–106.

Brown, S., Gillespie, A., Lugo, A.E., 1989. Biomass estimation methods for tropical forests with applications to forest inventory data. For. Sci. 35, 881–902.

Castel, J.R., Buj, A.C., 1992. Growth and evapotranspiration of young drip-irrigated Clementine-trees. In: Proceedings of the 7th International Citrus Congress, Acireale, Sicily. vol. 2. pp. 651–656.

Castle, W.S., 1980a. Citrus root systems: their structure, function, growth, and relationship of tree performance. In: Proceedings of the 4th International Society of Citriculture, Sydney, Australia, 15-23 Aug, 1978, pp. 62–69.

Castle, W.S., 1980b. Fibrous root distribution of "pineapple" orange trees on rough lemon root stock all three tree spacings. J. Am. Soc. Hortic. Sci. 105, 478–480.

Clark, D.A., Brown, S., Kicklighter, D.W., Chambers, J.Q., Thomlinson, J.R., Ni, J., 2001. Measuring net primary production in forests: concepts and field methods. Ecol. Appl. 11 (2), 356–370. https://doi.org/10.1890/1051-0761(2001)011[0356:MNPPIF]2.0.CO;2.

Djomo, A.N., Knohl, A., Gravenhorst, G., 2011. Estimations of total ecosystem carbon pool distribution and carbon biomass current annual increment of a moist tropical forest. For. Ecol. Manag. 261, 1448–1459. https://doi.org/10.1016/j.foreco.2011.01.031.

Doorenbos, J., Pruitt, W.O., 1977. Crop Water Requirements. FAO Irrigation and Drainage Paper N° 24, Rome.

Evanylo, G., Sherony, C., Spargo, J., Starner, D., Brosius, M., Haering, K., 2008. Soil and water environmental effects of fertilizer, manure, and compost-based fertility practices in an organic vegetable cropping system. Agric. Ecosyst. Environ. 127, 50–58.

Facini, O., Georgia, T., Nardino, M., Rossi, F., Maracchi, G., Motisi, A., 2007. (in Italian). Il contributo degli impianti da frutto all'assorbimento della CO2 atmosferica. In: Clima e Cambiamenti Climatici: le attività del CNR.

FAO, 2016. Citrus Fruits. Annual Statistics. Food and Agriculture Organization. viewed December 2018. Available from: http://www.fao.org/faostat/en/#data/QC.

FAO, 2017. Citrus Fruits. Annual Statistics. Food and Agriculture Organization. viewed December 2018. Available from: http://www.fao.org/faostat/en/#data/QC.

Grace, J., 2004. Understanding and managing the global carbon cycle. J. Ecol. 92, 189–202.

IGBP Terrestrial Carbon Working Group, 1998. The terrestrial carbon cycle: implications for the Kyoto Protocol. Science 280, 1393–1394.

Intergovernmental Panel on Climate Change, 2014. AR5: Impacts, Adaptation, and Vulnerability. https://www.ipcc.ch/report/ar5/wg2/.

Janiola, M.D., Marin, R.A., 2016. Carbon sequestration potential of fruit tree plantations in southern Philippines. J. Biol. Environ. Sci. 8 (5), 164–174.

Jasson, C.C., Wullschleger, S.D., Kalluri, U.C., Tuskan, G.A., 2010. Phytosequestration: carbon biosequestration by plants and the prospects of genetic engineering. Bioscience 60, 685–696.

Keller, J., Karmeli, D., 1974. Trickle irrigation design parameters. Trans. ASAE 17 (4), 678–684.

Leclerc, B., 1995. Guide des matières organiques. Institut Technique de l'Agriculture Biologique (ITAB), Paris (in French).

Liguori, G., Gugliuzza, G., Inglese, P., 2009. Evaluating carbon fluxes in orange orchards in relation to planting density. J. Agric. Sci. 147, 637–645. https://doi.org/10.1017/S002185960900882X.

Luyssaert, S., Inglima, I., Jung, M., Richardson, A.D., Reichstein, M., Papale, D., Piao, S.L., Schulze, D.E., et al., 2007. CO2 balance of boreal, temperate, and tropical forests derived from a global database. Glob. Change Biol. 13, 2509–2537. https://doi.org/10.1111/j.1365-2486.2007.01439.x.

Malhi, Y., Baldocchi, D.D., Jarvis, P.G., 1999. The balance of tropical, temperate and boreal forests. Plant Cell Environ. 22, 715–740. https://doi.org/10.1046/j.1365-3040.1999.00453.x.

Marín, Q.M.P., Andrade, H.J., Sandoval, A.P., 2016. Fijación de carbono atmosférico en la biomasa total de sistemas de producción de cacao en el departamento del Tolima, Colombia. Rev. U.D.C.A Act. Div. Cient. 19 (2), 351–360 (in Spanish).

McEwan, R.W., Lin, Y.C., Sun, I.F., Hsieh, C.F., Su, S.H., Chang, L.W., Song, G.Z.M., Wang, H.H., Hwong, J.L., Lin, K.C., Yang, K.C., Chiang, J.M., 2011. Topographic and biotic regulation of aboveground carbon storage in subtropical broadleaved forests of Taiwan. For. Ecol. Manag. 262, 1817–1825. https://doi.org/10.1016/j.foreco.2011.07.028.

Melero, S., Madejón, E., Ruiz, J.C., Herencia, J.F., 2007. Chemical and biochemical properties of a clay soil under dryland agriculture system as affected by organic fertilization. Eur. J. Agron. 26, 327–334.

Montanaro, G., Celano, G., Dichio, B., Xiloyannis, C., 2009. Effects of soil-protecting agricultural practices on soil organic carbon and productivity in fruit tree orchards. Land Degrad. Dev. https://doi.org/10.1002/ldr.917.

Nakicenovic, N., Alcamo, J., Davis, G., de Vries, B., Fenhann, J., Gaffin, S. et al., (Eds.), 2007. Working Group I, Report Summary for Policymakers, Special Report on Emissions Scenarios. A Special Report of Working Group III of the Intergovernmental Panel on Climate Change. Cambridge University Press, Cambridge.

Nasi, R., Mayaux, P., Devers, D., Bayol, N., Eba'a Atyi, R., Mugnier, A., Cassagne, B., Billand, A., Sonwa, D.J., 2009. Un aperç u des stocks de carbone et leurs variations dans les forêts du Bassin du Congo. In: Nasi, R., Mayaux, P., Devers, D., Bayol, N., Eba'a Atyi, R., Mugnier, A., Mayaux, P. (Eds.), Les Forêts du Bassin du Congo: etat des Forêts 2008. Office des publications de l'Union européenne, Luxembourg, pp. 199–216.

Navar, J., 2009. Allometric equations for tree species and carbon stocks for forests of northwestern Mexico. For. Ecol. Manag. 257, 427–434. https://doi.org/10.1016/j.foreco.2008.09.028.

Navarro, M.N.V., Jourdan, C., Sileye, T., Braconnier, S., Mialet-Serra, I., Saint-Andre, L., Dauzat, J., Nouvellon, Y., et al., 2008. Fruit development, not GPP, drives seasonal variation in NPP in a tropical palm plantation. Tree Physiol. 28, 1661–1674. https://doi.org/10.1093/treephys/28.11.1661.

Pernice, F., Motisi, A., Rossi, F., Georgiadis, T., Nardino, M., Fontana, G., Dimino, G., Drago, A., 2006. CO2 and water exchanges in an olive orchard: micrometeorological observations and agrometeorological application to sustainable water management. In: Biotechnology and Quality of Olive Tree Products Around the Mediterranean Basin. Marsala, 5-10 November. pp. 363–370.

Prentice, I.C., Farquhar, G.D., Fasham, M.J.R., Goulden, M.L., Heimann, M., Jaramillo, V.J., Kheshgi, H.S., LeQuéré, C., Scholes, R.J., Wallace, D.W.R., 2001. The carbon cycle and atmospheric carbon dioxide. In: Houghton, J.T., Ding, Y., Griggs, D.J., Noguer, M., van der Linden, P.J., Dai, X., Maskell, K., Johnson, C.A. (Eds.), Climate Change 2001: The Scientific Basis. Contributions of Working Group I to the Third Assessment Report of the Intergovernmental Panel on Climate Change. Cambridge University Press, Cambridge, UK, pp. 185–237.

Quiñones, A., Martínez-Alcántara, B., Primo-Millo, E., Legaz, F., 2011. Fertigation: concept and application in citrus. In: Srivastava, A.K. (Ed.), Advances in Citrus Nutrition. Springer Science+Business Media, Dordrecht.

Quiñones, A., Martínez-Alcántara, B., Font, A., Forner-Giner, M.A., Legaz, F., Primo-Millo, E., Iglesias, D.J., 2013. Allometric models for estimating carbon fixation in citrus trees. Agron. J. 105, 1355–1365.

Randerson, J.T., Chapin, F.S.I., Harden, J.W., Neff, J.C., Harmon, M.E., 2002. Net ecosystem production: a comprehensive measure of net carbon accumulation by ecosystems. Ecol. Appl. 12, 937–947. https://doi.org/10.1890/1051-0761(2002)012[0937:NEPACM]2.0.CO;2.

Rocuzzo, D., Zanotelli, M.S., Giuffriddaa, A., Torrisi, B.F., Leonardia, A., Quiñones, A., 2012. Assessing nutrient uptake by field-grown orange trees. Eur. J. Agron. 41, 73–80.

Rossi, F., Faccini, O., Georgiadis, T., Nardino, M., 2007. Seasonal CO_2 fluxes and energy balance in a kiwifruit orchard. Ital. J. Agrometeorol. 1, 44–56.

Ryan, M.G., Law, B.E., 2005. Interpreting, measuring, and modeling soil respiration. Biogeochemistry 73 (1), 3–27. https://doi.org/10.1007/s10533-004-5167-7.

Sanfeliu, I., 2016. Plan Star Citrus. Citrus Management Committee. Available from: http://www.agronegocios.es/digital/files/planstar/Sanfeliu_pstar_citricos_valencia.pdf.

Schulze, E.D., Ciais, P., Luyssaert, S., Schrumpf, M., Janssens, I.A., Thiruchittampalam, B., Thelokes, J., Saurat, M., et al., 2010. The European carbon balance, Part 4: integration of carbon and other trace-gas fluxes. Glob. Change Biol. 16, 1451–1469. https://doi.org/10.1111/j.1365-2486.2010.02215.x.

Shi, L., Wang, J., Liu, B., Nara, K., Lian, C., Shen, Z., Xia, Y., Chen, Y., 2017. Ectomycorrhizal fungi reduce the light compensation point and promote carbon fixation of *Pinus thunbergii* seedlings to adapt to shade environments. Mycorrhiza 27, 823–830.

Tagliavini, M., Panzacchi, P., Ceccon, C., Liguori, G., Bertolla, C., Meggio, F., Tonon, G., Corelli-Grappadelli, L., Celano, G., Gucci, R., Pitacco, A., Inglese, P., 2008. Fluxes of carbon in Italian orchards. In: First Symposium on Horticulture in Europe, Vienna, Austria, 17-20 February, 2008.

Valentini, R., Matteucci, G., Dolman, A.J., Schulze, E.D., Rebmann, C., Moors, E.J., Granier, A., Gross, P., Jensen, N.O., et al., 2000. Respiration as the main determinant of carbon balance in European forests. Nature 404, 861–865. https://doi.org/10.1038/35009084.

Watson, R.T., Noble, I.R., Bolin, B., Ravindranath, N.H., Verardo, D., Dokken, D., 2000. Land Use, Land Use Changes and Forestry. Cambridge Univ. Press, Cambridge.

Wauters, J.B., Coudert, S., Grallien, E., Jonard, M., Ponette, Q., 2008. Carbon stock in rubber tree plantations in Western Ghana and Mato Grosso (Brazil). For. Ecol. Manag. 255, 2347–2360. https://doi.org/10.1016/j.foreco.2007.12.038.

Webber, H.J., 1967. History and development of the citrus industry. In: Reuther, W., Webber, H.J., Batchelor, L.D. (Eds.), The Citrus Industry. In: vol. 1. Univ. of California, Riverside, CA, pp. 1–189.

Whittaker, R.H., 1975. Communities and Ecosystems, second ed. MacMillan, New York.

Wu, T., Wang, Y., Yu, C., Chiarawipa, R., Zhang, X., et al., 2012. Carbon sequestration by fruit trees—Chinese apple orchards as an example. PLoS One 7 (6), e38883. https://doi.org/10.1371/journal.pone.0038883.

Youkhana, A.H., Idol, T.W., 2011. Allometric models for predicting above- and belowground biomass of Leucaena-KX2 in a shaded coffee agroecosystem in Hawaii. Agrofor. Syst. 83 (3), 331–345. https://doi.org/10.1007/s10457-011-9403-6.

Zaragoza, S., 2007. (in Spanish). Aproximación a la historia de los cítricos. Origen, dispersión y evolución de su uso y cultivo. (Doctoral thesis) Polytechnic University of Valencia, Vegetable Production.

Zhang, M., Alva, A.K., Li, Y.C., Calvert, D.V., 1996. Root distribution of grapefruit trees under dry granular broadcast vs. fertigation method. Plant Soil 183, 79–84. https://doi.org/10.1007/BF02185567.

7

Effects of climate change on fruit nutrition

Alyssa L. Stewart, Selena Ahmed*

The Food and Health Lab, Department of Health and Human Development, Montana State University,
Bozeman, MT, United States
*Corresponding author. E-mail: selena.ahmed@montana.edu

1 Introduction

Climate change is impacting the quality and quantity of fruits produced in agricultural systems worldwide, with notable implications for farmer livelihoods and human nutrition (Ahmed and Stepp, 2016; Wheeler and Von Braun, 2013). Climate change refers to the multiple changes in global and regional climates, including increased global temperatures and carbon dioxide levels along with greater weather variability and more extreme weather conditions that have occurred primarily since the mid-20th century (IPCC, 2014). The Intergovernmental Panel on Climate Change (IPCC) has determined that climate change is largely attributed to increased anthropogenic greenhouse gas emissions since the preindustrial era, including from fossil fuel combustion and deforestation (IPCC, 2014). Changes in climate impact both natural and human systems, particularly agricultural systems, which use 37.5% of earth's land area (FAO, 2018a).

Numerous studies have highlighted how climate change has impacted crop yields that have direct and indirect effects on food supply, crop prices, and farmer livelihoods (Hertel et al., 2010). In addition to yields, studies indicate that climate change is impacting the crop quality (Ahmed and Stepp, 2016). Crop quality refers to the nutrition, phytochemical composition, health attributes, sensory properties, safety of a food crop (Ahmed and Stepp, 2016).

Assessing the effects of climate change on crop quality is particularly important for fruits where quality is determined in large part by the presence and concentrations of specific nutrients (i.e., vitamins, minerals, amino acids, and carbohydrates) and phytochemicals that benefit consumers (Ahmed et al., 2014). Fruit nutritional quality provides essential vitamins, minerals, and phytochemical compounds to the global population that are linked to the mitigation of diet-related chronic disease (Liu et al., 2000). Diet is the number one risk factor for disease (Achadi et al., 2016) and is directly linked to the inadequate consumption of fruits and vegetables that causes 2.635 million deaths annually (Lock et al., 2005). Consumption of fruits and vegetables is associated with decreased risk of coronary heart disease, diabetes, obesity, cataracts, and degenerative diseases (Liu et al., 2000).

A.K. Srivastava, Chengxiao Hu (eds.)
Fruit Crops: Diagnosis and Management of Nutrient Constraints
https://doi.org/10.1016/B978-0-12-818732-6.00007-1

The positive health effects of fruits are primarily due to the nutrients and phytochemicals present in plant tissues (Hounsome et al., 2008; Liu et al., 2000). Phytochemicals, also known as secondary metabolites, protect plant tissues from stresses, including climatic conditions, pathogens, and pests (Feeny, 1976; Fraenkel, 1959). Polyphenols are the most studied class of phytochemicals in fruits because of their prevalence with several thousand compounds and their benefits to human nutrition as antioxidant compounds with antiinflammatory properties (Hounsome et al., 2008; Liu et al., 2000). For example, anthocyanins, a subgroup of flavonoids found in red-purple pigmented fruits, help prevent neuronal diseases, cardiovascular disease, cancer, and diabetes (Yousuf et al., 2016). Each fruit contains a complex mixture of phytochemicals and nutrients that have been demonstrated to have synergistic effects in the human body including antioxidant, antimutagenic, antifungal, and antibacterial attributes (Liu, 2003). For example, although an apple only contains about 5.7 mg of vitamin C, its antioxidant effects in the body are equivalent to 1500 mg of vitamin C because of its phytochemical composition and interactions with vitamin C (Eberhardt et al., 2000).

This chapter provides a review of climate change effects on some of the most consumed fruits in the world. We start with a brief overview on climate change and its effects on agriculture. Next, we provide a review of the effects of climate change on the nutritional quality of the five most consumed fruit crops globally by weight including watermelons, bananas, apples, grapes, and oranges (FAO, 2018b). This chapter ends with a brief discussion on the implications of climate change on fruit nutrition and directions for further research.

2 Notable changes in climate for fruit nutrition

One of the most prominent climate change variables impacting agriculture is increased temperature (Parry et al., 2004). By 2065, the global mean surface temperature is expected to rise 0.5–2.6°C (IPCC, 2014). Temperature directly affects plant photosynthesis and growth. Each plant species has specific temperature thresholds in which it can grow and has an optimal temperature for yields and quality (Hatfield and Prueger, 2015). The increasing temperatures that are associated with climate change will result in fruit crops being cultivated at nonoptimal temperatures. This change will impact fruit nutrition and phytochemicals, with implications for human nutrition consuming these crops (Ketellapper, 1963).

Another global climate trend that is notably impacting fruit nutrition is the increased levels of atmospheric CO_2. Atmospheric levels of CO_2 have increased by 40% since the 1750s with the fasted recorded decadal rate of change between 2002 and 2011 (IPCC, 2014). The increased atmospheric levels of CO_2 and other greenhouse gases associated with climate change have impacted the growth of fruits and crops and their responses to stress (IPCC, 2014; Unsworth and Hogett, 1996). The uptake of atmospheric CO_2 into plants through photosynthesis is critical to plant metabolism. Increased levels of atmospheric CO_2 with climate change are impacting crop development, physiology, and chemistry including nutrition (Taub, 2010). Specifically, increased levels of CO_2 are impacting crop productivity and yield by promoting increased net photosynthetic rates of plants (Long et al., 2004). In addition, increased levels of CO_2 can also enhance plant tolerance to environmental stresses by increasing soluble sugars, antioxidants, and root exudates (Drake et al., 2011; Huang and Xu, 2015). The application of CO_2 has thus been used as a gas fertilizer for greenhouse cultivation in the past few decades (Bisbis et al., 2018).

Lastly, changes in the water cycle including changes in precipitation intensity, variability, and timing are having notable impacts on fruit nutrition. While the effects of changes in the water are predicted to vary by latitude and geography, modeling studies have identified these changes to include increased rainfall intensities, increased rainfall variabilities, changes in the timing of precipitation, and more extreme droughts and floods (Bates et al., 2008; IPCC, 2014; Zeppel et al., 2011). These effects impact fruits in agricultural systems through changes in soil water content and ecological processes (Zeppel et al., 2011). It is important to note that these climatic variables do not impact fruit crops in isolation but rather interact with one another and other abiotic variables along with biotic factors to impact fruit yields and quality.

3 Effects of climate change on crops

Numerous studies have shown that climate change has decreased agricultural productivity and shifted the geographic range of many crops (Porter and Semenov, 2005). Shifts in yields are important to understand because the amount of crops produced directly affects the food supply, crop prices, and farmer livelihoods (Hertel et al., 2010). Effects of climate change on yield vary by crop and geographic location. For example, many crops growing at higher latitudes, such as corn and soybeans in the United States, are predicted to increase in yields with increasing temperature (up to a certain temperature threshold (Schlenker and Roberts, 2009)). However, as global temperatures continue to increase, these crops are expected to see a 30%–80% decrease in yields by the end of the century, depending on the rate of warming (Schlenker and Roberts, 2009). In some regions, the increasing temperatures have been shown to decrease yields for specific crops (Peng et al., 2004). For example, rice grain yield has been decreasing by 10% for every 1°C increase in minimum temperature during the dry season at the International Rice Research Institute Farm in the Philippines (Peng et al., 2004). Globally, global maize and wheat production declined by 3.8% and 5.5%, respectively (Lobell et al., 2011). Droughts are further modeled to decrease the yield of major crops worldwide by 50% in 2050 (Li et al., 2009).

In addition to yields, climate change is impacting crop quality, which is critical to understand for its effects on consumer-buying choices and human nutrition and health (DaMatta et al., 2010; Grunert, 2005). Measures of crop quality include the concentrations of nutrients and secondary metabolites and sensory characteristics such as taste, aroma, texture, and color (Ahmed and Stepp, 2016; Mattos et al., 2014). Common parameters of crop quality include, but are not limited to, concentrations of ascorbic acid, calcium (Ca), carotenoids, chlorophyll a, chlorophyll b, copper (Cu), fructose, glucose, sucrose, total soluble sugar, titratable acidity, total protein, nitrate, total antioxidant capacity, total phenols, total flavonoids, lycopene, anthocyanins, phosphorus (P), potassium (K), magnesium (Mg), sulfur (S), iron (Fe), manganese (Mn), and zinc (Zn) (Dong et al., 2018). Many of these crop quality parameters have relevance for human nutrition; Table 7.1 describes the nutritional relevance of several nutrients and phytochemicals in fruits that are impacted by climate and will be discussed in this review.

Increased temperature affects the antioxidant content, sugar content, nutrient content, texture, and color in several fruits with a trend of reducing vitamin content in specific fruit crops (Mattos et al., 2014; McKeown et al., 2006). Increased atmospheric CO_2 has been shown to decrease the quality of iron, zinc, and protein levels in many staple crops, including wheat, rice, soybeans, barley, potatoes, and peas (Myers et al., 2014; Taub et al., 2008). A literature review on climate effects on specialty crops found a range of climate variables that impacts the nutrient, phytochemical, and sensory attributes of fruits, vegetables, and other crops (Ahmed and Stepp, 2016). Rainfall, temperature, and humidity were found to influence the terpenes and volatiles in apples (Vallat et al., 2005). The anthocyanidin concentration in bilberries was impacted by overall climatic factors (Åkerström et al., 2009). Eight individual flavor compounds in fig were found to statistically significantly shift with overall changes in climatic conditions (Darjazi and Larijani, 2012). In grapes, cooler temperatures were positively correlated to phenolic compounds and antioxidant properties (Xu et al., 2011). Another study in grapes found that a rise in carbon dioxide levels resulted in an increase in tartaric acid (Bindi et al., 2001). In pomegranate, seasonal temperature was inversely correlated to anthocyanin accumulation (Borochov-Neori et al., 2011). Warmer nights and days resulted in higher antioxidant activity and flavonoid composition in strawberries compared with cooler days (Wang and Zheng, 2001).

A recent metaanalysis on effects of CO_2 on vegetables synthesized 57 articles and found that increased CO_2 levels increased the concentrations of multiple quality parameters while decreasing the concentration of other quality parameters and not having an impact on others (Dong et al., 2018). Specifically, increased CO_2 levels were associated with increased fructose, glucose, total soluble sugar, total antioxidant capacity, total phenols, total flavonoids, ascorbic acid, and calcium in the edible part of vegetables included in their study by 14.2%, 13.2%, 17.5%, 59.0%, 8.9%, 45.5%, 9.5%, and 8.2%, respectively (Dong et al., 2018). At the same time, they found increased CO_2 levels were associated with decreased concentrations of protein, nitrate, magnesium, iron, and zinc in the edible part of vegetables included in their study by 9.5%, 18.0%, 9.2%, 16.0%, and 9.4%, respectively (Dong et al., 2018). The study further found that increased CO_2 levels had no significant impact on concentrations of titratable acidity, total chlorophyll, carotenoids, lycopene, anthocyanins, phosphorus, potassium, sulfur, copper, and manganese (Dong et al., 2018).

4 Effects of climate change on fruit nutrition

The following databases were searched to address the overall research question of this review "What are effects of climate change on fruit nutrition?" for the five most consumed fruit crops globally by weight including watermelons, bananas, apples, grapes, and oranges: Agricola, GreenFILE, Proquest, PubMed, and Web of Science. The searches were restricted to scientific articles published from 2000 to 2018 in English. A derivation of the following search terms was used in each database to identify the articles synthesized in this review: (Climate change OR Climate variability OR Weather OR Precipitation OR Temperature OR Carbon dioxide OR Rainfall) AND (Quality OR Nutrient* OR Secondary metabolite* OR Phytochemical* OR Antioxidant* OR Phenolic* OR Vitamin* OR Mineral* OR Micronutrient* OR Macronutrient*) AND (Apple OR *Malus pumila* OR *Malus sylvestris* OR *Malus communis* OR *Pyrus malus*). Derivations

TABLE 7.1 Examples of nutrients that are affected by climate change.

Nutrient	Relevance to human nutrition	Source
Anthocyanins	Anthocyanins are a subclass of flavonoid commonly found in red to blue pigmented plants. They act as antioxidants in the human body. Over time, their consumption is correlated with a decreased risk for many cancers and chronic diseases.	Crozier et al. (2008)
Antioxidants	Antioxidants reduce the negative effects of reactive oxygen species in the body. They are also critical to cell membrane stability. Many of the nutrients listed in this table function as antioxidants.	WHO and FAO (2005)
Calcium (Ca^{2+})	Calcium plays a vital role in providing strength to human bones, and inadequate calcium consumption can lead to bone loss and osteoporosis. Additionally, Ca^{2+} ions are involved in most metabolic processes, including neural conduction, muscle contraction, and blood clotting.	WHO and FAO (2005)
Carbohydrates	Carbohydrates are humans' primary energy source. Fruits provide carbohydrates in the forms of fiber and sugars. The carbohydrates are especially healthful when consumed in whole fruits because they provide a slow release of energy in the body. Additionally, adequate fiber intake can reduce the risk of colorectal cancer.	Mann et al. (2007)
Flavonoids	Flavonoids are a class of phenolic compounds that are commonly found in the skins of fruits. They act as antioxidants, and their sustained consumption over time is correlated with a decreased risk for many cancers and chronic diseases.	Crozier et al. (2008)
Iron (Fe)	Most of the iron in human bodies is found in the form of hemoglobin. Hemoglobin is a protein structure that transports oxygen from the lungs to the remainder of the body. Iron deficiency is the most common nutritional deficiency worldwide. Severe iron deficiency causes anemia, a condition in which oxygen delivery to tissues is diminished.	WHO and FAO (2005)
Magnesium (Mg^{2+})	Fifty to sixty percent of the human body's Mg^{2+} is located on the surface of bones. The remainder is found in muscles and soft tissues, where it helps maintain the electrical potentials of cell membranes in the muscles and the nervous system.	WHO and FAO (2005)
Phenolic compounds	Phenolic compounds are a type of secondary metabolite commonly found in food plants and closely linked to their nutritional quality. They act as antioxidants, and over time, their consumption is linked to a decreased risk for many cancers and chronic diseases.	Crozier et al. (2008)
Potassium (K^+)	Potassium is an essential player in maintenance of total body fluid and electrolyte balance. Potassium consumption at or above daily recommended values can help protect against heart disease and high blood pressure.	WHO (2012a)
Proteins	Protein is essential to human development, muscular function, and mental function. Proteins are made up of combinations of amino acids. Both quality and quantity are important when it comes to protein. Uncooked whole foods are good sources of high-quality bioavailable protein.	Lee et al. (2016)
Sodium (Na^+)	Sodium is the main ion present in extracellular fluid in the body. It is critical component in cell signaling. However, excessive consumption of sodium (primarily from processed foods) is associated with increased blood pressure and heart disease.	WHO (2012b)
Sugars	Sugars are one of the common carbohydrates found in fruits. "Free sugars" or sugars that have been refined to some extent and their consumption are correlated with obesity, diabetes, and other noncommunicable diseases. However, the sugars that are found in whole fruits are nutritious carbohydrates, which will be slowly absorbed by the body to provide energy. Their consumption is associated with lower body weights and can improve gut health.	WHO (2003)
Tannins	Tannins are a type of phenolic acid that are typically found in immature plants and create an astringent taste. Tannins can bind to dietary proteins during digestion, making them harder to digest. Therefore, overconsumption of tannins can have a negative effect on nutrition.	Crozier et al. (2008)
Vitamin C	In the human body, vitamin C acts as an electron donor. Electron donation has antioxidant effects, so vitamin C is critical in disease prevention. Additionally, inadequate vitamin C consumption can cause scurvy, a potentially lethal disease.	WHO and FAO (2005)

of these search terms replaced the "apple" search term with another fruit crop, as follows: "Banana OR *Musa sapientum* OR *Musa cavendishii* OR *Musa nana*", "Grape OR *Vitis vinifera*", "Orange OR *Citrus sinensis* OR *Citrus aurantium*", and "Watermelon OR *Citrullus vulgaris*." The scientific names of crops searched correspond to the species of each crop as defined by the FAO. The resulting articles were screened for relevance. For each fruit, a table is included, which summarizes relevant studies. Each table includes the full range of nutrients studied for that fruit, and, in the case of contradictory evidence, multiple studies may be shown for a single nutrient.

4.1 Effects of climate change on apple quality

Climate change is variably impacting the nutritional and phytochemical quality of apples (*Malus pumila*, *Malus sylvestris*, and *Malus communis*) around the world (Table 7.2). In the Himalayan states of India, apple farmers have reported decreases in yield, fruit size, and fruit quality and increases in pests and disease (Basannagari and Kala, 2013). In turn, apple farmers have reported responding to shifts in climate by switching varieties of apples or switching the crops they cultivate (Sen et al., 2015).

Over the past 50 years, climate change has been demonstrated to notably impact five major apple-producing regions in China (Qu and Zhou, 2016). Climate change has impacted different climate dimensions in each region, with mixed effects on apple quality (Qu and Zhou, 2016). In two regions, increased temperatures and decreased solar radiation improved the nutritional quality of apples by increasing vitamin C content, sugar-acid ratio, and anthocyanin concentration (Qu and Zhou, 2016). However, the other regions in China are observing decreased apple quality on several or all of these measures, due to temperatures above the optimal range, sunlight hours below the optimal range, and humidity levels either above or below the optimal range (Qu and Zhou, 2016).

While apple-producing regions in Japan have experienced increased temperatures with varying changes in solar radiation and precipitation over the past 40 years, each region has seen a steady increase in the soluble solid content of apples (Sugiura et al., 2013). Higher soluble solid content indicates a higher sugar content, which is generally associated with improved taste for human consumers. In Slovenia, a study demonstrated that increased solar radiation and temperature increased the total phenolic content of apples, indicating an improvement in nutritional quality (Zupan et al., 2014). However, extended periods of high temperatures and high solar radiation have been shown to result in sunburn in apples that is detrimental for yields (Wagner, 2010). Rates of sunburn have been increasing with climate changes and can cause large losses in crop yields (Wagner, 2010). A study in Germany that used hail nets to protect apples from sunburn found that uncovered fruits experienced sunburn while displaying high levels of vitamin C, anthocyanins, and soluble solid content (Solomakhin and Blanke, 2010). Covering apple trees with white or white-red hail nets slightly lowered the levels of temperature and solar radiation while increasing humidity with no effects on the nutrients measured (Solomakhin and Blanke, 2010). However, covering apple trees with colored hail nets reduced temperature and solar radiation while increasing humidity, which resulted in nutritionally inferior apples based on measures of vitamin C, anthocyanin content, and soluble solid content.

On the basis of this literature review, evidence demonstrates that climate change variables are improving specific parameters of nutritional quality of apples up to a certain threshold. When temperatures, solar radiation, and humidity extend beyond this threshold in different apple-producing regions, apple quality is reduced. It is thus critical to keep in mind that, while specific changes linked to the climate such as increased temperatures may be helping one apple-producing region at present, they could negatively impact the same region in the future if they continue to change at the same rate. Thus, apple-producing regions need to implement climate adaptation strategies to mitigate crop loss while preserving nutritional quality such as using white hail nets to prevent sunburn.

4.2 Effects of climate change on banana quality

Research studies on bananas (*Musa sapientum*, *Musa cavendishii*, and *Musa nana*) have demonstrated that changes in temperature, water supply, wind, and soil nutrients can both positively and negatively affect banana yields (Bugaud et al., 2009; Ndabamenye et al., 2012; Ramlall, 2014; Taulya, 2013). However, little research has focused on the effects of climate variation on the nutritional quality of bananas (Table 7.2). There is evidence that an increase in mean daily temperature results in a decrease in the average weight of banana fruits along with both a decrease and increase in specific nutrients, specifically a decrease in the concentrations of glucose and fructose and an increase in concentrations of phosphorus, magnesium, and calcium (Bugaud et al., 2009). The increase in micronutrients with a decrease in yields is suggestive of a dilution effect: with a smaller yield and approximately the same amount of total nutrients, the concentration of nutrients in bananas increases. While an increase in micronutrients may indicate a positive effect on human nutrition as global temperature increases, the decrease in carbohydrate concentration could have negative effects in regions where bananas are a staple carbohydrate. Many tropical and subtropical populations rely heavily on bananas for nutrition. For example, in rural areas of Uganda, Rwanda, and Cameroon, bananas often provide up to 25% of the daily caloric intake (FAO, 2017). For this reason, climate impacts on carbohydrates, sugars, and overall yields of bananas could have negative effects on human nutrition.

TABLE 7.2 Climate effects on apple and banana nutritional quality

Title	Author (year)	Location	Climate variable(s)	Nutritional quality variable(s)	Observations
APPLE					
Possible impact of climate change on the quality of apples from the major producing areas of China	Qu and Zhou (2016)	China	Temperature, humidity, and solar radiation	Vitamin C content, anthocyanin content, and sugar-acid ratio	In two Chinese regions, increased temperatures and decreased sunlight have improved the nutritional quality of apples by increasing vitamin C content, sugar-acid ratio, and anthocyanin concentration. However, other regions are observing decreased fruit quality on some or all of these measures, due to temperatures above the optimal range, sunlight hours below the optimal range, and humidity levels either above or below the optimal range.
Can colored hail nets improve taste (sugar-acid ratio), consumer appeal (coloration), and nutritional value (anthocyanin and vitamin C) of apple fruit?	Solomakhin and Blanke (2010)	Germany	Temperature, humidity, solar radiation	Vitamin C content, anthocyanin content, SSC (° Brix)	Covering trees with variously colored hail nets affected the solar radiation, temperature, and humidity of the apples. Apples under white netting experienced slightly lower temperatures and solar radiation than uncovered apples but higher humidity. These apples had similar levels of vitamin C, anthocyanins, and SSC to uncovered apples. All other colors of netting decreased solar radiation, decreased temperature, increased humidity, and yielded nutritionally inferior apples, based on these variables.
Changes in the taste and textural attributes of apples in response to climate change	Sugiura et al. (2013)	Japan	Temperature, precipitation, solar radiation	SSC (°Brix)	Two apple-producing locations in Japan have experienced increased average temperatures, while changes in precipitation and solar radiation have been variable. In both regions, apples have consistently increased in SSC over the past 40 years.
Leaf micromorphology and sugar may contribute to differences in drought tolerance for two apple cultivars	Wu et al. (2014)	China	Water deficit	Nonstructural carbohydrate levels	The leaves of both drought-tolerant and drought-sensitive apple trees experiencing a water deficit display increased levels of nonstructural carbohydrates (especially sorbitol).
Individual phenolic response and peroxidase activity in peel of differently sun-exposed apples in the period favorable for sunburn occurrence	Zupan et al. (2014)	Slovenia	Solar radiation and temperature	Total phenolic content and content of individual phenols	As apples experienced increased levels of solar radiation and higher temperature, the total phenolic content in their peels increased. However, levels of sunburn also increased. Individual phenols displayed varied results.

TABLE 7.2 Climate effects on apple and banana nutritional quality—cont'd

Title	Author (year)	Location	Climate variable(s)	Nutritional quality variable(s)	Observations
BANANA					
Relative importance of location and period of banana bunch growth in carbohydrate content and mineral composition of fruit	Bugaud et al. (2009)	Martinique	Temperature, rainfall, solar radiation	Micronutrient concentrations and carbohydrate content	As mean daily temperature increased, the average weight of fruits decreased, the carbohydrate content of fruits decreased, and the concentrations of phosphorus, magnesium, and calcium increased. Temperature, rainfall, and soil type exhibited varying effects on these quality variables.

4.3 Effects of climate change on grape quality

Most research about the effects of climate on grape (*Vitis vinifera*) nutritional quality has centered on the levels of phenolic compounds, including anthocyanins and flavanols (Table 7.3). In addition to providing antioxidant benefits to humans, these compounds have important effects on the overall flavor of wine. The overall evidence of the effects of climate change on grape nutrition indicates variation on the basis of region. Experimental and observational studies in several regions have elucidated an association between increased temperature and increased anthocyanin, phenol, and flavanol composition (Cozzolino et al., 2010; Del-Castillo-Alonso et al., 2016; Fernandes de Oliveira et al., 2015; Ramos et al., 2015; Torres et al., 2017). However, a few studies have also demonstrated the opposite effect of decreased phenolic, anthocyanin, and flavanol contents under high temperatures (Fourment et al., 2017; Pastore et al., 2017; Xu et al., 2011). Xu et al. (2011) highlighted that the likely reason they observed that grapes grown in winter in China display higher phenolic compound contents and antioxidant properties than grapes grown in the summer was because summer temperatures were often above optimal for anthocyanin synthesis in the grapes. This finding suggests that as global temperatures increase, the antioxidant properties of grapes may increase, up to a specific threshold. Once temperatures become too hot, the antioxidant properties of grapes could be negatively impacted (Xu et al., 2011).

Solar radiation may also impact the nutrient content of grapes. One study found that solar radiation has an even greater effect on phytochemical profiles than temperature (Del-Castillo-Alonso et al., 2016). Specifically, a positive correlation was found between increased levels of solar radiation and increased flavanols, flavanonols, and cinnamic acids (Del-Castillo-Alonso et al., 2016). Likewise, another study found that increasing sunlight exposure and UV exposure of grapes increased anthocyanin content, phenolic content, tannin content, and soluble solid content of the resulting wine (Song et al., 2015). In contrast, another study found that the intensity of UV light did not impact total anthocyanin content but did impact the proportions of anthocyanin derivatives present in grapes, indicating the effects of radiation on anthocyanin synthesis (Fernandes de Oliveira et al., 2015). Overall, changing levels of solar radiation will likely impact some of the antioxidant properties of grapes, though different grape cultivars and different regions could see varying impacts (Fernandes de Oliveira et al., 2015).

Water stress associated with climate change is also impacting grape nutritionally quality. Most studies have observed a positive effect of water deficit during any period of growth on anthocyanin and flavanol content in grapes (Bucchetti et al., 2011; Castellarin et al., 2007; Ollé et al., 2011; Ramos et al., 2015; Zarrouk et al., 2012). However, the timing of the water deficit could impact which specific anthocyanins are affected (Ollé et al., 2011). It appears that water stress may play a smaller role in the nutrient content of grapes than temperature and solar radiation (Fourment et al., 2017). This could explain why one study found a correlation between increased rainfall during ripening and increased phenolic content (Del-Castillo-Alonso et al., 2016). In this study, increased rainfall was also correlated with increased temperature and solar radiation, which may have played a larger role in increasing the phenolic content (Del-Castillo-Alonso et al., 2016).

Additionally, several studies simulated the overall expected effects of climate change on grapes. In one study, increased CO_2, temperature, and humidity resulted in reduced anthocyanin concentration and acid concentrations

TABLE 7.3 Climate effects on grape nutritional quality

Title	Author (year)	Location	Climate variable(s)	Nutritional quality variable(s)	Observations
Grape (*Vitis vinifera*) compositional data spanning 10 successive vintages in the context of abiotic growing parameters	Cozzolino et al. (2010)	Australia	Temperature, rainfall, CO_2 level	Anthocyanin content	In several wine growing regions of Australia, data from several years show that increases in temperature generally increase anthocyanin content in grapes. Relationships between grape quality and rainfall and CO_2 levels were not clear.
Environmental factors correlated with the metabolite profile of *Vitis vinifera* cv. Pinot Noir berry skins along a European latitudinal gradient	Del-Castillo-Alonso et al. (2016)	Europe	Solar radiation, temperature, rainfall, aridity	Metabolite composition	Of the measured variables, solar radiation had the greatest effect on metabolite profiles of grapes. Specifically, increasing levels of solar radiation particularly increased flavanols, flavanonols, and cinnamic acids. Additionally, increased temperature and rainfall during fruit ripening were correlated with increased phenolic content, though this relationship is weaker than that of solar radiation.
Tannat grape composition responses to spatial variability of temperature in Uruguay's coastal wine region	Fourment et al. (2017)	Uruguay	Temperature, rainfall	Phenolic content, anthocyanin content	Phenolic and anthocyanin content in grapes displayed a negative correlation with increasing temperatures. When rainfall was below average, temperature had increased effects on phenolic content.
Grape yield and quality responses to simulated year 2100 expected climatic conditions under different soil textures	Leibar et al. (2017)	Spain	CO_2, temperature, relative humidity, water deficit	Anthocyanin concentration	Simulated conditions of future climate change (increased CO_2, temperature, and humidity) caused reduced anthocyanin concentration and acid concentrations in grapes. Water deficit further reduced acid concentrations but did not affect anthocyanin concentration.
Growing season climate variability and its influence on Sauvignon Blanc and Pinot Gris berries and wine quality: study case in Romania (2005–15)	Nistor et al. (2018)	Romania	Temperature, rainfall	SSC (°Brix)	Over 10 years, researchers observed that higher sugar content in grapes was correlated with higher temperatures and lower rainfall levels.
Effect of pre- and postveraison water deficit on proanthocyanidin and anthocyanin accumulation during Shiraz berry development	Ollé et al. (2011)	Australia	Water deficit	Anthocyanin and proanthocyanidin contents	Water deficit before the onset of ripening and after the onset of ripening both increased anthocyanin content overall. However, water deficits at different times increased different anthocyanins, suggesting that the timing of water deficit affects the biosynthesis of anthocyanins.
Effects of climate change scenarios on Tempranillo grapevine (*Vitis vinifera* L.) ripening: response to a combination of elevated CO_2 and temperature and moderate drought	Salazar Parra et al. (2010)	Spain	CO_2, temperature, water deficit	Anthocyanin concentration	Elevated CO_2 concentration and temperature, as is predicted in the future, decreased anthocyanin concentration, compared with current conditions. Additionally, water deficit decreased anthocyanin extractability under climate change conditions.

TABLE 7.3 Climate effects on grape nutritional quality—cont'd

Title	Author (year)	Location	Climate variable(s)	Nutritional quality variable(s)	Observations
Effect of grape bunch sunlight exposure and UV radiation on phenolics and volatile composition of *Vitis vinifera* L. cv. Pinot Noir wine	Song et al. (2015)	China	Sunlight exposure, UV exposure	Anthocyanin content, total phenolic content, tannin content, SSC (°Brix)	As sunlight exposure and UV exposure for grapes increased, the anthocyanin content, phenolic content, tannin content, and SSC of the resulting wine also increased.
Flavonoid and amino acid profiling on *Vitis vinifera* L. cv Tempranillo subjected to deficit irrigation under elevated temperatures	Torres et al. (2017)	Spain	Temperature, water deficit	Anthocyanin content, flavanol content, amino acid concentration	In general, elevated temperatures caused increased concentrations of flavanols, hexoses, and amino acids in grapes. Experiencing a water deficit during the latter half of production increased anthocyanin and flavanol content, compared with no water deficit. However, a water deficit during the earlier half of production decreased anthocyanin and flavanol content, compared with no water deficit.
Influence of growing season on phenolic compounds and antioxidant properties of grape berries from vines grown in subtropical climate	Xu et al. (2011)	China	Precipitation, temperature,	Phenolic compound and antioxidant properties, including anthocyanins	In a subtropical climate, grapes grown in winter (mild temperature and drier conditions) displayed higher phenolic compound contents and antioxidant properties than grapes grown in the summer (high temperature and wetter conditions). Summer temperatures were often above optimal for anthocyanin synthesis in the grapes.
Impact of irrigation regime on berry development and flavonoids composition in Aragonez (Syn. Tempranillo) grapevine	Zarrouk et al. (2012)	Portugal	Water deficit	Total flavanol content, proanthocyanidin and anthocyanin content	Water deficit at all stages of growth decreased proanthocyanidins and flavanols in grapes. Changes in anthocyanin content were only apparent in maturing fruits and varied between years, indicating that variables other than water may have had an influence.

in grapes (Leibar et al., 2017). Adding water deficit to the simulation did not affect anthocyanin concentration (Leibar et al., 2017). In another study, increased CO_2 concentration, increased temperature, and water deficit decreased anthocyanin concentration, compared with current conditions (Salazar Parra et al., 2010). In these simulations, predicted impacts of climate change reduced the nutritional quality of grapes.

4.4 Effects of climate change on orange quality

Based on the literature focused on the effects of climate variables on oranges (*Citrus aurantium* and *Citrus sinensis*), climate change could have a variety of impacts on the nutritional quality of oranges (Table 7.4). As global temperature increases over time, the fruit size of oranges may increase at the expense of nutritional quality based on the phenolic compounds and vitamin C concentrations (Hussain et al., 2017). However, an increase in atmospheric CO_2, as is expected over the course of the next century, may cause an increase in the number of oranges produced, size of the oranges, and vitamin C content of oranges (Idso et al., 2002). Additionally, if soil salinity increases due to decreased

TABLE 7.4 Climate effects on orange nutritional quality.

Title	Author (year)	Location	Climate variable(s)	Nutritional quality variable(s)	Findings
Salinity reduces growth, gas exchange, chlorophyll, and nutrient concentrations in diploid sour orange and related allotetraploid somatic hybrids	Garcia-Sanchez et al. (2002)	Florida, United States	Soil salinity	Micronutrient content in leaves	As soil salinity increased, the concentrations of micronutrients present in the leaves of sour orange trees on a variety of rootstocks varied. Na^+ concentration increased, Ca^{2+} concentration decreased, K^+ concentration decreased, Mg^{2+} concentration increased, N concentration increased, and P concentration did not change.
Physicochemical profiling of promising sweet orange cultivars grown under different agroclimatic conditions of Pakistan	Hussain et al. (2017)	Pakistan	Temperature and precipitation levels	Phenolic compounds and vitamin C content	In a variety of sweet orange cultivars grown in Pakistan, a warmer, drier, low-altitude site produced oranges that were larger and heavier. As altitude increased and both temperature and precipitation decreased, oranges generally decreased in size but became juicer and contained higher vitamin C and phenolic compound concentrations.
The effect of elevated atmospheric CO_2 on the vitamin C concentration of (sour) orange juice	Idso et al. (2002)	Arizona, United States	Atmospheric CO_2 level	Vitamin C content	Sour oranges were grown for 12 years in controlled atmospheric CO_2 conditions, at levels considered ambient ($400\,\mu L/L$) or increased by 75% ($700\,\mu L/L$). Eight years of data revealed that a 75% increase in atmospheric CO_2 concentration generally increased the number of fruits produced by 65%–83%, increased the weight of fruits by 2%–6%, and increased the vitamin C concentration of the fruit juice by 4%–6%.
Effects of sodium chloride stress on gas exchange, chlorophyll content, and nutrient concentrations of nine citrus rootstocks	Khoshbakht et al. (2015)	Iran	Soil salinity	Micronutrient content in leaves and roots	Increased soil salinity decreased the overall photosynthetic rates for orange trees and affected the concentrations of micronutrients in their leaves and roots. As salinity increased, Na^+ and Cl^- concentrations in roots and leaves increased, Ca^{2+} and K^+ concentrations decreased in roots and leaves, and Mg^{2+} concentrations decreased in leaves but increased in roots.
Proteomic and metabolomic profiling of Valencia orange fruit after natural frost exposure	Perotti et al. (2015)	Argentina	Frost	Proteomic and metabolic profiles	Oranges were harvested before a natural frost event, directly after the event, and 1 week after the event. Levels of proteins, primary metabolites, and secondary metabolites were affected by the frost. Due to the frost, some components of the fruit exhibited increased expression, some exhibited decrease expression, and some varied as time passed after the frost.

TABLE 7.5 Climate effects on watermelon nutritional quality

Title	Author (year)	Location	Climate variable(s)	Nutritional quality variable(s)	Observations
Dynamic changes in the leaf proteome of a C3 xerophyte, *Citrullus lanatus* (wild watermelon), in response to water deficit	Akashi et al. (2011)	Japan	Water deficit	Protein expression	The proteome of watermelon leaves was examined under water-deficit conditions (simulating drought), compared with control conditions. Under a water deficit, researchers identified 23 proteins that were increased in concentration and 6 proteins that decreased in concentration. The upregulated proteins were commonly heat shock proteins, proteins related to antioxidative defense, and proteins related to carbohydrate metabolism.
Effects of temperature around the fruit on sugar accumulation in watermelon during the latter half of fruit developmental period	Fukuoka et al. (2009)	Japan	Temperature	Sugar content (sucrose, fructose, and glucose)	During the second half of fruit development, the temperature around watermelons was raised, lowered, or left unaffected. As temperature increased, fruits were larger overall, displayed larger cell size, and displayed decreased glucose and fructose concentrations.
Effects of heating bearing shoot near fruit on cell size, sucrose metabolizing enzymes, and sugar accumulation in watermelon fruit	Kano et al. (2012)	Japan	Nighttime temperature	Sugar content (sucrose, fructose, and glucose)	Bearing shoots near fruits were heated from 12 p.m. to 6 a.m. each day starting 5 days after anthesis. Compared with control fruits, the surface and flesh of the heated fruits was hotter. Although there was not difference in fruit size, heated fruits had higher sucrose and fructose content, due to high sucrose phosphate synthase activity. Additionally, heated fruits displayed accelerated pigmentation, ending with more red flesh.
Effect of preharvest-deficit irrigation on second crop watermelon grown in an extremely hot climate	Kirnak et al. (2009)	Turkey	Water deficit	Soluble solid content (°Brix)	In a semiarid region, as water stress increased, watermelon yield decreased linearly. Water stress at a rate of less than 75% full irrigation had severe negative impacts on soluble solids content (SSC). However, 75% of standard irrigation can be used to conserve water while obtaining high-quality fruit.
Effect of heat treatment around the fruit set region on growth and yield of watermelon (*Citrullus lanatus* [Thunb.] Matsum. and Nakai)	Noh et al. (2013)	Japan	Temperature	Soluble solid content (°Brix)	Temperature was adjusted around the fruiting regions of watermelon plants grow during winter season. While a higher temperature did not affect fruit size, it did increase the SSC of fruits. It also increased calcium and magnesium concentrations in leaves.
Resistance to cold and heat stress: accumulation of phenolic compounds in tomato and watermelon plants	Rivero et al. (2001)	Spain	Temperature	Total phenolic content	Watermelon plants were grown for the first 30 days at an optimal temperature; then, plants were transferred to chambers at varying temperatures. Since watermelons are sensitive to lower temperatures, as temperature

Continued

TABLE 7.5   Climate effects on watermelon nutritional quality—cont'd

Title	Author (year)	Location	Climate variable(s)	Nutritional quality variable(s)	Observations
					decreased below optimal (33°C), the total phenol levels in the melons increased.
Influence of temperature on biomass, iron metabolism, and some related bioindicators in tomato and watermelon plants	Rivero et al. (2003)	Spain	Temperature	Iron levels	Watermelon plants were grown for the first 30 days at an optimal temperature; then plants were transferred to chambers at varying temperatures. At day 60, Fe levels in plant leaves and roots were measured. As temperature decreased below optimal (35°C), Fe levels decreased.
Yield, mineral composition, water relations, and water use efficiency of grafted miniwatermelon plants under deficit irrigation	Rouphael et al. (2008)	Italy	Water deficit	Soluble solid content (°Brix) and micronutrient levels	Watermelons were given optimal irrigation for 3 weeks after planting. Then, irrigation was either maintained at the optimal level or decreased to 0.75 or 0.5 of the optimal level. Decreased irrigation decreased yield linearly. Irrigation rate had no effect on the SSC or the content of P or Ca in fruits, but it did impact K and Mg content in fruits. K content was highest at the 0.75 irrigation rate. Mg content increased linearly as irrigation level decreased.

precipitation with climate change, the nutrient content of leaves and roots of orange trees will change (Garcia-Sanchez et al., 2002; Khoshbakht et al., 2015), which may impact the micronutrient composition of fruits. Overall, increased temperature, increased atmospheric CO_2, and increased soil salinity might all impact the nutritional quality of oranges in the future. However, some of these climate factors may improve the quality, while others decrease the quality.

4.5 Effects of climate change on watermelon quality

As droughts become more frequent and severe, they may impact the nutritional quality of watermelons (*Citrullus vulgaris*; Table 7.5). Evidence suggests that water stress increases the content of the micronutrients potassium and magnesium in watermelon fruits (Rouphael et al., 2008). While these changes indicate that watermelons grown during periods of drought may be more nutritious, there is evidence that water stress will reduce the yield of watermelons and may make them less palatable (Kirnak et al., 2009; Rouphael et al., 2008). Palatability of watermelons is related to their soluble solid content. There is mixed evidence as to whether drought conditions could decrease soluble solid content or have a negligible effect (Kirnak et al., 2009; Rouphael et al., 2008). Another effect of water deficit that has been observed in watermelons is increased protein levels in watermelon leaves (Akashi et al., 2011). Although this effect has only been observed in leaves, it could possibly translate into fruits, which may positively affect the nutritional quality of watermelons.

Increased temperatures over time could also have a variety of impacts on the nutritional quality of watermelons. While one group of researchers found that lower daytime temperatures and higher nighttime temperatures can increase the soluble solid content of watermelons (Kano, 2004; Kano et al., 2012), others reported that higher daytime and nighttime temperatures will increase the soluble solid content of watermelons (Noh et al., 2013). It appears that the effects of increased global temperatures may combine with water stress and other factors to have an overall impact on soluble solid content, whether positive of negative. In addition to soluble solid content, temperature affects the nutrients in watermelons. Watermelons grown below the ideal temperature of 35°C display higher total phenolic compound levels, indicating more nutritious fruit (Rivero et al., 2001). However, other studies have shown that the micronutrients calcium, magnesium, and iron increase in watermelon leaves as temperature increases (Noh et al., 2013; Rivero et al., 2003). While it is unknown whether these increased micronutrients would also occur in fruits, it is possible that micronutrient levels could increase in watermelons as temperature increases. Overall, the literature available on the impacts of climate variables on the nutritional quality of watermelons indicates that as global temperatures rise and droughts become more common, there may be increased levels of several micronutrients and proteins in watermelons, with the trade-off of fewer phenolic compounds. There may also be changes in the flavor of watermelon, based upon the sugar contents of the fruits.

5 Conclusion

This chapter highlights that multiple environmental variables linked to climate change significantly impact fruit nutrition. Findings from our literature review suggest that climate change is impacting the nutritional quality of the five most produced fruits in the world and that these changes vary based on the type of crop and geographic area. The evidence reviewed on apples suggests that the effects of climate in several apple-producing regions around the world are improving the sugar content, phenolic content, and vitamin content up to certain plant thresholds. However, when temperatures, solar radiation, and humidity extend beyond these thresholds, apple quality will likely be reduced. The studies reviewed showed a similar finding for grapes, where flavanols, anthocyanins, and other phenols increased due to changes in climate variables until a certain threshold, beyond which they are expected to decline. This review further demonstrated that predicted changes in climate may impact bananas by decreasing sugar and overall carbohydrate levels and decreasing yield. Decreases in yields and quality of bananas have the potential to negatively impact diets and nutrition of people around the world that depend on bananas as a staple crop. This review further found that increased temperature, increased atmospheric CO_2, and increased soil salinity have both positive and negative effects on vitamin C and phenolic content of oranges. Specific changes in orange quality may also vary by region, depending on the specific changes in climate that each region incurs. Lastly, rising global temperatures and increasing frequency of droughts have been shown to increase the levels of several micronutrients and proteins in watermelons, with the trade-off of lower phenolic content. Climate change is also impacting the flavor of watermelons by changing the sugar contents of the fruits.

The findings synthesized in this review highlight that the impacts of climate change on the nutritional quality of fruits will be diverse and will vary by region. Temperature, solar radiation, water availability, atmospheric CO_2, and other climate variables are impacting several nutritional variables of fruits, including soluble solids, vitamin content, phenolic compound content, and the levels of several micronutrients. However, it is important to note that climate variables interact with one another, creating synergistic or antagonistic effects within agricultural systems and within fruit crops. Therefore, the expected changes in fruit nutritional quality will vary depending on the specific interaction of climate changes incurred in each region with specific agricultural systems and their associated management practices (Ahmed and Stepp, 2016). Additionally, as indicated by the literature on apples and grapes, some fruit-producing regions may experience improved fruit quality until climate variables exceed plant thresholds, after which there may then be a sharp decrease in crop quality.

While studies in this synthesis focused on examining the effects of climate change on fruit nutrition, findings have implications for farmer livelihoods and human well-being. Market prices and associated farmer livelihoods may be compromised if humans are able to perceive flavor shifts in crops (Ahmed et al., 2010, 2014). In addition, climate factors that resulted in a decrease in specific nutrients and phytochemicals can compromise the therapeutic effect associated with these compounds.

6 Future research

On the basis of the importance of fruit consumption to human nutrition, future research is called for to understand the effects of future climate change on high-volume fruit crops in various regions around the world and to identify strategies to mitigate climate risk. The current literature review suggests that there are multiple studies examining the effects of climate on grape quality due to worldwide demand for high-quality value-added grape products, namely, wine. However, we found a lack of relevant research on the nutritional quality of bananas, a crop which is relied on as a staple food in many tropical regions. Future research can help address how climate change is impacting the nutritional quality of bananas and other culturally and nutritionally relevant fruits around the world that support farmer livelihoods and diets.

In addition, further research is called for to measure, monitor, and mitigate the negative effects of climate change on fruit nutrition. Collecting evidence on the effects of climate change on fruit nutrition for a wide array of crops will enable identification of trends over time including specific climate challenges that different crops will face in different regions. After understanding climate challenges for different crops in different regions, management strategies can be identified to mitigate climate risks on fruit production. For example, in the case of apples, researchers have determined that covering apple orchards with hail nets is an effective strategy in reducing the surface temperature of apples and the amount of UV radiation they receive (Kalcsits et al., 2017; McCaskill et al., 2016; Solomakhin and Blanke, 2010). In areas that will incur above-optimal temperatures and solar radiation with climate change, the use of hail nets can reduce sunburn losses, improve flavor, and improve consumer desirability of apples (Kalcsits et al., 2017; Solomakhin and Blanke, 2010).

Another management strategy to mitigate against decreased precipitation that emerged in our literature review is the application of mulching in vineyards (Timbal et al., 2006). Viticulturists in Australia have reported using mulch to build soil carbon so that their soil will better retain water (Galbreath, 2014). Researchers in Southeast Australia identified that applying composted mulch in vineyards can mitigate the effects of drought by reducing evaporation from the soil surface and reducing soil temperature, both of which increase soil moisture levels (Campell and Sharma, 2003). Mulching and setting up hail nets are two examples of many examples of how farmers are responding to climate change to protect their crops. Ongoing and future research is expected to reveal many other management strategies for fruit crops, which can ensure the ongoing production of nutritious fruits over time. Multiple other agricultural, physiological, and molecular innovations have been identified for mitigating climate risk in fruit production systems including agricultural diversification, tree planting and maintaining vegetative cover, management of soil organic matter and carbon sequestration, water management, controlling pests and disease, and migration and relocating agricultural systems to more suitable locations (Ahmed and Stepp, 2016). Further research is needed to evaluate the effectiveness of these management strategies for mitigating the effects of climate change on fruit production.

In addition to identifying management strategies for climate mitigation, research is needed to identify climate-resilient fruit cultivars in different geographic areas and for different farming methods. Such research should focus on evaluating both of crop yields and quality of various varieties, varietals, and cultivars in response to multiple climate factors. Furthermore, such research should focus on the identification of crop yield and quality thresholds for different fruit cultivars.

References

Achadi, E., Ahuja, A., Bendech, M.A., Bhutta, Z.A., De-Regil, L.M., Fanzo, J., et al., 2016. Global nutrition report 2016: From promise to impact: Ending malnutrition by 2030. International Food Policy Research Institute.

Ahmed, S., Stepp, J.R., 2016. Beyond yields: climate effects on specialty crop quality and agroecological management. Elementa 4, 92.

Ahmed, S., Unachukwu, U., Stepp, J.R., Peters, C.M., Long, C., Kennelly, E., 2010. Pu-erh tea tasting in Yunnan, China: correlation of drinkers' perceptions to phytochemistry. J. Ethnopharmacol. 132 (1), 176–185.

Ahmed, S., Stepp, J.R., Orians, C., Griffin, T., Matyas, C., Robbat, A., et al., 2014. Effects of extreme climate events on tea (Camellia sinensis) functional quality validate indigenous farmer knowledge and sensory preferences in tropical China. PLoS One 9 (10), e109126.

Akashi, K., Yoshida, K., Kuwano, M., Kajikawa, M., Yoshimura, K., Hoshiyasu, S., et al., 2011. Dynamic changes in the leaf proteome of a C_3 xerophyte, Citrullus lanatus (wild watermelon), in response to water deficit. Planta 233 (5), 947–960. https://doi.org/10.1007/s00425-010-1341-4.

Åkerström, A., Forsum, Å., Rumpunen, K., Jaderlund, A., Bång, U., 2009. Effects of sampling time and nitrogen fertilization on anthocyanidin levels in Vaccinium myrtillus fruits. J. Agric. Food Chem. 57 (8), 3340–3345.

Basannagari, B., Kala, C.P., 2013. Climate change and apple farming in Indian Himalayas: a study of local perceptions and responses. PLoS One 8 (10), e77976.

Bates, B., Kundzewicz, Z., Wu, S., 2008. Climate Change and Water. Intergovernmental Panel on Climate Change Secretariat.

Bindi, M., Fibbi, L., Miglietta, F., 2001. Free Air CO2 Enrichment (FACE) of grapevine (Vitis vinifera L.): II. Growth and quality of grape and wine in response to elevated CO2 concentrations. Eur. J. Agron. 14 (2), 145–155.

Bisbis, M.B., Gruda, N., Blanke, M., 2018. Potential impacts of climate change on vegetable production and product quality—a review. J. Clean. Prod. 170, 1602–1620.

Borochov-Neori, H., Judeinstein, S., Harari, M., Bar-Ya'akov, I., Patil, B.S., Lurie, S., Holland, D., 2011. Climate effects on anthocyanin accumulation and composition in the pomegranate (Punica granatum L.) fruit arils. J. Agric. Food Chem. 59 (10), 5325–5334.

Bucchetti, B., Matthews, M.A., Falginella, L., Peterlunger, E., Castellarin, S.D.J.S.H., 2011. Effect of water deficit on merlot grape tannins and anthocyanins across four seasons. Sci. Hortic. 128 (3), 297–305.

Bugaud, C., Daribo, M.-O., Beauté, M.-P., Telle, N., Dubois, C., 2009. Relative importance of location and period of banana bunch growth in carbohydrate content and mineral composition of fruit. Fruits 64 (2), 63–74.

Campell, A., Sharma, G., 2003. Composted Mulch for Sustainable and Productive Viticulture. Recycled Organics Unit, Sydney.

Castellarin, S.D., Matthews, M.A., Di Gaspero, G., Gambetta, G.A., 2007. Water deficits accelerate ripening and induce changes in gene expression regulating flavonoid biosynthesis in grape berries. Planta 227 (1), 101–112.

Cozzolino, D., Cynkar, W.U., Dambergs, R.G., Gishen, M., Smith, P., 2010. Grape (Vitis vinifera) compositional data spanning ten successive vintages in the context of abiotic growing parameters. Agric. Ecosyst. Environ. 139 (4), 565–570.

Crozier, A., Clifford, M.N., Ashihara, H., 2008. Plant Secondary Metabolites: Occurrence, Structure and Role in the Human Diet. John Wiley & Sons.

DaMatta, F.M., Grandis, A., Arenque, B.C., Buckeridge, M.S., 2010. Impacts of climate changes on crop physiology and food quality. Food Res. Int. 43 (7), 1814–1823.

Darjazi, B.B., Larijani, K., 2012. The effects of climatic conditions and geographical locations on the volatile flavor compounds of fig (Ficus carica L.) fruit from Iran. Afr. J. Biotechnol. 11 (38), 9196–9204.

Del-Castillo-Alonso, M.A., Castagna, A., Csepregi, K., Hideg, E., Jakab, G., Jansen, M.A., et al., 2016. Environmental factors correlated with the metabolite profile of *Vitis vinifera* cv. Pinot Noir berry skins along a European latitudinal gradient. J. Agric. Food Chem. 64 (46), 8722–8734.

Dong, J., Gruda, N., Lam, S.K., Li, X., Duan, Z., 2018. Effects of elevated CO2 on nutritional quality of vegetables: a review. Front. Plant Sci. 9 (924).

Drake, J.E., Gallet-Budynek, A., Hofmockel, K.S., Bernhardt, E.S., Billings, S.A., Jackson, R.B., et al., 2011. Increases in the flux of carbon belowground stimulate nitrogen uptake and sustain the long-term enhancement of forest productivity under elevated CO2. Ecol. Lett. 14 (4), 349–357.

Eberhardt, M.V., Lee, C.Y., Liu, R.H., 2000. Nutrition: antioxidant activity of fresh apples. Nature 405 (6789), 903.

FAO, 2017. Banana Statistical Compendium. http://www.fao.org/fileadmin/templates/est/COMM_MARKETS_MONITORING/Bananas/Documents/Banana_Statistical_Compendium_2017.pdf. (Accessed 25 January 2019).

FAO, 2018a. Crop Production and Natural Resource Use. http://www.fao.org/docrep/005/y4252e/y4252e06.htm. (Accessed 25 January 2019).

FAO, 2018b. FSOStat: Crops. http://www.fao.org/faostat/en/#data/QC. (Accessed 25 January 2019).

Feeny, P., 1976. Plant apparency and chemical defense. In: Biochemical Interaction Between Plants and Insects. Springer, pp. 1–40.

Fernandes de Oliveira, A., Mercenaro, L., Del Caro, A., Pretti, L., Nieddu, G., 2015. Distinctive anthocyanin accumulation responses to temperature and natural UV radiation of two field-grown (*Vitis vinifera* L.) cultivars. Molecules 20 (2), 2061–2080.

Fourment, M., Ferrer, M., González-neves, G., Barbeau, G., Bonnardot, V., Quénol, H., 2017. Tannat grape composition responses to spatial variability of temperature in an Uruguay's coastal wine region. Int. J. Biometeorol. 61 (9), 1617–1628.

Fraenkel, G.S., 1959. The raison d'etre of secondary plant substances. Sci. Total Environ., 1466–1470.

Fukuoka, N., Masuda, D., Kanamori, Y., 2009. Effects of temperature around the fruit on sugar accumulation in watermelon (Citrullus lanatus (Thunb.) Matsum. & Nakai) during the latter half of fruit developmental period. J. Japanese Soc. Hort. Sci. 78 (1), 97–102.

Galbreath, J., 2014. Climate change response: evidence from the Margaret River wine region of Australia. Bus. Strateg. Environ. 23 (2), 89–104.

Garcia-Sanchez, F., Martinez, V., Jifon, J., Syvertsen, J.P., Grosser, J.W., 2002. Salinity reduces growth, gas exchange, chlorophyll and nutrient concentrations in diploid sour orange and related allotetraploid somatic hybrids. J. Hortic. Sci. Biotechnol. 77 (4), 379–386.

Grunert, K.G., 2005. Food quality and safety: consumer perception and demand. Eur. Rev. Agric. Econ. 32 (3), 369–391.

Hatfield, J.L., Prueger, J.H., 2015. Temperature extremes: effect on plant growth and development. Weather Clim. Extrem. 10, 4–10.

Hertel, T.W., Burke, M.B., Lobell, D.B., 2010. The poverty implications of climate-induced crop yield changes by 2030. Glob. Environ. Chang. 20 (4), 577–585.

Hounsome, N., Hounsome, B., Tomos, D., Edwards-Jones, G., 2008. Plant metabolites and nutritional quality of vegetables. J. Food Sci. 73 (4), R48–R65.

Huang, B., Xu, Y., 2015. Cellular and molecular mechanisms for elevated CO2–regulation of plant growth and stress adaptation. Crop Sci. 55 (4), 1405–1424.

Hussain, S.B., Maqsood, A., Muhammad Akbar, A., Sajjad, H., Shaghef, E., 2017. Physico-chemical profiling of promising sweet orange cultivars grown under different agro-climatic conditions of Pakistan. Erwerbs-Obstbau 59 (4), 315–324.

Idso, S.B., Kimball, B.A., Shaw, P.E., Widmer, W., Vanderslice, J.T., Higgs, D.J., et al., 2002. The effect of elevated atmospheric CO2 on the vitamin C concentration of (sour) orange juice. Agric. Ecosyst. Environ. 90 (1), 1.

IPCC, 2014. In: Pachauri, R.K., Allen, M.R., Barros, V.R., Broome, J., Cramer, W., Christ, R., … Dasgupta, P. (Eds.), Climate change 2014: Synthesis report. Contribution of Working Groups I, II and III to the fifth assessment report of the Intergovernmental Panel on Climate Change. IPCC.

Kalcsits, L., Musacchi, S., Layne, D.R., Schmidt, T., Mupambi, G., Serra, S., et al., 2017. Above and below-ground environmental changes associated with the use of photoselective protective netting to reduce sunburn in apple. Agric. For. Meteorol. 237, 9–17.

Kano, Y., 2004. Effects of summer day-time temperature on sugar content in several portions of watermelon fruit (*Citrullus lanatus*). J. Hortic. Sci. Biotechnol. 79 (1), 142–145.

Kano, Y., Matsumoto, J., Aoki, Y.-S., Madachi, T., 2012. Effects of heating bearing shoot near fruit on cell size, sucrose metabolizing enzymes and sugar accumulation in watermelon fruit. J. Jap. Soc. Hortic. Sci. 81 (2), 171–176.

Ketellapper, H., 1963. Temperature-induced chemical defects in higher plants. Plant Physiol. 38 (2), 175.

Khoshbakht, D., Ramin, A.A., Baninasab, B., 2015. Effects of sodium chloride stress on gas exchange, chlorophyll content and nutrient concentrations of nine citrus rootstocks. Photosynthetica 53 (2), 241–249.

Kirnak, H., Berakatoglu, K., Bilgel, L., Dogan, E., 2009. Effect of preharvest deficit irrigation on second crop watermelon grown in an extremely hot climate. J. Irrig. Drain. Eng. 135 (2), 141–148.

Lee, W.T., Weisell, R., Albert, J., Tomé, D., Kurpad, A.V., Uauy, R., 2016. Research approaches and methods for evaluating the protein quality of human foods proposed by an FAO Expert Working Group in 2014. J. Nutr. 146 (5), 929–932.

Leibar, U., Pascual, I., Morales, F., Aizpurua, A., Unamunzaga, O., 2017. Grape yield and quality responses to simulated year 2100 expected climatic conditions under different soil textures. J. Sci. Food Agric. 97 (8), 2633–2640.

Li, Y., Ye, W., Wang, M., Yan, X., 2009. Climate change and drought: a risk assessment of crop-yield impacts. Clim. Res. 39 (1), 31–46.

Liu, R.H., 2003. Health benefits of fruit and vegetables are from additive and synergistic combinations of phytochemicals. Am. J. Clin. Nutr. 78 (3), 517S–520S.

Liu, S., Manson, J.E., Lee, I.-M., Cole, S.R., Hennekens, C.H., Willett, W.C., Buring, J., 2000. Fruit and vegetable intake and risk of cardiovascular disease: the Women's Health Study. Am. J. Clin. Nutr. 72 (4), 922–928.

Lobell, D.B., Bänziger, M., Magorokosho, C., Vivek, B., 2011. Nonlinear heat effects on African maize as evidenced by historical yield trials. Nat. Clim. Chang. 1 (1), 42.

Lock, K., Pomerleau, J., Causer, L., Altmann, D.R., McKee, M., 2005. The global burden of disease attributable to low consumption of fruit and vegetables: implications for the global strategy on diet. Bull. World Health Organ. 83, 100–108.

Long, S.P., Ainsworth, E.A., Rogers, A., Ort, D.R., 2004. Rising atmospheric carbon dioxide: plants FACE the future. Annu. Rev. Plant Biol. 55, 591–628.

Mann, J., Cummings, J., Englyst, H., Key, T., Liu, S., Riccardi, G., et al., 2007. FAO/WHO scientific update on carbohydrates in human nutrition: conclusions. Eur. J. Clin. Nutr. 61 (S1), S132.

Mattos, L.M., Moretti, C.L., Jan, S., Sargent, S.A., Lima, C.E.P., Fontenelle, M.R., 2014. Climate changes and potential impacts on quality of fruit and vegetable crops. In: Emerging Technologies and Management of Crop Stress Tolerance. vol. 1. Elsevier, pp. 467–486.

McCaskill, M.R., Lexie, M., Ian, G., Steve, G., Debra, L.P., 2016. How hail netting reduces apple fruit surface temperature: a microclimate and modelling study. Agric. For. Meteorol. 226–227, 148–160.

McKeown, A., Warland, J., McDonald, M., 2006. Long-term climate and weather patterns in relation to crop yield: a minireview. Can. J. Bot. 84 (7), 1031–1036.

Myers, S.S., Zanobetti, A., Kloog, I., Huybers, P., Leakey, A.D., Bloom, A.J., et al., 2014. Increasing CO2 threatens human nutrition. Nature 510 (7503), 139.

Ndabamenye, T., Swennen, R., Annandale, J.G., Barnard, R.O., Van Asten, P.J.A., Vanhoudt, N., Blomme, G., 2012. Ecological characteristics influence farmer selection of on-farm plant density and bunch mass of low input East African highland banana (Musa spp.) cropping systems. Field Crop Res. 135, 126–136.

Nistor, E., Dobrei, A.G., Dobrei, A., Carmen, D., 2018. Growing season climate variability and its influence on Sauvignon Blanc and Pinot Gris berries and wine quality: study case in Romania (2005–2015). S. Afr. J. Enol. Vitic. 39 (2), 196–207.

Noh, J., Kim, J.M., Sheikh, S., Lee, S.G., Lim, J.H., Seong, M.H., Jung, G.T., 2013. Effect of heat treatment around the fruit set region on growth and yield of watermelon [*Citrullus lanatus* (Thunb.) Matsum. and Nakai]. Physiol. Mol. Biol. Plants 19 (4), 509–514.

Ollé, D., Guiraud, J.-L., Souquet, J.M., Terrier, N., Ageorges, A., Cheynier, V., Verries, C., 2011. Effect of pre-and post-veraison water deficit on proanthocyanidin and anthocyanin accumulation during Shiraz berry development. Aust. J. Grape Wine Res. 17 (1), 90–100.

Parry, M.L., Rosenzweig, C., Iglesias, A., Livermore, M., Fischer, G., 2004. Effects of climate change on global food production under SRES emissions and socio-economic scenarios. Glob. Environ. Chang. 14 (1), 53–67.

Pastore, C., Movahed, N., Allegro, G., Valentini, G., Zenoni, S., Santo, S.D., et al., 2017. Phenolic contents and genome-wide expression profiling of grapevine berries (*Vitis vinifera* L. 'Sangiovese') ripened under two different temperature regimes. Acta Hortic. 1172, 289–294.

Peng, S., Huang, J., Sheehy, J.E., Laza, R.C., Visperas, R.M., Zhong, X., et al., 2004. Rice yields decline with higher night temperature from global warming. Proc. Natl. Acad. Sci. 101 (27), 9971–9975.

Perotti, V.E., Moreno, A.S., Tripodi, K.E., Meier, G., Bello, F., Cocco, M., Vazquez, D., Anderson, C., Podesta, F.E., 2015. Proteomic and metabolomic profiling of Valencia orange fruit after natural frost exposure. Physiol. Plant 153 (3), 337–354.

Porter, J.R., Semenov, M.A., 2005. Crop responses to climatic variation. Phil. Trans. R. Soc. Lond. B: Biol. Sci. 360 (1463), 2021–2035.

Qu, Z., Zhou, G., 2016. Possible impact of climate change on the quality of apples from the major producing areas of China. Atmosphere 7 (9), 113.

Ramlall, I., 2014. Gauging the impact of climate change on food crops production in Mauritius. Int. J. Clim. Change Strateg. Manag. 6 (3), 332–355.

Ramos, M.C., Jones, G.V., Yuste, J., 2015. Spatial and temporal variability of cv. Tempranillo phenology and grape quality within the Ribera del Duero DO (Spain) and relationships with climate. Int. J. Biometeorol. 59 (12), 1849–1860.

Rivero, R.M., Ruiz, J.M., Garcia, P.C., Lopez-Lefebre, L.R., Sanchez, E., Romero, L., 2001. Resistance to cold and heat stress: accumulation of phenolic compounds in tomato and watermelon plants. Plant Sci. 160 (2), 315–321.

Rivero, R.M., Sanchez, E., Ruiz, J.M., Romero, L., 2003. Influence of temperature on biomass, iron metabolism and some related bioindicators in tomato and watermelon plants. J. Plant Physiol. 160 (9), 1065–1071.

Rouphael, Y., Cardarelli, M., Colla, G., Rea, E., 2008. Yield, mineral composition, water relations, and water use efficiency of grafted mini-watermelon plants under deficit irrigation. HortScience 43 (3), 730–736.

Salazar Parra, C., Aguirreolea, J., Sánchez-díaz, M., Irigoyen, J.J., Morales, F., 2010. Effects of climate change scenarios on Tempranillo grapevine (*Vitis vinifera* L.) ripening: response to a combination of elevated CO2 and temperature, and moderate drought. Plant Soil 337 (1–2), 179–191.

Schlenker, W., Roberts, M.J., 2009. Nonlinear temperature effects indicate severe damages to US crop yields under climate change. Proc. Natl. Acad. Sci. 106 (37), 15594–15598.

Sen, V., Rana, R.S., Chauhan, R.C., Aditya, 2015. Impact of climate variability on apple production and diversity in Kullu valley, Himachal Pradesh. Ind. J. Hortic. 72 (1), 14–20.

Solomakhin, A., Blanke, M.M., 2010. Can coloured hailnets improve taste (sugar, sugar:acid ratio), consumer appeal (colouration) and nutritional value (anthocyanin, vitamin C) of apple fruit? Food Sci. Technol. 43 (8), 1277–1284.

Song, J., Smart, R., Wang, H., Dambergs, B., Sparrow, A., Qian, M.C., 2015. Effect of grape bunch sunlight exposure and UV radiation on phenolics and volatile composition of *Vitis vinifera* L. cv. Pinot noir wine. Food Chem. 173, 424–431.

Sugiura, T., Ogawa, H., Fukuda, N., Moriguchi, T., 2013. Changes in the taste and textural attributes of apples in response to climate change. Sci. Rep. 3, 2418.

Taub, D.R., 2010. Effects of rising atmospheric concentrations of carbon dioxide on plants. Nat. Educ. Knowl. 3 (10), 21.

Taub, D.R., Miller, B., Allen, H., 2008. Effects of elevated CO2 on the protein concentration of food crops: a meta-analysis. Glob. Chang. Biol. 14 (3), 565–575.

Taulya, G., 2013. East African highland bananas (Musa spp. AAA-EA) 'worry' more about potassium deficiency than drought stress. Field Crop Res. 151, 45–55.

Timbal, B., Arblaster, J.M., Power, S., 2006. Attribution of the late-twentieth-century rainfall decline in southwest Australia. J. Clim. 19 (10), 2046–2062.

Torres, N., Hilbert, G., Luquin, J., Goicoechea, N., Carmen Antolin, M., 2017. Flavonoid and amino acid profiling on *Vitis vinifera* L. cv Tempranillo subjected to deficit irrigation under elevated temperatures. J. Food Compos. Anal. 62, 51–62.

Unsworth, M.H., Hogsett, W.E., 1996. Combined effects of changing CO2, temperature, UV-B radiation and O3 on crop growth. In: Global Climate Change and Agricultural Production: Direct and Indirect Effects of Changing Hydrological, Pedological and Plant Physiological Processes. pp. 171–197.

Vallat, A., Gu, H., Dorn, S., 2005. How rainfall, relative humidity and temperature influence volatile emissions from apple trees in situ. Phytochemistry 66 (13), 1540–1550.

Wagner, J., 2010. Sunburn on apples. Obst-und Weinbau 146 (11), 10–12.

Wang, S.Y., Zheng, W., 2001. Effect of plant growth temperature on antioxidant capacity in strawberry. J. Agric. Food Chem. 49 (10), 4977–4982.

Wheeler, T., Von Braun, J., 2013. Climate change impacts on global food security. Science 341 (6145), 508–513.

WHO, 2003. Diet, nutrition, and the prevention of chronic diseases: Report of a Joint WHO/FAO Expert Consultation. vol. 916 World Health Organization.

WHO, 2012a. Guideline: Potassium Intake for Adults and Children. World Health Organization.

WHO, 2012b. Guideline: Sodium Intake for Adults and Children. World Health Organization.

WHO, FAO, 2005. Vitamin and Mineral Requirements in Human Nutrition. World Health Organization and Food and Agriculture Organization of the United Nations.

Wu, S., Liang, D., Ma, F., 2014. Leaf micromorphology and sugar may contribute to differences in drought tolerance for two apple cultivars. Plant Physiol. Biochem. 80, 249–258.

Xu, C., Zhang, Y., Zhu, L., Huang, Y., Lu, J., 2011. Influence of growing season on phenolic compounds and antioxidant properties of grape berries from vines grown in subtropical climate. J. Agric. Food Chem. 59 (4), 1078–1086.

Yousuf, B., Gul, K., Wani, A.A., Singh, P., 2016. Health benefits of anthocyanins and their encapsulation for potential use in food systems: a review. Crit. Rev. Food Sci. Nutr. 56 (13), 2223–2230.

Zarrouk, O., Francisco, R., Pinto-Marijuan, M., Brossa, R., Santos, R.R., Pinheiro, C., et al., 2012. Impact of irrigation regime on berry development and flavonoids composition in Aragonez (Syn. Tempranillo) grapevine. Agric. Water Manag. 114, 18–29.

Zeppel, M.J., Adams, H.D., Anderegg, W.R., 2011. Mechanistic causes of tree drought mortality: recent results, unresolved questions and future research needs. New Phytol. 192 (4), 800–803.

Zupan, A., Mikulic-Petkovsek, M., Slatnar, A., Stampar, F., Veberic, R., 2014. Individual phenolic response and peroxidase activity in peel of differently sun-exposed apples in the period favorable for sunburn occurrence. Plant Physiol. 171 (18), 1706–1712.

8

NIR spectroscopy and management of bioactive components, antioxidant activity, and macronutrients in fruits

Jasenka Gajdoš Kljusurić, Tamara Jurina, Davor Valinger, Maja Benkovi,
Ana Jurinjak Tušek*

University of Zagreb, Faculty of Food Technology and Biotechnology, Zagreb, Croatia
*Corresponding author. E-mail: jasenka.gajdos@pbf.hr

1 Daily fruit consumption and composition

The conceptual models of daily recommended presence of certain food groups known as Food Guide Pyramid (USDA, 1996) or MyPlate (USDA, 2019) are educational tools reminding consumers of a healthy eating style. Conceptual models (Fig. 8.1) are used to present an idea with colors and shapes (like in the food guidelines—pyramid and table). But the important factor is the recommendation of daily fruit consumption when consumers are educated about proper and varied nutrition. Consumption of fruits in a fresh, frozen, or dried form is more than welcomed and is encouraged because they serve as sources of functional components whose intake encourages human well-being due to their macronutrients (carbohydrate content), micronutrients (minerals and vitamins), richness of fibers, and a number of bioactive components (Naderi et al., 2018). The main macronutrients found in fruits are carbohydrates, which provide energy for the body in general, including the nervous system and brain. Dietary fibers, as complex carbohydrates, are needed to improve digestive function, maintain healthy body weight (Resman et al., 2019), and lower the cholesterol levels and risks of heart disease and type 2 diabetes (Manach et al., 2004). Polyphenols, as secondary plant metabolites, are also present in fruits and act as a major source of antioxidants with an important physiological and morphological role for growth and reproduction of plants contributing to the color and sensory characteristics of fruits and vegetables (Alasalvar et al., 2001; Belščak-Cvitanović et al., 2018). Fruit antioxidant activity derives from the cumulative and synergistic effects of antioxidants such as vitamin C, vitamin E, and other various phenolic compounds (Georgieva et al., 2013) whose roles are inversely proportional to the role of free radicals that

A.K. Srivastava, Chengxiao Hu (eds.)
Fruit Crops: Diagnosis and Management of Nutrient Constraints
https://doi.org/10.1016/B978-0-12-818732-6.00008-3

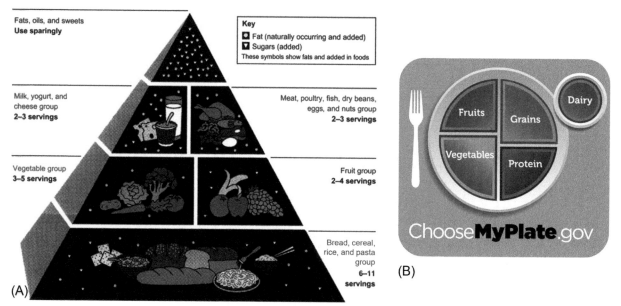

FIG. 8.1   Conceptual models (A) Food Guide Pyramid (USDA, 1996) and (B) MyPlate (USDA, 2019).

are unstable and reactive molecules. Free radicals in a given concentration on a local level are essential for functions of human immune system (Cervellati et al., 2001), but an excess of free radicals, which are not neutralized in the body, could increase the oxidative stress and cause damage to proteins, lipids, and carbohydrates. Consequences of such damages include chronic health problems such as diabetes, atherosclerosis, cancer, and/or other chronic diseases (Resman et al., 2019). So, antioxidant capacity can be defined as the sum of the present endogenous and exogenous defense mechanisms that ensure the oxidative balance and neutralize free radicals and reduce oxidative stress (Cervellati et al., 2001). This is the main reason why fruits are an important part of a daily diet leading to well-being (Bravo, 1998).

Daily recommendations of fruit consumption depend on the age, gender, and daily physical activity of the individual (Table 8.1). High content of carbohydrates (mostly sugar) in fruit is the limiting factor in daily consumption.

TABLE 8.1   Daily recommendations of fruit consumption (USDA, 2019).

Group	Age (years)	Serving size (cup[a])
Children	2–3	1
	4–8	1–1½
Girls	9–13	1½
	14–18	1½
Boys	9–13	1½
	14–18	2
Women	19–30	2
	31–50	1½
	>50	1½
Men	19–30	2
	31–50	2
	>50	2

[a] 1 cup = 1 cup raw fruits or ½ cup chopped, cooked, or canned fruit or vegetable, ¾ cup of fruit juice.

TABLE 8.2 Energy and macronutrient composition of fruits (USDA, 2016).

Fruit	Water (g)	Energy		Prot. (g)	Fat (g)	PUFA (g)	CHO (g)	M+Ds (g)	Ps (g)	Fibers (g)
		kJ	kcal							
Grapefruit	91.0	117	28	0.0	0.0	0.0	7.0	6.0	1.0	1.4
Lemon	91.0	50	12	0.0	0.0	0.0	3.0	2.0	1.0	1.8
Mandarin	87.0	142	34	0.9	0.0	0.0	8.0	8.0	0.0	1.9
Orange	86.0	146	35	0.8	0.0	0.0	8.5	8.5	0.0	2.0
Orange, red	87.0	172	41	0.8	0.3	0.3	9.3	9.3	0.0	1.9
Blueberries	85.0	234	56	0.6	0.6	0.6	14.3	14.3	0.0	0.0
Cranberries	87.0	63	15	0.4	0.0	0.0	3.5	3.5	0.0	4.2
Mulberry	85.0	151	36	1.3	0.0	0.0	8.1	8.1	0.0	1.7
Grapes, white	79.0	251	60	0.6	0.0	0.0	15.8	15.8	0.0	0.9
Grapes, red	84.0	224	54	0.5	0.1	0.1	13.5	13.5	0.0	1.5
Strawberries	89.0	109	26	0.6	0.0	0.0	6.2	6.2	0.0	2.2
Blackberries	82.0	121	29	1.3	0.0	0.0	6.4	6.4	0.0	7.3
Raspberries	83.0	105	25	0.9	0.0	0.0	5.6	5.6	0.0	7.4
Gooseberries	84.0	155	37	0.6	0.0	0.0	9.2	9.2	0.0	3.5
Apple	87.0	167	40	0.0	0.0	0.0	10.0	10.0	0.0	2.3
Pear	60.0	121	29	0.2	0.0	0.0	7.6	7.6	0.0	0.0
Peach	86.0	155	37	0.6	0.0	0.0	9.1	9.1	0.0	1.4
Apricot	86.0	117	28	0.6	0.0	0.0	6.7	6.7	0.0	2.1
Plum	84.0	159	38	0.6	0.0	0.0	9.6	9.6	0.0	2.1
Pineapple	84.0	192	46	0.5	0.0	0.0	11.6	11.6	0.0	1.2
Banana	71.0	331	79	1.1	0.3	0.2	19.2	16.2	3.0	3.4
Kiwi	83.0	167	40	1.0	0.0	0.0	9.0	7.0	2.0	1.0
Mango	83.0	247	59	0.5	0.0	0.0	15.3	15.3	0.0	1.5
Pomegranate	80.0	314	75	1.0	0.3	0.3	17.0	12.0	5.0	1.1

Prot., proteins; *PUFA*, polyunsaturated fatty acids; *M+Ds*, mono+disaccharides; *Ps*, polysaccharides.

One gram of monosaccharides will release ≈15.7 kJ (3.75 kcal/g), disaccharides around 16.5 kJ (3.95 kcal/g), and polysaccharides 17.4–17.6 kJ (4.15–4.20 kcal/g). The energy of hydrolysis is very small, and these values are essentially equivalent when calculated on a monosaccharide basis. Thus, 100-g sucrose gives 105.6-g monosaccharide after hydrolysis, and 100-g starch gives 110-g glucose after hydrolysis (FAO/WHO, 1997). Despite high sugar content, fruits are considered to be beneficial because of a low fat contents and high fiber content (black currant, 8.7-g/100-g fruit) (Table 8.2).

Unfortunately, fruit has not been spared from adulteration and false statements of its geographic origin with an aim of selling those products for much higher prices than its actual worth. Studies of adulteration define it as the addition of a compound or material that is cheaper or of inferior quality with a similar chemical profile. So, it was necessary to develop and apply measurement methods, which will detect the chemical composition. One such method is near-infrared (NIR) spectroscopy. Applications using Fourier transform mid-infrared (FT-MIR) spectroscopy and near-infrared (NIR) spectroscopy have been developed to detect adulteration in all foods and feeds (Cen and He, 2007; Sarraguca and Lopes, 2009; Bureau et al., 2019).

2 Near-infrared spectroscopy

Agricultural application of NIR spectroscopy was first used to measure moisture in grain by Norris (1964). In the past 55 years, NIR spectroscopy has been imposed as a rapid analysis mainly used in the detection of moisture, protein,

FIG. 8.2 Electromagnetic spectrum range.

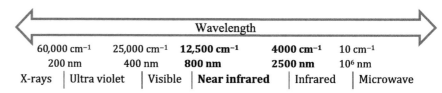

and fat content of a wide variety of food products (Nicolaï et al., 2007; Pasquini, 2018). The near-infrared range of the electromagnetic spectrum is between 780 and 2500 nm, and the product is irradiated with NIR radiation, and the reflected or transmitted radiation is measured (Fig. 8.2).

Radiation penetrates in the product causing spectral characteristics change depending on the chemical composition of the product. Infrared spectroscopy is based on the absorbance of radiation at molecular vibrational frequencies occurring for the O—H, N—H, S—H, and C—H groups (Chadha and Haneef, 2015). Depending on the equipment, a spectrum of the scanned product will result with a large data set, so multivariate techniques are used to extract significant information from the spectra.

3 Chemometrics

Data analysis of spectra includes spectral data preprocessing, calibration modeling, and model transfer (Cortés et al., 2019). Preprocessing will minimize noises and undesirable factors in spectra, which are subjected to construction of calibration models. Unfortunately, the preprocessing (spectra pretreatment) is not unique (Oliveri et al., 2018) and is often based on the experience of the researcher. Preprocessing can include just one pretreatment method, or a combination of them applied simultaneously. Most frequent pretreatment methods conducted on the NIR spectra include smoothing methods (e.g., moving average filter, Gaussian filter, Savitzky-Golay filter), derivation methods (first and second), multiplicative scatter correction (MSC), orthogonal signal correction (OSC), standard normal variate (SNV), wavelet transformation, normalization and/or scaling, and detrending to eliminate the baseline drift in the spectrum (Wang et al., 2015). Implementation of calibration on such treated spectra data is the basis of a good prediction model.

Multivariate tools mostly used are factor analysis (FA), principal component analysis (PCA), partial least squares regression (PLSR), multiple linear regression (MLR), principal component regression (PCR), quadratic discriminant analysis (QDA), etc. (Cortés et al., 2019). As in the case of pretreatments, one or more multivariate tools can be used in the calibration, which implies quantitative or qualitative analysis. First will be used the PCA, to detect the patterns and potential outliers (Cozzolino et al., 2011; Herceg et al., 2016). Qualitative models are used to classify samples, and a part of the data is used for model training, which is followed with model evaluation and its testing on unknown samples. Qualitative modeling tools are used in linear discriminant analysis (LDA) (Baranowski et al., 2012), partial least squares-discriminant analysis (PLS-DA) (Liu et al., 2011), soft independent modeling of class analogy (SIMCA) (Pontes et al., 2006), and support vector machine (SVM) (Chen et al., 2007). If the quantitative model is the aim, factor analysis is used to identify important wavelengths based on the lined contribution weight. Then follow application of methods such as: partial least square (PLS) regression, principal component regression (PCR), multiple linear regression (MLR), or broadly used artificial neural networks (ANN) (Kumaravelu and Gopal, 2015).

Model efficacy is evaluated using the coefficient of determination (R^2) and the regression point displacement that is the ratio of the standard error of performance (RPD, Eq. 8.5). Performance of the calibration models was calculated as the ratio of the range of reference chemistry values to standard error of prediction (RER, Eq. 8.6). Miloš and Bensa (2017) gave an overview of the most common parameters used in the evaluation of model performance (Eqs. 8.1–8.6):

$$R^2 = 1 - \frac{SS_{res}}{SS_{tot}} = 1 - \frac{\sum_{i=1}^{N}(y_i - f_i)^2}{\sum_{i=1}^{N}(y_i - \overline{y})^2} \tag{8.1}$$

$$RMSEP = \sqrt{\sum_{i=1}^{N} \frac{(\overline{y}_i - y_i)^2}{N}} \tag{8.2}$$

$$SEP = \sqrt{\frac{1}{N}\sum_{i=1}^{N}\left(\overline{y_i} - y_i - Bias\right)^2} \qquad (8.3)$$

$$Bias = \frac{1}{N}\sum_{i=1}^{N}\left(\overline{y_i} - y_i\right) \qquad (8.4)$$

$$RPD = SD_v / SEP \qquad (8.5)$$

$$RER = (range) / SEP \qquad (8.6)$$

where the parameters used are

SS_{res}, sum of squares of residuals, also called the residual sum of squares
SS_{tot}, total sum of squares (proportional to the variance of the data)
f_i, defined as the part of the residuals as $e_i = y_i - f_i$ (forming a vector e)
N, number of samples
$\overline{y_i}$ and y_i are the observed and predicted values of sample i
SD_v, standard deviation of the validation data set
SEP, standard error of prediction
$range$, min and max, minimum and maximum values of the validation set.

4 Near-infrared spectroscopy measurement of fruits

An overview of the last 15 years of applied near-infrared spectroscopy in the assessment of fruit quality is appended (Table 8.3).

Apples are the most investigated fruit for which are by use of NIR spectroscopy evaluated and predicted attributes as Streif index, respiratory maturity, physiological maturity, flavonol content, bruise detection, internal flesh and tissue browning, soluble solids content, softening index, moldy core, vitamin C, total polyphenols, titratable acidity, firmness starch, influence of packaging on apples, defect level, bitter pit detection, total antioxidant capacity, dry mater, total phenol content and origin (Peirs et al., 2005; Merzlyak et al., 2005; Xing et al., 2005; McGlone et al., 2005; Nicolaï et al., 2007; Xiaobo et al., 2007; Xing and De Baerdemaeker, 2007; Shenderey et al., 2010; Pissard et al., 2013, 2018; Ignat et al., 2014; Guo et al., 2016; Beghi et al., 2017; Khatiwada et al., 2016; Kafle et al., 2016; Schmutzler and Huck, 2016; Eisenstecken et al., 2019; Xia et al., 2019). For watermelons are available just five researches in the last 15 year investigating soluble solids content (Abebe, 2006; Jie et al., 2013, 2014; Tamburini et al., 2017). Lycopene and β-carotene were investigated by Tamburini et al. (2017), and Jie and Wei (2018) investigated the internal quality of watermelons. In the overview, nectarines (Golic and Walsh, 2006; Pérez-Marín et al., 2009, 2011; Sánchez et al., 2011; Munera et al., 2017, 2018; Cortés et al., 2017a,b), pears (Sirisomboon et al., 2007; Nicolaï et al., 2008; Li et al., 2013; Travers et al., 2014; Xu et al., 2012, 2014; Sun et al., 2016; Yu et al., 2018; Tian et al., 2018), and olives are also presented (Salguero-Chaparro et al., 2012, 2013; Salguero-Chaparro and Peña-Rodríguez, 2014; Giovenzana et al., 2018). The olive fruits are not so often investigated as olive oils or leaves (Valinger et al., 2018) are mostly used for oil production (Giovenzana et al., 2018). There are numerous examples of food adulteration studies using NIR spectroscopy: (i) Kemsley et al. (1996) and Holland et al. (1997) investigated adulteration of fruit purees; (ii) adulteration of apple juices with different types of sugar were investigated by Kelly and Downey (2005); (iii) the amount of grape juice concentrate as an adulterant in pomegranate juice was studied by Vardin et al. (2008); (iv) quality of processed agricultural crops (Walsh and Kawano, 2009); and (v) the percentage of added aqueous solution of glucose, fructose, and sucrose in orange juice were studied by Ellis et al. (2016). Near-infrared spectra and stability of phenols in pomegranate juices under different process conditions were investigated by Herceg et al. (2016), in chokeberry juices by Bursać Kovačević et al. (2016), while the amount of added steviol glycosides in juices was studied by Kujundžić et al. (2017).

In this paper, the PLSR model effectiveness parameters (R_v^2, R_p^2, $RMSEP$, and SEP) with calculated parameters RPD and RER (Eqs. 8.5, 8.6) were used to assess the efficacy of the predictions according to the guidelines proposed by Malley et al. (2004). Potential outliers in the fruit composition parameter sets are investigated using the Grubbs test (Dan and Ijeomao, 2013).

TABLE 8.3 Applications of NIR spectroscopy in the assessment of fruit quality.

Sample	Multivariate tool	Spectral range used (nm)	Analyzed attributes	Reference
Apples	PLS	380–2000	Streif index Respiratory maturity Physiological maturity	Peirs et al. (2005)
	Conceptual model	400–800	Flavonol content	Merzlyak et al. (2005)
	PLS-DA	400–1700	Bruise detection	Xing et al. (2005)
	PLS	650–950	Internal tissue browning	McGlone et al. (2005)
	Kernel PLS regression	800–1690	Soluble solid content	Nicolaï et al. (2007)
	PLS	910–2632	Soluble solid content	Xiaobo et al. (2007)
	PLS-DA	500–1600	Softening index	Xing and De Baerdemaeker (2007)
	PLS	400–1000	Moldy core	Shenderey et al. (2010)
	LS-SVM	400–2500	Vitamin C Total polyphenols Soluble solid content	Pissard et al. (2013)
	PLS	340–1014 and 850–1888	Soluble solid content Titratable acidity Firmness Starch	Ignat et al. (2014)
	ICA-SVM	500–1100	Soluble solid content	Guo et al. (2016)
	PLS-DA	400–1000 1100–2100	Influence of packaging on apples	Beghi et al. (2017)
	PLS PLS-DA, LDA, and SVM	302–1150 and 600–973	Defect level Internal flesh browning	Khatiwada et al. (2016)
	QDA, SVM	800–2500	Bitter pit detection	Kafle et al. (2016)
	PLS	1596–2397	Total antioxidant capacity	Schmutzler and Huck (2016)
	PLS	408–2498	Dry matter Total phenol content	Pissard et al. (2018)
	PCA-DA PCA-QDA	1000–2500	Origin	Eisenstecken et al. (2019)
	PLS	550–950	Soluble solid content	Xia et al. (2019)
Nectarine	PLS	735–930	Soluble solid content	Golic and Walsh (2006)
	MPLS	1600–2400 400–1700	Soluble solid content Flesh firmness Weight Diameter	Pérez-Marín et al. (2009)
	PLS-DA	1600–2400	Shelf-life discrimination	Pérez-Marín et al. (2011)
	MPLS:LOCAL algorithm	1600–2400	Weight Diameter Flesh firmness Soluble solid content	Sánchez et al. (2011)
	PLS	450–1040	Ripening index Internal quality index	Munera et al. (2017)
		360–1760		Cortés et al. (2017a)

TABLE 8.3 Applications of NIR spectroscopy in the assessment of fruit quality—cont'd

Sample	Multivariate tool	Spectral range used (nm)	Analyzed attributes	Reference
	PLS PLS-DA and LDA		Internal quality index Varietal discrimination	
	PLS-DA and LDA	600–1100	Varietal discrimination	Cortés et al. (2017b)
	PLS-DA	450–1040	Quality Similarity (fraud)	Munera et al. (2018)
Pears	MLR	1100–2500	Pectin constituents	Sirisomboon et al. (2007)
	PLS	780–1700 875–1030	Soluble solid content Firmness	Nicolaï et al. (2008)
	EW-LS-SVM	380–1800	Soluble solid content pH Firmness	Li et al. (2013)
	PLS	300–1100 and 100–2500	Dry meter Soluble solid content	Travers et al. (2014)
	PLS	465–1150	Soluble solid content	Xu et al. (2014)
	SMLR GA-PLS iPLS GA-SPA MLR	200–1100	Soluble solid content	Xu et al. (2012)
	PLS	200–1100	Brown core Soluble solid content	Sun et al. (2016)
	SAE-FNN	380–1030	Firmness Soluble solid content	Yu et al. (2018)
	Separate location model Global locations model Average spectra model	550–950, 550–780, 780–950, and 550–700	Soluble solid content	Tian et al. (2018)
Olives	ANOVA and LSD	380–1690	Focal distance and integration time	Salguero-Chaparro et al. (2012)
	PLS	380–1690	Free acidity Moisture content Fat content	Salguero-Chaparro et al. (2013)
	PLS LS-SVM	380–1690	Fat content Free acidity Moisture content Fat content	Salguero-Chaparro and Peña-Rodríguez (2014)
	PLS	400–1650	Oil content prediction on intact olives entering the mill	Giovenzana et al. (2018)
Watermelons	PLS	700–1100	Soluble solid content	Abebe (2006)
	MC-UVE-GA-PLS	690–950	Soluble solid content	Jie et al. (2013)
	MC-UVE-SMLP	200–1100	Soluble solid content	Jie et al. (2014)
	PLS	900–1700	Lycopene β-Carotene Soluble solid content	Tamburini et al. (2017)
	Review	Review	Internal quality	Jie and Wei (2018)

4.1 Practical application of NIR spectroscopy related to multivariate analysis

How and why NIR spectroscopy can be applied in fruit adulteration studies is explained in detail as follows.

5 Near-infrared scan of different fruits

Near-infrared spectra recorded in the wavelength range from 904 to 1699 nm were related to analytical measurements of 12 different fruits (Valinger et al., 2017, 2018; Georgieva et al., 2013). NIR spectra of fruits are presented in Fig. 8.3.

Scanned NIR spectra (Fig. 8.3) presented in the range of $\lambda = 904$–1699 nm showed molecule vibration of C—H, O—H, and N—H bonds, which are bond characteristic of macronutrients (N—H, for proteins; C—H, for carbohydrates; and HOH and OH for water and fats). Overlapping of NIR scans is visible with specific differences in the range of $\lambda = 904$–925 nm indicating absorption detection of the third overtone region and detecting different vibration of C—H bonds. Second specific region where the spectral differences are visible starts with the HOH region at $\lambda = 1400$ nm (water $\lambda = 1400$–1460 nm) and continues to the end of the recorded spectra indicating differences in the vibrations of the first overtone of C—H$_3$, Ar—CH, C—H, and C—H$_2$ bonds and second overtone of O—H, N—H, Ar—CH, and R—OH. All mentioned bonds can be found in observed composition parameters: water, carbohydrates, mono- and disaccharides, fibers, proteins, fats, PUFA, polyphenols, and antioxidant activity. Composition of investigated fruits taken from literature data (USDA, 2018; Neveu et al., 2010; Rothwell et al., 2012, 2013) and the parameters determined in this study are presented (Table 8.4).

Twelve different fruits were chosen: avocado, banana, blackberry, blueberry, cherry, clementine, currant, dried currant, mixed berries, raspberry, strawberry, and tangerine. Avocado was chosen because of its high fat content (10 g/ 100 g) and banana because of its high content of carbohydrates (19.2 g/100 g). Clementine and tangerine were chosen as representatives of the citrus group. Different berries were chosen as fruits, which contain PUFA (e.g., blueberry 0.6 g/ 100 g), and currant as a fruit, which not only is a rich source of dietary fibers (7.5 g/100 g) but also contains proteins (1.4 g/100 g rising to 4.1 g/100 g in dried currant). The water content of the chosen fruits ranged from 71% in banana to 89% in the strawberries, excluding the dried fruit samples. Phenol explorer was used to identify the polyphenol content (Neveu et al., 2010; Rothwell et al., 2012, 2013). The antioxidant activity (AOA) was measured using the Briggs-Rauscher (BR) oscillating reaction according to the method used by Gajdoš Kljusurić et al. (2016), and the AOA was presented with two reaction parameters: (i) the inhibition time and (ii) the Briggs-Rauscher antioxidant index (BRAI) calculated as suggested by Cervellati et al. (2001) and Aljović and Gojak-Salimović (2017). For the identification of the macronutrient contents, food composition databases were used (Kaić-Rak and Antonić, 1990; USDA, 2018). Fruit composition (content of macronutrients, content of polyphenols, and the antioxidant activity, which were used in the data processing) is further presented (Table 8.4).

BRAI is the equivalent of gallic acid showing the extent to which the antioxidant effect of phenolic molecules is stronger or weaker than the antioxidant effect of gallic acid (Bečić et al., 2014; Gajdoš Kljusurić et al., 2016). The greatest value of IT and BRAI was detected for blueberries (1377.7 s and 2.9 GAE), and the minimal AOA was detected for banana and avocado (BRAI = 0.1). In the study of Pantelidis et al. (2007), the highest AOA value was detected for

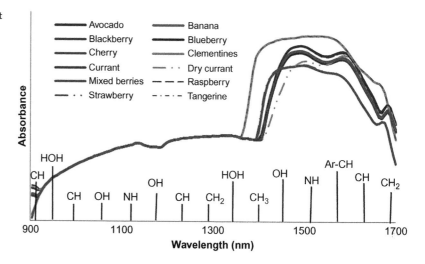

FIG. 8.3 Near-infrared (NIR) spectra of 12 different fruits.

TABLE 8.4 Fruit composition (experimental data: polyphenolic content (PC), inhibition time (IT), and Briggs-Rauscher antioxidant index (BRAI)).

Fruit	Water[a] g	CHO[a] g	M+Ds[a] g	Fibers[a] g	P[a] g	F[a] g	PUFA[a] g	PC[b] mg/100g FW	IT s	BRAI GAE
Avocado	81.0	7.0	5.0	2.0	2.0	10.0	2.0	140.0	50.0	0.1
Banana	71.0	19.2	16.2	3.4	1.1	0.3	0.2	154.7	67.0	0.1
Blackberry	82.0	6.4	6.4	7.3	1.3	0.0	0.0	302.0	1328.1	2.8
Blueberry	85.0	14.3	14.3	0.0	0.6	0.6	0.6	560.0	1377.7	2.9
Cherry	86.0	11.7	11.7	1.3	0.8	0.1	0.1	352.0	1100.0	2.3
Clementine	87.0	12.0	9.2	1.7	0.9	0.2	0.0	150.0	250.0	0.5
Currant	77.0	15.0	15.0	7.5	1.4	0.4	0.2	132.0	172.9	0.2
Dry currant	19.2	74.0	67.0	6.8	4.1	0.3	0.2	180.0	329.2	0.6
Mixed berries	83.2	9.5	9.5	4.9	1.0	0.2	0.2	250.0	894.0	1.9
Raspberry	83.0	5.6	5.6	7.4	0.9	0.0	0.0	215.0	1328.1	2.8
Strawberry	89.0	6.2	6.2	2.2	0.6	0.0	0.0	235.0	569.0	1.1
Tangerine	87.0	8.0	8.0	1.9	0.9	0.0	0.0	192.0	394.0	0.8

[a] Data from nutridata.
[b] Phenol explorer (Neveu et al., 2010; Rothwell et al., 2012; Rothwell et al., 2013).

the blackberry sample, which was also confirmed by our results. The lowest value of AOA was determined for the currant sample ($PC = 132 \, mg/100 \, g$ FW), which is also in agreement with the literature (Pantelidis et al., 2007; Lugasi et al., 2011). The mix of berry fruits analyzed in this study contained equal shares of strawberries, raspberries, and currant, and this sample was tested to confirm the findings of Hidalgo et al. (2010).

The matrix of spectral data was imported into the Unscrambler software 10.0 (CAMO, Trondheim, Norway).

5.1 Qualitative analysis

The recorded NIR spectra for each measurement contained 796 data points, and when all composition data were included in the matrix, it summed up to a data set of 38.688 points arranged in 48 rows and 806 columns. Such large data set is appropriate for multivariate data analysis (MVA). In this study, the MVA was used to identify the similarity patterns in the experimental data (Kujundžić et al., 2017). First, the principal component analysis (PCA) was performed. PCA is used to examine the interrelationships among the observed set of fruits to identify the underlying structure and to identify patterns based on their similarities and/or differences (species, content of macronutrients, polyphenolic content, or AOA). The first used multivariate tool was the principal component analysis, with the results further presented (Fig. 8.4).

The PCA biplot has grouped fruits in different quadrants showing their main similarities and differences. Water is dominant parameter in fruits that are positioned in the second quadrant (clementine, tangerine, and strawberries), while the sample containing the minimal water amount is in the opposite—fourth quadrant. All berries are positioned in the third quadrant together with the parameters such as polyphenolic content (PC) and the parameters of AOA (IT and BRAI). Avocado, positioned in the first quadrant, shows the high contents of fats and PUFA, while banana and currant contain some amounts of fats and PUFA, but their position close to the axis indicates higher values of carbohydrates, mono- and disaccharides, fibers, and proteins than in other observed fruits.

The biplot (Fig. 8.5) represents the qualitative analysis of fruits, and now, the only the question that remains is whether the NIR spectra can be quantitatively related to the parameters of fruit composition. The answer yes indicates the possibility of development of quantitative models, which allow prediction of the observed fruit parameters just by recording the NIR spectra of an unknown fruit extract without additional chemical analyses leading to low time-consuming analyses with additional environmental benefits (Belščak-Cvitanović et al., 2018).

If the pattern and relationship are successfully identified, the application of the next multivariate tool, the partial least squares regression (PLSR), follows.

FIG. 8.4 Principal component analysis (PCA) of 12 fruits and their chemical composition.

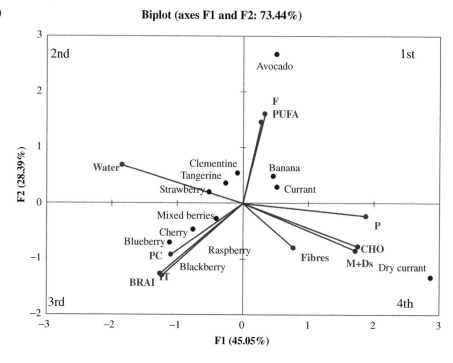

Biplot (axes F1 and F2: 73.44%)

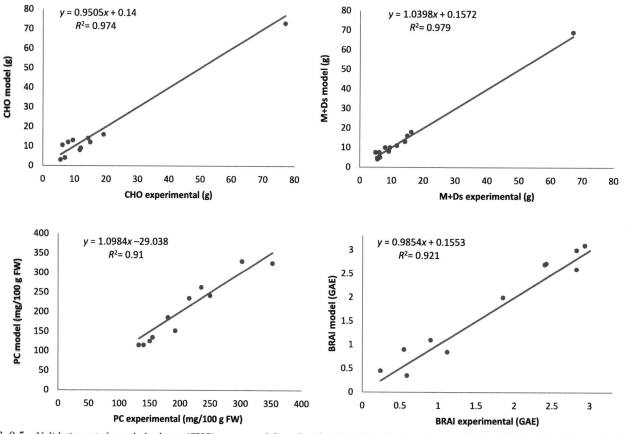

FIG. 8.5 Validation sets for carbohydrates (CHO), mono- and disaccharides (M+Ds), polyphenol content (PC), and Briggs-Rauscher antioxidant index (BRAI).

5.2 Quantitative analysis

Partial least squares regression feature is used to model multivariate data set which contains noisy and redundant data, such as spectral data and other chemical characteristics (such as total phenols and antioxidant activity and/or species, origin, and freshness) (Herceg et al., 2016; Bursać Kovačević et al., 2016). If the PLSR model efficiency is high, then the NIR spectra allow quantitative prediction of observed parameters (such as macronutrient contents, polyphenols, fibers, and/or AOA).

Samples were divided into calibration (2/3 of data) and validation sets (1/3 of data). Partial least squares regression (PLS) was used to develop validation models for each parameter (mentioned in Table 8.5). According to Kim et al. (2007), wavelengths from 1100 nm to higher were chosen for modeling. For the carbohydrate and protein content, the sample spectra were preprocessed (Savitzky-Golay first derivation), and for proteins, one additional pretreatment (multiplicative scatter correction (MSC)) was used. For the prediction of fats, the spectra were preprocessed with MSC + Savitzky-Golay second derivative treatment (Kim et al., 2007). For the other parameters, raw spectra of the same wavelengths were used (Valinger et al., 2018). PLS regression modeling was used for the quantitative analysis. The model used two-thirds of the data for calibration and one-third for validation (Table 8.5).

Efficient models that could be used as quantitative should have R^2s close to one and $RMSEP$ values as low as possible (Kujundžić et al., 2017), as well as models with $R^2 > 0.95$, $RPD > 4$, and $RER > 20$. Successful calibrations bind up R^2 in the range 0.9–0.95, RPD values in the range 3–4, and RER values between 15 and 20. Moderately successful calibration is characterized by R^2 form 0.8 to 0.9, RPD from 2.25 to 3, and the RER from 10 to 15. Moderately useful models have R^2 in the range from 0.7 to 0.8, RPD between 1.75 and 2.25, and RER values between 8 and 10. Less reliable models are the ones with $R^2 < 0.7$, $RPD < 1.75$, and $RER < 8$ (Malley et al., 2004; Miloš and Bensa, 2017).

The prediction in the range of $RPDs$ 2–2.5 and 2.5–3 shows possible approximate quantitative prediction and prediction that is classified as good, while models with the $RPD > 4$ are classified as excellent (Gajdoš Kljusurić et al., 2016).

Model parameters showed excellent prediction possibilities for all observed parameters, with the exception of fibers and proteins, while for the aforementioned two parameters, the prediction quality is successful (Miloš and Bensa, 2017). Results of the applied Grubbs test indicated that the PC content of blueberries as an outlier (560 mg/100 g FW) and the PLRS modeling was therefore conducted without the data for blueberries. Regarding the prediction models of IT and BRAI, it was important to exclude fruits containing high amounts of vitamin E (which cannot be detected using the standard Briggs-Rauscher reaction) and the validation efficiency plots (for carbohydrates, mono and disaccharides, polyphenols and the BRAI index) are presented in Fig. 8.5.

Effective validation was confirmed in the results presented in Fig. 8.5, presenting the results of the PLSR for carbohydrates, mono- and disaccharides, polyphenol content, and Briggs-Rauscher antioxidant index (BRAI). Coefficients of determination (R^2) confirmed the results from Table 8.5. The model effectiveness is obvious, where better model ($R^2 = 0.979$ for M+Ds) had experimental results more closely to the model line than models with lower coefficient of determination ($R^2 = 0.91$ for PC).

Although the content of polyunsaturated fatty acids (PUFA) was not the primary parameter that was aimed to be predicted, the results showed high correlations obtained for this parameter ($R^2 = 0.9$; $RPD > 4$ and $RER > 14$).

TABLE 8.5   Model statistics for prediction of compositional parameters of fruits based on the NIR spectra.

Parameter	R_v^2	RMSEP	R_p^2	SEP	RPD	RER
Water	0.984	3.185	0.969	3.326	5.717	20.983
CHO	0.987	2.931	0.974	3.060	6.150	22.353
M+Ds	0.989	2.376	0.979	2.482	6.836	24.980
Fibers	0.996	0.167	0.897	0.926	2.979	8.099
P	0.965	0.242	0.933	0.254	3.792	13.780
F	0.995	0.401	0.891	0.483	5.876	20.704
PUFA	0.983	0.058	0.900	0.139	4.074	14.388
PC	0.999	29.835	0.910	35.088	3.463	12.198
IT	0.909	65.437	0.889	60.329	8.618	22.008
BRAI	0.963	0.355	0.921	0.368	8.046	20.677

Validation data indicate that the developed models are capable of accurate predictions, and in general, this means that all developed PLS models could replace the time-consuming and laborious chemical procedures. Adding the recommendations of five or more servings of fruits and vegetables per day, our findings support the studies that fruit consumption is a good way to decrease the oxidative stress and increase human well-being (Kujundžić et al., 2017; Sirisomboon, 2018).

6 Conclusion and future directions

Presented approach where near-infrared spectroscopy was used in qualitative and quantitative analysis and model development can replace time-consuming chemical methods and can be used as a useful tool in food composition analysis. NIR spectroscopy has become a powerful tool for the nondestructive monitoring and prediction of multiple quality and safety attributes of fruits and fruit products. This technique, combined with chemometric methods, has proven to be an alternative to destructive analysis due to its fast detection speed, no need for sample disposal, relative lower cost, and potential to predict multiple quality parameters at the same time and therefore to distinguish the products according to different characteristics.

Although fruits differ substantially based on their nutritional composition (carbohydrates and fats), their bioactive content and effect on the well-being in general are the reason why nutritionists recommend their consumption, regardless of the health status. Food composition can be followed and successfully predicted for each fruit variety we tested, because the composition is based on the chemical bonds that can be detected and scanned using near-infrared spectroscopy coupled with multivariate analyses. Although NIR spectra must be coupled with multivariate analysis, it allows quantitative separation of fruit samples based on their composition (using the PCA analysis) and quantitative prediction of the chemical parameters observed in fruits, as macronutrients and their bioactivity.

The future researches have to be devoted to the unification of spectrum pretreatments and optimization of the quantitative and qualitative modeling, as well as increase of sensitivity to low concentrations undesirable supplements used in adulteration.

References

Abebe, A.T., 2006. Total sugar and maturity evaluation of intact watermelon using near infrared spectroscopy. J. Near Infrared Spectrosc. 14 (1), 67–70.

Alasalvar, C., Grigor, J.M., Zhang, D., Quantick, P.C., Shahidi, F., 2001. Comparison of volatiles, phenolics, sugars, antioxidant, vitamins and sensory quality of different colored carrot varieties. J. Agric. Food Chem. 49, 1410–1416.

Aljović, I., Gojak-Salimović, S., 2017. Evaluation of the antioxidant activity of ferulic, homovanillic and vanillic acids using the Briggs-Rauscher oscillating reaction method. Bull. Chem. Technol. Bosnia Herzegovina 49, 35–38.

Baranowski, T., Islam, N., Baranowski, J., Martin, S., Beltran, A., Dadabhoy, H., Adame, S.H., Watson, K., Thompson, D., Cullen, K.W., Subar, A., 2012. Comparison of a web-based versus traditional dietary recall among children. J. Am. Diet. Assoc. 112 (4), 527–532.

Bečić, I., Polović, I., Djaković, S., Lapić, J., Gajdoš Kljusurić, J., Kurtanjek, Ž., 2014. Nutritional contribution of berries for fighting oxidative stress. J. Hyg. Eng. Des. 8 (451), 44–52.

Beghi, R., Buratti, S., Giovenzana, V., Benedetti, S., Guidetti, R., 2017. Electronic nose and visible-near infrared spectroscopy in fruit and vegetable monitoring. Rev. Anal. Chem.. 36(4). https://doi.org/10.1515/revac-2016-0016.

Belščak-Cvitanović, A., Valinger, D., Benković, M., Jurinjak Tušek, A., Jurina, T., Komes, D., Gajdoš Kljusurić, J., 2018. Integrated approach for bioactive quality evaluation of medicinal plant extracts using HPLC-DAD, spectrophotometric, near infrared spectroscopy and chemometric techniques. Int. J. Food Prop. 20, 1–18.

Bravo, L., 1998. Polyphenols: chemistry, dietary sources, metabolism and nutritional significance. Nutr. Rev. 56, 317–333.

Bureau, S., Cozzolino, D., Clarkc, C.J., 2019. Contributions of Fourier-transform mid infrared (FT-MIR) spectroscopy to the study of fruit and vegetables: a review. Postharvest Biol. Technol. 148, 1–14.

Bursać Kovačević, D., Gajdoš Kljusurić, J., Putnik, P., Vukušić, T., Herceg, Z., Dragović-Uzelac, V., 2016. Stability of polyphenols in chokeberry juice treated with gas phase plasma. Food Chem. 212 (1), 323–331.

Cen, H., He, Y., 2007. Theory and application of near infrared reflectance spectroscopy in determination of food quality. Trends Food Sci. Technol. 18 (2), 72–83.

Cervellati, R., Höner, K., Furrow, S.D., Neddens, C., Costa, S., 2001. The Briggs-Rauscher reaction as a test to measure the activity of antioxidants. Helv. Chim. Acta 84, 3533–3547.

Chadha, R., Haneef, J., 2015. Near-infrared spectroscopy: effective tool for screening of polymorphs in pharmaceuticals. Appl. Spectrosc. Rev. 50 (7), 565–583. https://doi.org/10.1080/05704928.2015.1044663.

Chen, Q., Jiewen Zhao, J., Fang, C.H., Wang, D., 2007. Feasibility study on identification of green, black and Oolong teas using near-infrared reflectance spectroscopy based on support vector machine (SVM). Spectrochim. Acta A 66 (3), 568–574.

Cortés, V., Blasco, J., Aleixos, N., Cubero, S., Talens, P., 2017a. Visible and near-infrared diffuse reflectance spectroscopy for fast qualitative and quantitative assessment of nectarine quality. Food Bioprocess Technol. 10 (10), 1755–1766.

Cortés, V., Cubero, S., Aleixos, N., Blasco, J., Talens, P., 2017b. Sweet and nonsweet taste discrimination of nectarines using visible and near-infrared spectroscopy. Postharvest Biol. Technol. 133, 113–120.

Cortés, V., Blasco, J., Aleixosc, N., Cubero, S., Talens, P., 2019. Monitoring strategies for quality control of agricultural products using visible and near-infrared spectroscopy: a review. Trends Food Sci. Technol. 85, 138–148.

Cozzolino, D., Cynkar, W.U., Shah, N., Smith, P., 2011. Multivariate data analysis applied to spectroscopy: potential application to juice and fruit quality. Food Res. Int. 44 (7), 1888–1896.

Dan, E.D., Ijeomao, O.A., 2013. Statistical analysis/methods of detecting outliers in a Univariate data in a regression analysis model. Int. J. Educ. Res. 1 (5), 1–24.

Eisenstecken, D., Stürz, B., Robatscher, P., Lozano, L., Zanella, A., Oberhuber, M., 2019. The potential of near infrared spectroscopy (NIRS) to trace apple origin: study on different cultivars and orchard elevations. Postharvest Biol. Technol. 147, 123–131.

Ellis, D.I., Ellis, J., Muhamadali, H., Xu, Y., Horn, A.B., Goodacre, R., 2016. Rapid, high-throughput, and quantitative determination of orange juice adulteration by Fourier Transform infrared spectroscopy. Anal. Methods 8, 5581–5586.

FAO/WHO, 1997. The role of carbohydrates in nutrition. In: Carbohydrates in Human Nutrition. WHO, Geneva (FAO Food and Nutrition Paper—66). (Chapter 1).

Gajdoš Kljusurić, J., Mihalev, K., Bečić, I., Polović, I., Georgieva, M., Djaković, S., Kurtanjek, Ž., 2016. Near-infrared spectroscopic analysis of total phenolic content and antioxidant activity of berry fruits. Food Technol. Biotechnol. 54 (2), 236–242.

Georgieva, M., Nebojan, I., Mihalev, K., Yoncheva, N., Gajdoš Kljusurić, J., Kurtanjek, Ž., 2013. Application of NIR spectroscopy in quality control of wild berry fruit extracts during storage. Croat. J. Food Technol. Biotechnol. Nutr. 3–4, 67–73.

Giovenzana, V., Beghi, R., Romaniello, R., Tamborrino, A., Guidetti, R., Leone, A., 2018. Use of visible and near infrared spectroscopy with a view to on-line evaluation of oil content during olive processing. Biosyst. Eng. 172, 102–109.

Golic, M., Walsh, K.B., 2006. Robustness of calibration models based on near infrared spectroscopy for the in-line grading of stonefruit for total soluble solids content. Anal. Chim. Acta 555 (2), 286–291.

Guo, Z., Huang, W., Peng, Y., Chen, Q., Ouyang, Q., Zhao, J., 2016. Color compensation and comparison of shortwave near infrared and long wave near infrared spectroscopy for determination of soluble solids content of 'Fuji' apple. Postharvest Biol. Technol. 115, 81–90.

Herceg, Z., Bursać Kovačević, D., Gajdoš Kljusurić, J., Režek Jambrak, A., Zorić, Z., Dragović-Uzelac, V., 2016. Gas phase plasma impact on phenolic compounds in pomegranate juice. Food Chem. 190 (1), 665–672.

Hidalgo, M., Sánchez-Moreno, C., Pascual-Teresa, S., 2010. Flavonoid-flavonoid interaction and its effect on their antioxidant activity. Food Chem. 121, 691–696.

Holland, J.K., Kemsley, E.K., Wilson, R.H., 1997. Transfer of spectral data between Fourier-transform infrared spectrometers for use in discriminant analysis of fruit purees. J. Sci. Food Agric. 75, 391–400.

Ignat, T., Lurie, S., Nyasordzi, J., Ostrovsky, V., Egozi, H., Hoffman, A., Friedman, H., Weksler, A., Schmilovitch, Z.E., 2014. Forecast of apple internal quality indices at harvest and during storage by VIS-NIR spectroscopy. Food Bioprocess Technol. 7 (10), 2951–2961.

Jie, D., Wei, X., 2018. Review on the recent progress of non-destructive detection technology for internal quality of watermelon. Comput. Electron. Agric. 151, 156–164.

Jie, D., Xie, L., Fu, X., Rao, X., Ying, Y., 2013. Variable selection for partial least squares analysis of soluble solids content in watermelon using near-infrared diffuse transmission technique. J. Food Eng. 118 (4), 387–392.

Jie, D., Xie, L., Rao, X., Ying, Y., 2014. Using visible and near 548 infrared diffuse transmittance technique to predict soluble solids content of watermelon in an on-line detection system. Postharvest Biol. Technol. 90, 1–6.

Kafle, G.K., Khot, L.R., Jarolmasjed, S., Yongsheng, S., Lewis, K., 2016. Robustness of near infrared spectroscopy based spectral features for non-destructive bitter pit detection in honeycrisp apples. Postharvest Biol. Technol. 120, 188–192.

Kaić-Rak, A., Antonić, K., 1990. Tablice o sastavu namirnica i pića. Zavod za zaštitu zdravlja SR Hrvatske.

Kelly, J.F.D., Downey, G., 2005. Detection of sugar adulterants in apple juice using Fourier-transform infrared spectroscopy and chemometrics. J. Agric. Food Chem. 53, 3281–3286.

Kemsley, E.K., Holland, J.K., Defernez, M., Wilson, R.H., 1996. Detection of adulteration of raspberry purees using infrared spectroscopy and chemometrics. J. Agric. Food Chem. 44, 3864–3870.

Khatiwada, B.P., Subedi, P.P., Hayes Jr., C., Cunha-Carlos, L.C., Walsh, K.B., 2016. Assessment of internal flesh browning in intact apple using visible-short wave near infrared spectroscopy. Postharvest Biol. Technol. 120, 103–111.

Kim, Y., Singh, M., Kays, S.E., 2007. Near-infrared spectroscopic analysis of macronutrients and energy in homogenized meals. Food Chem. 105, 1248–1255.

Kujundžić, D., Režek Jambrak, A., Vukušić, T., Stulić, V., Gajdoš Kljusurić, J., Banović, M., Herceg, Z., 2017. Near-infrared spectroscopic characterization of steviol glycosides extracted from Stevia rebaudiana Bertoni using high-power ultrasound and gas-phase plasma. J. Food Nutr. Res. 56 (2), 109–120.

Kumaravelu, C., Gopal, A., 2015. A review on the applications of near-infrared spectrometer and chemometrics for the agro-food processing industries. In: 567 Technological Innovation in ICT for Agriculture and Rural Development (TIAR), IEEE, pp. 8–12.

Li, J., Huang, W., Zhao, C., Zhang, B., 2013. A comparative study for the quantitative determination of soluble solids content, pH and firmness of pears by Vis/NIR spectroscopy. J. Food Eng. 116 (2), 324–332.

Liu, X., Yanrong Ren, Y., Peng Zhou, P., Shang, Z., 2011. Prediction of protein 13Cα NMR chemical shifts using a combination scheme of statistical modeling and quantum-mechanical analysis. J Mol. Struct. 995 (1–3), 163–172.

Lugasi, A., Hóvári, J., Kádár, G., Dénes, F., 2011. Phenolics in raspberry, blackberry and currant cultivars grown in Hungary. Acta Aliment. 40, 52–64.

Malley, D.F., Martin, P.D., Ben-Dor, E., 2004. Application in analysis of soils. In: Roberts, C.A., Workman, J., Reeves, J.B. (Eds.), Near-Infrared Spectroscopy in Agriculture, Agronomy, vol. 44. ASA-CSSA-SSSA, Madison, WI, USA, pp. 729–784.

Manach, C., Scalbert, A., Morand, C., Rémésy, C., Jiménez, L., 2004. Polyphenols: food sources and bioavailability. Am. J. Clin. Nutr. 79 (5), 727–747. https://doi.org/10.1093/ajcn/79.5.727.

McGlone, V.A., Martinsen, P.J., Clark, C.J., Jordan, R.B., 2005. On-line detection of brown heart in Braeburn apples using near infrared transmission measurements. Postharvest Biol. Technol. 37 (2), 142–151.

Merzlyak, M.N., Solovchenko, A.E., Smagin, A.I., Gitelson, A.A., 2005. Apple flavonols during fruit adaptation to solar radiation: spectral features and technique for non-destructive assessment. J. Plant Physiol. 162 (2), 151–160.

Miloš, B., Bensa, A., 2017. Prediction of soil organic carbon using VIS-NIR spectroscopy: application to Red Mediterranean soils from Croatia. Eur. J. Soil Sci. 6 (4), 365–373.

Munera, S., Amigo, J.M., Blasco, J., Cubero, S., Talens, P., Aleixos, N., 2017. Ripeness monitoring of two cultivars of nectarine using VIS-NIR hyperspectral reflectance imaging. J. Food Eng. 214, 29–39.

Munera, S., Amigo, J.M., Aleixos, N., Talens, P., Cubero, S., Blasco, J., 2018. Potential of VIS-NIR hyperspectral imaging and chemometric methods to identify similar cultivars of nectarine. Food Control 86, 1–10.

Naderi, A., Rezaei, S., Moussa, A., Levers, K., Earnest, C.P., 2018. Fruit for sport. Trends Food Sci. Technol. 74, 85–98.

Neveu, V., Perez-Jiménez, J., Vos, F., Crespy, V., du Chaffaut, L., Mennen, L., Knox, C., Eisner, R., Cruz, J., Wishart, D., Scalbert, A., 2010. Phenol-Explorer: an online comprehensive database on polyphenol contents in foods. Database. https://doi.org/10.1093/database/bap024.

Nicolaï, B.M., Theron, K.I., Lammertyn, J., 2007. Kernel PLS regression on wavelet transformed NIR spectra for prediction of sugar content of apple. Chemom. Intell. Lab. Syst. 85 (2), 243–252.

Nicolaï, B.M., Verlinden, B.E., Desmet, M., Saevels, S., Saeys, W., Theron, K., Cubeddu, R., Pifferi, A., Torricelli, A., 2008. Time-resolved and continuous wave NIR reflectance spectroscopy to predict soluble solids content and firmness of pear. Postharvest Biol. Technol. 47 (1), 68–74.

Norris, K.H., 1964. A simple spectroradiometer for the 0.4 to 1.2 micron region. Trans. Am. Soc. Agric. Eng. 7, 240–242.

Oliveri, P., Malegori, C., Simonetti, R., Casale, M., 2018. The impact of signal pre-processing on the final interpretation of analytical outcomes—a tutorial. Anal. Chim. Acta. https://doi.org/10.1016/j.aca.2018.10.055. (in press).

Pantelidis, G.E., Vasilakakis, M., Manganaris, G.A., Diamantidis, G., 2007. Antioxidant capacity, phenol, anthocyanin and ascorbic acid contents in raspberries, blackberries, red currants, gooseberries and Cornelian cherries. Food Chem. 102, 777–783.

Pasquini, C., 2018. Near infrared spectroscopy: a mature analytical technique with new perspectives—a review. Anal. Chim. Acta 1026 (5), 8–36.

Peirs, A., Schenk, A., Nicolaï, B.M., 2005. Effect of natural variability among apples on the accuracy of VIS-NIR calibration models for optimal harvest date predictions. Postharvest Biol. Technol. 35 (1), 1–13.

Pérez-Marín, D., Sánchez, M.T., Paz, P., Soriano, M.A., Guerrero, J.E., Garrido-Varo, A., 2009. Non-destructive determination of quality parameters in nectarines during on-tree ripening and postharvest storage. Postharvest Biol. Technol. 52 (2), 180–188.

Pérez-Marín, D., Sánchez, M.T., Paz, P., González-Dugo, V., Soriano, M.A., 2011. Postharvest shelf-life discrimination of nectarines produced under different irrigation strategies using NIR-spectroscopy. Food Sci. Technol. 44 (6), 1405–1414.

Pissard, A., Fernández Pierna, J.A., Baeten, V., Sinnaeve, G., Lognay, G., Mouteau, A., Dupont, P., Rondia, A., Lateur, M., 2013. Non-destructive measurement of vitamin C, total polyphenol and sugar content in apples using near-infrared spectroscopy. J. Sci. Food Agric. 93 (2), 238–244.

Pissard, A., Baeten, V., Dardenne, P., Dupont, P., Lateur, M., 2018. Use of NIR spectroscopy on fresh apples to determine the phenolic compounds and dry matter content in peel and flesh. Biotechnol. Agron. Soc. Environ. 22 (1), 3–12.

Pontes, M.J.C., Santos, S.R.B., Araujo, M.C.U., Almeida, L.F., Lima, R.A.C., Gaiao, E.N., Souto, T.C.P., 2006. Classification of distilled alcoholic beverages and verification of adulteration by near infrared spectrometry. Food Res. Int. 39, 182–189.

Resman, B., Rahelić, D., Gajdoš Kljusurić, J., Martinis, I., 2019. Food composition compliance and reliability in calculations of diabetes and chronic pancreatitis diet offers. J. Food Compos. Anal. https://doi.org/10.1016/j.jfca.2019.01.013. (in press).

Rothwell, J.A., Urpi-Sarda, M., Boto-Ordoñez, M., Knox, C., Llorach, R., Eisner, R., Cruz, J., Neveu, V., Wishart, D., Manach, C., Andres-Lacueva, C., Scalbert, A., 2012. Phenol-Explorer 2.0: a major update of the Phenol-Explorer database integrating data on polyphenol metabolism and pharmacokinetics in humans and experimental animals. Database. https://doi.org/10.1093/database/bas031.

Rothwell, J.A., Pérez-Jiménez, J., Neveu, V., Medina-Ramon, A., M'Hiri, N., Garcia Lobato, P., Manach, C., Knox, K., Eisner, R., Wishart, D., Scalbert, A., 2013. Phenol-Explorer 3.0: a major update of the Phenol-Explorer database to incorporate data on the effects of food processing on polyphenol content. Database. https://doi.org/10.1093/database/bat070.

Salguero-Chaparro, L., Peña-Rodríguez, F., 2014. On-line versus off-line NIRS analysis of intact olives. Food Sci. Technol. 56 (2), 363–369.

Salguero-Chaparro, L., Baeten, V., Abbas, O., Peña-Rodríguez, F., 2012. On-line analysis of intact olive fruits by vis–NIR spectroscopy: optimisation of the acquisition parameters. J. Food Eng. 112 (3), 152–157.

Salguero-Chaparro, L., Baeten, V., Fernández-Pierna, J.A., Peña-Rodríguez, F., 2013. Near infrared spectroscopy (NIRS) for on-line determination of quality parameters in intact olives. Food Chem. 139 (1–4), 1121–1126.

Sánchez, M.T., De la Haba, M.J., Guerrero, J.E., Garrido-Varo, A., Pérez-Marín, D., 2011. Testing of a local approach for the prediction of quality parameters in intact nectarines using a portable NIRS instrument. Postharvest Biol. Technol. 60 (2), 130–135.

Sarraguça, M.C., Lopes, J.A., 2009. The use of net analyte signal (NAS) in near infrared spectroscopy pharmaceutical applications: interpretability and figures of merit. Anal. Chim. Acta 642 (1–2), 179–185.

Schmutzler, M., Huck, C.W., 2016. Simultaneous detection of total antioxidant capacity and total soluble solids content by Fourier transform near-infrared (FT-NIR) spectroscopy: a quick and sensitive method for on-site analyses of apples. Food Control 66, 27–37.

Shenderey, C., Shmulevich, I., Alchanatis, V., Egozi, H., Hoffman, A., Ostrovsky, V., Lurie, S., Arie, R.B., Schmilovitch, Z.E., 2010. NIRS detection of moldy core in apples. Food Bioprocess Technol. 3 (1), 79.

Sirisomboon, P., 2018. NIR spectroscopy for quality evaluation of fruits and vegetables. Mater. Today Proc. 5 (10), 22481–22486.

Sirisomboon, P., Tanaka, M., Fujita, S., Kojima, T., 2007. Evaluation of pectin constituents of Japanese pear by near infrared spectroscopy. J. Food Eng. 78 (2), 701–707.

Sun, X., Liu, Y., Li, Y., Wu, M., Zhu, D., 2016. Simultaneous measurement of brown core and soluble solids content in pear by on-line visible and near infrared spectroscopy. Postharvest Biol. Technol. 116, 80–87.

Tamburini, E., Costa, S., Rugiero, I., Pedrini, P., Marchetti, M.G., 2017. Quantification of lycopene, β-carotene, and total soluble solids in intact red-flesh watermelon (Citrullus lanatus) using on-line near-infrared spectroscopy. Sensors 17 (4), 746.

Tian, X., Wang, Q., Li, J., Peng, F., Huang, W., 2018. Non-destructive prediction of soluble solids content of pear based on fruit surface feature classification and multivariate regression analysis. Infrared Phys. Technol. 92, 336–344.

Travers, S., Bertelsen, M.G., Petersen, K.K., Kucheryavskiy, S.V., 2014. Predicting pear (cv. Clara Frijs) dry matter and soluble solids content with near infrared spectroscopy. Food Sci. Technol. 59 (2), 1107–1113.

USDA, 1996. The Food Guide Pyramid. Home and Garden Bulletin No. 252, pp. 1–29.

USDA, 2016. USDA National Nutrient Database for Standard Reference. Agricultural Research Service, Nutrient Data Laboratory. Release 28 (Slightly revised). Version Current: May 2016, http://www.ars.usda.gov/ba/bhnrc/ndl.

USDA, 2018. National Nutrient Database for Standard Reference Legacy Release, April.

USDA, 2019. https://www.choosemyplate.gov/fruit. (Accessed 21 January 2019).

Valinger, D., Benković, M., Jurina, T., Jurinjak Tušek, A., Belščak-Cvitanović, A., Gajdoš Kljusurić, J., Bauman, I., 2017. Use of NIR spectroscopy and 3D principal component analysis for particle size control of dried medicinal plants. J. Process. Energy Agric. 21 (1), 17–22.

Valinger, D., Kušen, M., Jurinjak Tušek, A., Panić, M., Jurina, T., Benković, M., Radojčić Redovniković, I., Gajdoš Kljusurić, J., 2018. Development of Near Infrared Spectroscopy models for the quantitative prediction of olive leaves bioactive compounds content. Chem. Biochem. Eng. Q. 32 (4), 535–543.

Vardin, H., Tay, A., Ozen, B., Mauer, L., 2008. Authentication of pomegranate juice concentrate using FTIR spectroscopy and chemometrics. Food Chem. 108, 742–748.

Walsh, K., Kawano, S., 2009. Near infrared spectroscopy. In: Zude, M. (Ed.), Optical Monitoring of Fresh and Processed Agricultural Crops. CRC Press, Boca Raton, FL, pp. 192–239.

Wang, H., Peng, J., Xie, C., Bao, Y., He, Y., 2015. Fruit quality evaluation using spectroscopy technology: a review. Sensors 15 (5), 11889–11927.

Xia, Y., Huang, W., Shuxiang Fan, S., Li, J., Chen, L., 2019. Effect of spectral measurement orientation on online prediction of soluble solids content of apple using Vis/NIR diffuse reflectance. Infrared Phys. Technol. https://doi.org/10.1016/j.infrared.2019.01.012. (in press).

Xiaobo, Z., Jiewen, Z., Xingyi, H., Yanxiao, L., 2007. Use of FT-NIR spectrometry in non-invasive measurements of soluble solid contents (SSC) of 'Fuji' apple based on different PLS models. Chemom. Intell. Lab. Syst. 87 (1), 43–51.

Xing, J., De Baerdemaeker, J., 2007. Fresh bruise detection by predicting softening index of apple tissue using VIS/NIR spectroscopy. Postharvest Biol. Technol. 45 (2), 176–183.

Xing, J., Van Linden, V., Vanzeebroeck, M., De Baerdemaeker, J., 2005. Bruise detection on Jonagold apples by visible and near-infrared spectroscopy. Food Control 16 (4), 357–361.

Xu, H., Qi, B., Sun, T., Fu, X., Ying, Y., 2012. Variable selection in visible and near infrared spectra: application to on-line determination of sugar content in pears. J. Food Eng. 109 (1), 142–147.

Xu, W., Sun, T., Wu, W., Hu, T., Hu, T., Liu, M., 2014. Determination of soluble solids content in Cuiguan pear by Vis/NIR diffuse transmission spectroscopy and variable selection methods. In: Wen, Z., Li, T. (Eds.), Knowledge Engineering and Management. Advances in Intelligent Systems and Computing, vol. 278. Springer, Berlin, Heidelberg, pp. 269–276.

Yu, X., Lu, H., Wu, D., 2018. Development of deep learning method for predicting firmness and soluble solid content of postharvest Korla fragrant pear using Vis/NIR hyperspectral reflectance imaging. Postharvest Biol. Technol. 141, 39–49.

9

Role of sensors in fruit nutrition

Daniel Cozzolino[a],*, Madeleine F. Dupont[a], Aaron Elbourne[a],
Vi Khanh Truong[a], Aoife Power[b], James Chapman[b]

[a]School of Science, RMIT University, Melbourne, VIC, Australia
[b]School of Medical and Applied Sciences, Central Queensland University, Rockhampton, QLD, Australia
*Corresponding author. E-mail: daniel.cozzolino@rmit.edu.au

1 Introduction

The definition of fruit quality generally refers to the physical and mechanical properties (e.g., mass, volume, sphericity, density, and firmness), the sensory properties (texture, taste, and aroma), and its relationship with organoleptic and/or attributes of the fruit, such as appearance, defects, or even diseases (e.g., insects and fungus) (Nicolai et al., 2007; Ayvaz et al., 2015, 2016a,b; Nicola et al., 2009; Hertog, 2016; Su and Sun, 2018; Minas et al., 2018; Roberts et al., 2018).

Fruit quality parameters, including inherent parameters such as sugar and acid content, ripening and shelf-life, and other fruit quality chemical and physical properties, such as color, shape, stage of growth and firmness, have been reported to be closely correlated with the uptake nutrients such as nitrogen (N), phosphorus (P), potassium (K), calcium (Ca), and magnesium (Mg) by the plant. Therefore, sensing nutrients in the plant is of vital importance not only to produce high-quality fruits but also to better monitor nutrients for sustainable production systems (Muñoz-Huerta et al., 2013; Zude-Sasse et al., 2016; Roberts et al., 2018).

The chapter is organized to provide a general introduction about infrared (IR) and multivariate data analysis (MVA) methods and to give examples of recent applications of these methods to measure and monitor fruit nutrition (e.g., macro- and micronutrients and water) (see Fig. 9.1).

2 Sensors based in vibrational spectroscopy

2.1 Near- and mid-infrared spectroscopy

Methods based on vibrational spectroscopy (near [NIR] and mid [MIR]) have been widely used to measure nutrients in crops and plants. These methods are based on the measurement of chemical bonds present in the sample that

A.K. Srivastava, Chengxiao Hu (eds.)
Fruit Crops: Diagnosis and Management of Nutrient Constraints
https://doi.org/10.1016/B978-0-12-818732-6.00009-5

FIG. 9.1   Applications of optical sensors (vibrational spectroscopy) in fruit and plant analysis.

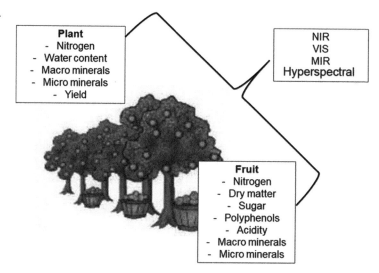

vibrate at specific frequencies, which are directly associated with the mass of the constituent atoms and the shape of the molecule (Karoui et al., 2010; Rodriguez-Saona and Allendorf, 2011). Specific vibrational bonds absorb in the infrared (IR) spectral region where diatomic molecules have only one bond that may stretch (e.g., the distance between two atoms may increase or decrease) (McClure, 2003; Blanco and Villaroya, 2002).

The IR radiation belongs to the electromagnetic spectrum between the visible (VIS) and the microwave wavelengths (McClure, 2003; Blanco and Villaroya, 2002; Cozzolino and Roberts, 2016; Clark et al., 2019). The nominal range of wavelengths for NIR is between 750 and 2500 nm (13,400–4000 cm^{-1}) while for the MIR is from 2500 to 25,000 nm (4000–400 cm^{-1}) (McClure, 2003; Blanco and Villaroya, 2002; Roggo et al., 2007; Cozzolino and Roberts, 2016).

Near-infrared spectroscopy is characterized by low molar absorptivities (McClure, 2003; Blanco and Villaroya, 2002; Roggo et al., 2007; Cozzolino and Roberts, 2016; Clark et al., 2019). More broadly NIR spectroscopy is capable to measure organic chemical molecules containing O-H (oxygen-hydrogen), N-H (nitrogen-hydrogen), and C-H (carbon-hydrogen) bonds through the absorption of energy in the NIR region of the spectrum (Huang et al., 2008; McClure, 2003; Blanco and Villaroya, 2002; Roggo et al., 2007; Cozzolino and Roberts, 2016; Clark et al., 2019).

On the other hand, absorptions in the MIR region result from the fundamental stretching, bending, and rotating vibrations of the sample molecules, while NIR spectra result from complex overtones and high-frequency combinations at shorter wavelengths (McClure, 2003; Blanco and Villaroya, 2002; Roggo et al., 2007; Cozzolino and Roberts, 2016; Clark et al., 2019). Although NIR intensities are 10–1000 times lower than for those in the MIR region, highly sensitive spectrometers can be built through several means including the use of efficient detectors and brighter light sources (McClure, 2003; Blanco and Villaroya, 2002; Cozzolino and Roberts, 2016; Clark et al., 2019).

Spectral peaks in the MIR range are often sharper and better resolved than in the NIR domain; the higher overtones of the O-H, N-H, C-H, and S-H (sulfur-hydrogen) bonds from the MIR wavelengths are still observed in the NIR region, although much weaker than the fundamental frequencies in the MIR region. In addition, the existence of combination bonds (e.g., CO stretch and NH bend in protein) gives rise to a crowded NIR spectrum with strongly overlapping bonds (McClure, 2003; Blanco and Villaroya, 2002; Roggo et al., 2007; Cozzolino and Roberts, 2016; Clark et al., 2019). A major disadvantage of this characteristic overlap and complexity in the NIR spectra has been the difficulty of quantification and interpretation of data from NIR spectra (e.g., need for calibration). However, the broad overlapping bonds can diminish the need for using many wavelengths in calibration and analysis routines. In recent years, new instrumentation and computer algorithms have taken advantage of this complexity and have made the technique much more powerful and simple to use. Yet, the advent of inexpensive and powerful computers has contributed to the surge of new NIR applications (Nicolai et al., 2007; Gishen et al., 2010; Cozzolino and Roberts, 2016; Clark et al., 2019) (see Table 9.1).

2.2 Hyperspectral spectroscopy

Most of the spectroscopy applications have been primarily relying on the spot measurements using either the NIR or MIR regions of the electromagnetic spectrum (Roberts et al., 2018; Clark et al., 2019; Kamruzzaman and Sun, 2016;

TABLE 9.1 Characteristics of optical sensors used to monitor nutrients in fruits.

	Wavelength range	Sample collection	Type of samples	Instrument portability
HYPER	VIS and NIR range	• Spot and spatial resolution	Solid samples	Broad overtones
MIR		• Spot, small samples • Excellent wavelength resolution	Liquids (transmission) and solids (ATR and diffuse reflectance)	Fundamental stretching, bending, and rotating
NIR	750–2500 nm	• Spot, small and large samples • Minimal sample preparation	Liquids (transmission) and solids (diffuse reflectance)	Broad overtones
VIS	400–750 nm	• Small samples, liquids	Liquids (transmission)	

ATR, attenuated total reflectance; *HYPER*, hyperspectral; *NIR*, near infrared; *MIR*, mid infrared; *VIS*, visible.

Pallone et al., 2018). In the last 20 years, a diverse range of hyperspectral devices, including cameras and spectral imaging devices, now readily available in the market provide with exciting new possibilities for the analysis of samples. The availability of hyperspectral (HSI) and multispectral imaging (MSI) systems allowed for obtaining spatial, spectral, and multiconstituent information about the sample being analyzed (Roberts et al., 2018; Clark et al., 2019; Pallone et al., 2018; Wang et al., 2016; Kamruzzaman and Sun, 2016). In the last decade, both HSI and MSI have started to feature with increasing importance as rapid and efficient methods for the assessment different samples (e.g., food, microbial contamination, and soil) (Roberts et al., 2018; Clark et al., 2019; Pallone et al., 2018; Wang et al., 2016; Kamruzzaman and Sun, 2016). Spectral imaging can be classified as either HSI or MSI (Roberts et al., 2018; Clark et al., 2019; Pallone et al., 2018; Wang et al., 2016; Kamruzzaman and Sun, 2016). MSI involves the acquisition of spectral images at few discrete and narrow wavebands (bandwidths of between 5 and 50 nm), and it is an improvement of HSI as this technology is cost effective (Roberts et al., 2018; Clark et al., 2019; Pallone et al., 2018; Wang et al., 2016; Kamruzzaman and Sun, 2016). The capability of MSI to simultaneously predict multiple components provides this technique with a key advantage and rendering a promising outlook with a single, automated image acquisition compared with other methods (Roberts et al., 2018; Clark et al., 2019; Pallone et al., 2018; Wang et al., 2016; Kamruzzaman and Sun, 2016). HSI relies on one of the two sensing modes as in-line scanning (push broom) mode or as filter-based imaging mode (Roberts et al., 2018; Clark et al., 2019; Pallone et al., 2018; Wang et al., 2016; Kamruzzaman and Sun, 2016). HSI methods require a high-performance digital camera that can cover the spectral region of interest, in a wide dynamic range with high signal-to-noise ratio, and good quantum efficiency (Roberts et al., 2018; Clark et al., 2019; Pallone et al., 2018; Wang et al., 2016; Kamruzzaman and Sun, 2016). HSI is particularly attractive for crop, food, and plant applications because of the integration of imaging ability with spectroscopy, which enables the simultaneous acquisition of both spectral and spatial information from the sample (Roberts et al., 2018; Clark et al., 2019; Pallone et al., 2018; Wang et al., 2016; Kamruzzaman and Sun, 2016).

3 Data and information: Multivariate data analysis

The development of methods based on sensors requires the creation of a model or calibration (Brereton, 2008; Naes et al., 2002; Walsh and Kawano, 2009; Westad and Marini, 2015; Williams et al., 2017). Several algorithms are used to interpret the data generated and to develop the so-called calibration models where principal component analysis (PCA), partial least-square (PLS) regression, and discriminant analysis are the most commonly used. Once a model or calibration was developed, the predictive ability of these models to predict new samples must be tested using an independent set of samples (e.g., samples not used when developing the calibration) (Brereton, 2008; Naes et al., 2002; Walsh and Kawano, 2009; Westad and Marini, 2015; Williams et al., 2017). To test the robustness of the model, some aspects need to be considered during the selection of independent samples. For example, samples should be gathered from other similar experiments, batches, or conditions (e.g., harvests, different temperatures, moisture content, and origins) (Brereton, 2008; Naes et al., 2002; Walsh and Kawano, 2009; Westad and Marini, 2015; Williams et al., 2017).

Several statistics and acronyms are in use to report the results obtained during model development and validation (Brereton, 2008; Naes et al., 2002; Walsh and Kawano, 2009; Westad and Marini, 2015; Williams et al., 2017). These statistics include the prediction error of a calibration model, which can be defined as the root-mean-square error for cross validation (RMSECV), when cross validation is used, or the root-mean-square error for prediction (RMSEP), when internal or external validation is used (Brereton, 2008; Naes et al., 2002; Walsh and Kawano, 2009; Westad and Marini, 2015; Williams et al., 2017). The standard error of prediction (SEP) can be reported instead of the RMSEP (Brereton, 2008; Naes et al., 2002; Walsh and Kawano, 2009; Westad and Marini, 2015; Williams et al., 2017). The residual predictive deviation (RPD) value has also been proposed to evaluate the ability of a calibration model to predict unknown samples (Naes et al., 2002; Walsh and Kawano, 2009; Williams et al., 2017). The RPD value is defined as the ratio of the standard deviation of the response variable to the RMSEP or RMSECV (other authors use the term SDR) (Naes et al., 2002; Walsh and Kawano, 2009; Williams et al., 2017; Roberts and Cozzolino, 2016). A common and general statistic often used is the coefficient of determination (R^2). This statistic represents the proportion of explained variance of the response variable in either the training or test sets (Brereton, 2008; Naes et al., 2002; Walsh and Kawano, 2009; Westad and Marini, 2015; Williams et al., 2017; Roberts and Cozzolino, 2016).

After the calibration model is finalized, the fit-for-purpose criterion needs to be considered to judge the applicability of the model developed for routine use. These models (calibrations) need to be understood in the overall context of the application and not only on the cold interpretation of the statistics (Brereton, 2008; Naes et al., 2002; Walsh and Kawano, 2009; Westad and Marini, 2015; Williams et al., 2017). Therefore, the loadings or coefficients of regression need to be interpreted in the context of the analyte that is being analyzed (Brereton, 2008; Naes et al., 2002; Walsh and Kawano, 2009; Westad and Marini, 2015; Williams et al., 2017; Roberts and Cozzolino, 2016).

4 Applications

The following sections present recent applications of sensor-based techniques (NIR, MIR, and hyperspectral) combined with multivariate data analysis to predict and measure nutrients in fruits.

4.1 Nitrogen status in plants and fruits

Nitrogen (N) is one of the essential elements in plants due to its key role in chlorophyll production, fundamental for the photosynthesis process (Muñoz-Huerta et al., 2013). Nitrogen is also an important building block of several enzymatic proteins that catalyze and regulate plant-growth processes and contributes to the production of chemical components that protect against parasites and plant diseases (Muñoz-Huerta et al., 2013; Parks et al., 2012; Timmer et al., 2005; Tremblay et al., 2011).

One of the most important issues in agriculture is related with the optimization of farming practices to achieve environmental friendly yields and to reduce the cost of production through the application of precision agriculture (sustainability agriculture) (Zude-Sasse et al., 2016). Precision agriculture requires the efficient supply of water and nutrients where the use of sensors has an important role to play (e.g., monitor plant nutrients, soil properties, and yield) (Zude-Sasse et al., 2016). Various methods have been proposed by several scientists to optimize fertilization strategies in crops and fruits. Most of these methods are based on routine analytical methods such as Kjeldahl digestion and Dumas combustion (Muñoz-Huerta et al., 2013; Parks et al., 2012; Timmer et al., 2005; Tremblay et al., 2011). Although these methods and techniques have been widely applied to the analysis of plants, they are considered time-consuming and destructive and impracticable for precision agriculture (Muñoz-Huerta et al., 2013; Parks et al., 2012; Timmer et al., 2005; Tremblay et al., 2011). Therefore, the development of methods based on sensors (e.g., optical and electrochemical) including those based on the use of vibrational spectroscopy becomes very attractive in precision agriculture (Muñoz-Huerta et al., 2013; Parks et al., 2012; Timmer et al., 2005; Tremblay et al., 2011).

Several studies have been reported for the rapid and nondestructive analysis of plant N status (Muñoz-Huerta et al., 2013; Parks et al., 2012; Timmer et al., 2005; Tremblay et al., 2011). However, most of these method were based in the use of leaf optical properties where pigments such as chlorophyll and polyphenols were used as proxies to estimate the N status of the plant (Muñoz-Huerta et al., 2013; Parks et al., 2012; Timmer et al., 2005; Tremblay et al., 2011). An extensive review by Muñoz-Huerta et al. (2013) discussed these applications and the use of different sensors to measure nitrogen in plants.

The use of VIS-NIR spectroscopy combined with PLS regression was proposed to determine the N status of pear (*Pyrus communis* L.) leaves during different growing season, to monitor fertilization, and to predict fruit quality

(Jie et al., 2014). The results by these authors showed that the model developed using full spectra performed better than the developed using specific wavelength regions (Jie et al., 2014). Calibration results reported by the authors indicated that PLS regression yielded a R^2 of 0.96 and 0.95 for calibration and cross validation, respectively (Jie et al., 2014). The validation of the method resulted in R^2 of 0.847 and low mean relative error (4.48%) (Jie et al., 2014). The authors concluded that the combination of VIS-NIR spectroscopy could provide a rapid and more reliable method to determine the total N concentration in fresh pear leaves and could be useful for fertilization management in pear orchards (Jie et al., 2014).

An infield method based on the combination of VIS-NIR spectroscopy with different calibration methods such as principal component regression (PCR), PLS, stepwise multiple linear regression (SMLR), and back propagation neural network (BPNN) was reported to monitor N status in an orchard (Wang et al., 2017). The results reported by these authors showed that the best modeling and prediction accuracy were found in the model established by PLS regression (Wang et al., 2017). The randomly separated subsets of calibration ($n = 1000$) and validation ($n = 420$) of this model resulted in high R^2 values of 0.86 and 0.85, respectively, and a low mean relative error (<6%) (Wang et al., 2017).

The use of NIR spectroscopy was reported to measure N content in citrus leaves by Min and Lee (2005). These authors identified several spectral regions in the VIS and NIR region associated with N content in wavelengths around 448, 669, 719, 1377, 1773, and 2231 nm. The best prediction model resulted in an R^2 of 0.84 with a residual mean standard deviation (RMSD) of 0.122% (Min and Lee, 2005). The same authors also investigated the use of VIS-NIR spectroscopy to predict N content in Chinese cabbage seedlings. In this study, the authors reported that absorbances at specific wavelengths such as 710 nm (VIS) and 1467, 1910, and 1938 nm (NIR) were important to develop the PLS model to measure N content (Min et al., 2006).

Menesatti et al. (2010) evaluated the use of VIS-NIR spectroscopy to measure N and other nutrients in citrus leaves such as P, K, Ca, Mg, Fe, Zn, and Mn. According to these authors, the best model was reported for K using PLS regression, R^2 0.99 (calibration) and 0.99 (validation) (Menesatti et al., 2010). Correlation for other parameters ranged between 0.883 (Mg) and 0.481 (P) with a standard error in prediction ranging from 0.01 to 12.418 for P and Fe, respectively (Menesatti et al., 2010).

The concentration of chlorophyll in cucumber leaves was monitored using NIR-HSI as proxy to determine N deficiency in the plant (Ji-Yong et al., 2012). Fresh leaves ($n = 120$) were analyzed, and multiple linear regression (MLR) was used to build calibration models where chlorophyll concentration was estimated by high-performance liquid chromatography (HPLC) (Ji-Yong et al., 2012). Chlorophyll concentration was predicted with an R of 0.87 (Ji-Yong et al., 2012). The predicted chlorophyll values were used to develop maps of N deficiency. The results reported by the authors indicated that NIR-HSI can be used as diagnostic tool to determine N deficiency in cucumber plants (Ji-Yong et al., 2012).

Nitrate is an important component of the nitrogen cycle and is therefore present in all plants (Yang et al., 2017). However, excessive nitrogen fertilization results in a high nitrate content in vegetables, which is unhealthy for humans (Yang et al., 2017). Understanding the spatial distribution of nitrate in leaves is beneficial for improving nitrogen assimilation efficiency and reducing its content in vegetables. The use of NIR-HSI was evaluated as nondestructive method to measure nitrate content in spinach (*Spinacia oleracea* L.) leaves (Yang et al., 2017). Calibrations developed using PLS regression yield a high R^2 0.74 and root-mean-square error of cross validation 710.16 mg/kg (Yang et al., 2017). The NIR data were used by the authors to map dynamic changes in the nitrate content in intact leaf samples under different storage conditions, showing the value of this nondestructive tool for future analyses of the nitrate content in vegetables (Yang et al., 2017).

The influence of fertilization on the N content of *Physalis peruviana* (cape gooseberry) and *Physalis angulata* (camapú) was evaluated using attenuated total reflectance in the MIR range (da Silva Leite et al., 2018). Leaves, stems, and roots of both species were characterized by ATR-FTIR spectroscopy where specific compounds related to the functional groups present in the fractions of the analyzed species, such as cellulose, pectin, and phenolic compounds, were analyzed as function of N fertilization (da Silva Leite et al., 2018). The authors concluded that the study offered a precise means to identify functional groups present in the species of the genus and with possible pharmacological use (da Silva Leite et al., 2018).

In a previous study, Schmidtke et al. (2012) utilized ATR-FTMIR spectroscopy to monitor N and starch reserves in the perennial tissues of the vine to better predict grapevine yield and to monitor sugar accumulation and secondary metabolites during berry ripening. The authors reported that ATR-FTMIR offers significant advantages over the standard traditional method of N concentration determination in plant tissue combustion and glucose quantification by enzymatic analysis due to the ease, reproducibility, and speed of analysis (Schmidtke et al., 2012). Best predictive models for starch were developed using PLS regression ($R^2 = 0.95$ and RSMEP = 1.43% DW) compared with SVM ($R^2 = 0.95$ and RMSEP = 1.56% DW) with the RMSEP reported being less than the recorded seasonal flux for these analytes in the samples. The authors concluded that ATR-FTMIR could provide grape growers with a rapid method for monitoring vine reserve status within commercial grape production systems (Schmidtke et al., 2012).

4.2 Water-dry matter

Water is an important nutrient associated with several properties and reactions in the plant (Rodriguez-Perez et al., 2007; Serrano et al., 2010, 2012; De Pascale et al., 2018; Facini et al., 2004). The determination of this nutrient is of importance to schedule irrigation and to determine optimum harvest times (Serrano et al., 2010, 2012; Santos and Kaye, 2009; de Bei et al., 2011; Schmidtke et al., 2012). Grapevines require a more rapid, precise, and reliable tool for determining their leaf water potential (ψ). Portable NIR spectroscopy tools coupled with multivariate analysis have been demonstrated to accurately predict (exhibiting R^2 values ranging from 0.87 to 0.95) the leaf ψ of grapevines, with the most significant leaf absorption bands occurring between 1440 and 1950 nm (Santos and Kaye, 2009). de Bei et al. (2011) developed NIR calibration using the spectra of the abaxial surface of the Chardonnay leaves compared with that of the spectra from the adaxial surface yielded a better correlation with the leaf ψ of the vine (de Bei et al., 2011). It was concluded by these authors that NIR spectroscopy can be used to nondestructively measure the leaf ψ of grapevines when appropriate calibration models are incorporated and the spectra collected were dependent on the leaf surface analyzed, which in turn impacted on the accuracy of the calibration statistics (de Bei et al., 2011). The capability of noninvasive, portable NIR spectroscopy to assess the grapevine water status in diverse varieties, grown under different environmental conditions, in a fast and reliable way was reported recently (Tardaguila et al., 2017). Both PCA and modified PLS regression were used to interpret the spectra and to develop reliable prediction models for stem water potential with a R^2 in cross validation ranged from 0.77 to 0.93 and SECV ranged from 0.10 to 0.23 (Tardaguila et al., 2017). For leaf, relative water content (RWC) R^2 ranged from 0.66 to 0.81 and SECV between 1.93 and 3.20 (Tardaguila et al., 2017). Shao et al. (2017) reported the use of VIS-NIR-HIS spectroscopy to predict water content in grape leaves yielding an R^2, RMSEP, bias, and RPD of 0.900, 0.826, $-2.213e-04$, and 2.084, respectively (Shao et al., 2017). Facini et al. (2004) reported the measurement of leaf and canopy reflectance as indicators of plant water status in apple trees (cv *Golden Delicious*) grown in pots under progressive water stress (Facini et al., 2004). These authors observed an increase in reflectance in the wavelength range between 900 and 1000 nm associated with the RWC decreased from 90.4% to 85.4% (Facini et al., 2004).

The combination of VIS and NIR spectroscopy to monitor plant water stress was evaluated on satsuma mandarin (*Citrus unshiu* Marc. var. Satsuma) and peach (*Persica vulgaris* Mill.) orchards (Kriston-Vizi et al., 2008). Differences in reflectance were found between water-stressed and nonwater-stressed mandarin and peach plants both at the individual leaf level and at the field-scale canopy level (Kriston-Vizi et al., 2008). Moderately good correlations were found between reflectance and leaf water potential (Kriston-Vizi et al., 2008).

4.3 Other macro and trace elements

Phosphorus (P) is an important element to litchi yield and fruit quality in addition to N and potassium (K) content (Li et al., 2018). Infrared spectroscopy was used to correlate optical with reference data for the measurement of N, P, and K in plants where the range in R^2 0.54–0.98 and RMSECV 0.02–0.03 were reported (Li et al., 2018). The most relevant wavelengths used by the model were found in the VIS and short NIR (SWNIR) regions (Li et al., 2018). These wavelengths were related to the absorption features of pigments (e.g., anthocyanin and chlorophyll), proteins, nitrogen, starch, sugar, oil, cellulose, and lignin in the plant tissues analyzed (Li et al., 2018). Overall, these authors concluded that the use of SWNIR in foliar nutrient monitoring will be very important for precision agriculture (Li et al., 2018).

The estimation of macro- and micronutrients in citrus leaves using NIR spectroscopy was reported (Galvez-Sola et al., 2015). Both NIR spectral data and chemical data for N, K, Ca, Mg, B, Fe, Cu, Mn, and Zn concentration were obtained using leaf samples from different citrus tree species (Galvez-Sola et al., 2015). Calibration models were developed using PLS regression and different preprocessing treatments (Galvez-Sola et al., 2015). The authors reported a high R 0.99 and R 0.98 for N and Ca, respectively, and acceptable estimation for K, Mg, Fe, and Zn. However, no successful calibrations were obtained for the estimation of B, Cu, and Mn (Galvez-Sola et al., 2015).

NIR spectroscopy was used to predict essential mineral composition in the skin and flesh of summer squash fruits (*Cucurbita pepo* subsp. pepo) using a large data set ($n = 200$) from diverse morphotypes (Martínez-Valdivieso et al., 2014). The R^2 reported using an external validation for the skin and flesh of the fruit was for total mineral content, 0.84 and 0.70; for P, 0.74 and 0.62; for K, 0.83 and 0.67; for Ca, 0.57 and 0.60; for Mg, 0.78 and 0.45; for Fe, 0.78 and

0.65; for Cu, 0.67 and 0.66; for Mn, 0.67 and 0.64; for Zn, 0.80 and 0.79; and for Na, 0.33 and 0.33; respectively (Martínez-Valdivieso et al., 2014). Both NIR and X-ray fluorescence (XRF) spectroscopy were investigated to predict the concentration of Ca, K, Fe, Mg, Mn, and Zn in artichoke samples (Mir-Marqués et al., 2016). NIR and XRF spectra, combined with PLS regression, were used to develop calibration models where inductively coupled plasma optical emission spectrometry (ICP-OES) was used to obtain the reference data (Mir-Marqués et al., 2016). The coefficients of determination obtained for the regression between predicted values and reference ones for calcium, potassium, magnesium, iron, manganese, and zinc were The R^2 reported were 0.61, 0.79, 0.53, 0.77, 0.54, and 0.60 for Ca, K, Mg, Fe, Mn, and Zn, respectively (Mir-Marqués et al., 2016).

5 Final considerations

A review of the literature on the application of optical sensors reveals a significant disparity in the numbers of papers published in relation with the measurement of nutrients in fruits, with the vast number of papers reported applications in fruit quality.

Adapting and using advanced technologies are a promising way to efficiently and reliably improve management farming toward sustainability. These sensors will allow to move forward novel applications that can deal with best management practices in the process and commercialization of agricultural products and commodities. In recent years, different researchers have shown the important role of proximal sensors based on NIR spectroscopy in the analysis of crops and their use on harvesters. This approach could reduce the manpower and cost required for the determination of different properties in the crop associated with nutrition and quality.

The accuracy and robustness of the models reported by different authors were considered acceptable, considering that several of these nutrients are components of large molecules or they do not have a specific absorption characteristic in the IR region (e.g., inorganic compounds). It is important to note that calibration models to be used in practice should be based on large datasets, encompassing several sources including climate conditions, seasons, and operational settings (e.g., temperature and moisture). These models should be also optimized toward robustness by incorporating appropriate spectral preprocessing methods.

Overall, the use of these techniques will allow for potential savings in relation to reduction of analysis time and cost placing the use of sensors as a very attractive technique in farm applications to monitor nutrient status in both crops and plants. Recent development in instrumentation (e.g., HSI and portable spectrophotometers) and algorithms (topics not covered in this chapter) will place these technologies as one of the most useful tools in crop/plant nutrient monitoring and as the preferred in field technology in the future.

Acknowledgments

The financial support of RMIT University is acknowledged.

References

Ayvaz, H., Santos, A.M., Moyseenko, J., Kleinhenz, M., Rodriguez-Saona, L.E., 2015. Application of a portable infrared instrument for simultaneous analysis of sugars, asparagine and glutamine levels in raw potato tubers. Plant Foods Hum. Nutr. 70, 215–220.

Ayvaz, H., Sierra-Cadavid, A., Aykas, D.P., Mulqueeney, B., Sullivan, S., Rodriguez-Saona, L.E., 2016a. Monitoring multicomponent quality traits in tomato juice using portable mid-infrared (MIR) spectroscopy and multivariate analysis. Food Control 66, 79–86.

Ayvaz, H., Bozdogan, A., Giusti, M.M., Mortas, M., Gomez, R., Rodriguez-Saona, L.E., 2016b. Improving the screening of potato breeding lines for specific nutritional traits using portable mid-infrared spectroscopy and multivariate analysis. Food Chem. 211, 374–382.

Blanco, M., Villaroya, I., 2002. NIR spectroscopy: a rapid-response analytical tool. Trends Anal. Chem. 21, 40–50.

Brereton, R.G., 2008. Applied Chemometrics for Scientist. John Wiley & Sons Ltd, Chichester.

Clark, C., Cozzolino, D., Bureau, S., 2019. Contributions of Fourier-transform mid infrared (FT-MIR) spectroscopy to the study of fruit and vegetables: a review. Postharvest Biol. Technol.

Cozzolino, D., Roberts, J.J., 2016. Applications and developments on the use of vibrational spectroscopy imaging for the analysis, monitoring and characterisation of crops and plants. Molecules 21, 755–763.

da Silva Leite, R., Hernandéz-Navarro, S., Neves do Nascimento, M., Ruiz Potosme, N.M., Carrión-Prieto, P., dos Santos Souza, E., 2018. Nitrogen fertilization affects Fourier Transform Infrared spectra (FTIR) in Physalis L. species. Comput. Electron. Agric. 150, 411–417.

de Bei, R., Cozzolino, D., Sullivan, W., Cynkar, W., Fuentes, S., Dambergs, R., Tyerman, S., 2011. Non-destructive measurement of grapevine water potential using near infrared spectroscopy. Aust. J. Grape Wine Res. 17, 62–71.

De Pascale, S., Rouphael, Y., Gallardo, M., Thompson, R.B., 2018. Water and fertilization management of vegetables: state of art and future challenges. Eur. J. Hortic. Sci. 83 (5), 306–318.

Facini, O., Loreti, S., Rossi, F., Bignami, C., 2004. Canopy and leaf light reflectance features in relation to water content in apple. Acta Hortic. (664), 217–224.

Galvez-Sola, L., García-Sánchez, F., Pérez-Pérez, J., Gimeno, V., Navarro, J.M., Moral, R., Martínez-Nicolás, J.J., Nieves, M., 2015. Rapid estimation of nutritional elements on citrus leaves by near infrared reflectance spectroscopy. Front. Plant Sci. 6571.

Gishen, M., Cozzolino, D., Dambergs, R.G., 2010. In: Li-Chan, E., Chalmers, J., Griffiths, P. (Eds.), Applications of Vibrational Spectroscopy in Food Science. John Wiley and Sons.

Hertog, M.L.A.T.M., 2016. Modeling Fruit Quality. Reference Module in Food Science. Elsevier.

Huang, H., Yu, H., Xu, H., Ying, Y., 2008. Near infrared spectroscopy for on/in-line monitoring of quality in foods and beverages: a review. J. Food Eng. 87, 303–313.

Jie, W., Hua-bing, Z., Chang-wei, S., qiao-wei, C., Cai-xia, D., Yang-chun, X., 2014. Determination of nitrogen concentration in fresh pear leaves by visible/near-infrared reflectance spectroscopy. Agron. J. 106, 1867–1872.

Ji-Yong, S., Xiao-Bo, Z., Jie-Wen, Z., Kai-Liang, W., Zheng-Wei, C., Xiao-Wei, H., De-Tao, Z., Holmes, M., 2012. Non-destructive diagnostics of nitrogen deficiency by cucumber leaf chlorophyll distribution map based on near infrared hyperspectral imaging. Sci. Hortic. 138, 190–197.

Kamruzzaman, M., Sun, D.-W., 2016. Introduction to hyperspectral imaging technology. In: Computer Vision Technology for Food Quality Evaluation. (Chapter 5).

Karoui, R., Downey, G., Blecker, C., 2010. Mid-infrared spectroscopy coupled with chemometrics: a tool for the analysis of intact food systems and the exploration of their molecular structure – quality relationships—a review. Chem. Rev. 110, 6144–6168.

Kriston-Vizi, J., Umeda, M., Miyamoto, K., 2008. Assessment of the water status of mandarin and peach canopies using visible multispectral imagery. Biosyst. Eng. 100 (3), 338–345.

Li, D., Wang, C., Jiang, H., Peng, Z., Yang, J., Su, Y., Song, J., Chen, S., 2018. Monitoring litchi canopy foliar phosphorus content using hyperspectral data. Comput. Electron. Agric. 154, 176–186.

Martínez-Valdivieso, D., Font, R., Gómez, P., Blanco-Díaz, T., Del Río-Celestino, M., 2014. Determining the mineral composition in *Cucurbita pepo* fruit using near infrared reflectance spectroscopy. J. Sci. Food Agric. 94, 3171–3180.

McClure, W.F., 2003. Review: 204 years of near infrared technology: 1800–2003. J. Near Infrared Spectrosc. 11, 487–518.

Menesatti, P., Antonucci, F., Pallottino, F., Roccuzzo, G., Allegra, M., Stagno, F., Intrigliolo, F., 2010. Estimation of plant nutritional status by Vis–NIR spectrophotometric analysis on orange leaves [*Citrus sinensis* (L) Osbeck cv Tarocco]. Biosyst. Eng. 105, 448–454.

Min, M., Lee, W.S., 2005. Determination of significant wavelengths and prediction of nitrogen content for citrus. Trans. ASAE 48 (2), 455–461.

Min, M., Lee, W.S., Kim, Y.H., Bucklin, R.A., 2006. Non-destructive detection of nitrogen in Chinese cabbage leaves using VIS–NIR spectroscopy. HortScience 41 (1), 162–166.

Minas, I.S., Tanou, G., Molassiotis, A., 2018. Environmental and orchard bases of peach fruit quality. Sci. Hortic. 235, 307–322.

Mir-Marqués, A., Martínez-García, M., Garrigues, S., Cervera, M.L., de la Guardia, M., 2016. Green direct determination of mineral elements in artichokes by infrared spectroscopy and X-ray fluorescence. Food Chem. 196, 1023–1030.

Muñoz-Huerta, R., Guevara-Gonzalez, R., Contreras-Medina, L., Torres-Pacheco, I., Prado-Olivarez, J., Ocampo-Velazquez, R., 2013. A review of methods for sensing the nitrogen status in plants: advantages, disadvantages and recent advances. Sensors 13, 10823–10843.

Naes, T., Isaksson, T., Fearn, T., Davies, T., 2002. A User-Friendly Guide to Multivariate Calibration and Classification. NIR Publications, Chichester. 420 p.

Nicola, S., Tibaldi, G., Fontana, E., 2009. fresh-cut produce quality: implications for a systems approach. In: Florkowski, W.J., Shewfelt, R.L., Brueckner, B., Prussia, S.E. (Eds.), Postharvest Handling, second ed. Academic Press, San Diego, CA, pp. 247–282 (Chapter 10).

Nicolai, B.M., Beullens, K., Bobelyn, E., Peirs, A., Saeys, W., Theron, K.I., Lammertyn, J., 2007. Non-destructive measurement of fruit and vegetable quality by means of NIR spectroscopy: a review. Postharvest Biol. Technol. 46, 99–118.

Pallone, J.A., dos Santos Carames, E.M., Domingues Alamar, P., 2018. Green analytical chemistry applied in food analysis: alternative techniques. Curr. Opin. Food Sci. 22, 115–121.

Parks, S.E., Irving, D.E., Milham, P.J., 2012. A critical evaluation of on-farm rapid tests for measuring nitrate in leafy vegetables. Sci. Hortic. 134, 1–6.

Roberts, J.J., Cozzolino, D., 2016. An overview on the application of chemometrics in food science and technology—an approach to quantitative data analysis. Food Anal. Methods 9, 3258–3267.

Roberts, J., Power, A., Chapman, J., Chandra, S., Cozzolino, D., 2018. Vibrational spectroscopy methods for agro-food product analysis. Compr. Anal. Chem.

Rodriguez-Perez, J.R., Riano, D., Carlisle, E., Ustin, S., Smart, D.R., 2007. Evaluation of hyperspectral reflectance indexes to detect grapevine water status in vineyards. Am. J. Enol. Vitic. 58, 302–317.

Rodriguez-Saona, L.E., Allendorf, M.E., 2011. Use of FTIR for rapid authentication and detection of adulteration of food. Annu. Rev. Food Sci. Technol. 2, 467.

Roggo, Y., Chalus, P., Maurer, L., Lema-Martinez, C., Edmond, A., Jent, N., 2007. A review of near infrared spectroscopy and chemometrics in pharmaceutical technologies. J. Pharm. Biomed. Anal. 44, 683–690.

Santos, A.O., Kaye, O., 2009. Grapevine leaf water potential based upon near infrared spectroscopy. Sci. Agric. 66, 287–292.

Schmidtke, L.M., et al., 2012. Rapid monitoring of grapevine reserves using ATR–FT-IR and chemometrics. Anal. Chim. Acta 732, 16–25.

Serrano, L., Gonzalez-Flor, C., Gorchs, G., 2010. Assessing vineyard water status using the reflectance-based water index. Agric. Ecosyst. Environ. 139, 490–499.

Serrano, L., Gonzalez-Flor, C., Gorchs, G., 2012. Assessment of grape yield and composition using the reflectance-based water index in Mediterranean rain fed vineyards. Remote Sens. Environ. 118, 249–258.

Shao, Y., Zhou, H., Jiang, L., et al., 2017. Using reflectance and gray-level texture for water content prediction in grape vines. Trans. ASABE 60, 207–213.

Su, W.-H., Sun, D.-W., 2018. Multispectral imaging for plant food quality analysis and visualization. Compr. Rev. Food Sci. Food Saf. 17 (1), 220–239.

Tardaguila, J., Fernandez-Novales, J., Gutierrez, S., Diago, M.P., 2017. Non-destructive assessment of grapevine water status in the field suing a portable NIR spectrophotometer. J. Sci. Food Agric. 97, 3772–3780.

Timmer, B., Olthuis, W., van den Berg, A., 2005. Ammonia sensors and their applications: a review. Sensors Actuators B Chem. 107, 666–677.

Tremblay, N., Fallon, E., Ziadi, N., 2011. Sensing of crop nitrogen status: opportunities, tools, limitations, and supporting information requirements. Hortic. Technol. 21, 274–281.

Walsh, K.B., Kawano, S., 2009. Near infrared spectroscopy. In: Zude, M. (Ed.), Optical Monitoring of Fresh and Processed Agricultural Crops. CRC Press, Boca Raton, FL, pp. 192–239.

Wang, N.-N., et al., 2016. Recent advances in the application of hyperspectral imaging for evaluating fruit quality. Food Anal. Methods 9, 178–191.

Wang, J., Shen, C., Liu, N., Jin, X., Fan, X., Dong, C., Xu, Y., 2017. Non-destructive evaluation of the leaf nitrogen concentration by in-field visible/near-infrared spectroscopy in pear orchards. Sensors 17, 538.

Westad, F., Marini, F., 2015. Validation of chemometric models: a tutorial. Anal. Chim. Acta 893, 14–23.

Williams, P., Dardenne, P., Flinn, P., 2017. Tutorial: items to be included in a report on a near infrared spectroscopy project. J. Near Infrared Spectrosc. 25 (2), 85–90.

Yang, H.-Y., Inagaki, T., Ma, T., Satoru, T., 2017. High-resolution and non-destructive evaluation of the spatial distribution of nitrate and its dynamics in spinach (*Spinacia oleracea* L.) leaves by near-infrared hyperspectral imaging. Front. Plant Sci. 81937.

Zude-Sasse, M., Fountas, S., Gemtos, T.A., Abu-Khalaf, N., 2016. Applications of precision agriculture in horticultural crops. Eur. J. Hortic. Sci. 81 (2), 78–90.

Further reading

Huang, Y., Lu, R., Chen, K., 2017. Development of a multichannel hyperspectral imaging probe for property and quality assessment of horticultural products. Postharvest Biol. Technol. 133, 88–97.

10

Omics in fruit nutrition: Concepts and application

*Jeanette M. Van Emon**

EVE Sciences, Henderson, NV, United States
*Corresponding author. E-mail: jmvanemon@gmail.com

1 Introduction

Food science in the United States in the early 1900s was geared toward providing a steady food supply to support World War I soldiers. These efforts included developing safe food preservation techniques and expanding agronomic efforts for food exportation. The 1920s saw advances in vitamin research in foods and linking vitamin deficiencies to health issues. Vitamin C was discovered, and the vitamin C content of fruits and vegetables was extensively studied. About this time, chemists isolated other vitamins, and linkages to nutrition and good health were made. Food science in the 1940s had to deal with food rationing and finding substitutes for fats that were in short supply because of World War II. In 1941, US President Franklin Roosevelt convened the National Nutrition Conference for Defense in response to the rising complaints that enlistees were not found physically fit for military duty, which was attributed to poor nutrition. This led to the first nutrition study based on human subjects and introduced the actions of making food policy. Food enrichment activities of adding vitamins and iron to diet staples such as white bread, corn meal, and yeast products began to address health concerns. The idea that protein deficiencies could be addressed through hybrid wheat and corn was being pursued. These enrichment activities were not without their detractors. Up until this time, nutrition was not considered a major selling point for the food industry. Finally, in the 1950s, the idea of nutritional food became a concern. The USDA Handbook No. 8 listed the nutritional contents of over 75 foods, including the many varieties of frozen foods, which were rapidly gaining in popularity. Books also began to appear espousing the nutritional value of individual "health" foods. In 1958, the enrichment of rice was mandated by the congress. However, it wasn't until the White House Conference in December 1969 that nutrition was emphasized with the goal of providing wholesome nutritional food to reduce hunger and

A.K. Srivastava, Chengxiao Hu (eds.)
Fruit Crops: Diagnosis and Management of Nutrient Constraints
https://doi.org/10.1016/B978-0-12-818732-6.00010-1

malnutrition. Interestingly, there was a concern that consumers may become too involved in nutrition leading to nutrition wars within the food industry. Decades later, savvy consumers began to emerge that wanted new and exotic fresh produce from global markets, all year long. This shift in eating patterns spurred the movement for safe fresh produce with more of an emphasis on health. Food products were now definitely being developed with an eye toward supporting wellness and addressing certain health conditions. Specialized food products to address specific concerns such as the nutritional aspects of aging began to emerge. Today, the advancements in nutraceuticals and functional foods are widely recognized as research goals in food science. If food is to be used to improve health, our knowledge of the many aspects of nutrition must be expanded, and the mechanisms of bioactive food components in the body must be determined. Dietary biomarkers need to be identified and quantitated particularly in reference to fluctuations during disease or one's health status. The impact of various nutritional regiments on biological responses must be untangled. These types of studies at the molecular level require the application of multiomic techniques, advanced analytical methods, and in depth data analysis and integration across omic platforms. The term foodomics has become popularized as omics begin to play a key role in understanding the bioactivity of food at the molecular level.

2 Biofortified and nutritionally enhanced food crops

Biofortified crops are poised to make a tremendous impact on populations that are underfed and malnourished (Hefferon, 2015). Maize and cassava have been biofortified with β-carotene and other micronutrients (Mugode et al., 2014). Dietary studies indicated that these biofortified foods increased vitamin A and could be useful in preventing vitamin deficiency. Wheat with increased lysine and reduce gliadins that are responsible for celiac disease has been developed using RNAi-techniques. Products made from this enhanced wheat may help gluten sensitive individuals (Gil-Humanes et al., 2014).

Biofortification can occur through conventional plant breeding or through biotechnology. Using omics, specific genes of interest are identified for either silencing or introducing into a plant of interest from another plant type or even from another species making a transgenic plant. Transgenic plants contain a gene or genes (i.e., transgenes) that have been artificially inserted rather than the plant acquiring them through pollination. Plants that are both nutritionally enhanced and designed to grow under adverse conditions are particularly attractive for areas that have poor agricultural conditions. Fig. 10.1 shows the development of such a transgenic plant.

Much of the omic research to date for biofortification and nutritional enhancements has been for field crops rather than specialty crops such as fruits and vegetables. This is more to do with economic reasons rather than science issues. Examples are given here specifically regarding fruit omic nutrition studies, while examples from other plants illustrate what is possible for fruit given the resources.

3 Omics

The broad discipline of omics encompasses the four major areas of genomics, transcriptomics, proteomics, and metabolomics (Table 10.1). Each omic discipline has been applied to fruit crops. Omics can probe mechanisms in plant physiology and metabolic pathways in plants providing fundamental biological pathways and biochemical processes for cultivar development (Shiratake and Suzuki, 2016). The development of enhanced cultivars requires the manipulation and introduction of traits to improve or enhance plants. Nutritional changes in a plant proteome can be monitored during growth, time of harvest, and postharvest processing to optimize growing conditions and minimize nutrient losses. Proteins have the important functions of being enzymes, receptors, and structural components; thus, changes in the proteome are important indicators of plant health.

Omics can help conventional plant breeders by determining if a desired trait is sufficiently heritable before attempting breeding experiments (McGhie and Rowan, 2012). Metabolomic studies with tomatoes showed that a metabolome analysis can be used for the discovery and monitoring of phenotypic traits including nutritional value (Schauer et al., 2006). Metabolomics will help to discover if processes throughout the food production and commercialization chain affect nutritional qualities (Castro-Puyana et al., 2017). The metabolome can describe the nutritional quality of the crop as it changes during food processing (Kim et al., 2016). Metabolomics can provide information indicative of an individual's current health status, making it applicable for supporting nutritional intervention studies and looking at dietary exposures (Scalbert et al., 2014). The effects of black and green tea consumption on human metabolism were studied using ^{1}H NMR metabolomic profiling on human urine and plasma samples. Green tea drinkers exhibited

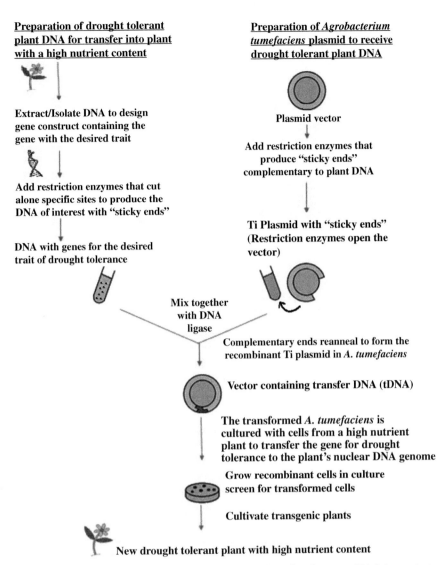

Preparation of drought tolerant plant DNA for transfer into plant with a high nutrient content

Extract/Isolate DNA to design gene construct containing the gene with the desired trait

Add restriction enzymes that cut alone specific sites to produce the DNA of interest with "sticky ends"

DNA with genes for the desired trait of drought tolerance

Preparation of *Agrobacterium tumefaciens* plasmid to receive drought tolerant plant DNA

Plasmid vector

Add restriction enzymes that produce "sticky ends" complementary to plant DNA

Ti Plasmid with "sticky ends" (Restriction enzymes open the vector)

Mix together with DNA ligase

Complementary ends reanneal to form the recombinant Ti plasmid in *A. tumefaciens*

Vector containing transfer DNA (tDNA)

The transformed *A. tumefaciens* is cultured with cells from a high nutrient plant to transfer the gene for drought tolerance to the plant's nuclear DNA genome

Grow recombinant cells in culture screen for transformed cells

Cultivate transgenic plants

New drought tolerant plant with high nutrient content

FIG. 10.1 A simplified description for making a transgenic plant that is both drought tolerant and high in nutrient content. In agriculture, the bacteria Agrobacterium tumefaciens is often used as a vector to deliver genes into plants. The bacteria are parasites with the natural ability to transfer their genes into plants.

TABLE 10.1 Omic terms associated with plant omic nutrition studies.

Omic terms encountered in plant nutrition studies		
Analytical approaches	Omics	Comprehensive analysis of a biological system
	Genomics	Study of the way genes and genetic information are organized within the genome
	Proteomics	Study of protein structure and function
	Transcriptomics	The analysis of gene expression patterns across a wide array of cellular responses, phenotypes, and conditions
	Metabolomics	Large-scale study of metabolites at the cellular, tissue or organism level
	Epigenomics	Genome-wide characterization of reversible modifications of DNA (e.g., DNA methylation)
Applications of omics	Foodomics	Study of food and nutrition through integrated omic approaches
	Nutrigenomics	Systems approach to understand the relationship between diet and health
	Nutriomics	An omics approach for studying molecular links between diet and health OR for the improvement of plant nutrient efficiency
	Microbiomics	Microbial communities associated with various plant structures
	Thiolomics	Molecular mechanisms of the thiol-cascade in plant growth and nutrition

The table differentiates between actual omic techniques and the application of these techniques to further the understanding of plant nutrition.

a stronger increase in urinary excretion of several citric acid cycle intermediates, suggesting a greater effect of green tea on oxidative energy metabolism (Van Dorsten et al., 2006). Metabolomics was used to evaluate Korean black raspberry as an intervention against oxidative stress and inflammation (Kim et al., 2017).

Transcriptomics can be used to look at changes in gene transcription in response to dietary changes. RNA microarrays and sequencing can be used to evaluate the interactions between diet and genes as measured in changes in gene expression. Transcriptomics can help identify and characterize pathways regulated by nutrients (Norheim et al., 2012). Human dietary intervention studies have successfully used transcriptomics to show that diet induces alterations in gene expression (Crujeiras et al., 2008). Transcriptomics puts proteomic and metabolomic nutrition marker data into a larger biological perspective and provides a good first assessment of regulatory networks within plants and humans (Kussmann et al., 2006). Omic data can help the food industry in providing new approaches and opportunities for nutrient-rich foods using nutritional biomarkers (Ordriozola and Corrales, 2015). Omics is providing a deeper understand of plant biology, which initially was obtained on model plant species. However, as the omic fields matures, and particularly as costs decrease, research is moving to food crops including fruits, building upon the successes of omic approaches (Gapper et al., 2014).

4 Transgenic fruit crops

Transgenic crops, commonly referred to as genetically modified (GM) crops, enable breeders to bring favorable genes, often previously inaccessible, into desired cultivars improving their value and providing the advantages for controlling insects and other pathogens (Dias and Ortiz, 2014). Several transgenic fruit crops have been designed for slow ripening to extend shelf life and nutritional value. Other transgenic fruit shows a higher resistance to pests and pathogens and tolerance to herbicides. Many fruit crops like papaya, banana, mango, apple, plum, citrus, and strawberry have increased yields and nutrition as transgenic plants (Rajan et al., 2018). Fruit is widely recognized as contributing to a healthy diet providing vitamins, minerals, dietary fiber, and various phytochemicals. However, it is not clear if some classes of fruit may be more beneficial. Furthermore, it is not known if the response to fruit consumption is dependent upon age, gender, health status, genotype, or gut microbiota (Ulaszewska et al., 2018). An omic approach can help to answer these questions.

An international research project between the United States, France, Poland, Romania, Spain, and the Czech Republic has been directed toward the development of a transgenic plum with resistance to the Plum pox virus (Dias and Ortiz, 2014). A specific gene for the virus protein coat was isolated, sequenced, cloned, and used in a Agrobacterium-mediated transformation of a plum cultivar through genetic engineering (Ravelonandro et al., 1992; Dias and Ortiz, 2014). The transgenic plum is resistant to the Plum pox virus as verified through several years of field testing. An added benefit is that the resistance can be transferred among seedlings through cross breeding. This resistance did not hamper yield or fruit quality. More large cooperative projects will lead to further advancements in transgenic plants including nutritional enhancements and assist with worldwide acceptance.

Fresh tomatoes and processed tomato products are the principal dietary sources of carotenoids including lycopene and β-carotene (Dias and Ortiz, 2014). The levels of carotenoids may be increased through genetic engineering. Transgenic plants containing bacterial genes that affect carotenoid production show increased levels of these compounds without effecting yield or quality. An interesting observation was the change in color of the tomato ranging from orange to red depending on the ratio of lycopene to β-carotene content (Dias and Ortiz, 2014).

Transgenic tomatoes have been developed that can provide the daily adult requirement of folate in one serving. Folate deficiency is a global health issue leading to disease and birth defects. The folate content of the transgenic tomatoes was 25 times higher than control nonenhanced varieties (Díaz de la Garza et al., 2004).

Flavonoid synthetic pathway genes (e.g., stilbene synthase, chalcone synthase, chalcone reductase, chalcone isomerase, and flavone synthase) were introduced into tomatoes (Schijlen et al., 2006). The resulting fruit showed increased levels of flavonoids-flavones and flavonols increasing the antioxidant qualities and nutritional value of the fruit.

Transgenic papaya trees resistant to papaya ringspot virus (PRSV) have led to a significant increase in papaya production in Hawaii (Susuki et al., 2007). Prior to the introduction of PRSV-resistant transgenic papaya, growing papaya in Hawaii was no longer viable despite massive efforts to destroy infected trees. The cultivation of PRSV-resistant transgenic papaya cultivars has reduced the incidence of PRSV to the extent that some areas are able to return to growing nontransgenic papaya for export to countries not allowing GM crops.

Food safety can also be enhanced through transgenic approaches. The first transgenic cassava plants were developed to reduce their cyanogenic content (Siritunga and Sayre, 2004). This has provided health benefits for populations that use this root crop as the basis for their diet as it reduces the amount of exposure to toxic cyanogenic compounds.

Plants are good synthetic chemists and can be genetically engineered to produce biologically active compounds including vaccines and "plantibodies" (Van Emon, 2016). Bioactive compounds produced in transgenic plants to fight cancer and heart disease and to reduce the effects of chronic conditions continue to be developed. Plants have also been coaxed into synthesizing omega-3 fatty acids for dietary benefits (Hefferon, 2015). None of these efforts would be possible without omics.

4.1 Methods of analysis

Paper chromatography was used in the 1950s to determine the concentrations of sucrose, glucose, fructose and other sugars in citrus juices and other foods. This simple technique was augmented in the 1960s by gas chromatography, which conferred a much wider utility for compound detection. Today, food science is supported by much more sophisticated instrumental methods (Castro-Puyana et al., 2017; Herrero et al., 2010, 2012; Yates et al., 2009). Intact proteins can be analyzed in a top-down approach by mass spectrometry (MS). Alternatively, a bottom-up approach starts with a proteolytic digestion followed by separation of the resulting peptides with tandem MS analysis. Search engines are used to match the spectra for peptide identification. These approaches have been successfully applied to the analysis of plant proteomes.

Matrix-assisted laser desorption ionization (MALDI) imaging provides a spatial distribution of proteins and metabolites. This approach has been used for fruit analysis to determine nutrient stability when coupled to two-dimensional gel electrophoresis. Electrophoresis provides a very efficient high-resolution separation capability based on migration differences under the effect of an electric field. The analytical capability of MS coupled with highly resolving separation techniques including high-performing liquid chromatography enables the characterization at the molecular level of a complex biological system (Gallo and Ferranti, 2016).

MS can analyze several thousand proteins in a single sample for use in determining and characterizing posttranslational modifications and protein interactions. Advancements in analytical methods including sample preparation, chromatography, and MS can provide additional capabilities for future nutritional research (Norheim et al., 2012). Advances in MS-based metabolomics methods are driving new applications in omic studies (Hu and Xu, 2013).

MS-based metabolomic studies can be either targeted or nontargeted. Basically, nontargeted measurements use generic sample preparations and chromatography, combined with full-scan mass spectra detection. The MS acquires and stores a massive amount of data acquired using a nontargeted approach. All ions are detected during the entire chromatography run time so there are no limitations to the number of substances that can be detected. Untargeted methods have been developed to monitor hundreds of metabolites at a given time (Gallo and Ferranti, 2016). An untargeted approach using UPLC-ESI-QTOF-MS on lettuce postcutting and 5 days in storage provided a metabolic profile to look at nutrient changes (Garcia et al., 2016). Targeted approaches use analytical standards for the identification of analytes in the sample. This approach is useful to monitor the concentration or appearance of an analyte of interest. Nuclear magnetic resonance (NMR), DNA, RNA, and gene expression arrays and immunochemical methods have found application to omic studies.

4.2 Bioinformatics

Bioinformatics is an indispensable interdisciplinary field that uses computer science and information technology to develop and enhance techniques to make it easier to organize and retrieve and use biological data in an integrated approach. Databases are necessary to gain the full benefits of bioinformatics (Ohlhorst et al., 2013). Omic applications for nutritional enhancements of fruits depend on the application of in depth high-throughput analyses that generate a tremendous amount of data, which can be overwhelming. Many challenges remain in applying a multiomics approach to plant nutrition. Improvements need to be made in bioinformatic capability to accommodate large data sets and present the data such that it can be integrated over several different omic levels (García-Cañas et al., 2012). Fortunately, sequencing technologies, computer infrastructure, and computational biology have been developing at a rapid rate to support omic studies (Gapper et al., 2014).

4.3 Databases

GenBank is an annotated collection of publicly available DNA sequences that is frequently updated. Databases are essential to track and observe trends related to nutrition and health (Ohlhorst et al., 2013). Placing metabolites into metabolic pathways and networks for assignment into functional roles will be enhanced as more metabolites are chemically identified, and the information is added to publicly accessible databases (Scalbert et al., 2011). Databases must be routinely expanded and updated as data become available. Data on phytochemicals including their health-promoting effects have been constructed and reviewed with an emphasis on needs and recommendations for additional databases or improvements to existing ones (Scalbert et al., 2011). Construction of databases are clearly needed that address omic information for nutritional enhancements of fruit.

5 Omics in nutrition

Omics can also play a key role in personalizing nutritional needs of individuals based, in part, on their genetic information (Betts and Gonzalez, 2016). Omics may answer the question of how nutrients modulate gene expression, protein synthesis, and metabolism. This information can be used for plant nutritional enhancements to address specific health and well-being concerns. Omic studies indicate that an individual's gene expression profile is related to dietary patterns (Bouchard-Mercier et al., 2013). Omic-based nutrition research helps to understand the relationships between diet and disease states, malnutrition, and health and may ultimately lead to personalized nutrition and personalized medicine to improve health and combat disease based on a person's genome (Bayram and Gökirmakli, 2018). Eventually, it may be possible to manage dietary intake such that the genes responsible for a particular disorder are not expressed (de Toro-Martín et al., 2017). It may also be possible to determine what nutrients are recommended for an individual based on their genome, or alternatively, what nutrients should be avoided. The proteome can vary due to diet induced posttranslational modifications of proteins, providing an indication of health (Norheim et al., 2012). Gut microbiota have an important role in health and disease. Omics can provide key information on the gut microbiome to study disease onset, progression, and perhaps eventual treatment and cure (Guirro et al., 2018; Khalsa et al., 2017; Portune et al., 2017; Putignani and Dallapiccola, 2016).

Foodomics includes the many rapidly appearing omic terms or subdisciplines put forward by research groups. Nutrigenomics, nutritranscriptomics, nutritional genomics, nutriproteomics, nutrimetabolomics, nutriomics, and many other descriptors are terms found in the literature, which need to be further justified with experimental data. Many of these terms represent conceptual approaches, which overlap, leading to confusion. What is needed are real data to substantiate these approaches and perhaps streamline the everexpanding omic lexicon. For example, nutriomics has been described as an omic approach for studying molecular links between diet and health using an integrated omic approach in the context of systems biology (Manivannan et al., 2013). It is known that foods can alter cellular functions and that there is an association between nutritional status and certain diseases such as diabetes and cardiovascular disorders (Manivannan et al., 2013). The goal of these studies is to understand the response of the body to diet and food components with eventual application to the prevention and management of disease or health disorders (Kato et al., 2011). Another descriptive study for nutriomics is the improvement of plant nutrient efficiency resulting in higher crop yields through genetic and molecular approaches (Yan et al., 2006). Genetic variations can result in differences in nutrient efficiencies within plant species. Studies were conducted regarding the essential minerals of N, P, K, Ca, Fe, Zn, and B looking at the effect of these minerals on plant growth development, reproducibility, and adaptability to harsh environmental conditions (Yan et al., 2006). The goal of these studies is to increase crop production using reduced rates of fertilizer. More efficient use of fertilizers by plants can help preserve water quality and result in more economical crops. As can be seen, the term nutriomics has been applied to different research initiatives. However, the overarching term of foodomics encompasses both avenues of research.

Metabolomics can further the understanding of human responses to dietary phytochemicals (McGhie and Rowan, 2012). Comprehensive metabolomic studies have been used to characterize the metabolic responses of humans to the intake of various food substances. Metabolomic data are useful for determining an in vivo response in near real-time as metabolite changes are typically the first to appear in response to exposure to xenobiotics. Metabolomics is also important in studying adverse outcome pathways and changes induced by nutrients, identifying biomarkers for exposure and food safety (Gibney et al., 2005; Lodge, 2010).

Humans and animals are unable to synthesize essential amino acids such as the branched chain amino acids of methionine, lysine, and tryptophan (Le et al., 2016). Furthermore, several essential amino acids are deficient or completely lacking among many crops used for human and animal consumption. Nutritional improvements using a biotechnology approach of overexpressing genes encoding for proteins that are rich in the lacking amino acids are being explored through transgenic plants (Le et al., 2016). Maize genetically altered to double the lysine content without changing the protein content is recognized by the USDA to be as safe as other varieties bred through conventional methods and has been approved in several countries for commercial use (Le et al., 2016).

Omics will help to determine how specific nutrients interact with genes, proteins, and metabolites to determine or even predict an individual's health. However, connecting these interactions is still far in the future (Capozzi and

Bordoni, 2013). Omics provide information on individualized nutrient requirements, including how nutrients are digested, absorbed, and metabolized, as well as their function in the body. Although the practice of omics in the food industry is not routine, the growing integration of omic disciplines in the field of systems biology will continue to move nutritional studies forward (D'Allessandro and Zolla, 2013).

5.1 Foodomics

An omic approach in studying food science and nutrition has led to a new discipline termed foodomics (Cifuentes, 2009, 2012; León et al., 2018). This relatively new application of omics has already impacted several areas of food science including improving nutrient content in plants to combat health issues (Braconi et al., 2018). Foodomics studies the food and nutrition domains through the application and integration of advanced omics technologies (Herrero et al., 2012). This omic approach enables studies linking food components to health and disease (Cifuentes, 2009). Nutrigenomics support foodomics by addressing the impact of specific nutrients or diets on gene expression. Nutrigenomics is a broad field to study the genome-wide influence of nutrition and the time-dependent response in transcriptomics, proteomics, and metabolomics to describe the phenotype of a biological system (Muller and Kersten, 2003). As with other omic subdisciplines, advances in nutrigenomics are dependent on analytical approaches to study the interrelationship of bioactivity and food (García-Cañas et al., 2010). Nutrigenetics addresses how genetic variability influences the body's response to a nutrient (Norheim et al., 2012). Both omic subdisciplines study the interplay of diet and genes but from different starting points.

Multiomic approaches encompass a holistic approach to address biological problems. Multiomic approaches weave together the answer to complex biological interactions occurring at several levels (e.g., genes, proteins, and metabolites) to study biological mechanisms such as the effect of nutrients on cell health. Outcomes from these studies at a system biology level can further the impact of nutrition on health and wellness. The integration of data from various levels of omics may also differentiate the impact of environmental factors and nutrition in the development of disease or health disorders. Multiomic approach enables the study of the chemical composition of a plant at the molecular level, which may allow an in depth "fingerprint" to assess nutritional value (Gallo and Ferranti, 2016).

Foodomics has also been used for quality assurance and identification of adulterants (Andjelković et al., 2017; Davies and Shepherd, 2012). Where foodomics leads depend not only on advancements in biotechnology, analytical chemistry, and data analysis but also on environmental issues, social implications, and an appetite for change among consumers to embrace GM foods for nutritional improvements.

5.2 Acceptance issues

Genetically modified crops with an increased vitamin or mineral content have a great potential to improve public health, but their availability for consumers is still hampered by negative public opinion. Various GMO crops designed to provide specific health benefits have been developed in which genes, mostly originating from other organisms, have been added. A notable example is Golden Rice, a rice enriched with β-carotene. Golden Rice is a biofortified transgenic crop, which has been genetically modified with extra genes that make a precursor to vitamin A (Ye et al., 2000). According to its developers at the International Rice Research Institute (IRRI), a single helping can supply half the recommended daily intake of vitamin A to prevent blindness and early death in undernourished children. Golden rice was recently judged safe to eat by the US Food and Drug Administration (FDA), as well as by Canada, Australia, and New Zealand. Ironically, the variety is still waiting on regulatory approvals in countries that need the nutrient the most before it can be cultivated there. The IRRI believes that the FDA's stance brings the approval of Golden Rice a step closer to the people that desperately need the nutrient-enriched crop (https://irri.org/news-and-events/news/golden-rice-meets-food-safety-standards-three-global-leading-regulatory-0).

6 Future research trends and needs

Additional research in the development of transgenic plants using more targeted and directed techniques is warranted. Targeted gene editing has become a popular method in genetic engineering. The relatively new gene editing biotechnique of clustered regularly interspaced short palindromic repeats (CRISPR) enables genome editing in plants based on precise incisions, mutations, and substitutions (Gaj et al., 2013). CRISPR is a family of DNA sequences found within the genomes of prokaryotes such as bacteria used in an antiviral defense system. This technique may prove to be revolutionary in the biofortification of plants to improve nutritional qualities. CRISPR has already been applied to a number of crops (e.g., sweet orange, rice, and soybean) resulting in a targeted genome manifesting in plants showing selected enhanced traits.

As the technique is more controlled and precise, then the procedure for developing transgenic plants the acceptance of biofortification through this route may be more acceptable. Since CRISPR is very precise, it can be used to manipulate feedback inhibition mechanisms of biosynthetic pathways such as increasing amino acid production. When combined with conventional plant breeding techniques, the transgenes encoding the editing machinery can be removed, and the now "edited plant" with improved amino acid synthesis can be considered as not being genetically modified (Le et al., 2016). There are other gene editing technologies; however, CRISPR is typically recognized as being more accurate and more efficient at a lower cost.

Another approach for the development of GM plants is cis-genesis. In this process, genes from the plant itself or a close plant relative capable of cross-pollination is used to enhance or introduce the desired traits. As these genes could also be introduced via conventional cross breeding, this procedure may alleviate concerns regarding genetically engineered plants.

Another area that needs to be addressed is the harmonization of terms and the research areas represented by them. There is a tendency to add the omic suffix to exploratory research efforts that do not fully represent an omic approach. The compilation of data in new areas of omic nutrition research will delineate concepts from actual integrated omic approaches supported by omic data. These data should be included in public databases enabling access to other researchers to further plant omic nutrition studies.

7 Summary

Food science is being constantly challenged to optimize crop productivity under increasing environmental stresses. The main challenge is to develop crops with enhanced nutritional value to feed a hungry world. Omics can help with the development of transgenic crops that have high nutrient content and are resistant to biotic and abiotic stresses (Sun and Liu, 2004). The biofortification of staple food crops of rice, maize, and wheat and the development of crops to combat heart disease and cancer can positively impact public health. A biotechnology approach is typically more cost-effective, sustainable, and reliable, and development of certain enhanced plants may not be possible using conventional plant breeding (Hefferon, 2015). However, conventional plant breeding that utilizes nontransgenic approaches will continue to be the mainstay of fruit nutritional enhancements. Transgenic crop cultivars though should be viewed as being capable of contributing to a more nutritious and healthy food supply (Dias and Ortiz, 2014). The amount of transgenic crop production continues to increase in spite of concerns for their effects on human health and the environment. Skepticism among consumers regarding genetically modified food needs to be overcome (Apel, 2010). Approval is especially needed for plants that are developed to be highly nutritious and able to grow in harsh environmental conditions. Such acceptance would particularly help those who have basic food needs.

Several studies show that GM plants may have positive impacts on human health. Through genetic modification, several nutrients may be enhanced in the same crop resulting in a broad positive impact on public health at a relatively low cost. Although GM plants with health benefits are not a panacea for eliminating malnutrition and health disorders, they offer a complementary and cost-effective alternative to other health strategies particularly for developing countries. Vaccines and other bioactive compounds produced in plants can provide relief to populations in developing countries in a cost-effective and efficient manner (Hefferon, 2013).

The differences in food concerns between developed and developing nations have become emphasized by global conditions of drought and famine. Today, consumers in developed nations are concerned with nutritional benefits to improve their health and lifestyle, while those in developing nations are concerned with enough food to eat, as well as basic nutritional needs. A large part of the world's population suffers from nutrient deficiencies, which in part could be remedied by fruit and other plants with a higher nutrient profile, the development of which could be aided through omics.

References

Andjelković, U., Gajdošik, M.S., Gašo-Sokač, D., Martinović, T., Josić, D., 2017. Foodomics and food safety: where we are. Food Technol. Biotechnol. 55 (3), 290–307.
Apel, A., 2010. The costly benefits of opposing agricultural biotechnology. New Biotechnol. 27, 635–640.

Bayram, M., Gökirmakli, Ç., 2018. Horizon scanning: how will metabolomics applications transform food science, bioengineering, and medical innovation in the current era of foodomics? OMICS 22 (3), 177–183.

Betts, J.A., Gonzalez, J.T., 2016. Personalized nutrition: what makes you so special? Nutr. Bull. 41, 353–359.

Bouchard-Mercier, A., Paradis, A.M., Rudkowska, I., Lemieux, S., Couture, P., Vohl, M.C., 2013. Associations between dietary patterns and gene expression profiles of healthy men and women: a cross-sectional study. Nutr. J. 12, 24.

Braconi, D., Bernardini, G., Millucci, L., Santucci, A., 2018. Foodomics for human health: current status and perspectives. Expert Rev. Proteomics 15 (2), 153–164.

Capozzi, F., Bordoni, A., 2013. Foodomics: a new comprehensive approach to food and nutrition. Genes Nutr. 8, 1–4.

Castro-Puyana, M., Pérez-Míguez, R., Montero, L., Herrero, M., 2017. Application of mass spectrometry-based metabolomics approaches for food safety, quality and traceability. TrAC Trend. Anal. Chem. 93, 102–118.

Cifuentes, A., 2009. Food analysis and foodomics. J. Chromatogr. A 1216 (43), 7109.

Cifuentes, A., 2012. Food analysis: present, future, and foodomics. Anal. Chem. 2012, 1–16.

Crujeiras, A.B., Parra, D., Milagro, F.I., et al., 2008. Differential expression of oxidative stress and inflammation related genes in peripheral blood mononuclear cells in response to a low-calorie diet: a nutrigenomics study. OMICS 12, 251–261.

D'Allessandro, A., Zolla, L., 2013. Meat science: from proteomics to integrated omics towards system biology. J. Proteome 78, 558–577.

Davies, H., Shepherd, L., 2012. Integrating omics in food quality and safety assessment. In: Seed Development: OMICS Technologies Toward Improvement of Seed Quality and Crop Yield, Springer, Dordrecht, pp. 555–567.

de Toro-Martín, J., Arsenault, B.J., Després, J.-P., Vohl, M.-C., 2017. Precision nutrition: a review of personalized nutritional approaches for the prevention and management of metabolic syndrome. Nutrients 9, 913.

Dias, J.S., Ortiz, R., 2014. Advances in transgenic vegetable and fruit breeding. Agric. Sci. 5, 1448–1467.

Díaz de la Garza, R., Quinlivan, E.P., Klaus, S.M.J., Basset, G.J.C., Gregory, J.F., Hanson, A.D., 2004. Folate biofortification in tomatoes by engineering the pteridine branch of folate synthesis. Proc. Natl. Acad. Sci. U. S. A. 101 (38), 13720–13725.

Gaj, T., Charles, A., Gersbach, C.A., Barbas, C.F., 2013. ZFN, TALEN, and CRISPR/Cas-based methods for genome engineering. Trends Biotechnol. 31, 397–405.

Gallo, M., Ferranti, P., 2016. The evolution of analytical chemistry methods in foodomics. J. Chromatogr. A 1428, 3–15.

Gapper, N.E., Giovanni, J.J., Watkins, C.B., 2014. Understanding development and ripening of fruit crops in an 'omics' era. Hortic. Res. 1, 14034–14043.

Garcia, C.J., García-Villalba, R., Garrido, Y., Gil, M.I., Tomás-Barberán, F.A., 2016. Untargeted metabolomics approach using UPLC-ESI-QTOF-MS to explore the metabolome of fresh-cut iceberg lettuce. Metabolomics 12, 138.

García-Cañas, V., Simó, C., León, C., Cifuentes, A., 2010. Advances in nutrigenomics research: novel and future analytical approaches to investigate the biological activity of natural compounds and food functions. J. Pharm. Biomed. 51, 290–304.

García-Cañas, V., Simó, C., Herrero, M., Ibáñez, E., Cifuentes, A., 2012. Present and future challenges in food analysis: foodomics. Anal. Chem. 84, 10150–10159.

Gibney, M.J., Walsh, M., Brennan, L., Roche, H.M., German, B., van Ommen, B., 2005. Metabolomics in human nutrition: opportunities and challenges. Am. J. Clin. Nutr. 82 (3), 497–503.

Gil-Humanes, J., Pistón, F., Barro, F., Rosell, C.M., 2014. The shutdown of celiac disease-related gliadin epitopes in bread wheat by RNAi provides flours with increased stability and better tolerance to over-mixing. PLoS One 9, e91931.

Guirro, M., Costa, A., Gual-Grau, A., et al., 2018. Multi-omics approach to elucidate the gut microbiota activity: metaproteomics and metagenomics connection. Electrophoresis 39 (13), 1692–1701.

Hefferon, K., 2013. Plant-derived pharmaceuticals for the developing world. Biotechnol. J. 8, 1193–1202.

Hefferon, K., 2015. Nutritionally enhanced food crops; progress and perspectives. Int. J. Mol. Sci. 16, 3895–3914.

Herrero, M., Garcia-Cañas, V., Simó, C., Cifuentes, A., 2010. Recent advances in the application of capillary electromigration methods for food analysis and foodomics. Electrophoresis 31 (1), 205–228.

Herrero, M., Simó, C., Garcia-Cañas, V., Ibáñez, E., Cifuentes, A., 2012. Foodomics: MS-based strategies in modern food science and nutrition. Mass Spectrom. Rev. 31, 49–69.

Hu, C., Xu, G., 2013. Mass-spectrometry-based metabolomics analysis for foodomics. TrAC Trend. Anal. Chem. 52, 36–46.

Kato, H., Takahashi, S., Saito, K., 2011. Omics and integrated omics for the promotion of food and nutrition science. J. Tradit. Complement. Med. 1 (1), 25–30.

Khalsa, J., Duffy, L.C., Riscuta, G., Starke-Reed, P., Hubbard, V.S., 2017. Omics for understanding the gut-liver-microbiome axis and precision medicine. Clin. Pharmacol. Drug Dev. 6 (2), 176–185.

Kim, S., Kim, J., Yun, E.J., Kim, K.H., 2016. Food metabolomics: from farm to human. Curr. Opin. Biotechnol. 37, 16–23.

Kim, Y.J., Huh, I., Kim, J.Y., Park, S., Ryu, S.H., Kim, K.-B., Kim, S., Park, T., Kwon, O., 2017. Integration of traditional and metabolomics biomarkers identifies prognostic metabolites for predicting responsiveness to nutritional intervention against oxidative stress and inflammation. Nutrients 9, 233–246.

Kussmann, M., Raymond, F., Affolter, M., 2006. OMICS-driven biomarker discovery in nutrition and health. J. Biotechnol. 124, 758–787.

Le, D.T., Chu, H.D., Le, N.Q., 2016. Improving nutritional quality of plant proteins through genetic engineering. Curr. Genomics 17, 220–229.

León, C., Cifuentes, A., Valdés, A., 2018. Foodomics applications. Compr. Anal. Chem. 82 (22), 643–685.

Lodge, J.K., 2010. Symposium 2: modern approaches to nutritional research challenges: targeted and non-targeted approaches for metabolite profiling in nutritional research. Proc. Nutr. Soc. 69 (1), 95–102.

Manivannan, J., et al., 2013. Nutriomics-systems biology of nutrition. In: Essa, M.M., Memon, M.A. (Eds.), Food as Medicine. Nova Publishers, Pullman, WA, pp. 441–456 (Chapter 24).

McGhie, T.K., Rowan, D.D., 2012. Metabolomics for measuring phytochemicals, and assessing human and animal responses to phytochemicals, in food science. Mol. Nutr. Food Res. 56, 147–158.

Mugode, L., Há, B., Kaunda, A., Sikombe, T., Phiri, S., Mutale, R., Davis, C., Tanumihardjo, S., de Moura, F.F., 2014. Carotenoid retention of biofortified provitamin a maize (*Zea mays* L.) after Zambian traditional methods of milling, cooking and storage. J. Agric. Food Chem. 62, 6317–6325.

Muller, M., Kersten, S., 2003. Nutrigenomics: goals and strategies. Nat. Rev. Genet. 4, 315–322.

Norheim, F., Gjelstad, I.M.F., Hjorth, M., et al., 2012. Molecular nutrition research—the modern way of performing nutritional science. Nutrients 4, 1898–1944.

Ohlhorst, S.D., Russell, R., Bier, D., Klurfeld, D.M., Li, Z., Mein, J.R., Milner, J., Ross, A.C., Stover, P., Konopka, E., 2013. Nutrition research to affect food and a healthy lifespan. Adv. Nutr. 4, 579–584.

Ordriozola, L., Corrales, F.J., 2015. Discovery of nutritional biomarker: future directions based on omics technologies. Int. J. Food Sci. Nutr. 66 (Suppl. 1), S31–S40.

Portune, K., Benítez-Páez, A., Del Pulgar, E., Cerrudo, V., Sanz, Y.E., 2017. Gut microbiota, diet, and obesity-related disorders—the good, the bad, and the future challenges. Mol. Nutr. Food Res. 61 (1), 1600252.

Putignani, L., Dallapiccola, B., 2016. Foodomics as part of the host-microbiota-exposome interplay. J. Proteome 147, 3–20.

Rajan, R., Chavda, J.K., Joshi, C.J., Gotur, M., 2018. Prospects of transgenic fruit crops: a review paper. J. Pharmacogn. Phytochem. 7 (2), 2820–2823.

Ravelonandro, M., Monsion, M., Teycheney, P.Y., Delbos, R., Dunez, J., 1992. Construction of a chimeric viral gene expressing plum pox virus coat protein. Gene 120, 167–173.

Scalbert, A., Andres-Lacueva, C., Arita, M., Kroon, P., et al., 2011. Databases on food phytochemicals and their health-promoting effects. J. Agric. Food Chem. 59, 4331–4348.

Scalbert, A., Brennan, L., Manach, C., Andres-Lacueva, C., Dragsted, L.O., Draper, J., Rappaport, S.M., van der Hooft, J.J., Wishart, D.S., 2014. The food metabolome: a window over dietary exposure. Am. J. Clin. Nutr. 99 (6), 1286–1308.

Schauer, N., Semel, Y., Roessner, U., Gur, A., et al., 2006. Comprehensive metabolic profiling and phenotyping of interspecific introgression lines for tomato improvement. Nat. Biotechnol. 24, 447–454.

Schijlen, E., Ric de Vos, C.H., Jonker, H., van den Broeck, H., Molthoff, J., van Tunen, A., Martens, S., Bovy, A., 2006. Pathway engineering for healthy phytochemicals leading to the production of novel flavonoids in tomato fruit. Plant Biotechnol. J. 4, 433–444.

Shiratake, K., Suzuki, M., 2016. Omics studies of citrus, grape and rosaceae fruit trees. Breed. Sci. 66, 122–138.

Siritunga, D., Sayre, R.T., 2004. Engineering cyanogen synthesis and turnover in cassava (Manihot esculents). Plant Mol. Biol. 56, 661–669.

Sun, S.S.M., Liu, Q., 2004. Transgenic approaches to improve the nutritional quality of plant proteins. In Vitro Cell Dev. Biol. Plant 40, 155.

Susuki, Y.I., Tripathi, S., Gonsalves, D., 2007. Virus-resistant transgenic papaya: commercial development and regulatory and environmental issues. In: Punka, Z.K., De Boer, S.H., Sanfaçon, H. (Eds.), Biotechnology and Plant Disease Management. CAB International, Wallingford, pp. 436–461.

Ulaszewska, M., Vázquez-Manjarrez, N., Carcia-Aloy, M., et al., 2018. Food intake biomarkers for apple, pear, and stone fruit. Genes Nutr. 13, 29.

Van Dorsten, F.A., Daykin, C.A., Mulder, T.P.J., Van Duynhoven, J.P.M., 2006. Metabonomics approach to determine metabolic difference between green tea and black tea consumption. J. Agric. Food Chem. 54, 6929–6938.

Van Emon, J.M., 2016. The omics revolution in agricultural research. J. Agric. Food Chem. 64, 36–44.

Yan, X., Wu, P., Ling, H., Xu, G., Xu, F., Zhang, Q., 2006. Plant nutriomics in China: an overview. Ann. Bot. 98, 473–482.

Yates, J.R., Ruse, C.I., Nakorchevsky, A., 2009. Proteomics by mass spectrometry: approaches, advances, and applications. Annu. Rev. Biomed. Eng. 11, 49–79.

Ye, X., Al-Babili, S., Klöti, A., Zhang, J., Lucca, P., Beyer, P., Potrykus, I., 2000. Engineering the provitamin A (beta-carotene) biosynthetic pathway into (carotenoid-free) rice endosperm. Science 287, 303–305.

Further reading

Pérez-Massot, E., Banakar, R., Gómez-Galera, S., et al., 2013. The contribution of transgenic plants to better health through improved nutrition: opportunities and constraints. Genes Nutr. 8, 29–41.

Remely, M., Stefanska, B., Lovrecic, L., Magnet, U., Haslberger, A.G., 2015. Nutriepigenomics: the role of nutrition in epigenetic control of human diseases. Curr. Opin. Clin. Nutr. Metab. Care 18 (4), 328–333.

Zhang, Y., Xu, X., Zhou, X., Chen, R., Yang, P., Meng, Q., Meng, K., Luo, H., Yuan, J., Yao, B., et al., 2013. Overexpression of an acidic endo-ß-1,3-1,4-glucanase in transgenic maize seed for direct utilization in animal feed. PLoS One 8, e81993.

11

Leaf analysis as diagnostic tool for balanced fertilization in tropical fruits

Renato de Mello Prado[a],*, *Danilo Eduardo Rozane[b]*

[a]São Paulo State University—UNESP, Jaboticabal, Brazil
[b]São Paulo State University, UNESP, Registro, Brazil
*Corresponding author. E-mail: rmprado@fcav.unesp.br

1 Introduction

For any culture to be able to manifest its full genetic potential by producing any food, it is necessary to optimize the vital factors such as climate, genotype, soil, light, water, temperature, and nutrients. Plants require that adequate amounts of nutrients fulfill their functions in the plant metabolism throughout their life cycle. It is noteworthy that the determining factors for the profitability of the agricultural enterprise are product price; production cost; and, the most important, productivity/yield, which in turn is mostly determined by the nutritional status of the crop. It becomes necessary to know whether the plant or the crop is well nourished or not because the adequate nutritional state is a determining factor on the production/yield of any crop. Therefore, to achieve optimal plant nutrition by avoiding either the lack or excess of nutrients, soil analysis is used as a criterion for recommending correctives and fertilizers, as well as the plant itself as a diagnosis subject. Thus, the nutritional status of plants can be evaluated using soil and/or plant chemical analysis (leaves) as well as nonroutine alternative methods such as sap analysis and biochemical tests. The usefulness of leaf diagnosis depends on leaf sampling, chemical analysis of plant tissue, and the correct interpretation of the results. The interpretation step consists of comparing the standards obtained from productive plants with adequate nutritional balance with the sample to be evaluated using uni-, bi-, or multivariate methods to generate the diagnosis indicating sufficiency, deficiency, or toxicity of a nutrient. The obtained data are then used to correct fertilization management, thus promoting the increase of crop productivity.

To this end, the objective of this study was to conduct a literature review focusing on the importance of leaf diagnosis and its main aspects for the adequate diagnosis aiming at better fertilization management of fruit crops.

A.K. Srivastava, Chengxiao Hu (eds.)
Fruit Crops: Diagnosis and Management of Nutrient Constraints
https://doi.org/10.1016/B978-0-12-818732-6.00011-3

2 Basic aspects of foliar diagnosis

Leaf diagnosis is performed in two steps, first by analyzing the plant visual symptoms and second via a chemical analysis of the plant itself (Prado, 2008).

2.1 Visual diagnosis

The visual method requires ensuring that the problem in the field is caused by the lack or excess of a nutrient, since the incidence of pests and diseases, among others, can be "masked" by the fact that the disease symptoms are sometimes very similar to nutritional problems. Thus, nutritional disorder symptoms usually have the following characteristics:

2.1.1 Dispersion

The "nutritional problem" occurs homogeneously in the field, whereas diseases and/or pests, for example, occur in isolated plants or seedlings.

2.1.2 Symmetry

In a pair of leaves, the nutritional disorder occurs in both leaves.

2.1.3 Gradient

The disease symptoms in a branch or plant respect a gradient, that is, the symptoms aggravate from the old to the new leaves or vice versa.

Thus, the symptoms of lack/excess observed in the visual diagnosis may vary with cultures. Typically, the lack symptoms occur in old or new leaves (for mobile nutrients) or in shoots (poorly mobile nutrients), but it can still be visualized at the root, characterizing different types of symptoms. Thus, the visual symptoms of nutritional deficiency can be grouped into six categories: (a) reduced growth, (b) uniform or stained chlorosis in leaves, (c) interveinal chlorosis, (d) necrosis, (e) purplish coloration, and (f) deformations.

Although with certain limitations, the visual diagnosis allows evaluating the symptoms of the lack or excess of nutrients quickly, becoming possible to correct fertilization problems right away. However, this method is criticized for several limitations:

- In the field, the plant is susceptible to interference from agents (pests and pathogens) that may mask the detection accuracy of the nutrient problem, as previously stated.
- In the field, the lack of nutrients' symptoms may be different from those described in specialized publications since such studies illustrate "severe" symptoms of nutritional disorder, whereas such symptoms may be "mild" in the field.
- The lack of a certain nutrient symptom may differ in different cultures, so that the knowledge for a given species may not be valid for another. For example, the lack of Zn symptoms in fruit can appear as small leaves, while in corn the new leaves show bleaching.
- The lack symptoms of different nutrients may be/appear the same in different crops.
- There may be a level of nutrient deficiency that reduces production without the plant developing any symptoms.
- The simultaneous lack of two or more nutrients makes it impossible to identify the deficient nutrients.
- The excess of one given nutrient can be confused with the lack of another nutrient.
- Proper use of the visual diagnosis technique requires technicians with significant experience in the culture of the region.
- In addition, the visual diagnosis does not quantify the deficiency or excess level of the studied nutrient.

It is noteworthy that the visual symptoms characteristic of the lack or excess of nutrients manifest themselves clearly and can be differentiated only when the plant presents an acute nutritional disorder. However, at this point, a significant part of the production (about 40%–50%) is already compromised, as a series of physiological damages have already been triggered and the visualization of the symptoms shows irreversible tissue damage at this stage. Therefore, the use of visual diagnosis should not be the rule, but rather a complement to the diagnosis.

2.2 Foliar diagnosis

A method with fewer limitations than the visual diagnosis and the most common method used to monitor the nutritional status of cultures is the leaf diagnosis, which consists of the chemical analysis of the leaves. Lagatu and Maume (1934) first proposed using the plant nutrient content to indicate their nutritional status in the 1930s, and although the chemical analysis technique is relatively old, it is still rarely used by Brazilian farmers (Prado, 2008).

The leaf diagnosis consists of evaluating the nutritional status of a plant by comparing a sample of the plant tissue with a preestablished standard. The standard is a plant that has all nutrients in adequate proportions and, therefore, capable of providing favorable conditions for the plant to express its maximum genetic potential for production.

The leaf tissue is the most commonly used for the analysis because this organ is the site of plant metabolism. Even though other plant tissues may be used, such as part of the leaf (petiole) or fruits, the nutrients do not accumulate evenly in these tissues, especially the immobile or slightly mobile nutrients, so it is unlikely that this organ adequately reflects the nutritional status of all macro- and micronutrients in the crop.

The leaf chemical analysis can be interpreted using a single nutrient by means of the critical level method or the sufficiency range or alternatively, based on the relationship between nutrients given by the diagnosis and recommendation integrated system (DRIS). Therefore, several tools can be used and, preferably, an integration of two or more tools to provide enough data on the soil-plant system and a nutrition diagnosis so that, when appropriate, fertilization practices can be corrected and/or become more efficient.

Leaf diagnosis identifies the nutritional status of a plant via chemical analysis of the plant leaf tissue, which is more sensitive to showing plant nutrient variations and the center of the physiological activities. The leaf analysis as a diagnostic criterion assumes that there is a correlation between nutrient supply and leaf nutrient levels, whose increase or decrease in concentrations is related to higher or lower yields. The nutrient levels in the plant result from all factors that have interacted to affect it so that, to interpret the results of chemical analysis of plants, it is necessary to know the factors affecting nutrient concentrations, standardized sampling procedures, and pertinent relationships (basic assumptions for using leaf diagnosis):

(a) Soil nutrient supply × production so that production should be higher for a more fertile soil compared with less fertile soil.
(b) Soil nutrient supply × leaf content, as the nutrient supply in the soil increases, the plant leaf content also increases.
(c) Leaf content × production/yield, the increase of leaf nutrient would explain the increasing production.

Specifically, the relationship between leaf content and production has several phases or areas such as the following:

2.2.1 Deficient zone or range

This stage shows clear symptoms of deficiency and occurs in soils (or substrates) with high deficiency of the nutrient supplied by the doses (still insufficient).

2.2.2 Transition zone or range

Although there are nonvisible deficiency symptoms (hidden hunger) at this stage, there is a direct relationship between leaf content and yield.

2.2.3 Luxury consumption zone or range

At this stage, increasing nutrient concentration does not result in increased yield, a fact that is observed in nutrient-deficient soils that receive nutrient doses.

2.2.4 Toxicity range

At this stage, the nutrient content increases significantly while decreasing yield. Thus, when a given nutrient content results in a drop between 5% and 20% of the maximum production, it is interpreted as a toxic level.

Research studies in the literature investigate critical deficiency level the most because it corresponds to the concentration below which the growth rate (production/yield or quality) decreases significantly and production is not economical above it.

After reaching maximum yield, a further increase in nutrient concentration in the leaf no longer results in increased production, and thus, the plant reaches the stage of "luxury consumption." It is noteworthy that, in the luxury consumption phase, nutrients accumulate in the vacuoles of the cells and may be released gradually to meet any nutritional requirements of the plants. After luxury consumption range, the nutrient increase in the vegetal tissue can lead to decreasing production/yield, characterizing the zone of toxicity.

It is emphasized that the leaf diagnosis should be performed when the plant is at peak physiological activity, such as during flowering or beginning of fruiting. Therefore, this requirement for performing the analysis at the plant developmental peak makes leaf diagnosis of little use for correcting a nutrient deficiency in annual crops during the crop production cycle. However, leaf diagnosis has a high potential for diagnosing the nutritional status of perennial crops such as citrus, since it is possible to make the correction with satisfactory efficiency in the same agricultural year.

It is also emphasized that foliar diagnosis has the advantage of using the plant itself as an extractor while having several applications:

- To assess nutrient requirements and nutrients exported in crops
- To identify deficiencies that cause similar symptoms, making visual diagnosis difficult or impossible
- To evaluate the nutritional status and assist the fertilization management programs

Foliar/leaf diagnosis is a direct method of evaluating the nutritional status of the crops since it uses the nutrient content present in the plant. On the other hand, there are indirect methods to evaluate the organic compound level or enzyme activity, in which the nutrient is present in the organic compound or is activator of this enzyme, that is, N-deficient plant should present low amount of chlorophyll or low nitrate reductase activity (NO_3^- induces the enzyme as it is the substrate thereof). In this sense, Malavolta et al. (1997) described biochemical tests for several nutrients to evaluate the nutritional status of the plant, for example, N (reductase activity, glutamine synthetase, N-amide, and asparagine), P (fructose-1,6-2P and photosynthesis, phosphatase activity), K (amide and pipecolic acid contents and putrescine content), Mg (pipecolic acid), S (glutaraldehyde reaction; free amino acids), Mn (peroxidases, chlorophyll a/b ratio), B (ATPase activity), and Zn (ribonuclease, carbonic anhydrase, and arginine content). In the case of P, other studies indicate that P in vacuole cells may indicate the nutritional status of the plant (Bollons and Barraclough, 1997).

3 Criteria for leaf sampling

It should be reminded that, for obtaining valid results, the leaf analysis requires comparing the studied leaf sample with a previously determined standard. There are variations between species and intraspecies, making it difficult to generalize the standards. Thus, the strict control of these criteria is the guarantee of a valid result for the chemical analysis of the leaf and its interpretation so that future fertilizations can correct the detected nutrient deficiency. Also, most errors that can compromise a fertilization management program occur at the sampling stage, such as poor sampling, and not from laboratory analytical problems or using inappropriate recommendation tables.

For the correct sampling of the leaf diagnosis itself, it is necessary to consider a few criteria specific to each culture:

- Leaf type
- Sampling time/season
- The number of leaves per field/plot

Usually, the most appropriate **leaf type** is newly mature and fully developed since it should be able to reflect the true nutritional state of the plant without being affected by the redistribution of nutrients. This leaf is known as the diagnosis or index leaf. The leaf needs to be standardized because the older leaf has a high concentration of little mobile nutrients (Ca, S, and micros), while younger leaf has a higher concentration of mobile nutrients (N, P, and K). Chadha et al. (1980) evaluated the leaves of mango trees and reported greater stability or balance between macro and micronutrients in leaves aged 6–8 months (leaves that were neither young nor old, Table 11.1) while the application of nitrogen fertilizer increased N in older leaves only.

The diagnostic leaf should clearly discriminate the nutrient concentrations in the plant tissue as to express deficient, adequate, and toxic levels. In addition, nutrients cannot change abruptly with the sampling time. Thus, the diagnostic leaf must be the one that provided the critical level/appropriate range for the standards of the Tables for the respective crop.

The **sampling time** should occur at defined physiological stages since the nutrients may vary with plant age. This means that plant analysis data should be calibrated with plant age.

Normally, it is assumed that the sampling time should coincide with the higher physiological activity of the plant (or photosynthetic activity), often occurring during the plant reproductive phase when the nutrient concentration is higher. Natale et al. (1994) reported that the best correlation (R^2 value) between nitrogen fertilizer and leaf N doses was obtained at flowering compared with fruiting.

TABLE 11.1 Effect of age on the nutrient concentration of mango tree leaf.

Leaf age Month	N g/kg	P	K	Ca	Mg	S	Zn mg/kg	Cu	Mn	Fe
1	12.8	1.52	11.07	9.1	2.0	0.88	20	12	27	105
2	11.8	1.18	9.8	10.8	2.9	0.81	28	11	32	153
3	11.9	0.98	8.1	12.2	3.2	1.05	28	11	46	171
4	11.7	0.90	7.7	13.1	3.4	0.88	14	8	46	129
5	12.0	0.84	8.1	14.0	3.5	1.14	15	12	54	193
6	11.7	0.73	7.0	15.9	3.3	1.13	13	11	63	156
7	11.7	0.73	6.4	16.7	3.3	1.14	13	10	63	154
8	11.7	0.73	5.8	17.2	3.3	1.15	12	12	78	169
9	11.6	0.66	5.7	18.8	3.1	1.13	17	21	100	143
10	12.8	0.73	4.8	19.1	3.4	1.19	22	22	87	108
11	12.9	0.70	5.4	20.7	3.3	1.39	15	14	112	145
12	13.0	0.77	4.2	21.2	3.7	1.32	50	17	100	182

In any case, the leaf sampling of a particular crop should coincide with the time of the diagnostic leaf collection, the one used to establish the critical level/appropriate range or the standard for the study.

Usually, between 25 and 100 leaves per sample are used. Sampling error usually decreases when using a high number of leaves together with a high number of plants (Holland et al., 1967). In the literature, there are recommendations of leaf sampling for various crops; however, research-supported sampling standards to reflect the nutritional status of the plant adequately have not yet been defined for some crops.

Quaggio et al. (1997) stated the criteria for sampling leaves from the following crops:

Acerola—Fully expanded young leaves should be collected from the fruiting branches around the perimeter of 50 plants per trees.

Apple—Collect 4–8 freshly expanded leaves. Sampling 25 plants per field, totaling 100 leaves.

Avocado—Newly expanded leaves aged between 5 and 7 months should be sampled/collected from the average height of the crowns of 50 trees in February/March.

Banana—Remove the central 5–10 cm of the third leaf from the inflorescence, eliminating the midrib and peripheral halves. Sample 30 plants.

Cashew—Collect 4 mature leaves (fourth leaf) from new growths per tree on 10 plants/trees in productive orchards.

Citrus—Collect the third leaf from the fruit, produced in the spring (aged 6 months), in branches with fruits (2–4-cm diameter). Sample 4 leaves per plant (25 plants per field).

Fig—Collect freshly and fully expanded leaves from the medial portion of the branches 3 months after budding. Sample 25 plants per field, totaling 100 leaves.

Grape—Collect the youngest newly mature leaf, counted from the apex of the vine branches, sampling a total of 100 leaves.

Guava[a]—Collect the third pair of leaves with the petiole, from the end of the branch, 1.5 m from the ground. Sample 4 pairs of leaves per tree, in 25 plants per plot.

Mango—Collect leaves from the middle of the last flow of vegetation branches with flowers at the end. Sample 4 leaves per tree, 20 plants per field.

Papaya—Collect 15 petioles of young leaves, fully expanded and mature (17th to 20th leaves from the apex), with a visible flower in the axil.

Passionfruit—Collect the third leaf from the apex of unshaded branches or leaf with a floral button about to open in the axil, in autumn. Sample 20 plants.

Peach—Collect 26 freshly expanded leaves from the middle portion of the branches. Sample 25 plants per field.

Pineapple—Fresh ripe "*D*" leaf (fourth leaf from the apex) should be collected before the floral induction. Cut the leaves (1 cm wide) as to eliminate the basal portion without chlorophyll.

    Starfruit[b]—Collect the 6th leaf with the petiole, in branches with flowers, between August and October, sampling 30 leaves per plot.

    Strawberry—Collect the third or fourth newly developed leaf (without petiole), at the beginning of flowering. Sample 30 plants.

    Watermelon/melon—Collect the 5th leaf, excluding the apical tuft from the tip, in the period between 1/2 and 2/3 of the plant life cycle. Sample 15 plants.

[a]Natale et al. (1996).
[b]Prado and Natale (2004).

To obtain representative samples, the leaf sampling in the field requires sampling techniques like those followed for soil sampling to circumvent the heterogeneity that may occur on the farm. Below are some general guidelines for leaf sampling:

- Zigzag trekking
- Level trekking
- Avoid nearby plants or roads

In addition, leaf sampling should not be conducted under the following conditions:

- Plants showing signs of pests and diseases.
- Plots that were fertilized or treated with agrochemicals (herbicide, pesticide, etc.) less than 30 days before.
- Different varieties, because the nutritional status is influenced by the genetic factor, a fact widely reported in the literature for several cultures.
- In the case of perennial grafted crops, do not mix leaves of plants that had a different crown or rootstock/rhizome, as they influence nutritional status, such as the yellow passion fruit crown (Prado et al., 2005).
- Leaves of different ages should never be mixed.
- In the case of perennial crops, the same sample must not contain leaves of productive and nonproductive branches.
- Dead or damaged (mechanical) tissues.
- Avoid collecting leaves after high rainfall since some nutrients may have undergone leaching, such as N and K.

Finally, when dealing with an isolated problem in a given culture such as the need to evaluate which nutrient is causing a particular symptom of a deficiency in any plant; it is recommended to remove and store separately leaf samples with heavy, mild, and no symptoms. The sampled leaves should be of the same age and collected from the same position in the plant.

4 Sampling preparation and chemical analysis

Sample preparation and chemical analysis constitute an important step in foliar diagnosis studies and are usually performed in the Plant Nutrition laboratory.

Field leaf sampling should be followed by a few immediate procedures such as (Malavolta, 1992) the following:

(a) If the sample can be in the laboratory within 2 days after the collection, place the leaves in paper bags and send to the laboratory.

(b) If the sample takes longer than 2 days after collection to reach the laboratory, the leaves need to be washed in "clean" running water, using detergent solution (0.1%) and water; dried in a regulated oven at approximately 70°C or in full sun (to interrupt leaf breathing); placed in paper bags; and sent to the laboratory.

Then, upon arriving in the laboratory, the leaf sample is treated as follows:

1. The sample is tagged with an identification number.
2. Fresh leaves should be washed with "distilled" running water, detergent solution (0.1%), hydrochloric acid solution (0.3%), and deionized water and dried. The detergent solution is used to eliminate the remaining soil and Fe contamination, while the acid should remove metals previously applied in foliar fertilization (e.g., Zn).
3. Samples should be dried as soon as possible to minimize changes, both biological and chemical. Excess water is removed by dripping; place the samples into paper bags to dry in temperatures ranging from 65°C to 70°C (Bataglia et al., 1983), in a forced air circulation oven, until constant weight, approximately 48–72 h.

4. The dried leaves are then pulverized in a mill as to obtain a thin and homogeneous material for analysis. The used mill should have stainless steel or plastic chambers to avoid contamination of the plant material with micronutrients, such as Fe and Cu.
5. The ground leaves should be stored in properly labeled paper bags until further analysis.
6. Still, in the laboratory, the sample to be analyzed is submitted to different procedures for the chemical analysis as weighing, obtaining the extract, and identification of the nutrient.

5 Leaf diagnosis of fruit trees

As previously mentioned, different tools such as foliar diagnosis (critical level/adequate range or DRIS/CND), visual diagnosis, and soil chemical analysis can be used to evaluate the nutritional status of the crops.

5.1 Conventional method (critical level or adequate range)

Leaf diagnosis to evaluate the nutritional status of plants requires following a few procedures, such as leaf sampling, material/sample preparation, chemical analysis in the laboratory, and obtaining the analytical results. The results may be used for defining the critical nutrient levels and preparing the tables of adequate levels or the standards. In the case, the standard tables have already been prepared, the sampled leaves are compared with the standards, and the appropriate interpretations should indicate whether the nutrient levels are adequate, deficient, or excessive. Therefore, a diagnosis of the nutritional status of the crops is presented and should be used to recommend either a fertilization management or adjustment, with is expected to reflect on the productivity/yield and profitability of the studied farm.

There are Tables in the literature indicating the adequate nutrient levels for cultures of high growth and production worldwide. Table 11.2 presents the adequate macro- and micronutrient leaf contents for several crops (Quaggio et al., 1997).

Therefore, the results of the chemical analysis of a particular sample are compared with the standards (Tables), and there are three possible outcomes (Prado, 2008):

1. The sample content is less than the standard, indicating possible deficiency.
2. The sample content is equal to the standard, indicating neither deficiency nor toxicity.
3. The sample content is higher than the standard, indicating possible toxicity.

The adequate levels or standards shown in the Tables are sometimes obtained by extrapolating the values found for plants grown in other countries to Brazil, and therefore, errors may occur due to different soil and climatic conditions. Another possibility is to obtain the standards in local experiments conducted in Brazil, which are the most accurate to guarantee the highest yields when using as reference for the leaf diagnosis.

The standards are determined in field experiments, in which more than three doses of a given nutrient are commonly used, in the presence of sufficient doses of the other elements. Thus, we obtain the relationships between the following:

1. Dose (nutrient supplied by soil) × production/yield
2. Dose (nutrient supplied by soil) and/or soil nutrient content × leaf content (or production/yield)

5.2 Leaf content × production/yield

Thus, the relationship between foliar content versus yield allows to establish the adequate levels of nutrients in plants, and again, the literature has these relationships for various nutrients and cultures. A coefficient of variation or experimental error, common in field experiments, is associated with increased nutrient (or foliar content) and yield ratios, assuming adequate levels slightly above the critical level according to a sufficiency range (SR $\sim$ 100% of maximum production), so that the future fertilization recommendation guarantees the expected productivity. Thus, depending on the author, the interpretation of low, average, adequate, and high foliar nutrient content is associated with relative production <70%, 70%–90%, 90%–100%, and >100%, respectively.

Furthermore, the sufficiency range is the most adequate to interpret the results of the plant nutritional state compared with the point value such as the critical level since, in addition to the coefficient of variation obtained in the calibration curve, the cultures show genetic variations due to diversity of variety and/or hybrids/clones. Therefore,

TABLE 11.2 Adequate macro- and micronutrient ranges in leaves of several fruits grown in Brazil.

Crop	N (g/kg)	P	K	Ca	Mg	S
Pineapple	15–17	0.8–1.2	22–30	8–12	3–4	–
Avocado	16–20	0.8–2.5	7–20	10–30	2.5–8	2.0–6.0
Acerola	20–24	0.8–1.2	15–20	15–25	1.5–2.5	4.0–6.0
Banana	27–36	1.8–2.7	35–54	3–12	3–6	2.5–8.0
Fig	20–35	1.0–3.0	10–30	30–50	7.5–10	1.5–3.0
Guava[a]	20–23	1.4–1.8	14–17	7–11	3.4–4.0	2.5–3.5
Orange	23–27	1.2–1.6	10–15	35–45	2.5–4.0	2.0–3.0
Apple	19–26	1.4–4.0	15–20	12–16	2.5–4.0	2.0–4.0
Papaya	10–25	2.2–4.0	33–55	10–30	4.0–12.0	–
Mango	12–14	0.8–1.6	5–10	20–35	2.5–5.0	0.8–1.8
Passion fruit	42–52	1.5–2.5	20–30	17–27	3.0–4.0	3.2–4.0
Peach	30–35	1.4–2.5	20–30	18–27	3.0–8.0	1.5–3.0
Grape	30–35	2.4–2.9	15–20	13–18	4.8–5.3	3.3–3.8

Crop	B (mg/kg)	Cu	Fe	Mn	Mo	Zn
Pineapple	20–40	5–10	100–200	50–200	–	5–15
Avocado	50–100	5–15	50–200	30–100	0.05–1.0	30–100
Acerola	25–100	5–15	50–100	15–50	–	30–50
Banana	10–25	6–30	80–360	200–2000	–	20–50
Fig	30–75	2–10	100–300	100–350	–	50–90
Guava[a]	20–25	20–40	60–90	40–80	–	25–35
Orange	36–100	4–10	50–120	35–300	0.1–1.0	25–100
Apple	25–50	6–50	50–300	25–200	0.1–1.0	20–100
Papaya	20–30	4–10	25–100	20–150	–	15–40
Manga	50–100	10–50	50–200	50–100	–	20–40
Passion fruit	40–60	5–20	100–200	100–250	1.0–1.2	50–80
Peach	20–60	5–16	100–250	40–160	–	20–50
Grape	45–53	18–22	97–105	67–73	–	30–35

Obs.: These values are valid only for cultures whose leaf sampling procedures were performed according to Quaggio et al. (1997). Shaded value 20–23 mg/kg indicates adequate nutritional status for nitrogen.
[a] Refers to cv. Paluma (Natale et al., 2002).
Modified from Quaggio, J.Á., Van, R.B., Piza Jr., C.T., 1997. Frutíferas. In: Van, R.B., Cantarella, H., Quaggio, J.Á., Furlani, A.M.C. (Eds.), Recomendações de adubação e calagem para o Estado de São Paulo. second ed. rev. IAC, Campinas, pp. 121–125 (Boletim Técnico, 100).

the leaf nutrient content is best interpreted using the sufficiency range method because it is less affected by the varying environmental conditions and the variations of the plant itself.

Another important aspect of adequate nutrition, besides increasing yield, is fruit quality. Normally, fruit quality can be influenced by genetic aspects, cultural traits, climatic conditions, and especially plant nutrition.

There are a few works on mango trees in the literature that investigated the relationship between nutritional imbalance and problems in fruit quality. In a literature review on the internal collapse of mango fruits, Prado (2004) concluded that the nutritional factor was important to explain the problem, reporting that an adequate Ca level provided by liming, irrigation, and moderate nitrogen application can reduce nutritional imbalances and improve fruit quality.

Schaffer et al. (1988) evaluated the nutrition of the "Tommy Atkins" mango trees that displayed symptoms of decline (a not-yet-defined disease) using the DRIS diagnostic method and reported higher nutrient imbalance in orchards with a high percentage of declining trees compared with those without the symptoms. The nutritional imbalance was associated with low Mn and Fe and high Mg contents in the leaves, while DRIS also indicated P deficiency; however, P concentration was above the critical value.

Finally, a balanced nutrient management has also been shown to be a valid and efficient alternative to control certain plant diseases; however, further research under the Brazilian conditions is necessary, to know/understand better the nutritional requirements, disease behavior in different levels, and nutrient sources and combination (Zambolim and Ventura, 1996).

5.3 DRIS and CND methods

The calibration experiments provided a range of responses between deficiency and sufficiency treatments, which facilitated establishing nutrient critical limits. Even in well-conducted repeated experiments, it is very rare that all nutrients are in optimal concentration or even close to that, so when establishing the relationship between productivity and nutrient variability the limiting nutrient is modeled more accurately when all other nutrients are enough and toxicity is avoided (Munson and Nelson, 1990).

Generally, a higher coefficient of variation is observed when developing standards for nutrients from research results compared with the data collected in surveys of commercial fields in the same place. Another disadvantage of the experiments is that they include luxury consumption cases, which influence optimal ranges calculated as confidence intervals over averages or using boundary lines to determine nutrient ranges (Webb, 1972).

Brunetto et al. (2007, 2008) conducted experiments with vines and warned that leaf analysis has limitations as a diagnostic tool for nutritional status since the N content in the leaves increased with increasing N doses applied to the soil but this N increase did not increase grapevine productivity. Therefore, the foliar concentration exceeded the value required by its metabolic functions, contributing to the occurrence of interpretation errors regarding the culture nutritional requirements. Parent and Dafir (1992) point out that the nutrient content in the plant tissue, which expresses the values necessary for the diagnosis, can be affected by the dilution and concentration effects caused by the alteration in the dry matter produced by the plants. Therefore, models for fertilization prediction, based only on total plant nutrient contents, must be interpreted with caution.

An alternative to the calibration experiments would be the use of nutritional monitoring information obtained from commercial plots since these data aggregate information from a wide variety of environments and, therefore, cannot be used for the determination of response curves, such as those obtained in the calibration experiments. However, there is the possibility of using bivariate methods, such as the diagnosis and recommendation integrated system (DRIS), proposed by Beaufils (1973), and the multivariate compositional nutrient diagnosis (CND) presented by Parent and Dafir (1992).

The DRIS method assumes that the binary relationships (between two nutrients) are a good indicator of the nutritional balance; however, in the literature, nutrient interrelationships are often reported by mistakenly assuming that pairs of elements are linearly correlated (Kenworthy, 1967). Nutrient relationships are also affected by interactions with other factors such as genotype, growth phase, planting date, seasonal fluctuations, and variations in fertilization (Geraldson et al., 1973). Optimal nutrient relationships alone are considered insufficient since they can be obtained in the deficiency or toxicity context (Marschner, 1986). Considering all dual relationships simultaneously for diagnostic purposes is impractical due to their large number: $(D \times (D-1))/2$, the dual relations can be calculated from a composition of D-parts. Most dual relations are redundant, such as $(K/Ca) \times (Ca/Mg) = (K/Mg)$. However, it is possible to use proportions such as the Redfield N/P (Redfield, 1934; Güsewell, 2004) and the stoichiometric ratios C-N-S-P for soil organic matter stabilization (Stevenson, 1986). The DRIS method is considered to be less affected by dilution and concentration effects because it takes into account the balance between nutrients in the nutritional diagnosis process. As a result, the DRIS standards are less affected by the local conditions compared with the standards generated by calibration curves.

The advantage of the DRIS and CND methods over the univariate and bivariate methods is the possibility of identifying the level of nutritional limitation (Bataglia et al., 1996; Parent and Dafir, 1992) while allowing to group the nutrients in classes varying from the most limiting due to deficiency to those in excess (toxicity). The DRIS method defines the reference population broadly (Rozane et al., 2016), with some arbitrariness when choosing the high-yield subpopulation. The DRIS interpretation is based on each nutrient index, obtained as a function of the ratios (quotients) between nutrient contents so that positive and negative values indicate excess and deficiency, respectively, whereas values equal or close to zero indicate nutritional balance.

Holland (1966) observed that diagnosis consistency increases as the number of examined nutrients also increases simultaneously until the method is extended to all nutrients, thus indicating multivariate analysis as a tool for diagnosing plant nutritional status. The CND technique evolved and was improved by using the relationships between one nutrient content and the geometric mean of other dry matter components (multivariable relations), indicating a more complete nutritional balance. The method uses the log-centered ratio transformation (Aitchison, 1986) and compositional data analyses, such as nutrient contents, being one of the most recent methods for interpreting plant tissue analysis, including those not analytically determined, and considered the best expression of plant tissue balance (Parent and Dafir, 1992; Egozcue and Pawlowsky-Glahn, 2005).

The CND and DRIS concepts differ regarding the correction factor. CND determines one correction factor for any nutrient by considering all nutrients in the analysis (multinutrient), while DRIS determines a simple correction factor for the nutrient ratios between pairs. Thus, the CND expands the DRIS concept from a two-dimensional analysis to a multidimensional one (Rozane et al., 2016). Parent et al. (2012) also indicated that the isometric log ratio (*ilr*) representation of composite data, such as coordinates in Euclidean space, must be used in contrasts, among the geometric means of the groups of components to obtain a coefficient of orthogonality.

The CND is an evolution of the univariate and bivariate methods since, because it is based on the compositional data and main component analyses, it can potentially improve the diagnosis of the studied vegetal tissue. Another great advantage is that the available modern computational tools reduced the analysis effort since the development of CND standards resulted in numerous free software that can be assessed on http://www.registro.unesp.br/sites/cnd (Parent and Dafir, 1992; Khiari et al., 2001 and Parent et al., 2005, 2009). These programs allow determining the nutritional balance, indicating deficiency or excess compared with other elements based on leaf analysis. Also, there is a practical guide for elaborating the CND from a databank (Traspadini et al., 2018).

It is noteworthy that the standards used by the CND or DRIS methods require regular updates over the individual crop cycles because agriculture is a dynamic system that changes with management, climate, and genotypes. These updates are important to maintain the high accuracy of the standards so that correct diagnoses and the tool efficacy to monitor the nutritional status of crops increase as well.

6 Leaf diagnosis using alternative methods

6.1 Sap analysis

Sap analysis is considered a precise and sensitive technique for determining and quantifying plant nutritional status, especially for nitrogen and potassium, at the time of sampling, as demonstrated in vines by Nagarajah (1999) and Tecchio et al. (2011) and in citrus by Souza et al. (2012).

The sap nutrient contents are monitored using portable ion-sensitive devices, which are very effective when associated with fertigation since the sap nutritional control allows evaluating the demand for nutrients in the different plant developmental stages that can be immediately corrected by changing the fertilization management.

The sap is usually collected from the conductive tissues of leaves intended for foliar analysis, such as petioles and veins (Oliveira et al., 2003). There are several methodologies for extracting sap: among them, press extraction using ethyl ether and subsequent sample freezing, pressure chamber, and collection of exudates. The sap nutrients are measured by argon plasma or atomic emission spectrometers and flame photometers, according to the applied methodology. Colorimeters can also be used, but since it requires staining the sap sample, the reading becomes more difficult in this equipment. There is still portable and versatile equipment with specific electrodes capable of measuring the NO_3^-, K^+, Na^+, and pH values in the sap (Faquin, 2002).

Souza et al. (2015) evaluated the relationship between NO_3^- and K^+ concentrations in the sap of petioles and in the soil solution of a vineyard to assist the fertigation management. The authors reported a direct relationship between sap and the soil solution since the results showed that both nutrients were very sensitive to the increased nitrogen and potassium contents that resulted from the fertilization.

Therefore, Souza et al. (2012) and Cadahía and Lucena (2005) reported that sap analysis should be applied only as a nutritional evaluation for a determined instant over time since sap nutrient levels are susceptible to variation caused by several factors, such as time of day, irrigation, crop phase, organ sampled and its position in the plant, fertilization, and climatic conditions. Thus, these authors consider this high number of interferences as a disadvantage of the method because there is no local calibration for the most important crops, as well as on the appropriate levels or critical nutrient levels. Likewise, Tecchio et al. (2011) made the same suggestions for grape culture.

6.2 Biochemical tests

Biochemical tests determine the levels (deficiency, excess, and standard) of a specific plant mineral nutrient by measuring the changing enzyme activity (enzymatic tests), accumulation or disappearance of certain metabolites, immunological response, and profile or still in the expression of certain genes, events that are related to metabolic pathways directly or indirectly dependent on the studied nutrient. This mineral deficiency can trigger several of these events in the plants at the same time since they are generally interconnected by the metabolic pathways.

Pioneering studies in Israel added equipment and techniques of chemical and biochemical analysis so that many studies in the literature have investigated the biochemical responses of plants as a function of nutritional status. It is emphasized that biochemical tests do not directly provide the concentration of mineral nutrients in plants; however, the activity of a certain enzyme or the changing content of certain substances allows indicating sufficiency or deficiency of a related nutrient. The test may be performed on the defective tissue or on the tissue where the element/nutrient to be evaluated has been infiltrated to reactivate the enzyme system, being performed on leaf tissue extracts or by incubating parts of the tissue itself to provide faster results (Bar-akiva, 1984).

Tests involving peroxidase were successfully used to distinguish iron and manganese deficiency in citrus (*Citrus* spp. L.) (Bar-akiva, 1961; Bar-akiva et al., 1967). In this case, iron deficiency inhibits peroxidase activity, while manganese deficiency may increase peroxidase activity, a result explained by the fact that iron is a component of peroxidase, while manganese is not.

Also, there is evidence of the relationship between carbonic anhydrase activity with zinc deficiency in citrus (Bar-akiva and Lavon, 1969) and pecan (*Carya illinoinensis* Koch) (Snir, 1983). Polle et al. (1992) reported that the activity of the superoxide dismutase and some other protective enzymes increased in manganese-deficient leaves of perennial plants such as Norway spruce (*Picea abies* L.).

In the last 50 years, several studies on the possibility of using enzymes for diagnosing nutritional status of plants have been conducted (Malavolta et al., 1997), and the best perspectives of practical application are the enzymes presented in the extensive review by Cazetta et al. (2010).

6.3 Chlorophyll concentration

The development of the portable chlorophyll meter for instant readings of leaf content became a tool for the indirect evaluation of nitrogen in plants (Varvel et al., 1997) since 50%–70% of the total N in the leaves is an integral part of the enzymes that make up the chloroplasts present in chlorophylls (Chapman and Barreto, 1997) and responsible for capturing the light used in photosynthesis, being essential to convert light radiation to chemical energy, in the form of ATP and NADPH. Thus, chlorophylls are related to the photosynthetic efficiency of plants and, consequently, their growth and adaptability to the different environments (Jesus and Marenco, 2008).

The chlorophyll meter has diodes that emit radiation at 650 nm (red light) and 940 nm (infrared radiation), providing indirect readings of the chlorophyll content in the leaf. During the measurement, the light travels through the leaf and is received by a silicone photodiode where it is converted first into an analog electrical signal and then into a digital signal, which is converted by a microprocessor into values proportional to the chlorophyll content in the leaf (Minolta, 1989). It is pertinent to add that the index values are susceptible and change according to the model/brand of the equipment used; therefore, index values obtained by different models should not be compared.

However, the chlorophyll meter readings and, consequently, the indirect estimation of the N content of the leaf used in the reading are not always representative of the nutritional status of the plant especially due to the dilution effect of N in the plant as the aerial shoots grow (Pocojeski, 2007), which may be related to the several factors affecting the reading of the green color index and, not only the nitrogen, causing the yellowing of the leaves and varying indices. The low levels of magnesium (Neals, 1956); iron (Barton, 1970); and other nutrients such as sulfur, calcium, manganese, and zinc can also affect chlorophyll formation, inducing different degrees of chlorosis (Mengel and Kirkby, 1987). Also, the effects of genotypic variation, diagnostic leaf, and time of sampling cannot be neglected, as indicated by Rozane et al. (2009) for guava trees (*Psidium guajava* L.) and Jesus and Marenco (2008) for cupuaçu (*Theobroma grandiflorum* Schum., Sterculiaceae), lemon (*Citrus limon* L., Rutaceae), urucum (*Bixa orellana* L., Bixaceae), and araçá-boi (*Eugenia stipitata* Mac Vaugh, Myrtaceae).

Shaahan et al. (1999) stated that the chlorophyll meter is a simple and fast tool to predict the nitrogen nutritional status in mango, guava, and mandarin trees and vines in the field. Prado and Vale (2008) and Vale and Prado (2009) investigated leaf rootstocks of clove and citrumelo, respectively, and observed that applied nitrogen influences the dark green color index reading (SPAD) while having a high positive correlation with the nitrogen content and dry matter production, which indicates the method feasibility for evaluating this nutrient status. The green color index may also be useful to indicate the best time to apply the nitrogen fertilizer in the crop.

The use of the green color index (SPAD) for determining the nutritional status of nitrogen in fruit tree seedlings could shorten plant developmental time and, consequently, lower production costs (Rozane et al., 2009; Vale and Prado, 2009).

Finally, future perspective on the nutritional status assessment of fruit trees requires improving the available reference standards so that they are reliable and can be updated frequently (every 4 years), given how genotypes and environment change over time and crop cycle. Therefore, an interesting research line for investing further would be to obtain low-cost nutritional reference standards without requiring experimentation, such as the interpretation methods of leaf analysis using the DRIS and CND methods.

References

Aitchison, J., 1986. The Statistical Analysis of Compositional Data. Chapman Hall, p. 416.

Bar-akiva, A., 1961. Biochemical indications as a means of distinguishing between iron and manganese deficiency symptoms in citrus plants. Nature 190, 647–648.

Bar-akiva, A., 1984. Substitutes for benzidine as H-donors in the peroxidase assay for rapid diagnosis of iron deficiency in plants. Commun. Soil Sci. Plant Anal. 15, 929–934.

Bar-akiva, A., Lavon, R., 1969. Carbonic anhydrase activity as an indicator of zinc deficiency in citrus leaves. J. Hortic. Sci. Biotechnol. 44, 359–362.

Bar-akiva, A., Kaplan, M., Lavon, R., 1967. The use of biochemical indicator for diagnosing micronutrient deficiencies of grapefruit trees under field conditions. Agrochemical 11, 283–288.

Barton, R., 1970. The production and behavior of phytoferritin particles during senescence of Phaseolus leaves. Planta 94, 73–77.

Bataglia, O.C., Furlani, A.M.C., Furlani, P.R., Teixeira, J.P.F., Gallo, J.R., 1983. Métodos de análise química de plantas. Instituto Agronômico, Campinas, p. 48 (Boletim Técnico, 78).

Bataglia, O.C., Dechen, A.R., Santos, W.R., 1996. Princípios da diagnose foliar. In: Alvarez, V.H.V., Fontes, L.E.F., Fontes, M.P.F. (Eds.), O solo nos grandes domínios morfoclimáticos do Brasil e o desenvolvimento sustentado. SBCS-UFV, Viçosa, pp. 647–660.

Beaufils, E.R., 1973. Diagnosis and Recommendation Integrated System (DRIS). University of Natal, p. 132 (Soil Science Bulletin, 1).

Bollons, H.M., Barraclough, P.B., 1997. Inorganic orthophosphate for diagnosing the phosphorus status of wheat plants. J. Plant Nutr. 20 (6), 641–655.

Brunetto, G., Ceretta, C.A., Kaminski, J., Melo, G.W.B., Lourenzi, C.R., Furlanetto, V., Moraes, A., 2007. Aplicação de nitrogênio em videiras na Campanha Gaúcha: produtividade e características químicas do mosto da uva. Ciênc. Rural 37, 389–393.

Brunetto, G., Borgignon, C., Mattias, J.L., Deon, M., Melo, G.W.B., Kaminski, J., Ceretta, C.A., 2008. Produção, composição da uva e teores de nitrogênio na folha e no pecíolo em videiras submetidas à adubação nitrogenada. Ciênc. Rural 38, 2622–2625.

Cadahía, C., Lucena, J.J., 2005. Diagnostico de nutrición y recomendaciones de abonado. In: Adahía, C. (Ed.), Fertirrigación: cultivos hortícolas, frutales y ornamentales, third ed. Ediciones Mundi-Prensa, Madrid, pp. 183–257.

Cazetta, J.O., Fonseca, I.M., Prado, R.M., 2010. Perspectivas de uso de métodos diagnósticos alternativos: testes bioquímicos. In: Prado, R.M., Cecilio Filho, A.B., Correia, M.A.R., Puga, A.P. (Eds.), Nutrição de Plantas: Diagnose foliar em hortaliças. first ed. FCAV/FAPESP/CAPES/FUNDU-NESP, Jaboticabal, pp. 109–134.

Chadha, K.L., Samra, J.S., Thakur, R.S., 1980. Standardization of leaf-sampling technique for mineral composition of leaves of mango cultivar Chausa. Sci. Hortic. 13, 323–329.

Chapman, S.C., Barreto, H.J., 1997. Using a chlorophyll meter to estimate specific leaf nitrogen of tropical maize during vegetative growth. Agron. J. 89, 557–562.

Egozcue, J.J., Pawlowsky-Glahn, V., 2005. Groups of parts and their balances in compositional data analysis. Math. Geol. 37, 795–828.

Faquin, V., 2002. Diagnose do estado nutricional das plantas. UFLA/FAEPE, Lavras, p. 227.

Geraldson, C.M., Klacan, G.R., Lorenz, O.A., 1973. Plant analysis as an aid in fertilizing vegetable crops. In: In Walsh, L.M., Beaton, J.D. (Eds.), Soil Testing and Plant Analysis. Soil Science Society of America, Madison, WI, pp. 365–379.

Güsewell, S., 2004. N:P ratios in terrestrial plants: variation and functional significance. New Phytol. 164, 243–266.

Holland, D.A., 1966. The interpretation of leaf analysis. J. Hortic. Sci. 41, 311–329.

Holland, D.A., Little, R.C., Allen, M., Dermott, W., 1967. Soil and leaf sampling in apple orchards. J. Hortic. Sci. 42, 403–417.

Jesus, S.V., Marenco, R.A., 2008. O SPAD-502 como alternativa para a determinação dos teores de clorofila em espécies frutíferas. Acta Amazon. 38, 815–818.

Kenworthy, A.L., 1967. Plant analysis and interpretation of analysis for horticultural crops. In: In Stelly, M., Hamilton, H. (Eds.), Soil Testing and Plant Analysis. Part II. Soil Science Society of America, Madison, WI, pp. 59–75.

Khiari, L., Parent, L.E., Tremblay, N., 2001. Selecting the high-yield subpopulation for diagnosing nutrient imbalance in crops. Agron. J. 93, 802–808.

Lagatu, H., Maume, L., 1934. Le diagnostic foliare de la pomme de terre. Ann. Ecole Natl. Agric. 22, 150–158.

Malavolta, E., 1992. ABC da análise de solos e folhas: Amostragem, interpretação e sugestões de adubação. Agronômica Ceres, São Paulo. 124 p.

Malavolta, E., Vitti, G.C., Oliveira, A.S., 1997. Avaliação do estado nutricional das plantas: Princípios e aplicações, second ed. Associação Brasileira para Pesquisa da Potassa e do Fosfato, Piracicaba, p. 319.

Marschner, H., 1986. Mineral Nutrition of Higher Plant. Academic Press, London. 674 p.

Mengel, K., Kirkby, E.A., 1987. Principles of Plant Nutrition. International Postash Institute, Worblaufen-Bern. 687 p.

Minolta, 1989. Chlorophyll meter SPAD-502. Instruction Manual. Minolta Co., Osaka. 22 p.

Munson, R.D., Nelson, W.L., 1990. Principles and practices in plant analysis. In: Westerman, R.L. (Ed.), Soil Testing and Plant Analysis, third ed. Soil Science Society of America Book Series No. 3, Madison, WI, pp. 359–387.

Nagarajah, S.A., 1999. Petiole sap test for nitrate and potassium in *Sultana grapevines*. Aust. J. Grape Wine Res. 5, 56–60.

Natale, W., Coutinho, E.L.M., Boaretto, A.E., Banzatto, D.A., 1994. Influência da época de amostragem na composição química das folhas de goia-beira (*Psidium guajava* L.). Rev. Agric. 69, 247–255.

Natale, W., Coutinho, E.L.M., Boaretto, A.E., Pereira, F.M., 1996. Goiabeira: Calagem e adubação. Funep. Jaboticabal. 22 pp.

Natale, W., Coutinho, E.L.M., Boaretto, A.E., Pereira, F.M., 2002. Nutrients foliar content for high productivity cultivars of guava in Brazil. Acta Hortic. 594, 383–386.

Neals, T.F., 1956. Components of total magnesium content within the leaves of white clover and perennial rye grass. Nature 177, 388–389.

Oliveira, M.N.S., Oliva, M.A., Martinez, C.A., Silva, M.A.P., 2003. Variação diurna e sazonal do pH e composição mineral da seiva do xilema em tomateiro. Hortic. Bras. 21, 10–14.

Parent, L.E., Dafir, M., 1992. A theorical concept of compositional nutrient diagnosis. J. Am. Soc. Hortic. Sci. 117, 239–242.

Parent, L.E., Khiari, L., Pettigrew, A., 2005. Nitrogen diagnosis of Christmas needle greenness. Can. J. Plant Sci. 85, 939–947.

Parent, L.E., Natale, W., Ziadi, N., 2009. Compositional nutrient diagnosis of corn using the Mahalanobis distance as nutrient imbalance índex. Can. J. Soil Sci. 89, 383–390.

Parent, S.E., Parent, L.E., Rozane, D.E., Hernandes, A., Natale, W., 2012. Nutrient balance as paradigm of soil and plant chemometrics. In: Issaka, R.N (Ed.), Soil Fertility. InTech, pp. 83–114.

Pocojeski, E., 2007. Estimativa do estado nutricional de arroz irrigado por alagamento. Dissertação (Mestrado), Universidade Federal de Santa Maria, Santa Maria. 97 p.

Polle, A., Chakrabarti, K., Chakrabarti, S., Seifert, F., Schramel, P., Rennenberg, H., 1992. Antioxidants and manganese deficiency in needles of Nor-way spruce (*Picea abies* L.) trees. Plant Physiol. 99, 1.084–1.089.

Prado, R.M., 2004. Nutrição e desordens fisiológicas na cultura da manga. In: Rozane, D.E., Darezzo, R.J., Aguiar, R.L., Aguilera, G.H.A., Zambolim, L. (Eds.), Manga: Produção integrada, industrialização e comercialização. UFV, Viçosa, pp. 199–231.

Prado, R.M., 2008. Nutrição de Plantas, first ed. vol. 1. Editora UNESP, São Paulo. 407 p.

Prado, R.M., Natale, W., 2004. Leaf sampling in carambola trees. Fruits 52, 281–289.

Prado, R.M., Vale, D.W., 2008. Nitrogênio, fósforo e potássio na leitura SPAD em porta-enxerto de limoeiro cravo. Pesqui. Agropecu. Trop. 38, 227–232.

Prado, R.M., Natale, W., Braghirolli, L.F., Ragonha, E., 2005. Estado nutricional do maracujazeiro-amarelo "FB 200" sobre cinco porta-enxertos, cul-tivado em um Latossolo Vermelho distrófico. Rev. Agric. 80, 388–399.

Quaggio, J.Á., Van, R.B., Piza Jr., C.T., 1997. Frutíferas. In: Van, R.B., Cantarella, H., Quaggio, J.Á., Furlani, A.M.C. (Eds.), Recomendações de adu-bação e calagem para o Estado de São Paulo, second ed. rev. IAC, Campinas, pp. 121–125 (Boletim Técnico, 100).

Redfield, A.C., 1934. On the proportions of organic derivations in sea water and their relation to the composition of plankton. In: Daniel, R.J. (Ed.), James Johnstone Memorial. University Press of Liverpool, pp. 177–192.

Rozane, D.E., Souza, H.A., Prado, R.M., Franco, C.F., Leal, R.M., 2009. Influência do cultivar, do tipo de folha e do tempo de cultivo na medida indireta da clorofila (SPAD) em mudas de goiabeira. Ciênc. Agrotec. 33, 1538–1543.

Rozane, D.E., Parent, L.E., Natale, W., 2016. Evolution of the predictive criteria for the tropical fruit tree nutritional status. Cientifica 44, 102–112.

Schaffer, B., Larson, K.D., Snyder, G.H., Sanchez, C.A., 1988. Identification of mineral deficiencies associated with mango decline by DRIS. HortScience 23, 617–619.

Shaahan, M.M., El-Sayed, A.A., Abou El-Nour, E.A.A., 1999. Predicting nitrogen, magnesium and iron nutritional status in some perennial crops using a portable clorophyll meter. Sci. Hortic. 82, 339–348.

Snir, I., 1983. Carbonic anhydrase activity as an indicator of zinc deficiency in pecan leaves. Plant Soil 74, 287–289.

Souza, T.R., Villas Bôas, R.L., Quaggio, J.Á., Salomão, L.C., 2012. Nutrientes na seiva de plantas cítricas fertirrigadas. Rev. Bras. Frutic. 34, 482–492.

Souza, T.R., Bardiviesso, D.M., Andrade, T.F., Villas Boas, R.L., 2015. Nutrientes no solo e na solucao do solo na citricultura fertirrigada por gote-jamento. Eng. Agr. 35, 484–493.

Stevenson, F.J., 1986. Cycles of Soil Carbon, Nitrogen, Phosphorus, Sulfur, Micronutrients. Wiley, New York.

Tecchio, M.A., Moura, M.F., Paioli-Pires, E.J., Terra, M.M., Teixeira, L.A.J., Smarsi, R.C., 2011. Teores foliares de nutrientes, índice relativo de clorofila e teores de nitrato e de potássio na seiva do pecíolo na videira 'Niagara Rosada'. Rev. Bras. Frutic. 33, 649–659.

Traspadini, E.I.F., Prado, R.M., Vaz, G.J., Silva, F.C., Mancini, A.L., Silva, G.P., Santos, E.H., Wadt, P.G.S., 2018. Guia prático para aplicação do método da diagnose da composição nutricional (CND): Exemplo de uso na cultura da cana-de-açúcar. Embrapa Informática Agropecuária, Campinas, p. 32.

Vale, D.W., Prado, R.M., 2009. Adubação NPK e o estado nutricional de 'Citrumelo' por medida indireta da clorofila. Rev. Ciênc. Agron. 40, 266–271.

Varvel, G.E., Schepers, J.S., Francis, D.D., 1997. Ability for in-season correction of nitrogen deficiency in corn using chlorophyll meters. Soil Sci. Soc. Am. J. 61, 1233–1239.

Webb, R.A., 1972. Use of the boundary line in the analysis of biological data. J. Hortic. Sci. 47, 309–319.

Zambolim, L., Ventura, J.A., 1996. Resistência a doenças induzidas pela nutrição mineral das plantas. Inf. Agron. 75, 1–16.

Further reading

Lavon, R., Goldshmith, E.E., Salomon, R., Frank, A., 1995. Effect of potassium, magnesium, and calcium deficiencies on carbohydrate pools and metabolism in citrus leaves. J. Am. Soc. Hortic. Sci. 120, 54–58.

Parent, L.E., 2011. Diagnosis of the nutrient compositional space of fruit crops. Rev. Bras. Frutic. 33, 321–334.

Srivastava, A.K., Singh, S., Huchche, A.D., Ram, L., 2000. Yield based leaf and soil test interpretations for Nagpur mandarin (*Citrus reticulata* Blanco) in central India. Commun. Soil Sci. Plant Anal. 32, 585–599.

12

Diagnosis of nutrient composition in fruit crops: Major developments

Léon Etienne Parent[a],*, Danilo Eduardo Rozane[b], José Aridiano Lima de Deus[c], William Natale[d]

[a]Department of Soil and Agri-food Engineering, Université Laval, Québec, QC, Canada
[b]São Paulo State University, UNESP, Registro, Brazil
[c]Institute of Technical Assistance and Rural Extension of Paraná (EMATER-PR), Curitiba, Brazil
[d]Federal University of Ceará, Fortaleza, Brazil
*Corresponding author. E-mail: leon-etienne.parent@fsaa.ulaval.ca

1 Introduction

Because the roots of fruit trees can explore soil layers well below the sampled layers, plant tissue analysis is a useful means to diagnose the nutrient status of fruit crops (Smith et al., 1997). Tissue nutrient concentrations are thought to integrate growth factors under the prevailing ecological conditions affecting crop response to fertilization (Munson and Nelson, 1990). Where one nutrient is supplied in insufficient amounts or in inappropriate combination with other nutrients to support plant growth, agroecosystems cannot be sustainable. At the other extreme, the overuse of fertilizers and inefficient nutrient uptake leads to luxury consumption (Nowaki et al., 2017; Deus et al., 2018), biochemical and physiological disorders, yield losses (Prado and Caione, 2012), wastage of fertilizers, and environmental damage (Das and Mandal, 2015).

A.K. Srivastava, Chengxiao Hu (eds.)
Fruit Crops: Diagnosis and Management of Nutrient Constraints
https://doi.org/10.1016/B978-0-12-818732-6.00012-5

FIG. 12.1 Representation of the change in NPK balance in the potato diagnostic leaf in response to P fertilization, projected into a Lagatu and Maume (1934) center-scaled ternary diagram (N + P + K = 100%). Nutrients move along the line as P dosage increased with spot size. The *gray blob* encloses the NPK composition of high-yielding crops, and the *central spot* is blob centroid. *Reproduced with permission from Parent, S.-É., Parent, L.E., Rozane, D.E., Hernandes, A., Natale, W., 2012. Nutrient balance as paradigm of soil and plant chemometrics. In: Issaka, R.N. (Ed.), Soil Fertility. IntechOpen Ltd., London, pp. 83–114. https://doi.org/10.5772/53343.*

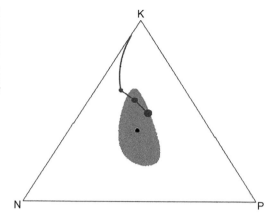

To diagnose plant nutrient status, total, soluble, or extractable nutrients are quantified in a selected tissue collected at a specific developmental stage (Bould et al., 1960). Cost-effective sampling procedures have been developed for each species (Benton Jones et al., 1991). Tissue analysis includes sample preparation, digestion, and quantification. The basis for nutrient expression is generally tissue dry matter but could be also fresh matter (Clements, 1957), petiole sap (Vitosh and Silva, 1994), or nitrogen (Ingestad, 1987). Quantified plant nutrients and undetermined tissue components are intrinsically multivariate and interrelated parts. Tissue compositions should thus be diagnosed holistically as nutrient combinations rather than as collections of nutrients addressed separately (Deus et al., 2018).

In this chapter, we examine the science behind plant nutrient diagnosis from earlier methods that diagnosed nutrients separately under the laws of the minimum and the optimum to modern tools of compositional nutrient diagnosis that address nutrient combinations impacting fruit yield and quality.

2 Diagnostic methods

2.1 Early evidence of physiological balances between nutrients

In the 1830s, Jean-Baptiste Boussingault advanced the idea that nutrient balances were more important for plant growth than nutrients examined separately (Epstein and Bloom, 2005). Lagatu and Maume (1934) represented Boussingault's inkling as a blob or domain delineating optimum NPK combinations in potato leaves. To validate their theory, they used a ternary diagram, an old concept dating 1704 but widely used in many fields of science to synthesize three variables adding up to 100% (Howarth, 1996). Lagatu and Maume (1934) showed (Fig. 12.1) that Carl Sprengel's Law of the Minimum popularized by Justus von Liebig (Epstein and Bloom, 2005) was based on overoptimistic assumptions for not accounting for nutrient interactions. Their brilliant demonstration of nutrient interactions remained unattended for nearly 50 years.

2.2 Critical nutrient concentrations and concentration ranges

Macy (1936) was the first to define adequate concentration ranges in line with Georg Liebscher's Law of the optimum stating that a growth-limiting factor contributes more to production the closer other production factors are to their optimum (De Wit, 1992). Ulrich (1952) set critical concentration values at 90%–95% maximum yield. Percentage of maximum yield was designed to remove the location-to-location variability due to factors other than the nutrient being studied, but this has serious agronomic (absolute profitability vs nutrient dosage) and statistical (variance heterogeneity and negligible interactions assumed) limitations (Nelson and Anderson, 1984). There is luxury consumption, accumulation, or contamination without yield loss above sufficiency range (Ulrich, 1952). At the far end of the response curve, a slow yield decrease indicates nutrient excess due to nutrient antagonism or toxicity.

To establish nutrient concentration ranges, a gradient of nutrient doses must be planned in controlled experiments where the effect of target nutrient on yield is not altered by any harmful effect of another factor (Ulrich and Hills, 1967). Nutrient concentration ranges can also be computed from crop surveys as confidence intervals of nutrient concentrations at high-yield level. However, the assumptions behind the law of the optimum are rarely met under field conditions where factor levels vary widely.

Critical concentration ranges are easy to interpret (Benton Jones et al., 1991). However, confidence intervals diagnose nutrients in one dimension and two nutrients in 2-D rectangles or squares delineated by the intervals (Nowaki et al., 2017). The 2-D scheme expands to hyperrectangles or hypercubes as more nutrients are added. Assuming normal (ellipsoidal) data distribution at high-yield level, the success across nutrient-by-nutrient diagnosis is conceptually low (Nowaki et al., 2017). Indeed, the volume of a hypersphere is small compared with that of a hypercube across several intervals.

2.3 Weighted nutrient diagnosis

Running fertilizer experiments to delineate nutrient concentration ranges is expensive. Using both crop surveys and experimental data, Kenworthy (1967) suggested to standardize nutrient concentrations as percentages of mean values for "normal" crops and then to weight the result by the coefficient of variation V to account for variation in the data. A balance index B relative was computed as follows:

$$\text{If } X < Std, (X/Std) \times 100 = P; (100 - P) \times (V/100) = I; P + I = B$$
$$\text{If } X > Std, (X/Std) \times 100 = P; (P - 100) \times (V/100) = I; P - I = B$$

where X is tissue concentration of nutrient X, Std is standard nutrient concentration for "normal" crops, P is percentage of standard value, and I reflects the variation in standard values. Because Kenworthy's equations were set symmetrical about 100, they allowed assessing relative shortage (<100) or excess (>100) of nutrients. However, the Kenworthy model did not address nutrient interactions.

3 Nutrient balances as dual or stoichiometric ratios

3.1 Nutrient ratios

Interactions are important where one nutrient competes with or dilutes another one close to its deficiency threshold (Marschner, 1986). Nutrient interactions are reported as dual (Bergmann, 1988; Walworth and Sumner, 1987; Wilkinson, 2000) or stoichiometric (Ingestad, 1987) ratios. Selected ratios must reflect physiological functions such as protein synthesis (N and S) and plant-available energy (P) (Epstein and Bloom, 2005) or guide nutrient applications relative to nitrogen (Ingestad, 1987). Nutrient ratios assume sometimes erroneously that pairs of elements are linearly related (Kenworthy, 1967). Ratios can return similar values within deficiency, sufficiency, or toxicity ranges, confusing the interpretation of dual ratios if concentration values are not examined concomitantly (Marschner, 1986).

3.2 Diagnosis and recommendation integrated system

The DRIS is an appealing diagnostic method that integrates dual ratios and functions into nutrient indices and arranges nutrients in the order of their limitation to yield (Beaufils, 1973). Dual ratios used to compute DRIS functions are as follows (Walworth and Sumner, 1987):

$$f(A/B) = \left[\frac{(A/B)}{(a/b)} - 1 \right] \frac{\kappa}{cv}, \quad \text{if } (A/B) > (a/b),$$
$$f(A/B) = \left[1 - \frac{(a/b)}{(A/B)} \right] \frac{\kappa}{cv}, \quad \text{if } (A/B) < (a/b),$$
$$f(A/B) = 0, \quad \text{if } (A/B) = (a/b),$$

where A/B and a/b are dual nutrient ratios in diagnosed and reference compositions, respectively, and cv is coefficient of variation (standard deviation divided by mean) of the reference dual ratio. Factor κ accounts for differential measurement units. The ratio expressions (X/Y or Y/X) are selected from the highest variance ratio between the low- and high-yielding subpopulations. However, such selection procedure is questionable because false-positive specimens (high yield despite nutrient imbalance) may result from luxury consumption or contamination.

The DRIS indices I_A, I_B, and I_C are computed across nutrients A, B, and C by averaging DRIS functions after multiplying DRIS functions by $(+1)$ if the nutrient is at numerator or (-1) otherwise, as follows (Walworth and Sumner, 1987):

$$I_A = \frac{f(A/B) + f(A/C)}{2}; \quad I_B = \frac{-f(A/B) - f(C/B)}{2}; \quad I_C = \frac{-f(A/C) + f(C/B)}{2}$$

$$I_A + I_B + I_C = 0$$

$$|I_A| + |I_B| + |I_C| = NII$$

where NII is nutrient imbalance index. Because DRIS indices are symmetrical, their sum is constrained to zero (Beaufils, 1973).

The DRIS was claimed wrongly to diagnose nutrient status irrespective of plant age and location (Walworth and Sumner, 1987; Epstein and Bloom, 2005). The DRIS has been modified empirically at several occasions, including nutrient products to account for nutrients accumulating in opposite directions with time and a dry matter index to delineate nutrient shortage from excess (Walworth and Sumner, 1987). The symmetry of DRIS indices was lost using nutrient products. The dry matter basis was viewed erroneously as a component rather than a scale of measurement.

Beverly (1987a,b) suggested to log transform nutrient ratios to facilitate deriving DRIS norms. Indeed, the advantages of log transformations are as follows:

- It is common to use a logarithmic scale where ratios are greater than 10^4 (Budhu, 2010).
- $\ln(A/B)$ and $\log(B/A)$ are reflective because $\ln(A/B) = -\ln(B/A)$; therefore, the variance is the same between both expressions.
- The geometric mean is the most appropriate centroid to conduct statistical analyses on ratios (Fleming and Wallace, 1986).

DRIS was a major step forward combining nutrients for diagnostic purposes. Nonetheless, the assumed additivity of DRIS functions and indices led to conceptual errors compared with the unbiased compositional data analysis methods (Parent and Dafir, 1992; Parent et al., 2012b).

4 Compositional nutrient diagnosis

4.1 Centered log ratios

Aitchison (1986) used a ternary diagram to illustrate the closure problem of three-part geochemical compositions, where any change in the proportion of one part must resonate on others. The negative covariance inherent to resonance distorts the results of linear multivariate analysis and may lead to wrong conclusions. Nutrient interactions (synergism and antagonism), luxury consumption, nutrient excess or toxicity, and contamination are sources of resonance specific to the closed space of the tissue measurement unit (e.g., 1000 g/kg on a dry mass basis).

To control resonance, Aitchison (1986) derived log-ratio transformations from the multinomial Dirichlet distribution function across proportions and the logistic function $\ln\left(\frac{p}{1-p}\right)$ closed to one for probability p, $(1-p)$ being the filling value to one. Because $\ln(A/B) = \ln(A) - \ln(B)$, a log ratio is a log contrast between two quantities. To close system space to 1000 g/kg in tissue dry matter, the filling value F_v in a D-part composition is computed by difference between measurement unit and the sum of proportions or concentrations as follows:

$$F_v = 1000 - \sum_{i=1}^{D-1} X_i$$

where X_i is the ith proportion or concentration. A compositional simplex comprising 12 components of which 11 are nutrients is described as follows on dry matter basis:

$$S_{12} = \left\{ \begin{array}{c} [N, S, P, K, Mg, Ca, B, Fe, Mn, Zn, Cu, F_v]; \\ \left[\begin{array}{c} N > 0; S > 0; P > 0; K > 0; Mg > 0; Ca > 0; \\ B > 0; Fe > 0; Mn > 0; Zn > 0; Cu > 0; F_v > 0 \end{array} \right]; \\ [N + S + P + K + Mg + Ca + B + Fe + Mn + Zn + Cu + F_v = 1000 \, \text{g kg}^{-1}] \end{array} \right\}$$

Aitchison (1986) proposed to express relationships among components as centered log ratios scaled on the geometric mean, as follows:

$$g(c_i) = \sqrt[12]{N \times S \times P \times K \times Mg \times Ca \times B \times Fe \times Mn \times Zn \times Cu \times F_v}$$

TABLE 12.1 Centered log-ratio (*clr*) standards (mean and standard deviation (SD)) of Brazilian fruit crops.

	Guava		Mango		Grape		Atemoia	
	Psidium guajava		*Mangifera indica*		*Vitis vinifera*		*Annona squamosa*	
clr	Mean	SD	Mean	SD	Mean	SD	Mean	SD
N	2.831	0.190	2.541	0.129	2.761	0.160	3.294	0.164
P	0.435	0.236	−0.075	0.227	0.652	0.164	0.070	0.616
K	2.689	0.186	2.175	0.198	1.981	0.208	2.606	0.117
Ca	2.074	0.164	2.978	0.203	2.085	0.178	2.497	0.232
Mg	0.744	0.182	0.620	0.178	0.531	0.147	0.947	0.285
S	0.870	0.174	0.253	0.145	0.718	0.111	0.452	0.220
B	−3.499	0.330	−3.718	0.273	−3.953	0.257	−2.934	0.215
Cu	−3.310	0.911	−4.072	0.650	−4.956	0.152	−4.795	0.463
Fe	−2.827	0.386	−2.807	0.287	−2.719	0.228	−2.549	0.342
Mn	−2.790	0.259	−0.782	0.344	−1.263	0.246	−2.596	0.540
Zn	−3.972	0.188	−3.950	0.336	−2.148	0.322	−3.677	0.624

Courtesy from Rozane, D.E., Natale, W., Parent, L.E., Parent, S.-É., Santos, E.M.H., 2013. Programa de computador: Instituto Nacional da Propiedade Industrial—INPI: BR5120130003806. Universidade Estadual Paulista "Júlio de Mesquita Filho"; Université Laval. 2013. http://www.registro.unesp.br/%23!/sites/cnd/ (Accessed 10 December 2018). (in Portuguese, Spanish, English and French).

The *clr* is much easier to compute than DRIS indices, as follows:

$$clr_{X_i} = ln\left(\frac{X_i}{G}\right) = ln(X_i) - ln(G)$$

where X_i is the *i*th nutrient concentration, G is geometric mean, and *clr* is a log contrast between X_i and G. The *clr* is a linear combination of dual ratios as follows:

$$clr_N = ln\frac{N}{G} = ln\left(\frac{N^{12}}{N \times P \times K \times Mg \times Ca \times B \times Fe \times Mn \times Zn \times Cu \times F_v}\right)^{\frac{1}{12}}$$

$$= ln\left(\frac{N}{N} \times \frac{N}{P} \times \frac{N}{K} \times \frac{N}{Mg} \times \frac{N}{Ca} \times \frac{N}{B} \times \frac{N}{Fe} \times \frac{N}{Mn} \times \frac{N}{Zn} \times \frac{N}{Cu} \times \frac{N}{F_v} \times\right)^{\frac{1}{12}}$$

$$= ln\frac{N}{P} + ln\frac{N}{K} + ln\frac{N}{Mg} + ln\frac{N}{Ca} + ln\frac{N}{B} + ln\frac{N}{Fe} + ln\frac{N}{Mn} + ln\frac{N}{Zn} + ln\frac{N}{Cu} + ln\frac{N}{F_v}$$

As a result, each component is adjusted to every other, avoiding optimistic assumptions on optimum levels of other nutrients under the law of the optimum or equal levels of other nutrients under the law of the minimum.

Parent and Dafir (1992) proposed to compute *clr* indices as follows:

$$I_{clr_i} = \frac{clr_i - clr_i^*}{s_{clr_i}^*},$$

where clr_i is *clr* value of the *i*th component of the composition being diagnosed, clr_i^* is *clr* mean of the *i*th component of the reference composition, and $s_{clr_i}^*$ is standard deviation of the *i*th component of the reference composition. Positive and negative I_{clr_i} values can be ranked on a histogram to illustrate nutrient limitations as shortage or excess. The *clr* standards (clr_i^* and $s_{clr_i}^*$) of some Brazilian fruit crops are presented in Table 12.1.

The difference between any two compositions X and Y can be computed as Euclidean distance $\in$ as follows (Aitchison, 1986):

$$\in = \sum_{i=1}^{D} (clr_{X_i} - clr_{Y_i})^2 = (clr_{X_i} - clr_{Y_i})I^{-1}(clr_{X_i} - clr_{Y_i})$$

where I is the identity matrix and X and Y refer to compositions X and Y, respectively. The *clr* and the ordinary log transformation return identical Euclidean distances if and only if geometric means are identical.

As a means to evaluate intercorrelations between *clr* variates, Badra et al. (2006) used the measure of sampling adequacy (MSA), allowing to compute a chi-square (χ^2) variable across "independent" (low MSA value) *clr* variates, as follows:

$$\chi^2 = \sum_{i=1}^{D} \left(I_{clr_i}\right)^2 = (clr_{X_i} - clr_{Y_i})VAR^{-1}(clr_{X_i} - clr_{Y_i})$$

where *VAR* is the variance matrix.

4.2 Isometric log ratios

Because there are *D clr* variates computable from a *D*-part composition, one *clr* variate must be discarded (generally F_v that is barely interpretable) to avoid computing a singular matrix when running multivariate analysis. Indeed, there are $D-1$ degrees of freedom available to run multivariate analysis of compositions (Aitchison and Greenacre, 2002). To reduce dimensionality, a useful case of linear independence is orthogonality whereby vectors are at perfectly right angles to each other (Rodgers et al., 1984). Egozcue et al. (2003) had the idea to partition *D* components into $D-1$ orthonormal balances hierarchically arranged among the *D* components of the system under study to compute isometric log ratios (*ilr*), as follows:

$$ilr_k = \sqrt{\frac{r_k s_k}{r_k + s_k}} \ln\left(\frac{\sqrt[r_k]{\prod_{i=1}^{r_k} x_i}}{\sqrt[s_k]{\prod_{j=1}^{s_k} x_j}}\right)$$

where r_k and s_k are numbers of components in subsets at numerator and denominator, respectively; $\sqrt{\frac{r_k s_k}{r_k + s_k}}$ is a normalization coefficient; *i* and *j* refer to components at numerator and denominator, respectively; and $\sqrt[r_k]{\prod_{i=1}^{r_k} x_i}$ and $\sqrt[s_k]{\prod_{j=1}^{s_k} x_j}$ are geometric means of component concentrations at numerator and denominator, respectively.

The balance design is an illustration allowing the researcher to describe how components and their related functions are connected in a closed system under study. Subsets of components are arranged into balances following a sequential binary partition (SBP). There are $D!(D-1)!/2^{D-1}$ possible balance designs in a *D*-part composition (Pawlowsky-Glahn et al., 2011). Multivariate distances remain the same whatever the SBP because switching from a balance design to another just rotates orthogonal axes. Concentration values in buckets do not change, but *ilr* values at fulcrum change where the balance design has been changed.

The SBP can be supported by theory, biplot analysis, or management objectives such as fertilizer source and balance, expected synergistic effect, or selection of liming material based on Ca and Mg contents (Parent, 2011; Hernandes et al., 2012; Parent et al., 2012a,b; Montes et al., 2016). The dual ratios most relevant to the study are assigned first. Examples of balance design are presented in Table 12.2 and Fig. 12.2. Computations can be performed using freeware such as CoDaPack (Comas-Cufí and Thió-Henestrosa, 2011) and R (van den Boogaart et al., 2014). Log ratio means including that of the filling value can be back-transformed to familiar concentration values using the *clrinv* or *ilrinv* procedures in R. Several R codes are available to handle compositional data and run statistical analyses.

TABLE 12.2 Sequential binary partition and balance formulation.

ilr	r	s	ilr formula	Balance arrangement
1	2	3	$\sqrt{6/5}\ln(\sqrt{N \times P}/\sqrt[3]{K \times Mg \times Ca})$	Anions versus cations
2	1	1	$\sqrt{1/2}\ln(N/P)$	Protein synthesis versus energy
3	1	1	$\sqrt{1/2}\ln(\sqrt{K/Mg})$	K-Mg antagonism
4	2	1	$\sqrt{2/3}\ln(\sqrt[2]{K \times Mg}/Ca)$	Cationic antagonisms
5	5	1	$\sqrt{5/6}\ln(\sqrt[5]{N \times P \times K \times Mg \times Ca}/F_v)^a$	Nutrient dilution in biomass

[a] F_v = *filling value between sum of analyzed nutrients and measurement unit.*

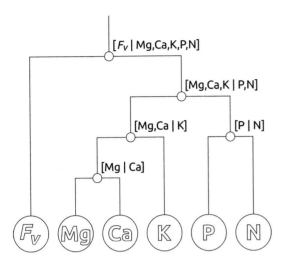

FIG. 12.2 Example of balance design for macronutrients. The *ilrs* are computed at fulcrums using concentration values located in buckets.

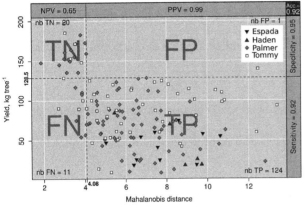

FIG. 12.3 Partitioning of 156 Brazilian mango (*Mangifera indica*) specimens returning an accuracy of 92% and critical Mahalanobis distance of 4.08. *FN*, false negative; *FP*, false positive; *NPV*, negative predictive value; *PPV*, positive predictive value; *TN*, true negative; *TP*, true positive; accuracy = 92%. *Reproduced with permission from Parent, S.-É., Parent, L.E., Rozane, D.E., Natale, W., 2013. Plant ionome diagnosis using sound balances: case study with mango* (Mangifera indica). *Front. Plant Sci. 4, 449. https://doi.org/10.3389/fpls. 2013.00449.*

5 Data partitioning

Based on research in soil fertility (Nelson and Anderson, 1984) and clinical biology (Swets, 1988), Parent et al. (2013b) proposed to partition data into four categories: true negative (TN), false negative (FN), true positive (TP), and false positive (FP). Crude data are first transformed into *ilr* variates. Mahalanobis distance (covariance matrix) is iterated across *ilr* values of healthy crops (true-negative specimens) as reference *ilr* variates (Fig. 12.3), starting with the compositions of preselected high-yielding crops. Accuracy, computed as the sum of TN and TP specimens over total, is generally >80% (Parent et al., 2012a, 2013a; Rozane et al., 2015; Souza et al., 2016). Lower accuracy is due to a high number of false-negative (Marchand et al., 2013) or true-positive (Deus et al., 2018) specimens.

Imbalanced specimens are those showing Mahalanobis distance exceeding the critical multivariate distance. The *clr* values of TN specimens are computed as standards (means and standard deviations). The Mahalanobis distance squared is assumed to be distributed like a chi-square variable.

To guide fertilizer recommendations, specimens are diagnosed using indices and nutrient ranking in the order of nutrient limitations to yield.

Steps to diagnose nutrients are as follows:

1. Compute the Mahalanobis distance using *ilr* mean and covariance matrix of TN specimens as reference values at fulcrum (Fig. 12.2). The *ilr* means of TN specimens are back-transformed into reference concentration centroids (*ilrinv*) assigned to buckets that equilibrate at fulcrum (circles in Fig. 12.2) equilibrating at fulcrum. Compare specimen's composition to the reference composition in buckets.
2. If the specimen is imbalanced (Mahalanobis distance larger than the critical distance), compute *clr* indices using mean and standard deviation of TN specimens as reference values (e.g., Table 12.2), and then order *clr* indices from the most negative to the most positive or provide a histogram of *clr* indices.
3. Make a recommendation to reestablish nutrient balance.

6 New lines of research

Compositional data analysis provides sound modern tools to address old issues and face challenges related to plant tissue diagnosis. New knowledge could be developed on the following topics:

1. Based on literature and experimentation, develop and animate customized nutrient balance schemes picturing nutrient relationships specific to the problem under study.
2. Look beyond total elemental analysis to improve diagnostic accuracy.
3. Test coefficient (exponent) assignment to concentration values to normalize log ratios or account for nutrient mobility.
4. Validate numerical order of nutrient limitations against the order of plant responses to nutrient stress.
5. Calibrate nutrient indices against crop response.
6. Relate ionomics to fruit quality, nutrient management, and biofortification to enhance the contribution of fruit crops to human nutrition and health.
7. Acquire and store big data to support artificial intelligence models and optimize nutrient management in fruit agroecosystems worldwide.

6.1 Customized nutrient balance designs

A customized balance (e.g., Fig. 12.2) is a comprehensive representation of the system designed managerially, phenomenologically, or heuristically to facilitate its interpretation. A catalog of customized balance designs could be elaborated to visualize and animate nutrient relationships relevant to a study or a diagnosis and monitor specific imbalances driven by treatments. For example, K fertilization, liming, and lime composition could be managed based on the interactive cationic subsystem (K, Ca, and Mg) operating in both soils and plant tissues (Parent and Parent, 2017). The $N \times P$ synergism (protein synthesis vs energy supply) could be contrasted with subsystems of other macronutrients or cationic micronutrients to assess the impact of NP fertilization on other subsystems.

The Cu, Zn, and Mn subsystem could be balanced with other subsystems to conduct global nutrient diagnosis. In some situations, fungicide applications can lead to excessive accumulations of Cu, Zn, and Mn on leaves that impact positively on yield by controlling foliar diseases. Nutrients contained in fungicides could be set apart as a specific subsystem of microcationic balances to assess fungicide capacity to sustain plant health with useful "contaminants" (Yamane, 2018).

6.2 Total analysis versus extracted forms

Tissue diagnosis is commonly conducted using total elemental analysis, disregarding the fact that just a portion of total concentration may be available to accomplish biological functions. If some extraction methods allow partitioning total elemental concentration into more or less available or unavailable forms, total concentration can be split into several contrasts. If the new partition impacts on yield, diagnostic accuracy could be increased.

6.3 Box-Cox coefficients

It is generally assumed that *ilr* variates are normally distributed, but this is not always the case. Non normal distribution can be recovered where appropriate after discarding outliers or by assigning Box-Cox coefficients varying

TABLE 12.3 Relative phloem mobility of essential elements (Epstein and Bloom, 2005).

Mobile	Intermediate	Immobile
Potassium	Sodium	Calcium
Nitrogen	Iron	Silicon
Magnesium	Manganese	Boron[a]
Phosphorus	Zinc	
Boron[a]	Copper	
Sulfur	Molybdenum	
Chlorine		

[a] *Immobile except for phloem-mobile boron-sorbitol complexes in fruit trees of genera* Pyrus, Malus, *and* Prunus *(Brown and Hu, 1996).*

between 0 and 1 to raw concentration values (Box and Cox, 1964). On the other hand, nutrients move differently from leaf to fruit through phloem transport, while xylem transport to low-transpiring organs may be limited. Assigning coefficients close to zero to the most immobile nutrients, coefficients close to one to most mobile ones, and intermediate coefficient values to partially mobile nutrients before *ilr* transformations may account for nutrient mobility (Table 12.3). While skewness of *ilr* values decreased using Box-Cox coefficients in a Brazilian banana (*Musa* spp.) data set, diagnostic accuracy was not improved (Deus et al., 2018). Nevertheless, the Box-Cox transformation may be an option for other crops.

6.4 How physiologically meaningful is nutrient numerical ordering?

It is a truism that growers are more concerned about negative than positive nutrient indices. While nutrient needs are prioritized according to the order of nutrient limitations to crop yield and quality, such ordering should bear some physiological significance. For the diagnosis to be reliable and reproducible, the addition of the most limiting nutrients in the order of their limitation should influence tissue composition and crop yield in the same order. This could only be validated by local fertilizer trials.

6.5 What is the minimum nutrient dosage to recover from nutrient imbalance?

By how much should present fertilization regimes be increased or decreased through soil or leaf applications in the case of nutrient imbalance? Plant response to applied nutrients must depend on whether and to what extent nutrient status is impacted by nutrient dilution in tissue mass, antagonism, synergism, luxury consumption, or contamination. As for soil testing where compositional models were also applied (Parent et al., 2012a,b; Parent and Parent, 2017), field calibration of tissue tests are needed to guide site-specific dosage of soil- and foliar-applied nutrients.

6.6 Ionomics, biofortification, and fruit quality

The ionome is the mineral nutrient composition of living organisms (Huang and Salt, 2016). Ionomics is a tool to target healthier food through genetics and biofortification (Welch, 2002; Zuo and Zhang, 2008; Gang et al., 2018). Baxter (2015) raised the problem of how to solve myriads of known and unknown interconnections between nutrients in ionomics and viewed a possible solution in the nutrient balance concept proposed by SE Parent et al. (2013b). Understanding nutrient interactions involving Zn and Fe absorption by plant and human might support plant breeding programs aimed at improving food quality and human health.

Fruit quality depends on genetics, management, and the environment. Fruit quality is often measured as Brix index, phenolics, and total anthocyanins (TAcy), as well as fruit acidity, firmness, and supply of minerals and vitamins essential to human health. Because organic acids and their salts (e.g., ascorbic acid, fumarate, malate, and citrate) can increase Zn and Fe bioavailability in fresh fruits (Bouis and Welch, 2010), such components could be balanced against each other (human nutrition forms also a compositional simplex). Some undesirable minerals could also be addressed.

The leaf ionome at high fruit yield may differ from that at high fruit quality, and this may impact on nutrient management. Compositional data analysis provides a means (1) to optimize leaf ionomes at high fruit quality level and (2) to balance fruit concentrations of sugars, antioxidants, acids, and minerals to reach high fruit yield and quality. Such models should be developed.

6.7 Big data, machine learning, and artificial intelligence

Plant nutrient diagnosis developed over the past centuries relied on overoptimistic assumptions. It was not until the theory of compositional data analysis was established on strong theoretical basis in 1986 using log ratios as data transformation techniques that plant tissue diagnosis gained a new momentum. Nutrient interactions, deficiency, luxury consumption, and contamination were viewed as numerical interplays between nutrient concentrations within the closed space of diagnostic tissues. Centered log ratios adjust every nutrient to the levels of other components. Isometric log ratios arrange selected nutrient interactions hierarchically. Data classification methods borrowed from soil fertility and clinical biology allowed delineating a hyperellipsoid within a critical Mahalanobis distance.

There is no reason to believe that the healthy state of ionomes has regular geometry such as hyperellipsoidal. Data distribution could be of any shape (blob) as impacted by several factors (Fig. 12.1). The many interacting growth factors such as genetics (e.g., cultivar), soil (pH, texture, genesis, etc.), and weather documented in big data sets can be handled using a wide spectrum of machine learning methods. Nutrient imbalance can be defined as the shortest distance between an observation and the nearest centroids of a hyperblob portion or blob islands delineating highly performing crops (Yamane, 2018). The k-nearest neighbor technique assumes that an observation will respond in line with its closest highly performing neighbors. An international initiative to acquire and store big data from different fruit agroecosystems worldwide is required to develop models of artificial intelligence.

In future research, machine learning methods and other tools of artificial intelligence will process higher-level models including an array of growth-impacting factors and metadata collected in big data sets. International collaboration will be needed to build such large data sets across fruit agroecosystems. Progress in compositional data analysis and artificial intelligence will undoubtedly foster the development of progressively more robust diagnostic tools to support high agronomic and environmental performance of fruit crops worldwide at lowest cost of inputs.

References

Aitchison, J., 1986. The Statistical Analysis of Compositional Data. Chapman and Hall, London.

Aitchison, J., Greenacre, M., 2002. Biplots of compositional data. J. R. Stat. Soc. Ser. C Appl. Stat. 51, 375–392. https://doi.org/10.1111/1467-9876.00275.

Badra, A., Parent, L.E., Allard, G., Tremblay, N., Desjardins, Y., Morin, N., 2006. Effect of leaf nitrogen concentration versus CND nutritional balance on shoot density and foliage colour of an established Kentucky bluegrass (Poa pratensis L.) turf. Can. J. Plant Sci. 86, 1107–1118. https://doi.org/10.4141/p05-242.

Baxter, I., 2015. Should we treat the ionome as a combination of individual elements, or should we be deriving novel combined traits? J. Exp. Bot. 66, 2127–2131. https://doi.org/10.1093/jxb/erv040.

Benton Jones, J., Wolf, B., Mills, H.A., 1991. Plant Analysis Handbook: A Practical Sampling, Preparation, Analysis, and Interpretation Guide. Micro Macro Intl., Athens, GA.

Bergmann, W., 1988. Ernährungs-störungen bei Kulturpflanzen. In: Auflage 2. Gustav Fischer Verlag, Stuttgart, New York.

Beverly, R.B., 1987a. Comparison of DRIS and alternative nutrient diagnostic methods for soybean. J. Plant Nutr. 10, 901–920.

Beverly, R.B., 1987b. Modified DRIS method for simplified nutrient diagnosis of "Valencia" oranges. J. Plant Nutr. 10, 1401–1408.

Bouis, H.E., Welch, R.M., 2010. Biofortification—a sustainable agricultural strategy for reducing micronutrient malnutrition in the Global South. Crop Sci. 50, S-20–S-32. https://doi.org/10.2135/cropsci2009.09.0531.

Bould, C., Bradfield, E.G., Clarke, G.M., 1960. Leaf analysis as a guide to the nutrition of fruit crops. I. General principles, sampling techniques and analytical methods. J. Sci. Food Agric. 11, 229–242. https://doi.org/10.1002/jsfa.2740110501.

Box, G.E.P., Cox, D.R., 1964. An analysis of transformations. J. R. Stat. Soc. Ser. B 26 (2), 211–252.

Brown, P.H., Hu, H., 1996. Phloem mobility of boron is species dependent: evidence for phloem mobility in sorbitol-rich species. Ann. Bot. 77, 497–505.

Budhu, M., 2010. Soil Mechanics and Foundations. University of Arizona, Tucson, AR. https://doi.org/10.1017/CBO9781107415324.004.

Clements, H.F., 1957. Crop-logging of sugar cane: the standard nitrogen index and the normal nitrogen index. In: Hawaii Agric. Exp. Sta. Tech. Bull. 35. University of Hawaii, Honolulu, HI.

Das, D.K., Mandal, M., 2015. Advanced technology of fertilizer uses for crop production. In: Shishir, S., Pant, K.K., Bajpai, S. (Eds.), Fertilizer Technology I: Synthesis. Studium Press LLC, Mumbai, pp. 18–68.

De Wit, C.T., 1992. Resource use efficiency in agriculture. Agric. Syst. 40, 125–151. https://doi.org/10.1016/0308-521x(92)90018-j.

Deus, J.A.L.d., Neves, J.C.L., Corréa, M.C.M., Parent, S.-É., Natale, W., Parent, L.E., 2018. Balance design for robust foliar nutrient diagnosis supervising the fertigation of banana "Prata" (Musa spp.). Nat. Sci. Rep. https://doi.org/10.1038/s41598-018-32328-y.

Egozcue, J.J., Pawlowsky-Glahn, V., Mateu-Figueras, G., Barceló-Vidal, C., 2003. Isometric logratio transformations for compositional data analysis. Math. Geol. 35, 279–300. https://doi.org/10.1023/A:1023818214614.

Epstein, E., Bloom, A.J., 2005. Mineral Nutrition of Plants: Principles and Perspectives. Sinauer Associates Inc., Sunderland, MA.

Fleming, P.J., Wallace, J.J., 1986. How not to lie with statistics: the correct way to summarize benchmark results. Commun. ACM 29, 218–221. https://doi.org/10.1145/5666.5673.

Gang, M., Sharma, N., Sharma, S., Kapoor, P., Kumar, A., Chunduri, V., Arora, P., 2018. Biofortified crops generated by breeding, agronomy, and transgenic approaches are improving lives of millions of people around the world. Front. Nutr. https://doi.org/10.3389/fnut.2018.00012.

Hernandes, A., Parent, S.-É., Natale, W., Parent, L.E., 2012. Balancing guava nutrition with liming and fertilization. Rev. Bras. Frutic. 34, 1224–1234. https://doi.org/10.1590/S0100-29452012000400032.

Howarth, R.J., 1996. Sources for a history of the ternary diagram. Br. Soc. Hist. Sci. 29 (3), 337–356.

Huang, X.Y., Salt, D.E., 2016. Plant ionomics: from elemental profiling to environmental adaptation. Mol. Plant 9, 787–797.

Ingestad, T., 1987. New concepts on soil fertility and plant nutrition as illustrated by research on forest trees and stands. Geoderma 40, 237–252. https://doi.org/10.1016/0016-7061(87)90035-8.

Kenworthy, A.L., 1967. Plant analysis and interpretation of analysis for horticultural crops. In: Stelly, M., Hamilton, H. (Eds.), Soil Testing and Plant Analysis, Part II. Soil Science Society of America, Madison, WI, pp. 59–75.

Lagatu, H., Maume, L., 1934. Le diagnostic foliaire de la pomme de terre. Ann. Éc. Natl. Agron. Montp. 22, 50–158 (in French).

Macy, P., 1936. The quantitative mineral nutrient requirements of plants. Plant Physiol. 11, 749–764. https://doi.org/10.1104/pp.11.4.749.

Marschner, H., 1986. Mineral Nutrition of Higher Plants. Academic Press, London/Orlando, FL. https://doi.org/10.1146/annurev.es.11.110180.001313.

Marchand, S., Parent, S.-É., Deland, J.P., Parent, L.E., 2013. Nutrient signature of Quebec (Canada) cranberry (*Vaccinium macrocarpon* Ait.). Rev. Bras. Frutic. 35 (1), 199–209.

Montes, R.M., Parent, L.E., de Amorim, D.A., Rozane, D.E., Parent, S.-É., Natale, W., Modesto, V.C., 2016. Nitrogen and potassium fertilization in a guava orchard evaluated for five cycles: effects on the plant and production. Rev. Bras. Ciênc. Solo. https://doi.org/10.1590/18069657rbcs20140532.

Munson, R.D., Nelson, W.L., 1990. Principles and practices in plant analysis. In: Westerman, R.L. (Ed.), Soil Testing and Plant Analysis. Soil Science Society of America, Madison, WI, pp. 359–387.

Nelson, L.A., Anderson, R.L., 1984. Partitioning of soil test-crop response probability. In: Stelly, M. (Ed.), Soil Testing: Correlating and Interpreting the Analytical Results. American Society of Agronomy, Madison, WI, pp. 19–28.

Nowaki, R.H.D., Parent, S.-É., Cecilio Filho, A.B., Rozane, D.E., Meneses, N.B., da Silva, J.A.D.S., Natale, W., Parent, L.E., 2017. Phosphorus over-fertilization and nutrient misbalance of irrigated tomato crops in Brazil. Front. Plant Sci. https://doi.org/10.3389/fpls.2017.00825.

Parent, L.E., 2011. Diagnosis of the nutrient compositional space of fruit crops. Rev. Bras. Frutic. 33, 321–334. https://doi.org/10.1590/S0100-29452011000100041.

Parent, L.E., Dafir, M., 1992. A theoretical concept of compositional nutrient diagnosis. J. Am. Soc. Hortic. Sci. 117, 239–242.

Parent, S.-É., Parent, L.E., 2017. Balance designs revisit indices commonly used in agricultural science and eco-engineering. In: Hron, K., Tolosana-Delgado, R. (Eds.), CodaWork 2017. The 7th International Workshop on Compositional Data Analysis. Abbadia San Salvatore, Italy, June 5-9, pp. 195–227. Proceedings book, http://www.compositionaldata.com/codawork2017/proceedings/ProceedingsBook2017_May30.pdf.

Parent, L.E., Parent, S.-É., Hébert-Gentile, V., Naess, K., Lapointe, L., 2013a. Mineral balance plasticity of cloudberry (*Rubus chamaemorus*) in Quebec-Labrador. Am. J. Plant Sci. 4, 1509–1520.

Parent, S.-É., Parent, L.E., Egozcue, J.J., Rozane, D.E., Hernandes, A., Lapointe, L., Hébert-Gentile, V., Naess, K., Marchand, S., Lafond, J., Mattos Jr., D., Barlow, P., Natale, W., 2013b. The plant ionome revisited by the nutrient balance concept. Front. Plant Sci. 4, 1–10. https://doi.org/10.3389/fpls.2013.00039.

Parent, L.E., Parent, S.-É., Rozane, D.E., Amorim, D.A., Hernandes, A., Natale, W., 2012a. Unbiased approach to diagnose the nutrient status of red guava (*Psidium guajava*). In: Santos, C.A.F. (Ed.), 3rd International Symposium on Guava and Other Myrtaceae, Petrolina, Brazil, April 23–25, 2012, pp. 145–159. https://doi.org/10.17660/ActaHortic.2012.959.18 ISHS Acta Horticulturae, Paper #959.

Parent, S.-É., Parent, L.E., Rozane, D.E., Hernandes, A., Natale, W., 2012b. Nutrient balance as paradigm of soil and plant chemometrics. In: Issaka, R.N. (Ed.), Soil Fertility. IntechOpen Ltd., London, pp. 83–114. https://doi.org/10.5772/53343

Prado, R.M., Caione, G., 2012. Plant analysis. In: Issaka, R.N. (Ed.), Soil Fertility. IntechOpen Ltd., London, pp. 115–134.

Rodgers, J.L., Nicewander, W.A., Toothaker, L., 1984. Linearly independent, orthogonal, and uncorrelated variables. Am. Stat. 38, 133. https://doi.org/10.2307/2683250.

Rozane, D.E., Parent, L.E., Natale, W., 2015. Evolution of the predictive criteria for the tropical fruit tree nutritional status. Cientifica 44, 102–112. https://doi.org/10.15361/1984-5529.2016v44n1p102-112.

Smith, G., Asher, G.J., Clark, C.J., 1997. Kiwifruit Nutrition Diagnosis of Nutritional Disorders. AG Press Communications Ltd, Wellington.

Souza, H.A., Parent, S.-É., Rozane, D.E., Amorim, D.A., Modesto, V.C., Natale, W., Parent, L.E., 2016. Guava waste to sustain guava (*Psidium guajava*) agroecosystem: nutrient "balance" concepts. Front. Plant Sci. 7, 1–13. https://doi.org/10.3389/fpls.2016.01252.

Swets, J.A., 1988. Measuring the accuracy of diagnostic systems. Science 240, 1285–1293.

Ulrich, A., 1952. Physiological bases for assessing the nutritional requirements of plants. Annu. Rev. Plant Physiol. 3, 207–228. https://doi.org/10.1146/annurev.pp.03.060152.001231.

Ulrich, A., Hills, F.J., 1967. Principles and practices of plant analysis. In: Stelly, M., Hamilton, H. (Eds.), Soil Testing and Plant Analysis. Part II. Soil Science Society of America, Madison, WI, pp. 11–24.

Vitosh, M.L., Silva, G.H., 1994. A rapid petiole sap nitrate-nitrogen test for potatoes. Commun. Soil Sci. Plant Anal. 25 (3–4), 183–190.

Walworth, J.L., Sumner, M.E., 1987. The diagnosis and recommendation integrated system (DRIS). Adv. Soil Sci. 6, 149–188. https://doi.org/10.1007/978-1-4612-4682-4.

Welch, R.M., 2002. The impact of mineral nutrients in food crops on global human health. Plant Soil 247, 83–90.

Wilkinson, S.R., 2000. Nutrient interactions in soil and plant nutrition. In: Sumner, M.E. (Ed.), Handbook of Soil Science. CRC Press, Boca Raton, FL, pp. D89–D112.

Yamane, D.R., 2018. Nutrient Diagnosis of Orange Crops Applying Compositional Data Analysis and Machine Learning Methods. (Ph.D. thesis). Universidade Estadual Paulista (UNESP), Jaboticabal.

Zuo, Y., Zhang, F., 2008. Iron and zinc biofortification strategies in dicot plants by intercropping with gramineous species. Agron. Sustain. Dev. 29, 63–71.

Web references

Beaufils, E.R., 1973. Diagnosis and recommendation integrated system (DRIS). Soil Sci. Bull., 1–132. Available from: http://www.worldcat.org/title/diagnosis-and-recommendation-integrated-system-dris/oclc/637964264?ht=edition&referer=di. (Accessed 2 May 2013).

Comas-Cufí, M., Thió-Henestrosa, S., 2011. CoDaPack 2.0. Universitat de Girona. Available from: http://ima.udg.edu/codapack/. (Accessed 10 December 2018).

Pawlowsky-Glahn, V., Egozcue, J.J., Tolosana-Delgado, R., 2011. Principal balances. In: Egozcue, J.J., Tolosana-Delgado, R., Ortego, M.I. (Eds.), 4th International Workshop on Compositional Data Analysis (Codawork 2011). San Feliu de Guixols, Spain. Available from: http://congress.cimne.com/codawork11/Admin/Files/FilePaper/p55.pdf. (Accessed 10 December 2018).

Rozane, D.E., Natale, W., Parent, L.E., Parent, S.-É., Santos, E.M.H., 2013. Programa de computador: Instituto Nacional da Propiedade Industrial—INPI: BR5120130003806. Universidade Estadual Paulista "Júlio de Mesquita Filho"; Université Laval. http://www.registro.unesp.br/#!/sites/cnd/. (Accessed 10 December 2018) (in Portuguese, Spanish, English and French).

van den Boogaart, K.G., Tolosana-Delgado, R., Bren, M., 2014. "Compositions": Compositional Data Analysis in R Package. Available from: http://cran.r-project.org/package=compositions. (Accessed 10 December 2018).

Further reading

Parent, S.-É., Barlow, P., Parent, L.E., 2015. Nutrient balances of New Zealand kiwifruit (*Actinidia deliciosa* cv. Hayward) at high yield level. Commun. Soil Sci. Plant Anal. 46, 256–271. https://doi.org/10.1080/00103624.2014.989031.

Parent, S.-É., Parent, L.E., Rozane, D.E., Natale, W., 2013c. Plant ionome diagnosis using sound balances: case study with mango (*Mangifera indica*). Front. Plant Sci. 4, 449. https://doi.org/10.3389/fpls.2013.00449.

13

Floral analysis in fruit crops: A potential tool for nutrient constraints diagnosis

Chengxiao Hu[a,b,*], Zhihao Dong[a,b], Yuanyuan Zhao[a,b], Wei Jia[a,b], Miaomiao Cai[a,b], Ting Zhan[a,b], Qiling Tan[a], Jinxue Li[a,b]

[a]College of Resources and Environment/Micro-element Research Center/Hubei Provincial Engineering Laboratory for New Fertilizers, Huazhong Agricultural University, Wuhan, People's Republic of China
[b]Key Laboratory of Horticultural Plant Biology (HZAU), MOE, Wuhan, People's Republic of China
*Corresponding author. E-mail: hucx@mail.hzau.edu.cn

1 Plant response to nutrient concentration

As we know, inadequate fertilizer particularly chemical fertilizer use usually results in yield losses, lower product quality, weaker resistance to stresses (such as low temperature, drought, disease, pests, etc.), and adverse environmental impact (such as nonpoint contamination). Scientific fertilizer recommendation for crop is becoming increasingly important due to: (i) growing demand in the global world market for food quality, (ii) increasing fertilizer particularly chemical fertilizer costs, (iii) increasing awareness of environmental problems caused by agriculture, and (iv) increasing climate changes such as drought, flooding, and heavy storms. For scientific fertilizer recommendation, correct diagnosis of plant nutrient deficiency is important and should be an integrative approach to crop production (Petra Marschner, 2012).

Plant growth response curve (Fig. 13.1) shows the relationship between plant growth or production yield and nutrient concentration in tissue (such as leaf); the diagnosis of plant nutrition status depends on this curve of diverse growing plant. There are four phases as follows: (i) Under severe deficiency, rapid increase in yield with added nutrient can cause a small decrease in nutrient concentration; this is called the *Steenberg effect*, which results from dilution of the nutrient in the plant by rapid plant growth. (ii) The nutrient concentration in the plant below which a yield response to added nutrient occurs means that *critical levels or ranges* vary among plants and nutrients but occur somewhere in the transition between nutrient deficiency and sufficiency. (iii) The nutrient concentration range in which added nutrient will not increase yield but can increase nutrient concentration means *luxury consumption* is often used

FIG. 13.1　Plant growth response curve to concentration of nutrient in tissue.

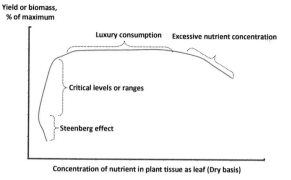

to nutrient absorption by plant that does not influence yield. (iv) When the concentration of essential or other elements is high enough to reduce plant growth and yield means *excessive nutrient concentration* can cause an imbalance in other essential nutrients that can also reduce yield. Depending on careful sampling, analysis and tests that are correlated with plant response, the roles of soil and plant analysis in quantifying crop nutrient requirement are that of high yield, high profit and good quality, friendly environment.

2　Nutrient deficiency symptom diagnosis

Several techniques are commonly employed to assess the nutrient status of plant: (i) nutrient deficiency symptoms of plants, (ii) analysis of tissue from plants growing on the soil, (iii) soil analysis, and (iv) biological tests in which the growth of either higher plants or certain microorganisms is used as a measure of soil fertility. Growing plants act as integrators of all ecosystem factors, visual characteristic symptoms may appear while a plant is lacking a particular nutrient, or plants exhibit a host of symptoms reflecting various disorders that can impact their health, vigor, and productivity to varying degrees. Identifying symptoms correctly is also an important aspect of nutrient management, as inappropriate fertilizer applications or other actions can be highly costly and sometimes detrimental to the ecosystem or themselves. Nutrient deficiency symptom diagnosis of plants or visual evaluation of nutrient stress requires systematic approach described as you can see in Table 13.1.

TABLE 13.1　Nutrient deficiency symptom diagnosis of crop.

1. Occurs only on new growth while persist in mature growth	1.1 Leaves with uniform color, growth decreasing with shorter internodes and bushy appearance	1.1.1 Leaves are large and dark green; shoots are long and willowy at early stages but have short and bushy secondary growth after long shoots dieback; gum blisters form along vigorous shoots at base of each petiole; multiple buds or sprouts form at the nodes; fruits show gum in tips of locules and brownish eruption on peel surface (exanthema)	Cu
		1.1.2 New leaves are pale green but turning yellow green with expanding; growth is sparse	N
		1.1.3 New growth is drab green, lusterless, sparse with misshapen leaves; fruit has gum deposits in the albedo peel layer	B
	1.2 Leaves with chlorosis patterns	1.2.1 Leaves are pointed and narrow with reduced size and sharply contrasting bright yellow mottling	Zn
		1.2.2 Approximately normal leaves in size and shape with pale green mottle over whole leaf, or marbled mottle with dark green color, and crooked veins network with light color in between	Mn
		1.2.3 Approximately normal leaves with feather-like straight veins on light green or yellow background, or leaves turning totally yellow and reduced size with twigs dying on the outer end of branches	Fe

TABLE 13.1 Nutrient deficiency symptom diagnosis of crop.—cont'd

2. Occurs on mature leaves with normal or nearly normal young leaves	2.1 Fading of chlorophyll in localized area and gradual enlargement with time	2.1.1 Chlorophyll fading starts from the leaf basal between midrib and lateral leaf margin, spreads usually outward with green "wedge" at the base of leaf, and turns inward and yellow wedge, or all leaf fade to golden bronze color	Mg
		2.1.2 Chlorophyll fading starts along lateral leaf margin, spreads inward about halfway to midrib, and forms irregular front margin	Ca
		2.1.3 Chlorophyll fading starts as blotches in distal half of leaf, blotches spread and coalesce with pale yellow firstly and deepened to bronze, foliage is drab, and fruit becomes smaller but of good quality	K
		2.1.4 Chlorophyll fading occurs in spots distributed randomly over whole leaf; brown centers with yellow or orange halo develop from spots that range from one-quarter to one-half inch in diameter but appear only in the fall	Mo
	2.2 Fading of chlorophyll not localized	2.2.1 Leaves fade to dull green then orange yellow eventually and turn burned tips and spots extremely. Fruits become coarse, spongy, and hollow-centered with thickened peel and above-normal acid	P
		2.2.2 Leaves fade to pale green then yellow color with whitish veins; fruits become sparse and pale-colored extern and internally with good quality but low juice concentration	N

Nutrient-deficiency symptoms appear only when the nutrient supply or concentration is so low that the plant can no longer function properly, so it should be used only as a supplement to other diagnostic techniques, the reasons are that the visual symptom may be caused by more than one nutrient, deficiency of one nutrient may be related to an excessive quantity of another, deficiency symptoms are difficult to distinguish among disease, insect, or herbicide damage which resemble micronutrients deficiency sometime in field, somehow a visual symptom may be caused by more than one factor. In contrast, symptoms resulted from diseases and pests are nonsymmetric or randomly positioned for individual plants and in a field particularly at the advanced phase of infections. Some differences are shown in Table 13.2.

Visual diagnosis of plant nutrient deficiency provides sometimes enough information for fertilization recommendation as fertilizer type and amount, especially applying foliar spraying micronutrient (B, Zn, Fe, and Mn) or Mg. But sometimes, the visual diagnosis is inadequate for making fertilizer recommendation, it is necessary to acquire additional information including soil pH, water and soil testing, and application of fertilizers, pesticides, etc., especially further chemical and biochemical leaf analysis.

3 Floral analysis in fruit crops: Sampling and analysis

The use of chemical analysis of plant material for nutrient diagnosis scientifically depends on the hypothesis that the plant growth or yield is related closely to nutrient concentration in shoot or tissue dry or fresh matter. The plant nutritional status is oftentimes better reflected by the element concentration in leaves than the other organs; thus, the leaves are usually used for plant analysis. In fact, the leaf or floral analysis is more important and efficacious for fruit crops than field crops or annual crops; the tissue sap analysis and soil test can be used for annual crops nutrient diagnosis and

TABLE 13.2 Distinguish between pathological diseases and nutrient deficiency.

Difference	Pathological disease	Nutrient deficiency
Process	Developing from a center	Developing in a scattered way
Relation to the soil type	Easily appeared in high fertility soil	Related to soil type significantly, such as acid soil
Relation to the climate	Easily appeared as it's cloudy and wet	Be relevant to drought, water logging, and low temperature

fertilizer recommendation because of crop nutrient mainly relying on soil fertility or soil supply, while soil test cannot reflect but only floral analysis does reflect the nutritional status of fruit tree because of the prior year nutrient storage in plant being the main sources for the following growth. Numerous researches have been carried out in the past to develop and improve leaf analysis for identifying nutritional constrains and subsequently the fertilizer recommendation in fruit trees (Srivastava and Singh, 2004).

Leaf analysis is a useful tool to detect problems and adjust fertilizer programs for citrus trees because leaf nutrient concentrations are the most accurate indicator of fruit crop nutritional status. Of all nutrient diagnosis methods, leaf tissue testing is useful to evaluate tree nutritional status with respect to most nutrients but is particularly effective for (i) macronutrients, primarily nitrogen (N) and potassium (K), that readily move with soil water and (ii) the micronutrients copper (Cu), manganese (Mn), zinc (Zn), and iron (Fe). Leaf tissue analysis is a much better indicator of the effectiveness of soil-applied fertilizer for these elements than soil analysis. Because citrus is a perennial plant, it is its own best indicator of appropriate fertilization. In addition, if particular elements have not been applied as fertilizer, leaf tissue analysis indicates the availability of those nutrients in the soil, leaves reflect nutrient accumulation and redistribution throughout the plant, so the deficiency or excess of an element in the soil is often reflected in the leaf. An annual leaf tissue sampling program can establish trends in tree nutrition resulting from fertilizer practices carried for several years (Obreza and Morgan, 2008). Considerable research involving citrus leaf testing has established its reliability as a management tool, but sampling guidelines should be followed precisely to insure that analytical results are meaningful. Leaf analysis integrates all the factors that might influence nutrient availability and uptake. Tissue analysis shows the relationship of nutrients to each other. For example, K deficiency may result from a lack of K in the soil or from excessive Ca, Mg, and/or Na. Similarly, adding N when K is low may result in K deficiency because the increased growth requires more K.

3.1 Sampling

Samples from field-growing plants are usually contaminated by dust or spraying and need to be washed; however, washing leaf samples may result in different loss among different elements. For example, washing leaves with water for a few minutes resulted in high boron loss due to boron passive diffusion across plasma membranes (Brown et al., 2002), while washing with diluted acids or chelating reagents did not remove the leaves surface contaminants of Fe, Zn, and Cu completely. The greatest challenge in utilizing leaf analysis for nutrient diagnosis purpose is the short-term fluctuations in nutrient concentration frequently; it means the leaf nutrient concentration reflecting deficiency, sufficiency, and toxicity ranges may change with climate factors, environmental factors and plant genotype, and growth stage of whole plant and single leaf. For example, the K concentration on DW basis declines with plant age, but the K concentration in plant cell sap remains relatively constant during growth (Petra Marschner, 2012). Procedures for proper sampling, preparation, and analysis of leaves have been standardized to achieve meaningful comparisons and interpretations. If done correctly, the reliability of the chemical analysis, data interpretation, fertilization recommendations, and adjustment of fertilizer programs will be sound. Therefore, considerable care should be taken from the time leaves are selected for sampling to the time they are received at the laboratory for analysis.

3.1.1 Leaf sample timing

Leaf samples must be taken at the correct time of year because nutrient concentrations within leaves continuously change. For citrus, as leaves age from spring to fall, N, P, and K concentrations decrease; Ca increases; and Mg first increases and then decreases. However, leaf mineral concentrations are relatively stable from 4 to 6 months after emergence in the spring. The best time to collect 4–6-month-old spring flush leaves is July and August (Obreza and Morgan, 2008). If leaves are sampled later in the season, summer leaf growth can easily be confused with spring growth. Leaf sampling time adopted for K analysis on different citrus-growing countries (Srivastava and Singh, 2004) can be seen in Table 13.3.

3.1.2 Leaf sampling technique

Leaves are collected from nonfruiting spring branches (6 months old) of the citrus trees in mid-September (Obreza et al., 1992) each year. A sampled citrus grove block or management unit should be no larger than 20 acres. The sampler should make sure that the selected leaves represent the block being sampled. Each leaf sample should consist of about 100 leaves taken from nonfruiting twigs of 15–20 uniform trees of the same variety and rootstock that have received the same fertilizer program. Use clean paper bags to store the sample, and label the bags with an identification number that

TABLE 13.3 Leaf sampling time (months) adopted for K analysis on different citrus-growing countries (Srivastava and Singh, 2004).

Months	Position and cultivar	Country
3–7	Spring cycle leaves from behind the green fruits	United States
4–7	Spring cycle leaves from fruit-bearing terminals	United States
4–7	Spring cycle leaves from nonfruiting terminals	California, United States
4–10	Spring cycle leaves from fruiting terminals	United States
4–10	Nonfruiting terminals of Washington Naval and Valencia	United States
5–7	Nonfruiting terminals of Naval	United States
9–10	Nonfruiting terminals of Valencia	Israel
5–7	Nonfruiting terminals of Valencia and Washington	Morocco
6–7	Fruiting terminals of Clementine mandarin	Corsica
6–7	Fruiting terminals of Valencia	Brazil
4–5	Nonfruiting terminals of Valencia	Carmer, Ivory C.
8–10	Nonfruiting terminals of Valencia	Israel
8–10	Nonfruiting terminals of Valencia	South Africa
5.5–7.0	Fruiting terminals of Valencia	Morocco
7–9	Leaves from fruiting terminals of grapefruit and lemon	South Africa
7–9	Nonfruiting terminals of Valencia	South Africa
4–6	Middle leaves from nonfruiting spring flush terminal of Valencia	Australia
4–7.5	Nonfruiting terminals of kinnow mandarin	India
6–8	Middle leaves from nonfruiting terminals of Nagpur mandarin	India
3–5	Middle leaves from nonfruiting terminals of Acid lime in central India	India
4.5–6.5	Middle leaves from nonfruiting terminals of Acid lime in north-west India	India
5–6	Fruiting terminals of Eureka lemon on *Citrus volkameriana*	South Africa

can be referenced when the analytical results are received. Avoid immature leaves due to their rapidly changing composition. Do not sample abnormal-appearing trees, such as at the edge of the block or at the end of rows, because they may be coated with soil particles and dust. Do not include diseased, insect-damaged, or dead leaves in a sample. Select only one leaf from a shoot and remove it with its petiole (leaf stem).

To minimize soil- and tree-type variability of the sampling grove, the sampling scheme is one area of the nutritional testing process controlled by the individual taking the sample. Thus, the individual needs to ensure that its leaf sample is representative of a particular area trees. For sampling purposes, partition the grove into management units of not more than 20 acres; each unit contains similar soil and scion/rootstock types. For small groves, the entire grove is partitioned into these units with a sample taken from each. For large groves, it is unfeasible to sample the entire grove; indicator block is used as a designated zone within a uniform span of grove from which the sample is taken (e.g., a 20-acre block within a uniform 100-acre span of grove). Aerial photos are useful for designing indicator blocks. The sample results obtained from the indicator block are assumed to represent the entire span, and management decisions made from the sample data are applied to the entire span; the same block should be sampled repeatedly in the succeeding years. A more elaborate approach to citrus leaf tissue sampling is to use the global positioning satellites (GPS) and geographic information system (GIS). Groves sample in a regular, grid-like pattern and record the geographic position of each sample using GPS technology. The analyzed results of the samples are processed with GIS, and contour maps are drawn. According to the maps, the spatial variation of tree nutritional level and areas of high or low nutrition in whole grove can be determined and identified. This method is more expensive than the traditional one but may provide a higher level and more information that can improve management decisions.

The leaf position on the tree twig has distinctly significant effect on mineral element distribution; to avoid these positional effects, sampled leaves should be collected from the north, east, south, and west directions and the same height of the tree's periphery.

In special cases as sampling for diagnosing disorders of trees growth, leaf samples should be collected from both affected trees and normal trees. Trees selected for comparison sampling should be of the same age, scion type, and rootstock. If possible, confine the sampling area to trees that are close to each other.

3.2 Handling and analysis of leaf samples

Protect leaves from heat and keep them dry, place them in a refrigerator for overnight storage if they cannot be washed and oven dried the day of collection. For macronutrient analysis, leaves do not need to be washed; if accurate micronutrient analysis is desired, the leaves will need to be washed. Leaf samples are firstly inactivated at 105°C for 30 min and then drying at 70°C to constant weight in a ventilated oven. The oven-dried leaves should be ground to fine powder with an agate mortar or ground prototype.

Leaves that have been sprayed with micronutrients for fungicidal (Cu) or nutritional (Mn and Zn) purposes should not be analyzed for those elements because it is almost impossible to remove all surface contamination from sprayed leaves. For accurate Fe and B or other micronutrient determinations, leaf samples require handwashing that is best done shortly after collection before they dehydrate. For micronutrient determinations, rub the leaves between the thumb and forefinger while soaking them in a mild detergent solution, and then, thoroughly rinse with pure water. It is difficult to remove all surface residues, but this procedure removes most of them.

If samples require handwashing for accurate element determination, it is best done when the leaves are still in a fresh condition. Laboratories do not normally handwash leaves, so washing should be finished by the person who collected the sample at that time. When the sample arrives at the laboratory, the following steps are typically taken: (1) Dry and finely grind the leaves; (2) weighted sample is either digested in strong acid (for N analysis) or ashed in furnace (for all other elements); (3) the element concentration of solution originated from the digested or ashed are measured; (4) nutrient concentrations are expressed as either percentage or milligram per kilogram in the tissue. Procedures for plant tissue analysis are usually the same among laboratories because the entire amount of each nutrient in the leaves is measured, thus results from different laboratories can be directly compared.

4 Floral analysis in fruit crops: Results interpretation and nutrient constraints diagnosis

4.1 Result interpretation

As total, each nutrient concentration of leaf sample is measured in laboratory, with standard sample simultaneously, there should be no difference in leaf analysis results among laboratories.

To interpret analysis results, compare the values with the leaf analysis standards, for example, citrus tree leaf shown in Tables 13.4 and 13.5; these standards are based on long-term field observations and experiments conducted in different countries with different citrus varieties, rootstocks, and management practices and are used to gauge citrus tree nutrition level throughout the world. Well-defined categories of classification for citrus leaf tissue analysis values are "deficient," "low," "optimum," "high," and "excess." As shown in Tables 13.4 and 13.5; remember that this classification applies only to the standard age leaf sample taken from mature trees as described earlier and is not valid for young, nonbearing trees.

4.2 Nutrient constraint diagnosis and recommendation

Actually, the first commercial citrus or fruit tree growers understood some needs of the macronutrients nitrogen, phosphorus, and potassium for citrus growth. Meanwhile, worldwide researches proved gradually that plants needed nutrients in addition to nitrogen, phosphorus, and potassium to grow properly. In 1939, A.F. Camp and B.R. Fudge showed that secondary nutrients and micronutrients were needed to grow citrus. Included examples were deficiency symptoms of copper, zinc, manganese, magnesium, boron, and iron. As early as in 1908, the yellow spot disease in citrus was first reported in Florida, which was rather widespread and caused extensive defoliation and tree death, and till 1951, it was Ivan Stewart and C.D. Leonard who reported that the yellow spot disease in citrus was due to molybdenum deficiency and could be corrected by spraying as little as 1 oz of sodium molybdate per acre. Then,

TABLE 13.4 Guidelines for interpretation of orange tree leaf analysis based on 4–6-month-old spring flush leaves from nonfruiting twigs (Koo et al., 1984).

Element	Unit of measure	Deficient	Low	Optimum	High	Excess
N	%	<2.2	2.2–2.4	2.5–2.7	2.8–3.0	>3.0
P	%	<0.09	0.09–0.11	0.12–0.16	0.17–0.30	>0.30
K	%	<0.7	0.7–1.1	1.2–1.7	1.8–2.4	>2.4
Ca	%	<1.5	1.5–2.9	3.0–4.9	5.0–7.0	>7.0
Mg	%	<0.20	0.20–0.29	0.30–0.49	0.50–0.70	>0.70
Cl	%	–	–	<0.2	0.2–0.7	>0.70
Na	%	–	–	–	0.15–0.25	>0.25
Mn	mg/kg	<18	18–24	25–100	101–300	>300
Zn	mg/kg	<18	18–24	25–100	101–300	>300
Cu	mg/kg	<3	3–4	5–16	17–20	>20
Fe	mg/kg	<35	35–59	60–120	121–200	>200
B	mg/kg	<20	20–35	36–100	101–200	>200
Mo	mg/kg	<0.05	0.06–0.09	0.10–2.0	2.0–5.0	>5.0

TABLE 13.5 Leaf analysis standards for mature, bearing citrus trees, which exist from years of experimentation in Florida and California, based on 4–6-month-old spring cycle leaves from nonfruiting terminals.

Element	Unit of measure	Deficient	Low	Optimum	High	Excess
N	%	<2.2	2.2–2.4	2.5–2.7	2.8–3.0	>3.0
P	%	<0.09	0.09–0.11	0.12–0.16	0.17–0.30	>0.30
K	%	<0.7	0.7–1.1	1.2–1.7	1.8–2.4	>2.4
Ca	%	<1.5	1.5–2.9	3.0–4.9	5.0–7.0	>7.0
Mg	%	<0.20	0.20–0.29	0.30–0.49	0.50–0.70	>0.70
Cl	%	–	–	0.05–0.10	0.11–0.25	>0.25
Na	%	–	–	–	0.15–0.25	>0.25
Mn	mg/kg	<17	18–24	25–100	101–300	>300
Zn	mg/kg	<17	18–24	25–100	101–300	>300
Cu	mg/kg	<3	3–4	5–16	17–20	>20
Fe	mg/kg	<35	35–59	60–120	121–200	>200
B	mg/kg	<20	20–35	36–100	101–200	>200
Mo	mg/kg	<0.05	0.06–0.09	0.10–2.0	2.0–5.0	>5.0

the fruit leaf analysis and fertilizer recommendation were globally accepted and developed rapidly while particularly focused on the diagnosis standard or norm of leaf analysis.

The leaf nutrient norms are developed with great variety of diagnostic methods using different citrus cultivars; the differences in diagnostic methods apart from climate condition of growing area and nutrient uptake characteristic of citrus cultivars are the major contributory factors toward variation in reference values being recommended in relation to yield.

It's not easy to establish the diagnosis standard or norm of fruit tree leaf analysis. Embleton published "Citrus fertilization" and "Leaf analysis as a diagnostic tool and guide to fertilization" in *The Citrus Industry* (Vol. 3) edited by Reuther as early as 1973, particularly discussed the relationship among soil-nutrient-plant and explained the leaf analysis with detail. What Embleton had introduced was rapidly cited in *Soil Testing and Plant Analysis* edited by Walsh and Soil Science Society of American in the same year. After that, lots of scientists have done researches and setup diagnosis standards or norms of citrus leaf analysis as we can see in Table 13.6.there are some variations in standards

TABLE 13.6 Leaf macronutrient norms (%) for diagnosis in different countries and area of citrus.

Resources		N	P	K	Ca	Mg
Chapman (1949a,b)		2.20–3.16	0.09–0.182	0.38–1.12	–	–
Chapman (1960)		2.20–2.70	0.12–0.18	1.0–1.7	3.0–6.0	0.3–0.6
Rodriguez and Gallo (1961)		2.20	0.12	1.00	3.00	0.30
Smith (1966)		2.5–2.7	0.12–0.16	1.2–1.7	3.0–4.5	0.30–0.49
Jorgensen and Price (1978)		2.4–2.6	0.14–0.16	0.9–1.2	3.0–6.0	0.23–0.60
Takidze (1981)		2.5–2.8	0.19–0.20	1.5–1.7	2.5–3.0	0.30–0.35
Koo et al. (1984)		2.5–2.7	0.12–0.16	1.2–1.7	3.0–4.9	0.30–0.49
Japan (Yu, 1984)	Early Satsuma	2.71–3.2	0.16–0.20	1.01–1.6	2.01–4.5	0.21–0.30
	General Satsuma	2.91–3.4	0.16–0.20	1.01–1.6	2.01–4.5	0.21–0.30
	Summer orange, ponkan	2.51–3.0	0.16–0.20	0.71–1.3	2.01–4.5	0.21–0.30
	Navel orange	3.11–3.6	0.16–0.20	1.41–2.0	2.01–4.5	0.21–0.30
Wang (1985)		3.0–3.5	0.15–0.18	1.0–1.6	2.5–5.0	0.30–0.60
Zhuang et al. (1986)		2.7–3.3	0.12–0.15	1.0–1.8	2.3–2.7	0.25–0.38
Ko and Kim (1987)		2.97	0.15	1.48	3.57	0.34
Wang (1987)		3.0–3.5	0.15–0.18	1.0–1.6	2.5–5.0	0.30–0.60
Wang et al. (1988)		2.7–3.3	0.12–0.15	1.0–1.8	2.3–2.7	0.25–0.38
Gallasch and Pfeiler (1988)		2.4–2.69	0.14–0.17	0.70–1.49	–	–
New Zealand, 1989		2.4–2.6	0.14–0.16	0.9–1.2	3.0–6.0	0.25–0.6
Koto et al. (1990)		2.4–2.6	0.13–0.15	–	3.4–4.8	0.20–0.29
Zhuang et al. (1991)		2.5–3.1	0.14–0.18	1.4–2.2	2.0–3.8	0.32–0.47
Zhou et al. (1991)		2.75–3.25	0.14–0.17	0.7–1.5	3.2–5.5	0.2–0.5
Wang et al. (1992a,b)		2.5–3.0	0.12–0.18	1.0–2.0	2.0–3.5	0.22–0.40
Wang et al. (1992a,b)	Ponkan	2.9–3.5	0.12–0.16	1.0–1.7	2.5–3.7	0.25–0.50
	Guanxi pomelo	2.5–3.1	0.14–0.18	1.4–2.2	2.0–3.8	0.32–0.47
	Sweet orange	2.5–3.3	0.12–0.18	1.0–2.0	2.0–3.5	0.22–0.40
Du Plessis et al (1992)	Valencia orange	2.0–2.4	0.11–0.16	0.95–1.5	3.5–5.5	0.30–0.55
	Navel	2.4–2.8	0.11–0.16	0.9–1.1	3.5–5.5	0.30–0.55
	Grapefruit	2.3–2.6	0.11–0.16	0.9–1.6	3.5–5.5	0.30–0.55
	Lemon	1.9–2.2	0.11–0.15	1.1–1.4	3.5–5.5	0.30–0.55
Dettori et al. (1996)		2.15–2.55	0.10–0.12	0.9–1.3	4.4–5.5	0.3–0.4
Perez (1996)		2.5–2.7	0.15–0.18	–	3.0–4.0	0.4–0.5
Beridze (1986)		2.1–2.2	0.15–0.16	0.95–1.23	5.74–6.94	0.27–0.34
Quaggio et al. (1996)		2.3–2.7	0.12–0.16	1.0–1.5	3.5–4.5	0.25–0.40
Zhuang et al. (1997)		2.8–3.3	0.14–0.18	1.4–2.1	2.0–4.0	0.25–0.45
Xie et al. (2014)		2.7–3.2	0.13–0.20	1.2–1.9	2.5–4.0	0.22–0.45
Recommendation norm		2.5–2.9	0.13–0.17	1.0–1.6	2.8–4.5	0.28–0.45

Data from Srivastava, A.K., Singh, S., 2004. Diagnosis methods, Nutrient Diagnosis and Management in Citrus. National Research Center for Citrus, Nagpur, pp. 3–19.

or norms among citrus cultivars and planting areas, but it is difficult to distinguish particularly significantly. We use the average or high frequency value as the recommended standards or norms showed as the last row as showed in Tables 13.7 and 13.8; the macronutrient (N, P, K, Ca, and Mg) concentration range of the standards is relatively narrower than the micronutrients (Fe, Mn, Cu, Zn, B, and Mo); the upper limit/lower limit of macronutrients is 1.16, 1.31, 1.60, 1.61, and 1.61 and of micronutrient is 2.00, 4.50, 3.00, 2.80, 3.33, and 10.0, respectively; the lower ratio of the micronutrient means the narrower range; the higher ratio of macronutrient means the wider range and a wider range concentration for plant to tolerate. The lower limit of the nutrient diagnosis norm is often the critical concentration of the deficiency for the plant, while the upper limit is of the excess or toxicity for plant. According to this diagnosis standard, the leaf diagnosis results of nearly whole China's citrus orchard are showed in Table 13.8; the citrus leaves of more than half sampled orchards are deficient of Ca, Mg, and Zn; all these deficient nutrients shall be suggested to apply.

Leaf analysis shows promise of being a reliable guide for nitrogen fertilization practices. An 8-year study in northern San Diego County was started with 11-year-old Fuerte trees. Statistical analyses of the data collected show that the curvature in the graph was highly significant at the 1% level. The trees with nitrogen leaf values below the most productive range were deficient of nitrogen and weakly vegetative, with small sparse and light green to yellow leaves and less new shoot growth. The trees with nitrogen leaf values higher than the most-productive-range were highly vegetative with large, dense, deep green leaves and abundant long new shoot growth. It also indicates that from 100 to 150 pounds of actual nitrogen per acre annually will generally be adequate unless a volunteer or planted cover crop exists in an orchard (Embleton et al., 1960).

The interpretation of leaf nutrient levels is based on the promise that there is a significant relationship between the nutrient concentration in leaf, plant growth, and fruit yield with the aim at predicting fertilizer requirement of fruit

TABLE 13.7 Leaf micronutrient norms (mg/kg) for diagnosis in different countries and areas of citrus.

Resources		Fe	Mn	Cu	Zn	B	Mo
Chapman (1949a,b)		70–200	20–80	4–100	20–80	20–100	0.1–0.2
Chapman (1960)		60–150	25–100	5.1–15.0	25–100	50–200	0.10–3.0
Smith (1966)		50–120	25–49	5–12	25–49	36–100	0.10–1.0
Jorgensen and Price (1978)		12–60	10–25	5–10	25–100	–	–
Koo et al. (1984)		60–120	25–100	5–16	25–100	36–100	0.10–1.0
Wang (1985)		50–120	25–100	4–10	25–100	–	–
Wang (1987)		50–120	25–100	4–10	25–100	30–100	–
Gallasch and Pfeiler (1988)		50–129	–	6–15	25–60	–	–
New Zealand, 1989		60–120	25–100	5–10	25–100	30–100	
Koto et al. (1990)		40–46	14–23	3.7–10	23–30	17–19	–
Zhuang et al. (1991)		60–140	15–140	8–17	24–44	15–50	–
Zhou et al. (1991)		60–170	20–40	4–8	13–20	40–110	–
Wang et al. (1992a,b)		90–160	20–100	4–18	25–70	25–100	–
Wang et al. (1992a,b)	Ponkan	50–140	20–150	4–16	20–50	20–60	–
	Guanxi pomelo	60–140	15–140	8–17	24–44	15–50	–
	Sweet orange	90–160	20–100	4–18	25–70	25–100	–
Dettori et al. (1996)		100–150	17–37	5–7	19–43	68–85	–
Quaggio et al. (1996)		50–120	35–50	–	35–50	36–100	0.10–1.0
Perez (1996)		30–80	25–90	–	30–90	–	–
Zhuang et al. (1997)		50–160	20–100	5–18	20–50	25–100	–
Xie et al. (2014)		50–200	20–150	5–25	20–50	35–150	0.05–1.0
Recommendation norm		60–120	20–90	5–15	25–70	30–100	0.1–1.0

Data from Srivastava, A.K., Singh, S., 2004. Diagnosis methods, Nutrient Diagnosis and Management in Citrus. National Research Center for Citrus, Nagpur, pp. 3–19.

TABLE 13.8 The leaf diagnosis and soil testing results of China's citrus orchards (not published, more than 2000 leaf and soil samples were tested).

Nutrient	Deficient low (%)		High excess (%)	
	Soil	Leaf	Soil	Leaf
N	58.80	35.89	4.43	20.04
P	39.77	13.32	20.74	24.92
K	36.78	43.32	28.08	19.90
Ca	57.51	54.06	25.98	12.47
Mg	65.10	61.49	20.31	4.96
Fe	22.96	7.61	79.06	49.53
Mn	16.93	12.45	47.53	51.19
Zn	22.72	72.66	25.86	0.96
Cu	22.63	20.78	60.75	43.25
B	81.02	32.51	0.00	9.14

tree. An extensive survey of 108 Khasi mandarin orchards located in humid tropical climate of northeast India covering 590 km^2 from 50 locations of 8 states was carried out during 2001–05. The optimum values of leaf nutrient concentration in relation to fruit were fixed through multivariate quadratic correlation and regression based on pooled data of both the states considering the similarity in growing conditions and cultural practices. Linear coefficient of correlation and regression analysis were used to test the soil properties governing the fruit yield and quality. The different nutrient concentration in leaf showed significant difference when divided in various yield levels, except Fe and Mn, while fruit yield was more strongly correlated with N ($r = 0.708$, $P = .01$), P ($r = 0.697$, $P = .01$), Ca ($r = 0.817$, $P = .01$), Cu ($r = 0.519$, $P = .01$), and Zn ($r = 0.793$, $P = .01$) than K ($r = 0.436$, $P = .01$), Fe ($r = -0.309$, $P = .05$), and Mn ($r = -0.296$, $P = .05$). The leaf nutrient optimum values using multivariate quadratic regression analysis were estimated as 2.32% N, 0.10% P, 1.92% K, 2.26% Ca, 0.30% Mg, 158.4 mg/kg Fe, 82.0 mg/kg Mn, 12.2 mg/kg Cu, and 30.3 mg/kg Zn in relation to fruit yield of 46.3 kg/tree according to regression equation (Singh et al., 2006), and you can see it in Table 13.9 (Srivastava and Singh, 2002, 2006). Based on the relationship between leaf nutrient content and fruit yield or using multivariate quadratic regression analysis, the leaf nutrient optimum values for diagnosis can be used not only to identify the nutrients constrains but also to predict the fruit yield or sometimes fruit quality.

TABLE 13.9 Leaf nutrient optimum values for diagnosis in citrus related to fruit yield.

Nutrients	Nagpur mandarin	Khasi mandarin	Kinnow mandarin	Mosambi sweet orange	Sathgudi sweet orange
N (%)	1.70–2.81	1.97–2.56	2.28–2.53	1.98–2.57	2.01–2.42
P (%)	0.09–0.15	0.09–0.10	0.10–0.13	0.091–0.17	0.09–0.12
K (%)	1.02–2.59	0.99–1.93	1.28–1.63	1.33–1.72	1.12–1.82
Ca (%)	1.80–3.28	1.97–2.49	2.12–3.12	1.73–2.98	1.93–2.73
Mg (%)	0.43–0.92	0.24–0.48	0.32–0.53	0.32–0.69	0.36–0.53
Fe (mg/kg)	74.9–113.4	84.6–249.0	52.3–89.4	69.5–137.1	53.5–82.1
Mn (mg/kg)	54.8–84.6	41.6–87.6	41.7–76.3	42.2–87.0	48.7–79.3
Cu (mg/kg)	9.8–17.6	2.13–14.4	6.1–10.3	6.6–15.8	3.7–8.9
Zn (mg/kg)	13.6–29.6	16.3–26.6	21.3–28.5	11.6–28.7	16.5–23.2
Yield (kg/tree)	47.7–117.2	31.6–56.3	61.8–140.3	76.6–137.9	81.2–145.3

Data from Srivastava and Singh (2002) and Srivastava, A.K., Singh S., 2006. Diagnosis of nutrient constraints in citrus orchards of humid tropical India. J. Plant Nutr. 29 (6), 1061–1076.

TABLE 13.10 Macronutrient norms (%) for leaf diagnosis of the other fruit tree.

Resources		N	P	K	Ca	Mg
DECIDUOUS FRUIT TREE						
Pear	Li et al. (1987)[a]	2.0–2.4	0.12–0.25	1.0–2.0	1.0–2.5	0.25–0.80
Peach		2.8–4.0	0.15–0.29	1.5–2.7	1.5–2.2	0.30–0.70
Grape		0.6–2.4	0.14–0.41	0.44–3.0	0.7–2.0	0.26–1.50
	Zhu et al. (2008)	1.0–1.7	0.31–0.54	2.3–2.9	0.6–1.1	0.28–0.41
Grape ، Zhang et al. (2003)	Ontario, Canada	0.7–1.3	0.15–0.40	1.0–3.0	1.0–3.0	0.50–1.50
	New Zealand	0.8–1.0	0.21–0.50	1.5–3.5	1.4–2.5	0.31–0.38
	Michigan, United States	0.8–1.2	0.16–0.30	1.5–2.5	0.5–1.0	0.30–0.50
	New York, United States	0.8–1.2	0.14–0.30	1.5–2.5	1.2–2.0	0.35–0.50
Grape	Recommendation	0.8–1.5	0.18–0.41	1.4–2.9	0.9–2.0	0.33–0.80
Apple	Li et al. (1987)[a]	2.0–2.6	0.15–0.23	1.0–2.0	1.0–2.0	0.22–0.35
	Jiang et al. (1995)	2.24	0.184	1.40	1.39	0.288
	Guiyang (2004)	2.31–2.50	0.14–0.17	0.73–0.98	1.7–2.2	0.37–0.43
	New Zealand	2.2–2.5	0.15–0.20	1.0–1.4	1.2–1.6	0.25–0.35
	American	2.0–2.25	0.2–0.3	1.25–1.75	1.25–1.75	0.25–0.40
	France	2.3–2.5	0.16–0.18	1.8–2.0	1.49–2.0	0.22–0.26
Apple[b]	Australia	2.0–2.4	0.15–0.20	1.2–1.5	1.1–2.0	0.21–0.25
	American	1.8–3.0	0.15–0.40	1.3–2.5	1.5–2.0	0.24–0.40
	Italy	2.0–2.6	0.16–0.24	1.3–1.9	1.4–2.0	0.24–0.36
	Canada	2.0–2.7	0.15–0.30	1.4–2.2	0.8–1.5	0.24–0.40
	Japan	3.4–3.6	0.17–0.19	1.3–1.5	0.8–1.3	0.27–0.40
	Hebei of China	2.2–2.9	0.09–0.13	0.85–1.04	1.29–1.55	0.11–0.12
	Henan of China	2.0–2.6	0.21–0.27	0.79–1.07	–	–
	Shanxi of China	2.3–2.5	0.14–0.17	0.7–1.0	1.7–2.3	0.37–0.43
	Shandong of China	2.7–3.2	0.11–0.25	0.6–0.9	0.9–1.4	0.19–0.27
	Liaoning of China	2.5–2.9	0.17–0.25	0.95–1.29	0.95–1.75	0.28–0.52
	Jiangsu of China	1.5–1.8	0.08–0.10	1.03–1.34	1.25–1.79	0.25–0.40
Apple	Recommendation	2.2–2.6	0.15–0.22	1.1–1.5	1.2–1.9	0.25–0.35
EVERGREEN FRUIT TREES[c]						
Litchi	Wang et al. (1988)	1.5–2.2	0.12–0.18	0.7–1.4	0.3–0.8	0.18–0.38
Longan		1.5–2.0	0.10–0.17	0.4–0.8	0.7–1.7	0.14–0.30
	Liu et al. (1986)	>1.7	0.12–0.20	0.6–0.8	1.5–2.5	0.20–0.30
	Zhuang et al. (1997)	1.4–1.9	0.10–0.18	0.5–0.9	0.9–2.0	0.13–0.30
	Recommendation	1.5–1.9	0.10–0.18	0.5–0.8	1.0–2.1	0.15–0.30
Banana	Jamaica	2.5	0.20	2.74	–	–
	India, 1981	2.8	0.35	3.10	–	–
	Australia, 1986	2.8–4.0	0.20–0.25	3.1–4.0	0.8–1.2	0.30–0.46
	Zhang (2001)	3.5–4.3	0.23–0.27	3.7–4.4	0.28–0.44	0.15–0.37
	Recommendation	2.8–4.2	0.20–0.30	3.1–4.2	0.8–1.2	0.20–0.40

[a] The sampling positions. **Apple**, leaves in the middle part of new shoot on mid-July to mid-August. **Pear**, leaves in the middle part of new shoot or brachyplast on mid-July to mid-August. **Peach**, leaves in the middle part of new shoot on mid-July to mid-August. **Grape**, petiole of leaf locates in the middle part of new shoot-bearing fruit on July to August.

[b] Wang et al. (2018).

[c] The sampling position. **Banana**, the third leaves from the plant roof at heading period. **Longan**, **Litchi**, the second to third lobule grows at the second compound leaf from the normal annual shoot top or roof, with 5–7 months and 3–5 months growth, respectively.

TABLE 13.11   Micronutrient norms (%) for leaf diagnosis of the other fruit tree.

Resources		Fe	Mn	Cu	Zn	B	Mo
DECIDUOUS FRUIT TREE							
Pear	Li et al. (1987)[a]	100	30–60	6–50	20–60	20–50	–
Peach		100–250	35–280	7–25	12–60	25–60	–
Grape		30–100	30–65	10–50	25–50	25–60	–
	Zhu et al. (2008)	26–52	230–453	78–206	19–47	–	
Grape, Zhang et al. (2003)	Ontario, Canada	15–100	20–200	–	–	–	
	New Zealand	31–100	25–200	5–20	25–50	31–50	
	Michigan, United States	20–100	30–60	10–20	30–60	25–50	
	New York, United States	30–100	50–1000	10–50	30–60	25–50	
Grape	Recommendation	25–100	30–200	10–35	25–60	25–50	
Apple	Li et al. (1987)[a]	150–290	25–150	5–15	15–80	20–60	–
	Jiang et al. (1995)	210	78.1	–	39.6	37.5	–
	Guiyang (2004)	120–150	52–80	20–50	24–45	33–37	–
	New Zealand	90–150	30–90	6–20	20–50	20–50	–
	United States	100–300	50–100	5–20	20–50	25–50	–
	France	60–240	50–120	5–12	9–53	25–35	–
Apple[b]	Australia	>100	50–100	6–20	20–50	–	–
	American	0–300	–	5–20	25–50	–	–
	Italy	40–150	>8	>1	>15	–	–
	Canada	25–200	20–200	–	15–100	–	–
	Japan	–	50–200	10–30	30–50	–	–
	Hebei of China	87–110	65–80	15–22	8.4–9.8	–	–
	Henan of China	89–120	28–40	3–4	25–35	–	–
	Shanxi of China	120–150	52–80	20–50	24–45	–	–
	Shandong of China	217–353	121–351	–	24–45	–	–
	Liaoning of China	114–176	4–20	60–216	25–49	–	–
	Jiangsu of China	274–494	41–99	85–354	40–84	–	–
Apple	Recommendation	112–218	45–112	18–60	22–50	25–45	–
EVERGREEN FRUIT TREES[c]							
Longan	Zhuang et al. (1997)	30–100	40–200	4–10	10–40	15–40	–
Banana	Australia, 1986	70–2200	25–1000	7–20	21–35	20–80	–
	Zhang (2001)	145–253	567–2339	12–26	30–86	8–18	–
	Recommendation	70–250	25–1000	7–26	21–86	8–80	–

[a] *The sampling positions.* **Apple,** *leaves in the middle part of new shoot on mid-July to mid-August.* **Pear,** *leaves in the middle part of new shoot or brachyplast on mid-July to mid-August.* **Peach,** *leaves in the middle part of new shoot on mid-July to mid-August.* **Grape,** *petiole of leaf locates in the middle part of new shoot-bearing fruit on July to August.*
[b] *Wang et al. (2018).*
[c] *The sampling position.* **Banana,** *the third leaves from the plant roof at heading period.* **Longan,** **Litchi,** *the second to third lobule grows at the second compound leaf from the normal annual shoot top or roof, with 5–7 months and 3–5 months growth, respectively.*

The nutrient norms for leaf diagnosis of the other fruit crops and recommendation norms are also raised as you can see in Tables 13.10 and 13.11; the norms of several fruit trees are gathered as you can see in Table 13.12. It shows the differences in nutrient requirement among several fruit crops: the banana leaves need higher N, P, and K concentrations; the citrus leaves require higher N, Ca, and Mg contents; the Mg content of grape leaves and the

TABLE 13.12 Macronutrient (%) and micronutrient (mg/kg) norms for several fruit tree leaf diagnosis.

	Evergreen fruit tree			Deciduous fruit tree	
Nutrient	Citrus	Longan	Banana	Apple	Grape
N	2.5–2.9	1.5–1.9	2.8–4.2	2.2–2.6	0.8–1.5
P	0.13–0.17	0.10–0.18	0.20–0.30	0.15–0.22	0.18–0.41
K	1.0–1.6	0.5–0.8	3.1–4.2	1.1–1.5	1.4–2.9
Ca	2.8–4.5	1.0–2.1	0.8–1.2	1.2–1.9	0.9–2.0
Mg	0.28–0.45	0.15–0.30	0.20–0.40	0.25–0.35	0.33–0.80
Fe	60–120	30–100	70–250	112–218	25–100
Mn	20–90	40–200	25–1000	45–112	30–200
Cu	5–15	4–10	7–26	18–60	10–35
Zn	25–70	10–40	21–86	22–50	25–60
B	30–100	15–40	8–80	25–45	25–50
Mo	0.1–1.0	–	–	–	–

TABLE 13.13 The relationship between fruit quality and yield of citrus (not published, five sweet orange orchards for each yield level were tested in the same area in Yunnan Province).

Fruit (kg/ha)	g/fruit	Peel (g/fruit)	TSS (%)	TA (%)	TSS/TA	Edible rate (%)
>1500	188.1	47.85	10.18	0.67	15.2	74.6
1000–1500	185.9	52.62	9.38	1.05	8.9	71.7
<1000	171.3	43.64	9.31	0.92	10.1	74.5

Cu contents of deciduous fruit trees are higher than the other trees. But it is difficult to distinguish the variations in P and particularly the micronutrient contents among fruit crops, because of the wide range of leaf micronutrient norms, so some more detailed researches are necessary in future.

Based on the field investigation, most of the standards were calculated from leaf sample analysis of the higher fruit yield tree group, meaning that the nutrient concentration of the higher yield tree leaves was suitable or the critical range or concentration. How it related to the fruit quality? You can see in Table 13.13 that the higher yield group fruit had higher TSS (total soluble solid in juice) concentration and TSS/TA ratio but lower TA (total acid in juice) content than the lower yield group. It seems that the high fruit yield and quality of fruit crop is consistent, the nutrient concentration of the high yield tree leaves is also suitable for high fruit quality tree, and the leaf analysis or nutrient diagnosis is the available tool for nutrient constraints of fruit yield and quality.

The goal in nutrition management is to maintain leaf nutrient concentrations within the optimum range every year. If the interpretation for a particular nutrient is not optimum, various strategies can be used to address the situation as you can see in Table 13.14.

The optimum or critical nutrient concentration (CNC) of leaf is one approach of leaf analysis and nutrient diagnosis; due to the problems arising from different CNCs during plant development and due to the importance of nutrient ratios or interactions, the Diagnosis and Recommendation Integrated System (DRIS) was developed by Beaufils in 1973. The system bases on large amount of analysis data on plant nutrient concentrations, which can be used to calculate the optimal nutrient ratios of N/P, N/K, K/(Ca + Mg), etc.; Table 13.15 shows the difference between CNC and DRIS briefly. However, the recommendations of DRIS are not always accurate; particularly, it is not the choosing method for cropping system with wide variations of annual crop species, cultivars, rotations, input, and farming scale.

5 Floral analysis in fruit crops: Combining use of the soil testing

Soil testing has been practically utilized in agriculture and horticulture for many years successfully. Its availability is closely related to (i) the extent to the data that can be calibrated with field trials of fertilizer and (ii) the interpretation of

TABLE 13.14   Adjusting a citrus fertilization program based on leaf tissue analysis.

Nutrient	What if it is less than optimum in the leaf? Option could be:	What if it is greater than optimum in the leaf? Option could be:
N	1. Check yield 2. Check tree health 3. Review water management 4. Review and increase N fertilizer rate	1. Check soil organic matter 2. Review and decrease N fertilizer rate 3. Grow some plants
P	1. Apply P fertilizer	1. Do nothing or grow green manure
K	1. Increase K fertilizer rate 2. Apply foliar K fertilizer 3. Apply crop straw	1. Decrease K fertilizer rate 2. Check leaf Mg and Ca status
Ca	1. Check soil pH and soil moisture 2. Check tested soil Ca status 3. Consider applying lime or soluble Ca fertilizer depending on soil pH 4. Apply foliar Ca fertilizer	1. Do nothing
Mg	1. Check tested soil Mg status 2. Check soil pH 3. Consider applying dolomitic lime or soluble Mg fertilizer depending on pH 4. Apply foliar Mg fertilizer	1. Do nothing
Micronutrients	1. Check soil pH and adjust if needed 2. Apply foliar micronutrients 3. Include micronutrients in soil-applied fertilizer	1. Check for spray residue on tested leaves 2. Do nothing

Revised from Obreza, T.A., Morgan, K.T., 2008. Soil and leaf tissue testing. In: Nutrition of Florida Citrus Trees, second ed. Florida Cooperative Extension Service, Institute of Food and Agricultural Sciences, University of Florida, pp. 24–32.

TABLE 13.15   The difference between CNC and DRIS.

CNC	DRIS
Nutrient concentration varied with growth stage of plant	Optimum value could obtained for each growth stage of plant
Identifying the deficiency or sufficiency of different nutrients, without evaluation for interaction of nutrients	Identifying the demand order of different nutrients
Critical concentration varied with environmental factors	Optimum value may not be affected by environmental factors

the analysis. Soil testing measures organic matter content, pH, and extractable nutrient concentrations that is useful in formulating and improving an enriching soil fertility program, particularly useful when conducted for several consecutive years so that variation trends can be observed.

Soil testing mainly reveals the potential or capacity of tested soil supplying nutrients to plant, but does not characterize efficiently or sufficiently the nutrient mobility in the soil. Especially, it commonly fails to display the information of soil structure, microorganism, root development, architecture, etc., which are very important factors affecting plant nutrient uptake under field condition. Similar to leaf analysis, methods to determine organic matter and soil pH are universal, so results should not differ among laboratories. However, soil nutrient extraction procedures can vary from lab to lab. Soil testing uses a wide range of conventional extractive reagents such as dilute acid, salts, complex agents, and water. Several accepted chemical procedures exist that remove different amounts of nutrients from the soil because of the extraction strength or ability of the reagents. To draw useful information from soil tests, consistency in the use of a single extraction procedure from year to year is important to avoid confusion when interpreting the amount of nutrients extracted (Petra Marschner, 2012).

Soil extraction procedure does not measure the total nutrient amount in soil, nor does it measure the quantity actually available to fruit trees. A perfect extractive reagent would extract nutrient from the soil nearly the same as the amount available to the plant; it means that the utility of a soil testing procedure depends on how well the extractable values correlate with the nutrient amount that the plant takes up; the process of relating these two quantities is called calibration. Calibration means that nutrient availability to plants increases foreseeing with soil testing value increasing. Low soil testing values imply that crop will respond to fertilization with particular nutrient; in turn, high soil testing values indicate the soil can supply most of the plant needs, and up to no fertilization required, the soil testing value is called the critical or sufficiency soil testing value, respectively. A soil testing is only useful if it is calibrated with plant response.

It's all known that there is a long-term argument as whether soil testing or plant analysis is more suitable for making fertilization recommendations. As you can see in Table 13.8, only the nutrients of Ca, Mg, Mn, and Cu exhibit the consistency between leaf analysis and soil testing diagnosis. In fruit crops, soil analysis alone is not a satisfactory guide for fertilization recommendation; a citrus grower cannot rely on soil analysis alone to formulate a fertilizer program or diagnose a nutritional problem in a grove, mainly because of the difficulty in determining accurately enough the nutrient availability in root zones where deep-rooted plants take up most nutrients. In reality, the fruit crop as perennial plant stands in a settled site for a long time of several years to more than hundreds years sometimes, the year after year's inconformity between root absorbing nutrients and farmer applying fertilizers, and the depth and width of tree roots distribution, are more deeply affecting the accuracy of soil testing and availability of soil nutrients.

Both leaf tissue and soil testing can be valuable, but leaf analysis provides more useful information about fruit tree nutrition than soil analysis. With the results of a soil test, one tries to predict how much of a particular nutrient will be available to plants in the future. Obviously, the further the future prediction is made, the less accurate it will be. Predictive soil testing works best with (1) short-term crops, and (2) nutrients are not very mobile in soil. Thus, for long-term crops such as fruit tree, predictive sampling should be used for only those nutrients that have slight mobility in soils, such as phosphorus (P), calcium (Ca), copper (Cu), and magnesium (Mg). Soil testing has limited value for the more mobile nutrients such as N and K (Obreza and Morgan, 2008).

The single most useful soil test in citrus grove is for pH. Soil pH greatly influences nutrient availability, while some nutrient deficiencies can be avoided by maintaining soil pH between 6.0 and 6.5. In some cases, soil tests can determine the best way to correct a deficiency identified by leaf analysis. For example, Mg deficiency may result from low soil pH or excessively high soil Ca. Dolomitic lime applications are advised if the pH is too low, but magnesium sulfate is preferred if soil Ca is very high and the soil pH is in the desirable range. If soil Ca is excessive and soil pH is relatively high, then foliar application of magnesium nitrate is recommended. A poor relationship may exist between soil test values and leaf nutrient concentrations in perennial crops like citrus. Often, fruit trees contain sufficient levels of a nutrient even though the soil test is low. On the other hand, a high soil test does not assure a sufficient supply to the trees. Tree nutrient uptake can be hindered by problems like drought or flooding stress, root damage, and cool weather. Tissue analysis combined with soil tests can help identify the problem.

References

Beridze, Z.A., 1986. Interrelation between ions in the soil and mandarin tree. Subtropicheskie Kul'tury 4, 109–113.

Brown, P.H., Bellaloui, N., Wimmer, M.A., Bassil, E.S., Ruiz, Z., Hu, H., Pfeffer, H., Dannel, F., Römheld, V., 2002. Boron in plant biology. Plant Biol. 4, 205–223.

Chapman, H.D., 1949a. Tentative leaf analysis standards. Calif. Citrog. 34, 518.

Chapman, H.D., 1949b. Citrus leaf analysis: nutrient deficiencies, excesses and fertilizer requirements of soil indicated by diagnostic aid. Calif. Agri. (11), 11, 12, 14.

Chapman, H.D., 1960. Leaf and soil analysis in citrus orchards. Univ. Calif. Div. Agri. Sci. Ext. Ser. Manual 25, 53.

Dettori, A., 1996. S.E.M. Studies on Penicillium italicum—'Star Ruby' grapefruit interaction as affected by hot water dripping. In: Proc. Int. Soc. Citriculture , vol. 2. 1288–1289.

Du Plessis, S.F., Koen, T.J., Odendaal, W.J., 1992. Interpretation of leaf analysis by means of N/K ratio approach. Proc. 7th Int. Citrus Congr. 2, 553–555.

Embleton, T.W., Jones, W.W., Garber, M.J., 1960. Leaf analysis as a guide to nitrogen. Calif. Agric., 1, 12.

Gallasch, P.T., Pfeiler, G.R., 1988. Develop a unique leaf analysis service, including a computer programme. In: Goren, R., Mendel, K. (Eds.), Proc. Sixth Int. Citrus Congress. Balaban Publ, Rehovot, Israel.

Guiyang, A., 2004. Study on the Standard Range and Effective Factors of Nutritional Elements in Apple Leaves. (Master Degree thesis). Northwest Agriculture & Forestry University, Xian.

Jiang, Y., Gu, M., Shu, H., 1995. Nutrient diagnosis of 'Starking Delicious' apple. Acta Hortic. Sin. 22 (3), 215–220 (in Chinese with English abstract).

Jorgensen, K.R., Price, G.H., 1978. The citrus leaf and soil analysis system in Queensland. Proc. Int. Soc. Citriculture 1, 297–299.

Ko, K.D., Kim, S.K., 1987. Chemical properties of soils and leaf mineral contents in Cheju citrus orchards. J. Korean Soc. Hort. Sci 28 (1), 45–52.

Koo, R.C.J., Anderson, C.A., Stewart, I., Tucker, D.P.H., Calvert, D.V., Wutscher, H.K., 1984. Recommended fertilizers and nutritional sprays for citrus. In: Fla. Coop. Extension Serv. Bulletin 536D.

Koto, M., Shunrokuro, F., Shengzhi, S., 1990. Leaf analysis standards for determining fertilizer needs and its computerisation. In: Bangyan, H., Qian, Y. (Eds.), Int. Symp. Citriculture, Gunazhou, China 383–389.

Li, G., Su, R., Shen, T., 1987. Studies on the nutritional ranges in some deciduous fruit trees. Acta Hortic. Sin. 14 (2), 81–89 (in Chinese with English abstract).

Liu, X., Zheng, J., Pan, D., Xie, H., 1986. An investigation on the leaf nutritional diagnosis criteria of Longan (Dimocarpus longan Lour.). J. Fujian Agric. Coll. 15 (3), 237–245 (in Chinese with English abstract).

Obreza, T.A., Morgan, K.T., 2008. Soil and leaf tissue testing. In: Nutrition of Florida Citrus Trees, second ed. Florida Cooperative Extension Service, Institute of Food and Agricultural Sciences, University of Florida, pp. 24–32.

Obreza, T.A., Alva, A.K., Hanlon, E.A., Rouse, R.E., 1992. Citrus Grove Leaf Tissue and Soil Testing: Sampling, Analysis, and Interpretation. Cooperative Extension Service, University of Florida, Institute of Food and Agricultural Sciences.

Perez, G.C., 1996. Argentina. Proc. Int. Soc. Citri., vol. 2.

Petra Marschner, J., 2012. Diagnosis of deficiency and toxicity of nutrients. In: Marschner's Mineral Nutrition of Higher Plants. Elsevier Ltd, pp. 299–312.

Quaggio, J.A., Cantarella, H., Mattos Jr., D., 1996. Soil testing and leaf analysis in Brazil—recent developments. Proc. Int. Soc. Citriculture 2, 1269–1275.

Rodriguez, O., Gallo, J.R., 1961. Pela analise foliar. Bragantia, Campinas, SP 20, 184–202.

Singh, S., Shivankar, V.J., Gupta, S.G., Singh, I.P., Srivastava, A.K., Das, A.K., 2006. Methods of indentifying nutrientional disorder. In: Citrus in Neh Region. Mudrashilpa Offset Printers, pp. 74–80.

Smith, P.F., 1966. Citrus nutrition. In: Childers, N.F. (Ed.), Temperate to Tropical Fruit Nutrition. Horticultural Publications, Rutgers, The State University, New Brunswick, NJ, USA, pp. 208–228.

Srivastava, A.K., Singh, S., 2002. Soil analysis based diagnostic norms for Indian citrus cultivar. Commun. Soil Sci. Plant Anal. 33 (11&12), 1689–1706.

Srivastava, A.K., Singh, S., 2004. Diagnosis methods. In: Nutrient Diagnosis and Management in Citrus. National Research Center for Citrus, Nagpur, pp. 3–19.

Srivastava, A.K., Singh, S., 2006. Diagnosis of nutrient constraints in citrus orchards of humid tropical India. J. Plant Nutr. 29 (6), 1061–1076.

Takidze, R.M., 1981. Determination of the requirement of citrus crops in inorganic fertilizers by the foliar diagnostics. Subtropicheskie Kul'tury 1, 61–68.

Wang, T.C., 1985. Study on guiding fertilization for satsuma orange by leaf analysis. J. Soil Sci. (Turang. Tongbo) 16 (6), 275–277.

Wang, T.C., 1987. Study on citrus leaf analysis for guiding fertilization. J. Guizhou Agri. Sci. 2, 36–37 (in Chinese).

Wang, R., Zhuang, Y., Chen, L., Xie, Z., Li, L., 1988. Study on leaf optimum nutrients concentration of main subtropical fruit trees in Fujian. Subtrop. Plant Res. Commun. 2, 1–5 (in Chinese).

Wang, R., Zhuang, Y., Chen, L., Xu, W., Li, L., Xie, Z., 1992a. Optimum range of mineral element contents in leaves of sweet orange. Subtrop. Plant Res. Commmun. 21 (1), 11–19 (in Chinese with English abstract).

Wang, R., Zhuang, Y., Chen, L., Xu, W., Xie, Z., 1992b. Optimum range of mineral element contents in leaves of citrus. Zhejiang Citrus 4, 1–4 (in Chinese).

Wang, L.-B., Chen, X.-W., Li, T.-Y., Shi, Y., Wang, S.-H., Gao, Z.-H., Qu, S.-C., 2018. Appropriate content of leaf mineral element in 'Fuji' apple orchards of Fengxian, Jiangsu Province. Acta Agric. Univ. Jiangxiensis 10 (1), 56–65 (in Chinese with English abstract).

Xie, W.-L., Li, J., Shi, Q., Li, M.-G., Xie, Z.-C., 2014. Studies on the optimum parameters for mineral nutrition in Newhall navel orange leaves. Acta Hortic. Sin. 41 (6), 1069–1079 (in Chinese with English abstract).

Yu, L., 1984. Introduction of citrus nutrient diagnosis of leaf analysis in Japan. Zhejiang Citrus (4), 15–17. 47 (in Chinese).

Zhang, H., 2001. Studies on Reverse Season Banana Nutritive Peculiarity and Nutritional Diagnostic Ranges in Hainan Province. Huanan Tropical Agricultural University, Hainan (Master Degree thesis).

Zhang, Z.-Y., Qiu, H.-L., Zhang, Y.-F., Ma, W.-Q., 2003. Application of leaf analysis in nutrition diagnosis and grape fertilization. Sino-Overseas Grapevine Wine 5, 17–21 (in Chinese with English abstract).

Zhou, X., Cheng, C., Lu, B., Z., L., Hui, L., 1991. Studies on mineral nutrient indices for sweet orange leaves. J. Southwest Agric. Univ. 13 (1), 15–20 (in Chinese with English abstract).

Zhu, X.-P., Liu, W., Zhang, J.-Z., Wang, T.-K., Qi, Y.-S., 2008. Study on the nutritional ranges of Cabernet Sauvignon grape by leaf analysis. North. Hortic. (10), 51–52 (in Chinese with English abstract).

Zhuang, Y., Li, L., Jiang, Y., Su, M., 1986. Preliminary study on nutrition status of Ponkan orchard in Fujian. Fujian Citrus 1, 13–16 (in Chinese).

Zhuang, Y., Wang, R., Chen, L., Xie, Z., Xu, W., Huang, Y., Zhou, Z., 1991. Optimum range of mineral element contents in the leaves of Guanxi Honey Pomelo (Citrus grandis). J. Fujian Acad. Agric. Sci. (2), 52–58 (in Chinese with English abstract).

Zhuang, Y., Wang, R., Xie, Z., Xu, W., 1997. Optimum range of mineral element contents in the leaves of parent navel orange. Subtrop. Plant Res. Commmun. 26 (2), 1–6 (in Chinese with English abstract).

14

Calcium nutrition in fruit crops: Agronomic and physiological implications

Fulya Gulbagca, Hakan Burhan, Fatima Elmusa, Fatih Sen*

Sen Research Group, Department of Biochemistry, Faculty of Arts and Science, Dumlupınar University, Evliya Çelebi Campus, Kütahya, Turkey

*Corresponding author. E-mail: fatih.sen@dpu.edu.tr

1 Introduction

The first stage of crop cultivation starts with the germination of agricultural production in appropriate conditions. Due to adverse ecological conditions, technical errors (such as low soil temperature or the formation of a slime layer in soil) and negativity caused by crop structure negatively affect germination and seedling output. Many studies have been conducted to determine the relationship between crop dormancy and germination. Ripe crops cannot be germinated, even if environmental conditions such as temperature, humidity, oxygen, and light are appropriate. Due to the impermeability of the crop shell water and oxygen deficiency, water and expanding embryo cannot penetrate the crop crust, incomplete embryos and embryo or growth in the shell due to the growth of the crop germination of the crop is not germinated. This event, which we call dormancy, is seen in the crops of fruit species such as plums, almonds, peaches, apricots, and cherries. The dormancy of the crop embryo and the outer shell is different. Removal of the layers surrounding the embryo removes the dormancy of the outer shell. Germination starts in crops that are stronger than the mechanical limitation of the outer shell. Dormant crops cannot germinate because the conditions for germination are not sufficient. If the necessary conditions for germination are provided, nondormant crops germinate. Water is adequate for the crops of some plant species. In some other species, light is essential in other factors such as soil conditions and temperature fluctuations. If these factors are not involved, germination is prevented, and the crops enter

A.K. Srivastava, Chengxiao Hu (eds.)
Fruit Crops: Diagnosis and Management of Nutrient Constraints
https://doi.org/10.1016/B978-0-12-818732-6.00014-9

rest (resting). Dormancy status is associated with essential gibberellic acid (GA) and abscisic acid (ABA). It has been determined that GA and ABA applications have significant effects on germination in mutant plants showing an endogenous hormone deficiency (Fidler et al., 2018; Tuan et al., 2018).

In order to break the dormancy and ensure proper germination and output of crops sown under inappropriate conditions, some applications are made before planting and after harvesting. These applications called priming include crops, folding, classification according to size, wetting preplanting, growth regulators, acid etching, vitamins, nutrients or osmotic solutions, planting in a gel after germination, coating, and banding (Fidler et al., 2018; Jaime et al., 2018; Li et al., 2018; Blancaflor et al., 2014).

2 Environmental factors affecting crop germination

2.1 Water

It is one of the most critical factors to start crop germination and to maintain the formation of the seedlings. The osmotic potential in the soil for germination depends on the presence of salts in the water. If the humidity in the environment is low, the presence of high salt in the germination environment can have an adverse effect (Mattana et al., 2018; Yin et al., 2018; Ishibashi et al., 2018).

2.2 Temperature

One of the most important factors regulating the germination time is temperature. It is very related to the control of dormancy. The germination rate is generally low at low temperatures. Temperate climate crops have the ability to germinate at an optimum temperature range of 4.5–40°C while they grow at an optimum temperature of 24–30°C. The temperature, which varies depending on the species, but increases to an optimum temperature usually accelerate the development of the product through the phases of life cycles (phenological) (Mattana et al., 2018; Yin et al., 2018; Ishibashi et al., 2018).

2.3 Oxygen

The gas exchange between the germination medium and the embryo is vital for rapid and uniform germination. Oxygen plays a role in the respiration process of germinated crops. Oxygen uptake increases as the amount of metabolic activity increases. Oxygen deposition is limited when excessive water is present in the medium (Mattana et al., 2018; Yin et al., 2018; Ishibashi et al., 2018).

2.4 Light

The light-stimulated dormancy in some plants has been determined while some plants don't experience this phenomenon. It has been decided that the underlying mechanism of reaction to light in crops is related to phytochrome, which is a chemically active pigment. It was discovered that red and infrared rays had an effect on GA biosynthesis in lettuce and Arabidopsis crops. In the studies, when the crops are kept in water for a second while being exposed to red light, germination rates have increased. It was determined that it had an inhibitory effect when exposed to infrared light. It was determined that the crop shell and embryo were sensitive to light in the plants and the impact of light was lost when they were removed. Light is a stimulating factor in crop germination in some species. The effect of light is on the GA3-oxidase enzyme in m-RNA, which catalyzes the final step of the bioactive GA3 synthesis. The predicted GA3 biosynthesis site appears to be related to the light-sensitive region determined by using microwaves in lettuce crops. Nongerminated mutants with GA deficiency have been useful in studying how GA stimulates crop germination (Mattana et al., 2018; Yin et al., 2018; Ishibashi et al., 2018).

3 Nutrient feeding of crops

As in all living things in nature, the presence of at least 18 elements is necessary for the growth of plants and crops. These elements are examined in two groups as macronutrients and micronutrients. The elements taken from air and water are generally seen as nonmineral plant nutrients. Although they cover approximately 95% of the plant content, they are not seen as necessary in plant nutrition (Wierzbowska et al., 2018; Kacar and Katkat, 2010).

For plants, Ca is one of the essential elements taken directly from the soil. Ca is the third most-used plant nutrient. The plant is an integral part of the cell wall and is therefore known as the plant nutrient that regulates the cell wall structure. Polygalacturonase, which accumulates in plant tissues in the absence of Ca, causes the breakdown of Ca-pectates. As a result, the cell walls are broken and the tissues are affected. Symptoms of this phenomenon are seen especially at the leaves and in the upper parts of the trunk. Ca pectates in cell walls also protect plant tissues and fruits against fungal and bacterial infections. Ca, which has many more functions besides its mentioned functions, also has essential functions in fruit formation, development, and quality (Dong et al., 2018). Although there are various symptoms of Ca deficiency in green parts, the effect on fruit quality is remarkable. The minerals such as anorthite, plagioclase, pyroxenes, amphiboles, augite, hornblende, apatite, calcite, limestone, dolomite, gypsum, marl, and Ca phosphates are the Ca sources of the soil (Thangella et al., 2018) (Fig. 14.1).

The majority of Ca ions, which are released as a result of the cleavage and degradation of these minerals, are absorbed by the exchange complexes. Ca ions enhance the soil structure by increasing granulation and adjusts the soil pH. More products are provided than soils with good structure. Ca in the intake of nutrients plays a role in the deposition of toxic substances in plants and soil. Ca is effective on root secretion in plants and protects plant tissues against freezing-thawing stress. In case of sufficient Ca, the plants are more resistant to diseases. Ca plays a vital role in protein formation and transport of carbohydrates in plants. In the soils that are poor in Ca, less product is obtained, and the protein content in the product is very low. Ca deficiency in plants slows the growth of meristem tissues. Some of these are some of the nutritional problems encountered in Ca deficiency, such as flower nose rot, decay, cracking, deformations, shortening of shelf life. and reduced storage quality. Ca is taken directly in the Ca^{2+} form through the ion channels that can pass through the CaPu in the epidermal cell walls of the root hairs in contact with the soil solution, and is carried to xylem conduction bundles (Dong et al., 2018).

Ca, which has a significant effect on plant nutrition, is usually taken up by the plant's endodermis cells. A well-developed root and root hair at the root end of the root significantly expands the absorption surface area, especially in the intake of nutrients such as Ca provides significant convenience (Alcock et al., 2017). Water is a significant factor in the uptake and transport of the plant. The mass flow, which is the primary mechanism for Ca transportation to the root domain, occurs only in the presence of water. At the same time, water is the main factor in the transportation of Ca taken by the root. Therefore, it is inevitable that plants show Ca deficiency in conditions where there is no water movement. Water movement in the plant is closely related to transportation. Even if there is sufficient Ca in the soil under the conditions where the transcription rate decreases, the plants cannot benefit from this and the symptoms of Ca deficiency appear (Alomari et al., 2017). Due to the fact that the mineral nutrition of the fruits is mostly through the phloem, it is tough to transfer elements such as Ca, which is immobile in phloem, to fruits. When the presence of conditions preventing the xylem motion is added to this situation, it is inevitable that the most affected organ is the fruits. This becomes even more important in greenhouses where natural transpiration is largely hindered.

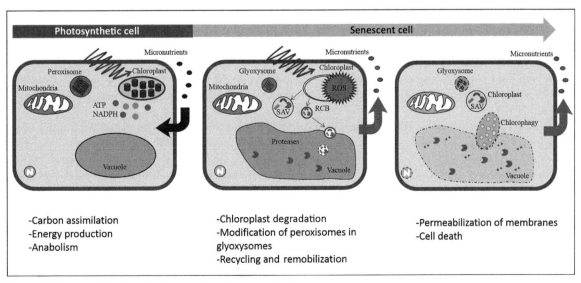

FIG. 14.1 Utilization of nutrients in photosynthetic and senescent cell.

As mentioned above, the application of Ca from foliage to eliminate the nutritional problems of plants in the presence of various factors that prevent the negative intake of Ca from the soil has been one of the frequently used methods (García-Gaytán et al., 2018). Ca in arid regions has an antagonistic effect on the uptake of other nutrients, especially micronutrients. For example, potassium, iron, phosphorus, and other elements are transformed into forms that cannot be used by plants (Foyer et al., 2018).

Ca, in the form of primary minerals in soils and bound to organic and inorganic soil colloids as Ca^{2+}, increases the coagulation of soil colloids, improves soil structure, and increases the durability of soil aggregates (Alori and Babalola, 2018). Ca is an essential element for plant growth and development. In the process of cell growth and development, the permeability of the membrane, the stabilization of tissues, and the quality criteria of the plant to gain the requirements are macro elements (Karkanis et al., 2018). According to the soil science encyclopedia, Ca, essential for fauna, microflora, plants, and soil, is a vital element for the function and structural properties of plant cell plasma membranes, which have significant effects on the physical and chemical properties of soil (García-Jiménez et al., 2018).

For Ca, which makes up about 3.5% of the Earth's crust; photosynthesis, hormone metabolism, enzyme activation and the role of plants in ensuring effective water use are still under discussion (Achari and Kowshik, 2018). Several studies have reported that Ca is an essential element in plant nutrition and that it has an effect on yield and quality. Reductions in the Ca content of soils due to various reasons manifest themselves in the plant, especially in the generative circuit, and adversely affect the development. Therefore, starting from the introduction of the plant into the generative phase after basic fertilization, calcium fertilization studies are available in the literature considering the plant type and Ca requirement. Ca is a motionless element in the plant and the response is generally faster than the calcium given by the leaf (Artyszak, 2018).

4 Applications of calcium nutrition in fruit crops

Fruit trees are biological and economically significant. The development of treatments that accelerate crop production is significant. Proper food management is necessary to optimize production. Fruit trees that are exposed to more chemical fertilizer rather than the soil it needs cause environmental pollution. The fertilization method, which is one of the methods applied, is a more preferred and environmentally friendly method (Thavarajah et al., 2015).

Product yields of perennial fruit trees that are frequently exposed to various abiotic stresses are limited. The solution focuses on the stimulation of plant cell development programs for modern fruit tree physiology using stress-related tolerance mechanisms and biostimulants. Biological stimulants are used in the production of fruits due to the fact that fruit intake increases nutrient uptake, promotes plant growth, and minimizes fertilizer use (Kunicki et al., 2010). Although the precise mechanisms of biostimulators are unknown, they are widely used in the fruit industry. In addition, the use of priming techniques compared to natural or synthetic compounds in plants has aroused considerable interest in recent years to awaken habits against environmental stresses (Tanou et al., 2012).

A thorough review of the molecular aspects of primate in plants, including fruit crops, has recently been reviewed (Conrath, 2011; Tanou et al., 2012; Molassiotis et al., 2016). In literature reviews, it has been reported that Ca, given in gypsum form from the soil, significantly increases the yield of marketable fruit with the Ca contents of the soil and the leaves of the obtained plant (Zhang et al., 2017). In another study, it was reported that Ca, K, and P contents increased in fruit leaves with gypsum and calcium nitrate applications and Ca fertilization had a positive effect on tuber quality (Sharma et al., 2017).

In a study in which the effects of different levels of calcium nitrate fertilizer on the yield and quality of strawberry plants were investigated, significant increases were reported in the Ca content of the plant with their yield and quality characteristics (Marwein et al., 2017). Thanks to the application of Ca leaf, a lot of research has been done in fruits and vegetables that report that quality criteria can be increased and marketing value will be increased. Another common finding obtained in these studies is that preharvest and postharvest Ca applications allow for a more healthy plant (Tian et al., 2016).

One of the essential features of Ca, which increases the quality criteria in plants, is the ratio of the Ca pectate compound to the total and cell walls. Studies have shown that preharvest or postharvest application of Ca increases the amount of this compound (Kurt et al., 2017). In many plants, mainly tomatoes, watermelons, melons, and peppers, physiological disorders occur with Ca deficiency. Ca/N, Ca/Mg, Ca/K imbalances can be caused if there is not enough Ca in the soil. Therefore, emphasis should be placed on balanced fertilization (Berkovich et al., 2016).

5 Calcium mineral evaluation after migration in fruits

In studies, lime and dolomite applied to the soil show that the soil has changed the Ca content and the tuber yield significantly in the potato plant. Salinity, germination, seedling struggle, the establishment of the crop, and yield stages have harmful effects on agriculturally essential plants. After the plants are exposed to salinity stress, osmotic effects cause cell division in stoma closure (Geilfus, 2018). Decreasing the photosynthetic area required to achieve optimal growth leads to aging of the leaves. Plant tolerance to environmental stress generally includes responses (Gao et al., 2018; Saini and Keum, 2018).

As a very important element (Kim et al., 2016), silicon may support the polymerization of SiO_3^- gas, a potential chelating agent for toxic elements (Debona et al., 2017; Etesami and Jeong, 2018). As a result, crop germination, which is the first stage of plant development, is prevented or does not occur as a result of the crop inhibitor in its structure, the crop being hard and impermeable, and the various technical errors and adverse environmental factors made during crop cultivation. Therefore, taking into account the crop characteristics and ecological conditions that vary according to the plant species and varieties, some preapplication to optimize crop germination can directly affect crop germination and indirectly affect plant growth (Fig. 14.2).

In another study, the Ca^{2+} and other nutrient concentrations of soil, leaves, and fruits were studied in the application of Ca for grapefruit. The 2-year field experiment evaluated the yield and fruit quality. The application of calcium on the soil, leaf, or fruit Ca^{2+} content did not have an effect on the mulberry sugar concentration, similar to the control process. In the treatments where $CaCl_2$ was applied to the soil, the concentration of chlorite in the soil, leaves, and fruits has been found to be higher. In the second season, berry tightness was higher in these same treatments, but this effect was lost after storage simulation. In a similar study, data such as calcium responses, mineral nutrient concentrations, quality and postharvest changes, mineral nutrient contents of grapes obtained with different application modes are given in Table 14.1–14.4 (Bonomelli and Ruiz, 2010).

TABLE 14.1 Response of grapes to calcium supplied through different modes of application (Pooled data: 2005–07).

Treatments	EC (dS/m)	Ca (cmol/kg)	Mg (cmol/kg)	K (cmol/kg)	Cl (mmol/L)
Control	0.45	24.95	5.15	0.95	0.49
Ca foliar	0.45	25.20	5.05	0.95	0.44
Ca soil	0.75	26.45	5.20	1.05	1.26
Ca foliar + Ca soil	0.95	26.6	5.25	0.80	1.26
P	0.019	0.373	0.237	0.308	0.005

EC, electric conductivity; P: probabilities from ANOVA. $P < 0.05$.
Based on Bonomelli, C., Ruiz, R., 2010. Effects of foliar and soil calcium application on yield and quality of table grape cv. 'thompson seedless'. J. Plant Nutr. 33, 299–314. https://doi.org/10.1080/01904160903470364.

TABLE 14.2 Mineral nutrient concentration for Ca application during the growing seasons of grapes (Pooled data: 2005–07).

Treatments	N	P	K	Ca	Mg	Cl
Control	1.66	0.14	0.90	2.20	0.38	0.34
Ca foliar	1.68	0.16	0.91	2.36	0.39	0.37
Ca soil	1.69	0.15	0.90	2.31	0.42	0.56
Ca foliar + Ca soil	1.68	0.15	0.90	2.42	0.41	0.57
P	0.568	0.325	0.465	0.397	0.353	0.003

Based on Bonomelli, C., Ruiz, R., 2010. Effects of foliar and soil calcium application on yield and quality of table grape cv. 'thompson seedless'. J. Plant Nutr. 33, 299–314. https://doi.org/10.1080/01904160903470364.

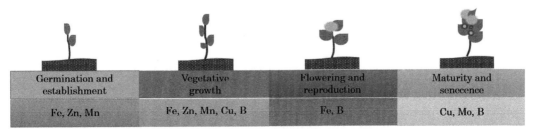

FIG. 14.2 Utilization of micronutrients in different growth phases of plants.

TABLE 14.3　Response of different methods of calcium application on quality and postharvest changes in grapes (Pooled data: 2005–07).

Treatments	Yield (t/ha)	Weight (g)	Sugar (°Brix)	Dry matter (%)	Firmness (gFx/mm) Harvest	Firmness (gFx/mm) 30 DAH[a]
Control	23.7	845.5	15.95	17.2	239	213
Ca foliar	23.95	852.5	15.9	17.6	250.5	225
Ca soil	24.7	881.5	15.75	16.1	266.5	228
Ca foliar + Ca soil	23.85	849	15.75	16.25	264.5	226
P	0.712	0.8655	0.816	0.0405	0.025	0.334

$P < 0.05$.
[a] DAH: Days after harvest (cold storage).
Based on Bonomelli, C., Ruiz, R., 2010. Effects of foliar and soil calcium application on yield and quality of table grape cv. 'thompson seedless'. J. Plant Nutr. 33, 299–314. doi: https://doi.org/10.1080/01904160903470364.

TABLE 14.4　Mineral nutrient contents after Ca application for growing seasons in grapes (Pooled data: 2005–07).

Treatments	N	P	K	Ca	Mg	Cl
Control	0.57	0.12	0.90	0.05	0.04	0.029
Ca foliar	0.55	0.12	0.51	0.05	0.05	0.028
Ca soil	0.61	0.12	0.92	0.06	0.04	0.060
Ca foliar + Ca soil	0.61	0.12	0.92	0.06	0.05	0.061
P	0.724	0.307	0.094	0.451	0.156	0.001

$P < 0.05$.
Based on Bonomelli, C., Ruiz, R., 2010. Effects of foliar and soil calcium application on yield and quality of table grape cv. 'thompson seedless'. J. Plant Nutr. 33, 299–314. doi: https://doi.org/10.1080/01904160903470364.

In a study by Madani et al., different sources and concentrations of foliar-treated Ca, nutrient concentration, and growth of plants of Exotica II papaya (*Carica papaya*) were studied. Papaya seedlings, four prehydric sprays, and three different Ca {calcium chloride in four different concentrations [$CaCl_2$] were applied as leaf treatments with calcium nitrate [$Ca(NO_3)_2$] and calcium propionate [$Ca(C_2H_5COO)_2$]} (0, 60, 120, and 180 mg/L). It was observed that the concentration of plant Ca was not affected by different Ca sources. However, increased Ca concentration in the leaves increased phosphorus and Ca plant accumulation in the plant, but decreased concentrations of K and Mg in the tissues. It was also observed that the plants given Ca at 180 mg/L had a larger size and diameter than the control plants. In a field trial with mature trees, preharvest applications of Ca (0, 4000, and 5400 mg/L) in the form of $CaCl_2$ showed that increasing concentrations improved fruit Ca concentration, texture, and flavor while decreasing weight loss, Mg content, and apparent disease incidence of the fruit. Similarly, four Ca concentrations were studied in the Exotika II papaya plant and leaf nutrient concentrations are given in Table 14.5 1 month after transplantation at 2-week intervals (Madani et al., 2015).

TABLE 14.5　Effects of three sources of calcium (Ca) at four Ca concentrations sprayed on the foliage 1 month after transplanting at 2-week intervals on leaf nutrient concentrations and stem growth of Exotika II papaya (Madani et al., 2015).

Treatment		Macronutrients (mg/gL dry weight) Ca	Mg	N	P	K	Stem growth Ht (cm)	Diam (mm)
Source	Calcium chloride	11.7	8.3	15.6	5.0	60.8	145.8	28.1
	Calcium nitrate	12.3	8.5	18.6	6.1	71.9	148.9	27.5
	Calcium propionate	12.4	8.2	16.3	5.9	58.5	144.5	27.7
Concentration (mg/L)	0	11.1	7.4	15.8	2.6	64.9	131.2	25.9
	60	11.6	9.5	16.5	5.9	72	148.4	27.7
	120	12.4	8.5	17.1	6.6	62.7	151.4	27.5
	180	13.4	7.7	18.3	7.5	55.5	154.6	29.8
Significance		Q***	Q***	Q*	Q***	Q*	Q**	Q**

Q*, Q*** = significant quadratic response at $P < 0.05$ or $P < 0.001$, respectively.

6 Nanotechnological nutrition in fruit crops

The literature indicates that nanotechnology can be a solution to the problem of lack of food, rather than transforming it into nonbiologically available forms. Nanoparticles (NPs) are materials that display sizes between 1 and 100 nm. They can be in the form of atoms and molecules as well as in zero, one, two, or three dimensions (Sen et al., 2018a,b,c,d, e,f) (Fig. 14.3). There are also hybrid quantum effects between biological activities and cast minerals. NPs can represent the properties of the particles in the 1–250 nm dimensions (Sen et al., 2018a,b,c,d,e,f; Sahin et al., 2018; Eris et al., 2018). By forming of nanoparticles, fertilizer is prevented from decomposing and volatilizing, and wasting manure is reduced. Plants carry NPs to all plant parts through pores on the leaves and root epidermis. When applied to plants as a spray, NPs enter into fruits and vegetables, thereby increasing their nutritional quality (Sen et al., 2018a,b,c,d,e,f; Gunbatar et al., 2018; Yang et al., 2018; Talan et al., 2018; Mukhopadhyay, 2014). The findings of the researchers indicated that NPs showed both positive and negative effects on plant growth and development and that the effect of the nanoparticles applied to the plants was related to the composition, concentration, size, and physical and chemical properties of NPs as well as the plant species. The activity of NPs depends on their concentration and varies from plant to plant (Table 14.6).

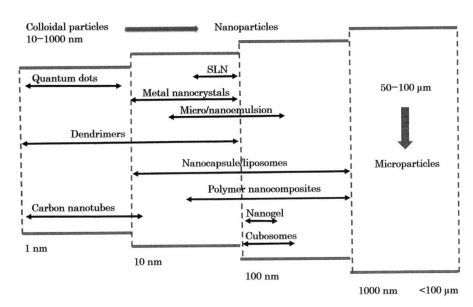

FIG. 14.3 Deciphering the size of a nanoparticle.

TABLE 14.6 Effectiveness of nanoparticles.

Nanoparticle(s)	Beneficiary concentration(s)	Plant	Part of plant/process	Reference(s)
Ca(NO$_3$)$_2$, CaCl$_2$	0.0125 M	*Malus domestica* Borkh	Fruit color, quality, fruit firmness	Wojcik and Borowik (2013)
CaCl$_2$	40 mM	*Malus domestica*	Flesh firmness, contents, cell wall properties	Siddiqui and Bangerth (1995)
CaCl$_2$	0.34%	*Malus domestica*	Fruit color, quality, fruit firmness	Kadir (2004)
CaCl$_2$	0.75%	*Pyrus communis*	Fruit color, quality, organoleptic rate, fruit firmness	Bhat et al. (2012)
CaCl$_2$	2%	*Diospyros kaki* Thunb	Antioxidative enzyme activities, fruit firmness, weight, quality	Bagheri et al. (2015)
CaCl$_2$	862 mM, 9°C	*Solanum betaceum* Cav	Fruit firmness, weight, quality	Pinzón-Gómez et al. (2014)
CaCl$_2$	3%	*Eriobotrya japonica* Lindl	Fruit firmness, weight, quality	Babu et al. (2015)
CaCl$_2$	1.5%	*Raphanus sativus* L.	Antioxidative enzyme activities, quality	Devi et al. (2018)

Continued

TABLE 14.6   Effectiveness of nanoparticles.—cont'd

Nanoparticle(s)	Beneficiary concentration(s)	Plant	Part of plant/process	Reference(s)
Nano-Ca	4 g/L	*Capsicum annum*	Antioxidative enzyme activities, quality	Amini et al. (2016)
CaCl$_2$	4%	*Prunus armeniaca* L.	Fruit firmness, weight loss, total soluble solids, sugar estimation	Ali et al. (2013)
Ca(NO$_3$)$_2$	20 g/L	*Diospyros kaki*	Fruit firmness	Agusti et al. (2004)
CaCl$_2$, CaO	15%	*Malus domestica* Borkh	Root growth, uptake of nutrients	Torres et al. (2017)
CaCl$_2$, nano-Ca	2 g/L	*Malus domestica*	Fruit firmness, weight loss, antioxidant activity	Ranjbar et al. (2018)
CaCO$_3$	15.1 g	*Mushroom*	Yield and growth	Royse and Sanchez-Vazquez (2003)
Graphene oxide	400 and 800 mg/L	*Vicia faba* L.	Germination	Anjum et al. (2014)
CNTs	40 µg/mL	*Lycopersicum esculentum*	Germination and seedling growth	Morla et al. (2011)
	75 wt% CNTs	*Medicago saliva, Triticum aestivum*	Root elongation	Miralles et al. (2012)
	75 wt% CNTs impurities	*Medicago saliva, Triticum aestivum*	Root elongation	Miralles et al. (2012)
SWCNTs	9, 56, 315, and 1750 mg/L	*Allium cepa, Cucumis sativus*	Root elongation	Cañas et al. (2008)
MWCNTs	25–100 µg/mL	*Hordeum vulgare* L., *Glycine max, Zea mays*	Germination	Lahiani et al. (2013)
	50 and 200 µg/mL	*Lycopersicon esculentum* Mill	Plant height and number of flowers	Khodakovskaya et al. (2013)
	5 up to 500 µg/mL	*Nicotiana tabacum*	Growth	Khodakovskaya et al. (2012)
o-MWCNTs	10–160 µg/mL	*Triticum aestivum*	Root growth, vegetative biomass	Wang et al. (2012a)
MWCNTs, dMWCNT	40 µg/mL	*Lycopersicon esculentum* Mill	Uptake nutrients (K, Ca, Fe, Mn, and Zn)	Tiwari et al. (2013)
Pristine MW CNTs	20 mg/L	*Zea mays*	Nutrient transport, biomass	Tiwari et al. (2014)
ZnO NPs	400 mg/kg	*Cucumis sativus* fruit	Micronutrients (Cu, Mn, and Zn)	Zhao et al. (2014)
	1.5 ppm (foliar spray)	*Cicer arietinum* L.	Shoot dry weight	Burman et al. (2013)
	20 ppm (suspension, foliar spray)	*Vigna radiata*	Biomass	Dhoke et al. (2013)
	1000 ppm	*Arachis hypogaea*	Germination	Prasad et al. (2012)
	1000 ppm	*Arachis hypogaea*	Stem, root growth, and yield	Prasad et al. (2012)
	500, 1000, 2000, and 4000 ppm	*Vigna radiate* L. Wilczek	Dry weight	Patra et al. (2013)
GNPs	10 and 80 µg/mL	*Arabidopsis thaliana*	Germination	Kumar et al. (2013)
	10 and 80 µg/mL	*Arabidopsis thaliana*	Root length	Kumar et al. (2013)
	10 µg/mL	*Arabidopsis thaliana*	Shoot and root system (longer), early flowering, yield	Kumar et al. (2013)

TABLE 14.6 Effectiveness of nanoparticles.—cont'd

Nanoparticle(s)	Beneficiary concentration(s)	Plant	Part of plant/process	Reference(s)
AgNPs	10–30 µg/mL	*Boswellia ovalifoliolata*	Germination and seed ling growth	Savithramma et al. (2012)
	60 ppm	*Phaseolus vulgaris* L., *Zea mays* L.	Root length	Salama (2012)
	60 ppm	*Phaseolus vulgaris* L., *Zea mays* L.	Shoot length	Salama (2012)
	60 ppm	*Phaseolus vulgaris* L., *Zea mays* L.	Dry weight of root and shoot	Salama (2012)
	100 µM	*Vigna radiata*	Antagonize inhibition by 2,4-dichlorophenoxyacetic acid (2,4-D) at 500 µM of plant growth	Karuppanapandian et al. (2011) Contd..
Sulfur NPs	500, 1000, 2000, and 4000 ppm	*Vigna radiata*	Dry weight	Patra et al. (2013)
SiO_2NPs	15 kg/ha	*Zea mays* L.	Growth parameters	Yuvakkumar et al. (2011), Suriyaprabha et al. (2012)
TiO_2 NPs	400 mg/L	*Arabidopsis thaliana*	Root length	Lee et al. (2010)
	60 ppm	*Foeniculum vulgare*	Germination	Feizi et al. (2013)
	lower than 200 mg/L	*Lemna minor*	Plant growth	Song et al. (2012)
	1000 mg/L	*Triticum aestivum*	Chlorophyll content	Mahmoodzadeh et al. (2013)
	0.05–0.2 g/L	*Lycopersicon esculentum* Mill	Net photosynthetic rate, conductance to H_2O, and transpiration rate, regulation of photosystem II (PSII)	Qi et al. (2013)
Nano-anatase TiO_2	0.25% (foliar spray)	*Spinacia oleracea*	Rubisco activase (rca) mRNA expressions	Ma et al. (2008)
	0.25% (foliar spray)	*Spinacia oleracea*	Oxygen evolution, Rubisco carboxylation, Rubisco activase, rate of photosynthetic carbon reaction	Gao et al. (2006), Zheng et al. (2007), Gao et al. (2008), Ma et al. (2008)
	0.25%	*Spinacia oleracea*	Several enzymes activities induction	Yang et al. (2006)
Aluminum oxide NPs	400–4000 mg/L	*Arabidopsis thaliana*	Root length	Lee et al. (2010)
Alumina NPs	10 mg/L	*Lemna minor*	Root length	Juhel et al. (2011)
	0.3 g/L	*Lemna minor*	Biomass accumulation	Juhel et al. (2011)
nZVI (nanoscale Zero-Valent Iron particles) Iron oxide NPs	0.5 g/L	*Arabidopsis thaliana*	Root elongation	Kim et al. (2014)
Iron oxide NPs	0.5–0.75 g/L	*Glycine max*	Yield and quality	Sheykhbaglou et al. (2010)
	50 ppm (foliar spray)	*Vigna radiata*	Biomass	Dhoke et al. (2013)
ZnFeCu oxide NPs (suspension)	50 ppm (foliar spray)	*Vigna radiata*	Biomass	Dhoke et al. (2013)
CeO_2 NPs	250 ppm	*Arabidopsis thaliana*	Biomass	Ma et al. (2013)
CO_3O_4 NPs	5 g/L	*Raphanus sativus* L.	Root elongation	Wu et al. (2012)
CuO NPs	500 mg/kg (sand culture)	*Triticum aestivum*	Biomass	Dimkpa et al. (2012)
Hydroxyapatite suspension	100–2000 mg/L	*Lactuca sativa*	Root length	Wang et al. (2012b)

7 Foliar application

NPs applied to the leaves enter through the cuticle or stoma of the leaves. Cuticulitis of leaves, the first obstacle, restricts NPs to the size of <5nm. NPs from the stoma can pass through the vascular system of the cell through apoplastic or sympathetic paths >10nm. While NPs having a size of 10–50nm prefer the sympathetic pathway in the cell, translocation of large NPs with a size of 50–200nm prefer apoplast (passing through neighboring cells). There are effective factors for this purpose. In addition, there are important factors such as the chemical composition and morphology of the leaves. Also, the microsphere of the photosphere, sunlight, temperature, and humidity also affect NP transformation. Photosynthesis may also affect its efficacy (Varshney et al., 2018; Yang et al., 2018; Talan et al., 2018).

According to the literature, Citrus tankan, California red scale, and Ca concentration in *Ziziphus mauritiana* plants have increased the controlled invasion of fruit flies, thus encouraging the growth of plants. Again, according to the literature, N, P, and K minerals on the leaves of the NP application have not been studied. Due to some urea transformations of urea fertilizer applied to soil, N loss has been caused. Because the ureas (NH_4^+ and NO_3) formed considerable damage to the leaves, leaf applications were a preferred N source for NP application. In addition, positive results were obtained by applying N, P, and K fertilizers on the leaves. Thanks to the leaf application that will overcome the limitations of traditional fertilizers used today, the requirements for N, P, and K minerals will also be eliminated in the future. Zn is an essential cofactor for enzymes in the synthesis of chlorophyll. (Joshi et al., 2018; Jabeen et al., 2017; Kole et al., 2013).

An increase in the photosynthesis ratio, stomatal conduction, and stomatal patency in *Glycine max* was observed. In *Cucurbita maxima*, Fe_2O_3 NPs have been observed to have a positive effect on leaf application. However, the Fe concentration applied to the plants plays a significant role. The optimal concentration for increasing the amount of biomass and chlorophyll in Nc is 250–500mg. In addition, when SiO_2 NPs are applied to leaves, it has been observed that rice decreases cadmium toxicity (Kole et al., 2013).

The effects of some of the transition nanoparticles such as manganese (Mn) NPs on the application of leaves were not investigated. The efficacy of leaf sprays of Mn, Mo, and Ni fertilizers has been proven to eliminate nutritional deficiencies and increase crop yield. Due to the holding capacity of the surface areas of the leaves, applications on the leaves may be limited. It has been reported that plants act faster than leaf fertilization compared to the application on leaves (Prasad et al., 2017). For this reason, it would be advantageous to modify the surface of hydrophobic/hydrophilic molecules for micronutrient NPs to remain on the surface of the leaves for longer and better absorption. In recent years, researchers have shown interest in carbon-based nanomaterials as they can improve plant biomass. Carbon-based NPs can pass through the cell wall and membrane to mediate small molecules and potential carriers for DNA75. Thus, functionalized NPs can act as nanocarriers of specific DNA regulatory molecules and regulate the gene expression of plants (Varshney et al., 2018; Mukhopadhyay, 2014).

In a study conducted on Luhua 4, Nano-$CaCO_3$ application coordinated with humus acid and organic fertilizers was performed. Compared to the single nitric acid calcium application, it has been observed that this significantly improves peanut growth and development. The application of nano-$CaCO_3$ (coordinated) with humic acid and organic fertilizer could improve the physiological state of peanuts. It has been observed that the content of soluble sugar and protein is increased, especially in the stems and leaves of peanuts. Nano-$CaCO_3$ combined with humus acid and peanut nitric acid compared to calcium, nitrogen, phosphorus, and potash was found to be able to absorb nutrients. The content of nutrients in the shoot and root of the peanut increased Ca by 0.72% and 0.32%, N by 1.3% and 0.43%, P by 0.08% and 0.04%, and K by 0.49% and 0.014%, respectively (Liu et al., 2005).

8 Plant growth responses of nanoparticles applied to soil

The use of chemical fertilizers in agriculture that started in the twentieth century has increased. However, applying chemical fertilizers more than needed without doing soil analysis causes negative effects such as salinization in soil, heavy metal accumulation, deterioration of nutrient element balance, damage of the microorganism population and activity, nitrate accumulation, and formation of eutrophication in water. The pH balance of the soil deteriorates, the acidity rate is increased, some nutrients such as P and Ca are prevented from being taken by plants, and growth deficiencies are observed by improper chemical fertilizer application. Especially, applying phosphorus fertilizers in large amounts with Ca in soil disturbs the nutrient element balance by preventing the plants from getting Zn and Fe. While nanotechnology provides innovative opportunities in many areas, the use of nanomaterials, especially in food and agriculture, is extremely important to meet the needs of the growing population and to obtain the highest amount of top-quality products with minimum input from the unit area (Duhan et al., 2017).

Nanofertilizers can be defined as nanomaterials that provide one or more nutrients to the plant and increase the growth and development of the plant. These fertilizers are slow soluble and highly effective fertilizers. Slow-release nanofertilizers that are easily gotten by plants without being washed or changed from the soil, without binding to organic material, clay, and lime, and that are easily gotten by the plants without forming compounds with other elements have the potential to be much preferred. The particle sizes of nanofertilizers are smaller than plant pores but have a large surface area. These properties can increase the penetration of nanofertilizers from the surface to the plant as well as the intake and utilization efficiency of the nutrients. On the other hand, it also increases the specific surface area and the number of particles per unit area of nanofertilizers by decreasing the particle size. Thus, plants will be able to benefit more from the applied nutrients by increasing in the contact area of the encapsulated nanogubers in the nanoparticles. Nanofertilizers transmit plant nutrients to the plant in one of three ways; (i) the nutrient can be encapsulated in nanomaterials such as nanotubes or nanoparticles, (ii) coated with a thin protective polymer film, or (iii) transported to the plant in nanoscale-sized particles or emulsions. Nanofertilizers provide healthy growth and development of the plant. Thus, the healthy plant gains more resistance against severe and variable atmospheric conditions and diseases. The amount of microelement that can be sufficiently taken in soil increases the resistance of plants to pathogens. But, some factors limit the uptake of nutrients by plants in the soil. For example, as the pH of the soil approaches the alkaline nutrient, like and Fe, Mn, and Zn bioavailability in the soil decreases and the roots are limited by the intake of these elements. However, contrary to macro and microdimensions, nano-sized fertilizers can be easily passed through the stoma opening of the plant, allowing the plant to easily make use of minerals while fertilizer yields are close to 100%. Thanks to the use of nanotechnological fertilizers, the maximum yield from sowing is avoided by preventing fertilizer waste. In addition, existing and economic losses can be invested. Thanks to nanotechnological fertilizer, it is possible to obtain higher yields by using less fertilizer. Nanofertilizers will vary depending on many factors such as type, content, dose of application, time of administration, and default plant variety. Nanofertilizers that are produced and used extensively and expected to be used more intensively in the future will vary depending on many factors such as the type, content, dose of application, time of application, and type of plant being applied. For all these reasons, nanofertilizers supported by environmentally friendly macro- and micronutrients must be developed for sustainable agriculture. Nanofertilizers can be classified as macro- and micronutrient nanofertilizers (Vishwakarma et al., 2018).

8.1 Nanoparticles of macronutrients

Macronutrient materials are mostly used to increase production. There are some environmental problems as well as waste and leakage of fertilizers due to the low intake of macronutrients by the plant. Therefore, the use of a slow-release macronutrient is necessary to reduce manure waste and control environmental problems. The extruded urea-montmorillonite nanocomposites encapsulated in *Gliricidia sepium* wood chips and the urea-modified hydroxyapatite NPs have been shown to have a controlled release capacity for 30 days. With the continuous release of urea-hydroxyapatite NPs (urea:hydroxyapatite ratio 6:1 by weight), 50% less urea fertilizer could improve the rice yield, resulting in reduced N waste. The soil leaching test was tested with urea-hydroxyapatite-KCl. As in paddy cultivation, nitrates pass into the atmosphere as they turn to N_2, which leads to the generation of greenhouse gases such as N_2O. The side effects of crop nitrogen germination and seedling growth have been documented by reports (Apodaca et al., 2018). These reports suggest that nitrogen fertilizers with sustained release should be applied to the soil.

Furthermore, there are a few works on the effectiveness of nanofertilizers with the N-based sustained release. The development of novel macronutrient NPs that exhibit controlled release and respond to plant signals is necessary to increase the uptake of macronutrients by the plant. There is no doubt that under greenhouse/field conditions, it will be necessary to test these new and intelligent NPs to determine their release in natural environmental conditions and absorption by plants.

In another study, a 1:1 peat-perlite mixture was used as the seed mortar and 1400 g $Ca(H_2PO_4)_2$ (42%–44% P_2O_5), 800 g K_2SO_4 (50% K_2O), 1200 g NH_4NO_3 (33.5% N), and 1000 g $Mg(NO_3)_2 \cdot 6H_2O$ were added. Accordingly, 402 g N, 616 g P_2O_5, 400 g K_2O, and 95 Mg were added to the mixture as pure material. The prepared mortar was filled into seedling cultivation containers (viol) and the seeds were sown at two to three times depth. Seeds were germinated and covered with newspaper on top of the soil until they reached the soil surface at 18–21°C as the ambient temperature. After germination, necessary day and night temperatures were provided and irrigation, fertilization and protective spraying were carried out. As a result, increased harvest rates and increased nutrient content were reported (Khandaker et al., 2017).

In a study on apple fruits, the effects of nanocalcium and calcium chloride spraying on harvest quality and cell wall enzyme activities were investigated. Nanocalcium (0%, 1.5%, 2%, and 2.5%) and calcium chloride (0%, 1%, 1.5%, and 2%) solutions were applied five times over 2 weeks. The measurements included fruit hardness, weight loss, titratable acidity (TA), total soluble solids (TSS), total phenolic content (TPC), browning, total antioxidant activity (TAA), and fiber content; enzymatically, polygalacturonase (PG), pectinmethylesterase (PME), and β-galactosidase (β-Gal) were investigated. The results showed that the hardness, TA, TPC, TAA, and fiber content were increased with both nanocalcium and calcium chloride compared to the control fruit, but these parameters decreased with the prolongation of storage time. Also, during storage, lower activities of PG, PME, and β-Gal enzymes were observed in fruits sprayed with calcium fertilizers. In addition, it was observed that the quality of the apple fruit treated with nanocalcium was better than that for all parameters treated with calcium chloride (Ranjbar et al., 2018).

In another study, nitrogen, phosphorus, potassium, calcium, and magnesium in peanut calcium fertilizer applications were investigated under salt stress. Using Huayu 25 as the test material, the four Ca levels (T1 [0], T2 [75], T3 [150], and T4 [225] kg/hm CaO) were adjusted to 0.3% salt stress in a pot. The results showed that the nutrient content of the peanut was followed by nitrogen > potassium > calcium > phosphorus > magnesium. In the seedling stage, the leaves are the absorption center of nitrogen and calcium while the stems are the center of phosphorus, potassium, and magnesium, and about half of the nutrient accumulation is dispersed in the corresponding growth center. At the mature stage, nitrogen, phosphorus, and potassium absorption centers were transferred to the pod. Nitrogen and phosphorus accumulation in the seed kernel reached 72.3%–78.9%. The absorption centers of calcium and magnesium were still 49.8% and 32.6% in the leaves, respectively. Salt stress significantly inhibited the absorption and distribution of nutrients in peanuts, especially reducing nitrogen accumulation in leaves and seed nuclei. However, salt stress increased the accumulation of magnesium in the pods. The application of exogenous calcium has shown a significant positive effect on the absorption and accumulation of nitrogen, phosphorus, calcium, and magnesium in different organs of peanuts under salt stress. It is also mentioned that there is a significant adjustment in phosphorus accumulation in the seed nucleus, which increases more than 50%. Proper calcium content can significantly promote peanut nutrient absorption and accumulation under salt stress, and improve the nitrogen, phosphorus, and potassium distribution rate in mature pods of peanuts. According to the reactions of nutrient absorption and distribution, the optimized application amount for calcium fertilizer under 0.3% salt stress was transferred as 150 kg/hm CaO (Shi et al., 2018).

In another study, nutrient problems from the soil, especially for P farmland, were studied. Calcium sulfate or gypsum is a common soil change and has a strong complication for orthophosphates. The results showed that calcium sulfate reduced the amount of P (WEP) removable through soil incubation tests, with less P loss than farmland. A greater reduction in WEP occurred with greater amounts of calcium sulfate. Compared with calcium sulfate, nanocalcium sulfate further reduced WEP by providing a much more specific surface area, higher solubility, better contact with fertilizer and soil particles, and superior dispersibility. Increasing the nanocalcium sulfate for WEP reduction is more pronounced than a powdered fertilizer for a pellet. WEP with a Ca/P weight ratio of 2.8 was reduced by $31 \pm 5\%$ with coarse calcium sulfate and nanocalcium sulfate during pellet fertilizer, and $20 \pm 5\%$ with coarse calcium sulfate. Calculation of the chemical equilibrium specification indicates that calcium hydroxyapatite has the lowest solubility. However, other mineral phases, such as hydroxydicalcium phosphate, dicalcium phosphate dihydrate, octacalcium phosphate, and tricalcium phosphate, may form before calcium hydroxyapatite. Because calcium sulfate is the main product of the flue gas desulfurization (FGD) process, this study demonstrates potentially useful reuse and reduction of solid FGD waste (Chen et al., 2016).

8.2 Nanoparticles of micronutrients

The main cationic micronutrient that plants need is zinc. The products in the soil respond efficiently to Zn fertilizers. Zn deficiency results in poor quality of growth, chlorosis, yield, and malnutrition. Zinc oxide NPs as nanofertilizers have been shown as nontoxic compared to bulk materials. ZnO NPs are the status of antioxidant and growth parameters. Zinc oxide NPs also reduce the toxicity induced by copper oxide NPs. Iron is required for cellular redox reactions and chloroplast synthesis. In case of deficiency, the leaf becomes white. The role of ferrihydrite NPs ($5Fe_2O_3 \cdot 9H_2O$) and hematite NPs (α-Fe_2O_3 NPs) in increased chlorophyll content in hydroponically grown *G. max* and *Z. mays* has been reported. When Fe_2O_3 NPs are applied to roots in amounts of 2–50 mg/L, Fe_2O_3 NPs have been transported to leafy parts, crop germination, and growth of seedlings. Indole-3-acetic acid, transgenic and nontransgenic abscisic acid, and malondialdehyde levels were increased by Fe_2O_3 NPs (Jabeen et al., 2017; Kole et al., 2013).

Although it has been reported that copper oxide NPs in plants cause oxidative stress, a recent study has shown that redox status, leaf area, and photosynthetic pigments increase with the combination of Cu-Zn NPs. The concentration of CuO-NPs between 0.25 and 5 mg/L is nontoxic. It was observed that when applied to *L. sativa* at a concentration of 0.26 g/L, it increased germination and growth (Calderón et al., 2017).

Manganese is another vital micronutrient of plants. The inadequacy of plants causes chlorosis (Calderón et al., 2017). Si is an essential material and usually accumulates in plant intracellular cavities. SiO_2 NPs obtained by *Oryza sativa, L. esculentum*, and *Z. mays* have been reported to be nontoxic for the development of plants. The deposition of Si as a result of the use of SiO_2 NPs contributed to the drought tolerance in the plants of *Z. mays* and *L. esculentum*. SiO_2 NPs are related to the structure of leaves, photosynthetic activity, the population of beneficial rhizobacteria (Prasad et al., 2017; Soukoulis and Bohn, 2018).

Numerous studies have shown significant advances in various micronutrient NPs. The response of plants to these NPs does not show a general feature as they may vary depending on the type of plants and NPs. In addition, the positive effects of these micronutrient NPs on plant growth vary according to the environment in which the study is performed. Most of the tests and studies conducted so far have been performed using hydroponic systems or under laboratory conditions. Their behavior as growth-enhancing agents may be affected by NPs under natural environmental conditions. For laboratory results, the growth-stimulating effects of NPs should also be investigated under field conditions. However, during the fieldwork, the effects of NPs on flora and fauna should also be considered. In short, it should be assured that micronutrient NPs do not harm animals/microorganisms by performing multiple experiments before determining their role as a positive effector on plants and labeling them as micronutrient fertilizers.

Besides, even if there are some problems related to NPs in the first generation, in the following generations, growth and physiological parameters were reduced in repeated applications of NPs. These parameters are particularly chlorophyll content, ferritin, photosynthesis, and Fe and Ca content. Doses of $CaCO_3$ NPs above 26% induce K. However, it is difficult to interpret whether these adverse effects are caused by NPOs or only by ionic toxicity.

Capitata is 2–45 times more than the tolerable daily intake determined by the US Environmental Protection Agency, which suggests that leaf application of copper oxide NPs may lead to the contamination of vegetables. Studies show that the doses that cause toxicity of NPs vary with host plants. For example, in one study, γ-Fe_2O_3-NPs above 100 mg/L concentration had a toxic effect on *C. maxima*. These studies have limited our understanding of the potential risks of NPs applied to leaves (Jabeen et al., 2017; Kole et al., 2013; Prasad et al., 2017; Mukhopadhyay, 2014). A recent study has demonstrated the long-term effects of NPs on nuclei, the accumulation of ions in corn kernels throughout generations, and the risks associated with it. Some NPs have adverse effects on animals. In order to do this, it is necessary to carry out extensive nanotoxicological studies in plant models and to establish standard test procedures for the safe use of nanofertilizers.

9 Nanoparticles and side effects applied to soil

Hydroxyapatite-based materials formed larger particles that influenced the entry of cells. The adverse effects of ZnO NPs are due to the aggregation of NPs on roots and the release of Zn^{2+} ions to the soil.

ZnO NPs damage some of the cells by increasing the permeability of cell walls in the *Lolium perenne*. Plants exposed to CuO and Ag NPs are prevented from ingestion and growth of minerals. While the CuO NPs in *O. sativa* and *Brassica juncea* caused an increase of antioxidative and photosynthetic enzymes, inhibition of the production of phytohormone (absicic acid, indoleacetic acid, and glycemic acid) in transgenic *Gossypium hirsutum* was observed. When TiO_2 NPs were administered at a concentration of 4307.5 mg, they caused a decrease. The use of Al_2O_3, Co, and Ni NPs adversely affected the biomass of the above and below *L. esculentum* (Jabeen et al., 2017; Kole et al., 2013; Prasad et al., 2017; Du et al., 2015; Mukhopadhyay, 2014; Soukoulis and Bohn, 2018; Achari and Kowshik, 2018).

The information available so far shows that the larger size of NPs cannot be attracted to the smaller one, thus eliminating the toxic effect. In the studies, it has been reported that NPs contained chitosan and montmorillonite while Ca, Mo, Mg, and mineral nanoconjugates negatively affect the plants in soil application.

10 Future research

In this review, recent advances in nanotechnological calcium feed for seeds have been reported, and it is thought that this will be an inspiring study to make new studies more efficient. Research on NP-based crop feeding for agricultural systems is still at an early stage. Briefly, the studies in the literature show that the use of NPs varies to stimulate growth and physiological processes while inducing toxic effects. For this purpose, specific tests must be performed to determine whether micronutrient NPs are useful for the growth and productivity of a plant. It is clear that the removal of dissolved NPs and NPs from the ions using the leaf or soil will be carried out successfully in later times. However, researchers should focus on the production of nanomaterials with sustained release properties, which allow greater efficacy. Biopolymer-mineral nanoconjugates that are biodegradable with reduced toxicity and higher stability for agricultural applications can be used as new nanofertilizers.

Further research on the synthesis of new nano-sized plant-friendly materials requires further enlightening of the importance of these nanomaterials in plant systems. There are very few studies on the applications and effects of NPs on leaves. Therefore, the research should be directed toward this. Gene expression and physiology studies are required. It is also necessary to examine the performance of these NPs in field conditions by combining them with hydroponic studies in the laboratory to increase their acceptability and to confirm their efficacy compared to bulk mineral fertilizers. Therefore, in order to achieve full efficiency in the future, it is necessary to develop NP structures with intelligent design, good molecular specificity, and precise targeting in live systems. These nanomaterials will provide superior solutions for the improvement of nanotechnology crops, the prevention of diseases, and the protection of the environment.

References

Achari, G.A., Kowshik, M., 2018. Recent developments on nanotechnology in agriculture: plant mineral nutrition, health, and interactions with soil microflora. J. Agric. Food Chem. 66 (33), 8647–8661. https://doi.org/10.1021/acs.jafc.8b00691.

Agusti, M., Juan, M., Martinez-Fuentes, A., et al., 2004. Calcium nitrate delays climacteric of persimmon fruit. J. Appl. Biol. Res. 144, 65–69.

Alcock, T.D., Havlickova, L., He, Z., et al., 2017. Identification of candidate genes for calcium and magnesium accumulation in Brassica napus L. by association genetics. Front. Plant Sci. 8, 1968. https://doi.org/10.3389/fpls.2017.01968.

Ali, S., Masud, T., Abbasi, K.S., et al., 2013. Influence of CaCl2 on physico-chemical, sensory and microbial quality of Apricot cv. Habi at ambient storage. J. Chem. Bio. Phys. Sci. Sec. 3, 2744–2758.

Alomari, D.Z., Eggert, K., von Wirén, N., et al., 2017. Genome-wide association study of calcium accumulation in grains of European wheat cultivars. Front. Plant Sci. 8, 1797. https://doi.org/10.3389/fpls.2017.01797.

Alori, E.T., Babalola, O.O., 2018. Microbial inoculants for improving crop quality and human health in Africa. Front. Microbiol. 9, 2213. https://doi.org/10.3389/fmicb.2018.02213.

Amini, F., Bayat, L., Hosseinkhani, S., 2016. Influence of preharvest nano-calcium applications on postharvest of sweet pepper (Capsicum annum). Nusant. Biosci. 8 (2), 215–220.

Anjum, N.A., Singh, N., Singh, M.K., et al., 2014. Single-bilayer graphene oxide sheet impacts and underlying potential mechanism assessment in germinating faba bean (Vicia faba L.). Sci. Total Environ. 472, 834–841. https://doi.org/10.1016/j.scitotenv.2013.11.018.

Apodaca, S.A., Medina-Velo, I.A., Lazarski, A.C., et al., 2018. Different forms of copper and kinetin impacted element accumulation and macromolecule contents in kidney bean (Phaseolus vulgaris) seeds. Sci. Total Environ. 636, 1534–1540. https://doi.org/10.1016/j.scitotenv.2018.04.360.

Artyszak, A., 2018. Effect of silicon fertilization on crop yield quantity and quality—a literature review in Europe. Plants (Basel) 7 (3), 54. https://doi.org/10.3390/plants7030054.

Babu, I., Ali, M.A., Shamim, F., et al., 2015. Effect of calcium chloride application on quality characteristics and post harvest performance of loquat fruit during storage. IJAR 3 (1), 602–610.

Bagheri, M., Esna-Ashari, M., Ershadi, A., 2015. Effect of postharvest calcium chloride treatment on the storage life and quality of persimmon fruit (Diospyros kaki Thunb.) cv. 'Karaj'. Int. J. Hortic. Sci. Technol. 2, 15–26.

Berkovich, Y.A., Krivobok, N.M., Krivobok, A.S., et al., 2016. Advanced nutrient root-feeding system for conveyor-type cylindrical plant growth facilities for microgravity. Life Sci. Space Res. 14, 1–21. https://doi.org/10.1016/j.lssr.2015.12.002.

Bhat, M.Y., Ahsan, H., Banday, F.A., et al., 2012. Effect of harvest dates, pre harvest calcium sprays and storage period on physico-chemical characteristics of pear CV Bartlett. J. Agric. Res. Dev. 2 (4), 101–106.

Blancaflor, E.B., Kilaru, A., Keereetaweep, J., et al., 2014. N-Acylethanolamines: lipid metabolites with functions in plant growth and development. Plant J. 79 (4), 568–583. https://doi.org/10.1111/tpj.12427.

Bonomelli, C., Ruiz, R., 2010. Effects of foliar and soil calcium application on yield and quality of table grape cv. 'thompson seedless'. J. Plant Nutr. 33, 299–314. https://doi.org/10.1080/01904160903470364.

Burman, U., Saini, M., Kumar, P., 2013. Effect of zinc oxide nanoparticles on growth and antioxidant system of chickpea seedlings. Toxicol. Environ. Chem. 95 (4), 605–612.

Calderón, R., Godoy, F., Escudey, M., et al., 2017. A review of perchlorate (ClO(4)(-)) occurrence in fruits and vegetables. Environ. Monit. Assess. 189 (2), 82. https://doi.org/10.1007/s10661-017-5793-x.

Cañas, J.E., Long, M., Nations, S., et al., 2008. Effects of functionalized and nonfunctionalized single walled carbon nanotubes on root elongation of select crop species. Environ. Toxicol. Chem. 27 (9), 1922–1931.

Chen, D., Szostak, P., Wei, Z., et al., 2016. Reduction of orthophosphates loss in agricultural soil by nano calcium sulfate. Sci. Total Environ. 539, 381–387. https://doi.org/10.1016/j.scitotenv.2015.09.028.

Conrath, U., 2011. Molecular aspects of defense priming. Trends Plant Sci. 16 (10), 524–531. https://doi.org/10.1016/j.tplants.2011.06.004.

Debona, D., Rodrigues, F.A., Datnoff, L.E., 2017. Silicon's role in abiotic and biotic plant stresses. Annu. Rev. Phytopathol. 55, 85–107. https://doi.org/10.1146/annurev-phyto-080516-035312.

Devi, J., Bhatia, S., Alam, M.S., et al., 2018. Effect of calcium and salicylic acid on quality retention in relation to antioxidative enzymes in radish stored under refrigerated conditions. J. Food Sci. Technol. 55 (3), 1116–1126. https://doi.org/10.1007/s13197-017-3027-4.

Dhoke, S.K., Mahajan, P., Kamble, R., et al., 2013. Effect of nanoparticles suspension on the growth of mung (*Vigna radiata*) seedlings by foliar spray method. Nanotechnol. Dev. 3 (1), e1. https://doi.org/10.4081/nd.2013.e1.

Dimkpa, C.O., McLean, J.E., Latta, D.E., et al., 2012. CuO and ZnO nanoparticles: phytotoxicity, metal speciation, and induction of oxidative stress in sand-grown wheat. J. Nanopart. Res. 14 (9), 1–15.

Dong, J., Gruda, N., Lam, S.K., et al., 2018. Effects of elevated CO(2) on nutritional quality of vegetables: a review. Front. Plant Sci. 9, 924. https://doi.org/10.3389/fpls.2018.00924.

Du, W., Gardea-Torresdey, J.L., Ji, R., et al., 2015. Physiological and biochemical changes imposed by CeO2 nanoparticles on wheat: a life cycle field study. Environ. Sci. Technol. 49 (19), 11884–11893. https://doi.org/10.1021/acs.est.5b03055.

Duhan, J.S., Singh, J., Kumar, R., et al., 2017. Nanotechnology: the new perspective in precision agriculture. Biotechnol. Rep. 15, 11–23. https://doi.org/10.1016/j.btre.2017.03.002.

Eris, S., Dasdelen, Z., Sen, F., 2018. Enhanced electrocatalytic activity and stability of monodisperse Pt nanocomposites for direct methanol fuel cells. J. Colloid Interface Sci. 513, 767–773.

Etesami, H., Jeong, B.R., 2018. Silicon (Si): review and future prospects on the action mechanisms in alleviating biotic and abiotic stresses in plants. Ecotoxicol. Environ. Saf. 147, 881–896. https://doi.org/10.1016/j.ecoenv.2017.09.063.

Feizi, H., Kamali, M., Jafari, L., et al., 2013. Phytotoxicity and stimulatory impacts of nanosized and bulk titanium dioxide on fennel (*Foeniculum vulgare* Mill). Chemosphere 91 (4), 506–511. https://doi.org/10.1016/j.chemosphere.2012.12.012.

Fidler, J., Grabowska, A., Prabucka, B., et al., 2018. The different ability of grains to synthesize and catabolize ABA is one of the factors affecting dormancy and its release by after-ripening in imbibed triticale grains of cultivars with different pre-harvest sprouting susceptibilities. J. Plant Physiol. 226, 48–55. https://doi.org/10.1016/j.jplph.2018.03.021.

Foyer, C.H., Siddique, K.H.M., Tai, A.P.K., et al., 2018. Modelling predicts that soybean is poised to dominate crop production across Africa. Plant Cell Environ.. https://doi.org/10.1111/pce.13466.

Gao, F.Q., Hong, F.S., Liu, C., et al., 2006. Mechanism of nano-anatase TiO2 on promoting photosynthetic carbon reaction of spinach: inducing complex of Rubisco–Rubisco activase. Biol. Trace Elem. Res. 111, 286–301. https://doi.org/10.1385/BTER:111:1:239.

Gao, F.Q., Liu, C., Qu, C.X., et al., 2008. Was improvement of spinach growth by nano-TiO2 treatment related to the changes of rubisco activase? Biometals 21, 211–217. https://doi.org/10.1007/s10534-007-9110-y.

Gao, Y., Cui, Y., Long, R., et al., 2018. Salt-stress induced proteomic changes of two different alfalfa cultivars during the germination stage. J. Sci. Food Agric.. https://doi.org/10.1002/jsfa.9331.

García-Gaytán, V., Hernández-Mendoza, F., Coria-Téllez, A.V., et al., 2018. Fertigation: nutrition, stimulation, and bioprotection of the root in high performance. Plants (Basel) 7 (4), 88. https://doi.org/10.3390/plants7040088.

García-Jiménez, A., Trejo-Téllez, L.I., Guillén-Sánchez, D., et al., 2018. Vanadium stimulates pepper plant growth and flowering, increases concentrations of amino acids, sugars and chlorophylls, and modifies nutrient concentrations. PLoS One 13 (8), e0201908. https://doi.org/10.1371/journal.pone.0201908.

Geilfus, C.M., 2018. Review on the significance of chlorine for crop yield and quality. Plant Sci. 270, 114–122. https://doi.org/10.1016/j.plantsci.2018.02.014.

Gunbatar, S., Aygun, A., Karatas, Y., et al., 2018. Carbon-nanotube-based rhodium nanoparticles as highly-active catalyst for hydrolytic dehydrogenation of dimethylamineborane at room temperature. J. Colloid Interface Sci. 530, 321–327.

Ishibashi, Y., Yuasa, T., Iwaya-Inoue, M., 2018. Mechanisms of maturation and germination in crop seeds exposed to environmental stresses with a focus on nutrients, water status, and reactive oxygen species. Adv. Exp. Med. Biol. 1081, 233–257. https://doi.org/10.1007/978-981-13-1244-1_13.

Jabeen, N., Maqbool, Q., Bibi, T., et al., 2017. Optimised synthesis of ZnO-nano-fertiliser through green chemistry: boosted growth dynamics of economically important *L. esculentum*. IET Nanobiotechnol. 12 (4), 405–411. https://doi.org/10.1049/iet-nbt.2017.0094.

Jaime, R., Alcántara, J.M., Manzaneda, A.J., et al., 2018. Climate change decreases suitable areas for rapeseed cultivation in Europe but provides new opportunities for white mustard as an alternative oilseed for biofuel production. PLoS One 13 (11), e0207124. https://doi.org/10.1371/journal.pone.0207124.

Joshi, A., Kaur, S., Dharamvir, K., et al., 2018. Multi-walled carbon nanotubes applied through seed-priming influence early germination, root hair, growth and yield of bread wheat (*Triticum aestivum* L.). J. Sci. Food Agric. 98 (8), 3148–3160. https://doi.org/10.1002/jsfa.8818.

Juhel, G., Batisse, E., Hugues, Q., et al., 2011. Alumina nanoparticles enhance growth of *Lemna minor*. Aquat. Toxicol. 105 (3), 328–336.

Kacar, B., Katkat, V., 2010. Plant Nutrition, fifth ed. Nobel Publications, Kızılay-Ankara.

Kadir, S.A., 2004. Fruit quality at harvest of "Jonathan" apple treated with foliar-applied calcium chloride. J. Plant Nutr. 27, 1991–2006. https://doi.org/10.1081/PLN-200030102.

Karkanis, A., Ntatsi, G., Lepse, L., et al., 2018. Faba bean cultivation—revealing novel managing practices for more sustainable and competitive European cropping systems. Front. Plant Sci. 9, 1115. https://doi.org/10.3389/fpls.2018.01115.

Karuppanapandian, T., Wang, H.W., Prabakaran, N., et al., 2011. 2, 4-dichlorophenoxyacetic acid-induced leaf senescence in mung bean (*Vigna radiata* L. Wilczek) and senescence inhibition by co-treatment with silver nanoparticles. Plant Physiol. Biochem. 49 (2), 168–177. https://doi.org/10.1016/j.plaphy.2010.11.007.

Khandaker, M.M., Rohani, F., Dalorima, T., et al., 2017. Effects of different organic fertilizers on growth, yield and quality of *Capsicum annuum* L. Var. Kulai (Red Chilli Kulai). Biosci. Biotechnol. Res. Asia 14 (1), 185–192. https://doi.org/10.13005/bbra/2434.

Khodakovskaya, M.V., de Silva, K., Biris, A.S., et al., 2012. Carbon nanotubes induce growth enhancement of tobacco cells. ACS Nano 6 (3), 2128–2135. https://doi.org/10.1021/nn204643g.

Khodakovskaya, M.V., Kim, B.S., Kim, J.N., et al., 2013. Carbon nanotubes as plant growth regulators: effects on tomato growth, reproductive system, and soil microbial community. Small 9 (1), 115–123. https://doi.org/10.1002/smll.201201225.

Kim, J.H., Lee, Y., Kim, E.J., et al., 2014. Exposure of iron nanoparticles to *Arabidopsis thaliana* enhances root elongation by triggering cell wall loosening. Environ. Sci. Technol. 48 (6), 3477–3485. https://doi.org/10.1021/es4043462.

Kim, Y.H., Khan, A.L., Lee, I.J., 2016. Silicon: a duo synergy for regulating crop growth and hormonal signaling under abiotic stress conditions. Crit. Rev. Biotechnol. 36 (6), 1099–1109.

Kole, C., Kole, P., Randunu, K.M., et al., 2013. Nanobiotechnology can boost crop production and quality: first evidence from increased plant biomass, fruit yield and phytomedicine content in bitter melon (*Momordica charantia*). BMC Biotechnol. 13, 37. https://doi.org/10.1186/1472-6750-13-37.

Kumar, V., Guleria, P., Kumar, V., et al., 2013. Gold nanoparticle exposure induces growth and yield enhancement in *Arabidopsis thaliana*. Sci. Total Environ. 461, 462–468. https://doi.org/10.1016/j.scitotenv.2013.05.018.

Kunicki, E., Grabowska, A., Sękara, A., et al., 2010. The effect of cultivar type, time of cultivation, and biostimulant treatment on the yield of spinach (*Spinacia oleracea* L.). Folia Hortic. Ann. 22 (2), 9–13. https://doi.org/10.2478/fhort-2013-0153.

Kurt, A., Torun, H., Colak, N., et al., 2017. Nutrient profiles of the hybrid grape cultivar 'Isabel' during berry maturation and ripening. J. Sci. Food Agric. 97 (8), 2468–2479. https://doi.org/10.1002/jsfa.8061.

Lahiani, M.H., Dervishi, E., Chen, J., et al., 2013. Impact of carbon nanotube exposure to seeds of valuable crops. ACS Appl. Mater. Interfaces 5, 7965–7973. https://doi.org/10.1021/am402052x.

Lee, C.W., Mahendra, S., Zodrow, K., et al., 2010. Developmental phytotoxicity of metal oxide nanoparticles to *Arabidopsis thaliana*. Environ. Toxicol. Chem. 29, 669–675. https://doi.org/10.1002/etc.58.

Li, Y., Yu, Z., Jin, J., et al., 2018. Impact of elevated CO(2) on seed quality of soybean at the fresh edible and mature stages. Front. Plant Sci. 9, 1413. https://doi.org/10.3389/fpls.2018.01413.

Liu, X.M., Zhang, F.D., Zhang, S.Q., et al., 2005. Responses of peanut to nano-calcium carbonate. Plant Nutr. Fertilizer Sci. 11 (3), 385–389.

Ma, L., Liu, C., Qu, C., et al., 2008. Rubisco activase mRNA expression in spinach: modulation by nanoanatase treatment. Biol. Trace Elem. Res. 122 (2), 168–178.

Ma, C., Chhikara, S., Xing, B., et al., 2013. Physiological and molecular response of *Arabidopsis thaliana* (L.) to nanoparticle cerium and indium oxide exposure. ACS Sustain. Chem. Eng. 1 (7), 768–778. https://doi.org/10.1021/sc400098h.

Madani, B., Wall, M., Mirshekari, A., Bah, A., et al., 2015. Influence of calcium foliar fertilization on plant growth, nutrient concentrations, and fruit quality of papaya. HortTechnology 25 (4), 496–504.

Mahmoodzadeh, H., Nabavi, M., Kashefi, H., 2013. Effect of nanoscale titanium dioxide particles on the germination and growth of canola (*Brassica napus*). J. Orna. Hortic. Plants 3, 25–32.

Marwein, M.A., Choudhury, B.U., Chakraborty, D., et al., 2017. Response of water deficit regime and soil amelioration on evapotranspiration loss and water use efficiency of maize (*Zea mays* l.) in subtropical northeastern Himalayas. Int. J. Biometeorol. 61 (5), 845–855. https://doi.org/10.1007/s00484-016-1262-4.

Mattana, E., Sacande, M., Bradamante, G., et al., 2018. Understanding biological and ecological factors affecting seed germination of the multipurpose tree *Anogeissus leiocarpa*. Plant Biol. 3, 602–609. https://doi.org/10.1111/plb.12702.

Miralles, P., Johnson, E., Church, T.L., et al., 2012. Multiwalled carbon nanotubes in alfalfa and wheat: toxicology and uptake. J. R. Soc. Interface 9 (77), 3514–3527. https://doi.org/10.1098/rsif.2012.0535.

Molassiotis, A., Job, D., Ziogas, V., et al., 2016. Citrus plants: a model system for unlocking the secrets of NO and ROS-inspired priming against salinity and drought. Front. Plant Sci. 7, 229. https://doi.org/10.3389/fpls.2016.00229.

Morla, S., Ramachandra Rao, C.S.V., Chakrapani, R., 2011. Factors affecting seed germination and seedling growth of tomato plants cultured in vitro conditions. J. Chem. Bio. Phys. Sci. B 1, 328–334.

Mukhopadhyay, S.S., 2014. Nanotechnology in agriculture: prospects and constraints. Nanotechnol. Sci. Appl. 7, 63–71. https://doi.org/10.2147/NSA.S39409.

Patra, P., Choudhury, S.R., Mandal, S., et al., 2013. Effect sulfur and ZnO nanoparticles on stress physiology and plant (*Vigna radiata*) nutrition. In: Advanced Nanomaterials and Nanotechnology. Springer, Berlin Heidelberg, pp. 301–309.

Pinzón-Gómez, L.P., Deaquiz, Y.A., Álvarez-Herrera, J.G., 2014. Postharvest behavior of tamarillo (*Solanum betaceum* Cav.) treated with CaCl2 under different storage temperatures. Agron. Colomb. 32 (2), 238. https://doi.org/10.15446/agron.colomb.v32n2.42764.

Prasad, T.N.V.K.V., Sudhakar, P., Sreenivasulu, Y., et al., 2012. Effect of nanoscale zinc oxide particles on the germination, growth and yield of peanut. J. Plant Nutr. 35 (6), 905–927.

Prasad, T.N.V.K.V., Adam, S., Visweswara Rao, P., et al., 2017. Size-dependent effects of antifungal phytogenic silver nanoparticles on germination, growth and biochemical parameters of rice (*Oryza sativa* L), maize (*Zea mays* L) and peanut (*Arachis hypogaea* L). IET Nanobiotechnol. 11 (3), 277–285. https://doi.org/10.1049/iet-nbt.2015.0122.

Qi, M., Liu, Y., Li, T., 2013. Nano-TiO2 improve the photosynthesis of tomato leaves under mild heat stress. Biol. Trace Elem. Res. 156 (1–3), 323–328. https://doi.org/10.1007/s12011-013-9833-2.

Ranjbar, S., Rahemi, M., Ramezanian, A., 2018. Comparison of nano-calcium and calcium chloride spray on postharvest quality and cell wall enzymes activity in apple cv. Red Delicious. Sci. Hortic. 240, 57–64. https://doi.org/10.1016/j.scienta.2018.05.035.

Royse, D.J., Sanchez-Vazquez, J.E., 2003. Influence of precipitated calcium carbonate (CaCO3) on shiitake (*Lentinula edodes*) yield and mushroom size. Bioresour. Technol. 90 (2), 225–228.

Sahin, B., Aygun, A., Gündüz, H., et al., 2018. Cytotoxic effects of platinum nanoparticles obtained from pomegranate extract by the green synthesis method on the MCF-7 cell line. Colloids Surf. B Biointerfaces 163, 119–124.

Saini, R.K., Keum, Y.S., 2018. The significance of genetic, environmental, and pre- and postharvest factors affecting carotenoid contents in crops: a review. J. Agric. Food Chem. 66 (21), 5310–5324. https://doi.org/10.1021/acs.jafc.8b01613.

Salama, H.M.H., 2012. Effects of silver nanoparticles in some crop plants, common bean (*Phaseolus vulgaris* L.) and corn (*Zea mays* L.). Int. Res. J. Biotech. 3 (10), 190–197.

Savithramma, N., Ankanna, S., Bhumi, G., 2012. Effect of nanoparticles on seed germination and seedling growth of Boswellia ovalifoliolata an endemic and endangered medicinal tree taxon. Nano Vis. 2, 61–68.

Sen, B., Savk, A., Sen, F., 2018a. Highly efficient monodisperse nanoparticles confined in the carbon black hybrid material for hydrogen liberation. J. Colloid Interface Sci. 520, 112–118.

Sen, B., Demirkan, B., Savk, A., et al., 2018b. Trimetallic PdRuNi nanocomposites decorated on graphene oxide: a preferred catalyst for the hydrogen evolution reaction. Int. J. Hydrog. Energy 43, 17984–17992.

Sen, B., Demirkan, B., Simsek, B., et al., 2018c. Monodisperse palladium nanocatalysts for dehydrocoupling of dimethylamineborane. Nano-Struct. Nano-Objects 16, 209–214.

Sen, B., Akdere, E.H., Savk, A., et al., 2018d. A novel thiocarbamide functionalized graphene oxide supported bimetallic monodisperse Rh-Pt nanoparticles (RhPt/TC@GO NPs) for Knoevenagel condensation of aryl aldehydes together with malononitrile. Appl. Catal. B Environ. 225 (5), 148–153.

Sen, B., Demirkan, B., Levent, M., et al., 2018e. Silica-based monodisperse PdCo nanohybrids as highly efficient and stable nanocatalyst for hydrogen evolution reaction. Int. J. Hydrog. Energy. https://doi.org/10.1016/j.ijhydene.2018.07.080.

Sen, B., Aygun, A., Savk, A., et al., 2018f. Bimetallic palladium-iridium alloy nanoparticles as a highly efficient and stable catalyst for the hydrogen evolution reaction. Int. J. Hydrog. Energy. https://doi.org/10.1016/j.ijhydene.2018.07.081.

Sharma, D., Jamra, G., Singh, U.M., et al., 2017. Calcium biofortification: three pronged molecular approaches for dissecting complex trait of calcium nutrition in finger millet (*Eleusine coracana*) for devising strategies of enrichment of food crops. Front. Plant Sci. 7, 2028. https://doi.org/10.3389/fpls.2016.02028.

Sheykhbaglou, R., Sedghi, M., Shishevan, M.T., et al., 2010. Effects of nano-iron oxide particles on agronomic traits of soybean. Not. Sci. Biol. 2 (2), 112–113.

Shi, X.L., Zhang, Z.M., Dai, L.X., et al., 2018. Effects of calcium fertilizer application on absorption and distribution of nutrients in peanut under salt stress. Ying Yong Sheng Tai Xue Bao 29 (10), 3302–3310. https://doi.org/10.13287/j.1001-9332.201810.026.

Siddiqui, S., Bangerth, F., 1995. Differential effect of calcium and strontium on flesh firmness and properties of cell walls in apples. J. Hortic. Sci. 70, 949–953. https://doi.org/10.1080/14620316.1995.11515370.

Song, G., Gao, Y., Wu, H., et al., 2012. Physiological effect of anatase TiO2 nanoparticles on *Lemna minor*. Environ. Toxicol. Chem. 31 (9), 2147–2152.

Soukoulis, C., Bohn, T., 2018. A comprehensive overview on the micro- and nano-technological encapsulation advances for enhancing the chemical stability and bioavailability of carotenoids. Crit. Rev. Food Sci. Nutr. 58 (1), 1–36. https://doi.org/10.1080/10408398.2014.971353.

Suriyaprabha, R., Karunakaran, G., Yuvakkumar, R., et al., 2012. Silica nanoparticles for increased silica availability in maize (*Zea mays* L) seeds under hydroponic conditions. Curr. Nanosci. 8, 902–908.

Talan, A., Mishra, A., Eremin, S.A., et al., 2018. Ultrasensitive electrochemical immuno-sensing platform based on gold nanoparticles triggering chlorpyrifos detection in fruits and vegetables. Biosens. Bioelectr. 105, 14–21. https://doi.org/10.1016/j.bios.2018.01.013.

Tanou, G., Fotopoulos, V., Molassiotis, A., 2012. Priming against environmental challenges and proteomics in plants: update and agricultural perspectives. Front. Plant Sci. 3, 216. https://doi.org/10.3389/fpls.2012.00216.

Thangella, P.A.V., Pasumarti, S.N.B.S., Pullakhandam, R., et al., 2018. Differential expression of leaf proteins in four cultivars of peanut (*Arachis hypogaea* L.) under water stress. 3 Biotech (3), 157. https://doi.org/10.1007/s13205-018-1180-8.

Thavarajah, D., Thavarajah, P., Vial, E., et al., 2015. Will selenium increase lentil (*Lens culinaris* Medik) yield and seed quality? Front. Plant Sci. 6, 356. https://doi.org/10.3389/fpls.2015.00356.

Tian, S., Xie, R., Wang, H., et al., 2016. Calcium deficiency triggers phloem remobilization of cadmium in a hyperaccumulating species. Plant Physiol. 172 (4), 2300–2313.

Tiwari, D.K., Dasgupta-Schubert, N., Villaseñor, L.M., et al., 2013. Interaction of carbon nanotubes with mineral nutrients for the promotion of growth of tomato seedlings. Nano Stud. 7, 87–96.

Tiwari, D.K., Dasgupta-Schubert, N., Villaseñor-Cendejas, L.M., et al., 2014. Interfacing carbon nanotubes (CNT) with plants: enhancement of growth, water and ionic nutrient uptake in maize (*Zea mays*) and implications for nanoagriculture. Appl. Nanosci. 4, 577–591.

Torres, E., Recasens, I., Lordan, J., Alegre, S., 2017. Combination of strategies to supply calcium and reduce bitter pit in 'Golden Delicious' apples. Sci. Hortic. 217, 179–188.

Tuan, P.A., Kumar, R., Rehal, P.K., et al., 2018. Molecular mechanisms underlying abscisic acid/gibberellin balance in the control of seed dormancy and germination in cereals. Front. Plant Sci. 9, 668. https://doi.org/10.3389/fpls.2018.00668.

Varshney, R.K., Pandey, M.K., Chitikineni, A., 2018. Plant genetics and molecular biology: an introduction. Adv. Biochem. Eng. Biotechnol. 164, 1–9. https://doi.org/10.1007/10_2017_45.

Vishwakarma, K., Upadhyay, N., Kumar, N., et al., 2018. Potential applications and avenues of nanotechnology in sustainable agriculture. In: Nanomaterials in Plants, Algae, and Microorganisms. vol. 1, pp. 473–500. (Chapter 21). https://doi.org/10.1016/B978-0-12-811487-2.00021-9.

Wang, X., Han, H., Liu, X., et al., 2012a. Multi-walled carbon nanotubes can enhance root elongation of wheat (*Triticum aestivum*) plants. J. Nanopart. Res. 14 (6), 1–10.

Wang, M., Chen, L., Chen, S., et al., 2012b. Alleviation of cadmium-induced root growth inhibition in crop seedlings by nanoparticles. Ecotoxicol. Environ. Saf. 79, 48–54.

Wierzbowska, J., Kovačik, P., Sienkiewicz, S., et al., 2018. Determination of heavy metals and their availability to plants in soil fertilized with different waste substances. Environ. Monit. Assess. 190 (10), 567. https://doi.org/10.1007/s10661-018-6941-7.

Wojcik, P., Borowik, M., 2013. Influence of preharvest sprays of a mixture of calcium formate, calcium acetate, calcium chloride and calcium nitrate on quality and 'Jonagold' apple storability. J. Plant Nutr. 36, 2023–2034. https://doi.org/10.1080/01904167.2013.816730.

Wu, S.G., Huang, L., Head, J., et al., 2012. Phytotoxicity of metal oxide nanoparticles is related to both dissolved metals ions and adsorption of particles on seed surfaces. J. Pet. Environ. Biotechnol. 3, 126.

Yang, F., Hong, F., You, W., et al., 2006. Influence of nano-anatase TiO2 on the nitrogen metabolism of growing spinach. Biol. Trace Elem. Res. 110 (2), 179–190.

Yang, J., Jiang, F., Ma, C., et al., 2018. Alteration of crop yield and quality of wheat upon exposure to silver nanoparticles in a life cycle study. J. Agric. Food Chem. 66 (11), 2589–2597.

Yin, Z., Guo, W., Xiao, H., et al., 2018. Nitrogen, phosphorus, and potassium fertilization to achieve expected yield and improve yield components of mung bean. PLoS One 13 (10), e0206285. https://doi.org/10.1371/journal.pone.0206285.

Yuvakkumar, R., Elango, V., Rajendran, V., et al., 2011. Influence of nanosilica powder on the growth of maize crop (*Zea mays* L.). Int. J. Green Nanotechnol. 3 (3), 80–190.

Zhang, R., Sun, Y., Liu, Z., et al., 2017. Effects of melatonin on seedling growth, mineral nutrition, and nitrogen metabolism in cucumber under nitrate stress. J. Pineal Res.. 62(4). https://doi.org/10.1111/jpi.12403.

Zhao, L., Peralta-Videa, J.R., Rico, C.M., et al., 2014. CeO2 and ZnO nanoparticles change the nutritional qualities of cucumber (*Cucumis sativus*). J. Agric. Food Chem. 62 (13), 2752–2759.

Zheng, L., Su, M., Liu, C., et al., 2007. Effects of nanoanatase TiO2 on photosynthesis of spinach chloroplasts under different light illumination. Biol. Trace Elem. Res. 119 (1), 68–76.

Further reading

Farooq, M., Hussain, M., Usman, M., et al., 2018. Impact of abiotic stresses on grain composition and quality in food legumes. J. Agric. Food Chem. 66 (34), 8887–8897. https://doi.org/10.1021/acs.jafc.8b02924.

Sahin, B., Demir, E., Aygun, A., et al., 2017. Investigation of the effect of pomegranate extract and monodisperse silver nanoparticle combination on MCF-7 cell line. J. Biotechnol. 260C, 79–83.

15

Boron deficiency in fruit crops

Seyed Majid Mousavi[a],*, Babak Motesharezadeh[b]

[a]Soil and Water Research Institute, Agricultural Research, Education and Extension Organization (AREEO), Karaj, Iran
[b]Department of Soil Science, University of Tehran, Tehran, Iran
*Corresponding author. E-mail: majid62mousavi@gmail.com

1 Introduction

More than 80 years ago, boron was known as one of the essential elements for plant growth. The importance of boron in plant nutrition was first reported by Warington (1923) and later on by Brandenburg (1931) who found heart and dry rot in sugar beet and mangolds, respectively. Boron is also regarded as a hazardous element (Uluisik et al., 2018), because of its high potency, even small quantities gave damages, such as germination inhibition, root growth inhibition, shoot chlorosis, and necrosis to plants (Bergmann Jena, 1984). Ahmad et al. (2012) stated that boron is probably the trace element, which most commonly limits crops yield and consequently is most widely used in agriculture. It is also well known that monocotyledons require less boron than dicotyledons, which refers to the lower capacity of roots in monocotyledons for boron absorption compared with roots of dicotyledons (Tariq and Mott, 2007). Considering the role of boron in plant nutrition and the occurrence of boron disorder symptoms, Bergmann Jena (1984) stated that boron inhibits the dropping of grapes and frost resistance of fruit trees presumably due to the beneficial effect of B on carbohydrate and protein metabolism. He also showed that the scab disease of potatoes should also be reduced by boron application, through improvement of skin resistance. Not only the boron status of soils and plants varies with soil type, plant species, and environmental conditions, but also its excess or deficiency may affect the plant growth, yield, and quality. This is due to the small concentration range between its deficiency and toxicity in soil-plant systems (Tariq and Mott, 2007).

A.K. Srivastava, Chengxiao Hu (eds.)
Fruit Crops: Diagnosis and Management of Nutrient Constraints
https://doi.org/10.1016/B978-0-12-818732-6.00015-0

2 Chemical behavior and nature of boron

Boron (B), a member of the subgroup III of metalloids, is considered as an element with properties intermediate between metals and metalloids. It is also an essential nutrient for vascular plants, and its limited bioavailability affects growth, yield, and quality of plants producing considerable economic losses (Tanaka and Fujiwara, 2008). In spite of its low content in nature, B is significantly distributed in lithosphere and hydrosphere. Its concentration is reported from 5 to 10 mg/kg in rocks (Shorrocks, 1997), 3–30 mg/kg in rivers (Power and Woods, 1997), and 4.5 mg/L in ocean (Lemarchand et al., 2000). The total boron concentration of most agricultural soils changes from 1 to 467 mg/kg, with an average concentration of 9 to 85 mg/kg. Such wide-range variations of soils in the total boron concentration are generally due to the parent rock types and soil types falling under divergent geographical and climatic areas. It is mostly high in derived soils from marine sediments (Dar, 2017). Boron exists in soil solution in several forms but, at public soil pH values (5.5–7.5), the most abundant form is the soluble undissociated boric acid ($B(OH)_3$), which it can be significantly leached under drastic rainfall conditions (Yan et al., 2006) and consequently leading to deficiencies in plants that grow there. In contrast, under low rainfall conditions, B cannot be adequately leached and hence may accumulate to toxic levels for plant growth (Reid, 2007). This status is usual in arid and semiarid regions with high-boron groundwater. In these areas, accumulation of B in topsoil because of the evaporation of groundwater reaches toxic levels that reduce plant growth and yield (Tanaka and Fujiwara, 2008).

Although the B requisite is varied in different plants but the optimum concentration of B for most plants is about 20 mg/kg, the B availability in right rate in soil is needed for plant growth, development, yield, and quality. In the earth crust B availability varies from 20 to 200 μg/g, and B available in soils varies between 0.4 and 5 μg/g. The average of B concentration in plant tissues is to be determined about 2 μmol/g (20 μg/g) of dry weight (Shah et al., 2017). It is well known that plants take up B from soil in the boric acid form. Absorption of boric acid by roots can happened by three different molecular mechanisms depending on B availability: (a) passive diffusion through lipid bilayer; (b) aided transport by major intrinsic protein (MIP) channel; and (c) energy-dependent high-affinity transport system encouraged in react to low B supply, which is facilitated via BOR transporters (Tanaka and Fujiwara, 2008).

Under sufficient or excessive B availability conditions, uptake of boric acid by roots is facilitated through a passive process that includes mainly B diffusion across the lipid bilayer (Tanaka and Fujiwara, 2008). As a matter of fact, the lipid absorptivity coefficient for boric acid, calculated both theoretically (Raven, 1980) and experimentally (Stangoulis et al., 2001), confirms the idea that B can cross membranes by a passive procedure to meet plant B requirements (Brown et al., 2002).

It was reported that B absorption may be facilitated by MIP channels, which can transfer small neutral molecules (Dannel et al., 2002). The first experimental confirmation suggesting the participation of channel proteins in B transport was belonged to Dordas et al. (2000), who show that B permeation across plasma membrane vesicles gained from squash roots was somewhat restrained by channel blockers such as mercuric chloride and phloretin. These findings were consequently certified in some studies carried out with intact squash roots (Dordas and Brown, 2001). Physiological researches also have reported the existence of an active B uptake by roots under insufficient B conditions (Dannel et al., 2002). This active uptake of B is confirmed by the fact that B absorption was restrained by both metabolic inhibitors and cold treatment in roots (Dannel et al., 2000). But, so far, only one BOR transporter (OsBOR1 in rice) has been proposed to be involved in the effective B uptake into root cells under B deficiency (Tanaka and Fujiwara, 2008).

3 Factors affecting availability of boron in fruit crops

Optimum plant yield depends on different factors including nutrient uptake capacity and distribution to other growing tissues of the plant (Brown et al., 2002). Boron is widely distributed in the earth's crust, and it is accessible to plants in the form of uncharged boric acid or borate that depends on local soil conditions, such as soil-water content, soil temperature, soil acidity, salinity, organic matter content, and climatic conditions such as rainfall (Shorrocks, 1997). The B availability in many areas of the world is limited due to its high solubility and leaching off by irrigation or rainfall in shallow or coarse-textured soils (Zhou et al., 2014). Furthermore, the probabilities of B bioavailability under drought conditions or in soils with low organic matter are also lowered as a result of the alkalization and organic matter degradation, respectively (Shorrocks, 1997). The maintenance of optimum B concentrations in the soil solution is important for maximum production, which can be attained by applying numerous beneficial and eco-friendly systems. These systems not only improve the B uptake and transfer to other plant parts but also increase the soil fertility and crop production. Some important factors and procedures to increase the B acquisition are explained as follows:

3.1 Soil properties

Soil properties are considered as crucial factor on mobility and bioavailability of nutrients. The soil properties such as soil pH, soil texture, soil temperature/moisture, soil calcium carbonate content, and soil organic matter have a significant effect on boron mobility and availability to plants;

Soil pH; With increasing pH, B becomes less mobile and therefore less bioavailable to plants due to the increase of its adsorption on soil particles (Lehto, 1995). Boron adsorption by soils increased in the range of soil solution pH 3–9 and decreased in the range of pH 10–11.5 (Goldberg and Glaubig, 1986).

Soil texture; Deficiency of B in plants often happens in sandy soils (Mahmoud et al., 2006). It was reported that with increasing clay content soil adsorbed B increased (Mezuman and Keren, 1981).

Soil temperature/moisture; Boron mobility and bioavailability reduced in the range of soil temperature between 5°C and 40°C (Ye et al., 2003). Influence of temperature may be an interactive effect between soil moisture and soil temperature, because it was shown that B deficiency is associated with dry summer conditions (Goldberg, 1997).

Soil calcium carbonate content; Calcium carbonate can reduce B mobility and availability to plants on the one hand by its effect on increasing of soil pH (Mahmoud et al., 2006) and on the other hand by its role as a sink for B in the soil, where it adsorbs a large portion of the soluble B on its particles surfaces (Goldberg and Forster, 1991).

Soil organic matter; With increasing of soil organic matter, content boron adsorption increased as well (Yermiyahu et al., 1995). Considering equilibrium between soluble boron and boron adsorbed by organic matter, soil application of reasonable contents of organic matter to soil can increase boron bioavailability to plant roots (Lehto, 1995).

3.2 Change of root behaviors

Maximum of nutrients absorption by plants is obtained by changing the postembryonic development of the root system through changing the dynamic root structure, such as division of root meristematic; formation of root hair and lateral root; and enlarged root length, diameter, and surface area, which intermediates its adjustment to low nutrient availability. Improved root length and dense root hairs promote root system exploratory capacity by making capable plants to obtain nutrients deeper in the root zone that are commonly distributed extensively and unevenly in the soil under inadequate ion conditions. Indeed, there is an important relationship between B absorption and root architecture (Miwa and Fujiwara, 2010). In different plants, the ion uptake is specific to root units and categories. For example, in wheat, lateral root uptake and transport more water with maximum ion uptake compared with nodal roots (Navara, 1987). Also, fine roots are more dependable in water and ion uptake compared with the coarse and middle roots of *Gossypium hirsutum* (Zhang et al., 2009).

3.3 Grafting

Another important and environmentally friendly approach to decrease limitation of B availability is using of appropriate rootstocks for different plants (Bie et al., 2017) that can uptake a large amount of B from the soil and transport to upper plant tissues to be used for good physiological functioning. Certainly, rootstocks affect plant nutritional status in different plants as a result of their efficacious water- and chemical-absorbing capabilities from soil solution compared with self-rooted plants (Huang et al., 2016). In addition, rootstocks improve the tolerance of scion cultivars to deficiency and toxicity of B (Wojcik et al., 2003). The influence of B on Carrizo citrange (*Citrus sinensis* Osb. × *Poncirus trifoliata* [L.] Raf.) and trifoliate orange (*P. trifoliata* [L.] Raf.) rootstock grafted on orange plants was studied. An increase in B absorption and recently absorbed B concentration in younger and older leaves of Carrizo citrange grafted plants happened than trifoliate-orange-grafted plants (Liu et al., 2011). In pistachio (*Pistacia vera* cv. Kerman), *Pedicypraedia atlantica* rootstock significantly absorbed B and other elements from the soil solution with maximum concentration (1.2–2.4 times more) in leaves followed by PG-II than other rootstocks (Brown et al., 1994).

The inarching of Carrizo citrange (*C. sinensis* Osb. × *P. trifoliata* [L.] Raf.) on Newhall orange (*C. sinensis* Osb.) budded onto trifoliate orange (*P. trifoliata* [L.] Raf.) was studied and was observed that plants sustained better growth under insufficient B availability in root medium and reacted positively to an increased B level in new leaves, twigs, scion, and rootstock stem, verifying that grafting is an ideal technique for ion absorption from the soil solution (Shireen et al., 2018). Therefore, this procedure may be used to meet the B deficiency of citrus plants already growing in orchards.

3.4 Biostimulators

Biostimulants are compounds other than fertilizers, soil conditioners, or pesticides that affect plant metabolic processes, including cell division, respiration, photosynthesis, and ion absorption, when used in small quantities (Kauffman et al., 2007). They react with the plant signaling cascade to lower the negative plant responses under stress conditions, leading to optimal plant growth and yield (Brown and Saa, 2015). Among organic biostimulants, humic substances (HS) are familiar for improving soil structure and root architecture through promoting the root H^+-ATPase activity; thus, depending on the concentration, plant species, and environmental conditions, they are widely used for ion absorption (Halpern et al., 2015). The HS creates a complex with micronutrients, and the plasma membrane creates the proton motive force to help active and passive transport of ions through the symplastic pathway, consequently enhancing trace metal availability to plants (Canellas et al., 2015). Plant nutrition is improved by amino acids through affecting soil microbial activity by the production of a useful microbial community and mineralization of nutrients in the soil solution, consequently improving micronutrients mobility (Halpern et al., 2015). The change in absorption of one ion affects the availability of another ion, maintaining the theory of B absorption through application of biostimulator; however, this needs further studies (Yakhin et al., 2017).

3.5 Mycorrhizal fungi

Fungi that make associations with roots of plant are well known as mycorrhizas; arbuscular mycorrhizal fungi (AMF), as a group of them, is greatly important in soil fertility and plant nutrition (Sarkar et al., 2015). These ectomycorrhizas are obviously common in the mundane ecosystem and are associated with more than 80% of vascular plants, with the exception of some members of the families *Chenopodiaceae, Proteaceae, Cyperaceae*, and *Cruciferae* (Newman and Reddell, 1987). This linkage between two symbiotic partners is dependent upon bidirectional reflections, and under contradictory conditions, mycorrhizal fungi (MF) entraps soil nutrients and acts as carrier of these nutrients from soil to plant roots by gaining plant sugar compounds in return for their metabolism (Lavola et al., 2011). Such as, in the Arctic tundra, AMF contributes 61%–86% of N uptake and transports to plants and obtain 8%–17% of plant photosynthetic carbon in response (Hobbie and Hobbie, 2006).

The capacity of nutrient absorption and also retaining or excluding capacity are due to the fungal mycelium that makes a hyphal sheath around the fine roots of vascular plants or hyphal spirals inside root cortical cells; these structures permit the beneficial plant elements to enter and impede the toxic metals (such as cadmium, chromium, nickel, lead, and arsenic) absorption, from the soil solutions, thus guarding host plants from environmental stresses (Hildebrandt et al., 2007). Total concentration of macro- and micronutrients in different tissues of *Miscanthus sacchariflorus* plants inoculated with mycorrhiza which were grown under natural sterilized soil conditions was increased. The bioavailability of these nutrients to plants enhanced due to the effect of exuded organic compounds from the plant roots and fungi (Sarkar et al., 2015). AMF inoculation affects the concentration of B in plant tissues. Reducing, unchanging, and increasing B acquisition in shoots of the MF-inoculated plants were reported (respectively, Clark et al., 1999; Lu and Miller, 1989; Kothari et al., 1990), but the real effect of mycorrhizal functioning for B has not been verified yet and needs further studies. It is assumed sucrose is a main carbohydrate that is mostly responsible for mobilization of B due to its little B attraction in vascular plants (Lewis, 1980). Conversely, fungal carbohydrates, especially mannitol, have great affinity to readily form a complex with B, causing slight B mobility from fungal symbiotic partner to the host. But this mannitol-B complex mobility has been detected in some mycelia, which provides the continuous absorption and long-distance transport of B in plant tissues (Hu and Brown, 1997) (Table 15.1).

4 Sensitivity of fruit crops, boron deficiency and diseases

Sufficient boron nutrition is vital not only for high yields but also for high quality of plants. Deficiency of B is a nutritional disorder which differs from high B soils and high rainfall areas all over the world (Marschner, 2012). Plants which cultivate in B-deficient conditions must display B deficiency symptoms in first stage, such as unusual growth of the apical growing section. Root growth is also affected by B deficiency, mainly due to disturbing cell wall plasticity and furthermore as result of the inhibition of root meristem functioning (Herrera-Rodríguez et al., 2010). Some important and usual symptoms of B deficiency in different fruit crops are presented in Table 15.4.

Symptoms of a harsh boron deficiency are most usual in fully grown plants. Dissimilar to nitrogen, boron is not easily transported throughout the plant therefore its deficiency symptoms first seem in the shoot apical meristem and in fruits (Nelson, 2012). Boron deficiency may arise on a soil with sufficient boron level if its absorption is limited

TABLE 15.1 Influence of grafting and mycorrhizal fungi on B uptake of plants.

Plants	Rootstocks	Grafting Plant parts utilized	Percent of increase in B concentration	Reference
	Carrizo citrange (*Citrus sinensis* [L.] Osb. × *Poncirus trifoliata* [L.] Raf.)	New leaves Middle leaves Basal leaves Roots New twig Old twig	24–51 53 66–149 63 25 24	Sheng et al. (2009), Wang et al. (2016)
Navel orange (*Citrus sinensis* Osb.)	Cleopatra mandarin (*Citrus reticulata*)	Leaves	29	Brown et al.(1994)
	Sweet cherry (*Prunus avium*)	Leaves	15	Taylor and Dimsey (1993)
Sweet Cherry (*Prunus avium* L.), cultivar Rita	GiSelA 6 (*Prunus cerasus* × *Prunus canescens*, Gi 148/1)	Leaves	13	Taylor and Dimsey (1993)
Grapes (*Vitis vinifera* L.)	Riparia Gloirede Montpellier (RGM) and 1103 Paulsen (*Vitis vinifera* L.)	Leaves	93	Lecourt et al. (2015)
	Pistacia atlantica	Leaves	3	Brown et al. (1999)
Pistachio (*Pistacia atlantica* L.) cultivar Kerman	*Pistacia integerrima*	Leaves	68	Brown et al. (1999)
	UCBI (*Pistacia atlantica* × *Pistacia integerrima*)	Leaves	3	Borgognone et al. (2013)

Plant	AMF species	Mycorrhizal fungi		
		Plant part utilized	Percent of increase in B concentration	Reference
Silver Birch (*Betula pendula*)	*Paxillus involutus*	Roots Stem Leaves	78 5–15 14	Lehto et al. (2004), Ruuhola and Lehto (2014)
Rough lemon (*Citrus jambhiri* Lush)	*Glomus fasciculatum*	Leaves	11–18	Dixon et al. (1989)

by over liming, dry or wet soil conditions, and a low partial pressure of oxygen in soil (Wojcik et al., 2008). Based on Loomis and Durst (1992) B has a vital role in generative processes affecting another development, pollen germination, and pollen tube growth. Hence, in boron-sensitive plants abortion of flower initials and poor set of fruit or seeds are recorded under boron deficiency conditions (Mozafar, 1989). The main effect of boron deficiency looks to be the interruption of the customary functioning of the apical meristems with changes in membrane structure, metabolisms of auxin, carbohydrate, ascorbate, RNA, lignifications, cell wall synthesis, phenol accumulation and sucrose transport being secondary effects (Brown et al., 2002).

As well as other plants, deficiency and excess of B also affect citrus growth and productivity. Deficiency of B is known as "hard fruit," since the fruit is hard and dry due to lumps in the rind making happen by gum impregnations. The main fruit signs contain premature shedding of young fruits. Such fruits have brownish discolorations in the white zone of the rind (albedo), which are defined as gum pockets or impregnations of the tissue with gum and abnormally thick albedo. Older fruit are undersized, unsmooth, and deformed with an extraordinarily thick albedo including gum deposits. Seeds fail to mature, and gum deposits are usual surrounding the axis of the fruit (Fig. 15.1A). Boron is rather immobile in plant tissues; therefore, the major visual symptoms of B deficiency are usually the death of the terminal growing point of the main stem (Zekri and Obreza, 2014).

More symptoms contain brittle and partly condensed leaves, vein splitting, affinity for the leaf blade to curl descending, and occasionally chlorosis (Fig. 15.1B). Deficiency of B also tends to create corking and extension of the upper

FIG. 15.1 Boron deficiency symptoms in (A) fruit and (B) leaves; (C) leaf chlorosis affected by Boron deficiency. *Based on Zekri, M., Obreza, T., 2014. Boron (B) and Chlorine (Cl) for Citrus Trees. A part of a series about understanding nutrient requirements for citrus trees. For the rest of the series, visit http://edis.ifas.ufl.edu/ topic_series_citrus_tree_nutrients.*

surface of the main veins (Fig. 15.1C). Fruit symptoms are the most consistent and credible tool to detect B deficiency. Boron deficiency is also related to citrus greening (HLB) disease. It is seemingly produced by limitations in nutrient uptake and/or transport (Zekri and Obreza, 2014).

Apple (*Malus domestica* Borkh.) has been identified to have considerable requirements of boron (Shorrocks, 1997). Boron deficiency can cause rosette symptoms in leaves. Leaves become dwarfed, thickened, and brittle and are found on much shortened twigs. Zinc deficiency can also lead to leaf rosettes, but the presence of corky lesions in fruit will recognize between the two (Gürel and Başar, 2016). Low fruit-set and in turn lowered yielding are initial visual symptoms of boron deficiency since this nutrient has a key role in the reproductive growth (Loomis and Durst, 1992). Role of boron on fruit-set, yield, and quality of apple is presented in Table 15.2. Under boron deficiency conditions, apple trees are small, misshapen, corked, and sensitive to cracking and have yellow skin with a weak red color (Peryea, 1994). Apple fruits may also have low contents of soluble solids and acids (Shear and Faust, 1980). Boron deficiency symptoms in fruits can occur as either internal or external cork (Dart, 2004). Fruits with internal cork in many cases show no external symptoms. Internal cork is the more usual disorder of deficiency in New South Wales (NSW). It is supposed to happen when deficiency happens later than 8 weeks after petal fall. Dry climatic conditions are typically the trigger in NSW. Small, round water-soaked spots are observed throughout the flesh. The brown tissue usually happens nearer to the core than the skin. As the corky tissue dies, fruit becomes deformed. Fruit with severe internal cork may also have a bumpy appearance.

External cork is more likely to happen if deficiency happens early in the growing season. It results in irregular depressions to seem as fruit matures. Raised brown or reddish spots may happen. In apples, the fruit symptoms of boron deficiency can often be confused with calcium deficiency, mainly where cracking has not advanced. But the two disorders can be recognized by the features in Table 15.3.

TABLE 15.2 Fruit-set, yield, and quality of apple affected by B.

Treatment of B	Fruit-set (%) 14	28	42	Yield (kg/tree)	Average of fruit weight (g)	Total soluble solids (%)
Soil application	36.2	15.3	7.2	4.3	226	13.6
Foliar application	40.2	25.3	15.2	6.8	191	12.5
Control	39.4	11.4	6.9	3.1	188	12.4

Based on Wojcik, P., Wojcik, M., Klamkowski, K., 2008. Response of apple trees to boron fertilization under conditions of low soil boron availability. Sci. Hortic. 116 (1), 58–64; Dar, G.A., 2017. Impact of boron nutrition in fruit crops. Int. J. Curr. Microbiol. App. Sci. 6 (12), 4145–4155.

TABLE 15.3 Comparison between symptoms of boron deficiency and calcium deficiency in fruit.

Boron	Calcium
Occurring the brown tissue near the core	Appearing the brown tissue nearer to the skin
Cracks may happen in fruit	Without cracking
Symptoms do not exacerbate in storage	Symptoms often seem worse, or increase after storage

Based on Dart, J., 2004. Boron deficiency (cork) in pome fruits. AgFact H4. AC 2, p. 2.

Besides pomaceous fruits, stone fruits are also severely affected by boron deficiency, like boron-deficient cherry shoots grow for some time and thereafter tips die. Leaves are deformed in shape, with rough serration, and may cup or roll downward. Under B deficiency conditions, splitting of the bark frequently happens. Some buds of B-deficient plants may fail to open in the spring, but others shrivel and die. Some symptoms of B deficiency in cherry fruit are cracking, shriveling, deforming, internal and external browning, and corking around pit and in flesh (Wojcik and Wojcik, 2006).

Based on different reports, peach trees are not very susceptible to show boron deficiency symptoms compared with other fruit trees. However, in some areas were reported the initial deficiency symptoms at the shoot tips, which were caused terminal dieback. This may then result in the expansion of many side branches. New leaves that follow are small, deformed, thick, and fragile (Wojcik et al., 2008). Defoliation often happens, beginning with terminal leaves and moving down the shoot. Sunken, necrotic spots have been reported on apricot fruit, but peach and Japanese plum are obviously less sensitive to this disorder. Cracking, shriveling, deformation, internal and external browning, and corking around pit and in flesh are signs of cherry fruit under B deficiency conditions (Wojcik et al., 2008).

Brown rot, produced by *Monilinia laxa* (Aderh and Ruhland Honey), is the most important fruit rot disease of peaches that its control is obtained mainly through the use of fungicides to protect blossoms and ripening fruit (Adaskaveg et al., 2011). Though, with recent public concerns concerning pesticide residues on fruit, no postharvest fungicide treatment is permitted in Europe, and there is a necessity for alternative disease management practices that will decrease risk to consumers (Thomidis and Exadaktylou, 2010). Results of Plich and Wójcik (2002) have shown the role of boron on the resistance of fruits to this disorder and fruit quality.

Grapevines reproductive tissues are susceptible to temporary boron deficiency of developing tissues during drought periods, suggesting limited boron availability. Based on Christensen (Christensen et al., 2006), B deficiency in grapevine results in reduced fruit-set, small "shot berries" that are round to pumpkin shaped, and flower and fruit cluster necrosis (Fig. 15.2). Boron deficiency can have a severe impact on fruit quality and yield, even when there are only mild-to-moderate foliar symptoms (Christensen et al., 2006).

FIG. 15.2 Boron deficiency in Thompson Seedless cluster and leaf. *Based on Christensen, L., Beede, R., Peacock, W., 2006. Fall foliar sprays prevent boron-deficiency symptoms in grapes. Calif. Agric. 60 (2), 100–103.*

TABLE 15.4 Symptoms of B deficiency in different fruit crops.

Fruit crop	Symptom of B deficiency
Prunes	Extreme multiple branching in tree tops, bushy branch, significant reduction in fruit-set, brown sunken zones in fruit
Apricot	Pitting, skin discolored, cracking, and corking
Apple	Pitting, skin discolored, cracking, and corking
Almond	Flowers fall and nuts abort or gummy
Walnut	Dieback from shoot tips, leaf fall
Strawberry	Pale chlorotic skin of fruit, cracking, and die back
Pistachio	Fruit-set reduces, and blanks and nonsplit nuts enhances
Peanut	Dark, hollow zone in center of nut, called "hollow heart"
Pear	Blossom blast, pitting, internal corking, and bark cankers
Grape	"Hen and chicken" symptom, dead central shoots

Based on Dar, G.A., 2017. Impact of boron nutrition in fruit crops. Int. J. Curr. Microbiol. App. Sci. 6 (12), 4145–4155.

In nonfruiting papaya plants, the young leaves become deformed, brittle, bunchy, and clawlike. Matured plants are dwarfed, and fruit-set is significantly decreased. Affected fruits are susceptible to be seedless and poorly developed or absent, to ripen irregularly and to have low sugar content. The surface of fruits on harshly affected plants may be covered with lumps. Most of the seeds in the seed cavity fail to germinate. If symptoms seem on young fruits, the fruits may not grow to full size (Wang and Ko, 1975). Under undesirable conditions, farmers in Hawaii in the past have experienced losses of close to 100% of their crop because of this deficiency disease (Nishina, 1991) (Table 15.4).

5 Importance of boron with reference to sink-source relationship

Relationship between source organs, exporters of photoassimilates and sink organs, and importers of fixed carbon has a key role in carbohydrate assimilation and distributing during plant growth and development. Plant productivity is improved by sink strength (highly competitive capacity for import of photoassimilates or biosynthetic rate of macromolecules, such as proteins) and source activity (high photosynthesis or nutrient remobilization rates), which are controlled by a complex signaling network including sugars, hormones, and environmental factors (Yu et al., 2015). Accordingly, an imbalance condition happened by insufficient sink strength, or slow sugar export leading to the enhancement of carbohydrates contents in source organs will cause feedback down controlling of photosynthetic efficiency in leaves (Rossi et al., 2015).

Boron is passively absorbed and transported through the transpirational stream; therefore, its deficiency may be temporary (Brown et al., 1996). Such deficiencies usually happen during periods of quick plant growth, particularly during flowering and seed set. Premature flower and fruit drop of tree crops has been attributed to B deficiency, proposing that B movement to reproductive organs is limited or that growth and enlargement of floral structures have a higher requisite for B than the vegetative structures (Dell and Huang, 1997).

Theoretically, boron might contribute in foliar transport procedures such as veinal loading or superiority of phloem filter plates in the basipetal flow of sucrose. Whether active in source or sink regions, an important problem has been to display that a chemical union of boron and sugar can form in physiological conditions, withstand transport through a barrier membrane, and dissociate easily at the contrary membrane surface. Boron is normally regarded phloem immobile in some fruit trees, spray application of B in the fall impermanently enhanced boron content of leaves, but during late fall and winter boron transported to the bark. In spring, the boron transported from the bark into flowers and caused improved fruit-set. Boron is an essential nutrient and plays a key role in the synthesis and stability of nuclear membranes, in the ribonucleic acid metabolism, and in the transport of assimilates, mostly sugars (Marschner, 2011). Some researchers also reported borate influences the amount of sucrose diffusion from storage tissues bathed in distilled water (Acin-Diaz and Alexander, 1972).

6 Boron functionality; Functionality (hormonal properties of boron) of boron

There is growing evidence that B is necessary for the protection of the structure and functions of membranes and, particularly, plasma membrane (Brown et al., 2002). Boron deficiency changes the membrane potential and lowers the activity of proton-pumping ATPase in *Helianthus annuus* (Ferrol and Donaire, 1992) and *Daucus carota* (Blaser-Grill et al., 1989) roots. Moreover, it has been also explained that B deficiency changes plasma membrane permeability for ions and other solutes (Wang et al., 1999). In spite of the obvious and rapid effects of B deficiency, the fundamental mechanisms by which the structure and function of plasma membrane are affected by B deficiency are still unknown (Goldbach and Wimmer, 2007). Hence, it has been proposed that some membrane molecules including hydroxylated ligands such as glycoproteins and glycolipids are suitable candidates for a potential B function in membranes (Goldbach and Wimmer, 2007). But, so far, the happening of these B complexes has not been verified yet. A latest study reported that at least three possibly B-binding membrane glycoproteins did not discover in B-deficient pea nodules and also in other B-deficient plant tissues, which could show that B and certain membrane glycoproteins have important roles in membrane processes associated with public cell growth (Redondo-Nieto et al., 2007). Furthermore, surface proteins bounded to the membrane via a glycosylphosphatidylinositol anchor such as arabinogalactan proteins (AGP) have been recommended to be putative B-binding structures (Goldbach and Wimmer, 2007). Remarkably, a similar change in cell wall pectins has been proved in pollen tubes affected by B deficiency (Yang et al., 1999) or exposed to Yariv reagent (Roy et al., 1998), a complex that cross-links plasma membrane–associated AGPs.

Newly, it has been reported that B deficiency results in a fast decrease in the expression of several AGP genes in *Arabidopsis* roots (Camacho-Cristóbal et al., 2008). Thus, it is essential to indicate that B might exert its function in membranes not only by stabilizing membrane molecules with *cis*-diol groups but also by controlling the expression of genes which have role in membrane structure and function. Different studies reported the necessity of B for N_2 fixation in the heterocyst of the cyanobacterium *Anabaena* PCC 7119 (Garcia-González et al., 1990) and in the vesicles of actinomycetes of the genus *Frankia* (Bolaños et al., 2002). Both types of microorganisms need B for the resistance of the envelopes that keep nitrogenase from passivation by oxygen under N_2-fixing conditions. Besides, it has been defined that there exists a limited number of developed nodules with the capacity of N_2 fixation in legumes under B deficiency conditions (Yamagishi and Yamamoto, 1994), which could be referred to the potential role of B in *Rhizobium*-legume cell surface interaction. Specially, B is necessary for the targeting of nodule certain plant-derived glycoproteins (Bolaños et al., 2002) that are vital as signals for bacteroid differentiation into a N_2-fixing form (Bolaños et al., 2004). Also, the cell walls of B-deficient nodules have low contents of hydroxyproline-/proline-rich protein, for example, $ENOD_2$, which causes higher oxygen diffusion into the nodules and the following inactivation of nitrogenase (Bonilla et al., 1997). There are many reports on the potential involvement of B in assimilation of nitrogen. For example, a lowered nitrate reductase (NR) activity and heightened accumulation of nitrate have been shown in B-deficient plants (Shen et al., 1993). It is reported that B has a key role in changes of concentration and metabolism of phenolic compounds in vascular plants. Indeed, it is concluded that B deficiency results in an accumulation of phenolics due to the motivation of the enzyme phenylalanine ammonium-lyase (PAL) (Camacho-Cristóbal et al., 2002). Other findings have shown that B deficiency not only resulted in quantitative changes but also qualitative changes in the phenolic pool of plants (Karioti et al., 2006). Hence, B deficiency resulted in an accumulation of two polyamine-phenolic conjugates that were not identified under B-sufficient conditions.

Boron deficiency also improved the activity of polyphenoloxidase (PPO) enzyme that catalyzes the oxidation of phenolic compounds into quinones (Camacho-Cristóbal et al., 2002). Even though it has been suggested that the loss of membrane integrity under B deficiency may be because of accumulated phenolics and their oxidation products (Cakmak and Römheld, 1997), it has been proven that resupply of B to deficient concentrations does not recover plasma membrane unity throughout complexing phenols or restraining PPO activity (Cara et al., 2002). The ascorbate/glutathione cycle plays an important role in the oxygen toxic species detoxification mechanisms in cells, and many studies have shown that B deficiency plays an important role on this cycle. Indeed, both ascorbate and glutathione levels have been proven to reduce in root and leaves under B deficient (Cakmak and Römheld, 1997). For instance, based on Lukaszewski and Blevins (1996), a decrease in ascorbate concentration in root tips of squash suffering from B deficiency that was not related to ascorbate oxidation happened. Remarkably, the ascorbate level decreased in proportion to growth rate under low B supply conditions, and the external addition of ascorbate to the low B medium increased root growth. Hence, it is concluded that root growth inhibition in B-deficient squash may be due to the impaired ascorbate metabolism.

B is involved in protein and enzymatic functioning of the cell membrane, resulting in improved membrane unity (Goldbach et al., 2001). Optimum concentration of B increases the plasma membrane hyperpolarization, while

B deficiency changes the membrane potential and decreases H^+-ATPase activity (Goldbach and Wimmer, 2007). Limitation in B availability decreases ATPase in plasmalemma-enriched vesicles of chickpea roots compared with the control (Lawrence et al., 1995). The direct effect of B on plasma membrane–bound proton-pumping ATPase affects ion flux; the change of H^+, K^+, PO_4^{3-}, Rb^+, and Ca^{2+} ions across the membrane was detected in *Vicia faba* under B deficiency (Robertson and Loughman, 1973). The increased B requirement of young growing tissues verifies its vital role primarily in cell division and elongation (Dell and Huang, 1997). B deficiency dramatically inhibits root elongation, with misshapen flower and fruit formation due to impaired cell division in the meristematic region, whereas enough B supply increases beneficial root development (Gupta and Solanki, 2013). Boron is involved in phenolic metabolism, and phenol accumulation is a usual feature of B-deficient plants (Marschner, 2012). B-sugar *cis*-diol complex formation is essential to lowering phenol accumulation. Nevertheless, plants fail to form this complex due to the shifting of the pathway from glycolysis to phosphate under B deficiency conditions, causing phenolic compound production and accumulation (Mengel and Kirkby, 2001). Boron deficiency activates enzymatic and nonenzymatic oxidation by using phenol as substrate, causing increased polyphenol oxidase and quinine concentrations, which are harmful for plant growth and development (Hajiboland et al., 2013). B deficiency may stimulate reactive oxygen species generation which significantly decreases ascorbic acid and glutathione metabolism (Marschner, 2012). Even though B-deficient leaves of citrus displayed antioxidant enzymatic activity against ascorbate peroxidase, ascorbate, and superoxide dismutase, they were not strong enough to maintain against oxidative damage (Han et al., 2008).

B plays a vital role in nitrogen (N) metabolism. It increases nitrate levels and decreases nitrate reductase activity under limited B conditions (Shen et al., 1993). Bolaños et al. (2004) reported the effect of B in rhizobial N fixation, actinomycete symbiosis, and cyanophyceae heterocyst formation in leguminous crops. Until now, there is no study on the direct involvement of B in photosynthesis. Photosynthesis indirectly is affected by B deficiency through weakening vascular tissues responsible for ion transport (Wang et al., 2015). Goldbach and Wimmer recommended that the disturbance in chloroplast membranes, stomatal apparatus, the energy gradient across the membrane, and thylakoid electron transport are the main reasons for photosynthetic reduction under B deficiency (Goldbach and Wimmer, 2007). During pollen tube growth and germination, B increases the probabilities of fruit setting and promotes seed production, leading to increased crop productivity. A sufficient supply of B decreases the incidence of empty grains, increases the yield by up to 5.5% in barley (*Hordeum vulgare* L.) (El-Feky et al., 2012), enhances spike length and plant pigment content, hinders the probabilities of sterility in wheat (*Triticum aestivum* L.) (Abdel-Motagally and El-Zohri, 2018), and increases the shelf life and quality of tomato (*Lycopersicon esculentum* L.) (Salam et al., 2011). It is also reported that flowering and seed setting in *Arabidopsis thaliana* are supported by overexpressing efflux B transporter BOR1, which enhances not only the yield but also the mineral transport under B-deficient conditions (Miwa et al., 2006). Boron affects the bioavailability and absorption of other plant nutrients from the soil. An obvious increase in the absorption and translocation of P, N, K, Zn, Fe, and Cu in leaves, buds, and seeds was observed after B application in cotton (Ahmed et al., 2011). An enhanced or restricted B supply reduces the nitrate levels by changing the nitrate transporter activity and preventing PMA2 transcript level in the roots, resulting in decreased plasma membrane H^+-ATPase activity (Camacho-Cristóbal and González-Fontes, 2007). The functions of B in different parts of plants are summarized in Fig. 15.3.

7 Interrelationship of boron with other nutrients

Not only the response of plants to B differs with plant species, soil type, and environmental conditions, but also its excess and deficiency may affect the availability and uptake of the other plant nutrients. Some researchers have found that B supply may affect, as a regulator or inhibitor, the accumulation and utilization of other essential nutrients (Alvarez-Tinaut et al., 1979a,b), because excessive concentration of B may interfere with metabolic processes, thus affecting the absorption of other nutrients by plants (Corey and Schulte, 1973). On the other hand, B deficiency may also decrease the levels of plant nutrients (Carpena-Artes and Carpena-Ruiz, 1987). Usually, it can be concluded from the findings that B has important role in physiological and biochemical processes inside the plant cell, changing the concentration and translocation of nutrients. Singh concluded that high levels of applied B had an antagonistic effect on the absorption of nutrients by wheat plants (Singh et al., 1990). But in contrast, a synergistic effect was also observed due to excess B supply on the absorption of nutrients by tomato plants (Carpena-Artes and Carpena-Ruiz, 1987). However, there is a lack of agreement among researchers in respect to the specific role of B on the behavior of any given nutrient, either present in the nutrient substrate or in the plants. Some studies show that B effects are related to all the cation and anion contents in plant. Wallace reported that increase in B uptake caused a reduce cation-anion equivalent ratio in the leaves and enhance in the roots of alfalfa plant (Wallace and Bear, 1949). Correspondingly, other researchers concluded that an increase in B concentrations in the nutrient solution leads to the decrease in cation-anion

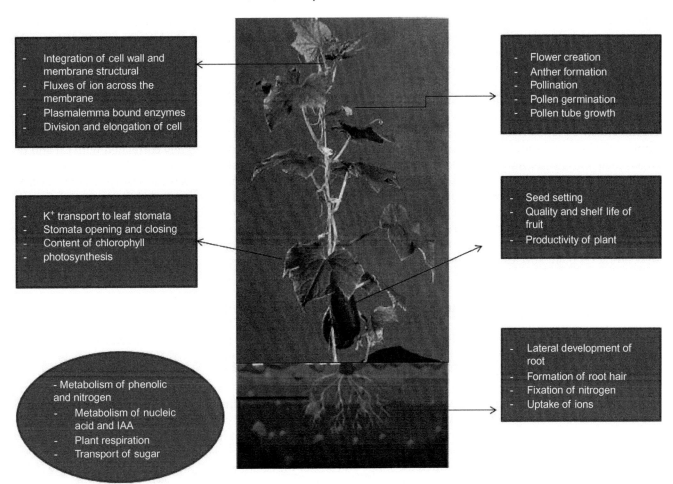

Integration of cell wall and membrane structural
- Fluxes of ion across the membrane
- Plasmalemma bound enzymes
- Division and elongation of cell

- Flower creation
- Anther formation
- Pollination
- Pollen germination
- Pollen tube growth

- K+ transport to leaf stomata
- Stomata opening and closing
- Content of chlorophyll
- photosynthesis

- Seed setting
- Quality and shelf life of fruit
- Productivity of plant

- Metabolism of phenolic and nitrogen
- Metabolism of nucleic acid and IAA
- Plant respiration
- Transport of sugar

- Lateral development of root
- Formation of root hair
- Fixation of nitrogen
- Uptake of ions

FIG. 15.3 Boron functions in different parts of plant. *Based on Shireen, F., Nawaz, M., Chen, C., Zhang, Q., Zheng, Z., Sohail, H., Sun, J., Cao, H., Huang, Y., Bie, Z., 2018. Boron: functions and approaches to enhance its availability in plants for sustainable agriculture. Int. J. Mol. Sci. 19 (7), 1856.*

contents in leaves of sugar beet (Vlamis and Ulrich, 1971). These findings propose that the B effect is related to the type of ions or valence of ions in the plants. Additionally, a further strand in soil B research has been the consideration of nutrient ratios, rather than simple amounts of individual elements in yield analysis. Based on other reports, a high ratio of Ca + K/Mg and Ca + Mg/K in plants is related to B toxicity, but high K + Mg/Ca ratio has little effect on B toxicity (Wallace and Bear, 1949). Also, it was reported that in the presence of B, P/Zn, Fe/Zn, Cu/Zn, and Mn/Zn ratios enhanced in maize crop (Leece, 1978). But these results suggest that B may be indirectly involved in the nutrient balance of plants, resulting in either increase or decrease of plant growth and production.

It was observed that B serves to control the accumulation of ions even from nutrient solutions (Lal and Rao, 1954). Correspondingly, Santra et al. showed that B acts not only within the plant but also in the nutrient medium, thus affecting the uptake of nutrients (Santra et al., 1989). These findings proposed that B may also serve to control or delay the bioavailability of ions form soils. It was also reported that greater than normal levels of B would lessen Al toxicity, superficially by promoting the root growth of crops (Belvins, 1995). These findings show that B may help in the improvement of soil acidity and neutralize the Al toxicity in acid soils. The influences of B deficiency, sufficiency, and toxicity on the chemical composition of plants are not well proven. The investigations of these relationships appear to be contradictory, but the dissimilarities observed is most likely the result of using different plant species (Lombin and Bates, 1982) and varieties (Mozafar, 1989). Similarly, the use of hydroponic (Rodriguez et al., 1981) or sand culture approaches and different soils (Agbenin et al., 1991), analysis of different plant tissues at different growth stages (Carpena-Artes and Carpena-Ruiz, 1987) were used for clarifying these relationships. Because B has important roles in many physiological and biochemical processes, probably affects the utilization of other plant nutrients. Hence, one might imagine the relationship between B and other element utilization to be very complex. In the scientific reports, quantitative information on the role of B on the availability and absorption of other nutrients is slightly found. Even when information is existed, reasons are by no means obvious, because no physiological and chemical mechanisms were suggested. Based on Parks et al. (1944) with graded B levels, the contents of NH_4-N, NO_3-N, Org-N, P, K, Ca, Mg, Na, Zn, Cu, Fe, Mn, Mo, and B were changed in the tomato leaflets as much as several times. Furthermore, they reported that B supply had particular effects, and the trends found were entirely dissimilar with respect to different nutrients.

Boron through changing the permeability of plasmalemma at the root surface, increases P uptake (Morsey and Taha, 1986). Absorption of K enhanced due to their mutual synergistic relationship, but Ca lowered because of antagonistic effect. Uptake of Fe and Cu was positively correlated, whereas Mn and Zn negatively correlated with the applied B. Correspondingly, it was reported that the tip burn percentage was negatively correlated with Ca and positively with N, P, K, Fe, and Cu contents in the leaves of lettuce (Abd-El-Fattah and Agwah, 1987).

Studies on nutrient interactions reveal that boron deficiency in stem adversely affects the absorption of Ca, Fe, Mn, Cu, and Zn in the stem of pomegranate plant (Marathe et al., 2016). Generally, it is concluded that a great deficient or toxic level of B may be responsible for secondary effects on the account of the decline in plant growth and resulting in change of nutrient absorption because of direct or indirect interactions of B with other plant nutrients. However, the interactions of B with other plant nutrients are greatly complex, and the effects can be antagonistic or synergistic depending on plant species and varieties, growth medium, and environmental conditions.

8 Anatomical (structural), physiological, and molecular response of boron

The effect of B as a structural element in plant cell walls and its roles for plant growth have been well studied. It has been revealed that most of the B in the wall exists as a borate diol diester, which cross-links two chains of rhamnogalacturonan II (RG-II), proposing that a physiologically vital function of B is to cross-link cell wall pectin (Ishii and Matsunaga, 1996). Such a cross-linked pectic network is probably to play a role in regulating the mechanical and biochemical properties of the wall (Fleischer et al., 1999). Boron cross-linking with RG-II in cell walls enhances the mechanical strength of the cell wall and support turgor-driven plant cell growth (O'Neill et al., 2001). The studies have shown that B deficiency causes changes in cell wall structure, for example, swelling of the cell walls and the formation of small and irregularly shaped cells (Ishii et al., 2001). Cell walls are vital for the structural unity of the cell and, in fact, for the whole plant. They concluded the shape and size of the cells and also other important properties such as texture, mechanical strength, resistance to pathogenic microorganisms, and the ability to bind and sequester toxic elements and molecules are related to the quality of cell walls (Hayot et al., 2012). The plant cell wall is a greatly complex and active structure made up of cellulose, hemicelluloses, pectins, lignin, and incrusting materials such as cutin, suberin, wax, and various proteins and chemical molecules (Showalter, 1993).

It is commonly well known that B deficiency causes the formation of abnormal cell wall with changed physical properties (Ryden et al., 2003). The short-term influence of B deficiency on structural alterations in cell walls was studied (Yu et al., 2002). Besides, it has been shown that the expression of numerous cell wall–modifying enzymes is downregulated after 6 and 24 h of B deficiency (Camacho-Cristóbal et al., 2008), which could change the cell wall loosening, due to cell elongation. Even though it has usually been shown that boron deficiency results in abnormal cell wall structure based on microscopic observation, what exactly happens in the architecture of cell wall under this condition remains unidentified.

Some studies have reported a potential role of B in cytoskeleton structure and associated processes (Yu et al., 2003; Bassil et al., 2004). Hence, B deficiency enlarged the levels of actin and tubulin proteins in *Arabidopsis* roots and altered the cytoskeletal polymerization patterns in the cells of maize root apices (Yu et al., 2003). The accumulation of cytoskeletal proteins has been imagined to be an adaptive reaction for contributing to mechanical strengthening of cells of root periphery under B deficiency (Yu et al., 2003). With the aim of adapting their self in severe environmental conditions, citrus plants evolved several processes to cope with various stresses including B deficiency and excess. Antioxidant enzymes are one of these approaches that endure plant against multiple stresses including B deficiency and toxicity. As in most abiotic stresses, B toxicity causes the formation of reactive oxygen species (ROS), which results in oxidative destruction of cellular membranes, and lastly the cell death may happen (Cervilla et al., 2007). In order to lower cellular injury caused by ROS, plants have established the scavenging systems composed of antioxidant enzymes, such as peroxidase (POD), superoxide dismutase (SOD), catalase (CAT) and ascorbate peroxidase (APX) (Chen et al., 2005), and nonenzymatic scavenging system similar to alpha-tocopherol, ascorbate, glutathione (Miller et al., 2008), and osmoprotective solutes like proline (Cervilla et al., 2012). It has been well known that under B deficiency conditions plants, show antioxidant enzymes to scavenge ROS (Tewari et al., 2009).

9 Response of boron to fruit yield and quality

Fruit quality is considered as an important parameter in determining shelf life and purchasing choice for customers. Fruit quality is significantly affected by practices prior to harvest that can affect both postharvest quality and storage life. One factor that is crucial in plants is an adequate supply of essential nutrients during growth. Optimum nutrition during the production season and prior to fruit harvest can improve fruit quality and reduce the nutrient deficiencies (Muengkaew et al., 2018).

Boron deficiency is a usual micronutrient problem in agriculture, which causes yield reductions and weakened crop quality (Barker and Pilbeam, 2015). Boron has known functions on the germination of pollen grains, the elongation of pollen tube, fruit-set, and yield and is also indirectly responsible for the activation of dehydrogenase enzymes, sugar translocation, nucleic acids, and plant hormones (Marschner, 2012). Boron application before or after harvest can inhibit physiological disorders, delay ripening, and enhance the concentrations of B in fruit tissues, which all increase fruit quality (Dong et al., 2009). The positive roles of boron on sugar transport, cell membrane synthesis, ascorbate metabolism, nitrogen fixation, indole acetic acid metabolism, respiration, lignification, cell wall structure, carbohydrate, metabolisms of RNA and phenol, membrane biology and functional characteristics, and fruit growth and development were reported (Camacho-Cristóbal et al., 2008; Ahmed et al., 2009).

Boron has important role in calcium transportation in the plant and controls the ratio of potassium and calcium in plant tissues. Boron application also improves fruit firmness, total concentration of soluble solids and sugar while lowering total acidity, weight loss percentage, and fruit rot in fruit crops such as apple and peach (Thomidis and Exadaktylou, 2010). Boron can also decrease pitting disorder of mango (Sharma and Singh, 2009). In order to investigate the effects of B on yield, quality, and leaf nutrients of Isabella (*Vitis labrusca* L.), a study of grape cultivar exhibited that foliar B treatments may support plant growth and development and 0.3% boric acid treatment is recommended for high quality and quantity yields (Ekbic et al., 2018).

The effects of foliar applications of boron nanofertilizers on pomegranate (*Punica granatum* cv. Ardestani) fruit yield and quality were studied (Davarpanah et al., 2016). Their results showed that a single foliar spray with relatively low dose of B (34 mg B/tree) results in improvement of pomegranate fruit yield, mainly because of increase in the number of fruits per tree. Also, they reported the improvement in tree nutrient status and fruit quality affected by B. The last sections, especially "Sensitivity of fruit crops, boron deficiency, and diseases in fruit crops," explain the importance of B on yield and fruit quality of citrus, apple, grapevines, pomaceous fruits, stone fruits, and papaya.

10 Boron nutrition and shelf life of fruits

Shelf life of fruits is controlled by the nutritional status of fruit trees (Krishna et al., 2012), and the nutrient balance is the key factor in postharvest quality of fruits. Boron is a vital element that influences cell wall formation to maintain the firmness of fruit (Khalaj et al., 2017), and its deficiency in agriculture is a common challenge that lessens yield and fruit quality (Barker and Pilbeam, 2015). Boron deficiency also reduces pollen germination and growth of pollen tubes, which subsequently causes the development of malformed fruits, which lowers crop yield and deteriorates fruit quality (Sharma and Sharma, 2004). Based on Khalaj et al., fruit firmness was considerably increased by B spray (Khalaj et al., 2017). Increasing pectin levels in the cell wall also likely play a role in this regard (Veltman et al., 1999). Studies on red beet hybrids and cultivars showed that boron nutrition positively affect the yield and reduce scab prevalence and intensity (Bundinienė, 2009). The variation of fruit produce quality and yield depends on the produce variety, growing environment, and nutrient solution provided and used on the produce as it grows (Ercisli and Orhan, 2007).

Boron and calcium generally have similar uptake, translocation, and correlated physiological roles. Cell strength, cell wall thickness, and tissue firmness are greatly dependent upon sufficient amounts of calcium, mostly forming pectin in the middle lamella (Subburamu et al., 1990). For example, deficiency of boron and calcium in apple causes early fruit softening, high incidence of physiological disorders such as internal breakdown, bitter pit, high respiration, ethylene production, and fruit decay (Conway et al., 2001). Furthermore, B deficiency in fruits results in sensitivity of the fruits to sunburn, splitting, and rotting, and also these fruits have a short storage life (Sánchez and Righetti, 2005). Studies confirm that boric acid applications increase fruit firmness and decrease weight losses percentages and fruit rot decay percentage of apples induced by *Botrytis cinerea* during cold storage at 5°C for 60 days (Omaima and Karima, 2007). Boron can also affect the cell wall and plasma membrane cell wall interface, metabolism, reproductive growth and development, and root elongation and shoot growth of a growing plant structure (Marschner, 2012). Boron supports maintain cell membrane structures integrity and is also known as an antioxidant that hinders fruit internal browning (Khalaj et al., 2017).

B. cinerea Pers. Fr., a present everywhere fungal pathogen, results in gray mold rot on a considerable number of economically important agricultural and horticultural productions (Keller et al., 2003). It is the most usual postharvest pathogen of table grapes in most areas of the world, resulting in severe losses of table grapes after harvest (Cappellini et al., 1986). Findings of Qin et al. (2010) confirmed the inhibitory effect of boron against *B. cinerea* on table grapes. They reported that the mechanisms by which boron reduced gray mold decay of table grapes may be directly associated with the disruption effect of boron on cell membrane of the fungal pathogen that resulted in the breakdown of the cell membrane and loss of cytoplasmic materials from the hyphae.

Based on other reports, the most important factor in postharvest quality is firmness, which can maintain fruit quality and prolong shelf life by increasing the fruit's cell wall thickness or cell compactness (Islam et al., 2016). They reported that B + Ca treatment improves cell wall compactness, decreases the respiration rate and fresh weight loss, increases shelf life, and maintains flesh firmness of cherry tomato.

11 Future research

Boron is an essential plant micronutrient to optimum plant growth and fruit quality. The sources, behavior, and fate of B in the environment remain unclear. The information on B fractionation/speciation in natural compartments, that is, the atmosphere, soils, natural waters, sediments, and biota, is incomplete and often contradictory. Remarkably, until now, B geochemistry in the biosphere was usually defined in terms of inorganic chemistry, and boric acid, borates, and borosilicate minerals were considered the main B compounds and acting agents in the environmental cycle. Boron has been also used extensively in agriculture for control of fungi, bacteria, and many insects. Treatment of crops with boron induces an effective, economical, and environmentally safe management strategy to control different physiological disorders and pathogens. The influence of boron in combination with other salts for integrated control of postharvest pathogens deserves further work which needs to be considered in future researches.

The capabilities of crops to efficiently utilize B resources vary significantly. Accordingly, from an agricultural point of view, there is a necessity to identify the important cultivars of agronomic and horticultural crops with vigorous root systems to utilize the available B and that can thrive best under B deficiency. Understanding of the rootstock and scion relationship may be useful in recognizing excellent root systems of crops that are tolerant to B deficiency or toxicity conditions. Besides, the mechanistic study at the molecular level for B in plants opens novel views to increase B stress tolerance in crops.

Grafting and mycorrhiza (AMF) inoculation increase plant physiological and nutritional properties, and based on last sections, a number of investigations have verified their vital role in B uptake. A stressful environment induces a negative effect on plant growth and development. But this stress can be alleviated by the use of plant growth promoting rhizobacteria (PGPR) and AMF. Most researches on AMF and rhizobacteria inoculation focused on increasing B availability and plant growth under normal and stress conditions (Lehto et al., 2004; Ruuhola and Lehto, 2014). Definitely, these applications improve the water and B content of plants, but the combined inoculation of AMF and PGPR could be more suitable in B assimilation compared with their individual application. In some plant species, the dual inoculation of AMF and PGPR increases growth by increasing water and macronutrient contents. Though to the best of our knowledge, no data are accessible for micronutrients, particularly B. The mechanism of nutrient acquisition by these microorganisms is also slightly understood. Hence, the role of combined inoculation of these microorganisms for efficient B acquisition and its molecular mechanism needs to be studied further to gain better results and to increase B absorption and utilization in crops.

References

Abd-El-Fattah, M.A., Agwah, E.M.R., 1987. Physiological studies on lettuce tipburn. Egypt. J. Hortic. 14, 143–153.

Abdel-Motagally, F.M.F., El-Zohri, M., 2018. Improvement of wheat yield grown under drought stress by boron foliar application at different growth stages. J. Saudi Soc. Agric. Sci. 17 (2), 178–185.

Acin-Diaz, N., Alexander, A.G., 1972. Studies on sugar diffusion from sugarcane storage tissue. PR Univ. J. Agric. 51 (3), 253–266.

Adaskaveg, J.E., Gubler, W.D., Michailides, T.J., Holtz, B.A., 2011. Efficacy and Timing of Fungicides, Bactericides, and Biologicals for Deciduous Tree Fruit, Nut, Strawberry, and Vine Crops 2011. UC Davis.

Agbenin, J.O., Lombin, G., Owonubi, J.J., 1991. Direct and interactive effect of boron and nitrogen on selected agronomic parameters and nutrient uptake by cowpea (*Vigna unguiculata* L. Walp) under glass house conditions. Trop. Agric. 68, 356–362.

Ahmad, W., Zia, M., Malhi, S.S., Niaz, A., Ullah, S., 2012. Boron deficiency in soils and crops: a review. In: Crop Plant., p. 39. https://doi.org/10.5772/36702.

Ahmed, W., Niaz, A., Kanwal, S., Rahmatullah, A., Rasheed, M.K., 2009. Role of boron in plant growth. A review. Agric. Res. 47 (3), 329–338.

Ahmed, N., Abid, M., Ahmad, F., Ullah, M.A., Javaid, Q., Ali, M.A., 2011. Impact of boron fertilization on dry matter production and mineral constitution of irrigated cotton. Pak. J. Bot. 43 (6), 2903–2910.

Alvarez-Tinaut, M.C., Leal, A., Agui, I., Recalde-Martinez, L., 1979a. Physiological effects of B-Mn interaction in tomato plants. II. The uptake and translocation of macro elements. Anal. Edafol. Agrobiol. 38, 991–1012.

Alvarez-Tinaut, M.C., Leal, A., Agui, I., Recalde-Martinez, L., 1979b. Physiological effects of B-Mn interaction in tomato plant. III. The uptake and translocation of macro elements. Anal. Edafol. Agrobiol. 38, 1013–1029.

Barker, A.V., Pilbeam, D.J., 2015. Handbook of Plant Nutrition, second ed. CRC Press.

Bassil, E., Hu, H., Brown, P.H., 2004. Use of phenylboronic acids to investigate boron function in plants. Possible role of boron in transvacuolar cytoplasmic strands and cell-to-wall adhesion. Plant Physiol. 136 (2), 3383–3395.

Belvins, D.G., 1995. Boron deficiency. In: Shorrocks, V.M. (Ed.), Boron in Agriculture. In: A Quarterly Bulletin, 15, Borax Consolidated Limited, London, pp. 2–3.

Bergmann Jena, W., 1984. The significance of the micronutrient boron in agriculture. In: Symposium Held by the Borax Group in the International Trade Center of GDR. Borax House, Berlin, p. 7.

Bie, Z., Nawaz, M.A., Huang, Y., Lee, J.-M., Colla, G., 2017. Introduction of vegetable grafting. In: Vegetable Grafting: Principles and Practices. CABI Publishing, Wallingford, pp. 1–21.

Blaser-Grill, J., Knoppik, D., Amberger, A., Goldbach, H., 1989. Influence of boron on the membrane potential in *Elodea densa* and *Helianthus annuus* roots and H+ extrusion of suspension cultured *Daucus carota* cells. Plant Physiol. 90 (1), 280–284.

Bolaños, L., Redondo-Nieto, M., Bonilla, I., Wall, L.G., 2002. Boron requirement in the *Discaria trinervis* (Rhamnaceae) and Frankia symbiotic relationship. Its essentiality for Frankia BCU110501 growth and nitrogen fixation. Physiol. Plant. 115 (4), 563–570.

Bolaños, L., Lukaszewski, K., Bonilla, I., Blevins, D., 2004. Why boron? Plant Physiol. Biochem. 42 (11), 907–912.

Bonilla, I., Mergold-Villasenor, C., Campos, M.E., Sanchez, N., Perez, H., Lopez, L., Castrejón, L., Sanchez, F., Cassab, G.I., 1997. The aberrant cell walls of boron-deficient bean root nodules have no covalently bound hydroxyproline-/proline-rich proteins. Plant Physiol. 115 (4), 1329–1340.

Borgognone, D., Colla, G., Rouphael, Y., Cardarelli, M., Rea, E., Schwarz, D., 2013. Effect of nitrogen form and nutrient solution pH on growth and mineral composition of self-grafted and grafted tomatoes. Sci. Hortic. 149, 61–69.

Brandenburg, E., 1931. Heart and dry rot of beet—causes and treatment. Angew. Bot. 14, 194–228.

Brown, P., Saa, S., 2015. Biostimulants in agriculture. Front. Plant Sci. 6, 671.

Brown, P.H., Zhang, Q., Ferguson, L., 1994. Influence of rootstock on nutrient acquisition by pistachio. J. Plant Nutr. 17 (7), 1137–1148.

Brown, P.H., Hu, H., Nyomora, A., Freeman, M., 1996. Foliar application enhances almond yields. In: Better Crops With Plant Food, No. 1. Potash and Phosphate Inst. Ref. No. 323054/95223.

Brown, P.H., Bellaloui, N., Hu, H., Dandekar, A., 1999. Transgenically enhanced sorbitol synthesis facilitates phloem boron transport and increases tolerance of tobacco to boron deficiency. Plant Physiol. 119 (1), 17–20.

Brown, P.H., Bellaloui, N., Wimmer, M.A., Bassil, E.S., Ruiz, J., Hu, H., Pfeffer, H., Dannel, F., Römheld, V., 2002. Boron in plant biology. Plant Biol. 4 (02), 205–223.

Bundinienė, O., 2009. Influence of boron fertilizer and meteorological conditions on red beet infection with scab and productivity. Sodinink. daržinink, 28 (3), 29–40.

Cakmak, I., Römheld, V., 1997. Boron deficiency-induced impairments of cellular functions in plants. Plant Soil 193 (1–2), 71–83.

Camacho-Cristóbal, J.J., González-Fontes, A., 2007. Boron deficiency decreases plasmalemma H+-ATPase expression and nitrate uptake, and promotes ammonium assimilation into asparagine in tobacco roots. Planta 226 (2), 443–451.

Camacho-Cristóbal, J.J., Anzellotti, D., González-Fontes, A., 2002. Changes in phenolic metabolism of tobacco plants during short-term boron deficiency. Plant Physiol. Biochem. 40 (12), 997–1002.

Camacho-Cristóbal, J.J., Herrera-Rodríguez, M.B., Beato, V.M., Rexach, J., Navarro-Gochicoa, M.T., Maldonado, J.M., González-Fontes, A., 2008. The expression of several cell wall-related genes in Arabidopsis roots is down-regulated under boron deficiency. Environ. Exp. Bot. 63 (1–3), 351–358.

Canellas, L.P., Olivares, F.L., Aguiar, N.O., Jones, D.L., Nebbioso, A., Mazzei, P., Piccolo, A., 2015. Humic and fulvic acids as biostimulants in horticulture. Sci. Hortic. 196, 15–27.

Cappellini, R., Ceponis, M., Lightner, G., 1986. Disorders in table grape shipments to the New York Market, 1972-1984. Plant Dis. 70, 1075–1079.

Cara, F.A., Sánchez, E., Ruiz, J.M., Romero, L., 2002. Is phenol oxidation responsible for the short-term effects of boron deficiency on plasma-membrane permeability and function in squash roots? Plant Physiol. Biochem. 40 (10), 853–858.

Carpena-Artes, O., Carpena-Ruiz, R., 1987. Effects of boron in tomato plant. Leaf evaluations. Agrochimica 31, 391–400.

Cervilla, L.M., Blasco, B., Ríos, J.J., Romero, L., Ruiz, J.M., 2007. Oxidative stress and antioxidants in tomato (*Solanum lycopersicum*) plants subjected to boron toxicity. Ann. Bot. 100 (4), 747–756.

Cervilla, L.M., Blasco, B., Rios, J.J., Rosales, M.A., Sánchez-Rodríguez, E., Rubio-Wilhelmi, M.M., Romero, L., Ruiz, J.M., 2012. Parameters symptomatic for boron toxicity in leaves of tomato plants. J. Bot. https://doi.org/10.1155/2012/726206.

Chen, L., Qi, Y., Liu, X., 2005. Effects of aluminum on light energy utilization and photoprotective systems in citrus leaves. Ann. Bot. 96 (1), 35–41.

Christensen, L., Beede, R., Peacock, W., 2006. Fall foliar sprays prevent boron-deficiency symptoms in grapes. Calif. Agric. 60 (2), 100–103.

Clark, R.B., Zobel, R.W., Zeto, S.K., 1999. Effects of mycorrhizal fungus isolates on mineral acquisition by *Panicum virgatum* in acidic soil. Mycorrhiza 9 (3), 167–176.

Conway, W.S., Sams, C.E., Hickey, K.D., 2001. Pre- and postharvest calcium treatment of apple fruit and its effect on quality. In: International Symposium on Foliar Nutrition of Perennial Fruit Plants 594, pp. 413–419.

Corey, R.B., Schulte, E.E., 1973. Factors affecting the availability of nutrients to plants. In: Walsh, L.M., Beaton, J.D. (Eds.), Soil Testing and Plant Analysis (Revised). Soil Science Society of America, Madison, WI, pp. 23–33.

Dannel, F., Pfeffer, H., Römheld, V., 2000. Characterization of root boron pools, boron uptake and boron translocation in sunflower using the stable isotopes 10 B and 11 B. Funct. Plant Biol. 27 (5), 397–405.

Dannel, F., Pfeffer, H., Römheld, V., 2002. Update on boron in higher plants—uptake, primary translocation and compartmentation. Plant Biol. 4 (02), 193–204.

Dar, G.A., 2017. Impact of boron nutrition in fruit crops. Int. J. Curr. Microbiol. App. Sci. 6 (12), 4145–4155.

Dart, J., 2004. Boron deficiency (cork) in pome fruits. In: AgFact H4. AC 2, p. 2.

Davarpanah, S., Tehranifar, A., Davarynejad, G., Abadía, J., Khorasani, R., 2016. Effects of foliar applications of zinc and boron nano-fertilizers on pomegranate (Punica granatum cv. Ardestani) fruit yield and quality. Sci. Hortic. 210, 57–64.

Dell, B., Huang, L., 1997. Physiological response of plants to low boron. Plant Soil 193 (1–2), 103–120.

Dixon, R.K., Garrett, H.E., Cox, G.S., 1989. Boron fertilization, vesicular-arbuscular mycorrhizal colonization and growth of citrus jambhiri lush. J. Plant Nutr. 12 (6), 687–700.

Dong, T., Xia, R., Xiao, Z., Wang, P., Song, W., 2009. Effect of pre-harvest application of calcium and boron on dietary fibre, hydrolases and ultrastructure in 'Cara Cara' navel orange (Citrus sinensis L. Osbeck) fruit. Sci. Hortic. 121 (3), 272–277.

Dordas, C., Brown, P.H., 2001. Evidence for channel mediated transport of boric acid in squash (Cucurbita pepo). Plant Soil 235 (1), 95–103.

Dordas, C., Chrispeels, M.J., Brown, P.H., 2000. Permeability and channel-mediated transport of boric acid across membrane vesicles isolated from squash roots. Plant Physiol. 124 (3), 1349–1362.

Ekbic, H.B., Gokdemir, N., Erdem, H., 2018. Effects of boron on yield, quality and leaf nutrients of Isabella (Vitis labrusca L.) grape cultivar. Acta Sci. Pol. Hortorum Cultus 17 (1), 149–157.

El-Feky, S.S., El-Shintinawy, F.A., Shaker, E.M., Shams El-Din, H.A., 2012. Effect of elevated boron concentrations on the growth and yield of barley ('Hordeum vulgare' L.) and alleviation of its toxicity using different plant growth modulators. Aust. J. Crop Sci. 6 (12), 1687.

Ercisli, S., Orhan, E., 2007. Chemical composition of white (Morus alba), red (Morus rubra) and black (Morus nigra) mulberry fruits. Food Chem. 103 (4), 1380–1384.

Ferrol, N., Donaire, J.P., 1992. Effect of boron on plasma membrane proton extrusion and redox activity in sunflower cells. Plant Sci. 86 (1), 41–47.

Fleischer, A., O'Neill, M.A., Ehwald, R., 1999. The pore size of non-graminaceous plant cell walls is rapidly decreased by borate ester cross-linking of the pectic polysaccharide rhamnogalacturonan II. Plant Physiol. 121 (3), 829–838.

Garcia-González, M., Mateo, P., Bonilla, I., 1990. Effect of boron deficiency on photosynthesis and reductant sources and their relationship with nitrogenase activity in Anabaena PCC 7119. Plant Physiol. 93 (2), 560–565.

Goldbach, H.E., Wimmer, M.A., 2007. Boron in plants and animals: is there a role beyond cell-wall structure? J. Plant Nutr. Soil Sci. 170 (1), 39–48.

Goldbach, H.E., Yu, Q., Wingender, R., Schulz, M., Wimmer, M., Findeklee, P., Baluška, F., 2001. Rapid response reactions of roots to boron deprivation. J. Plant Nutr. Soil Sci. 164 (2), 173–181.

Goldberg, S., 1997. Reactions of boron with soils. Plant Soil 193 (1–2), 35–48.

Goldberg, S., Forster, H.S., 1991. Boron sorption on calcareous soils and reference calcites. Soil Sci. 152 (4), 304–310.

Goldberg, S., Glaubig, R.A., 1986. Boron adsorption on California soils 1. Soil Sci. Soc. Am. J. 50 (5), 1173–1176.

Gupta, U.C., Solanki, H.A., 2013. Impact of boron deficiency on plant growth. Int. J. Bioassays 2 (7), 1048–1050.

Gürel, S., Başar, H., 2016. Effects of applications of boron with iron and zinc on the contents of pear trees. Not. Bot. Horti Agrobot. Cluj-Napoca 44 (1), 125–132.

Hajiboland, R., Bahrami-Rad, S., Bastani, S., 2013. Phenolics metabolism in boron-deficient tea [Camellia sinensis (L.) O. Kuntze] plants. Acta Biol. Hung. 64 (2), 196–206.

Halpern, M., Bar-Tal, A., Ofek, M., Minz, D., Muller, T., Yermiyahu, U., 2015. The use of biostimulants for enhancing nutrient uptake. In: Advances in Agronomy. Elsevier, pp. 141–174.

Han, S., Chen, L.-S., Jiang, H.-X., Smith, B.R., Yang, L.-T., Xie, C.-Y., 2008. Boron deficiency decreases growth and photosynthesis, and increases starch and hexoses in leaves of citrus seedlings. J. Plant Physiol. 165 (13), 1331–1341.

Hayot, C.M., Forouzesh, E., Goel, A., Avramova, Z., Turner, J.A., 2012. Viscoelastic properties of cell walls of single living plant cells determined by dynamic nanoindentation. J. Exp. Bot. 63 (7), 2525–2540.

Herrera-Rodríguez, M.B., González-Fontes, A., Rexach, J., Camacho-Cristóbal, J.J., Maldonado, J.M., Navarro-Gochicoa, M.T., 2010. Role of boron in vascular plants and response mechanisms to boron stresses. Plant Stress 4 (2), 115–122.

Hildebrandt, U., Regvar, M., Bothe, H., 2007. Arbuscular mycorrhiza and heavy metal tolerance. Phytochemistry 68 (1), 139–146.

Hobbie, J.E., Hobbie, E.A., 2006. 15N in symbiotic fungi and plants estimates nitrogen and carbon flux rates in Arctic tundra. Ecology 87 (4), 816–822.

Hu, H., Brown, P.H., 1997. Absorption of boron by plant roots. Plant Soil 193 (1–2), 49–58.

Huang, Y., Jiao, Y., Nawaz, M.A., Chen, C., Liu, L., Lu, Z., Kong, Q., Cheng, F., Bie, Z., 2016. Improving magnesium uptake, photosynthesis and antioxidant enzyme activities of watermelon by grafting onto pumpkin rootstock under low magnesium. Plant Soil 409 (1–2), 229–246.

Ishii, T., Matsunaga, T., 1996. Isolation and characterization of a boron-rhamnogalacturonan-II complex from cell walls of sugar beet pulp. Carbohydr. Res. 284 (1), 1–9.

Ishii, T., Matsunaga, T., Hayashi, N., 2001. Formation of rhamnogalacturonan II-borate dimer in pectin determines cell wall thickness of pumpkin tissue. Plant Physiol. 126 (4), 1698–1705.

Islam, M.Z., Mele, M.A., Baek, J.P., Kang, H.-M., 2016. Cherry tomato qualities affected by foliar spraying with boron and calcium. Hortic. Environ. Biotechnol. 57 (1), 46–52.

Karioti, A., Chatzopoulou, A., Bilia, A.R., Liakopoulos, G., Stavrianakou, S., Skaltsa, H., 2006. Novel secoiridoid glucosides in Olea europaea leaves suffering from boron deficiency. Biosci. Biotechnol. Biochem. 70 (8), 1898–1903.

Kauffman, G.L., Kneivel, D.P., Watschke, T.L., 2007. Effects of a biostimulant on the heat tolerance associated with photosynthetic capacity, membrane thermostability, and polyphenol production of perennial ryegrass. Crop Sci. 47 (1), 261–267.

Keller, M., Viret, O., Cole, F.M., 2003. Botrytis cinerea infection in grape flowers: defense reaction, latency, and disease expression. Phytopathology 93 (3), 316–322.

Khalaj, K., Ahmadi, N., Souri, M., 2017. Improvement of postharvest quality of Asian pear fruits by foliar application of boron and calcium. Horticulturae 3 (1), 15.

Kothari, S.K., Marschner, H., Römheld, V., 1990. Direct and indirect effects of VA mycorrhizal fungi and rhizosphere microorganisms on acquisition of mineral nutrients by maize (*Zea mays* L.) in a calcareous soil. New Phytol. 116 (4), 637–645.

Krishna, H., Das, B., Attri, B.L., Kumar, A., Ahmed, N., 2012. Interaction between different pre-and postharvest treatments on shelf life extension of 'Oregon Spur' apple. Fruits 67 (1), 31–40.

Lal, K., Rao, M., 1954. Microelement Nutrition of Plants, first ed. Plant Physiology Laboratory College of Agriculture, Banarus Hindu University, Banaras.

Lavola, A., Aphalo, P.J., Lehto, T., 2011. Boron and other elements in sporophores of ectomycorrhizal and saprotrophic fungi. Mycorrhiza 21 (3), 155–165.

Lawrence, K., Bhalla, P., Misra, P.C., 1995. Changes in NAD (P) H-dependent redox activities in plasmalemma-enriched vesicles isolated from boron- and zinc-deficient chick pea roots. J. Plant Physiol. 146 (5–6), 652–657.

Lecourt, J., Lauvergeat, V., Ollat, N., Vivin, P., Cookson, S.J., 2015. Shoot and root ionome responses to nitrate supply in grafted grapevines are rootstock genotype dependent. Aust. J. Grape Wine Res. 21 (2), 311–318.

Leece, D.R., 1978. Effects of boron on the physiological activity of zinc in maize. Aust. J. Agric. Res. 29 (4), 739–747.

Lehto, T., 1995. Boron retention in limed forest mor. For. Ecol. Manag. 78 (1–3), 11–20.

Lehto, T., Lavola, A., Kallio, E., Aphalo, P.J., 2004. Boron uptake by ectomycorrhizas of silver birch. Mycorrhiza 14 (3), 209–212.

Lemarchand, D., Gaillardet, J., Lewin, E., Allegre, C.J., 2000. The influence of rivers on marine boron isotopes and implications for reconstructing past ocean pH. Nature 408 (6815), 951.

Lewis, D.H., 1980. Boron, lignification and the origin of vascular plants—a unified hypothesis. New Phytol. 84 (2), 209–229.

Liu, G., Jiang, C., Wang, Y., Peng, S.A., Zhong, B., Ceng, Q., Yuan, S., 2011. Changes in mineral element contents of 'Newhall' navel orange (*Citrus sinensis* Osb.) grafted on two different rootstocks under boron deficiency. Plant Nutr. Fertil. Sci. 1, 180–185.

Lombin, G.L., Bates, T.E., 1982. Comparative responses of peanuts, alfalfa, and soybeans to varying rates of boron and manganese on two calcareous Ontario soils. Can. J. Soil Sci. 62 (1), 1–9.

Loomis, W.D., Durst, R.W., 1992. Chemistry and biology of boron. BioFactors 3 (4), 229–239.

Lu, S., Miller, M.H., 1989. The role of VA mycorrhizae in the absorption of P and Zn by maize in field and growth chamber experiments. Can. J. Soil Sci. 69 (1), 97–109.

Lukaszewski, K.M., Blevins, D.G., 1996. Root growth inhibition in boron-deficient or aluminum-stressed squash may be a result of impaired ascorbate metabolism. Plant Physiol. 112 (3), 1135–1140.

Mahmoud, M., Shaaban, A.F., El-Sayed, A., El-Nour, E.A.M., Aly, E.S., Mohamed, A.K., 2006. Boron/nitrogen interaction effect on growth and yield of faba bean plants grown under sandy soil conditions. Int. J. Agric. Res. 1 (4), 322–330.

Marathe, R.A., Murkute, A.A., Babu, K.D., 2016. Mineral nutrient deficiencies and nutrient interactions in pomegranate. Natl. Acad. Sci. Lett. 39 (6), 407–410.

Marschner, H., 2011. Mineral Nutrition of Higher Plants, second ed. Academic Press, San Diego, CA.

Marschner, H., 2012. Marschner's Mineral Nutrition of Higher Plants. .

Mengel, K., Kirkby, E., 2001. Principles of Plant Nutrition, fifth ed. Dordrecht, Kluwer Academic Publishers.

Mezuman, U., Keren, R., 1981. Boron adsorption by soils using a phenomenological adsorption equation 1. Soil Sci. Soc. Am. J. 45 (4), 722–726.

Miller, G., Shulaev, V., Mittler, R., 2008. Reactive oxygen signaling and abiotic stress. Physiol. Plant. 133 (3), 481–489.

Miwa, K., Fujiwara, T., 2010. Boron transport in plants: co-ordinated regulation of transporters. Ann. Bot. 105 (7), 1103–1108.

Miwa, K., Takano, J., Fujiwara, T., 2006. Improvement of seed yields under boron-limiting conditions through overexpression of BOR1, a boron transporter for xylem loading, in *Arabidopsis thaliana*. Plant J. 46 (6), 1084–1091.

Morsey, M.A., Taha, E.M., 1986. Effect of boron, manganese and their combination on sugar beet under El-Minia conditions. 2: Concentration and uptake of N, P, K, B and Mn. Ann. Agric. Sci. Ain Shams Univ. Cairo 31, 1241–1259.

Mozafar, A., 1989. Boron effect on mineral nutrients of maize. Agron. J. 81 (2), 285–290.

Muengkaew, R., Whangchai, K., Chaiprasart, P., 2018. Application of calcium–boron improve fruit quality, cell characteristics, and effective softening enzyme activity after harvest in mango fruit (*Mangifera indica* L.). Hortic. Environ. Biotechnol. 59 (4), 537–546.

Navara, J., 1987. Participation of individual root types in water uptake by maize seedlings. Biologia 42, 17–26.

Nelson, S., 2012. Boron Deficiency of Papaya.

Newman, E.I., Reddell, P., 1987. The distribution of mycorrhizas among families of vascular plants. New Phytol. 106 (4), 745–751.

Nishina, M.S., 1991. Bumpy Fruit of Papaya as Related to Boron Deficiency. University of Hawaii, Honolulu, HI. 4 p. (Commodity Fact Sheets; CFS-PA-4B).

O'Neill, M.A., Eberhard, S., Albersheim, P., Darvill, A.G., 2001. Requirement of borate cross-linking of cell wall rhamnogalacturonan II for Arabidopsis growth. Science 294 (5543), 846–849.

Omaima, M.H., Karima, H.E.H., 2007. Quality improvement and storability of apple cv. Anna by pre-harvest applications of boric acid and calcium chloride. J. Agric. Biol. Sci. 3, 176–183.

Parks, R.Q., Lyon, C.B., Hood, S.L., 1944. Some effects of boron supply on the chemical composition of tomato leaflets. Plant Physiol. 19 (3), 404.

Peryea, F., 1994. Boron nutrition in deciduous tree fruits. In: Peterson, A.B., Stevens, R.G. (Eds.), Tree Fruit Nutrition. Good Fruit Grower, Yakima, Washington.

Plich, H., Wójcik, P., 2002. The effect of calcium and boron foliar application on postharvest plum fruit quality. Acta Hortic, 445–452.

Power, P.P., Woods, W.G., 1997. The chemistry of boron and its speciation in plants. Plant Soil 193 (1–2), 1–13.

Qin, G., Zong, Y., Chen, Q., Hua, D., Tian, S., 2010. Inhibitory effect of boron against *Botrytis cinerea* on table grapes and its possible mechanisms of action. Int. J. Food Microbiol. 138 (1–2), 145–150.

Raven, J.A., 1980. Short-and long-distance transport of boric acid in plants. New Phytol. 84 (2), 231–249.

Redondo-Nieto, M., Pulido, L., Reguera, M., Bonilla, I., Bolanos, L., 2007. Developmentally regulated membrane glycoproteins sharing antigenicity with rhamnogalacturonan II are not detected in nodulated boron deficient *Pisum sativum*. Plant Cell Environ. 30 (11), 1436–1443.

Reid, R., 2007. Update on boron toxicity and tolerance in plants. In: Xu, et al., (Eds.), Advances in Plant and Animal Boron Nutrition. Springer Publications, pp. 83–90.

Robertson, G.A., Loughman, B.C., 1973. Rubidium uptake and boron deficiency in *Vicia faba* L. J. Exp. Bot. 24 (6), 1046–1052.

Rodriguez, M.V.G., Gomez-Ortega, M., Alvarez-Tinaut, M.C., 1981. Boron, copper, iron, manganese and zinc contents in leaves of flowering sunflower plants (*Helianthus annuus*, L.), grown with different boron supplies. Plant Soil 62 (3), 461–464.

Rossi, M., Bermudez, L., Carrari, F., 2015. Crop yield: challenges from a metabolic perspective. Curr. Opin. Plant Biol. 25, 79–89.

Roy, S., Jauh, G.Y., Hepler, P.K., Lord, E.M., 1998. Effects of Yariv phenylglycoside on cell wall assembly in the lily pollen tube. Planta 204 (4), 450–458.

Ruuhola, T., Lehto, T., 2014. Do ectomycorrhizas affect boron uptake in *Betula pendula*? Can. J. For. Res. 44 (9), 1013–1019.

Ryden, P., Sugimoto-Shirasu, K., Smith, A.C., Findlay, K., Reiter, W.-D., McCann, M.C., 2003. Tensile properties of Arabidopsis cell walls depend on both a xyloglucan cross-linked microfibrillar network and rhamnogalacturonan II-borate complexes. Plant Physiol. 132 (2), 1033–1040.

Salam, M.A., Siddique, M.A., Rahim, M.A., Rahman, M.A., Goffar, M.A., 2011. Quality of tomato as influenced by boron and zinc in presence of different doses of cowdung. Bangladesh J. Agric. Res. 36 (1), 151–163.

Sánchez, E.E., Righetti, T.L., 2005. Effect of postharvest soil and foliar application of boron fertilizer on the partitioning of boron in apple trees. HortScience 40 (7), 2115–2117.

Santra, G.H., Das, D.K., Mandal, B.K., 1989. Relationship of boron with iron, manganese, copper and zinc with respect to their availability in rice soil. Environ. Ecol. 7 (4), 874–877. Kalyani.

Sarkar, A., Asaeda, T., Wang, Q., Rashid, M.H., 2015. Arbuscular mycorrhizal influences on growth, nutrient uptake, and use efficiency of *Miscanthus sacchariflorus* growing on nutrient-deficient river bank soil. Flora Morphol. Distrib. Funct. Ecol. Plants 212, 46–54.

Shah, A., Wu, X., Ullah, A., Fahad, S., Muhammad, R., Yan, L., Jiang, C., 2017. Deficiency and toxicity of boron: alterations in growth, oxidative damage and uptake by citrange orange plants. Ecotoxicol. Environ. Saf. 145, 575–582.

Sharma, R., Sharma, V., 2004. The Strawberry. ICAR, New Delhi.

Sharma, R.R., Singh, R., 2009. The fruit pitting disorder—a physiological anomaly in mango (*Mangifera indica* L.) due to deficiency of calcium and boron. Sci. Hortic. 119 (4), 388–391.

Shear, C.B., Faust, M., 1980. Nutritional ranges in deciduous tree fruits and nuts. Hortic. Rev. 2, 142–163.

Shen, Z., Liang, Y., Shen, K., 1993. Effect of boron on the nitrate reductase activity in oilseed rape plants. J. Plant Nutr. 16 (7), 1229–1239.

Sheng, O., Song, S., Peng, S., Deng, X., 2009. The effects of low boron on growth, gas exchange, boron concentration and distribution of 'Newhall'-navel orange (*Citrus sinensis* Osb.) plants grafted on two rootstocks. Sci. Hortic. 121 (3), 278–283.

Shireen, F., Nawaz, M., Chen, C., Zhang, Q., Zheng, Z., Sohail, H., Sun, J., Cao, H., Huang, Y., Bie, Z., 2018. Boron: functions and approaches to enhance its availability in plants for sustainable agriculture. Int. J. Mol. Sci. 19 (7), 1856.

Shorrocks, V.M., 1997. The occurrence and correction of boron deficiency. Plant Soil 193 (1–2), 121–148.

Showalter, A.M., 1993. Structure and function of plant cell wall proteins. Plant Cell 5 (1), 9.

Singh, J.P., Dahiya, D.J., Narwal, R.P., 1990. Boron uptake and toxicity in wheat in relation to zinc supply. Fertil. Res. 24 (2), 105–110.

Stangoulis, J.C.R., Reid, R.J., Brown, P.H., Graham, R.D., 2001. Kinetic analysis of boron transport in Chara. Planta 213 (1), 142–146.

Subburamu, K., Singaravelu, M., Nazar, A., Irulappan, I., 1990. Pre-harvest spray of calcium in grapes (*Vitis vinifera*). South Indian Hortic. 38 (5), 268–269.

Tanaka, M., Fujiwara, T., 2008. Physiological roles and transport mechanisms of boron: perspectives from plants. Pflügers Arch. Eur. J. Physiol. 456 (4), 671–677.

Tariq, M., Mott, C.J.B., 2007. The significance of boron in plant nutrition and environment—a review. J. Agron. 6 (1), 1.

Taylor, B.K., Dimsey, R.T., 1993. Rootstock and scion effects on the leaf nutrient composition of citrus trees. Aust. J. Exp. Agric. 33 (3), 363–371.

Tewari, R.K., Kumar, P., Sharma, P.N., 2009. Morphology and oxidative physiology of boron-deficient mulberry plants. Tree Physiol. 30 (1), 68–77.

Thomidis, T., Exadaktylou, E., 2010. Effect of boron on the development of brown rot (*Monilinia laxa*) on peaches. Crop Prot. 29 (6), 572–576.

Uluisik, I., Karakaya, H.C., Koc, A., 2018. The importance of boron in biological systems. J. Trace Elem. Med. Biol. 45, 156–162.

Veltman, R.H., Sanders, M.G., Persijn, S.T., Pemppelenbos, H.W., Oosterhaven, J., 1999. Decreased ascorbic acid levels and brown core development in pears (*Pyrus communis* L. cv. Conference). Physiol. Plant. 107 (1), 39–45.

Vlamis, J., Ulrich, A., 1971. Boron nutrition in the growth and sugar content of sugarbeets. Am. Soc. Sugar Beet Technol. J. 16, 428–439.

Wallace, A., Bear, F.E., 1949. Influence of potassium and boron on nutrient-element balance in and growth of Ranger alfalfa. Plant Physiol. 24 (4), 664.

Wang, D.-N., Ko, W.H., 1975. Relationship between deformed fruit disease of papaya and boron deficiency. Phytopathology 65 (4), 445–447.

Wang, Z.Y., Tang, Y.L., Zhang, F.S., Wang, H., 1999. Effect of boron and low temperature on membrane integrity of cucumber leaves. J. Plant Nutr. 22 (3), 543–550.

Wang, N., Yang, C., Pan, Z., Liu, Y., Peng, S., 2015. Boron deficiency in woody plants: various responses and tolerance mechanisms. Front. Plant Sci. 6, 916.

Wang, N., Wei, Q., Yan, T., Pan, Z., Liu, Y., 2016. Improving the boron uptake of boron-deficient navel orange plants under low boron conditions by inarching boron-efficient rootstock. Sci. Hortic. 199, 49–55.

Warington, K., 1923. The effect of boric acid and borax on the broad bean and certain other plants. Ann. Bot. 37 (148), 629–672.

Wojcik, P., Wojcik, M., 2006. Effect of boron fertilization on sweet cherry tree yield and fruit quality. J. Plant Nutr. 29 (10), 1755–1766.

Wojcik, P., Wojcik, M., Treder, W., 2003. Boron absorption and translocation in apple rootstocks under conditions of low medium boron. J. Plant Nutr. 26 (5), 961–968.

Wojcik, P., Wojcik, M., Klamkowski, K., 2008. Response of apple trees to boron fertilization under conditions of low soil boron availability. Sci. Hortic. 116 (1), 58–64.

Yakhin, O.I., Lubyanov, A.A., Yakhin, I.A., Brown, P.H., 2017. Biostimulants in plant science: a global perspective. Front. Plant Sci. 7, 2049.

Yamagishi, M., Yamamoto, Y., 1994. Effects of boron on nodule development and symbiotic nitrogen fixation in soybean plants. Soil Sci. Plant Nutr. 40 (2), 265–274.

Yan, X., Wu, P., Ling, H., Xu, G., Xu, F., Zhang, Q., 2006. Plant nutriomics in China: an overview. Ann. Bot. 98 (3), 473–482.

Yang, X., Sun, S., Li, Y., 1999. Boron deficiency causes changes in the distribution of major polysaccharides of pollen tube wall. Acta Bot. Sin. 41 (11), 1169–1176.

Ye, Z., Huang, L., Bell, R.W., Dell, B., 2003. Low root zone temperature favours shoot B partitioning into young leaves of oilseed rape (*Brassica napus*). Physiol. Plant. 118 (2), 213–220.

Yermiyahu, U., Keren, R., Chen, Y., 1995. Boron sorption by soil in the presence of composted organic matter. Soil Sci. Soc. Am. J. 59 (2), 405–409.

Yu, Q., Hlavacka, A., Matoh, T., Volkmann, D., Menzel, D., Goldbach, H.E., Baluška, F., 2002. Short-term boron deprivation inhibits endocytosis of cell wall pectins in meristematic cells of maize and wheat root apices. Plant Physiol. 130 (1), 415–421.

Yu, Q., Baluška, F., Jasper, F., Menzel, D., Goldbach, H.E., 2003. Short-term boron deprivation enhances levels of cytoskeletal proteins in maize, but not zucchini, root apices. Physiol. Plant. 117 (2), 270–278.

Yu, S.-M., Lo, S.-F., Ho, T.-H.D., 2015. Source–sink communication: regulated by hormone, nutrient, and stress cross-signaling. Trends Plant Sci. 20 (12), 844–857.

Zekri, M., Obreza, T., 2014. Boron (B) and Chlorine (Cl) for Citrus Trees. A part of a series about understanding nutrient requirements for citrus trees. For the rest of the series, visit http://edis.ifas.ufl.edu/topic_series_citrus_tree_nutrients.

Zhang, Z.-Y., Qing-Lian, W., Zhao-Hu, L.I., Liu-Sheng, D., Xiao-Li, T., 2009. Effects of potassium deficiency on root growth of cotton seedlings and its physiological mechanisms. Acta Agron. Sin. 35 (4), 718–723.

Zhou, G.F., Peng, S.A., Liu, Y.Z., Wei, Q.J., Han, J., Islam, M.Z., 2014. The physiological and nutritional responses of seven different citrus rootstock seedlings to boron deficiency. Trees 28 (1), 295–307.

Further reading

Ruiz, J.M., Baghour, M., Bretones, G., Belakbir, A., Romero, L., 1998. Nitrogen metabolism in tobacco plants (*Nicotiana tabacum* L.): role of boron as a possible regulatory factor. Int. J. Plant Sci. 159 (1), 121–126.

16

Boron toxicity in fruit crops: Agronomic and physiological implications

Christos Chatzissavvidis*, Chrysovalantou Antonopoulou

Department of the Agricultural Development, School of Agricultural and Forestry Sciences, Democritus University of Thrace, Orestiada, Greece

*Corresponding author. E-mail: cchatz@agro.duth.gr

1 Introduction

Boron (B) is an essential micronutrient for plants, and plant requirements for this nutrient are lower than the requirements for all the other nutrients except for molybdenum and copper. B is involved in a number of metabolic pathways and functions such as cell wall synthesis and structure, carbohydrate metabolism, phenol metabolism, lignification, and plasma membrane integrity (Marschner, 1995). However, despite its essentiality, it is harmful to plants when present in excess.

Although the requirement of B varies in different crops, the optimum value of B content for most crops is about 20 mg/kg (Shah et al., 2017). The availability of B concentration in proper amount in soil is necessary for plant growth, development, and yield (Shah et al., 2017). Plants uptake B as an undissociated boric acid (Mengel and Kirkby, 2001). Total B content in soils is generally much greater than the available B required for optimum plant growth, and as with other micronutrients, it frequently bears little relationship with plant available B (Ryan et al., 2013). Boron is unique as a micronutrient since the range between its deficient and toxic levels is relatively narrow and both situations produce toxic symptoms that affect crop yield and production (Camacho-Cristóbal et al., 2008; Yau and Ryan, 2008).

Toxicity in crops can be defined as a situation in which a potentially harmful substance accumulates in the plant tissue to a level affecting its optimal growth and development. Certain essential plant nutrients and nonessential elements can be absorbed by plants in sufficient quantities to be toxic. Essential trace elements that are more frequent to produce toxicity in plants are Mn, Cu, B, Mo, and Cl (Bennett, 1993).

Although B toxicity is less common than B deficiency (a widespread problem in several fruit crops), it is still a severe nutritional disorder that can limit fruit crop yield and the quality of fruit production in many agricultural regions across the world, including Egypt, Turkey, Iraq, Jordan, Syria, Morocco, Chile, the United States, Spain, and South Australia (Miwa et al., 2007; Landi et al., 2012; Ayvaz et al., 2013) (Table 16.1). However, it is noticeable that nowadays B toxicity has also become an important problem, especially in the soils of arid and semiarid regions throughout the

A.K. Srivastava, Chengxiao Hu (eds.)
Fruit Crops: Diagnosis and Management of Nutrient Constraints
https://doi.org/10.1016/B978-0-12-818732-6.00016-2

TABLE 16.1 Cases of boron toxicity in some fruit crops worldwide.

Fruit crop	Country	Conditions	References
Apricot (*Prunus armeniaca*)	New Zealand	Excessive supply of B fertilizers	Dye et al. (1983)
Kiwifruit (*Actinidia* spp.)	Greece	Well water high in boron	Sotiropoulos et al. (1999)
Orange (*Citrus sinensis*)	Spain	Irrigation water polluted by ceramic industrial wastewater	Aucejo et al. (1997)
Peach (*Prunus persica*)	Canada	Soils with high B	Nielsen et al. (1985)
Prickly pear (*Opuntia ficus-indica*)	United States	Naturally high levels of B in soils	Bañuelos (2014)
Pomelo (*Citrus grandis*)	China	Overuse of B fertilizers (foliar applications)	Li et al. (2015)

world. Under Mediterranean climatic conditions (dry and warm during the growing seasons of most of the high-value crops), and where rainfall is not enough to leach the element from the root zone, B tends to accumulate and, in many cases, to exceed toxic levels. Moreover, in such areas, irrigation water is often high in B (Dionysiou et al., 2006). For example, in some areas of northern Greece, well water with high B concentration has been used for irrigation of olive and kiwifruit orchards (Chatzissavvidis et al., 2004; Sotiropoulos et al., 2006a).

A concentration of B in the irrigation water that exceeds 1 mg/L can affect the yield of most traditional Mediterranean crops such as orange trees (*Citrus sinensis*) and avocados (*Persea americana*) (Gimeno et al., 2012). These species can start to show toxicity symptoms when the leaf concentration of B is higher than 50 mg/kg dry weight (DW) (Hakki et al., 2007). However, olive is more resistant to an excess of B than other species such as peach, apple, plum, or apricot (Hansen, 1945). It was suggested that the differences are due to minimal accumulation of B in olive leaves when grown on soils with high concentrations of this element.

Generally, B toxicity occurs under the following conditions: (i) in B-enriched soils originating from marine rock materials, (ii) in soils overfertilized with minerals high in B, (iii) in soils receiving fossil combustion residues, (iv) in soils being used as disposal sites of B-containing waste materials, and (v) in soils being irrigated with water high in B content (Gupta et al., 1985; Leyshon and James, 1993; Nable et al., 1997; Papadakis et al., 2004a,b; Chatzissavvidis and Therios, 2011).

As farmers have been approached about potentially transferring to recycled wastewater (RWW) to conserve precious water resources for urban, recreational, and environmental users, concerns have been raised about the utilization of treated wastewater for irrigation of agricultural crops since it involves the risk of yield reduction resulting from elevated salt and B concentrations (Tsadilas et al., 1997; Grattan et al., 2015). Furthermore, in recent years, B toxicity has attracted increasing interest owing to the greater demand for desalinated water, in which the B concentration may be too high for healthy irrigation (Parks and Edwards, 2005). The addition of water from seawater desalinating plants and urban wastewater treatment plants is common in areas with a Mediterranean climate, where good-quality water resources are scarce. These nonconventional water sources can contain excessive amounts of B for sensitive crops, such as in the case of citrus plants, where a concentration above 0.3 mg/L is considered to be the threshold of B toxicity (Grattan, 2013).

2 Boron tolerance mechanism

Even though research over the past few decades has improved our understanding of B toxicity, basic fundamental mechanisms of why excess B is toxic are still lacking (Chatzissavvidis and Therios, 2011). In general terms, B is relatively unreactive, but it can form strong complexes with a handful of metabolites that have multiple hydroxyl groups in the *cis* conformation. Of these, ribose appears to be the most likely candidate for toxicity-related effects (Reid, 2010). Additionally, in leaves, accumulation of high concentrations of B at the end of the transpiration stream might lead to osmotic imbalances (Reid et al., 2004).

On the basis of our understanding of B chemistry, there appear to be three main possible sites for B toxicity: (i) disruption of cell wall development; (ii) metabolic disruption by binding to the ribose moieties of ATP, NADH, or NADPH; and (iii) disruption of cell division and development by binding to ribose, either as the free sugar or within RNA (Stangoulis and Reid, 2002).

A possible biochemical mechanism for B tolerance could include the internal compartmentation of B in vacuoles (Torun et al., 2002), an antioxidant defense response to oxidative stress (Aftab et al., 2010), or by counteracting the B-induced upregulation of the cell wall-associated proteins pectin methylesterases (PME) and expansins (Wang et al., 2010).

Given that amelioration of B-rich soils in orchards is very difficult, the use of rootstocks is an important factor in salinity and in B tolerance, to limit the uptake and translocation of toxic ions to the shoots (El-Motaium et al., 1994; Papadakis et al., 2004a,b; López-Gómez et al., 2007; Chatzissavvidis et al., 2008). On that basis, it has been found that cherry rootstocks can greatly affect the scion's tolerance to salinity or high B concentrations (Perry, 1987) and the proper selection of rootstocks may help growers to alleviate specific problems.

3 Detrimental effects of B toxicity

Boron excess results in plant toxicity and damages the plant growth and productivity. Generally, the primary symptoms are burning of older leaves margins or tips, that is, chlorosis and necrosis and reduced vigor (Brown and Shelp, 1997; Nable et al., 1997). Leaf chlorosis is observed first at the leaf tips and then expands to the margins and finally whole leaves (Shah et al., 2017). Afterward, it delays development, inhibits plant growth, and decreases weight, number, and size of fruits (Reid and Fitzpatrick, 2009; Herrera-Rodríguez et al., 2010). Sensitivity to B toxicity apparently involves several metabolic processes (Reid et al., 2004) including reduction of meristem expansion, development of necrotic areas in mature tissues (limiting photosynthetic capacity), and reduction of photosynthate supply to developing plant regions (Gimeno et al., 2012).

In higher plants, B excess causes impairments of cellular functions and physiological and biochemical processes, such as plant growth, hormone balance, uptake, and use efficiency of other mineral elements; photosynthetic rate; leaf photosynthetic electron transport chain; photosynthetic enzyme activities; reactive oxygen metabolism; chlorophyll (Chl) and carotenoid levels; cell wall biosynthesis; carbohydrate, energy, nucleic acid, and nitrogen metabolisms; and leaf and root structure (Guo et al., 2016). It is well known from the previous studies that B toxicity increases oxidative stress (Cervilla et al., 2007; Landi et al., 2014).

3.1 Visible symptoms

Boron toxicity is crop specific, and the characteristics of B injury are related to plant's ability to mobilize this element (Brown and Shelp, 1997) (Table 16.2). Boron-immobile plants such as *Citrus* spp., pistachio, and walnut do not have high concentrations of polyols and B concentrates in older leaves, where injury first develops (Grattan et al., 2015). In *Citrus* spp. trees, B toxicity usually appears in late summer or early fall, as apical leaf mottling and chlorosis (yellowing; light gray in the print version) between the leaf veins particularly closer to the leaf margins (Chapman, 1968a,b; Gimeno et al., 2012) (Fig. 16.1). Excess B is known to make *Citrus* spp. leaves thinner, to reduce chlorophyll content, and to reduce photosynthesis by damaging thylakoid structure in the chloroplasts (Papadakis et al., 2004a,b;

TABLE 16.2 Symptoms of boron toxicity in some fruit crops.

Fruit crop	Boron toxicity symptoms	References
Almond	Stem dieback, gum exudation in leaf axils and buds	Brown et al. (1999)
Banana	Continuous irregular chlorosis and necrosis from the edge toward the internal part of the leaf blade leaving the central portion green	Vargas et al. (2007)
Blueberry, cranberry	Mottled interveinal chlorosis, scorching of leaf margins	Caruso and Ramsdell (1995)
Cherry	Yellow-green interveinal and marginal chlorosis	Sotiropoulos et al. (2006c)
Citrus	Apical leaf mottling and chlorosis (yellowing) between the leaf veins particularly closer to the leaf margins	Chapman (1968a,b), Gimeno et al. (2012)
Kiwifruit	Yellow-green leaf interveinal chlorosis, necrotic spots between the minor veins	Sotiropoulos et al. (2002)
Olive	Reduced shoot growth, defoliation, leaf marginal chlorosis, necrosis of the apical third of the limp area	Benlloch et al. (1991)
Pear	Leaves small, developed late, curled and elongated or cup-shaped. Delayed blossoms, few and small flowers. Early fruit maturation and lower storage ability	Crandall et al. (1981), Wojcik and Wojcik (2003)
Pistachio	Reduced plant height, leaf area, root length, and total plant biomass	Arzani and Hokmabadi (2015)

FIG. 16.1 Boron injury in Ponkan mandarin leaves. *Data from Grattan, S.R., Díaz, F.J., Pedrero, F., Vivaldi, G.A., 2015. Assessing the suitability of saline wastewaters for irrigation of* Citrus *spp: emphasis on boron and specific-ion interactions. Agric. Water Manag. 157, 48–58.*

Chen et al., 2012). As injury continues to advance, similar symptoms begin to show up progressively on younger leaves. Injury may also progress from tip chlorosis and interveinal mottling to the formation of tan-colored, resinous blisters on the underside of the leaves (Wutscher et al., 1996), particularly in orange, grapefruit, and mandarins (Chapman, 1968a,b). In severe cases, B toxicity can cause leaves to prematurely abscise and the entire branch and the tree can eventually die (Grattan et al., 2015). However, B toxicity symptoms in *Citrus* spp. may resemble other nutritional toxicities or disorders. Therefore, leaf samples should be collected to confirm the visual diagnosis.

In blueberry and cranberry, B toxicity has also been associated with mottled interveinal chlorosis and scorching of leaf margins (Caruso and Ramsdell, 1995). On the other hand, the symptoms of B toxicity in B-mobile species (e.g., *Prunus* spp., *Malus* spp., and *Pyrus* spp.) firstly appear in the meristematic regions and fruits, but not in the mature leaves (Brown and Hu, 1996; Yau and Ryan, 2008).

Since B content in roots remains relatively low compared with that in leaves, even at very high level of B supply, visible symptoms of B toxicity do not appear to develop in roots (Nable et al., 1997), suggesting that B distribution is related to the transpiration stream (Papadakis et al., 2004a,b; Shah et al., 2017). In citrange (*Poncirus trifoliata* × *C. sinensis*) B is mainly accumulated in leaves, but in relatively low concentrations, it also remains in the roots, as well. Although there are no visible symptoms in roots under B excess, overall, its length, surface area, and volume decrease significantly (Shah et al., 2017).

Kiwifruit is a very sensitive species to excess B (Sotiropoulos et al., 1999). When the B concentration in the nutrient solutions exceeded 0.5 mg/L of B, characteristic symptoms of B toxicity appeared (Sotiropoulos et al., 1999). The first sign of B toxicity was a yellow-green interveinal chlorosis, which first developed on the oldest leaves and progressed to the youngest. Later, small patches of necrotic tissue appeared between the minor veins and extended to the midribs resulting in a reduction of the photosynthetic leaf area (Sotiropoulos et al., 2002).

In comparison with peach, apple, prune, and apricot trees, some species, such as those of the genus *Punica* and *Olea*, are very tolerant of B toxicity. For example, Brown et al. (1998) reported that pomegranate (*Punica granatum*) plants were very tolerant of B toxicity, and no clear B toxicity symptoms were observed even when treated with a nutrient solution containing 25 mg/L B for 5 months. On the other hand, in olive, high B nutrient solution did not cause visible B toxicity symptoms in leaves, since olive accumulates small quantity of B due to B phloem mobility (Chatzissavvidis et al., 2005). Moreover, B excess usually reduces crop load, since the percentage of perfect flowers was reduced (Chatzissavvidis et al., 2005). However, in "Manzanillo" and "Picual" olive trees B excess reduced shoot growth, induced the appearance of toxic symptoms in the leaves and caused defoliation. Toxic symptoms in mature olive leaves began with marginal chlorosis that extended to the limb, necrosis appearing mostly on the apical third of the limb area (Benlloch et al., 1991).

Cherry rootstocks irrigated with 0.2 mM B in the solution developed B toxicity symptoms (Sotiropoulos et al., 2006c). The first sign of B toxicity was a yellow-green interveinal and marginal chlorosis, which developed on the older leaves and progressed to the younger ones. Plants exposed to the highest levels of B and salinity became progressively wilted and suffered from leaf burn. CAB 6P (*Prunus cerasus* L.) plants maintained most of their leaves until the end of the experiment, whereas Gisela 5 (*P. cerasus* × *Prunus canescens*) plants showed higher leaf shedding (Sotiropoulos et al., 2006c).

High levels of B nutrition can also affect flowering and fruit crop (Crandall et al., 1981). Blossoms of pears were delayed by 7–10 days. The flowers were small and few, and practically none of the flowers set fruit. Leaves developed late and were small, curled, and elongated or cup-shaped. Those that developed later in the season appeared to be nearly normal. There was some dieback of fruit spurs. These dead spurs were quite evident during the later years of the study (Crandall et al., 1981).

3.2 Plant growth

Kiwifruit plants irrigated with solutions containing 100 mM of B showed a reduction in growth and characteristic B toxicity symptoms (Sotiropoulos et al., 1997). The strong influence of B toxicity on growth parameters of kiwifruit vines is in agreement with other reports (Smith and Clark, 1989; Tsadilas et al., 1997). Similar results were reported in several fruit and nut crops including citrus (*C. sinensis* L.) (Chapman and Vanselow, 1955), pecan (*Carya illinoinensis* Wangenh. C. Koch) (Picchioni and Miyamoto, 1991), olive (*Olea europaea* L.) (Benlloch et al., 1991), and sour cherry (*P. cerasus* L.) (Hanson, 1991).

In some citrus species, high B concentrations in the nutrient solution inhibited plant growth (plant height, root volume, and the dry masses of various tissues) and resulted in symptoms of toxicity in the old leaves (Papadakis et al., 2003; Sheng et al., 2008; Gimeno et al., 2012). In addition, the growth inhibition was reduced in trees that were grafted on certain rootstocks (Gimeno et al., 2012).

In other experiments with citrus trees, Sheng et al. (2009) observed that B toxicity in the navel orange mainly reduced the root growth, while, in both the "Navelina" orange and the "Clementine" mandarin, Papadakis et al. (2004a,b) did not observe any effect on the growth parameters when they were treated with 2.5 mg/L B for 204 days. Oxidative damage by hydrogen peroxide and lipid peroxidation may have negatively affected the photosynthetic system and consequently affect plant growth including plant height, root volume, and biomass of citrange plants (Shah et al., 2017). Decreased transpiration rate, stomatal conductance, photosynthetic rate, and intercellular CO_2 were also observed.

The decline in vegetative growth due to B excess in olive has been also reported by several researchers (Benlloch et al., 1991; Chatzissavvidis et al., 2008). The observed growth decline due to B toxicity could be ascribed to decreased photosynthetic rate and water use efficiency. Water transport through the roots is very sensitive to various environmental factors, including nutrient stress. Therefore, B toxicity may reduce water flow and affect nutrient uptake. A decline in root hydraulic conductance could be caused by a reduction in new root growth, which was observed in B-treated plants (Dannel et al., 1998; Apostol et al., 2002). Reduced hydraulic conductance and nutrient uptake could be partly responsible for growth reduction observed in B-treated plants (Nable et al., 1997; Dannel et al., 1998).

Increasing the amount of added B in acidic soil enhanced fresh matter weight of apple plants. In the limed soil, there were not significant effects, and this may be attributed to lower B availability in that soil (Paparnakis et al., 2013). Toxic levels of B have also been shown to cause root growth inhibition (Reid et al., 2004). B excess led to cell death in pomelo (*C. grandis*) roots (Huang et al., 2014). The most possible mechanism for reduction Li in root can be interference with mitosis (Reid, 2007).

3.3 Boron content and distribution

Except in B-deficient soil, B can cross root membranes by passive diffusion and through aquaglyceroporins, channel-like proteins that permit bidirectional movements of small molecules (Camacho-Cristóbal et al., 2008; Reid, 2010). When soils have abundant B, the high permeability of B may lead to excessive B accumulation by plants leading to B toxicity (Ryan et al., 2013).

The typical pattern of B accumulation in plants is for a higher concentration of B to be found in oldest leaves with progressively lower B concentrations in younger tissues and fruits. This distribution of B is typical of an element with limited phloem mobility and results in the development of leaf tip and edge burn of the oldest leaves because of the high B concentrations in these tissues (Brown et al., 1998). Toxicity effects appear to be loosely correlated with the accumulation of high concentrations of B in old leaves, especially at the margin of leaves. This is because B introduced into the transpiration stream accumulates at the end of this stream.

The concentration of B and its distribution in the different plant tissues depends on the mobility of B inside the plant. In plants that transport sugar alcohols such as mannitol and sorbitol, the rapid and significant retranslocation of B occurs. In plant species that do not produce significant amounts of sugar alcohols, B has been thought to be immobile in the phloem, and its distribution in the shoots primarily follows the transpiration streams (Brown and Shelp, 1997;

Takano et al., 2008). The results of Gimeno et al. (2012) showed that lemon cv. Verna trees belong to the group of plants in which B has a restricted mobility and where the transpiration flow determines the total B concentration and distribution.

Eaton et al. (1941) reported, however, that the leaf symptoms of B toxicity, so characteristic in many plants, are frequently absent in stone fruit trees. In support of these observations, others have also demonstrated that species of the genus *Malus*, *Prunus*, and *Pyrus* do not accumulate high levels of B in their leaves and that B toxicity is expressed as twig dieback and gum exudation in leaf axils and buds (Hansen, 1948, 1955; Woodbridge, 1955; Maas, 1984; Choe et al., 1986; El-Motaium et al., 1994). These "unusual" symptoms of B toxicity are the result of the high phloem mobility of B in these species, which occurs as a result of the formation and phloem transport of a B-sugar alcohol complex (Brown and Hu, 1996; Hu et al., 1997). Since sugar alcohols such as sorbitol have a very high affinity to bind B (Makkee et al., 1985), and as they represent primary products of photosynthesis, it is possible that the B-sugar alcohol complex will be readily transported to active sinks such as apical tissues (Brown and Hu, 1996; Hu et al., 1997). As a result, B will be accumulated in the meristematic regions or fruit but not in the mature leaves, of species in which sugar alcohols are present. In these species, B toxicity will likely appear in apical and meristematic tissues, while the "typical" symptoms of B toxicity such as leaf tip/edge burn will not be observed (Brown et al., 1998).

Generally, B contents of about 100–130 mg/kg DW of leaves are usually considered the limit at which slight to moderate leaf symptoms begin to appear, while contents in the range of 200–250 mg/kg DW indicate pronounced B excess (Aucejo et al., 1997). However, there are also exceptions: In two citrus genotypes, the B concentration in leaves with visible toxicity symptoms was higher than 419 mg/kg DW (Papadakis et al., 2003). In another experiment on apples (Antoniadis et al., 2013)—soil B concentration was relatively high—with additional B treatment, leaf B concentrations of apple plants did not reach the toxicity threshold of 200 mg/kg (Brady and Weil, 2002; Antoniadis et al., 2013).

In two kiwifruit genotypes, increasing B concentration in the nutrient solution from 20 to 500 mM, B concentration was nearly doubled in the leaf margin and quadrupled in the remaining leaf blade (Sotiropoulos et al., 2002).

It is well known in plants that an antagonism between B, P, K, and Ca can occur when the nutrient solution contains a high concentration of B. This antagonism between B and P has been reported in kiwifruit (Sotiropoulos et al., 1999). Moreover, the interaction of B with other nutrient elements present at a certain plant tissue can determine the occurrence of B excess symptoms. In kiwifruit, an important determining factor for B toxicity seems to be the ratio of Ca/B in a particular tissue (Sotiropoulos et al., 1999). At the highest (12 mM) Ca concentration in solution, the Ca/B ratio in leaves was significantly higher than at 4 and 8 mM of Ca. Specifically, when this ratio in the leaf margin was less than 106, B toxicity symptoms developed, while at greater ratios no toxicity symptoms were observed (Sotiropoulos et al., 1999).

It is well established that an antagonistic relationship exists between B and Ca in plants. Therefore, the toxic effect of B in plants may be reduced by adding Ca to soils. As an example, the inclusion of Ca in nutrient solutions significantly diminished B levels in kiwifruit plants and alleviated B toxicity symptoms (Sotiropoulos et al., 1999).

Another nutrient with great effect on B uptake is nitrogen (N). Thus, application of N fertilizers seems to alleviate B toxicity symptoms in citrus trees (Chapman and Vanselow, 1955). What is more, Chatzissavvidis et al. (2007) suggested that supply of N in the form of NH_4^+ to olives grown under high B was more beneficial for growth than NO_3^- or the combination NO_3^-/NH_4^+.

Although the content of leaves in nutrient elements are usually a good indicator of the nutritional status of fruit trees, experiments have shown that also other plant organs can be used for B. The results of Crandall et al. (1981) indicate that toxicity symptoms can be expected in pear cv. Bartlett when B levels exceed 115 mg/kg in the blossom at full bloom and 45 mg/kg in the cortex of the fruit at harvest, while toxic B levels for pear cv. d'Anjou are in excess of 90 mg/kg in the blossoms and 55 mg/kg in the fruit (Crandall et al., 1981).

3.4 Interactions with other elements

By elevating B concentration in the nutrient solution up to 500 mM, Ca and Mn concentrations in the leaves of two *Actinidia* species were reduced. In detail, Ca concentration was reduced in the leaf margin and in the remaining part of the leaf blade, whereas it was increased in the petioles of kiwifruit (Sotiropoulos et al., 1999). Boron toxicity exerted a strong influence in Ca and Mn absorption and/or transport, did not affect the concentration of the other mineral nutrients, and increased Zn and B contents (Sotiropoulos et al., 2002, 2005). The decrease in B content from the basal part of the fruit toward the apical could be ascribed to the low mobility of B within the plant. The higher B content in the basal part of flesh suggests that this tissue is the most sensitive indicator of B toxicity in fruit (Sotiropoulos et al., 2005).

In pistachio (*Pistacia vera* L.), the toxic Ca/B ratio was the lowest (45) for a semitolerant cultivar and the highest (263) for a very sensitive cultivar (Sepaskhah et al., 1988).

Significantly lower concentrations of K, Mn, and Cl and higher concentrations of Ca and Mg were found in the leaves of control banana plants than in those treated with 400 μM B. Concerning the nutrient status of pseudostem, B excess resulted in an increase of the concentrations of K, Mn, Cl, and Na. Finally, the concentrations of all studied nutrients in plant roots were similar in both treatments (25 and 400 μM B), aside from B and Cl that were significantly increased under B excess conditions (Karantzi et al., 2015).

The levels of N concentration in apple leaves decreased as the amount of applied B increased suggesting an antagonism between B and N, which is likely to have occurred due to the preferred B uptake in the expense of N (Paparnakis et al., 2013). Potassium levels showed an increase disproving the expected B-K antagonism, as suggested by some researchers who reported that increasing B levels in the nutrient solution led to a decline in K concentration of olive and pomegranate (Chatzissavvidis et al., 2007; Sarafi et al., 2017). Other works have also reported enhanced K uptake with added B (Chatzissavvidis and Therios, 2010).

Leaf N in several olive cultivars was reduced by increasing B levels, while it did not affect leaf Mg, Fe, Zn, and Mn contents (Chatzissavvidis and Therios, 2003; Chatzissavvidis et al., 2005). Also, the increased B concentration in nutrient solution led to a decline in root K concentration of olive cv. Oblonga plantlets (Chatzissavvidis and Therios, 2003).

3.5 Antioxidant activity

According to the relative literature, B acts as a regulatory factor of the antioxidant protective system in plants (Heidarabadi et al., 2011; Landi et al., 2012; Singh et al., 2012). Boron toxicity has been associated with oxidative stress by B accumulation in leaf cell walls causing imbalances in cytoplasmic metabolism (Herrera-Rodríguez et al., 2010; Landi et al., 2012). Boron excess induced oxidative damage in leaves, as indicated by the increased content of 2-thiobarbituric acid (TBA) reactive compounds in citrus trees and in an apple rootstock (*Malus domestica* Borkh) (Dell and Huang, 1997; Han et al., 2009). Likewise, B excess (400 and 800 μM) caused oxidative stress in leaves and roots of highbush blueberry (*Vaccinium corymbosum* L.) (Meriño-Gergichevich et al., 2017).

In apple, it has been reported that B toxicity induces oxidative damage by lipid peroxidation and hydrogen peroxide accumulation (Gunes et al., 2006). Molassiotis et al. (2006) and Sotiropoulos et al. (2006c) found that B toxicity increased nonenzymatic antioxidant activity. According to Herrera-Rodríguez et al. (2010), excessive B in tissues decreased plant capacity to avoid the photooxidative damage arising from the accumulation of ROS that trigger lipid peroxidation. However, Tewari et al. (2009) reported that lipid peroxidation was not influenced significantly in leaves of mulberry (*Morus alba* L.) grown at three B rates.

When three citrus genotypes used as rootstocks were grown under B excess conditions, the A_{CO_2}/Φ_{PSII} ratio was decreased, so that reactive oxygen species were created, and these plants responded to this increase by elevating the enzymatic activities related to the antioxidant systems (Simón-Grao et al., 2018). Moreover, sour orange (*Citrus aurantium* L.) had a combination of different responses to B excess: (i) it has a very potent antioxidant system that is based on the high activity of the enzymes superoxide dismutase, ascorbate peroxidase, and catalase, and (ii) the overproduction of quaternary ammonium compounds could contribute, to a certain degree, to the avoidance of cellular damage in its photosynthetic machinery (Simón-Grao et al., 2018).

3.6 Photosynthesis

The lower CO_2 assimilation observed in B-stressed leaves of three citrus genotypes was primary caused by nonstomatal factors as the lower assimilation coincided with an increase of the intercellular CO_2 concentration. This is in agreement with the results obtained for B-excess kiwifruit (Sotiropoulos et al., 2002). The reduction of CO_2 assimilation by B excess is probably caused by a combination of factors such as oxidative damage, reduced photosynthetic enzyme activities, and impaired electron transport capacity (Keles et al., 2004; Papadakis et al., 2004a; Han et al., 2009; Han et al., 2009). High irradiance appears to increase the harmful effects of B toxicity, probably because elevated B concentrations may impair plant mechanisms to cope with photooxidation stress (Reid et al., 2004).

In citrus, CO_2 assimilation was lower in toxic leaves than in control leaves, while stomatal conductance was not lower in the former (Guo et al., 2014; Simón-Grao et al., 2018). The reduction of net assimilation of CO_2 that was found in lemon plants grown as a response to an excess of B was mainly due to nonstomatal factors (Gimeno et al., 2012). A_{CO_2} depletion due to nonstomatal factors includes changes such as to carboxylation efficiency, the photochemical efficiency of photosystem II, chlorophyll concentration, and the activity of photosynthetic enzymes (Pezeshki, 1994; Nishiyama et al., 2006; Yordanova and Popova, 2007). An excess of B in the leaf tissue damaged the photochemical machinery of the lemon leaves, as the proportion of absorbed energy being used in photochemistry (Φ_{PSII}) decreased. Moreover, changes in the Φ_{PSII} were due only to changes in F_v'/F_m', indicating that the loss of the quantum efficiency of PSII was because the light harvesting complex was damaged (Guerfel et al., 2009). This damage could be because an excess of B in the leaves can injure the chloroplast structure (Papadakis et al., 2004a). Likewise, in "Navelina" orange, both chlorophyll fluorescence parameters (F_v/F_m and F_v/F_0) and leaf chlorophyll concentration decreased, as B concentration in the nutrient solution increased (Papadakis et al., 2004b). Also, B toxicity impaired the whole photosynthetic electron transport from PSII donor side up to the reduction of end acceptors of PSI in pomelo leaves (Han et al., 2009).

Boron toxicity can cause a decrease in chlorophyll content and structural damage of chloroplasts, which leads to a decrease in CO_2-assimilatory capacity of the mesophyll cells (Papadakis et al., 2004a,b), ultimately affecting plant growth. Boron toxicity may indirectly affect photosynthesis by decreasing the photosynthetic leaf area and by altering leaf constituents (Lee and Aronoff, 1966). In the case of B toxicity in peach (*Prunus persica* L. Batsch) leaves, shrinkage and deorganization of the cells of the spongy parenchyma were observed (Kamali and Childers, 1966).

High B concentration in leaves can cause a decrease in photosynthesis, as has been reported for olives (Chatzissavvidis and Therios, 2010) and kiwifruit (Sotiropoulos et al., 1999). A reason for this decline could be the reduction of N concentration in leaves because most of this element in green tissues is a constituent of chlorophyll molecule and it is utilized in the photosynthetic machinery.

Chlorophyll (especially total Chl and Chl *b*) of apple leaves tended to decrease with added B (Mouhtaridou et al., 2004; Paparnakis et al., 2013). An explanation given by authors was that increased B levels may be causing thinner cell walls and thus lower Chl content. The increased Chl *b/a* ratio caused by the added B may be attributed to the decreased thickness of cell walls and, consequently, to enhanced penetration of a wider spectrum of light (Paparnakis et al., 2013).

Boron toxicity may cause severe impairments in the metabolism and as a result to growth and productivity of fruit crops. However, the physiological basis for B toxicity is not clear. Interestingly, there is a great diversity among genotypes that offers opportunities to utilize various genomic resources and technologies in an attempt to manipulate B in plants. Moreover, as concerns research priorities on this topic, assessment for B should be made for the following:

- The need to evaluate the inherent ability to tolerate excessive levels of B of the existing cultivars/rootstocks
- The determination of the exact B needs of fruit crops for conducting more precise fertilization programs
- Evaluation of methods and techniques, such as grafting, to limit B levels into fruit trees and shrubs

References

Aftab, T., Khan, M.M.A., Idree, M., Naeem, M., Ram, M., 2010. Boron induced oxidative stress, antioxidant defense response and changes in artemisinin content in *Artemisia annua* L. J. Agron. Crop Sci. 196, 423–430.

Antoniadis, V., Chatzissavvidis, C., Paparnakis, A., 2013. Boron behavior in apple plants in acidic and limed soil. J. Plant Nutr. Soil Sci. 176, 267–272.

Apostol, K.G., Zwiazek, J.J., MacKinnon, M.D., 2002. NaCl and Na_2SO_4 alter responses of jack pine (*Pinus banksiana*) seedlings to boron. Plant Soil 240, 321–329.

Arzani, K., Hokmabadi, H., 2015. Growth response of two young pistachio (*Pistacia vera* L.) rootstock seedlings to boron excess in irrigation water under a soilless culture system. Acta Hortic. 1062, 67–70.

Aucejo, A., Ferrer, J., Gabaldón, C., Marzal, P., Seco, A., 1997. Diagnosis of boron, fluorine, lead, nickel and zinc toxicity in citrus plantations in Villarreal, Spain. Water Air Soil Pollut. 94, 349–360.

Ayvaz, M., Avci, M.K., Yamaner, C., Koyuncu, M., Guven, A., Fagerstedt, K., 2013. Does excess boron affect malondialdehyde levels of potato cultivars? Eurasia J. Biosci. 7, 47–53.

Bañuelos, G.S., 2014. Coping with naturally-high levels of soil salinity and boron in the westside of central California. Perspect. Sci. 3, 4–6.

Benlloch, M., Arboleda, F., Barranco, D., Fernandez-Escobar, R., 1991. Response of young olive trees to sodium and boron excess in irrigation water. HortScience 26, 867–870.

Bennett, W.F., 1993. Plant nutrient utilization and diagnostic plant symptoms. In: Bennett, W.F. (Ed.), Nutrient Deficiencies and Toxicities in Crop Plants. APS Press, St Paul, MN, pp. 1–7.

Brady, N.C., Weil, R.R., 2002. The Nature and Properties of Soils, thirteen ed. Pearson Prentice Hall, New Jersey.

Brown, P.H., Hu, H., 1996. Phloem mobility of boron is species dependent: evidence for phloem in sorbitol-rich species. Ann. Bot. 77, 497–505.

Brown, P.H., Shelp, B.J., 1997. Boron mobility in plants. Plant Soil 193, 85–101.

Brown, P.H., Hu, H., Roberts, W.G., 1998. Redefining boron toxicity symptoms in some ornamentals. Slosson Report 95, pp. 1–7.

Brown, P.H., Hu, H., Roberts, W.G., 1999. Occurrence of sugar alcohols determines boron toxicity symptoms of ornamental species. J. Am. Soc. Hortic. Sci. 124, 347–352.

Camacho-Cristóbal, J.J., Rexach, J., González-Fontes, A., 2008. Boron in plants: deficiency and toxicity. J. Integr. Plant Biol. 50, 1247–1255.

Caruso, F.L., Ramsdell, D.C., 1995. Compendium of blueberry and cranberry diseases. Am. Phytopathol. Soc. 87.

Cervilla, L.M., Blasco, B., Ríos, J.J., Romero, L., Ruiz, J.M., 2007. Oxidative stress and antioxidants in tomato (*Solanum lycopersicum*) plants subjected to boron toxicity. Ann. Bot. 100, 747–756.

Chapman, H.D., 1968a. The mineral nutrition of Citrus. In: Reuther, W., Batchelor, L.D., Webber, H.J. (Eds.), The Citrus Industry II. University of California, Division of Agricultural Sciences, pp. 127–274.

Chapman, H.D., 1968b. Salinity and alkali. In: Reuther, W., Batchelor, L.D., Webber, H.J. (Eds.), The Citrus Industry II. University of California, Division of Agricultural Sciences, pp. 243–266.

Chapman, H.D., Vanselow, A.P., 1955. Boron deficiency and excess. Calif. Citrogr. 40, 92–94.

Chatzissavvidis, C., Therios, I., 2003. The effect of different B concentrations on the nutrient concentrations of one olive (*Olea europaea* L) cultivar and two olive rootstocks. In: Proceedings of the International Symposium on the Olive Tree and the Environment, pp. 214–220.

Chatzissavvidis, C., Therios, I., 2010. Response of four olive (*Olea europaea* L) cultivars to six B concentrations: growth performance, nutrient status and gas exchange parameters. Sci. Hortic. 127, 29–38.

Chatzissavvidis, C., Therios, I., 2011. Boron in higher plants. In: Perkins, G.L. (Ed.), Boron: Compounds, Production and Application. Nova Science Publishers Inc, pp. 147–176.

Chatzissavvidis, C.A., Therios, I., Antonopoulou, C., 2004. Seasonal variation of nutrient concentration in two olive (*Olea europaea* L) cultivars irrigated with high boron water. J. Hortic. Sci. Biotechnol. 79, 683–688.

Chatzissavvidis, C.A., Therios, I.N., Molassiotis, A.N., 2005. Seasonal variation of nutritional status of olive plants as affected by boron concentration in nutrient solution. J. Plant Nutr. 28, 309–321.

Chatzissavvidis, C.A., Therios, I.N., Antonopoulou, C., 2007. Effect of nitrogen source on olives growing in soils with high boron content. Aust. J. Exp. Agric. 47, 1491–1497.

Chatzissavvidis, C., Therios, I., Antonopoulou, C., Dimassi, K., 2008. Effects of high boron concentration and scion-rootstock combination on growth and nutritional status of olive plants. J. Plant Nutr. 31, 638–658.

Chen, L., Han, S., Qi, Y., Yang, L., 2012. Boron stresses and tolerance in citrus. Afr. J. Biotechnol. 11, 5961–5969.

Choe, J.S., Lee, J.C., Kim, S.B., Moon, J.Y., 1986. Studies on the cause of shoot die-back in pear trees (*Pyrus serotina* Rehder). J. Korean Soc. Hortic. Sci. 27, 149–156.

Crandall, P.C., Chamberlain, J.D., Garth, J.K.L., 1981. Toxicity symptoms and tissue-levels associated with excess boron in pear trees. Commun. Soil Sci. Plant Anal. 12, 1047–1057.

Dannel, F., Pfeffer, H., Römheld, V., 1998. Compartmentation of boron in roots and leaves of sunflower as affected by boron supply. J. Plant Physiol. 153, 615–622.

Dell, B., Huang, L., 1997. Physiological response of plants to low boron. Plant Soil 193, 103–120.

Dionysiou, N., Matsi, T., Misopolinos, N.D., 2006. Use of magnesia for boron removal from irrigation water. J. Environ. Qual. 35, 2222–2228.

Dye, M.H., Buchanan, L., Dorofaeff, F.D., Beecroft, F.G., 1983. Die-back of apricot trees following soil application of boron. N. Z. J. Exp. Agric. 11, 331–342.

Eaton, F.M., McCallum, R.D., Mayhugh, M.S., 1941. Quality of irrigation waters of the Hollister area of California. In: Tech Bull No 746. United States Department of Agriculture, Washington, DC.

El-Motaium, R., Hu, H., Brown, P.H., 1994. The relative tolerance of six *Prunus* rootstocks to boron and salinity. J. Am. Soc. Hortic. Sci. 119, 1169–1175.

Gimeno, V., Simón, I., Nieves, M., Martínez, V., Cámara-Zapata, J.M., García, A.L., García-Sánchez, F., 2012. The physiological and nutritional responses to an excess of boron by Verna lemon trees that were grafted on four contrasting rootstocks. Trees 26, 1513–1526.

Grattan, S.R., 2013. Evaluation of the impact of boron on citrus orchards in riverside county riverside county water task force. Report submitted to Riverside County Water Task Force.

Grattan, S.R., Díaz, F.J., Pedrero, F., Vivaldi, G.A., 2015. Assessing the suitability of saline wastewaters for irrigation of *Citrus* spp: emphasis on boron and specific-ion interactions. Agric. Water Manag. 157, 48–58.

Guerfel, M., Baccouri, O., Boujnah, D., Chaïbi, W., Zarrouk, M., 2009. Impacts of water stress on gas exchange, water relations, chlorophyll content and leaf structure in the two main Tunisian olive (*Olea europaea* L) cultivars. Sci. Hortic. 119, 257–263.

Gunes, A., Soylemezoglu, G., Inal, A., Bagci, E.G., Coban, S., Sahin, O., 2006. Antioxidant and stomatal responses of grapevine (*Vitis vinifera* L) to boron toxicity. Sci. Hortic. 110, 279–284.

Guo, P., Qi, Y.P., Yang, L.T., Ye, X., Jiang, H.X., Huang, J.H., Chen, L.S., 2014. cDNA-AFLP analysis reveals the adaptive responses of citrus to long-term boron-toxicity. Plant Biol. 14, 284.

Guo, P., Qi, Y.P., Yang, L.T., Ye, X., Huang, J.H., Chen, L.S., 2016. Long-term boron-excess-induced alterations of gene profiles in roots of two citrus species differing in boron-tolerance revealed by cDNA-AFLP. Front. Plant Sci. 7, 898.

Gupta, U.C., James, Y.W., Campbell, C.A., Leyshon, A.J., Nicholaichuk, W., 1985. Boron toxicity and deficiency: a review. Can. J. Soil Sci. 65, 381–409.

Hakki, E.E., Atalay, E., Harmankaya, M., Babaoglu, M., Hamurcu, M., Gezgin, S., 2007. Determination of suitable maize (*Zea mays* L.) genotypes to be cultivated in boron-rich central Anatolian soil. In: Xu, F., et al. (Ed.), Advances in Plant and Animal Boron Nutrition. Springer, pp. 231–247.

Han, S., Tang, N., Jiang, H.X., Yang, L.T., Li, Y., Chen, L.S., 2009. CO_2 assimilation, photosystem II photochemistry, carbohydrate metabolism and antioxidant system of citrus leaves in response to boron stress. Plant Sci. 176, 143–153.

Hansen, C.J., 1945. Boron content of olive leaves. Proc. Am. Soc. Hortic. Sci. 46, 78–80.

Hansen, C.J., 1948. Influence of the rootstock on injury from excess boron in French (Agen) prune and President plum. In: Proceedings of the American Society for Horticultural Science. vol. 51, No. JUN. Amer Soc Horticultural Science, Alexandria, VA, pp. 239–244.

Hansen, C.J., 1955. Influence of rootstock on injury from excess boron in Nonpareil almond and Elberta peach. Proc. Am. Soc. Hortic. Sci. 65, 128–132.

Hanson, E.J., 1991. Sour cherry trees respond to foliar boron applications. Hortic. Sci. 26, 1142–1145.

Heidarabadi, M.D., Ghanati, F., Fujiwara, T., 2011. Interaction between boron and aluminum and their effects on phenolic metabolism of *Linum usitatissimum* L roots. Plant Physiol. Biochem. 49, 1377–1383.

Herrera-Rodríguez, M.B., González-Fontes, A., Rexach, J., Camacho-Cristóbal, J.J., Maldonado, J.M., Navarro-Gochicoa, M.T., 2010. Role of boron in vascular plants and response mechanisms to boron stresses. Plant Stress 4, 115–122.

Hu, H., Penn, S.G., Lebrilla, C.B., Brown, P.H., 1997. Isolation and characterization of soluble boron complexes in higher plants (the mechanism of phloem mobility of boron). Plant Physiol. 113, 649–655.

Huang, J.H., Cai, Z.J., Wen, S.X., Guo, P., Ye, X., Lin, G.Z., Chen, L.S., 2014. Effects of boron toxicity on root and leaf anatomy in two *Citrus* species differing in boron tolerance. Trees 28, 1653–1666.

Kamali, A., Childers, N., 1966. Effect of boron nutrition on peach anatomy. J. Am. Soc. Hortic. Sci. 90, 33–38.

Karantzi, A.D., Papadakis, I.E., Psychoyou, M., Ioannou, D., 2015. Nutrient status of the banana cultivar 'FHIA-01' as affected by boron excess. In: III Balkan Symposium on Fruit Growing 1139, pp. 399–404.

Keles, Y., Öncel, I., Yenice, N., 2004. Relationship between boron content and antioxidant compounds in *Citrus* leaves taken from fields with different water source. Plant Soil 265, 345–353.

Landi, M., Degl'Innocenti, E., Pardossi, A., Guidi, L., 2012. Antioxidant and photosynthetic responses in plants under boron toxicity: a review. Am. J. Agric. Biol. Sci. 7, 255–270.

Landi, M., Guidi, L., Pardossi, A., Tattini, M., Gould, K.S., 2014. Photoprotection by foliar anthocyanins mitigates effects of boron toxicity in sweet basil (*Ocimum basilicum*). Planta 240, 941–953.

Lee, S.G., Aronoff, S., 1966. Investigations on the role of boron in plants. III Anatomical observations. Plant Physiol. 41, 1570–1577.

Leyshon, J.A., James, W.Y., 1993. Boron toxicity and irrigation management. In: Gupta, U. (Ed.), Boron and Its Role in Crop Production. CRC Press Inc., Boca Raton, FL.

Li, Y., Han, M.Q., Lin, F., Ten, Y., Lin, J., Zhu, D.H., et al., 2015. Soil chemical properties, 'Guanximiyou' pummelo leaf mineral nutrient status and fruit quality in the southern region of Fujian province, China. J. Soil Sci. Plant Nutr. 15, 615–628.

López-Gómez, E., San Juan, M.A., Diaz-Vivancos, P., Beneyto, J.M., Garcia-Legaz, M.F., Hernández, J.A., 2007. Effect of rootstocks grafting and boron on the antioxidant systems and salinity tolerance of loquat plants (*Eriobotrya japonica* Lindl). Environ. Exp. Bot. 60, 151–158.

Maas, E.V., 1984. Salt tolerance of plants. In: Christie, B.R. (Ed.), Handbook of Plant Science in Agriculture. In: Vol. 2. CRC Press, Bota Raton, FL, pp. 57–75.

Makkee, M., Kieboom, A.P.G., Van Bekkum, H., 1985. Studies on borate esters. III Borate esters of D-mannitol, D-glucitol, D-fructose and D-glucose in water. Recl. Trav. Chim. Pays-Bas 104, 230–235.

Marschner, M., 1995. Mineral Nutrition of Higher Plants, second ed. Academic Press, London.

Mengel, K., Kirkby, E.A., 2001. Principles of Plant Nutrition, fifth ed. Kluwer Academic Publishers, Dordrecht.

Meriño-Gergichevich, C., Reyes-Díaz, M., Guerrero, J., Ondrasek, G., 2017. Physiological and nutritional responses in two highbush blueberry cultivars exposed to deficiency and excess of boron. J. Soil Sci. Plant Nutr. 17, 307–318.

Miwa, K., Takano, J., Omori, H., Seki, M., Shinozaki, K., Fujiwara, T., 2007. Plants tolerant of high boron levels. Science 318, 1417.

Molassiotis, A., Sotiropoulos, T., Tanou, G., Diamantidis, G., Therios, I., 2006. Boron induced oxidative damage and antioxidant and nucleolytic responses in shoot tips culture of the apple rootstock EM9 (*Malus domestica* Borkh). Environ. Exp. Bot. 56, 54–62.

Mouhtaridou, G.N., Sotiropoulos, T.E., Dimassi, K.N., Therios, I.N., 2004. Effects of boron on growth, and chlorophyll and mineral contents of shoots of the apple rootstock MM 106 cultured *in vitro*. Biol. Plant. 48, 617–619.

Nable, R.O., Banuelos, G.S., Paull, J.G., 1997. Boron toxicity. Plant Soil 193, 181–198.

Nielsen, G.H., Hout, P., Yorston, J., Lierop, W.V., 1985. Relationships between leaf and soil boron and boron toxicity of peaches in British Colombia. Can. J. Soil Sci. 65, 213–217.

Nishiyama, Y., Allakhverdiev, S.I., Murata, N., 2006. A new paradigm for the action of reactive oxygen species in the photoinhibition of photosystem II. Biochim. Biophys. Acta Biomembr. Bioenerg. 1757, 742–749.

Papadakis, I.E., Dimassi, K.N., Therios, I.N., 2003. Response of two citrus genotypes to six boron concentrations: concentration and distribution of nutrients, total absorption, and nutrient use efficiency. Aust. J. Agric. Res. 54, 571–580.

Papadakis, I.E., Dimassi, K.N., Bosabalidis, A.M., Therios, I.N., Patakas, A., Giannakoula, A., 2004a. Boron toxicity in 'Clementine' mandarin plants grafted on two rootstocks. Plant Sci. 166, 539–547.

Papadakis, I.E., Dimassi, K.N., Bosabalidis, A.M., Therios, I.N., Giannakoula, A., 2004b. Effects of B excess on some physiological and anatomical parameters of 'Navelina' orange plants grafted on two rootstocks. Environ. Exp. Bot. 51, 247–257.

Paparnakis, A., Chatzissavvidis, C., Antoniadis, V., 2013. How apple responds to boron excess in acidic and limed soil. J. Soil Sci. Plant Nutr. 13, 787–796.

Parks, J.L., Edwards, M., 2005. Boron in the environment. Crit. Rev. Environ. Sci. Technol. 35, 81–114.

Perry, R., 1987. Cherry rootstocks. In: Rom, R.C., Carlson, R.F. (Eds.), Rootstocks for Fruit Crops. John Wiley, New York, pp. 145–183.

Pezeshki, S.R., 1994. Responses of baldcypress (*Taxodium distichum*) seedlings to hypoxia: leaf protein content, ribulose-1,5-biosphophate carboxylase/oxygenase activity and photosynthesis. Photosynthetica 30, 59–68.

Picchioni, G.A., Miyamoto, S., 1991. Growth and boron uptake of five pecan cultivar seedlings. Hortic. Sci. 26, 386–388.

Reid, R., 2007. Update on boron toxicity and tolerance in plants. In: Xu, F., et al. (Ed.), Advances in Plant and Animal Boron Nutrition. Springer, pp. 83–90.

Reid, R., 2010. Can we really increase yields by making crop plants tolerant to boron toxicity? Plant Sci. 178, 9–11.

Reid, R., Fitzpatrick, K., 2009. Influence of leaf tolerance mechanisms and rain on boron toxicity in barley and wheat. Plant Physiol. 151, 413–420.

Reid, R.J., Hayes, J.E., Post, A., Stangoulis, J.C.R., Graham, R.D., 2004. A critical analysis of the causes of boron toxicity in plants. Plant Cell Environ. 25, 1405–1414.

Ryan, J., Rashid, A., Torrent, J., Yau, S.K., Ibrikci, H., Sommer, R., Erenoglu, E.B., 2013. Micronutrient constraints to crop production in the Middle East–west Asia region: significance, research, and management. Adv. Agron. 122, 1–84.

Sarafi, E., Chatzissavvidis, C., Therios, I., 2017. Response of two pomegranate (*Punica granatum* L.) cultivars to six boron concentrations: growth performance, nutrient status, gas exchange parameters, chlorophyll fluorescence, proline and carbohydrate content. J. Plant Nutr. 40, 983–994.

Sepaskhah, A.R., Maftoun, M., Yasrebi, J., 1988. Seedling growth and chemical composition of three pistachio cultivars as affected by soil applied boron. J. Hortic. Sci. 63, 743–749.

Shah, A., Wu, X., Ullah, A., Fahad, S., Muhammad, R., Yan, L., Jiang, C., 2017. Deficiency and toxicity of boron: alterations in growth, oxidative damage and uptake by citrange orange plants. Ecotoxicol. Environ. Saf. 145, 575–582.

Sheng, O., Song, S.W., Chen, Y.J., Peng, S.A., Den, X.X., 2008. Effects of exogenous B supply on growth, B accumulation and distribution of two navel orange cultivars. Trees 23, 59.

Sheng, O., Song, S., Peng, S., Deng, S., 2009. The effects of low boron on growth, gas exchange, boron concentration and distribution of 'Newhall' navel orange (*Citrus sinensis* Osb) plants grafted on two rootstocks. Sci. Hortic. 121, 278–283.

Simón-Grao, S., Nieves, M., Martínez-Nicolás, J.J., Cámara-Zapata, J.M., Alfosea-Simón, M., García-Sánchez, F., 2018. Response of three citrus genotypes used as rootstocks grown under boron excess conditions. Ecotoxicol. Environ. Saf. 159, 10–19.

Singh, D.P., Beloy, J., McInerney, J.K., Day, L., 2012. Impact of boron, calcium and genetic factors on vitamin C, carotenoids, phenolic acids, anthocyanins and antioxidant capacity of carrots (*Daucus carota*). Food Chem. 132, 1161–1170.

Smith, G.S., Clark, C.J., 1989. Effect of excess boron on yield and post-harvest storage of kiwifruit. Sci. Hortic. 38, 105–115.

Sotiropoulos, T.E., Therios, I.N., Dimassi-Theriou, K.N., 1997. Boron toxicity of kiwifruit orchards in Northern Greece. Acta Hortic. 444, 243–247.

Sotiropoulos, T.E., Therios, I.N., Dimassi, K.N., 1999. Calcium application as a means to improve tolerance of kiwifruit (*Actinidia deliciosa* L) to boron toxicity. Sci. Hortic. 81, 443–449.

Sotiropoulos, T.E., Therios, I.N., Dimassi, K.N., Bosabalidis, A., Kofidis, G., 2002. Nutritional status, growth, CO$_2$ assimilation, and leaf anatomical responses in two kiwifruit species under boron toxicity. J. Plant Nutr. 25, 1249–1261.

Sotiropoulos, T.E., Therios, I.N., Dimassi, K.N., 2005. Seasonal accumulation and distribution of nutrient elements in fruit of kiwifruit vines affected by boron toxicity. Aust. J. Exp. Agric. 4612, 1639–1644.

Sotiropoulos, T.E., Dimassi, K.N., Tsirakoglou, V., 2006a. Effect of boron and methionine on growth and ion content in kiwifruit shoots cultured *in vitro*. Biol. Plant. 50, 300–302.

Sotiropoulos, T.E., Therios, I.N., Almaliotis, D., Papadakis, I., Dimassi, K.N., 2006c. Response of cherry rootstocks to boron and salinity. J. Plant Nutr. 29, 1691–1698.

Stangoulis, J.C., Reid, R.J., 2002. Boron toxicity in plants and animals. In: Boron in Plant and Animal Nutrition. Springer, Boston, MA, pp. 227–240.

Takano, J., Miwa, K., Fujiwara, T., 2008. Boron transport mechanisms: collaboration of channels and transporters. Trends Plant Sci. 13, 451–457.

Tewari, R.K., Kumar, P., Sharma, P.N., 2009. Morphology and oxidative physiology of boron-deficient mulberry plants. Tree Physiol. 30, 68–77.

Torun, B., Kalayci, M., Ozturk, L., Torun, A., Aydin, M., Cakmak, I., 2002. Differences in shoot boron concentrations, leaf symptoms, and yield of Turkish barley cultivars grown on boron-toxic soil in field. J. Plant Nutr. 26, 1735–1747.

Tsadilas, C.D., Dimoyiannis, D., Samaras, V., 1997. Methods of assessing boron availability to kiwifruit plants growing on high boron soils. Commun. Soil Sci. Plant Anal. 28, 973–987.

Vargas, A., Arias, F., Serrano, E., Arias, M., 2007. Toxicidad de boro en plantaciones de banano (Musa AAA) en Costa Rica. Agron. Costarric. 31, 21–29.

Wang, B.L., Shi, L., Li, Y.X., Zhang, W.H., 2010. Boron toxicity is alleviated by hydrogen sulfide in cucumber (*Cucumis sativus* L) seedlings. Planta 231, 1301–1309.

Wojcik, P., Wojcik, M., 2003. Effects of boron fertilization on 'Conference' pear tree vigor, nutrition, and fruit yield and storability. Plant Soil 256, 413–421.

Woodbridge, C.G., 1955. The boron requirements of stone fruit trees. Can. J. Agric. Sci. 35, 282–286.

Wutscher, H.K., Smith, P.F., Bennett, W.F. (Eds.), 1996. Citrus. In: Nutrient Deficiencies and Toxicities in Crop Plants. The Am Phytopath Soc, St Paul, MN, pp. 165–170.

Yau, S.K., Ryan, J., 2008. Boron toxicity tolerance in crops: a viable alternative to soil amelioration. Crop Sci. 48, 854–865.

Yordanova, R.Y., Popova, L.P., 2007. Flooding-induced changes in photosynthesis and oxidative status in maize plants. Acta Physiol. Plant. 296, 535–541.

Further reading

Haydon, G.F., 1981. Boron toxicity of strawberry. Commun. Soil Sci. Plant Anal. 12, 1085–1091.

Nyomora, A.M.S., Brown, P.H., Krueger, B., 1999. Rate and time of boron application increase almond productivity and tissue boron concentration. HortScience 34, 242–245.

Sotiropoulos, T.E., Molassiotis, A., Almaliotis, D., Mouhtaridou, G., Dimassi, K., Therios, I., Diamantidis, G., 2006b. Growth, nutritional status, chlorophyll content, and antioxidant responses of the apple rootstock MM 111 shoots cultured under high boron concentrations *in vitro*. J. Plant Nutr. 29, 575–583.

17

Aluminum toxicity and fruit nutrition

Li-Song Chen[a,*], *Lin-Tong Yang*[a], *Peng Guo*[a,b], *Huan-Xin Jiang*[a], *Ning Tang*[a,c]

[a]Institute of Plant Nutritional Physiology and Molecular Biology, College of Resources and Environment, Fujian Agriculture and Forestry University, Fuzhou, China
[b]College of Life Sciences, Henan Agricultural University, Zhengzhou, China
[c]Research Institute for Special Plants, Chongqing University of Arts and Sciences, Chongqing, China
*Corresponding author. E-mail: lisongchen2002@hotmail.com

1 Introduction

Acid soils are very prevalent in many regions of the world throughout the tropics and subtropics and accounts for ~30% of the world's ice-free land and ~12% of the global area used for arable crops (von Uexküll and Mutert, 1995). Moreover, soil acidity is becoming an increasingly serious problem due to the improper farming practices (the overuse of nitrogen [N] fertilizers), environmental deterioration (acid rain), intensive agriculture, and monoculture (Guo et al., 2010; Yang et al., 2013; Long et al., 2017). Aluminum (Al), the most abundant metallic element and the third most abundant element in the earth's crust after oxygen (O) and silicon (Si), comprises 7% of its mass (Foy et al., 1978). In neutral or mildly acidic soils, Al exists primarily as silicate or oxide forms and is biologically inactive. In acidic soils (pH < 5), however, phytotoxic Al^{3+} was released from the nontoxic silicate or oxide forms into solution (Kinraide, 1991). Micromolar concentration of Al^{3+} can rapidly inhibit root elongation. The consequent impairments on the uptake of water and nutrients can cause poor growth and yield loss of crops (Kochian, 1995; Guo et al., 2018a). Thus, Al toxicity is a major constraint for crop productivity worldwide (von Uexküll and Mutert, 1995; Kochian et al., 2004; Reyes-Díaz

A.K. Srivastava, Chengxiao Hu (eds.)
Fruit Crops: Diagnosis and Management of Nutrient Constraints
https://doi.org/10.1016/B978-0-12-818732-6.00017-4

et al., 2015; Chen et al., 2010b; Guo et al., 2018a). Many important fruit trees (viz., *Citrus*, longan, pineapple, banana, litchi, and blueberry) are commercially planted in humid and subhumid of tropical, subtropical, and temperate regions of the world mainly on acidic soils and are prone to Al toxicity. In acidic soils with high active Al, *Citrus* often has a poor growth and a shortened life span (Lin and Myhre, 1990). During 1998–99, Huang et al. (2001) measured the pH of 200 soil samples from Pinghe pummelo (*Citrus grandis*) orchards, situated in Zhangzhou, China, and found that 85.5% of the soils displayed a pH < 5.0, with an average pH value of 4.63 (3.57–7.25). In 2011, Li et al. (2015) assayed the pH of 319 soil samples from Pinghe pummelo orchards and found that over 90.0% of the soils had a pH < 5.0, with an average pH value of 4.34 (3.26–6.22). Obviously, soil pH is rapidly decreasing in pummelo orchards in last decades. In this chapter, we summarized the recent progress in our understanding of Al toxicity and the mechanisms of Al tolerance in fruit trees.

2　Al toxic effects on growth and structure

Aluminum toxicity–induced inhibition of root and shoot growth have been observed in many fruit trees including *Citrus* (Arunakumara et al., 2012; Chen, 2006, 2012; Jiang et al., 2008, 2009c), apple (Wojcik, 2003), litchi, longan, wampee (Xiao et al., 2002; Chen et al., 2005c), and peach (Reyes-Díaz et al., 2015). Al toxicity decreased root, stem, leaf, shoot, and whole plant dry weight (DW) and increased the ratio of root DW to shoot DW in Al-intolerant *C. grandis* seedlings, but all the six parameters were not altered in Al-tolerant *Citrus sinensis* seedlings (Yang et al., 2011; Guo et al., 2018a). Chen et al. (2005a) observed that Al-treated "Cleopatra" tangerine (*Citrus reshni*) plants had decreased root, stem, and leaf fresh and dry weight, with a greater decrease in leaf and stem fresh and dry weight than that in root fresh and dry weight, but increased leaf DW per area. However, Al-treated "Swingle" citrumelo (*Citrus paradisi* × *Poncirus trifoliata*) seedlings had increased leaf DW per area but unaltered ratio of shoot DW/root DW (dos Santos et al., 2000). Jiang et al. (2009c) reported that Al toxicity led to increased ratio of root DW/shoot DW in *Citrus reticulata* cv. Chachiensis, *C. reticulata* cv. Fuju, *C. reticulata* cv. Sichuanhongju, sour orange (*Citrus aurantium*) from Hubei, and *P. trifoliata* seedlings, but not in *C. reticulata* cv. Ponkan, *C. aurantium* cv. Guotoucheng from Zhejiang, *Citrus limon* cv. Eureka, *Citrus limonia* cv. Red limonia, *C. grandis* cv. Wendanyou, and citrumelo seedlings. In a study with five *Citrus* rootstocks (viz., "Cleopatra" mandarin, rough lemon, sour orange, "Swingle" citrumelo, and "Carrizo" citrange), Lin and Myhre (1991a) observed that, at high Al concentration, seedlings had fewer new lateral and fibrous roots and thickened root tips and that some older roots rotted at the highest level. Near the end of experiment (60 days), seedlings of some rootstocks submitted to high level concentration displayed yellow, mottled, and withered new leaves.

In a study with longan seedlings, Xiao et al. (2002) observed that root, stem, and leaf DW and root vitality were decreased by 13.8%–31.7%, 5.4%–32.6%, 14.5%–58.2%, and 22.0%–66.0%, respectively, as Al concentration in the nutrient solution increased from 0 to 1.850 mM, indicating that the relative sensitivity of shoots and roots to Al depended on Al concentration. Xiao et al. (2003a) examined Al toxic effects on longan root and leaf ultrastructure. The most striking Al toxic effect on leaf cell ultrastructure was the chloroplast injury. Al toxicity caused the loss of chloroplast integrity, as indicated by the disappearance of chloroplast membrane and the distortion of grana lamellae. Plasmolysis occurred in the Al-treated mesophyll cells. Al toxicity led to the degradation of cell wall, the decrease of intercellular space, and the breakdown of tonoplast in leaves. Al toxicity also impaired mitochondrion membranes and decreased the number of cristae in leaves. Al-treated longan seedlings had shorter and thicker root tips. In root elongating zone, Al toxicity caused the degradation of cell wall, the decrease of intercellular space, and the breakdown of nuclear and mitochondrial membranes. Al toxicity decreased greatly the number of starch grains in the root elongating zone, and even no starch grain was observed in the elongating zone of some Al-treated roots. 4'-6-Diamidino-2-phenylindole (DAPI) staining showed that Al toxicity caused the deformation and decrease of nuclei in the root apical meristem zone. Chen et al. (2005c) observed that Al toxicity decreased the number of fibrous roots and the length of roots in litchi and wampee seedlings. Under high Al concentration, root tips became brown and thicker, and leaves became wilted due to loss of water.

In "Elberta" peach (*Prunus persica*) seedlings, leaf symptoms included marginal chlorosis that later developed into necrotic areas at the early stage of Al toxicity, and collapse of the midrib, terminal dieback, and defoliation are typical symptoms of calcium (Ca) deficiency in peaches at the advanced stage of Al toxicity. Root symptoms at high Al concentration included dieback of roots and abnormal development of new roots (Edwards and Horton, 1977; Edwards et al., 1976). Graham (2001) showed that Al-treated "Nemaguard" peach seedlings had decreased total leaf number, total leaf area, and total vegetative growth but similar leaf, stem, and root DW. In a study with 4-week-old "Flordaguard" peach rootstock seedlings, Kongsri et al. (2014) observed that shoot fresh weight (shoot DW and leaf size) linearly (quadratically) decreased in response to Al toxicity, but leaf burn quadratically increased.

Schuch et al. (2010) investigated Al toxic effects on growth and morphogenesis (in leaf explants) in in vitro cultures of "BA 29" quince (*Cydonia oblonga*), one of the most widely used pear rootstocks in Europe. Al strongly decreased shoot growth in the proliferation (0.5-mM Al) and rooting (2.2-mM Al) phases, inhibited shoot proliferation (1.1-mM Al), and caused conspicuous tissue browning (2.2-mM Al). Lin (2010) observed visible Al toxic symptoms (viz., chlorosis on young leaves and short and coarse roots) in Al-sensitive "Tainung No. 17" pineapple grown in 200- and 300-μM AlCl$_3$.

3 Physiological and biochemical responses of fruit trees to Al toxicity

3.1 Uptake of mineral nutrients and water

Aluminum toxicity can inhibit root growth and impair their function, thus affecting the uptake of nutrient elements and water (Chen, 2012; Kochian, 1995; Guo et al., 2018a). A potted sand culture experiment with "Cleopatra" tangerine seedlings showed that Al toxicity decreased the concentrations of phosphorus (P), potassium (K), magnesium (Mg), zinc (Zn), copper (Cu), and molybdenum (Mo) in leaves, but did not significantly alter the concentrations of N, Ca, iron (Fe), boron (B), and manganese (Mn) (Chen et al., 2010a). In a field experiment with 30-year-old trees of *C. sinensis* cv. Hamlin on sour orange rootstock, Lin and Myhre (1990) observed that Al-treated fibrous roots had decreased Mn, Fe, and Zn concentrations but unaltered P, K, Ca, Mg, and B concentrations. In a solution-culture experiment with five *Citrus* rootstocks: "Cleopatara" tangerine, rough lemon, "Swingle" citrumelo, sour orange, and "Carrizo" citrange, Lin and Myhre (1991b) found that the concentrations of Ca, Mg, Fe, and Cu in shoots and Ca, Fe, Cu, Zn, and Mn in roots decreased as Al in the nutrient solution increased from 4 to 178 μM, but the reverse was the case for the concentrations of K and P in shoots and K, P, and Mg in roots. Recently, Guo et al. (2018a) found that in *C. grandis* seedlings, Al toxicity greatly decreased leaf and root relative water content (RWC) and K, N, Ca, Mg, P, and S uptake per plant but, in *C. sinensis* seedlings, Al toxicity only slightly decreased K, P, and N uptake per plant; did not alter leaf and root RWC and Mg uptake per plant; and increased sulfur (S) uptake per plant. Evidently, the uptake of water and macroelements were less affected by Al in Al-tolerant *C. sinensis* seedlings than those in Al-intolerant *C. grandis* seedlings. Similarly, N, P, K, Ca, and Mg concentrations were less affected in Al-treated *C. sinensis* leaves, roots, and stems than those in Al-treated *C. grandis* ones. To conclude, Al-treated *C. sinensis* seedlings had higher capacity to maintain the homeostasis of water and nutrients.

Edwards and Horton (1977) investigated Al toxic effects on total nutrient content, uptake rates, and distribution in peach seedlings. Generally viewed, total nutrient content per seedlings decreased with increasing Al concentration. The uptake rates for P, Ca, Mg, Zn, and Mn decreased and for K and Fe increased with increasing Al concentration. Al toxicity had no influence on the translocation of most nutrient elements, but more of the absorbed Ca was accumulated in the leaves than that in the roots or stems.

Xiao et al. (2005a) investigated the effects of 0–1.850-mM Al in the nutrient solution on the concentrations of N, P, K, Ca, and Mg in longan roots, stems, and leaves. The concentrations of N and K in roots and stems increased as Al supply increased from 0 to 0.185 mM and then decreased with further increasing Al. The concentrations of N, P, and K in leaves and Ca and Mg in roots, stems, and leaves decreased with increasing Al supply. By contrast, the concentration of P in roots and stems increased with increasing Al supply. A study with longan, litchi, and wampee seedlings showed that the uptake of N, K, Ca, and water decreased with increasing Al supply (Chen et al., 2005c).

In a sand culture experiment with "Nemaguard" peach seedlings, Chibiliti and Byrne (1990) showed that Al toxicity decreased the concentrations of Ca, Cu, Fe, Zn, and Mo in shoots and K, Ca, Cu, Zn, and Mo in roots, but had no influence on the concentrations of K, Mg, and Mn in shoots and Fe and Mn in roots. In another study, Graham (2001) observed that Al-treated "Nemaguard" peach seedling had decreased concentrations of Ca, Mg, P, Fe, and Mo in leaves and roots, Mg and P in stems, and Mn in leaves.

In a study with two apple rootstocks (Polish rootstock [P22] and Malling 26 [M.26]), Wojcik (2003) found that the concentrations of P, Ca, and Mg and their uptake rates and translocation rates to shoots decreased in response to Al toxicity.

Lin (2010) investigated the effects of 0, 100, 200, and 300 μM Al in the nutrient solution on root growth and nutrient uptake by two pineapple cultivars—namely, Al-tolerant "Cayenne" and Al-sensitive "Tainung No.17." Root elongation of "Cayenne" increased with increasing Al concentration, while root elongation of "Tainung No.17" was inhibited at 200- and 300-μM Al. In "Tainung No.17," Ca, Mg, and K uptake was decreased at 200 μM Al, and Fe, Mn, and Cu uptake was decreased at 300 μM Al. However, Ca, Mg, and K uptake of "Cayenne" was increased at 200-μM Al.

A study with grape rootstocks showed that elevating soil Al level decreased the concentrations of Ca and S in roots and Ca in stems and leaves but increased the concentrations of Cu and Zn in roots and Zn and Mn in stems and leaves (Alvarenga et al., 2004).

Al treatment increased the concentrations of Fe and Mn both in the soil and in the raspberry plants (Sikirić et al., 2009).

To conclude, Al toxic effects on nutrient uptake depend on time of exposure to Al, Al concentration in growth medium, and the species or cultivar of fruit trees.

3.2 Leaf gas exchange and chlorophyll *a* florescence

The Al-induced inhibition of leaf photosynthesis is a very common phenomenon in many higher plants including some fruit trees (*Citrus*, longan, and blueberry) (Chen, 2012; Reyes-Díaz et al., 2011, 2015; Xiao et al., 2005b). Several studies from our laboratory showed that the Al-induced decrease in *Citrus* leaf CO_2 assimilation was primarily due to nonstomatal limitation, because Al toxicity increased or had no influence on intercellular CO_2 concentration (Chen et al., 2005a; Jiang et al., 2008, 2009a,b; Guo et al., 2017b, 2018a; Liao et al., 2015; Yang et al., 2012a). This is also supported by the study of Pereira et al. (2000), who observed that the Al-induced decrease in leaf CO_2 assimilation of four *Citrus* rootstock ("Cravo" lemon, "Volkamer" lemon [*Citrus volkameriana*], "Cleopatra" tangerine, and "Sunki" tangerine) seedlings was accompanied by increased or unaltered stomatal conductance and intercellular CO_2 concentration increased with increasing Al concentration in the nutrition solution. They concluded that the decline in CO_2 assimilation in response to Al toxicity might be caused by the structural damage to the thylakoids, as shown by the decreased ratio of variable fluorescence to initial fluorescence (F_v/F_o). Because Al toxicity affected CO_2 assimilation more than chlorophyll (Chl) concentration in *Citrus* leaves (Chen et al., 2005a; Jiang et al., 2008, 2009a,b; Guo et al., 2017b, 2018b; Liao et al., 2015; Yang et al., 2012a), the Al-induced decrease in CO_2 assimilation was not explained alone by the decreased Chl level. This is also supported by the data that Al-treated *Citrus* leaves had increased dissipated energy flux per reaction center (RC) at $t=0$ (DI_o/RC), dissipated energy flux per cross section (CS) at $t=0$ (DI_o/CS_o), and quantum yield at $t=0$ for energy dissipation (DI_o/ABS; Jiang et al., 2008, 2009a,b; Liao et al., 2015; Yang et al., 2012a; Guo et al., 2018a). Al toxicity did not significantly alter the concentration of malondialdehyde (MDA, a marker of peroxidative damage) in "Cleopatra" tangerine (Chen et al., 2005a,b), and 10-μM sodium nitroprusside (SNP, an NO donor) (Yang et al., 2012a) and 0.5-mM $MgSO_4$ + 0.5-mM Na_2SO_4 (Guo et al., 2017b) treated *C. grandis* leaves, implying that the Al did not affect the membrane lipid in chloroplast. Thus, the lower CO_2 assimilation in these leaves could not be caused by Al-induced photo-oxidative damage.

In a sand culture experiment with "Cleopatra" tangerine seedlings, Chen et al. (2005a,b) observed that Al-treated leaves had decreased CO_2 assimilation, increased or unaltered activities of enzymes involved in the Calvin cycle, and unaltered or slightly decreased concentrations of nonstructural carbohydrates. They concluded that the Al-induced decline in leaf CO_2 assimilation was not caused by feedback inhibition due to carbohydrate accumulation, but was probably caused by a combination of factors such as decreased electron transport rate through PSII (ETR); increased closure of PSII RCs, as indicated by decreased photochemical quenching coefficient (qP); and elevated photorespiration, as indicated by increased activities of catalase (CAT) and ribulose bisphosphate carboxylase/oxygenase (Rubisco).

Several studies (Jiang et al., 2008, 2009a,b; Yang et al., 2012a; Liao et al., 2015; Guo et al., 2017b, 2018a) from our laboratory showed that Al-treated *Citrus* leaves had altered Chl *a* fluorescence (OJIP) transients, as indicated by the increased O-step, the depressed P-step, and the positive ΔL, ΔK, ΔJ, and ΔI bands around 100 μs, 300 μs, 2 ms, and 30 ms, respectively, decreased maximum amplitude of IP phase, maximum fluorescence (F_m), F_v, F_v/F_o, fraction of oxygen evolving complex (OEC) in comparison with control, normalized total complementary area above the OJIP or total electron carriers per RC (S_m or EC_o/RC), electron carriers per absorption (ABS) at $t=0$ (EC_o/ABS), total performance index ($PI_{abs,total}$), performance index (PI_{abs}), efficiency with which an electron can move from the reduced intersystem electron acceptors to the photosystem I (PSI) end electron acceptors (δ_{Ro}), efficiency with which a trapped exciton can move an electron into the electron transport chain from Q_A^- to the PSI end electron acceptors (ρ_{Ro}), probability (at $t=0$) that a trapped exciton moves an electron into the electron transport chain beyond Q_A^- (ψ_{Eo}), amount of active PSII RCs per CS at $t=0$ (RC/CS_o), amount of active PSII RCs per ABS (RC/ABS), electron transport flux per CS at $t=0$ (ET_o/CS_o), reduction of end acceptors at PSI electron acceptor side per CS at $t=0$ (RE_o/CS_o), maximum PSII efficiency of dark-adapted leaves (F_v/F_m), quantum yield for the reduction of end acceptors of PSI per photon absorbed (φ_{Ro}), quantum yield for electron transport at $t=0$ (φ_{Eo}), reduction of end acceptors at PSI electron acceptor side per RC at $t=0$ (RE_o/RC), electron transport flux per RC at $t=0$ (ET_o/RC) and overall grouping probability (P_{2G}), and increased F_o, relative variable fluorescence at J-step (V_J) and I-step (V_I), approximated initial slope (in m/s^1) of the fluorescence transient $V = f(t)$ (M_o), absorption flux per CS at $t = 0$ (ABS/CS$_o$), DI$_o$/CS$_o$, DI$_o$/ABS, DI$_o$/RC, trapped energy flux per RC at $t = 0$ (TR$_o$/RC), trapped energy flux per CS at $t = 0$ (TR$_o$/CS$_o$), absorption flux per RC (ABS/RC) and CS per RC at $t = 0$ (CS$_o$/RC). Regressive analysis showed that leaf CO_2 assimilation decreased with increasing leaf Al concentration, F_o, V_J, V_I, M_o, DI$_o$/ABS, ABS/RC, or DI$_o$/RC, respectively, but it increased with increasing maximum amplitude of IP phase, PI$_{abs,total}$, F_m, F_v/F_m, φ_{Eo}, or φ_{Ro}, respectively. Based on these results, we concluded that Al toxicity impaired the whole photosynthetic electron transport chain from PSII donor side up to the reduction of PSI end acceptors, thus decreasing ETR and photosynthesis.

Xiao et al. (2005b) investigated the effects of 0-, 0.185-, 0.740-, or 1.850-mM Al in the nutrition solution on gas exchange in longan leaves and observed that Al-treated leaves had decreased CO_2 assimilation, light saturation point, CO_2 saturation point, apparent quantum yield, carboxylation efficiency (CE), and photophosphorylation activity but increased dark respiration, light compensation point, CO_2 compensation point, and glycolate oxidase activity. Al at 0.740 and 1.850 mM decreased stomatal conductance, while Al at 0.185 mM increased stomatal conductance. Based on these results, they concluded that the Al-induced decrease in leaf CO_2 assimilation might be caused by a combination of factors such as decreased CE and photophosphorylation and increased photorespiration.

3.3 Reactive oxygen species and methylglyoxal detoxification systems

Aluminum toxicity can trigger reactive oxygen species (ROS) and methylglyoxal (MG) production, thus disturbing redox homeostasis and leading to lipid peroxidation in plant cells. Lipid peroxidation may result in the loss of plasma membrane integrity, hence increasing electrolyte leakage (Yamamoto et al., 2002; Guo et al., 2018a; Dawood et al., 2012; Xiao et al., 2003b). Guo et al. (2018a) observed that the production of superoxide anion and H_2O_2 was increased in Al-treated C. sinensis and C. grandis roots, with greater increases in Al-intolerant C. grandis roots but only in Al-treated C. grandis leaves. In addition, Al toxicity increased the concentration of MG in C. grandis leaves and roots, but not in C. sinensis leaves and roots. Al-induced increases of superoxide anion and H_2O_2 production have been obtained on longan and "Cleopatra" leaves (Xiao et al., 2003b; Chen et al., 2005b) and C. grandis roots and leaves (Guo et al., 2017b). Increased concentration of H_2O_2 was also observed in Al-treated blueberry plantlets cultivated in vitro (Manquián-Cerda et al., 2018). Plants have evolved diverse enzymatic and nonenzymatic detoxification systems to protect cells against oxidative stress due to increased production and accumulation of ROS and MG (Chen and Cheng, 2003; Guo et al., 2017b, 2018a; Yadav et al., 2005). Antioxidant enzymes are the first line of defense against oxidative damage (Manna et al., 2008). Xiao et al. (2003b) investigated the effects of 0-, 0.185-, 0.370-, 0.740-, 1.110-, and 1.850-mM Al in the nutrition solution on ROS metabolism and membrane system in longan leaves and found that Al treatments increased the concentration of MDA and the leakages of K$^+$ and macromolecules, increased or did not affect the activities of glutathione (GR), ascorbate (ASC) peroxidase (APX), superoxide dismutase (SOD), and guaiacol peroxidase (GuPX) activities and the concentration of reduced glutathione (GSH) except for a slight decrease in the activity of SOD and the concentration of GSH at 1.850-mM Al, but the activity of CAT and the concentration of ASC decreased with increasing Al concentration in the nutrient solution. Although the activities of most antioxidant enzymes and the concentration of GSH were kept high, the antioxidant system as a whole did not provide sufficient protection to these leaves against oxidative damage. Schuch et al. (2010) found that shoot SOD activity was increased by 2.2-mM Al at proliferation phase, shoot MDA concentration was increased by 1.7-mM Al or higher at proliferation phase and by 1.1-mM Al or higher at rooting phase, but Al toxicity had no significant influence on shoot CAT activity at proliferation and rooting phases and shoot SOD activity at rooting phase in in vitro cultures of "BA 29" quince (C. oblonga), one of the most widely used pear rootstocks in Europe. Chen et al. (2005b) observed that the activities of antioxidant enzymes—namely, CAT, APX, SOD, GR, monodehydroascorbate reductase (MDAR), and dehydroascorbate reductase (DHAR)—and the concentrations of antioxidant metabolites, namely, GSH, oxidized glutathione (GSSG), ASC, and dehydroascorbate (DHA), in "Cleopatra" tangerine leaves increased in response to Al toxicity, but the concentration of MDA kept unchanged. They concluded that the Al-induced upregulation of the antioxidant system provided considerable protection to Al-treated leaves against photo-oxidative damage. Guo et al. (2018a) reported that Al toxicity decreased CAT, MDHAR, and DHAR activities; increased GuPX and SOD activities; and had no significant influence on GR and APX activities in C. sinensis leaves but it decreased the activities of all the seven enzymes in C. grandis leaves. Al toxicity decreased MDHAR, APX, GR, and SOD activities; increased GuPX activity; and did not alter DHAR and CAT activities in C. sinensis roots, while it decreased GuPX, CAT, GR, MDHAR, and SOD activities, and did not affect DHAR and APX activities in C. grandis roots. Thiol-based antioxidant system is the second line of defense against oxidative damage (Manna et al., 2008). Al-treated C. sinensis leaves had increased

γ-glutamylcysteine synthetase (γGCS) and sulfite reductase (SiR) activities, decreased glutathione S-transferase (GST) activities, and unaltered adenosine 5'-phosphosulfate reductase (APR), ATP sulfurylase (ATPS), glutathione peroxidase (GlPX), cysteine (Cys) synthase, and glutamine synthetase (GS) activities, while Al-treated *C. grandis* leaves had decreased Cys synthase, ATPS, SiR, and APR activities; increased GlPX and GST activities; and unchanged γGCS and GS activities. Al toxicity decreased APR, GS, SiR, and Cys synthase activities; increased γGCS and GST activities; and did not affect GlPX and ATPS activities in *C. sinensis* roots, but it increased GlPX activity; decreased ATPS, Cys synthase, APR, SiR, and GS activities; and did not alter γGCS and GST activities in *C. grandis* roots (Guo et al., 2018a). MG (a cytotoxic compound) is mainly detoxified by the glyoxalase (Gly) system, where GSH serves as a cofactor (Hossain et al., 2012). The activities of Gly I and Gly II in *C. sinensis* and *C. grandis* roots and leaves decreased in response to Al toxicity. Al toxicity increased ASC, DHA, and ASC + DHA concentrations in *C. grandis* and *C. sinensis* leaves and roots; decreased GSH, GSSG, and GSH + GSSG concentrations in *C. grandis* leaves and roots; but had no influence on GSH, GSSG, and GSH + GSSG concentrations in *C. sinensis* leaves and roots. Both the ratios of GSH/(GSH + GSSG) and ASC/(ASC + DHA) were decreased in Al-treated *C. grandis* roots and leaves but unaltered or slightly elevated in Al-treated *C. sinensis* roots and leaves. Al toxicity increased the concentration of MDA and electrolyte leakage in *C. sinensis* and *C. grandis* roots, but it only increased electrolyte leakage in *C. grandis* leaves. Obviously, the detoxification systems of ROS and MG did not provide sufficient protection to Al-treated *C. sinensis* and *C. grandis* roots. The Al-induced increase of electrolyte leakage in *C. grandis* leaves might be caused by the increased concentration of MG, because the concentration of MDA did not differ between Al-treated leaves and controls (Guo et al., 2018a).

Schuch et al. (2010) investigated Al toxic effects on growth and morphogenesis (in leaf explants) in in vitro cultures of "BA 29" quince. Al strongly decreased shoot growth in the proliferation (0.5-mM Al) and rooting (2.2-mM Al) phases, inhibited shoot proliferation (1.1-mM Al), and caused conspicuous tissue browning (2.2-mM Al). Manquián-Cerda et al. (2018) reported that exposure to 100-µM Al for 14–30 days increased SOD activity and MDA concentration; decreased total phenolic concentration; increased or did not alter gallic acid, ellagic acid, and chlorogenic acid concentrations; and increased or did not affect antioxidant capacity in blueberry plantlets cultivated in vitro.

3.4 Nonstructural carbohydrates

Chen et al. (2005a) investigated Al toxic effects on the concentrations of glucose, fructose, sucrose, total soluble sugars (the summation of glucose, fructose, and sucrose), starch, and total nonstructural carbohydrates (TNC; the summation of glucose, fructose, sucrose, and starch) in "Cleopatra" tangerine leaves at dusk and predawn and observed that Al had little effect on the concentrations of all these nonstructural carbohydrates despite greatly decreased CO_2 assimilation. Decreased utilization of carbohydrates for growth and the translocation of carbohydrates from the leaves might be responsible for the little alteration in the concentrations of nonstructural carbohydrates. In another study with *C. grandis* seedlings, Yang et al. (2012a) found that Al-treated leaves had increased concentrations of glucose and fructose and decreased concentrations of sucrose, starch, and TNC, while Al-treated roots had increased concentrations of glucose, fructose, and sucrose; decreased concentration of sucrose; and unaltered concentration of TNC.

Reyes-Díaz et al. (2010) reported that under long-term Al toxicity, the concentration of total soluble carbohydrates in leaves decreased in the Al-tolerant "Legacy" blueberry, but not in the Al-sensitive "Bluegold" blueberry. In a greenhouse experiment with "Nemaguard" peach seedlings, Graham (2002) observed that 8 weeks of treatment with 1-mM Al increased the concentrations of sucrose in roots and leaves, starch in stems, and prunasin in roots; decreased the concentrations of glucose in stems and roots, fructose in roots and stems, sorbitol in leaves, and total soluble sugars in roots and leaves; but had no influence on the concentration of TNC (the summation of glucose, fructose, sucrose, sorbitol, and starch) in roots, stems, and leaves. In a study with longan seedlings, Wan (2007) observed that total C level in roots, stems, and leaves decreased and the concentrations of total soluble sugar, reducing sugar, and sucrose concentrations in roots and stems increased as Al concentration in the nutrient solution increased from 0 to 1.85 mM. Interestingly, the concentrations of total soluble sugar, reducing sugar, and sucrose concentrations in leaves were increased by Al concentration up to 0.37 mM and then decreased with increasing Al concentration.

3.5 Organic acids

Chen et al. (2009) investigated the responses of organic acid metabolism to Al and P interactions in *C. grandis* roots and leaves. Leaf concentrations of malate and citrate decreased with increasing P supply under Al stress, but they kept unchanged in response to P supply without Al stress. Al toxicity increased leaf concentrations of malate under 50-µM

P and citrate under 50- and 100-µM P but decreased leaf concentration of malate under 500-µM P. Root concentrations of malate and citrate did not significantly differ among P and Al combinations except for a slight decrease in malate concentration under 50-µM P + 1.2-mM Al and an increase in malate and citrate concentrations under 50-µM P + 0-mM Al. In most cases, the activities of aconitase (ACO), citrate synthase (CS), phosphoenolpyruvate carboxylase (PEPC), phosphoenolpyruvate phosphatase (PEPP), NADP-isocitrate dehydrogenase (NADP-IDH), pyruvate kinase (PK), NAD-malate dehydrogenase (NAD-MDH), and NADP-malic enzyme (NADP-ME) were more affected by P and Al interactions in the leaves relative to the roots. This might be associated with the greater changes in leaf Al *versus* root Al. Similarly, organic acid metabolism was more affected by nitric acid (NO)-Al (B-Al) interactions in leaves than that in roots (Tang et al., 2011; Yang et al., 2012c).

Wan (2007) reported that, when Al concentration in the nutrient solution increased from 0 to 1.85 mM, the concentrations of malate, citrate, and oxalate increased in longan roots and decreased in longan stems and leaves.

4 Factors affecting Al tolerance of fruit trees

4.1 Species and cultivars

In a sand culture experiment, Jiang et al. (2009c) examined Al toxic effects on seedling growth of 12 *Citrus* species and cultivars and observed that "Xuegan" (*C. sinensis*), "Chachiensis," "Ponkan," "Fuju," "Sichuanhongju," "Eureka," and *P. trifoliata* were relatively Al tolerant; "Goutoucheng" from Zhejiang, sour orange from Hubei, Red limonia, and citrumelo were relatively Al sensitive; and "Wendanyou" was the most Al sensitive. Magalhães (1987) reported that "Rugoso da Florida" lemon was the most tolerant to Al, followed by "Cleopatra" tangerine and "Rangpur" lime. Based on new-growth fresh weight of whole plants, the Al tolerance for five *Citrus* rootstock seedlings was as follows: "Cleopatra" tangerine > rough lemon > sour orange > "Swingle" citrumelo > "Carrizo" citrange (Lin and Myhre, 1991a). Pereira et al. (2003) used hydroponic culture to identify the Al tolerance of "Rangpur" lime, "Volkamer" lemon, "Cleopatra" tangerine, and "Sunki" tangerine and found that "Rangpur" lime ("Cleopatra" tangerine) was the most sensitive (tolerant) rootstock to Al. A hydroponic culture approach was used to test the Al tolerance of five *Citrus* species/cultivars—namely, *Citrus nobilis* var. *microcarpa*; "Volkamer" lemon; and "Ngot," "Banhxe," and "Ngang" pummelo. "Banhxe" pummelo was the least Al tolerant, and "Volkamer" lemon was the most tolerant among the five *Citrus* species/cultivars (Toan et al., 2003). Liao et al. (2015) investigated the Al tolerance of four scion–rootstock combination (i.e., X/X and X/SP, "Xuegan" grafted on "Xuegan" and "Sour" pummelo, respectively, and SP/X and SP/SP, "Sour" pummelo grafted on "Xuegan" and "Sour" pummelo, respectively) plants and concluded that the Al tolerance of grafted *Citrus* plants depended on the scion and rootstock genotype and the scion-rootstock interaction.

Pineapple is especially well adapted to acid soils and can tolerate high levels of soluble soil Al and Mn. Le Van and Masuda (2004) reported that all seven pineapple cultivars—namely, "Cayenne," "Queen," "Soft Touch," "Honey Bright," "Bogor," "Red Spanish," and "Cream Pine"—could tolerate 100-µM $AlCl_3$ in the nutrient solution. Expose of pineapple roots to 300-µM $AlCl_3$ indicated that "Soft Touch" and "Cayenne" were the most Al-sensitive and Al-tolerant cultivars, respectively. After exposure to 300-µM $AlCl_3$ for 72 h, the root length of "Tainung No. 17," "Tainung No. 6," "Tainung No. 13," and "Cayenne" pineapple were 73%, 85%, 93%, and 115% of those without Al, respectively. Obviously, "Tainung No. 17" and "Cayenne" were the most Al-sensitive and Al-tolerant cultivars, respectively (Lin and Chen, 2011). Al treatments decreased Chl index; root length; root DW; and leaf, stem, and root accumulation of N, P, K, Ca, and Mg in "Vitória" pineapple plants, but not in "IAC Fantástico" ones. Obviously, "IAC Fantástico" was more tolerant to Al than "Vitória" (Mota et al., 2016).

Rufyikiri et al. (2002) investigated Al toxic effects on root and shoot yield of five banana cultivars—namely, "Grande Naine," "Agbagba," "Obino l'Ewaï," "Kayinja," and "Igitsir." The Al-induced relative decrease in root and shoot DW was greater in "Kayinja" than that in the other cultivars.

Naing et al. (2009) evaluated the Al tolerance of "Carabao," "Pahutan," "Pico," and "Kachamitha" mango seedlings. "Kachamitha" was the most tolerant to high soil Al level.

Reyes-Diaz et al. (2009) investigated the effects of short-term Al toxicity on root growth, root and leaf Al concentrations, and leaf pigments and Chl a fluorescence parameters in three blueberry cultivars and found that "Brigitta" was the most Al tolerant, followed by "Legacy," and "Bluegold" was highly sensitive to Al toxicity.

Wojcik (2003) observed that 100-µM $AlCl_3$ decreased the biomass production of "M.26" and "P22" apple rootstock by 41% and 22%, respectively, at 20 µM B, indicating that "P22" was more tolerant to Al than "M.26."

Cançado et al. (2009) used the phenotypic indexes of relative root growth, root fresh and dry weight, area, hematoxylin staining profile, and Al concentration to evaluate the Al tolerance of six grape rootstock genotypes. "Kober 5BB," "Gravesac," "Paulsen 1103," and "IAC 766" were the most Al-tolerant genotypes and "IAC 572" and "R110" the most Al-sensitive ones.

To conclude, wide differences in Al tolerance exist in various fruit trees species and/or cultivars (Table 17.1).

TABLE 17.1 Aluminum tolerance of fruit trees.

Fruit trees	System	Al tolerance	References
Citrus spp.	Sand culture	**Al-tolerant species/cultivars:** *C. reticulata* cv. Chachiensis, *C. reticulata* cv. Ponkan, *C. reticulata* cv. Fuju, *C. reticulata* cv. Sichuanhongju, *C. limon* cv. Eureka, and *Poncirus trifoliata* **Relatively Al-sensitive species/cultivars:** *C. aurantium* cv. Goutoucheng from Zhejiang, sour orange (*C. aurantium*) from Hubei, *C. limonia* cv. Red limonia, and citrumelo (*P. trifoliata* × *C. grandis*) **The most Al-sensitive cultivar:** *C. grandis* cv. Wendanyou	Jiang et al. (2009c)
	Soil culture	"Rugoso da Florida" lemon (*C. jambhiri*) was the most tolerant to Al, followed by "Cleopatra" tangerine (*C. reshni*) and "Rangpur" lime (*C. limonia*)	Magalhães (1987)
	Solution culture	"Cleopatra" tangerine > rough lemon (*C. jambhiri*) > sour orange > "Swingle" citrumelo > "Carrizo" citrange (*C. sinensis* × *P. trifoliata*)	Lin and Myhre (1991a)
	Solution culture	Among the four *Citrus* rootstocks—namely, "Rangpur" lime, "Volkamer" lemon (*C. volkameriana*), "Cleopatra" tangerine, and "Sunki" tangerine (*C. sunki*), "Rangpur" lime and "Cleopatra" tangerine were the most sensitive and tolerant rootstocks to Al, respectively.	Pereira et al. (2003)
	Solution culture	Among the five *Citrus* species/cultivars—namely, *C. nobilis* var. *microcarpa*; "Volkamer" lemon; and "Ngot," "Banhxe," and "Ngang" pummelo (*C. grandis*), "Banhxe" pummelo was the least Al tolerant, and "Volkamer" lemon was the most tolerant	Toan et al. (2003)
	Sand culture	Al-induced decrease of photosynthesis was less pronounced in X/X ("Xuegan" [*C. sinensis*] grafted on "Xuegan") and X/SP ("Xuegan" grafted on "Sour pummelo" [*C. grandis*]) leaves than in SP/SP ("Sour" pummelo grafted on "Sour" pummelo) and SP/X ("Sour" pummelo grafted on "Sour" pummelo) leaves	Liao et al. (2015)
Pineapple (*Ananas comosus*)	Solution culture	Among "Cayenne," "Queen," "Soft Touch," "Honey Bright," "Bogor," "Red Spanish," and "Cream Pine," "Soft Touch" and "Cayenne" were the most Al-sensitive and Al-tolerant cultivars, respectively	Le Van and Masuda (2004)
	Solution culture	Among "Tainung No. 17," "Tainung No. 6," "Tainung No. 13," and "Cayenne," "Tainung No. 17" and "Cayenne" were the most Al-sensitive and Al-tolerant cultivars, respectively	Lin and Chen (2011)
	Solution culture	"IAC Fantástico" > "Vitória"	Mota et al. (2016)
Banana (*Musa* spp.)	Solution culture	Al-tolerance was greater in "Kayinja" than that in "Grande Naine," "Agbagba," "Obino l'Ewaï," and "Igitsir"	Rufyikiri et al. (2002)
Mango	Soil culture	Among "Carabao," "Pahutan," "Pico," and "Kachamitha," "Kachamitha" was the most tolerant to Al	Naing et al. (2009)
Blueberry (*Vaccinium corymbosum*)	Medium (1 oat:1 shell sawdust: pine needles by volume) culture	"Brigitta" > "Legacy" > "Bluegold"	Reyes-Diaz et al. (2009)
Apple (*Malus* spp.) rootstocks	Medium (perlite) culture	"P22" > "M.26"	Wojcik (2003)
Grape (*Vitis* spp.) rootstocks	Solution culture	"Kober 5BB," "Gravesac," "Paulsen 1103," and "IAC 766" were the most Al-tolerant genotypes, and "IAC 572" and "R110" the most Al-sensitive ones	Cançado et al. (2009)

4.2 Aluminum uptake and distribution

Tolerance of higher plants to Al toxicity is associated not only with low Al uptake but also with relatively little Al translocation from roots to shoots. There is evidence showing that some Al-tolerant plant species can accumulate more Al in roots and less Al in shoots than Al-sensitive ones (Roy et al., 1988). In a study with longan seedlings, Xiao (2005b) reported that the distribution of Al in roots increased from 58.5% to 75.1% as Al concentration in the nutrient solution increased from 0 to 1.850 mM, since the Al-induced increase in Al concentration was greater in roots than that in stems and leaves, and that over 86% of Al accumulated in leaf cell walls. This might be an adaptive strategy of longan seedlings to Al toxicity. Le Van and Masuda (2004) showed that the higher Al tolerance of "Cayenne" pineapple might be associated with the lower Al concentration in root apices. Lin and Chen (2011) found that, at 200- or 300-μM AlCl$_3$, Al uptake by cell wall of Al-tolerant "Cayenne" root tips was less than that by cell wall of Al-sensitive "Tainung No. 17" root tips. Subsequent study showed that "Tainung No. 17" pineapple, which contained the highest functional (carboxylic and phenolic) groups with strong Al adsorption in its cell wall of root apices, had the most uptake of Al by the cell wall of root apices, followed by "Tainung No. 6" and "Tainung No. 13," and that "Cayenne," which contained the lowest carboxylic and phenolic groups, had the lowest uptake of Al (Lin and Chen, 2013). Al-tolerant "IAC 572" and "Gravesac" grape rootstocks accumulated more Al in roots when soil Al level increased (Alvarenga et al., 2004). However, there were no major differences in Al root concentration among six grape rootstock genotypes with contrasting Al tolerance in the absence or presence of Al (Cançado et al., 2009). Lin and Myhre (1991b) observed that the concentration of Al in roots was higher in Al-tolerant *Citrus* rootstocks that that in Al-sensitive ones, but no systematic relation existed between the concentration of Al in shoots and the Al tolerance of shoot growth. Under Al toxicity, the Al-tolerant *C. sinensis* seedlings accumulated more Al in roots and less Al in shoots than the Al-sensitive *C. grandis* seedlings (Yang et al., 2011, 2018). However, under Al toxicity, leaf Al concentration was higher in the Al-tolerant "P22" apple rootstock than that in the less Al-tolerant "M.26" apple rootstock, and root Al concentration was not higher in the former (Wojcik, 2003). Reyes-Diaz et al. (2009) observed that, under short-term Al toxicity (24 h), the concentration of Al in roots and leaves was higher in the most Al-tolerant "Brigitta" blueberry cultivar than that in the other two cultivars ("Legacy" and "Bluegold"). Similarly, the concentration of Al concentration was much higher in Al-tolerant "Legacy" than that in Al-sensitive "Bluegold" after exposure to Al for 48 h (Ulloa-Inostroza et al., 2017). However, in another experiment, Reyes-Díaz et al. (2010) reported that, under long-term Al toxicity (7–20 days), the concentration of Al in roots (leaves) was less (higher) in the Al-tolerant "Legacy" than that in the Al-sensitive "Bluegold."

4.3 Secretion of organic acid anions

The Al-induced secretion of organic acid anions, mainly citrate, malate, and oxalate, from roots is the best-characterized mechanism of Al tolerance in higher plants (Kochian et al., 2004; Yang et al., 2013). Yang et al. (2011) observed that the Al-tolerant *C. sinensis* seedling roots secreted more citrate and malate than the Al-sensitive *C. grandis* seedling roots in response to Al toxicity. Liao et al. (2015) found that *C. sinensis* rootstock alleviated the Al toxic effects on leaf photosynthetic electron transport chain and CO$_2$ assimilation in SP/X plants through increasing the Al-induced secretion of organic acid anions, while the revers was the case for *C. grandis* rootstock in X/SP plants. Using the rapid amplification of cDNA ends and the Y-shaped adaptor-dependent extension methods, Deng et al. (2008) cloned a mitochondrial dicarboxylate/tricarboxylate carrier gene, *CjDTC*, from Yuzu (*Citrus junos*). qRT-PCR analysis showed that *CjDTC* was induced by Al. They suggested that CjDTC protein might play a role in the Al-induced excretion of organic acid anions and rhizotoxic Al tolerance. Deng et al. (2009) observed that Al activated the secretion of citrate from *Yuzu* roots. In addition, Al increased mitochondrial CS (CjCS) gene expression level, CjCS activity, and citrate concentration in roots. *Nicotiana benthamiana* plants overexpressing a Yuzu *CjCS* showed enhanced Al tolerance accompanied by increased CjCS activity and citrate concentration in roots and Al-induced secretion of citrate from roots. They concluded that the Al-induced secretion of citrate from roots played a role in the Al tolerance of Yuzu. Zhang et al. (2008) observed that transgenic tobacco (*Nicotiana tabacum* cv. Xanthi) plant overexpressing a Yuzu malate dehydrogenase gene displayed enhanced Al tolerance.

Cançado et al. (2009) observed that Al-resistant "Kober 5BB" grape rootstock had a three higher constitutive exudation of citrate than that of Al-sensitive "R110" grape rootstock. The exudation of citrate in "Kober 5BB" was independent from the absence or presence of Al. They suggested that the exudation of citrate in "Kober 5BB" might confer Al tolerance/amelioration in grape. Xiao (2005a) reported that citrate supply greatly decreased the concentration of Al in Al-treated longan roots, stems, and leaves. Le Van and Masuda (2004) observed that Al-tolerant "Cayenne" pineapple roots grown in the nutrient solution without Al excluded malate and a large amount of succinate but undetectable citrate and that Al induced the secretion of citrate from roots, but had no significant influence on succinate and malate. They suggested that the Al-induced secretion of citrate might be involved in the higher Al tolerance of "Cayenne." In another study, Lin and Chen (2011) found that Al uptake by cell walls of "Cayenne" root tips were reduced by 18% and 31%, respectively, when its root tips were pretreated with 1 and 10 mM of malate.

4.4 Mineral nutrients

Increasing evidence shows that mineral nutrients can alleviate Al toxicity in many higher plants including some fruit trees (*Citrus*, apple, longan, blueberry, and raspberry) (Table 17.2; Li et al., 2018; Riaz et al., 2018a; Rahman et al., 2018; Wojcik, 2003).

TABLE 17.2   Mineral nutrient–induced alleviation of Al toxicity in fruit trees

Fruit trees	Nutrients	Nutrient-induced alleviation of Al toxicity	References
"Sour" pummelo (*C. grandis*)	50-, 100-, 250-, and 500-μM KH_2PO_4	Alleviating Al-induced inhibition of growth and photosynthesis via increasing Al immobilization in roots and P level in roots and shoots	Jiang et al. (2009a)
	0-, 50-, 100-, and 200-μM KH_2PO_4	Alleviating Al toxicity by increasing immobilization of Al in roots and P level in seedlings rather than by increasing organic acid anion secretion	Yang et al. (2011)
Longan (*Dimocarpus longan*)	0- and 2-mM KH_2PO_4	Decreasing Al concentrations in roots, stems, and leaves	Xiao (2005a)
"Sour" pummelo	2.5-, 10-, 25-, and 50-μM H_3BO_3	(a) The sequence of the ameliorative effect of B on Al toxicity was 25 μM B > 10 μM B ≥ 50 μM B > 2.5 μM B (b) B-induced amelioration of root inhibition was probably caused by B-induced changes in Al speciation and/or subcellular compartmentation (c) B-induced amelioration of shoot growth and photosynthesis inhibition could be caused by less Al accumulation in the stems and leaves	Jiang et al. (2009b)
Poncirus trifoliata	2.5- and 20-μM H_3BO_3	(a) Isolating 127 differentially expressed genes in leaves (b) Activating multiple signal transduction pathways (c) Inducing genes related to transport and energy production (d) Improving amino acid accumulation and protein degradation (e) Genes related to nucleic acid metabolism, stress responses, and cell wall modification might play a role in the B-induced alleviation of Al toxicity	Wang et al. (2015)
	2.5- and 20-μM H_3BO_3	(a) Isolating 100 differentially expressed genes in roots (b) Genes related to detoxification of ROS and aldehydes, cellular transport, gene regulation, cell wall modification, Ca signal, and hormone might play a role in the B-induced alleviation of Al toxicity	Zhou et al. (2015)
	2.5- and 20-μM H_3BO_3	(a) Identifying 61 differentially abundant proteins in roots (b) Mediating cell wall remodeling and modification (c) Elevating the abundances of proteasomes and stress response proteins (d) Reinforcing signal transduction	Yang et al. (2018)
	0-, 10-, and 50-μM H_3BO_3	Promoting root elongation via (a) reducing Al accumulation to cell wall and the distribution of homogalacturonan (HG) epitopes in the roots and (b) mitigating Al-induced alterations of cell wall structure	Riaz et al. (2018b)
	0- and 10-μM H_3BO_3	Alleviating Al-induced inhibition of root elongation through decreasing Al uptake, cell wall modification, and oxidative stress	Riaz et al. (2018c)

TABLE 17.2 Mineral nutrient–induced alleviation of Al toxicity in fruit trees—cont'd

Fruit trees	Nutrients	Nutrient-induced alleviation of Al toxicity	References
	0- and 10-μM H_3BO_3	(a) Decreasing ROS and MDA levels and increasing antioxidant enzyme activities (b) Reducing Al concentration in roots and leaves (c) Decreasing root cell damage (d) Promoting root growth	Riaz et al. (2018d)
	0- and 10-μM H_3BO_3	(a) Alleviating Al-induced accumulation of MDA in roots through regulating antioxidant enzyme activities and decreasing Al and H_2O_2 concentrations (b) Enhancing root elongation, plant height, and biomass	Yan et al. (2018a)
	0.1- and 10-μM H_3BO_3	Alleviating Al toxicity through (a) shielding potential Al binding sites and (b) decreasing Al-induced alterations of cell wall cellulose and pectin components	Yan et al. (2018b)
"P22" (apple rootstock)	20-, 40-, and 60-μM H_3BO_3	(a) Decreasing the uptake of Al and increasing the uptake of P, Mg, and Ca (b) Increasing root and leaf DW	Wojcik (2003)
Blueberry	0-, 2.5-, and 5-mM $CaSO_4$	(a) Increasing Ca and S levels, decreasing Al and MDA levels in roots and leaves (b) Improving leaf photosynthesis	Reyes-Díaz et al. (2011)
	0- and 5-mM $CaSO_4$	Increasing antioxidant capacity	Meriño-Gergichevich et al. (2015)
Malus domestica cv. Gala/MM 106	Calcitic lime (Ca-lime), phosphogypsum, $CaCl_2$, and magnesite (Mg-lime)	(a) Decreasing exchangeable Al in soils (b) Increasing Ca (Mg) in soils and leaves (c) Increasing root development and yielding more and larger fruits	Pavan et al. (1987)
Raspberry (*Rubus* spp.)	9 t/ha lime	(a) Elevating soil pH (b) Decreasing Al, Fe, Mn, and Fe levels both in soils and raspberry plants	Sikirić et al. (2009)
Longan	0- and 2-mM $CaCl_2$ and $MgSO_4$	Lowering Al levels in roots, stems, and leaves	Xiao (2005a)
"Sour" pummelo	0.5-mM $MgSO_4$ and 0.5-mM $MgSO_4$+0.5-mM Na_2SO_4	(a) Reducing the transport of Al from roots to shoots and the accumulation of Al in leaves (b) Promoting the Al-induced secretion of citrate (c) Enhancing leaf and root RWC and P, Ca, and Mg contents per plant (d) Alleviating the Al-induced increases in MDA concentration and electrolyte leakage through decreasing ROS production and increasing antioxidant capacity	Guo et al. (2017b)

4.4.1 Phosphorus

Using *C. grandis* seedlings, Jiang et al. (2009a) and Yang et al. (2011) found that P alleviated the Al-induced decreases in root, stem, and leaf DW and P concentration, leaf CO_2 assimilation, Chl concentration and Rubisco activity, F_v/F_m, φ_{Eo}, φ_{Ro}, ET_o/RC, RE_o/RC, ψ_{Eo}, δ_{Ro}, maximum amplitude of IP phase, and $PI_{abs,total}$; increases in leaf F_o, relative variable fluorescence at K- and I-steps, ABS/RC, and TR_o/RC; and alterations in OJIP transients. Under Al stress, P increased root Al concentration but decreased stem and leaf Al concentration. Al-induced secretion of citrate and malate from Al-treated excised roots increased with decreasing P supply. Based on these results, they concluded that P alleviated Al-induced inhibition of growth and impairment of the whole photosynthetic electron transport chain from PSII donor side up to the reduction of end acceptors of PSI, thus preventing photosynthesis inhibition through increasing Al immobilization in roots and P level in roots and shoots rather than through increasing organic acid anion secretion. Yang et al. (2012b) used qRT-PCR to examine the effects of Al and P interactions on the expression of genes involved in alternative glycolytic pathways, P scavenging and recycling in *Citrus* roots, and concluded that the tolerance of *Citrus* to Al and/or P deficiency might be caused by a complex coordinated regulation of gene expression involved in these pathways. Xiao (2005a) reported that K_2PO_4 alleviated the Al-induced increase in Al concentrations in roots, stems, and leaves of longan seedlings.

4.4.2 Boron

Jiang et al. (2009b) observed that B alleviated the Al toxicity–induced inhibition of growth and decreases in leaf Chl concentrations, CO_2 assimilation, Rubisco activity, and alterations of OJIP transients and related fluorescence parameters in *C. grandis* seedlings. The ameliorative effect of B was not caused by increased concentration of B in the roots, stems, and leaves, because B concentration increased in response to Al toxicity. Generally speaking, the sequence of the ameliorative effect of B on all these parameters was 25 µM B > 10 µM B ≥ 50 µM B > 2.5 µM B. At the absence of Al, root DW was slightly less at 50-µM B than at 10- and 25-µM B, implying that 50-µM B-treated plants received excess B. This might be the cause why the ameliorative effect of 50-µM B was lower than that of 25-µM B. The B-induced amelioration of root inhibition was probably caused by B-induced changes in Al speciation and/or subcellular compartmentation because root Al concentration did not differ among B treatments. By contrast, the B-induced amelioration of shoot growth and photosynthesis inhibition, impairment of photosynthetic electron transport chain, and decreases of Rubisco activity and Chl concentrations could be caused by less Al accumulation in the stems and leaves. Recently, several studies showed that B alleviated Al toxicity of *P. trifoliata* seedlings through keeping cell wall stability, decreasing Al uptake and root impairment, stimulating antioxidant enzyme activities and preventing oxidative damage, and improving root growth (Riaz et al., 2018b,c,d; Yan et al., 2018a,b). Wang et al. (2015) and Zhou et al. (2015) used a cDNA-amplified fragment length polymerization (cDNA-AFLP) approach to examine the molecular mechanisms of the alleviative actions of B against Al toxicity in *C. grandis* roots and leaves. The ameliorative effects of B on Al toxicity might be associated with (a) inducing multiple signal transduction pathways, (b) upregulating the expression of cell transport–related genes, (c) activating genes related to lipid and energy metabolisms, (d) increasing the capacity of ROS and aldehyde detoxification, and (e) enhancing protein degradation and amino acid accumulation. Also, genes related to nucleic acid and cell wall metabolisms might play a role in B-induced alleviation of Al toxicity. In another study, Yang et al. (2018) used a two-dimensional electrophoresis (2-DE)–based MS approach to investigate the mechanisms underlying the B-induced alleviation of Al toxicity in *C. grandis* roots. This occurred through (a) mediating cell wall remodeling and modification, (b) promoting turnover of dysfunctional proteins, (c) enhancing the abundance of proteins involved in stress response, and (d) reinforcing signal transduction.

Wojcik (2003) reported that supraoptimal B could alleviate Al toxicity in "P22" apple rootstock through decreasing the concentration of Al in leaves and shoots, the uptake rate of Al, and the translocation rates of Al to shoots and increasing the concentrations of P in roots and leaves; Ca in roots, stems, and leaves; and Mg in leaves and the uptake rates of P, Mg, and Ca, but not in "M.26" apple rootstock.

4.4.3 Calcium and magnesium

Aluminum can compete with Mg and Ca in the root exchange sites, which has been regarded as a possible cause for Al toxicity to higher plants. Rufyikiri et al. (2002) assayed the cation exchange capacity of roots (CECR) of five banana cultivars—namely, "Grande Naine," "Agbagba," "Igitsiri," "Obino l'Ewaï," and "Kayinja" in two tropical soils and in nutrient solutions with and without Al. The CECR of the five cultivars were similar in the two soils and in the nutrient solution without Al, but it was higher in the nutrient solution with Al. There was a positive between root total Al concentration and CECR. The most Al-sensitive "Kayinja" had a greater CECR than the others. The Cu-extractable Mg was substantially decreased in the nutrient solution with Al, whereas the Cu-extractable Ca was little altered. They proposed that Al/Mg ratio in root exchange sites was a better indicator of Al toxicity than Al/Ca ratio. In a factorial experiment with "Nemaguard" peach seedlings at Al concentrations of 0, 50, or 100 mg/L and a Ca concentration of 0, 50, and 100 mg/L in the nutrient solution, Chibiliti and Byrne (1990) observed that an antagonism existed between Al and Ca uptake, and that shoot and root DW, shoot length, number of branches, stem caliper, and number of leaves decreased in response to Al toxicity. Interestingly, Ca only alleviated toxicity at 50-mg/L Al. In a solution-culture experiment with three blueberry cultivars (Al-tolerant "Legacy" and "Brigitta," and Al-sensitive "Bluegold"), Reyes-Díaz et al. (2011) found that $CaSO_4$ could alleviate Al toxicity through increasing Ca and S concentrations in leaves and roots and CO_2 assimilation, actual quantum yield of PSII electron transport (Φ_{PSII}) and ETR in leaves, and decreasing Al and MDA concentrations in leaves and roots. A study with two blueberry cultivars ("Elliot" and "Jersey") indicated that $CaSO_4$ supply might provide an important approach for alleviating Al toxic effect through enhancing the antioxidant capacity in fruit trees planted in acid soils with high active Al (Meriño-Gergichevich et al., 2015). In a field experiment with apple (*Malus domestica* cv. Gala/MM 106) plants, Pavan et al. (1987) found that Ca (Mg) supply could enhance Ca (Mg) in soils and leaves and decrease exchange Al in soils, thus increasing root density and producing more and larger fruits. In a solution culture with longan seedlings, Xiao (2005a) found that $CaCl_2$ and

$MgSO_4$ decreased the concentration of Al in Al-treated roots, stems, and leaves. Lime supply (9 t/ha) increased the pH of acid soils (pH 3.79), thus lowering Al, Fe, Mn, and Fe concentrations both in soils and raspberry plants (Sikirić et al., 2009).

4.4.4 Sulfur

In a study with *C. grandis* seedlings, Guo et al. (2017b) observed that S supply could alleviate the Al-induced decreases of growth and photosynthesis in *C. grandis* seedlings via (a) decreasing the transport of Al from roots to shoots and leaf accumulation of Al; (b) increasing the Al-induced secretion of citrate by roots, leaf and root RWC, and P, Ca, and Mg contents per plant; and (c) decreasing the Al-stimulated ROS production and increasing the activities of S metabolism-related enzymes and antioxidant enzymes and the concentrations of ASC and GSH in roots and leaves, thus lowering the Al-induced increases in MDA concentration and electrolyte leakage.

4.5 Nitric oxide and methyl jasmonate

4.5.1 Nitric oxide

There was a study with *C. grandis* seedlings showing that NO alleviated the Al toxicity–induced decreases in root, stem, and leaf DW, leaf CO_2 assimilation, and concentrations of P and pigments; increases in root and leaf concentration of MDA and Al, and alterations of ASC, DHA, GSH, and GSSH concentrations and antioxidant enzyme activities in roots and leaves; and OJIP transients and related fluorescence parameters in leaves, but it increased root Al and the Al-induced secretion of malate and citrate from roots. The ameliorative effect of NO on Al toxicity occurred through (a) enhancing Al immobilization and P concentration in roots and the Al-induced secretion of malate and citrate from roots and decreasing Al accumulation in shoots and (b) preventing the Al-induced accumulation of MDA in roots and leaves.

4.5.2 Methyl jasmonate

Ulloa-Inostroza et al. (2017) examined the effects of 5-, 10-, and 50-μM methyl jasmonate (MeJA) applied prior to or simultaneously with Al on biochemical and physiological properties in roots and leaves of two blueberry cultivars (Al-tolerant "Legacy" and Al-sensitive "Bluegold"). Low MeJA concentration applied simultaneously with Al decreased root and leaf concentrations of Al, MDA, and H_2O_2 of the two cultivars. Low MeJA concentration decreased oxidative damage of "Legacy" mainly via increasing nonenzymatic compounds and "Bluegold" by increasing SOD activity. The higher MeJA concentration might be potentially toxic.

4.6 Efficient maintenance of redox homeostasis via reactive oxygen species and methylglyoxal detoxification systems

Reyes-Díaz et al. (2010) observed that, under long-term Al toxicity, the highly Al-sensitive "Bluegold" blueberry had lower relative growth rate (RGR) and radical scavenging activity (RSA) in roots and leaves and higher MDA concentration in roots in contrast to Al-tolerant "Legacy," indicating that RSA played an important in the long-term acclimation response to Al toxicity. Inostroza-Blancheteau et al. (2011b) observed that the Al-tolerant "Brigitta" blueberry had a higher capacity to control Al toxicity–induced oxidative damage, as indicated by several enhanced antioxidant and other physiological properties—namely, RSA, activities of SOD and CAT, F_v/F_m, ETR, Φ_{PSII}, and nonphotochemical quenching (NPQ), than the Al-sensitive "Bluegold" blueberry.

Guo et al. (2018a) observed that Al toxicity decreased the ratios of GSH/(GSH+GSSG) and ASC/(ASC+DHA) in Al-intolerant *C. grandis* roots and leaves, but not in Al-tolerant *C. sinensis* ones, and that the activities of most enzymes involved in the detoxification of ROS and MG and the concentrations of S-containing compounds (GSH, GSH+GSSG, phytochelatins, and metallothioneins) were higher in Al-treated *C. sinensis* roots and leaves than those in Al-treated *C. grandis* ones. They concluded that the higher detoxification capacity of ROS and MG might contribute to the higher Al tolerance of *C. sinensis*.

4.7 Other mechanisms

At 308-μM Al or higher, some *Citrus* rootstock root caps were covered with black gelatinous material. Rootstocks with higher Al tolerance had more of this kind of root cap. It appeared that the black gelatinous material might play a role in Al tolerance of *Citrus* rootstocks (Lin and Myhre, 1991a).

There is evidence showing that callose concentration in root tips positively relate with Al-induced inhibition of root elongation and Al-induced callose formation can be used as a selection criterion for Al sensitivity (Horst et al., 1997; Massot et al., 1999). Le Van and Masuda (2004) found that, up to 2 days of Al treatment, callose was not detected in the root apices of Al-sensitive "Soft Touch" and Al-tolerant "Cayenne" pineapple treated with or without 300-μM AlCl$_3$ and that, after exposure to 300-μM AlCl$_3$ for 3 days, callose concentration was fivefold more in "Soft Touch" root tips than that in "Cayenne" root tips. Similarly, no callose was detected in the root tips of Al-tolerant "Cayenne" and Al-sensitive "Tainung No. 17" pineapple grown in the absence of Al. When treated with 300-μM AlCl$_3$, callose concentration was much higher in "Tainung No. 17" root tips than that in "Cayenne" root tips (Lin and Chen, 2011).

Rootstock grapevine "Gravesac" and "Paulsen 1103" had mechanisms of Al exclusion by raising the pH in the culture medium (Burkhardt et al., 2008).

5 Molecular aspects of Al tolerance in fruit trees

Using a cDNA-AFLP method, Inostroza-Blancheteau et al. (2011a) isolated 70 Al-responsive transcript-derived fragments (TDFs) from 2 cDNA libraries established using Al-resistant "Brigitta" and Al-sensitive "Bluegold" blueberry roots. Several genes involved in stress responses—namely, vacuolar H$^+$-pyrophosphatase, GST, aldehyde dehydrogenase, and S-adenosylmethionine decarboxylase genes—were identified. In another study, Inostroza-Blancheteau et al. (2013) cloned a calmodulin gene, VcCaM1 (Vaccinium corymbosum Calmodulin 1) induced by Al toxicity. VcCaM1 might indirectly improve plant Al tolerance via regulating Ca^{2+} homeostasis and antioxidant systems in leaves.

Using a 2-DE-based MS approach, Chen and Lin (2010) identified 17 Al-responsive proteins from Al-tolerant pineapple cultivar ("Cayenne") roots. Most of these differentially abundant proteins were involved in the production of organic acids and energy and the maintenance of redox homeostasis and root morphology. This might be an adaptive mechanism of pineapple to Al toxicity via providing more energy and organic acids for the maintenance of root morphology and the chelation of Al under Al stress.

Guo et al. (2017a, 2018b) used RNA-Seq to reveal Al-responsive genes in roots and leaves of two Citrus species (Al-tolerant C. sinensis and Al-intolerant C. grandis) with contrasting Al tolerance. They isolated 496 downregulated and 116 upregulated and 224 downregulated and 118 upregulated genes in Al-treated C. grandis and C. sinensis leaves, respectively. Most of these Al-responsive genes were isolated only in C. grandis leaves or C. sinensis leaves. Only 112 Al-responsive genes with the same accession number were isolated simultaneously in C. grandis and C. sinensis leaves. Similar results have been obtained in Al-treated C. grandis and C. sinensis roots. Further analysis indicated that the higher Al tolerance of C. sinensis might be associated with the following several aspects: (a) Al-treated C. sinensis roots and leaves had higher capacity to keep the homeostasis of cellular energy and P, the integrity of cell wall, and the stability of lipid metabolism than did Al-treated C. grandis ones; (b) Al-induced generation of ROS and the other cytotoxic compounds was greater in Al-treated C. grandis leaves and roots than that in Al-treated C. sinensis ones; accordingly, more upregulated genes related to the detoxifications of ROS, aldehydes, and MG were detected in Al-treated C. grandis leaves; however, more upregulated genes involved in antioxidation were identified in Al-treated C. sinensis roots; (c) Al-treated C. sinensis seedlings had a higher concentration of S in leaves and roots possibly by increasing S uptake and decreasing S export; and (d) protein phosphorylation and dephosphorylation and the equilibrium of hormones and hormone-mediated signal transduction were less affected in Al-treated C. sinensis roots and leaves than those in Al-treated C. grandis ones.

Li et al. (2016) and Jiang et al. (2015) used isobaric tags for relative and absolute quantification (iTRAQ) to investigate Al-induced alterations of proteome in C. sinensis and C. grandis roots and leaves. They isolated 134 (131) proteins with decreased abundance and 176 (120) proteins with increased abundance and 11 (101) proteins with decreased and 6 (44) proteins with increased abundance from Al-treated C. sinensis and C. grandis leaves (roots), respectively. Most of these Al-responsive proteins were only presented in C. sinensis or C. grandis leaves (roots). Only 3 (49) Al-responsive proteins with the same accession number were shared by C. sinensis and C. grandis leaves (roots). The higher Al tolerance of C. sinensis might involve in following several aspects: (a) activation of S metabolism in roots and leaves; (b) decreasing the production of ROS and other toxic compounds and enhancing the detoxifying capacity of ROS and other toxic compounds in leaves and roots; (c) better maintenance of photosynthesis and energy balance by upregulating photosynthesis, carbohydrate, and energy metabolism-related proteins in leaves and/or roots; and (d) increasing the abundances of low-P-responsive proteins in leaves and roots. Also, Al-responsive proteins related to cellular transport, signal transduction (i.e., protein phosphorylation and dephosphorylation and hormone biosynthesis), and nucleic acid and protein metabolisms might contribute to the higher Al tolerance of C. sinensis.

Understanding of the mechanisms underlying Al tolerance is crucial not only to allow us to screen Al-tolerant fruit species/cultivars but also to provide us opportunities to improve fruit tree production on acidic soils with high active Al level via using genetic engineering. However, many physiological and molecular mechanisms for Al tolerance of fruit trees remain to be answered. It is important to remember that the responses of fruit trees to Al toxicity are highly dependent on the species/cultivars. However, most of studies regarding Al toxicity of fruit trees have been focused on *Citrus*, blueberry, longan, banana, and pineapples; more fruit trees need to be systematically investigated. It is well known that mineral nutrients, NO, MeJA, and organic acids can alleviate Al toxicity in some fruit trees, but their potential roles are not understood fully. Many plants have been known to detoxify Al internally and/or externally. The Al-induced secretion of organic acid anions is the major mechanism for Al tolerance of higher plants including some fruit trees. However, internal detoxification of Al at the molecular level is not clear and requires further investigation in fruit trees. So far, not a few Al-tolerant genes involved in Al-induced secretion of organic acid anions, cell wall modification, sequestration of Al, and other processes have been isolated from model plants including *Arabidopsis*, wheat, rice, and barley, but little information is available on the complete sequence of genes related to Al tolerance in fruit trees. Obviously, it is of great importance to carry out more researches to characterize the genome of fruit trees at the molecular level. Last but not least, fruit tree breeders have to develop novel and more efficient approaches of fruit tree genetic transformation to improve Al tolerance using transgenic technique. Fortunately, many new techniques including genome-wide association study (GWAS), RNA-Seq, miRNA-Seq, proteomics, and metabolomics will speed up our clarification of Al toxicity and Al tolerance of fruit trees in the physiological and molecular levels.

Acknowledgments

This study was financially supported by the National Key Research and Development Program of China (2018YFD1000305) and the National Natural Science Foundation of China (31772257).

References

Alvarenga, A.A., Regina, M.D., Fraguas, J.C., Da Silva, A.L., Chalfun, N.N.J., 2004. Aluminum effect on nutrition and development of grapevine rootstocks (*Vitis* spp.). J. Int. Sci. Vigne Vin 38 (2), 119–129.

Arunakumara, K.K.I.U., Walpola, B.C., Yoon, M.H., 2012. How do *Citrus* crops cope with aluminum toxicity? Korean J. Soil Sci. Fert. 45 (6), 928–935.

Burkhardt, S.L., Villa, F., da Silva, A.L., Comim, J.J., Pasqual, M., 2008. Evaluation of rootstock grapevine *in vitro* in conditions of stress per aluminum. Ciênc. Téc. Vitiv. 23 (1), 21–27.

Cançado, G.M.A., Ribeiro, A.P., Piñeros, M.A., Miyata, L.Y., Alvarenga, Â.A., Villa, F., et al., 2009. Evaluation of aluminium tolerance in grapevine rootstocks. Vitis 48 (4), 167–173.

Chen, L.S., 2006. Physiological responses and tolerance of plant shoot to aluminum toxicity. J. Plant Physiol. Mol. Biol. 32, 143–155.

Chen, L.S., 2012. Physiological responses and tolerance of *Citrus* to aluminum toxicity. In: Srivastava, A.K. (Ed.), Advances in *Citrus* Nutrition. Springer-Verlag, New York, pp. 435–452.

Chen, L.S., Cheng, L., 2003. Both xanthophyll cycle-dependent thermal dissipation and the antioxidant system are up-regulated in grape (*Vitis labrusca* L. cv. Concord) leaves in response to N limitation. J. Exp. Bot. 54, 2165–2175.

Chen, J.H., Lin, Y.H., 2010. Effect of aluminum on variations in the proteins in pineapple roots. Soil Sci. Plant Nutr. 56 (3), 438–444.

Chen, L.S., Qi, Y.P., Smith, B.R., Liu, X.H., 2005a. Aluminum-induced decrease in CO_2 assimilation in *Citrus* seedlings is unaccompanied by decreased activities of key enzymes involved in CO_2 assimilation. Tree Physiol. 25, 317–324.

Chen, L.S., Qi, Y.P., Liu, X.H., 2005b. Effects of aluminum on light energy utilization and photoprotective systems in *Citrus* leaves. Ann. Bot. 96, 35–41.

Chen, Z.C., Chen, H.Z., Huang, H., Zhang, N., Zhu, S.Y., 2005c. Effect of aluminum on seedling growth of semi-tropical fruit tree. J. Agro-Environ. Sci. 24 (Sp.), 34–37.

Chen, L.S., Tang, N., Jiang, H.X., Yang, L.T., Li, Q., Smith, R.B., 2009. Changes in organic acid metabolism differ between roots and leaves of *Citrus grandis* in response to phosphorus and aluminum interactions. J. Plant Physiol. 166, 2023–2034.

Chen, L.S., Tang, N., Li, Q., Zheng, S., Wang, Y.N., 2010a. Effects of aluminum on leaf water relation and nutritional elements of *Citrus* seedlings. In: Deng, X.X., Xu, J., Lin, S.M., Guan, G. (Eds.), Proceedings of the International Society of Citriculture. In: vol. I. China Agriculture Press, Beijing, pp. 650–653.

Chen, L.S., Qi, Y.P., Jiang, H.X., Yang, L.T., Yang, G.H., 2010b. Photosynthesis and photoprotective systems of plants in response to aluminum toxicity. Afr. J. Biotechnol. 9, 9237–9247.

Chibiliti, G.B., Byrne, D.H., 1990. Interaction of aluminum and calcium on 'Nemaguard' peach seedling nutrient contents and growth in sand culture. Sci. Hortic. 43, 29–36.

Dawood, M., Cao, F., Jahangir, M.M., Zhang, G., Wu, F., 2012. Alleviation of aluminum toxicity by hydrogen sulfide is related to elevated ATPase, and suppressed aluminum uptake and oxidative stress in barley. J. Hazard. Mater. 20, 121–128.

Deng, W., Luo, K., Li, Z., Yang, Y., 2008. Molecular cloning and characterization of a mitochondrial dicarboxylate/tricarboxylate transporter gene in *Citrus junos* response to aluminum stress. Mitochondrial DNA 19 (4), 376–384.

Deng, W., Luo, K., Li, Z., Yang, Y., Hu, N., Wu, Y., 2009. Overexpression of *Citrus junos* mitochondrial citrate synthase gene in *Nicotiana benthamiana* confers aluminum tolerance. Planta 230, 355–365.

dos Santos, C.H., Filho, H.G., Rodrigues, J.D., de Pinho, S.Z., 2000. Influence of different levels of aluminum on the development of *Citrus* rootstock 'Swingle' citrumelo (*Citrus paradisi* Mcf. × *Poncirus trifoliata* Raf.) in nutrient solution. Braz. Arch. Biol. Technol. 43, 27–33.

Edwards, J.H., Horton, B.D., 1977. Aluminum-induced calcium deficiency in peach seedlings. J. Am. Soc. Hort. Sci. 102, 459–461.

Edwards, J.H., Horton, B.D., Kirkpatrick, H.C., 1976. Aluminum toxicity symptoms in peach seedlings. J. Am. Soc. Hortic. Sci. 101, 139–142.

Foy, C.D., Chaney, R.L., White, M.C., 1978. The physiology of metal toxicity in plants. Ann. Rev. Plant Physiol. 29, 511–566.

Graham, C.J., 2001. The influence of nitrogen source and aluminum on growth and elemental composition of nemaguard peach seedlings. J. Plant Nutr. 24 (3), 423–439.

Graham, C.J., 2002. Nonstructural carbohydrate and prunasin composition of peach seedlings fertilized with different nitrogen sources and aluminum. Sci. Hortic. 94, 21–32.

Guo, J.H., Liu, X.J., Zhang, Y., Shen, J.L., Han, W.X., Zhang, W.F., et al., 2010. Significant acidification in major Chinese croplands. Science 327, 1008–1010.

Guo, P., Qi, Y.P., Yang, L.T., Lai, N.W., Ye, X., Yang, Y., et al., 2017a. Root adaptive responses to aluminum-treatment revealed by RNA-Seq in two *Citrus* species with different aluminum-tolerance. Front. Plant Sci. 8, 330.

Guo, P., Li, Q., Qi, Y.P., Yang, L.T., Ye, X., Chen, H.H., Chen, L.S., 2017b. Sulfur-mediated-alleviation of aluminum-toxicity in *Citrus grandis* seedlings. Int. J. Mol. Sci. 18, 2570.

Guo, P., Qi, Y.P., Cai, Y.T., Yang, T.Y., Yang, L.T., Huang, Z.R., et al., 2018a. Aluminum effects on photosynthesis, reactive oxygen species and methylglyoxal detoxification in two *Citrus* species differing in aluminum tolerance. Tree Physiol. 38, 1548–1565.

Guo, P., Qi, Y.P., Huang, W.L., Yang, L.T., Huang, Z.R., Lai, N.W., et al., 2018b. Aluminum-responsive genes revealed by RNA-Seq and related physiological responses in leaves of two *Citrus* species with contrasting aluminum-tolerance. Ecotoxicol. Environ. Saf. 158, 213–222.

Horst, W.J., Püschel, A.K., Schmohl, N., 1997. Induction of callose formation is a sensitive marker for genotypic aluminium sensitivity in maize. Plant Soil 192, 23–30.

Hossain, M.A., Piyatida, P., da Silva, J.A.T., Fujita, M., 2012. Molecular mechanism of heavy metal toxicity and tolerance in plants: central role of glutathione in detoxification of reactive oxygen species and methylglyoxal and in heavy metal chelation. J. Bot. 2012, 872–875.

Huang, Y.Z., Li, J., Wu, S.H., Pang, D.M., 2001. Nutrition condition of the orchards in the main production areas of Guanxi honeypomelo trees (Pinhe county). J. Fujian Agric. Univ. 30, 40–43.

Inostroza-Blancheteau, C., Aquea, F., Reyes-Díaz, M., Alberdi, M., Arce-Johnson, P., 2011a. Identification of aluminum-regulated genes by cDNA-AFLP analysis of roots in two contrasting genotypes of highbush blueberry (*Vaccinium corymbosum* L.). Mol. Biotechnol. 49 (1), 32–41.

Inostroza-Blancheteau, C., Reyes-Díaz, M., Aquea, F., Nunes-Nesi, A., Alberdi, M., Arce-Johnson, P., 2011b. Biochemical and molecular changes in response to aluminium-stress in highbush blueberry (*Vaccinium corymbosum* L.). Plant Physiol. Biochem. 49, 1005–1012.

Inostroza-Blancheteau, C., Aquea, F., Loyola, R., Slovin, J., Josway, S., Rengel, Z., et al., 2013. Molecular characterisation of a calmodulin gene, VcCaM1, that is differentially expressed under aluminium stress in highbush blueberry. Plant Biol. 15 (6), 1013–1018.

Jiang, H.X., Chen, L.S., Zheng, J.G., Han, S., Tang, N., Smith, B.R., 2008. Aluminum-induced effects on photosystem II photochemistry in *Citrus* leaves assessed by the chlorophyll a fluorescence transient. Tree Physiol. 28, 1863–1871.

Jiang, H.X., Tang, N., Zheng, J.G., Li, Y., Chen, L.S., 2009a. Phosphorus alleviates aluminum-induced inhibition of growth and photosynthesis in *Citrus grandis* seedlings. Physiol. Plant. 137, 298–311.

Jiang, H.X., Tang, N., Zheng, J.G., Chen, L.S., 2009b. Antagonistic actions of boron against inhibitory effects of aluminum toxicity on growth, CO$_2$ assimilation, ribulose-1,5-bisphosphate carboxylase/oxygenase, and photosynthetic electron transport probed by the JIP-test, of *Citrus grandis* seedlings. BMC Plant Biol. 9, 102.

Jiang, H.X., Chen, L.S., Han, S., Zhang, L.F., Lin, J.Q., 2009c. Effects of aluminum on the growth of young *Citrus* seedlings. Chin. Agric. Sci. Bull. 25 (4), 167–170.

Jiang, H.X., Yang, L.T., Qi, Y.P., Lu, Y.B., Huang, Z.R., Chen, L.S., 2015. Root iTRAQ protein profile analysis of two *Citrus* species differing in aluminum-tolerance in response to long-term aluminum-toxicity. BMC Genomics 16, 949.

Kinraide, T.B., 1991. Identity of the rhizotoxic aluminium species. Plant Soil 134, 167–178.

Kochian, L.V., 1995. Cellular mechanisms of aluminum toxicity and resistance in plants. Ann. Rev. Plant Physiol. Mol. Biol. 46, 237–260.

Kochian, L.V., Hoekenga, O.A., Pineros, M.A., 2004. How do crop plants tolerate acid soils? Mechanisms of aluminum tolerance and phosphorous efficiency. Ann. Rev. Plant Biol. 55, 459–493.

Kongsri, S., Boonprakob, U., Byrne, D.H., 2014. Assessment of morphological and physiological responses of peach rootstocks under drought and aluminum stress. Acta Hortic. 1059, 229–236.

Le Van, H., Masuda, T., 2004. Physiology and biological studies on aluminum tolerance in pineapple. Aust. J. Soil Res. 42, 699–707.

Li, Y., Han, M.Q., Lin, F., Ten, Y., Lin, J., Zhu, D.H., et al., 2015. Soil chemical properties, 'Guanximiyou' pummelo leaf mineral nutrient status and fruit quality in the southern region of Fujian province, China. J. Soil Sci. Plant Nutr. 15, 615–628.

Li, H., Yang, L.T., Qi, Y.P., Guo, P., Lu, Y.B., Chen, L.S., 2016. Aluminum-toxicity-induced alterations of leaf proteome in two *Citrus* species differing in aluminum-tolerance. Int. J. Mol. Sci. 17, 1180.

Li, X., Li, Y.L., Mai, J., Tao, L., Qu, M., Liu, J.Y., et al., 2018. Boron alleviates aluminum toxicity by promoting root alkalization in transition zone *via* polar auxin transport. Plant Physiol. 177, 1254–1266.

Liao, X.Y., Yang, L.T., Lu, Y.B., Ye, X., Chen, L.S., 2015. Roles of rootstocks and scions in aluminum-tolerance of *Citrus*. Acta Physiol. Plant. 37, 1743.

Lin, Y.H., 2010. Effects of aluminum on root growth and absorption of nutrients by two pineapple cultivars [*Ananas comosus* (L.) Merr.]. Afr. J. Biotechnol. 9, 4034–4041.

Lin, Y.H., Chen, J.H., 2011. Behavior of aluminum adsorption on cell wall of pineapple root apices. Afr. J. Agric. Res. 6 (4), 949–955.

Lin, Y.H., Chen, J.H., 2013. Aluminum resistance and cell-wall characteristics of pineapple root apices. J. Plant Nutr. Soil Sci. 176 (5), 795–800.

Lin, Z., Myhre, D.L., 1990. *Citrus* root growth as affected by soil aluminum level under field conditions. Soil Sci. Soc. Am. J. 54, 1340–1344.

Lin, Z., Myhre, D.L., 1991a. Differential response of *Citrus* rootstocks to aluminum levels in nutrient solutions: I. Plant growth. J. Plant Nutr. 14, 1223–1238.

Lin, Z., Myhre, D.L., 1991b. Differential response of *Citrus* rootstocks to aluminum levels in nutrient solutions: II. Plant mineral concentrations. J. Plant Nutr. 14, 1239–1254.

Long, A., Zhang, J., Yang, L.T., Ye, X., Lai, N.W., Tan, L.L., et al., 2017. Effects of low pH on photosynthesis, related physiological parameters and nutrient profile of *Citrus*. Front. Plant Sci. 8, 185.

Magalhães, A.F.J., 1987. Tolerância de porta-enxertos de *Citrus* ao alumínio. Rev. Bras. Frutic. 9, 51–55.

Manna, P., Sinha, M., Sil, P.C., 2008. Arsenic-induced oxidative myocardial injury: protective role of arjunolic acid. Arch. Toxicol. 82, 137–149.

Manquián-Cerda, K., Cruces, E., Escudey, M., Zúñiga, G., Calderón, R., 2018. Interactive effects of aluminum and cadmium on phenolic compounds, antioxidant enzyme activity and oxidative stress in blueberry (*Vaccinium corymbosum* L.) plantlets cultivated in vitro. Ecotoxicol. Environ. Saf. 150, 320–326.

Massot, N., Llugany, M., Poschenrieder, C., Barceló, J., 1999. Callose production as indicator of aluminum toxicity in bean cultivars. J. Plant Nutr. 22 (1), 1–10.

Meriño-Gergichevich, C., Ondrasek, G., Zovko, M., Šamec, D., Alberdi, M., Reyes-Díaz, M., 2015. Comparative study of methodologies to determine the antioxidant capacity of Al-toxified blueberry amended with calcium sulfate. J. Soil Sci. Plant Nutr. 15 (4), 965–978.

Mota, M.F., Pegoraro, R.F., Batista, P.S., Pinto, V.D.O., Maia, V.M., Silva, D.F.D., 2016. Macronutrients accumulation and growth of pineapple cultivars submitted to aluminum stress. Rev. Bras. Eng. Agríc. Ambient. 20 (11), 978–983.

Naing, K.W., Angeles, D.E., Protacio, C.M., 2009. Tolerance of mango (*Mangifera indica* L. Anacardiaceae) seedlings to different levels of aluminum. Philipp. J. Crop Sci. 34 (3), 33–42.

Pavan, M.A., Bingham, F.T., Peryea, F.J., 1987. Influence of calcium and magnesium salts on acid soil chemistry and calcium nutrition of apple. Soil Sci. Soc. Am. J. 51, 1526–1530.

Pereira, W.E., de Siqueira, D.L., Martinez, C.A., Puiatti, M., 2000. Gas exchange and chlorophyll fluorescence in four *Citrus* rootstocks under aluminum stress. J. Plant Physiol. 157, 513–520.

Pereira, W.E., de Siqueira, D.L., Puiatti, M., Martínez, C.A., Salomão, L.C.C., Cecon, P.R., 2003. Growth of *Citrus* rootstocks under aluminium stress in hydroponics. Sci. Agric. 60, 31–41.

Rahman, M., Lee, S.H., Ji, H., Kabir, A., Jones, C., Lee, K.W., 2018. Importance of mineral nutrition for mitigating aluminum toxicity in plants on acidic soils: current status and opportunities. Int. J. Mol. Sci. 19, 3073.

Reyes-Diaz, M., Alberdi, M., de la Luz Mora, M., 2009. Short-term aluminum stress differentially affects the photochemical efficiency of photosystem II in highbush blueberry genotypes. J. Am. Soc. Hort. Sci. 134, 14–21.

Reyes-Díaz, M., Inostroza-Blancheteau, C., Millaleo, R., Cruces, E., Wulff-Zottele, C., Alberdi, M., de la Luz Mora, M., 2010. Long-term aluminum exposure effects on physiological and biochemical features of highbush blueberry cultivars. J. Am. Soc. Hort. Sci. 135, 212–222.

Reyes-Díaz, M., Meriño-Gergichevich, C., Alarcón, E., Alberdi, M., Horst, W.J., 2011. Calcium sulfate ameliorates the effect of aluminum toxicity differentially in genotypes of highbush blueberry (*Vaccinium corymbosum* L.). J. Soil Sci. Plant Nutr. 11 (4), 59–78.

Reyes-Díaz, M., Inostroza-Blancheteau, C., Rengel, Z., 2015. Physiological and molecular regulation of aluminum resistance in woody plant species. In: Panda, S.K., Baluška, F. (Eds.), Aluminum Stress Adaptation in Plants. Springer, Cham, pp. 187–202.

Riaz, M., Yan, L., Wu, X., Hussain, S., Aziz, O., Jiang, C., 2018a. Mechanisms of organic acids and boron induced tolerance of aluminum toxicity: a review. Ecotoxicol. Environ. Saf. 165, 25–35.

Riaz, M., Yan, L., Wu, X., Hussain, S., Aziz, O., Jiang, C., 2018b. Boron increases root elongation by reducing aluminum induced disorganized distribution of HG epitopes and alterations in subcellular cell wall structure of trifoliate orange roots. Ecotoxicol. Environ. Saf. 165, 202–210.

Riaz, M., Yan, L., Wu, X., Hussain, S., Aziz, O., Imran, M., et al., 2018c. Boron reduces aluminum-induced growth inhibition, oxidative damage and alterations in the cell wall components in the roots of trifoliate orange. Ecotoxicol. Environ. Saf. 153, 107–115.

Riaz, M., Yan, L., Wu, X., Hussain, S., Aziz, O., Wang, Y., et al., 2018d. Boron alleviates the aluminum toxicity in trifoliate orange by regulating antioxidant defense system and reducing root cell injury. J. Environ. Manag. 208, 149–158.

Roy, A.K., Sharma, A., Talukder, G., 1988. Some aspects of aluminum toxicity in plants. Bot. Rev. 54, 145–178.

Rufyikiri, G., Dufey, J.E., Achard, R., Delvaux, B., 2002. Cation exchange capacity and aluminum–calcium–magnesium binding in roots of bananas cultivated in soils and in nutrient solutions. Commun. Soil Sci. Plant Anal. 33, 991–1009.

Schuch, M.W., Cellini, A., Masia, A., Marino, G., 2010. Aluminium-induced effects on growth, morphogenesis and oxidative stress reactions in *in vitro* cultures of quince. Sci. Hortic. 125, 151–158.

Sikirić, B., Mrvić, V., Stevanović, D., Maksimović, S., Stajković, O., Bogdanović, D., 2009. The effects of calcification, urea and Al salts on Fe, Mn and Al contents in the soil and raspberry leaves. Agrochimica 53 (4), 250–259.

Tang, N., Jiang, H.X., Yang, L.T., Li, Q., Yang, G.H., Chen, L.S., 2011. Boron-aluminum interactions affect organic acid metabolism more in leaves than in roots of *Citrus grandis* seedlings. Biol. Plant. 55, 681–688.

Toan, N.B., Debergh, P., Ve, N.B., Bornman, C.H., 2003. Aluminium tolerance of *Citrus* seedlings in the Mekong Delta, Vietnam. South Afr. J. Bot. 69 (4), 526–531.

Ulloa-Inostroza, E.M., Alberdi, M., Meriño-Gergichevich, C., Reyes-Díaz, M., 2017. Low doses of exogenous methyl jasmonate applied simultaneously with toxic aluminum improve the antioxidant performance of *Vaccinium corymbosum*. Plant Soil 412, 81–96.

von Uexküll, H.R., Mutert, E., 1995. Global extent, development and economic impact of acid soils. Plant Soil 171, 1–15.

Wan, Q., 2007. Effect of aluminum stress on the content of carbohydrate in *Dimocarpus longan* Lour seedlings. Chin. J. Trop. Crops 28 (4), 10–14.

Wang, L.Q., Yang, L.T., Guo, P., Zhou, X.X., Ye, X., Chen, E.J., et al., 2015. Leaf cDNA-AFLP analysis reveals novel mechanisms for boron-induced alleviation of aluminum-toxicity in *Citrus grandis* seedlings. Ecotoxicol. Environ. Saf. 120, 349–359.

Wojcik, P., 2003. Impact of boron on biomass production and nutrition of aluminum-stressed apple rootstocks. J. Plant Nutr. 26 (12), 2439–2451.

Xiao, X.X., 2005a. Responses of longan to aeration under aluminum stress and rectification of aluminum toxicity. J. Nanjing Forest. Univ. (Nat. Sci. Ed.) 29 (5), 41–44.

Xiao, X.X., 2005b. Characteristics of aluminum absorption by longan (*Dimocarpus longan*) seedlings. Sci. Silvae Sin. 41 (3), 40–47.

Xiao, X.X., Liu, X.H., Zhang, X.W., Yang, Z.W., Chen, L.S., Xie, Y.Q., 2002. Effect of aluminium stress on the growth of young longan seedling. Fujian J. Agric. Sci. 17, 182–185.

Xiao, X.X., Liu, X.H., Yang, Z.W., Zheng, R., Chen, L.S., 2003a. Effects of aluminum stress on cell ultra-structure of leaf and root of longan (*Dimoscarpus longan*). Sci. Silvae Sin. 39 (Sp. 1), 58–61.

Xiao, X.X., Liu, X.H., Yang, Z.W., Xiao, H., Xie, Y.Q., 2003b. Effects of aluminum stress on active oxygen metabolism and membrane system of longan (*Dimoscarpus longan*) leaves. Sci. Silvae Sin. 39 (Sp. 1), 52–57.

Xiao, X.X., Chen, L.S., Cai, Y.H., Huang, Y., Xie, Y.Q., 2005a. The effect of aluminum stress on the absorption of nutrient elements in longan (*Dimocarpus longana*) seedlings. Acta Agric. Univ. Jiangxiensis 27 (2). 230–233, 316.

Xiao, X.X., Liu, X.H., Yang, Z.W., Chen, L.S., Cai, Y.H., 2005b. Effect of aluminum stress on the photosynthesis of longan seedlings. Chin. J. Trop. Crops 26 (1), 63–69.

Yadav, S.K., Singla-Pareek, S.L., Ray, M., Reddy, M.K., Sopory, S.K., 2005. Transgenic tobacco plants overexpressing glyoxalase enzymes resist an increase in methylglyoxal and maintain higher reduced glutathione levels under salinity stress. FEBS Lett. 579, 6265–6271.

Yamamoto, Y., Kobayashi, Y., Devi, S.R., Rikiishi, S., Matsumoto, H., 2002. Aluminum toxicity is associated with mitochondrial dysfunction and the production of reactive oxygen species in plant cells. Plant Physiol. 128, 63–72.

Yan, L., Riaz, M., Wu, X., Du, C., Liu, Y., Lv, B., et al., 2018a. Boron inhibits aluminum-induced toxicity to *Citrus* by stimulating antioxidant enzyme activity. J. Environ. Sci. Health C 36 (3), 145–163.

Yan, L., Riaz, M., Wu, X., Du, C., Liu, Y., Jiang, C., 2018b. Ameliorative effects of boron on aluminum induced variations of cell wall cellulose and pectin components in trifoliate orange (*Poncirus trifoliata* (L.) Raf.) rootstock. Environ. Pollut. 240, 764–774.

Yang, L.T., Jiang, H.X., Tang, N., Chen, L.S., 2011. Mechanisms of aluminum-tolerance in two species of *Citrus*: secretion of organic acid anions and immobilization of aluminum by phosphorus in roots. Plant Sci. 180, 521–530.

Yang, L.T., Qi, Y.P., Chen, L.S., Sang, W., Lin, X.J., Wu, Y.L., et al., 2012a. Nitric oxide protects 'Sour' pummelo (*Citrus grandis*) seedlings against aluminum-induced inhibition of growth and photosynthesis. Environ. Exp. Bot. 83, 1–13.

Yang, L.T., Jiang, H.X., Qi, Y.P., Chen, L.S., 2012b. Differential expression of genes involved in alternative glycolytic pathways, phosphorus scavenging and recycling in response to aluminum and phosphorus interactions in *Citrus* roots. Mol. Biol. Rep. 39, 6353–6366.

Yang, L.T., Chen, L.S., Peng, H.Y., Guo, P., Wang, P., Ma, C.L., 2012c. Organic acid metabolism in *Citrus grandis* leaves and roots is differently affected by nitric oxide and aluminum interactions. Sci. Hortic. 133, 40–46.

Yang, L.T., Qi, Y.P., Jiang, H.X., Chen, L.S., 2013. Roles of organic acid anion secretion in aluminium tolerance of higher plants. Biomed. Res. Int. 2013, 173682.

Yang, L.T., Liu, J.W., Wu, Y.M., Qi, Y.P., Lai, N.W., Ye, X., et al., 2018. Proteome profile analysis of boron-induced alleviation of aluminum-toxicity in *Citrus grandis* roots. Ecotoxicol. Environ. Saf. 162, 488–498.

Zhang, M., Luo, X.Y., Bai, W.Q., Li, Y.R., Hou, L., Pei, Y., et al., 2008. Characterization of malate dehydrogenase gene from *Citrus junos* and its transgenic tobacco's tolerance to aluminium toxicity. Acta Hortic. Sin. 35, 1751–1758.

Zhou, X.X., Yang, L.T., Qi, Y.P., Guo, P., Chen, L.S., 2015. Mechanisms on boron-induced alleviation of aluminum-toxicity in *Citrus grandis* seedlings at a transcriptional level revealed by cDNA-AFLP analysis. PLoS One 10, e0115485.

Further reading

Schaedle, M., Thornton, F.C., Raynal, D.J., Tepper, H.B., 1989. Response of tree seedlings to aluminum. Tree Physiol. 5, 337–356.

18

The importance of selenium in fruit nutrition

Pavel Ryant[a,*], Jiří Antošovský[a], Vojtěch Adam[b], Ladislav Ducsay[c], Petr Škarpa[a], Eva Sapáková[d]

[a]Department of Agrochemistry, Soil Science, Microbiology and Plant Nutrition, Faculty of AgriSciences, Mendel University in Brno, Brno, Czech Republic
[b]Department of Chemistry and Biochemistry, Faculty of AgriSciences, Mendel University in Brno, Brno, Czech Republic
[c]Department of Agrochemistry and Plant Nutrition, The Faculty of Agrobiology and Food Resources, Slovak University of Agriculture in Nitra, Nitra, Slovakia
[d]Department of Languages, Faculty of Regional Development and International Studies, Mendel University in Brno, Brno, Czech Republic
*Corresponding author. E-mail: pavel.ryant@mendelu.cz

1 Introduction

Selenium is a solid, red or gray water-insoluble element, which was randomly discovered by the Swedish chemist Jons Jacob Berzelius in 1817. Selenium is classified as a metalloid because of the position between sulfur and tellurium in VI.A group and between arsenic and bromine in the fourth period of the periodic table. Ionic radius of Se and S are closer; therefore, physicochemical properties of both elements are similar to each other because of the same group of the periodic table (Bodnar et al., 2012). The trace element of selenium can be considered as an essential nutrient for human and animal health, although the demand of selenium has not been observed in higher plants (Rayman, 2000; Combs, 2001; Finley, 2005; Zhu et al., 2009). In physiology, the importance of selenium is confirmed in the occurrence in Se-containing amino acids such as selenomethionine (SeMet) and selenocysteine (SeCys) classified as the 21st amino acid due to a necessity in a small set of selenoproteins working for redox functions (Cone et al., 1976). Selenium compounds have proved a key role in many important biochemical processes such as antioxidant activity, antitumor effects, or effective antiviral, antibacterial, or antimycotic agents (Mugesh et al., 2001). Appropriate selenium intake levels show a relatively narrow range between deficiency and toxicity. An intake of selenium up to 30 µg per day can be considered as an inappropriate dose for humans, whereas the intakes exceeding 900 µg per day could be

A.K. Srivastava, Chengxiao Hu (eds.)
Fruit Crops: Diagnosis and Management of Nutrient Constraints
https://doi.org/10.1016/B978-0-12-818732-6.00018-6

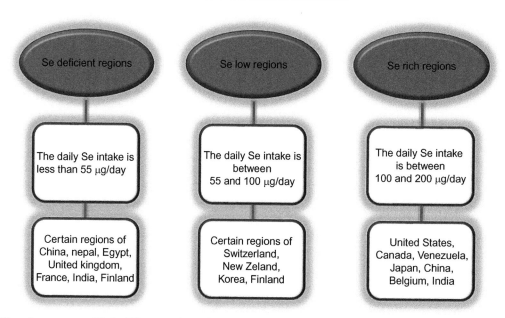

FIG. 18.1   Outline of occurrence of Se in different regions of world as Se-deficient, Se-low, and Se-rich regions (Gupta and Gupta, 2016).

potentially harmful (Fairweather-Tait et al., 2011). The intake levels of dietary selenium (Fig. 18.1) are dependent on the total concentration and bioavailability of selenium in food sources influenced by the multiple factors such as selenium content in soil, irrigation, and animal feed. Selenium content may vary considerably according to food sources and geographical origin. Selenium deficiency is more widespread than selenium excess, but both factors can be found out in agricultural and natural areas (Rayman et al., 2008). The chemical form of selenium also plays an important role because of affecting the bioavailability and nutritional value of selenium. The various chemical forms of selenium prove different properties in sorption, bioavailability, mobility, and toxicity (Hamilton, 2004; Fernandez-Martinez and Charlet, 2009; Wallschlaeger and Feldmann, 2010; Zwolak and Zaporowska, 2012).

In areas with high selenium content in soil, an increased transfer of selenium to plants and subsequently to livestock has been detected. Selenium content in food sources can vary from plant to plant depending on selenium uptake and accumulating capacity of plant and soil selenium content differing from geographical locations and the occurrence of other elements in soil (Dumont et al., 2006; Bodnar et al., 2012; Mehdi et al., 2013). Fruits generally contain lower selenium amount in comparison with vegetables. Brazil nuts, Brassica species, and garlic can effectively accumulate selenium to be rich sources of selenium into diet (Dumont et al., 2006; Bodnar et al., 2012). The high content of bioavailable selenium is also contained in yeast produced by *Saccharomyces cerevisiae* grown in a selenium-rich medium (Schrauzer, 2006). Food supplement, based on inorganic selenium (usually in the form of sodium selenite), proves lower bioavailability, and part of it is uselessly excluded. But there is no clear evidence that selenium is essential element for higher plants. Sulfur (S)-loving plants often show trace levels of selenium (Ellis and Salt, 2003).

2 Selenium in soil

Selenium may occur naturally in sedimentary rocks formed from the carboniferous to quaternary period (White et al., 2004). The occurrence of selenium in soil can depend on type of soil, organic matter, and rainfall (Sors et al., 2005b). The concentration of Se in the soil ranges from less than 0.1 to more than 100 $\mu g/g$ but in most parts of the world the content ranges from 0.1 to 1.5 $\mu g/g$ (Lyons et al., 2003). Worldwide, the average selenium concentration in soils can reach up to 0.4 $\mu g/g$: however, elevated levels of selenium (>2–5000 $\mu g/g$) are found out in seleniferous soils (Hartikainen, 2005). Generally, the total content of 0.1–0.6 $\mu g/g$ is deficient (Lyons et al., 2003). Selenium occurs in soil in both organic and inorganic forms. The organic form (SeCys and SeMet) comes mainly from degraded vegetation with a high selenium content. Inorganic selenium in the soil is in the form of metal selenides, elemental selenium, selenites, or selenates often occurring as stable salts (ferric selenite and calcium selenium) (Adriano, 1986). The distribution of the species depends on soil properties such as aeration, the mineral and organic matter acidity, and the microbiological activity (Nakamaru and Altansuvd, 2014). Selenium is bound in acidic soils as ferrous selenite with very low solubility reducing the availability of selenium for vegetation. In the alkaline soils, more selenium is present, preparing it more available for the plant (Merian, 1991). In well-aerated alkaline soils from semidry areas, selenium is included in the form of selenium. However, under acidic and reducing conditions, which are in the wet regions, the selenium may be selectively present. The rate of conversion of selenites to selenates and vice versa is very low, and the rate of conversion of selenites to elemental selenium is even slower (Adriano, 1986). The sources of Se in the soils are lithogenic (parent rocks), pedogenic (through organic matter), atmospheric (precipitation and volcanic emission), phytogenetic (volatilization from plants) and anthropogenic (agriculture and industry) (Kabata-Pendias, 2001).

3 Bioavailability of selenium from soil

Selenium is transferred to plants from the surrounding environment particularly from soil and water received into the bodies of herbivores and omnivores from the plant mass. As follows, in full compliance with the general principles in food chains, the selenium intake of the organism can occur at higher levels of the food chain including humans. The amount of selenium of the soil does not necessarily have to be a clear and decisive indicator of the sufficiency or deficiency in the given area. The bioavailability of selenium is dependent on pH (Fig. 18.2), oxidation-reduction conditions, hydroxide content, organic matter content, and content of clayey fractions. When pH decreases, selenium adsorption increases due to fewer negatively charged clay surfaces and sesquioxide edges resulting in stronger electrostatic interactions (Eich-Greatorex et al., 2007). Thus, the mobility and bioavailability of inorganic selenium in the environment increases with increasing pH and with decreasing clay and iron oxide content in the soil (Peak and Sparks, 2002). Moreover, some microorganisms can convert relatively chemically inert forms of selenium into much more bioavailable organic forms or, in some cases, soluble inorganic forms, that is, SeO_3^{2-} selenites and optionally SeO_4^{2-} selenates (Sarathchandra and Watkinson, 1981). However, the opposite effects of microorganisms, that is, the reduction of oxidized inorganic forms (Se^{IV} and Se^{VI}) on elemental selenium or Se^{-II} compounds (Javed et al., 2015; Hageman et al., 2013), are more frequently described. The intensity of intake of inorganic forms of selenium by plants varies. SeO_4^{2-} can enter to the plant organism more actively than SeO_3^{2-}. At pH <6, the soils prove predominantly Se^{4+} and Se^{6+} in alkaline soils. The mobility and availability of various forms of selenium depend on processes in the soil such as adsorption and precipitation. The selenites can form adsorption complexes with iron oxide in the soil rather than reacting in the form of ferric selenite. The precipitation of selenite with ferric hydroxide takes place through the adsorption process. Many studies have shown that Se^{4+} is bound to soil much stronger than Se^{6+} (Adriano, 1986).

In soils with a high content of calcium and magnesium, $CaSeO_4$ and $MgSeO_4$ are mainly formed. In acidic soils, especially $KHSe$, NH_4HSe, and $MnSe$ are created. Selenates (Se^{6+}) are mobile in inorganic forms of alkaline soils and are not adsorbed to hydrated sesquioxides. Selenite (Se^{4+}) is poorly mobile in neutral or acidic soils in wet regions and is readily adsorbed to hydrated sesquioxides and organic matter. Selenides (Se^{2-}) are low mobile in acidic soils because of formation stable mineral and organic compounds. The conversion of selenium forms into soils is based on the following

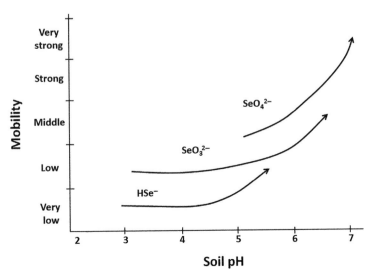

FIG. 18.2 Effect of soil pH on selenium mobility (edited according Kabata-Pendias, 2001).

processes: (1) from selenates (Se^{6+}) to selenite (Se^{4+}), a slow process and (2) from selenite (Se^{4+}) to selenides (Se^{2-}) and to elemental selenium and to organic compounds—a very slow process (Fig. 18.2, Kabata-Pendias, 2001).

4 Selenium in plants

The effect of selenium addition was performed in stress condition in several experiments (Proietti et al., 2013; Irmak, 2017; Huang et al., 2018; Zhu and Ma, 2018; Zhao et al., 2011) on different plants (*Vitis vinifera*, *Olea europaea*, *Fragaria ananassa*, *Arachis hypogaea*, and *Actinidia chinensis*). Obviously, selenium always enables to alleviate the stress condition (drought, water stress, copper toxicity, and abiotic stress) using the antioxidant function.

The sufficient selenium content can also reduce the disintegration of the chlorophyll. Foliar spraying with selenium alleviated the decline in the net photosynthetic rate. Selenium can induce a partial stomatal closure, as evidenced by the values of stomatal conductance, resulting in a reduction in net assimilation, and thus a decrease in dry-matter production. The addition of selenium to the plants proves also a positive effect on increased tolerance against heavy metals in soil. Some authors (Long et al., 2016a,b) describe that the application of exogenous selenium in irrigation water or in organic fertilizer may increase selenium content, but Cd and Pb can be decreased in accumulation in kiwifruit (*Actinidia deliciosa*). The reduced accumulation of heavy metals Pb, Cr, Cd, As, and Ni in grapes after foliar spraying with selenium fertilizers was also observed in the experiment with *V. vinifera* (Zhu et al., 2017c). Selenium proves similar chemical properties to sulfur taken up inside the plants using sulfate transporters and assimilated by sulfur assimilating pathway (Sors et al., 2005b; Dumont et al., 2006). This is especially true for selenan (Se^{VI}), but phosphate transport mechanism is also used for selenite intake (Se^{IV}) (Li et al., 2008a).

The absorption of both organic and inorganic forms of selenium can be passive (nonmetabolic) that means ion diffusion from soil solution to root endoderm, although the role of simple diffusion is limited (Terry et al., 2000; Li et al., 2008b). Some plants can accumulate selenium against the concentration gradient (active/metabolic) requiring metabolic energy. The accumulation of selenium against the concentration gradient may be successfully used for both bioremediation and for increase the selenium content in food in areas with selenium deficiency (Neuhierl and Bock, 1996; Freeman et al., 2012). Because of this reason, selenium concentration in different plant species differs in the same locality leading to a different selenium intake—for example, at selective grazing of specific plant species by herbivores or at consumption of specific plant crops originating from a given site.

Obviously, selenium has also been found out to modify the transport or absorption of other elements (Pilon-Smits and LeDuc, 2009). Depending on the selenium concentration, selenium may cause synergistic or antagonistic effects. The antagonistic effect of Se-selenite on P detected in the leaves can be probably the result of the competition between these ions (Hopper and Parker, 1999), because both are absorbed through phosphate transporters (Zhao et al., 2010). S and Se generally follow the same molecular pathways for their uptake, metabolism, enzymatic reduction, incorporation into protein, and volatilization in plant leaves and roots (Sors et al., 2005a). After the uptake into the roots, SeO_4^{2-} (similar to SO_4^{2-}) is transported in most cases to the shoot chloroplasts and often reduced and incorporated into different organic selenium compounds (Ellis and Salt, 2003).

Most of the S molecules found out in plants proved to have Se analogs. Most plant species contain less than 25 μg of Se per g of dry weight in their natural environment and are also unable to tolerate the increased concentrations of selenium (White et al., 2004). As selenium is integrated into SeMet and SeCys, which can be misincorporated, and nonspecifically may replace Met and Cys in proteins frequently resulting in selenium toxicity to plants (Brown and Shrift, 1982). SeMet may be used to form selenoproteins or methylated to form methyl-SeMet (Me-SeMet). Me-SeCys or Me-SeMet may be further volatilized into the atmosphere as nontoxic dimethylselenide (DMSe) in or dimethyldiselenide (DMDSe). Both DMSe and DMDSe are unstable, and several days after the volatilization, they return to the soil as organic selenium (Kubachka et al., 2007; Martens and Suarez, 1999). The brief outline of beneficial selenium effect in plants is described in Fig. 18.3. The specific effect of sulfur and selenium interactions in strawberry cultivation is described by Santiago et al. (2018). The results showed that the concentration and accumulation of selenium were significantly reduced in the shoots and fruits of strawberry with the application of the sulfated fertilization, while absorption of S was maximized with increased levels of selenium. Generally, selenium concentrations are found out to be higher in younger leaves in comparison with older ones during seedling growth. The content of selenium in seeds is increasing during the reproduction stage, whereas the content of selenium in leaves is dramatically decreased. Plants have been classified as hyperaccumulators, secondary accumulators, and nonaccumulators according to selenium accumulation inside of their cells (Galeas et al., 2007; Bodnar et al., 2012). Hyperaccumulators accumulate higher amounts of selenium in their cells, that is, >1000 μg/g Se DW (dry weight) and suitably prosper in selenium-rich regions of the world.

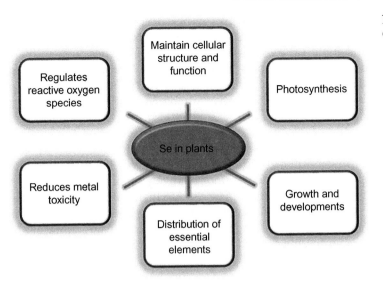

5 Selenium toxicity in plants

Selenosis or selenium toxicity can occur in plants in the conditions when the optimum concentration of selenium is exceeded. Selenium may cause toxicity using two mechanisms such as malformation of selenoproteins or inducing oxidative stress. Both mechanisms have been known to be harmful to plants. Malformed selenoproteins are formed because of the misincorporation of SeCys/SeMet in the place of Cys/Met in protein chain. In comparison with SeMet, the substitution of SeCys can be more reactive and detrimental to protein functioning (Hondal et al., 2012). Sulfur substitution in enzymes can affect their activity such as nitrogenase activity which may be decreased after the replacement of Fe-S cluster with Fe-Se (Hallenbeck et al., 2009). Selenium works as prooxidant at high doses and can generate reactive oxygen species causing the oxidative stress in plants. The level of glutathione is decreased in selenium stress (Hugouvieux et al., 2009), but Se-tolerant plants proved higher levels (Grant et al., 2011). In supplementing types of fruit with selenium, an increase can be achieved in fruit content, but parallelly, necrotic spots may occur on the leaves, as reported by Licina et al. (1998). Differential sensitivity to selenium toxicity is also described in different peach rootstocks demonstrating genotypic variation in selenium uptake and the accumulation (Pezzarossa et al., 2009).

6 Methods of selenium determination in soil and plants

Trace concentrations (milligram/gram) of selenium in plants and soil samples are determined using various analytical detection methods such as fluorometer, neutron activation analysis (NAA), inductively coupled plasma mass spectrometry (ICP-MS), gas chromatography (GC), spectrophotometry, X-ray, atomic absorption spectroscopy (AAS), hydride generation atomic absorption spectroscopy (HGAAS), inductively coupled plasma atomic emission spectroscopy (ICP-AES), and fluorescence analysis (Krishnaiah et al., 2003). Enzymatic formation of volatile selenium compounds should be reduced by freezing the biological samples immediately after the collection.

The sample pretreatment procedures need to be in attention. The flow injection systems can provide the advantage of full quick analysis, automation, minimum sample and less reagent consumption, and low risk of contamination. Online sorption and preconcentration of selenium enable to exceed the matrix effect, the insufficient detection power of the analytical techniques, and other shortcomings of batchwise operation. This procedure is based on solid or solvent-phase extraction and can provide the possibility for a differential determination of selenium oxidation states (Yan et al., 1999; Ko and Yang, 1996). Online reduction using HGAAS method in a closed system (Cobo et al., 1994) enable to prevent losses of selenium due to the formation of volatile selenium chloride. The use of microwave irradiation can perform the same process. The effectiveness of the electrothermal vaporization is used as a thermochemical reactor. Citric acid may be used as a chemical modifier converting selectively selenium into a volatile molecular form and, therefore, preparing relatively easy determination by ICP-MS at ultra-trace levels (Sturgeon and Lam, 1999). Subcritical water extraction procedure of solid samples was combined to a continuous flow system for succeeding treatment, derivatization, and finally transport to the detector (Varadé and de Castro, 1998). Speciation analysis of plants has usually been carried out by HPLC/ICP-MS method using the extraction process not to modify and alter the chemical forms of the analyte or to disturb the equilibrium between the present various species. Leaching with hot water has often been used for selenoamino acids because of water-soluble feature. Classical flame AAS techniques do not provide sufficiently low detection limits for selenium detection in biological samples (Koirtyohann and Morris, 1986). Selenium species are difficult to determine by inductively coupled plasma atomic emission spectrometry (ICP-AES) because of the inadequate sensitivity in comparison with other elements, but graphite furnace atomic absorption spectrophotometry (GFAAS) can be used for the detection (Watson and Isaac, 1990).

HGAAS has been carried out instead of the determination of selenium in biological samples (Aldea and Constantin, 2010). Graphite furnace atomic absorption spectroscopy (GFAAS) can prove a high sensitivity of 5×10^{-11} g selenium/g sample, but the interference from the matrix can confirm significant difficulties (Lewis, 1988). GFAAS may depend based on selenium compounds and to react to numerous metal compounds relatively forming refractory metal selenides (Oster and Prellwitz, 1982). Molybdenum, nickel, and platinum can be used to thermally stabilize the selenium. Really high temperature in the furnace is used prior to atomization of the sample so the destruction of organic material can happen. The graphite sample cell can be chemically (in situ) treated to reduce chemical interference leading to an advantage of GFAAS, and so one limitation requiring the correction for background absorption may exist.

The Zeeman splitting of the absorption line (Koirtyohann and Morris, 1986) and the deuterium continuum light source method (Hoenig and Hoeyweghen, 1986) are included between modified techniques. A Zeeman-effect system requires a magnetic field to the atomizer allowing the background correction performed at the exact analyte wavelength without the use of auxiliary light sources. ICP-AES with hydride vapor generation can be suitable to analyze small sample enabling to determine the total selenium in biological samples. The samples should be stored in sulfuric, nitric, and perchloric acids at temperatures up to 310°C. The detection limit of this method is carried out down hundreds of nanogram per liter (Marchante-Gayón et al., 2000; Weekley and Harris, 2013). The determination of selenium content in natural fruit juices can be caused by using atomic absorption spectrometry, spectrophotometry, and spectrofluorimetry (Gawloska-Kamocka, 2008).

6.1 Biofortification

The agronomic biofortification of selenium is generally created by increasing concentration of the nutrient in agricultural crops using fertilization or eventually applying different chemical forms (Graham et al., 2007). This strategy of selenium biofortification of food crops has been carried out especially in selenium-deficient regions by adding inorganic selenium-containing fertilizers to soils or using organic sources of selenium. The most commonly added form of selenium such as selenate and to a lesser extent selenite used as sodium or barium salts have been included in these studies. Granular or blended forms can be directly applied to the soil or using high volume liquid drenches getting into the soil in a large majority (Broadley et al., 2010; Iwashita and Nishi, 2004; Rayman et al., 2008; Shrestha et al., 2006). The application of high quantities of selenium fertilizers may not always be included between the most sustainable strategies. Potential leaching of excessive selenium may occur, and therefore, regular applications can increase the costs (Hirschi, 2009; White and Broadley, 2009; Winkler, 2011). The implementation efficiency of selenium fertilizer strategy may be increased in Se biofortification using organic acids (Morgan et al., 2005), organic forms of selenium (Schiavon et al., 2013; Pezzarossa et al., 2014), or microbes (Duran et al., 2013, 2014), enhancing the chances of selenium to be available for plants.

Disadvantages of selenium application to the soil can lead to a low selenium recovery rates in edible portions of crops (Brown and Shrift, 1982) or excessive selenium accumulation in the soil with long-term application of selenium fertilizer becoming toxic for nearby ecosystems (Feugang et al., 2006).

Hence, an effective method of biofortifying food crops with selenium is the foliar application. Both forms of selenite and selenate have been proved to be more bioavailable for plants applied directly to leaf surfaces opposed to soils (Combs et al., 2011). The effect of direct foliar exposure route can ensure a high efficiency of selenium assimilation into the plant because of the independence from root-to-shoot translocation (Winkel et al., 2015). The rate of biotransformation from absorbed inorganic selenium to different organic forms of selenium in a plant can be species specific depending on specific biochemical pathways (Zou et al., 2005). A selenium-containing solution is sprayed onto the leaf surface of the crop using foliar application. In this respect, soil chemistry and microbiological processes prove less impact on selenium ensuring a higher uptake efficiency with low volumes of applied selenium solution. The factors such as the amount of applied selenium, leaf area and surface structure, and differences in plant-specific metabolism of selenium can differ between the crops and should be considered. Ros et al. (2016) discovered that selenate foliar fertilization seems to be the most effective fertilizer strategy to increase selenium uptake in most arable crops.

Foliar application strategies demand careful consideration of the following factors related to spraying the selenium solution onto plants: (1) selenium solutions must be carefully prepared and delivered using well-calibrated spraying equipment; (2) windy and/or rainy days should be avoided, and plants must prove the adequate leaf surface area to ensure the absorption of selenium; and (3) growth stages need to be determined in timing applications to prove the highest selenium absorption.

Another effective alternative soil treatment may be the use of Se-enriched organic or green fertilizers to produce selenium biofortified crops. One of the first studies of Ajwa et al. (1998), Bañuelos et al. (1992), and more recently the results of Bañuelos et al. (2015) have confirmed that selenium applied in organic matter (green manure) could be taken up by various plant species. As the first one, Freeman and Bañuelos (2011) suggested the possibility of using Se-hyperaccumulating plant materials as an organic Se-enriched fertilizer for biofortifying food crops. Plant uptake of selenium from Se-laden organic matter containing big proportions of organic compounds (e.g., SeMet and SeCys) may supposedly occur in higher rates compared with inorganic sources of selenium (Abrams et al., 1990; Kikkert and Berkelaar, 2013), although the uptake mechanisms for organic selenium compounds have been poorly known but probably amino acid transporters can be involved (Kikkert and Berkelaar, 2013). The application of selenium-enriched organic/green fertilizers has proved potential advantages of inorganic selenium sources because of selenium gradual release from organic matter into soil solution. The bioaccessibility of selenium may be influenced by the formation of indigestible Se-containing polysaccharides (Bhatia et al., 2013) and/or association of selenium with chitin-containing structures in cell walls (Serafin Muñoz et al., 2006). The specific content of selenium in fruit crops can highly vary. According to the reported literature, the highest selenium content of fruits was determined in the range of 0.5–10 µg/g. The highest amount up to 10 µg/g was detected in kiwifruit. The lowest amount with less than 1 µg/g was found out in peach (Wang and Huang, 2007). Most of the selenium-content fruits such as pear, apple, jujube, and grape proved that the selenium concentration below 5 µg/g (Wu and Tian, 2009). As we can see from the example of strawberries grown in Mexico containing an average of 6.35 µg/g of selenium, after the application of 4 mg/L, the average selenium concentration reached up to 31.2 µg/g of DW in fruits that is lower amount compared with the results of strawberries treated with selenate in the concentration of 7.9 mg/L (46.04 µg/g DW) (Mimmo et al., 2017). Similarly, selenium concentrations were detected in higher levels in other types of fruit crops fertilized with selenium reaching up to 46.7 µg/g in Opuntia (Bañuelos et al., 2011) and up to 35.8 µg/g in tomatoes (Castillo-Godina et al., 2016). Blinnikova and Eliseeva (2016) carried out the experiment focused on the selenium enrichment in several fruit crops. The research crops included the experiment of the fruits of edible honeysuckle, garden strawberry, Actinidia kolomikta, and field ash and black chokeberry. Selenium was applied in the form of aqueous sodium selenate (the concentration of 1, 2, and 3 mg/L of solution) during the formation of fruits and berries. The application of selenium in a dose of 3 mg/L proved the most influencing effect on the selenium content in fruits. The selenium content in this dose was increased by 5.2-fold in honeysuckle, by 3.9-fold in strawberry, by 3-fold in Actinidia, and by 2.7-fold in the field ash and black chokeberry's fruits. In the fruits of the individual types of fruits, the selenium content was raised up to the values in micrograms per 100 in the range of 32.3–35.2 in honeysuckle, from 11.7 to 13.1 in strawberry, from 3.5 to 4.1 in A. kolomikta, from 4.6 to 5.1 in fruits of the field ash, and from 2.8 to 3.2 in black chokeberry's fruits.

The apples are well-available and globally widespread fruits. Jakovljevic et al. (1998) mentioned that unfertilized apple trees reached up to 0.03 µg/g of the selenium concentration in leaves. The fruits showed up to 0.04 µg/g of the selenium concentration. The application of selenium (up to 1.5 kg/ha Se as Na_2SeO_4) using special root-soaking application can significantly increase the selenium concentration up to 3.22 µg/g or up to 2.81 µg/g, respectively. Apparently, based on published results, the foliar application of selenium proved to be less effective especially the effect on the selenium content in fruit (up to 0.20 µg/g).

The fertilizers, containing selenium, have been extended in China to produce selenium-enriched fruits such as apple, pear, jujube, kiwifruits, grape, oranges, and watermelon (Wu and Tian, 2009). Some researchers injected inorganic Se liquid into the trunk during the period of growth to improve the organic Se content of fruits. The results showed that this method could quantitatively improve the Se content up to 30–200 µg/g of fresh fruit without disrupting the normal growth of the fruit trees (Wu et al., 2004). Three major methods of production selenium-rich fruits such as foliage spraying, root irrigating, and injection with drip transport were tested. The results of Hu et al. (1998) detected that the injecting and drip transport of the trunk of the apple tree proved better results compared with other methods. Hu et al. (1998) injected and drip transported the trunk of 10-year-old tree in spring (in the middle of April) using up to 500-mL liquid with up to 0.5 g of sodium selenite resulting in up to 25–40 µg/g of the Se content of apple fruit. Wu et al. (2009) showed similar results in 8-year-old apple trees. The selenium content was increased from 0.96- to 8.08-fold after the application of selenium. The optimum time for the dripping injection of selenium is stated in the young fruit period. During fertilizing, the optimum concentration reached up to 600 mg/L. Similar experiments were carried out in peer trees. After foliage spraying of the pear tree using the Se fertilizer, the selenium content was increased from 2- to 4-fold in fruit and vegetable organs achieved up to 7–60-fold. The selenium content of fruit was enhanced with the prolonged period of growth and the increased repetition of spraying (Meng, 2006). The nonpolluted, Se-enriched watermelon (*Citrullus lanatus*) achieved the selenium content up to 19.3 µg/g in a greenhouse using foliage spraying and root irrigating (Liu, 2006). Foliage spraying may also produce enriched selenium kiwifruit (*A. deliciosa*). The enriched kiwifruit proved an average of organic selenium content up to 0.108 µg/g (Chen et al., 2001). Foliage spraying and root irrigating with the Se fertilizer should also improve the selenium content in sarcocarp and pericarp of satsuma mandarin (*Citrus unshiu*) during fruit setting. The foliar application is included between the effective approaches of producing selenium-enriched oranges. The difference between the effect of root irrigating and foliage spraying can exist. The roots were sprayed with sodium selenite up to 4 g per tree so the selenium content of sarcocarp was increased up to 43.6% compared with the control group.

However, foliage spraying with sodium selenite in a dose of 1–2 g per orange tree means that the selenium content of sarcocarp increased from 43.6% to 369.2% compared with the control group (Zhang et al., 1995; Cheng et al., 2007). The same effects were detected in the papaya (Liu et al., 2007).

7 Effect of selenium application on yield and product quality

An increase of selenium content in plant organs can be ensured by various application technologies (see earlier). Foliar application of selenium may appear as an easy and conventional method. The literature describes the experiments where, for example, peaches (*Prunus persica*, cv. Batch) and pears (*Pyrus communis* L.) were sprayed with selenate solution (Pezzarossa et al., 2012; Meng, 2006) or table grapes (*V. vinifera* L.) sprayed with organic selenium (Zhu et al., 2017b), pears (*Pyrus bretschneideri* cv. "Huangguan"), grapes (*V. vinifera*, cv. "Kyoho"), and peaches (*P. persica*, cv. Jinliuzaohong) sprayed with amino acid–chelated selenium solution (Feng et al., 2015). Selenium is not an essential element for plants, but the effects can be reflected in better growth and thus higher yields of fruit. Selenium fertilization can affect the synthesis of amino acids and proteins (Schiavon et al., 2013). An increased net photosynthetic rate and the decreased stomatal conductance and transpiration rate were found out in pears, grapes, and peach foliar sprayed with amino acid–chelated selenium (Feng et al., 2015). The available studies, dealing with strawberries (*F. ananassa*) describe either an increase of the fresh weight of the fruits after the selenium application (Santiago et al., 2018; Narváez-Ortiz et al., 2018) or yield of strawberries not affected on selenium application (Mimmo et al., 2017). On the other hand, the foliar application of selenium resulted in the increased yield and higher selenium content in the fruit at the stages of first flowering and early fruiting of watermelon (Yang et al., 2008). The application of selenium without affecting the yield of fruit trees is described in studies of Pezzarossa et al. (2012) and Zhu et al. (2017b). The evidence of positive effects of selenium on the quality parameters such as soluble solid content (Zhu et al., 2016, 2017a; Lee et al., 2007), titratable acidity (Zhu et al., 2016; Lee et al., 2007), glucose, fructose, and the total sugar content (Lee et al., 2007) and firmness (Zhu et al., 2016) of tomato fruit has been reported. Selenium can affect the soluble solid content in peach and pear (Pezzarossa et al., 2012). The content of glucose, fructose, organic acid, and protein in grape is also positively influenced (Zhu et al., 2017b). Additionally, selenium delayed the decreasement of firmness (Zhu et al., 2016; Pezzarossa et al., 2012) and titratable acidity, as well as the weight loss in tomato fruits during storage (Zhu et al., 2016). A delayed reduction positively influences the shelf life of fruits in flesh firmness. Selenium can also stimulate the formation of ascorbic acid and polyphenolics in fruits (D'Amato et al., 2014; Zhao et al., 2013; Schiavon et al., 2013). The bioactive substances can substantially contribute to the health value of fruits (Boeing et al., 2012). In tomato fruits, selenium may also induce an increased content of pigments and antioxidant compounds (Pezzarossa et al., 2013, 2014; Zhu et al., 2017a; Schiavon et al., 2013; Andrejiová et al., 2016). Foliar application of sodium selenite was applied in different concentrations on jujube (*Ziziphus jujube*) with similar results (Zhao et al., 2013; Jing et al., 2017; Wu and Ning, 2013). Obviously, foliar application of selenium increased the selenium content in fruits during the fruitlet stage. Higher selenium concentration in fruit led to the increased fruit quality in terms of fruit firmness, vitamin C, soluble protein and total soluble sugar. The average fruit weight and yield of jujube showed an increasing trend together with higher selenium addition. The higher selenium content in fruits can cause the reduction of the total titratable acid in fruits, thus, to improve the fruit flavor and quality.

The same results were also observed in the experiments carried out in mandarin (*Citrus reticulata*), raspberry (*Rubus idaeus*), strawberry (*Fragaria vesca*), grape fruits (*Citrus paradisi*), grape (*V. vinifera*), or pears (*P. bretschneideri*) (DeXiang and Xin Qi, 2000; Licina et al., 1998; Li et al., 2018b; Wang et al., 2016; Liu et al., 2015; Carvalho et al., 2003). Acid invertase activity, total soluble sugar and the selenium content in berries of grape (*V. vinifera* L.), and photosynthetic rate in leaves were found out higher using selenium fertilizer compared with the control variant without selenium application (Zhu et al., 2017a,c). These authors confirmed the significant correlation between the selenium content and sugar content. Selenium-enriched kiwifruit is also produced using foliage spraying. The enriched kiwifruit proved a high quality fruit containing increased fiber (Chen et al., 2001). The content of vitamin C, soluble sugar, soluble solid, and dry matter was increased with higher amount of selenium supplement. The titratable acid was decreased (Long et al., 2016a,b). Li et al. (2018a) detected the significant increase of anthocyanin concentration and the intact fruit rate of blueberry.

An **antioxidant effect** of selenium has been determined in Se-enriched fruit crops due to an improved antioxidative status and the reduction of biosynthesis of ethylene, which is the hormone with a primary role in plant senescence and fruit ripening. This fact highlights the possible positive effect of selenium in preserving a longer shelf life and longer-lasting quality (Puccinelli et al., 2017; Luksic and Germ, 2017). Selenium can increase the antioxidant enzyme activity in tomatoes (Zhu et al., 2017a) or strawberries (Narváez-Ortiz et al., 2018). Various reactive oxygen species may by decreased during storage (Zhu et al., 2017a). The experiments, carried out by Pezzarossa et al. (2014), detected a delay in the onset of fruit ripening in Se-enriched tomato showing a reduced rate in color change and an earlier harvesting of control plants compared with Se-treated plants. The delay in fruit ripening can depend on a reduced ethylene production. This reduction could be due to the higher cellular concentration of selenomethionine (SeMet) than methionine (Met), which is a precursor of ethylene in the ethylene biosynthesis pathway (Yang and Hoffman, 1984). Another hypothesis showed the genes of the enzymes involved in the ethylene biosynthesis pathway. Zhu et al. (2017a) detected that selenium can suppress the transcription of genes, controlling the ethylene production during ripening, and, thus, the reduction of the production in tomato fruits. The testing of changes of quality and physiology using preharvest application of selenium was examined in stored nectarine (*P. persica*). The results confirmed that the flesh firmnesses and the total acid content of treated fruits were detected in higher levels compared with the control variant. The content of soluble solids was not affected by the applications. In treated fruits, peroxidase (POD) and polyphenoloxidase (PPO) activities were found out in lower levels compared with the control variant (Guo et al., 2005). Selenium proves an indirect positive effect on the postharvest storage of vegetables by reducing germination and the mycelial growth of some harmful fungi such as *Botrytis cinerea* (Wu et al., 2016; Zhu et al., 2016) and *Penicillium expansum* (Wu et al., 2014). *B. cinerea* can cause gray mold decay, one of the main pre- and postharvest diseases in fruit and vegetables (Quinn et al., 2010; Youssef and Roberto, 2014), leading to economic losses (Soylu et al., 2010; Cabot et al., 2013). Selenium also counteracts to the infections caused by *Alternaria brassicicola* and *Fusarium* sp. and enhances the resistance to fungal diseases (Hanson et al., 2003) reducing the damage due to *Fusarium* wilt infection in tomato (Companioni et al., 2012). Selenium can also contribute to longer stability of plant fruits observed in the experiments performed in apricot (*Prunus armeniaca*) and nectarine (*P. persica*) (Gao et al., 2016; Yan et al., 2014). These findings stated that selenium can prove potential application in postharvest disease control because of the effective inhibition of *Alternaria alternata* inoculated on the apricot fruit.

8 Relationship between selenium and silicon

Synergistic effect of selenium with silicon can be emphasized from the interaction of selenium with other elements in the plant. The individual roles of selenium and silicon are the same as the synergistic effects of Si-Se examined in the experiment with their different concentrations in the foliar application used on date palm (*Phoenix dactylifera* L., cv. Zaghloul) (El-Kareem et al., 2014). As we can see from the results, the leaf area and the content of the total chlorophylls and N, P, and K were significantly stimulated after the continuous cyclic treatment by silicon and selenium. After the comparison of the observation, we can conclude that the use of silicon and selenium together are more effective compared with the use of the individual compound. Palm fruit can be increased in weight and dimension in four repetitions of treatment. A significant higher fruit quality (the increased content of the total soluble sugar, the total and reduced sugars, the decreased percentage of seed, the total acidity, content of the total soluble tannins, and the total crude fiber) was detected in this combined treatment. Ibrahim and Al-Wasfy (2014) showed similar result in the experiment with orange (*Citrus sinensis*, cv. Valencia). The detected orange fruits also provided high quality of fruit yield in combined treatment of selenium and silicon. It is evident from their results, that the combination of selenium and silicon (with Potassium and Boron) can be very useful for stimulating growth aspects, nutritional status of the trees, yield, and both physical and chemical characteristics of fruits. The positive influence on yield and quality of early sweet grapevines is also described by Uwakiem (2015) in a 2-year trial with foliar application of silicon, selenium, and humic acid. Many studies show the positive joint action of selenium and silicon on crops under cadmium stress (Liu et al., 2017; Gao et al., 2018) or salinity (Bybordi et al., 2018; Sattar et al., 2017).

9 Future research

A significant increase has developed in agricultural land production during the last century. It is also very important to screen Se-hyperaccumulator plants and those plant species that can accumulate selenium in edible parts within the safer limits for human consumption (Gupta and Gupta, 2016). Genetic engineering is also a useful strategy to obtain Se-biofortified food products focused on manipulation of Se-related enzymes for selenium uptake, assimilation, and volatilization. Generally, in association with selenium, we should target on (1) increasing plant tolerance of high soil selenium concentration, (2) increasing selenium uptake and the transport to shoot, (3) enhancing of selenium accumulation in shoot tissues, and (4) improving of selenium volatilization. Numerous opportunities are opening including not only "simple" plant genome modifications but also introducing of nonplant-based genes in CRISPR times to fulfill the aforementioned conditions.

References

Abrams, M., Shennan, C., Zasoski, R., Burau, R., 1990. Selenomethionine uptake by wheat seedlings. Agron. J. 82, 1127–1130.

Adriano, D.C., 1986. Trace Elements in the Terrestrial Environment. Springer-Verlag, New York, p. 1986. 533 p.

Ajwa, H.A., Bañuelos, G.S., Mayland, H.F., 1998. Selenium uptake by plants from soils amended with inorganic and organic materials. J. Environ. Qual. 27, 1218–1227.

Aldea, M.M., Constantin, L., 2010. An analytical method for chemical speciation of selenium in soil. Sci. Study Res. Chem. Chem. Eng. 11 (3), 323–328.

Andrejiová, A., Hegedűsová, A., Mezeyová, I., 2016. Effect of genotype and selenium biofortification on content of important bioactive substances in tomato (Lycopersicon esculentum Mill.) fruits. J. Int. Sci. Publ. 4, 8–18.

Bañuelos, G., Mead, R., Wu, L., Beuselinck, P., Akohoue, S., 1992. Differential selenium accumulation among forage plant species grown in soils amended with selenium-enriched plant tissue. J. Soil Water Conserv. 47, 338–342.

Bañuelos, G.S., Fakra, S.C., Walse, S.S., Marcus, M.A., Yang, S.I., Pickering, I.J., Pilon-Smits, E.A.H., Freeman, J.L., 2011. Selenium accumulation, distribution, and speciation in spineless prickly pear cactus: a drought-and salt-tolerant, selenium-enriched nutraceutical fruit crop for biofortified foods. Plant Physiol. 155, 315–327.

Bañuelos, G.S., Arroyo, I., Pickering, I.J., Yang, S.I., Freeman, J.L., 2015. Selenium biofortification of broccoli and carrots grown in soil amended with Se-enriched hyperaccumulator Stanleya pinnata. Food Chem. 166, 603–608.

Bhatia, P., Aureli, F., D'Amato, M., Prakash, R., Cameotra, S.S., Nagaraja, T.P., Cubadda, F., 2013. Selenium bioaccessibility and speciation in biofortified Pleurotus mushrooms grown on selenium-rich agricultural residues. Food Chem. 140, 225–230.

Blinnikova, O.M., Eliseeva, L.G., 2016. Enrichment of fruits and berries with selenium and prospects for their using in the preventive nutrition. Vopr. Pitan. 85 (1), 85–91.

Bodnar, M., Konieczka, P., Namiestnik, J., 2012. The properties, functions and use if selenium compound in living organisms. J. Environ. Sci. Health C Environ. Carcinog. Exotoxicol. Rev. 30 (3), 225–252.

Boeing, H., Bechthold, A., Bub, A., Ellinger, S., Haller, D., Kroke, A., Leschik-Bonnet, E., Müller, M.J., Oberritter, H., Schulze, M., et al., 2012. Critical review: vegetables and fruit in the prevention of chronic diseases. Eur. J. Nutr. 51, 637–663.

Broadley, M.R., Alcock, J., Alford, J., Cartwright, P., Foot, I., Fairweather-Tait, S.J., Hart, D.J., Hurst, R., Knott, P., McGrath, S.P., 2010. Selenium biofortification of high-yielding winter wheat (Triticum aestivum L.) by liquid or granular Se fertilisation. Plant Soil 332, 5–18.

Brown, T.A., Shrift, A., 1982. Selenium: toxicity and tolerance in higher plants. Biol. Rev. Camb. Philos. Soc. 57, 59–84.

Bybordi, A., Saadat, S., Zargaripour, P., 2018. The effect of zeolite, selenium and silicon on qualitative and quantitative traits of onion grown under salinity conditions. Arch. Agron. Soil Sci. 64 (4), 520–530.

Cabot, C., Gallego, B., Martos, S., Barceló, J., Poschenrieder, C., 2013. Signal cross talk in Arabidopsis exposed to cadmium, silicon, and Botrytis cinerea. Planta 237, 337–349.

Carvalho, K.M., Gallardo-Williams, M.T., Benson, R.F., Martin, D.F., 2003. Effects of selenium supplementation on four agricultural crops. J. Agric. Food Chem. 0021-8561. 51 (3), 704–709. [cit. 2019-01-18]. Available from: http://pubs.acs.org/doi/abs/10.1021/jf0258555.

Castillo-Godina, R.G., Foroughbakhch-Pournavab, R., Benavides-Mendoza, A., 2016. Effect of selenium on elemental concentration and antioxidant enzymatic activity of tomato plants. J. Agric. Sci. Technol. 18, 233–244.

Chen, L., Zhang, B., Zhang, Y., 2001. The hypolipidemic effect of selenium rich kiwifruits and influence on hemorrheology. Acta Nutr. Sin. 1, 448–449.

Cheng, Z., Long, J., Song, L., 2007. The produce technique of enriched-Se oranges. J. Mod. Hortic. 8, 15–16.

Cobo, M.G., Palacios, M.A., Cámara, C., Reis, F., Quevauviller, P., 1994. Effect of physicochemical parameters on trace inorganic selenium stability. Anal. Chim. Acta 286, 371–379.

Combs, G.F., 2001. Selenium in global food systems. Brit. J. Nutr. 85, 517–547.

Combs, G.F., Watts, J.C., Jackson, M.I., Johnson, L.K., Zeng, H.W., Scheett, A.J., Uthus, E.O., Schomburg, L., Hoeg, A., Hoefig, C.S., Davis, C.D., Milner, J.A., 2011. Determinants of selenium status in healthy adults. Nutr. J. 10, 10(75).

Companioni, B., Medrano, J., Torres, J.A., Flores, A., Rodríguez, E., Benavides, A., 2012. Protective action of sodium selenite against Fusarium wilt in tomato: total protein contents, levels of phenolic compounds and changes in antioxidant potential. Acta Hortic. (947), 321–328.

Cone, J.E., Del Río, R.M., Davis, J.N., Stadtman, T.C., 1976. Chemical characterization of the selenoprotein component of clostridial glycine reductase: identification of selenocysteine as the organoselenium moiety. Proc. Natl. Acad. Sci. U. S. A. 73, 2659–2663.

D'Amato, R., Proietti, P., Nasini, L., Del Buono, D., Tedeschini, E., Businelli, D., 2014. Increase in the selenium content of extra virgin olive oil: quantitative and qualitative implications. Grasas Aceites 65, 1–9.

DeXiang, X., Xin Qi, W., 2000. Effect of foliar spray of selenic fertilizer on the fruit and leaf content of selenium of Ponggan mandarin variety. South China Fruits 29 (3), 4.

Dumont, E., Vanhaecke, F., Cornelis, R., 2006. Selenium speciation from food source to metabolites: a critical review. Anal. Bioanal. Chem. 385, 1304–1323.

Duran, P., Acuna, J.J., Jorquera, M.A., 2013. Enhanced selenium content in wheat grain by co-inoculation of selenobacteria and arbuscular mycorrhizal fungi: a preliminary study as a potential Se biofortification strategy. J. Cereal Sci. 57, 275–280.

Duran, P., Acuna, J.J., Jorquera, M.A., 2014. Endophytic bacteria from selenium-supplemented wheat plants could be useful for plant-growth promotion, biofortification and *Gaeumannomyces graminis* biocontrol in wheat production. *Biol. Fertil*. Soils 50, 983–990.

Eich-Greatorex, S., Sogn, T.A., Øgaard, A.F., Aasen, I., 2007. Plant availability of inorganic and organic selenium fertiliser as influenced by soil organic matter content and pH. Nutr. Cycl. Agroecosyst. 79, 221–231.

El-Kareem, M.R.G., Abdel-Aal, A.M.K., Mohamed, A.Y., 2014. The synergistic effects of using silicon and selenium on fruiting of Zaghloul date palm (*Phoenix dactylifera* L.). Int. J. Agric. Biosyst. Eng. 8, 1–4.

Ellis, D.R., Salt, D.E., 2003. Plants, selenium and human health. Curr. Opin. Plant Biol. 6 (3), 273–279.

Fairweather-Tait, S.J., Bao, Y., Broadley, M.R., Collings, R., Ford, D., Hesketh, J.R., Hurst, R., 2011. Selenium in human health and disease. Antioxid. Redox Signal. 14, 1337–1383.

Feng, T., Chen, S.S., Gao, D.Q., Liu, G.Q., Bai, H.X., Li, A., Peng, L.X., Ren, Z.Y., 2015. Selenium improves photosynthesis and protects photosystem II in pear (*Pyrus bretschneideri*), grape (*Vitis vinifera*), and peach (*Prunus persica*). Photosynthetica 53, 609–612.

Fernandez-Martinez, A., Charlet, L., 2009. Selenium environmental cycling and bioavailability: a structural chemist point of view. Rev. Environ. Sci. Biotechnol. 8, 81–110.

Feugang, J.M., Konarski, P., Zou, D.M., Stintzing, F.C., Zou, C.P., 2006. Nutritional and medicinal use of Cactus pear (*Opuntia spp.*) cladodes and fruits. Front. Biosci. 11, 2574–2589.

Finley, J.W., 2005. Selenium accumulation in plant foods. Nutr. Rev. 63, 196–202.

Freeman, J.L., Bañuelos, G.S., 2011. Selection of salt and boron tolerant selenium hyperaccumulator Stanleya pinnata genotypes and characterization of Se phytoremediation from agricultural drainage sediments. Environ. Sci. Technol. 45, 9703–9710.

Freeman, J.L., Marcus, M.A., Fakra, S.C., Devonshire, J., McGrath, S.P., Quinn, C.F., Pilon-Smits, E.A.H., 2012. Selenium hyperaccumulator plants *Stanleya pinnata* and *Astragalus bisulcatus* are colonized by Se-resistant, Se-excluding wasp and beetle seed herbivores. PLoS One 7 (12), 1–13.

Galeas, M.L., Zhang, L.H., Freeman, J.L., Wegner, M., Pilon-Smits, E.A.H., 2007. Seasonal fluctuations of selenium and sulfur accumulation in selenium-hyperaccumulators and related non-accumulators. New Phytol. 173, 517–525.

Gao, C., Li, Y., Yang, B., Xiao, L., Lan, Y., Wenjing, Q., Di, W., Ying, T., 2016. Effects of sodium selenite treatment on black spot disease and storage quality of postharvest apricot fruit. Food Sci. 37 (14), 258–263.

Gao, M., Zhou, J., Liu, H., Zhang, W., Yuanmei, H., Liang, J., Zhou, J., 2018. Foliar spraying with silicon and selenium reduces cadmium uptake and mitigates cadmium toxicity in rice. Sci. Total Environ. 631/632, 1100–1108.

Gawloska-Kamocka, A., 2008. The determination of content of selenium in natural fruit juices by spectral methods. Rocz. Panstw. Zakl. Hig. 59 (2), 173–178.

Graham, R.D., Welch, R.M., Saunders, D.A., Ortiz-Monasterio, I., Bouis, H.E., Bonierbale, M., de Haan, S., Burgos, G., Thiele, G., Liria, R., Meisner, C.A., Beebe, S.E., Potts, M.J., Kadian, M., Hobbs, P.R., Gupta, R.K., Twomlow, S., 2007. Nutritious subsistence food systems. In: Sparks, D.L. (Ed.), Advances in Agronomy.In: vol. 92, pp. 1–74.

Grant, K., Carey, N.M., Mendoza, M., Schulze, J., Pilon, M., Pilon-Smits, E.A.H., et al., 2011. Adenosine 5-phosphosulfate reductase (APR2) mutation in *Arabidopsis* implicates glutathione deficiency in selenate toxicity. Biochem. J. 438, 325–335.

Guo, Y., Wang, Y., Han, T., Yu, J., 2005. Changes of physiology and fruit quality in stored nectarine treated with selenium and boron. J. Beijing Agric. Coll. 20 (2), 1–4.

Gupta, M., Gupta, S., 2016. An overview of selenium uptake, metabolism and toxicity in plants. Front. Plant Sci. 7, 2074.

Hageman, P.W., van der Weijden, R.G., Weijma, J., Buisman, C.J.N., 2013. Microbial selenite to selenite conversion for selenium removal. Water Res. 47, 2118–2128.

Hallenbeck, P.C., George, G.N., Prince, R.C., Thorneley, R.N., 2009. Characterization of a modified nitrogenase Fe protein from *Klebsiella pneumoniae* in which the 4Fe4S cluster has been replaced by a 4Fe4Se cluster. J. Biol. Inorg. Chem. 14, 673–682.

Hamilton, S.J., 2004. Review of selenium toxicity in the aquatic food chain. Sci. Total Environ. 326, 1–31.

Hanson, B., Garifullina, G.F., Lindblom, S.D., Wangeline, A., Ackley, A., Kramer, K., Norton, A.P., Lawrence, C.B., Pilon-Smits, E.A.H., 2003. Selenium accumulation protects *Brassica juncea* from invertebrate herbivory and fungal infection. New Phytol. 159, 461–469.

Hartikainen, H., 2005. Biogeochemistry of selenium and its impact on food chain quality and human health. J. Trace Elem. Med. Biol. 18, 309–318.

Hirschi, K.D., 2009. Nutrient biofortification of food crops. Annu. Rev. Nutr. 29, 401–421.

Hoenig, M., Hoeyweghen, P.W., 1986. Determination of selenium and arsenic in animal tissues with platform furnace atomic absorption spectrometry and deuterium background correction. Int. J. Environ. Anal. Chem. 24, 193–202.

Hondal, R.J., Marino, S.M., Gladyshev, V.N., 2012. Selenocysteine in thiol/disulfide-like exchange reactions. Antioxid. Redox Signal. 18, 1675–1689.

Hopper, J.L., Parker, D., 1999. Plant availability of selenite and selenate as influenced by the competing ions phosphate and sulfate. Plant Soil 210, 199–207.

Hu, S., Feng, G., Zhao, X., Li, J., Xue, C., 1998. Researches on the Se absorption and accumulation characteristics of apple. Acta Botan. Boreali-Occiden. Sin. 18 (1), 110–115.

Huang, C., Qin, N., Sun, L., Yu, M., 2018. Selenium improves physiological parameters and alleviates oxidative stress in strawberry seedlings under low-temperature stress. Int. J. Mol. Sci. 19 (7), 1913.

Hugouvieux, V., Dutilleul, C., Jourdain, A., Reynaud, F., Lopez, V., Bourguignon, J., 2009. *Arabidopsis* putative selenium-binding protein1 expression is tightly linked to cellular sulfur demand and can reduce sensitivity to stresses requiring glutathione for tolerance. Plant Physiol. 151, 768–781.

Ibrahim, H.I.M., Al-Wasfy, M.M., 2014. The promotive impact of using silicon with potassium and boron of fruiting of Valencia orange trees grown under Minia region conditions. World Rural Observations 2, 28–36.

Irmak, S., 2017. Effects of selenium application on plant growth and some quality parameters in peanut (*Arachis hypogaea*). Pak. J. Biol. Sci. 20 (2), 92–99.

Iwashita, Y., Nishi, K., 2004. Cultivation of selenium-enriched vegetables in large scale. Biomed. Res. Trace Elem. 15, 72–75.

Jakovljevic, M., Licina, V., Velickovic, M., 1998. The effects of selenium application on replant soil and its content in apple leaves and fruits. Acta Hortic. (477), 163–166.

Javed, S., Sarwar, A., Tassawar, M., Faisal, M., 2015. Conversion of selenite to elemental selenium by indigenous bacteria isolated from polluted. Chem. Speciat. Bioavailab. 27, 162–168.

Jing, D., Zhen-Yu, D., Hai-Lin, M., Bing-Yao, M., Fang-Chun, L., Song, Y., Xu, Y., Li, L., 2017. Selenium enrichment, fruit quality and yield of winter jujube as affected by addition of sodium selenite. Sci. Hortic. 225, 1–5.

Kabata-Pendias, A., 2001. Trace Elements in Soils and Plants, third ed. CRC Press, Boca Raton, London, New York, Washington, D.C. 413 p.

Kikkert, J., Berkelaar, E., 2013. Plant uptake and translocation of inorganic and organic forms of selenium. Arch. Environ. Contam. Toxicol. 65, 458–465.

Ko, F.H., Yang, M.H., 1996. On-line removal of interferences via anion-exchange column separation for the determination of germanium, arsenic and selenium in biological samples by inductively coupled plasma mass spectrometry. J. Anal. At. Spectrom. 11, 413–420.

Koirtyohann, S.R., Morris, J.S., 1986. General review of analytical methods. Some metals: As, Bc, Cd, Cr, Ni, Pb, Se, Zn. IARC Sci. Publ. 8, 159–190.

Krishnaiah, L., Kumar, K.S., Suvardhan, K., Chiranjeevi, P., 2003. Simple spectrophotometric determination of traces of selenium in environmental samples. In: Proceedings of the Third International Conference on Environment and Health, Chennai, India, pp. 15–17.

Kubachka, K.M., Meija, J., LeDuc, D.L., Terry, N., Caruso, J.A., 2007. Selenium volatiles as proxy to the metabolic pathways of selenium in genetically modified Brassica juncea. Environ. Sci. Technol. 41, 1863–1869.

Lee, G.J., Kang, B.K., Kim, T.I., Kim, T.J., Kim, J.H., 2007. Effects of different selenium concentrations of the nutrient solution on the growth and quality of tomato fruit in hydroponics. Acta Hortic. 761, 443–448.

Lewis, S.A., 1988. Determination of selenium in biological matrices. Methods Enzymol. 158, 391–402.

Li, Y.H., Wang, W., Luo, K.L., Li, H.R., 2008a. Environmental behaviors of selenium in soil of typical selenosis area. China J. Environ. Sci. 20, 859–864.

Li, H.F., McGrath, S.P., Zhao, F.J., 2008b. Selenium uptake, translocation and speciation in wheat supplied with selenate or selenite. New Phytol. 178, 92–102.

Li, M., Zhao, Z., Zhou, J., Zhou, D., Chen, B., Huang, L., Zhang, Z., Liu, X., 2018a. Effects of a foliar spray of selenite or selenate at different growth stages on selenium distribution and quality of blueberries. J. Sci. Food Agric. 98 (12), 4700–4706.

Li, X., Liu, J., Zhang, X., Sheng, Y., Wang, Y., 2018b. Effects of spraying selenium on character and yield of grape fruit. J. North. Agric. 46 (1), 115–117.

Licina, V., Jakovljevic, M., Antic-Mladenovic, S., Oparnica, C., 1998. The content of selenium in raspberry plant and its improvement by Se-fertilization. Acta Hortic. 477, 167–171.

Liu, S., 2006. Pollution-free cultivation technique of Se-enriched in watermelons. J. Anhui Agric. Sci. 34 (15), 3660–3662.

Liu, D., Chen, X., Wei, J., Zeng, X., Chen, Z., 2007. Study on the selenium content and effects of selenium application on papaya in Hainan province. South China Fruits 36 (6), 34–36.

Liu, Q., Hao, Y., Wu, G., Hao, G., Zhang, P., 2015. Effect of spraying exogenous selenium on fruit quality and selenium content of Dangshan Suli. J. Henan Agric. Sci. 44 (8), 113–117.

Liu, Y.X., Pan, L.P., Huang, Y.F., Nong, M.L., Lu, S., Zhao, Y.Y., Liang, P.X., Xiong, L.M., Li, K.B., Lan, X., 2017. Effects of selenium or silicon foliar fertilizer on cadmium accumulation in rice. Southwest China J. Agric. Sci. 30 (7), 1588–1592.

Long, Y., Cheng, Z., Gong, F., Wu, X., Yin, X., 2016a. Effects of foliar application of selenium fertilizer on selenium content, accumulation of cadmium and lead, and fruit quality of kiwifruit. Food Sci. 37 (11), 74–78.

Long, Y., Cheng, Z., Wu, X., Li, M., Yao, X., Shao, S., 2016b. Application of Se-fertilizer affects selenium content, cadmium and lead accumulation and fruit quality in kiwifruits. Food Sci. 37 (13), 82–88.

Luksic, L., Germ, M., 2017. Selenium in water and in terrestrial plants. Folia Biol. Geol. 58 (2), 165–174.

Lyons, G., Graham, R., Stanghoulis, J., 2003. Nutriprevention of disease with high-selenium wheat. J. Australas. Coll. Nutr. Environ. Med. 22 (3), 3–9.

Marchante-Gayón, M.J., Thomas, C., Feldmann, I., Jakubowski, N., 2000. Comparison of different nebulisers and chromatographic techniques for the speciation of selenium in nutritional commercial supplements by hexapole collision and reaction cell ICP-MS. J. Anal. At. Spectrom. 15, 1093–1102.

Martens, D.A., Suarez, D.L., 1999. Transformations of volatile methylated selenium in soil. Soil Biol. Biochem. 31, 1355–1361.

Mehdi, Y., Hornick, J.L., Istasse, L., Dufrasne, I., 2013. Selenium in the environment, metabolism and involvement in body functions. Molecules 18, 3292–3311.

Meng, D., 2006. Se-enriched pear in Yangxin country, Shandong province of China. China Fruit News 23 (9), 35.

Merian, E., 1991. Metals and Their Compounds in the Environment. VCH Verlagsgesellschaft, Weinheim. 1438 p.

Mimmo, T., Tiziani, R., Valentinuzzi, F., et al., 2017. Selenium biofortification in Fragaria × ananassa: implications on strawberry fruits quality, content of bioactive health beneficial compounds and metabolomic profile. Front. Plant Sci. 8, [cit. 2019-01-18]. Available from: http://journal.frontiersin.org/article/10.3389/fpls.2017.01887/full.

Morgan, J.A., Bending, G.D., White, P.J., 2005. Biological costs and benefits to plant-microbe interactions in the rhizosphere. J. Exp. Bot. 56, 1729–1739.

Mugesh, G., du Mont, W.W., Sies, H., 2001. Chemistry of biologically important synthetic organoselenium compounds. Chem. Rev. 101, 2125–2179.

Nakamaru, Y.M., Altansuvd, J., 2014. Speciation and bioavailability of selenium and antimony in non-flooded and wetland soils: a review. Chemosphere 111, 366–371.

Narváez-Ortiz, W.A., Martínez-Hernández, M., Fuentes-Lara, L.O., Benavides-Mendoza, A., Valenzuela-García, J.R., González-Fuentes, J.A., 2018. Effect of selenium application on mineral macro- and micronutrients and antioxidant status in strawberries. J. Appl. Bot. Food Qual. 91, 321–331.

Neuhierl, B., Bock, A., 1996. On the mechanism of selenium tolerance in selenium-accumulating plants. Eur. J. Biochem. 239, 235–238.

Oster, O., Prellwitz, W., 1982. A methodological comparison of hydride and carbon furnace atomic absorption spectroscopy for the determination of selenium in serum. Clin. Chim. Acta 124, 277–291.

Peak, D., Sparks, D.L., 2002. Mechanisms of selenate adsorption on iron oxides and hydroxides. Environ. Sci. Technol. 36, 1460–1466.

Pezzarossa, B., Remorini, D., Piccotino, D., Malagoli, M., Massai, R., 2009. Effects of selenate addition on selenium accumulation and plant growth of two Prunus rootstock genotypes. J. Plant Nutr. Soil Sci. 172 (2), 261–269.

Pezzarossa, B., Remorini, D., Gentile, M.L., Massai, R., 2012. Effects of foliar and fruit addition of sodium selenate on selenium accumulation and fruit quality. J. Sci. Food Agric. 92, 781–786.

Pezzarossa, B., Rosellini, I., Malorgio, F., Borghesi, E., Tonutti, P., 2013. Effects of selenium enrichment of tomato plants on ripe fruit metabolism and composition. Acta Hortic. 247–251.

Pezzarossa, B., Rosellini, I., Borghesi, E., Tonutti, P., Malorgio, F., 2014. Effects of Se-enrichment on yield, fruit composition and ripening of tomato (*Solanum lycopersicum*) plants grown in hydroponics. Sci. Hortic. 165, 106–110.

Pilon-Smits, E.A.H., LeDuc, D.L., 2009. Phytoremediation of selenium using transgenic plants. Curr. Opin. Biotechnol. 20, 207–212.

Proietti, P., Nasini, L., Buono, D., del D'Amato, R., Tedeschini, E., Businelli, D., 2013. Selenium protects olive (*Olea europaea* L.) from drought stress. Sci. Hortic. 164, 165–171.

Puccinelli, M., Malorgio, F., Pezzarossa, B., 2017. Selenium enrichment in horticultural crops. Molecules 4 (22), 6.

Quinn, C.F., Freeman, J.L., Reynolds, R.J.B., Cappa, J.J., Fakra, S.C., Marcus, M.A., Lindblom, S.D., Quinn, E.K., Bennett, L.E., Pilon-Smits, E.A.H., et al., 2010. Selenium hyperaccumulation offers protection from cell disruptor herbivores. BMC Ecol. 10, 19.

Rayman, M.P., 2000. The importance of selenium to human health. Lancet 356, 233–241.

Rayman, M.P., Infante, H.G., Sargent, M., 2008. Food-chain selenium and human health: spotlight on speciation. Br. J. Nutr. 100, 238–253.

Ros, G., van Rotterdam, A., Bussink, D., Bindraban, P., 2016. Selenium fertilization strategies for biofortification of food: an agro-ecosystem approach. Plant Soil 1, 99–112.

Santiago, F.E.M., de Souza, M.L., Ribeiro, F.O., Cipriano, P.E., Guilherme, L.R.G., 2018. Influence of sulfur on selenium absorption in strawberry. Acta Sci. Agron. 1807-8621. 40.

Sarathchandra, S.U., Watkinson, J.H., 1981. Oxidation of elemental selenium to selenite by *Bacillus megaterium*. Science 211, 600–601.

Sattar, A., Cheema, M.A., Abbas, T., Sher, A., Ijaz, M., Hussain, M., 2017. Separate and combined effects of silicon and selenium on salt tolerance of wheat plants. Russ. J. Plant Physiol. 64 (3), 341–348.

Schiavon, M., Dall'Acqua, S., Mietto, A., Pilon-Smits, E.A.H., Sambo, P., Masi, A., Malagoli, M., 2013. Selenium fertilization alters the chemical composition and antioxidant constituents of tomato (*Solanum lycopersicon* L.). J. Agric. Food Chem. 61, 10542–10554.

Schrauzer, G.N., 2006. Selenium yeast: composition, quality, analysis, and safety. Pure Appl. Chem. 78, 105–109.

Serafin Muñoz, A.H., Kubachka, K., Wrobel, K., Gutierrez Corona, J.F., Yathavakilla, S.K., Caruso, J.A., Wrobel, K., 2006. Se-enriched mycelia of Pleurotus ostreatus: distribution of selenium in cell walls and cell membranes/cytosol. J. Agric. Food Chem. 54, 3440–3444.

Shrestha, B., Lipe, S., Johnson, K., Zhang, T., Retzlaff, W., Lin, Z.-Q., 2006. Soil hydraulic manipulation and organic amendment for the enhancement of selenium volatilization in a soil–pickleweed system. Plant Soil 288, 189–196.

Sors, T.G., Ellis, D.R., Na, G.N., Lahner, B., Lee, S., Leustek, T., et al., 2005a. Analysis of sulfur and selenium assimilation in *Astragalus* plants with varying capacities to accumulate selenium. Plant J. 42, 785–797.

Sors, T.G., Ellis, D.R., Salt, D.E., 2005b. Selenium uptake, translocation, assimilation and metabolic fate in plants. Photosynth. Res. 86, 373–389.

Soylu, E.M., Kurt, Ş., Soylu, S., 2010. In vitro and in vivo antifungal activities of the essential oils of various plants against tomato grey mould disease agent *Botrytis cinerea*. Int. J. Food Microbiol. 143, 183–189.

Sturgeon, R.E., Lam, J.W., 1999. The ETV as a thermochemical reactor for ICP-MS sample introduction. J. Anal. At. Spectrom. 14, 785–791.

Terry, N., Zayed, A.M., de Souza, M.P., Tarun, A.S., 2000. Selenium in higher plants. Annu. Rev. Plant Physiol. 51, 401–432.

Uwakiem, M.K.H., 2015. Effect of spraying silicon, selenium and humic acid on fruiting of early sweet grapevines. Egypt. J. Hortic. 42 (1), 333–343.

Varadé, R.C.M., de Castro, M.D.L., 1998. Determination of selenium in solid samples by continuous subcritical water extraction, flow injection derivatisation and atomic fluorescence detection. J. Anal. At. Spectrom. 13, 787–791.

Wallschlaeger, D., Feldmann, J., 2010. Formation, occurrence, significance, and analysis of organoselenium and organotellurium compounds in the environment. In: Sigel, A. (Ed.), Organometallics in Environment and Toxicology. In: vol. 7. Royal Society of Chemistry, Cambridge, pp. 319–364.

Wang, Q., Huang, W., 2007. Application and development of selenium accumulating food in China. J. Guangdong Trace Elem. Sci. 15 (3), 7–10.

Wang, X., Zhang, L., Wan, Y., Wang, Q., Sun, H., Guo, Y., Li, H., 2016. Effects of foliar-applied selenite and selenate on selenium accumulation in strawberry. J. Agric. Resour. Environ. 33 (4), 334–339.

Watson, M.E., Isaac, R.A., 1990. Analytical instruments for soil and plant analysis. In: Westerman (Ed.), Soil Testing and Plant Analysis, third ed. SSSA Book Series No. 3, Soil Science Society of America, Madison, WI, pp. 691–740.

Weekley, C.M., Harris, H.H., 2013. Which form is that? The importance of selenium speciation and metabolism in the prevention and treatment of disease. Chem. Soc. Rev. 42, 8870–8894.

White, P.J., Broadley, M.R., 2009. Biofortification of crops with seven mineral elements often lacking in human diets—iron, zinc, copper, calcium, magnesium, selenium and iodine. New Phytol. 182, 49–84.

White, P.J., Bowen, H.C., Parmaguru, P., Fritz, M., Spracklen, W.P., Spiby, R.E., Meacham, M.C., Mead, A., Harriman, M., Trueman, L.J., Smith, B.B., Thomas, B., Broadley, M.R., 2004. Interactions between selenium and sulphur nutrition in *Arabidopsis thaliana*. J. Exp. Bot. 55, 1927–1937.

Winkel, L.H., Vriens, B., Jones, G.D., Schneider, L.S., Plon-Smith, E., Banuelos, G.S., 2015. Selenium cycling across soil-plant-atmosphere interfaces: a critical review. Nutrients 7 (6), 4199–4239.

Winkler, J., 2011. Biofortification: improving the nutritional quality of staple crops. In: Pasternak, C. (Ed.), Access Not Excess. Smith-Gordon Publishing, London.

Wu, G.L., Ning, C.J., 2013. Effects of applying selenium to Chinese jujube on the fruit quality and mineral elements content. Acta Hortic. (993), 199–207.

Wu, G.L., Tian, J.B., 2009. Progress of fruit plants in selenium-enriched research in China. In: Patil, B. (Ed.), II. International Symposium on Human Health Effects of Fruits and Vegetables: Favhealth 2007, pp. 599–602.

Wu, G., Liu, Q., Liu, H., Ji, L., Yan, H., Wang, Y., 2004. Effects of dripping injection of microelements on fruit tree physiology and fruit quality. Chin. J. Appl. Environ. Biol. 10, 2154–2157.

Wu, G.L., Jian, Z.H., Liu, Q.L., Tian, J.B., 2009. Dynamic absorption by stem dripping injection of selenium element in 'Red Fuji' apple trees. In: Patil, B. (Ed.), II. International Symposium on Human Health Effects of Fruits and Vegetables: Favhealth 2007, pp. 595–598.

Wu, Z., Yin, X., Lin, Z., Bañuelos, G.S., Yuan, L., Liu, Y., Li, M., 2014. Inhibitory effect of selenium against *Penicillium expansum* and its possible mechanisms of action. Curr. Microbiol. 69, 192–201.

Wu, Z., Yin, X., Bañuelos, G.S., Lin, Z.Q., Zhu, Z., Liu, Y., Yuan, L., Li, M., 2016. Effect of selenium on control of postharvest gray mold of tomato fruit and the possible mechanisms involved. Front. Microbiol. 6, 1–11.

Yan, X.P., Sperling, M., Welz, B., 1999. On-line coupling of flow injection microcolumn separation and preconcentration to electrothermal atomic absorption spectrometry for determination of (ultra)trace selenite and selenate in water. Anal. Chem. 71, 4353–4360.

Yan, G., Liu, H., JiZhou, Y., Wang, Y., 2014. Effects of selenium and boron on refrigerated nectarine fruit cell membrane performance. Guizhou Agric. Sci. 42 (1), 170–173.

Yang, S.F., Hoffman, N.E., 1984. Ethylene biosynthesis and its regulation in higher plants. Annu. Rev. Plant Phys. 35, 155–189.

Yang, H., ZiChong, W., ShenPu, Z., Liang, X., 2008. Effects of spraying selenium (Se) on watermelon. Soil Fertil. Sci. China 1, 37–39.

Youssef, K., Roberto, S.R., 2014. Salt strategies to control Botrytis mold of "Benitaka" table grapes and to maintain fruit quality during storage. Postharvest Biol. Technol. 95, 95–102.

Zhang, L., Zhang, P., Huang, G., 1995. Effect of spraying sodium selenite on selenium content of fruit in Citrus. Zhejiang Agric. Sci. 2, 70–72.

Zhao, X.Q., Mitani, N., Yamaji, N., Shen, R.F., Ma, J.F., 2010. Involvement of silicon influx transporter OsNIP2;1 in selenite uptake in rice. Plant Physiol. 153, 1871–1877.

Zhao, W., Xi, Z., Lin, G., Tang, J.F., 2011. Effect of selenium on physiological and biochemical indexes of *Vitis vinifera* cv. Cabernet Sauvignon leaves under water stress. J. Fruit Sci. 28 (6), 984–990.

Zhao, Y., Wu, P., Wang, Y., Feng, H., 2013. Different approaches for selenium biofortification of pear-jujube (*Ziziphus jujuba* M. cv. Lizao) and associated effects on fruit quality. J. Food Agric. Environ. 11 (2), 529–534.

Zhu, J.J., Ma, H.J., 2018. Effects of selenium on physiological characteristics of wine grape seedling under copper stress. J. South. Agric. 49 (1), 91–97.

Zhu, Y.G., Pilon-Smits, E.A., Zhao, F.J., Williams, P.N., Meharg, A.A., 2009. Selenium in higher plants: understanding mechanisms for biofortification and phytoremediation. Trends Plant Sci. 14, 436–447.

Zhu, Z., Chen, Y., Zhang, X., Li, M., 2016. Effect of foliar treatment of sodium selenate on postharvest decay and quality of tomato fruits. Sci. Hortic. 198, 304–310.

Zhu, S., Liang, Y., An, X., Kong, F., Gao, D., Yin, H., 2017a. Changes in sugar content and related enzyme activities in table grape (*Vitis vinifera* L.) in response to foliar selenium fertilizer. J. Sci. Food Agric. 0022-5142. 97 (12), 4094–4102. [cit. 2019-01-16]. Available from: http://doi.wiley.com/10.1002/jsfa.8276.

Zhu, Z., Chen, Y., Shi, G., Zhang, X., 2017b. Selenium delays tomato fruit ripening by inhibiting ethylene biosynthesis and enhancing the antioxidant defense system. Food Chem. 219, 179–184.

Zhu, S., Liang, Y., Gao, D., An, X., Kong, F., 2017c. Spraying foliar selenium fertilizer on quality of table grape (*Vitis vinifera* L.) from different source varieties. Sci. Hortic. 218, 87–94.

Zou, D.M., Brewer, M., Garcia, F., Feugang, J.M., Wang, J., Zang, R.Y., Liu, H.G., Zou, C.P., 2005. Cactus pear: a natural product in cancer chemoprevention. Nutr. J. 4 (25), 75–95.

Zwolak, I., Zaporowska, H., 2012. Selenium interactions and toxicity: a review. Cell Biol. Toxicol. 28, 31–46.

Further reading

WHO, 2009. Global Health Risks: Mortality and Burden of Disease Attributable to Selected Major Risks. [cit. 2019-01-18] Available from: https://apps.who.int/iris/bitstream/handle/10665/44203/9789241563871_eng.pdf?sequence=1&isAllowed=y.

Zhu, L., Wei, Q., Xu, X., 2007. Selenium absorption, distribution and accumulation in grapevine. Acta Hortic. Sin. 34 (2), 325–328.

19

Importance of silicon in fruit nutrition: Agronomic and physiological implications

Hassan Etesami[a],*, Byoung Ryong Jeong[b]

[a]Agriculture & Natural resources Campus, Faculty of Agricultural Engineering & Technology, Department of Soil Science, University of Tehran, Tehran, Iran

[b]Horticulture Major, Division of Applies Life Science (BK21 Plus Program), Graduate School, Gyeongsang National University, Jinju, Republic of Korea

*Corresponding author. E-mail: hassanetesami@ut.ac.ir

1 Introduction

Silicon accounts for 28% of the earth's crust and is the second most abundant element in the lithosphere after oxygen, where it is found in the form of 0.1–2.0-mM silicic acid (or $Si[OH]_4$) at pH <9 (Epstein, 1994). Silicon as silicate or aluminum silicates is a major soil constituent. Plants generally uptake silicon as monomeric or monosilicic acid (H_4SiO_4) from soils, which is typically present in the soil solution from 0.1 to 0.6 mM (Ma and Yamaji, 2006). Monosilicic acid absorbed by plants is polymerized into silica gel or biogenetic opal as amorphous $SiO_2 \bullet nH_2O$, called phytoliths, in cell walls, bracts, and intercellular spaces between the leaf and root cells (Ma and Yamaji, 2006; Mitani et al., 2005). Although silicon contents considerably differ among species, all plants contain the element in their tissues. The silicon contents in a plant's aerial parts vary from 0.1% to 10% of the plant dry matter (Liang et al., 2007). The silicon content in a plant is typically equal to or greater than the amount of each of the three main macronutrients, namely, potassium (K), nitrogen (N), and phosphorus (P) (Meena et al., 2014). Silicon is essential for animals, ubiquitous in earth's soil, and present in all plants, but there is little evidence that the element is essential to vascular plants (Epstein, 1994; Liang et al., 2015e; Yan et al., 2018), as plants can fulfill their life cycles without it. However, it is well known that silicon is still beneficial and useful for the healthy growth of many plant species due to its role in the metabolism, physiological and/or structural activities, promotion of plant mechanical strength and light interception, and improvement in survival of plants exposed to abiotic (aluminum toxicity, drought, flooding, freezing, elevated

A.K. Srivastava, Chengxiao Hu (eds.)
Fruit Crops: Diagnosis and Management of Nutrient Constraints
https://doi.org/10.1016/B978-0-12-818732-6.00019-8

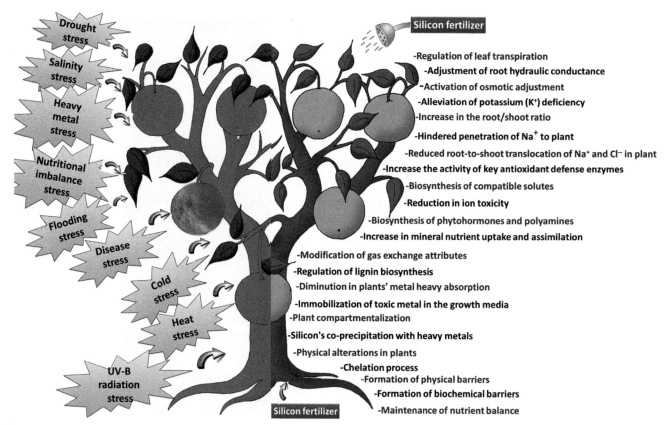

FIG. 19.1   The multiple action mechanisms of silicon in alleviating biotic and abiotic stresses in plants. For details, see the reference Etesami and Jeong (2018).

temperature, heavy metal toxicity, lodging, nutrient deficiency and imbalance, UV-B radiation, salinity, wounding, etc.) and biotic (insect pests, pathogenic diseases, plant maladies, etc.) stresses (Fig. 19.1) (Adrees et al., 2015; Bakhat et al., 2018; Coskun et al., 2016; Debona et al., 2017; Etesami and Jeong, 2018; Guo et al., 2016; Hernandez-Apaolaza, 2014; Imtiaz et al., 2016; Liang et al., 2015d; Meharg and Meharg, 2015; Rizwan et al., 2015; Tripathi et al., 2017; Wang et al., 2017; Zhu and Gong, 2014), which render the element "quasiessential" (Epstein and Bloom, 2005) to plants, helping them improve quality and yield. Laboratory experiments and field trials have verified the benefits of silicon in promoting plant growth. Silicon is anticipated to finally be recognized as essential to vascular plants, pursuant to the newly suggested definition of essentialness by Epstein and Bloom (2005).

Silicon is mostly present in soils as insoluble crystalline aluminosilicates, which are not directly available to plants (Liang et al., 2015e), and therefore, despite its abundance in the soil, it makes silicon absorption by plants low (Zhu and Gong, 2014). Silicon does not harm the environment and is not destructive, which makes silicon fertilizers highly suitable for ecologically sustainable agriculture. Indeed, agricultural field applications of silicon fertilizers, especially those with low available silicon, are becoming a common farming practice in many countries and regions (Yan et al., 2018) for high productivity and sustainable production (Liang et al., 2015e). Silicon is well documented to positively affect plant growth and yield (Etesami and Jeong, 2018; Liang et al., 2015a). Silicon promotes the progression of plants and biomass accumulation, improves yield, and enhances the quality of various plants, ranging from monocotyledonous crops such as barley (*Hordeum vulgare* L.), maize (*Zea mays* L.) (Liang et al., 2006; Liu et al., 2011; Mitani et al., 2009; Wang et al., 2001), millet (*Setaria italica* L.), rice (*Oryza sativa* L.) (Alvarez and Datnoff, 2001; Correa-Victoria et al., 2001; Korndörfer and Lepsch, 2001; Prabhu et al., 2001), sorghum (*Sorghum bicolor* L.), sugarcane (*Saccharum officinarum* L.) (Ashraf et al., 2009; Huang et al., 2011; Jiang et al., 2011; Meyer and Keeping, 2001; Wang et al., 2001; Zeng et al., 2007), and wheat (*Triticum aestivum* L.) (Liu et al., 2011; Montpetit et al., 2012; Wang et al., 2001; Yu and Gao, 2012), to dicotyledonous crops like cotton (*Gossypium arboreum* L.) and soybean (*Glycine max* L.), vegetables, and fruits (Al-Wasfy, 2012, 2013; Babini et al., 2012; El-Kareem et al., 2014; Gaofeng et al., 2012; Kamenidou et al., 2010; Liang et al., 2015a; QuanYu et al., 2009; Savvas et al., 2002; Stamatakis et al., 2003; Su et al., 2011; Voogt and Sonneveld, 2001; Xiyan et al., 2007; Yu and Gao, 2012; Zhang et al., 2007), which readily absorb and retain high silicon levels in their organs.

Many studies have demonstrated that silicate fertilizers significantly increase the crop yields (Etesami, 2018; Etesami and Jeong, 2018; Liang et al., 2015e) and also largely improve crop quality parameters (i.e., fruit sugar and vitamin C levels) (Dehghanipoodeh et al., 2016; Jarosz, 2013; Liu et al., 2014). Silicon has also been used to prevent biotic and abiotic stresses from adversely affecting various fruit crops (Kaluwa et al., 2010). Silicate fertilizers have been reported to prolong fruit shelf life and increase the quality and yield parameters of fruit trees (Al-Wasfy, 2012, 2013; Zhang et al., 2017). Improving silicon nutrition also helps initiate root and fruit formation in vascular plants (Matichenkov et al., 2001). In apple, citrus, or mango, for example, silicon application of various formulations at different concentrations led to a decreased level of diseases caused by bacterial (Cazorla et al., 2006; Gutiérrez-Barranquero et al., 2012) or fungal pathogens (Helaly et al., 2017; Marais, 2015), enhanced resistance to insect pests (Vieira et al., 2016) and physiological disorders (Edgerton et al., 1976), and increased yield (Helaly et al., 2017), fruit size (Costa et al., 2015), and biomass (Matichenkov et al., 2001). Compared with those in agronomic crops, the effects of applying silicate fertilizers to fruit crops have been rarely reported, while using silicon to improve the quality and yield parameters of fruit crops enhances plant tolerance to abiotic and biotic stresses and extends shelf life of fruits that is predicted to become an emerging trend in agriculture in the near future. Thus, this chapter reviews the recent progress in research regarding the mechanisms by which silicon improves growth and fruit quality in various plants.

2 Silicon helps improve plants' uptake of minerals

Plants require a sufficient supply of essential nutrients for natural growth and development. Proper nutrition of fruit trees is an important tool to achieve the maximum yield and fruit quality and improve the agricultural products quantitatively and qualitatively (Brunetto et al., 2015). Nutritional imbalance hampers plant growth, development and yield (Paul and Lade, 2014). Environmental stresses, such as heavy metal toxicity, salinity, and water stresses, affect the absorption and translocation of elemental macronutrients (Parida and Das, 2005), including calcium (Ca), potassium (K), magnesium (Mg), nitrogen (N), and phosphorus (P) and elemental micronutrients boron (B), copper (Cu), iron (Fe), manganese (Mn), and zinc (Zn) in plants (Etesami and Jeong, 2018; Wang and Han, 2007). Salinity may cause nutritional imbalances by affecting nutrient availability, intraplant transport or partitioning, competitive intake, and physiological inactivation of a nutrient that results in an increased requirement for the said nutrient in plants (Grieve and Grattan, 1999). A high proportion of nutrients is unavailable for plant roots to uptake; as soil nutrients are generally bound to soil constituents, both organic and inorganic, or alternatively, are present as insoluble precipitates. Managing the nutritional balance is the most effective and easiest way of combating plant stresses (Abbas et al., 2015). The mineral contents in plants greatly influence plants' ability to cope with adverse environmental conditions, in particular with abiotic stresses; plants with impaired mineral nutrition experience amplified adverse effects from abiotic stresses (Baligar et al., 2001; Grieve and Grattan, 1999; Khoshgoftarmanesh et al., 2010). Many studies have demonstrated that plants under environmental stresses required supplementary provisions of mineral nutrients to mollify detrimental influences of stresses (Endris and Mohammad, 2007; Heidari and Jamshid, 2010; Kaya et al., 2002). In general, nutritional imbalances can limit plant growth by affecting plant fitness and nutritional status (toxic ions).

Past studies well document the positive role of silicon in increasing plant growth and the amount of nutrients in plants grown under nutritional imbalances (nutrient deficiency or excess) (Bloemhard and Van Moolenbroek, 1994; Datnoff et al., 2001; Etesami and Jeong, 2018; Guntzer et al., 2012). Silicon can improve, via different mechanisms, plants' capability to survive and produce in an environment with nutritional imbalances (Etesami and Jeong, 2018).

Since roots are used to take up essential plant nutrients from the soil (Mills et al., 1996), good root growth is crucial to enhanced plant development. Silicon has been observed to stimulate the elongation of lateral roots in plants (Isa et al., 2010). Alterations in the root architecture brought forth by silicon might lead to increased total root surface area and consequently to enhanced nutrient and water uptake, which is likely to positively affect plants as a whole. Silicon treatments enhance the levels of long-distance transport molecules, such as citrate, in plants, which can contribute to the root-to-shoot metal transport, and will diminish deficiency symptoms (Hernandez-Apaolaza, 2014). For example, silicon was observed to increase the translocation of 11 elemental nutrients—boron, calcium, copper, iron, potassium, magnesium, manganese, molybdenum, nitrogen, phosphorus, and zinc—to growing plant parts, which could enhance the growth of stressed plants (Ji et al., 1992). In another study, silicon amendment was also shown to reduce phosphate, nitrate, and potassium leaching (Matichenkov and Bocharnikova, 2010). In general, silicon treatments to plants under various stresses can help maintain nutrient balances it may improve plant growth and biomass (Etesami and

Jeong, 2018). In this section, the silicon application methods to improve crop health and productivity under micronutrient and macronutrient deficiencies are discussed.

2.1 Nitrogen

Nitrogen is an important nutrient required for crop production. Nitrogen is directly related to the amino acid composition of proteins and therefore also influences the nutritional quality of the economic produce. Nitrogen influences several fruit quality parameters such as color, flavor, nutritional composition, shape, size, and texture (Maheswari et al., 2017). For example, nitrogen fertilization resulted in minimal physiological losses in weight and high levels of total soluble solids (TSS) and total sugars in guava (Sharma et al., 2013). Nitrogen application was shown to affect the fruit quality and anthocyanin composition in grapes (Hilbert et al., 2015). Silicon fertilization significantly increased the nitrogen levels in plants and, as a result, led to increased growth and higher biomass yield. Thus, plant yield affected by silicon fertilization may be pertinent to an enhanced uptake of nitrogen (Huang et al., 1997; Ji et al., 1992; Li et al., 1999). Plants also displayed improved calcium and nitrogen absorption when supplied with increasing levels of sodium metasilicate (50–800 mg silicon/kg) (Mali, 2008; Mali and Aery, 2008). It was discovered that silicon improves the use efficiency of nitrogen and stimulates amino acid remobilization to alter the primary plant metabolism (Detmann et al., 2012).

Excessive nitrogen concentration is toxic and can negatively affect produce quality, as demonstrated by delayed fruit maturing, diminished fruit firmness, poor fruit colors, reduced storage life, and diminutive cold acclimation abilities (Maheswari et al., 2017). For example, high nitrogen levels induced by fertilization led to poor fruit colors, reduced firmness, decreased TSS levels, and bad texture in apple (Nava et al., 2007; Sharma, 2016). Reduced fruit color and firmness and increased ethylene evolution, respiration rate, and bitter pit were observed in apple after a higher rate of soil nitrogen application (Neilsen et al., 2008). High-level nitrogen fertilization in berries delayed fruit maturation, increased the arginine content, and reduced anthocyanin levels with relatively higher acylated anthocyanins compared with berries of vines supplied with low nitrogen levels (Hilbert et al., 2015). Silicon has shown to also mitigate stresses from excessive nitrogen (Liang et al., 2015b). Previous studies report that decreased plant leaf erectness can be mitigated by supplying silicon to the nutrient solution (Okamoto, 1969; Sadanandan and Varghese, 1970; Singh et al., 2006; Yoshida et al., 1969).

2.2 Phosphorus

It has been known that applications of phosphorus (P) can increase the vigor and accelerate flowering in newly planted trees and increase cropping and leaf and fruit tissue phosphorus concentration in mature trees. In orange, foliar application of soluble phosphorus compounds has increased phosphorus levels in fruits and reduced the vulnerability to low-temperature breakdown (LTB), while, in apple, it increased fruit firmness and resistance to LTB (Neilsen et al., 2008). It has been known for a while that phosphorus availability rises after silicon fertilization in various plant species (Liang et al., 2015b; Ma, 2004; Owino-Gerroh and Gascho, 2005; Singh et al., 2006). How silicon influences phosphorus uptake by plants was one of the earliest researches on the effects of silicon (Anderson et al., 1991; Eneji et al., 2008; Fisher, 1929; Ji et al., 1992; Li et al., 1999). An *in planta* mechanism may be responsible for the response of phosphorus-deficient plants to silicon, implying an improved phosphorus utilization, likely via increased phosphorylation (Cheong and Chan, 1973) or decreased uptake of excess iron and manganese (Ma and Takahashi, 1990b; Ma, 2004). Silicon's positive effects on phosphorus availability have also been attributed to a competitive exchange and interaction of the two elements (Smyth and Sanchez, 1980), an increased soil pH that resultantly enhances the soil phosphorus availability (Owino-Gerroh and Gascho, 2005), decreased metal uptake (Al, Cd, Fe, and Mn), interactions with cationic metals such as iron and manganese (Ma and Takahashi, 1990a), and increased water-soluble phosphorus concentration (Eneji et al., 2008), which indirectly improves plants' utilization of phosphorus (Liang et al., 2005; Ma and Takahashi, 1990a). In acidic soils, phosphorus sorption is the chief cause of phosphorus deficiency. Phosphate ($H_2PO_4^-$) anions replace OH groups from the coordination spheres of the metals and are specifically adsorbed onto hydrated iron and aluminum oxides. The chemical competition for sorption sites between $H_2PO_4^-$ and silicate ($H_3SiO_4^-$) has been demonstrated in soils resulting in the mutual displacement of each other between silicon and phosphorus (Smyth and Sanchez, 1980). It has also been reported that silicon mitigates excessive phosphorus stresses (Liang et al., 2015b). Excessive phosphorus supply resulted in silicon limiting phosphorus uptake and the appearance of chlorosis, possibly by reducing the transpiration rate (Chinnasami and Chandrasekaran, 1978; Ma et al., 2001; Silva, 1971; Singh et al., 2006).

2.3 Potassium

Potassium constitutes about 2%–10% of the dry mass in vascular plants and is a very important cation for them (Leigh and Wyn Jones, 1984). Potassium plays various roles in plant metabolism and physiology (Cruz et al., 2017). It is essential for stomatal movements, photosynthesis, enzyme activation, protein synthesis, and for modulating osmotic regulation during cell expansion, tropism (Mäser et al., 2002), and transport of photoassimilates of fabric sources for fabric drains (Marschner, 2012a). A sufficient potassium supply increases fruit size and yield, ascorbic acid, and soluble solids levels and improves the fruit color and shelf life (Cruz et al., 2017; Kanai et al., 2007; Lester et al., 2006, 2005). Furthermore, potassium is related to the quality of harvested vegetables and fruits; it affects the postharvest attributes, which include acidity, color, size, industrial qualities, nutritional value, and resistance to transportation, handling, and storage, which naturally makes it relevant to the quality of products of plant origin (Cruz et al., 2017; Javaria et al., 2012).

Silicon application can influence the soil potassium availability and nutrient concentrations in plants (Chen et al., 2016, 2011; Huang et al., 1997; Ji et al., 1992; Kaya et al., 2006; Pei et al., 2010). Miao et al. (2010) have shown that silicon application to potassium-deficient plants improved both the internal potassium status and plant growth. Silicon modulates antioxidant enzymes to alleviate oxidation and membrane lipid peroxidation induced by potassium deficiency in plants. Mali and Aery (2008) demonstrated that even low-level silicon application improves the hydroponic and soil potassium uptake through the activation of H^+-ATPase. Chen et al. (2016) recently demonstrated that silicon improves the water status in plants to moderate potassium deficiency.

2.4 Calcium and magnesium

Calcium (Ca) is vital to plant growth and fruit development (McLaughlin and Wimmer, 1999; Terraza et al., 2008). Calcium protects the cell wall from the disintegrating enzymes secreted by pathogens to significantly improve plant resistance to diseases (Torres-Olivar et al., 2014). In a study evaluating how calcium protects pumpkin fruit tissues from *Botrytis cinerea* infection (Chardonnet and Doneche, 1995), it was found that calcium application to fruits increased calcium concentration in the cell walls and thus decreased the pectin digestion by the fungus pectinolytic enzymes. There is evidence that calcium controls physiological disorders and delays senescence in plants (Torres-Olivar et al., 2014). The element helped overcome the negative impacts of high salinity (Nedjimi and Daoud, 2009) and reduced the incidence of bitter pit in apple (Lötze and Theron, 2007; Peryea et al., 2007). Calcium is primarily associated with pectins in plants, which majorly influences the cell wall's rheological properties and therefore the storage life and texture of fruits (Rose, 2003).

Silicon application increased the calcium and magnesium levels in soils and plants (Chen et al., 2011; Huang et al., 2011; Ji et al., 1992; Kaya et al., 2006; Mali, 2008; Mali and Aery, 2008; Pei et al., 2010). Silicate fertilizers were reported to improve soil fertility and influence crops: silicate fertilization at a soil dose of 1.41 g silicon/kg led to the greatest changes in soil acidity, which suggests an improved nutrient availability, especially of calcium, to plants (Matichenkov and Calvert, 2002). Such nutritional improvements may also be accountable for the improved performance of plants treated with silicon. The increased uptake of calcium may be due to increased plasma membrane H^+-ATP activity and decreased plasma membrane permeability as a result of silicon addition (Kaya et al., 2006; Liang, 1999).

2.5 Iron

Iron deficiency chlorosis in plants have been reported to be mitigated by adding silicon to the nutrient solution (Bityutskii et al., 2010; Gonzalo et al., 2013; Ji et al., 1992; Pavlovic et al., 2013). One hypothesis is that the added silicon increases the root oxidation capacity and therefore iron oxidation to insoluble ferric compounds, which are from the iron plaque (You-Qiang et al., 2012). Ions bound to oxides are typically inaccessible by plants, but they may become available for uptake when plants exude phytosiderophores, which either chelate or dissolve the iron oxide to make iron soluble in the immediate vicinity of roots (Zhang et al., 1998). It was determined that silicon increases the root-to-shoot iron transport (You-Qiang et al., 2012). It was also suggested that the increased expression of silicon transporters that follows the silicon addition to the nutrient solution might influence iron uptake and translocation, which improves iron nutrition in iron-deficient plants (You-Qiang et al., 2012). Plants treated with silicon were observed to have a high iron accumulation in roots (Bityutskii et al., 2014; Gonzalo et al., 2013; Pavlovic et al., 2013), which is attributed to high levels of iron in the root apoplastic pools (Pavlovic et al., 2013) or to root surface iron precipitation. It is also hypothesized that silicon helps maintain other micronutrient balances, such as the iron/manganese ratio (Pich et al., 1994), which helps enhance chlorophyll synthesis. This provides a plausible explanation for the growth stimulation of iron-deficient plants treated with silicon (Bityutskii et al., 2014; Gonzalo et al., 2013; Pavlovic et al., 2013).

Silicon application also facilitated iron mobility and its xylem translocation toward the shoot and promoted the accumulation of iron-mobilizing compounds such as citrate in the root and shoot tissues and the xylem sap and catechins in roots (Bityutskii et al., 2014; Pavlovic et al., 2013). This leads to hypothesizing that an increased level of citrate in silicon-treated plants and the resultant mechanisms should be considered as beneficial to overcoming iron deficiency. Citrate joins iron on its transport through the xylem (Rellán-Álvarez et al., 2010). The increase in leaf, root tissue, and xylem sap citrate concentration in plants treated with silicon, especially on the first day after eliminating iron from the nutrient solution (Bityutskii et al., 2014; Pavlovic et al., 2013), should facilitate long-distance iron transport and utilization in leaves. Studies have been conducted with cucumber to see how silicon affects iron chelate reductase and the expression profiles of strategy I—related *FRO2*, *HA1*, and *IRT1* genes (Pavlovic et al., 2013). Generally, silicon improves iron distribution in plants and consequently mediates iron deficiency. Silicon application to plants seems to promote root pool (apoplastic or plaque) iron storage, and remobilization of the stored iron seems to be the major driver of how silicon helps mollify iron deficiency. Additionally, new findings suggest the mechanisms with which silicon alleviates iron deficiency are indirect in nature, affecting the activation of genes associated with iron deficiency to enhanced iron acquisition via the root and mobilization through tissues (Pavlovic et al., 2013).

Ma and Takahashi (2002) observed that, under excess iron concentrations, silicon appears to enhance the oxidizing capacity of roots, which converts ferric iron from ferrous iron, thus avoiding a large iron uptake and limiting its toxicity. It has been hypothesized that roots of plants treated with silicon release OH^- to regulate iron uptake from acidic soils (Wallace, 1992). In rice, silicon application effectively alleviated iron toxicity (Ma, 2004; You-Qiang et al., 2012), largely by increasing iron precipitation at root surfaces (iron plaque) or in the growth media. Silicon addition likely increases pH and may lead to iron precipitation. Silicon may precipitate as negatively charged silica particles, with which the iron cations may bind, or iron and silicon could also coprecipitate, as Fe(III) silicates (Currie and Perry, 2007), which lead to iron depletion from the solution.

2.6 Manganese

The interaction of silicon and manganese in certain plants has been researched (Horst and Marschner, 1978; Ji et al., 1992; Li et al., 1999). Silicon helps enhance the oxidizing capacity of rice roots, which yields an increased rhizospheric manganese oxidation rate and increases manganese precipitation outside the rice plant (Okuda and Takahashi, 1962). Said deposits may be used in manganese deficiency and temporarily ameliorate manganese deficiency symptoms. Silicon helps homogeneously distribute manganese in leaves, rather than concentrating it in necrotic spots. A lower apoplastic manganese level was observed in cowpea where silicon was added to the medium. This could be explained by silicon deposits mediating metal adsorption on cell walls (Horst et al., 1999). Soluble apoplastic silicon may affect the oxidation state of manganese and promote its precipitation (Iwasaki et al., 2002). It was determined that silicon indirectly leads to decreased leaf apoplastic OH^- by reducing the amount of free apoplastic Mn^{2+}, hence controlling the Fenton reaction and protecting plants against manganese toxicity (Dragišić Maksimović et al., 2007).

2.7 Zinc

Silicon affects zinc (Zn) nutrition in plants under deficiency stresses (Bityutskii et al., 2014; Li et al., 1999). Rhizospheric zinc solubility is regulated through root excretion of low-molecular-weight chelating agents and growth media acidification (Sinclair and Krämer, 2012; von Wirén et al., 1996). Several studies report that addition of silicon to hydroponics and soil affected the zinc distribution (Bityutskii et al., 2014; Gu et al., 2011, 2012) and that both elements were located at similar sites within plants (Gu et al., 2012). Silicon treatments help strengthen cell walls and delay Zn^{2+} transport to the shoots (Currie and Perry, 2007; Gong et al., 2006; Huang et al., 2009; Peleg et al., 2010). Silicon and zinc were mainly found near the root endodermis (Gu et al., 2011), precipitated as zinc silicates, which could hinder xylem loading and transport through the xylem (da Cunha and do Nascimento, 2009); these zinc silicates slowly degraded to SiO_2, and zinc was subsequently accumulated in an unknown form in vacuoles (Neumann and zur Nieden, 2001). Cell wall silicate precipitation (Currie and Perry, 2007) augmented the number of Zn^{2+}-binding sites, which could improve Zn^{2+} adsorption on the silicate deposits (Wang et al., 2000). This implies avoiding zinc transport to more sensitive plant organs and signifies a detoxifying action (da Cunha and do Nascimento, 2009; Shi et al., 2005). Root zinc deposits could be used in zinc deficiency. It has been reported that silicon treatments enhance citrate levels in plants under iron deficiency (Bityutskii et al., 2014; Pavlovic et al., 2013); citrate could also significantly impact zinc distribution in plants.

Silicon addition to plants in zinc deficiency could enhance the mobility of root zinc pools, similarly as it does for iron and manganese, to help better distribute the element and mollify deficiency symptoms. Silicon and zinc deposits were detected in leaves, which indicates remobilization through the phloem of the two elements, which in turn could lead to higher zinc contents in the fruit and seeds; this points to a more efficient use of the available zinc, especially in zinc deficiency (Bityutskii et al., 2014; Hernandez-Apaolaza, 2014).

2.8 Copper

Silicon-copper interaction in plants has not been extensively studied (Frantz et al., 2011; Ji et al., 1992; Khandekar and Leisner, 2011; Li et al., 2008). Adding silicon to the nutrient solution decreased copper toxicity symptoms, such as leaf chlorosis, and reduced shoot and root biomass, in *Arabidopsis thaliana* (Khandekar and Leisner, 2011; Li et al., 2008). Similar results were found in wheat (Nowakowski and Nowakowska, 1997). It is known that silicon affects the bioavailability and the distribution of copper within leaves of plants under copper stress (Li et al., 2008). It has been proposed that cell wall silicon accumulation, which increases the number of binding sites for copper and reduces the impact of high copper levels on plant cells, is responsible for the aforementioned redistribution of copper by silicon, similar to those proposed for other micronutrient toxicities (Frantz et al., 2011; Liang et al., 2007; Ma and Yamaji, 2006; Rogalla and Römheld, 2002; Wang et al., 2000). In addition, high levels of molecules that bind to copper, which minimizes copper toxicity effects, were maintained or even increased when silicon was added, which indicates that more than one response is activated to cope with copper toxicity (Khandekar and Leisner, 2011). Researchers proposed that silicon sequesters copper and reduces its toxic effects by helping generate additional apoplastic binding sites for copper (Hernandez-Apaolaza, 2014; Samuels et al., 1991). Silicon also promotes apoplastic and cell wall copper binding sites, and their incidence on some copper deficiency induced oxidative mechanisms (Samuels et al., 1991).

With respect to the toxicity of the rest of micronutrients, formation of root metal deposits similar to iron root deposits is reported in the literature. For example, Liang and Shen (1994) reported that, in oilseed rape, silicon supply decreased boron uptake at adequate or excess boron levels while increased the boron uptake under boron-deficient conditions. Silicon application to soils with a high boron concentration decreased the root-to-shoot boron translocation to effectively mitigate boron toxicity in barley, spinach, and wheat and furthermore modulated the antioxidant enzyme activities to prevent ROS membrane damages (Gunes et al., 2007a,b; Inal et al., 2009; Karabal et al., 2003).

Studies featured earlier lead to the conclusion that silicon can improve the uptake of micronutrients and macronutrients by plants grown under nutrient-deficient conditions. The role of silicon in improved uptake of micronutrients and macronutrients by fruit crops needs further research, as the topic has not been well investigated in comparison with that in agronomic crops.

3 Silicon and improved growth and quality parameters of fruit crops

Studies demonstrate that silicon is critical to fruit trees, as it can promote leaf chlorophyll content to improve photosynthesis; prevent root rot and premature aging; promote root growth and development; regulate fruit absorption of potassium, nitrogen, phosphorus, and micronutrients; effectively prevent cracking, early defoliation, and other physiological diseases; inhibit moisture evaporation in leaves and fruits; enhance resilience of fruit trees; increase fruit firmness; and improve storage and transportation (Dehghanipoodeh et al., 2018; Fang and Ma, 2005; Jana and Jeong, 2014; Meena et al., 2014; Patil et al., 2017; Wang et al., 2016). In this section, the role of silicon in improving the growth and quality of some fruit trees explored until now is discussed.

It is well known that grape (*Vitis vinifera* L.) is highly perishable during fresh fruit storage due to fungal decay, mechanical injury, physiological deterioration, and water loss (Sogvar et al., 2016); as a result, grapes have a short postharvest life. This majorly affects the local grape farmer's income and the stability of the fresh grape industry. The application of silicate fertilizers (600 kg SiO_2/ha) from two sources of water-cooling slag and steel slag significantly increased the fruit yield (24.7–26.0 Mg/ha, 10.9%–16.9% higher than that of the control), berry weight, berry size (i.e., length and width), and cluster weight in two table grape cultivars (Monukka and Red Globe) (Zhang et al., 2017). The yield of table grape fruit was 13.5% higher for plants treated with either water-cooling slag fertilizer or steel slag fertilizer compared with the control, for the 2 years of testing. Silicon fertilization considerably enhanced the total soluble solids to titratable acidity (TSS/TA) ratio (11.4% increase over the control), the fruit TSS, and fruit firmness (11.5%–15.0% increase over the control). These studies report that silicate fertilization decreased the fruit respiratory intensity, decay incidence (22.2%–27.0% decrease over the control), and weight loss to significantly extend the shelf life of fruits. Silicate fertilization probably led to reduced fruit weight loss by having silicon accumulate beneath the cuticle to form a cuticle-silicon double layer, strengthening the cell wall structures (Kader, 1992; Kim et al., 2002), and consequently suppressing fruit transpiration and respiration (Vogler and Ernst, 1999). Silicon could regulate plant metastasis, resultantly facilitating nitrogen and potassium assimilation in crops to significantly improve the quality of berries (Datnoff et al., 2001; Liu et al., 2014). A considerable increase in the root and green mass of silicon-treated germinated marsh grapefruit seedlings has also been reported (Matichenkov et al., 2001). Bangalore blue grapevines' responses to foliar silicic acid and boron sprays were studied (Bhavya, 2010). Foliar silicic acid application of 4 and 6 mL at 10-day intervals (six sprays) and 6 mL/L at 20-day intervals (three sprays) significantly affected the growth parameters over those of the control. The effects of silicon on the antioxidant and stomatal responses of grapevine rootstocks grown in boron-toxic saline solution and boron-toxic saline soil were investigated by Soylemezoglu et al. (2009). These studies suggest that rootstocks supplied with 4-mM silicon responded to the harmful effects of salinity and boron toxicity on shoot growth by lowering boron, chlorine, and sodium accumulation in rootstocks.

The response of passion fruit seedlings to calcium silicate application was studied (Prado and Natale, 2005). The maximum dry matter of shoot and root, number of leaves, plant height, and stem diameter were recorded in plants treated with silicon.

In another study with Zaghloul date palms (*Phoenix dactylifera* L.), applications of silicon at 0.05%–0.1% effectively enhanced the total chlorophyll level; potassium, nitrogen, and phosphorus contents in leaves, leaf area; yield; bunch weight; and physiochemical characteristics of fruits compared with the control (El-Kareem et al., 2014).

Application of silicon fertilizers can enhance citrus growth and fruit yield. Matichenkov et al. (1999) reported that silicon may directly play an important role in the growth and development of citrus trees. Previous studies investigated the relationship between the available soil silicon contents, the leaf contents of silicon, and the health of citrus trees (Matichenkov et al., 1999, 2000). These studies were further supplemented with an investigation of citrus trees grown in a greenhouse with silicon supplementation, and it was observed that silicon nutrition was attributed to significantly increased root mass of citrus trees (Matichenkov et al., 2001). The application of Pro-Sil silica slag (2, 4, and 8 t/ha) to young orange trees considerably augmented both the total tree height (14%–41% increase over the control) and the length of tree branches (31%–48% increase over the control) over a 6-month period (Matichenkov et al., 2001). Mathaba et al. (2009) observed that 0.5 mg/kg silicon had a greater potential in mitigating chilling injury in citrus fruits with less weight loss and membrane damage. Matichenkov et al. (1999) also reported that silicon may play a very important part in citrus tree growth and development.

In a study (Kaluwa et al., 2010), postharvest applications of 2940 mg/kg silicon in the form of potassium silicate (K_2SiO_3) improved avocado fruit quality, probably by suppressing respiration, reducing ethylene evolution, increasing catalase enzyme activity, maintaining fruit moisture, and increasing phenolic compound accumulation, thereby enhancing the shelf life.

A more direct impact of phenolic phytochemicals on the fruit quality is their ability to modulate biochemical, physiological, and molecular cellular physiology via the response pathways of antioxidant enzymes (ascorbate, catalase, glutathione, glutathione-transferase interface, and superoxide dismutase) (Serdula et al., 1996). Phenols can participate in the activation/deactivation of proteins, enzymes, and key metabolic pathways and the induction/repression of gene expression due to their structural similarities with several key signal molecules and biological effectors. The physiochemical properties of phenols also enable them to participate in moderating the cellular homeostasis (Vattem et al., 2005). Phenolics are important for plants in resisting and defending against microbial infections, which are closely related with the occurrence of reactive oxygen species (ROS). Free phenols provide protection against pathogens to contribute to the overall fruit fitness (Beckman, 2000). Phenolic compounds are present in plants in both the cell membrane–bound form and free form (Renger and Steinhart, 2000). When the concentration of free phenolic compounds is low, phenolic compounds are less capable to function as antioxidants. The availability of free hydroxyl groups on the phenol ring is vital to neutralizing free radicals. Therefore, releasing cell membrane–bound phenolics into the free-form phenolics enhances the tissue antioxidant capacity. Enzymatic hydrolysis of these free-form phenolics could enhance the levels of free phenolic compounds in food products, to enrich the aroma and flavor and potentially augment their nutritional value (Kapasakalidis et al., 2009; Vattem et al., 2005; Zheng and Shetty, 2000).

Optimal preharvest and postharvest management practices result in the maintenance of quality over longer storage periods for fruits and increase consumer confidence (Tesfay et al., 2011). It is known that silicon can increase the pool of free phenols. In other words, silicon functions as a major elicitor in increasing free polyphenol levels. For instance, in a study with avocado, silicon applications increased the pool of free phenols in the mesocarp, thereby increasing the fruit quality (Tesfay et al., 2011). There is a low concentration of free phenols in the avocado mesocarp but a higher level of cell membrane–bound phenols (Tesfay, 2009). Vattem et al. (2005) report that phenols help protect fruits and seeds against oxidation and ensure a healthy propagation of plants. Silicon, by increasing the free phenols released from the membrane-bound form and, as a result, increasing the antioxidant pool in fruits, could be a major factor in improving the postharvest fruit quality (Tesfay et al., 2011). Fruits treated with silicon had lower mass loss than the control. Silicon possibly contributes to maintaining fruit moisture. Silicon accumulation has been reported to induce impregnation of fruit peels' intercellular parts. Silicon treatments envelop the fruit stomata with a silicon layer, which reduces fruit respiration, and lead to a reduced weight loss (Hammash and Assi, 2007). Reduced activity levels of polyphenol oxidase limit/prevent mesocarp browning of cut avocados (Bower and Dennison, 2005). Decreased polyphenol oxidase activity induces membrane-bound phenols to leak and act as antioxidants with no oxidant interferences to reduce browning. This allows cut fruits to be kept fresh for a longer time. Silicon could be used as it functions to bind cellular oxygen, reducing the accumulation of oxidants (Řezanka and Sigler, 2008). Silicon oxidizes to form solid silicon dioxide, where a lattice is formed with one silicon atom surrounded by four oxygen atoms (Bekker, 2007). The results propose that silicon application could maintain membrane integrity and suppress lipid peroxidation under stresses. Silicon treatments were observed to improve firmness and reduce mass loss in "Hass" avocado, which can be attributed to reduced electrolyte leakage, lipid peroxidation, and increased catalase expression and activity (Tesfay et al., 2011). Silicon deposits could furthermore be related to how the element affects membrane integrity. Electrolyte leakage is related to the breakdown of the cell membrane and is therefore indicative of reduced membrane integrity (Thompson, 1988). Avocado fruits treated with silicon had a lower electrolyte leakage than the control, which may be due to silicon accumulation between the cell membrane and the cell wall, upholding a barrier against solute leakage (Tesfay et al., 2011).

Hanumanthaiah et al. (2015) reported in a field study of banana that foliar silicon application in 15-day intervals effectively improved the quality parameters such as TSS (26.67 °Brix), shelf life (6.33 days), pulp/peel ratio (7.44), acidity (0.26%), reducing sugars (19.93%), and nonreducing sugars (2.24%). How different in vitro silicon sources affect the leaf anatomy of banana "Maca" plants was investigated (Magno Queiroz Luz et al., 2012), where it was discovered that calcium silicate addition induced thicker epidermis (upper and lower), mesophyll, and palisade parenchyma and enhanced the photosynthetic rate. Kumbargire et al. (2016) also reported the greatest composition of macronutrients like nitrogen (3.48%), phosphorus (0.37%), potassium (4.23%), calcium (0.75%), and magnesium (0.35%) and micronutrients like zinc (27 mg/kg) and copper (12.37 mg/kg) in banana leaves in recommended fertilizer dose of +750 kg/ha of diatomaceous earth (as a silicon source), which indirectly reflects on the growth, development, and yield of banana. Different in vitro silicon sources and their effects on the development of micropropagated banana seedlings were investigated (Asmar et al., 2011). The study demonstrated increased chlorophyll a and b and total chlorophyll levels in the following calcium silicate application. Weak neck is observed in *Musa* sp. cv. Rastali was solved by spraying boron, magnesium, and silicon (Putra et al., 2010).

In field experiments conducted by Lalithya et al. (2013) and Lalithya et al. (2014), it was reported that sapota trees (cv. Kalipatti) sprayed with K_2SiO_3 saw improvements in the number of shoots (23.96 per m^2), total chlorophyll content (8.47 mg/g of fresh weight), number of flowers (250.16 per m^2), and a minimal number of mummified fruits (34 per tree[1]). In addition, sapota trees sprayed with K_2SiO_3 saw improvements in fruit weight (99.96 g), fruit length (5.55 cm), fruit diameter (5.85 cm), volume of fruit (102.38 g), maximum shelf life (10.90 days), and yield (12.48 t/ha) compared with the control.

Costa et al. (2015) observed a quadratic increase in mango diameter from trees treated with 1600 kg/ha of agrosilicon. More et al. (2015) found that application of stabilized silicic acid with five sprays during the initial stages of fruit growth of Alphonso mango (*Mangifera indica* L.) (before flowering and 15, 30, 45, and 60 days after flowering) improved the yield and quality of mango. Helaly et al. (2017) reported indole-3-acetic acid (IAA), gibberellins (GA), and cytokinins (CK) endogenous levels were improved in mango trees treated with silicon, and, in contrast, abscisic acid (ABA) level was lower in these trees. In addition, silicon supplementation reduced the levels of antioxidative enzymes of peroxidase (POX), catalase (CAT), and superoxide dismutase (SOD), compared with those of the control. Silicon application resulted in improved fruit quality for three successive seasons. According to these researchers, silicon protects plants against destructive oxidative reactions by scavenging reactive oxygen species and helped mango trees better withstand the environmental stresses in arid regions.

In the experiment carried out by El-Rhman (2010), it was observed that pomegranate plants irrigated every 2 days with foliar application of kaolin at 6% saw reduced fruit cracking (27.98%), increased fruits per plant (57.53), fruit weight (418.5 g), yield (23.98 kg/plant), pulp (60.69%), total sugars (13.03%), and reduced sugars (12.02%). Melgarejo et al. (2004) reported that four sprays of kaolin at 2.5% to pomegranate plants resulted in the greatest ratio of premium quality fruits (90.6%) and minimum unmarketable fruits (9.4%). Yazici and Kaynak (2006) found that plants covered with foliar applications of kaolin at 3% had decreased sunburn damage in pomegranate (cv. Hicaznar) compared with the control.

How chitosan/nanosilica coating affects the physiochemical characteristics of longan fruits was investigated (Shi et al., 2013). It was discovered that an excellent semipermeable chitosan/nanosilica film on fresh longan fruits considerably reduces browning, increases the shelf life, slows weight loss, and inhibits increase of malondialdehyde levels and polyphenol oxidase activities.

Silicon fertilization is known to enhance the growth and quality of apples (Cai and Qian, 1995). Silicon treatments could significantly enhance the silicon content in apple leaf, peel, flesh, and whole fruit. This positively impacts apple fruits, because silicon is the main component of the cell wall, depositing in the plant cell wall and root cortical cells, forming a silicide layer with host cells that acts as a barrier against pathogen invasion. Thus, silicon plays an effective role in improving fruit firmness and pest resistance (Wang et al., 2016). Polygalacturonase (PG) is the dominant factor of the cell wall and pectin degradation and is a key enzyme in a variety of fruit softening. PG is an enzyme that can degrade pectin; however, pectin plays a pivotal role in sustaining the internal and external environments and maintaining the firmness of fruit peels. During apple storage process, a lower PG content keeps a better apple quality and creates a longer storage time in apple trees treated with silicon fertilizer (100 g/tree). Apple yield was 19.7% higher in trees treated with silicon compared with the control. Silicon is the major component to maintain fruit firmness by tightening the cell wall arrangement. Therefore, a high amount of silicon fertilization could enhance the ability of cell wall to defend against the activity of malondialdehyde (MDA) and PG and then extend the storage time of harvested apple (Wang et al., 2016). Gao et al. (2006) studied how silicon application affects apple internal bark necrosis (IBN) induced by high manganese levels. The results showed that simultaneous application of silicon and manganese, both at 400 mg/kg, effectively prevented Fuji apple trees from developing IBN. Su et al. (2011) also studied how silicon affected the quality of apple fruits grown in acidic soils. It was found that silicon could considerably reduce the titratable acid content in fruits and increase TSS content and vitamin C levels, but do not seem to affect the firmness. The results of the studies earlier demonstrate that silicate fertilization during fruit production can help effectively improve fruit quality and increase yield.

4 Silicon and control of diseases of fruit crops

Pests and plant diseases poses a major threat to agricultural production, causing serious crop yield losses and quality degradation throughout the world (Etesami and Alikhani, 2017). Chemicals are commonly used to control pests and plant diseases around the world, which, in addition to imposing a lot of costs, cause environmental pollution and are unusable, not efficient or applicable according to host plant–microbe pathosystems (Cooksey, 1990). Other methods of controlling plant pathogens such as breeding for host resistance are not available to growers in some cases, and owing to the local environmental conditions and the genetic diversity of the microbial pathogens, a failure of usable resistance has often been reported (Lindgren, 1997). Antibiotics and copper fungicides have been adopted to control diseases but were met with limited success in curbing diseases and subsequent yield losses. Therefore, other strategies to combat the diseases are necessary. A practical alternative could be using mineral nutrition to increase the disease resistance in plants (Marschner, 1995).

Silicon stands out among minerals due to its effectiveness in reducing the severity of several plant diseases (Datnoff et al., 2007; Epstein, 1999). With the emergence of serious physiological diseases and reduction of quality and storage issues, the use of silicon gradually attracted people's attention. The use of silicon is known as one of the most environmentally friendly and sustainable ways to battle plant diseases and pests. Silicon's role in alleviating biotic stresses was first identified in the modification of plant cell wall properties (Lux et al., 2002). Silicon's role in effectively controlling fungal, bacterial, viral, and nematode diseases in different plant species and in increasing plant resistance to diseases has been documented for decades (Etesami and Jeong, 2018; Fauteux et al., 2005; Fortunato et al., 2015b; Rodrigues et al., 2015; Rodrigues and Datnoff, 2015; Sahebi et al., 2016; Sakr, 2016; Van Bockhaven et al., 2013). Silicic acid polymerization within the apoplast forms an amorphous silica barrier (Exley, 2015) and helps deter pathogen infection (Guerriero et al., 2016). The improvement of overall mechanical strength and an additional outer layer of enhance protection for plants account for most of the reported benefits in crop quality and yield after silicon fertilization (Epstein, 1999). Successful infection requires plant pathogens to penetrate physical barriers including wax, cuticles, and cell walls and enter the host plant (Łaźniewska et al., 2012; Nawrath, 2006). Physical barrier formation is a mechanism to control plant diseases (Guerriero et al., 2016; Kim et al., 2002; Liang et al., 2015d). Physical barriers make plant cells less vulnerable to fungal pathogen invasion and the subsequent enzymatic degradation (Fauteux et al., 2005; Van Bockhaven et al., 2013). Silicon promotes plant growth by forming an outer protective layer and improving the mechanical strength of plants (Epstein, 1999; Sun et al., 2010). Silicon is typically cross-linked to hemicellulose, which improves the mechanical properties and regeneration of cell walls (Guerriero et al., 2016; He et al., 2015). Silicon plays a role in cell wall rigidity and reinforcement and helps increase its elasticity during extension growth (Marschner, 2012b). In primary cell walls, silicon contributes to increased elasticity during extension growth by interacting with cell wall constituents, such as polyphenols and pectins (Emadian and Newton, 1989). Increased resistance of plants treated with silicon to diseases is associated with the thick silica layer under the cuticle, double cuticular layer, density of silicified short and long epidermal cells, papilla formation, thickened cellulose membrane, and complexes formed with organic compounds in epidermal cell walls that mechanically strengthen plants (Fauteux et al., 2005; Van Bockhaven et al., 2013).

Silicon accumulation in the cell wall, cuticle, papillae, and wax layer helps increase the physical resistance of plants against pathogen penetration. Although the silicon polymerization theory was placed under skepticism, the hypothesis of cell wall fortification from silicon treatment to explain the augmented resistance of plants against pathogenic fungi has recently been strongly challenged (Liang et al., 2015d). Recent studies report that biochemical mechanisms with silicon play a bigger part in increasing disease resistance in plants than physical mechanisms with silicon do (Etesami and Jeong, 2018; Ratnayake et al., 2016; Song et al., 2016), which needs verification from future research. Silicon promotes defense-related enzyme activities in plant-pathogen interactions, which is closely connected to disease resistance (Fauteux et al., 2005; Van Bockhaven et al., 2013). Such defense-related enzymes include **chitinases**, a PR protein that contributes to hydrolyzing many phytopathogenic fungi cell walls; **polyphenol oxidases**; **catalase**; **lipoxygenase**; **phenylalanine ammonia lyase** (aid to total soluble phenolic and lignin-thioglycolic acid derivative accumulation); **superoxide dismutase**; **ascorbate peroxidase**; **glutathione reductase**; **β-1,3-glucanases**; **peroxidase** (contribute to the host defense response by producing antimicrobial quantities of H_2O_2 and phenolic polymerization that increases tissue lignification and also involved in the final stages of lignin biosynthesis, cell wall protein cross-linking, and cell wall reinforcement); and **polyphenoloxidase**, which mainly exists in a free form in the cytoplasm or bound in chloroplasts, mitochondria, and other subcellular organelles; helps oxidize phenolic compounds to quinines, usually more toxic than the original phenols to plant pathogens; and aids lignin biosynthesis. Its activity has been positively correlated with disease resistance in plants. Polyphenoloxidase is involved in lignin synthesis and helps increase the antibacterial defense of plants (Brisson et al., 1994; Cruz et al., 2013; Etesami, 2018; Etesami and Jeong, 2018; Keen and Yoshikawa, 1983; Piperno, 2006; Quarta et al., 2013; Shewry and Lucas, 1997; Silva et al., 2010; Song et al., 2016; Torres et al., 2006; Waewthongrak et al., 2015). Silicon supply helps raise total soluble phenolic compound contents in plants and delays the growth of invasive pathogens to enhance disease resistance of plants (Dallagnol et al., 2011; Fortunato et al., 2015a).

Ethylene (ET), jasmonic acid (JA), and salicylic acid (SA) regulate plant defense responses and play key roles in plant immunity networks (Clarke et al., 2000; Devadas et al., 2002). It is reported in several studies that silicon may modulate phytohormone homeostasis and signaling pathways to regulate plant stress responses (De Vleesschauwer et al., 2008; Fauteux et al., 2006; Ghareeb et al., 2011; Reynolds et al., 2016; Zhang et al., 2004). There is much evidence that the biochemical modifications and the incurred increase in resistance to environmental stresses and diseases from silicon application also help plants deal with pests, and it has recently been confirmed that silicon helps prime plants for a better pest infestation defense response. JA signaling pathways seem to mediate the primed state (Etesami and Jeong, 2018; Liang et al., 2015c).

The change in antimicrobial substances is a substantial response to defense-related enzymes; silicon application and the resultant lower plant disease occurrences are generally associated with a higher level of defense-related enzyme activities, which stimulates the production and accumulation of antimicrobial compounds, such as flavonoids, phenols, phytoalexins, and PR proteins following pathogen penetration in plants (Chérif et al., 1994; Fawe et al., 2001; Rémus-Borel et al., 2005; Rodrigues et al., 2004). Several studies have documented that silicon stimulates the accumulation of antimicrobial compounds, such as flavonoids, phenols, and phytoalexins during pathogen infection. Antimicrobial compounds help higher plants to combat diseases (Datnoff et al., 2007; Dodds and Rathjen, 2010; Fauteux et al., 2005; Van Bockhaven et al., 2013). Phytoalexins are recognized to be critical in plant defense against pathogen infection.

Well-nourished plants are known to be more resistant to diseases. Another mechanism with which silicon increases plant disease resistance is by affecting the plant mineral nutrition. As previously mentioned, silicon can increase the essential nutrient concentrations in plants. Additionally, silicon contributes to the uptake of important nutrients in plants (Pozza et al., 2015). The silicate anion competes for adsorption sites in soils, increasing the sulfate, nitrate, and phosphate availability in soil and the ability for plants to retain these anions (Etesami and Jeong, 2018; Pozza et al., 2015). For example, phosphorus becomes available in the soil solution after being desorbed from the reactive surfaces of the soil components, which can be incorporated into plants to reinforce plant disease resistance by going into the metabolic pathways and using the given nutrients to form defense walls. Marschner and Rimmington (1988) report that macronutrient and micronutrient imbalances affect the vigor and defense responses of plants and influence host susceptibility to diseases. They can directly work on secondary metabolic pathways, where cell wall defense system expansion and fungistatic phenolic compound formations occur, either by increasing middle lamella resistance or by forming physical and chemical barriers, as in calcium conditions.

In a study, Ferreira et al. (2015) found that silicon clearly did not directly affect *Acidovorax citrulli*. Instead, silicon played an important part in further raising plant calcium and magnesium contents, which could also have been involved in suppressing bacterial fruit blotch symptoms in melon. Soil silicates behave similarly to calcium and magnesium carbonates by correcting acidity, reducing Al^{3+} and H^+ levels, and neutralizing exchangeable aluminum. Silicates are also associated with increased levels of available soluble silicon and exchangeable calcium and magnesium (Epstein, 1999). Matichenkov and Calvert (2002) found that although higher Si doses did not significantly increase silicon accumulation in the shoots, elevated pH values were observed that made Ca and Mg more plant available thus increasing their shoot levels in the plant.

Silicon application to plants susceptible to diseases has been found to increase the resistance of these cultivars to the resistance level of innately resistant cultivars (Debona et al., 2017; Etesami and Jeong, 2018). The role of silicon in alleviating pathogen-induced stresses has been comprehensively reviewed (Chanchal Malhotra et al., 2016; Cooke and Leishman, 2011; Etesami and Jeong, 2018; Liang et al., 2015c,d; Ma, 2004; Meena et al., 2014; Pozza et al., 2015; Song et al., 2016; Van Bockhaven et al., 2013; Wang et al., 2017). Silicon employs many mechanisms to alleviate plant disease stresses or increase plant resistance to diseases (Etesami, 2018; Etesami and Jeong, 2018; Wang et al., 2017). Generally, silicon prevents penetration of different pathogens into plants by (i) accumulating in the epidermal tissue, beneath the cuticle and acting as a barrier, resulting in an increase of plant tissue rigidity; (ii) strengthening plant structures; (iii) inducing phenolic compounds; (iv) producing phytoalexin, glucanase, and peroxidase; (v) stimulating systemic acquired resistance to inhibit pathogen colonization; (vi) manufacturing antimicrobial compounds; (vii) forming organic compound complexes in cell walls, thereby increasing the resistance of plant cell walls to degradation; (viii) activating gene expressions related to defense and multiple signaling pathways to increase plant resistance; (ix) limiting pathogen invasion and colonization by regulating stress-related gene expression or pathogenicity; and (x) improving plant mineral nutrition. Due to many intricacies surrounding silicon properties, absorption, and efficiency (Liang et al., 2015e), the complete picture of silicon mechanisms in plant disease regulation has yet to be understood. However, silicon can employ one or a combination of the aforementioned mechanisms to increase plant disease resistance.

Many studies indicate that silicon increases plant resistance to diseases with biochemical, molecular, and physical mechanisms (Etesami, 2018; Etesami and Jeong, 2018; Wang et al., 2017), and more research is needed to unearth the mechanisms that silicon uses to regulate plant-microbe interactions, such as plant signal transmission and transcriptome regulation of defense-related pathways (Etesami, 2018; Etesami and Jeong, 2018). There is also little information on how silicon and bacterial diseases interact, compared with the vast information available on how silicon enhances plant resistance to fungal diseases (Etesami and Jeong, 2018; Song et al., 2016), which needs further research. Studies on the effects of soluble silicon to horticultural tree crops are scarce, compared with those on agricultural crops (mostly cereal grains and dicotyledonous annual crops like cucumber and tomato) (Pozza et al., 2015; Tubana et al., 2016). Table 19.1 summarizes studies in this area.

5 Necessity of using silicon fertilizers

Since the amount of silicon taken up by crops from the soil usually exceeds the amount of silicon released via gradual degradation of silicate clay minerals (the regular removal of silicon-rich litter during harvest), the soluble silicon removed from the soil by plants cannot usually be compensated, and the soil silicon supply declines (Song et al., 2015; Vandevenne et al., 2015). This results in increasingly worsening soil silicon deficiency, which may effectively be addressed with silicon fertilization to increase the yield and quality of many agricultural crops.

Various sources of silicon-containing fertilizers have been widely used throughout the world. Silicon can be applied preharvest or postharvest, to the root or to the leaves. Silicon is the only element reported to increase the resistance to multiple stresses in plants. Even in excess, silicon does not harm plants (Epstein, 1994) since it is noncorrosive and free from pollution (Zhu and Gong, 2014). Therefore, silicon fertilizers can be utilized to cultivate an ecologically sound agronomy (Etesami and Jeong, 2018; Zhu and Gong, 2014). Economically, silicon fertilizers are affordable, priced at 10%–20% levels of other fertilizers (Feng, 2000). The economic and ecological benefits make silicon fertilizers very favorable (Etesami and Jeong, 2018). The industrial waste materials, such as blast furnace slag, electric furnace steel slag, stainless steel slag, and other sources like sodium silicate, potassium silicate, calcium silicate, and rice hull (raw or partially burnt) are frequently used as the main sources of silicon fertilizers (Datnoff et al., 2001; Haynes, 2014;

TABLE 19.1 Studies conducted on the role of silicon in reducing the fungal disease incidence in fruit crops.

Pathogen	Disease	Fruit crop	Effect of silicon (Si)	Reference
Alternaria alternata	Brown spot	Citrus	Reduced disease severity	Asanzi et al. (2015)
Penicillium digitatum	Green mold	Citrus	Controlled green mold	Liu et al. (2010)
P. digitatum	Green mold	Citrus	Controlled green mold	Abraham (2010)
P. digitatum	Green mold	Citrus	Reduced the severity of green mold	Moscoso-Ramírez and Palou (2014)
P. digitatum	Green mold	Lemon	Increased the resistance of lemon to *P. digitatum*	Mkhize et al. (2012)
Cylindrocladium spathiphylli	Root rot disease	Banana	A reduction of about 50% of root necrosis after 14 days inoculation for the Si-supplied (2 mM of soluble Si) plants compared with those not supplied with Si and alleviation of growth reduction caused by the pathogen	Vermeire et al. (2011)
Fusarium oxysporum f. sp. *cubense*	Fusarium wilt	Banana	Reduced the intensity of Fusarium wilt concomitantly with more concentrations of hydrogen peroxide, total soluble phenolics and lignin-thioglycolic acid derivatives and greater activities of phenylalanine ammonialyases, polyphenoloxidases, peroxidises, β-1,3-glucanases, and chitinases	Fortunato et al. (2012b)
Mycosphaerella fijiensis	Black sigatoka	Banana	The disease developed more rapidly and more severely on banana plants grown without Si than on plants supplied with Si	Kablan et al. (2012)
Fusarium oxysporum f. sp. *cubense*	Fusarium wilt	Banana	Reduced the intensity of Fusarium wilt	Fortunato et al. (2012a)
Fusarium oxysporum f. sp. *cubense*	Fusarium wilt	Banana	Reduced the intensity of Fusarium wilt	Kidane and Laing (2008)
Uncinula necator	Powdery mildew	Grapevine	Reduced numbers of colonies of fungus	Bowen et al. (1992)
Colletotrichum gloeosporioides	Anthracnose	Avocado	Decreased the severity and incidence of anthracnose	Anderson et al. (2005)
C. gloeosporioides	Anthracnose	Avocado	Decreased presence of anthracnose in stored fruit	Bosse et al. (2011)
Calonectria ilicicola	Black root rot	Avocado	Reduced root necrosis of avocado seedlings	Dann and Le (2017)
Phytophthora cinnamomi	Phytophthora root rot	Avocado	Controlled the disease	Bekker et al. (2007)
Podosphaera xanthii	Powdery mildew	Pumpkin	Reduced disease incidence and severity	Lepolu Torlon et al. (2016)
Sphaerotheca aphanis	Powdery mildew	Strawberry	Reduced disease incidence and severity	Kanto et al. (2006)
Ganoderma boninense	Basal stem rot	Oil palm	Suppressed basal stem rot and provided a 53% disease reduction compared with other treatments and reduced the numbers of primary roots infected and of stem tissues that developed lesions	Najihah et al. (2015)
Fusarium oxysporum f. sp. *vasinfectum*	Fusarium wilt	Cotton	Reduced disease incidence and severity by affecting cellular defense responses in cotton roots	Whan et al. (2016)
Hemileia vastatrix	Leaf rust	Coffee	Controlled the coffee leaf rust development	Carré-Missio et al. (2014)
Penicillium expansum and *Monilinia fructicola*	Fruit decay	Cherry	Controlled disease by inducing a significant increase in the activities of phenylalanine ammonia lyase, polyphenoloxidase, and peroxidase in sweet cherry fruit and activating a cytochemical reaction and causing tissue browning near the site of wounding	Qin and Tian (2005)

Continued

TABLE 19.1   Studies conducted on the role of silicon in reducing the fungal disease incidence in fruit crops.—cont'd

Pathogen	Disease	Fruit crop	Effect of silicon (Si)	Reference
Erysiphe sp.	Powdery mildew	Bitter gourd	Strengthened resistance in bitter gourd plants against powdery mildew by stimulating expression of several biochemical defense reactions	Ratnayake et al. (2016)
Uncinula necator (Schwein) *Burrill*	Powdery mildew	Grape	Reduced disease incidence and severity and numbers of colonies of pathogen on leaves	Bowen et al. (1992)
Penicillium expansum	Blue mold	Apple	Controlled apple blue mold	Etebarian et al. (2013)
P. expansum	Blue mold	Apple	Controlled apple blue mold	Farahani et al. (2012)
P. expansum	Blue mold	Apple	Inhibited the mycelial growth of *P. expansum* and reduced the lesion diameter of blue mold decay of apples	Ebrahimi et al. (2012)
Monilinia fructicola	Brown rot	Apple	Controlled disease caused by *M. fructicola*	Yang et al. (2010)
Colletotrichum gloeosporioides	Anthracnose	Mango	Reduced disease incidence and severity	Umaña-Rojas (2009)
Penicillium expansum	Blue mold	Jujube fruit (*Ziziphus jujube*)	Controlled apple blue mold	Tian et al. (2005)
Alternaria alternate	Alternaria fruit rot	Jujube fruit (*Z. jujube*)	Controlled the diseases caused by *A. alternata*	Tian et al. (2005)

Hossain et al., 2001; Liang et al., 2015e; Pereira et al., 2004) for fruits and vegetables (Weerahewa and Somapala, 2016). Calcium silicate treatment also supplies Ca^{2+}, which contributes to the defense block formations (Pozza et al., 2015). As previously mentioned, a balanced nutrition is crucial to resistance to biotic and abiotic stresses in plants. Complementary cations from silicon fertilizers—Ca^{2+} and K^+ from calcium and potassium silicates—can disrupt the nutritional balance in plants. For example, Ca^{2+} may compete for the same absorption sites with Mg^{2+}, K^+, and NH_4^+, leading to deficiencies in these minerals (Pozza et al., 2015). Therefore, the soil composition, regarding cation concentrations, should be considered in choosing the source of silicon application—as calcium silicate, potassium silicate, etc. On the other hand, calcium, like silicon, plays a role in forming resistance barriers. The role of calcium in increasing plant disease resistance should be considered along with the role of silicon.

6 Future research

The use of silicon in the agricultural sector has attracted many researchers due to the many benefits of this element for fruit crops. The use of silicon has reduced the severity of certain diseases in very important economic products. In organic farming, silicon application may pave the way for increasing the quality and yield of fruits while reducing the use of chemical fertilizers, pesticides, and fungicides and may provide an environmentally friendly alternative to pesticides for the integrated control of important fruit crop disease. Previous studies have not looked at the use of pesticides and fungicides as a positive control or the use of a combination of pesticides and fungicides along with silicon to evaluate silicon's effects on controlling plant diseases. Using these treatments with silicon in future studies is recommended, to determine whether silicon can completely control diseases in comparison with pesticides and fungicides. Silicon can be used to manage diseases, not as the sole solution to problems caused by pathogens and pests, but as an important component of integrated pest management systems (Pozza et al., 2015).

Foliar silicon application is considerably more cost-effective and more convenient compared with soil fertilization. Foliar silicon application is biostimulative, and the best results are observed in stressed plants exposed to water deficiency, water excess, salinity, extreme temperatures, and the prevalence of diseases, pests, etc. Many studies indicate that foliar nutrition should become a standard crop management practice for many species of agricultural plants (Artyszak, 2018). Foliar nutrition can help farmers increase the yield of fruit crops and is environmentally safe.

Silicon can easily be fixed in the soil, similar to phosphorus. The amendment of organic materials, such as humic acid, can improve the availability of soil silicon/phosphorus. This is due to the fact that phosphorus adsorption sites are occupied by organic anions, which are decomposed from organic materials, thereby increasing phosphorus availability in soil (Wang et al., 2016). Hence, the combined use of silicon and organic materials to prevent fixation of phosphorus in soil is suggested for future studies. Further investigation of the action mechanisms by which silicon relieves various stresses in fruit crops, as well as the biochemical and molecular bases of how silicon improves the quality and yield of fruit crops, is necessary.

Acknowledgment

We wish to thank the University of Tehran and Gyeongsang National University for providing the necessary facilities for this study.

References

Abbas, T., Balal, R.M., Shahid, M.A., Pervez, M.A., Ayyub, C.M., Aqueel, M.A., Javaid, M.M., 2015. Silicon-induced alleviation of NaCl toxicity in okra (*Abelmoschus esculentus*) is associated with enhanced photosynthesis, osmoprotectants and antioxidant metabolism. Acta Physiol. Plant. 37, 1–15.

Abraham, A.O., 2010. Integrated Use of Yeast, Hot Water and Potassium Silicate Treatments for the Control of Postharvest Green Mould of Citrus and Litchi. (Doctoral dissertation).

Adrees, M., Ali, S., Rizwan, M., Zia-ur-Rehman, M., Ibrahim, M., Abbas, F., Farid, M., Qayyum, M.F., Irshad, M.K., 2015. Mechanisms of silicon-mediated alleviation of heavy metal toxicity in plants: a review. Ecotoxicol. Environ. Saf. 119, 186–197.

Alvarez, J., Datnoff, L.E., 2001. The economic potential of silicon for integrated management and sustainable rice production. Crop Prot. 20, 43–48.

Al-Wasfy, M.M., 2012. Trails for improving water use efficiency and improving productivity in Williams banana orchards by spraying salicylic acid. Minia J. Agric. Res. Dev. 32, 139–160.

Al-Wasfy, M.M., 2013. Response of Sakkoti date palms to foliar application of royal jelly, silicon and vitamins B. J. Am. Sci. 9, 315–321.

Anderson, D., Synder, G., Martin, F., 1991. Multi-year response of sugarcane to calcium silicate slag on Everglades Histosols. Agron. J. 83, 870–874.

Anderson, J.M., Pegg, K.G., Dann, E.K., Cooke, A.W., Smith, L.A., Willingham, S.L., Giblin, F.R., Dean, J.R., Coates, L.M., 2005. New strategies for the integrated control of avocado fruit diseases, pp. 1–6.

Artyszak, A., 2018. Effect of silicon fertilization on crop yield quantity and quality—a literature review in Europe. Plan. Theory 7, 54.

Asanzi, N.M., Taylori, N.J., Vahrmeijer, J.T., 2015. Can silicon be used to prevent *Alternaria alternata* in citrus trees? Technology, 48–51.

Ashraf, M., Ahmad, R., Afzal, M., Tahir, M., Kanwal, S., Maqsood, M., 2009. Potassium and silicon improve yield and juice quality in sugarcane (*Saccharum officinarum* L.) under salt stress. J. Agron. Crop Sci. 195, 284–291.

Asmar, S.A., Pasqual, M., Rodrigues, F.A., Araujo, A.G.d., Pio, L.A.S., Silva, S.d.O., 2011. Sources of silicon in the development of micropropagated seedlings of banana 'Maçã'. Ciência Rural 41, 1127–1131.

Babini, E., Marconi, S., Cozzolino, S., Ritota, M., Taglienti, A., Sequi, P., Valentini, M., 2012. Bio-available silicon fertilization effects on strawberry shelf-life. Acta Hortic.

Bakhat, H.F., Bibi, N., Zia, Z., Abbas, S., Hammad, H.M., Fahad, S., Ashraf, M.R., Shah, G.M., Rabbani, F., Saeed, S., 2018. Silicon mitigates biotic stresses in crop plants: a review. Crop Prot. 104, 21–34.

Baligar, V.C., Fageria, N.K., He, Z.L., 2001. Nutrient use efficiency in plants. Commun. Soil Sci. Plant Anal. 32, 921–950.

Beckman, C.H., 2000. Phenolic-storing cells: keys to programmed cell death and periderm formation in wilt disease resistance and in general defence responses in plants? Physiol. Mol. Plant Pathol. 57, 101–110.

Bekker, T.F., 2007. Efficacy of Water Soluble Silicon for Control of *Phytophthora cinnamomi* Root Rot of Avocado. University of Pretoria, Pretoria.

Bekker, T.F., Labuschagne, N., Aveling, T., Kaiser, C., 2007. Efficacy of water soluble potassium silicate against Phytophthora root rot of avocado under field conditions. S. Afr. Avocado Grower Assoc. Yearb. 30, 39–48.

Bhavya, H.K., 2010. Effect of Foliar Silicic Acid and Boron in Bangalore Blue Grapes.

Bityutskii, N.P., Yakkonen, K.L., Zlotina, M.M., 2010. Vliyanie kremniya na proyavlenie khlorosa rasteniy v usloviyakh defizita zheleza i marganza. Agrokhimiia 2, e51.

Bityutskii, N., Pavlovic, J., Yakkonen, K., Maksimović, V., Nikolic, M., 2014. Contrasting effect of silicon on iron, zinc and manganese status and accumulation of metal-mobilizing compounds in micronutrient-deficient cucumber. Plant Physiol. Biochem. 74, 205–211.

Bloemhard, C., Van Moolenbroek, J., 1994. Management of mineral elements of roses grown in closed rockwool systems. In: International Symposium on Growing Media & Plant Nutrition in Horticulture. 401, pp. 481–492.

Bosse, R.J., Bower, J.P., Bertling, I., 2011. Pre- and post-harvest treatments on 'Fuerte' avocados to control anthracnose (*Colletotrichum gloeosporioides*) during ripening. S. Afr. Avocado Grower Assoc. Yearb. 34, 65–69.

Bowen, P., Menzies, J., Ehret, D., Samuels, L., Glass, A.D.M., 1992. Soluble silicon sprays inhibit powdery mildew development on grape leaves. J. Am. Soc. Hortic. Sci. 117, 906–912.

Bower, J.P., Dennison, M.T., 2005. A process to prevent browning of frozen avocado halves and chunks. S. Afr. Avocado Growers' Assoc. Yearb. 28, 40–41.

Brisson, L.F., Tenhaken, R., Lamb, C., 1994. Function of oxidative cross-linking of cell wall structural proteins in plant disease resistance. Plant Cell 6, 1703–1712.

Brunetto, G., Melo, G.W.B.D., Toselli, M., Quartieri, M., Tagliavini, M., 2015. The role of mineral nutrition on yields and fruit quality in grapevine, pear and apple. Rev. Bras. Frutic. 37, 1089–1104.

Cai, D.L., Qian, F.J., 1995. Effect of Si fertilization on yield and quality of apples. J. Reg. Res. Dev. 14, 64–66.

Carré-Missio, V., Rodrigues, F.A., Schurt, D.A., Resende, R.S., Souza, N.F.A., Rezende, D.C., Moreira, W.R., Zambolim, L., 2014. Effect of foliar-applied potassium silicate on coffee leaf infection by *Hemileia vastatrix*. Ann. Appl. Biol. 164, 396–403.

Cazorla, F.M., Arrebola, E., Olea, F., Velasco, L., Hermoso, J.M., Pérez-García, A., Tores, J.A., Farre, J.M., de Vicente, A., 2006. Field evaluation of treatments for the control of the bacterial apical necrosis of mango (*Mangifera indica*) caused by *Pseudomonas syringae* pv. syringae. Eur. J. Plant Pathol. 116, 279–288.

Chanchal Malhotra, C., Kapoor, R., Ganjewala, D., 2016. Alleviation of abiotic and biotic stresses in plants by silicon supplementation. Scientia 13, 59–73.

Chardonnet, C., Doneche, B., 1995. Influence of calcium pretreatment on pectic substance evolution in cucumber fruit (*Cucumis sativus*) during *Botrytis cinerea* infection. Phytoparasitica 23, 335–344.

Chen, W., Yao, X., Cai, K., Chen, J., 2011. Silicon alleviates drought stress of rice plants by improving plant water status, photosynthesis and mineral nutrient absorption. Biol. Trace Elem. Res. 142, 67–76.

Chen, D., Cao, B., Wang, S., Liu, P., Deng, X., Yin, L., Zhang, S., 2016. Silicon moderated the K deficiency by improving the plant-water status in sorghum. Sci. Rep. 6.

Cheong, Y.W.Y., Chan, P.Y., 1973. Incorporation of P 32 in phosphate esters of the sugar cane plant and the effect of Si and Al on the distribution of these esters. Plant Soil 38, 113–123.

Chérif, M., Asselin, A., Bélanger, R.R., 1994. Defense responses induced by soluble silicon in cucumber roots infected by Pythium spp. Phytopathology 84, 236–242.

Chinnasami, K.N., Chandrasekaran, S., 1978. Silica status in certain soils of Tamil Nadu. Madras Agric. J. 65, 743–746.

Clarke, J.D., Volko, S.M., Ledford, H., Ausubel, F.M., Dong, X., 2000. Roles of salicylic acid, jasmonic acid, and ethylene in cpr-induced resistance in Arabidopsis. Plant Cell 12, 2175–2190.

Cooke, J., Leishman, M.R., 2011. Is plant ecology more siliceous than we realise? Trends Plant Sci. 16, 61–68.

Cooksey, D.A., 1990. Genetics of bactericide resistance in plant pathogenic bacteria. Annu. Rev. Phytopathol. 28, 201–219.

Correa-Victoria, F., Datnoff, L., Okada, K., Friesen, D., Sanz, J., Snyder, G., 2001. Effects of silicon fertilization on disease development and yields of rice in Colombia. Stud. Plant Sci. 8, 313–322.

Coskun, D., Britto, D.T., Huynh, W.Q., Kronzucker, H.J., 2016. The role of silicon in higher plants under salinity and drought stress. Front. Plant Sci. 7.

Costa, I.J.S., Pereira, M.C.T., Mizobutsi, G.P., Maia, V.M., Silva, J.F., Oliveira, J.A.A., Oliveira, M.B., Souza, V.N.R., Nietsche, S., Santos, E.F., 2015. Influence of silicon fertilization on 'Palmer' mango tree cultivation. Acta Hortic. 1075, 229–234.

Cruz, M.F.A.d., Rodrigues, F.Á., Polanco, L.R., Curvêlo, C.R.d.S., Nascimento, K.J.T., Moreira, M.A., Barros, E.G., 2013. Inducers of resistance and silicon on the activity of defense enzymes in the soybean-*Phakopsora pachyrhizi* interaction. Bragantia 72, 162–172.

Cruz, F.J.R., de Mello Prado, R., Felisberto, G., Santos, Á.S., Barreto, R.F., 2017. Potassium Nutrition in Fruits and Vegetables and Food Safety through Hydroponic System, Potassium-Improvement of Quality in Fruits and Vegetables Through Hydroponic Nutrient Management. IntechOpen.

Currie, H.A., Perry, C.C., 2007. Silica in plants: biological, biochemical and chemical studies. Ann. Bot. 100, 1383–1389.

da Cunha, K.P.V., do Nascimento, C.W.A., 2009. Silicon effects on metal tolerance and structural changes in maize (*Zea mays* L.) grown on a cadmium and zinc enriched soil. Water Air Soil Pollut. 197, 323–330.

Dallagnol, L.J., Rodrigues, F.A., DaMatta, F.M., Mielli, M.V.B., Pereira, S.C., 2011. Deficiency in silicon uptake affects cytological, physiological, and biochemical events in the rice–*Bipolaris oryzae* interaction. Phytopathology 101, 92–104.

Dann, E.K., Le, D.P., 2017. Effects of silicon amendment on soilborne and fruit diseases of avocado. Plan. Theory 6, 51.

Datnoff, L.E., Snyder, G.H., Korndörfer, G.H., 2001. Silicon in Agriculture. Elsevier.

Datnoff, L.E., Elmer, W.H., Huber, D.M., 2007. Mineral Nutrition and Plant Disease. American Phytopathological Society (APS Press).

De Vleesschauwer, D., Djavaheri, M., Bakker, P.A.H.M., Höfte, M., 2008. *Pseudomonas fluorescens* WCS374r-induced systemic resistance in rice against *Magnaporthe oryzae* is based on pseudobactin-mediated priming for a salicylic acid-repressible multifaceted defense response. Plant Physiol. 148, 1996–2012.

Debona, D., Rodrigues, F.A., Datnoff, L.E., 2017. Silicon's role in abiotic and biotic plant stresses. Annu. Rev. Phytopathol.

Dehghanipoodeh, S., Ghobadi, C., Baninasab, B., Gheysari, M., Bidadabi, S.S., 2016. Effects of potassium silicate and nanosilica on quantitative and qualitative characteristics of a commercial strawberry (Fragaria × ananassa cv.'camarosa'). J. Plant Nutr. 39, 502–507.

Dehghanipoodeh, S., Ghobadi, C., Baninasab, B., Gheysari, M., Shiranibidabadi, S., 2018. Effect of silicon on growth and development of strawberry under water deficit conditions. Hortic. Plant J. 4, 226–232.

Detmann, K.C., Araújo, W.L., Martins, S.C., Sanglard, L.M., Reis, J.V., Detmann, E., Rodrigues, F.Á., Nunes-Nesi, A., Fernie, A.R., DaMatta, F.M., 2012. Silicon nutrition increases grain yield, which, in turn, exerts a feed-forward stimulation of photosynthetic rates via enhanced mesophyll conductance and alters primary metabolism in rice. New Phytol. 196, 752–762.

Devadas, S.K., Enyedi, A., Raina, R., 2002. The Arabidopsis hrl1 mutation reveals novel overlapping roles for salicylic acid, jasmonic acid and ethylene signalling in cell death and defence against pathogens. Plant J. 30, 467–480.

Dodds, P.N., Rathjen, J.P., 2010. Plant immunity: towards an integrated view of plant–pathogen interactions. Nat. Rev. Genet. 11, 539.

Dragišić Maksimović, J., Bogdanović, J., Maksimović, V., Nikolic, M., 2007. Silicon modulates the metabolism and utilization of phenolic compounds in cucumber (*Cucumis sativus* L.) grown at excess manganese. J. Plant Nutr. Soil Sci. 170, 739–744.

Ebrahimi, L., Aminian, H., Etebarian, H.R., Sahebani, N., 2012. Control of apple blue mould disease with *Torulaspora delbrueckii* in combination with Silicon. Arch. Phytopathol. Plant Protect. 45, 2057–2065.

Edgerton, L.J., Veinbrants, N., Hutchinson, J.F., 1976. Foliar sprays of silicon dioxide-containing compound reduce russeting in 'Golden Delicious' apple fruits. HortScience.

El-Kareem, M.R.G., Aal, A.M.K.A., Mohamed, A.Y., 2014. The synergistic effects of using silicon and selenium on fruiting of Zaghloul date palm (*Phoenix dactylifera* L.). World Acad. Sci. Eng. Technol. Int J. Biol. Biomol. Agric. Food Biotechnol. Eng. 8, 259–262.

El-Rhman, I., 2010. Physiological studies on cracking phenomena of pomegranates. J. Appl. Sci. Res. 6, 696–703.

Emadian, S.F., Newton, R.J., 1989. Growth enhancement of loblolly pine (*Pinus taeda* L.) seedlings by silicon. J. Plant Physiol. 134, 98–103.

Endris, S., Mohammad, M.J., 2007. Nutrient acquisition and yield response of barley exposed to salt stress under different levels of potassium nutrition. Int. J. Environ. Sci. Technol. 4, 323–330.

Eneji, A.E., Inanaga, S., Muranaka, S., Li, J., Hattori, T., An, P., Tsuji, W., 2008. Growth and nutrient use in four grasses under drought stress as mediated by silicon fertilizers. J. Plant Nutr. 31, 355–365.

Epstein, E., 1994. The anomaly of silicon in plant biology. Proc. Natl. Acad. Sci. 91, 11–17.

Epstein, E., 1999. Silicon. Annu. Rev. Plant Biol. 50, 641–664.

Epstein, E., Bloom, A.J., 2005. Mineral Nutrition of Plants: Principles and Perspectives, second ed. Sinauer, Sunderland, MA.

Etebarian, H.R., Farhangian-Kashani, S., Ebrahimi, L., 2013. Combination of silicon and hot water to control of postharvest blue mould caused by *Penicillium expansum* in apple. Int. J. Agric. 3, 72.

Etesami, H., 2018. Can interaction between silicon and plant growth promoting rhizobacteria benefit in alleviating abiotic and biotic stresses in crop plants? Agric. Ecosyst. Environ. 253, 98–112.

Etesami, H., Alikhani, H.A., 2017. Evaluation of Gram-positive rhizosphere and endophytic bacteria for biological control of fungal rice (*Oryza sativa* L.) pathogens. Eur. J. Plant Pathol. 147, 7–14.

Etesami, H., Jeong, B.R., 2018. Silicon (Si): review and future prospects on the action mechanisms in alleviating biotic and abiotic stresses in plants. Ecotoxicol. Environ. Saf. 147, 881–896.

Exley, C., 2015. A possible mechanism of biological silicification in plants. Front. Plant Sci. 6, 853.

Fang, J., Ma, X., 2005. Progress of silicon improving plant resistance to stress. Chin. Agric. Sci. Bull. 21, 304–306.

Farahani, L., Etebarian, H.R., Sahebani, N., Aminian, H., 2012. Effect of two strains of antagonistic yeasts in combination with silicon against two isolates of *Penicillium expansum* on apple fruit. Int. Res. J. App. Basic Sci. 3, 18–23.

Fauteux, F., Rémus-Borel, W., Menzies, J.G., Bélanger, R.R., 2005. Silicon and plant disease resistance against pathogenic fungi. FEMS Microbiol. Lett. 249, 1–6.

Fauteux, F., Chain, F., Belzile, F., Menzies, J.G., Bélanger, R.R., 2006. The protective role of silicon in the Arabidopsis–powdery mildew pathosystem. Proc. Natl. Acad. Sci. 103, 17554–17559.

Fawe, A., Menzies, J.G., Chérif, M., Bélanger, R.R., 2001. Silicon and disease resistance in dicotyledons. Stud. Plant Sci. 8, 159–169.

Feng, Y., 2000. Siliceous fertilizer to become a new fertilizer product in expansion of agriculture in China. J. Chem. Fertil. Ind. 27, 9–11.

Ferreira, H.A., do Nascimento, C.W.A., Datnoff, L.E., de Sousa Nunes, G.H., Preston, W., de Souza, E.B., Mariano, R.d.L.R., 2015. Effects of silicon on resistance to bacterial fruit blotch and growth of melon. Crop Prot. 78, 277–283.

Fisher, R.A., 1929. A preliminary note on the effect of sodium silicate in increasing the yield of barley. J. Agric. Sci. 19, 132–139.

Fortunato, A.A., Rodrigues, F.Á., Baroni, J.C.P., Soares, G.C.B., Rodriguez, M.A.D., Pereira, O.L., 2012a. Silicon suppresses Fusarium wilt development in banana plants. J. Phytopathol. 160, 674–679.

Fortunato, A.A., Rodrigues, F.Á., do Nascimento, K.J.T., 2012b. Physiological and biochemical aspects of the resistance of banana plants to Fusarium wilt potentiated by silicon. Phytopathology 102, 957–966.

Fortunato, A.A., Debona, D., Bernardeli, A.M.A., Rodrigues, F.A., 2015a. Defence-related enzymes in soybean resistance to target spot. J. Phytopathol. 163, 731–742.

Fortunato, A.A., Rodrigues, F.A., Datnoff, L.E., 2015b. Silicon control of soil-borne and seed-borne diseases. In: Silicon and Plant Diseases. Springer, pp. 53–66.

Frantz, J.M., Khandekar, S., Leisner, S., 2011. Silicon differentially influences copper toxicity response in silicon-accumulator and non-accumulator species. J. Am. Soc. Hortic. Sci. 136, 329–338.

Gao, Y.M., Xu, J., Gao, S.Q., 2006. Effects of silicon application on apple internal bark necrosis induced by high content of manganese. Plant Nutr. Fertil. Sci. 12, 571–577.

Gaofeng, X., Guilong, Z., Yanxin, S., Shangqiang, L., Yanhua, C., 2012. Influences of spraying two different forms of silicon on plant growth and quality of tomato in solar greenhouse. Chin. Agric. Sci. Bull. 16, 051.

Ghareeb, H., Bozsó, Z., Ott, P.G., Repenning, C., Stahl, F., Wydra, K., 2011. Transcriptome of silicon-induced resistance against *Ralstonia solanacearum* in the silicon non-accumulator tomato implicates priming effect. Physiol. Mol. Plant Pathol. 75, 83–89.

Gong, H., Randall, D., Flowers, T., 2006. Silicon deposition in the root reduces sodium uptake in rice (*Oryza sativa* L.) seedlings by reducing bypass flow. Plant Cell Environ. 29, 1970–1979.

Gonzalo, M.J., Lucena, J.J., Hernández-Apaolaza, L., 2013. Effect of silicon addition on soybean (*Glycine max*) and cucumber (*Cucumis sativus*) plants grown under iron deficiency. Plant Physiol. Biochem. 70, 455–461.

Grieve, C.M., Grattan, S.R., 1999. Mineral nutrient acquisition and response by plants grown in saline environments. In: Handbook of Plant and Crop Stress, second ed. CRC Press, pp. 203–229.

Gu, H.-H., Qiu, H., Tian, T., Zhan, S.-S., Chaney, R.L., Wang, S.-Z., Tang, Y.-T., Morel, J.-L., Qiu, R.-L., 2011. Mitigation effects of silicon rich amendments on heavy metal accumulation in rice (*Oryza sativa* L.) planted on multi-metal contaminated acidic soil. Chemosphere 83, 1234–1240.

Gu, H.-H., Zhan, S.-S., Wang, S.-Z., Tang, Y.-T., Chaney, R.L., Fang, X.-H., Cai, X.-D., Qiu, R.-L., 2012. Silicon-mediated amelioration of zinc toxicity in rice (*Oryza sativa* L.) seedlings. Plant Soil 350, 193–204.

Guerriero, G., Hausman, J.-F., Legay, S., 2016. Silicon and the plant extracellular matrix. Front. Plant Sci. 7, 1–8

Gunes, A., Inal, A., Bagci, E., Coban, S., Sahin, O., 2007a. Silicon increases boron tolerance and reduces oxidative damage of wheat grown in soil with excess boron. Biol. Plant. 51, 571–574.

Gunes, A., Inal, A., Bagci, E.G., Pilbeam, D.J., 2007b. Silicon-mediated changes of some physiological and enzymatic parameters symptomatic for oxidative stress in spinach and tomato grown in sodic-B toxic soil. Plant Soil 290, 103–114.

Guntzer, F., Keller, C., Meunier, J.-D., 2012. Benefits of plant silicon for crops: a review. Agron. Sustain. Dev. 32, 201–213.

Guo, B., Liu, C., Ding, N., Fu, Q., Lin, Y., Li, H., Li, N., 2016. Silicon alleviates cadmium toxicity in two cypress varieties by strengthening the exodermis tissues and stimulating phenolic exudation of roots. J. Plant Growth Regul. 35, 420–429.

Gutiérrez-Barranquero, J.A., Arrebola, E., Bonilla, N., Sarmiento, D., Cazorla, F.M., De Vicente, A., 2012. Environmentally friendly treatment alternatives to Bordeaux mixture for controlling bacterial apical necrosis (BAN) of mango. Plant Pathol. 61, 665–676.

Hammash, F., Assi, N.E., 2007. influence of pre-storage waxing and wrapping on quality attributes of stored 'Shamouti' oranges. Acta Hortic.

Hanumanthaiah, M.R., Kulapatihipparagi, R.C., Renuka, D.M., Kumar, K.K., Santhosha, K.V., 2015. Effect of soil and foliar application of silicon on fruit quality parameters of banana cv. Neypoovan under hill zone. Plant Arch. 15, 221–224.

Haynes, R.J., 2014. A contemporary overview of silicon availability in agricultural soils. J. Plant Nutr. Soil Sci. 177, 831–844.

He, C., Ma, J., Wang, L., 2015. A hemicellulose-bound form of silicon with potential to improve the mechanical properties and regeneration of the cell wall of rice. New Phytol. 206, 1051–1062.

Heidari, M., Jamshid, P., 2010. Interaction between salinity and potassium on grain yield, carbohydrate content and nutrient uptake in pearl millet. ARPN J. Agric. Biol. Sci. 5, 39–46.

Helaly, M.N., El-Hoseiny, H., El-Sheery, N.I., Rastogi, A., Kalaji, H.M., 2017. Regulation and physiological role of silicon in alleviating drought stress of mango. Plant Physiol. Biochem. 118, 31–44.

Hernandez-Apaolaza, L., 2014. Can silicon partially alleviate micronutrient deficiency in plants? A review. Planta 240, 447–458.

Hilbert, G., Soyer, J.P., Molot, C., Giraudon, J., Milin, M., Gaudillere, J.P., 2015. Effects of nitrogen supply on must quality and anthocyanin accumulation in berries of cv. Merlot. VITIS J. Grapevine Res. 42, 69.

Horst, W.J., Marschner, H., 1978. Effect of silicon on manganese tolerance of bean plants (*Phaseolus vulgaris* L.). Plant Soil 50, 287–303.

Horst, W.J., Fecht, M., Naumann, A., Wissemeier, A.H., Maier, P., 1999. Physiology of manganese toxicity and tolerance in *Vigna unguiculata* (L.) Walp. J. Plant Nutr. Soil Sci. 162, 263–274.

Hossain, K.A., Horiuchi, T., Miyagawa, S., 2001. Effects of silicate materials on growth and grain yield of rice plants grown in clay loam and sandy loam soils. J. Plant Nutr. 24, 1–13.

Huang, X., Zhang, Z., Ke, Y., Xiao, C., Peng, Z., Wu, L., Li, Y., Jian, H., Cen, J., Zhang, Y., 1997. Effects of silicate fertilizer on nutrition of leaves, yield and sugar of sugarcanes. Trop Subtrop. Soil Sci. 6, 242–246.

Huang, C.-F., Yamaji, N., Nishimura, M., Tajima, S., Ma, J.F., 2009. A rice mutant sensitive to Al toxicity is defective in the specification of root outer cell layers. Plant Cell Physiol. 50, 976–985.

Huang, H.-R., Xu, L., Bokhtiar, S., Srivastav, M.K., Li, Y.-R., Yang, L.-T., 2011. Effect of calcium silicate fertilizer on soil characteristics, sugarcane nutrients and its yield parameters. J. South. Agric. 42, 756–759.

Imtiaz, M., Rizwan, M.S., Mushtaq, M.A., Ashraf, M., Shahzad, S.M., Yousaf, B., Saeed, D.A., Rizwan, M., Nawaz, M.A., Mehmood, S., 2016. Silicon occurrence, uptake, transport and mechanisms of heavy metals, minerals and salinity enhanced tolerance in plants with future prospects: a review. J. Environ. Manag. 183, 521–529.

Inal, A., Pilbeam, D.J., Gunes, A., 2009. Silicon increases tolerance to boron toxicity and reduces oxidative damage in barley. J. Plant Nutr. 32, 112–128.

Isa, M., Bai, S., Yokoyama, T., Ma, J.F., Ishibashi, Y., Yuasa, T., Iwaya-Inoue, M., 2010. Silicon enhances growth independent of silica deposition in a low-silica rice mutant, lsi1. Plant Soil 331, 361–375.

Iwasaki, K., Maier, P., Fecht, M., Horst, W.J., 2002. Leaf apoplastic silicon enhances manganese tolerance of cowpea (*Vigna unguiculata*). J. Plant Physiol. 159, 167–173.

Jana, S., Jeong, B.R., 2014. Silicon: the most under-appreciated element in horticultural crops. Trends Hortic. Res. 4, 1–19.

Jarosz, Z., 2013. The effect of silicon application and type of substrate on yield and chemical composition of leaves and fruit of cucumber. J. Elem. 18, 403–414

Javaria, S., Khan, M.Q., Bakhsh, I., 2012. Effect of potassium on chemical and sensory attributes of tomato fruit. J. Anim. Plant Sci. 22, 1081–1085.

Ji, M., Huang, X., Yu, J., 1992. Study on behavior mechanism of increasing production and sugar of cane caused by silicon. VI Effect of silicon on activities of invertases in blades of cane. Sugarc. Canesugar 4, 24.

Jiang, Z., Liao, Q., Wei, G., Tan, Y., Chen, G., Liu, B., Wang, Y., 2011. Effects of optimum fertilization model between silicon and phosphorous on yield and quality of sugarcane in lateristic red earth area of Guangxi. J. Anhui Agric. Sci. 39, 22233–22236.

Kablan, L., Lagauche, A., Delvaux, B., Legrve, A., 2012. Silicon reduces black sigatoka development in banana. Plant Dis. 96, 273–278.

Kader, A.A., 1992. Postharvest biology and technology: an overview. In: Postharvest Technology of Horticultural Crops.

Kaluwa, K., Bertling, I., Bower, J.P., Tesfay, S.Z., 2010. Silicon application effects on 'Hass' avocado fruit physiology. S. Afr. Avocado Grower Assoc. Yearb. 33, 44–47.

Kamenidou, S., Cavins, T.J., Marek, S., 2010. Silicon supplements affect floricultural quality traits and elemental nutrient concentrations of greenhouse produced gerbera. Sci. Hortic. 123, 390–394.

Kanai, S., Ohkura, K., Adu-Gyamfi, J.J., Mohapatra, P.K., Nguyen, N.T., Saneoka, H., Fujita, K., 2007. Depression of sink activity precedes the inhibition of biomass production in tomato plants subjected to potassium deficiency stress. J. Exp. Bot. 58, 2917–2928.

Kanto, T., Miyoshi, A., Ogawa, T., Maekawa, K., Aino, M., 2006. Suppressive effect of liquid potassium silicate on powdery mildew of strawberry in soil. J. Gen. Plant Pathol. 72, 137–142.

Kapasakalidis, P.G., Rastall, R.A., Gordon, M.H., 2009. Effect of a cellulase treatment on extraction of antioxidant phenols from black currant (*Ribes nigrum* L.) pomace. J. Agric. Food Chem. 57, 4342–4351.

Karabal, E., Yücel, M., Öktem, H.A., 2003. Antioxidant responses of tolerant and sensitive barley cultivars to boron toxicity. Plant Sci. 164, 925–933.

Kaya, C., Higgs, D., Saltali, K., Gezerel, O., 2002. Response of strawberry grown at high salinity and alkalinity to supplementary potassium. J. Plant Nutr. 25, 1415–1427.

Kaya, C., Tuna, L., Higgs, D., 2006. Effect of silicon on plant growth and mineral nutrition of maize grown under water-stress conditions. J. Plant Nutr. 29, 1469–1480.

Keen, N.T., Yoshikawa, M., 1983. β-1, 3-Endoglucanase from soybean releases elicitor-active carbohydrates from fungus cell walls. Plant Physiol. 71, 460–465.

Khandekar, S., Leisner, S., 2011. Soluble silicon modulates expression of *Arabidopsis thaliana* genes involved in copper stress. J. Plant Physiol. 168, 699–705.

Khoshgoftarmanesh, A.H., Schulin, R., Chaney, R.L., Daneshbakhsh, B., Afyuni, M., 2010. Micronutrient-efficient genotypes for crop yield and nutritional quality in sustainable agriculture. A review. Agron. Sustain. Dev. 30, 83–107.

Kidane, E.G., Laing, M.D., 2008. Integrated control of Fusarium wilt of banana (Musa spp.). pp. 315–321.

Kim, S.G., Kim, K.W., Park, E.W., Choi, D., 2002. Silicon-induced cell wall fortification of rice leaves: a possible cellular mechanism of enhanced host resistance to blast. Phytopathology 92, 1095–1103.

Korndörfer, G., Lepsch, I., 2001. Effect of silicon on plant growth and crop yield. Stud. Plant Sci. 8, 133–147.

Kumbargire, G.A., Swamy, G.S.K., Kalatippi, S.A., 2016. Influence of diatomaceous earth as source of silicon on leaf nutrient status and yield attributing characters of banana cv. Grand Naine. Bioscan 11, 435–438.

Lalithya, K.A., Hipparagi, K., Thippeshappa, G.N., 2013. Effect of silicon and micronutrients on growth and yield attributes of sapota cv. Kalipatti under hill zone. Crop Res. 0970-4884. 46.

Lalithya, K.A., Bhagya, H.P., Choudhary, R., 2014. Response of silicon and micro nutrients on fruit character and nutrient content in leaf of sapota. Biolife 2, 593–598.

Łaźniewska, J., Macioszek, V.K., Kononowicz, A.K., 2012. Plant-fungus interface: the role of surface structures in plant resistance and susceptibility to pathogenic fungi. Physiol. Mol. Plant Pathol. 78, 24–30.

Leigh, R.A., Wyn Jones, R.G., 1984. A hypothesis relating critical potassium concentrations for growth to the distribution and functions of this ion in the plant cell. New Phytol. 97, 1–13.

Lepolu Torlon, J., Heckman, J.R., Simon, J.E., Wyenandt, C.A., 2016. Silicon soil amendments for suppressing powdery mildew on pumpkin. Sustainability 8, 293.

Lester, G.E., Jifon, J.L., Rogers, G., 2005. Supplemental foliar potassium applications during muskmelon fruit development can improve fruit quality, ascorbic acid, and beta-carotene contents. J. Am. Soc. Hortic. Sci. 130, 649–653.

Lester, G.E., Jifon, J.L., Makus, D.J., 2006. Supplemental foliar potassium applications with or without a surfactant can enhance netted muskmelon quality. HortScience 41, 741–744.

Li, C., Chu, T., Liu, X., Yang, Q., 1999. Silicon nutrition effects and its study and application development in China. In: Proceedings of Symposium of Plant Nutrition. Shaanxi Science and Technology Press, China, pp. 329–333.

Li, J., Leisner, S.M., Frantz, J., 2008. Alleviation of copper toxicity in *Arabidopsis thaliana* by silicon addition to hydroponic solutions. J. Am. Soc. Hortic. Sci. 133, 670–677.

Liang, Y., 1999. Effects of silicon on enzyme activity and sodium, potassium and calcium concentration in barley under salt stress. Plant Soil 209, 217.

Liang, Y., Shen, Z., 1994. Interaction of silicon and boron in oilseed rape plants. J. Plant Nutr. 17, 415–425.

Liang, Y., Sun, W., Si, J., Römheld, V., 2005. Effects of foliar-and root-applied silicon on the enhancement of induced resistance to powdery mildew in *Cucumis sativus*. Plant Pathol. 54, 678–685.

Liang, Y., Hua, H., Zhu, Y.G., Zhang, J., Cheng, C., Römheld, V., 2006. Importance of plant species and external silicon concentration to active silicon uptake and transport. New Phytol. 172, 63–72.

Liang, Y., Sun, W., Zhu, Y.-G., Christie, P., 2007. Mechanisms of silicon-mediated alleviation of abiotic stresses in higher plants: a review. Environ. Pollut. 147, 422–428.

Liang, Y., Nikolic, M., Bélanger, R., Gong, H., Song, A., 2015a. Effect of silicon on crop growth, yield and quality. In: Liang, Y., Nikolic, M., Bélanger, R., Gong, H., Song, A. (Eds.), Silicon in Agriculture: From Theory to Practice. Springer Netherlands, Dordrecht, pp. 209–223.

Liang, Y., Nikolic, M., Bélanger, R., Gong, H., Song, A., 2015b. Silicon-mediated tolerance to other abiotic stresses. In: Silicon in Agriculture. Springer, pp. 161–179.

Liang, Y., Nikolic, M., Bélanger, R., Gong, H., Song, A., 2015c. Silicon and insect pest resistance. In: Liang, Y., Nikolic, M., Bélanger, R., Gong, H., Song, A. (Eds.), Silicon in Agriculture: From Theory to Practice. Springer Netherlands, Dordrecht, pp. 197–207.

Liang, Y., Nikolic, M., Bélanger, R., Gong, H., Song, A., 2015d. Silicon and plant–pathogen interactions. In: Liang, Y., Nikolic, M., Bélanger, R., Gong, H., Song, A. (Eds.), Silicon in Agriculture: From Theory to Practice. Springer Netherlands, Dordrecht, pp. 181–196.

Liang, Y., Nikolic, M., Bélanger, R., Gong, H., Song, A., 2015e. Silicon in Agriculture: From Theory to Practice. Springer.

Lindgren, P.B., 1997. The role of hrp genes during plant-bacterial interactions. Annu. Rev. Phytopathol. 35, 129–152.

Liu, J., Zong, Y., Qin, G., Li, B., Tian, S., 2010. Plasma membrane damage contributes to antifungal activity of silicon against *Penicillium digitatum*. Curr. Microbiol. 61, 274–279.

Liu, J., Han, C., Sheng, X., Liu, S., Qi, X., 2011. Potassium-containing silicate fertilizer: its manufacturing technology and agronomic effects. In: Oral Presentation at 5th International Conference on Silicon in Agriculture, pp. 13–18.

Liu, H., Li, J.M., Zheng, G., Du, Q.J., Pan, T.H., Chang, Y.B., 2014. Effects of silicon on plant growth and fruit quality of cucumber. Acta Agric. Boreali-occident. Sin. 23, 117–121.

Lötze, E., Theron, K.I., 2007. Evaluating the effectiveness of pre-harvest calcium applications for bitter pit control in 'Golden Delicious' apples under South African conditions. J. Plant Nutr. 30, 471–485.

Lux, A., Luxová, M., Hattori, T., Inanaga, S., Sugimoto, Y., 2002. Silicification in sorghum (*Sorghum bicolor*) cultivars with different drought tolerance. Physiol. Plant. 115, 87–92.

Ma, J.F., 2004. Role of silicon in enhancing the resistance of plants to biotic and abiotic stresses. Soil Sci. Plant Nutr. 50, 11–18.

Ma, J., Takahashi, E., 1990a. The effect of silicic acid on rice in a P-deficient soil. Plant Soil 126, 121–125.

Ma, J., Takahashi, E., 1990b. Effect of silicon on the growth and phosphorus uptake of rice. Plant Soil 126, 115–119.

Ma, J.F., Takahashi, E., 2002. Soil, Fertilizer, and Plant Silicon Research in Japan. Elsevier.

Ma, J.F., Yamaji, N., 2006. Silicon uptake and accumulation in higher plants. Trends Plant Sci. 11, 392–397.

Ma, J.F., Miyake, Y., Takahashi, E., 2001. Silicon as a beneficial element for crop plants. Stud. Plant Sci. 8, 17–39.

Magno Queiroz Luz, J., Abreu Asmar, S., Pasqual, M., Gomes de Araujo, A., Pio, L.A.S., Ferreira Resende, R., 2012. Modifications in leaf anatomy of banana plants cultivar 'Maca' subjected to different silicon sources in vitro, pp. 239–243.

Maheswari, M., Murthy, A.N.G., Shanker, A.K., 2017. Nitrogen Nutrition in Crops and Its Importance in Crop Quality. In: The Indian Nitrogen Assessment: Sources of Reactive Nitrogen, Environmental and Climate Effects, Management Options, and Policies, p. 175.

Mali, M., 2008. Silicon effects on nodule growth, dry-matter production, and mineral nutrition of cowpea (*Vigna unguiculata*). J. Plant Nutr. Soil Sci. 171, 835–840.

Mali, M., Aery, N., 2008. Influence of silicon on growth, relative water contents and uptake of silicon, calcium and potassium in wheat grown in nutrient solution. J. Plant Nutr. 31, 1867–1876.

Marais, L.J., 2015. Efficacy of water soluble silicon in managing Fusarium dry root rot of citrus. Acta Hortic. 1065, 993–999.

Marschner, H., 1995. Mineral Nutrition of Higher Plants, second ed. Academic Press, London, 889 pp.

Marschner, H., 2012a. Marschner's Mineral Nutrition of Higher Plants. vol. 89 Academic Press.

Marschner, P., 2012b. Rhizosphere biology. In: Marschner's Mineral Nutrition of Higher Plants, third ed. Elsevier, pp. 369–388.

Marschner, H., Rimmington, G., 1988. Mineral nutrition of higher plants. Plant Cell Environ. 11, 147–148.

Mäser, P., Gierth, M., Schroeder, J.I., 2002. Molecular mechanisms of potassium and sodium uptake in plants. Springer, pp. 43–54.

Mathaba, N., Tesfay, S., Bower, J.P., Bertling, I., 2009. The potential of postharvest silicon dips to mitigate chilling injury in citrus fruit with special emphasis on lemon (cv Eureka). In: Proceedings of the 1st Annual Research Day, Silicon in Agriculture.

Matichenkov, V.V., Bocharnikova, E.A., 2010. Technology for natural water protection against pollution from cultivated areas.

Matichenkov, V.V., Calvert, D.V., 2002. Silicon as a beneficial element for sugarcane. J. Am. Soc. Sugarcane Technol. 22, 21–30.

Matichenkov, V., Calvert, D., Snyder, G., 1999. Silicon fertilizers for citrus in Florida, pp. 5–8.

Matichenkov, V.V., Calvert, D.V., Snyder, G.H., 2000. Prospective of Silicon fertilization for citrus in Florida. Soil and Crop Science Society of Florida, pp. 137–141.

Matichenkov, V., Bocharnikova, E., Calvert, D., 2001. Response of citrus to silicon soil amendments, pp. 94–97.

McLaughlin, S.B., Wimmer, R., 1999. Tansley Review No. 104 Calcium physiology and terrestrial ecosystem processes. New Phytol. 142, 373–417.

Meena, V.D., Dotaniya, M.L., Coumar, V., Rajendiran, S., Kundu, S., Rao, A.S., 2014. A case for silicon fertilization to improve crop yields in tropical soils. Proc. Natl. Acad. Sci. India Sect. B Biol. Sci. 84, 505–518.

Meharg, C., Meharg, A.A., 2015. Silicon, the silver bullet for mitigating biotic and abiotic stress, and improving grain quality, in rice? Environ. Exp. Bot. 120, 8–17.

Melgarejo, P., Martínez, J.J., Hernández, F.C.A., Martínez-Font, R., Barrows, P., Erez, A., 2004. Kaolin treatment to reduce pomegranate sunburn. Sci. Hortic. 100, 349–353.

Meyer, J., Keeping, M., 2001. Past, present and future research of the role of silicon for sugarcane in southern Africa. Stud. Plant Sci. 8, 257–275.

Miao, B.-H., Han, X.-G., Zhang, W.-H., 2010. The ameliorative effect of silicon on soybean seedlings grown in potassium-deficient medium. Ann. Bot. 105 (6), 967–973

Mills, H.A.J.J., Cottenie, A., Faithfull, N.T., Larrahondo, J.P., Fj Ramirez, J., Lopez, A.E., J Vargas, A., Malavolta, E., Sa Vitti, G.C., Ritas, L., 1996. Plant Analysis Handbook II: A Practical Preparation, Analysis, and Interpretation Guide. Potash and Phosphate Institute.

Mitani, N., Ma, J.F., Iwashita, T., 2005. Identification of the silicon form in xylem sap of rice (Oryza sativa L.). Plant Cell Physiol. 46, 279–283.

Mitani, N., Yamaji, N., Ma, J.F., 2009. Identification of maize silicon influx transporters. Plant Cell Physiol. 50, 5–12.

Mkhize, N., Bower, J.P., Bertling, I., Mathaba, N., 2012. Response of citrus physiology to phosphorus acid and silicon as elicitors of induced disease resistance, pp. 135–141.

Montpetit, J., Vivancos, J., Mitani-Ueno, N., Yamaji, N., Rémus-Borel, W., Belzile, F., Ma, J.F., Bélanger, R.R., 2012. Cloning, functional characterization and heterologous expression of TaLsi1, a wheat silicon transporter gene. Plant Mol. Biol. 79, 35–46.

More, S.S., Gokhale, N.B., Shinde, S.E., Korake, G.N., 2015. Effect of different sources of silica on nutrient content of leaves and fruit at different stages of alphonso mango (Mangifera indica L.) in lateritic soil. J. Progress. Agric. 6, 84–88.

Moscoso-Ramírez, P.A., Palou, L., 2014. Preventive and curative activity of postharvest potassium silicate treatments to control green and blue molds on orange fruit. Eur. J. Plant Pathol. 138, 721–732.

Najihah, N.I., Hanafi, M.M., Idris, A.S., Hakim, M.A., 2015. Silicon treatment in oil palms confers resistance to basal stem rot disease caused by Ganoderma boninense. Crop Prot. 67, 151–159.

Nava, G., Dechen, A.R., Nachtigall, G.R., 2007. Nitrogen and potassium fertilization affect apple fruit quality in southern Brazil. Commun. Soil Sci. Plant Anal. 39, 96–107.

Nawrath, C., 2006. Unraveling the complex network of cuticular structure and function. Curr. Opin. Plant Biol. 9, 281–287.

Nedjimi, B., Daoud, Y., 2009. Effects of calcium chloride on growth, membrane permeability and root hydraulic conductivity in two Atriplex species grown at high (sodium chloride) salinity. J. Plant Nutr. 32, 1818–1830.

Neilsen, G.H., Neilsen, D., Peryea, F.J., Fallahi, E., Fallahi, B., 2008. Effects of mineral nutrition on fruit quality and nutritional disorders in apples, pp. 49–60.

Neumann, D., zur Nieden, U., 2001. Silicon and heavy metal tolerance of higher plants. Phytochemistry 56, 685–692.

Nowakowski, W., Nowakowska, J., 1997. Silicon and copper interaction in the growth of spring wheat seedlings. Biol. Plant. 39, 463–466.

Okamoto, Y., 1969. Effect of silicic acid upon rice plants (9) On growth under high and low temperature of the culture solution (10) On growth under high and low air temperatures, pp. 743–752.

Okuda, A., Takahashi, E., 1962. Effect of silicon supply on the injuries due to excessive amounts of Fe, Mn, Cu, As, Al, Co of barley and rice plant. Jpn. J. Soil Sci. Plant Nutr. 33, 1–8.

Owino-Gerroh, C., Gascho, G., 2005. Effect of silicon on low pH soil phosphorus sorption and on uptake and growth of maize. Commun. Soil Sci. Plant Anal. 35, 2369–2378.

Parida, A.K., Das, A.B., 2005. Salt tolerance and salinity effects on plants: a review. Ecotoxicol. Environ. Saf. 60, 324–349.

Patil, H., Tank, R.V., Patel, M., 2017. Significance of silicon in fruit crops—a review. Plant Arch. 17, 769–774.

Paul, D., Lade, H., 2014. Plant-growth-promoting rhizobacteria to improve crop growth in saline soils: a review. Agron. Sustain. Dev. 34, 737–752.

Pavlovic, J., Samardzic, J., Maksimović, V., Timotijevic, G., Stevic, N., Laursen, K.H., Hansen, T.H., Husted, S., Schjoerring, J.K., Liang, Y., 2013. Silicon alleviates iron deficiency in cucumber by promoting mobilization of iron in the root apoplast. New Phytol. 198, 1096–1107.

Pei, Z., Ming, D., Liu, D., Wan, G., Geng, X., Gong, H., Zhou, W., 2010. Silicon improves the tolerance to water-deficit stress induced by polyethylene glycol in wheat (Triticum aestivum L.) seedlings. J. Plant Growth Regul. 29, 106–115.

Peleg, Z., Saranga, Y., Fahima, T., Aharoni, A., Elbaum, R., 2010. Genetic control over silica deposition in wheat awns. Physiol. Plant. 140, 10–20.

Pereira, H.S., Korndörfer, G.H., Vidal, A.d.A., Camargo, M.S.d., 2004. Silicon sources for rice crop. Sci. Agric. 61, 522–528.

Peryea, F.J., Neilsen, G.H., Faubion, D., 2007. Start-timing for calcium chloride spray programs influences fruit calcium and bitter pit in 'Braeburn' and 'Honeycrisp' apples. J. Plant Nutr. 30, 1213–1227.

Pich, A., Scholz, G., Stephan, U.W., 1994. Iron-dependent changes of heavy metals, nicotianamine, and citrate in different plant organs and in the xylem exudate of two tomato genotypes. Nicotianamine as possible copper translocator. Plant Soil 165, 189–196.

Piperno, D.R., 2006. Phytoliths: A Comprehensive Guide for Archaeologists and Paleoecologists. Rowman Altamira.

Pozza, E.A., Pozza, A.A.A., Botelho, D.M.d.S., 2015. Silicon in plant disease control. Rev. Ceres 62, 323–331.

Prabhu, A.S., Barbosa Filho, M.P., Filippi, M.C., Datnoff, L.E., Snyder, G.H., 2001. Silicon from rice disease control perspective in Brazil. Stud. Plant Sci. 8, 293–311.

Prado, R.d.M., Natale, W., 2005. Effect of application of calcium silicate on growth, nutritional status and dry matter production of passion fruit seedlings. Rev. Bras. Eng. Agríc. Ambient. 9, 185–190.

Putra, E.T.S., Zakaria, W., Abdullah, N.A.P., Saleh, G., 2010. Crop nutrition and post harvest. J. Agron. 9, 45–51.

Qin, G.Z., Tian, S.P., 2005. Enhancement of biocontrol activity of Cryptococcus laurentii by silicon and the possible mechanisms involved. Phytopathology 95, 69–75.

QuanYu, S., WenZhong, Z., YaDong, H., Rong, R., Hai, X., ZhensJin, X., WenFu, C., 2009. Effect of silicon fertilizer application on yield and grain quality of japonica rice from Northeast China. Chin. J. Rice Sci. 23, 661–664.

Quarta, A., Mita, G., Durante, M., Arlorio, M., De Paolis, A., 2013. Isolation of a polyphenol oxidase (PPO) cDNA from artichoke and expression analysis in wounded artichoke heads. Plant Physiol. Biochem. 68, 52–60.

Ratnayake, R., Daundasekera, W.A.M., Ariyarathne, H.M., Ganehenege, M.Y.U., 2016. Some biochemical defense responses enhanced by soluble silicon in bitter gourd-powdery mildew pathosystem. Australas. Plant Pathol. 45, 425–433.

Rellán-Álvarez, R., Giner-Martínez-Sierra, J., Orduna, J., Orera, I., Rodríguez-Castrillón, J.Á., García-Alonso, J.I., Abadía, J., Álvarez-Fernández, A., 2010. Identification of a tri-iron (III), tri-citrate complex in the xylem sap of iron-deficient tomato resupplied with iron: new insights into plant iron long-distance transport. Plant Cell Physiol. 51, 91–102.

Rémus-Borel, W., Menzies, J.G., Bélanger, R.R., 2005. Silicon induces antifungal compounds in powdery mildew-infected wheat. Physiol. Mol. Plant Pathol. 66, 108–115.

Renger, A., Steinhart, H., 2000. Ferulic acid dehydrodimers as structural elements in cereal dietary fibre. Eur. Food Res. Technol. 211, 422–428.

Reynolds, O.L., Padula, M.P., Zeng, R., Gurr, G.M., 2016. Silicon: potential to promote direct and indirect effects on plant defense against arthropod pests in agriculture. Front. Plant Sci. 7, 744.

Řezanka, T., Sigler, K., 2008. Biologically active compounds of semi-metals. In: Studies in Natural Products Chemistry. Elsevier, pp. 835–921.

Rizwan, M., Ali, S., Ibrahim, M., Farid, M., Adrees, M., Bharwana, S.A., Zia-ur-Rehman, M., Qayyum, M.F., Abbas, F., 2015. Mechanisms of silicon-mediated alleviation of drought and salt stress in plants: a review. Environ. Sci. Pollut. Res. 22, 15416–15431.

Rodrigues, F.A., Datnoff, L.E., 2015. Silicon and Plant Diseases. Springer.

Rodrigues, F.Á., McNally, D.J., Datnoff, L.E., Jones, J.B., Labbé, C., Benhamou, N., Menzies, J.G., Bélanger, R.R., 2004. Silicon enhances the accumulation of diterpenoid phytoalexins in rice: a potential mechanism for blast resistance. Phytopathology 94, 177–183.

Rodrigues, F.A., Dallagnol, L.J., Duarte, H.S.S., Datnoff, L.E., 2015. Silicon control of foliar diseases in monocots and dicots. In: Silicon and Plant Diseases. Springer, pp. 67–108.

Rogalla, H., Römheld, V., 2002. Role of leaf apoplast in silicon-mediated manganese tolerance of Cucumis sativus L. Plant Cell Environ. 25, 549–555.

Rose, J.K.C., 2003. The Plant Cell Wall. CRC Press.

Sadanandan, A.K., Varghese, E.J., 1970. Role of silicate in the uptake of nutrients by rice plants in the laterite soils of Kerala. Agric. Res. J. Kerala 7, 91–96

Sahebi, M., Hanafi, M.M., Azizi, P., 2016. Application of silicon in plant tissue culture. In Vitro Cell Dev. Biol. Plant 52, 226–232.

Sakr, N., 2016. Silicon control of bacterial and viral diseases in plants. J. Plant Prot. Res. 56, 331–336.

Samuels, A., Glass, A., Ehret, D., Menzies, J., 1991. Mobility and deposition of silicon in cucumber plants. Plant Cell Environ. 14, 485–492.

Savvas, D., Manos, G., Kotsiras, A., Souvaliotis, S., 2002. Effects of silicon and nutrient-induced salinity on yield, flower quality and nutrient uptake of gerbera grown in a closed hydroponic system. J. Appl. Bot. 76, 153–158.

Serdula, M.K., Byers, T., Mokdad, A.H., Simoes, E., Mendlein, J.M., Coates, R.J., 1996. The association between fruit and vegetable intake and chronic disease risk factors. Epidemiology, 161–165.

Sharma, L.K., 2016. Effect of nutrients sprays on growth, yield and fruit quality of apple under cold desert condition of Himachal Pradesh. J. Appl. Nat. Sci. 8, 297–300.

Sharma, A., Wali, V.K., Bakshi, P., Jasrotia, A., 2013. Effect of integrated nutrient management strategies on nutrient status, yield and quality of guava. Indian J. Hortic. 70, 333–339.

Shewry, P.R., Lucas, J.A., 1997. Plant proteins that confer resistance to pests and pathogens. In: Advances in Botanical Research. Elsevier, pp. 135–192.

Shi, Q., Bao, Z., Zhu, Z., He, Y., Qian, Q., Yu, J., 2005. Silicon-mediated alleviation of Mn toxicity in Cucumis sativus in relation to activities of superoxide dismutase and ascorbate peroxidase. Phytochemistry 66, 1551–1559.

Shi, S., Wang, W., Liu, L., Wu, S., Wei, Y., Li, W., 2013. Effect of chitosan/nano-silica coating on the physicochemical characteristics of longan fruit under ambient temperature. J. Food Eng. 118, 125–131.

Silva, J.A., 1971. Possible mechanisms for crop response to silicate applications.

Silva, I.T., Rodrigues, F.Á., Oliveira, J.R., Pereira, S.C., Andrade, C.C.L., Silveira, P.R., Conceição, M.M., 2010. Wheat resistance to bacterial leaf streak mediated by silicon. J. Phytopathol. 158, 253–262.

Sinclair, S.A., Krämer, U., 2012. The zinc homeostasis network of land plants. Biochim. Biophys. Acta Mol. Cell Res. 1823, 1553–1567.

Singh, K., Singh, R., Singh, J.P., Singh, Y., Singh, K.K., 2006. Effect of level and time of silicon application on growth, yield and its uptake by rice (Oryza sativa). Indian J. Agric. Sci. 76, 410–413.

Smyth, T.J., Sanchez, P.A., 1980. Effects of lime, silicate, and phosphorus applications to an Oxisol on phosphorus sorption and ion retention. Soil Sci. Soc. Am. J. 44, 500–505.

Sogvar, O.B., Saba, M.K., Emamifar, A., 2016. Aloe vera and ascorbic acid coatings maintain postharvest quality and reduce microbial load of strawberry fruit. Postharvest Biol. Technol. 114, 29–35.

Song, A., Ning, D., Fan, F., Li, Z., Provance-Bowley, M., Liang, Y., 2015. The potential for carbon bio-sequestration in China's paddy rice (Oryza sativa L.) as impacted by slag-based silicate fertilizer. Sci. Rep. 5, 17354.

Song, A., Xue, G., Cui, P., Fan, F., Liu, H., Yin, C., Sun, W., Liang, Y., 2016. The role of silicon in enhancing resistance to bacterial blight of hydroponic- and soil-cultured rice. Sci. Rep. 6.

Soylemezoglu, G., Demir, K., Inal, A., Gunes, A., 2009. Effect of silicon on antioxidant and stomatal response of two grapevine (*Vitis vinifera* L.) rootstocks grown in boron toxic, saline and boron toxic-saline soil. Sci. Hortic. 123, 240–246.

Stamatakis, A., Papadantonakis, N., Savvas, D., Lydakis-Simantiris, N., Kefalas, P., 2003. Effects of silicon and salinity on fruit yield and quality of tomato grown hydroponically. In: International Symposium on Managing Greenhouse Crops in Saline Environment 609, pp. 141–147.

Su, X.W., Wei, S.C., Jiang, Y.M., Huang, Y.-y., 2011. Effects of silicon on quality of apple fruit and Mn content in plants on acid soils. Shandong Agric. Sci. 6, 018.

Sun, W., Zhang, J., Fan, Q., Xue, G., Li, Z., Liang, Y., 2010. Silicon-enhanced resistance to rice blast is attributed to silicon-mediated defence resistance and its role as physical barrier. Eur. J. Plant Pathol. 128, 39–49.

Terraza, S.P., Romero, M.V., Pena, P.S., Madrid, J.L.C., Verdugo, S.H., 2008. Effect of calcium and osmotic potential of the nutritive solution on the tomato blossom-end rot, mineral composition and yield. Interciencia 33, 449–456.

Tesfay, S.Z., 2009. Special Carbohydrates of Avocado: Their Function as 'Sources of Energy' and 'Anti-Oxidants'.

Tesfay, S.Z., Bertling, I., Bower, J.P., 2011. Effects of postharvest potassium silicate application on phenolics and other anti-oxidant systems aligned to avocado fruit quality. Postharvest Biol. Technol. 60, 92–99.

Thompson, J.E., 1988. The molecular basis for membrane deterioration during senescence. In: Senescence and Aging in Plants, Academic Press, New York, pp. 51–83.

Tian, S., Qin, G., Xu, Y., 2005. Synergistic effects of combining biocontrol agents with silicon against postharvest diseases of jujube fruit. J. Food Prot. 68, 544–550.

Torres, M.A., Jones, J.D.G., Dangl, J.L., 2006. Reactive oxygen species signaling in response to pathogens. Plant Physiol. 141, 373–378.

Torres-Olivar, V., Villegas-Torres, O.G., Domínguez-Patiño, M.L., Sotelo-Nava, H., Rodríguez-Martínez, A., Melgoza-Alemán, R.M., Valdez-Aguilar, L.A., Alia-Tejacal, I., 2014. Role of nitrogen and nutrients in crop nutrition. J. Agric. Sci. Technol. B 4, 29.

Tripathi, D.K., Shweta, S.S., Yadav, V., Arif, N., Singh, S., Dubey, N.K., Chauhan, D.K., 2017. Silicon: a potential element to combat adverse impact of UV-B in plants. In: UV-B Radiation: From Environmental Stressor to Regulator of Plant Growth.vol. 1. pp. 175–195.

Tubana, B.S., Babu, T., Datnoff, L.E., 2016. A review of silicon in soils and plants and its role in US agriculture: history and future perspectives. Soil Sci. 181, 393–411.

Umaña-Rojas, G., 2009. Control of Rots of Tropical Fruits With Generally Regarded as Safe (GRAS) Compounds. pp. 189–196.

Van Bockhaven, J., De Vleesschauwer, D., Hofte, M., 2013. Towards establishing broad-spectrum disease resistance in plants: silicon leads the way. J. Exp. Bot. 64, 1281–1293.

Vandevenne, F.I., Barão, L., Ronchi, B., Govers, G., Meire, P., Kelly, E.F., Struyf, E., 2015. Silicon pools in human impacted soils of temperate zones. Glob. Biogeochem. Cycles 29, 1439–1450.

Vattem, D.A., Ghaedian, R., Shetty, K., 2005. Enhancing health benefits of berries through phenolic antioxidant enrichment: focus on cranberry. Asia Pac. J. Clin. Nutr. 14, 120.

Vermeire, M.-L., Kablan, L., Dorel, M., Delvaux, B., Risède, J.-M., Legrève, A., 2011. Protective role of silicon in the banana-*Cylindrocladium spathiphylli* pathosystem. Eur. J. Plant Pathol. 131, 621.

Vieira, D.L., de Oliveira Barbosa, V., de Souza, W.C.O., da Silva, J.G., Malaquias, J.B., de Luna Batista, J., 2016. Potassium silicate-induced resistance against blackfly in seedlings of Citrus reticulata. Fruits 71, 49–55.

Vogler, B.K., Ernst, E., 1999. *Aloe vera*: a systematic review of its clinical effectiveness. Br. J. Gen. Pract. 49, 823–828.

von Wirén, N., Marschner, H., Romheld, V., 1996. Roots of iron-efficient maize also absorb phytosiderophore-chelated zinc. Plant Physiol. 111, 1119–1125.

Voogt, W., Sonneveld, C., 2001. Silicon in horticultural crops grown in soilless culture. Stud. Plant Sci. 8, 115–131.

Waewthongrak, W., Pisuchpen, S., Leelasuphakul, W., 2015. Effect of Bacillus subtilis and chitosan applications on green mold (*Penicillium digitatum* Sacc.) decay in citrus fruit. Postharvest Biol. Technol. 99, 44–49.

Wallace, A., 1992. Participation of silicon in cation-anion balance as a possible mechanism for aluminum and iron tolerance in some gramineae. J. Plant Nutr. 15, 1345–1351.

Wang, X.S., Han, J.G., 2007. Effects of NaCl and silicon on ion distribution in the roots, shoots and leaves of two alfalfa cultivars with different salt tolerance. Soil Sci. Plant Nutr. 53, 278–285.

Wang, L., Wang, Y., Chen, Q., Cao, W., Li, M., Zhang, F., 2000. Silicon induced cadmium tolerance of rice seedlings. J. Plant Nutr. 23, 1397–1406.

Wang, H., Li, C., Liang, Y., 2001. Agricultural utilization of silicon in China. Stud. Plant Sci. 8, 343–358.

Wang, M., Wang, X., Wang, J.J., 2016. Effect of silicon application on silicon contents in "Fuji" apple in Loess Plateau. Commun. Soil Sci. Plant Anal. 47, 2325–2333.

Wang, M., Gao, L., Dong, S., Sun, Y., Shen, Q., Guo, S., 2017. Role of silicon on plant–pathogen interactions. Front. Plant Sci. 8, 1–14.

Weerahewa, D., Somapala, K., 2016. Role of Silicon on Enhancing Disease Resistance in Tropical Fruits and Vegetables: A Review.

Whan, J.A., Dann, E.K., Aitken, E.A.B., 2016. Effects of silicon treatment and inoculation with *Fusarium oxysporum* f. sp. vasinfectum on cellular defences in root tissues of two cotton cultivars. Ann. Bot. 118, 219–226.

Xiyan, W., Yulong, Z., Na, Y., Qinglong, M., Jun, L., 2007. Effect of silicon fertilizer on cucumber photosynthesis and yield in protected field. J. Changjiang Veg. 2, 027.

Yan, G.-c., Nikolic, M., Ye, M.-j., Xiao, Z.-x., Liang, Y.-c., 2018. Silicon acquisition and accumulation in plant and its significance for agriculture. J. Integr. Agric. 17, 2138–2150.

Yang, L., Zhao, P., Wang, L., Filippus, I., Meng, X., 2010. Synergistic effect of oligochitosan and silicon on inhibition of *Monilinia fructicola* infections. J. Sci. Food Agric. 90, 630–634.

Yazici, K., Kaynak, L., 2006. Effects of kaolin and shading treatments on sunburn on fruit of Hicaznar cultivar of pomegranate (*Punica granatum* L. cv. Hicaznar), pp. 167–174.

Yoshida, S., Navasero, S.A., Ramirez, E.A., 1969. Effects of silica and nitrogen supply on some leaf characters of the rice plant. Plant Soil 31, 48–56.

You-Qiang, F.U., Hong, S., Dao-Ming, W.U., Kun-Zheng, C.A.I., 2012. Silicon-mediated amelioration of Fe2+ toxicity in rice (*Oryza sativa* L.) roots. Pedosphere 22, 795–802.

Yu, L., Gao, J., 2012. Effects of silicon on yield and grain quality of wheat. J. Triticeae Crops 32, 469–473.

Zeng, X., Liang, J., Tan, Z., 2007. Effects of silicate on some photosynthetic characteristics of sugarcane leaves. J. Huazhong Agric. Univ. 26, 330.

Zhang, X., Zhang, F., Mao, D., 1998. Effect of iron plaque outside roots on nutrient uptake by rice (*Oryza sativa* L.). Zinc uptake by Fe-deficient rice. Plant Soil 202, 33–39.

Zhang, H., Zhang, D., Chen, J., Yang, Y., Huang, Z., Huang, D., Wang, X.-C., Huang, R., 2004. Tomato stress-responsive factor TSRF1 interacts with ethylene responsive element GCC box and regulates pathogen resistance to *Ralstonia solanacearum*. Plant Mol. Biol. 55, 825–834.

Zhang, G., Wang, J., Zhang, H., Huo, Z., Ling, L., Wang, X., Zhang, J., 2007. Effects of Silicon Fertilizer Rate on Yield and Quality of Japonica Rice Wuyujing 3.

Zhang, M., Liang, Y., Chu, G., 2017. Applying silicate fertilizer increases both yield and quality of table grape (*Vitis vinifera* L.) grown on calcareous grey desert soil. Sci. Hortic. 225, 757–763.

Zheng, Z., Shetty, K., 2000. Solid-state bioconversion of phenolics from cranberry pomace and role of *Lentinus edodes* β-glucosidase. J. Agric. Food Chem. 48, 895–900.

Zhu, Y., Gong, H., 2014. Beneficial effects of silicon on salt and drought tolerance in plants. Agron. Sustain. Dev. 34, 455–472.

Further reading

Chen, D., Wang, S., Yin, L., Deng, X., 2018. How does silicon mediate plant water uptake and loss under water deficiency? Front. Plant Sci. 9, 281.

Cover cropping for increasing fruit production and farming sustainability

M. Ângelo Rodrigues*, Margarida Arrobas

Centro de Investigação de Montanha (CIMO), Instituto Politécnico de Bragança, Bragança, Portugal
*Corresponding author. E-mail: angelor@ipb.pt

1 Introduction

Weeds are a chronic phytosanitary problem appearing every year in any cultivated field. They must be kept under control since they compete for resources with cultivated species. In spite of several tools that have been developed to deal with weeds in cultivated fields, their control is still one of the most important tasks the farmers face. In fact, in fruit growing, if man ceases its care, the ground is firstly invaded by grass, followed by shrubby and arboreal vegetation in an ecological succession that leads to the failure of the cultivated trees. The competition is mainly for water and nutrients, although in young plantations and in some shrubby and creeping species, competition for light may also occur.

In fruit growing, the need to eliminate or keep under control herbaceous vegetation is not only due to the reduction of direct competition with the trees. The vegetation can hinder the trafficability of equipment and people and several ordinary operations, such as pruning, phytosanitary treatments, and harvesting. Additionally, the herbaceous vegetation can harbor pests and diseases that may require more phytosanitary treatments increasing monetary and environmental cost. It can also be emphasized that weeds make a property appear neglected reflecting on the competence of the grower. Nowadays, however, major benefits are recognized of herbaceous vegetation. Not only the reduction of soil erosion but also carbon sequestration in the soil with increased organic matter and enhanced general soil fertility has been highlighted. Weeds may also have a role in pest and disease biological control by harboring beneficial

A.K. Srivastava, Chengxiao Hu (eds.)
Fruit Crops: Diagnosis and Management of Nutrient Constraints
https://doi.org/10.1016/B978-0-12-818732-6.00020-4

organisms and providing several other advantages that arise from the increase of functional biodiversity. At present, the major challenge for fruit growers is to assess what advantages and drawbacks are relevant in their agroecological contexts to take the best options for weed management.

To compare methods of weed management, a diverse range of aspects should be equated. The role of the ground management in soil protection from erosion is of crucial importance given its relationship to the sustainability of the agrosystem. Also, the influence of soil management on soil organic matter content must be taken into account due to its role on the overall soil fertility. However, a number of other aspects deserve attention in the assessment of a soil management system, such as physical, chemical, and biological soil properties that influence the soil's ability to retain water, drainage, nutrient cycling, and biodiversity. The way the soil is managed can also be associated with relevant environmental impacts that must be taken into account, especially eutrophication and silting of watercourses and lakes, resulting from leaching, erosion and runoff, and the emission of harmful gases into the atmosphere. Finally, the way the soil is managed influences the nutritional status of trees and fruit yield. It will never be too much to point out that without profit the farmer does not maintain the activity. Thus, an advised soil management system cannot ignore an economistic view. Unfortunately, crop productivity has not yet deserved due attention because it is very laborious and time consuming to assess.

2 Soil management systems

The options to manage the orchard floor are diverse. The most common in regions of developed agriculture probably is weed control in the rows by herbicides and a sod strip between the rows. Mechanical cultivation is still widespread in regions of less intensive farming and in rainfed managed orchards of arid and semiarid environments. Several other methods are often used, such as integral floor management with herbicides, cover cropping, or mulching with mineral and organic materials. The topic maintains high scientific and technical importance insofar as it relates to such relevant aspects as erosion control, soil organic matter and carbon sequestration, functional biodiversity, fruit yield, and farmer profit, aspects that have been highlighted in recent book chapters on the subject (Rodrigues and Cabanas, 2009; Jackson, 2011; Penman and Chapman, 2011; Velarde, 2015; Alcántara et al., 2017; Rodrigues and Arrobas, 2017).

2.1 Soil tillage

Soil cultivation is the original method of weed management in fruit growing (Fig. 20.1). Although the importance of tillage as a method of weed management has progressively been reduced in several parts of the world, it maintains widespread use in orchards of arid and semiarid regions, especially when the orchards are rainfed managed, and in regions of least developed agriculture (Palese et al., 2014; Slatnar et al., 2014; Velarde, 2015; Ping et al., 2018). Tillage is easy to perform, and growers can do it without advising. At the installation of the orchard, tillage is also the preferred method of ground management. Weeds are very competitive, and herbicides should be used with caution around young trees and particularly on sandy soils, where herbicides may leach into the rooting zone where they are available for root uptake (Jackson, 2011; Penman and Chapman, 2011).

FIG. 20.1 Conventional tillage in a young rainfed managed orchard of almond.

Soil tillage is done with the main objective of reducing the competition of weeds to the trees for water and nutrients, although the incorporation of fertilizers, organic amendments, and green manures subsists as a complementary objective. Tillage is often also performed to improve the trafficability of equipment, to facilitate harvesting, to destroy superficial crusts, to facilitate water infiltration, to reduce the incidence of diseases, or to install cover crops. However, the effectiveness of soil tillage in achieving some of those objectives has been progressively questioned (Winkler et al., 1974; Keesstra et al., 2016; Rodrigues and Arrobas, 2017). Soil tillage is in turn susceptible to cause diverse problems in the physical, chemical, and biological soil properties, aspects that will be discussed in more detail in Section 3.

Soil tillage can be done with different farm implements, depending on the nature of the soil, the purpose of the operation, and the region of the globe. Cultivator is probably the widespread used implement, but soil tillage can also be achieved by harrowing, disking, or plowing (Alcántara et al., 2017).

Soil management was originally carried out with animal traction. With the mechanization of tillage operations, it became possible to transform a system based on a single to two tillage trips per year in three to four and in some situations up to six. Tillage is held in spring to destroy vegetation, in the summer to eliminate crusts, and in the autumn to prepare the ground for harvest (Pastor, 2008; Velarde, 2015). With more mechanical power available, it is also possible to till the soil when dry or excessively wet, which increase the potential damage on soil structure and the exposure of the organic substrates to the action of soil microorganisms, reducing soil organic matter and favoring the erosive process, especially when associated with sloping terrains (Bárberi, 2006; Alcántara et al., 2017). Tillage may also destroy the root system. Although this aspect has not been yet properly quantified, there is a common consensus that it can be problematic in shallow soils, which causes the plants to spend valuable resources in its regeneration (Cockroft and Wallbrink, 1996; Rodrigues and Cabanas, 2009; Arquero and Serrano, 2013; Lisek and Buler, 2018). In addition, tillage prevents the plants to explore the surface layers, usually of greater biological activity and higher availability of most nutrients (Rodrigues and Cabanas, 2009; Velarde, 2015; Rodrigues and Arrobas, 2017).

2.2 Bare soil managed with herbicides

The use of herbicides as a new tool in the control of weeds in orchards was widespread from the 1970s. Keeping a bare free-weed soil throughout the year can be achieved by using residual or preemergence herbicides or herbicides that combine residual and postemergence active ingredients applied several times a year. Residual herbicides prevent seed germination, while postemergence herbicides kill already established vegetation but do not control new emergencies. To achieve both effects, several commercial herbicides combine pre- and postemergence active ingredients (Rodrigues and Cabanas, 2009). The most common dates of residual herbicide application coincide with periods of high seed germination, such as after the first rains following a long dry period or after a cold period when temperatures start to raise (Rodrigues and Arrobas, 2017).

Postemergence herbicides can be separated into those that are absorbed but not translocated among tissues, called contact herbicides, and those that translocate to shoot or root meristems, called systemic herbicides. The latter are particularly useful since they are able to destroy perennial weeds with reproductive subterranean structures such as rhizomes, stolons, regenerating roots, bulbils, and nutlets (Rodrigues and Cabanas, 2009; Penman and Chapman, 2011). In orchards, it is also possible to use nonselective herbicides, those able to kill any species, being used just before bud burst in spring in deciduous species or using protective tools to avoid spraying the shoots (Jackson, 2011).

In comparison with soil cultivation, herbicide use in the entire floor has the main advantages of not destroying the root system. However, maintaining the ground free from weeds with herbicides keeps the soil exposed throughout the year and then also susceptible to erosion (Martínez et al., 2016). In addition, it does not allow the deposition of plant debris decreasing the organic matter and the biological activity of the soil (Rodrigues et al., 2015a). The use of herbicides may allow the reversal of vegetation. In orchards managed with herbicides, the vegetation tends to evolve to a reduced number of difficult-to-combat species (Zimdahl, 1993; Rodrigues et al., 2009). The strategy can become unsustainable, since it can be necessary to use more expensive herbicides, those that combine different active ingredients or increasingly expensive new molecules. The use of herbicides implies greater knowledge and training than tillage. The farmer should select an active ingredient, prepare a spray in a concentration adequate to the weed population present, and select the best timing of application (Velarde, 2015). The use of herbicides may also be associated with damage to cultivated plants (Cañero et al., 2011) (Fig. 20.2). Some molecules are quite toxic to operators or hazardous to the

FIG. 20.2 Severe damage in a young olive tree after the application of a nonselective herbicide.

FIG. 20.3 Mulching with blanket black plastic in a blueberry plantation.

environment, especially to aquatic ecosystems (Celis et al., 2007), being nowadays the reduction of pesticide use a priority in the European Union (European Union, 2009).

2.3 Mulching

Mulching consists of covering the soil surface with organic or mineral products, such as gravel, bark, straw, hay, sawdust, and pruning wood, or a diverse range of synthetic materials, such as polyethylene plastics (Lisek and Buler, 2018). The use of living vegetation is also usually seen as a mulching technique, but in this work, the topic is dealt in the next section as cover cropping. Mulching is not common in large-scale fruit growing due to the high costs and/or availability of the materials, but important in particular species or ecological conditions (Fig. 20.3). The objectives of mulching are diverse, such as weed control, soil temperature reduction, and water saving (Hoagland et al., 2008). Despite the numerous advantages of mulches, some can also be a source of problems, since they can be a fire hazard, and may harbor rodents, and frost can be more severe (Jackson, 2011). Furthermore, mulching can make it difficult to spread or incorporate manures and fertilizers (Velarde, 2015).

Mulching can be used as a method of weed control since it prevents light from reaching recently germinated seedlings. However, excluding the use of weed blanket black plastic ground cover, for the treatment to be effective, a layer of 10–30 cm of a natural mulching materials may be required (Jackson, 2011), which greatly limits their potential for expansion. Sometimes, mulches are used in restricted spaces, such as under the canopy, or in the rows of the trees. They are also more frequent in shrub fruit species such as blueberries and raspberries than in tree crop species (Penman and Chapman, 2011).

2.4 Cover cropping

Cover cropping consists of allowing the development of herbaceous vegetation in the orchards, seeking to take advantage of its presence. It is assumed that the benefits will be higher than the loss associated with competition for resources. The soil is covered with vegetation during a relevant part or all year round. Cover cropping has the potential to solve problems of erosion and to increase soil organic matter, nutrient cycling, and functional biodiversity.

Cover cropping may consist on the management of spontaneous vegetation in the orchard or on sowing selected species having in mind a particular objective. Sown cover crops can be managed as green manures or in a similar manner of the natural vegetation where once sown it is expected that they may self-reseed without supplemental work or cost.

2.4.1 Cover cropping with spontaneous vegetation

Unsown vegetation is the most widespread and natural method of cover cropping in orchards. It does not require sowing or the incorporation of the residues of the seeded species into the soil, but vegetation needs to be mown frequently to reduce the competition with the trees (Fig. 20.4).

In a first approach, cover cropping aims to achieve a primary objective consisting on the reduction of soil erosion. The herbaceous vegetation protects the soil from the direct impact of the raindrops and sustains the soil preventing erosion as long as the cover is abundant (Keesstra et al., 2016; Torres et al., 2018). Since mowing replaces tillage, the rooting system of the trees is not damaged (Rodrigues and Arrobas, 2017; Lisek and Buler, 2018). The development of vegetation increases the deposition of organic residues, which, together with the reduction of soil aeration, allows the increase of organic matter and the biological activity of the soil with great influence on nutrient cycling (Hoagland et al., 2008; Rodrigues et al., 2015a; Almagro et al., 2016). Moreover, the presence of herbaceous vegetation favors the trafficability of equipment and people and reduces the risks of soil compaction. Herbaceous vegetation that grows during the winter can also act as a catch crop preventing the loss of N by leaching and/or denitrification (Rodrigues et al., 2002; Gabriel et al., 2012). Continuous root development improves soil structure, which increases water infiltration and storage (Blanco-Canqui et al., 2017).

Cover cropping with natural vegetation increases biodiversity with richer food chains, which may reduce the incidence of some pests by favoring beneficial organisms (Arquero and Serrano, 2013; Irvin et al., 2016; Muscas et al., 2017). Unfortunately, sometimes, spontaneous vegetation can also be an alternative host of pests and diseases and aggravate particular phytosanitary problems in the orchard (Penman and Chapman, 2011; Arquero and Serrano, 2013).

The management of the cover crops can be done by mowing. The number of cuts should be adjusted to the need for more or less tight vegetation control. As a principle, the greater the permissiveness to the herbaceous vegetation, the better the soil protection and the greater the amount of carbon that will be sequestered in the soil. However, the risk of competition for water and nutrients increases, which may reduce crop yield (Ferreira et al., 2013). In hot and dry climates with high grass fire risk, as occurs in the Mediterranean region, the tolerance to herbaceous vegetation should also be adjusted accordingly (Rodrigues and Arrobas, 2017). The management of natural vegetation by mowing tends to select the vegetation for perennial species that better withstand cutting, which is progressively difficult to control (Alcántara et al., 2017).

Natural vegetation can alternatively be managed with nonselective postemergence herbicides. In this situation, the farmer must allow the vegetation to develop in winter, to protect the soil from erosion, and to act as an N catch crop. Thereafter, the vegetation is killed in spring at the beginning of the dry season with herbicides to reduce the competition for water and nutrients when the trees restart the biological activity. Dead vegetation may act as mulch protecting the soil during the summer period (Rodrigues et al., 2011).

The application of herbicides, even nonselective ones, becomes progressively inefficient as some weeds acquire resistance (Wakclin et al., 2004). This may lead to a reversal of the flora, where few species of difficult control tend to dominate the cover (Zimdahl, 1993; Rodrigues et al., 2009), which requires the use of different molecules sometimes very expensive or alternating the use of herbicides with mechanical weed control. Some species of the genus *Conyza* are an interesting example of tolerance to glyphosate, which is causing great difficulties of control in rainfed managed orchards of the Mediterranean region (Rodrigues and Arrobas, 2017).

FIG. 20.4   Cover cropping with natural vegetation in a chestnut plantation.

Natural vegetation can also be geared toward a specific goal. If the soil is clayey, for instance, with deficient trafficability and high risk of compaction, an antibroadleaf herbicide can be used to favor the development of grasses, which better helps fulfill those objectives (Pastor, 2008).

It should always be taken into account that cover cropping reduces productivity through the competition of herbaceous vegetation with trees. The problem may be residual and negligible if the orchard is not exposed to relevant nutritional or drought stresses to very problematic in rainfed fruticulture and poor fertility soils. In Mediterranean-rainfed agriculture, the introduction of poorly controlled herbaceous vegetation can significantly reduce fruit yield (Rodrigues et al., 2011; Ferreira et al., 2013).

Herbaceous vegetation can also be used to reduce soil fertility and plant vigor with potential advantages in fruit quality. This is a frequent goal in viticulture (Lopes et al., 2011; Giese et al., 2014; Muscas et al., 2017). However, the reason of excessive vigor should firstly be sought, often occurring in the lowland parts of the vineyards, and control measures should be implemented, such as reduction of fertilizer application.

2.4.2 Green manures and short-term cover crops

Green manures are crops sown with a specific purpose of being incorporated into the soil in an immature stage (Niemsdorff and Kristiansen, 2006). Although various groups of plants, such as grasses and cruciferous species can be used, legumes are the most frequently grown for this purpose since they are able to access atmospheric dinitrogen afforded by the ability to establish symbiotic relationships with N-fixing organisms living in their roots (Cooper and Scherer, 2012). N fixed by these diazotroph microorganisms can meet N needs of legumes, being thereafter some N transferred to a nonlegume crop after their N-rich tissues have been mineralized (Snoeck et al., 2000; Pirhofer-Walzl et al., 2012). This aspect is particularly important in organic farming since legumes provide N from a natural source taken into account that commercial synthetic fertilizers are not allowed (see Section 5.3).

The availability of other nutrients is less affected, since green manures release, by mineralization, the nutrients that were previously taken up from the soil. However, some plants, also mostly those of the Fabaceae family, can uptake P from sparingly soluble P sources, through the development of cluster roots or releasing organic acids and acid phosphatases into the rhizosphere (Veneklaas et al., 2003; Wang et al., 2011; Sepehr et al., 2012). Phosphorus taken up by green manures may thereafter be used by other species, such as companion crops, after the mineralization of debris of the legume species (Arrobas et al., 2015).

Some plants have also been studied for their biofumigation potential. Biofumigation is the suppression of soilborne pests and pathogens resulting from biocidal compounds released from cover crops, particularly brassicaceous crops, following the hydrolysis of glucosinolates present in their tissues (Meng et al., 2018). Alcántara et al. (2017) reported studies carried out in Spain where the use of *Sinapis alba* as green manure was particularly effective in reducing the incidence of *Verticillium dahliae* in olive (*Olea europaea* L.).

The great inconvenience and difficulty of using green manures in commercial orchards is the requirement of soil tillage for sowing, which can coincide with a period of high risk of soil erosion, and a second tillage for the incorporation of plant debris into the soil, which can also damage the root system. The need of tillage for sowing and incorporation of green manures is perhaps the main reason for the low adoption of green manures in commercial orchards. Rodrigues et al. (2013) tried to overcome the problem of the destruction of the root system mowing the aboveground biomass and leaving a mulch of the dead vegetation on the ground. The technique was partially unsuccessful since most of N present in the mulched materials seems to disappear without entering the soil. The results of this experiment recommend some caution in the management of pure legume cover crops as mulch due to the reduced transfer of N to the trees (Fig. 20.5).

Several annual plants, such as grass and cruciferous species, have also been tested as cover crops with diverse objectives. The use of annual plants as cover crops has a major drawback requiring soil tillage for sowing and eventually incorporation, and unlike legumes, they do not supply N to companion crops. However, they can reduce soil erosion in the period where the soil is covered by vegetation and increase soil organic matter (Alcántara et al., 2017).

2.4.3 Self-reseeding legumes and continuous cover cropping

Cover cropping with self-reseeding annual legumes or possibly perennial species is a management approach close to natural vegetation since it goes without soil tillage, except for the first installation. The principle is similar to improved pastures where once installed sown species must be managed to maintain their persistence in the cover without the need for resowing in a period of time as long as possible.

FIG. 20.5 Cover cropping with vetch (*Vicia villosa* Roth). The above-ground biomass is left on the ground as mulch.

When sown cover crops are used, it is assumed that sown species bring a benefit over the spontaneous vegetation, since they also have supplementary cost of installation and management. It is also assumed that sown species are less competitive to the trees than spontaneous vegetation (Rodrigues and Arrobas, 2017).

Legume species are the most frequently used vegetation since they tend to be less competitive for water and nutrients and are able to incorporate significant amounts of N into the soil. Legumes also have a greater facility in establishing in poor fertility soils as they can access atmospheric N. If annual species are used, which is the most common situation, the sward should be managed to ensure the maturing of seeds to allow a successful reseeding in the next season. Previous studies with pasture legumes as cover crops have shown very promising results (Driouech et al., 2008; Mauromicale et al., 2010; Rodrigues et al., 2015a,b).

Orchard floor managed by sown species is also characterized by high biodiversity. In addition to the cultivated species, native vegetation always has the opportunity to persist in the cover even if their relative abundance is reduced due to the increased competition for resources with the seeded species. Over time, the relative abundance of the seeded species may recede with native vegetation recovering its relative abundance. Rodrigues et al. (2015b) observed that a cover of self-reseeding legumes suffers a false break, which reduced their presence in the cover in the third year after sowing. However, seeded species recover the dominance of the cover in the fourth growing season, since the bank of hard seeds established in the previous seasons ensured the persistence of the sown legumes. The dynamic between sown species and spontaneous vegetation may also occur due to fluctuation on soil-available N in a similar manner as occurs in pastures (Rodrigues et al., 2015b).

2.5 Mixed systems

The aforementioned soil management systems are not often used on the whole ground. Sometimes, vegetation is managed in strips, with different methods of control. It is also possible to implement management strategies that combine different methods in different timings. Nowadays, methods of mixed management of orchard floor are among the most widespread.

In wet climates or irrigated orchards, the most frequent soil management system consists of weed control by chemicals in the rows and sod alleys between the rows (Sirrine et al., 2008; Giese et al., 2014). The vegetation in the row is controlled with residual herbicides often in combination with postemergence herbicides. The vegetation developed between rows is usually mowed (Fig. 20.6). When a sown cover is used, it can be composed of a great diversity of species according to the objective. In intensive fruticulture, grasses are the most frequent type of vegetation used since they consume more water in the wet period, favoring the trafficability of equipment and people and reducing soil compaction. In these systems, the fertilizers must be applied in the rows beneath the trees where there is no vegetation (Jackson, 2011).

Although less frequent, a great diversity of other methods can be implemented. In grapevine (*Vitis vinifera* L.), for instance, it seems quite ingenious to till and grow vegetation alternatively between lines, with people and equipment circulating along the lines managed with sward (Fig. 20.7). Techniques combining tillage-herbicides or tillage-mulching in different timings all year round are also common as well as techniques of reduced tillage in the interrow combined with herbicides in the row (Velarde, 2015; Alcántara et al., 2017).

FIG. 20.6   Soil management using herbicides in a narrow strip along the row and natural vegetation managed by mowing between rows.

FIG. 20.7   Interrows of a vineyard alternately managed by soil tillage and cover cropping with a commercial mixture containing legume and grass species. *Courtesy of Henrique Chia.*

3 Impact of weed management on soil and environment

The original problem of soil management in fruit growing is the elimination of the competition of weeds. In this perspective, the more effective the elimination of the herbaceous vegetation, the more reputed is the method of control. Nowadays, the issue of weed management in orchards is known to be rather more complex, with interactions that go beyond the original problem of eliminating weed competition with the trees. In addition to produce fruit, a set of other aspects must be taken into account to ensure the sustainability of the agrosystem. The way the ground is maintained significantly influences the loss of soil through erosion, the organic matter content and nutrient cycling, water infiltration and conservation, and biodiversity with diverse interactions with organisms that can be pests, diseases, or auxiliary organisms. Together, the trees and the herbaceous community can provide a wide range of ecosystem services that should be taken into account at the time of making a decision (Salomé et al., 2016; Montanaro et al., 2017; Daryanto et al., 2018).

3.1 Soil erosion

Soil can be lost from an agricultural field by the action of rainwater (water erosion) or wind (wind erosion). Soil erosion is a major worldwide problem. In fruit growing, water erosion is particularly important (Fig. 20.8). Water erosion occurs when raindrops strike bare soil detaching particles and overland flow carries the detached soil from the field (Singer and Munns, 2002).

FIG. 20.8   Soil erosion in an olive grove. The loss of soil is particularly detectable near the tree trunks.

Soil erosion is probably the primary reason that has intensified the research on cover cropping as an alternative to conventional tillage or bare soil managed with herbicides. It represents an irreversible loss of soil fertility and a great threat to the sustainability of the production systems as it selectively removes fine particles with higher content in organic matter and clay. As a result, soil erosion reduces effective soil depth and the volume of water and available nutrients that can be stored and used by plants (Gómez et al., 2009; Palese et al., 2014; López-Vicente et al., 2016).

Cover crops protect the soil against the high energy of raindrops, avoiding soil particle disaggregation and allowing water infiltration. A soil covered with herbaceous vegetation develops a better structure due to root activity and presents increased biological activity. Vegetation is a physical barrier to water runoff, which is associated to the improved structure, and the canicula created by roots may increase water infiltration (Palese et al., 2014; Ruiz-Colmenero et al., 2013).

Soil erosion is a particularly important problem in tropical and subtropical areas due to the high erosivity of rainwater and susceptibility of soils to erosion. Also in the Mediterranean region, soil erosion is a huge problem in orchards due to high rainfall intensity, steep slopes, and erodible parent material (Keesstra et al., 2016). It is abundantly demonstrated that the introduction of a permanent cover in an orchard can greatly reduce soil erosion in comparison with conventional tillage or bare soil managed with herbicides (Gómez et al., 2009; Marquez-Garcia et al., 2013; Keesstra et al., 2016; Torres et al., 2018).

3.2 Organic matter and soil microbiology

Organic matter is a complex entity involving humic substances, relatively stable to decomposition and nonhumic materials that are poorly decomposed and much more reactive. The microbial biomass also takes part of the organic matter; although small, it is an important reservoir of nutrients. Organic matter plays a major role in the general soil fertility, because it determines important physical, chemical, and biological soil properties. In cultivated soils, it is commonly accepted that, for a given agroecological context, the higher the organic matter content, the better the soil quality.

The organic matter content of a soil depends on the balance of gains and losses of organic carbon. Usually, the gains are due to the deposition of plant debris, such as the leaves of the trees that fall and also the roots and shoots of the herbaceous community that can be incorporated into the soil. For a given soil and climate, the greater the development of annual herbaceous vegetation, the greater the deposition of these debris, and the higher the organic matter content of the soil. The losses of organic matter or soil carbon emissions are due to the microbial decomposition of the organic substrates. The microorganisms most effective in degrading the organic substrate are aerobic, which means that soil aeration accelerates the mineralization of the organic residues. Thus, soil management systems more tolerant to the development of herbaceous vegetation and those reducing excessive soil aeration are able to increase the organic matter content. In accordance, several studies have shown that cover cropping may increase soil organic matter in comparison with systems based on conventional tillage or on bare soil managed with herbicides (Montanaro et al., 2010; Ramos et al., 2010; Marquez-Garcia et al., 2013; Herencia, 2015; Torres et al., 2018).

Organic matter, particularly recently incorporated fresh organic residues, supports a trophic chain with relevance in soil biological processes. Soil management systems allowing the entry of more organic substrates into the soil increase enzymatic and microbiological activity, normally used as indicators of soil quality. On the contrary, systems that favor aeration, such as tillage, accelerate the mineralization of organic debris. Excessive tillage can even destroy soil aggregates exposing the native and stable soil organic matter to the action of microorganisms. These systems are usually associated with worse indicators of soil quality (Yao et al., 2005; Moreno et al., 2009; Ramos et al., 2010; Wang et al., 2011; Calabrese et al., 2015; Herencia, 2015; Almagro et al., 2016; Cucci et al., 2016).

Arbuscular mycorrhizal (AM) fungi are symbiotically associated with roots of most plant species. Plants provide photosynthetic carbon needed for fungal growth. AM fungi, via increasing the absorbing area associated to an extended hyphal network, can, in turn, improve plant growth by increasing nutrient uptake and biotic and abiotic resistance (Duc et al., 2018; Ouledali et al., 2018; Zhu et al., 2018; Hack et al., 2019). Plants can also promote microbial growth by releasing carbohydrates into the rhizosphere. Phosphorus-solubilizing bacteria and free-living and facultative endophytic diazotrophs are some of the microbes benefiting from root exudates. In turn, they can become a source of P and N to plants particularly in low-fertility soils (Cooper and Scherer, 2012; Bhanwariya et al., 2013; Ansari et al., 2015). Tillage, through a continuous disruption of these energetically expensive symbiotic relationships and partially destroying the rhizosphere, may difficult the tree's access to nutrients and water with a great negative impact in fruit tree crops growth and yield although currently poorly assessed.

3.3 Water conservation

Cover crops built a physical barrier to water transport on soil surface enhancing soil hydraulic properties related to the ability of the soil to capture and store precipitation or irrigation water (Blanco-Canqui et al., 2017). The physical action of the root systems and the release of root exudates improve soil aggregation, which is also a favorable condition to rainwater infiltration. On the contrary, a bare soil managed with herbicides is more prone to compaction, which reduces infiltration and accelerates surface runoff (Moreno et al., 2009; Montanaro et al., 2017). In soils subjected to mechanical disturbance, the decline in organic matter unavoidably leads to a reduction in natural porosity and to an increase in soil bulk density, which may decrease infiltration capacity and hydraulic conductivity (Cucci et al., 2016).

Cover crops in turn transpire water that may counter the beneficial effects on water infiltration and storage. Cover crops should be managed during the growing season. Exuberant development should not be allowed as the dry season approaches or as the mean annual precipitation reduces (Alcántara et al., 2017).

Soil tillage by exposing a moist soil can accelerate the loss of water by evaporation. On the contrary, tillage of a dehydrated surface layer can cause a discontinuity that hinders the transfer of energy to deep layers reducing the temperature and creating a physical barrier to the rise of water by capillarity (Ozpinar and Cay, 2006: Keesstra et al., 2016). However, farmers continue to till their orchards indifferent to soil water status, which leads to an uncertain outcome (Pastor, 2008).

3.4 Environmental contamination

Major aspects of environmental contamination regarding soil management systems are water eutrophication and silting of shallow waters and lakes as a result of soil erosion and runoff. Soil management systems that favor soil erosion, such as tillage and a bare free-weed soil with herbicides, are the most critical. On the contrary, cover cropping preventing the loss of soil may reduce the arrival of sediments to watercourses. Furthermore, cover crops uptake water and nutrients, acting as a catch crop, which may reduce nitrate leaching and denitrification during the wet season.

Agricultural activities may contribute to the emission of greenhouse gases into the atmosphere. The practices promoting the mineralization of soil organic matter can contribute to the release of carbon dioxide (CO_2) into the atmosphere, while the practices promoting soil carbon sequestration have the inverse effect. Once again, cover cropping is more favorable, while soil management systems promoting the degradation of the organic substrates, such as tillage, and those restricting the entry of organic substrate into the soil, such as bare free-weed soil by herbicides, accentuate the level of CO_2 emissions (Montanaro et al., 2012; Tribouillois et al., 2017).

Other important greenhouse gases such as N oxides can be lost from the soil in different rates depending on the soil management system. The increase of the organic pool may accelerate the turnover of organic matter with a positive balance on the emission of N oxides. However, cover cropping is not expected to increase nitrous oxide (N_2O) emission, since they can uptake N (NH_4^+ and NO_3^-) reducing nitrification and denitrification, soil processes that increase N_2O emissions (Guardia et al., 2016; Tribouillois et al., 2017). Cover cropping with grasses can be more effective than with legumes due to their higher ability to absorb water and reduce inorganic N in the soil (Kallenbach et al., 2010).

3.5 Biodiversity and relation with pests, diseases, and auxiliary organisms

More organic substrate means more food for macrofauna and heterotrophic soil microorganisms. Thus, systems that allow incorporating more plant debris, such as cover crops or green manures, greatly increase soil microbial biodiversity. Furthermore, soil management systems more permissive with the development of herbaceous vegetation also tend to exhibit more floristic biodiversity, which, in turn, supports a richer and more diverse trophic chain of arthropods, insects, reptiles, birds, and mammals. The food chain supported on cover crops may favor pests or diseases, but it is usually expected that it can enhance auxiliary organisms. Relevant literature has demonstrated the role of local flora or sown species in the food or refuge of natural enemies of key pest of important crops (Nicholls et al., 2000; Berndt and Wratten, 2005; Irvin et al., 2016; Muscas et al., 2017).

4 Tree crop nutrition and fruit yield

Cover crops play a fundamental role in nutrient cycling and, consequently, in the nutrition of trees and crop productivity. Despite the abundant literature on soil management dealing with aspects related to soil fertility, such as soil erosion, organic matter content, enzymatic activity, and soil microbiology, studies assessing the tree crop nutritional status and productivity, major components of the sustainability of the agrosystem and directly associated to farmer profit, are comparatively less frequent.

The herbaceous vegetation competes with trees for nutrient uptake. Some of the nutrients may return to the soil being recycled through the decay process of plant debris. When the covers are dominated by grasses, they tend to depress the N nutritional status of cultivated plants, as observed by Giese et al. (2014) after comparing a grass cover with an undertrellis herbicide strip treatment in a vineyard. When the covers are dominated by legumes, a positive net N balance may occur due to the ability of legumes to access atmospheric N. Silvestri et al. (2018) recorded a return of 70-kg N/ha to the soil at the end of the growing season of a cover crop of clover (*Trifolium subterraneum* L.). Rodrigues et al. (2015b) using a mixture of self-reseeding annual legumes in an olive orchard recorded values of apparently fixed N in aboveground biomass ranging from 50 to 115 kg N/ha depending on the length of the growing cycle, with the late-maturing species fixing higher amount of N. Annual legumes, such as lupine (*Lupinus albus* L.) and vetch (*Vicia villosa* Roth), are able to recover more than 150 kg N/ha in the aboveground biomass at the end of flowering even when cultivated in a poor soil of a rainfed olive grove (Rodrigues et al., 2013). N previously fixed by legume cover crops can thereafter be used by the trees. Rodrigues et al. (2015a) found a consistent increase in leaf N concentration after growing legume species in olive groves. In an apple (*Malus domestica* Borkh.) orchard, Sofi et al. (2018) found that N fixing cover crops gave rise to a significant increase in N, K, S, Mg, Fe, and Mo, whereas no discernible differences were observed for B, Cu, P, and Zn.

Some legume species, in addition to their ability to access atmospheric N, appear also to be efficient in using P from soils where P is only sparingly available to most plants. The interest in these species has increased tremendously, since they can grow without application of N and P fertilizers. The P contained in plant tissues may become available in the soil after mineralization of the residues and be used by companion crops (Damon et al., 2014; Dube et al., 2014; Silvestri et al., 2018). White lupine has been shown to be an effective cover crop in retaining P available for the subsequent crop (Arrobas et al., 2015; Soltangheisi et al., 2018) and is considered a P-mobilizing species regardless of the source of applied P (Soltangheisi et al., 2018).

Cover crops compete for resources with the trees, which can lead to a reduction in crop growth and yield as has been shown in almost all the studies where tree crop performance was analyzed. In apple, Hoagland et al. (2008) observed that tree growth under cover cropping was less than in standard tillage weed control likely in response to competition with the living cover understory for space and water. Gucci et al. (2012) recorded a reduced trunk section and fruit yield in plots of natural flora in comparison with conventional tillage in an irrigated olive orchard. Ferreira et al. (2013) recorded a significant increase in olive yield in plots where the soil was managed with herbicide in comparison with cover cropping in rainfed olive orchards. In poplar (*Populus deltoides* W. Bartran ex Marshall), perennial ryegrass (*Lolium perenne* L.) reduced growth compared with less competitive legume cover crops (*T. subterraneum* L.) with natural vegetation registering intermediate values (Silvestri et al., 2018). Lopes et al. (2011) consider cover cropping particularly hazardous for low-vigor vineyards in dry areas of the Mediterranean basin.

Properly managed, cover crops can, in turn, be compatible with good tree crop performance. Palese et al. (2014) did not record significant differences between conventional tillage and cover crops in a mature rainfed olive orchard due to the prompt removal of the vegetation. Lisek and Buler (2018) also did not found differences in plum (*Prunus domestica* L.) productivity after comparing several different floor management systems. Rodrigues et al. (2015a) reported improved N nutritional status and olive yield using early-maturing self-reseeding annual legumes in rainfed olive groves in comparison with conventional tillage and natural flora managed by mowing. Rodrigues et al. (2011) reported higher olive yield with a cover of natural vegetation managed with a nonselective herbicide compared with conventional tillage or bare soil performed by residual herbicides (Fig. 20.9). In a second field trial, Rodrigues et al. (2011) also reported increased yield by managing a natural cover with nonselective herbicide in comparison with plots managed by soil tillage or as a sheepwalk (Fig. 20.10).

The competition of herbaceous vegetation may also help to achieve particular benefits. Several studies have reported the role of cover crops in regulating the excessive vigor of the grapevine (Lopes et al., 2008, 2011; Giese et al., 2014; Muscas et al., 2017). Must quality may also be improved with increased competition by cover crops. The results of Monteiro and Lopes (2007) have shown that a sward treatment did not affected grapevine yield but reduced must acidity and increased berry skin total phenols and anthocyanins. Bouzas-Cid et al. (2016) reported an increase in wine anthocyanin concentrations in plots managed under cover crops. Muscas et al. (2017) also observed that utilizing competitive cover crops on grapevine, while reducing yield, would improve must quality and reduced pest development. In an apple orchard, an increase in phenolic compounds was observed under living mulch in comparison with plots managed with herbicides or a black polyethylene strip application (Slatnar et al., 2014). Otherwise, in situations where it is beneficial to reduce the vigor of the grapevine, legume cover crops may be inadequate, since they may supply more N than desired (Patrick-King and Berry, 2005).

FIG. 20.9 (A) Cumulative olive yield after eight consecutive harvests and (B) cumulative trunk circumference increase between two consecutive years (02; 2002–2001) from a field trial carried out in rainfed olive grove with three ground management treatments. Error bars are the mean standard deviations.

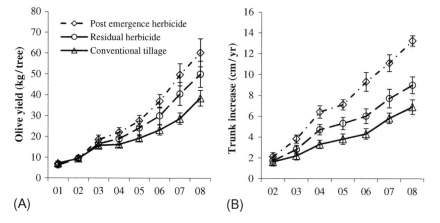

FIG. 20.10 Cumulative olive yield after two consecutive periods of harvests (2001–11; 2012–18) from a field trial carried out in rainfed olive grove with three ground management treatments: Plot 1, vegetation managed with a postemergence herbicide between 2001 and 2011 and as a grazed pasture between 2012 and 2018; Plot 2, vegetation managed with conventional tillage between 2001 and 2018; and Plot 3, vegetation managed as a grazed pasture between 2001 and 2011 and with a postemergence herbicide between 2012 and 2018 (in spring 2012, the plot 1, previously managed with a postemergence herbicide, started to be managed as a grazed pasture, and in plot 3, the inverse was followed). Error bars are the mean standard deviations.

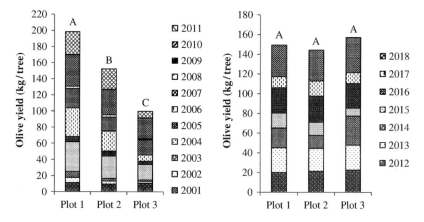

5 Adequacy of the soil management system to local agroecological conditions

There is no soil management system that should be recommended for all agroecological situations. The adoption of a soil management system should take into account the fertility of the soil; the growing conditions related to water availability, in particular if the orchard is or not irrigated; and the farming system. In organic farming, for instance, the fertility of the soil deserves particular attention due to the restricted use of industrial synthetic fertilizers.

5.1 Soil management in irrigated orchards

Under conditions where the trees do not suffer from significant water stress during the growing season, the competition between herbaceous vegetation and trees is restricted to soil nutrients. In irrigated orchards, for instance, water conservation is of less importance as it can be supplemented as irrigation water. Furthermore, the nutritional status of the orchard can be adjusted by performing additional applications of fertilizers to sustain the performance of the tree. Nutrients can be supplied as fertigation or foliar sprays, which favor the trees in detriment of weeds. In any case, without water restriction, it can be allowed a more luxuriant development of herbaceous vegetation, trying to obtain the benefits of a better soil protection from erosion, increased carbon sequestration in the soil, and better trafficability of people and equipment.

In fruit growing of wet climates or in irrigated orchards, there is a method of ground management widespread among farmers consisting on the establishment of a narrow strip herbicide managed along the rows with the herbaceous vegetation developing between rows and managed by mowing as previously mentioned. The cutoff frequency is intended to prevent vegetation from interfering to the remaining cropping techniques (Alcántara et al., 2017; Rodrigues and Arrobas, 2017).

5.2 Cover crops for rainfed fruticulture

In rainfed fruticulture, the abandonment of the conventional soil tillage has been particularly slow, and when it happens, the vegetation is managed with herbicides. In rainfed managed orchards, cover cropping is a problematic strategy due to the competition for water and the risk of yield reduction. The problem is more severe as the aridity of the region increases. However, cover cropping is possible in rainfed fruit growing. Cover crops for rainfed orchards should display a set of specific particularities, where the most important is the reduced competition for water. Furthermore, is of great importance to ensure adequate soil protection and the other benefits that cover crops provide.

Olive probably is the fruit tree crop where most of the respective acreage is managed in rainfed conditions. Olive is extensively cultivated throughout the Mediterranean basin, a region highly vulnerable to soil erosion. Furthermore, scenarios for climate change seem to be not favorable, with substantial shift in precipitation patterns with higher winter rains and drier summers (Goubanova and Li, 2007; Cramer et al., 2018). Rodrigues et al. (2015a,b) have developed a very promising model of cover cropping for this scenario consisting of growing early-maturing self-reseeding annual legumes, in particular some cultivars of subterranean clover. These plants germinate early in the fall, following the first autumn rains. They are very short and less competitive for water but provide a good protection of soil and sequester significant amounts of C and N in the soil (Fig. 20.11). With particular importance, they develop asynchronously with the trees, that is, their vegetative growth occurs during the winter resting period of olive, from October to April. By spring, when the trees resume vegetative growth, the growing cycle of these plants finishes. During summer, a mulch of dead vegetation persists in the ground keeping the soil protected (Fig. 20.12). Rodrigues et al. (2015a) have shown that a cover of early-maturing self-reseeding annual legumes improved the N nutritional status of the trees and olive yield over a treatment of natural vegetation fertilized with 60 kg N/ha (Fig. 20.13). This kind of cover crops can also be extended to other important crop of the Mediterranean basin, such as almond and grapevine. Results from other studies have also indicated that, in the Mediterranean environment, annual self-reseeding legumes are the most desirable species for cover cropping (Driouech et al., 2008; Mauromicale et al., 2010).

Unfortunately, the use of covers of early-maturing self-reseeding annual legumes is not yet common among farmers likely due to the low dissemination of the technique and the difficulty of implementation. Firstly, commercial mixtures with the right seeds are not yet available, contrarily to several other commercial mixtures of legumes and grasses. On the other hand, these seeds are expensive, and the installation technique is sensitive. Sowing must occur early in the autumn to benefit from mild temperatures for seed germination. Due to the high seed price, if the seeded species do not persist in the cover for several years, it can be little stimulating for farmer to make the investment.

5.3 Soil management in organic fruticulture

One of the major obstacles to the expansion of organic farming is the need to provide the agrosystems with enough N to balance crop removals (Hoagland et al., 2008). Haber-Bosch N fertilizers are not allowed and available organic amendments insufficient for large-scale agriculture or expensive in view of their fertilizing value (Rodrigues et al.,

FIG. 20.11 Cover cropping with self-reseeding early-maturing annual legumes in an olive grove.

FIG. 20.12 Mulching of dead vegetation late in spring from a cover of self-reseeding annual legumes.

FIG. 20.13 Cumulative olive yield after five consecutive harvests from a field trial carried out in rainfed olive grove with three ground management treatments. Error bars are the mean standard deviations.

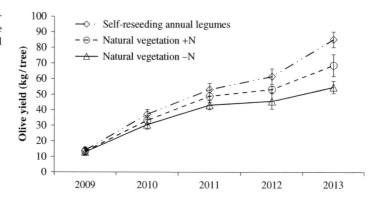

2006). Thus, legume species are a fundamental component of organic farming systems, since they reduce or eliminate the need for external N fertilizers providing they are effectively nodulated to fix atmospheric N. Furthermore, a strategy is needed to combat weeds, and this is also a major challenge in organic crop husbandry because the use of herbicides is not allowed (Niemsdorff and Kristiansen, 2006; Hoagland et al., 2008). Legume cover crops can contribute to meet both objectives.

In organic farming, legume cover crops can be managed in two different ways: cultivated as green manures or as self-reseeding annual legumes. Legume green manures, with the incorporation of the aboveground biomass, have a particularly high potential to introduce N into the soil. They can be recommended for orchards of high yield potential and when water stress is not a problem. The technique requires that the soil be tilled to sow the cover and to incorporate the biomass into the soil. Thus, it should only be recommended to deep soils to reduce the potential damage on the root system and to flat soils where there is a reduced risk of soil erosion. Rodrigues et al. (2013) tried to manage green manures of white lupine and vetch without soil incorporation, by leaving a mulch of dead vegetation on the ground. The authors observed that most of the N was lost without entering the soil, probably due to the volatilization of ammonia from decaying mulch deposited on the ground.

In rainfed orchards, often installed in shallow soils and steep slopes, where the risk of soil erosion is high, tillage is always avoided. The best option probably is cover cropping with early-maturing self-reseeding annual legumes. Although the early-maturing legumes fix less N than late-maturing ones (Rodrigues et al., 2015b), the rainfed managed orchards usually present low N requirements due to the lower yield potential (Rodrigues et al., 2012; Arrobas et al., 2014; Fernández-Escobar et al., 2015).

6 Future research

The management of orchard floor by cover cropping has not been an easy task in rainfed fruticulture. Farmers using cover crops are still rare exceptions. The problem is to keep the level of the tree crop production and the farmer profit when herbaceous vegetation is allowed to compete for resources with the trees. Future research should start from this point; profit is a determinant factor to a new tool of soil management to be adopted. Future work should be focused on the demonstration of the beneficial effects of herbaceous vegetation on soil protection, carbon sequestration, or functional biodiversity, keeping in mind that proposed models of soil management should also ensure fruit yield and quality.

References

Alcántara, C., Soriano, M.A., Saavedra, M., Gómez, J.A., 2017. Sistemas de manejo del suelo. In: Barranco, D., Fernández-Escobar, R., Rallo, L. (Eds.), El Cultivo del Olivo, seventh ed. Mundi-Prensa, Madrid, pp. 335–417.

Almagro, M., de Vente, J., Boix-Fayos, C., García-Franco, N., Aguilar, J.M., González, D., Solé-Benet, A., Martínez-Mena, M., 2016. Sustainable land management practices as providers of several ecosystem services under rainfed Mediterranean agroecosystems. Mitig. Adapt. Strateg. Glob. Chang. 21, 1029–1043.

Ansari, M.F., Tipre, D.R., Dave, S.R., 2015. Efficiency evaluation of commercial liquid biofertilizers for growth of *Cicer arietinum* (chickpea) in pot and field study. Biocatal. Agric. Biotechnol. 4, 17–24.

Arquero, O., Serrano, N., 2013. Manejo dei suelo. In: Arquero, O. (Ed.), Manual del Cultivo dei Almendro. Junta de Andalucía, Consejería de Agricultura, Pescas y Desarrollo Rural, Sevilla.

Arrobas, M., Ferreira, I.Q., Freitas, S., Verdial, J., Rodrigues, M.A., 2014. Guidelines for fertilizer use in vineyards based on nutrient content of grapevine parts. Sci. Hortic. 172, 191–198.

Arrobas, M., Ferreira, I.Q., Claro, A.M., Rodrigues, M.A., 2015. The effect of legume species grown as cover crops in olive orchards on soil phosphorus bioavailability. J. Plant Nutr. 38, 2294–2311.

Bárberi, P., 2006. Tillage: how bad is it in organic agriculture? In: Kristiansen, P., Taji, A., Reganold, J. (Eds.), Organic Agriculture, a Global Perspective. Comstock Publishing Association Press, Ithaca, NY, pp. 295–303.

Berndt, L.A., Wratten, S.D., 2005. Effects of alyssum flowers on the longevity, fecundity, and sex ratio of the leafroller parasitoid *Dolichogenidea tasmanica*. Biol. Control 32, 65–69.

Bhanwariya, B., Manohar Ram, M., Kumawat, N., Kumar, R., 2013. Influence of fertilizer levels and biofertilizers on growth and yield of linseed (*Linum usitatissimum* L.) under rainfed condition of South Gujarat. Madras Agric. J. 100 (4–6), 403–406.

Blanco-Canqui, H., Wienhold, B.J., Jin, V.L., Schmer, M.R., Kibet, L.C., 2017. Long-term tillage impact on soil hydraulic properties. Soil Tillage Res. 170, 38–42.

Bouzas-Cid, Y., Portu, J., Pérez-Álvarez, E.P., Gonzalo-Diago, A., Garde-Cerdán, T., 2016. Effect of vegetal ground cover crops on wine anthocyanin content. Sci. Hortic. 211, 384–390.

Calabrese, G., Perrino, E.V., Ladisa, G., Aly, A., Solomon, M.T., Mazdaric, S., Benedetti, A., Ceglie, F.G., 2015. Short-term effects of different soil management practices on biodiversity and soil quality of Mediterranean ancient olive orchards. Org. Agric. 5, 209–223.

Cañero, A.I., Cox, L., Redondo-Gómez, S., Mateos-Naranjo, E., Hermosín, M.C., Cornejo, J., 2011. Effect of the herbicides terbuthylazine and glyphosate on photosystem II photochemistry of young olive (*Olea europaea*) plants. J. Agric. Food Chem. 59, 5528–5534.

Celis, R., Trigo, C., Facenda, G., Hermosín, M.C., Cornejo, J., 2007. Selective modification of clay minerals for the adsorption of herbicides widely used in olive groves. J. Agric. Food Chem. 55, 6650–6658.

Cockroft, B., Wallbrink, J.C., 1996. Root distribution of orchard trees. Aust. J. Agric. Res. 17, 49–54.

Cooper, J., Scherer, H., 2012. Nitrogen fixation. In: Marschner, P. (Ed.), Marschner's Mineral Nutrition of Higher Plants. Academic Press, London, pp. 229–332.

Cramer, W., Guiot, J., Fader, M., Garrabou, J., Gattuso, J.P., Iglesias, A., Lange, M.A., Lionello, P., Llasat, M.C., Paz, S., Peñuelas, J., Snoussi, M., Toreti, A., Tsimplis, M.N., Xoplaki, E., 2018. Climate change and interconnected risks to sustainable development in the Mediterranean. Nat. Clim. Chang. 8, 972–980.

Cucci, G., Lacolla, G., Crecchio, C., Pascazio, S., Giorgio, D., 2016. Impact of long term soil management practices on the fertility and weed flora of an almond orchard. Turk. J. Agric. For. 40, 194–202.

Damon, P.M., Bowden, B., Rose, T., Rengel, Z., 2014. Crop residue contributions to phosphorus pools in agricultural soils: a review. Soil Biol. Biochem. 74, 127–137.

Daryanto, S., Fu, B., Wang, L., Jacinthe, P.-A., Zhao, W., 2018. Quantitative synthesis on the ecosystem services of cover crops. Earth-Sci. Rev. 185, 357–373.

Driouech, N., Abou Fayad, A., Ghanem, A., Al-Bitar, L., 2008. Agronomic performance of annual self-reseeding legumes and their self-establishment potential in the Apulia region of Italy. In: 16th IFOAM Organic World Congress, Modena, Italy, June, pp. 16–20.

Dube, E., Chiduza, C., Muchaonyerwa, P., 2014. High biomass yielding winter cover crops can improve phosphorus availability in soil. S. Afr. J. Sci 110 (3/4), 1–4.

Duc, N.H., Csintalan, Z., Posta, K., 2018. Arbuscular mycorrhizal fungi mitigate negative effects of combined drought and heat stress on tomato plants. Plant Physiol. Biochem. 132, 297–307.

European Union, 2009. Directive 2009/128/EC of the European Parliament and of the Council of 21 October 2009 establishing a framework for Community action to achieve the sustainable use of pesticides. Off. J. Eur. Commun. L 309, 71–86.

Fernández-Escobar, R., Sánchez-Zamora, M.A., García-Novelo, J.M., Molina-Soria, C., 2015. Nutrient removal from olive trees by fruit yield and pruning. HortScience 50, 474–478.

Ferreira, I.Q., Arrobas, M., Claro, A.M., Rodrigues, M.A., 2013. Soil management in rainfed olive orchards may result in conflicting effects on olive production and soil fertility. Span. J. Agric. Res. 11 (2), 472–480.

Gabriel, J.L., Muñoz-Carpena, R., Quemada, M., 2012. The role of cover crops in irrigated systems: water balance, nitrate leaching and soil mineral nitrogen accumulation. Agric. Ecosyst. Environ. 155, 50–61.

Giese, G., Velasco-Cruz, C., Roberts, L., Heitman, J., Wolf, T.K., 2014. Complete vineyard floor cover crops favorably limit grapevine vegetative growth. Sci. Hortic. 170, 256–266.

Gómez, J.A., Sobrinho, T.A., Giráldez, J.V., Fereres, E., 2009. Soil management effects on runoff, erosion and soil properties in an olive grove of Southern Spain. Soil Tillage Res. 102 (1), 5–13.

Goubanova, K., Li, L., 2007. Extremes in temperature and precipitation around the Mediterranean basin in an ensemble of future climate scenario simulations. Glob. Planet. Chang. 57, 27–42.

Guardia, G., Abalos, D., García-Marco, S., Quemada, M., Alonso-Ayuso, M., Cárdenas, L.M., Dixon, E.R., Vallejo, A., 2016. Effect of cover crops on greenhouse gas emissions in an irrigated field under integrated soil fertility management. Biogeosciences 13, 5245–5257.

Gucci, R., Caruso, G., Bertolla, C., Urbani, S., Taticchi, A., Esposto, S., Servili, M., Sifola, M.I., Pellegrini, S., Pagliai, M., Vignozzi, N., 2012. Changes of soil properties and tree performance induced by soil management in a high-density olive orchard. Eur. J. Agron. 41, 18–27.

Hack, C.M., Porta, M., Schäufele, R., Grimoldi, A.A., 2019. Arbuscular mycorrhiza mediated effects on growth, mineral nutrition and biological nitrogen fixation of *Melilotus alba* Med. in a subtropical grassland soil. Appl. Soil Ecol. 134, 38–44.

Herencia, J.F., 2015. Enzymatic activities under different cover crop management in a Mediterranean olive orchard. Biol. Agric. Hortic. 31, 45–52.

Hoagland, L., Carpenter-Boggs, L., Granatstein, D., Mazzola, M., Smith, J., Peryea, F., Reganold, J.P., 2008. Orchard floor management effects on nitrogen fertility and soil biological activity in a newly established organic apple orchard. Biol. Fertil. Soils 45, 11–18.

Irvin, N.A., Bistline-East, A., Hoddle, M.S., 2016. The effect of an irrigated buckwheat cover crop on grape vine productivity, and beneficial insect and grape pest abundance in southern California. Biol. Control 93, 72–83.

Jackson, D., 2011. Soils, nutrients and water. In: Jackson, D., Looney, N., Morley-Bunker, M., Thiele, G. (Eds.), Temperate and Subtropical Fruit Production, third ed. CAB International, Cambridge, MA, pp. 101–118.

Kallenbach, C.M., Rolston, D.E., Horwath, W.R., 2010. Cover cropping affects soil N_2O and $CO2$ emissions differently depending on type of irrigation. Agric. Ecosyst. Environ. 137, 251–260.

Keesstra, S., Pereira, P., Novara, A., Brevik, E.C., Azorin-Molina, C., Parras-Alcántara, L., Jordán, A., Cerdà, A., 2016. Effects of soil management techniques on soil water erosion in apricot orchards. Sci. Total Environ. 551–552, 357–366.

Lisek, J., Buler, Z., 2018. Growth and yield of plum trees in response to in-row orchard floor management. Turk. J. Agric. For. 42, 97–102.

Lopes, C.M., Monteiro, A., Machado, J.P., Fernandes, N., Araujo, A., 2008. Cover cropping in a slopping non-irrigated vineyard: II—effects on vegetative growth, yield, berry and wine quality of 'Cabernet Sauvignon' grapevines. Cienc. Tec. Vitiviníc. 23, 37–43.

Lopes, C.M., Santos, T.P., Monteiro, A., Rodrigues, M.L., Costa, J.M., Chaves, M.M., 2011. Combining cover cropping with deficit irrigation in a Mediterranean low vigor vineyard. Sci. Hortic. 129, 603–612.

López-Vicente, M., García-Ruiz, R., Guzmán, G., Vicente-Vicente, J.L., Van Wesemael, B., Gómez, J.A., 2016. Temporal stability and patterns of runoff and runon with different cover crops in an olive orchard (SW Andalusia, Spain). Catena 147, 125–137.

Marquez-Garcia, F., Gonzalez-Sanchez, E.J., Castro-Garcia, S., Ordoñez-Fernandez, R., 2013. Improvement of soil carbon sink by cover crops in olive orchards under semiarid conditions. Influence of the type of soil and weed. Span. J. Agric. Res. 11 (2), 335–346.

Martínez, J.R.F., Zuazo, V.H.D., Raya, A.M., 2016. Environmental impact from mountainous olive orchards under different soil-management systems (SE Spain). Sci. Total Environ. 358, 46–60.

Mauromicale, G., Occhipinti, A., Mauro, R.P., 2010. Selection of shade-adapted subterranean clover species for cover cropping in orchards. Agron. Sustain. Dev. 30, 473–480.

Meng, L., Yao, X., Yang, Z., Zhang, R., Zhang, C., Wang, X., Xu, N., Li, S., Liu, T., Zheng, C., 2018. Changes in soil microbial diversity and control of Fusarium oxysporum in continuous cropping cucumber greenhouses following biofumigation. Emir. J. Food Agric. 30 (8), 644–653.

Montanaro, G., Celano, G., Dichio, B., Xiloyannis, C., 2010. Effects of soil-protecting agricultural practices on soil organic carbon and productivity in fruit tree orchards. Land Degrad. Dev. 21, 132–138.

Montanaro, G., Dichio, B., Bati, C.B., Xiloyannis, C., 2012. Soil management affects carbon dynamics and yield in a Mediterranean peach orchard. Agric. Ecosyst. Environ. 161, 46–54.

Montanaro, G., Xiloyannis, C., Nuzzo, V., Dichio, B., 2017. Orchard management, soil organic carbon and ecosystem services in Mediterranean fruit tree crops. Sci. Hortic. 217, 92–101.

Monteiro, A., Lopes, C.M., 2007. Influence of cover crop on water use and performance of vineyard in Mediterranean Portugal. Agric. Ecosyst. Environ. 121, 336–342.

Moreno, B., Garcia-Rodriguez, S., Cañizares, R., Castro, J., Benítez, E., 2009. Rainfed olive farming in south-eastern Spain: long-term effect of soil management on biological indicators of soil quality. Agric. Ecosyst. Environ. 131, 333–339.

Muscas, E., Cocco, A., Mercenaro, L., Cabras, M., Lentini, A., Porqueddu, C., Nieddu, G., 2017. Effects of vineyard floor cover crops on grapevine vigor, yield, and fruit quality, and the development of the vine mealybug under a Mediterranean climate. Agric. Ecosyst. Environ. 237, 203–212.

Nicholls, C.I., Parrella, M.P., Altieri, M.A., 2000. Reducing the abundance of leafhoppers and thrips in a northern California organic vineyard through maintenance of full season floral diversity with summer crops. Agric. For. Entomol. 2, 107–113.

Niemsdorff, P.F., Kristiansen, P., 2006. Crop agronomy in organic agriculture. In: Kristiansen, P., Taji, A., Reganold, J. (Eds.), Organic Agriculture, a Global Perspective. Comstock Publishing Association Press, Ithaca, New York, pp. 53–82.

Ouledali, S., Ennajeh, M., Zrig, A., Gianinazzi, S., Habib Khemira, H., 2018. Estimating the contribution of arbuscular mycorrhizal fungi to drought tolerance of potted olive trees (Olea europaea). Acta Physiol. Plant. 40, 81.

Ozpinar, S., Cay, A., 2006. Effect of different tillage systems on the quality and crop productivity of a clay–loam soil in semi-arid north-western Turkey. Soil Tillage Res. 88, 95–106.

Palese, A.M., Vignozzi, N., Celano, G., Agnelli, A.E., Pagliai, M., Xiloyannis, C., 2014. Influence of soil management on soil physical characteristics and water storage in a mature rainfed olive orchard. Soil Tillage Res. 144, 96–109.

Pastor, M., 2008. Sistemas de manejo del suelo. In: Barranco, D., Fernández-Escobar, R., Rallo, L. (Eds.), El Cultivo del Olivo, sixth ed. Ediciones Mundi-Prensa, Madrid, pp. 239–295.

Patrick-King, A.P., Berry, A.M., 2005. Vineyard d15N, nitrogen and water status in perennial clover and bunch grass cover crop systems of California's central valley. Agric. Ecosyst. Environ. 109, 262–272.

Penman, D., Chapman, B., 2011. Crop protection. In: Jackson, D., Looney, N., Morley-Bunker, M., Thiele, G. (Eds.), Temperate and Subtropical Fruit Production, third ed. CAB International, Cambridge, MA, pp. 119–129.

Ping, X.Y., Wang, T.M., Yao, C.Y., Lu, X.S., 2018. Impact of floor management practices on the growth of groundcover species and soil properties in an apple orchard in northern China. Biol. Rhythm. Res. 49 (4), 597–609.

Pirhofer-Walzl, K., Rasmussen, J., Høgh-Jensen, H., Eriksen, J., Søegaard, K., Rasmussen, J., 2012. Nitrogen transfer from forage legumes to nine neighbouring plants in a multi-species grassland. Plant Soil 350, 71–84.

Ramos, M.E., Benítez, E., García, P.A., Robles, A.B., 2010. Cover crops under different managements vs. frequent tillage in almond orchards in semi-arid conditions: effects on soil quality. Appl. Soil Ecol. 44, 6–14.

Rodrigues, M.A., Arrobas, M., 2017. Manutenção do solo. In: Rodrigues, M.A. (Ed.), Amendoeira: Estado da Produção. CNCFS, pp. 185–231.

Rodrigues, M.A., Cabanas, J.E., 2009. Manutenção do solo. In: Rodrigues, M.A., Correia, C. (Eds.), Manual da Safra e contra Safra do Olival. Instituto Politécnico de Bragança, Portugal, pp. 41–57.

Rodrigues, M.A., Coutinho, J., Martins, F., 2002. Efficacy and limitations of triticale as nitrogen catch crop in a Mediterranean environment. Eur. J. Agron. 17 (3), 155–160.

Rodrigues, M.A., Pereira, A., Cabanas, J.E., Dias, L., Pires, J., Arrobas, M., 2006. Crops use-efficiency of nitrogen from manures permitted in organic farming. Eur. J. Agron. 25, 328–335.

Rodrigues, M.A., Cabanas, J.E., Lopes, J.I., Pavão, F., Aguiar, C., Arrobas, M., 2009. Grau de cobertura do solo e dinâmica da vegetação em olivais de sequeiro com a introdução de herbicidas. Rev. Ciênc. Agrár. XXXII (2), 30–42.

Rodrigues, M.A., Lopes, J.I., Pavão, F.M., Cabanas, J.E., Arrobas, M., 2011. Effect of soil management on olive yield and nutritional status of trees in rainfed orchards. Commun. Soil Sci. Plant Anal. 42, 993–2011.

Rodrigues, M.A., Ferreira, I.Q., Claro, A.M., Arrobas, M., 2012. Fertiliser recommendations for olive based upon nutrients removed in crop and pruning. Sci. Hortic. 142, 205–211.

Rodrigues, M.A., Correia, C.M., Claro, A.M., Ferreira, I.Q., Barbosa, J.C., Moutinho-Pereira, J.M., Bacelar, E.A., Fernandes-Silva, A.A., Arrobas, M., 2013. Soil nitrogen availability in olive orchards after mulching legume cover crop residues. Sci. Hortic. 156, 45–51.

Rodrigues, M.A., Dimande, P., Pereira, E., Ferreira, I.Q., Freitas, S., Correia, C.M., Moutinho-Pereira, J., Arrobas, M., 2015a. Early-maturing annual legumes: an option for cover cropping in rainfed olive orchards. Nutr. Cycl. Agroecosyst. 103, 153–166.

Rodrigues, M.A., Ferreira, I.Q., Freitas, S., Pires, J., Arrobas, M., 2015b. Self-reseeding annual legumes for cover cropping in rainfed managed olive orchards. Span. J. Agric. Res. 13 (2), e0302. 13 pages.

Ruiz-Colmenero, M., Bienes, R., Eldridge, D.J., Marques, M.J., 2013. Vegetation cover reduces erosion and enhances soil organic carbon in a vineyard in the central Spain. Catena 104, 153–160.

Salomé, C., Coll, P., Lardo, E., Metay, A., Villenave, C., Marsden, C., Blanchart, E., Hinsinger, P., Cadre, E., 2016. The soil quality concept as a framework to assess management practices in vulnerable agroecosystems: a case study in Mediterranean vineyards. Ecol. Indic. 61, 456–465.

Sepehr, E., Rengel, Z., Fateh, E., Sadaghiani, M.R., 2012. Differential capacity of wheat cultivars and white lupin to acquire phosphorus from rock phosphate, phytate and soluble phosphorus sources. J. Plant Nutr. 35, 1180–1191.

Silvestri, N., Giannini, V., Antichi, D., 2018. Intercropping cover crops with a poplar short rotation coppice: effects on nutrient uptake and biomass production. Ital. J. Agron. 13, 934.

Singer, M.J., Munns, D.N., 2002. Soil: An Introduction, fifth ed. Prentice Hall, New Jersey.

Sirrine, J.R., Letourneau, D.K., Shennan, C., Sirrine, D., Fouch, R., Jackson, L., Mages, A., 2008. Impacts of groundcover management systems on yield, leaf nutrients, weeds, and arthropods of tart cherry in Michigan, USA. Agric. Ecosyst. Environ. 125, 239–245.

Slatnar, A., Licznar-Malanczuk, M., Mikulic-Petkovsek, M., Stampar, F., Veberic, R., 2014. Long-term experiment with orchard floor management systems: influence on apple yield and chemical composition. J. Agric. Food Chem. 62, 4095–4103.

Snoeck, D., Zapata, F., Domenach, A.-M., 2000. Isotopic evidence of the transfer of nitrogen fixed by legumes to coffee trees. Biotechnol. Agron. Soc. Environ. 4, 95–100.

Sofi, J.A., Dar, I.H., Chesti, M.H., Bisati, I.A., Mir, S.A., Sofi, K.A., 2018. Effect of nitrogen fixing cover crops on fertility of apple (*Malus domestica* Borkh) orchard soils assessed in a chronosequence in North-West Himalaya of Kashmir valley, India. Legum. Res. 41 (1), 87–94.

Soltangheisi, A., Rodrigues, M., Coelho, M.J.A., Gasperini, A.M., Sartor, L.R., Pavinato, P.S., 2018. Changes in soil phosphorus lability promoted by phosphate sources and cover crops. Soil Tillage Res. 179, 20–28.

Torres, M.A.R.-R., Ordóñez-Fernández, R., Giráldez, J.V., Márquez-García, J., Laguna, A., Carbonell-Bojollo, R., 2018. Efficiency of four different seeded plants and native vegetation as cover crops in the control of soil and carbon losses by water erosion in olive orchards. Land Degrad. Dev. 29, 2278–2290.

Tribouillois, H., Constantin, J., Justes, E., 2017. Cover crops mitigate direct greenhouse gases balance but reduce drainage under climate change scenarios in temperate climate with dry summers. Glob. Chang. Biol. 24, 2513–2529.

Velarde, F.G.-A., 2015. El cultivo de las plantaciones frutales. Ediciones Mundi-Prensa, Madrid.

Veneklaas, E.J., Stevens, J., Cawthray, G.R., Turner, S., Grigg, A.M., Lambers, H., 2003. Chickpea and white lupin rhizosphere carboxylates vary with soil properties and enhance phosphorus uptake. Plant Soil 248, 187–197.

Wakclin, A.M., Lorraine-Colwill, D.F., Preston, C., 2004. Glyphosate resistance in four different populations of *Lolium rigidum* in associated reduced translocation of glyphosate to meristematic zones. Weed Res. 44, 453–459.

Wang, P., Liu, J.-H., Xia, R.-X., Wu, Q.-S., Wang, M.-Y., Dong, T., 2011. Arbuscular mycorrhizal development, glomalin-related soil protein (GRSP) content, and rhizospheric phosphatase activity in citrus orchards under different types of soil management. J. Plant Nutr. Soil Sci. 174, 65–72.

Winkler, A.J., Cook, J.A., Kliewer, W.M., Lider, L.A., 1974. General Viticulture. University of California Press, Los Angeles, CA.

Yao, S., Merwin, I.A., Bird, G.W., Abawi, G.S., Thies, J.E., 2005. Orchard floor management practices that maintain vegetative or biomass groundcover stimulate soil microbial activity and alter soil microbial community composition. Plant Soil 271, 377–389.

Zhu, X., Song, F., Liu, S., Liu, F., Li, X., 2018. Arbuscular mycorrhiza enhances nutrient accumulation in wheat exposed to elevated CO2 and soil salinity. J. Plant Nutr. Soil Sci. 181, 836–846.

Zimdahl, R., 1993. Fundamentals of Weed Science. Academic Press, California.

21

Rootstock: Scion combinations and nutrient uptake in grapevines

*Antonio Ibacache, Nicolás Verdugo-Vásquez, Andrés Zurita-Silva**

Instituto de Investigaciones Agropecuarias (INIA), Centro Regional de Investigación Intihuasi, La Serena, Chile

*Corresponding author. E-mail: andres.zurita@inia.cl

1 Introduction

The use of rootstocks in viticulture was initially intended to overcome phylloxera infections that infected European vineyards in the late 19th century (Corso and Bonghi, 2004). However, the recognition of the benefits of rootstocks has since expanded to include nematode control, nutrient absorption, water uptake, vine vigor, yield, and fruit quality (Satisha et al., 2010; Walker et al., 2000; Keller, 2001; Tambe and Gawade, 2004; Ibacache and Sierra, 2009; Ibacache et al., 2016). Also, in some arid or semiarid areas, rootstocks are used to replace old or unproductive vineyards (Ibacache and Sierra, 2009; Satisha et al., 2010).

Most of the world viticulture is based on grafted grapevines, where the scion is a cultivar of *Vitis vinifera* and the rootstock is either an American *Vitis* species or an interspecific *Vitis* hybrid. The different effects of the rootstocks on scions take place in a more or less indirect manner and are consequences of interactions between environmental factors and the physiology of both scion and rootstock cultivars employed.

Grapevine cultivars are known to exhibit wide differences in mineral nutrient status. Likewise, the use of rootstocks with the resistance and tolerance to phylloxera and/or nematodes can also have a major influence on the mineral nutrient status of scion cultivar (Garcia et al., 2001; Bavaresco et al., 2003; Fisarakis et al., 2004; Ibacache and Sierra, 2009). This influence has important implications for decisions involving soil adaptability, grapevine fertilizer requirements, canopy management, yield, and fruit quality. Rootstocks are also known to affect the uptake of mineral nutrients that can be damaging to grapevines, including sodium, chloride, and boron (Stevens and Harvey, 1995; Walker et al., 2004). This should be considered when choosing a rootstock for soils, which have potentially damaging levels of salts.

Although information about the interactions among the cultivar, the rootstock, and nutrient content in grapevines is reduced, several studies have shown that rootstocks differ in their effect on the nutrient levels in the grafted cultivar

A.K. Srivastava, Chengxiao Hu (eds.)
Fruit Crops: Diagnosis and Management of Nutrient Constraints
https://doi.org/10.1016/B978-0-12-818732-6.00021-6

(Grant and Matthews, 1996; Nikolaou et al., 2000; Garcia et al., 2001; Bavaresco et al., 2003; Fisarakis et al., 2004; Robinson, 2005).

To select the type of rootstock to be used in a specific edaphoclimatic condition, it is important to carry out local long-term studies, given the interactions that exist among the rootstock, the cultivar, and the environment. This implies that the results obtained with a particular cultivar-rootstock combination in a particular environment cannot be extrapolated to other situations (Keller et al., 2001).

2 Characteristics of rootstocks associated with the uptake of water and nutrients

Root system is the interface between the grapevine and the soil. It provides anchorage for grapevine in the soil and is responsible for water uptake and nutrient mining, therefore enabling combination of genotypes and cultivated scion to express the productive potential and attributes of the target cultivar. The roots also serve as storage organs for carbohydrates and other nutrients, which support the initial growth of scion and roots in spring, and for water. In addition, they are a source of plant hormones (cytokinins and ABA), which modify shoot physiology. Commonly used rootstocks are either individual *Vitis* species or crosses of two or more species, which harbors traits of agronomic relevance in grapevine rootstocks for grapevine production (Table 21.1, adapted after Keller, 2015). The practice of rootstocks on grapevines is a technological tool widely used, where about 10 rootstocks occupy 90% of grafted grapevines in the world (Keller, 2015).

Despite this limited genetic diversity, new rootstock breeding programs have been developed in recent years with the aim of not only expanding frontiers of grapevine production, using marginal soils, dealing with biotic constrains (root parasites, such as phylloxera and nematodes), and higher efficiency in the use of water and nutrients (Tortosa et al., 2016; Bianchi et al., 2018; Romero et al., 2018), but also for their ability to influence crop maturity or their tolerance of adverse abiotic soil conditions such as waterlogging, lime, acid or saline soils, and excess of toxic mineral elements (Walker et al., 2018). Indeed, a major goal is to develop rootstocks that can influence scion growth and productivity under drought, particularly those that can increase water conservation through reducing the need for irrigation while ameliorating negative impacts on yields (Zhang et al., 2016). Therefore, grapevine is grown in a great diversity of geographical areas in the world, with a wide variety of cultivars and productive purposes (table grapes, wine, spirits, raisins, etc.).

Notwithstanding grafting does not directly affect the quality-relevant traits of the grapevines produced by the scion cultivar, because the rootstock to which a scion cultivar is grafted may alter water and nutrient uptake and distribution, plant growth, and yield formation, it seems logical to expect the rootstock to influence fruit composition too (Keller, 2015). Thus, an indirect effect on fruit composition, especially on acidity, is possible due to the potential influence of the rootstock on scion vigor, canopy configuration, and yield components (Keller et al., 2001; Ruhl and Alleweldt, 1990). Nevertheless, the rootstock may alter amino acids, especially arginine, in the berries of its grafting partner. Reports had determined that 140 Ruggeri and 101-14 Mgt may sometimes lead to considerably lower amino acid concentrations in fruits of their grafting partner than some other rootstocks and own-rooted vines (Treeby et al., 1998).

3 Rootstocks differences on the uptake of macro and micronutrients

The vineyard fertilization is a key management within the productive cycle, determining the yield and quality of the production. To carry out an adequate fertilization of macro- and micronutrients, information about the soil and climatic conditions, quality of the irrigation water, and the combination of cultivar rootstock should be considered. As already mentioned, rootstocks differ in the capacity to capture water and nutrients, and this can be affected by the cultivar used and environmental conditions (Ibacache and Sierra, 2009; Pachnowska and Ochmian, 2018).

Although there are rootstock comparative tables in the literature and their effect on the absorption of few nutrients (Table 21.1; Keller, 2015), the aim is to present an update of the recent studies, which have been carried out in new geographical areas and the incorporation of new rootstocks and cultivars. This information can serve as a guide to establish guidelines for fertilization, considering the rootstock employed, increasing efficiency in the fertilizer use, reducing costs, and avoiding environmental contamination.

TABLE 21.1 Traits of agronomic relevance in grapevine rootstocks.

Rootstock	Parent species	Resistance				Tolerance						Susceptibility		Influenced traits			
		Phyll	Nem	CGa	Phyt	Droug	Lime	Salty	AcidS	ClayS	SandS	Mg-Def	K-Def	SciV	Mat	B-Graft	RootA
Riparia Gloire	*V. riparia*	E	P			P	L	L		L	P	N		L	A		M
Rupestris St. George	*V. rupestris*	H	P		L	P	P	L	P	L	L		N	H	D	P	M
Rupestris du Lot	*V. rupestris*		M			M	P	L		P	L			H	D	E	
420A Millardet et de Grasset	*V. berlandieri × V. riparia*	H	P		L	P	M			M	P		Y	P	D	P	L
5BB Kober	*V. berlandieri × V. riparia*	H	P			L	M	P		P	L	Y	Y	P	D	P	P
SO4	*V. berlandieri × V. riparia*	H	P	P	L	L	P	L	M	P	L	Y	N	M		P	P
8B	*V. berlandieri × V. riparia*	H	P			P	P							M		L	
5C Teleki	*V. berlandieri × V. riparia*	H	H			L	P	P		M	P			H	A	P	P
161-49 Couderc	*V. berlandieri × V. riparia*	E	P			L	M	L		P	P			M		L	L
99 Richter	*V. berlandieri × V. rupestris*	H	P	L		M	P	L	L	M	P	Y	Y	H	D	E	P
110 Richter	*V. berlandieri × V. rupestris*	H	P	L	L	E	P	P	M	M	M	Y	Y	M	D	H	L
1103 Paulsen	*V. berlandieri × V. rupestris*	H	P	P	P	M	P	H	P	M	M	N	Y	M	D	M	M
140 Ruggeri	*V. berlandieri × V. rupestris*	H	L	L	P	H	M	H	H	M	H	N	Y	H	D	M	P

Continued

TABLE 21.1 Traits of agronomic relevance in grapevine rootstocks.—cont'd

Rootstock	Parent species	Resistance				Tolerance						Susceptibility		Influenced traits			
		Phyll	Nem	CGa	Phyt	Droug	Lime	Salty	AcidS	ClayS	SandS	Mg-Def	K-Def	SciV	Mat	B-Graft	RootA
44-53 Malègue	*V. riparia × V. rupestris*	H	M			H	L		M	M	L	Y	N	M	A	E	H
3309 Couderc	*V. riparia × V. rupestris*	H	L	H	L	P	L	L		P	P	N		P	A	P	M
101-14 Millardet et de Grasset	*V. riparia × V. rupestris*	H	P	H	P	P	L	M	M	P	L			M	A	P	M
Schwarzmann	*V. riparia × V. rupestris*	H	M			P	P	M						M		H	M
Gravesac	*V. berl. × V. rip. × V. rupestris*	H	L			M	P		H					P		M	M
1616 Couderc	*V. solonis × V. riparia*	M	L		L	M	L	M		L	P			P	A	L	M
Salt Creek (Ramsey)	*V. champini*	P	H	L	H	P		H	L	L	H			H		L	L
DogRidge	*V. champini*	P	H		H	P		H		L	H			H		L	L
Harmony	*V. champini × V. sol. × V. riparia*	P	H	H		P		H		L	P			P		M	H
Freedom	*V. champini × V. sol. × V. riparia*	P	H	H		P		H		L	M			P		M	M

Scale: *A*, advanced; *D*, delayed; *E*, excellent; *H*, high; *L*, low; *M*, intermediate; *N*, absent; *P*, poor; *Y*, present.
AcidS, acid soil; *B-Graft*, bench grafting; *CGa*, crown gall; *ClayS*, clay soil; *Droug*, drought; *K-Def*, K deficiency; *Lime*, lime; *Mat*, maturation; *Mg-Def*, Mg deficiency; *Nem*, nematode; *Phyll*, phylloxera; *Phyt*, phytophthora; *RootA*, rooting ability; *Salty*, salinity; *SandS*, sandy soil; *SciV*, scion vigor.
Modified from Keller, M., 2015. The Science of Grapevines, second ed. Elsevier. 491 p.

3.1 Macronutrients

Nitrogen is the most abundant soil-derived macronutrient in grapevine, influencing plant growth and playing a major role in most of the biological functions and processes of both grapevine and fermentative microorganisms (Bell and Henschke, 2005). On the other hand, the nutritional status of the grapevine with respect to nitrogen is associated with the quality of wine or must and physiological disorders, mainly due to problems of vegetative balance. Viticultural practices aimed at reaching nitrogen requirements are of special relevance, because interactions between rootstock and vineyard supply strongly influence scion mineral nutrient status and shoot vigor and, via those processes, fruit composition (Holzapfel and Treeby, 2007). In this regard, it is important to carry out an adequate nitrogen fertilization to each productive system (table grapes, wine, etc.), considering from the vineyard design the rootstock to stablish. Table 21.2 shows rootstock-scion combinations that have displayed a higher absorption of nitrogen and phosphorus, compared with other rootstocks in each cited study.

In addition, the rootstock parentals and type of study and tissue analyzed are mentioned. There are nine rootstocks that presented a higher nitrogen absorption compared with own-rooted plants and other rootstocks (statistically significant), according to field studies that have been conducted in the last 18 years. Frequently, more vigorous rootstocks result in higher N levels on the grafted variety. Thus, when a cultivar is grafted onto rootstocks that exhibit high nitrogen uptake, fertilization programs for this element should be adjusted. Excessive applications of nitrogen can have negative effects on the environment (contaminating ground water with nitrates) and increasing pest and diseases damage, physiological disorders, and reducing fruit quality (Keller et al., 2001).

Phosphorus (P) is considered as a major limiting factor of crop production. Phosphorus deficiency is currently mitigated by P fertilizer application. However, P fertilizers are mainly produced from nonrenewable phosphate rock, and concerns have been expressed that this natural resource will be exhausted in the near future (Gautier et al., 2018). In Table 21.2, seven rootstocks that showed a higher absorption of phosphorus compared with own-rooted plants and other rootstocks are shown (statistically significant); thus, most of studies were carried out in the field. It is important to underline that some authors showed that own-rooted plants have a higher absorption of this element with respect to rootstocks (Vijaya and Rao, 2015).

Information regarding the macronutrients, potassium (K), magnesium (Mg), and calcium (Ca), is presented in Table 21.3. Potassium corresponds to major cation in grape juice (Mpelasoka et al., 2003). A high concentration of K in grape juice can lead to high juice pH (e.g., >3.8) and, in turn, lower quality wines, for example, reduced color stability and poor taste (Kodur, 2011). Potential vineyard management options to manipulate berry K accumulation include selective use of rootstock/scion combination, canopy management, and irrigation strategies. With respect to rootstocks, the use of low K accumulating rootstocks can be an alternative to avoid quality problems in wine, associated with a high K concentration. Sixteen rootstocks that have shown a higher K absorption compared with own-rooted plants and other rootstocks are shown (Table 21.3).

It is important to remark that potassium-magnesium and potassium-calcium cations may present antagonism in their uptake, and therefore, rootstocks with a high capacity to absorb potassium may have a low absorption of Mg-Ca or vice versa. This aspect should be considered since a normal fertilization, without differentiating by established rootstock can increase deficiencies of magnesium and/or calcium. For example, SO4 rootstock presents a high absorption of K, but low Mg and Ca (Table 21.3). In this fashion, it is important to consider the balance between these cations to avoid deficiencies. According to Delas and Pouget (1979), an adequate range for K/Mg in grape petioles at veraison is between 3 and 7. Values below 1 indicate K deficiency and above 10, Mg deficiency.

Leaf symptoms of cv. Muscat of Alexandria grafted on 110 Richter with magnesium deficiency are shown (Fig. 21.1). The leaves with symptoms had 45% less petiole magnesium content than normal leaves. A yellowing of the basal leaves is observed, keeping the green vein (intervening chlorosis) together with burns in the leaf margin (Fig. 21.1). On the other hand, magnesium deficiency is related to the development of the physiological disorder called "bunch stem necrosis" that may develop during ripening (Keller, 2015).

Information regarding micronutrients is presented (Table 21.4). Micronutrients such as zinc, boron, and molybdenum and the macronutrient calcium are essential for the process of pollination and fertilization in grapevines. Kidman et al. (2014) found that rootstocks affect the sequestration of nutrients, which affected reproductive performance on the cv. Syrah. Specifically, Kidman et al. (2014) showed that 1103 Paulsen rootstock had a significantly higher amount of boron and a lower number of seedless berries and a lower millerandage index. On the other hand, Zn deficiency was observed for 110 Richter and 140 Ruggeri. Finally, it is important to remark that tables presented should be used with caution, since site-cultivar-specific responses can be generated in each study. Therefore, the results may not be strictly extrapolated to other edaphic-climatic conditions or cultivars.

TABLE 21.2 Rootstock–scion combination that has presented a higher absorption of nitrogen and phosphorus, compared with other rootstocks in each mentioned study.

Nutrient	Rootstock	Parentage	Cultivars	Reference	Type of study/analysis
Nitrogen (N)	- Salt creek[a]	- *V. champini*	- Flame Seedless - Thompson Seedless - Superior Seedless - Red Globe	Ibacache and Sierra (2009)	Field/petioles at flowering

	- 420A
	- 110 Richter[b]
	- 1103 Paulsen[b]
	- 99 Richter[b]
		- *V. riparia* ×
		  V. berlandieri
		- *V. berlandieri* ×
		  V. rupestris
			- Thompson Seedless	Nikolaou et al. (2000)	Field/blades at veraison

	- 161-49 C
	- Sori
		- *V. riparia* ×
		  V. berlandieri
		- *V. solonis* ×
		  V. riparia
			- Regent	Pachnowska and Ochmian (2018)	Field/leaves and berries at fruit set

	- 140 Ruggeri	- *V. berlandieri* × *V. rupestris*	- Chardonnay - Pinot Noir	Wooldridge et al. (2010)	Field/petioles at fruit set
	- IAC 572	-*V. caribaea* × 101-14	- Niagara Rosada	Tecchio et al. (2014)	Field/branches at pruning
Phosphorus (P)	- Salt Creek	- *V. champini*	- Flame Seedless - Thompson Seedless - Superior Seedless - Red Globe	Ibacache and Sierra (2009)	Field/petioles at flowering

	- 125 A Kober
	- Börner
		- *V. berlandieri* ×
		  V. riparia
		- *V. riparia* 183
		  GM ×
		  V. cinerea
			- Regent	Pachnowska and Ochmian (2018)	Field/leaves at fruit set

	- Harmony	- *Couderc 1613* × *V. champini*	- Pinot Noir	Candolfi and Castagnoli (1997)	Field/petioles at flowering
			- Thompson Seedless	Morales et al. (2014)	Glasshouse/petioles at veraison
	- R99	- *V. berlandieri* × *V. rupestris*	- Niagara Rosada - Concord	Dalbó et al. (2011)	Field/petioles at veraison
	- 140 Ruggeri	- *V. berlandieri* × *V. rupestris*	- Chardonnay - Pinot Noir	Wooldridge et al. (2010)	Field/petioles at fruit set
	- IAC 572	- *V. caribaea* × 101-14	- Niagara Rosada	Tecchio et al. (2014)	Field/branches at pruning

[a] *Syn. Ramsey.*
[b] *Corresponding to V. berlandieri × V. rupestris.*

4 Rootstocks differences in chloride (Cl), sodium (Na), and boron (B) accumulation

High yields and grapevine quality are best achieved with soils and irrigation water that have optimal levels of salinity and the correct composition of salts. In regions with lower rainfall, where leaching of soluble salts is often incomplete, soil salinity can be a serious constraint to grape production. Field problems often associated with salinity include decreased soil-water availability, and accumulation of specific elements like chloride, sodium and boron that lead to toxic levels in plant tissues. Grapevines are classified as moderately sensitive to salinity (Walker, 1994) and sensitive to boron excess (Christensen et al., 1978).

Depending on concentration, salt accumulation in the medium may lead to a poor growth and yield performance of the vines (Walker et al., 1997, 2014; Zhang et al., 2002), and high concentrations may even cause death of plants

TABLE 21.3 Rootstock-scion combination that has presented a higher absorption of potassium, magnesium, and calcium, compared with other rootstocks in each mentioned study.

Nutrient	Rootstock	Parentage	Cultivars	Reference	Type of study/analysis
Potassium (K)	- Harmony - 1613C	- *Couderc 1613 × V. champini* - (*V. solonis × V. vinifera*) × (*V. labrusca × V. riparia*)	- Flame Seedless - Thompson Seedless - Red Globe	Ibacache and Sierra (2009)	Field/petioles at flowering
	- 420A - 110 Richter[a] - 1103 Paulsen[a] - 99 Richter[a]	- *V. riparia × V. berlandieri* - *V. berlandieri × V. rupestris*	- Thompson Seedless	Nikolaou et al. (2000)	Field/petioles at veraison
	- Dogridge	- *V. champini*	- Thompson Seedless	Vijaya and Rao (2015)	Field/petioles at flowering
	- SO4[b]	- *V. berlandieri × V. riparia*	- Négrette	Garcia et al. (2001)	Greenhouse/blades at veraison
			- Chardonnay - Pinot Noir	Wooldridge et al. (2010)	Field/petioles at fruit set
	- Sori	- *V. solonis × V. riparia*	- Regent	Pachnowska and Ochmian (2018)	Field/leaves and berries at fruit set
	- 44-53[c]	- *Riparia grand glabre × 144M*	- Pinot Noir	Candolfi and Castagnoli (1997)	Field/petioles at flowering
	- 140 Ruggeri - 101-14 - Saint George	- *V. berlandieri × V. rupestris* - *V. riparia × V. rupestris* - *V. rupestris*	- Chardonnay - Shiraz	Walker and Blackmore (2012)	Field/petioles at flowering
	- 043-43[c]	- *V. vinifera × V. rotundifolia*	- Niagara Rosada - Concord	Dalbó et al. (2011)	Field/petioles at veraison
	- IAC 766	- *V. caribaea × Traviú (106-8)*	- Niagara Rosada	Dalbó et al. (2011)	Field/petioles at veraison
	- IAC 572	- *V. caribaea × 101-14*		Tecchio et al. (2014)	Field/branches at pruning
Magnesium (Mg)	- 3309 C	- *V. riparia × V. rupestris*	- Négrette	Garcia et al. (2001)	Greenhouse/blades at veraison
	- Sori - 125A Kober - Börner - 5BB Kober	- *V. solonis × V. riparia* - *V. berlandieri × V. riparia* - *V. riparia 183 GM × V. cinerea* - *V. riparia × V. berlandieri*	- Regent	Pachnowska and Ochmian (2018)	Field/leaves at fruit set
	- Harmony	- *Couderc 1613 × V. champini*	- Pinot Noir	Candolfi and Castagnoli (1997)	Field/petioles at flowering
	- 140 Ruggeri	- *V. berlandieri × V. rupestris*	- Chardonnay - Pinot Noir	Wooldridge et al. (2010)	Field/petioles at fruit set
	- -IAC 572	- *V. caribaea × 101-14*	- Niagara Rosada	Tecchio et al. (2014)	Field/branches at pruning
	- Isabella[d]	- *V. labrusca*	- Niagara Rosada	Dalbó et al. (2011)	Field/petioles at veraison
	- IAC766 - 420-A[d]	- *V. caribaea × Traviú (106-8)* - *V. berlandieri × V. riparia*	- Concord		
	- -420A - 110 Richter[a] - 1103 Paulsen[a] - 99 Richter[a]	- *V. riparia × V. berlandieri* - *V. berlandieri × V. rupestris*	- Thompson Seedless	Nikolaou et al. (2000)	Field/petioles at veraison
Calcium (Ca)	- 3309 C	- *Riparia tomenteuse × Rupestris Martin*	- Négrette	Garcia et al. (2001)	Greenhouse/blades at veraison
	- Sori	- *V. solonis × V. riparia*	- Regent		Field/leaves at fruit set

Continued

TABLE 21.3 Rootstock-scion combination that has presented a higher absorption of potassium, magnesium, and calcium, compared with other rootstocks in each mentioned study.—cont'd

Nutrient	Rootstock	Parentage	Cultivars	Reference	Type of study/analysis
	- 125A Kober - Börner - 5BB Kober - 161-49 C	- *V. berlandieri* × *V. riparia* - *V. riparia* 183 GM × *V. cinerea* - *V. riparia* × *V. berlandieri* - *V. riparia* × *V. berlandieri*		Pachnowska and Ochmian (2018)	
	- 140 Ruggeri	- *V. berlandieri* × *V. rupestris*	- Chardonnay - Pinot Noir	Wooldridge et al. (2010)	Field/petioles at fruit set
	- IAC 572	- *V. caribaea* × 101-14	- Niagara Rosada	Tecchio et al. (2014)	Field/branches at pruning
	- 420A - 110 Richter[a] - 1103 Paulsen[a] - 99 Richter[a]	- *V. riparia* × *V. berlandieri* - *V. berlandieri* × *V. rupestris*	- Thompson Seedless	-Nikolaou et al. (2000)	Field/blades at veraison

[a] *Corresponding to* V. berlandieri × V. rupestris.
[b] *Less magnesium and calcium absorption were recorded.*
[c] *Less magnesium absorption was recorded.*
[d] *Less potassium absorption was recorded.*

(Troncoso et al., 1999). Physiological effects to salinity exposure in grapevine include reduced stomatal conductance and photosynthesis systemic disturbances, which can lead to reductions in growth, biomass accumulation, and yield (Downton, 1977a; Ben-Asher et al., 2006; Walker et al., 2004). The inhibition of grapevine growth and CO_2 assimilation by high salinity is mainly due to changes in stomatal conductance, electron transport rate, leaf water potential, chlorophyll, fluorescence, osmotic potential, and leaf ion concentration (Stevens and Harvey, 1995; Cramer et al., 2007).

Among the strategies adopted for sustaining growth and productivity of vine cultivars under salinity, the use of tolerant rootstocks is widely accepted (Walker, 1994). Moreover, in areas affected by salinity constrains one of the more important uses of rootstocks is for modifying fruit composition, because rootstocks can exclude much of the salt dissolved in the soil solution from root uptake and xylem transport, the scions grafted to them accumulate less Na^+ and Cl^- in the fruit (Walker et al., 2000, 2010).

Susceptibility or rootstock tolerance to high salinity is a coordinated action of multiple factors. Salinity tolerance mediated by rootstocks is attributed to root system restricting the movement and/or limiting absorption and accumulation of toxic ions from saline soils (Hepaksoy et al., 2006; Walker et al., 2002). Fisarakis et al. (2001) showed that there is a great variability in the uptake and accumulation of Na^+ and Cl^- exclusion among rootstocks. Specifically, they demonstrated that *Vitis berlandieri* species had a great ability for Cl^- and/or Na^+ exclusion, although this ability is reduced in hybrids having *V. vinifera* as parent. On the other hand, Walker et al. (2004) working with field-grown vines concluded that a high innate vigor of a rootstock combined with moderate to high chloride and sodium exclusion ability represents the best combination for salt tolerance in Sultana grapevines as measured by yield at moderate to high salinity.

4.1 Chloride toxicity

The visible symptoms that develop on vines growing under saline conditions are, in most cases, due to accumulation of chloride toxic concentrations in leaves. The symptoms appear first as marginal chlorosis on leaves, followed by necrosis developing inward from leaf margins. This effect is frequently referred to as "leaf burn" (Walker, 1994). Petiole Cl concentrations that exceed 1.0% in spring are considered excessive (Robinson, 2005).

Petiole analysis has been shown to provide a good tool to assess Cl concentration in Cl-stressed vines (Christensen et al., 1978). In a long-term field study carried out under semiarid conditions of northern Chile, we determined the Cl, Na, and B petiole concentration in two cultivars, Flame Seedless (table grape) and Muscat of Alexandria (syn. Moscatel de Alejandría and Muscat Gordo *Blanco*), a cultivar that is used to produce a distilled spirit called Pisco in Chile. Both varieties were grafted onto 10 different rootstocks, 1613 Couderc, Freedom, Harmony, 1103 Paulsen, 110 Richter, 99 Richter, 140 Ruggeri, SO4, Salt Creek (syn. Ramsey), and Saint George, which were compared with own-rooted vines. The study was carried out in a slightly saline soil (electrical conductivity 1.0 dS/m in saturated paste) located at the Vicuña Experimental Center (30°02′S, 70°44′W) of the Instituto de Investigaciones Agropecuarias (INIA).

FIG. 21.1 Leaf of Muscat of Alexandria grafted on 110 Richter with magnesium deficiency.

The rootstock effect on chloride levels in Flame Seedless and Muscat of Alexandria cultivars are shown (Figs. 21.2 and 21.3), respectively. Own-rooted vines accumulated the highest amount of petiole chloride compared with grafted vines in both cultivars. Among the rootstocks, Harmony accumulated more chloride than the others rootstock studied.

The higher chloride concentration in petioles at full flowering stage of cv. Flame Seedless and cv. Muscat of Alexandria on own roots, relative to cultivars on the various rootstocks, reflects the poor capacity of *V. vinifera* vines for chloride exclusion (Downton, 1977). A number of studies have been undertaken comparing chloride exclusion ability of grapevine species and varieties (Walker, 1994). Downton (1977) ranked them as follows: *Vitis rupestris* < *V. berlandieri*, *Vitis riparia* < *Vitis champini* < *V. vinifera*. Also, Fisarakis et al. (2001) demonstrated that *V. berlandieri* species had a greater ability for chloride exclusion than hybrids having *V. vinifera* as parent. In our study, *V. berlandieri* × *V. rupestris* rootstocks, 99 Richter, 110 Richter, 140 Ruggeri, and 1103 Paulsen were all comparable with *V. rupestris* Saint George rootstock performance. According to Walker et al. (2010), 1103 Paulsen and 140 Ruggeri rootstocks were among the lowest petiole chloride concentration in cv. Shiraz at full flowering. 140 Ruggeri is also mentioned as a good chloride excluder (Walker et al., 2018). The greater chloride exclusion capacity of this rootstock appears to be associated with restricted entry of Cl to xylem and lower root-to-shoot Cl transport (Tregeagle et al., 2010).

While there is uncertainty as to whether high chloride accumulation by certain cultivars is linked to growth reductions, there is a general agreement that grapevine chloride content is an important factor for vine health under saline conditions (Walker et al., 2004). Chloride exclusion has subsequently been regarded as an important rootstock trait and therefore commonly used as a screening test for salt tolerance in grapevine rootstock breeding programs (Antcliff et al., 1983; Newman and Antcliff, 1984; Sykes, 1985; Walker, 1994). Studies on grapevine salt tolerance have shown that high uptake and root-to-shoot transport of chloride, resulting in excessive accumulation in leaf tissues, is a major factor in impaired leaf function and damage (Walker et al., 1997). Chloride-excluding rootstocks have similarly been thought to protect against excessive chloride accumulation and therefore contribute to salt tolerance (Walker et al., 2002).

TABLE 21.4 Rootstock-scion combination that has presented a higher absorption of micronutrients, compared with other rootstocks in each mentioned study.

Nutrient	Rootstock	Parentage	Cultivars	Reference	Type of study/analysis
Iron (Fe)	- 161-49 C	- *V. riparia* × *V. berlandieri*	- Regent	Pachnowska and Ochmian (2018)	Field/leaves at fruit set
	- 41B - 8B Teleki	- *V. vinifera* × *V. berlandieri* - *V. riparia* × *V. berlandieri*	- Thompson Seedless	Nikolaou et al. (2000)	Field/petioles at veraison
	- 1613 C - Salt Creek	- (*V. solonis* × *V. vinifera*) × (*V. labrusca* × *V. riparia*) - *V. champini*		Vijaya and Rao (2015)	Field/petioles at flowering
Zinc (Zn)	- 420A - 110 Richter[a] - 1103 Paulsen[a] - 99 Richter[a]	- *V. riparia* × *V. berlandieri* - *V. berlandieri* × *V. rupestris*	- Thompson Seedless	Nikolaou et al. (2000)	Field/blades and petioles at veraison
	- Sori - Börner - 5BB Kober	- *V. solonis* × *V. riparia* - *V. riparia* 183 GM × *V. cinerea* - *V. riparia* × *V. berlandieri*	- Regent	Pachnowska and Ochmian (2018)	Field/leaves and berries at fruit set
	- IAC 572	- *V. caribaea* × 101-14	- Niagara Rosada	Tecchio et al. (2014)	Field/branches at pruning
Copper (Cu)	- 420A - 110 Richter[a] - 1103 Paulsen[a] - 99 Richter[a]	- *V. riparia* × *V. berlandieri* - *V. berlandieri* × *V. rupestris*	- Thompson Seedless	Nikolaou et al. (2000)	Field/petioles at veraison
	- 125AA Kober - 5BB Kober	- *V. berlandieri* × *V. riparia* - *V. riparia* × *V. berlandieri*	- Regent	Pachnowska and Ochmian (2018)	Field/leaves at fruit set
Boron (B)	- 44-53 - 3309 C	- *Riparia grand glabre* × 144M - *V. riparia* × *V. rupestris*	- Pinot Noir	Candolfi and Castagnoli (1997)	Field/petioles at flowering
	- 1103 Paulsen - 110 Richter	- *V. berlandieri* × *V. rupestris*	- Shiraz	Kidman et al. (2014)	Field/petioles at flowering
	- IAC 572	- *V. caribaea* × 101-14	- Niagara Rosada	Tecchio et al. (2014)	Field/branches at pruning
Manganese (Mn)	- 41B - 8B Teleki	- *V. vinifera* × *V. berlandieri* - *V. riparia* × *V. berlandieri*	- Thompson Seedless	Nikolaou et al. (2000)	Field/blades and petioles at veraison
	- 1613 C	- (*V. solonis* × *V. vinifera*) × (*V. labrusca* × *V. riparia*)		Vijaya and Rao (2015)	Field/petioles at flowering
	- 5BB Kober	- *V. riparia* × *V. berlandieri*	- Regent	Pachnowska and Ochmian (2018)	Field/leaves at fruit set

[a] *Corresponding to V. berlandieri × V. rupestris.*

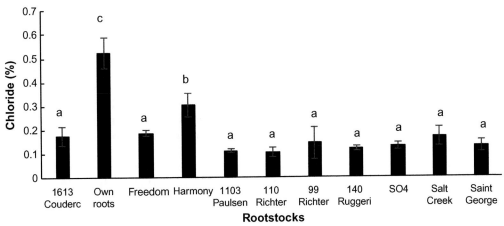

FIG. 21.2 Chloride concentration in petioles from cv. Flame Seedless on various rootstocks. Each bar represents the average of 4 years (2006–09) standard error. *Different letters* denote significant differences (*P* < .05).

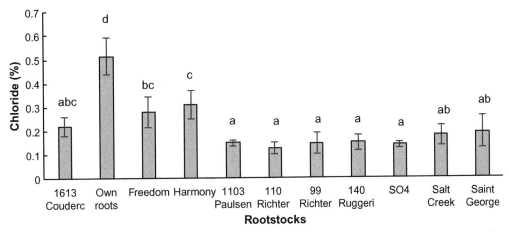

FIG. 21.3 Chloride concentration in petioles from cv. Muscat of Alexandria on various rootstocks. Each bar represents the average of 4 years (2006–09) standard error. *Different letters* denote significant differences (*P* < .05).

4.2 Sodium toxicity

The principal effects of sodium (Na) excess in grapevines are driven by soil physical and permeability difficulties. However, grapevines can also accumulate fairly high levels of Na from highly sodic soils. The direct effects of Na excess in plant tissue are not always clear, because excess Na is most commonly associated with excess chloride uptake as well (Christensen et al., 1978). Dag et al. (2015) showed that more sodium than chloride was accumulated in woody tissue of 5-year-old Cabernet Sauvignon grapevines, evaluated under three irrigation salinity levels. As a result, mortality rates as high as 17.5% were found for poor salt-excluding rootstocks irrigated with the highest salty water. The apparent collapse of tolerance mechanisms, leading to salt damage and vine mortality, might be due to sodium reaching critical levels in woody tissues. This contradicts earlier findings of Ehlig (1960), who concluded that chloride was the dominant cause of salt toxicity in grapevines. The prevalence of sodium as the governing factor in salinity damage has been suggested previously for Fisarakis et al. (2001) and Stevens et al. (2011), who similarly found stronger long-term accumulation of sodium than chloride in Colombard vines grafted onto the chloride-excluding Ramsey rootstock.

It is well known that an excess of Na causes a decrease of other cations due to an ionic antagonism (Troncoso et al., 1999). Downton (1985) showed a Na-Ca antagonism in grapevines. Garcia and Charbaji (1993) related that Cabernet Sauvignon vines grown hydroponically and using a standard nutrient solution with various NaCl doses brought about an increase in the NaCl content in all plant vegetative organs. Thus, Na-K antagonism was shown by the decrease in K content, even at low NaCl doses. The Ca and Mg contents of the different plant organs were also decreased along with NaCl content that was increased in the nutrient solution. Also, Upreti and Murti (2010) concluded that grape rootstocks exhibited considerable variations in salinity tolerance as evidence from changes in Na and K contents and their corresponding ratios.

In our study carried out in northern Chile, sodium concentration in petioles of cv. Flame Seedless (Fig. 21.4) and cv. Muscat of Alexandria (Fig. 21.5) were comparatively higher in own-rooted vines compared with grafted vines on various rootstocks. There were nonsignificant differences in Na concentration among the rootstocks examined. These results are in agreement with Walker et al. (2004, 2010), who showed that own-rooted vines contained higher petiole Na concentrations than grafted vines. Rootstocks may also affect berry sodium concentrations. Berries from own-rooted Shiraz vines contain two- to threefold Na concentrations higher than grafted vines, but the difference was much less for Cabernet Sauvignon (Downton, 1977). Concentrations in petioles at flowering greater than 0.5% are regarded as excessive (Robinson, 2005).

Regarding Cl-Na accumulation ratio in grapevines, Kuiper (1968) reported an inverse relation between chloride and sodium transport capacities by grape root lipids, suggesting that transport chloride readily should restrict sodium transport and vice versa. Walker et al. (2004) found that chloride concentrations in petioles of Sultana vines were higher than sodium concentrations by 1.3–22.1-fold at flowering stage. In the same way, chloride concentrations in laminae were higher than sodium concentrations by 1.6–25.2-fold at harvest stage. The same trend was found by Dag et al. (2015) who showed that Na concentrations in leaves of 5-year-old cv. Cabernet Sauvignon were much lower than Cl concentrations. There is often a negative correlation between Na and Cl concentration in the leaves. Walker et al. (2004) say that for any given treatment combination, Cl-ion concentration in plant parts are generally higher than

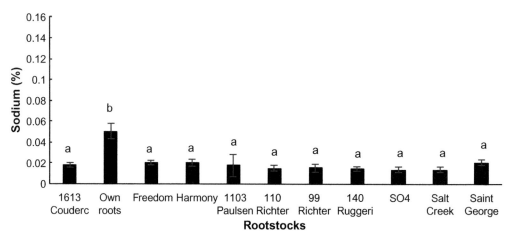

FIG. 21.4   Sodium concentration in petioles of cv. Flame Seedless on various rootstocks. Each bar represents the average of 4 years (2006–09) standard error. *Different letters* denote significant differences ($P < .05$).

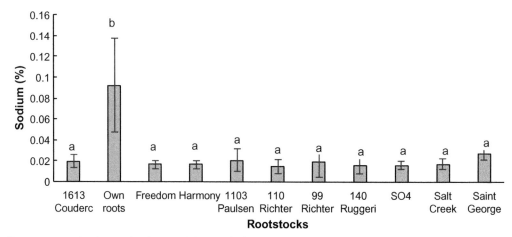

FIG. 21.5   Sodium concentration in petioles from cv. Muscat of Alexandria on various rootstocks. Each bar represents the average of 4 years (2006–09) standard error. *Different letters* denote significant differences ($P < .05$).

the corresponding Na-ion concentrations. In making those comparisons, note the ordinate scale differences between Cl and Na in (Figs. 21.2–21.5).

4.3 Boron toxicity

Boron toxicity is an important nutritional disorder in arid and semiarid environments. Boron excess can be avoided simply by not planting in affected areas. Grapevines should never be planted where the irrigation water source contains 1 ppm B and above. Even with supply of good water quality, vineyard establishment should be delayed until soil B levels are lowered by leaching to near 1 ppm or below (Christensen et al., 1978).

Symptoms of B toxicity are characterized as chlorosis and necrosis of leaves beginning at their margins, reduced leaf size, and reduced internodal distance between adjacent leaves (Yermiyahu and Ben-Gal, 2006). Boron tends to accumulate along the leaf margins until the concentration becomes sufficiently high to be toxic to leaf tissues (Christensen et al., 1978) (Fig. 21.6). Boron toxicity symptoms are commonly observed during the vegetative growing periods of grapevines, especially at the end of the seasons (Yermiyahu and Ben-Gal, 2006). Christensen et al. (1978) and Robinson (2005) mention that values of boron greater than 150 mg/kg in blades taken at flowering are indicative of B toxicity.

There is little information about the influence of rootstocks on B uptake in grapevines. Vines of cv. Sugraone grafted on Ramsey and Ruggeri rootstocks and irrigated at two salinity levels did not show a significant difference in the accumulation of boron in leaves (Yermiyahu et al., 2007). Table grapes are commonly grafted onto rootstocks developed for their hardiness or tolerance to environmental conditions. Both Ramsey (*V. champini*) and 140 Ruggeri (*V. berlandieri* × *V. rupestris*) are rootstocks that have shown relatively high tolerance to conditions of root-zone salinity (Walker et al., 2002; Zhang et al., 2002) and are found in commercial vineyards. Data comparison for two rootstocks, 41B (*V. vinifera* × *V. berlandieri*) and 1103 Paulsen (*V. berlandieri* × *V. rupestris*), showed no differences in boron content in leaves (Soylemezoglu et al., 2009).

The influence of rootstocks on B accumulation in the leaf laminae in cv. Flame Seedless and cv. Muscat of Alexandria grown in the arid zone of northern Chile is shown (Figs. 21.7 and 21.8, respectively). The leaf laminae are used to test B concentration in vines because B accumulates more in the blades (Christensen et al., 1978; Pech et al., 2013). Boron is considered highly immobile in most plants in that it is restricted to transpiration flow and accumulates in leaves (Nable et al., 1997), especially in leaf edges where necrotic specks may also develop along the leaf margins (Christensen et al., 1978). Own-rooted vines of cv. Flame Seedless accumulated less B than 1103 Paulsen, 110 Richter, and Salt Creek rootstocks (Fig. 21.7), and B content in vines of Muscat of Alexandria grafted on Salt Creek was significantly lower than vines grafted onto SO4 rootstock (Fig. 21.8).

In general, B concentration in leaves of cv. Flame Seedless variety was higher than in Muscat of Alexandria vines, independent of the rootstock (Figs. 21.7 and 21.8) effect that was also demonstrated by Walker et al. (2014) for chloride and sodium accumulation in trunk wood and grape juice of cv. Chardonnay and cv. Shiraz over own roots and also over a range of rootstocks.

FIG. 21.6 Boron toxicity symptoms along the leaf margins (at harvest stage, cv. Thompson Seedless, own rooted).

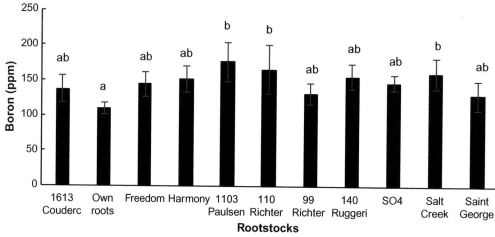

FIG. 21.7 Boron concentration in petioles from cv. Flame Seedless on various rootstocks. Each bar represents the average of 4 years (2006–09) ± standard error. *Different letters* denote significant differences ($P < .05$).

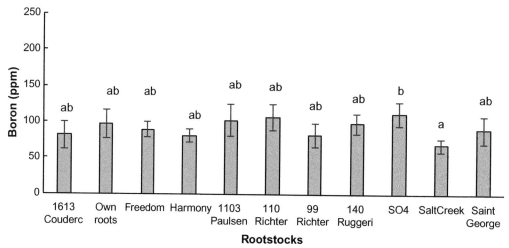

FIG. 21.8 Boron concentration in petioles from cv. Muscat of Alexandria on various rootstocks. Each bar represents the average of 4 years (2006–09) ± standard error. *Different letters* denote significant differences ($P < .05$).

To determine the tolerance to boron, Pech et al. (2013) evaluated various *Vitis* species, including *V. acerifolia*, *V. berlandieri*, *V. caribaea*, *V. champini*, *V. labrusca*, *V. rupestris*, *V. vinifera*, and hybrids. They determined that concentration of B in leaves from boron treatments was equivalent within comparisons. However, genotypes *V. acerifolia* × (*V. vinifera* × [*V. riparia* × *V. labrusca*]) (1613 Couderc), *V. acerifolia*, and *V. vinifera* × *V. rupestris* (1202 Couderc) showed tolerance to excess B in terms of absolute and/or relative growth.

5 Carbohydrates and nitrogen reserves

Both reserve carbohydrates and nitrogen are essential for the initial growth and development of grapevine in spring, as they provide energy and building blocks for the new growth before any net carbon assimilation and significant root uptake of nitrogen takes place (Cheng and Xia, 2004). The majority of the stored assimilates that are needed in the subsequent year for successful growth in the vineyard is accumulated in the roots (Zapata et al., 2004; Vrsic et al., 2009). According to Zapata et al. (2004), roots contained more than 90% of the carbohydrates and 75% of the nitrogen stored in the dormant Pinot noir vines. In the same fashion, Bates et al. (2002) found that roots were the major storage organ for carbohydrates and nutrients, accounting for 84% of starch and 75% of nitrogen stored in the vines at the season start. Ruhl and Alleweldt (1990) mention that in late summer, starch is major carbohydrate in roots and those concentrations of sugars other than glucose are low at that stage. In their study, root starch

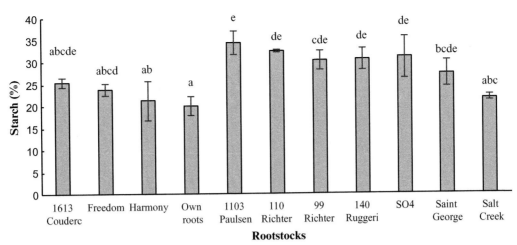

FIG. 21.9   Starch concentration in roots from cv. Flame Seedless on various rootstocks. Each bar represents the average of 4 years (2006–09) ± standard error. *Different letters* denote significant differences ($P < .05$).

concentrations ranged from 11.0% to 26.3%. Zapata et al. (2004) found that root starch concentration at dormancy was 29% dry weight, four times higher than in other perennial tissues. Along with carbohydrates, nitrogen reserves play a crucial role in supporting early season growth in woody plants (Conradie, 1980). In grapevines, nitrogen reserves are made mainly by amino acids, mostly arginine (Kliewer, 1991). Total nitrogen in the form of proteins are free amino acids typically accounts for less than 2% of dry matter of a dormant vine (Xia and Cheng, 2004). Compared with starch storage, nitrogen was far less represented at dormancy (33% vs 1.6% of root dry weight, respectively), in cv. Pinot noir grafted onto SO4 rootstock (Zapata et al., 2004).

In our long-term study conducted in arid zone of northern Chile, we determined the levels of reserve carbohydrates and nitrogen accumulated in roots of cv. Flame Seedless and cv. Muscat of Alexandria grafted onto various rootstocks (Fig. 21.9–21.12).

The results of our study show that in cv. Flame Seedless 1103 Paulsen, 110 Richter, 99 Richter, and SO4 rootstocks accumulated more starch in roots than own-rooted vines and Harmony rootstock. However, there was no significant effect of rootstocks in starch root accumulation in cv. Muscat of Alexandria. Significant differences in arginine root accumulation were found in both grapevine cultivars. Salt Creek rootstock had almost three times more arginine in roots than 110 Richter in cv. Flame Seedless (3.23% and 1.11%, respectively). In this cultivar, neither of the rootstocks accumulated more arginine than the own-rooted vines. On the other hand, in cv. Muscat of Alexandria SO4, Salt Creek and Harmony rootstocks accumulated arginine 1.7-fold more, in average, than own-rooted vines and Saint George rootstock. Because there are scarce available data about the rootstock influences in reserves accumulation under field conditions, it was not possible to compare our information. In general, high yield and low leaf area per vine decreased the nitrogen concentration in roots. The leaf-fruit ratio, expressed as the "light-exposed leaf area per kg fruit," substantially influenced the nitrogen and starch concentration in the roots (Zufferey et al., 2015).

6 Rootstock strategies to cope with salinity

In semiarid and arid regions, the competition for scarce water resources inevitably reduces the supplies of fresh water for irrigation; thus, agriculture is forced to utilize low-quality water for irrigation, increasing the risks of soil salinization where about 6% of the world's land is already affected by salinity and 20% of irrigated land (Ollat et al., 2016). Considering the context of global change, precipitation patterns are also likely to change with a reduction of the balance between precipitation and evapotranspiration, leading to an acceleration of salinization in dry (and drying) regions (Keller, 2010).

As moderately sensitive to salinity, grapevine responses depend on several factors, such as the rootstock-scion combination, irrigation system, soil type, and climate (Ollat et al., 2016). It is well known that variations in salt exclusion exist between grapevine species and cultivars, and rootstocks are considered as one important means for improving grapevine salt tolerance (Sykes, 1985; Walker et al., 2010; Ollat et al., 2016).

Physiological effects of exposure to salinity in grapevines include reduced stomatal conductance and photosynthesis (Ben-Asher et al., 2006; Downton, 1977b); systemic disturbances can lead to reduced growth, vegetative biomass

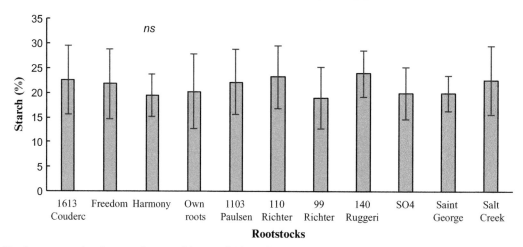

FIG. 21.10 Starch concentration in roots from cv. Muscat of Alexandria on various rootstocks. Each bar represents the average of 4 years (2006–09) ± standard error. *Different letters* denote significant differences (P < .05).

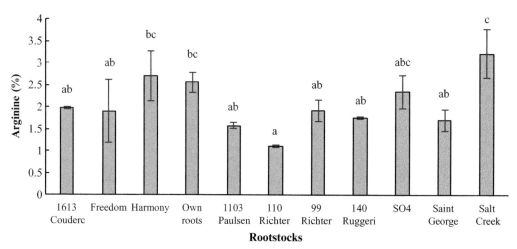

FIG. 21.11 Arginine concentration in roots from cv. Flame Seedless on various rootstocks. Each bar represents the average of 4 years (2006–09) ± standard error. *Different letters* denote significant differences (P < .05).

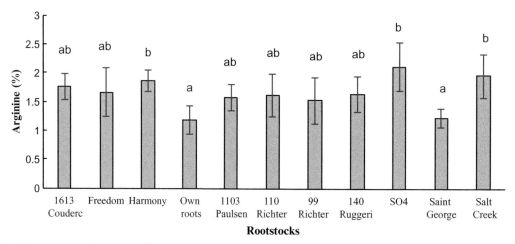

FIG. 21.12 Arginine concentration in roots from cv. Muscat of Alexandria on various rootstocks. Each bar represents the average of 4 years (2006–09) ± standard error. *Different letters* denote significant differences (P < .05).

(Shani and Ben-Gal, 2005), and yield (Walker et al., 2010; Zhang et al., 2002). Similar to other plants, grapevines have osmotic and toxicity-related responses to salinity, and tolerance mechanisms involve exclusion of salt uptake, accumulation and storage of ions in cellular compartments, and restriction of toxic ions in leaves (Shani and Ben-Gal, 2005).

The differences in nutrient uptake and distribution may be explained in several ways. First, rootstocks may have different absorption capability or affinity for some specific nutrients and minerals. In this concern, Bavaresco et al. (1991) pointed out that rootstocks with lime tolerance have a strategy to overcome chlorosis with high root iron uptake and reducing capacity. Grant and Matthews (1996) observed that different rootstocks might have different abilities to absorb phosphorus. Ruhl (2000) also found a high potassium acquisition mechanism on some rootstocks, which would affect pH of fruit and wines. Second, translocation and nutrient distribution may differ among rootstocks. Giorgessi et al. (1997) found differences in number and size of the xylem vessels between rootstocks and own-rooted grapevines. Third, some nutrients might be assimilated mostly by roots, thus reducing the amount translocated to the shoots. Ruhl (1993) points out that rootstocks delivering low amounts of potassium to the cultivar accumulate majority of absorbed cations in the vacuole of root cells, and Keller et al. (2001) discovered that over 85% of nitrogen was assimilated by means of vine root metabolism. In all cases, further studies are required to understand the influence of the root system in mineral absorption (Nikolaou et al., 2000).

Salt exclusion by roots is considered a main mechanism to contribute to tolerance, since previous studies have shown that high uptake and root-to-shoot transport of chloride resulted in its excessive accumulation in leaf tissues, causing impaired leaf function and damage (Walker et al., 2002). Chloride exclusion by roots may prevent chloride accumulation in leaves contributing to salt tolerance (Walker et al., 1997, 2000, 2002; Stevens et al., 2011; Fisarakis et al., 2001; Zhang et al., 2002). Indeed, a much lower chloride concentration was found in xylem sap and shoot tissue of potted grape vines when grafted on to a salt tolerant (140 Ru) than on to a salt-sensitive rootstock (K51-40).

The ability of rootstock-scion combinations to inhibit salt accumulation in leaf tissue is used to categorize sensitivity to salinity. One management strategy for vineyards with saline irrigation water is to select rootstocks based on their ability to prevent salt uptake and accumulation. There is significant variation in scion Cl content among rootstocks (Walker et al., 2004, 2007). Dag et al. (2015) determined that apparent breakdown of tolerance mechanisms, leading to salt damage and vine mortality, might be due to Na-reaching critical levels in woody tissues. The ability to exclude Na and Cl from shoots and fruit was found to (a) increase wine quality by reducing concentrations of salt ions in must and wine and (b) reduce mortality rates that result from long-term exposure to salt, since *Vitis* spp. rootstocks can mediate salt exclusion from grafted *V. vinifera* scions enabling higher grapevine yields and production of superior wines with lower salt content.

Recently, by using a cross between two *Vitis* interspecific hybrid rootstocks, Henderson et al. (2017) mapped a dominant quantitative trait locus (QTL) associated with leaf Na+exclusion (NaE) under salinity stress. The NaE locus encodes six high-affinity potassium transporters (HKT). Transcript profiling and functional characterization in heterologous systems identified *VisHKT1;1* as the best candidate gene for controlling leaf Na+exclusion. The origin of the recessive *VisHKT1;1* alleles was traced to *V. champini* and *V. rupestris* and therefore is possible to assist breeding Na+-tolerant grapevine rootstocks through the genetic and functional data generated (Henderson et al., 2017).

In the field, tolerance to salt may be defined as the ability to maintain shoot growth and yield under high-salinity conditions. Over time, it appears that vine vigor and intrinsic ability for salt tolerance are both required to maintain yield performance in saline environments (Walker et al., 2010). Moreover, as grapevine rootstocks differ widely in their ability to exclude chloride from the shoot and in their salinity tolerance, field experiments showed that Ramsey was one of the best performing salt-tolerant rootstocks combined with various scion varieties, soils, and climate conditions (summarized by Ollat et al., 2016). 140Ru was also considered as highly salinity tolerant, and differences among extreme genotypes are maintained in the field in a range of environments and salinities (Walker et al., 2010); nevertheless, large genotype–environment interactions have been reported (Gong et al., 2011), suggesting that chloride exclusion ability of certain rootstocks can vary with long-term exposure to salinity (Tregeagle et al., 2010). These variations could be related to the volume of irrigation, the salt concentration in the soil, the leaching conditions, and the evapotranspiration levels in each site. Finally, Sivritepe et al. (2010) observed a strong effect of scion varieties on the salt tolerance of grafted plants, reinforcing the notion of a terroir based cultivar/rootstock responses.

Rootstocks have a key role in the grapevine response to the environment and in grape berry composition as they represent a physical and biological link between the soil and the aboveground part of the plant. Nevertheless, as they are the hidden half of the grapevine, our knowledge about their functioning is still very scarce (Ollat et al., 2016). The features involved in the influence of grapevine rootstocks on scion growth and nutrient uptake and the interactions between rootstock and scion in a grafted system are gathering renewed interest from scientific community.

Since grafting is required in the cultivation of grapevine in most areas in the world, rootstocks have a wide range of impacts on scion behavior (summarized from Keller, 2015; Ollat et al., 2016). The study of rootstock-scion interaction is incredibly complex integrating structural changes at the graft interface, hydraulic integration, hormonal communication, and even exchange of genetic materials (Zhang et al., 2016). It has been described that the effect of the interaction between the two genotypes is, in general, larger than the rootstock effect. By analogy with genotype-environment studies, the concept of plasticity was introduced to characterize the phenotypic variations produced by a genotype in response to grafting partners. There is a good amount of data describing the variability among rootstocks, but the mechanisms underlying their responses to the environment are still unknown. Also, further research is still needed to elucidate the genetic determinism of these traits and how rootstocks can be used to mitigate plant stress in the context of the ongoing climate changes.

Indeed, Tandonnet et al. (2018) have identified key QTLs for traits assessed on field-grown grafted grapevines, where root number and section had the largest phenotypic variance explained. Genetic control of root and aerial traits was independent, when analyzing genetic architecture of root traits in a segregating progeny from an interspecific cross between *V. vinifera* cv. Cabernet Sauvignon × *V. riparia* cv. Gloire de Montpellier. Identified QTLs for aerial-to-root biomass ratio suggested that aerial and root traits are controlled independently, opening new venues for breeding rootstocks with improved root development capacities.

Moreover, studies concerning the influences of rootstocks on scion growth, fruit composition, or wine quality do not always produce consistent results, possibly due to experimental conditions (e.g., potted vs field and young vines vs old vines), soil type and/or climatic conditions, scion variety, etc. Metaanalysis studies could be useful to better understand and integrate the studies that have already been carried out. Further studies aimed at understanding the physiology and traits responsible for rootstock control over scion behavior might benefit by integrating different approaches (genetic, transcriptomic, metabolic, hydraulic, etc.) in the experimental approach. There is still a lot to be gained from investigations in ungrafted material creating a foundation of understanding regarding the differences between the rootstock genotypes themselves (Zhang et al., 2016).

Finally, future research demands the application of new technologies in the vineyard and the integration of multidisciplinary approaches. One of the most common goals pursued by grapevine researchers is the development of new rootstock varieties that meet growers' demands, especially in the context of climate change (Zhang et al., 2016). The complexity of the responses, putative mechanisms, and interactions with environment present significant challenges, but the rootstock breeding is critical for the development of new sustainable approaches to adapt and mitigate climate change effects in viticulture.

Acknowledgments

The authors are grateful to their technical staff who perform a lot of experimental work related to rootstock studies, including María Isabel Rojas, Elizabeth Pastén, Carmen Jopia, and Cristián González for their valuable contribution in maintaining the experimental trials and technical support. They also thank the financial support of Instituto de Investigaciones Agropecuarias (INIA) (long-term rootstocks studies), FONDECYT Regular Grant no. 1140039 (to AZS and AI), and Postdoctoral FONDECYT 2018 N°3180252 (to NVV).

References

Antcliff, A.J., Newman, H.P., Barrett, H.C., 1983. Variation in chloride accumulation in some American species of grapevine. Vitis 22, 357–362.

Bates, T.R., Dunst, R.M., Joy, P., 2002. Seasonal dry matter, starch, and nutrient distribution in "Concord" grapevines roots. HortScience 37 (2), 313–316.

Bavaresco, L., Fregoni, M., Fraschini, P., 1991. Investigations on iron uptake and reduction by excised roots of different grapevine rootstocks and *V. vinifera* cultivar. Plant Soil 130, 109–113.

Bavaresco, L., Giachino, E., Pezutto, S., 2003. Grapevine rootstock effects on lime-induced chlorosis, nutrient uptake, and source-sink relationships. J. Plant Nutr. 26 (7), 1451–1465.

Bell, S., Henschke, P., 2005. Implications of nitrogen nutrition for grapes, fermentation and wine. Aust. J. Grape Wine Res. 11 (3), 242–295.

Ben-Asher, J., Tsuyuki, B., Bravdo, B., Sagih, M., 2006. Irrigation of grapevines with saline water: I. Leaf area index, stomatal conductance, transpiration and photosynthesis. Agric. Water Manag. 83, 13–21.

Bianchi, D., Grossi, D., Tincani, D., Simone Di Lorenzo, G., Brancadoro, L., Rustioni, L., 2018. Multi-parameter characterization of water stress tolerance in *Vitis* hybrids for new rootstock selection. Plant Physiol. Biochem. 132, 333–340.

Candolfi, M., Castagnoli, S., 1997. Grape rootstocks and nutrient uptake efficiency. In: Proceeding of the Oregon Horticultural Society. vol. 88, pp. 221–228.

Cheng, L., Xia, G., 2004. Growth and fruiting of young "Concord" grapevines in relation to reserve nitrogen and carbohydrates. J. Am. Soc. Hortic. Sci. 129 (5), 660–666.

Christensen, L.P., Kasimatis, A.N., Jensen, F.L., 1978. Grapevine Nutrition and Fertilization in the San Joaquin Valley. Publication 4097, Division of Agricultural Sciences, University of California 40 p.

Conradie, W.J., 1980. Seasonal uptake of nutrients by Cherin blanc in sand culture. I. Nitrogen. S. Afr. J. Enol. Vitic. 1, 59–65.

Corso, M., Bonghi, C., 2004. Grapevine rootstock effects on abiotic stress tolerance. Plant Sci. Today 1 (3), 108–113.

Cramer, G., Ergul, A., Grimplet, I., Tillet, R., Tattersall, E.R., Bohlman, M., Crushman, J., 2007. Water and salinity stress in grapevines: early and late changes in transcript and metabolite profiles. Funct. Integr. Genomics 7, 11–134.

Dag, A., Ben-Gal, A., Goldberger, S., Yermiyahu, U., Zipori, I., Or, E., David, I., Netzer, Y., Kerem, Z., 2015. Sodium and chloride distribution in grapevines as a function of rootstock and irrigation water salinity. Am. J. Enol. Vitic. 66 (1), 80–84.

Dalbó, M., Schuck, E., Basso, C., 2011. Influence of rootstock on nutrient content in grape petioles. Rev. Bras. Frutic. 33 (3), 941–947.

Delas, J., Pouget, R., 1979. Influence du greffage sur la nutrition minerale de la vigne. Consequences sur la fertilization. Oeno One 13 (4), 241–261.

Downton, W.J., 1977a. Chloride accumulation in different species of grapevine. Sci. Hortic. 7, 249–253.

Downton, W.J.S., 1977b. Influence of rootstocks on the accumulation of chloride, sodium and potassium in grapevines. Aust. J. Agric. Res. 28, 879–889.

Downton, W.J., 1985. Growth and mineral composition of the Sultana grapevine as influenced by salinity and rootstock. Aust. J. Agric. Res. 36, 425–434.

Ehlig, C.F., 1960. Effects of salinity of four varieties of table grapes grown in sand culture. Proc. Am. Soc. Hortic. Sci. 76, 323–331.

Fisarakis, I., Chartzoulakis, K., Stavrakas, D., 2001. Response of Sultana vines (*V. vinifera* L.) on six rootstocks to NaCl salinity exposure and recovery. Agric. Water Manag. 51, 13–27.

Fisarakis, I., Nikolaou, N., Tsikalas, P., Therios, I., Stavrakas, D., 2004. Effect of salinity and rootstock on concentration of potassium, calcium, magnesium, phosphorus, and nitrate-nitrogen in Thompson Seedless grapevine. J. Plant Nutr. 27 (12), 2117–2134.

Garcia, M., Charbaji, T., 1993. Effect of sodium chloride salinity on cation equilibria in grapevine. J. Plant Nutr. 16 (11), 2225–2237.

Garcia, M., Gallego, P., Daveride, C., Ibrahim, H., 2001. Effect of three rootstocks on grapevine (*Vitis vinifera* L.) cv. Négrette, grown hydroponically. I. Potassium, calcium and magnesium nutrition. S. Afr. J. Enol. Vitic. 22 (2), 101–103.

Gautier, A., Cookson, S., Mollier, A., Hevin, C., Vivin, P., Lauvergeat, V., 2018. Phosphorus acquisition efficiency and phosphorus remobilization mediate genotype-specific differences in shoot phosphorus content in grapevine. Tree Physiol. https://doi.org/10.1093/treephys/tpy074.

Giorgessi, F., Bortolin, C., Sansone, L., Giulino, C., 1997. Stock and scion relationships in *Vitis vinifera*. Acta Hortic. 427, 311–318.

Gong, H., Blackmore, D., Clingeleffer, P., Sykes, S., Jha, D., Tester, M., Walker, R., 2011. Contrast in chloride exclusion between two grapevine genotypes and its variation in their hybrid progeny. J. Exp. Botany 62, 989–999.

Grant, R.S., Matthews, M.A., 1996. The influence of phosphorus availability, scion, and rootstock on grapevine shoot growth, leaf area, and petiole phosphorus concentration. Am. J. Enol. Vitic. 47, 217–224.

Henderson, S.W., Dunlevy, J.D., Wu, Y., Blackmore, D.H., Walker, R.R., Edwards, E.J., et al., 2017. Functional differences in transport properties of natural HKT1;1 variants influence shoot Na$^+$ exclusion in grapevine rootstocks. New Phytol. 217 (3), 1113–1127.

Hepaksoy, S., Ben-Asher, I., De Malach, Y., David, I., Sagih, M., Bravdo, B.A., 2006. Grapevine irrigation with saline water: effect of rootstocks on quality and yield of Cabernet Sauvignon. J. Plant Nutr. 29, 783–795.

Holzapfel, B.P., Treeby, M.T., 2007. Effects of timing and rate of N supply on leaf nitrogen status, grape yield and juice composition from Shiraz grapevines grafted to one of three different rootstocks. Aust. J. Grape Wine Res. 13, 14–22.

Ibacache, A., Sierra, C., 2009. Influence of rootstocks on nitrogen, phosphorus and potassium content in petioles of four table grape varieties. Chil. J. Agric. Res. 69 (4), 503–508.

Ibacache, A., Albornoz, F., Zurita-Silva, A., 2016. Yield responses in Flame Seedless, Thompson Seedless and Red Globe table grape cultivars are differentially modified by rootstocks under semiarid conditions. Sci. Hortic. 204, 25–32.

Keller, M., 2001. Reproductive growth of grapevines in response to nitrogen supply and rootstock. Aust. J. Grape Wine Res. 7, 12–18.

Keller, M., 2010. Managing grapevines to optimise fruit development in a challenging environment: a climate change primer for viticulturists. Aust. J. Grape Wine Res. 16, 56–69.

Keller, M., 2015. The Science of Grapevines, second ed. Elsevier. 491 p.

Keller, M., Kummer, M., Carmo Vasconcelos, M., 2001. Soil nitrogen utilization for growth and gas exchange by grapevines in response to nitrogen supply and rootstock. Aust. J. Grape Wine Res. 7, 2–11.

Kidman, C.M., Dry, P.R., McCarthy, M.G., Collins, C., 2014. Effect of rootstock on nutrition, pollination and fertilisation in 'Shiraz' (*Vitis vinifera* L.). Vitis 53 (3), 39–145.

Kliewer, W.M., 1991. Methods for determining the nitrogen status of vineyards. In: Rantz, J.M. (Ed.), Proceeding of the International Symposium on Nitrogen in Grapes and Wine. The American Society for Enology and Viticulture, Seattle, Washington, pp. 133–147.

Kodur, S., 2011. Effects of juice pH and potassium on juice and wine quality, and regulation of potassium in grapevines through rootstocks (*Vitis*): a short review. Vitis 50, 1–6.

Kupier, P.J.C., 1968. Lipids in grape roots in relation to chloride transport. Plant Physiol. 43, 1367–1371.

Morales, M., Toro, G., Riquelme, A., Sellés, G., Pinto, M., Ferreyra, R., 2014. Effect of different rootstocks on photosynthesis and nutritional response of grapevines cultivar 'Sultanina' under flooding stress. Acta Hortic. (1045), 123–131.

Mpelasoka, B., Schachatman, D., Treeby, M., Thomas, M., 2003. A review of potassium nutrition in grapevines with special emphasis on berry accumulation. Aust. J. Grape Wine Res. 9, 154–168.

Nable, R.O., Banuelos, G.S., Paull, J.G., 1997. Boron toxicity. Plant Soil 198, 181–198.

Newman, H.P., Antcliff, A.J., 1984. Chloride accumulation in some hybrids and backcrosses of *Vitis berlandieri* and *Vitis vinifera*. Vitis 23, 106–112.

Nikolaou, N., Koukourikou, M.A., Karagiannidis, N., 2000. Effects of various rootstocks on xylem exudates cytokinin content, nutrient uptake and growth patterns of grapevine *Vitis vinifera* L. cv. Thompson Seedless. Agronomie 20, 363–373.

Ollat, N., Peccoux, A., Papura, D., Esmenjaud, D., Marguerit, E., Tandonnet, J.-P., Bordenave, L., Cookson, S.J., Barrieu, F., Rossdeutsch, L., Lecourt, J. L., Lauvergeat, V., Vivin, P., Bert, P.-F., Delrot, S., 2016. Rootstocks as a component of adaptation to environment. In: Gerós, H., Chaves, M.M., Gil, H.M., Delrot, S. (Eds.), Grapevine in a Changing Environment. John Wiley & Sons, Ltd., Chichester, pp. 68–108.

Pachnowska, K., Ochmian, I., 2018. Influence of rootstock on nutrients and heavy metals in leaves and berries of the vine cultivar 'Regent' grown in North-Western Poland. J. Appl. Bot. Food Qual. 91, 180–186.

Pech, J.M., Stevens, R.M., Grigson, G.J., Cox, J.W., Schrale, G., 2013. Screening the *Vitis* genus for tolerance to boron with and without salinity. Aust. J. Grape Wine Res. 19, 446–456.

Robinson, J.B., 2005. Critical plant tissue values and application of nutritional standards for practical use in vineyards. In: Christensen, L.P., Smart, D. R. (Eds.), Proceedings of the Soil Environment and Vine Mineral Nutrition Symposium. The American Society for Enology and Viticulture, Davis, CA, pp. 61–68.

Romero, P., Botía, P., Navarro, J.M., 2018. Selecting rootstocks to improve vine performance and vineyard sustainability in deficit irrigated Monastrell grapevines under semiarid conditions. Agric. Water Manag. 209, 73–93.

Ruhl, E.H., 1993. Effect of K supply on ion uptake and concentration in expressed root sap and xylem sap of several grapevine rootstock varieties. Wein-Wiss 48, 61–68.

Ruhl, E.H., 2000. Effect of rootstock and K+ supply on pH and acidity of grape juice. Acta Hortic. (512), 31–37.

Ruhl, E.H., Alleweldt, G., 1990. Effect of water stress on carbohydrate accumulation in root and stem of four different grapevine varieties. Wein-Wiss 45, 156–159.

Satisha, J., Somkuwar, R.J., Sharma, J., Upadhyay, A.K., Adsule, P.G., 2010. Influence of rootstocks on growth yield and fruit composition of Thompson Seedless grapes grown in the Pune region of India. S. Afr. J. Enol. Vitic. 31 (1), 1–8.

Shani, U., Ben-Gal, A., 2005. Long-term response of grapevines to salinity: osmotic effects and ion toxicity. Am. J. Enol. Vitic. 56, 148–154.

Sivritepe, N., Sivritepe, H.O., Celik, H., Katkat, A.V., 2010. Salinity responses of grafted grapevines: effects of scion and rootstock genotypes. Not. Bot. Horti Agrobot. Cluj-Napoca 38, 193–201.

Soylemezoglu, G., Demir, K., Inal, A., Gunes, A., 2009. Effect of silicon on antioxidant and stomatal response of two grapevine (Vitis vinifera L.) rootstocks grown in boron toxic, saline and boron toxic-saline soil. Sci. Hortic. 123, 240–246.

Stevens, R.M., Harvey, G., 1995. Effects of waterlogging, rootstock and salinity on Na, Cl, K concentrations of the leaf and root, and shoot growth of Sultanina grapevines. Aust. J. Agric. Res. 46, 541–551.

Stevens, R.M., Harvey, G., Partington, D.L., 2011. Irrigation of grapevines with saline water at different growth stages: effects on leaf, wood and juice composition. Aust. J. Grape Wine Res. 17, 239–248.

Sykes, S.R., 1985. Variation in chloride accumulation by hybrid vines from crosses involving the cultivars Ramsey, Villard Blanc and Sultana. Am. J. Enol. Vitic. 36, 30–37.

Tambe, T.B., Gawade, M.H., 2004. Influence of rootstocks on vine vigour, yield and quality of grapes. Acta Hortic. 662, 259–263.

Tandonnet, J.-P., Marguerit, E., Cookson, S.J., Ollat, N., 2018. Genetic architecture of aerial and root traits in field-grown grafted grapevines is largely independent. Theor. Appl. Genet. 33, 1–13.

Tecchio, M., Moura, M., Teixeira, L., Pires, E., Leonel, S., 2014. Influence of rootstocks and pruning times on yield and on nutrient content and extraction in 'Niagara Rosada' grapevine. Pesq. Agrop. Brasileira 49 (5), 340–348.

Tortosa, I., Escalona, J.M., Bota, J., Tomás, M., Hernández, E., Escudero, E.G., Medrano, H., 2016. Exploring the genetic variability in water use efficiency: evaluation of inter and intra cultivar genetic diversity in grapevines. Plant Sci. 251, 1–27.

Treeby, M.T., Holzapfel, B.P., Walker, R.R., Nicholas, P.R., 1998. Profiles of free amino acids in grapes of grafted Chardonnay grapevines. Aust. J. Grape Wine Res. 4 (3), 121–126.

Tregeagle, J.M., Tisdall, J.M., Tester, M., Walker, R.R., 2010. Cl-uptake, transport and accumulation in grapevine rootstocks of differing capacity for Cl-exclusion. Funct. Plant Biol. 37, 665–673.

Troncoso, A., Matte, C., Cantos, M., Lavee, S., 1999. Evaluation of salt tolerance of in-vitro-grown grapevine rootstock varieties. Vitis 38 (2), 55–60.

Upreti, K.K., Murti, G.S.R., 2010. Response of grape rootstocks to salinity: changes in root growth, polyamines and abscisic acid. Biol. Plant. 54 (4), 730–734.

Vijaya, D., Rao, B., 2015. Effect of rootstocks on petiole mineral nutrient composition of grapes (Vitis vinifera L. cv. Thompson Seedless). Curr. Biot. 8 (4), 367–374.

Vrsic, S., Pulko, B., Valdhuber, J., 2009. Influence of defolation on carbohydrates reserves of young grapevines in the nursery. Eur. J. Hortic. Sci. 74 (5), 218–222.

Walker, R.R., 1994. Grapevine responses to salinity. Bull. l'O.I.V. 67, 634–661.

Walker, R., Blackmore, D., 2012. Potassium concentration and pH inter-relationships in grape juice and wine of Chardonnay and Shiraz from a range of rootstocks in different environments. Aust. J. Grape Wine Res. 18 (2), 183–193.

Walker, R.R., Blackmore, D.H., Chingelffer, P.R., Jacono, F., 1997. Effect of salinity and Ramsey rootstock on ion concentration and carbon dioxide assimilation in leaves of drip-irrigated, field-grown grapevines (Vitis vinifera L. cv. Sultana). Aust. J. Grape Wine Res. 3, 66–74.

Walker, R.R., Read, P.E., Blackmore, D.H., 2000. Rootstock and salinity effects on rates of berry maturation, ion accumulation and colour development in Shiraz grapes. Aust. J. Grape Wine Res. 6, 227–239.

Walker, R.R., Blackmore, D.H., Clingeleffer, P.R., Correl, R.L., 2002. Rootstock effects on salt tolerance of irrigated field-grown grapevines (Vitis vinifera L. cv. Sultana). 1. Yield and vigour inter-relationships. Aust. J. Grape Wine Res. 8 (1), 3–14.

Walker, R.R., Blackmore, D.H., Clingeleffer, P.R., Correl, R.L., 2004. Rootstock effects on salt tolerance of irrigated field-grown grapevines (Vitis vinifera L.cv. Sultana). 2. Ion concentrations in leaves and juice. Aust. J. Grape Wine Res. 10, 90–99.

Walker, R.R., Blackmore, D., Clingeleffer, P., Tarr, C., 2007. Rootstock effects on salt tolerance of irrigated field-grown grapevines (Vitis vinifera L. cv. Sultana). 3. Fresh fruit composition and dried grape quality. Aust. J. Grape Wine Res. 13, 130–141.

Walker, R.R., Blackmore, D.H., Clingeleffer, P.R., 2010. Impact of rootstock on yield and ion concentrations in petioles, juice and wine of Shiraz and Chardonnay in different viticultural environments with different irrigation water salinity. Aust. J. Grape Wine Res. 16, 243–257.

Walker, R.R., Blackmore, D.H., Clingeleffer, P.R., Emanuelli, D., 2014. Rootstock type determines tolerance of Chardonnay and Shiraz to long-term saline irrigation. Aust. J. Grape Wine Res. 20, 496–506.

Walker, R.R., Blackmore, D.H., Gong, H., Henderson, S.W., Gilliham, M., Walker, A.R., 2018. Analysis of the salt exclusion phenotype in rooted leaves of grapevine (Vitis spp.). Aust. J. Grape Wine Res. 24, 317–326.

Wooldridge, J., Louw, P., Conradie, W.J., 2010. Effects of rootstock on grapevine performance, petiole and must composition, and overall wine score of Vitis vinifera cv. Chardonnay and Pinot Noir. S. Afr. J. Enol. Vitic. 31, 45–48.

Xia, G., Cheng, L., 2004. Foliar urea application in the fall affects both nitrogen and carbon storage in young "Concorde" grapevines grown under a wide range of nitrogen supply. J. Am. Soc. Hortic. Sci. 129 (5), 653–659.

Yermiyahu, U., Ben-Gal, A., 2006. Boron toxicity in grapevine. HortScience 41 (7), 1698–1703.

Yermiyahu, U., Ben-Gal, A., Sarig, P., Zipilevitch, E., 2007. Boron toxicity in grapevine (Vitis vinifera L.) in conjunction with salinity and rootstock effects. J. Hortic. Sci. Biotechnol. 82 (4), 547–554.

Zapata, C., Deliens, E., Cahillou, S., Magné, C., 2004. Partitioning and mobilization of starch and N reserves in grapevine (Vitis vinifera L.). J. Plant Physiol. 161, 1031–1040.

Zhang, X., Walker, R.R., Stevens, R.M., Prior, L.D., 2002. Yield-salinity relationships of different grapevine (Vitis vinifera L.) scion-rootstock combinations. Aust. J. Grape Wine Res. 8, 150–156.

Zhang, L., Marguerit, E., Rossdeutsch, L., Ollat, N., Gambetta, G.A., 2016. The influence of grapevine rootstocks on scion growth and drought resistance. Theor. Exp. Plant Physiol. 28 (2), 143–157.

Zufferey, V., Murisier, F., Belcher, S., Lorenzini, F., Vivin, P., Spring, J.L., Viret, O., 2015. Nitrogen and carbohydrate reserves in the grapevine (Vitis vinifera L. "Chasselas"): the influence of the leaf to fruit ratio. Vitis 54, 183–188.

22

Microbial ecology in sustainable fruit growing: Genetic, functional, and metabolic responses

Adriano Sofo[a],*, Patrizia Ricciuti[b], Alba N. Mininni[a], Cristos Xiloyannis[a], Bartolomeo Dichio[a]

[a]Department of European and Mediterranean Cultures: Architecture, Environment and Cultural Heritage (DiCEM), University of Basilicata, Matera, Italy
[b]Department of Soil, Plant and Food Sciences (DiSSPA), University of Bari 'Aldo Moro', Bari, Italy
*Corresponding author. E-mail: adriano.sofo@unibas.it

1 Introduction

Microorganisms comprise the majority of soil biological diversity (Riesenfeld et al., 2004). Soil microbial communities influence soil fertility and plant growth, and changes in their structure and dynamics in response to different soil management practices can give information about soil status, in terms of its quality and biological complexity (Bai et al., 2018) (Fig. 22.1). Particularly, under Mediterranean climates, a sustainable soil management aimed at increasing soil organic carbon and microbiological fertility is of key importance. In these environments, the need for a new approach in orchard management is urgent for improving or maintaining soil quality, health, and fertility (Sofo et al., 2019a). Particularly, in fruit crops, the positive influence of sustainable management systems on soil physical, biochemical, and microbiological characteristics has been described (Sofo et al., 2019b). Sustainable soil management (S_{mng}) of fruit orchards can have positive effects on both soils and crop yields due to increases in microbial biomass, activity, and complexity. In Mediterranean orchards, a S_{mng} aimed at increasing soil organic carbon (SOC) stocks (e.g., by no-tillage, increased C and N inputs through recycling of pruning residuals, cover crops, and compost addition) can affect many soil parameters, such as physical characteristics (Palese et al., 2014), chemical parameters (Sofo et al., 2010a; Montanaro et al., 2010), water content (Celano et al., 2011), and CO_2 fluxes (Montanaro et al., 2012).

Soil management, if not well planned and conducted, can provoke decreases in soil organic matter (SOM), mainly due to SOM mineralization (Montanaro et al., 2010). For instance, in Mediterranean olive orchards, high temperature and water shortage, via stimulated microbial metabolism and respiration, cause decreases of SOC and other nutrients (primarily soil N; Pascazio et al., 2018). In these agroecosystems, a S_{mng} that includes a localized and evapotranspiration-based irrigation can improve plant physiological status, yield, and fruit quality in the midterm (Palese et al., 2009; Sofo et al., 2010a). The application of a S_{mng} also gives considerable economic advantages to the farmers and provides efficient ecosystemic and sociocultural services (Montanaro et al., 2017). In addition, the S_{mng} applied to fruit crops can reduce the negative repercussions on the environment linked to nutrient leaching/runoff, particularly weighty for nitrates (Palese et al., 2015).

A.K. Srivastava, Chengxiao Hu (eds.)
Fruit Crops: Diagnosis and Management of Nutrient Constraints
https://doi.org/10.1016/B978-0-12-818732-6.00022-8

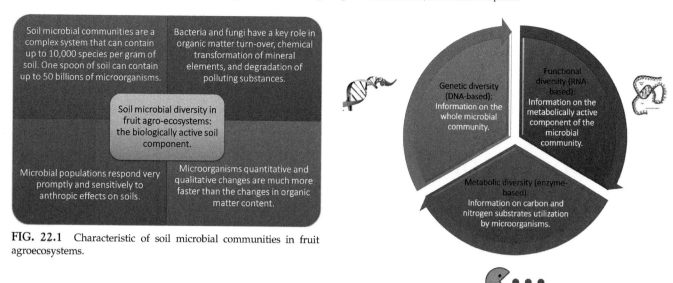

FIG. 22.1 Characteristic of soil microbial communities in fruit agroecosystems.

FIG. 22.2 The three major components of the microbial diversity.

In view of circular economy principles and to capitalize on natural potential of soils, sustainable strategies have to be developed for land use practices that optimize nutrient and energy use in fruit groves. The application of these strategies can not only reduce soil organic matter decline, soil erosion, and degradation but also promote ecosystem services and foster all the components of microbial biodiversity (Fig. 22.2). Through an innovative multidisciplinary approach, with a particular emphasis on microbial ecology, region-specific land management practices should be proposed. In this chapter, we present a survey on the researches carried out in several experimental fruit groves in the last two decades by our research group. In particular, the Metapontino area of Basilicata Region (Italy), where most of the researches and experimental trials here presented have been conducted, is listed as a "nitrate vulnerable zone" (NVZ), and for this reason, a steady microbiological and chemical monitoring is continuously needed. We here discuss the changes in the structure, dynamics, complexity, and genetic diversity of soil microbial communities and their relationships with soil health status and soil fertility.

2 Soil microbial dynamics in orchard agrosystems

Soil microorganisms' dynamics (e.g., mobility, growth, nutrient absorption, and respiration), the major responsible of soil fertility and quality (Bünemann et al., 2018), are strongly affected both by the type of soil management and irrigation, and this has been widely demonstrated in fruit crops (Palese et al., 2009; Sofo et al., 2012, 2014a,b; Pascazio et al., 2018). Soil microorganisms influence and, at the same time, are influenced by the soil C and N contents, being bacteria an essential part of C and, even more, of N-cycling processes (Zhang et al., 2014; Li et al., 2018).

Because of the complexity and site specificity of soils, defining soil quality is not an easy task (Bünemann et al., 2018). Qualitative soil profile descriptions and the analysis of C/N ratios and pH values both in topsoil and subsoil can be very useful when interpreting soil C and N data for understanding the general context of the environment soil microorganisms live in (Li et al., 2018). Particularly, the characteristics of the litter layer are a very informative addition to the soil profile description, as they reflect the equilibrium between litter production, litter microbial decomposition, and interaction with the mineral soil (Zanella et al., 2018). Among microbiological techniques, one of the easiest and reliable techniques for defining soil microbiological status is the determination of microbial metabolic diversity by the Biolog method, having a high discriminating power between microbial soil communities from different soil environments and from soils subjected to various agronomic treatments (Sofo et al., 2010a,b). Culture-based and genetic techniques have been used successfully in olive orchards to ascertain the presence of C- and N-cycling functional bacteria (Sofo et al., 2010a,b, 2014a). The study of the soil C- and N-cycling bacteria and of the C and N dynamics could help to understand how soil management can affect soil status (Pascazio et al., 2018; Li et al., 2018). This is particularly important for Mediterranean orchards, where N is often a limiting factor, even when soils are managed sustainably and have high organic C inputs (Montanaro et al., 2010; Celano et al., 2011).

Besides microbiological and genetic analysis, nowadays, next-generation sequencing (NGS), coupled with bioinformatic tools and metagenomic approach, has made it easier to comprehensively analyze microbial communities on any type of matrix, including soils (Jansson and Hofmockel, 2018). Using a metagenomic approach, Sofo et al. (2019a,b) evaluated the possible persistence of potential human pathogenic bacteria (HPB) in olive orchards sustainably irrigated with treated urban wastewater. Indeed, under suitable conditions, low-quality, urban wastewater is an additional water resource for irrigation in water-scarce environments, but its use in agriculture requires a careful monitoring of a range of hygiene parameters, including HPB. On the basis of the results obtained and from the general analysis of previous researches, the authors concluded that irrigation with urban wastewater, if adequately treated and applied, does not constitute health risks for consumers and farmers.

3 Sustainable systems versus conventional systems

Several works published by research group have evaluated the medium-term effect (approximately 20 years) of two different soil management systems, so-called sustainable (S_{mng}) and conventional (C_{mng}), on bacterial and fungal genetic, functional, and metabolic diversity in soils of different types of orchards. The S_{mng} system included no-tillage and endogenous and polygenic organic matter inputs deriving from spontaneous cover crops, pruning material left on the field and/or compost, while C_{mng} soils were tilled, without cover crops and pruning residues were removed. In many cases, the microbial analyses were carried out by culture-based (plating, spectrophotometry, and Biolog) and molecular-based approaches (Denaturing Gradient Gel Electrophoresis [DGGE] 16S DNA cloning/sequencing and metagenomic analysis) (Fig. 22.3).

Significant differences between S_{mng} and C_{mng} fruit systems, such as olive, peach, and kiwifruit groves (Sofo et al., 2010a,b, 2014a, 2019b), have generally been observed, in terms of number of the activities of microbial soil enzymes, Biolog carbon source utilization patterns and related indices, DNA abundance, and presence of the bacteria involved in soil dynamics, such as C and N biogeochemical cycles, lignin degradation, humification, and organic matter mineralization. In detail, in S_{mng} soils, significant increases were observed for the following parameters: (a) functional groups of culturable bacteria (e.g., actinomycetes, ammonifying, proteolytic, and free N-fixing bacteria); (b) activities of microbial enzymes (β-glucosidase and protease); (c) diversity indices (Shannon's diversity index, evenness, and richness) obtained from Biolog assay; and (d) number and type of bacterial taxa involved in C and N dynamics,

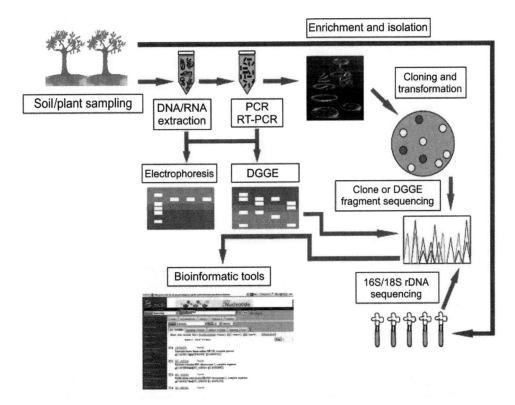

FIG. 22.3 An overview of the methods used for studying plant and soil microorganisms.

particularly symbiont N-cycling bacteria and the abundance of their related genes (*nifH* and *amoA*). Generally, the data revealed a significantly higher microbial abundance, activity, diversity, and complexity in S_{mng} soils. This higher biodiversity could lead to greater soil stability and multifunctionality, positively affecting also plant status and product quality. The results obtained in these papers highlighted that in Mediterranean orchards, under semiarid climatic conditions, the application of endogenous and/or exogenous organic matter can be a key factor to enhance soil quality/ fertility and produce in a sustainable way, preserving natural resources and avoiding detrimental effects on the environment. The composition of the microbiota of S_{mng} and C_{mng} soils resulted to be significantly different, particularly for the bacteria involved in soil N cycle, lignin degradation, and humification, more abundant in the S_{mng} system.

The aim of the study of Casacchia et al. (2010) was to explore the effects of the application of S_{mng} system on the diversity of important groups of microorganisms (fungi and actinomycetes), bacterial species (*Pseudomonas* spp., *Bacillus* spp., and *Azotobacter* spp.), and functional groups (proteolytic and ammonifying bacteria) related to soil fertility. The culture-based analysis revealed that higher populations of total bacteria, actinomycetes, and proteolytic bacteria were induced by the S_{mng} application, whereas *Pseudomonas* spp., *Azotobacter* spp., and ammonifying bacteria showed increased populations in the C_{mng} system. Furthermore, the higher inputs of organic matter of the S_{mng} plot caused an increase in the microbial groups responsible for N metabolism in well-watered zones and higher activities of decomposer and humus-forming microorganisms in the drier areas. Similarly, in a recent paper, Pascazio et al. (2018) applied some microbial indicators of soil quality in drip-irrigated olive and peach orchards managed with sustainable agricultural practices, finding that irrigation plays a key role in the determination of soil quality. This latter was evaluated by the authors by measuring and following two parameters: (a) a biochemical soil indicator (N_c/N_k ratio) based on soil N/C turnover and soil enzyme activities and (b) the abundance of three important N-cycling genes (*nifH*, *amoA*, and *nosZ*). The N_c/N_k ratio exhibited all the attributes of a reliable soil fertility indicator. Both N_c/N_k and gene abundances gave a precise idea on N and C soil dynamics in the olive orchard examined.

The results of all these trials demonstrated that soil microorganisms respond significantly to a sustainable orchard management characterized by periodic applications of locally (cover crops and pruning residues) and externally (compost) derived organic matter. These sustainable practices caused an improvement in soil organic matter in the topsoil layers of the S_{mng} system, compared with the C_{mng}, with consequent increases in the abundance of soil N-cycling bacteria and of the values of the indices related to total microbial metabolic activity and diversity.

4 Microbial translocation from soil to plant

Microorganism-plant interactions, where both profit from each other, play an important role in orchards, positively affecting plant status and improving product quality (Pascazio et al., 2015). Indeed, microbial endophytes colonizing plant do not cause apparent damage and contribute to host plant's protection and survival. Plant microbiota forms a complex network with microbial diversity and community dynamically changing throughout the plant life cycle (Ying-Ning et al., 2017). The studies focused on epiphytic and endophytic microbiota of fruit trees grown under different agronomic systems can be useful for the promotion of plant growth and a higher crop quality (Pascazio et al., 2015). Many studies demonstrated that plant-associated microbes live either inside plant tissue or on the surface of plant organs (e.g., Berendsen et al., 2012; Bulgarelli et al., 2012). Particularly, the microbial inhabitants of the soil and phyllosphere (plant aerial surface) are considered epiphytes, whereas those residing within plant tissues (xylem sap) are the so-called endophytes. The relative abundance of microorganisms in different soil and plant compartments varies, as some ecological niches, like xylem vessels, are very selective and restrictive, whereas others, such as rhizospheric soils, are very rich of microbes (Fig. 22.4).

The microbial colonization of plants depends on some key factors, such as plant genotype, tissue, growth stage and physiological status, and on soil environmental conditions, as well as on some agricultural practices (Singh et al., 2009). It is known that the taxonomic diversity of plant-associated bacteria is tissue specific and that microorganisms exhibit a particular spatial compartmentalization within plants. Experimental supports provided that microorganisms reach the rhizosphere by chemotaxis toward root exudate components and that the preferred site of attachment and subsequent entry is the apical root zone (Bulgarelli et al., 2012). The differentiation zone and the intercellular spaces in the epidermis have been suggested to be preferential sites for colonization (Berendsen et al., 2012). Soil microenvironment, particularly in terms of temperature and humidity, affects the colonization of endophytes and their community structure.

Studies on olive plants revealed that soil sustainable management practices have positive effects on soil fertility and on fruit and leaves as concerns microbiological genetic diversity and metabolic activity (Sofo et al., 2014a; Pascazio et al., 2015). Particularly, the phyllosphere represents a niche with great agricultural and environmental importance,

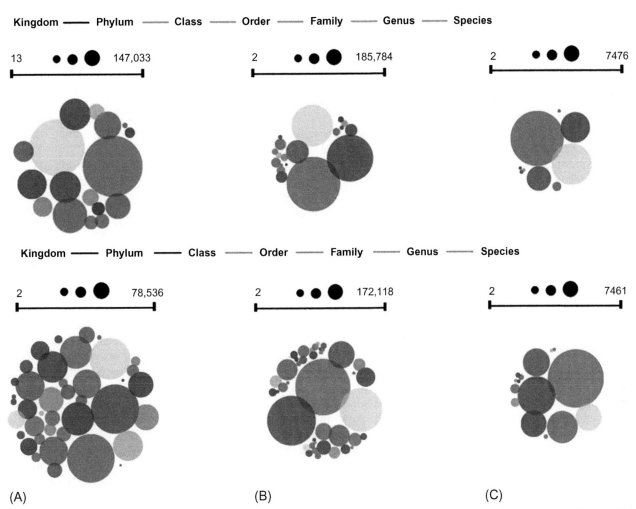

FIG. 22.4 Bubble charts of the values of bacterial relative abundance in (A) soil, (B) leaves, and (C) xylem sap at the phylum (row above) and class level (row below) in an olive orchard. *Each bubble* represents a bacterial taxon filled with a specific color, the size of which is proportional to the summary level of this taxon in the examined samples. The increasing scale of the total reads, represented by the *smaller bubbles* to the *bigger ones*, is displayed above each bubble chart. Xylem sap is the poorest in terms of microbial abundance of diversity.

as there is growing evidence that phyllosphere microbial inhabitants could affect the quality and productivity of agricultural crops, as they promote plant growth and contribute to host plant protection and survival (Rasche et al., 2006). In the study of Pascazio et al. (2015), phyllosphere and carposphere bacterial communities of olive trees subjected to two different soil management systems (S_{mng} and C_{mng}) were characterized in a mature olive grove. From a genetic point of view, the 16S ribosomal RNA eubacterial genes (16S *rDNA*) separated by DGGE and belonging to the bacteria living on leaf and fruit surface, and in fruit pulp, clustered separately in the two systems. Moreover, a clone library of 16S *rDNA* amplicons extracted from the bacteria living in pulp homogenates and a phylogenetic analysis were performed. The medium-term sustainable orchard management resulted in a higher number of bacterial species from olive fruit pulp. Finally, phyllosphere and carposphere communities evaluated by DGGE were affected by the type of the agricultural practices adopted.

These results are of key importance, as many endophytic microorganisms have been appreciated for their capacity to protect their hosts against insects-pests and pathogens (Azevedo et al., 2000). Furthermore, they could confer other important characteristics to plants, such as greater resistance to stress conditions, alteration in physiological properties, and production of phytohormones and compounds of biotechnological interest (Azevedo et al., 2000). Epiphytes and endophytes establish a symbiotic relationship with the plant and colonize an ecological niche like that of phytopathogens and are possible candidates for biocontrol agents. Thus, the identification and quantification of endophytic microflora present in and on plants provide information and data that may also have potential practical implications for disease management in fruit trees, which can deeply impact crop quality and production.

5 Future perspectives

A sustainable orchard management is a key factor for increasing the functionality and genetic diversity of soil microbiota that in turn enhances soil health, quality, and fertility (Fig. 22.5). This amelioration positively affects plant physiological status and productivity. The studies mentioned here confirm the need for Mediterranean orchards to encourage farmers to practice soil management based on organic matter inputs associated with zero/minimum tillage to improve soil functionality and the genetic diversity of soil microorganisms. The results deriving from these researches could promote the development of new approaches for optimizing soil carbon cycling, managing nutrient transport, and sustaining and improving olive yield and quality.

The amount and type of C and N inputs, the composition of N-cycling bacterial community, the study of C/N and pH imbalances, and the kinetic of soil C dynamics and N cycling have a major role in the genetic, functional, and metabolic responses of soil microorganisms, according to the Sofo et al. (2012, 2014b, 2019b). Among the agronomic sustainable practices, the input of soil organic matter as compost is another relevant factor affecting soil fertility in orchards. All these issues should be taken into account for planning and establishing new orchards based on sustainable agronomic techniques, or for fruit growing conversion from conventional/traditional to sustainable/innovative. Particular attention should be paid on the most important groups of microorganisms, including many plant-growth-promoting fungi/bacteria and biocontrol agents. Finally, a better understanding of phyllosphere and carposphere microbiota of cultivated trees could be useful for the promotion of plant growth, a better plant protection and a higher crop quality.

Although studies of microbial community based on isolation/culture techniques have been conducted for a long time in fruit groves, in the immediate future, next-generation sequencing coupled with bioinformatic tools and metagenomic approach will made it easier to comprehensively analyze microbial communities of plant tissues. One of the main advantages of metagenomics compared with culture-dependent methods is the ability to theoretically detect all microorganisms that cannot or are extremely difficult to culture, which represent about 99% of the total estimated microbial diversity, as well as rare taxa that are usually missed by culturing techniques. For this reason, studies aimed at characterizing the microbial communities of groves and at determining the effects of different management systems on microbial diversity of soil and plant through a metagenomic approach will be needed.

The importance of soil microorganisms to ecosystem services of fruit groves is often overlooked and must be taken into account in future fruit growing management strategies. Nature-based solutions are required to facilitate sustainable use and conservation of soils, including adaptation and resilience to climate change. Scientific synthesis of the current understanding of soil microorganisms is needed, and guidelines for future experimentation and best regenerative practices to exploit soil multifunctionality have to be developed, tested, and validated. The application of sustainable agricultural practices in fruit groves can positively affect the variability and composition of soil and plant

FIG. 22.5 The definition of soil health, soil quality, and soil fertility. Soil fertility is a subset of soil quality, in turn a subset of soil health.

microbial communities in orchards, promoting plant growth and plant protection, enhancing fruit production and quality, restoring soil fertility, improving resource use efficiency, and providing orchard multifunctionality, with evident benefits to the whole agroecosystem stability. For sure, a sustainable orchard management is a key factor for increasing the genetic diversity of soil microbiota that in turn enhances soil microbiological fertility. The final target of these researches should be to transfer science to farmers, enterprises, policy makers, and related end users in the domain of microbial ecology in fruit productive systems, to promote cost-effective sustainable land use and soil restoration strategies.

References

Azevedo, J.L., Maccheroni Jr., W., Pereira, J.O., 2000. Endophytic microorganisms: a review on insect control and recent advances on tropical plants. Electron. J. Biotechnol. 3, 40–65. https://doi.org/10.2225/vol3-issue1-fulltext-4.

Bai, Z., Caspari, T., Ruiperez Gonzalez, M., Batjes, N.H., Mäder, P., Bünemann, E.K., de Goede, R., Brussaard, L., Xu, M., Santos Ferreira, C.S., Reintam, E., Fan, H., Mihelič, R., Glavan, M., Tóth, Z., 2018. Effects of agricultural management practices on soil quality: a review of long-term experiments for Europe and China. Agric. Ecosyst. Environ. 265, 1–7. https://doi.org/10.1016/j.agee.2018.05.028.

Berendsen, R.L., Pieterse, C.M.J., Bakker, P.A.H.M., 2012. The rhizosphere microbiome and plant health. Trends Plant Sci. 17, 478–486. https://doi.org/10.1016/j.tplants.2012.04.001.

Bulgarelli, D., Rott, M., Schlaeppi, K., van Themaat, E.V.L., Ahmadinejad, N., Assenza, F., Rauf, P., Huettel, B., Reinhardt, R., Schmelzer, E., et al., 2012. Revealing structure and assembly cues for *Arabidopsis* root-inhabiting bacterial microbiota. Nature 488, 91–95. https://doi.org/10.1038/nature11336.

Bünemann, E.K., Bongiorno, G., Bai, Z., Creamer, R.E., De Deyn, G., de Goede, R., Fleskens, L., Geissen, V., Kuyper, T.W., Mäder, P., Pulleman, M., Sukkel, W., van Groenigen, J.W., Brussaard, L., 2018. Soil quality—a critical review. Soil Biol. Biochem. 120, 105–125. https://doi.org/10.1016/j.soilbio.2018.01.030.

Casacchia, T., Briccoli Bati, C., Sofo, A., Dichio, B., Motta, F., Xiloyannis, C., 2010. Long-term consequences of tillage, organic amendments, residue management and localized irrigation on selected soil micro-flora groups in a Mediterranean apricot orchard. Acta Hortic. 862, 447–452. https://doi.org/10.17660/ActaHortic.2010.862.69.

Celano, G., Palese, A.M., Ciucci, A., Martorella, E., Vignozzi, N., Xiloyannis, C., 2011. Evaluation of soil water content in tilled and cover-cropped olive orchards by the geoelectrical technique. Geoderma 163, 163–170. https://doi.org/10.1016/j.geoderma.2011.03.012.

Jansson, J.K., Hofmockel, K.S., 2018. The soil microbiome—from metagenomics to metaphenomics. Curr. Opin. Microbiol. 43, 162–168. https://doi.org/10.1016/j.mib.2018.01.013.

Li, Z., Zhao, B., Olk, D.C., Jia, Z., Mao, J., Cai, Y., Zhang, J., 2018. Contributions of residue-C and -N to plant growth and soil organic matter pools under planted and unplanted conditions. Soil Biol. Biochem. 120, 91–104. https://doi.org/10.1016/j.soilbio.2018.02.005.

Montanaro, G., Xiloyannis, C., Nuzzo, V., Dichio, B., 2017. Orchard management, soil organic carbon and ecosystem services in Mediterranean fruit tree crops. Sci. Hortic. 217, 92–101. https://doi.org/10.1016/j.scienta.2017.01.012.

Montanaro, G., Dichio, B., Briccoli Bati, C., Xiloyannis, C., 2012. Soil management affects carbon dynamics and yield in a Mediterranean peach orchard. Agric. Ecosyst. Environ. 161, 46–54. https://doi.org/10.1016/j.agee.2012.07.020.

Montanaro, G., Celano, G., Dichio, B., Xiloyannis, C., 2010. Effects of soil-protecting agricultural practices on soil organic carbon and productivity in fruit tree orchards. Land Degrad. Dev. 21, 132–138. https://doi.org/10.1002/ldr.917.

Palese, A.M., Ringersma, J., Baartman, J.E.M., Peters, P., Xiloyannis, C., 2015. Runoff and sediment yield of tilled and spontaneous grass-covered olive groves grown on sloping land. Soil Res. 53, 542–552. https://doi.org/10.1071/SR14350.

Palese, A.M., Vignozzi, N., Celano, G., Agnelli, A.E., Pagliai, M., Xiloyannis, C., 2014. Influence of soil management on soil physical characteristics and water storage in a mature rainfed olive orchard. Soil Tillage Res. 144, 96–109. https://doi.org/10.1016/j.still.2014.07.010.

Palese, A.M., Pasquale, V., Celano, G., Figliuolo, G., Masi, S., Xiloyannis, C., 2009. Irrigation of olive groves in Southern Italy with treated municipal wastewater: effects on microbiological quality of soil and fruits. Agric. Ecosyst. Environ. 129, 43–51. https://doi.org/10.1016/j.agee.2008.07.003.

Pascazio, S., Crecchio, C., Scagliola, M., Mininni, A.N., Dichio, B., Xiloyannis, C., Sofo, A., 2018. Microbial-based soil quality indicators in dry and wet soil portions of olive and peach Mediterranean orchards. Agric. Water Manag. 195, 172–179. https://doi.org/10.1016/j.agwat.2017.10.014.

Pascazio, S., Crecchio, C., Ricciuti, P., Palese, A.M., Xiloyannis, C., Sofo, A., 2015. Phyllosphere and carposphere bacterial communities in olive plants managed with different cultivation practices. Int. J. Plant Biol. 6, 6011. (15-19)https://doi.org/10.4081/pb.2015.6011.

Rasche, F., Marco-Noales, E., Velvis, H., van Overbeek, L.S., Lopez, M.M., van Elsas, J.D., Sessitsch, A., 2006. Structural characteristics and plant-beneficial effects of bacteria colonizing the shoots of field grown conventional and genetically modified T4-lysozyme producing potatoes. Plant Soil 289, 123–140. https://doi.org/10.1007/s11104-006-9103-6.

Riesenfeld, C.S., Schloss, P.D., Handelsman, J., 2004. Metagenomics: genomic analysis of microbial communities. Annu. Rev. Genet. 38, 525–552. https://doi.org/10.1146/annurev.genet.38.072902.091216.

Singh, G., Singh, N., Marwaha, T.S., 2009. Crop genotype and a novel symbiotic fungus influences the root endophytic colonization potential of plant growth promoting rhizobacteria. Physiol. Mol. Biol. Plants 15, 87–92. https://doi.org/10.1007/s12298-009-0009-7.

Sofo, A., Mininni, A.N., Fausto, C., Scagliola, M., Crecchio, C., Xiloyannis, C., Dichio, B., 2019a. Evaluation of the possible persistence of potential human pathogenic bacteria in orchards irrigated with treated urban wastewater. Sci. Total Environ. 658, 763–767. https://doi.org/10.1016/j.scitotenv.2018.12.264.

Sofo, A., Ricciuti, P., Fausto, C., Mininni, A.N., Crecchio, C., Scagliola, M., Malerba, A.D., Xiloyannis, C., Dichio, B., 2019b. The metabolic and genetic diversity of soil bacterial communities is affected by carbon and nitrogen dynamics: a qualitative and quantitative comparison of soils from an olive grove managed with sustainable or conventional approaches. Appl. Soil Ecol. 137, 21–28. https://doi.org/10.1016/j.apsoil.2018.12.022.

Sofo, A., Ciarfaglia, A., Scopa, A., Camele, I., Curci, M., Crecchio, C., Xiloyannis, C., Palese, A.M., 2014a. Soil microbial diversity and activity in a Mediterranean olive orchard using sustainable agricultural practices. Soil Use Manag. 30 (1), 160–167. https://doi.org/10.1111/sum.12097.

Sofo, A., Palese, A.M., Casacchia, T., Xiloyannis, C., 2014b. Sustainable soil management in olive orchards: effects on telluric microorganisms. In: Parvaiz, A., Rasool, S. (Eds.), Emerging Technologies and Management of Crop Stress Tolerance. In: A Sustainable Approach, vol. 2. Academic Press, San Diego, CA, ISBN: 978-0-12-800875-1, pp. 471–484. https://doi.org/10.1016/B978-0-12-800875-1.00020-X.

Sofo, A., Palese, A.M., Casacchia, T., Dichio, B., Xiloyannis, C., 2012. Sustainable fruit production in Mediterranean orchards subjected to drought stress. In: Ahmad, P., Prasad, M.N.V. (Eds.), Abiotic Stress Responses in Plants. Metabolism, Productivity and Sustainability. Springer, New York, ISBN: 978-1-4614-0633-4, pp. 105–129. https://doi.org/10.1007/978-1-4614-0634-1_6.

Sofo, A., Celano, G., Ricciuti, P., Curci, M., Dichio, B., Xiloyannis, C., Crecchio, C., 2010a. Changes in composition and activity of soil microbial communities in peach and kiwifruit Mediterranean orchards under an innovative management system. Soil Res. 48 (3), 266–273. https://doi.org/10.1071/SR09128.

Sofo, A., Palese, A.M., Casacchia, T., Celano, G., Ricciuti, P., Curci, M., Crecchio, C., Xiloyannis, C., 2010b. Genetic, functional, and metabolic responses of soil microbiota in a sustainable olive orchard. Soil Sci. 175 (2), 81–88. https://doi.org/10.1097/SS.0b013e3181ce8a27.

Ying-Ning, H., Dony, C.M., Chieh-Chen, H., 2017. Plant-microbe ecology: interactions of plants and symbiotic microbial communities. In: Yousaf, Z. (Ed.), Plant Ecology—Traditional Approaches to Recent Trends. Intech, Rijeka, pp. 93–119. https://doi.org/10.5772/intechopen.69088.

Zanella, A., Ponge, J.F., Topoliantz, S., Bernier, N., Juilleret, J., 2018. Humusica 2, Article 15: agro humus systems and forms. Appl. Soil Ecol. 122, 204–219. https://doi.org/10.1016/j.apsoil.2017.10.011.

Zhang, X., Xu, S., Li, C., Zhao, L., Feng, H., Yue, G., Ren, Z., Cheng, G., 2014. The soil carbon/nitrogen ratio and moisture affect microbial community structures in alkaline permafrost-affected soils with different vegetation types on the Tibetan plateau. Res. Microbiol. 165, 128–139. https://doi.org/10.1016/j.resmic.2014.01.002.

23

Mycorrhizosphere of fruit crops: Nature and properties

Jia-Dong He, Bo Shu, Qiang-Sheng Wu*

College of Horticulture and Gardening, Yangtze University, Jingzhou, China
*Corresponding author. E-mail: wuqiangsh@163.com

1 Introduction

The emergence of chemical fertilizers and balanced fertilization technology is an important symbol of the first agricultural industry revolution. The application of chemical fertilizers has made indelible achievements in agricultural production, whereas, as a high energy-consuming product, chemical fertilizers seriously damage the ecological environment and endanger human health. At the same time, it is limited by the lack of phosphorus and potassium mineral resources and the safety of agricultural products. More and more countries will gradually reduce the output and use of chemical fertilizers such as developed countries in Europe and the United States in the future. Microbial fertilizers refer to a type of fertilizers that contains the life activities of microorganisms as the core and enables crops to obtain specific fertilizer effects. The traditional concept of microbial fertilizer is inoculant or solid living microbial preparation made by the microbial solution adsorbed by aseptic carrier, which contains a large number of beneficial microorganisms. When applied to the soil, microbial fertilizers can fix nitrogen in the air, activate nutrients in the soil, improve the nutrient status of plants, and/or produce active substances in the process of microbial life activities to stimulate plant growth. Among microbial fertilizers, mycorrhizal biofertilizers have been developed and used in field and are attracting more and more attention from all countries.

On the basis of fungus hyphal arrangement within cortical tissues, mycorrhizae can be simply classified as ectomycorrhizae, endomycorrhizae, and ectendomycorrhizae. Ectomycorrhizae modify plant roots into swollen and

A.K. Srivastava, Chengxiao Hu (eds.)
Fruit Crops: Diagnosis and Management of Nutrient Constraints
https://doi.org/10.1016/B978-0-12-818732-6.00023-X

profusely branched as compared with nonmycorrhizal roots. The fungal hyphae form a "fungus mantle" around the feeder roots. Fungi penetrate the roots but remain around the cortical cells and form a Hartig net. Most of the ecto-mycorrhizae belong to *Basidiomycetes*, especially *Tricholoma*. In fruit trees, only chestnut of *Fagaceae* and hazelnut of *Betulaceae* has ectomycorrhizae. Endomycorrhizae are mostly formed by fungi belonging to the genera of *Acaulospora*, *Archaeospora*, *Entrophospora*, *Funneliformis*, *Gigaspora*, *Glomus*, *Sclerocystis*, and *Scutellospora*. However, some basidio-mycetes also can form endomycorrhizae. Ectendomycorrhizae form external mantle with hyphae (ectomycorrhizal character) and penetrate host cells for intracellular growth (endomycorrhizal character).

Arbuscular mycorrhizal fungi (AMF), a kind of beneficial soil microorganisms in plant rhizosphere, widely exist in natural ecosystems, which can form a symbiotic association, namely, arbuscular mycorrhizas (AMs), with about 80% of terrestrial plant roots (Jiang et al., 2018). In general, AMs include hyphae, entry points, inter- and extrahyphae, arbuscules, vesicles, and spores. External mycelium colonizes root systems through entry points and then establishes symbiotic association between roots and AMF, where the typical structure of arbuscule is formed in cortical cells. When plant roots establish symbiosis with AMF, AMF has the capacity to absorb soil nutrients and water for plant partners, and in return, host plants provide fatty acids and carbohydrates to maintain AMF growth (Jiang et al., 2017).

Mycorrhizosphere, formed by the changes in root exudates that affect the microbial communities around the roots, includes roots, hyphae of the directly connected mycorrhizal fungi, associated microorganisms, and the soil within their direct influence (Rambelli, 1973; Mukerji et al., 1997). Thus, mycorrhizosphere comprises both rhizosphere and hyphosphere. Mycorrhizosphere is responsible for the majority of mycorrhizal functions, because it contains various ecoenvironments, ecosystems, and substances released by mycorrhizas into the zone.

2 Mycorrhizosphere

The mycorrhizosphere is a soil zone where the physical, chemical, and microbiological processes are affected by plant roots and their associated mycorrhizal fungi (Timonen and Marschner, 2006). The mycorrhizosphere microbiota is different from the rhizosphere of nonmycorrhizal plants qualitatively and quantitatively. The soil microfauna affects the mycorrhiza formation and the host growth response (Fitter and Garbaye, 1994). The distinct difference between nonmycorrhizal rhizosphere and mycorrhizosphere is the existence of extracellular hyphae of mycorrhizal fungi (Giri et al., 2005). More specifically, mycorrhizosphere is a microecosystem in which plant roots, mycorrhizal fungi, other organisms and soil interact with each other (Bending et al., 2006). In this system, microorganisms colonized in root system, root surface, and mycorrhizosphere soil play an important role in activating nutrient elements, nutrient cycling, plant growth, and soil physical and chemical properties (Liu and Chen, 2007). Among them, the interaction between AMF and other microorganisms is an important aspect of the complex relationship of mycorrhizosphere. The unique microbial flora protective layer formed by mycorrhizosphere is the outermost defensive system to prevent pathogens from invading. Hence, AM symbiosis alters the number, and proportion of bacteria resistant to pathogens increased greatly (Fig. 23.1). The number of actinomycetes resistant to *Fusarium* and *Pseudomonas* spp. was significantly higher in the mycorrhizosphere of pure cultured plant of *Glomus fasciculatum* than in the non-AM one (Secilia and Bagyaraj, 1987). Meyer and Linderman (1986) also observed that the production of both sporangia and free spores of *Phytophthora cinnamomi* was inhibited in the mycorrhizosphere in chrysanthemum (Fig. 23.1).

AMF can selectively change the balance of microorganisms in mycorrhizosphere, stimulate the activity of micro-organisms antagonistic to soilborne pathogenic bacteria, and increase the number of some beneficial microorganisms (Liu and Chen, 2007). For example, AMF has mutually reinforcing effects on other antagonistic microorganisms such as *Trichoderma* spp., *Gliocladium* spp., and *Streptomyces* spp. and plant growth-promoting bacteria such as *Pseudomonas fluorescens*, *Bacillus*, *Rhizobium*, phosphorus-solubilizing bacteria, *Azotobacter*, and *Actinomycetes*. These beneficial microorganisms can be indirectly counteracted and resist the attack of pathogens (Harrier and Watson, 2004).

2.1 Arbuscular mycorrhizas in fruit crops

The roots of fruit trees living in the soil environment are infected by various soil microorganisms, such as AMF, nematodes, and *Rhizobium*. In the 1950s, China began to investigate fruit trees that can be symbiotic with AMF. The fruit trees with AMs were found in many fruits, such as citrus (Fig. 23.2A), ginkgo (Fig. 23.2B), plum (Fig. 23.2C), grape (Fig. 23.2D), apple, pear, peach, apricot, pomegranate, persimmon, loquat, hawthorn, kiwi, walnut, cherry, strawberry, banana, lychee, pineapple, longan, and mango. After AMs establish in the roots of fruit trees, it can develop a large hypha system in the soil surrounding the roots. The spreading mycelium system is the main absorption

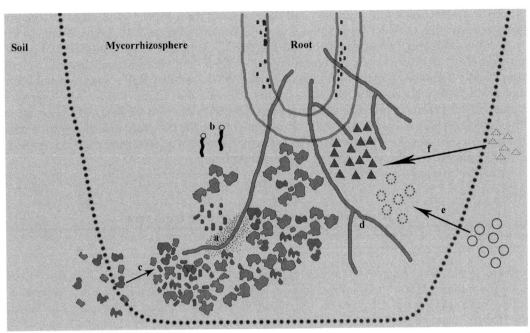

FIG. 23.1 Schematic view of different components in mycorrhizosphere. (a) Glomalin-related soil protein secreted by arbuscular mycorrhizal fungi (AMF) mycelia, (b) spore germination of AMF, (c) modification of soil structure and better soil fertility in mycorrhizosphere, (d) growth and development of extraradical hyphae, (e) increase of soil bacteria in mycorrhizosphere, and (f) decrease of pathogenic fungi in mycorrhizosphere.

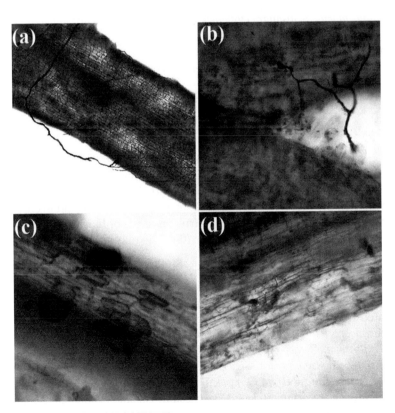

FIG. 23.2 Root mycorrhizal colonization of citrus (a), ginkgo (b), plum (c), and grape (d) in field.

organ of AMs. It can expand the absorption range of water and nutrients of fruit trees, activate mineral nutrients in soil, and promote the absorption of nutrient elements, especially easily fixed mineral elements such as phosphorus (P), zinc (Zn), iron (Fe), and copper (Cu). On the other hand, AMs can branch and grow between cortical tissues and cells in roots and invade cortical cells at the same time. After branching, AMF forms arbuscules. Through the arbuscule structure, mycorrhizal fungi can transfer water and nutrients absorbed by external hyphae from soil to fruit trees and exchange carbohydrates needed for their growth and development from fruit trees. A considerable portion of these carbohydrates is released into the rhizosphere in the form of glomalin (Wright et al., 1996). Glomalin is characterized by its stable performance and high preserved in the soil. In general, glomalin is defined as glomalin-related soil protein (GRSP) in soils according to the Bradford protocol (Rillig, 2004). GRSP can bind soil particles with a "super glue" ability, which can promote the soil aggregate formation and stability. Therefore, GRSP is seen as a stabilizer of soil structure formation, which can change soil moisture status (Spohn and Giani, 2010). AMF secretes GRSP into mycorrhizosphere of fruit crops to improve soil structure (Fig. 23.1) (Wu et al., 2008). AMF plays an important role in the growth of fruit crops. AMs can promote the absorption of specific mineral nutrients by fruit trees and replace chemical fertilizers. At the same time, mycorrhizas increase the content of organic matter in orchard soil and improve the structure and quality of soil through forming mycorrhizosphere.

3 Mycorrhizal fungal diversity in fruit crops

3.1 Fruit crops forming mycorrhizas

The mycorrhizal germplasm resources of horticultural plants are extremely abundant. According to incomplete statistics, there are more than 300 fruit trees known to be colonized by AMF: *Actinidia chinensis, Anacardium occidentale, Ananas comosus, Aronia melanocarpa, Artocarpus heterophyllus, Averrhoa carambola, Carica papaya, Cerasus tomentosa, Chaenomelis fructus, Citrus, Cocos nucifera, Coffea arabica, Crataegus pinnatifida, Dimocarpus longan, Diospyros kaki, Eriobotrya japonica, Ficus carica, Fortunella, Fragaria ananassa, Ginkgo biloba, Hippophae rhamnoides, Juglans regia, Litchi chinensis, Lonicera edulis, Macadamia integrifolia, Malus komarovii, Malus pumila, Mangifera indica, Mangifera persiciformis, Morella rubra, Musa supientum, Olea europaea, Padus asiatica, Passiflora edulis, Persea americana, Phoenix dactylifera, Pistacia vera, Poncirus, Prunus armeniaca, Prunus avium, Prunus persica, Prunus salicina, Psidium guajava, Punica granatum, Pyrus bretschneideri, Ribes nigrum, Rosa roxburghii, Rubus alleghaniensis, Rubus idaeus, Schisandra chinensis, Syzygium jambos, Theobroma cacao, Vitis amurensis, Vitis* spp., *Ziziphus jujuba*, etc. (Wu, 2010).

In short, the berry fruit trees except the bilberry and cranberry belong to the ericoid mycorrhizal type, and the others are arbuscular mycorrhizal types. Nut fruit trees are arbuscular mycorrhizal plants except pecans, chestnuts, and hazelnuts, which belong to ectomycorrhizal fungi.

3.2 Mycorrhizal fungal diversity in fruit crops

The investigation of AMF resources in evergreen fruit crops can be traced back to the investigation of 79 citrus orchards and nurseries in California and 66 citrus orchards and nurseries in Florida in 1981 (Nemec et al., 1981). Meanwhile, *Glomus fasciculatus, Glomus macrocarpus, Glomus etunicatus, Glomus constrictus*, and *Sclerocystis sinuosa* were found in citrus orchards of California and Florida, and *Glomus microcarpus* and *Glomus monosporus* were found in citrus orchards of California. *Gigaspora margarita* and *Glomus mosseae* were also found in citrus orchards of Florida. An interesting result was that *G. fasciculatus* could generally coexist with young citrus trees (0–30 years), while *G. constrictus* coexisted with old trees (30–70 years).

The earliest investigation of AMF resources on deciduous fruit crops was done by Miller et al. (1985). They investigated apple rhizosphere in Arkansas, California, Georgia, Illinois, Indiana, Iowa, Kansas, Kentucky, Massachusetts, Michigan, New York, Ohio, Oregon, Pennsylvania, Virginia, Washington, and Wisconsin. And 9 apple rootstocks from 18 apple orchards (*Malus domestica*) were investigated for AMF diversity. A total of 43 species of AMF were found. Among them, 36 species were identified, namely, *Acaulospora bireticulata, Acaulospora scrobiculata, Acaulospora spinosa, Gigaspora calospora, Gigaspora coralloidea, Gigaspora erythropa, Gigaspora gigantea, Gigaspora gilmorei, Gigaspora heterogama, G. margarita, Gigaspora pellucida, Gigaspora reticulata, Gigaspora rosea, Gigaspora verrucosa, Glomus albidum, Glomus bitunicatum, Glomus caledonium, Glomus constrictum, Glomus etunicatum, G. fasciculatum, Glomus fragile, Glomus geosporum, Glomus hyalosporum, Glomus invermaium, Glomus macrocarpum, Glomus maculosum, Glomus manihotis, Glomus melanosporum, Glomus microcarpum, Glomus monosporum, G. mosseae, Glomus occultum, Glomus stramentotinctum, Glomus versiforme, Sclerocystis rubiformis*, and *S. sinuosa*. Seven species were not identified, including *Acaulospora* sp., *Gigaspora* sp., and *Sclerocystis* sp. and four *Glomus* sp. In addition, *A. spinosa, G. coralloidea, G. erythropa, G. gilmorei, G. margarita, G. pellucida, G. reticulata, G. verrucosa, G. albidum, G. bitunicatum, G. caledonium, G. etunicatum, G. hyalosporum, G. invermaium, G. maculosum, G. manihotis, G. mosseae, G. occultum, Glomus* sp. 2, *G. stramentotinctum*, and *G. versiforme* could successfully propagate in pots with apples and sorghum.

There are 18 kinds of AMF in rhizosphere of Troyer citrange orchards in India (Vinayak and Bagyaraj, 1990). They were *G. fasciculatum* (three different ecological species), *G. mosseae* (two different ecological species), *G. albidum*, *G. macrocarpum*, *Glomus merredum*, *Glomus velum*, *G. monosporum*, *G. caledonium*, *G. calospora* (two different ecological species), *G. margarita* (two different ecological species), *Acaulospora* sp., *Acaulospora laevis*, and *Sclerocystis dussi*.

In October 2001, Cheng and Baumgartner (2004) investigated the AMF diversity in 10 vineyards in northern California. A total of 19 species of AMF were found in rhizosphere. Among them, 16 identified species were *Glomus aggregatum*, *Glomus claroideum*, *Glomus clarum*, *G. etunicatum*, *G. fasciculatum*, *Glomus fistulosum*, *G. geosporum*, *Glomus leptotichum*, *Glomus microaggregatum*, *G. mosseae*, *Glomus pulvinatum*, *Glomus scintillans*, *Glomus sinuosum*, *Glomus spurcum*, *Entrophospora infrequens*, and *Paraglomus occultum*. In addition, there were three species of AMF that have not been identified, all of which belong to *Glomus*.

Ananthakrishnan et al. (2004) separated AMF species from five cashew nut plantations in Tamil Nadu, southern Indian. A total of 23 AMF resources were found, belonging to 4 genera, including 14 species of *Glomus* (*G. aggregatum*, *G. clarum*, *G. constrictum*, *Glomus deserticola*, *G. fasciculatum*, *G. geosporum*, *Glomus intraradices*, *G. macrocarpum*, *G. microcarpum*, *G. mosseae*, *Glomus multisubstensum*, *G. occultum*, *Glomus radiatum*, and *G. scintillans*), 4 species of *Acaulospora* (*Acaulospora elegans*, *Acaulospora foveata*, *A. laevis*, and *Acaulospora morrowiae*), 3 species of *Gigaspora* (*Gigaspora candida*, *Gigaspora decipiens*, and *G. margarita*), and 2 species of *Scutellospora* (*Scutellospora calospora* and *Scutellospora nigra*). Among them, *Glomus* is the most abundant, especially *G. aggregatum*, *G. fasciculatum*, and *G. mosseae* appeared most frequently in orchards.

There were 35 species of AMF resources in the rhizosphere of 14-year-old apple trees (*M. domestica*) in traditional and organic apple orchards in southwestern Brazil, of which 23 were identified as *A. bireticulata*, *Acaulospora denticulata*, *A. foveata*, *A. laevis*, *Acaulospora mellea*, *A. morrowiae*, *Acaulospora rugosa*, *A. scrobiculata*, *A. spinosa*, *Acaulospora tuberculata*, *Archaeospora trappei*, *E. infrequens*, *Entrophospora kentinensis*, *G. claroideum*, *G. etunicatum*, *G. fasciculatum*, *G. geosporum*, *G. invermaium*, *G. mosseae*, *G. microaggregatum*, *G. spurcum*, *Paraglomus brasilianum*, and *P. occultum*; 12 species were not identified, including 4 species from *Acaulospora*, 6 species from *Glomus*, 1 species from *Gigaspora*, and 1 species from *Scutellospora* (Purin et al., 2006).

Trindade et al. (2006) collected 67 soil samples from 47 papaya commercial orchards in the southwest of Salvador and north of San Espiritu, Brazil, to investigate the AMF resources. There were 24 AMF strains identified, including *Acaulospora delicata*, *Acaulospora dilatata*, *Acaulospora longula*, *A. mellea*, *A. morrowiae*, *A. rugosa*, *A. scrobiculata*, *Entrophospora colombiana*, *P. brasilianum*, *P. occultum*, *G. etunicatum*, *G. intraradices*, *G. macrocarpum*, *G. mosseae*, *Glomus tortuosum*, *Scutellospora cerradensis*, *Scutellospora heterogama*, *Scutellospora pellucida*, and *Scutellospora persica*. Four species were not identified, belonging to *Gigaspora* (one species), *Glomus* (one species), *Acaulospora* (one species), and *Entrophospora* (one species).

In a word, mycorrhizal fungal diversity in fruit crops is relatively abundant in mycorrhizosphere. It can provide superior AMF community in rhizosphere of fruit crops than nonmycorrhizosphere. As a result, mycorrhizosphere possesses more potential in improving tree growth and maintaining soil health of fruit crops.

4 Mycorrhizosphere roles in fruit crops

In fruit crops, mycorrhizosphere can increase nutrient absorption of host plants, improve rhizospheric environments, and enhance stress resistance (drought stress, salt stress, temperature stress, and environmental pollution), disease and pest. Some abiotic stresses such as drought stress, salt stress, and temperature stress negatively affect fruit yield and quality. Many studies had shown that mycorrhizal inoculation can prove useful for promoting plant growth and development, and the use of AMF has a significant effect on improving the tolerance of fruit trees to abiotic stresses (Table 23.1).

4.1 Increase in nutrient absorption

A role of mycorrhizosphere in fruit crops mainly promotes the absorption of P, nitrogen (N), potassium (K), Zn, Fe, Cu, calcium (Ca), and other elements, especially P. Interestingly, under drought stress, the effect of mycorrhizal fungi on nutrient uptake is more important than under sufficient water conditions. As reported by Wu and Zou (2009), AMF-increased mineral nutrient concentrations were higher under soil water deficit than under ample water in trifoliate orange. Xie et al. (2010) inoculated AMF (*G. mosseae* and *G. margarita*) during grape plantlets to study the colonization process of AMF during micropropagation seedling transplantation and analyzed the changes of root

TABLE 23.1   Impacts of mycorrhizae on fruit crops under different stress conditions.

Type of stress	Host plants	AMF	Effects	References
Drought stress	Papaya (*C. papaya* var. Solo)	*Gigaspora margarita*	Leaf water potential ↑; plant fresh weight ↑; ethylene concentration ↑	Cruz et al. (2000)
	Pistachio (*Pistacia vera* L. cv. Akbari)	*Glomus etunicatum*	Growth performance ↑; soluble sugars ↑; proteins ↑; flavonoid ↑; proline ↑; P, K, Zn, and Cu ↑	Abbaspour et al. (2012)
	Trifoliate orange (*Poncirus trifoliata* L. Raf.)	*Funneliformis mosseae*	Plant growth performance ↑; leaf moisture status ↑; diverse responses of seven aquaporins to drought	He et al. (2019a)
Waterlogging stress	Peach (*Prunes persica* L. Batsch)	*Gigaspora margarita*	Ethanol accumulation ↓	Rutto et al. (2002)
	Sclerocarya birrea	*Gigaspora margarita*	N, P, Ca, and Mg ↑	Muok and Ishii (2006)
	Citrus (*Citrus junos*)	*Diversispora spurca*	Plant growth ↑; root system architecture ↑; antioxidant enzyme activities ↑	Wu et al. (2013)
	Trifoliate orange (*Poncirus trifoliata*)	*Diversispora spurca*	Superoxide dismutase ↑; catalase ↑	Zou et al. (2014)
	Peach (*Prunes persica* L. Batsch)	*Funneliformis mosseae*	Chlorophyll *a*, *b*, *a + b* ↑; proline ↑	Tuo et al. (2015)
Salt stress	Citrus (*Citrus tangerina* Hort. ex Tanaka)	*Glomus mosseae* and *Paraglomus occultum*	Plant growth, root morphology, photosynthesis, and ionic balance ↑	Wu et al. (2010)
	Strawberry (*Fragaria × ananassa* Duch.)	*Glomus irregulare*	Root morphology ↑	Fan et al. (2011)
	Beach plum (*Prunus maritima*)	*Funneliformis mosseae*	K^+/Na^+ ↑	Zai et al. (2014)
	Trifoliate orange (*Poncirus trifoliata* L. Raf.)	*Diversispora versiformis*	Root biomass, length, surface area, volume, and the overground growth ↑	Zhang et al. (2017)
	Trifoliate orange (*Poncirus trifoliata* L. Raf.)	*Rhizoglomus intraradices*	Root brassinosteroids, abscisic acid, methyl jasmonate, and indole-3-acetic acid levels under nonsalt stress ↑; root abscisic acid, indole-3-acetic acid, and methyl jasmonate concentrations under salt stress ↑	He et al. (2019b)
Temperature stress	Strawberry (*Fragaria × ananassa*, Duch., cv. Nohime)	*Gigaspora margarita*, *Glomus fasciculatum*, *G. mosseae*, *Glomus* sp. R10 and *G. aggregatum*	Tolerance against high temperature ↑; browning of leaves and roots ↓	Matsubara et al. (2004)
	Citrus (*Citrus tangerina* Hort. ex Tanaka)	*Glomus mosseae*	Plant growth ↑, photosynthesis ↑, root morphology ↑ and part nutrient uptake ↑ at moderate temperature (25°C), but not at low temperature (15°C)	Wu and Zou (2010)
	Blueberry (*Vaccinium* spp.)	*Scleroderma* sp., *Boletus* sp., *Coprinus* sp. and *Tricholoma* sp.	Superoxide dismutase, peroxidase and catalase under low-temperature stress ↑	Luo et al. (2017)

activity during mycorrhizal process and the promotion effect of AM on the growth and development of micropropagation seedlings. The results showed that AMF played an important role in promoting the growth and development of grape micropropagation seedlings and could promote root activity. There was a significant correlation between the activity of phosphatase and the colonization rate of AMF in the root of micropropagation seedlings. Hence, AMF inoculation could enhance root activity, promote root uptake and accumulation of mineral nutrients such as N and P, and promote plant photosynthesis, so as to improve the growth potential of seedlings. As stated by Wu et al. (2011), mycorrhizal mycelium also sustained better nutritional (especially P) uptake and water absorption in trifoliate orange (*Poncirus trifoliata*) seedlings though drought stress seriously decreased the active, functional, and total hyphae activities. Hence, the key physiological mechanism of mycorrhizal fungi in improving drought resistance of host plants is that AMF extraradical mycelium enhances nutrient uptake of host plants. In a pot experiment, Liu et al. (2018) found that P content and P uptake of trifoliate orange seedlings increased significantly when inoculated with *Funneliformis mosseae*. In 2011, a study of Khalil et al. showed that the inoculation of AMF could significantly increase the levels of P, K, Mg, and Zn in leaves of sour orange and Volkamer lemon rootstocks, while, in the leaf, N, Ca, and Na significantly decreased. *Pyrus pashia* Buch.-Ham. seedlings were used in pot experiment and inoculated with *G. intraradices, G. mosseae*, and *G. versiforme* in four stages of seedlings (germination, three leaves, five leaves, and seven leaves) (Yang et al., 2016). The results showed that all the three AMF used in the experiment could effectively colonize the seedlings, and mycorrhizal colonization rate of *G. intraradices* was significantly higher than that of *G. mosseae* and *G. versiforme*. The formation of AM significantly increased the longest lateral root length, the number of first-order new roots, and the volume of roots. The root activity and soluble protein content in roots of *P. pashia* were significantly increased by AM presence (Yang et al., 2016). Also in 2016, Cheng and Zeng studied the effects of AMF on growth, mineral nutrient absorption, photosynthesis, and protective enzyme systems of kiwifruit (*Actinidia deliciosa*) seedling by inoculating *G. mosseae* and *G. versiforme*. The results showed that AMF could significantly increase the absorption of N, K, and P. And the inoculation of AMF significantly improved the photosynthesis and antioxidant enzyme system of kiwifruit. The study also found that *G. mosseae* treatment was superior to *G. versiforme* in P absorption of kiwifruit seedlings. However, compared with *G. mosseae*, the effect of *G. versiforme* on N and K concentrations of kiwifruit seedlings was more significant. As reported by Liu et al. (2017), *F. mosseae* inoculation considerably increased root biomass, number of first- and second-order lateral roots, root acid phosphatase activity, and the transcript level of root acid phosphatase gene *PtPAP1* of trifoliate orange under 1 mmol L P conditions. Under 0.1 mmol L and 1 mmol L P, mycorrhizal symbiosis could markedly increase the relative expression of root P transporter gene (*PtaPT3, PtaPT5,* and *PtaPT6*) (Liu et al., 2017). It concludes that the increase in nutrient absorption by mycorrhizosphere might be due to the root morphological improvement, the release of acid phosphatase into soil, and the enhancement of root acid phosphatase gene and root P transporter genes.

4.2 Enhancement in drought tolerance

Globally, drought stress is the most serious and widespread abiotic stress. The effect on mycorrhizas on drought tolerance of fruit crops is well understood. Water deficit can cause root development of fruit trees to be blocked and water and nutrient supply to be insufficient, thus reducing fruit quality. In Cruz et al., 2000, Cruz et al. inoculated AMF (*G. margarita*) into papaya (*C. papaya*) in a pot experiment. They found that the leaf water potential decreased during the water-stress treatment and this decrease was more severe in the non-AM plants. Plant fresh weight was higher for AM than both non-AM plants. Under well-irrigated conditions, the ethylene concentration in the roots was increased by the presence of AMs, although there was no significant difference between AM and non-AM roots in 1-aminocyclopropane-1-carboxylic acid levels. 1-Aminocyclopropane-1-carboxylic acid increased in both AM and non-AM roots under water-stress conditions. The drought treatment resulted in a significant increase in ethylene concentration in non-AM roots, but the concentration in AM roots was slightly lower than under normal conditions. A study of Abbaspour et al. (2012) in 3-month-old pistachio (*P. vera* L. cv. Akbari) showed that, regardless of water status, the inoculated seedlings with *G. etunicatum* had a greater growth performance and higher contents of soluble sugars, proteins, flavonoid, proline, P, K, Zn, and Cu than non-AM seedlings. The inoculation of AMF increased the activities of peroxidase enzyme in shoot and root of pistachio in both well-watered conditions and drought stress conditions. Such enhancement in physiological activities by mycorrhizosphere confers greater drought tolerance in mycorrhizal pistachio. Recently, a potted experiment was conducted to investigate the effects of an AMF, *F. mosseae*, on plant growth, leaf water status, root abscisic acid, and relative expression level of root tonoplast intrinsic protein (TIP) genes of trifoliate orange seedlings under well-watered and drought stress (He et al., 2019a). They had a conclusion that, in both well-watered and drought stress conditions, AMF significantly enhanced leaf relative water content, leaf water potential, plant growth performance, and root abscisic acid levels. Meanwhile, under well-watered condition, AMF treatment upregulated *PtTIP1;2, PtTIP2;1, PtTIP4;1,* and *PtTIP5;1* and downregulated *PtTIP1;1* and *PtTIP2;2*. Under drought stress condition, AMF seedlings showed higher expression levels of root *PtTIP1;2, PtTIP1;3,* and *PtTIP4;1* and lower expression levels of root *PtTIP2;1* and *PtTIP5;1*. This implies that mycorrhizas had two responsive mechanisms in terms of *TIP* expression pattern: drought tolerance by upregulation of *TIPs* and drought resistance by downregulation of *TIPs*.

4.3 Enhancement in waterlogged tolerance

Waterlogging, an abiotic stress, often results in anoxic respiration as its hypoxic conditions (Elzenga and van Veen, 2010; Tanaka et al., 2011). As a result, fruit crops grown in waterlogging have bad root hydraulic conductivity, stomatal aperture, photosynthetic capacity, and nutrient availability (Ashraf, 2012; Yin et al., 2012). Earlier studies indicated that AM citrus and peach plants presented greater plant growth performance and plant biomass than non-AM plants under waterlogging stress (Rutto et al., 2002; Wu et al., 2013; Zou et al., 2014). Rutto et al. (2002) found that AMF (*G. margarita*) could significantly reduce ethanol accumulation in peach (*P. persica* L. Batsch) roots under flooding conditions, showing a healthier growth condition of flooded roots. However, the flooding tolerance of peach seedlings to AMF colonization may be limited. Muok and Ishii (2006) discovered that *G. margarita* was able to increase the tolerance of *Sclerocarya birrea* to drought stress, salt stress, and flooding stress. Further, under waterlogging stress, plant height of potted *Citrus junos* (Wu et al., 2013) and potted *P. trifoliata* (Zou et al., 2014) seedlings was significantly increased by *Diversispora spurca* inoculation. After 12 days of flooding treatment, AMF (*F. mosseae*) significantly increased the concentrations of chlorophyll *a*, *b*, *a* + *b*, and proline accumulation, thus improving the waterlogging resistance of potted peach (*P. persica*) (Tuo et al., 2015). As a result, mycorrhizosphere in fruit crops can provide superior physiological activity in plants to tolerate waterlogging, whereas more molecular studies need to be conducted.

4.4 Enhancement in salt tolerance

Salt stress, one of the important abiotic stresses in plants, heavily declines fruit crop growth and yields. The response of fruit plants to salt stress depends on a variety of factors, in which phytohormones are considered to be the most important substances involved in the tolerance or susceptibility (Mahmud et al., 2016). The mechanism of phytohormones in regulating the salt responses can be divided into two groups, positive-related hormones and stressed hormones. The former includes auxin, cytokinins, gibberellins, and brassinosteroids (BRs). Abscisic acid (ABA), jasmonic acid, and salicylic acid belong to the latter (Kosová et al., 2012). In trifoliate orange, He et al. (2019b) found that mycorrhizal (*Rhizoglomus intraradices*) plants had significantly higher root BRs, ABA, methyl jasmonate, and indole-3-acetic acid levels under nonsalt stress and higher root ABA, indole-3-acetic acid, and methyl jasmonate concentrations under salt stress. Wu et al. (2010) conducted an experiment on *Citrus tangerina* Hort. ex Tanaka under salt stress and observed greater stomatal conductance, net photosynthetic rate, and transpiration rate of AM citrus. In the study of Fan et al. (2011), the inoculation of *Glomus irregulare* significantly increased root length percentages of medium and coarse roots, shoot and root biomass, root-to-shoot ratio and specific root length, and the root morphology of three elite strawberry cultivars. A study of Zai et al. (2014) found that the synergic effect of *F. mosseae* and a phosphate-solubilizing fungus could enhance the salt tolerance of beach plum (*Prunus maritima*) by regulating the stability of Na$^+$ contents, K$^+$ contents, and K$^+$/Na$^+$ contents. Zhang et al. (2017) found that AMF (*Diversispora versiformis*) significantly increased the parameters of root biomass, length, surface area, volume, and the overground growth in trifoliate orange seedlings under salt stress.

4.5 Enhancement in temperature stress tolerance

Among many abiotic stresses, temperature is also one of the important environmental factors affecting growth and productivity development of fruit crops. In growing season, temperature stress, including high temperature and low temperature, can negatively affect crop growth and play a decisive role in yield (Wahid et al., 2007). AM symbiosis represents potential functionings on tolerance of temperature stress in host plants (Ruotsalainen and Kytöviita, 2004; Zhu et al., 2015). After 10 weeks of inoculation with five different AMF species (*G. margarita*, *G. fasciculatum*, *G. mosseae*, *Glomus* sp. R10, and *G. aggregatum*), Matsubara et al. (2004) found that the browning on strawberry (*Fragaria* × *ananassa*, Duch., cv. Nohime) leaves and roots under high temperature stress was significantly relieved. Among them, *G. mosseae* and *G. aggregatum* had the best ability to alleviate the browning of strawberries under high temperature. Such alleviation in the strawberry browning caused by high temperature under mycorrhization is dependent on AMF species. However, opposite results were appeared on *C. tangerina* (Wu and Zou, 2010) under low temperature (15°C) stress. AMF-colonized *C. tangerina* seedlings showed lower net photosynthetic rate and plant growth responses, compared with non-AM seedlings (Wu and Zou, 2010). However, mycorrhizal functionings on mitigating high-temperature stress were positively shown in citrus plants (Wu and Zou, 2010). Luo et al. (2017) selected four mycorrhizal fungi (*Scleroderma* sp., *Boletus* sp., *Coprinus* sp., and *Tricholoma* sp.) to inoculate blueberry seedlings by bacterial solution and investigated the mycorrhizal colonization after 90 days of inoculation subjected to low-temperature stress treatment. The results showed that the four mycorrhizal fungi could colonize blueberry seedlings, and the highest inoculation rate and the lowest inoculation rate occurred in plants inoculated with *Coprinus* sp. and *Tricholoma* sp. separately. The mycorrhizal blueberry seedlings possessed higher activities of superoxide dismutase, peroxidase, and catalase under low-temperature stress. Among them, the efficacy was *Boletus* sp. > *Scleroderma* sp. = *Coprinus* sp. > *Tricholoma* sp. in the decreasing order.

4.6 Mitigation of soil replant disease

Replant disease refer to the abnormal growth and development of crops caused by continuous cultivation of the same crop or related crops on the same soil. Symptoms are generally poor growth and development and yield and quality decline. Most of the injured plants have browning, branching reduction, low vigor, and narrow distribution range, resulting in a decline in the ability to absorb water and nutrients. The disease has a serious impact on the yield and quality of fruit trees worldwide. Especially with the large-scale intensive development of modern agriculture, the cost of orchard reconstruction and planting is huge, which has become an important bottleneck for sustainable development. Generally, replant disease of fruit trees is mainly caused by allelopathy, autotoxicity, and the imbalance of both soil physical-biochemical traits and soil microflora.

Early studies conducted by Čatská (1994) found that AMF application could replace soil chemical treatment to mitigate replant disease of apple (*M. domestica*). Waschkies et al. (1994) found that AMF intervention could significantly reduce soil pathogenic fungi (*Phytophthora parasitica*, *Pythium ultimum*, and *Fusarium oxysporum*) and soil pathogenic bacteria (*Pseudomonas syringae* and *P. solanacearum*) in the soils of replant disease grapevine. In addition, AMF could increase the number of bacteria and actinomycetes in the soils of replant disease watermelon (*Citrullus lanatus*) and decrease the number of harmful fungi (Zhao et al., 2010). Similar results were obtained in apple (*M. domestica*) under replanting conditions, in which AMF inoculation increased the number of fungi, bacteria, and actinomycetes in the soil, thus enabling host plants to have better pH and nutrient levels (Mehta and Bharat, 2013). Guo et al. (2009) reported that the inoculation of *G. versiforme* in grape (*Vitis vulpina* × *Vitis labrusca* "Beta") could increase the released nerolidols in root exudates, but not in non-AMF grape. Furthermore, Hu (2016) found that *G. mosseae* could change the contents of allelochemicals (especially phenolic acid allelochemicals) of strawberry (*Fragaria* × *ananassa* "Sweet Charlie"). Based on previous studies, Lü and Wu (2018) summarized the potential mechanisms: (1) regulating the root and soil microflora balance; (2) enhancement of enzyme activities in the soil and plants; (3) changes in root exudates; (4) greater soil aggregate stability by GRSP; (5) balanced nutrient levels in plants, and (6) greater root system architecture.

4.7 Potential roles in mitigating heavy metal pollution

Since Bradley et al. reported in 1981 that the mycorrhiza could reduce the absorption of excessive heavy metals Cu and Zn, the role of mycorrhizal symbionts in plant adaptation to heavy metal pollution has received increasing attention. A large number of studies have confirmed that mycorrhizal fungi could help plants absorb mineral nutrients and promote their growth in heavy metal–contaminated soil, thereby enhancing the tolerance of plants to heavy metal pollution (Chen et al., 2003; Meier et al., 2012; Wu et al., 2014). On the other hand, mycorrhizal fungi could absorb and immobilize heavy metals themselves, thus reducing the toxicity of heavy metals to plants to a certain extent (Zhang et al., 2015).

Heavy metal contamination could affect the diversity of mycorrhizal fungi (Colpaert et al., 2011). It is generally believed that indigenous mycorrhizal fungi in heavy metal–contaminated soil have strong tolerance to heavy metals, and mycorrhizal fungi that can survive in heavy metal–contaminated habitats have certain adaptability to contaminated environment (Weissenhorn et al., 1993; Redon et al., 2009; Vallino et al., 2011). When spore germination experiments were carried out in the culture medium containing $Cd(NO_3)_2$, Weissenhorn et al. (1993) found that the spore germination rate and mycelial growth ability of two *F. mosseae* isolated from the Cd-contaminated environment were better than those from the uncontaminated environment. Similarly, Sudová et al. (2007) also confirmed that *Rhizophagus irregularis* from heavy metal–contaminated soil was more tolerant to Pb than it from uncontaminated soil. Under the same concentration of Pb treatment, *R. irregularis* isolated from the polluted environment was significantly higher in mycorrhizal infection rate, arbuscular branch, and vesicle abundance than it from uncontaminated environment. And studies showed that *Gigaspora* and *Glomus* occurred more frequently in heavy metal–contaminated soils, which might be related to their higher sporulation rates (Lin et al., 2013). Among the ectomycorrhizal fungi, *Ascomycete* and *Basidiomycetes* were usually found in heavy metal–contaminated habitats (Baar et al., 1999; Trowbridge and Jumpponen, 2004). Therefore, mycorrhizal fungi originated from heavy metal–contaminated soils are considered to be highly adaptable to adversity.

Though the information of mycorrhizas on fruit crops is scarce, we believe that mycorrhizosphere in fruit crops has potential capacity to enhance heavy metal pollution tolerance. At present, more and more fruit crops are being planted in cities to make it easier for residents to visit and enjoy nature. Under such conditions, fruit crops possibly grow on heavy metal–contaminated soils. This provides a new route for the application of mycorrhizal fungi.

5 Mycorrhizosphere effects on soil ecology of orchards

The ecological effects of AMF on orchards play an important role in the material circulation of orchard ecosystems and the protection and restoration of orchard ecosystems. AMF is unable to perform photosynthesis, and it is inevitable to obtain carbohydrates from host plants and convert them into their own biomass, making mycorrhizosphere as an important carbon pool in the soil. The carbohydrates obtained from the host of AMF are used for their own synthesis and metabolism on the one hand and the mycorrhizosphere in the form of the secretion GRSP on the other hand. GRSP is the most important source of soil active organic carbon pool, and its content is 2–24 times higher than that of humus, which is considered to be one of the most important sources of soil organic carbon (Rillig et al., 2010). Mycorrhizosphere is involved in the circulation of inorganic elements and water. Extracellular hyphae increase the contacted area of roots to the soil. Hyphae can reach areas where roots cannot reach, which promote the absorption of water and nutrients from the soil. It is documented that extracellular hyphae absorb more than 80% of P, 60% of Cu, 25% of N, 25% of Zn, and other elements to host plants (Marschner and Dell, 1994).

AMF can increase the effectiveness of poorly soluble elements in the soil by secreting organic acids (Liang et al., 2002). It can also exhibit synergistic effects with nitrogen-fixing bacteria, phosphate-dissolving bacteria, and plant growth–promoting bacteria in soil; increase soil biological activity; and jointly increase the total amount of inorganic elements in the orchard ecosystem cycle. Compared with well-watered condition, drought stress could significantly stimulate AMF releases GRSP into rhizosphere soil for better soil structure and aggregate stability (Wu et al., 2008). Meanwhile, under drought stress condition, the total GRSP levels in rhizosphere soil could be increased by the inoculation of *G. mosseae*, *Glomus diaphanum*, and *G. versiforme* (Wu et al., 2008). As a result, in mycorrhizal soils, extraradical mycelia secrete the GRSP to maintain superior soil structure under drought stress, which results in higher soil available water content than poorly structured nonmycorrhizal soils (Augé, 2001). GRSP has a chelation effect on heavy metals in soil, inhibiting their transfer from roots to shoots. Higher concentrations of GRSP in mycorrhizosphere can reduce the damage of soil structure and the growth and development of fruit trees to some extent.

6 Production of mycorrhizal fertilizers

Since the 1960s, the propagation of AMF has attracted much attention. So far, pure culture of arbuscular mycorrhizal fungi has not been successful. On the one hand, AMF are symbiotic fungi, which cannot be separated from host plants. On the other hand, it is difficult to avoid infection of other microorganisms in the process of pure culture. Therefore, the reproduction of AMF can only depend on living plants for propagation.

Nowadays, many countries in the world are actively carrying out research on aseptic cultivation of AMF. There are already many effective propagation techniques that have been created so far, such as pot culture, medium culture (Mosse, 1962; Mosse and Hepper, 1975), resting hydroponic culture (Crush and Hay, 1981), nutrition flow culture (Elmes and Mosse, 1984; Mosse and Thompson, 1984), dual axenic culture (St-Arnaud et al., 1996), field culture (Liu and Chen, 2007), and aeroponic culture (Jarstfer and Sylvia, 1995). Following is a detailed introduction of several viable methods of AMF propagation on fruit trees.

At present, the reproductive techniques of AMF in horticultural plants are mainly concentrated on the soil with AMF or on the indirect host roots that form AMF to propagate AMF. The commercial production of citrus arbuscular mycorrhizal fungi should be the most successful in horticultural plants. Menge (1983) reported a protocol of mycorrhizal production in citrus crops. He explained that the original AMF could be obtained from live roots in the field or inoculated on plant roots cultured in sterilized soil in greenhouse. The Sudan grass (*Sorghum vulgare* Pers.) was the most commonly used culture plant. Other plants such as tomato, soybean, corn, peanut, and asparagus could also be used as culture plants. The soil for cultivating plants should be low-nutrient sandy soil, irrigated with 1/2 Hoagland nutrient solution once a week. When a large number of spores were produced in potted Sudan grass, the spores were separated by wet sieve-centrifugation method and disinfected with sodium hypochlorite and streptomycin to ensure that the spores were free from harmful pathogens. These sterilized spores were then inoculated on the root segments of plants cultured under sterile conditions. The container for the root segment could be a plastic petri dish containing low-nutrient sand. After 1–4 weeks, when AMF had invaded the root segments, the root segments were cut into small segments and inoculated into potted host plants in sterilized soil of greenhouse. If there was no pathogen in the root segment, then potted Sudan grass in greenhouse can be used as the "mother plant" of inoculant in the production field.

In China, Shen and Wan (1994) proposed a technology to produce mycorrhized citrus plants in nursery. The specific operation is as follows: First, the biological host was used to carry out a simple propagation of the fungus agent, and then, the hotbed that had been used for the simple propagation of the fungus agent was used to sow trifoliate oranges. The roots of biological hosts could also be used as the source of fungi to colonize trifoliate oranges. After the seedlings were infected by AMF, they were moved out of the hotbed. Because the spores of AMF still existed in the hotbed, the seedlings with AMF could be obtained without reinoculation treatment when the seedlings were sown again. In this way, many batches of trifoliate oranges seedlings infected by AMF could be obtained.

Single-spore inoculation technique was first reported by Selvakumar et al. (2018).

They collected the spores of AMF from salt-affected soil in South Korea and inoculated single, healthy spores on filter paper wrapping a *Sorghum bicolor* L. seedling. And then, the successfully inoculated seedlings were selected and transferred into 1-kg capacity pots containing sterilized soil. The pots were maintained under controlled conditions (25°C, 12 h day/night, and relative humidity ∼70%) in a greenhouse for 120 days for the first plant cycle. Subsequently, seedlings and sterilized soil were transferred to 2.5-kg capacity pots containing sterilized soil for 120 days. Each pot received 0.5 × Hoagland's nutrient solution every week. The ingredients for Hoagland's nutrient solution were specifically used for AMF mass production (Habte and Osorlo, 2001). The potted plants were treated with drought stress by withholding water 15 days before the end of the experiment to induce spore germination. At the end of 120 days, the roots and soil were collected as inoculants.

In a word, all the earlier three methods of reproduction can obtain AMF agents to a certain extent, but the spore infectivity and genetic stability of these reproductions are low, and the reproductive coefficient is not high enough to meet the field application. Developing an efficient and stable reproductive system of mycorrhizal agents is an urgent task in the research of AMF. In a recent study of Wu et al. (2019), indigenous AMF were propagated in fresh root segments ($\Phi < 2$ mm), fresh rhizospheric soil (<4-mm size), and air-dried rhizospheric soil (<4-mm size) as the source collected from *Citrus unshiu* cv. Guoqing 1 grafted on trifoliate orange, in which white clover was used as host plants to propagate indigenous AMF. The results showed that the spore propagation rate of fresh root segments as a propagator was significantly higher than that of other treatments. After inoculated with AMF agents from fresh root segments, the effects of AMF agents on promoting growth and nutrient uptake and enhancing antioxidant enzyme activity of leaves were better than those of fresh rhizospheric soil and air-dried rhizospheric soil. And they got a conclusion that rapid proliferation of indigenous AMF from fresh root segments is a good choice. This provides a new idea for rapid and efficient propagation and application of fruit tree indigenous AMF.

7 Conclusions and outlook

Functionings of mycorrhizosphere have been summarized in Fig. 23.1. Hence, mycorrhizosphere has become a topic in international mycorrhizal researches. At present, many countries have established companies to produce mycorrhizal fertilizers and apply them to a certain extent, such as the United States, European countries, Japan, and Taiwan. They are mainly used for transplanting fruit crops, such as strawberry and citrus. China began to develop mycorrhizal biological agents in the 1980s, which developed rapidly in the 1990s and has made significant progress. Three efficient strains, *G. mosseae* 32, *G. versiforme*, and *G. margarita*, have been obtained by the Institute of Mycorrhizal Biotechnology of Qingdao Agricultural University. These efficient strains can increase the yield of watermelon and other crops by 15%–60% under field conditions. However, the current screening and domestication of strains in China is far from being able to adapt to the needs of production, especially the development of AMF agents specifically for fruit crops. In the future, efforts should be made to screen high-efficiency strains for different fruit crops and different environmental orchards and to create conditions for the commercial production and application of arbuscular mycorrhizal agents for fruit trees.

The screening and domestication of high-efficiency strains and the commercial production of mycorrhizal fungal agents are a key and difficult point in the current research of mycorrhizosphere. There is no substantial breakthrough in the pure culture technique of arbuscular mycorrhizal fungi, which is an important limiting factor for the production and application of AMF. At present, it can only be produced by the pure pot culture method, static liquid culture method, flowing liquid culture method, spray liquid culture method, etc., but the cost is high, which severely limits the large-area application of AM fertilizer. Therefore, the establishment of AM fungal agent production line suitable for Chinese national conditions and its application supporting technology is an urgent task.

Mycorrhizosphere is a microzone providing a microenvironment, which makes the difficulty to study and observe in situ. Possibly, there is no destructive root box that can be used for mycorrhizosphere researches.

Acknowledgements

This study was supported by the Hubei Agricultural Science and Technology Innovation Action Project, the Hubei Agricultural Major Technical Cooperation Project, and the Plan in Scientific and Technological Innovation Team of Outstanding Young Scientists, Hubei Provincial Department of Education (T201604).

References

Abbaspour, H., Saeidi-Sar, S., Afshari, H., Abdel-Wahhab, M.A., 2012. Tolerance of mycorrhiza infected pistachio (*Pistacia vera* L.) seedlings to drought stress under glasshouse conditions. J. Plant Physiol. 169, 704–709.

Ananthakrishnan, G., Ravikumar, R., Girija, S., Ganapathi, A., 2004. Selection of efficient arbuscular mycorrhizal fungi in the rhizosphere of cashew and their application in the cashew nursery. Sci. Hortic. 100, 369–375.

Ashraf, M.A., 2012. Waterlogging stress in plants: a review. Afr. J. Agric. Res. 7, 1976–1981.

Augé, R.M., 2001. Water relations, drought and vesicular-arbuscular mycorrhizal symbiosis. Mycorrhiza 11, 3–42.

Baar, J., Horton, T.R., Kretzer, A.M., Bruns, T.D., 1999. Mycorrhizal colonization of *Pinus muricata* from resistant propagules after a stand-replacing wildfire. New Phytol. 143, 409–418.

Bending, G.D., Aspray, T.J., Whipps, J.M., 2006. Significance of microbial interactions in the mycorrhizosphere. Adv. Appl. Microbiol. 60, 97–132.

Bradley, R., Burt, A.J., Read, D.J., 1981. Mycorrhizal infection and resistance to heavy metal toxicity in *Calluna vulgaris*. Nature 292, 335–337.

Čatská, V., 1994. Interrelationships between vesicular arbuscular mycorrhiza and rhizosphere microflora in apple replant disease. Biol. Plant. 36, 99–104.

Chen, B.D., Li, X.L., Tao, H.Q., Christie, P., Wong, M.H., 2003. The role of arbuscular mycorrhiza in zinc uptake by red clover growing in a calcareous soil spiked with various quantities of zinc. Chemosphere 50, 839–846.

Cheng, X.M., Baumgartner, K., 2004. Survey of arbuscular mycorrhizal fungal communities in northern California vineyards and mycorrhizal colonization potential of grapevine nursery stock. HortScience 39, 1702–1706.

Cheng, X.Y., Zeng, M., 2016. Effect of arbuscular mycorrhizal fungi on physiological metabolism and protective enzymes system of kiwifruit. South China Fruits 45, 56–60 (in Chinese with English abstract).

Colpaert, J.V., Wevers, J.H.L., Krznaric, E., Adriaensen, K., 2011. How metal-tolerant ecotypes of ectomycorrhizal fungi protect plants from heavy metal pollution. Ann. For. Sci. 68, 17–24.

Crush, J.R., Hay, M.K., 1981. A technique for growing mycorrhizal clover in solution culture. N. Z. J. Agric. Res. 24, 371–372.

Cruz, A.F., Ishii, T., Kadoya, K., 2000. Effects of arbuscular mycorrhizal fungi on tree growth, leaf water potential, and levels of 1-aminocyclopropane-1-carboxylic acid and ethylene in the roots of papaya under water-stress conditions. Mycorrhiza 10, 121–123.

Elmes, R.P., Mosse, B., 1984. Vesicular arbuscular endomycorrhizal inoculum production. II. Experiments with maize (*Zea mays*) and other hosts in nutrient flow culture. Can. J. Bot. 62, 1531–1536.

Elzenga, J.T.M., van Veen, H., 2010. Waterlogging and plant nutrient uptake. In: Stefano, M., Shabala, S. (Eds.), Waterlogging Signalling and Tolerance in Plants. Springer, Berlin Heidelberg, pp. 23–35.

Fan, L., Dalpé, Y., Fang, C., Dubé, C., Khanizadeh, S., 2011. Influence of arbuscular mycorrhizae on biomass and root morphology of selected strawberry cultivars under salt stress. Botany 89, 397–403.

Fitter, A.H., Garbaye, J., 1994. Interaction between mycorrhizal fungi and other soil organisms. Plant Soil 159, 123–132.

Giri, B., Giang, P.H., Kumari, R., Prasad, R., Sachdev, K., Garg, A.P., Oelmüller, R., Varma, A., 2005. Mycorrhizosphere: strategies and functions. In: Varma, A., Buscot, F. (Eds.), Microorganisms in Soils: Roles in Genesis and Functions. Springer Nature Singapore Pte Ltd, pp. 213–252.

Guo, X.W., Li, K., Guo, Y.S., Zhang, L.H., Sun, Y.N., Xie, H.G., 2009. Effect of arbuscular mycorrhizal fungi (AMF) strains on growth and root exudation characteristics of grapevine. J. Shenyang Agric. Univ. 40, 392–395 (in Chinese with English abstract).

Habte, M., Osorlo, N.W., 2001. Arbuscular Mycorrhizas: Producing and Applying Arbuscular Mycorrhizal Inoculum. College of Tropical Agriculture and Human Resources, University of Hawaii at Manoa, Hawaii.

Harrier, L.A., Watson, C.A., 2004. The potential role of arbuscular mycorrhizal (AM) fungi in the bioprotection of plants against soil-borne pathogens in organic and/or other sustainable farming systems. Pest Manag. Sci. 60, 149–157.

He, J.D., Dong, T., Wu, H.H., Zou, Y.N., Wu, Q.S., Kuča, K., 2019a. Mycorrhizas induce diverse responses of root *TIP* aquaporin gene expression to drought stress in trifoliate orange. Sci. Hortic. 243, 64–69.

He, J.D., Li, J.L., Wu, Q.S., 2019b. Effects of *Rhizoglomus intraradices* on plant growth and root endogenous hormones of trifoliate orange under salt stress. J. Anim. Plant Sci. 29, 245–250.

Hu, X.P., 2016. Study on Regulation of Arbuscular Mycorrhizal Fungi in Strawberry Root Allelopathic Stress. (Master dissertation). Southwest Univ, Chongqing (in Chinese with English abstract).

Jarstfer, A.G., Sylvia, D.M., 1995. Aeroponic culture of VAM fungi. In: Varma, A., Hock, B. (Eds.), Mycorrhiza. Springer-Verlag, Berlin Heidelberg, pp. 427–441.

Jiang, Y.N., Wang, W.X., Xie, Q.J., Liu, N., Liu, L.X., Wang, D.P., Zhang, X.W., Yang, C., Chen, X.Y., Tang, D.Z., Wang, E.T., 2017. Plants transfer lipids to sustain colonization by mutualistic mycorrhizal and parasitic fungi. Science 356, 1172.

Jiang, Y., Xie, Q.J., Wang, W.X., Yang, J., Zhang, X.W., Yu, N., Zhou, Y., Wang, E.T., 2018. Medicago AP2-domain transcription factor *WRI5a* is a master regulator of lipid biosynthesis and transfer during mycorrhizal symbiosis. Mol. Plant 11, 1344–1359.

Khalil, H.A., Eissa, A.M., El-Shazly, S.M., Aboul Nasr, A.M., 2011. Improved growth of salinity-stressed citrus after inoculation with mycorrhizal fungi. Sci. Hortic. 130, 624–632.

Kosová, K., Prasil, I., Vítámvás, P., Dobrev, P., Motyka, V., Floková, K., Novak, O., Turečková, V., Rolčík, J., Pešek, B., Trávníčková, A., Gaudinová, A., Galiba, G., Janda, T., Vlasáková, E., Prášilová, P., Vanková, R., 2012. Complex phytohormone responses during the cold acclimation of two wheat cultivars differing in cold tolerance, winter Samanta and spring Sandra. J. Plant Physiol. 169, 567–576.

Liang, Y., Guo, L.D., Ma, K.P., 2002. The role of mycorrhizal fungi in ecosystems. Acta Phytoecol. Sin. 26, 739–745 (in Chinese with English abstract).

Lin, S.S., Sun, X.W., Wang, X.J., Li, Y.Y., Luo, Q.Y., Sun, L., et al., 2013. Mechanism of plant tolerance to heavy metals enhanced by arbuscular mycorrhizal fungi. Pratacultural Sci. 30, 365–374.

Liu, R.J., Chen, Y.L., 2007. Mycorrhizology. China Science Publishing & Media Ltd., Beijing.

Liu, C.Y., Wu, Q.S., Zou, Y.N., 2017. Effects of arbuscular mycorrhizal fungi on phosphorus uptake and phosphatase release in trifoliate orange seedlings. Mycosystema 36, 942–949 (in Chinese with English abstract).

Liu, C.Y., Wang, P., Zhang, D.J., Zou, Y.N., Kuča, K., Wu, Q.S., 2018. Mycorrhiza-induced change in root hair growth is associated with IAA accumulation and expression of *EXPs* in trifoliate orange under two P levels. Sci. Hortic. 234, 227–235.

Lü, L.H., Wu, Q., 2018. Mitigation of replant disease by mycorrhization in horticultural plants: a review. Folia Hortic. 30, 269–282.

Luo, X., Huang, G.H., Yao, P., Wang, L.D., Ju, F.C., Zhao, F.J., 2017. Influence of arbuscular mycorrhizal fungi on antioxidant systems in the *in vitro* branch of blueberry under low temperature stress. Jiangsu J. Agric. Sci. 33, 903–913 (in Chinese with English abstract).

Mahmud, S., Sharmin, S., Chowdhury, B.L.D., Hossain, M.A., Bhuiyan, M.J.H., 2016. Mitigation of salt stress in rice plant at germination stage by using methyl jasmonate. Asian J. Med. Biol. Res. 2, 74–81.

Marschner, H., Dell, B., 1994. Nutrient uptake in mycorrhizal symbiosis. Plant Soil 159, 89–102.

Matsubara, Y.I., Hirano, I., Sassa, D., Koshikawa, K., 2004. Alleviation of high temperature stress in strawberry plants infected with arbuscular mycorrhizal fungi. Environ. Control. Biol. 42, 105–111 (in Japanese with English abstract).

Mehta, P., Bharat, N.K., 2013. Effect of indigenous arbuscular-mycorrhiza (*Glomus* spp.) on apple (*Malus domestica*) seedlings grown in replant disease soil. Indian J. Agric. Sci. 83, 1173–1178.

Meier, S., Borie, F., Bolan, N., Cornejo, P., 2012. Phytoremediation of metal-polluted soils by arbuscular mycorrhizal fungi. Crit. Rev. Environ. Sci. Technol. 42, 741–775.

Menge, J.A., 1983. Utilization of vesicular–arbuscular mycorrhizal fungi in agriculture. Can. J. Bot. 61, 1015–1024.

Meyer, J.R., Linderman, R.G., 1986. Selective influence on populations of rhizosphere-rhizoplane bacteria and actinomycetes by mycorrhizas formed by *Glomus fasciculatum*. Soil Biol. Biochem. 18, 191–196.

Miller, D.D., Domoto, P.A., Walker, C., 1985. Mycorrhizal fungi at eighteen apple rootstock plantings in the united states. New Phytol. 100, 379–391.

Mosse, B., 1962. The establishment of vesicular-arbuscular mycorrhiza under aseptic conditions. J. Gen. Microbiol. 27, 509–520.

Mosse, B., Hepper, C.M., 1975. Vesicular-arbuscular mycorrhizal infections in root organ cultures. Physiol. Plant Pathol. 5, 215–223.

Mosse, B., Thompson, J.P., 1984. Vesicular-arbuscular endomycorrhizal inoculum production. I. Exploratory experiments with beans (*Phaseolus vulgaris*) in nutrient flow culture. Can. J. Bot. 62, 1523–1530.

Mukerji, K.G., Chamola, B.P., Sharma, M., 1997. Mycorrhiza in control of plant pathogens. In: Agnihotri, V.P., Sarbhoy, A., Singh, D.V. (Eds.), Management of Threatening Plant Diseases of National Importance. MPH, New Delhi, pp. 297–314.

Muok, B.O., Ishii, T., 2006. Effect of arbuscular mycorrhizal fungi on tree growth and nutrient uptake of *Sclerocarya birrea* under water stress, salt stress and flooding. J. Jpn. Soc. Hortic. Sci. 75, 26–31.

Nemec, S., Menge, J.A., Platt, R.G., Johnson, E.L.V., 1981. Arbuscular mycorrhizal fungi associated with citrus in Florida and California notes on their distribution and ecology. Mycologia 73, 112–127.

Purin, S., Filho, O.K., Sturmer, S.L., 2006. Mycorrhizae activity and diversity in conventional and organic apple orchards from Brazil. Soil Biol. Biochem. 38, 1831–1839.

Rambelli, A., 1973. The rhizosphere of mycorrhizae. In: Marks, G.L., Koslowski, T.T. (Eds.), Ectomycorrhizae. Academic Press, New York, pp. 299–343.

Redon, P.O., Béguiristain, T., Leyval, C., 2009. Differential effects of am fungal isolates on *Medicago truncatula* growth and metal uptake in a multi-metallic (Cd, Zn, Pb) contaminated agricultural soil. Mycorrhiza 19, 187–195.

Rillig, M.C., 2004. Arbuscular mycorrhizae, glomalin, and soil aggregation. Can. J. Soil Sci. 84, 355–363.

Rillig, M.C., Hernández, G.Y., Newton, P.C.D., 2010. Arbuscular mycorrhizae respond to elevated atmospheric CO_2 after long-term exposure: evidence from a CO_2 spring in New Zealand supports the resource balance model. Ecol. Lett. 3, 475–478.

Ruotsalainen, A.L., Kytöviita, M., 2004. Mycorrhiza does not alter low temperature impact on *Gnaphalium norvegicum*. Oecologia 140, 226–233.

Rutto, K.L., Mizutani, F., Kadoya, K., 2002. Effect of root-zone flooding on mycorrhizal and non-mycorrhizal peach (*Prunus persica* Batsch) seedlings. Sci. Hortic. 94, 285–295.

Secilia, J., Bagyaraj, D.J., 1987. Bacteria and actinomycetes associated with pot cultures of vesicular-arbuscular mycorrhizas. Can. J. Microbiol. 33, 1069–1073.

Selvakumar, G., Shagol, C.C., Kang, Y., Chung, B.N., Han, S.G., Sa, T.M., 2018. Arbuscular mycorrhizal fungi spore propagation using single spore as starter inoculum and a plant host. J. Appl. Microbiol. 124, 1556–1565.

Shen, T.H., Wan, S.L., 1994. A practical study on the propagation and inoculation technique of VA mycorrhizal fungi in citrus. Acta Agric. Jiangxi 6, 25–30 (in Chinese with English abstract).

Spohn, M., Giani, L., 2010. Water-stable aggregates, glomalin-related soil protein, and carbohydrates in a chrono-sequence of sandy hydromorphic soils. Soil Biol. Biochem. 42, 1505–1511.

St-Arnaud, M., Hamel, C., Vimard, B., Caron, M., Fortin, J.A., 1996. Enhanced hyphal growth and spore production of the arbuscular mycorrhizal fungus *Glomus intraradices* in an in vitro system in the absence of host roots. Mycol. Res. 100, 328–332.

Sudová, R., Jurkiewicz, A., Turnau, K., Vosatka, M., 2007. Persistence of heavy metal tolerance of the arbuscular mycorrhizal fungus *Glomus intraradices* under different cultivation regimes. Symbiosis 43, 71–81.

Tanaka, K., Masumori, M., Yamanoshita, T., Tange, T., 2011. Morphological and anatomical changes of *Melaleuca cajuputi* under submergence. Trees 25, 695–704.

Timonen, S., Marschner, P., 2006. Mycorrhizosphere concept. In: Mukerji, K.J. (Ed.), Microbial Activity in the Rhizosphere. Springer Berlin Heidelberg Pte Ltd, pp. 155–172.

Trindade, A.V., Siqueira, J.O., Sturmer, S.L., 2006. Arbuscular mycorrhizal fungi in papaya plantations of Espirito Santo and Bahia, Brazil. Braz. J. Microbiol. 37, 283–289.

Trowbridge, J., Jumpponen, A., 2004. Fungal colonization of shrub willow roots at the forefront of a receding glacier. Mycorrhiza 14, 283–293.

Tuo, X.Q., Li, S., Wu, Q.S., Zou, Y.N., 2015. Alleviation of waterlogged stress in peach seedlings inoculated with *Funneliformis mosseae*: changes in chlorophyll and proline metabolism. Sci. Hortic. 197, 130–134.

Vallino, M., Zampieri, E., Murat, C., Girlanda, M., Picarella, S., Pitet, M., et al., 2011. Specific regions in the *Sod1* locus of the ericoid mycorrhizal fungus *Oidiodendron maius* from metal-enriched soils show a different sequence polymorphism. FEMS Microbiol. Ecol. 75, 321–331.

Vinayak, K., Bagyaraj, D.J., 1990. Vesicular-arbuscular mycorrhizae screened for Troyer citrange. Biol. Fertil. Soils 9, 311–314.

Wahid, A., Gelani, S., Ashraf, M., Foolad, M.R., 2007. Heat tolerance in plants: an overview. Environ. Exp. Bot. 61, 199–223.

Waschkies, C., Schropp, A., Marschner, H., 1994. Relations between grapevine replant disease and root colonization of grapevine (*Vitis* sp.) by fluorescent pseudomonads and endomycorrhizal fungi. Plant Soil 162, 219–227.

Weissenhorn, I., Leyval, C., Berthelin, J., 1993. Cd-tolerant arbuscular mycorrhizal (AM) fungi from heavy-metal polluted soils. Plant Soil 157, 247–256.

Wright, S.F., Franke-Snyder, M., Morton, J.B., Upadhyaya, A., 1996. Time-course study and partial characterization of a protein on hyphae of arbuscular mycorrhizal fungi during active colonization of roots. Plant Soil 181, 193–203.

Wu, Q.S., 2010. Arbuscular Mycorrhizal Research and Application of Horticultural Plants. China Science Publishing & Media Ltd., Beijing.

Wu, Q.S., Zou, Y.N., 2009. Mycorrhizal influence on nutrient uptake of citrus exposed to drought stress. Philipp. Agric. Sci. 92, 33–38.

Wu, Q.S., Zou, Y.N., 2010. Beneficial roles of arbuscular mycorrhizas in citrus seedlings at temperature stress. Sci. Hortic. 125, 289–293.

Wu, Q.S., Xia, R.X., Zou, Y.N., 2008. Improved soil structure and citrus growth after inoculation with three arbuscular mycorrhizal fungi under drought stress. Eur. J. Soil Biol. 44, 122–128.

Wu, Q.S., Zou, Y.N., He, X.H., 2010. Contributions of arbuscular mycorrhizal fungi to growth, photosynthesis, root morphology and ionic balance of citrus seedlings under salt stress. Acta Physiol. Plant. 32, 297–304.

Wu, Q.S., Zou, Y.N., He, X.H., 2011. Differences of hyphal and soil phosphatase activities in drought stressed mycorrhizal trifoliate orange (*Poncirus trifoliata*) seedlings. Sci. Hortic. 129, 294–298.

Wu, Q.S., Zou, Y.N., Huang, Y.M., 2013. The arbuscular mycorrhizal fungus *Diversispora spurca* ameliorates effects of waterlogging on growth, root system architecture and antioxidant enzyme activities of citrus seedlings. Fungal Ecol. 6, 37–43.

Wu, S.L., Chen, B.D., Sun, Y.Q., Ren, B.H., Zhang, X., Wang, Y.S., 2014. Chromium resistance of dandelion (*Taraxacum platypecidum* Diels.) and bermudagrass (*Cynodon dactylon* [Linn.] Pers.) is enhanced by arbuscular mycorrhiza in Cr (Vi)-contaminated soils. Environ. Toxicol. Chem. 33, 2105–2113.

Wu, Q.S., He, J.D., Srivastava, A.K., Zhang, F., Zou, Y.N., 2019. Development of propagation technique of indigenous AMF and their inoculation response in citrus. Ind. J. Agric. Sci. 89, 1190–1194.

Xie, L.Y., Zhang, Y., Xiong, B.Q., Zeng, M., 2010. Study on the effects of AMF on the growth and development of micropropagated grape plantlets. North. Hortic. 5, 18–23 (in Chinese with English abstract).

Yang, Y.T., Zhang, N.N., Zhang, F., Wang, L.Q., Zeng, M., 2016. Effect of AMF strains and time of inoculating on root growth and development of *Pyrus pashia* seedlings. J. Fruit Sci. 33, 114–120 (in Chinese with English abstract).

Yin, D., Zhang, Z., Luo, H., 2012. Anatomical responses to waterlogging in *Chrysanthemum zawadskii*. Sci. Hortic. 146, 86–91.

Zai, X.M., Hao, Z.P., Zai, Y., Zhang, H.S., Qin, P., 2014. Arbuscular mycorrhizal fungi (AMF) and phosphate-solubilizing fungus (PSF) on tolerance of beach plum (*Prunus maritima*) under salt stress. Aust. J. Crop. Sci. 8, 945–950.

Zhang, X., Ren, B.H., Wu, S.L., Sun, Y.Q., Lin, G., Chen, B.D., 2015. Arbuscular mycorrhizal symbiosis influences arsenic accumulation and speciation in *Medicago truncatula* L. in arsenic-contaminated soil. Chemosphere 119, 224–230.

Zhang, Y.C., Wang, P., Wu, Q.H., Zou, Y.N., Bao, Q., Wu, Q.S., 2017. Arbuscular mycorrhizas improve plant growth and soil structure in trifoliate orange under salt stress. Arch. Agron. Soil Sci. 63, 491–500.

Zhao, M., Li, M., Liu, R.J., 2010. Effects of arbuscular mycorrhizae on microbial population and enzyme activity in replant soil used for watermelon production. Int. J. Eng. Sci. Technol. 2, 17–22.

Zhu, Y., Xiong, J.L., Lü, G.C., Asfa, B., Wang, Z.B., Li, P.F., Xiong, Y.C., 2015. Effects of arbuscular mycorrhizal fungi and plant water relation and its mechanism. Acta Ecol. Sin. 35, 2419–2427 (in Chinese with English abstract).

Zou, Y.N., Srivastava, A.K., Wu, Q.S., Huang, Y.M., 2014. Increased tolerance of trifoliate orange (*Poncirus trifoliata*) seedlings to waterlogging after inoculation with arbuscular mycorrhizal fungi. J. Anim. Plant Sci. 24, 1415–1420.

24

Mycorrhizas in fruit nutrition: Important breakthroughs

İbrahim Ortaş*

University of Cukurova, Faculty of Agriculture, Department of Soil Science and Plant Nutrition, Adana, Turkey
*Corresponding author. E-mail: iortas@cu.edu.tr

1 Introduction

The main symbiotic organisms in the rhizosphere are mycorrhizae, N_2-fixing bacteria, phosphorus-solubilizing bacteria, and phytostimulator hormone–producer bacteria. This beneficial symbiosis including mycorrhizae has significant benefits to plant health and soil quality (Table 24.1). In terrestrial ecosystem, mycorrhizal fungi are surrounded by a number of complex microbial communities such as bacteria and actinomycetes. Possibly, the other rhizosphere organisms may modulate the mycorrhizal symbioses in the rhizosphere area as well.

It is estimated from the fossil record, approximately 480 million years ago, arbuscular mycorrhiza (AM) has an associations in the earliest land plant fossils (Delaux, 2017). The majority of plant species are naturally arbuscular mycorrhiza (AM) dependent. It is obvious that many plant species depend on these symbionts for growth and survival under several different soil and climate conditions. Nearly 90% of plants have symbiotic relation arbuscular mycorrhiza, and this soil fungus belongs to *Glomeromycota* (Schussler et al., 2001). Arbuscular mycorrhizal fungi (AMF) are obligate symbionts of many plants that biotrophically colonize the root cortex and develop an extramatrical mycelium, which helps the plant to acquire water and mineral nutrients from the soil (Elsen et al., 2003; Marschner, 1995) and increases plant resistance against stress factors. Mycorrhizae are providing a critical linkage between the plant root and soil particles. AM fungi provide a range of important ecological services, in particular by improving

TABLE 24.1 Type of microorganisms and their role on plant and rhizosphere soil systems.

Type of microorganisms	Effects of interaction on plant life
N_2-fixing bacteria (biofertilizers)	N_2 fixation, N cycling for some fruit trees
Mycorrhiza fungi (biofertilizers)	Nutrient and water uptake and improved soil quality for nearly 90% of plant species
Phosphorus solubilizes (biofertilizers)	P cycling, use of rock and organic phosphates
Rhizobacteria (PGPR)	Plant growth and health promotion and regulation of diversity of the microbial population in the rhizosphere
Plant hormone producers (phytostimulators)	Rooting and establishment of seedlings
Biological control agents of plant diseases (bioprotectors and biopesticides)	Increased resistance/tolerance to root diseases and soilborne disease

stress resistance and tolerance, soil structure, and fertility. According to Rouphael et al. (2015), AMF interfere with the phytohormone balance of host plants, thereby influencing plant development (bioregulators) and inducing tolerance to soil and environmental stresses (bioprotector) factors.

2 Facultative and obligatory mycorrhizal plants

The microorganisms play an important role in the plant, soil, and environmental sustainability. Most plant species are facultative symbionts, and they get to benefit from AM fungi. Also, numerous microorganisms including AMF cannot survive without plant roots. Some plant species cannot grow without undergoing association with friendly AMF. Also, some at the same time can survive without AMF. Plant taxa such as *Brassicaceae* and *Chenopodiaceae* are symbiotic, and they lost the capacity to interact with AM fungi (Table 24.2). Those group of taxa devolved an alternative strategy to get their mineral nutrient demands (Brundrett, 2004).

Some plant species cannot grow without mycorrhizal associations, but others do not because they have other strategies for accessing nutrients (Lambers et al., 2011). Plants benefit from AMF associations only in some of the least fertile soils in which they naturally occur (Janos, 1980, 2007). In ecosystem, some herbaceous plant species has low levels of AMF colonization (less than 25%) (Brundrett and Kendrick, 1988). Plants with coarsely branched roots

TABLE 24.2 Plant species and their relation with mycorrhizal symbioses.

Mycorrhizal type	Plants	Number of plant species hosting mycorrhizal fungi	Fungi	Fungal colonization	Total estimated number of fungal taxa
Arbuscular mycorrhiza (AM)	Most herbs, grasses, many flowering plants and several trees (mainly fruit tree), shrublands, hornworts and liverworts	200,000	*Glomeromycota*	Endo	300–1600
Ectomycorrhiza	*Pinaceae* and angiosperms (mostly shrubs and trees, mostly temperate), tropical forests, tundra, and agroforestry	6000	*Basidiomycota* and *Ascomycota*	Ecto	20,000
Orchid mycorrhiza	Orchids species	20,000–35,000	*Basidiomycota*	Endo	25,000
Ericoid mycorrhiza	Members of the *Ericaceae* family	3900	Mainly *Ascomycota*, some *Basidiomycota*	Endo	>150
Nonmycorrhizal plant species	*Brassicaceae, Crassulaceae, Orobanchaceae, Proteaceae*, etc.	51,500			0

Reproduced with based on Brundrett, M., 1991. Mycorrhizas in natural ecosystems. Advances in Ecological Research. Elsevier, pp. 171–313; Brundrett, M.C., 2009. Mycorrhizal associations and other means of nutrition of vascular plants: understanding the global diversity of host plants by resolving conflicting information and developing reliable means of diagnosis. Plant Soil 320, 37–77; Van Der Heijden, M.G., Martin, F.M., Selosse, M.A., Sanders, I.R., 2015. Mycorrhizal ecology and evolution: the past, the present, and the future. New Phytol. 205, 1406–1423.

and with few or no root hairs are expected to be more dependent on AMF than plants with finely branched or fibrous roots (Smith and Read, 2008).

Rhizosphere soil of the many plants revealed the presence of many species of mycorrhizal fungi. Anand and Reenu (2009) reported that in the rhizosphere of two medicinal plants (*Centella asiatica* and *Ocimum sanctum*) have 16–17 different species of AM fungi. In another study, the results of Shi et al. (2013) shown that 31 AM fungus species belonging to 3 genera (*Glomus* [21], *Acaulospora* [7], and *Scutellospora* [3]) were identified in the rhizospheric soil. In the same rhizosphere soils, other soil beneficial organism richness is supposed to be higher than noninoculated plants.

3 Mycorrhizal dependency

3.1 Mycorrhizal dependency and dependent plant species

Baylis (1975) hypothesized that mycorrhizal dependency is largely controlled by the root system architecture. It was defined mycorrhiza dependency "host inability for growth without AMF at given soil fertility." Mycorrhizal dependency (MD) of the host crop is dependent on soil fertility and fertilizer levels (Ortas, 2012a,c). So far, many hypotheses were suggested to measure the mycorrhizal response on plant growth and nutrient uptake. Gerdemann (1975) and Baylis (1975) were the first researchers who defined the mycorrhizal dependency as the degree to which a plant is dependent on AMF to produce maximum growth or yield at a given level of soil fertility. Baylis (1975) also postulated that magnolioid roots were most dependent upon an AMF for phosphorus uptake, indicating the existence of species differences in the changes in P uptake associated with mycorrhizal colonization. Later on, Plenchette et al. (1983) have determined the "relative mycorrhizal dependency (RMD)" by expressing the difference between the dry weight of the mycorrhizal plant and the dry weight of the nonmycorrhizal plant as a percentage of the dry weight of the mycorrhizal inoculated plant.

Plant benefits from AM fungal colonization in large degree depend on the environmental conditions (Chen et al., 2018a) such as nutrient concentrations and other stress factors (Azcón et al., 2003). Plants will not survive to reproductive maturity without being associated with AMF in the soils (or at the fertility levels) of their natural habitats (Janos, 1980). Graham et al. (2017) reported that some plant species have turned to obligate parasites on the AM fungus and they became fully dependent on fungal nutrition and most probably they lost their photosynthetic capacity. The orchid plant is an example. In different studies, the positive influence of the AM symbiosis on horticultural production was provided by many researchers (Gianinazzi et al., 1989; Lovato et al., 1999). Almost all fruit tree plant species are mycorrhizae dependent for growth and nutrient uptake (Fitter et al., 2011; Kungu et al., 2008). Some species are strongly mycorrhizal dependent than others. The AMF also have many advantages on inducing plant that can tolerate biotic stress (Barea et al., 1996) and abiotic stress factors. Species variation in AMF may cause a difference in the crop growth due to the mycorrhiza effect on certain key processes of plant physiology. AMF influence on optimizing the uptake of mineral nutrients in the rhizosphere is one of a very important regulator of plant development.

In this context, it is natural that many plant species and genus are AMF obligate with good infection. Mycorrhiza species may be selectively preferred by some plant species than the others. For citrus plant, it has been clearly shown that *Glomus clarium* inoculum significantly inoculated sour orange seedling better than the other mycorrhiza species (Ortas, 2012b). It has been reported that some efficient AMF species such as *Glomus aggregatum* are helping plant growth than nonefficient AMF species such as *Glomus intraradices* (Ba et al., 2000; Guissou et al., 1998). Since AMF are obligate symbionts, they may have preferred plant species and ecologically growth conditions.

4 Advantages of mycorrhizal dependency on plant

Mycorrhizal dependency and mineral nutrition potential have been focused by many researchers, and it has been indicated that the benefits of AM fungi on plant growth could vary widely among plant species and even among cultivars or species from different geographic locations (Ortas, 2012b; Plenchette et al., 2005; da Sousa et al., 2013).

The degree of plant dependence is of great practical and ecological interest for plant nutrition. Fungi are better than plants at acquiring mineral nutrition (P, K, N, Zn, Cu, and Ca) from the soil. AMF improve a plant's access to water. Guissou et al. (2016) indicated that AM inoculation significantly improved the N, P, and K absorption compared with non-AM fruit trees. They indicated that AM-inoculated plant leaves have more P concentration than noninoculated tree plants. Plant gets nutrients P, Zn, Cu, N, and water, and the fungus gets carbohydrates. Carbohydrates are moved from the leaves to the rhizosphere the benefit of both sides. The most efficient method of carbon mitigation is photosynthesis mechanism, which can pull the atmospheric carbon to the rhizosphere soil. However, still unknown mechanisms are regulating AMF colonizing such as carbon expenditure, the toxic effect of high P, nutrients, and different other ecological factors. The benefit is provided by AMF, and plant will depend on the relative contribution of root and mycorrhizal mediated nutrient uptake to plants (Janos, 2007). Mycorrhizal dependency has often been quantified by calculating the yield between mycorrhizal and nonmycorrhizal control plants grown in a particular soil at a single soil P level (Hetrick et al., 1992; Koide, 1991; Manjunath and Habte, 1991).

5 Mycorrhiza-dependent horticultural plant species

As can be seen in Table 24.3, many horticultural tree plants are mycorrhiza dependents. Jaizmevega and Azcon (1995) indicated that RMD of tropical horticultural fruit crops such as avocado, papaya citrus, banana, vineyard, cherry, fig, pistachio, and pineapple is highly responsive to AMF when the inoculum consists of *Glomus* spp. Plant species give different response to the different mycorrhiza species. Under similar growth conditions, by using the data of Fig. 24.1, mycorrhizal dependency (MD) of horticultural plant was calculated. Plant species mycorrhiza dependency was significantly differed without considering the growth mediums. It seems that citrus plant is strongly mycorrhizal dependent than the other plant species (Table 24.3). So far in our different research works, it has been observed that sour orange seedlings are a strongly mycorrhizal dependent under several soil and nutrition conditions (Ortas, 2012b). After soil sterilization, without mycorrhizal inoculation, sour orange seedlings are nearly not grown (Ortas et al., 2018). On the other hand, mycorrhiza-inoculated seedling is grown in several times bigger than noninoculated one (Fig. 24.2).

Many studies have been done to determine mycorrhizal dependency by using Plenchette et al. (1983) equation. Hetrick et al. (1996) indicated that, when the soil available phosphate concentration is the growth-limiting factor of the plant, MD is positively correlated with the MD in phosphorus uptake. The degree of AMF dependency varied according to phosphorus levels and fungal inoculum (Cardoso et al., 2008). The results of Costa et al. (2005) shown that the mycorrhizal dependency varied according to the AMF and soil condition. In another research, Sharma et al. (2001) reported that the values of MD were negatively correlated with soil P levels. Host plants' phosphate acquisition ability and phosphate utilization efficiency can determine the dependency of mycorrhizae. Plant ability of phosphate acquisition mainly depends on morphological and physiological characteristics of plant roots (Koide, 1991; Schachtman et al., 1998).

AMF must be considered a necessary factor for promoting horticulture plant productivity and health. Since horticultural plants are produced in nursery beds, containers, or by tissue culture, AM biotechnology or soil biotechnology is feasible and rewarding mainly for crops, which involve a transplant stage application. With efficient AMF species and a careful selection of compatible host/fungus/substrate combinations, maximum benefits can be obtained from AMF inoculation. In general, the earlier AMF inoculation or seedling produced with AMF infection at the appropriate stage of seedling production can provide better yield and benefit.

Under greenhouse conditions with similar growth medium, different mycorrhizal species were used. Generally, in comparison with noninoculation, mycorrhizal inoculation increased banana, vineyard, cherry, citrus, fig, and pistachio plant total dry weight (Fig. 24.1). In general, selected mycorrhiza species produce more total dry weight than

TABLE 24.3 Effects of several mycorrhiza species on different plant mycorrhiza dependencies.

Treatments	Banana	Vineyard	Cherry	Citrus	Fig	Pistachio	Mean
G. mosseae	32	27	59	91	24	48	47
G. intraradices	28	27	40	89	38	34	43
G. caledonium	33	25	24	92	39	31	41
G. clarium	36	23	37	82	38	41	43
G. etunicatum	34	27	24	81	33	38	40
Cocktail	25	26	10	91	28	30	35
Indigenous Myco.	28	23	37	49	17	36	32
Mean	31	26	33	82	31	37	40

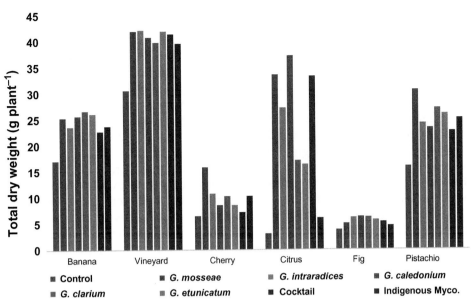

FIG. 24.1 Effect of different mycorrhizal species on plant growth.

FIG. 24.2 Effects of mycorrhizal inoculation on sour orange citrus seedling growth.

cocktail and indigenous inoculation. Indigenous mycorrhiza-inoculated citrus seedlings produce less DW than other mycorrhiza species inoculation.

Generally, indigenous mycorrhizal inoculation effect on mycorrhiza dependency is less than selected mycorrhiza species. Cocktail mycorrhiza inoculation also has fewer effects on MD. This may be related to with a number of mycorrhizal propagules and mycorrhizal spore quality.

6 Mycorrhizae and plant nutrition

6.1 Mycorrhizae reduce the nutrient deficiency stress

Increased nutrient uptake made possible by the AM symbiosis results in more vigorous plants. When AMF-inoculated young plantlets raised in nurseries are planted in field, they quickly adapt to the dry climate conditions (Chen et al., 2018b). Similarly, Sellal et al. (2017) reported that AMF inoculation significantly increases the growth and health of young argan trees, and they indicated that inoculated seedlings increase their fitness and survival after planting. Mycorrhizal symbiosis results in reduced nutrient stress and also can reduce other environmental stresses such as soil drought; they can benefit plant growth.

These fungi can benefit plants by enhancing the nutrient-absorbing ability of roots especially important in facilitating uptake of phosphorus. In addition to element uptake via mycorrhizal mycelia, AMF has also shown to affect root morphology and functioning, as well as mycorrhizosphere soil properties. With their thin diameter ($<10\,\mu m$), AM hyphae might be able to access smaller soil pores and better compete with soil microbes for nutrient resources, compared with plant roots. Neumann and George (2010) indicated that just like plant root systems, AM hyphae seem to differ considerably in their architecture and physiological activities depending on their genotype. This enhancement of nutrient uptake is a result of the extensive system of hyphae and mycelia that pervade soils. Mycorrhizal hypha length is also controlled by nutrient level mainly by soil phosphorus levels. Recent works are strongly indicating that phosphate delivery is among the most important benefits for the host in AM symbiosis (Karandashov and Bucher, 2005), and all results showed that the arbuscules are the site of transfer of phosphate from the fungus to the plant cells (MacLean et al., 2017).

All physiological evidence indicates that the P uptake pathway by AM is often accompanied by reduction in P absorbed directly by root hairs and epidermis (Smith and Read, 2008). The reduction is usually related to the depletion of P in the rhizosphere area. Also, the operation of AM P uptake pathway is believed to be under the reduced expression of Pht1 transporters in root epidermal cells. Also, mycorrhizal inoculation may transfer the nutrient elements such as nitrogen, sulfur, and microminerals such as copper and zinc via the arbuscules as well. The relationship of this fungus with plants is a mutually beneficial one, with the fungi receiving energy in the form of carbohydrates from the host plant. One of the most dramatic effects of mycorrhizal inoculation on the host plant is the increase in P uptake and Zn mainly due to the capacity of the AMF to absorb phosphate from soil and transfer it to the host roots. Even under abiotic stress conditions, mycorrhizal inoculation increases the nutrient concentration.

7 Contribution of mycorrhizae on P uptake and plant growth

The AMF can utilize soil nutrient efficiently. Mycorrhizal fungi may be making significant contributions to ecosystem nutrient cycling in a large scale on mineral nutrients. AM fungi affect plant growth only via an increased nutrient supply under a well-inoculated condition. AMF can secrete phosphatases to hydrolyze phosphate from organic P compounds (Marschner, 2012), thus improving crop productivity under low input conditions (Smith et al., 2011). The extensive hyphal network of mycorrhizal fungi and extension of mycorrhizosphere influence the physicochemical and biochemical properties of the soil and directly or indirectly contribute to the release of phosphate from inorganic complexes of low solubility (Parniske, 2008). Phosphorus exists in the natural soil ecosystems as inorganic orthophosphate, primarily involved in inert complexes with cations such as iron phosphate ($FePO_4$), aluminum phosphate ($AlPO_4$), and calcium phosphate ($CaPO_4$) and in organic molecules such as lecithin and phytate, the latter of which can account up to 50% of total soil organic phosphate. P uptake mechanisms in between plant species are so complex. Since P is highly immobile in the soil, its acquisition by the roots generates a depletion zone surrounding the epidermis and the root hairs of rhizosphere.

Some plant increases the root-soil interface to maximize access to available P in rhizosphere (Fig. 24.3). Some plants are using root secrete organic acids such as malate and citrate to compete with P cation binding. Nearly 90% of plant species get an important part of P thorough mycorrhizal external hyphae (fungal hyphae grow up to 100 times longer than root hairs).

As shown in Fig. 24.3, the depletion zone around roots is very narrow, and depletion zone of AM-inoculated root area is much larger than root alone. At the same time, AM fungal colonization induces expression and secretion of a plant-derived acid phosphatase in the rhizosphere, which further liberates Pi (Ezawa et al., 2005). The characterization of a high-affinity Pi transporter (PT) through extraradical hyphae is a milestone in the definition of AM fungi as biofertilizers. All research work data showed that Pi, taken up by the extraradical mycelium from soil solutions, is translocated through the AM fungal hyphae as polyphosphate (poly-Pi). Also, research clearly shows that Pi uptake is under gene control. Benedetto et al. (2005) have reported that tomato Pi transporter genes are consistently expressed

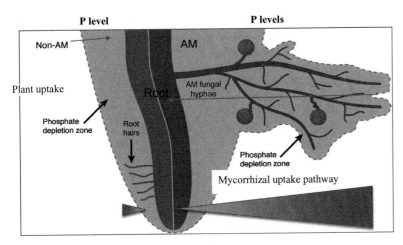

FIG. 24.3 Phosphate depletion zones that develop around a nonmycorrhizal root hairs and the root colonized by AM fungi. Plant P and other nutrient uptake and mycorrhizal inoculation uptake pathways are completely different from each other. *Modified from Hodge, A., 2017. Accessibility of inorganic and organic nutrients for mycorrhizas. Mycorrhizal Mediation of Soil. Elsevier, pp. 129–148.*

inside the arbusculated cells, suggesting that plants guarantee maximum Pi uptake through the activation of the whole gene family. According to Javot et al. (2007) after poly-Pi hydrolysis in the arbuscule, the phosphorous is then released as Pi into the periarbuscular space. Mycorrhiza-specific Pi transporters are responsible for plant Pi uptake in arbuscule-containing cells to plant tissue.

8 Contribution of mycorrhizae on other nutrient uptake

In both major inorganic N sources, nitrate and ammonium are regarded as relatively mobile in soil, transported to roots by mass flow in the soil solution. Since ammonium is less mobile than nitrate AMF, mainly ectomycorrhizae absorb ammonium from the soil solution. The results of Chalot et al. (2006) indicated that NH_3/NH_4^+ is thought to be the preferential form of N released by the fungus and uptake by the plant. Guether et al. (2009) reported that a mycorrhizal-specific lotus AMT gene has recently been identified to be responsible for N uptake. It seems that nitrogen is then transferred from the fungus to the plant as ammonium form. Govindarajulu et al. (2005) indicated that with isotope-labeling experiments in combination with gene expression data have demonstrated that inorganic N (nitrate and ammonium) is taken up by the extraradical mycelium, incorporated into amino acids, and translocated to the intraradical mycelium, mainly as arginine. Van Der Heijden et al. (2015) indicated that, up to 80% of plant, N and P are provided by mycorrhizal fungi.

Nutrient deficiency is a common nutritional problem in crop production in many arid and semiarid soil conditions. Under such soil conditions, also, there are salt and alkalization problems. Latef and Chaoxing's (2011) results show that mycorrhiza-inoculated tomato plants grown under nonsaline and saline conditions have higher P and K concentrations than compared with non-AM-inoculated plants. Mycorrhizal inoculation or indigenous potential of mycorrhizae in such soil is a critical factor in crop production for P, Cu, and Zn. The results of Navarro and Morte (2019) showed that, under lower P supply conditions, Zn or Fe increased more in the mycorrhiza inoculation of *Citrus macrophylla* Wester seedling plants. In addition, mycorrhizal infection results in an increase in the uptake of other macro- and micronutrient and water uptake. The results of Grimaldo-Pantoja et al. (2017) revealed that, under salt conditions, mycorrhiza inoculation significantly increases pepper plant P concentration.

Boron deficiency is also common in many soil conditions, especially under semiarid and low-fertile soil conditions. Boron is mainly accumulated in the fruits and upper leaves of plants. And translocation can only occur from other plant parts toward the fruits. Mycorrhizae also have effects on boron uptake of apple seedling (Gastol et al., 2016) and translocation (Shireen et al., 2018; Watts-Williams and Cavagnaro, 2014). Roots and hyphae were the first structure to respond to boron uptake and translocation in plant fruits.

9 Mycorrhizae and other rhizosphere organisms

9.1 Mycorrhiza is an efficient mechanism in rhizosphere for plant health

The physiology of the plant is highly affected by the presence of the fungal symbionts (Smith et al., 1994). The effects of AMF on the growth and development of horticultural plants have been studied and described by many research papers (George and Marschner, 1996; Lovato et al., 1996, 1999, 2006). In general, fruit crops have received more attention than vegetable and ornamental crops. Obviously, the interest of horticulturists in AM technology is due to the easy inoculum application. The ability of AMF to increase the uptake of phosphorus and other nutrients and to increase resistance to biotic and abiotic stress. Fruit tree plants are having benefit through AMF by increasing uptake of P and other nutrients and also increasing resistance to biotic and abiotic stresses. Plant species, which can use organic phosphate by root exudation of phosphatase, may not need to depend on mycorrhizal colonization, and they are known as nonmycorrhizal plants. The AM fungi can alter rhizosphere of plant through the profile of volatile organic carbon released by roots that can affect the number and richness of rhizosphere organisms. Also with mycorrhizal inoculation, the root morphology of plants can increase, and root morphology increases the plant adaption to the soil environments. Flasinski and Rogozinska (1988) reported that root exudation increases for phosphatase under P-deficient condition in no host plant species.

9.2 Relationship between plant-bacteria-mycorrhizae

Rhizobacteria, referred as plant growth–promoting rhizobacteria (PGPR), are agronomically useful and active root-colonizing microbes that are involved in N fixation, nutrient solubilization, water and nutrient uptake, salinity and drought tolerance, enzyme production against soilborne pathogens, and phytohormone production. Bacteria are the most abundant microorganisms in the rhizosphere. Bacteria mostly collaborate with mycorrhiza fungi in the rhizosphere. Both mycorrhizae and PGPR can influence the plant physiology to a greater extent, especially considering their collaboration or competitiveness in root colonization; PGPRs act as biofertilizers having efficient symbiotic relationship with mycorrhiza species on many plant roots and promote plant growth. AMF and PGPR have potential to increase plant growth and also control soilborne diseases including plant-parasitic organisms. Numerous plant growth–promoting rhizobacteria are well known to exhibit beneficial effects with mycorrhizal inoculation on plenty of horticultural crops. The results of Sharma and Sharma (2017) showed that tomato plants treated with dual or individual inoculation of AMF and PGPR showed significantly enhanced plant growth.

When plant-bacteria-mycorrhizae work together, it is termed as tripartite relationship, which can promote plant growth, nutrients, and water uptake and make plant defense system stronger against stress factors. The establishment of tripartite relationship depends on plant-soil-microorganism feedback system, and all the contributors are involved in multidirectional exchange of goods and services, which induce changes in the aboveground and belowground interactions (Ortaş et al., 2017). Positive and negative interaction between tripartite organisms mainly depends on energy resource limitation. There are several different models to explain the way of tripartite interaction where the optimal resource allocation model is emphasize under resource-limited condition (Revillini et al., 2016) (Fig. 24.4).

The interaction between mycorrhiza and plant root has several benefits irrespective of their contribution to host plant growth and nutrient uptake. Also, mycorrhizal inoculation is a strong sink for photosynthesis. Mycorrhizae demand more carbon from plant to survive and consequently leeched more carbon to the rhizosphere. Finally, when there is a good infection, mycorrhizal fungi compensate the carbon drainage and carbon use.

It is believed that AM fungi and *Rhizobium* stimulate the photosynthetic process in their host plant as a C sink. The calculation of Miransari (2014) indicated that the amounts of plant photosynthesized C supplied to AM fungi and *Rhizobium* range from 4% to 16% of the total fixed C.

10 Compatibility and collaboration of AMs with other rhizosphere microbes

In the rhizosphere, the complex relationship between mycorrhiza and other organisms is still under extensive search by several advance laboratory conditions. Recently, the use of DNA-based research methods has increased our understanding of the specificity of mycorrhizal fungi toward their host plants and interaction with other soil rhizosphere organisms.

Maybe, more efficient nutrient cycling by mycorrhiza fungi would relieve some plant stress, thereby leading to improved nodulation and N_2 fixation by *Rhizobium*. Under stress conditions, AM and *Rhizobium* may influence the efficiency by improving the nutrient and water uptake, enhance the plant morphology and physiology, and improve hormone signaling between host and microorganism to establish better symbiosis, and AM controls the pathogen and disease factors. In our long-term observation, some plant such as soybeans under sterile soil conditions without mycorrhizal inoculation plant nodulation is not well developed. It seems that still there are several unknown

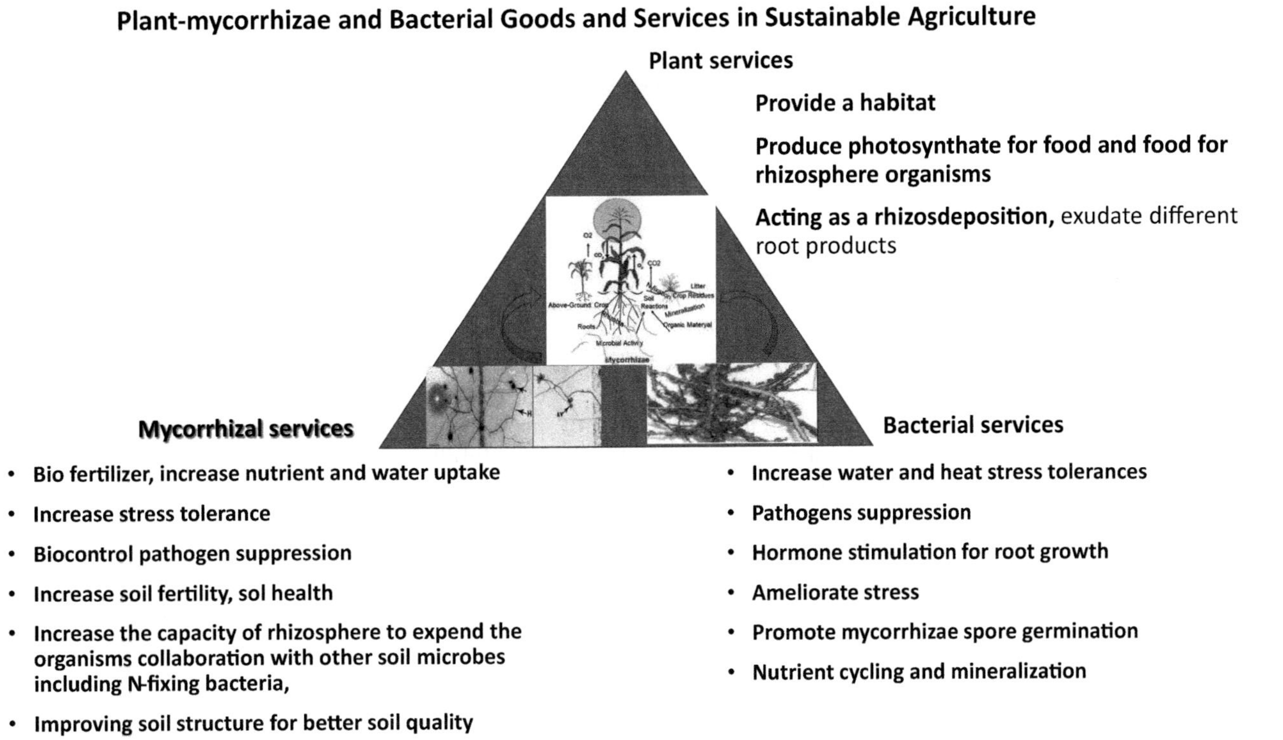

Plant-mycorrhizae and Bacterial Goods and Services in Sustainable Agriculture

Plant services

Provide a habitat

Produce photosynthate for food and food for rhizosphere organisms

Acting as a rhizosdeposition, exudate different root products

Mycorrhizal services

- Bio fertilizer, increase nutrient and water uptake
- Increase stress tolerance
- Biocontrol pathogen suppression
- Increase soil fertility, sol health
- Increase the capacity of rhizosphere to expend the organisms collaboration with other soil microbes including N-fixing bacteria,
- Improving soil structure for better soil quality

Bacterial services

- Increase water and heat stress tolerances
- Pathogens suppression
- Hormone stimulation for root growth
- Ameliorate stress
- Promote mycorrhizae spore germination
- Nutrient cycling and mineralization

FIG. 24.4 Multidirectional exchanges of goods and services among plants, mycorrhizal fungi, and rhizobacteria relation. Exchanges presented here show mycorrhizae and bacteria include plants benefits to plant growth, mycorrhizal fungi and rhizobacteria in several ways. Bacteria and mycorrhizal inoculation also increase soil fertility and soil quality directly and indirectly.

mechanisms between AM and *Rhizobium* interaction. Also, the N_2-fixing rhizobium bacteria and mycorrhiza fungi are the most relevant representatives of beneficial plant symbionts as well (Barea, 2015), which are directly related to nutrient cycling and uptake. Under different ecological conditions, dual inoculation of both mycorrhizal fungi and N-fixing bacteria can benefit the host plant. Both *Rhizobium* and mycorrhizal fungi are interacting with their host plant like lentils and green bean. It is estimated that *Rhizobium* and mycorrhizal fungus well-infected plant provide up to 80% and 75% of necessary N and P for plant use and growth (Miransari, 2014). Dual inoculation of both AM-plant symbiosis and the bacteria-legume symbiosis is of agricultural and environmental implications, and they can substantially contribute to available P and N production and utilization by plant and decreasing the amount of N and P chemical fertilization. The results of Kuang et al. (2005) indicated that the positive interactions between P and N, which are resulted by mycorrhizal and N-fixing bacterial symbioses, significantly increase plant growth. Also, it has been reported by Smith et al. (2010) that AM species increase the efficiency of legume plants, through enhancing the uptake of nutrients such as P, Cu, and Zn, which are important for nodulation and N fixation (Smith et al., 2010).

It seems that N-fixing capability of *Rhizobium* may be enhanced when host plant roots are infected with optimum mycorrhizal species. Our early research showed that mycorrhizal inoculation improved soybean nodulation and root morphology (unpublished data). The results of Ames et al. (1987) indicated that preestablishment of mycorrhizae improved cowpea nodule activity and root dry weight. Long time ago, Hayman (1986) reported that the *Rhizobium*-AM-leguminous plant interaction exists, and some legumes grew so poorly without mycorrhizas as to be ecologically obligate mycorrhizae. Many studies also showed that AM fungal communities in legume roots are different from those in no legume roots. This mean mycorrhizae and rhizobia relationship and interaction may be selective depending on host plant. Root colonization of AM may be different with rhizobia inoculation for each host plants. In an experiment, N-fixing plant roots densely mycorrhizae colonized >70%–80% of its root system (Ortas, 2008).

For example, dual inoculation of mycorrhizal fungus and *Thiobacillus* on maize crop with sulfur application on alkaline soil has decreased rhizosphere soil pH, which made nutrients available to the plant roots (Ansori and Gholami, 2015). The symbiotic effectivity of dual and tripartite symbiotic agent bacteria and AM fungi was investigated in two pot culture experiments on different soybean cultivars by Takacs et al. (2018). The results indicated that both

TABLE 24.4 Density of total bacteria, nitrogen-fixing bacteria, and *Actinomycetes* populations in rhizosphere soil inoculated with different AM fungal species (Secilia and Bagyaraj, 1987).

AM species	Rhizosphere soil population density (colony-forming units/g soil)		
	Bacteria ($\times 10^6$)	N$_2$ fixers ($\times 10^5$)	Actinomycetes ($\times 10^4$)
Control (non-AM)	14.7	12.4	13.4
Glomus fasciculatum	41.9	42.0	26.1
Gigaspora margarita	34.0	87.9	17.7

microbial-inoculated cultivars were better than that of control, proving that even drought-tolerant genotypes. The results of Efthymiou et al. (2018) suggest that AM and phosphate-solubilizing microorganism *Penicillium aculeatum* may possibly act synergistically without showing and antagonistic interactions to improve the uptake of P in wheat. In general, plant-mycorrhizae-bacteria association gets the benefit in terms of three core functions of bacteria, that is, mineral nutrient mobilization from soil minerals, fixation of atmospheric nitrogen, and increased plant resistance against stress factors and pathogens.

Until now, the following bacterial genera have been reported for the association with mycorrhizal fungi such as gram-negative *Proteobacteria* and gram-positive *Firmicutes* and gram-positive *Actinomycetes* (Frey-Klett et al., 2007; Secilia and Bagyaraj, 1987). Since AM spores are obligate in terms of their growth, development, and multiplication, they are grown on host plant root medium. Through mycorrhiza multiplication, also rhizosphere is rich of viable useful bacterial species (Table 24.4). Mycorrhizal spores' application as inoculum is also rich of bacteria medium is applied to the soil medium. As a result of dual inoculation, plants are getting benefit from mycorrhiza spores and bacteria as well in several ways (Fig. 24.3). Also, microbial communities such as the activity of mycorrhizal helper bacteria may increase mycorrhizal activity to uptake less mobile nutrient from soil. The mycorrhizal helper bacteria assist the AM fungi up to 3.8-fold increase in root colonization (Duponnois and Plenchette, 2003). Also, bacteria help plant to increase the number of mycorrhizal spores and mycorrhizal colonization. The results of Secilia and Bagyaraj (1987) showed that different AMF species inoculated plant rhizosphere soil populations of total bacteria, nitrogen-fixing bacteria, and *Actinomycetes* were significantly higher than noninoculated control treatments (Table 24.4). Also, their results clearly suggest that different AMF species affect the physiology of the host in different ways. According to Marschner (2012) effect of rhizosphere microorganisms on root morphology and nutrient availability, this alteration may affect nutrient acquisition and plant growth.

11 Future aspects and concluding remarks

In this review, the role of mycorrhizae in plant growth and nutrient uptake clearly and extensively was determined. It is obvious that mycorrhiza is an important part of plant symbiosis for sustainable life. However, still there is more need to work on the role of mycorrhizae under field conditions for better sustainable ecological agriculture. Under field conditions, several experiments were performed to understand the potential contribution of mycorrhizae on field and horticultural plant growth and nutrient uptake. After long-term evaluation, it is sound to use mycorrhizal inoculation for horticultural plant, and for field crop plant, it is sound to manage the soil and crop systems. Since horticultural plant is grown as a seedling and transplanted to the field conditions, it is sound to produce mycorrhiza-inoculated seedlings. And it is sound to suggested using mycorrhiza for mycorrhizae dependent plant species before transplanting to field conditions. The future challenge of the use of mycorrhizal fungi in the production of horticultural fruit plants will be to optimize combinations of plant species and mycorrhizal fungi species, inoculation methods, and soil or substrate properties for mycorrhizal use and effectiveness. These facts show that mycorrhizal inoculation is necessary for healthy, effective, and well-grown quality seedling production. Mycorrhizal inoculation is going to be the key mechanism for healthy food for plant physiology. To manage indigenous mycorrhizae under greenhouse and field conditions, the effect of soil and crop management will be important. For future, our research direction will be focusing on soil and crop management systems and using mycorrhizal inoculated horticultural seedlings for large agricultural practice.

References

Ames, R.N., Mihara, K.L., Bethlenfalvay, G.J., 1987. The establishment of microorganisms in vesicular-arbuscular mycorrhizal and control treatments. Biol. Fertil. Soils 3, 217–223.

Anand, S., Reenu, K., 2009. Fungal associates of *Centella asiatica* and *Ocimum sanctum*. J. Pure Appl. Microbiol. 3, 243–248.

Ansori, A., Gholami, A., 2015. Improved nutrient uptake and growth of maize in response to inoculation with Thiobacillus and Mycorrhiza on an alkaline soil. Commun. Soil Sci. Plant Anal. 46, 2111–2126.

Azcón, R., Ambrosano, E., Charest, C., 2003. Nutrient acquisition in mycorrhizal lettuce plants under different phosphorus and nitrogen concentration. Plant Sci. 165, 1137–1145.

Ba, A.M., Plenchette, C., Danthu, P., Duponnois, R., Guissou, T., 2000. Functional compatibility of two arbuscular mycorrhizae with thirteen fruit trees in Senegal. Agrofor. Syst. 50, 95–105.

Barea, J.M., 2015. Future challenges and perspectives for applying microbial biotechnology in sustainable agriculture based on a better understanding of plant-microbiome interactions. J. Soil Sci. Plant Nutr. 15, 261–282.

Barea, J.M., Calvet, C., Estaun, V., Camprubi, A., 1996. Biological control as a key component in sustainable agriculture. Plant Soil 185, 171–172.

Baylis, G.T.S., 1975. The magnolioid mycorrhiza and mycotrophy in root systems derived from it. In: Sanders, F.E., Mosse, B., Tinker, P.B. (Eds.), Endomycorrhizas. Academic Press, New York, pp. 373–389.

Benedetto, A., Magurno, F., Bonfante, P., Lanfranco, L., 2005. Expression profiles of a phosphate transporter gene (GmosPT) from the endomycorrhizal fungus *Glomus mosseae*. Mycorrhiza 15, 620–627.

Brundrett, M., 2004. Diversity and classification of mycorrhizal associations. Biol. Rev. 79, 473–495.

Brundrett, M.C., Kendrick, B., 1988. The mycorrhizal status, root anatomy, and phenology of plants in a sugar maple forest. Can. J. Bot. 66, 1153–1173.

Cardoso, J.A., de Lemos, E.E.P., dos Santos, T.M.C., Caetano, L.C., Nogueira, M.A., 2008. Mycorrhizal dependency of mangaba tree under increasing phosphorus levels. Pesq. Agrop. Brasiliera 43, 887–892.

Chalot, M., Blaudez, D., Brun, A., 2006. Ammonia: a candidate for nitrogen transfer at the mycorrhizal interface. Trends Plant Sci. 11, 263–266.

Chen, M., Arato, M., Borghi, L., Nouri, E., Reinhardt, D., 2018a. Beneficial services of arbuscular mycorrhizal fungi—from ecology to application. Front. Plant Sci. 9, 14.

Chen, M., Arato, M., Borghi, L., Nouri, E., Reinhardt, D., 2018b. Beneficial services of arbuscular mycorrhizal fungi—from ecology to application. Front. Plant Sci. 9, 1–14.

Costa, C.M.C., Cavalcante, U.M.T., Goto, B.T., dos Santos, V.F., Maia, L.C., 2005. Arbuscular mycorrhizal fungi and phosphorus supply on seedlings of mangabeira. Pesq. Agrop. Brasiliera 40, 225–232.

da Sousa, C.S., Menezes, R.S.C., de Sampaio, E.V.S.B., de Lima, F.S., Oehl, F., Maia, L.C., 2013. Arbuscular mycorrhizal fungi within agroforestry and traditional land use systems in semi-arid Northeast Brazil. Acta Sci. Agron. 35, 307–314.

Delaux, P.M., 2017. Comparative phylogenomics of symbiotic associations. New Phytol. 213, 89–94.

Duponnois, R., Plenchette, C., 2003. A mycorrhiza helper bacterium enhances ectomycorrhizal and endomycorrhizal symbiosis of Australian Acacia species. Mycorrhiza 13, 85–91.

Efthymiou, A., Jensen, B., Jakobsen, I., 2018. The roles of mycorrhiza and Penicillium inoculants in phosphorus uptake by biochar-amended wheat. Soil Biol. Biochem. 127, 168–177.

Elsen, A., Baimey, H., Sweenen, R., De Waele, D., 2003. Relative mycorrhizal dependency and mycorrhiza-nematode interaction in banana cultivars (Musa spp.) differing in nematode susceptibility. Plant Soil 256, 303–313.

Ezawa, T., Hayatsu, M., Saito, M., 2005. A new hypothesis on the strategy for acquisition of phosphorus in arbuscular mycorrhiza: up-regulation of secreted acid phosphatase gene in the host plant. Mol. Plant-Microbe Interact. 18, 1046–1053.

Fitter, A., Helgason, T., Hodge, A., 2011. Nutritional exchanges in the arbuscular mycorrhizal symbiosis: implications for sustainable agriculture. Fungal Biol. Rev. 25, 68–72.

Flasinski, S., Rogozinska, J., 1988. Acid phosphatase activity in oilseed rape depending on phosphorus content in the medium and the plant. Acta Physiol. Plant 10, 3–10.

Frey-Klett, P., Garbaye, J.A., Tarkka, M., 2007. The mycorrhiza helper bacteria revisited. New Phytol. 176, 22–36.

Gastol, M., Domagala-Swiatkiewicz, I., Bijak, M., 2016. The effect of mycorrhizal inoculation and phosphorus application on the growth and mineral nutrient status of apple seedlings. J. Plant Nutr. 39, 288–299.

George, E., Marschner, H., 1996. Nutrient and water uptake by roots of forest trees. Z. Pflanzenernahr. Bodenkd. 159, 11–21.

Gerdemann, J.W., 1975. VA mycorrhizae. In: Torrey, J.G., Clarkson, D.T. (Eds.), Development and Function of Roots. Academic Press, London, pp. 575–591.

Gianinazzi, S., Gianinazzi-Pearson, V., Trouvelot, A., 1989. Potentialities and procedures for the use of endomycorrhizas with special emphasis on high value crops. In: Whipps, J.M., Lumsden, R.D. (Eds.), Biotechnology of Fungi for Improving Plant Growth. Cambridge University Press, pp. 41–54.

Govindarajulu, M., Pfeffer, P.E., Jin, H., Abubaker, J., Douds, D.D., Allen, J.W., Bücking, H., Lammers, P.J., Shachar-Hill, Y., 2005. Nitrogen transfer in the arbuscular mycorrhizal symbiosis. Nature 435, 819.

Graham, S.W., Lam, V.K., Merckx, V.S., 2017. Plastomes on the edge: the evolutionary breakdown of mycoheterotroph plastid genomes. New Phytol. 214, 48–55.

Grimaldo-Pantoja, G.L., Niu, G.H., Sun, Y.P., Castro-Rocha, A., Alvarez-Parrilla, E., Flores-Margez, J.R., Corral-Diaz, B., Osuna-Avila, P., 2017. Negative effect of saline irrigation on yield components and phytochemicals of pepper (*Capsicum annuum*) inoculated with arbuscular mycorrhizal fungi. Rev. Fitotec. Mex. 40, 141–149.

Guether, M., Neuhäuser, B., Balestrini, R., Dynowski, M., Ludewig, U., Bonfante, P., 2009. A mycorrhizal-specific ammonium transporter from *Lotus japonicus* acquires nitrogen released by arbuscular mycorrhizal fungi. Plant Physiol. 150, 73–83.

Guissou, T., Bâ, A., Ouadba, J.-M., Guinko, S., Duponnois, R., 1998. Responses of Parkia biglobosa (Jacq.) Benth, *Tamarindus indica* L. and *Zizyphus mauritiana* Lam. to arbuscular mycorrhizal fungi in a phosphorus-deficient sandy soil. Biol. Fertil. Soils 26, 194–198.

Guissou, T., Babana, A.H., Sanon, K.B., Ba, A.M., 2016. Effects of arbuscular mycorrhizae on growth and mineral nutrition of greenhouse propagated fruit trees from diverse geographic provenances. Biotechnol. Agron. Soc. Environ. 20 (3), 417–426.

Hayman, D., 1986. Mycorrhizae of nitrogen-fixing legumes. MIRCEN J. Appl. Microbiol. Biotechnol. 2, 121–145.

Hetrick, B.A.D., Wilson, G.T., Cox, T.S., 1992. Mycorrhizal dependence of modern wheat varieties, land races, and ancestors. Can. J. Bot. 70, 2032–2040.

Hetrick, B.A.D., Wilson, G.W.T., Todd, T.C., 1996. Mycorrhizal response in wheat cultivars: relationship to phosphorus. Can. J. Bot. 74, 19–25.

Jaizmevega, M.C., Azcon, R., 1995. Responses of some tropical and subtropical cultures to endomycorrhizal fungi. Mycorrhiza 5, 213–217.

Janos, D.P., 1980. Vesicular-arbuscular mycorrhizae affect lowland tropical rain-forest plant-growth. Ecology 61, 151–162.

Janos, D.P., 2007. Plant responsiveness to mycorrhizas differs from dependence upon mycorrhizas. Mycorrhiza 17, 75–91.

Javot, H., Pumplin, N., Harrison, M.J., 2007. Phosphate in the arbuscular mycorrhizal symbiosis: transport properties and regulatory roles. Plant Cell Environ. 30, 310–322.

Karandashov, V., Bucher, M., 2005. Symbiotic phosphate transport in arbuscular mycorrhizas. Trends Plant Sci. 10, 22–29.

Koide, R.T., 1991. Nutrient supply, nutrient demand and plant-response to mycorrhizal infection. New Phytol. 117, 365–386.

Kuang, R.B., Liao, H., Yan, X.L., Dong, Y.S., 2005. Phosphorus and nitrogen interactions in field-grown soybean as related to genetic attributes of root morphological and nodular traits. J. Integr. Plant Biol. 47, 549–559.

Kungu, J.B., Lasco, R.D., Dela Cruz, L.U., Dela Cruz, R.E., Husain, T., 2008. Effect of vesicular arbuscular mycorrhiza (VAM) fungi inoculation on coppicing ability and drought resistance of senna spectabilis. Pak. J. Bot. 40, 2217–2224.

Lambers, H., Brundrett, M.C., Raven, J.A., Hopper, S.D., 2011. Plant mineral nutrition in ancient landscapes: high plant species diversity on infertile soils is linked to functional diversity for nutritional strategies. Plant Soil 348, 7.

Latef, A.A.H.A., Chaoxing, H., 2011. Effect of arbuscular mycorrhizal fungi on growth, mineral nutrition, antioxidant enzymes activity and fruit yield of tomato grown under salinity stress. Sci. Hortic. 127, 228–233.

Lovato, P.E., Gianinazzi-Pearson, V., Trouvelot, A., Trouvelet, A., Gianinazzi, S., 1996. The state of art of mycorrhizas and micropropagation. Adv. Hortic. Sci., 10, 46–52.

Lovato, P.E., Schüepp, H., Trouvelot, A., Gianinazzi, S., 1999. Application of arbuscular mycorrhizal fungi (AMF) in orchard and ornamental plants. In: Mycorrhiza. Springer, pp. 443–467.

Lovato, P.E., Trouvelot, A., Gianinazzi-Pearson, V., Gianinazzi, S., 2006. Enhanced growth of wild cherry using micropropagated plants and mycorrhizal inoculation. Agron. Sustain. Dev. 26, 209–213.

MacLean, A.M., Bravo, A., Harrison, M.J., 2017. Plant signaling and metabolic pathways enabling arbuscular mycorrhizal symbiosis. Plant Cell 29, 2319–2335.

Manjunath, A., Habte, M., 1991. Root morphological-characteristics of host species having distinct mycorrhizal dependency. Can. J. Bot. 69, 671–676.

Marschner, H., 1995. Mineral Nutrition of High Plants. Academic Press, London.

Marschner, P., 2012. Marschner's Mineral Nutrition of Higher Plants. Academic Press, London.

Miransari, M., 2014. The interactions of soil microbes, arbuscular mycorrhizal fungi and N-fixing bacteria, rhizobium, under different conditions including stress. In: Use of Microbes for the Alleviation of Soil Stresses. Springer, pp. 1–21.

Navarro, J.M., Morte, A., 2019. Mycorrhizal effectiveness in *Citrus macrophylla* at low phosphorus fertilization. J. Plant Physiol. 232, 301–310.

Neumann, E., George, E., 2010. Nutrient uptake: the arbuscular mycorrhiza fungal symbiosis as a plant nutrient acquisition strategy. In: Arbuscular Mycorrhizas: Physiology and Function. Springer, pp. 137–167.

Ortas, I., 2008. The effect of mycorrhizal inoculation on forage and non forage plant growth and nutrient uptake under the field conditions. In: Sustainable Mediterranean Grasslands and Their Multi-Functions. Options Mediterraneennes, CIHEAM, Zaragoza, pp. 463–469.

Ortas, I., 2012a. Do maize and pepper plants depend on mycorrhizae in terms of phosphorus and zinc uptake? J. Plant Nutr. 35, 1639–1656.

Ortas, I., 2012b. Mycorrhiza in citrus: growth and nutrition. In: Advances in Citrus Nutrition. Springer, pp. 333–351.

Ortas, I., 2012c. Mycorrhiza in citrus: growth and nutrition. In: Srivastava, A.K. (Ed.), Advances in Citrus Nutrition. Springer-Verlag, The Netherlands.

Ortaş, I., Rafique, M., Ahmed, İ.A., 2017. Application of arbuscular mycorrhizal fungi into agriculture. In: Arbuscular Mycorrhizas and Stress Tolerance of Plants. Springer, pp. 305–327.

Ortas, I., Demirbas, A., Akpinar, C., 2018. Time period and nutrient contents alter the mycorrhizal responsiveness of citrus seedlings. Eur. J. Hortic. Sci. 83, 72–80.

Parniske, M., 2008. Arbuscular mycorrhiza: the mother of plant root endosymbioses. Nat. Rev. Microbiol. 6, 763.

Plenchette, C., Fortin, J.A., Furlan, V., 1983. Growth responses of several plant species to mycorrhizae in a soil of moderate P-fertility. Plant Soil 70, 199–209.

Plenchette, C., Clermont-Dauphin, C., Meynard, J.M., Fortin, J.A., 2005. Managing arbuscular mycorrhizal fungi in cropping systems. Can. J. Plant Sci. 85, 31–40.

Revillini, D., Gehring, C.A., Johnson, N.C., 2016. The role of locally adapted mycorrhizas and rhizobacteria in plant-soil feedback systems. Funct. Ecol. 30, 1086–1098.

Rouphael, Y., Franken, P., Schneider, C., Schwarz, D., Giovannetti, M., Agnolucci, M., De Pascale, S., Bonini, P., Colla, G., 2015. Arbuscular mycorrhizal fungi act as biostimulants in horticultural crops. Sci. Hortic. 196, 91–108.

Schachtman, D.P., Reid, R.J., Ayling, S.M., 1998. Phosphorus uptake by plants: from soil to cell. Plant Physiol. 116, 447–453.

Schussler, A., Schwarzott, D., Walker, C., 2001. A new fungal phylum, the Glomeromycota: phylogeny and evolution. Mycol. Res. 105, 1413–1421.

Secilia, J., Bagyaraj, D., 1987. Bacteria and actinomycetes associated with pot cultures of vesicular–arbuscular mycorrhizas. Can. J. Microbiol. 33, 1069–1073.

Sellal, Z., Mouden, N., Selmaoui, K., Dahmani, J., Benkirane, R., Douira, A., 2017. Effect of an endomycorrhizal inoculum on the growth of argan tree. Int. J. Environ. Agric. Biotechnol 2, 928–939.

Sharma, I.P., Sharma, A.K., 2017. Physiological and biochemical changes in tomato cultivar PT-3 with dual inoculation of mycorrhiza and PGPR against root-knot nematode. Symbiosis 71, 175–183.

Sharma, M.P., Bhatia, N.P., Adholeya, A., 2001. Mycorrhizal dependency and growth responses of Acacia nilotica and *Albizzia lebbeck* to inoculation by indigenous AM fungi as influenced by available soil P levels in a semi-arid Alfisol wasteland. New For. 21, 89–104.

Shi, Z.Y., Chen, Y.L., Hou, X.G., Gao, S.C., Wang, F.Y., 2013. Arbuscular mycorrhizal fungi associated with tree peony in 3 geographic locations in China. Turk. J. Agric. For. 37, 726–733.

Shireen, F., Nawaz, M.A., Chen, C., Zhang, Q.K., Zheng, Z.H., Sohail, H., Sun, J.Y., Cao, H.S., Huang, Y., Bie, Z.L., 2018. Boron: functions and approaches to enhance its availability in plants for sustainable agriculture. Int. J. Mol. Sci. 19, 20.

Smith, S.E., Read, D.J., 2008. Mycorrhizal Symbiosis. Academic Press, San Diego, CA.

Smith, S.E., Gianinazzipearson, V., Koide, R., Cairney, J.W.G., 1994. Nutrient transport in mycorrhizas—structure, physiology and consequences for efficiency of the symbiosis. Plant Soil 159, 103–113.

Smith, S.E., Facelli, E., Pope, S., Smith, F.A., 2010. Plant performance in stressful environments: interpreting new and established knowledge of the roles of arbuscular mycorrhizas. Plant Soil 326, 3–20.

Smith, S.E., Jakobsen, I., Grønlund, M., Smith, F.A., 2011. Roles of arbuscular mycorrhizas in plant phosphorus nutrition: interactions between pathways of phosphorus uptake in arbuscular mycorrhizal roots have important implications for understanding and manipulating plant phosphorus acquisition. Plant Physiol. 156, 1050–1057.

Takacs, T., Cseresnyes, I., Kovacs, R., Paradi, I., Kelemen, B., Szili-Kovacs, T., Fuzy, A., 2018. Symbiotic effectivity of dual and tripartite associations on Soybean (*Glycine max* L. Merr.) cultivars inoculated with *Bradyrhizobium japonicum* and AM Fungi. Front. Plant Sci. 9, 14.

Van Der Heijden, M.G., Martin, F.M., Selosse, M.A., Sanders, I.R., 2015. Mycorrhizal ecology and evolution: the past, the present, and the future. New Phytol. 205, 1406–1423.

Watts-Williams, S.J., Cavagnaro, T.R., 2014. Nutrient interactions and arbuscular mycorrhizas: a meta-analysis of a mycorrhiza-defective mutant and wild-type tomato genotype pair. Plant Soil 384, 79–92.

Further reading

Brundrett, M., 1991. Mycorrhizas in natural ecosystems. In: Advances in Ecological Research. Elsevier, pp. 171–313.

Brundrett, M.C., 2009. Mycorrhizal associations and other means of nutrition of vascular plants: understanding the global diversity of host plants by resolving conflicting information and developing reliable means of diagnosis. Plant Soil 320, 37–77.

Hodge, A., 2017. Accessibility of inorganic and organic nutrients for mycorrhizas. In: Mycorrhizal Mediation of Soil. Elsevier, pp. 129–148.

25

Microbial consortia: Concept and application in fruit crop management

Gloria Padmaperuma, Thomas O. Butler, Faqih A.B. Ahmad Shuhaili,
Wasayf J. Almalki, Seetharaman Vaidyanathan*

Department of Chemical and Biological Engineering, The University of Sheffield, Sheffield, United Kingdom
*Corresponding author. E-mail: s.vaidyanathan@sheffield.ac.uk

1 Introduction

With the rising world population, the consumption of fruits and vegetables is expected to rise, and the need for sustainable food systems for human health and well-being is ever more pressing (Lindgren et al., 2018; Martin, 2018). To meet rising demands and to ensure most of the harvest reaches the consumer's plate, various agricultural practices, sometimes detrimental to the environment, have been adopted. Currently, fertilizers, insecticides, and herbicides have been used liberally to maximize crop yields, to protect them from pathogens and environmental factors, and to enhance the flavor profile. The liberal use of harsh chemicals to protect plants from pests and diseases, to develop resistance to environmental constraints, and to increase yields has all contributed to soil and water pollution (Agostini et al., 2010). The accumulation of compounds such as nitrates, heavy metals, and toxins has also destabilized soil microbiota, contributing to the prevalence of plant pathogens (Babin et al., 2019).

Innovative crop management practices are constantly needed to bridge yield gaps between what is theoretically possible and what is realistic. In addition, sustainable methods are required to mitigate pollution while ensuring that yield losses from crop pathogens are minimized. This is especially true in the tropics, where crop losses are double those of temperate zones (Drenth and Guest, 2016). Furthermore, the increased resistance of pests and pathogens to pesticides further poses problems in regard to crop storage, especially during transportation. Food handling and

A.K. Srivastava, Chengxiao Hu (eds.)
Fruit Crops: Diagnosis and Management of Nutrient Constraints
https://doi.org/10.1016/B978-0-12-818732-6.00025-3

inappropriate storage practices have been shown to provide breeding environments for pathogenic bacteria, fungi, and viruses.

In the pursuit of obtaining novel and eco-friendly strategies, research has transitioned to the use of microbial consortia. Learning from nature and from ancient practices that rely on the use of specific microbial assemblages, consortium application can be used in fruit crop management. The approach aims to construct microbial consortia that can possess varied functionalities including the ability to breakdown nutrients for plant uptake, to eliminate contaminating microorganisms, to release molecules to protect the outer shell of crops, and to reduce crop losses, and in the long-term leads to an increase in crop yields. In this chapter, we highlight the concept of developing microbial consortia, which is increasingly finding its way into different fields, and discuss the suitability of applying the approach as an innovative tool in different areas of fruit crop management.

2 Natural consortia

Microbial consortia are found in abundance in the natural world (Fig. 25.1), for example, in the mammalian gut (Clavel et al., 2017), foods (Montel et al., 2014), soils (Van Der Heijden et al., 2008), aquatic ecosystems (Paerl and Pinckney, 1996), and biological wastes (Bayer et al., 2007). These are populated by various types of microbes, composed of balanced communities for the benefit of the host and the consortium members themselves.

Ancient civilizations used various combinations of microbes to ferment food and beverages. Unlike today, these practices were carried out without defined knowledge of the concentration of a single taxon within the starter culture. The role of microbial consortia over the past years has become a focal point in medical and biotechnological research (Hays et al., 2015). Investigating the social interactions between various microorganisms, including bacteria, yeast, and microalgae, has proven to be essential in the pursuit of improving human health, production of food and beverages (Santos et al., 2014), ecology (Kouzuma and Watanabe, 2014), and bioproductions (Jones and Wang, 2018). Natural microbial communities have been used for bioremediation, wastewater treatment, methanogenic digestion, and microbial fuel cells. Some examples of natural consortia are observed in biofilms and lichens. The growth environment also influences the type of microbial consortium required. In recent years, the communities involved in these applications have been under scrutiny by scientists and engineers alike to better elucidate the role of individual microbes within the group. This knowledge is believed to aid in the synthesis of more targeted artificial assemblages to be deployed in agriculture, bioremediation, and bioproduction (Kouzuma and Watanabe, 2014).

The longevity of a microbial community is dictated by how each member interacts with the other. The interactions within the communities have broadly been categorized into symbiotic (mutualism and commensalism) (Santos and Reis, 2014; Wilmes and Bond, 2009), and antagonistic (parasitic, predator-prey (Giannone et al., 2015), amensalism, and competitive) (Fig. 25.1). There is a reason to believe that microorganisms live in clusters to withstand adversities encountered in their habitat, such as extreme environments and incursions from other species into the niche, to

FIG. 25.1 Natural consortia are encountered in various habitats and play a major role in ecology, bioremediation, production of fermented foods and beverages, and biofilms and within mammalian guts. These consortia consist of various microorganisms whose relationships can fall into the following categories: commensalism, amensalism, mutualism or cooperation, competition, parasitism, and predator-prey.

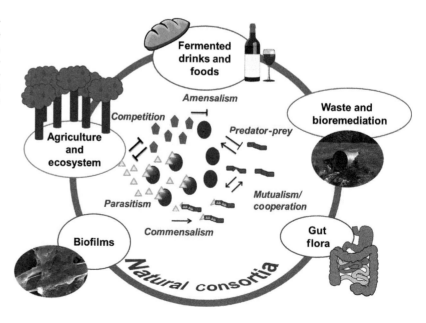

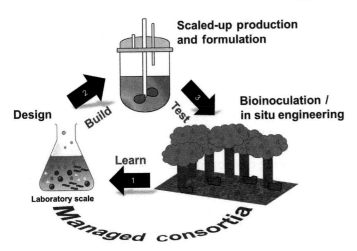

FIG. 25.2 Managed consortia: From flask to fruit crops.

scavenge nutrients, to evolve (Brenner et al., 2008; Hays et al., 2015; Loeuille and Leibold, 2008), and to become more robust (Stenuit and Agathos, 2015). Within a microbial consortium, a network is established through communication, usually in the form of metabolites, such as lactones involved in quorum sensing (Stringlis et al., 2018). The metabolite exchanges dictate the nature of their interaction, population dynamics, division of labor, screening, and aiding in evolution (Beech and Sunner, 2004; Delaux et al., 2015).

Fig. 25.2 shows managed consortia from flask to fruit crops, the design-build-test-learn model to illustrate how an artificial consortium conceived within the laboratory can be applied to the fruit crop field. It could involve (1) isolation; (2) selection, characterization, and construction; and (3) application. Isolated candidates from nature are selected based on their capacity to effect the desired outcome, consortia constructed and applied in the field as bioinoculant. Alternatively, community structure in nature is characterized and engineered in situ to affect the desired outcome. The process sequence could involve laboratory testing, scale-up, and subsequent field application. A cycle of build → test → learn → build… is established until the desired outcome is reached. Data are acquired for improving the initial design in an evolutionary process.

3 Concept of managed consortia

A managed consortium consists of an assemblage of microbes. The microbes are either isolated from a natural environment that can be manipulated, maintained, or selectively evolved to achieve a desired trait or constructed by "scouting" based on metabolic, genotypic, or phenotypic profiles and cultured together in the laboratory (Goers et al., 2014; Padmaperuma et al., 2018) toward end use application (Fig. 25.2). The aim is to create a consortium that would have a specific purpose, for example, improving product yields in biomanufacturing. The construction of an artificial consortium within the laboratory faces several challenges. The consortia found in nature have evolved over tens of thousands of years. Within a laboratory setting, the goal may be to replicate some of these adaptations in a fraction of the time through artificial adaptive evolution. There are many variables that need to be considered, such as priority effects, community backgrounds, ratio of component partners, competition, overyielding, and underyielding effects (Schmidtke et al., 2010). In addition, the efficacy over varying environmental conditions poses a further problem, as artificial consortia may not behave the same way when used outside of the laboratory or the experimental rigs (Preininger et al., 2018). The limitations within the experimental setups may only allow for artificial consortia to be used within indoor greenhouses or within process plants; however, with deeper understanding, these practices could come to fruition and have the potential to be used in conventional agricultural settings for fruit crops (Enebe and Babalola, 2019). There is thus an element of designing with a cycle of learn-build-test (Pham et al., 2017) (Fig. 25.2) that needs to be undertaken with each scenario to arrive at functional consortia that achieve the desired effects.

4 Consortia types, approaches, and applications in agriculture

The construction of a symbiotic artificial consortium for increased bioproduction has been demonstrated using combinations of bacteria, yeast, microalgae, and other microorganisms. The artificial consortia predominantly consist of two interacting members and are commonly referred to as cocultures. By working with two key microorganisms, researchers have been able to address some of the challenges encountered when establishing a symbiosis within the laboratory (Xu et al., 2018). Within the constructs, one main microorganism, usually the producer of the desired product, is equipped with an aiding microorganism that would instigate the primary producer to increase bioproduction output. These consortia have been constructed with an end product in mind, for example, between two bacteria (*Weissella confusa* 11GU1 and *Propionibacterium freudenreichii* JS15) for the production of acetate and exopolysaccharide (Tinzl-Malang et al., 2015), between two yeasts (*Rhodutorula glutinis* DBVPG 3853 and *Debaryomyces castellii* DBVPG 3503) to increase the yield of carotenoids (Buzzini, 2001), and between microalgae and yeast for increasing lipid production for biofuel applications (e.g., *Chlorella vulgaris* TISTR 826 [alga] and *Trichosporonoides spathulata* JU4-57 [yeast]) (Kitcha and Cheirsilp, 2014). There is also a wealth of knowledge being developed in understanding and managing consortia to benefit human health, such as with the gut microbiome (Clavel et al., 2017).

Artificial consortia have shown potential in the laboratory, and it has been hypothesized that this can be replicated in the field to benefit fruit crops. The benefits that agricultural practices can gain from the use of carefully studied consortia are immense. Research has shown that the use of fertilizers, which pollute air, soil, water, and disrupt the original biochemical composition of the rhizosphere, can be decreased by introducing microbial consortia. Pesticide residues in the environment harm humans and mammals upon consumption, as they enter the food chain, and these can be degraded by microbial consortia (Arya and Sharma, 2016). Microbial consortia have also been shown to be beneficial in aiding plants in extreme crop conditions (Bilal et al., 2018). Managing a microbial consortium is still a challenge. Various studies have been conducted, demonstrating that one of the main bottlenecks is finding a "master" consortium that could be used as the basis for agricultural plant growth. This is because each crop and perhaps each crop lineage/variant seem to need a different microbial consortium recipe to be fruitful as the effect of the microbial consortia has great impact on the genetic makeup of the plant (Balestrini et al., 2017).

5 Applications in fruit crop management

The characterization of microbial consortia and its application in different areas of crop management are not new, and there is a considerable body of literature on maximizing crop yields and warding against plant diseases with the help of microorganisms. The application of managed consortia is increasingly finding relevance in fruit crop management. Some of the areas of influence where managed consortia can make a difference are discussed in the succeeding text, and evidence from literature on application to fruit corps or other allied crops is highlighted.

5.1 Soil management and nutrient mobilization: Effect on the rhizosphere

Obtaining a healthy, high yielding crop is vital for a successful agricultural business. In pursuit of improving soil quality for healthier crops and higher yields, crop rotation, fertilizers, and pesticides are frequently employed. Although these methods offer benefits, mitigate crop spoilage, increase crop yields, and supplement essential nutrients to the soil, malpractices and misuses have resulted in ecological issues. Mixed fertilizers are rich in phosphorus, ammonium nitrate, and potassium; however, they also contain heavy metals (Ni, Pb, Cd, and Hg), which accumulate in the soil and harm the local ecosystem (Sharma and Pathak, 2014). Pollution by excess fertilizer additions destabilizes the soil environment, spoils crops, and results in eutrophication and bioaccumulation of toxins throughout the trophic web. High doses of heavy metals have been shown to eliminate insects and microorganisms that can be helpful for the soil. There is, therefore, a necessity for natural and sustainable alternatives to chemical fertilizers.

Microbial products, such as biofertilizers, consist of mixtures of living microbes that can participate to enhance crop yields, and potentially conventional fertilizers (Trivedi et al., 2017). Upon inoculation, the microbial mixture provides essential nutrients to the plant by releasing compounds into the surrounding environment or increase nutrient availability by breaking down complex compounds into simple ones, which can be, for example, easily assimilated through the roots of the plant. A combination of *Azospirillum*, *Glomus mosseae*, *Pseudomonas striata*, and blue-green algae (composed of *Aulosira fertilissima*, *Anabaena variabilis*, *Tolypothrix tenuis*, and *Nostoc muscorum*) has been used as a rice biofertilizer and shown to result in better growth and grain yield production (Chinnusamy et al., 2006). Applying the consortia of *Punica granatum*, *Azospirillum brasilense*, and *G. mosseae* to custard apples and bananas resulted in greater biomass production compared with single microbe treatments (Aseri et al., 2008).

Soil microorganisms have been considered as an integral part of ecosystems because they are able to provide essential nutrients such as N, P, and C for plant growth through the turnover of organic substances in the soil and mobilization of plant nutrients (Francioli et al., 2018). Phosphorus is an important nutrient for plant growth and crops. The reduction of phosphorus supplies decreases crop yields. Thus, there is attention to develop sustainable technologies to provide phosphorus for plant uptake (Baas et al., 2016). Bacteria and fungi are able to provide soluble phosphorus molecules to the plant, by transforming insoluble phosphate to soluble phosphorus via secreting organic acids to lower the pH of the soil and release restricted phosphate forms. Soil bacteria have a key role in soil phosphorus concentration and improving phosphorus uptake through stimulating plant root growth (Alori et al., 2017; Suleman et al., 2018). A consortium of four bacterial taxa (*Enterobacter cloacae*, *Citrobacter freundii*, *Pseudomonas putida*, and *Comamonas testosterone*) has been shown to be effective in mobilizing phosphate and improve productivity by twofold in crops including tomatoes (Baas et al., 2016). It should also be noted that several of the plant growth–promoting rhizobacteria (PGPR) release minute amounts of auxin indole-3-acetic acid (IAA), which encourages plant root growth (Raja et al., 2006).

Microbial consortia act as strongly structured networks in the rhizosphere and can provide support to the host and may survive better than individual microorganisms (Trivedi et al., 2017). Resident microbes in the rhizosphere can enhance plant growth by stimulating the root, releasing regulators of growth, providing nutrients to plant, and controlling soilborne disease suppression (Saha et al., 2016). For example, an association between the mycorrhizal fungus *Rhizophagus irregularis* and the diazotrophic nitrogen-fixing bacterium *Azotobacter vinelandii* has been shown to benefit wheat cultivation, especially by improving root exploration in field conditions with improved nutrient uptake profiles (Dal Cortivo et al., 2018). Microbial consortia may be able to carry out nitrogen fixation and denitrification on the different phase of their cultivation. The dual behavior of microbial consortia in the rhizosphere has shown positive results with the inoculation of nitrogen-fixing microbes on plant growth and yield production. The advantages of inoculating rhizobacteria and cyanobacteria with crops not only improves fixation of biological nitrogen but also enhances root permeability and production of extracellular antibiotics and increases some plant hormones such as gibberellins, auxin, and cytokinin (Subhashini et al., 2016).

5.2 Stress tolerance

Microbial consortia help to increase tolerance toward several stresses such as drought, salinity, heavy metal, and also biotic stress. Microbial community in consortia responds to stresses through different mechanisms such as production of enzymes, hormones, and metabolites (Sultan et al., 2016). For example, the consortium containing *P. putida* KT2440, *Sphingomonas* sp. OF178, *A. brasilense* Sp7, and *Acinetobacter* sp. EMM02 bacteria has been shown to bring improvement in drought tolerance of blue maize (Molina-Romero et al., 2017). This strengthens the premise that inoculation with multiple beneficial microbes promotes more growth and control compared with single bacterial species. This synergistic design and formulation between these four desiccation-tolerant strains was also shown to benefit weight, height, and plant diameter. Another example is inoculation of pepper with *Microbacterium* sp. 3J1 and *Arthrobacter koreensis* 5J12A, which has been shown to increase tolerance of the crops toward drought. This protection from drought is achieved through the accumulation of xeroprotectants, which help to withstand extreme abiotic stresses. The plants were found to be more tolerant to drought when they were inoculated with this microbial consortium. Another improvement in drought tolerance was observed in a consortium of Arbuscular Mycorrhizal Fungi (AMF) and PGPR in maize crops (Zoppellari et al., 2014). While both fungi and rhizobacteria are known to improve drought tolerance in crops, inoculation of AMF helped the establishment of the rhizobacteria. In this consortium, higher tolerance to water deficiency stress, higher content of mineral nutrients in leaves, and more biomass yield were all recorded.

Consortia can also be used to aid the crop to improve heavy metal tolerance. One example is the interaction between the fungus *Paecilomyces formosus* LHL10 and a bacterium *Sphingomonas* sp. LK11 in soybean (Bilal et al., 2018). Synergistic interactions between the fungus and bacterium were observed during combined Al and Zn stress that not only resulted in lower metal uptake and transport in roots but also enhanced macronutrient uptake and modulated soil enzyme activities. Similarly, the consortium UHasselt Sofie 3 with seven members belonging to *Burkholderia*, *Variovorax*, *Bacillus*, *Pseudomonas*, and *Ralstonia* species was capable of successfully enhancing root length in *Arabidopsis* (flowering plant) under 2,4-DNT stress (Thijs et al., 2014).

5.3 Disease prevention and role in plant defense

It is important for the early diagnosis of fruit crop pathogens to prevent further crop losses. The development of natural resistance through selective breeding is often difficult in fruit crops because many have been domesticated from limited germplasm diversity (Drenth and Guest, 2016). There are a range of microbes that can cause disease in fruit crops including fungi, bacteria, and viruses. Chemical fungicides are in wide use, but these have several issues including low efficacy in the wet season and single mode of action allowing fungicide resistance to rapidly develop, and they cause environmental issues including pathogenicity, bioaccumulation in the food chain of toxic substances, and the displacement of nontarget organisms (Drenth and Guest, 2016; Enebe and Babalola, 2019; Mcdonald and Linde, 2002).

One of the best studied examples of disease in fruit crops is in bananas. Bananas exported from Central America and parts of Asia are grown as monocultures of the Cavendish variety and are susceptible to black leaf streak (black sigatoka) caused by the ascomycete fungus, *Mycosphaerella fijiensis* (Pasberg-Gauhl et al., 2000). Control of this pathogen requires sprays of curative and protective fungicides with applications up to 70 times annually (Drenth and Guest, 2016). In some regions, 25% of the production cost of bananas is attributed to the control of the disease alone (Ploetz et al., 2015). These banana monocultures are prone to soilborne diseases such as Panama wilt caused by *Fusarium oxysporum f.* sp. *cubense* (Ploetz, 2015). Recently, there has been an alarming increase in TR4 spread worldwide by the fungus *F. oxysporum f.* sp. *cubense*, which causes *Fusarium* wilt in Cavendish bananas. Control of these diseases with consortia including endophytic and rhizospheric bacterial isolates is possible (Kavino and Manoranjitham, 2018).

Endophytic bacteria like *Alpha-, Beta-,* and *Gammaproteobacteria, Firmicutes, Actinobacteria* (*Streptomyces* spp.) isolated from mesocarp of *Cucumis melo reticulatus* group "dulce" fruit show antimicrobial activity against major cucurbit pathogens like *Macrophomina phaseolina* (Mac), *F. oxysporum f.* sp. *melonis* races 1 and 2 (FOM 1 and FOM 2, respectively), *F. oxysporum f.* sp. *radicis-cucumerinum* (Forc), and *Pseudomonas syringae* (P.s.) (Glassner et al., 2015). In citrus fruits, a major disease is citrus black spot disease caused by the fungus *Phyllosticta citricarpa* (Enebe and Babalola, 2019). Endophytic bacteria have been identified as having a biocontrol effect against this fungus, and they have been found to inhibit pathogens through direct competition and indirect competition for the same niche. A range of microbes with potential antagonistic effects against pathogens in melons including actinomycetes, which remain relatively unexplored against fruit crop pathogens, have also been shown (Glassner et al., 2015).

Another alternative to treating fruit crop diseases is to prime the plant itself, preparing the plant for a faster and stronger resistance only once a pathogen attacks (Jebakumar and Selvarajan, 2018; Le Mire et al., 2016). Priming has been found to be more cost-effective than elicited defense because the energy cost of induced resistance is optimized. The defense signals in plants are complex and are initiated by microbes at the site of infection. The first lines of defense in plants are the production of oxygen-derived free radicals such as hydrogen peroxide and superoxide molecules (Nanda et al., 2010). Other compounds implicated in these phytodefense strategies are hormones such as salicylic acid and jasmonic acid (Enebe and Babalola, 2019). Microbes can induce systemic resistance within the plant. Antagonistic microbes can be applied to increase the immunity of fruits through the upregulation of pathogenesis-related genes (Enebe and Babalola, 2019).

Plant growth–promoting microbes (PGPM) aid in the sensitization and priming of the plant immune defense arsenal for it to conquer invading pathogens (Enebe and Babalola, 2019). PGPM have been found to suppress plant pathogens by producing antagonistic metabolites and enhancing the immunity of potential crops. Applying *Bacillus subtilis* OSU-142 to apricot trees when they were in full bloom has been shown to result in a reduction of shot-hole disease severity and incidence of the disease caused by the fungus *Coryneum blight* (Esitken et al., 2002). The success of this strain is potentially caused by the ability to inhibit fungal spore germination and penetration.

To avoid pathogens in the first place, the material used for planting should be free of pathogens. Using microbial consortia as a protective layer on seeds could be a way of preventing the establishment of these pathogenic organisms combined with a regular spraying application. *Pseudomonas* spp. have been used for seed coatings but with mixed degrees of success (O'callaghan, 2016). The fungal entomopathogen *Metarhizium anisopliae* has been identified as an effective colonizer of the plant rhizosphere and could be applied as a seed agent. However, the most success has been attained with applying *Bacillus firmus* as a seed treatment as opposed to a drench product, resulting in a 1000-fold reduction in bacteria load (O'callaghan, 2016). To use microbial consortia for this purpose, it is essential that it is inexpensive, is non time consuming, has a similar shelf life to normal treatments, and has the ability to be used with traditional farming practices. To date, the molecular mechanisms for priming remain to be elucidated. Further studies need to be on documenting the shelf life with particular reference to the long-term viability of the inoculant and its efficacy because to date these reports are scarce.

5.4 Biostimulants

A biostimulant can be defined as a factor separate from a plant nutrient that enhances plant growth and development and is soon to be included in the Fertilizers Regulation EC 2003/2003 in early 2019, which means the biostimulant products will require evidence for any claims made (Caradonia et al., 2018). It has been reported that they can increase crop yields by 5%–10% while improving fertilizer use efficiency by 5%–25%. The market for biostimulants is expected to reach US $3.68 billion by 2022 with a compound annual growth rate of 13.58% with Europe retaining 42% of the global market (Caradonia et al., 2018). Many scientists consider biostimulants lacking peer-reviewed scientific evaluation, and their composition and modes/mechanisms of action are generally unknown. There are currently five categories of biostimulants: microbial inoculants, humic acids, fulvic acids, protein hydrolysates and amino acids, and seaweed extracts (Calvo et al., 2014). It is hypothesized that one of the modes of success of biostimulants is attributed to the changes in the microbial flora. It has been identified that the optimal PGPR consist of local strains with specificity to the host plant and a good capacity for adaptation and that can coevolve with the native strains (Le Mire et al., 2016).

Most studies have focused on single strains of bacteria. It was observed that applying *B. subtilis* OSU-142 to apricot trees in full bloom resulted in increases in average fruit weight and yield in field trials (Esitken et al., 2002). It was hypothesized that auxins from *B. subtilis* were responsible for the increases as this has been well documented (Perez-Garcia et al., 2011). It was found that *Pseudomonas* BA-8 and *Bacillus* OSU-142 alone or in combination as a coculture had a great potential to increase the yield, nutrition, and growth in the sweet cherry plant in field trials in Turkey (Esitken et al., 2006). Using an indigenous consortium of arbuscular mycorrhizal fungi (AMF) consisting of three species of *Glomus* and *Scutellospora* sp. resulted in increased plant/root biomass total yield in green peppers and tomatoes investigated in greenhouses and farm trials (Regvar et al., 2003). A series of factors for enhancing mycorrhizal symbiosis has been shown that include soil fertility management (organic manure), pest and disease management (reduced chemical pesticides), establishing cropping systems, tillage management, crop choice, and landscape management (Oruru and Njeru, 2016).

There are several commercial formulations of biostimulants involving microbes available on the market: SumaGrow (AgriBiotic Products), Salavida (Sourcon Padena) containing *Pseudomonas trivialis*, and RhizoVital(R) 42 (ABiTEP) containing *Bacillus* sp., which use different microbial additions and concentrations ranging from 1×10^8 to 10^{17} CFU/g (Preininger et al., 2018). For commercial use as a biostimulant only, SumaGrow from AgriBiotics appears to exist, which contains a consortium of microbes with >30 microbe species (containing nitrogen fixers, phosphate solubilizers, and growth hormone producers) but is applied at a high rate of 10^{17} CFU/mL (Preininger et al., 2018). These microbials are typically applied as a seed treatment, soil additive, foliar spray, or root dip.

One challenge is in figuring how to treat seeds. Conventionally, seeds are inoculated with 10^8 cells, and the seeds are coated with a carrier with or without adhesives, but the inoculum is affected by temperature and humidity affecting viability (Le Mire et al., 2016). More research is needed on biostimulants compared with conventional chemical inputs in field trails with a variety of fruit crops and over varying climates and conditions before farmers will be willing to accept them. With the regulatory framework coming into enforcement in early 2019, this will help ensure a gold standard biostimulant product.

5.5 Biopesticides

Chemical pesticides are effective, but they are costly and have issues with health and safety, tougher regulatory conditions, and concerns of environmental damage and disposal. The use of pesticides is an issue because it lowers the yield of the crop, they can cause secondary outbreaks of pests, the chemicals can be harmful to higher organisms, they can cause environmental pollution, and resistance in pests has been observed (Enebe and Babalola, 2019). Furthermore, some weeds, pests, and organisms responsible for diseases have become resistant to agrochemicals. Foodborne bacteria such as *Escherichia coli*, *Salmonella enteritidis*, and *Shigella sonnei* not only were found to survive in some pesticide solutions but could also increase in number and may contribute a potential health hazard to fruit crops. Consumer pressure, stricter regulations such as lowering the maximum residue levels, and the cost of development for registering synthetic pesticides have resulted in the withdrawal of many synthetic pesticides (O'callaghan, 2016).

There has been an emphasis on developing biobased solutions, which avoid environmental damage and health effects for humans. Biopesticides are a broad group of agents that are mass produced and biologically based agents for controlling plant pests (Greaves and Grant, 2010). The biopesticide market was around $3.5 billion in 2015 (8% of the global pesticides trade) and is projected to reach $4.5 billion in 2023 (7% of the total pesticide market) (Jouzani et al., 2017). From a grower's perspective, biopesticides have several advantages over conventional pesticides: organic (more environmentally friendly), useful in resistance management, and effective at low doses. Biopesticides are often multifunctional: promote growth, cause antibiosis, induce parasitism on pest organisms, and induce systemic resistance in host plants. However, there are also several challenges to overcome: not broad spectrum, not as efficacious, slower acting, unreliable, inconsistent, and have a short shelf life.

The number of genera used for pest and disease control is small. *Bacillus thuringiensis* is the active ingredient in 90% of microbial biopesticides, and 200 products are sold in the United States and 60 in the EU (Preininger et al., 2018). *B. thuringiensis* is known as the most successful microbial bioinsecticide of the last century (Jouzani et al., 2017). It has been found to be present in >98 ($424 million) formulated bacterial pesticides (Jouzani et al., 2017). *Bacillus* is host specific and fast acting, and the endospores are heat resistant. Conventional spraying can be used, and the production process is easy and cheap.

Another microbe that has seldom been reported as a biopesticide is the virus. Baculoviruses can be effective for the control of insect pests including polyhedrosis viruses and granulosis viruses, which are highly specific, but are constraining by having slow bacterial activity and are vulnerable to UV damage (Preininger et al., 2018). *Cydia pomonella* granulovirus (CpGV) is arguably the most important and widely used in organic apple production against the codling moth whereby it is ingested by the larvae and begins replicating, thus causing the larva to cease feeding (Preininger et al., 2018).

There are at least 15 microbial commercial products on the market with varying inoculums (1×10^8 to 1×10^{12} CFU/g) with applications against a variety of pests such as insects, fungi, and bacteria (Preininger et al., 2018). However, to date, no known commercial microbial consortia have been developed against fruit crop pathogens. Vestaron produces a wide range of insecticides/miticides derived from naturally occurring peptides sourced from *B. thuringiensis* and other microbes, which have been designed to be nontoxic to the environment and in particular pollinators.

5.6 Postharvest control of fruit spoilage

Fruit spoilage is a major issue in the food industry, resulting in high levels of wastage and economic losses for consumers and manufacturers. Postharvest handling can result in 20%–25% of harvested fruits and vegetables being degraded by pathogens (Sharma et al., 2009). Fruits are particularly susceptible to spoilage because of their natural nutrient-rich composition. Fungi are primarily responsible for food spoilage, and some genera such as *Aspergillus*, *Penicillium*, and *Fusarium* produce mycotoxins, which have been found to be resistant to food processing steps such as heating and acidification (Sanzani et al., 2016). Bacteria have also been known to be responsible for fruit spoilage. *Alicyclobacillus acidoterrestris* and *Alicyclobacillus acidocaldarius* isolated from orchard soil (pears and apples) in Western Cape, South Africa, are bacteria known for spoilage, but the role of these spoilage organisms has not been elucidated (Groenewald et al., 2008).

Concerns over environmental damage and health effects of fungicides have created demand for natural methods to control fruit spoilage. For protection against fruit spoilage, the main organisms investigated have been bacteria (primarily the *Bacillus* genus, the lactic acid bacteria: *Pediococcus*, *Lactobacillus*, and *Leuconostoc*) and various yeast species (*Candida fructus* and *Issatchenkia orientalis*) (Salas et al., 2017). *Lactobacillus plantarum* TK9 isolated from naturally fermented congee was active against *Penicillium* in apples and citrus fruits (Zhang et al., 2016). *B. subtilis* strains produce a range of bioactives and in particular antimicrobial peptides and nonpeptides (e.g., the antibiotic iturin) (Sharma et al., 2009). Lactobacilli are highly efficacious against fungal spoilage organisms that are attributable to antibiosis and a pH decrease (acetic, propionic, and succinct acids) resulting in the "weak acid theory"–inducing cell death (Salas et al., 2017). Typically when microbial antagonists are applied for controlling postharvest decay on fruits, they are applied at high concentrations ranging from 10^7 to 10^8 CFU/mL (Sharma et al., 2009).

The most well-studied fruit in terms of microbial characterization has been grapes. In unblemished grapes, microbial flora are dominated by basidomycetous and ascomycete yeasts (Kántor et al., 2017). Acetic acid bacteria are often at controllable levels, but in grape sour rot, there is exaggerated production of acetic acid that can cause wine spoilage. It has been proposed that yeasts are effective against fruit spoilage organisms because they compete for nutrients, acidify the medium, and produce mycocins, which affect fungal growth (Salas et al., 2017). Typically, there has been an emphasis on a single species against spoilage organisms in fruit (tomatoes, jackfruit, wild cherries, apples, citrus, peaches, grapes, and pomegranates) rather than consortia (Salas et al., 2017; Sharma et al., 2009). More research is needed to investigate the mechanism of action from antifungal microbes in consortia. For developing microbes, they have to be efficacious in cold conditions as that is usually where they are stored. To date, only single species microbes have been tested for fruit spoilage, and a variety of commercial products are on the market, but to date, there is no microbial consortium that has been developed in the laboratory against fruit spoilage. A major challenge to consider for the future development of microbes against fruit crop spoilage is the possibility of microbial gene transfer to avoid the transfer of virulent genes. Microbial communities can be used to preserve ancient artifacts, by releasing chemical cues that help protect the surfaces (Pinna, 2014). This same concept can be used in protecting fruits from damage, during shipping and transportation.

5.7 Ecomanagement

In addition to direct influence on fruit crop growth, microbial consortia can play important roles in managing the ecosystem that enables the fruit crops to thrive. Synergistic interactions between microbes indirectly contribute to aspects needed in fruit crop production such as pollination, buffering system against pathogens, and also good soil structure. An example is a consortium of two taxonomically different bacteria, that is, *Acetobacter pomorum* and *L. plantarum*, that modulate insulin signaling and target of rapamycin (TOR) pathway of honeybee (Crotti et al., 2013). Insulin, which is modulated by the acetic acid produced by the action of *A. pomorum*, controls several honeybee homeostatic programs such as energy metabolism and cell activity. Meanwhile, *L. plantarum* regulates amino acid that activates TOR signaling, which involves in promoting growth rate. Together, this consortium helps the honeybee to protect their health and resist pathogens or parasite attacks and enhances their well-being. This enables honeybees to carry out pollination for the crops for the latter to produce fruits.

Another application of microbial consortium with a role in ecomanagement is by secondary microbes that exist in fungiculture associated with ants, termites, and beetles (Mueller et al., 2005). These insects cultivate fungi as their source of food. Secondary microbes such as *Escovopsis* fungus are managed by insect farmers to provide antibiotic-producing and disease-suppressing action. It appears that the fungiculture helps in providing the ecosystem a buffering against disease organisms, which can influence the production of fruits. A similar report on fungus-bacteria consortia that interact with leafcutter ants has also been reported (Aylward et al., 2012). In this consortium, bacteria help to synthesize amino acids, B vitamins, and other nutrients needed by basidiomycete fungi (e.g., *Leucoagaricus gongylophorus*) to enhance the growth of the latter and biomass-processing efficiency. The ants provide the leaves as food for the fungi, which in turn act as food source for the ants to survive and proliferate. In return, leafcutter ants contribute to the ecosystem by improving soil aeration and breaking down plants for soil nourishment.

Pesticides have been used to protect crops from pest infestations. However, one of the many disadvantages is its accumulation on the earth surface that can reach the water table. Consequently, the pesticide contamination can transfer from one region to another and destabilize the biological and biochemical processes, important in water and soil ecosystems. Bioremediation has received much interest as an active biotechnological process to decontaminate environmental pollution (Jariyal et al., 2018). Bioremediation is the use of living microorganisms for degradation of environmental contaminants. These microorganisms may be native to polluted sites or be isolated from other sites to be then introduced to the polluted site to enhance pollution degradation. During the biodegradation process, microorganisms convert pesticides into less complex compounds that in the end get converted to carbon dioxide, water, oxides, and mineral salts. Diuron, a pollutant herbicide, has been shown to be degraded by a consortium of three microbes, which was not possible by the individual components of the consortium when tested alone (Villaverde et al., 2017). The consortium was composed of three diuron-degrading strains, *Arthrobacter sulfonivorans*, *Variovorax soli*, and *Advenella* sp. JRO. Combined use of microbial consortia isolated from different agricultural soils and cyclodextrin, a biodegradable organic enhancer of pollutant bioavailability, has been shown to work well in degrading diuron (Villaverde et al., 2018).

Hydrocarbons, pesticides, herbicides, and heavy metals (Cu, Hg, Cd, Pb, and Zn) are the most widespread pollution sources in the soil. Seed germination can decline significantly in heavy metal–polluted soils. While soil microbes can degrade organic contaminants, heavy metals require physical removal (Sharma and Pathak, 2014). Microorganisms are important for heavy metal remediation from polluted areas because they can tolerate metal toxicity in different ways. Applying microbial consortia in heavy metal bioremediation is better than using an individual culture. For example, a consortium of *Sporosarcina soli* B-22, *Viridibacillus arenosi* B-21, *E. cloacae* KJ-47, and *E. cloacae* KJ-46 has shown the synergistic effect to remove Cu, Pb, and Cd from polluted soil, the performance of the consortium observed to be better than the individual components (Ojuederie and Babalola, 2017). Symbiotic bacteria can influence crop management indirectly by controlling disease suppression pests, as has been shown of a soil-dwelling wireworm agricultural pest (Kabaluk et al., 2017). This would imply potential pest control by engineering associated microbiome.

6 Microbiome engineering

In many areas of fruit crop management, as has been highlighted earlier, there is a drive to move toward the use of environmentally sustainable approaches where microbial consortia can play a crucial role in replacing existing practices that are environmentally unsustainable. These include roles in the rhizosphere and phyllosphere, including carposphere and spermosphere, such as in soil conditioning, nutrient mobilization, stress tolerance, disease prevention, pathogen control, control of fruit spoilage, ecomanagement, biostimulants, biopesticides, and biofertilizers (Fig. 25.3).

FIG. 25.3 Microbiome engineering in fruit crop management.

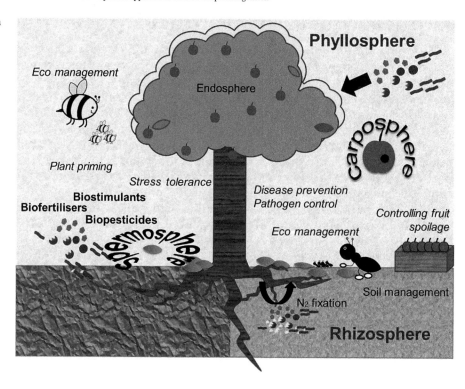

A concerted effort is required at defining the composition of consortia that can be developed for use in fruit crop management, in bespoke applications, or for a specific type of fruit crop. Bacteria, fungi, algae, archaea, and protists can all contribute to construction of microbial consortia that can have a positive influence on fruit crop management. Endophytic and epiphytic microbes can be characterized and community composition engineered to cater to specific applications. Endophytic and epiphytic fungal communities have been studied, for example, in olive cultivars with different susceptibilities to anthracnose (Preto et al., 2017). Differences in size and composition have been observed, perhaps attributable to the chemical nature of the habitats that could potentially be put to use in biocontrol of anthracnose in olives. Despite the potential diversity in habitats and the resulting combinatorial possibilities of community composition, dominant genera can be identified, as can be seen in a recent study on citrus rhizosphere microbiome (Xu et al., 2018). Samples from distinct biogeographical regions of the world showed remarkable similarities in the core composition of the rhizosphere communities. Although bacterial genotypes have been predominantly characterized, the increasing knowledge of eukaryotic genomes should see more eukaryotic populations appear in such characterizations. Potential partners could, for example, be a combination of *Trichoderma* that have been shown to be effective biofertilizers, biostimulants, and bioenhancers and PGP bacteria such as nonpathogenic *Pseudomonas*, *Bacillus*, *Azotobacter*, *Serratia*, and *Azospirillum*, which have been shown to increase nutrient assimilation and nitrogen fixation of the crop (Woo and Pepe, 2018).

Managed consortia in fruit crop management could influence the phyllosphere, carposphere, spermosphere, and/or rhizosphere associated with the desired crop. The application of a well-devised microbial consortium as a biofertilizer, biostimulant, and biopesticide to fruit crops would further help the crops withstand disease, stress tolerance, plant priming, pathogen control, disease prevention, controlling fruit spoilage, etc.

Equally, nonendogenous microbes with enhanced capabilities could be mixed with endogenous communities to engineer microbiomes with optimal benefits for fruit crop management. It might even be possible to generate microbiomes entirely composed of nonendogenous microbes toward the desired outcome. This would, however, carry the caution of careful tracking of genetic variations, as horizontal gene transfer can lead to antibiotic resistance in plant pathogens (Sundin and Wang, 2018), and would require development of knowledge of the host-microbiome dynamics, with respect to not only preventing evolution of antibiotic resistance in plant pathogens, but also the overall ecosystem at large.

Clearly, comprehensive characterization of the natural microbiome will help in defining managed strategies to achieve the desired outcomes in a tractable manner. Amplicon-based community profiling is a common approach used to characterize community structure at the molecular level, but this has been augmented by deep shotgun metagenomics sequencing, for example, to get the community structure in citrus rhizosphere microbiome, in distinct biogeographical regions from six continents (Xu et al., 2018). Changes at the genetic, metabolic, or gross phenotypic levels could be characterized and used to engineer microbiomes for specific end use. The strategies developed could be based on host-mediated microbiome engineering, several options for which have been elaborated (Mueller and Sachs, 2015), where the development is centered around host genotype and/or phenotype and evolving host-microbiome relationships to desired outcome. Inclusion of community coalescence and incorporation of information gathered in different stages of host-microbe interactions into the development of strategies for constructing microbiomes has also been mooted (Rillig et al., 2016). Development of synthetic communities through social-interaction engineering in the laboratory has been demonstrated (Kong et al., 2018), as well as in situ engineering of microbiomes at the community level, with possible precise alteration of metagenomics content for desired effects in the native environments (in situ) (Sheth et al., 2016).

The process of microbiome engineering would require meaningful strategies for isolation of indigenous community partners and proper characterization of microbiome dynamics in specific host-pathogen interactions. Interpartner relationships such as amensalism, commensalism, mutualism, and parasitism (Fig. 25.2) require to be investigated with respect to their evolution vis-à-vis host-microbiome interactions. A deeper understanding of not only microbial community partners and host-microbe interactions but also the evolving relationships over time need to be established for each scenario and host species to enable functional and tractable strategies for microbiome engineering to be developed. Given the multimodality of the data and the variety of subjects and environments that are required to be monitored over time, mobile imaging and sensing technologies can be particularly helpful and cost-effective (Ballard et al., 2018). In addition, computational tools and statistical models (Garcia-Jimenez et al., 2018) will be required, and development of predictive models (Stenuit and Agathos, 2015) will enable more meaningful translation of data. This will lead to more meaningful design and construction of relevant microbiome structure and strategies geared to the desired outcomes with respect to host-specific functionalities. Strategies to construct microbiomes to potentially benefit as many of the desired outcomes in a host-specific or a regioselective manner require systematic development of knowledge at different levels, the tools for which are increasingly becoming available. Regulatory frameworks need to evolve to catch up with the technology available for effective translation of the technology to field applications.

7 Conclusions and future prospects

Microbial consortia can be useful in a wide range of application in fruit crop management. While inoculation of plants with microbes is a well-known practice, construction of managed consortia and its application for specific purposes require a systematic approach with a design based on the learn-build-test cycle. Several areas of fruit crop management can be influenced by appropriate microbiome engineering. This will require proper characterization of indigenous microbiomes and potential exogenous partners with tractable host-microbiome relationships. Both laboratory construction and in situ engineering of consortia can be employed, and engineering at community level and of metagenomes can be effected. There is increasing interest in developing microbial consortia for fruit crop management, but this needs further developments at several stages for it to be effectively implemented in the field. This would also require appropriate regulatory developments.

Acknowledgments

GP and TOB acknowledge financial assistance from UK-EPSRC (DTA 1623367 and 1912024), FAS from Malaysian government, and WJA from Saudi Arabian Cultural Bureau.

References

Agostini, F., Tei, F., Silgram, M., Farneselli, M., Benincasa, P., Aller, M.F., 2010. Decreasing nitrate leaching in vegetable crops with better N management. In: Genetic Engineering, Biofertilisation, Soil Quality and Organic Farming. vol. 4, Springer Science+Business Media B.V., pp. 147–200.

Alori, E.T., Glick, B.R., Babalola, O.O., 2017. Microbial phosphorus solubilization and its potential for use in sustainable agriculture. Front. Microbiol 8, 971.

Arya, R., Sharma, A.K., 2016. Bioremediation of carbendazim, a benzimidazole fungicide using *Brevibacillus borstelensis* and *Streptomyces albogriseolus* together. Curr. Pharm. Biotechnol. 17, 185–189.

Aseri, G.K., Jain, N., Panwar, J., Rao, A.V., Meghwal, P.R., 2008. Biofertilizers improve plant growth, fruit yield, nutrition, metabolism and rhizosphere enzyme activities of pomegranate (*Punica granatum* L.) in Indian Thar Desert. Sci. Hortic. 117, 130–135.

Aylward, F.O., Burnum, K.E., Scott, J.J., Suen, G., Tringe, S.G., Adams, S.M., Barry, K.W., Nicora, C.D., Piehowski, P.D., Purvine, S.O., Starrett, G.J., Goodwin, L.A., Smith, R.D., Lipton, M.S., Currie, C.R., 2012. Metagenomic and metaproteomic insights into bacterial communities in leaf-cutter ant fungus gardens. ISME J. 6, 1688–1701.

Baas, P., Bell, C., Mancini, L.M., Lee, M.N., Conant, R.T., Wallenstein, M.D., 2016. Phosphorus mobilizing consortium Mammoth P (TM) enhances plant growth. PeerJ 4.

Babin, D., Deubel, A., Jacquiod, S., Sorensen, S.J., Geistlinger, J., Grosch, R., Smalla, K., 2019. Impact of long-term agricultural management practices on soil prokaryotic communities. Soil Biol. Biochem. 129, 17–28.

Balestrini, R., Salvioli, A., Dal Molin, A., Novero, M., Gabelli, G., Paparelli, E., Marroni, F., Bonfante, P., 2017. Impact of an arbuscular mycorrhizal fungus versus a mixed microbial inoculum on the transcriptome reprogramming of grapevine roots. Mycorrhiza 27, 417–430.

Ballard, Z.S., Brown, C., Ozcan, A., 2018. Mobile technologies for the discovery, analysis, and engineering of the global microbiome. ACS Nano 12, 3065–3082.

Bayer, E.A., Lamed, R., Himmel, M.E., 2007. The potential of cellulases and cellulosomes for cellulosic waste management. Curr. Opin. Biotechnol. 18, 237–245.

Beech, W.B., Sunner, J., 2004. Biocorrosion: towards understanding interactions between biofilms and metals. Curr. Opin. Biotechnol. 15, 181–186.

Bilal, S., Shahzad, R., Khan, A.L., Kang, S.M., Imran, Q.M., Al-Harrasi, A., Yun, B.W., Lee, I.J., 2018. Endophytic microbial consortia of phytohormones-producing fungus Paecilomyces formosus LHL 10 and bacteria Sphingomonas sp LK11 to Glycine max L. regulates physio-hormonal changes to attenuate aluminum and zinc stresses. Front. Plant Sci 9, 1273.

Brenner, K., You, L.C., Arnold, F.H., 2008. Engineering microbial consortia: a new frontier in synthetic biology. Trends Biotechnol. 26, 483–489.

Buzzini, P., 2001. Batch and fed-batch carotenoid production by Rhodotorula glutinis-Debaryomyces castellii co-cultures in corn syrup. J. Appl. Microbiol. 90, 843–847.

Calvo, P., Nelson, L., Kloepper, J.W., 2014. Agricultural uses of plant biostimulants. Plant Soil 383, 3–41.

Caradonia, F., Battaglia, V., Righi, L., Pascali, G., La, A., 2018. Plant biostimulant regulatory framework: prospects in Europe and current situation at international level. J. Plant Growth Regul 38 (2), 438–448.

Chinnusamy, M., Kaushik, B.D., Prasanna, R., 2006. Growth, nutritional, and yield parameters of wetland rice as influenced by microbial consortia under controlled conditions. J. Plant Nutr. 29, 857–871.

Clavel, T., Lagkouvardos, I., Stecher, B., 2017. From complex gut communities to minimal microbiomes via cultivation. Curr. Opin. Microbiol. 38, 148–155.

Crotti, E., Sansonno, L., Prosdocimi, E.M., Vacchini, V., Hamdi, C., Cherif, A., Gonella, E., Marzorati, M., Balloi, A., 2013. Microbial symbionts of honeybees: a promising tool to improve honeybee health. New Biotechnol. 30, 716–722.

Dal Cortivo, C., Barion, G., Ferrari, M., Visioli, G., Dramis, L., Panozzo, A., Vamerali, T., 2018. Effects of field inoculation with VAM and bacteria consortia on root growth and nutrients uptake in common wheat. Sustainability 10, 3286.

Delaux, P.M., Radhakrishnan, G., Oldroyd, G., 2015. Tracing the evolutionary path to nitrogen-fixing crops. Curr. Opin. Plant Biol. 26, 95–99.

Drenth, A., Guest, D.I., 2016. Fungal and oomycete diseases of tropical tree fruit crops. Annu. Rev. Phytopathol. 54, 373–395.

Enebe, M.C., Babalola, O.O., 2019. The impact of microbes in the orchestration of plants' resistance to biotic stress: a disease management approach. Appl. Microbiol. Biotechnol. 103, 9–25.

Esitken, A., Karlidag, H., Ercisli, S., Sahin, F., 2002. Effects of foliar application of Bacillus subtilis Osu-142 on the yield, growth and control of shot-hole disease (Coryneum blight) of apricot. Gartenbauwissenschaft 67, 139–142.

Esitken, A., Pirlak, L., Turan, M., Sahin, F., 2006. Effects of floral and foliar application of plant growth promoting rhizobacteria (PGPR) on yield, growth and nutrition of sweet cherry. Sci. Hortic. 110, 324–327.

Francioli, D., Schulz, E., Buscot, F., Reitz, T., 2018. Dynamics of soil bacterial communities over a vegetation season relate to both soil nutrient status and plant growth phenology. Microb. Ecol 75 (1), 216–227.

Garcia-Jimenez, B., De La Rosa, T., Wilkinson, M.D., 2018. MDPbiome: microbiome engineering through prescriptive perturbations. Bioinformatics 34, 838–847.

Giannone, R.J., Wurch, L.L., Heimerl, T., Martin, S., Yang, Z.M., Huber, H., Rachel, R., Hettich, R.L., Podar, M., 2015. Life on the edge: functional genomic response of Ignicoccus hospitalis to the presence of Nanoarchaeum equitans. ISME J. 9, 101–114.

Glassner, H., Zchori-Fein, E., Compant, S., Sessitsch, A., Katzir, N., Portnoy, V., Yaron, S., 2015. Characterization of endophytic bacteria from cucurbit fruits with potential benefits to agriculture in melons (Cucumis melo L.). FEMS Microbiol. Ecol 91 (7), 1–13.

Goers, L., Freemont, P., Polizzi, K.M., 2014. Co-culture systems and technologies: taking synthetic biology to the next level. J. R. Soc. Interface 11https://doi.org/10.1098/rsif.2014.0065.

Greaves, J., Grant, W., 2010. Underperforming policy networks: the biopesticides network in the United Kingdom. Br. Polit. 5, 14–40.

Groenewald, W.H., Gouws, P.A., Witthuhn, R.C., 2008. Isolation and identification of species of Alicyclobacillus from orchard soil in the Western Cape, South Africa. Extremophiles 12, 159–163.

Hays, S.G., Patrick, W.G., Ziesack, M., Oxman, N., Silver, P.A., 2015. Better together: engineering and application of microbial symbioses. Curr. Opin. Biotechnol. 36, 40–49.

Jariyal, M., Jindal, V., Mandal, K., Gupta, V.K., Singh, B., 2018. Bioremediation of organophosphorus pesticide phorate in soil by microbial consortia. Ecotoxicol. Environ. Saf. 159, 310–316.

Jebakumar, R.M., Selvarajan, R., 2018. Biopriming of micropropagated banana plants at pre- or post-BBTV inoculation stage with rhizosphere and endophytic bacteria determines their ability to induce systemic resistance against BBTV in cultivar Grand Naine. Biocontrol Sci. Tech. 28, 1074–1090.

Jones, J.A., Wang, X., 2018. Use of bacterial co-cultures for the efficient production of chemicals. Curr. Opin. Biotechnol. 53, 33–38.

Jouzani, G.S., Valijanian, E., Sharafi, R., 2017. Bacillus thuringiensis: a successful insecticide with new environmental features and tidings. Appl. Microbiol. Biotechnol. 101, 2691–2711.

Kabaluk, T., Li-Leger, E., Nam, S., 2017. Metarhizium brunneum—an enzootic wireworm disease and evidence for its suppression by bacterial symbionts. J. Invertebr. Pathol. 150, 82–87.

Kántor, A., Mareček, J., Ivanišová, E., Terentjeva, M., Kačániová, M., 2017. Microorganisms of grape berries. Proc. Latv. Acad. Sci. Sect. B Nat. Exact Appl. Sci. 71, 502–508.

Kavino, M., Manoranjitham, S.K., 2018. In vitro bacterization of banana (Musa spp.) with native endophytic and rhizospheric bacterial isolates: novel ways to combat Fusarium wilt. Eur. J. Plant Pathol. 151, 371–387.

Kitcha, S., Cheirsilp, B., 2014. Enhanced lipid production by co-cultivation and co-encapsulation of oleaginous yeast *Trichosporonoides spathulata* with microalgae in alginate gel beads. Appl. Biochem. Biotechnol. 173, 522–534.

Kong, W.T., Meldgin, D.R., Collins, J.J., Lu, T., 2018. Designing microbial consortia with defined social interactions. Nat. Chem. Biol. 14, 821–829.

Kouzuma, A., Watanabe, K., 2014. Microbial ecology pushes frontiers in biotechnology. Microbes Environ. 29, 1–3.

Le Mire, G., Nguyen, M.L., Fassotte, B., Du Jardin, P., Verheggen, F., Delaplace, P., Jijakli, M.H., 2016. Review: Implementing plant biostimulants and biocontrol strategies in the agroecological management of cultivated ecosystems. Biotechnol. Agron. Soc. Environ. 20, 299–313.

Lindgren, E., Harris, F., Dangour, A.D., Gasparatos, A., Hiramatsu, M., Javadi, F., Loken, B., Murakami, T., Scheelbeek, P., Haines, A., 2018. Sustainable food systems—a health perspective. Sustain. Sci. 13, 1505–1517.

Loeuille, N., Leibold, M.A., 2008. Evolution in metacommunities: on the relative importance of species sorting and monopolization in structuring communities. Am. Nat. 171, 788–799.

Martin, C., 2018. A role for plant science in underpinning the objective of global nutritional security? Ann. Bot. 122, 541–553.

Mcdonald, B.A., Linde, C., 2002. Pathogen population genetics, evolutionary potential, and durable resistance. Annu. Rev. Phytopathol. 40, 349–379.

Molina-Romero, D., Baez, A., Quintero-Hernández, V., Castañeda-Lucio, M., Fuentes-Ramírez, L.E., Bustillos-Cristales, M.D.R., Rodríguez-Andrade, O., Morales-García, Y.E., Munive, A., Muñoz-Rojas, J., 2017. Compatible bacterial mixture, tolerant to desiccation, improves maize plant growth. PLoS One 12, 1–21.

Montel, M.C., Buchin, S., Mallet, A., Delbes-Paus, C., Vuitton, D.A., Desmasures, N., Berthier, F., 2014. Traditional cheeses: rich and diverse microbiota with associated benefits. Int. J. Food Microbiol. 177, 136–154.

Mueller, U.G., Sachs, J.L., 2015. Engineering microbiomes to improve plant and animal health. Trends Microbiol. 23, 606–617.

Mueller, U.G., Gerardo, N.M., Aanen, D.K., Six, D.L., Schultz, T.R., 2005. The evolution of agriculture in insects. Annu. Rev. Ecol. Evol. Syst. 36, 563–595.

Nanda, A.K., Andrio, E., Marino, D., Pauly, N., Dunand, C., 2010. Reactive oxygen species during plant-microorganism early interactions. J. Integr. Plant Biol. 52, 195–204.

O'callaghan, M., 2016. Microbial inoculation of seed for improved crop performance: issues and opportunities. Appl. Microbiol. Biotechnol. 100, 5729–5746.

Ojuederie, O.B., Babalola, O.O., 2017. Microbial and plant-assisted bioremediation of heavy metal polluted environments: a review. Int. J. Environ. Res. Public Health 14, 1504.

Oruru, M.B., Njeru, E.M., 2016. Upscaling arbuscular mycorrhizal symbiosis and related agroecosystems services in smallholder farming systems. BioMed Res. Int 2016, 4376240.

Padmaperuma, G., Kapoore, R.V., Gilmour, D.J., Vaidyanathan, S., 2018. Microbial consortia: a critical look at microalgae co-cultures for enhanced biomanufacturing. Crit. Rev. Biotechnol. 38, 690–703.

Paerl, H.W., Pinckney, J.L., 1996. A mini-review of microbial consortia: their roles in aquatic production and biogeochemical cycling. Microb. Ecol. 31, 225–247.

Pasberg-Gauhl, C., Lockhart, B.E., Duran, S., 2000. First outbreak of banana streak badnavirus infection in commercial export bananas in Costa Rica. Plant Dis. 84, 1152.

Perez-Garcia, A., Romero, D., De Vicente, A., 2011. Plant protection and growth stimulation by microorganisms: biotechnological applications of Bacilli in agriculture. Curr. Opin. Biotechnol. 22, 187–193.

Pham, H.L., Ho, C.L., Wong, A., Lee, Y.S., Chang, M.W., 2017. Applying the design-build-test paradigm in microbiome engineering. Curr. Opin. Biotechnol. 48, 85–93.

Pinna, D., 2014. Biofilms and lichens on stone monuments: do they damage or protect? Front. Microbiol. 5, 1–3.

Ploetz, R.C., 2015. Fusarium wilt of banana. Phytopathology 105, 1512–1521.

Ploetz, R.C., Kema, G.H.J., Ma, L.J., 2015. Impact of diseases on export and smallholder production of banana. Annu. Rev. Phytopathol. 53, 269–288.

Preininger, C., Sauer, U., Bejarano, A., Berninger, T., 2018. Concepts and applications of foliar spray for microbial inoculants. Appl. Microbiol. Biotechnol. 102, 7265–7282.

Preto, G., Martins, F., Pereira, J.A., Baptista, P., 2017. Fungal community in olive fruits of cultivars with different susceptibilities to anthracnose and selection of isolates to be used as biocontrol agents. Biol. Control 110, 1–9.

Raja, P., Uma, S., Gopal, H., Govindarajan, K., 2006. Impact of bio inoculants consortium on rice root exudates, biological nitrogen fixation and plant growth. J. Biol. Sci 6 (5), 815–823.

Regvar, M., Vogel-Mikus, K., Severkar, T., 2003. Effect of AMF inoculum from field isolates on the yield of green pepper, parsley, carrot, and tomato. Folia Geobot. 38, 223–234.

Rillig, M.C., Tsang, A., Roy, J., 2016. Microbial community coalescence for microbiome engineering. Front. Microbiol 7, 1967.

Saha, S., Loganathan, M., Rai, A.B., Singh, A., 2016. Role of Microbes in Soil Health Improvement Introduction. Genetic improvement of tomato through resistance breeding for EB View project gcp project View project.

Salas, M.L., Mounier, J., Valence, F., Coton, M., Thierry, A., Coton, E., 2017. Antifungal microbial agents for food biopreservation—a review. Microorganisms 5, 37.

Santos, C.A., Reis, A., 2014. Microalgal symbiosis in biotechnology. Appl. Microbiol. Biotechnol. 98, 5839–5846.

Santos, C.C.A.D., Libeck, B.D., Schwan, R.F., 2014. Co-culture fermentation of peanut-soy milk for the development of a novel functional beverage. Int. J. Food Microbiol. 186, 32–41.

Sanzani, S.M., Reverberi, M., Geisen, R., 2016. Mycotoxins in harvested fruits and vegetables: insights in producing fungi, biological role, conducive conditions, and tools to manage postharvest contamination. Postharvest Biol. Technol. 122, 95–105.

Schmidtke, A., Gaedke, U., Weithoff, G., 2010. A mechanistic basis for underyielding in phytoplankton communities. Ecology 91, 212–221.

Sharma, S., Pathak, H., 2014. Basic techniques of phytoremediation. Int. J. Sci. Eng. Res 5, 584–605.

Sharma, R.R., Singh, D., Singh, R., 2009. Biological control of postharvest diseases of fruits and vegetables by microbial antagonists: a review. Biol. Control 50, 205–221.

Sheth, R.U., Cabral, V., Chen, S.P., Wang, H.H., 2016. Manipulating bacterial communities by in situ microbiome engineering. Trends Genet. 32, 189–200.

Stenuit, B., Agathos, S.N., 2015. Deciphering microbial community robustness through synthetic ecology and molecular systems synecology. Curr. Opin. Biotechnol. 33, 305–317.

Stringlis, I.A., Zhang, H., Pieterse, C.M.J., Bolton, M.D., De Jonge, R., 2018. Microbial small molecules—weapons of plant subversion. Nat. Prod. Rep. 35, 410–433.

Subhashini, D.V., Anuradha, M., Damodar Reddy, D., Vasanthi, J., 2016. Development of bioconsortia for optimizing nutrient supplementation through microbes for sustainable tobacco production. Int. J. Plant Prod 10 (4), 479–490.

Suleman, M., Yasmin, S., Rasul, M., Yahya, M., Atta, B.M., Mirza, M.S., 2018. Phosphate solubilizing bacteria with glucose dehydrogenase gene for phosphorus uptake and beneficial effects on wheat. PLoS One 13(9) e0204408.

Sultan, A., Andersen, B., Svensson, B., Finnie, C., 2016. Exploring the plant-microbe interface by profiling the surface-associated proteins of barley grains. J. Proteome Res. 15, 1151–1167.

Sundin, G.W., Wang, N., 2018. Antibiotic resistance in plant-pathogenic bacteria. Annu. Rev. Phytopathol. 56, 161–180.

Thijs, S., Weyens, N., Sillen, W., Gkorezis, P., Carleer, R., Vangronsveld, J., 2014. Potential for plant growth promotion by a consortium of stress-tolerant 2,4-dinitrotoluene-degrading bacteria: isolation and characterization of a military soil. Microb. Biotechnol. 7, 294–306.

Tinzl-Malang, S.K., Rast, P., Grattepanche, F., Sych, J., Lacroix, C., 2015. Exopolysaccharides from co-cultures of *Weissella confusa* 11GU-1 and *Propionibacterium freudenreichii* JS15 act synergistically on wheat dough and bread texture. Int. J. Food Microbiol. 214, 91–101.

Trivedi, P., Schenk, P.M., Wallenstein, M.D., Singh, B.K., 2017. Tiny microbes, big yields: enhancing food crop production with biological solutions. Microb. Biotechnol. 10, 999–1003.

Van Der Heijden, M.G.A., Bardgett, R.D., Van Straalen, N.M., 2008. The unseen majority: soil microbes as drivers of plant diversity and productivity in terrestrial ecosystems. Ecol. Lett. 11, 296–310.

Villaverde, J., Rubio-Bellido, M., Merchán, F., Morillo, E., 2017. Bioremediation of diuron contaminated soils by a novel degrading microbial consortium. J. Environ. Manag 188, 379–386.

Villaverde, J., Rubio-Bellido, M., Lara-Moreno, A., Merchan, F., Morillo, E., 2018. Combined use of microbial consortia isolated from different agricultural soils and cyclodextrin as a bioremediation technique for herbicide contaminated soils. Chemosphere 193, 118–125.

Wilmes, P., Bond, P.L., 2009. Microbial community proteomics: elucidating the catalysts and metabolic mechanisms that drive the Earth's biogeochemical cycles. Curr. Opin. Microbiol. 12, 310–317.

Woo, S.L., Pepe, O., 2018. Microbial consortia: promising probiotics as plant biostimulants for sustainable agriculture. Front. Plant Sci 9, 1801.

Xu, J., Zhang, Y., Zhang, P.F., Trivedi, P., Riera, N., Wang, Y.Y., Liu, X., Fan, G.Y., Tang, J.L., Coletta, H.D., Cubero, J., Deng, X.L., Ancona, V., Lu, Z.J., Zhong, B.L., Roper, M.C., Capote, N., Catara, V., Pietersen, G., Verniere, C., Al-Sadi, A.M., Li, L., Yang, F., Xu, X., Wang, J., Yang, H.M., Jin, T., Wang, N.A., 2018. The structure and function of the global citrus rhizosphere microbiome. Nat. Commun 9, 4894.

Zhang, N., Liu, J.H., Li, J.J., Chen, C., Zhang, H.T., Wang, H.K., Lu, F.P., 2016. Characteristics and application in food preservatives of *Lactobacillus plantarum* TK9 isolated from naturally fermented congee. Int. J. Food Eng. 12, 377–384.

Zoppellari, F., Malusà, E., Chitarra, W., Lovisolo, C., Spanna, F., Bardi, L., 2014. Improvement of drought tolerance in maize (*Zea mays* L.) by selected rhizospheric microorganisms. Ital. J. Agrometeorol. 19 (1), 5–18.

26

Biofertigation in fruit crops: Concept and application

Juan Manuel Covarrubias-Ramírez[a],*, Maria Del Rosario Jacobo-Salcedo[b], Erika Nava-Reyna[b], Víctor Manuel Parga-Torres[a]

[a]CESAL-INIFAP, Saltillo, México
[b]CENID RASPA-INIFAP, Gómez Palacio, México
*Corresponding author. E-mail: covarrubias.juan@inifap.gob.mx

1 Introduction

Biofertigation was created in 2002 (Covarrubias-Ramírez et al., 2002), with three areas of agricultural knowledge that are involved: soil fertility, soil microbiology, and irrigation, as you can see in Fig. 26.1. Soil microbiology studies beneficial microorganisms that live in symbiosis, associated or in free life in soil and plants as one of its aspects (Covarrubias-Ramírez and Hernández, 2005). The addition of microorganisms to fertilizers is known as biofertilization (Díaz-Moreno et al., 2007). Biofertilizers are microorganisms that can fix atmospheric nitrogen, symbiotically as the genus *Rhizobium* or in the form of free life as the genus *Azospirillum*, and the rest of the essential elements must obtain it from the soil reserve with microorganisms of the genus *Bacillus* by solubilization of phosphorus and heavy metals, production of growth stimulators, and reduction of fungal and bacterial diseases (Covarrubias-Ramírez et al., 2005; Pérez-Pérez and Espinosa-Victoria, 2014; León-López and Peña-Cabriales, 2015). There are microorganisms that produce phytohormones such as auxins and gibberellins that increase the area of plant absorption (Hernández-Flores and Covarrubias-Ramírez, 2012). By increasing of absorption areas, the plant could tolerate water stress, nutrimental and pathogenic (Ruscitti et al., 2007). On the other hand, the benefits of the application of bioliquid and compost derived from the vermicompost are the following: improves the structure of the soil, increases the infiltration of water, increases the permeability, favors the capacity of water retention in soils, strengthens the capacity of cation exchange, provides organic matter, increases the presence of beneficial microorganisms, stimulates the development of plants, provides organic nutrients, expands the adsorption of nutrients in the soil, promotes the remediation of soil structure, amplifies the aggregation of soil particles, and neutralizes the pH by its buffer capacity (Cornell composting, 2012). The efficiency of the application of water by the type of irrigation and the quality of the water during irrigation influence the efficiency of application of fertilizers and microorganisms by the plugging of the emitters due to the precipitation of nutrients (Boštjan et al., 2014). A lack of analysis of water and its adequate treatment makes this one of the most important reasons in the failure of irrigation systems, in addition to the type of irrigation system (Subbaiah, 2013).

A.K. Srivastava, Chengxiao Hu (eds.)
Fruit Crops: Diagnosis and Management of Nutrient Constraints
https://doi.org/10.1016/B978-0-12-818732-6.00026-5

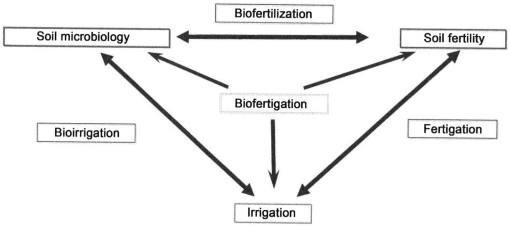

FIG. 26.1   Conceptualization of biofertigation.

Bioirrigation is the application of beneficial microorganisms to plants through irrigation water and the specific concepts of the irrigation and fertilizer technology disciplines, and by integrating the discipline of plant nutrition and crop physiology, they make fertigation a technology derived from science and defined with the following conceptual framework: "Fertigatión is to provide to the plant its food in quantity and opportunity through the irrigation system" (Covarrubias-Ramírez and Vázquez-Ramos, 2014). The natural fertility of soil does not normally provide enough nutrients to meet the demand of the crops, necessary to obtain maximum yield and production (Etchevers, 1997). The fertilization needs of a crop depend on the demand of the crop minus the contribution made by the soil based on its natural fertility, considering the efficiency of the fertilizer used (Alcántar-González et al., 2016). The natural fertility of the soil depends of the processes between supplies and extract of the nutrients by the crops; the processes are properties and characteristic of soil, type of crop, climatic conditions and the action of the microorganisms (Rojas et al., 2007).

Composting is a process of aerobic decomposition (which is supplied with oxygen) of organic matter under controlled conditions of temperature, humidity, and aeration in which different groups of microorganisms participate (Paul and Clark, 1996). Compost is the final product of the composting process, and if we add a worm, it will be the final process of vermicomposting, which should have an earthy appearance, soft consistency, and dark color with adequate physical properties and excellent attributes as organic soil fertilizer; the quality as fertilizer depends on the material from which it comes from (Etchevers, 1999).

The vermicompost is the use of worms that have the ability to feed on organic matter, which is to be composted (Basal and Kapoor, 2000); there are four types of worms used in vermicomposting: the red worm (*Eisenia andrei*), the tiger worm (*Eisenia foetida*), the African worm (*Eudrilus eugeniae*), and the eastern worm (*Perionyx excavatus*); the best is the red worm. The vermicompost can be made with cattle, horse, pig, goat, sheep, rabbit, etc. produced in the same year of production, in addition to any crop residue from healthy crops (Wagner and Wolf, 1998). The composition of the plant organic matter depends on the natures of the source, which could be perennial herbaceous legumes, perennial grasses, deciduous trees, conifers, bacteria, mosses, and algae; on average, they have 0.8%–7.7% of waxes, fats, and resins; 30%–35% of proteins; 3%–5% of cellulose; 5%–12% of hemicellulose; 30%–50% of lignins, and 8%–15% of C/N (Labrador, 1996; Velazco, 2002).

The decomposition of organic matter depends on two processes: the depolymerization of large organic molecules and the mineralization or transformation of organic molecules into inorganic ions (Kögel-Knabner, 2002). In the first case, the microorganisms release enzymes to the soil solution, to break and solubilize complex molecules and thus assimilate them. In mineralization, the nutrients contained in the depolymerized molecules are absorbed and incorporated into the microbial biomass (immobilized) and subsequently discarded to the soil solution in the form of inorganic ions that act as nutrients to the plants (Havlin et al., 2016).

Microorganisms produce specific enzymes for each stage of composting to easily break large molecules into structurally simple molecules with a high content of nitrogen and phosphorus (Kögel-Knabner, 2002). The enzymatic activity regulates the decomposition, the availability of dissolved organic compounds, the mineralization of nutrients, and the humification. This activity varies in relation to environmental conditions and the characteristics of the organic matter to decompose. In the presence of labile organic compounds, the cellulase enzymes, phosphatases, and some hydrolases dominate. In contrast, in the presence of more complex compounds, phenol oxidases (laccases) and peroxidases (lignin peroxidase and manganese peroxidase) dominate (Sinsabaugh et al., 2008).

Other enzymes that participate in the decomposition of organic matter are those related to the depolymerization of C compounds (cellobiohydrolases and β-glucosidases), in the degradation of chitin (*N*-acetyl-glucosaminidases) and in the depolymerization of lignin and phenolic compounds (polyphenol oxidases) and also for the mineralization of P (phosphomonoesterases), in addition to dehydrogenase, acid phosphatase, alkaline phosphatase, and urease, the latter for N mineralization (Fioretto et al., 2009).

The following bacteria are involved: *Bacillus, Clostridium, Pseudomonas, Cytophaga,* and *Corynebacterium* in degrading cellulose (Miller, 1993); *Bacillus, Cytophaga, Erwinia,* and *Streptomyces* in degrading hemicellulose (Alexander, 1994); *Azotobacter, Arthrobacter, Flavobacterium, Micrococcus,* and *Pseudomonas,* which depolarize lignin and reduce its molecular size (Mathews and Van Holde, 2001).

2 Application

In a commercial orchard with more than 30 years in production with standard Golden apple variety, with a spacing of 4 m between trees and 6 m between rows in the Ejido Rancho Nuevo, in Arteaga, Coahuila, located at 25° 23′5.25″ N, 100° 31′44.84″ W and 2522 masl. The orchard is under organic management certified by OTCO since December 2016.

The evaluation of the efficiency of water application is done under the methodology described by Covarrubias-Ramírez et al. (2006); the orchard has a drip irrigation system with emitters spaced every 1 m and placed at 0.5 m from the trunk of the tree as a minimum, and the capacity of the emitter was 4.0 LPH. Due to the slope of more than 20% in the orchard, the infiltration measured with concentric rings exceeds 6 cm/h, which indicates a soil with more than 1% organic matter and with a horizontal flow similar to the slope.

Vermicompost channel is trapezoidal 2.5 m wide and up to 1.0 m high, depending on the ease of handling and the organic matter that is being used as seen in Fig. 26.2. The channel must have good drainage and a firm floor with a slope of 0.5% to drain off the wet organic matter, in Fig. 26.3 (Jacobo-Salcedo et al., 2017). The raw material of the vermicomposter can be manure, and the material is transferred in covered containers to avoid disperse; to perform the vermicomposting, the manure is milled in a hammer mill to be less than 5 cm, as you can see in Fig. 26.4, and the process can be accelerated; this is done by placing the ground manure into the channel and adding the red or Californian worm (*Eisenia andrei*) at the rate of 1000 earthworms for a channel of 2.5 m wide, 1.0 high, and 1.5 m long, that

FIG. 26.2 Vermicompost channel trapezoidal.

FIG. 26.3   The channel with good drainage and a firm floor with a slope of 0.5% to drain.

FIG. 26.4   A trough of vermicompost industrial.

is, 267 earthworms/m^3, as seen in the Fig. 26.5. For example, if the channel is 24 m long, 16,000 earthworms will be needed.

If more or less surface is measured depending on the amount of compost to be used, the calculation is made to determine how many worms are needed. In the vermicompost, it is not required to turn the substrate because the worm performs the aeration process. When the product is mature, we proceed to remove the worms by placing a plastic mesh on the channel and covering it with the same compost with a layer of approximately 0.20 m, wet it, and then leave it overnight. The next day the worms will be on the mesh and must be moved, and the substrate is ready to use as a compost, a soil improver.

Bacillus subtilis was added to vermicompost at 1 L per 10 m^3 with a concentration of 1×10^7 CFU/mL; this microorganism can live in the worm's digestive system or can be transported. For the bioliquid, quantification was performed at the beginning, the middle, and the end of the vermicomposting by counting on trypticase soy agar plates, incubated 24 h at 37°C in laboratory.

The temperature is determined with a rod thermometer daily in the substrate, as you can see in Fig. 26.6, to know the temperature in the mesophilic, thermophilic, cooling, and maturation processes. When the temperature is the minimum and stabilizes for 5 days and the vermicompost has an earthy appearance, soft consistency, and dark color, it indicates that it was already formed. In the mesophilic and thermophilic stage, the temperature in the manure under vermicomposting can reach up to 40°C, which causes the elimination of weed seeds and pathogens sensitive to tem-

FIG. 26.5   Red or California worm (*Eisenia andrei*).

FIG. 26.6   Rod thermometer to measure in the substrate.

perature. The microbiological specifications of quality in the vermicompost must comply with the microbiological specifications established in the Official Mexican Standards of the Ministry of Health NMX-FF-109-SCFI-2007, valid for *Salmonella* and *Escherichia coli*.

3 Crop application

The biofertigation was applied by means of a drip irrigation system from the irrigation head with a venturi injector during the irrigation time. The biofertigation was the levels of 0, 1, 2, and 3 L/ha of the bioliquid of the vermicompost. These levels represent four treatments. The arrangement was in the fringe of 50 apple trees each, which is an irrigation line with a design in random blocks with 4 treatments and with 3 repetitions, to have 12 experimental units. Each experimental unit will consist of 1/3 of the fringe with trees. The evaluated variables were fruit yield in kilogram per tree and by tons/ha and the nutritional content of the bioliquids in the three stages of the vermicompost, as well as the measurement of their effect on nutrition by foliar analysis to determine the sufficiency levels into the apple tree with the applied doses (Covarrubias-Ramírez and Vázquez-Ramos, 2014).

In addition, root colonization at 0.2 m depth of soil was determined by *B. subtilis* in CFU/g of the root, applied in biofertigation. The compost generated from the vermicomposting was used to determining the effect as a soil improvement, the population of *B. subtilis* in CFU/g of compost was defined.

The yield in tons/ha was analyzed with the response surface method to determine if there are differences between levels, and after determining the optimal dose with the Ridge regression, the analysis was performed with the SAS package.

The efficiency of water application was 90.2%, which is acceptable for drip irrigation systems (Covarrubias-Ramírez et al., 2006). The 16 emitters evaluated in the irrigation unit had a behavior with 9.8% deficiency; applying a sheet (Li) in 4 h of irrigation had expenses higher than 4 LPH and lower than these, as you can see in Fig. 26.7.

The value of the irrigation depth (Zi) shows 90.2% efficiency with brown lines (light gray in print version) and 9.8% deficiency with red lines (dark gray in print version). The models used to represent this condition were not significant and have the same coefficient of determination. The importance of knowing the efficiency of water application allows the rest of the inputs to be applied efficiently (Espinosa, 2013).

The vermicomposting process started on July 20, 2017, and lasted 59 days, of which the mesophilic process lasted 4 days in the beginning, the thermophilic process of 31 days, the cooling process of 15 days, and maturation in 9 days as you can see in Fig. 26.8.

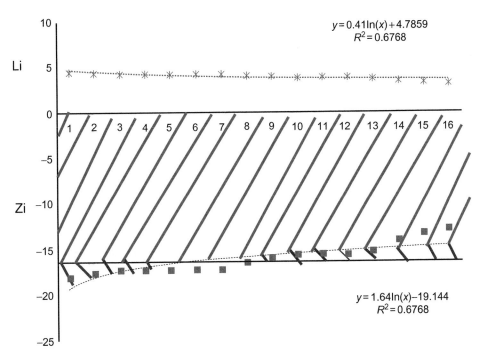

$$y = 0.41\ln(x) + 4.7859$$
$$R^2 = 0.6768$$

$$y = 1.64\ln(x) - 19.144$$
$$R^2 = 0.6768$$

FIG. 26.7 Water application´s uniformity in drip irrigation system from apple orchard.

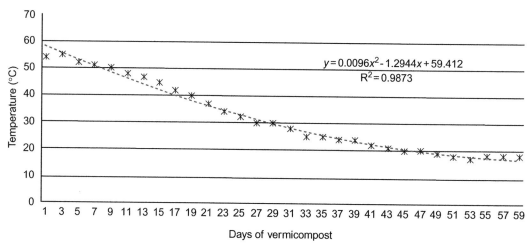

FIG. 26.8   Vermicompost of sheep manure.

In the stage of the mesophilic process after moistening the substrate, the temperature reached up to 55°C at a depth of 0.30 m, temperature at which the process of elimination of weeds and pathogenic microorganisms begins, and they begin to degrade amines, amino acids, and proteins; the worm to escape this temperature is transferred to the first 10 cm where moisture is greater by irrigation (Sinsabaugh, et al., 2008). The thermophilic process is most important for the degradation of organic matter; it starts at a temperature of 52°C until it reaches 25°C. In this process, the waxes, fats, and resins are decomposed, also the polymerization of lignin and phenolic compounds (Fioretto et al., 2009).

The initial cooling process with temperatures of 24°C until reaching 19°C initiates the decomposition of lignin, pectin, chitin, cellulose, and hemicellulose, which, being long-chain polymers, have a slower degradation process, and reduces the temperature to allow the microorganisms to act; once that through the high temperatures eliminated most of the pathogens, the beneficial microorganisms act and activate their defense mechanisms against pathogens with antibiosis, which is the production of antibiotics to eliminate pathogenic bacteria (León-López and Peña-Cabriales, 2015).

The ripening process was realized when the temperature was stabilized at 18°C for more than 5 days; at this stage, the microorganisms already acted to degrade most of the organic matter; the initial volume of 1.5 m in height is reduced to 0.5 m; the substrate becomes friable organic matter; and the density is reduced, which facilitates its management as a greenhouse substrate or soil improver (Domínguez-Villaseñor and Covarrubias-Ramírez, 2017). The temperatures in the stages were low for manure (18°C) considering the altitude of the locality and the incidence of rain in the crop cycle.

The measurement of *B. subtillis* has been made at the beginning of the thermophilic process (BTP) and water applications have been made to the vermicompost, the first sampling was 5 days later (BTP), when there was already enough bioliquid to perform the biofertigation, the second sampling was at the end of the thermophilic process that corresponds to 28 days later (BTP) and the third sampling was done 50 days later (BTP).

The bioliquid has a salt concentration much greater than 16 dS/m, all manure supplies salts, and the apple tree tolerates up to 1.7 dS/m in the soil solution; the nitrogen content does not exceed the level that a fertilizer must have, only 0.61% (6100 mg/kg); therefore, the level is deficient in this nutrient, even though, with the level of biofertigation with 3 L/ha, this element was applied at the highest concentration, with respect to the rest (Table 26.1).

The content of phosphorus and potassium was deficient as fertilizer with 0.034% (337 mg/kg) and 0.35% (3500 mg/kg), respectively, as well as the levels of calcium and magnesium (SPACI, 2000; Havlin et al., 2016), but the advantage of biofertigation is to apply low and frequent doses to supply the crop requirements (Irrigation Association, 1999).

With respect to microelements, all levels have a deficient supply except sodium, which is the microelement with the highest concentration in biofertigation. Iron and boron (29.7 and 17.4 mg/kg) are followed in abundance, and also, silicon and selenium are added, which are beneficial elements (Alcántar-González et al., 2016). This mineral concentration with most essential nutrients makes the bioliquid a complete fertilizer in low concentrations but is more efficient due to the contribution of humic and fulvic acids, which both do not exceed 0.2%.

The concentration in the bioliquid of *B. subtilis* is shown in Table 26.2. The inoculum was applied at 1×10^7 CFU/mL; at the beginning and at the end of the vermicomposting process, the concentration was lower than this value, but at half of the process, the value increased with respect to the applied concentration that corresponds to the thermophilic stage of vermicomposting. *B. subtilis* inhabits and reproduces in the intestine of the earthworm (*Eisenia andrei*) and,

TABLE 26.1 The nutrimental concentration of the bioliquid and applied biofertigation dose.

Element	Units	Concentration	Biofertigation		
			1 L/ha	2 L/ha	3 L/ha
Potential hydrogen		7.77			
Saturated extract conductivity	dS/m	16.51	10.6[a]	21.1[a]	31.7[a]
Nitrogen	mg/kg	6100	6.1	12.2	18.3
Phosphorus-PO$_4$	mg/kg	337	0.337	0.674	1.011
Potassium	mg/kg	3500	3.5	7	10.5
Calcium	mg/kg	120	0.12	0.24	0.36
Magnesium	mg/kg	50	0.05	0.1	0.15
Sulfur	mg/kg	20.16	0.02	0.04	0.12
Chlorine	mg/kg	1.6	0.0016	0.0032	0.0096
Iron	mg/kg	29.73	0.02973	0.05946	0.08919
Copper	mg/kg	4.78	0.00478	0.00956	0.01434
Zinc	mg/kg	2.58	0.00258	0.00516	0.00774
Manganese	mg/kg	1.2	0.0012	0.0024	0.0036
Boron	mg/kg	17.44	0.01714	0.03428	0.05142
Sodium	mg/kg	367	0.367	0.734	1.101
Silicon	mg/kg	1.95	0.0019	0.0039	0.0058
Selenium	mg/kg	0.20	0.0002	0.00041	0.00061
Humic acids	%	0.19	1.9	3.8	5.7
Fulvic acids	%	0.18	1.8	3.6	5.4

[a] Grams applied.

TABLE 26.2 The population of *Bacillus subtilis* in bioliquids in biofertigation, colonization in the root of the apple tree, and content in the compost elaborated.

Form	Mean	Standard deviation
Bioliquids at the beginning	6.77×10^6 CFU/mL	6.35×10^5 CFU/mL
Bioliquids in the middle	28.3×10^6 CFU/mL	4.04×10^5 CFU/mL
Bioliquids at the end	6.83×10^6 CFU/mL	6.11×10^5 CFU/mL
Apple root	1.07×10^4 CFU/g	1.15×10^3 CFU/g
Compost	1.77×10^6 CFU/g	2.89×10^5 CFU/g

through the production of different molecules, promotes the growth of lateral roots and root length and increases the absorption of phosphorus, microelements, and the biomass of the plant (Covarrubias-Ramírez et al., 2005; Pérez-Pérez and Espinosa-Victoria, 2014; León-López and Peña-Cabriales, 2015).

B. subtilis is a bacterium with variable tolerance to salinity, and when performing symbiosis with the tree, it is verified that colonization was done in the root of the tree and the compost elaborated and applied as a soil improver has a concentration of 1.77×10^6 CFU/g (Table 26.2). The compost generated is a solid organic fertilizer. The effect of microorganisms in biofertigation has already been detected with other crops even in high concentrations of salts when applied in nutrient solution (Rojas et al., 2007). Although the bioliquid as a fertilizer does not have the concentration of an inorganic fertilizer, it has the advantage of being a compound fertilizer, with most of the essential elements, and supplemented with beneficial microorganisms and organic acids that facilitate the absorption of nutrients, because

TABLE 26.3   Apple tree and orchard yield with biofertigation and vermicompost bioliquids.

Doses L/ha	Yield		Distinctiveness
	kg/tree	tons/ha[a]	%
0	10.3	4.33 d	64
1	19.7	8.17 c	33
2	26.0	10.83 b	11
3	29.2	12.17 a	0

[a] *Difference of the highest value with respect to the lowest.* a = Tukey (P ≤ .05).

these polymers of long chain are recognized by transmembrane proteins in the cell, which allows their absorption with nutrients increasing their absorption efficiency (Paredes, 2010; Marschner, 2012).

Thus, the fruit yield increased with increasing the biofertigation dose (Table 26.3), the maximum yield with 3 L/ha significantly higher than the rest of the levels, with 2 L/ha in 11%, and with 1 L/ha in 33%. The yield of the level of 0 L/ha, although 64% different from the highest level of biofertigation, could have been due to the supply of soil and the contribution of nitrogen by rain (Paz and Etchevers, 2016; Domínguez-Villaseñor and Covarrubias-Ramírez, 2017). The maximum yield is considered average with respect to the management of an orchard with 50% mineral fertilization and biofertigation, where yields vary from 21 to 40 tons/ha from the 2 years of application (Orozco-Corral et al., 2016).

In the response of the yield of apple tree to the bioliquids applied as you can see in Fig. 26.9, a maximum was obtained with the maximum level of biofertigation, which indicates that the dose can be increased, the model is highly significant ($R^2 = 0.9925$), and the function an answer is

$$yield = 4.325 + 4.49\,bio - 0.625\,bio^2$$

where yield = yield in tons/ha and bio = dose of biofertigation in L/ha; applying the model when increasing the dose to 4 L/ha, the yield reaches 12.3 tons/ha of fruit, 0.13 tons/ha higher than the level of 3 L/ha; therefore, this level is the maximum for the management with the same level of nutrients in the bioliquid.

All sustainable management focuses on the yield, but the tree as a living being is subject to its nutrient concentration to demonstrate the effect of biofertigation, as you can see in Table 26.4. The nitrogen content exceeds 2% of the level of sufficiency; therefore, the treatments are efficient in this nutrient, except the level of 0 L/ha, with which it is observed that, if the bioliquids do not contain the concentration adequate, the recovery efficiency of the element is high, due to the effect of organic acids and the beneficial microorganism and the contributions of soil and rain (Covarrubias-Ramírez, 2016).

Regarding the phosphorus content, no level reached the level of sufficiency; the level of potassium is high (greater than 1.5%) at 3 L/ha, while at the rest of the levels it is in the range of sufficiency even in the control; this content is due to the contribution of the soil. In calcium, all treatments are at the level of sufficiency (greater than 1.2%) for the contribution of soil; in addition, magnesium is at the level of sufficiency (0.15%–0.22%) for the same situation (Reuter and Robinson, 1986).

FIG. 26.9   The optimal dose of bioliquids in the biofertigation of an apple orchard.

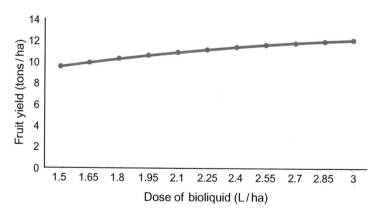

TABLE 26.4 The nutrimental concentration of the apple tree under levels of biofertigation.

| Foliar concentration | 0 L/ha | Biofertigation doses | | | Sufficient Range[a] |
		1 L/ha	2 L/ha	3 L/ha	
Total nitrogen (N, %)	1.84	2.1	2.12	2.41	2.0–2.5
Total phosphorus (P, %)	0.09	0.1	0.09	0.11	0.2–0.3
Total potassium (K, %)	1.02	1.33	1.38	1.73	0.7–1.4
Calcium (Ca, %)	1.38	1.6	1.71	1.99	0.7–1.2
Total magnesium (Mg, %)	0.17	0.19	0.21	0.22	0.15–0.22
Sulfur (S, %)	0.02	0.06	1.10	1.5	1.0–3.0
Iron (Fe, mg/kg)	103	123.1	128.2	129.9	150–250
Copper (Cu, mg/kg)	9.6	10.5	13.8	13.9	6–20
Zinc (Zn, mg/kg)	12.7	12.9	15.9	16.6	20–50
Manganese (Mn, mg/kg)	18.7	24.6	29.5	31.3	30–120
Boron (B, mg/kg)	2.8	3.2	4.2	5.7	21–40
Sodium (Na, mg/kg)	100	100	200	800	200–500
Chlorine (Cl, mg/kg)	1.6	2.1	2.9	3.2	2–10
Nickel (Ni, mg/kg)	0.77	0.89	0.99	1.10	<1.5
Silicon (Si, mg/kg)	2.75	3.19	4.62	5.66	<100
Selenium (Se, mg/kg)	0.72	0.97	1.00	1.46	<10

[a] *Covarrubias-Ramírez and Vázquez-Ramos (2014).*

With respect to microelements, all treatments do not exceed the level of iron sufficiency (150 mg/kg); with zinc in all the treatments, they presented deficiency (less than 20 mg/kg); in the content of manganese, only the level of 3 L/ha exceeds the level of sufficiency (30 mg/kg); the rest of the treatments are under deficiency of this element. In the case of copper, all levels are in the sufficiency range (6–20 mg/kg); for the boron level, it is an element that is considered toxic at levels higher than 4 mg/kg in the soil solution for most crops (Cadahia, 2005). If the bioliquid contained 17.4 mg/kg, this level no presented toxicity, because the range of sufficiency is 21–40 mg/kg, so the low level of boron in the tree can be due to an antagonism with copper (Covarrubias-Ramírez, 1999a,b).

Although the apple tree requires sodium, its levels of sufficiency are 0.02–0.05 mg/kg (Petersen and Stevens, 1994), and those observed are very high because they exceed 100 mg/kg. The apple tree from 1.7 dS/m of salts in the soil solution begins to reduce its yield, so in some studies of bioliquids, its analysis is not included, to determine the effect of soil salinization (Preciado-Rangel et al., 2014).

In the case of nickel, silicon, and selenium, nickel is considered essential, but the others are beneficial. Nickel was found at concentrations of 0.89–1.1 mg/kg, suitable concentrations for tomato that has sufficiency up to 1.5 mg/kg, so, the apple tree, which is a culture more sensitive to salts than tomato. The concentration found is considered adequate (Covarrubias-Ramírez., 2003; Alcántar-González et al., 2016).

The silicon in plants was a concentration of 0.01%; so that the contributions of this element are not within the level of sufficiency (2.7–5.6 mg/kg) and selenium in this case organic, the range is at an adequate concentration of 0.97–1.46 mg/kg for this fruit, with respect to pasture and forage that require up to 50 mg/kg (Marschner, 2012). The manures derived from livestock emit greenhouse gases (GHG) during the decomposition process, so this sector contributes 52% of CH_4 and 44% of N_2O; these gases have a greater harmful effect on the atmosphere than CO_2 (Ramírez, 2010; Saynes-Santillán et al., 2016). So, the use of manure in agriculture can contribute to environmental deterioration. Even with the addition of salts by the bioliquid, vermicomposting reduces GHG volatilization such as CO_2, N_2O, and CH_4, by applying water when generating the bioliquid (Ozores-Hampton, 2017), like the use of manure in agriculture, and it must be prevented.

On the other hand, the prices of the most used fertilizers in the agricultural sector of Mexico have an upward trend, due to the behavior of the price at international level, as well as the increase of the logistic costs and the demand of the agricultural products (INIFAP, 2016). The bioliquid produced can replace the fertilizer, since it contains all the elements that a mixture of fertilizers would have. The implementation of biofertigation in organic or sustainable agriculture in an apple tree is not achieved with a recommendation, because the crop is subject to the severity of changes between cycles to achieve organic production (Williams et al., 2015; Orozco-Corral et al., 2016).

4 Conclusions

The biofertigation dose of 3 L/ha was the best response in yield, but it can be increased because the apple tree is a perennial crop.

The nutritional content of the bioliquid contributed in supply the most essential nutrients. The bioliquid has nutrient recovery efficiency; it's a sustainable and economical option to fertilize the crops.

The application of *B. subtilis* as a beneficial microorganism responded in the symbiosis and colonization with the apple crop.

The association of the bioliquid and *B. subtilis* contributed some nutrients in a sufficient way, others in excess and others deficient, so it is necessary to adjust the management of biofertigation.

To make biofertigation first, you must have an efficient irrigation system in the application of water.

References

Alcántar-González, G., Trejo-Téllez, L.I., Gómez-Merino, F.C., 2016. Nutrición de cultivos, second ed. Mundi Prensa México. S. A. de C. V., México, D. F. (454 p.).

Alexander, M., 1994. Introducción a la microbiología del suelo. 2ª Reimpresión en español. AGEDITOR, S. A. México. (210 p).

Basal, S., Kapoor, K.K., 2000. Vermicomposting of crop residues and cattle dung with Eisenia fetida. Bioresour. Technol. 73, 95–98.

Boštjan, N., Kechavarzi, C., Coulon, F., Pintar, M., 2014. Numerical investigation of the influence of texture, surface drip emitter discharge rate and initial soil moisture condition on wetting pattern size. Irrigation Sci. 32, 421–436.

Cadahia, L.C., 2005. Fertirrigación. Cultivos hortícolas y ornamentales. 3a ed. Mundi-Prensa, Barcelona, Esp. (475 p).

Cornell composting. 2012. The Science and Engineering of Composting. En línea: compost.css.cornell.edu/science.html. (Consultado el 22 de junio de 2012).

Covarrubias-Ramírez, J. M. 1999a. Corrección de deficiencia de cobre en manzano. En: 500 Tecnologías Llave en Mano. Tomo II. SAGAR-INIFAP. México, D. F. p 34.

Covarrubias-Ramírez, J. M. 1999b. Corrección de la deficiencia de zinc en manzano. En: 500 Tecnologías Llave en Mano. Tomo II. SAGAR-INIFAP. México, D. F. p 33.

Covarrubias-Ramírez, J. M. 2016. Factores del suelo que afectan la eficiencia de recuperación del fertilizante. *In:* Peña Cabriales, J. J y Vera Núñez, A. (Compilación y Edición). Memorias del XLI Congreso Nacional de la Ciencia del Suelo. Sociedad Mexicana de la Ciencia del Suelo. León, Gto. 41:194–200.

Covarrubias-Ramírez, J. M. y Hernández F.L. 2005. Biofertirrigación en maíz, manzano y papa en México. *In:* Memorias del VIII Simposio Internacional de Agricultura Sostenible. SOMAS. Cd. Victoria, Tam. CD-ROM. p 37.

Covarrubias-Ramírez, J. M. y J. A. Vázquez-Ramos. 2014. Guía de fertirrigación del manzano en Coahuila y Nuevo León. INIFAP-CIRNE. Campo Experimental Saltillo. Folleto Técnico Núm. MX-0-310391-28-03-15-09-60. Coahuila, México. (77 p).

Covarrubias-Ramírez, J.M., Núñez, E.R., Hernández, F.L., 2002. Memories of IV International Symposium on Cleaner Bioprocesses and Sustainable Development. Veracruz, Ver, (Ed.), Biofertigation: a new sustainable technology, p. 47.

Covarrubias-Ramírez, J. M., S. Castillo-Aguilar, J. A. Vera-Núñez, R. Núñez-Escobar, P. Sánchez-García, R. Aveldaño-Salazar y J. J. Peña-Cabriales. 2005. Absorción y eficiencia en uso de fósforo en papa cultivar alpha con 32P. Agrociencia 39:127–136.

Covarrubias-Ramírez, J. M., F. J. Contreras de la Ree. y I. Sánchez C.I. 2006. Como evaluar un sistema de riego por goteo en huertas de manzano. Secretaria de Agricultura, Ganadería, Desarrollo Rural, Pesca y Alimentación. Instituto Nacional de Investigaciones Forestales, Agrícolas y Pecuarias, Campo Experimental Saltillo. Saltillo, Coah., México. (2 p). (Desplegable Técnica Núm. 3).

Covarrubias-Ramírez., J. M, 2003. Guía para preparar mezclas de fertilizantes en sistemas de fertirrigación. Folleto técnico Num. 6. INIFAP-CIRNE. Campo experimental Saltillo. Saltillo, Coah. (27 p).

Díaz-Moreno, R., A. Díaz-Franco, I. Garza-Cano y A. Ramírez-De León. 2007. Brassinoesteroides e inoculación con *glomus intraradices* en el crecimiento y la producción de sorgo en campo. Terra Latinoamericana 25: 77–83.

Domínguez-Villaseñor, M. A. y J. M. Covarrubias-Ramírez. 2017 El uso de materia orgánica diversa para generar vermicomposta. *In:* Vázquez Navarro, J. M.; López Calderón, M. J.; Soto González, R. y Camacho Rodríguez, Y. (eds). Memoria XXIX semana internacional de agronomía. UJED-FAZ-DEP. Venecia Durango, México, 29:721–726.

Espinosa, E.J.L., 2013. Tecnologías para ofrecer servicios de asesoramiento en riego. Tesis M.C. Hidrociencias. Colegio de Postgraduados, Montecillo, Edo de México. (85 p)

Etchevers, B.J.D., 1997. Evaluación del estado nutrimental del suelo y de los cultivos ferti-irrigados. In: Memoria, 2° Simposio internacional de fertirrigación. Querétaro, México, pp. 51–60.

Etchevers B.J.D 1999, Análisis para asegurar la calidad en el proceso y el producto de los materiales orgánicos degradados. En: Martínez C., R. Romero, L. Corlay, A. Trinidad y L. F. Santoyo (Eds). I Simposium internacional y reunión nacional de Lombricultura y abonos orgánicos. SAGAR-CP. Montecillo, Texcoco, México.

Fioretto, A., A. Papa, A. Pellegrino y A. Ferrigno. 2009. Microbial activities in soils of a Mediterranean ecosystem in different successional stages. Soil Biol. Biochem. 41:2061–2068.

Havlin, J.L., Tisdale, S.L., Nelson, W.L., Beaton, J.D., 2016. Soil fertility and fertilizers. In: eight ed. Prentice Hall, Inc., Englewood Cliffs, NJ (499 p).

Hernández-Flores, L. y J.M. Covarrubias-Ramírez. 2012. Diversidad genética de *Azospirillum brasilense* en papa (*Solanum tuberosum* L.), manzano (*Malus domestica*) y maíz (*Zea mays* L.). Agrofaz 12(1):27–37.

Instituto Nacional de Investigaciones Forestales, Agrícolas y Pecuarias (INIFAP), 2016. Convenio de Administración por Resultados del INIFAP 2016–2019. Progreso No. 5, Barrio de Santa Catarina. Delegación Coyoacán, C.P. 04010, Ciudad de México, MEX. (144 p).

Irrigation Association, 1999. Chemigation. Irrigation Association, Fairfax, Va (192 p).

Jacobo-Salcedo, M. D. R.; J. M. Covarrubias-Ramírez: J. A. Hernández-Maruri y V. M. Parga Torres. 2017. Diseño y manejo de una pila de vermi-composteo. In: Vázquez Navarro, J. M; López Calderón, M. J.; Soto González, R. y Camacho Rodríguez, Y. (eds). Memoria XXIX semana internacional de agronomía. UJED-FAZ-DEP. Venecia Durango, México. 29:741–745.

Kögel-Knabner, I., 2002. The macromolecular organic composition of plant and microbial residues as input to soil organic matter. Soil Biol. Biochem. 34, 139–162.

Labrador, M.J., 1996. La materia orgánica en los agrosistemas. Ministerio de Agricultura, Pesca y Alimentación. Mundi-Prensa, Madrid, España. (123 p).

León-López, L. y J. J. Peña-Cabriales. 2015. La producción de compuestos volátiles por rizobacterias y su relación con la promoción del crecimiento vegetal. In: Loredo Osti, C., Lara Mireles, J.L., Beltrán López, S. y Valdez Cepeda, R.D. (Compilación y Edición). Memorias del XL Congreso Nacional de la Ciencia del Suelo. Sociedad Mexicana de la Ciencia del Suelo. 40:250–253.

Marschner, H., 2012. Mineral Nutrition of Higher Plants, second ed. Elsevier Science & Technology (672 p).

Mathews C. K. y K. E. Van Holde. 2001. Bioquímica. 3ª Reimpresión McGraw-Hill interamericana de España. Madrid, España. (1283 p).

Miller, C.F., 1993. Composting as a process base on the control of ecologically selective factors. In: Blaine-Meeting Jr., F. (Ed.), Soil Microbial Ecology. Marcel Dekker, Inc., New York, NY, pp. 515–544

Orozco-Corral, A. L., M. I. Valverde-Flores, R. Martínez-Téllez, C. Chávez-Bustillos y R. Benavides-Hernández. 2016. Propiedades físicas, químicas y biológicas de un suelo con biofertilización cultivado con manzano. Terra Latinoamericana 34: 441–456.

Ozores-Hampton, M., 2017. Guidelines for assessing compost quality for safe and effective utilization in vegetable production. HortTechnology 27, 162–165.

Paredes, M.M., 2010. Aislamiento y caracterización bioquímica de metabolitos producidos por rizobacterias que solubilizan fosfato. Doctor en Ciencias. Colegio de postgraduados. Montecillo, Edo. de Mex. (125 p).

Paul, E.A., Clark, F.E., 1996. Soil microbiology and biochemistry, second ed. Academic Press, San Diego, CA. (340 p).

Paz P. F., y Etchevers B.J.D. 2016. Distribución a profundidad del carbono orgánico en los suelos de México. Terra Latinoamericana 34: 339–355.

Pérez-Pérez, J.A. y D. Espinosa-Victoria. 2014. Caracterización microbiológica de las bacterias residentes en el tracto digestivo de la lombriz de tierra *eisenia foetida*. In: Flores Margez, J.P. *et al*. (eds). Memoria XXXIX Congreso Nacional de la Sociedad Mexicana de la Ciencia del Suelo. Cd. Juárez, Chih. 39:128–133.

Petersen, A.B., Stevens, R.G., 1994. Tree Fruit Nutrition. Good Fruit Grower, Yakima, WA (211 p).

Preciado-Rangel, P.; J. L. García Hernández, M. A. Segura-Castruita, L. Salas Pérez; A.V. Ayala Garay, J. R. Esparza Rivera. y E. Troyo-Diéguez. 2014. Efecto del lixiviado de vermicomposta en la producción hidropónica de maíz forrajero. Terra Latinoamericana 32: 333–338.

Ramírez, H.I.F., 2010. Emisiones de metano generadas por excretas de animales de granja y contenido ruminal de bovino. Tesis Doctor en ciencias. Colegio de Postgraduados. Montecillo, Texcoco, Edo. de México. (166 p).

Reuter, D.J., Robinson, J.B., 1986. Plant Analysis. An interpretation manual. Inkata Press, Melbourne, Australian (219 p).

Rojas P.L., J. M. Covarrubias-Ramírez, R. Mendoza V., L.F. Hernández y V. M. Parga Torres. 2007. Biofertirrigacion con solución nutritiva Steiner y *Azospirillum brasilense* en el cultivo de papa. *In*: Vásquez Alarcón, A. y I. Aaimers-de A. (eds). Memoria del XVII Congreso Latinoamericano de la Ciencia del Suelo. León, Gto. 17:1071–1074.

Ruscitti, M. F.; M.C. Arango, M.G. Ronco, O. Peluso, y A. Beltrano. 2007. Efecto del estrés hídrico simulado y la inoculación con esporas de *glomus mosseae* sobre el crecimiento y la partición de biomasa en avena. Terra Latinoamericana 25: 135–143.

Saynes-Santillán, V., J. D.B. Etchevers, F.P.P. Paz y L.O.C. Alvarado. 2016. Emisiones de gases de efecto invernadero en sistemas agrícolas de México. Terra Latinoamericana 34: 83–96.

Sinsabaugh, R., Lauber, C., Weintraub, M., Ahmed, B., Allison, S., Crenshaw, C., Contosta, A., Zeglin, L., 2008. Stoichiometry of soil enzyme activity at global scale. Ecol. Lett. 11 (11), 1252–1264.

Soil and Plant Analysis Council Inc. (SPACI), 2000. Soil Analysis: Handbook of Reference Methods. CRC Press LLC, Boca Raton, FL (247 p).

Subbaiah, R., 2013. A review of models for predicting soil water dynamics during trickle irrigation. Irrigation Sci. 31, 225–258.

Velazco, V.J., 2002. Alternativa tecnológica del reciclaje de los desechos orgánicos del Colegio de Postgraduados. Tesis de Maestría. Colegio de Postgraduados. Montecillos, Texcoco, Edo. de México. (95 p).

Wagner, G.H., Wolf, D.C., 1998. Carbon transformations and soil organic matter formation. In: Sylva, D.M., Fuhrmann, J.J., Hartel, P.G., Zuberer, D. A. (Eds.), Principles and Applications of Soil Microbiology. Prentice-Hall, Upper Saddle River, NJ, pp. 412–478 (Chapter 11).

Williams, M.A., Strang, J.G., Bessin, R.T., Law, D., Scott, D., Wilson, N., Witt, S., Archbold, D.D., 2015. An assessment of organic apple production in Kentucky. HortTechnology 25, 154–161.

27

Nutrient management in fruit crops: An organic way

Moreno Toselli*, Elena Baldi, Luciano Cavani, Giovambattista Sorrenti

Department of Agricultural and Food Sciences, Alma Mater Studiorum University of Bologna, Bologna, Italy
*Corresponding author. E-mail: moreno.toselli@unibo.it

1 Introduction

According to the principles of organic farming, the main issue of tree nutrient management is the buildup and maintenance of soil fertility, defined as the ability of the soil to supply water and nutrients for plant growth and reproduction (Bünemann et al., 2018). Fruit crops are thus supposed to find, in soil, the nutrients required for their growth and reproduction cycle. This approach makes the soil, not the tree, the target of the fertilization practice. Therefore, the satisfactory fertility of soil should be achieved and maintained in advance of fruit crop demands. Once trees are planted, the efficiency of the fertilization is higher the more synchronized is nutrient availability with root uptake. This goal is achieved when the amount of available nutrients equalizes tree crop requests. However, synchronicity is not easy to be established, since the demand of nutrients of fruit crops varies during the growing season, according to the genetic combination (e.g., rootstock and scion), phenological stage, environmental conditions, tree age, soil management, etc. Moreover, plant nutrient availability depends on biological (e.g., soil microbiota interaction with roots), chemical (e.g., pH and cation exchange capacity (CEC)) and physical (e.g., texture and soil particle aggregation) soil factors. An excessive nutrient availability increases the risk of environmental pollution (nutrient leaching, volatilization, etc.), induces nutritional unbalances and promotes antagonisms among elements. These factors contribute to reduce nutrient use efficiency (NUE). Since organic fertilization management promotes the whole soil fertility (biological, chemical, and physical), it should be included routinely, in fertilization management.

2 Soil organic matter

Soil organic matter (OM) includes heterogeneous organic residues such as cells and tissues of animal and vegetal organisms as well as substances synthesized by soil organisms in different stages of decomposition that cohabit in the

A.K. Srivastava, Chengxiao Hu (eds.)
Fruit Crops: Diagnosis and Management of Nutrient Constraints
https://doi.org/10.1016/B978-0-12-818732-6.00027-7

379

soil and that are subjected to complex processes to build up nonhumic primary and humic secondary components. The former includes carbohydrates, amino acids, proteins, lipids, nucleic acids, and lignin and contributes to modify physical characteristics of the soil such as structure, water holding capacity (WHC), thermic capacity, bulk density, and biological activity (Sequi, 1989). Primary compounds in aerobic environments, with a relatively high-temperature (such as in temperate and tropical ecosystems) feed mineralization process to form carbon dioxide (CO_2), water, and ammonia (NH_3) and release nutrients under mineral forms. In anoxic environments, with low temperatures, nonhumic primary components proceed toward reduction (carbonification) reactions. In intermediate conditions, humification prevails leading to the biosynthesis of humic compounds. Degradation processes are replaced by condensation, resynthesis, and polymerization that bring about the production of organic secondary compounds named humic substances. Humic compounds are responsible for chemical and physical reactions that involve soil CEC, production of metal-organic and mineral-organic compounds that show molecular retention and pesticide deactivation capacity. Thereby, during its modification, OM in soils is responsible for a number of benefits including the increase of: (1) biological activity, (2) biodiversity, (3) plant nutrient availability, (4) moisture retention, (5) CEC, (6) tolerance to lime, (7) tolerance to salinity, (8) tolerance to replanting disease, and (9) ion solubility. Organic matter also maintains pH near to neutrality, prevents soil erosion, etc.

Mineralization and humification balance affects the release of nutrients (i.e., N), whose rate depends on environmental conditions and type of OM, in particular on its carbon:nitrogen (C:N) ratio. If C:N is <20, the soil microbial biomass increases its NUE for C (the C:N ratio of soil microbial biomass ranges between 5 and 8), and a release of N is expected. On the other hand, with a C:N ratio >20, soil microbe growth is stimulated by C addition, and N may become a limiting factor for bacteria growth; as a matter of fact, soil N immobilization is expected (increase of microbial NUE for N), with potential competition with agricultural crops.

The knowledge of C:N ratio allows the prediction of the nutrient releasing rate, a crucial factor for the rational management of fruit crop fertilization. In particular, if the release of nitrate (NO_3^-)-N does not match with crop N uptake, a decrease of the N use efficiency and a potential increase of the risk of pollution are expected. These implications contribute to the lower yield usually observed in organic compared with conventionally managed farming. Soil mineral N determination is therefore important to estimate N availability and consequently modulate N supply to tree demand at each phenological stages. To this end, the measurements of the KCl-extractable soil NO_3^-- and ammonium (NH_4^+)-N fractions provide an accurate estimation of the amount of mineral N available for root uptake (Tagliavini et al., 1996).

For its multiple beneficial effects, soil OM content should be increased toward sustainable fertilizer management that includes: (1) application of organic fertilizers, such as animal manure, compost, digestate, and sewage sludges, either fresh or mature, after a chemical and/or biological stabilization process; (2) green manure; (3) floor management with temporary (cover crops) or permanent grass; (4) mulching with organic residues; (5) recycling of abscised leaves, pruned wood, etc.

2.1 Temporary and permanent cover crops

Temporary cover crops are 1-year-cycle crops, cultivated during orchard lifetime with the aim of, along with the increase of soil OM, trapping nutrients, reducing NO_3^--N and other nutrient leaching, limiting nutrient run off and soil erosion, improving soil structure and aeration, and reducing nematode infestations. Cover crops can be established by either natural or selected species. Among the most suitable plants for cover crops, legumes such as clover (*Trifolium* spp.), broad bean (*Vicia faba minor*) and vetch (*Vicia sativa*) are the most used and effective in fixing atmospheric N. The magnitude of this ability ranges between 20 and 200 kg N/ha (Adjei-Nsiah et al., 2008; Rahman et al., 2004). Because of their deep root system, most of leguminous plants show competition with tree crops, and their cultivation should be limited to the tree dormancy phase.

Graminaceous species, such as oat (*Avena sativa*), barley (*Hordeum vulgare*), rye grass (*Lolium perenne*), and sorghum (*Sorghum vulgare*), are characterized by a high-density shallow root system, produce a high biomass, and are particularly suitable in sloping orchards where they can prevent erosion and nutrient runoff.

Most species of the Brassicaceae family (i.e., *B. napus*, *B. nigra*, *B. alba*, etc.) present a glucosinolate-*myrosinase* defensive system, active in pest (i.e., *Pratylenchus penetrans*) and diseases (i.e., *Rhizoctonia* spp.) control of many horticultural crops (Baldi et al., 2015; Lazzeri et al., 2013). Once glucosinolates and *myrosinase* come in contact (they are physically separated in the plant cells), isothiocyanates, volatile compounds toxic for nematode and pathogens, are produced (Brown et al., 2008; Wentzell and Kliebenstein, 2008). In addition, isothiocyanates have been found to inhibit nitrification of NH_4^+ (Bending and Lincoln, 2000) with beneficial effect on N leaching. Other species, like *Tagetes patula* are known for their toxic effect on nematodes (Marahatta et al., 2012) and can be used in temporary cover crops. A mix of grasses and legumes may be preferred to combine the tolerance to low winter temperature of graminaceous with the positive effect on N_2 fixation of legumes. Cover crops can be sowed in the fall, late winter, or early spring (according to the temperature required) and tilled into the soil in spring or early summer, usually few days before full bloom. If cover crops are mowed later in the season, the relative C:N ratio of the biomass increases. The mowed biomass should remain a couple of days on the ground to partially dehydrate, before soil incorporation.

The advantages of temporary cover crop are similar to permanent cover crop floor management and include: the addition of OM to the soil, fixing and trapping nutrients, reducing NO_3^--N leaching, nutrient run off, and soil erosion, increasing availability in the upper soil profile of nutrients with a low mobility such as potassium (K), phosphorus (P), and magnesium (Mg) (Schliemann et al., 1983). It is also well known that the consociation of fruit trees with graminaceous can produce positive effect on the prevention of lime-induced Fe-deficiency symptoms (Zuo and Zhang, 2011).

The amount of organic material provided yearly by the understory grass is usually twice than the aboveground biomass, since root turnover and consequent rhizodeposition have the same magnitude of aboveground organs. However, permanent grass management on the whole soil surface (row and interrow) may reduce yield compared with the temporary cover crop.

2.2 Mulching

Mulching is a traditional weed control method (Neilsen et al., 2003) that consists in the coverage of the soil of the tree row (1.5–2-m wide) with either organic or synthetic materials. When organic residues are adopted as mulching material, beside benefits related to weed control and soil moisture content, they promote biological activity and the release of nutrients and organic compounds. Compared with other practices, the impact of the mulching on soil OM is much lower; however, the release of nutrients can be significant in terms of tree nutrition. For instance, when grass littering is used as mulching materials, the release of K, completed in 5 weeks, is faster than other macronutrients such as N, calcium (Ca), and P (Tagliavini et al., 2007). This response has implications on the availability of nutrients and their effect on fruit quality (i.e., bitter pit represents a postharvest disorder related to K:Ca ratio imbalance in fruits).

2.3 Organic fertilizers

The most effective mean to increase soil OM is the continuous application of organic material to the soil. There are a number of products that can be used for this purpose identified with different names like amendments, manures, biosolids, organic by-products, and organic fertilizers. Any kind of organic material can be recycled in agriculture once its composition and its behavior in soil are known. Stabilized organic materials should be preferred to fresh ones, since stabilization includes an intense biological oxidation process operated by aerobic microorganisms, which improves the quality of the amendments. If this oxidation occurs in the soil, anoxia can be detrimental for the root system. As a matter of fact, to achieve all the beneficial effect of OM, the organic material should be tilled into the soil, where it can provide benefits on soil structure, microbiota, and nutrient availability.

Among organic compounds suitable as soil conditioner, compost is the by-product of the controlled biological decomposition of a large range of organic materials (Senesi, 1989). The stabilization process lasts around 90 days, during which the fresh material loses CO_2, NH_3, and water while producing heat and consuming O_2. At the end of the composting process, the initial weight is reduced by more than 50%, and consumption and release of O_2 are in equilibrium. The chemical composition of compost depends on the starting material: that is, a predominance of the vegetal fraction leads to a higher C:N ratio compared with a compost obtained from a large portion of animal manure or municipal solid wastes (MSW). This makes it possible to define the final composition of the compost and choose it according to its prevalent use. Organic materials from agroindustry (i.e., wastes from fruit and vegetable processing, winery, etc.), MSW, management of the urban landscape (e.g., pruning and trimming), and livestock are potentially organic fertilizers with a precious agronomical value at relatively low cost. These fertilizers allow the closure of the "nutrient cycling" defined as the breakdown of organic substances, release of energy, and matter captured by life processes, used to stimulate the new growth. By incorporating recycled organic material into the soil, a sequestration of C, which otherwise would follow disposal processes, which potentially release CO_2 in the atmosphere, is achieved (Table 27.1).

The use of compost in agriculture has many advantages including: (1) increase availability of mineral nutrients (Bhattacharya et al., 2016) through mineralization of OM (Table 27.1), chelation of micronutrients (Sorrenti et al., 2012); (2) enlargement of soil biodiversity (Valarini et al., 2009) and consequent decrease of pest and disease incidence; (3) stimulation of microbial biomass (Baldi et al., 2018a) associated to an increase of the use of nutrients by agricultural crop roots; (4) amelioration of soil structure, by increasing of micro- and macroporosity (Bronick and Lal, 2005); (5) increase WHC (Ahn et al., 2008) and reduction of water irrigation rate (Sorrenti and Toselli, 2016); and (6) increase

TABLE 27.1 Soil organic carbon, average mineral N, leaf N, and yield of nectarine trees after 14 years of fertilization with either organic or mineral sources.

Treatment	Soil organic C (Mg/ha)	Soil mineral N (Mg/kg DW)	Leaf N (g/kg DW)	Yield (kg/Tree)
Mineral	99 bc	11.1 a	3.11	48.2 b
Compost 5 Mg/ha/year	115 b	9.1 b	3.04	47.3 b
Compost 10 Mg/ha/year	149 a	12.4 a	3.06	50.3 a
Significance	***	***	n.s.	*

In the same column, values followed by the same letter are not statistically different according to Student Neuman Keul test ($P \leq .05$). n.s., *, and ***: effect of the fertilization management not significant or significant at $P \leq .05$ and $P \leq .001$, respectively.
Mineral fertilization included the application of 130 kg N/ha/year as ammonium nitrate.
Soil mineral N (NO_3-N and NH_4-N): average from 2006 to 2014 (4 sampling times per year, January, May, July, and September).
Leaf N: average from 2002 to 2014 (1 sampling per year).
Yield: average from 2004 to 2014.

the soil buffer capacity (Latifah et al., 2018). These benefits contribute to increase soil fertility and net primary production (NPP), with a positive effect on crop yield and orchard C sequestration in soil, which is of relevant importance as a strategy to mitigate the rise of anthropogenic emission of CO_2 in the atmosphere. Table 27.1 shows the positive effect of 14 years of continuous applications of compost on organic C sequestration, increase availability of soil mineral N, and yield of a commercial nectarine orchard.

Soil compost application modifies the bacterial community profile and may contribute to suppress soil-borne disease (Postma et al., 2003). In container-grown bearing nectarine trees, application of compost promoted *Nitrosomonas* spp. and *Nitrobacter* spp. relative abundance (Sorrenti et al., 2017). In a walnut orchard treated with compost for 14 years, Mazzon et al. (2018) observed significantly difference between untreated control and mineral fertilization, in several biochemical indicators, such as microbial biomass, soil basal respiration, metabolic quotient, hydrolytic and phenol oxidase enzymatic activity, indicating that the compost applications induce large differences in soil microbial functionality.

In general, an increase of plant growth promoting rhizobacteria has been found in response to compost application (Viti et al., 2010), along with a biocidal effect against nematodes. Unlike noncomposted raw material (e.g., manure), compost application reduced pathogen indicator bacteria like fecal coliform, *Escherichia coli*, *Enterococci*, and *Staphylococci* (Smiciklas et al., 2013), as these bacteria are in raw materials but disappear, with few exceptions (i.e., *Staphylococci*), after composting process.

The effects of compost application on fruit and vegetable quality, in terms of fruit composition and nutraceutical effectiveness, are still unclear with contrasting results (Prange, 2015) related to different environment conditions, species, and composts used in the experiments. According to some reports (Winter and Davis, 2006), different fertilization practices have different impacts on plant metabolism. Indeed, the amount of available N released by synthetic fertilizers frequently exceeds that of organic-based fertilizers and may accelerate plant growth and development. Therefore, plants allocate resources mainly for growth purposes, with a decrease in the biosynthesis of plant secondary metabolites (organic acids, polyphenols, chlorophyll, and amino acids). In addition, plants respond to stressful environments such as nutrient limitation. Organic fertilization may limit nutrient availability and poses greater stresses on plants that devote greater resources toward antioxidants such as polyphenols. Consequently, if organic fertilization is managed correctly, in the absence of biotic and/or abiotic stresses, plants usually do not respond with a modification of their physiological behavior, that is, synthesis of secondary metabolism compounds such as antioxidants in fruits (Winter and Davis, 2006).

For its relatively slow mineralization rate and release of nutrients, mature compost is particularly suitable for perennial crops such as fruit trees that alternate periods of high nutrient requirements with periods of dormancy used to build up reserves. The continuous application of compost allows the stimulation of microbial populations (Baldi et al., 2010) that degrade organic C according to temperature oscillations: a release of nutrients is then expected in spring, summer, and early fall, while a reduced release occurs in cold seasons. In the temperate area of the Po valley (Italy), a continuous application of compost at 10 Mg/ha/year, was found to promote a release of mineral N that peaked in summer, during the intense shoot growth and fruit cell enlargements, with a low soil NO_3^--N in winter, during tree dormancy. From an environmental point of view, this trend allows an optimal synchronicity of the availability of mineral N and root uptake, an increase in soil chemical fertility, and NPP to levels similar to the chemical fertilization commonly used in the area (Baldi et al., 2018a).

The correct compost application rate depends on the goals to be achieved, keeping in mind to avoid excess of available N, which would lead to negative effects: worse production in qualitative-quantitative terms and risk of N-mineral form leaching (Sorrenti and Toselli, 2016). In normal conditions, we suggest an annual application rate ranging between 10 and 20 Mg/ha of a high-quality and stabilized compost.

3 Biochar

Biochar is the high porous and recalcitrant C-rich residue generated by the thermal decomposition of organic materials (e.g., wood, agroindustrial residues, energetic crops, manures, and MSW) under depleted O_2 supply and at relatively low (<700°C) temperatures (pyrolysis), intentionally used for agronomical purposes as a soil conditioner (Lehmann et al., 2006). Other than a long-term strategy to sequester atmospheric CO_2 (biochar in soil may last several hundreds of years) and increase the stable soil C pools (source of OM), once incorporated in soil, biochar might yield several cobenefits: enhancement of soil health (Ameloot et al., 2013), increase in plant-available nutrients (e.g., K and P) and soil WHC, decrease in bulk density (increase of soil aeration and reduce anoxic microsites), enhancement of CEC, stimulation of microbial activity and biodiversity (Atkinson et al., 2010; Spokas et al., 2012; Verheijen et al., 2010), decrease in heavy metal contamination risks (Namgay et al., 2010), and decrease in greenhouse gas emissions (Singh et al., 2010).

Biochar added to soil acts as a sponge, soaking up different forms of OM as well as water and nutrients (Glaser et al., 2002) through its porous structure, which reflects the cellular arrangement of the original feedstock. Even though it is accepted that biochar in soil interacts with microbes, plant roots, water, and minerals, the extent, rates, and implications of these interactions are not fully understood yet. However, to explain how biochar might benefit plant growth and crop yield, generally, four mechanisms are proposed (Lehmann and Joseph, 2009; Sohi et al., 2010): alteration of soil chemistry (direct source of nutrients or improvement of nutrient efficiency); mitigation and/or removal of soil constraints (e.g., low pH, aluminum (Al) toxicity, and contaminants), which may limit plant growth; modification of the nutrient dynamics in soil and/or altering soil reactions by providing chemically active surfaces; and change soil physical parameters that benefit root growth and/or nutrient, water retention, and uptake.

An overall increase in crop yield has been reported (Genesio et al., 2015; Jeffery et al., 2011) following biochar application, although some studies stressed neutral or even negative effects (Sorrenti et al., 2016a; Spokas et al., 2012). Indeed, literature reports benefits mostly obtained in tropical and subtropical environments, in acid, weathered, and poorly fertile soils (Jeffery et al., 2011). In contrast, scarcely pronounced effects on soil properties, yield, fruit and berry quality were described by Ventura et al. (2013), Schmidt et al. (2014), and Sorrenti et al. (2016b) who incorporated biochar at a rate of 10, 8, and up to 30 Mg/ha in apple, grapevine, and nectarine orchards, respectively, grown in temperate environments. These contrasting results suggest that plant response to biochar application varies with crop species, environmental conditions, soil type, biochar properties, and application rates.

Due to its high surface area and to the presence of polar and nonpolar surface sites through electrostatic forces, biochar has the ability to bind and desorb cations (Alling et al., 2014). This property may have important environmental (reduce leaching) and agronomical implications (increase NUE). The application of biochar at 15 and 30 Mg/ha in a sandy soil significantly increased NO_3^--N concentration in the top soil (0–0.15 m), while it decreased in the 0.90-m-deep layers (Kammann et al., 2016). Similarly, NO_3-N (Ventura et al., 2013) and NH_4-N (Sorrenti et al., 2016b) leaching was reduced by the incorporation in field conditions of biochar in the first 0.2-m soil, indicating a retention effect in the soil top layers.

Biochar usually promotes synergistic interactions when incorporated into soils either with stabilized organic residues or used as a cocomposting agent. However, direction and extent of these interactions are not always granted (Bonanomi et al., 2017). Mixing biochar with compost increases compost stability and nutrient availability, alters soil water retention potential, and stimulates biomass biodiversity (Fischer and Glaser, 2012; Liu et al., 2012; Schulz et al., 2013; Sorrenti et al., 2017). The mixture of biochar with compost was more effective in increasing soil water retention capacity than the mere addition of the two amendments separately (Sorrenti and Toselli, 2016). On the other hand, Bonanomi et al. (2017) reported that antagonistic responses were more frequent than synergistic interactions when biochar was mixed with organic amendments, highlighting that synergistic interactions occurred mostly when N-rich and lignin-poor organic amendments were mixed with biochars.

The main concern related with the use of biochar is that, if not properly produced, it has the potential to contaminate the environment through the releasing of polluting compounds such as polycyclic aromatic hydrocarbons, *polychlorobiphenyls*, dioxins, furans, and heavy metals, depending on the precursor biomass that is pyrolyzed.

To date, biochar in agriculture has been legally regulated and included as a soil amendment in few countries (e.g., Italy) and even less permit its adoption, if produced according to stringent protocols, in organic farming (e.g., France). In this sense, the voluntary memberships to the internationally recognized certification systems (e.g., IBI, EBC, and BQM-UK) attesting the biochar quality in terms of properties and pollutant concentration are increasing worldwide.

4 Tree nutrient balance

Two factors mainly drive a sustainable fertilization schedule at each phenological stage: tree nutrient requirement and nutrient availability in the soil.

4.1 Tree nutrient requirement

The best way to measure the yearly nutrient demand of a tree is the determination of the amount of nutrients that the tree uptakes. This amount includes: (1) the portion accumulated into fruits; (2) the portion that goes into woody perennial organs such as roots, trunk, and branches; and (3) the portion accumulated into organs that at the end of the season return to the soil like leaves, pruned wood, and thinned fruits. While the first two portions are actually removed from the soil, the last fraction will be available during the commercial lifetime of the orchard. Although only portions 1 and 2 negatively impact the nutritional balance, among the tree nutrient requirement, we consider also portion 3, because NUE is usually lower than 100%. For example, 50% of the N-based fertilizers are worldwide lost in the environment, and less than 50% is estimated to be removed by crops (Galloway et al., 2004), and this percentage is lower in fruit trees (Bravo et al., 2017). The magnitude of portions 1 and 2 of N equalizes portion 3, consequently we can assume that portion 3 compensate for the low N use efficiency. For each element, the three portions can be easily estimated knowing its concentration and the mass of a given organ. Multiplying the mass by the concentration, the amount of nutrients used during the growing season will be estimated. For the permanent structures, such as skeleton and root system, an estimation of the nutrients used during tree lifetime will be obtained, while the annual growth of these organs is correlated with the increase of trunk diameter. A large body of literature deals with the amount of nutrient removed by the different species of fruit trees (Table 27.2). These values depend on genetic (i.e., grafting combination), environment (soil and climate), agricultural practices (soil management), soil fertility, etc. Beside the yearly amount of nutrients used by a tree, the kinetics of uptake is also important particularly for nutrients that are considered mobile in the tree and, for this reason, stored in perennial organs at the end of the season and remobilized in spring, to promote growth resumption. Some of these nutrients, such as N, K, and boron (B), are well studied, while for others the information is still scarce. This is the case of zinc (Zn), considered a low-mobility nutrient. However, in a field experiment, the Zn concentration increased from midsummer to senescent and finally in abscised leaves of mature Abbé Fétel pear trees, suggesting negligible remobilization. However, its content in nectarine leaves decreased before leaf abscisions (Baldi et al., 2014), as for in the mobilization to the storage wood.

TABLE 27.2 Amount of the yearly macro- and micronutrients removed by several fruit tree crops.

Species	Macronutrients (kg/ha)					Note	Literature
	N	P	K	Ca	Mg		
Apple	55–60	9–11	83–100	40–74	12–19	6-years-old Gala/M26 on sandy soil; 6-years-old Gala/M9 on silty-clay-loam	Cheng and Raba (2009) and Scandellari et al. (2010)
Pear	80–126	11–25	69–74	104–142	15–43	9-years-old Forelle/BP1, Forelle/Quince A	Stassen and North (2005)
Orange	83	9	33	187	13	Tarocco/sour orange Tarocco/Citrange Carrizo, on sandy-loam soil	Roccuzzo et al. (2012)
Almond	20–45	1–8	19–43	1–4	1–3	Nonpareil/Nemaguard on sandy loam soil	Muhammad et al. (2015)
Kiwi	200	30	220	184	29	9-years-old Hort 16A/Hayward	Boyd et al. (2010)
Cherry	39–65	6–11	16–47	26–55	5–9	Unbearing Bing/Gisela 6; 12-years-old Bigaerrau Morau, Ferrovia, Stella, Colafemmina, Droganova, Montagnola/Mazzard	Bonomelli et al. (2010) and Roversi and Monteforte (2005)

TABLE 27.2 Amount of the yearly macro- and micronutrients removed by several fruit tree crops.—cont'd

Species	Macronutrients (kg/ha)					Note	Literature
	N	P	K	Ca	Mg		
Grapevine	16–59	3–16	38–118	33–80	6–14	41-years-old Concord, sandy loam soil	Pradubsuk and Davenport (2010) and Schreiner et al. (2006)
Peach	49–170	6–26	60–212	94–259	14–37	BabyGold 5/GF677 Catherina/GF677, on calcareous, clay-loam soil	Rufat and DeJong (2001) and El-Jendoubi et al. (2013)
Plum	117–154	10	112	95	17	Dabrowica Plume and Stanley/Myrobalan on sandy loam soil; 4-years-old Red-Beaut	Plich and Wójcik (2002) and Alcaraz-Lopez (2003)

Species	Micronutrients (g/ha)					Note	Literature
	B	Cu	Fe	Mn	Zn		
Apple	122–261	70–130	415–1547	217–516	125–170	6-years-old Gala/M26 on sandy soil; 6-years-old Gala/M9 on silty-clay-loam	Cheng and Raba (2009) and Scandellari et al. (2010)
Peach	–	134–400	603–2050	134–350	134–450	BabyGold 5/GF677 Catherina/GF677, on calcareous, clay-loam soil	El-Jendoubi et al. (2013)
Pear	224–786	197–802	2911–4545	457–1630	547–786	Forelle/BP1, Forelle/Quince A	Stassen and North (2005)

4.2 Nutrient availability

The nutrient soil availability is a priority information to define the fertilization management. In clay and loamy soils, most of the essential nutrients for tree crops are held by soil properties (CEC, aggregation with metal hydroxides, organic matter bounding, etc.). In fact, a little fraction of nutrients are soluble in soil solution; most of them can be solubilized by roots before uptake; and consequently, they have a high use efficiency and low risk of leaching. Under aerobic conditions, N (i.e., ammonium, organic N) is rapidly oxidized to nitrate (via nitrite), an anion scarcely held by soil. The N fraction not absorbed by plants is thus lost through soil profile, with the leaching water. A sustainable fertilization should consider the amount of nutrients available that can be calculated by multiplying concentration by the mass of soil explored by roots (soil unit). The soil unit on a hectare (1 ha = 10,000 m^2) basis depends on root depth and soil bulk density. Supposing a root depth of 0.8 m and an apparent soil weight of 1.3 Mg/m^3, the soil volume and mass explored by roots accounts for approximately 8000 m^3/ha and 10,400 Mg/ha, respectively. This value is based on the assumption that the root system explores all the volume of soil in which it grows. However, this is far from reality. According to our knowledge, root density of fruit crops ranges between less than 1 to 10 kg/m^3 (Govi et al., 1996), meaning roughly from 0.1% to 1% of soil mass and lower probability to intercept the nutrient.

4.2.1 Nitrogen

Among nutrients, N is the most abundant in fruit trees, with the largest impact on shoot and root growth and fruit quality. Average N concentration in fruit trees ranges from 0.1% in pome fruits, 0.5% in woody organs, 1% in roots, and up to 5% in leaves at the beginning of their development. Nitrogen deficiency reduces shoot growth, bud fertility (Baldi et al., 2017), yield, and fruit size and stimulates the synthesis of secondary metabolite compounds such as organic acids and polyphenols, which may accumulate in fruits. Contrarily, excess of N promotes a higher protein synthesis and vegetative growth; delays fruit ripening; and hinders fruit skin color, fruit size, and sugar content. Excess of N may also increase the susceptibility of the tree to pest and disease infections (Pfeiffer and Burts, 1983).

Mineral N is the final product of the soil N cycle in orchard ecosystem; the amount of N available for root absorption can be calculated by multiplying the mineral N (sum up both mineral N forms) by the soil mass explored by roots. In agricultural soil, the mineral N concentration ranges from few units to hundreds of mg/kg that means an amount of mineral N from 10 to 1000 kg/ha. To optimize the fertilization efficiency, the availability of mineral N has to meet tree request. According to Table 27.2, nectarine trees require approx. 100 kg of N/ha/year; hence, with a soil unit of ~10,000 Mg, the availability of mineral N should be of 10 mg/kg. If the request is of 200 kg N/ha/year (e.g., kiwifruit), the optimal mineral N should be maintained at 20 mg/kg. If the root depth changes (e.g., 0.5 m for apple trees when grafted on dwarfing rootstocks), then the soil unit to be considered would be reduced to approx. 6500 Mg. If an apple orchard request is of 90 kg/ha, then soil available mineral N is expected to be 14 mg/kg. The low root density is

FIG. 27.1 Relationship between soil nitrate-N (NO$_3$-N) measured in summer and leaf N in an organically managed pear variety (Abbé Fétel) grafted on different root-stocks and grown in northern Italy. Optimal leaf N concentration for Abbé Fétel in the area is 2%–2.45% (Toselli et al., 2002). $r =$ Pearson correlation coefficient.

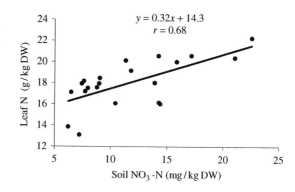

compensated by the relatively steady concentration of NO$_3^-$-N during the season. Our experience indicates that oscillations of mineral N may occur between warm (high N availability) and cold season (low availability), when the activity of bacteria slows down. The relationship between mineral N in soil and leaf N concentration in mature pear trees shows that to reach the optimal level of leaf N (i.e., 2.2% for Abbé Fétel pear variety) a soil concentration of mineral N of 20 mg/kg is expected (Fig. 27.1).

Spring bud sprouting, blooming, fruit set, and the first stages of shoot growth usually lay on the remobilization of N stored the previous year. In fact, we calculated in nectarine trees an autumnal remobilization of N from abscised leaves to skeleton ranging from 20 to 50 kg/ha/year, according to the soil fertility (Baldi et al., 2014). This amount accounts for almost 20%–35% of the total N yearly required by tree. In this view, the efficiency of root N uptake depends on the time of the year, and it increases with the progress of the season, reaching the maximum at the end of the summer (Bravo et al., 2017).

Soil N availability should be low until fruit set and then increase according to tree requirement (Table 27.3). Immediately after fruit set, N is polarized mainly by fruits, in their cytokinesis step. Later on (late spring), N goes to the shoots and again (in summer) to fruits in their cell enlargement stage. Finally, before fall leaf abscission, N is partitioned preferentially to the woody organs to build up new reserves. In kiwifruit and grapevine species, berries continue to accumulate N until harvest. Consequently, high mineral N availability in the weeks before harvest should be avoided, as it can turn into an excess of N in the fruits with detrimental effects on kiwifruit storability and higher incidence of postharvest diseases. On the other hand, a high N concentration in grapevine berries promotes the development of yeasts and the start of alcoholic fermentation.

A sustainable N management should monitor the soil mineral N concentration at least three times a year: (1) late spring, beginning of summer when N reserves stored in winter are exhausted, tree requirements account roughly for 25% of the total estimated N requirement; (2) at fruit cell enlargement (tree requirement of 50%); and (3) late summer when N is requested to build up new reserves (tree requirement of 25%). At a given phenological stages, if mineral N meets the crop demand, no fertilization is necessary; if mineral N is lower than the established threshold, then it should be applied according to the phenological stage previously mentioned. A continuous monitoring of soil mineral N should be available for precision farming practices to evaluate the most appropriate N application rate. In the case of fertigation, NUE is expected to increase.

TABLE 27.3 Optimal soil concentration of mineral N (mg/kg) in different phenological stages of some temperate fruit crops.

Tree crop	Phenological stage					
	Bud sprout	Full bloom	Fruit set	Fruit growth	Harvest	Leaf fall
Pome fruit	<5	5–10	5–10	10–20	10–20	<5
Stone fruit	<5	<5	5–10	20–30	20–30	<5
Kiwifruit	<5	10–20	10–20	20–30	<5	<5
Grapevine	<5	10–20	10–20	10–20	10–20	<5
Walnut	<5	<5	10–20	20–30	20–30	<5

TABLE 27.4 Chemical composition of selected organic fertilizers admitted in organic farming with different mineralization rate.

Mineralization rate	Fertilizer	N (% DW)	P (% DW)	K (% DW)
Fast	Blood meal	5–12	–	–
	Molasses	1–3	–	0–4
	Fish meal	5	1.3	–
	Poultry manure	3.7	–	–
Medium	Oil seed panel	3	–	–
	Meat meal	8.1	–	–
	Hydrolized hides and skins	5	–	–
	Compost	2–3	0.5	1–4
Slow	Residual leather processing	5	–	–
	Wool waste	8	–	–
	Bone meal	2	8	
	Cow manure	1.5–2	0.6	2.3
Very slow	Ground feather	10	–	–
	Horn and hoof meals	9	–	–

Modified from Carli, G., 1998. I prodotti per la fertilizzazione. In: CRPV (Eds.) Linee Guida per l'agricoltura Biologica, Fruttiferi e Fragola (Fertilizer for Organic Farming of Fruit Crop and Strawberry, in Italian). Edagricole, Bologna, pp. 131–155.

During the season, temporary N deficiencies may appear in response to cold and wet weather, high fruit crop load, and the use of an organic fertilizer with high C:N ratio. To overcome these deficiencies, organic fertilizers with a prompt N release (e.g., blood meal and poultry manure) may be conveniently employed (Table 27.4) (Carli, 1998). At the same time, if the mineral N concentration is considerably higher than tree request (i.e., >100 mg/kg), a temporary intercropping can be established, along with application of organic fertilizer with high C:N ratio.

4.2.2 Phosphorus

Phosphorous concentration in vegetal tissues ranges between 0.2% and 0.5% (in leaves at the beginning of the vegetative season). Below this range, deficiency symptoms may appear, including a reduction of photosynthetic efficiency, leaf area, and leaf number. If the P deficiency is severe, a precocious leaf abscission may occur, bud flower differentiation can be impaired, and a number of flowers and seeds can be also reduced. Organic P accounts for 20%–80% in soil, and phytate can represent up to 50% of organic P (Hocking, 2001). Roots exudation of organic acids (citric, malic, oxalic, etc.) can solubilize P at the rhizosphere level. Addition of P, above the sufficiency threshold, promotes root growth in grapevines (Baldi et al., 2018b). In natural and cultivated soils, P is mainly present, in soil solution, as phosphate (PO_4^{3-}) in very low concentration (<0.1 mg/L). The phosphate concentration in soil solution is at equilibrium with the sorbed phosphate on the surface of iron (Fe) and Al oxides and clay minerals. Furthermore, the phosphate has high affinity with Ca^{2+}, Mg^{2+}, Al^{3+}, Fe^{3+}, and Fe^{2+} ions, with which they can form poorly soluble phosphates: in acid and subacid soils with Fe^{2+}, Fe^{3+}, and Al^{3+} and in alkaline soils with Ca^{2+} and Mg^{2+}. As a matter of fact, P is almost immobile in soil, and its availability is highest at neutral pH. Fruit tree request of P is approximately 10–20 kg/ha/year; consequently, an annual fertilization with organic matter (i.e., 10–20 Mg/ha) allows a proper P availability. We found that, after 14 years of repeated applications of organic matter, Olsen extractable P doubled from 20 mg/kg to more than 40 mg/kg. In case of P deficiency, some fertilizers used in organic farming include: (1) meal bone (22%–28% of P and 2%–4% of N); (2) soft ground rock phosphate, a phosphatic rock powder (phosphorite), with low solubility in subalkaline soils, where it should be mixed with an organic fertilizer to decrease soil pH; (3) aluminum calcium phosphate suitable in subalkaline soils, while in acid soils Al increases solubility and can be toxic; and (4) Thomas slag, suitable for acid soils.

4.2.3 Potassium

Potassium is highly mobile in tree where it moves in the vascular system toward both acropetal (via xylem) and basipetal (via phloem) directions. Potassium is the most abundant element in cell vacuole where it regulates osmotic potential and tissue hydration. Along with N is the most used nutrient by tree, accounting for a removal rate of 100–200 K/ha/year. In fruit trees, K is partitioned preferentially to the fruits. For this reason, in case of deficiency, K is mobilized from mature leaves to fruits. Our studies demonstrate a mobilization of K from senescent leaves to skeleton ranging from 4 to 15 kg K/ha in nectarine, persimmon, and pear. If the deficiency is severe, tree becomes more susceptible to drought stress, cold damages, and diseases. Excess of exchangeable K in soil decreases Ca and Mg uptake. Potassium is usually present in soil as cation and is adsorbed on negative charged surfaces of soil colloids. Potassium deficiencies may occur in sandy, low-CEC soils or in heavy, clay soils, where the bounds between K^+ and clay are strong enough to prevent roots from K uptake. Potassium deficiency symptoms include leaf chlorosis, followed by lamina curling, browning, and eventually necrosis. In calcareous, loam, and clay-loam soils, K management is quite easy, since K can be applied at any phenological stage, while in sandy soils K should be applied upon tree request. Potassium concentration in organic fertilizers ranges from 1% in green compost, to 2% in cow manure, to 3% in compost; thereby, they provide a large amount of K when applied to the soil (Table 27.4). To overcome K deficiency, in organic farming, application of raw salts of K obtained from kainite (potassium chloride and magnesium sulfate) or sylvinite (potassium chloride + sodium chloride) can be effective. Rock powder (from orthoclase rocks) has a lower K concentration and solubility than K raw salts. Foliar application and fertigation with soluble salts (e.g., KCl and KNO_3) are not allowed in organic farming. However, molasses, a by-product of sugar industry (with a K concentration up to 4%), can be used as an organic fast-release K fertilizer, also in fertigation.

4.2.4 Calcium

Calcium is associated to fruit firmness and storability because of its cross-linking function between two molecules of galacturonic acid (pectins) at wall cell level (Fisher and Bennet, 1991). Calcium seems to inhibit the activity of the enzyme polygalacturonase (pectinase), contributing to delay fruit softening.

While in calcareous soils Ca^{2+} is not a limiting factor, in sandy, acid soils, it can be one of the most important nutrients in fertilization management of pome fruits, which suffer from postharvest disorders. Calcium deficiency symptoms in pome fruits include pre- and postharvest disorders, such as bitter pit and senescent breakdown. In neutral to subacid soils, Ca^{2+} competes with other cations such as NH_4^+, K^+, Mg^{2+}, and copper (Cu^{2+}) that may inhibit its uptake and worsen deficiency symptoms. Foliar applications of Ca are frequently used in fruit trees fertilization management and are recognized as the most effective tool to prevent Ca deficiency. However, the effectiveness of such treatments is often species dependent, as are more effective in apple and kiwifruit than pear or stone fruit. Organic farming Italian regulation allows the use of calcium chloride in apple variety susceptible to storage disorders.

4.2.5 Magnesium

In fruit trees, Mg is considered a mobile nutrient. Nevertheless, our experience on nectarine (Baldi et al., 2016) and persimmon indicates that Mg is not remobilized from senescent leaves to tree skeleton, while scarce mobilization was measured in pear trees. In soil, Mg is present as cation and competes with other ions such as K^+, Ca^{2+}, manganese (Mn^{2+}), and NH_4^+ root uptake. In particular, K^+ and Mg^{2+} leaf concentrations are negatively correlated, and excess of Mg may contribute to originate K deficiency symptoms and. In organic farming, kieserite (magnesium sulfate monohydrate) is the most common fertilizer.

4.2.6 Iron

Although Fe is enumerated among the most abundant elements of the cultivated soils, in some environment, plant Fe acquisition is often impaired, which in turn develops in Fe chlorosis. This nutritional disorder occurs mainly as a consequence of a scarce solubility of mineral Fe sources in soil and of a reduced Fe uptake by the symplast (Pestana et al., 2003; Röemheld and Nikolic, 2007). The active calcium carbonate soil fraction and the alkaline soil pH are the major constrains for Fe uptake, although other factors such as genotypes (rootstock and scion combination), fruit load and climate conditions (e.g., reiterate rainy events in spring or late frosts) may accentuate Fe-deficiency occurrence (Tagliavini and Rombolà, 2001). Peach, pear, quince, olive, citrus, and kiwifruit are the most susceptible species to Fe chlorosis. Among grapevine genotypes, *Vitis vinifera* and *Vitis rupestris* are considered tolerant, while *Vitis riparia* is susceptible. Typical symptoms of Fe chlorosis are the interveinal leaf yellowing (in susceptible pear varieties, such as Abbé Fétel) the entire leaf lamina becomes yellow), starting from the youngest ones usually at growth resumption, when tree Fe requirement is highest. Unlikely, Fe chlorosis will appear in mid or late summer.

Fe chlorosis negatively affects leaf water relations (Eichert et al., 2010), canopy light absorption, leaf photosystem II, and Rubisco carboxylation; thus, leaf photosynthetic and transpiration rates are impaired (Larbi et al., 2006). Thereby, Fe deficiency may reduce tree growth and yield, depress fruit quality, and shorten orchard lifetime (Álvarez-Fernández et al., 2006; Sorrenti et al., 2012).

Synthetic Fe-containing compounds such as Fe chelates supplied either to the soil (e.g., Fe-EDDHA) or to the canopy (e.g., Fe-EDTA and Fe-DPTA) are effective to overcome Fe deficiency (Lucena, 2006). If there are no suitable alternatives, the use of these compounds is legally permitted in organic farming, under the supervision of the certification bodies. However, costs and environmental implications induced by Fe chelates are incompatible with a sustainable approach. Indeed, soil-applied Fe chelates are scarcely degradable and easily leached out of the root zone and may induce toxic effects in soil microorganisms (i.e., mycorrhiza) (Tagliavini and Rombolà, 2001).

Among alternatives, foliar-applied Fe salts (i.e., $FeSO_4$) were generally as effective as Fe synthetic chelates (i.e., Fe-DTPA) in kiwifruit (Tagliavini et al., 2000), orange trees (Pestana et al., 2001), and pear (Álvarez-Fernández et al., 2004).

Blood meal, the by-product of industrial slaughter houses, is a natural source of Fe (20–30 g Fe/kg), chelated under the ferrous from (Fe^{2+}) available for plants. The application of 70 g/tree of blood meal in a pear orchard alleviated Fe-chlorosis symptoms (Tagliavini et al., 2000). The use of blood meal is allowed in organic farming, not only as a Fe source but also as a fast-release N fertilizer.

It is well proved that soil OM improves soil Fe availability and contributes to alleviate the risk of Fe chlorosis. The complex with organic ligands (i.e., organic acids, amino acids, phenols, hydroxamates, catechols, and microbial siderophores) increases the Fe availability for plant uptake to a greater extent than inorganic insoluble pools. Humic substances are able to solubilize, mobilize Fe from hydroxides, and establish a strong complex with Fe. These properties were also recognized in the organic compounds contained in amendments, whose ability as Fe chelators have been demonstrated to serve as Fe source for several crops (Wang et al., 1993; Miller et al., 1985).

Our studies indicate that the yearly incorporation (26.3 Mg/ha FW) of a compost derived from MSW and pruning residues in a commercial pear orchard was a successful strategy to alleviate Fe chlorosis in calcareous soils. Table 27.5 shows that, in a 3-year field study, compost improved Fe nutrition of pear trees with benefits on tree yield and fruit weight (Sorrenti et al., 2012). In a 14-year field study, the organic fertilization with compost (10 Mg/ha/year) increased soil available DTPA-extractable Fe in comparison with mineral fertilization (Table 27.6).

Furthermore, results obtained in controlled and in field conditions highlight the potential of the soil-applied *Amaranthus retroflexus* aqueous extract in improving Fe nutritional status of pear trees, likely as a consequence of the natural Fe chelating capacity of the compounds released by its tissues (Sorrenti et al., 2011).

TABLE 27.5 Effect of the compost soil incorporation on the leaf chlorophyll content, tree yield, and fruit weight in a 3-year study in commercial pear orchard (Abbé Fétel/quince BA29).

Fertilization strategy	Leaf chlorophyll (Spad units)	Yield (kg/tree)	Fruit weight (g/fruit)
Control	18.2 b	12.6 b	249 a
Fe chelate	37.8 a	17.3 a	236 b
Compost	28.0 ab	14.4 ab	255 a
Significance	***	*	*

*, *** = effect of the fertilization significant at $P < .05$ or $P < .001$, respectively. In the same column, means followed by the same letter are not statistically different ($P < .05$).

TABLE 27.6 Effect of 14-year-long fertilization strategy on the average concentration of DTPA-extractable metals in top soil samples (0–0.25 m).

Treatment	DTPA extractable (mg/kg DW)					
	Fe	Mn	Zn	Cu	Ni	Pb
Mineral	16.4 c	11.9 b	2.21 c	7.23	0.76 b	0.96 b
Compost 5 Mg/ha/year	24.6 b	14.2 a	8.00 b	8.15	1.07 a	1.61 a
Compost 10 Mg/ha/year	32.8 a	15.0 a	10.8 a	6.83	1.04 a	2.11 a
Significance	***	***	***	n.s.	***	***

Mineral fertilization included the application of 130 kg N/ha/year as ammonium nitrate.
n.s. and ***: effect of fertilization management not significant or significant at $P < .001\%$. Within the same column, values with the same letter are not statistically different.

4.2.7 Manganese and zinc

Manganese deficiency resembles Fe-chlorosis symptoms as it appears not easy to distinguish them. The best ways to identify a Mn deficiency are the analyses of the leaf tissues or to supply a Mn source to leaves and see if symptoms disappear.

Zinc deficiency symptoms include a reduction of shoot length and leaf area and may develop to leaf necrosis. Zinc deficiency often appears in soils with high P concentration, as a result of a decrease of Zn solubility. Manganese and Zn chelates are the most effective tools to supply these nutrients to the trees. However, fertilization with organic fertilizers is effective in increasing soil availability of micronutrients. Table 27.6 shows the effect of 14 years of application of MSW compost on availability (DTPA extractable) of trace nutrients and heavy metals.

4.2.8 Boron

Boron is involved in fruit set, anther fertility, pollen germinability, and fruit cell division at cytokinesis stage; its deficiency reduces fruit set and yield. Boron deficiencies are frequent in sandy soils and in calcareous, subalkaline conditions and include chlorosis uniformly distributed on young leaf lamina, while, in the older, chlorosis is localized between veins. Boron deficiency can be overcome through foliar sprays with products based on boric acid or sodium borate. Late-summer sprays are more effective than spring application because B is mobile in the tree of Rosaceae family and can be rapidly mobilized in spring, no matter the efficiency of the root system.

References

Adjei-Nsiah, S., Kuyper, T.W., Leeuwis, C., Abekoe, M.K., Cobbinah, J., Sakyi-Dawson, O., et al., 2008. Farmers' agronomic and social evaluation of productivity, yield and N_2-fixation in different cowpea varieties and their subsequent residual N effects on a succeeding maize crop. Nutr. Cycl. Agroecosys 80 (3), 199–209.

Ahn, H.K., Richard, T.L., Glanville, T.D., 2008. Laboratory determination of compost physical parameters for modeling of airflow characteristics. Waste Manag. 28 (3), 660–670.

Alcaraz-Lopez, C., Botia, M., Alcaraz, C.F., Riquelme, F., 2003. Effects of foliar sprays containing calcium, magnesium and titanium on plum (*Prunus domestica* L.) fruit quality. J. Plant Physiol. 160 (12), 1441–1446.

Alling, V., Hale, S.E., Martinsen, V., Mulder, J., Smebye, A., Breedveld, G.D., et al., 2014. The role of biochar in retaining nutrients in amended tropical soils. J. Plant Nutr. Soil Sci. 177 (5), 671–680.

Álvarez-Fernández, A., García-Laviña, P., Fidalgo, C., Abadía, J., Abadía, A., 2004. Foliar fertilization to control iron chlorosis in pear (*Pyrus communis* L.) trees. Plant Soil 263 (1), 5–15.

Álvarez-Fernández, A., Abadía, J., Abadía, A., 2006. Iron deficiency, fruit yield and fruit quality. In: Barton, L.L., Abadía, J. (Eds.), Iron Nutrition in Plants and Rhizospheric Microorganisms. Springer, Dordrecht, The Netherlands, pp. 85–101.

Ameloot, N., Graber, E.R., Verheijen, F.G., De Neve, S., 2013. Interactions between biochar stability and soil organisms: review and research needs. Eur. J. Soil Sci. 64 (4), 379–390.

Atkinson, C.J., Fitzgerald, J.D., Hipps, N.A., 2010. Potential mechanisms for achieving agricultural benefits from biochar application to temperate soils: a review. Plant Soil 337 (1–2), 1–18.

Baldi, E., Toselli, M., Marcolini, G., Quartieri, M., Cirillo, E., Innocenti, A., et al., 2010. Compost can successfully replace mineral fertilizers in the nutrient management of commercial peach orchard. Soil Use Manage. 26 (3), 346–353.

Baldi, E., Marcolini, G., Quartieri, M., Sorrenti, G., Toselli, M., 2014. Effect of organic fertilization on nutrient concentration and accumulation in nectarine (*Prunus persica* var. *nucipersica*) trees: the effect of rate of application. Sci. Hortic. 179, 174–179.

Baldi, E., Lazzeri, L., Malaguti, L., Toselli, M., 2015. Evaluation of the biocidal effects of Brassica seed meal on *Armillaria mellea*. Ann. Appl. Biol. 167 (3), 364–372.

Baldi, E., Marcolini, G., Quartieri, M., Sorrenti, G., Muzzi, E., Toselli, M., 2016. Organic fertilization in nectarine (*Prunus persica* var. nucipersica) orchard combines nutrient management and pollution impact. Nutr. Cycl. Agroecosys 105, 39–50.

Baldi, E., Colucci, E., Gioacchini, P., Valentini, G., Allegro, G., Pastore, C., Filippetti, I., Toselli, M., 2017. Effect of post-bloom foliar nitrogen application on vines under two level of soil fertilization in increasing bud fertility of 'Trebbiano Romagnolo' (*Vitis vinifera* L.) vine. Sci. Hortic. 218, 117–124.

Baldi, E., Cavani, L., Margon, A., Quartieri, M., Sorrenti, G., Marzadori, C., Toselli, M., 2018a. Effect of compost application on the dynamics of carbon in a nectarine orchard ecosystem. Sci. Total Environ. 637–638, 918–925.

Baldi, E., Miotto, A., Ceretta, C.A., Brunetto, G., Muzzi, E., Sorrenti, G., Quartieri, M., Toselli, M., 2018b. Soil application of P can mitigate the copper toxicity in grapevine: physiological implications. Sci. Hortic. 238, 400–407.

Bending, G.D., Lincoln, S.D., 2000. Inhibition of soil nitrifying bacteria communities and their activities by glucosinolate hydrolysis products. Soil Biol. Biochem. 32, 1261–1269.

Bhattacharya, S.S., Kim, K.H., Das, S., Uchimiya, M., Hun Jeon, B., Kwon, E., et al., 2016. A review on the role of organic inputs in maintaining the soil carbon pool of the terrestrial ecosystem. J. Environ. Manag. 167, 214–227.

Bonanomi, G., Ippolito, F., Cesarano, G., Nanni, B., Lombardi, N., Rita, A., et al., 2017. Biochar as plant growth promoter: better off alone or mixed with organic amendments? Front. Plant Sci., 1–13. https://doi.org/10.3389/fpls.2017.01570.

Bonomelli, C., Gilabert, H., Ayala, M., 2010. Nitrogen uptake, growth and biomass accumulation in the first growing season of cherry trees on Gisela® 6. Acta Hortic. 868, 177–183.

Boyd, L., Barnett, A., Civolani, C., Fini, E., 2010. Whole plant excavations to determine nutritional requirements in 'Hort16A' kiwifruit vines. Acta Hortic. (868), 171–176.

Bravo, K., Marcolini, G., Sorrenti, G., Baldi, E., Quartieri, M., Toselli, M., 2017. Effect of time of application on nitrogen uptake, partitioning, and remobilization in walnut trees. J. Plant Nutr. 40 (5), 719–725.

Bronick, C.J., Lal, R., 2005. Soil structure and management: a review. Geoderma 124 (1–2), 3–22.

Brown, B.D., Gibson, R.C., Geary, B., Morra, M.J., 2008. Biofumigant biomass, nutrient content and glucosinolate response to phosphorus. J. Plant Nutr. 31 (4), 743–757.

Bünemann, E.K., Bongiorno, G., Bai, Z., Creamer, R.E., De Deyn, G., Goede, R., et al., 2018. Soil quality—a critical review. Soil Biol. Biochem. 120, 105–125.

Carli, G., 1998. I prodotti per la fertilizzazione. In: CRPV, (Ed.), Linee guida per l'agricoltura biologica, fruttiferi e fragola (Fertilizer for organic farming of fruit crop and strawberry, in Italian). Edagricole, Bologna, pp. 131–155.

Cheng, L., Raba, R., 2009. Accumulation of macro- and micronutrients and nitrogen demand-supply relationship of "gala"/'Malling 26' apple trees grown in sand culture. J. Am. Soc. Hortic. Sci. 134, 3–13.

Eichert, T., Peguero-Pina, J.J., Gil-Pelegrín, E., Heredia, A., Fernández, V., 2010. Effects of iron chlorosis and iron resupply on leaf xylem architecture, water relations, gas exchange and stomatal performance of field-grown peach (Prunus persica). Physiol. Plant. 138, 48–59.

El-Jendoubi, H., Abadía, J., Abadía, A., 2013. Assessment of nutrient removal in bearing peach trees (Prunus persica L. Batsch) based on whole tree analysis. Plant Soil 369 (1–2), 421–437.

Fischer, D., Glaser, B., 2012. Synergisms Between Compost and Biochar for Sustainable Soil Amelioration. INTECH Open Access Publisher, pp. 167–198.

Fisher, R.L., Bennet, A.B., 1991. Role of cell wall hydrolases in fruit cells. Annu. Rev. Plant Physiol. Plant Mol. Biol. 42, 675–703.

Galloway, J.N., Dentener, F.J., Capone, D.G., Boyer, E.W., Howarth, R.W., Seitzinger, S.P., et al., 2004. Nitrogen cycles: past, present, and future. Biogeochemistry 70, 153–226.

Genesio, L., Miglietta, F., Baronti, S., Vaccari, F.P., 2015. Biochar increases vineyard productivity without affecting grape quality: results from a four years field experiment in Tuscany. Agric. Ecosyst. Environ. 201, 20–25.

Glaser, B., Lehmann, J., Zech, W., 2002. Ameliorating physical and chemical properties of highly weathered soils in the tropics with charcoal—a review. Biol. Fertil. Soils 35 (4), 219–230.

Govi, G., Toselli, M., Gaspari, N., Scudellari, D., 1996. Effetti dell'irrigazione localizzata sull'apparato radicale dell'albicocco (Effect of Localized Irrigation on Apricot Root System, in Italian). Irrigazione e Drenaggio (in Italian) 43 (4), 13–19.

Hocking, P.J., 2001. Organic acids exuded from roots in phosphorus uptake and aluminum tolerance of plants in acid soils. Adv. Agron. 74, 63–97. Academic Press, Canberra, Australia.

Jeffery, S., Verheijen, F.G.A., Van Der Velde, M., Bastos, A.C., 2011. A quantitative review of the effects of biochar application to soils on crop productivity using meta-analysis. Agric. Ecosyst. Environ. 144 (1), 175–187.

Kammann, C., Glaser, B., Schmidt, H.P., 2016. Combining biochar and organic amendments. In: Shackley, S., Ruysschaert, G., Zwar, K., Glaser, B. (Eds.), Biochar in European Soils: Science and Practice. Routledge, London, pp. 136–164.

Larbi, A., Abadía, A., Abadía, J., Morales, F., 2006. Down co-regulation of light absorption, photochemistry, and carboxylation in Fe deficient plants growing in different environments. Photosynth. Res. 89, 113–126.

Latifah, O., Ahmed Haruna, O., Abdul Majid, N.M., 2018. Soil pH buffering capacity and nitrogen availability following compost application in a tropical acid soil. Compost Sci. Util. 26 (1), 1–15.

Lazzeri, L., Malaguti, L., Cinti, S., Ugolini, L., De Nicola, G.R., Bagatta, M., et al., 2013. The Brassicaceae biofumigation system for plant cultivation and defense. An Italian twenty-year experience of study and application. Acta Hortic. 1005, 331.

Lehmann, J., Joseph, S., 2009. Biochar for environmental management: an introduction. In: Lehmann, J., Joseph, S. (Eds.), Biochar for Environmental Management: Science and Technology. Earthscan, London, pp. 1–12.

Lehmann, J., Gaunt, J., Rondon, M., 2006. Bio-char sequestration in terrestrial ecosystems—a review. Mitig. Adapt. Strat. Gl. 11, 403–427.

Liu, J., Schulz, H., Brandl, S., Miehtke, H., Huwe, B., Glaser, B., 2012. Short-term effect of biochar and compost on soil fertility and water status of a dystric Cambisol in NE Germany under field conditions. J. Plant Nutr. Soil Sci. 175, 698–707.

Lucena, J.J., 2006. Synthetic iron chelates to correct iron deficiency in plants. In: Barton, L.L., Abadía, J. (Eds.), Iron Nutrition in Plants and Rhizospheric Microorganisms. Springer, Dordrecht, The Netherlands, pp. 103–128.

Marahatta, S.P., Wang, K.H., Sipes, S., Cerruti, R., Hooks, R., 2012. Effects of Tagetes patula on active and inactive stages of root-knot nematodes. J. Nematol. 44 (1), 26–30.

Mazzon, M., Cavani, L., Margon, A., Sorrenti, G., Ciavatta, C., Marzadori, C., 2018. Changes in soil phenol oxidase activities due to long-term application of compost and mineral N in a walnut orchard. Geoderma 316, 70–77.

Miller, G.W., Pushnik, J.C., Brown, J.C., Emery, T.E., Jolley, V.D., Warnick, K.Y., 1985. Uptake and translocation of iron from ferrated rhodotorulic acid in tomato. J. Plant Nutr. 8, 249–264.

Muhammad, S., Sanden, B.L., Lampinen, B.D., Saa, S., Siddiqui, M.I., Smart, D.R., et al., 2015. Seasonal changes in nutrient content and concentrations in a mature deciduous tree species: studies in almond (Prunus dulcis (mill.) DA Webb). Eur. J. Agron. 65, 52–68.

Namgay, T., Singh, B., Singh, B.P., 2010. Influence of biochar on the availability of As, Cd, Cu, Pb and Zn to maize (Zea mays L.). Aust. J. Soil Res. 48, 638–647.

Neilsen, G.H., Hogue, E.J., Forge, T., Neilsen, D., 2003. Mulches and biosolids affect vigor, yield and leaf nutrition of fertigated high density apple. HortScience 38 (1), 41–45.

Pestana, M., Correia, P.J., Varennes, A., Abadía, J., Faria, E.A., 2001. The use of floral analysis to diagnose the nutritional status of orange trees. J. Plant Nutr. 24 (12), 1913–1923.

Pestana, M., de Verennes, A., Araujo Faria, E., 2003. Diagnosis and correction of iron chlorosis in fruit trees: a review. Food Agric. Environ. 1 (1), 46–51.

Pfeiffer, D.G., Burts, E.C., 1983. Effect of tree fertilization on numbers and development of pear psylla (Homoptera: Psyllidae) and on fruit damage. Environ. Entomol. 12, 895–901.

Plich, H., Wójcik, P., 2002. The effect of calcium and boron foliar application on postharvest plum fruit quality. Acta Hortic. (594), 445–451.

Postma, J., Montanari, M., van den Boogert, P.H.J.F., 2003. Microbial enrichment to enhance the disease suppressive activity of compost. Eur. J. Soil Biol. 39 (3), 157–163.

Pradubsuk, S., Davenport, J.R., 2010. Seasonal uptake and partitioning of macronutrients in mature 'Concord' grape. J. Am. Soc. Hortic. Sci. 135, 474–483.

Prange, R.K., 2015. Fruit and vegetable quality as affected by the use of compost and other organic amendments. Acta Hortic. (1076), 127–136.

Rahman, M.M., Amano, T., Inoue, H., Matsumoto, Y., 2004. Nitrogen accumulation and recovery from legumes and N fertilizer in rice-based cropping systems. In: Fischer, T., Turner, N., Angus, J., McIntyre, L., Robertson, M., Borrell, A., Lloyd, D. (Eds.), Proceedings for the 4th International Crop Science Congress, vol. 26. Brisbane, Australia.

Roccuzzo, G., Zanotelli, D., Allegra, M., Giuffrida, A., Torrisi, B.F., Leonardi, A., et al., 2012. Assessing nutrient uptake by field-grown orange trees. Eur. J. Agron. 41, 73–80.

Röemheld, V., Nikolic, M., 2007. Iron. In: Barker, A.V., Pilbeam, D.J. (Eds.), Handbook of Plant Nutrition. CRC Press, Taylor & Francis Group, Boca Raton, FL, pp. 329–350.

Roversi, A., Monteforte, A., 2005. Preliminary results on the mineral uptake of six sweet cherry varieties. Acta Hortic. 721, 123–128.

Rufat, J., DeJong, T.M., 2001. Estimating seasonal nitrogen dynamics in peach trees in response to nitrogen availability. Tree Physiol. 21, 1133–1140.

Scandellari, F., Ventura, M., Malaguti, D., Ceccon, C., Menarbin, G., Tagliavini, M., 2010. Net primary productivity and partitioning of absorbed nutrients in field-grown apple trees. Acta Hortic. (868), 115–122.

Schliemann, G.K., Terblanche, J.H., De Koch, I.S., 1983. Preparation and cultivation of plum soil. Deciduous Fruit Grower 1, 20–27.

Schmidt, H.P., Kammann, C., Niggli, C., Evangelou, M.W.H., Mackie, K.A., Abiven, S., 2014. Biochar and biochar-compost as soil amendments to a vineyard soil: influences on plant growth, nutrient uptake, plant health and grape quality. Agric. Ecosyst. Environ. 191, 117–123.

Schreiner, R.P., Scagel, C.F., Baham, J., 2006. Nutrient uptake and distribution in a mature pinot noir vineyard. HortScience 41 (2), 336–345.

Schulz, H., Dunst, G., Glaser, B., 2013. Positive effects of composted biochar on plant growth and soil fertility. Agron. Sustain. Dev. 33 (4), 817–827.

Senesi, N., 1989. Composted materials as organic fertilizers. Sci. Total Environ. 81–82, 521–542.

Sequi, P., 1989. Chimica del Suolo (Soil Chemistry, in Italian). Patron, Bologna, Italy.

Singh, B.P., Hatton, B.J., Singh, B., Cowie, A., Kathuria, A., 2010. Influence of biochars on nitrous oxide emission and nitrogen leaching from two contrasting soils. J. Environ. Qual. 39, 1224–1235.

Smiciklas, K.D., Walker, P.M., Kelley, T.R., 2013. Evaluation of compost for use as a soil amendment in corn and soybean production. Compost Sci. Util. 16 (3), 183–191.

Sohi, S.P., Krull, E., Lopez-Capel, E., Bol, R., 2010. A review of biochar and its use and function in soil. Adv. Agron. 105, 47–82.

Sorrenti, G., Toselli, M., 2016. Soil leaching as affected by the amendment with biochar and compost. Agric. Ecosyst. Environ. 226, 56–64.

Sorrenti, G., Toselli, M., Marangoni, B., 2011. Effectiveness of *Amaranthus retroflexus* L. aqueous extract in preventing iron chlorosis of pear trees (*Pyrus communis* L.). Soil Sci. Plant Nutr. 57 (6), 813–822.

Sorrenti, G., Toselli, M., Marangoni, B., 2012. Use of compost to manage Fe nutrition of pear trees grown in calcareous soil. Sci. Hortic. 136, 87–94.

Sorrenti, G., Masiello, C.A., Toselli, M., 2016a. Biochar interferes with kiwifruit Fe-nutrition in calcareous soil. Geoderma 272, 10–19.

Sorrenti, G., Ventura, M., Toselli, M., 2016b. Effect of biochar on nutrient retention and nectarine tree performance: a three-year field trial. J. Plant Nutr. Soil Sci. 179 (3), 336–346.

Sorrenti, G., Buriani, G., Gaggia, F., Baffoni, L., Spinelli, F., Di Gioia, D., Toselli, M., 2017. Soil CO_2 emission partitioning, bacterial community profile and gene expression of *Nitrosomonas* spp. and *Nitrobacter* spp. of a sandy soil amended with biochar and compost. Appl. Soil Ecol. 112, 79–89.

Spokas, K.A., Cantrell, K.B., Novak, J.M., Archer, D.W., Ippolito, J.A., Collins, H.P., et al., 2012. Biochar: a synthesis of its agronomic impact beyond carbon sequestration. J. Environ. Qual. 41 (4), 973–989.

Stassen, P.J.C., North, M.S., 2005. Nutrient distribution and requirement of 'Forelle' pear trees on two rootstocks. Acta Hortic. (671), 493–500.

Tagliavini, M., Rombolà, A.D., 2001. Iron deficiency and chlorosis in orchard and vineyard ecosystems: a review. Eur. J. Agron. 15, 71–92.

Tagliavini, M., Scudellari, D., Marangoni, B., Toselli, M., 1996. Nitrogen fertilization management in orchards to reconcile productivity and environmental aspects. Fertil. Res. 43, 93–102.

Tagliavini, M., Abadìa, J., Rombolà, A.D., Tsipouridis, C., Marangoni, B., 2000. Agronomic means for the control of iron deficiency chlorosis in deciduous fruit trees. J. Plant Nutr. 23 (11&12), 2007–2022.

Tagliavini, M., Tonon, G., Scandellari, F., Quiñones, A., Palmieri, S., Menarbin, G., et al., 2007. Nutrient re cycling during the decomposition of apple leale (*Malus domestica*) and mowed grasses in an orchard. Agric. Ecosyst. Environ. 118, 191–200.

Toselli, M., Mazzanti, F., Marangoni, B., Tagliavini, M., Scudellari, D., 2002. Determination of leaf standards for mineral diagnosis in pear orchards in the Po Valley, Italy. Acta Hortic. (596), 665–669.

Valarini, P.J., Curaqueo, G., Seguel, A., Manzano, K., Rubio, R., Cornejo, P., et al., 2009. Effect of compost application on some properties of a volcanic soil from central South Chile. Chil. J. Agric. Res. 69, 416–425.

Ventura, M., Sorrenti, G., Panzacchi, P., George, E., Tonon, G., 2013. Biochar reduces short-term nitrate leaching from a horizon in an apple orchard. J. Environ. Qual. 42 (1), 76–82.

Verheijen, F., Jeffery, S., Bastos, A.C., Van der Velde, M., Diafas, I., 2010. Biochar Application to Soils. Institute for Environment and Sustainability, Luxembourg.

Viti, C., Tatti, E., Decorosi, F., Lista, E., Rea, E., Tullio, M., et al., 2010. Compost effect on plant growth-promoting rhizobacteria and mycorrhizal fungi population in maize cultivations. Compost Sci. Util. 18 (4), 273–281.

Wang, Y., Brown, H.N., Crowley, D.E., Szaniszlo, P.J., 1993. Evidence for direct utilization of a siderophore, ferrioxamine B, in axenically grown cucumber. Plant Cell Environ. 16 (5), 579–585.

Wentzell, A.M., Kliebenstein, D.J., 2008. Genotype, age, tissue, and environment regulate the structural outcome of glucosinolate activation. Plant Physiol. 147, 415–428.

Winter, C.K., Davis, S.F., 2006. Organic food. J. Food Sci. 71 (9), 117–124.

Zuo, Y., Zhang, F., 2011. Soil and crop management strategies to prevent iron deficiency in crops. Plant Soil 339 (1–2), 83–95.

28

Biodynamic soil fertility management in fruit crops

Matjaž Turinek*

University of Maribor, Maribor, Slovenia
*Corresponding author. E-mail: matjazturinek@gmail.com

1 Introduction

Biodynamic (BD) agriculture, as one of the sustainable agricultural systems, is a holistic approach toward farming. It was proposed by Rudolf Steiner in 1924 (Turinek et al., 2009), who emphasized that we should be perceiving the world around us as a complex system, where next to the material dimension there is also the invisible world of energies and the spiritual dimension. One of the core ideas is also the perception of the farm as an organism or individuality, where the development of the farmer and the farm as such through time is taken into consideration. For the farms to be independent and self-supporting, BD farmers strive to have as-closed-as-possible nutrient and energy cycles.

The main principles of modern organic farming, such as composting, green manures, closed nutrient, and life cycles, were taken after BD farming principles (Conford, 2001) and are at the core of BD farm management. BD farming also encompasses a broader idea of the farm placement in its surroundings, the involvement of the people connected to a farm, a balance between the subsystems or "organs" of the farm (arable crops, pastures, livestock, horticulture, etc.) and the elements of nature, such as forests, heaths, moors, and watercourses (Vereijken et al., 1997).

The methods and proposals developed from the agricultural course present the basis for most BD farmers. However, as emphasized in the agricultural course, it is important that each farmer develops his/her personal approach and understanding toward the farm, plants, animals, and broader natural setting in which the farm is placed and act upon that. Most of research done and literature published (scientific and professional) is done in Germany, since the German-speaking countries present the origin of BD farming and also have many of the farms and research institutions working on BD agriculture. However, BD agriculture is practiced on every farmable continent of the Earth and is thus practically applicable in every climatic, economic, cultural, and social environment. The nonprofit Demeter International association is responsible for coordinating and publishing BD standards, also known under the private trademark "Demeter." There are more than 5300 BD farms in 54 countries, whose area of approximately 187,000 ha is certified according to Demeter standards (Demeter, 2019). Most of that area lies in Germany (85,000 ha), as the development of BD agriculture began there and has the longest tradition, followed by France (13,600 ha) and Netherlands (6300 ha). Next to that, there are many farmers, who apply BD methods, but do not apply for Demeter certification. In Slovenia, for example, there are 39 certified Demeter farms, joined in the Slovene Demeter Association; however, over

A.K. Srivastava, Chengxiao Hu (eds.)
Fruit Crops: Diagnosis and Management of Nutrient Constraints
https://doi.org/10.1016/B978-0-12-818732-6.00028-9

2000 small gardeners, farmers, and beekeepers have joined the BD "Ajda" associations, which promote BD farming practices and organize lectures, workshops, and training for those interested in learning this way of farming.

2 Biodynamic approach to soil fertility

As mentioned before, a BD farm should ideally be composed of different parts, each one of them contributing its share to the whole farm organism. This is of great importance especially for soil fertility, as it depends on the functioning of the whole farm system. Thus, animal husbandry is a requirement for BD farms. There can be exemptions for specialized fruit, vegetable, or wine production farms. However, a cooperation or an agreement with a BD or organic animal production farm must be in place to ensure the input of organic fertilizers (in the form of animal manure), which, of course, must be composted with other organic materials from the farm and treated with BD compost preparations before it is used on the fields (Fig. 28.1). There are also an increasing number of BD fruit growers that incorporate animal husbandry in their orchards. Usually, small ruminants and poultry are the preferred entry point, whereas some farmers strive to create quite a diverse plant and animal ecosystem to improve soil fertility, plant nutrition, and control pests and diseases in their fruit crops. Another important measure is the use of diverse crop rotations (especially for field crops and vegetables). However, crop rotations are also strongly encouraged and promoted in fruit and vine production. In practice, this means that one alternates the cultivation of the soil between the rows of trees and seed diverse mixtures of different plant species from different families.

One of the measures to improve agriculture, presented in the agricultural course, are the BD preparations (Table 28.1, Fig. 28.2), which are being further developed and tested for the last 95 years. They are made from various medicinal plants, cow dung, and minerals using respective preparation procedures (von Wistinghausen et al., 2005, 2007). For a detailed description on preparing and using the preparations, the interested reader is kindly referred to the abundant scientific and professional literature published worldwide. The ideas behind the preparations are unconventional and sometimes difficult to understand, as they are to be used as homeopathic preparations for the soil and plants. For example, BD 500, which is made out of cow dung and buried over winter in the soil, is regarded as a soil fertility preparation that enhances soil biological functions through the stimulation of soil life. BD 501, made out of finely ground quartz or pure silica and buried in the soil over summer, is supposed to enhance resistance and plant resilience by supporting a physiological balance in a plant, making it more active and at the same time "aware" of its surroundings. Preparations BD 502–507 have a similar function in establishing well balanced and mature compost fertilizers for the soils, aiding to build long-term humus complexes and contributing to the long-term soil fertility. Effects of BD preparations are difficult to explain from the natural science mechanistic principle point of view. Attempts were made to explain it as a normalization (normalizing yields under low-yielding conditions) or compensation (BD preparations compensating for lower N fertilization) effect (Raupp and Konig, 1996), where there are many

FIG. 28.1 BD compost is essential for soil fertility improvement.

TABLE 28.1 BD preparations, main ingredients, type of use, and mentioned areas of influence.

Number of preparation	Main ingredient[a]	Use	Mentioned in the agricultural course in connection with
BD 500	Cow manure	Field spray	Soil biological activity
BD 501	Silica	Field spray	Plant resilience
BD 502	Yarrow flowers (*Achillea millefolium* L.)	Compost preparation	K and S processes
BD 503	Chamomile flowers (*Matricaria recutita* L.)	Compost preparation	Ca and K processes
BD 504	Stinging nettle shoots (*Urtica dioica* L.)	Compost preparation	N management
BD 505	Oak bark (*Quercus robur* L.)	Compost preparation	Ca processes
BD 506	Dandelion flowers (*Taraxacum officinale* Web.)	Compost preparation	Si management
BD 507	Valerian extract (*Valeriana officinalis* L.)	Field spray, compost preparation	P and warmth processes

[a] *The procedure of preparation and fermentation is in detail described in various sources (Steiner, 1924; von Wistinghausen et al., 2005, 2007). BD preparations are designed to be used together on a farm/farming system.*

FIG. 28.2 Storage of BD compost preparations in a specially designed box.

open questions remaining. A system response and adaptation model suggests a possible explanation, where the effects of BD preparations do not depend only on their properties and how they are applied. Properties of soils, plants, and environmental conditions and how they interact are supposed to be factors that to the greatest extent determine the effects of BD preparations (Raupp and Konig, 1996). It has to be noted that BD spray preparations are applied in small quantities of 4–160 g/ha, where physical or biological effects are less likely to occur (Reganold, 1995). However, BD preparations were also shown to have hormone-like effects (Goldstein et al., 2004).

One of the possible, although indirect, explanations on their mode of action was given by Montagnier et al. (2010), who have shown that some bacterial and viral DNA sequences induce low-frequency electromagnetic waves in high aqueous dilutions, which then invoke the "creation" of the same DNA sequences in a "clean" medium, although none of the original DNA sequence exists in the medium as such. In other words, this means that water acts as a transmitter of information, which is dissolved in it through electromagnetic waves and this information is then transmitted to the recipient of this aqueous dilution—in the case of BD preparations to the soil or plants—and is then responsible for the creation of substances in it (although there is little or none of the original preparation in the water solution). However, the information to be transferred from the BD preparations to the aqueous solution has to be qualitative, in balance, and adapted to soil/plant conditions—thus, the active preparation procedures with burying them in biologically active soils and/or fermentation are very important and have to be done with utmost care and craftsmanship.

Some regional differences in applications of BD preparations can be seen on BD farms reported in studies; for example, in Australia, only preparation BD 500 was applied 1–2 times each year (Ryan and Ash, 1999). Also, Nguyen and Haynes (1995) report only preparation BD 500 being used on a studied BD farm in New Zealand. However, the preparations were designed to be used together (Steiner, 1924) and only used together on a farm they can have the desired, wholesome effect in helping to create balance in soils and plants.

Another aspect often followed on BD farms is the influence of the moon and its rhythms on the processes in soils, growth and quality of plants, and the health and behavior of animals. Steiner (1924) mentioned this in the agricultural course; Spiess (1990a,b) scientifically researched the effects of lunar rhythms and concluded that several rhythms have an influence on growth, yield, and quality of little radish and rye. Thun (1994), however, found one single rhythm to be the most important one on a number of other plants. Kollerstrom and Staudenmaier (2001) argue that Spiess' experimental results (after a reanalysis) comply with the findings and recommendations of Thun, thus confirming a lunar influence on crop development and growth. It has to be added that this area of research is highly interesting but difficult to research and measure, for example, the influence of other astronomical bodies on crop growth and development. We cannot live without a certain astronomical body for a year and then see the influence this body had. Furthermore, with our current knowledge, it is also impossible to shield areas of planet Earth from just certain, not all, influences coming from the universe. An innovative and at the same time trustworthy idea would be needed to conduct scientific research in an interdisciplinary way. It must be added that Thun and her coworkers conducted field research on experimental fields for more than 50 years and an "astronomical planting calendar" is published each year, which is nowadays translated into over 40 languages worldwide. Some BD organizations also publish their own, locally adapted astronomical calendars.

3 Do BD preparations affect soil fertility?

Experimental results show effects of BD preparations not only on yields but also on some processes in compost piles and in the long-term in the soil. Carpenter-Boggs et al. (2000) report 3.4°C higher temperatures in BD compost piles throughout the active composting period, whereas Zaller (2007) measured no significant differences in BD and conventional (CON) compost piles. BD treated compost also respired carbon dioxide (CO_2) at a 10% lower rate, contained 65% more nitrate in the final samples (Carpenter-Boggs et al., 2000), and had a larger dehydrogenase enzyme activity to CO_2 production ratio. Carpenter-Boggs et al. (2000) argue that BD preparations caused these effects through their bioactive ingredients or by serving as microbial inoculants, which is in line with the aforementioned explanations. In addition, Rupela et al. (2003) found a substantial microbial population in BD preparations, where bacteria population ranged from 3.45 to 8.59 $\log_{10} g^{-1}$. Moreover, populations of fungi were found in the preparations 502 and 506 (5.30 and 4.26 $\log_{10} g^{-1}$, respectively). Authors also report that several bacteria and fungi strains found in BD preparations showed a potential for suppressing fungal plant pathogens (Rupela et al., 2003). This population could also be the reason for the significant difference in dehydrogenase, protease, and phosphatase activity between farming systems in the DOK (biodynamic, organic, and conventional agriculture long-term comparison) trial, where highest values were measured in the BD system (Maeder et al., 2002). Similarly, differences were found in another long-term trial (Raupp, 2001), whereas Zaller and Köpke (2004) found no differences between the BD and organic plots.

In addition, higher soil organic matter contents were found on BD plots in two separate field trials (Raupp, 2001; Maeder et al., 2002) and in two studies on farms in New Zealand and in the Netherlands after using BD farming practices for several years (Reganold et al., 1993; Droogers and Bouma, 1996). Microbial biomass nitrogen also differed significantly in the DOK trial and was highest in the BD system with 59% more than in the CON farmyard manure (FYM) system (Fließbach et al., 2007). Furthermore, the microbial biomass carbon was 35% higher in the BD system, compared with the CON-FYM system (Maeder et al., 2002; Oehl et al., 2004). In contrast, Zaller and Köpke (2004) report no differences between treatments in regard to microbial biomass carbon, with untreated FYM and FYM treated with BD preparations, respectively. In both cases, microbial biomass carbon was significantly higher on BD plots compared with the control plots (Zaller and Köpke, 2004), which can lead to the conclusion that FYM has an important effect on the soil microbial biomass buildup. Wada and Toyota (2007) discovered that FYM applications play a role in stabilizing soil biological functions, where microbial and fungal populations are more resilient and resistant against disinfection. Moreover, FYM can change soil nitrogen composition and contributes to higher rates of protein amino acids, which bind nitrogen in the soil (Scheller and Raupp, 2005). However, differences between treatments do not seem to depend solely on amino acid supply from manure. An altered amino acid metabolism in the soil also influences

FIG. 28.3 A mixed farming approach also in intensive orchards by integrating small ruminants into the farm organism.

soil amino acid composition and contents. Soils receiving FYM with BD preparations have a lower catabolism:anabolism ratio than soils receiving nonprepared FYM, which results also in a more intensive humification process. The explanation for the influence of BD preparations on anabolism has yet to be found (Scheller and Raupp, 2005). However, these results add to the importance of including or at least incorporating animal husbandry on biodynamic farms, as well as on intensive vegetable, wine, or fruit-growing farms (Fig. 28.3).

3.1 Importance of the plant-soil interaction

Moreover, when science makes progress in researching the interconnectedness of the Earth's living systems and their complexity, numerous of the ideas presented in the "agricultural course" can be perceived in a different way or at least give impulses for future research. In the last lecture of this "course," it was suggested that "...if you work the soil as just explained, then the plant will be ready to attract "things" in its wider surroundings. The plant can take benefit not just from the contents of the field, where it grows, but also of the contents of the soil in the neighboring pasture, if the plant needs it. The plant can also benefit from the soil in the neighboring forest, if it is made sensible in the described way..." (Steiner, 1924, p. 160). And indeed, with today's knowledge on the existence of extensive mycelial networks in soils, which have been shown to connect individual species, genera, and even families of plants (He et al., 2003), in connection with results indicating the improvement of arbuscular mycorrhizal fungi by BD preparations (Ryan and Ash, 1999; Maeder et al., 2002), we can better understand and confirm this more than 95 years old statement. Arbuscular mycorrhizal fungi are not only responsible for nutrient mobilization from soil minerals.

Frey-Klett et al. (2007) have reported that arbuscular mycorrhizal fungi and mycorrhiza helper bacteria play an important role in the fixation of atmospheric nitrogen and also protect plants against root pathogens. In addition, higher density of roots on plots treated with BD preparations were reported (Raupp, 2001). Mycorrhiza helper bacteria could be the possible reason for this effect, as they have been shown to stimulate lateral root formation and increase potential root-mycorrhiza interaction points (Frey-Klett et al., 2007). An active and ever-present interaction between the soil and plants is often mentioned in the "agricultural course." BD farming also advocates "feeding" (fertilizing) the soil, so the plant can feed/nourish itself from it. This is however sometimes hard to understand in all its complexity and scale. To add one piece of the puzzle, we came to some interesting results in trials on red beet (Bavec et al., 2010), where samples from BD and control plots had higher levels of malic acid. Rudrappa et al. (2008) present one of the possible reasons for this phenomena, which was observed in other studies as well. Namely, it was demonstrated that malic acid, selectively excreted through roots, signals beneficial rhizobacteria and encourages their interaction with plants. Beneficial soil bacteria have been found to confer immunity against a wide range of foliar diseases by activating plant defenses. Organic acids and phenolic compounds have also been found to participate in leveling out P deficiency by being excreted through plant roots (Badri and Vivanco, 2009). Moreover, it has been demonstrated that BD systems have an influence on microbial structure, enzyme activity, or amino acid metabolism (Turinek et al., 2009). Plant-microbe interactions and plant-soil interactions seem to play an important role in providing plants with nutrients and activating resilience against pests and diseases and are thus increasingly being researched, where as a consequence some beneficial constituents/compounds can also be found in food products (Badri and Vivanco, 2009). Reganold et al. (2010), for example, have found more than 200 unique strains of microorganisms in organic soils, as compared to only two in conventional soils for strawberry production.

3.2 Effects on perennial crops

Most of the research on perennial crops in BD farming has been done on vines, since BD wine grape production is increasingly attracting attention, as well-known and well-established wine producers are converting to BD production practices. Experimental results show an effect of BD practices on wine grape canopy and chemistry, where a more balanced canopy and wine composition was measured for BD vineyards and wines, respectively. No significant effects on soil fertility parameters were shown in the same 6 years on farm comparison trial between ORG and BD practices in an organic vineyard in California (Reeve et al., 2005). Probst et al. (2008), however, measured significant differences in soil fertility between CON and BD soils in vineyards with a long history of BD (since 1981) and CON cultivation in Germany. An influence of BD farming was measured on microbial communities in the soil, on plant structures, and on the developing crop (Morrison-Whittle et al., 2017). It was also shown that not only the use of but also the timing in plant growth stage and the kind of spray preparation (BD 500 or BD 501) have an influence on grape and consequently wine quality (Meissner, 2011), that is, overuse of the BD 500 preparation resulted in unbalanced and unripe, "green" wines.

Moreover, BD soils were found to contain 12 times as much earthworms and had significantly higher water infiltration rates, porosity, organic C, and soil respiration and lower bulk density and penetration resistance than CON farmed soils (Reganold, 1995). Pergola et al. (2017) report of higher amounts of CO_2 sequestered in BD apricot orchards, where also the energy balance is favorable compared with orchards managed in an integrated way. We also came to similar results comparing BD, organic, CON, and integrated production of various field crops (Bavec et al., 2012), where BD plots had 7–11 times lower energy consumption than CON plots. This is important also in the face of climatic changes and the role of agriculture in lowering CO_2 emissions or even acting as a sink for carbon in the soils.

There are some professional/research associations dealing with BD fruit production. The most organized and visible one is the International Working Group on Biodynamic Fruit Production (https://www.sektion-landwirtschaft.org/en/thematic-areas/fruit-production/), which meets twice a year to exchange experiences, research results, and insights into BD fruit production. The group was also part of a research project with a good cooperation between farmers, researchers, and advisors, dealing with the question of rootstocks, spacing and origin of fruit trees, and their influence on the quality of the fruit. The results are published online (http://www.louisbolk.org/downloads/2811.pdf) and give many interesting insights into the researched interactions. It has to be added that the work of this group is mostly focused on Europe, with emphasis on apple trees. Another group, dealing with the production of olives, was started in 2017. Therefore, regional and/or continental groups in other parts of the world are strongly encouraged and can bring all stakeholders a step forward in understanding the complexity of BD farming.

4 Summarized biodynamic recommendations and concluding remarks

A biodynamic farmer strives to help create balance on his farm, soils, fields, meadows, orchards, animals, and himself. As manifold demonstrated in nature and also in research presented in this chapter, a balanced organism is the basis for long-term stability and successful organic and biodynamic production. What is then the added value of the BD method? Certainly, results show better yields and healthier plants with the use of BD preparations—be it in compost, soils, or plants (Table 28.2). Another important aspect is the personal development of the farmer through time and space through detailed observation and reflection, also called the "Goethean phenomenological approach," since it was successfully practiced and introduced into natural sciences by Johann Wolfgang Goethe. Through dedicated observation of a phenomenon in the time span of several months or years one can, of course only with the subsequent reflection of the observed, eventually understand the phenomena itself. This approach may not be something new, but it is certainly something that is not done consciously and with discipline anymore; be it by farmers, researchers, teachers, or others. Raising awareness about and understanding for the processes in the world around us is another important aspect and recommendation for practical and research work, which is central in BD farming. Since only with the understanding of the specific conditions each farm is positioned in, one can choose and use the right

TABLE 28.2 Summarized effects of BD preparations found in research trials compared with organic practices.

Researched issue	Increase/decrease
Yields	↗
Produce quality	↗
Soil quality/fertility	↗
Compost quality	↗
Soil biodiversity	↗
Financial viability	↗
Problems with pests/diseases	↘
Work intensity (work hours used)	↗

measures at the right time in the right way at that specific farm. This brings us to another important aspect, already mentioned at the beginning of this chapter. The BD "method" is not a "one-size-fits-all" recipe. The beauty and advantage of it is its adaptability to local conditions all over the world. Therefore, the BD farming practice can also be regarded as a path of personal development for those engaged with it. However, one is not obliged or forced to take this path. The positive effects of the BD approach, mentioned in this chapter, are not conditioned with any personal development.

The interested reader is kindly referred to the numerous professional literatures on the topic of BD agriculture, where more details on the use and practice of BD practices can be found.

But what are some of the future research challenges we are faced with? What about the energy efficiency or ecological impact of BD production on a wider scale (production to consumption)? Do we need to include economic feasibility into our studies? Then, there are some more detailed questions regarding BD preparations. Does it make a difference if they are made on-farm or bought from a distant location? Does this affect the effectiveness of the preparations? What about research on farm animals? Moreover, is there a difference between BD prepared compost of animal and plant origin? How does this affect soil fertility and health? How do we need to change and adapt our soil fertilization and tillage systems to get balanced soils? Clearly, there is a need for more research on the effects and use of BD practices in perennial crops. However, how to approach this matter? Do we need to make more production systems comparison trials for that matter? If yes, how well defined are the systems to be compared? And what are the areas of interest to compare? Soil quality and long-term fertility are certainly of high interest, as they present the basis for healthy plants and high-quality produce. Also, the question of rootstocks, varieties, and management systems needs attention. As a continuation, food quality is still a highly discussed and debated area, which would also deserve more attention on this account.

A worldwide network of farmers, researchers, advisors, teachers, and others interested in BD farming, joined under Demeter International, the Section for Agriculture and the International BD Association, contributes toward naming and addressing questions from everyday practice. For it to work efficiently, however, every stakeholder needs to be motivated and dedicated to actively participate in the creative process.

References

Badri, D.V., Vivanco, J.M., 2009. Regulation and function of root exudates. Plant Cell Environ. 32 (6), 666–681. https://doi.org/10.1111/j.1365-3040.2009.01926.x.

Bavec, M., Turinek, M., Grobelnik-Mlakar, S., Slatnar, A., Bavec, F., 2010. Influence of industrial and alternative farming systems on contents of sugars, organic acids, total phenolic content, and the antioxidant activity of red beet (Beta vulgaris L. ssp. vulgaris Rote Kugel). J. Agric. Food Chem. 58 (22), 11825–11831. https://doi.org/10.1021/jf103085p.

Bavec, M., Narodoslawsky, M., Bavec, F., Turinek, M., 2012. Ecological impact of wheat and spelt production under industrial and alternative farming systems. Renew. Agric. Food Syst. 27 (3), 242–250. https://doi.org/10.1017/S1742170511000354.

Carpenter-Boggs, L., Reganold, J.P., Kennedy, A.C., 2000. Effects of biodynamic preparations on compost development. Biol. Agric. Hortic. 17 (4), 313–328.

Conford, P., 2001. The Origins of the Organic Movement. Floris Books, Edinburgh.

Demeter, International e.V, 2019. Demeter-International e. V.—A World-Wide Network. http://www.demeter.net/.

Droogers, P., Bouma, J., 1996. Biodynamic vs. conventional farming effects on soil structure expressed by simulated potential productivity. Soil Sci. Soc. Am. J. 60 (5), 1552–1558.

Fließbach, A., Oberholzer, H.-R., Gunst, L., Mäder, P., 2007. Soil organic matter and biological soil quality indicators after 21 years of organic and conventional farming. Agric. Ecosyst. Environ. 118 (1–4), 273–284. https://doi.org/10.1016/j.agee.2006.05.022.

Frey-Klett, P., Garbaye, J., Tarkka, M., 2007. The mycorrhiza helper bacteria revisited. New Phytol. 176 (1), 22–36. https://doi.org/10.1111/j.1469-8137.2007.02191.x.

Goldstein, W., Barber, W., Lynn, C.-B., Dalsoren, D., Koopmans, C., 2004. Comparisons of Conventional, Organic, and Biodynamic Methods. Michael Fields Agricultural Institute, East Troy. http://www.michaelfieldsaginst.org/education/comparison.pdf.

He, X.-H., Critchley, C., Bledsoe, C., 2003. Nitrogen transfer within and between plants through common mycorrhizal networks (CMNs). Crit. Rev. Plant Sci. 22 (December), 531–567. https://doi.org/10.1080/0735268031878237.

Kollerstrom, N., Staudenmaier, G., 2001. Evidence for lunar-sidereal rhythms in crop yield: a review. Biol. Agric. Hortic. 19 (3), 247–259.

Maeder, P., Fliessbach, A., Dubois, D., Gunst, L., Fried, P., Niggli, U., 2002. Soil fertility and biodiversity in organic farming. Science 296 (5573), 1694–1697. https://doi.org/10.1126/science.1071148.

Meissner, G., 2011. Grundbetrachtung der Rebe. In: Bio-Dynamische Landwirtschaft III. Lehr- und Forschungsgemeinschaft für bio-dynamische Lebensfelder, Edelschrott, pp. 61–72.

Montagnier, L., Aissa, J., Del Giudice, E., Lavallee, C., Tedeschi, A., Vitiello, G., 2010. DNA Waves and Water. ArXiv:1012.5166, December, http://arxiv.org/abs/1012.5166.

Morrison-Whittle, P., Lee, S.A., Goddard, M.R., 2017. Fungal communities are differentially affected by conventional and biodynamic agricultural management approaches in vineyard ecosystems. Agric. Ecosyst. Environ. 246 (August), 306–313. https://doi.org/10.1016/j.agee.2017.05.022.

Nguyen, M.L., Haynes, R.J., 1995. Energy and labour efficiency for three pairs of conventional and alternative mixed cropping (pasture-arable) farms in Canterbury, New Zealand. Agric. Ecosyst. Environ. 52 (2–3), 163–172. https://doi.org/10.1016/0167-8809(94)00538-P.

Oehl, F., Frossard, E., Fliessbach, A., Dubois, D., Oberson, A., 2004. Basal organic phosphorus mineralization in soils under different farming systems. Soil Biol. Biochem. 36 (4), 667–675. https://doi.org/10.1016/j.soilbio.2003.12.010.

Pergola, M., Persiani, A., Pastore, V., Palese, A.M., Arous, A., Celano, G., 2017. A comprehensive life cycle assessment (LCA) of three apricot orchard systems located in Metapontino area (southern Italy). J. Clean. Prod. 142 (January), 4059–4071. https://doi.org/10.1016/j.jclepro.2016.10.030.

Probst, B., Schüler, C., Joergensen, R., 2008. Vineyard soils under organic and conventional management—microbial biomass and activity indices and their relation to soil chemical properties. Biol. Fertil. Soils 44 (3), 443–450. https://doi.org/10.1007/s00374-007-0225-7.

Raupp, J., 2001. Research issues and outcomes of a long-term fertilization trial over two decades: a contribution to assessing long-term agronomic trials. Berichte Uber Landwirtschaft 79 (1), 71–93.

Raupp, J., Konig, U.J., 1996. Biodynamic preparations cause opposite yield effects depending upon yield levels. Biol. Agric. Hortic. 13 (2), 175–188.

Reeve, J.R., Carpenter-Boggs, L., Reganold, J.P., York, A.L., McGourty, G., McCloskey, L.P., 2005. Soil and winegrape quality in biodynamically and organically managed vineyards. Am. J. Enol. Vitic. 56 (4), 367–376.

Reganold, J.P., 1995. Soil quality and profitability of biodynamic and conventional farming systems: a review. Am. J. Altern. Agric. 10, 36–45.

Reganold, J.P., Palmer, A.S., Lockhart, J.C., Neil Macgregor, A., 1993. Soil quality and financial performance of biodynamic and conventional farms in New Zealand. Science 260 (5106), 344.

Reganold, J.P., Andrews, P.K., Reeve, J.R., Carpenter-Boggs, L., Schadt, C.W., Alldredge, J.R., Ross, C.F., Davies, N.M., Zhou, J., 2010. Fruit and soil quality of organic and conventional strawberry agroecosystems. PLoS One. 5(9) e12346. https://doi.org/10.1371/journal.pone.0012346.

Rudrappa, T., Czymmek, K.J., Paré, P.W., Bais, H.P., 2008. Root-secreted malic acid recruits beneficial soil bacteria. Plant Physiol. 148 (3), 1547–1556. https://doi.org/10.1104/pp.108.127613.

Rupela, O.P., Gopalakrishnan, S., Krajewski, M., Sriveni, M., 2003. A novel method for the identification and enumeration of microorganisms with potential for suppressing fungal plant pathogens. Biol. Fertil. Soils 39 (2), 131–134. https://doi.org/10.1007/s00374-003-0680-8.

Ryan, M., Ash, J., 1999. Effects of phosphorus and nitrogen on growth of pasture plants and VAM fungi in SE Australian soils with contrasting fertiliser histories (conventional and biodynamic). Agric. Ecosyst. Environ. 73 (1), 51–62. https://doi.org/10.1016/S0167-8809(99)00014-6.

Scheller, E., Raupp, J., 2005. Amino acid and soil organic matter content of topsoil in a long term trial with farmyard manure and mineral fertilizers. Biol. Agric. Hortic. 22 (4), 379–397.

Spiess, H., 1990a. Chronobiological investigations of crops grown under biodynamic management. 1. Experiments with seeding dates to ascertain the effects of lunar rhytms on the growth of winter Rye (Secale Cerale, cv Nomaro). Biol. Agric. Hortic. 7 (2), 165–178.

Spiess, H., 1990b. Chronobiological investigations of crops grown under biodynamic management. 2. Experiments with seeding dates to ascertain the effects of lunar rhytms on the growth of little Radisch (Raphanus Sativus, cv Parat). Biol. Agric. Hortic. 7 (2), 179–189.

Steiner, R., 1924. Geisteswissenschaftliche Grundlagen Zum Gedeihen Der Landwirtschaft: Landwirtschaftlicher Kurs. Rudolf Steiner Verlag, Dornach.

Thun, M., 1994. Hinweise Aus Der Konstellationsforschung Für Bauern, Weinbauern, Gärtner Und Kleingärtner, 8., Aufl. Biedenkopf, Germany, Aussaattage Thun-Verlag.

Turinek, M., Grobelnik-Mlakar, S., Bavec, M., Bavec, F., 2009. Biodynamic agriculture research progress and priorities. Renew. Agric. Food Syst. 24 (2), 146–154. https://doi.org/10.1017/S174217050900252X.

Vereijken, J.F.H.M., van Gelder, T., Baars, T., 1997. Nature and landscape development on organic farms. Agric. Ecosyst. Environ. 63 (2–3), 201–220. https://doi.org/10.1016/S0167-8809(97)00013-3.

von Wistinghausen, C., Scheibe, W., Heilmann, H., Wistinghausen, E., König, U.J., 2005. Anleitung Zur Anwendung Der Biologisch-Dynamischen Feldspritzpräparate Und Düngerpräparate, third ed. Lebendige Erde, Darmstadt.

von Wistinghausen, C., Scheibe, W., Wistinghausen, E., König, U.J., 2007. Anleitung Zur Herstellung Der Biologisch-Dynamischen Präparate, fourth ed. Lebendige Erde, Darmstadt.

Wada, S., Toyota, K., 2007. Repeated applications of farmyard manure enhance resistance and resilience of soil biological functions against soil disinfection. Biol. Fertil. Soils 43 (3), 349–356. https://doi.org/10.1007/s00374-006-0116-3.

Zaller, J.G., 2007. Seed germination of the weed Rumex obtusifolius after on-farm conventional, biodynamic and vermicomposting of cattle manure. Ann. Appl. Biol. 151 (2), 245–249. https://doi.org/10.1111/j.1744-7348.2007.00172.x.

Zaller, J.G., Köpke, U., 2004. Effects of traditional and biodynamic farmyard manure amendment on yields, soil chemical, biochemical and biological properties in a long-term field experiment. Biol. Fertil. Soils 40 (4), 222–229. https://doi.org/10.1007/s00374-004-0772-0.

29

Manipulating fruit quality through foliar nutrition

Vasileios Ziogas[a], Michail Michailidis[b], Evangelos Karagiannis[b],
Georgia Tanou[c], Athanassios Molassiotis[b],*

[a]Institute of Olive Tree, Subtropical Plants and Viticulture, Hellenic Agricultural Organization (H.A.O.)—Demeter, Chania, Greece
[b]Laboratory of Pomology, Department of Agriculture, Aristotle University of Thessaloniki, Thessaloniki, Greece
[c]Institute of Soil and Water Resources, ELGO-DEMETER, Thessaloniki, Greece
*Corresponding author. E-mail: amolasio@agro.auth.gr

1 Introduction

Due to life-supporting role of fruits in human diet, the study of fruit quality undergoes a great burst of knowledge (Fotopoulos et al., 2010). Fruit quality has been associated with various factors like genetics, microclimate conditions, cultivation practices, plant nutrition, proper pollination, harvesting method, and maturation index at harvest (Coêlho de Lima and Alves, 2011). Schreiner et al. (2013) described quality as "the sum of characteristics, properties, and attributes of a product or commodity which aims in fulfilling the established or presumed customer requirements" (by the International Organization of Standardization, ISO 8402, 1989). The external fruit parameters include color, shape, size, and lack of defects, while the internal are related with taste, texture, aroma, nutritional value, sweetness, acidity, and postharvest shelf life (Shewfelt, 1999; Kingston, 2010). Attributes like skin color, titratable acidity, and ratio of total solid content to titratable acidity are being considered important fruit quality factors at harvest time (Oosthuyse and Westcott, 2005). Meanwhile, total soluble solids, sugar content, firmness, shape, and external fruit appearance are closely related to fruit nutrient levels. The relation between the adequate nutritional status of the fruit tree and the overall fruit quality has been well documented (Dris et al., 1999). Several nutrients like nitrogen (N), phosphorus (P), potassium (K), calcium (Ca), and magnesium (Mg) exert a notable influence upon several fruit quality parameters (Fallahi and Simons, 1996). The sufficient supply of macro- and micronutrients can be achieved via the use of soil surface application of fertilizers, fertigation, or even via foliar spray application (Nicola et al., 2009). Particularly, foliar

A.K. Srivastava, Chengxiao Hu (eds.)
Fruit Crops: Diagnosis and Management of Nutrient Constraints
https://doi.org/10.1016/B978-0-12-818732-6.00029-0

application has been characterized as the most appropriate method of nutrient application in fruit trees, providing rapid plant response and uniform distribution upon the foliage (Umar et al., 1999; Mengel, 2002). The most common macronutrients applied as foliar fertilizers are N (as urea, ammonium nitrate, and ammonium sulfate), P (as H_3PO_4, KH_2PO_4, $NH_4H_2PO_4$, $Ca(H_2PO_4)_2$, and phosphites), K (as K_2SO_4, KCl, KNO_3, K_2CO_3, and KH_2PO_4), Mg (as $MgSO_4$, $MgCl_2$, $Mg(NO_3)_2$), and Ca (as $CaCl_2$, Ca propionate, and Ca acetate). Also, the most commonly foliar applied micronutrients belong the B (as boric acid ($B(OH)_3$), borax ($Na_2B_4O_7$), Na octaborate ($Na_2B_8O_{13}$), B-polyols, Fe (as $FeSO_4$, Fe(III) chelates, and Fe complexes), Mn (as $MnSO_4$, Mn(II) chelates), and Zn (as $ZnSO_4$, Zn(II)-chelates, ZnO, and Zn organic complexes) (Tanou et al., 2017).

In addition to this, foliar application has the potential to provide a higher status of nutrient bioavailability compared with soil application (Li et al., 2018). Micronutrient absorption has been proven of being a fruit species-dependent factor related with the thickness and the composition of the plant cuticle (Eichert and Goldbach, 2008). Great body of experimental data has focused upon the beneficial effect of foliar application of nutrients toward fruit yield and the improvement of the overall fruit quality (Eichert and Goldbach, 2008; Pacheco et al., 2014). This chapter outlines up-to-date progress in terms of the regulation of nutrient-mediated fruit quality with an emphasis upon foliar nutrition.

2 Nitrogen (N)

Nitrogen is an essential ingredient for plant survival, as it participates in the formation of amino acids, and possesses a multiple principal role in plant metabolism (proteins, enzymes, storage compounds of nitrogen, etc.). Exogenous supply of N to trees via foliar spray application, to improve the fruit quality and yield, is mainly achieved by the use of urea and in combination with potassium, in the form of potassium nitrate (KNO_3). Although N is an important element for plant nutrition, current knowledge is not able to fully support its role in fruit quality when applied in the form of foliage-sprayed nitrogen.

Nitrogen prolonged the vegetative phase at the expense of ripening. In mango, N foliar spray prolonged green coloring of mature fruits reducing fruit quality (Nguyen et al., 2004). Evidence suggested that the increasing rates of foliar applied urea induced the soluble protein and ascorbic acid, total amino acids, titratable acidity, and alcohol acyltransferase activity, whereas it decreased soluble sugars in apples (Zhao et al., 2013) and in mandarins (Al-Obeed et al., 2018). Application of nitrogen via the foliar spray of KNO_3 resulted also in the increase of titratable acidity in pomegranate (Pulla Reddy et al., 2011). In regard to productive parameters, urea foliar spray increased fruit size and yield of apples, pears, and mandarins (Dong et al., 2005; Sánchez et al., 2008; Curetti et al., 2013; Al-Obeed et al., 2018), but when it was applied postharvestly at apple trees, it didn't affect the next year's yield (Wojcik, 2006a). Previous data support the beneficial effects of applied foliage N in almond nuts on fruit weight, width, percentage of hard shells, protein content, and percentage of green shells (Bybordi and Malakouti, 2006).

3 Phosphorus (P)

Phosphorus is a macronutrient that regulates a variety of cellular functions due to its participation to the ATP molecular structure. Foliar application of P at sweet persimmon increased fruit yield and decreased fruit dropping. On the other hand, in guava fruits, P did not affect fruit production, but it has been associated with disease control (Natale et al., 2002). Foliar P application apart from alleviating P deficiency may also contribute to the overall fruit quality at harvest and during postharvest period since a positive impact on fruit hardness and sugar content with parallel acidity decrease has been reported at sweet persimmon (Hossain and Ryu, 2009). The combined P, Fe, and Zn foliar application has been reported to exert significant influence on the morphophysical and qualitative parameters of guava fruits as it increased volume, weight, yield, sugars, acidity, ascorbic acid, and pectin content (Kanpure et al., 2016).

4 Potassium (K)

Potassium is an essential macronutrient and exerts an important role in plant physiology since it provides the correct ionic environment for metabolic processes in plant cells (Leigh and Wyn Jones, 1984). There are several evidences showing that foliar K application has a positive impact in fruit quality and yield.

Citrus, apples, pears, peach, plum, mango, and date palm trees sprayed with K exhibited a significant increase in yield (Abdi and Hedayat, 2010; Ashraf et al., 2013; Ben Mimoun and Marchand, 2013; Baiea and Moneim, 2015;

Jawandha et al., 2017; Dbara et al., 2018) sometimes due to decreased fruit drop phenomenon after flowering (Baiea and Moneim, 2015; Oosthuyse, 2015). In olive trees, a notable effect in yield increase was observed only after 5 years of systemic K application (Ben Mimoun and Marchand, 2013). In terms of fruit quality parameters, many reports support the beneficial role of K foliar application as it increased fruit size/weight of orange (Hafez et al., 2017), pear (Gill et al., 2012), peach (Dbara et al., 2018), plum (Jawandha et al., 2017), and mango (Baiea and Moneim, 2015); fruit firmness in apples (Solhjoo et al., 2017), pear (Gill et al., 2012), but not in peach (Dbara et al., 2018); the coloring percentage in orange (Hafez et al., 2017) and pear (Gill et al., 2012); the content of total soluble solids in orange (Hafez et al., 2017), apple (Javaid et al., 2016), pear (Gill et al., 2012), peach (Ghanem and Ben Mimoun, 2010b; Dbara et al., 2018), plum (Ghanem and Ben Mimoun, 2010a), and mango (Baiea and Moneim, 2015); the ascorbic acid in orange (Hafez et al., 2017) and mango (Baiea and Moneim, 2015), and also olive soluble carbohydrates and anthocyanins (Zivdar et al., 2016). On the contrary, acidity was reduced at orange (Hafez et al., 2017), peaches (Ghanem and Ben Mimoun, 2010b), and plum (Jawandha et al., 2017) or remained unaffected at pear (Gill et al., 2012) and peach (Dbara et al., 2018). It has been suggested that potassium may exert a beneficial effect toward physiological disorders as its foliar application has been shown to reduce fruit cracking and improve fruit quality of lemons (Devi et al., 2018), whereas oranges did not demonstrate any significant effect on fruit quality parameters (Ramezanian et al., 2018).

5 Calcium (Ca)

Calcium is an essential macroelement for plant growth, and its absence, in certain content, leads to plant and cell malfunctions and eventually to cell death. On-tree Ca foliar spray is an effective way to increase Ca content in leaves and fruits. Calcium seems to be associated with fruit quality as reports indicate a beneficial impact on flesh firmness, accumulation of bioactive compounds, and prevention of physiological disorders, which are related to cell-wall consistency. Quality attributes, such as soluble solid content, acidity, mean weight, weight loss during storage, respiration, and ethylene production, undergo a controversial impact due to Ca foliar application, and this can be observed under several species of fruit trees.

Several reports demonstrate that Ca foliar spray increases fruit flesh firmness at harvest and during postharvest period in a variety of fruits, including peaches (Alcaraz-López et al., 2004; Val and Fernández, 2011; Ekinci, 2018), apples (Wojcik, 2002; Kadir, 2005; Domagala-Światkiewicz and Blaszczyk, 2009; Wójcik et al., 2016; Ghorbani et al., 2017; Solhjoo et al., 2017), oranges (Ramezanian et al., 2018), strawberries (Wójcik and Lewandowski, 2003; Singh et al., 2007a,b, 2009; Bieniasz et al., 2012), plums (Plich and Wójcik, 2002; Alcaraz-Lopez et al., 2004), sweet cherries (Brown et al., 1996; Tsantili et al., 2007; Michailidis et al., 2017) kiwifruits (Shiri et al., 2016), papaya (Madani et al., 2015) olives (Tsantili et al., 2008), guava (Goutam et al., 2010), and pears (Frances et al., 1999). No significant shifts to fruit firmness were observed due to Ca foliar application, for some species like apples (Brown et al., 1998; Wojcik and Szwonek, 2002; Val et al., 2008), peaches (Crisosto et al., 2000), and strawberries (Toivonen and Stan, 2001), while a decrease of fruit firmness was observed for peaches (Val et al., 2010). It has been suggested that fruits treated with Ca solution at preharvest time exhibited a significant accumulation of pectin (Ekinci, 2018).

On-tree Ca spray of fruits alleviated physiological disorders; skin cracking in sweet cherries (Brown et al., 1996; Michailidis et al., 2017), lemons (Devi et al., 2018), and pomegranates (Bakeer, 2016; Davarpanah et al., 2018); and scald incidents in apples (Kadir, 2005). During postharvest period, Ca foliar application alleviated the internal browning in peaches (Val and Fernández, 2011) and apples (Wojcik, 2002), the sensitivity of apples to bitter pit (Wojcik, 2002; Wojcik and Szwonek, 2002; Lötze and Theron, 2006), and the severity of chilling injury in mandarins (D'Aquino et al., 2005).

Application of Ca via the foliage is considered a potent method to increase total phenolic content of fruits at harvest and during postharvest storage for sweet cherries (Michailidis et al., 2017), oranges (Ramezanian et al., 2018), and strawberries (Xu et al., 2014). Furthermore, an increase in ascorbic acid and anthocyanin content in strawberries (Singh et al., 2007a,b, 2009; Xu et al., 2014), in ascorbic acid in guava fruits (Goutam et al., 2010), and in carotenoid content in kiwifruits (Shiri et al., 2016) was observed.

The soluble solid concentration (°Brix) at harvest and during postharvest period was decreased following Ca foliar spray in peaches (Val an d Fernández, 2011), apples (Wójcik et al., 2016; Ghorbani et al., 2017), strawberries (Singh et al., 2007a,b, 2009), and papaya fruits (Mirshekari and Madani, 2018); by contrast, it was increased in strawberries (Wójcik and Lewandowski, 2003), apples (Solhjoo et al., 2017), kiwifruits (Shiri et al., 2016), mango (El-Razek et al., 2017), jujube (Mao et al., 2016), and guava (Goutam et al., 2010). In addition to this observation, an increase in fruit acidity was observed in peaches (Val and Fernández, 2011), strawberries (Wójcik and Lewandowski, 2003; Singh et al.,

2007a,b, 2009), papaya fruits (Mirshekari and Madani, 2018), and apples (Domagala-Światkiewicz and Blaszczyk, 2009) in Ca-sprayed fruits at harvest and after cold storage.

It has been reported that foliar spray of Ca decreases weight loss during postharvest storage in peaches (Ekinci, 2018) and kiwifruits (Shiri et al., 2016). Moreover, Ca application resulted in an increase of fruit weight during harvest in apples (Kadir, 2005; Solhjoo et al., 2017) and mango (El-Razek et al., 2017). Fruit skin redness improved in apples as a result of Ca applications (Kadir, 2005), while the respiration rate and ethylene production were decreased in papaya fruits (Mirshekari and Madani, 2018).

6 Magnesium (Mg)

Macroelement magnesium (Mg) has a multifunctional role in plants, but its function in fruit quality has not been fully elucidated. Reports indicated that foliar spray of Mg had no impact on firmness and color index of peaches and plums (Alcaraz-Lopez et al., 2003; Alcaraz-López et al., 2004), while a lower fruit juice pH was reported with no parallel effect upon the acidity of peaches (Ekinci, 2018) at harvest and during postharvest period.

Recent studies demonstrate that Mg has a positive function upon litchi pericarp color index. At molecular level, these reports indicate an increase of anthocyanin content in the pericarp that contributes to the red color of the pericarp and boosted the overcome of the stay-green phenomenon. Moreover, Mg foliar spray is linked with several transcripts that are associated with flavonoid biosynthesis, anthocyanin biosynthesis, and the ABA signaling pathway. Magnesium could contribute to the accumulation of anthocyanin content via the increase of ABA concentration in fruits (Wang et al., 2017a,b).

7 Sulfur (S)

The effect of foliar sulfur (S) application in fruit quality is difficult to study, since S exists in its anionic form ($^{-2}SO_4$) and participates as a stabilizer of cationic elements, such as magnesium ($MgSO_4$), manganese ($MnSO_4$, (Ekinci, 2018)), iron (Fe_2SO_4, (Song et al., 2016)), calcium, and nitrogen. Elemental sulfur with proven fungicidal activity could be used to clarify the effect of sulfur in fruit quality. Indeed, foliar sprays with S increased pecan nut weight under perennial open field experiments (Wells et al., 2014), thus amplifying the need for a deeper understanding of S-controlled metabolic processes of fruit at harvest and postharvest period.

8 Boron (B)

Boron (B) has long being considered an essential nutrient for higher plants (Marschner, 1995), since it has been associated with reproductive processes affecting flower development, pollen germination, and pollen tube elongation (Loomis and Durst, 1992). B deficiency is associated with malformed fruits, which reduces overall fruit yield and negatively affects fruit quality (Sharma et al., 2004). Cracking, shriveling, deformation, internal and external browning, and corking near the pit or into the flesh are notable signs of B deficiency (Wojcik and Wojcik, 2006). Application of boron via the soil may cause phytotoxicity, since the limit between deficiency and toxicity is rather narrow in many fruit crops (Gupta, 1979). Therefore, it has been proposed that foliar application is more effective than soil application (Singh et al., 2007a,b).

Several studies have demonstrated that B foliar application could have a positive outcome to the overall fruit yield of strawberry (Singh et al., 2007a,b), olive (Larbi et al., 2011), raspberry (Wojcik, 2005), and pistachio (Acar et al., 2016). However, B application in sweet cherry (Wojcik and Wojcik, 2006) and hazelnut (Silva et al., 2003) did not have any significant physiological effect. The foliar application of B has a variable effect in total fruit yield, and there is no clear rule between the applied concentration and yield response (Sotiropoulos et al., 2006).

Foliar application of B could contribute to the qualitative characteristic of fruit firmness. Studies indicate that there was a positive correlation of B foliar spray and fruit firmness in raspberry (Wojcik, 2005) and pear (Wojcik and Wojcik, 2003), while B spray in apples (Peryea and Drake, 1991; de SÁ et al., 2014) had no effect. It has been proposed that total firmness is a marker of fruit ripeness, and foliar application of B sustained the firmness of the applied fruit (Wojcik and Wojcik, 2003).

Fruit quality traits are linked and determined by the content of various attributes like vitamins, phenols, anthocyanins, starch, and the antioxidant capacity of the fruit extract (Molassiotis et al., 2013). Foliar application of B increased the concentration of vitamin C in mandarins (Al-Obeed et al., 2018). Also, the foliar application of B increases the anthocyanin content of sweet cherry fruits (Wojcik and Wojcik, 2006), while when applied upon pomegranate plants the fruit concentration of anthocyanins remained unaffected (Davarpanah et al., 2016). Furthermore, foliar spray of B in pomegranate plants did not exhibit any significant difference in the concentration of total phenols and total antioxidant activity of fruit extracts (Davarpanah et al., 2016). Moreover, external B applied in apple trees (de SÁ et al., 2014) reduced the starch content of fruits. A similar trend of reduction was also observed in olive fruits (Saadati et al., 2013), where the foliar application of B also resulted in a reduction of soluble carbohydrates.

The foliar B application increased total soluble solid content in sweet cherry (Wojcik and Wojcik, 2006), raspberry (Wojcik, 2005), pomegranate (Davarpanah et al., 2016), and mandarins (Al-Obeed et al., 2018), while the TSS content remained unaffected in apples (Peryea and Drake, 1991; de SÁ et al., 2014) and strawberry (Wójcik and Lewandowski, 2003). Several reports proposed that foliar B treatment could increase the titratable acidity of pear fruits (Wojcik and Wojcik, 2003) but could provoke a decrease in apples (de SÁ et al., 2014) and pomegranate fruits (Davarpanah et al., 2016) while had no significant effect concerning this attribute in sweet cherry (Wojcik and Wojcik, 2006), apples (Peryea and Drake, 1991), tart cherries (Wojcik, 2006a,b), and strawberry (Wójcik and Lewandowski, 2003). The foliar application of B was positively correlated with the increase of total sugars in mandarin fruits (Al-Obeed et al., 2018), but not in pomegranate (Davarpanah et al., 2016).

It has been shown that B element is linked with internal fruit characteristics, like fruit malformations, cracking, split nut ratio, internal browning, length and diameter values, internal tissue structure, and dietary fiber content. In this regard, foliar application of B exhibited an effect toward these quality parameters through a significant increase in B concentration into the core and inner cortex of apples (Peryea and Drake, 1991) and increase in volume, length, and diameter when applied to mandarins (Al-Obeed et al., 2018). Also, the foliar application of B could increase the mean nut mass and kernel mass of hazelnuts (Silva et al., 2003) and decrease internal cracking events in mango fruits (Saran and Kumar, 2011), decrease internal membrane permeability in pear (Wojcik and Wojcik, 2003), decrease of the split nut ratio and split kernel ratio in pistachio fruits (Acar et al., 2016), and reduce the appearance of fruit malformation in strawberry fruits (Singh et al., 2007a,b). However, foliar application of B in strawberry fruits could not influence albinism appearance or the appearance of gray mold (Singh et al., 2007a,b). Also, foliar application of B was not effective toward the sensitivity to fruit cracking in cherry fruits (Wojcik and Wojcik, 2006).

A body of scientific data indicate that B foliar application could pose a positive effect toward the oil content from olive fruits and increase the ratio of monounsaturated fatty acids to saturated ones during olive oil extraction (Saadati et al., 2013). Moreover, foliar B application did not have any effect to the fat content in avocado fruit (Abdel-Karim et al., 2015).

Scientific data also pinpoint the fact that B foliar application was able to decrease the activity and expression level of certain enzymes related to the postharvest shelf life. In detail, the activity and expression level of polygalacturonase, pectinesterase, and β-galactosidase were decreased in B-treated orange fruits (Dong et al., 2009).

9 Cobalt (Co)—copper (Cu)—iron (Fe)

Cobalt (Co) application plays a regulatory role in plant metabolic cascades. Application of Co significantly inhibited ethylene biosynthesis in plants (Yang and Hoffman, 1984). The alleviated floral malformation syndromes were attributed to the ability of Co to exert an inhibitory effect in ethylene biosynthesis (Singh et al., 1991) via the prevention of the oxidation of 1-aminocyclopropane-1-carboxylic acid to ethylene (Singh et al., 1994). In the work of Singh et al. (1994), Co foliar feeding prior to flower bud differentiation was evaluated for its effect in mango fruit quality. Data indicated that total and nonreducing sugars were adversely affected by the foliar application of high levels of Co; however, total soluble solids or reducing sugars were not affected.

Scientific data highlight the role of copper (Cu) toward disease resistance due to its ability to participate in enzymatic activities, reactive oxygen species (ROS) production, regulation of gene expression, and processes linked with the biosynthesis of pathogen-related proteins and phytoalexins (Evans and Solberg, 2007). The effect of Cu foliar spray in cherry and apple fruit quality was studied by Brown et al. (1996). Post bloom application of Cu, in the form of copper hydroxide, was applied to sweet cherry trees after 6 weeks of flowering and in apple trees following 8 weeks of flowering in the presence or in the absence of Ca treatments. Results indicated that only when Cu was applied in tandem with calcium the formula was able to improve fruit resistance to cracking and to increase fruit firmness. The combination of Cu and Ca was also able to improve the flesh firmness of "Golden Delicious" apples (Brown et al., 1996) and minimize significantly the rain-induced fruit cracking in sweet cherry (Brown et al., 1995).

In fruit tree cultivation, iron (Fe) is an indispensable cation required for mineral nutrition (Chouliaras et al., 2004). Iron deficiency consists a major nutritional problem when trees are cultivated to calcareous or alkaline soils, affecting fruit quality parameters and overall yield (Chouliaras et al., 2004). Foliar spray of compounds containing Fe favorably improved fruit quality in several tree cultivations (Song et al., 2016). Foliar application of compounds containing Fe, under calcareous soils, could favor the fruit yield, the Brix index, and the titratable acidity in pomegranate fruits (Mirzapour and Khoshgoftarmanesh, 2013). Also, the beneficial effect of Fe-containing compounds in foliar spray was studied in the work of Song et al. (2016), where foliar spray in peach fruits increased fruit firmness, total soluble solid content, and Fe tissue concentration and enhanced the activity of succinate dehydrogenase (SDH) and aconitase (ACO).

10 Zinc (Zn)

Zinc (Zn) in plant nutrition is considered a microelement that generally linked with important roles in fruit set and retention, fruit yield, and fruit quality. Notably, Zn is essential for the activation of different enzymes, including dehydrogenases, aldolases, isomerases, transphosphorylases, and DNA and RNA polymerases and to its implication to the biosynthesis of tryptophan, cellular division, maintenance of cell membrane integrity, and photosynthesis (Marschner, 2012).

In response to Zn foliar application, the overall fruit yield was increased in pomegranate (Davarpanah et al., 2016) while remained unaffected in mango fruits (Bahadur et al., 1998). Specific physical parameters of fruits were also influenced after foliar application of Zn solutions in several types of fruit trees. In this sense, Zn foliar applications provoked a significant increase of fruit weight, pulp weight, fruit juice, fruit volume, and circumference/diameter ratio in mandarin fruits (Ashraf et al., 2013; Al-Obeed et al., 2018) while remained unaffected in pomegranate (Davarpanah et al., 2016) and mango fruits (Bahadur et al., 1998).

Zinc foliar treatments altered the quality characteristics of the fruit related with acidity and soluble solids. In detail, foliar sprays with Zn increased the content of total soluble solids and titratable acidity in pomegranate (Davarpanah et al., 2016), mango (Bahadur et al., 1998), and mandarin fruits (Ashraf et al., 2013; Al-Obeed et al., 2018), while these ripening features remained unaffected in apples (Rasouli and Koushesh Saba, 2018). Foliar Zn feeding increased total sugars in mandarins (Al-Obeed et al., 2018) while reduced the levels of starch in apples (Rasouli and Koushesh Saba, 2018) and the soluble carbohydrates in olive fruits (Saadati et al., 2013). However, no effect of Zn in sugars, total phenols, and total anthocyanins was observed in pomegranate (Davarpanah et al., 2016); in addition, total phenolic content in apple (Rasouli and Koushesh Saba, 2018) and sugar in mango fruits (Bahadur et al., 1998) were unaffected under Zn exposure.

The ability of Zn foliar application to interfere with antioxidative fruit parameters was evaluated by several groups. Foliar spray of Zn-containing solutions increased vitamin C in apple fruits (Rasouli and Koushesh Saba, 2018) and mandarins (Ashraf et al., 2013; Al-Obeed et al., 2018), while the antioxidative activity of the fruit extract was increased in apples with parallel increase of the enzymatic activity of superoxide dismutase (SOD) (Rasouli and Koushesh Saba, 2018). This effect could be attributed to the role of Zn as a constituent of SOD enzyme, thus facilitating its inhibitory effect upon free radical generation (Cackam and Marschner, 1988). Foliar sprays of Zn failed to provoke a significant increase to the antioxidant activity of fruit extracts from pomegranate (Davarpanah et al., 2016) and decreased enzymatic activity of polyphenol oxidase in apples, resulting to the decreased levels of enzymatic browning (Rasouli and Koushesh Saba, 2018).

Furthermore, foliar spray of Zn in olive fruits increased the oil content of the drupe and favored the ratio of monounsaturated fatty acids to the saturated ones (Saadati et al., 2013). This increase was estimated that could be attributed to the continued activity of the triglyceride-forming biosynthesis pathway until the fruit reaches full maturation (Dag et al., 2011). Also, Zn foliar spray increased the maturation index of apples (Rasouli and Koushesh Saba, 2018) and the pH of fruit juice extracted from pomegranate (Davarpanah et al., 2016) and mandarins (Al-Obeed et al., 2018).

11 Manganese (Mn)—nickel (Ni)—selenium (Se)

Scientific data related with the impact of manganese (Mn) foliar application on various quality parameters from fruit species are rather limited. In the work of Ekinci (2018), Mn foliar application in peach trees increased polygalacturonic acid (pectin) content followed by a decrease of fruit weight loss. Additionally, foliar spray of Mn in citrus plants (grape fruits and oranges) had no significant effect to the overall fruit yield, fruit number, or the average fruit weight (Swietlik and LaDuke, 1991).

Experimental data have highlighted the fact that Ni is part of the active center of urease (Dixon et al., 1975), and the activation of urease is the most important attributed function of Ni in the plant metabolism (Römheld, 2011). Scientific reports indicate the positive effect of Ni application in plant growth, fruit yield, and seed germination (Jamali et al., 2013). In the work of Eshghi and Ranjbar (2015), foliar application of Ni-containing compounds in strawberries increased the crown number at the studied fruits.

Selenium (Se) is not an essential element for plants, although it can benefit their growth and survival in some environments (Kápolna et al., 2009; Pacheco et al., 2014). The concentration of Se in plants mainly depends upon the vegetal species and the form that Se is supplied to the plant (Kápolna et al., 2009). Data indicate that foliar Se application has a higher bioavailability than soil sprayed (Li et al., 2018). In the work of Jing et al. (2017), foliar application of various Se compound in winter jujube resulted to the increase of vitamin C content, soluble sugars, and total flavonoid content and favored the sugar-to-acid ratio and had a beneficial effect in the fruit weight and overall yield. Beneficial effects of foliar applied Se were also recorded by Li et al. (2018) in blueberries. The foliar spray of Se-containing compounds resulted in augmentation of Se concentration to the pomace of the fruit, the internal anthocyanin content, and favored the intact rate fruit rate, thus resulting to fruits with improved quality characteristics.

12 Titanium (Ti)

Titanium (Ti) is considered a beneficial element for plant growth. Ti applied via roots or leaves at low concentrations has been documented to improve crop performance; however, the physiological role of Ti in higher plants has not been clarified. In the work of Pais (1983), it was stated that Ti has an impact upon tissue development similar to auxins, while the foliar application of Ti upon paprika pepper improved the photosynthetic activity of the plant (Carvajal and Alcaraz, 1998). Wócik et al. (2010) found that foliar application of Ti in apple trees resulted to a significant increase of overall apple fruit yield. Furthermore, the foliar Ti spray did not provoke any significant change to the mean apple weight, firmness of the pomace, total soluble solid concentration, and titratable acidity. Although Ti application in apples did not alter the coloring of the fruits (Wócik et al., 2010), however, foliar Ti increased the color formation in plum (Alcaraz-Lopez et al., 2003) and peach (Alcaraz-López et al., 2004). Also, Ti foliar application was correlated positively with the increased ability of plum (Alcaraz-Lopez et al., 2003) and peach (Alcaraz-López et al., 2004) and with the ability to store increased quantities of Fe, Cu, Zn, and Ca into the peel of the fruit. This work further indicated that Ca assimilation to the flesh and skin of the fruits after Ti foliar spray application is attributed to the enhanced absorption, translocation, and assimilation procedures into the fruit of plum (Alcaraz-Lopez et al., 2003) and peach (Alcaraz-López et al., 2004). Furthermore, Ti foliar application was related with the ability of plum and peach fruits to delay ripening procedures and exert an overall decreased ripening status (Alcaraz-Lopez et al., 2003; Alcaraz-López et al., 2004).

13 Conclusions and future perspectives

Scientific data indicate and support the fact that foliar application of macro- or micronutrients can act as a beneficial factor toward the improvement of fruit quality, an action that is closely related with the chemical formula of the element, the applied dosage, and the number of repeated applications (Tables 29.1 and 29.2). Summarizing current scientific knowledge of foliar applied nutrients, data indicate that the most profound effect was established after the application of Ca, K, and B. The foliar spray of these elements exerted significant positive effect and beneficial effect toward fruit quality parameters. Lack of specific scientific data linked with the mode of action of other macro- or micronutrients upon fruit quality attributes does not allow the extraction of clear and specific results for the rest of the elements. Also, another factor that needs attention is the diversity that is related with the interpretation of the results. Specific quality attributes do not always have the same meaning in all types of fruits. For example, in some cases like citrus, the increase of acidity might be a positive attribute, but to other species like fig, a decrease of acidity is called a positive result. An in-depth molecular, metabolomic analysis could shed light to the metabolic cascades that are controlled and influenced by the foliar application of each macro- or microelement, providing a holistic approach toward the understanding and manipulation of fruit quality via the usage of foliar nutrition.

TABLE 29.1   Effect of macroelement foliar application in fruit quality characteristics.

Formula	Fruit	Dose	Sprays	Increase	Decrease	Ref.
$CaCl_2 \cdot 2H_2O$	Papaya	0.5%, 1%, 1.5%, 2%	6	Firmness, trititable acidity	Respiration rate, ethylene production, soluble solid concentration	Madani et al. (2014)
	Apple	0.5%	6 in 4 seasons	Fresh weight, firmness, starch concentration, soluble solid concentration		Ghorbani et al. (2017)
		0.5%	5	Fresh weight, firmness, anthocyanin concentration, soluble solid concentration		Solhjoo et al. (2017)
		8.971 kg Ca/ha	1–8	Fruit size, fresh weight	Scald	Kadir (2005)
		6–9 kg/ha	6	Firmness	Soluble solid concentration, starch concentration	Wójcik et al. (2016)
	Pomegranate	1%, 2%	4	Yield, fruit quality	Fruit cracking	Bakeer (2016)
		1%, 2%	2		Fruit cracking	Davarpanah et al. (2018)
	Kiwifruit	1.5%	3	Trititable acidity	Weight loss of fruit, decay	Shiri et al. (2016)
	Strawberry	10, 20, 50 mM	Every 4 days AP	Total phenol concentration, anthocyanin concentration		Xu et al. (2014)
		1.5 kg Ca/ha	5	Firmness	Decay	Wójcik and Lewandowski (2003)
		2 kg Ca/ha	5	Yield, firmness, trititable acidity, ascorbic acid concentration	Decay, physiological disorders, soluble solid concentration	Singh et al. (2007a,b)
		0.4%	3	–	–	Toivonen and Stan (2001)
	Olive	58.5 mM	3	Firmness		Tsantili et al. (2008)
	Guava	0.5%, 1%, 1.5%	1	Firmness, soluble solid concentration, ascorbic acid concentration		Goutam et al. (2010)
	Sweet cherry	0.5%	2	Skin penetration, stem removal, total phenol concentration	Fruit cracking	Michailidis et al. (2017)
		0.5%	3	Firmness	Fruit cracking	Erogul (2014)
		22.5, 45, 58.5 mM	2	Firmness, stem removal		Tsantili et al. (2007)
	Peach	1.5%	4	Firmness	Weight loss of fruit	Ekinci (2018)
		0.5%, 1%	5	Firmness	Physiological disorders,	Val and Fernández (2011)
	Pear	1.1%	3		decay	Sugar and Basile (2011)

TABLE 29.1 Effect of macroelement foliar application in fruit quality characteristics—cont'd

Formula	Fruit	Dose	Sprays	Increase	Decrease	Ref.
$Ca(NO_3)_2$	Sweet cherry	0.5%	3		Fruit cracking	Erogul (2014)
	Apple	0.4%, 0.8%	4	Firmness, trititable acidity	Physiological disorders, decay	Domagala-⊠wiatkiewicz and Blaszczyk (2009)
		5–8 kg/ha	6	Firmness	Physiological disorders	Wojcik (2002)
$Ca(OH)_2$	Sweet cherry	0.2 M	3		Fruit cracking	Erogul (2014)
		3%	2	Firmness	Fruit cracking	Brown et al. (1996)
Calcium propionate	Peach	0.5%, 1%	5	Firmness	Physiological disorders	Val and Fernández (2011)
Calcium caseinate	Sweet cherry	0.5%	3		Fruit cracking	Erogul (2014)
—	Orange	0.3%	5	Cellulose, protopectin	Hemicellulose, total pectin, water-soluble pectin	Dong et al. (2009)
—	Plum	4 mg Ca/L	3	Compression resistance, penetration resistance		Alcaraz-Lopez et al. (2004)
—	Jujube	0.2%, 0.4%	4	Soluble solid concentration		Mao et al. (2016)
$MgSO_4$	Peach	2%	4	Firmness	Weight loss of fruit	Ekinci (2018)
$MgCl_2$	Litchi	1.5%	1	Flavonoid biosynthesis transcripts, anthocyanin biosynthesis transcripts, ABA signal pathway transcripts		Wang et al. (2017b)
	Lychee	1.5%	2	Anthocyanins concentration, ABA concentration		Wang et al. (2017a)
—	Plum	0.103 mM	3	Penetration resistance		Alcaraz-Lopez et al. (2003)
—	Peach	0.103 mM	3		Weight loss of fruit	Alcaraz-López et al. (2004)
KNO_3	Apple	0.25%	3	Fresh weight, anthocyanin concentration, soluble solid concentration, trititable acidity		Solhjoo et al. (2017)
	Mango	2%, 4%	1–2	Yield		Oosthuyse (2015)
		2%	4	Yield, fresh weight, soluble solid concentration, ascorbic acid concentration	Trititable acidity	Baiea and Moneim (2015)
	Plum	1%, 1.5%, 2%	2	Yield, fresh weight, soluble solid concentration	Trititable acidity	Jawandha et al. (2017)
	Clementine	5%, 8%	2–3	Fresh weight, fruit size		Hamza et al. (2015)
	Pear	1%, 1.5%, 2%	1–3	Fresh weight, fruit size, firmness, soluble solid concentration		Gill et al. (2012)

Continued

TABLE 29.1 Effect of macroelement foliar application in fruit quality characteristics—cont'd

Formula	Fruit	Dose	Sprays	Increase	Decrease	Ref.
K_2SO_4	Apple	0.25%	3	Fresh weight, anthocyanin concentration, soluble solid concentration, trititable acidity		Solhjoo et al. (2017)
	Peach	143 g/tree	3	Yield, fruit size, soluble solid concentration		Dbara et al. (2018)
	Citrus	1%	4	Yield, fresh weight		Ashraf et al. (2013)
	Olive	0.1%, 0.2%	2	Soluble solid concentration, anthocyanin concentration		Zivdar et al. (2016)
	Clementine	2.5%, 4%	2–3	Fresh weight, fruit size		Hamza et al. (2015)
	Pear	1%, 1.5%, 2%	1–3	Fresh weight, fruit size, firmness, soluble solid concentration		Gill et al. (2012)
KCl	Apple	0.25%	3	Fresh weight, firmness, anthocyanin concentration, soluble solid concentration, trititable acidity		Solhjoo et al. (2017)
KH_2PO_4	Mango	2%	4	Yield, fresh weight, soluble solid concentration, ascorbic acid concentration	Trititable acidity	Baiea and Moneim (2015)
K_2HPO_4	Mango	2%	4	Yield, fresh weight, soluble solid concentration, ascorbic acid concentration	Trititable acidity	Baiea and Moneim (2015)
$B(OH)_3$	Strawberry	160 g B/ha	3	–	–	Wójcik and Lewandowski (2003)
		150 g B/ha	3	Yield		Singh et al. (2007a,b)
–	Jujube	0.2%, 0.4%	4	Soluble solid concentration		Mao et al. (2016)
$MnSO_4$	Peach	1%	4	Polygalacturonic acid content	Fruit weight loss	Ekinci (2018)
NH_4NO_3	Mango	0.75%	4	Physiological disorders, green color		Nguyen et al. (2004)
$CO(NH_2)_2$	Apple	0.5%	7	Yield, fruit size		Dong et al. (2005)
		0.2%, 0.5%, 0.8%		Ascorbic acid concentration	Soluble solid concentration	Zhao et al. (2013)
	Pear	5%	1	Fruit size		Curetti et al. (2013)
	Mandarin	0.1%	2	Yield, fresh weight, fruit size, soluble solid concentration, pH	Trititable acidity	Al-Obeed et al. (2018)
$Ca(H_2PO_4)_2 \cdot 2H_2O$	Sweet persimmon	7.5, 10, 20 ppm P		Yield, soluble solid concentration, firmness	Trititable acidity	Hossain and Ryu (2009)

TABLE 29.2 Effect of microelement foliar application in fruit quality characteristics.

Formula	Fruit	Dose	Sprays	Increase	Decrease	Ref.
B(OH)$_3$	Strawberry	8%	3	Fruit yield	Fruit deformation	Singh et al. (2007a,b)
	Sweet cherry	8%	2	Titratable acidity, total anthocyanins		Wojcik and Wojcik (2006)
	Raspberry	8%	4	Fruit yield, firmness, titratable acidity		Wojcik (2005)
	Tart cherry	8%	3			Wojcik (2006a,b)
	Apple	0.3%, 0.6%	2		Starch content, titratable acidity	de SÁ et al. (2014)
	Pear	0.2, 0.8 B/ha	10	Ca content, firmness, titratable acidity, internal browning	Membrane permeability	Wojcik and Wojcik (2003)
	Olive tree	0.25%	2	Oil content, MUFA/PUFA ratio	Soluble carbohydrates	Saadati et al. (2013)
	Mandarin	300 mg/L	2	Volume length, fruit diameter, total soluble solids, juice pH, total sugars, ascorbic acid content		Al-Obeed et al. (2018)
Solubor	Olive tree	300 mg/L	2	Fruit yield		Larbi et al. (2011)
Solubor	Hazelnut	300, 600, 900 mg/L	4	Nut mass, kernel mass		Silva et al. (2003)
Polybor	Apple	0.30, 0.60 g B/L	1	Core boron, cortex boron, fruit color		Peryea and Drake (1991)
Tarimbor fertilizer	Pistachio	0.1% 0.2% 0.3%	2	Fruit yield	Split nut ratio, split kernel ratio	Acar et al. (2016)
B	Orange	3 g/kg	5	Tissue structure, segment membrane, dietary fiber content	Polygalacturonase activity, Pectinesterase activity, β-galactosidase activity	Dong et al. (2009)
Nano-B-chelate fertilizer	Pomegranate	3.25, 6.5 mg B/L	1	Total soluble solids, maturation index, fruit juice pH	Titratable acidity	Davarpanah et al. (2016)
Bormit	Strawberry	160 g/ha	3	Microelement concentration		Wójcik and Lewandowski (2003)
Na$_2$B$_8$O$_{13}$·4H$_2$O	Mango	0.05%, 0.0075%, 0.1% tree^{-1}	3		Internal cracking	Saran and Kumar (2011)
CoSO$_4$	Mango	250, 500, 1000, 1500 ppm	1		Nonreducing sugars	Singh et al. (1994)
Cu(OH)$_2$	Sweet cherry	50% Cu 100 L/tree	1	Resistance to cracking, firmness		Brown et al. (1995)
FeSO$_4$·7H$_2$O·FeSO$_4$·7H$_2$O	Pomegranate	2 g/L	2	Fruit yield, total soluble content		Mirzapour and Khoshgoftarmanesh (2013)
Aminoacid-Fe compound fertilizer	Peach	1000 mg/kg	2	Firmness, total soluble solids, Fe concentration, succinate		Song et al. (2016)

Continued

TABLE 29.2 Effect of microelement foliar application in fruit quality characteristics—cont'd

Formula	Fruit	Dose	Sprays	Increase	Decrease	Ref.
				dehydrogenase activity, aconitase activity		
$MnSO_4$	Peach	1%	4	Polygalacturonic acid content	Fruit weight loss	Ekinci (2018)
$NiSO_4$	Strawberry	150, 300, 400 mg/L	2	Crown number		Eshghi and Ranjbar (2015)
$SeSO_4$	Winter jujube	25, 50, 100, 200 mg/L	3	Ascorbic acid content, soluble sugars, total flavonoids, sugar/acid ratio, fruit weight, fruit yield		Jing et al. (2017)
Na_2SeO_3-Na_2SeO_4	Blueberry	200 mg/L	2	Se concentration, pomace storage, anthocyanin content, intact fruit rate		Li et al. (2018)
$C_{24}H_{32}O_{24}Ti$	Apple	3 g Ti/ha	6	Fruit yield		Wócik et al. (2010)
$C_{24}H_{32}O_{24}Ti$	Plum	0.0042 mM, 5 L/tree	3	Resistance to compression, resistance to penetration, color formation, peel and flesh Fe-Cu-Zn concentration, absorption-translocation and assimilation of Ca	Weight, ripening status	Alcaraz-Lopez et al. (2004)
$C_{24}H_{32}O_{24}Ti$	Peach	0.0042 mM, 2 mg Ti L/tree	3	Resistance to compression, color formation, peel Fe-Cu-Zn, Peel and flesh Ca absorption, translocation and assimilation process	Ripening status	Alcaraz-Lopez et al. (2003)
$ZnSO_4$	Olive tree	0.25%	2	Oil content, UFA/SFA ratio	Soluble carbohydrates	Saadati et al. (2013)
$ZnSO_4$ Nano-Zn-chelated fertilizer	Mango	0.25%, 0.50%, 1.0%	1	Zn pulp concentration, total soluble solid content		Bahadur et al. (1998)
	Mandarin	0.5 g L/tree	2	Fruit weight, fruit pulp, fruit juice, fruit volume, fruit length-diameter, total soluble solid content, acidity, fruit juice pH, total sugars, ascorbic acid content		Al-Obeed et al. (2018)
	Mandarin	1%	4	Circumference/diameter, fruit juice content, juice total soluble solids, ascorbic acid content, total soluble solids/acid ratio	Peel thickness, citric acid content	Ashraf et al. (2013)

TABLE 29.2 Effect of microelement foliar application in fruit quality characteristics—cont'd

Formula	Fruit	Dose	Sprays	Increase	Decrease	Ref.
	Pomegranate	60, 120 mg Zn L/tree	1	Fruit yield, titratable acidity, total soluble solids, maturation index, fruit juice pH		Davarpanah et al. (2016)
Nanochelated Zn	Apple	0.13%	1	Ascorbic acid content, total antioxidative activity, superoxide dismutase activity	Starch content, polyphenol oxidase activity	Rasouli and Koushesh Saba (2018)

References

Abdel-Karim, H.A., et al., 2015. Effect of foliar application of boron and zinc on fruit set, yield and some fruit characteristics of Fuerte avocado. Res. J. Pharm. Biol. Chem. Sci. 6, 443–449. Available at: https://scholar.google.gr/scholar?hl=el&as_sdt=0%2C5&q=Abdel-Karim+et+al.+%282015+ +Effect+of+foliar+application+of+Boron+and+zinc+on+fruit+set%2C+yiled+an+some+fruit+chracteristics+of+Fuert+avocado&btnG=. (Accessed 2 February 2019).

Abdi, G.H., Hedayat, M., 2010. Yield and fruit physiochemical characteristics of "Kabkab" date palm as affected by methods of potassium fertilization. Adv. Environ. Biol. 4 (3), 437–442.

Acar, İ., et al., 2016. Boron affects the yield and quality of nonirrigated pistachio (Pistacia vera L.) trees. Turk. J. Agric. For. 40, 664–670. https://doi.org/10.3906/tar-1511-80.

Alcaraz-Lopez, C., et al., 2003. Effects of foliar sprays containing calcium, magnesium and titanium on plum (Prunus domestica L.) fruit quality. J. Plant Physiol. 160 (12), 1441–1446. Available at: http://www.ncbi.nlm.nih.gov/pubmed/14717435%5Cnhttp://www.sciencedirect.com/science/article/pii/S0176161704705404.

Alcaraz-Lopez, C., et al., 2004. Effects of calcium-containing foliar sprays combined with titanium and algae extract on plum fruit quality. J. Plant Nutr. 27 (4), 713–729. https://doi.org/10.1081/PLN-120030377.

Alcaraz-López, C., et al., 2004. Effect of foliar sprays containing calcium, magnesium and titanium on peach (*Prunus persica* L) fruit quality. J. Sci. Food Agric. 84 (9), 949–954. https://doi.org/10.1002/jsfa.1703.

Al-Obeed, R.S., et al., 2018. Improvement of "Kinnow" mandarin fruit productivity and quality by urea, boron and zinc foliar spray. J. Plant Nutr. 41 (5), 609–618. https://doi.org/10.1080/01904167.2017.1406111. Taylor & Francis.

Ashraf, M.Y., et al., 2013. Modulation in yield and juice quality characteristics of citrus fruit from trees supplied with zinc and potassium foliarly. J. Plant Nutr. 36 (13), 1996–2012. https://doi.org/10.1080/01904167.2013.808668.

Bahadur, L., Malhi, C.S., Singh, Z., 1998. Effect of foliar and soil applications of zinc sulphate on zinc uptake, tree size, yield, and fruit quality of mango. J. Plant Nutr. 21 (3), 589–600. https://doi.org/10.1080/01904169809365426.

Baiea, M.H.M., Moneim, E.A., 2015. Effect of different forms of potassium on growth, yield and fruit quality of mango cv. Hindi. Int. J. Chemtech Res. 8 (4), 1582–1587.

Bakeer, S.M., 2016. Effect of ammonium nitrate fertilizer and calcium chloride foliar spray on fruit cracking and sunburn of Manfalouty pomegranate trees. Sci. Hortic. 209, 300–308. https://doi.org/10.1016/j.scienta.2016.06.043. Elsevier B.V.

Ben Mimoun, M., Marchand, M., 2013. Effects of potassium foliar fertilization on different fruit tree crops over five years of experiments. Acta Hortic. 984, 211–218. Available at: https://www.actahort.org/books/984/984_23.htm. (Accessed 3 December 2018).

Bieniasz, M., Małodobry, M., Dziedzic, E., 2012. The effect of foliar fertilization with calcium on quality of strawberry cultivars 'Luna' and 'Zanta'. Acta Hortic. https://doi.org/10.1016/j.scitotenv.2013.10.104.

Brown, G., et al., 1995. Effects of copper-calcium sprays on fruit cracking in sweet cherry (Prunus avium). Sci. Hortic. 62 (1–2), 75–80. https://doi.org/10.1016/0304-4238(94)00746-3.

Brown, G., et al., 1996. The effects of copper and calcium foliar sprays on cherry and apple fruit quality. Sci. Hortic. 67 (3–4), 219–227. Available at: http://www.sciencedirect.com/science/article/pii/S0304423896009375. (Accessed 24 October 2016).

Brown, G.S., Kitchener, A.E., Barnes, S., 1998. Calcium hydroxide sprays for the control of black spot on apples—treatment effects on fruit quality. Acta Hortic. 513 (513), 47–52. https://doi.org/10.17660/ActaHortic.1998.513.4.

Bybordi, A., Malakouti, M.J., 2006. Effects of foliar applications of nitrogen, boron and zinc on fruit setting and quality of almonds. Acta Hortic. 726, 351–357.

Cackam, I., Marschner, H., 1988. Enchanced superoxide radical production in roots of zinc-deficient plants. J. Exp. Bot. 207 (39), 1449–1460.

Carvajal, M., Alcaraz, C., 1998. Titanium as a beneficial element for Capsicum annuum L. plants. Recent Res. Dev. Phytochem. 2 (1), 83–94. Available at: https://scholar.google.gr/scholar?hl=el&as_sdt=0%2C5&q=.+Titanium+as+a+beneficial+element+for+Capsicum+annuum+&btnG. (Accessed 2 February 2019).

Chouliaras, V., et al., 2004. Iron chlorosis in grafted sweet orange (Citrus sinensis L.) plants: physiological and biochemical responses. Biol. Plant. 48 (1), 141–144. https://doi.org/10.1023/B:BIOP.0000024292.51938.aa.

Coêlho de Lima, M.A., Alves, R.E., 2011. Soursop (Annona muricata L.). In: Postharvest Biology and Technology of Tropical and Subtropical Fruits. Woodhead Publishing, pp. 363–392e. https://doi.org/10.1533/9780857092618.363.

Crisosto, C.H., et al., 2000. Influence of in-season foliar calcium sprays on fruit quality and surface discoloration incidence of peaches and nectarines. J. Am. Pomol. Soc. 54 (3), 118–122. https://doi.org/10.1057/crr.2011.9.

Curetti, M., et al., 2013. Foliar-applied urea at bloom improves early fruit growth and nitrogen status of spur leaves in pear trees, cv. Williams Bon Chretien. Sci. Hortic. 150, 16–21. https://doi.org/10.1016/j.scienta.2012.10.022 Elsevier B.V.

D'Aquino, S., et al., 2005. Effect of preharvest and postharvest calcium treatments on chilling injury and decay of cold stored "fortune" mandarins. Acta Hortic. 682 (682), 631–638. https://doi.org/10.17660/ActaHortic.2005.682.81.

Dag, A., et al., 2011. Influence of time of harvest and maturity index on olive oil yield and quality. Sci. Hortic. 127 (3), 358–366. https://doi.org/10.1016/j.scienta.2010.11.008.

Davarpanah, S., et al., 2016. Effects of foliar applications of zinc and boron nano-fertilizers on pomegranate (*Punica granatum* cv. Ardestani) fruit yield and quality. Sci. Hortic. 210, 57–64. https://doi.org/10.1016/j.scienta.2016.07.003.

Davarpanah, S., et al., 2018. Foliar calcium fertilization reduces fruit cracking in pomegranate (*Punica granatum* cv. Ardestani). Sci. Hortic. 230, 86–91. https://doi.org/10.1016/j.scienta.2017.11.023. Elsevier.

Dbara, S., Lahmar, K., Ben Mimoun, M., 2018. Potassium mineral nutrition combined with sustained deficit irrigation to improve yield and quality of a late season peach cultivar (*Prunus persica* L. cv "Chatos"). Int. J. Fruit Sci. 18 (4), 369–382. https://doi.org/10.1080/15538362.2018.1438329. Taylor & Francis.

de SÁ, A.A., et al., 2014. Forms of boron application and its influence on quality and yield of apples (Malus dosmetica). Rev. Bras. Frutic.. 36(2) https://doi.org/10.1590/0100-2945-171/13.

Devi, K., et al., 2018. Effect of foliar nutrition and growth regulators on nutrient status and fruit quality of Eureka lemon (Citrus limon). Indian J. Agric. Sci. 88 (5), 704–708.

Dixon, N.E., et al., 1975. Jack bean urease (EC 3.5.1.5). A Metalloenzyme. A simple biological role for nickel? J. Am. Chem. Soc. 97 (14), 4131–4133. https://doi.org/10.1021/ja00847a045.

Domagala-Światkiewicz, I., Blaszczyk, J., 2009. Effect of calcium nitrate spraying on mineralcontents and storability of "elise" apples. Pol. J. Environ. Stud. 18 (5), 971–976.

Dong, S., et al., 2005. Foliar N application reduces soil NO_3^--N leaching loss in apple orchards. Plant Soil 268 (1), 357–366. https://doi.org/10.1007/s11104-004-0333-1.

Dong, T., et al., 2009. Effect of pre-harvest application of calcium and boron on dietary fibre, hydrolases and ultrastructure in "Cara Cara" navel orange (Citrus sinensis L. Osbeck) fruit. Sci. Hortic. 121 (3), 272–277. https://doi.org/10.1016/j.scienta.2009.02.003.

Dris, R., Niskanen, R., Fallahi, E., 1999. Relationships between leaf and fruit minerals and fruit quality attributes of apples grown under northern conditions. J. Plant Nutr. 22 (12), 1839–1851. https://doi.org/10.1080/01904169909365760.

Eichert, T., Goldbach, H.E., 2008. Equivalent pore radii of hydrophilic foliar uptake routes in stomatous and astomatous leaf surfaces—further evidence for a stomatal pathway. Physiol. Plant. 132 (4), 491–502. https://doi.org/10.1111/j.1399-3054.2007.01023.x.

Ekinci, N., 2018. Foliar spray of nutrients affects fruit quality, polygalacturonic acid (pectin) content and storage life of peach fruits in Turkey. Appl. Ecol. Environ. Res. 16 (1), 749–759. https://doi.org/10.15666/aeer/1601_749759.

El-Razek, E.A., et al., 2017. Effect of foliar application of biosimulated nanomaterials (calcium/yeast nanocomposite) on yield and fruit quality of 'Ewais' mango trees. Ann. Res. Rev. Biol. 18 (3), 1–11. https://doi.org/10.9734/ARRB/2017/36395.

Erogul, D., 2014. Effect of preharvest calcium treatments on sweet cherry fruit quality. Not. Bot. Horti. Agrobot. Cluj-Na. 42 (1), 150–153.

Eshghi, S., Ranjbar, R., 2015. Vegetative growth, yield and leaf mineral composition in strawberry (Fragaria × Ananassa DUCH. CV. Pajaro) as influenced using nickel sulfate and urea sprays. J. Plant Nutr. 38 (9), 1336–1345. https://doi.org/10.1080/01904167.2014.983121.

Evans, I., Solberg, E., 2007. Copper and plant disease. In: Mineral Nutrition and Plant Disease. American Phytopathological Society (APS Press), St. Paul, USA, pp. 177–188.

Fallahi, E., Simons, B.R., 1996. Interrelations among leaf and fruit mineral nutrients and fruit quality in "delicious" apples. J. Tree Fruit Prod. 1 (1), 15–25. https://doi.org/10.1300/J072v01n01_02.

Fotopoulos, V., et al., 2010. Involvement of AsA/DHA and GSH/GSSG ratios in gene and protein expression and in the activation of defence mechanisms under abiotic stress conditions. In: Ascorbate-Glutathione Pathway and Stress Tolerance in Plants. Springer Netherlands, Dordrecht, pp. 265–302. https://doi.org/10.1007/978-90-481-9404-9_10.

Frances, J., et al., 1999. Minimised post-harvest chemical treatments, fruit density per tree, and calcium sprays affect the storability of "passe crassane" and "conference" pears in Girona (Spain). Acta Hortic. 485 (485), 161–166. https://doi.org/10.17660/ActaHortic.1999.485.21.

Ghanem, M., Ben Mimoun, M., 2010a. Effect of potassium foliar spray on two plum trees cultivars: "strival" and "black star". Acta Hortic. 874 (874), 83–90. https://doi.org/10.17660/ActaHortic.2010.874.10.

Ghanem, M., Ben Mimoun, M., 2010b. Effects of potassium foliar sprays on Royal Glory peach trees. Acta Hortic. 868, 261–265. Available at:https://www.actahort.org/books/868/868_34.htm. (Accessed 3 December 2018).

Ghorbani, E., et al., 2017. Evaluation of pre-harvest foliar calcium applications on "Fuji" apple fruit quality during cold storage. Aust. J. Crop Sci. 11 (2), 228–233. https://doi.org/10.21475/ajcs.17.11.02.p5853.

Gill, P.P.S., et al., 2012. Effect of foliar sprays of potassium on fruit size and quality of "Patharnakh" pear. Indian J. Hortic. 69 (4), 512–516.

Goutam, M., Dhaliwal, H.S., Mahajan, B.V.C., 2010. Effect of pre-harvest calcium sprays on post-harvest life of winter guava (*Psidium guajava* L.). J. Food Sci. Technol. 47 (5), 501–506. https://doi.org/10.1007/s13197-010-0085-2. Springer-Verlag.

Gupta, U., 1979. Boron nutrition of crops. Adv. Agron. 31, 273–307.

Hafez, O.M., et al., 2017. Enhancement yield and fruit quality of Washington navel orange by application of spraying potassium microencapsulated biodegradable polylactic acid. Agric. Eng. Int. CIGR J. Special Issue, 101–110.

Hamza, A., et al., 2015. Response of "cadoux" clementine to foliar potassium fertilization: effects on fruit production and quality. Acta Hortic. 1065 (1065), 1785–1794. https://doi.org/10.17660/ActaHortic.2015.1065.228.

Hossain, M.B., Ryu, K.S., 2009. Effect of foliar applied phosphatic fertilizer on absorption pathways, yield and quality of sweet persimmon. Sci. Hortic. 122 (4), 626–632. https://doi.org/10.1016/j.scienta.2009.06.035.

Jamali, B., Eshghi, S., Taffazoli, E., 2013. Vegetative growth, yield, fruit quality and fruit and leaf composition of strawberry cv. "Pajaro" as influenced by salicylic acid and nickel sprays. J. Plant Nutr. 36 (7), 1043–1055. https://doi.org/10.1080/01904167.2013.766803.

Javaid, K., et al., 2016. Influence of crop load and foliar nutrient sprays on plant nutrient content and post harvest quality of apple (Malus x domestica Borkh.) cv. Red delicious during storage. Ecol. Environ. Conserv. 22 (3), 1457–1463.

Jawandha, S.K., et al., 2017. Effect of potassium nitrate on fruit yield, quality and leaf nutrients content of plum. Vegetos 30 (Special Issue 2), 325–328. https://doi.org/10.5958/2229-4473.2017.00090.8.

Jing, D.W., et al., 2017. Selenium enrichment, fruit quality and yield of winter jujube as affected by addition of sodium selenite. Sci. Hortic. 225, 1–5. https://doi.org/10.1016/j.scienta.2017.06.036.

Kadir, S.A., 2005. Fruit quality at harvest of "Jonathan" apple treated with Foliarly-applied calcium chloride. J. Plant Nutr. 27 (11), 1991–2006. https://doi.org/10.1081/PLN-200030102.

Kanpure, R.N., et al., 2016. Growth, yield and quality of guava (Psidium guajava I.) augmented by foliar application of phosphorus and micronutrients. Ecol. Environ. Conserv. 22, S7–S11.

Kápolna, E., et al., 2009. Effect of foliar application of selenium on its uptake and speciation in carrot. Food Chem. 115 (4), 1357–1363. https://doi.org/10.1016/J.FOODCHEM.2009.01.054.

Kingston, C.M., 2010. Maturity indices for apple and pear. In: Horticultural Reviews. John Wiley & Sons, Inc., Oxford, pp. 407–432. https://doi.org/10.1002/9780470650509.ch10

Larbi, A., et al., 2011. Effect of foliar boron application on growth, reproduction, and oil quality of olive trees conducted under a high density planting system. J. Plant Nutr. 34 (14), 2083–2094. https://doi.org/10.1080/01904167.2011.618570.

Leigh, R.A., Wyn Jones, R.G., 1984. A hypothesis relating critical potassium concentrations for growth to the distribution and functions of this ion in the plant cell. New Phytol. 97 (1), 1–13. https://doi.org/10.1111/j.1469-8137.1984.tb04103.x. Wiley/Blackwell.

Li, M., et al., 2018. Effects of a foliar spray of selenite or selenate at different growth stages on selenium distribution and quality of blueberries. J. Sci. Food Agric. 98 (12), 4700–4706. https://doi.org/10.1002/jsfa.9004.

Loomis, W.D., Durst, R.W., 1992. Chemistry and biology of boron. Biofactors 3 (4), 229–239. Available at: http://www.ncbi.nlm.nih.gov/pubmed/1605832. (Accessed 1 February 2019).

Lötze, E., Theron, K.I., 2006. Dynamics of calcium uptake with pre-harvest sprays to reduce bitter pit in 'Golden delicious'. Acta Hortic. 721, 313–319.

Madani, B., et al., 2014. Preharvest calcium chloride sprays affect ripening of Eksotika II'papaya fruits during cold storage. Sci. Hortic. 171, 6–13. https://doi.org/10.1016/j.scienta.2014.03.032. Elsevier B.V.

Madani, B., et al., 2015. Influence of calcium foliar fertilization on plant growth, nutrient concentrations, and Fruit Quality of Papaya. Horttechnology 25 (August), 496–504.

Mao, Y.M., et al., 2016. Effects of foliar applications of boron and calcium on the fruit quality of "Dongzao" (Zizyphus jujuba mill.). Acta Hortic. (1116), 105–108. https://doi.org/10.17660/ActaHortic.2016.1116.18.

Marschner, H., 1995. Mineral Nutrition of Higher Plants. Academic Press, London.

Marschner, P., 2012. Rhizosphere Biology, Marschner's Mineral Nutrition of Higher Plants. Available at:https://www.sciencedirect.com/science/article/pii/B9780123849052000157. (Accessed 2 February 2019).

Mengel, K., 2002. Alternative or complementary role of foliar supply in mineral nutrition. Acta Hortic. (594), 33–47. https://doi.org/10.17660/ActaHortic.2002.594.1.

Michailidis, M., et al., 2017. Metabolomic and physico-chemical approach unravel dynamic regulation of calcium in sweet cherry fruit physiology. Plant Physiol. Biochem. 116, 68–79. https://doi.org/10.1016/j.plaphy.2017.05.005.

Mirshekari, A., Madani, B., 2018. Effects of calcium spraying to the leaves and fruits on postharvest physiological characteristics of papaya fruits. Acta Hortic. (1208), 409–416. https://doi.org/10.17660/ActaHortic.2018.1208.56.

Mirzapour, M.H., Khoshgoftarmanesh, A.H., 2013. Effect of soil and foliar application of iron and zinc on quantitative and qualitative yield of pomegranate. J. Plant Nutr. 36 (1), 55–66. https://doi.org/10.1080/01904167.2012.733049.

Molassiotis, A., et al., 2013. Proteomics in the fruit tree science arena: new insights into fruit defense, development, and ripening. Proteomics 13 (12−13), 1871–1884. https://doi.org/10.1002/pmic.201200428.

Natale, W., et al., 2002. Phosphorus foliar fertilization in guava trees. Acta Hortic. 594 (594), 171–177. https://doi.org/10.17660/ActaHortic.2002.594.17.

Nguyen, H., et al., 2004. Effect of nitrogen on the skin colour and other quality attributes of ripe "Kensington pride" mango (Mangifera indica L.) fruit. J. Hortic. Sci. Biotechnol. 79 (2), 204–210. https://doi.org/10.1080/14620316.2004.11511749.

Nicola, S., Tibaldi, G., Fontana, E., 2009. Fresh-cut produce quality: implications for a systems approach. In: Postharvest Handling. Elsevier, pp. 247–282. https://doi.org/10.1016/B978-0-12-374112-7.00010-X.

Oosthuyse, S.A., 2015. Spray application of KNO$_3$, low biuret urea, and growth regulators and hormones during and after flowering on fruit retention, fruit size and yield of mango. Acta Hortic.

Oosthuyse, S.A., Westcott, D.J., 2005. Determination of harvest maturation stages appropriate for export of Madras litchi as identified by measurable fruit attributes. In: South African Litchi Growers' Association Yearbook.vol. 17, pp. 29–33.

Pacheco, P., Hanley, T., Landero Figueroa, J.A., 2014. Identification of proteins involved in Hg-Se antagonism in water hyacinth (Eichhornia crassipes). Metallomics 6 (3), 560–571. https://doi.org/10.1039/c3mt00063j.

Pais, I., 1983. The biological importance of titanium. J. Plant Nutr. 6 (1), 3–131. https://doi.org/10.1080/01904168309363075.

Peryea, F.J., Drake, S.R., 1991. Influence of mid-summer boron sprays on boron content and quality indices of "delicious" apple1. J. Plant Nutr. 14 (8), 825–840. https://doi.org/10.1080/01904169109364245.

Plich, H., Wójcik, P., 2002. The effect of calcium and boron foliar application on postharvest plum fruit quality. Acta Hortic. 594 (594), 445–451. https://doi.org/10.17660/ActaHortic.2002.594.57.

Pulla Reddy, C., et al., 2011. Effect of ethylene inhibiting chemicals on yield and quality of pomegranate. Acta Hortic. 890, 353–358.

Ramezanian, A., Dadgar, R., Habibi, F., 2018. Postharvest attributes of "Washington navel" orange as affected by preharvest foliar application of calcium chloride, potassium chloride, and salicylic acid. Int. J. Fruit Sci. 18 (1), 68–84. https://doi.org/10.1080/15538362.2017.1377669. Taylor & Francis.

Rasouli, M., Koushesh Saba, M., 2018. Pre-harvest zinc spray impact on enzymatic browning and fruit flesh color changes in two apple cultivars. Sci. Hortic. 240, 318–325. https://doi.org/10.1016/j.scienta.2018.06.053.

Römheld, V., 2011. Diagnosis of deficiency and toxicity of nutrients. In: Marschner's Mineral Nutrition of Higher Plants, third ed. Academic Press, pp. 299–312. https://doi.org/10.1016/B978-0-12-384905-2.00011-X.

Saadati, S., et al., 2013. Effects of zinc and boron foliar application on soluble carbohydrate and oil contents of three olive cultivars during fruit ripening. Sci. Hortic. 164, 30–34. https://doi.org/10.1016/j.scienta.2013.08.033.

Sánchez, E.E., Curetti, M., Sugar, D., 2008. Foliar application of urea during bloom increases fruit size in "Williams" pears. Acta Hortic. (800), 583–586. https://doi.org/10.17660/ActaHortic.2008.800.77.

Saran, P.L., Kumar, R., 2011. Boron deficiency disorders in mango (*Mangifera indica*): field screening, nutrient composition and amelioration by boron application. Indian J. Agric. Sci. 81 (June), 506–510. Available at: https://www.researchgate.net/profile/Parmeshwar_L_Saran/publication/266341676_Boron_deficiency_disorders_in_mango_Mangifera_indica_Field_screening_nutrient_composition_and_amelioration_by_boron_application/links/5b8e0b5b92851c6b7eba8f7f/Boron-deficiency-d. (Accessed 2 February 2019).

Schreiner, M., et al., 2013. Current understanding and use of quality characteristics of quality characteristics of horticulture products. Sci. Hortic. 163, 63–69. Available at:https://www.sciencedirect.com/science/article/pii/S030442381300486X. (Accessed 1 February 2019).

Sharma, R.R., Sharma, V.P., Pandey, S.N., 2004. Mulching influences plant growth and albinism disorder in strawberry under subtropical climate. Acta Hortic. 662 (1), 187–191. https://doi.org/10.17660/ActaHortic.2004.662.25.

Shewfelt, R.L., 1999. What is quality? Postharvest Biol. Technol. 15 (3), 197–200. https://doi.org/10.1016/S0925-5214(98)00084-2.

Shiri, M.A., et al., 2016. Effect of CaCl2 sprays at different fruit development stages on postharvest keeping quality of "Hayward" kiwifruit. J. Food Process. Preserv. 40 (4), 624–635. https://doi.org/10.1111/jfpp.12642.

Silva, A.P., Rosa, E., Haneklaus, S.H., 2003. Influence of foliar boron application on fruit set and yield of hazelnut. J. Plant Nutr. 26 (3), 561–569. https://doi.org/10.1081/PLN-120017665.

Singh, Z., Dhillon, B.S., Arora, C.L., 1991. Nutrient levels in malformed and healthy tissues of mango (Mangifera indica L.). Plant Soil 133 (1), 9–15. https://doi.org/10.1007/BF00011894.

Singh, Z., et al., 1994. Effect of cobalt, cadmium, and nickel as inhibitors of ethylene biosynthesis on floral malformation, yield, and fruit quality of mango. J. Plant Nutr. 17 (10), 1659–1670. https://doi.org/10.1080/01904169409364838.

Singh, R., Sharma, R.R., Tyagi, S.K., 2007a. Pre-harvest foliar application of calcium and boron influences physiological disorders, fruit yield and quality of strawberry (*Fragaria × ananassa* Duch.). Sci. Hortic. 112 (2), 215–220. https://doi.org/10.1016/j.scienta.2006.12.019.

Singh, R., Sharma, R.R., Tyagi, S.K., 2007b. Pre-harvest foliar application of calcium and boron influences physiological disorders, fruit yield and quality of strawberry (Fragaria Â ananassa Duch.). Sci. Hortic. 112, 215–220. https://doi.org/10.1016/j.scienta.2006.12.019.

Singh, R., et al., 2009. Foliar application of calcium and boron influences physiological disorders, fruit yield and quality of strawberry (*Fragaria × ananassa* Duch.). Acta Hortic. 842 (842), 835–838. https://doi.org/10.17660/ActaHortic.2009.842.184.

Solhjoo, S., Gharaghani, A., Fallahi, E., 2017. Calcium and potassium foliar sprays affect fruit skin color, quality attributes, and mineral nutrient concentrations of "red delicious" apples. Int. J. Fruit Sci. 17 (4), 358–373. https://doi.org/10.1080/15538362.2017.1318734. Taylor & Francis.

Song, Z., et al., 2016. Differential expression of iron-sulfur cluster biosynthesis genes during peach fruit development and ripening, and their response to iron compound spraying. Sci. Hortic. 207, 73–81. https://doi.org/10.1016/j.scienta.2016.05.024. Elsevier B.V.

Sotiropoulos, T.E., et al., 2006. Growth, nutritional status, chlorophyll content, and antioxidant responses of the apple rootstock MM 111 shoots cultured under high boron concentrations in vitro. J. Plant Nutr. 29 (3), 575–583. https://doi.org/10.1080/01904160500526956. Taylor & Francis Group.

Sugar, D., Basile, S.R., 2011. Orchard calcium and fungicide treatments mitigate effects of delayed postharvest fungicide applications for control of postharvest decay of pear fruit. Postharvest Biol. Technol. 60 (1), 52–56. https://doi.org/10.1016/j.postharvbio.2010.11.007. Elsevier B.V.

Swietlik, D., LaDuke, J.V., 1991. Productivity, growth, and leaf mineral composition of orange and grapefruit trees foliar-sprayed with zinc and manganese. J. Plant Nutr. 14 (2), 129–142. https://doi.org/10.1080/01904169109364189.

Tanou, G., Ziogas, V., Molassiotis, A., 2017. Foliar nutrition, biostimulants and prime-like dynamics in fruit tree physiology: new insights on an old topic. Front. Plant Sci. 8. https://doi.org/10.3389/fpls.2017.00075.

Toivonen, P.M.A., Stan, S., 2001. Effect of preharvest CaCl2 sprays on the postharvest quality of "Rainier" and "Totem" strawberries. Acta Hortic. 564 (564), 159–163. https://doi.org/10.17660/ActaHortic.2001.564.18.

Tsantili, E., et al., 2007. Effects of two pre-harvest calcium treatments on physiological and quality parameters in "vogue" cherries during storage. J. Hortic. Sci. Biotechnol. 82 (4), 657–663. https://doi.org/10.1080/14620316.2007.11512287.

Tsantili, E., et al., 2008. Texture and other quality attributes in olives and leaf characteristics after preharvest calcium chloride sprays. HortScience 43 (6), 1852–1856.

Umar, S., et al., 1999. Effect of foliar fertilization of potassium on yield, quality, and nutrient uptake of groundnut. J. Plant Nutr. 22 (11), 1785–1795. https://doi.org/10.1080/01904169909365754.

Val, J., Fernández, V., 2011. In-season calcium-spray formulations improve calcium balance and fruit quality traits of peach. J. Plant Nutr. Soil Sci. 174 (3), 465–472. https://doi.org/10.1002/jpln.201000181.

Val, J., et al., 2008. Effect of pre-harvest calcium sprays on calcium concentrations in the skin and flesh of apples. J. Plant Nutr. 31 (11), 1889–1905. https://doi.org/10.1080/01904160802402757.

Val, J., et al., 2010. The effects of Ca applications on peach fruit mineral content and quality. Acta Hortic. 868 (868), 405–408. https://doi.org/10.17660/ActaHortic.2010.868.55.

Wang, Z., Yuan, M., et al., 2017a. Applications of magnesium affect pericarp colour in the Feizixiao lychee. J. Hortic. Sci. Biotechnol. 92 (6), 559–567. https://doi.org/10.1080/14620316.2017.1322922. Taylor & Francis.

Wang, Z., Li, S., et al., 2017b. De novo transcriptome assembly for pericarp in Litchi chinesis Sonn. cv. Feizixiao and identification of differentially expressed genes in response to Mg foliar nutrient. Sci. Hortic. 226 (58), 59–67. https://doi.org/10.1016/j.scienta.2017.08.023. Elsevier.

Wells, L., Brock, J., Brenneman, T., 2014. Effects of foliar sulfur sprays on pecan independent of pecan scab control. HortScience 49 (4), 434–437.

Wócik, P., et al., 2010. Response of "granny smith" apple trees to foliar titanium sprays under conditions of low soil availability of iron, manganese, and zinc. J. Plant Nutr. 33 (13), 1914–1925. https://doi.org/10.1080/01904167.2010.512051.

Wojcik, P., 2002. Yield and "Jonagold" apple fruit quality as influenced by spring sprays with commercial Rosatop material containing calcium and boron. J. Plant Nutr. 25 (5), 999–1010. https://doi.org/10.1081/PLN-120003934.

Wojcik, P., 2005. Response of primocane-fruiting "Polana" red raspberry to boron fertilization. J. Plant Nutr. 28 (10), 1821–1832. https://doi.org/10.1080/01904160500251191.

Wojcik, P., 2006a. Effect of postharvest sprays of boron and urea on yield and fruit quality of apple trees. J. Plant Nutr. 29 (3), 441–450. https://doi.org/10.1080/01904160500524894.

Wojcik, P., 2006b. "Schattenmorelle" tart cherry response to boron fertilization. J. Plant Nutr. 29 (9), 1709–1718. https://doi.org/10.1080/01904160600853813.

Wójcik, P., Lewandowski, M., 2003. Effect of calcium and boron sprays on yield and quality of "Elsanta" strawberry. J. Plant Nutr. 26 (3), 671–682. https://doi.org/10.1081/PLN-120017674 Taylor & Francis Group.

Wojcik, P., Szwonek, E., 2002. The efficiency of different foliar-applied calcium materials in improving apple quality. Acta Hortic. 594 (594), 563–567. https://doi.org/10.17660/ActaHortic.2002.594.75.

Wojcik, P., Wojcik, M., 2003. Effects of boron fertilization on "conference" pear tree vigor, nutrition, and fruit yield and storability. Plant Soil 256 (2), 413–421. https://doi.org/10.1023/A:1026126724095.

Wojcik, P., Wojcik, M., 2006. Effect of boron fertilization on sweet cherry tree yield and fruit quality. J. Plant Nutr. 29 (10), 1755–1766. https://doi.org/10.1080/01904160600897471.

Wójcik, P., Skorupińska, A., Gubbuk, H., 2016. Impacts of pre- and postbloom sprays of tryptophan on calcium distribution within "red jonaprince" apple trees and on fruit quality. HortScience 51 (12), 1511–1516. https://doi.org/10.21273/HORTSCI11216-16.

Xu, W., et al., 2014. Effect of calcium on strawberry fruit flavonoid pathway gene expression and anthocyanin accumulation. Plant Physiol. Biochem. 82, 289–298. https://doi.org/10.1016/j.plaphy.2014.06.015. Elsevier Masson SAS.

Yang, S.F., Hoffman, N.E., 1984. Ethylene biosynthesis and its regulation in higher plants. Annu. Rev. Plant Physiol. 35 (1), 155–189. https://doi.org/10.1146/annurev.pp.35.060184.001103.

Zhao, Y., et al., 2013. Influence of girdling and foliar-applied urea on apple (Malus domestica L.) fruit quality. Pak. J. Bot. 45 (5), 1609–1615.

Zivdar, S., et al., 2016. Physiological and biochemical response of olive (Olea europaea L.) cultivars to foliar potassium application. J. Agric. Sci. Technol. 18, 1897–1908.

Open field hydroponics in fruit crops: Developments and challenges

José S. Rubio-Asensio[a,*], *Margarita Parra*[a], *Diego S. Intrigliolo*[a,b]

[a]Irrigation Deparment, CEBAS-CSIC, Murcia, Spain
[b]CSIC Associated Unit "Riego en la agricultura Mediterránea," Instituto Valenciano de Investigaciones Agrarias, Moncada, Spain
*Corresponding author. E-mail: jsrubio@cebas.csic.es

1 Introduction

The most immediate agriculture challenge for the 21st century, to meet global food demand, is the increasing of the crops yield per surface unit while increasing the water use efficiency (WUE) and the nutrients use efficiency (NUE). Increasing WUE is mandatory in countries where water is limited, and irrigation water competes heavily with other more competitive economic activities and human uses (Morison et al., 2008). Increasing NUE is likewise mandatory to reduce nutrient losses that can be cause of pollution (Fageria et al., 2008). Fruit crops—temperate fruit crops (apple, pear, and prunus species), small fruits (strawberry, grapevine, ribes species, rubus species, and vaccinium species), tropical and subtropical fruit crops (avocado, banana, citrus species, kiwifruit, mango, papaya, persimmon, and pineapple)—are a major part of agricultural production, both in terms of human diet and economic importance. In a scenario where the fruit crop industry faces higher productions costs (labor, energy, and agrichemicals) and growers have to deal with climate change, water availability, and sustainability (Retamales, 2011), all the developments that increase yields and quality and are respectful of the environment should be pursued.

In arboriculture, a large body of research has made important advances in increasing water use efficiency, for example, regulated deficit irrigation strategies (De La Rosa et al., 2015; Gonzalez-Dugo et al., 2012; Intrigliolo et al., 2013; Laribi et al., 2013; Phogat et al., 2013), irrigation scheduling (Intrigliolo and Ramon Castel, 2010), partial root-zone drying (Faber and Lovatt, 2014; Yang et al., 2011), and different irrigation systems (Maris et al., 2015; Panigrahi et al., 2012). However, increasing NUE in fruit crops is complex (Srivastava and Malhotra, 2017). This complexity

A.K. Srivastava, Chengxiao Hu (eds.)
Fruit Crops: Diagnosis and Management of Nutrient Constraints
https://doi.org/10.1016/B978-0-12-818732-6.00030-7

is due to its large life span, extended and diverse physiological stages during its development with preferential requirement of some nutrients over others, differential root distribution pattern (root volume distribution), and special nutrients requirement by a specific fruit crop. This means a large variation in its nutrient preferences and therefore a challenge to increase of the efficient use of nutrients. In this context, novel approaches to cultivate tree fruits crops aimed to keep fruit trees in good nutritional and water conditions are still needed. Open field soilless system (OFSS) adapts the principles of commercial soilless culture to soil based production. Professor Rafael Martinez Valero (University Miguel Hernández, Alicante, Spain) was the first to bring all of the concepts together to develop OFH in the early 1990s. The original system that does not incorporate substrate to the soil was developed for the citrus industry, and the bibliography is still limited to citrus (Kruger et al., 2000; Kuperus et al., 2002; Martínez-Valero and Fernández, 2004; Stover et al., 2008), although recent reports suggest that this technique can be applied to other crops (Falivene et al., 2015; Morgan and Kadyampakemi, 2012). The key principles OFH have been described by Morgan and Kadyampakemi (2012) (i) reducing the size of the root zone by reducing the wetted soil volume, (ii) maintaining the soil volume always near field capacity, and (iii) the continuous application of a balanced nutrient solution. To achieve these principles, the irrigation system design, usually drip irrigation, provides water and nutrients on a short timescale (daily supply) rather than using the soil as a storage medium to release nutrients as required at particular physiological stages throughout time (monthly or quarterly). This system is an integrated set of practices that combine grove design, size limiting rootstocks, irrigation and nutrient management, and mechanical harvesting. In citrus, OFH has proven to increase productivity, early growth, water, and nutrient use efficiency and reduce nutrient leaching (Morgan et al., 2010). Its application has been proven to address low-fertility gravel base soils and saline water (Falivene et al., 2015). The inclusion of a substrate other than soil in this system, where roots can grow and develop and easily obtain water and nutrients, is an important advance in the OFH system. Besides, whenever soil conditions are unfavorable, soilless culture in open field can be a solution. Here, we present recent advances in incorporating the soilless technique into the open field hydroponics for fruit trees crops.

2 What is new in this system

Soilless culture is a method of growing plants in any medium other than soil. Soilless system techniques have been integrated into horticultural production, maximizing the efficiency in the use of water and nutrients (Gorbe and Calatayud, 2010; Urrestarazu, 2013; Van Kooten et al., 2004). The new system that uses soilless techniques and is named open field soilless system (OFSS) where substrate bags are incorporate to the field is also an insight that comes from soilless horticulture, and it is an important advancement for keeping the roots in a controlled environment (Rubio-Asensio et al., 2018). To be strict this is a near-soilless system since soil is also used by the growing tree. This system reduces the influence and interaction of the soil as a media to store water and nutrients by concentration in the active roots in the bag. The adaptation of soilless systems to tree orchards and its success depend almost entirely on the proportion of the root system colonizing the substrate with regard to the total amount of tree roots. Therefore, the time lapse until the roots colonized the substrate is one of the major factors affecting the efficiency and success of this system. In common with OFH, this system attempts to concentrate some roots, the more the better, but into the substrate bags, leaving these active roots in a balanced nutrient mixture after each irrigation event, inducing the tree to use water and nutrients more efficiently.

3 Implementation of the system in the field: Our experience

The use of substrate bags in the field introduces the possibility of changes in the arrangement of the substrate bags according to the crop size and the soil. The type of the substrate, the number of bags per tree, the size of the bag and its disposition in the ground and with respect to the trunk, the inclusion and the design of the irrigation system, and the water and nutrient management are important aspects to consider when OFSS are using in the field.

3.1 Choosing the substrate type

The election of the growing media, inorganic (e.g., perlite, rockwool, etc.) or organic (peat, coconut coir dust, etc.), is important since its physical, chemical, and biological characteristics will determine water and nutrient management (Urrestarazu, 2013; Savvas and Gruda, 2018). Since fruit trees at certain phenological periods could have a huge demand of water and nutrients (rapid phase of leaf growth) and also at a given point could have a very low or null demand of water and nutrients (dormancy), it is mandatory to have a medium with high buffering capacity, where occasional irrigation events during the tree dormancy keeps the roots healthy without disrupting water and nutrient plant status. For this purpose, organic growing media are the most suitable due to its high water retention capacity and high cation exchange capacity (CEC) (Dannehl et al., 2015; Domeno et al., 2009; Quintero et al., 2009).

An important consideration of using organic material is its "instability" (Verhagen, 2009). Organic materials are subjected to variable rates of microbial decomposition that can cause undesirable physical and chemical changes in the resultant growing medium (Jackson et al., 2009). The physical changes of the organic media may result in shrink (or "slump") within the plastic bag (Särkkä et al., 2008). Our experience with coconut substrate bags shows that after 4 years in the field there is an important change in the volume of substrate in the bag (Fig. 30.1). This physical change may lead to reduced air-holding capacity and excessive water retention (Nash and Pokorny, 1990). In cases where the vigor of the root (rootstock) is low it may not be a good idea to use the bags until the tree root system is well developed, thus reducing the time it takes for the roots to colonize the substrate. The chemical changes, as microbes decompose carbon compounds in organic material, consume plant available nutrients. This microbial uptake of nutrients, for example, nitrogen (Handreck, 1992) and phosphate (Handreck, 1996), can detrimentally affect plant performance if not appropriately compensated for (Handreck, 1993).

When analyzing nutrient input and output from a coconut fiber substrate bag without plant, we distinguished two groups of nutrients that differ in the balance between the concentration of nutrients in the irrigation water and amount of nutrients drained (Rubio-Asensio et al., 2019). In these conditions, the drained volume was similar to that irrigated, and therefore, if there is no absorption/consumption of nutrients in the substrate, the input and output amounts of nutrients should be similar. In the first group were NO_3^-, K^+, and Ca^{2+}; the concentration of these nutrients in the drainage of the substrate decreased with respect to that of entry into the irrigation water, indicating that there was a net consumption or adsorption by the substrate. The second group consisted of PO_4^{3-}, SO_4^{2-}, Mg^{2+}, Na^+, and Cl^-; the concentration of these nutrients in the drainage of the substrate was similar to that of entry into the irrigation water, indicating that there was no consumption or adsorption by the substrate. Moreover, our results comparing the electrical conductivity (EC) and pH of the irrigation solution with those of the drainage solution of a coconut coir dust bag without plant (Fig. 30.2) showed that there was consumption and/or transformation of nutrients by the substrate and its microorganism since EC decrease and pH decrease. To build up their own body protein components, the microorganism present in the substrate need nutrients, for example, mineral nitrogen (N), which they gain from the available N content in the substrate or supply by fertigation. Consequently, EC decrease and N may not be readily available for the plants. The decrease in the pH of the drainage in regard to the irrigation water may due to decomposition of organic N and the release of H^+ ions when soil bacteria transform $N-NH_4^+$ to $N-NO_3^-$, since there is no plant and therefore no absorption of $N-NH_4^+$.

Until relatively recently, the main drivers for the selection of the component materials in growing media were largely based on performance and economic considerations (Barrett et al., 2016). Nowadays, the use of renewable materials and ease to dispose them are also important drivers (Raviv, 2013). For this purpose, in relative terms, organic materials are renewable and easier to dispose of, making them a more environmentally sustainable option at field-scale

FIG. 30.1   External appearance of the substrate bags in the second (A) and fourth (B) year of experimentation in young mandarin trees.

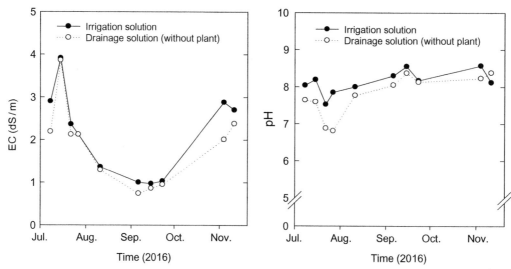

FIG. 30.2 Changes in the EC and pH of the irrigation solution and the drainage solution in a substrate coconut bag without plant.

FIG. 30.3 Two-year-old mandarin trees planted directly in coconut substrate bags (left line) and directly in the soil with coconut substrate bag on the soil (right line).

basis, and except the plastic the organic substrates can be left in the field. At present, few organic materials dominate soilless cultivation worldwide; these are primarily peat, coir, wood, and composted materials. We used coconut coir dust, a waste product of the coconut (*Cocos nucifera*) industry, consisting of the dust and short fibers derived from the mesocarp of the fruit (Barrett et al., 2016). They are from renewable, high-volume waste streams and, through various degrees of secondary processing, can provide growers with consistent, predictable results. Coconut coir dust substrate provides a favorable balance of air and water to plant roots and has both, high water holding capacity and high CEC (Blok and Wever, 2008). Under identical irrigation and fertilization conditions, mandarin trees planted directly in the coconut substrate performed better, in terms of nutritional leaf status, than trees where the bags were on the soil surface (Fig. 30.3), which demonstrate the suitability of this substrate for growing fruit trees. As a waste product of coconuts grown in coastal areas or washed in saline water (during primary processing), its physical, chemical, and biological properties can vary widely (Abad et al., 2005). Consequently, in addition to a period of aging to stabilize the material (Carlile et al., 2015), coir requires several washings in fresh water and a "buffering" treatment (in which calcium nitrate is added to the material to displace harmful concentrations of sodium and potassium) before it is suitable for use as a growing medium (Poulter, 2014).

By using coconut fiber, the OFSS relies on the specific characteristics of this substrate in terms of water and nutrient holding capacity to attain a high flexibility in the irrigation scheduling (Rubio et al., 2010, 2011). This will facilitate both the adaptation to conditions where the water for irrigation is uncertain or unreliable throughout the growing season and the adaptation to deciduous tree crops that need more spacing between irrigation events. Also, as much of the root system is in the soil, it serves as an anchor for the tree at the same time that root performance is buffered by the soil water and nutrient content.

3.2 Irrigation system and design according with the OFSS

The irrigation system that best suit the requirements of the OFSS is drip irrigation. Should be convenient to have drippers a short distance (20 cm) to make sure that the substrate is uniformly moistened and the nutrients are uniformly distributed, avoiding areas where salts can concentrate. If drippers go inserted in the pipeline, it should be place inside the bag and in the upper part. This has the inconvenience that drippers cannot be checked for obstructions. The design of the irrigation system and equipment should follow the indications described by Falivene et al. (2015). In the OFSS necessary is a precise control of the root environment once that roots colonize the substrate, for keeping water, nutrient concentration, and nutrient balance, at optimum. We introduce in the OFSS the "sampling station" for the substrate when it goes under the soil surface (Fig. 30.4). The objective is to monitor CE, pH, and nutrient concentration in the drainage solution and compare it with the irrigation solution. From our experience, the EC value is a good indicator of the substrate colonization by the roots (Fig. 30.5), better than the amount of solution drainage from the substrate since inevitably there are losses of nutrient solution from the sampling station. If the substrate goes on the soil surface, taking a sample of water with a syringe at the end of the irrigation event will be enough.

3.3 When, where, and how implement the substrate bag

Answering when, where, and how (number of bags per tree, size, and position of windows) to incorporate the bags of substrate is an important decision facing the farmer when it comes to deciding to apply OFSS. Firstly, when the bags have been placed in young plantations at the same time as the plantation, in persimmon trees (Rojo brillante/Lotus), the bags were colonized by the roots in the first year of the plantation, whereas in mandarin trees (Clemenrubi/Citrange carrizo) the roots have taken 3 years to colonize the substrate (Parra et al., unpublished results). However, in adult plantation of nectarine (Viowhite 15/GF 677) or peach (Baby gold 9/GF 677), the bags were colonized immediately (15–45 days) (Rubio-Asensio et al., 2018; Parra et al., unpublished results). Secondly, the desirable position of the substrate in regard to the trunk for young tress will be close to the trunk to facilitate the substrate colonization by the roots, whereas in adult trees the desirable position will be near to the current pipeline with drippers (Fig. 30.6).

FIG. 30.4 Sampling station for OFSS when the substrate is under the soil. Position of the station in the field in regard to the mandarin tree (left) and sketch of the sampling station (right).

FIG. 30.5 Volume values of irrigation solution and drainage solution and its EC in young persimmon trees at its second and third year after the plantation. The amount of root per bag harvested during the winter of each year is also indicated.

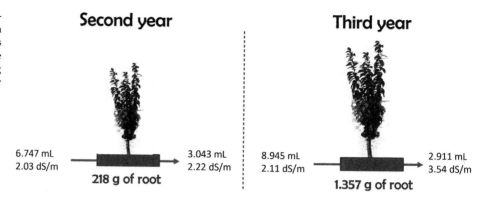

Second year

6.747 mL
2.03 dS/m → 3.043 mL
2.22 dS/m

218 g of root

Third year

8.945 mL
2.11 dS/m → 2.911 mL
3.54 dS/m

1.357 g of root

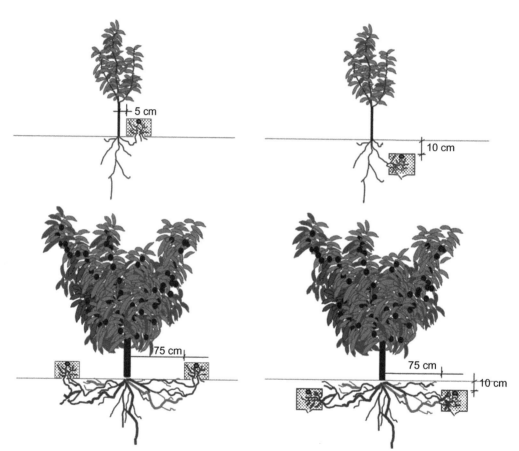

5 cm

10 cm

75 cm

75 cm

10 cm

FIG. 30.6 Position of the bags containing the substrate in regard to the tree, in young and adult trees.

With regard to the soil surface, we had compared substrate place on the soil surface and substrate place 10 cm under the soil surface (Fig. 30.6). On the soil surface, the cost of installing the substrate is much lower than under the soil surface, where heavy machinery is needed. Under the soil surface, there is also an important damage on the roots in adult trees, when digging the soil where the bags will be placed. The water losses by evaporation are lower when the substrate is under the soil surface; however, even when the substrate is on the soil surface, the plastic bags prevent water losses through evaporation. On the contrary, the higher daytime temperate in the root environment when the substrate is on the soil surface may enhance nutrient uptake (Tindall et al., 1990). To answer how to implement the system, a higher number of bags per tree or higher size of individual bags will put more substrate per tree and make the system more efficient. Finally, it remains to be decided where to make the windows in the substrate and its size, as well as the drainage openings (Fig. 30.6). This decision is important because it is a matter of perforating the plastic and letting in the roots or letting out the drainage. The greater the surface that we remove from plastic, the more possibility the roots will have of accessing the substrate but less control of the water and nutrients that are incorporated in the substrate by fertigation since the higher surface in contact with the soil.

3.4 Water and nutrient management

Extensive information on water and nutrient management in OFH and soilless system exist in the literature (Falivene et al., 2015; Morgan and Kadyampakemi, 2012) that is valid for the OFSS, and it is not the purpose of this chapter to go through these bases again. The combination of soilless culture techniques with the OFH techniques is a natural step for improving both water and nutrient efficiency in fruit crops. Firstly, OFSS relies on daily drip irrigation or pulse fertigation during daytime hours to maintain enough available water content through the day to satisfice tree demand of water for transpiration and evaporation. By incorporating the new growing media, the plastic bags, and the drip irrigation to the OFH management, the new system reduces the losses and inefficiencies due to evaporation. Therefore, it is important to note that the evaporation compound should be minimized since the water is applied inside the bag. In nectarine trees under water-deficit irrigation (65% of ETc), the use of substrate on the soil surface (80 L/tree of coconut substrate) reduced water-deficit effects on fruit yield by increasing the number of fruits per tree, which resulted in increasing the WUE by 60%, proving that the substrate favor water use efficiency, especially under deficit irrigation conditions (Rubio-Asensio et al., 2018). The irrigation event frequency is another important factor. Through the irrigation frequency, the system should keep a wealthy environment in the substrate bag in terms of available water, available nutrient, pH, and electrical conductivity. If any of these factors is under nonoptimal conditions, plants may suffer from stress leading to a decline of yields and product qualities. Secondly, by using fertigation—the application of water and nutrients at the same time in a nutrient solution—proper timing, rates, and balances among ions will make fulfilling fruit tree nutrient demand more easy at certain growth stages or development phase that will reduce the inefficiencies due to leaching and volatilization.

Greenhouse growers generally use high nutrient concentrations in an attempt to maximize crop yield (Rouphael and Colla, 2005), but this relationship is not necessarily straightforward in our system, since there are also roots outside the bag, and there is not any possibility of recovery of the drainage solution. So, with high nutrient concentration in the nutrient solution exist the risk of increase the EC of the soil to toxic values or having toxicity by specific ions such as Na^+ or Cl^-. In horticulture, several studies have documented the advantage of using lower concentrations than the standard. Zheng et al. (2005) and Rouphael et al. (2008) proved that nutrient solution concentration used by growers can be reduced by 50% without any adverse effect on biomass and quality parameters in geranium and gerbera, respectively. Similarly, pepper plants irrigated eight times per day and with half-strength Hoagland nutrient solution perform the same in terms of yield as plants irrigated with full-strength Hoagland nutrient solution (Rubio et al., 2011). Multiple applications of N in relatively small amounts with drip irrigation increased N uptake efficiency in citrus and contributed to minimized leaching of $N-NO_3^-$ (Alva and Paramasivam, 1998; Alva et al., 2006). These studies indicate that, with proper irrigation scheduling, soil N can be maintained in citrus tree root zones and not leached to groundwater. Therefore, as a proxy, it will be desirable to use nutrient solution with lower concentration than the conventional.

The efforts should be made from an environmental standpoint to find and use the less concentrated but optimum nutrient solution possible. A combination of low concentration with high frequency and low duration of the irrigation events may be desirable for trees crops. In general, increasing the frequency of irrigation reduces the variations in nutrient concentration, thereby increasing their availability to plants (Silber and Bar-Tal, 2008). In addition, long irrigation events increase the leaching fraction, which reduces the availability of nutrients to the roots (Lieth and Oki, 2008).

Under not restricted water and nutrient conditions, the performance of the OFSS has been studied during four seasons in a 6-year-old peach orchard and two new plantations of persimmon and mandarin (Parra et al., unpublished results). We kept trees without substrate (control trees) and trees with substrate on the soil surface or under the soil surface (substrate trees) (Fig. 30.7). In all crops, water and nutrients were applied by drip irrigation, to fulfill the 100% of ETc or the whole nutrient requirements of N, P_2O_5, and K_2O (peach (Espada-Carbó, 2010), persimmon (Pomares, 2014), and mandarin (Quinones et al., 2010)). In both trees growing in soil (standard conditions) and trees with substrate, the daily water dose and fertilizer dose were kept the same during the 4 years' study. The irrigation frequency was higher for trees with substrate (ranged from 2 to 4 events per day) to restore water and nutrients into the substrate at optimum levels for tree growth after each irrigation event. The results of using substrate on the different fruit crops anticipated that (i) persimmon trees roots colonized the substrate (on surface and under surface) at the first year, with high amounts of roots at the fourth year (Fig. 30.8), (ii) persimmon trees increased mean fruit weight, commercial yield, and tree growth (Table 30.1); (iii) mandarin trees delayed for 3 years the root colonization of the substrate, and did not

FIG. 30.7 Position of the coconut substrate bag in an adult peach orchard. (A) The substrate is on the soil surface with two windows in the bottom, and (B) the substrate is under the soil surface with two windows in a lateral.

FIG. 30.8 Persimmon roots inside a coconut substrate bag after 4 years with substrate on the soil surface.

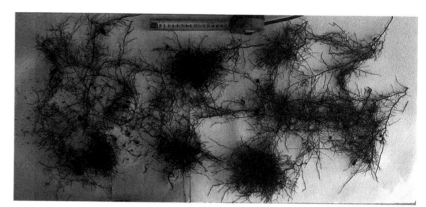

TABLE 30.1 Yield and mean fruit weight of different experiment with different fruit tree crops.

Tree crop	Growth conditions[a]	Treatments	Yield (kg/tree)	Mean fruit weight (g)	Effect of the substrate in the yield components
Persimmon (Parra et al., unpublished results)	3-year-old trees Two bags per tree Four drippers per bag	No substrate	26.4	241.5	Increases mean fruit weight
		Substrate on the soil surface	28.7	278.5	
		Substrate under the soil surface	27.8	288.6	
Peach (Parra et al., unpublished results)	15-year-old trees Three bags per tree Four drippers per bag	No substrate	79.7	223.7	Increases mean fruit weight
		Substrate on the soil surface	72.2	270.8	
		Substrate under the soil surface	70.3	249.4	
Nectarine (Rubio-Asensio et al., 2018)	5-year-old trees Two bags per tree Bags on the soil surface Two drippers per bag	100% ETc, not substrate	41.5	145.8	Increases the numbers of fruits per tree in the 65% ETc treatment
		65% ETc, not substrate	49.1	148.9	
		100% ETc, substrate	30.5	136.3	
		65% ETc, substrate	50.3	138.5	

[a] In all the experiments, the substrate bags were 40 L volume filled with coconut coir fiber.

colonize the substrate on the soil surface; (iv) mandarin trees after the second year increased tree growth and canopy volume; and finally, (iv) adult peach roots colonized the substrate immediately (days) and increased mean fruit weight (Table 30.1). These evidences support our argument that substrate incorporated to the OFH technique, said OFSS, will contribute to increase water and nutrient use efficiency in a wide range of fruit crops.

When using the OFSS under restricted water and nutrient conditions in a 5-year-old nectarine orchard, the results showed that the colonization of the substrate by the root system, the suitable substrate water content profiles, and values of the midday stem water potential indicated that the supportive substrate could be a good strategy for irrigating tree orchards in conditions of deficit irrigation (Rubio-Asensio et al., 2018). In these conditions, the presence of the substrate increased the number of fruits per tree and increased fruit yield (Table 30.1). Open field hydroponics relies on providing water and nutrients directly to the tree's roots without the interference of the soil's role as a reservoir for water and nutrients in a demand-based system (Morgan and Kadyampakemi, 2012), and this may be a risk in periods and places where water is not a secure resource. For these reasons, just as the OFH is suitable for perennial trees, the near-soilless system could be more suitable for deciduous trees that experience long periods without irrigation supply.

3.5 Advantages

- *Greater efficiency in the combined use of water and nutrients.* Once that the root system has colonized the substrate, the joint application of water and nutrients into the confined space of the bag will make the use of water and nutrients by the trees more efficient. In peach trees under different irrigation and nitrogen levels, WUE was related to the combined effects of irrigation and nitrogen, and a rational N application improved WUE (Pascual et al., 2016).
- *Cultivation in unsuitable soils.* Adding substrate into the fields will let to cultivate in soils that due its low fertility, salinity or sodic conditions are inappropriate for specific tree crops. Besides, in soils with high variability in fertility, the substrate will standardize the medium where the roots thrive, which will have a positive impact on the achievement of more homogeneous fruit yields and homogeneous fruit quality.
- *Environmentally friendly substrate.* As compared with other substrates, such as rockwool or perlite, coconut fiber is biodegradable and environmentally friendly.
- *Suitable for diverse techniques aimed to save water.* By using two lines with substrate per tree file, OFSS could be adapted to include partial root-zone drying (PRD), aimed to increase WUE without detrimental effect on fruit quality, fruit yield, or fruit size (Yang et al., 2011). Partial root-zone drying involves approximately half of the root system exposes to drying soil, while the remaining half is irrigated as conventional irrigation, and the wetting and drying sides of the root system are repeatedly alternated with certain frequency. Partial root-zone drying reduces transpiration rate significantly and maintained higher level of photosynthesis rate, which leads to the increase of leaf water use efficiency (WUE) (Kirda et al., 2007).

3.6 Disadvantages

- *High initial investment.* These systems imply a higher initial capital investment for the grower. The cost of the substrate, the implantation of the substrate in the field, and the implantation of the irrigation system with the substrate need to be considered. Even if the substrate bag is on the soil surface, the soil under the bag had to be cleared of stones and leveled to ensure an adequate contact between the coconut fiber and the soil surface.
- *Qualified staff.* The personnel in charge of soilless systems should be trained to maintain the substrate conditions optimum for root growth. This implies periodic controls of the EC, pH, and stability of the substrate. The high cation exchange capacity of the coconut fiber could lead to high concentrations of toxic cations such Na^+, which could be a problem even when Na^+ concentration in the irrigation water is low (Rubio et al., 2011). To avoid this, a leaching fraction is required to reduce the salt concentration in the substrate.
- *Keep soil-substrate interface.* The bags could not be moved to avoid disrupting the soil-substrate interface and let the roots colonize the substrate.
- *Drip emitters cannot be checked.* The in-line drip emitters inside the bag could not be checked for obstruction, which made the periodic use of acidic solutions imperative to clean the pipes and to avoid the obstruction in the drip emitters as much as possible.

- *Weeds.* The use of the substrate plastic bags was suitable in an agricultural system where the use of herbicides was allowed for maintaining the weeds controlled; otherwise, weed roots could colonize the substrate, and even some herbs such as *Cyperus rotundus* could perforate the bag.
- *Plastic stability.* The stability of the plastic surrounding the substrate over time may be another disadvantage, but in our conditions, this was not a problem in the fourth year after placing the bags in the field. In adult trees, during the months of greater sun exposure (spring and summer), the bags were partially protected by the shade provided by the trees.
- *Substrate stability.* The stability of the substrate, as it is an organic substrate and is therefore chemically active, needs close monitoring, especially for changes in the pH (Abad et al., 2004) and water holding capacity. In the Mediterranean conditions, the substrate started losing the volume in the third year after placing the bags in the field.

4 Conclusions and future research

Fruit crop cultivation is expanding quickly due to its importance in the human diet. To keep this tendency, an arboriculture sustainable, efficient, and able to cope with climate change and water and nutrients restrains is necessary. Previous and recent results of the implementation of the OFSS at field-scale basis for fruit crops demonstrated that soilless system techniques that maximized water and/or nutrient use by the plant, such as the incorporation of substrate at field-scale basis, could be successfully applied in fruit trees in the field. Coconut fiber bags have proven to adapt well to a diverse fertigation strategies, and its use in the field is a natural step to maximize the water and nutrient use efficiency when advanced production techniques are applied, in both deciduous trees and perennial trees. The fact that water and nutrients are provided inside the substrate bag and roots successfully colonized the bag in a short-medium term will help to manipulate tree physiology to maximize yield and fruit quality. Incorporating high tree density in combination with size-controlling rootstock selection to OFSS will help to obtain higher yield per square meter and probably earlier harvests. This work also left many open questions that will need to be addressed before the system could be implemented at commercial field crop scale: (i) it is still necessary go deeper in how different tree crops and different tree ages and substrate positioning will affect the colonization of the substrate by the root system, (ii) how this system performs to adjust the nutrients application with plant nutrient demand, (iii) how to adapt the fertigation protocols to the system to attract roots to the substrate and keep them in good condition, (iv) how this system influences the fruit quality, for example, with proper management of the salt concentration of the nutrient solution (v) explores substrates other than coconut fiber and explore biodegradable plastics, (vi) how this system will perform in conditions of water and nutrient scarcity, (vii) how to plan the implantation of substrate in time and space, so that there were always roots in the substrate in tree plantations that last several years in the field, and (viii) study the effect of root temperature in tree physiology and fruit quality. It will also be important to complement the studies, where it is possible, of a cost-benefit study to know the economic viability of this system in real field conditions.

Acknowledgments

This research was funded by a grant agreement between "Comercial Projar" and the "Instituto Valenciano de Investigaciones Agrarias" (IVIA) and carried out within the framework of the MINEICO funded Project Hipofrut RTC-2014-2141-2 in collaboration with the Sistema AZUD company.

References

Abad, M., Noguera, P., Carrión, C.B., 2004. Los sustratos en los cultivos sin suelo. In: Urrestarazu, M.G. (Ed.), Tratado de Cultivo sin Suelo. Mundi-Prensa, España, Madrid, pp. 113–158.

Abad, M., Fornes, F., Carrión, C., Noguera, V., Noguera, P., Maquieira, A.N., Puchades, R., 2005. Physical properties of various coconut coir dusts compared to peat. Hortscience 40, 2138–2144.

Alva, A.K., Paramasivam, S., 1998. Nitrogen management for high yield and quality of citrus in sandy soils. Soil Sci. Soc. Am. J. 62, 1335–1342.

Alva, A.K., Paramasivam, S., Obreza, T.A., Schumann, A.W., 2006. Nitrogen best management practice for citrus trees—I. Fruit yield, quality, and leaf nutritional status. Sci. Hortic. 107, 233–244.

Barrett, G.E., Alexander, P.D., Robinson, J.S., Bragg, N.C., 2016. Achieving environmentally sustainable growing media for soilless plant cultivation systems—a review. Sci. Hortic. 212, 220–234.

Blok, C., Wever, G., 2008. Experience with selected physical methods to characterize the suitability of growing media for plant growth. Acta Hortic 779, 239–250.

Carlile, W.R., Cattivello, C., Zaccheo, P., 2015. Organic growing media: constituents and properties. Vadose Zone J. 14.

Dannehl, D., Suhl, J., Ulrichs, C., Schmidt, U., 2015. Evaluation of substitutes for rock wool as growing substrate for hydroponic tomato production. J. Appl. Bot. Food Qual. 88, 68–77.

De La Rosa, J.M., Domingo, R., Gomez-Montiel, J., Perez-Pastor, A., 2015. Implementing deficit irrigation scheduling through plant water stress indicators in early nectarine trees. Agric. Water Manag. 152, 207–216.

Domeno, I., Irigoyen, N., Muro, J., 2009. Evolution of organic matter and drainages in wood fibre and coconut fibre substrates. Sci. Hortic. 122, 269–274.

Espada-Carbó, J.L., 2010. Abonado De Los Frutales Caducifolios. In: Ministerio De Medio Ambiente Y Medio Rural Y Marino (Ed.), Guía Práctica De La Fertilización Racional De Los Cultivos En España. Ministerio De Medio Ambiente Y Medio Rural Y Marino, Madrid, España.

Faber, B.A., Lovatt, C.J., 2014. Effects of applying less water by partial root zone drying versus conventional irrigation on navel orange yield. In: Braun, P., Stoll, M., Zinkernagel, J. (Eds.) VII International Symposium on Irrigation of Horticultural Crops.

Fageria, N.K., Baligar, V.C., Li, Y.C., 2008. The role of nutrient efficient plants in improving crop yields in the twenty first century. J. Plant Nutr. 31, 1121–1157.

Falivene, S.G., Navarro, J.M., Connolly, K., 2015. Open hydroponics of citrus compared to conventional drip irrigation best practice: first three years of trialling and Australian experience. In: Sabatermunoz, B., Moreno, P., Pena, L., Navarro, L. (Eds.) XII International Citrus Congress—International Society of Citriculture.

Gonzalez-Dugo, V., Suarez, L., Ruz, C., Berni, J.A.J., Zarco-Tejada, P.J., Soriano, M.A., Fereres, E., 2012. Responses of nectarine to regulated deficit irrigation at the field scale. In: VII International Peach Symposium. vol. 962. pp. 349–353.

Gorbe, E., Calatayud, A., 2010. Optimization of nutrition in soilless systems: a review. In: Kader, J.C., Delseny, M. (Eds.) Advances in Botanical Research.In: vol. 53.

Handreck, K.A., 1992. Rapid assessment of the rate of nitrogen immobilisation in organic components of potting media: I. Method development. Commun. Soil Sci. Plant Anal. 23, 201–215.

Handreck, K.A., 1993. Use of the nitrogen drawdown index to predict fertilizer nitrogen requirements in soilless potting media. Commun. Soil Sci. Plant Anal. 24, 2137–2151.

Handreck, K.A., 1996. Phosphorus immobilization in wood waste-based potting media. Commun. Soil Sci. Plant Anal. 27, 2295–2314.

Intrigliolo, D.S., Ramon Castel, J., 2010. Response of grapevine cv. 'Tempranillo' to timing and amount of irrigation: water relations, vine growth, yield and berry and wine composition. Irrig. Sci. 28, 113–125.

Intrigliolo, D.S., Bonet, L., Nortes, P.A., Puerto, H., Nicolas, E., Bartual, J., 2013. Pomegranate trees performance under sustained and regulated deficit irrigation. Irrig. Sci. 31, 959–970.

Jackson, B.E., Wright, R.D., Seiler, J.R., 2009. Changes in chemical and physical properties of pine tree substrate and pine bark during long-term nursery crop production. Hortscience 44, 791–799.

Kirda, C., Topcu, S., Cetin, M., Dasgan, H.Y., Kaman, H., Topaloglu, F., Derici, M.R., Ekici, B., 2007. Prospects of partial root zone irrigation for increasing irrigation water use efficiency of major crops in the Mediterranean region. Ann. Appl. Biol. 150, 281–291.

Kruger, J.A., Britz, K., Tolmay, C.D., Du Plessis, S.F., 2000. Evaluation of an open hydroponic system (OHS) for citrus in South Africa: preliminary results. Proc. Int. Soc. Citric. 9, 239–242.

Kuperus, K.H., Combrink, N., Britz, K., Ngalo, J., 2002. Evaluation of an open hydroponics system (OHS) for citrus in South Africa. Annual Researh Report. Citrus Research International (Pty) Ltd.

Laribi, A.I., Palou, L., Intrigliolo, D.S., Nortes, P.A., Rojas-Argudo, C., Taberner, V., Bartual, J., Perez-Gago, M.B., 2013. Effect of sustained and regulated deficit irrigation on fruit quality of pomegranate cv. 'Mollar De Elche' at harvest and during cold storage. Agric. Water Manag. 125, 61–70.

Lieth, J.H., Oki, L.O., 2008. Irrigation in soilless production. In: Raviv, M., Lieth, J.H. (Eds.) Soilless Culture: Theory and Practice. Elsevier, Amsterdam.

Maris, S.C., Teira-Esmatges, M.R., Arbones, A., Rufat, J., 2015. Effect of irrigation, nitrogen application, and a nitrification inhibitor on nitrous oxide, carbon dioxide and methane emissions from an olive (Olea europaea L.) orchard. Sci. Total Environ. 538, 966–978.

Martínez-Valero, R., Fernández, C., 2004. Preliminary results in citrus groves grown under the Moht system. In: El-Otmani, M., Ait-Oubahou, A. (Eds.) 10th International Citrus Congress. Morocco: International Society Of Citriculture.

Morgan, K., Kadyampakemi, D., 2012. Open field hydroponics: concept and application. In: Srivastava, A.K. (Ed.), Advances in Citrus Nutrition. Springer, Netherlands.

Morgan, K.T., Zotarelli, L., Dukes, M.D., 2010. Use of irrigation technologies for Citrus trees in Florida. HortTechnology 20, 74–81.

Morison, J.I.L., Baker, N.R., Mullineaux, P.M., Davies, W.J., 2008. Improving water use in crop production. Philos. Trans. R. Soc. Lond B, Biol. Sci. 363, 639–658.

Nash, M.A., Pokorny, F.A., 1990. Shrinkage of selected two-component container media. Hortscience 25, 930–931.

Panigrahi, P., Srivastava, A.K., Huchche, A.D., 2012. Effects of drip irrigation regimes and basin irrigation on Nagpur mandarin agronomical and physiological performance. Agric. Water Manag. 104, 79–88.

Pascual, M., Villar, J.M., Rufat, J., 2016. Water use efficiency in peach trees over a four-years experiment on the effects of irrigation and nitrogen application. Agric. Water Manag. 164, 253–266.

Phogat, V., Skewes, M.A., Mahadevan, M., Cox, J.W., 2013. Evaluation of soil plant system response to pulsed drip irrigation of an almond tree under sustained stress conditions. Agric. Water Manag. 118, 1–11.

Pomares, F., 2014. Necesidades Nutricionales Del Cultivo Del Caqui. Vida Rural 375, 14–18.

Poulter, R., 2014. Quantifying differences between treated and untreated coir substrate. In: International Society for Horticultural Science (Ishs), Leuven, Belgium, pp. 557–564.

Quinones, A., Martinez-Alcántara, B., Primo-Millo, E., Legaz, F., 2010. Abonado De Los Cítricos. In: Ministerio De Medio Ambiente Y Medio Rural Y Marino (Ed.), Guía Práctica De La Fertilización Racional De Los Cultivos En España. Ministerio De Medio Ambiente Y Medio Rural Y Marino, Madrid, España.

Quintero, M.F., Gonzalez-Murillo, C.A., Florez, V.J., Guzman, J.M., 2009. Physical evaluation of four substrates for cut-rose crops. In: Rodriguezdelfin, A., Martinez, P.F. (Eds.) International Symposium on Soilless Culture and Hydroponics.

Raviv, M., 2013. Composts in growing media: what's new and what's next? In: International Society for Horticultural Science (Ishs), Leuven, Belgium, pp. 39–52.

Retamales, J.B., 2011. World temperate fruit production: characteristics and challenges. Rev. Bras. Frutic. 33, 121–130.

Rouphael, Y., Colla, G., 2005. Growth, yield, fruit quality and nutrient uptake of hydroponically cultivated zucchini squash as affected by irrigation systems and growing seasons. Sci. Hortic. 105, 177–195.

Rouphael, Y., Cardarelli, M., Rea, E., Colla, G., 2008. The influence of irrigation system and nutrient solution concentration on potted Geranium production under various conditions of radiation and temperature. Sci. Hortic. 118, 328–337.

Rubio, J.S., Rubio, F., Martinez, V., Garcia-Sanchez, F., 2010. Amelioration of salt stress by irrigation management in pepper plants grown in coconut coir dust. Agric. Water Manage. 97, 1695–1702.

Rubio, J.S., Pereira, W.E., Garcia-Sanchez, F., Murillo, L., Garcia, A.L., Martinez, V., 2011. Sweet pepper production in substrate in response to salinity, nutrient solution management and training system. Hortic. Bras. 29, 275–281.

Rubio-Asensio, J.S., Franch, V., Lopez, F., Bonet, L., Buesa, I., Intrigliolo, D.S., 2018. Towards a near-soilless culture for woody perennial crops in open field conditions. Sci. Hortic. 240, 460–467.

Rubio-Asensio, J.S., Parra, M., Abrisqueta, I., Hortelano, D., Intrigliolo, D.S., 2019. Balance De Nutrientes En Árboles Jóvenes De Caqui Mediante Técnicas De Cultivo Semi-Hidropónico. In: Sech, (Ed.), VII Jorndas Fertilización Sech, Valencia, Spain.

Särkkä, L.E., Tuomola, P., Reinikainen, O., Herranen, M., 2008. Long-term cultivation of cut gerbera in peat-based growing media. In: International Society for Horticultural Science (Ishs), Leuven, Belgium, pp. 423–430.

Savvas, D., Gruda, N., 2018. Application of soilless culture technologies in the modern greenhouse industry—a review. Eur. J. Hortic. Sci. 83, 280–293.

Silber, A., Bar-Tal, A., 2008. Nutrition of substrate-grown plants. In: Raviv, M., Lieth, J.H. (Eds.) Soilless Culture: Theory and Practice. Elsevier, Amsterdam.

Srivastava, A.K., Malhotra, S.K., 2017. Nutrient use efficiency in perennial fruit Cropsa review. J. Plant Nutr. 40, 1928–1953.

Stover, E., Castle, W.S., Spyke, P., 2008. The Citrus grove of the future and its implications for Huanglongbing management. Proc. Fla State Hortic. Soc. 121, 155–159.

Tindall, A.J., Mills, H., Radcliffe, D.E., 1990. The effect of root zone temperature on nutrient uptake of tomato. J. Plant Nutr. 13, 939–956.

Urrestarazu, M., 2013. State of the art and new trends of soilless culture in Spain and in emerging countries. In: Martinez, F.X., Carlile, W.R., Bures, S. (Eds.) International Symposium on Growing Media, Composting and Substrate Analysis.

Van Kooten, O., Heuvelink, E., Stanghellini, C., 2004. Nutrient supply in soilless culture: on-demand strategies. In: Cantliffe, D.J., Stoffella, P.J., Shaw, N.L. (Eds.) Proceedings of the VIIth International Symposium on Protected Cultivation in Mild Winter Climates: Production, Pest Management and Global Competition, Vols 1 and 2.

Verhagen, J.B.G.M., 2009. Stability of growing media from a physical, chemical and biological perspective. Acta Hortic. 819, 135–141.

Yang, Q., Zhang, F., Li, F., 2011. Effect of different drip irrigation methods and fertilization on growth, physiology and water use of young apple tree. Sci. Hortic. 129, 119–126.

Zheng, Y., Graham, T.H., Richard, S., Dixon, M., 2005. Can low nutrient strategies be used for pot Gerbera production in closed-loop subirrigation? In: International Society for Horticultural Science (Ishs), Leuven, Belgium, pp. 365–372.

31

Role of biochars in soil fertility management of fruit crops

Raffaella Petruccelli[a],*, *Sara Di Lonardo*[b]

[a]Institute of BioEconomy—Italian National Research Council (IBE-CNR), Sesto Fiorentino, Italy
[b]Research Institute on Terrestrial Ecosystems—Italian National Research Council (IRET-CNR), Sesto Fiorentino, Italy
*Corresponding author. E-mail: petruccelli@ivalsa.cnr.it

1 Introduction

The development and growth, as well as the high yield and the good quality of agricultural crops, depend on the relationships among atmosphere, plant, and soil (White and Brown, 2010; McGrath et al., 2014). In particular, soil fertility plays a central role in crop production. "Agricultural soil fertility" is the soil ability to provide physical, chemical, and biological characteristics (e.g., organic matter content, acidity, texture, depth, and water retention capacity) needed for production of the "crops desired" (Cooke, 1967; Diacono and Montemurro, 2010). These characteristics provide a support to root growth, a habitat for soil microorganisms and, above all, offer water and nutrient supplies to plants (Larson and Pierce, 1994). There are 17 "essential nutrients" for plants of which 6 are major nutrients (nitrogen, phosphorus, potassium, calcium, magnesium, and sulfur), 8 are micronutrients (boron, chlorine, cobalt, copper, iron, manganese, molybdenum, and zinc), and 3 are structural nutrients (carbon, hydrogen, and oxygen). These elements are involved in all the physiological and metabolic processes of the plant life cycle (Marschner, 1995; Epstein and Bloom, 2005; McGrath et al., 2014). In general, fertilization (synthetic mineral fertilizers and organic amendments), decomposition (via mineralization by soil microorganisms) of soil organic matter, and nutrients buffering capacity of soils supply these nutrients. Plants often face significant challenges in obtaining an adequate supply of these nutrients to meet the demands of basic cellular processes due to their relative immobility. The success of soil management and the high agricultural productivity could be determined by maintaining a sufficient level of organic matter in soil and ensuring an efficient biological cycle of nutrients (McGrath et al., 2014). Mineral fertilization is the major physical input into agricultural production and, in particular, the use of organic amendments to restore C stocks and soil biological functions is particularly important (Lal, 2004). The regular addition of exogenous organic matter has the additional key role of supplying the essential nutrients to maintain crop growth and yield. Although the use of fertilizers is

A.K. Srivastava, Chengxiao Hu (eds.)
Fruit Crops: Diagnosis and Management of Nutrient Constraints
https://doi.org/10.1016/B978-0-12-818732-6.00031-9

necessary for food production, potentially detrimental effects on long-term soil fertility and environmental pollution (water, soil, and air pollution) have also been recognized (Savci, 2010; Bitew and Alemayehu, 2017).

Traditionally, organic amendments such as manures and composts have been applied in agricultural soils (Albiach et al., 2000). Moreover, other strategies have been used to increase fertilizer-use efficiency and to reduce negative impacts on the environment, such as precision fertilization and environmentally friendly fertilizers (Albiach et al., 2000; Diacono and Montemurro, 2010; Lü et al., 2016; Chen et al. 2018). More recently, residues from bioenergy production have been used as "fertilizers" (Galvez et al., 2012; Cayuela et al., 2014), and among all bioenergy by-products, the use of biochar has recently captured the most attention within the soil science community due to its positive effects.

Biochar is a predominantly stable, recalcitrant organic carbon compound, created by heating biomass (feedstock) under temperatures usually between 300°C and 1000°C with low (preferably zero) oxygen concentrations (Verheijen et al., 2010). In addition to its action in removing carbon from the atmosphere as biochar is a potential carbon sequestration matrix (Lehmann, 2007; Matovic, 2011; Roberts et al., 2010; Whitman et al., 2011a,b, 2013), it could also increase production and soil fertility (Hossain et al., 2010; Liu et al., 2013; Jeffery et al., 2015) especially by improving low-quality soils (Chen et al., 2008; Cao et al., 2009; Lu et al., 2012), diminish disease incidence in crops (Elad et al., 2010; Elmer and Pignatello, 2011), reduce nutrient leaching loss by reducing fertilizer needs (Liang et al., 2006; Laird et al., 2010), and provide an appropriate habitat for beneficial microorganisms (Schmalenberger and Fox, 2016; Zhu et al., 2017). These results have been achieved in experiments where herbaceous or annual plants were involved; there are no many data on the management of trees/plants or fruit crops, in general. The knowledge on fruit crops are at the moment not so long explored. In this review, we critically discussed the role of biochar for the soil fertility in the management of fruit crops. Moreover, the general mechanisms of biochar in the improvement of soil fertility were also reviewed.

2 Biochar: Production and chemical and physical characteristics

Biochar, the solid material formed during the thermochemical decomposition of biomasses, is defined, by the International Biochar Initiative (http://www.biochar-international.org/biochar), as "a solid material obtained from the carbonization of biomass." Biochar is the product of thermochemical biomass conversion processes such as pyrolysis (Chan et al. 2007), gasification, hydrothermal carbonization, flash carbonization (Antal Jr and Gronli, 2003; Wade et al. 2006; Chen et al. 2016), and torrefaction (Benavente and Fullana 2015; Chioua et al. 2016). In all these processes, temperature could be different, and so, the resulted biochar has different characteristics despite the use of the same initial feedstock (Cha et al. 2016). It has been shown that biochar characteristics and the yield are influenced also by the production variables such as highest treatment temperature and holding time at highest treatment temperature.

In general, feedstock properties (both physical and chemical) and treatment temperature are considered to be among the main factors influencing biochar characteristics (Novak et al., 2009; Bird et al., 2010; Enders et al., 2012) (Fig. 31.1). The use and the utility of a specific biochar depend on its properties. Biochar rich in available nutrients and minerals and/or showing high water-holding capacity could be better used as soil amendments to improve fertility (Graber et al., 2010). For example, increases of pH, CEC, and trace metals concentration occur with increasing production temperature (Hossain et al., 2011; Yuan et al., 2011; Mukherjee et al., 2011). Biochars derived from wood biomasses often have higher surface area than grass biochar (Mukherjee et al., 2011; Kloss et al., 2012). However, most previous studies focused on the impact of production parameters on physicochemical properties of biochar used either a narrow range of feedstock materials, often falling into one or two categories such as agricultural residues, wood derivatives, or manures, or a narrow range of production temperatures. For example, Cantrell et al. (2012) studied the impact of pyrolysis temperature and manure source on physicochemical characteristics of five manures biochar made at only two temperatures.

Pereira et al. (2011) investigated the labile fraction of C in biochar derived from three trees (pine, poplar, and willow) at two temperatures. Biochars of the feedstock with the same category might show similar properties compared with those made from parent material of very different types. If the benefits offered by biochar have to be maximized, it is important to develop an understanding of its physiochemical variations and their relations to functions in soil for a broad range of biochar types. The biochar parameters most affected by feedstock properties are total organic carbon (TOC), fixed carbon, and mineral elements of biochar. On the other hand, biochar surface area and pH is mainly, influenced by highest treatment temperature. Biochar recalcitrance is mainly determined by production temperature, while the potential total C sequestration depends more on feedstock.

Overall, the work sheds some light on the relative importance of different biochar production process parameters on the final biochar product, which is an important step toward "designed" biochar (Zhao et al., 2013). In fact, optimizing

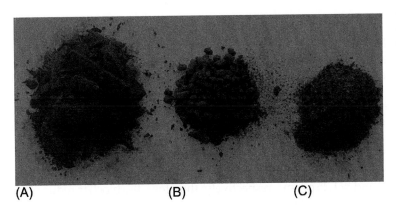

FIG. 31.1 Different biochar types. (A) Biochar from hardwood; (B) biochar from maize silage; (C) biochar from wheat straw.

biochar for a specific application may require a purposeful selection of a feedstock and production technique process, within constraints given by particular scenario, to produce biochar with desired characteristics (Ippolito et al., 2012). Considering this evaluation, moreover, a recent meta-analysis confirmed that the greatest positive impacts of biochar may occur in acidic and neutral pH soils, suggesting that one of the main mechanisms for yield increase may be a liming effect of biochar in soil (Jeffery et al., 2011).

3 Effects of biochar on the soil-plant system

In the recent past, many reviews (Lehmann, 2007; Jeffery et al., 2011; Spokas et al., 2012; Biederman and Harpole, 2013; Solaiman and Anawar, 2015; Ding et al., 2016; Srivastava, 2017; Subedi et al., 2017; Shaaban et al., 2018) highlighted that biochar amendment application to cropland can affect, directly or indirectly, soil characteristics by interdependent modification of chemical, physical, and biological properties (Fig. 31.2). In addition, the effective management of these properties could cause an increase in soil fertility with the ability to promote plant growth and increase production (Srivastava, 2017). Given the intrinsic heterogeneity of biochars, the complexity of the soil-biochar interaction, and the scarcity of long-term studies on biochar effects on soil properties (Zhang et al., 2016), attention must be paid to generalizing on the biochar potential role of soil-plant system. Indeed, results with "no effect" and even "negative effects" have also been reported, both in soil fertility and yield improvements.

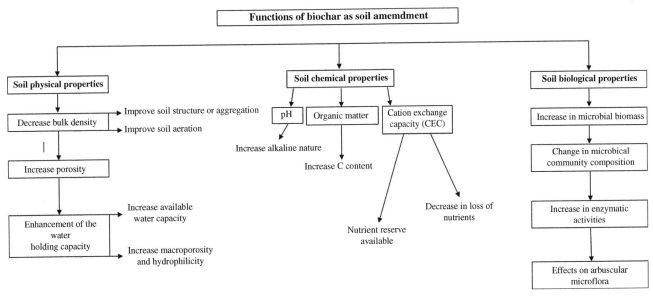

FIG. 31.2 Physical, chemical and biological soil properties that can be modified after biochar addition.

3.1 Effects on soil physical properties

The addition of biochar to soil can potentially alter its physical properties (i.e., texture, structure, surface area, particle size, total porosity, bulk density, soil moisture content, and water-holding capacity), although discordant results have been observed. Biochar addition can significantly decrease bulk density (from 3% to 31%) (Omondi, 2016) and increase the porosity of the soil (from 2% to 41%) modifying field crops behavior and production. Recently, Blanco-Canqui (2017) in an exhaustive review analyzed the influence of biochar on the mechanical structural and hydraulic soil properties, highlighting the limits and opportunities of the use and biochar management in different soils. Both chemical and physical properties of biochar, such as its high surface area and a highly porosity (Keech et al., 2005; Suliman et al., 2017), are key factors controlling the many physical properties of the soil system.

Blanco-Canqui (2017) has hypothesized two mechanisms responsible for bulk density reduction: (1) the bulk density of the biochar is lower than that of the soil, <0.6 and $\approx 1.25\,g/cm^3$, respectively; therefore, the biochar reduces the density of the bulk soil through dilution or mixing effect, and (2) the reduction could be due to the aggregation that occurs, in the long term, between the biochar and the soil particles. Several authors (Oguntunde et al., 2008; Ding et al., 2016) reported that the decrease in bulk density depends on the biochar rate and soil type. Rogovska et al. (2016) reported that amounts of biochar below 60 Mg/ha had greater effects on reducing bulk density. Furthermore, the effect of biochar is more evident in sandy soil than in clayey soil, and in coarse-textured soils than in fine-textured soils (Blanco-Canqui, 2017). As a final point, recent studies (Mukherjee and Lal, 2014) have also shown that the application of biochar did not affect bulk density.

Evidence suggests that biochar addition can increase the soil total porosity (Liu et al., 2014, 2017; Omondi, 2016). That is due to the following: (1) the interaction of the biochar particles with the soil particles, (2) the decrease in the bulk density, and (3) the increase in the state of soil aggression (Hardie et al., 2014; Blanco-Canqui, 2017). The increase in porosity has positive implications for the movement of water and gases in the soil. Biochar appears to increase porosity more in coarse-textured soils than in fine-textured soils (Blanco-Canqui, 2017).

Biochar application can also alter the soil pore size distribution and the soil's aggregate structure. Different mechanisms have been proposed to explain the action of the biochar on the soil pore size distribution particles (e.g., the intrinsic porosity of the biochar (Fig. 31.3), the increase of the aggregation between the particles of the soil, and the biochar and their rearrangement (Hardie et al., 2014; Blanco-Canqui, 2017)). Some authors reported that biochar can reduce the amount of drainable pores (Petersen et al., 2016) and increase the amount of mesopores (Lu et al., 2014); these effects could increase available water into soil (Blanco-Canqui, 2017).

FIG. 31.3 SEM image of pores of biochar produced by a woody plant feedstock.

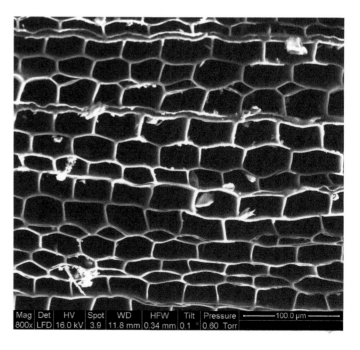

The specific surface area and high porosity of biochar have positive effects on soil water-holding capacity or plant available water content observing an increase of up to 30% in the soil water-holding capacity (Novak et al., 2009; Jeffery et al., 2011; Kinney et al., 2012; Hardie et al., 2014; Laghari et al., 2016). The effect on water-holding capacity could be particularly positive in sandy soils, where the abundance of macropores limits the capacity for holding water. Therefore, it can be neutral in medium-textured soil and potentially detrimental in clay soil (Sohi et al., 2010). The application of biochar to croplands and the subsequent increase in plant available water content can contribute to the rational use of water, thus reducing the frequency of irrigation, in particular in water-limited or semiarid regions. Nevertheless, it is necessary to underline that an increase of water-holding capacity does not always correspond to an increase in the water availability for the plant (Sohi et al., 2010).

In a field experiment for two consecutive seasons, Baronti et al. (2014) and Genesio et al. (2015) reported that the addition of a large volume of biochar (22–44 Mg/ha) on vineyard determined an increase in soil water content and in soil bulk density. However, other authors reported no significant changes in soil water-holding capacity. For example, Ventura et al. (2013) showed inconsistencies into soil water retention capability when biochar was applied to soil in apple orchard. Also, Hardie et al. (2014) observed that biochar application in apple orchard had no significant effects on soil hydrological properties and Jeffery et al. (2015) in a sandy soil. Different mechanisms have been proposed to explain the neutral results, for example, the replacement of the clay particles by biochar ones, increase of leaching flows, and/or decrease in water infiltration caused by hydrophobicity of biochar (Verheijen et al., 2010; Major et al., 2010). Furthermore, it has been observed from some authors that the biochar can relieve soil compaction (Laird et al., 2010), reduce tensile strength and penetration resistance (Chan et al., 2008; Busscher et al., 2010), reduce runoff, decrease erosion (Asai et al., 2009), and affect soil aggregation (Fungo et al., 2017) through interactions with soil microorganisms, organic matter, and minerals (Wang et al., 2017). Others, such as Busscher et al. (2010), reported that biochar-amended soil decreased aggregation in loamy sand soil when compared with the control, or it has no effect on a coarse-textured soil (Wang et al., 2017). The influence of the combination of biochar and cultivation practices on soil properties has not been understood yet. Several authors (Ahmed et al., 2016; Darusman et al., 2017) evaluated the action of tillage and cultivation practices on the degradation of biochar particles; this may result in an increase in bulk density caused by the biochar-soil mixture. However, the excessive destruction of biochar fractions may cause the formation of granules very smooth and determine the closure of soil pores.

Summarizing, evidence suggests that the effect of biochar on soil physical properties can be soil and biochar specific. Soil could have beneficial or neutral by biochar application depending on the biochar feedstocks, conditions of production, biochar application rates and on different soil types, and environmental and agricultural conditions (Ahmed et al., 2016), as before discussed, and more positive results can be observed in nonfertile, degraded, nutrient depleted, or interfertile soils (Mukherjee and Lal, 2013; Agegnehu et al., 2017; Blanco-Canqui, 2017).

3.2 Effects on soil chemical properties

Biochar-soil additions can alter soil chemical properties including pH, electrical conductivity, cation exchange capacity (CEC), organic carbon, and nutrient levels (Liang et al., 2006; Singh et al., 2010; Bera et al., 2016; Agegnehu et al., 2017). Although biochar pH values can range from acid to basic (pH from <3 to >12) depending on the nature of feedstock and of pyrolysis temperature, generally, the biochars used as amendment showed a neutral or alkaline pH value. Therefore, the addition of biochar can increase soil pH, resulting in a liming effect. Several studies reported that the greatest positive impacts of biochar may occur in acidic and neutral pH soils (Jeffery et al., 2011; Mosley et al., 2015), where the liming effect of the biochar also reduced the exchangeable acidity and toxicity of Al^{3+} (Rajakumar and Sankar, 2016). On the contrary, some authors reported no or negative effects on soil pH, when biochar was added to alkaline or saline soil or soil with low buffering capacity (Scott et al., 2014; Blanco-Canqui, 2017).

Biochar has the potential to improve soil CEC generally increasing from 4% to 30% (Laird et al., 2010; Jien and Wang, 2013). Although the effectiveness and duration of this effect after addition to soils remains poorly understood, the increase in CEC in soil may be attributed to the specific surface area and the number of carboxylic groups of the biochar (Cheng et al., 2006). It has been speculated that improvement in CEC is related to the following: the oxidation of aromatic carbon and formation of carboxyl and phenolic types functional groups on the biochar surface (Liang et al., 2006); conversion and or addition, by abiotic and biotic process, of surface functional groups containing oxygen (Mukherjee et al., 2011); and microbial colonization on soil/biochar surfaces (Mukherjee et al., 2011). Granatstein et al. (2009) studied the effect of biochar on soils with different textures and concluded that in both sandy and silt loam soils, the CEC was increased with an increased rate of biochar. The increase in soil CEC increases soil fertility by providing a nutrient reserve available to plant roots and preventing soil nutrients from leaching. However, no significant changes in CEC were observed in the different soil types to which biochar had been added (Mukherjee and Lal, 2014; Nelissen et al., 2015). The application of biochar in apple orchard increased TOC and available phosphorus and potassium contents but not CEC and pH (Safaei Khorram et al., 2018).

Moreover, biochar application can affect native soil organic matter (Ventura et al., 2015; Dong et al., 2016) although contrasting evidence was reported on the effect of biochar in the stability of soil organic matter (Kuzyakov et al., 2000). Some studies reported a stimulation (Cross and Sohi, 2011; Zimmerman et al., 2011) and others no effect or inhibition (Spokas et al., 2010; Singh and Cowie, 2014; Keith et al., 2015) of native soil organic matter decomposition after the addition of biochar to soil. Biochar, directly or indirectly, can also modify soil organic carbon dynamics increasing soil organic carbon stabilization processes and contributing to soil organic carbon sequestration or influencing net primary production with an increase in soil organic carbon (Oshunsanya and Aliku, 2016; Wang et al., 2016). However, the results obtained are mostly based on short-term laboratory studies and therefore not representative of the impacts over time (Zhang et al., 2016).

3.3 Effects on soil biological properties

Soil microorganisms (e.g., bacteria, fungi, actinomycetes, and microalgae) are responsible for several biological soil properties, such as organic matter decomposition, nutrient cycling, immobilizing inorganic nutrients, and other chemical transformation into the soil (Aislabie and Deslippe, 2013). The nature and activity of soil microbial communities are very sensitive to any change in management practices. Biochar, as a soil amendment, has the potential to alter the composition and the microbial biomass (Kolb et al., 2009), to promote different microbial communities (Thies and Rillig, 2009; Anderson et al., 2011; Lehmann et al., 2011) and to stimulate soil microbial activity (Lehmann et al., 2011; Ajema, 2018). Although it is difficult to distinguish between direct and indirect effects of biochar on the behavior of soil microorganisms, several mechanisms are proposed to explain its effects: (1) Biochar can provide a habitat for soil microorganisms. The porous structure of biochar and its high internal surface area can provide an ideal microhabitat for colonization and growth and reproduction for bacteria, actinomycetes, and arbuscular mycorrhizal fungi (AMF), by protecting them from predators (Pietikäinen et al., 2000). (2) Biochar modifies soil characteristics (aeration conditions, water content, and pH) promoting the microbial biomass population. (3) Biochar supplies nutrients and promotes carbon availability nutrients for microbes growth. (4) Biochar reduces the bioavailability of various soil toxins through absorption by its particles. (5) Biochar induces changes in enzyme activities that affect soil elemental cycles (Smith et al., 2010; Lehmann et al., 2011; Quilliam et al., 2013; Zhang et al., 2014; Stefaniuk and Oleszczuk, 2016; Sun et al., 2016; Yang et al., 2016).

Biochar can also alter the synergism between AMF and plants (Warnock et al., 2007; Thies and Rillig, 2009; Lehmann et al., 2011; Ajema, 2018). Warnock et al. (2007) proposed several mechanisms to explain the action of biochar on activity of AMF in soils:

- Biochar can serve as a refuge from hyphal grazers.
- Biochar can alter plant-AMF signaling processes.
- Biochar may adsorb compounds that are toxic to AMF.
- Biochar can modify nutrient availability.
- The positive or negative influence of biochar on other soil microbes can have indirect consequences on AMF.

Generally, the microbiological biomass, the different biota groups, and activity of microorganisms are influenced by the quantity and quality of biochar and by soil intrinsic properties. Several authors (Warnock et al., 2007; Gell et al., 2011; Ennis et al., 2012; Mukherjee and Lal, 2014; Ding et al., 2016) reported a significant decrement in soil biota. Although the reasons for this decrease are not known, it has been hypothesized that the presence of organic pyrolytic products (phenolic compounds and polyphenols), the retention of toxic substances (high levels of salt, heavy metals) and/or the release of pollutants from biochar (bio-oil and polycyclic aromatic hydrocarbons) may decrease microbial abundance and activities.

Moreover, in several soil types, biochar can also determine a decrease in the population of AMF (Gaur and Adholeya, 2000; Birk et al., 2009; Warnock et al., 2010). These results can be determined by modifications in soil conditions (e.g., pH or water relations), by the high contents of mineral elements or organic compounds detrimental and by high P levels into soil (Mukherjee and Lal, 2014).

3.4 Effects of biochar on plant nutrition and crop yield

The application of biochar to agricultural soil, affecting the nutrient dynamics, may represent a potential source of macro- and micronutrients for plants. Biochar is more effective than other organic amendments in retaining and making nutrients available to plants for long time (Chen et al., 2007). Nutrient contents in biochar-amended soils could be affected by different specific factors such as feedstock, pyrolysis conditions, soil properties, and management practices (Downie et al., 2009; Ding et al., 2016; Laghari et al., 2016). For example, animal-derived biochar contained higher content of available nutrient than plant-derived biochar (apple wood, pine chips, and magnolia leaves) (Zheng et al., 2013; Laghari et al., 2016).

Scott et al. (2014) have proposed several mechanisms to explain biochar effects on soil nutrient availability:

- Biochar can directly release nutrients to the soil due to its composition and structure. Beside C, biochar can contain important plant nutrients (N, P, K, Ca, Mg, S, and micronutrient), which can be released and assimilated by plants, although nutrient release is often negligible and reduced over time (1–2 years after biochar application) (Lehmann et al., 2003, 2009).
- Biochar can affect the availability of nutrients present in the soil by modifying the pH.
- Biochar can retain nutrients through adsorption and desorption mechanisms (Ding et al., 2016).
- Biochar can influence N cycle and P availability (Deluca et al., 2015).
- Physical biochar properties, such as porosity and surface area, may control the release progressive of nutrients from charred biomass.
- Biochar reduces nutrient leaching (Deluca et al., 2015; Laghari et al., 2016).

3.4.1 Effects of biochar on crop yield

Reviews (Sohi et al., 2010; Ahmed et al., 2016; Zhang et al., 2016; Agegnehu et al., 2017; Shareef and Zhao, 2017; El-Naggar et al., 2018) and statistical metaanalysis (Jeffery et al., 2011; Biederman and Harpole, 2013; Crane-Droesch et al., 2013; Baidoo et al., 2016; Yuan et al., 2018) have reported the effects of biochar on crop productivity and plant growth. Different scenarios, both increases and decreases, as well as no significant changes, have been observed in crop yield and agronomic productivity, after the application of biochar to soil (Asai et al., 2009; Feng et al., 2014; Mukherjee and Lal, 2014; Subedi et al., 2017; Laghari et al., 2016; Bass et al., 2016; Cornelissen et al., 2018; Moland et al., 2018). Spokas et al. (2012) observed that the biochar had positive impact on crop and plant growth in 50% of the analyzed studies, in 30% the differences that were not significant, and in 20% of the cases the responses that were negative. Several metaanalyses provided a general average estimate of an increase of 10%–20% in crop productivity, although higher values were reported in the literature (Park et al., 2011; De la Rosa et al., 2014; Kammann et al., 2015). No significant increase of crop yield or negative responses to biochar application have, also, been reported (Mukherjee and Lal, 2014; Maroušek et al., 2016). Decreases were observed in different crops with variable percentages, for example, 18%–22% in sorghum (Laghari et al., 2016), 3%–28% in radish (Nelissen et al., 2015), and 18%–24% in banana (Bass et al., 2016).

Various mechanisms have been suggested to explain how biochar might affect crop production both positively and negatively. Positive mechanisms can be

- modification of the physical-chemical characteristics of the soil favoring root growth and/or nutrient and water retention and acquisition (pH, CEC),
- changes in the biological activity,
- reduction of phytotoxic phenolic compounds,
- detoxification of allelochemicals (Hussain et al., 2017).

Negative impacts can be

- immobilization of N due to high C/N ratio,
- micronutrient deficiency,
- hydrophobic biochar characteristics,
- excessive application rate (>50 t/ha),
- high volatile matter content (Mukherjee and Lal, 2014; Subedi et al., 2017),
- ethylene evolution from freshly produced biochars (Fulton et al., 2013) and nutrient immobilization (Ahmed et al., 2016).

Tests on the agronomic effect of biochar on crop productivity were conducted for cereal crops (57%), noncereal crops (35%), and pastures or grasslands (8%) (Zhang et al., 2016). Few studies have been devoted to horticulture or perennial crops (Table 31.1). The perennial wood nature of fruit trees, their physiological stages of growth, and the large and differential root system required specific predictions of biochar applications (biochar type, biochar rate, time of application, etc.).

Grape yield per plant has been increased from 16% to 66% on average in respect to control, with 22 t/ha of commercial biochar derived from orchard pruning feedstock applied over 4 years (Genesio et al., 2015). Schmidt et al. (2014), on the contrary, investigated the agronomic performance of biochar on grape vines (*Vitis vinifera* cv. Pinot Noir), showing that biochar did not show relevant effects on plant growth parameters of vine. Wang et al. (2014) reported that the height and fresh weight of *Malus hupehensis* seedlings increased after the application of biochar into replant soil. The authors hypothesized that biochar amendment application can decrease the concentration of phytotoxic phenolic compounds in soil, favoring the development of plants. In a 5-month glasshouse pot trial of M26 apple rootstocks, Street et al. (2014) observed that biochar has positive effects on apple rootstock nutrition and growth only in a sandy soil and not in a sandy loam soil. In a 4-year trial on an apple orchard, Eyles et al. (2015) found that the trunk girth of trees in the biochar treatment was positively impacted. However, the positive effect of biochar was only evident in the last year, hypothesizing that the impact of amendments may require a longer timescale for study. In those few time of study, the authors observed that biochar had a limited effects on tree physiological processes and no effect on yield efficiency and average fruit weight.

Atuka and Litus (2015) reported the application of biochar to replant soils had beneficial effects on biomass production, on root carbon content, and on leaf magnesium concentration in peach rootstock "Lovell." Sorrenti et al. (2016) carried out a 3-year field trial on a commercial nectarine orchard and found that biochar application at different concentrations (0, 5, 15, and 30 t/ha) had no effect on yield and fruit weight.

Sánchez-García et al. (2016) have evaluated the effects of biochar on soil quality in drip-irrigated organic crop olives. In the semiarid Mediterranean agroecosystem, the biochar application did not play a significant role in N_2O mitigation, although it resulted in a consistent and persistent increase of TOC into the soil.

Few studies have analyzed the effect of biochar on qualitative characteristics of agricultural product. In particular, in fruit trees, the quality, such as their content in antioxidant compounds, is sometimes more important than yield. However, both no effect responses and positive responses have been reported for a wide variety of crop. For example, Akhtar et al. (2014) noted an improvement in tomato quality under reduced irrigation, while changes were observed in the qualitative characteristics of tomato fruits grown on three different biochars (Petruccelli et al., 2015). On the contrary, biochar had no significant effect on in apple tree (Eyles et al., 2015) and in grapevine (Schmidt et al., 2014; Genesio et al., 2015).

4 Conclusions

The examined literature pointed out that the benefits of biochar on soil fertility and crop productivity depends by many factors determined by the production technology of the biochar (feedstock type and pyrolysis process conditions), by the characteristics of the soil and the environment, and by the biochar management (application rate and time, application methods, etc.). Furthermore, it has been suggested that biochar prefers a "specific agroecological niche" (Crane-Droesch et al., 2013), maximizing its effects in nutrient-poor and acid soil and in tropical soils than in temperate ones.

At authors' knowledge, there are only few papers that reveal what happens to fruit crops when biochar is added. In fruit orchard, characterized by high inputs of nutrients and water, the results of biochar application may be of minor magnitude. Accurate biochar management is necessary to control effectively plant development and fruit quality. In particular, field studies and long-term studies are needed; in fact many effects, for example, carbon sequestration and release of nutrients, evolve slowly, and it is necessary to monitor the action of the biochar over time. Furthermore, long-term studies can provide realistic information on the use of biochar in maintaining or increasing fertility of soils and in yields of fruit tree orchards.

Acknowledgments

Authors would really thank Dr. Emanuela Pusceddu from Institute of Biometeorology—Italian National Research Council (IBIMET-CNR) for her valuable work at SEM: we really appreciated the provided SEM picture for this review. We would also thank Dr. Alessandro Pozzi from Advanced Gasification Technology (AGT) to book this chapter his picture on different biochar.

TABLE 31.1 References paper on biochar application on fruit crops and its effects on soil and plants.

Reference	Time and type/years of plants	Main results
Apple		
Eyles et al. (2015)	A 4-year trial on 1-year-old nursery plants (~1 m tall) in nonlimiting conditions (e.g., water availability and soil fertility)	Crop yield and fruit quality parameters were unaffected by the soil amendment treatments. Trunk girth was significantly higher than the control in the biochar treatments, in the first year and fourth year, respectively, while compost had no effect in any year. Neither photosynthetic capacity nor leaf nutrient concentration was influenced by treatment. Seasonal daily tree water use was similar between biochar and control treatments
Safaei Khorram et al. (2018)	A 4-year field study on replanted 3-year-old (~2 m tall) apple trees	Biochar and compost were beneficial to improve the soil quality, mainly by increasing soil nutrient contents and decreasing soil bulk density, and to increase plant growth at early growth stages of apple orchards. However, they failed to enhance overall yield and fruit quality most likely due to their limited ability to suppress apple replant diseases
Ventura et al. (2013)	A 3-year field experiment on 13-year-old apple trees	Biochar addition increased soil organic carbon-derived respiration and reduced rhizosphere-derived respiration, even if the root length intensity increased in biochar-treated soils relative to that in the control. A decrease in root metabolic activity was postulated to explain these contrasting results
Wang et al. (2014)	A 3-month pot experiment on about 4 months plants	The addition of biochar can alleviate apple replant disease by activating antioxidant enzymes, decreasing lipid peroxidation, and significantly reducing the phenolic acids content of replant soil through the sorption of biochar
Pear		
Lyu et al. (2016)	A 15-day drought stress study on potted 2-year-old plants	Biochar could significantly retard the loss of soil efficient moisture and effectively change the chlorophyll fluorescence parameters avoiding serious damage caused by drought stress
Olive		
Sánchez-García et al. (2016)	A 2-year field experiment in N-limited and deficit irrigated semiarid agroecosystem on 20-year-old plants	Compost amendment has an impact on soil microbiological activity with a link to N availability. Biochar applied alone does not alter the N dynamics, but markedly builds-up soil C. In both cases, only during the first year, these effects were statistically significant. When applied in combination, a synergistic effect was observed and the highest values of denitrifying enzyme activity and N_2O emissions were detected
Peach		
Atuka and Litus (2015)	A 33-week trial on rootstock liners of "Lovell" planted in 18.9-L containers and placed in a glass greenhouse with temperature ranging from 15°C at night to 27°C during daytime	Total aboveground and belowground biomass production was significantly higher in the biochar and sterilized soil treatment compared with the control. Root carbon content was significantly greater in the 20% (v/v) with soil mixture treatment compared with the control. Soil nitrate-N was significantly greater in this treatment by the third harvesting date, and foliar magnesium concentrations were significantly higher in both biochar treatments (10% and 20% (v/v) with the soil mixture, equivalent to an application of 16 and 32 Mg/ha to the top 5 cm of soil) for all harvesting dates. The results from this study provide evidence that biochar could reduce the detrimental effects of replant disease in peach trees

Continued

TABLE 31.1 References paper on biochar application on fruit crops and its effects on soil and plants—cont'd

Reference	Time and type/years of plants	Main results
Sorrenti et al. (2016)	A 3-year field trial on 12-year-old commercial nectarine peach in nonlimiting conditions (e.g., water availability and soil fertility)	Biochar reduced the leached amount of NH_4-N in the top 0.25-m soil layer over 13 months; nevertheless, it did not affect soil pH, soil N mineral availability, soil moisture, tree nutritional status, yield, and fruit quality
Grapevine		
Baronti et al. (2014)	A two-growing season field experiment on a not irrigated Merlot vineyard in a shallow acids sandy-clay-loam soil	The treatments did not show a significant increase in soil hydrophobicity. Moreover, soil analysis and ecophysiological measurements indicated a substantial relative increases in available soil water content compared with control soils (from 3.2% to 45% in the 22 and 44 t/ha application rates, respectively) and in leaf water potential (24%–37%) during droughts
Genesio et al. (2015)	A 4-year field study on a not irrigated Merlot vineyard in a shallow acids sandy-clay-loam soil	Results from four harvest-years showed a higher productivity, up to 66%, of treated plots ([A] 22 t/ha of biochar applied in 2009; [B] 22 t/ha in 2009 and further 22 t/ha in 2010) with respect to control, while no significant differences were observed in grape quality parameters. The observed increase in productivity was inversely correlated with rainfall in the vegetative period, confirming the role of biochar in regulating plant water availability
Schmidt et al. (2014)	A 3-year field study on a Pinot Noir vineyard in poor fertility, alkaline, temperate soil	Biochar treatments ([A] 8 t/ha of biochar; [B] 8 t/ha of biochar +55 t/ha of compost, mixed before the composting process) did not show relevant effects on plant growth parameters over 3 years. There were effects on grape quality only in the first year.

References

Agegnehu, G., Srivastava, A.K., Bird, M.I., 2017. The role of biochar and biochar-compost in improving soil quality and crop performance: a review. Appl. Soil Ecol. 119, 156–170.

Ahmed, A., Jiby, K., Vijaya, R., 2016. Biochar influences on agricultural soils, crop production and the environment—a review. Environ. Rev. 24, 495–502.

Aislabie, J., Deslippe, J.R., 2013. Soil microbes and their contribution to soil services. In: Dymond, J.R. (Ed.), Ecosystem services in New Zealand—conditions and trends. Manaaki Whenua Press, Lincoln, New Zealand, pp. 143–161.

Ajema, L., 2018. Effects of biochar application on beneficial soil organism review. Int. J. Res. Stud. Sci. Eng. Technol. 5, 9–18.

Akhtar, S.S., Li, G., Andersen, M.N., Liu, F., 2014. Biochar enhances yield and quality of tomato under reduced irrigation. Agric. Water Manag. 138, 37–44.

Albiach, R., Canet, R., Pomares, F., Ingelmo, F., 2000. Microbial biomass content and enzymatic activities after the application of organic amendments to a horticultural soil. Bioresour. Technol. 75, 43–48.

Anderson, C.R., Condron, L.M., Clough, T.J., Fiers, M., Stewart, A., Hill, R.A., Sherlock, R.R., 2011. Biochar induced soil microbial community change: implications for biogeochemical cycling of carbon, nitrogen and phosphorus. Pedobiologia 54, 309–320.

Antal Jr., M.J., Gronli, M., 2003. The art, science, and technology of characol production. Ind. Eng. Chem. Res. 42, 1619–1640.

Asai, H., Samson, B.K., Stephan, H.M., Songyikhangsuthor, K., Homma, K., Kiyono, Y., Inoue, Y., Shiraiwa, T., Horie, T., 2009. Biochar amendment techniques for upland rice production in Northern Laos 1. Soil physical properties, leaf SPAD and grain yield. Field Crop Res. 111, 81–84.

Atuka, A., Litus, G., 2015. Effect of biochar amendments on peach replant disease. HortScience 50, 863–868.

Baidoo, I., Sarpong, D.B., Bolwig, S., 2016. Biochar amended soils and crop productivity: a critical and meta-analysis of literature. Int. J. Dev. Sustain. 5 (9), 414–432.

Baronti, S., Vaccari, F.P., Miglietta, F., Calzolari, C., Lugato, E., Orlandini, S., Pini, R., Zulian, C., Genesio, L., 2014. Impact of biochar application on plant water relations in Vitis vinifera L. Eur. J. Agron. 53, 38–44.

Bass, A.M., Bird, M.I., Kay, G., Muirhead, B., 2016. Soil properties, greenhouse gas emissions and crop yield under compost, biochar and co-composted biochar in two tropical agronomic systems. Sci. Total Environ. 550, 459–470.

Benavente, V., Fullana, A., 2015. Torrefaction of olive mill waste. Biomass Bioenerg. 73, 186–194.

Bera, T., Collins, H.P., Alva, A.K., Purakayastha, T.J., Patra, A.K., 2016. Biochar and manure effects on soil biochemical properties under corn production. Appl. Soil Ecol. 107, 360–367.

Biederman, L.A., Harpole, W.S., 2013. Biochar and its effects on plant productivity and nutrient cycling: a meta-analysis. Glob. Change Biol. Bioenergy. 5, 202–214.

Bird, M.I., Wurster, C.M., de Paula Silva, P.H., Bass, A.M., de Nys, R., 2010. Algal biochar–production and properties. Bioresour. Technol. 102, 1886–1891.

Birk, J.J., Steiner, C., Teixiera, W.C., Zech, W., Glaser, B., 2009. Microbial response to charcoal amendments and fertilization of a highly weathered tropical soil. In: Woods, W.I., Teixeira, W.G., Lehmann, J., Steiner, C., WinklerPrins, A.M.G.A., Rebellato, L. (Eds.), Amazonian Dark Earths: Wim Sombroek's Vision. Berlin Springer, pp. 309–324.

Bitew, Y., Alemayehu, M., 2017. Impact of crop production inputs on soil health: a review. Asian J. Plant Sci. 16, 109–131.

Blanco-Canqui, H., 2017. Biochar and soil physical properties. Soil Sci. Soc. Am. J. 81, 687–711.

Busscher, W.J., Novak, J.M., Evans, D.E., Watts, D.W., Niandou, M.A.S., Ahmedna, M., 2010. Influence of pecan biochar on physical properties of a Norfolk loamy sand. Soil Sci. 175, 10–14.

Cantrell, K.B., Hunt, P.G., Uchimiya, M., Novak, J.M., Ro, K.S., 2012. Impact of pyrolysis temperature and manure source on physicochemical characteristics of biochar. Bioresour. Technol. 107, 419–428.

Cao, X.D., Ma, L., Gao, B., Harris, W., 2009. Dairy-manure derived biochar effectively sorbs lead and atrazine. Environ. Sci. Technol. 43, 3285–3291.

Cayuela, M.L., Zwieten, L.V., Singh, B.P., Jeffery, S., Roig, A., Sánchez-Monedero, M.A., 2014. Biochar's role in mitigating soil nitrous oxide emissions: a review and metanalysis. Agric. Ecosyst. Environ. 191, 5–16.

Cha, J.S., Park, S.H., Jung, S.C., Ryu, C., Jeon, J.K., Shin, M.C., Park, K., 2016. Production and utilization of biochar: a review. J. Ind. Eng. Chem. 40, 1–15.

Chan, K.Y., Zwieten, L.V., Meszaros, I., Downie, A., Joseph, S., 2007. Agronomic values of green waste biochar as a soil amendment. Aust. J. Soil Res. 45, 629–634.

Chan, K.Y., Van Zwieten, Y., Meszaros, I., Downie, A., Joseph, S., 2008. Using poultry litter biochar as soil amendments. Aust. J. Soil Res. 46 (5), 437–444.

Chen, B.L., Zhou, D.D., Zhu, L.Z., 2008. Transitional adsorption and partition of nonpolar and polar aromatic contaminants by biochars of pine needles with different pyrolytic temperatures. Environ. Sci. Technol. 42, 5137–5143.

Chen, W.H., Zhuang, Y.Q., Liu, S.H., Juang, T.T., Ysai, C.M., 2016. Product characteristics from the torrefaction of oil palm fiber pellets in inert and oxidative atmospheres. Bioresour. Technol. 199, 367–374.

Chen, J., Lü, S., Zhang, Z., Zhao, X., Li, X., Ning, P., Liu, M., 2018. Environmentally friendly fertilizers: a review of materials used and their effects on the environment. Sci. Total Environ. 613–614, 829–839.

Cheng, C.H., Lehmann, J., Thies, J.E., Burton, S.D., Engelhard, M.H., 2006. Oxidation of black carbon by biotic and abiotic processes. Org. Geochem. 37, 1477–1488.

Chioua, B.S., Medina, D.V., Sainz, C.B., Klamczynski, A.P., Bustillos, R.J.A., Milczarek, R.R., Du, W.X., Glenn, G.M., Orts, W.J., 2016. Torrefaction of almond shells: effects of torrefaction conditions on properties of solid and condensate products. Ind. Crop. Prod. 86, 40–48.

Cooke, G.W., 1967. The Control of Soil Fertilizer. London Crosby Lock Wood Ltd, London, pp. 15–18.

Cornelissen, G., Jubaedah, N.N.L., Hale, S.E., Martinsen, V., Silvania, L., Mulder, J., 2018. Fading positive effect of biochar on crop yield and soil acidity during five growth seasons in an Indonesian Ultisol. Sci. Total Environ. 634, 561–568.

Crane-Droesch, A., Abiven, S., Jeffery, S., Torn, M.S., 2013. Heterogeneous global crop yield response to biochar: a meta-regression analysis. Environ. Res. Lett. 80, 440–449.

Cross, A., Sohi, S.P., 2011. The priming potential of biochar products in relation to labile carbon contents and soil organic matter status. Soil Biol. Biochem. 43, 2127–2134.

Darusman, D., Syahruddin, S., Syakur, S., Manfarizah, M., 2017. Biochar and tillage systems influenced on soil physical properties. Aceh. Int. J. Sci. Technol. 6 (2), 68–74.

De la Rosa, J.M., Paneque, M., Miller, A.Z., Knicker, H., 2014. Relating physical and chemical properties of four different biochars and their application rate to biomass production of *Lolium perenne* on a Calcic Cambisol during a pot experiment of 79 days. Sci. Total Environ. 499, 175–184.

Deluca, T.H., Gundale, M.J., Mackenzie, M.D., Jones, D.L., 2015. Biochar effects on soil nutrient transformations. In: Lehmann, J., Joseph, S. (Eds.), Biochar for Environmental Management: Science, Technology and Implementation, second ed. Earthscan, London, pp. 421–454.

Diacono, M., Montemurro, F., 2010. Long-term effects of organic amendments on soil fertility. A review. Agron. Sustain. Dev. 30, 401–422.

Ding, Y., Liu, Y., Liu, S., Li, Z., Tan, X., Huang, X., Zeng, G., Zhou, L., Zheng, B., 2016. Biochar to improve soil fertility. A review. Agron. Sustain. Dev. 36 (2), 36.

Dong, X.L., Guan, T.Y., Li, G.T., Lin, Q.M., Zhao, X.R., 2016. Long-term effects of biochar amount on the content and composition of organic matter in soil aggregates under field conditions. J. Soils Sediment. 16, 1481–1497.

Downie, A., Crosky, A., Munroe, P., 2009. Physical properties of biochar. In: Lehmann, J., Joseph, S. (Eds.), Biochar for Environmental Management. Science and Technology. Earthscan, London, pp. 13–32.

Elad, Y., David, D.R., Harel, Y.-M., Borenshtein, M., Kalifa, H.B., Silber, A., Graber, E.R., 2010. Induction of systemic resistance in plants by biochar, a soil-applied carbon sequestering agent. Phytopathol 100, 913–921.

Elmer, W.H., Pignatello, J.J., 2011. Effect of biochar amendments on mycorrhizal associations and Fusarium crown and root rot of asparagus in replant soils. Plant Dis. 95, 960–966.

El-Naggar, A., Lee, S.S., Rinklebe, J., Farooq, M., Song, H., Sarmah, A.K., Zimmerman, A.R., Ahmad, M., Shaheen, S.M., Yong Sik, O.K., 2018. Biochar application to low fertility soils: A review of current status, and future prospects. Geoderma 337, 536–554.

Enders, A., Hanley, K., Whitman, T., Joseph, S., Lehmann, J., 2012. Characterization of biochars to evaluate recalcitrance and agronomic performance. Bioresour. Technol. 114, 644–653.

Ennis, C.J., Evans, A.G., Islam, M., Ralebitso-Senior, K., Senior, E., 2012. Biochar: carbon sequestration, land remediation, and impacts on soil microbiology. Crit. Rev. Environ. Sci. Technol. 42, 2311–2364.

Epstein, E., Bloom, A.J., 2005. Mineral Nutrition of Plants: Principles and Perspectives, second ed. Sinauer, Sunderland, MA.

Eyles, A., Bound, S., Oliver, G., Corkrey, R., Hardie, M., Green, S., Close, D.C., 2015. Impact of biochar amendment on the growth, physiology and fruit of a young commercial apple orchard. Trees 29, 1817–1826.

Feng, L., Gui-Tong, L., Qi-mei, L., Xiao-Rong, Z., 2014. Crop yield and soil properties in the first 3 years after biochar application to a calcareous soil. J. Integr. Agric. 13, 525–532.

Fulton, W., Gray, M., Prahl, F., Kleber, M., 2013. A simple technique to eliminate ethylene emissions from biochar amendment in agriculture. Agron. Sustain. Dev. 33 (3), 469–474.

Fungo, B., Lehmann, J., Kalbitz, K., Thiongo, M., Okeyo, I., Tenywa, M., Neufeldt, H., 2017. Aggregate size distribution in a biochar-amended tropical Ultisol under conventional hand-hoe tillage. Soil Tillage Res. 165, 190–197.

Galvez, A., Sinicco, T., Cayuela, M.L., Mingorance, M.D., Fornasier, F., Mondini, C., 2012. Short term effects of bioenergy by-products on soil C and N dynamics, nutrient availability and biochemical properties. Agric. Ecosyst. Environ. 160, 3–14.

Gaur, A., Adholeya, A., 2000. Effects of the particle size of soil-less substrates upon AM fungus inoculum production. Mycorrhiza 10, 43–48.

Gell, K., Van Groenigen, J., Cayuela, M.L., 2011. Residues of bioenergy production chains as soil 17 amendments: immediate and temporal phytotoxicity. J. Hazard. Mater. 186, 2017–2025.

Genesio, L., Miglietta, F., Baronti, S., Vaccari, F.P., 2015. Biochar increases vineyard productivity without affecting grape quality: results from a four years field experiment in Tuscany. Agric. Ecosyst. Environ. 201, 20–25.

Graber, E.R., Harel, Y.M., Kolton, M., Cytryn, E., Silber, A., David, D.R., Tsechansky, L., Borenshtein, M., Elad, Y., 2010. Biochar impact on development and productivity of pepper and tomato grown in fertigated soilless media. Plant Soil 337 (1–2), 481–496.

Granatstein, D., Kruger, C.E., Collins, H., Galinato, S., Garcia-Perez, M., Yoder, J., 2009. Use of Biochar From the Pyrolysis of Waste Organic Material as a Soil Amendment. Final project report, Center for Sustaining Agriculture and Natural Resources, Washington State University, Wenatchee, WA, p. 168.

Hardie, M., Clothier, B., Bound, S., Oliver, G., Close, D., 2014. Does biochar influence soil physical properties and soil water availability? Plant Soil 376, 347–361.

Hossain, M.K., Strezov, V., Chan, K.Y., Nelson, P.F., 2010. Agronomic properties of wastewater sludge biochar and bioavailability of metals in production of cherry tomato (Lycopersicon esculentum). Chemosphere 78, 1167–1171.

Hossain, M.K., Strezov, V., Chan, K.Y., Ziolkowski, A., Nelson, P.F., 2011. Influence of pyrolysis temperature on production and nutrient properties of wastewater sludge biochar. J. Environ. Manag. 92, 223–228.

Hussain, M., Farooq, M., Nawaz, A., Al-Sadi, A.M., Solaiman, Z.M., Alghamdi, S.S., Ume, A., Yong, S.O., Siddique, K.H.M., 2017. Biochar for crop production: potential benefits and risks. J. Soil Sediment. 17 (3), 685–716.

Ippolito, J.A., Laird, D.A., Busscher, W.J., 2012. Environmental benefits of biochar. J. Environ. Qual. 41, 967–972.

Jeffery, S., Verheijen, F.G.A., Van der Velde, M., Bastos, A.C., 2011. A quantitative review of the effects of biochar application to soils on crop productivity using meta-analysis. Agric. Ecosyst. Environ. 144 (1), 175–187.

Jeffery, S., Bezemer, T.M., Cornelissen, G., Kuyper, T.W., Lehmann, J., Mommer, L., Sohi, S.P., van de Voorde, T.F.J., Wardle, D.A., van Groenigen, J.W., 2015. The way forward in biochar research: targeting trade-offs between the potential wins. Glob. Change Biol. Bioenergy. 7, 1–13.

Jien, S.H., Wang, C.S., 2013. Effects of biochar on soil properties and erosion potential in a highly weathered soil. Catena 110, 225–233.

Kammann, C.I., Schmidt, H.P., Messerschmidt, N., Linsel, S., Steffens, D., Müller, C., Koyro, H.W., Conte, P., Joseph, S., 2015. Plant growth improvement mediated by nitrate capture in co-composted biochar. Sci. Rep. 511080.

Keech, O., Carcaillet, C., Nilsson, M.C., 2005. Adsorption of allelopathic compounds by wood derived charcoal: the role of wood porosity. Plant Soil 272, 291–300.

Keith, A., Singh, B., Dijkstra, F.A., 2015. Biochar reduces the rhizosphere priming effect on soil organic carbon. Soil Biol. Biochem. 88, 372–379.

Kinney, T.J., Masiello, C.A., Dugan, B., Hockaday, W.C., Dean, M.R., Zygourakis, K., Barnes, R.T., 2012. Hydrologic properties of biochars produced at different temperatures. Biomass Bioenergy 41, 34–43.

Kloss, S., Zehetner, F., Dellantonio, A., Hamid, R., Ottner, F., Liedtke, V., Schwanninger, M., Gerzabek, M.H., Soja, G., 2012. Characterization of slow pyrolysis biochars: effects of feedstocks and pyrolysis temperature on biochar properties. J. Environ. Qual. 41, 990–1000.

Kolb, S.E., Fermanich, K.J., Dornbush, M.E., 2009. Effect of charcoal quantity on microbial biomass and activity in temperate soils. Soil Sci. Soc. Am. J. 73, 1173–1181.

Kuzyakov, Y., Friedel, J.K., Stahr, K., 2000. Review of mechanisms and quantification of priming effects. Soil Biol. Biochem. 32, 1485–1498.

Laghari, M., Naidu, R., Xiao, B., Hu, Z., Mirjat, M.S., Hu, M., Kandhro, M.N., Chen, Z., Guo, D., Jogi, Q., Abudi, Z.N., Fazal, S., 2016. Recent developments in biochar as an effective tool for agricultural soil management: a review. J. Sci. Food Agric. 96 (15), 4840–4849.

Laird, D., Fleming, P., Davis, D.D., Horton, R., Wang, B., Karlen, D., 2010. Impact of biochar amendments on the quality of a typical midwestern agricultural soil. Geoderma 158, 443–449.

Lal, R., 2004. Carbon sequestration in dryland ecosystems. Environ. Manag. 33, 528–544.

Larson, W.E., Pierce, F.J., 1994. The dynamics of soil quality as a measure of sustainable management. In: Doran, J.W. (Ed.), Defining Soil Quality for a Sustainable Environment. Soil Science Society of America and American Society of Agronomy, Madison, WI, pp. 37–51. Special Publication No. 35.

Lehmann, J., 2007. Bio-energy in the black. Front. Ecol. Environ. 5, 381–387.

Lehmann, J., da Silva Jr., J.P., Steiner, C., Nehls, T., Zech, W., Glaser, B., 2003. Nutrient availability and leaching in an archaeological Anthrosol and a Ferralsol of the Central Amazon basin: fertilizer, manure and charcoal amendments. Plant Soil 249, 343–357.

Lehmann, J., Czimczik, C., Laird, D., Sohi, S., 2009. Stability of biochar in soil. In: Lehmann, J., Joseph, S. (Eds.), Biochar for Environmental Management: Science and Technology. Earthscan, London, pp. 183–205.

Lehmann, J., Rillig, M.C., Thies, J., Masiello, C., Hockaday, W., Crowley, D., 2011. Biochar effects on soil biota—a review. Soil Biol. Biochem. 43, 1812–1836.

Liang, B., Lehmann, J., Solomon, D., Kinyangi, J., Gross-man, J., O'Neill, B., Skjemstad, J.O., Luizao, T.J., FJ, P.J., Neves, E.G., 2006. Black carbon increases cation exchange capacity in soils. Soil Sci. Soc. Am. J. 70, 1719–1730.

Liu, X., Zhang, A., Ji, C., Joseph, S., Bian, R., Li, L., Pan, G., Paz-Ferreiro, J., 2013. Biochar's effect on crop productivity and the dependence on experimental conditions—a meta-analysis of literature data. Plant Soil 73 (12), 583–594.

Liu, X., Zheng, J., Zhen, J., 2014. Sustainable biochar effects for low carbon crop production: a 5-crop season field experiment on a low fertility soil from Central China. Agric. Syst. 129, 22–29.

Liu, Z., Dugan, B., Masiello, C.A., Gonnermann, H.M., 2017. Biochar particle size, shape, and porosity act together to influence soil water properties. PLoS One. 12(6). e0179079.

Lu, H.L., Zhang, W.H., Yang, Y.X., Huang, X.F., Wang, S.Z., Qiu, R.L., 2012. Relative distribution of Pb^{2+} sorption mechanisms by sludge-derived biochar. Water Res. 46, 854–862.

Lu, S., Sun, F., Zong, Y., 2014. Effect of rice husk biochar and coal fly ash on some physical properties of expansive clayey soil (vertisol). Catena 114, 37–44.

Lü, S., Feng, C., Gao, C., Wang, X., Xu, X., Bai, X., Gaoì, N., Liu, M., 2016. Multifunctional environmental smart fertilizer based on L-aspartic acid for sustained nutrient release. J. Agric. Food Chem. 64, 4965–4974.

Lyu, S., Du, G., Liu, Z., Zhao, L., Lyu, D., 2016. Effects of biochar on photosystem function and activities of protective enzymes in *Pyrus ussuriensis* Maxim. under drought stress. Acta Physiol. Plant. 38, 220.

Major, J., Lehmann, J., Rondon, M., Goodale, C., 2010. Fate of soil-applied black carbon: downward migration, leaching and soil respiration. Glob. Chang. Biol. 16 (4), 1366–1379.

Maroušek, J., Vochozka, M., Plachý, J., Zak, J., 2016. Glory and misery of biochar. Clean Techn. Environ. Policy 19 (2), 311–317.

Marschner, H., 1995. Mineral Nutrition of Higher Plants, second ed. Academic Press, Amsterdam.

Matovic, D., 2011. Biochar as a viable carbon sequestration option: global and Canadian perspective. Energy 36, 2011–2016.

McGrath, J.M., Spargo, J., Penn, C.J., 2014. Soil fertility and plant nutrition. In: Van Alfen, N. (Ed.), Encyclopedia of agriculture and food systems. In: vol. 50. Elsevier, San Diego, CA, pp. 166–184.

Moland, S., Brent, R., Robin, B., Ruth, N., Allison, W., 2018. Determining the effects of biochar and an arbuscular mycorrhizal inoculant on the growth of fowl mannagrass (*Glyceria striata*) (Poaceae). Facets 3, 441–454.

Mosley, L., Willson, P., Hamilton, B., Butler, G., Seaman, R., 2015. The capacity of biochar made from common reeds to neutralise pH and remove dissolved metals in acid drainage. Environ. Sci. Pollut. Res. 22, 15113–15122.

Mukherjee, A., Lal, R., 2013. Biochar impacts on soil physical properties and greenhouse gas emissions. Agronomy 3 (2), 313–339.

Mukherjee, A., Lal, R., 2014. The biochar dilemma. Soil Res. 52, 217–230.

Mukherjee, A., Zimmerman, A.R., Harris, W., 2011. Surface chemistry variations among a series of laboratory-produced biochars. Geoderma 163, 247–255.

Nelissen, V., Ruysschaert, G., Kumar Saha, B., Boeckx, P., Manka'Abusi, D., D'Hose, T.K., Al-Barri, B., Cornelis, W., Boeckx, P., 2015. Impact of a woody biochar on properties of a sandy loam soil and spring barley during a two-year field experiment. Eur. J. Agron. 62, 65–78.

Novak, J.M., Busscher, W.J., Laird, D., Ahmedna, M., Watts, D.W., Niandou, M.A.S., 2009. Impact of biochar amendment on fertility of a southeastern coastal plain soil. Soil Sci. 174, 105–112.

Oguntunde, P.G., Abiodun, B.J., Ajayi, A.E., van de Giesen, N., 2008. Effects of charcoal production on soil physical properties in Ghana. J. Plant Nutr. Soil Sci. 171, 591–596.

Omondi, M., 2016. Quantification of biochar effects on soil hydrological properties using meta-analysis of literature data. Geoderma 274, 28–34.

Oshunsanya, S.O., Aliku, O.O., 2016. Biochar technology for sustainable organic farming. In: Organic Farming. IntechOpen. https://doi.org/10.5772/61440.

Park, J.H., Choppala, G.K., Bolan, N.S., Chung, J.W., Chuasavathi, T., 2011. Biochar reduces the bioavailability and phytotoxicity of heavy metals. Plant Soil 348, 439–451.

Pereira, R.C., Kaal, J., Arbestain, M.C., Lorenzo, R.P., Aitkenhead, W., Hedley, M., Macías, F., Hindmarsh, J., Maciá-Agulló, J.A., 2011. Contribution to characterisation of biochar to estimate the labile fraction of carbon. Org. Geochem. 42, 1331–1342.

Petersen, C.T., Hansen, E., Larsen, H.H., Hansen, L.V., Ahrenfeldt, J., Hauggaard-Nielsen, H., 2016. Pore-size distribution and compressibility of coarse sandy subsoil with added biochar. Eur. J. Soil Sci. 67, 726–736.

Petruccelli, R., Bonetti, A., Traversi, M.L., Faraloni, C., Valagussa, M., Pozzi, A., 2015. Influence of biochar application on nutritional quality of tomato (*Lycopersicon esculentum*). Crop Pasture Sci. 66 (7), 747–755.

Pietikäinen, J., Kiikkila, O., Fritze, H., 2000. Charcoal as a habitat for microbes and its effects on the microbial community of the underlying humus. Oikos 89, 231–242.

Quilliam, R.S., Glanville, H.C., Wade, S.C., Jones, D.L., 2013. Life in the 'charosphere'—does biochar in agricultural soil provide a significant habitat for microorganisms? Soil Biol. Biochem. 65, 287–293.

Rajakumar, R., Sankar, J., 2016. Biochar for sustainable agriculture. A review. IJAPSA 2 (9), 173–184.

Roberts, K.G., Gloy, B.A., Joseph, S., Scott, N.R., Lehmann, J., 2010. Life cycle assessment of biochar systems: estimating the energetic, economic, and climate change potential. Environ. Sci. Technol. 44, 827–833.

Rogovska, N., Laird, D.A., Karlen, D.L., 2016. Corn and soil response to biochar application and stover harvest. Field Crop Res. 187, 96–106.

Safaei Khorram, M., Zhang, G., Fatemi, A., Kiefer, R., Maddah, K., Baqar, M., Zakaria, M.P., Li, G., 2018. Impact of biochar and compost amendment on soil quality, growth and yield of a replanted apple orchard in a 4-year field study. J. Sci. Food Agric. https://doi.org/10.1002/jsfa.9380.

Sánchez-García, M., Sánchez-Monedero, M.A., Roig, A., López-Cano, I., Moreno, B., Benitez, E., Cayuela, M.L., 2016. Compost vs biochar amendment: a two-year field study evaluating soil C build-up and N dynamics in an organically managed olive crop. Plant Soil 408, 1–14.

Savci, S., 2010. An agricultural pollutant: chemical fertilizer. Int. J. Environ. Sci. Dev. 3, 77–80.

Schmalenberger, A., Fox, A., 2016. Bacterial mobilization of nutrients from biochar-amended soils. Adv. Appl. Microbiol. 94, 109–159.

Schmidt, H.P., Kammann, C., Niggli, C., Evangelou, M.W.H., Mackied, K.A., Abiven, S., 2014. Biochar and biochar-compost as soil amendments to a vineyard soil: influences on plant growth, nutrient uptake, plant health and grape quality. Agric. Ecosyst. Environ. 191, 117–123.

Scott, H., Ponsonby, D., Atkinson, C.J., 2014. Biochar: an improver of nutrient and soil water availability—what is the evidence? CAB Rev. 9 (19), 1–19.

Shaaban, M., Van Zwieten, L., Bashir, S., Younas, A., Núñez-Delgado, A., Chhajro, M.A., Kubar, K.A., Ali, U., Rana, M.S., Mehmood, M.A., Hu, R., 2018. A concise review of biochar application to agricultural soils to improve soil conditions and fight pollution. J. Environ. Manag. 228, 429–440.

Shareef, T.M.E., Zhao, B., 2017. Review paper: the fundamentals of biochar as a soil ammendmend tool and management in agriculture scope: an overview for farmer and gardners. J. Agric. Chem. Environ. 6, 38–61.

Singh, B.P., Cowie, A.L., 2014. Long-term influence of biochar on native organic carbon mineralisation in a low-carbon clayey soil. Sci. Rep. 4, 3687.

Singh, B., Singh, B.P., Cowie, A.L., 2010. Characterisation and evaluation of biochars for their application as a soil amendment. Aust. J. Soil Res. 48, 516–525.

Smith, J.L., Collins, H.P., Bailey, V.L., 2010. The effect of young biochar on soil respiration. Soil Biol. Biochem. 42, 2345–2347.

Sohi, S.P., Krull, E., Lopez-Capel, E., Bol, R., 2010. A review of biochar and its use and function in soil. In: Sparks, D.L. (Ed.), Advances in Agronomy. Elsevier Academic Press Inc., San Diego, CA, pp. 47–82.

Solaiman, Z., Anawar, H., 2015. Application of biochars for soil constraints: challenges and solutions. Pedosphere 25, 631–638.

Sorrenti, G., Ventura, M., Toselli, M., 2016. Effect of biochar on nutrient retention and nectarine tree performance: a three-year field trial. J. Plant Nutr. Soil Sci. 179, 336–346.

Spokas, K.A., Baker, J.M., Reicosky, D.C., 2010. Ethylene: potential key for biochar amendment impacts. Plant Soil 333, 443–452.

Spokas, K.A., Cantrell, K.B., Novak, J.M., Archer, D.A., Ippolito, J.A., Collins, H.P., Boateng, A.A., Lima, I.M., Lamb, M.C., McAloon, A.J., Lentz, R. D., Nichols, K.A., 2012. Biochar: a synthesis of its agronomic impact beyond carbon sequestration. J. Environ. Qual. 41, 973–989.

Srivastava, A.K., 2017. The role of biochar and biochar-compost in improving soil quality and crop performance: a review. Appl. Soil Ecol. 119, 156–170.

Stefaniuk, M., Oleszczuk, P., 2016. Addition of biochar to sewage sludge decreases freely dissolved PAHs content and toxicity of sewage sludge-amended soil. Environ. Pollut. 218, 242–251.

Street, T.A., Doyle, R.B., Close, D.C., 2014. Biochar media addition impacts apple rootstock growth and nutrition. HortScience 49 (9), 1188–1193.

Subedi, R., Bertora, C., Zavattaro, L., Grignani, C., 2017. Crop response to soils amended with biochar: expected benefits and unintended risks. Ital. J. Agron. 12, 161–173.

Suliman, W., Harsh, J.B., Abu-Lail, N.I., Fortuna, A.M., Dallmeyer, I., Garcia-Pérez, M., 2017. The role of biochar porosity and surface functionality in augmenting hydrologic properties of a sandy soil. Sci. Total Environ. 574, 139–147.

Sun, D., Hale, L., Crowley, D., 2016. Nutrient supplementation of pinewood biochar for use as a bacterial inoculum carrier. Biol. Fertil. Soil 52 (4), 1–8.

Thies, J.E., Rillig, M., 2009. Characteristics of biochar: biological properties. In: Lehmann, J., Joseph, S. (Eds.), Biochar for Environmental Management: Science and Technology. Earthscan, London, pp. 85–105.

Ventura, M., Sorrenti, G., Panzacchi, P., George, E., Tonon, G., 2013. Biochar reduces short-term nitrate leaching from a horizon in an apple orchard. J. Environ. Qual. 42, 76–82.

Ventura, M., Alberti, G., Viger, M., Jenkins, J.R., Girardin, C., Baronti, S., Zaldei, A., Taylor, G., Rumpel, C., Miglietta, F., Tonon, G., 2015. Biochar mineralization and priming effect on SOM decomposition in two European short rotation coppices. Glob. Change Biol. Bioenergy. 7 (5), 1150–1160.

Verheijen, F., Jeffery, S., Bastos, A.C., Van der Velde, M., Diafas, I., 2010. Biochar Application to Soils. A Critical Scientific Review of Effects on Soil Properties Processes and Functions. European Communities, Luxembourg. 24099:162. ISBN 9789279142932.

Wade, S.R., Nunoura, T., Antal, M.J., 2006. Studies of the flash carbonization process. 2. Violent ignition behavior of pressurized packed beds of biomass: a factorial study. Ind. Eng. Chem. Res. 45, 3512–3519.

Wang, Y., Pan, F., Wang, G., Zhang, G., Wang, Y., Chen, X., Mao, Z., 2014. Effects of biochar on photosynthesis and antioxidative system of *Malus hupehensis* Rehd. seedlings under re-plant conditions. Sci. Hortic. 175, 9–15.

Wang, J., Xiong, Z., Kuzyakov, Y., 2016. Biochar stability in soil: meta-analysis of decomposition and priming effects. GCB Bioenergy 8, 512–523.

Wang, D., Fonte, S.J., Parikh, S.J., Six, J., Scow, K.M., 2017. Biochar additions can enhance soil structure and the physical stabilization of C in aggregates. Geoderma 303, 110–117.

Warnock, D.D., Lehmann, J., Kuyper, T.W., Rillig, M.C., 2007. Mycorrhizal responses to biochar in soil–concepts and mechanisms. Plant Soil 300, 9–20.

Warnock, D.D., Mummey, D.L., McBride, B., Major, J., Lehmann, J., Rillig, M.C., 2010. Influences of non-herbaceous biochar on arbuscular mycorrhizal fungal abundances in roots and soils: results from growth-chamber and field experiments. Appl. Soil Ecol. 46, 450–456.

White, P.J., Brown, P.H., 2010. Plant nutrition for sustainable development and global health. Ann. Bot. 105, 1073–1080.

Whitman, T., Hanley, K., Enders, A., Lehmann, L., 2011a. Predicting pyrogenic organic matter mineralization from its initial properties and implications for carbon management. Org. Geochem. 64, 76–83.

Whitman, T., Nicholson, C.F., Torres, D., Lehmann, J., 2011b. Climate change impact of biochar cook stoves in western Kenyan farm households: system dynamics model analysis. Environ. Sci. Technol. 45, 3687–3694.

Whitman, T., Hanley, K., Enders, A., Lehmann, J., 2013. Predicting pyrogenic organic matter mineralization from its initial properties and implications for carbon management. Org. Geochem. 64, 76–83.

Yang, X., Liu, J., Mc Grouther, K., Huang, H., Lu, K., Guo, X., He, L., Lin, X., Che, L., Ye, Z., Wang, H., 2016. Effect of biochar on the extractability of heavy metals (Cd, Cu, Pb, and Zn) and enzyme activity in soil. Environ. Sci. Pollut. Res. 23, 974–984.

Yuan, J.H., Xu, R.K., Zhang, H., 2011. The forms of alkalis in the biochar produced from crop residues at different temperatures. Bioresour. Technol. 102, 3488–3497.

Yuan, W., Meng, X., UJames, A., 2018. The role of biochar in sustainable agriculture and environment: promising but inconsistent. Adv. Biotech. Micro.. 9 (4). AIBM.MS.ID.555770.

Zhang, Q.-Z., Dijkstra, F.A., X-R, L., Y-D, W., Huang, J., Lu, N., 2014. Effects of biochar on soil microbial biomass after four years of consecutive application in the north China plain. PLoS One. 9(7). e102062.

Zhang, D., Yan, M., Niu, Y., Liu, X., van Zwieten, L., Chen, D., Bian, R., Cheng, K., Li, L., Joseph, S., Zheng, J., Zhang, X., Zheng, J., Crowley, D., Fillet, T.R., Pan, G., 2016. Is current biochar research addressing global soil constraints for sustainable agriculture? Agric. Ecosyst. Environ. 226, 25–32.

Zhao, L., Cao, X., Mašek, O., Zimmerman, A., 2013. Heterogeneity of biochar properties as a function of feedstock sources and production temperatures. J. Hazard. Mater. 256–257, 1–9.

Zheng, H., Wang, Z., Deng, X., Zhao, J., Luo, Y., Novak, J., Herbert, S., Xing, B., 2013. Characteristics and nutrient values of biochars produced from giant reed at different temperatures. Bioresour. Technol. 130, 463–471.

Zhu, X., Chen, B., Zhu, L., Xing, B., 2017. Effects and mechanisms of biochar-microbe interactions in soil improvement and pollution remediation: a review. Environ. Pollut. 227, 98–115.

Zimmerman, A.R., Gao, B., Ahn, M.-Y., 2011. Positive and negative carbon mineralization priming effects among a variety of biochar-amended soils. Soil Biol. Biochem. 43, 1169–1179.

Further reading

Novak, J.M., Busscher, W.J., 2009. Evaluation of designer biochars to ameliorate select chemical and physical characteristics of degraded soils. In: AIChE Annual Meeting, Nashville TN, pp. 152–158.

32

Physiological and molecular basis of salinity tolerance in fruit crops

Riaz Ahmad, Muhammad Akbar Anjum*

Department of Horticulture, Bahauddin Zakariya University, Multan, Pakistan
*Corresponding author. E-mail: akbaranjum@bzu.edu.pk

1 Introduction

Salinity is one of the most critical menaces for horticultural crops to increase their area and attain higher productions all over the globe. The world population is growing hurriedly that may exceed from 6 to 9.3 billion by the year 2050, while the crop yield is reducing due to negative effects of several environmental stresses (Hussain et al., 2018). Agricultural development and crop growth are suppressed due to excessive salt concentrations in soil and becoming challenging aspect worldwide. All over the world, more than 950 million hectares of agricultural land is affected by soil salinity (Fu et al., 2019). So, it is imperative to produce stress-tolerant cultivars to cope with problems of food security in the future. Salinization is a most serious environmental stress among all abiotic stresses and damaging at least 20% crop production of world's arable land. Mineral weathering, irrigation with salty water, low rainfall, and high evaporation are the responsible factors for the accumulation of excessive salts in soils

Study of physiological and genetic basis of salt tolerance in fruit crops is of vital importance to find out more effective solutions of practical complications caused by salinity. It is possible to reduce production losses due to high salinity stress through grafting of salt-sensitive cultivars on those rootstocks, which have salt tolerance capability. This approach can also produce plants with good shoot and root characters. However, salt tolerance is a polygenic phenomenon composed of a complicated network of molecular and physiological processes, strongly influenced by climatic factors (Munns et al., 2006). Therefore, the identification and evaluation of physiological attributes are considered as more effective markers to evaluate salt tolerance in many fruit crops.

The proficient use of limited water resources is becoming more essential due to fast increase of irrigated farming in arid and semiarid regions. Existences of high amount of soluble salts in irrigation water and in the soil solution are the limiting factors reducing plant growth and production, particularly in arid and semiarid environments (Hussain et al., 2009). Salinity problem has been increasing with passage of time due to mishandling of canal irrigation system and also due to saline groundwater irrigation. Leaf burn and leaf defoliation are associated with accumulation of Na^+ and Cl^- at

A.K. Srivastava, Chengxiao Hu (eds.)
Fruit Crops: Diagnosis and Management of Nutrient Constraints
https://doi.org/10.1016/B978-0-12-818732-6.00032-0

toxic levels in leaf cells. Alterations in different physiological processes, that is, photosynthesis, stomatal conductance, osmotic adjustment, ion uptake, protein synthesis, nucleic acid production, enzymatic activity, and hormonal stability, are main causes of plant growth reduction under salinity conditions. Reduction of plant growth is also caused by osmotic effect or due to high amount of ion accumulation in plant tissues resulting in ion toxicity and nutritional imbalance (Patane et al., 2013)

Stress-induced conditions mainly depend on numerous factors including plant species, cultivar, growth stage, composition of soluble salts, stress intensity and its duration, and natural climatic conditions. Several physiological and molecular mechanisms are involved in plant resistance against several abiotic stresses including salt stress. Fruit plants store a huge number of metabolites that are very similar and compatible to solutes and do not inhibit plant metabolism during salt stress. Proline increase is one of the most frequently described changes in salt-stressed plants. A greater involvement of proline is observed in stress resistance mechanism of different fruit crops (Abdallah et al., 2017). There is urgent need of time to reduce salinity problems for higher fruit production all over the globe. Several factors, that is, proteomics, different proteins, environmental pressures, molecular influences, and genotypic variations, can be utilized to reduce salt stress in fruit crops (Miranda et al., 2018)

Development and identification of salt-tolerant cultivars are very slow and worrying in fruit crops. In these conditions, it is very necessary to understand the importance of development and identification of salt-tolerant cultivars. Traditional approaches do not have great potential to evolve salt-tolerant cultivars. Therefore, it is important to use marker-assisted selection (MAS), QTL mapping, association mapping, and whole genome sequencing used to enhance salt tolerance in salt-sensitive cultivars

2 Effects of salt stress on fruit crops

Plants growing under salinity conditions may suffer from three different but interconnected stresses, that is, osmotic stress, salt-specific stress, and oxidative stress. Plants are also affected from high root zone pH, nutritional differences, structural harms, oxygen deficit, and poor respiration of roots in saline soils due to osmotic and ionic stresses. Annual and perennial plants are affected with salinity in a similar manner, primarily disturbing the osmotic balance by producing salt-specific effects. Accumulation of salts in leaves and shoots may vary in both annual and perennial plants. Salt accumulation in leaves may reach to a high toxicity level within few days or weeks in annual salt-stressed crops, while the effects of salt stress in perennial crops may appear after few months or even in years (Munns, 2005). Plants are tolerant or susceptible to either osmotic stress or salt-specific stress even separately or both stresses depending upon the species or cultivars. Carrizo citrange seedlings exposed to 50-mM NaCl resulted in growth reduction and net gaseous exchange that were not caused by leaf turgor loss, but caused by ion toxicity in leaves because salt stress decreased leaf water potential and osmotic potential so leaf turgor was increased (Sánchez and Syvertsen, 2009). Sodium and chloride ions in seedlings of Citrus limonia and rooted cuttings of olive cv. "Arbequina" were enhanced due to salt stress resulting in growth reduction, decrease in plant dry weight, and alterations in net photosynthesis with high ion toxicity level (Melgar et al., 2008). Therefore, it has been concluded that cultivars can react differentially under osmotic and salt-specific mechanisms of salinity. Salt-tolerant rootstocks may significantly reduce salt stress by decreasing salt transport to shoots. Moreover, salt rejection by roots can also reduce energy usage for osmotic balance (Fig. 32.1)

Annual crops and forage species remain comparatively salt tolerant throughout their life cycle, while fruit crops loss their resistance against salt stress after few years of growth. In fruit crops, salts are stored in basal stem and root tissues when they are exposed to high salinity stress during initial few years, while accumulated salts move slowly toward the leaves due to longtime salinity exposure. Fruit crops can tolerate excessive amount of salts during vegetative growth while harmfully affected during reproductive growth and vice versa. QTL information confirmed that intraphase variability was found in salt tolerance mechanism, so several species exhibiting salt resistance at germination and seedling stage can become salt susceptible during later growth stages (Yamaguchi and Blumwald, 2005). In fruit crops, salt toxicity is more damaging and harmful at reproductive stage, and harvestable produce or edible portion may also be affected

2.1 General signs

Salt stress reduces uptake of K+ and Ca^{2+}, and nutrient deficiency may occur in fruit crops. Morphology of the plants also exhibits their susceptibility to salt stress. Reduction in fresh and dry weights in all plant tissues exposed to salt stress resulting in yield losses is visible characteristics in aerial parts of plants (Sharma et al., 2011). Depending

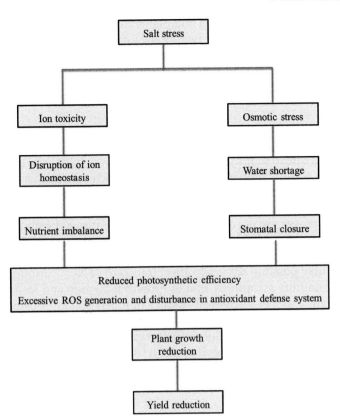

FIG. 32.1 Effects of salt stress on fruit plants.

on salt stress tolerance, plants have been categorized into two major groups; halophytes and glycophytes. Halophytes are those plants that have better ability to survive and reproduce under high salt concentration nearly 200-mM NaCl. These plants have nearly 1% portion of the world's flora. Halophytes are mostly naturally grown plants, which can tolerate salt stress (Patane et al., 2013). Glycophytes are those plants that are not able to survive and reproduce under salinity conditions. Fruit crops are grouped in glycophytes. Salt stress is one of the most critical environmental problems, which can damage fruit crops. Considerable decrease in total leaf area and leaf growth is the initial response of glycophytes showing salt stress. Some other signs are marginal and tip scorching of leaves, followed by yellowing and bronzing. In fruit crops, main symptoms of salt stress are early leaf drop, twig dieback, blackening, necrosis, and leaf burning (Bernstein et al., 2004)

2.1.1 Osmotic stress

High amount of salts primarily causes soil water-deficit conditions reducing plant root capacity to uptake sufficient amount of water and nutrients for several metabolic activities. Water-deficit stress conditions can also be increased due to water loss from leaves. Harmful effects of osmotic stress in citrus are significantly based on salt accumulation pattern. Under salt stress, abscisic acid and ethylene levels increase in leaves and accelerate leaf fall. Water availability in saline soils is suppressed due to salt ions that encourage plants to store organic and inorganic osmolytes to reduce leaf water potential for producing negative pump that helps to keep water flux through xylem sap (Arbona et al., 2003). $Na+$ and $Cl-$ reduce enzymatic functions at 100 mM or exceeding concentrations. However, greater energy is required for salt elimination and process of intracellular compartmentation of ions that might be the cause of ionic injury in salt-sensitive crops. Moreover, it has been confirmed that plants consume 10 times more energy for proline buildup as compared to accumulate an equal amount of $Na+$ (Munns, 2002)

Plants are either salt sensitive or salt tolerant. Halophytes are the plant that can tolerate high salinity levels by keeping the water uptake from the roots through the accumulation of Na^+ and Cl^- ions. Glycophytes are salt-sensitive plants, and their growth and yield are greatly affected when exposed to salt stress. These plants produce certain compounds such as proline, sugars, and polyols even at moderate level of salinity (Ferguson and Grattan, 2005). The compatible solutes can be grouped into three, namely, sugars/polyols (e.g., manitol), amino acids (e.g., proline), and quaternary ammonium compound (glycine betaine). The process of osmotic adjustment involves accumulation of these compatible solutes in cytoplasm so that osmotic pressure can be maintained and plants can uptake water. Furthermore, few of these solutes may reduce ROS-induced injury by protecting cell membrane and also some enzymes (Nuccio et al., 1999). Osmotic adjustment strategy in fruit plants mainly depends on salt tolerance, genotype, and management practices. In salt-sensitive plants, proline is the major compatible solute produced in stressed conditions. Olive plants tend to increase mannitol synthesis in leaf mesophyll cells as the salinity level is enhanced in root zone (Gucci et al., 1998)

2.1.2 Salt-specific effects

Salt ions remain still outside the plant during osmotic stress due to salinity, while salt-specific stress initiates when ions enter inside the plants. In salt-sensitive cultivars, salts are accumulated in stem and leaves, which are rapid cause of physiological disorders in fruit crops. Citrus, grapevines, avocado, and stone fruits are considered as highly sensitive to chloride injury. Twofold concentration of Cl^- was accumulated in citrus treated with NaCl and KCl salts as compared with controlled plants (Brumos et al., 2010). Moreover, excess of Na+ and Cl^- may reduce accessibility of important cations such as NH^{4+}, K+, and Ca^{2+} and anions, that is, NO^{3-}. Excess of Na+ is the cause of depolarization in plasma membrane that reduces the passive uptake of K+. Under saline soils, $NO3^-$ uptake decreases due to equal size and similar charge of Cl^- and $NO3^-$. Hence, $Cl-$ may possibly adversely affect transport of NO^{3-} through membranes. In citrus plants, it has been observed that some salinity symptoms, that is, growth reduction and lower fruit yield, are correlated with nitrogen deficiency demonstrating that excessive salts reduce nitrogen uptake. Mostly, Cl^- ions harshly affect the absorption of $NO3^-$ to a greater extent in Troyer citrange and in Cleopatra seedlings (Cerezo et al., 1997). Under salinity stress, plants need 5–10-mM calcium in external solution for better growth, while more than 10-mM calcium can retard the plant growth. The optimum ratio of Na/Ca is ranged 10–20 in fruit crops (Cramer et al., 2007). Application of supplemental calcium has greater capability to reduce the harmful Na+ effects in numerous fruit crops worldwide

2.2 Physiological responses to salt stress

Accumulation of hormones, osmotic balance, preferential accumulation of potassium, maintenance of photosynthesis, antioxidant enzymes, different ion channels, salt avoidance, salt exclusion, salt excretion, and numerous genetic strategies is involved to manage the adverse effects of salt stress in fruit crops. These responses may influence different growth stages of plants.

2.2.1 Water relations

In salinized olive trees, reduction in water uptake has been recorded mainly due to the reduction of osmotic potential in solutions having NaCl. Under salt stress, water relations in olive leaves have been studied in salt-tolerant "Frantoio" and in salt-sensitive "Leccino" cultivar (Chartzoulakis, 2005). Relatively, more Cl^- in leaves and less in roots were accumulated in trifoliate orange as compared with Cleopatra mandarin, while relatively lower Na+ concentration was recorded in trifoliate orange as compared with Cleopatra mandarin. Application of 30-mM L−1 calcium nitrate significantly reduced Cl^- and Na+ in leaves of rootstock and scion (Banuls et al., 1997). The decrease of leaf water potential and relative water content is early responses of salinized olive plants as in many woody crops. These differences in leaf water potential, relative water content, and water uptake might occur due to higher amount of salts in several fruit species including olive. Calcium acetate can be used to reduce Cl^- and Na+ ions and enhance photosynthesis mechanism and stomatal conductance in citrus leaves. Moreover, it is also involved to neutralize the decrease of leaf water potential and osmotic capability during high salts. So, it has been found that chloride accumulation in citrus leaves is a major cause of adverse effects under salt stress (Banuls and Millo, 1992).

Due to high salt concentration in soil solution, reduction in relative water content was recorded, which caused osmotic stress and dehydration at cellular level. The highest leaf dehydration was a major cause of extensive drop in leaf water potential during salt stress and its capability to recover upon relief of stress. The salt-induced reduction of leaf water potential also resulted in osmotic potential reduction by producing turgor potential (Yadav et al., 2011). Therefore, the maximum turgor pressure was recorded in salinized plants as compared with control plants. Moreover, osmotic adjustment is mostly attained by the increase of inorganic ions with lower energy costs in olive trees. The increase of soluble carbohydrates mostly glucose and mannitol in response to enhanced NaCl concentration of the external medium supports their influence to osmotic adjustment, whereas other ions, that is, nitrate, sulfate, phosphate, and amino compounds contribute very little in this process. The variations in leaf water relations between salt-tolerant cultivars Frantoio, Hamed, and Picual, and salt-sensitive cultivars Leccino, Nabal, and Meski reflect their differences in the exclusion capabilities for Na+ and Cl⁻ (Chartzoulakis, 2005).

2.2.2 Gaseous exchange

Rootstock-scion interaction has great impact on plant performance under saline conditions. Very little work has been carried out on the interaction of rootstock/scion for ion uptake and gaseous exchange in leaves under salt stress. In mango, the highest reduction in CO2 assimilation and transpiration rate was recorded in "turpentine" grafted on "13-1" rootstock. The highest Cl— was found in rootstock–scion combination, and the highest decrease in shoot growth was also recorded. Therefore, it has been found that "13-1" as a rootstock was totally unable to protect the "turpentine" as a salt-sensitive scion. Moreover, it has been noted that "13-1" as a scion has greater capability to keep higher K+/Na+ ratio in the leaves (Schmutz and Lüdders, 1999).

Olive has hard leaves due to thick cuticle layer and a packed mesophyll that may reduce CO2 entrance in chloroplasts and photosynthesis. When leaf thickness enhances, it might reduce mesophyll conductance through extending and producing more complex CO2 pathway toward the chloroplasts. In fact, all available works for olive crop shows reduction in photosynthesis process under salt stress. Although salinity effects on CO2 assimilation rate, but it varies depending on salt concentrations at which different type of cultivars are exposed. At 200-mM NaCl, 1-year-old plants of 6 olive cultivars showed a significant reduction in net assimilation rate, which may be 20% for salt-tolerant "Kalamata," while 62% for moderately sensitive "Amphissis" in young leaves (Chartzoulakis, 2005). Application of 30-mM/L calcium nitrate significantly reduced Cl⁻ and Na+ uptake and also enhanced the growth attributes and gaseous exchange mechanism in citrus plants (Banuls et al., 1997).

A reduction in stomatal conductance may alter the rate of photosynthesis in salt-stressed olive plants. At 100–200-mM NaCl, a clear decrease in photosynthesis of olive plants was recorded. A complete recovery of photosynthesis in 50- and 100-mM salt-stressed plants occurred during the relief period in salt-tolerant "Frantoio," accompanied by an increase in stomatal conductance and transpiration (Tabatabaei, 2006). So, during the initial stages of salinization, stomatal limits of photosynthesis are found to be more prevalent. The major limitations of photosynthesis in moderately salt-stressed olive plants are low stomatal conductance and mesophyll conductance, which fix low chloroplast CO2 concentration.

Two wild species of Indian jujube, that is, Ziziphus rotundifolia and Ziziphus nummularia, were found to be more resistant to saline conditions due to higher CO2 assimilation, higher stomatal conductance, higher hormonal regulation, and accumulation of several antioxidants. Moreover, these two species had better translocation of nutrients through roots. Therefore, these two species are widely used as rootstocks (Meena et al., 2003). An excess of Na⁺ is found to be more toxic for plants due to negative effects on photosynthesis and metabolism (Chakraborty et al., 2012). So, regulation of intercellular ion homeostasis is very vital for normal plant growth and development. In Indian jujube, the highest accumulation of Na⁺ was recorded in roots of Gola cultivar and in leaves of Umran cultivar. Na⁺ accumulated in Gola roots through restricted translocation to keep ionic balance in leaves. Moreover, it has been found that cultivar Gola had better performance as compared with other cultivars due to its better photosynthetic mechanism, antioxidant defense mechanism, and high K+/Na+ ratio (Agrawal et al., 2013).

2.2.3 Hormonal regulation

Salinity greatly affects photosynthetic metabolism in leaves, and it indirectly reduces growth of fruits and other sink organs. Growth hormones, that is, indole acetic acid (auxin), aminocyclopropane-1-carboxylic acid (cytokinin), and abscisic acid, have capability to adjust the synthesis of assimilates and also control their partitioning and assimilation in salt-affected plants. Under osmotic stress, these hormones may cause to increase the growth process by increasing ion homeostasis and root development and also affect source-sink relationship (Alfocea et al., 2010). It can also depress accumulation of Na⁺ and Cl⁻ ions to a toxic level. Chloride accumulation may reach toxic level that can initiate production of ethylene precursor ACC. In prolonged salt stress, leaf abscission starts due to ethylene synthesis. In counteract, plants trigger the synthesis of ABA to reduce ethylene production. The pretreatment of citrus plants with ABA decreases the leaf fall that may prevent the Cl⁻ accumulation in leaves. Similarly, 8'-methylene ABA and 8'-acetylene ABA also function like ABA that may reduce the harmful effects of salts in citrus plants (Arbona et al., 2006). Silver thiosulfate and phosgene are also ethylene inhibitors and may reduce leaf fall in salt-affected plants (Cadenas et al., 2002). Approximately, threefold increase in ABA levels was noted in roots of some grape cultivars such as "Salt Creek," "Dogridge," "1613," and "St. George" at 250-mM NaCl. ABA synthesis was adversely linked with Na⁺ and K⁺ ratio, which signals that Na⁺ uptake was inhibited by ABA accumulation (Upreti and Murti, 2010).

Preferential accumulation of potassium

Several fruit cultivars favorably accumulate K^+ in leaf and in stem tissues to tolerate harmful effects of salinity. The lowest Cl^- and the highest K^+ were accumulated in leaves of grafted lemon cv. "Fino 49" during application of 50-mM NaCl with 10-mM potassium nitrate (Gimeno et al., 2009). Trifoliate orange seedlings inoculated with two arbuscular mycorrhiza fungi, *Glomus mosseae* and *G. versiforme*, exhibited considerably lower levels of Na^+ and Ca^{2+} in leaves while higher K^+, carbohydrates, and proline concentrations under 100-mM NaCl. Under 60-mM NaCl, mango rootstock (*Microhyla zeylanica*) continued greater photosynthetic adjustment as compared with *Mangifera indica* due to a greater K^+/Na^+ ratio in roots and lesser in leaf/root Na^+ (Schmutz, 2000). Salt stress considerably enhanced K^+ concentration in leaves and in shoots of salt-sensitive fig cultivars. The highest cytosolic K^+ ion may involve in better appropriation of Na^+ ion in vacuoles. Therefore, it is very helpful to manage large Na^+ content in leaves (Zarei et al., 2016). In pistachio rootstocks, *Pistacia atlantica* gave better performance as compared with *Polycythemia vera* by controlling Na^+ and Cl^- uptake under salinity stress (Tounsi et al., 2017). Regulation of sodium and potassium toxic ions plays a major role in salinity tolerance mechanism of glycophyte salinity (Assaha et al., 2017).

Maintenance of photosynthesis

Salt stress significantly affects the photosynthesis mechanism for short or long time. During short time, salt stress can greatly affect photosynthesis through stomatal restrictions (stomatal factor) resulting in the reduction of carbon assimilation. During the long time, salt stress can adversely affect the photosynthetic mechanism due to accumulation of salts in young leaves resulting in photosynthetic pigments loss and Calvin cycle enzymes inhibition (nonstomatal factors). Hence, salt resistance crops have good potential to maintain or enhance their chlorophyll content (Motos et al., 2017).

Under salt stress, decrease in stomatal conductance is initial response of plants to reduce transpiration. Stomatal closure reduces CO_2 circulation into the leaves causing inner CO_2 reduction, limited pressure, and ultimately little photosynthesis. In several fruit crops, nonstomatal problems also affect the photosynthesis. Scions of Taylor lemon, Ellendale tangor, and Valencia orange were grafted on Cleopatra mandarin rootstock. Induced salt level up to 75-mM NaCl caused photosynthesis reduction, stomatal closure, Na^+ toxicity, and chlorophyll denaturation. Sodium ions have large regulating influence on photosynthesis and transpiration as compared with chloride ions (Behboudian et al., 1986). Salinity tolerance mainly depends on vine strength, regulation of photosynthesis, and maintenance of Na^+ in shoots and old basal leaves in grape cultivars (Mohammadkhani, 2018).

Under salt stress, photosynthetic efficiency is reduced due to stomatal and nonstomatal limitations. Stomatal limitations are diffusional resistance toward CO_2 transportation, while nonstomatal limitations are physiological and molecular factors. Better understanding is available on stomatal factors, while nonstomatal limitations are still unclear. The evaluation of stomatal and nonstomatal limitations is necessary, while investigation on diffusional and nondiffusional limits of photosynthesis can offer a good indication of photosynthesis reduction in salt-stressed plants.

Structural disorders were produced due to chloride salts, that is, KCl, CaCl2, and NaCl in both sensitive (Carrizo citrange) and tolerant (Cleopatra mandarin) cultivars of citrus. Structural disorders observed were increase of leaf thickness and decrease in area of mesophyll cells. Hence, increase in leaf thickness, excessive chlorides, lower Mg^{2+} ions, interruption in stomatal closure, and loss of chlorophyll significantly hamper the photosynthesis mechanism (Aranda et al., 1998).

Antioxidative metabolism

Oxygen is considered as one of the most essential constituents in plants. Generally, it is more useful in metabolism, oxidative phosphorylation, and mitochondrial respiration to generate energy, though oxygen is stimulated into reactive oxygen species (ROS) through metabolism. ROS had a strong oxidative potential causing cytoplasmic membrane injury, permanent metabolic disruption, and finally cell decease. Oxidative stress is considered as one of the consequences of salt stress. Moreover, oxidative stress might be liable for further injury. Oxidative stress may also occur due to continuous accumulation of ROS in salt-stressed plants. Oxidative stress and accumulation of excessive Na^+ in cytosol reduce stomatal conductance and photosynthesis processes resulting in partial consumption of light through photosynthetic pigments. So, enhanced production of ROS in leaves causes lipid peroxidation, DNA denaturation, and protein degradation and ion like K^+ efflux (Shabala and Munns, 2017).

Enzymatic antioxidant activity

Numerous enzymatic reactions such as superoxide dismutase (SOD), peroxidase (POD), ascorbate peroxidase (APX), glutathione peroxidase (GPX), catalase (CAT), and glutathione reductase (GR) are activated in plants under salt stress as a defense mechanism. SOD is first antioxidant enzyme in plant defense system that has potential to remove excessive amount of superoxide anions from the cells. In higher plants based on several metal ions bounded by its auxiliary positions, SOD is categorized into three major groups, that is, manganese SOD (Mn-SOD), iron SOD (Fe-SOD), and copper/zinc SOD (Cu/Zn-SOD). SOD can imbalance the O_2 to form H_2O_2 and remove the harmfulness of superoxide anion. H_2O_2 and its derivatives are quickly reduced or sometimes reduced through APX or GPX (Wang et al., 2016). Ascorbate peroxidase (APX) is famous as one of the major enzymes that decrease H_2O_2 level in plants. Chloroplast APX chiefly eliminates the H_2O_2 production. The expression of APX gene is encouraged through ozone. Increasing the APX activity, plants tolerance to ozone might be enhanced (Sofo et al., 2015).

Catalase (CAT) is largely found in plant peroxisomes and glyoxylic acid-circulating bodies, which is one of the important enzymes that directly convert H_2O_2 into H_2O. Generally, CAT is found to be different from APX because it needs a reductive substrate and the highest affinity for H_2O_2. CAT primarily decreases the amount of H_2O_2, which is manufactured in light respiration. Numerous studies have revealed that CAT is a basic enzyme for the elimination of H_2O_2 in C_3 plants and is also important for H_2O_2 tolerance of these plants (Willekens et al., 1994; Ashraf, 2009).

The increase of ROS in plants causes lipid peroxidation in cell membranes under high salt stress. Malondialdehyde (MDA) is the major constitute of membrane lipid peroxidation in plants under high salinity, and its content characterizes the level of cell membrane injury. Therefore, MDA content can identify the plant salt stress and salt tolerance. At 200-mM NaCl application for 6 days, MDA content enhanced in TaSP transgenic *Arabidopsis* and control plants. The increase of MDA content in transgenic plants was lesser as compared with control plants. TaSP transgenic Arabidopsis plants exhibited higher salt tolerance mechanism through reduction in lipid peroxidation under salt stress (Ashraf, 2009).

Glutathione is one of the most critical metabolites in fruit crops. It has a capability to scavenge H_2O_2 and ROS like hydroxyl radical. Moreover, glutathione involves in antioxidant defense mechanism via stimulating water-soluble antioxidants and ascorbic acid through ascorbate-glutathione cycle. Glutathione is the substrate of glutathione S-transferase, which shows an essential role in detoxification of dehydroascorbate reductase and xenobiotics (Zhang and Shi, 2013). Therefore, glutathione is significant involved to maintain normal cellular redox system either in normal or even in stressed environments.

Nonenzymatic antioxidant activity

Numerous nonenzymatic reactions such as ascorbic acid, tocopherols, flavonoids, proline, soluble sugars, and phenols are efficient antioxidant systems for ROS scavenging. These antioxidants may vary in gene expression, also accomplish as redox buffers and as a metabolic interface to control the optimum stimulation of acclimation reactions (Miller et al., 2010).

Tocopherols are present in all parts of plants. These have high capability to scavenge the ROS and lipid radicals in plants. These are vital constitutes of biological membranes for antioxidant and nonantioxidant functions. There are four isomers, that is, α^-, β^-, γ^-, and δ^- of tocopherols found in plants. The highest antioxidant activity was calculated in α-tocopherol (vitamin E). Chloroplast membranes comprise a considerable amount of α-tocopherols in higher plants. Therefore, these are well secure against photooxidative injury. Thylakoid membranes are protected through these tocopherols as do the carotenoids. The α-tocopherols have greater potential to directly repair oxidizing radicals by inhibiting chain transmission period during lipid autooxidation (Sairam and Tyagi, 2004).

Flavonoid compounds are extensively found in plant kingdom and very frequent in plant foliage, floral portion, and pollens. These are commonly stored in plant vacuole by means of glycosides. The concentration of flavonoids in plant cells or tissues is nearly 1 mM (Potapovich and Kostyuk, 2003). These flavonoids have greater potential to inhibit the lipoxygenase.

Proline may act as a metal chelator, an antioxidant compound and a signaling molecule to maintain osmotic regulation. Plants produce proline mainly under salt-stressed conditions. Hence, it may be used as a biochemical marker to differentiate the sensitive and tolerant species. Studies revealed that, in date palm, proline synthesis largely increased in leaves and roots under drought, salinity, extreme temperatures, ABA treatment, and other stress conditions. Precursor of proline is different under normal and stress conditions. It is produced by ornithine under normal conditions, while it is synthesized from glutamate in stress conditions. Glutamate and proline are also involved in chlorophyll synthesis in salt-stressed plants that may possibly lead to greater loss of chlorophyll as recorded in Yaghooti cultivar of grapes. Proline might be accumulated under salt stress conditions, and its osmoprotection activity is well known. The accumulation of metabolites is considered as one of the most common responses of plants that act as compatible solutes to change in external osmotic potential. Proline is considered as antioxidant that improves salt tolerance in pistachio plants (Shamshiri and Fattahi, 2014). Highly significant increases were recorded in pistachio cv. "Badami-Rize-Zarand" (Karimi and Kuhbanani, 2015). Proline can act as free radical scavenger for subcellular structures and cell redox balancer. Thus, pistachio cv. "Badami-Rize-Zarand" was found to be more resistant under high salinity due to more proline accumulation as compared to cv. "Badami-e-Sefid" (Rahneshan et al., 2018). Proline exogenously applied to *Citrus sinensis* cv. Valencia late plants significantly minimized the salt damage (Costa et al., 2010).

Plants tend to produce more soluble sugars under salt stress, low temperature, water deficit, or even water logging conditions. Sugar level decreases due to nutrient deficiency, heavy metal toxicity, and also in high light intensity. However, soluble sugar level in plants may vary with the intensity of stress and genotype to be studied. Glucose and sucrose act as osmolyte and also considered as substrates for cellular respiration. However, fructose can be more helpful in secondary metabolites production in salt-stressed plants (Rosa et al., 2009). In olive cultivars "Zard" and "Roghani," soluble sugars gradually increase with the increase of salt level up to 80-mM NaCl, and then, decline starts when the salinity level was further increased (Mousavi et al., 2008). Moreover, the level of soluble sugars (glucose, fructose, and sucrose) increased in leaves of Cleopatra mandarin and Troyer citrange under salinity level of 0–80-mM NaCl, which indicated that salt-tolerant plants showed more sugar accumulation as compared with salt-sensitive plants (Anjum, 2008).

Ascorbic acid plays a significant role to reduce the hypoxia-induced oxidative injury in salt-tolerant apple rootstock (*Malus hupehensis*) and salt-sensitive rootstock (*M. sieversii*) through controlling antioxidants activities, that is, SOD, POD, APX, CAT, and GR (Bai et al., 2013). At 50-mM NaCl, citrus rootstock Jatti Khatti has considerably greater potential for SOD and CAT activities as compared with Attani-1 and Attani-2 rootstocks. At 25-mM NaCl, POD levels might be enhanced in several citrus rootstocks (Singh et al., 2014). Salinity also enhances production of phenolic compounds in plants. Oleuropein is one of the most identified phenolics that mainly protect olive trees from salt stress possibly through acting as a glucose reservoir for osmoregulation and a constituent of antioxidant defense mechanism. Under salt stress, enzymatic and nonenzymatic antioxidants were enhanced in Carrizo citrange seedlings, while lipid peroxidation continued at an adequate level demonstrating that damaging effects of salinity in Carrizo citrange were due to toxic effects of Cl⁻ ions (Arbona et al., 2003).

Another important osmolyte is glycine betaine, which is mostly produced in salt-stressed plants. Glycine betaine is very famous to be used in maintaining photosynthetic process and protein stability. The grape cultivars "Yaghooti" and "1103P" are salt tolerant and accumulate more glycine betaine than proline (Sohrabi et al., 2017). In these cultivars, the functioning of photosynthetic mechanism and chlorophyll synthesis is normal due to less proline production.

Under salt stress, antioxidant enzymes and nonenzymatic antioxidants levels are upregulated and highly differ among plant species or cultivars. Salt-tolerant and salt-sensitive cultivars vary in their antioxidant level that also differs based on crop growth stage and several management practices (Table 32.1). In apple, Dongbeihuanghaitang, Daguohongsanyehaitang, and Qiuzi are salt-tolerant rootstocks, while Yingyehaitang and Lushihongguo are salt-sensitive rootstocks. It has been observed that salt-tolerant rootstocks showed the highest ROS scavenging activity as compared with salt-sensitive rootstocks (Yin et al., 2010).

2.2.4 Salt avoidance

In some fruit crops, reduction in the canopy area results in minimum water losses, when stomata are closed during transpiration. Canopy management by the plants is an important salt avoidance mechanism.

Salt exclusion

Salt exclusion refers to inhibiting the access of salts into vascular system of fruit crops. Salt exclusion, that is, Na⁺ and Cl⁻, by roots makes sure that these salts do not store to toxic levels in leaves. Mostly, salt exclusion is the capability of rootstocks to keep salts in the roots and basal portion of stem, and these are unable to translocate in photosynthetically active leaves and other parts of fruit crops. Several rootstocks of fruit crops have been found salt excluder with excellent properties of Na⁺ and Cl⁻ exclusion. Moreover, it has been evaluated that Cleopatra mandarin showed good adaptation as compared with Troyer citrange under high salt stress (Anjum, 2008). Ploidy level is an important genetic feature in fruit crops that can affect the salt tolerance of several rootstocks. It has been examined that tetraploid citrus seedlings exhibit higher salinity tolerance as compared with diploid ones (Ruiz et al., 2016).

Genome duplication has greater ability to eliminate salts by changing physiological properties of citrus plants. Tetraploid seedlings of *Citrus macrophylla* accumulate lower amount of Na⁺ and Cl⁻ in leaves as compared with diploid seedlings at 40- and 80-mM NaCl salinity levels. Moreover, K⁺ concentrations in leaf may reduce in diploid seedlings under salinity, and significant variations were observed in mineral uptake (Ruiz et al., 2016). In citrus, *Poncirus trifoliata* has distinctive root system and confirmed that fine roots are continuously produced from stressed plants to eliminate the excess amount of salts. In mango, polyembryonic rootstocks have good capability to withstand under salt stress conditions than monoembryonic rootstocks. Salt exclusion capability can be a failure due to excessive amount of salts

TABLE 32.1 Role of different antioxidants in salt tolerance mechanism of fruit crops.

Antioxidant	Function	Reference
SOD	It can imbalance the O_2 to form H_2O_2 and remove the harmfulness of superoxide anion	Wang et al. (2016)
POD	POD level was enhanced in citrus rootstocks under high salinity level	Singh et al. (2014)
Ascorbates	It chiefly eliminates the H_2O_2 production in plants. Increasing the ascorbate activity, plant tolerance might be enhanced	Sofo et al. (2015)
Glutathione	It is involved to maintain normal cellular redox system either in normal or even in stressed conditions	Zhang and Shi (2013)
	H_2O_2 and its derivatives are quickly reduced through glutathione	Wang et al. (2016)
Catalases	It also primarily decreases the amount of H_2O_2, which is manufactured in light respiration	Ashraf (2009)
Malondialdehyde	It is used to reduce membrane lipid peroxidation	Miranda et al. (2018)
	Higher salt tolerance mechanism recorded through reduction in lipid peroxidation under salt stress	
Tocopherols	These have high capability to scavenge the ROS and lipid radicals in plants	Sairam and Tyagi (2004)
	These have greater potential to directly repair oxidizing radicals by inhibiting chain transmission period during lipid autooxidation	
Flavonoids	These flavonoids have greater potential to inhibit the lipoxygenase	Potapovich and Kostyuk (2003)
Proline	Proline may act as a signaling molecule to maintain osmotic regulation	Ashraf and Foolad (2007)
	Date palm proline synthesis largely increased in leaves and roots under salt stress conditions	Rabey et al. (2016)
	Proline is considered as antioxidant that improves salt tolerance in pistachio plants	Shamshiri and Fattahi (2014)
Ascorbic acid	Ascorbic acid plays a significant role to reduce the hypoxia-induced oxidative injury in salt-tolerant apple rootstock (*Malus hupehensis*)	Bai et al. (2013)
Glycine betaine	It is very famous in maintaining photosynthesis and protein stability	Sohrabi et al. (2017)
	Grape cultivars "Yaghooti" and "1103P" are salt tolerant and accumulate more glycine betaine than proline	
Soluble sugars	Soluble sugars increased in leaves of Cleopatra mandarin and Troyer citrange indicated that salt-tolerant plants showed more sugar accumulation as compared with salt-sensitive plants	Anjum (2008)
Phenolic content	Oleuropein is one of the most identified phenolics that mainly protect olive trees from salt stress by acting as a glucose reservoir for osmoregulation and a constituent of antioxidant defense mechanism	Mohammadkhani (2018)

than critical threshold. Dogridge is a famous grape rootstock that was initially considered as a salt tolerant approximately up to 6.5 dS/m NaCl. When Thompson Seedless scion is grafted on dogridge rootstock, then, it does not withstand under continued saline irrigation (Sharma et al., 2011). Salt exclusion is a cultivar-specific trait in fruit crops, and the highest salt exclusion potential of a particular cultivar does not necessarily characterize the salt tolerance of species. Cleopatra mandarin and shekwasha exhibited negligible leaf drop and chlorosis symptoms, while fuzhu, beauty, willow leaf, nasnaran, and King of Siam were highly affected due to translocation of Cl− from roots to leaves. Even with heavy leaf abscission, numerous leaves on salt-affected plants remained green due to escape mechanism whereby leaves holding too much toxic ions were dropped. However, young leaves maintain the photosynthesis (Yahmed et al., 2015).

Two olive cultivars, that is, "Arvanitolia Serron" and "Lefkolia Serron" were found to be tolerant against NaCl stress among nine studied cultivars. These cultivars translocated Na+ and Cl− to leaves very slowly by holding them in roots. Under salinity, cultivar "Arvanitolia Serron" also retains the highest level of K+ in younger leaves (Assimakopoulou et al., 2017). The highest rate of growth suppression was recorded in olive cultivar "Barnea" as compared with "Arbequina." Lower level of Na+ and Cl− and the highest photosynthesis were recorded in Japanese pear "Akibae" when grafted on *Pyrus betulaefolia* rootstock under 100-mM NaCl. Moreover, their scions were also further grafted on *P. calleryana* and *P. pyrifolia* rootstocks evaluating that salt exclusion ability of *P. betulaefolia* rootstock remains functional and purposeful even after grafting (Matsumoto et al., 2007). The suppression in plant growth was measured in fig cultivars under salinity stress. The cultivar "S × P" is most sensitive, while "S × K" is very tolerant. The accumulation of $Cl-$ concentrations was enhanced in different plant parts but was higher in roots as compared with leaves showing salinity tolerance in fig based on Na^+ and $Cl-$ elimination from shoots (Zarei et al., 2016). F_1 hybrids of grapevine A15 and A17 under salt stress conditions exhibited greater salt tolerance. High Na^+ was obstructed in roots. They had strong capability to release efficient Na^+ efflux and retain K^+, which decreased Na^+ buildup in leaves and even in whole plant. It also retained greater K^+ levels in plant, particularly in leaves (Fu et al., 2019).

2.2.5 Salt excretion

Halophyte plants have greater capability to survive under 30–500-mM concentrations of NaCl. These plants developed good salt tolerance mechanism by salt-secreting glands and root hairs that vigorously reduce salts. Thus, salt concentrations in leaves remain under a threshold level (Soni et al., 2017).

2.2.6 Ion channels involvement

Sodium and potassium channels

Under optimum soil conditions, plants sustain the highest cytosolic K^+ and Na^+ ratio, while, under salt stress, ionic stability is disturbed due to excess of Na^+ that may be helpful for passive transport of Na^+ in excess of K^+. Sodium and potassium channels are mostly unable to differentiate between K^+ and Na^+ ions because these have powerful similar hydrated ionic ranges. Low- and high-affinity K^+ transporters can be used to transport Na^+ ions into the cell. Three different low-affinity K^+ channels have been well known in the plants such as voltage-dependent inward rectifying channels (KIRCs), voltage-dependent outward rectifying channels (KORCs), and voltage-independent cation channels (VICs). Under high salinity levels, KIRCs and KORCs show a high K^+/Na^+ differentiation ratio at functional K^+ and Na^+ concentrations, while VICs have a comparatively high Na^+/K^+ discrimination and seem to reduce Na^+ uptake (Blumwald, 2000). Low-affinity K^+ channels have low Na^+/K^+ discrimination and are comparatively highly affected by Na^+. Therefore, it is necessary to keep high-affinity K^+ channels for K^+ uptake in salinized plants.

Solute movements can be controlled through integrated membrane proteins (IMPs). High-affinity K^+ transporters (HKTs) are considered as one of the most important classes of IMPs, which are usually present only in plants. These HKTs can vary in Na^+/K^+ differentiation because few are highly specific for Na^+ and others for K^+ (Waters et al., 2013). Transcriptome analysis played an important role for identification and documentation of 387 unique ion transporter transcripts in salt-stressed olive cv. "Kalamon" including 9 Na^+ transporters, 30 K^+ transporters, and numerous other transcripts associated with proton transport system (Bazakos et al., 2015).

Up to 35-mM NaCl, reduction in plant growth, fruit yield, plant water use, and membrane permeability was observed in strawberry plants. Moreover, high NaCl level may enhance the calcium deficiency in leaves, and these growth and yield attributes can be enhanced by calcium application (Karlidag et al., 2011).

Chloride channels

Under salinity stress, plant response and $Cl-$ transport processes are linked with salt tolerance of plants that can vary from species to species and even within cultivars. In plants, chloride ions catalyze numerous enzymatic activities and known as cofactor in photosynthesis. These are also helpful to regulate membrane potential, stabilize cell turgor pressure, and maintain pH. High level of $Cl-$ is responsible for toxic symptoms at 4–7 mg/g concentration in salt-sensitive crops and at 15–50 mg/g in salt-tolerant crops. $Cl-$ exclusion through roots and slower passage to shoot or leaves may decrease or inhibit $Cl-$-induced damage signs in salt-sensitive fruit crops like citrus, strawberry, and grapes (Tavakkoli et al., 2010). Moreover, it has been studied that excess of $Cl-$ in many plants is more damaging than Na^+. Recently, it has been reported that strawberry cultivars have potential to limit the accumulation of Na^+ in leaves even under high salinity stress, while $Cl-$ was accumulated in every plant organ. Therefore, old leaves can be damaged due to exudation of chlorides by guttation on leaf margins (Ferreira et al., 2019). There is little work available on absorption and transport mechanisms of $Cl-$. Several major processes, that is, slower loading into xylem, intracellular compartmentation, and root efflux, are important to control $Cl-$ entry and its translocation in plants similar to Na^+. However, reduced loading into xylem inhibits the buildup of excess $Cl-$ in shoots, and preferential accumulation of $Cl-$ in leaf epidermis can reduce the possible injury of mesophyll cells carrying out photosynthesis. Previously, it was thought that $Cl-$ exclusion in shoots was regulated by a single gene, while, recently, it was found that it is controlled by several genes. Aquaporins (AQP) are essential membrane proteins that may be used to control water flow in roots. ATP-binding cassette (ABC) transporters are several distinguished genes of plants, which are openly or incidentally involved in $Cl-$ flux. It has been confirmed that various candidate genes have been identified for

Cl— transporters and channels that might be involved in salt tolerance. Moreover, candidate genes in two citrus cultivars for Cl— transport mechanisms were identified through transcriptome analysis (Teakle and Tyerman, 2010). Therefore, efficient characterization will be very helpful for evaluating the physiological behavior of these genes in salt tolerance.

---

3 Response of different pathways under salinity

Apoplastic, symplastic, and transmembrane pathways are very useful for water and ion movement in plant roots. Direct apoplastic transportation without any functional discrimination for ion passage may become predominant under high salinity stress. Grapevine is reasonably sensitive to salinity stress due to accumulation of Cl— at toxic levels in leaves. Therefore, it is the main reason for salt-induced symptoms in grapevine. Two grape rootstocks "1103 Paulsen," a salt tolerant, and "K 51-40," a salt sensitive, were evaluated that revealed that Cl— has no relationship with apoplastic tracer but represents alike bypass salts flow in xylem of both rootstocks. Moreover, differences in membrane transport clarify the variations in Cl— transport toward the shoot (Abbaspour et al., 2014). On the basis of a proposed model for symplastic control of Cl— homeostasis in salt includer rootstock (Carrizo citrange) and excluder rootstock (Cleopatra mandarin), the highest affinity Cl—/H$^+$ symporters control Cl— uptake of Cl— concentrations. Symplastic regulation showed differences in the includer rootstock and excluder rootstock depending upon Cl— translocation rate toward xylem. So, symplastic regulation is very low in excluder Cleopatra as compared with includer Carrizo rootstock (Brumos et al., 2010). NRT1 and NRT2 are membrane transporter genes that are clearly expressed by showing their good potential in Cl— homeostasis in salt-stressed Cleopatra and Carrizo plants (Brumos et al., 2009). Chloride channels (CLCs) are also involved in plant survival under salt stress by Cl— homeostasis regulation. Moreover, chloride channels encoding gene CsCLCc expressed in trifoliate orange leaves and roots are controlled by low temperature, ABA, and NaCl (Wei et al., 2013).

3.1 Salt stress signaling

A single abiotic stress can drastically reduce plant growth and yield. In fact, plants frequently face several stresses in a particular time. Immediate happening of abiotic stresses, that is, drought and excessive salts, which suggested that stress sensing and signaling, are very complicated mechanism. Moreover, underground and aboveground portions of plants can be affected through a single or more stresses occurring simultaneously. It is essential to understand plant sense and their response under stress particularly in grafted plants comprising two diverse cultivars. Therefore, it is known that plant signaling is not a lonely mechanism but basically comprises the interaction between numerous transduction measures in a difficult way. Further, the complex transduction chain among several biotic and abiotic stresses and their interaction were also detected (Singh and Sharma, 2018).

---

4 Strategies for improvement of salt tolerance

The maintenance of excessive amount of salts at favorable level is excellent way to control harmful salt accumulation in soil. Optimum salt level can be attained through excessive water to leach down salts below the root zone and not into groundwater reserves. Quality of irrigated water may have a greater influence on fruit production. Dissolved minerals are found in irrigated water, but salt level and compositions mainly depend on irrigation water source. Dissolved salts are also found in natural rainfall. Therefore, irrigation water and rainfall are two major causes of salinity in fruit crops. Water is absorbed by plant roots, and salts remain left behind in soil and start to accumulate. Soil salinity is a cause of difficulty for plants to absorb soil moisture. However, these salts can be leached down below root zone through additional water supply (Soni et al., 2017).

Development of salt tolerance cultivars is very slow due to the lack of knowledge about various complex tolerance mechanisms, which perform totally diverse at cellular and whole plant levels. Earlier work reported that osmotic adjustment and ion transportation attained greater attention because plants exhibit greater salt tolerance due to greater osmotic adjustment capacity by eliminating toxic ions from the cells and tissues by accumulating low-molecular-weight organic osmotic contents, that is, proline, soluble sugars, and several minerals. Moreover, photosynthesis plays an important role during salt stress by synthesizing organic solutes (Yaish et al., 2017). However, further search work is needed to elaborate the physiological changes in photochemical, gaseous exchange, and metabolic occurrences during photosynthesis under salt stress.

Diverse photochemical, gaseous exchange, and metabolic developments of photosynthesis are strongly associated with each other. Therefore, considerable changes in any process can change the series of mechanisms that eventually affect the overall photosynthesis rate. Under saline conditions, plant physiologists determined a relationship of physiological traits for plant growth and yield potential. The evaluation of such essential traits and their potential use for plant breeders is very costly mainly in segregating populations. However, genetic improvements can be made for salt tolerance by isolating natural variations through direct selection or with QTL mapping in fruit crops. Earlier studies mentioned that direct selection under field conditions may have very limited success due to environmental influence on whole plant response to salt tolerance. Plant response to salt stress may vary depending on different developmental stages (Yamaguchi and Blumwald, 2005). Therefore, molecular interests have been increased for identification of genes and gene products that can be transmitted to well-known cultivars using marker-assisted breeding and genetic transformation.

4.1 Genetic basis of salt tolerance

Seed germination, root development, seedling growth, vegetative growth, flower formation, fruit set, and its development are mainly affected under salt stress. This occurrence eventually causes reduction in crop yield. The application of different molecular technologies to examine crop tolerance mechanism, evaluation of resistance genes, and evolve tolerant cultivars through genetic transformation is an imperative constituent of advanced agricultural research. Molecular markers and their applications in salt tolerance studies of fruit crops have been listed (Table 32.2). Under stress situations, plants express specific stress-related genes that can be categorized into three groups:

a. Genes encoding proteins involved in metabolic processes, that is, detoxification and osmolyte biosynthesis
b. Genes encoding proteins having unidentified utilities
c. Monitoring class genes that regulate signaling flows and transcriptional mechanism

Salt tolerance is very complicated characteristic, and its polygenic nature is a basic limitation for the development of salt-tolerant cultivars using traditional strategies. Salt tolerance through physiological and molecular basis remains still unclear in many fruit crops. However, accurate description of even physiological and biochemical characters exhibiting salt tolerance and compound gene linkages does not remain a critical challenge in woody perennials. In citrus plants, a huge number of traits were involved in evaluating the whole plant reaction under salt stress (Raga et al., 2016). Limited work was accompanied to know the molecular basis of salt tolerance in fruit crops. Salt tolerance is a monogenic or polygenic character that is still unclear even in more studied fruit crops such as grapevine (Gong et al., 2010). For salt stress screening, broad field and laboratory trials are very expensive and time consuming. In spite of cost and resource limitations, accurate evaluation of salt tolerance under filed conditions is not promising due to complicated relationships between the plants and environmental conditions. Thus, pot studies play a major role to evaluate salt tolerance capability in plants. Recently, in vitro culture studies also proved helpful for evaluation of salt-tolerant cultivars providing short time and lesser space under controlled conditions. In vitro screening practices are more helpful for evaluation of large populations. However, a poor correlation was observed between in vivo and in vitro observations in plants regenerated from salt-tolerant lines (Vijayan et al., 2003).

4.1.1 Marker-assisted selection

MAS is a most favorable breeding strategy for the improvement of fruit production. A lot of time is required for the development of salt-tolerant cultivars through traditional techniques. Conventionally, genetically diverse germplasm of a species can be outcrossed and then selected on the basis of their phenotypic characteristics. Now, this mechanism is efficiently using QTL analysis combined with MAS (Ashraf and Foolad, 2007). Therefore, the critical basis for effective breeding with MAS requires huge information of genomic variations in the preferred fruit crops. Molecular markers and different sequencing techniques such as whole genome sequencing are prospective and more critical to facilitate the monitoring of further advanced breeding programs (Song et al., 2012). Hence, the significant role of MAS is evident in the development of salt-tolerant cultivars. Several studies were conducted to enhance salt tolerance level through genomic influence of certain genes. Few genes may not have an excellent potential to enhance crop capability under saline conditions (Song et al., 2012). A traditional breeding practice for stacking of traits is a promising method depending on the introduction of numerous useful and valuable genes for improvement of plant performance. This method is

TABLE 32.2 Molecular markers and their applications in salt tolerance studies of fruit crops.

Molecular marker	Function	Reference
MAS	MAS facilitate pyramiding traits of interest to attain significant improvement in crop salt tolerance	Yamaguchi and Blumwald (2005)
	Zinc finger nucleases are successfully implemented for producing trait landing pads	Belhaj et al. (2013)
QTLs	Nearly, 98 QTLs were strongly linked and detected salt tolerant in a hybrid citrus population (Cleopatra mandarin [salt tolerant] × trifoliate orange [salt sensitive])	Raga et al. (2016)
	Biparental QTL mapping can be used for separating genomic regions related to salt tolerance. Such QTLs may enclose thousands of the recognized genes	Long et al. (2013)
Genome sequencing	Availability of complete genome sequences has allowed the determination of the number and structure of genes in a particular genome and their organization on the chromosome	Pichersky and Gerats (2011)
Functional genomics	RAPDs (OPC-02) were used to reveal DNA alleles that can distinguish gene coding for salt tolerance in date palm genome regions	Kurup et al. (2009)
	Sometimes, a specific function is controlled with more than one gene; then, reverse genetic practices are the only method to accomplish a step-by-step investigation of such terminated functions	Holtorf et al. (2002)
Transcription factors	In olive, 209 transcripts in salt tolerant cv. "Kalamon" and 36 transcripts in salt sensitive cvs. "Chondrolia" and "Chalkidikis" were expressed	Bazakos et al. (2012)
	Huge number of transcripts was associated with sugar breakdown and transport and aquaporin genes during the recovery of salt-stressed grapevines	Perrone et al. (2012)
Gene expression	A set of 2630 genes in leaves and 4687 genes in roots were differentially expressed in date palm	Yaish et al. (2017)
	Nearly, 21 HAK/KUP/KT genes were detected in pear genome, which involved in K^+ deficient	Li et al. (2018)
Proteomics	A total of 271 S-nitrosylated proteins were detected in "Chardonnay" and in "Cabernet Sauvignon" grapevines due to salt stresses	Vincent et al. (2007)
	Moreover, 35 protein spots were detected in salinized date palm plants out of 55 protein spots	Rabey et al. (2016)
Metabolites	In grapes, 32 metabolites measured exhibited tissue-specific variations in abundance, and the accumulation of seven of these compounds was disturbed under water stress conditions	Grimplet et al. (2009)
Genetic transformation	Overexpression of Musa DHN1 gene enhanced salt tolerance by increasing accumulation of proline and decreasing lipid peroxidation in leaves of transgenic banana lines	Shekhawat et al. (2011)

inadequate due to the segregation of attributes, which disturb the breeding approaches (Ainley et al., 2013). However, zinc finger nucleases are successfully implemented for producing trait landing pads (Belhaj et al., 2013).

4.1.2 QTL mapping

QTL mapping and molecular marker usage are becoming an urgent need and are still a challenge for prediction of salinity tolerance in fruit crops. Single marker assisted, simple interval mappings, and composite interval mappings are three basic methods for QTL detection. Among three basic methods, single marker assisted is very simple and widely used method because single marker is involved in QTL detection (Collard et al., 2005). Salinity tolerance in fruit crops have been examined through different molecular markers, but few QTLs and genes were identified for several physiological traits under salt stress. The development of salt-resistant cultivars in fruit crops through QTLs and molecular markers is very difficult and slow.

The importance of gene identification and cloning to develop tolerance against one or more abiotic stresses has progressively enhanced due to limitations of conventional breeding. Conventional breeding strategies do not provide actual or real outcomes. Further, marker-assisted breeding is significantly based on the identification of genomic regions through QTLs, which are strongly related to desired traits. Subsequent introgression into well-known cultivars continues to suffer from linkage drag problem because insertion of such QTLs frequently brings along unwanted characters from donor parents.

QTLs are genetic stretches interrelated with variations of a quantitative trait of interest such as salt resistance. In plants, such quantitative variations are resulting from mutual accomplishment of numerous segregating genes and climatic conditions. QTLs are involved in crossing of two diverse parents opposing in one or more quantitative characters. Hybrids are examined to connect QTLs with known DNA marker for indirect selection of desired variable. Numerous QTL mapping studies were conducted with aim to achieve marker-assisted selection, gene cloning, and characterization of germplasm (Asins, 2002). Nearly, 98 QTLs were strongly linked and detected salt tolerance in a hybrid citrus population (Cleopatra mandarin [salt tolerant] × trifoliate orange [salt sensitive]). A cluster of QTLs leading plant vigor and leaf boron concentration pointed a genomic region in linkage group 3. So, this linkage group is most appropriate to enhance salt tolerance using the Cleopatra parent as donor (Raga et al., 2016). Under salt stress, QTL mapping revealed 70 potential QTLs in 16 regions of citrus genome, and six of them were involved in growth and dry weight production (Tozlu et al., 1999). Moreover, an apparent relationship was also identified between genomic regions, which control salt and cold tolerance in citrus genome. RAPD markers are codominant markers that have been used to reveal DNA alleles, which can distinguish gene coding for salt tolerance in date palm genome regions. RAPD marker (OPC-02) was applied, which amplified fragments at 1400 bp in salt-tolerant date palm cv. Bugal White (Kurup et al., 2009).

4.1.3 Association mapping

Biparental QTL mapping can be used for separating genomic regions related to salt tolerance, which offers very slight information as how allelic variability found in complete gene pool interrelate and affect the salt tolerance in fruit crops. Moreover, such QTLs may enclose thousands of the recognized genes (Long et al., 2013). Therefore, it is very necessity to develop more appropriate strategies for understanding the genomic control of measurable attributes. Advances in DNA fingerprinting, genome sequencing, association mapping, and development of different statistical analyses are very interesting for the utilization of genetic diversity in fruit crops. Hence, association mapping is also known as disequilibrium mapping. Association mapping is also used to detect variations in complex trait by manipulating historical and evolutionary recombination measures at population level. However, QTL mapping detects differences between two characters or individuals, while association mapping exhibits phenotypic and genetic variability across a natural population (Raga et al., 2016).

4.1.4 Genome sequencing

Recently, whole genome sequencing in fruit crops, that is, date palm, citrus, strawberry, peach, olive, grape, apple, and papaya, is available, while, in several other fruits, whole genome mapping is under progress (Lor et al., 2012). High-throughput sequencing techniques reduced the developmental cost of sequencing, and genotyping might be helpful to improve current genome sequences. So, these developments are more useful for genome sequences and also for references to identify single nucleotide polymorphisms and duplicate number variants. However, unique growth and genomic appearances of fruit crops such as long juvenile phase and high level of heterozygosity must be measured during phenotyping and genotyping data to confirm the selection of suitable association mapping approach (Khan and Korban, 2012).

4.1.5 Gene-based markers

Generally, plant genome has more number of genes than other organisms. Therefore, the information described by sessile and autotrophic habit of plants is exciting them to produce a group of compounds compulsory for defense against harsh environments. The availability of DNA sequencing information in fruit crops has encouraged the attention of researchers in exploration of gene functioning. Accessibility of whole genome sequencing is very helpful for evaluation of gene number and structure and also their association on the chromosome in a specific genome (Pichersky and Gerats, 2011). Now, "quantitative" genome facts existing in many species are being adapted into "qualitative" facts with a procedure of value addition to the nucleotide sequence information through allocating novel tasks to the anonymous genes. Regrettably, knowing the overall function often does not offer proper understanding into particular role in the organism, thus requiring extensive functional genomics to allocate clearly stated roles to the genes (Holtorf et al., 2002).

The role of particular genes might be attained by "forward" or "reverse" genetic strategies. Genetic basis of an attribute can be exposed through investigating the improved or mutated phenotype such as phenotype to genotype in forward genetics. However, when a specific gene is improved, the resulting phenotype is studied as genotype to phenotype in reverse genetics. The studies on classical forward genetics are carried out through phenotypic screening of mutant populations attained with chemical or physical mutagenesis. Map-based cloning is time consuming and is conducted to detect the sequence change essential for the mutant phenotype. Such forward genetic approaches are mainly inappropriate for high-throughput functional analysis even if saturated genetic maps are developed. Tagging through transposable elements or T-DNA constructs may have greater potential to detect the essential function of a particular gene by uncovering a particular phenotype. Sometimes, a specific function is controlled with more than one gene; then, reverse genetic practices are the only method to accomplish a step-by-step investigation of such terminated functions. Moreover, it has been examined that insertional knockout mutants of *Arabidopsis* do not exhibit useful phenotype that may possibly due to functional severance among genes or the fact that mutations are uncertain and do not freely change plant phenology. Therefore, in such circumstances, genetic function has to be concluded from capacities of gene expression and gene action (Peters et al., 2003).

Previous studies recommended that successful achievement of genome-based markers is mainly based on appropriate high-throughput systems, that is, transcriptional regulation, metabolites, genomics, and proteomics, which are mutually known as "omic methods." Omic techniques are genome-wide or system-wide experimental methodologies to evaluate the functions of gene worldwide. Gene-based markers are extremely related to "functional biology" in that it severely depends on complete profiling of nearly every gene expression product to concurrently monitor all the biological mechanisms functioning as an integrated method (Sumner et al., 2003). Moreover, 21 HAK/KUP/KT genes were detected in pear genome, which revealed their involvement in K^+ deficient and abiotic stress responses (Li et al., 2018).

4.1.6 Transcriptional regulation

Plant genome remains constant, while transcriptome expression changes depending upon the plant growth stage and ecological stimuli. Several practices were applied to calculate changes in transcriptome in response to stress environments. A recent significant revolution in high-throughput next-generation sequencing is of vital importance for comprehensive transcriptional linkages studies. Polyamines such as putrescence, spermine, and spermidine exhibited a high protecting role under salt stress conditions. Exogenous application of polyamines can be used for partial recovery of salt-induced injuries in plants. Gene expression strongly involved in polyamine biosynthesis and catabolism exhibited that biosynthetic genes except ODC were enhanced through NaCl usage in sour orange (Tanou et al., 2014).

Approximately 1900 genes were identified and expressed in the roots of salt-treated plants of date palm cv. "Deglet Beida." These genes are involved in DNA, RNA, and protein biosynthesis and also in signaling pathways. Hence, it has been believed that these genes play a functional role in salt tolerance (Radwan et al., 2015). A set of 2630 genes in leaves and 4687 genes in roots were differentially expressed in date palm cv. "Khalas" under salt stress. A total of 194 genes were usually expressed in leaves and roots (Yaish et al., 2017). A combined examination of miRNA and mRNA expression proved that these were involved in ABA-activated signaling pathway and ROS metabolism in citrus roots under salt stress (Xie et al., 2017). In grapevines, the huge number of transcripts was associated with sugar breakdown and transport, flavonoid biosynthesis, and aquaporin genes during the recovery of stressed plants (Perrone et al., 2012).

Under salt stress, many transcripts connected to transcription, protein synthesis, and metabolism have been identified (Vincent et al., 2007). Grapes cultivars, that is, "H6" and "Gharashani" are salt tolerant, while "Shirazi" and "Ghezel Uzum" cultivars are salt sensitive. Under salinity, it has been confirmed that salt-tolerant and salt-sensitive cultivars may possibly differ in gene expression. Expression profile of VvNHL1 and VvEDS1 was similar in leaves of all cultivars and in roots of salt-tolerant cultivars, while VvChS and VvPAL transcripts considerably enhanced only in leaves of salt-tolerant cultivars (Mohammadkhani et al., 2016). In olive cultivars, 209 transcripts in salt-tolerant cv. "Kalamon" and 36 transcripts in salt-sensitive cvs. "Chondrolia" and "Chalkidikis" were expressed under salinity stress. Under salt stress, transcriptomic profiles were produced, and target transcripts were also identified in two olive cultivars. Transcriptional regulatory networks indicated similar regulatory TF homologues in olive and *Arabidopsis* (Bazakos et al., 2012).

Genomics and proteomics

Under abiotic stresses, physiological adjustment may occur in fruit plants through variations in gene expression and alterations in plant genomics and proteomics. However, alterations in gene expression do not continuously relate to the variations at protein level. Therefore, proteomic change studies are required because proteins directly stimulate plant stress response that can be identified. Proteomic approaches encode the whole set of proteins, which are found in biological sample. So, they permit a comprehensive understanding on how different proteins are involved in stress tolerance regulation in fruit crops. Improvements in proteomics have played an important role in identification and evaluation of diverse salt stress receptive proteins, which generally are involved in signal transduction, maintenance of photosynthesis, ion homeostasis, osmotic regulation, and stimulation of antioxidant defense mechanism. Those genes that encode such essential proteins are being cloned and transferred in numerous glycophytes (Kumari et al., 2015).

Hydrogen peroxide and nitric oxide are also helpful for signal transduction pathways, which stimulate plant defense mechanism under stress conditions. In a proteomic study, 85 leaf proteins were recorded with significant quantitative variability in *Citrus aurantium* L. plants under salinity stress. When these salinized plants were pretreated with hydrogen peroxide and sodium nitroprusside, these protein differences were not noticed (Tanou et al., 2009).

Under salt stress, polyamines reduced the protein carbonylation and tyrosine nitration, while S-nitrosylation protein was produced by all polyamines in sour orange plants. A total of 271 S-nitrosylated proteins were detected, which normally were affected by salt stress and polyamines. Proteins are successfully involved in photosynthetic adjustment and protein production dropped in "Chardonnay" and in "Cabernet Sauvignon" grapevines due to water and salt stresses. Many proteins controlled in tolerant cv. Chardonnay were of an unknown function, while proteins involved in protein breakdown/absorption were upregulated in cv. "Cabernet Sauvignon" (Vincent et al., 2007). Moreover, 35 protein spots were detected in salinized date palm plants out of 55 protein spots (Rabey et al., 2016).

4.1.7 Metabolites

Recently, greater advancements were recorded in transcriptome and proteome analysis under stress conditions in fruit crops, but the information regarding gene function is still confusing. Approximately 30%–40% of the open reading frames have unknown function(s) emphasizing the basic need for evaluating the biological function(s) of these pretended orphan genes (Hollywood et al., 2006). Metabolomics refers to wide-ranging high-throughput analysis of complicated metabolites manufactured by the plants. However, plant metabolic alignment is more complex and demands very high expenses and technical skills. Attention in metabolomic activities was developed by the fact that biological response of an organism can be characterized by its influence on difference in buildup of individual metabolites. Phenotype of an organism is mainly influenced by genetic make of cultivar and its interaction with environmental factors. Phenotypic expression is also controlled by several subcellular physiological mechanisms considering that exact description of metabolic fluxes involved in biological pathways is completely vital for clear understanding (Bino et al., 2004).

Several metabolites, that is, proline, glycinebetaine, mannitol, antioxidant molecules, and chlorophyll content, are consistently calculated to assess plant response under salinity stress in fruit crops. Hence, the understanding of plant adaptation to salt stress as a function of the whole metabolite network still remains very poor. Metabolite profiling showed that higher concentrations of glucose, proline, and malate were found in water-stressed plants as compared with salinized "Cabernet Sauvignon" grape vines. Metabolite changes were associated to the difference in transcript richness of numerous genes involved in energy absorption and nitrogen integration practices such as photosynthesis and photorespiration (Cramer et al., 2007). In grapes, approximately half of the 32 metabolites measured exhibited tissue-specific variations in abundance, and the accumulation of seven of these compounds was disturbed under water stress conditions (Grimplet et al., 2009).

4.1.8 Genetic transformation

Genetic improvement of fruit crops with traditional techniques is very laborious and time consuming. The major constrains that obstruct the success rate are long juvenile period, big tree size, and prerequisite to screen huge number of seedlings. For example, the development of good apple cultivar requires about 20 years and huge cost. Further, a lot of time is required to transfer most wanted character from a diverse or wild relative to the desired cultivar. Time is also needed for removal of unwanted fragments produced from the wild relatives, that is, 5–6 generation cycles each of 4–10 years. Approximately, five decades will be more vital to achieve new cultivar with good traits and one other decade to commercialize the cultivar to meet consumer demand (Flachowsky et al., 2011). Introgression of genes brings a great revolution in tree breeding by using Agrobacterium mediated or numerous other direct transfer techniques such as electroporation and microinjection. These methods reduce time and costs as compared with conventional selection and hybridization programs.

Currently, genetic transformation has been efficiently used in more than 100 species. Genetically improved fruit crops are also present in the market and have demonstrated catalytic to the increase of transgenic cultivars in fruit crops. However, numerous natural, monitoring, and public opinion limitations are needed to be effectively addressed before actual uses of such transgenic materials in orchards. Overexpression of chemically developed grape gene VvbHLH1 considerably improved salt tolerance in transgenic Arabidopsis thaliana plants via controlling the genes involved in flavonoid biosynthesis, proline biosynthesis, ABA signaling paths, and ROS scavenging (Wang et al., 2016). Arabidopsis vacuolar Na^+/H^+ antiporter gene (AtNHX1) in transgenic kiwifruit plants resulted in better capability to tolerate NaCl level more than 200 mM contributing to a greater osmotic regulation and the highest antioxidant activities (Tian et al., 2011). Moreover, introgressive hybridization of Arabidopsis gene CBF3/DREB1A into Citrus macrophylla genome expressively enhanced salt tolerance mechanism as compared with wild relatives (Gerding et al., 2015). Similarly, overexpression of Musa DHN1 gene enhanced salt tolerance by increasing accumulation of proline and decreasing lipid peroxidation in leaves of transgenic banana lines (Shekhawat et al., 2011).

In recent times, transgrafting is also being popular as an innovative practice to relieve several uncertainties interrelated to transgenic cultivars. Genetically engineered scion or rootstock both had excellent capability to enhance tolerance under different stress conditions during transgrafting. Therefore, transgrafting is considered as one of the most important graft components. Moreover, transgenic product(s) movement through a graft union is alarmed, because long distance is extremely unlike for transmission of genomic or organelle DNA. However, several genetic alterations are induced by epigenetic changes of DNA that are transmissible (Haroldsen et al., 2012).

5 Conclusion and future prospects

During the last few years, a positive change has been observed in employing methods and approaches to resolve the complicated physiological and genetic basis of salt tolerance in fruit crops. Therefore, marker-based techniques and genomic methods are progressively employed for the identification of genes related to salt tolerance in fruit crops. The introduction of next-generation sequencing techniques and plant omic methods is the beginning of a new era in the field of functional genomics. Fruit physiologists and plant breeders are selecting such genetic technologies for identification and evaluation of transcripts, genes, proteins, and metabolites that are strongly associated with fundamental salt tolerance attributes. Thus, these can be transferred to salt-sensitive high yielding cultivars. Most of the present information on the employment of these high-end technologies has been derived from two major fruit crops, that is,. citrus and grapes, showing their enormous marketable value. Moreover, it is probable that frequent improvements would further reduce the expenses and increase effectiveness of presently employed genomic techniques leading to their extensive usage in several other fruit crops for evolving high yielding salt tolerance cultivars.

References

Abbaspour, N., Kaiser, B., Tyerman, S., 2014. Root apoplastic transport and water relations cannot account for differences in Cl$^-$ transport and Cl$^-$/ NO^{3-} interactions of two grapevine rootstocks differing in salt tolerance. Acta Physiol. Plant. 36, 687–698. https://doi.org/10.1007/s11738-013-1447-y.

Abdallah, B.M., Trupiano, D., Ben Youssef, N., Scippa, S.G., 2017. An efficient method for olive leaves proteins extraction and two-dimensional electrophoresis. Nat. Prod. J. 7, 12–17.

Agrawal, R., Gupta, S., Gupta, N.K., Khandelwal, S.K., Bhargava, R., 2013. Effect of sodium chloride on gas exchange, antioxidative defense mechanism and ion accumulation in different cultivars of Indian jujube (*Ziziphus mauritiana* L.). Photosynthetica 51, 95–101. https://doi.org/10.1007/s11099-013-0003-8.

Ainley, W.M., Dent, S.L., Welter, M.E., Murray, M.G., Zeitler, B., Amora, R., Corbin, D.R., Miles, R.R., Arnold, N.L., Strange, T.L., Simpson, M.A., 2013. Trait stacking via targeted genome editing. Plant Biotechnol. J. 11, 1126–1134. https://doi.org/10.1111/pbi.12107.

Alfocea, P.F., Albacete, A., Ghanem, M.E., Dodd, I.C., 2010. Hormonal regulation of source–sink relations to maintain crop productivity under salinity: a case study of root-to-shoot signaling in tomato. Funct. Plant Biol. 37, 592–603. https://doi.org/10.1071/FP10012.

Anjum, M.A., 2008. Effect of NaCl concentrations in irrigation water on growth and polyamine metabolism in two citrus rootstocks with different levels of salinity tolerance. Acta Physiol. Plant. 30, 43–52. https://doi.org/10.1007/s11738-007-0089-3.

Aranda, R., Moya, J.L., Tadeo, F.R., Legaz, F., Millo, P.E., Talon, M., 1998. Physiological and anatomical disturbances induced by chloride salts in sensitive and tolerant citrus: beneficial and detrimental effects of cations. Plant Cell Environ. 21, 1243–1253. https://doi.org/10.1046/j.1365-3040.1998.00349.x.

Arbona, V., Flors, V., Jacas, J., García-Agustín, P., Cadenas, A., 2003. Enzymatic and non-enzymatic antioxidant responses of Carrizo citrange, a salt-sensitive citrus rootstock, to different levels of salinity. Plant Cell Physiol. 44, 388–394. https://doi.org/10.1093/pcp/pcg059.

Arbona, V., López-Climent, M.F., Mahouachi, J., Clemente, P.R.M., Abrams, S.R., Cadenas, G.A., 2006. Use of persistent analogs of abscisic acid as palliatives against salt-stress induced damage in citrus plants. J. Plant Growth Regul. 25, 1–9. https://doi.org/10.1007/s00344-005-0038-6.

Ashraf, M., 2009. Biotechnological approach of improving plant salt tolerance using antioxidants as markers. Biotechnol. Adv. 27, 84–93. https://doi.org/10.1016/j.biotechadv.2008.09.003.

Ashraf, M., Foolad, M.R., 2007. Roles of glycine betaine and proline in improving plant abiotic stress resistance. Environ. Exp. Bot. 59, 206–216. https://doi.org/10.1016/j.envexpbot.2005.12.006.

Asins, M.J., 2002. Present and future of quantitative trait locus analysis in plant breeding. Plant Breed. 121, 281–291. https://doi.org/10.1046/j.1439-0523.2002.730285.x.

Assaha, D.V., Ueda, A., Saneoka, H., Al-Yahyai, R., Yaish, M.W., 2017. The role of Na+ and K+ transporters in salt stress adaptation in glycophytes. Front. Physiol. 8, 509. https://doi.org/10.3389/fphys.2017.00509.

Assimakopoulou, A., Salmas, I., Roussos, P.A., Nifakos, K., Kalogeropoulos, P., Kostelenos, G., 2017. Salt tolerance evaluation of nine indigenous Greek olive cultivars. J. Plant Nutr. 40, 1099–1110.

Bai, T., Ma, P., Li, C., Yin, R., Ma, F., 2013. Role of ascorbic acid in enhancing hypoxia tolerance in roots of sensitive and tolerant apple rootstocks. Sci. Hortic. 164, 372–379. https://doi.org/10.1016/j.scienta.2013.10.003.

Banuls, J., Millo, P.E., 1992. Effects of chloride and sodium on gas exchange parameters and water relations of citrus plants. Physiol. Plant. 86, 115–123. https://doi.org/10.1111/j.1399-3054.1992.tb01319.x.

Banuls, J., Serna, M.D., Legaz, F., Talon, M., Primo-Millo, E., 1997. Growth and gas exchange parameters of citrus plants stressed with different salts. J. Plant Physiol. 150, 194–199. https://doi.org/10.1016/S0176-1617(97)80202-7.

Bazakos, C., Manioudaki, M.E., Therios, I., Voyiatzis, D., Kafetzopoulos, D., Awada, T., Kalaitzis, P., 2012. Comparative transcriptome analysis of two olive cultivars in response to NaCl-stress. PLoS One. 7e42931 https://doi.org/10.1371/journal.pone.0042931.

Bazakos, C., Manioudaki, M.E., Sarropoulou, E., Spano, T., Kalaitzis, P., 2015. 454 pyrosequencing of olive (*Olea europaea* L.) transcriptome in response to salinity. PLoS One 10, e0143000. https://doi.org/10.1371/journal.pone.0143000.

Behboudian, M.H., Törökfalvy, E., Walker, R.R., 1986. Effects of salinity on ionic content, water relations and gas exchange parameters in some Citrus scion-rootstock combinations. Sci. Hortic. 28, 105–116. https://doi.org/10.1016/0304-4238(86)90130-5.

Belhaj, K., Garcia, C.A., Kamoun, S., Nekrasov, V., 2013. Plant genome editing made easy: targeted mutagenesis in model and crop plants using the CRISPR/Cas system. Plant Methods 9, 39. https://doi.org/10.1186/1746-4811-9-39.

Bernstein, N., Meiri, A., Zilberstaine, M., 2004. Root growth of avocado is more sensitive to salinity than shoot growth. J. Am. Soc. Hortic. Sci. 129, 188–192.

Bino, R.J., Hall, R.D., Fiehn, O., Kopka, J., Saito, K., Draper, J., Nikolau, B.J., Mendes, P., Roessner-Tunali, U., Beale, M.H., Trethewey, R.N., 2004. Potential of metabolomics as a functional genomics tool. Trends Plant Sci. 9, 418–425. https://doi.org/10.1016/j.tplants.2004.07.004.

Blumwald, E., 2000. Sodium transport and salt tolerance in plants. Curr. Opin. Cell Biol. 12, 431–434. https://doi.org/10.1016/S0955-0674(00)00112-5.

Brumos, J., Flores, C.J.M., Conesa, A., Izquierdo, P., Sánchez, G., Iglesias, D.J., López-Climent, M.F., Cadenas, G.A., Talón, M., 2009. Membrane transporters and carbon metabolism implicated in chloride homeostasis differentiate salt stress responses in tolerant and sensitive Citrus rootstocks. Funct. Integr. Genomics 9, 293. https://doi.org/10.1007/s10142-008-0107-6.

Brumos, J., Talon, M., Bouhlal, R.Y.M., Flores, C.J.M., 2010. Cl-homeostasis in includer and excluder citrus rootstocks: transport mechanisms and identification of candidate genes. Plant Cell Environ. 33, 2012–2027. https://doi.org/10.1111/j.1365-3040.2010.02202.x.

Cadenas, G.A., Arbona, V., Jacas, J., Millo, P.E., Talon, M., 2002. Abscisic acid reduces leaf abscission and increases salt tolerance in citrus plants. J. Plant Growth Regul. 21, 234–240. https://doi.org/10.1007/s00344-002-0013-4.

Cerezo, M., Agustin, G.P., Serna, M.D., Millo, P.E., 1997. Kinetics of nitrate uptake by citrus seedlings and inhibitory effects of salinity. Plant Sci. 126, 105–112. https://doi.org/10.1016/S0168-9452(97)00095-2.

Chakraborty, K., Sairam, R.K., Bhattacharya, R.C., 2012. Differential expression of salt overly sensitive pathway genes determines salinity stress tolerance in Brassica genotypes. Plant Physiol. Biochem. 51, 90–101.

Chartzoulakis, K.S., 2005. Salinity and olive: growth, salt tolerance, photosynthesis and yield. Agric. Water Manag. 78, 108–121. https://doi.org/10.1016/j.agwat.2005.04.025.

Collard, B.C.Y., Jahufer, M.Z.Z., Brouwer, J.B., Pang, E.C.K., 2005. An introduction to markers, quantitative trait loci (QTL) mapping and marker-assisted selection for crop improvement: the basic concepts. Euphytica 142, 169–196. https://doi.org/10.1007/s10681-005-1681-5.

Costa, L.M.E., Ferreira, S., Duarte, A., Ferreira, A.L., 2010. Alleviation of salt stress using exogenous proline on a citrus cell line. In: VI International Symposium on Mineral Nutrition of Fruit Crops 868, pp. 109–112.

Cramer, G.R., Ergül, A., Grimplet, J., Tillett, R.L., Tattersall, E.A., Bohlman, M.C., Vincent, D., Sonderegger, J., Evans, J., Osborne, C., Quilici, D., 2007. Water and salinity stress in grapevines: early and late changes in transcript and metabolite profiles. Funct. Integr. Genom. 7, 111–134. https://doi.org/10.1007/s10142-006-0039-y.

Ferguson, L., Grattan, S.R., 2005. How salinity damages citrus: osmotic effects and specific ion toxicities. HortTechnology 15, 95–99.

Ferreira, J.F., Liu, X., Suarez, D.L., 2019. Fruit yield and survival of five commercial strawberry cultivars under field cultivation and salinity stress. Sci. Hortic. 243, 401–410. https://doi.org/10.1016/j.scienta.2018.07.016.

Flachowsky, H., Roux, L.P.M., Peil, A., Patocchi, A., Richter, K., Hanke, M.V., 2011. Application of a high-speed breeding technology to apple (Malus × domestica) based on transgenic early flowering plants and marker-assisted selection. New Phytol. 192, 364–377. https://doi.org/10.1111/j.1469-8137.2011.03813.x.

Fu, Q.Q., Tan, Y.Z., Zhai, H., Du, Y.P., 2019. Evaluation of salt resistance mechanisms of grapevine hybrid rootstocks. Sci. Hortic. 243, 148–158. https://doi.org/10.1016/j.scienta.2018.07.034.

Gerding, A.X., Espinoza, C., Blancheteau, I.C., Johnson, A.P., 2015. Molecular and physiological changes in response to salt stress in Citrus macrophylla plants overexpressing Arabidopsis CBF3/DREB1A. Plant Physiol. Biochem. 92, 71–80. https://doi.org/10.1016/j.plaphy.2015.04.005.

Gimeno, V., Syvertsen, J.P., Nieves, M., Simón, I., Martínez, V., Sánchez, G.F., 2009. Additional nitrogen fertilization affects salt tolerance of lemon trees on different rootstocks. Sci. Hortic. 121, 298–305. https://doi.org/10.1016/j.scienta.2009.02.019.

Gong, H., Blackmore, D., Clingeleffer, P., Sykes, S., Jha, D., Tester, M., Walker, R., 2010. Contrast in chloride exclusion between two grapevine genotypes and its variation in their hybrid progeny. J. Exp. Bot. 62, 989–999. https://doi.org/10.1093/jxb/erq326.

Grimplet, J., Wheatley, M.D., Jouira, H.B., Deluc, L.G., Cramer, G.R., Cushman, J.C., 2009. Proteomic and selected metabolite analysis of grape berry tissues under well-watered and water-deficit stress conditions. Proteomics 9, 2503–2528. https://doi.org/10.1002/pmic.200800158.

Gucci, R., Moing, A., Gravano, E., Gaudillère, J.P., 1998. Partitioning of photosynthetic carbohydrates in leaves of salt-stressed olive plants. Funct. Plant Biol. 25 (5), 571–579. https://doi.org/10.1071/PP98003.

Haroldsen, V., Szczerba, M.W., Aktas, H., Lopez, J., Odias, M.J., Chi-Ham, C.L., Labavitch, J., Bennett, A.B., Powell, A.L., 2012. Mobility of transgenic nucleic acids and proteins within grafted rootstocks for agricultural improvement. Front. Plant Sci. 3, 39. https://doi.org/10.3389/fpls.2012.00039.

Hollywood, K., Brison, D.R., Goodacre, R., 2006. Metabolomics: current technologies and future trends. Proteomics 6, 4716–4723. https://doi.org/10.1002/pmic.200600106.

Holtorf, H., Guitton, M.C., Reski, R., 2002. Plant functional genomics. Sci. Nat. 89, 235–249. https://doi.org/10.1007/s00114-002-0321-3.

Hussain, K., Majeed, A., Nawaz, K., Nisar, M.F., 2009. Effect of different levels of salinity on growth and ion contents of black seeds (Nigella sativa L.). Curr. Res. J. Biol. Sci. 1, 135–138.

Hussain, M., Ahmad, S., Hussain, S., Lal, R., Ul-Allah, S., Nawaz, A., 2018. Rice in saline soils: physiology, biochemistry, genetics, and management. Adv. Agron. 148, 231–287. https://doi.org/10.1016/bs.agron.2017.11.002.

Karimi, H.R., Kuhbanani, A.M., 2015. The evaluation of inter-specific hybrid of P. atlantica × P. vera cv. 'Badami Zarand' as a pistachio rootstock to salinity stress. J. Nuts 6, 113–122.

Karlidag, H., Yildirim, E., Turan, M., 2011. Role of 24-epibrassinolide in mitigating the adverse effects of salt stress on stomatal conductance, membrane permeability, and leaf water content, ionic composition in salt stressed strawberry (Fragaria × ananassa). Sci. Hortic. 130, 133–140. https://doi.org/10.1016/j.scienta.2011.06.025.

Khan, M.A., Korban, S.S., 2012. Association mapping in forest trees and fruit crops. J. Exp. Bot. 63, 4045–4060. https://doi.org/10.1093/jxb/ers105.

Kumari, A., Das, P., Parida, A.K., Agarwal, P.K., 2015. Proteomics, metabolomics, and ionomics perspectives of salinity tolerance in halophytes. Front. Plant Sci. 6, 537. https://doi.org/10.3389/fpls.2015.00537.

Kurup, S.S., Hedar, Y.S., Al Dhaheri, M.A., El-Heawiety, A.Y., Aly, M.A., Alhadrami, G., 2009. Morpho-physiological evaluation and RAPD markers-assisted characterization of date palm (Phoenix dactylifera L.) varieties for salinity tolerance. J. Food Agric. Environ. 7, 3–50.

Li, Y., Peng, L., Xie, C., Shi, X., Dong, C., Shen, Q., Xu, Y., 2018. Genome-wide identification, characterization, and expression analyses of the HAK/KUP/KT potassium transporter gene family reveals their involvement in K^+ deficient and abiotic stress responses in pear rootstock seedlings. Plant Growth Regul. 85, 187–198. https://doi.org/10.1007/s10725-018-0382-8.

Long, N.V., Dolstra, O., Malosetti, M., Kilian, B., Graner, A., Visser, R.G., Linden, V.D.C.G., 2013. Association mapping of salt tolerance in barley (*Hordeum vulgare* L.). Theor. Appl. Genet. 126, 2335–2351. https://doi.org/10.1007/s00122-013-2139-0.

Lor, G.A., Luro, F., Navarro, L., Ollitrault, P., 2012. Comparative use of InDel and SSR markers in deciphering the interspecific structure of cultivated citrus genetic diversity: a perspective for genetic association studies. Mol. Gen. Genomics. 287, 77–94. https://doi.org/10.1007/s00438-011-0658-4.

Matsumoto, K., Tamura, F., Chun, J.P., Ikeda, T., Imanishi, K., Tanabe, K., 2007. Enhancement in salt tolerance of Japanese pear (*Pyrus pyrifolia*) by using *Pyrus betulaefolia* rootstock. Hort. Res. 6, 47–52.

Meena, S.K., Gupta, N.K., Gupta, S., Khandelwal, S.K., Sastry, E.V.D., 2003. Effect of sodium chloride on the growth and gas exchange of young Ziziphus seedling rootstocks. J. Hortic. Sci. Biotechnol. 78 (4), 454–457.

Melgar, J.C., Syvertsen, J.P., Martínez, V., Sánchez, G.F., 2008. Leaf gas exchange, water relations, nutrient content and growth in citrus and olive seedlings under salinity. Biol. Plant. 52, 385–390. https://doi.org/10.1007/s10535-008-0081-9.

Miller, G.A.D., Suzuki, N., Yilmaz, C.S., Mittler, R.O.N., 2010. Reactive oxygen species homeostasis and signalling during drought and salinity stresses. Plant Cell Environ. 33, 453–467. https://doi.org/10.1111/j.1365-3040.2009.02041.x.

Miranda, P.J., Achalandabaso, Y.A., Aguirresarobe, A., Canto, D.A., López, P.U., 2018. Similarities and differences between the responses to osmotic and ionic stress in quinoa from a water use perspective. Agric. Water Manag. 203, 344–352. https://doi.org/10.1016/j.agwat.2018.03.026.

Mohammadkhani, N., 2018. Effects of salinity on phenolic compounds in tolerant and sensitive grapes. Agric. Forest. 64, 73–86.

Mohammadkhani, N., Heidari, R., Abbaspour, N., Rahmani, F., 2016. Salinity effects on expression of some important genes in sensitive and tolerant grape genotypes. Turk. J. Biol. 40, 95–108. https://doi.org/10.3906/biy-1501-67.

Motos, A.J.R., Ortuño, M.F., Vicente, B.A., Vivancos, D.P., Blanco, S.M.J., Hernandez, J.A., 2017. Plant responses to salt stress: adaptive mechanisms. Agronomy 7, 18. https://doi.org/10.3390/agronomy7010018.

Mousavi, A., Lessani, H., Babalar, M., Talaei, A.R., Fallahi, E., 2008. Influence of salinity on chlorophyll, leaf water potential, total soluble sugars, and mineral nutrients in two young olive cultivars. J. Plant Nutr. 31, 1906–1916.

Munns, R., 2002. Comparative physiology of salt and water stress. Plant Cell Environ. 25, 239–250. https://doi.org/10.1046/j.0016-8025.2001.00808.x.

Munns, R., 2005. Genes and salt tolerance: bringing them together. New Phytol. 167, 645–663. https://doi.org/10.1111/j.1469-8137.2005.01487.x.

Munns, R., James, R.A., Läuchli, A., 2006. Approaches to increasing the salt tolerance of wheat and other cereals. J. Exp. Bot. 57, 1025–1043. https://doi.org/10.1093/jxb/erj100.

Nuccio, M.L., Rhodes, D., McNeil, S.D., Hanson, A.D., 1999. Metabolic engineering of plants for osmotic stress resistance. Curr. Opin. Plant Biol. 2, 128–134.

Patane, C., Saita, A., Sortino, O., 2013. Comparative effects of salt and water stress on seed germination and early embryo growth in two cultivars of sweet sorghum. J. Agron. Crop Sci. 199, 30–37. https://doi.org/10.1111/j.1439-037X.2012.00531.x.

Perrone, I., Pagliarani, C., Lovisolo, C., Chitarra, W., Roman, F., Schubert, A., 2012. Recovery from water stress affects grape leaf petiole transcriptome. Planta 235, 1383–1396. https://doi.org/10.1007/s00425-011-1581-y.

Peters, J.L., Cnudde, F., Gerats, T., 2003. Forward genetics and map-based cloning approaches. Trends Plant Sci. 8, 484–491. https://doi.org/10.1016/j.tplants.2003.09.002.

Pichersky, E., Gerats, T., 2011. The plant genome: an evolutionary perspective on structure and function. Plant J. 66, 1–3. https://doi.org/10.1111/j.1365-313X.2011.04564.x.

Potapovich, A.I., Kostyuk, V.A., 2003. Comparative study of antioxidant properties and cytoprotective activity of flavonoids. Biochemist 68, 514–519. https://doi.org/10.1023/A:1023947424341.

Rabey, E.H.A., Al-Malki, A.L., Abulnaja, K.O., 2016. Proteome analysis of date palm (*Phoenix dactylifera* L.) under severe drought and salt stress. Int. J. Genomics 2016, 1–8. https://doi.org/10.1155/2016/7840759.

Radwan, O., Arro, J., Keller, C., Korban, S.S., 2015. RNA-Seq transcriptome analysis in date palm suggests multi-dimensional responses to salinity stress. Trop. Plant Biol. 8, 74–86. https://doi.org/10.1007/s12042-015-9155-y.

Raga, V., Intrigliolo, D.S., Bernet, G.P., Carbonell, E.A., Asins, M.J., 2016. Genetic analysis of salt tolerance in a progeny derived from the citrus rootstocks Cleopatra mandarin and trifoliate orange. Tree Genet. Genomes 12, 34. https://doi.org/10.1007/s11295-016-0991-1.

Rahneshan, Z., Nasibi, F., Moghadam, A.A., 2018. Effects of salinity stress on some growth, physiological, biochemical parameters and nutrients in two pistachio (*Pistacia vera* L.) rootstocks. J. Plant Interact. 13, 73–82.

Rosa, M., Prado, C., Podazza, G., Interdonato, R., González, J.A., Hilal, M., Prado, F.E., 2009. Soluble sugars: metabolism, sensing and abiotic stress: a complex network in the life of plants. Plant Signal. Behav. 4, 388–393. https://doi.org/10.4161/psb.4.5.8294.

Ruiz, M., Quiñones, A., Martínez-Alcántara, B., Aleza, P., Morillon, R., Navarro, L., Primo-Millo, E., Cuenca, M.M.R., 2016. Effects of salinity on diploid (2x) and doubled diploid (4x) *Citrus macrophylla* genotypes. Sci. Hortic. 207, 33–40. https://doi.org/10.1016/j.scienta.2016.05.007.

Sairam, R.K., Tyagi, A., 2004. Physiological and molecular biology of salinity stress tolerance in plants. Curr. Sci. 86, 407–420.

Sánchez, G.F., Syvertsen, J.P., 2009. Substrate type and salinity affect growth allocation, tissue ion concentrations, and physiological responses of Carrizo citrange seedlings. HortScience 44, 1432–1437.

Schmutz, U., 2000. Effect of salt stress (NaCl) on whole plant CO_2 gas exchange in mango. Acta Hortic. 509, 269–276. https://doi.org/10.17660/ActaHortic.2000.509.29.

Schmutz, U., Lüdders, P., 1999. Effect of NaCl salinity on growth, leaf gas exchange, and mineral composition of grafted mango rootstocks (var.'13-1'and 'Turpentine'). Gartenbauwissenschaft 64 (2), 60–64.

Shabala, S., Munns, R., 2017. Salinity stress: physiological constraints and adaptive mechanisms. In: Plant Stress Physiology. CABI, Wallingford, pp. 24–63.

Shamshiri, M.H., Fattahi, M., 2014. Evaluation of two biochemical markers for salt stress in three pistachio rootstocks inoculated with arbuscular mycorrhiza (*Glomus mosseae*). J. Stress Physiol. Biochem. 10, 335–346.

Sharma, J., Upadhyay, A.K., Bande, D., Patil, S.D., 2011. Susceptibility of Thompson seedless grapevines raised on different rootstocks to leaf blackening and necrosis under saline irrigation. J. Plant Nutr. 34, 1711–1722.

Shekhawat, U.K.S., Srinivas, L., Ganapathi, T.R., 2011. MusaDHN-1, a novel multiple stress-inducible SK3-type dehydrin gene, contributes affirmatively to drought-and salt-stress tolerance in banana. Planta 234, 915. https://doi.org/10.1007/s00425-011-1455-3.

Singh, A., Sharma, P.C., 2018. Recent insights into physiological and molecular regulation of salt stress in fruit crops. Adv. Plants Agric. Res. 8, 171–183.

Singh, A., Prakash, J., Srivastav, M., Singh, S.K., Awasthi, O.P., Singh, A.K., Chaudhari, S.K., Sharma, D.K., 2014. Physiological and biochemical responses of citrus rootstocks under salinity stress. Indian J. Hortic. 71, 162–167.

Sofo, A., Scopa, A., Nuzzaci, M., Vitti, A., 2015. Ascorbate peroxidase and catalase activities and their genetic regulation in plants subjected to drought and salinity stresses. Int. J. Mol. Sci. 16, 13561–13578. https://doi.org/10.3390/ijms160613561.

Sohrabi, S., Ebadi, A., Jalali, S., Salami, S.A., 2017. Enhanced values of various physiological traits and VvNAC1 gene expression showing better salinity stress tolerance in some grapevine cultivars as well as rootstocks. Sci. Hortic. 225, 317–326. https://doi.org/10.1016/j.scienta.2017.06.025.

Song, Y., Ji, D., Li, S., Wang, P., Li, Q., Xiang, F., 2012. The dynamic changes of DNA methylation and histone modifications of salt responsive transcription factor genes in soybean. PLoS One. 7e41274 https://doi.org/10.1371/journal.pone.0041274.

Soni, A., Dhakar, S., Kumar, N., 2017. Mechanisms and strategies for improving salinity tolerance in fruit crops. Int. J. Curr. Microbiol. App. Sci. 6, 1917–1924. https://doi.org/10.20546/ijcmas.2017.608.226.

Sumner, L.W., Mendes, P., Dixon, R.A., 2003. Plant metabolomics: large-scale phytochemistry in the functional genomics era. Phytochemistry 62, 817–836. https://doi.org/10.1016/S0031-9422(02)00708-2.

Tabatabaei, S.J., 2006. Effects of salinity and N on the growth, photosynthesis and N status of olive (*Olea europaea* L.) trees. Sci. Hortic. 108, 432–438. https://doi.org/10.1016/j.scienta.2006.02.016.

Tanou, G., Job, C., Rajjou, L., Arc, E., Belghazi, M., Diamantidis, G., Molassiotis, A., Job, D., 2009. Proteomics reveals the overlapping roles of hydrogen peroxide and nitric oxide in the acclimation of citrus plants to salinity. Plant J. 60, 795–804. https://doi.org/10.1111/j.1365-313X.2009.04000.x.

Tanou, G., Ziogas, V., Belghazi, M., Christou, A., Filippou, P., Job, D., Fotopoulos, V., Molassiotis, A., 2014. Polyamines reprogram oxidative and nitrosative status and the proteome of citrus plants exposed to salinity stress. Plant Cell Environ. 37, 864–885. https://doi.org/10.1111/pce.12204.

Tavakkoli, E., Rengasamy, P., McDonald, G.K., 2010. High concentrations of Na^+ and Cl^- ions in soil solution have simultaneous detrimental effects on growth of faba bean under salinity stress. J. Exp. Bot. 61, 4449–4459. https://doi.org/10.1093/jxb/erq251.

Teakle, N.L., Tyerman, S.D., 2010. Mechanisms of Cl^- transport contributing to salt tolerance. Plant Cell Environ. 33, 566–589. https://doi.org/10.1111/j.1365-3040.2009.02060.x.

Tian, N., Wang, J., Xu, Z.Q., 2011. Overexpression of Na^+/H^+ antiporter gene AtNHX1 from *Arabidopsis thaliana* improves the salt tolerance of kiwifruit (*Actinidia deliciosa*). South Afr. J. Bot. 77, 160–169. https://doi.org/10.1016/j.sajb.2010.07.010.

Tounsi, M.H., Chelli-Chaabouni, A., Boujnah, M.D., Boukhris, M., 2017. Long-term field response of pistachio to irrigation water salinity. Agric. Water Manag. 185, 1–12. https://doi.org/10.1016/j.agwat.2017.02.003.

Tozlu, I., Guy, C.L., Moore, G.A., 1999. QTL analysis of morphological traits in an intergeneric BC1 progeny of *Citrus* and *Poncirus* under saline and non-saline environments. Genome 42, 1020–1029.

Upreti, K.K., Murti, G.S.R., 2010. Response of grape rootstocks to salinity: changes in root growth, polyamines and abscisic acid. Biol. Plant. 54, 730–734. https://doi.org/10.1007/s10535-010-0130-z.

Vijayan, K., Chakraborti, S.P., Ghosh, P.D., 2003. In vitro screening of mulberry (*Morus* spp.) for salinity tolerance. Plant Cell Rep. 22, 350–357. https://doi.org/10.1007/s00299-003-0695-5.

Vincent, D., Ergül, A., Bohlman, M.C., Tattersall, E.A., Tillett, R.L., Wheatley, M.D., Woolsey, R., Quilici, D.R., Joets, J., Schlauch, K., Schooley, D.A., 2007. Proteomic analysis reveals differences between *Vitis vinifera* L. cv. Chardonnay and cv. Cabernet Sauvignon and their responses to water deficit and salinity. J. Exp. Bot. 58, 1873–1892. https://doi.org/10.1093/jxb/erm012.

Wang, F., Zhu, H., Chen, D., Li, Z., Peng, R., Yao, Q., 2016. A grape bHLH transcription factor gene, VvbHLH1, increases the accumulation of flavonoids and enhances salt and drought tolerance in transgenic *Arabidopsis thaliana*. Plant Cell Tissue Organ Cult. 125, 387–398. https://doi.org/10.1007/s11240-016-0953-1.

Waters, S., Gilliham, M., Hrmova, M., 2013. Plant high-affinity potassium (HKT) transporters involved in salinity tolerance: structural insights to probe differences in ion selectivity. Int. J. Mol. Sci. 14, 7660–7680. https://doi.org/10.3390/ijms14047784.

Wei, Q., Liu, Y., Zhou, G., Li, Q., Yang, C., Peng, S.A., 2013. Overexpression of CsCLCc, a chloride channel gene from *Poncirus trifoliata*, enhances salt tolerance in Arabidopsis. Plant Mol. Biol. Rep. 31, 1548–1557. https://doi.org/10.1007/s11105-013-0592-1.

Willekens, H., Langebartels, C., Tire, C., Montagu, V.M., Inze, D., Camp, V.W., 1994. Differential expression of catalase genes in *Nicotiana plumbaginifolia* (L.). Proc. Natl. Acad. Sci. 91, 10450–10454. https://doi.org/10.1073/pnas.91.22.10450.

Xie, R., Zhang, J., Ma, Y., Pan, X., Dong, C., Pang, S., He, S., Deng, L., Yi, S., Zheng, Y., Lv, Q., 2017. Combined analysis of mRNA and miRNA identifies dehydration and salinity responsive key molecular players in citrus roots. Sci. Rep. 7, 42094.

Yadav, S., Irfan, M., Ahmad, A., Hayat, S., 2011. Causes of salinity and plant manifestations to salt stress: a review. J. Environ. Biol. 32, 667.

Yahmed, J.B., Novillo, P., Garcia-Lor, A., Salvador, A., Mimoun, M.B., Luro, F., Talon, M., Ollitrault, P., Morillon, R., 2015. Salt tolerance traits revealed in mandarins (*Citrus reticulata* Blanco) are mainly related to root-to-shoot Cl^- translocation limitation and leaf detoxification processes. Sci. Hortic. 191, 90–100. https://doi.org/10.1016/j.scienta.2015.05.005.

Yaish, M.W., Patankar, H.V., Assaha, D.V., Zheng, Y., Al-Yahyai, R., Sunkar, R., 2017. Genome-wide expression profiling in leaves and roots of date palm (*Phoenix dactylifera* L.) exposed to salinity. BMC Genomics 18, 246. https://doi.org/10.1186/s12864-017-3633-6.

Yamaguchi, T., Blumwald, E., 2005. Developing salt-tolerant crop plants: challenges and opportunities. Trends Plant Sci. 10, 615–620. https://doi.org/10.1016/j.tplants.2005.10.002.

Yin, R., Bai, T., Ma, F., Wang, X., Li, Y., Yue, Z., 2010. Physiological responses and relative tolerance by Chinese apple rootstocks to NaCl stress. Sci. Hortic. 126, 247–252. https://doi.org/10.1016/j.scienta.2010.07.027.

Zarei, M., Azizi, M., Rahemi, M., Tehranifar, A., 2016. Evaluation of NaCl salinity tolerance of four fig genotypes based on vegetative growth and ion content in leaves, shoots, and roots. HortScience 51, 1427–1434. https://doi.org/10.21273/HORTSCI11009-16.

Zhang, J.L., Shi, H., 2013. Physiological and molecular mechanisms of plant salt tolerance. Photosynth. Res. 115, 1–22. https://doi.org/10.1007/s11120-013-9813-6.

Salt stress alleviation through fertilization in fruit crops

Rui Machado[a,b,*], Ricardo Serralheiro[a,c]

[a]ICAAM—Mediterranean Institute for Agricultural and Environmental Sciences, School of Science and Technology, University of Évora, Évora, Portugal
[b]Crop Sciences Department, School of Science and Technology, University of Évora, Évora, Portugal
[c]Agricultural Engineering Department, School of Science and Technology, University of Évora, Évora, Portugal
*Corresponding author. E-mail: rmam@uevora.pt

1 Introduction

Salt stress is one of the most widespread abiotic constraints in food production a tendency to increase due to climate change, increasing of the use of low-quality water in irrigation, and massive introduction of irrigation associated with intensive farming and low leaching fraction (Machado and Serralheiro, 2017).

Salts affecting soil root zone can be a natural process if they are originated in deeper layers of the soil, as often occurs in soils formed upon calcareous, limestones, and other calcic rocks. However, soil and water salinity is irrevocably associated with irrigated agriculture. With rainfed agriculture, the water used by the crops comes from the rain, free from salts, leaching from the root zone the salts eventually in excess. In nonirrigated lands as a result of water losses, through evapotranspiration situations of poor drainage, with water accumulating over slowly permeable soil layers containing salts, these can rise up to the root zone and accumulate there creating salt stress.

On the contrary, the irrigation water contains naturally dissolved minerals, which are usually added with salts to provide nutrients to the plants, which can affect plant growth and contribute to salt accumulation within the root zone. The use of chemical fertilizers to supplement soil fertility and to high-yielding crop requirements has been increasing rapidly due to the increase in intensive farming.

If drainage is limited, every excess of applied irrigation water will result in uprising of the soil water table, carrying salts from bottom layers to the root zone. Therefore, throughout the world, irrigation requires that salt management becomes an integral part of the production system (Ayars, 2003). As a major concern in irrigation water management, some efficient drainage system is required in parallel to the irrigation water delivery system, the availability being needed of some excess water above irrigation requirements, for leaching excess salts from the soil profile.

Irrigation water quality can have a deep impact on crop yield (Grattan, 2002) and in soil fertility. Water quality will in the next future be increasingly stressed, as population grows increasing the needs for irrigation, available water becoming more scarce. Water scarcity is caused not only by the physical exiguity of the resource but also by the progressive deterioration of water quality in many basins, reducing the quantity of water that is safe to use (Mateo-Sagasta et al., 2018).

Irrigation of olives with saline water will inevitably increase in the future in the Mediterranean due to negative effects of population growth and climate change on the availability and quality of existing freshwater supplies (Chartzoulakis, 2005). To attenuate water scarcity, the use of recycled waters is increasing in many countries. Huge amounts of low-quality water, such as that coming from urban and industrial wastewater treatment plants, could be recovered and reused for irrigation, attenuating the demand of high-quality water. However, the use of wastewaters in irrigation may result in various problems such as toxicity for crops, damage to soil quality, diffusion of parasites, and drawbacks in irrigation systems (Raveh and Ben-Gal, 2016). These waters typically have higher salinity than potable water (Nackley et al., 2015; Becerra-Castro et al., 2015) and cannot be used in crop vegetables grown for raw eating, but can be used for irrigating fruit crops, namely, fruit trees with high salt tolerance. On the other hand, wastewaters can contain essential nutrients for plant growth, which can reduce the need for input of chemical fertilizers (Magesan et al., 1999).

Irrigation water requirements are generally projected to escalate with higher global mean temperature (Haddeland et al., 2014) and are likely to have an even higher salt content, due to solute concentration following evaporation. This phenomenon in fruit crops can be more intense than in other crops, due to the long growing cycle period with irrigation and respective water requirements. Moreover, due to the reduction in water resources, the farmers are stimulated to increase water-use efficiency as the way to maintain crop yields using less water in irrigation. Improving water-use efficiency is important, but it is often achieved by reducing the leaching fraction, which can result in the increase of soil salinity. Among the various sources of soil salinity, irrigation combined with poor drainage is the most serious, because it represents losses of once productive agricultural land (Zhu, 2007). Moreover, the farmers may use the saved water for irrigating new areas or other crops, also deficiently and risking the increase of soil salinity. Therefore, in that case there is no true save of water with the increase of application efficiency. On the other hand, reduced flow through natural drainage channels and waterways will result in harmful higher salt concentration in the natural drainage network. Therefore, from the environmental point of view, the increase of irrigation efficiency may be contradictory. Some authors (Grafton et al., 2018) refer to this problem as "the paradox of irrigation efficiency."

Irrigation water quality, depending on the concentration and also on the nature of the ions present, can have adverse effects on physiochemical properties of the soil and affect plant growth due to water salinity (EC_w) and specific ion accumulation of toxicity ions. The physical and mechanical properties of the soil, for example, soil structure (stability of aggregates) and permeability, are very sensitive to the type of exchangeable ions present in irrigation water. Fruit crops are less tolerant to irrigation water salinity (EC_w) than to soil salinity (EC_e). Salinity is one of the major factors limiting avocado yield, primarily due to the high concentration of ions in irrigation water (Bonomelli et al., 2018). Salinity tolerance of fruit crops has increasing importance, as a result of the decreasing availability of high-quality water for irrigation. Therefore, the main objective of this chapter is to analyze how fertilization can contribute to the alleviation of salt stress associated with the irrigation water.

2 Irrigation water quality

All irrigation waters contain dissolved minerals, which concentration and composition varies with the geological and chemical origin and in case of use recycled water depends their origin and treatment (Magen, 2005) and the nature of the wastes added during use (Asano and Pettygrove, 1987).

The most frequent ions in irrigation water are nitrate, chloride, sulfate, carbonates, and bicarbonates, which are anions, and sodium, calcium, magnesium and sometimes potassium, which are cations. The composition and concentration of ions in irrigation water can affect plant growth due to water salinity, since that the ions present have different conductances (through their charges and mobilities) (Table 33.1), water pH and alkalinity, and specific ion toxicity. Moreover, it also can affect plant growth due to soil structure degradation and decreasing of the nutrient availability. Physical and mechanical properties of the soil, for example, soil structure (stability of aggregates) and permeability, are very sensitive to the type of exchangeable ions present in irrigation water. The chemical composition of the irrigation water also affects irrigation water uniformity due to emitter clogging caused by chemical precipitation when pressure

TABLE 33.1 Ionic equivalent conductance at infinite dilution in aqueous solution (µS/cm per meq./L) at 25°C for the most common ions in agriculture.

Cations	λ°_+	Anions	λ°_-
H^+	349.7	OH^-	198.0
Na^+	50.1	Cl^-	76.3
K^+	73.5	NO_3^-	71.4
NH_4^+	73.5	HCO_3^-	44.5
Mg^{+2}	53.0	CO_3^{-2}	69.3
Ca^{+2}	59.5	SO_4^{-2}	80.0
		HPO_4^{-2}	33.0
		$H_2PO_4^-$	33.0

Based on Lide, D.R., 1993. CRC Handbook of Chemistry and Physics: A Ready-Reference Book of Chemical and Physical Data: 1993–1994. CRC Press.

irrigation systems are used. This problem can increase when reclaimed water is applied (Zhou et al., 2018). In irrigated fruit crops, the assessment of the water chemical characteristics is vital, to improve crop productivity, to avoid soil salinization and alkalinity, and to conserve the water nutritional capability, and their importance is raising due to the increase the use of low-quality water.

In the recycled waters, the concentrations of Na, Cl, and B are the main concern. Therefore, the use of recycled waters should be carried out with caution, to avoid salt accumulation in soil and excessive salt specific toxic accumulation in plant tissues (Kafkafi and Tarchizky, 2011). When two sources of water, saline and not saline, are available, they can be mixed to meet the crop's needs. Dilution is the only economical way to reduce the salt concentration in irrigation water.

3 Fruit crops salt tolerance

Among the factors determining the fruit crop production, the concentration of salts (salinity) present in the soil plays a critical role. Salinity can affect crop growth due to osmotic stress, nutrient deficiencies and nutritional imbalances, specific toxic ions, soil alkalinity, and interactions between these factors. The effect of salinity on crops is modeled according to the effect of the electrical conductivity (EC) on their productive response, which reaches a threshold at a given EC, above that plant growth, and the yield is affected. Salinity, as salt concentration in water or soil water, can be measured through electrical conductivity (EC). It is a measure of total salinity, but it does not give any indication of the salt composition.

The tolerance of any crop to salinity is defined as the ability to endure the effects of excess salt in the root zone. Most fruit crop tolerance guidelines for salinity are related to electrical conductivity, which is denoted as EC_e for saturated soil extracts and EC_w for irrigation water. Salt tolerance can be characterized on the basis of a salinity threshold (EC_t), which is the maximum salinity within root zone that the crops can tolerate, above which yield decrease, and the decreasing slope, which describes the rate of yield decline with soil salinity. Each fruit crop salt sensitivity is consequently a function of the threshold and the slope. Therefore, salt tolerance is higher for crops with high thresholds and low slope. The relative yield of a crop growing in a soil with average salinity within the root zone above the threshold value can be calculated through Eq. (33.1) (based on Maas and Hoffman, 1977):

$$Yr = 100 - S(EC_e - EC_t) \tag{33.1}$$

where Yr is the relative yield of the crop (%); S is the crop decreasing slope in yield per unit increase in soil salinity; EC_e is the average root-zone salinity, expressed of the saturated soil past, in dS/m; and EC_t is the salt tolerance threshold of the fruit crop. Parameters relative to fruit crops for use in Eq. (33.1) are listed in Table 33.2. Based on this model, fruit crops can be rated from tolerant to sensitive being the majority of the fruit tree species rated as sensitive to salinity (Table 33.2). In peach (Tattini, 1990), almond (Zrig et al., 2011), and kiwi fruit (Chartzoulakis et al., 1995), shoot growth was suppressed by relatively low concentrations NaCl of salt in the soil solution. Pistachio is one of the most tolerant, followed by olive and date palm (Table 33.2). The ratings are useful in predicting how a crop may perform relative to

TABLE 33.2 Salt tolerance of different fruit crops to soil (EC_e) and irrigation water salinity (EC_W) and tolerance to boron in irrigation water.

Crops	Soil		Rating[a]	Irrigation water	
	Threshold[b] EC_e (dS/m)	Slope (% per dS/m)		Threshold[b] EC_W (dS/m)	Boron[b] (mg/L)
Almond	1.5	19.1	S	1.0	0.3
Apple	1.7	24.0	S	1.0	
Apricot	1.6	24.0	S	1.1	
Avocado	2.8	9.2	S	0.9	0.5–0.75
Blackberry	1.5	22	S	–	0.5
Blueberry[c]	2	–	S	–	1.0
Date palm	4.0	9.6	T	2.7	2–4
Fig	2.7	14.0	MT	1.8	0.5–0.75
Grape	1.5	9.6	MS	1.0	0.5–75
Grapefruit	1.2	13.5	S	1.2	–
Guava	4.7	9.8	MT	–	–
Kiwifruit[d]	–	–	S	–	–
Lemon	1.7	16.0	S	1.1	<0.5
Mango	–	–	S	–	–
Olive	4.0	–	MT	2.0	1–2
Orange	1.3	13.1	S	1.1	0.5–0.75
Peach	1.7	21.0	S	1.1	0.5–0.75
Pear	1.7	–	S	1.0	–
Pistachio[e]	9.4	8.4	T	8.0	–
Plum	1.5	12.0	MS	1.0	0.5–0.75
Pomegranate	2.7	14.0	MT	1.8	–
Raspberry	4.0	12	S	0.7	–
Walnut	1.5	–	S	1.1	0.5

–, data not available; EC_e, electrical conductivity (EC) of saturated paste extract of soil and EC_W, electrical conductivity (EC) of irrigation water. Slope = yield reduction per unit increase in EC beyond threshold.

[a] Tolerance to soil salinity is rated as tolerant (T), moderately tolerant (MT), moderately sensitive (MS), and sensitive (S).

[b] Based on Maas and Hoffman (1977), Ayers and Westcot (1984), Maas and Grattan (1999), Grattan (2002), Tanji and Kielen (2002), and Hanson et al. (1999).

[c] Retamales and Hancock (2012).

[d] Chartzoulakis et al. (1995).

[e] Sanden et al. (2004, 2005).

another in saline conditions of the soil and/or of irrigation water. However, fruit crop salt tolerance is a dynamic process and is not influenced only by soil salinity and species. Salt stress effect is a combination of osmotic stress, specific ion toxicities, and nutrient imbalances each affecting the trees at different intensities dependent upon soil salinity, scion/rootstock characteristics, and the presence of the biotic and abiotic factors and the physiological condition of fruit trees (Srivastava, 2012) and interactions between them. In fruit crops, the rootstocks and cultivars of a species assume a high influence in salt tolerance. Fruit crops and different cultivars of same crop vary considerably in their tolerance to salinity. Avocado is specifically sensitive to chloride toxicity with some rootstocks, notably Mexican, being more sensitive than others Guatemalan and West Indian (Shalhevet, 1999). Citrus trees are sensitive to excess of salt; however, the tolerance to soil salinity is correlated with its ability to restrict the entry of toxic ions as chloride, sodium, and boron (Srivastava and Ram, 2000). Salt tolerance is influenced by several abiotic factors as the following: soil type and drainage characteristics within the root zone, soil temperature, levels of *carbon dioxide* (CO_2) in the *atmosphere*, time of exposure to salt stress and agronomical practices (e.g., fertilization and irrigation method and leaching fraction), and irrigation water chemical characteristics. Salt tolerance of perennial crops such as olive is influenced by additional detrimental effects caused by specific ion toxicities derived from the progressive toxic accumulation of Cl and Na in the leaves (Benlloch et al., 1991; Bongi and Loreto, 1989; Maas and Grattan, 1999).

Most fruit crops with the exception of pistachio, guava, date palm, and olive presented a salinity threshold (EC_t), lower than the EC_e of saline soil ($EC_e > 4 dS/m$). Moreover, the EC threshold of the fruit crops decreases with the water salinity, when that is used continuously for irrigation, and with the exception of the pistachio, date palm, and olive, they have low water salinity threshold (EC_w) ($EC_w \leq 1.1 dS/m$) (Table 33.2). Therefore, waters moderately salty (with EC_W ranging from 0.8 to 2.5 dS/m) affected the growth of the majority of the fruit crops. However, fruit crop salt tolerance to EC_w is also influenced by the type of salts and their concentration in irrigation water. Olives can tolerate high values of EC_w when the concentration of NaCl in irrigation water is low (Chartzoulakis, 2005). In olive, the exposure to moderate levels of salinity in irrigation water did not negatively affect yields and additionally appears to contribute to positive attributes of olive oil (Ben-Gal, 2011). Three-year old irrigated olives, which water had an EC_w of 8 dS/m, did not suffer salt stress at NaCl concentrations lower than 80 mM during a period of 90 days (Therios and Misopolinos, 1988). For olive trees, sodium is more deleterious than chloride (Chartzoulakis, 2005; Aragüés et al., 2005). On the contrary, on citrus, the Cl ion is considered to be a more important limitation than Na on growth and yield (Bañuls et al., 1997; Lopez-Climent et al., 2008).

3.1 pH and alkalinity

pH and alkalinity are two important factors in determining the suitability of water for irrigating plants since they can affect plant growth, soil pH, and structure. The water pH expresses the concentration of hydronium (H_3O^+) in aqueous solution affecting plant nutrition by influencing the solubility of the ionic species and microorganism's activity.

The optimum pH of irrigation water ranges from 6.5 to 7.5. Soil pH increase was observed after long-term irrigation with wastewater, in soil with different management regimes and irrigated with different types of wastewater (Becerra-Castro et al., 2015). The alkalinity of the water, which is the relative measurement of the capacity of the water to resist a change in pH, increases with the rising amount of dissolved bicarbonates (HCO_3^-) and carbonates (CO_3^{2-}) and hydroxide ion (OH^-). Applying water that's high in bicarbonate and carbonates can increase soil pH and decrease the permeability of the soil. Moreover, the bicarbonate in irrigation water can contribute to the clogging of the emitters affecting irrigation uniformity and impacting the water availability and soil salinity. Bicarbonate concentrations exceeding about 2 meq./L and pH exceeding about 7.5 can cause emitter clogging due to calcium carbonate precipitation.

3.2 Specific ion toxicity

Specific ion toxicities are a consequence of chloride, boron, and sodium ion accumulation in plant tissues, either from root uptake and/or leaf absorption due to high concentrations in irrigation water and/or soil solution. Specific ion toxicities are more common in woody plants since they have a longer growing season. Recycled water usually has higher concentrations of sodium, chloride, and boron ions than the water from which it was made. Boron is typically found in wastewater from domestic sources due to its use in detergents (NRMMC et al., 2006).

Despite the essentiality of chloride and boron as micronutrients for fruit crops, they can become toxic to plants even at very low concentrations in irrigation water (Table 33.3). In Table 33.3, chloride classification of the water shows that values ranging from 70 to 140 mg/L can be injurious to sensitive plants.

TABLE 33.3 Chloride classification of irrigation water[a].

Chloride		Effect on crops
(mg/L)	(meq./L)	
<70	<2	Generally safe for all plants
70–140	2.1–4	Sensitive plants show injury
141–350	4.1–10	Moderately tolerant plants show injury
<350	<10	Can cause severe problems

[a] Maas (1990).

However, tolerance to chloride, boron (Table 33.4), and sodium in irrigation water vary between species and rootstocks within species, climate, and soil and level of salinity and water deficit. In citrus, the chloride tolerance is highly dependent of rootstock and can range from 250 to 600 mg/L in irrigated water (Hanson et al., 1999). The threshold of olive trees sensitivity to chloride is 350 mg/L (10 meq./L) (Chartzoulakis, 2005) and of almonds is 140 mg/L (4 meq./L). Irrigation water containing chloride concentration of less than 150 mg/L of chloride is safe for most fruit crops. In apple trees, chloride and sodium increase in leaves with exposure to salinity stress and with soil matrix potential stress (West, 1978). Common symptoms of chloride toxicity in plants include necrosis of leaf margins and tips, which typically occurs first in older leaves. Excessive leaf burn might eventually result in leaf drop.

In addition to affecting the growth of plants due to the accumulation of Cl^- in plant tissues, chloride high concentrations in irrigation water can interfere with absorption of nitrates. This effect of chloride may be due to ion competition between Cl^- and NO_3^- at the soil-root interface or, in the case of young trees, due to a dilution effect linked to increased growth (Iglesias et al., 2004). The continuous application of nitrates, under saline conditions, might reduce Cl accumulation in scions grafted on susceptible rootstocks of citrus (Bar et al., 1997; Levy et al., 2000) and in avocado (Bar et al., 1987). However, Lea-Cox and Syvertsen (1993) observed that in citrus (Cleopatra mandarin and lemon) it appears that reductions in NO_3^- uptake are more strongly related to reduction in water availability for use than to Cl competition from salt stress. In kiwi fruit, the severity of leaf necrosis following potassium chloride (KCl) application was attributed not to Cl toxicity but rather to N deficiency, enhanced by competition between Cl and NO_3 (Buwalda and Smith, 1991). In tomato, it also appeared that the inhibition of NO_3-N absorption was more strongly related to reduced water uptake than to Cl competition from salt stress (Abdelgadir et al., 2005).

Chloride is difficult to remove from water, advanced treatments being necessary, as the reverse osmosis. When irrigation water with a low concentration of chloride is available, dilution is the best practice.

Boron in arid regions is considered the most harmful element in irrigation water (Phocaides, 2007). Boron toxicity on many crops caused by high B concentration in irrigation water has been reported in different parts of the world: India, Egypt, the United States, Greece, the Philippines (Chauhan and Asthana, 1981; Elseewi, 1974; Salinas et al., 1981; Connery, 2011; Sotiropoulos et al., 1995; Dobermann and Fairhurst, 2000). Safe concentrations of B in irrigation water range from 0.3 to 1.0 mg/L for sensitive plants (avocado, apple, grape, etc.), 1–2 mg/L for semitolerant plants (olive), and 2–4 mg/L for tolerant plants (date palm) (Table 33.3).

Boron toxicity visual symptoms are dependent on boron mobility in phloem. Boron mobility in plants varies with species, B is phloem immobile in pecan and walnut, while B is phloem mobile in apple, apricot, pear, grape, peach, olive, and pomegranate (Brown and Hu, 1998). Plants like walnut, in which B is immobile, accumulate B in the tip and edge of old leaves like walnut. B toxicity symptoms in those species are always exhibited as leaf tip and edge burn. On the other hand, for those plants in which B is mobile, instead of the marginal leaf burn, these species exhibit B toxicity as die back in young shoots profuse gumming in the leaf axil and the appearance of brown, corky lesions along stems and petioles.

TABLE 33.4   Critical levels of specific ions in leaf tissue of walnut and olives (July samples) and pistachio (August—prior to harvest).

Crop		Degree of toxicity		
		None	Increasing	Severity
Walnut[a]	Chloride (%)	<0.3	0.3–0.35	>0.5
	Boron (ppm)	<36	36–200	>200
	Sodium (%)	<0.10	0.10–0.30	>0.30
Pistachio[b]	Chloride (%)	<0.2	0.2–0.3	>0.3
	Boron (ppm)	<300	300–700	>800
	Sodium (%)			
Olives trees[c]	Chloride (%)			>0.20
	Boron (ppm)			>185
	Sodium (%)			>0.5

[a] Fulton et al. (1988).
[b] Sanden et al. (2005).
[c] Fernandez-Escobar (2004).

Sodium is not essential (Marschner, 2012) for the most plants, but their growth of most plants is stimulated at low Na concentrations (Cramer, 2002). Like chloride and boron, sodium accumulation in leaves is progressive. It is important to diagnosticate ion accumulation before levels become elevated in the woody and leaf tissue (Table 33.4) once ions accumulate there, the trees have no mechanism to expel them (Sanden et al., 2005). Tissue analysis gives the best indicator of the toxic element hazard. Correcting the toxicity in the root zone may require several seasons of tree growth, production, and management. Therefore, analyses of concentration of these ions in soil and irrigation water are essential. These specific salts can be leached as total salts. However, boron and sodium are more difficult to remove, especially when compared with chloride.

3.3 Sodium effects on soil

Sodium content in irrigation water can also harm fruit crop growth through its deleterious effect on the physico-chemical properties of the soil. Sodium is attracted by negatively charged soil particles (clay minerals and long molecules of organic acids), replacing the cohesive ion calcium and magnesium in the soil microaggregates, which cause the dispersion of these soil particles due to the low flocculating power of the sodium, blocking soil pores, decreasing water infiltration, and causing poor aeration. The increase in the concentration of exchangeable sodium may cause an increase in the soil pH designated as alkalinity, which reduces nutrient availability (e.g., iron and phosphorus) and decreases calcium and potassium uptake due to increase ion competition Ca/Na and K/Na. The permeability problem is mainly related to a relatively high sodium or very low calcium content in the soil or in the applied water. The sodium problem is reduced if the amount of calcium plus magnesium is high compared with the amount of sodium. This relation is called the sodium adsorption ratio (SAR) and is a calculated value from the equation:

$$SAR = \frac{[Na^+]}{\sqrt{\frac{[Ca^{++}] + [Mg^{++}]}{2}}}$$

(33.2)

in which the concentrations (meq./L) of sodium, calcium, and magnesium are represented in brackets. In general, sodium hazard increases with SAR. The quality of water for irrigation may be classified according to SAR values. In general, sodium hazard increases as SAR increases and EC decreases. For water containing a significant concentration of bicarbonate, it is more useful to use the adjusted sodium adsorption ratio (SAR_{adj}). It is the SAR modified to include the effects of bicarbonates and carbonates, in addition to Ca and Mg. If SAR_{adj} is less than 6.0, there should be no problems with either sodium or permeability.

4 Fertilization

Fertilization can contribute to alleviate fruit crop salt stress and prevent soil salinization. For achieving that it is necessary to correctly choose the fertilization method and schedule, as well as the fertilizer source and rate of application. Moreover, the salts contained in the irrigation water should be considered as a nutrient source (Machado and Serralheiro, 2017). On the other hand, fertilization management must incorporate the application of organic composts, biostimulants and/or biofertilizers, and antioxidants.

4.1 Nutrient stewardship

4.1.1 Fertilization method

Fertigation combines fertilization and irrigation by injecting fertilizers and other water-soluble products into an irrigation system. Fertigation combined with drip irrigation compared with other methods of irrigation and fertilization promotes increases in the water and nutrient use efficiency, in the yield, and permits to use saline irrigation water and reduce the volume of the soil to be reclaimed.

Drip irrigation is characterized by localized and uniform application of the water or nutrient solution through emitters, in a small moistened volume of soil (wet bulb) where the root system is concentrated. Therefore, in dry climates, as the fruit crops are mainly installed in rows, the major part of the soil between rows is not moistened. The plant nutrition is being assured by the part of the root system in the wet bulb. This can allow the use of saline water since the salts concentrate on the borders of the wet bulb, the interior remains moist and relatively free of salts, creating a suitable root-zone salinity ($EC_e < EC_t$). Due to the frequent irrigation and evaporation cycles, the ions in wet bulb tend to accumulate in the interface between the dry soil and the wetting front, mainly next to soil surface. However, to maintain on the wet bulb the suitable root-zone salinity in the wet bulb the irrigation frequency must be adjusted to maintain always moist the wet bulb, irrigation water uniformity must be high. In case of rain, the irrigation must continue, to avoid displacement of salts from the salinized soil into the bulb.

In drip-fertigated fruit crops on soil affected by excess salts, the amendments required for soil remediation should mainly focus on the crop row. The application of gypsum ($CaSO_4$) and elemental sulfur ($S°$) to reduce soil sodium content and pH can be made in a more or less narrow strip of the crop row, where the roots are concentrated. This procedure has the advantage of decreasing the amount of the amendments and of water with high quality for leaching the salts of the soil. The application of micronized elemental sulfur sprayed over the soil in different times for avoiding a high increase in EC_e also decreasing soil pH and reducing sodium concentration in the soil (Machado, unpublished). However, soil remediation also can be achieved by applying sulfuric acid and micronized elemental sulfur in fertigation (Almutairi et al., 2017). Moreover, gypsum application either through the irrigation water or by direct application to the soil, when the irrigation water has a high content in sodium, contributes to reducing the SAR. Besides the improvement of the physical and chemical conditions of the soil, it is also important to promote the proliferation of microorganisms in the soil by applying humic acids in fertigation and organic composts.

4.1.2 Fertilization scheduling

Fertigation, the application of fertilizers with the irrigation water, allows applying small amounts of fertilizer in each fertigation event, that contributes to high flexibility in fertigation frequency. Small and frequent applications of fertilizers reduce of potential for plant salt injury since the application of fertilizers can take into account the EC_w, the nutrients contained in the irrigation water, the type of fertilizer, and the EC_t of fruit crops. For example, for the irrigation water used (Fig. 33.1), the maximum value of calcium nitrate in irrigation water of almonds must be around 1.0 g/L, since EC_t of this crop is 1.5 dS/m.

4.1.3 Fertilizer source and rate

With regard to fertilizer source and rate of its application, the following guidelines are priority: the irrigation water must be considered as a nutrient source; the fertilizers used in fertigation should have low salinity, high purity, acid reaction, and free of chloride, sodium, and boron; the micronutrients must be applied in the chelated form; and the EC_W after addition of fertilizer should be less than the EC_t of the fruit crop.

Irrigation waters can contain considerable concentrations of nutrients such as nitrogen, phosphorus, potassium, calcium, and magnesium that must be discounted in the amounts to be applied in fertigation. The plants are able to utilize these nutrients very efficiently because the nutrients are directly in wet bulb where roots are concentrated, regularly, according to water requirements, which closely follow the production of dry matter. High levels of nitrate in irrigation water from groundwater aquifers has been observed in different parts of the world (Hudak, 2000; Machado et al., 2008; Mohamed et al., 2003; Ramos et al., 2002), which may be due to the fact that nitrate from agriculture is the most common chemical contaminant in the world (WWAP, 2013). Nitrate leaching losses in citrus orchards in Valencia Spain were, in general, lower than 100 kg N/ha/year, representing about 33% of the total N input (Ramos et al., 2002). The recycled waters typically contain nitrogen, phosphorus, and potassium as a result of prior use. In vineyards, in California, nutrients in the recycled water (18.8 mg K/L, 0.9 mg P/L, and 53.5 mg NO_3/L) applied by drip irrigation

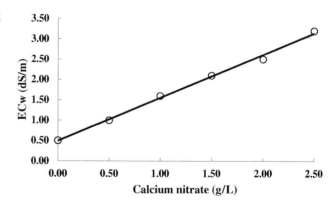

FIG. 33.1 Effect of calcium nitrate [$Ca(NO_3)_2$] concentration on electrical conductivity of irrigation water (EC_W) (Machado, 2002).

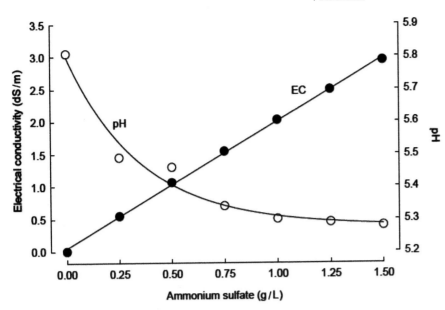

FIG. 33.2 Effects of ammonium sulfate [$(NH_4)_2SO_4$] concentration on EC_w (●) and pH (○) in distilled water (Machado et al., 2008).

may be beneficial to vineyards, though the levels of nitrogen may need to be reduced by planting cover crops (Weber et al., 2014).

The fertilizer source can contribute to decreasing osmotic stress and the Na, Cl, and B addition and the increase of nutrient availability in soil. For this purpose, the fertilizers applied in the fertigation must cause a low increase in EC_W and decreased water pH (Fig. 33.2). Irrigation water pH must be adjusted when pH or alkalinity of irrigation water is high, to prevent soil pH increase and also to contribute to reducing soil pH. The maintenance of pH of irrigation water within the optimal range of most crops (5.5–6.5) is necessary for fertigation systems to allow optimal nutrient uptake, mainly micronutrients and to keep the irrigation system free from clogging. However, the decrease of water pH depends on the alkalinity of the water, which is the relative measurement of the capacity of the water to resist a change in pH. When the alkalinity is high due to the high concentration of bicarbonate and/or carbonates, this must be neutralized by acidification of water through injection of acids (e.g., nitric acid [HNO_3], phosphoric acid [H_3PO_4], and sulfuric acid [H_2SO_4]]. Citric acid ($H_3C_6H_5O_7$), a weak organic acid, also can be used, as it is safer than the others; but it is much less effective. Acidification of the irrigation water reduces the concentration of bicarbonates and carbonates in water resulting in the formation of carbon dioxide (CO_2) and water (H_2O) [($HCO_3^- + H_3O^+ \leftrightarrow H_2O + CO_2$) ($CO_3^= + 2H_3O^+ \leftrightarrow H_2O + CO_2$)]. The nitrate or phosphate must be discounted in the amount of total N or P to be applied to satisfy plant nutrient requirements. The acidifying of the irrigation water reducing the bicarbonate can reduce the SAR and SAR_{adj}.

The addition of K, Ca, and NO_3, can prevent Na and Cl accumulation in tissues of the fruit crops. In kiwifruit, the severity of leaf necrosis caused by Cl toxicity increase with nitrogen deficiency and is enhanced by competition between Cl and NO_3 (Smith et al., 1988; Buwalda and Smith, 1991). Addition of NO_3 to the irrigation water reduced Cl accumulation in the plant and alleviated its adverse effects in citrus and avocado provided that nitrate is supplied continuously at a molar concentration equivalent to half that of chloride (Bar et al., 1997). The addition of calcium, which tends to replace exchangeable Na, and reducing the SAR is a valuable strategy to saline soil. The ability of vines to tolerate the Na is influenced by Ca nutrition (Weber et al., 2014) being Na toxicity is often reduced or completely overcome if sufficient Ca is made available to roots (Ayers and Westcot, 1984). When drip fertigating with P under saline water conditions, precautions should be taken to increase the SAR by either increasing Na or reducing Ca and Mg ions in soil solution (Mmolawa, and Or, 2000).

Saline soils have usually high pH, so the occurrence of deficiencies of iron, calcium, zinc, magnesium, boron, and molybdenum are common. The most effective way to correct these deficiencies is to apply fertilizers with these nutrients in chelated form to the soil or in foliar solution. The chelates to be used in these conditions must be stable under conditions of high pH. In case of Zn, Mn, Ca, and Mg, this can be assured by EDTA whether applied through foliar fertilization, fertigation, and over the soil. The application of magnetic iron to a natural row rock (magnetite) that has very high iron content increased salt tolerance in different fruit crops (Table 33.5). However, iron chelates made with HEDTA and EDDHA are the most effective iron fertilizers on high pH soils.

TABLE 33.5 Nutrients and substances that improved salt tolerance in different fruit crops.

Crop	Substance	References
Almond	Humic acid	Hatami et al. (2018)
Apple	Salicylic acid	Shaaban et al. (2011)
Avocado[a]	Seaweed extract	Bonomelli et al. (2018)
	Nitrate	Bar et al. (1987)
Citrus	Nitrate	Bar et al. (1997) and Iglesias et al. (2004)
Date palm	Salicylic acid Humic acids Acetic acid	El-Khawaga (2013)
	Amino acids Yeast	Darwesh (2013)
	Calcium	Jasim et al. (2016)
Grapevine	Humic acid Magnetic iron[b] Uni-sal[c]	Ali et al. (2013)
Lime	Humic acid	Ennab (2016)
Olive tree	K	Karimi et al. (2012) and Chartzoulakis (2005)
	Ascorbic acid Jasmonic acid Proline	El-Sayed et al. (2014)
	Proline	Ben Ahmed et al. (2010)
Orange	Calcium	Zekri and Parsons (1990)
	K-humate Active dry yeast Amino acids	Mustafa and El-Shazly (2015)
	K-humate Magnetic iron Ascorbic acid	Abobatta (2015)
	Proline Salicylic acid	Aly et al. (2015)
Peach	Humic acid Magnetic iron Uni-sal	Soliman et al. (2017)
Pistachio	Humic acid	Moghaddam and Soleimani (2012) and Javanshah and Nasab (2016)
	Silicon	Habibi et al. (2014) and Nooghi and Mozafari (2012)

[a] *Seaweed extract reduced the effects of salt stress only at an early stage.*
[b] *Is a natural row rock (Magnetite) that has very high iron content.*
[c] *Uni-sal contains polyethylene glycol (PEG), some elements and amino acids.*

5 Organic amendments

Organic residues as manure and composts are widely used as sources of organic matter and nutrients that improve soil physiochemical proprieties and promote the microbial soil activity. However, their use in saline conditions can contribute to increasing the soil pH and salinity, since organic composts presented usually a high EC and pH values. However, these composts can also be used for soil reclamation, since their application to saline soils can accelerate Na leaching, decrease exchangeable sodium percentage and electrical conductivity, and increase water infiltration and water-holding capacity and aggregate stability (El-Shakweer et al., 1998). The addition of calcium, which tends to replace exchangeable Na and reduce the SAR value, is a strategy to reduce soil sodicity. The addition of organic matter to the soil can also contribute to this objective (García et al., 2000). One of the major elements in MSW is the Ca (He et al., 1995; Stoffela et al., 2001).

An increase in organic matter content of saline soils increase soil stability, soil bulk density, and therefore soil microbial biomass (Tejada and Gonzalez, 2005). The addition of two organic wastes improved soil structure and reduced exchangeable sodium percentage, but the effect can be affected by the nature of the compost (Tejada et al., 2006) and rate of the application the compost. The benefits of compost application at low rates in lettuce, tomato, and blueberry outweighed the possible negative influence of compost salts (Reddy and Crohn, 2012). Nowadays, the offer of composted municipal solid wastes (MSW) is increasing due to the pressure to reduce the amount of refuse going to landfills and the biodegradable fraction of wastes, as food residues, represents a high percentage of the total urban waste collection. The municipal composts beyond the problem of high EC and pH still can present high levels of Na and Cl (Hargreaves et al., 2008). Despite this, MSW can be used to remediate saline soils (Lakhdar et al., 2009), since organic matter increases soil cation exchange capacity and decreases soil Na, EC, and pH likely due to the high supplies of Ca, Mg, and K. Municipal solid waste increased total and extractable soil Ca concentrations (Warman et al., 2009). MSW minimizes Na adsorption and increases Na leaching during precipitation or irrigation events (Ouni et al., 2014). However, MSW application must be combined with a posterior application of water to percolate of the exchanged Na and subsequent reduction in soil sodicity. Under greenhouse conditions, the use of municipal solid wastes (MSW) reduced the adverse effects of salinity in *Hordeum maritimum* (Lakhdar et al., 2008). The use of composted MSW associated with saline irrigation applications to the soil seems to be a practical and effective way to improve soil fertility and enhance the salt tolerance and growth of crops and could substitute the mineral fertilizer (Leogrande et al., 2016). The addition of vineyard compost supplemented with the beneficial microorganism *Trichoderma harzianum* isolate T78 to saline soils (NaCl) represent a promising approach for the treatment and improvement of saline soil properties (Mbarki et al., 2017)

The organic composts can be inserted in the soil or used as much. In this case, it can contribute to reduce water losses by evaporation and soil temperature in the first centimeter of the soil, which is important in warm climates to avoiding that root are exposed to supraoptimal temperatures (Machado and Bryla, unpublished). However, further research is needed to determine how MSW incorporated in the soil or as mulch affect fruit crop salt tolerance.

5.1 Bioestimulants and biofertilizers

Despite there has been a phenomenal growth over the last decade of biostimulants and biofertilizer products, there are no formally agreements on definitions, sometimes being difficult to find the differences between them. However, their use in saline conditions assumes high importance. In Europe, the EBIC defined plant biostimulants as follows: "Plant biostimulants contain substance(s) and/or microorganisms whose function when applied to plants or the rhizosphere is to stimulate natural processes to enhance/benefit nutrient uptake, nutrient efficiency, tolerance to abiotic stress, and crop quality. Biofertilizer can be defined as biological products containing living microorganisms that, when applied to seed, plant surfaces, or soil, promote growth by several mechanisms such as increasing the supply of nutrients, increasing root biomass or root area and increasing nutrient uptake capacity of the plant (Vessey, 2003). Therefore, the application of biostimulants and/or biofertilizers can contribute to alleviate salt stress and prevent the soil salinization since increasing fruit crops salt tolerance and nutrition and reduce the fertilizer application.

Biostimulants and/or biofertilizers include microbial inoculants, humic acids, fulvic acids, protein hydrolysates, and amino acids, silicon, and seaweed extracts. Despite the recognized potential of biostimulants, most studies presented on their influence in salt tolerance of fruit crops have been carried under controlled conditions with young trees, although usually more sensitive to salinity results cannot be easily extrapolated to field conditions.

5.2 Microbial inoculants

Arbuscular mycorrhizal fungi (AMF) and plant growth-promoting rhizobacteria (PGPR) could play a significant role in the increase of the salt tolerance of the fruit crops. However, the effects of AMF and PGPR application in increasing of salt tolerance of the fruit crops in field conditions are limited. AMF symbiosis can increase plant growth and nutrient uptake, improve fruit quality, and alleviate salt stress (Miransari, 2010; Porcel et al., 2012). AMF symbiosis is particularly important for enhancing the uptake of the relatively immobile and insoluble phosphate ions in the soil, due to interactions with soil bi- and trivalent cations, principally Ca, Fe, and Al (Rouphael et al., 2015). Arbuscular mycorrhizal fungi increased salt tolerance of apple (Yang et al., 2014), citrus (Wu et al., 2010), seedlings, and peach trees (Soliman et al., 2017). The effect of AMF on salt tolerance is influenced by the species. In inoculating olive plantlets under nursery conditions, *Glomus mosseae* was most efficient in the protection offered against the detrimental effects of salinity than *G. intraradices* and *G. claroideum* (Porras-Soriano et al., 2009).

Plant growth-promoting bacteria (PGPR) can be a tool to improve salinity tolerance (Shrivastava and Kumar, 2015) since that promote plant growth and development through diverse mechanisms such as enhanced nutrient assimilation by biological nitrogen fixation, phosphorous solubilization, or iron acquisition (Jin et al., 2014; Kuan et al., 2016). Plant growth-promoting rhizobacteria (PGPR) can alleviate salinity stress and improve water-use efficiency by removal of ethylene from the rhizosphere (Yang et al., 2009). Ethylene is produced by the plant roots in response to drought and salinity stress and results in the cessation of root growth.

5.3 Humic substances

Humic substances increase nutrient availability, nutrient uptake, and fruit crop salt tolerance. Its application through the drip irrigation system without clogging the irrigation system directly in the wet bulb is very effective and easy. The general category of humic substances includes humins (insoluble, in both alkali and acid media), humic acids (soluble in basic media), and fulvic acids (soluble in both alkali and acid media). The humic substances improve soil physiochemical characteristics and stimulate proliferation of microorganisms and plant growth. They improve soils' water-holding capacity, cation exchange capacity, aggregation, aeration, permeability, and buffering soil capacity, and they are a good source of energy for beneficial soil organisms. Humic substances neutralize the pH and are also involved in chelation of mineral elements (i.e., Cu, Ca, Zn, Mg, and Mn). The chelation process also increases the mass flow of the micronutrients (Pettit, 2004). Humic substances improve micronutrient transport and availability (Tan, 2014). Humic substances increase micronutrient uptake even under unfavorable conditions of high alkalinity (Bidegain et al., 2000; Khaled and Fawy, 2011). Therefore, their application in saline conditions usually with high pH can be very advantageous since they can reduce ion toxicity and leaching and increase their uptake by plants.

Humic acids improved fruit crop salt tolerance in different fruit crops (Table 33.5). Its application increased salt tolerance, by improving plant growth, increasing uptake of P and Fe, and decreasing Na uptake and proline concentration. The accumulation of osmolytes such as proline (Pro) is a well-known adaptive mechanism in plants against salt stress conditions. Proline concentration is closely related to salinity and water stress (Pérez-López et al., 2018). It has also been suggested that Pro accumulation can serve as a selection criterion for the tolerance of most species to stressed conditions (Parida and Das, 2005; Ashraf and Foolad, 2007). Humic acids increase lateral root and induce the production of smaller but ramified secondary roots (Ouni et al., 2014) increasing the volume of depletion. In avocado, root growth was more affected by salinity than shoot growth (Bernstein et al., 2004). In a field study, with pistachio, humic acid ameliorated negative effects on plant growth resulting from irrigation with low to moderate rates of NaCl (Moghaddam and Soleimani, 2012). The application of potassium humate (K-humate) the potassium salt of humic acid enhanced the salt tolerance of the oranges (Table 33.5).

5.4 Others biostimulants

Foliar, fertigation, or soil applications of amino acids, seaweed extract, and silicon (Si) alleviated salt stress in some fruit crops (Table 33.5). The amino acids can improve salt tolerance due to improving micronutrients uptake by plant due to their action of the chelation and plant growth. Plants can produce amino acids, but this synthesis is highly energy consuming. Therefore, the application of amino acids allows plants to save energy and increase plant growth. In avocado, seaweed extract reduced the effects of salinity stress only at an early stage and increased potassium and calcium concentrations in leaves (Bonomelli et al. 2018). Silicon can alleviate salt stress, since it decreases Na uptake and toxicity and increases K uptake (Rizwan et al. 2015). Salt stress alleviation by silicon has been reported in crop vegetables (Machado and Serralheiro, 2017) and cereals (Xie et al., 2015; Kim et al., 2014). In young pistachio plants grown in pots, Si alleviated salt stress due to increasing photosynthesis and water use efficiency, along with decreasing Na and increasing stimulation of the plant's antioxidative defense capacity (Habibi et al., 2014) (Table 33.5). The application of different antioxidants as ascorbic and salicylic acid and proline increased salt tolerance of fruit crops (Table 33.5). Proline supplements seem to improve olive salt tolerance by amelioration of some antioxidative enzyme activities, photosynthetic activity, and, so, plant growth as well as the preservation of a suitable plant water status under salinity conditions (Ben Ahmed et al., 2010).

6 Conclusions

The irrigation water quality has a deep impact on fruit crop growth and soil salinization, and this phenomenon can be more intense in irrigated fruit crops due to the long growing season and respective high water requirements. The majority of the fruit crops are rated as sensitive to salt stress, and growth is impaired a low EC_w. Fertigation combined with drip irrigation can alleviate fruit crop salt stress and reduce soil salinization. It reduces the input of salts due to increases in water and nutrient use efficiency and the use of the nutrients contained in irrigation water and creates a suitable root-zone salinity ($EC_e < EC_t$). Moreover, decreasing the soil volume to be reclaimed, soil reclamation should focus on row crop, in a more or less narrow strip, where the roots are concentrated. Fertilization management must incorporate the best nutrient stewardship with the application of biostimulants (humic acids, microbial inoculants, etc.) and organic composts, including municipal solid wastes, which have high potential to improve physicochemical soil properties, reducing soil salinization, and to enhance fruit crop salt tolerance. That potential is high. However, further research is needed to analyze the effects of the application of the biostimulants under field conditions and of the incorporation in the soil or as a mulch of organic composts. Moreover, further research that incorporates the different ways (mineral fertilization, biostimulants, and organic amendments) of the prevention and mitigation of the effects of the salinity of soil and/or the irrigation water is necessary.

References

Abdelgadir, E.M., Oka, M., Fujiyama, H., 2005. Characteristics of nitrate uptake by plants under salinity. J. Plant Nutr. 28 (1), 33–46.

Abobatta, W.F., 2015. Influence of magnetic iron and k-humate on productivity of Valencia orange trees (*Citrus sinensis* L.) under salinity conditions. Int. J. Sci. Res. Agric. Sci. 2, 108–119 (Proceedings).

Ali, M.A., El-Gendy, R.S., Ahmed, O.A., 2013. Minimizing adverse effects of salinity in vineyards. J. Hort. Sci. Ornamen. Plants 5 (1), 12–21.

Almutairi, K.F., Machado, R.M., Bryla, D.R., Strik, B.C., 2017. Chemigation with micronized sulfur rapidly reduces soil pH in a new planting of Northern Highbush Blueberry. HortScience 52 (10), 1413–1418.

Aly, M.A., Ezz, M., Osama, S.S.M., Mazek, A.A., 2015. Effect of magnetic irrigation water and some anti-salinity substances on the growth and production of Valencia orange. Middle East J. Agric. Res 4 (1), 88–98.

Aragüés, R., Puy, J., Royo, A., Espada, J.L., 2005. Three-year field response of young olive trees (*Olea europaea* L., cv. Arbequina) to soil salinity: trunk growth and leaf ion accumulation. Plant Soil 271 (1–2), 265–273.

Asano, T., Pettygrove, G.S., 1987. Using reclaimed municipal wastewater for irrigation. Calif. Agric. 41 (3), 15–18.

Ashraf, M.F.M.R., Foolad, M., 2007. Roles of glycine betaine and proline in improving plant abiotic stress resistance. Environ. Exp. Bot. 59 (2), 206–216.

Ayars, J.E., 2003. Field crop production in areas with saline soils and shallow saline groundwater in the San Joaquin Valley of California. J. Crop. Prod. 7 (1-2), 353–386.

Ayers, R.S., Westcot, D.W., 1984. Water Quality for Agriculture. vol. 29. Food and Agriculture Organization of the United Nations, Rome.

Bañuls, J., Serna, M.D., Legaz, F., Talon, M., Primo-Millo, E., 1997. Growth and gas exchange parameters of citrus plants stressed with different salts. J. Plant Physiol. 150 (1-2), 194–199.

Bar, Y., Kafkafi, U., Lahav, E., 1987. Nitrate nutrition as a tool to reduce chloride toxicity in avocado. In: Yearbook, South African Avocado Growers Association, 10.pp. 47–48.

Bar, Y., Apelbaum, A., Kafkafi, U., Goren, R., 1997. Relationship between chloride and nitrate and its effect on growth and mineral composition of avocado and citrus plants. J. Plant Nutr. 20 (6), 715–731.

Becerra-Castro, C., Lopes, A.R., Vaz-Moreira, I., Silva, E.F., Manaia, C.M., Nunes, O.C., 2015. Wastewater reuse in irrigation: a microbiological perspective on implications in soil fertility and human and environmental health. Environ. Int. 75, 117–135.

Ben Ahmed, C., Ben Rouina, B., Sensoy, S., Boukhriss, M., Ben Abdullah, F., 2010. Exogenous proline effects on photosynthetic performance and antioxidant defense system of young olive tree. J. Agric. Food Chem. 58, 4216–4222.

Ben-Gal, A., 2011. Salinity and olive: from physiological responses to orchard management. Isr. J. Plant Sci. 59 (1), 15–28.

Benlloch, M., Arboleda, F., Barranco, D., Fernandez-Escobar, R., 1991. Response of young olive trees to sodium and boron excess in irrigation water. HortScience 26, 867–870.

Bernstein, N., Meiri, A., Zilberstaine, M., 2004. Root growth of avocado is more sensitive to salinity than shoot growth. J. Am. Soc. Hortic. Sci. 129 (2), 188–192.

Bidegain, R.A., Kaemmerer, M., Guiresse, M., Hafidi, M., Rey, F., Morard, P., Revel, J.C., 2000. Effects of humic substances from composted or chemically decomposed poplar sawdust on mineral nutrition of ryegrass. J. Agric. Sci. 134 (3), 259–267.

Bongi, G., Loreto, F., 1989. Gas-exchange properties of salt-stressed olive (*Olea europea* L.) leaves. Plant Physiol. 90 (4), 1408–1416.

Bonomelli, C., Celis, V., Lombardi, G., Mártiz, J., 2018. Salt stress effects on Avocado (*Persea americana* Mill.) plants with and without seaweed extract (*Ascophyllum nodosum*) application. Agronomy 8 (5), 64.

Brown, P.H., Hu, H., 1998. Boron mobility and consequent management in different crops. Better Crops 82 (2), 28–31.

Buwalda, J.G., Smith, G.S., 1991. Influence of anions on the potassium status and productivity of kiwifruit (*Actinidia deliciosa*) vines. Plant Soil 133, 209–218.

Chartzoulakis, K.S., 2005. Salinity and olive: growth, salt tolerance, photosynthesis and yield. Agric. Water Manag. 78 (1–2), 108–121.

Chartzoulakis, K.S., Therios, I.N., Misopolinos, N.D., Noitsakis, B.I., 1995. Growth, ion content and photosynthetic performance of salt-stressed kiwifruit plants. Irrig. Sci. 16 (1), 23–28.

Chauhan, R.P.S., Asthana, A.K., 1981. Tolerance of lentil, barley and oats to boron in irrigation water. J. Agric. Sci. 97 (1), 75–78.

Connery, E., 2011. Boron in Irrigation Water. University of California. Available from: http://ccmg.ucanr.edu/files/83816.pdf.

Cramer, G.R., 2002. Sodium-calcium interactions under salinity stress. In: Salinity: Environment-Plants-Molecules. Springer, Dordrecht, pp. 205–227.

Darwesh, R.S., 2013. Improving growth of date palm plantlets grown under salt stress with yeast and amino acids applications. Ann. Agric. Sci. 58 (2), 247–256.

Dobermann, A., Fairhurst, T.H., 2000. Nutrient Disorders and Nutrient Management. Potash and Phosphate Institute, Potash and Phosphate Institute of Canada and International Rice Research Institute, Singapore.

El-Khawaga, A.S., 2013. Effect of anti-salinity agents on growth and fruiting of different date palm cultivars. Asian J. Crop Sci. 5, 65–80.

El-Sayed, O.M., El-Gammal, O.H.M., Salama, A.S.M., 2014. Effect of ascorbic acid, proline and jasmonic acid foliar spraying on fruit set and yield of Manzanillo olive trees under salt stress. Sci. Hortic. 176, 32–37.

Elseewi, A.A., 1974. Some observations on boron in water, soils and plants at various locations in Egypt. Alex. J. Agric. Res. 22, 463–473.

El-Shakweer, M.H.A., El-Sayad, E.A., Ejes, M.S.A., 1998. Soil and plant analysis as a guide for interpretation of the improvement efficiency of organic conditioners added to different soils in Egypt. Commun. Soil Sci. Plant Anal. 29, 2067–2088.

Ennab, H.A., 2016. Effect of humic acids on growth and productivity of Egyptian lime trees (*Citrus aurantifolia* Swingle) under salt stress conditions. J. Agric. Res. Kafr. El-Sheikh Univ. 42 (4), 494–505.

Fernandez-Escobar, R., 2004. Fertilizacion. In: Fernández-Escobar, R., Barranco, D., Rallo, L. (Eds.), El cultivo del olivo (2001), 5ª ed Mundi prensa, Madrid.

Fulton, A.E., Oster, J., Nanson, B., 1988. Salinity Management of Walnut. Walnut Production Manual. pp. 54–65.

García-Gil, J.C., Plaza, C., Soler-Rovira, P., Polo, A., 2000. Long-term effects of municipal solid waste compost application on soil enzyme activities and microbial biomass. Soil Biol. Biochem. 32 (13), 1907–1913.

Grafton, R.Q., Williams, J., Perry, C.J., Molle, F., Ringler, C., Steduto, P., Allen, R.G., 2018. The paradox of irrigation efficiency. Science 361 (6404), 748–750.

Grattan, S., 2002. Irrigation Water Salinity and Crop Production. vol. 9. UCANR Publications.

Habibi, G., Norouzi, F., Hajiboland, R., 2014. Silicon alleviates salt stress in pistachio plants. Prog. Biol. Sci. 4 (2), 189–202.

Haddeland, I., Heinke, J., Biemans, H., Eisner, S., Flörke, M., Hanasaki, N., Konzmann, M., Ludwig, F., Masaki, Y., Schewe, J., Stacke, T., Tessler, Z.D., Wada, Y., Wisser, D., 2014. Global water resources affected by human interventions and climate change. Proc. Natl. Acad. Sci. 111, 3251–3256.

Hanson, B., Grattan, S.R., Fulton, A., 1999. Agricultural Salinity and Drainage. University of California Irrigation Program, University of California, Davis.

Hargreaves, J.C., Adl, M.S., Warman, P.R., 2008. A review of the use of composted municipal solid waste in agriculture. Agric. Ecosyst. Environ. 123 (1–3), 1–14.

Hatami, E., Shokouhian, A.A., Ghanbari, A.R., Naseri, L.A., 2018. Alleviating salt stress in almond rootstocks using of humic acid. Sci. Hortic. 237, 296–302.

He, X., Logan, T.J., Traina, S.J., 1995. Physical and chemical characteristics of selected U.S. municipal solid waste composts. J. Environ. Qual. 24, 543–551.

Hudak, P.F., 2000. Regional trends in nitrate content of Texas groundwater. J. Hydrol. 228 (1–2), 37–47.

Iglesias, D.J., Levy, Y., Gòmez-Cadenas, A., Tadeo, F.R., Primo-Millo, E., Talon, M., 2004. Nitrate improves growth in salt-stressed citrus seedlings through effects on photosynthetic activity and chloride accumulation. Tree Physiol. 24 (9), 1027–1034.

Jasim, A.M., Abbas, M.F., Shareef, H.J., 2016. Calcium application mitigates salt stress in Date Palm *Phoenix dactylifera* L. (offshoots cultivars of Berhi and Sayer). Acta Agric. Slov. 107 (1), 103–112.

Javanshah, A., Nasab, S.A., 2016. The effects of humic acid and calcium on morpho-physiological traits and mineral nutrient uptake of pistachio seedling under salinity stress. J. Nuts 7 (2), 125–135.

Jin, C.W., Ye, Y.Q., Zheng, S.J., 2014. An underground tale: contribution of microbial activity to plant iron acquisition via ecological processes. Ann. Bot. 113, 7–18.

Kafkafi, U., Tarchizky, J., 2011. Fertigation: A Tool for Efficient Fertilizer and Water Management. International fertilizer industry association, Paris, France.

Karimi, E., Abdolzadeh, A., Sadeghipour, H.R., 2012. Increasing salt tolerance in Olive Olea europaea L. plants by supplemental potassium nutrition involves changes in ion accumulation and anatomical attributes. Int. J. Plant Prod. 3 (4), 49–60.

Khaled, H., Fawy, H.A., 2011. Effect of different levels of humic acids on the nutrient content, plant growth, and soil properties under conditions of salinity. Soil Water Res. 6 (1), 21–29.

Kim, Y.H., Khan, A.L., Waqas, M., Shim, J.K., Kim, D.H., Lee, K.Y., Lee, I.J., 2014. Silicon application to rice root zone influenced the phytohormonal and antioxidant responses under salinity stress. J. Plant Growth Regul. 33, 137–149.

Kuan, K.B., Othman, R., Rahim, K.A., Shamsuddin, Z.H., 2016. Plant growth-promoting rhizobacteria inoculation to enhance vegetative growth, nitrogen fixation and nitrogen remobilisation of maize under greenhouse conditions. PLoS One 11(3)e0152478.

Lakhdar, A., Hafsi, C., Rabhi, M., Debez, A., Montemurro, F., Abdelly, C., Ouerghi, Z., 2008. Application of municipal solid waste compost reduces the negative effects of saline water in *Hordeum maritimum* L. Bioresour. Technol. 99 (15), 7160–7167.

Lakhdar, A., Rabhi, M., Ghnaya, T., Montemurro, F., Jedidi, N., Abdelly, C., 2009. Effectiveness of compost use in salt-affected soil. J. Hazard. Mater. 171 (1-3), 29–37.

Lea-Cox, J.D., Syvertsen, J.P., 1993. Salinity reduces water use and nitrate-N-use efficiency of citrus. Ann. Bot. 72 (1), 47–54.

Leogrande, R., Lopedota, O., Vitti, C., Ventrella, D., Montemurro, F., 2016. Saline water and municipal solid waste compost application on tomato crop: effects on plant and soil. J. Plant Nutr. 39 (4), 491–501.

Levy, Y., Raveh, E., Lifshitz, J., 2000. The effect of rootstock and nutrition on the response of grapefruit trees to salinity. In: Proc. Int. Soc. Citricult.-vol. 1, pp. 334–337.

Lopez-Climent, M.F., Arbona, V., Pérez-Clemente, R.M., Gòmez-Cadenas, A., 2008. Relationship between salt tolerance and photosynthetic machinery performance in citrus. Environ. Exp. Bot. 62 (2), 176–184. Louise Ferguson (Ed.) Pistachio Production Manual. 2008. UC Fruit & Nut Research and Information Center, University of California, Davis (276 p.).

Maas, E.V., 1990. Crop salt tolerance. In: Tanji, K.K. (Ed.), Agricultural Salinity Assessment and Management, ASCE Manual Reports on Engineering Practices. In: vol. 71. ASCE, New York, pp. 262–304.

Maas, E.V., Grattan, S.R., 1999. Crop yields as affected by salinity. Agronomy 38, 55–110.

Maas, E.V., Hoffman, G.J., 1977. Crop salt tolerance—current assessment. J. Irrig. Drain. Div. 103 (2), 115–134.

Machado, R.M.A., 2002. Estudos Sobre a Influência da Rega-Gota-a-Gota Subsuperficial na Dinamica de Enraizamento, no Rendimento Físico e na Qualidade da Matéria-Prima do Tomate de Indústria. Ph.D. thesis. Universidade de Évora, Évora.

Machado, R., Serralheiro, R., 2017. Soil salinity: effect on vegetable crop growth. Management practices to prevent and mitigate soil salinization. Horticulturae 3 (2), 30.

Machado, R.M.A., Bryla, D.R., Verissimo, M.L., Sena, A.M., Oliveira, M.R.G., 2008. Nitrogen requirements for growth and early fruit development of drip-irrigated processing tomato (*Lycopersicon esculentum* Mill.) in Portugal. J. Food Agric. Environ. 6, 215–218.

Magen, H., 2005. Potential development of fertigation and its effect on fertilizer use. In: International Potash Institute (IPI) International Symposium on Fertigation: Optimizing the Utilization of Water and Nutrients, Beijing, September Available from: http://www.ipipotash.org/udocs/Potential%20development%20of%20fertigation.pdf.

Magesan, G.N., Williamson, J.C., Sparling, G.P., Schipper, L.A., Lloyd Jones, A.R., 1999. Hydraulic conductivity in soils irrigated with 269 wastewaters of differing strengths: field and laboratory studies. Aust. J. Soil Res. 37, 391–402.

Marschner, H., 2012. Mineral Nutrition of Higher Plants, second ed. Academic Press, New York.

Mateo-Sagasta, J., Zadeh, S.M., Turral, H. (Eds.), 2018. More People, More Food, Worse Water?: A Global Review of Water Pollution from Agriculture. International Water Management Institute (IWMI). CGIAR Research Program on Water, Land and Ecosystems (WLE), Rome, Italy/FAO Colombo, Sri Lanka.

Mbarki, S., Cerdà, A., Brestic, M., Mahendra, R., Abdelly, C., Pascual, J.A., 2017. Vineyard compost supplemented with Trichoderma harzianum T78 improve saline soil quality. Land Degrad. Dev. 28 (3), 1028–1037.

Miransari, M., 2010. Contribution of arbuscular mycorrhizal symbiosis to plant growth under different types of soil stress. Plant Biol. 12, 563–569.

Mmolawa, K., Or, D., 2000. Root zone solute dynamics under drip irrigation: a review. Plant Soil 222 (1-2), 163–190.

Moghaddam, A.R., Soleimani, A., 2012. Compensatory effects of humic acid on physiological characteristics of pistachio seedlings under salinity stress. Acta Hortic. 940, 253–255.

Mohamed, M.A., Terao, H., Suzuki, R., Babiker, I.S., Ohta, K., Kaori, K., Kato, K., 2003. Natural denitrification in the Kakamigahara groundwater basin, Gifu prefecture, central Japan. Sci. Total Environ. 307 (1-3), 191–201.

Mustafa, N.S., El-Shazly, S.M., 2015. Impact of some biostimulant substances on growth performance of Washington navel orange trees. Acta Hortic. 1065, 1795–1800.

Nackley, L.L., Barnes, C., Oki, L.R., 2015. Investigating the impacts of recycled water on long-lived conifers. AoB Plants. 7.

Nooghi, F., Mozafari, V., 2012. Effects of calcium on eliminating the negative effects of salinity in pistachio (*Pistacia vera* L.) seedlings. Aust. J. Crop Sci. 6 (4), 711.

NRMMC, EPHC, AHMC, 2006. Australian guidelines for water recycling: managing health and environmental risks (phase 1). In: National Water Quality Management Strategy, Natural Resource Management Ministerial Council, Environment Protection and Heritage Council and Australian Health Ministers' Conference, November, 149.

Ouni, Y., Ghnaya, T., Montemurro, F., Abdelly, C., Lakhdar, A., 2014. The role of humic substances in mitigating the harmful effects of soil salinity and improve plant productivity. Int. J. Plant Prod. 8 (3), 353–374.

Parida, A.K., Das, A.B., 2005. Salt tolerance and salinity effects on plants: a review. Ecotoxicol. Environ. Saf. 60 (3), 324–349.

Pérez-López, U., Sgherri, C., Miranda-Apodaca, J., Micaelli, F., Lacuesta, M., Mena-Petite, A., Muñoz-Rueda, A., 2018. Concentration of phenolic compounds is increased in lettuce grown under high light intensity and elevated CO2. Plant Physiol. Biochem. 123, 233–241.

Pettit, R.E., 2004. Organic Matter, Humus, Humate, Humic Acid, Fulvic Acid and Humin: Their Importance in Soil Fertility and Plant Health. CTI Research.

Phocaides, A., 2007. Handbook on Pressurized Irrigation Techniques. Food & Agriculture Org.

Porcel, R., Aroca, R., Ruiz-Lozano, J.M., 2012. Salinity stress alleviation using arbuscular mycorrhizal fungi. Agron. Sustain. Dev. 32, 181–200.

Porras-Soriano, A., Soriano-Martín, M.L., Porras-Piedra, A., Azcón, R., 2009. Arbuscular mycorrhizal fungi increased growth, nutrient uptake and tolerance to salinity in olive trees under nursery conditions. J. Plant Physiol. 166 (13), 1350–1359.

Ramos, C., Agut, A., Lidon, A.L., 2002. Nitrate leaching in important crops of the Valencian Community region (Spain). Environ. Pollut. 118 (2), 215–223.

Raveh, E., Ben-Gal, A., 2016. Irrigation with water containing salts: evidence from a macro-data national case study in Israel. Agric. Water Manag. 170, 176–179.

Reddy, N., Crohn, D.M., 2012. Compost induced soil salinity: a new prediction method and its effect on plant growth. Compost Sci. Util. 20 (3), 133–140.

Retamales, J.B., Hancock, J.F., 2012. Blueberries. vol. 28. CABI.

Rizwan, M., Ali, S., Ibrahim, M., Farid, M., Adrees, M., Bharwana, S.A., Abbas, F., 2015. Mechanisms of silicon-mediated alleviation of drought and salt stress in plants: a review. Environ. Sci. Pollut. Res. 22 (20), 15416–15431.

Rouphael, Y., Franken, P., Schneider, C., Schwarz, D., Giovannetti, M., Agnolucci, M., Colla, G., 2015. Arbuscular mycorrhizal fungi act as biostimulants in horticultural crops. Sci. Hortic. 196, 91–108.

Salinas, M.R., Cerda, A., Romero, M., Caro, M., 1981. Boron tolerance of pea (*Pisum sativum*). J. Plant Nutr. 4 (2), 205–215.

Sanden, B.L., Ferguson, L., Reyes, H.C., Grattan, S.R., 2004. Effect of salinity on evapotranspiration and yield of San Joaquin Valley pistachios. Acta Hortic. 664, 583–589.

Sanden, B., Fulton, A., Ferguson, L., 2005. Managing salinity, soil and water amendments. In: Pistachio Production Manual, pp. 129–146.

Shaaban, M.M., El-Aal, A.A., Ahmed, F.F., 2011. Insight into the effect of salicylic acid on apple trees growing under sandy saline soil. Res. J. Agric. Biol. Sci. 7 (2), 150–156.

Shalhevet, J., 1999. Salinity and water management in avocado. In: Proceedings of Avocado Brainstorming, 99. pp. 27–28.

Shrivastava, P., Kumar, R., 2015. Soil salinity: a serious environmental issue and plant growth promoting bacteria as one of the tools for its alleviation. Saudi J. Biol. Sci. 22, 123–131.

Smith, G.S., Clark, C.J., Buwalda, J.G., Gravett, I.M., 1988. Influence of light and form of nitrogen on chlorine requirement of kiwifruit. New Phytol. 110 (1), 5–12.

Soliman, M.A.M., Abo-Ogiela, H.M., El-Saedony, N.A., 2017. In: Soliman, M.A.M., Abo-Ogiela, H.M., El-Saedony, N.A. (Eds.), Reducing Adverse Effects of Salinity in Peach Trees Grown in Saline Clay Soil. Deciduous Fruit Trees Research Department, Hort. Res. Instit. ARC, Giza, Egypt.

Sotiropoulos, T., Therios, I., Dimassi-Theriou, K., 1995. Boron toxicity of kiwifruit orchards in northern Greece. In: III International Symposium on Kiwifruit 444, pp. 243–248.

Srivastava, A.K. (Ed.), 2012. Advances in Citrus Nutrition. Springer Science & Business Media.

Srivastava, A.K., Ram, L., 2000. Irrigation water quality of Nagpur mandarin orchards of central India. Indian J. Agric. Sci. 70 (10), 679–681.

Stoffella, P.J., Kahn, B.A., 2001. Compost Utilization in Horticultural Cropping Systems. CRC Press.

Tan, K.H., 2014. Humic Matter in Soil and the Environment: Principles and Controversies. CRC Press.

Tanji, K.K., Kielen, N.C., 2002. Agricultural Drainage Water Management in Arid and Semi-arid Areas. FAO.

Tattini, M., 1990. Effect of increasing nitrogen concentration on growth and nitrogen uptake of container-grown peach and olive. In: van Beusichem, M.L. (Ed.), Plant Nutrition-Physiology and Application. Kluwer Academic Publishers, Dordrecht, The Netherlands, pp. 515–518.

Tejada, M., Gonzalez, J.L., 2005. Beet vinasse applied to wheat under dryland conditions affects soil properties and yield. Eur. J. Agron. 23 (4), 336–347.

Tejada, M., Garcia, C., Gonzalez, J.L., Hernandez, M.T., 2006. Use of organic amendment as a strategy for saline soil remediation: influence on the physical, chemical and biological properties of soil. Soil Biol. Biochem. 38 (6), 1413–1421.

Therios, I.N., Misopolinos, N.D., 1988. Genotypic response to sodium chloride salinity of four major olive cultivars (Olea europea L.). Plant Soil 106 (1), 105–111.

Vessey, J.K., 2003. Plant growth promoting rhizobacteria as biofertilizers. Plant Soil 255, 571–586.

Warman, P.R., Burnham, J.C., Eaton, L.J., 2009. Effects of repeated applications of municipal solid waste compost and fertilizers to three lowbush blueberry fields. Sci. Hortic. 122 (3), 393–398.

Weber, E., Grattan, S., Hanson, B., Vivaldi, G., Meyer, R., Prichard, T., Schwankl, L., 2014. Recycled water causes no salinity or toxicity issues in Napa vineyards. Calif. Agric. 68 (3), 59–67.

West, D.W., 1978. Water use and sodium chloride uptake by apple trees. Plant Soil 50 (1–3), 51–65.

Wu, Q.S., Zou, Y.N., He, X.H., 2010. Contributions of arbuscular mycorrhizal fungi to growth, photosynthesis, root morphology and ionic balance of citrus seedlings under salt stress. Acta Physiol. Plant. 32 (2), 297–304.

WWAP, (United Nations World Water Assessment programme), 2013. The United Nations Worldwater Report 2013. UNESCO, Paris.

Xie, Z., Song, R., Shao, H., Song, F., Xu, H., Lu, Y., 2015. Silicon improves maize photosynthesis in saline-alkaline soils. Sci. World J. 2015.

Yang, J., Kloepper, J.W., Ryu, C.M., 2009. Rhizosphere bacteria help plants tolerate abiotic stress. Trends Plant Sci. 14 (1), 1–4.

Yang, S.J., Zhang, Z.L., Xue, Y.X., Zhang, Z.F., Shi, S.Y., 2014. Arbuscular mycorrhizal fungi increase salt tolerance of apple seedlings. Bot. Stud. 55 (1), 70.

Zekri, M., Parsons, L.R., 1990. Calcium influences growth and leaf mineral concentration of citrus under saline conditions. HortScience 25 (7), 784–786.

Zhou, B., Wang, D., Wang, T., Li, Y., 2018. Chemical clogging behavior in drip irrigation systems using reclaimed water. Trans. ASABE 61 (5), 1667–1675.

Zhu, J.K., 2007. Plant salt stress. In: Encyclopedia of Life Sciences. John Wiley & Sons, Ltd, www.els.net.

Zrig, A., Tounekti, T., Vadel, A.M., Mohamed, H.B., Valero, D., Serrano, M., Khemira, H., 2011. Possible involvement of polyphenols and polyamines in salt tolerance of almond rootstocks. Plant Physiol. Biochem. 49 (11), 1313–1322.

Further reading

Bauder, T.A., Waskom, R.M., Sutherland, P.L., Davis, J.G., Follett, R.H., Soltanpour, P.N., 2011. Irrigation Water Quality Criteria. Service in Action; no. 0.506. .

El Shakweer, M.H.A., El Sayad, E.A., Ewees, M.S.A., 1998. Soil and plant analysis as a guide for interpretation of the improvement efficiency of organic conditioners added to different soils in Egypt. Commun. Soil Sci. Plant Anal. 29 (11–14), 2067–2088.

Khemira, H., 2011. Possible involvement of polyphenols and polyamines in salt tolerance of almond rootstocks. Plant Physiol. Biochem. 49, 1313–1320.

Lide, D.R., 1993. CRC Handbook of Chemistry and Physics: A Ready-reference Book of Chemical and Physical Data: 1993-1994. CRC Press.

Zrig, A., Tounekti, T., BenMohamed, H., Abdelgawad, H., Vadel, A.M., Valero, D., Khemira, H., 2016. Differential response of two almond rootstocks to chloride salt mixtures in the growing medium. Russ. J. Plant Physiol. 63 (1), 143–151.

34

Trunk nutrition in fruit crops: An overview

Paula Alayón Luaces*, Melanie D. Gomez Herrera, José E. Gaiad

Fruticultura, Facultad de Ciencias Agrarias, Universidad Nacional del Nordeste, Corrientes, Argentina
*Corresponding author. E-mail: palayonluaces@yahoo.com

1 Introduction

The nutrition is one of the bases for fruit tree production. Superior plants can contain until 60 elements, but only 17 are considered essential for normal development. If one of them is deficient, the plant cannot complete the normal vegetative cycle. The essential elements are carbon (C), oxygen (O), hydrogen (H), nitrogen (N), phosphorus (P), potassium (K), calcium (Ca), sulfur (S), magnesium (Mg), iron (Fe), boron (B), manganese (Mn), copper (Co), zinc (Zn), molybdenum (Mo), nickel (Ni), and chlorine (Cl) (Mengel and Kirkby, 2001). The first 3 are supplied mainly from the air and water, while the remaining 14 are provided from the soil. The nutritive elements delivered from the soil are classified as macroelements (N, P, K, Ca, Mg, and S) and microelements (Fe, B, Mn, Co, Zn, Mo, and Cl) according to the amount needed by the plant.

The role of these nutrients involves different characteristics of the metabolism and plant structure, and their deficiencies are detected at different levels, depending on the depth of the deficiency. The principal nutrient disorder techniques are based on foliar symptoms, soil and plant tissue tests, and crop growth responses.

The nutrient supplying methods depend on different factors. The conventional method is to apply fertilizers to the soil, in particular, those essential plant nutrients that are required in higher amounts. The soil fertilization is mainly done according to the results of soil and plant tissue tests. The soil application method is more common and most effective for nutrients required in higher amounts. Although soil fertilization is the most widespread, it presents some disadvantages. All the nitrogenous fertilizer sources are eventually susceptibility to N loss mechanisms as leaching volatilization, denitrification, or immobilization. Nitrogen, as nitrate, does not remain strongly bound to soil particles, especially in soils with lighter textures, and can be leached from the soil profile, for example, with excessive rainfall, with potential economic and environmental consequences. In different areas of fruit production, there are registered different levels of nitrate pollution associated with excess fertilization and particular environmental condition as high rainfall and sandy soils.

Fertilizer N use efficiency by trees increased from the first to the third year but was generally small (6%, 14%, and 33%), and the estimated N losses were large (89%, 46%, and 53%), respectively, in the first, second, and third years (Srivastava and Malhotra, 2017).

The P applied on the soil surface during fertilization can be adsorbed with high binding energy with the mineral fraction surface of the soil (Schmitt et al., 2013). However, successive applications of P onto the soil surface may cause

A.K. Srivastava, Chengxiao Hu (eds.)
Fruit Crops: Diagnosis and Management of Nutrient Constraints
https://doi.org/10.1016/B978-0-12-818732-6.00034-4

the occupation of adsorption surfaces, decreasing their adsorption energy, which can enhance desorption and therefore migration in the soil profile, increasing the P amount in deeper layers.

According to soil availability for plants, the potassium is classified as relatively unavailable, easily available K-forms and slowly available. Since in many types of soil the amount of interchangeable K is low, it is necessary to preserve it with potassium fertilization. The exchange sites in the interlayers of clay minerals and primary minerals buffer the K supply. K migration occurs when the quantity added is greater than that used by the crops. Its adsorption to the functional groups of inorganic and organic particles takes place with low binding energy, facilitating its migration in the soil profile (Kaminski et al., 2007). Moraes and Dynia (1992) considered K as the most easily leached cation, due to its displacement to the soil solution and to its percolation, especially in sandy soils. However, because the soil properties have a strong influence on nutrient availability, clay soils typically have a high K-fixing capacity and, thus, often show little response to soil-applied K fertilizers (Brady and Weil, 2002).

Micronutrients play many complex roles in plant nutrition and crop production, but very few of each is needed. The soil application of micronutrients, especially inorganic salts, is often not so effective due to the immediate reaction of added micronutrient cations with the mineral portion of soil through various processes like adsorption, fixation, and chemical precipitation (Srivastava and Singh, 2008a).

In perennial fruit crops, nutrient doses need to be recommended in tandem with the level of fruit yield targeted; a nutrient dose optimum for one fruit yield target will become suboptimum for the higher targeted fruit yield level in combination in the subsequent years (Srivastava and Singh, 2008b; Srivastava, 2013a,b).

Environmental contamination as a result of nutrients leached, especially N, from orchards is a major concern for some local communities; new strategies are required in orchard management to reduce the potential of nutrient leaching loss (Dong et al., 2005). Previous studies about soil fertilization efficiency proved that a small amount of the fertilizers is taken up by the plant roots while the great portion (62%–85% of N and 80%–95% of P and K) is lost by leach, volatilization, and fixation (Shaaban, 2012).

Nowadays, the techniques and management of perennial fruit production are directed to the need to conserve resources and energy and a commitment to the environment. In this sense, fertigation (open hydroponics and intensive fertigation programs) has become a valuable tool in recent years that has spread throughout the world. Many studies have provided a large amount of information with unanimous results that fertigation reduced irrigation and nutrient requirements by 30%–50% compared with conventional fertilization (Srivastava and Malhotra, 2017). Although these techniques are very efficient systems, they require a very exhaustive control of the soil and the levels of foliar nutrients and the threshold values of leaf nutrient diagnostics (Shirgure and Srivastava, 2014). In addition to the costs, the irrigation system must be well balanced and maintained, and the water quality must be good.

Another very common system to supply nutrients to the plants is the foliar fertilization. Foliar nutrition cannot replace or eliminate the natural way of nutrient uptake by plant roots, but this method is most commonly used for supplying the nutrient deficiency, improving the nutritional status of plants, and thus increasing crop yield and its quality. However, depending on plant species and environmental factors, this treatment can be applied for other purposes, like mitigation of the negative effects of stress conditions—drought, frost damages, and others (Smolen, 2012).

Srivastava and Singh (2003) presented a list of the most important exogenous and environmental factors that affect foliar fertilization: light, temperature, wind, time of day, photoperiod, moisture, amount and intensity of precipitation, drought, osmotic potential of growing medium (or soil water), and nutrient stress. Foliar fertilization requires a higher leaf area index to absorb the nutrient solution. For this reason, it may be necessary to have more than one application depending on the severity of nutrient deficiency. Nutrient concentration and day temperature should be optimal to avoid leaf burning, and fertilizer source should be soluble in water to be more effective (Fageria et al., 2009).

For both commercial and environmental reasons, fertilizers should be used with caution, and crop production for future food security will require sustainable fertilizer management, which might include more sophisticated decision support tools, improved agronomic practices, and crops or cropping systems that require less fertilizer input (White and Brown, 2010).

The trunk injection is a novel alternative to conventional or widely used fertilization methods, to supply nutrients to fruit tree crops. This method inserts substances (nutrients, pesticides, and hormones) directly to the vascular system of the tree. Injecting compounds into the vascular system of a tree requires knowledge of the woody plant system and the know-how of these compounds moves throughout the vascular system.

2 Trunk injection history

Many author mentions several events in relation to the trunk injection history (Robbins, 1982; Sánchez-Zamora and Fernández-Escobar, 2004; Doccola and Wild, 2012); according to them, this method was first investigated by Leonardo da Vinci, but some of the most early tree injection experiments were not recorded until early in the 20th century (Roach, 1939).

Concerning to the advances in this techniques through the history, a deep description is presented by Robbins (1982). Many researchers investigated the use of trunk injections for different purposes associated to trunk conservation, protection, tree pathology, entomological threats, and nutritional and physiological aspects, even to kill trees.

Also, the evolution of this technique refers to the injection process itself and improved formulations. Some examples of the development of this technique are: powder substances (Wallace, 1935; Duggan, 1943), chemical tablets (Roach and Roberts, 1945), gelatin capsules (Neely, 1973, 1976), metal pieces (May 1941), bark banding (Backhaus et al., 1976; Koehler and Rosenthal, 1967), or plugs of solidified polyethylene glycol (Miller et al., 2001); the trunk injection of solids is usually done through one or more holes in a plant and by inserting water-soluble substances.

The liquid formulation was extended more widely, and many technological alternatives were developed around this possibility. One of the possible alternatives is that the solution dispensation is passive or active. A passive alternative is the trunk infusion where the solution reservoir is placed above the injection site (Schreiber, 1969) and it consists of an unpressurized container suspended from the tree above the injector assembly so that the liquid is introduced by gravity into the tree.

In active injection methods, the supply of the solution can be high pressure (Filer Jr., 1973; Helburg et al., 1973; Reil and Beutel, 1976; Sachs et al., 1977) or low pressure. For high-pressure injection, basic components of a system involve at least a solution reservoir, pressure source, pressure regulator, supply hoses, and injector heads. These pressurized injections force solutions into the xylem, so all the system must be prepared for pressure tolerance that usually operates in the $0.7–14\,kg/cm^2$ pressure range providing rapid injection, and the volume of solution can be around of 1 or more liters. This technology was widely used for experimental purposes; however, this system is limited because of high labor costs and availability of the equipment. Low-pressure injections were developed as an alternative method with cost and labor advantages. The apparatus consists of two elements: a plastic injector for insertion into the tree trunk after a hole has been mechanically drilled and a latex tube that contains the solution to be injected. The solution was placed into the tube under pressure by means of a syringe connected to the tube. The tube was closed at both ends through ties or clamps. After removing one of the ties or clamps, the pressurized tube provides a pressure of 60–80 kPa with the use of a 9-mm-diameter latex tube (Navarro et al., 1992).

Examples of the new technologies are the TREE I.V. microinfusion system and Air/Hydraulic microinjector (Arborjet, Inc., Woburn, Massachusetts, United States) and the EcoJect Microinjection System (Bioforest Technologies, Canada).

Trunk injection technology can be classified according to the targeted trunk issue (xylem or cambium) and type of injection port (drill based or needle insertion based). The drill based creates a cylindrical injection port, and the part of the wood tissue is removed while administering the injection solution. The needle insertion based creates a lenticular injection port, and only a small portion of wood bark tissue can be removed (Acimovic, 2014). Some examples of drill based are ChemJet Tree Injector (ChemJet Trading PTY Ltd., Bongaree, Queensland, Australia) and Mauget Tree Injection Capsules (Mauget Inc., Arcadia, California, United States), and examples of needle based are Blade for Infusion in Trees (BITE) (University of Padova, Italy) and Wedgle Direct-Inject System (ArborSystems LLC., Omaha, Nebraska, United States) (Acimovic, 2014).

In addition to new injection technology, formulations are being designed for injecting into trees that improve plant safety and reduce application time. Actually, tree injection is an alternative of chemical application with certain advantages: efficient use of chemicals that reduced potential environmental exposure and useful when soil and foliar applications are either ineffective or difficult to apply (Doccola and Wild, 2012).

3 Factors that influence trunk injections

The injection technology or system, injection pressure, injection site characteristics, substance characteristics, plant species, plant environment, and persistence of the operator are some factors that determine the time and effort expended on the operation of trunk injection.

It is desirable that the solution injected into the trunk reaches the leaves in an optimal dosage and in different stages of growth. Many known factors that influence the absorption of compounds must be considered for the success of the injection. The proportional distribution of compounds along the tree depends on the translocation capacity of the vascular system. Different species have different physiologies, and injection techniques must differ according to these characteristics. This makes injection techniques specific to each species; according to Perry et al. (1991), appropriate treatments for trees with different wood, bark, and leaf anatomies are not likely to be the same.

In their work, Sánchez Zamora and Fernández (2000) quoted authors that investigated the numerous factors that influence the uptake and distribution of substances injected in tree trunks. Technical application-related factors include injection pressure of the treatment solution, the type of substance injected, hole diameter, the number of injections and hole depth, the speed and type of drill used, and the injection site in the tree trunk or branch. Other factors that affect uptake are specie and tree transpiration rate. The last factor varies according to the treatment timing, stress condition, wind speed, soil water content, tree size, tree health, and phenological state.

Hole location and the hole depth and the number of holes play a large role in translocation. Sánchez Zamora and Fernández (2000) results suggest that, in adult trees with thick trunks, it may not be necessary to drill holes right into the pith, requiring the use of an especially long bit as suggested by Navarro et al. (1992), since 4.5- to 5-cm-deep holes made in the trunk would reach the active xylem, the part of the wood where active water flow occurs. Furthermore, the internal part of the wood of some adult trees has normally lost its conductive capacity. Injectors would thus have to be inserted to a maximum depth of 1.5cm in the hole, thus leaving an internal air chamber some 3 ± 3.5 cm in length, which is essential for ensuring injection uptake.

The number of injections necessary to uniformly distribute chemicals throughout the tree depends on tree size (Nyland and Moller, 1973; Navarro et al., 1992).

Doccola et al. (2011) examined wound responses in *Fraxinus pennsylvanica* for 2 years following trunk injection, to look for evidence of infection associated with injection sites. All healthy trees successfully compartmentalized injection wounds without any signs of infection, decay, or structural damage. Wound closure was positively correlated with tree health as measured by annual radial growth.

Wallace and Wallace (1986) suggested that the hole is covered with an asphalt-based tree-sealing compound after the Fe solution has been applied and the sterilization of the drill bit, either by alcohol or flame, would be recommended in those areas where bacterial or viral diseases are problems.

The drill speed and drill bit type also affect the rate of uptake. Brad point drills make a cleaner cut of the xylem cells, improving the rate of uptake and uniformity. Regardless of the type, a sharp cutting edge on the drill bit is essential for injection. High-speed drills may cause excessive friction to the conductive tissue of the plant, impeding uptake. This may relate directly to the poor wounding response by the tree (Orr et al., 1988).

The type of substance injected and concentration is important in the distribution. Many authors have shown that the chemical characteristics of an injected chemical affect distribution (Rumbold, 1920; Collison, et al., 1932; Morris, 1951).

Systemic compounds should be liquid formulated for proper distribution. If the compound is not liquid formulated and is a dry mixture, it must be highly water soluble. The more water soluble the compound is, the more efficient its distribution will be (VanWoerkom, 2012).

Sánchez Zamora and Fernández (2000) described that uptake rates varied according to injection timing. In sweet orange, grapefruit, and mandarin (perennial species), uptake rates during the vegetative rest period were generally lower than in full growth period, due to the reduced physiological activity of the trees. In peach trees, the presence or absence of leaves between the phenological visible calyx and recently set fruit stages only influenced the uptake rate shortly after injection; no significant differences were observed in terms of the final volumes taken up. This aspect is of agronomic interest since uptake speed is less important than the uptake of the volume of liquid applied within an acceptable time (48 or 72h).

Fernández-Escobar et al. (1993) suggest the possibility of injecting chemical formulations into trees before the initial period of leaf development when some chemical products might cause leaf burning by trunk injection.

Environmental characteristics influence the uptake rate of trunk injections. The rate of transpiration of the trees is one of the factors that determine the rate of compound translocation and the level of its accumulation in the tree canopy. Thus, for fast translocation and satisfactory distribution of trunk-injected solutions into the canopy, ideal conditions would be medium to high temperatures, low relative air humidity, high vapor pressure deficit in the air, sunny and windy weather, and substantial water supply in the soil (Acimovic, 2014). On clear days, olive trees took up the injections more quickly (higher uptake rate) than on rainy days, and this difference increased with smaller diameter injectors. This variation in the injection rate was due to the abundant rain and relative air humidity that lead to greater water accumulation in the soil and thus reducing tree transpiration (Sánchez Zamora and Fernández, 2000). Reil and Beutel (1976) report that injection is faster when a tree is under slight water stress; they found that it took longer to inject in the early morning or at night than in the afternoon and that cloudy humid days increased injection time.

Different species and their trunk anatomy play an important role in the injection efficiency. Trees with diffuse porous xylem, tracheids and vessels dispersed among the annual rings, need deeper injection points. While ring porous xylem types with tracheids and vessels located only in the annual ring are benefited from superficial injections. Therefore, injections past the current annual ring are not necessary for this type of wood. Proper translocation throughout a diffuse-porous tree requires an injection through the bark, cambium, and phloem and into the active xylem tissue. With vessels and tracheids being distributed fairly even throughout each xylem growth ring, injecting into more than one xylem layer is optimal. When the compound reaches the phloem and xylem, it travels up the trunk out to the branches and leaves. It also travels downward through the phloem to the root system (VanWoerkom, 2012).

3.1 Trunk anatomy

Due to the fact that trunk injections supply the nutritional solution directly to the vascular system of the tree (Fig. 34.1), it is important to have a competent understanding of the trunk anatomy.

Wood types and structure vary among tree species, which plays a large role in the translocation characteristics of injected compounds. A noticeable difference is between dicot and monocot species. While the first has a vascular system produced by a secondary meristematic called cambium, the monocot system does not have secondary growth and do not present cambium. In dicot, due to the cambium, new phloem is produced toward the outside of the tree and xylem or wood toward the inner portion. The trunk vascular systems are organized in xylem, cambium, and phloem tissues. In a cross section of the dicot tree trunk, from outside to inside, there are: bark, phloem, and cambium. The xylem is organized in concentric circles with sapwood next to cambium and heartwood near the trunk center (Fig. 34.2A).

The xylem is composed of several cell types, with different functions. Among the elements qualified for the movement of water and solutes, mainly from the region of absorption (roots) to the region of use (leaf and reproductive organs), are the vessels and tracheids. Both cell types have lignified walls and lack cytoplasm, facilitating the apoplastic movement. Vessels and tracheids are arranged vertically moving the water upward and connected by pits in the lateral walls and in the vessels; there are also perforation plates in the terminal walls. Vessels are the main conductive element in the wood of the dicot. Tracheids, a type of cell with conifer characteristics, present in a few dicotyledonous genera, are imperforated elements and less evolved as conductive elements. As these cells mature, they turn into dead transport cells. Other dead, nonconductive cells are the libriform fibers and fiber tracheids; they are the background tissue in which the conductive elements are embedded, constituting the structural and supporting elements. Finally, there are parenchymal cells, organized in a vertical and horizontal system or rays; they represent the only system of living cells in the wood and store carbohydrates needed for growth and also move water laterally and radially throughout the xylem (Fig. 34.3).

The sapwood is the portion of the wood that in the living tree contains living cells and reserve materials. Meanwhile, the heartwood is defined as the inner layers of wood that, in the growing tree, have ceased to contain living cells; it is generally darker in color than sapwood, though not always clearly differentiated (IAWA, 1964).

The trunk of monocot is enveloped by an outer cortex made of parenchyma cells and fibers, and the center consists of vascular bundles (Darvis et al., 2013; Fathi and Frühwald, 2014; Fathi et al., 2014) embedded in parenchymatous ground tissue (Ramle et al., 2012) (Fig. 34.2B).

The anatomy of the monocot stem is organized through scattered vascular bundles, surrounded each one by a ring of cell called bundle sheath, and inside is the cell arranging the xylem and the phloem (Fig. 34.4). These species usually

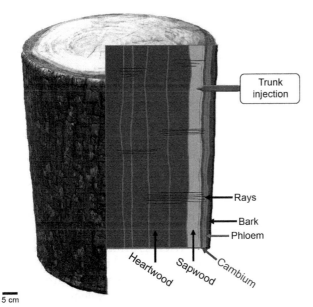

FIG. 34.1 Scheme of trunk injections to the vascular system of the tree. *Courtesy of Ana Maria Gonzalez—www.biologia.edu.ar/botanica.*

Trunk injection

Rays

Bark

Phloem

Heartwood Sapwood Cambium

5 cm

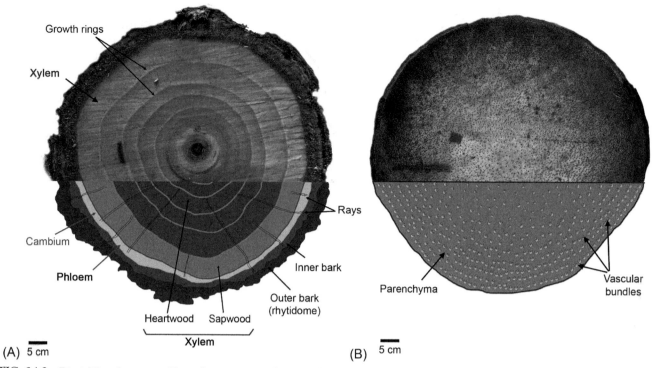

FIG. 34.2 Dicot (A) and monocot (B) trunk cross section. *Courtesy of Ana Maria Gonzalez—www.biologia.edu.ar/botanica.*

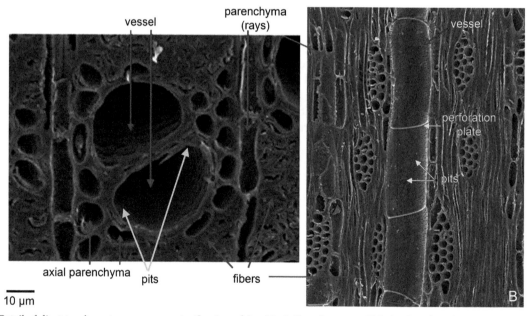

FIG. 34.3 Detail of dicot trunk anatomy components. *Courtesy of Ana Maria Gonzalez—www.biologia.edu.ar/botanica.*

lack secondary growth from a cambium but may develop substantial stems (e.g., palms) thickening growth resulting from division and enlargement of parenchyma cells of the ground tissue (Evert, 2006).

The xylem functions are water transport and structural support of the plant; the xylem is primarily responsible for the upward movement of water and mineral nutrients and phytohormonal signals from the soil to the leaves. Otherwise, the phloem is responsible for the bidirectional transport of photosynthates, amino acids, and plant growth regulators and electrolytes mainly from the leaves to the rest of the plant (Myburg et al., 2013). Mostly xylem elements are dead when functional and movement is apoplastic, whereas phloem translocation occurs in living cells and is symplastic.

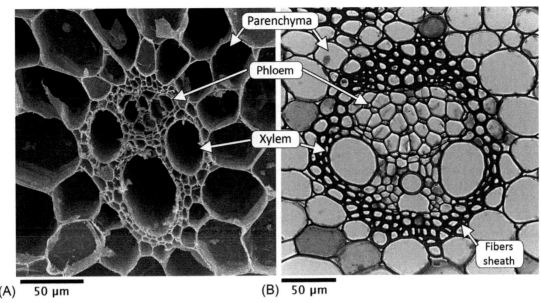

FIG. 34.4 Detail of monocot bundle sheath. (A) Scanning electron microscopy and (B) Optic microscope. *Courtesy of Ana Maria Gonzalez—www. biologia.edu.ar/botanica.*

In 1894, Dixon and Joly proposed the cohesion tension theory, which exposes that the water is pumped by transpiration from the roots through the xylem to the leaves. This widely accepted theory about long-distance flow in plants indicates that the tension on the surface of the leaves is generated by transpiration and this produces a pressure difference between the root and the leaves, which is the force driving the transport of xylem. The upward movement of the water column is achieved by its weight, the adhesion to the cell walls of the tracheary elements, and the adhesion of water to the soil particles (Myburg et al., 2013). Actually, this theory has been reviewed with new findings and controversies appear (Zimmermann et al., 1993; Tyree, 1997; Wei et al., 1999; Windt et al., 2006).

The rate of water flow differs among tree species, plant characteristics, and particularly the structure of the vascular system. The anatomical characteristics of the conduction system in plants can have a profound impact on the hydraulic conductivity of the tree (Tyree and Zimmerman, 2002).

Diffuse-porous woods are those in which the vessels have more or less the same diameter throughout the growth ring. The vast majority of tropical species and most temperate species show this pattern. In some temperate diffuse-porous woods, the latest formed vessels in the latewood may be considerably smaller than those of the earlywood of the next ring, but vessel diameter is more or less uniform throughout most of the growth ring, while in ring-porous wood the vessels in the earlywood are distinctly larger than those in the latewood of the previous growth ring and of the same growth ring and form a well-defined zone or ring, in which there is an abrupt transition to the latewood of the same growth ring (IAWA, 1989). Pérez Olvera et al. (2008) described the wood anatomy of some species of the Rosaceae family: pear and apple have a ring-porous distribution, while peach has a diffuse-porous wood. Ring-porous hardwoods move injected liquid at a faster rate than diffuse-porous hardwoods (Coder, 1999). For water movement, diffuse-porous hardwoods tend to use a larger proportion of sapwood than the ring-porous hardwoods. Sterrett and Creager (1977) have stated that diffuse-porous species are more receptive to injection than ring-porous species.

In ring-porous species, water movement is generally confined to a thin cylinder of the outer sapwood and often to the outermost annual ring. The path of ascent in diffuse-porous trees usually occupies a greater portion of the sapwood (Kozlowski, 1961).

4 Previous studies in trunk nutrition

In 1925, Lipman and Gordon studied passive injections in 10-year-old pear trees, with various salts of Ca, Mg, and K. Three liters of 0.3N solutions were injected in early spring, and observations were made throughout the growing season. Calcium salts of $Ca(NO_3)_2$, $CaHPO_4$, $CaCl_2$, and $CaSO_4$ were found to be somewhat toxic and reduced growth. Magnesium nitrate and $MgHPO_4$ injections were stimulating to tree growth, resulting in longer, greener leaves and flower buds of greater vigor than other treatments. The K salts, KNO_3, KH_2PO_4, and K_2HPO_4, were much more toxic than Mg salts and in no case stimulated tree growth. They assume that the concentration of the solutions of Ca and K were too high and that lower concentrations should stimulate tree growth. On the other hand, Lipman and Gordon eliminated chlorosis caused by Fe deficiency with $FeSO_4$ injections. Doses of 0.1–7 g of $FeSO_4$ produced green foliage and renewed shoot growth in lemon trees that had been chlorotic for some time prior to treatment.

Temporary elimination of chlorosis in orange trees was achieved by passive injection of solutions containing 3–7 g of $FeSO_4$ or 1–2 g of Fe tartrate (Thomas and Haas, 1928). They found retreatment necessary with each new cycle of tree growth.

Chandler et al. (1933) had obtained remission of Zn deficiency symptoms for two growing seasons in fruit trees from solid injections of either $ZnSO_4$, ZnO, or Zn dust. Drilled holes, 0.96 cm in diameter and 3.81 cm deep, spaced 7.62–10.16 cm apart around the tree trunk were filled with the Zn material. New growth on trees did not display "little leaf" or "rosette" symptoms typically associated with Zn deficiency of fruit trees. Zinc-coated nails pounded 2.54 cm apart in a spiral around the trunk were also an effective treatment.

In 1933, Roach suggested that trunk injections of K could possibly be used to speed up the slow response to soil-applied K fertilizers. He applied a KNO_3 injection of a 15-year-old apple tree with potassium deficiency. Ten liters of 1% KNO_3 solution was passively injected in the early part of the growing season. The following day, leaf scorch was visible on the fully expanded leaves, particularly on branches above the injection hole, but leaves expanding after this time were undamaged. After 1 month, these leaves were darker green, thicker, larger, and apparently healthier than before. By autumn, strong healthy shoot growth was evident, exceeding growth of the previous year by 50%–70%. The branches that showed the most damage from injection had the greatest amount of shoot growth.

In 1934, Roach described a test of the tree injection of nutrients on a larger scale in eight 21-year-old apple trees that were passively injected with a solution containing 0.25% K_2HPO_4 plus 0.25% urea. Shoot growth increased and was positively correlated with the amount of injected material. The number of pruning from injected trees was 1.9 times more than from uninjected trees. The injected trees also showed the best-looking foliage and lower infestations of leaf hopper insects and red spider mites. No effect was measured on crop size; however, fruit from injected trees had inferior color and finish to those from control trees. At the same year, Roach reported that whole-tree injection of 0.05% $FeCl_3$ solution for 2 days during the growing season cured chlorosis in a peach tree within 2 weeks without any ill effects.

Manganese deficiency in cherry trees was effectively treated with solid $MnSO_4$, injections by Duggan in 1943. One-centimeter-diameter holes spaced 2.54 cm apart around the trunk were filled with 2–5 g of $MnSO_4$ in early spring. This treatment resulted in complete remission of chlorosis and an increase in tree growth and cropping without apparent tree injury.

Southwick (1945) used pressure injection to effectively treat Fe deficiency in 15-year-old orange trees. He reported that 3.5–4.2 kg/cm^2 of pressure was employed to inject trees with solutions containing 30–200 g of $FeSO_4$. All treatments corrected chlorosis and were effective for 2–4 years. Because of injury associated with dosages above 100 g, the resulting recommended dosage range was 50–100 g. Some injury of small shoots was associated with injections of 70–100 g of $FeSO_4$, but rapid recovery with normal foliage, shoot growth, and cropping was observed.

In 1946, Levy injected K-deficient 6-year-old dwarf apple trees in midsummer with a nutrient solution containing 0.125% urea, 0.125% K_2SO_4, and 0.25% KH_2PO_4 by branch injection method. The dosage administered was 0.5 g of total salt per 1 cm of trunk cross-sectional area. The same growing season "incipient" flower buds were stimulated into vegetative growth.

Successful spring time solid injections of K salts alone or in combination with Fe or Mn salts were described by Roberts and Landau in 1947. Potassium sulfate and K_2HPO_4 were injected into apple trees at the rate of 6–12 g per 2.54 cm of trunk diameter depending on tree size, although it was found that maximum safe doses were 24–144 g for K_2SO_4 and 18–108 g for K_2HPO_4. Ferrous sulfate and $MnSO_4$ were injected at the rate of 2–4 g of salt per 2.54 cm of trunk diameter. In general, trees injected with iron had deeper green foliage than untreated trees. All trees injected with K plus Fe were even deeper green than the Fe injected trees.

Water-insoluble FeO was effective as a solid injection compound for correcting lime-induced chlorosis of several ornamental trees (Brown and Hildreth, 1960). Drilled holes, 0.9 cm in diameter and 5.1 cm deep, spaced 7.6–10.2 cm around the trunk in a spiral were filled with FeO. Response to this treatment was reported to be nearly as rapid as with water-soluble iron salts and lasted nearly three times longer.

In 1960, Tanaka investigated that labeled ammonium phosphate was applied to the soil or by injection to one side only of 20-year-old Satsuma orange trees in the field and in pots. The absorbed P accumulated on the treated side, little or none being found on the other side, and was greatest in the upper part of the tree. No difference was found due to age of leaves, and no radioactivity appeared in the fruit.

Worley et al. (1976) reported their preliminary results of pressure injecting Zn compounds into pecan trees. Pressures of 2.1–2.5 kg/cm^2 were used to inject 8 L of solution into the tree trunk. Injection of 2270 g/tree of $ZnSO_4$ killed all first flush, leaves, and stem tips in a tree 201 cm in circumference. A larger tree of 270 cm in circumference had some uninjured small limbs. Damaged foliage contained over 5000-ppm Zn, and within 3 weeks, new shoots emerged with normal foliage.

Reil et al. (1978) reported successful results with pressure injection of Fe and Zn compounds into trees. Pressure of 14 kg/cm^2 was used to inject pear trees with Fe and Zn sulfates and Fe and zinc chelates. The sulfates were more effective than the chelates in correcting Fe and Zn deficiencies. A volume of 0.95 L of 1% or 2% $FeSO_4$, solution corrected severe chlorosis in mature trees for at least 1 year. Larger trees were not injured by 1.9 L of 2% $FeSO_4$ solution and remained free of chlorosis for 2 years. An injection of 0.95 L of 1% $ZnSO_4$ solution controlled Zn deficiency for at least 1 year.

Barney et al. (1984) evaluated chlorotic "Red Delicious" apple trees that were treated for iron deficiency through pressure injection and soil treatment. Injection treatments were either ferrous sulfate ($FeSO_4 \cdot 7H_2O$) or ferric citrate ($FeC_6H_5O_7 \cdot H_2O$) at rates of 100 mL of 1% solution per year of tree age. Chlorotic conditions were completely eliminated by the following spring by ferrous sulfate injection. Ferric citrate injection decreased, but did not completely eliminate iron chlorosis. The improvement in chlorotic conditions gave correspondingly heavier fruit set, higher chlorophyll levels, and greater shoot growth for trees injected with both ferrous sulfate and ferric citrate. Some injury at the injection site was observed for both injected compounds.

Apple and pear trees were treated with various iron (Fe) and sulfur (S) compounds in an attempt to correct or reduce severe Fe chlorosis. Fall plus spring trunk injections of ferrous sulfate solution (1%) were the most effective and long-lasting (3 or 4 years) method for reducing Fe chlorosis (Raese et al., 1986).

Injection of Fe salts (mainly Fe^{2+} sulfate and Fe ammonium citrate) in liquid form into xylem vessels has been reported to alleviate Fe chlorosis symptoms in citrus plants (Wallace and Wallace, 1986); they report that about 5 mL of the ferric ammonium citrate increased the leaf Fe concentration by 50% in trees mature about 20 years old.

To study the response of date palm trees to Fe fertilization, two date palm cultivars Khlas and Ruzaiz were fertilized with different levels of FeEDDHA (trunk injected and through soil) and $FeSO_4 \cdot 7H_2O$ (trunk injected). Leaves were collected 100 days after iron application and analyzed for micronutrients. The injection with 100-g $FeSO_4 \cdot 7H_2O$ tree increased the Fe contents in the leaves of both cultivars and with 100-g FeEDDHN tree in the leaves of Ruzaiz cultivar (Abo-Rady et al., 1987).

Peach trees grown in soil of pH 8.1 and with poor water drainage were treated with various rates of $FeSO_4$ to overcome iron chlorosis. Two methods to reduce iron deficiency consisted of limb scoring and tree injection. Results of 2 years of experimentation demonstrated that the treatments significantly regreened the trees, enhanced bloom, regenerated vegetative growth, increased fruit set, and improved the yield. Fruit quality (i.e., size, color, and total soluble solids) was also significantly improved. Leaf analysis showed treated trees had significantly more Ca but significantly less P and K than the control (Yoshikawa, 1988).

Fernández-Escobar et al. (1993) injected tress of chlorotic "Manzanillo" olive trees and "Maycrest" peach with Fe solutions. All treatments increased chlorophyll content compared with that of the control. Ferrous sulfate was the most effective Fe compound in alleviating chlorosis; its effect lasted for two seasons in peach and for at least three seasons in olive. Also, ferrous sulfate increased vegetative growth and affected cropping the year following injections. Ferrous sulfate at 0.5%–1% was recommended to reduce the risk of foliar burning. The injection method effectively introduced Fe compounds into olive and peach trees.

Cui et al. (2005) stated that high-pressure trunk injection of iron fertilizer was effective to correct chlorosis, but the mechanism involved was unknown. Regreening in iron deficiency chlorosis in apple trees was slower, but the effect of correcting iron deficiency chlorosis lasted longer when iron fertilizer was injected than when it was root supplied.

Tian et al. (2007) studied the effects of infusion fertilizer on the fruit quality of citrus. He applied the infusion fertilizer with different concentrations in the branches and near the root to investigate their effects on citrus quality. He found that applying infusion fertilizer promotes effects on the growth of plants and the ratio of sugar to acid was enlarged and vitamin C content was increased. He also found that the effect of applying 6% infusion fertilizer in the branches was the best.

Shaaban (2009) studied trunk fertilization through injections for mango and grapevine trees. He stated that dicotyledonous vascular trees can be fully fertilized by injection through xylem. Trunk injection plants compared with soil-fertilized plants presented 20%–25% higher growth in mango (*Mangifera indica* var. "Sukkary white") trees and 32%–49% higher fruit yield in grapevine (*Vitis vinifera* vars. "White Riesling" and "Spätburgunder"). In the same study, fruit quality of grapevine clusters was assessed (juice °Brix, pH, reduced sugars, total acidity, grape vinegar, apple vinegar, ethanol, and glycerin content); the plants fertilized through injection were better than those fertilized through soil. Grapevine fresh juice content of the reduced sugars and ethanol increased by 7.5%–11.9% and 41.4%–50%, respectively, while the total acidity decreased by 6.2%–19.7%.

Abdi and Hedayat (2010) evaluated the effects of different potassium fertilization method on yield and chemical, physical characteristics of date palm var. "Kabkab." Treatments were control treatment (no fertilization), soil surface application of 3 kg/palm as potassium sulfate (48% K_2O), foliar spray of potassium sulfate (2%), and potassium sulfate injection into the trunk of tree (2%). Higher and lower yield were obtained from trunk injection method and control, respectively. The greater amount of flash weight, fruit weight, and fruit size was resulted from injection method.

Injections of iron into the trunk help Mohebi et al. (2010) to resolve the problem of absorption and transmission of Fe in date palm "Sayer." To study the effects of iron fertilization on yield and yield components, different treatments were assessed: control treatment, soil surface application of Fe in two levels, application of Fe as localized placement method in two levels, and Fe injection into the trunk of tree in four doses. Results showed that, in most cases, injection of 25-g $FeSO_4$/tree into the trunk showed the highest yield.

Shaaban (2012) found that the yield of guava and kaki trees fully fertilized through trunk increased by 84% and 89%, respectively, compared with soil-fertilized trees. He attributed the increase in yield achieved by trunk nutrition, to the optimal phosphorus and micronutrient supplement.

Tian et al. (2015) investigated the effects of different compound fertilizers on the photosynthesis characteristics and the growth and fruiting of chestnut by trunk injection method with $(NH_2)_2CO$ and KH_2PO_4 being mixed at the different ratios. Results showed that compound fertilizer of $(NH_2)_2CO$ and KH_2PO_4 induced positive synergistic effects to enhance photosynthetic capacity, yield, and quality of chestnut. The content of chlorophyll was decreased by $(NH_2)_2CO$ and increased by KH_2PO_4, but increased by the compound fertilizer. The 0.3% $(NH_2)_2CO + 0.3\%$ KH_2PO_4 treatment significantly increased the photosynthetic rate, the maximum net photosynthesis, apparent quantum yield, carboxylation efficiency, instantaneous water use efficiency, and nitrogen use efficiency.

Prieto Martinez et al. (2015) supplied zinc salt tablets to coffee trees. In Brazil, there are two usual forms of supplying zinc (Zn) to coffee plants: in the soil with macronutrient fertilizers or by foliar sprayings. Both have limitations that compromise the Zn availability to the plant. This study examined the plant responses to the foliar sprays with $ZnSO_4$, and insertion of Zn salt tablets, made with a mix of Zn sources, into the orthotropic branches of *Coffea arabica* L. Zinc supplied by foliar sprays or tablets inserted into the trunk both increased leaf Zn content, although the tablets provided more consistent Zn supply. Coffee production was higher with Zn treatments, regardless of the method of supply. The supply of Zn by the insertion of tablets into coffee tree trunks thus proved to be a promising way to supply Zn.

Alayón Luaces et al. (2015) evaluated the effect of nutritional trunk injections (NI) of N, P_2O_5, K_2O, CaO, and MgO (Arbolesanos) on quantitative and qualitative productivity. Experimental trials were conducted in an orchard of orange "Valencia Late" in Corrientes, Argentina. NI were applied to each side of the trunk at 40 cm from the soil (Figs. 34.5–34.7).

FIG. 34.5   Drill-based injection.

FIG. 34.6   Nutritional injection applied to one side of the trunk.

FIG. 34.7 Nutritional injection applied to each side of the trunk.

The treatments consisted in T1, 1.5 kg/plant of 15-6-15-6 fertilization to the soil (half dose, HF); T2, HF treatment plus NI spring; T3, HF plus NI spring and summer; T4, HF plus NI spring, summer, and autumn; and T5, 3 kg/plant of 15-6-15-6 fertilization to the soil (Table 34.1).

Leaf concentrations of N, P, K, Ca, Mg, Fe, and Zn were determined in leaves of fruiting branches obtained in autumn (Table 34.2).

At harvest, total fruit production was measured (YIELD) (Fig. 34.8). Diameter (DIAM), percentage of juice (JUI), soluble solid contents (SST), acidity (ACI), and ratio (RAT) were determined on a sample of 40 fruits per plot (Table 34.3).

However, none of the treatments significantly affected the fruit quality (Table 34.3).

The principal component analysis allows to relate random variables between them and associate it with a particular factor (in this case are the treatments). In this analysis, three groups were defined: the first consisting of T1, JUI, DIAM, and Ca; a second conformed by T2, T5, SST, Fe, ACI, and Mg; and the third composed by T3, T4, N, P, K, ZN, Yield, and RAT (Fig. 34.9).

TABLE 34.1 Description of the fertilization treatments applied to orange trees "Valencia Late," in Corrientes, Argentina.

		Trunk injections			
Treatments	Soil fertilization (kg/plant)	Spring	Summer	Autumn	TVNS (mL)
T1	1.5	No	No	No	0
T2	1.5	Yes	No	No	250
T3	1.5	Yes	Yes	No	500
T4	1.5	Yes	Yes	Yes	750
T5	3	No	No	No	0

TVNS = total volume of the nutrient solution.
Reproduced from Alayón Luaces, P., Yfran Elvira, M.M., Chabbal Monzón, M.D., Mazza Jeandet, S.M., Rodríguez Da Silva Ramos, V.A., Martínez Bearzzotti, C.G., 2015. Efecto de inyecciones nutritivas al tronco en la productividad de naranja valencia. Cultivos Trop. 36(2), 142–147.

TABLE 34.2 Nutrient foliar concentrations found as a result of fertilization treatments to the soil and nutritional trunk injections in orange trees "Valencia Late," in Corrientes, Argentina.

Treatments	N (%)	P (%)	K (%)	Ca (%)	Mg (%)	Fe (ppm)	Zn (ppm)
Optimal values according to Quaggio et al. (2005)	2.3–2.7	0.12–0.15	1.0–1.5	3.5–4.5	0.3–0.4	50–100	35–50
T1	2.12	0.12	0.90	3.10	0.15	92.13	30.71
T2	2.22	0.13	0.95	3.05	0.15	110.92	33.88
T3	2.27	0.13	1.08	3.07	0.14	89.57	36.35
T4	2.22	0.13	0.99	2.94	0.14	93.42	36.13
T5	2.24	0.13	0.89	2.89	0.15	103.14	31.04

T1, 1.5 kg/plant of 15-6-15-6 fertilization to the soil (half dose, HF); T2, HF treatment plus NI spring; T3, HF plus NI spring and summer; T4, HF plus NI spring, summer, and autumn; and T5, 3 kg/plant of 15-6-15-6 fertilization to the soil.
Modified from Alayón Luaces, P., Yfran Elvira, M.M., Chabbal Monzón, M.D., Mazza Jeandet, S.M., Rodríguez Da Silva Ramos, V.A., Martínez Bearzzotti, C.G., 2015. Efecto de inyecciones nutritivas al tronco en la productividad de naranja valencia. Cultivos Trop. 36(2), 142–147.

FIG. 34.8 Box plot of fruit production as a result of fertilization to the soil and of nutritional trunk injection treatments in orange trees "Valencia Late," in Corrientes, Argentina. *Modified from Alayón Luaces, P., Yfran Elvira, M.M., Chabbal Monzón, M.D., Mazza Jeandet, S.M., Rodríguez Da Silva Ramos, V.A., Martínez Bearzzotti, C.G., 2015. Efecto de inyecciones nutritivas al tronco en la productividad de naranja valencia. Cultivos Trop. 36(2), 142–147.*

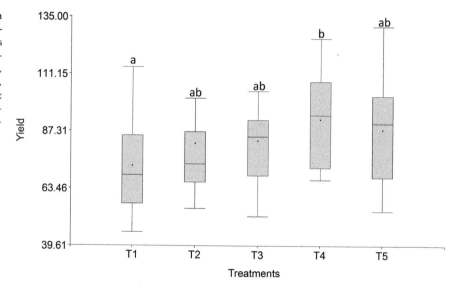

TABLE 34.3 External and internal fruit quality parameters as a result of fertilization treatments to the soil and of nutritional trunk injections in orange trees "Valencia Late," in Corrientes, Argentina.

Treatments	Yield (kg plant)	Diameter (mm)	Juice (%)	SST °Brix	Acidity (%)	SST/acidity
T1	72.90	62.45	52.99	11.91	1.46	8.21
T2	82.23	60.36	50.63	13.10	1.54	8.48
T3	83.30	60.97	51.93	12.59	1.46	8.68
T4	92.25	61.52	51.25	12.55	1.54	8.23
T5	87.88	60.72	50.08	13.21	1.53	8.71

T1, 1.5 kg/plant of 15-6-15-6 fertilization to the soil (half dose, HF); T2, HF treatment plus NI spring; T3, HF plus NI spring and summer; T4, HF plus NI spring, summer, and autumn; and T5, 3 kg/plant of 15-6-15-6 fertilization to the soil.
Modified from Alayón Luaces, P., Yfran Elvira, M.M., Chabbal Monzón, M.D., Mazza Jeandet, S.M., Rodríguez Da Silva Ramos, V.A., Martínez Bearzzotti, C.G., 2015. Efecto de inyecciones nutritivas al tronco en la productividad de naranja valencia. Cultivos Trop. 36(2), 142–147.

The application of trunk injections as a supplement to the soil fertilization increased the Valencia fruit production. This work shows the utility of NI as a complement to conventional fertilization in commercial fields, considering that the use of NI could increase fruit production without modifying the quality parameters. In addition, the decrease in the amount of fertilizers applied to the soil with the NI as a supplement could benefit the environment.

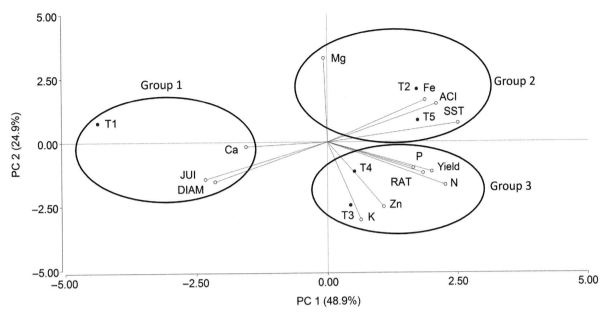

FIG. 34.9 Principal component biplot analysis as a result of fertilization treatments to the soil and of nutritional trunk injections in orange trees "Valencia Late," in Corrientes, Argentina. *Modified from Alayón Luaces, P., Yfran Elvira, M.M., Chabbal Monzón, M.D., Mazza Jeandet, S.M., Rodríguez Da Silva Ramos, V.A., Martínez Bearzzotti, C.G., 2015. Efecto de inyecciones nutritivas al tronco en la productividad de naranja valencia. Cultivos Trop. 36(2), 142–147.*

5 Concluding remarks

Trunk nutrition method regulates the mineral fertilization of the fruit tree, which is quite fast, highly efficient, and economical.

It is a helpful technique used as a complement to conventional fertilization in commercial fields. The application of these techniques includes the correction of deficiencies, particularly supplementation for special situations and research applications, among others. In addition, the decrease in the amount of fertilizers applied to the soil due to the nutritional supplement of the trunk benefits the environment. Future lines of research should be related to the nutritional solutions availability, solubility, and reaction in the vascular system of different fruit crops.

References

Abdi, G.H., Hedayat, M., 2010. Yield and fruit physiochemical characteristics of 'Kabkab' date palm as affected by methods of potassium fertilization. Advances in environmental. Biology 4 (3), 437–443.

Abo-Rady, M.D.K., Ahmed, H.S., Ghanem, M., 1987. Response of date palm to iron fertilization by trunk injection and soil application. Z. Pflanzenernähr. Bodenkd. 150 (4), 197–200.

Acimovic, S.G., 2014. Disease Management in Apples Using Trunk Injection Delivery of Plant Protective Compounds. Thesis plant pathology—Doctor of Philosophy, Michigan State University.

Alayón Luaces, P., Yfran Elvira, M.M., Chabbal Monzón, M.D., Mazza Jeandet, S.M., Rodríguez Da Silva Ramos, V.A., Martínez Bearzzotti, C.G., 2015. Efecto de inyecciones nutritivas al tronco en la productividad de naranja valencia. Cultivos Trop. 36 (2), 142–147.

Backhaus, R.A., Hield, H., Sachs, R.M., 1976. Tree growth inhibition by bark applications of morphactins. HortScience 11, 578–580.

Barney, D., Walser, R.H., Nelson, S.D., Williams, C.F., Jolley, V.D., 1984. Control of iron chlorosis in apple trees with injections of ferrous sulfate and ferric citrate and with soil-applied iron-sul. J. Plant Nutr. 7 (1–5), 313–317.

Brady, N.C., Weil, R.R., 2002. The Nature and Properties of Soils, thirteenth ed. Prentice Hall, Upper Saddle River, NJ. 960 p.

Brown, C.B., Hildreth, A.C., 1960. Injections of reduced iron for control of lime induced chlorosis of trees. Arborist's News, 71–72.

Chandler, W.H., Hoagland, D.R., Hibbard, P.L., 1933. Little leaf or rosette of fruit trees. III. Proc. Am. Soc. Hortic. Sci. 30, 70–86.

Coder, K., 1999. Water Movement in Trees. Daniel B. Warnell School of Forest Resources, University of Georgia. Extension Publication. FOR99-007. 4 p.

Collison, R.C., Harlan, J.D., Sweeney, M.P., 1932. Direct tree injection in the study of tree nutrition problems. N.Y. State Agric. Exp. Sta. Tech. Bull. 192, 3–36.

Cui, M.X., Xue, J.J., Wang, X.R., Tai, S.Z., Zhang, F.S., Li, S.H., 2005. Application of iron fertilizer by use high-pressure trunk-injection to remedy the iron deficiency chlorosis in apple trees and its mechanism. J. Plant Nutr. Fertil. Sci. 1.

Darvis, A., Nurrochmat, D.R., Massijaya, Y., Nugroho, N., Alamsyah, E.M., Bahtiar, E.T., Safe'i, R., 2013. Vascular bundle distribution effect on density and mechanical properties of oil pal trunk. Asian J. Plant Sci. 12, 208–213.

Dixon, H.H., Joly, J., 1894. On the ascent of sap. Philos. Trans. R. Soc. Lond. Ser. B 186, 563–576.

Doccola, J.J., Wild, P.M., 2012. Tree injection as an alternative method of insecticide application. In: Soloneski, S., Larramendy, M. (Eds.), Insecticides—Basic and Other Applications. InTech, p. 268. (Chapter 4). Available from: http://www.intechopen.com/books/insecticides-basic-and-other-applications/tree-injection-asan-alternative-method-of-insecticide-application.

Doccola, J.J., Smitley, D.R., Davis, T.W., Aiken, J.J., Wild, P.M., 2011. Tree wound responses following systemic insecticide trunk injection treatments in green ash (Fraxinus pennsylvanica marsh.) as determined by destructive autopsy. Arboricult. Urban For. 37 (1), 6–12.

Dong, S., Neilsen, D., Neilsen, G.H., Fuchigami, L.H., 2005. Foliar N application reduces soil NO-3-N leaching loss in apple orchards. Plant Soil 268, 357–366.

Duggan, J.B., 1943. A promising attempt to cure chlorosis, due to manganese deficiency, in a commercial cherry orchard. J.Pomol. 20, 69–79.

Evert, R.F., 2006. Esau's Plant Anatomy, Meristems, Cells, and Tissues of the Plant Body: Their Structure, Function, and Development, third ed. John Wiley & Sons, Inc., New Jersey (601 p).

Fageria, N.K., Barbosa Filho, M.P., Moreira, A., Guimarães, C.M., 2009. Foliar fertilization of crop plants. J. Plant Nutr. 32 (6), 1044–1064.

Fathi, L., Frühwald, A., 2014. The role of vascular bundles on the mechanical properties of coconut palms wood. Wood Mater. Sci. Eng. 9 (4), 214–223.

Fathi, L., Frühwald, A., Koch, G., 2014. Distribution of lignin in vascular bundles of coconut wood (Cocos nucifera) by cellular UV-spectroscopy and relationship between lignification and tensile strength in single vascular bundles. Holzforschung 68, 915–925.

Fernández-Escobar, R., Barranco, D., Benlloch, M., 1993. Overcoming iron chlorosis in olive and peach trees using a low-pressure trunk-injection method. HortSci. 28 (3), 192–194.

Filer Jr., T.H., 1973. Pressure apparatus for injecting chemicals into trees. Plant Dis. Rep. 57, 338–340.

Helburg, L.B., Schomaker, M.E., Morrow, R.A., 1973. A new trunk injection technique for systemic chemicals. Plant Dis. Rep. 57, 513–514.

IAWA, 1964. Multilingual Glossary of Terms Used in Wood Anatomy. (46 p.). Available from:http://www.iawa-website.org/uploads/soft/Abstracts/IAWA_glossary.pdf.

IAWA Committee, 1989. IAWA list of microscopic features for hardwood identification. IAWA Bull. n.s. 10, pp. 219–332.

Kaminski, J., Brunetto, G., Moterle, D., Rheiheimer, D.S., 2007. Depleção de formas de potássio do solo afetada por cultivos sucessivos. Revista Brasileira de Ciência do Solo, Viçosa 31, 1003–1010.

Koehler, C.S., Rosenthal, S.S., 1967. Bark vs. foliage applications of insecticides for control of Psylla uncatoides on acacia. J. Econ. Entomol. 60, 1554–1558.

Kozlowski, T.T., 1961. The movement of water in trees. For. Sci. 7 (2), 177–192.

Levy, B.F.G., 1946. Tree Injection III—Re-Invigoration of Debilitated Trees. E. Mailing Res. Sta. Ann. Rpt, pp. 107–118.

Lipman, C.B., Gordon, A., 1925. Further studies on new methods in the physiology and pathology of plants. J. Gen. Physiol. 7, 615–623.

May, C., 1941. Methods of tree injection. Trees 4, 7–16.

Mengel, K., Kirkby, E.A., 2001. Principles of Plant Nutrition, fifth ed. Kluwer, Dordrecht.

Miller, G.W., Awada, S., Salam, A., 2001. Dissolving Polymer Plug for Introducing Nutrients and Medicinal Materials into Tree Trunks. US Patent No. 6216388B1.

Mohebi, A.H., Nabhani, L., Dialami, H., 2010. Yield and yield components of 'Sayer' date palm as affected by levels and methods of iron fertilization. In: IV International Date Palm Conference 882, pp. 131–136.

Moraes, J.F.V., Dynia, J.F., 1992. Alterações nas características químicas e físico-químicas de um solo Gley Pouco Húmico sob inundação e após a drenagem. Pesq. Agrop. Brasileira 27, 223–235.

Morris, R.F., 1951. Tree injection experiments in the study of birch dieback. For. Chron. 27 (4), 313–329.

Myburg, A.A., Lev-Yadun, S., Sederoff, R.R., 2013. Xylem Structure and Function. In eLS, John Wiley & Sons, Ltd. (Ed.).

Navarro, C., Fernández-Escobar, R., Benlloch, M., 1992. A low-pressure, trunk-injection method for introducing chemical formulations into olive trees. J. Am. Soc. Hortic. Sci. 117 (2), 357–360.

Neely, D., 1973. Pin oak chlorosis, trunk implants correct iron deficiency. J. For. 71 (6), 340–342.

Neely, D., 1976. Iron deficiency chlorosis of shade trees. J. Aboricul. 2 (7), 128–130.

Nyland, G., Moller, W.J., 1973. Control of pear decline with a tetracycline. Plant Dis. Rep. 57, 634–637.

Orr, J.W., Leonard, S., Lentz, J., 1988. Field observations of tree injection. J. Arboric. 14, 269–273.

Pérez Olvera, C.P., Aguirre, M.M., Ceja Romero, J., Pacheco, L., 2008. Wood anatomy of five specie of the Rosaceae family. Madera y Bosques 14 (1), 81–105.

Perry, T.O., Santamour Jr., F.S., Stipes, R.J., Shear, T., Shigo, A.L., 1991. Exploring alternatives to tree injection. J. Arboric. 17, 217–226.

Prieto Martinez, H.E., Poltronieri, Y., Cecon, P.R., 2015. Supplying zinc salt tablets increased zinc concentration and yield of coffee trees. J. Plant Nutr. 38, 1073–1082.

Quaggio, J.A., Mattos Jr., D., Cantarella, H., 2005. In: Citros, D., Mattos Jr., J.D., De Negri, R.M., Pio, J., Pompeu Jr., (Eds.), Manejo da fertilidade do solo na citricultura. In: Campinas: Instituto Agronomico (Chapter 17).

Raese, J.T., Parish, C.L., Staiff, D.C., 1986. Nutrition of apple and pear trees with foliar sprays, trunk injections or soil applications of iron compounds. J. Plant Nutr. 9 (3–7), 987–999.

Ramle, S.F.M., Sulaiman, O., Hashim, R., Arai, T., Kosugi, A., Abe, H., Murata, Y., Mori, Y., 2012. Characterization of parenchyma and vascular bundle of oil palm trunk as function of storage time. Lignocellulose 1, 33–44.

Reil, W.O., Beutel, J.A., 1976. A pressure machine for injecting trees. Calif. Agric. 30 (12), 4–5.

Reil, W., Beutel, J., Hemstreet, C., Seyman, W., 1978. Trunk injection corrects iron and zinc deficiency in pear trees. Calif. Agric. 32 (10), 22–24.

Roach, W.A., 1933. Tree Injection—A Progress Report. E. Mailing Res. Sta. Ann. Rpt. 6, pp. 137–141.

Roach, W.A., 1934. Injection for the diagnosis and cure of physiological disorders of fruit trees. Ann. Appl. Biol. 21, 333–343.

Roach, W.A., 1939. Plant injection as a physiological method. Ann. Bot. 3 (9), 155–227.

Roach, W.A., Roberts, W.O., 1945. Further work on plant injection for diagnostic and curative purposes. J. Pomol. 21, 108–119.

Robbins, S.H., 1982. Pressure trunk injections of potassium as a possible short term corrective measure for potassium deficiency in sweet cherry (Prunus avium L.) and prune (Prunus domestica L.). Masters thesis, pp. 103. Avaliable from: https://ir.library.oregonstate.edu/concern/graduate_thesis_or_dissertations/12579w35p.

Roberts, W.O., Landau, N., 1947. Multiple mineral deficiencies in fruit trees: injections as a first aid treatment. J. Pomol. 23, 80–91.

Rumbold, C., 1920. Effect on chestnuts of substances injected into their trunks. Am. J. Bot. 7 (2), 45–57.

Sachs, R.M., Nyland, G., Hackett, W.P., Coffelt, J., Debie, J., Giannini, G., 1977. Pressurized injection of aqueous solutions into tree trunks. Sci. Hortic. 6, 297–310.

Sánchez Zamora, M.A., Fernández, R., 2000. Escobar injector-size and the time of application affects uptake of tree trunk-injected solutions. Sci. Hortic. 84, 163–177.

Sánchez-Zamora, M.A., Fernández-Escobar, R., 2004. Uptake and distribution of trunk injections in conifers. J. Arboric. 30 (2), 73–79.

Schmitt, D.E., Comin, J.J., Gatiboni, L.C., Tiecher, T., Lorensini, F., Melo, G.W.B., Girotto, E., Guardini, R., Heinzen, J., Brunetto, G., 2013. Phosphorus fractions in sandy soils of vineyards in southern Brazil. Revista Brasileira de Ciência do Solo 37, 472–481.

Schreiber, L.R., 1969. A method for the injection of chemicals into trees. Plant Dis. Rep. 53, 764–765.

Shaaban, M.M., 2009. Injection fertilization: a full nutritional technique for fruit trees saves 90-95% of fertilizers and maintains a clean environment. Fruit Veg. Cereal Sci. Biotechnol. 3 (1), 22–27.

Shaaban, M.M., 2012. Trunk nutrition. In: Advances in Citrus Nutrition. Springer, Dordrecht (Chapter 8).

Shirgure, P.S., Srivastava, A.K., 2014. Fertigation in perennial fruit crops: major concerns. Agrotechnol 3(1).

Smolen, S., 2012. Foliar nutrition: current state of knowledge and opportunities. In: Srivastava, A.K. (Ed.), Advances in Citrus Nutrition. In: 4, Springer Science, pp. 41–58.

Southwick, R.W., 1945. Pressure injection of iron sulfate into citrus trees. Proc. Am. Soc. Hortic. Sci. 46, 27–31.

Srivastava, A.K., 2013a. Nutrient deficiency symptomology in citrus: an effective diagnostic tool or just an aid for postmortem analysis. Agric. Adv. 2, 177–194.

Srivastava, A.K., 2013b. Recent developments in diagnosis and management of nutrient constraints in acid lime. J. Agric. Sci. 2, 86–96.

Srivastava, A.K., Malhotra, S.K., 2017. Nutrient use efficiency in perennial fruit crops—a review. J. Plant Nutr. https://doi.org/10.1080/01904167.2016.1249798.

Srivastava, A.K., Singh, S., 2003. Foliar fertilization in citrus—a review. Agric. Rev. 24 (4), 250–264.

Srivastava, A.K., Singh, S., 2008a. DRIS norms and their field validation in Nagpur mandarin (*Citrus reticulata* Blanco). J. Plant Nutr. 31, 1091–1107.

Srivastava, A.K., Singh, S., 2008b. Analysis of citrus orchard efficiency in relation to soil properties. J. Plant Nutr. 30, 2077–2090.

Sterrett, J.P., Creager, R.A., 1977. A miniature pressure injector for deciduous woody seedlings and branches. HortScience 12 (2), 156–158.

Tanaka, Y., 1960. Effects of phosphorus fertilization on Citrus unshiu Marc. I. a study on translocation and distribution of phosphorus applied to the soil or injected into the trunk of citrus trees using P32 as a tracer. J. Hortic. Assoc. Jpn 29, 63–69.

Thomas, E.E., Haas, A.R.C., 1928. Injection method as a means of improving chlorotic orange trees. Bot. Gaz. 86, 355–362.

Tian, J., Zhabg, C., Xiong, Y., Kong, C., 2007. Effects of infusion fertilizer on the fruit quality of citrus. J. Anhui Agric. Sci. 34, 98.

Tian, S.L., Sun, X.L., Shen, G.N., Xu, L., 2015. Effects of compound fertilizer of $(NH_2)2CO$ and KH_2PO_4 on the chestnut photosynthesis characteristics, growth and fruiting. Ying yong sheng tai xue bao. J. Appl. Ecol. 26 (3), 747–754.

Tyree, M.T., 1997. The cohesion-tension theory of sap ascent: current controversies. J. Exp. Bot. 48, 1753–1765.

Tyree, M.T., Zimmerman, M.H., 2002. Xylem Structure and the Ascent of Sap. Springer Verlag, Berlin.

VanWoerkom, A.H., 2012. Trunk Injection: A New and Innovative Technique for Pesticide Delivery in Tree Fruits. Thesis to for the degree of Master of Science. Entomology, Michigan State University.

Wallace, T., 1935. Investigations on chlorosis of fruit trees. V. The control of lime-induced chlorosis by injection of iron salts. J.Pomol. 13, 54–67.

Wallace, G.A., Wallace, A., 1986. Correction of iron deficiency in trees by injection with ferric ammonium citrate solutions. J. Plant Nutr. 9, 981–986.

Wei, C.F., Steudle, E., Tyree, M.T., 1999. Water ascent in plants: do ongoing controversies have a sound basis? Trends Plant Sci. 4, 372–375.

White, P.J., Brown, P.H., 2010. Plant nutrition for sustainable development and global health. Ann. Bot. 105 (7), 1073–1080.

Windt, C.W., Vergeldt, F.J., De Jager, P.A., Van As, H., 2006. MRI of long-distance water transport: a comparison of the phloem and xylem flow characteristics and dynamics in poplar, castor bean, tomato and tobacco. Plant Cell Environ. 29, 1715–1729.

Worley, R.E., Littrell, R.H., Polles, S.G., 1976. Pressure trunk injection promising for pecan and other trees. Hortic. Sci. 11 (6), 590–591.

Yoshikawa, F.T., 1988. Correcting iron deficiency of peach trees. J. Plant Nutr. 11 (6–11), 1387–1396.

Zimmermann, U., Haase, A., Langbein, D., Meinzer, F., 1993. Mechanisms of long-distance water transport in plants: a re-examination of some paradigms in the light of new evidence. Philos. Trans. Biol. Sci. 341 (1295), 19–31.

CHAPTER

35

Importance of nanofertilizers in fruit nutrition

Dámaris Leopoldina Ojeda-Barrios[a], Isidro Morales[b], Antonio Juárez-Maldonado[c], Alberto Sandoval-Rangel[d], Laura Olivia Fuentes-Lara[e], Adalberto Benavides-Mendoza[d],*

[a]Autonomous University of Chihuahua, Laboratory of Plant Physiology, Chihuahua, Mexico
[b]National Polytechnic Institute, CIIDIR-Oaxaca, Oaxaca, Mexico
[c]Autonomous Agricultural University Antonio Narro, Department of Botany, Saltillo, Mexico
[d]Autonomous Agricultural University Antonio Narro, Department of Horticulture, Saltillo, Mexico
[e]Autonomous Agricultural University Antonio Narro, Department of Animal Nutrition, Saltillo, Mexico
*Corresponding author. E-mail: abenmen@gmail.com

1 Fruit production and fertilizers

Fertilization in crops is directly related to the replenishment of nutrients that a plant requires to achieve its highest biological expression. Mineral nutrition is the process that allows plants to absorb, process, and use the essential elements for optimal development and fruiting. The chemical elements considered essential for plants are 17, divided into macro- and micronutrients. The concentrations of these elements vary for each plant species and growth medium (Benton Jones, 2012). Fertilization constitutes one of the main factors that limit the production of the crops since plants absorb only a fraction (about 10%–60%) of the applied fertilizer (Peña Cabriales et al., 2002).

Based on the volume of production, tropical fruits of greater importance are mango, papaya, pineapple, and avocado. On the other hand, based on the volume of production, some important deciduous fruit crops are apple, grape, pears, peaches and nectarines, and nuts (FAO, 2017) (Tables 35.1 and 35.2).

The fertilizers most used in fruit crops are those that provide N, P, K, Ca, Mg, and S (Tables 35.3 and 35.4). Furthermore, depending on the species, there are nutrients for specific needs and problems. For example, Ca has a significant influence on the quality of the apple fruit, Fe on the productive capacity of the peach, and Zn on the quality of the pecan tree (Parra Quezada et al., 2011; Tarango Rivero et al., 2009).

Regarding the development of nanofertilizer technology, the strategy to enter the market should be directed toward the development of high-efficiency products for the major elements N, P, K, Ca, Mg, and S, which are applied in higher

A.K. Srivastava, Chengxiao Hu (eds.)
Fruit Crops: Diagnosis and Management of Nutrient Constraints
https://doi.org/10.1016/B978-0-12-818732-6.00035-6

TABLE 35.1 Worldwide production of the most important tropical fruits.

Crop	Worldwide production (tons)
Mango	46,508,697
Papaya	25,809,038
Pineapple	13,050,749
Avocado	5,567,044

Data from FAO, 2017. FAOSTAT: Production Crops [WWW Document]. http://www.fao.org/faostat/en/#data/QC.

TABLE 35.2 Worldwide production some important deciduous fruit trees (FAO, 2017).

Crop	Worldwide production (tons)
Apple	89,329,179
Grape	77,438,929
Pears	27,345,930
Peaches and nectarines	24,975,649
Nuts	936,783

TABLE 35.3 Fertilizer sources for tropical fruit species.

Crop	Most common sources of fertilizers
Mango	Calcium nitrate, ammonium sulfate, calcium superphosphate, potassium chloride, magnesium sulfate, and magnesium nitrate
Papaya	Urea, ammonium nitrate, phosphoric acid, triple 17, potassium chloride, potassium nitrate, calcium single superphosphate, magnesium sulfate, and magnesium nitrate
Pineapple	Ammonium sulfate, urea, calcium nitrate, potassium nitrate, phosphoric acid, potassium chloride, potassium sulfate, and magnesium sulfate
Avocado	Ammonium nitrate, ammonium sulfate, urea, diammonium phosphate, triple 17, potassium chloride, potassium sulfate, calcium single superphosphate, calcium triple superphosphate, magnesium sulfate, and magnesium nitrate

TABLE 35.4 Fertilizer sources for deciduous fruit trees of temperate climate.

Crop	Most common sources of fertilizers
Apple	Ammonium nitrate, ammonium sulfate, urea, diammonium phosphate, triple 17, potassium chloride, potassium sulfate, calcium single superphosphate, calcium triple superphosphate, magnesium sulfate, and magnesium nitrate
Pecan	Urea, ammonium nitrate, phosphoric acid, triple 17, potassium chloride, potassium nitrate, calcium single superphosphate, magnesium sulfate, magnesium nitrate, zinc nitrate, and zinc sulfate
Peach	Ammonium sulfate, urea, diammonium phosphate, triple 17, phosphoric acid, calcium nitrate, potassium nitrate, potassium chloride, potassium sulfate, magnesium sulfate, and chelated iron
Grape vine	Ammonium nitrate, ammonium sulfate, urea, diammonium phosphate, triple 17, potassium chloride, potassium sulfate, calcium single superphosphate, calcium triple superphosphate, magnesium sulfate, magnesium nitrate

volume and that are currently less efficient and more environmentally impactful. If it is possible to apply less kg per hectare and fewer applications per season with the use of nanofertilizers, the cost of fertilization per unit of fruit yield will be significantly reduced. More profits make the use of nanomaterials attractive.

In the case of microelements, especially Fe, Zn, B, and Cu, the world market has a wide range of highly effective specialty products. For this reason, the strategy to introduce nanofertilizers of these elements should consider the cost that the agricultural producer will pay. Additional advantages of the nanoelements are induced stress tolerance and increase in the nutraceutical quality of the fruit (Juárez-Maldonado et al., 2018).

Climate change is imposing significant challenges on production systems in both tropical and temperate regions. Compared with annual species, perennial species show an increased risk of exposure to adverse biotic and abiotic factors. Because of the greater unpredictability in the temperature, atmospheric humidity, and precipitation regimes, the development patterns based on the likelihood of seasonal variation are sometimes in a situation of lag, excess, or deficiency (Ramírez-Legarreta et al., 2011). In these cases, the use of nanofertilizers of microelements such as Cu, Zn, or Fe could be useful. In addition to covering the need for the element or elements contained in its formulation, the nanofertilizers of metallic microelements could mitigate the adverse effect of environmental variability, through the induction of antioxidants and other metabolites (Juárez-Maldonado et al., 2018).

The tropical and deciduous fruit trees, as well as most of the agricultural species, are cultivated in open field production systems. The surface of species of fruit trees grown under protected production systems such as greenhouses, tunnels, shade mesh, and plastic mulching is minimal in comparison (Hernández-Hernández et al., 2018). The information presented in this chapter, therefore, refers to the advantages of nutrition with nanofertilizers in plants grown in soil under open field conditions.

2 Common sources of fertilizers for fruit production

The sources of fertilizers for crops depend on the type of soil, quality and application system of irrigation water, the characteristics of the species or plant variety, and the age of the plant.

The different types of fertilizers used are variable in their characteristics, cost, and availability in the market. Granulated fertilizers are the most economical and accessible to apply on soil and provide the primary elements N-P-K and secondary Ca-Mg-S. They are followed by fertilizers of high solubility and that are more expensive and relatively less available in the market. Its application is made with pressurized irrigation systems or by foliar spray and provides the primary elements, secondary elements, and microelements Cl, Fe, Zn, Mn, B, Cu, Mn, and Ni. Finally, the most expensive are slow-release fertilizers (SRF) or controlled-release fertilizers (CRF), chelated fertilizers, and organic fertilizers; their use is minimal compared with the first two categories (Morgan et al., 2009).

However, for fertilizers, not only the cost but also the ease of access for acquisition should be considered. Another critical point is the efficiency of the fertilizer, that is, the quotient between the amount of plant absorption and the amount applied to the soil. Different physicochemical and biological factors of the soil such as pH, ORP, composition of the soil parent material, extremes of temperature and humidity, organic matter and inorganic soil colloids, cation exchange capacity, porosity or compaction, and the edaphic microbiome make this efficiency very low, approximately 50% for N, 10% or less for P, and around 40% for K (Baligar et al., 2001).

The different edaphic factors mentioned transform the elements contained in the fertilizers into nonbioavailable chemical forms or induce leaching or volatilization. These last two processes, especially for the N and P elements, constitute an important source of water pollution for the subsoil and the atmosphere (Foley et al., 2011).

The less efficient fertilizers are granulated, followed by those with high solubility. The low efficiency of nutrient supply of granular fertilizers causes the need to apply large quantities, spread over different moments to mitigate losses due to fixation, leaching, and volatilization. Besides the increase in the economic costs due to greater use of machinery and labor, the excess of fertilizers and excessive tillage cause soil degradation, mainly due to the loss of organic matter (Baligar et al., 2001).

High solubility fertilizers are more efficient than granulates due to the use of high-technology and high-pressure irrigation systems or the use of foliar spray (Baligar et al., 2001). Pressurized irrigation techniques allow constant application of small amounts of fertilizers according to the needs of the plants, thus reducing the exposure to the aforementioned edaphic factors that make the elements nonbioavailable. Water is used efficiently, and nutrients can be applied at the right time and in the right amount, reducing manual labor. The fertilizer is distributed evenly, and there is less risk of environmental pollution. The result is higher productivity and better fruit quality (Pinto et al., 2002).

However, there are some disadvantages for the use of pressurized irrigation, such as the high initial cost of irrigation systems, the nonuniform application of fertilizer when the system is poorly designed, and the possibility of precipitation due to incompatibility of the fertilizer formulations, causing blockage of the emitters and significant economic losses to farmers (Silva et al., 1996).

The fertilizers that currently have the highest efficiency are SRF, CRF, organic fertilizers, and chelates. The basis of the functioning of the SRF is the contribution of the nutrient element in a form not immediately soluble but released through a process of microbial decomposition or chemical hydrolysis (as with organic fertilizers). CRF are soluble fertilizers coated with or embedded in low-solubility materials such as sulfur or polymers. Both SRF and CRF provide necessary elements for plants slowly, which increases the opportunity for plants to absorb dissolved elements, decreasing fixation, leaching, and volatilization. The main disadvantages of slow release fertilizers are the cost and dependence on temperature and microbial activity for the release of nutrients (Morgan et al., 2009). This temperature dependence may limit its use in regions with low temperature and soils with a low amount of organic matter. In the case of chelated fertilizers, some elements form a complex (called a chelate) with the ligands of a natural or synthetic organic molecule. This chelate is chemically stable against the losses associated with edaphic conditions and can be absorbed by the roots or foliar surfaces of plants. The main disadvantage of chelates is the high cost and availability only for some elements, mainly microelements (Souri, 2016).

3 Use of nanofertilizers in fruit production

Except for a few studies (Table 35.5), the literature does not report works with fruit crops. In most of the agronomically oriented studies, results are presented on annual crop species. However, even in these cases, a considerable part of the studies has a focus on toxicity or biostimulation, rather than on fertilizer value. This situation occurs even in the case of NPs and NMs of essential elements such as Fe or Zn.

In many cases reported by literature, NPs and NMs were applied to plants in concentrations above the recommended levels for each element. For example, it is recommended that a nutrient solution for the production of guava seedlings contain 1.8 mg/L of Fe and 0.11 mg/L of Zn (Monteiro Corrêa et al., 2012), while in many studies it is

TABLE 35.5 Classification of nanofertilizers (NFs), examples, and comparative advantages compared with conventional fertilizers.

NFs categories	Examples and references	Comparative advantages
(1) Nanostructured materials adding the bulk ionic form of essential elements	Nano chitosan + NPK (Abdel-Aziz et al., 2016)	Superior efficiency in fertilization, increase in yield
	Nanozeolite/nanohydroxyapatite + P (Mikhak et al., 2017)	Higher bioavailability of P by decreasing P fixation in the soil
	Nano hollow core-shells of polymers with Zn^{+2} (Yuvaraj and Subramanian, 2014)	Controlled release of Zn^{2+}
(2) Nanostructured essential elements incorporated in a carrier material (which in turn may or may not be a nanomaterial)	Cu NPs encapsulated in chitosan (Juárez-Maldonado et al., 2016)	Cu NPs increased yield and fruit quality
	Cu NPs encapsulated in chitosan-PVA (Hernández-Hernández et al., 2017)	Cu NPs increased yield and fruit quality
	Nano-hydroxyapatite (Liu and Lal, 2015a)	Higher efficiency of nanohydroxyapatite versus $Ca(H_2PO_4)_2$
(3) Nanostructured essential elements for direct application in soil or leaves	Nano-NH_4NO_3 (Davarpanah et al., 2017)	Nanofertilizer applied by foliar spraying increased the quality of the fruit compared with bulk urea
	Nano-Ca (Ranjbar et al., 2018)	Higher effectiveness of nano-Ca to increase the quality and postharvest life of the fruits compared with $CaCl_2$
	Zn and B NPs (Davarpanah et al., 2016)	Foliar application of nano-Zn and nano-B increased yield and quality compared with conventional fertilizer
	Nano size calcite with Ca, Mg, and Fe (Sabir et al., 2014)	Higher quality and yield of berries in grapevines

common to find concentrations of up to 60 mg/L of NP of Fe or Zn (Juárez-Maldonado et al., 2018). Thus, the results described are difficult to extrapolate to situations encountered in the field (Morales-Díaz et al., 2017).

There is not yet a consensus on the definition of nanofertilizer (Raliya et al., 2018); both essential and beneficial elements are grouped in the same concept (El-Ramady et al., 2018). In this chapter, we have considered that a nanofertilizer is any material with at least a dimension between 1 and 100 nm and that contains and directly contributes at least one of the following 14 elements: N, P, K, Ca, Mg, S, Fe, B, Cl, Zn, Mn, Cu, Ni, and Mo. This list includes the elements considered essential that the plant takes from the soil or that can be applied as leaf or fruit spray.

Raliya et al. (2018) mention that nanofertilizers can provide the nutrients directly or indirectly. The direct contribution occurs if at least one essential element is included in the composition of the material so that the reaction with the soil will make it available in ionic form (e.g., Cu^{2+}) or as a nanoparticle (e.g., Cu or CuO). On the other hand, the indirect contribution occurs under two situations: (i) one or more components of the nanofertilizer cause a physicochemical reaction in the soil or the soil solution, resulting in a higher bioavailability of an essential element; (ii) one or more components of the nano-fertilizer modify the metabolic capacities of the plants and their microbiome, resulting in added capacity to absorb essential nutrients from the soil (Morales-Díaz et al., 2017; Raliya et al., 2018).

Indirect inputs of essential nutrients can occur even in the presence of NPs from elements considered nonessential for plants (such as TiO_2 and silica) or by carbon NMs (Raliya et al., 2018). However, this type of action corresponds more to a biostimulant or soil improver action, rather than a nanofertilizer.

Mineral nutrients in the form of nanofertilizers can be provided in three ways that could correspond to a primary classification (Fig. 35.1 and Table 35.5). The first way of use is through the bulk ionic form of elements absorbed or adsorbed in a nanostructured complex such as hydroxyapatite, zeolites, or nanochitosan (NCs) that regulates the release and availability of fertilizer.

For example, the urea-hydroxyapatite nanofertilizer has a high N content and a slow release rate (Kottegoda et al., 2017). In the case of nanochitosan, the functional groups of NCs (amino and hydroxyl) form complexes with the ions of metals such as Mg, Ca, Cu, and Fe and constitute a potential alternative for the development of agricultural chelates (Saharan and Pal, 2016).

The second form of application of the nanofertilizers used the essential elements in nanoform but incorporated in a carrier substance, which in turn may or may not be a nanomaterial. Some examples are the NPs of hydroxyapatite (source of P) stabilized in carboxymethyl cellulose (Liu and Lal, 2015a) and the NPs of Cu and Fe incorporated by absorption or adsorption in materials such as clays, zeolites, chitosan, or polymeric hydrogels (Golbashy et al., 2016; González-Gómez et al., 2017).

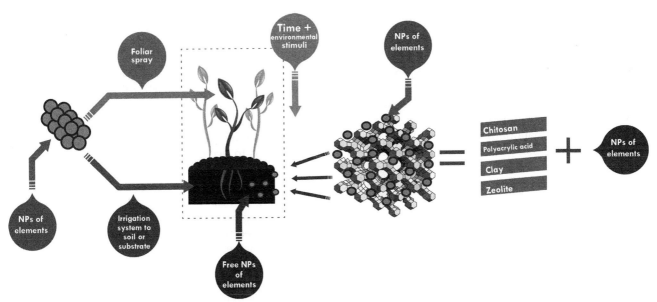

FIG. 35.1 Forms of application of nanofertilizers. The nanoparticles of essential elements that are applied directly in the soil, irrigation water, or over the leaf surfaces and the fruits in pre- or postharvest are represented on the left side. On the right side, carrier materials are schematized. These can be nanomaterials and contain essential elements in bulk or nanostructure that are released in a prolonged time depending on the chemical properties of soil or plant surface where they were applied. *From Morales-Díaz, A.B.A.B., Ortega-Ortíz, H., Juárez-Maldonado, A., Cadenas-Pliego, G., González-Morales, S., Benavides-Mendoza, A., 2017. Application of nanoelements in plant nutrition and its impact in ecosystems. Adv. Nat. Sci. Nanosci. Nanotechnol. 8, 013001.*

In the third form of use, the essential element as NPs in suspension is applied in a nanostructured form directly to the soil and substrate and by sprinkling in the leaves or fruits. Examples of this last form refer to the use of NPs of Fe, Cu, Zn, and their oxides (Fedorenko et al., 2015; López-Vargas et al., 2018; Morales-Díaz et al., 2017).

In the previous section of this chapter advantages and disadvantages were listed for the different kinds of fertilizers. In this sense, the proposals for the use of nanofertilizers should point to at least one advantage over conventional fertilizers to solve the problems of agricultural producers. There are numerous studies where NPs and NMs have been tested in crop plants. Some reviews where this information can be consulted are (Chhipa, 2017; Das and Brar, 2018; Ditta and Arshad, 2016; Ghormade et al., 2011; Guerriero and Cai, 2018; Khan et al., 2016; Khot et al., 2012; Liu and Lal, 2015b; Raliya et al., 2018; Vázquez-Núñez et al., 2018; Verma et al., 2018; Zuverza-Mena et al., 2017).

Different studies (Table 35.5) indicate some properties that represent advantages of nanofertilizers over conventional fertilizers. These properties increase the efficiency of the nanofertilizers, promoting higher yield and quality and decreasing the amount of the element applied in the form of nanofertilizer, as well as the number of applications. Some examples of increased fertilization efficiency when using nanofertilizers will be mentioned later. However, not all examples refer to fruit species, because there are few studies conducted on the use of nanofertilizers in fruit crops. Nevertheless, the physicochemical processes that are modified by nanofertilizers in soils and annual crops are equally applicable in fruit crops, both in tropical and temperate regions.

4 Lower loss from leaching and volatilization

Volatilization and leaching are processes that lead to the loss of nutrients in soils. Volatilization is crucial for nitrogen fertilizers and increases with N application rate, does not decrease with split applications, but does with urease or nitrification inhibitors, deep placement of fertilizers, and use of controlled release coatings (Pan et al., 2016). Not only does the leaching occur in fertilizers with N and K, but also it occurs with all elements contained in fertilizers when applied in light textured soils (Lal, 2015). Leaching decreases due to the recycling of nutrients by plants, the presence of inorganic and organic ion exchange complexes, and chemical speciation phenomena that change the solubility of the elements. In agricultural systems, leaching can accompany excess application of water and fertilizers; however, it is also a phenomenon that occurs in soils subject to degradation (Lal, 2015).

To mitigate volatilization and leaching losses, nanofertilizers can provide some of the following advantages: (i) decreased solubility, allowing the slow release or controlled release of the nutrient, and (ii) increased capacity to stabilize by aggregation or adsorption in soil components (Raliya et al., 2018).

To mitigate volatilization and leaching, it has been proposed to decrease the release rate of nutrients by combining them in NMs such as hydrogels, films, or other biopolymers such as chitosan (Hasaneen et al., 2014; Kashyap et al., 2015), montmorillonite-urea clay nanocomposites (Golbashy et al., 2016) or other types of ceramic materials (Choy et al., 2007), C nanotubes (Liu et al., 2009), NPs of SiO_2, and natural or synthetic zeolites (Ditta and Arshad, 2016; Le et al., 2014; Lehman and Larsen, 2014). Zeolites are low-cost and innocuous materials that positively regulate the availability of nutrients and decrease leachate losses (Ming and Allen, 2001).

Another alternative to reduce the leaching of essential nutrients such as Fe and Zn is to apply them directly to the soil in the form of nanoparticles (NPs). Due to their large surface/volume ratio (Fig. 35.2), NPs have high surface energy. In soil and soil solution, the natural tendency of the system is to decrease the surface energy, which leads to the aggregation of the NPs or their adsorption with the soil components and the soil solution.

Aggregation in the soil stabilizes the NPs and decreases their transfer to other ecosystem components by leaching or mass flow, but in turn, decreases the immediate absorption by the roots and can even induce adverse effects by excess aggregation in the roots. To decrease the aggregation process, the engineered NPs are functionalized before use, although when exposed to natural fluids the surface of the NPs is covered with a new corona of organic components that substantially modify their stability and bioavailability (Morales-Díaz et al., 2017).

On the other hand, adsorption on soil colloids stabilizes NPs against leaching while keeping them available for plants. The balance between adsorption/desorption is dynamic and depends positively on the concentration of salts in the soil solution and negatively on the amount of dissolved organic matter (Bae et al., 2013). In sandy soils, the adsorption of NPs represents an advantage as reported by Rui et al. (2016) with the highest availability of Fe when NPs of Fe_2O_3 are used, in comparison with Fe-EDTA chelate. In loam or clayey soils, greater adsorption of the NPs on the surfaces of the colloids would be expected; mainly as the concentration of salts in the soil solution increases, the opposite effect is exerted by the presence of humic substances (Bae et al., 2013) and possibly exudates from the roots.

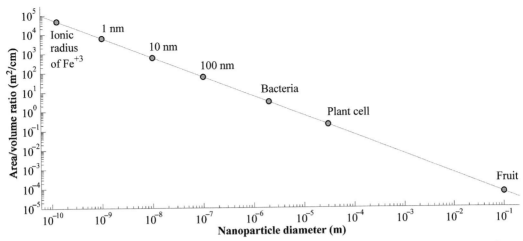

FIG. 35.2 Effect of decreasing the diameter of a geometric body (in this case, a sphere) on the area/volume ratio. Fruit, plant cell, and bacteria represent spheres equivalents in size to these objects. In nanoparticles (1, 10, and 100 nm), 1 cm³ of nanoparticles with 100 nm diameters can contain a display area of 60 m²; with the same volume of 1 cm³, the nanoparticles of 10 nm in diameter have a display area of 600 m². Both scales are logarithmic. *From Juárez-Maldonado, A., González-Morales, S., Cabrera-De la Fuente, M., Medrano-Macías, J., Benavides-Mendoza, A., 2018. Nanometals as promoters of nutraceutical quality in crop plants. In: Grumezescu, A.M., Holban, A.M. (Eds.), Impact of Nanoscience in Food Industry. Academic Press, London, pp. 277–310.*

5 Decrease in fixation, precipitation, or speciation in nonbioavailable forms

Fertilizers with N, P, S, Fe, Zn, and other trace metals are susceptible to fixation in the soil. This complexation process with inorganic and organic soil components results in nonbioavailable element forms for plants (Pilbeam, 2015).

Compared with conventional fertilizers, the nanofertilizers lower the rate of nutrient release and, on the other hand, increase their stability by aggregation or adsorption but without causing changes in chemical speciation due to pH-dependent phenomena, redox potential, and availability of ligands in the soil (Raliya et al., 2018).

Examples of the aforementioned are (i) the use of nano rock phosphate that increases the bioavailability of P for plants, since it prevents the fixation of P by Fe, Ca, or silicic acid, and (ii) the application of modified zeolites with high internal surface for ion exchange for the regulation of N, P, and K for plants (Singh et al., 2017).

6 Higher growth and yield

Nanofertilizers provide nutrients in two forms: by releasing ions from NPs or NMs or by direct absorption of NPs in plant cells (Morales-Díaz et al., 2017). The essential elements released in ionic form from the NMs in the soil solution are absorbed by the usual mechanisms that involve integral proteins that act as channels and transporters. On the other hand, the NPs applied to the soil can be absorbed by the radical cells and mobilized by the apoplastic or symplastic route, directly through preexisting pores in the cell walls or by the formation of new pores (Juárez-Maldonado et al., 2019).

It is also possible to apply nanofertilizers by foliar spray. Depending on the larger or smaller size of the NPs or NMs used, absorption can occur through stomata or transport through the cuticle, through the interaction between the epidermal surfaces and the free energy of the surface of NPs (Juárez-Maldonado et al., 2019). Once the nanofertilizers enter the plant through the stomata or the epidermal surface, they are transported to the different tissues (Eichert et al., 2008).

For both soil and foliar applications, nanofertilizers present a faster dissolution, absorption, and assimilation by the plants compared with traditional fertilizers. The aforementioned advantages have been shown for the elements N, P, K, Ca, Mg, Fe, Mn, Zn, Cu, and Mo (Ditta and Arshad, 2016) and probably explain why the application of nanofertilizers generally shows superior results in crop growth and yield when compared with conventional fertilizers (Abdel-Aziz et al., 2016; Kabir et al., 2018; Liu and Lal, 2015a).

7 Biostimulation with nanofertilizers: Stress tolerance and nutraceutical quality

The nutritional quality of the crops and the stress tolerance of the plants are characteristics that go together: the adequate balance of mineral nutrients, antioxidants or osmolytes, and metabolites associated with signaling and response results on the one hand in more ability to adapt to the environment and, on the other hand, in greater health for the consumer.

The nanofertilizers improve stress tolerance and the nutraceutical quality of fruit crops, in addition to providing essential nutrients. This response occurs through two mechanisms: (i) the adequate supply of essential nutrients that are associated with a vigorous plant and with the adequate density of vitamins, minerals, and other nutrients (Cabrera-De la Fuente et al., 2018), and (ii) the effect of biostimulation of the components of the nanofertilizers (Juárez-Maldonado et al., 2018).

The biostimulant effect is supposed to occur due to interactions between NPs and NMs with proteins, membranes, nucleic acids, and various metabolites; or due to the presence in the nanostructures of unpaired electrons that induce redox signals; and by the release of ions of transition metals able to induce a controlled oxidative stress (Juárez-Maldonado et al., 2018, 2019; Morales-Díaz et al., 2017). When the concentration of the NPs or NMs is adequate, the level of oxidative stress does not exceed the toxic threshold, and induces in the plant cells a phenomenon of defense induction accompanied by the expression of resistance genes (Van Aken, 2015), synthesis of metabolites and defense proteins, and a higher concentration of enzymatic and nonenzymatic antioxidants (Jiang et al., 2012; Juárez-Maldonado et al., 2013).

This effect of biostimulation accompanied by the promotion of nutraceutical quality and stress tolerance has been described in different studies that were reviewed by Juárez-Maldonado et al. (2018). For example, a higher yield of essential oil was obtained in *Anethum graveolens* (Gholinezhad, 2017), as well as the increased shelf life of ornamental species such as *Gerbera jamesonii* (Mohammadbagheri and Naderi, 2017). The use of NPs of ZnO induced a higher tolerance to water deficit (Sedghi et al., 2013), while NPs of ZnO and Fe_3O_4 increased tolerance to salinity (Amira et al., 2015).

The use of nanocalcium in "red delicious" apple has also been reported, improving the quality and postharvest life of the fruits in comparison with the conventional fertilizer calcium chloride ($CaCl_2$) (Ranjbar et al., 2018). In *Punica granatum*, the foliar application of nanofertilizers of Zn and B (Davarpanah et al., 2016) and nano-N (Davarpanah et al., 2017) produced higher yield and fruit quality when compared with conventional fertilizers.

8 The environmental fate of nanofertilizers

In conventional granulated fertilizers and in high-solubility, slow-release, controlled-release, or organic fertilizers, the essential elements are applied in ionic forms whose fate in ecosystems is relatively well known; therefore, there is no concern about the health of the consumer. In the case of nanofertilizers, the issue is not well defined, since questions prevail about the effect in both the short and long term of the NPs and the NMs used in their composition (Dubey and Mailapalli, 2016; Sekhon, 2014). For example, it is not known with certainty if the nanofertilizers are entirely transformed to ionic forms in the plants or if a part is conserved as NPs that could be transferred to the consumers (Morales-Díaz et al., 2017). In the case of gold NPs, it was observed that they are transported in their nanoform from the leaves to the rest of the plant (Raliya et al., 2016).

The information presented on the use of nanofertilizers in the previous sections is confident regarding the nutrition and composition of the plants. In contrast to the NPs and NMs from industrial and domestic sources, nanofertilizers (both for the relatively small amounts that are applied to the soil or by foliar spray and the fact that they are applied in a timely manner on the sites where they will have their effect) do not seem to represent a threat to the health of humans or ecosystems (Tolaymat et al., 2017). However, not much information is available about the fate of nanofertilizers regarding their transfer between different trophic levels, including human consumers (Morales-Díaz et al., 2017).

Because of its potential economic importance, the use of nanofertilizers should be subject to a careful analysis to review the possible trophic transfer and its consequences (Du et al., 2017). Different works indicate the absorption of NPs that make up nanofertilizers by the roots of plants and by soil microorganisms and sediments. Once in the roots, the NPs move and accumulate in other organs of the plant (Gardea-Torresdey et al., 2014). When NPs are applied by foliar spray, mobility also occurs to other organs via phloem (Raliya et al., 2016). The transfer to the

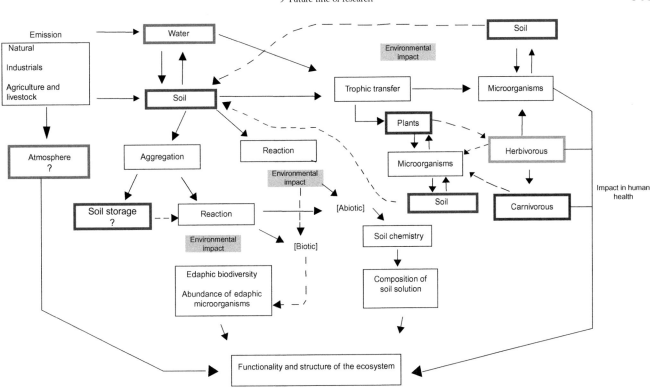

FIG. 35.3 The dynamic flow of nanoparticles and nanomaterials in ecosystems. The left side includes emissions that flow into water, soil, and atmosphere. From that point, they are incorporated into the plants and microorganisms of the soil and water. The transfer to other trophic components of the ecosystem occurs by herbivores or carnivores or by the decomposition of the remains of the organisms by the microorganisms of the soils, water, or sediments. *From Morales-Díaz, A.B.A.B., Ortega-Ortíz, H., Juárez-Maldonado, A., Cadenas-Pliego, G., González-Morales, S., Benavides-Mendoza, A., 2017. Application of nanoelements in plant nutrition and its impact in ecosystems. Adv. Nat. Sci. Nanosci. Nanotechnol. 8, 013001.*

next trophic level (Fig. 35.3) occurs when microorganisms, plant structures, or wastes thereof serve as food for the soil micro- and mesofauna and from there possibly transferred to vertebrates (Morales-Díaz et al., 2017).

The use of nanofertilizers shows great functional and environmental advantages compared with conventional fertilizers. According to the presented information, the benefits have been observed for practically all the essential elements for the plants. Despite the small number of studies carried out on perennial fruit crops, the very encouraging results obtained in other crop species, mainly annual crops, indicates the possibility of successful application in tree species.

9 Future line of research

The toxicological impact of NPs and NMs has been well studied and constitutes an extensive and relatively well-established area of knowledge. On the other hand, less information is available about nanofertilizers, almost all referring to annual species. In fruit crops and in general in perennial species, there are very few published reports.

Some future lines of research that are suggested about the use of nanofertilizers in fruit crops are as follows:

(a) Determine the impact of repeated applications of nanofertilizers on the soil microbiome and the metabolism and development of different species of fruit crops in tropical or temperate regions.

(b) Review the transfer of NPs and NMs from nanofertilizers to soil components, the food chain, and human consumers.

(c) Carry out comparisons of the effectiveness of the application of the elements contained in the nanofertilizers against those of conventional fertilizers. The comparison should include the differences in cost, in ecological impact (i.e., on the soil microbiome), and in the capacity to induce higher productivity, yield, fruit quality, and tolerance to stress in fruit crops.

References

Abdel-Aziz, H.M.M., Hasaneen, M.N.A., Omer, A.M., 2016. Nano chitosan-NPK fertilizer enhances the growth and productivity of wheat plants grown in sandy soil. Spanish. J. Agric. Res. 14, 1–9.

Amira, S.S., Souad, A.E.f., Essam, D., 2015. Alleviation of salt stress on Moringa peregrina using foliar application of nanofertilizers. J. Hortic. For. 7, 36–47.

Bae, S., Hwang, Y.S., Lee, Y.-J., Lee, S.-K., 2013. Effects of water chemistry on aggregation and soil adsorption of silver nanoparticles. Environ. Health Toxicol. 28. e2013006.

Baligar, V.C., Fageria, N.K., He, Z.L., 2001. Nutrient use efficiency in plants. Commun. Soil Sci. Plant Anal. 32, 921–950.

Benton Jones, J.J., 2012. Plant Nutrition and Soil Fertility Manual, second ed. CRC Press, Boca Raton, FL.

Cabrera-De la Fuente, M., González-Morales, S., Juárez-Maldonado, A., Leija-Martínez, P., Benavides-Mendoza, A., 2018. Plant nutrition and agronomic management to obtain crops with better nutritional and nutraceutical quality. In: Holban, A.M., Grumezescu, A.M. (Eds.), Therapeutic Foods. Academic Press, London, pp. 99–140.

Chhipa, H., 2017. Nanofertilizers and nanopesticides for agriculture. Environ. Chem. Lett. 15, 15–22.

Choy, J., Choi, S., Oh, J., Park, T., 2007. Clay minerals and layered double hydroxides for novel biological applications. Appl. Clay Sci. 36, 122–132.

Das, R.K., Brar, S.K., 2018. Plant-derived metallic nanoparticles: environmental safety and colloidal behaviour. Curr. Sci. 114, 2018–2019.

Davarpanah, S., Tehranifar, A., Davarynejad, G., Abadía, J., Khorasani, R., 2016. Effects of foliar applications of zinc and boron nano-fertilizers on pomegranate (Punica granatum cv. Ardestani) fruit yield and quality. Sci. Hortic. 210, 57–64.

Davarpanah, S., Tehranifar, A., Davarynejad, G., Aran, M., Abadía, J., Khorassani, R., 2017. Effects of foliar nano-nitrogen and urea fertilizers on the physical and chemical properties of pomegranate (Punica granatum cv. Ardestani) fruits. HortScience 52, 288–294.

Ditta, A., Arshad, M., 2016. Applications and perspectives of using nanomaterials for sustainable plant nutrition. Nanotechnol. Rev. 5, 209–229.

Du, W., Tan, W., Peralta-Videa, J.R., Gardea-Torresdey, J.L., Ji, R., Yin, Y., Guo, H., 2017. Interaction of metal oxide nanoparticles with higher terrestrial plants: physiological and biochemical aspects. Plant Physiol. Biochem. 110, 210–225.

Dubey, A., Mailapalli, D.R., 2016. Nanofertilisers, nanopesticides, nanosensors of pest and nanotoxicity in agriculture. In: Lichtfouse, E. (Ed.), Sustainable Agriculture Reviews. vol. 19. Springer, Cham, pp. 307–330.

Eichert, T., Kurtz, A., Steiner, U., Goldbach, H.E., 2008. Size exclusion limits and lateral heterogeneity of the stomatal foliar uptake pathway for aqueous solutes and water-suspended nanoparticles. Physiol. Plant. 134, 151–160.

El-Ramady, H., El-Ghamry, A., Mosa, A., Alshaal, T., 2018. Nanofertilizers vs. biofertilizers: new insights. Environ. Biodivers. Soil Secur. 2, 40–50.

FAO, 2017. FAOSTAT: Production Crops [WWW Document]. http://www.fao.org/faostat/en/#data/QC.

Fedorenko, V.F., Buklagin, D.S., Golubev, I.G., Nemenushchaya, L.A., 2015. Review of Russian nanoagents for crops treatment. Nanotechnol. Russ. 10, 318–324.

Foley, J.A., Ramankutty, N., Brauman, K.A., Cassidy, E.S., Gerber, J.S., Johnston, M., Mueller, N.D., O'Connell, C., Ray, D.K., West, P.C., Balzer, C., Bennett, E.M., Carpenter, S.R., Hill, J., Monfreda, C., Polasky, S., Rockström, J., Sheehan, J., Siebert, S., Tilman, D., Zaks, D.P.M., 2011. Solutions for a cultivated planet. Nature 478, 337–342.

Gardea-Torresdey, J.L., Rico, C.M., White, J.C., 2014. Trophic transfer, transformation, and impact of engineered nanomaterials in terrestrial environments. Environ. Sci. Technol. 48, 2526–2540.

Gholinezhad, E., 2017. Effect of drought stress and Fe nano-fertilizer on seed yield, morphological traits, essential oil percentage and yield of dill (Anethum graveolens L.). J. Essent. Oil Bear. Plants 20, 1006–1017.

Ghormade, V., Deshpande, M.V., Paknikar, K.M., 2011. Perspectives for nano-biotechnology enabled protection and nutrition of plants. Biotechnol. Adv. 29, 792–803.

Golbashy, M., Sabahi, H., Allahdadi, I., Nazokdast, H., Hosseini, M., 2016. Synthesis of highly intercalated urea-clay nanocomposite via domestic montmorillonite as eco-friendly slow-release fertilizer. Arch. Agron. Soil Sci. 0, 1–12.

González-Gómez, H., Ramírez-Godina, F., Ortega-Ortiz, H., Benavides-Mendoza, A., Robledo Torres, V., Cabrera-De la Fuente, M., 2017. Use of chitosan-PVA hydrogels with copper nanoparticles to improve the growth of grafted watermelon. Molecules 22, 1031.

Guerriero, G., Cai, G., 2018. Interaction of nano-sized nutrients with plant biomass: a review. In: Faisal, M., Saquib, Q., Alatar, A.A., Al-Khedhairy, A.A. (Eds.), Phytotoxicity of Nanoparticles. Springer International Publishing, Cham, pp. 135–149.

Hasaneen, M., Abdel-Aziz, H., El-Bialy, D., Omer, A., 2014. Preparation of chitosan nanoparticles for loading with NPK fertilizer. Afr. J. Biotechnol. 13, 3158–3164.

Hernández-Hernández, H., Benavides-Mendoza, A., Ortega-Ortiz, H., Hernández-Fuentes, A.D., Juárez-Maldonado, A., 2017. Cu nanoparticles in chitosan-PVA hydrogels as promoters of growth, productivity and fruit quality in tomato. Emir. J. Food Agric. 29.

Hernández-Hernández, H., Pérez-Labrada, F., Enciso, E.L.G., Leija-Martínez, P., López-Pérez, M.C., Medrano-Macías, J., González-Morales, S., Maldonado, A.J., Dávila, L.R.G., Mendoza, A.B., 2018. Tolerance-induction techniques and agronomical practices to mitigate stress in extensive crops and vegetables. In: Andjelkovic, V. (Ed.), Plant, Abiotic Stress and Responses to Climate Change. InTech, Rijeka, Croatia, pp. 145–181.

Jiang, H.-S., Li, M., Chang, F.-Y., Li, W., Yin, L.-Y., 2012. Physiological analysis of silver nanoparticles and AgNO3 toxicity to Spirodela polyrhiza. Environ. Toxicol. Chem. 31, 1880–1886.

Juárez-Maldonado, A., Rosales-Velázquez, J., Ortega-Ortiz, H., Cabrera-De-la-Fuente, M., Ramírez, H., Benavides-Mendoza, A., 2013. Accumulation of silver nanoparticles and its effect on the antioxidant capacity in Allium cepa L. Phyton 82, 91–97.

Juárez-Maldonado, A., Ortega-Ortíz, H., Pérez-Labrada, F., Cadenas-Pliego, G., Benavides-Mendoza, A., 2016. Cu nanoparticles absorbed on chitosan hydrogels positively alter morphological, production, and quality characteristics of tomato. J. Appl. Bot. Food Qual. 89, 183–189.

Juárez-Maldonado, A., González-Morales, S., Cabrera-De la Fuente, M., Medrano-Macías, J., Benavides-Mendoza, A., 2018. Nanometals as promoters of nutraceutical quality in crop plants. In: Grumezescu, A.M., Holban, A.M. (Eds.), Impact of Nanoscience in Food Industry. Academic Press, London, pp. 277–310.

Juárez-Maldonado, A., Ortega-Ortíz, H., Morales-Díaz, A.B., González-Morales, S., Morelos-Moreno, Á., Cabrera-De la Fuente, M., Sandoval-Rangel, A., Cadenas-Pliego, G., Benavides-Mendoza, A., 2019. Nanoparticles and nanomaterials as plant biostimulants. Int. J. Mol. Sci. 20, 162.

Kabir, E., Kumar, V., Kim, K.H., Yip, A.C.K., Sohn, J.R., 2018. Environmental impacts of nanomaterials. J. Environ. Manag. 225, 261–271.

Kashyap, P.L., Xiang, X., Heiden, P., 2015. Chitosan nanoparticle based delivery systems for sustainable agriculture. Int. J. Biol. Macromol. 77, 36–51.

Khan, M.N., Mobin, M., Abbas, Z.K., AlMutairi, K.A., Siddiqui, Z.H., 2016. Role of nanomaterials in plants under challenging environments. Plant Physiol. Biochem. 110, 194–209.

Khot, L.R., Sankaran, S., Maja, J.M., Ehsani, R., Schuster, E.W., 2012. Applications of nanomaterials in agricultural production and crop protection: a review. Crop Prot. 35, 64–70.

Kottegoda, N., Sandaruwan, C., Priyadarshana, G., Siriwardhana, A., Rathnayake, U.A., Berugoda Arachchige, D.M., Kumarasinghe, A.R., Dahanayake, D., Karunaratne, V., Amaratunga, G.A.J., 2017. Urea-hydroxyapatite nanohybrids for slow release of nitrogen. ACS Nano 11, 1214–1221.

Lal, R., 2015. Restoring soil quality to mitigate soil degradation. Sustainability 7, 5875–5895.

Le, V., Rui, Y., Gui, X., Li, X., Liu, S., Han, Y., 2014. Uptake, transport, distribution and bio-effects of SiO_2 nanoparticles in Bt-transgenic cotton. J. Nanobiotechnol. 12, 50.

Lehman, S.E., Larsen, S.C., 2014. Zeolite and mesoporous silica nanomaterials: greener syntheses, environmental applications and biological toxicity. Environ. Sci. Nano 1, 200.

Liu, R., Lal, R., 2015a. Synthetic apatite nanoparticles as a phosphorus fertilizer for soybean (Glycine max). Sci. Rep. 4, 5686.

Liu, R., Lal, R., 2015b. Potentials of engineered nanoparticles as fertilizers for increasing agronomic productions. Sci. Total Environ. 514, 131–139.

Liu, Q., Chen, B., Wang, Q., Shi, X., Xiao, Z., Lin, J., Fang, X., 2009. Carbon nanotubes as molecular transporters for walled plant cells. Nano Lett. 9, 1007–1010.

López-Vargas, E., Ortega-Ortíz, H., Cadenas-Pliego, G., de Alba Romenus, K., Cabrera de la Fuente, M., Benavides-Mendoza, A., Juárez-Maldonado, A., 2018. Foliar application of copper nanoparticles increases the fruit quality and the content of bioactive compounds in tomatoes. Appl. Sci. 8, 1020.

Mikhak, A., Sohrabi, A., Kassaee, M.Z., Feizian, M., 2017. Synthetic nanozeolite/nanohydroxyapatite as a phosphorus fertilizer for German chamomile (Matricariachamomilla L.). Ind. Crop. Prod. 95, 444–452.

Ming, D.W., Allen, E.R., 2001. Use of natural zeolites in agronomy, horticulture and environmental soil remediation. Rev. Mineral. Geochem. 45, 619–654.

Mohammadbagheri, L., Naderi, D., 2017. Effect of growth medium and calcium nano-fertilizer on quality and some characteristics of gerbera cut flower. J. Ornam. Plants 7, 205–212.

Monteiro Corrêa, R., Pinto, C., do, S.I., Soares, É., Mendes de Carvalho, V.A., 2012. Hydroponic production of fruit tree seedlings in Brazil. In: Asao, T. (Ed.), Hydroponics—A Standard Methodology for Plant Biological Researches. InTech, Rijeka, Croatia, pp. 225–244.

Morales-Díaz, A.B.A.B., Ortega-Ortíz, H., Juárez-Maldonado, A., Cadenas-Pliego, G., González-Morales, S., Benavides-Mendoza, A., 2017. Application of nanoelements in plant nutrition and its impact in ecosystems. Adv. Nat. Sci. Nanosci. Nanotechnol. 8, 013001.

Morgan, K.T., Cushman, K.E., Sato, S., 2009. Release mechanisms for slow- and controlled-release fertilizers and ttrategies for their use in vegetable production. HortTechnology 19, 10–12.

Pan, B., Lam, S.K., Mosier, A., Luo, Y., Chen, D., 2016. Ammonia volatilization from synthetic fertilizers and its mitigation strategies: a global synthesis. Agric. Ecosyst. Environ. 232, 283–289.

Parra Quezada, R.A., Ávila Marioni, M.R., Amado Álvarez, J.P., 2011. Perspectivas del sistema de producción de manzano en Chihuahua, ante el cambio climático. Rev. Mex. Cienc. Agríc. 2 (spe2), 265–279.

Peña Cabriales, J.J., Grageda Cabrera, O.A., Vera Núñez, J.A., 2002. Manejo de los fertilizantes nitrogenados en México: uso delas técnicas isotópicas (15n). Terra Latinoam. 20, 51–56.

Pilbeam, D.J., 2015. Breeding crops for improved mineral nutrition under climate change conditions. J. Exp. Bot. 66, 3511–3521.

Pinto, J.M., Silva, D.J., Borges, A.L., Coelho, E.F., Feitosa, F.J., 2002. Fertirrigação. In: Genu, P.J.C., Pinto, A.C. (Eds.), A Cultura Da Mangueira. Embrapa Informação Tecnológica, Brasília, pp. 225–241.

Raliya, R., Franke, C., Chavalmane, S., Nair, R., Reed, N., Biswas, P., 2016. Quantitative understanding of nanoparticle uptake in watermelon plants. Front. Plant Sci. 7, 1288.

Raliya, R., Saharan, V., Dimkpa, C., Biswas, P., 2018. Nanofertilizer for precision and sustainable agriculture: current state and future perspectives. J. Agric. Food Chem. 66, 6487–6503.

Ramírez-Legarreta, M.R., Ruiz-Corral, J.A., Medina-García, G., Jacobo-Cuéllar, J.L., Retes-López, R., Palafox, A.R.N., Medina, S.M., Ballesteros, F.G.D., Rivera, M.M., 2011. Perspectives on the apple production system in Chihuahua facing climate change. Rev. Mex. Cien. Agrícolas 2, 265–279.

Ranjbar, S., Rahemi, M., Ramezanian, A., 2018. Comparison of nano-calcium and calcium chloride spray on postharvest quality and cell wall enzymes activity in apple cv. Red delicious. Sci. Hortic. 240, 57–64.

Rui, M., Ma, C., Hao, Y., Guo, J., Rui, Y., Tang, X., Zhao, Q., Fan, X., Zhang, Z., Hou, T., Zhu, S., 2016. Iron oxide nanoparticles as a potential iron fertilizer for peanut (Arachis hypogaea). Front. Plant Sci. 7, 815.

Sabir, A., Yazar, K., Sabir, F., Kara, Z., Yazici, M.A., Goksu, N., 2014. Vine growth, yield, berry quality attributes and leaf nutrient content of grapevines as influenced by seaweed extract (Ascophyllum nodosum) and nanosize fertilizer pulverizations. Sci. Hortic. 175, 1–8.

Saharan, V., Pal, A., 2016. Current and future prospects of chitosan-based nanomaterials in plant protection and growth. In: Saharan, V., Pal, A. (Eds.), Chitosan Based Nanomaterials in Plant Growth and Protection. Springer, New Delhi, pp. 43–48.

Sedghi, M., Hadi, M., Toluie, S.G., 2013. Effect of nano zinc oxide on the germination parameters of soybean seeds under drought stress. Ann. West Univ. Timişoara, ser. Biol. 16, 73–78.

Sekhon, B., 2014. Nanotechnology in agri-food production: an overview. Nanotechnol. Sci. Appl. 7, 31.

Silva, E.M., Pinto, A.C., Azevedo, J.A., 1996. Manejo da irrigação e fertirrigação na cultura da mangueira. Brasília.

Singh, M.D., Chirag, G., Prakash, P.O., Mohan, M.H., Prakasha, G., Vishwajith, 2017. Nano fertilizers is a new way to increase nutrients use efficiency in crop production. Int. J. Agric. Sci. 9, 3831–3833.

Souri, M.K., 2016. Aminochelate fertilizers: the new approach to the old problem: a review. Open Agric. 1.

Tarango Rivero, S.H., Nevárez Moorillón, V.G., Orrantia Borunda, E., 2009. Growth, yield, and nutrient status of pecans fertilized with biosolids and inoculated with rhizosphere fungi. Bioresour. Technol. 100, 1992–1998.

Tolaymat, T., Genaidy, A., Abdelraheem, W., Dionysiou, D., Andersen, C., 2017. The effects of metallic engineered nanoparticles upon plant systems: an analytic examination of scientific evidence. Sci. Total Environ. 579, 93–106.

Van Aken, B., 2015. Gene expression changes in plants and microorganisms exposed to nanomaterials. Curr. Opin. Biotechnol. 33, 206–219.

Vázquez-Núñez, E., López-Moreno, M.L., Álvarez, G.d.l.R., Fernández-Luqueño, F., 2018. Incorporation of nanoparticles into plant nutrients: the real benefits. In: López-Valdez, F., Fernández-Luqueño, F. (Eds.), Agricultural Nanobiotechnology. Springer, Cham, pp. 49–76.

Verma, S.K., Das, A.K., Patel, M.K., Shah, A., Kumar, V., Gantait, S., 2018. Engineered nanomaterials for plant growth and development: a perspective analysis. Sci. Total Environ. 630, 1413–1435.

Yuvaraj, M., Subramanian, K.S., 2014. Controlled-release fertilizer of zinc encapsulated by a manganese hollow core shell. Soil Sci. Plant Nutr. 61, 319–326.

Zuverza-Mena, N., Martínez-Fernández, D., Du, W., Hernandez-Viezcas, J.A., Bonilla-Bird, N., López-Moreno, M.L., Komárek, M., Peralta-Videa J.R., Gardea-Torresdey, J.L., 2017. Exposure of engineered nanomaterials to plants: insights into the physiological and biochemical responses-a review. Plant Physiol. Biochem. 110, 236–264.

4R nutrient stewardship in fruit crops

*David R. Bryla**

U.S. Department of Agriculture, Agricultural Research Service, Horticultural Crops Research Unit,
Corvallis, OR, United States
*Corresponding author. E-mail: david.bryla@ars.usda.gov

1 Introduction

The 4R concept represents the "four rights" of nutrient stewardship and is defined as using the right fertilizer sources at the right rate, right time, and right place (Roberts, 2007). This concept, championed by the International Plant Nutrition Institute (IPNI), was developed through many years of research and has been supported by a long history of cooperation between agronomists and the fertilizer industry (Bruulsema et al., 2012). The goal is to provide farmers and other stakeholders a science-based framework on how to best utilize fertilizers to produce the most economical outcomes for a given crop and, at the same time, provide desirable social and environmental benefits essential to sustainable agriculture. Potential benefits to implementing 4R management can include better crop quality and yields, reduced input costs, less soil erosion and nutrient leaching, and financial incentives through nutrient trading and other sustainability markets.

In this chapter, the utility of the 4R nutrient concept is illustrated to discuss its application in fruit crops. Asking whether a particular crop was given the right nutrient source at the right rate, time, and place helps farmers and advisers identify opportunities for improving their fertilizer programs. While each of these principles are interlinked, grouping them under specific headings guarantees balanced effort among the four "rights" and ensures that no critical steps in fertilizer management are overlooked. These headings also provide a platform for outlining specific nutrient requirements to the fertilizer industry, and help farmers, crop advisers, and scientists to clearly communicate with stakeholders less familiar with agriculture.

2 Goals and performance indicators

The 4R nutrient stewardship concept provides a framework to achieve cropping system goals, such as increased production, increased farmer profitability, enhanced environmental protection, and improved sustainability.

A.K. Srivastava, Chengxiao Hu (eds.)
Fruit Crops: Diagnosis and Management of Nutrient Constraints
https://doi.org/10.1016/B978-0-12-818732-6.00036-8

Performance indicators such as yield and quality assessments, net profits, and soil productivity and nutrients losses are selected by stakeholders and serve as valuable tools for evaluating each goal.

2.1 Productivity

Yield is the primary indicator of productivity in many cropping systems and, for most fruit crops, is defined as the total mass of fruit produced per unit land area in a given season. Quality is also important and is an essential component of handling, storage, and marketability. Together, yield and quality are affected by nutrient management decisions, which ultimate impact profitability through volume and value (Sanchez et al., 1995). Depending on the interest of the stakeholders, multiple efficiencies can also be calculated to evaluate productivity, including nutrient and water use efficiency, labor use efficiency, and energy use efficiency.

2.2 Profitability

Profitability is determined by the difference between the volume and value of the crop produced and the total costs of production (net profit). Capital investment and amortization should also be considered and used to determine the return on investment associated with each practice decision that is related to nutrient management. Performance can also be expressed in terms economic efficiency, that is, increase in production in response to the cost of each management decision.

2.3 Environmental health

Crop production has a wide range of effects on the environment and can impact the quality of soil, water, and air resources directly and indirectly. Management decisions that improve nutrient and water use efficiency generally help reduce resource losses and improve quality. Some impacts such as nutrient loading into watersheds or the atmosphere are extremely important but difficult to monitor at the farm level. Biodiversity and ecosystem services such as natural insect predators and pollinators or links to hunting, fishing, and outdoor activities are likewise difficult to quantify, and usually, assistance from multiple stakeholders is required for assessment.

2.4 Sustainability

A sustainable cropping system is one in which good productivity and profitability continue for many years without any long-term impacts on the environment. Many management decisions have impacts that extend beyond a single growing season. A few local measures of sustainability include soil productivity, consistent crop yields and quality, stable farm income, and favorable working conditions. Broader measures are more difficult to define and often intertwined with environmental issues such as acid rain and greenhouse gases.

3 The 4Rs

3.1 Right source

Some common products used to apply nutrients to fruit and other crops are listed in Table 36.1. These products include both dry and liquid sources of conventional and organic materials. The use of liquid fertilizers has increased steadily over the last 50 years, particularly in high-value crops such as fruit, which are often irrigated by drip or other types of microirrigation systems. Dry materials are generally less expensive than liquid fertilizers and easier to handle, particularly when large quantities are used. Dry fertilizers are most commonly applied to fruit crops during months with abundant rainfall, or to fields or orchards that rely on sprinklers for irrigation or have no irrigation systems.

When developing a nutrient management plan, fruit growers must first choose what form of nutrients to apply and consider the characteristics of the specific product or combination of products they decide to use. Are the nutrients readily available to the plant or must they be converted into a plant-available form in the soil? For example, under conditions of good plant growth, NH_4^+ ions in the soil are rapidly converted to NO_3^- by bacteria (*Nitrosomonas* and *Nitrobacter* sp.) through a natural process called nitrification. While both forms can be taken up and utilized by plants, most fruit crops respond more quickly to NO_3 applications because of its immediate movement to the leaves. Furthermore, NH_4 uptake can interfere with uptake and movement of K and Ca in the plants, resulting in reduced production and fruit quality. Therefore, a balance of NH_4 and NO_3 nutrition is often recommended. Young plants (<3 weeks of age) or acid-loving

TABLE 36.1 Common fertilizers used in fruit crops (Bruulsema et al., 2012).

Fertilizer	Chemical formula	Composition	Comments
NITROGEN MATERIALS			
Ammonium nitrate	NH_4NO_3	33%–34% (dry) or 20% (liquid) N	Highly soluble and efficient source of N. NO_3 portion is easily taken up by the plants, while the NH_4 portion provides a delayed supply of N. Due to high crop recovery, ease of use, and suitability for in-season top dressing, ammonium nitrate is widely used in many European countries. However, because of concerns over its illegal use for explosives, availability is highly controlled and restricted in many parts of the world. The fertilizer is available without restrictions as a soluble fertilizer and is often mixed with P and K to produce NPK compound fertilizers
Urea	$CO(NH_2)_2$	46% (dry) or 20% (liquid) N	Moves freely in soil solution, and potential for movement and leaching losses are dependent upon the rate of hydrolysis and subsequent conversion to immobile NH_4-N. Urea can sometimes contain excessive amounts of biuret (toxic to plants) due to faulty manufacturing. Biuret is most toxic when urea is sprayed on leaves and immature fruit during foliar applications of the fertilizer
Urea-ammonium nitrate (UAN) solution	$CO(NH_2)_2 + NH_4NO_3$	28%–32% N	Contains the highest concentration of N of all the available N solution products and is among the most commonly used fertilizer in fruit crops
Ammonium sulfate	$(NH_4)_2SO_4$	21% N, 24% S	Widely used as a dry or liquid fertilizer in many fruit crops and is particularly useful in soils that are deficient in both N and S. Compared with urea, it is much more resistant to NH_3 volatilization. Has a higher salt index than most fertilizers, and extra caution is needed when using it in single or split applications as a dry product. Soil acidification is also pronounced and, therefore, is often used to help reduce soil pH
Ammonium thiosulfate (ATS)	$(NH_4)_2S_2O_3$	12% N, 26% S	Like ammonium sulfate, can be applied in liquid mixes or by itself, and when added to soil, *Thiobacillus* bacteria breakdown the thiosulfate portion of the fertilizer and oxidize it to sulfuric acid (acidifying agent)
Calcium nitrate	Solid: 5 $Ca(NO_3)_2 \cdot NH_4NO_3 \cdot 10\ H_2O$ Liquid: $Ca(NO_3)_2 \cdot 4\ H_2O$	Solid: 15.5% N, 19%–22% Ca, <1.5% NH_4 Liquid: 8%–9% N, 11%–12% Ca, <1% NH_4	Popular for providing both N and Ca. Highly susceptible to leaching and denitrification losses. Careful attention to water management is needed. To avoid precipitating insoluble salts, it should not be mixed in solutions with soluble phosphate or sulfate fertilizers
Calcium ammonium nitrate (CAN-17)	$Ca(NO_3)_2 \cdot NH_4NO_3$	17% N, 8.8% Ca	A popular liquid source of N and Ca. Less susceptible to leaching than calcium nitrate
Sodium nitrate	$NaNO_3$	16% N	Mined from natural rock deposits found in the Atacama Desert in Chile and, therefore, is often referred to as *Chilean nitrate*. Unlike other inorganic sources of N, Chilean nitrate is approved by the US National Organic Program for use as a supplemental N source. Concern is sometimes expressed over Na in the fertilizer, but the risk is minimal at the typical rates used in organic production
PHOSPHORUS MATERIALS			
Monoammonium phosphate (MAP) and diammonium phosphate (DAP)	MAP: $NH_4H_2PO_4$ DAP: $(NH_4)_2HPO_4$	MAP: 10%–12% N, 48%–61% P_2O_5 DAP: 18% N, 46% P_2O_5	Both fertilizers contain N, but their primary value is as a P source. They are readily soluble in soil and easily applied in concentrated bands in the vicinity of the roots. Application of DAP to the soil increases pH temporarily until the NH_4^+ ions are converted to NO_3^-, while MAP is moderately acidic. Most field comparisons between these two P sources show only minor or no differences in growth and yield under most conditions

Continued

TABLE 36.1 Common fertilizers used in fruit crops (Bruulsema et al., 2012)—cont'd

Fertilizer	Chemical formula	Composition	Comments
Ammonium polyphosphate (APP)	$[NH_4PO_3]_n(OH)_2$	Typically 10% N, 34% P_2O_5 or 11% N, 37% P_2O_5	Roughly one-quarter to one-half of the P is orthophosphate and available for immediate plant uptake, while the remaining polymer phosphate chains will hydrolyze to orthophosphates within a couple of weeks and breakdown into simple phosphate molecules by enzymes produced by plant roots and soil microorganisms. When mixed in hard water containing a lot of Ca, highly insoluble calcium pyrophosphates will form and can plug the emitters in drip and other types of irrigation systems. Under such circumstances, sulfuric acid or similar should be injected to reduce the pH of water to <7.0. A number of nutrients are mixed well with polyphosphate fertilizers. However, micronutrients should only be added to acidified solutions made with ammonium polyphosphate
Mineral phosphate materials: rock phosphate, rock dust (crushed granite), and colloidal (soft) Bone sources: steamed bone meal, fish bone meal	Variable	3%–33% P	Inorganic P sources approved for organic production. Mineral phosphate materials release P over a period of months or years in neutral and acidic soils but breakdown slowly or not at all in alkaline soils. Bone sources, on the other hand, are readily available for plant uptake
POTASSIUM MATERIALS			
Potassium chloride (muriate of potash)	KCl	60%–62% K_2O	Usually less expensive than the other sources of K, but because most fruit crops are sensitive to high amounts of Cl^-, use should be minimized or avoided in many cases, particularly on sites with sandy or sandy loam soils
Potassium nitrate (nitrate of potash)	KNO_3	16% N, 44%–46% K_2O	Good source of K and N. Synergistic effect between K^+ and NO_3^- facilitates the uptake of both ions by the plant roots
Potassium sulfate (sulfate of potash)	K_2SO_4	48%–53% K_2O, 17%–18% S	Good source of K and S. Only one-third as soluble as KCl and, therefore, is not usually dissolved for application through an irrigation system unless S is also needed. Liquid potassium thiosulfate (25% K_2O and 17% S) solution is available as an alternative, and, much like ATS, is a good product for reducing soil pH
Potassium magnesium sulfate (sulfate of potash magnesia, KMag, or langbeinite)	$K_2SO_4 \cdot 2\,MgSO_4$	21%–22% K_2O, 10%–11% Mg, 21%–22% S	A popular choice for situations where Mg is needed in addition to K; however, this fertilizer is even slower to dissolve than potassium sulfate and, therefore, is usually only applied as a dry product
SECONDARY NUTRIENTS (CALCIUM, MAGNESIUM, AND SULFUR)			
Calcium carbonate (limestone) and calcium magnesium carbonate (dolomitic limestone)	Limestone: $CaCO_3$ Dolomite: $Ca \cdot Mg(CO_3)_2$	Limestone: 40% Ca Dolomite: 2%–13% Mg	Calcium is often added as a result of using limestone to raise soil pH. Secondary nutrient benefits of adding lime to acid soils include increased P availability and enhanced N mineralization and nitrification. When liming is needed and Mg is also deficient, dolomitic limestone (a mixture of Ca and Mg carbonates) should be used. Liming with Ca only can also provoke Mg deficiency
Gypsum	$CaSO_4 \cdot 2H_2O$	23% Ca, 18% S	No effect on soil pH and is often used to supply Ca to fruit crops without raising the pH. Also used to remediate high Na (sodic) soils, replacing Na with Ca on soil exchange sites and using irrigation to leach the soluble Na from the root zone. The process results in enhanced properties in sodic soils, such as lower bulk density and increased permeability and water infiltration
Magnesium sulfate (kieserite or Epsom salt)	$MgSO_4$	15%–16% Mg, 20%–22% S	Good source of Mg and S. No effect on soil pH
Elemental sulfur	S		Must be oxidized to sulfate (SO_4^{2-}) by soil bacteria (*Thiobacillus* sp.) to become available for plant uptake. This process produces acid as a by-product and is often used to reduce soil pH

plants such as blueberry (*Vaccinium* sp.) and cranberry (*V. macrocarpon* Aiton) are exceptions, lacking N reductase in the leaves and relying primarily on NH_4-N.

Nutrient sources must also suit the physical and chemical properties of the soil. This is particularly important when choosing N fertilizers. Since NO_3^+ is much more mobile than NH_4^+ in soil, NH_4 forms of fertilizer are recommended when N applications are made in early spring and prior to the time of greatest need (i.e., during rapid growth and fruit development in the summer). This practice minimizes the potential for N loss by leaching. However, even under acidic soil conditions, most of the NH_4^+ that is applied as a fertilizer is nitrified to NO_3^- within a matter of weeks. The process produces two hydrogen ions (H^+) for each molecule of NH_4^+ converted to NO_3^-, quickly acidifying the soil and driving down pH. Continued use of NH_4 fertilizers for many years can eventually lower soil pH enough to require liming. The rate in which soil pH will fall depends on the kind and amount of fertilizer added and the buffering capacity (i.e., cation-exchange capacity) of the soil. Since clay soils or those high in organic matter tend to have more buffering capacity, they are usually more resistant to pH change than sandy soils.

Urea is one of the least expensive and most widely used N fertilizer and is available in dry and liquid forms. Once applied to the soil, urea is hydrolyzed within a matter of days into ammonium bicarbonate by the naturally occurring enzyme, urease. The transformation process is basic and results in a net increase of only one H^+ for each molecule of NH_4^+ converted to NO_3^-. Consequently, urea is generally less acidifying than NH_4 fertilizers. Sometimes when urea is applied to the soil surface, as much as 50% or more of the NH_4^+ produced by urease, will convert to gaseous NH_3 and volatilize; however, this is seldom a problem during cooler and wetter months of the year such as in early spring. The most favorable conditions for volatile N loss from surface-applied urea are alkaline soils, warm temperatures, and sandy soils with low organic matter content and low cation-exchange capacities. Alternative sources of N should be used under such circumstances. Urease inhibitors can also be used to reduce activity of the enzyme and slow NH_3 volatilization.

Manures and composts are also good sources of N and other nutrients and are commonly used for organic production of fruit crops (Nelson and Mikkelsen, 2008). Nutrient mineralization rates are relatively slow in these materials and typically provide <20% of the total N during the first year after application (Hartz et al., 2000). High-N animal by-products, including fish emulsions, blood and feather meals, corn steep liquors, and bat and seabird guano, are also available and mineralize much more rapidly than manures and composts, releasing the majority of available N within several weeks of application. Generally, these high-N products are more expensive but useful for quickly improving nutrition of organic fruit crops. Some are also fairly soluble and can be added through the irrigation system. However, a number of long-term studies have shown that losses of NO_3 and P due to runoff and leaching are sometimes an issue when organic manures are used as the sole source of nutrients (Kirchmann and Bergström, 2001).

Other questions that need to be considered is whether there are any synergisms between or among the nutrients applied, and if using more than one fertilizer, are they compatible with each other (e.g., positive interactions between N and P or NO_3-N and K, negative interactions between N and B or P and Zn). Furthermore, accompanying ions in the fertilizer could have benefits or detriments in certain crop. The chloride in muriate of potash (KCl), for example, is beneficial to corn (*Zea mays* L.) but can be detrimental to the production and quality of many fruit crops. Growers must also consider their application equipment at hand and the availability and cost of the product.

Nanofertilizers are a new potential source of nutrients worth considering for 4R management of high-value crops, including many fruit species. Three classes of nanofertilizers have been proposed, including nanoparticles that contain nutrients, traditional fertilizers coated or loaded with nanoparticles, and traditional fertilizers with nanoscale additives. The technology is still in the early stages of development but has many unique properties that are now being explored for new opportunities in agriculture (Dimkpa and Bindraban, 2018). Nanoparticles are extremely small (1–100 nm) and can be engineered to release nutrients under very specific circumstances as follows: upon contact with a leaf surface or with a specific chemical or enzyme, in the presence of water, when the temperature reaches a set point, at specified acid or alkaline conditions, or by using an external ultrasound frequency or a magnetic field (Manjunatha et al., 2016). Each method could be used potentially to improve plant nutrition and nutrient efficiency relative to traditional fertilizers. Recently, Davarpanah et al. (2017) examined the use of foliar nanonitrogen in pomegranate (*Punica granatum* L.) and found that yield and fruit quality was increased with only 2.7 g N/tree or 1.8 kg N/ha. In contrast, they found that at least 10 times as much N was needed to produce a similar yield response from the trees using foliar applications of urea.

3.2 Right rate

The right rate is defined simply as applying just enough nutrients to meet production and quality goals in a given cropping system. To do so, an assessment of how much nutrients the plants actually need is required. One way to do this is to measure the total amount of nutrients removed during harvest, as well as pruning and senescence. However, unless the measurements are taken on the farm of interest, these will only be average values and will vary depending on region, cultivars, and cultural practices. Furthermore, these data only reflect the amount of nutrients the crop loses over the growing season and not what should be applied. Growers need to factor in nutrients supplied from the soil and irrigation water (through soil and water testing) and adjust for fertilizer use efficiency and resulting unavoidable loses due to soil sorption and leaching, runoff, and volatilization. Atmospheric deposition and estimates of nutrient release from organic sources such as manures, composts, and crop residues must also be considered (Forge et al., 2013; Thompson and Peck, 2017).

Another approach is to evaluate plant performance over different doses of fertilizer and develop response curves to specific nutrients. The goal is to identify the range of nutrient supply needed to achieve the desired crop performance (Fig. 36.1). Initially, plant growth, yield, and/or quality are limited by nutrient supply and increase as more is added (deficient). Then, growth or production reaches a maximum and remains unaffected by nutrient supply (adequate). Finally, growth and production falls with additional nutrient supply (toxic). The results must be calibrated and verified under a wide range of growing conditions. Again, results vary among locations, cultivars, and practices and may be affected by interactions with other nutrients.

Experiments designed to measure plant response to nutrients are time-consuming and expensive. Consequently, they are usually complemented with easier, faster tests that are based on data from representative plant-response experiments. These tests include soil, soil solution, plant tissue, plant sap, and irrigation water. Each test provides different insights into the nutritional needs of the crop.

FIG. 36.1 Fertilizer rate trials in northern highbush blueberry (*V. corymbosum* L.) and pineapple [*Ananas comosus* (L.) Merr.]. Blueberry was fertilized using dry N fertilizer or by fertigation with a liquid N source. Pineapple was fertilized using different K sources, including potassium sulfate (K_2SO_4), potassium chloride (KCl), or both. *Modified from Bryla, D.R., Machado, R.M.A., 2011. Comparative effects of nitrogen fertigation and granular fertilizer application on growth and availability of soil nitrogen during establishment of highbush blueberry. Front. Plant Sci. 2, 46. doi:10.3389/fpls.2011.00046 and Teixeira, L.A.J., Quaggio, J.A., Cantarella, H., Mellis, E.V., 2011. Potassium fertilization for pineapple: effects on plant growth and fruit yield. Rev. Bras. Frutic. 33, 618–626.*

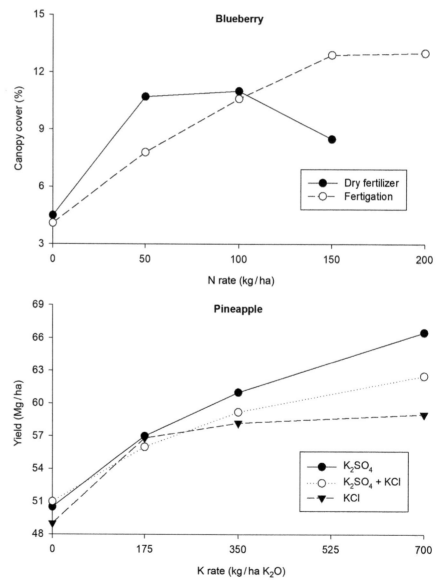

Soil testing measures what is available or potentially available in the field or orchard for plant uptake. Laboratories routinely test soils for pH, electrical conductivity (EC), cation-exchange capacity (CEC), NH_4 and NO_3 concentrations, total N, and other essential nutrients (total N, P, K, Ca, Mg, SO_4-S, B, Cu, Fe, Mn, Zn, Mo, and Cl). Sodium and carbonates/bicarbonates can also be analyzed upon request and provide useful information for correcting problems in sodic and saline soils. Additionally, on-farm quick tests are becoming increasingly popular and allow farmers and consultants to immediately measure soil properties such as pH, EC, NO_3, P, and K (Hartz et al., 1994).

Soil sampling should represent one uniform area or soil condition within a field and is usually done near or at the end of the growing season every year or two. Sampling depth should be adjusted to fit laboratory suggestions, the crops being grown, and the cultivation practices used, and care should be taken to avoid weedy areas, mulches and composts on the soil surface, and old fertilizer bands. Alternatively, soil solution samplers may be used to monitor nutrient changes throughout the growing season and are well suited for situations in which nutrients are applied by fertigation through a drip irrigation system (Bryla et al., 2010; Bryla and Machado, 2011).

Like other test procedures, soil and soil solution tests are calibrated against field fertilizer trials. Hue and Fox (2010) determined the P requirements for numerous crops in Hawaii, including banana [*Musa × paradisiaca* L. (pro sp.)], by (1) obtaining soil solution P levels using sorption isotherms for different soil series and families, (2) identifying critical soil solution P levels for each crop, (3) regressing soil solution against extractable P, and (4) establishing relationships between extractable P and fertilizer P. This technique provides the user with information on whether a particular soil needs additional P, and if so, how much P is needed to produce optimal growth and production.

Plant tissue analysis allows growers to detect nutritional issues in the plants before they become a problem and, like soil tests, can be used to adjust application of fertilizers for better production and fruit quality (Fallihi, 1985, 1988). Nutrient concentrations change throughout the plant and growing season, and therefore, meaningful results require that plant tissues are sampled consistently and in the same manner used as the standard for comparison in making the nutrient recommendation. The most common tissues sampled are mature, recent fully expanded leaves or leaf petioles. Concentration of nutrients that are immobile within the plant, such as Ca, S, B, Fe, Mn, and Zn, will decrease in new leaves when soil nutrients are low, while mobile nutrients, such as N, P, K, Mg, and Cu, will be lower in older leaves than in newer leaves. Sufficiency ranges are perhaps the most common method used to interpret plant tissue tests and are usually developed from either single-factor experiments to determine the optimum level for a particular nutrient or from analysis of large data sets from field surveys.

More involved methods are also used occasionally to interpret the results of plant and soil tests, including nutrient ratios, which are used to overcome problems with tissue age and nutrient interactions (Walworth and Sumner, 1988; Yang et al., 2015), and the diagnosis and recommendation integrated system (DRIS) approach, which is used to rank deficient nutrients and provide a recommendation as to which nutrients should be added first (e.g., Beverly et al., 1984).

Under- and overapplication of nutrients often hampers crop production and has economic and environmental consequences (Weng and Cheng, 2011). In addition to reductions in yield and fruit quality, nutrient deficiencies can lead to more problems with insects and plant pathogens and thereby result in more pesticide use. Attempts to alleviate nutrient deficiencies could also require frequent applications of more expensive foliar fertilizers. Overapplication, on the other hand, may result in excessive vegetative growth, which can reduce fruit quality [e.g., wine grapes (*Vitis vinifera* L.)] and increase pruning costs, and lead to contamination of groundwater and surface water sources. Efficient water management is a key factor for improving availability and uptake of nutrients by the plants and reducing runoff and leaching. Contamination of water sources are sometimes dire, leading to extreme environmental and social problems, such as eutrophication of major water ways from N and P runoff, unsafe levels of NO_3^- in the drinking water, or emission of ozone-depleting nitrous oxides and other greenhouse gases.

3.3 Right time

Timing applications properly ensures that nutrients are available during peak periods of plant uptake and are in adequate supply at critical stages of development in the crop. Knowing how nutrient uptake pattern vary by stage of growth is very useful for establishing a schedule for fertilizer applications. A good example of this can be seen by examining the seasonal N demands in floricane-fruiting raspberries (*Rubus idaeus* L.) (Fig. 36.2). These plants produce biennial shoots (canes) on a perennial root system. New shoots on the plants are completely vegetative in the first year (primocanes) and overwinter and develop fruit the following year (floricanes). Both types of canes exist simultaneously on the plant, but the floricanes senesce after harvest each year. As result, the plants have two major peaks in N uptake. The first peak occurs in early spring and coincides with the development of fruiting laterals on the

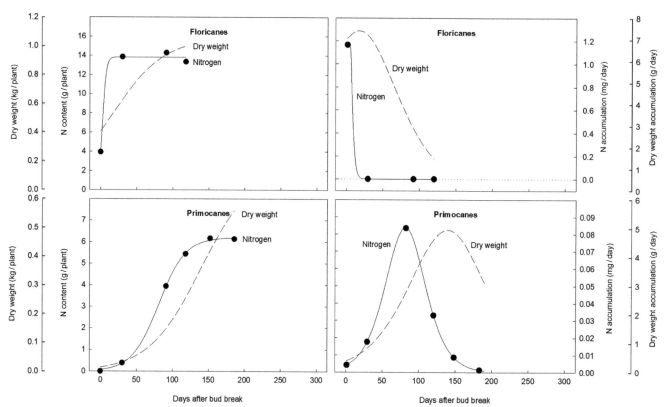

FIG. 36.2 Changes in total content and accumulation of N *(symbols)* and plant dry weight *(dashed lines)* in floricanes (bud break-harvest) and primocanes (bud break-leaf senescence) of red raspberry. To maximum N availability during peaks in N uptake, a split-application of N fertilizer should be applied at bud break (early spring) for the floricanes and at ≈80 days after bud break (early summer) for the primocanes. *Modified from Rempel, H.G., Strik, B.C., Righetti, T.L., 2004. Uptake, partitioning and storage of fertilizer nitrogen in red raspberry as affected by rate and timing of application. J. Am. Soc. Hortic. Sci. 129, 439–448.*

floricanes. The second peak occurs during summer and coincides with production and growth of the new primocanes. Rempel et al. (2004) determined that split applications of N fertilizer in spring and early summer were more efficient in terms of N recovery than a single application in the spring. Information on nutrient uptake is fairly limited in many fruit crops, and acquiring more information of this sort would be very useful for developing more efficient nutrient management strategies.

Once peak demands for nutrients have been identified, a variety of products and practices can be employed to better coordinate availability with demand. Among the simplest is using multiple applications of dry fertilizer over the season. Work with stable isotopes indicates that split application of dry N fertilizer increases N recovery in a number of crops, including blueberry (Bañados et al., 2012). Timing may be further improved by using soluble fertilizers and applying them frequently by fertigation or exact times using foliar applications (Melgar et al., 2010).

Controlled-release fertilizers are also available to improve timing of nutrient availability. The fertilizers are coated with S or other specialized materials (polymers) that reduce their solubility and slow down the rate at which they are released into the soil. Prolonged release can provide more uniform plant nutrition and reduce leaching losses (Shaviv and Mikkelsen, 1993). Nutrient release from these fertilizers is controlled by a variety of environmental factors such as increased soil moisture and temperature and may be dependent on microbial activity. These fertilizers are relatively expensive but may be cost-effective under certain circumstances for a fruit crop. For example, many strawberry (*Fragaria × ananassa* Duch.) growers use controlled-release fertilizers routinely (Bottoms et al., 2013). Increasing concerns with nutrient contamination of water sources and development of less expensive products may stimulate more use of controlled-release fertilizers in other fruit crops. Fertilizer additives such as nitrification inhibitors and urease inhibitors are also available and can be used to maintain N in the NH_4 form longer, releasing NO_3^- slowly from the fertilizer and minimizing the potential for N leaching. These inhibitors tend to be most effective on sandy soils and soils low in organic matter.

In many fruit crops, demands for nutrients coincide with fruiting. Fruiting often varies among cultivars and can differ by weeks or months. Application of nutrients should be scheduled accordingly and adjusted for differences in weather conditions from year to year. However, such adjustments are sometimes unfeasible due to practical or economic considerations. For example, applications of dry fertilizers are weather dependent and require sufficient rain or irrigation to dissolve and wash into the root zone. In parts of China, farmers must apply all of the N to crops at planting due to a shortage of labor for applying additional N later in the season (Johnston and Bruulsema, 2014). Soils exhibit varying capacities to retain plant-available nutrients. Furthermore, most fruit crops are capable of storing nutrients (luxury consumption) and reallocating them to new growth at a later time. Nitrogen, phosphorus, potassium, magnesium, and sulfur are readily transported in the phloem from older tissues to actively growing parts of the plant (Epstein and Bloom, 2005). Therefore, it might be unnecessary to perfectly synchronize application of certain nutrients with crop demands. For examples, many growers find that a single application of K in the spring is sufficient to meet crop demands for the nutrient. Furthermore, K is relatively immobile in many soils and concerns with leaching are minimal.

Mineralization of organic amendments and fertilizers can supply large quantities of N and other nutrients, but if the needs of the crop precede the release of these nutrients, deficiencies may limit productivity. In a study on organic blueberries, plants fertilized with a dry feather meal required additional N from fish emulsion during the first year after planting, which increased costs by \$573/ha (Julian et al., 2012). This likely happened because the feather meal was not washed into the soil thoroughly as a result of little rainfall that spring. Additional fish emulsion was not required in subsequent years when the first application of feather meal was done earlier in the season.

3.4 Right place

To be effective, nutrients must be placed in areas where they are accessible to the plant. This could include application to the soil in the vicinity of the roots, application through the irrigation system (fertigation), or spraying them directly on the leaves of the plants. Fertilizers placed correctly ensure that nutrients are readily available for plant uptake when needed and reduce potential losses that may result in negative economic and environmental consequences. Proper placement evolves continuously during growth and development of a plant and is affected by a host of factors, including, but not limited to, plant spacing, crop rooting patterns, weather, soil conditions, nutrient placement technologies, and tillage practices. There is still much to learn about what constitutes the right place and how well it can be predicted when management decisions need to be made.

Generally, soil applied fertilizers should be placed as close to the plant roots as possible. Nutrient uptake is highest in young roots and declines sharply as roots age (Bouma et al., 2001; Volder et al., 2005). New roots emerge from seeds, bare roots, or the root ball of newly transplanted plants and develop in response to temperature (Kasper and Bland, 1992; McMichael and Burke, 1998) and availability of soil water and nutrients (Abbott and Gough, 1987; Dong et al., 2001; Hogue and Neilsen, 1986; Jackson and Bloom, 1990; Tagliavini et al., 1991). In young plants, fertilizers are usually applied in the vicinity of the roots by banding them with the seeds during sowing or by mixing them with the soil at the bottom of the planting hole prior to transplanting (Watson, 1994). In mature, woody fruit trees and shrubs, new roots extend a distance equivalent to the spread of the branch tips or beyond (Gilman, 1988), and therefore, most dry fertilizers are applied on the soil near the drip line (outer edge of the canopy). More research is needed to determine optimum placement of fertilizers in mature fruit trees and shrubs.

Rooting depth also increases with plant age and varies considerably among crop species and cultivars (Sponchiado et al., 1989; Thorup-Kristensen and van den Boogaard, 1998). It is likely that shallow-rooted plants are more sensitive to fertilizer placement than those that are deeply rooted, since NO_3^- and other mobile nutrients are more easily leached below the roots. Care is needed when applying fertilizers soon after planting or to crops with shallow root systems.

Fertilizers are often banded or broadcast on the soil surface in many crops, but this practice provides limited benefit to plants irrigated by drip. Therefore, most growers using drip will apply nutrients by fertigation (Kafkafi and Tarchitzky, 2011). With fertigation, placement of fertilizers can be controlled precisely with emitter spacing and the number and location of the drip laterals in each row. Because most roots are located near the drip emitters, fixed and diffusion-limited ions such as NH_4^+, PO_4^{-2}, and Fe^{+2} can be placed directly where needed and reduce potential for soil sorption (Bañuls et al., 2003). Fertigation also facilitates efficient use of mobile nutrients, including NO_3^- and urea, provided irrigation is managed properly (Ehaliotis et al., 2010; Quiñones et al., 2003). Overwatering by drip quickly leaches mobile nutrients below the root zone (Hanson et al., 2006).

Foliar sprays are widely used to apply micronutrients to many fruit crops (Mortdevt, 2011). Sticker-spreader agents are usually recommended when applying micronutrient solutions and help the spray adhere to leaves. Macronutrients including urea can also be applied to the leaves, but repeated applications are needed (Dong et al., 2005; Johnson et al., 2001; Khayyat et al., 2012), and in many situations, the fertilizer drips off and is ultimately absorbed by the roots (Reickenberg and Pritts, 1996; Strik et al., 2004).

Ideally, whether fertilizers are wet or dry and applied by broadcasting, banding, fertigation, or as a foliar spray, the method should maximize nutrient uptake by the crop and minimize nutrient losses. Actual methods used are compromises between these considerations and suitability to existing equipment, cultivation practices, and costs. However, all good methods apply the right source and rate uniformly and should be selected appropriately. GPS and other precision practices are now available to assist with fertilizer placement and help improve application uniformity.

References

Abbott, J.D., Gough, R.E., 1987. Seasonal development of highbush blueberry roots under sawdust mulch. J. Am. Soc. Hortic. Sci. 112, 60–62.

Bañados, M.P., Strik, B.C., Bryla, D.R., Righetti, T.L., 2012. Response of highbush blueberry to nitrogen fertilizer during field establishment. I: accumulation and allocation of fertilizer nitrogen and biomass. HortScience 47, 648–655.

Bañuls, J., Quiñones, A., Martín, B., Primo-Millo, E., Legaz, F., 2003. Effects of the frequency of iron chelate supply by fertigation on iron chlorosis in citrus. J. Plant Nutr. 26, 1985–1996.

Beverly, R.B., Stark, J.C., Ojala, J.C., Embleton, T.W., 1984. Nutrient diagnosis of Valencia oranges by DRIS. J. Am. Soc. Hortic. Sci. 109, 649–654.

Bottoms, T.G., Hartz, T.K., Cahn, M.D., Farrara, B.F., 2013. Crop and soil nitrogen dynamics in annual strawberry production in California. HortScience 48, 1034–1039.

Bouma, T.J., Yanai, R.D., Elkin, A.D., Hartmond, U., Flores-Alva, D.E., Eissenstat, D.M., 2001. Estimating age-dependent costs and benefits of roots with contrasting life span: comparing apples and oranges. New Phytol. 150, 685–695.

Bruulsema, T.W., Fixen, P.E., Sulewski, G.D. (Eds.), 2012. 4R Plant Nutrition. A Manual for Improving the Management of Plant Nutrition. International Plant Nutrition Institute, Norcross, GA.

Bryla, D.R., Machado, R.M.A., 2011. Comparative effects of nitrogen fertigation and granular fertilizer application on growth and availability of soil nitrogen during establishment of highbush blueberry. Front. Plant Sci. 2, 46. https://doi.org/10.3389/fpls.2011.00046.

Bryla, D.R., Machado, R.M.A., Shireman, A.D., 2010. Effects of method and level of nitrogen fertilizer application on soil pH, electrical conductivity, and availability of ammonium and nitrate in blueberry. Acta Hortic. (868), 95–102.

Davarpanah, S., Tehranifar, A., Davarynejad, G., Aran, M., Abadía, J., Khorassani, R., 2017. Effects of foliar nano-nitrogen and urea fertilizers on the physical and chemical properties of pomegranate (Punica granatum cv. Ardestani) fruits. HortScience 52, 288–294.

Dimkpa, C.O., Bindraban, P.S., 2018. Nanofertilizers: new products for the industry. J. Agric. Food Chem. 66, 6462–6473.

Dong, S., Scagel, C.F., Cheng, L., Fuchigami, L.H., Rygiewicz, P.T., 2001. Soil temperature and plant growth stage influence nitrogen uptake and amino acid concentration of apple during early spring growth. Tree Physiol. 21, 541–547.

Dong, S., Neilsen, D., Neilsen, G.H., Fuchigami, L.H., 2005. Foliar N application reduces soil NO_3-N leaching loss in apple orchards. Plant Soil 268, 357–366.

Ehaliotis, C., Massas, I., Pavlou, G., 2010. Efficient urea-N and KNO_3-N uptake by vegetable plants using fertigation. Agron. Sustain. Dev. 30, 763–768.

Epstein, E., Bloom, A.J., 2005. Mineral Nutrients of Plants: Principles and Perspectives. Sinauer Assoc., Inc., Sunderland, MA.

Fallihi, E., 1985. Prediction of quality by preharvest fruit and leaf mineral analyses in Starkspur Golden delicious apple. J. Am. Soc. Hortic. Sci. 110, 524–527.

Fallihi, E., 1988. Ranking tissue mineral analyses to identify mineral limitation quality in fruit. J. Am. Soc. Hortic. Sci. 113, 282–289.

Forge, T., Neilsen, G., Neilsen, D., Hogue, E., Faubion, D., 2013. Composted dairy manure and alfalfa hay mulch affect soil ecology and early production of 'Braeburn' apple on M.9 rootstock. HortScience 48, 645–651.

Gilman, E.F., 1988. Predicting root spread from trunk diameter and branch spread. J. Arboric. 14, 85–89.

Hanson, B., O'Connell, N., Hopmans, J., Simunek, J., Beede, R., 2006. Fertigation with Microirrigation. University of California, Agriculture and Natural Resources, Communication Services Publication. 21620.

Hartz, T.K., Smith, R.F., Schulbach, K.F., LeStrange, M., 1994. On-farm nitrogen tests improve fertilizer efficiency, protect groundwater. Calif. Agric. 48, 29–32.

Hartz, T.K., Mitchell, J.P., Giannini, C., 2000. Nitrogen and carbon mineralization dynamics of manures and composts. HortScience 35, 209–212.

Hogue, E.J., Neilsen, G.H., 1986. Effect of root temperature and varying cation ratios on growth and leaf cation concentration of apple seedlings grown in nutrient solution. Can. J. Plant Sci. 66, 637–645.

Hue, N.V., Fox, R.L., 2010. Predicting plant phosphorus requirements for Hawaii soils using a combination of phosphorus sorption isotherms and chemical extraction methods. Commun. Soil Sci. Plant Anal. 41, 133–143.

Jackson, L.E., Bloom, A.J., 1990. Root distribution in relation to soil nitrogen availability in field grown tomatoes. Plant Soil 128, 115–126.

Johnson, R.S., Rosecrance, R.C., Weinbaum, S.A., Andris, H., Wang, J., 2001. Can we approach complete dependence on foliar-applied urea nitrogen in an early-maturing peach? J. Am. Soc. Hortic. Sci. 126, 364–370.

Johnston, A.M., Bruulsema, T.W., 2014. 4R nutrient stewardship for improved nutrient use efficiency. Process. Eng. 83, 365–370.

Julian, J.W., Strik, B.C., Larco, H.O., Bryla, D.R., Sullivan, D.M., 2012. Costs of establishing organic northern highbush blueberry: impacts of planting method, fertilization, and mulch type. HortScience 47, 866–873.

Kafkafi, U., Tarchitzky, J., 2011. Fertigation: A Tool for Efficient Fertilizer and Water Management. Intl. Fert. Industry Assoc, Paris, France.

Kasper, T.C., Bland, W.L., 1992. Soil temperature and root growth. Soil Sci. 154, 290–299.

Khayyat, M., Tehranifar, A., Zaree, M., Karimian, Z., Aminifard, M.H., Vazifeshenas, M.R., Amini, S., Noori, Y., Shakeri, M., 2012. Effects of potassium nitrate spraying on fruit characteristics of 'Malas Yazdi' pomegranate. J. Plant Nutr. 35, 1387–1393.

Kirchmann, H., Bergström, L., 2001. Do organic farming practices reduce nitrate leaching? Commun. Soil Sci. Plant Anal. 32, 997–1028.

Manjunatha, S.B., Biradar, D.P., Aladakatti, Y.R., 2016. Nanotechnology and its applications in agriculture: a review. J. Farm Sci. 29, 1–13.

McMichael, B.L., Burke, J.J., 1998. Soil temperature and root growth. HortScience 33, 947–951.

Melgar, J.C., Schumann, A.W., Syvertsen, J.P., 2010. Fertigation frequency affects growth and water and nitrogen use efficiencies of Swingle citrumelo citrus rootstock seedlings. HortScience 45, 1255–1258.

Mortdevt, J., 2011. Micronutrients. In: Efficient Fertilizer Use Manual. http://www.back-to-basics.net/efu/pdfs/Micronutrients.pdf. (19.04.07).

Nelson, N., Mikkelsen, R., 2008. Meeting the phosphorus requirement on organic farms. Better Crops 92, 12–14.

Quiñones, A., Bañuls, J., Primo-Millo, E., Legaz, F., 2003. Effects of ^{15}N application frequency on nitrogen uptake efficiency in *Citrus* trees. J. Plant Physiol. 160, 1429–1434.

Reickenberg, R.L., Pritts, M.P., 1996. Dynamics of nutrient uptake from foliar fertilizers in red raspberry (*Rubus idaeus* L.). J. Am. Soc. Hortic. Sci. 121, 158–163.

Rempel, H.G., Strik, B.C., Righetti, T.L., 2004. Uptake, partitioning and storage of fertilizer nitrogen in red raspberry as affected by rate and timing of application. J. Am. Soc. Hortic. Sci. 129, 439–448.

Roberts, T.L., 2007. Right product, right rate, right time and right place: the foundation of best management practices for fertilizer. In: Fertilizer Best Management Practices. Intl. Fert. Industry Assoc, Paris, France, pp. 29–32.

Sanchez, E.E., Khemira, H., Sugar, D., Righetti, T.L., 1995. Nitrogen management in orchards. In: Bacon, P.E. (Ed.), Nitrogen Fertilization in the Environment. Marcel Dekker, New York, NY, pp. 327–380.

Shaviv, A., Mikkelsen, R.L., 1993. Controlled-release fertilizers to increase efficiency of nutrient use and minimize environmental degradation—a review. Fert. Res. 35, 1–12.

Sponchiado, B.N., White, J.W., Castillo, J.A., Jones, P.G., 1989. Root growth of four common bean cultivars in relation to drought tolerance in environments with contrasting soil types. Exp. Agric. 25, 249–257.

Strik, B., Righetti, T., Buller, G., 2004. Influence of rate, timing, and method of nitrogen fertilizer application on uptake and use of fertilizer nitrogen, growth, and yield of June-bearing strawberry. J. Am. Soc. Hortic. Sci. 129, 165–174.

Tagliavini, M., Hogue, E.J., Neilson, G.H., 1991. Influence of phosphorus nutrition and root zone temperature on growth and mineral uptake of peach seedlings. J. Plant Nutr. 14, 1267–1276.

Thompson, A.A., Peck, G.M., 2017. Alternative fertility management for establishing new apple orchards in the mid-Atlantic. HortScience 52, 1313–1319.

Thorup-Kristensen, K., van den Boogaard, R., 1998. Temporal and spatial root development of cauliflower (*Brassica oleracea* L. var. botrytis L.). Plant Soil 201, 37–47.

Volder, A., Smart, D.R., Bloom, A.J., Eissenstat, D.M., 2005. Rapid decline in nitrate uptake and respiration with age in fine lateral roots of grape: implications for root efficiency and competitive effectiveness. New Phytol. 165, 493–502.

Walworth, J.L., Sumner, M.E., 1988. Foliar diagnosis: a review. Adv. Plant Nutr. 3, 193–241.

Watson, G.W., 1994. Root growth response to fertilizers. J. Arboric. 20, 4–8.

Weng, H., Cheng, L., 2011. Differential effects of nitrogen supply on skin pigmentation and flesh starch breakdown of 'gala' apple. HortScience 46, 1116–1120.

Yang, B., Li, G., Yang, S., He, Z., Zhou, C., Yao, L., 2015. Effect of application ratio of potassium over nitrogen on litchi fruit yield, quality, and storability. HortScience 50, 916–920.

Further reading

Teixeira, L.A.J., Quaggio, J.A., Cantarella, H., Mellis, E.V., 2011. Potassium fertilization for pineapple: effects on plant growth and fruit yield. Rev. Bras. Frutic. 33, 618–626.

Thorup, J.T., Stewart, J.W.B., 1988. Optimum fertilizer use with differing management practices and changing government policies. In: Proceedings of the 25th Anniversary Symposium of Division S-8 Soil Science Society of America, Advances in Fertilizer Technology and Use, Anaheim, CA, November 28. Potash & Phosphate Institute, Atlanta, GA.

37

Climate-smart integrated soil fertility management in fruit crops: An overview

A.K. Srivastava*

Indian Council of Agricultural Research-Central Citrus Research Institute, Nagpur, India
*Corresponding author. E-mail: aksrivas2007@gmail.com

1 Introduction

Fruit crops represent hardly 1% of the global agricultural land area, while Mediterranean region covers maximum of 11%, which is of great economic importance in world trade and tariff. Approximately 1.7 million (2.8%) deaths worldwide are attributable to micronutrient deficiency induced through suboptimum consumption of fruits and vegetables and regarded as top 10 selected risk factors for global mortality. In the 21st century, nutrient-efficient plants will play a major role in increasing crop yields compared with the 20th century, mainly due to limited land and water resources available for crop production, higher cost of inorganic fertilizer inputs, declining trends in crop yields globally, and increasing environmental concerns. According to one estimate, at least 60% of the world's arable lands have mineral deficiencies or elemental toxicity problems, and on such soils, fertilizers and lime amendments are essential for achieving improved crop yields. Perennial fruit trees play an important role in carbon cycle of terrestrial ecosystems and sequestering atmospheric CO₂ (Wu et al., 2012; Lakso, 2010; Bwalya, 2013). Fruit crops by the virtue of their perennial nature of woody framework (nutrients locked therein), extended physiological stages of growth, differential root distribution pattern (root volume distribution), growth stages from the point of view of nutrient requirement, and preferential requirement of some nutrients by specific fruit crop collectively make them nutritionally more efficient than the annual crops (Srivastava and Singh, 2008a; Scholberg and Morgan, 2012).

A guesstimate proclaims over 900 million people in the world are undernourished and malnutrition alone is responsible for 3.5 million deaths annually (Srivastava, 2010a, b). In the backdrop of demography-driven diminishing per capita availability of land (more so in fruit crops), sustaining soil fertility management has gained a phenomenal

A.K. Srivastava, Chengxiao Hu (eds.)
Fruit Crops: Diagnosis and Management of Nutrient Constraints
https://doi.org/10.1016/B978-0-12-818732-6.00037-X

significance in meeting the multipronged challenges of sustaining the quality production on the one hand and ensuring the carrying capacity of soil health on the other hand (Srivastava and Singh, 2003c). Considering the economics of fruit production, fertilizers alone on an average constitutes about 20%–30% of total cost of production, which is a significant recurring expenditure that a grower needs to invest every year (Srivastava and Singh, 2003a, b, c, 2005).

Renewed and intensified efforts are in progress during the past 10–15 years using alternative management practices, ever since the depleting soil fertility has become a serious concern with regard to the practice of high-density orcharding coupled with heavy use of chemical fertilizers (Kohli et al., 1998; Srivastava, 2013c, d), bringing unprecedented reduction in soil organic matter (Intrigliolo and Stagno, 2001). However, the fertilizers act in exactly the same way as nutrient from organic resources in the soil, since they are chemically the same (Srivastava and Kohli, 1997). The plant itself cannot tell where the nutrient is coming from (Srivastava et al., 2014). In recent years the nutrient additions have been exclusively in favor of mineral fertilizers due to demographic pressure and of demands related to lifestyles and trade involvement. While the quick and substantial response to fruit yield due to mineral fertilizers eclipsed the use of organic manures, the inadequate supply of the latter sources exacerbated this change (Ghosh, 2000). Integrated soil fertility management (ISFM) with emphasis on the use of bioorganics is a comparatively recent concept that needs to be vigorously pursued to achieve the sustainability in fruit crop production trend spaced over the years. Additionally, crop nutrition, therefore, must respect the prescriptions of ISFM (Srivastava and Ngullie, 2009). The merits of ISFM-based practices also take into account the mobilization of unavailable nutrients, which could be effected by speeding up the rate of mineralization of various organic substrates.

A dynamic concept of nutrient management through ISFM is looked upon the economic yield in terms of fruit yield coupled with quality on the one hand and soil physicochemical and microbiological health on other hand as a marker of resistance against the nutrient mining that arises because of failure to strike a balance between annual nutrient demand versus quantum of nutrients applied (Srivastava and Singh, 2008a). Soils under citrus differ from other cultivated soils, which remain fallow for 3–6 months every year forcing depletion of soil organic matter (Bhargava, 2002). On the contrary, biological oxidation of existing C continues in soil covered under crop (Srivastava et al., 2002). Multiple nutrient deficiencies are considered to have triggering effect on potential source of atmospheric CO_2. Soil carbon stock is, hence, considered as an important criterion of determining the impact of ISFM in the longer version of impact assessment (He et al., 1997; Joa et al., 2006). The amount of accumulated C within the rhizosphere soil does not continue to increase with time with increasing C outputs. An upper limit of C saturation level occurs, which governs the ultimate limit of soil C sink and rate of C sequestration in mineral soils, independent of C input rate. An understanding of mechanism involved in C stabilization in soils is needed for controlling and enhancing soil C sequestration (Goh, 2004) with fruit crop-based land uses.

Over the years the concepts of integrated nutrient management (INM) and integrated soil fertility management (ISFM) have been gaining acceptance, moving away from a more sectoral and input-driven approach (Srivastava et al., 2012, 2019). INM advocates the careful management of nutrient stocks and flows in a way that leads to profitable and sustained production. ISFM not only emphasizes the management of nutrient flows but also highlights other important aspects of soil complex such as maintaining organic matter content, soil structure, moisture, and microbial biodiversity (Srivastava et al., 2015b). Still, more attention is needed toward integrated soil biological management as a crucial aspect of soil fertility management since providing protection to citrus rhizosphere against the nutrient depletion is of utmost importance for sustained orchard production in which the objectivity of ISFM could have far reaching consequences (Srivastava et al., 2008). Exploring microbial diversity perspectives in citrus crop is, therefore, important and equally useful to arrive at measures that can act as indicators of soil quality and sustainable orchard productivity using biological soil management to be intimately integrated with ISFM (Srivastava et al., 2015b). Diagnosis of nutrient constraints and their efficient management has, therefore, now shifted in favor of ISFM through collective use of organic manures, inorganic fertilizers, and beneficial microorganisms that becomes all the more difficult.

2 Soil fertility and functions: Important facts

Soil is a distinct living entity that is one of the core building blocks of land. Land consists of soil, rocks, rivers, and vegetation (Lal, 2010). Soil contributes five principal functions within a landscape: (i) nutrient cycling; (ii) water retention; (iii) biodiversity and habitat; (iv) storing, filtering, buffering, and transforming compounds; and (v) provision of physical stability and support (Blum, 1993). A fully functioning soil lies at the heart of solving the big issue of food security, biodiversity, climate change, and freshwater regulation, but to date, there has been no easy way to communicate these linkages. The narrative on soils must be improved, and its voice must be raised if the required response is to be achieved. The key aim in securing soil is to maintain and optimize its functionality; its diverse and complex ecosystems of soil biota; its nutrient cycling capacity; and its roles as substrate for growing plants, as a regulator, filter, and holder of freshwater, and as a potential mediator of climate change through the sequestration of atmospheric carbon dioxide (Koch et al., 2013). Soil security is explored as a conceptual framework that could be used as the basis for a soil policy framework with soil carbon as an exemplar indicator.

Soil function fertility refers to the ability of soil to support and sustain plant growth, which relates to making all the essential nutrients available for root uptake. This is facilitated by their storage in soil organic matter, nutrient recycling from organic to plant available mineral forms, and physicochemical processes that control their fixation and release (Srivastava and Kohli, 1997). On the other hand, managed soils are highly dynamic system that makes the soil work and supply ecosystem services to humans. Overall the fertility and functioning of soils strongly depend on interactions between soil mineral matrix, plants, and microbes. These are responsible for both building and decomposing soil organic matter, and therefore, for the preservation and availability of nutrients in soils, cycling of nutrients in soils must be preserved (Srivastava et al., 2008).

Soil health considers the physical, chemical, and biological properties of the soil and the disturbance and ameliorative responses of land managers. Soil health also describes the capacity of the soil to meet performance standards relating to nutrient and water storage and supply, biological diversity and function, and resistance to degradation. The most important of these manageable services include biological nitrogen fixation (BNF), other symbiotic and beneficial organisms, nutrient and moisture supply, carbon storage, and protection from erosion (Srivastava et al., 2014). Let us look at soil nutrient imbalances, nutrient mining, and sustainability of nutrient management practices. Managing soil carbon for multiple benefits addresses to enhance a range of ecological services. Increasing the soil organic matter of degraded soils can boost crop productivity, sequester CO_2, enhance soil microbial growth and activities, and improve water capture and retention. Soil carbon stocks, highly vulnerable to human activities, decrease significantly in response to changes in land capability and land use such as deforestation and increased tillage continues (Srivastava and Singh, 2009b; Tagaliavini et al., 2007). Opportunity to use existing mechanisms encouraging active management of soil carbon—land use planning that excludes vulnerable soils from land uses eventually lead to soil organic carbon losses. Promotion of proper management practices to protect and enhance soil organic matter as an essential element of good soil and environmental quality. Promotion of sources of plant nutrients (e.g., cover crops, legumes, crop diversification that enhance soil organic carbon stocks. Integration of several crops in a field at the same time to increase soil organic matter, soil biodiversity and soil health. Decline in soil fertility is the major constraint limiting the productivity of fruit crops. Continuous reduction in nutrient density of different fruit crops is an indication of nutrient mining-induced decline in fruit crop productivity (Srivastava and Ngullie, 2009).

3 Fruit crops as carbon sink and response at elevated CO_2

Pedospheric, atmospheric, and biotic carbon pools contain 2400, 750, and 550 Gt of carbon, respectively (Brady and Weil, 2004). In particular, about 80% of the biotic pool of carbon is fixed in plants and fungi (Kimmins, 1997). Modern agricultural practices convert the pedosphere, which is normally a carbon sink, into a significant carbon source, a process that is resulting in significant repercussions on the total amount of CO_2 in the atmosphere. This is the case in modern fruit orchards, especially in areas where rainfall is infrequent during the growing season and the soil is managed with shallow and repeated tillage (Xiloyannis et al., 2002).

Fruit production and quality depend on adequate source-sink relationships. Carbohydrates (CH) translocated from leaves or reserve organs are the most important for the growth and development of sink organs (mainly fruits). Up to 60% of CH produced daily can be lost through respiration. Carbohydrates constitute over 65% of the dry matter of tree crops. Increasing the leaf-fruit ratio generally increases fruit growth and CH content. Photosynthesis increases with fruit load, and the leaves next to fruits are strong sources for CH. The leaf-fruit ratio is species, cultivar, and geographic location dependent. The optimal leaf area in various species is 200 cm2 per 100 g of fruit (Fischer et al., 2012). Several studies have recently documented the effects of elevated atmospheric CO_2 concentrations on photosynthesis in various fruiting trees. In a longer 3-month study, Keutgen and Chen (2001) noted that cuttings of Citrus madurensis exposed to 600-ppm CO_2 displayed rates of photosynthesis that were more than 300% greater than rates observed in control cuttings exposed to 300-ppm CO_2. Likewise, Schaffer et al. (1997) reported that atmospheric CO_2 enrichment enhanced the rates of net photosynthesis in various tropical and subtropical fruit trees, including avocado, banana, citrus, mango, and mangosteen.

There is a little information in the literature on the effects of atmospheric CO_2 enrichment on mineral element concentrations in tissues of tropical C_3 plants (Hocking and Meyer, 1991a, b). Conroy (1992) observed that atmospheric CO_2 enrichment alerts foliar nutrient concentrations required to maintain maximum productivity (critical concentrations). Critical concentrations in leaves are routinely used to evaluate nutrient status of crops and manage fertilizer programs (Conroy, 1992). Since atmospheric CO_2 is expected to increase steadily (Ehleringer and Cerling, 1995; Houghton and Skole, 1990), knowledge of changes in foliar nutrient concentrations in response to CO_2 enrichment is important for diagnosing nutrient deficiencies that are based on critical concentrations (Conroy, 1992; Hocking and Meyer, 1991b). Leaf mineral element concentrations were generally lower for trees grown at the higher ambient CO_2 concentration, presumably due to a dilution effect from an increased growth rates (Schaffer et al., 1997). Under changing pattern of climatic condition, there is a need to classify the plant species for their efficient response to enhance CO_2 to climate change. Nutrient concentration in various tree components varies in accordance to their utilization for regulating different physiological processes. These nutrients are also translocated to various components as and when required. Most of the macronutrients (N, P, K, Mg, S, and Na) are highly mobile and leachable except for Ca. Ca being an immobile element can be used as indicator of carbon content in different tree components (Negi et al., 2003).

The total carbon emission and the heterotrophic carbon emission in the grape orchard were estimated to 422.7-g C m sup(−2)y sup(−1) and 222.5-g C m sup(−2) y sup(−1), respectively, and both values were one-half of those in the peach orchard. The total carbon supply was 401.0-g C m sup(−2)y sup(−1) in the grape orchard (litter from floor vegetation 54.5%, litter from grapevine 34%, and fertilizer 11.5%) and was only one-third of the value in the peach orchard. It was determined that carbon from floor vegetation is the largest input to the soil in the grape orchard, which is similar to that in the peach orchard. These results indicate that the grape orchard sequestered carbon of 178.5-g m sup(−2)y sup(−1), which is one-third of that in the peach orchard suggesting that soil in orchards acts as a carbon sink owing to a large amount of carbon input from the floor vegetation (Sekikawa et al., 2003).

Trials were carried out in southern Italy on olive (Olea europaea L.) and peach orchards (Prunus persica L.) at different age and plant densities. At the end of each vegetative season, values of fixed atmospheric CO_2 were calculated by measuring dry matter accumulation and partitioning in the different plant organs. In the early years sequestered CO_2 was primarily distributed in the permanent structures and in the root system, while in mature orchards the fixed CO_2 was distributed in leaves, pruning materials, and fruit. Significant differences in amounts of fixed CO_2 were observed in peach orchards cultivated using different planting and training strategies. The results underline the importance of training system, plant density, and cultivation techniques in the absorption of atmospheric CO_2 and its storage as organic matter in the soil (Sofo et al., 2005). Of all the component parts, that is, fruits, stem, bark, branches, twigs, and roots, the carbon storage was maximum in fruits followed by roots. The present studies indicate the innate potential of Nagpur mandarin as CO_2 sequester being dependent on stem, leaf, and root biomass under different diameter classes (Bhatnagar et al., 2016).

Elevated CO_2 enhances the photosynthetic rates of fruiting trees; it should also lead to increased biomass production in them. And it does. In the 2-year study of Centritto et al. (1999a), cherry seedlings grown at 700-ppm CO_2 exhibited photosynthetic rates that were 44% greater than those displayed by seedlings grown in ambient air, independent of a concomitant soil moisture treatment. In the 2-year study of Centritto et al. (1999b), for example, well-watered and water-stressed seedlings growing at twice-ambient CO_2 concentrations displayed basal trunk areas that were 47% and 51% larger than their respective ambient controls. Similarly, in a study spanning more than 13 years, Idso and Kimball (2001) demonstrated that the aboveground wood biomass of mature sour orange trees growing in air enriched with an additional 300 ppm of CO_2 was 80% greater than that attained by control trees growing in ambient air. However, elevated CO_2 did not improve plant-water relations (e.g., bulk leaf-water potential, osmotic potentials at full and zero turgor, relative water content at zero turgor, and bulk modulus of elasticity of the cell) and thus did not increase water-stress tolerance of cherry seedlings (Centritto et al., 1999a) but accelerated ontogenic development, irrespective of water status (Centritto et al., 1999b). According to Pan et al. (1998), elevated CO_2 enhanced the photosynthesis of apple plants and altered carbohydrate accumulation in mature leaves in favor of starch and sorbitol over sucrose, while Idso and Kimball (2001) observed that CO_2-enriched sour orange trees may have reached an equilibrium condition with respect to the CO_2-induced enhancement of wood biomass and fruit production and that they will not substantially depart from these steady-state responses over the remainder of their lifespan.

In the light of climate change-related issues, perennial fruit trees play an important role in carbon cycle of terrestrial ecosystems and sequestering atmospheric CO_2 (Lobell et al., 2005; Guimarães et al., 2014). According to Wu et al. (2012), net C sink and C storage in biomass of apple orchard ranged from 19- to 32-Tg C, respectively, and from 230- to 475-Tg C in 20-year period, amounting to 4.5% of total net C sink in the terrestrial ecosystems in China. In an estimate, Lakso (2010) observed that an acre of apple orchard fixed about 20 tons of CO_2 from the air each season and provided over 15 tons of O_2, equivalent to over 5 billion BTUs of cooling power, while Bwalya (2013) showed that citrus tree carbon sequestration in biomass ranged from 23.9-ton CO_2/ha for young trees to 109-ton CO_2/ha for mature trees. Perennial fruit trees act as strong carbon sink by sequestering the atmospheric carbon (Sugiura et al., 2007). Studies in the past have shown increase in the yield of fruit crops like apple (Wu et al., 2012), grape (Bindi et al., 1997), Japanese pears (Ito et al., 1999), mango (Goodfellow et al., 1997), and citrus (Peng et al., 2000) in response to elevated

CO_2 concentration. However, conversion of forest land into fruit orchard cultivation led to a reduction of 5%–23% and 4%–21% reductions in soil organic carbon and N-stock, respectively. Compared with other soil uses, fruit crops like guava, mango, and sapota contributed to improving soil organic carbon stratification index (de Campos Bernardi et al., 2007).

4 Nutrient removal versus nutrient requirement

Fruit crops by the virtue of their perenniality, longer growth period, and developing fruits acting as major sink, are considered nutrient extracting in nature and hence so nutrient responsive. Interestingly, nutrient uptake pattern (Table 37.1) by major fruit crops displays the fact that K removal is far higher than N or K; however, P removal is nearly half of N removal. Crops like banana, citrus, grape, kiwifruit, and mango are considered highly nutrient-exhaustive crops, warranting their replenishment to ensure long-term sustainability in production (Tandon and Muralidharadu, 2010; Srivastava and Singh, 2004a, b).

TABLE 37.1 Nutrient removal pattern by major fruit crops.

Crops	Nutrients removal (kg/ha)			Yield (t/ha)
	N	P_2O_5	K_2O	
Apple (*Malus pumila*)	100.0	45.0	180.0	25.0
Avocado (*Persea americana*)	11.3	3.9	23.4	10.0
Banana (*Musa acuminata* L.)	250.0	60.0	1000.0	40.0
Citrus spp.				
Nagpur mandarin (*C. reticulata* Blanco)	3.4	0.37	3.72	1.0
Khasi mandarin (*C. reticulata* Blanco)	1.81	0.16	2.18	1.0
Coorg mandarin (*C. reticulata* Blanco)	3.04	0.27	3.31	1.0
Kinnow mandarin (*C. nobilis* T. × *C. deliciosa* L.)	2.40	0.57	2.35	1.0
Acid lime (*C. aurantifolia* Swingle)	1.22	0.21	4.06	1.0
Durian (*Durio zibethinus* L.)	16.1	6.6	33.5	6.7
Grapes (*Vitis vinifera* L.)	170.0	60.0	220.0	20.0
Guava (*Psidium guajava* L.)	120.0	50.0	150.0	20.0
Ber (*Ziziphus mauritiana* L.)	80.0	35.0	125.0	20.0
Kiwifruit (*Actinidia deliciosa* Chev.)	165.0	50.0	273.0	40.0
Mango (*Mangifera indica* L.)	100.0	25.0	110.0	15.0
Papaya (*Carica papaya* L.)	90.0	25.0	130.0	50.0
Pecan (*Carya illinoinensis* W.)	9.7	5.3	5.4	1.2
Pistachio (*Pistacia vera* L.)	30.0	12.0	15.0	0.057
Passion fruit (*Passiflora* spp. F.)	60.0	15.0	75.0	15.0
Pineapple (*Ananas comosus* L.)	185.0	55.0	350.0	50.0
Sapota (*Achras zapota* L.)	130.0	50.0	170.0	80.0
Strawberry (*Fragaria ananassa* D.)	200.0	100.0	900.0	0.75
Walnut (*Juglans regia* L.)	37.9	10.7	21.2	2.7

Note: In the case of citrus, figures are expressed as grams per ton fruits.

Sources: Tandon, H.L.S., Muralidharadu, Y., 2010. Nutrient Uptake, Removal and Recycling by Crops. Fertilizer Development and Consultation Organization, New Delhi. pp. 1–140; Srivastava, A.K., Singh, S., Albrigo, L.G., 2008. Diagnosis and remediation of nutrient constraints in Citrus. Hortic. Rev. 34, 277–64; Chadha, K.L., Bhargava, B.S., 1997. Plant nutrient supply, needs, efficiency and policy issues for fruit crops in Indian agriculture from 2000 to 2025 AD. In: Kanwar, J.S., Katyal, J.C. (Eds.). Plant Nutrient Needs, Supply, Efficiency and Policy Issues: 2000–2025. NAAS, New Delhi, pp. 1134–132; Kemmler, G., Hobt, H., 1985. Potassium a Product of Nature. Kali and Salz, Kassel; Chadha, K.L., 2007. Handbook of Horticulture. Indian Council of Agricultural Research ICAR, New Delhi, p. 1031; Sparks, D., 1989. Pecan nutrition—a review. Proc. S.E. Pecan Grow. Assoc. 82, 101–122.

5 Nutrient-microbe synergy in unlocking productivity

Recognition of the importance of soil microorganisms has led to increased interest in measuring the quantum of nutrients held in their biomass (Srivastava et al., 2002). An increase in the microbial biomass often goes along with increased nutrient immobilization. Plant growth-promoting microorganisms play an important role exerting various mechanisms such as biological nitrogen fixation, growth hormone production, phosphate solubilization siderophore production, hydrolytic enzymes production, and antagonistic activity, individually or collectively leading to improved nutrient use efficiency (Srivastava et al., 2015a, b; Wu and Srivastava, 2015). Activation of each mechanism implies the production of specific compounds and metabolites, such as plant growth factors, hydrolytic enzymes, siderophores, antibiotics, and carbon and nitrogen permeases. These metabolites can be either overproduced or combined with appropriate biocontrol strains to obtain new formulations for their more effective applications Studies have demonstrated that *Azotobacter* inoculation alone can substitute up to 50% nitrogen requirement of banana and 25% phosphorus requirement of papaya (Singh and Varu, 2013). Arbuscular mycorrhizal fungi has also been reported to substantially improve nutrient acquisition capacity of host plant and fruit yield in addition to enriching the rhizosphere biologically in a much activated form. Mineral fertilizers on the other hand have limited direct effects, but their application can enhance soil biological activity via increases in system productivity, crop residue return, and soil organic matter (Wu et al., 2016a). Another important indirect effect especially of nitrogen fertilization is the soil acidification, with considerable negative effects on soil organisms (Wu et al., 2019). However, the outcome of a long-term fertilizer experiment in rice established that a balanced application of nutrients promoted microbial biomass through improved diversity of the microbial community (Wu et al., 2016b; Liu et al., 2018).

There are ample evidences accrued through worldwide research that nutrient-microbe synergy is the launching pad for any perennial plant to mobilize and accumulate the required nutrients as per the metabolic nutrient demand, a prerequisite to improved NUE (Wu and Srivastava, 2015). Many genes play a central role in the acquisition and distribution of nutrients, including many protein-coding genes and microRNAs (miR395, miR398, miR397, and miR408) (Chiou, 2007; Sunkar et al., 2007). Oustric et al. (2019) reported that higher tolerance to nutrient deficiency could be explained by better activation of their antioxidant system. However, for the other genotypes, tetraploidization did not induce greater tolerance to nutrient deficiency. Rengel et al. (1996) observed that the total number of bacterial colony-forming units increased in the rhizosphere of Zn-efficient genotypes of wheat under Zn deficiency and in Mn-efficient genotypes under conditions of Mn deficiency. In contrast a Zn deficiency treatment acted synergistically with the number of fluorescent *Pseudomonas* in the rhizospheres. A variety of fruit crops have displayed an excellent synergy with a large number of microbes that could play an important role in improving use efficiency of applied nutrients (Table 37.2).

A still bigger question emerges, whether rhizosphere-competent microbes could collectively contribute toward improved resilience of plant's rhizosphere. And if those microbes are so successful in promoting growth response, addition of starter nutrients in such combination may further magnify the magnitude of response called nutrient-microbe synergy. Our earlier studies have shown that rhizosphere effective microbes have the tendency to play multiple roles to overcome various biotic and abiotic stresses while interacting with an environment (Huang et al., 2014). Rhizosphere modification through roots by soil microorganism exudation is an important attribute that regulates not only the availability of nutrients in the soil but also their acquisition by plants (Zou et al., 2014). A number of studies have suggested that the whole range of microorganisms including arbuscular mycorrhizal fungi (AMF) have helped to alleviate different abiotic stresses (Wu et al., 2013, 2014; Wang et al., 2014; Chi et al., 2018; Zou et al., 2015) in fruit crops and aid in improving the efficient use of applied nutrients.

5.1 Microbial consortium: A novel concept

The most common objective of developing microbial consortium is to capitalize both the capabilities of individual microbes and their interactions to create useful systems in tune with enhanced productivity and soil health improvements through efficient metabolic functionality (Brenner et al., 2008). Two major underlying principles are applied in the whole process of development of microbial consortium. The first one is resource ratio theory, which uses both qualitatively and quantitatively to assess the outcomes between component microorganisms competing for shared limiting resources. This permits coexistence of multiple microbes or the competitive exclusion of all but a single microbe (Brauer et al., 2012). And the second principle theory relevant to microbial consortium is maximum power principle initially proposed by Lotka (1992) and, later modified at various levels, is value for analyzing consortial interactions. It also dictates that biological systems that maximize fitness by maximizing power are analogous to metabolic

TABLE 37.2 Response of different microbes on various growth, yield and nutrient uptake of fruit crops.

Fruit crop	Microbes involved	Response parameters	Reference
Pomegranate (*Punica granatum* L.)	*Azotobacter chroococcum* *Glomus mosseae*	Plant canopy, pruned material, rhizosphere changes, and fruit yield	Mir and Sharma (2012)
Grape (*Vitis vinifera* L.)	*Pseudomonas fluorescens*	Root development	Wange and Ranawade (1998)
Quince (*Cydonia oblonga* Mill.)	*Bacillus subtilis*	Fruit firmness, soluble dry matter and fruit yield	Arikan et al. (2013)
Navel orange (*Citrus sinensis* (L.) Osbeck)	*Pseudomonas fluorescens* (843) *Azospirillum brasilense* (W24)	Canopy volume and soil fertility	Shamseldin et al. (2010)
Apple (*Malus domestica* Borkh.)	*Pseudomonas striata* *Azotobacter chroococcum* *Trichoderma viride*	Germination, root growth, and pest incidence	Raman (2012)
Peach (*Prunus persica* (L.) Stokes)	*Azospirillum brasilense* *Bacillus megatarium*	Plant height, girth, and canopy growth	Mahmoud and Mahmoud (1999)
Peach (*Prunus persica* (L.) Stokes)	*Glomus fasciculatum* *Azotobacter chroococcum*	Plant height, girth, and micronutrient concentration	Godara et al. (1996)
Pomegranate (*Punica granatum* L.)	*Azotobacter chroococcum* *Glomus mosseae*	Plant height, pruned material, and fruit yield	Aseri et al. (2008)
Apple (*Malus domestica* Borkh.)	*Bacillus* (OSU 142,M-3) *Pseudomonas* (BA-8)	Tree growth and fruit yield	Aslantas et al. (2007)
Mango (*Mangifera indica* L.)	*Azotobacter chroococcum*	Seedling diameter and number of leaves	Kerni and Gupta (1986)
Passion (*Passiflora edulis* Sims.)	*Azotobacter* sp. *Azospirillum* sp. *Trichoderma* sp.	Improved plantlet growth and yield	Quiroga-Rojas et al. (2012)
Sweet orange (*Citrus sinensis* Osbeck)	*Azospirillum brasilense* *Glomus fasiculatum*	Growth, fruit yield, and nutrient uptake	Singh and Sharma (1993)
Banana (*Musa acuminata* L.)	*Azospirillum* sp.	Height and girth of pseudo-mostem leaf area and yield	Jeeva et al. (1988)
Banana (*Musa acuminata* L.)	*Azotobacter chroococcum* *Azospirillum brasilense*	Number of fingers, bunch weight, and leaf area	Tiwari et al. (1999)
Banana (*Musa acuminata* L.)	*Azospirillum brasilense* *Pseudomonas fluorescens*	Fruit weight and finger size	Suresh and Hasan (2001)
Sweet cherry (*Prunus avium* L.)	*Pseudomonas* (BA-8) *Bacillus* (OSU-142)	Growth, yield, and leaf nutrient composition	Esitken et al. (2006)
Apple (*Malus domestica* Borkh.)	*Bacillus* (M3) *Bacillus* (OSU-143) *Microbacterium*	Growth, yield, and plant nutrition	Karlidag et al. (2007)
Apricot (*Prunus armeniaca* (L.))	*Bacillus* (OSU-142)	Shoot length, yield, and leaf nutrient concentration	Esitken et al. (2003)
Apricot (*Prunus armeniaca* (L.))	*Bacillus* (OSU-142) *Pseudomonas* (BA-8)	Growth, yield, and leaf nutrient composition	Pirlak et al. (2007)
Nagpur mandarin (*Citrus reticulata* Blanco)	*Bacillus mycoides* *Bacillus polymyxa*	Shoot weight, root weight, and rhizosphere microbial properties	Keditsu and Srivastava (2014)

Continued

TABLE 37.2   Response of different microbes on various growth, yield and nutrient uptake of fruit crops—cont'd

Fruit crop	Microbes involved	Response parameters	Reference
	Trichoderma harzianum *Azotobacter chroococcum* *Pseudomonas fluorescens*		
Walnut (*Juglans hegia* L.)	*Pseudomonas chlororaphis* *Pseudomonas fluorescens* *Bacillus cereus*	Plant height, shoot, and root dry	Xuan et al. (2011)
Papaya (*Carica papaya* L.)	*Glomus mosseae* *Glomus fasciculatum* *Gigaspora margarita*	Growth and nutrient uptake	Padma and Kandasamy (1990)

rate or the capacity to capture and utilize energy (Sciubba, 2011). The microbial consortium is classified (Handelsman et al., 1998; Kim et al., 2008; Klitgord and Segré, 2011) as artificial (carrying two or more wild-type microbes whose interactions do not typically occur naturally), synthetic (carrying microbes that are modified through manipulations of genetic content), and natural (carrying microbes having much wider applications like bioremediation, wastewater treatment, and biogas synthesis).

In the past a number of studies have suggested the coinoculation of different microbes, which can be summarized as follows: *A. brasilense—P. striata / B. polymyxa, A. lipoferum, Agrobacterium radiobacter / A. lipoferum-Arthrobacter mysorens, A. brasilense—Rhizobium, A. brasilense—A. chroococcum—Klebsiella pneumoniae—R. meliloti, A. brasilense—R. leguminosarum,* and *A. brasilense / Streptomyces mutabilis—A. chroococcum* (Alagawadi and Gaur, 1992; Belimov et al., 1995; Yadav et al., 1992; Fabbrie and Gallo Del, 1995; Hassouma et al., 1994; Neyra et al., 1995; Elshanshoury, 1995). Microbes involving AM fungi and bacteria have also been suggested for improvement in both yield and quality. These include *A. brasilense-G. fasciculatum* in wheat (Gori and Favilli, 1995), strawberry (Bellone and Bellone de, 1995), *A. brasilense-Pantoea dispersa* in sweet pepper, and *A. chroococcum-G. mosseae* in pomegranate (Aseri et al., 2008).

Studies were carried out with an aim to develop rhizosphere-specific microbial consortium. Growth-promoting microbes were isolated from rhizosphere (0–20 cm) for development of MC through extensive soil sampling (from the rhizosphere of as many as 110 plants) at the experimental site. The microbial diversity existing within rhizosphere soil was isolated following standard procedures and characterized the promising microbes for their nutrient mobilizing capacity through laboratory-based incubation study using the same experimental soil. The efficient microbes, namely, *Aspergillus flavus* (MF113270) (P solubilizer), *Bacillus pseudomycoides* (MF113272) (K solubilizer), *Acinetobacter radioresistens* (MF113273) (N solubilizer), *Micrococcus yunnanensis* (MF113274) (P solubilizer), and *Paenibacillus alvei* (MF113275) (P solubilizer) were finally identified. Pure culture of these microbes in value-added form was developed in broth and prepared a mixture called microbial consortium. The compatibility among these microbes was tested thoroughly by their population dynamics in consortium mode, which showed no antagonism among them up to 90 days of laboratory-oriented incubation study (Srivastava et al., 2014). The response of microbial consortium on rough lemon seedlings (*C. jambhiri* Lush) showed a significant increase in various growth parameters over control. Further, we tested the microbial consortium through various modules of ISFM using different levels of recommended doses of fertilizers (RDF).

6   Climate resilient approach of ISFM: Our case study

Our long-term study entitled "Integrated nutrient management in relation of sustained quality production of Nagpur mandarin (*C. reticulata* Blanco)" was evaluated under field condition during 2007–16 initiated in 2007 with a total of five treatments, namely, Module I (T_1), 100% RDF (600-g N, 200-g P, 300-g K, 200-g ZnSO$_4$, 200-g FeSO$_4$, 200-g MnSO$_4$/tree/year); Module II (T_2), 75% RDF + 25% vermicompost; Module III (T_3), 75% RDF + 25% vermicompost + microbial consortium; Module IV (T_4), 50% RDF + 50% vermicompost; and Module V (T_5), 50% RDF + 50% vermicompost + microbial consortium. These modules were evaluated in terms of growth parameters, fruit yield, fruit quality (including micronutrient composition of fruit juice), plant available pool of nutrients in soil, soil carbon stock, soil microbial load, CO_2 emission rate, and spectral properties of soil on a black clay soil (typic ustochrept). The pooled results obtained during 2007–16 (10 years) are summarized in the succeeding text.

6.1 ISFM: Plant growth, fruit yield, and quality response

All the ISFM-based treatments displayed significant response on changes in canopy volume (Table 37.3). Maximum cumulative increase in canopy volume was observed with Module III (18.67 m^3) followed by Module IV (17.16 m^3), Module V (16.93 m^3), on part with Module II (16.70 m^3), and canopy volume with either Module III or Module V, which was better than either Module II, or Module IV compared with Module I, which is composed of only inorganic fertilizers.

Module I: T_1 (11.50 m^3). Incorporation of microbial consortium either with Module III or with Module V invariably induced higher canopy volume suggesting better response on fruit yield was observed to be significantly affected by different ISFM-based treatments (Table 37.4). The maximum fruit yield of 88.8 kg/tree was observed with treatment module V, which was better than 83.2 kg/tree with T_4 or 80.5 kg/tree with Module III than 71.0 kg/tree with treatment module II. However, all these treatment modules were far superior in magnitude of response when compared with 100% RDF as Module I (67.7 kg/tree). Hence, different ISFM-based modules (71.0–88.8 kg/tree) were much better than exclusive inorganic fertilizer treatment-like module (67.7 kg/tree).

Different fruit quality parameters, except peel thickness (Table 37.6) displayed significant response in relation to different treatments. There was much high fruit weight with different ISFM-based treatment modules I–V 101.4–114.6 g compared with inorganic RDF Module I (101.4), showing that incorporation of both organic manure and microbial cultures improved the efficiency of organic fertilizers. Similar observations were obtained with respect to other three fruit quality-related parameters such as juice content, acidity, and TSS. On the other hand, acidity observed a significant reduction with those superior treatment, for example, Module III (0.86%) and Module V (0.80%) compared with Module II (0.93%) and Module IV (0.91%) highlighting the favorable changes in different fruit quality changes in response to different ISFM-based treatments.

6.2 ISFM: Response on soil health

Rational soil use practices must allow economically and environmentally sustainable yields, which will only be reached with the maintenance or recovery of the soil health. Thus a healthy soil has been defined as "the continued capacity of soil to function as a vital living system, within ecosystem and land-use boundaries, to sustain biological

TABLE 37.3 Growth attributing parameters in response to different vermicompost-based modules of ISFM (Pooled data: 2007–16).

ISFM modules	Plant height (m)	Tree spread (m)		Canopy volume (m^3)	Cumulative increase in canopy volume over 2007–08 (m^3)
		E-W	N-S		
Module I	3.82 (2.02)	2.52 (1.27)	2.50 (1.20)	13.04 (1.54)	11.50
Module II	5.09 (2.10)	2.98 (1.39)	2.89 (1.25)	18.51 (1.81)	16.70
Module III	3.87 (2.13)	3.00 (1.27)	3.22 (1.23)	20.28 (1.61)	18.67
Module IV	3.76 (2.10)	3.73 (1.30)	2.93 (1.30)	19.78 (1.62)	17.16
Module V	3.86 (1.87)	2.97 (1.18)	2.96 (1.22)	18.33 (1.40)	16.93
CD (P=0.05)	0.49	0.38	0.51	1.18	1.32

RDF stands for recommended doses of fertilizer (600-g N, 200-g P, 300-g K, 200-g ZnSO$_4$, 200-g FeSO$_4$, 200-g MnSO$_4$ tree/year).
Vm stands for vermicompost (nutrient composition: 2.38% N, 0.09% P, 1.42% K, 1072 ppm Fe, 116 ppm Mn, 39 ppm Cu, and 46 ppm Zn).
MC stands for microbial consortium developed by isolating the native microbes as a mixture of Bacillus pseudomycoides (MF113272), *Acinetobacter radioresistens* (MF113273), *Micrococcus yunnanensis* (MF113274), Paenibacillus *alvei* (MF113275), and *Aspergillus flavus* (MF113270).
Figures in parenthesis indicates the value obtained in 2007–08.
Source: Srivastava, A.K., Singh, S., Huchche, A.D., 2015b. Evaluation of INM in citrus on Vertic Ustochrept: biometric response and soil health. J. Plant Nutr 38(5), 1–15.

TABLE 37.4   Fruit yield and quality parameter in response to different vermicompost-based modules of ISFM treatments (Pooled data: 2007–16).

| ISFM modules | Yield (kg/tree) | Fruit weight (g/fruit) | Peel thickness (mm) | Fruit quality parameters | | | |
				Juice content (%)	TSS (°Brix)	Acidity (%)	TSS/acid ratio
Module I	67.7	101.4	4.1	40.2	8.4	0.96	8.92
Module II	71.0	104.1	3.0	39.6	8.6	0.93	9.23
Module III	80.5	106.2	3.4	41.5	9.5	0.86	10.05
Module IV	83.2	109.1	3.4	42.7	8.4	0.91	9.56
Module V	88.8	114.6	3.0	44.2	9.3	0.80	11.62
CD (P=0.05)	2.1	1.4	0.50	2.1	0.30	0.10	1.10

RDF stands for recommended doses of fertilizer (600-g N, 200-g P, 300-g K, 200-g ZnSO$_4$, 200-g FeSO$_4$, 200-g MnSO$_4$ tree/year).

Vm stands for vermicompost (nutrient composition: 2.38% N, 0.09% P, 1.42% K, 1072 ppm Fe, 116 ppm Mn, 39 ppm Cu, and 46 ppm Zn).

MC stands for microbial consortium developed by isolating the native microbes (mixture of Bacillus pseudomycoides (MF113272), *Acinetobacter radioresistens* (MF113273), *Micrococcus yunnanensis* (MF113274), Paenibacillus *alvei* (MF113275), and *Aspergillus flavus* (MF113270)).

Sources: Srivastava, A.K., Malhotra, S.K., 2014. Nutrient management in fruit crops: issues and strategies. Indian J. Fert. 10(12), 72–88; Srivastava, A.K., Singh, S., Huchche, A.D., 2015b. Evaluation of INM in citrus on Vertic Ustochrept: biometric response and soil health. J. Plant Nutr. 38(5), 1–15.

productivity, promote the quality of air and water environments, and maintain plant, animal, and human health" (Doran and Safley, 1997). Soil health refers to the ecological equilibrium and the functionality of a soil and its capacity to maintain a well-balanced ecosystem with high biodiversity above and below surface and productivity. To understand and use soil health as a tool for sustainability, physical, chemical, and biological properties must be employed to verify which respond to the soil use and management within a desired timescale. Attributes with a rapid response to natural or anthropogenic actions are considered good indicators of soil health. However, most of soil health-related parameters generally have a slow response, when compared with the biological ones, such as microbial biomass C and N, biodiversity, soil enzymes, and soil respiration, in addition to macro- and mesofauna. Thus a systemic approach based on different kinds of indicators (physical, chemical, and biological) in assessing soil health would be safer than using only one kind of attribute (Cardoso et al., 2013). The soil microbial load in terms of total bacterial count and fungal count in response to different ISFM-based treatments showed variable higher than soil microbial load with Module III and Module V compared with either Module II or Module IV including I. In our study, changes in soil fertility indices with regard to available macro- and micronutrients were observed highly significant, but of variable nature in response to different treatments (Table 37.5). Among macronutrients, KMnO$_4$-N and Olsen-P showed significant responses, without any significant response on NH$_4$OAc-K. The treatments such as Module III (155.7- and

TABLE 37.5   Changes in soil fertility status in response to different modules of ISFM-based treatments (Pooled data: 2007–16).

| ISFM modules | Available nutrients (mg/kg) | | | | | | |
| | Macronutrients | | | DTPA micronutrients | | | |
	KMnO$_4$-N	Olsen-P	NH$_4$OAc-K	Fe	Mn	Cu	Zn
Module I	140.0	9.32	185.2	10.35	10.54	2.2	0.98
Module II	1435	9.27	196.6	12.57	11.54	2.3	1.00
Module III	155.7	9.15	203.6	15.18	12.45	2.7	1.10
Module IV	161.0	9.57	209.2	16.56	11.67	2.8	1.18
Module V	178.5	9.55	212.6	19.50	12.93	2.5	1.32
CD (P=0.05)	6.2	NS	1.3	1.32	1.01	NS	0.46

RDF stands for recommended doses of fertilizer (600-g N, 200-g P, 300-g K, 200-g ZnSO$_4$, 200-g FeSO$_4$, 200-g MnSO$_4$/tree/year).

Vm stands for vermicompost (nutrient composition: 2.38% N, 0.09% P, 1.42% K, 1072-ppm Fe, 116-ppm Mn, 39-ppm Cu, and 46-ppm Zn).

MC stands for microbial consortium developed by isolating the native microbes (mixture of Bacillus pseudomycoides (MF113272), *Acinetobacter radioresistens* (MF113273), *Micrococcus yunnanensis* (MF113274), Paenibacillus *alvei* (MF113275) and *Aspergillus flavus* (MF113270)).

Sources: Srivastava, A.K., 2010a. Development of INM to harmonize with improved citrus production: changing scenario. J. Adv. Plant Physiol. 12, 294–68; Srivastava, A.K., Singh, S., Huchche, A.D., 2015b. Evaluation of INM in citrus on Vertic Ustochrept: biometric response and soil health. J. Plant Nutr. 38(5), 1–15; Srivastava, A.K., Paithankar, D.H., Venkataramana, K.T., Hazarika, B., Patil, P., 2019. INM in fruit crops sustainable quality production and soil health. Indian J. Agric. Sci. 89(3), 379–395.

TABLE 37.6 Soil carbon fractions and C:N ratio in response to different modules of ISFM (Pooled data: 2007–16).

ISFM modules	pH	EC (dS/m)	Soil carbon (g/kg)			Soil total N (g/kg)	Soil C:N ratio	Soil microbial load (×10³ cfu/g)	
			SOC	SiC	TC			Bacterial count	Fungal count
Module I	7.7	0.168	6.61	1.71	8.32	0.721	11.54	26	12
Module II	7.8	0.174	6.70	1.71	8.41	0.732	11.48	36	18
Module III	7.7	0.163	7.02	1.74	8.76	0.741	11.82	54	22
Module IV	7.6	0.173	7.11	1.75	8.86	0.738	12.00	44	26
Module V	7.8	0.174	7.43	1.81	9.24	0.748	12.32	86	42
CD ($P=0.05$)	NS	NS	0.08	NS	0.09	0.006	NS	8.2	4.8

RDF stands for recommended doses of fertilizer (600-g N, 200-g P, 300-g K, 200-g ZnSO$_4$, 200-g FeSO$_4$, 200-g MnSO$_4$/tree/year).

Vm stands for vermicompost (nutrient composition: 2.38% N, 0.09% P, 1.42% K, 1072-ppm Fe, 116-ppm Mn, 39-ppm Cu, and 46-ppm Zn).

MC stands for microbial consortium developed by isolating the native microbes (mixture of *Bacillus pseudomycoides* (MF113272), *Acinetobacter radioresistens* (MF113273), *Micrococcus yunnanensis* (MF113274), *Paenibacillus alvei* (MF113275) and *Aspergillus flavus* (MF113270)).

SOC, SiC, and TC stand for soil organic carbon, soil inorganic carbon, and total carbon, respectively.

Source: Srivastava, A.K., Paithankar, D.H., Venkataramana, K.T., Hazarika, B., Patil, P., 2019. INM in fruit crops sustainable quality production and soil health. Indian J. Agric. Sci. 89(3), 379–395.

203.6-mg/kg KMnO$_4$-N and Olsen-P, respectively) and Module V (178.5- and 212.6-mg/kg KMnO$_4$-N and Olsen-P, respectively) was comparatively higher than Module II (193.5- and 196.6-mg/kg KMnO$_4$-N and Olsen-P, respectively) and Module IV (161.0 and 209.2 mg/kg KMnO$_4$-N and NH$_4$OAc-K, respectively). These observations suggested superiority of INM-based treatments than those treatments without microbial consortium. Module I involving exclusive inorganic fertilizers registered lowest test values as against treatment-like module V registering maximum values, validating the supremacy of those treatments that carry all the three components of ISFM.

The soil properties such as soil pH and soil EC were not affected by any of the ISFM-based treatments (Table 37.6), while, among different fractions of soil carbon, namely, organic-C (SOC), inorganic-C (SiC), and total-C (TC), only SOC and TC were significantly affected. These observations showed that changes in soil carbon stock are more governed by organic fraction than inorganic fraction. Maximum SOC and TC of 7.43 and 9.14 g/kg were observed with Module V. Likewise, Module III has much SOC and TOC than Module II, displaying the significant role of microbial consortium in improving the carbon sink capacity of soil. Different ISFM-based modules were also observed to aid in improving the total soil N stock, being maximum with treatment-like modules III–V (0.741%–0.748%) compared with the rest of the treatments like Module I, Module II, or Module IV (0.721%–0.738%). However, soil C:N ratio in the range of 12.00–12.32, without displaying significant changes in response to different treatments.

6.3 ISFM: Response on enzyme profiling

Rhizopshere enzymes displayed significant changes in response to different modules of ISFM. Acid phosphatase (EC:3.1.3.2) showed no significant variation in response to different ISFM-based modules (Fig.37.1), while alkaline phosphatase (EC:3.1.3.1) varied from 752.31 units/g soil (Module IV) to as high as 6666.66 units/g soil (Module V), although inorganic fertilizers as 100% RDF also registered significantly higher alkaline phosphatase (EC:3.1.3.1) activity (4444.44 units/g) with both the combinations of inorganic fertilizers and vermicompost as Module III or Module V. The peroxidase (EC:1.11.1.7) enzyme on the other hand was lowest with treatment module I (65.14 unit/g) where inorganic fertilizers was used, while maximum peroxidase (EC:1.11.1.7) activity was observed with Module III (704.62 units/g soil). The modules where no microbial consortium was involved either with Module II or Module IV were far less effective. The another soil enzyme, namely, succinate dehydrogenase (EC:1.3.99.1) also followed the similar pattern of response being minimum (0.21-µmol TTZ reduced/g) with inorganic fertilizers (Module I), but the same activity was higher with Module II (0.38-µmol TTZ reduced/g) involving replacement of 25% of RDF with vermicompost. And further replacement of RDF up to 50% with vermicompost further improved the enzyme activity with Module IV (0.59-µmol TTZ reduced/g); by incorporating microbial consortium, it was maximum with Module V (0.72-µmol TTZ reduced/g) based on enzyme profile of rhizosphere.

In endosphere, both the forms of phosphatase enzyme as acid phosphatase (EC:3.1.3.2) and alkaline phosphatase (EC:3.1.3.1) expressed significant changes in response to different ISFM-based treatments (Fig. 37.2). Acid phosphatase (EC:3.1.3.2) enzyme activity was statistically on par with each other when compared with inorganic fertilizers (Module I) versus (Module II) and Module III. But, with Module II or Module V vermicompost, acid phosphatase (EC:3.1.3.2)

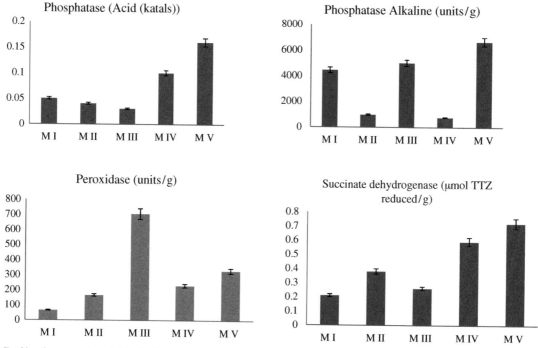

FIG. 37.1   Profile of enzymatic activities in rhizosphere in response to different modules of ISFM.

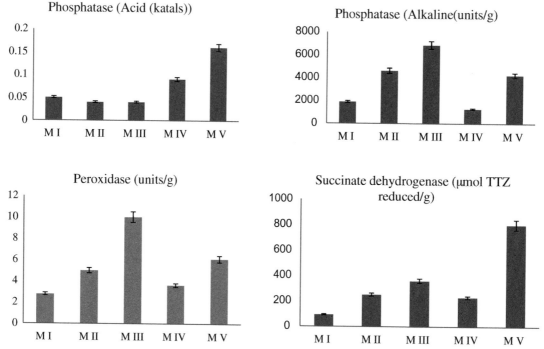

FIG. 37.2   Profile of enzymatic activities in endosphere in response to different modules of ISFM.

enzyme displayed significant changes, while alkaline phosphatase (EC:3.1.3.1) activity was lowest in inorganic fertilizer treatment as Module I (1888.88 units/g) and highest with Module III (6861.11 units/g). Peroxidase (EC:1.11.1.7) activity was minimum with Module I (2.76 units/g), and it increased further with Module II (4.97 units/g) and Module III (9.99 units/g). Replacing further 50% RDF with vermicompost either without microbial consortium (Module IV) or with microbial consortium induced no further significant changes. The succinate dehydrogenase (EC:1.3.99.1) enzyme activity was observed minimum with Module I (91.66-μmol TTZ reduced/g) and maximum with Module V (799.99-μ mol TTZ reduced/g). It was interestingly to observe the concurrent increase in the activity of all the three enzymes with addition of vermicompost and microbial consortium under different ISFM-based modules.

6.4 ISFM: Changes in rhizosphere CO_2 emission

All the ISFM-based modules influenced the CO_2 emission to varying proportions across various seasons in a year (Fig. 37.3). The appraisal on CO_2 emission rate in response to differential treatments showed comparatively higher CO_2 emission in the morning and was observed over evening time, irrespective of the season. While, much higher CO_2 emission was observed during summer season (3127.2–4530.6 mg C^2/h) compared with either rainy season (1858.9–2371.4 mg C/m^2/h) or winter season (1212.1–2052.9 mg C/m^2/h) with Module I involving exclusive use of inorganic fertilizers. With replacement of 25% of RDF with vermicompost (Module II), CO_2 emission rate was slowed down to 1297.0–1959.7 mg C/m/h, 1381.7–2075.8 mg C/m/h, and 746.3–1189.8 mg C/m/h during summer, rainy, and winter seasons, respectively, irrespective of time of sampling. Incorporation of microbial brought down the CO_2 emission rate to 980.5–1030.6 mg C/m/h, 408.6–528.8 mg C/m^2/h, and 988.3–1061.6 mg/C/m^2/h during summer, rainy, and winter seasons, respectively, irrespective of time of sampling. Combined use of 50% RDF+50% vermicompost+microbial consortium (Module V) further brought down the CO_2 emission rate to 724.6–938.2 mg C/m^2/h, 685.5–864.9 mg C/m/h, and 697.2–1007.1 mg C/m^2/h during summer, rainy and winter seasons, respectively. On the other hand, Module IV showed comparatively hither CO_2 emission, 2205.0–2635.0 mg C/m^2/h, 1108.4–1670.0 mg C/m/h, and 831.3–1402.7 mg C/m2/h during summer, rainy, and winter seasons, respectively. These observations showed better carbon accreditation with those ISFM modules having all the three components.

6.5 ISFM: Spectral behavior of soil

The spectral properties of the soil were evaluated in response to different treatments (Fig. 37.4). Although pattern spectral response is same, segregation of various wavelength peaks seemed to have some marginal improvements in

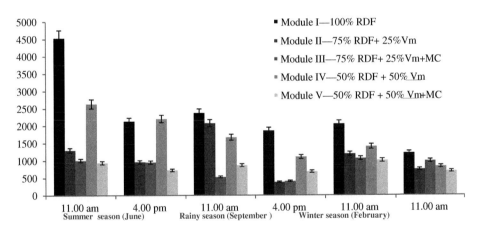

FIG. 37.3 ISFM: Reduction in CO_2 emission rate (mg C/m^2/h) in response to different modules. *Source: Srivastava, A.K., Paithankar, D.H., Venkataramana, K.T., Hazarika, B., Patil, P., 2019. INM in fruit crops sustainable quality production and soil health. Indian J. Agric. Sci. 89(3), 379–395.*

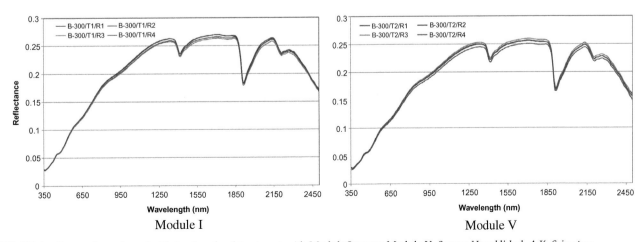

FIG. 37.4 Comparison of spectral behavior of soil treatment with Module I versus Module V. *Source: Unpublished, A.K. Srivastava.*

spectral properties of soil. These observations highlight the soils treated with ISFM-based treatments in improving the liveliness of the soil.

We have attempted to review the published literature on various modules of ISFM in different fruit crops (Table 37.7). The results accured have also confirmed that any mode of ISFM that supplies nutrients at differential rates can sustain the production coupled with quality, besides soil health resilience (Srivastava and Singh, 2004c), which are quite synonymous to climate-smart ISFM system.

TABLE 37.7 Different modules of INM/ISFM recommended for important fruit crops.

Crop	Nutrient-microbe combination	Reference
Guava (*Psidium guajava* L.)	FYM 50 kg/plant, *Azotobacter sp.* 50 g/plant, *Azospirillum sp.* 50 g/plant, *Sesbania* sp. as green manure	Ram and Rajput (2000)
Pomegranate (*Punica granatum* L.)	400-g N, 100-g P_2O_5, 300-g K_2O/plant, FYM 20 kg/plant	Ghosh et al. (2012)
Papaya (*Caria papaya* L.)	Vermicompost 20 kg/plant, rhizosphere culture 50 g/plant, 150 N, 200 P_2O_5, 200 K_2O g/plant (75% RDF)	Kirad et al. (2010)
Banana (*Musa acuminata* L.)	FYM 12 kg/plant, *Azospirillum sp.* 50 g/plant, phosphate solubilizing bacteria 50 g/plant, *T. harzianum* 50 g/plant	Hazarika and Ansari (2010)
Banana (*Musa acuminata* L.)	50% RDF, FYM 20 kg/plant, *Azotobacter sp.* 50 g/plant, phosphate solubilizing bacteria 50 g/plant, VAM 250 g/plant	Patil and Shinde (2013)
Guava (*Psidium guajava* L.)	488-g N, 244-g P_2O_5, 281-g K_2O/plant, FYM 50 kg/plant, *Azotobacter* 250 g/plant, phosphate solubilizing bacteria 25 g/plant	Barne et al. (2011)
Strawberry (*Fragaria ananassa* Duches)	75% N as RDF, 25% N as FYM, *Azotobacter sp.*	Umer et al. (2009)
Pomegranate (*Punica granatum* L.)	300-g N/plant, neem cake 1 kg/plant	Ray Dutta et al. (2014)
Banana (*Musa acuminata* L.)	100% RDF, 40% Wellgrow organic manure	Kuttimani et al. (2013)
Peach (*Pr. persica* (L.) Stokes)	75% RDF, 25% N equivalent FYM	Shah et al. (2014)
Lemon (*Citrus limon* (L.) Burm.f.)	N 525 g/plant, FYM 150 kg/plant, *Azotobacter sp.* 18 g/plant	Khehra and Bal (2014)
Apricot (*Pr. armeniaca* (Linn.))	75% RDF, 25% FYM	Shah et al. (2006)
Papaya (*Caria papaya* L.)	50% RDF (100 N, 100 P_2O_5, 125 K_2O g/plant), *Azotobacter sp.* 50 g/plant, phosphate solubilizing bacteria 2.5 g/m^2	Singh and Varu (2013)
Guava (*Psidium guajava* L.)	50% RDF (250-g N, 100-g P_2O_5, 250 K_2O g/plant), FYM 25 kg/plant, vermicompost 5 kg/plant	Dwivedi (2013)
Sapota (*Ach. zapota* L.)	75% RDF + 25% RDF equivalent vermicompost	Hebbarai et al. (2006)
Mango (*Mangifera indica* L.)	500-g N, 250-g P_2O_5, 250 K_2O g/plant, 50-kg FYM/plant, *Azospirillum sp.* 250 g/plant	Singh and Banik (2011)
Mango (*Mangifera indica* L.)	250 N, 425 P_2O_5, 1000 K_2O, *Azospirillum sp.* 250 g/plant, PSB, 250 g/plant, $ZnSO_4$ 100 g/plant, Borax 100 g/plant	Hasan et al. (2012)
Banana (*Musa acuminata* L.)	100% RDF, FYM 10 kg/plant, *Azospirillum sp.* 25 g/plant, phosphate solubilizing bacteria 250 g/plant	Bhalerao et al. (2009)
Guava (*Psidium guajava* L.)	236-g N, 66-g P_2O_5, *Azospirillum sp.* 30 g/plant, VAM 30 g/plant	Dutta et al. (2009)
Mosambi (*Citrus sinensis* Osbeck)	300-g N, 250-g P_2O_5, 300-g K_2O, AMF 10 g/plant, *Azospirillum sp.* 25 g/plant	Patel et al. (2009)

TABLE 37.7 Different modules of INM/ISFM recommended for important fruit crops—cont'd

Crop	Nutrient-microbe combination	Reference
Guava (*Psidium guajava* L.)	250-g N, 100-g P, 250-g K$_2$O/plant, *Azotobacter sp.* 250 g/plant	Shukla et al. (2009)
Litchi (*Litchi chinensis* Sonn.)	500-g N, 250-g P$_2$O$_5$, 500-g K$_2$O/plant, FYM 50 kg/plant, *Azotobacter sp.* 150 g/plant, VAM 100 g/plant	Dutta et al. (2010)
Aonla (*Emblica officinalis* Gaertn.)	50% NPKS (105-kg N, 7.20-kg P$_2$O$_5$, 125.25-kg K$_2$O/ha), biofertilizers (*Azotobacter sp.*, *Azospirillum sp.*, phosphate solubilizing bacteria), FYM (2 tons/ha)	Yadav et al. (2007)
Aonla (*Emblica officinalis* Gaertn.)	100-g N, 25-g P$_2$O$_5$, 150-g K$_2$O/plant, FYM 10 kg/plant, phosphate solubilizing bacteria 50 g/plant	Mandal et al. (2013)
Sapota (*Ach. zapota* L.)	1500-g N, 1000 P$_2$O$_5$, 500-g K$_2$O/plant, 75-kg FYM, 12.5 g/plant PSB	Dalal et al. (2004)
Guava (*Psidium guajava* L.)	50% RDF (225-g N, 195-g P$_2$O$_5$, 150-g K$_2$O/plant), FYM 50 kg/plant, *Azospirillum* 250 g/plant	Goswami et al. (2012)

RDF and PSB stand for recommended doses of fertilizers and phosphate solubilizing bacteria predominantly (*Pseudomonas fluorescens*), respectively.
FYM and AMF stand for farmyard manure and arbuscular mycorrhizal fungi, respectively.

7 Suggested future viewpoints

A cultivar evaluated under both intensive and organic farming system may not perform with similar magnitude of success. The major difference lies between the nutrient availability pattern and form in both the cases. The plants suitable for intensive (conventional) farming get high amount of nutrients at its peak stage, whereas in organic farming the manure applied is needed to be decomposed first by microorganisms followed by mineralization process to be converted to available form like NO$_3^-$ and NH$_4^+$; hence its availability was low when it was highly required. Considering the mentioned problem, it is highly desirable to breed the plants for organic conditions such that it can change the plants' nutrient absorption pattern, increase nutrient absorption capacity, reduced root losses due to pathogens, ability to maintain a high mineralization activity in rhizosphere via root exudates, increased rooting depth, and associated ability to recover N leached from the topsoil. A considerable approach is urgently required to sustain the rising organic food requirement (Sharma and Bardhan, 2017). Despite many cutting-edge technologies addressing a variety of core issues of nutrient management, many more issues are yet to be attempted with respect to ISFM-based production vis-à-vis rhizosphere dynamics.

Nutrient dynamics is another virgin area where limited attempts have been made using citrus as test crop (Srivastava and Ngullie, 2009). Among different nutrients, Zn has attracted worldwide investigation from various angles (Srivastava and Singh, 2009a, b). The changes in rhizosphere bring different simultaneous changes in microbial diversity vis-a-vis Cmic, Nmic, and Pmic, and nutrient regime especially for diffusion-limited nutrients like P, Zn, Fe, and Mn has to find serious considerations in any nutrient management program that involves ISFM-based corrective treatments (Srivastava et al., 2007a, b). Additionally, the conditions under which citrus trees are most likely to respond to corrective Zn treatments are still not fully understood. The role of Zn in flowering, fruit set, fruit quality (external and internal), and juice shelf life; models defining the critical periods of Zn supply to assure sustained response and its uptake for helping the management decision under different fruit crop-based cropping systems; and devising means for improved Zn uptake efficiency need to be attempted to unravel many of the complexities involved with Zn nutrition under ISFM-based production management (Srivastava and Singh, 2006).

Out of different soil properties, the microbial biomass is one of the biological properties of soil that undergoes immediate change in response to fertilizer-like input (Srivastava and Singh, 2007, 2008b). Studies, therefore, need to be undertaken with a view to explore the possibility whether microbial properties could be used as a potential tool for finding out soil fertility constraint instead of available supply of nutrients in soil. Simultaneously an eye should be kept on long-term changes in total carbon pool of soil to arrive at the logistic conclusion that sequestration of carbon through improved production level could rejuvenate the lost productivity potential of nutritionally depleted soil (Srivastava and Singh, 2015). However, it remains to be further established that any change in microbial diversity within the rhizosphere is brought about with different sources of substrate and, if there is any, how the nutrient dynamics is associated with orchard productivity.

The microbial biomass is one of the biological properties of soil that undergoes immediate change in response to fertilizer-like input. Studies, therefore, need to be undertaken with a view to explore the possibility of which soil microbial property could be used as a potential tool for finding out soil health-related constraint instead of concentration of available nutrients in soil using some indicator fruit crops as bioaccumulators (Marshner et al., 2004). While the genetic, functional, and metabolic diversity of soil microorganisms within the rhizosphere of a wide range of fruit crops is important, the capacity of soil microbial communities to maintain functional diversity of those critical soil processes could ultimately be more important to ecosystem productivity and stability than mere taxonomic diversity. In this context, it remains to be assessed how nutrient-microbe synergism is associated with productivity of perennial fruits.

With the availability of more technical know-how on combined use of organic manures, prolonged shelf life of microbial biofertilizers, and inorganic chemical fertilizers, an understanding on nutrient acquisition and regulating the water relations would help switch orchards to CO_2 sink (expanding carbon capturing capacity of rhizosphere) with emphasis on nutrient-use efficiency so that a more sustainable fruit-based integrated crop production system under biotic and abiotic stress could evolve. The molecular approach to breeding of mineral deficiency resistance and mineral efficiency would facilitate produce nutritionally efficient biotypes to maximize the quality production of fruit crops on sustained basis. The role of AMF in providing an additional resilience to rhizosphere's ability of carbon accreditation through within rhizosphere and associated development of plant's antioxidant profile as a defense mechanism should divert the research studying strong mycorrhizal dependency of fruit crops. Rhizosphere-specific AMF-based microbial consortium would add a new dimension in providing newer options for raising the productivity potential of fruit crops. The framework on soil biodiversity effects from field to fork comprises (i) recognizing both direct and indirect mechanisms of soil biodiversity effects on crops properties, (ii) identifying postharvest processes that affect biodiversity legacy effects on crop properties, and (iii) pinpointing biodiversity-related crop properties that influence the efficacy and success of operations occurring in the agrifood chain (Rillig et al., 2018).

Impacts due to environmental changes and anthropogenic activity are the potential threats to the conservation of soil quality while expanding citriculture to marginal soils having a wide range of limitations. With the availability of more technical know-how on efficient use of bulky organic manures, prolonged shelf life of microbial biofertilizers, and better understanding on citrus-mycorrhiza symbiosis with regard to nutrient acquisition and regulating the water relations, a more effective integrated citrus production system could evolve in the future. The molecular approach for the breeding of mineral deficiency resistance and mineral efficiency would facilitate to produce nutritionally efficient biotypes to maximize the quality production on sustained basis. However, it remains to be established whether or not polyploidy could render a given fruit crop better tolerance against mineral deficiencies. Fertilizer applications are currently managed to protect environmentally sensitive areas by using controlled release fertilizers (the use of organic manures, a befitting option), frequent low concentration fertigation, multiple applications, and variable rate application technology to improve fertilizer use efficiency. However, using newly emerging techniques of nutrient management and site-specific management on the principles of ISFM could be worked out accommodating soil's nature and properties. Simultaneously, concerted efforts would be required to develop ISFM-based yield monitors and soil quality indicators to develop a comprehensive system, whereby the concept of soil security could be effectively brought into reality with an emphasis on the development of minimum data set to define crop-based soil health card—the efforts on these lines are still in infancy. The efforts such as these are likely to transform climate-smart ISFM like a common conventional management system using fruit crop-based land uses.

References

Alagawadi, A.R., Gaur, A.C., 1992. Inoculation of *Azospirillum brasilense* and phosphate-solubilizing bacteria on yield of sorghum [*Sorghum bicolor* (L.) Moench] in dry land. Trop. Agric. 69, 347–350.

Arikan, S., Ipek, M., Pirlak, L., 2013. Effect of plant growth promoting rhizobacteria (PGPR) on yield and fruit quality of Quince. IPCBEE 60, 97–106. https://doi.org/10.7763/IPCBEE.2013.V60.19.

Aseri, G.K., Jain, N., Panwar, J., Rao, A.V., Meghwal, P.R., 2008. Biofertilizers improve plant growth, fruit yield, nutrition, metabolism and rhizosphere enzyme activities of pomegranate (*Punica granatum* L.) in Indian Thar desert. Sci. Hortic. 117, 130–135.

Aslantas, R., Cakmakci, R., Sahin, F., 2007. Effect of plant growth promoting rhizobacteria on young apple tree growth and fruityield under orchard conditions. Sci. Hortic. 111, 371–377.

Barne, V.G., Bharad, S.G., Dod, V.N., Baviskar, M.N., 2011. Effect of integrated nutrient management on yield and quality of guava. Asian J. Hort. 6, 546–548.

Belimov, A.A., Kojemiakov, A.P., Chuvarliyeva, C.V., 1995. Interaction between barley and mixed cultures of nitrogen fixing and phosphate solubilizing bacteria. Plant Soil 173, 29–37.

Bellone, C.H., Bellone de, S.C., 1995. Morphogenesis of strawberry roots infected by *Azospirillum brasilense* and VA mycorrhiza. NATO ASI Ser. G 37, 251–255.

Bhalerao, N.M., Patil, N.M., Badgujar, C.O., Patil, D.R., 2009. Studies on integrated nutrient management for tissue cultured Grand Naine banana. Indian J. Agric. Res. 43, 107–112.

Bhargava, B.S., 2002. Leaf analysis for nutrient diagnosis recommendation and management in fruit crops. J. Indian Soc. Soil Sci. 50 (4), 352–373.

Bhatnagar, P., Singh, J., Chauhan, P.S., Sharma, M.K., Meena, C.B., Jain, M.C., 2016. Carbon assimilation potential of Nagpur mandarin (*Citrus reticulata* Blanco). Int. J. Environ. Sci. Technol. 5, 1402–1409.

Bindi, M., Fibbi, L., Gozzini, B., Orlandini, S., Seghi, L., Poni, S., 1997. The effect of elevated CO_2 concentration on grapevine growth under field conditions. Acta Hortic. 42, 325–330.

Blum, W.E.H., 1993. Soil protection concept of the council of Europe and integrated soil research. In: Eijsackers, H.J.P., Hamer, T. (Eds.), Integrated Soil and Sediment Research: A Basis for Proper Protection, Soil and Environment. In: vol. 1. Kluwer Academic Publishers, Dordrecht, pp. 37–47.

Brady, N.C., Weil, R.R., 2004. Elements of the Nature and Properties of Soils, second ed. Pearson Prentice Hall, New Jersey.

Brauer, V.S., Stomp, M., Huisman, J., 2012. The nutrient-load hypothesis: patterns of resource limitation and community structure driven by competition for nutrients in light. Am. Nat. 179, 721–740.

Brenner, K., You, L., Arnold, F.H., 2008. Engineering microbial consortia: a new frontier in synthetic biology. Trends Biotechnol. 26, 483–489.

Bwalya, J.M. 2013. Estimation of Net Carbon Sequestration Potential of Citrus Under Different Management Systems Using the Life Cycle Approach. Research Thesis. University of Zambia Research Repository Online. http://dspace.unza.zm:8080/xmlui/handle/123456789/2202.

Cardoso, E.J.B.N., Vasconcellos, R.L.F., Bini, D., Miyauchi, M.Y.H., Santos, C.A.D., Alves, P.R.L., Paula, A.M., Nakatani, A.S., Pereira, J.M., Nogueira, M.A., 2013. Soil health: looking for suitable indicators. What should be considered to assess the effect of use and management on soil health? Sci. Agric. 70 (4), 274–289.

Centritto, M., Magnani, F., Lee, H.S.J., Jarvis, P.G., 1999a. Interactive effects of elevated [CO_2] and drought on cherry (*Prunus avium*) seedlings. II. Photosynthetic capacity and water relations. New Phytol. 141, 141–153.

Centritto, M., Lee, H.S.J., Jarvis, P.G., 1999b. Interactive effects of elevated [CO_2] and drought on cherry (*Prunus avium*) seedlings. I. Growth, whole-plant water use efficiency and water loss. New Phytol. 141, 129–140.

Chi, G.G., Srivastava, A.K., Qiang-Sheng, W., 2018. Exogenous easily extractable glomalin-related soil protein improves drought tolerance of trifoliate orange. Arch. Agron. Soil Sci. https://doi.org/10.1080/03650340.2018.1432854.

Chiou, T.J., 2007. The role of microRNAs in sensing nutrient stress. Plant Cell Environ. 30, 323–332. https://doi.org/10.1111/j.1365-3040.2007.01643.x.

Conroy, J., 1992. Influence of elevated atmospheric CO_2 on concentrations of plant nutrients. Aust. J. Bot. 40, 445–456.

Dalal, S.R., Gonge, V.S., Jogdande, N.O., Moharia, A., 2004. Response of different levels of nutrients and PSB on fruit yield and economics of sapota. PKV Res. J. 28, 126–128.

de Campos Bernardi, A.C., de Almeida Machdo, P.L.O., Madari, B.E., de Lucena Tavares, S.R., de Campos, D.V.B., de Arabújo Crisóstomo, L., 2007. Carbon and nitrogen stocks of an Arenosol under irrigated fruit orchards in semiarid Brazil. Sci. Agric. 64. https://doi.org/10.1590/S0103-90162007000200010.

Doran, J.W., Safley, M., 1997. Defining and assessing soil health and sustainable productivity. In: Pankhurst, C.E., Doube, B.M., Gupta, V.V.S.R. (Eds.), Biological Indicators of Soil Health. CAB International, Wallingford, pp. 1–28.

Dutta, P., Maji, S.B., Das, B.C., 2009. Studies on the response of bio-fertilizer on growth and productivity of guava. Indian J. Hort. 66, 39–42.

Dutta, P., Kundu, S., Biswas, S., 2010. Integrated nutrient management in litchi cv Bombai in new alluvial zone of West Bengal. Indian J. Hort. 67, 181–184.

Dwivedi, V., 2013. Effect of integrated nutrient management on yield, quality and economics of guava. Ann. Plant Soil Res. 15, 149–151.

Ehleringer, J., Cerling, T.E., 1995. Atmospheric CO_2 and the ratio of intercellular to ambient CO_2 concentrations in plants. Tree Physiol. 15, 105–111.

Elshanshoury, A.R., 1995. Interactions of *Azotobacter chroococcum, Azospirillum brasilense* and *Streptomyces mutabilis* in relation to their effect on wheat development. J. Agron. Crop Sci. 175, 119–127.

Esitken, A., Karlidag, H., Ercisli, S., Turan, M., Sahin, F., 2003. The effect of spray a growth promoting bacterium on the yield, growth and nutrient element composition of leaves of apricot (*Prunus armeniaca* L. cv. Hacihaliloglu). Aust. J. Agric. Res. 54, 377–380.

Esitken, A., Pirlak, L., Turan, M., Sahin, F., 2006. Effect of floral and foliar application of plant growth promoting rhizobacteria (PGPR) on yield, growth and nutrition of sweet cherry. Sci. Hortic. 110, 324–327.

Fabbrie, P., Gallo Del, M., 1995. Specific interaction between chickpea (*Cicer arietinum*) and three chickpea-Rhizobium strains inoculated singularly and in combination with *Azospirillum brasilense*. NATO ASI Ser. G 37, 267–277.

Fischer, G., Almanza-Marchan, P.S., Ramirez, F., 2012. Source-sink relationship in fruit species: a review. Rev. Colomb. Cienc. Hortic. 6, 238–253.

Ghosh, S.P., 2000. Nutrient management in fruit crops. Fertilizer News 45, 71–76.

Ghosh, S.N., Roy, S.N., Bora, B., 2012. Nitrogen and potassium nutrition in sapota grown in laterite soil. J. Crop Weed 8, 152–154.

Godara, R.K., Awasthi, R.P., Kaith, N.S., 1996. Interaction effect of VA-mycorrhizae and *Azotobacter* inoculation on micronutrient content of peach seedlings. J. Hill Res. 9, 5–10.

Goh, K.M., 2004. Carbon sequestration and stabilization in soils: implications for soil productivity and climate change. J. Soil Sci. Plant Nutr. 50, 467–476.

Goodfellow, J., Eamus, D., Duff, G., 1997. Diurnal and seasonal changes in the impact of CO_2 enrichment on assimilation, stomatal conductance and growth in long term study of *Mangifera indica* in the wet-dry tropics of Australia. Tree Physiol. 17, 291–299.

Gori, A., Favilli, F., 1995. First results on individual and dual inoculation with Azospirillum—Glomus on wheat. NATO ASI Ser. G 37, 245–249.

Goswami, K.A., Shant, L., Misra, K.K., 2012. Integrated nutrient management improves growth and leaf nutrient status of guava cv. Pant Prabhat. Indian J. Hort. 69, 168–172.

Guimarães, D.V., Gonzaga, M.I.S., de Melo Neto, J.O., 2014. Management of soil organic matter and carbon storage in tropical fruit crops. Rev. Bras. Eng. Agríc. Ambient. 18, 301–306.

Handelsman, J., Rondon, M.R., Brady, S.F., Clardy, J., Goodman, R.M., 1998. Molecular biological access to the chemistry of unknown soil microbes: a new frontier for natural products. Chem. Biol. 5, R245–R249.

Hasan, M.A., Manna, M., Dutta, P., Bhattacharya, K., Mandal, S., Banerjee, H., Ray, S.K., Jha, S., 2012. Integrated nutrient management in improving fruit quality of Mango 'Himsagar'. In: ISHS Acta Horticulturae, IX International Mango Symposium vol. 992.

Hassouma, M.G., Hassan, M.T., Madkour, M.A., 1994. Increased yield of alfalfa (*Medicago sativa*) inoculated with N2-fixing bacteria and cultivated in a calcareous soil of Northwestern Egypt. Arid Soil Res. Rehabil. 8, 389–393.

Hazarika, B.N., Ansari, S., 2010. Effect of integrated nutrient management on growth and yield of banana cv. Jahaji. Indian J. Hort. 67, 270–273.

He, Z.L., Yao, H., Chen, G., Zhu, J., Huang, C.Y., 1997. Relationship of crop yield to microbial biomass in highly weathered soils of China. In: Ando, T. et al., (Eds.), Plànt Nutrition Sustainable Food Production and Environment. Kluwer Academic Publishers, Tokyo, pp. 745–746.

Hebbarai, M., Ganiger, V.M., Masthana, R.B.G., Joshi, V.R., 2006. Integrated nutrient management in sapota (*Manikara* zapota) using vermicompost to increase yield and quality. Indian J. Agric. Sci. 76, 587–590.

Hocking, P.J., Meyer, C.P., 1991a. Effects of CO_2 enrichment and nitrogen stress on growth, and partitioning of dry matter and nitrogen in wheat and maize. Aust. J. Plant Physiol. 18, 339–356.

Hocking, P.J., Meyer, C.P., 1991b. Carbon dioxide enrichment decreases critical nitrate and nitrogen concentration in wheat. J. Plant Nutr. 14, 571–584.

Houghton, R.A., Skole, D.L., 1990. Carbon. In: Turner II, B.L., Clark, W.C., Kates, R.W., Richards, J.F., Mathews, J.T., Meyer, W.B. (Eds.), The Earth as Transformed by Human Action, Global and Regional Changes in the Biosphere Over the Past 300 Years. Cambridge University Press, Cambridge, pp. 393–408.

Huang, Y.-M., Srivastava, A.K., Ying-Ning, Z., Qiu-DanNi, Y.H., Wu, Q.S., 2014. Mycorrhizal-induced calmodulin mediated changes in antioxidant enzymes and growth response of drought-stressed trifoliate orange. Front. Microbiol. 5, 682–688.

Idso, S.B., Kimball, B.A., 2001. CO_2 enrichment of sour orange trees: 13 years and counting. Environ. Exp. Bot. 46, 147–153.

Intrigliolo, F., Stagno, F., 2001. Organic fertilizers in citrus crops. Riv. Frutticolt. Ortofloricolt. 63 (11), 75–79.

Ito, J., Hasegawa, S., Fujita, K., Ogasawara, S., Fujiwara, T., 1999. Effect of CO_2 enrichment on fruit growth and quality in Japanese pear (*Pyrus serotina* Reheder cv Kousui). Soil Sci. Plant Nutr. 45, 385–393.

Jeeva, S., Kulasekaran, M., Shanmugavelu, K.G., Oblisami, G., 1988. Effect of Azospirillum on growth and development of banana cv. Poovan (AAB). South Indian Hort. 36, 1–4.

Joa, J.-H., Lim, H.C., Han, S.G., Moon, K.H., Jeon, S.J., 2006. Microbial activities in dark brown volcanic ash soils was affected by different fertilizer application methods in citrus orchard. Abstract 138-84, In: 18th World Congress on Soil Science, Pennsylvania, USA.

Karlidag, H., Estiken, A., Turan, M., Shain, F., 2007. Effect of root inoculation of plant growth promoting rhizobacteria (PGPR) on yield, growth and nutrient element contents of leaves of apple. Sci. Horticult. 114 (2), 16–20.

Keditsu, R., Srivastava, A.K., 2014. Substrate dynamics: developments and issues. Ann. Plant Soil Res. 16 (1), 1–8.

Kerni, P.N., Gupta, A., 1986. Growth parameters affected by azotobacterization of mango seedlings in comparison to different nitrogen doses. Res. Dev. Rep. 3, 77–79.

Keutgen, N., Chen, K., 2001. Responses of citrus leaf photosynthesis, chlorophyll fluorescence, macronutrient and carbohydrate contents to elevated CO_2. J. Plant Physiol. 158, 1307–1316.

Khehra, S., Bal, J.S., 2014. Influence of organic and inorganic nutrient sources on growth of lemon (*Citrus limon* (L.) Burm.) cv Baramasi. J. Exp. Biol. Agric. Sci. 2, 126–129.

Kim, H.J., Boedicker, J.Q., Choi, J.W., Ismagilov, R.F., 2008. Defined spatial structure stabilizes a synthetic multispecies bacterial community. Proc. Natl. Acad. Sci. U. S. A. 105 (18), 188–193.

Kimmins, J.P., 1997. Forest Ecology. Pearson Prentice Hall, New Jersey.

Kirad, K.S., Barche, S., Singh, D.B., 2010. Integrated nutrient management in papaya cv. Surya. Acta Hortic. 1, 377–380.

Klitgord, N., Segré, D., 2011. Ecosystems biology of microbial metabolism. Curr. Opin. Biotechnol. 22, 541–546.

Koch, A., Bratney, M.C., Adams, M., Field, D., Hill, R., Crawford, J., Minasny, B., Lal, R., Abbott, L., O'Donnell, A., Angers, D., Baldock, J., Barbier, E., Binkley, D., Parton, W., Wall, D.H., Bird, M., Bouma, J., Chenu, C., Flora, C.B., Goulding, K., Grunwald, S., Hampel, J., Jastrow, J., Lehmann, J., Lorenz, K., Morgan, C.L., Rice, C.W., Whitehead, D., Young, I., Zimmerman, M., 2013. Soil security: solving the global soil crisis. Glob. Pol. https://doi.org/10.1111/1758-5899.12096https://www.researchgate.net/publication/318652369.

Kohli, R.R., Srivastava, A.K., Shivankar, V.J., 1998. Organic culture in citrus cultivation. Indian Hort. 43, 12–15.

Kuttimani, R., Velayudham Somasundram, E., Muthukrishnan, P., 2013. Effect of integrated nutrient management on yield and economics of banana. Global J. Biol. Agric. Health Sci. 2, 191–195.

Lakso, A.L., 2010. Estimating the environmental footprint of New York apple orchards. New York Fruit Quart. 181, 26–28.

Lal, R., 2010. Managing soil and ecosystems for mitigating Anthrogenic Carbon Emissions and advancing Global food security. Bioscience 60, 708–712.

Liu, C.-Y., Zhang, F., Zhang, D.J., Srivastava, A.K., Qiang-Sheng, W., Ying-Ning, Z., 2018. Mycorrhiza stimulates root-hair growth and IAA synthesis and transport in trifoliate orange under drought stress. Sci. Rep. 8, 1978. https://doi.org/10.1038/s41598-018-20456-4.

Lobell, D.B., Field, C.B., Chahill, K.N., Bonfils, C., 2005. Impacts of future climate changes on California perennial crop yields: model projections with climate and crop uncertainties. Agric. For. Meteorol. 141, 208–218.

Lotka, A.J., 1992. Contribution to the energetic of evolution. Proc. Natl. Acad. Sci. U. S. A. 8, 147–151.

Mahmoud, H.M., Mahmoud, F.A.F., 1999. Studies on effect of some biofertilizers on growth of peach seedlings and root rot disease incidence. Egypt. J. Hortic. 26, 7–18.

Mandal, K.K., Rajak, A., Debnath, S., Hasan, M.A., 2013. Integrated nutrient management in Aonla cv A-7 in the red lateritic region of West Bengal. J. Crop Weed 9, 121–123.

Marshner, P., Crowley, D., Yang, C.H., 2004. Development of specific rhizosphere bacterial communities in relation to plants species, nutrition and soil type. Plant Soil 261, 199–208.

Mir, M., Sharma, S.D., 2012. Influence of biofertilizers on plant growth, fruit yield, nutrition and rhizosphere microbial activity of pomegranate (*Punica granatum* L.) cv. Kandhari Kabuli. J. Appl. Hortic. 14, 129–136.

Negi, J.D.S., Manhas, R.K., Chauhan, P.S., 2003. Carbon allocation in different components of some tree species of India: a new approach of carbon estimation. Curr. Sci. 85, 1528–1531.

Neyra, C.A., Atkinson, A., Olubayi, O., 1995. Coaggregation of Azospirillum with other bacteria: basis for functional diversity. NATO ASI Ser. G 37, 429–439.

Oustric, J., Morillon, R., Luro, F., Herbette, S., Martin, P., Giannettini, J., Berti, L., Santini, G., 2019. Nutrient deficiency tolerance in citrus is dependent on Genotype or ploidy level. Front. Plant Sci. 10, 1–13.

Padma, T.M.R., Kandasamy, D., 1990. Effect of interaction between VA-mycorrhizae and graded levels of phosphorous on the growth of papaya (*Carica papaya*). In: Proceedings of the National Conference on Trends in Mycorrhizal Research. vols. 14–16. Haryana Agricultural University, Hisar, pp. 133–134.

Pan, Q., Wang, Z., Quebedeaux, B., 1998. Responses of the apple plant to CO_2 enrichment: changes in photosynthesis, sorbitol, other soluble sugars and starch. Aust. J. Plant Physiol. 25, 293–297.

Patel, V.B., Singh, S.K., Asrey, R., Nain, L., Singh, A.K., Laxman, S., 2009. Microbial and inorganic fertilizers application influenced vegetative growth, yield, leaf nutrient status and soil microbial biomass in sweet orange cv Mosambi. Indian J. Hort. 66, 163–168.

Patil, V.K., Shinde, B.N., 2013. Studies on integrated nutrient management on growth and yield of banana cv. Ardhapuri (Musa AAA). J. Hort. Forestry 5, 130–138.

Peng, L., Wang, C., He, S., Guo, C., Yan, C., 2000. Effects of elevation and climatic factors on the fruit quality of Navel orange. South China Fruits 29, 3–4.

Pirlak, L., Turan, M., Sahin, F., Esitken, A., 2007. Floral and foliar application of plant growth promoting rhizobacteria (PGPR) to apples increases yield, growth and nutrient element contents of leaves. J. Sustain. Agric. 30, 145–155.

Quiroga-Rojas, L.I., Ruiz-Quinones, N., Munoz-Motta, G., Lozano- Tovar, M.D., 2012. Rhizosphere microorganism, potential antagonists of *Fusarium* sp. causing agent of root rot in passion fruit (*Passiflora edulis* Sims). Acta Agron. 61, 244–250.

Ram, R.A., Rajput, M.S., 2000. Role of biofertilizers and manures on production of guava (Psidium *guajava* L.) cv. Allahabad Safeda. Haryana J. Hort. Sci. 29, 193–194.

Raman, J., 2012. Response of Azotobacter, Pseudomonas and Trichoderma on growth of apple seedling. In: International Conference on Biological and Life Sciences. vol. 40, pp. 83–90.

Ray Dutta, S.K., Takawale, P.V., Chatterjee, R., Hnamte, V., 2014. Yield and quality of pomegranate as influenced by organic and inorganic nutrients. Bioscan 9, 317–620.

Rengel, Z., Gutteridge, R., Hirsch, P., Hornby, D., 1996. Plant genotype, micronutrient fertilization and take-all infection influence bacterial populations in the rhizosphere of wheat. Plant Soil 183, 269–277.

Rillig, M.C., Lehmann, A., Lehmann, J., Camenzind, T., Corneli, R., 2018. Soil biodivercity effects from field to fork. Trends Plant Sci. 23 (1), 17–24.

Schaffer, B., Whiley, A.W., Searle, C., Nissen, R.J., 1997. Leaf gas exchange, dry matter partitioning, and mineral element concentrations in mango as influenced by elevated atmospheric carbon dioxide and root restriction. J. Am. Soc. Hortic. Sci. 122, 849–855.

Scholberg, J., Morgan, K.T., 2012. Nutrient use efficiency in citrus. In: Srivastava, A.K. (Ed.), Advances in Citrus Nutrition. Springer Verlag, The Netherlands, pp. 205–229.

Sciubba, E., 2011. What did lotka really say? A critical reassessment of the maximum power principle. Ecol. Model. 222 (8), 1347–1353.

Sekikawa, S., Kibe, T., Koizumi, H., Mariko, S., 2003. Soil carbon sequestration in a grape orchard ecosystem in Japan. J. Jpn. Agric. Syst. Soc. 19, 141–150.

Shah, M.S., Wisal, M., Azam, S., Haq, N., 2006. Integrated nitrogen management of young deciduous apricot orchard. Soil Environ. 25, 59–63.

Shah, A.S., Mohammad, W., Shah, M.S., Ali, E.R., Haroon, A.A.B., 2014. Integrated effect of organic and inorganic nitrogen on peach fruit yield and orchard fertility. Agric. Sci. Res. J. 4, 78–82.

Shamseldin, A., El-Sheikh Mohamad, H., Hassan, H.S.A., KabeilS, S., 2010. Microbial biofertilization approaches to improve yield and quality of Washington navel orange and reducing the survival of nematode in the soil. J. Am. Sci. 6, 264–271.

Sharma, K.M., Bardhan, K., 2017. Can high yielding varieties perform similarly in organic farms? Plant Arch. 17 (1), 675–680.

Shukla, A.K., Sarolia, D.K., Kumari, B., Kaushik, R.A., Mahawer, L.N., Bairwa, H.L., 2009. Evaluation of substrate dynamics for integrated nutrient management under high density planting of guava cv. Sardar. Indian J. Hort. 66, 461–464.

Singh, S.R., Banik, B.C., 2011. Response of integrated nutrient management on flowering, fruit setting, yield and fruit quality in mango cv. Himsagar (*Mangifera indica* L.). Asian J. Hort. 6, 151–154.

Singh, C., Sharma, B.B., 1993. Leaf nutrient composition of sweet orange as affected by combined use of bio and chemical fertilizers. South Indian Hort. 41, 131–134.

Singh, J.K., Varu, D.K., 2013. Effect of integrated nutrient management in Papaya (*Carica papaya* L.) cv Madhubindu. Asian J. Hort. 8, 667–670.

Sofo, A., Nuzzo, V., Palese, A.M., Xyloyannis, C., Cellano, G., 2005. Net CO_2 storage in Mediterranean olive and peach orchards. Sci. Hortic. 107, 17–24.

Srivastava, A.K., 2010a. Development of INM to harmonize with improved citrus production: changing scenario. J. Adv. Plant Physiol. 12, 294–368.

Srivastava, A.K., 2010b. Integrated nutrient management in citrus: frontier developments. Indian J. Fert. 6 (11), 34–44.

Srivastava, A.K., 2013c. Recent developments in diagnosis and management of nutrient constraints in acid lime. J. Agric. Sci. 2 (3), 86–96.

Srivastava, A.K., 2013d. Nutrient deficiency symptomology in citrus: an effective diagnostic tool or just an aid for post–mortem analysis. J. Agric. Sci. 2 (6), 177–194.

Srivastava, A.K., Kohli, R.R., 1997. Soil suitability criteria for citrus—an appraisal. Agric. Rev. 18 (3), 134–146.

Srivastava, A.K., Ngullie, E., 2009. Integrated nutrient management: theory and practice. Dyn. Soil Dyn. Plant 3 (1), 1–30.

Srivastava, A.K., Singh, S., 2003a. Diagnosis and Management of Nutrient Constraints in Citrus. Manual No. 2, National Research Centre for Citrus, Nagpur, pp. 1–70.

Srivastava, A.K., Singh, S., 2003b. Soil-plant nutrient limits in relation to optimum fruit yield of sweet orange (*Citrus sinensis* Osbeck) cultivar Mosambi. Indian J. Agric. Sci. 73 (4), 209–211.

Srivastava, A.K., Singh, S., 2003c. Plant and soil diagnostic norms for optimum productivity of Nagpur mandarin (*Citrus reticulata* Blanco). Fertilizer News 48 (2), 47–63.

Srivastava, A.K., Singh, S., 2004a. Soil and plant nutritional constraints contributing to citrus decline in Marathawada region, India. Commun. Soil Sci. Plant Anal. 35 (17/18), 2537–2550.

Srivastava, A.K., Singh, S., 2004b. Zinc nutrition, a global concern for sustainable citrus production. J. Sustain. Agric. 25 (3), 5–42.

Srivastava, A.K., Singh, S., 2004c. Zinc nutrition and citrus decline—a review. Agric. Rev. 25 (3), 173–188.

Srivastava, A.K., Singh, S., 2005. Diagnosis of nutrient constraints in citrus orchards of humid tropical. J. Plant Nutr. 29 (6), 1061–1076.

Srivastava, A.K., Singh, S., 2006. Nutrient diagnostics and management in citrus. In: Technical Bulletin No. 8, National Research Centre for Citrus, Nagpur, Maharashtra, India, pp. 1–130.

Srivastava, A.K., Singh, S., 2007. DRIS-based nutrient norms for Nagpur mandarin (*Citrus reticulata* Blanco). Indian J. Agric. Sci. 77 (6), 363–365.

Srivastava, A.K., Singh, S., 2008a. Citrus nutrition research in India: problems and prospects. Indian J. Agric. Sci. 78, 3–16.

Srivastava, A.K., Singh, S., 2008b. DRIS norms and their field validation in Nagpur mandarin (*Citrus reticulata* Blanco). J. Plant Nutr. 31 (6), 1091–1107.

Srivastava, A.K., Singh, S., 2009a. Citrus decline: soil fertility and plant nutrition. J. Plant Nutr. 32 (2), 197–245.

Srivastava, A.K., Singh, S., 2009b. Zinc nutrition in Nagpur mandarin on haplustert. J. Plant Nutr. 32 (7), 1065–1081.

Srivastava, A.K., Singh, S., 2015. Site-Specific nutrient management in Nagpur mandarin (*Citrus reticulata* Blanco) raised on contrasting soil types. Commun. Soil Sci. Plant Anal. 47 (3), 447–456.

Srivastava, A.K., Singh, S., Marathe, R.A., 2002. Organic citrus: soil fertility and plant nutrition. J. Sustain. Agric. 19 (3), 5–29.

Srivastava, A.K., Huchche, A.D., Lallan, R., Singh, S., 2007a. Yield prediction in intercropped versus monocropped citrus orchards. Sci. Hortic. 114, 67–70.

Srivastava, A.K., Singh, S., Tiwari, K.N., 2007b. Diagnostic tools for citrus: their use and implications in India. Better Crops—India 1 (1), 26–29.

Srivastava, A.K., Singh, S., Albrigo, L.G., 2008. Diagnosis and remediation of nutrient constraints in citrus. Hortic. Rev. 34, 277–364.

Srivastava, A.K., Singh, S., Huchche, A.D., 2012. Evaluation of INM in citrus (*Citrus reticulata* Blanco): Biometric response, soil carbon and nutrient dynamics. Int. J. Innov. Hort. 1 (2), 126–134.

Srivastava, A.K., Das, S.N., Malhotra, S.K., Majumdar, K., 2014. SSNM-based rationale of fertilizer use in perennial crops: a review. J. Agric. Sci. 84 (1), 3–17.

Srivastava, A.K., Malhotra, S.K., Krishna Kumar, N.K., 2015a. Exploiting nutrient-microbe synergy in unlocking productivity potential of perennial fruits: a review. Indian J. Agric. Sci. 85 (4), 459–481.

Srivastava, A.K., Singh, S., Huchche, A.D., 2015b. Evaluation of INM in citrus on Vertic Ustochrept: biometric response and soil health. J. Plant Nutr. 38 (5), 1–15.

Srivastava, A.K., Paithankar, D.H., Venkataramana, K.T., Hazarika, B., Patil, P., 2019. INM in fruit crops sustainable quality production and soil health. Indian J. Agric. Sci. 89 (3), 379–395.

Sugiura, T., Kuroda, H., Sugiura, H., 2007. Influence of the current state of global warming on fruit tree growth in Japan. Hortic. Res. 6, 257–263.

Sunkar, R., Chinnusamy, V., Zhu, J., Zhu, J.K., 2007. Small RNAs as big players in plant abiotic stress responses and nutrient deprivation. Trends Plant Sci. 12, 301–309. https://doi.org/10.1016/S0065-2660(08)6049-3.

Suresh, C.P., Hasan, M.A., 2001. Studies on the response of Dwarf Cavendish banana Musa AAA to biofertilizer inoculation. Hortic. J. 14, 35–41.

Tagaliavini, M., Tonon, G., Scandellari, F., Quinones, A., Palmieri, S., Menarbin, G., Gioacchini, P., Masia, A., 2007. Nutrient recycling during the decomposition of apple leaves (*Malus domestica*) and mowed grasses in an orchard. Agric. Ecosyst. Environ. 118, 191–200.

Tandon, H.L.S., Muralidharadu, Y., 2010. Nutrient Uptake, Removal and Recycling by Crops. Fertilizer Development and Consultation Organization, New Delhi, pp. 1–140.

Tiwari, D.K., Hasan, M.A., Chattopadhyay, P.K., 1999. Leaf nutrient and chlorophyll content in banana (Musa AAA) under influence of *Azotobacter* and *Azospirillum* inoculation. Environ. Ecol. 17, 346–350.

Umer, I., Wali, V.W., Kher, R., Jamwal, M., 2009. Effect of FYM, Urea and Azotobacter on growth, yield and quality of strawberry cv. Chandler. Not. Bot. Hort. Agrobot. Chuj. 37, 139–143.

Wang, S., Srivastava, A.K., Wu, Q.-S., Fokom, R., 2014. The effect of mycorrhizal inoculation on the rhizosphere properties of trifoliate orange (*Poncirus trifoliata* L. Raf.). Sci. Hortic. 170, 137–142.

Wange, S.S., Ranawade, D.B., 1998. Effect of microbial inoculants on fresh root development on grape var. Kishmis chorni. Recent Hortic. 4, 27–31.

Wu, Q.S., Srivastava, A.K., 2015. Effect of mycorrhizal symbiosis on growth behavior and carbohydrate metabolism of trifoliate orange under different substrate P levels. J. Plant Growth Regul. 34, 499–508.

Wu, T., Wang, Y., Changjilang, Y., Chiarawipa, R., Xinzhoing, Z., Zhenhai, H., Wu, L., 2012. Carbon sequestration by fruit trees—Chinese apple orchards as an example. PLoS One. 7, e38883. https://doi.org/10.1371/journal.pone.0038883, http://www.plosone.org/article/info%3Adoi%2F10.1371%2Fjournal.pone.0038883.

Wu, Q.S., Srivastava, A.K., Zou, Y.N., 2013. AMF-induced tolerance to drought stress in citrus: a review. Sci. Hortic. 164, 77–87.

Wu, Q.S., Srivastava, A.K., Cao, M.Q., Wang, J., 2014. Mycorrhizal function on soil aggregate stability in root zone and root-free hyphae zone of trifoliate orange. Arch. Agron. Soil Sci. 61 (6), 813–825.

Wu, Q.S., Wang, S., Srivastava, A.K., 2016a. Mycorrhizal hyphal disruption induces changes in plant growth, glomalin-related soil protein and soil aggregation of trifoliate orange in a core system. Soil Tillage Res. 160, 82–91.

Wu, Q.S., Srivastava, A.K., Cao, M.Q., 2016b. Systematicness of glomalin in roots and mycorrhizosphere of a split-root trifoliate orange. Plant Soil Environ. 62 (11), 508–514.

Wu, Q.S., He, J.D., Srivastava, A.K., Zou, Y.N., Kamil, K., 2019. Mycorrhizas enhance drought tolerance of citrus by altering root fatty acid composition and their saturation levels. Tree Physiol. 39, 1149–1158.

Xiloyannis, C., Montanaro, G., Sofo, A., 2002. Proposte per contenere i danni da siccita` alle piante da frutto. Frutticoltura 7–8, 19–27.

Xuan, Y., Liu, X., Zhu, T.H., Liu, G.H., Mao, C., 2011. Isolation and characterization of phosphate solubilising bacteria from walnut and their effect on growth and phosphorous mobilization. Biol. Fertil. Soils 47, 437–446.

Yadav, K., Prasad, V., Mandal, K., Ahmed, N., 1992. Effect of coinocuation (*Azospirillum* and *Rhizobium* strains) on nodulation, yield, nutrient uptake and quality of lentil [*Lens culinaris*]. Lens Newsletter 19, 29–31.

Yadav, R., Singh, B.H., Singh, H.K., Yadav, A.L., 2007. Effect of integrated nutrient management on productivity and quality of Aonla (*Emblica officinalis* Gaetm.) fruits. Plant Arch. 1&2, 881–883.

Zou, Y.N., Srivastava, A.K., Qiang-Sheng, W., Yong-Ming, H., 2014. Glomalin-related soil protein and water relations in mycorrhizal citrus (*Citrus tangerina*) during soil water deficit. Arch. Agron. Soil Sci. 60 (8), 1103–1104.

Zou, Y.N., Srivastava, A.K., Qiu-Dan, N., Qiang-Sheng, W., 2015. Disruption of mycorrhizal extraradical mycelium and changes in leaf water status and soil aggregate stability in rootbox-grown trifoliate orange. Front. Microbiol. 6, 20–31.

Further reading

Chadha, K.L. (Ed.), 2007. Handbook of Horticulture. Indian Council of Agricultural Research ICAR, New Delhi, p. 1031.

Chadha, K.L., Bhargava, B.S., 1997. Plant nutrient supply, needs, efficiency and policy issues for fruit crops in Indian agriculture from 2000 to 2025 AD. In: Kanwar, J.S., Katyal, J.C. (Eds.), Plant Nutrient Needs, Supply, Efficiency and Policy Issues: 2000–2025. NAAS, New Delhi. 1134–132.

Kemmler, G., Hobt, H., 1985. Potassium a Product of Nature. Kali and Salz, Kassel.

Sparks, D., 1989. Pecan nutrition—a review. Proc. S.E. Pecan Grow. Assoc. 82, 101–122.

Srivastava, A.K., Malhotra, S.K., 2014. Nutrient management in fruit crops: issues and strategies. Indian J. Fert. 10 (12), 72–88.

38

Evaluation of organic versus conventional nutrient management practices in fruit crops

Maciej Gąstoł*

Department of Pomology and Apiculture, Agricultural University in Kraków, Kraków, Poland
*Corresponding author. E-mail: rogastol@cyfronet.pl

1 Introduction

"You are what you eat"—this common phrase reflects our belief that our diet and its quality strongly influence human's behavior and health. However, during the last decade, consumers' trust in food quality has drastically decreased, mainly because of some food scandals (BSE, dioxins, and adulteration). There is a rapidly growing group of consumers unsatisfied with conventional agriculture products and thus seeking alternatives (Canavari and Olson, 2007; Hughner et al., 2007; Stolz et al., 2011). They look for authentic, better controlled foods, produced without artificial additives. Organic food is also associated by the general public as being more attractive with its improved nutritional properties, produced in a more environmentally friendly manner (Zanoli and Naspetti, 2002; Carlson and Jaenicke, 2016; Mie et al., 2016). Studies indicate that many consumers buy organic products because of the perceived taste, health, and nutrient benefits and avoided pesticides or genetically modified foods (Winter and Davis, 2006). As far as other organic food consumption motives are concerned, ethical issues, animal welfare along with one's care for the environment tends to play a certain role (Seyfang, 2008; Goetzke et al., 2014).

However, health-related issues seem to be the most important reasons (Huber et al., 2012; Kahl et al., 2012; Truong et al., 2012). Some authorities like the World Health Organization and Food and Agriculture Organization (WHO and FAO, 2004) have increasingly put emphasis on the relationship between food, nutrition, and health. In addition, some recent human epidemiological studies associated consumption of organic foods with lower risks of some civilization disease like allergies (Huber et al., 2011).

Therefore, the consumption of organic food has been growing all over the world (Raigón et al., 2010). The global organic food market has grown rapidly over the last decade (Sahota, 2018), passing EUR 80 billion in 2016 (Willer and Lernoud, 2018). Hence, organic agriculture has considerably increased from 200,000 producers in 1999 to 2.7 million

A.K. Srivastava, Chengxiao Hu (eds.)
Fruit Crops: Diagnosis and Management of Nutrient Constraints
https://doi.org/10.1016/B978-0-12-818732-6.00038-1

producers in 2016, being this tendency still forecasted (IFOAM, 2018). From 2004 to 2012, the size of the European organic food market doubled, reaching 22.8 billion EUR, which means organic food retail sales per capita at 35 EUR (Bryła, 2015). Organic food sales have been growing 10%–12% annually in the United States and 6% annually in Europe in recent years (Willer and Schaack, 2015).

At the category level, organic fruit tree share is quite small, with 1.8%, 0.9%, and up to 0.9% of all temperate, citrus, and tropical/subtropical tree fruit area worldwide, respectively. However, fruits and vegetables accounted for 34% of all retail organic food sale. Moreover, from 2008 to 2013, the area of production grew 109%, 42%, and 53% for organic temperate tree fruits, citrus, and tropical/subtropical fruits, respectively (FAO, 2015).

Mexico, Italy, and China are the top three countries in terms of organic tree fruit area, while Poland has seen the most rapid expansion of organic apple acreage anywhere in the world. Italy has led Europe for years in organic citrus and temperate fruit production (Granatstein, 2015) but has been replaced by both China and Poland for temperate fruits worldwide (CNCA, 2015).

Therefore, the organic food quality issues considerably attract not only customers but also many scientists (Woese et al., 1997; Heaton, 2001; Bonti-Ankomah and Yiridoe, 2006). There is a debate whether organic farming is better for both consumers and the environment (Hansen et al., 2001; Köpke, 2005; Hughner et al., 2007). However, the data on nutritional quality of organic produce in comparison with conventional produce are often inconclusive, the reports' results are contradictory, and they are the subject of much controversy (Woese et al., 1997; Brandt and Mølgaard, 2001; Bourn and Prescott, 2002; Williams, 2002; Magkos et al., 2003). Some research reports point to statistical differences (Worthington, 2001; Pussemier et al., 2006; Benbrook et al., 2008), while other studies do not (Dangour et al., 2009). Organic diets have been convincingly demonstrated to expose consumer to fewer pesticides (Lu et al., 2006) and antibiotic agents (Hamer and Gill, 2002). Recent reviews on vegetables, fruits, and milk showed significantly higher amounts of bioactive compounds in the organic food products (Palupi et al., 2012; Brandt and Mølgaard, 2001). Several of the studies reported that "bio"-products have lower nitrate content and higher dry matter and mineral content compared with conventionally grown alternatives (Benbrook et al., 2008; Mader et al., 1993; Rembiałkowska, 2000; Gąstoł et al., 2011). Some reviews of the comparative studies (Woese et al., 1997; Bourn and Prescott, 2002; Hunter et al., 2011) indicated contrasting conclusions. The limited number of studies and variable obtained results make difficult to compose a general conclusion (Matt et al., 2011).

Winter and Davis (2006) reviewed three meta-analyses carried out in the late 1990s and early 2000s, one reporting no major difference, one significant differences, and one contradictory results (differences in some cases and not in others). Recent reviews also provided contrasting results: Lairon (2010) reported that the nutritional properties of organic produce have never been inferior to conventional produce and in many cases have been superior (i.e., vegetables and fruit, milk, and meat); Dangour et al. (2009) conclude that there is no evidence for a difference in nutrient quality between organically and conventionally produced foodstuffs.

In 2009, the United Kingdom Food Standard Agency (FSA) published an extensive report on organic food that concluded that there was no evidence of a difference in a nutritional quality between organically and conventionally produced foodstuffs. The conclusion was based on a review of 55 relevant studies conducted over the past 50 years (Dangour et al., 2009). Scientists convened by The Organic Center (TOC) carried out the similar but more rigorous review of the same literature (Benbrook et al., 2008). The TOC team analyzed published research just on plant-based foods. Results differed significantly from the narrower FSA review included studies over a 50-year period (1958–2008). Since 2008, the cut-off date of the London study, some 15 new studies have been published most of which use superior design and analytical methods based on criticism of older studies (Benbrook et al. 2008). This indicated the great need to elaborate proper methodology for such research.

It is explained that crop variety, soil type, climate, duration of experiment, postharvest practices, and statistical design can all influence conclusions on the nutritive and sensory characteristics of a product (Hornick, 1992; Heaton, 2001; Woese et al., 1997; Bourn and Prescott, 2002). Environmental conditions likely to affect food quality include geographical area, soil type (differences in clay mineralogy), soil pH, soil moisture, soil health (humus content, fertility, microbial activity, etc.), weather and climatic conditions (temperature, rainfall, flooding, drought), and pollution (Heaton, 2001). Since soils can vary greatly from region to region, it is difficult to compare nutrient quality of crops from different sites.

Assessing the nutritional quality of fruits produced in different production systems is an even more challenging scientific exercise. This is due to long-term factorial field trials that must be carried out before valid conclusions could be made. Many, other than setup experimental factors, may influence the perennial crops. Therefore, any methodological insufficiencies, may affect the accuracy of the comparative studies' final results.

2 Methodology of organic versus conventional: Comparative studies

As mentioned earlier, the most important factor making the comparison between the effect of organic and conventional cultivation effects accurate is a methodological aspect (Vijver et al., 2007). The agriculture methodologists (Vetter et al., 1987) distinguish three types of the comparative experiments: field trials, on-farm research, and market surveys (shopping basket).

Of the three methods, cultivation tests (field trials) are assumed by food scientists as the best way for optimal control of agricultural practices and finally most accurate (Woese et al., 1997). However, due to complicated logistics, large sample size, and a high number of tests required to produce reliable information accessibility and costs, these experiments are not commonly used. The "on-farm surveys" are performed on existing neighboring (organic/conventional) farms. This kind of study is thought as more suitable for the practice. It seems to be the best alternative if all important practices/routines are recorded. It also allows to prepare study covering more fields and farms. Therefore, in our study, we decided to choose this model of experiment. For the experiment, 66 fruit fields (36 farms) producing organic and conventional crops were chosen. The detailed methodology is presented in our previous reports (Gąstoł et al., 2011; Domagała-Świątkiewicz and Gąstoł, 2012). In the studies more orientated on consumer, the market surveys (shopping basket) are also performed. However, this type of study is not suitable to compare the agricultural practices as we have no data on it.

2.1 Mineral content

Mineral compounds, including magnesium, phosphorus, and trace elements like iron and manganese are vital in healthy diet. However, some long-term studies comparing mineral content of fruits made from the 1930s to 1980s show continuous decreasing mineral content in respect of some essential elements like K, Mg, Fe, and Cu (Mayer, 1997). This is linked with an intensification of fruit production in the last decades.

While there are many factors that can influence the nutrient contents of crops, the method of farming is also shown to be a strong influence, with the valid scientific studies demonstrating a trend toward significantly higher mineral contents, in organically grown than nonorganically grown fruit and vegetables (Holden, 2001). However, analyzing results of our experiments (Gąstoł et al., 2011; Domagała-Świątkiewicz and Gąstoł, 2012), it should be concluded that the effect of farming system on mineral content of fruit is often scattered (Figs. 38.1 and 38.2). While for some investigated species (black currant), we obtained significant and consistent results, for some being nonsignificant (pears).

Many authors have reported similar trend (no differences) for plums (Lombardi-Boccia et al., 2004), grapefruit (Lester, 2007), strawberry (Conti et al., 2014), or jujube (Reche et al., 2019). On the contrary, Raigón et al. (2010) found a higher mineral content in organic eggplant.

2.2 Macronutrients

A comparative study by Worthington (2001) indicates a higher concentration of mineral elements (iron, magnesium, and phosphorus) in organic raw material that is linked to higher soil microorganism content and its activity in organically managed soils (Fliessbach and Mäder, 2000). Soil microorganisms produce active compounds that are adsorbed by soil minerals making them more available to plant roots (Rembiałkowska, 2007). Matt et al. (2011) report shows no clear conclusion in the case of fruits. In the Heaton (2001) review among 14 studies, 7 demonstrated a trend toward higher mineral content in organically grown crops, and 6 revealed inconsistent or no significant differences.

A range of studies suggest that organic cropping systems promote mineral content, but there are instances where differences were small or nonexistent. Rembiałkowska (2007) estimated that organic crops overall contain 29% more Mg than their conventional counterparts. Several studies have reported contrasting results. For instance, Amodio et al. (2007) reported higher concentrations of main minerals in organic kiwifruits compared with those grown under conventional production system. In the study of Benge et al. (2000), calcium was the only element which content was significantly different between organic and conventional production system. The authors did not provide information about the types and amounts of fertilizers in their investigation. Fruit harvested from conventional orchards has been linked to a lower phosphorus and potassium concentrations and higher nitrogen concentrations than those grown under organic production system (DeEll and Prange, 1993; Raigón et al., 2010). This could be attributed to nitrogen fertilization, which is known of inhibiting phosphorus levels in apples. On the other hand, the phosphorus level could be alleviated by higher mycorrhizal activity. Comparing the fertilization program between

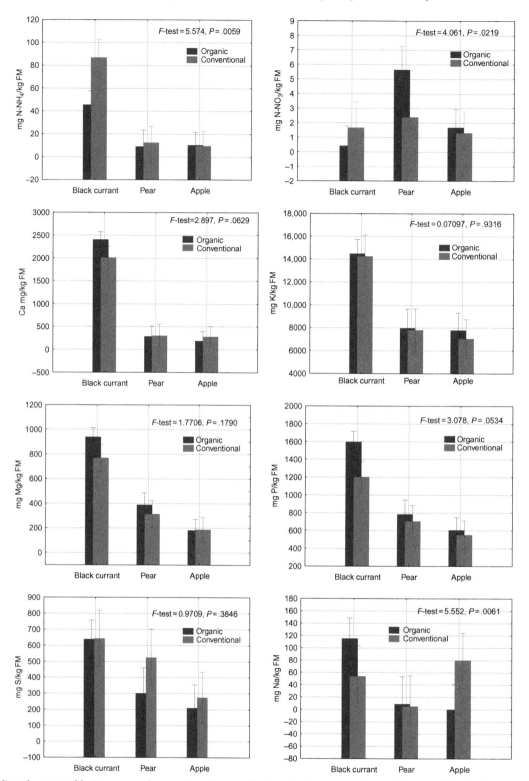

FIG. 38.1 Mineral content of fruit grown under organic and conventional methods—macroelements (*F*-test and *P* value for two-way interaction between variables: farming method and species, *P* = .05). *Data from Gąstoł, M, Domagała-Świątkiewicz, I. 2012. Comparative study on mineral content of organic and conventional apple, pear and black currant juices. Acta Sci. Pol. Hortorum Cultus 11(3), 3–14.*

production systems could explain the mineral composition of fruit at postharvest. For instance, Amarante et al. (2008) reported lower concentrations of K, Mg, and N in organic "Royal Gala" and "Fuji" apples compared with conventional ones.

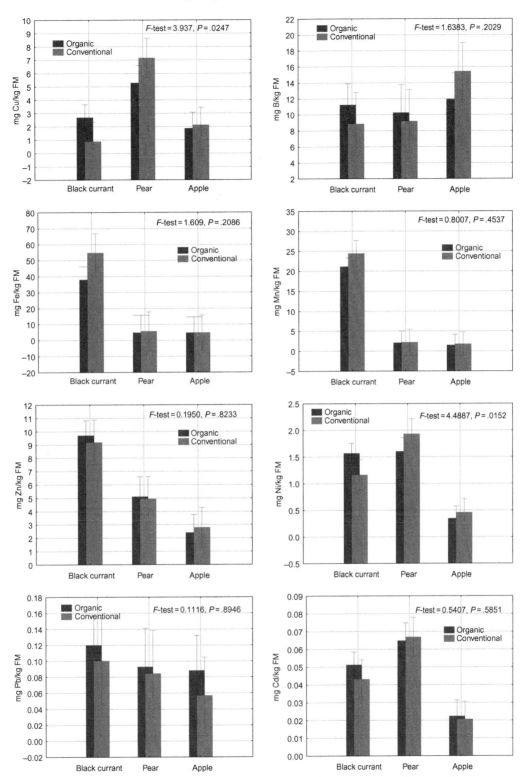

FIG. 38.2 Mineral content of fruits grown under organic and conventional methods—microelements (*F*-test and *P* value for two-way interaction between variables: farming method and species, *P* = .05 *Data from Gąstoł, M, Domagała-Świątkiewicz, I. 2012. Comparative study on mineral content of organic and conventional apple, pear and black currant juices. Acta Sci. Pol. Hortorum Cultus 11(3), 3–14.*

The organic fertilization program affects not only the trees' mineral content but also external quality fruit properties. DeEll and Prange (1993) and Jönsson (2007) reported reduced marketability and a higher incidence of storage rots caused by fungi in organically produced apples compared with conventionally produced ones. It has been hypothesized that higher pathogen pressure and lower mineral supply in organic orchards are responsible for the low microbiological quality in organic fruits. Studies by Amarante et al. (2008) also indicated a higher incidence of the moldy core in "Fuji" apples than fruit from conventional orchards.

2.2.1 Nitrogen

Worthington (2001) summarized the results of 41 studies comparing nitrate levels of organic and conventional foods and found 127 cases where nitrate levels were higher in conventional foods, 43 cases where nitrate levels were higher in organic foods, and 6 cases where no difference was observed. The ratio of nitrate levels in conventional foods relative to organic foods ranged from 97% to 819%. Heaton (2001) in his review reported that 16 valid studies comparing the nitrate content of organically and nonorganically grown fruit and vegetables revealed the following: 14 studies demonstrated a trend toward significantly lower nitrate content in organically grown crops, and two studies found inconsistent or not significant differences. The data obtained for food sold in Belgium showed a mean nitrate value of 1703 mg/kg for organic products and 2637 mg/kg for conventional products (Pussemier et al. 2006).

Also, Rembiałkowska (2007) concluded that nitrates and nitrites are unarguably higher in conventionally cultivated crops compared with organic ones. In their study, nitrate levels in conventionally grown fruits were almost twofold of the organically grown fruits. Lairon (2010) indicated that the use of slow-release nitrogen fertilizers such as compost in organically managed soils is key in explaining lower nitrate accumulation in organic produce. Nitrate concentrations of fruit are linked to nitrogen availability in the root system.

In our studies (Gąstoł et al. 2011; Domagała-Świątkiewicz and Gąstoł 2012), no differences were found between conventional and organic fruit in nitrate level (Table 38.1). The levels of fruit nitrates and nitrites were low. The similar results demonstrated Woese et al. (1997) in the review concerned with the summary and evaluation of the results from more than 150 investigations comparing the quality of conventionally and organically produced food. Many researchers (Dangour et al., 2009; Guadagnin et al., 2005; Lester, 2007; Lairon, 2010) have confirmed that the nitrate levels in plant crops are higher when nitrogen is supplied in high-dose inorganic fertilizers than when supplied from slowly mineralized organic manures.

The difference in soil management techniques in organic and conventional agriculture, especially nitrogen source, could be the main factor in fruit coloration. Studies on "Galaxy Gala," "Royal Gala," and "Fuji" apples (DeEll and Prange, 1993; Amarante et al., 2008), found that color intensity was significantly less with high nitrogen levels than with low levels. The reduced fruit coloration at high nitrogen levels was largely attributed to higher nitrogenous substances stored in fruits and reduced carbohydrate accumulation in fruit (Kliewer, 1977). The excessive growth of N rich trees and the reduced light within the canopy may decrease the activity of enzymes that regulate the synthesis of anthocyanins (Brunetto et al., 2015), which give color to fruits and vegetables.

Fruit firmness loss is another fruit quality predictor that is closely linked to water content loss and metabolic changes (García et al., 1998). A lower N:Ca ratio in organically managed soils, as opposed to higher ratio in conventional soils, led to higher fruit firmness in organic "Galaxy Gala," "Fuji," and "Royal Gala" apples (Peck et al., 2006; Amarante et al., 2008). The results were largely attributed to differences in nitrogen concentration between the two production systems. Therefore, maintaining lower N:Ca ratio is highly critical for fruit firmness and consequently better overall postharvest quality.

TABLE 38.1   The effect of farming system (organic vs conventional) on the mineral content and some quality parameters in fruit juices ($P = .05$).

	Black currant	Pear	Apple
DM	Organic > conventional	ns	ns
Total sugars	ns	ns	ns
TA	ns	ns	ns
Ascorbic acid	ns	ns	ns
Polyphenols	ns	ns	ns
FRAP	ns	ns	ns
N-NH$_4$	Organic < conventional	ns	ns
N-NO$_3$	ns	ns	ns

DM, dry matter content; FRAP, ferric reducing antioxidant power; ns, not significant; TA, titratable acidity.
Data from Gąstoł, M., Domagała-Świątkiewicz, I., Krośniak, M. 2011. Organic versus conventional—a comparative study on quality and nutritional value of fruit and vegetable juices. Biol. Agric. Hortic. 27(3), 310–319.

2.2.2 Phosphorus and sulfur

The effect of cultivation method on *phosphorus* content was recorded merely in black currant juices. This relationship was correlated with higher level of soil phosphorus in organic black current fields (Gąstoł, 2009). Weibel et al. (2000) compared the dry matter and mineral (P, K, Ca, Mg, and Se) content in apples from organic and conventional farming systems. The data showed higher phosphorus content for organic apples (increase of one-third), whereas no differences were found in the K, Ca, Mg, and Se levels. Phosphorus solubility is low in the most of soils. A varied crop rotation in organic farming and animal manures, green manures, and compost applied lead to optimal soil biological activity. Under organic management, colonization by mycorrhizal fungi is greatly increased, which is very important for feeding orchard plants. This increases the utilizable root space and enhances P uptake by plant (Hildermann et al., 2010). However, in conventional farming, soluble phosphorus fertilizers promote high phosphorus uptake of plants.

As far as *sulfur* content is concerned, in our study, the higher level was noted for conventional apple juices as compared with organic ones (Fig. 38.1). For the rest of fruits, no significant differences were found between orchard management systems. Many orchards have significant natural inputs of sulfur from mineralization of soil organic matter, with precipitation, irrigation, and unintended applications of S with sulfur-containing fertilizers and pesticides. Most organic amendments also have large sulfur content (Ferree and Warrington, 2003; Eriksson 2001), and S overdose due to excessive plant protection is concerned (Lind et al., 2003).

2.2.3 Calcium, magnesium, and potassium

The cultivation method influenced *calcium* accumulation in black currants. However, for the rest of investigated species, no significant differences in calcium concentration in juices were found. These results are in agreement with Benbrook et al. (2008) report of higher concentrations of minerals and other nutrients in organically grown crops. Also, the results of Raigón et al. (2010) showed that organic management and fertilization have a positive effect on the accumulation of certain beneficial mineral compounds like K, Ca, and Mg in eggplants. In contrast, DeEll and Prange (1993) in their farm survey found no significant differences in Ca and Mg content, whereas organically grown apples had higher concentration of P and K. Similarly, Amarante et al. (2008) found no impact of husbandry method on Ca content in "Royal Gala" apples. However, the author proved lower K and Mg levels for organic fruits. A lower N:Ca ratio in organically managed soils, as opposed to higher ratio in conventional soils, leads to higher fruit firmness in organic fruit (Peck et al., 2006; Amarante et al., 2008).

Organic fruits revealed higher *magnesium* amounts (Gąstoł et al., 2011; Domagała-Świątkiewicz and Gąstoł, 2012). It was proved for pears and currants. Weibel et al. (2000) found the similar for apples. However, in our study, there was no impact of farming system on Mg content in apple juices.

In the experiment (Domagała-Świątkiewicz and Gąstoł, 2012), we recorded some differences in *potassium* content caused by growing method; however, the differences were not statistically significant. Pither and Hall (1990) found higher levels of potassium content in organic apples. A sufficient release of nutrients from soil reserves causes a high soil microbial activity making them available for root uptake (Hildermann 2010). Probably, for this reason, some studies demonstrate a trend toward significantly higher nutrient mineral content in organically grown crops.

2.2.4 Sodium and trace elements

Studies on mineral content of these elements in fruits have not given any clear conclusion. Some have reported no differences between organic and conventional (Woese et al., 1997; Bourn and Prescott, 2002), whereas Smith (1993) reported higher micronutrient levels for organically grown apples and pears.

Although the measured *sodium* levels were low in our study, some differences between fruits, in respect of cultivation method, were observed. Conventional apple juices contained more sodium than organic, whereas the reverse was true for currants (Fig. 38.1).

The cultivation system had no impact on iron fruit content (Fig. 38.2). The study of Woese et al. (1997) showed organic growth of vegetables had higher Fe content. Organically produced currant juices contained more *zinc* and *copper* as compared with their conventional contrapartners. Several significant sources such as fertilizers, sewage sludge, manures, and agrochemicals (fungicides) have contributed to increased Cu levels in agricultural soils and plant foods.

Report of Matt et al. (2011) demonstrated that studies have not shown distinct differences in the heavy metal content between organic and conventional plant materials. The ambiguous results in Cd level indicate the need for further study of the agriculture factors determining the Cd level in plant crops (2000). The heavy metal concentration in measured samples was very low. *Cadmium* content for conventional apple was significantly higher than organic, while the reverse was noted black currants.

No differences in *lead* level in juices were found. The results of Rossi et al. (2008) study show organic tomatoes had higher Cd and Pb levels but a lower Cu content than conventionally ones. Kabata-Pendias (2011) concluded a relatively minor effect of the Pb soil concentration due to agricultural activity. Also, Rembiałkowska (2000) pointed out that many studies did not show distinct differences in heavy metals between "bio" and conventional foods.

Dry matter, soluble solid content, and organic acid content

In general, the differences between organic and conventional products are commonly justified by the different dry matter content of these products, with higher dry matter being expected for the organic ones. Huber et al. (2011) compared dry matter content of 19 organic and conventional fruits and vegetables, in which only 10 showed that organic products had 20% higher dry matter content.

Our own analyses have shown just a little increased concentration of dry matter in ecological fruits as compared with conventional ones. Only in respect of organic black currants, the fruit dry matter was significantly increased. Amarante et al. (2008) and Peck et al. (2006) reported that organic apples had higher soluble solid content than conventional fruit.

The total sugar content not only promotes better taste but is an important component of the technological quality. The results from the study of Wang et al. (2008) showed that blueberry fruit grown from organic culture had significantly higher sugar content (fructose and glucose) than fruit from conventional farms. Lester (2007) proved the sugar content in organic grapefruits was higher compared with conventional ones.

Studies on kiwifruits, plums, and strawberries have shown the small or nonexistent effect of organic and conventional production system on total sugars (Lombardi-Boccia et al., 2004; Amodio et al., 2007; Raigón et al., 2010). Based on the cultural practices, particularly fertilization program, between organic and conventional agriculture, total sugars should be higher in organically produced fruit. Between nitrogen concentration in the soil and content of total sugars in fruits (Elamin and Al-Wehaibi, 2005), Beinard et al. (2009) have proved that fruit sucrose, fructose, and glucose levels increased when nitrate levels were reduced. Organic sources of nitrogen and ammonium have previously been found superior to inorganic nitrate in increasing total sugars (Heeb et al., 2005).

Dangour et al. (2009) summarized the results of nutrient content comparison from 46 satisfactory quality-crop studies and found significantly higher titratable acidity in organically produced crops. Amarante et al. (2008) found lower titratable acidity in "Royal Gala" fruit from an organic orchard than in fruit from a conventional orchard. Peck et al. (2006) studied fruit quality of "Galaxy Gala" apples from organic, conventional, and integrated orchards and found no differences in the titratable acidity in fruit from the examined management systems.

However, in our studies, the titratable acidity did not differ significantly between organic and conventional production methods for any of the fruit tested.

These data agreed with those obtained in other organic fruits, such strawberry (Conti et al., 2014), passion fruit (De Oliveira et al., 2017), and grapefruit (Lester, 2007). Lester (2007) also found more acidity in organically than conventionally grown grapefruit, with greater amounts of ascorbic acid. Also, in a recent study of Cuevas et al. (2015) on 13 European plum cultivars (*Prunus domestica*), it demonstrated that organic production significantly promotes accumulation of organic acids. In their study, higher concentrations of malic, succinic, tartaric, and shikimic acid were reported. However, Lombardi-Boccia et al. (2004) in yellow plums did not find significant differences in the total content of organic acids.

The lower organic acid accumulation in conventional fruit is explained by higher amounts of nitrogen fertilization used. Previous research has shown that such fertilizers may retard the concentration of organic acids (Amodio et al., 2007). Additionally, some pesticides linked with conventional pest management program could be classified as shikimate pathway inhibitors (Lydon and Duke, 1989), and such pesticides may negatively affect the accumulation of organic acids in fruits. This was confirmed by Cuevas et al. (2015) who showed the lower content of shikimic acid in conventionally grown plums as compared with organic ones.

One of technical fruit parameters that is strongly correlated with our palatability perception is a soluble solid content-to-organic acid ratio (SSC:TA). Peck et al. (2006) reported that neither instrumental measurements nor sensory evaluations detected differences in SSC, TA, and SSC:TA of "Galaxy Gala" apples grown in organic or conventional production system. Similarly, Khalil and Hassan (2015) in their comparative study reported that TSS of organically grown oranges and strawberry was not significantly different from those conventionally grown. Amarante et al. (2008) reported no significant differences for fruit attributes of taste and flavor from organic and conventional production system. Also, DeEll and Prange (1993) found no significant differences perceived in juiciness, sweetness, tartness, and off-flavor of organically and conventionally produced "McIntosh" and "Cortland" apples at harvest or after storage.

Peck et al. (2006) reported clear trends on consumer preference for flavor and sweetness of organically and conventionally produced "Galaxy Gala" apples. Weibel et al. (2000) reported inconsistent results. On the first year of study, organic fruit received a 15% higher sensory score compared with conventional fruit. In the second and third year, however, panelists found no differences in the sensory quality between the production systems. Raigón et al. (2010) using hedonic/intensity ratings reported that sensory panels judged organic strawberries to be sweeter and have superior flavor and overall acceptance compared with conventional fruit. In the new studies on organic strawberries (Gąstoł, 2016), organically grown fruits were characterized by higher level of sugars and titratable acidity, also having better SSC:TA ratio and thus superior taste than conventionally grown ones.

The aforementioned debates may be due to the fact that different results may be associated with different crop types not equally influenced by different agroecosystems.

Ascorbic acid, polyphenols content, and antioxidant activity

Of the 13 studies comparing vitamin C content mentioned in the review of Heaton (2001), 7 demonstrated significantly higher vitamin C in organically grown crops. The increment of the content ranged from 6% to 100%. Six studies reported inconsistent or nonsignificant differences, while no studies found significantly higher vitamin C content in conventionally than in organically grown crops. In this respect, the results may be inconsistent due to several factors, not only the cultivation method but also time of harvest, the mode of transport, and storage. Hunter et al. (2011) found no statistically significant differences in organic and conventional foodstuff. In the presented study, the cultivation method did not influence the ascorbic content.

Cardoso et al. (2011) reported a higher content of vitamin C in organic acerola compared with its conventional production. The reverse was true for strawberries. The use of lower concentrations of N fertilizers (45 kg/ha) along with higher K fertilizer doses is linked to higher vitamin C content in fruits (Lee and Kader, 2000). Therefore, the higher vitamin C content reported in conventionally grown fruit could be associated with the use of the fertilizer program, which consisted of higher potassium fertilizer rates. On the other hand, Raigón et al. (2010) demonstrated that fresh organic strawberries would supply almost 10% more of daily vitamin C requirement of adults than the conventionally produced strawberries.

Numerous studies showed that organic fruits and vegetables contained a wide variety of phenolic compounds and antioxidants, which could help enhance human immunity, eliminate free radicals, and have positive functions in anticancer and immunomodulation (Brandt and Mølgaard, 2001).

According to the literature reviewed, organically produced fruits have higher phenolic content. There are many factors affecting the accumulation of secondary metabolites in plants. Fertilization management is one of the main factors affecting the chemical composition of plants, including the secondary metabolites. Previous research has shown that production systems have a tremendous effect on the content of phenolic compounds in fruits. The high phenolic content in organically produced fruit might be the result of an increase in endogenous phenolics that enrich plant defense mechanisms as the result of the absence of synthetic pesticides and highly soluble fertilizers commonly used in the conventional production system (Carbonaro and Mattera, 2001). The higher antioxidant levels in organically produced fruits are also linked to zero use of pesticides. Letaief et al. (2016) attributed the higher antioxidant levels in organic fruits to enhanced synthesis of active phytochemicals produced in response to biotic and abiotic stress.

Weibel et al. (2000) found that organic apples had 18.6% more phenolic compounds than nonorganic fruits. Wang (2008) measured higher anthocyanin content in organically cultivated blueberries. Worthington (2001) reported that organic fruits/vegetables contained 44.7% and 57.4%, respectively, more polyphenols than conventional ones. However, in our study (Gąstoł et al., 2011), there were no significant differences between organic and conventional fruits in this respect for any of the investigated species. The overall mean for FRAP was higher for organic than for conventional juices, although no significant differences were established for individual species.

A higher content of phenolics has been reported in "Regina bianca" peaches from organically managed farms compared with those from conventional production systems (Carbonaro and Mattera, 2001). This finding could be linked to the fact that, unlike organic fruits, conventionally grown fruits were applied with synthetic pesticides. Application of pesticides, particularly synthetic herbicides, may reduce carbon fixation by plants and consequently decrease the carbon available for the synthesis of phenolics (Lombardi-Boccia et al., 2004). Earlier studies by Lydon and Duke (1989) demonstrated that some herbicides block the shikimate pathway, reducing the synthesis of aromatic amino acids at the starting point of the synthesis of phenolic compounds.

In general, no significant effects of the farming type were observed on the antioxidant activity with any of the methods used. It was true for apples (Valavanidis et al., 2009) or passion fruit (De Oliveira et al., 2017). Faller and Fialho (2010) also found similar contents of total phenols in organic and conventional banana, orange, and apple pulp, although they found higher amounts in organic papaya and mandarin and smaller in organic mango. Similarly, no significant effect of the farming system was reported in eggplant fruits (Raigón et al., 2010). However, Lombardi-Boccia et al. (2004) found higher total phenols in conventional than in organic yellow plums. In the study of Anttonen and Karjalainen (2006) the profile of flavonoids in black currant is dependent more on where they have been grown than whether they were grown under organic or conventional farming conditions.

Phenolic compounds not only contribute to plant's resistance to fungal infections, insect wounds or mechanical damage (Ohazurike and Arinze, 1996); but also foster with potential antioxidant activity and epidemiological properties, including anticarcinogenic activity. Benbrook (2005) reported that 85% of studies have proved that organic farming increases antioxidant levels as compared with conventional farming. His results demonstrated that, on average, organically produced foods have 30% more antioxidant level compared with conventionally produced foods. Olsson et al. (2006) demonstrated that the effect of the production system may depend on the choice of cultivar. In the study by You et al. (2011), comparing the organic and conventional production of four blueberry cultivars, it was observed that not all the organic grown blueberry cultivars exhibit higher antioxidant capacity than those grown under conventional production systems. Such difference may be due to different response of cultivars to the environmental changes. Also, the variation of organic and conventional treatments from study to study may be contributing to such discrepancies. According to Tarozzi et al. (2006), fruits grown without pesticides contain higher levels of antioxidants. This is attributed to enhanced synthesis of active phytochemicals produced in response to biotic and abiotic stress (Letaief et al., 2016).

Brandt et al. (2011) found that the content of secondary metabolites in organic produce was 12% higher than in corresponding conventional samples. Different results were obtained by Smith-Spangler et al. (2012), who did not find strong evidence that organic food is significantly more nutritious than conventional food. Results show statistically meaningful differences in composition between organic and nonorganic crops/crop-based foods. Organic products had higher antioxidant activity, contained higher concentrations (18%–69%) of a wide range of nutritionally desirable antioxidants/(poly)phenolics and other plant secondary metabolites. However, in our study, the differences between investigated agroecosystems were nonsignificant. The overall mean for ferric reducing antioxidant power (FRAP) was higher for organically raised fruits, although no significant and valid differences were established for individual species.

3 Conclusions

The differences identified in a nutritional content of fruits could be largely attributed to the different fertilization/cultivation systems between organically and conventionally managed soils. These routines change to some extent the plants' gene expression and metabolism. As a result, plants grown under organic agroecosystem are forced to synthesize their own chemical defenses (e.g., secondary metabolites and antioxidants). On the other hand, high levels of fast released fertilizers and pesticides are often detrimental for desirable mineral elements and metabolite accumulation in fruits. Also, residues and nitrates in conventionally grown fruits are also a cause of concern, while some bacterial and fungal contamination of organic fruits could be an issue.

The aforementioned experiments indicate that there is no simple confirmation of nutritional superiority of organic fruit. In some aspects like the lack of pesticides and antibiotics, the beneficial effect of organic diet is evident. In some parameters (dry matter content, polyphenol content, and antioxidant activity), usually, the better results are noted for organic foods. However, in many cases, the differences are not so evident or inconclusive.

A key challenge for scientific research of food is to elaborate and unify the research methodologies on comparison organic and conventional foodstuffs. The research approaches should include the whole production chain, from "a field to fork." As the definition of food quality is continuously changing, there is a great need to not only investigate the analytical parameters but also use more systemic, holistic view. This holistic approach is postulated by some authors (Meier-Ploeger and Vogtmann, 1991; Kahl et al., 2010) indicate that food quality assessment should be more holistic than reductionistic. It should cover the nutritional and technological value and sensory quality to integrate all quality aspects of food. It opens new exciting fields to explore.

Acknowledgments

The presented results were elaborated within a project financed by the Polish Ministry of Agriculture and Rural Development (project no. RR-re-401-25-173/09).

References

Amarante, C.V.T., Steffens, C.A., Mafra, A.L., Albuquerque, J.A., 2008. Yield and fruit quality of apple from conventional and organic production systems. Pesq. Agrop. Brasileira. 43, 333–340.

Amodio, M.L., Colelli, G., Hasey, J.K., Kader, A.A., 2007. A comparative study of composition and postharvest performance of organically and conventionally grown kiwifruits. J. Sci. Food Agric. 87, 1228–1236.

Anttonen, M.J., Karjalainen, R.O., 2006. High-performance liquid chromatography analysis of black currant (*Ribes nigrum* L.) fruit phenolics grown either conventionally or organically. J. Agric. Food Chem. 54, 7530–7538.

Beinard, C., Gautier, H., Bourgaud, F., Grasselly, D., Navez, B., Caris-Veyrat, C., Weiss, M., Genard, M., 2009. Effects of low nitrogen supply on tomato (*Solanum lycopersicum*) fruit yield and quality with special emphasis on sugars, acids, ascorbate, carotenoids, and phenolic compounds. J. Agric. Food Chem. 57, 4112–4123.

Benbrook, C.M., 2005. Elevating Antioxidant Levels in Food Through Organic Farming and Food Processing. The Organic Center, Foster, RI, p. 81.

Benbrook, C., Xin, Z., Yanez, J., Davie, N., Andrews, P., 2008. New Evidence Confirms the Nutritional Superiority of Plant-Based Organic Foods. An Organic Center State of Science Review. Available from: http://www.organic-center.org.

Benge, J., Banks, N., Tillman, R., De Silva, H.N., 2000. Pairwise comparison of the storage potential of kiwifruit from organic and conventional production systems. HortScience 28, 147–152.

Bonti-Ankomah, S., Yiridoe, E.K., 2006. Organic and Conventional Food: A Literature Review of the Economics of Consumer Perceptions and Preferences. Final Report. Organic Agriculture Centre of Canada (OACC).

Bourn, D., Prescott, J., 2002. A comparison of the nutritional value, sensor qualities, and food safety of organically and conventionally produced foods. Crit. Rev. Food Sci. Nutr. 42, 1–34.

Brandt, K., Mølgaard, J.P., 2001. Organic agriculture: does it enhance or reduce the nutritional value of plant foods? J. Sci. Food Agric. 81 (9), 924–931.

Brandt, K., Leifert, C., Sanderson, R., Seal, C., 2011. Agroecosystem management and nutritional quality of plant foods: the case of organic fruits and vegetables. Crit. Rev. Plant Sci. 30, 177–197.

Brunetto, G., Melo, G.W.B., Toselli, M., Quartieri, M., Tagliavini, M., 2015. The role of mineral nutrition on yields and fruit quality in grapevine, pear and apple. Rev. Bras. Frutic. 37, 1089–1104.

Bryła, P., 2015. The role of appeals to tradition in origin food marketing: a survey among polish consumers. Appetite 91, 302–310.

Canavari, M., Olson, K.D. (Eds.), 2007. Organic Food. Consumers' Choices and Farmers' Opportunities. Springer, New York, USA.

Carbonaro, M., Mattera, M., 2001. Polyphenoloxidase activity and polyphenol levels in organically and conventionally grown peach (*Prunus persica* L., cv. Reginabianca) and pear (*Pyrus communis* L., cv. Williams). Food Chem. 72, 419–424.

Cardoso, P.C., Tomazini, A.P.B., Stringheta, P.C., Ribeiro, S.M., Pinheiro-Sant'Ana, H.M., 2011. Vitamin C and carotenoids in organic and conventional fruits grown in Brazil. Food Chem. 126, 411–416.

Carlson, A., Jaenicke, E., 2016. Changes in retail organic price premiums from 2004 to 2010. United States Department of Agriculture, Economic Research Report Number 209. Available online: https://www.ers.usda.gov/webdocs/publications/err209/59472_err209.pdf.

Chinese National Certification and Accreditation (CNCA), 2015. Country Report on the Development of the Chinese Organic Sector. Chinese National Certification and Accreditation Admin., Beijing, China

Conti, S., Villari, G., Faugno, S., Melchionna, G., Somma, S., Caruso, G., 2014. Effects of organic vs. conventional farming system on yield and quality of strawberry grown as an annual or biennial crop in southern Italy. Sci. Horticult. 180, 63–71.

Cuevas, F.J., Pradas, I., Ruiz-Moreno, M.J., Arroyo, F.T., Perez-Romero, L.F., Montenegro, J.C., 2015. Effect of Organic and Conventional Management on Bio-Functional Quality of Thirteen Plum Cultivars (Prunus salicina Lindl.). PLoS ONE 10 (8), e0136596. https://doi.org/10.1371/journal.pone.0136596.

Dangour, A., Dodhia, S.K., Hayter, A., Allen, E., Lock, K., Uauy, R., 2009. Nutritional quality of organic foods: a systematic review. Am. J. Clin. Nutr. 90, 680–685.

DeEll, J.R., Prange, R.K., 1993. Postharvest physiological disorders, diseases and mineral concentrations of organically and conventionally grown McIntosh and Cortland apples. Can. J. Plant Sci. 73 (1), 223–230.

De Oliveira, A.B., Lopes, M.M.A., Moura, C.F.H., Oliveira, L.S., de Souza, K.O., Filho, E.G., 2017. Effects of organic vs. conventional farming systems on quality and antioxidant metabolism of passion fruit during maturation. Sci. Horticult. 222, 84–89.

Domagała-Świątkiewicz, I., Gąstoł, M., 2012. Comparative study on mineral content of organic and conventional carrot, celery and red beet juices. Acta Sci. Pol. Hortorum Cultus 11 (2), 173–183.

Elamin, E., Al-Wehaibi, N., 2005. Alternate use of good and saline irrigation water (1:1) on the performance of tomato cultivar. J. Plant Nutr. 28, 1061–1072.

Eriksson, J.E., 2001. Concentration of 61 Trace Elements in Sewage Sludge, Farmyard Manure, Mineral Fertilizers, Precipitation and in Oil and Crops. Swedish EPA. Report nr 5159, Stockholm, Sweden.

Faller, A., Fialho, E., 2010. Polyphenol content and antioxidant capacity in organic and conventional plant foods. J. Food Comp. Anal. 23, 561–568.

Ferree, D.C., Warrington, I., 2003. Apples: Botany, Production and Uses. CABI Publishing, Cambridge.

Fliessbach, A., Mäder, P., 2000. Microbial biomass and size density factions differ between soils of organic and conventional agricultural systems. Soil Biol. Biochem. 32, 757–768.

Food and Agriculture Organization (FAO), 2015. FAOSTAT. Food and Agriculture Organization, Rome, Italy. http://faostat3fao.org/download/Q/QC/E.

García, M.A., Martinó, M.N., Zaritzky, N.E., 1998. Plasticized starch-based coatings to improve strawberry (*Fragaria ananassa*) quality and stability. J. Agric. Food Chem. 46, 3758–3767.

Gąstoł, M., 2009. Organic versus conventional—a comparative study on quality and nutritional value of fruit and vegetables and its preserves. The report from project financed by the Polish Ministry of Agriculture and Rural Development (grant decision RR-re-401-25-173/09), https://www.gov.pl/documents/912055/913531/Streszczenie+wyników+badań_2009.pdf/8da4af46-a91e-4545-30b0-52247e45f4d4.

Gąstoł, M., 2016. Innovative methods in organic strawberry protection. Ministry of Agriculture and Rural Development Annual Grant Report, Warsaw, Poland. Available online: http://www.zielen.ogr.ar.krakow.pl/Ksip/pliki/Raport_grant_eko_2016.pdf.

Gąstoł, M., Domagała-Światkiewicz, I., Krośniak, M., 2011. Organic versus conventional—a comparative study on quality and nutritional value of fruit and vegetable juices. Biol. Agric. Hortic. 27 (3), 310–319.

Goetzke, B., Nitzko, S., Spiller, A., 2014. Consumption of organic and functional food. A matter of well-being and health? Appetite 77, 96–105.

Granatstein, D., 2015. Another burst of organic fruit? Good Fruit Grower (May 1) 36–37. www.goodfruitgrower.com.

Guadagnin, S.G., Rath, S., Reyes, F.G., 2005. Evaluation of the nitrate content in leaf vegetables produced through different agricultural systems. Food Addit. Contam. 22 (12), 1203–1208.

Hamer, D.H., Gill, C.J., 2002. From the farm to the kitchen table: the negative impact of antimicrobial use in animals on humans. Nutr. Rev. 60 (8), 261–264.

Hansen, B., Alroe, H.J., Kristensen, E.S., 2001. Approaches to assess the environmental impact of organic farming with particular regard to Denmark. Agric. Ecosyst. Environ. 83, 11–26.

Heaton, S., 2001. Organic Farming Food Quality and Human Health. A Review of an Evidence. Soil Association, Bristol.

Heeb, A., Lundegårdh, B., Ericsson, T., Savage, G.P., 2005. Effects of nitrate-, ammonium-, and organic-nitrogen-based fertilizers on growth and yield of tomatoes. J. Plant Nutr. Soil Sci. 168, 123–129.

Hildermann, I., Messmer, M., Dubois, D., Boller, T., Wiemken, A., Mader, P., 2010. Nutrient use efficiency and arbuscular mycorrhizal root colonisation of winter wheat cultivars in different farming systems of the DOK long-term trial. J. Sci. Food Agric. 90, 2027–2038.

Holden, P., 2001. Organic Farming, Food Quality and Human Health: A Review of the Evidence. Soil Association, Bristol.

Hornick, S.B., 1992. Factors affecting the nutritional quality of crops. Am. J. Altern. Agric. 7, 63–68.

Huber, M., Rembiałkowska, E., Średnicka, D., Bügel, S., Vijver, L.P., 2011. Organic food and impact on human health: assessing the status quo and prospects of research. NJAS Wagening J. Life Sci. 58 (3–4), 103–109.

Huber, M., Bakker, M., Dijk, W., Prins, H., Wiegant, A.C., 2012. The challenge of evaluating health effects of organic food: operationalisation of a dynamic concept of health. J. Sci. Food Agric. 92, 2766–2773.

Hughner, R.S., McDonagh, P., Prothero, A., Shultz, C.J., Stanton, J., 2007. Who are organic food consumers? A compilation and review of why people purchase organic food. J. Consum. Behav. 6, 94–110.

Hunter, D., Foster, M., McArthur, J.O., Ojha, R., Petocz, P., Samman, S., 2011. Evaluation of the micronutrient composition of plant foods produced by organic and conventional agricultural methods. Crit. Rev. Food Sci. Nutr. 51 (6), 571–582.

IFOAM, 2018. The World of Organic Agriculture. Statistics and Emerging Trends. Fibl, Fricks, and IFOAM-Organics International. https://doi.org/10.4324/9781849775991.

Jönsson, A., 2007. Organic Apple Production in Sweden: Cultivation and Cultivars. Swedish University of Agricultural Sciences, Balsgård, (PhD thesis), p. 33.

Kabata-Pendias, A., 2011. Trace Elements in Soils and Plants. CRC Press, Taylor & Francis Group, London.

Kahl, J., Busscher, N., Ploeger, A., 2010. Questions on the validation of holistic methods of testing organic food quality. Biol. Agric. Hortic. 27, 81–94.

Kahl, J., Baars, T., Bügel, S., Busscher, N., Huber, M., Kusche, D., 2012. Organic food quality: a framework for concept, definition and evaluation from the European perspective. J. Sci. Food Agric. 92, 2760–2765.

Khalil, H.A., Hassan, S.M., 2015. Ascorbic acid, β-carotene, total phenolic compound and microbiological quality of organic and conventional citrus and strawberry grown in Egypt. Afr. J. Biotech. 14 (4), 272–277.

Kliewer, W.M., 1977. Influence of temperature, solar radiation and nitrogen on coloration and composition of Emperor grapes. Am. J. Enol. Vitic. 28, 96–103.

Köpke, U., 2005. Organic foods: do they have a role? Forum Nutr. 5 (57), 62–72.

Lairon, D., 2010. Nutritional quality and safety of organic food. A review. Agron. Sustain. Dev. 30, 33–41.

Lee, S.K., Kader, A.A., 2000. Preharvest and postharvest factors influencing vitamin C content of horticultural crops. Postharvest Biol. Technol. 20, 207–220.

Lester, G.E., 2007. Organic vs conventionally grown Rio red whole grapefruit and juice: comparison of production inputs, market quality, consumer, acceptance, and human health-bioactive compounds. J. Agric. Food Chem. 55 (11), 4474–4480.

Letaief, H., Zemni, H., Mlik, A., Chebil, S., 2016. Composition of *Citrus sinensis* (L.) Osbeck cv "Maltaise demi-sanguine" juice. A comparison between organic and conventional farming. Food Chem. 194, 290–295.

Lind, K., Lafer, G., Schloffer, K., Innerhofer, G., Meister, H., 2003. Organic Fruit Growing. CABI Publishing, Cambridge.

Lombardi-Boccia, G., Lucarini, M., Lanzi, S., Aguzzi, A., Cappelloni, M., 2004. Nutrients and antioxidant molecules in yellow plums (*Prunus domestica* L.) from conventional and organic productions: a comparative study. J. Agric. Food Chem. 52, 90–94.

Lu, C., Toepel, K., Irish, R., Fenske, R.A., Barr, D.B., Bravo, R., 2006. Organic diets significantly lower children's dietary exposure to organophosphorus pesticides. Environ. Health Perspect. 114 (2), 260–263.

Lydon, J., Duke, S.O., 1989. Pesticide effects on secondary metabolism of higher plants. Pestic. Sci. 25, 361–373.

Mader, P., Pfiffner, L., Niggli, U., Balzar, U., Balzar, F., Plochberger, K., Velimirov, A., Besson, J.M., 1993. Effect of three farming systems (biodynamic, bio-organic, conventional) on yield and quality of beetroot (*Beta vulgaris* L. var. *esculenta* L.) in a seven year crop rotation. Acta Hortic. 330, 10–31.

Magkos, F., Arvaniti, F., Zampelas, A., 2003. Organic food: nutritious food or food for thought? A review of the evidence. Int. J. Food Sci. Nutr. 54 (5), 357–371.

Matt, D., Rembiałkowska, E., Luik, A., Peetsmann, E., Pehme, S., 2011. Quality of Organic vs. Conventional Food and Effects on Health. Estonian University of Life Science.

Mayer, A.M., 1997. Historical changes in the mineral content of fruits and vegetables. Br. Food J. 99 (6), 207–211.

Meier-Ploeger, A., Vogtmann, H. (Eds.), 1991. Lebensmittelqualität—Ganzheitliche Methoden und Konzepte. Alternative Konzepte Nr. 66, Verlag C. F. Müller, Karlsruhe, Germany.

Mie, A., Kesse-Guyot, E., Kahl, J., Rembiałkowska, E., Raun Andersen, H., Grandjean, P., Gunnarsson, S., 2016. Human Health Implications of Organic Food and Organic Agriculture. European Parliament Research Service, Belgium. Available online: http://www.europarl.europa.eu/thinktank/en/document.html?reference=EPRS_STU(2016)581922.

Ohazurike, N., Arinze, A., 1996. Changes in phenol oxidase and peroxidase levels in cocoyam tubers of different postharvest ages infected by *Sclerotium rolfsii* sacc. Food Nahrung 40, 25–27.

Olsson, M.E., Andersson, C.S., Oredsson, S., Berglund, R.H., Gustavsson, K., 2006. Antioxidant levels and inhibition of cancer cell proliferation in vitro by extracts from organically and conventionally cultivated strawberries. J. Agric Food Chem. 54, 1248–1255.

Palupi, E., Jayanegara, A., Ploeger, A., Kahl, J., 2012. Comparison of nutritional quality between conventional and organic dairy products: a meta-analysis. J. Sci. Food Agric. 92 (14), 2774–2781. https://doi.org/10.1002/jsfa.5639.

Peck, G., Preston, A.K., Reganold, J.P., Fellman, J.K., 2006. Apple orchard productivity and fruit quality under organic, conventional, and integrated management. HortScience 41, 99–107.

Pither, R., Hall, M.N., 1990. Analytical Survey of the Nutritional Composition of Organically Grown Fruit and Vegetables. Technical Memorandum 597. MAFF Project 4350.

Pussemier, L., Larondelle, Y., Peteghem, C.V., Huyghebaert, A., 2006. Chemical safety of conventionally and organically produced foodstuffs: a tentative comparison under Belgian conditions. Food Control 17, 14–21.

Raigón, M.D., Rodríguez-Burruezo, A., Prohens, J., 2010. Effects of organic and conventional cultivation methods on composition of eggplant fruits. J. Agric. Food Chem. 58 (11), 6833–6840.

Reche, J., Hernández, F., Almansa, M.S., Carbonell-Barrachina, A.A., Leguab, P., Amorós, A., 2019. Effects of organic and conventional farming on the physicochemical and functional properties of jujube fruit. LWT Food Sci. Technol. 99, 438–444.

Rembiałkowska, E., 2000. The nutritive and sensory quality of carrots and white cabbage from organic and conventional farms. In: Proc.13th IFOAM Sci. Conf. 28–31 August 2000 Zurich: 297.

Rembiałkowska, E., 2007. Quality of plant products from organic agriculture. J. Sci.Food Agric. 87, 2757–2762.

Rossi, F., Godani, F., Bertuzzi, T., Trevisan, M., Ferrari, F., Gatti, S., 2008. Health-promoting substances and heavy metal content in tomatoes grown with different farming techniques. Eur. J. Nutr. 47 (5), 266–272.

Sahota, A., 2018. The global market for organic food & drink. In: Willer, H., Lernoud, J. (Eds.), The World of Organic Agriculture. Statistics and Emerging Trends. Research Institute of Organic Agriculture (FiBL) and IFOAM, Bonn, pp. 145–150.

Seyfang, G., 2008. Avoiding Asda? Exploring consumer motivations in local organic food networks. Local Environ. 13, 187–201.

Smith, B., 1993. Organic foods vs. supermarket foods: element levels. J. Appl. Nutr. 45, 35–39.

Smith-Spangler, C., Brandeau, M.L., Hunter, G.E., Bavinge, J.C., Pearson, M., Eschbach, P.J., Sundaram, V., Liu, H., Schirmer, P., Stave, C., 2012. Are organic foods safer or healthier than conventional alternatives?: a systematic review. Ann. Intern. Med. 157, 348–366.

Stolz, M., Stolze, M., Hamm, U., Janssen, M., Ruto, M., 2011. Consumer attitudes towards organic versus conventional food with specific quality attributes. NJAS – Wageningen J. Life Sci. 58, 67–72.

Tarozzi, A., Hrelia, S., Angeloni, C., Morron, F., Biagi, P., Guardigli, M., Cantelli-Forti, G., Hrelia, P., 2006. Antioxidant effectiveness of organically and non-organically grown red oranges in cell culture systems. Eur. J. Nutr. 45, 152–158.

Truong, T., Yap, M., Ineson, E., 2012. Potential Vietnamese consumers' perceptions of organic foods. Br. Food J. 114, 529–543.

Valavanidis, A., Vlachogianni, T., Psomas, A., Zovoili, A., Siatis, V., 2009. Polyphenolic profile and antioxidant activity of five apple cultivars grown under organic and conventional agricultural practices. Int. J. Food Sci. Technol. 44, 1167–1175.

Vetter, H., von Abercron, M., Bischoff, R., Kampe, W., Klasink, A., Ranfft, K., 1987. Qualitat p-anzlicher nah- rungsmittel- 'alternative' und 'modern' im Vergleich, Teil III (AID-Schriftenreihe Verbraucherdienst, 3100), 13 pp.

Vijver, L.P.L., van de Huber, M.A.S., Doesburg, P., Nielsen, J.H., Wyss, G., Bloksma, J.R. et al. (Eds.), 2007. Measuring Food Quality: Concepts, Methods and Challenges. Driebergen. 12–14 February 2007; Louis Bolk Institute, Driebergen, http://www.louisbolk.org/downloads/1887.pdf.

Wang, S.Y., Chen, C.T., Sciarappa, W., Wang, C.Y., Camp, M., 2008. Fruit quality, antioxidant capacity, and flavonoid content of organically and conventionally grown blueberries. J. Agric. Food Chem. 56 (14), 5788–5794.

Weibel, F.P., Bickel, R., Leuthold, S., Alfoldi, T., 2000. Are organically grown apples testier and healthier? A comparative field study using conventional and alternative methods to measure fruit quality. Acta Hortic. (517), 417–426.

Willer, H., Lernoud, J. (Eds.), 2018. The World of Organic Agriculture. Statistics and Emerging Trends. Research Institute of Organic Agriculture (FiBL) and IFOAM – Organics International, Bonn.

Willer, H., Schaack, D., 2015. Organic farming and market development in Europe. In: Willer, H., Lernoud, J. (Eds.), The World of Organic Agriculture: Statistics and Emerging Trends 2015. Research Institute of Organic Agriculture (FiBL), Frick, Switzerland and International Federation of Organic Agriculture Movements (IFOAM), Bonn, Germany, pp. 181–214. 300 pp. https://www.fibl.org/fileadmin/documents/shop/1663-organic-world-2015.pdf.

Williams, C.M., 2002. Nutritional quality of organic food: shades of grey or shades of green. Proc. Nutr. Soc. 61, 19–24.

Winter, C.K., Davis, S.F., 2006. Organic food. J. Food Sci. 71, 117–124.

Woese, K., Lange, D., Boess, C., Bögl, K.W., 1997. A comparison of organically and conventionally grown foods—results of a review of the relevant literature. J. Sci. Food Agric. 74, 281–293.

World Health Organization and Food and Agriculture Organization (WHO and FAO), 2004. Vitamin and Mineral Requirements in Human Nutrition, second ed. WHO, Geneva, Switzerland.

Worthington, V., 2001. Nutritional quality of organic versus conventional fruits, vegetables, and grains. J. Altern. Complement. Med. 7, 161–173.

You, Q., Wang, B., Chen, F., Huang, Z., Wang, X., Luo, P.G., 2011. Comparison of anthocyanins and phenolics in organically and conventionally grown blueberries in selected cultivars. Food Chem. 125, 201–208.

Zanoli, R., Naspetti, S., 2002. Consumer motivations in the purchase of organic food: a means-end approach. Br. Food J. 104 (8), 643–653.

Further reading

Cassandro, V., Mafra, Á.L., Albuquerque, J.A., 2008. Yield and fruit quality of apple from conventional and organic production systems. Pesq. Agrop. Brasileira 43, 333–340.

Ferrante, A., Spinardi, A., Maggiore, T., Testoni, A., Gallina, P.M., 2008. Effect of nitrogen fertilisation levels on melon fruit quality at the harvest time and during storage. J. Sci. Food Agric. 88, 707–713.

Forman, J., Silverstein, J., 2012. Organic foods: health and environmental advantages and disadvantages. Pediatrics 130 (5), 1406–1415. https://doi.org/10.1542/peds.2012-2579.

Gąstoł, M., Domagała-Świątkiewicz, I., 2012. Comparative study on mineral content of organic and conventional apple, pear and black currant juices. Acta Sci. Pol. Hortorum Cultus 11 (3), 3–14.

Nguyen, H., Hofman, P., Holmes, R., Bally, I., Stubbings, B., McConchie, R., 2004. Effect of nitrogen on the skin colour and other quality attributes of ripe 'Kensington Pride' mango (*Mangifera indica* L.) fruit. J. Hortic. Sci. Biotechnol. 79, 204–210.

39

Role of controlled and slow release fertilizers in fruit crop nutrition

Xiangying Wei[a,b,c], Jianjun Chen[c,*], Bin Gao[d], Zonghua Wang[a,b]

[a]Institute of Oceanography, Minjiang University, Fuzhou, China
[b]Fujian University Key Laboratory for Plant-Microbe Interaction,
Fujian Agriculture and Forestry University, Fuzhou, China
[c]Department of Environmental Horticulrture and Mid-Florida Research and Education Center,
University of Florida, Institute of Food and Agricultural Sciences, Apopka, FL, United States
[d]Department of Agricultural and Biological Engineering, University of Florida,
Institute of Food and Agricultural Sciences, Gainesville, FL, United States
*Corresponding author. E-mail: jjchen@ufl.edu

1 Introduction

Perennial, woody plants that produce edible fleshy or dry fruits are commonly referred to as fruit crops (Rieger, 2006). Depending on their origination, fruit crops are divided into temperate fruit crops, such as almond, apple, apricot, blueberry, cranberry, blackberry, cherry, grape, kiwifruit, pear, and plum, and tropical and subtropical fruit crops, such as avocado, banana, citrus, mango, papaya, persimmon, and pineapple. Fruit crops are also classified into small fruits, like blackberry, blueberry, cranberry, grape, and raspberry; tree fruits, such as apple, pear, apricot, cherry, peach, plum, citrus, and persimmon; tree nuts, such as almond, chestnut, pecan, pistachio, and walnut; and emerging fruits, such as blue honeysuckle, elderberry, fig, goji berry, mayhaws, pawpaw, pomegranate, and quince (Badenes and Byrne, 2012).

Fruits are rich in anthocyanin, flavonoid, phenolic compound, vitamin, and mineral nutrients. A large body of research suggests that consumption of fruit helps prevent nutrient deficiencies and reduces the risk of cardiovascular disease and cancer (Chen et al., 2013). As a result, fruit production has significantly increased. For example, worldwide fresh fruit production increased from 13.58 million metric tons in 1990 to 33.25 million metric tons in 2016 (FAO, 2018). During that time, blackberry acreage was doubled in the southeastern United States (US) (Studholme et al., 2011) and worldwide highbush blueberry production increased from 58,400 ha in 2007 to 110,800 ha in 2014 (Brazelton, 2015).

A.K. Srivastava, Chengxiao Hu (eds.)
Fruit Crops: Diagnosis and Management of Nutrient Constraints
https://doi.org/10.1016/B978-0-12-818732-6.00039-3

North America represented more than 50% of the production area and accounted for almost 60% of the global high-bush blueberry production in 2014 (Brazelton, 2015). Grape acreage almost doubled in China from 2002 to 2011 and then tripled in New Zealand from 2000 to 2011 (Gergerich et al., 2015). Additionally, the promotion of mandarins as healthy and easy-to-eat food resulted in acreage doubling in California from 2002 to 2012 (California Department of Food and Agriculture, 2012). Nearly 50% of the total global volume of fruit crops is produced by five countries: China, the United States, Brazil, Italy, and Spain (Retamales, 2011).

Fruit crops are often planted in orchards with soils differing considerably in biological, chemical, and physical properties (Srivastava and Malhotra, 2017). A desirable orchard soil is that with 1 m of cultivatable and well-drained one. To ensure longevity and support healthy growth, fertilizers must be applied. Fertilizers applied to fruit crops include organic and inorganic forms. Organic fertilizers contain plant- or animal-based materials that are either a byproduct or end product of naturally occurring processes, such as animal manure and composted organic materials. Inorganic fertilizers, also referred to as chemical fertilizers, are synthesized artificially and contain essential mineral elements. Plants require 14 essential mineral nutrients to complete their life cycle: six macronutrients, namely, nitrogen (N), phosphorus (P), potassium (K), calcium (Ca), magnesium (Mg), and sulfur (S) and eight micronutrients, namely, iron (Fe), zinc (Zn), manganese (Mn), copper (Cu), boron (B), molybdenum (Mo), nickel (Ni), and chlorine (Cl). Organic fertilizers used to be the main nutrition source applied to fruit crops; however, modern commercial production of fruit crops has primarily used chemical fertilizers. Chemical fertilizer types include single nutrient, such as ammonium nitrate, urea, and superphosphates; multinutrient, such as NPK fertilizers; and multiple micronutrients, such as a mix of Mo, Zn, Cu, and chelated Fe. Commercial chemical fertilizers are granulated without coating, which are soluble in water immediately, coated or encapsulated as controlled- or slow-release fertilizers (CRFs or SRFs).

Chemical fertilizers represent a major cost in fruit crop production, accounting for nearly 20% of total citrus production costs (Obreza and Morgan, 2008). Meanwhile, two pronouncing constraints in fruit crop production are also closely related to fertilization. NO_3-N leaching and the nutrient imbalance resulting in crop susceptible to pathogens, such as Huanglongbing. Accumulating evidence has pointed to the fact that CRFs and SRFs could be used to alleviate the two constraints. This chapter is intended to discuss the role of CRFs and SRFs in fruit crop production in general with an emphasis on the two problems.

2 Controlled- and slow-release fertilizers

CRFs are granules that are purposely designed for releasing nutrients in a controlled, delayed manner in synchrony with plant requirements for nutrients (Shaviv, 2005; Trenkel, 2010; Sempeho et al., 2014). CRFs are typically coated or encapsulated with inorganic or organic membranes that control the rate, pattern, and duration of plant nutrient release. The membrane technology differs among companies and can be resin-, plastic-, or polymer-based. Common CRFs are presented in Table 39.1 (Ozores-Hampton, 2017; Chen and Wei, 2018). A CRF must meet the following three criteria: (1) Less than 15% of the nutrients should be released in 24 h, (2) less than 75% should be released in 28 days, and (3) at least 75% should be released by the stated release time (40–360 days) (Trenkel, 2010).

SRFs are a group of fertilizers with a slow-release characteristic, which is due to either the permeability of coating material or limited solubility of fertilizer compounds (Table 39.2). SRFs include (1) sulfur-coated fertilizers, urea- or ammonium-based salt blends coated with elemental sulfur, resulting in individual fertilizer prills similar in size and appearance to CRFs; (2) methylene urea, a product synthesized by urea reacting with formaldehyde to produce monomethylol urea, which further reacts with urea to produce methylene urea that varies in chain length (2–5 urea), thus varying in solubility; (3) urea formaldehyde, a class of slowly soluble N fertilizers synthesized by combining formaldehyde with urea; (4) isobutylidenediurea (IBDU), a single slowly soluble N fertilizer synthesized by combining isobutyraldehyde with urea; (5) crotonylidene diurea, which is produced by the reaction of urea and acetic aldehyde; and (6) slowly soluble fertilizers, such as gypsum ($CaSO_4 \cdot 2H_2O$), triple superphosphate or 0-46-0 ($9Ca(H_2PO_4)_2 + CaF_2$), limestone, or micronutrient oxides (Shaviv, 2005; Trenkel, 2010).

Both CRFs and SRFs belong to enhanced-efficiency fertilizers (EEFs), which is defined as "fertilizer products with characteristics that allow increased plant uptake and reduce the potential of nutrient losses to the environment (e.g., gaseous losses, leaching, or runoff) when compared with an appropriate reference product" (AAPFCO, 2013). In addition to CRFs and SRFs, EEFs include stabilized N fertilizers, nitrification inhibitors, and urease inhibitors.

The terms CRF and SRF are generally considered analogous. However, Shaviv (2005) and Trenkel (2010) clearly defined their differences. In SRFs, the pattern of nutrient release is generally unpredictable and remains subject to changes by soil type, microbial activity, and climatic conditions. In contrary, the pattern, quantity, and time of release can be predicted, within limits, for CRFs.

TABLE 39.1 Commercially available controlled-release fertilizers (CRFs) commonly used in crop production.

Manufacturer	Trade name	Coating materials	Selected commercial products
Agrium, Inc.	TriKote	Urea coated with polymer and sulfur	Trikote 42-0-0
	Duration	Polymer-coated urea or urea coated with microthin polymer membrane	Duration CR, Duration 44-0-0, Duration 19-6-13, Duration 35-0-10
	ESN	Urea coated with flexible microthin polymer	ESN 44-0-0 (environmentally smart nitrogen)
	Polyon	Nutrients coated with patented "reactive layers coating" (ultrathin polyurethane coating)	Polyon 41-0-0, Polyon NPK 20-6-13, Polyon 16-5-10 with micronutrients
Everris, Inc.	Osmocote	Granule contains NPK coated with organic resin	Osmocote Exact, Osmocote Exact Mini, Osmocote Pro, Osmocote Start, Osmocote Bloom
	Agrocote	Nutrients Coated with polymer/sulfur and resin coatings	Agrocote 19-6-12, Agrocote 39-0-0+11% S
	Poly-S	Urea coated with sulfur followed by polymer	Poly-S 37-0-0
Haifa Group	Multicote	Nutrients encapsulated in a polymeric shell	Multicote Agri 6 22-8-13, Multicote Agri 6 34-0-7, Multicote Agri 8 34-0-7
JCAM Agri Co.	Nutricote	Nutrients coated with polymer along with a special chemical release agent	Nutricote NPK 20-7-10, Nutricote 18-6-8, Nutricote 13-13-13, Nutricote 13-11-11
	Meister	Granular urea coated with a polymer of natural products, resin, and additives	Meister 15-5-15, Meister 19-5-14
J.R. Simplot	Florikote	Nutrients coated with polymer by dual layer technology	Florikote 40-0-0, Florikote 12-0-40, Florikote 19-6-13

TABLE 39.2 Commercially available slow-release fertilizers (SRFs) commonly used in crop production.

Manufacturer	Trade name	Formulation	Selected commercial products
Aglukon	Azolon	Methylene urea	Azolon 39N, Azolon Green-Keeper (15-0-20-3MgO-13S), Azolon Fluid (28% N)
	PlantoDur	Methylene urea	PlantoDur
Agrium, Inc.	Nitroform	Urea formaldehyde (UF)	Nitroform 39-0-0, Nitroform 38-0-0
	Nutralene	Methylene urea	40-0-0 Standard
	IBDU nitrogen	Isobutylidenediurea formation	XCU
BASF	Triabon	NPK fertilizer based on crotonylidene diurea (CDU)	Triabon 16-8-12+(4.22)+micronutrients, Triabon 16-8-13+4+9
Georgia-Pacific	Nitamin	UF	Nitamin 42G, Nitamin 30L
Growth Products	Nitro-30	UF solution 30-0-0	Nitro-30 SRN 30-0-0
Helena Chemicals	CoRoN	UF solution	CoRoN 25-0-0-.05B, CoRoN 28-0-0
JCAM Agri Co.	CDU nitrogen	CDU	CM-CDU-4, CM-CDU-10
	UBER	Controlled-mineralization CDU	Hyper CDU
	IBDU nitrogen	Isobutylidenediurea	Good-IB, Super-IB

2.1 Common CRFs and SRFs used in crop production

The leading producers and/or suppliers of CRFs (shown in Tables 39.1 and 39.2) include Agrium Inc., Calgary, Alberta, Canada; Everris NA, Inc., a subsidiary of Israel Chemicals Ltds; Haifa Group, Haifa, Israel; JCAM Agri Co., Tokyo, Japan; and J.R. Simplot, Boise, Idaho, U.S. CRFs produced by Agrium include those with the following trade names: Duration, ESN, Polyon, and XCU in which urea is coated by polymer. Everris, Inc. produces Agrocote, Osmocote, and poly-S where urea is coated by sulfur/polymer and resin, resin, and sulfur and polymer, respectively. Urea in Mulitcote produced by Haifa group is coated by resin or polymer. Popular CRFs Nutricote and Meister are manufactured by JCAM Agri co, and urea is coated by resin. Florikote produced by J.R. Simplot is coated by polymer (Ozores-Hampton, 2017; Chen and Wei, 2018).

Nitrogen is a major component of most CRFs. N sources include ammonium nitrate, ammonium phosphate, potassium nitrate, and urea, of which urea is a major N source for formulation of CRFs. This is due to the fact that urea contains 46% of N, has the lowest transportation costs per unit of N, and allows ease of application (Schlesinger and Hartley, 1992; Trenkel, 2010). Additionally, urea is highly soluble in water and has much lower risk of causing fertilizer burn to crop. Other components include phosphorus and potassium. Commercially, CRF formulation is often presented as $N-P_2O_5-K_2O$ in different ratios as shown in Table 39.1, which are coated or encapsulated with different materials. CRFs with micronutrients are referred to micronutrients that are either encapsulated with $N-P_2O_5-K_2O$ or mixed with encapsulated $N-P_2O_5-K_2O$ granules.

Resins and polymers are commonly used to coat or encapsulate nutrients to produce CRFs. Each coated particle is known as a prill, and nutrient release is controlled by the chemical composition and thickness of the coating materials. Polymers include thermoplastic, thermosetting, or biodegradable ones. Thermoplastic resins are not widely used as they either are not soluble in a solvent or make a very viscous solution that is not suitable for spraying; however, polyolefin is used in the art of coating the fertilizer granules. Thermoset polymers include urethane resin, epoxy resin, alkyd resin, unsaturated polyester resin, phenol resin, urea resin, melamine resin, phenol resin, and silicon resin (Sempeho et al., 2014). Among them, urethane resin, or urethane, is most commonly used (Ukessays, 2015). Polyacrylamide is known to reduce soil erosion, and more studies should be conducted for its advanced use in CRFs (Shaviv, 2005; Subbarao et al., 2013). Biodegradable polymers are naturally available and are known to be environmentally friendly because they can be decomposed in bioactive environments and degraded by the enzymatic action of microorganisms, such as bacteria, fungi, and algae, and their polymer chains may also be broken down by nonenzymatic processes, such as chemical hydrolysis. Thus, polymers used for coating urea include alkyd resin (Osmocote), polyurethane (Polyon, Multicote, and Plantacote), and thermoplastic polymers.

Materials used for coating nutrients of SRFs are largely sulfur and also guanylurea, magnesium ammonium phosphate, oxalic acid diamide, potassium calcium phosphate, potassium poly-phosphate, and urea aldehyde (Trenkel, 2010). The Tennessee Valley Authority developed the basic production process for sulfur-coated urea more than 50 years ago (Azeem et al., 2014) in which preheated urea granules were coated with molten sulfur and wax. The sulfur coating is an impermeable layer that can be slowly degraded through microbial activities and soil chemical and physical processes. The uniformity in coating coverage and thickness of coating determine the speed and effectiveness of urea release. Incompletely coated or cracked prills are immediately amenable to dissolution in soil water and hydrolysis by urease. However, due to its amorphous nature, sulfur alone cannot be used to produce slow-release urea. Subsequently, many other materials, such as binders, plasticizers, and sealants, were evaluated for reducing the immediate burst effect. Some tested materials reduced the effect but increased the cost and complexity (Azeem et al., 2014). As a result, sulfur alone has not been used as a coating agent. If used, it is in combination with some polymers.

2.2 N release patterns from CRFs and SRFs

Nutrient release from CRFs is governed by diffusion mechanisms. Shaviv (2005) and Liu (2008) proposed a multistage diffusion model, in which irrigation water penetrates the coating to condense on the solid fertilizer core followed by partial nutrient dissolution. As osmotic pressure builds within the containment, the granule swells and results in two events: One could be "catastrophic release" when osmotic pressure surpasses threshold membrane resistance; the coating bursts and the entire core are spontaneously released. This is also referred to as the "failure mechanism." In the second, if the membrane withstands the developing pressure, core fertilizer is thought to be released slowly via diffusion for which the driving force may be a concentration, pressure gradient, or combination thereof called the "diffusion mechanism." The failure mechanism is generally observed in frail coatings (e.g., sulfur or modified sulfur), while polymer coatings (e.g., polyolefin) are expected to exhibit the diffusion release mechanism (Azeem et al., 2014). Nutrient release from CRFs is generally classified into linear and sigmoidal patterns (Trenkel, 2010; Zhang et al., 2005). In most cases, the energy of activation of the release, EA_{rel}, is calculated on the basis of estimates of the rate of the release (percentage release per day) during the linear period obtained from the release curves (Du et al., 2006). Nutrient release profiles are established in both laboratory and field tests. Laboratory tests include extraction of nutrients at 25°C, 40°C, and 100°C. Field tests include the placement of net bags in the plowed layer or soil in the actual production soil (Trenkel, 2010). Shaviv (2001) reported that nutrient release consists of three stages: the initial stage or lag period during which little release is observed, the constant release stage characterized with an increasing release, and the last or mature stage where nutrient release is gradually reduced.

The release of nutrients from SRFs is dependent on multiple factors, not just a single factor like temperature (Morgan et al., 2009a; Chen and Wei, 2018). Thus, the nutrient release from SRFs is less predictable than from CRFs. For sulfur-coated SRFs, nutrients are released when water penetrates the sulfur coating through pores or imperfections in the coating. Once water has penetrated the coating, nutrient release from the prill is rapid. Since the thickness of the sulfur coat will influence the time required for water penetration, sulfur-coated fertilizers usually contain a range of coating thicknesses to get an extended release duration. If sulfur-coated SRFs are waxed, then microbial activity is needed to break down the wax to allow the sulfur coat to be exposed before nutrient release can occur. If a wax coating is present, anything that affects microbial activity, such as temperature, soil moisture, pH, or aeration, will also influence nutrient release. The release of N from urea formaldehyde is a multistep process depending largely on microbial decomposition. Therefore, factors that influence microbial activity will affect the release of N from urea formaldehyde. On the other hand, the release of N from isobutylidenediurea (IBDU) is controlled by a process called hydrolysis and is affected by soil moisture level and pH, but not microbial activity. Nutrient release from slowly soluble fertilizers such as gypsum ($CaSO_4 \cdot 2H_2O$), triple superphosphate or 0-46-0 ($9Ca(H_2PO_4)_2 + CaF_2$), limestone, or micronutrient oxides is dependent on the solubility of the individual fertilizer salt. In addition, the particle size of the salt will influence the release duration.

3 Role of CRFs and SRFs in fruit crop production

CRFs and SRFs have been available since the 1950s. Significant advances in the development of these products were made in the 1980s and 1990s (Trenkel, 2010). Traditionally, CRFs and SRFs have been applied to ornamental crops (Chen and Wei, 2018) and turfgrass (LeMonte et al., 2016). Their use in field and fruit crops is limited because of the cost of per unit of N (Simonne and Hutchinson, 2005; Trenkel, 2010). However, due to increasing concerns on environmental problems, particularly NO_3-N leaching, NH_3 volatilization, and N_2O emission in crop production and nutrient imbalance related fruit production problems, CRFs and SRFs have been considered as a means to alleviate these problems. Increasing evidence suggests that application of CRFs or SRFs enhances fruit crop yield and fruit quality, reduces N leaching and N_2O emission, improves rhizosphere microbial community, and also alleviates Huanglongbing in citrus plants.

3.1 CRFs and SRFs improve fruit crop production

Application of CRFs or SRFs has been shown to improve apple, banana, citrus, grape, kiwifruit, papaya, pawpaw, and pear production and fruit quality (Table 39.3). Koo (1986) Reported that yields of total soluble solids (TSS) of mature "Valencia" and "pineapple" orange trees fertilized with an IBDU were greater than sulfur-coated urea or ammonium nitrate in 2 of 4 years, and the highest yields were obtained at 200 kg N/ha/yr. Slow-release N applied once a year produced equal or higher TSS yield than ammonium nitrate applied twice a year. In another study with six types of commercially available CRFs (Escote, Meister 9-month, Meister 12-month, Nutricote 360, Prokote Plus, and Sierra) and a conventional, water-soluble fertilizer applied to "Valencia" orange trees on Swingle rootstock, annual application rates were at full, ½, and 1/4 recommended N rates from planting through 6 years of age. CRFs were applied once a year, and conventional fertilizer was applied 6, 5, 4, and 3 times in years 1, 2, 3, and 4–6, respectively. Fruit was harvested the third through the sixth year. Averaged across N rates, Prokote Plus, Nutricote, and Sierra produced cumulative fruit yields of 4.9–4.8 boxes/tree compared with 4.3 boxes/tree for conventional fertilizer in 4 years. Prokote plus and sierra also produced higher TSS yield (12.4 kg/tree in 4 years) and gross dollar return ($28.58/tree in 4 years) than the conventional fertilizer (Rouse et al., 1999). A study conducted with a pear variety "Mit Ghamr" using three slow release N fertilizers resulted in increased shoot length, leaf area, percentage of leaf N, and nutritional status of trees and fruit quality than a urea fertilizer (Kandil et al., 2010). Furthermore, a recent study conducted in China showed that single fruit weight and fruit yield of apple fertilized with a controlled-release N fertilizer were 7.8% and 14% greater than plants fertilized with a common urea (Sha et al., 2018). An attribute for the increased productivity was higher and long-lasting N residue in 0–40-cm soil applied with the controlled-release N than that of the common urea fertilizer

The enhanced crop productivity is largely attributed to the continuous and long-lasting release of nutrients, specifically N. Such a release characteristic is particularly important to woody perennial plants, like fruit crops. This is because N can be bidirectionally retranslocated within tree canopy, i.e., from top-to-lateral, lateral-to-top, and lateral-to-lateral position among simultaneously flushed leaves (Ueda, 2012), which is different from the traditional

TABLE 39.3   Application of controlled- and slow-release fertilizers improves fruit crop production.

Plant	Species or background	Fertilizer	Effects	Reference
Apple	Variety Orin	A controlled-release N fertilizer	Increase individual fruit weight and fruit yield	Sha et al. (2018)
Banana	Banana	Multicote Agri (17-7-25 + 2MgO)	Enhance plant growth by increased perimeter, canopy height, and leaf number	Haifa (2017)
Citrus	*Citrus sinensis* Valencia and pineapple sweet orange	Two SRFs (isobutylidene diurea and sulfur-coated urea) compared with ammonium nitrate	Slow-release N applied once a year produced equal or higher TSS yield than ammonium nitrate applied twice a year	Koo (1986)
Citrus	Valencia	Sulfur-coated urea (SCU) Methylene urea (MU)	Increase leaf mineral concentration	Zekri and Koo (1991)
Citrus	Valencia on Swingle Citrumelo rootstock	Prokote Plus (20-3-10) Nutricote 360 (17-6-8) Sierra (16-6-10) Meister (17-6-12) Escote (19-6-12)	CRFs increased fruit production compared with the conventional fertilizer	Obreza et al. (2006)
Citrus	*Citrus* × *clementina*	SRF (17N-12P-18K + 2Mg) SRF (13N-5P-27K + 2Mg)	Increased fruit yield and quality	Bettaga and Ben Mimoun (2008)
Citrus	Sweet orange trees on Swingle rootstock, 15–18 years of age	Harrell's CRF + Tiger micronutrient mix	Decreased preharvest fruit drop and increase total soluble solids	Vashisth (2017)
Grape	Thompson seedless grapevines	Methylene urea Phosphorus coated urea Sulfur coated urea Urea	Improved plant growth, berry setting, and fruit quality	Refaai (2016)
Kiwifruit	*Actinidia deliciosa*	Controlled-release urea Urea	Use of controlled release urea reduced fertilizer application	Lu et al. (2018)
Papaya	*Carica papaya* L. "Formosa"	Coated urea Kimcoat N, coated with polymer layers, conventional urea	Coated urea promotes a higher growth and yield of "Formosa" papaya compared to the conventional urea	Silva Jr et al. (2016)
Pawpaw	*Asimina triloba*	Osmocote 14-14-14	Enhanced seedling growth	Pomper et al. (2002)
Pawpaw	*Asimina triloba*	Osmocote 14-14-14 (14N-6.1P-11.6K)	Enhance production of container-grown pawpaw seedlings	Pomper et al. (2002)
Pear	*Prunus persica* L "Mit Ghamr"	Formaldehyde urea Phosphorus coated urea Sulfur coated urea Urea	Increased shoot length, leaf area, leaf N content. Nutritional status of trees and fruit quality were greater than a urea fertilized trees	Kandil et al. (2010)

one-way process, that is, from older to newer leaves. N storage over winter is the primary N source for flushing. The effect of N fertility in the current growing season was larger in the late phase than in the early phase (Millard and Grelet, 2010; Ueda et al., 2009). The difference in N fertility in the current growing season is expected to affect retranslocation in the following spring (Ueda, 2012). This means that tree plants largely use retranslocated rather than currently absorbed N. Foliar N content is strongly related to leaf photosynthetic capacity and plant growth. As CRFs and SRFs are able to sustain N supply over a long period of time, more absorbed N can be either used or stored and retranslocated for immediate use in the following season, thus improving plant productivity.

The use of CRFs and SRFs has been shown to increase nutrient use efficiency and decrease fertilizer application. Trenkel (2010) Suggested that CRFs and SRFs can potentially decrease fertilizer use by 20%–30% of the recommended rate of a conventional fertilizer while obtaining the same yield. In several field trials in Florida, young or nonbearing citrus trees fertilized with CRFs at a 50% of the recommended rate performed equally well compared with 100% of the recommended rate with water-soluble fertilizers (Obreza and Sartain, 2010). Applying CRFs and SRFs generally reduces salt accumulation, thus minimizing the possibility of leaf burning. The use of CRFs reduces labor costs.

Depending on plant species, one or two applications of appropriate amount of CRFs will ensure healthy growth or expected fruit yield, while water-soluble fertilizers have to be applied several times. This is because CRFs or SRFs can meet the crop N demand for the entire season through one or two applications depending on fertilizer longevity, resulting in saving in fertilizer application and labor costs

3.2 CRFs and SRFs reduce nutrient loss

Both CRFs and SRFs have been documented to reduce NO_3^--N leaching in laboratory and field experiments (Broschat, 1995; Wang and Alva, 1996; Paramasivam and Alva, 1997; Chen et al., 2001; Mello et al., 2017; Chen and Wei, 2018). Some reports on N leaching from CRFs or SRFs are presented in Table 39.4. For example, Alva (1992) showed that near 40% and 50% of NO_3^{-1} and 80% and 100% NH_4^+ were leached from ammonium nitrate and calcium nitrate, respectively, after 1000 mL of water was penetrated through 8-cm soil columns, but NO_3^{-1} and NH_4^+ leaching from IBDU, IBDU + Escote, Nutralene, Osmocote, and Meister was much lower (less than 4%) except NH_4^+ that was leached by more than 40% and 20%, respectively, from IBDU and Nutralene fertilizers. In another laboratory leaching study, Wang and Alva (1996) reported that 100% of applied N was leached from ammonium nitrate. However, applied N leached from IBDU and Meister polyolefin resin-coated urea were 32% and 12%, respectively. These results demonstrated that the amount of N leached from SRFs or CRFs was substantially lower than granular fertilizers or uncoated urea in laboratory test.

Field experiments also showed that CRFs and SRFs can reduce NO_3^--N leaching in fruit crop production. In an experiment conducted with 21-year-old orange trees on "Cleopatra" mandarin rootstock in a fine sand soil in Florida, Paramasivam et al. (2001) reported that the amount of N leached below the rooting depth ranged from 5% to 12% for dry granular fertilizer, 10% to 16% for a liquid fertilizer through fertigation, and only 1% to 5% for a CRF. A number of similar studies were conducted in Florida as commercial citrus production in Florida used to occupy 342,200 ha. Citrus

TABLE 39.4 Nitrogen lost in leachates or runoff water or emitted into the atmosphere when controlled- and slow-release fertilizers used in fruit crop production or leaching experiments.

Experiments	Soil types	Fertilizers	N leached or N conc. in leachates/runoff	Comments	References
A leaching column experiment	Florida sandy soil	NH_4NO_3 $Ca(NO_3)_2$ IBDU (isobutylidene diurea) IBDU plus Escote Nutralene Osmocote Meister	37% and 88% of NO_3 and NH_4 were leached from NH_4NO_3; 48% and 100% of NO_3 and NH_4 were leached from $Ca(NO_3)_2$ but leaching loss of NO_3 and NH_4 from the rest less than 3% and 4%	Much less amount of NO_3 and NH_4 was leached from CRFs and SRFs than water-soluble fertilizers	Alva (1992)
A leaching column study	Florida sandy soil	NH_4NO_3 IBDU Meister polyolefin resin-coated urea	100% of applied N leached from NH_4NO_3 32% of applied N leached from IBDU 12% of applied N leached from polyolefin resin-coated urea	All N was leached from NH_4NO_3, but much less N was leached from IBDU and Meister	Wang and Alva (1996)
A leaching column study	Florida fine sandy soil	Urea Poly-S Meister polyolefin resin-coated urea Osmocote	28% of applied N leached from urea 12% of applied N leached from Poly-S 6% of applied N leached from Meister, and 5% leached from Osmocote	Meister and Osmocote leached much less N than urea	Paramasivam and Alva (1997)
Kiwifruit	Actinidia deliciosa	Controlled-release urea Urea	More than 28% N leached from urea	Plant growth was not affected	Lu et al. (2018)
Coffee	Dystroferric red latosol, equivalent to an oxisol	Granulated urea Polyblen extend Polyblen Montanha	83.4%, 60.2% and 47.4% of NH_3-N loss from urea, Polyblen Extend, Polyblen Montanha, respectively	Coffee growth was not affected	Chagas et al. (2016)
Turfgrass	A timpanogos loam soil	Polymer-coated urea Urea	1.25 mg N_2O-N/m^2/h 2.22 mg N_2O-N/m^2/h	Polymer-coated urea emitted significantly low amount of N_2O-N	LeMonte et al. (2016)

plants have been primarily produced in sandy soils, and granular chemical fertilizers are traditionally broadcast over the entire grove, which has been related to N contamination of ground and surface water. Results suggest that the benefits of using CRFs and SRFs to reduce N leaching depend on soil properties and rainfall regime or irrigation practices. Several studies conducted in Southwest Florida showed that maximum growth and early yield of "Hamlin" orange trees planted on typical Flatwoods soils (Alfisols and Spodosols) could be obtained at 50% of the recommended rate of ~220 kg N/ha (Marler et al., 1987; Matthews, 1988; Obreza and Rouse, 1993). Other studies in Central Florida, dominated by the well-drained Entisols ($\geq$95% sand), also reported higher leaching under fertigation and dry granular fertilization compared with CRFs and SRFs (Morgan et al., 2009b; Paramasivam et al., 2001; Wang and Alva, 1996)

Application of CRFs and SRFs reduces NH_3 volatilization and N_2O emission compared with the use of conventional fertilizers. In a leaching column experiment (Paramasivam and Alva, 1997), N transformation and leaching from three urea-based CRFs and uncoated urea from a sandy soil were monitored. After 18 leaching and dry cycles (3960 mL of water leached, equivalent to 90-cm rainfall), recovery of total N (sum of all forms in the leachate, residual soil N, and residual fertilizer N) represented 89%, 85%, 84%, and 45% of the total applied N for poly-S, Meister, Osmocote, and urea, respectively. Cumulative NO_3^--N in 3960 mL leachate was 80%, 98%, 97%, and 24% of the total N leached from poly-S, Meister, Osmocote, and urea amended soil columns, respectively. The deficit of N that was uncovered from urea could be lost due either to volatilization or emission. A field experiment conducted in a dystroferric red latosol, equivalent to an Oxisol soil with coffee plants in Brazil showed that accumulated NH_3 loss from urea, Polyblend extend, and Polyblend Montanha were 83.4, 60.2, and 47.4 kg/ha, respectively (Chagas et al., 2016). Application of CRFs can reduce N_2O emission. For example, application of urea in turfgrass production resulted in 127%–476% more N_2O emission into atmosphere compared with 45%–73% emission by using a CRF (LeMonte et al., 2016)

3.3 CRFs and SRFs improve rhizosphere microbial community

The rhizosphere harbors diverse microbes; many of them could be beneficial to plants by facilitating nutrient acquisition and reducing plant pathogen activities in the soil. CRF application has been shown to improve the rhizosphere microbial community. Controlled-release urea was reported to continuously affect the dehydrogenase activity over a short distance from roots, while conventional urea could greatly increase the enzyme activity for a shorter period of time (Chu et al., 2005). A study conducted in Japan showed that application of urea formaldehyde fertilizers to onion bulbs and main roots of sugar beet changed the diversity of the microbial community and the abundances of certain bacterial species (Hayatsu, 2014). Arbuscular mycorrhizal fungi (AMF) play important roles in the functioning of agricultural ecosystems. To gain information on fertilizer application on rhizosphere microbial population, Van Geel et al. (2016) used 454-pyrosequencing of the small subunit rRNA gene amplicons to quantify AMF diversity and community composition in apple roots. Plants were fertilized with two inorganic, three slow release P fertilizers, and a control treatment. Results showed that SRF treatments had significantly higher AMF richness and differed in community composition compared with the inorganic fertilizer treatments. The distribution of AMF operational taxonomic units (OTUs) showed a significantly nested pattern. Additionally, AMF communities in the inorganic fertilizer treatments were a subset of the communities in the SRF treatments. This study clearly demonstrates that application of SRFs promoted AMF diversity in the roots of cultivated apple trees in comparison with the other treatments.

Recently, bulk soil and rhizosphere samples of citrus were collected from distinct biogeographical regions of six continents, and microbial compositions of the samples were analyzed using both amplicon and deep shotgun metagenomic sequencing (Xu et al., 2018). Results showed that predominant taxa include Acidobacteria, Actinobacteria, Bacteroidetes, and Proteobacteria. The core citrus rhizosphere microbiome comprises Agrobacterium, Bradyrhizobium, Burkholderia, Cellvibrio, Cupriavidus, Mesorhizobium, Paraburkholderia, Pseudomonas, Rhizobium, Sphingomonas, and Variovorax, and some of which are potential plant beneficial microbes. This study also identified overrepresented microbial functional traits mediating plant-microbe and microbe-microbe interactions, nutrition acquisition, and plant growth promotion in citrus rhizosphere. How different CRFs and SRFs in different soil conditions affect citrus rhizosphere microbial composition is unclear. Understanding microbial composition and factors influencing the composition could allow manipulation the rhizosphere microbial community for improving plant growth and preventing disease.

3.4 CRFs and SRFs boosts health of HLB trees

Application of CRFs and SRFs has been shown to improve plant tolerance of certain pathogens. HLB is now recognized as the main reason for the decline in citrus yields worldwide. HLB is a bacterial disease caused by the fastidious, phloem-restricted bacterium *Candidatus* Liberibacter asiaticus (CLas), transmitted by the Asian psyllid (*Diaphorina citri* Kuwayama). Once CLas is transmitted to the citrus trees, the phloem plugging results in the disruption of vascular function, loss of roots, and an alteration in mineral nutrition, which results in the stunting of plants, fruit growth, and eventual death of the tree (Vashisth and Grosser, 2018). The typical symptoms of HLB in a tree include reduced plant height, leaf yellowing, stain spotting, and chlorotic leaf patterns that resemble those caused by zinc and iron deficiencies. Application of CRFs has been shown to alleviate HLB in citrus. A series of experiments conducted by faculty at the University of Florida showed that application of balanced nutrients, that is, CRFs along with micronutrients significantly improve overall plant health and reduced disease severity (Spann et al., 2011; Spyke et al., 2017a,b). For example, severely debilitated HLB-infected citrus trees that lost the ability to bear fruit started to bear fruit after 2-year application of complete nutrients including microelements three times a year (Spann et al., 2011). Application of CRFs in combination with conventional fertilizers, called a hybrid program, has been developed to alleviate HLB-infected plants, resulting in the recovery of fruit production (Spyke et al., 2017a,b). Vashisth and Grosser (2018) reported that yields of 4-year-old sweet orange "Valquarius" fertilized with five CRFs, Florikote (14 N-4 P-10 K), Citriblend (17 N-5 P-12 K), Harrell's (13 N-4 P-9 K), Citriblend (18 N-6 P-11 K), and Harrell's (16 N-5 P-10 K), were exceptionally high for 4-year-old tree under high disease pressure. Overall, all the CRFs yielded good production and fruit quality. Analysis of leaf tissue showed that all the nutrients were within the optimum to the high range of recommended levels by the University of Florida (Obreza and Morgan, 2008) with an exception of manganese.

Over the years of combating with HLB, it becomes clear that plant health is an important factor contributing to the well-being of citrus. Conventional ground application of granular fertilizers at several times a year did not provide citrus with desired nutrient compositions. First, the granular fertilizers are largely N compounds. Ground application of them only provides roots a peak of N and then diminished due to leaching, volatilization, or emission, resulting in environmental problems. Second, the fertilizers are not complete nutrients. Florida sandy soils that are poor in fertility in general, continuous application of single nutrient will result in imbalance of nutrients in plants, thus jeopardizing overall plant health and increasing plant susceptible to pest and pathogens. Third, there is possibility that the recommended nutrient range for citrus may not be appropriate. Citrus may need more micronutrients than previously thought. The current realization of the benefits of CRF application in remediation of HLB could be attributed to the following factors. One, CRFs, particularly those with micronutrients, provide plants with complete nutrients required for healthy growth and development. Remember Liebig's "Law of Minimum." If one of the essential plant nutrients is deficient, plant growth will be poor even when all other essential nutrients are abundant. Two, the controlled release characteristic may synchronize with plants' need for nutrients in a prolonging fashion, thus sustaining nutrient supply through roots. Three, as discussed above, N is retranslocated bidirectionally in plants, and plant growth in spring generally uses the existing, retranslocated N, rather than freshly applied N. Continuous supply of N through CRFs allows N being stored in plants for immediate use in next spring, which enhanced plant health and increase plant tolerance of abiotic and biotic stresses. Four, complete nutrients absorbed by roots and transported to shoot through xylem should be more efficient for nutrient delivery compared with foliar spray of micronutrients. It is likely that the application of CRFs or SRFs may facilitate the interactions of these factors in a synergistic way, thus improving plant overall health and enhancing plants in adaptation to stressful factors.

4 Concerns of using CRFs and SRFs in fruit crop production and perspectives

The benefits of using CRFs and SRFs in fruit crop production are documented in this chapter. However, concerns over their commercial application exist, including the cost, the limitation in release patterns and formulations for specific fruit crops, and application methods (Chen and Wei, 2018; Shaviv, 2001; Trenkel, 2010). The cost is probably the greatest hurdle limiting commercial use of CRFs and SRFs in fruit crop production. For example, 1 ton of a CRF (44% N) could be $650 compared with 1 ton of urea (46% N) at $481 (Ruark, 2012). More specifically, Obreza et al. (1999) reported that the cost of fertilizing citrus with CRFs at the full N rate was four times greater than conventional fertilization, but the return by using CRFs was only 15% greater. The authors concluded that the high cost of CRF products made their exclusively use to produce citrus economically unfeasible. This conclusion was drawn in the 1990s and was largely based on dollar value; it might not fully take conventional fertilizer application related environmental problems, application labor cost, and plant overall health into consideration. As the attention to fertilizer application related environmental problem mounting and the severity of HLB in the citrus industry continuing, the value of using CRFs and SRFs in fruit crop, particularly in citrus production, may offset the cost concern. The fact is that CRFs and SRFs have been used as a tool to combat HLB in Florida right now. CRFs have been used to rescue HLB-infected plants, and CRF or SRFs with conventional fertilizers have been used for citrus production (Spann et al., 2011; Spyke et al., 2017a,b). CRFs and SRFs are also used for production of other fruit crops as presented in Table 39.3.

Technologies for developing CRFs and SRFs have been evolving, transitioning from sulfur-coating to polymer-coating technologies. With the development of nanotechnology, future CRFs may need to integrate this technology for improving controlled-release characteristics (Zhang et al., 2014). Coating materials used to produce CRFs should be capable of decomposing naturally in most common environmental conditions. For development of SRFs, technologies should focus not only on reducing solubility of N and other nutrient elements but also on development of appropriate matrixes for carrying and release of nutrients. Among the promising matrixes, alginate (Wang et al., 2018), biochar (Yao et al., 2013a,b; Yu et al., 2019) and natural zeolites should be considered. Additionally, future SRFs should consider incorporating beneficial microbes, such as plant growth-promoting bacteria (Vessey, 2003) and mycorrhizal fungi (Wei et al., 2016a,b) into the matrixes to maximize nutrient use efficiency and minimize negative impact on the environment. Currently, there are different types of CRFs and SRFs with different compositions and longevities on the market; fruit crops also differ significantly in their growth and development and in nutrient requirements. Manufacturers, researchers, and producers should work together to design and develop CRFs and SRFs that are sustainable, economically affordable, and crop specific for the production of different fruit crops. Nevertheless, the increased nutrient-use efficiency with minimum nutrient losses is anticipated to fuel market growth of CRFs and SRFs in the future. The CRF and SRF market was projected to grow at a compound annual growth rate of 5.6% during the period 2018–22 (Research and Markets, 2018). The increase in quantity may potentially reduce the cost, further promoting the use of CRFs and SRFs in fruit crop production.

Acknowledgments

The authors would like to thank Caroline Roper Warwick and Terri A. Mellich for reviewing this chapter. The work was supported in part by the Scientific Research Foundation of Graduate School at the Fujian Agriculture and Forestry University (324-1122YB026) and the National Nature Science Foundation of China (Grant No. 31801897) to X.W.

References

AAPFCO, 2013. Official publication number 65. Association of American Plant Food Control Officials. http://www.aapfco.org/publications.html. (Accessed 8 November 2017).

Alva, A.K., 1992. Differential leaching of nutrients from soluble vs. controlled-release fertilizers. Environ. Manag. 16, 769–776.

Azeem, B., KuShaari, K., Man, Z.B., Basit, A., Thanh, T.H., 2014. Review on materials & methods to produce controlled release coated urea fertilizer. J. Control. Release 181, 11–21.

Badenes, M.L., Byrne, D.H., 2012. Fruit Breeding. vol. 8 Springer Science + Business Media, New York, NY.

Bettaga, N., Ben Mimoun, M., 2008. Effects of controlled-release fertilizer on fruit yield and quality of clementine citrus trees. Acta Hortic. 868, 429–432.

Brazelton, C., 2015. World Blueberry Statistics and Global Market Analysis. US Highbush Blueberry Council, Folsom, CA (105 p).

Broschat, T.K., 1995. Nitrate, phosphate, and potassium leaching from container-grown plants fertilized by several methods. HortScience 30, 74–77.

California Department of Food and Agriculture, 2012. California Agricultural Statistical Service. 2012 California Citrus Acreage Report. http://wwwnassusdagov/Statistics_by_State/California/Publications/Fruits_and_Nuts/201208citacpdf. (Accessed 8 January 2019).

Chagas, W.F.T., Guelfi, D.R., Caputo, A.L.C., Souza, T.L.D., Andrade, A.B., Faquin, V., 2016. Ammonia volatilization from blends with stabilized and controlled-released urea in the coffee system. Ciênc. Agrotec. 40, 497–509.

Chen, J., Wei, X., 2018. Controlled-release fertilizers as a means to reduce nitrogen leaching and runoff in container-grown plant production. In: Khan, A., Fahad, S. (Eds.), Nitrogen in Agriculture-Updates. InTechOpen, London, pp. 33–52. https://doi.org/10.5772/intechopen.73055.

Chen, J., Huang, Y., Caldwell, R.D., 2001. Best management practices for minimizing nitrate leaching from container-grown nurseries. Sci. World J. 1, 96–102.

Chen, L., Xin, X., Yuan, Q., Su, D., Liu, W., 2013. Phytochemical properties and antioxidant capacities of various colored berries. J. Sci. Food Agric. 94, 180–188.

Chu, H., Hosen, Y.Y., Yagi, K., Okada, K., Ito, O., 2005. Soil microbial biomass and activities in a Japanese Andisol as affected by controlled release and application depth of urea. Biol. Fertil. Soils 42, 89–96.

Du, C., Zhou, J., Shaviv, A., 2006. Release characteristics of nutrients from polymer-coated compound controlled release fertilizers. J. Polym. Environ. 14, 223–230.

FAO (Food Agricultural Organization), 2018. Fresh Fruit Production Worldwide 1990–2016. https://www.statista.com/statistics/264001/worldwide-production-of-fruit-by-variety/. (Accessed 8 January 2019).

Gergerich, R.C., Welliver, R.A., Gettys, S., Osterbauer, N.K., Kamenidou, S., Martin, R.R., Golino, D.A., Eastwell, K., Fuchs, M., Vidalakis, G., 2015. Safeguarding fruit crops in the age of agricultural globalization. Plant Dis. 99, 176–187.

Haifa, 2017. Crop Guide: Banana Fertilizer Recommendations. Haifa Group. https://www.haifa-group.com/banana-fertilizer/crop-guide-banana-fertilizer-recommendations-0. (Accessed 5 January 2019).

Hayatsu, M., 2014. A novel function of controlled-release nitrogen fertilizers. Microbes Environ. 29, 121–122.

Kandil, E., Fawzi, M., Shahin, F., 2010. The effect of some slow release nitrogen fertilizers on growth, nutrient status and fruiting of "Mit Ghamr" peach trees. J. Am. Sci. 6, 195–201.

Koo, R., 1986. Controlled-release sources of nitrogen for bearing citrus. In: Paper Presented at: Proceedings of Florida State Horticultural Society. vol. 99, pp. 46–48.

LeMonte, J.J., Jolley, V.D., Summerhays, J.S., Terry, R.E., Hopkins, B.G., 2016. Polymer coated urea in turfgrass maintains vigor and mitigates nitrogen's environmental impacts. PLoS One. 11. https://doi.org/10.1371/journal.pone.0146761.

Liu, L., Kost, J., Fishman, M., Hicks, K.B., 2008. A review: controlled release systems for agricultural and food applications. In: ACS Symposium Series 992, pp. 265–281. New delivery systems for controlled drug release from naturally occurring materials, https://doi.org/10.1021/bk-2008-0992.ch014.

Lu, Y., Kang, T., Gao, J., Chen, Z., Zhou, J., 2018. Reducing nitrogen fertilization of intensive kiwifruit orchards decreases nitrate accumulation in soil without compromising crop production. J. Integr. Agric. 17, 1421–1431.

Marler, T., Ferguson, J., Davies, F., 1987. Growth of young 'Hamlin' orange trees using standard and controlled-release fertilizers. Weed Sci. 33, 779–785.

Matthews, C., 1988. Controlled-release fertilizers and growth of young 'Hamlin' orange trees. In: Paper Presented at: Proceedings of Florida State Horticultural Society. vol. 101, pp. 17–20.

Mello, S.C., Li, Y.C., Migliaccio, K.W., Linares, E.P., Colee, J., Angelotti-Mendonça, J., 2017. Effects of polymer coated urea and irrigation rates on lantana growth and nitrogen leaching. Soil Sci. Soc. Am. J. 81, 546–555.

Millard, P., Grelet, G., 2010. Nitrogen storage and remobilization by trees: ecophysiological relevance in a changing world. Tree Physiol. 30, 1083–1095.

Morgan, K.T., Cushman, K.E., Sato, S., 2009a. Release mechanisms for slow- and controlled-release fertilizers and strategies for their use in vegetable production. HortTechnology 19, 10–12.

Morgan, K.T., Wheaton, T.A., Castle, W.S., Parsons, L.R., 2009b. Response of young and maturing citrus trees grown on a sandy soil to irrigation scheduling, nitrogen fertilizer rate, and nitrogen application method. HortScience 44, 145–150.

Obreza, T., Morgan, K., 2008. Nutrition of Florida citrus trees. In: University of Florida/Institute of Food and Agricultural Sciences, Electronic Database Information System (EDIS) SL253. https://edis.ifas.ufl.edu/pdffiles/SS/SS47800.pdf.

Obreza, T.A., Rouse, R.E., 1993. Fertilizer effects on early growth and yield of 'Hamlin' orange trees. HortScience 28, 111–114.

Obreza, T.A., Sartain, J.B., 2010. Improving nitrogen and phosphorus fertilizer use efficiency for Florida's horticultural crops. HortTechnology 20, 23–33.

Obreza, T., Rouse, R., Sherrod, J., 1999. Economics of controlled-release fertilizer use on young citrus trees. J. Prod. Agric. 12, 69–73.

Obreza, T., Rouse, R., Hanlon, E., 2006. Advancements with Controlled-Release Fertilizers for Florida Citrus Production: 1996-2006. Soil and Water Science Department, Florida Cooperative Extension Service, Institute of Food and Agricultural Sciences, University of Florida Fact Sheet SL-243. https://ucanr.edu/sites/nm/files/76682.pdf.

Ozores-Hampton, M., 2017. Description of enhanced-efficiency fertilizers for use in vegetable production. UF/IFAS Extension, Document HS1247. University of Florida, Immokalee, FL. http://edisifasufledu/pdffiles/HS/HS124700pdf.

Paramasivam, S., Alva, A., 1997. Leaching of nitrogen forms from controlled-release nitrogen fertilizers. Commun. Soil Sci. Plant Anal. 28, 1663–1674.

Paramasivam, S., Alva, A., Fares, A., Sajwan, K., 2001. Estimation of nitrate leaching in an Entisol under optimum citrus production. Soil Sci. Soc. Am. J. 65, 914–921.

Pomper, K.W., Layne, D.R., Reed, E.B., 2002. Determination of the optimal rate of slow-release fertilizer for enhanced growth of pawpaw seedlings in containers. HortTechnology 12, 397–402.

Refaai, M.M., 2016. Response of Thompson seedless grapevines to application of methylene urea and some slow release N fertilizers as apartial replacement of the fast release mineral urea fertilizers. J. Plant Prod. (Mansoura University) 7, 99–104.

Research and Markets, 2018. Global controlled-release fertilizer market 2018–2022. https://www.researchandmarkets.com/research/37tpg6/global?w=5. (Accessed 8 January 2019).

Retamales, J.B., 2011. World temperate fruit production: characteristics and challenges. Rev. Bras. Frutic. 33, 121–130.

Rieger, M., 2006. Introduction to Fruit Crops. CRC Press.

Rouse, R.E., Obreza, T.A., Sherrod, J.B., 1999. Yield and relative cost of controlled-release fertilizer on young bearing citrus trees. In: Proceedings of Florida State Horticultural Society, 112, pp. 46–50.

Ruark, M., 2012. Advantages and Disadvantages of Controlled-Release Fertilizers. Departement of Soil Science, WI FFVC, Wisconsin-Madison University. http://www.soils.wisc.edu/extension/materials/Overview_of_fertilizer_technologies_2012_WIFFVC.pdf.

Schlesinger, W.H., Hartley, A.E., 1992. A global budget for atmospheric NH$_3$. Biogeochemistry 15, 191–211.

Sempeho, S.I., Kim, H.T., Mubofu, E., Hilonga, A., 2014. Meticulous overview on the controlled release fertilizers. Adv. Chem. 2014. https://doi.org/10.1155/2014/363071.

Sha, J., Wang, F., Tian, G., Yu, B., Ge, S., Jiang, G.Y., 2018. Effects of controlled-release nitrogen fertilizer and bag-controlled release fertilizer on utilization of ^{15}N-urea in 'Orin' apple and its accumulation in soil. Chin. J. Appl. Ecol. 29, 1421–1428.

Shaviv, A., 2001. Advances in controlled-release fertilizers. Adv. Agron. 71, 1–49.

Shaviv, A., 2005. Controlled release fertilizers. In: Paper Presented at: IFA International Workshop on Enhanced-Efficiency Fertilizers, Frankfurt. International Fertilizer Industry Association, Paris, France.

Silva Jr., G.B.D., Santos, E.M.D., Silva, R.L., Cavalcante, Í.H., 2016. Nutritional status and fruit production of Carica papaya as a function of coated and conventional urea. Rev. Bras. Eng. Agrícola Ambient. 20, 322–328.

Simonne, E.H., Hutchinson, C.M., 2005. Controlled-release fertilizers for vegetable production in the era of best management practices: teaching new tricks to an old dog. HortTechnology 15, 36–46.

Spann, T.M., Schumann, A.W., Rouse, B., Ebel, B., Rouse, B., Ebel, B., 2011. Foliar nutrition for HLB. Citrus Ind. 92, 6–10.

Spyke, P., Sherrod, J., Grosser, J., 2017a. Controlled-release fertilizer boosts health of HLB trees (part 1). Citrus Ind., 12–18.

Spyke, P., Sherrod, J., Grosser, J., 2017b. Controlled-release fertilizer boosts health of HLB trees (part 1). Citrus Ind., 10–14.

Srivastava, A.K., Malhotra, S.K., 2017. Nutrient use efficiency in perennial fruit crops—a review. J. Plant Nutr. 40, 1928–1953.

Studholme, D.J., Glover, R.H., Boonham, N., 2011. Application of high-throughput DNA sequencing in phytopathology. Annu. Rev. Phytopathol. 49, 87–105.

Subbarao, C.V., Kartheek, G., Sirisha, D., 2013. Slow release of potash fertilizer through polymer coating. Int. J. Appl. Sci. Eng. 11, 25–30.

Trenkel, M.E., 2010. Slow-and Controlled-release and Stabilized Fertilizers: An Option for Enhancing Nutrient Use Efficiency in Agriculture. IFA (International Fertilizer Industry Association), Paris, France.

Ueda, M.U., 2012. Gross nitrogen retranslocation within a canopy of *Quercus serrata* saplings. Tree Physiol. 32, 859–866.

Ueda, M.U., Mizumachi, E., Tokuchi, N., 2009. Allocation of nitrogen within the crown during leaf expansion in *Quercus serrata* saplings. Tree Physiol. 29, 913–919.

Ukessays, 2015. Controlled release fertilizers and nanotechnology traces biology essay. https://wwwukessayscom/essays/biology/controlled-release-fertilizers-and-nanotechnology-traces-biology-essayphp.

Van Geel, M., De Beenhouwer, M., Ceulemans, T., Caes, K., Ceustermans, A., Bylemans, D., Gomand, A., Lievens, B., Honnay, O., 2016. Application of slow-release phosphorus fertilizers increases arbuscular mycorrhizal fungal diversity in the roots of apple trees. Plant Soil 402, 291–301.

Vashisth, T., 2017. An update on UF/IFAS-grower citrus nutrition trials. Citrus Ind. 98, 20–27.

Vashisth, T., Grosser, J., 2018. Comparison of controlled-release fertilizer (CRF) for newly planted sweet orange trees under Hunaglongbing prevalent conditions. J. Hortic. 5(3). https://doi.org/10.4172/2376-0354.1000244.

Vessey, J.K., 2003. Plant growth promoting rhizobacteria as biofertilizers. Plant Soil 255, 571–586.

Wang, F., Alva, A., 1996. Leaching of nitrogen from slow-release urea sources in sandy soils. Soil Sci. Soc. Am. J. 60, 1454–1458.

Wang, B., Wan, Y., Zheng, Y., Lee, X., Liu, T., Yu, Z., Huang, J., Ok, Y.S., Chen, J., Gao, B., 2018. Alginate-based composites for environmental applications: a critical review. Crit. Rev. Environ. Sci. Technol. 1–39. https://doi.org/10.1080/10643389.2018.1547621.

Wei, X., Chen, J., Zhang, C., Pan, D., 2016a. A new *Oidiodendron maius* strain isolated from *Rhododendron fortunei* and its effects on nitrogen uptake and plant growth. Front. Microbiol. 7, 1327. https://doi.org/10.3389/fmicb.2016.01327.

Wei, X., Chen, J., Zhang, C., Pan, D., 2016b. Differential gene expression in *Rhododendron fortunei* roots colonized by an ericoid mycorrhizal fungus and increased nitrogen absorption and plant growth. Front. Plant Sci. 7, 1594. https://doi.org/10.3389/fpls.2016.01594.

Xu, J., Zhang, Y., Zhang, P., Trivedi, P., Riera, N., Wang, Y., Liu, X., Fan, G., Tang, J., Coletta-Filho, H.D., 2018. The structure and function of the global citrus rhizosphere microbiome. Nat. Commun. 9, 4894. https://doi.org/10.1038/s41467-018-07343-2.

Yao, Y., Gao, B., Chen, J., Yang, L., 2013a. Engineered biochar reclaiming phosphate from aqueous solutions: mechanisms and potential application as a slow-release fertilizer. Environ. Sci. Technol. 47, 8700–8708.

Yao, Y., Gao, B., Chen, J., Zhang, M., Inyang, M., Li, Y., Alva, A., Yang, L., 2013b. Engineered carbon (biochar) prepared by direct pyrolysis of Mg-accumulated tomato tissues: characterization and phosphate removal potential. Bioresour. Technol. 138, 8–13.

Yu, H., Zou, W., Chen, J., Chen, H., Yu, Z., Huang, J., Tang, H., Wei, X., Gao, B., 2019. Biochar amendment improves crop production in problem soils: a review. J. Environ. Manag. 232, 8–21.

Zekri, M., Koo, R.C., 1991. Evaluation of controlled-release fertilizers for young citrus trees. J. Am. Soc. Hortic. Sci. 116, 987–990.

Zhang, M., Yang, Y., Song, F., Shi, Y., 2005. Study and industrialized development of coated controlled release rertilizers. J. Chem. Fertil. Ind. 32, 7–12.

Zhang, M., Gao, B., Chen, J., Li, Y., Creamer, A.E., Chen, H., 2014. Slow-release fertilizer encapsulated by graphene oxide films. Chem. Eng. J. 255, 107–113.

40

Diagnosis and management of nutritional constraints in berries

*Rolf Nestby[a], Jorge B. Retamales[b],**

[a]Division Food and Society (Horticulture), Norwegian Institute of Bioeconomy (NIBIO), Ås, Norway
[b]Head ISHS Division Vine and Berry Fruits, Viña del Mar, Chile
*Corresponding author. E-mail: jretamal@utalca.cl

1 Availability of nutrients

Today, growers have the possibility to highly modernize the irrigation and thereby the fertilization, by adding nutrients through the irrigation system (fertigation). Through modern fertigation technique using valves, sensors, timers, and IT support, growers avoid a lot of time-consuming manual work. However, data achieved from sensors have to be in a context to make the fertigation optimally adjusted at all developmental stages, also taking into account the variations in climate, mainly sunlight and temperature. Several companies deliver such equipment today. The competition is high, which is a driver for continuous improvements of the systems.

This chapter starts with a general section that describes the conditions for nutrient uptake by the roots, which will highlight soil conditions, interactions with soil organisms (the biome), and the importance of the rhizosphere (growth medium in the immediate proximity of the root hairs). In the following three main sections, we will focus on the diagnosis of nutrient constraints (stresses) in berry crops. We will show how the knowledge of these constraints is helpful to manage optimal growing conditions using modern fertilization technology and help the growers to achieve this. Since there is a wide range of berry crops, the focus will be on strawberry and highbush blueberry. There are similarities and differences between these species on nutrient demand and especially of soil preferences, discussed in two separate sections.

Throughout the world, strawberries tend to grow in light soils, but the result can be very good in heavy and nutrient-rich soils, properly drained. Blueberries prefer acid soils, typically with pH of ±4.5 to 5.5. We will discuss the general need of macronutrients and boron (B), at different developmental stages. This will vary between the two species and between cultivars of each species. We will concentrate on general demands of the species and a few cultivar-related differences.

A.K. Srivastava, Chengxiao Hu (eds.)
Fruit Crops: Diagnosis and Management of Nutrient Constraints
https://doi.org/10.1016/B978-0-12-818732-6.00040-X

2 The roots and the rhizosphere

Mineral nutrients are transported from the soil to the different plant organs (roots, crowns, leaves, runners (strawberry), and fruits), generally starting with the roots and transported from there to the other plant organs. Leaves can also absorb mineral nutrients, but that is of minor importance; however, this property is often used to alleviate acute deficiencies. When the mineral nutrients have entered the plant, they move between plant organs depending on plant signaling. A special plant organ will be of interest to the grower, which for the berries is a high fruit yield of good eating and postharvest quality. To achieve this, it is important that transport of minerals from the soil through the roots is sufficient and that all plant organs have optimal levels of macro- and micronutrients, at all developmental stages.

Unless unfavorable aboveground conditions are dramatically limiting for plant health, the plant health depends on interactions occurring at the root-soil-microorganism interface (Benbrook, 2017). Successful berry cultivation has to start with a healthy plant, and this chapter will deal with strawberry and blueberry as examples of berries. A healthy plant in strawberry could be a short-day plant where flowers are not yet initiated, or more commonly, a fully initiated short-day plant ready to sustain a proper yield shortly after planting (60-day plants). Another alternative of increasing interest is remontant (long-day) strawberry plants. In blueberry 1- to 2-year-old in vitro or plants from cuttings are used.

In strawberries, the plant is established in a bed normally mulched with one of the different types of polyethylene, and the planting system varies from one to several rows on the bed. Plant density (number of plants per square meter) will depend on the cultivar. However, the plants will compete for nutrients (from both soil and fertilizers), and the closer they are planted, the greater the competition. In the end, fruit yield, size, and quality will decide the density. There is also a limit to the sum of cations that can be simultaneously absorbed by the roots (Greenwood and Stone, 1998). The growth of strawberry roots, typically, looks like in Fig. 40.1, which shows a cross section of a bed mulched with black polyethylene film, supported with two drip lines for fertigation. The majority of root growth occurs just below the two drip lines where most of the water and nutrients end up. In this case, it was intentionally fertigated down to 30 cm, to avoid leaching of nutrients (Nestby and Guery, 2018).

The roots seek for water and mineral nutrients and grow between the soil particles. The roots active in uptake of nutrients—the root hairs, which are 5–20-μm long–are surrounded by the rhizosphere (Waisel et al., 1996). The rhizosphere microbiome (population of microbes) extends the functional repertoire of the plant (Bakker et al., 2013). In the rhizosphere, abundant organisms are found that help the plant to absorb nutrients, and organisms also enter into the root hairs; that is the case of the mycorrhiza that surround the roots and provide them with minerals and get assimilates in return. Additionally, beneficial microbes in the microbiome improve the conditions for root uptake of mineral nutrients. The rhizosphere is well described in several chapters of the classic book *Plant Roots: The Hidden Half* (Waisel et al., 1996). However, given the unraveling processes that drive selection and activities of the rhizosphere microbiome, it has open up new avenues to manipulate crop health and yield (Bakker et al., 2013). Later, several authors have shown that using biofumigation (green manure) and adding biofertilizers are beneficial to improve uptake of mineral nutrients from the soil (Koron et al., 2014; Chakraborty et al., 2017; Kapur et al., 2018;

FIG. 40.1 Root development of "Sonata" strawberry in a silt loam bed with two drip lines. Soil depth between tubes and plow sole is 30–35 cm.

Mihàlka et al., 2017; Tomic et al., 2018). Some amendments have reduced injury of *Fusarium* wilt in strawberry, which has been an important problem since the ban of methyl bromide (Borrero et al., 2017).

Highbush blueberries have two types of root: thick storage roots (<11 mm in diameter) and fine, threadlike roots (as small as 1 mm in diameter). The former anchor plants and perform a storage function, while the fine roots absorb water and nutrients. Blueberries do not have root hairs, and a specific type of endotropic mycorrhiza, ericoid mycorrhizae, inhabits the roots (Coville, 1910; Jacobs et al., 1982). These fungi form symbiotic associations with blueberry roots and help them prosper in soils with low pH and high organic matter (Vega et al., 2009). Mycorrhizal inoculation increases plant, root, and shoot dry weight without influencing shoot/root ratios (Yang et al., 2002). Mycorrhizae increase the uptake of soil nutrients and the efficiency of fertilizer application, improve water use, and protect the blueberry plant from toxic elements, such as Al (Scagel and Yang, 2005). Mycorrhizal colonization of blueberries varies significantly with cultivar, rate of fertilizer application, and the amount and type of soil organic matter present in the soil. In highbush blueberry fields in Oregon, large variations were found in mycorrhizal infection levels (0.5%–44% of total root length). Most colonization occurred in the upper 15 cm of the soil profile (Scagel and Yang, 2005).

In general, about 50% of blueberry roots are located within 30 cm of the crown, and 80%–85% are within 60 cm (Paltineanu et al., 2017). Mulching concentrates roots near the surface. Abbott and Gough (1987) found that high rates of irrigation tended to increase root depth and that mulched highbush plants had 83% of their roots in the top 15 cm of soil compared with 40% in nonmulched plants.

Vaccinium spp. absorptive roots can have diameters of less than 50 mm compared with a typical diameter of more than 200 mm in most other woody species. Valenzuela-Estrada et al. (2008) studied the root system of mature northern highbush "Bluecrop" plants with minirhizotrons and established that the ephemeral portion of the root system was mainly in the first three root orders. First- and second-order roots, despite being extremely fine, had median life spans of 115–120 days. The more permanent portion of the root system occurred in fourth- and higher-order roots. Roots in these orders had the lowest specific root length, nitrogen/carbon (N/C) ratios and levels of mycorrhizal colonization.

Young containerized highbush plants growing in sawdust had two peaks of root growth during the season (Abbott and Gough, 1987). The first (weaker) peak occurs near fruit set and extends to the immature green fruit stage. The second peak occurs once fruit harvest started and ends before plant dormancy.

3 Strawberry (*Fragaria* × *ananassa* Duch.)

3.1 Accumulation of minerals in the plant

A root takes up mineral nutrients by absorption of nutrient ions from the soil solution: These ions are readily available, but their concentrations in the soil solution are usually very low. The most abundant is NO_3^- (nitrate) in concentrations as high as 5–10 mM and followed by SO_4^{2-}, Mg^{2+}, and Ca^{2+} in concentrations up to 2–5 mM, K^+ up to 1–2 mM, and PO_4^- up to 4 µM, respectively. By releasing H^+ and HCO_3^- as dissociation products of respiratory CO_2, roots promote ion exchange at the surface of the clay minerals and humic particles, obtaining in return the nutrient ions (Larcher, 2003). From the roots, the minerals enter into the aboveground plant organs. The amount of mineral nutrients accumulated in the vegetative and generative part of the strawberry plant at the end of harvest season was published as early as 1978 for soil-growing plants (Albregts and Howard, 1978, 1980), and a little later in soilless growing systems (Lieten and Misotten, 1993).

These relative early examinations indicated that cultivated strawberries of different genotypes did not differ much in the amount of macroelements accumulated and in the balance between them, despite that soil type and large differences in mineral content and balance, pH, and cultivation system could create some deviations. In strawberries, ±31 kg N/ha was stored in the vegetative plant organs, and there was a similar amount harvested from the field as fruit yield at the end of harvest, if the fruit yield was approximately 30 t/ha. A higher or lower fruit yield would accordingly contain more or less N, respectively (Nestby et al., 2005). Later, the Haifa Group (Haifa, 2018a,b) published values that confirm the earlier findings of Albregts and Howard (1978). In addition, Haifa published a table showing removal of all macronutrients at different yields. Remontant strawberries produce higher yields than short-day strawberries and thereby remove more N from the field.

However, it is important to establish that the amount of mineral nutrients needed and the balance between them change at different developmental stages. Additionally, it is decisive to secure a proper pH, which, in soil, has to be optimal before planting, and, in soilless culture, must be balanced continuously to secure an optimal pH of the fertigation water. With this knowledge in mind, it should be relatively easy to set up a plan to fertilize a given cultivar in soil or in soilless culture. Lower correlations are reported in the literature for the microelements, possibly because of the differences in availability of nutrients in various growing systems (Nestby et al., 2005). Accordingly, we know the approximate amount of nutrients needed to build a well-functioning plant with potential to develop high-quality fruit.

In the next section, we will look further into diagnosis and management of nutrient constraints with focus mainly on macronutrients.

3.2 Causes of nutrient constraints

Setbacks may arise in the strawberry culture, because of problems with the growing medium being soil or soilless, lack of regularly monitoring of the field situation, climate, pest and insects, etc. If plants show symptoms of nutrient deficiencies, the problems have started long before these are visible. Sometimes, there are no clear deficiency symptoms, but the yield or the fruit size or quality does not reach their potentials. This could be because of a climatic situation, such as below optimal temperature for flower development in a biannual growing system, too high temperatures during fruit development that stresses the plants resulting in small fruits, and sometimes discoloring. However, it could also be that the rhizosphere is not well developed or that plants in spite of enough mineral nutrients in the soil have reduced availability of the nutrients. Too high or too low pH will also influence availability of mineral nutrients. Therefore, it is important to keep pH within an optimal range. Soilless production has increased dramatically in the last 25 years. However, in England (as in the rest of Europe), the essentials of the systems implemented have not changed much since a booklet on protected cultivation by Dennis Wilson was published in 1998 (Wilson, 1998). In the beginning, crops were mostly soil grown with moving polyethylene tunnels between fields, but because of constraints caused by soilborne diseases, more than 50% of the English strawberry production was soilless in 2015 (Moore, 2015). The situation is much the same in the rest of Europe and in varying degree in the rest of the world.

There have been reports of direct or indirect effects on fruit yield and quality for all mineral nutrients. However, the optimal values change between cultivars, and new cultivars are released every year, so there should be continuous adjustments placing new cultivars at least into a known nutritional group of cultivars.

3.3 Diagnosis of nutrient constraints

3.3.1 Nitrogen (N)

Nitrogen is absorbed in roots as NO_3^- (nitrate) and NH_4^+ (ammonium) and is an essential component of protoplasm and enzymes and is found as NO_3^- in vacuoles. During the transition from the vegetative to the reproductive phases and during the mobilization of storage proteins for new shoots of perennial plants, organic N compounds are transferred in large quantities. In leaves, growing shoots, and ripening fruits, the transport of N as organic compounds provides amino groups for the synthesis of amino acids and for transamination and serves as building blocks in protein synthesis and cell growth. Ammonium is metabolized in the roots, where it reacts with sugars. On the other hand, nitrate transported to the leaves is reduced to ammonium and then reacts with sugars. At high temperature, the plant's respiration increases, making the sugars less available for ammonium metabolism in the roots. The practical conclusion is that at higher temperatures applying a lower ammonium/nitrate ratio is advisable (Larcher, 2003; SFM, 2018). To avoid N deficiency, there are standards for appropriate level of N and other mineral nutrients in strawberry plants. For macroelements, these levels are expressed in the percentage of dry matter (DM) (and in parts per million for micronutrients) of young mature leaves as indication of the situation of the whole plant. Recommendations throughout the world vary; for example, Haifa (2018a), situated in Israel, recommends an optimal N range of 1.9%–2.8%, a little lower than the 2.5%–3.5% recommended in Australia (Lawrence, 2010), while a general recommendation in Norway is 1.8%–2.2% (Yara, 2018). The differences are probably a result of the nature of leaf sampling. Haifa (2018a) and Lawrence (2010) recommend leaf sampling preferably when the plants are actively growing at early harvest, while Yara based their recommendation on sampling after harvest when the growth is slowing down (Yara, 2018). The drop in N content of the leaves from start to end of harvest was confirmed by Bottoms et al. (2013). In addition, the optimal level may be different between cultivars, and cultivars that grow in the warm climate of Israel and Australia are different from those grown in the colder climate of Norway.

Fertilization affects root growth such that low availability of N improves root growth but restrains shoot growth (Vamerali et al., 2003; Arevalo et al., 2005), while high N supply enhances shoot growth more than root growth. A similar effect was achieved in strawberry seedlings using pure organic fertilizers, but the effect was stronger by adding a combined organic/inorganic fertilizer. Simultaneously, the concentrations of indole-3-acetic acid (IAA) and abscisic acid (ABA) in the roots decreased when seedlings were fertilized from the initial to the late growth phase, while isopentenyl adenosine (iPA) levels increased at all growth stages. This suggests that the concentrations of endogenous phytohormones in strawberry plants could be responsible for the morphological changes of roots due to fertilization (Wang et al., 2009). Adding N in strawberry at 58.8 kg N/ha in the spring of the fruiting year, in addition to N supply from the soil, resulted in slightly less firm fruits than in unfertilized strawberries (Shoemaker and Greve, 1930). This effect was confirmed by some researchers (Overholser and Claypool, 1932; Miner et al., 1997), but not by Darrow (1931) who showed that fruits receiving extra N were firmer than those in plants receiving no extra N and that plants with low or moderate growth tended to have firmer fruits. Others found no influence of N on fruit firmness (Cochran and Webster, 1931; Haut et al., 1935; Bell and Downes, 1961). The divergence in these early results concerning fruit firmness is probably due to the variable amount of readily available N in the soil and that N leaching was not considered. It is therefore difficult from these experiments to establish a maximum limit of added N in fertilizers for plants growing in soil for avoiding soft fruits. However, as mentioned earlier, short-day strawberry (plant + fruit), depending on cultivar, contains ±61 kg N/ha, important to keep in mind when managing fertilization of strawberries.

Nitrogen also has effects on the number of flowers initiated, fruit size and malformation. Adding 100 kg N/ha increased the number of flowers in the first inflorescence; however, fruit malformation and greater number of small fruits (<2.0 g) occurred, resulting in reduced yield (Yoshida et al., 1991; Kopanski and Kaweci, 1994). Others confirmed the effect of N on fruit malformation (Kirsch, 1959; Yoshida et al., 1991; Yoshida, 1992). However, fruit malformation was reduced by applying N after sepal differentiation (Yoshida, 1992). This amount of fertilizer, in short-day strawberry growing in a well-balanced soil, is usually more than necessary to give a good strawberry development and will rather reduce than improve yield and fruit quality. However, the results often vary by soil type and field age. Nestby (1998) showed that there were influences on yield parameters in a nutrient-rich silt loam supplied with a basal fertilizer at weekly intervals in the summer season of the planting year and from flowering to end of harvest in harvesting years. The yield increased only in the second fruiting year giving 124 kg N/ha extra in both fruiting years, but fruit size decreased. On the other hand, in a nutrient-poor sandy loam, fruit yield strongly increased in both years giving extra 62 kg N/ha, with a small significant increase above this by doubling the amount of fertilizer, but fruit size decreased. The percentage of malformed fruits increased in the first fruiting year using extra fertilization, but in the second fruiting year, malformation decreased. A level of fertilizer estimated according to Albregts and Howard (1978) was used by Gariglio et al. (2000) applying 53 kg N/ha, which increased strawberry yield, a response due to increased fruit number, but not fruit weight, while fruit weight increased in everbearing strawberries by adding 40 kg N/ha (Burgess, 1997).

3.3.2 Potassium (K)

Plants adsorb K as K^+, which has good transportability in plants. It regulates hydration and is synergistic to NH_4^+ and Na^+ and antagonistic to Ca^{2+}. Besides, it has electrochemical effect (membrane potential and osmoregulation) and activates enzymes in photosynthesis (Larcher, 2003; Raviv and Lieth, 2008). The recommendations of contents of K in leaf samples vary; however, for example, in Israel, Haifa (2018a) recommend an optional K range of 1.6%–2.5% similar to Lawrence (2010) in Australia, while Yara (2018) recommend 1.2%–1.8% for Norwegian growers. The same comments as for N are valid here. According to Haifa (2018a), the plant is deficient at 1.2% K in leaf DM, which is equal to the minimum of the optimum range recommended in Norway. Besides deficiency symptoms in leaves shown as discoloration, fruits from deficient plants failed to develop full color; their texture was pulpy and insipid in taste (Ulrich et al., 1980; Raviv and Lieth, 2008). In early stages of K deficiency, dead calyces appeared, and in advanced stages of deficiency, wilting and drying up of pedicles and peduncles resulted in shriveling of fruits (Lineberry and Burkhart, 1943). Too much K in the soil can restrict the uptake of Ca and Mg ions and may create physiological disorders such as tip burn of emerging leaves and flowers. Main causes of tip burn disorders is high K, low Ca, high salinity, and cation imbalance; K deficiency symptoms could be seen on susceptible cultivars and under unfavorable environmental conditions (Kaya et al., 2002; Trejo-Téllez and Gómez-Merinoand, 2012; Nestby et al., 2005; Raviv and Lieth, 2008).

Because of the limited substrate volume in soilless cultivation, an unbalanced nutrient solution has larger influence on the crop compared with soil cultivation (Adak, 2009). For optimum crop performance, the electrical conductivity (EC) of the nutrient solution was more important than the $K^+:Ca^+:Mg_2^+$ ratio (Neocleous and Savvas, 2013). In regard to interactions between cations, the ratio K:Ca:Mg (Neocleous and Savvas, 2013), the levels of K (Seyedi et al., 2014), and the pH (Kim et al., 2005) are particularly important in terms of yield and quality in various horticultural crops. Likewise, high levels of K^+ and high EC in the nutrient solution increased fruit dry matter, total soluble solid content (SSC), and lycopene concentration of tomato (Fanasca et al., 2006). A short application of high EC during ripening may be of practical interest for enhancing lycopene in fruit without affecting overall growth or yield (Wu and Kubota, 2008). Trejo-Téllez and Gómez-Merinoand (2012) reported that strawberry plants with a good supply of K can synthetize more sugar and thereby develop sweeter fruits. In contradiction, Pivot and Gilliéoz (2001) indicated that excessive K in a soilless closed system reduced fruit quality due to lower sugar content.

3.3.3 Calcium (Ca)

Ca enters the roots as Ca_2^+ and has very poor transportability in plants, where it is transported as ion and chelate, and binds in the tissue as pectates. It regulates hydration, activates amylase and ATPase, and regulates elongation growth. Further, it is a signaling substance via calmodulin and in the form of Ca pectate is responsible for stabilizing the cell walls of plants (Larcher, 2003; Raviv and Lieth, 2008). The recommendations of Ca content in leaf samples vary; for example, Haifa (2018a) recommend an optimal Ca range of 0.7%–1.7%; the range in Australia is much narrower, 1.0%–1.2% (Lawrence 2010), and the Norwegian recommendation is 1.0%–1.5% (Yara, 2018). The differences are smaller than for N and K. When Ca is deficient, new tissues such as root tips, young leaves, and shoot tips often exhibit distorted growth from improper cell wall formation. In addition, Ca^{2+} shows antagonism with K^+ and Mg^{2+}.

Calcium moves from the roots to the rest of the plant via evapotranspiration by the water-conducting elements of the plant (i.e., xylem). If the evapotranspiration is high, such as on a hot dry day, calcium will move from the roots and up through the plant. Conversely, lengthy spells of cool, humid weather will reduce evapotranspiration, and subsequently, calcium movement can be restricted. Plant organs such as the fruit and developing leaves do not transpire as much as a fully mature, expanded leaf and therefore would tend to be the first to express Ca deficiency (Larcher, 2003; Bolda, 2010).

3.3.4 Magnesium (Mg)

Mg enters the roots as Mg^{2+} and has partly good transportability. Mg regulates hydration and is antagonistic to Ca^{2+}, is important in basal metabolism (photosynthesis and phosphate transfer), and is synergistic with Mn^{2+} and Zn^{2+} (Larcher, 2003). The recommendations of Mg content in leaf samples vary; for example, Haifa (2018a) recommend an optimal Mg range of 0.30%–0.49 % of DM, the range in Australia is one step up to 0.40%–0.60% of DM (Lawrence, 2010), and the Norwegian recommendation of 0.20%–0.30% is on the lower end of the scale (Yara, 2018).

There are few reports of deficiency symptoms on fruits caused by low Mg, but Mg application increased fruit size in 1 of 3 years for "Tribute" strawberry (Lamarre and Lareau, 1997). According to Ulrich et al. (1980), in Mg-deficient plants, fruit appears nearly normal, except for a lighter red color and a tendency to albinism, and the upper margins of mature blades develop scorching, moving inward.

3.3.5 Phosphorus (P)

P enters the roots as HPO_4^{2-} or $H_2PO_4^-$ and has good transportability in organically bound form. It incorporates in the plant as free ions, in esteric compounds, nucleotides, phosphatides, and phytin and functions in plants in basal metabolism and syntheses (phosphorylation). The recommendations of P content in leaf samples vary; for example, Haifa (2018a) recommend an optimal P range of 0.25%–0.40% of DM, the range in Australia is not much different (0.30%–0.50% of DM) (Lawrence, 2010), and the Norwegian recommendation is 0.20%–0.30% of DM, a little lower than for the other two.

The effect of P on fruit quality has received little attention. However, Haut et al. (1935) found that P had no effect on fruit firmness, but flowers and fruits of P-deficient plants tended to be smaller than normal. Fruits of susceptible cultivars occasionally developed albinism (Ulrich et al., 1980) and had a lower SSC (Valentinuzzi et al., 2015). There was positive correlation between P and SSC in 24 strawberry cvs, and the concentration of P and SSC both increased from the lower part toward the tip of the fruit. After irrigating with 6.0-mM phosphoric acid, the SSC increased, and a KH_2PO_4 solution was the most effective (Zhang et al., 2017).

3.3.6 Boron (B)

B is a microelement that is noticed here since it is an essential element for vascular plants and influences fruit set and quality. It enters the roots as HBO_3^{2-} and $H_2BO_3^-$ and plays a role in carbohydrate transport and metabolism, in phenol metabolism, and in activation of growth regulators (Shkolnik, 1974; Larcher, 2003). The recommendations of B content in leaf samples vary; for example, Haifa (2018a) recommend optimal B range of 30–64 ppm, in Australia 30–50 ppm (Lawrence, 2010), and in Norway 20–40 ppm (Yara, 2018).

Several authors have reported influence of B on pollen germination and fruit set (Visser, 1955; Vasil, 1964; Guttridge and Turnbull, 1975; Ulrich et al., 1980). Since many years, blasting of flowers and distorted fruits and a tendency to produce fascinated fruits (Johanson, 1963; Willis, 1945) were reported. Vitamin C content of "Redcoat" strawberry grown on light soil deficient in B increased linearly with B application rates of 0–8 kg/ha and sugar for rates up to 4 kg/ha (Cheng, 1994). Later, Lieten (2000) showed that a nutrient solution without boron produced a strongly reduced number of fruits and a high percentage of malformed fruit.

3.4 Management of nutrient constraints

Growing conditions of strawberries in open field and in protected culture vary because of differences in growth medium and climate. In addition, pests and diseases may influence the crop. Strawberry plants have a shallow root system and are sensitive to water stress and salinity (Sauvageau et al., 2017). Earlier, we have dealt with the influence of macronutrients and boron on crop yield and fruit quality and focused on what the plant needs to grow optionally. The focus here will be to discuss management to achieve optimal plant growth and fruit quality when growing in soil and to some extent in soilless growing. However, this is an extensive topic, and only general recommendations are given. Growers should search for local information to manage strawberry fields (e.g., Haifa, 2018a,b; Queensland, 2008; Strik, 2013; NCDA, 2015. Bolda et al., 2010, local extension reports).

The level of organic matter in soil is important, and to bring the content to an optimal level should start early and be continuous. Soils with higher levels of fine silt and clay usually have higher levels of organic matter than those with a sandier texture. There are some general guidelines. For example, 2% organic matter in a sandy soil is very good and difficult to reach, while a soil with 50% clay would need 6% organic matter to reach an aggregation level similar to a soil with 16% clay and 2% organic matter (SARE, 2012). To maintain a good level of organic matter, green manure crops (cereals, millet, legumes, clover, etc.) in between strawberry crops are recommended (Queensland, 2008), and higher soil activity was shown in intercropping with beans and clover (Dane et al., 2016). Other amendments such as vermicompost, biochar, N_2-fixing bacterium (*Gluconacetobacter diazotrophicus*), seaweed extract (*Ascophyllum nodosum*), other microinoculants, and fermented food waste had positive influence on growth conditions of the soil (Chalanska et al., 2016; Yadav et al., 2016; Delaporte-Quintana et al., 2017; Hou et al., 2017; de Tender et al., 2016; Patil et al., 2016; Holden and Ross, 2017). In addition, pH has to be within the optimal range (±6.5). For strawberry, the general recommendation is to use lime or dolomite (if soil Mg is low) at least 6 months before planting (Lawrence, 2010). Soil analyses will give a good indication of the demand of preplanting fertilizer to add, at least 4–5 months before planting or earlier if there is a long period of cold winter and frozen soil in between. The soil should be prepared before planting to obtain a good structure.

Adding fertilizer and lime/dolomite should be compatible with modern methods, such as broadcasting using a tractor with Global Positioning System (GPS) before laying the plastic mulch. Together with information of variation in mineral nutrients and pH, the broadcasting equipment can deliver fertilizer or lime/dolomite in varying amount depending on soil analysis. The use of slow release fertilizers is an option, and they are best applied during bed formation in a band 10–15 cm below the surface. The benefit of slow release fertilizers is that they supply nutrients to the plants at a regular rate over 3–9 months, giving more even plant growth and reducing the risk of fertilizer burn of the roots (Lawrence, 2010). When the field is established, drones carrying modern photo equipment can be used to map the nutritional situation in the field, and again, GPS on a tractor (or a robot) can be used to add a balanced supplementary fertilization that may be necessary on a low fertile soil (Bossin, 2016). In open field, there is one more problem, and that is water availability. This will vary a lot between regions and states and time. In some places, the average rainfall can cover the need of the plants. However, the rain does not necessarily fall when it is needed, and when it falls, it may be too little or too much. Too high rainfall will increase nutrient leaching, and a fertilization plan has to be adapted, preferably after analyses of the runoff. To escape the effect of too much rain, especially during fruit growth and ripening, causing leaching and the infection of fruit rot (*Botrytis cinerea*), the field could be protected with polyethylene tunnels or rain roofs. This is frequently used, for example, in Huelva (Andalucia, Spain) where high tunnels are dominating (Guéry et al., 2018). This creates a dessert-like condition under the shelter, and watering becomes necessary, but now, it can be completely controlled preferably using drip irrigation. Again, modern techniques using sensors for analyzing volume water content (VWC) and EC are valuable tools. These sensors can connect to an automated sensor-based control that starts and stops watering depending on sensor readings (Guéry et al., 2018). The drip water is an excellent carrier of fertilizers. Combined with EC measurements and drainage lysimeters, the use of water and fertilizer can be programed to fit the actual needs of the plants (Nestby and Guery, 2018; Guéry et al., 2018; Garcia-Tejero et al., 2018). Moving strawberry into soilless culture has been more common. However, comparing soil-grown strawberries with soilless production showed that SSC, glucose, fructose, ascorbic acid, tocopherol, and total polyphenolic compounds were significantly higher in soil-grown strawberries. This should be a topic in future research (Treftz and Omaye, 2015).

4 Highbush blueberry (*Vaccinium* sp.)

4.1 Accumulation of minerals in the plant

The information provided in this review refers mainly to high chill or northern highbush (NHB) and low chill or southern highbush (SHB) blueberries. The nutrient demand of blueberries is low, compared with fruit trees (Table 40.1). However, in most situations, regular fertilizer applications are usually necessary for commercial fields (Hanson and Hancock, 1996; Krewer and NeSmith, 1999). There are various conditions, in both the plant and the soil, that explain the low nutritional requirements of blueberries compared with other fruit crops. Blueberries are said to be calcifuge plants, which means they are adapted to acidic soil conditions. Best growth and productivity are obtained when blueberries grow in soils with a pH in the range of 4.2–5.5. At this pH, the availability of most soil nutrients is limited, and this reduces the amount of mineral elements that are absorbed by the plant (Korcak, 1989; Hanson and Hancock, 1996). Blueberry roots are shallow and devoid of hairs (which limits the surface area in contact with the soil or substrate), and in natural habitats, they are colonized by a specialized type of fungus called ericoid mycorrhizae (ErM). Studies on NHB blueberries showed that increasing fertilization rates decreased ErM colonization of "Duke" but had little influence on colonization of "Reka" (Golldack et al., 2001). This type of cultivar-specific response in sensitivity to ErM colonization by nutrient availability may be responsible for some of the differences in the frequency and intensity of colonization detected among different highbush blueberry cultivars (Scagel, 2005). In addition, the fine root system of blueberries demands a loose soil, which makes sandy loams high in organic matter preferable for their cultivation.

4.2 Causes of nutrient constraints

To obtain optimum yields, plants must have sufficient nutrient levels during active growth. Nutrient imbalances will affect fruit yield and quality. The degree of the effect will depend on the magnitude, opportunity, and duration of the deviation of nutrient levels from the optimum (Marschner, 1986). The high yield and quality required in commercial plantings demand constant field monitoring to satisfy the nutrient requirements to avoid nutrient deficiencies or excesses (Hart et al., 2006).

Weather (temperatures, wind, and rainfall), fruit load, shoot growth, soil moisture, pruning intensity and timing, yield, and insect and disease load can affect plant functioning and the nutrient status of the plant (Stiles and Reid, 1991). In a 2-year study on the seasonal evolution of nutrients in NHB blueberries ("Aurora," "Bluecrop," "Draper," "Duke," "Legacy," and "Liberty"), Strik and Vance (2015) found that the pattern of nutrient changes was similar between organic and conventional sites, but they had fewer differences in nutrient concentrations among cultivars at the organic site. In addition, the cultivar had a significant effect on all fruit nutrients except for P at the conventional site.

Nitrogen is usually the mineral nutrient most frequently applied to blueberries (Hanson and Hancock, 1996). Fertilizer rates can be lower in soils high in organic matter since they have a higher N supply (Eck et al., 1990). Ca is another important nutrient because of its impacts on fruit quality. The goal of fertilization is to remove limitations to yield and quality by supplying the blueberry crop with ample nutrition in advance of demand. Fertilizer applications should be based on soil and plant analysis, information on environmental conditions, plant performance and management, and grower's experience. A fertilizer application should produce a measurable change in plant growth, plant performance, and/or nutrient status. Results from nutrient applications can vary from year to year and from field to field (Retamales and Hancock, 2018). Fertilizers are only one part of a complete management package. If some parts of the blueberry-growing system are not working properly, extra fertilization is no solution (Hart et al., 2006).

Plants interact with the environment (nutrients, light, water, and biotic factors) to generate growth. The amount of growth and the balance between reproductive and vegetative growth determine yield. Adequate nutrition is based on the soil/plant interaction (Marschner, 1986). Plant growth requires satisfaction of certain biological, physical, and chemical conditions by the soil (Retamales and Hancock, 2018).

To establish the nutritional status and pH soil analysis is important before planting. Once fields are planted, nutrient management should be based on leaf analysis and soil pH monitoring. To be useful, leaf sampling must follow strict procedures. Estimation of fertilizer needs is based on nutrient demand/supply and fertilizer use efficiency. Fertigation has expanded due to higher efficiency and ease of use. Organic nutrient management is also expanding for blueberries. Whatever the method, efficiency increases with greater number of applications and as the rate of each is reduced. Except for some micronutrients and on specific occasions, foliar feeding is usually inefficient and costly (Retamales and Hancock, 2018).

TABLE 40.1 Sufficient or normal foliar concentrations of macronutrients (%) and micronutrients (ppm) for northern highbush (NHB) blueberries (Hanson and Hancock, 1996), and apple (Hanson, 1998; Stiles and Reid, 1991) in Michigan, USA.

Nutrient	NHB blueberry	Apple
Macronutrients (%)		
N	1.70–2.10	2.00–2.16
P	0.08–0.40	0.16–0.30
K	0.40–0.65	1.30–1.50
Ca	0.30–0.80	1.10–1.60
Mg	0.15–0.30	0.30–0.50
S	0.12–0.20	–
Micronutrients (ppm)		
B	25–70	25–50
Cu	5–20	10–20
Fe		150–250
Mn		50–80
Zn		20–40

Data on dry-weight basis.

4.3 Diagnosis of nutrient constraints

4.3.1 Nitrogen (N)

Soil N is in constant flux, moving in the soil profile and having different chemical forms (Subbarao et al., 2006). The N cycle is mediated by microorganisms, whose activity depends on chemical and physical soil conditions. Some processes of the N cycle increase N availability for the plants (nitrification and mineralization), while others have the opposite effect (immobilization, denitrification, volatilization, and leaching).

Plants deficient in N are usually stunted, initiate fewer canes, and have low vigor and pale green to chlorotic (yellow) leaves. The chlorosis is uniform across the leaf. Symptoms appear first on older leaves and will eventually include the entire plant if no N is applied. Leaves drop early, and yields are usually reduced. Excessive N results in plants with numerous, vigorous shoots and large, dark green leaves. Growth occurring at the end of the season may not harden properly before winter. The tips of these shoots often suffer freezing injury by low winter temperatures. Plants with excessive N have lower yields and smaller berries that ripen later (Hanson and Hancock, 1996; Hart et al., 2006).

Several reports confirm the preference of blueberries for NO_3^- or NH_4^+. They concluded that differences in response to form of N may be due to variability in rhizosphere pH. Absorption of NO_3^- is accompanied by net release of excess OH^-, which will raise the rhizosphere pH, while NH_4^+ uptake requires net release of H^+ with a concomitant drop in rhizosphere pH (Merhaut and Darnell, 1995).

The recommended N rates vary greatly among various producing regions. While 73 kg N/ha is suggested for fields older than 7 years in Michigan (Hanson and Hancock, 1996), 185 kg N/ha is advised in Oregon (Hart et al., 2006). For NHB blueberries grown under mulch, rates of 158–170 kg N/ha in year 1 and 238–257 kg N/ha for later years have been estimated in Oregon. However, in recent trials in Oregon comparing N sources (urea vs $(NH_4)_2SO_4$) and mode of application (granular vs fertigation), the highest yields were obtained in plants fertigated with 63–93 kg N/ha/year (Vargas and Bryla, 2015).

4.3.2 Potassium (K)

Leaf K is rarely low in blueberries, except on sandy soils. As with other fruit crops, the first deficiency symptom is chlorosis of older leaves margins. Greater K deficit can cause scorching of leaf margins, cupping, curling, and necrotic spots and dieback of shoot tips (Hart et al., 2006). Low leaf K can be due to lower root function, flooding, poor drainage, high N, drought, and very acid soils (Stiles and Reid, 1991). Root growth is important for K nutrition. In clayey or compacted soils, root growth will be reduced; then, even though soil nutrient analysis can show high K, leaf K could be low (Shaw, 2008). Excessive K (leaf K > 0.9%) can generate Mg and Ca deficiencies (Stiles and Reid, 1991).

K availability for crops depends on soil exchange dynamics. There is an active K fraction for immediate uptake and another long-term passive fraction. The passive fraction does not supply K during a growth cycle. The active fraction is composed of K in the soil solution (0.1%–0.2% of total soil K), exchangeable K and nonexchangeable K (occluded within phyllosilicate clays). If Na^+ and NH_4^+ levels are high, K absorption is reduced. In contrast, optimum soil K levels improve uptake of Cu, Mn, and Zn. Release of exchangeable K is often slower than plant uptake, and thus, the K supply in some soils does not always satisfy crop K demand. Soil K application efficiency is 40%–60% and depends on the fertilizer form and dose, as well as the crop's absorption capacity (Guerrero-Polanco et al., 2017).

As fruit is an important sink for K in the plant, fruit load greatly influences leaf K levels. Fruit K increases strongly as fruits mature, approaching 60 mg per berry for ripe fruit (Hart et al., 2006). At harvest, 10 t of fruit removed 6.5, 7, and 8 kg K/ha in "Brigitta," "O'Neal," and "Duke" NHB blueberries, respectively (Hirzel, 2014). When leaf K is deficient, yields have been increased with K fertilization on various soil types. In Oregon, K fertilizers are not recommended if soil levels are >150 ppm and leaf K is >0.40%. If soil test readings are 101–150 ppm and tissue K is 0.21%–0.40%, up to 84 kg K_2O/ha is suggested for fertilization. Application of 84–112 kg K_2O/ha is advised when soil levels are 0–100 ppm and leaf K is <0.2% (Hart et al., 2006). Similar rates are recommended in Michigan, but it is suggested that if crop load is high, 0.35–0.40 leaf K would be adequate (Hanson and Hancock, 1996).

4.3.3 Calcium (Ca)

Blueberries are calcifuges that thrive in low pH, are efficient in Ca^{2+} uptake, and have low Ca requirements relative to other temperate fruit crops. Healthy bushes typically have 0.3%–0.8% Ca in leaves (Eck, 1988) compared with 1%–3% in temperate tree crops (Shear and Faust, 1980). Blueberries are seldom deficient in leaf Ca (Hanson and Hancock, 1996; Hart et al., 2006); however, even when leaf levels indicate adequate Ca supply to plants (Hanson et al., 1993), Ca can affect several fruit quality characteristics (fruit texture, firmness, and ripening rate). Fruit deficient in Ca can occur due to insufficient mobilization of Ca from internal stores, dilution due to tissue growth, or reduced Ca supply through the xylem (often due to low transpiration rates; Hocking et al., 2016). A high K and N supply, as well as wide fluctuations in moisture during the season, can reduce even further Ca supply to the fruit (Hirschi, 2004). Ca deficiency symptoms are most common in plants growing in lower pH soils, which also tend to have low Ca levels. Low leaf Ca levels can also occur in heavily fertilized, vigorously growing plants (Stiles and Reid, 1991).

Ca serves various roles within plant cells, including structural, defense, and communication among tissues. The Ca levels needed for each role are quite different; 10^{-4} M Ca is required for the structural role and 10^{-7} M for communication functions (a 1000-fold gradient). When Ca uptake exceeds plant needs, some Ca is sequestered to form Ca oxalates, which also helps in defense and detoxification caused by heavy metals (Franceschi and Nakata, 2005). This can in part explain the weak relationships usually found between Ca applications and changes in Ca-related processes such as firmness and decay prevention (Retamales and Hancock, 2018).

Ca is preferably absorbed by young roots. Soil NH^{4+}, K, and Mg interfere with Ca absorption by roots; hence, high levels of these nutrients will reduce fruit Ca levels. The fruit tends to accumulate most of its Ca in their early stages of development and when shoot growth is limited (White and Broadley, 2003). As Ca is a phloem-immobile nutrient, the fruit relies mainly on transpirational water flow for its accumulation. Fruit transpiration rate is highest at fruit set (e.g., in kiwifruit, it can be as high as 2.3 mmol/m^2/s, but this quickly declines to as little as 10% of this value later in development), whereas leaf transpiration remains at more than 10 mmol/m^2/s (Montanaro et al., 2015). As blueberry fruit grows near harvest, Ca levels decrease by dilution (Retamales and Hancock, 2018).

There are contradictory results reported from trials with Ca sprays on blueberries during the season; thus, while, some researchers reported positive effects on fruit quality, (Stückrath et al., 2008) others (Hanson, 1995; Vance et al., 2017) found no effect.

4.3.4 Magnesium (Mg)

Mg deficiencies have been reported in fields of many blueberry-growing areas (Eck et al., 1990). They occur occasionally in Georgia (Krewer and NeSmith, 1999) and periodically in Michigan (Hanson and Hancock, 1996). The high variation in Mg content among source materials reflects on the total soil Mg contents (0.05%–0.5%). Differences in soil silicate content also explain the higher Mg contents typically found in clay and silty soils than in sandy soils. Mg availability to plants depends on the distribution and chemical properties of the source rock material and its degree of weathering, site-specific climatic factors, and, to a high degree, the management practices of a specific field, including the cultivar used and the organic/mineral fertilization practices (Gransee and Führs, 2013).

Lower Mg levels are common in low-pH fields. Mg deficiencies are most frequent in rapid growing plants or in those with heavier fruit loads. Although the leaf deficiency level would occur <0.1%, Mg deficiency has been found in bushes with 0.2% (Hanson and Hancock, 1996). High Ca and/or K reduce Mg absorption and may indicate a need for Mg application. Desirable ranges of the percentage of bases (as proportion of the CEC) in soil samples are 60%–80% Ca, 15%–30% Mg, and 10%–15% K. Mg should be applied if Mg is <4% of the bases or if K exceeds Mg as a percentage of the bases (Hanson and Hancock, 1996). For fruit crops in general, a ratio in leaf samples of the percentages of K:Mg greater than or equal to 4:1 usually indicates that the Mg supply is inadequate (Stiles and Reid, 1991).

If foliar levels indicate a deficiency, soil pH should help decide which fertilizer to use. If soil pH is >4.5, magnesium sulfate ($MgSO_4$, Epsom salts) or Sul-Po-Mag (21%–24% K_2O, 21% S, and 10%–18% magnesium oxide (MgO)) should be chosen. Epsom salts can be fertigated. However, if the pH <4.5, dolomitic lime at 1 t/ha should be used. All soil applications should be in autumn (Hart et al., 2006).

4.3.5 Phosphorous (P)

As with other fruit crops, P deficiencies are infrequent (Stiles and Reid, 1991). However, P supply to the roots of perennial crops is particularly constrained in acid, calcareous/alkaline, and old, highly weathered soils (Plassard and Dell, 2010). P is by far the least mobile and least available nutrient to plants in most soil conditions. Most soil P is bound tightly to the surface of soil particles or tied up as organic P compounds and thus mostly unavailable for plant uptake (Kochian, 2012). In addition, organic materials in the soil (e.g., from manure, organic mulch, or crop debris) can bind phosphate, in particular phytate (inositol compounds). It is estimated that plants use 10%–25% of applied inorganic P.

The effects of excessive P are unusual in blueberry fields (Stiles and Reid, 1991; Krewer and NeSmith, 1999). Leaf P levels are highest early in the season and lowest at harvest. Tissue levels are little affected by crop load and moisture status. Threshold foliar P levels to establish deficiency vary. They are defined as 0.07% in Michigan and 0.09% in Wisconsin and Minnesota (Hart et al., 2006).

In Oregon, P applications are only recommended if soil test (Bray) readings are <50 ppm and leaf P is <0.10%. At 26–50 ppm, soil P and 0.08%–0.10% leaf P up to 45 kg P_2O_5/ha are suggested. When soil P is <25 ppm and leaf P is <0.07%, 45–67 kg/ha of P_2O_5 is suggested (Hart et al., 2006). In Michigan, although recommendations are based on soil P levels, they suggest to apply P only when foliar P levels are <0.08% (Hanson and Hancock, 1996).

4.3.6 Boron (B)

B deficiency is one of the most common plant micronutrient deficiencies worldwide (Gürel and Başar, 2016). Low B is particularly prevalent in light-textured soils, where water-soluble B readily leaches and becomes unavailable to the plants. Deficiency is aggravated with dry weather and heavy crop load. Low plant B may accentuate deficiencies of other nutrients because of impaired root function (Ganie et al., 2013). In fruit crops in general, low B is often associated with Ca deficiency problems (Stiles and Reid, 1991). Its incidence varies across the blueberry-producing regions: while common in Oregon (Hart et al., 2006), it has not been found in Michigan (Hanson and Hancock, 1996).

If B is deficient, 11–22 kg borax/ha (11% B) applied in the autumn or early spring prior to rain is beneficial. Alternatively, 0.9–2.7 kg Solubor (20% B) in 950 L water/ha can be sprayed before bloom or after harvest and before leaf senescence. An annual application of 560 g B/ha has been suggested (Hart et al., 2006). Foliar and soil B treatments (four applications of 0.2 kg B/ha between early bloom and 6-week post bloom) to "Bluecrop" NHB blueberries increased leaf and flower B levels, as well as fruit SSC, but had no effect on plant vigor, number of flowers per cane, fruit set, or yield (Wojcik, 2005). In contrast, a 4-year study in Missouri (United States) on the effect of autumn and spring foliar B sprays to "Blueray" and "Collins" NHB blueberries increased yield by an average of 10% for all seasons, mostly due to greater fruit number per plant. B sprays also reduced tip dieback symptoms (Blevins et al., 1998). Foliar B levels need frequent monitoring, as toxic B levels can rapidly appear (Hanson and Hancock, 1996).

4.4 Management of nutrient constraints

The soil pH range recommended for highbush blueberries is 4.5–5.5. Soil pH influences nutrient availability for plants. Blueberries grown in high soil pH have yellow leaves with green veins. These leaves are small and often turn brown and fall from the plant before the season is over. Little growth occurs, and some plants may die. Plants stunted by high soil pH usually do not recover, even if soil pH is reduced (Hart et al., 2006). Plants established in high-pH soils often require replanting. Fe, Mn, or Cu deficiencies are common in high-pH soils; thus, instead of applying these elements to the soil, correcting the pH will usually be more effective. Soils are acidified either with elemental S incorporated before planting or with sulfuric acid (H_2SO_4) applied through the irrigation system. When soil S is applied, its conversion into acid is mediated by microorganisms, which require time, moisture, and warm temperatures. Thus, soil pH should be corrected at least 1 year before planting. Two variables affect the amount of elemental S needed to drop soil pH: (1) the initial soil pH and (2) soil CEC. The higher the difference between the initial and the desired pH and the higher the CEC (or the soil buffer capacity), the more elemental S will be required to adjust soil pH. If more than 3.4t/ha of S is necessary, the dosage should be split. Elemental S has to be thoroughly blended and incorporated in the top 20 cm of the soil (Retamales and Hancock, 2018). The increasing popularity of blueberries all across the world has forced growing the crop outside its natural habit (acid soils, high organic matter, and loose soil) in different soil conditions. Besides modifying soil acidity, the maintenance of moisture near the soil surface is critical because of the shallow rooting of blueberries. The most successful and widespread practice to enhance soil moisture and reduce weed infestation has been the use of mulch (Retamales and Hancock, 2018). A mulch is defined as any covering applied to the soil surface. This broad definition includes crop residues, weeds and other plant material, and artificial materials such as paper and plastic (Kumar et al., 2013).

Mulches are common in blueberry production, and it is a prevalent approach to weed management in organic blueberry production (DeVetter et al., 2015). An Oregon survey showed that 58% of blueberry growers used some type of mulch (Scagel and Yang, 2005). Mulches can improve crop growth and productivity because they mitigate soil temperature fluctuations and reduce water loss from the soil surface (Clark and Moore, 1991; Burkhard et al., 2009; Cox, 2009). Within the soil profile, roots will grow primarily where organic matter is present. Mulching has improved root growth in the upper soil layer (Scagel and Yang, 2005), and a large part of the root system is developed at the soil-mulch interface (Cox et al., 2014).

Mulches from plant-derived materials supply organic matter and provide plant mineral nutrients (Clark and Moore, 1991; Himelrick et al., 1995). Sawdust or bark derived from Douglas fir is common for mulching throughout Washington, United States. Generally, compost is not recommendable for blueberry production, as they typically have high pH, EC, and K content, all of which are undesirable for blueberry growth and development (Sullivan et al., 2014; DeVetter et al., 2015).

Plastic mulches are used in many blueberry-growing regions. Most benefits from plastic mulches occur in the first years after planting, as this is the period when competition for water, light, and nutrients is strongest. Depending on the type of mulch, they can last from 2 to 7 years, with weed mats (black landscape fabric made from woven polypropylene or polyethylene) having the longest active life (Cox, 2009). In blueberries, plastic mulches should be combined with fertigation, as the fertilizer placed under the plastic often depletes after 1–2 years (Williamson et al., 2006). Plastic mulching reduces fluctuations of soil moisture and temperature in the top 25 cm where most blueberry roots grow (Kader et al., 2017). On heavy, wet, clay soils, plastic mulches restrict soil microbial activity, leading to anaerobic soil conditions (Retamales and Hancock, 2018).

Plastic mulching provides mechanical protection of the top soil, enhances root growth, stabilizes soil aggregates, increases mucilage production, and promotes soil fauna activity. The accelerated soil processes under plastic mulch can thus alter soil organic matter composition and quality (Steinmetz et al., 2016). Mulching has led to slight increases in total microbial diversity compared with nonmulched soil. The enhanced productivity under plastic mulches has often resulted in lower soil contents of Mg, K, P, and N than bare soils (Steinmetz et al., 2016). However, the effects of soil-mulch on microbial activity and soil nutrient contents would depend on both the type of mulch and the environmental conditions of the research site (Retamales and Hancock, 2018).

References

Strawberry

Adak, N., 2009. Effects of different growing media on the yield and quality of soilless grown strawberries. Akdeniz University, Faculty of Agriculture, Department of Horticulture, Turkey, p. 232. PhD thesis.

Albregts, E.E., Howard, C.M., 1978. Accumulation of nutrients by strawberry plants and fruit grown in annual hill culture. J. Am. Soc. Hortic. Sci. 105 (3), 386–388.

Albregts, E.E., Howard, C.M., 1980. Elemental composition of fresh strawberry fruit. J. Am. Soc. Hortic. Sci. 103 (3), 293–296.

Arevalo, C.B.M., Drew, A.P., Volk, T.A., 2005. The effect of common Dutch white clover (*Trifolium repens* L.) as a green manure, on biomass production, allometric growth and foliar nitrogen of two willow clones. Biomass Bioenergy 29 (1), 22–31.

Bakker, P.A.H.M., Berendsen, R.L., Doornbos, R.F., Wintermans, P.C.A., Pieterse, C.M.J., 2013. The rhizosphere revisited: root microbiomics. Front. Plant Sci. 4, 7. https://doi.org/10.3389/fpls.2013.00165.

Bell, H.K., Downes, J.D., 1961. Some cultural studies with Robinson strawberry in Michigan. Q. Bull. Michigan Agric. Exp. Station 44, 171–174.

Benbrook, C.M., 2017. The dollars and cents of soil health. In: Field, D.J., Morgan, C.L.S., Mcbratney, A.B. (Eds.), Global Soil Security. Progress on Soil Science. Springer, Cham, pp. 219–226.

Bolda, M., Cahn, M., Bottoms, T., Farrara, B., 2010. Strawberry Nutrient Management. Hartz Watsonville Berry, Santa Cruz county, USA. Available from: www.cesantacruz.ucanr.edu/files/73967.pdf.

Borrero, C., Bascón, J., Gallardo, M.A., Orta, M.S., Avilés, M., 2017. New foci of strawberry wilt in Huelva (Spain) and susceptibility of the most commonly used cultivars. Sci. Hortic. 226, 85–90.

Bossin, S., 2016. Drones in agriculture. European drone convention 2016. Available from: http://www.droneconvention.eu/agriculture.html.

Bottoms, T.G., Bolda, M.P., Gaskell, M.L., Hartz, T.K., 2013. Determination of strawberry nutrient optimum ranges through diagnosis and recommendation integrated system analysis. HortTechnology 23 (3), 312–318.

Burgess, C.M., 1997. Nutrition of new everbearing strawberry cultivars. Acta Hortic. (439), 693–700.

Chakraborty, S., Narsimhaiah, L., Kumar, G., Gupta, P.K., Pateel, L.V., Sathish, G., 2017. Empirical study on urease activity and nitrogen fixing capacity in organic and conventional status—a case of Nadia district, India. Environ. Ecol. 34 (3B), 2026–2031.

Chalanska, A., Malusa, E., Bogumil, A., 2016. The dynamic of nematodes community in organic strawberry plantations is affected by the kind of fertilizers and soil conditioners applied. Ecofruit. In: 17th International Conference on Organic Fruit-Growing: Proceedings, 15–17 February, Hohenheim, Germany, pp. 286–288.

Cheng, B.T., 1994. Ameliorating *Fragaria* ssp. and *Rubus idaeus* L. productivity through boron and molybdenum addition. Agrochemica 38 (3), 177–185.

Cochran, G.W., Webster, J.E., 1931. The effect of fertilizers on the handling qualities and chemical analysis of strawberries and tomatoes. Proc. Am. Soc. Hortic. Sci. 28, 236–243.

Dane, S., Sterne, D., Sterne, Z., 2016. Soil fertility in strawberry—legume intercrop. Vecauce 2016: Lauksaimnieci bas zina tne nozares atti bai, Vecauce, Latvia. 3 November 2016, pp. 16–19.

Darrow, G.M., 1931. Effect of fertilizers on firmness and flavor of strawberries in North Carolina. Proc. Am. Soc. Hortic. Sci. 28, 231–238.

Delaporte-Quintana, P., Grillo-Puertas, M., Lovaisa, N.C., Teixeira, K.R., Rapisarda, V.A., Pedraza, R.O., 2017. Contribution of *Gluconacetobacter diazottrophicus* to phosphorus nutrition in strawberry plants. Plant Soil 419 (1/2), 335–347.

Fanasca, S., Colla, G., Maiani, G., Venneria, E., Rouphael, Y.T., Azzini, E., Saccardo, F., 2006. Changes in antioxidant content of tomato fruits in response to cultivar and nutrient solution composition. J. Agric. Food Chem. 54, 4319–4325.

Gariglio, N.F., Pilatti, R.A., Baldi, B.L., 2000. Using nitrogen balance to calculate fertilization in strawberries. HortTechnology 10 (1), 147–150.

Garcia-Tejero, I.F., López-Borrallo, D., Miranda, L., Medina, J.J., Arriaga, J., Muriel-Fernández, J.L., Martínez-Ferri, E., 2018. Estimating strawberry crop coefficients under plastic tunnels in Southern Spain by using drainage lysimeters. Sci. Hortic. 231, 233–240.

Greenwood, D.J., Stone, D., 1998. Prediction and measurement of the decline in the critical-K, the maximum-K and total cation plant concentration during growth of field vegetables crops. Ann. Bot. 82, 871–881.

Guéry, S., Lea-Cox, J.D., Martinez Bastida, M.A., Belayneh, B.E., Ferrer-Alegre, F., 2018. Using sensor-based control to optimize soil moisture availability and minimize leaching in commercial strawberry production in Spain. Acta Hortic. (1197), 71–178.

Guttridge, C.G., Turnbull, J.M., 1975. Improving anther dehiscence and pollen germination in strawberry with boric acid and salts of divalent cations. Hortic. Res. 14, 73–79.

Haifa, 2018a. Nutritional Recommendations for Strawberry. Available from: https://www.haifa-group.com/crop-guide/vegetables/strawberry-fertilizer/crop-guide-strawberry-1.

Haifa, 2018b. Crop Guide: Strawberry Fertilizer Recommendations. Available from: https://www.haifa-group.com/print/1719.

Haut, I.C., Webster, J.E., Cockran, G.W., 1935. The influence of commercial fertilizers upon the firmness and chemical composition of strawberries and tomatoes. Proc. Am. Soc. Hortic. Sci. 33, 405–410.

Holden, D., Ross, R., 2017. Six years of strawberry trials in commercial fields demonstrate that an extract of the brown seaweed *Ascophyllum nodosum* improves yield of strawberries. Acta Hortic. 1156, 249–254.

Hou, J.Q., Li, M.X., Mao, X.H., Hao, Y., Ding, J., Liu, D.M., Xi, B.D., Liu, H.L., 2017. Response of microbial community of organic-matter-impoverished arable soil to long-term application of soil conditioner derived from dynamic rapid fermentation of food waste. PLoS One 12 (4), e0175715. https://doi.org/10.1371/journal.pone.0175715.

Johanson, F.D., 1963. Nutrient deficiency symptoms. In: Smith, C.R., Childers, N.F. (Eds.), The Strawberry. Rutgers the State University, New Brunswick, NJ, pp. 75–80.

Kapur, B., Celiktopuz, E., Sandas, M.A., Kargi, S.P., 2018. Irrigation regimes and bio-stimulant application effects on yield and morpho-physiological responses of Strawberry. Hortic. Sci. Technol. 36 (3), 313–325.

Kaya, C., Kirnak, H., Higgs, D., Saltali, K., 2002. Supplementary calcium enhances plant growth and fruit yield in strawberry cultivars grown at high (NaCl) salinity. J. Hortic. Sci. Biotechnol. 93, 65–74.

Kim, H.J., Cho, Y.S., Kwon, O.K., Cho, M.W., Hwang, J.B., Jeon, W.T., Bae, S.D., 2005. Effect of hydroponic solution pH on the growth of greenhouse rose. Asian J. Plant Sci. 4 (1), 17–22.

Kirsch, R.K., 1959. The importance of interaction effects in fertilizer and lime studies with strawberry. Proc. Am. Soc. Hortic. Sci. 73, 181–188.

Kopanski, K., Kaweci, Z., 1994. Nitrogen fertilization and growth and cropping of strawberries in the conditions of Zlawy. III. Cropping and fruit chemical composition. Acta Acad. Agric. Tech. Olsten. 58, 135–142.

Koron, D., Sonjak, S., Regvar, M., 2014. Effects of non-chemical soil fumigant treatments on root colonisation with arbuscular mycorrhizal fungi and strawberry fruit production. Crop Prot. 55, 35–41.

Lamarre, M., Lareau, M.J., 1997. Influence of nitrogen, potassium and magnesium fertilization on day-neutral strawberries in Quebec. Acta Hortic. 439, 701–704.

Larcher, W., 2003. Physiological Plant Ecology. Ecophysiology and Stress Physiology of Functional Groups. Springer Verlag, Berlin, Heidelberg, New York, p. 513. Printed in Germany.

Lawrence, U., 2010. Strawberry Fertilizer Guide. NSW Government. Industry & Investment. Primefacts 941, p. 9.

Lieten, F., 2000. The effect of nutrient prior to and during flower differentiation on phyllody and plant performance of short day strawberry Elsanta. Acta Hortic. 567, 345–348.

Lieten, F., Misotten, C., 1993. Nutrient uptake of strawberry plants (cv. Elsanta) grown on substrate. Acta Hortic. 348, 299–306.

Lineberry, R.A., Burkhart, L., 1943. Nutrient deficiency in the strawberry leaf and fruit. Plant Physiol. 18, 324–333.

Mihálka, V., Hüvely, A., Petö, J., Király, I., 2017. Effects of a Third Generation Microbial Inoculant on Strawberry Grown Under Organic Conditions. Faculty of Agricultural and food sciences, University of Sarajevo LXII (67/2), 10 pp.

Miner, G.S., Poling, E.B., Carroll, D.E., Nelson, L.A., Campbell, C.R., 1997. Influence of fall nitrogen and spring nitrogen-potassium applications on yield and fruit quality of 'Chandler' strawberry. J. Am. Soc. Hortic. Sci. 122 (2), 290–295.

Moore, G., 2015. Strawberry culture in the UK—managing without soil on a grand scale. In: Zbornik radova V savetovanja "Innovacije u vocarstvu", Beagrad 2015, pp. 47–55.

NCDA, 2015. Strawberry Fertility Management. Available from: https://www.ncagr.gov/agronomi/documents/StrawberryFertility-Feb2015.pdf.

Neocleous, D., Savvas, D., 2013. Response of hydroponically grown strawberry (Fragaria x ananassa Duch.) plants to different ratios of K:Ca:Mg in the nutrient solution. J. Hortic. Sci. Biotechnol. 88 (3), 293–300.

Nestby, R., 1998. Effect of N-fertigation on fruit yield, leaf N and sugar content in fruits of two strawberry cultivars. J. Hortic. Sci. Biotechnol. 73 (4), 563–568.

Nestby, R., Lieten, F., Pivot, D., Raynal-Lacroix, C., Tagliavini, M., 2005. Influence of mineral nutrients on strawberry fruit quality and their accumulation in plant organs: a review. Int. J. Fruit Sci. 5 (1), 139–156.

Nestby, R., Guery, S., 2018. Balanced fertigation and improved sustainability of June bearing strawberry cultivated three years in open polytunnel. J. Berry Res. 7, 203–216.

Overholser, E.L., Claypool, L.L., 1932. The relation of fertilizers to respiration and certain physical properties of strawberries. Proc. Am. Soc. Hortic. Sci. 28, 220–224.

Patil, N.N., Rao, V., Dimri, D.C., Sharma, S.K., 2016. Effect of mulching on soil properties, growth and yield of strawberry cv. Chandler under mid hill conditions of Uttarakhand. Progress. Hortic. 48 (1), 42–47.

Pivot, D., Gillioz, J.M., 2001. Mineral imbalance in strawberries grown in soilless closed system: influence of climate. Rev. Suisse Vitic. Arboric. Hortic. 33 (4), 217–221.

Queensland, 2008. Strawberry, Best Soil, Water and Nutrient Management Practices. Available from: http://era.daf.qld.gov.au/id/eprint/1441/1/strawb_final-web3.pdf.

Raviv, M., Lieth, J.H., 2008. In: Raviv, M., Lieth, J.H. (Eds.), Soilless Culture. Theory and Practice. Elsevier B.V, p. 587. https://doi.org/10.1016/B978-0-444-52975-6.X5001-1 Available from:.

SARE, 2012. In: Magdoff, F., Van Es, H. (Eds.), Amount of Organic Matters in Soil. Available from: https://www.sare.org/Learning-Center/Books/Building-Soils-for-Better-Crops-3rd-Edition/Text-Version/Amount-of-Organic-Matter-in-Soils/How-Much-Organic-Matter-Is-Enough.

Sauvageau, G., Pepin, S., Depardieu, C., Gendron, L., Anderson, L., Caron, J., 2017. Tension-based irrigation management and leaching monitoring using electrical conductivity in strawberry production. Acta Hortic. 1156, 309–314.

Seyedi, A., Ebadi, A., Babalar, M., 2014. Effect of potassium levels in nutrient solution, harvest season, and plant density on quantity and quality of strawberry fruit (cv. Selva) in hydroponic system conditions. Iran. J. Hortic. Sci. 44 (4), 423–429.

Shoemaker, J.S., Greve, E.W., 1930. Relation of N fertilizer to firmness and composition of strawberries. Ohio Agric. Exp. Station Bull. 466.

Shkolnik, M.Y., 1974. General conception of the physiological role of boron in plants. Sov. Plant Nutr. 21 (2), 137–150.

SFM, 2018. Ammonium/nitrate ratio. In: Smart Fertilizer Management. Available from: www.smart-fertilizer.com/articles/ammonium-nitrate-ratio.

Strik, B., 2013. Nutrient Management of Berry Crops in Oregon. Available from: www.oregon-strawberries.org/...Nutrient_Management_Berry_-Crops_OSU.pdf

de Tender, C.A., Debode, J., Vandecasteele, B., D'Hose, T., Cremelie, T., Haegeman, A., Ruttnik, T., Dawyndt, P., Maes, M., 2016. Biological, physiochemical and plant health responses in lettuce and strawberry in soil or peat amended with biochar. Appl. Soil Ecol. 107(1–12).

Tomic, J., Pesakovic, M., Milevojevic, J., Karaklajic-Stajic, Z., 2018. How to improve strawberry productivity, nutrients composition, and beneficial rhizosphere microflora by biofertilization and mineral fertilization. J. Plant Nutr., 12. https://doi.org/10.1080/01904167.2018.1482912.

Treftz, C., Omaye, T., 2015. Nutrient analysis of soil and soilless strawberries and raspberries grown in a greenhouse. Food Nutr. Sci. 6, 805–815.

Trejo-Téllez, I., Gómez-Merinoand, F.C., 2012. Asao, T. (Ed.), Nutrient Solutions for Hydroponic Systems. Hydroponics—A Standard Methodology for Plant Biological Researches. Available from: http://www.intechopen.com/books/hydroponics-a-standardmethodology-forplantbiological-researches/nutrient-solutions-for-hydroponic-systems doi:10.5772/37578.

Ulrich, A., Mostafa, M.A.E., Allen, W.W., 1980. Strawberry Deficiency Symptoms: A Visual and Plant Analysis Guide to Fertilization. Division of Agricultural Sciences, University of California, p. 58. 4098.

Valentinuzzi, F., Mason, M., Scampicchio, M., Andreotti, C., Cesco, S., Mimmo, T., 2015. Enhancement of the bioactive compound content in strawberry fruits grown under iron and phosphorus deficiency. J. Sci. Food Agric. 95 (10), 2088–2094.

Vamerali, T., Ganis, A., Bona, S., Mosca, G., 2003. Fibrous root turnover and growth in sugar beet (Beta vulgaris var. saccharifera) as affected by nitrogen shortage. Plant Soil 255 (1), 169–177.

Vasil, I.K., 1964. Effect of boron on pollen germination and pollen tube growth. In: Linskens, H.F. (Ed.), Pollen Physiology and Fertilization, pp. 107–119 North Holland, Amsterdam.

Visser, L.G., 1955. Germination and storage of pollen. Meded. Landbouwhog. Wageningen 55, 1–68.

Wang, B., Lai, T., Huang, Q.W., Yang, X.M., Shen, Q.R., 2009. Effect of N fertilizers on root growth and endogenous hormones in strawberry. Pedosphere 19 (1), 86–95.

Waisel, Y., Eshel, A., Kafkafi, U., 1996.In: Waisel, Y., Eshel, A., Kafkafi, U. (Eds.), Plant Roots: The Hidden Half, second ed, p. 1002 (several chapters).

Willis, L.G., 1945. Defective strawberry fruit corrected by borax. Better Crops Plant Food. 29 (2) 22, 39–40.

Wilson, D., 1998. Strawberry Under Protection. In: Grower Books. Nexus Media Ltd, Kent, GB, p. 84 Grower Guide No. 6, Second Series.

Wu, M., Kubota, C., 2008. Effects of high electrical conductivity of nutrient solution and its application timing on lycopene, chlorophyll and sugar concentrations of hydroponic tomatoes during ripening. Sci. Hortic. 116, 122–129.

Yara, 2018. Gjødslingsnormer for frukt og bær. Jordbær. Available from: https://www.yara.no/gjoedsel/frukt-og-bar/.

Yadav, S.K., Khokhar, U.U., Sharma, S.D., Pramod, K., 2016. Response of strawberry to organic versus inorganic fertilizers. J. Plant Nutr. 39 (2), 194–203.

Yoshida, Y., Ohi, M., Fujimoto, K., 1991. Fruits malformation, size and yield in relation to nitrogen and nursery plants in large fruited strawberry (Fragaria x ananassa Duch. cv. Ai-berry). J. Jpn. Soc. Hortic. Sci. 59 (4), 727–735.

Yoshida, Y., 1992. Studies of Flower and Fruit Development in Strawberry, With Special Reference to Fruit Malformation in 'Ai-Berry'. Memories of Faculty of Agriculture, Kagawa University, 57, 94 pp.

Zhang, Z., Cao, F., Li, H., Li, X., Dai, H., 2017. Correlation between phosphorus content and fruit quality in strawberries. Acta Hortic. 1156, 661–666.

Blueberries

Abbott, J.D., Gough, R.E., 1987. Seasonal development of highbush blueberry roots under sawdust mulch. J. Am. Soc. Hortic. Sci. 112, 60–62.

Blevins, D.G., Scrivner, C.L., Reinbott, T.M., Schon, M.K., 1998. Foliar boron increases berry number and yield of two highbush blueberry cultivars in Missouri. J. Plant Nutr. 19, 99–113.

Burkhard, N., Lynch, D., Percival, D., Sharifi, M., 2009. Organic mulch impact on vegetation dynamics and productivity of highbush blueberry under organic production. HortScience 44, 688–696.

Clark, J.R., Moore, J.N., 1991. Southern highbush blueberry response to mulch. HortTechnology 1, 52–54.

Coville, F.V., 1910. Experiments in Blueberry Culture. US Department of Agriculture Bulletin No. 193.

Cox, J., 2009. Comparison of plastic weedmat and woodchip mulch on low chill blueberry soil in New South Wales, Australia. Acta Hortic. 810, 475–482.

Cox, J.A., Morris, S., Dalby, T., 2014. Woodchip or weedmat? A comparative study on the effects of mulch on soil properties and blueberry yield. Acta Hortic. 1018, 369–374.

DeVetter, L., Granatstein, D., Kirby, E., Brady, M., 2015. Opportunities and challenges of organic highbush blueberry production in Washington State. HortTechnology 25, 796–804.

Eck, P., 1988. Blueberry Science. Rutgers University Press, New Brunswick, NJ, pp. 106–109.

Eck, P., Gough, R.E., Hall, I.V., Spiers, J.M., 1990. Blueberry management. In: Galletta, G.J., Himelrick, D.G. (Eds.), Small Fruit Crop Management. Prentice Hall, Upper Saddle River, NJ, pp. 273–333.

Franceschi, V.R., Nakata, P.A., 2005. Calcium oxalate in plants: formation and function. Annu. Rev. Plant Biol. 56, 41–71.

Ganie, M.A., Akhter, F., Bhat, M.A., Malik, A.R., Junaid, J.M., Shah, M.A., Bhat, A.H., Bhat, T.A., 2013. Boron—a critical nutrient element for plant growth and productivity with reference to temperate fruits. Curr. Sci. 104, 76–85.

Golldack, J., Schubert, R., Tauschke, M., Schwarzel, H., Hofflich, G., Lentzsch, P., Munzenberger, B., 2001. Mycorrhization and plant growth of high-bush blueberry (Vaccinium corymbosum L.) on arable land in Germany. In: Smith, S.E. (Ed.), Diversity and Integration in Mycorrhizas. Proceedings of the 3rd International Conference on Mycorrhiza, 8–13 July 2001, Adelaide, Australia. Springer, Dordrecht, The Netherlands.

Gransee, A., Führs, H., 2013. Magnesium mobility in soils as a challenge for soil and plant analysis, magnesium fertilization and root uptake under adverse growth conditions. Plant Soil 368, 5–21.

Guerrero-Polanco, F., Alejo-Santiago, G., Luna-Esquivel, G., 2017. Potassium fertilization in fruit trees. Rev. Bio Cienc. 4, 143–152.

Gürel, S., Başar, H., 2016. Effects of applications of boron with iron and zinc on the contents of pear trees. Notulae Bot. Horti Agrobot. 44, 125–132.

Hanson, E.J., 1995. Preharvest calcium sprays do not improve highbush blueberry (Vaccinium corymbosum L.) quality. HortScience 30, 977–978.

Hanson, E.J., Hancock, J.F., 1996. Managing the Nutrition of Highbush Blueberries. Extension Bulletin E-2011. Michigan State University Extension, East Lansing, MI.

Hanson, E.J., Beggs, J.L., Beaudry, R.M., 1993. Applying calcium chloride postharvest to improve highbush blueberry firmness. HortScience 28, 1033–1034.

Hart, J., Strik, B., White, L., Yang, W., 2006. Nutrient Management for Blueberries in Oregon. EM 8918. Oregon State University Extension Service, Corvallis, OR.

Himelrick, D.G., Powell, A.A., Dozier, W.A., 1995. Commercial Blueberry Production Guide for Alabama. Available from: http://www.aces.edu/pubs/docs/A/ANR0904/ANR-0904.pdf .

Hirschi, K.D., 2004. The calcium conundrum. Both versatile nutrient and specific signal. Plant Physiol. 136, 2438–2448.

Hirzel, J., 2014. Principios de fertilización en frutales y vides (Fertilization principles in fruits and grapes). In: Hirzel, J. (Ed.), Diagnóstico Nutricional y Principios de Fertilización en Frutales y Vides. [Nutritional diagnostic and fertilization principles in fruits and grapes.] Colección Libros INIA-31. second ed. Instituto de Investigaciones Agropecuarias, Chillán, pp. 225–293.

Hocking, B., Tyermn, S.D., Burton, R.A., Gilliham, M., 2016. Fruit calcium: transport and physiology. Front. Plant Sci. 7, 569.

Jacobs, L.A., Davies, F.S., Kimbrough, J.M., 1982. Mycorrhizal distribution in Florida rabbiteye blueberries. HortScience 17, 951–953.

Kader, M.A., Senge, M., Mojid, M.A., Ito, K., 2017. Recent advances in mulching materials and methods for modifying soil environment. Soil Tillage Res. 168, 155–166.

Kochian, L.V., 2012. Rooting for more phosphorus. Nature 488, 466–467.

Korcak, R.F., 1989. Variation in nutrient requirements of blueberries and other calcifuges. HortScience 24, 573–578.

Krewer, G., NeSmith, D.S., 1999. Blueberry Fertilization in Soil. No 01-1. University of Georgia Cooperative Extension, Athens, GA.

Kumar, D., Shivay, Y.S., Dha, S., Kumar, C., Prasad, R., 2013. Rhizospheric flora and the influence of agronomic practices on them: a review. Proc. Natl. Acad. Sci. India, Sect. B Biol. Sci. 83, 1–14.

Marschner, H., 1986. Mineral Nutrition of Higher Plants. Academic Press, London.

Merhaut, D.J., Darnell, R.L., 1995. Ammonium and nitrate accumulation in containerized southern highbush blueberry plants. HortScience 30, 1378–1381.

Montanaro, G., Dichio, B., Lang, A., Mininni, A.N., Xiloyannis, C., 2015. Fruit calcium accumulation coupled and uncoupled from its transpiration in kiwifruit. J. Plant Physiol. 181, 67–74.

Paltineanu, C., Coman, M., Nicolae, S., Ancu, I., Calinescu, M., Sturzeanu, M., Chitu, E., Ciucu, M., Nicola, C., 2017. Root system distribution of highbush blueberry crops of various ages in medium-textured soils. Erwerbs-obstbau. https://doi.org/10.1007/s10341-017-0357-3.

Plassard, C., Dell, B., 2010. Phosphorus nutrition of mycorrhizal trees. Tree Physiol. 30, 1129–1139.

Retamales, J.B., Hancock, J.F., 2018. Blueberries, second ed. CABI, Oxfordshire. 412 pp.

Scagel, C.F., 2005. Inoculation with ericoid mycorrhizal fungi alters fertilizer use of highbush blueberry cultivars. HortScience 40, 786–794.

Scagel, C.F., Yang, W.Q., 2005. Cultural variation and mycorrhizal status of blueberry plants in NW Oregon commercial production fields. Int. J. Fruit Sci. 5, 85–111.

Shaw, M., 2008. Soil pH is more important than fertilizer for blueberries. New York Fruit Q. 16, 25–28.

Shear, C.B., Faust, M., 1980. Nutritional ranges in deciduous tree fruits and nuts. Hortic. Rev. 2, 142–163.

Steinmetz, Z., Wollmann, C., Schaefer, M., Buchmann, C., David, J., Tröger, J., Muñoz, K., Frör, O., Schaumann, G.E., 2016. Plastic mulching in agriculture: trading short-term agronomic benefits for long-term soil degradation? Sci. Total Environ. 550, 690–705.

Stiles, W.C., Reid, W.S., 1991. Orchard Nutrition Management. Information Bulletin 219. Cornell Cooperative Extension, Ithaca, NY.

Strik, B.C., Vance, A.J., 2015. Seasonal variation in leaf nutrient concentration of northern highbush blueberry cultivars grown in conventional and organic production systems. HortScience 50, 1453–1466.

Stückrath, R., Quevedo, R., de la Fuente, L., Hernández, A., Sepúlveda, V., 2008. Effect of foliar application of calcium on the quality of blueberry fruits. J. Plant Nutr. 31, 1299–1312.

Subbarao, G.V., Ito, O., Sahrawat, K.L., Berry, W.L., Nakahara, K., Ishikawa, T., Watanabe, T., Suenaga, K., Rondon, M., Rao, I.M., 2006. Scope and strategies for regulation of nitrification in agricultural systems: challenges and opportunities. Crit. Rev. Plant Sci. 25, 303–335.

Sullivan, D.M., Bryla, D.R., Costello, R.C., 2014. Chemical characteristics of custom compost for highbush blueberry. In: He, Z., Zhang, H. (Eds.), Applied Manure and Nutrient Chemistry for Sustainable Agriculture and Environment. Springer Science+Business Media, Dordrecht, The Netherlands, pp. 293–311.

Vance, A.J., Jones, P., Strik, B.C., 2017. Foliar calcium applications do not improve quality or shelf life of strawberry, raspberry, blackberry, or blueberry fruit. HortScience 52, 382–387.

Vargas, O.L., Bryla, D.R., 2015. Growth and fruit production of highbush blueberry fertilized with ammonium sulfate and urea applied by fertigation or as granular fertilizer. HortScience 50, 479–485.

Valenzuela-Estrada, L.R., Vera-Caraballo, V., Ruth, L.E., Eissenstat, D.M., 2008. Root anatomy, morphology and longevity among root orders in Vaccinium corymbosum (Ericaceae). Am. J. Bot. 95, 1506–1514.

Vega, A.R., Garciga, M., Rodriguez, A., Prat, L., Mella, J., 2009. Blueberries mycorrhizal symbiosis outside the boundaries of natural dispersion for Ericaceous plants in Chile. Acta Hortic. (810), 665–671.

White, P.J., Broadley, M.R., 2003. Calcium in plants. Ann. Bot. 92, 487–511.

Williamson, J., Krewer, G., Pavlis, G., Mainland, C.M., 2006. Blueberry soil management, nutrition and irrigation. In: Childers, N.F., Lyrene, P.M. (Eds.), Blueberries for Growers, Gardeners and Promoters. Dr Norman F. Childers Publications, Gainesville, FL, pp. 60–74.

Wojcik, P., 2005. Response of 'Bluecrop' highbush blueberry to boron fertilization. J. Plant Nutr. 28, 1897–1906.

Yang, W.Q., Goulart, B.L., Demchak, K., Li, Y., 2002. Interactive effects of mycorrhizal inoculation and organic soil amendments on nitrogen acquisition and growth of highbush blueberry. J. Am. Soc. Hortic. Sci. 127, 742–748.

Soil fertility: Plant nutrition vis-à-vis fruit yield and quality of stone fruits

Tomo Milošević[a],*, Nebojša Milošević[b]

[a]Department of Fruit Growing and Viticulture, Faculty of Agronomy, University of Kragujevac, Čačak, Republic of Serbia
[b]Department of Pomology and Fruit Breeding, Fruit Research Institute, Čačak, Republic of Serbia
*Corresponding author. E-mail: tomomilosevic@kg.ac.rs

1 Introduction

Agriculture started in multiple locations in the world some 10,000 years ago. These locations (where crop domestications took place) are distributed in some 10 areas generally between 30-degree northern latitude and southern latitude. They tend to occur in areas with higher levels of biodiversity. Although wild ancestors tend to be much more widely distributed than domesticated crops (Smartt and Simmonds, 1995), many crops originate from distinct geographic regions, which have been called Vavilov centers (Vavilov, 1926, 1951; Meyer et al., 2012). So, crop domestication has been considered as one of the key developments that enabled the rise of major civilizations (Gepts, 2014).

Data on the timing of the domestication of fruit crops are different, as new discoveries arise. Generally, the origins of fruit culture occurred in the Fertile Crescent in the late Neolithic and Bronze Age, about 8000 years ago, a period known as the second Neolithic Revolution that involved the change from villages to urban communities (Janick, 2011). For example, in the archeological excavations of lake dwelling in Bosnia (former SFR Yugoslavia) from the late Bronze Age, stones of black thorn (sloe) and sweet cherry were found (Mišić, 2006).

A.K. Srivastava, Chengxiao Hu (eds.)
Fruit Crops: Diagnosis and Management of Nutrient Constraints
https://doi.org/10.1016/B978-0-12-818732-6.00041-1

Regarding domestication of fruit crops, in contrast with mere collection, between 6000 and 3000 BC (the late Neolithic and Bronze Ages), the ancient Mediterranean fruits (date, olive, grape, fig, sycamore fig, and pomegranate) were domesticated (Zohary and Spiegel-Roy, 1975). Some groups of fruits such as citrus, banana, various pome fruits (apple, pear, quince, and medlar), and stone fruits (almond, apricot, cherry, peach, and plum) were domesticated in Central and East Asia and reached the West in antiquity.

In ancient communities, the fruits were highly respected (Ebert, 2009). According to the same author, the Egyptians were renowned for the sycamore fig or fig-mulberry (mulberry fig, *Ficus sycomorus*) so named because of the resemblance of the leaves to those of the mulberry. In ancient Greece, the very harsh laws introduced by the Greek aristocrat Drakon (621 BC) included the death penalty for fruit thieves. A similar attitude toward fruit was registered much later. Namely, in 18th century, the Prussian King Friedrich II (also known as "Der Alte Fritz") committed fruit thieves to jail (Ebert, 2009).

On the other hand, a more number of interesting and popular fruits with high economic and social importance for the development of some countries worldwide were domesticated in the 18th to 20th centuries including various brambles, blueberry, cranberry, pecan, and kiwifruit, while many fruits, although extensively collected and marketed, are in the process of domestication (lingonberry, various cacti, durian, etc.) (Janick, 2011).

In the development of knowledge and understanding of the botany of fruit trees and their cultivation that increased rapidly, fruit production has reached a very high level in economic and social terms worldwide. Many of the common techniques and methods in use today such as pruning, grafting, artificial pollination, and irrigation were developed centuries ago (Ebert, 2009). This author also reported that, regarding mineral fertilization, the thinking of fruit growers of today is far removed from the old ideas that a high input of fertilizers per se would result in the production of high yields of quality fruit.

The modern concept of fruit nutrition is based on new principles that imply minimal but effective use of mineral fertilizers, of course other chemical agents, primarily pesticides. This will be discussed in more detail in the future sections of this chapter.

2 Distribution, production, and importance of stone fruits

2.1 Global distribution

Stone fruits are deciduous tree species originating from the temperate zone of the northern hemisphere. Plum, peach and nectarine, sweet and sour (tart) cherry, apricot, and cornelian cherry belong to the group of stone fruits according to pomological classification.

With the exception of cornelian cherry, all members of the group of stone fruits belong to the *Prunus* L. genera from Rosaceae family.

Cornelian cherry belongs to the *Cornus* genus from Cornaceae family. The Latin, that is, binomial, name is *Cornus mas* L.

The main temperate zone stone fruit production areas are located between 60- and 35-degree, latitude north and between 30- and 50-degree latitude south (Fig. 41.1A–D). In relation to the globalization process, especially to climate change, temperate fruits are also grown in mountainous areas of the tropics and subtropics, for example, in the Andean Cordilleras of South America or in the foothills of the Himalayas in India (Ebert, 2009).

2.2 Production statistics

In the last 30 years, world fruit production and harvested area have risen dramatically. According to relevant world services, total fruit production (without group of citrus) in 1986 were 387,685,120 t, in 1996 496,654,561 t, in 2006 686,919,388 t, and in 2016 865,876,405 t (FAOSTAT, 2018). Compared with 1986, world fruit production in 2016 increased by 2.23 times or by 55.23%.

Similar tendencies are valid to a group of stone fruits. Among them, in 2016, the highest production has peach and nectarine with 24,975,649 t, followed by plum with 12,050,800 t, apricot with 3,881,204 t, sweet cherry with 2,317,956 t, and sour cherry with 1,378,216 t (FAOSTAT, 2018).

From 2006 to 2016, the most stable production growth was found in peach and plum, partly in sweet cherry, whereas, in apricot and sour cherry, the increase in production is evident but with large drops and rises. In apricot, instability of production is mainly caused by climatic factors, while in sour cherry the main cause is situation on the world market, that is, hyperproduction in certain periods, which the market cannot accept (Milošević et al., 2010,

(A)

(B)

FIG. 41.1 Distribution map of plum and sloe (A), peach and nectarine (B), apricot

Continued

2012a). For the mentioned period, the production of peach and nectarine increased by 26.41%, plums for 12.66%, apricot for 10.79%, sweet cherry for 17.88%, and sour cherry for 17.11%.

Asia is the continent with the largest production of stone fruits, except sour cherry in 2016. In the case of this fruit species, Europe is the largest producer (FAOSTAT, 2018). According to same source, for peach and nectarine, the "top five" countries with the highest production, in descending order, are China, Spain, Italy, the United States, and Iran; for plum China, Romania, Serbia, the United States, and Turkey; for apricot Turkey, Uzbekistan, Iran, Algeria, and Italy; for sweet cherry Turkey, the United States, Iran, and Italy; and for sour (tart) cherry Russian Federation, Poland, Turkey, Ukraine, and the United States.

Cornelian cherry has a minor economic importance compared with other members of stone fruits. This plant is popular in southern Europe with the northern limit being southern Belgium and central Germany (Mamedov and Craker, 2004). They are grown predominantly at a small-scale level, and their production is focused on the preservation of genetic sources in gene pools (Dokoupil and Řezníček, 2012). Earlier, cornelian cherry was widely cultivated in southern regions

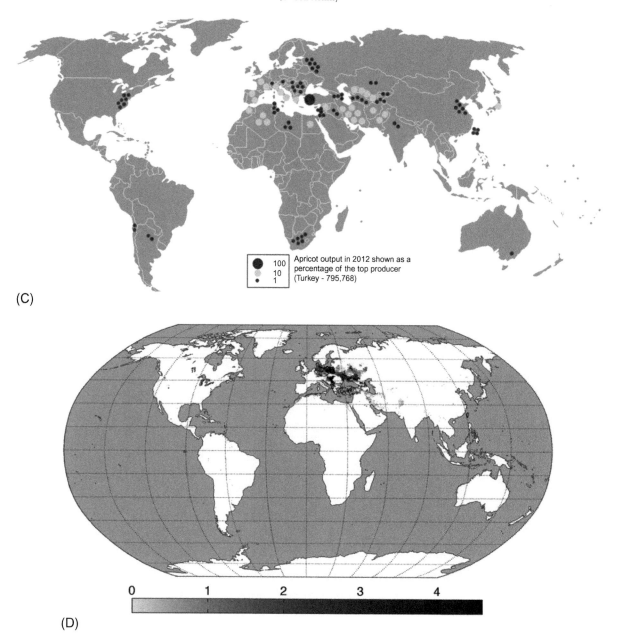

(C)

(D)

FIG. 41.1—Cont'd (C), and sour cherry (D). *From Wikimedia Commons.*

of the Czech Republic; nowadays, it is nearly forgotten, although it still wildly grows in this locality (Cetkovská et al., 2015). In the Czech and Slovak Republic, it is spread in the area of the White Carpathian Mountains (Tetera, 2006). Additionally, in Brusnica village (44°01′ N, 20°26′ E) near Gornji Milanovac City (western Serbia), small cornelian cherry orchard with selected genotypes was established several years ago (Milošević, unpublished data).

Therewith, some smallholder plantations exist in some countries in Europe and Minor Asia. According to data from the end of the 20th century, in Turkey, there were 1,585,000 cornelian cherry trees, in which yield is approximately 14,000 t/year. Also, 758 t fruit were harvested from the 77,060 cornelian cherry trees located in Artvin and Erzurum regions (Güleryüz et al., 1998). Nowadays, Turkey is an important center of cornelian cherries (Ercisli et al., 2008), especially in northern Anatolia (Ercisli, 2004). In Georgia, the total area of the commercial orchards of cornelian cherry is 135 ha (Maghradze et al., 2009). According to the same source, the number of the trees of this species in orchards makes up 60,204, and there are 146,534 scattered trees.

It must be acknowledged that in the last few decades intensification of fruit production, including stone fruits, has been primarily based on the application of irrigation and nutrition (fertilization), which has undoubtedly contributed to the increase in production costs, and when it comes to fertilizers, especially mineral, environmental problems, and health safety of fruits (Stefanelli et al., 2010).

3 Short overview about importance of stone fruits

Stone fruits are used for fresh (direct) consumption (table fruit) and for processing and some of them (plum, apricot) for drying. Their fruits are rich source of organic and inorganic compounds useful for human diet and health. The fruits of these plants contain water; carbohydrates, that is, sugars; organic acids; fiber; proteins; minerals (Ca, P, Fe, and K); and high amounts of nonnutritious or bioactive compounds, especially vitamins (A, B_1, B_2, K, E (α-tocopherol), niacin, and pantothenic acid), β-carotene, lutein + zeaxanthin, phenols, flavonoids, anthocyanins, etc. (Ferretti et al., 2010).

Plum fruit contains major macroelements. The total quantities of the macroelements accumulated in fully ripe plums followed a decreasing order: $K > N > Mg > Ca > P$ (Milošević and Milošević, 2012b), also contains some microelements i.e. Na and Fe (Milošević and Milošević, 2012c). Among single sugars, the predominant is glucose, followed by sucrose and fructose (Milošević and Milošević, 2012d).

Peach and nectarine, apricot, plum, and sweet cherry are among the most delectable fruits available, and they are cultivated for family consumption or for sale on local and/or world markets (Milošević et al., 2013a, 2014; Milošević and Milošević, 2018). The ripe fruit can reach the consumer only in local markets, while regarding to distant market needs they are mostly harvested immature (Bavcon Kralj et al., 2014). An earlier survey conducted by some researchers indicated that the main reasons why consumers do not eat more stone fruits are as follows: fruits lack taste, fail to ripen, and are hard and mealy (Crisosto, 2006).

Fruits of sour cherry and cornelian cherry are less suitable for direct consumption. In jargon, they are often named as "industrial" fruit crops due to its fruits that are primarily used for processing into juices, syrups, jams, jellies, marmalades, stewed fruit, yogurts, and liquors (Milošević et al., 2012a; Dokoupil and Řezníček, 2012; Cetkovská et al., 2015). Sour cherries have lower sugar content and a higher acid content than its sweet counterpart; however, sour cherries with higher sugar content are becoming more common on the fresh fruit market in recent decades.

Fruit of cornelian cherry is a rich source in primary and secondary metabolites. Data from Georgia revealed that fruit weighs 1.1–5.6 g and dry matter varies from 17.7% to 26.1%, total sugar from 8.5% to 9.2%, acidity from 1.7% to 2.3%, pectin from 0.98% to 1.12%, and vitamin C from 50.5 to 128.0 mg/100 g (Maghradze et al., 2009). According to the same authors, cornelian cherry in Georgia is used for preparing traditional dishes, dried products, and preserves.

Finally, the fruits of all members of these crops can be used to produce alcoholic beverages.

4 Soil fertility

It has been long known that soils are fundamental to life on Earth; however, human pressures on soil resources are reaching critical limits (FAO and ITPS, 2015). According this source, careful soil management is one essential factor of sustainable agriculture and also provides a valuable lever for climate regulation and a pathway for safeguarding ecosystem services and biodiversity.

The term "soil fertility" is defined as the ability of soil to sustain plant growth and optimize crop yield. This can be enhanced through application of organic and inorganic fertilizers to the soil. However, this agrotechnical measure is possible under usual conditions, but the reality is completely different. Soil degradation, devastation, and pollution are major factors that limited successful, economically justified, and sustainable crop production worldwide.

Many agricultural soils around the world are degraded, devastated, and polluted by the influence of natural factors but much more by the influence of human activities. Namely, rapid economic development and improved technologies have caused increase in the degradation, devastation, and contamination of soil, which is a basic substrate for crop cultivation and food production. At the same time, world population growth has brought about both food consumption increase and soil area decrease (Ovuka, 2000; Ramankutty et al., 2002; Lepers et al., 2005). In addition, soil structure deterioration, soil disturbance, frequent accumulation of heavy metals and other hazardous substances, destruction of beneficial soil macro- and microorganisms, and alteration of soil chemical properties are induced partly by the effect of nature, abundant precipitations, acid rains, and erosion (Kirkby, 2000; Evans, 2002; Lal, 2001, 2006).

On the other hand, the uncontrolled usage of mineral fertilizers and pesticides by humans contributes to these soil states (Popov et al., 2005; Hartemink, 2006). Not only do such soils have reduced fertility, but they also carry risks of producing food of unsuitable quality and questionable safety (Glisic et al., 2009).

Data from relevant literature showed that soil degradation is the physical, chemical, and biological decline in soil quality (Jie et al., 2002). It can be the loss of organic matter; decline in soil fertility; structural condition; erosion; adverse changes in salinity, acidity, or alkalinity; and the effects of toxic chemicals, pollutants, or excessive flooding.

Basically, soil degradation can involve (Oldeman, 1992) the number of problems like water erosion (includes sheet, rill, and gully erosion), wind erosion, salinity (includes dryland, irrigation, and urban salinity), loss of organic matter, fertility decline, soil acidity or alkalinity, structure decline (includes soil compaction and surface sealing), mass movement, and soil contamination, including effects of toxic chemicals and pollutants.

It seems that soil erosion, chemical deterioration, and physical degradation are the important parts among various types of soil degradation (Jie et al., 2002). These authors also stated that, as a result of soil degradation, it is estimated that about 11.9%–13.4% of the global agricultural supply has been lost in the past five decades of 20th century.

The following data is also disturbing:

- During the past 40 years, nearly one-third of the world's cropland (1.5 billion hectares) has been abandoned because of soil erosion and degradation.
- About 2 million hectares of rainfed and irrigated agricultural lands are lost to production every year due to severe land degradation, among other factors.
- It takes approximately 500 years to replace 2.5 cm of topsoil lost to erosion. The minimal soil depth for agricultural production is 15 cm. From this perspective, productive fertile soil is a nonrenewable, endangered ecosystem.

Regeneration of such soils is a long-term process that may, only with the effect of nature, take more than 20 years (Ruecker et al., 1998; Tiag et al., 2000).

For these reasons, improving food security and environmental sustainability in farming systems requires an integrated soil fertility management approach that maximizes crop production while minimizing the mining of soil nutrient reserves and the degradation of the physical and chemical properties of soil that can lead to land degradation, including soil erosion. Such soil fertility management practices include the use of fertilizers, organic inputs, crop rotation with legumes, and the use of improved germplasm, combined with the knowledge on how to adapt these practices to local conditions (Jie et al., 2002).

5 Fertilization of stone fruit frees

5.1 General overview

As mentioned earlier, total fruit production in the world has risen dramatically in the past 20–30 years. However, in many, fruitful advanced, countries, the land area is not enlarged; on the contrary, they are reduced, but fruit production, especially pome and stone fruits, has an upward trend. The reasons for such a phenomenon are numerous. This was largely related to changes in the choice of rootstocks, introduction of new high-quality and more productive cultivars, increased plant density, improved soil conditions, more effective disease and pest protection, weed control, and large-scale switchover to the grass strip system (Delver, 1982; Milošević et al., 2008, 2013a). Nevertheless, irrigation (Delver and Wertheim, 1988) and fertilization (Milošević et al., 2014) are the most important cultural practices that contribute to improving fruit quality and productivity of fruit species including stone fruits.

Although the positive impact of irrigation on the growth and development of fruit trees and other crops has been long known, the significance of this agrotechnical measure arises due to global climate change (drought) and more problematic due to water deficiency in nature. Namely, more number of regions of the world that now depend on rainfed agriculture may require irrigation, bringing higher costs and conflict over access to water (Hribar and Vidrih, 2015).

Fertilization also known as nutrition of crops is an important tool used by most farmers to boost crop yield and quality. However, excessive fertilization has been verified, especially on the horticultural enterprises, where the fertilizer costs represented less than 10% of the variable crop costs (Huett and Dirou, 2000). Except for an economic point of view, excessive fertilization has been connected to groundwater and stream water contamination (Cuquel et al., 2011) and causing an increment of pest (Marschner, 1995) and diseases occurrence (Tratch et al., 2010). Cuquel et al. (2011) reported that increasing the public concern about environment aspects caused by overfertilization renews the interest on evaluating the adequate fertilization recommended on field to maintain productivity and fruit quality with less environment impact. As known, the some fruit growing area is not suitable for intensive growing systems, excluding soil fertility maintenance through manure and other organic fertilizer usage. Thus, mineral fertilizer application needs to be investigated as a potential alternative. However, the addition of inorganic fertilizers (mostly N and/or NPK) to improve fruit tree and other crop yield produced highly erratic results and remains unreliable (O'Sullivan and Ernest, 2008; Milošević et al., 2013a, 2014). The reasons for the variability of fruit tree yield responses to mineral fertilizer application have not been fully investigated. Fertilizer effects may rely on other factors such as plant density, tillage, and the overall cropping system.

During the 20th century, mineral nutrition alone contributed significantly to increased crop yields. Namely, 50% of the increase in crop yields worldwide during the 20th century was due to application of chemical fertilizers (Borlaug and Dowswell, 1994). They also stated that, during the 21st century, essential plant nutrients would be the single most important factor limiting crop yields, especially in developing countries. Additionally, average percentage of yield attributable to fertilizer generally ranged from about 40% to 60% in England and the United States (Stewart et al., 2005). However, in developed countries, during the last 30–40 years, mineral fertilization in fruit orchards has changed considerably. More information about modern approach to the application of mineral fertilizers in stone fruit orchards will be in the next sections.

5.2 Response of stone fruit trees to fertilization

Fruit trees, including stone fruits, are perennial plants, have a deep root, and remain for several decades in the same place, that is, soil. To achieve high and stable yields, accompanied by good external and internal quality of the fruit, the reserves of nutrients in the soil fall from year to year and must be compensated by fertilization. It has been long known that fruit trees can be cultivated on different soil types. However, in highly fertile soils, fruit trees are very vigor, late started to produce and give uneven yields from year to year. The biggest problem for producers in such soils is the control of tree vigor. A syntagma is known: "who controls tree vigor controls the productiveness; who controls the productiveness controls the fruit quality." In paradox, very fertile soils should be avoided for the fruit tree growing.

Different soils contain unequal amounts of essential nutrients. On the other hand, their utilization by roots of different fruit species is different. Nutrient storage capacity and accessibility are influenced by soil texture, rooting depth, and organic matter content, but the availability is modified by soil moisture and pH (Keller, 2005). According to same author, plant-available nutrients are taken up as ions, dissolved in the soil solution, and their uptake depends also on the water flow through the soil-root-shoot pathway. A well-balanced nutrient supply is known to be crucial for all the crops to avoid excessive growth or mineral deficiency, since mineral elements affect fruit tree physiology and, thereafter, also fruit tree development (Bergmann, 1992).

5.3 Role of individual elements in stone fruit trees

Data from relevant literature indicate that between 16 and 22 elements are necessary for the normal fruit tree growth and development, including group of stone fruits (Bergmann, 1992). Their deficiency caused numerous physiological, morphological, and anatomical disorders that adversely affect the yield and fruit quality.

According to the presence in plants, the necessary elements are divided into macro- and microelements or macro- and micronutrients (Table 41.1).

Stone fruit trees require relatively large quantities of macroelements, whereas microelements are required in smaller quantities. Macronutrients are needed in the range (or more) of $1\,g/kg$ ($1000\,\mu g/g = 1000\,ppm = 0.1\%$) of plant's dry weight (DW), whereas micronutrients are required in plant tissue at level of $100\,mg/kg$ ($100\,\mu g/g = 100\,ppm$) DW or lower (Mehra and Farago, 1994).

Besides nutrient classified as macronutrients and micronutrients, other classifications were proposed in the literature. One of them divides the elements on the basis of physicochemical properties into "metals" (K, Ca, Mg, Fe, Mn, Zn, Cu, and Mo) and "nonmetals" (N, S, P, B, and Cl) (Marschner, 1995).

Other authors reported that successful growth and development of fruit trees are caused by six macroelements (N, P, K, Ca, Mg, and S) and eight microelements (Fe, Mn, Cu, Zn, B, Mo, Co, and Cl) (Ubavić et al., 1990). As for the pome and stone fruit species, it is usually manipulated with five macro- (N, P, K, Ca, and Mg) and five microelements (Fe, Mn, Cu, Zn, and B) as an essential nutrients (Milošević and Milošević, 2011a; Milošević et al., 2012b, 2013a, 2014). However, in 1- and 2-year-old shoots, flower buds, and flowers of some cultivars of European plum (*Prunus domestica* L.), important levels of sodium (Na) were detected (Milošević and Milošević, 2012a).

Unlike C, O, and H, which rarely limit the growth and development of fruit trees because they are naturally in sufficient quantities, other macro- and microelements, especially N, P, and K, are often missing and need to be added as fertilizers. In some sources of literature, C, O, H, N, P, and K are classified as primary macronutrients, whereas

TABLE 41.1 Elements those are essential for all higher plants.

Organic nutrients (basic elements in organic matter)	Mineral elements		
	Macroelements (macronutrients)	Microelements (micronutrients)	
Carbon (C)	Nitrogen (N)	Iron (Fe)	Aluminum (Al)
Hydrogen (H)	Phosphorus (P)	Manganese (Mn)	Cobalt (Co)
Oxygen (O)	Potassium (K)	Copper (Cu)	Sodium (Na)
	Calcium (Ca)	Zinc (Zn)	Nickel (Ni)
	Magnesium (Mg)	Boron (B)	Silicon (Si)
	Sulfur (S)	Molybdenum (Mo)	Vanadium (V)
		Chlorine (Cl)	

From Bergmann, W., 1992. Nutritional disorders of plants. Development, Visual and Analytical Diagnosis. Gustav Fischer/Verlag, Jena/Stuttgart/New York.

Ca, Mg and S are classified as secondary macroelements (author's remark). In general, all types of fertilizers (organic, organomineral, and mineral) as a source of nutrients should be used in preplanting fertilization, fertilization during juvenile stage, and, especially, fertilization for maintaining stone and other fruit orchards.

Nutrient deficiencies and/or toxicities are widespread and have been documented for various soils all over the world. Nutrient deficiencies occur when the plant cannot acquire sufficient amounts of nutrients for its internal needs, whereas an excessive supply of elements, especially trace metals (e.g., Zn, Cu, and Mn), results in toxicity to the plant (He et al., 2005).

Specific deficiency symptoms appear on all plant parts, but discoloration (chlorosis) of leaves is most commonly observed. Deficiency symptoms of low mobile nutrients (like Fe and Zn) appear initially and primarily on upper leaves or leaf tips, while deficiency symptoms of mobile nutrients (like Mg) appear primarily on lower fully expanded leaves (Marschner, 1995).

Although fruit quality is genetically fixed and a cultivar characteristic, cultural practice, including irrigation and fertilization, also influences fruit quality. Namely, balanced nutrient supply to trees is one of the most important prerequisites for optimal fruit growth and development (Ebert, 2009). Thus, nutrient deficiencies and imbalances impair the composition and appearance of the fruit and may reduce its internal and/or external quality.

Detailed overviews about the role of some essential elements in fruit trees and its deficiency/sufficiency/excess symptoms were made by Johnson and Uriu (1989), Hanson (1996), and Ebert (2009).

6 Relationship of soil conditions: Nutrients

6.1 Soil conditions

Soil conditions and their characteristics play important role in efficiency of fruit tree nutrition, that is, fertilization. The physical, chemical, and biological soil properties are fundamental. Shortly, the physical properties of soil are—to a great extent—fixed and will not change over the years. These properties play a vital role in the sustainable productivity of soils. Most of the physical properties can be evaluated visually on site. The most important are texture, structure, depth, layering or stratification, and aeration.

The chemical properties of soil are constantly in a state of flux (changing) and can be changed, although this may sometimes prove prohibitive because of high costs. The most relevant chemical properties of soils are soil reaction (pH value), resistance or electrical conductivity, salinity, fertility level, cation exchange capacity, and organic matter content.

Soil biological properties refer to the living organisms found in the soil and include both micro- and macroplants and/or animals. A group of microplants are bacteria, fungi, and actinomycetes, whereas a group of macroplants are roots of higher plants. The group of microanimals present in the soil consists of nematodes and protozoa. A group of macroanimals in the soil are numerous, and they are as follows: earthworms, rodents (prairie dogs, moles, gophers, etc.), arthropods (mites, insects, spiders, etc.), and gastropods (slugs, snails, etc.).

There is no doubt that the presence of microorganisms and their activity have vital importance for the normal and sustainable state of the soil. Namely, soil microbial activity that reflects microbiological processes of soil microorganisms is the potential indicator of soil quality, as plants rely on soil microorganisms to mineralize organic nutrients for fruit tree growth and development (Chen et al., 2003). Soil microorganisms also process plant litter and residues into soil organic matter, a direct and stable reservoir of C and N that consists of living and dead organic materials subject to rapid biological decomposition. Hence, in natural systems, the activity of soil microorganisms is a major determinant of efficient nutrient cycling (Chen et al., 2003).

6.2 Source of nutrients and their availability

In natural soils, the presence and distribution of plant mineral nutrients are extremely heterogeneous (Ebert, 2009). According to same author, nutrient amounts and availabilities in the soil of a fruit orchard may thus vary both spatially and temporally, which is of importance in their acquisition by tree roots.

For fruit trees, including group of stone fruits, several sources of nutrients can be distinguished (Ebert, 2009):

(a) Mineralization of organic matter (plant residues and microorganisms)
(b) Mineral fertilizers
(c) Organic fertilizers
(d) Release from parent rocks by weathering and release from fixed sites in clay minerals (e.g., K from illite, some macro- and micronutrients from natural clinoptilolite, also called natural zeolite) (Milošević and Milošević, 2010)
(e) Atmospheric deposition
(f) N-fixation by specialized microorganisms

Fertilizer supply and the biochemical release or fixation of mineral nutrients are the most important factors influencing the nutrient balance in the soil. Soil conditions and its biological activity are the key factor determining the availability of nutrients to tree roots. So, the roots of fruit tree only absorb nutrients that are dissolved in the soil solution and present in ionic form. According to Ebert (2009), two main processes are involved in this movement, mass flow and diffusion. Mass flow of nutrients to the root is brought about by the transpiration of the tree, which during the uptake of water literally sucks the nutrients in the soil to the root surface. Nutrients are presented in high amounts like NO_3^-, Ca^{2+}, and Mg^{2+}; mass flow is the means by which they are transported to the root surface prior to uptake. Contrary, nutrients that have low ionic concentrations in soil such as K, especially P, move to the roots by diffusion, that is, down a concentration gradient induced by the removal of these nutrients at the tree root surface during uptake. Namely, amounts of these ions transported by mass flow are not high enough to meet the nutrient requirements of the tree.

Soil conditions such as water content, soil type and structure, organic matter content, and soil pH determine the availability of nutrients to tree roots. These conditions are autonomous and affect individually the availability of nutrients. But, full positive influence is expressed together. Nonetheless, the content of organic matter and soil pH (Fig. 41.2) is the basis for determining the type, amount, and time of application of fertilizers (Ubavić et al., 1990).

Generally, macroelements are available to the root of fruit tree in a wider range of soil pH with exception of Mg, especially P. Broadly speaking, orchard soils should have pH within 6.0 and 7.3 (Wang et al., 2015). However, within the soil pH < 6.0, forms of Mg, Ca, P, and Mo become less available, and when pH falls below 5.5, the uptake of N, K, and S becomes hindered (Tkaczyk et al., 2018). In contrast, the availability of micronutrients relates closely to soil pH. Namely, most of micronutrients are poorly available to tree root at high values of soil pH. Hence, their accessibility increases when soil pH decreases. A classic example is the availability of Fe. Although soil can be very rich in this element, Fe deficiency occurs very frequently in some stone fruit species, such as sweet cherry, growing on calcareous

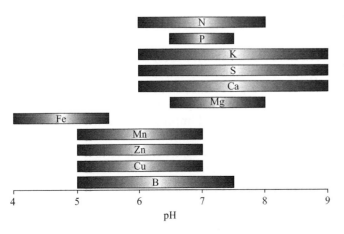

FIG. 41.2 Nutrient availability as a function of soil pH. (Bars represent pH ranges of nutrient availability to plant roots). *From Ebert, G., 2009. Fertilizing for High Yield and Quality: Pome and Stone Fruits of the Temperate Zone. IPI Bulletin No 19, Imprimerie de Saint-Louis, France.*

soils with a pH higher than 7 (Jiménez et al., 2007). However, we observed Fe, Mn, Zn, and B deficiency in same stone fruits at low soil pH (4.71), although soil contains these elements in sufficient quantities (Milošević et al., 2014). The reasons for this phenomenon are numerous, but the primary is antagonism between some nutrients, for example, Fe versus P.

On the other hand, this is probably related to the leaves of scion and roots of rootstock being the areas of the most complex physiological activity (Webster and Looney, 1996). Contrary, at pH > 7.3, Fe, Mn, B, and Cu become hardly available for plants, and as a result, fruit trees succumb to calcium chlorosis (Zia et al., 2006).

Otherwise, individual species differ in their soil pH requirements, for example, stone fruit trees prefer acidic and slightly acidic soils with the optimal pH between 5.6 and 7.7 (Ubavić et al., 1990; Mišić, 2006). Among others, rootstocks and interstocks for these species play an important role in the assimilation of macro- and micronutrients at different levels of soil pH (Milošević and Milošević, 2011b; Milošević et al., 2014).

As is also true for the macronutrients, total micronutrient concentrations provide little information concerning availability to root of fruit trees in orchards (Ebert, 2009). According to recent concepts, this process is strictly regulated by large groups of genes, which are specific for each nutrient (Mitra, 2017). These genes produce mRNA transcripts that translate sets of transporter proteins specific for each nutrient.

Regarding effect of soil organic matter content on availability and assimilation of nutrients, its positive role has long been known. In soil, naturally organic matter or organic fertilizers (manure, compost, etc.) when applied to the soil may be mineralized over time by soil microorganisms in the fruit orchards, which would increase the total organic C content, the microbial biomass C content, and the metabolic activity of the microorganisms (Adani et al., 2007). Also, organic matter improves soil physical, chemical, and biological properties and water, air, and temperature regimes that contribute to better assimilation and utilization of nutrients.

7 Stone fruit fertilization to improve yield and fruit quality

7.1 Fertilizers used in stone fruit production

In general, orchard nutrition is a preharvest factor that affects productivity and fruit quality and has to be performed very carefully since, after harvest, fruit quality cannot be improved, only maintained (Crisosto et al., 1997). Fruit trees are fertilized to ensure continued growth, fruit production, and quality.

Nutrition (fertilization) as an agrotechnical measure is based on the application of organic, organomineral, and mineral fertilizers. Representatives of organic fertilizers are manure, livestock slurry, artificial manure, compost, and green manure. Most commonly used are manure and compost. Horse manure is the best, followed by sheep and cattle, and the worst is swinish manure. Recently, vermicompost (so-called worm castings, worm humus, worm manure, and/or worm feces) has attracted increasing attention. This organic fertilizer is the product of the composting process using various species of worms, usually red wigglers, white worms, and other earthworms to create a mixture of decomposing vegetable or food waste, bedding materials, and vermicast. Vermicompost contains water-soluble nutrients and is an excellent, nutrient-rich organic fertilizer and soil conditioner (Zularisam et al., 2010). It is used in farming and small-scale sustainable, organic farming.

Mineral fertilizers can be individual (single) and composite. Single mineral fertilizers are nitrogenous, phosphoric, potassium, magnesium, and lime. The composite mineral fertilizers are divided into mixed and complex. According to the chemical reaction, mineral fertilizers are divided into acidic, basic, and neutral. A more detailed overview about the properties of mineral fertilizers used in fruit production is given by Ebert (2009). Nitrogen (N) and other macro- and micronutrients, with the exception of zinc (Zn), can be broadcast over the ground and watered in or applied in a band in the irrigation furrows prior to irrigation (Hanson, 1996).

Agrochemicals (mineral fertilizers, pesticides, and synthetic hormones) have been widely used in agronomic, also in horticultural, industry since its introduction in the 1960s during the "Green Revolution" and have since been proven to boost food productivity (Adhikary, 2012). However, the usage of these agrochemicals has been reported to produce harmful consequences to the environment and society, that is, humans and animals (Theunissen et al., 2010). Namely, fertilizers, especially mineral, can pollute lakes, streams, and groundwater if used improperly (Hanson, 1996). Also, fertilizers can contribute nitrate to groundwater and phosphorus to lakes and streams.

Excessive fertilization has been verified, especially on the horticultural enterprises, where the fertilizer costs represented less than 10% of the variable crop costs (Huett and Dirou, 2000). Beside economic aspects, excessive fertilization has been associated to ground and stream soil and water contamination and causing an increment of pest and disease (Marschner, 1995). Also, applications of aforementioned amounts needed by the fruit tree species (Fig. 41.3) are an

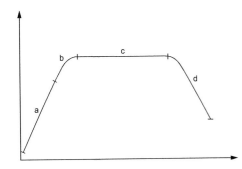

FIG. 41.3 The relationship between yield (y-axis) and content of elements (x-axis) in fruit trees: acute deficiency (a), latent deficiency (b), luxury nutrition (c), and toxic concentration (d). *From Ubavić, M., Kastori, R., Peić, A., 1990. Fertilization of Orchards and Vineyards. DP H.I. "Zorka", Subotica.*

unnecessary expense and may result in reduced fruit quality, toxicities, or deficiencies of other elements (Hanson, 1996).

Proper fertilizer rates and application practices minimize the adverse effects of nutrient movement out of fruit plantings and reduce fertilization costs (Hanson, 1996). Thus, efforts have been made by various scientific communities to look for alternative strategies to improve food production while ensuring environmental sustainability.

Since organic and mineral fertilizers used in fruit orchards have a number of advantages and disadvantages, the need for the development of a new fertilizer type has been imposed. On this line, many companies worldwide produce a combination of organic and mineral fertilizers, so-called organomineral fertilizers. Organomineral fertilizer is defined as a fertilizer obtained by blending, chemical reaction, granulation, or dissolution in water of inorganic fertilizers having a declarable content of one or more primary nutrients with organic fertilizers or soil improver (Paré et al., 2009). These products have been derived from a range of different organic and inorganic sources (Paré et al., 2010) and show a slow release of nutrients, which is the main advantage in relation to mineral fertilizers. The organic matter in liquid organomineral fertilizers is sourced from high-quality compost or other sources of organic matter, such as, poultry manure, sewage sludge, and food waste (Kominko et al., 2017), that contains high levels of organic matter (at least 65%) and humic-fulvic acids (at least 25%). This organic matter (a) increases soil nutrient retention capacity (cation exchange capacity), (b) increases soils water and air retention capacity, (c) increases soil trace mineral levels, (d) balances soils pH, and (e) increases soil beneficial microorganism populations. For information, the direct application of poultry manure or sewage sludge makes risk for humans, fruit trees, and environment, due to contain pathogenic microorganisms (*Escherichia coli*) and toxic compounds (Kominko et al., 2017) (Fig. 41.3).

One of the measures considered highly effective, biologically justified, and environmentally safe, especially on degraded and other soils having unfavorable productive traits for crop cultivation, including fruit trees, is the use of natural zeolite mineral (Polat et al., 2004). Natural zeolites, due to their structural, ion exchange, and sorption properties and many other properties, are well suited for agricultural uses—in animal and plant production (Reháková et al., 2004), especially on devastated, degraded, and acidic soils. The application of zeolites to soils increases their electrical conductivity (EC), and as a result, it increases nutrient retention capacity and usually increases soil pH (soil conditioner) (Glisic et al., 2009; Milošević and Milošević, 2010). It was verified that, when mixed with N, P, and K compounds, zeolite enhances the action of such compounds as slow-release fertilizers, both in horticultural and extensive crops (Reháková et al., 2004). Finally, zeolites improve growth and development of different fruit trees, and their application resulted in a yield and fruit quality increase (Milošević and Milošević, 2013; Milošević et al., 2013a,b). A detailed overview of usage of natural zeolites in agriculture, especially in fruit growing, was made by Polat (2004), Milošević and Milošević (2010), and Milošević et al. (2018).

7.2 Factors influencing stone fruit tree fertilization

As previously reported, for proper growth and large, high-quality yielding, fruit trees, including stone fruits, need appropriate amounts of water, nutrients, solar radiation, and CO_2 (Tkaczyk et al., 2018). According to the same authors, this means that the fruit yield and fruit quality are affected by a number of external factors over which the farmer has no control, such as the precipitation, temperature, insolation, or concentration of CO_2 in the atmosphere. On the other hand, among factors that have a decisive impact on orchard production that can be controlled by a fruit farmer are the level of macro- and microelements in soil and soil reaction (Zia et al., 2006).

Several factors influenced positive results of fertilizers, that is, macro- and micronutrient application. The most important factors are soil fertility, soil type, climatic conditions, fruit species and cultivars, rootstock and its vigor, developmental stage of the tree, planting density, and the scope to be achieved by fertilizing (Ubavić et al., 1990). We reported about influence and importance of soil fertility, soil type, and climatic conditions on stone fruit tree growth and development in the previous sections of this chapter.

The effect of the rootstock on the nutrition of the fruit trees, including stone fruits, is known for a long time. Rootstock affects, for example, vegetative growth, precocity, productivity, fruit quality, resistance to negative biotic and abiotic factors, physiological and phonological traits, increase or decrease of the nutrient uptake, and leaf and shoot nutrient composition (Jiménez et al., 2007; Milošević et al., 2014).

Rootstocks for stone fruit trees are genetically more different from the grafting partner i.e. component than are the apple and/or pear rootstocks. As a consequence, problems of incompatibility are more pronounced in *Prunus* spp. combinations than in apple.

The same is true for a great number of sweet cherry (*P. avium* L.) cultivars grafted on less vigorous *P. cerasus* L. or other *Prunus* rootstocks. In sweet cherry production, a great deal of effort has been made in recent years to find suitable, growth-reducing rootstocks. For an economically reasonable production, the tree's height should not exceed 4 m. However, as yet, there is still no *Prunus* rootstock that simultaneously reduces tree height and favors fruit quality as does the M.9 rootstock in apple production. Regarding rootstocks, similar phenomenon characterizes other stone fruits such as plum, peach, apricot, or sour cherry. Adequate choice of suitable rootstock, for cultivars of stone fruits, broadly speaking, harmonious rootstock/cultivar combination, of stone fruits for growing in soils with different physical, chemical, and biological traits to improve high yield and fruit quality is a decisive factor. In the studies of Ystaas and Frøynes (1998) and Milošević et al. (2014), rootstocks significantly changed leaf macro- and micronutrient levels in sweet cherry cultivars at midsummer, that is, 120 days after full bloom (DAFB). Also, macro- and micronutrient levels of plum, apricot, and/or sweet cherry leaves at 120 DAFB appeared to be significantly influenced by cultivars (Milošević et al., 2012b, 2013a,b, 2014). In general, low yielding cultivar trees, for example, requires less N than other heavy bearers.

Growing system also plays important role in the uptake of nutrients. Namely, trees in high-density planting orchards are more susceptible to nutrient deficiencies including micronutrient deficiencies because their small and shallow root systems do not allow the exploitation of nutrient reserves in deeper soil layers (Ebert, 2009).

7.3 Determination of adequate nutrient requirement of stone fruits

The fertilization practice lays mainly on two concepts, maintenance of the rooting medium fertility and balancing of the crop nutrient demand. Both benefited from recent scientific advances, which serve the purposes of diagnosis, crop modeling, and optimization of fertilizer use (Le Bot and Adamowicz, 2005).

The nutrient requirement of fruit trees is determined by the soil and tissue nutrient content, which is necessary for normal vegetative growth, development, productivity (Ebert, 2009), and acquirement of fruits with excellent external and internal quality, of course biologically valuable and health safe (Milošević et al., 2013a,b). Basically, requirement is closely related to the defined normal (optimum) range of nutrient amounts in soils (Hanson, 1996) and also in leaves, flowers, fruits, and/or 1- and 2-year-old shoots (Leece, 1975a,b; Leece and van den Ende, 1975; Bergmann and Neubert, 1976; Bergmann, 1992; Jiménez et al., 2004, 2007).

7.3.1 Soil testing

Generally, soil testing provides a means of monitoring soil pH and organic matter content and estimating nutrient supply (Hanson, 1996). This author also reported that inadequate relationship exists between soil and plant nutrient levels in perennial crops. Often, fruit trees contain sufficient levels of a nutrient even though soil test values are low. Conversely, high soil nutrient levels do not assure an adequate supply to the tree. Soil tests do give a reasonable estimate of the nutrient status of shallow-rooted crops such as strawberries, raspberries, currants, and/or other small fruit species (Ebert, 2009). Soil analysis is done every third or fourth year.

Soil tests are most useful in monitoring soil pH in established orchards because it greatly influences nutrient availability to plants, and many nutrient deficiencies can be avoided by maintaining soil pH between 6.0 and 7.0 (Hanson, 1996), that is, between 5.7 and 7.7 (Ubavić et al., 1990). Nutrient deficiencies or excesses (toxicities) are more likely when the pH is outside of this range. In general, soils too low in pH are more common worldwide than soils with an excessively high pH. For example, in Serbia, over 60% agricultural soils are acidic (Milošević and Milošević, 2012a; Milošević et al., 2012b); in southeastern Poland, 52% of soils are highly acidic and acidic, 23% slightly acidic, and 25% neutral or alkaline (Tkaczyk and Bednarek, 2011), whereas in Michigan soil pH varies widely, but soils with low pH values dominate (Hanson, 1996). According to same author, low pH may be due to the native soil acidity or repeated use of acidifying nitrogen fertilizers. The pH of many older plantings is low because of acidic fertilizer use. So, liming is an operation required for these soils or obligates application of basic (alkaline) mineral fertilizers (Milošević et al., 2013b).

Contrary, soil salinity is a rising problem in many areas of Mediterranean countries and of the subtropics and tropics (Ebert, 2009). It is often associated with drip irrigation systems. Sodium chloride (NaCl) is the dominant salt involved in the problem. In some Mediterranean countries (Spain), problem of calcareous soils, which show lime-induced chlorosis, is solved by applying an appropriate rootstocks for some stone fruits, such as SL 64 and Adara for sweet cherry (Moreno et al., 1996) or almond × peach hybrid rootstocks—GF 677, Felinem, Garnem, and Mongero for other members of *Prunus* genera (Felipe, 2009). These rootstocks are characterized by adaptation to calcareous soils and other Mediterranean agroecological conditions. On calcareous soils, application of acidic fertilizers is "conditio sine qua non."

To assess the levels of available plant nutrients in the soil, a standard classification scheme is useful, indicating ranges of nutrient concentrations relating to soil type (i.e., sandy, clay, and calcareous soils) (Tables 41.2 and 41.3).

For example, in South Tyrol, an area with very intensive apple, pear, and other fruit species production, soil orchards are also divided into five classes (Drahorad, 1998). Based on these classes, a recommendation for fertilization was made by available P, K, Mg, and B.

In Germany, orchard soils are divided into five classes according to their available nutrient contents (Ebert, 2009). Similar schemes are used elsewhere throughout the world where fruit tree crops are grown (Ubavić et al., 1990). The classification was based on the content of nutrients in the soil. For soil evaluation, soil pH, soil type, and content of organic matter are taken. The optimum class of supply (marked with C) should be sought (Table 41.4).

Numerous methods and equipments are used to determine the chemical composition of the soil. For determination of N content in soil, N_{min} method is used (Wehrmann and Scharpf, 1979). The method is based on the measurement of residual nitrogen in the soil, that is, the amount of nitrate (NO_3^-) and ammonium (NH_4^+) nitrogen form that is available for the tree. Soil samples are taken from two depths: from 0 to 30 cm and from 31 to 60 cm within the tree row shortly (approximately 2 weeks) before flowering. So, N_{min} method provides valuable data in estimating the appropriate N application.

For other macronutrients and micronutrients, standard and routine analyses were taken using different extraction methods (e.g., water, $CaCl_2$, and electroultrafiltration). For a meaningful interpretation of the soil data, classification schemes are used, which are regional or local as well as fruit crop specific and dependent on soil pH, soil

TABLE 41.2 Classification of soil according to total nitrogen (N_{TOT}) content.

N_{TOT} content (% of dry matter)	Soil class
<0.1	Poor
0.1–0.2	Medium
>0.2	Good

From Ubavić, M., Kastori, R., Peić, A., 1990. Fertilization of Orchards and Vineyards. DP H.I. "Zorka", Subotica.

TABLE 41.3 Classification of soil according to available P_2O_5 and K_2O contents and fertilization recommendation using the VDLUFA[a] scheme.

Soil class	Nutrient content	P_2O_5 mg 100/g dry soil	K_2O[b] mg 100/g dry soil	Fertilization recommendation
A	Very low	<3.5	<8	Very high
B	Low	3.5–7.4	8–16	Higher than C
C	Optimum	7.4–14	16–24	Equal to that removed by fruits
D	High	14–20	24–32	Less than C
E	Very high	>20	>32	None

[a] *Association of German Agricultural Research and Research Institutes (Verband Deutscher Landwirtschaftlicher Untersuchungs- und Forschungsanstalten) (Source: Kerschberger et al., 1997).*
[b] *For loamy soils.*

TABLE 41.4   Soil classification and nutrient requirements (kg/ha) in South Tyrol in accordance with soil analysis.

Soil class	P$_2$O$_5$	K$_2$O	MgO	B
A	40–70	140–180	50–70	1.0–1.5
B	20–40	100–140	30–50	0.7–1.0
C	10–20[a]	60–100	20–30	0.5–0.7
D	0[b]	20–60	0–20	0.5–0.7
E + S	0[b]	0[b]	0[b]	0[b]

[a] In 2–3 years.
[b] To the next soil analysis.
From Drahorad, W., 1998. Nutrition of fruit trees. In: Gvozdenović D (Ed.), Apple. Faculty of Agriculture, Novi Sad, pp. 62–73 (in Serbian).

type, and climatic conditions, as previously reported (Ebert, 2009) Soil nutrient levels and pH usually change gradually (Milošević and Milošević, 2010), so sampling every 3–5 years is adequate to monitor established plantings (Hanson, 1996).

Soil nutrient analysis does not always provide sufficient information about nutrient availability to trees. Therefore, additional analysis is recommended to assess the uptake of nutrients from the soil.

7.3.2 Plant analysis

Since the soil testing is insufficient, to obtain better information about real requirements of fruit trees for nutrients, it is necessary to carry out a chemical analysis of some plant parts. For this purpose, the most commonly used organs are leaf, fruit, flower, and 1- and 2-year-old shoots (Leece, 1975a,b; Marschner, 1995; Jiménez et al., 2004, 2007). In addition, it is necessary to integrate both techniques, chemical analysis of plants and soil testing, besides visual diagnosis to maximize fertilization efficiency in terms of cost and prevention of environmental damage (de Mello Prado and Caione, 2012). Apart from the content of individual macro- and microelements in the leaves, their mutual relationship is very important to avoid antagonism (Papp and Tamási, 1979).

Leaf is the most common organ used for chemical analysis. The use of plant analysis as a diagnostic tool has a history dating back to studies of plant ash content in the early 1800s, that is, 1804 by de Saussure (Ulrich, 1952). Chemist working on the composition of plant ash recognized that relationships existed between yield and the nutrient concentrations in plant tissues. Later, Liebig in 1852 suggested that the fertility of a soil could be maintained through the sample expedient of the returning to the soil as fertilizers the nutrients contained in the crops removed from the field.

Quantitative methods for interpreting these relationships in a manner that could be used for assessing plant nutrient status arose from the work of Macy (1936). Since then, much effort has been directed toward refining plant analysis as a diagnostic tool. Readers interested in following these developments are referred to the articles of Goodall and Gregory (1947), Ulrich (1952), Ulrich et al. (1959), Ulrich and Hills (1967), Bates (1971), Bouma (1983), Martin-Prével et al. (1984), Munson and Nelson (1990), Jones et al. (1991), and Bergmann (1992).

Deficient, low, normal (optimal and adequate), high, and excess or toxic levels of macro- and microelements in stone fruit leaves were proposed by several authors (Leece, 1975a,b; Leece and van den Ende, 1975; Bergmann and Neubert, 1976; Leece et al., 1971; Bergmann, 1992; Heckman, 2004). So, surveys of nutrient concentrations in "deficient" and "adequate" plants have been used to establish standard nutrient levels for some stone fruit species. Publications of these authors also report previously defined fruit tree organs, sample numbers, and the sampling period for diverse fruit cultures, including stone fruits. But to utilize such data as a standard, it is necessary to be careful about the physiological age of each plant and leaf as stated by the author (de Mello Prado and Caione, 2012).

Leaf analysis of stone fruit trees is restricted to midsummer (120 days after full bloom), which is the standard time for leaf sampling of fruit trees (Montañés and Sanz, 1994). In general, leaf contents of specific macro- and micronutrients vary during the growing period with similar tendencies in all stone fruit trees (Milošević and Milošević, 2011b, 2012a). The content of certain micronutrients in leaves may vary considerably that shows the foliar ranges in deciduous fruit trees. Within these ranges, growth and yield of stone fruit trees are usually satisfactory (Milošević et al., 2013a,b, 2014), but at the lower end of the range leaf, deficiency symptoms appear, whereas too high micronutrient concentrations may result in toxicity symptoms (Ebert, 2009).

Because of the dynamic nature of the leaf tissue composition, strongly influenced by leaf age, maturation stage, and the interactions involving nutrient absorption and translocation, the tissue diagnosis may be a practice of difficult understanding and utilization (Walworth and Sumner 1987). For these reasons, several methods for nutritional diagnosis using leaf tissue analysis have been proposed and used, including the critical value (CV), the sufficiency range approach (SRA), deviation from optimum percentage (DOP index), and the diagnosis and recommendation integrated system (DRIS). Between them, DOP index and DRIS system are the most important (Tables 41.5 and 41.6).

The DOP index is an alternative methodology for plant mineral analysis interpretation (Montañés et al., 1993). According to the same authors, it was estimated for the diagnosis of the leaf mineral status of the trees. The DOP index was calculated from the leaf analysis at 120 DAFB by the following mathematical expression:

$$DOP = \left(\frac{C_n}{C_o} - 1\right) \times 100$$

where C_n = foliar content of the tested major and/or minor nutrient and C_o = critical optimum nutrient content for stone fruit species; both values are given on a dry matter basis. C_o is the optimum (normal) value to be considered as optimum, both values given on a dry matter basis. The C_o or C_{ref} has been taken from optimum values, proposed by different prominent authors previously mentioned. An optimum nutritional situation for any element is defined by

TABLE 41.5 Leaf composition standards for stone fruits based on midshoot leaves sampled in midsummer expressed on a dry weight basis.

Element	Plum (1)	Peach (2)	Apricot (3)	Sweet cherry (4)	Sour cherry (5)
Macroelements (% on dry matter)					
N	2.40–3.00	3.00–3.50	2.40–3.00	2.20–2.60	2.60–3.00
P	0.14–0.25	0.14–0.25	0.14–0.25	0.14–0.25	0.16–0.22
K	1.60–3.00	2.00–3.00	2.00–3.50	1.60–3.00	1.60–2.10
Ca	1.50–3.00	1.80–2.70	2.00–4.00	1.40–2.40	1.50–2.60
Mg	0.30–0.80	0.30–0.80	0.30–0.80	0.30–0.80	0.30–0.75
Microelements (µg/g on dry matter)					
Fe	100–250	100–250	100–250	100–250	100–200
Mn	40–160	40–160	40–160	40–160	40–60
Cu	6–16	5–16	5–16	5–16	8–28
Zn	20–50	20–50	20–50	20–50	20–50
B	25–60	20–60	20–60	20–60	20–55

From (1 and 4) Leece, D.R., 1975. Diagnostic leaf analysis for stone fruit. 4. Plum. Aust. J. Exp. Agric. Anim. Husb. 15, 112–117; Leece, D.R., 1975. Diagnostic leaf analysis for stone fruit. 5. Sweet cherry. Aust. J. Exp. Agric. Anim. Husb. 15, 118–122; (2) Leece, D.R., Cradock, F.W., Carter, O.G., 1971. Development of leaf nutrient concentration standards for peach trees in New South Wales. J. Hortic. Sci. 46 (2), 163–175; (3) Leece, D.R., van den Ende, B., 1975. Diagnostic leaf analysis for stone fruit. 6. Apricot Aust. J. Exp. Agric. Anim. Husb. 15, 123–128; and (5) Heckman, J., 2004. Leaf Analysis for Fruit Trees. Rutgers Cooperative Research & Extension NJAES, Rutgers the State University of New Jersey, New Brunswick, NJ, pp. 1–2.

TABLE 41.6 Average contents and/or concentrations of essential mineral elements in shoots that are sufficient for optimal growth.

Macroelements	g/kg	Microelements	mg/kg
Nitrogen (N)	15	Chlorine (Cl)	100
Phosphorus (P)	2	Iron (Fe)	100
Potassium (K)	10	Manganese (Mn)	50
Calcium (Ca)	5	Boron (B)	20
Magnesium (Mg)	2	Zinc (Zn)	20
Sulfur (S)	1	Cooper (Cu)	6
		Molybdenum (Mo)	0.1

From Marschner, H., 1995. Mineral Nutrition of Higher Plants, second ed. Ademic Press, London, 889 p.

a DOP index equal to zero. Furthermore, the DOP indexes indicate whether an element is in defect (negative indexes) or excess (positive indexes) (Montañés et al., 1993). Hence, DOP index is a good relationship between the nutritive state of the plant and the soil.

The ΣDOP index is obtained by adding the values of DOP index irrespective of sign. Basically, this index is a supplement to the DOP index. The larger the ΣDOP index value, the greater the intensity of imbalances among nutrients and vice versa.

The DRIS system as a new interpretation for leaf analysis was firstly developed and proposed earlier by Beaufils (1957) with several modifications to increase accuracy in the nutritional diagnosis for several crops. Shortly, the DRIS method uses nutrient ratios instead of absolute and/or individual nutrient concentrations for the interpretation of tissue analysis (Mourão Filho, 2004). This method or system for plant nutritional diagnosis can basically be applied in two forms: DRIS graphs and DRIS indexes with adequate norms. In some countries such as the United States, Canada, and China, this method is being adopted as part of a representative diagnosis in selected areas (Walworth and Sumner, 1987).

Much more information about interpretation for leaf analysis, especially DOP index and DRIS methods, is provided by Montañés et al. (1993) and Mourão Filho (2004).

7.4 Improving stone fruits productivity and fruit quality using fertilizers

7.4.1 Organic fertilizers

There are different philosophies on how often fertilizer should be applied to stone fruits, but all agree that they prefer frequent small amounts rather than one or two large applications. Stone fruits are generally quite vigorous, but if the tree is overly vigorous, cut back the amount of fertilizer by 20%–25% the following year and observe the response. Excess vigor will produce vegetative growth and wood that has more leaf buds and few fruit buds.

Stone fruit trees have similar requirements for organic and mineral nutrients. In modern plantations, grass residues and crushed remains of branches after pruning and leaves are humified that increase the level of humus. If soil contains 3%–5% humus, the addition of expensive organic fertilizers makes no sense. In the case of low humus content (<2%), it is recommended to add organic fertilizers.

Many literature data clearly showed that, for example, cattle manure slightly improved vegetative growth and productivity but highly improved fruit size and some fruit internal quality attributes of apricot (Milošević and Milošević, 2013; Milošević et al., 2013a,b), of course, and other fruit crops such as apple (Milošević and Milošević, 2017) or blackberry (Milošević et al., 2018). Other organic fertilizers have similar effects on the growth and development of stone fruit trees (Hanson, 1996). Fresh materials are high in nutrients (poultry manure) and can injure tree roots if applied at excessive rates.

7.4.2 Macronutrients

Among macroelements, the nutrient with the single greatest effect on tree growth, productivity, and fruit quality of stone and other fruit crops is nitrogen (Crisosto et al., 1997). This element can be applied as NO_3, NH_4 ion, or other forms like amides in urea. Nitrate is immediately available to plant roots; however, due to its high mobility in the soil, it is also subject to leaching after rainfall or irrigation (Ebert, 2009).

Response of plum, peach and nectarine, apricot, and cherries trees to N fertilization is dramatic and different. According to new trends of N fertilization management, small doses are preferable. Namely, it is much easier to apply additional N than to manage excessive vigor caused by too high rates (Hanson, 1996). For example, Tagliavini and Marangoni (2002) recommend an application rate between 0- and 60-kg N/ha for pome fruits. Excessive vigor is particularly damaging in new, high-density stone fruit plantings grafted on dwarf and/or semidwarf rootstocks such as Pixy for European plums, Gisela 5, Gisela 6, PHL-A and PHL-C for sweet cherry, and Pumiselect for apricot or peach etc. (Milošević, unpublished data). Hence, high N levels stimulate vigorous vegetative growth, causing shading out of lower fruiting wood and its death, especially in peach and nectarine, and increase accumulation in fruit skin and leaves (Crisosto et al., 1997). Also, according to the same authors, excess N does not increase fruit size, production, or soluble solid content; delays stone fruit maturity; induces poor visual red color development; inhibits ground color change from green to yellow; and reduces tree hardiness. In contrast, N deficiency leads to small fruit with poor flavor and unproductive trees (Daane et al., 1995). Hence, judicious use of N fertilizers is an important issue in the present and future stone fruit production based on soil testing and leaf chemical analysis because optimum N rates vary considerably from site to site. In addition, the use of mineral fertilizers, especially N-containing mineral fertilizers, is limited by the EU Nitrates Directive and the regulations of members of EU and other countries.

Application of P, K, Ca, and Mg without real requirement is very harmful (Hanson, 1996).

Phosphorus (P) is the second major nutrient that is essential for plant healthy growth, development, and greater resistance to diseases. Because P moves very slowly in soil, small rates annually between 6 and 11 kg/ha will sustain most fruit crops, including stone fruits, for many years (Johnson and Uriu, 1989). On this line, high amount of P fertilizer applied to the soil induced excess P in leaves of apricot (Milošević et al., 2013a,b). In peach, P applications have increased flower formation (Fukuda and Kondo, 1957) but decreased yield. Chatzitheodorou et al. (2004) also reported that applications of P alone or mixed PK decreased yield of peach cv. "Red Haven." Some studies reported that cherries usually do not respond to P fertilization; however, certain circumstances might indicate otherwise (Westwood and Wann, 1966). The circumstances under which cherries respond to P fertilization occur when other nutrients become immobilized and are not moving through the tree. Cherries need P to maintain a balance of micronutrients within the tree. Research has shown that P fertilization in sweet cherries has also increased fruit firmness (Neilsen et al., 2007).

Deficiency of P is rarely seen due to little plant requirement but occurred in some peach plantations. Under these conditions, yield and fruit size are reduced although fruit quality properties were not directly affected by the P application. In addition, the fruit are more highly colored and ripen earlier but exhibit surface defects and poor eating quality (Johnson and Uriu, 1989). Foliar sprays of phosphorus improve yield and reduce the water core (Tripathi et al., 2017).

Potassium (K) is the third macronutrient; it is necessary for the development of strong plants. Unlike P, K is a nutrient that needs to be supplied to fruit trees in relatively large quantities. This element is the most abundant nutrient in the fruit where it affects positively the size, firmness, skin color, soluble solid content, acidity, juiciness, and aroma (Brunetto et al., 2015). Although excessive K is not directly toxic, high soil K amounts may inhibit Mg or Ca uptake and so induce deficiencies of these elements (Hanson, 1996). In peach, yield and fruit weight increased with increasing rate of application to up to 700-g K/tree. The lowest yield of "Red Haven" peach was recorded in P, PK, and control treatments (Chatzitheodorou et al., 2004). Also, in peaches, K fertilization resulted in an increase in titratable acidity in the fruit. In contrast, early research has shown that P and K added during the growing season had little to no effect on fruit quality in sour cherry (Reuther et al., 1958). However, recent research indicates that fruit size, acidity, and color are positively related to the application of K and P during the early growing season as well (Stiles and Reid, 1991). To keep the leaf major nutrients within the optimal range, sweet cherries have a requirement of 80-kg K/ha to produce a satisfactory crop on loamy sand soil high in organic matter (Ystaas, 1990).

Because K is antagonist to both Ca and Mg, if K increases in the soil over the normal range, both Ca and Mg uptake may decrease (Brunetto et al., 2015). Determining the amounts of P and K in the leaves and fruit under a variety of conditions allows development of an economical and effective nutrient management plan specific to concrete pedoclimatic conditions.

Calcium (Ca) plays very important role to maintain the cell membranes and also helpful to prevent the physiological disorders, which occur due to Ca deficiency in apple and pear production. However, little is known about the Ca requirements for either the European or Japanese plum, sweet and sour cherries, and other stone fruit species. Calcium deficiencies are rare if pH is maintained above 6.0. However, many of the low calcium-related problems in tree fruits are attributable to inadequate amounts of soil Ca. Low levels of soil Ca are usually associated with low soil pH and low cation exchange particularly in subsurface soils (Stiles and Reid, 1991). These authors also reported that this phenomenon usually reflects inadequate lime application prior to establish the orchard and/or failure to maintain an adequate liming program throughout the life of the orchard. For example, the application of 5000-kg ground limestone/ha significantly reduced fruit size in loamy sand soil with 7% organic matter, whereas application of 2500-kg limestone/ha every fifth year will be a safeguard against soil acidification and provide adequate supply of exchangeable Ca within a favorable pH regime of 5.5–6.5 (Ystaas, 1990).

In our earlier studies, a large deficit of Ca in apricot and sweet cherry leaves occurred and was somewhat smaller in plum leaves during their growing on acidic soil (pH 4.86) under Serbian conditions, but their yield and fruit quality were very good (Milošević et al., 2012b, 2013a,b, 2014). It seems that the requirements of apricot, plum, and sweet cherry to Ca are rather modest.

It has been long known that Ca deficiency causes cracking of cherries and plums. On this line, some authors reported that Ca sprays, due to its low mobility, caused a marked increase in the firmness of "Stanley" plum fruit at harvest and reduced softening during storage but had no effects on "Dabrowicka Prune" (Plich and Wojcik, 2002). It is possible that the benefits of supplementary Ca on "Stanley" plums may relate to its lower Ca status compared with the "Dabrowicka Prune" cultivar. Abdel-Hafeez et al. (2010) revealed that foliar application of $Ca(NO_3)_2$ increased N level in leaf and improved fruit quality and storage capacity of European plum cv. "Kelsey." On the other hand, as a possible method of reducing rain-induced fruit cracking on sweet cherry, a range of spray products containing Ca have been tested (Cline, 1995). However, detailed studies carried out under conditions of Great Britain led to the conclusion that it was questionable whether there were economic benefits of applying Ca sprays for this purpose (Cline, 1995).

Postharvest application of $CaCl_2$ increased the flesh Ca amount of cherries and improved the texture (firmness and bioyield) and the incidence of pitting resulting from impact damage (Lidster et al., 1978, 1979). In addition, fruit Ca accumulation is higher at the beginning of fruit development, that is, during fruit cytokinesis (Brunetto et al., 2015). Crisosto et al. (1997) suggests that $CaCl_2$ spray on peaches and nectarines should be treated with caution because their heavy metal content (Fe, Al, Cu, etc.) may contribute to fruit skin discoloration. Generally, Ca sprays may reduce some problems but rarely eliminate them (Hanson, 1996). From these reasons, a wider survey experiment would be needed to distinguish between stone fruit species and their cultivar effects and indirect effects of cultivar on tree calcium uptake.

Magnesium (Mg) uptake can be strongly depressed by K^+, NH_4^+, Ca^{2+}, and/or H^+ (Johnson and Uriu, 1989; Marschner, 1995). Levels of Mg in the tree are generally less than either Ca or K (Johnson and Uriu, 1989). This element accumulates linearly at a slow rate throughout the growing season (Ubavić et al., 1990). Generally, Mg is classified as mobile because symptoms tend to develop first on older leaves (Hanson, 1996). Deficiencies in plant tissues are common in acidic soils, particularly where the soil is high in plant-available K. Similarly to Ca, we also observed Mg deficiency in leaves of apricots, plums, and sweet cherries grown on acidic but different soil types with high K content (Milošević et al., 2012b, 2013a,b, 2014). So, it can be said that Ca and Mg deficiency in some sites worldwide such as California (Johnson and Uriu, 1989) or Serbia (Milošević et al., 2012b, 2013a,b, 2014) usually does not damage yield of some stone fruits, and therefore, no treatment is required. Nevertheless, when correction is desired, soil applications of magnesium sulfate, magnesium oxide, and dolomite have proved effective (Johnson and Uriu, 1989).

7.4.3 Micronutrients

Among micronutrients, boron (B) is one of the most critical in fruit orchards including stone fruit species. About 90 years ago, B was recognized as an essential plant nutrient (Warington, 1923). B is usually deficient in many soils because most of this nutrient is prone to leaching and adsorbed to clay minerals, hydrous metal oxides, and organic matter in soils (Johnson and Uriu, 1989).

Typical symptoms of B deficiency are the reduction of fruit set and yielding and small, deformed, cracked, and corked fruits in some pome fruit species (Wojcik and Wojcik, 2003). Under different soil types such as sandy loamy, heavy and shallow vertisol, and/or clay loamy with soil pH below 5, we observed medium to high B deficiency (<20-mg/kg DW) in leaves of plum (Milošević et al., 2012b), apricot (Milošević et al., 2013a,b), sweet cherry (Milošević et al., 2014), and blackberry (Milošević et al., 2018). However, these B deficiencies does not damage vegetative growth, productivity, and fruit quality attributes. Similarly to enumerated stone fruits, peaches, compared with other fruit trees, are not sensitive to B deficiency (Johnson and Uriu, 1989). Shortly thereafter, Stiles and Reid (1991) reported quite the opposite data. According to them, stone fruits such as peach are extremely sensitive to both deficiencies and excesses of B and should be treated only if the leaf analysis and soil tests indicate a need for B application.

Other experimental investigations indicated that foliar application of micronutrients (B, Cu, Fe, Mn, and Zn) alone and in combination were the most effective treatments in increasing chemical and physical parameters of peach fruit, vegetative parameters of tree and yield under silt loam, and calcareous and alkaline condition of Peshawar (Pakistan) soil (Ali et al., 2014). Taha and Sherif (2015) reported that B spraying improved fruit set and leaf nutrient content while decreased fruit drop herein reflected on yield and fruit quality in "Canino" apricot. As previously reported (Ebert, 2009), the availability of B and other extractable Zn, Cu, Fe, and Mn decreases with increasing soil pH (Fig. 41.2). Since B is mobile in the trees of the *Rosaceae* family (Brown et al., 1999), the most appropriate application period is the end of summer when B is stored in permanent organs and can be remobilized to the developing fruits the following spring (Sánchez et al., 1998). Other researchers reported that B may be applied at any time but is more effective in improving blossom quality and fruit set if applied shortly before full bloom (Taha and Sherif, 2015).

The opposite results of different studies can be explained with the fact that micronutrients, including B, are more soluble in the low soil pH and their availability in soils varies considerably with the seasonal changes in temperature and moisture.

Iron (Fe), one of the most abundant minerals in the soil, constitutes about 5% of the weight of the earth's crust (Johnson and Uriu, 1989). However, despite the abundance of this element in soils, Fe acquisition by fruit crops are often impaired, compromising yield and fruit quality. Fe deficiency occurs very frequently in stone fruit species, especially in peach and nectarine, growing on calcareous soils with a pH > 7 (lime-induced chlorosis), although the soil can be very rich in this element. So, Fe availability increases as the soil pH falls; the best is when pH ranged from 4 to 5.5. Otherwise, Fe chlorosis is a phenomenon that depends on many environmental and agronomic factors such as soil type, climatic conditions, grafting, and rootstock used (Brunetto et al., 2015). In our earlier studies, we established that leaves of European plum cultivars grafted on suckers of local (autochthonous) plum "Belošljiva" grown on sandy loam soil with low pH (4.86) had excessive Fe level (Milošević et al., 2012b), whereas leaves of apricots grafted on seedlings of Myrobalan (*P. cerasifera* Ehrh.) had deficiency level (Milošević and Milošević, 2013; Milošević et al., 2013a,b).

According to DOP index, leaves of sweet cherry cultivars grafted on seedlings of Mazzard (*P. avium* L.) and vegetative Colt rootstock, grown on heavy (vertisol) and acidic soil (pH 4.71), showed Fe deficiency but without chlorosis symptoms and damage on yield and fruit quality (Milošević et al., 2014). Our results support knowledge that soil type and buffer capacity of the rootstocks play an important role in the availability and uptake of Fe. Namely, although Fe levels overall were very deficient, the Fe present may have been quite active physiologically (Leece 1975a,b; Leece and van den Ende, 1975). Iron mostly is present in its oxidized form Fe^{3+}, but to be available to the roots of stone fruit trees, it has to be chemically reduced to Fe^{2+} (Ebert, 2009). The mobility of Fe in the tree is very low; therefore, young leaves and shoots show typical deficiency symptoms.

Corrections of Fe deficiency are difficult and very complex agrotechnical measure. According to Johnson and Uriu (1989), applying inorganic Fe to the soil seldom corrects the deficiency, since conversion to insoluble minerals occurs rapidly. These authors also reported that synthetic Fe chelates applied to soil or foliage have been effective, but results are often short lived and may not be cost-effective. In situations of lime-induced chlorosis (due to reduced chlorophyll synthesis), applications of sulfur (S) to lower the soil pH are somewhat effective.

Manganese (Mn) deficiencies are more likely on alkaline soils (pH > 7.0) and more prevalent during dry seasons (Hanson, 1996). Also, we detected leaf Mn deficiencies in acidic soils (pH < 4.8) in plum (Milošević et al., 2012b), apricot (Milošević and Milošević, 2011b; Milošević et al., 2013a,b), and sweet cherry (Milošević et al., 2014) orchards under western Serbian conditions but without damage and negative impact on the yield and fruit quality. Mixture of essential micronutrients, including Mn, increased chemical and physical parameters of peach fruit (Ali et al., 2014). Deficiency of leaf Mn associated with its low content in soil and also lack of solubility or losses in sandy loam and acidic soil (Milošević et al., 2013a,b). In general, the total amount of soil Mn varies widely from one soil to another, but there is usually an adequate amount to supply the limited requirements of fruit trees (Johnson and Uriu, 1989). According to the same authors, high levels of other cations in soil such as K, Ca, and especially Mg can inhibit uptake of Mn. In addition, Mn ions are effective in till cation competition. To correct deficiency, foliar sprays of $MnSO_4$ have proved effective, whereas soil Mn application is not recommended since manganese (Mn^{2+}) is quickly oxidized to less available forms (Johnson and Uriu, 1989).

Zinc (Zn) deficiency symptoms ("little leaf") are common in many calcareous or alkaline soils worldwide (Ebert, 2009). However, Zn deficiency occurred also in acidic soils. In each case, leaf P levels were high, which indicates P:Zn imbalance, that is, the high P levels having inactivated the Zn (Marschner, 1995). In our earlier works on plums, apricots, and sweet cherries, this phenomenon has been confirmed (Milošević and Milošević, 2011b; Milošević et al., 2013a,b, 2014). Stone fruit are particularly sensitive to Zn deficiency. However, reaction on Zn deficiency varies between rootstocks and cultivars, especially in plums (Johnson and Uriu, 1989).

Since soil applications of Zn have not proven effective, especially on high pH soils, the application of a Zn sprays prior to bud break in the spring is recommended. On this line, three sprays with mixture Zn + Ca prior the onset of vegetative cycle improved fruit set, yield, and fruit quality of apricot cv. "Canino" (Taha and Sherif, 2015). In general, applications made within 3 days before or after an application of dormant oil can cause injury and should be avoided. In this case, except for apricots, zinc sulfate used in a postharvest (September) (Hanson, 1996) or applied from mid-October through the dormant season is very effective (Johnson and Uriu, 1989). Zinc sulfate may also be used with equal amounts of hydrated lime (Hanson, 1996). Foliar sprays of Zn chelates (Zn EDTA) at recommended rates and timing are the preferred Zn sources. Some fungicides containing Zn can be effective sources of this element. The use of zinc sulfate or Zn EDTA during vegetative cycle damages the leaves and fruits.

Copper (Cu) insufficiency is extremely rare in plum (Shear and Faust, 1980) and/or peach and apricot orchards (Johnson and Uriu, 1989), although some cases of Cu deficiency have been reported (Milošević et al., 2012b, 2013a,b). Among stone fruits, plum and sweet cherry are more sensitive to Cu deficiency (Ubavić et al., 1990). In addition, the degree of sensitivity to Cu insufficiency varies between rootstocks and cultivars. In general, Cu deficiency symptoms occur most often on sandy soils, acidic sands and soils heavily fertilized with N or high in organic matter content (Milošević et al., 2013a,b). Excessive leaf Cu amount was observed in sweet cherry orchards and probably represented accumulated levels of inactive Cu (Leece, 1975b; Milošević et al., 2014). Generally, Cu is more available in low pH soils (Fig. 41.2), that is, between 5 and 7 pH units (Johnson and Uriu, 1989).

According to Johnson and Uriu (1989), correction of the Cu deficiency has been obtained with soil applications of copper sulfate at recommended rates. These authors reported that correction can also be expected with foliar chelate or Bordeaux sprays in early spring or soil applications of Cu chelate. Similarly to Zn, foliar application of Cu during vegetative cycle is not recommended (Ubavić et al., 1990).

There is a narrow range between microelement deficiency and toxicity. Because the difference between deficient and excessive levels is small, apply micronutrients only if needed, of course, according to both soil testing and leaf analysis.

8 Conclusions and future research

Fruit production, including stone fruits, increases rapidly in the world. In period 2006–16, the most intensive growth was observed in peach and nectarine (26.41%) and the smallest in apricot (10.79%). The reasons for the large increase of fruit production are numerous, but the most important are the raising of new plantations in countries that have not previously had significant production and improvement of growing technology in countries that are traditionally large fruit producers. However, new cultivation technologies, including irrigation and fertilization, are key factors that have contributed to the achievement of high yields of stone and other fruit species per unit area. Earlier, traditional fruit growing was based on the natural soil fertility and application of mineral, especially nitrogen, fertilizers in large quantities. This inevitably led to degradation and devastation of soil, reduction or loss of fertility, and contamination of soil, water, air, and of course fruits. Such fruits are not health safe. This should add to the high price of mineral fertilizers that significantly burden the costs of fruit production. More recently, agronomic professions and science are facing great challenges in fruit tree fertilization. The new philosophy of fertilizing fruit trees is based on mathematical models that contribute to the rational application of organic, organomineral, and mineral fertilizers to obtain high and regular yields and fruit quality without any risks to human health and environmental pollution and, of course, a drastic reduction of production costs. The new philosophy of fertilizing fruit trees is based exclusively on the conjugation of soil tests and leaf chemical analysis and, of course, also on the general physical and health condition of the trees. In short, only then one can consciously draw up a nutrient balance for a given stone fruit according to the principle "as much as necessary, as little as possible." Finally, we hope that our overview opens new prospective in the fertilization management in stone fruit orchards allowing reduction of input without negative influence on yield, fruit quality, human health, and the environment.

Acknowledgments

This very extensive, difficult, and responsible job was realized with the unselfish help of our family. We greatly thank the little granddaughter Milica, daughter Aleksandra, and my wife Gordana Milošević for moral and technical support.

References

Abdel-Hafeez, A.A., Mohamed, A.I., Taher, N.M., Mehaisen, S.M.A., 2010. Effects of some sources of potassium and calcium as a foliar spray on fruit quality and storability of "Kelsey" plums. Egyptian. J. Hort. 2, 151–168.

Adani, F., Genevini, P., Ricca, G., Tambone, F., Montoneri, E., 2007. Modification of soil humic after 4 years of compost application. Waste Manag. 27 (2), 319–324.

Adhikary, S., 2012. Vermicompost, the story of organic gold: a review. Agric. Sci. 3, 905–917.

Ali, A., Perveen, S., Shah, M.N.S., Zhang, Z., Wahid, F., Shah, M., Bibi, S., Majid, A., 2014. Effect of foliar application of micronutrients on fruit quality of peach. Am. J. Plant Sci. 5, 1258–1264.

Bates, T.E., 1971. Factors affecting critical nutrient concentrations in plants and their evaluation: a review. Soil Sci. 112, 116–130.

Bavcon Kralj, M., Jug, T., Komel, E., Fajt, N., Jarni, K., Živković, J., Mujić, I., 2014. Aromatic compound in different peach cultivars and effect of preservatives on the final aroma of cooked fruits. Hem. Ind. 68 (6), 767–779.

Beaufils, E.R., 1957. Research for rational exploitation of *Hevea* using a physiological diagnosis based on the mineral analysis of various parts of the plants. Fertilite 3, 27–38.

Bergmann, W., 1992. Nutritional disorders of plants. In: Development, Visual and Analytical Diagnosis. Gustav Fischer/Verlag, Jena/Stuttgart/New York.

Bergmann, W., Neubert, P., 1976. Pflanzendiagnose und Pflanzenanalyse flanzendiagnose und Pflanzenanalyse zur Ermittlung von Ernäh-rungsstörungen und des Gesundheitszustandes der Kulturpflanzen. Verlag VEB Gustav Fischer, Jena.

Borlaug, N.E., Dowswell, C.R., 1994. Feeding a human population that increasingly crowds a fragile planet. In: Paper Presented at the 15th World Congress of Soil Science, 10–16 July 1994, Acapulco, Mexico.

Bouma, D., 1983. Diagnosis of mineral deficiencies using plant tests. In: Läuchli, A., Bieleski, R.L. (Eds.), Inorganic Plant Nutrition. In: Encyclopedia of plant physiology (New Series), vol. 15. Springer, Berlin, Heidelberg.

Brown, P.H., Hu, H., Roberts, G.W., 1999. Occurrence of sugar alcohols determines boron toxicity symptoms of ornamental species. J. Am. Soc. Hort. Sci. 124 (4), 347–352.

Brunetto, G., de Melo, B.W.G., Toselli, M., Quartieri, M., Tagliavini, M., 2015. The role of mineral nutrition on yields and fruit quality in grapevine, pear and apple. Rev. Bras. Frutic. 37 (4), 1089–1104.

Cetkovská, J., Diviš, P., Vespalcová, M., Pořízka, J., Řezníček, V., 2015. Basic nutritional properties of cornelian cherry (*Cornus mas* L.) cultivars grown in the Czech Republic. Acta Aliment. 44 (3), 357–364.

Chatzitheodorou, T.I., Sotiropoulos, E.T., Mouhtaridou, I.G., 2004. Effect of nitrogen, phosphorus and potassium frtilisation and manure on fruit yield and fruit quality of the peach cultivars 'spring time' and red haven. Agron. Res. 2 (2), 135–143.

Chen, G., Zhu, H., Zhang, Y., 2003. Soil microbial activities and carbon and nitrogen fixation. Res. Microbiol. 154 (6), 393–398.

Cline, J.A., 1995. Physiological Investigations Into the Rain-Induced Cracking of Sweet Cherries (*Prunus avium* L.): Reasons for Cultivar Differences in Cracking Susceptibility and the Practical Use of Ameliorating Sprays and Protective Rain Covers to Alleviate Cracking in Britain. PhD thesis-University of London.

Crisosto, C.H., 2006. Peach quality and postharvest technology. Acta Hort. (713), 479–488.

Crisosto, C.H., Johnson, R.S., DeJong, T., Day, K.R., 1997. Orchard factors affectingpostharvest stone fruit quality. HortScience 32, 820–823.

Cuquel, F.L., Motta, A.C.V., Tutida, I., May de Mio, L.L., 2011. Nitrogen and potassium fertilization affecting the plum postharvest quality. Rev. Bras. Frutic. 33, 328–336 (Special issue).

Daane, K.M., Johnson, R.S., Michailides, T.J., Crisosto, C.H., Dlott, J.W., Ramirez, H.T., Yokota, G.T., Morgan, D.P., 1995. Excess nitrogen raises nectarine susceptibility to disease and insects. Calif. Agric. 49 (4), 13–17.

de Mello Prado, R., Caione, G., 2012. Plant Analysis. In: Issaka, N.R. (Ed.), Soil Fertility. IntechOpen. Available from: https://www.intechopen.com/books/soil-fertility/plant-analysis.

Delver, P., 1982. Changes in nitrogen fertilization in Dutch orchards. Comp. Fruit Tree 15, 57–72.

Delver, P., Wertheim, S.J., 1988. Promotion of early growth and cropping of apple by trickle irrigation and planting-hole treatments. Gartenbauwissenschaft 53, 128–132.

Dokoupil, L., Řezniček, V., 2012. Production and use of the cornelian cherry—Cornus mas L. Acta Univ. Agric. Silvic. Mendel. Brun. 60 (8), 49–58.

Drahorad, W., 1998. Nutrition of fruit trees. In: Gvozdenović, D. (Ed.), Apple. Faculty of Agriculture, Novi Sad, pp. 62–73 (in Serbian).

Ebert, G., 2009. Fertilizing for High Yield and Quality: Pome and Stone Fruits of the Temperate Zone. IPI Bulletin No 19, Imprimerie de Saint-Louis, France.

Ercisli, S., 2004. A short review of the fruit germplasm resources of Turkey. Genet. Resour. Crop. Evol. 51, 419–435.

Ercisli, S., Orhan, E., Esitken, A., Yildirim, N., Agar, G., 2008. Relationship among some cornelian cherry genotypes (*Cornus mas* L.) based on RAPD analysis. Genet. Resour. Crop. Evol. 55, 613–618.

Evans, R., 2002. An alternative way to assess water erosion of cultivated land-field based measurements and analysis of some results. Appl. Geogr. 22, 187–208.

FAO and ITPS, 2015. Status of the World's Soil Resources (SWSR)—Main Report. Food and Agriculture Organization of the United Nations and Intergovernmental Technical Panel on Soils, Italy.

FAOSTAT, 2018. FAOSTAT Crops. http://faostat.fao.org/beta/en/#data/QC. (Accessed 11 October 2018).

Felipe, J.A., 2009. "Felinem", "Garnem", and "Monegro" almond × peach hybrid rootstocks. HortScience 44 (1), 196–197.

Ferretti, G., Bacchetti, T., Belleggia, A., Neri, D., 2010. Cherry antioxidants: from farm to table. Molecules 15, 6993–7005.

Fukuda, A., Kondo, G., 1957. Growth and yield as related to the concentrations of nitrogen, phosphoric acid, and potassium in sand culture. Stud. Inst. Hort. Kyoto Univ. 8, 16–23.

Gepts, P., 2014. Domestication of plants. In: Van Alfen, N. (Ed.), Encyclopedia of Agriculture and Food Systems. In: vol. 2. Elsevier, San Diego, CA.

Glisic, P.I., Milosevic, M.T., Glisic, S.I., Milosevic, T.N., 2009. The effect of natural zeolites and organic fertilizers on the characteristics of degraded soils and yield of crops grown in Western Serbia. Land Degrad. Develop. 20 (1), 33–40.

Goodall, G.W., Gregory, F.G., 1947. Chemical Composition of Plants as an Index of Their Nutritional Status. Imperial Bureau of Horticulture and Plantation Crops, East Malling, Kent, p. 170.

Güleryüz, M., Bolat, I., Pirlak, L., 1998. Selection of table cornelian cherry (*Cornus mas* L.) types in Çoruh valley. Turk. J. Agric. For. 22, 357–364.

Hanson, E., 1996. Fertilizing Fruit Crops. In: Extension Bulletin E-8S2. Michigan State University Extension, East Lansing, MI.

Hartemink, A.E., 2006. Assessing soil fertility decline in the tropics using soil chemical data. Adv. Agron. 89, 179–225.

He, Z.L., Yang, X.E., Stoffella, P.J., 2005. Trace elements in agroecosystems and impacts on the environment. J. Trace Elem. Med. Biol. 19, 125–140.

Heckman, J., 2004. Leaf Analysis for Fruit Trees. Rutgers Cooperative Research & Extension NJAES, Rutgers the State University of New Jersey, New Brunswick, NJ, pp. 1–2.

Hribar, J., Vidrih, R., 2015. Impacts of climate change on fruit physiology and quality. In: Proceedings of 50th Croatian and 10th International Symposium on Agriculture, February 16–20, Opatija, Croatia, pp. 42–45.

Huett, D.O., Dirou, J.F., 2000. An evaluation of the rationale fertilizer management of tropical fruit crops. Aust. J. Exp. Agr. 40 (8), 1137–1143.

Janick, J., 2011. History of fruit breeding. In: Flachowsky, H., Hanke, V.-M. (Eds.), Methods in Temperate Fruit Breeding. Fruit, Vegetable and Cereal Science and Biotechnology, pp. 1–7. 5 (Special Issue 1).

Jie, C., Jing-Zhang, C., Man-Zhi, T., Zi-Tong, G., 2002. Soil degradation: a global problem endangering sustainable development. J. Geogr. Sci. 12 (2), 243–252.

Jiménez, S., Garín, A., Gogorcena, Y., Betrán, J.A., Moreno, M.A., 2004. Flower and foliar analysis for prognosis of sweet cherry nutrition: influence of different rootstocks. J. Plant Nutr. 27, 701–712.

Jiménez, S., Pinochet, J., Gogorcena, Y., Betrán, J.A., Moreno, M.A., 2007. Influence of different vigor cherry rootstocks on leaves and shoots mineral composition. Sci. Hortic. 112, 73–79.

Johnson, R.S., Uriu, K., 1989. Mineral nutrition. In: LaRue, J., Johnson, R.S. (Eds.), Peach, Plum and Nectarine: Growing and Handling for Fresh Market. University of California Cooperative Extension, Division of Agriculture and Natural Resources, Oakland, CA, pp. 68–81.

Jones, J.B., Wolf Jr., B., Mills, H.A., 1991. Plant Analysis Handbook. Micro-Macro Publishing, Inc, Athens, GA.

Keller, M., 2005. Deficit irrigation and vine mineral nutrition. Am. J. Enol. Vitic. 56, 267–283.

Kerschberger, M., Hege, U., Jungk, A., 1997. Phosphordüngung Nach Bodenuntersuchung Und Pflanzenbedarf (Phosphorus Fertilization Following Soil Analysis and Plant Requirement). VDLUFA-Standpunkt, VDLUFA, Darmstadt.

Kirkby, M.J., Bissonnais, Y., Coulthard, T.J., Daroussin, J., McMahon, M.D., 2000. The development of land quality indicators for soil degradation by water erosion. Agric. Ecosyst. Environ. 81, 125–136.

Kominko, H., Gorazda, K., Wzorek, Z., 2017. The possibility of Organo-mineral fertilizer production from sewage sludge. Waste Biomass Valori. 8, 1781–1791.

Lal, R., 2001. Soil degradation by erosion. Land Degrad. Develop. 12, 519–539.

Lal, R., 2006. Enhancing crop yields in the developing countries through restoration of the soil organic carbon pool in agricultural lands. Land Degrad. Develop. 17, 197–209.

Le Bot, J., Adamowicz, S., 2005. Nitrogen nutrition and use in horticultural crops. In: Goyal, S.S., Tischner, R., Basra, S.A. (Eds.), Enhancing the Efficiency of Nitrogen Utilization in Plants. Food Products Press, The Haworth Press, Inc., Philadelphia, PA, pp. 323–367

Leece, D.R., 1975a. Diagnostic leaf analysis for stone fruit. 4. Plum. Aust. J. Exp. Agric. Anim. Husb. 15, 112–117.

Leece, D.R., 1975b. Diagnostic leaf analysis for stone fruit. 5. Sweet cherry. Aust. J. Exp. Agric. Anim. Husb. 15, 118–122.

Leece, D.R., van den Ende, B., 1975. Diagnostic leaf analysis for stone fruit. 6. Apricot. Aust. J. Exp. Agric. Anim. Husb. 15, 123–128.

Leece, D.R., Cradock, F.W., Carter, O.G., 1971. Development of leaf nutrient concentration standards for peach trees in New South Wales. J. Hortic. Sci. 46 (2), 163–175.

Lepers, E., Lambin, E.F., Janetos, A.C., Defries, R., Achard, F., Ramankutty, N., Scholes, R.J., 2005. A syntesis of information on rapid land-cover change for the period 1981–2000. Bioscience 55, 115–124.

Lidster, P.D., Porrit, S.W., Tung, M.A., 1978. Texture modification of "Van" sweet cherries by postharvest calcium treatments. J. Am. Soc. Hort. Sci. 103 (4), 527–530.

Lidster, P.D., Tung, M.A., Yada, R.G., 1979. Effects of Preharvest and postharvest calcium treatments on fruit calcium content and the susceptibility of 'Van' cherry to impact damage. J. Am. Soc. Hort. Sci. 104 (6), 790–793.

Liebig, J., 1852. Nouvelles Lettres sur la Chimie. Masson, Paris.

Macy, P., 1936. The quantitative mineral nutrient requirements of plants. Plant Physiol. 2, 749–764.

Maghradze, D., Abashidze, E., Bobokashvili, Z., Tchipashvili, R., Maghlakelidze, E., 2009. Cornelian cherry in Georgia. Acta Hortic. (818), 65–72.

Mamedov, N., Craker, L.E., 2004. Cornelian cherry: a prospective source for phytomedicine. Acta Hortic. 629, 83–86.

Marschner, H., 1995. Mineral Nutrition of Higher Plants, second ed. Ademic Press, London, p. 889.

Martin-Prével, P., Gagnard, J., Gautier, P., 1984. L'analyse vegetale dans le controle de l'alimentation des plantes temperees et tropicales. Lavoisier TEC et DOC, Paris, p. 832.

Mehra, A., Farago, E.M., 1994. Metal ions and plant nutrition. In: Farago, E.M. (Ed.), Plants and the Chemical Elements: Biochemistry, Uptake, Tolerance and Toxicity. VCH Verlagsgesellschaft mbH, pp. 31–66.

Meyer, R.S., DuVal, A.E., Jensen, H.R., 2012. Patterns and processes in crop domestication: an historical review and quantitative analysis of 203 global food crops. New Phytol. 196 (1), 29–48.

Milošević, T., Milošević, N., 2010. The effect of organic fertilizer, composite NPK and clinoptilolite on changes in the chemical composition of degraded vertisol in Western Serbia. Carpath. J. Earth Env. 5 (1), 25–32.

Milošević, T., Milošević, N., 2011a. Diagnose apricot nutritional status according to foliar analysis. Plant Soil Environ. 57 (7), 301–306.

Milošević, T., Milošević, N., 2011b. Seasonal changes in micronutrients concentrations in leaves of apricot trees influenced by different interstocks. Agrochimica 55 (1), 1–14.

Milošević, N., Milošević, T., 2012a. Seasonal changes and content of sodium in main organs of European plum trees (*Prunus domestica* L.), fruit size and yield as affected by rootstocks on acidic soil. Semin-Cienc. Agrar. 33 (2), 605–619.

Milošević, T., Milošević, N., 2012b. Factors influencing mineral composition of plum fruits. J. Elem. 17 (3), 453–464.

Milošević, T., Milošević, N., 2012c. Main physical and chemical traits of fresh fruits of promising plum hybrids (*Prunus domestica* L.) from Cacak (Western Serbia). Rom. Biotech. Lett. 17 (3), 7358–7365.

Milošević, T., Milošević, N., 2012d. The physical and chemical attributes of plum influenced by rootstock. Acta Aliment. 41 (3), 293–303.

Milošević, T., Milošević, N., 2013. Response of young apricot trees to natural zeolite, organic and inorganic fertilizers. Plant Soil Environ. 59 (1), 44–49.

Milošević, T., Milošević, N., 2017. Influence of mineral fertiliser, farmyard manure, natural zeolite and their mixture on fruit quality and leaf micronutrient levels of apple trees. Commun. Soil Sci. Plant Anal. 48 (5), 539–548.

Milošević, T., Milošević, N., 2018. Plum (*Prunus* spp.) breeding. In: Al-Khayri, J.M., Jain, M.S., Johnson, D.V. (Eds.), Advances in Plant Breeding Strategies: Fruits. In: vol. 3. Springer International Publishing AG, Part of Springer Nature, pp. 165–215.

Milošević, T., Zornic, B., Glisic, I., 2008. A comparison of low-density and high-density plum plantings for differences in establishment and management costs, and in returns over the first three growing seasons. J. Hortic. Sci. Biotech. 83 (5), 539–542.

Milošević, T., Milošević, N., Glišić, I., Krška, B., 2010. Characteristics of promising apricot (*Prunus armeniaca* L.) genetic resources in Central Serbia based on blossoming period and fruit quality. Hortic. Sci. 37 (2), 46–55.

Milošević, T., Milošević, N., Glišić, I., 2012a. Changes of fruit size and fruit quality of sour cherry during ripening process. CR Acad. Bulg. Sci. 65 (12), 1751–1758.

Milošević, T., Milošević, N., Glišić, I., 2012b. Vegetative growth, fruit weight, yield and leaf mineral content of plum grown on acidic soil. J. Plant Nutr. 35 (5), 770–783.

Milošević, T., Milošević, N., Glišić, I., Bošković Rakočević, L., Milivojević, J., 2013a. Fertilization effect on trees and fruits characteristics and leaf nutrient status of apricots which are grown at Cacak region (Serbia). Sci. Hortic. 164 (16), 112–123.

Milošević, T., Milošević, N., Glišić, I., 2013b. Tree growth, yield, fruit quality attributes and leaf nutrient content of 'Roxana' apricot as influenced by natural zeolite, organic and inorganic fertiliser. Sci. Hortic. 156, 131–139.

Milošević, T., Milošević, N., Milivojević, J., Glišić, I., Nikolić, R., 2014. Experiences with mazzard and colt sweet cherry rootstocks in Serbia which are used for high density planting system under heavy and acidic soil conditions. Sci. Hortic. 176 (12), 261–272.

Milošević, M.T., Glišić, P.I., Glišić, S.I., Milošević, T.N., 2018. Cane properties, yield, berry quality attributes and leaf nutrient composition of blackberry as affected by different fertilization regimes. Sci. Hortic. 227, 48–56.

Mišić, P., 2006. Plum. Partenon, Belgrade, Serbia (in Serbian).

Mitra, G., 2017. Essential plant nutrients and recent concepts about their uptake. In: Naeem, M., Ansari, A., Gill, S. (Eds.), Essential Plant Nutrients. Springer, Cham.

Montañés, L., Sanz, M., 1994. Prediction of reference values for early leaf analysis for peach trees. J. Plant Nutr. 17 (10), 1647–1657.

Montañés, L., Heras, L., Abadía, J., Sanz, M., 1993. Plant analysis interpretation basedon a new index: deviation from optimum percentage (DOP). J. Plant Nutr. 16, 1289–1308.

Moreno, M.A., Montañés, L.L., Tabuenca, M.C., Cambra, R., 1996. The performance of Adara as a cherry rootstock. Sci. Hortic. 65, 58–91.

Mourão Filho, A.A.F., 2004. DRIS: concept and applications on nutritional diagnosis in fruit crops. Sci. Agric. 61 (5), 550–560.

Munson, R.D., Nelson, W.L., 1990. Principles and practices in plant analysis. In: Westerman, R.L. (Ed.), Soil-Testing and Plant Analysis, thirrd ed. Soil Science Society of America, Inc., Madison, WI, pp. 359–387.

Neilsen, G., Kappel, F., Neilsen, D., 2007. Fertigation and crop load affect yield, nutrition, and fruit quality of 'Lapins' sweet cherry on Gisela 5 rootstock. HortScience 42, 1456–1462.

O'Sullivan, J.N., Ernest, J., 2008. Yam Nutrition and Soil Fertility Management in the Pacific. Australian Centre for International Agricultural Research, Brisbane.

Oldeman, L.R., 1992. Global Extent of Soil Degradation. ISRIC Bi-Annual Report 1991–1992, Wageningen, pp. 19–36.

Ovuka, M., 2000. More people, more erosion. Land use, soil erosion and soil productivity in Muranga District, Kenya. Land Degrad. Develop. 11, 111–124.

Papp, J., Tamási, J., 1979. Gyümölcsösök talajművelése és tápanyagellátása. Mezőgazdasági Kiadó, Budapest, p. 372 (in Hungarian).

Paré, C.M., Allaire, E.S., Khiari, L., Nduwamungu, C., 2009. Physical properties of organo-mineral fertilizers—short communication. Can. Biosyst. Eng. 51, 321–327.

Paré, C.M., Allaire, E.S., Khiari, L., Parent, E.L., 2010. Improving physical properties of organo-mineral fertilizers: substitution of peat by pig slurry composts. Appl. Eng. Agric. 26 (3), 447–454.

Plich, H., Wojcik, P., 2002. The effect of calcium and boron foliar application on postharvest plum fruit quality. Acta Hortic. (594), 445–451.

Polat, E., Karaca, M., Demir, H., Naci-Onus, A., 2004. Use of natural zeolite (clinoptilolite) in agriculture. J. Fruit Ornam. Plant. Res. 12, 183–189.

Popov, V.H., Cornish, P.S., Sultana, K., Morris, E.C., 2005. Atrazine degradation in soils: the role microbial communities, atrazine application history, and soil carbon. Aust. J. Soil. Res. 43, 861–871.

Ramankutty, N., Foley, J.A., Olejniczak, N.J., 2002. People on the land: changes in global population and croplands during the 20th century. Ambio 31, 251–257.

Reháková, M., Čuvanová, S., Dzivák, M., Rimár, J., Gaval'ová, Z., 2004. Agricultural and agrochemical uses of natural zeolite of the clinoptilolite type. Curr. Opin. Solid. St. M. 8, 397–404.

Reuther, W., Embleton, T.W., Jones, W.W., 1958. Mineral nutrition of tree crops. Annu. Rev. Plant Physiol. 9, 175–206.

Ruecker, G., Schad, P., Alcubilla, M.M., Ferrer, C., 1998. Natural regeneration of degraded soils and site changes on abandoned agricultural terraces in mediterranean Spain. Land Degrad. Develop. 9, 179–188.

Sánchez, E., Righetti, T., Sugar, D., 1998. Partitioning and recycling of fall applied boron in Comice pears. Acta Hortic. (475), 347–354.

Shear, C.B., Faust, M., 1980. Nutritional ranges in deciduous tree fruits and nuts. Hortic. Rev. 2, 142–163.

Smartt, J., Simmonds, N.W., 1995. Evolution of Crop Plants. John Wiley & Sons, Inc., New York.

Stefanelli, D., Goodwin, I., Jones, R., 2010. Minimal nitrogen and water use in horticulture: effects on quality and content of selected nutrients. Food Res. Int. 43 (7), 1833–1843.

Stewart, W.M., Dibb, D.W., Johnston, A.E., Smyth, T.J., 2005. The contribution of commercial fertilizer nutrients to food production. Agron. J. 97, 1–6.

Stiles, W.C., Reid, W.S., 1991. Orchard Nutrition and Soil Management. Cornell University Extension, Ithaca, NY. Information Bulletin 219.

Tagliavini, M., Marangoni, B., 2002. Major nutritional issues in deciduous fruit orchards of northern Italy. HortTechnology 12, 26–31.

Taha, N.M., Sherif, H.M., 2015. Increasing fruit set, yield and fruit quality of "Canino" apricot trees under two different soil conditions. Br. J. Appl. Sci. Technol. 10 (2), 1–18.

Tetera, V., 2006. Ovoce Bilych Karpat. CSOP, Veseli na Moravou.

Theunissen, J., Nhakidemi, P., Laublsher, C.P., 2010. Potential of vermicompost produced from plant waste on the growth and nutrient status in vegetable production. Int. J. Phys. Sci. 5, 1964–1973.

Tiag, G., Olimah, J.A., Adeoye, G.O., Kang, B.T., 2000. Regeneration of earthworm populations in a degraded soil by natural and planted fallows humid tropical conditions. Soil Sci. Soc. Am. J. 64, 222–228.

Tkaczyk, P., Bednarek, W., 2011. Evaluation of soil reaction (pH) in the Lublin region. Acta Agrophysica 18 (1), 173–186 (in Polish).

Tkaczyk, P., Bednarek, W., Dresler, S., Krzyszczak, J., Baranowski, P., Brodowska, S.M., 2018. Content of certain macro- and microelements in orchard soils in relation to agronomic categories and reaction of these soils. J. Elem. 23 (4), 1361–1372.

Tratch, R., May de Mio, L.L., Serrat, B.M., Motta, A.C.V., 2010. Nitrogen and potassium fertilization influences on intensity of peach leaf rust. Acta Hortic. 872, 313–318.

Tripathi, A., Uniyal, S., Sajwan, P., Negi, S.S., 2017. A review on impact of preharvest foliar sprays of macronutrients on yield and quality improvement of fruit crops. Res J Recent Sci 6 (6), 43–56.

Ubavić, M., Kastori, R., Peić, A., 1990. Fertilization of Orchards and Vineyards. DP H.I. "Zorka", Subotica (in Serbian).

Ulrich, A., 1952. Physiological bases for assessing the nutritional requirements of plants. Annu. Rev. Plant Physiol. 3, 207–228.

Ulrich, A., Hills, F.J., 1967. Principles and practices of plant analysis. In: Hardy, G.W. (Ed.), Soil Testing and Plant Analysis, Part II, Plant Analysis, pp. 11–24. Soil Science Society of America, Special Publication Series 2, Madison, WI.

Ulrich, A., Ririe, D., Hills, F.J., George, A.G., Morse, M.D., 1959. Plant Analysis: A Guide for Sugar Beet Fertilization. Bulletin No 766, California Agricultural Experiment Station, Berkeley, CA.

Vavilov, N.I., 1926. Studies on the origin of cultivated plants. Bull. Appl. Bot. Plant Breed. 16, 139–245.

Vavilov, N.I., 1951. The origin, variation, immunity and breeding of cultivated plants. Chron. Bot. 13, 1–366.

Walworth, J.L., Sumner, M.E., 1987. The diagnosis and recommendation integrated system (DRIS). Adv. Soil Sci. 6, 149–188.

Wang, G., Zhang, X., Wang, Y., Xu, X., Han, Z., 2015. Key minerals influencing apple quality in Chinese orchard identified by nutritional diagnosis of leaf and soil analysis. J. Integr. Agric. 14 (5), 864–874.

Warington, K., 1923. The effect of boric acid and borax on the broad bean and certain other plants. Ann. Bot. 37, 457–466.

Webster, A.D., Looney, N.E., 1996. Cherries: Crop Physiology, Production and Uses. CAB International, Wallingford.

Wehrmann, J., Scharpf, H.C., 1979. Der Mineralstickstoffgehalt des Bodens als Massstab für den Stickstoffdüngungsbedarf (N_min-Methode). Plant Soil 52, 109–126.

Westwood, M.N., Wann, F.B., 1966. Cherry nutrition. In: Childers, F.N. (Ed.), Fruit Nutrition; Temperate to Tropical Fruit Nutrition. Rutgers University, New Brunswick, NJ, pp. 158–173.

Wojcik, P., Wojcik, M., 2003. Effects of boron fertilization on 'conference' pear tree vigor, nutrition and fruit yield and storability. Plant Soil 256, 413–421.

Ystaas, J., 1990. Nutritional requirement of sweet cherry. Acta Hortic. (274), 521–526.

Ystaas, J., Frøynes, O., 1998. The influence of eleven cherry rootstocks on the mineral leaf content of major nutrients in 'Stela' and 'Ulster' sweet cherries. Acta Hortic. (468), 367–372.

Zia, M.H., Ahmad, R., Khaliq, I., Ahmad, A., Irshad, M., 2006. Micronutrients status and management in orchards soils: applied aspects. Soil Environ. 25 (1), 6–16.

Zohary, D., Spiegel-Roy, P., 1975. Beginning of fruit growing in the Old World. Science 187 (4174), 319–327.

Zularisam, A.W., Siti Zahirah, Z., Zakaria, I., Syukri, M.M., Anwar, A., Sakinah, M., 2010. Production of biofertilizer from vermicomposting process of municipal sewage sludge. J. Appl. Sci. 10, 580–584.

Further reading

Malusà, E., Buffa, G., Ciesielska, J., 2001. Effect of different fertilisation management on photosynthesis, yield and fruit quality of peach. In: Horst, W.J., et al. (Ed.), Plant Nutrition: Developments in Plant and Soil Sciences. vol. 92. Springer, Dordrecht, pp. 332–333.

Diagnosis and management of nutrient constraints in papaya

Róger Fallas-Corrales[a],*, Sjoerd E.A.T.M. van der Zee[a,b]

[a]Soil Physics and Land Management Group, Wageningen University, Wageningen, The Netherlands
[b]School of Chemistry, Monash University, Melbourne, VIC, Australia
*Corresponding author. E-mail: rogerarmando.fallas@ucr.ac.cr

1 Introduction

Carica papaya L. is a particularly high yielding tropical crop, originally from Central America and the South of México; its domestication is attributed to the Aztecs and Mayas (Fuentes and Santamaría, 2014). As it originates and is grown in tropical and subtropical regions (Céccoli et al., 2013), the crop suffers from chill conditions in temperate regions. The crop is now found in different tropical regions around the world.

The *Caricaceae* family, which *C. papaya* L. is part, is composed of six genus (*Carica, Vasconcellea, Horovitzia, Jarilla, Jacaratia,* and *Cylicomorpha*); most of them with an American origin and as shown in Fig. 42.1 can be found growing in wild areas, for example, in Central America. Only one of the genus (*Cylicomorpha*) is found growing wildly in regions of Africa, even though it is hypothesized that *Carica* has an African ancestor that traveled by surface sea currents through the Atlantic Ocean and started to colonize in the American continent (Carvalho and Renner, 2014).

2 Uses and nutritional value

Nowadays, papaya is used for several purposes, for example, as a fresh fruit (green or ripe), for juice production, jams, and as crystallized fruit and canned (Paull, 1993). Also, more industrialized products such as the production of

A.K. Srivastava, Chengxiao Hu (eds.)
Fruit Crops: Diagnosis and Management of Nutrient Constraints
https://doi.org/10.1016/B978-0-12-818732-6.00042-3

FIG. 42.1   Genus of the *Caricaceae* family (*Carica* and *Vasconcellea*, respectively) found in wild areas of Acosta, Costa Rica, at approximately 600 m above sea level. (A) *C. papaya* L. (B) *Vasconcellea* fruits.

TABLE 42.1   Average mineral concentrations for the fresh flesh fruit of C. *papaya* L. from one farm located in the Atlantic Region of Costa Rica.

	g/100 g flesh						mg/100 flesh			
	N	**P**	**Ca**	**Mg**	**K**	**S**	**Fe**	**Cu**	**Zn**	**B**
Concentration	109.40	10.20	17.60	11.90	171.00	6.30	0.28	0.04	0.10	0.15
Conf. Interval[a]	0.23	0.02	0.05	0.01	0.42	0.02	0.01	0.001	0.003	0.001

[a] *Confidence interval alpha = 0.95, N = 10.*

latex for the candy industry and the production of the enzyme papain that is used for digestive disorders, treatment of ulcers, and diphtheria are prepared from papaya (Reddy and Dinesh, 2000).

Papaya is considered as one of the most nutritional and healthy tropical fruits; it has antioxidant properties derived from the presence of several phenols, flavonoids, carotenoids, and amino acids (Nugroho et al., 2017; Somanah et al., 2018). Papaya also has been used in the preparation of several nutraceutical products, for example, cancer prevention, to reestablish the immunological and metabolic functions in tube-fed patients (Fujita et al., 2017) and to diminish the levels of glucose in type 2 diabetes patients (Danese et al., 2006). The mashed fruit is used for the treatment of infected burns, showing antimicrobial activity and proteolytic activity over necrotic tissue (Starley et al., 1999).

Papaya is a good source of minerals and vitamins, which can vary depending on cultivars, management practices, and ripening stage. Wall and Tripathi (2014) compiled the nutritional value of fresh fruit from different sources, and between the most interesting compounds found in papaya, the high contents of vitamin A and C are notable and also the high content of potassium. An example of mineral concentrations for the flesh fruit of papaya cv. Pococí from one farm in the Caribbean region of Costa Rica is highlighted (Table 42.1).

The total content of carotenoids varies between cultivars. For instance, the cultivars golden and sunrise solo have a total carotenoid content around 734 and 1013 μg/100 g of FW, respectively (Martins et al., 2016): for maradol considering a 10% of dry matter in the fresh fruit, the carotenoid content is around 3500 μg/100 g of FW (Ovando-Martínez et al., 2018), and for the cultivar Pococí, Schweiggert et al. (2011) reported contents between 5414 and 6214 μg/100 g of FW. However, for a detailed review about the nutritional value of papaya, the reader is referred to Wall and Tripathi (2014).

3 Botany

Papaya is a giant herb, and it can reach 9-m height, but in commercial plantations, it is rarely found with more than 5–6 m (Jiménez et al., 2014), as in taller plants, agronomical management is difficult. It is a fast-growing crop, normally with a single stem canopy and palmately lobed leaves (Campostrini et al., 2018; Jiménez et al., 2014; Wang et al., 2014), with a spiral 3:8 phyllotaxis that consists of 3 leaves positioned clockwise or counterclockwise each 360 degrees and

FIG. 42.2 Exemplification of xylem discretization in papaya through the staining of the plant tissue by the application of acid fuchsine dye to a single root. Note that only part of the plant was stained with the (*purple* color; *light gray* in print version) of the dye. *Photo: R.A. Fallas-Corrales.*

required 8 leaves to have a new leave perfectly aligned vertically with the one present in the upper part of the canopy (Campostrini et al., 2018; Fisher, 1980). This phyllotaxis confers to papaya a high radiation use efficiency and gas exchange. Papaya plants can be present in one of three different sex, female, male, or hermaphrodite. For commercial plantations, the hermaphrodite sex is preferred because of the lower internal cavity of the fruit and its elongated shape that can be stored in a smaller volume. The root system in adult plants is composed of a principal tap root of 0.6 m that in some cases may be branched and a set of secondary roots developed on the first 15 cm of depth, which can achieve up to 4-m long in a 1-year-old plant (Fisher, 1980).

In some plants, the plant xylem is organized with separate connections, as mentioned by Gutiérrez (1997). Papaya is one of those plants, for which each root is connected directly only to a specific part of the plant (Fig. 42.2). This property has direct consequences for the water and nutrient management in the crop as it affects processes like transport of water and nutrients within the plant.

4 Habitat

Papaya production is restricted to tropical and frost-free subtropical regions. In subtropical regions, it sometimes is necessary to protect it from chilling in greenhouse structures. The principal problems that papaya faces in subtropical cold regions are the ceasing of crop growth and fruit production due to low temperatures (Allan, 2002). Allan and Biggs (1987) using controlled temperature in greenhouses found that optimal temperature for the crop is in the range of 25–30°C at daytime and between 11°C and 16°C during the night for subtropical regions. They also mention that high-temperature regimes (36/28°C (day/night)) promote the fast growth of the plant, but high temperatures have the disadvantage of other problems, such as low pollen viability and sooner maturation of fruits, resulting in small and poor quality fruits; Allan and Biggs (1987) also mention that, under cool regimes (20/12°C (day/night)), papaya shows slow growth; sex reversal in female plants may occur as well as failure in pollen germination. The optimum temperature for growth according to Jiménez et al. (2014) is between 21°C and 33°C. Regarding altitude, in tropical conditions, papaya is preferably grown between 0 and 600 m above sea level, where temperatures are more suitable for the crop.

As a pioneering plant, papaya prefers habitats with high radiation, and papaya has a high photosynthetic capacity under high radiation regimes. For example, Wang et al. (2014) found net CO_2 assimilation rates from 20 to 24.2 μmol/m^2/s in fully expanded and fully exposed leaves, and Chutteang et al. (2007) found assimilation rates of 22.5 and 27.9 μmol/m^2/s in hermaphrodite and female plants, respectively. The light saturation point of papaya is very high, and Marler and Mickelbart (1998) found values between 1279 and 1325 μmol/m^2/s for different varieties grown in field conditions.

Papaya prefers habitats with relatively abundant water, as the crop is very sensible to a water deficit; for example, Marler and Mickelbart (1998) report the reductions in 85% of the CO_2 assimilation with slight reductions in the soil water tension (more negative values). As will be explained later, papaya can adapt to several soil conditions, but it prefers high-fertility and well-drained soils. Khondaker and Ozawa (2007) mentioned that flooding papaya fields for a period of 48 h may kill the crop.

5 Growth behavior and management-related aspects

The commercial cycle of the crop starts with the plantlet production, which stage is characterized by slow growth of the plant and it last between 2.5 and 3.5 months (Bogantes et al., 2011). After being transplanted to the field, the cycle is characterized by a vegetative stage, which lasts up to 2½–3 months after transplantation (MAT). Afterward, a reproductive stage that starts at 2½ or 3 MAT with the flowering and initiation of fruit production follows. This stage continues during all the remaining life cycle of the plant, as this crop continuously produces flowers, fruits, and vegetative structures. The commercial plantation cycle is usually limited to around 2–2.5 years due to the difficulties of management in big plants and decreasing yields as the plants age. However, under unmanaged conditions, there are reports of plants living for up to 20 years (Marler et al., 1994) (see Fig. 42.3A).

High yielding in this crop is closely related to efficient water and nutrient management. Regarding water management, Campostrini et al. (2018) reviewed the factors that affect water use efficiency in papaya and also the economic impact of water scarcity. Papaya is considered a drought-tolerant crop, as it can survive to very dry soil conditions; according to Mahouachi et al. (2006), papaya can increase ion concentrations to adjust its osmotic balance. But on the other hand, stomatal conductance and photosynthesis are severely reduced under soil water tensions around -50 and -70 KPa (Lima et al., 2016; Marler and Mickelbart, 1998). It appears that an optimum soil water suction for papaya is around -10 KPa (de Lima et al., 2015); unpublished data for the cultivar Pococí are in accordance with this value. The suction of soil water is usually given as the (negative) pressure. The optimum soil water pressure is therefore 10 kPa (this is a suction of about 100-cm water or in terms of the soil water retention function: pF = 2), which agrees with the pressure at what is generally used as the value of the field capacity (the amount of water that can be held by soil with capillary forces). Exceeding the field capacity leads to drainage of water. The implication is that water availability is optimal at field capacity, which is difficult to maintain: slightly excessive irrigation immediately leads to drainage. In addition, the hydraulic conductivity of soils at about field capacity can be large. For instance, the unsaturated hydraulic conductivity at pF = 2 estimated by the Van Genuchten-Mualem model using the R package "SoilHyP" (Dettmann and Andrews, 2018; R Core Team, 2017), for a loamy sand soil from La Rita, Costa Rica, and a clayey soil from Cariari, Costa Rica, showed values of 0.80 and 1.35 cm/day, respectively, while, in another sandy clay loam, it was in the order of 0.07 cm/day.

To give an impression, we show Fig. 42.4 that describes the loss function of water in a root zone at some distance above groundwater. This loss function gives the relationship between the "loss" of root zone water due to evapotranspiration and drainage and the water saturation ($s \approx \frac{\theta}{\theta_s}$, with θ_s being the saturated volumetric water fraction). If the root zone is well above the groundwater level, the loss function is unaffected by capillary rise of water from groundwater

(A)   (B)   (C)

FIG. 42.3 (A) Very tall and old *C. papaya* L. plant growing spontaneously in an urban area of San José, Costa Rica. (B) 4½-month-old (MAT) *C. papaya* L. plant in San Carlos, Costa Rica. (C) 8½-month-old (MAT) *C. papaya* L. cv. Pococí plantation in La Rita, Costa Rica. *(C) Courtesy of Antonio Bogantes and Eric Mora.*

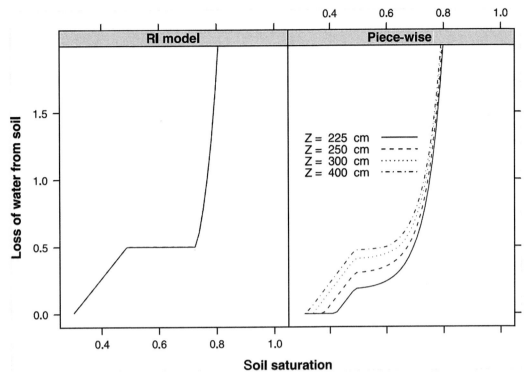

FIG. 42.4 Water loss function for a root zone in the absence (RI model) and in the presence (piecewise linear approximation) of groundwater that allows for capillary rise of water. *Reproduced from Vervoort, R.W., van der Zee, S.E.A.T.M., 2008. Simulating the effect of capillary flux on the soil water balance in a stochastic ecohydrological framework. Water Resour. Res. 44, W08425. doi:10.1029/2008WR006889.*

level into the root zone. In that case, it comprises a linearly increasing part, a horizontal part, and a curvilinear part, going from dry soil ($s = 0$) to complete saturation. The linearly increasing part represents the dry regime, where evapotranspiration is limited by the amount of water in soil. The horizontal part concerns optimal water availability to meet the evapotranspiration demand (which is controlled by incoming radiation, temperature, and water vapor pressure and the plant type, of course). For papaya, the curvilinear part is the most important. It shows that, as the water content approaches the so-called water field capacity of soil, the loss increases very rapidly with increasing soil wetness: root zone water loss at these saturations is predominantly drainage toward groundwater. The right panel of Fig. 42.4 shows that, if capillary rise can occur, this partly can compensate the evapotranspiration and drainage losses (making loss smaller). This effect is larger as groundwater depth (Z in Fig. 42.4) is smaller. As is well known, the soil water retention function, hence the field capacity, and the hydraulic conductivity depend on the soil type and texture. Accordingly, the precise shape of the loss function is also strongly dependent on soil type (Vervoort and van der Zee, 2008).

Campostrini and Glenn (2007) mentioned that papaya responds to irradiance changes with variations in stomatal conductance rates, and this way, under cloudy conditions, the plant reduces transpiration water losses. Additionally, the leaf to air vapor pressure deficit ($VPD_{leaf-air}$) is especially important for papaya; it has been observed that, when it increases above 1 KPa, there is a reduction in stomatal conductance and, consequently, net photosynthesis assimilation is also reduced (Campostrini et al., 2018). Whereas papaya favors a high soil moisture level for an optimum growth, it does not tolerate waterlogged soil conditions (Campostrini and Glenn, 2007; Khondaker and Ozawa, 2007). If soil becomes so anaerobic that oxygen concentrations in the root zone become too small, papaya is stressed, and its yield is quickly depressed, it reduces its net assimilation A and the stomatal conductance g_s (Thani et al., 2016), and it becomes more susceptible to papaya blight disease (*Phytophthora palmivora*). The need of a sufficiently aerated root zone implies that groundwater levels should not be too shallow and that the soil hydraulic conductivity must be sufficient to drain an excess of water at high saturation.

As far as experimental evidence is available, papaya may be considered as moderately sensitive to soil to salinity (Maas, 1993); then, a threshold of 3 dS/m is the maximum electrical conductivity to obtain 100% of the potential yield under otherwise optimum conditions. This factor may be of great importance in systems with application of high rates of fertilizer and low leaching of the nutrients. In summary, the agronomic experience with growing papaya is that it responds favorably to fertilization and good water supply under (sub)tropical climate conditions, but frost, inundation, and other causes of a water saturated soil that lead to poor soil aeration need to be avoided. This may give constraints with regard to rainfall and irrigation rates, and these constraints depend on the soil hydraulic properties (water retention and hydraulic conductivity). Though papaya is not highly sensitive to salt levels, the high fertilization rates that are needed to accomplish high yields may be constrained by papaya's salt tolerance. Moreover, the need to avoid large nutrient leaching losses to groundwater, which may deteriorate the quality of groundwater reserves, also constrains water and fertilizer applications. All these factors render the management of papaya growing a complicated process, with several trade-offs that need to be dealt with.

The scope of this chapter is to provide an inventory of the nutrient needs of papaya for the cultivation cases where large yields are aimed at. We particularly consider the perspective of soil fertility management. At this moment, a balance between soil fertility and environmentally sustainable management may not be well possible, due to gaps in knowledge. Therefore, we indicate the possible conflicting interests, and we suggest routes for improvement in this issue.

6 Diagnosis of nutrient deficiencies

The correct diagnosis of nutrient deficiencies is a critical task for the improvement of soil fertility and of the fertilization management in crop systems. Such a diagnosis can be done (including for papaya) looking for visible nutrient deficiency symptoms (morphological diagnosis) and by the interpretation of leaf or tissue analysis (leaf nutrient standards and chemical content-based methods).

6.1 Morphological diagnosis of nutrient deficiencies

The first mentioned approach has the drawback that the reduction of the yields may already occur when the deficiency symptoms are visible. Still, in the field practice where tissue analysis may not be simple, it is important to recognize deficiency symptoms as early as possible, to appropriately deal with the problem. Cibes and Gaztambide (1978) characterized the nutrient deficiency symptoms in papaya. In their results, nitrogen (N) appeared as the most limiting nutrient for the plant growth, affecting the growth of the complete plant under nitrogen omission. These symptoms have been also described by Thomas et al. (1995) and Kaisar et al. (2013). Kaisar et al. (2013) additionally reported a reduction in the number of fruits per plant and a lower fruit size under nitrogen scarcity. The reduction in the number of fruits per plant when nitrogen is deficient has also been observed by others in experiments where the applied nutrients have been varied (De Brito Neto et al., 2011; Sales Marinho et al., 2001).

For phosphorus, Cibes and Gaztambide (1978) concluded that it is the second most limiting nutrient, after nitrogen, for the growth of the vegetative parts of the plant. Nitrogen deficiency causes growth reduction of the entire plant, but only a reduced growth of the aboveground parts occurs in case of P-deficient conditions. Under such conditions, the papaya root system increases in mass, size, and density, as the plant becomes more efficient in extracting soil P. Vichiato et al. (2009) found that P deficiency can be related to lower plant dry matter production; additionally, Kaisar et al. (2013) found that P deficiency induced an earlier starting of the flowering process, reduction in the number of fruits per plant, lower yield, and a decrease in the quality (total soluble solids) of the harvested fruits.

Potassium deficiency also induces a reduction of plant growth, and in particular, the stem diameter and height become smaller (Cibes and Gaztambide, 1978; Kaisar et al., 2013). Potassium deficiency has been related to a reduction in the number of fruits per plant and with a lower postharvest quality (reduction in total soluble solids) (Kaisar et al., 2013). Besides its effect on yield and fruit quality, K plays an important role in reducing the frequency of occurrence and severity of some diseases, as documented by Dordas (2008), who also explained the underlying mechanisms. Papaya is not an exception; fruits of the cultivar Pococí submitted to very high potassium rates showed a reduction of the incidence and severity in postharvest diseases compared with normal fertilization (unpublished). Therefore, potassium deficiency symptoms can also be observed at the end of the commercialization process as more prompt to postharvest diseases.

Calcium deficiency has been associated with a decrease in the root growth of the plant, with impacts on fruit quality. For instance, Qiu et al. (1995) found postharvest problems (stiffened fruit) when low levels of Ca were present. For Mg and S deficiencies, Cibes and Gaztambide (1978) did not observe reductions in the root and stem growth, but the fruit set was reduced under its deficiency. On the other hand, Kaisar et al. (2013) reported a reduction in plant height as a consequence of S deprivation and agreed with the reduction of the fruit number per plant and its consequent lower yield.

Regarding micronutrients, boron deficiency is maybe the most common and most studied one. In vegetative stages, it is characterized by a reduction of the plant growth (Cibes and Gaztambide, 1978; Singh et al., 2010), reduction of leaf size, limited stem growth, reduced internode distance, and a distortion of the lamina and curvature of petioles (Nautiyal et al., 1986). Boron deficiency has a notorious effect on the fruit set and fruit deformation. Bogantes et al. (2011) showed the effect of boron deficiency in the early stages of fruit development. They describe it as a nonsealed apex of the fruit, which later can be invaded by pathogens that can cause the early drop of the fruit. Boron has been reported as the most limiting nutrient regarding the number of fruits per plant (Kaisar et al., 2013).

The most notorious symptom of boron deficiency is the fruit deformation, also known as a bumpy fruit (Wang and Ko, 1975), which is followed by exudation of latex. Using this symptom to diagnose boron deficiencies compromises the production because it is usually too late when detected. Even so, it is important to correct this deficiency because papaya has a continuous fruit production and vegetative tissue over its entire life (starting approximately 3 months after transplanting into the field), and commercial plot cycles normally last around 2 years.

Some of the symptoms reported as a boron deficiency in papaya can be confused with symptoms caused by viral diseases. For example, internode distance reduction, fruit deformation, leaf lamina distortion, the curvature of petioles, and fruit latex exudation have been reported as symptoms caused by virus pathogens (Becerra et al., 1999). Due to the aggressiveness of viral diseases in papaya production, it is recommendable to be aware and to distinguish when those symptoms are related to B deficiency or viral diseases.

Total boron uptake of the plant may depend on its availability in the soil solution, the demand of the hybrid or variety, and the attainable yield. Cunha and Haag (1980) reported a total B uptake of 74.2 mg per plant using a papaya plant not submitted to a breeding program; on the other hand, Fallas et al. (2014) investigating the nutrient demand of a high yielding hybrid (Pococí) found a total uptake around 190-mg B per plant.

6.2 Leaf nutrient standards and chemical content-based methods

A more effective method than the detection of nutrient deficiencies by morphological diagnosis is the use of chemical analysis of plant tissue samples. In this way, nutrient deficiencies can be detected and corrected early, but as it requires a correct interpretation of the results of the analyzed sampled, this approach must first have been developed sufficiently. A commonly used method for the interpretation of tissue analysis is the "sufficiency range method," which is based on critical concentrations for the crop determined under specific sampling conditions. The accurate interpretation of tissue analysis by conventional approaches as the sufficiency range method is limited by factors related to sampling conditions, plant varieties, and interactions between nutrients and their mobility in the plant. Therefore, for interpretation, it is important to consider the conditions in which the standard sufficiency ranges were developed.

A profound complication for tissue analysis interpretation arises when the deficiency is to be assessed in organs like fruits, which is quite common to diagnose for fruit deformations, fruit abortion, low growth and low weight of the fruit, or severe affection by diseases in fruit. At this moment, it appears that there is a lack of reference to compare fruit concentrations in diagnosing in these situations. In addition, it is not established yet if deficiency in petiole or leaf is well correlated with deficiency in the fruit. It is important to emphasize that when a deficiency appears in organs like fruits, the diagnosis might come too late to save this fruit, but it can help to avoid the same problem for the following fruits of the same plant. Nevertheless, plant tissue analysis is a powerful tool for nutrient constraint detection, and if complemented by alternative methods of interpretation, it can give valuable information for the nutrient management in production systems.

For papaya, mainly the macronutrients nitrogen (N), phosphorus (P), and potassium (K) restrict the growth and yield. Thus, it is common to detect symptoms of their deficiency by means of visual symptoms and by the use of leaf analysis. However, micronutrient deficiencies can also play an important role in papaya growth and the quality of harvested fruits, and their deficiency also has notorious effects on the plant growth. The sufficiency range standard method for papaya is used with two different plant tissues, that is, leaf and petiole, of which the analysis of petiole is the most commonly used. However, there is still a discussion about the suitability of leaf lamina, or petiole, in the diagnosis of nutrient deficiencies in papaya. Some results point to better predictions of the nutrient status of the plant based on petiole, but others based on leaf lamina (Marinho et al., 2002). The tissue used for the analysis normally corresponds to leaves or petioles from the most recent fully expanded leaf that contains the most recently open flower, which is sometimes called leaf F (Marinho et al., 2002; Mills and Jones, 1996).

The appropriate implementation of the nutrient deficiency diagnoses using the sufficiency range method requires to consider (i) a specific tissue, (ii) the season of sampling, (iii) the growth stage, and (iv) a representative number of

TABLE 42.2 Critical nutrient concentrations (sufficiency ranges) for C. *papaya* L., as compiled from different investigations.

Reference *Country*	Mills and Jones (1996) *United States*	Malavolta et al. (1989) *Brazil*	Malavolta et al. (1989) *Brazil*	Robinson et al. (1997) *Australia*	Chatterjee and Dube (2004)	Range of concentrations observed in Costa Rica (LSF)[a] *Costa Rica*
N %	1.01–2.50	1.0	4.5–5	1.3–2.5	1.01–2.5	0.99–4.60
P	0.22–0.40	0.3	0.5–0.7	0.2–0.4	0.22–0.40	0.11–0.32
K	3.30–5.50	2.5–3.0	2.5–3.0	3.0–6.0	3.3–5.5	2–4.4
Ca	1.00–3.00	1.5	2.0–2.2	1.0–2.5	1.0–3.0	0.68–1.90
Mg	0.40–1.20	0.4	1.0	0.5–1.5	0.40–1.50	0.21–0.40
S	–	–	0.4–0.6	0.3–0.8	0.20–0.40	0.1–0.34
Al mg/kg	–	–	–	–	–	8.0–39
Na	–	–	–	–	–	160–801
B	20–30	–	15	20–50	20–30	16–25
Cu	4–10		11	4–10	4–10	1.1–5.1
Fe	25–100		291	20–80	25–100	16–88
Mn	20–150		70	25–150	20–150	15–58
Zn	15–40		–	10–30	14–40	9.0–26
Sampling season					Six months after transplant	
Tissue	Leaf petioles from the most recent fully expanded leaf	Leaf petioles	Leaf lamina	Leaf petioles from the most recently opened flower	Sixth petiole from the apex	Leaf petioles

[a] *Range that includes the concentration of the 75% of the samples analyzed in Costa Rica.*
From Bogantes, A., Mora Newcomer, E., Umaña Rojas, G., Loría Quirós, C.L., 2011. Guía para el cultivo de papaya en Costa Rica. MAG/UCR/INTA, San José, Costa Rica. Retrieved from http://www.mag.go.cr/bibliotecavirtual/F01-10190.pdf.

samples for a statistical treatment and taking into account the variability that may be encountered in the data. The interpretation of tissue analysis by the sufficiency range method is normally done by the direct comparison of leaf or petiole nutrient concentration of the sample against the previously defined standards. For C. *papaya* L., different thresholds of nutrient concentrations have been proposed (Table 42.2).

6.3 Complementary methods for diagnosis of nutrient constraints

As the sufficiency range method has some limitations for the use and interpretation of tissue leaf analysis, it is necessary to evaluate alternative methods for the detection of nutrient constraints. The dilution effect method can be a useful concept for this purpose. The dilution effect concept (Jarrel and Beverly, 1981) states that the concentration of the plant at some time is a function of the nutrient uptake and its production of dry matter. Therefore, older organs with higher amounts of dry matter have lower concentrations of nutrients, as a consequence of the nutrient dilution in the new tissue produced.

Jarrel and Beverly (1981) summarized the dilution effect concept with examples of different factors such as light, water availability, temperature, and application of other nutrients, all affecting the dry matter production and consequently the concentration of nutrients in the plant. Jarrel and Beverly (1981) showed that, when one element is limiting the biomass production (i.e., phosphorus), there is an increase in the concentration of the other nutrients (i.e., nitrogen and potassium). When this limiting nutrient's availability is improved by application of fertilizer, this leads to an increase of dry matter. However, for the other (nonlimiting) nutrients, there is less reason to take up more of them by the plant; hence, a decrease in the concentration of these other nutrients occurs; a slightly higher uptake is overcompensated by the dry matter increase. This complexity of interactions between nutrient concentrations in the plant makes it very difficult to infer a real sufficiency status for optimal yields, as the interpretation of plant tissue analysis is valid only for the

TABLE 42.3 Increase in leaf nutrient concentration in a K-deficient solution treatment compared with a complete treatment.

Treatment	Leaf nutrient concentration (%)				
	K	N	P	Ca	Mg
Minus K	0.40	3.46	1.49	3.61	2.21
Complete	1.58	2.25	0.82	3.61	1.21

Data from Cibes, H. R., Gaztambide, S., 1978. Mineral-deficiency symptoms displayed by papaya plants grown under controlled conditions. J. Agr. U. Puerto Rico, 62(4), 413–423.

specific moment when the plant was sampled. Any measure that is taken, for example, fertilization for the limiting nutrient, changes the relative availability of all nutrients, and another nutrient may become growth limiting.

Taking into account the temporal specificity of the plant analysis and interactions between nutrients in the plant, the dilution effect concept can be applied for the identification of nutrient deficiencies, assuming that an organ developed under nutrient scarcity cannot develop its maximum potential size or growth. Nutrient-deficient plants have limited growth and consequently higher nutrient concentrations of all the nutrients when compared with plants without a nutrient restriction for the growth (without deficiency). This behavior is found for all the nutrients except that one(s) that is causing the growth restriction. So, it is possible to have a more customized (varieties, regions, etc.) interpretation and avoid some of the problems found with the interpretation with the sufficiency range method.

The interpretation of leaf analysis using the dilution effect concept can be exemplified using data of Cibes and Gaztambide (1978) for *C. papaya* cv. Solo (Table 42.3). In this example, it can be seen that, when a certain nutrient limits the growth of the crop (for this example is potassium), a larger concentration of all the other nutrients is observed because potassium limits the production of dry matter. If K is supplied in sufficient quantities, its concentration increases, and the concentrations of the other nutrients decrease (less additional uptake than additional dry matter production). From the comparison of these two treatments, it is immediately (and quantitatively) clear which element is limiting yield.

For the success of the interpretation of nutrient deficiencies using the dilution effect concept, the need of choosing a right reference sample to compare (i.e., a high yield plant, without deficiency symptoms, etc.) is crucial. As the dilution effect concept is useful for crops in general, this need to choose the correct reference is also needed for other crops than papaya.

As a final remark for the methods of chemical analysis of plant tissue, there are other methods for the interpretation that consider interactions between nutrients, for example, the Diagnosis and Recommendation Integrated System (DRIS). DRIS norms have been developed for several crops, and indeed, there are available for papaya (Bowen, 1992; Costa, 1995). However, due to the specificity of the norms to the conditions in which they were developed, their implementation for other regions or with other varieties requires calibration for specific conditions.

7 Nutrient management

This section identifies the high nutrient requirements of papaya for high yielding systems and addresses yield gaps and common empirical approaches to determine fertilizer applications. Where papaya can be a high yield crop, this may be associated with significant leaching and contamination of groundwater. Also, this conflict is elaborated in this section.

7.1 Nutrient requirement for *C. papaya* L

Reports about nutrient requirements for papaya are scarce. Cunha and Haag (1980), in Brazil, characterized the nutrient requirements for papaya. They used plants that were not previously submitted to any breeding program, and their potential yield is not comparable with modern hybrids that are used around the world. For their study, Cunha and Haag (1980) obtained a yield of 30,000 kg/ha. In Costa Rica, modern cultivars of *C. papaya* have been developed by a breeding program. As a result, the currently grown hybrids show a very high potential yield. For example, one grower mentioned to have achieved yields around 150,000 kg/ha in Costa Rica with the cultivar Pococí, and in

Brazil, (Campostrini et al., 2018) mentioned in a personal communication that papaya from the Formosa group can achieve the same (unofficial) yield. Official data for Costa Rica of the year 2014 show an (country specific) average yield around 100,000 kg/ha (FAO, 2018). Similar yields have been reported by Chan (2009) and by Gomes Oliveira and Correa Caldas (2004) in Brazil (93,000 kg/ha) and in Belize (113,000 kg/ha) (FAO, 2018). Despite the high potential yields of papaya, the average global yields in papaya for the year 2017 were around 29,540 kg/ha (FAO, 2018). This illustrates the huge yield gap in *C. papaya* production systems around the world and the high potential for the improvement of its production. The worldwide yield gap is in the order of 70%, considering a potential yield of 100,000 kg/ha or even higher as some yield estimates suggest. Some of the reports of low yields in papaya are related to diseases, principally viral diseases (Tennant et al., 2007), water management problems and water scarcity and related effects on the net photosynthesis of the crop (Campostrini and Glenn, 2007; Campostrini et al., 2018; Lima et al., 2016; Lopes Peçanha et al., 2017), and nutrient management. To obtain high yields, *C. papaya* L. requires a considerable amount of water and nutrients. Kumar et al. (2010) stated that the crop has a demand of 310, 105, 530, 332, and 185 kg/ha of N, P, K, Ca, and Mg, respectively. A similar study developed in Costa Rica, on volcanic ash-derived soil, showed a nutrient demand of 354, 44, 413, 124, 64, and 40 kg/ha of N, P, K, Ca, Mg, and S, respectively (Fallas et al., 2014). Comparison of these numbers reveals that they are different. There are several reasons that this may be the case. As already emphasized earlier in relation with the dilution effect, the relative availability of the different nutrients plays a role. This can be generalized toward other growing conditions, such as available light, temperature, and water availability. This underlines the need to characterize the important environmental conditions well enough, in publications such as these. The nutrient demand follows the development of dry matter of the crop. For a fast-growing crop like papaya, this results in a very high nutrient demand in short periods of time. Figs. 42.5 and 42.6 adapted from Fallas et al. (2014) show the total nutrient uptake and the nutrient demand per each stage of development. During the beginning of the growth period (from transplanting up to 2 months after transplanting), the nutrient requirements are lower than 4% of the total amount quantified at 8 months after transplantation (MAT). This behavior is in agreement with the low leaf area index of the plant at this stage and its consequently low capacity of light interception for the production of assimilates. The soil supply can play an important role during this period, but it is necessary to consider also that, during the first days of the establishment, the root system is small and its capacity to obtain the necessary nutrients may be a limiting factor. From the second month after transplantation and due to the fast growth rate of the crop, there is an increase in leaf area and light interception. Hence, the requirement for nutrients increases considerably, and fertilization and correction of nutrient constraints require even more attention from this period onward. The reason is that the quantities and proportions of nutrients required in this case may not be supplied by the soil system. Note that in Figs. 42.5 and 42.6, the sampling plant at the stage between 3 and 4 MAT was missing, so the quantity shown at 5 months corresponds to the uptake between 3 and 5 MAT.

As observed in Figs. 42.5 and 42.6, in general, the stage of maximum nutrient requirement in papaya is related to the period just before the beginning of the harvest (i.e., about 7–8 months after transplantation). In this stage, the plant is producing new vegetative growth, and at the same time, it has a big requirement for the fruit filling because fruits are a strong sink for nutrients in the plant. At this phenological stage, the plant has fruits in all the different stages of development, and the production of vegetative and reproductive structures (flowers and fruits) is continuous. Papaya fruit development is characterized by a low accumulation of sugars from the anthesis up to approximately 100 days after anthesis; after 100 up to the harvest (approx 140 days after anthesis), the sucrose content in the fruit increases from less than 5 to approximately 40 g kg of fresh weight; fructose and glucose content increases more than two times from day 100 to day 140 after anthesis (Zhou and Paull, 2001). This production implies the high demand for nutrients when papaya is approaching its harvesting stage (period from seven to eight MAT in Figs. 42.5 and 42.6).

It is important to emphasize that papaya has a hollow stem and does not accumulate starch; it is a plant with a low capacity to store assimilates. According to Jiménez et al. (2014), the production of assimilates that is needed to accommodate the high fruit load requires a steady flow from the leaves. Figs. 42.5 and 42.6 give an impression of the required amount of nutrient for each stage, which has implications for fertilizer recommendations.

7.2 Management aspects related to soil acidity

Papaya is native to tropical Central America. It is a crop that is preferably cultivated in high-fertility and well-drained soils. Nevertheless, it is adapted to soils with a broad range of chemical and physical properties. In Costa Rica, papaya is found under natural conditions in different regions, and it is also cultivated in soils that range from acid Ultisols in the region of Upala to more fertile Inceptisols with soil pH values near 7 and high Ca and Mg contents like the Central Pacific Region.

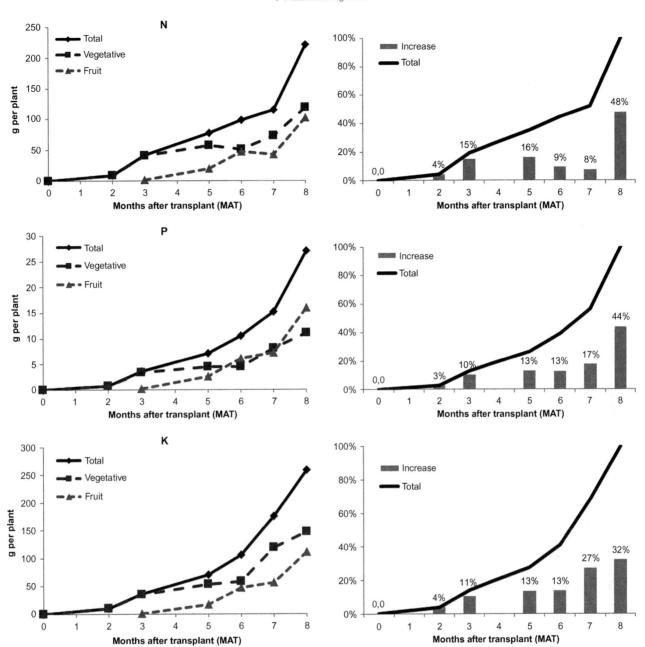

FIG. 42.5 Nitrogen, phosphorus, and potassium uptake by *C. papaya* L. cv. Pococí in the Atlantic Region of Costa Rica during a 8-month period. *Modified from Fallas, R., Bertsch, F., Barrientos, M. 2014. Curvas de absorción de nutrientes en papaya (Carica papaya L.) CV. "Pococí" en las fases de crecimiento vegetativo, floración e inicio de cosecha. Agronomía Costarricense, 38(2), 43–54.*

The soil pH does not appear to have a direct impact on papaya growth at the commonly found soil pH values. Marler (1998), studying the effect of the pH on dry matter production at the initial vegetative stage of papaya and varying the pH in the substrate solution from 3 to 9, did not find significant differences of dry matter production in the pH range from 4 to 9. Only when exposed to pH3, the plants showed growth problems, and lower dry matter yields were accomplished.

Although soil pH may not affect the dry matter production of papaya directly, its indirect effects can be more important. For example, soil pH can also play a role on papaya growth if it has a modifying effect on the solubility of nutrients in the soil. Therefore, pH is a necessary parameter to consider when aiming at a high nutrient use efficiency. The use of high quantities of fertilizer in intensive systems can also lead to the development of soil acidity problems if it causes a lowering in the soil pH values. Hence, fertilizer-pH interactions may be found in both directions. Additionally, low soil pH values commonly are correlated with low soil Ca and Mg contents, elements that are required in considerable

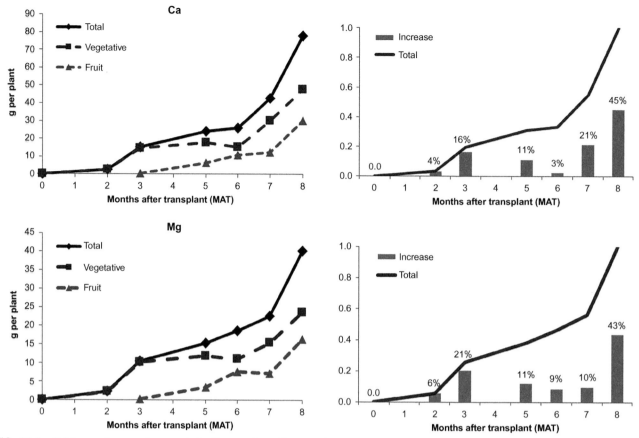

FIG. 42.6 Calcium and magnesium uptake by *C. papaya* L. cv. Pococí in the Atlantic Region of Costa Rica during an 8-month period. *Modified from Fallas, R., Bertsch, F., Barrientos, M. 2014. Curvas de absorción de nutrientes en papaya (Carica papaya L.) CV. "Pococí" en las fases de crecimiento vegetativo, floración e inicio de cosecha. Agronomía Costarricense, 38(2), 43–54.*

quantities by the crop (Fig. 42.6). Consequently, the need for liming low pH soils for papaya production is also related to the supply of Ca and Mg.

Regarding soil acidity, the high aluminum content in soil is a big concern for papaya producers. This cation has been related to the inhibition of root growth in several crops (Ryan et al., 1992), and it could have a notorious impact on the overall nutrient uptake and efficiency (Zhao and Shen, 2018). However, the effects of aluminum on papaya are not well established, for example, there is only one report about high Al concentrations that cause toxicity for papaya. (De la Fuente et al., 1997) found that, in nontransgenic plants, the root growth is inhibited at Al concentrations of 50 μM, while, on transgenic plants altered for the citrate synthesis, the roots grew even at concentrations up to 300 μM. Besides Al toxicity, also toxicity, for example, of Mn and Fe, may have to be taken into consideration, but information about Fe and Mn toxicity for papaya is not available.

When lime is applied to supply Ca and Mg for the crop, the commonly used methods to calculate the lime requirement (Cochrane et al., 1980; Quaggio et al., 1985; Shoemaker et al., 1961) are not very useful as they cannot predict the availability of those nutrients for the crop in the different stages, so alternative and preferably more mechanistic methods are necessary to improve the calculations of lime requirements for the crop; these tools can also help to improve the timing of application for papaya, as nowadays in some cases, lime is applied on acid soils only at the beginning of the crop cycle. Examples of alternative methods for liming calculation are discussed in the section of the mechanistic modeling approach.

7.3 Fertilizer dosage for papaya using an empirical approach

As explained earlier, papaya is a very high nutrient demanding crop, so it is predictable that the yield increased when the rates of fertilizer for the crop are increased. Several experiments have been conducted to evaluate the effect of nitrogen, phosphorus, and potassium on yield response, and recommendations of fertilizer rates for different regions have been based on the results of those experiments.

Kumar et al. (2010) recommended a fertilization with 300, 300, and 300 g of N, P_2O_5, and K_2O, respectively, per plant per year, based on an evaluation of fertilizer rates under different field locations. With a similar kind of experiment, Gomes Oliveira and Correa Caldas (2004) recommended a yearly fertilization dose of 347 and 360 kg/ha of N and K_2O, respectively, for the region of Cruz das Almas-BA, Brazil.

In Costa Rica, Fallas developed a fertilization experiment with N, P, and K rates based on the results of Fallas et al. (2014). For this experiment (unpublished), nitrogen and potassium rates were as high as 700 kg N/ha and 700 kg K_2O/ha, respectively, distributed in a period of 9 months and using a plant density of 1600 plants per hectare. The results of this fertilization experiment showed the high response of papaya to high rates of fertilization (up to 132 tonnes of fresh fruit per hectare in a dystric soil) and also a strong interaction between the nutrients on the yield of the crop (Fig. 42.7).

The recommendation of fertilization based on regional experiments of nutrient rates albeit practical and in some cases functional most of the time ignores the processes that govern the real availability, uptake, and movement of the nutrients in the soil-plant-atmosphere system, so there is an implicit risk to contaminate groundwater sources. Nevertheless, this kind of experiments in combination with methods that consider the relationships between nutrients and its relationship with environmental factors is a good option to detect limiting nutrients for specific conditions, for the correction of the nutrient constraints, and for the improvement of fertilization programs.

It is well known that different growth factors (light, temperature, water, and various nutrients) cannot be considered independently of fertilization, even though, for regions, their spatial variation may be small (e.g., light and temperature). The main reason is that different growing factors may affect the response of each other by the crop. Specifically for nutrients, interactions for uptake and crop growth have been recognized by Janssen et al. (1990). They developed and parameterized the model QUEFTS that accounts for such interactions, which for papaya are very necessary as shown in the Fig. 42.7.

The basic concept is related with Liebig's law of the minimum, according to which the nutrient that is lowest in supply determines the crop's yield and uptake of other better available nutrients. As the limiting nutrient is applied in the form of fertilizer, its uptake but also that of other nutrients increases. The ratios between uptake rates of different nutrients change in dependency of which nutrient is limiting and which rates can be accomplished, for example, very deficient or slightly deficient availability of the limiting factor (Janssen et al., 1990; Sattari et al., 2014). Accordingly, fertilization with the limiting nutrient may lead to synergies if at the same time also other nutrients are fertilized. This multidimensionality of critical nutrients and other growth factors (incoming radiation and water availability) complicates modeling of optimal conditions and sets high demands on the experimental basis of fertilization (taking cross factor effects into account).

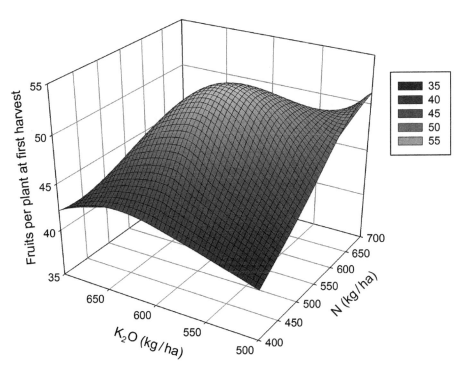

FIG. 42.7 Number of fruits per plant at the stage of the first harvest in *Carica papaya* L. cv. Pococí as a function of rates of nitrogen and potassium in field conditions, in the Atlantic Region of Costa Rica. Applied fertilizer rates correspond to the cumulative amount of a 9 months period.

As Sattari et al. (2014) conclude, other factors than N, P, and K (considered nowadays in QUEFTS) may be important and vary considerably from region to region. Major limitations of the QUEFTS approach are also the disregard of soil heterogeneity, weather variability, and the profound experimental effort to parameterize the model for a certain region and crop. Nevertheless, this latter effort is feasible if a region regarding those other factors (climate, hydrology, and soil type) is sufficiently homogeneous and of sufficient size to make extensive plot fertility experiments worthwhile. Then, this approach is no doubt far more appropriate than assessments that disregard interactions between growth factors.

On the other hand, there are some remarks to consider before the implementation of the QUEFTS approach, for example, in the Atlantic region of Costa Rica with volcanic soils, the model predictions for corn and grass differed substantially from the real data (Nieuwenhuyse, 1988), which was related to the behavior of allophanic soils regarding N mineralization and P retention. Also for papaya, as far as the authors are aware, no data are available about the simultaneous uptake of nutrients and its yield under variation in the supply of the major nutrients. These data are necessary to establish the potential supply and the actual uptake of N, P, and K as defined by Janssen et al. (1990).

At the moment, however, in Costa Rica and other countries, the interaction between different nutrients is not always taken into consideration. Then, instead, for each nutrient, the fertilizer requirement is based on known crop needs, selective extractions of soil (and not calibrated for papaya), and estimated fertilizer efficiency. For this approach, the calculation of nutrient requirement is based on the data of the uptake of nutrients by the crop, for example, according to information published by Cunha and Haag (1980) and Fallas et al. (2014). The calculation of the available quantity of nutrient is based on an empirical calibration of nutrient concentrations determined by a soil solution extractant, such as Mehlich I, II, III, KCl, and Olsen. This calibration is based as a response of the crop to fertilizer additions under different soil nutrient concentrations determined by the extractant. This kind of calibration requires a big quantity of experiments in different soil conditions. Therefore, the information of a specific calibration for papaya does not exist as the authors are aware.

If the solution concentrations in the soil are over the defined critical concentration, it is assumed that the soil can supply nutrients for the crop. If this is not the case, it is assumed that is necessary to apply all the plant requirement by means of fertilizer additions (organic and/or inorganic).

Finally, the fertilization efficiency could be estimated according to the environmental conditions (like precipitation and soil moisture) and the geochemical characteristics of the soil, which are principally related to the current content of the nutrient and its reactivity with the soil matrix. Most of the times, this calculation of the efficiency is a subjective decision, and it is assumed big losses to avoid the limitation of the crop growth and yield. This final aspect is reprehensible from an environmental perspective, and it could also lead to contamination of groundwater sources and higher emissions of greenhouse gases.

Having emphasized the procedure of this empirical approach and its limitations, the amount of required fertilizer for each stage of the crop can be calculated with

$$\text{kg nutrient/ha} = \frac{\text{Requirement of the crop (kg nutrient/ha)} - \text{Soil supply (kg nutrient/ha)}}{\%\text{Efficiency of fertilizer}} \times 100 \qquad (42.1)$$

Regarding fertilization management by means of organic fertilizers and biofertilizers, incidental investigations only are available. They can be useful in practice and can give an idea about the possible outcomes of using organic materials and microorganisms, but most of these studies lack explanations about the mechanisms that govern nutrient availability and uptake by papaya under different conditions of the soil-microorganisms-plant-atmosphere continuum. Consequently, it is difficult to establish general recommendations of organic amendments and microorganisms for the diverse and heterogeneous conditions in which papaya is grown around the world. Despite these issues, the beneficial effects of organic matter and beneficial microorganisms on the fertility, productivity, and sustainability of the soils are of great importance. About biofertilizer use in different conditions of climate, P availability, etc., Schütz et al. (2018) conducted a metaanalysis and gave some insight into the use of these products in several conditions. Because of the limited support of observations in the literature, a generalization as is the aim in this chapter is still not in reach.

7.4 A mechanistic modeling approach to management

In most countries, even nowadays, soil fertilization and fertility recommendations are strongly based on empirical approaches as were mentioned in the previous sections. However, models are finding an increasingly large position in soil-plant fertility research, and some approaches are shortly discussed here. One of these applications of modeling in papaya could be to improve liming practices for the crop. As already described, papaya requires big quantities of calcium and magnesium during the stages that do not necessarily coincide with the time of lime application. Chemical modeling can improve the assessment of the requirement of lime, regarding time and quantity, for example, in case of slow acid-base reactions. This may help avoid the need (e.g., Cochrane et al., 1980) to multiply the calculated quantity of lime per hectare by a factor of 2 to account for nondetermined Al sources. Considering the limitations of the empirical methods and its slow development during the last decades, an alternative (but knowledge demanding) method to improve Ca and Mg application practices for papaya is the use of multisurface adsorption models, which are based on mechanistic descriptions of element adsorption reactions.

Examples of multisurface models are implemented by the codes ORCHESTRA (Meeussen, 2003), MINTEQA2 (Allison et al., 1991), ECOSAT (Keizer and Van Riemsdijk, 2009), and PHREEQC (Parkhurst and Appelo, 1999). Multisurface adsorption models are capable to consider multicomponent interactions within reactants and with several adsorption surfaces such as organic matter, clay silicate, and crystallized and amorphous iron hydroxides (Peng et al., 2018; Weng et al., 2001). This kind of models considers the global adsorption surfaces as the addition of each individual soil component and calculates the equilibrium in the system.

The authors are aware that this kind of models have not been used for liming dosage calculations, but such models have been used to explain the effect of liming on C dynamics in the vadose zone using the HP-1 model (Thaysen et al., 2014). Also, predictions have been made of pH and Al, Ca, Mg, Zn, Cu, Cd, and Ni activities of a soil exposed to acidification treatments (Fest et al., 2005). Therefore, multisurface models can be useful to improve recommendations for the lime requirement for crops as papaya grown in acid soils, to estimate Al and Fe activities in the soil solution, as well as Ca and Mg availability, which are all the principal targets of liming practices in plant production systems. Another multicomponent model that might be useful for the improvement of liming management is UNSATCHEM-2D (Šimůnek and Suarez, 1994); this model considers major ion chemistry and calcite and dolomite solubility and precipitation.

An alternative approach for nutrient management, which commonly is much more demanding in terms of computer CPU demands, is the extension of agrohydrological modeling toward nutrients. Agrohydrological models are nowadays strongly focused on understanding water availability to crops and used to schedule irrigation requirements, to anticipate drought yield reductions, and to appreciate salinity stress by crops (Kuhlmann et al., 2012; Noory et al., 2011). This approach, for water availability, usually solves the unsaturated water flow equation (the well-known Richards equation) for a one-dimensional (vertical) water column. For the water flow, use can be made of several computer codes, such as SWAP (Kroes et al., 2017) and HYDRUS-1D (Šimůnek et al., 2008). The 3-D version of the Richards equation that combines Darcy's law with the continuity equation is

$$\frac{\partial \theta}{\partial t} = \nabla \cdot (K \nabla H) - S \tag{42.2}$$

where θ is the volumetric water fraction, t is time, K is the unsaturated soil hydraulic conductivity tensor, H is hydraulic head, and S the sink term for root water uptake. To account for RWU, evapotranspiration, and atmospheric forcing, the water flow models have to be equipped with plant root functionality, which is available in, for example, SWAP and HYDRUS-1D. Few models are available where the soil water model is interacting with a crop growth model (Vereecken et al., 2016), but they are available. A well-known example is the SWAP-WOFOST (Kroes and Supit, 2011) and HYDRUS-WOFOST combination (Zhou et al., 2012). Multidimensional models are also available, such as FUSSIM-3D (Heinen, 2014), PARFLOW (Kollet and Maxwell, 2008), and HYDRUS-3D (Šimůnek et al., 2016). Of great interest is also the advanced 3-D modeling by Vereecken et al. (2016). Examples where the next step, of combining nutrient transport with multi-D flow and RWU modeling, is OpenSimRoot model (Postma et al., 2017); also, pseudo multi-D flow and transport modeling have been undertaken.

The rapid development of software and CPU power is indicative of the advances that can be expected in the next decade. Whereas modeling for soil systems as done by Vereecken et al. (2016) may quickly be extended toward the fate of fertilizer derived nutrients, it is worthwhile to point out some major gaps in our understanding. These gaps concern both water flow and chemical fate (in this chapter, we will focus on water-dissolved chemicals, solutes, rather than gas phase (e.g., N_2O) or pure liquid (oil) compounds as studied by Rappoldt (1992) and Van Dijke and Van der Zee (1998), respectively).

Water uptake: It is evident that, if the roots take up soil water for the transpiration stream inside the plant, this uptake is easier if more roots are present: more roots in a soil volume imply a shorter distance that water needs to travel toward the roots, hence, less flow resistance. For this reason, it is obvious that the so-called root density is an important factor with regard to the uptake of water. The same is the case for the water content in soil: the more water, the easier it is for the plant uptake. The complexity arises if soil is heterogeneous with regard to the soil hydraulic functions (the hydraulic conductivity and the water retention function), and the water content varies in the root zone. In both cases, the plant needs to decide where to take the water. In our software, we need to program this decision: will the plant choose to use water from the place with highest water content, smallest water suction, or some combination of these? (Couvreur et al., 2014). The physical problem is that not only the local presence of water and its energetic status but also the supply rate of water by the soil to the root surface is important for RWU, which is controlled by the hydraulic conductivity function. In their study of RWU in a heterogeneous soil, Kuhlmann et al. (2012) noticed that a wrong choice in these issues could result in a reasonably wet soil with extremely desiccated parts, where the root zone had extracted all of its water: an unrealistic situation.

The complexity of soil-plant RWU relationships is that not only soil controls this interaction, if it is in control to begin with. For instance, we are just beginning to understand plant physiological responses to water stress. Davies and Zhang (1991) gave evidence about plant hormones and root shoot signaling in this context. This suggests an active response of plants to stress that is completely ignored at the soil side of plant–soil research, to our best knowledge. Whereas it is commonly recognized that different plant species have a different susceptibility to drought and indeed to the impact of soil salinity, the different responses of one-species' different genotypes have been recognized for both drought and salinity. Yet, it is not accounted for in any soil water model. These observations, which are (for genotypic differences) sometimes based on incidental evidence, are indicative that there is too little interaction between plant physiology and soil physics.

Whereas we therefore are confronted with essential gaps in our understanding of water uptake by the plant root system, this is even more the case for nutrient uptake. Partly, this is inherent to the very different chemical properties of the nutrient components. Each component can be present in the soil system in the form of a range of species. An example is nitrogen (as NO_3^-, NH_4^+, NO_2, organic N, and so on), for which the different species show different sorption behavior and redox conditions under which they are present.

The source of most nutrients is applied fertilizer, immisions from atmosphere or groundwater into the unsaturated soil and its root zone (RZ), and mineralization in situ. Just as with water, nutrient availability depends on the local quantity present in soil and the transport rate toward the root surfaces (or, adversely, away from these surfaces or out of soil). The multi-D version of the solute transport equation (the convection dispersion equation or CDE) combines the flux equation with the continuity equation and is given by

$$\frac{\partial \theta C}{\partial t} + \frac{\partial f(C)}{\partial t} = \nabla \cdot (\theta D \nabla C - qC) - S_d \tag{42.3}$$

where C is the solute concentration in the water phase, $f(C)$ is a function that describes solute present in a solid form (i.e., matrix), D is the hydrodynamic dispersion tensor, q is the Darcy water flow rate, and S_d represents sinks related with root uptake, volatilization, and degradation/transformation.

Focused on nutrient availability for papaya, several issues of bioavailability in relation to the CDE are apparent, and these are now briefly introduced.

1. For most nutrients, chemical reactions occur both in the aqueous solution and between the solution and the soil matrix. These reactions may be adsorption/desorption and precipitation/dissolution, as well as complexation, the latter in solution. These reactions are typically dependent on the concentrations or activities of more than one solute; hence, for each of these so-called species, a version of the CDE can be developed. In fact, it is also possible to combine those CDEs for one CDE for the component, where the component is dependent on all species of, for example, nitrogen, or zinc, or another element. Models are available to deal with this "multicomponent" approach of solute transport, such as the well-known HP models (HYDRUS-PHREEQC combinations) (Jacques et al., 2018) and ORCHESTRA (Meeussen, 2003).

2. Root nutrient uptake (RNU) is, just as root water uptake (RWU), dependent on the availability at different places in RZ and the transport of nutrients in the root zone (the replenishment). Accordingly, RNU depends on where nutrients enter the root zone and how often this happens (comprehensively called the "fertilization" of RZ) and the retardation due to chemical reactions. In the case of, for example, nitrogen, also, the other factors comprising S_d than root uptake are important. In modeling with the CDE, assumptions are required in the form of "boundary conditions" on how the concentrations are at the root zone and at the plane "halfway" between individual roots. But because the root architecture is spatiotemporally very variable, these boundary conditions change a lot during a growing season. Modeling can hardly (at this time) keep track of these changes, and therefore, it is necessary to find reasonable approximations from which we can still learn on the practicalities of how to fertilize papaya. An example

is the axially symmetric approach of the CDE by Roose et al. (2001) for the transport rate of a nutrient to the single plant root. They assumed that the uptake kinetics at the root surface is given by Michaelis-Menten (MM) kinetics. Whereas MM kinetics are flexible, the switch between limiting cases of MM, such as zero sink, linear concentration dependency of uptake rate, and constant uptake rate, is generally not well known. Moreover, it is likely dependent on water transpiration rate and other complications (see Hinsinger et al., 2011).

3. It has been well recognized that the root architecture is strongly spatiotemporally dependent. Indeed, models have been developed for important crops, to describe such dependencies throughout the growing season. For a crop as papaya, such a model has not yet been parameterized as far as the authors are aware.

4. Nutrient applications nowadays occur regularly for papaya, because their availability needs to be high at all times, to ascertain top yields, and because the soils with commonly not only a high hydraulic permeability but also a large rainfall excess are prone to have much drainage water and therefore much leaching of nutrients. For fertilization efficiency and sustainability of papaya management, it is important that the nutrients are taken up by the crop and not leached into groundwater. To attain this goal, a careful balance is needed for application and atmospheric (rainfall) forcing.

Despite the complexity that was pictured so far, the application of "distributed modeling" is promising. The reasons for that are diverse. For instance, Rappoldt (1992) already showed that geometric complexity (in his case of aggregates, in our case of the root system) can often be captured well in simplified equivalent behavior. It is plausible that this is also the case for RWU and RNU, as Couvreur et al. (2014) showed. They simplified the complex root system to elementary building blocks and showed that it is possible to separate the impacts of (i) water content and (ii) root architecture with appropriate cases that are simulated or perhaps measured (see Vereecken et al., 2016). If field data are available, the soil hydraulic parameters and the root water uptake parameters can also be estimated by inverse modeling using the Couvreur's approach (Cai et al., 2017). The necessary data to accomplish this kind of modeling for papaya include time series with data of soil water contents, its potential, and root distribution at different spatial scales.

We have to admit that, at this moment, the complexity of the root zone water and nutrient dynamics under temporally variable fertilization and rainfall regimes can only be modeled in broad features. Still, such modeling may result in important benefits. For instance, the recognition that most of the nutrients show relatively simple behavior can be approximated by (i) no to little adsorption and high mobility (nitrogen N), (ii) linear adsorption with distinct smaller mobility (Zn, P, and B), and (iii) some intermediary cases (Ca and Mg). These stereotypes can be confronted with dynamic aspects such as irregular (but high rate) rainfall and irrigation, which are likely to leach the more mobile nutrients. For a sustainability analysis, these issues of availability to crop and hazard of leaching need to be combined. With modeling, this is possible if RWU and RNU in dependency of quantities in the root zone are sufficiently known. A combination of such issues is a challenge, but not without precedence. For instance, for water, the impact of highly variable inputs has been investigated by Rodríguez-Iturbe and Porporato (2004), for salt accumulation by Suweis et al. (2010) and Shah et al. (2011), and for sodicity by Van Der Zee et al. (2014). A translation of such "ecohydrological" model approaches toward adsorbing nutrients, as considered by Boekhold and Van der Zee (1991) for metals, seems perfectly feasible.

In Fig. 42.8, the accumulation of "a nutrient" in soil is simulated with a model similar as used by Van Der Zee et al. (2014), where the application regime is varied. In one case, fertilizer is applied once a year, and in the other case, the same annual quantity is applied in 12 equal monthly applications. Furthermore, a season with leaching (i.e., rainfall and irrigation excess) is followed by one without leaching. We see that the fluctuation of the increasing concentration in the root zone is much larger for the annual application. This is logical, and in practice, it depends on many factors (amount of rainfall/irrigation, nutrient adsorption by soil, degradation, and uptake by plant) how large this fluctuation will be. Implicit to Fig. 42.8, also, the leaching will show a sawtooth pattern. This pattern will depend on whether the crop growth (and annual applications of fertilizer) occurs in the dry or in the wet season. Whereas this figure is only intended as an illustration, relatively simple models as used to prepare this figure can help in assessing trade-offs and optimize growing conditions and fertilization strategies.

The development of such models for nutrient-crop combinations requires their parameterization with experimental pot, greenhouse, or field fertilization research, but the rationale can be illustrated with very basic considerations. For instance, consider a well-developed papaya stand, in a high yield cultivation scheme. To accomplish those yields, fertility research may identify which application rates and frequency are needed. The risk of leaching of nutrients depends on the soil hydraulic functions (i.e., the hydraulic conductivity and the retention curves). These functions reflect how much water will drain depending on the irrigation management. In combination with an uptake rate, drainage rates and nutrient concentration levels can be translated into nutrient leaching rates. Accordingly, fertilization and rainfall/irrigation rates can be translated into nutrient leaching rates. With a model, options for good

FIG. 42.8 Simulation of soil nutrient accumulation (concentration) in the root zone (A) and the cumulative mass leached (B) as a function of annual or monthly application regimes.

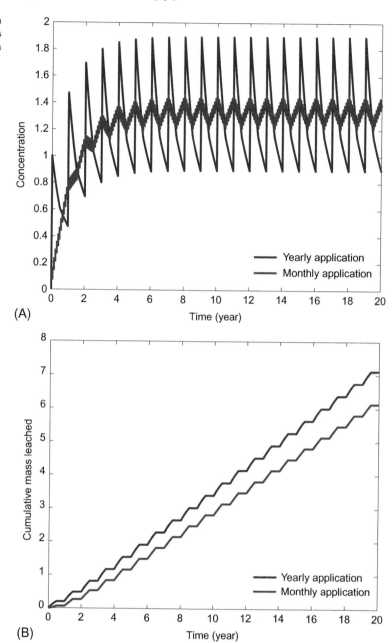

(A)

(B)

fertilization and also little leaching can then be identified. In stating this, we recognize that other aspects play a role: For instance, a 24/7 irrigation/fertilization scheme of every 30 min is unfeasible, but every 12 h might be feasible and every 7 days even more.

8 Conclusions and perspectives

With mechanistic modeling, experiments can be interpreted and designed. Such experiments, to characterize the soil or its reaction to interventions such as fertilization, are crucial to provide the ground truth for the modeling. The models, though, combine the knowledge we have about the soil-plant system and, if well-parameterized, enable the search for optimal conditions in a cost-effective way: fast and with less demands on time and labor.

A necessary step from experimentation toward interpretation is to recognize the interrelationships that have been formalized with the QUEFTS approach. So far, it has only been considered for nutrients. To account for water availability, a necessary step if this availability is highly variable is to use either "bucket" type of ecohydrological modeling or really distributed models. Despite that the later models are also confronted with several crucial gaps in knowledge, it is irrefutably true that, for water availability and irrigation management investigations, such models are already used. In fact, drought- or salinity-induced yield reductions are evaluated with such models and sometimes considered in liability or compensation measures. Based on the use of water modeling for crop yield improvement, it is clear that this can also be done for nutrient availability, despite shortcomings in our knowledge. Just as with experiments in the field, modeling can give us understanding of interdependencies between management, different crop growing factors, and yields.

Future advances in modeling of water and nutrient uptake for papaya require the parameterization, calibration, and evaluation of solute transport models for soils in which papaya is grown. These solute transport models are necessary to account for nutrient availability and for the leaching potential of the applied fertilizer. The use of multicomponent solute transport models is interesting to account for the interactions between nutrients and their reactions in the soil, as for papaya, commonly, a mixture of elements in each fertilization is being applied.

Parameterization and calibration of root water and nutrient uptake models are still pending. For the most common nutrient uptake models, basic parameters that are lacking are the concentration threshold that divides the nutrient uptake between the passive or the active uptake mechanism and parameters that describe the kinetics of the active nutrient uptake (Michaelis-Menten parameters).

Acknowledgments

We gratefully acknowledge support and funding of this work by Ministry of Science and Technology of Costa Rica (MICITT), University of Costa Rica (UCR), and "Fundación FITTACORI." We also appreciate that Pavan Cornelissen (Wageningen University & Research) made Fig. 42.8 and the permission of Antonio Bogantes (MAG-INTA, Costa Rica) and Eric Mora-Newcomer (University of Costa Rica) for using their figures and tables. This work was partially funded by the Dutch Science Foundation (NWO) project RUST under NWO-contract number ALWGK.2016.16.

References

Allan, P., 2002. Carica papaya responses under cool subtropical growth conditions. Acta Hortic. 575, 757–763. https://doi.org/10.17660/ActaHortic.2002.575.89.

Allan, P., Biggs, D., 1987. Environmental effects on clonal female and male Carica papaya L. plants. Sci. Hortic. 32, 221–232.

Allison, J.D., Brown, D., Novo-Gradac, K., 1991. MINTEQA2/PRODEFA2, A Geochemical Assessment Model for Environmental Systems: User Manual Supplement for Version 3.0. .

Becerra, E.N., Cárdenas, E., Lozoya, H., Mosqueda, R., 1999. Rhabdovirus en papayo (Carica papaya L.) en el sureste de MéxicoEN EL. Agron. Mesoam. 10 (2), 85–90.

Boekhold, A.E., Van der Zee, S.E.A.T.M., 1991. Long-term effects of soil heterogeneity on cadmium behaviour in soil. J. Contam. Hydrol. 7, 371–390.

Bogantes, A., Mora Newcomer, E., Umaña Rojas, G., Loría Quirós, C.L., 2011. Guía para el cultivo de papaya en Costa Rica. MAG/UCR/INTA, San José, Costa Rica. Retrieved from: http://www.mag.go.cr/bibliotecavirtual/F01-10190.pdf.

Bowen, J.E., 1992. Comparative DRIS (diagnostic recommendation and integrated system) and critical concentration interpretation of papaya tissue analysis data. Trop. Agric. 69 (1), 63–67.

Cai, G., Vanderborght, J., Couvreur, V., Mboh, C.M., Vereecken, H., 2017. Parameterization of root water uptake models considering dynamic root distributions and water uptake compensation. Vadose Zone J. 21. https://doi.org/10.2136/vzj2016.12.0125.

Campostrini, E., Glenn, D.M., 2007. Ecophysiology of papaya: a review. Braz. J. Plant Physiol. 19 (4), 413–424. Retrieved from: http://www.scielo.br/pdf/bjpp/v19n4/a10v19n4.pdf.

Campostrini, E., Schaffer, B., Ramalho, J.D.C., González, J.C., Rodrigues, W.P., da Silva, J.R., Lima, R.S.N., 2018. Environmental factors controlling carbon assimilation, growth, and yield of papaya (Carica papaya L.) under water-scarcity scenarios. In: Water Scarcity and Sustainable Agriculture in Semiarid Environment. Elsevier Inc.https://doi.org/10.1016/B978-0-12-813164-0.00019-3 (Chapter 19).

Carvalho, F.A., Renner, S.S., 2014. The phylogeny of the Caricaceae. In: Ming, R., Moore, P.H. (Eds.), Genetics and Genomics of Papaya. Springer New York, New York, NY, pp. 81–92. https://doi.org/10.1007/978-1-4614-8087-7_5.

Céccoli, G., Panigo, E.S., Gariglio, N., Favaro, J.C., Bouzo, C., 2013. Fruit yield and growth parameters of several Carica papaya L . genotypes in a temperate climate. Rev. FCA Uncuyo 45 (2), 299–310.

Chan, Y.-K., 2009. Breeding papaya (Carica papaya L.). In: Breeding Plantation Tree Crops, pp. 121–160. Tropical species.

Chatterjee, C., Dube, B.K., 2004. Nutrient deficiency disorders in fruit trees and their management. In: Mukerji, K.G. (Ed.), Fruit and Vegetable Diseases. Kluwer Academic Publishers, pp. 3–39.

Chutteang, C., Yingjajaval, S., Wasee, S., 2007. Leaf photosynthetic potential of female and hermaphrodite papaya (Carica papaya cv. Khaeg nuan). In: Acta Horticulturae. International Society for Horticultural Science (ISHS), Leuven, pp. 197–202. https://doi.org/10.17660/ActaHortic.2007.740.23.

Cibes, H.R., Gaztambide, S., 1978. Mineral-deficiency symptoms displayed by papaya plants grown under controlled conditions. J. Agr. U. Puerto Rico 62 (4), 413–423.

Cochrane, T.T., Salinas, J.G., Sánchez, P.A., 1980. An equation for liming acid mineral soils to compensate crop aluminium tolerance. Trop. Agric. 57 (2), 133–140.

Costa, A., 1995. Uso do Sistema Integrado de Diagnose e Recomendação (DRIS) na Avaliação do Estado Nutricional do Mamoeiro (Carica papaya L.) no Estado do Espírito Santo. Universidade Federal de Viçosa.

Couvreur, V., Vanderborght, J., Draye, X., Javaux, M., 2014. Dynamic aspects of soil water availability for isohydric plants: focus on root hydraulic resistances. Water Resour. Res. 50, 8891–8906. https://doi.org/10.1002/2014WR015608.

Cunha, J.R., Haag, H.P., 1980. Nutrição mineral do mamoeiro. Marcha de absorção de nutrientes em condições de campo. Anais Da E S A Luiz de Queiróz 37, 631–668.

Danese, C., Esposito, D., D'Alfonso, V., Cirene, M., Ambrosino, M., Colotto, M., 2006. Plasma glucose level decreases as collateral effect of fermented papaya preparation use. Clin. Ter. 157 (3), 195–198.

Davies, W.J., Zhang, J., 1991. Drying soil regulation of growth and developement of plants in drying soils. Annu. Rev. Plant. Physiol. Plant. Mol. Biol. 42, 55–76.

De Brito Neto, J.F., Pereira, W.E., Cavalcanti, L.F., Da Costa Araujo, R., De Lacerda, J.S., 2011. Produtividade e qualidade de frutos de mamoeiro "sunrise solo" em função de doses de nitrogênio e boro. Semina:Ciencias Agrarias 32 (1), 69–80.

De la Fuente, J., Ramírez-Rodríguez, V., Cabrera-Ponce, J., Herrera-Estrella, L., 1997. Aluminum tolerance in transgenic plants by alteration of citrate synthesis. Science 276 (1), 1566–1568. https://doi.org/10.1126/science.276.5318.1566.

de Lima, R.S.N., de Assis Figueiredo, F.A.M.M., Martins, A.O., de Deus, B.C., Ferraz, T.M., Gomes, M., de Sousa, E.F., … Campostrini, E., 2015. Partial rootzone drying (PRD) and regulated deficit irrigation (RDI) effects on stomatal conductance, growth, photosynthetic capacity, and water-use efficiency of papaya. Sci. Hortic. 183, 13–22. https://doi.org/10.1016/j.scienta.2014.12.005.

Dettmann, U., Andrews, F., 2018. SoilHyP. Retrieved from: https://cran.r-project.org/package=SoilHyP.

Dordas, C., 2008. Role of nutrients in controlling plant diseases in sustainable agriculture. A review. Agron. Sustain. Dev. 28, 33–46. https://doi.org/10.1007/978-3-319-67349-3_8.

Fallas, R., Bertsch, F., Barrientos, M., 2014. Curvas de absorción de nutrientes en papaya (*Carica papaya* L.) CV. "Pococí" en las fases de crecimiento vegetativo, floración e inicio de cosecha. Agronomía Costarricense 38 (2), 43–54.

FAO, 2018. FAOSTAT. Retrieved from: http://www.fao.org/faostat/en/#data/QC. (Accessed 19 February 2018).

Fest, E.P.M.J., Temminghoff, E.J.M., Griffioen, J., Van Riemsdijk, W.H., 2005. Proton buffering and metal leaching in sandy soils. Environ. Sci. Technol. 39 (20), 7901–7908. https://doi.org/10.1021/es0505806.

Fisher, J., 1980. The vegetative and reproductive structure of papaya. Lyonia 1 (4), 191–208.

Fuentes, G., Santamaría, J., 2014. Papaya (*Carica papaya* L.): origin, domestication, and production. In: Genetics and Genomics of Papaya, p. 438.

Fujita, Y., Tsuno, H., Nakayama, J., 2017. Fermented papaya preparation restores age-related reductions in peripheral blood mononuclear cell Cytolytic activity in tube-fed patients. PLoS One 12 (1), 1–18. https://doi.org/10.1371/journal.pone.0169240.

Gomes Oliveira, M.A., Correa Caldas, R., 2004. Produção do mamoeiro em função de adubação com nitrogênio, fósforo e potássio. Rev. Bras. Frutic. 26 (1), 160–163.

Gutiérrez, M., 1997. Nutrición mineral de las plantas: avances y aplicaciones. Agronomía Costarricense 21 (1), 127–137.

Heinen, M., 2014. Compensation in root water uptake models combined with three-dimensional root length density distribution. Vadose Zone J. 13 (2). https://doi.org/10.2136/vzj2013.08.0149.

Hinsinger, P., Brauman, A., Devau, N., Gérard, F., Jourdan, C., Laclau, J.-P., … Plassard, C., 2011. Acquisition of phosphorus and other poorly mobile nutrients by roots. Where do plant nutrition models fail? Plant and Soil 348 (1–2), 29–61. https://doi.org/10.1007/s11104-011-0903-y.

Jacques, D., Šimůnek, J., Mallants, D., Van Genuchten, M.T., 2018. The HPx software for multicomponent reactive transport during variably-saturated flow: recent developments and applications. J. Hydrol. Hydromech. 66 (2), 211–226. https://doi.org/10.1515/johh-2017-0049.

Janssen, B.H., Guiking, F.C.T., van der Eijk, D., Smaling, E.M.A., Wolf, J., van Reuler, H., 1990. A system for quantitative evaluation of the fertility of tropical soils (QUEFTS). Geoderma 46 (4), 299–318. https://doi.org/10.1016/0016-7061(90)90021-Z.

Jarrel, W.M., Beverly, R.B., 1981. The dilution effect in plant nutrition studies. Adv. Agron. 34, 197–224.

Jiménez, V., Mora-Newcomer, E., Gutiérrez-Soto, M., 2014. Biology of the papaya plant. In: Ming, R., Moore, P.H. (Eds.), Genetics and Genomics of Papaya. Springer Science + Business Media, New York, pp. 17–34. https://doi.org/10.1007/978-1-4614-8087-7.

Kaisar, M.O., Sadat, M.A., Khalequzzaman, K.M., 2013. Growth and yield of papaya subjected to nutrient deprivation. Int. J. Sustain. Crop Prod 8 (1), 25–27.

Keizer, M.G., Van Riemsdijk, W.H., 2009. ECOSAT: A Computer Program for the Calculation of Speciation and Transport in Soil-Water Systems. https://doi.org/10.1007/SpringerReference_28001.

Khondaker, N.A., Ozawa, K., 2007. Papaya plant growth as affected by soil air oxygen deficiency. In: Acta Horticulturae. International Society for Horticultural Science (ISHS), Leuven, pp. 225–232. https://doi.org/10.17660/ActaHortic.2007.740.27.

Kollet, S.J., Maxwell, R.M., 2008. Capturing the influence of groundwater dynamics on land surface processes using an integrated, distributed watershed model. Water Resour. Res. 44 (2), 1–18. https://doi.org/10.1029/2007WR006004.

Kroes, J.G., Supit, I., 2011. Impact analysis of drought, water excess and salinity on grass production in the Netherlands using historical and future climate data. Agric. Ecosyst. Environ. 144 (1), 370–381. https://doi.org/10.1016/j.agee.2011.09.008.

Kroes, J.G., van Dam, J.C., Bartholomeus, R.P., Groenendijk, P., Heinen, M., Hendriks, R.F.A., … van Walsum, P.E.V., 2017. SWAP Version 4: Theory Description and User Manual. Wageningen Environmental Research, Report 2780, Wageningen. Retrieved from: http://library.wur.nl/WebQuery/wurpubs/fulltext/416321.

Kuhlmann, A., Neuweiler, I., Van Der Zee, S.E.A.T.M., Helmig, R., 2012. Influence of soil structure and root water uptake strategy on unsaturated flow in heterogeneous media. Water Resour. Res. 48 (2), 1–16. https://doi.org/10.1029/2011WR010651.

Kumar, N., Soorianathasundaram, K., Meenakshi, N., Manivannan, M.I., Suresh, J., Nosov, V., 2010. Balanced fertilization in papaya (Carica). Horticulture 851, 357–362.

Lima, R.S.N., García-Tejero, I., Lopes, T.S., Costa, J.M., Vaz, M., Durán-Zuazo, V.H., … Campostrini, E., 2016. Linking thermal imaging to physiological indicators in *Carica papaya* L. under different watering regimes. Agric Water Manag 164, 148–157. https://doi.org/10.1016/j.agwat.2015.07.017.

Lopes Peçanha, A., Rangel da Silva, J., Pereira Rodrigues, W., Massi Ferraz, T., Torres Netto, A., Samara Nunes Lima, R., … Campostrini, E., 2017. Leaf gas exchange and growth of two papaya (*Carica papaya* L.) genotypes are affected by elevated electrical conductivity of the nutrient solution. Sci. Hortic. 218, 230–239. https://doi.org/10.1016/j.scienta.2017.02.018.

Maas, E.V., 1993. Testing crops for salinity tolerance. In: Maranville, J.W., Baligar, B.V., Duncan, R.R., Yohe, J.M. (Eds.), Workshop on Adaptation of Plants to Soil Stresses, pp. 234–247. https://doi.org/10.1109/ACC.1995.529272.

Mahouachi, J., Socorro, A.R., Talon, M., 2006. Responses of papaya seedlings (*Carica papaya* L.) to water stress and re-hydration: growth, photosynthesis and mineral nutrient imbalance. Plant and Soil 281 (1–2), 137–146. https://doi.org/10.1007/s11104-005-3935-3.

Malavolta, E., Vitti, G., Oliveira, S., 1989. Avaliação do estado nutricional das plantas. Principios e aplicações. Associação Brasileira para Pesquisa da Potassa e do Fosfato, Piracicaba.

Marinho, C.S., Monnerat, P.H., De Carvalho, A.J.C., Martins, S.L.D., Vieira, A., 2002. Análise química do pecíolo e limbo foliar como indicadora do estado nutricional dos mamoeiros "solo" e "formosa.". Sci. Agric. 59 (2), 373–381. https://doi.org/10.1590/S0103-90162002000200025.

Marler, T.E., 1998. Solution pH influences on growth and mineral element concentrations of "Waimanalo" papaya seedlings. J. Plant Nutr. 21 (12), 2601–2612. Retrieved from: http://www-tandfonline-com.ezproxy.library.wur.nl/doi/pdf/10.1080/01904169809365591?needAccess=true.

Marler, T., Mickelbart, M., 1998. Drought, leaf gas exchange, and chlorophyll fluorescence of field-grown papaya. J. Amer. Soc. Hort. Sci. 123 (4), 714–718.

Marler, T., George, A., Nissen, R., Andersen, P., 1994. Miscellaneous tropical fruits. Schaffer, B., Andersen, P.C. (Eds.), Handbook of Environmental Physiology of Fruits Crops. In: VII Subtropical and Tropical Crops, CRC Press, Florida, pp. 199–224.

Martins, G.F., Fabi, J.P., Mercadante, A.Z., de Rosso, V.V., 2016. The ripening influence of two papaya cultivars on carotenoid biosynthesis and radical scavenging capacity. Food Res. Int. 81, 197–202. https://doi.org/10.1016/j.foodres.2015.11.027.

Meeussen, J.C.L., 2003. Orchestra: an object-oriented framework for implementing chemical equilibrium models. Environ. Sci. Technol. 37 (6), 1175–1182. https://doi.org/10.1021/es025597s.

Mills, H., Jones, B., 1996. Plant Analysis Handbook II: A Practical Sampling Preparation Analysis, and Interpretation Guide. Micro-Macro Pub.

Nautiyal, B.D., Sharma, C.P., Agarwala, S.C., 1986. Iron, zinc and boron deficiency in papaya. Sci. Hortic. 29 (1–2), 115–123. https://doi.org/10.1016/0304-4238(86)90037-3.

Nieuwenhuyse, A., 1988. Application of WOFOST, DUET and QUEFTS for Modelling Maize and Grass Production Using Data From the Atlantic Zone of Costa Rica. IICA/CATIE.

Noory, H., Van Der Zee, S.E.A.T.M., Liaghat, A.M., Parsinejad, M., van Dam, J.C., 2011. Distributed agro-hydrological modeling with SWAP to improve water and salt management of the Voshmgir irrigation and drainage network in northern Iran. Agric Water Manag 98 (6), 1062–1070. https://doi.org/10.1016/j.agwat.2011.01.013.

Nugroho, A., Heryani, H., Choi, J.S., Park, H.-J., 2017. Identification and quantification of flavonoids in Carica papaya leaf and peroxynitrite-scavenging activity. Asian Pac. J. Trop. Biomed. 7 (3), 208–213. https://doi.org/10.1016/j.apjtb.2016.12.009.

Ovando-Martínez, M., López-Teros, V., Tortoledo-Ortiz, O., Astiazarán-García, H., Ayala-Zavala, J., Villegas-Ochoa, M., Gonzáles Aguilar, G., 2018. Effect of ripening on physico-chemical properties and bioactive compounds in papaya pulp, skin and seeds. Indian J. Nat. Prod. Resour. 9 (1), 47–59.

Parkhurst, D., Appelo, C.A.J., 1999. User's Guide to PHREEQC (Version 2)—A Computer Program for Speciation, Batch-Reaction, One-Dimensional Transport, and Inverse Geochemical Calculations. Water Investigations Report 99-4259, 312, U.S. Geological Survey: Earth Science Information Center.

Paull, R., 1993. Pineapple and papaya. In: Seymour, G.B., Taylor, J.E., Tucker, G.A. (Eds.), Biogeochemistry of Fruit Ripening, first ed. Springer Science+Business Media, Dordrecht, pp. 291–315. https://doi.org/10.1007/978-94-011-1584-1.

Peng, S., Wang, P., Peng, L., Cheng, T., Sun, W., Shi, Z., 2018. Predicting heavy metal partition equilibrium in soils: roles of soil components and binding sites. Soil Sci. Soc. Am. J. 82, 839–849. https://doi.org/10.2136/sssaj2014.07.0299.

Postma, J.A., Kuppe, C., Owen, M.R., Mellor, N., Griffiths, M., Bennett, M.J., … Watt, M., 2017. OpenSimRoot: widening the scope and application of root architectural models. New Phytol. 215 (3), 1274–1286. https://doi.org/10.1111/nph.14641.

Qiu, Y., Nishina, M.S., Paull, R.E., 1995. Papaya fruit growth, calcium uptake, and fruit ripening. J. Amer. Soc. Hort. Sci. 120 (2), 246–253.

Quaggio, J.A., Raij, B.v., Malavolta, E., 1985. Alternative use of the Smp-buffer solution to determine lime requirement of soils. Commun. Soil Sci. Plant Anal. 16 (3), 245–260. https://doi.org/10.1080/00103628509367600.

R Core Team, 2017. R: A Language and Environment for Statistical Computing. R Foundation for Statistical Computing, Vienna. Retrieved from: https://www.r-project.org/.

Rappoldt, C., 1992. Diffusion in Aggregated Soil. Wageningen University.

Reddy, Y.T.N., Dinesh, M.R., 2000. Papaya Cultivation. Indian Institute of Agricultural Research.

Robinson, J.B., Treeby, M.T., Stephenson, R.A., 1997. Fruits, vines and nuts. In: Reuter, D.J., Robinson, J.B., Dutkiewicz, C. (Eds.), Plant Analysis. An Interpretation Manual. second ed. CSIRO, p. 572.

Rodríguez-Iturbe, I., Porporato, A., 2004. Ecohydrology of Water-Controlled Ecosystems: Soil Moisture and Plant Dynamics, first ed. Cambridge University Press.

Roose, T., Fowler, A., Darrah, P., 2001. A mathematical model of plant nutrient uptake. J. Math. Biol. 42 (4), 347–360. https://doi.org/10.1007/s002850000075.

Ryan, P.R., Shaff, J.E., Kochian, L.V., 1992. Aluminum toxicity in roots 1. Plant Physiol. 99 (3), 1193–1200. Retrieved from: http://www.pubmedcentral.nih.gov/articlerender.fcgi?artid=1080602&tool=pmcentrez&rendertype=abstract.

Sales Marinho, C., Borges de Oliveira, M.A., Monnerat, P.H., Vianni, R., Maldonado, J.F., 2001. Fontes e doses de nitrogênio e a qualidade dos frutos do mamoeiro. Sci. Agric. 58 (2), 345–348. Retrieved from: http://www.scielo.br/pdf/sa/v58n2/4426.pdf.

Sattari, S.Z., van Ittersum, M.K., Bouwman, A.F., Smit, A.L., Janssen, B.H., 2014. Crop yield response to soil fertility and N, P, K inputs in different environments: testing and improving the QUEFTS model. Field Crop Res 157, 35–46. https://doi.org/10.1016/j.fcr.2013.12.005.

Schütz, L., Gattinger, A., Meier, M., Müller, A., Boller, T., Mäder, P., Mathimaran, N., 2018. Improving crop yield and nutrient use efficiency via biofertilization—a global meta-analysis. Front. Plant Sci. 8, 2204. https://doi.org/10.3389/fpls.2017.02204.

Schweiggert, R.M., Steingass, C.B., Mora, E., Esquivel, P., Carle, R., 2011. Carotenogenesis and physico-chemical characteristics during maturation of red fleshed papaya fruit (*Carica papaya* L.). Food Res. Int. 44 (5), 1373–1380. https://doi.org/10.1016/j.foodres.2011.01.029.

Shah, S.H.H., Vervoort, R.W., Suweis, S., Guswa, A.J., Rinaldo, A., Van Der Zee, S.E.A.T.M., 2011. Stochastic modeling of salt accumulation in the root zone due to capillary flux from brackish groundwater. Water Resour. Res. 47 (9), 1–17. https://doi.org/10.1029/2010WR009790.

Shoemaker, H.E., McLean, E.O., Pratt, P.F., 1961. Buffer methods for determining lime requirement of soil with appreciable amounts of extractable aluminum. Soil Sci. Soc. Am. J. 25 (4), 274–277.

Šimůnek, J., Suarez, D.L., 1994. Two-dimensional transport model for variably saturated porous media with major ion chemistry. Water Resour. Res. 30 (4), 1115–1133. https://doi.org/10.1029/93WR03347.

Šimůnek, J., van Genuchten, M.T., Šejna, M., 2008. Development and applications of the HYDRUS and STANMOD software packages and related codes. Vadose Zone J. 7 (2), 587. https://doi.org/10.2136/vzj2007.0077.

Šimůnek, J., van Genuchten, M.T., Šejna, M., 2016. Recent developments and applications of the HYDRUS computer software packages. Vadose Zone J. 15(7). https://doi.org/10.2136/vzj2016.04.0033.

Singh, D.K., Ghosh, S.K., Paul, P.K., Suresh, C.P., Technology, P.H., Banga, U., … Bengal, W., 2010. Effect of different micronutrients on growth, yield and quality of papaya Carica. Methods, 351–356.

Somanah, J., Putteeraj, M., Aruoma, O.I., Bahorun, T., 2018. Discovering the health promoting potential of fermented papaya preparation—its future perspectives for the dietary management of oxidative stress during diabetes. Fermentation. 4(4). https://doi.org/10.3390/fermentation4040083.

Starley, I.F., Mohammed, P., Schneider, G., Bickler, S.W., 1999. The treatment of paediatric burns using topical papaya. Burns 25 (7), 636–639. https://doi.org/10.1016/S0305-4179(99)00056-X.

Suweis, S., Rinaldo, A., Van Der Zee, S.E.A.T.M., Daly, E., Maritan, A., Porporato, A., 2010. Stochastic modeling of soil salinity. Geophys. Res. Lett. 37 (7), 1–5. https://doi.org/10.1029/2010GL042495.

Tennant, P.F., Fermin, G.A., Roye, M.E., 2007. Viruses infecting papaya (Carica papaya L.): etiology, pathogenesis, and molecular biology. Plant Viruses 1 (2), 178–188. Retrieved from: http://www.globalsciencebooks.info/Online/GSBOnline/images/0712/PV_1(2)/PV_1(2)178-188o.pdf.

Thani, Q.A., Schaffer, B., Liu, G., Vargas, A.I., Crane, J.H., 2016. Chemical oxygen fertilization reduces stress and increases recovery and survival of flooded papaya (Carica papaya L.) plants. Sci. Hortic. 202, 173–183. https://doi.org/10.1016/j.scienta.2016.03.004.

Thaysen, E.M., Jessen, S., Postma, D., Jakobsen, R., Jacques, D., Ambus, P., … Jakobsen, I., 2014. Effects of lime and concrete waste on Vadose zone carbon cycling. Vadose Zone J. 13(11). https://doi.org/10.2136/vzj2014.07.0083.

Thomas, M.B., Ferguson, J., Crane, J.H., 1995. Identification of N, K, Mg, Mn, Zn and Fe deficiency symptoms of carambola, lychee, and papaya grown in sand culture. Proc. Fla. State Hort. Soc. 108, 370–373.

Van Der Zee, S.E.A.T.M., Shah, S.H.H., Vervoort, R.W., 2014. Root zone salinity and sodicity under seasonal rainfall due to feedback of decreasing hydraulic conductivity. Water Resour. Res. 50 (12), 9432–9446. https://doi.org/10.1002/2013WR015208.

Van Dijke, M.I.J., Van der Zee, S., 1998. Modeling of air sparging in a layered soil: numerical and analytical approximations. Water Resour. Res. 34 (3), 341–353.

Vereecken, H., Schnepf, A., Hopmans, J.W., Javaux, M., Or, D., Roose, T., … Lamorski, K., 2016. Modeling soil processes: review, key challenges and new perspectives. Vadose Zone J. 15 (5), 1–57. https://doi.org/10.2136/vzj2015.09.0131.

Vervoort, R.W., van der Zee, S.E.A.T.M., 2008. Simulating the effect of capillary flux on the soil water balance in a stochastic ecohydrological framework. Water Resour. Res. 44, W08425. https://doi.org/10.1029/2008WR006889.

Vichiato, M., de Carvalho, J.G., Vichiato, M. R. de M., & Silva, C. R. de R. e., 2009. Interações fósforo-magnésio em mudas de mamoeiros tainung no. 1 e improved sunrise solo 72/12. Ciencia e Agrotecnologia 33 (5), 1265–1271.

Wall, M.M., Tripathi, S., 2014. Papaya nutritional analysis. In: Ming, R., Moore, P.H. (Eds.), Genetics and Genomics of Papaya. Springer New York, New York, NY, pp. 377–390. https://doi.org/10.1007/978-1-4614-8087-7_20.

Wang, D., Ko, W., 1975. Relationship between deformed fruit disease of papaya and boron deficiency. Phytopathology 65, 445–447. https://doi.org/10.1094/Phyto-65-445.

Wang, R.H., Chang, J.C., Li, K.T., Lin, T.S., Chang, L.S., 2014. Leaf age and light intensity affect gas exchange parameters and photosynthesis within the developing canopy of field net-house-grown papaya trees. Sci. Hortic. 165, 365–373. https://doi.org/10.1016/j.scienta.2013.11.035.

Weng, L., Temminghoff, E.J.M., Van Riemsdijk, W.H., 2001. Contribution of individual sorbents to the control of heavy metal activity in sandy soil. Environ. Sci. Technol. 35 (22), 4436–4443. https://doi.org/10.1021/es010085j.

Zhao, X.Q., Shen, R.F., 2018. Aluminum–nitrogen interactions in the soil–plant system. Front. Plant Sci. 9, 1–15. https://doi.org/10.3389/fpls.2018.00807.

Zhou, L., Paull, R., 2001. Sucrose metabolism during papaya (Carica papaya) fruit growth and ripening. J. Amer. Soc. Hort. Sci. 126 (3), 351–357.

Zhou, J., Cheng, G., Li, X., Hu, B., Wang, G., 2012. Numerical modeling of wheat irrigation using coupled HYDRUS and WOFOST models. Soil Sci. Soc. Am. J. 76 (2), 648–662. https://doi.org/10.2136/sssaj.

CHAPTER

43

Diagnosis and management of nutrient constraints in mango

R.A. Ram[a],*, M.A. Rahim[b], M.S. Alam[c]

[a]ICAR-Central Institute for Subtropical Horticulture, Lucknow, India
[b]Department of Horticulture, Bangladesh Agriculture University, Mymensingh, Bangladesh
[c]Horticulture Division, Bangladesh Institute of Nuclear Agriculture, Mymensingh, Bangladesh
*Corresponding author. E-mail: raram_cish@yahoo.co.in

1 Introduction

Mango (*Mangifera indica* L.) is an evergreen tree of the family Anacardiaceae, grown extensively for its edible fruit. Mango trees grow to 35–40 m tall, with a crown radius of 10–15 m. The trees are long lived, as some specimens still fruit after 300 years. The genus *Mangifera* is one among the 73 genera belonging to the family Anacardiaceae in the order Sapindales. Mango (*M. indica* L.) has been cultivated for more than 5000 years, and a wide genetic diversity exist in this crop. The largest numbers of *Mangifera* species are found in the Malay Peninsula, the Indonesian archipelago, Thailand, Indochina, and the Philippines. Edible fruits are produced by at least 27 species in the genus.

Mangoes are native to southern Asia, especially eastern India, Burma, and Andaman Islands. Its production predominates in dry and wet tropical low land areas 23°26′ North and South of the equator, on the Indian subcontinent, Southeast Asia, and Central and South America (Litz, 1997) (Fig. 43.1). Mango the "King of Fruits" is the most popular fruit among millions of people in India and abroad. In terms of total fruit production on a global basis, mango is next to banana. For mango, India ranks first in area and production in the world. Mango covers an area of 4,946,000 ha with a production of 46.5 million tons in the world during the year of 2016 (Brandon Gaille, 2018). India occupies top position among mango-growing countries of the world and produces 40.48% of the total world mango production (NHB, 2017). India's

A.K. Srivastava, Chengxiao Hu (eds.)
Fruit Crops: Diagnosis and Management of Nutrient Constraints
https://doi.org/10.1016/B978-0-12-818732-6.00043-5

629

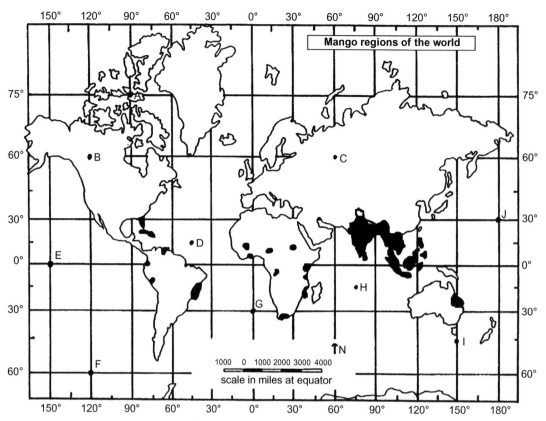

FIG. 43.1 Mango growing regions of the world. *Data from Bains, G., Pant, R.C., 2003. Mango malformation: etiology and preventive measures. Physiol. Mol. Biol. Plants 9, 41–61.*

contribution to the world's mango production is highest, that is, 197.87 lakh metric tons from 22.63 lakh ha, and about 53,177.26 metric tons of mango is exported of approx. value of INR 44,554.54 lakhs (NHB, 2017). In spite of the highest area (22.63 lakh ha) and production of mango in the world, the mango productivity is very low (8.70 metric tons/ha).

2 Nutritional value of fruit

Mango is rich in a variety of phytochemicals and nutrients that qualify it as a super fruit, a term used to highlight potential health value of certain edible fruits. A 100 g of mango has following composition: water 83.46 g, energy 60 kcal, protein 0.82 g, total lipid (fat) 0.38 g, carbohydrate 14.98 g, fiber, total dietary 1.6 g, total sugar13.66 g, Ca 11 mg, Fe 0.16 mg, Mg 10 mg, P 14 mg, K 168 mg, Na 1 mg, Zn 0.09 mg, vitamin C 36.4 mg, thiamin 0.028 mg, riboflavin 0.038 mg, niacin 0.669 mg, vitamin B6 0.119 mg, folate 43 mg, vitamin B12 0.00 mg, vitamin A 54 mg, vitamin A 1082 IU, vitamin E 0.90 mg, vitamin K 4.2 mg, saturated fatty acids 0.092 g, total monounsaturated fatty acids 0.140 g, and total polyunsaturated fatty acids 0.071 g (Source: USDA—National Nutriment database, 2019). Mango is considered highly nutrient exhaustive fruit crop. Nutrient removal by mango fruits has also been studied by Singh (1978) (Table 43.1). Krishnakumar et al. (1998) studied the nutrient removal through harvested fruits. The maximum removal of most of

TABLE 43.1 Nutrient removal in kg by mango cv. Dashehari (Singh et al., 1962).

Fruit's part	N	P₂O₅	K₂O	CaO	MgO	Ash
Whole fruit/1000 kg	1.23	0.60	2.14	0.91	0.52	6.27
Flesh/668 Kg	0.66	0.31	1.12	0.16	0.22	2.57
Peel/204 Kg	0.25	0.14	0.68	0.47	0.20	2.57
Kernel/54 kg	0.25	0.12	0.25	0.055	0.066	0.60
Seed coat/74 kg	0.067	0.025	0.10	0.22	0.029	0.53

the nutrients was recorded in Khas-ul-khas, Goamankur, and Alphonso varieties. The minimum removal was from Totapuri and Willard varieties. The bulk of the nutrients were removed through the edible portion of epicarp-mesocarp followed by kernel and endocarp portions. The average removal of the different nutrients from every tonne of harvested fruits was found to be 1363.64, 100.54, 1454.41, 1143.70, 123.10, 28.03, 8.31, 24.03, and 12.44 g each of N, P, K, Ca, Mg, Fe, Mn, Zn, and Cu, respectively. According to studies by Fowomola (2010), mango seed contained sodium 21.0 mg/100 g), potassium (22.3 mg/100 g), calcium (113.3 mg/100 g), magnesium (94.8 mg/100 g), iron (11.9 mg/ 100 g), and zinc (1.1 mg/100 g), nutritionally so promising indeed.

3 Geographical distribution

Gradually, mango moved from its center of origin in Indo-Burma (Myanmar) region to all the mango-growing countries. The mango seeds traveled with humans from Asia to the Middle East, East Africa, and South America in beginning around AD 300–400. In the beginning, it moved to the southeast Asia and the Malay Archipelago, and by the 7th century, it reached China. The Persians are said to have carried it to East Africa around the 10th Century AD. Muslim missionaries took mango to the Mindanao and the Sulu of the Philippines in the beginning of the 15th century. During the end of the 15th century and beginning of the 16th century, the mango entered into the Philippines from India through Spanish. The mango spread from south and southeast Asia over the tropical and subtropical areas of the world as a result of the Portuguese and Spanish voyages from the end of the 15th century onward. Mango was carried to Africa during the 16th century and later found their way abroad through Portuguese ships to Brazil in the 1700s. Portuguese brought mango from Goa to South Africa and from South Africa to Brazil in the beginning of the 18th century. From Africa, the Portuguese brought it to the West Indies also; however, it was commonly grown in the East Indies before the earliest visits of the Portuguese. The mango was taken to Santo Domingo of Barbados in the West Indies in the middle of the 18th century (Singh et al., 1991) and from Bourbon to Santo Domingo in 1782. In 1742, the mango was successfully introduced to Barbados from Rio de Janeiro by the traders from Portugal. Early in the 19th century, it reached Mexico from the Philippines and the West Indies. The mango reached Jamaica from Barbados in 1782 and Hawaii from Mexico in 1809. Afterward, it began popping up all over the world. In the 16th century, a special technique employing grafting was developed for propagating the mango. During 17th century, the Portuguese planted mango in coastal areas of both East and West Africa, but the acceptance by the Africans was slow and spread into the interior was erratic. Mango trees were present in a few interior market towns in West Africa, when European explorers arrived in the late 19th century, but most of the spread came later.

The earliest known successful introduction of mango to the new world was to Bahia in Brazil in about 1700 with plantings elsewhere along the Brazilian coast soon after. With the first permanent plantation, the mango reached Florida from India in 1861 as cv. Mulgoa. In 1882, this cultivar was subsequently introduced in Florida from India. From Bombay (now Mumbai, India), the mango was introduced into Egypt in 1825 from where it was introduced into Israel in 1929, and thereafter, the commercial varieties were introduced in Israel from South Africa, Indonesia, Florida (United States), and India. The most important mango cultivars of India, like Alphonso, Dashehari, Langra, and Chausa, are selections that were made at the time of Akbar (AD 1542–1605) and therefore have been propagated vegetatively for several hundred years.

4 Major cultivars

Mango, a tropical fruit, is now largely cultivated under subtropical conditions. In other words, it is considered as a pan tropical fruit. Presently, the mango is being cultivated commercially or in the backyard or as a mixed plantation in 89 countries of the world. The major mango-growing countries are India, Pakistan, Bangladesh, Myanmar, Sri Lanka, Florida and Hawaii (United States), Australia, Brazil, Thailand, Philippines, Malaysia, Vietnam, Indonesia, Fiji Islands, Egypt, Israel, South Africa, Sudan, Somalia, Kenya, Uganda, Tanzania, Niger, Nigeria, Zaire, Madagascar, Mauritius, Venezuela, Mexico, West Indies Islands, Cambodia, etc.

Reportedly, in India alone, there are more than 1000 types of mango, out of which only 30 are well known. The US Department of Agriculture (USDA) facility on Old Cutler Road in Coral Gables, Florida, has about 400 varieties of mango and is one of the largest depositories of mango plant cultures in the world. The USDA collection was originally believed to have over 500 varieties of mango germplasm, but genetic testing showed several duplicates. Reviews of older varieties of mangoes were based upon the competition available at the time. Most of the varieties of mango available in grocery stores in the United States can trace their lineage to the Haden mango tree, a tree planted by Jack Haden in 1902 in Coral Gables, Florida (Haden itself can be traced back to Mulgoba, which is misspelled Mulgoba in the United States and has its origin in Tamil Nadu, India). The BAU-GPC, the largest fruit repository in Bangladesh, also preserved about 350 varieties of mango (Rahim, 2018). Popular cultivars grown in different parts of the world are given later (Source: https://en.wikipedia.org/wiki/List_of_mango_cultivars):

1. *India:* Alphonso, Amrapali, Banganapalli, Himsagar, Langra, Fazli, Bangalora, Dashehari, Chausa, Mallika, Bombay Green, Badami, Raspuri, Kesar, Neelam, Swarnarekha, Baneshan, Lakshmanbhog, Vanraj, Mankurad, Mulgoa, Kishen Bhog, Ambika, Arunika, Arunima, Arka Aruna, Arka Anmol, Arka Neelkiran, Arka Puneet, Neeleshan, Neelgoa, Neeluddin, Swarn Jehangir, Ratna, Khasulkhas, Totapuri Red Small, Zardalu, Rajapuri, Jamadar, Swaranarekha, Gulab Khas, Pairi, Janardhan Pasand, Sukul, Mlada, Zardalu, etc.
2. *China:* Ivory, Dajhonmonag, Carabao, Nang Klangwan, Dashehari, Okrong, Irwin, Haden.
3. *Thailand:* Nam Doc Mai, Ngar Chan, Ok Rong, Keow Savoey, Pimsen mum.
4. *Pakistan:* Almaas, Anwar Ratol, Chok Anan, Sindhri, Fajri Kalan, Dashehari, Gulab Khas, Langra, Siroli, Zafran, Sindhri, etc.
5. *Mexico:* Ataulfo, Haden, Irwin, Keitt, Manila, Palmer, Sensation, Tommy Atkins, Van Dyke, Ostien.
6. *Indonesia:* Arumanis, Dodol, Gedong, Golek, Madu, Manalagi, Haden, Tommy Atkins, Keitt.
7. *Brazil:* Bourban, Carlota, Coracao, Espada, Itamaraca, Maco, Magoda, Rosa, Tommy Atkins, Pravin, Dixon, Florigon, Davis Haden.
8. *Bangladesh:* Khirsapat, Langra, Ashwina, Fazli, Gopalbhog, etc.
9. *Philippines:* Carabao, Manila Super, Pico.
10. *Nigeria:* Cotonou, Opioro, Benue, Sheri, Julie, Peter, Kerosene, Calabar.
11. *United State:* Anderson, Angie, Bennet Alphonso, Beverly, Coconut Cream, Cogshall, Edward, Fascell, Ford, Glenn, Jakarta, Julie, Mahachanok, Momi K, Pina Colada, Ruby, Sunset, Van Dyke, Fairchild, Fascell, Florigon, Ford, Gary, Gold Nugget, Golden Lippens, Graham, Hatcher, Van Duke, Palmer, Irwin, Golden Nuggets, Brooks, Davis Haden, Glen, Fascell, Lipens, Parvin, Smith, Zill, etc.
12. *Australia:* Brooks, Haden, Irwin, Keitt, Kensington Pride, Kent, Palmer.
13. *Sri Lanka:* Green Willard, Malwana.
14. *South Africa:* Heidi, Sensation, Tommy Atkins, Valencia Pride, Zill, Amelie, Haden, Fascell.
15. *Kenya:* Boubo, Ngowe, Batawi.
16. *Egypt:* Alphonso, Bullock's Heart, Hindi Be Sennara, Langra, Mabrouka, Pairie, Taimur, Zebda.

5 Physiological disorders and plant nutrition

A number of physiological disorders in mango have been reported. An attempt has been made to analyze different physiological disorders of mango from plant nutrition point of view. These are as follows:

5.1 Mango malformation: Causes and management

Malformation is one of the most important problems of mango and a serious threat to mango production in northern India and other mango-growing countries (tropical and subtropical) of the world (Crane and Campbell, 1994). It is described as a disease (Varma et al., 1974; Misra and Singh, 2002) and a physiological disorder (Majumdar et al., 1970). It was first reported about 120 years back from Darbhanga district of Bihar, India, by Maries (Watt, 1891), though it remained a debatable issue for researchers over the last 6–7 decades.

Broadly three distinct types of symptoms are described by various workers. These are bunchy top of seedlings, vegetative malformation, and floral malformation. Later, these were grouped under two broad categories, namely, vegetative and floral malformation (Fig. 43.2) (Varma, 1983). The malformed tissues contain more carbohydrates and nitrogen than normal ones (Mallik, 1963) and higher C/N ratio (Pandey et al., 1973), leading to greater percentage of staminate flowers on malformed panicles (Majumder and Sinha, 1972) and suppression in development of flower and fruit set (Pandey et al., 1973).

Vegetative malformation is more commonly observed on young seedlings (Nirvan, 1953). The seedlings produce small shootlets bearing small scaly leaves with a bunch like appearance on the shoot apex. Hence, the apical dominance is lost, and the seedling remains stunted, and numerous vegetative buds sprout producing hypertrophied growth, which constitutes vegetative malformation. According to Pandey et al. (1973), vegetative malformation is supported by internal metabolic process in the shoots of two varieties of mango, namely, Bombay Green (highly susceptible) and

FIG. 43.2 Showing vegetative (left) normal panicle (below) and floral mango malformation (right) and normal panicle (below).

Bhadauran (free from malformation). Like healthy shoots, malformed shoots also showed high C:N ratio, conducive toward the synthesis of floral stimulus, resulting in initiation and further development of flowers and fruits.

Floral malformation, on the other hand, is the malformation of panicles. The primary, secondary, and tertiary rachises become short, thickened, and hypertrophied. Such panicles are greener and heavier with increased crowded branching. These panicles have numerous flowers that remain unopened and are predominantly male and rarely bisexual (Singh et al., 1961; Schlosser, 1971; Hiffny et al., 1978).

5.1.1 Macronutrients nutrition

Shoots carrying malformed panicles had lower (Pandey et al., 1973, 1977) and higher (Mishra, 1976) levels of nitrogen than the healthy tissues, due to varietal response and varying soil conditions, as the former report pertains to the cv. Bombay Green under Delhi conditions and later to the cv. Dashehari under Ludhiana conditions. Enhanced nitrogen application was observed to curtail malformation, whereas addition of P and K increased the incidence significantly (Kanwar and Kahlon, 1987). However, Bindra and Bakhetia (1971) did not find any reduction in the malady by N, P, and K (9:3:3) treatment and concluded that the disorder is not associated with nutritional imbalance. However, these may influence the incidence of the malady. Singh et al. (1983) observed lower amount of Ca in malformed tissue than healthy ones and suggested that calcium deficiency may not be directly responsible, although it may predispose the tissue to become more prone to the incidence of this disorder. This however needs further investigations. The combined effect of potassium sulfate as soil application and monocrotophos as trunk injection has managed the incidence quite effectively. Results of 10-year-old trial of NPK fertilization on panicle malformation in mango cv. Dashehari indicated that increasing nitrogen doses reduced panicle malformation, whereas the effect of phosphorus and potassium was just the reverse (Minessey et al., 1971; Abou El-Dahab, 1975). Partial control of the disease was achieved in India by spraying the malformed parts with mangiferin-Zn^{2+} and mangiferin-Cu^{2+} chelates (Chakrabarti and Ghosal, 1989).

5.1.2 Micronutrient nutrition

A very little difference has been observed in the mineral constituent of healthy and malformed tissue, although micronutrient deficiency, particularly iron and zinc, has been reported to be associated with the causation of malformation (Abou El-Dahab, 1975; Martin-Preveli et al., 1975; Minessey et al., 1971; Singh and Rajput, 1976). However, Pandey and Pandey (1997) observed that zinc sulfate alone or in combination with NAA could not reduce malformation in cv. Amrapali. Soil treatment with Bayfolan (containing N, P, K, Mn, Fe, Cu, Mg, B, Zn, Ca, and Mo) at 100 mL/tree in 22-year-old trees of mango cv. Taimour also did not affect the extent of malformation (El-Beltagy et al., 1980). Malformation could not be managed with B, Ca, Cu, Fe, Mg, Mn, K, Na, and Zn or their chelates (Rajan, 1986; Ram, 1991; Saeed and Schlosser, 1972). Chander et al. (1998) observed 56.7% reduction in malformation with foliar spray of cobalt sulfate at 1000 ppm followed 49%–36% reduction with cobalt sulfate at 1250 ppm 43.28% reduction with and zinc sulfate at 12,000 ppm (43.28%). The nutrient deficiency was, therefore, not considered as the primary cause of mango malformation (Singh and Rathore, 1983).

5.2 Spongy tissue: Causes and management

Spongy tissue (ST) in mango cannot be detected, unless the mango is cut open. The area affected will appear white and soft, like a sponge. Several studies have shown that spongy tissue formation in mangoes is caused by the shift of the seed to germination mode inside the fruit where it absorbs nutrients from the surrounding fleshy part of the fruit, making it appear white and spongy. Hence, as the fruit ripens, the patch affected by this malady can be seen clearly. Severaj (1998) indicated that spongy tissues had low total carotene, sugars, sucrose, sugar-acid ratio, K, Ca, and Na and higher acidity; glucose-fructose ratio; and N, P, Mg, Zn, and Fe contents compared with healthy tissues. Harvesting the fruit at three-fourth matured stage and storage low temperature reduces the disorder. Mulching in basin has shown promising results in the prevention of spongy tissue effect. Shivashankar and Sumathi (2014) reported the preharvest application of the nutrient mix containing macro- and micronutrients or sea water on developing fruits between 60% and 70% maturity stage effectively reduced the incidence of spongy tissue to <5% compared with 54.3% in control, besides improving the shelf life and fruit quality of Alphonso mango. The study also demonstrated that treatment of fruits before reaching 60% maturity or after crossing 70% maturity was not as effective in reducing spongy tissue incidence.

5.3 Fruit drop: Causes and management

The low fruit set in mango is ascribed to self incompatibility and numerous problems encountered in pollination, fertilization, and low temperature during flowering. A significant advantage was found in hand pollination with foreign pollen over self-pollination. Recurring flowering is the emergence of new flowering panicle at the base of old flowering panicle due to prolonged cold climate. This results in fruit drop from old panicle. As GA inhibits the flowering in mango, use of GA has been suggested to control recurring flowering. The deficiency of nutrients and growth regulators also causes flower and fruit drop of varying degrees at various stages. Auxin is known to play an important role in control of flower and fruit drop. Boron deficiency impairs the floral development, pollen germination, and pollen tube growth. Potassium is known to improve yield and quality of mango. The foliar spray of GA_3 (50 ppm) at full-bloom stage significantly reduced the recurring flowering in Alphonso mango. Foliar spray of urea (2%), NAA (20 ppm), and micronutrient mixture (50 ppm) at peanut stage and 10 days after first spray produced significantly higher fruit yield of Alphonso mango over untreated control. The soil application of recommended dose of yard farm manure (50 kg/tree), N (1 kg/tree), P (0.5 kg P_2O_5/tree), and K (1.0 kg K_2O/tree) through sulfate of potash, with three foliar sprays of 0.9% K_2SO_4 at peanut, marble, and egg stages increased flowering, fruit set, yield, and quality of "Alphonso" mango. The foliar application of Alar (B-nine) 100 ppm or NAA 20 ppm at pea stage of fruit was observed effective in controlling fruit drop in mango. Management of fruit drop was effectively done in cv. Banganapalli with the application of micronutrient mixture consisting of zinc, boron, manganese, and molybdenum (Rameshwar and Kulkarni, 1979).

5.4 Biennial bearing: Causes and management

The term biennial, alternate, or irregular bearing generally signifies the tendency of mango trees to bear a heavy crop in 1 year (on-year) and very little or no crop in the succeeding year (off-year). Most of the commercial varieties of north India, namely, Dashehari, Langra, and Chausa, are biennial bearers, while south Indian varieties like Totapuri Red Small, Bangalora, and Neelum are known to be regular bearers. When a tree produces heavy crop in one season, it gets exhausted nutritionally and is unable to put forth new flush, thereby failing to yield in the following season. Sahithya et al. (2018) described the flowering phenomenon in mango as a complex one. Normally, it crops heavily in 1 year (on-year) and bears less or no crop the following year (off-year). Thus, the rhythm of bearing in mango is not strictly "alternate" but "irregular" or "erratic." Draining out of CHO and N reserves during on-year is known to lead to a lean crop in the off-year, as they are important for fruit bud initiation, that is, high C/N ratio helps for fruit bud initiation. As per physiological studies in regular and irregular varieties of mango, it was observed that mesophyll cells in leaves of the regular mango varieties are more uniform and highly efficient in CO_2 utilization.

The problem has been attributed to the causes like genetical, physiological, environmental, and nutritional in nature. For overcoming biennial bearing, deblossoming is recommended to reduce the crop load in the on-year, so that it is balanced in the off-year. Rao (1998) used four approaches to manage the problem with the intention of promoting flowering directly on fruited shoots. All these methods helped in developing appreciable crop in varying degree in the expected "off" year." The different methods were as follows: (i) application of spray of 1% each of KH_2PO_4 and urea after harvesting of on-year crop helped in the prevention of occurrence of dormancy in fruited shoots and consequently promoted development of new fruiting shoots and flowering on them, (ii) sprays of 2000 ppm paclobutrazol or 100 ppm TIBA during October-November period following harvest of on-year crop resulted in the promotion of flowering directly on fruited shoots, (iii) moderate NAA-deblossoming in on-year reflected on development of moderate cropping in off-year, and (iv) pruning of fruited shoots promoted lateral flowering in off-year.

5.5 Black tip: Causes and management

Black tip is a serious disorder, particularly in the cultivar Dashehari. This disorder of mango is caused by gases (i.e., SO_2, ethylene, and carbon monoxide) emanating from nearby brick kilns in northern India (Ram, 1989). The affected fruits become unmarketable and reduce the yield to a considerable extent. The damage to the fruit gets initiated right at marble stage with a characteristic yellowing of tissues at distal end. Gradually, the color intensifies into brown and finally black. At this stage, further growth and development of the fruit are retarded, and black ring at the tip extends toward the upper part of the fruit. Black tip disorder has generally been detected in orchards located in the vicinity of brick kilns.

The black-tip disorder in Dashehari mango (*M. indica* L.) generally occurs in the vicinity of brick kilns (Pal and Chadha, 1980). Hence, an index was worked out to quantify the intensity of black tip in a given orchard or locality, considering the degree of damage on individual tree. The results were interpreted with special reference to the number of brick kilns and their distance and direction from orchards. Khader et al. (1988) studied the intensity of the disorder (>75%) with a black tip index of 2.78–2.84. The disease intensity was severe as long as the distance remained within 2 km distance of mango orchards in north-south direction, and 5–6 km away from the brick kilns may reduce incidence of black tip to a greater extent (Srivastava, 1963). The incidence of black tip can be minimized by spraying borax (1%) or other alkaline solutions like caustic soda (0.8%) or washing soda (0.5%). The first spray of borax should be done positively at pea stage followed by two more sprays at 15 days' interval.

6 Diagnosis of nutrient constraints

The problems of mineral nutrition of fruit trees are different than those of field crops due to a variety of reasons: (i) large size of the tree, (ii) deep and extensive root system that penetrates the subsoil, (iii) methods of propagation, (iv) soil factors, and (v) its long perennial life. Moreover, the fruits are harvested, while the vegetative parts continue to grow. Mango is reported to recover from long period of undernourishment of mango trees due to suboptimum fertilization.

6.1 Nutrient deficiency symptoms

In field conditions, appearance of deficiency symptoms is a complex phenomenon that changes with age, season, rootstock, reproductive stage of the shoot, and type of soil and fertilizer management (Rajput et al., 1987). Deficiency symptoms in mango are the result of complex phenomenon associated with many factors, namely, age of tree, position of shoots, season, rootstock, reproductive stage of the shoots, soil type, nutrient management practices, and moisture level in the soil (Rajput et al., 1987). Some previous studies (Sen et al., 1947; Krishnamurthi and Randhawa, 1959) attempted to develop deficiency symptoms through pot studies. Agarawala et al. (1989) suggested that, for determination of B concentration, top young leaves should be sampled and, for Zn and Mn, young and middle leaves should be taken for analysis. They developed the deficiency of Mn, Zn, and B in 2-year-old seedlings of Dashehari in sand culture. The N, P, K, Ca, Mg, and S concentration in seedling mango were 0.63%, 0.03%, 0.24%, 0.81%, 0.10%, and 0.32%, respectively, displaying deficiency symptoms (Kumar and Nauriyal, 1977). A 27.5 mg/kg Zn in leaves was designated as critical limit for Zn deficiency (Nijjar et al., 1976). A range of various symptoms (Fig. 43.3) involved and management strategies (Table 43.2) are further described.

FIG. 43.3 Morphologically defined nutritional disorders in mango. *Source: http://agritech.tnau.ac.in/horticulture/plant_nutri/mango_cal.html; ICAR-Central Institute of Subtropical Horticulture, Lucknow, Uttar Pradesh, India.*

TABLE 43.2 Description of deficiency symptoms, management, and response parameters in mango.

Nutrient deficiency symptom	Management strategy	Response parameters	References
N deficiency: Irregular bearing, yellowing of leaves, stunted shoot growth, yellow twigs, smaller fruit size, and immature fruit drop	Dose of N fertilizer (100 g/plant/year of age) or urea 2% spray at fortnightly interval	Ascorbic acid, total soluble solids, and fruit acidity in Dashehari and Langra	Singh et al. (1979) and Chandra (1988)
P deficiency: Reduced plant growth, premature dropping of leaves, partial dieback at tip of younger leaves, necrosis, and premature abscission of leaves	Single superphosphate or foliar application 0.5% orthophosphoric acid thrice or spray of 0.5% orthophosphoric acid with 2% urea	Fruit yield, fruit size, fruit set acidity, ascorbic acid, sugar, and total soluble solids in Chausa	Reddy and Majumdar (1983) and Singh (1975)
K deficiency: Scorching of leaf margin from tip to downward; reduced growth; vigor of leaves; white, yellow, or orange chlorotic spots on leaves; poor root growth, fruit quality; dieback along with tip burn on small leaves	Foliar spray 2% KCL at fortnightly, 1 kg muriate of potash, sulfate of potash along with 2 kg urea and 6 kg superphosphate, spray 0.50% $KNO_3 + 1.0\%$ NaH_2PO_4	Fruit size and quality in Banarsi Langra and Alphonso	Singh and Tripathi (1978) and Ravishankar et al. (1989)
Ca deficiency: Abnormal growth of leaves, some limes dipping death of bud, tip pulp, distinguished by softening of tip of fruit soft nose in Tommy Atkins	Gypsum 50 kg/ha, preharvest dip in $CaCl_2$, spray of 0.60% $CaCl_2$	More calcium in the peel and flesh, reduction in weight loss and lower respiration in Tommy Atkins, reduced internal breakdown in Alphonso, delayed ripening and improved fruit quality in Dashehari	Young (1957), Malo and Campbell (1978), Galan et al. (1984), Gunjate et al. (1979), and Singh et al. (1993)
Mg deficiency: Premature defoliation of leaves, yellowish brown chlorosis, green wedge down the central port of leaf, leaf rounded margin	$MgSO_4$ 5–10 kg/ha, foliar spray 2% $MgSO_4$ 2% at fortnightly	Reduction in fruit drop, better fruit size and fruit quality	Singh et al. (1962)
Zn deficiency: Twigs die back, flower panicles with little leaf symptoms, leaves irregular shape drooping spikes, small and narrow leaf with margins bent upward or downward, reduced internodal length, twigs with crowded leaves like rosette appearance	0.50% Foliar spray and zinc sulfate 0.50 kg/tree or foliar spray $ZnSO_4$ 0.5% plus 10 g urea/1 water twice at 15 days	Fruit quality, sugars, ascorbic acid, total soluble solids in Langra, Dashehari	Rajput et al. (1975), Nijjar et al. (1976), Bahadur et al. (1998), Singh and Rajput (1976), and Daulta et al. (1981)
Fe deficiency: Plant leaves losses green color and turn white (bleaching), size of leaves reduced, leaves dry from tip downward, calcium-induced iron chlorosis is very common	Two sprays at fortnight interval with ferrous sulfate 2.5 g/L or foliar application 0.2% $FeSO_4$ solution or 0.2% $Fe(NO_3)_2$ and 1–5 mg/L iron sequestrene	Increase in chlorophyll fruit set, fruit yield, and fruit quality	Kadrnan and Gazit (1984)
B deficiency: Fruits become brown in color, soft flesh, and watery, which cracks down to the center, lusterless leathery leaves with thickened veins	Application of 0.40%–0.60% boric acid at bud swelling stage	Vegetative growth, length and breadth of the panicle, fruit retention, total sugar, ascorbic acid, acidity and total soluble solids, in Dashehari varieties and Langra varieties	Singh (1977), Rath et al. (1980), Singh and Dhillon (1987), and Coetzer et al. (1991)
Cu deficiency: Weak terminal shoots followed by defoliation, dieback of branches, on the top of long drooping or shaped branches	Spray of copper sulfate (250 g/ 10 years old tree) or 0.30% copper oxychloride at fortnightly interval	Improvement in fruit set, fruit drop, and fruit yield in Dashehari variety	Agarwala et al. (1991)

6.2 Soil and leaf nutrient analysis

The assessment of nutritional status of mango has some obstacles, as this species has phonological variable cycles that influence the absorption and translocation of minerals. On the other hand, there are few studies that address the nutritional requirements or quantifying nutrient uptake by mango tree in different stages of development (Faria et al., 2016). Soil or leaf analysis-based nutrient constraint diagnosis is comparatively better option to identify nutritional disorders and determine the nutritional requirement in perennial fruit crops including mango. Conventionally, diagnostic tools of identifying nutrient constraints comprise leaf analysis, soil analysis, juice analysis, and to some extent metalloenzyme-based biochemical analysis all have been under continuous use and refinement (Srivastava, 2013a,b). Combined use of leaf and soil analysis provides comparatively better diagnosis for nutrient requirement (Srivastava et al., 2001, 2006).

Productivity of the plant depends essentially on the nutrient balance and the biological activity. There are definite limitations with the leaf analysis application that is largely dependent upon composition of index leaves or any other plant parts. Faria et al. (2016) reported that nutrient content with lower mobility in the plant, namely, Ca, B, Fe, and Mn, increased in Tommy Atkins mango leaves from flowering to fruiting with the return of irrigation at 100% crop evapotranspiration, while the levels of N, P, K, and Mg known as high mobility nutrients in plant decreased. Oosthuyse (1997) sampled leaves on fruit-bearing terminal shoots on trees of the mango cultivars Zill, Tommy Atkins, Sensation, Heidi, and Kent in the months of October, November, and December 1996 and January 1997, which were taken for analysis of N, P, K, Ca, Mg, Cu, Fe, Mn, and Zn. These nutrients increased during the rapid phase of fruit growth and decreased during the later stages of fruit development. The variation in Ca, K, N, and Mn was most marked. Based on survey of different bearing orchards in Lucknow, India, Thakur et al. (1980) observed deficiency of Zn in mango cv. Dashehari at the time of flowering in most of the orchards. Thakur et al. (1981a) surveyed micronutrient concentrations of Dashehari in Lucknow, India, and observed that most of the orchards were Zn deficient, while Cu, Mn, and Fe concentrations of mango leaves were in optimum range and significantly higher in leaves sampled during the off-year (bearing a lean crop) compared with the on-year (bearing a heavy crop) of alternately bearing trees. Singh et al. (1998) surveyed 50 orchards representing different mango-growing areas in Surguja district of Madhya Pradesh (Northern hill zone), India. It was observed that the deficiency of phosphorus and calcium was more widespread than nitrogen and potassium. However, magnesium and iron were not found deficient in any of the mango orchards surveyed.

Rao and Mukherjee (1989) surveyed high and low yielding orchards of varieties, namely, Fazli, Himsagar, Langra, Gopalbhog, Bombai, and Aswina in West Bengal, India, which showed deficient level of leaf nitrogen, phosphorus, and potassium. Reddy et al. (1998) surveyed 25 mango orchards in states of Andhra Pradesh, Karnataka, and Tamil Nadu with Banganpally, Alphonso, and Totapuri varieties indicated higher fruit yield of these orchards associated with higher leaf N, P, and K concentration. Survey of 22 orchards representing four cultivars of mango (Dashehari, Langra, Chaura, and Fazli) was reported by Chaudhary et al. (1985), which showed large-scale potassium deficiency symptoms.

6.2.1 Leaf nutrient indices

The nutrient contents in mango leaves vary throughout the growth cycle and may be grouped into two distinct stages, the first between harvesting and initiation of new flowering, period in which there are nutrients accumulation, and second during the period of development of fruit to the harvest, when there is reduction in nutrient levels in leaves (Srivastava, 2013c). The knowledge of periods of greater demands in the plants shows the stage at which nutrients must be supplied or be available for them (Faria et al., 2016). Establishment of absolute figures of normal, deficient, or excess nutrient level is not real, unless the dynamic aspect of leaf nutrient concentration is considered, especially when various nutrient interactions produce resonance within close space of tissue composition (Srivastava et al., 2014). Tertiary diagrams and nutrient ratios are early representation of interacting nutrients in the tissue compositional space (Srivastava, 2013d). Time of leaf sampling suitable for leaf analysis in mango has been reported to vary a great deal, from 6 to 8 months in Dashehari (Devrani and Ram, 1978), 8–10 months in Atkins (Koo and Young, 1972), 5–12 months from nonfruiting branches in Langra (Kumar and Nauriyal, 1977; Thakur et al., 1981a,b), and 6–7-month-old leaves from nonfruiting branches in Chausa (Chadha et al., 1980). Joshi et al. (2016) observed a decline in soil nutrients and rise of leaf nutrients during preflowering, flowering, and fruit developmental stages indicating that, during reproductive stages, supply of nutrient through leaves would be more effective and efficient than soil. Nachegowda et al. (1998) studied the leaf nutrient status of mango varieties, namely, Mallika, Alphonso, Dashehari, Langra, Bombay Green, Fazli, Khas-ul-khas, Pairi, Goamankur, Totapuri, and Willard grown on Nekkare rootstock. Varieties differed

TABLE 43.3 Critical limits of nutrients based pot culture.

	Nutrients									
	Macronutrients (%)						Micronutrients (mg/kg)			
Sr. No.	N	P	K	Ca	M	S	Fe	Mn	Zn	Cu
1.	1.0	0.10	0.50	1.50	0.15	0.50	–	–	–	–
2.	1.23	0.06	0.54	1.71	0.91	0.12	171	66	25	12

Source: (1) Kumar, S., Nauriyal, J.P., 1977. Nutritional studies on mango—tentative leaf analysis standards. Indian J. Hortic. 34, 100–106; (2) Bhargava, B.S., Chadha, K.L., 1988. Leaf nutrient guide for fruit and plantation crops. Fert. News 33, 21–29.

significantly in their leaf nutrient status. The leaf N, P, and K concentrations were highest in Mallika (1.62%), Willard (0.11%), Dashehari (0.57%), and Khas-ul-khas (0.47%) varieties.

Raghupati et al. (2005) reported that nutrient concentration in "Totapuri" cultivar of mango leaf as affected by the application of varying levels of N, P, or K. The Diagnosis and Recommendation Integrated System (DRIS) ratio norms were developed from high yielding population, while diagnosis of nutrient imbalance was made through low yielding plants. Involvement of several nutrients in a single PC indicated that it was not possible to diagnose nutrient imbalance of any particular nutrient in isolation in fruit crops like mango. The nutrient concentration variation in mango leaf appeared to be an overall orchard phenomenon rather than individual tree phenomenon. DRIS indices suggested by Naik and Bhatt (2017) revealed a relative deficiency of Mg, Zn, and B corresponding to relative sufficiency of Ca, N, P, S, and K detected in 9-year-old mango orchards. For the younger orchards (6-year-old ones), the order of requirement of nutrient was observed as Ca > S > K > B > N > P > Mg > Zn. Parent et al. (2013) suggested plant ionome (the mineral nutrient and trace element composition of an organism) diagnosis using sound boundaries in mango considering the biasness between critical nutrient concentration ranges and DRIS.

Biswas et al. (1986) determined critical limits based on critical leaf nutrient levels of N, P, and K for bearing orchards on soil test values. Critical values for N and K were observed higher than those for seedlings, indicating that younger plants are more sensitive than older ones. However, this approach has some limitations, because plants are analyzed at only juvenile phase with restricted root system and media culture does not represent the root soil interface in the field conditions. These limitations were revealed by Samra et al. (1978) through survey for assessment of leaf nutrient status of 30 bearing orchards, relating results to sand culture limits for 1–2-year-old seedlings of mango from India and Florida. The studies have fixed critical limits of N, P, K, Mg, and S based on pot culture experiment. The critical limits (Table 43.3) fixed through pot culture may not be genuine because of juvenile phase of seedlings with limited root system compared with bearing trees with vast tree structure and extensive root system. And in pot culture, such growing conditions do not exist similar to field, with result; the reproducibility of such results is very difficult to obtain under varied orchard conditions. Optimum leaf nutrient concentration worked out for different mango cultivar has been further summarized (Table 43.4).

7 Optimum fertilizer requirement

Field experiments are being laid out for making fertilizers recommendation schedule in mango in different parts of mango-growing region of the world. In a field trail, soil application of 50-kg farmyard manure, 1.5-kg N, 0.5-kg P_2O_5, and 1.0-kg K_2O/tree in the first fortnight of June and application of paclobutrazol at 0.75 g ai/m canopy diameter during second fortnight of July to first fortnight of August with three foliar sprays of 1% KNO_3 at peanut, marble, and egg stage increased the yield and quality of Alphonso mango coupled with reduced spongy tissue incidences (Shinde et al., 2006). Sujatha and Reddy (2006) recommended drip irrigation at 0.75 Ep and NPK (100 g each/tree) + farmyard manure (75 kg/tree) in mango var. Keshar on Alfisols. According to Krishnakumar et al. (1998) mean nutrient removal by every kilogram of dried senescent leaves was 6.63, 0.39, 2.48, 35.81, 0.84, 0.276, 0.65, 0.053, and 0.059 g each N, P, K, Ca, Mg, Fe, Mn, Zn, and Cu, respectively. Sharma et al. (1998) investigated three levels of N (0, 400, and 800 g) in combination with three levels each of P_2O_5 (2, 200, and 400 g) and P_2O_5 (0, 300, and 600 g)/plant were applied to 25-year-old Dashehari mango plants. The study revealed optimum fertilizer requirement of 800 g N + 200 g P_2O_5 + 300 g K_2O/tree/year. The fertilizer recommendations for mango in India were reviewed by Tandon (1987) and presented (Rahim et al., 2018).

TABLE 43.4 Optimum leaf nutrient concentration suggested for different commercial mango cultivars.

	Mango cultivars				
Nutrient	Totapuri (1)	Kent, Parvin, Haden, Zill (2)	Dashehari (3)	Owaise (4)	Alfonso (5)
N (%)	0.92–1.37	1.0–1.5	–	0.74–1.43	0.77–1.65
P (%)	0.08–0.16	0.08–0.17	–	0.07–0.14	0.08–0.18
K (%)	0.21–0.44	0.30–0.80	–	0.54–1.04	0.77–1.73
Ca (%)	1.71–3.47	2.0–3.5	–	1.4–2.6	0.76–1.63
Mg (%)	0.15–0.37	0.15–0.40	–	0.15–0.30	0.13–0.35
Fe (mg/kg)	63–227	–	177–19	389–1148	657–961
Mn (mg/kg)	87–223	–	69–79	23–60	13–408
Cu (mg/kg)	1–6	–	14–26	2–12	14–18
Zn (mg/kg)	11–19	–	28–38	28–56	17–18

Data from (1) Hundal, H.S., Dhanwinder, S., Brar, J.S., 2005. Diagnosis and recommendation integrated system for monitoring nutrient status of mango trees in sub-mountainous area of Punjab, India, Commun. Soil Sci. Plant Anal., 36, 2085–2099; (2) Young, T.W., Koo, R.C.J., 1969. Mineral composition of Florida mango leaves. In: Florida State Horticultural Society. Florida Agricultural Experiment Station Journal Series, vol. 3368. pp. 324–328; (3) Kumar, P., Rehalia, A.S., 2007. Standardization of micronutrients range in mango (Mangifera indica L.) by orchard surveys. Asian J. Hortic. 2(1), 218–221; (4) Ali, A.M., 2018. Nutrient sufficiency ranges in mango using boundary-line approach and compositional nutrient diagnosis norms in El-Sahiya, Egypt. Commun. Soil Sci. Plant Anal. 49(2), 188–201; (5) Raghupathi, H.B. and Bhargava, B.S. 1999. Preliminary nutrient norms for 'alphonso' mango using diagnosis and recommendation integrated system. Indian J. Agric. Sci. 69, 648–650.

7.1 Time and method of fertilizer application

Timing of fertilizers application depends upon the intensity and period of rainfall, fruiting time, vegetative growth, and irrigation water availability. Chadha and Rajput (1993) suggested that, at the time of plantation of new mango orchard, 50-kg FYM, 2.5-kg single superphosphate, and 1-kg muriate of potash should be added in the pits along with insecticide for the management of soil pests. Foliar feeding of nutrients should be done as per need. Micronutrient application should be done by foliar spray for maximum recovery. Placement of manures and fertilizers bearing mango trees should be restricted to active root zone. Bhojappa and Singh (1974) observed that about 82%–88% active roots were located within a radius of 300 cm and the highest activity occurred at a distance of 120 cm from the trunk. During the rainy season, 65%–70% of root activity was observed within a depth of 60 cm, while most of it was confined to a depth of 30 cm from February to March.

Several studies have been carried out for finding appropriate method and time of placement of fertilizers for maximum recovery. In a long-term fertilizer experiment on full-grown Dashehari mango trees, no significant differences were found on increase in growth parameters (viz., plant height, girth of rootstocks, and scion and canopy spread) due to differential time of fertilizer application. Yield data showed significant difference due to various treatments. The highest yield (96.66 kg/tree) was obtained from the trees that received full dose of N, P, and K (1 kg/tree each) in the month of July, which was significantly higher than all treatments, except the one in which three split parts of full dose of N, P, and K were given in April, August, and October. Fertilizer application in April, August, and October gave significantly higher yield that treatment of March and July, August and October, and control. Munniswami (1970) reported that manures are to be applied in 15 cm deep and 60 cm wide trench 30 cm away from the trunk but 60 cm away from trunk in older trees.

7.2 Organic nutrient management

Indiscriminate use of chemical fertilizers, herbicides, and pesticides especially in commercial mango production has resulted in various environmental and health hazards along with socioeconomic problems. Organic mango cultivation of late is claimed in climate mitigation, carbon fixation, soil fertility improvement, and water conservation. In present scenario shifting from conventional to organic, maintenance of soil health and insect paste management are consider major challenges, which auger so well with a variety of practices associated with organic cultivation (Ram and Kumar, 2019). The on-farm organic input production with locally available materials normally leads to a reduction in variable

TABLE 43.5 Doses of fertilizer for different ages of mango plants.

Plant age (year)	Decomposed cowdung (kg)	Urea	TSP	MOP	Gypsum	Zinc	Boron
		Doses of fertilizer (g/plant)					
2	12	650	620	260	90	25	10
3	16	755	825	290	120	20	15
4	20	860	950	390	150	30	15
5	25	900	1050	450	180	40	20
6–8	50	1255	1260	750	220	50	25
9–11	60	1550	1575	1100	250	50	25
Above 12	80	1850	1875	1300	300	60	30

Aforementioned fertilizers are applied in two splits, that is, first 75% in September-October and second 25% in April-May.
Data from Rahim, M.A., Quddus, Islam, M.A., Karim, F., Kabir, M.M., Sarker, M.A., Naher, B.C.N., Alam M.S., Fatema, K., 2018. Present Status, future, prospect of growing temperate fruits in Bangladesh a—tropical environment–with special reference to contribution in nutrition and poverty alleviation. VIII Symposium of Temperate Zone Fruits in the Tropics and Subtropics Certifies, Florianopolis, SC.

input costs under organic management (Pathak et al., 2010; Ram and Pathak, 2016); in few cases, higher input costs due to purchase of composed and organic manures have been reported (Sellen et al., 1993). The diversity on organic farms provides many ecological services that significantly enhance farm resilience (Bengtsson et al., 2005; Hole et al., 2005). Interestingly, yields from organic production under conditions where water is limited during the growing period and under subsistence farming are equal or significantly higher than those from conventional production. A comparison of 133 studies from developing countries concluded that organic crops and livestock yields were 80% higher than conventional (increase in yield was 74% in crops according to Badgley et al. (2007). Ram et al. (2017) reported increase in production and benefit-cost ratio with biodynamic package of practice compared with recommended dose of fertilizers (Table 43.5). Singh et al. (2004) observed the different organic ammendments on soil, biophysical attributes, and yield of mango cv. Langra. The study showed that an maximum increase in organic carbon content was observed with vermicompost 0.495% treatment followed by cow pat pit (0.465%) as compared with control (0.307%). The fruit retension per panicle, fruit number, and fruit weight were also obtained with BD-500+biodynamic compost (10.35). The maximum fruit number (309) and fruit yield (68.4 kg) per tree were obtained with BD-500+biodynamic compost.

Ram et al. (2011) reported that highest number of fruit (260/tree) and fruit yield (30.1 kg/tree) with the treatment involving rhizosphere soil of *Ficus benghalensis* (250 g/tree)+5% *Amritpani*+organic mulching, After 2 years of treatment, maximum organic carbon (1.00%) was recorded in mango basin soil with vermicompost (45 kg/tree)+ *Azospirillum* culture (250 g/tree)+PSB (50 g/tree) in addition to Olsen-P (61.2 ppm), NH_4OAc-K (280.50 ppm), DTPA-Zn (11.68 ppm), and DTPA-Cu (9.60 ppm) in rhizosphere soil of *F. benghalensis* (250 g/tree). Ram and Verma (2015) estimated energy input, output, and economic analysis in organic mango production system. Study revealed that irrigation consumed the least energy (63 MJ) and diesel oil ranked first (3040.74 MJ) (Table 43.6) in energy consumption. Maximum input energy was consumed (8121.49 MJ) with application of 50-kg vermicompost +250-g *Azospirillum* +PSB culture per tree + vermiwash spray compared with 6232.24 MJ with the application of 50 kg FYM/tree (Table 43.7). Chavan et al. (2018) studied the different organic sources on yield of mango cv. Alphonso, and 6 years of pooled data revealed highest fruit yield 5.17 t/ha with the application of 100-kg vermicompost+20 kg of glyricidia green leaf manuring followed by 50 kg of vermicompost+20 kg of glyricidia green leaf manuring (2.55 t/ha) and 150-kg FYM+20-kg glyricidia green leaf manuring (2.20 t/ha). These studies strongly warranted the high-utility organic cultural practices.

7.3 Integrated nutrient management

The continuous use or excess supply of inorganic fertilizers as source of nutrient in imbalanced proportion is also a problem, causing economic inefficiency, damage to the environment, and in certain situations harm to the plants themselves and also to human beings who consume them. On the other hand, increase in productivity of mango removes large amounts of essential nutrients from the soil nutrient reserves, which unduly impose a variety of stresses on soil as growing medium. Another issue of great concern is the sustainability of soil productivity, as soil as growing medium began to be intensively exhausted to produce higher yields. Overtime, cumulative depletion decreased and yield and

TABLE 43.6 Budgeting of energy (MJ) consumption under different organic treatments in mango cv. Dashehari.

Particular	50-kg FYM/tree	50-kg FYM +250g *Azospirillum* +PSB culture/tree	50-kg FYM +250g *Azotobacter* +PSB culture/tree	50-kg Vermicompost/ tree	50-kg Vermicompost +250g *Azospirillum* +PSB culture/ tree	50-kg Vermicompost +250g *Azotobacter* +PSB culture/tree	50-kg Vermicompost +250g *Azospirillum* +PSB culture/tree + vermiwash spray
Human labor	397.45	478.73	532.52	494.26	536.58	534.08	486.70
Machinery	931.05	931.05	931.05	931.05	931.05	931.05	931.05
Organic manures	1500	2000	2000	2500	3000	3000	3300
Diesel oil	3040.74	3040.74	3040.74	3040.74	3040.74	3040.74	3040.74
Irrigation	63	63	63	63	63	63	63
Biopesticides	300	300	300	300	300	300	300
Total energy input	6232.24	6813.52	6867.31	7329.05	7871.37	7868.87	8121.49

Data from Ram, R.A., Verma, A.K., 2015. Energy input, output and economic analysis of organic production of mango (Mangifera Indica L.) cv. 'Dashehari'. Indian J. Agric. Sci. 85(6), 827–832.

TABLE 43.7 Economic analysis of an organic mango production.

	Fruit yield		Fruit quality parameters			
Treatments	Fruit yield (kg/tree)	Fruit yield (tons/ha)	TSS (°Brix)	Titrable acidity (%)	Total carotenoids (mg/ 100g)	Benefit-cost ratio
T_1	60.68	6.068	22.00	0.30	4.24	1.67
T_2	108.98	10.899	23.50	0.24	5.50	3.19
T_3	81.75	8.175	21.83	0.22	4.41	1.60
T_4	61.56	6.156	21.00	0.31	5.06	1.90
T_5	57.95	5.795	21.53	0.32	3.73	1.74
T_6	82.18	8.218	22.33	0.28	3.38	2.26
T_7	46.18	4.686	20.50	0.38	4.98	1.44
CD at 5%	27.20	2.72	1.74	NS	0.87	0.78

T_1, FYM (40kg/tree) + *Azotobacter* + *Azospirillum* + PSB (10^8 cfu/g) + VAM; T_2, biodynamic compost (30kg/tree) + bioenhancers (CPP 100g, BD – 500 and BD 501 as soil and foliar spray); T_3, Neem cake + farmyard manure (20kg each/tree) + *Azotobacter* + *Azospirillum* + PSB (10^8 cfu/g); T_4, vermicompost (30kg/tree) + *Azotobacter* + *Azospirillum* + PSB (10^8 cfu/g); T_5, farmyard manure (40kg/tree) + bioenhancer (Amritpani 5% soil application); T_6, FYM (40kg/tree) + green manuring (sun hemp) *Azotobacter* + *Azospirillum* + PSB (10^8 cfu/g); T_7, 1000g N P K/tree.

soil fertility coupled with soil degradation. The new approach to farming often referred to as sustainable agriculture, seeks to introduce agricultural practices that are eco-friendly and maintain the long-term ecological balance of soil as a dynamic ecosystem.

The judicious use of organic inputs with inorganic is considered as an alternative strategy to meet the nutrient requirement of mango. Haldavnekar et al. (2018a,b) studied the substrate dynamics for IPNM in mango cv. Alphonso with recommended dose of fertilizers (RDF) and different combinations of organic manure (Farmyard manure and vermicompost), organic mulching, inorganic fertilizer, and biofertilizers (*Azotobacter, Azospirillium, Pseudomonas fluorescence,* and *Trichoderma*). The treatment consisting of half RDF + 50-kg farmyard manure + 250-g *Trichoderma* + 250-g *Pseudomonas* proved to be the most effective with regard to maximum number of fruits per tree (289.67) and yield (72.30 kg/tree and 7.23 t/ha) over rest of the treatments. Bauri and Ghosh (1998) reported maximum fruit yield (by weight) in trees treated with organic matter and 300 g N + 200 g P_2O_5 + 200 g K_2O/tree/year, in addition to leaf nutrient composition (2.03% N, 0.08% P, and 0.72% K).

Fruit crops including mango have an excellent synergy with a variety of microbes. Sukhada (1998) isolated major species of vesicular arbuscular mycorrhizal (VAM) fungi belonging to genus *Glomus* from the rhizosphere of mango rootstocks Totapuri and Vellaikulumban in the natural habitat, which was multiplied in a glasshouse on Rhodes grass from single spore. The rootstocks took 6–8 months to colonize 90% roots up to and showed significant visible increase in plant height (70%–80%). Dry-matter accumulation was significantly higher in all VAM-inoculated plants compared with uninoculated plants. Gupta and Bhriguvansi (1998) studied the effects of VAM fungi (*Glomus fasciculatum* and *G. mosseae*) and humacil (an organic source) on growth and mineral nutrition of mango seedlings under pot culture conditions. At initial growth (3–9 months) of mango seedlings, all the plants grew with no significant difference irrespective of the mycorrhizal infection, but, at later stage of growth (9–15 months), soil inoculated with *G. fasciculatum* and half dose of P_2O_5 followed by humacil with full dose of P_2O_5 showed higher plant height over other treatments.

In a long-term experiment on organic production of mango cv. Dashehari, maximum yield (72 kg/tree) and average fruit number (305 fruits/tree) along with fruit quality was recorded with 50-kg vermicompost + 250-g *Azospirillum* + PSB culture + vermiwash per tree during the third year of experimentation. This treatment was associated with maximum production cost (INR 22,002.26/ha), productivity (0.33 kg per INR), net return INR (49,997.74/ha), and benefit-cost ratio (3.27) according to (Ram and Verma, 2018). In another experiment, application of different amendments on 35-year-old trees of mango cv. Mallika, highest fruit yield (160.30 kg/tree) was recorded with the application of biodynamic compost (30 kg/tree) + CPP 100 g, BD-500 and BD-501 as soil and foliar spray (Ram et al., 2016). Raut et al. (2018) studied the effect of manures and biofertilizers on soil microbial population of Alphonso mango. The results revealed that treatment involving FYM (50 kg/tree) + *Azospirillum* culture (250 g/tree) + PSB (250 g tree) recorded the highest population count of *Azospirillum* (23.7 cfu/g), while the highest PSB count (110.0 cfu/g) was recorded with the treatment having vermicompost (50 kg/tree) + *Azospirillum* culture (250 g/tree) + PSB (250 g/tree). In a long-term field trail in 17-year-old mango cv. Dashehari plants, growth parameters remained unaffected due to the application of various manures, fertilizers, and biofertilizers. Maximum yield (83.66 kg/tree) was obtained from plants receiving 300-g culture of *Azospirillum,* significantly higher than all other treatments except the treatments with 10-kg neem cake per tree. The lowest yield (39.66 kg/tree) was obtained from the trees under which Sesbania was applied as green manure (Ram and Rajput, 1998).

7.4 Foliar application of nutrients

Foliar application of nutrients basically micronutrients is an age-old practice to rectify the deficiency symptoms. Soil application of micronutrients is far less effective than foliar application (Srivastava and Singh, 2004), although combined application of Zn (0.4%) and urea (1%) produced the highest number of fruits and yield of mango, whereas the fruit quality was improved by the application of B (0.4%) + urea (1%) to younger plants (Banik et al., 1997). Nijjar et al. (1976) recommended 27.5 mg/kg Zn in deficient leaves and 35.2 mg/kg for healthy leaves to be maintained with three sprays of $ZnSO_4$ to correct the Zn deficiency.

Arora et al. (2018) studies on foliar sprays of potassium nitrate at 2%, 3%, and 4% as a single (10 days after fruit set), double (10 days after fruit set and superimposed 20 days after fruit set), and triple spray (10 days after fruit set and superimposed 20 and 30 days after fruit set), respectively, on Amrapali mango trees. The fruit retention (%) was significantly highest (2.41%) in double spray of KNO_3 3%, followed by 2% triple spray (2.16%) and 4% single day (2.22%), respectively, compared with control (1.59%). The fruit was significantly recorded highest with double-spray spread trees with KNO_3 3% (49.80 kg/tree). The study concluded that potassium nitrate as foliar spray significantly improves the fruit retention quality and yield the double spray of KNO_3 at 3% was the best when applied on Amrapali cultivar of mango. Hiray et al. (2018) carried out an experiment to find out the response of boiler fertilizers, namely, SOP 1%, SOP 2.0%, and KNO_3 1.5% and 2% at different fruit developmental stages, namely, 50% marble stages, 50% egg stage, and twice at 50% marble stage as along with control. Results revealed that maximum fruit set (%) and minimum fruit set to maturity (125.10 days). While maximum fruit yields (14.57 t/ha) were recorded with foliar application of 2% SOP twice spray at 50% marble stage and 50% egg stage. Some other recommendations of foliar application of micronutrients cultivar wise are further summarized (Table 43.8).

7.5 Fertigation

Fertigation is an effective means of controlling the timing and placement of fertilizers to root zone of crop (Shirgure and Srivastava, 2014). Fertigation is considered synonymous to nutrient use efficiency (Srivastava et al., 2014) that can further be fine-tuned with nitrification inhibitors (restrict the microbial conversion of ammonium to nitrate that it is

TABLE 43.8　Recommendations on foliar application of micronutrients in different commercial mango cultivars.

Treatments	Response parameters	References
Fazli: 0.40% Zn 0.40%B + 1% urea	Improvement in growth flowering intensity, fruit set, fruit yield, and fruit quality	Banik et al. (1997)
Chausa: 0.80% Zn	Fruit weight, fruit yield, and fruit quality	Singh and Rajput (1976)
Langra: 0.40% Zn + 0.20% Cu + 0.20% B	Fruit yield and fruit quality	Bhriguvanshi and Gupta (1998)
Dashehari: 0.50% $FeSO_4$ + 0.50% $MnSO_4$ + 0.50% $ZnSO_4$ + 0.20% borax	Fruit yield, pulp peel-stone ratio, fruit quality	Malhi and Joshi (1998)
Dashehari: 1000 ppm cobalt sulfate + 4000 ppm zinc sulfate + 100 ppm NAA	Flowering, fruit length, fruit breath, fruit set, fruit retention, fruit weight, and fruit yield	Chander and Singh (1998)
Dashehari and Langra: RDF + 100 g/tree $ZnSO_4$ + 50 g/tree $CuSO_4$ + 50 g/tree boric acid + foliar spray of 0.20% $ZnSO_4$ + 0.10% $CuSO_4$ + 0.10% boric acid	Fruit yield and quality	Mukesh et al. (2018)
Alphonso: 0.50% (urea, single superphosphate and potash) + 0.25%($ZnSO_4$, borax, and $CuSO_4$ each) + 0.01% sodium molybdate	Flowering, fruit set, fruit yield, quality, and benefit-cost ratio	Patil et al. (2018)
Dashehari: 0.50% $FeSO_4$ + 0.50% $ZnSO_4$ + 0.50% $MnSO_4$ + 1.0% Sodium tetraborate	Fruit quality and fruit yield	Joshi and Malhi (1998)
Alphanso: RDF + foliar spray of 0.40% $ZnSO_4$ + 0.20% boric acid	Flowering, fruit set, fruit yield, and fruit quality	Haldavnekar et al. (2018a, b)

mobile in soils) or plant growth-promoting bioeffectors (microorganisms and active natural compounds involved in plant growth). Bhriguvanshi et al. (2014) evaluated the effect of different soil moisture and nitrogen levels on the yield, quality, and water use efficiency (WUE) of mango cv. Dashehari. The results revealed that the drip irrigation irrespective of fertigation levels significantly increased fruit yield over basin irrigation. Irrigation given to the replenishment of 60% of United States Weather Bureau (USWB) Open Pan Evaporation was found to be statistically superior to 70% and 90% replenishments. The increase in fruit yield in 0.9, 0.7, and 0.6 OPE was 19.4%, 40.5%, and 57.5%, respectively, as compared with basin irrigation. The yield improvement in 0.6 OPE was higher to the tune of 32.0% and 15.5% as compared with 0.9 and 0.7 OPE, respectively. It was found that, under the condition of limited water supply, drip irrigation proved consistently superior to the basin application, which significantly saved 15.0%, 31.4%, and 40.5% of irrigation water in 0.9, 0.7, and 0.6 OPE, respectively, as compared with control. The water use efficiency ranged from 9.9- to 18.8-kg fruits/m^3 of water applied in the drip irrigation system as compared with 7.0-kg fruits/m^3 of water applied in basin irrigation.

Increased mobility of P dissolved in irrigation water was observed, especially for sandy soils (O'Neil et al., 1979) and fine-textured clay loams (Rauschkolb et al., 1976). The improved mobility has been attributed to movement of P in mass flow with irrigation waters after saturation of reaction sites near the zone of P application. The extent of P mobility was illustrated from a fertigation study carried out by Neilsen et al. (1997). Trees in an orchard on a loamy sand soil received annual applications of 40 g N and 17.5 g P as mixtures of ammonium nitrate (34 N-0 P-0 K) and ammonium polyphosphate (10 N-15 P-0 K), through a single drip emitter. Increases in extractable soil P were measurable as far as 30 cm from the emitter, regardless of whether the same amount of P was applied as a single dose, four weekly doses, or thirty daily doses in May. Changes were most apparent directly beneath the emitter (0 distance), with an elevated extractable P concentration at 30 cm depth immediately after application. This was in contrast with the behavior of surface broadcast P fertilizer, which failed to penetrate below 7 cm depth for sandy soil after 16 irrigation applications totaling 16 cm. This is a clear indication of the mobility of fertigated P.

The mobility of fertigated K is well documented and illustrated by changes in extractable soil K at selected depths and distances from an emitter through which K was added (Uriu et al., 1980). At the end of a single growing season in a clay loam soil, expected to inhibit downward movement of fertilizer K, increased in NH_4OAc-extractable K could be observed to depths of 60–75 cm and lateral distances as great as 60 cm. Thus, the distribution of soil K was similar in depth but greater in lateral extent than similar changes previously cited after broadcasting K directly beneath the emitter at rates that exceeded by >50 times normal unit area surface broadcast rates under flood irrigation.

8 Plant nutrition and shelf life of fruits

Studies on mango dealing with the factors that determine the final quality of fruit at the consumer level have generally focused on maturity at harvest (Jacobi et al., 1995; Lalel et al., 2003) and on postharvest management (Hoa et al., 2002; Nunes et al., 2007). However, as is the case with other stone fruits, preharvest cultural practices, which affect the environmental conditions of fruit development, profoundly influence postharvest performance and final quality (Crisosto et al., 1995; Hewett, 2006). Mango is a fleshy fruit containing more than 80% water (Lakshimnarayana et al., 1970). Its size depends on the accumulation of water and dry matter in the various compartments during fruit growth. The skin, the flesh, and the stone have specific compositions that appear to accumulate water and dry matter at different rates, depending on environmental conditions (Léchaudel and Joas, 2006). Fruit growth after cell division consists in the enlargement of fruit cells characterized by a large accumulation of water that results from the balance between incoming fluxes such as phloem and xylem and outgoing fluxes such as transpiration (Ho et al., 1987). Changing the balance between these various fluxes, which have elastic and plastic components, leads to large variations in fruit volume.

Another quality trait for mango is its shelf life, which can vary with postharvest conditions, the best known of which is temperature. However, this attribute can be influenced by conditions during fruit growth that affect the supply of minerals to the fruit. In fact, relationships between minerals (the main one in mango is potassium, followed by magnesium and calcium; Simmons et al. (1998)) and shelf life are often studied. In particular, it appears those variations in calcium content or the ratio between calcium and the other two minerals delays ripening and senescence (Ferguson, 1984) or reduces storage disorders (Bangerth, 1979). Calcium is supplied by the xylem (Jones et al., 1983), and potassium and magnesium are phloem-mobile nutrients (Nooden, 1988). Special attention must also be paid to the influence of environmental factors on ingoing fruit fluxes and on the balance between mineral ions in mango, as has been done for other fruits (Ferguson et al., 1999). Moreover, shelf life can be discussed in terms of dry matter content, which is directly affected by carbohydrate and water fluxes at the fruit level during its growth. Hence, improved plant nutrition is supposed to elevate the postharvest life of mango fruits.

9 Future line of research

Various studies on field's trials, surveys of orchards, and tissue analyses have been accomplished throughout the world, for fertilizer recommendations in mango plantation. These recommendations are restricted to specific climatic and soil conditions. An in-depth study is needed to develop target yield-based fertilizer recommendation using soil test value of leaf analysis values. At the same time, a database evidence as needed to testify how depletion in soil fertility is adversely affecting nutrient density of mango, besides yield and quality. Foliar application of nutrients particularly micronutrients has become a popular practice for correction of short-term deficiency symptoms. Besides micronutrients, foliar application of macronutrients is also cost effective and well suited for maximum nutrient recovery. There is need to develop cost-effective package of practice of nutrients for sustainable production in different agroclimatic conditions, with an eye on developing nutrient-dense mango.

Mango is a perennial crop; efforts should be directed toward proper in situ recycling of orchard organic wastes for reduction of cost of cultivation. Some of the other issues like on-farm input generation to make it cost effective; quantum production equal or higher what is expected from optimum combination of agrochemicals; continuous improvement in physicochemical and biological properties of soil; and improvement in produce quality with respect to nutrition, essential constituents, therapeutic value and storability, and eco-friendly and cost-effective technology with emphasis on quality production of mango need an immediate attention.

Role of rootstocks in nutrient absorption from the soil is equally important for better growth and development of plant. Therefore, there is a need to standardize various rootstocks for different agroclimatic and soil conditions with emphasis on salinity and drought. Study should also be carried out for the identification of proper source of nutrients, time, method of nutrient application, and source of nutrients (4R Nutrient Stewardship) in combination with plant protection measures for cost-effective production of mango. Despite all these efforts, we need a comprehensive study to streamline mango trees as a carbon sink, as the carbon footprint of mango plantation will further reveal the role of perennial fruit crops in climate change mitigation and soil health resilience. There is also a stronger need to develop equality index of mango as a comprehensive system of evaluating the nutrient quality of mango. Such studies should be duly supported by neutraceutical and phytonutrient profiling of mango germplasm to be later utilized in identifying nutrient-efficient genotypes of mango, the efforts on which are really limited. These studies will perch the mango industry on a different dimension to ensure sustained quality production of mango completed with soil health.

Mango, a tropical fruit of great economic importance, is generally harvested green and then commercialized after a period of storage. Unfortunately, the final quality of mango batches is highly heterogeneous, in fruit size as well as in gustatory quality and postharvest behavior. A large amount of knowledge has been gathered on the effects of the maturity stage at harvest and postharvest conditions on the final quality of mango. Considerably less attention has been paid to the influence of environmental factors on mango growth, quality traits, and postharvest behavior. It remains to be experimentally seen how preharvest conditions affect postharvest life of mango fruits, besides growth and quality (Léchaudel and Joas, 2007).

References

Abou El-Dahab, M.K., 1975. Correcting malformation symptoms on mango trees in Egypt by soil application of iron chelates. Egypt. J. Phytopath. 7, 97–99.

Agarawala, S.C., Naurial, B.D., Chatterji, C., Sharma, P.C., 1989. Manganese, zinc and boron deficiency in mango. Sci. Hortic. 35, 99–107.

Agarwala, S.C., Nautiyal, B.D., Sharma, C.P., Chatterjee, C., 1991. Studies on copper deficiency in mango, guava and jackfruit. Indian J. Hortic. 48, 192–200.

Arora, R., Singh, N., Gill, M., Kaur, S., 2018. Yield and quality attributes of mango cv. 'Amrapali' as influenced by foliar application of potassium nitrate. In: International Mango Conference, 2018 May 8–10, Vengurla, India, pp. 91–92.

Badgley, C., Moghtader, J., Quintero, E., Zakem, E., Jahi, Chappell, M., Aviles Vazquez, K., Samulon, A., Perfecto, I., 2007. Organic agriculture and the global food supply. Renewable Agric. Food Syst. 22 (2), 86–108.

Bahadur, L., Malhi, C.S., Chander, P., 1998. Zinc content of pulp, rag and stone as affected by ZnSO₄ application. In: National Symposium on Mango Production and Export, 25–27 June, ICAR-C.I.S.H., Lucknow, India, p. 36.

Bangerth, F., 1979. Calcium-related physiological disorders of plants. Annu. Rev. Phytopathol. 17, 97–122.

Banik, B.C., Sen, S.K., Bose, T.K., 1997. Effect of zinc, iron and boron in combination with urea on growth, flowering, fruiting and fruit quality of mango cv. Fazli. Environ. Ecol. 15 (1), 122–125.

Bauri, F.K., Ghosh, S.N., 1998. Growth, yield and physio-chemical properties of mango cv. 'Himsagar' grown in response to organic and inorganic fertilizers under rainfed laterite soil. In: National Symposium on Mango Production and Export, 25–27 June, ICAR-C.I.S.H., Lucknow, India, p. 36.

Bengtsson, J., Ahnström, J., Weibull, A.C., 2005. The effects of organic agriculture on biodiversity and abundance: a meta-analysis. J. Appl. Ecol. 42, 261–269.

Bhojappa, K.M., Singh, R.N., 1974. Root activity of mango by radioactive technique using P³². Indian J. Agric. Sci. 44, 175–180.

Bhriguvanshi, S.R., Gupta, M., 1998. Effect of various micronutrient sprays on yield and quality of mango (Mangifera indica L.) fruit. In: Symposium Mango Production and Export, 25–27 June, ICAR-C.I.S.H., Lucknow, India, p. 37.

Bhriguvanshi, S.R., Adak, T., Kumar, K., Singh, A., Sing, K., 2014. Impact of varying soil moisture regimes on growth and soil nutrient availability in mango. Indian J. Soil Conserv. 42 (1), 68–73.

Bindra, O.S., Bakhetia, D.R.C., 1971. Investigations on etiology and control of mango malformation. Indian J. hortic. 28, 80–85.

Biswas, P.P., Joshi, O.P., Rajput, M.S., 1986. Potassium supplying capacity of some mango orchards. J. Potassium Res. 2 (3), 83–89.

Brandon Gaille, 2018. 21 Mango Industry Statistics. Trends & Analysis. https://brandongaille.com/21-mango-industry-statistics-trends-analysis/.

Chadha, K.L., Rajput, M.S., 1993. Mango nutrition. In: Chadha, K.L., Pareek, O.P. (Eds.), Advance in Horticulture. In: vol. 2. Malhotra Publishing House, New Delhi, pp. 813–826.

Chadha, K.L., Samra, J.S., Thakur, R.S., 1980. Standardization of leaf sampling technique for mineral composition of leaves of mango cultivar 'Chausa'. Sci. Hortic. 13, 323–329.

Chakrabarti, D.K., Ghosal, S., 1989. The disease cycle of mango malformation induced by Fusarium moniliforme var. subglutinans and the curative effects of mangiferin metal chilates. J. Phytopathol. 125, 238–246.

Chander, P., Singh, A., 1998. Effect of cobalt, zinc and napthaline acetic acid (NAA) on yield and fruit quality in mango cv. 'Dashehari'. In: National Symposium on Mango Production and Export, 25–27 June, ICAR-C.I.S.H., Lucknow, India, p. 21.

Chander, P., Singh, A., Ohiman, J.S., 1998. Effect of exogenous application of cobalt, zinc and napthaline acetic acid on mango malformation. In: National Symposium on Mango Production and Export, 25–27 June, ICAR-C.I.S.H., Lucknow, India, p. 73.

Chandra, A., 1988. Note on control of pre-harvest fruit drop in mango. Curr. Agric. 12, 91–92.

Chaudhary, S.K., Nijjar, G.S., Rehalia, A.S., 1985. Nutritional status of mango orchards of Gurdaspur district in Punjab. Acta Hortic. 231, 282–285.

Chavan, L.S., Mhaskar, N.V., Karle, S.S., Jondhale, D.G., Haldankar, P.M., 2018. Studies on comparative effeciency of organic manures on fruit yield of alphonso mango in konkan region of Maharashtra. In: International Mango Conference, 8–11 May, 2018, RFRS, Vengurla, Maharashtra, p. 37.

Coetzer, L.A., Robbertse, P., Wet, E.D., 1991. The influence of boron application on fruit production and cold storage of mangoes. In: Year Book of the South African Mango Growers Association. vol. 11, pp. 29–31.

Crane, J.H., Campbell, C.W., 1994. Fact Sheet HS-2, Horticultural Science Department, Florida Cooperative Extension Service. Institute of Food and Agricultural Sciences, University of Florida, Gainesville, FL.

Crisosto, C.H., Mitchell, F.G., Johnson, S., 1995. Factors in fresh market stone fruit quality. Postharvest News Inf. 6, 171–211.

Daulta, B.S., Singh, H.K., Chauhan, K.S., 1981. Effect of zinc and CCC sprays on flowering, fruiting and physico-chemical composition of fruit in mango (Mangifera indica L.) cv. 'Dashehari'. Haryana J. Hortic. Sci. 10, 161–165.

Devrani, H.B., Ram, S., 1978. Studies on micro-nutrient in relation to age of mango leaves. Plant Sci. 10, 73–75.

El-Beltagy, M.S., El-Ghandour, M.A., El-Hanawi, 1980. Effects of Bayfolan and some growth regulators on modifying flowering and the incidence of flowering malformation of mango (Magnifera indica L.). Egyptian J. Hortic. 6, 125–139.

Faria, L.N., Donato, S.L.R., Dos Santos, M.R., Castro, L.G., 2016. Nutrient contents in Tommy Atkins mango leaves at flowering and fruiting stages. J. Braz. Assoc. Agric. Eng. 36 (6), 1073–1085.

Ferguson, I.B., 1984. Calcium in plant senescence and fruit ripening. Plant Cell Environ. 7, 397–405.

Ferguson, I.B., Volz, R., Woolf, A., 1999. Preharvest factors affecting physiological disorders of fruit. Postharvest Biol. Technol. 15, 255–262.

Fowomola, M.A., 2010. Some nutrients and antinutrients contents of mango (*Mangifera indica*) seed. Afr. J. Food Sci. 4 (8), 472–476.

Galan, S.V., Galvan, D.F., Calvo, R., 1984. Incidence of soft-nose on mangoes in the Canary Islands. Proc. Florida State Hortic. Soc. 97, 358–360.

Gunjate, R.T., Tare, S.J., Rangwala, A.D., Limaye, V.P., 1979. Effect of pre-harvest and post-harvest calcium treatments on calcium content and occurrence of spongy tissue in 'alphonso' mango fruits. Indian J. Hortic. 37, 140–144.

Gupta, M., Bhriguvansi, S.R., 1998. Effect of vesicular arbuscular mycorrhizal (VAM) fungi and humacil on growth and mineral nutrition of mango seedlings (*Mangifera indica* L.). In: National Symposium on Mango Production and Export, 25–27 June, ICAR-C.I.S.H., Lucknow, India, p. 22.

Haldavnekar, P.C., Salvi, B.R., Munj, A.Y., Raut, R.A., Baviskar, S.B., Satelkar, A.G., Shedge, M.S., 2018a. Integrated plant nutrient management in 'Alphonso' mango under konkan agro climatic conditions. In: International Mango Conference, May 8–10, Vengurla, India, p. 87.

Haldavnekar, P.C., Salvi, B.R., Satelkar, A.G., Munj, A.Y., Shedge, M.S., Baviskar, S.B., Raut, R.A., Sanas, M.P., Sawant, S.N., 2018b. Micronutrient effect on yield and physico-chemical composition of 'alphonso' mango under Konkan agro-climatic conditions. In: International Mango Conference, May 8–10, Vengurla, India, pp. 87–88.

Hewett, E.W., 2006. An overview of preharvest factors influencing postharvest quality of horticultural products. Int. J. Postharvest Technol. Innov. 1, 4–15.

Hiffny, H.A.A., El-Barkouki, M., El-Banna, G.S., 1978. Morphological and physiological aspects of the floral malformation of mangoes. Egypt. J. Hortic. 5, 43–53.

Hiray, S.A., Patel, B.N., Tandel, B.M., Patil, S., 2018. Effect of foliar fertilizers sprays on fruit set, fruit drop and yield of mango (*Mangifera indica* L.) cv. 'Kesar'. In: International Mango Conference, 2018 May 8–10, Vengurla, India, p. 89.

Ho, L.C., Grange, R.I., Picken, A.J., 1987. An analysis of the accumulation of water and dry matter in tomato fruit. Plant Cell Environ. 10, 157–162.

Hoa, T.T., Ducamp, M.N., Lebrun, M., Baldwin, E.A., 2002. Effect of different coating treatments on the quality of mango fruit. J. Food Qual. 25, 471–486.

Hole, D.G., Perkins, A.J., Wilson, J.D., Alexander, I.H., Grice, P.V., Evans, A.D., 2005. Does organic farming benefit biodiversity? Biol. Conserv. 122, 113–130.

Jacobi, K.K., Wong, L.S., Giles, J.E., 1995. Effect of fruit maturity on quality and physiology of high-humidity hot air-treated 'Kensington' mango (*Mangifera indica* Linn.). Postharvest Biol. Technol. 5, 149–159.

Jones, H.G., Higgs, K.H., Samuelson, T.J., 1983. Calcium uptake by developing apple fruits. I. Seasonal changes in calcium content of fruits. J. Hortic. Sci. 58, 173–182.

Joshi, A., Malhi, C.S., 1998. Effect of micronutrient sprays on growth, yield and fruit quality of mango cv. 'Dashehari'. In: National Symposium on Mango Production and Export, 25–27 June, Held at ICAR-C.I.S.H., Lucknow, India, p. 38.

Joshi, N.S., Prabhudesai, S.S., Burondarkar, M.M., Gokhale, N.B., Pujari, K.H., Dhekale, J.S., 2016. Stage-wise nutrient status of leaf and soil of alphonso mango grown in Ratnagiri district of Maharashtra, India. Indian. J. Agric. Res. 50 (4), 318–324.

Kadrnan, A., Gazit, S., 1984. The problem of iron deficiency in mango trees and experiments to cure it in Israel. J. Plant Nutr. 7, 283–290.

Kanwar, J.S., Kahlon, G.S., 1987. Effect of nitrogen, phosphorus and potassium fertilizer on panicle malformation in mango, Mangifera indica, L. cv. Dashehari. Punjab Hort. J. 27, 12–16.

Khader, S.E.S.A., Rajput, M.S., Biswas, P.P., 1988. Index to quantify black-tip disorder of mango (*Mangifera indica* L.) in orchards and its relationship with brick kilns. Indian J. Agric. Sci. 58 (7), 573–575.

Koo, R.C.J., Young, T.W., 1972. Effect of age and position on mineral composition of mango leaves. J. Am. Soc. Hortic. Sci. 97, 792–794.

Krishnakumar, B.G., Nachegowda, V., Bhargawa, B.S., 1998. Nutrient removal through harvested mango fruit. In: National Symposium on Mango Production and Export, 25–27 June, ICAR-C.I.S.H., Lucknow, India, p. 40.

Krishnamurthi, S., Randhawa, G.S., 1959. Role of potassium in nutrition of fruit crops. Indian J. Hortic. 16, 175–182.

Kumar, S., Nauriyal, J.P., 1977. Nutritional studies on mango—tentative leaf analysis standards. Indian J. Hortic. 34, 100–106.

Lakshimnarayana, S., Subhadra, N.V., Subramanyam, H., 1970. Some aspects of developmental physiology of mango fruit. J. Hortic. Sci. 45, 133–142.

Lalel, H.J.D., Singh, Z., Tan, S.C., 2003. Maturity stage at harvest affects fruit ripening, quality and biosynthesis of aroma volatile compounds in 'Kensington Pride' mango. J. Hortic. Sci. Biotechnol. 78, 225–233.

Léchaudel, M., Joas, J., 2006. Quality and maturation of mango fruits of cv. Cogshall in relation to harvest date and carbon supply. Aust. J. Agr. Res. 57, 419–426.

Léchaudel, M., Joas, J., 2007. An overview of preharvest factors influencing mango fruit growth, quality and postharvest behaviour. Braz. J. Plant Physiol. 19 (4), 287–298.

Litz, R.E., 1997. The Mango: Botany, Production and Uses. CAB International, Wallingford.

Majumdar, P.K., Singa, G.C., Singh, R.N., 1970. Effect of exogenous application of alfa-naphthylacetic acid on mango (*Magnifera indica* L.) malformation. Indian J. Hortic. 20, 130–131.

Majumder, P.K., Sinha, G.C., 1972. Seasonal variation in incidence of malformation in *Mangifera indica* L. Acta Hortic. 24, 221–223.

Malhi, C.S., Joshi, A., 1998. Effect of micro nutrient spray on the nutrient uptake and their distribution in various parts of fruits. In: Symposium Mango Production and Export, 25–27 June, ICAR-C.I.S.H., Lucknow, Indiapp. 37–38.

Mallik, P.C., 1963. Mango malformation-symptoms, causes and cure. Punjab Hortic. J. 3, 292–299.

Malo, S.E., Campbell, C.W., 1978. Studies on mango fruit breakdown in Florida. Proc. Am. Soc. Hortic. Sci. Trop. Reg. 22, 1–15.

Martin-Preveli, P., Marchal, J., Furon, V., 1975. A case of zinc deficiency in mango. Fruits 30, 201.

Minessey, F.A., Biely, M.P., El-Fahal, A., 1971. Effect of iron chelates in correcting malformation of terminal bud growth in mango. Sudan Agric. J. 6, 71–74.

Mishra, K.A., 1976. Studies on Bearing Behaviour of *Mangifera indica* L. and its Malformation. PhD thesis, PAU, Ludhiana.

Misra, A.K., Singh, V.K., 2002. *Fusarium subglutinans* (*F. moniliforme* var. *subglutinans*) in relation to mango malformation. Indian J. Plant Pathol. 20, 81–83.

Mukesh, K.M., Kaushik, R.A., Laxman, J., Sarolia, D.K., 2018. Response of mango (*Mangifera indica* L.) cultivars to micronutrients on fruit quality and yield attributes. In: International Mango Conference, May 08–10, Vengurla, India, p. 68.

Munniswami, D., 1970. Application of manures and fertilizers for mango crop. Lal Baugh 15, 7–8.

Nachegowda, V., Krishnakumar, B.G., Bhargawa, B.S., 1998. Leaf nutrient status of mango varieties. In: National Symposium on Mango Production and Export, 25-27 June, ICAR-C.I.S.H., Lucknow, India, p. 23.

Naik, S.K., Bhatt, B.P., 2017. Diagnostic leaf nutrient norms and identification of yield—Limiting nutrients of mango in eastern plateau and hilly region of India. Commun. Soil Sci. Plant Anal. 48, 1574–1583.

Neilsen, G.H., Parchomchuk, P., Neilsen, D., 1997. Distribution of soil P and K as affected by NP-fertigation in high density apple orchards. Acta Hortic. 451, 439–447.

NHB, 2017. http://nhb.gov.in/statistics/Publication/Horticulture%20At%20a%20Glance%202017%20for%20net%20uplod%20(2).pdf.

Nijjar, G.S., Arora, J.S., Singh, G., Dwivedi, R.S., 1976. Symptoms of zinc deficiency in mango. Punjab Hortic. J. 16, 113–114.

Nirvan, R.S., 1953. Bunchy top of young mango seedlings. Sci. Cult. 18, 335–336.

Nooden, L.D., 1988. Whole plant senescence. In: Nooden, L.D., Leopold, A. (Eds.), Senescence and Aging in Plants. Academic Press, San Diego, CA, pp. 391–439.

Nunes, C.N., Emond, J.P., Brecht, J.K., Dea, S., Proulx, E., 2007. Quality curves for mango fruit (cv. Tommy Atkins and palmer) stored at chilling and non-chilling temperatures. J. Food Qual. 30, 104–120.

O'Neil, M.K., Gardner, B.R., Roth, R.L., 1979. Orthophosphoric acid as phosphorus fertilizer in trickle irrigation. Soil Sci. Soc. Am. J. 43, 283–286.

Oosthuyse, S.A., 1997. Variation of leaf nutrition status in relation to fruit growth in mango. In: Mango Growers' Association Year Book.vol. 17, pp. 25–28.

Pal, R.N., Chadha, K.L., 1980. Black tip disorder of mango: a review. Punjab Hortic. J. 20, 112–121.

Pandey, D., Pandey, S.N., 1997. Effect of growth regulators and zinc sulphate on floral malformation in mango (*Mangifera indica* L.). In: Proceeding of the National Seminar ISPP, March, 19–21, IARI, New Delhi, p. 94.

Pandey, R.M., Sinha, G.C., Singh, R.N., Majumder, P.K., 1973. Some biochemical aspects of vegetative malformation in mango (*Mangifera indica* L.) carbohydrate reserve and nitrogenous fractions. Indian J. Hortic. 3–4, 475–480.

Pandey, R.M., Rao, M.M., Pathak, R.A., 1977. Biochemical changes associated with floral malformation in mango. Sci. Hortic. 6, 37–44.

Parent, S.E., Parent, L.E., Rozane, D.E., Natale, W., 2013. Plant ionome diagnosis using sound balances: case study with mango (*Mangifera indica*). Front. Plant Sci. 4, 1–12. Article 449.

Pathak, R.K., Ram, R.A., Garg, N., Kishun, R., Bhriguvanshi, S.R., Sharma, S., 2010. Critical review of indigenous technologies for organic farming in horticultural crops. Org. Farming News Lett. 6 (2), 3–16.

Patil, K.D., Salvi, B.R., Chavan, S.A., Khobragade, N.H., 2018. Supplementation of nutrients through foliar application at critical stages for increasing the yield of 'alphonso' mango. In: International Mango Conference, 2018 May 8–10, Vengurla, India, p. 78.

Raghupati, H.B., Reddy, Y.T.N., Reju, M.K., Bhargava, B.S., 2005. Diagnosis of nutrient imbalance in mango by DRIS and PCA approaches. J. Plant Nutr. 27 (7), 1131–1148.

Rahim, M.A., 2018. Horticultural crops and agroforestry in Bangladesh. In: Mirza, M.J.A. (Ed.), Agroforestry. GTI Publications, p. 302.

Rahim, M.A., Quddus, Islam, M.A., Karim, F., Kabir, M.M., Sarker, M.A., Naher, B.C.N., Alam, M.S., Fatema, K., 2018. Present Status, future, prospect of growing temperate fruits in Bangladesh a—tropical environment–with special reference to contribution in nutrition and poverty alleviation. In: VIII Symposium of Temperate Zone Fruits in the Tropics and Subtropics Certifies, Florianopolis, SC.

Rajan, S., 1986. Biochemical Basis of Mango (*Mangifera indica* L.), Malformation. PhD thesis, IARI, New Delhi.

Rajput, C.B.S., Singh, B.P., Singh, S.B., 1975. Effect of foliar application of zinc sulphate on vegetative growth of mango. Bangladesh Hortic. 4, 23–24.

Rajput, M.S., Srivastava, K.C., Singh, B.P., 1987. Factors affecting mineral composition of mango leaves—a review. Harayana J. Hortic. Sci. 16, 1–10.

Ram, S., 1989. Factors associated with black tip and internal necrosis in mango and their control. Acta Hortic. (231), 797–804.

Ram, S., 1991. Horticultural aspects of mango malformation. Acta Hortic. 291, 235–252.

Ram, R.A., Kumar, A., 2019. Growing fruit crop organically: challenges and opportunities. Curr. Hortic. 7 (1), 3–11.

Ram, R.A., Pathak, R.K., 2016. Organic approaches for sustainable production of horticulture crops: a review. Prog. Hortic. 48, 1–16.

Ram, R.A., Rajput, M.S., 1998. Studies on the effect of organic and bio-fertilizers on growth and yield of mango cv. 'Dashehari'. In: National Symposium on Mango Production and Export, 25–27 June, ICAR-C.I.S.H., India, Lucknow, p. 39.

Ram, R.A., Verma, A.K., 2015. Energy input, output and economic analysis of organic production of mango (*Mangifera Indica* L.) cv. 'Dashehari'. Indian J. Agric. Sci. 85 (6), 827–832.

Ram, R.A., Verma, A.K., 2018. Response of various organic amendments on yield, quality and economics of production in mango cv. 'Dashehari'. In: International Mango Conference, 8–11 May, 2018, RFRS, Vengurla, Maharashtra, Indiap. 173.

Ram, R.A., Singha, A., Bhriguvanshi, S.R., 2011. Response of different organic inputs on soil, plant nutrient status, yield and quality of mango cv. 'Amrapali'. In: Global Conference on Augmenting Production and Utilization of Mango: Biotic and Abiotic Stresses, 21–24 June, ICAR-CISH, Lucknow, India, pp. 100–101.

Ram, R.A., Verma, A.K., Vaish, S., 2016. Development of cost effective package of practice for organic production of mango cv. 'Mallika'. In: International Seminar on Recent Trends and Experimental Approaches in Science, Technology and Nature, 23–24 December, 2016 by Society for Science and Nature, IISR, Lucknow, India, p. 2.

Ram, R.A., Verma, A.G., Vaish, S., 2017. Studies on yield, fruit quality and economics of organic production of mango cv. 'Mallika'. In: Organic e Prints, pp. 485–488.

Rameshwar, A., Kulkarni, B., 1979. Effect of micronutrient spray on mango. In: Proceedings of mango workers meeting, Panaji, Goa, Horticultural Society of India, Bangalore, pp. 177–178.

Rao, M.M., 1998. Approaches for managing the problem of biennial bearing in 'alphonso' mango trees. In: National Symposium on Mango Production and Export, 25–27 June, ICAR-C.I.S.H., Lucknow, India, p. 23.

Rao, D.P., Mukherjee, S.K., 1989. Nutrient status in leaf and soil of some cultivars of mango in relation to yield. Acta Hortic. 231, 286–289.

Rath, S., Singh, R.L., Singh, B., Singh, D.B., 1980. Effect of boron and zinc sprays on the physico-chemical composition of mango fruits. Punjab Hortic. J. 20, 33–35.

Rauschkolb, R.S., Rolston, D.E., Miller, R.J., Carlton, A.B., Burau, R.G., 1976. Phosphorus fertilization with drip irrigation. Soil Sci. Soc. Am. J. 40, 68–72.

Raut, R.A., Patil, P.D., Munj, A.Y., Haldavnekar, P.C., Baviskar, S.B., Sanas, M.P., Shedge, M.S., 2018. Effect of organic manures and bio-fertilizers on soil microbial population of 'Alphonso' mango orchard. In: International Mango Conference; May 08–10, Vengurla, India, p. 112.

Ravishankar, H., Nalawadi, U.G., Hulmani, N.C., 1989. Effects of spraying chemicals and nutrients on growth and flowering in off year mango trees. J. Maharashtra Agric. Univ. 14, 319–322.

Reddy, S.E., Majumdar, A.M., 1983. Response of mango (Mangifera indica L.) to foliar application of phosphorus. Fert. Res. 4, 281–285.

Reddy, Y.T.N., Kurian, R.M., Sujatha, N.T., Srinivasa, M., 1998. Leaf and soil nutrient status of some cultivars of mango in relation to yield. In: National Symposium on Mango Production and Export, 25–27 June, ICAR-C.I.S.H., Lucknow, India, pp. 33–34.

Saeed, A., Schlosser, R., 1972. Effects of some cultural practices on the incidence of mango malformation. Z. Pflanzenkrankh Pflanzen Zchuz 79, 349–351.

Sahithya, B.R., Raju, B., Sagar, B.S., 2018. Alternate bearing of mango. In: International Mango Conference, May 08–10, Vengurla, India, p. 17.

Samra, J.S., Thakur, R.S., Chadha, K.L., 1978. Evaluation of existing critical limits of leaf nutrient standards in mango. Sci. Hortic. 8, 349–355.

Schlosser, S.E., 1971. Mango malformation: symptoms, occurrence and varietal susceptibility. Plant Protect. Bull. FAO 19, 12–14.

Sellen, D., Tollamn, J.H., McLeod, D., Weersink, A., Yiridoe, E., 1993. In the Economics of Organic and Conventional Horticulture Production. Department of Agricultural Economics and Business, University of Guelph, Guelph, ON.

Sen, P.K., Roy, P.K., De, B.N., 1947. Hunger signs on mango. Indian J. Hortic. 5, 35–44.

Severaj, Y., 1998. Investigations on the biochemistry of internal breakdown of 'Alphonso' mango (Mangifera indica L.) fruit. In: National Symposium on Mango Production and Export, 25–27 June, ICAR-C.I.S.H., Lucknow, India, p. 71.

Sharma, R.C., Mahajan, B.V.C., Dhillon, B.S., 1998. Fertilizer studies on mango cv. 'Dashehari'. In: National Symposium on Mango Production and Export, 25–27 June, ICAR-C.I.S.H., Lucknow, India, pp. 34–35.

Shinde, A.K., Dabke, B.B., Jadhav, M.N., Kandalkar, M.P., Burondkar, M.M., 2006. Effect of dose and source of potassium on yield quality 'alphonso' mango (Mangifera Indica L.). Indian J. Agric. Sci. 76 (4), 213–217.

Shirgure, P.S., Srivastava, A.K., 2014. Fertigation in perennial fruit crops: major concerns. Agrotechnol. 3 https://doi.org/10.4172/2168-9881-1000e109.

Shivashankar, S., Sumathi, M., 2014. A protocol for the control of spongy tissue disorder in 'alphonso' mango by pre-harvest treatments. J. Plant Sci. Res. 1 (4), 1–5.

Simmons, S.L., Hofman, P.J., Whiley, A.W., Hetherington, S.E., 1998. Effects of leaf: fruit ratios on fruit growth, mineral concentration and quality of mango (Mangifera indica L. cv. Kensington pride). J. Hortic. Sci. Biotechnol. 73, 367–374.

Singh, R.R., 1975. Effect of foliar spray of urea and superphosphate on the physico chemical composition of mango (Mangifera indica L.) fruits of cultivar 'Chausa'. South Indian Hortic. 23, 126–129.

Singh, R.R., 1977. Effect of various concentration of boron on growth characters and chemical composition of leaves of mango (Mangifera indica L.) cv. 'Langra'. Bangladesh Hortic. 5, 30–34.

Singh, R.N., 1978. Mango: Low Priced Book Series no 3. Indian Council of Agricultural Research, New Delhi, pp. 18–22.

Singh, Z., Dhillon, B.S., 1987. Effect of foliar application of boron on vegetative and panicle growth, sex expression, fruit retention and physico-chemical characters of fruits of mango (Mangifera indica L.) cv. 'Dashehari'. Trop. Agric. 64, 305–308.

Singh, R.R., Rajput, C.B.S., 1976. Effect of various concentrations of zinc on vegetative growth, characters, flowering, fruiting and physiochemical composition of fruits in mango. Haryana J. Hortic. Sci. 5, 10–15.

Singh, R.N., Rathore, V.S., 1983. Changes in moisture content, dry matter accumulation, chlorophyll and nutrients in mango malformation. Indian J. Hortic. 40, 21–25.

Singh, U.R., Tripathi, J.S., 1978. Effect of foliar spray of potassium nitrate and sodium dihydrogen orthophosphate on physico-chemical quality of mango fruits. Punjab Hortic. J. 18, 39–40.

Singh, L.S., Singh, S.M., Nirvan, R.S., 1961. Studies on mango malformation review, symptoms, extent, intensity and cause. Hortic. Adv. 5, 197–207.

Singh, K.K., Grewal, J.S., Jawanda, J.S., 1962. Improved practices for citrus and mango. Punjab Hortic. J. 1, 135–140.

Singh, R.L., Singh, B., Singh, R., 1979. Effect of foliar application of urea on the chemical composition of mango (Mangifera indica L.) fruits of cv. 'Langra'. Plant Sci. 11, 94–95.

Singh, D.S., Pathak, P.A., Singh, R.D., 1983. Studies on control of malformation of mango cv. Bombay green. Punjab Hortic. J. 23, 220–221.

Singh, Z., Dhillon, B.S., Arora, C.L., 1991. Nutrient levels in malformed and healthy tissue of mango (Mangifera indica L.). Plant Sci. 133, 9–15.

Singh, B.P., Tandon, D.K., Kalra, S.K., 1993. Changes in post-harvest quality of mangoes affected by pre-harvest application of calcium salts. Sci. Hortic. 54, 211–219.

Singh, V.K., Saini, J.P., Misra, A.K., 1998. Changes in ammonia assimilating enzymes associated with floral malformation in mango. In: National Symposium on Mango Production and Export, 25–27 June, ICAR-C.I.S.H., Lucknow, India, p. 71.

Singh, V.K., Ram, R.A., Garg, N., Pathak, R.K., 2004. Influence of organics on biophysical parameters and quality production of mango. In: National Conference on Role of Bio-pesticides, Bio-agents and Biofertilizers for Sustainable Agriculture and Horticulture, February 13–16, 2004, Lucknow, India, p. 101.

Srivastava, R.P., 1963. The black-tip disease of mango. Cause and control. Punjab Hortic. J. 5, 226–228.

Srivastava, A.K., 2013a. Recent developments in diagnosis and management of nutrient constraints in acid lime. J. Agric. Sci. 2, 86–96.

Srivastava, A.K., 2013b. Nutrientmanagement in Nagpur mandarin: frontier developments. J. Agric. Sci. 2, 1–14.

Srivastava, A.K., 2013c. Nutrient deficiency symptomology in citrus: an effective diagnostic tool or just an aid for post-mortem analysis. Agri. Adv. 2 (1), 77–94.

Srivastava, A.K., 2013d. Nutrient diagnostics in citrus: are they applicable to current season crop. Agrotechnology 2, 104–105. https://doi.org/10.4172/2168P-9881.1000e104.

Srivastava, A.K., Singh, S., 2004. Zinc nutrition, a global concern for sustainable citrus production. J. Sustain. Agri. 25, 5–42.

Srivastava, A.K., Singh, S., Huchche, A.D., Ram, L., 2001. Yield based leaf and soil test interpretations for Nagpur mandarin in Central India. Commun. Soil Sci. Plant Anal. 32, 585–599.

Srivastava, A.K., Singh, S., Tiwari, K.N., 2006. Site specific nutrient management for Nagpur mandarin (*Citrus reticulata* Blanco). Better Crops 88, 22–25.

Srivastava, A.K., Das, S.N., Malhotra, S.K., Kaushik, M., 2014. SSNM-based rationale of fertilizer uses in perennial crops: a review. Indian J. Agric. Sci. 84, 3–17.

Sujatha, S., Reddy, N.N., 2006. Economic feasibility of drip irrigation and nutrient management in mango (*Mangifera Indica* L). Indian J. Agric. Sci. 76 (5), 312–314.

Sukhada, M., 1998. Inoculation response of mango to VAM fungi in glass house and field trials. In: National Symposium on Mango Production and Export, 25–27 June, ICAR-C.I.S.H., Lucknow, India, p. 21.

Tandon, H.L.S., 1987. Fertilizer Recommendations for Horticultural Crop in India: A Guide Book. Fertilizer Development and Consultation Organization, New Delhi.

Thakur, R.S., Samra, J.S., Chadha, K.L., 1980. Assessment of micro-nutrient status in the foliage of mango trees around Malihabad, Lucknow, India. Indian J. Hortic. 37, 120–123.

Thakur, R.S., Rao, G.S.P., Chadha, K.L., Samra, J.S., 1981a. Variation in mineral composition of mango leaves as contributed by leaf sampling factors. Soil Sci. Plant Anal. 12, 331–341.

Thakur, R.S., Samra, J.S., Chadha, K.L., 1981b. The nutrient levels in fruiting and non-fruiting terminals of three mango cultivars. Sci. Hortic. 15, 355–361.

Uriu, K., Carlson, R.M., Henderson, D.W., Schulbach, H., Aldrich, T.M., 1980. Potassium fertilization of prune trees under drip irrigation. J. Am. Soc. Hortic. Sci. 105, 508–510.

USDA—National Nutriment database, 2019. http://nhb.gov.in/pdf/fruits/mango/man003.pdf.

Varma, A., 1983. Mango malformation. In: Singh, K.G. (Ed.), Exotic Plant Quarantine Pests and Procedures for Introduction of Plant Materials. Asian Plant Quarantine Centre and Training Institutes, Serdang, Malaysia, pp. 173–188.

Varma, A., Lele, V.C., Raychaudhuri, S.P., Ram, A., Sang, A., 1974. Mango malformation: a fungal disease. Phytopathol. J. 79, 254–257.

Watt, G., 1891. A Dictionary of Economic Products of India vol. 5. Govt. Printing Press, Calcutta, p. 149.

Young, T.W., 1957. 'Soft-nose', a physiological disorder in mango fruits. Proc. Florida State Hortic. Soc. 70, 280–283.

Further reading

Ali, A.M., 2018. Nutrient sufficiency ranges in mango using boundary-line approach and compositional nutrient diagnosis norms in El-Sahiya, Egypt. Commun. Soil Sci. Plant Anal. 49 (2), 188–201.

Bains, G., Pant, R.C., 2003. Mango malformation: etiology and preventive measures. Physiol. Mol. Biol. Plants 9, 41–61.

Bhargava, B.S., Chadha, K.L., 1988. Leaf nutrient guide for fruit and plantation crops. Fert. News 33, 21–29.

Hundal, H.S., Dhanwinder, S., Brar, J.S., 2005. Diagnosis and recommendation integrated system for monitoring nutrient status of mango trees in sub-mountainous area of Punjab, India. Commun. Soil Sci. Plant Anal. 36, 2085–2099.

Kumar, P., Rehalia, A.S., 2007. Standardization of micronutrients range in mango (*Mangifera indica* L.) by orchard surveys. Asian J. Hortic. 2 (1), 218–221.

Raghupathi, H.B., Bhargava, B.S., 1999. Preliminary nutrient norms for 'alphonso' mango using diagnosis and recommendation integrated system. Indian J. Agric. Sci. 69, 648–650.

Young, T.W., Koo, R.C.J., 1969. Mineral composition of Florida mango leaves. In: Florida State Horticultural Society.Florida Agricultural Experiment Station Journal Series, vol. 3368, pp. 324–328.

44

Diagnosis and management of nutrient constraints in bananas (*Musa* spp.)

*Kenneth Nyombi**

Makerere University, College of Agricultural and Environmental Sciences, Kampala, Uganda
*Corresponding author. E-mail: knyombi@yahoo.co.uk

1 Introduction

Bananas (*Musa* spp.) are major food crop globally, ranking fourth in the world after rice, wheat, and maize (FAO, 2017). They are large perennial herbs growing to a height of 2–9 m at maturity depending on variety. Native to the Indo-Malaysian, Asian, and Australian tropics, bananas are currently grown throughout the tropics and subtropics between latitudes 33°N and 33°S (Stover and Simmonds, 1987). They have developed secondary loci of genetic diversity in Africa (the great Lakes region), Latin America, and the Pacific (Papua New Guinea and the Solomon Islands).

Bananas grow in a wide altitude range from 0 to 2000 m.a.s.l., but lower temperatures at high altitudes reduce the rate of leaf emergence or the growth in leaf area index (LAI), the rate of fruit growth, and length of the crop cycle duration. With increasing altitude from 1174 to 1405 m.a.s.l, Taulya et al. (2014) noted that phenological development of East Africa highland bananas (Musa AAA-EA) involved trade-offs between physiological and chronological age. Generally, bananas require a base temperature of 14°C and the temperate range for optimal growth is 25°C–30°C (Stover and Simmonds, 1987). Under good growth conditions (if soil moisture and nutrients are not limiting), the net assimilation rate (NAR) depends on the total photosynthetically active radiation (PAR), but shading up to 50% in tropical regions may not reduce yields. At higher shading such as in plantations with leaf area index (LAI) greater than 4, growth of ratoon crops is delayed by a couple of months due to the low PAR penetration. Bananas take up large amounts of water owing to the large broad leaves and high leaf area index. Water use is a function of supply, demand (potential evaporation), and soil water holding capacity (Lahav and Turner, 1989). In the tropics, bananas are produced in areas receiving about 1000 mm of rainfall per annum, with no irrigation resulting into moisture stress, reduced nutrient uptake during the dry months, and consequently low yields. In the Caribbean, rainfall may exceed

A.K. Srivastava, Chengxiao Hu (eds.)
Fruit Crops: Diagnosis and Management of Nutrient Constraints
https://doi.org/10.1016/B978-0-12-818732-6.00044-7

2000 mm per annum, which coupled intensive mineral fertilizer and pesticide use results into leaching and pollution (Godefroy and Dormoy, 1990).

Global banana production is estimated at 120 million tons per year (FAOSTAT). Most of the export bananas are desert bananas—Cavendish (AAA)—and export volumes are estimated at 16.741 million tons per year (FAO, 2017), generating about USD 8 billion per year. Latin America and the Caribbean export 84%, Asia 12%, and Africa 4%. The main banana exporters are Ecuador (33%), the Philippines (14%), Costa Rica (12%), Guatemala (11%), and Colombia (11%), with organic bananas mainly exported by Colombia, Peru, and the Dominican Republic. Export destinations are the North American markets, Western Europe, Japan, Russia, and the Asian market, which mainly takes exports from the Philippines. The rest (about 85% of total production) is consumed locally, most especially in large producing countries such as India, China, Brazil, the Philippines, and in some African countries such as Uganda and Rwanda, where per capita consumption exceeds 200 kg of banana. Bananas provide potassium (358 mg/100 g), vitamin B6 (20%), and vitamin C. Potassium reduces the risk of heart disease and lowers blood pressure. Ripe bananas provide up to 90 cal, 12 g of sugars (mainly sucrose, fructose, and glucose), 0.3 g of fat, and 2.7 g of fiber per 100 g. Its nutritional content, health benefits coupled with increasing global population, have kept the banana demand strong.

2 Classification, geographical distribution, and major cultivars

There are over 1000 varieties of bananas produced globally, which are grouped according to the ploidy levels and the uses (cooking and desert). Ploidy level refers to the number of chromosome sets and the relative proportion of *Musa acuminata* (A) and *M. balbisiana* (B) in their genome (Karamura, 1998). The common cultivated cultivars globally are triploid hybrids: AAA (Cavendish, East Africa highland banana), AAB (Silk, Mysore), ABB (Saba, Bluggoe, Monthan, Simoi), and BBB (Saba). Diploids AA, AB (beer bananas), and BB, which are starchier, are also grown. Tetraploids AAAA (FHIAs), AAAB (PA12.03 and SH-3640), AABB (FHIA03), and ABBB are grown but are not so common.

In Southeast Asia, cooking bananas (ABB) and dessert bananas (AAB) are common. In central and South America and China, Cavendish banana (AAA) are grown in commercial plantations and account for about 47% of global production (FAO, 2017). This is attributed to their dwarf stems, implying reduced damage to storms and the ability to recover quickly from natural disasters. In Africa, there are over 200 cultivars of East African highland cooking bananas (AAA-EA) grown in the great Lakes region for home consumption with surplus sold to urban centers. In this region, a number of beer cultivars (AB, ABB, and AAA) and desert types (AAA and AB) are grown. Plantains (AAB) are starchier and are common in West and Central Africa and Central and South America (Stover and Simmonds, 1987).

3 Biotic production constraints and yields

The major biotic constraints to banana production, which differ from one region to another (given the soils, temperature, altitude, management, and production orientation), are pests: weevil (*Cosmopolites sordidus*), nematodes (burrowing nematode, *Radopholus similis*; root lesion nematode, *Pratylenchus goodeyi*; spiral nematode, *Helicotylenchus multicinctus*; reniform nematode, *Rotylenchulus reniformis*; stunt nematode, *Tylenchorhynchus* spp.; and ring nematode, *Criconemoides sphaerocephalum*), moths (scab moth, *Nacoleia octasema*), and diseases (black Sigatoka, *Mycosphaerella fijiensis*; bunchy top, banana bunchy top virus (BBTV); Panama disease or Fusarium wilt, *Fusarium oxysporum*; anthracnose, *Colletotrichum musae*; and banana bacterial wilt, BBW).

The banana weevil is the best known banana insect pest in the great Lakes region. In Uganda, it was first reported in 1908 and mainly affects bananas at altitudes below 1600 m.a.s.l. (Sseguya et al., 1999). The feeding larva eats through the rhizome reducing water and nutrient uptake. Commonly reported yield losses due range between 14% and 20% (Rukazambuga et al., 2002), but heavy attack can lead to 100% yield loss. It is controlled through field sanitation, better agronomic practices, and the use of chemicals (organophosphates and carbamates). Nematodes are a major yield reducing factor in all banana-producing regions, and their populations increase with age of the plantation, if not controlled. They attack roots and rhizomes causing necrosis of root tissues and corm, thus impeding water and nutrient uptake. *R. similis* is common at altitudes below 1500 m.a.s.l, while the root lesion nematodes are common at altitudes higher than 1500 m.a.s.l. Nematodes are controlled through cultural practices, using nematode-free planting materials, soil treatment before planting or fumigation, crop rotation, fallowing, varietal resistance, and nematicides (Risede et al., 2010). The scab moth larvae enter the banana flower and destroy the growing fruits, causing scars and deformed fingers. It is common in the Southwest Pacific and is controlled by injecting the banana flower after it emerges with an insecticide.

Globally, black sigatoka is the most important fungal banana disease. An attack leads to the development of lesions on leaves significantly reducing the number of green leaves, leaf area to intercept radiation, and the bunch yield (Stover and Simmonds, 1987). Usually, a combination of cultural practices (field sanitation, canopy aeration, pruning, cutting off affected leaves, and resistant varieties) and chemicals (fungicides) are important for controlling this disease. The bunchy top disease is a major problem for banana plantations in Hawaii and the Pacific. It is controlled by using clean planting materials and disinfecting tools used on plantations. The panama disease caused by a soil borne fungus is a major banana disease worldwide. Fungal attack results into stem, root, and rhizome necrosis. Plant death mainly occurs at or during flowering and periods of soil moisture stress. Since the fungus lasts long in soil, control is largely preventive, through planting resistant varieties or clean materials in clean soil. Banana anthracnose disease mainly damages fruits and is controlled by removing affected leaves that are close to the bunches. Banana bacterial wilt (BBW) devastated banana plantations in the great Lakes region, with yield losses up to 100% (Kagezi et al., 2006). Affected plants wilt when the leaves are green and die off. It is controlled by sanitation (planting disease free materials, disinfecting farm tools, and removing male buds).

Yields on banana farms are a complex interaction of both the biotic and abiotic (soil moisture stress and soil fertility) constraints. The biotic constraints and moisture stress reduce the recovery efficiency of applied fertilizer and the fertilizer-use efficiencies for farmers who use fertilizers. In Africa, banana production is mainly by smallholder farmers on fields <0.5 ha, with medium size commercial farms (>3 ha) gaining prominence due to increasing demand. In the great Lakes region, actual banana yields on smallholder farms are low (5–20 Mg/ha/yr FW) due the constraints mentioned earlier, far below the estimated potential yield (100 Mg/ha/yr FW) (Nyombi, 2010). Under average farmer management, Smithson et al. (2001) reported yields of 67 Mg/ha FW from Southwestern Uganda. This shows that the yield gap can actually be reduced with proper management. For example, Costa Rica, a leading banana producer and exporter have been able to increase yields from 10 to 20 Mg/ha/yr FW before 1960 to 50–80 Mg/ha/yr FW today for Cavendish (AAA) through better agronomic management and cultivar choice (Stover and Simmonds, 1987). In general, commercial production of the Cavendish variety gives yields ranging 40–60 Mg/ha FW in other regions (India and China), while small producers report yields of about 30 Mg/ha FW. In the Philippines, actual yields are low, with varieties such as Saba (cooking—ABB) and Lakatan (dessert—AAA), yielding 11–20 Mg/ha FW. It is important to note that yields are generally low for locally consumed bananas. Population growth coupled with increasing demand, improved infrastructure, market access, and value addition are likely to increase prices, and compel farmers invest more into management thus raising yields (Bagamba, 2007).

4 Major soil types and nutrient requirements for banana production

Globally, bananas are grown on a wide range of soils from the fertile and young Inceptisols (with a large capacity for stocking potassium) and Andisols, moderate K releasing capacity (Delvaux et al., 1987, 1989; Delvaux, 1989) to the old and low fertility Ferralsols and Acrisols (Fig. 44.1). In many areas especially in Africa, banana production systems are built on old soils with soil pH, extractable phosphorus (P), calcium (Ca), and potassium (K) below critical levels and nutrient availability largely depends on mineralization of soil organic matter. Generally, soils must be deep, free draining loams, or light clays, because bananas can't tolerate extreme water logging especially for clay soils (>60% clay).

Rooting depth of bananas determines the ease of root movement through the soil; however, maximum rooting depth of 1.5 m has been reported (Lahav and Turner, 1989). A higher clay content >50% depresses yields due to compaction, poor root movement, and reduced available water (Robinson, 1996) resulting poor emergence of fruit (Fig. 44.2).

The best pH range for good banana growth is 5.5–8.0. Low pH (4.5) reduces yield by 50% as compared with the optimal range due to low availability of important nutrients such as phosphorus especially in P-fixing old tropical soils. For acid soils, 600–750 kg of lime per hectare can be plowed into soil at land preparation to raise the pH. Bananas are heavy feeders (Jones, 1998). Nutrient requirements are in order as follows: potassium > nitrogen > phosphorus (Nyombi et al., 2010). In addition, bananas take up significant quantities of other nutrients such as calcium and magnesium.

Typical N, P, and K uptake levels amount to 388, 52, and 1438 kg/ha in a crop cycle for AAA Cavendish dessert banana plants yielding 50 Mg/ha FW at a density of 2400 plants/ha (Lahav and Turner, 1983). Due to the large quantities of potassium taken up by bananas, apparent K recovery fractions are high (75%) as reported by Lopez and Espinosa (2000), if soil moisture is not limiting. Potassium plays a crucial role in regulating the transfer of nutrients to the xylem, regulating water in the plant and the functioning of the stomata. When potassium supply is low, the transfer of nitrogen, phosphorus, calcium, magnesium, and other nutrients across the xylem is restricted

FIG. 44.2   Moisture stress from the floral initiation stage to flowering often results into choked bunches and flower abortions in East Africa highland bananas (*Musa* AAA-EA).

FIG. 44.1   A soil profile pit dug on a haplic ferralsol in central Uganda showing the shallow A horizon, the iron and aluminum oxide-rich B horizon and the stone layer.

(Turner, 1987). Low potassium supply reduces total dry matter production and its allocation within the plant, most especially the bunch. This attributed to the reduced photosynthesis, respiration, and total green leaf area of the plant (Fig. 44.3). Potassium deficiency also impairs protein synthesis, sugar transportation, and N utilization leading to N accumulation in plants (Nyombi et al., 2010). Potassium deficiency also leads into shortening of internodes, delayed flower initiation, choking of bunches, and stunted growth.

Nitrogen is component of enzymes and chlorophyll and give leaves a green color. It is very important in green leaf area formation, photosynthetically active radiation interception, and determining its use efficiency (Marschner, 2003). Low nitrogen supply results into pale green leaves, with midribs, leaf sheaths, and petioles turning pink. Phosphorus is crucial for root growth, energy transfer, and fruit formation. Good root growth is important due to the large fresh mass (stem and leaves) and to support the bunch after flowering. Low P supply leads to curling of leaves, and a bluish green color on young leaves. Magnesium uptakes in banana plants can be as high as 55 kg/ha/yr for Grand Naine on an Ultisol (Irizarry et al., 1988). Low magnesium supply results into yellow discoloration of midblade and midrib portion and purple mottling of the petioles and marginal necrosis. Calcium is important in cell wall structure and ensuring the integrity of the banana peel during ripening. Calcium deficiency results into the thickening of veins and the deformation of the leaf lamina. Sulfur is a component of chlorophyll molecule and is crucial for protein synthesis. Sulfur deficiency results into white or yellow patches on young leaves and on leaf margins, thickening of veins, and reduced bunch size. Copper deficiency results into chlorosis of leaves. Zinc deficiencies mostly occur in limed or alkaline soils.

FIG. 44.3   Potassium deficiency in a newly established banana plantation on a Lixic Ferralsol at Ntungamo, Southwestern Uganda. Potassium-deficient old leaves turn orange-yellow in color due to K relocation to the young leaves; show scorching along the margins, later curl inward; and die resulting into a reduction in total leaf area.

Leaves show a high amount of anthocyanin on the underside, with fruits small and twisted. Manganese deficiency results into light greening of leaf margins of younger leaves. Deficiency of boron results into thickening of secondary veins, delayed flowering, curling of leaves, and lamina deformation.

Potassium is sensitive to cation balance, and the assessment of soil K critical levels is dependent on the K/Ca/Mg balance. Therefore, soil analysis is important for monitoring K fertilization to determine the relationship between crop response to the added K, K concentration in soil, and the K/Ca/Mg balance. Soil critical K content was reported at 1.4 meq/100 g of soil, with a K/Mg ratio of 0.28 (Lahav and Turner, 1989). Optimal calcium absorption would require a soil balance K:Mg:Ca of 1:3.5:10.7 (Turner et al., 1989).

5 Diagnosis of banana nutrient constraints (nutrient deficiency symptoms, soil fertility norms for different Banana cultivars used globally, and leaf nutrient standards)

Plant nutrient deficiencies are diagnosed using foliar analysis (Hallmark and Beverly, 1991), which helps to accurately assess the need for fertilization. Foliar analysis is based on the assumption that there is always a positive relationship between the quantities of nutrient applied, leaf nutrient status, and consequently the yield. This assumption however may not hold if pests and moisture supply limit uptake of the applied fertilizers. Timely leaf analysis helps to prevent deficiencies rather than correcting them. In most banana-producing countries, not only laminar structure of third leaf is sampled, but also samples of the central vein of third leaf and the petiole of seventh leaf can also be used. Usually, a strip of tissue 10-cm wide, on both sides of the central vein, is taken (Lopez and Espinosa, 2000) and analyzed using standard laboratory methods for assessing leaf nutrient contents. The critical levels of foliar nutrients of Dwarf Cavendish (AAA) dessert banana are presented in Table 44.1.

However, as already discussed, nutrient uptake and distribution are affected by interactions within the plant; therefore, multinutrient approaches have been derived. There are two common methods used to diagnose nutritional imbalances: diagnosis and recommendation integrated system, DRIS (Walworth and Sumner, 1987) and compositional nutrient diagnosis, CND (Parent and Dafir, 1992). DRIS norms are considered as best and are commonly used. They envisage that the ratios N:P, N:K, P:K, Ca:Mg, and their reciprocals remain constant in the leaf irrespective age. Therefore, nutrient concentration ratios rather than concentrations of single nutrients themselves are used (Beaufils, 1971, 1973) to identify the imbalances. These norms are usually developed for specific locations (climate and soils), cultivar, and under a given crop and soil management regime.

For example, Wortmann et al. (1994) developed DRIS norms for AAA-EA cooking banana cultivars in the Kagera region of Northwest Tanzania as N (3.15), P (0.25), K (3.04), and Ca (1.13). Wairegi and van Asten (2011) using foliar analysis data from different banana growing zones in Uganda for similar cooking bananas reported that foliar N, P, K, Ca, and Mg ranged between 1.35%–3.89%, 0.10%–0.38%, 1.54%–5.83%, 0.23%–1.74%, and 0.25%–0.85%. This shows that on average, Wortmann et al. (1994) had higher N (3.15% vs 2.79%), P (0.25% vs 0.21%), Ca (1.13% vs 0.91%), and lower K (3.04% vs 3.84%) as compared with the findings of Wairegi and van Asten (2011). Smithson et al. (2001) in a study at Rubale, Southwest Uganda, suggested that the norm for K in AAA-EA cooking banana was lower than the 4.49 suggested by Angeles et al. (1993) but higher than 3.04 obtained by Wortmann et al. (1994). This emphasizes the

TABLE 44.1 Critical levels of foliar nutrients of Dwarf Cavendish (AAA) dessert banana.

Nutrient	Leaf 3 Lamina %	Leaf 3 Central vein %	Leaf 7 Petiole %
Nitrogen	2.6	0.65	0.4
Phosphorus	0.2	0.08	0.07
Potassium	3.0	3.0	2.1
Calcium	0.5	0.5	0.5
Magnesium	0.3	0.3	0.3
Sodium	0.005	0.005	0.005
Sulfur	0.23	–	0.35

656

44. Diagnosis and management of nutrient constraints in bananas (*Musa* spp.)

TABLE 44.2 Diagnosis and recommendation integrated system (DRIS) and critical nutrient levels in the 3rd leaf lamina.

Nutrient (%)	DRIS	Critical value range	Average of published values
Nitrogen	3.04	1.81–4.00	3.03
Phosphorus	0.23	0.12–0.41	0.22
Potassium	4.49	1.66–5.40	3.40

Critical levels of micronutrients: boron (10 mg/kg), iron (100 mg/kg), manganese (160 mg/kg), zinc (20 mg/kg), and copper (9 mg/kg).

Based on Angeles, D.E., Sumner, M.E., Lahav, E., 1993, Preliminary DRIS norms for banana. J. Plant Nutr. 16, 1059–1070.

location and management specificity of DRIS norms even for the same cultivar. Angeles et al. (1993) reported higher K (4.49%), N (3.04%), and P (0.23%) in a high yielding banana plantation—70 Mg/ha (Table 44.2).

6 Physiological disorders due to nutritional and environmental constraints

In most fruit crops, physiological disorders associated with micronutrients as compared with macronutrients are more common given their perennial nature and the focus mainly on macronutrients (N, P, and K) during fertilizer applications (Hanson, 1987). In bananas, high K:Mg ratios above 0.6–0.7 induce typical mottling of the petiole (blue), which reduces crop yields. For example, Delvaux (1988) working on an old and highly leached acid Ultisols in Cameroon reported that a K:Mg ratio of over 0.7 induces blue, only when root depth is shallow (<20 cm) and when both exchangeable Mg concentration and total Mg soil reserve are below 1.3 and 45 meq/100 g of soil, respectively. However, blue can also be associated with a low K:Mg ratio in the field caused by potassium deficiency and was not related to high magnesium supply.

Micronutrient deficiencies (Zn, Mn, and B) are more common in banana (Robinson, 1996), and these must be detected before the visual symptoms and addressed to reduce their effect on yields and fruit quality. These are also attributed to expansion of banana production to poor soils, limited use of organic manures, imbalanced NPK fertilizers, and high intensity crop production (>2500 plants/ha). To correct the micronutrient deficiencies, appropriate foliar, and soil nutrient applications are necessary. Disorders caused by micronutrient deficiencies include the following: bronzing (the development of bronze or copper color on leaves), chlorosis (the loss of green color resulting into pale yellow tissues), tip dieback (collapse and death of the growing tip or the younger leaves), lesions (small wounds on the leaf or stem tissue followed by discoloration), and necrosis or death of plant tissues.

Under field conditions, disturbances in the plant metabolic and physiological activities can occur resulting from an excess or deficit of environmental variables like temperature, light, and aeration, coupled with micronutrient deficiencies. Banana choke throat is due to low temperature, which turns the leaves yellow and in severe cases, lead to tissue necrosis. At flowering, the bunch is unable to emerge from the pseudostem properly and is deformed, and bunch maturity is delayed by 1–2 months. Tree shelter belts can be planted in the wind direction, or low-temperature tolerant cultivars can be planted at that location. Another common disorder is chilling injury. This occurs when preharvest or postharvest temperatures fall below 14°C for various time periods. Banana peels become dark, and the fruits do exhibit uneven ripening. In the commercial Cavendish and Gros Michel banana, chilling injury manifests its self at temperatures below 12.5°C. To avoid this problem, the major banana exporting companies ship bananas at a controlled temperature of 13–14°C.

Other physiological disorders happen during ripening, thus reducing postharvest fruit quality, quantity, and marketability. Peel cracking is common disorder, but it largely depends on genetic, environmental, and physiological factors before harvest and environmental conditions of storage after harvest. This exposes the banana pulp to pathogen infections. Peel cracking can be controlled by managing the humidity levels in the ripening, storage, and transportation chambers. Peet (1992) argued that splitting may also be attributed to preharvest conditions, as calcium and potassium deficiency. Most of the export bananas are sold as clusters of fingers in supermarkets. Finger drop usually occurs during ripening, with individual fingers dropping off from the cluster, reducing market value, and increasing susceptibility to pathogens. A weak finger neck may have three possible causes: genetics, nutrition, and postharvest conditions. Balanced banana nutrition is very important to deter neck breakage. Usually, the application of boron, calcium, magnesium, or silicon enforces cell structure and deters this disorder (Putra et al., 2010). It is important to ensure balance nutrition to have good quality banana fruits (Raghupathi et al., 2002).

7 Management of nutrient constraints

In Africa, few banana farmers use fertilizers; therefore, nutrient supply largely depends on the mineralization of soil organic matter, but the size the active pool is always small. Organic fertilizers (grasses, composts, crop residues, and animal manures) were proposed as the first line remedy to soil fertility decline from the 1930s in Africa. In the great Lakes region, research done in the 1940s and 1950s focused on the quantification of organic fertilizers applied and yield data collection on farms. But the yield increases above the control reported were low—<10 Mg/ha/yr (Nyombi, 2013). In the Americas and the Caribbean, from the 1960s, the focus was on integrated pest management and the use of mineral fertilizers to increase banana yield to meet the growing demand for the fruits. These efforts have largely paid off with yields of more than 70 Mg/ha FW in high yielding banana plantations today. In Africa, from the 1980s, studies focused principally on diagnosing plant nutrition status, the use of mineral fertilizers, and the combined application of organic and mineral fertilizers. The integrated plant nutrition approach was advocated for giving the range of nutrients from organic materials and other benefits such as soil structure improvement and soil water retention. But increasing land pressure and shrinking farm sizes have reduced farmers capacity to obtain off-farm organic resources from communal areas (Baijukya et al., 2005), with additional supply of N, P, and K expected to come from mineral sources (Hallmark et al., 1987).

Increasing commercialization of banana trade from the Americas from the 1960s led to increases in mineral fertilizer use and efforts to make production more efficient and profitable to maintain and attract new markets. Today, these efforts have led to a number of companies that are famous for trade in banana such as Chiquita, Del Monte, Dole Foods, and Fyffes, with a net worth valued at billions of dollars. Science has played a crucial role to determine the optimal amount of fertilizer or fertilizer amounts to apply for a target yield. Janssen et al. (1990) developed the quantitative evaluation of the fertility of tropical soil (QUEFTS) model originally to calculate maize yield as a function of N, P, and K supply from soil and fertilizer while accounting for interactions among the nutrients (N, P, and K). Using experimental data from Uganda, this static model was applied to East Africa highland bananas (AAA) with interesting results (Nyombi et al., 2010). With the developed model parameters, it is relatively easy to calculate the amount of fertilizer required for a target yield, provided moisture supply is not limiting. In South and Central America (Costa Rica and Honduras), N, P, and K apparent recovery fraction are estimated at 50%, 30%, and 75% (Lopez and Espinosa, 2000). Using these and other derived banana parameters, the QUEFTS model predicts that for a fresh banana yield of 70 Mg/ha, 426 kg N/ha + 55 kg P/ha + 1576 kg K/ha. These values are not so different from fertilizer application rates in high yielding plantations in the banana growing regions (Robinson, 1996).

With the fertilizer amount determined, it is important to identify the sources of nutrients, methods of application, and when to apply. The types of fertilizers to use as sources of nutrients depend on availability, the soil pH, and the texture (sandy or clayey). However, to achieve the above yields, special attention must be taken for potassium, which is the most important nutrient for banana growth. The common sources of potassium are potassium chloride (KCl) (60% K_2O and 45% Cl), potassium sulfate (50% K_2O and 17%–18% S), and potassium magnesium sulfate (21%–22% K_2O, 10%–11% Mg, and 21%–22% S). Potassium is susceptible to leaching losses in light textured (sandy) soils and fixation in clay soils; it is thus better to apply the granular form or to split applications. In most banana-producing regions, usually three six (six splits) per year are used for experimentation and are recommended to farmers. This improves the recovery efficiency of the applied fertilizer. Bananas are also produced on saline soils in drier environments (with irrigation) resulting into an increase in Na content in roots reducing K uptake and banana yield (Israeli et al., 1986). In the Canary Islands, the optimal value for the soil K:Na ratio is considered to be 2.5. For soils with high Na content, the combined application of gypsum, manure, and potassium (Jeyabaskaran et al., 2000).

Nitrogen sources include the following: calcium ammonium nitrate (CAN) (26% N, 3.5% CaO, and 4% MgO), urea (46% N), and ammonium sulfate (21% N and 24% S). Urea should be covered or applied in solution to minimize volatilization losses (Chan, 1986). Prasertsak et al. (2001) reported that 25% of applied N was lost from the system by runoff, leaching, ammonia volatilization, or denitrification in the wet tropics of Queensland, Australia (annual rainfall 1800–4000 mm and plant density 1300 mats/ha). Godefroy and Dormoy (1990) reported leaching losses from soils with a low cation exchange capacity in Martinique to be 165 kg N/ha/yr. Nitrogen must be applied in splits. CAN is preferred on acid soils to raise the pH. Urea and ammonium fertilizers are preferably used on soils with a high pH. Mono-ammonium phosphate (10%–12% N and 48%–61% P_2O_5) and diammonium phosphate (18% N and 46% P_2O_5) are also used. Other sources of phosphorus include the following: triple super phosphate (20% P) and single super phosphate (12%–18% P_2O_5, 18%–21% Ca, and 11%–12% S). Usually two applications per year are adequate.

The other source of magnesium is magnesium sulfate (13%–16% Mg and 13% S). Micronutrient deficiencies are more common now (especially Zn and B). Zinc is supplied by application of zinc sulfate (22% Zn and 11% S) and boron through application of borax (11% B). Fertigation is more efficient with less losses, but it is only possible for large-scale farmers. For small scale farmers, fertilizers are applied per plant by hand.

8 Conclusions

Bananas are and will remain an important source of income and nutrition to current and future generations. However, to increase the current yields and ensure future fruit supplies, the causes of low yields especially in Africa and Asia, where population increase is fast, must be addressed.

Low yields are attributed to biotic (pests and diseases) and abiotic constraints (moisture stress and nutrients). Biotic constraints reduce the uptake and utilization efficiency of water and nutrients and should be controlled using the different methods as discussed in this chapter.

Bananas take up large amounts of K, N, P, Mg, and Ca, if soil moisture is not limiting. Where the indigenous nutrient supply is low, these must be supplied from both organic manures and mineral fertilizers. Given the fact that nutrient deficiencies show a strong spatial and temporal variability in different banana systems due to soils, soil and foliar analysis are important if static models such as QUEFTS are to be successful applied to calculate fertilizer requirements for target yield levels. Balanced nutrition is important to minimize growth and postharvest physiological disorders.

Variation in soils (cation exchange capacity), rainfall received, and the need to synchronize nutrient demand by the plant and supply dictate that fertilizers supplying the macronutrients are applied in splits. This reduces losses due to runoff, leaching, and volatilization, hence improving the recovery efficiency of the applied fertilizers. Owing to the perennial nature of banana plantations, micronutrients have gained importance, and these must be applied on foliage or in soil before deficiency symptoms are seen.

9 Future research

Biotic constraints are a major yield reducing factor. Research should focus on identifying and inserting genes of resistance into bananas without changing the quality and taste of the different varieties. Such genes can be identified in other plants or in wild bananas.

Since bananas are also grown by resource poor farmers in areas receiving about 1000 mm of rainfall per annum, research should focus on soil moisture conservation and assess the potential for rain water harvesting and irrigation. Investments in irrigation and rain water harvesting need cash that must come from new market opportunities, improved fruit quality, and value addition.

Banana production systems may be susceptible to the impacts of climate change, due to the shallow rooting depth, rainfall seasonal changes, and the high LAI. The potential impacts of elevated temperatures, increased or reduced rainfall on the banana yield, must be fully understand, and mitigation and adaptation measures put in place in the different banana regions.

Soil fertility management is still a major problem in Africa. The potential for banana-livestock integration needs to be thoroughly investigated. Some key questions must be answered; how many cows can adequately supply manure for 0.5 ha, and what are the sources of feed for livestock. This may improve productivity of the smallholder plantations (through supply of nutrients and moisture conservation), especially where farmers use little or cannot afford mineral fertilizers.

A lot of research information has been generated over the past decades, but the yield is still low. Our scientific understanding of banana systems has greatly improved. It is thus imperative to put together all this systems knowledge and establish some banana trials. In such trials, pests and diseases are controlled adequate water and nutrients supplied. The costs and benefits can then be assessed. This may aid adoption of technologies fronted to the farmers.

References

Angeles, D.E., Sumner, M.E., Lahav, E., 1993. Preliminary DRIS norms for banana. J. Plant Nutr. 16, 1059–1070.

Bagamba, F., 2007. Market Access and Agricultural Production. The Case of Banana Production in Uganda. PhD thesis, Wageningen University, The Netherlands.

Baijukya, F.P., de Ridder, N., Masuki, K.F., Giller, K.E., 2005. Dynamics of banana based farming systems in Bukoba District, Tanzania: changes in land use, cropping and cattle keeping. Agric. Ecosyst. Environ. 106, 395–406.

Beaufils, E.R., 1971. Physiological diagnosis—a guide for improving maize production based on principles developed for rubber trees. Fertil. Soc. S. Afr. J. 1, 1–28.

Beaufils, E.R., 1973. Diagnosis and recommendation integrated system (DRIS) a general scheme for experimentation and calibration based on principle developed from research in plant nutrition. South Africa. Soil Sci. Bull. 1, 1–132.

Chan, K.S., 1986. The simple open soil method of measuring urea volatilization losses. Plant Soil 92, 73–79.

Delvaux, B., 1988. Constituents and Surface Properties of Soils Derived From Basalt Pyroclastic Materials in Western Cameroon-Genetic Approaches to Their Fertility. PhD thesis, University of Catholique de Louvain, Belgium.

Delvaux, B., 1989. Role of constituents of volcanic soils and their charge properties in the functioning of the banana agrosystem in Cameroon. Fruits 44 (6), 309–319.

Delvaux, B., Lassoudiere, A., Perrier, X., Marchal, J., 1987. A methodology for the study of soil-plant cultivation technique relations—results for banana-growing in Cameroon. In: Galindo, J.J. (Ed.), ACORBAT 85, pp. 351–357. Memorias VII Reunion.

Delvaux, B., Dufey, J.E., Vielvoye, L., Herbillion, A.J., 1989. Potassium exchange behaviour in a weathering sequence of volcanic ash soils. Soil Sci. Soc. Am. J. 53, 1679–1684.

FAO, 2017. Banana Statistical Compendium 2015–2016. Food and Agriculture organization, Rome, Italy.

Godefroy, J., Dormoy, M., 1990. The dynamics of mineral fertilizers in a ferrisol in Martinique used for banana growing. Fruits 45, 93–101.

Hallmark, W.B., Beverly, R.B., 1991. Review—an update in the use of the diagnosis and recommendation integrated system. J. Fertil. Issues 8, 74–88.

Hallmark, W.B., Walworth, J.L., Sumner, M.E., de Mooy, C.J., Pesek, J., Shao, K.P., 1987. Separating limiting from non-limiting nutrients. J. Plant Nutr. 10, 1381–1390.

Hanson, E.J., 1987. Fertilizer use efficiency: integrating soil tests and tissue analysis to manage the nutrition of highbush blueberries. J. Plant Nutr. 10, 1419–1427.

Irizarry, H., Rivera, E., Rodriguez, J., 1988. Nutrient uptake and dry matter composition in the plant crop and first ratoon of the grand Naine banana grown on an Ultisol. J. Agrie. Univ. P.R. 72, 337–351.

Israeli, Y., Lahav, E., Nameri, N., 1986. The effect of salinity and sodium absorption ratio in the irrigation water on growth and productivity of banana under drip irrigation conditions. Fruits 41, 297–301.

Janssen, B.H., Guiking, F.C.T., Van der Eijk, D., Smaling, E.M.A., Wolf, J., Van Reuler, H., 1990. A system for quantitative evaluation of the fertility of tropical soils (QUEFTS). Geoderma 46, 299–318.

Jeyabaskaran, K.J., Pandey, S.D., Laxman, R.H., 2000. Studies on reclamation of saline sodic soil for banana (cv. Nendran). In: Proceedings of the International Conference on Managing Natural Resources for Sustainable Agricultural Production in the 21st century, New Delhi, vol. 2, pp. 357–360.

Jones, J.B., 1998. Plant Nutrition Manual. CRC Press, Boca Raton, FL.

Kagezi, G.H., Kangire, A., Tushemereirwe, W., Bagamba, F., Kikulwe, E., Muhangi, J., Gold, C.S., Ragama, P., 2006. Banana bacterial wilt incidence in Uganda. Afr. Crop. Sci. J. 14 (2), 83–91.

Karamura, D.A., 1998. Numerical Taxonomic Studies of the East African Highland Bananas *Musa* AAA-East Africa in Uganda. PhD thesis, University of Reading, England.

Lahav, E., Turner, D.W., 1983. Fertilizing for High Yield—Banana. International Potash Institute, Berne, Switzerland, p. 62 Bulletin no. 7.

Lahav, E., Turner, D.W., 1989. Banana Nutrition, Bull. 12. International Potash Institute, Worblaufen-Bern, Switzerland.

Lopez, A., Espinosa, J., 2000. Manual on the Nutrition and Fertilization of Banana. National Banana Corporation and Potash and Phosphate Institute Office for Latin America, Limon, Costa Rica and Quito, Ecuador.

Marschner, H., 2003. Mineral Nutrition of Higher Plants, second ed. Academic Press, London/San Diego, CA.

Nyombi, K., 2010. Understanding Growth of East Africa Highland Bananas: Experiments and Simulation. PhD thesis, Wageningen University, Netherlands.

Nyombi, K., 2013. Towards sustainable highland banana production in Uganda: opportunities and challenges. Afr. J. Food Agric. Nutr. Dev. 13, 7545–7561.

Nyombi, K., van Asten, P.J.A., Corbeels, M., Taulya, G., Leffelaar, P.A., Giller, K.E., 2010. Mineral fertilizer response and nutrient use efficiencies of east African highland banana (Musa spp., AAA-EAHB, cv. Kisansa). Field Crop Res. 117, 38–50.

Parent, L.E., Dafir, M., 1992. A theoretical concept of compositional nutrient diagnosis. J. Am. Soc. Hortic. Sci. 117, 239–242.

Peet, M.M., 1992. Fruit cracking in tomato. Hortic. Technol. 2, 216–223.

Prasertsak, P., Freny, J.R., Saffigna, P.G., Denmead, O.T., Prove, B.G., 2001. Fate of urea nitrogen applied to a banana crop in wet tropics of Queensland. Nutr. Cycl. Agroecosyst. 59, 65–73.

Putra, E., Zakaria, W., Abdullah, N., Saleh, G., 2010. Weak neck of *Musa* sp. cv. Rastali: a review on it's genetic, crop nutrition and post-harvest. J. Agrotechnol. 9 (2), 45–51.

Raghupathi, H.B., Reddy, B.M.C., Srinivas, K., 2002. Multivariate diagnosis of nutrient imbalance in banana. Commun. Soil Sci. Plant Anal. 33, 2131–2143.

Risede, J.M., Chabrier, C., Dorel, M., Dambas, T., Achard, R., Queneherve, P., 2010. Integrated Management of Banana Nematodes: Lessons from a Case Study in the French West Indies. CIRAD, France.

Robinson, J.C., 1996. Bananas and Plantains. CAB International, England.

Rukazambuga, N.D.T.M., Gold, C.S., Gowen, S.R., 2002. The influence of crop management on banana weevil, *Cosmoplites sordidus* (Coleoptera: Curculionidae) populations and yield of highland banana (cv. Atwalira) in Uganda. Bull. Entomol. Res. 92, 413–421.

Smithson, P.C., McIntyre, B.D., Gold, C.S., Ssali, H., Kashaija, I.N., 2001. Nitrogen and potassium fertilizer vs. nematode and weevil effects on yield and foliar nutrient status of banana in Uganda. Nutr. Cycl. Agroecosyst. 59, 239–250.

Sseguya, H., Semana, A.R., Bekunda, M.A., 1999. Soil fertility management in the banana-based agriculture of Central Uganda: farmer's constraints and opinions. Afr. Crop. Sci. J. 7, 559–567.

Stover, R.H., Simmonds, N.W., 1987. Bananas, third ed. Longman Scientific and Technical, Harlow Essex.

Taulya, G., van Asten, P.J., Leffelaar, P.A., Giller, K.E., 2014. Phenological development of east African highland banana involves trade-offs between physiological age and chronological age. Eur. J. Agron. 60, 41–53.

Turner, D.W., 1987. Nutrient supply and water use of bananas in a subtropical environment. Fruits 42, 89–93.

Turner, D.W., Korawis, C., Robson, A.D., 1989. Soil analysis and its relationship with leaf analysis and banana yield with special reference to a study of Carnarvan, Western Australia. Fruits 44, 193–203.

Wairegi, L.W.I., van Asten, P.J.A., 2011. Norms for multivariate diagnosis of nutrient imbalance in east African highland cooking banana (Musa spp. AAAEA). J. Plant Nutr. 34 (10), 1453–1472.

Walworth, J.L., Sumner, M.E., 1987. The diagnosis and recommendations integrated system (DRIS). Adv. Soil Sci. 6, 149–188.

Wortmann, C.S., Bosch, C.H., Mukandala, L., 1994. Foliar nutrient analysis in bananas grown in the highlands of East Africa. J. Agron. Crop Sci. 172, 223–226.

45

Diagnosis and management of nutrient constraints in litchi

Lixian Yao*, Cuihua Bai, Donglin Luo

College of Natural Resources and Environment, South China Agricultural University, Guangzhou,
People's Republic of China
*Corresponding author. E-mail: lyaolx@scau.edu.cn

1 Introduction

Litchi (*Litchi chinensis* Sonn.), a tropical and subtropical woody perennial fruit tree with lifespans from decades to centuries, is cultivated in the region from 17 to 25°N in latitude and from 98 to 117°E in longitude (Qi et al., 2016). Commercial plantation of litchi is developed in Africa (e.g., South Africa, Mauritius, Madagascar, Gabon, and Congo), America (e.g., the United States, Brazil, Mexico, Panama, Cuba, Honduras, Puerto Rico, Trinidad, and Tobago), Asia (China, India, Thailand, Vietnam, Bengal, and Nepal), Europe (e.g., Spain, France, and Italy) (Farina et al., 2017; Huang, 2007), and Oceania (Australia) (Diczbalis, 2007). Among them, litchi cultivation area and fruit yield in China account for 69.2% and 62.9% of the total litchi cultivation area and total fruit yield worldwide, respectively (Qi et al., 2016). However, low and unstable fruit yield confuses litchi growers and researchers. Lack of suitable nutrient management has been recognized as one of the major constraints for litchi production. Understanding the nutrient requirement, diagnosis, and management in litchi is indispensable to achieve profitable litchi production.

2 Nutrient concentration in litchi plant

Foliar nutrient content reflects the nutritional status in a crop. Litchi grows perennially, and foliar nutrient concentration varies with the nutrient, the growth stage, and fruiting (Luo et al., 2019; Menzel et al., 1992; Yang et al., 2015). A recent study reports the tissue nutrient contents in litchi plants of 10 major litchi cultivars (cv. Feizixiao, Guiwei, Dadingxiang, Ziniangxi, Heiye, Lanzhu, Baitangying, Baila, Huaizhi, and Shuangjianyuhebao) with age of about 15 years from southern China when the fruits were harvested (Table 45.1), which yielded from 38.4 to 101.8 kg/tree (Yao et al., 2020). Generally, litchi leaves contain the highest N level and the lowest P value, whereas the trunks have the maximum Ca or N and the minimum S. N or K is detected with the upmost contents in the epicarp, endocarp, pulp,

TABLE 45.1 Nutrient concentrations in various parts of 10 major litchi cultivars at fruit harvest in South China (DW).

Part	Item	N (g/kg)	P (g/kg)	K (g/kg)	Ca (g/kg)	Mg (g/kg)	S (g/kg)	Si (g/kg)	Fe (mg/kg)	Mn (mg/kg)	Cu (mg/kg)	Zn (mg/kg)	B (mg/kg)	Mo (mg/kg)
Leaf	Range	15.5–19.0	0.8–1.8	5.7–36.6	5.9–14.2	1.3–4.1	0.8–2.2	1.4–7.0	100.0–222.1	44.7–482.7	4.4–10.4	18.9–43.8	11.5–24.9	ND[a]–2.98
	Mean	17.4 (1.4)	1.0 (0.3)	11.5 (9.2)	9.5 (2.7)	2.5 (0.9)	1.5 (0.4)	4.0 (2.1)	160.3 (33.7)	187.8 (154.9)	7.4 (2.3)	29.9 (9.4)	17.5 (3.9)	0.66 (1.14)
Trunk	Range	2.6–8.9	0.3–1.3	3.1–7.8	4.3–12.7	0.4–1.4	0.3–0.8	0.2–1.6	43.1–286.3	6.2–65.4	0.8–16.9	3.9–23.0	3.8–11.7	ND–0.50
	Mean	5.7 (1.9)	0.7 (0.3)	4.3 (1.6)	7.7 (3.0)	0.8 (0.3)	0.5 (0.1)	0.7 (0.5)	102.5 (70.8)	26.5 (18.1)	6.8 (6.9)	10.3 (5.6)	7.8 (2.7)	0.09 (0.18)
Root	Range	2.4–4.6	0.1–0.4	0.9–5.5	2.2–5.9	0.4–0.8	0.3–0.8	0.5–2.3	89.2–618.4	4.9–26.8	1.2–6.5	1.9–27.5	3.8–12.6	ND–1.06
	Mean	3.7 (0.6)	0.3 (0.1)	1.9 (1.4)	3.9 (1.2)	0.5 (0.2)	0.5 (0.2)	1.1 (0.7)	217.0 (150.0)	14.5 (7.8)	2.9 (1.5)	6.8 (7.4)	7.6 (3.0)	0.22 (0.39)
Epicarp	Range	8.4–16.3	0.6–1.3	5.5–14.0	4.1–10.3	1.1–2.8	0.5–1.0	0.1–1.9	19.1–70.8	15.8–152.5	3.2–89.2	13.0–29.7	9.7–25.5	ND–0.36
	Mean	11.5 (2.4)	0.9 (0.2)	9.3 (2.9)	6.2 (1.6)	2.0 (0.6)	0.7 (0.2)	0.8 (0.6)	41.3 (15.1)	70.6 (43.2)	19.3 (28.1)	20.2 (5.3)	15.9 (4.7)	0.11 (0.12)
Endocarp	Range	9.7–17.9	1.1–3.0	3.5–23.1	2.8–10.1	1.7–3.2	0.4–1.3	ND–2.8	26.2–86.9	14.8–176.0	3.2–28.3	16.7–34.0	17.4–26.7	ND–0.30
	Mean	13.0 (2.6)	1.9 (0.6)	13.9 (5.7)	5.5 (2.3)	2.2 (0.5)	0.9 (0.3)	1.1 (1.0)	49.3 (19.0)	74.5 (48.9)	11.2 (7.5)	23.4 (5.4)	22.1 (3.6)	0.09 (0.10)
Pulp	Range	7.9–13.1	1.2–1.7	6.6–15.6	0.0–3.1	0.7–1.1	0.4–1.0	ND–2.8	13.5–143.5	3.0–13.3	4.5–12.0	10.8–25.3	2.6–13.4	0.00–0.12
	Mean	10.7 (2.0)	1.5 (0.2)	10.7 (2.9)	0.6 (0.9)	0.8 (0.1)	0.7 (0.2)	1.0 (1.0)	39.5 (38.2)	7.6 (3.6)	9.0 (2.8)	15.7 (4.5)	7.3 (3.1)	0.03 (0.04)
Seed	Range	9.3–13.5	1.1–1.7	3.9–12.6	0.2–3.3	0.9–1.7	0.5–1.4	ND–1.9	22.5–60.0	9.9–31.4	5.2–14.1	13.7–35.7	5.8–39.2	0.00–0.28
	Mean	11.1 (1.2)	1.4 (0.2)	7.3 (2.7)	1.3 (1.0)	1.3 (0.3)	0.9 (0.3)	0.8 (0.8)	35.8 (14.1)	19.2 (7.8)	10.5 (2.9)	24.4 (7.1)	14.2 (9.4)	0.12 (0.09)

[a] ND refers to not detectable.
Data in the parenthesis are standard deviation.

and seed, but Ca or Si is present at the bottom level. Ca dominates in the roots, whereas P or S stays with the lowest value. Mo is commonly undetectable in various parts of litchi trees, which suggests that Mo might be an inhibiting nutrient for litchi production in China (Yao et al., 2020). Nutrient contents in litchi fruits (cv. Bengal and Tai So) are ranked in the order of K > N > P > Mg > Ca > Na > Fe > Zn > Cu > Mn > B in Australia (Menzel et al., 1988b). P, K, Ca, Mg, S, Fe, Mn, Zn, Cu, and B contents in the fruits of three litchi cultivars (cv. Bosworth, Groff, and Kaimana) from Hawaii are reported in terms of fruit nutrient for eating (Wall, 2006).

The annual variances of leaf nutrients in 15-year-old fruit bearing plants (cv. Feizixiao) are illustrated in Fig. 45.1 (Yang et al., 2015). Foliar N slowly decreases during the emergence and maturation of the postharvest vegetative flush (from June to October) and significantly rises to the peak before floral initiation (December) and then gradually declines during the development of panicle, flower, and fruit, and further falls to the bottom level at fruit harvest (from February to May) (Fig. 45.1A). Foliar K reaches the lowest value at fruit harvest and then sharply rises during

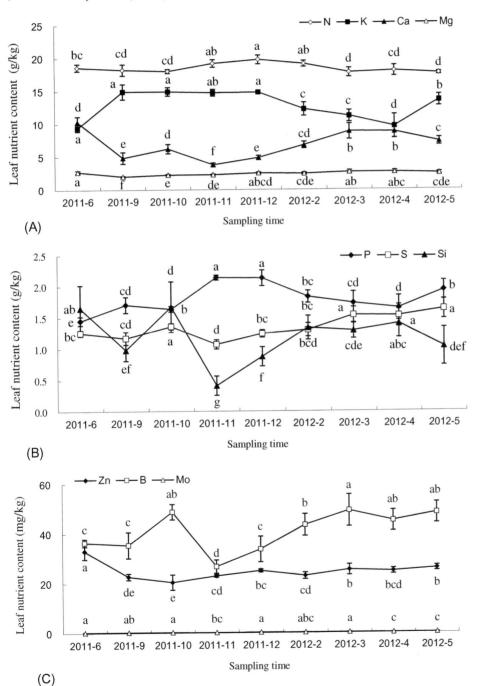

FIG. 45.1 Annual variations of foliar N, K, Ca, Mg (A), P, S, Si (B), Zn, B, and Mo (C) contents in litchi (cv. Feizixiao).

FIG. 45.2 Serious fruit crack in litchi (cv. Baitangying) trees over-dosed with potassium fertilizer at one application (upper), as compared with those with routine fertilization (down). *Photographed by Lixian Yao.*

the development of vegetative flush, maintains at a relatively constant value during vegetative flush maturation and floral initiation (from September to December), and continuously falls to the bottom level during panicle, flower, and fruit development (from February to April) and significantly rises again prior to fruit harvest. Contrary to foliar K, foliar Ca stays at the highest level in June and varies with an opposite trend to foliar K, indicating antagonism between leaf K and Ca during the growth period. Balanced application of K and Ca should be cautiously considered due to the interaction between them in practice. For example, in contrast to the litchi (cv. Baitangying) trees with routine fertilization, serious fruit crack was observed in trees overdosed with potassium chlorite at one application in Maoming Guangdong, southern China, in 2011 (Fig. 45.2), which is ascribed to suppressed fruit Ca caused by excessive K. Foliar Mg changes with a similar seasonal pattern to foliar Ca during the growth period.

Foliar P arrives at the bottom concentration after fruit harvest and gradually rises, then reduces during the emergence and maturation of the vegetative flush, significantly increases to the maximum value and stays at similar high level before floral initiation (from November to December), and then steadily declines during the emergence of panicle, fruit-set, and development (from December to April) and increases again prior to fruit harvest (Fig. 45.1B). Foliar S gradually varies with an alternative decrease and increase pattern during the emergence and development of vegetative flush and panicle (from June to February) and then continuously rises to the highest value and maintains constant till fruit harvest. Foliar Si significantly fluctuates during the emergence and development of vegetative flush and panicle, remains at a relatively constant level during fruit-set and development, and then sharply declines till fruit harvest.

Foliar Zn reaches to the upmost level after fruit harvest and then steadily declines to the lowest value during the emergence and maturation of vegetative flush and then alternatively and moderately rises and declines till fruit harvest (Fig. 45.1C). Foliar B remains at low concentration during the emergence and development of vegetative flush, however, significantly rises while the flush is completely mature in October and markedly declines to the minimum value before floral initiation, then continuously increases to the maximum during the emergence and development of panicle and fruit set (from November to March), and slightly decreases during fruit development and then increases again till fruit harvest. Foliar Mo remains at trace level as compared with all the other nutrients, however, significantly differs with growth stage as well.

3 Nutrient accumulation and distribution in litchi plant

Investigation on nutrient requirement in litchi is limited. Menzel and Simpson (1987) summarized the nutrient removal by a 100-kg plant (cv. Late Large Red, Bengal, and Tai so) was 90–250 g N, 35–50 g P, 240–320 g K, 20–60 g Ca, 2.0–2.5 g Cl, 1.0–1.4 g Na, 0.6–1.3 g Fe, 0.4–0.7 g Mn, 0.7–1.0 g Zn, 0.5–1.0 g Cu, and 0.3–0.7 g B, with unspecified tree age and yield level.

In southern China, the nutrient requirement by litchi plants (Table 45.2) of 10 major cultivars (cv. Feizixiao, Guiwei, Dadingxiang, Ziniangxi, Heiye, Lanzhu, Baitangying, Baila, Huaizhi and Shuangjianyuhebao) is reported (Yao et al., 2020). To nourish a litchi plants capable of yielding 50 kg fruit, 811.9 ± 298.3 g N, 86.4 ± 34.5 g P, 586.0 ± 259.2 g K, 792.5 ± 421.3 g Ca, 112.8 ± 54.7 g Mg, 66.0 ± 14.7 g S, 117.6 ± 76.0 g Si, 11.58 ± 7.41 g Fe, 4.79 ± 3.18 g Mn, 0.98 ± 0.99 g Cu, 1.44 ± 0.44 g Zn, 1.02 ± 0.42 g B, and 24.41 ± 37.50 mg Mo are needed in the aboveground parts of the plants, with the ratios of N:P:K:Ca:Mg:S being 1:0.11:0.72:0.98:0.14:0.08. And a total of 114.5 ± 12.8 g N, 14.4 ± 1.6 g P, 105.1 ± 31.5 g K, 21.6 ± 8.2 g Ca, 12.5 ± 1.6 g Mg, 7.7 ± 2.3 g S, 10.4 ± 10.1 g Si, 0.16 ± 0.13 g Cu, 0.27 ± 0.24 g Zn, 0.55 ± 0.50 g Fe, 0.33 ± 0.29 g Mn, 0.14 ± 0.08 g B, and 0.63 ± 0.58 mg Mo is removed by harvesting 50 kg fruit of the 10 major litchi cultivars, with the ratios of N:P:K:Ca:Mg:S being 1:0.13:0.92:0.19:0.11:0.07.

At fruit harvest, 42.1%–78.7% of the total nutrient requirement for all nutrients except Mo is allocated in the trunk, while 15.8 ± 5.3% of N, 18.9 ± 11.0% of P, 20.2 ± 7.7% of K, 3.4 ± 1.7% of Ca, 12.6 ± 6.3% of Mg, 11.1 ± 5.0% of S, 10.3 ± 12.5% of Si, 8.6 ± 8.8% of Cu, 8.2 ± 5.5% of Zn, 23.8 ± 21.5% of Fe, 20.3 ± 17.7% of Mn, 15.1 ± 11.3% of B, and 23.8 ± 24.5% of Mo are taken away by fruit harvest (Table 45.3) (Yao et al., 2020). The highly variable distribution of Mo in litchi plant supports the observation of Jongruaysup et al. (1994) who reported that Mo is readily remobilized in Mo-adequate plants, but not remobilized in Mo-deficient plants.

Feizixiao is the most widely planted cultivar in China. Approximately 259.5 ± 28.4 g of N, 28.3 ± 2.6 g of P, 186.5 ± 19.6 g of K, 41.6 ± 9.2 g of Ca, 36.1 ± 4.7 g of Mg, 12.4 ± 36.1 g of S, 316.8 ± 53.4 mg of Zn, 201.1 ± 29.0 mg of B, and 1.4 ± 0.3 mg of Mo are needed to produce 55.2 ± 7.8 kg of fruit, 39.78 ± 2.60 kg of matured vegetative flush, and 8.44 ± 1.96 kg of panicle for 10-year-old Feizixiao plants in 1 year, with the nutrient ratios of N:P:K:Ca:Mg being 1:0.11:0.72:0.16:0.14 for the vegetative flush, 1:0.12:0.75:0.22:0.12 for the panicle, and 1:0.13:1.06:0.16:0.12 for the fruit (Yao et al., 2017b). Moreover, all of the N, P, K, Mg, S, Mo, 67.5% of Zn, and 20.2% of B accumulated in litchi panicle are transferred from the terminal vegetative flush (the last autumn flush). Litchi plant hardly absorbs nutrients after the maturation of the terminal shoot and before flowering, with the exception of Ca, Zn, and B absorption during this stage. Almost all of the N, K, Ca, Zn, and S required by fruit development are newly taken up by the plant during fruit swelling, and partial P, Mg, B, and Mo are translocated from the first and the second vegetative flushes (Yao et al., 2017b).

TABLE 45.2 Nutrient accumulated in the aboveground parts of 10 major litchi cultivars to yield 50 kg fruit in South China.

Part	Item	N (g)	P (g)	K (g)	Ca (g)	Mg (g)	S (g)	Si (g)	Fe (g)	Mn (g)	Cu (g)	Zn (g)	B (g)	Mo (mg)
Fruit	Range	95.4–135.1	12.4–17.2	68.4–169.8	12.7–42.0	9.3–14.1	4.5–11.3	0.5–28.0	0.23–1.76	0.08–1.10	0.05–0.48	0.13–0.94	0.07–0.37	0.0–1.61
	Mean	114.5 (12.8)	14.4 (1.6)	105.1 (31.5)	21.6 (8.2)	12.5 (1.6)	7.7 (2.3)	10.4 (10.1)	0.55 (0.50)	0.33 (0.29)	0.16 (0.13)	0.27 (0.24)	0.14 (0.08)	0.63 (0.58)
Leaf	Range	124.9–428.3	6.1–41.5	48.3–829.8	54.3–322.3	10.1–79.1	9.9–22.2	11.2–158.6	0.84–5.04	0.60–5.76	0.06–1.32	0.21–0.89	0.12–0.38	0.0–35.55
	Mean	237.2 (106.9)	14.6 (10.0)	176.2 (232.2)	135.1 (86.5)	35.7 (23.0)	18.7 (3.7)	61.4 (51.6)	2.33 (1.30)	2.22 (1.75)	0.34 (0.51)	0.40 (0.22)	0.23 (0.08)	8.36 (13.39)
Trunk	Range	126.6–825.1	14.6–89.1	117.0–619.3	170.0–1124.4	24.3–147.9	22.1–57.5	4.9–121.6	1.41–29.25	0.47–5.21	0.10–1.76	0.19–1.58	0.295–0.15	0.0–60.34
	Mean	460.2 (225.0)	57.4 (27.4)	304.7 (146.2)	635.8 (354.8)	64.6 (37.1)	39.6 (11.4)	45.8 (32.7)	8.70 (7.75)	2.24 (1.84)	0.48 (0.54)	0.77 (0.37)	0.65 (0.41)	15.42 (25.14)
Total	Range	403.0–1156.4	38.5–145.9	233.7–1140.7	257.2–1318.2	48.1–218.8	37.2–82.0	36.1–281.4	3.08–31.04	1.24–10.51	0.26–3.37	0.74–2.27	0.54–1.85	0.0–81.96
	Mean	811.9 (298.3)	86.4 (34.5)	586.0 (259.2)	792.5 (421.3)	112.8 (54.7)	66.0 (14.7)	117.6 (76.0)	11.58 (7.41)	4.79 (3.18)	0.98 (0.99)	1.44 (0.42)	1.02 (0.42)	24.41 (37.50)

Data in the parenthesis are standard deviation.

TABLE 45.3 Nutrient distribution (%) in various parts of 10 major litchi cultivars in South China.

Part	Item	N	P	K	Ca	Mg	S	Si	Fe	Mn	Cu	Zn	B	Mo
Fruit	Range	10.1–23.9	1.3–39.7	8.1–34.1	1.4–5.8	1.1–20.8	0.4–17.0	0.6–37.4	6.4–75.7	9.1–69.9	0.9–26.7	3.2–20.8	5.5–44.1	0–64.5
	Mean	15.8 (5.3)	18.9 (11.0)	20.2 (7.7)	3.4 (1.7)	12.6 (6.3)	11.1 (5.0)	10.3 (12.5)	23.8 (21.5)	20.3 (17.7)	8.6 (8.8)	8.2 (5.5)	15.1 (11.3)	23.8 (24.5)
Leaf	Range	18.5–44.7	10.7–28.4	14.4–72.7	8.9–28.1	17.1–51.4	22.4–36.1	22.0–94.4	6.2–66.5	16.043.4	4.8–46.7	13.9–63.4	11.3–36.8	0–72.6
	Mean	30.0 (9.0)	16.9 (6.3)	25.6 (17.4)	18.0 (6.9)	30.4 (10.3)	28.7 (4.0)	48.2 (21.3)	27.3 (16.8)	27.1 (8.8)	25.3 (14.8)	47.1 (16.0)	24.2 (8.5)	45.0 (24.2)
Trunk	Range	31.4–71.3	33.8–74.1	19.1–72.0	66.1–88.3	40.6–67.7	51.9–70.1	5.1–75.6	15.4–85.8	14.1–71.1	45.4–94.2	15.9–81.8	35.0–83.2	0–80.1
	Mean	54.1 (11.7)	63.4 (13.2)	54.3 (15.8)	78.7 (7.4)	56.3 (10.1)	59.4 (5.6)	42.1 (20.6)	48.0 (19.4)	53.5 (16.5)	68.5 (17.5)	45.0 (17.9)	60.5 (15.4)	21.0 (31.6)

Data in the parenthesis are standard deviation.

As a heavily trimmed cultivar with high growth speed, the nutrient removal by harvesting 55.2 ± 7.8 kg fruit and trimming 39.78 ± 2.60 kg vegetative flush from Feizixiao plants is summed up to 320.6 ± 49.8 g N, 33.7 ± 5.3 g P, 273.2 ± 39.3 g K, 143.5 ± 28.6 g Ca, 38.5 ± 8.4 g Mg, 16.1 ± 2.5 g S, 544.2 ± 91.3 mg Zn, 357.9 ± 45.9 mg B, and 1.2 ± 0.3 mg Mo, which is the minimum nutrient requirement for the growth and development of vegetative flush, panicle and fruit, and the maintenance of soil fertility during the next growth period (Yao et al., 2017b). This implies that fertilization soon after fruit harvest is the key for emergence and development of healthy vegetative flush in heavily trimmed litchi cultivars like Feizixiao.

4 Nutrient imbalance in litchi

Foliar nutrient deficient symptoms in litchi are not easy to be clearly discriminated in field condition worldwide because the symptoms associated with insects, pathogens, pesticides, herbicides, growth regulators, climate, and so on might be mixed with those caused by nutrient deficiency. The functions of nutrients in litchi were reviewed by Menzel and Simpson (1987), and the deficiency symptoms of some nutrients including N, P, K, Fe, Cu, Zn, and B were described or illustrated in South Africa (de Villiers and Joubert, 2010). Visual symptoms of N, K, Mg, Fe, and Zn in litchi (cv. Brewster and Mauritius) grown in gravel culture and irrigated with modified Hoagland solution or nutrient solutions minus N, K, Mg, Mn, Zn, or Fe were recorded in Florida of the United States (Thomas et al., 1995). Digital pictures of typical foliar nutrient deficiency symptoms in litchi (cv. Huaizhi) nourished with optimum nutrient (complete nutrient) solution or nutrient-omitted solutions are recently recorded in China (Figs. 45.3–45.13).

FIG. 45.3 Foliar N deficiency symptom in litchi. *Photographed by Yongyuan Zhang.*

FIG. 45.4 Foliar P deficiency symptom in litchi. *Photographed by Yongyuan Zhang.*

FIG. 45.5 Foliar K deficiency symptom in litchi. *Photographed by Yongyuan Zhang.*

FIG. 45.6 Foliar Mg deficiency symptom in litchi. *Photographed by Yongyuan Zhang.*

FIG. 45.7 Foliar S deficiency symptom in litchi. *Photographed by Yongyuan Zhang.*

FIG. 45.8 Foliar Fe deficiency in litchi tree cultivated in calcareous soil in Luzhou Sichuan, southwest China. *Photographed by Lixian Yao.*

FIG. 45.9 Foliar Fe deficiency symptom in litchi. *Photographed by Minqiu Sun.*

FIG. 45.10 Foliar Mn deficiency symptom in litchi. *Photographed by Minqiu Sun.*

FIG. 45.11 Foliar Cu deficiency symptom in litchi. *Photographed by Minqiu Sun.*

FIG. 45.12 Foliar Zn deficiency symptom in litchi. *Photographed by Minqiu Sun.*

FIG. 45.13 Foliar B deficiency symptom in litchi. *Photographed by Minqiu Sun.*

N deficient litchi plant is severely stunted, and obvious symptoms first occur in the older leaves, characterized by even chlorosis and following marginal scorch. While N starvation is prolonged, the younger leaves fade green (gray in print version) and wither in the leaf apices as well (Fig. 45.3).

The typical P deficient symptoms include smaller, narrower, and longer young leaves as compared with the old leaves, and all the leaves grow dark green (black in print version) as P starvation prolongs (Fig. 45.4).

While a litchi plant is deficient in K, small rufous patches first appear in the apices of old leaves and spread to the margins and base and then merge into obvious patches, and the residue areas change to yellow (light gray in print version) from healthy green (dark gray in print version). The rufous patches also occur in the young leaves, and the whole leaf fades green (gray in print version) while in severe K deficiency (Fig. 45.5).

Although symptoms of Ca deficiency in litchi were documented as small leaflets, leaf necrosis along margins, and severe leaf drop by Goldweber (1959), litchi seedlings grow healthy and do not show visual symptoms even after being subjected to Ca-omitted nutrient solutions for 4 months. Leaf drop was recorded for all litchi plants separately grown in gravel and nourished with complete nutrient solution, minus N, P, K, Ca, and Mg solutions over 6 months (Goldweber, 1959), which is likely contributed to the secondary salinity stress caused by nutrient solutions. Ca deficiency is seldom reported and rare in field condition worldwide.

Large irregular necrotic patches first occur in the older leaves of Mg deficient litchi plants, with the remaining areas of the older leaves staying green. The patches progress along leaf apices and margins toward the base, followed by a burned appearance (Fig. 45.6).

Symptoms of S deficiency include serious stunting growth, considerably smaller young leaflets with light green (light gray in print version) to yellow (gray in print version) color, and older leaves with green (dark gray in print version) color (Fig. 45.7).

Fe is the micronutrient with the maximum requirement by litchi as shown in Table 45.2. However, symptoms of Fe deficiency are not common in acid soils of the main production region worldwide, with the exception in calcareous soils or heavily limed soils. Fe-deficient litchi tree is characterized by obvious interveinal chlorosis of young leaves, poor flowering and fruit-set and extremely low fruit yield in field condition, as found in calcareous soil with pH of 7.5 in Luzhou Sichuan, southwest China (Fig. 45.8). Litchi seedlings grown in minus Fe nutrient solution are significantly stunted, and serious interveinal chlorosis is also observed in the young leaflets. Some seedlings are subjected to dieback, whereas some of them sprout again, but the young leaflets appear pale, almost white in color (Fig. 45.9).

Mn deficient symptoms begin as interveinal chlorosis of young leaves, followed by randomized occurred tiny rufous patches on the leaves. Leaf size is not affected by Mn deficiency, and the older leaves stay green (dark gray in print version) (Fig. 45.10).

Symptoms of Cu deficiency are similar to those of Mn deficiency, with the exception figure of small young leaflets and bright orange-red (gray in print version) patches between the midribs and the margins (Fig. 45.11).

Zn deficient litchi plants are characterized by long and narrow young leaves with curled margins and young leaflets emerging from the axillae of the older leaves, with a caespitose appearance. Brown (gray in print version) patches occur in apices of the young leaflets as the starvation continues, while the old leaves stay healthy (Fig. 45.12).

Symptoms of B deficiency start as interveinal chlorosis in the young leaflets, followed by chlorosis of the veins, and brown (gray in print version) stains along the margins as the starvation prolongs (Fig. 45.13).

Although foliar Ca deficiency symptoms are not found, fruit abnormality caused by Ca or Ca and B deficiencies associated with bad weather including continuous drought, wet, coldness, or torridity is recently concerned (Yao et al., 2017a). Litchi fruit is significantly stunted during the fruitlet stage and grows abnormally at harvest, and more seriously, the endocarp, superior to the epicarp, turns brown (gray in print version), or the pericarp is cracked and the seed becomes necrosis after being subjected to continuous wet and coldness during fruitlet stage (Fig. 45.14). And, fruit matures with small and abnormal size or cannot turn red (light gray in print version) prior to harvest under xerothermic climate.

Comparably, investigation on nutrient toxicity in litchi is scarce up to date. Even fruit yield is significantly suppressed by overuse of boron, no visual symptoms are found in litchi plants in the field condition (Yang et al., 2016).

5 Nutrient diagnosis in litchi

Leaf analysis or foliar nutrient diagnosis has been recognized as an important tool to improve nutrient management in crop. As a perennial woody fruit tree, leaf age, phenological stage, fruit status, and yield level, vegetative flush control, promotion of floral initiation, fertilization, pest and disease control, and so on greatly affect the relationship between fruit yield and leaf nutrient concentration (Luo et al., 2019; Menzel et al., 1992; Roy et al., 1984). Therefore, a universal leaf nutrient standard might not be suitable for litchi production worldwide. Leaf nutrient standards are proposed in several countries/regions (Table 45.4), and Dai (1999) summarized the leaf nutrient norms for various litchi cultivars in China (Table 45.5). It is difficult to compare the practicability of various leaf nutrient norms due to discrepancies of soil type, cultivar, fruit level, diagnosis time, and even the approaches used to build the norms. For example, after comparing the reliability and diagnosis accuracy of four approaches including critical value approach, sufficiency range approach (also named as standard value approach), modified diagnosis and recommendation integrated system (M-DRIS), and compositional nutrient diagnosis, sufficiency range approach is selected to build the foliar nutrient norms for litchi (cv. Feizixiao) capable of yielding approximately 20 t/ha in southern China, based on leaf nutrient concentrations and fruit yields of 538 samples gathered from 22 orchards within 2 years (Luo et al., 2019). However, it is deemed that variable leaf nutrient standards for different growth stages are necessary because there are significant differences for leaf nutrient concentrations among phenological stages (Luo et al., 2019; Menzel et al., 1992; Yang et al., 2015).

FIG. 45.14 Fruit abnormality (cv. Caomeili and Xianjinfeng) caused by Ca deficiency due to bad weather in southern China. *Photographed by Yaliang Gu and Lixian Yao.*

Besides leaf analysis, soil analysis is deemed as a useful aid for proper fertilization in litchi as well. Menzel and Simpson (1987) suggested soil attribute standards from trees, nut, vine, and fruit crops for the evaluation of litchi soils in Australia, and litchi soil norms are proposed in Fujian, China, and South Africa as well (Table 45.6) (de Villiers and Joubert, 2010; Zhuang et al., 1994). The variable soil attribute norms for litchi indicate that soil properties differ greatly with regions and might considerably affect nutrient management in litchi grown in different soil types.

However, soil nutrient status might not act as reliable indicators for leaf or tissue nutrients in litchi plants. No close relation is observed between tissue elements and soil nutrients in 10 major litchi cultivars from southern China, with the

TABLE 45.4 Litchi foliar diagnosis norms from different countries/regions.

Location	Cultivar	Diagnosis stage (sampling time)	N (g/kg)	P (g/kg)	K (g/kg)	Ca (g/kg)	Mg (g/kg)	S (g/kg)	Fe (mg/kg)	Mn (mg/kg)	Cu (mg/kg)	Zn (mg/kg)	B (mg/kg)	Reference
Agrilink			15–18	1.4–2.2	7–11	6–10	3–5		50–100	100–250	10–25	15–30	40–60	Diczbalis (2007)
Australia	Tai So, Haak Yip, Wai Chee	1–2 weeks after panicle emergence	15.0–18.0	1.4–2.2	7.0–11.0	6.0–10.0	3.0–5.0	–	50–100	100–250	10–25	15–30	25–60	Menzel et al. (1992)
China Mainland	Feizixiao	Fruit bulking (4–5 weeks before harvest)	16.7–19.2	1.06–1.25	5.1–6.7	7.7–11.0	2.5–3.7	1.51–1.81	–	–	–	19.6–32.6	11.5–19.2	Luo et al. (2019)
	Feizixiao	Terminal shoot maturation	19.7–22.0	1.69–1.95	10.8–12.7	3.0–4.1	2.5–2.9	1.38–1.57	–	–	–	15.0–18.9	10.8–16.8	Luo et al. (2019)
China Taiwan	Haak Yip		16–19	1.2–2.7	3.9–11.5	6–10	3–5		50–100	100–250	10–25	15–30	25–60	Diczbalis (2007)
India														Diczbalis (2007)
Low yield population			13.9	1.8	7.8									
High yield population			15.1	2.0	7.7									
South Africa	Not specified	Mid-September to mid-November (6–8 month old leaves)	14.6–16.2	1.5–2.0	9.0–10.6	8.0–25.0	2.0–7.0	–	50–200	50–200	10–15	15–40	20–75	de Villiers and Joubert (2010)
Thailand			19.5–20.3	1.8–1.9	10.0	2.9–3.5	2.3–2.5		29–30	49–62	12–14	15–16	14	Diczbalis (2007)

TABLE 45.5 Foliar nutrient standards for litchi in China (Dai, 1999).

Cultivar	N	P	K	Ca	Mg
Lanzhu	1.5–2.2	0.12–0.18	0.7–1.4	0.3–0.8	0.18–0.38
Dazao	1.5–2.0	0.11–0.16	0.7–1.2	0.3–0.5	0.12–0.25
Heli	1.6–2.3	0.12–0.18	0.8–1.4	0.5–1.35	0.2–0.4
Nuomici	1.5–1.8	0.13–0.18	0.7–1.2	–	–
Huaizhi	1.4–1.6	0.11–0.15	0.6–1.0	–	–
Chenzi	1.4–1.8	0.12–0.17	0.8–1.2	–	–

TABLE 45.6 Soil attribute norms for litchi.

Soil attribute	Norms for trees, nut, vine, and fruit crops (Menzel and Simpson, 1987)			Norms for litchi (de Villiers and Joubert, 2010)	Norms for litchi (Zhuang et al., 1994)
	Low	No action required	High		
pH	<5.0	5.0–5.5	>5.5	5.7–6.8	–
	Low	Medium	High	–	–
Organic C (%)	<1.0	1–3	>3.0	–	–
NO₃-N (mg/kg)	<20	20–40	>40	–	–
P (mg/kg)	<20	20–60	>60	5–10 (resin), 15–25 (bray 1)	–
K (mg/kg)	<78	78–195	>195	60–80 (sandy soil), 80–200 (clay soil)	–
Ca (mg/kg)	<1200	1200–2000	>2000	–	–
Mg (mg/kg)	<192	192–384	>384	–	–
Ca/Mg				2.5–5.0	–
	No action required			–	–
Cl (mg/kg)	<250			–	–
Na (mg/kg)	<390			–	–
Electrical conductivity (ms/cm)	<0.4			–	–
Cu (mg/kg)	0.3–10.0			–	1.0–5.0
Zn (mg/kg)	2–15			–	1.5–5.0
Mn (mg/kg)	2–50			–	1.5–5.0
Fe (mg/kg)	2–50			–	
B (mg/kg)	1.0–5.0			–	0.4–1.0
Al (mg/kg)	<540			0–30	
Mo (mg/kg)	–			–	0.15–0.32

exception that foliar K, Ca, and Mg significantly correlate with soil available K and exchangeable Ca and Mg (Table 45.7) (Yao et al., 2020). Moreover, no close relation is computed between fruit yield and soil attributes including soil pH, organic matter, alkaline hydrolyzable N, Oslen-P, and available K, Ca, Mg, S, Fe, Mn, Cu, Zn, B, and Mo (Yao et al., 2020). Therefore, the utilization of soil nutrient standards might be limited due to the lack of linkage between soil properties versus tissue nutrients and fruit yield in litchi, as compared with the foliar nutrient norms in litchi.

TABLE 45.7   Pearson coefficients between soil nutrients and tissue nutrients in various parts of 10 major litchi cultivars in southern China (Yao et al., 2020).

Soil nutrient	Leaf	Trunk	Root	Epicarp	Endocarp	Pulp	Seed
N[a]	0.525	0.394	0.428	−0.035	−0.093	0.114	−0.073
P[b]	0.143	0.580	0.491	0.162	0.450	−0.634*	−0.602*
K	0.636*	0.071	−0.162	0.086	0.308	−0.106	−0.046
Ca	0.627*	0.733*	0.380	−0.206	−0.224	−0.329	−0.246
Mg	0.669*	0.566	0.562	0.049	−0.259	−0.288	0.351
S	0.137	−0.452	−0.245	−0.077	−0.142	−0.032	−0.011
Si	0.416	0.129	−0.328	0.067	−0.003	−0.033	0.223
Fe	−0.111	−0.250	−0.368	0.605*	0.800**	0.126	0.732*
Mn	−0.480	0.391	0.535	−0.390	−0.327	−0.345	−0.328
Cu	0.242	0.107	0.689*	0.064	−0.017	−0.634*	−0.505
Zn	−0.180	0.269	0.044	−0.366	−0.173	0.540	0.213
B	−0.131	0.040	0.249	0.173	0.261	−0.024	0.204
Mo	−0.274	−0.281	−0.151	−0.294	0.025	−0.305	0.399

* and ** denote significant at 0.05 level.
[a] *Soil N refers to alkaline hydrolyzable N.*
[b] *Soil P refers to Olsen-P.*

6 Nutrient management in litchi

Systematic nutrient management program in litchi is inadequately studied up to date. Fertilization practices to obtain greater tree size and bearing canopy for young tree in several countries were summarized in Australia, India, South Africa, and the United State (Table 45.8) (Menzel and Simpson, 1987; de Villiers and Joubert, 2010). The total amount of nutrients and timing differ greatly in different regions or even in different orchards in the same region. A tentative nutrition program, illustrating the fertilizer choices based on phenological stage of litchi, is also suggested in Australia (Menzel and Simpson, 1987).

In China, the total amount of nutrient input per year in litchi is commonly determined by target fruit yield, rather than by tree age. It maybe be linked with the fact that besides the old trees have been cultivated for decades or even for hundreds of years, most of the commercial litchi trees were planted in the late 1980s and the early 1990s in China and now are at fruit bearing stage (Yao, 2009). Although the nutrient takeaway by litchi fruit and flush is documented, the suitable fertilizer doses cannot be precisely calculated solely by nutrient removal because use efficiencies of fertilizers are not available in litchi. However, the reasonable fertilizer rates might be estimated by field experiments involving in fruit yield response to fertilizers.

Chen et al. (1998) suggested that 0.84 kg N, 0.5 kg P_2O_5, and 1.2 kg K_2O (N:P_2O_5:K_2O = 1:0.6:1.43) are applied for litchi canopy capable of yielding 50-kg fruit (cv. Guiwei) by a 5-year field trial. Dai et al. (1998) proposed 0.8 kg N, 1.0 kg P_2O_5, and 1.5 kg K_2O (N:P_2O_5:K_2O = 1:1.25:1.875) for producing 43.1 kg fruits (cv. Lanzhu) from a 5-year experiment. 0.25–0.5 kg N is recommended to generate 23.4–30.1 kg fruit (cv. Chenzi) through a 6-year experiment (Liang and Dai, 1984), with the ratios of N:P_2O_5:K_2O being 1:2:6 or 1:1:3. Excessive N promotes the development of vegetative flushing and leads to poor flowering and fruit setting in litchi (Liang and Dai, 1984; Menzel et al., 1988a), whereas overdose of K postpones fruit development and harvest and increases titratable acid content in fruit, leading to lower ratio of soluble sugar over acid in the pulp (Su et al., 2015; Yang et al., 2015).

As compared with the N:P:K ratio of nutrient accumulation and removal in litchi (Yao et al., 2017b, 2020), both P and K might be overused in the previous studies (Chen et al., 1998; Dai et al., 1998; Liang and Dai, 1984). He et al. (2003) examined the effect of application ratio of K over N fertilizer (K_2O/N 0.8, 1.0, 1.2, 1.4, and 1.6, with the ratio of P_2O_5/N = 0.6) on litchi fruit (cv. Sanyuehong) yield in three orchards in Guangxi, southern China, over 3 years and suggested that when K and N fertilizers are applied at the ratio of K_2O/N = 1.2, litchi yield the maximum. After investigating the response of fruit yield, quality and harvest time to K over N fertilization ratio (K_2O/N 0.6, 0.8, 1.0, 1.2, and 1.4) in litchi

TABLE 45.8 Fertilizer practices for litchi trees in Australia, India, South Africa, and the United States.

Country/region	Tree age	Amount of nutrients (g/tree/year)			Timing of fertilization (northern hemisphere equivalent)	Reference
		N	P	K		
Australia	1	60	12	100	July, November, March	Menzel and Simpson (1987)
	2	90	18	150		
	3	150	30	220		
	4	180	60	300		
Australia	1	190	10	25	N monthly, P, K trimonthly, organic in March in Years 2 and 3	Menzel and Simpson (1987)
	2	350	40	100		
	3	530	70	150		
	4	390	10	50		
India	1–3	175–350	20–60	105–235	December, February, April	Menzel and Simpson (1987)
	4–6	650–1200	75–125	300–470		
South Africa	1	28	13	25		de Villiers and Joubert (2010)
	2	84	21	50		
	3	196	26	75		
	4–5	280	26	100		
	6–7	420	53	150		
	8–9	560	53	200		
	10–11	700	79	250		
	12–13	840	79	375		
	14–15	980	105	500		
	15 and older	1120	105	500		
Florida, United States	1	10	15	15	Monthly February, May, August	Menzel and Simpson (1987)
	2	14	20	20		
	3	28	40	40		
	4	56	80	80		
	5	112	160	160		
Hawaii, United States	1	34	34	34	Quarter-monthly	Menzel and Simpson (1987)
	2	68	66	68		
	3	136	136	136		
	4	272–410	272–410	272–410		

(cv. Feizixiao) within 4 years, Yang et al. (2015) recommended that use ratio of $K_2O/N = 1.0–1.2$ is suitable for typical red soils with low N and K and medium to abundant P in southern China, based on the supplement of proper P ($P_2O_5/N = 0.3$), Ca, Mg, B, and Mo nutrients.

Further, fertilization programs in China are commonly recommended in terms of phenological stage (Table 45.9). The total amount of N, P, K, Ca, and Mg is divided into three to four splits after fruit harvest, before flowering, and/or after flower abscission and fruit bulking for fruit-bearing trees. Forty-five percent to 65% of N, 30%–45% P, 22.2%–30% of K, and 30%–40% Ca of the total nutrient dosage are used after fruit harvest; 15%–25% of N, 25%–35% of P, 35%–47.8% of K, and 40% of Ca and Mg are applied during fruit bulking stage, and the residue is added before flowering and/or after flower abscission.

TABLE 45.9 Fertilization programs for litchi in China.

Cultivar	Total amount of nutrients (g/tree)					Timing	Reference
	N	P_2O_5	K_2O	Ca	Mg		
Guiwei (14 a, 62.7 kg/tree), Nuomici (16 a, 62.7 kg/tree)	709	284	887		71	After fruit harvest: N 55%, P 30%, K 30% Before flowering: N 20%, P 35%, K 35% Fruit bulking: N 25%, P 35%, K 35%	Zhou et al. (2001)
Guiwei (21 a, 50 kg/tree)	840	500	1200			After fruit harvest: N 65%, P 33.3%, K 25% Before flowering: N 20%, P 33.3%, K 40% Fruit bulking: N 15%, P 33.3%, K 35%	Chen et al. (1998)
Feizixiao (5 a, 18.1 kg/tree)	450	130	540	80	40	After fruit harvest: N 45%, P 45%, K 22.2%, Ca 30%, Mg 30% Before flowering: N 10%, P 10%, K 7.4%, Ca 10%, Mg 10% After flower abscission: N 10%, P 10%, K 7.4%, Ca 10%, Mg 10% Fruit bulking: N 25%, P 25%, K 42.6%, Ca 40%, Mg 40%	Li et al. (2011)
Guiwei (10 a, 5.2 kg/tree)	350	120	300	150	60	After fruit harvest: N 45%, P 45%, K 26.1%, Ca 30%, Mg 30% Before flowering: N 10%, P 10%, K 8.7%, Ca 10%, Mg 10% After flower abscission: N 10%, P 10%, K 8.7%, Ca 10%, Mg 10% Fruit bulking: N 25%, P 25%, K 47.8%, Ca 40%, Mg 40%	

Besides land application, spraying macro- and secondary nutrient fertilizers is adopted to improve fruit set and quality as well. Foliar urea application at the rate of 120 g/plant after blossom in 6-year Bengal litchi significantly boosts fruit set and size in Brazil (Goncalves et al., 2016). Foliar spray of mixture of Ca and Mg (0.3% $CaCl_2$ + 1.5% $MgCl_2$, w/w) improves fruit pigmentation and maturation in Sanyuehong and Feizixiao litchi in Hainan, China (Gao et al., 2015; Zhou et al., 2015), which might overcome the retarded fruit maturation linked with K overuse due to antagonism between K and Ca and Mg. Meanwhile, Mg application enhances anthocyanin synthesis in the pericarp via promoting the ratio of abscisic acid/gibberellin and then stimulating the activity of flavonoid transferase (Zhou et al., 2016). Foliar application of 3% $CaCl_2$ + 1.5% borax ($Na_2B_4O_7 \cdot 10H_2O$) not only increases single fruit weight but also enhances total sugars, reducing sugars and nonreducing sugars in fruits (cv. Gola) in Pakistan (Haq et al., 2013). However, it is demonstrated that 1% urea spray in early May, 4% KNO_3 spray in mid-May, 2% KNO_3 during flowering, 2% $Ca(NO_3)_2$ foliar spray at fruit set, and 2% KNO_3 spray at fruitlet stage cannot increase fruit (cv. HLH Mauritius) yield in South Africa (Cronje and Mostert, 2009).

Additionally, foliar spray of Ca or Ca/B is beneficial to avoid fruit abnormality caused by Ca and Ca/B deficiencies due to bad weather during fruit development. Spraying 0.1 mmol/L $Ca(NO_3)_2$ or 1 mmol/L gibberellin three times at flowering, fruitlet, and fruit bulking stages, effectively alleviates fruit crack in litchi (cv. Nuomici) (Peng et al., 2001). Fruit crack is reduced by spraying chelated calcium solution containing 180 mmol/L Ca in Nuomici litchi as well (Li et al., 1999). Spraying 0.2% or 0.5% $CaCl_2$ after full blossom does not effectively enhance structural Ca in litchi pericarp, however, generally reduces fruit cracking rate in Nuomici litchi (Huang et al., 2008).

Foliar spray of micronutrients, a common practice to prevent or alleviate micronutrient deficiencies in litchi production, is summarized in Table 45.10. Micronutrients are suggested as the concentrations of either elements or compounds in various countries. Spraying concentrations of Zn, B, Cu, Mn, and Fe are proposed in Australia (Menzel and Simpson, 1987), and those of Zn, B, Cu, and Mn are recommended in South Africa (de Villiers and Joubert, 2010). Boron deficiency is a worldwide nutrient constraint in crop production, and the suitable B concentrations in litchi differ with countries. For example, 0.11–0.45 g B/L and 1 g solubor/L are recommended in Australia and South Africa, respectively; however, it is reported that spraying solubor ($Na_2B_4O_7 \cdot 10H_2O$)

TABLE 45.10 Supplement of micronutrients in litchi.

Country	Nutrient	Concentration	Reference
Australia	Zn	0.45–2.70 g Zn/L as $ZnSO_4 \cdot 7H_2O$ or ZnO	Menzel and Simpson (1987)
	B	0.11–0.45 g/B/L as H_3BO_3 or $Na_2B_4O_7$	
	Cu	0.51–3.05 g Cu/L as $CuSO_4 \cdot 4H_2O$	
	Mn	0.59 g Mn/L as $MnSO_4 \cdot 5H_2O$	
	Fe	0.48 g Fe/L as $FeSO_4 \cdot 7H_2O$	
South Africa	Zn	2 g ZnO/L or 1.5 mL nitrozinc/L	de Villiers and Joubert (2010)
	B	1 g solubor/L	
	Cu	2 g copper oxychloride/L	
	Mn	2 g $MnSO_4$/L	
China	B	0.5 g $Na_2B_4O_7 \cdot 10H_2O$/L	Yang et al. (2016)

at the rate of 0.5 g/L (equalling to 0.056 g B/L) prior to flowering, after flower drop and fruitlet stage, improves - fruit set and increases the yield by 98.8% as compared with the control, whereas spraying solubor at 1 and 2 g/L significantly reduces fruit yield by 81.4% and 14.7% in southern China (Yang et al., 2016). Supplement of Fe and Mn is not reported in China due to abundant Fe and Mn in the acid soils of most litchi orchard in the main production area.

7 Future research

Systematic nutrient management in litchi is under investigation, probably due to its huge biomass and longevity, in contrast to other perennial fruit crops such as apple and citrus. Some constraints, worthy to be focused on in the near future, are suggested as follows:

New litchi cultivars or the immigrated varieties are increasingly planted worldwide (Chen et al., 2016; Froneman et al., 2016; Marboh et al., 2018). The nutritional demand and response to nutrients of the new/immigrated species are urgently needed to be illuminated.

New fertilizers such as slow release fertilizer and controlled release fertilizer have been utilized in crops (Cheng et al., 2015; Ke et al., 2018; Saha et al., 2019), which are particularly suitable for litchi owing to its yearly growth period. Investigations are required to find how the new fertilizers enhance nutrient use efficiency and promote plant growth in litchi. Meanwhile, fertigation is commonly adopted in versatile crops. Fertilization schemes including right fertilizer combinations with right rates under fertigation in litchi are impending requirements as well.

With regard to nutrient diagnosis, time efficiency of leaf analysis is the bottleneck for nutrient supplement in the key growth stage. Canopy reflectance spectra have been used to estimate leaf N contents for field crops in terms of timeliness (Hansen and Schjoerring, 2003; Li et al., 2014). For example, leaf N levels can be evaluated by the optimized reflectance models at autumn shoot maturation stage, flower spike stage, fruit maturation stage, and flowering stage (Li et al., 2016). Hence, instant nutrient diagnosis approach is likely to be developed to timely ameliorate nutritional imbalance in litchi.

Additionally, most of the litchi is cultivated in Southeast Asia, where the soils are commonly acid. Overdose of N fertilizer (Guo et al., 2010; Yang et al., 2018), intensive tillage (Gui et al., 2018), and atmospheric deposition (Zhu et al., 2016) further acidify soils. Plant growth is inhibited by soil acidification per se (Thomas Raese, 1998; Zhao et al., 2009). Moreover, soil chemical and biological attributes are deteriorated by low soil pH (Goulding, 2016; Sato and Takahashi, 1996), which further suppresses plant growth. And, low soil pH enhances bioavailability of aluminum, leading to aluminum toxicity to litchi seedling (Chen et al., 2005). Though litchi originates from South China and adjusts itself to local soil condition, the suitable soil pH for litchi is still in vague up to date and needs to be revealed. Then, soil pH amelioration may be focused on in litchi orchard.

References

Chen, X.D., Chen, J.S., Zhang, F.B., Yang, S.H., 1998. Effect of nitrogen and potassium fertilizers on litchi. Guangdong Agric. Sci. (2), 27–29 (in Chinese).

Chen, Z.C., Chen, H.Z., Huang, H., Zhang, N., Zhu, S.Y., 2005. Effect of aluminum on seedling growth of semi-tropical fruit tree. J. Agro-Environ. Sci. 24 (S1), 34–37 (in Chinese).

Chen, Z., Zhao, J.T., Qin, Y.H., Hu, G.B., 2016. Study on the graft compatibility between 'Jingganghongnuo' and other litchi cultivars. Sci. Hortic. 199, 56–62.

Cheng, D.D., Wang, Y., Zhao, G.Z., Liu, Y.Q., 2015. Effects of polymeric slow release fertilizer on Chinese cabbage growth and soil nutrients. Arch. Agron. Soil Sci. 61, 959–968.

Cronje, R.B., Mostert, P.G., 2009. A management program to improve yield and fruit size of litchi, cv. HLH Mauritius—final report. In: South Africa Litchi Grower's Association Yearbook, 21, pp. 46–55.

Dai, L.Z. (Ed.), 1999. New Fertilization Technique in Litchi and Longan. China Agricultural Press, Beijing.

Dai, L.Z., Lin, C.X., Liu, L.R., Su, M.L., 1998. Effect of applying fertilizer on yield and leaf mineral nutrient of *Litchi chinensis*. Fujian J. Agric. Sci. (1), 52–55 (in Chinese).

de Villiers, E.A., Joubert, P.H. (Eds.), 2010. The Cultivation of Litchi, second ed. ARC-Institute for Tropical and Subtropical Crops.

Diczbalis, Y., 2007. Final Report: Unlocking Litchi Research Australia. Queensland Department of Primary Industries and Fisheries. FR01033, 63.

Farina, V., Gianguzzi, G., D'Asaro, A., Mazzaglia, A., Palazzolo, E., 2017. Fruit production and quality evaluation of four litchi cultivars (*Litchi chinensis* Sonn.) grown in Mediterranean climate. Fruits 72 (4), 203–211.

Froneman, I.J., Nonyane, D., Severn-Ellis, A., Cronje, R.B., Sippel, A.D., 2016. Expanding genetic diversity of the south African litchi germplasm collection to promote plant improvement. In: Onus, N., Currie, A. (Eds.), Xxix International Horticultural Congress on Horticulture: Sustaining Lives, Livelihoods and Landscapes. Int Soc Horticultural Science, Leuven, pp. 365–372.

Gao, D., Zhou, X.C., Su, Y., Zhou, K.B., 2015. Effect of spraying K and Ca and Mg foliar fertilizers on pericarp coloring of *Litchi chinensis* Sonn. cv Sanyuehong. J. South. Agric. 46 (10), 1849–1855 (in Chinese).

Goldweber, S., 1959. Observations on lychees grown in pot culture. Proc. Fla State Hortic. Soc. 72, 353–356.

Goncalves, V.D., Pires, M.C., Yamanishi, O.K., 2016. Foliar application of nitrogen at petal fall stage increases fruit set and size of 'Bengal' lychee. In: Milatovic, D., Milivojevic, J., Nikolic, D. (Eds.), III Balkan Symposium on Fruit Growing. Int Soc Horticultural Science, Leuven, pp. 591–596.

Goulding, K.W.T., 2016. Soil acidification and the importance of liming agricultural soils with particular reference to the United Kingdom. Soil Use Manag. 32, 390–399.

Gui, R., Wu, W., Zhuang, S., Zhong, Z., 2018. Intensive management increases soil acidification and phytotoxic Al content in Phyllostachys praecox stands in Southeast China. J. Sustain. Forest. 37, 46–55.

Guo, J.H., Liu, X.J., Zhang, Y., Shen, J.L., Han, W.X., Zhang, W.F., Christie, P., Goulding, K.W.T., Vitousek, P.M., Zhang, F.S., 2010. Significant acidification in major Chinese croplands. Science 327, 1008–1010.

Hansen, P.M., Schjoerring, J.K., 2003. Reflectance measurement of canopy biomass and nitrogen status in wheat crops using normalized difference vegetation indices and partial least squares regression. Remote Sens. Environ. 86, 542–553.

Haq, I., Rab, A., Sajid, M., 2013. Foliar application of calcium chloride and borax enhance the fruit quality of litchi cultivars. J. Anim. Plant Sci. 23, 1385–1390.

He, Y.Q., Long, S.Z., Wei, C.B., Li, Z., Liang, R.M., 2003. Experiments of applying various ratio of N and K fertilizers to litchi. Guangxi Agric. Sci. (1), 32–33 (in Chinese).

Huang, X.J., 2007. Global litchi production and trade: a review. World Tropic. Agric. Inf. (5), 1–4 (in Chinese).

Huang, X.M., Wang, H.C., Zhong, W.L., Yuan, W.Q., Lu, J.M., Li, J.G., 2008. Spraying calcium is not an effective way to increase structural calcium in litchi pericarp. Sci. Hortic. 117, 39–44.

Jongruaysup, S., Dell, B., Bell, R.W., 1994. Distribution and redistribution of molybdenum in black gram (*Vigna mungo* L. Hepper) in relation to molybdenum supply. Ann. Bot. 73, 161–167.

Ke, J., He, R.C., Hou, P.F., Ding, C., Ding, Y.F., Wang, S.H., Liu, Z.H., Tang, S., Ding, C.Q., Chen, L., Li, G.H., 2018. Combined controlled-released nitrogen fertilizers and deep placement effects of N leaching, rice yield and N recovery in machine-transplanted rice. Agric. Ecosyst. Environ. 265, 402–412.

Li, J.G., Gao, F.F., Huang, H.B., Tan, Y.W., Luo, J.T., 1999. Preliminary studies on the relationship between calcium and fruit cracking in litchi fruit. J. South China Agric. Univ. 20 (3), 45–49 (in Chinese).

Li, G.L., Yao, L.X., Yang, B.M., He, Z.H., Zhou, C.M., Huang, L.X., Guo, B., Tu, S.H., 2011. Effect of fertilizer applicaiton strategy on the growth, quality and yield of litchi. Chin. J. Trop. Crops 32, 15–20 (in Chinese).

Li, F., Miao, Y., Feng, G., Yuan, F., Yue, S., Gao, X., Liu, Y., Liu, B., Ustin, S.L., Chen, X., 2014. Improving estimation of summer maize nitrogen status with red edge-based spectral vegetation indices. Field Crop Res. 157, 111–123.

Li, D., Wang, C., Liu, W., Peng, Z., Huang, S., Huang, J., Chen, S., 2016. Estimation of litchi (*Litchi chinensis* Sonn.) leaf nitrogen content at different growth stages using canopy reflectance spectra. Eur. J. Agron. 80, 182–194.

Liang, Z.J., Dai, L.Z., 1984. Nutrient diagnosis and fertilization in litchi. Fujian Fruit (Z1), 6–13 (in Chinese).

Luo, D.L., Wang, W., Zhu, L.W., Bai, C.H., Li, H., Zhou, C.M., Zhou, Q.M., Yao, L.X., 2019. Establishment of foliar nutrient diagnosis norms for litchi (*Litchi chinensis* Sonn.) in South China. J. Plant Nutr. Fertil. 25, 859–870 (in Chinese).

Marboh, E.S., Gupta, A.K., Gyanesh, K., Singh, M., Singh, A., Nath, V., 2018. Genetic variability, heritability and genetic advance in litchi (*Litchi chinensis*). Indian J. Agric. Sci. 88, 1510–1514.

Menzel, C.M., Simpson, D.R., 1987. Lychee nutrition: a review. Sci. Hortic. 31, 195–224.

Menzel, C.M., Carseldine, M.L., Simpson, D.R., 1988a. Crop development and leaf nitrogen in lychee. Aust. J. Exp. Agric. 28, 793–800.

Menzel, C.M., Carseldine, M.L., Simpson, D.R., 1988b. The effect of fruiting status on nutrient composition of litchi (*Litchi chinensis* Sonn.) during the flowering and fruiting season. J. Hortic. Sci. 63, 547–556.

Menzel, C.M., Carseldine, M.L., Haydon, G.F., Simpson, D.R., 1992. A review of existing and proposed new leaf nutrient standards for lychee. Sci. Hortic. 49, 33–53.

Peng, J., Xi, J.B., Tang, X.D., Wang, Y.G., Si, X.M., Chen, J.S., 2001. The effect of Ca(NO₃)₂ and GA sprayed on leaves on the fruit cracking of 'Nuomici' litchi. Acta Hortic. Sin. 28 (4), 348–350 (in Chinese).

Qi, W.E., Chen, H.B., Li, W.W., Zhan, H.J., 2016. Development situation, trend and suggestions of Chinese litchi industry. Guangdong Agric. Sci. 43 (6), 173–179 (in Chinese).

Roy, R.N., Rao, D.P., Mukherjee, S.K., 1984. Orchard efficiency analysis of litchi. Indian J. Hortic. 41 (1–2), 16–21.

Saha, B.K., Rose, M.T., Wong, V.N.L., Cavagnaro, T.R., Patti, A.F., 2019. A slow release brown coal-urea fertiliser reduced gaseous N loss from soil and increased silver beet yield and N uptake. Sci. Total Environ. 649, 793–800.

Sato, K., Takahashi, A., 1996. Acidity neutralization mechanism in a forested watershed in Central Japan. Water Air Soil Pollut. 88, 313–329.

Su, Y., Zhou, X.C., Gao, D., Zhou, K.B., 2015. Studies on the relationship between the main flavor components and the contents of K, Ca, and Mg in flesh of Feizixiao litchi (Litchi chinensis Sonn. cv Feizixiao). Chin. J. Trop. Crops 36 (6), 1131–1135 (in Chinese).

Thomas Raese, J., 1998. Response of apple and pear trees to phosphate fertilization: a compendium. Commun. Soil Sci. Plant Anal. 29, 1799–1821.

Thomas, M.B., Ferguson, J., Crane, J.H., 1995. Identification of N, K, Mg, Mn, Zn, and Fe deficiency symptoms of carambola, lychee, and papaya grown in sand culture. Proc. Fla. State Hortic. Soc. 108, 370–373.

Wall, M.M., 2006. Ascorbic acid and mineral composition of longan (Dimocarpus longan), lychee (Litchi chinensis) and rambutan (Nephelium lappaceum) cultivars grown in Hawaii. J. Food Compos. Anal. 19, 655–663.

Yang, B.M., Yao, L.X., Li, G.L., He, Z.H., Zhou, C.M., 2015. Dynamic changes of nutrition in litchi foliar and effects of potassium-nitrogen fertilization ratio. J. Soil Sci. Plant Nutr. 15, 98–110.

Yang, B.M., Li, G.L., He, Z.H., Zhou, C.M., Xu, P.Z., Yang, S.H., Yao, L.X., 2016. Effect of boron on litchi yield and Ca-B forms. Guangdong Agric. Sci. (2), 71–76 (in Chinese).

Yang, X.D., Ni, K., Shi, Y.Z., Yi, X.Y., Zhang, Q.F., Fang, L., Ma, L.F., Ruan, J.Y., 2018. Effects of long-term nitrogen application on soil acidification and solution chemistry of a tea plantation in China. Agric. Ecosyst. Environ. 252, 74–82.

Yao, L.X., 2009. Survey on nutrient management in litchi of China. Litchi Newsl. (3), 41–54 (in Chinese).

Yao, L.X., Zhou, C.M., He, Z.H., Jiang, Z.D., Bai, C.H., 2017a. Abnormality of litchi and longan fruits and diagnosis on fruit miernal nutrition. South China Fruits 46 (4), 49–54 (in Chinese).

Yao, L.X., Zhou, C.M., He, Z.H., Li, G.L., Bai, C.H., 2017b. Nutrient demand of annual growth of autumn branch, spica and fruit in litchi. J. Plant Nutr. Fertil. 23 (4), 1128–1134 (in Chinese).

Yao, L.X., Zhou, C.M., He, Z.H., Li, G.L., Yang, B.M., Bai, C.H., 2020. Biological attributes and nutrient requirement characteristics of the main litchi cultivars. J. South China Agric. Univ.. 41, (in press) (in Chinese).

Zhao, J., Li, X., Zhang, X.X., Zhang, C., Sheng, X., Wang, H.Q., Chen, L.B., 2009. Effect of soil pH on fruit quality of Golden pearl. Anhui Agric. Sci. 37 (27), 13037–13040 (in Chinese).

Zhou, X.C., Portch, S., Xie, F., Mo, W.X., Yao, L.X., 2001. Nutritional characteristics of quality litchi and effect of potassium, magnesium and sulfur fertlizers on litchi growth. Guangdong Agric. Sci. (5), 31–33 (in Chinese).

Zhou, X.C., Su, Y., Zhang, R., Zhou, K.B., 2015. Effect of K, Ca and Mg applied in foliar nutrients on pericarp's coloring of Litchi chinensis Sonn. cv. Feizixiao, Southwest China. Southwest China J. Agric. Sci. 28 (4), 1713–1718 (in Chinese).

Zhou, K.B., Zhou, X.C., Su, Y., Gao, D., 2016. The physiological causes of foliar Mg application enhance the fruit pigmentation of Litchi chinensis Sonn. cv Feizixiao. Chin. J. Trop. Crops 37 (9), 1752–1758 (in Chinese).

Zhu, Q.C., De Vries, W., Liu, X.J., Zeng, M.F., Hao, T.X., Du, E.Z., Zhang, F.S., Shen, J.B., 2016. The contribution of atmospheric deposition and forest harvesting to forest soil acidification in China since 1980. Atmos. Environ. 146, 215–222.

Zhuang, Y.M., Wang, R.J., Xie, Z.N., Xu, W.B., 1994. Studies on the contents of trace elements in productive orchard soil of the south tropical zone in Fujian. Subtrop. Plant Res. Commun. 23 (1), 1–7 (in Chinese).

46

Diagnosis and management of nutrient constraints in pomegranate

John M. Chater[a],*, *Donald J. Merhaut*[a], *John E. Preece*[b]

[a]Department of Botany and Plant Sciences, University of California, Riverside, CA, United States
[b]National Clonal Germplasm Repository, USDA-ARS, University of California, Davis, CA, United States
*Corresponding author. E-mail: jchat004@ucr.edu

1 Commercial pomegranate cultivation

Pomegranate is an ancient fruit tree crop that has been domesticated for thousands of years and cultivated for its prized fruit, juice, and aril-coated seeds (Preece and Moersfelder, 2016). In the United States, the majority of commercial orchards consist of trees with one single trunk on its own rootstock (Fig. 46.1), but growing trees with multiple trunks is commonplace and can be advantageous in regions that experience hard frost conditions during winter and fall. Pomegranates are rarely propagated by grafting, although research in Iran, the United States, and other countries is underway to determine the efficacy, benefits, and disadvantages of this cultural practice, which is common in citriculture, viticulture, and many other fruit and nut cropping systems to improve tree performance (Vazifeshenas et al., 2009; Karimi and Nowrozy, 2017). There is at least one nonsuckering rootstock that is grown in Italy (Preka et al., 2016). Despite its long history of cultivation, limited consistent and reliable fertilizer recommendation information on pomegranate is available to growers, although a recent publication provided leaf nutrient ranges consistent with the information contained in this chapter (Kahramanoglu and Usanmaz, 2016). Extension materials are needed to help farmers so that yield and profitability can be optimized for farm sustainability. It is just as important that fertilizers are used responsibly so that the ecological harm, such as groundwater pollution, often associated with agricultural systems can be minimized.

This grove consisted of single-trunked trees grown on their own rootstocks and averaged more than 250 fruit per tree. This represents what is possible when appropriate cultivars are grown under best management practices, including proper plant nutrition.

A.K. Srivastava, Chengxiao Hu (eds.)
Fruit Crops: Diagnosis and Management of Nutrient Constraints
https://doi.org/10.1016/B978-0-12-818732-6.00046-0

FIG. 46.1 Nine-year-old "Wonderful" pomegranate tree in July, grown under conventional management practices in Kern County, California, United States.

2 Where pomegranates grow

To understand how to properly manage pomegranate nutrition in a commercial orchard, it is important to understand that if the environment is not conducive to pomegranate production, no matter how well the tree is managed nutritionally, the plant may not be able to provide salable fruit. Environmental conditions that are not conducive to pomegranate yield are the lack of sunlight (including shading), high humidity during bloom and fruit development, and extreme temperatures, especially freezing during bloom. Pomegranates are a Mediterranean crop and grow best in climates with mild, dry summers with rainfall during the winter. Cultivar selection is also crucial; if the wrong cultivar is selected, pests, diseases, and fruit defects leading to crop loss are to be expected.

Pomegranate is grown on every continent except for Antarctica, but that doesn't mean that pomegranates grow well, or at all, in all regions. Growing pomegranates in unfit climates can result in total crop loss and, even worse, total tree loss. When choosing any crop, consider climatic conditions, soil type, resource availability, and cultivar. Pomegranate is best commercially grown in Mediterranean climates (Stover and Mercure, 2007) that have no hard freeze events until late fall or winter, after the fruit are harvested (Glozer and Ferguson, 2008). This also means that there should be no rainy season during bloom, during fruit set, or during fruit development to avoid fruit rot and the summer should be dry and warm (Levin, 2006). Pomegranate trees are quite hardy when it comes to temperature extremes, and some cultivars can withstand temperatures of $-12°C$ (Levin, 2006; Westwood, 1988). From personal experience in Riverside, California, the trees of many cultivars from the USDA National Clonal Germplasm Repository, Davis, California, are also tolerant of extreme outdoor and greenhouse temperatures as high as 49°C. However, the sun-exposed fruit on the outer side of the canopy are likely to be scorched by sunburn that will cause reduction in internal and external color and flavor quality, thus making the fruit unmarketable. Pomegranates require full sun for best growth and production. It has been reported that pomegranate trees need a minimum of 6–8h of full sun (Glozer and Ferguson, 2008). Optimal yields are realized when the photosynthetic photon flux (PPF) of photosynthetically active radiation (PAR) ranges from 1500 to 1900 μmol/m^2/s (Chater et al., 2018a).

3 The effects of pomegranate cultivar on the cropping system

The cultivars that are grown will influence crop management. For example, some cultivars are more vigorous or grow taller than others, which means they will be spaced further apart than smaller-growing cultivars, altering the number of trees per hectare, thus influencing fertilizer applications. There has been limited published work on the influence of cultivar on tree nutrition, so this chapter will focus primarily on the nutrition of "Wonderful" pomegranate because it is the industry standard in many countries and it represents a productive cultivar in a commercial production setting. In warmer climates, temperature extremes can destroy quality and thus salability of fruit. To mitigate fruit sunburn, growers in California, United States, will spray formulations of kaolinite clay (diatomaceous

FIG. 46.2 Applications of kaolinite (kaolin clay) can affect foliar applications of fertilizers, so it is important to supplement fertilization before treatments of crop protectant. Similarly, using screen cover over trees or bagging fruit may also influence effects of foliar applications on pomegranate.

earth), and this is believed to increase the albedo of the fruit (Fig. 46.2), which reduces the temperature and protects the fruit from the sun's rays. If a grower uses kaolinite, they may influence fertilizer uptake when applying foliar nutrients. Kaolinite applications do not affect uptake of N and K with 2% and 3% KNO_3 applications with Activator 90 (Chater and Garner, 2018).

4 The pomegranate cropping system

A cropping system is the crop and its management techniques that are applied in a given horticultural field or setting. This includes decisions made before and during planting, such as tree spacing, and, subsequently, including pruning, pest control, and plant mineral nutrition. In commercial production, fertilizer scheduling is designed for the needs of any given cropping system. Each cropping system design can have very different characteristics that dictate how, when, and which fertilizers are applied. Some of the important characteristics of pomegranate cropping systems that should be considered when developing a fertilizer program include cultivar, tree spacing, tree height, cultural practices such as the use of kaolinite for crop protection, irrigation system type, trellising or other tree-support materials, pruning, canopy size, number of trunks, cultivar, climate and soil type. Pomegranate is a deciduous tree (Fig. 46.3) in most growing regions. In tropical regions, cultivars can remain evergreen, and the use of differing

FIG. 46.3 Pomegranate is a deciduous tree crop, and it is important to apply fertilizers at the proper time. Pruning of trees into desired forms is also performed during tree dormancy. The photographs depict about 1 ha of pomegranate trees in mid-January surrounded by citrus in Riverside, California, United States.

FIG. 46.4 Container-grown pomegranate struggling in tropical Hawaii in mid-September. Note how few fruit and the thinness for the tree canopy. The lack of chilling and wet humid tropical conditions is not ideal for pomegranate production.

irrigation schedules (Phadnis, 1974) or ethylene applications is required to defoliate the tree to restart the production cycle (personal communication). Otherwise, the tree will produce little fruit throughout the year in overlapping cycles (Fig. 46.4). The largest pomegranate producers in the United States use conventional fertilizers; however, smaller farms may use conventional and/or organic fertilizers. Globally, pomegranates are grown both conventionally or organically, with most production done conventionally.

Fertilizer choice can be influenced by time of year, weather, soil pH, soil type, and soil fertility status and plant age. A pomegranate grove that is planted in the proper climatic and soil conditions, sufficiently fertilized, and is of a commercial-grade cultivar should produce a minimum of between 50 and 60 fruit per tree (Fig. 46.1). Larger trees can have more fruit per tree (up to 1000), but large trees are typically not in commercial groves because they are difficult to manage and harvest. Tree height should be 3.5 m at a maximum to reduce the efforts of hired labor and the dangerous use of harvesting ladders. Tree heights of 2.5–3 m are more desirable because harvest and pruning can be done much easier. Pomegranate can be grown successfully on many different soil types (Blumenfeld et al., 2000). Pomegranate is considered moderately salt tolerant (Blumenfeld et al., 2000; Holland et al., 2009), with the production of some cultivars unaffected by soil salinity with conditions as high as 4 dS/m (Holland et al., 2009).

Pomegranate tree spacing for a commercial pomegranate grove is variable. Often, tree spacing in Californian orchards is 4 × 6 m. Tree spacing also varies based on cultivation equipment and cropping system practices, with in-row spacing ranging from 3.6 to 5.8 m and drive-row spacing ranging from 1.5 to 5.2 m (Day and Wilkins, 2009).

Cultural practices that affect leaf tissue nutrition can significantly influence yield. Crop load should be managed so that fruit do not contact the orchard floor because fruit can be nonmarketable because of blemishes, decay, and soil-borne microorganisms that may cause food safety problems. Ways to accomplish this task are to tie up drooping branches with rope (Fig. 46.5) and install stakes to prop up branches or to install a trellis system. Some growers will have limited yield in their orchards and will not require these management practices, and others will lose fruit, branches, and entire trees if mitigation practices are not utilized. Some growers will thin fruit to reduce the heavy crop loads that cause tree or branch breakage. Sloping orchard floor conditions may exacerbate the problems associated with heavy crop load.

5 Pomegranate nutrition and fertilization

5.1 Conventional versus organic production

The difference between conventional and organic production systems is explained in the other chapters of this book. Conventional production will utilize all appropriate agricultural products that are legally available, and organic production utilizes inputs that are not synthetic products. There are benefits and drawbacks of each method of crop production; however, most pomegranates are farmed conventionally, and conventional orchard managers may utilize organic products.

FIG. 46.5 Nine-year-old "Wonderful" pomegranate tree in mid-September, grown under conventional management practices in Kern County, California, United States. A nylon rope is used to keep the branches and fruit up away from the orchard floor to avoid food safety problems.

(A)   (B)

FIG. 46.6 A pomegranate irrigation pond (Panel A) and fertilizer reservoir (Panel B). These components are commonplace in the Central Valley of California, United States, for fertigation of pomegranate groves and many other fruit and nut tree crops.

In the United States, some commercial growers fertilize their pomegranates with rotted manure (Blumenfeld et al., 2000), but the vast majority of hectares in the United States do not use rotted manure and instead use inorganic fertilizer. The largest orchards in the United States utilize large irrigation ponds to pull water from and inject fertilizer into the irrigation lines being fed from the pond (Fig. 46.6). In the United States, most apply inorganic fertilizer by fertigation through the irrigation system (Blumenfeld et al., 2000), which is an efficient method compared with broadcast. Fertilizer applications and rates depend on conditions in the field. Some factors to consider include orchard age and tree spacing, soil properties, leaf tissue mineral nutrition status, and fertilizer delivery system.

5.2 Leaf tissue nutrition

Pomegranates need to have adequate levels of macronutrients (such as nitrogen, phosphorus, and potassium, N-P-K) in their leaves for the tree to produce commercial yields of fruit (Fig. 46.7). Because pomegranate fruit has such high concentrations of potassium (K) (about the same amount as a banana), it is important to add enough

FIG. 46.7 With proper nutrition, climate, and cultivar selection, pomegranate trees can produce flushes that have several flowers that produce up to five fruit at harvest. Having multiple fruit located on the same branch or floral cluster does not negatively usually impact flavor of "Wonderful" pomegranate.

TABLE 46.1 Leaf tissue mineral nutrient sufficiency ranges of nitrogen (N), phosphorus (P), potassium (K) in %, and zinc (Zn) in mg/L for a high yielding, 9-year-old commercial "Wonderful" pomegranate orchard in Kern County, California, United States.

		Nutrient		
Cultivar	N	P	K	Zn
"Wonderful"	>2%	0.13%–0.15%	1.0%–1.2%	12–14 mg/L

Each tree had over 200 fruit on average.

potassium to the cropping system so that the tree can have enough of this essential mineral nutrient for healthy leaves and high quality fruit (Chater and Garner, 2018). Additionally, it is very important that all crop plants have enough nitrogen (N) available for healthy growth, development, and flower and fruit production. From previous investigations involving the most successful pomegranate growers in the United States, "Wonderful" pomegranate leaves should have approximately 2% N at a minimum (Table 46.1). This value is much higher than what is recommended in Savita et al. (2016), which demonstrates much inconsistency on this topic in the literature.

For leaf tissue K, it is important to keep concentrations in the leaves between 1% and 1.2%. This is in agreement with what was reported in Savita et al. (2016). Phosphorus (P) is a mineral nutrient that is often associated with reproductive growth in plants and should be in the range of 0.130%–0.150% in the leaf tissues of fully developed, newly hardened off leaves. These values are lower than what was reported in Savita et al. (2016), again highlighting the need for a greater understanding of the needs of pomegranate trees and the differences among cultivars and growing conditions. The recommendations herein are based on field experiments involving foliar nutrient applications on 9-year-old "Wonderful" pomegranate, with each tree producing over 200 fruit. The recommended leaf tissue concentration of zinc (Zn) is between 12 and 14 mg/L. Foliar application of zinc sulfate to 9-year-old "Wonderful" pomegranate trees increased the concentration of Zn in the leaf tissues by up to an order of magnitude (to over 100). Specific fertilizer formulations are dependent on orchard conditions and some growers utilize the services of fertilizer companies or professional consultants to calculate dosages for applications.

5.3 The big four

Most experts will recommend three and sometimes four of the essential mineral nutrients be considered when managing pomegranate in commercial production settings. The main three are N, K, and Zn and sometimes P. This is not to say that the other nutrients are not important, it is just that applying these other nutrients in pomegranate groves has not consistently influenced yield. This chapter therefore focuses on N, P, K, and Zn. These values apply to "Wonderful" pomegranate but could likely be of use to growers of other cultivars.

5.3.1 Nitrogen

The most common symptom of N deficiency is chlorosis (a general yellowing) of the oldest leaves. A measure of plant health is relative chlorophyll content, which can be measured by a SPAD 502 chlorophyll meter (Konica-Minolta, Japan). Some cultivars naturally have more relative chlorophyll content than others (Chater et al., 2018b). In terms of fertilizing pomegranate groves with N by conventional means, there are no reliably consistent reports on nitrogen fertilizer requirements of pomegranate in the literature. Some have reported requirements of N are 0.20-kg or even 0.63-kg N/tree/year (Firake and Deolankar, 2000; Stover and Mercure, 2007), which is over a threefold difference in application rates for mature trees. These rates are for sandy loam soils; clayey soils may require more or less application and in areas with higher rainfall. Despite this inconsistency in the amount of N applied per tree per year, when fertilizing with N, the grower may apply fertilizer at one time during the season, or they could choose to do two applications per year (Glozer and Ferguson, 2008), with the best time to apply being in spring, during vegetative flush and bloom periods. Others have reported that multiple applications are a standard practice, with the final N application just weeks before harvest (Holland et al., 2009). Some growers have mentioned that applying too much N may cause excessive vegetative growth and negatively affect yield, harvest date, pruning needs, and fruit quality. There are multiple sources of N that are used in pomegranate cropping systems. Some of these sources include urea ($CO(NH_2)_2$), ammonium sulfate (($NH_4)_2SO_4$), and ammonium nitrate (NH_4NO_3) (Glozer and Ferguson, 2008). Potassium nitrate (KNO_3) is also a useful source of N and should be applied if foliar tests also call for K. The only issue with fertigating with KNO_3 is that, if soil pH is too high, the nitrate will contribute to increasing soil pH. In the United States, ammonium nitrate is highly regulated and difficult to obtain in large enough amounts for farming due to security concerns of it being used as an explosive.

5.3.2 Phosphorus

Forms of P used in pomegranate cropping systems include phosphoric acid (H_3PO_4) and phosphorus pentoxide (P_2O_5) (Holland et al., 2009). Application of these P-containing materials could be done once per season or in split applications or in multiple sets. Phosphoric acid is useful in fertigation applications because it is less likely to cause problems with irrigation emitters because other forms of phosphorus will often precipitate out of solution. There have been reports of trees requiring 250 g of phosphorus per plant per year. However, one study reported that 187.5-g P/tree/year produced yields no different than the 250-g rate (Firake and Deolankar, 2000). Others have suggested that P fertilization is not required in commercial pomegranate production because it does not improve yields (Glozer and Ferguson, 2008), but of course, that would depend on the available P in the soil, the soil type, and other factors.

5.3.3 Potassium

Potassium nutrition of pomegranate is not well understood (Khayyat et al., 2012). Some report that the amount of K applied should be equal or similar to the amount of N applied (Blumenfeld et al., 2000; Stover and Mercure, 2007). Others have reported that K applied should be similar to, but less than, the quantity of N applied. According to Firake and Deolankar (2000), the amount of K applied via fertigation per tree per year is 250 g in India. These authors also reported that reducing this K dosage by 25% did not significantly affect yield, and they also reported that yield was highest if micronutrients were coupled with 250-g K/tree/year. However, it is unknown what would happen to the system in the long term if reduced amounts of K were applied. Pomegranate fruit is very high in K, and it is important that enough is available over the long term for plant health and fruit quality. Visual symptoms of K deficiency in leaves have not yet been developed by scientific means (e.g., via traditional hydroponics-based methodologies using experimentation). Potassium is important because the exocarp and the arils contain large concentrations of K and are thus removed from the cropping system each harvest (Miklavčič Višnjevec et al., 2017).

5.3.4 Zinc

Zinc nutrition is considered important in pomegranate horticulture by some experts, but not by others. This micronutrient is one of the four nutrients that experts on pomegranate production will recommend considering in a

TABLE 46.2   Leaf tissue mineral macronutrient sufficiency ranges for calcium (Ca), magnesium (Mg), and sulfur (S) (in %) for a high yielding, 9-year-old commercial "Wonderful" pomegranate orchard in Kern County, California, United States.

Cultivar	Calcium	Nutrient Magnesium	Sulfur
"Wonderful"	4.5%–4.9%	0.38%–0.42%	0.18%–0.20%

Each tree had over 200 fruit on average, and higher concentrations than reported herein did not significantly increase yields.

TABLE 46.3   Leaf tissue mineral micronutrient sufficiency ranges for boron (B), copper (Cu), iron (Fe), and manganese (Mn) (in mg/L) for a high yielding, 9-year-old commercial "Wonderful" pomegranate orchard in Kern County, California, United States.

Cultivar	Boron	Nutrient (mg/L) Copper	Iron	Manganese
"Wonderful"	20.0–22.0	4.5–7.0	70.0–85.0	30.0–45.0

Each tree had over 200 fruit on average, and higher concentrations than reported herein did not significantly increase yields.

fertilization program. Pomegranate Zn leaf tissue deficiency symptoms include chlorosis and delayed flushing of vegetative and reproductive shoots. Zinc sulfate ($ZnSO_4$) can be used during the growing season to correct Zn deficiency (Glozer and Ferguson, 2008; Stover and Mercure, 2007). Dormant sprays of Zn, applied when the tree has no leaves, can also be utilized in pomegranate (Glozer and Ferguson, 2008), but there is limited information on the efficacy of this practice. Dormant sprays should be applied after winter pruning to maximize efficiency and cost-effectiveness because, if nutrients are taken up by biomass that will later be removed, there is a loss of efficiency. A foliar application dosage of 0.1% $ZnSO_4$ solution sprayed until runoff once during spring or early summer with a proper adjuvant would likely be sufficient to remedy a leaf tissue zinc deficiency.

5.4 The other macro- and micronutrients

Taking soil and leaf samples in the field is a neglected best management practice in horticulture. It is often neglected because the costs associated with testing or the technology is unavailable. Understanding what is in the soil and knowing the nutritional status of the leaves is extremely useful when planning fertilizer scheduling. While most labs have not developed standardized nutritional guidelines or sufficiency ranges, this section includes tables that can serve as a guidance for growers who have no standard for comparison. The other three macronutrients calcium, magnesium, and sulfur are typically of no importance unless there is a deficiency in the system. Calcium has been associated with fruit split in pomegranate, but the mechanism is not understood and has not been adequately demonstrated in the scientific literature. Potential sufficiency ranges for these macronutrients are reported (Table 46.2). With the exception of magnesium, these values are very different compared with another review (Savita et al., 2016), which again demonstrates disagreement in the literature on this subject. Similarly, potential sufficiency ranges for the micronutrients boron, copper, iron, and manganese are reported (Table 46.3). These potential sufficiency ranges are from 9-year-old "Wonderful" trees that were being grown by the largest pomegranate grower in the United States, and these trees had a yield of over 200 pomegranates per plant. It is important that these ranges are documented for growers to see how these leaf tissue nutrient statuses affect growers in other countries, in differing climates, and with other cultivars beyond "Wonderful."

6 Foliar applications

Foliar applications involve spraying fertilizers on the foliage (leaves) of the trees to increase levels of nutrients in the leaves and sometimes even the fruit. Spraying 2% or 3% potassium nitrate was successful in increasing both potassium

FIG. 46.8 Pomegranate leaf tissue necrosis from micronutrient foliar applications of ZnSO$_4$ on "Wonderful" pomegranate in Kern County, California, United States. The dosage of this application was 4000 mg/L and was only applied once in July when the fruit were about the size of a baseball.

and nitrogen in N- and K-deficient leaves (Chater and Garner, 2018). Use of an adjuvant or other surfactant (such as Tween 20 or Activator 90) when applying foliar fertilizers is recommended to increase uptake. Many plants have waxy leaves with thick cuticles that can reduce penetration of the fertilizer if a surfactant is not used. It is known that foliar applications of conventional fertilizers are utilized in Californian pomegranate groves, with some growers known to sample trees for leaf tissue nutrition analysis before such applications. Leaf tissue sufficiency ranges for macro- and micronutrients are unknown due to a lack of solid rigorous scientific inquiry on the subject (Holland et al., 2009), but there is some useful literature on N, P, K, and Zn nutrition of pomegranate (Firake and Deolankar, 2000; Glozer and Ferguson, 2008).

Foliar applications of potassium have been reported to increase internal redness color of pomegranates that have not colored up yet in production as there have been reports of K applications increase anthocyanin content versus control (Hamouda et al., 2015). Foliar applications of KNO$_3$ have also been reported to increase total phenolics in pomegranate (Chater and Garner, 2019). There have been studies on various foliar nutrient applications reducing fruit split in pomegranate, but the mechanism of fruit split is poorly understood, and the methods and results are not consistent.

Foliar applications of Zn have been reported to improve marketable yield, even without symptoms of Zn deficiency (Afria et al., 1999; Khorsandi et al., 2009). Foliar applications of Zn can also increase the amount of Zn in the fruit, which may have human health implications (Chater and Garner, 2019). When Zn applications are rather high for a micronutrient foliar spray (e.g., in the range of 3000–5000 mg/L), leaf damage can result, causing leaf staining and necrosis (Fig. 46.8).

6.1 Cultivar considerations

There has been limited work characterizing differences in fertilizer requirement among cultivars. According to one study, there were significant differences among cultivars for N, K, and Ca, which the authors implicated in fruit split (Hepaksoy et al., 2000). This particular study had leaf tissue mineral nutrient concentrations very different from what is reported in other sources, but they did identify differences in these three nutrients. More research is needed to characterize differences in fertilizer requirements among cultivars.

7 Future lines of research

Much is unknown about pomegranate physiology, including exact fertilizer requirements and visual leaf tissue nutrient deficiency symptoms. Pomegranate horticulture will benefit greatly from a deeper understanding of leaf nutrition. The traditional hydroponic experiments are critical to develop visual leaf mineral nutrient deficiency symptoms, and this work has yet to be done as it has been with most of the major fruit and nut tree crops. Being able to identify deficiencies is critically important to growers, who often apply a standard quantity of fertilizer every year regardless of leaf tissue or soil nutrient status. The effect of cultivar on fertilizer requirements is a research direction that also must be addressed to gain a better understanding of how genotype affects fertilizer needs. Another important line of research that is needed is investigation into how the use of rootstocks affects soil fertilizer applications and scheduling. Rootstocks are not yet commonly used in pomegranate orchard production systems, but new research is exploring the use of rootstocks in this crop. Finally, larger experiments must be designed in many locations (climates, soil types, and sites) over many years with different tree ages and cultivars to gain a systems-based understanding of pomegranate's fertilizer requirements in commercial orchard production systems and how leaf tissue nutrient status affects bloom and yield.

For pomegranates to produce enough marketable fruit for profitable returns, they must be planted in the right environment and be given the right care. The most important nutrients to consider and apply as fertilizers in pomegranate production are N, P, K and Zn. Minding these nutrients in commercial production is recommended. Best management practices include taking soil and leaf tissue samples for nutritional analysis at least once a season so that growers know important horticultural factors including soil pH, soil fertility, and leaf nutrient status. Every site is different, and knowing what is in the leaves and soil is the first step to correcting any nutritional deficiency in the system. Sometimes, fertigation is not the best cultural practice for a given system due to soil status or weather, and this is when foliar applications are best utilized. Application time in the season depends on the nutrient, but typically, N, P, and K are applied either in spring and/or early summer. Zinc is often applied during dormancy, and zinc sulfate in high concentrations can damage leaves. Foliar applications are typically a more efficient use of the fertilizer, and typically, it will have less of an ecological impact. This chapter includes potential sufficiency ranges for the "big four" nutrients in pomegranate horticulture and methods of mitigating nutrient deficiencies in leaf tissue. With proper cultivar selection, climate, and fertilizer management, healthy pomegranate trees grown conventionally should produce 60 or more fruit per tree in a commercial setting.

References

Afria, B.S., Pareek, C.S., Garg, D.K., Singh, K., 1999. Effect of foliar spray of micronutrients and their combinations on yield of pomegranate. Ann. Arid Zone 38 (2), 189–190.

Blumenfeld, A., Shaya, F., Hillel, R., 2000. Cultivation of pomegranate. Opt. Médit. Sér. A Sémin. Médit. 42, 143–147.

Chater, J.M., Garner, L.C., 2018. Foliar nutrient applications to 'Wonderful' pomegranate (Punica granatum L.). II. Effects on leaf nutrient status and fruit split, yield and size. Sci. Hortic. 242, 207–213.

Chater, J.M., Garner, L.C., 2019. Foliar nutrient applications to 'Wonderful' pomegranate (Punica granatum L.). I. Effects on fruit mineral nutrient concentrations and internal quality. Sci. Hortic. 244, 421–427.

Chater, J.M., Santiago, L.S., Merhaut, D.J., Preece, J.E., Jia, Z., 2018a. Diurnal patterns of photosynthesis and water relations for four orchard-grown pomegranate (Punica granatum L.) cultivars. J. Am. Pomol. Soc. 72 (3), 157–165.

Chater, J.M., Santiago, L.S., Merhaut, D.J., Jia, Z., Mauk, P.A., Preece, J.E., 2018b. Orchard establishment, precocity, and eco-physiological traits of several pomegranate cultivars. Sci. Hortic. 235, 221–227.

Day, K.R., Wilkins, E.D., 2009. Commercial pomegranate (Punica granatum L.) production in California. Acta Hortic. 890, 275–285.

Firake, N., Deolankar, K., 2000. Response of pomegranate to soluble fertilizers through drip. J. Maharashtra Agric. Univ. 25 (2), 196–197.

Glozer, K., Ferguson, L., 2008. Pomegranate Production in Afghanistan. University of California, Davis, College of Agricultural and Environmental Sciences, 32.

Hamouda, H.A., Elham, Z.A.M., Nagwa, G., 2015. Nutritional status and improving fruit quality by potassium, magnesium and manganese foliar application in pomegranate shrubs. Int. J. ChemTech Res. 8 (6), 858–867.

Hepaksoy, S., Aksoy, U., Can, H.Z., Ui, M.A., 2000. Determination of relationship between fruit cracking and some physiological responses, leaf characteristics and nutritional status of some pomegranate varieties. Cah. Opt. Médit. 87–92.

Holland, D., Hatib, K., Bar-Ya'akov, I., 2009. Pomegranate: botany, horticulture, breeding. Hortic. Rev. 35, 127–191.

Kahramanoglu, I., Usanmaz, S., 2016. Pomegranate Production and Marketing. CRC Press.

Karimi, H.R., Nowrozy, M., 2017. Effects of rootstock and scion on graft success and vegetative parameters of pomegranate. Sci. Hortic. 214, 280–287.

Khayyat, M., Tehranifar, A., Zaree, M., Karimian, Z., Aminifard, M., Vazifeshenas, M., Amini, S., Noori, Y., Shakeri, M., 2012. Effects of potassium nitrate spraying on fruit characteristics of 'Malas Yazdi' pomegranate. J. Plant Nutr. 35 (9), 1387–1393.

Khorsandi, F., Yazdi, F.A., Vazifehshenas, M.R., 2009. Foliar zinc fertilization improves marketable fruit yield and quality attributes of pomegranate. J. Agric. Biol. 11, 766–770.

Levin, G.M., 2006. Pomegranate. Third Millennium Publishing, Arizona.

Miklavčič Višnjevec, A., Ota, A., Skrt, M., Butinar, B., Smole Možina, S., Gunde Cimerman, N., Nečemer, M., Baruca Arteiter, A., Hladnik, M., Krapac, M., Ban, D., Bučar-Miklavčič, M., Poklar Ulrih, N., Bandelj, D., 2017. Genetic, biochemical, nutritional and antimicrobial characteristics of pomegranate (Punica granatum L.) grown in Istria. Food Technol. Biotechnol. 55 (2), 151–163.

Phadnis, N.A., 1974. Pomegranate for dessert and juice. Indian Hortic. 19 (3), 9–13.

Preece, J.E., Moersfelder, J., 2016. Pomegranate: the grainy apple. J. Am. Pomol. Soc. 70 (4), 187–193.

Preka, P., Cherubini, S., De Salvador, R., Engel, P., 2016. Register of new fruit and nut cultivars list 49: pomegranate rootstock. HortScience 51 (6), 645.

Savita, S., Krishnappa, R., Ngangom, B., Devi, M.T., Mishra, G., Rawat, D., Srivastava, P.C., 2016. Diagnosis and recommendation integrated system (DRIS) approach on nutritional diagnosis in fruit crops—a review. J. Appl. Nat. Sci. 8 (4), 2337–2345.

Stover, E., Mercure, E.W., 2007. The pomegranate: a new look at the fruit of paradise. HortScience 42 (5), 1088–1092.

Vazifeshenas, M., Khayyat, M., Jamalian, S., Samadzadeh, A., 2009. Effects of different scion-rootstock combinations on vigor, tree size, yield and fruit quality of three Iranian cultivars of pomegranate. Fruits 64 (6), 343–349.

Westwood, M.N., 1988. Temperate-Zone Pomology. Timber Press, Oregon.

Further reading

Chater, J.M., Merhaut, D.J., Preece, J.E., Blythe, E.K., 2017. Rooting and vegetative growth of hardwood cuttings of 12 pomegranate (*Punica granatum* L.) cultivars. Sci. Hortic. 221, 68–72.

CHAPTER

47

Diagnosis and management of nutrient constraints in grape

Gustavo Brunetto[a,*], Felipe Klein Ricachenevsky[b], Lincon Oliveira Stefanello[a], Betânia Vahl de Paula[a], Matheus Severo de Souza Kulmann[a], Adriele Tassinari[a], George Wellington Bastos de Melo[c], William Natale[d], Danilo Eduardo Rozane[e], Marlise Nara Ciotta[f], Alberto Fontanella Brighenti[g], Jucinei José Comin[h], Cledimar Rogério Lourenzi[h], Arcângelo Loss[i], Djalma Eugênio Schmitt[j], Jovani Zalamena[k], Lessandro De Conti[l], Tadeu Luis Tiecher[m], André Luiz Kulkamp de Souza[n], Betina Pereira de Bem[o]

[a]Universidade Federal de Santa Maria (UFSM), Campus Universitário, Centro de Ciências Rurais, Departamento de Solos, Camobi, Santa Maria, Brazil
[b]Universidade Federal de Santa Maria (UFSM), Departamento de Biologia, Santa Maria, Brazil
[c]Embrapa Uva e Vinho, Bento Gonçalves, Brazil
[d]Federal University of Ceará, Fortaleza, Brazil
[e]São Paulo State University, UNESP, Registro, Brazil
[f]Empresa de Pesquisa Agropecuária e Extensão de Rural de Santa Catarina (Epagri), Lages, Brazil
[g]Empresa de Pesquisa Agropecuária e Extensão de Rural de Santa Catarina (Epagri), São Joaquim, Brazil
[h]Universidade Federal de Santa Catarina (UFSC), Departamento de Engenharia Rural, Florianópolis, Brazil
[i]Universidade Federal de Santa Catarina (UFSC), Centro de Ciencias Agrarias, Florianopolis, Brazil
[j]Universidade Federal de Santa Catarina (UFSC), Curitibanos, Brazil
[k]Instituto Federal do Rio Grande do Sul (IFRS)—Campus Restinga, Porto Alegre, Brazil
[l]Instituto Federal de Educação, Ciência e Tecnologia Farroupilha—Campus Santo Augusto, Santo Augusto, Brazil
[m]Instituto Federal Farroupilha, Campus Alegrete, Alegrete, Brazil
[n]Empresa de Pesquisa Agropecuária e Extensão Rural de Santa Catarina (Epagri), Videira, Brazil
[o]Instituto Federal de Santa Catarina (IFSC), Urupema, Brazil
*Corresponding author. E-mail: brunetto.gustavo@gmail.com

A.K. Srivastava, Chengxiao Hu (eds.)
Fruit Crops: Diagnosis and Management of Nutrient Constraints
https://doi.org/10.1016/B978-0-12-818732-6.00047-2

1 Introduction

Viticulture is an agricultural activity that is typically more profitable per area than annual crops. Several cultivars of red and white grapes are grown for fresh consumption or for producing wines, juice, and raisins. Grape yield and quality are determined by soil fertility and plant nutritional status. Grapevines grown on acid soils with low fertility may present symptoms of nutrient deficiency and/or excess. Therefore, adequate technical procedures should be used in sampling soils where grapevines will be grown and those of producing vineyards and in analyzing plant tissue (typically leaf tissue). It is possible to accurately estimate the need for and the doses of nutrient application in vineyard soils by soil and tissue chemical analyses. We can also consider parameters such as vegetative growth and expected yield. We can use results interpreted by several methods, especially univariate methods (critical level and sufficiency range) or bivariate methods (diagnosis and recommendation integrated system and compositional nutrient diagnosis).

Nutrient application can affect plant vegetative growth, yield, and quality of the grape. However, grapevine response to fertilization is not always observed. This is sometimes due to nutrient losses (e.g., nitrogen, N; phosphorus, P; and potassium, K) in vineyards, which may result not only in the contamination of surface waters adjacent to vineyards but also of subsurface waters, especially in sandy soils. Thus, it is necessary to establish adequate strategies for nutrient supply in vineyards by fertilization (correction, growth, and maintenance), especially more mobile nutrients in soil, such as N forms. In this chapter, we will focus on the main grapevine cultivars grown in the most traditional wine regions of the world. We will also address the origin of excess and deficiency symptoms of nutrients such as N, P, K, calcium (Ca), magnesium (Mg), iron (Fe), zinc (Zn), and boron (B), the criteria for fertilization, lime and gypsum application, and possible alternative methods to better estimate plant nutritional status. Then, we will report on the main management practices used to increase nutrient use in grapevines, thus reducing soil contamination of vineyards and surrounding areas.

2 Major cultivars

Approximately 10,000 grapevine varieties are grown commercially in the world, but DNA analyses suggest that 5000 is a more accurate number (This et al., 2006). A great number of cultivated grapevines are closely related and known by several names or even synonyms (different names for the same variety) or homonyms (identical names for different varieties). Most varieties belong to *Vitis vinifera*, and analysis of chloroplast DNA suggests that they may originate from at least two geographically distinct populations of *Vitis sylvestris*: one in the Near and Middle East, and the other in a region comprising the Iberian Peninsula, Central Europe, and North Africa (Arroyo-Garcia et al., 2006). In this chapter, we have chosen to highlight some of the grape varieties typically grown in various wine regions of the world.

Chardonnay: Originally from Burgundy, France. It is most likely a product of a cross between Pinot Noir and Gouais Blanc (Bowers et al., 1999). It is a variety of early budding (therefore, it can be exposed to spring frost), medium fertility and yield, and early-to-medium ripening (Galet, 1990). In France, it is used for the production of wine and liqueur wines in the regions of Burgundy and Champagne. In other countries, it is used for the production of varietal wines, sparkling wines, and wines fermented in barrels, with or without malolactic fermentation.

Sauvignon Blanc: Most likely originates from Central or Southeastern France. It is a descendant of Fié (Fiét) grown in the Loire Valley (Galet, 1990; Robinson, 1996). It is a medium budding, vigorous variety with large sprouting of secondary buds. It is of very early ripening, and it is used to produce dry white and natural sweet wines. It is largely grown in different locations in France, especially and most importantly in the Sauternais region. Its aroma has citrus characteristics and something herbaceous of medium intensity. It is also used for the production of "late harvest" wines (Entav et al., 1995).

Merlot: Originally from the French region of Bordeaux. It is most likely a product of a cross between Pinot Noir and Gouais Blanc (Entav et al., 1995; Bowers et al., 1999). It is a medium-to-early budding variety, which is exposed to winter and spring frosts (Galet, 1990; Entav et al., 1995). It is vigorous, very fertile, but of irregular yields, because it is prone to fruiting problems generally intensified by low temperatures during flowering. Great wines such as Pomerol are produced in France. The wine has a smaller body and structure than Cabernet Sauvignon and relatively mild tannins. The aromas are complex and elegant. The palate has a good structure, lots of fruit, and sometimes dry herbaceous characteristics, reminiscent of raisins. The wines are rich in alcohol, with a distinctive, slightly fruity, and herbaceous flavor (Entav et al., 1995).

Cabernet Sauvignon: Originally from the Bordeaux region of France, product of a cross between Cabernet Franc and Sauvignon Blanc (Entav et al., 1995). It is a late budding variety, which makes it less prone to damage by spring frost. Terminal or apical buds tend to sprout first, which causes the central bus of the long branches to sprout with difficulty. Wines can feature pronounced herbaceous or green bell pepper flavors along with high acidity and lighter color and body. A variety used for the production of red wines for aging (Entav et al., 1995; Robinson, 1996).

Pinot Noir: Very old variety of Burgundy, France, which has many biotypes and great heterogeneity. Early ripening variety, medium vigor, low yield, and fertility (Kasimatis et al., 1979; Galet, 1990) adapted to temperate zones (Entav et al., 1995). A quality grape used to produce white, rose, and outstanding sparkling wines. It is also used to produce red wines for aging of great quality, intensity, and aromatic complexity. The wine may present lighter color, intense, sweet, fruity aromas, cherry, and violet notes (Galet, 1990).

Other grape varieties: Riesling, Gewurztraminer, Glera (Prosecco), Moscato, Viognier, and Pinot Gris (white). Sangiovese, Cabernet Franc, Zinfandel, Syrah, Nebbiolo, Tempranillo, and Malbec (red). More information on cultivars can be found in Porro et al. (2016).

3 Symptoms of nutrient deficiency and excess in grapevines

3.1 Symptom origin

Nutritional deficiency and toxicity are common in grapevine and often reduce plant health and yields. For several years, plant physiologists have described changes in plants exposed to nonideal nutrition, including visual symptoms, changes in growth, photosynthesis, root architecture, and mineral composition, all focusing on easily diagnosing nutritional problems (Rustioni et al., 2018; Giehl and von Wirén, 2014; Tiecher et al., 2018). However, our knowledge on the underlying molecular processes and key genes regulating the responses to these conditions is only beginning to emerge, with the characterization of genome-wide gene expression pattern changes regulated by nutrients, mapping of quantitative trait loci involved in differences in nutrient use efficiency, and characterization of transporters involved in nutrient uptake and partitioning (Leng et al., 2015; Vannozzi et al., 2017). With the availability of high-quality reference genome assemblies (Jaillon et al., 2007; Velasco et al., 2007), the feasibility of functional characterization of nutritional homeostasis in grapevine was greatly improved. Here, we review what is known at the molecular level for grapevine nutritional homeostasis.

The molecular mechanisms of N (mainly nitrate) sensing and how plants respond to varying nitrate levels are relatively well known for model plant species (reviewed in O'Brien et al., 2016). N absorption from the soil by plant roots is performed by nitrate, nitrite, and ammonium transporters at root cell plasma membranes. Transport within the plant, including into specific cellular organelles (i.e., mitochondria, plastids, and vacuoles), and long distance transport through xylem and phloem also rely on transporters for every membrane-crossing step. Thus, transporters of N forms in plant are usually found in large gene families (Tsay et al., 2007). Since grapevine is a fruit species in which scion is grafted onto rootstocks, which in turn have to be tolerant to environmental stresses, it is key to identify rootstocks that are more efficient in nutrient use. Studies have shown that N supply and growth of the grafted plant are controlled by the rootstock (Lecourt et al., 2015). Indeed, a recent study evaluated two different rootstock/scion combinations in a split-root experiment, using transcriptomics. Clearly, local N supply regulates core N–related genes, but there is an extensive difference between rootstocks, which may be related to how they control scion vigor (Cochetel et al., 2017). Moreover, VvNRT2.4A and VvNRT2.4B high-affinity transporters were induced upon nitrate resupply, together with the gene encoding the accessory protein VvNAR2.2, although the transporters have not been functionally characterized so far (Pii et al., 2013).

The only well-characterized N transporter in grapevine is a low-affinity nitrate/nitrite transporter named VvNPF3.2 (nitrate transporter1/peptide transporter family, NPF, formerly NRT1/PTR family). VvNPF3.2 is induced upon inoculation with *Erysiphe necator*, the causal agent of powdery mildew, in leaves of susceptible but not resistant genotypes. VvNPF3.2 promoter was shown to be induced by *E. necator* and is mainly expressed in leaf veins. It is known that overfertilization with nitrate increases the severity of mildews in crop plants and that high nitrogen supply increased powdery mildew infection in several grapevine cultivars (Marschner, 2012). Thus, these results indicate that the low-affinity nitrate/nitrite transporter VvNPF3.2 is induced by *E. necator* as part of the pathogen's strategy to increase nitrogen availability by manipulating the plant.

P-efficient rootstock genotypes and strategies to improve P extraction by plant are the key. One such strategy is the use of arbuscular mycorrhizal fungi (AMF), symbiotic organisms that improve plant P uptake. In fact, high P availability has been demonstrated to inhibit mycorrhiza formation in other plants (Balzergue et al., 2011). Interestingly, two phosphate transporter (PHT) genes from grapevine that are highly similar to genes specific to arbuscules in other plants were found to have increased expression when plants are inoculated with AMF (Valat et al., 2018).

Potassium is one the best studied nutrient at the molecular level, given its abundance in the grape berry, which accounts for ~80% of cation present in the fruit (Rogiers et al., 2017). Interestingly, grapevines can also suffer from K deficiency, especially in alkaline soils where Ca and Mg can be absorbed at higher levels (Hannan, 2011). Thus, some K transporters were molecularly described in detail. Low-affinity K transporters from the Shaker gene family, named VvK1.1, similar to AKT transporters of *Arabidopsis thaliana*, are expressed in root cortex cells, being thus involved in K uptake from the soil. Expression was also observed in berries, indicating a role in K berry accumulation. Two other genes from the same family, specific to berries, were also described, named SIRK and VvK1.2 (Cuéllar et al., 2013). Two K transporters from the KUP/HAK/KT gene family, also likely to be involved in K accumulation in fruits, were shown to transport K in heterologous systems (Davies et al., 2006). Finally, VvNHX1, a K^+/H^+ and Na^+/H^+ antiporter, was suggested to be important for berry vacuole expansion (Hanana et al., 2007). However, the physiological roles of these transporters are poorly understood to date, mainly due to the lack of genetic variants to access their effects on mutants and/or natural variants.

Although nutrition is clearly connected to fruit quality, we know little about how nutrients are transported into the grape berry. Ca can cross-link cell wall pectins and fruit cell wall (which is pectin rich) and determine the physical and structural properties of the fruit. Fruit cell wall layers can also be important pathogen susceptibility determinants, and spraying fruits with Ca can improve cell integrity and disease resistance (Hocking et al., 2016). Therefore, proper Ca homeostasis in fruit is key for both development and stress response. Interestingly, Ca has low mobility in the phloem, and thus, its delivery to developing berries in grape occurs early, when xylem conductivity is higher. Once berries are maturing, xylem conductivity drops, and Ca accumulations slows (Hocking et al., 2016). This is in contrast to K, a xylem and phloem mobile element that is delivered throughout berry formation. It is hypothesized that longer periods of maturation (i.e., increased water and Ca delivery) could contribute to cell integrity in berries, making them less susceptible to berry shrivel (Fuentes et al., 2010).

In grapevine, a Ca^{2+} transporter from the CAX (proton-cation exchangers) family was molecularly characterized. CAXs are low-affinity transporters that use the H^+ gradient to export cations from the cytoplasm into the vacuole, Golgi, or to the extracellular space (Pittman and Hirschi, 2016). VvCAX3 was shown to be highly similar to AtCAX1, a well-known Ca^{2+} vacuolar transporter from *Arabidopsis* (Conn et al., 2011). VvCAX3 was shown to be localized to the tonoplast and was able to transport Ca into vacuoles when expressed in yeast cell, but only when an auto inhibitory domain was truncated, a feature that is common to other CAX proteins (Martins et al., 2014). Expression was detected throughout the plant during the vegetative development, especially in stems. Moreover, VvCAX3 expression decreased in berries from green stage until maturity, which is consistent with a role in berry Ca loading (Martins et al., 2014). Thus, VvCAX3 might also be a great tool to improve Ca loading into berries and possibly reduce susceptibility to shrivel.

Zinc deficiency is a common problem that affects plant development, including grapevine, which shows smaller leaves and shortened internodes if Zn is lacking. Another interesting phenotype associated with low Zn concentration is the observation of berry clusters that can vary in size, from normal to very small. Thus, if Zn is low, grapevines should be corrected by Zn application (foliar or in the soil) to increase set fruit. VvZIP3 is a plasma membrane–localized Zn transporter highly expressed in reproductive tissues, especially in developing flowers, and inducible by Zn, indicating its role in Zn loading in flowers and likely in normal berry development. However, it should be noted that Zn-based fungicides are also being used, which can lead to Zn excess in soil and plant stress (Tiecher et al., 2018). Thus, Zn homeostasis should be carefully characterized in grapevine to fine-tune uptake and distribution depending on growth conditions.

Copper is a key micronutrient involved in photosynthesis, respiration, hormone signaling, and antioxidant activity. However, when in excess, Cu can lead to oxidative stress due to the generation of harmful reactive oxygen species via Fenton and Haber-Weiss reactions. Cu homeostasis and Cu concentration regulation in plants have been reviewed but are largely unknown in plants other than the model species *A. thaliana*. In grapevine, Cu toxicity is a particular problem, since a common fungicide used in grapevines is the Cu-based Bordeaux mixture, which leads to Cu accumulation in plant leaves and soils (Tiecher et al., 2016a,b). Grape cells exposed to Cu excess rely on plasma membrane and vacuolar transport for Cu detoxification (Martins et al., 2014). Two transporters from the copper transporter (COPT) gene family, usually involved in high-affinity transport at plasma membrane, were upregulated in grapevine left tissues after Cu treatment, namely, VvCTR1 and VvCTR2 (Leng et al., 2015). Interestingly, VvCTR1 molecular characterization showed tonoplast localization and indicated a role in intracellular Cu transport (Martins et al., 2014). The transcriptional changes observed in grapevine leaves after Cu treatment indicated that other uncharacterized transporters and enzymes are involved in Cu toxicity response (Leng et al., 2015). Moreover, little is known about the molecular players in root responses to high Cu, and thus, Cu homeostasis should be a focus of the grapevine community in the future.

Iron deficiency–induced leaf chlorosis is a common nutritional disorder in grapevine, especially when cultivated in soils with alkaline pH or high in lime, which reduces plant growth and productivity (Tagliavini and Rombolà, 2001). Mild Fe deficiency, however, can positively change wine quality due to improvement in traits such as soluble solids, pH, anthocyanins, and resveratrol. Grapevines are strategy I plants, relying on proton and organic compound release to the rhizosphere to increase Fe^{3+} solubility and Fe reductase activity to reduce Fe^{3+} to Fe^{2+}. Indeed, transcriptomic analyses of rootstock responses to Fe deficiency showed regulation of many orthologous genes to *A. thaliana* "ferrome," or the Fe regulon, confirming that grapevines are strategy I plants. Interestingly, bicarbonate presence in the soil, suggested as an important factor leading to Fe deficiency–induced chlorosis when in the soil, and thus mimicking low Fe, has only a partially overlapping transcriptional response compared to Fe deficiency (Vannozzi et al., 2017). Mapping of loci linked to Fe deficiency tolerance (using lime-induced chlorosis) in rootstocks indicated that (1) there is a QTL with strong effect in chromosome 13, (2) rootstocks are important for the chlorotic phenotype in the scion, and (3) the genetic architecture of the phenotype is polygenic in nature. However, no clear candidate gene associated with the interval was found. Thus, the causative genes involved in low Fe availability to grapevine remain to be described. Moreover, no VvIRT1-like gene has been characterized to date.

Boron is an essential element for normal plant development, being key for plant cell wall integrity. Although essential, B can become toxic in excessive amounts, and plants can only tolerate a narrow range of concentrations within their tissues. Its main role at the cellular level is to form borate-diol ester bonds to link two rhamnogalacturonan II (RGII) chains, a pectic polysaccharide found in the cell wall. B deficiency is associated with reproductive development impairment, having a negative effect on flowering, fruit, and seed set, and with pollen tube growth. In grapevine, a condition known as "shot berries," in which some berries in a cluster are smaller and seedless, might be linked to pollen tube defects caused by B deficiency. A B transporter, named VvBOR1, has recently been characterized. VvBOR1 is highly similar to the known *A. thaliana* AtBOR1, which is involved in B efflux from root endodermis and pericycle cells. VvBOR1 was localized to the plasma membrane and is able to at least partially complement *atbor1* mutants. Interestingly, VvBOR1 expression is correlated with B concentration in normal grape berries, suggesting it could be involved in B loading into fruits (Pérez-Castro et al., 2012).

3.2 Symptoms of nutrient deficiency and excess

Common symptoms of N deficiency in grapevines include decreased plant growth (dwarfism), reduced leaf size, chlorosis, shortening of internodes, reduced development of the root system, and low fertilization of clusters. On the other hand, conditions of N excess increase the vigor of the grapevines (Fig. 47.1A), prolonging of the period of vegetative growth, delaying the ripening of the clusters, reducing anthocyanin content in clusters, and increasing total titratable acidity (Duchene et al., 2011).

P-deficient grapevines exhibit red-violet coloring (dark gray in the print version) in marginal and interveinal leaf areas (especially in old leaves). There is red coloring (light gray in the print version) in the petioles and in the primary and secondary veins of the old leaves. Excess P promotes delayed ripening of the clusters and induces Fe, Zn, and Cu deficiency (Baldi et al., 2018).

K deficiency symptoms are first observed in the older leaves of the grapevines (Fig. 47.1B). The leaves of white grape cultivars exhibit an interveinal yellowing, followed by necrosis of the peripheral zone of the lamina, which progresses into the interveinal tissue. Leaves of red grape cultivars initially exhibit a purplish color (dark gray in the print version) between the veins, followed by progressive necrosis of lamina tissues. Excess of K not only causes stem desiccation, due to lower Ca and Mg uptake by the grapevine, but also favors the increase in wine pH, especially in red wines (Mpelasoka et al., 2003; Dalbó et al., 2015).

Leaves of Mg-deficient grapevines have yellowish margins (light gray in the print version), which are directed toward the center of the leaf, between the veins (Fig. 47.1C and D). The veins remain green (dark gray in the print version). Mg deficiency may cause reduced sugar content in the must (Marschner, 2012) and stem desiccation, due

FIG. 47.1   Symptoms of N excess (A), K deficiency (B), Mg deficiency (C and D), B deficiency (E), and B toxicity (F). *From Melo, G.W., 2003. Correção de deficiência de boro em videira—Circular Técnica (INFOTECA-E). Embrapa Uva e Vinho, Bento Gonçalves.*

to the imbalance of the K/Mg ratio. Ca deficiency affects cell permeability and causes new leaf margins to cup and interveinal and marginal chlorosis, followed by necrosis of lamina margins. Excess Ca in the environment may decrease K uptake.

Excess Cu and Zn are typically found in vineyards due to the intensive use of products for phytosanitary control. Grapevines grown in soils with excess of both metals may exhibit decreased root volume and dry matter (Ambrosini et al., 2016). The young leaves may present chlorosis, due to decreased chlorophyll and photosynthetic pigment content (Cambrollé et al., 2015). Excess Zn in the root system inhibits root lengthening, and the roots may have dark coloring, followed by necrosis (Tiecher et al., 2018).

Symptoms of Fe toxicity are bronzing of the leaves and browning of the roots (Siqueira-Silva et al., 2012). Grapevines grown on limestone soils may exhibit symptoms of Fe deficiency. Leaves may be yellowish with green veins (Tecchio et al., 2012). Symptoms of S deficiency are similar to those of N but appear first in younger leaves. B deficiency causes stunted growth with shortening of internodes, death of bud tips, and interveinal chlorosis of older leaves (Fig. 47.1E). It may also cause fruit abortion and affect pollen tube growth, resulting in poor fruit set. Symptoms of B toxicity include the formation of a "shell" on new shoots, followed by brown necrotic spots (light-gray spots in the print version) on the leaf margin and yellow stripes (light gray in the print version) between the veins (Fig. 47.1F).

4 Diagnosis of nutrient content in vineyard soils

A key issue in soil sampling is a representation in which a small amount of soil needs to adequately represent the cultivation area. The first step of soil sampling is subdividing the area into homogeneous plots, considering soil type, relief, vegetation, and previous use. Cultivation time, variety, rootstock, and yield are also considered in separating plots for sampling existing vineyards. In each homogeneous plot, 10–20 soil subsamples are randomly collected at a depth of 0–20 cm at the installation of new vineyards and in existing vineyards (Comissão de Química e Fertilidade do Solo-RS/SC, 2016). Soil sampling at a depth of 0–40 cm is also recommended in areas that will be grown with grapevines to diagnose the deeper soil layers (Comissão de Química e Fertilidade do Solo-RS/SC, 2016). However, depths may vary according to specific recommendations for the grapevine of each region.

For the installation of a new vineyard, the samples should be collected in the entire area 2–3 months prior to the planting of the seedlings. This allows time to interpret the results and apply and observe the reaction of the inputs in soil. At the installation of a vineyard, liming and fertilization are recommended on the soil surface, followed by incorporation through soil tillage operations. Correction of acidity and fertilization of the entire area are important, because they may favor the development of cover crop species between the rows of the grapevines. Cover crop species can protect the soil surface against the impact of raindrops, reduce erosive processes, and contribute to nutrient cycling (Oliveira et al., 2018). In producing vineyards, soil sampling should be carried out after harvesting, and the sampling locations may vary depending on fertilizer distribution (planting rows, crown projection area, total area, or random collections throughout the area). The soils of the subsamples should be broken down and mixed well. Approximately 500 g should be used for chemical analysis.

The tools used for soil sampling are typically the shovel and soil augers (Dutch, gouge, and spiral), which depend on soil texture and moisture content. The risk of errors due to the loss of the surface layer is greater in sampling with augers. Therefore, it is preferable to use a shovel to collect soil, although it is the tool with the lowest operating efficiency.

In the laboratory, the soil samples are air-dried or dried in an oven with forced air at approximately 45°C until constant weight. Available P and K can be extracted by Mehlich-1-extracting solution (Mehlich, 1953). P is determined by molecular absorption spectrometry (colorimetry), while K is determined by flame atomic emission spectrometry (flame photometry). Some methods suggest the determination of micronutrients such as Cu, Zn, and Mn in Mehlich-1-extracting solution by atomic absorption spectrometry (AAS) (Comissão de Química e Fertilidade do Solo-RS/SC, 2016). The use of anion and cation exchange resins is another method used to extract available P and K, especially for soil P after the application of natural phosphate and other less-soluble inputs. This is because extraction with Mehlich-1 tends to overestimate P availability, by making unavailable P fractions more soluble.

In recent years, multielemental methods that allow the determination of macro- and micronutrients in a single extraction have emerged, such as Mehlich-3 (Bortolon et al., 2009). Although this method is still in the calibration phase in southern Brazilian soils, it optimizes execution with the use of inductively coupled plasma optical emission spectrometry (ICP-OES). When a multielemental method is not used, the exchangeable contents of Ca, Mg, and Al are typically extracted with a concentrated solution of potassium chloride (KCl), with subsequent determination of Ca and Mg in AAS and Al by titration with sodium hydroxide. Soil buffering potential (SMP index) and pH can be determined with potentiometer, which evaluates the activity of H^+ (Tedesco et al., 1995). Organic matter contents are usually determined by sulfochromic digestion followed by titration (Walkley-Black). Soil texture can be determined with the physical separation of the sand fraction, the sedimentation of the clay fraction, and the difference in relation to 100 of the silt fraction, after the application of a dispersing agent in the soil sample (Klein, 2008). Then, cation exchange capacity (CEC), base, and Al saturation are calculated using the chemical attributes determined previously.

The results of the analytical methods need to be calibrated with yield or relative yield of the grapevines or even with grape quality indicators in long-term field experiments.

5 Diagnosis of nutrient contents in grapevines

Plant tissue analysis (typically leaf tissue) allows us to evaluate total nutrient content, which makes it possible to diagnose the nutritional status of fruit trees, such as the grapevine. However, the maximum yield depends on nutrient balance within the plant, characterized by ratios (relationships) among nutrients. Thus, establishing total nutrient content is not always enough to achieve high yields. Assuming that individual yield parameter performs better when the others are close to ideal, interpreting the results of leaf analysis alone has limited applicability for recommending fertilization. Thus, it would be more appropriate to interpret the results of plant tissue analysis in combination with those determined by soil analysis and with other criteria such as expected yield or plant growth parameters.

The diagnosis of nutritional status in grapevines typically aims to detect nutritional problems that are not expressed visually, to identify the cause of the visual symptoms observed in the field, to verify if a given nutrient was absorbed by the plant, to characterize the specific cause of a nutritional problem, and, along with soil analysis, to guide a rational soil fertilization and correction program (Carmo et al., 2000).

However, leaf nutrient content does not depend exclusively on the availability in the soil, as it also reflects the uptake rate by the roots and transport and redistribution to the various organs of the plant (Brunetto et al., 2006). In some cases, the amount of nutrients in soil may be enough to meet plant demand. However, other factors, such as water deficiency or excess (Rozane et al., 2009) or soil compaction (de Souza et al., 2018), may cause low absorption of one or more nutrients, resulting in nutritional imbalance. Moreover, there are situations in which nutrients absorbed by plants are redistributed to organs that act as sinks (e.g., fruits), which results in low leaf contents.

The studies of Brunetto et al. (2007) and Brunetto et al. (2008) on grapevines found that leaf analysis is limited as a tool to diagnose nutritional status, as the authors found leaf N contents increased with the application of increasing N doses, but this was not reflected in increased grape yields. The lack of relationship between leaf nutrient content and yield may arise from the compartmentalization of part of N in cell organelles. This occurs when the nutrient is absorbed in an amount greater than the physiological demand of the plant. It is stored in the vacuole, as is also the case with P and K (Brunetto et al., 2011).

When one must use leaf analysis to diagnose nutritional status, leaves should be collected following the official technical recommendations for grapevine. In general, these recommendations establish the best time of the year or phenological stage for leaf collection, position in the plant and branches where the leaves should be collected, and number of leaves per plant or area. For instance, the Soil Chemistry and Fertility Commission of the states of Rio Grande do Sul (RS) and Santa Catarina (SC) (Comissão de Química e Fertilidade do Solo-RS/SC, 2016) establishes leaf collection at veraison for the largest grape-producing region in Brazil, which is when about 50% of the fruits reach their final color. The complete leaf (leaf + petiole) should be collected opposite the first cluster of a fruit branch. Each plot should be represented by a composite leaf sample consisting of the grouping of 30 single samples. Each simple sample consists of the grouping of four pairs of complete leaves per plant.

If one must use petiole analysis, these should also be removed at the same physiological stage and in the same amounts indicated for leaf collection. However, it should be noted that after collection, leaves and even petioles should be placed in paper bags and sent to a laboratory as soon as possible for analysis. If possible, leaves and/or petioles sampled for analysis should be prepared as follows: washed with neutral detergent solution 1 mL/L, then in running water, followed by immersion in 30 mL/L HCl solution for 15 s and finally rinsed with distilled water. Afterward, samples are placed in paper bags to dry in an oven with forced circulation at $65 \pm 5°C$ until constant weight is reached. Samples are then ground in a Wiley mill (2-mm mesh) and sent to the laboratory, where they are typically subjected to wet digestion (e.g., sulfuric or nitric-perchloric digestion) to estimate total nutrient content.

The results of leaf or petiole analysis can be interpreted with the aid of interpretation tables found in liming and fertilization manuals for the grapevine around the world. Examples are shown in Table 47.1.

TABLE 47.1   Concentrations and nutrients ranges (adequate or normal) in complete leaves of vines in regions of Brazil.

References	g/kg					
	N	P	K	Ca	Mg	S
Comissão de Química e Fertilidade do Solo-RS/SC (2016)	16–24	1.2–4.0	8–16	16–24	2.0–6.0	–
Raij et al. (1997)	30–35	2.4–2.9	15–20	13–18	4.8–5.3	3.3–3.8
Malavolta et al. (1997)	25–27	2.0–3.0	15–20	30–40	3.0–4.0	2.0–3.0
Martinez et al. (1999)	25	2.0	15	4	4	–
Pauletti and Motta (2017)[a]	30–35	2.4–2.9	15–20	13–18	4.8–5.3	3.3–3.8
Pauletti and Motta (2017)[b]	28	2.0	12	20	4.0	2.0

References	mg/kg				
	B	Cu	Fe	Mn	Zn
Comissão de Química e Fertilidade do Solo-RS/SC (2016)	30–65	–	60–180	20–300	25–60
Raij et al. (1997)	45–53	18–22	97–105	67–73	30–35
Malavolta et al. (1997)	30–40	–	–	40–100	25–40
Martinez et al. (1999)	100	15	–	40–100	25–40
Pauletti and Motta (2017)[a]	45–53	18–22	97–105	67–73	30–35
Pauletti and Motta (2017)[b]	30	14	60	30	20

[a] Flowering.
[b] Veraison.

6 Alternative methods of predicting nutritional status in grapevines (DRIS, CND)

The results of plant tissue analysis can be interpreted by several methods, especially univariate methods such as critical level (CL) and sufficiency range (SR).

The interpretation of leaf analysis was proposed by Lagatu and Maume (1934a,b) in Montpellier (France). It is important to emphasize that these authors began their work with grapevines and, that in addition to establishing the basis for foliar diagnosis (chemically determining individual elements), they stressed the importance of the relationship between nutrients.

Among bivariate methods, diagnosis and recommendation integrated system (DRIS) proposed by Beaufils (1973) assumes that binary relationships (between two nutrients) provide a good indication of nutritional balance. The method uses nutritional relationships to explain interactions between elements in plant tissue analysis. The interpretation of DRIS is based on indexes for each nutrient, obtained according to the relationships (ratios) between nutrient contents. However, DRIS has some limitations. It does not allow exclusion of outliers using mathematical procedures; there is some arbitrariness in selecting high-yield subpopulation and incompatibility with multivariate analysis methods; there is a need for considerable computational calculations; results are expressed in indexes that are not independent (therefore may generate inadequate diagnoses) (Parent, 2011).

To obtain more effective tools for assessing plant nutritional status, Parent and Dafir (1992) developed the compositional nutrient diagnosis (CND) method, which is based on the application of principal component analysis to the data, allowing the detection of nutritional imbalances. This method considers that data of plant tissue nutrient concentration belong to the compositional data class (Aitchison, 1986) (i.e., multivariate), where each part cannot be interpreted in isolation, without being related to the others (Parent et al., 2013). This is because it constitutes a structure of dependence in which the variation in any part of the whole affects the relative values of the other parts.

Using CND, we evaluated 81 vineyards of commercial production of wine grapes in the state of Rio Grande do Sul, located in Southern Brazil. Vineyards had an average age of 8 years of planting. The database consisted of yield and leaf contents determined at the beginning of fruit development and showed variation in yields between 69 and 0.4 t/ha, with an average of 14 t/ha and standard deviation of 12 t/ha. Exploring the data before and after excluding 18 outliers by Mahalanobis distance (Parent et al., 2009), 63 vineyards were left in the database. Correlation matrix was used to verify the existence of significant correlations and an adequate coefficient of determination between nutrient contents and yield. We were unable to find any significant correlation with a high coefficient of determination using univariate analysis (Table 47.2). By dividing the reference population as indicated by Khiari et al. (2001b), the diagnosis of the nutritional composition of the 63 vineyards presented a yield of 11,111 kg/ha as the average inflection point in the cumulative function. This value was used as the basis for determining the high-yield (reference) subpopulation ($n = 29$). With the separation of the high- and low-yield subpopulations, we found that none of these subpopulations exceeded the amount of correlations observed in the full database (Tables 47.3 and 47.4). This was expected due to the

TABLE 47.2 Correlation between foliar concentrations of nutrients and grapevine yield ($n = 63$).

	N	P	K	Ca	Mg	S	B	Cu	Fe	Mn	Zn
Yield	−0.10	0.00	0.04	0.05	0.00	−0.10	0.05	−0.15	0.07	0.01	−0.03
N		0.52*	−0.17	0.08	0.13	0.56*	0.05	0.25	0.07	0.04	0.23
P			0.03	0.34*	0.53*	0.61*	0.37*	0.40*	0.27*	0.06	0.07
K				−0.03	0.06	0.08	0.22	0.14	−0.02	0.28*	0.02
Ca					0.77*	0.44*	0.42*	0.27*	0.01	0.10	0.05
Mg						0.45*	0.42*	0.25	0.18	0.11	0.13
S							0.40*	0.40*	−0.02	0.35*	0.35*
B								0.33*	0.05	0.07	0.09
Cu									0.08	0.22	−0.08
Fe										0.14	0.08
Mn											0.04

* Significant ($P < .05$) by Tukey's test.
From Rozane, D.E., Brunetto, G., Melo, G.W.B., et al., 2016. Avaliação do estado nutricional de videiras pela Diagnose da Composição Nutricional-CND. In: Melo, G.W.B., Zalamena, J., Brunetto, G., et al. (Eds.), Calagem, adubação e contaminação em solos cultivados com videiras, 100 ed. Embrapa-Uva e Vinho, Bento Gonçalves, pp. 45–60.

partition of the observations and, thus, a smaller number of occurrences. Also, the main relationships highlighted in the full database ($n = 63$) remained significant, in addition to complementing the S-Cu relationship in the high-yield subpopulation and the Ca-S relationship in the low-yield subpopulation. Minimum, maximum, and average nutrient contents evaluated in the high- and low-yield subpopulations did not differ as to their classification in the content range considered adequate by the Comissão de Química e Fertilidade do Solo-RS/SC (2016). Also, the effect of the correlations between nutrients and yield showed the need to implement bivariate or multivariate methods to diagnose plant nutritional status.

The CND norms developed according to Parent and Dafir (1992), Khiari et al. (2001a), and Parent et al. (2009) resulted in CND-Uva software. Based on foliar analysis of the vineyard, this tool allows us to determine whether nutrient levels in plots are adequate, deficient, or excessive, in relation to other elements. (The software is available for free use at http://www.registro.unesp.br/sites/cnd or www.gepaces.com.br.)

TABLE 47.3   Correlation between foliar concentrations of nutrients and grapevine yield (high yield) ($n = 29$).

	N	P	K	Ca	Mg	S	B	Cu	Fe	Mn	Zn
Yield	−0.07	−0.05	−0.21	0.07	0.06	−0.12	−0.05	0.14	0.04	−0.12	−0.14
N		0.39*	−0.11	0.02	−0.03	0.64*	−0.10	0.30	−0.11	0.19	0.24
P			0.07	0.29	0.51*	0.64*	0.38*	0.71*	0.28	0.01	−0.10
K				−0.19	−0.05	−0.13	0.22	0.04	0.05	0.04	−0.07
Ca					0.78*	0.39*	0.47*	0.27	−0.15	0.09	0.10
Mg						0.34	0.43*	0.27	0.16	−0.03	0.08
S							0.35	0.61*	−0.06	0.43*	0.20
B								0.48*	−0.16	0.11	−0.13
Cu									0.12	0.27	−0.08
Fe										0.10	0.18
Mn											0.05

* Significant ($P < .05$) by Tukey's test.
From Rozane, D.E., Brunetto, G., Melo, G.W.B., et al., 2016. Avaliação do estado nutricional de videiras pela Diagnose da Composição Nutricional-CND. In: Melo, G.W.B., Zalamena, J., Brunetto, G., et al. (Eds.), Calagem, adubação e contaminação em solos cultivados com videiras, 100 ed. Embrapa-Uva e Vinho, Bento Gonçalves, pp. 45–60.

TABLE 47.4   Correlation between foliar concentrations of nutrients and grapevine yield (low yield) ($n = 29$).

	N	P	K	Ca	Mg	S	B	Cu	Fe	Mn	Zn
Yield	−0.10	−0.29	0.09	−0.21	−0.22	−0.12	−0.05	−0.09	0.12	0.18	−0.11
N		0.66*	−0.20	0.14	0.27	0.50*	0.20	0.20	0.18	−0.06	0.23
P			−0.01	0.39*	0.56*	0.60*	0.35*	0.23	0.27	0.10	0.19
K				0.10	0.15	0.25	0.22	0.23	−0.07	0.44*	0.07
Ca					0.75*	0.51*	0.36*	0.33	0.12	0.12	0.00
Mg						0.57*	0.40*	0.26	0.20	0.22	0.17
S							0.47*	0.26	0.00	0.29	0.46*
B								0.26	0.19	0.05	0.25
Cu									0.07	0.19	−0.06
Fe										0.16	0.03
Mn											0.03

* Significant ($P < .05$) by Tukey's test
From Rozane, D.E., Brunetto, G., Melo, G.W.B., et al., 2016. Avaliação do estado nutricional de videiras pela Diagnose da Composição Nutricional-CND. In: Melo, G.W.B., Zalamena, J., Brunetto, G., et al. (Eds.), Calagem, adubação e contaminação em solos cultivados com videiras, 100 ed. Embrapa-Uva e Vinho, Bento Gonçalves, pp. 45–60.

TABLE 47.5 Diagnosis standards of the nutritional composition CND-clr (centered log ratio) in grapevines.

	IN	IP	IK	ICa	IMg	IS	IB	ICu	IFe	IMn	IZn	CND-r^2
LMa	2.13	2.47	2.00	1.86	1.84	1.85	2.17	2.42	2.53	2.52	2.22	20.26
LMi	−1.73	−2.08	−1.69	−1.84	−1.73	−2.33	−1.81	−1.82	−1.96	−1.65	−2.19	2.41
SP_m	0.79	0.77	0.84	0.81	0.85	0.80	0.82	0.77	0.73	0.80	0.74	3.94
D	0.36	0.30	0.25	0.04	−0.01	−0.11	−0.01	0.40	0.41	0.32	−0.25	0.16
V	9.07	8.81	9.67	9.27	9.72	9.19	9.38	8.81	8.38	9.22	8.48	100.00

	N	P	K	Ca	Mg	S	B	Cu	Fe	Mn	Zn
Mean	2.761	0.652	1.981	2.085	0.531	0.718	−3.953	−4.956	−2.719	−1.263	−2.148
SD	0.16	0.164	0.208	0.178	0.147	0.111	0.257	0.152	0.228	0.246	0.322

D, distortion; *LMa*, maximum limit; *LMi*, minimum limit; *SD*, standard deviation; *SP_m*, average standard deviation; *V*, mean percentage variation of the nutritional indexes that make up the average variation of CND-r^2 = 10.62.
From Rozane, D.E., Brunetto, G., Melo, G.W.B., et al., 2016. *Avaliação do estado nutricional de videiras pela Diagnose da Composição Nutricional-CND. In: Melo, G.W.B., Zalamena, J., Brunetto, G., et al. (Eds.), Calagem, adubação e contaminação em solos cultivados com videiras, 100 ed. Embrapa-Uva e Vinho, Bento Gonçalves, pp. 45–60.*

Based on average contents considered adequate for grapevines by the Comissão de Química e Fertilidade do Solo-RS/SC (2016) and analyzing average normal contents of leaf samples for comparison with standards evaluated by the CND norms, the IN, IP, IK, ICa, IMg, IB, IFe, IMn, and IZn indexes were −0.79, −0.35, 0.69, 3.08, 3.36, 2.12, 1.04, −3.78, and −4.26, respectively, where CND-r^2 = 63.72. The conditions under which the research was carried out and using the norms established by the CND method, we found that Mn and Zn contents are underestimated in the largest grape-growing region of Brazil (Table 47.5).

7 Correction of acidity and fertilization in vineyards

7.1 Limestone and gypsum in vineyards

Some of the soils cultivated with grapevines are acidic, especially in tropical regions. This restricts plant development, and therefore, liming is essential to achieve satisfactory yields. Because of low cost of liming compared with other inputs and high benefits of this agricultural practice in correcting acidity, it provides the greatest return in comparison with other technologies such as irrigation, phytosanitary treatment, and even fertilization.

Neutralizing acidity is essential in viticulture, as the crop requires the application of high doses of nitrogen fertilizers, organic fertilizers, and deposition of pruning residues, all of which intensify soil acidity. It is also important to highlight the extent of the root system of grapevine. If there is chemical restriction to root development, water and nutrient uptake will be impaired. The SMP index can be used to indicate the amount of limestone to be added to the soil (e.g., to reach a pH [water] = 6.0, which corresponds to base saturation of about 75%). The inputs used to correct acidity are limestones (typically calcium carbonate + magnesium carbonate), which are ground rocks whose constituents neutralize acidity while still providing Ca and Mg. Common limestones typically have low solubility, hence the need for incorporation into the soil. In the installment of the vineyard, the use of lime with thicker particle size (effective calcium carbonate equivalent [ECCE] < 60%) is recommended. It should be applied uniformly to the entire area 90 days prior to planting the seedlings, incorporating it into a depth of 0–20 cm (if possible, up to 30 cm, increasing the dose by 50%). The coarser particles of common limestone ensure a prolonged residual effect, which lasts for several years after installment of the vineyard. Uniform distribution and incorporation are critical to avoid the appearance over or under calcareous areas. When the amount of limestone exceeds 4 t/ha, it is suggested that half the dose be applied prior to plowing and harrowing and the other half applied subsequently, repeating the incorporation procedures once again.

Incorporation should be avoided in existing and producing vineyards, because it causes severe damages to the root system of fruit trees, in addition to disseminating existing pathogens in the area. In this case, liming could be done in the entire area and/or in the fertilization area of the grapevine. This decision should be made based on the results of the soil analysis of each site. In either case, application should be done without incorporation. The use of fine limestone (ECCE > 90%) is recommended, due to the greater ease of particle movement in the soil profile.

The topsoil (i.e., 0–20 cm) is typically used for soil sampling. However, the grapevine explores a much larger volume of soil. Thus, it is important to understand the conditions of the subsurface layers, especially with respect to Ca and Al concentrations, supporting the application of gypsum to soil. Gypsum ($CaSO_4 \cdot 2H_2O$) is a by-product from the production of phosphoric acid. Due to its chemical nature, it does not correct soil acidity, but it is a source of Ca and S to plants. The application of gypsum favors the increase of the Ca/Al ratio, neutralizes toxic Al, and promotes increases in Ca and S concentrations in depth, given the greater mobility of gypsum in comparison with limestone. The presence of Ca and the elimination of toxic Al are critical for root system growth and exploration of deeper soil layers. However, the effect of gypsum varies with the type of soil and application dose.

Gypsum application should only be used under specific conditions, such as those defined in the state of São Paulo (Brazil): when soil analysis of the 20–40-cm layer reveals Ca concentrations lower than $4\,mmol_c/dm^3$ and Al^{3+} greater than $5\,mmol_c/dm^3$ and/or Al saturation (m%) above 40%. The need for gypsum is estimated according to the following equation (Raij et al., 1997) (Eq. 47.1):

$$NG = clay \times 6 \tag{47.1}$$

where NG is the need for gypsum, given in kg/ha; clay content, given in g/kg.

An alternative to gypsum application is the use of single superphosphate in P fertilization. This is because this source of P has a large amount of gypsum in in its composition, adding Ca and S to soil, in addition to neutralizing toxic Al.

7.2 Fertilization in vineyards

Fertilization prior to the planting of seedlings and typically associated to soil preparation practices is characterized by correction fertilization. This type of fertilization aims to increase nutrient availability to sufficiency levels to further conduct the vineyard with seasonal surface fertilization, depending on the demand of the variety, soil chemical characteristics, and yield. Correction fertilization consists of adding P and K, typically according to the availability class of each nutrient in soil, whose requirements can be found in fertilization manuals of each wine region. P and K fertilizers should preferably be distributed on the soil surface and incorporated into a depth of 0–20 cm. The dose should be increased proportionally with deeper incorporation (e.g., up to 30 or 40 cm). Deep soil fertilization is desired in viticulture, because it will enable the root system to explore greater soil volume, enhancing water and nutrient uptake capacity.

Due to the high mobility of N in soil, N fertilization is not generally recommended in correction fertilization. In addition, rootstocks or seedlings have a small volume of roots exploring the soil, which limits nutrient uptake in deeper soil layers. When soil analysis diagnoses the need for micronutrients, addition should be done simultaneously with the incorporation of P and K fertilizers. Due to the high demand and the characteristics of the soils, B is typically added in correction fertilization of Brazilian vineyards (Melo, 2003).

Fertilization between the planting of the seedlings and the beginning of plant production is called growth fertilization. At this stage, fertilization aims to stimulate root and shoot growth. Considering adequate amounts of P and K were added in correction fertilization, N is usually the only or most important nutrient recommended in growth fertilization. However, in conditions of visual symptoms of deficiency, other nutrients may also be applied to soil, although this will occur as a result of failure in correction fertilization. Mineral and organic fertilizers can be used during this growth period. Application should be done on the planting row or crown projection area and on the soil surface without incorporation to avoid physical damage to the roots. Application of mineral N sources (e.g., urea) should split to increase use and minimize losses, whereas organic sources should be applied at the beginning of budding period (single dose). When organic compounds are available in the region, preference should be given to this type of fertilizer, which should be distributed throughout the area, thus benefiting cover crops and grapevines. Maintenance (or yield) fertilization aims to maintain soil fertility by replenishing nutrients exported from the soil through harvests, in addition to the natural losses caused by water movement in soil (Brunetto et al., 2011). P, K, and N are commonly applied to soil, but other micronutrients may be included depending on specific soil conditions or management system, although generally based on leaf or soil analysis. As in growth fertilization, fertilizers should be applied to the soil surface without incorporation. The application should be carried out on the crown projection area or on the entire area of the vineyard. Chemical or organic sources may be used to supply nutrients. Organic sources may present greater efficiency in relation to mineral fertilizers, due to the gradual nutrient release. This allows better synchronicity with plant uptake, in addition to the possibility of positively impacting physical and biological properties of the soil.

8 Practices to increase nutrient efficiency in vineyards

Some of the nutrients applied in vineyards may be lost. However, there are strategies to reduce nutrient losses, such as defining the best fertilizer application time, dose, source, etc. In general, grapevines are expected to absorb higher amounts of nutrients such as N during budding and flowering (Araújo et al., 1995; Brunetto et al., 2006; Schreiner and Scagel, 2006), because there is greater emission of roots and cell division and elongation of shoot organs. In a study on sandy soil, Brunetto et al. (2016) evaluated the application of 10 kg N/ha at budding + 10 kg N/ha at full flowering, 20 kg N/ha at budding + 20 kg N/ha at full flowering, 20 kg N/ha at budding, and 40 kg N/ha at full flowering. These authors found that Cabernet Sauvignon grapevines absorbed the highest amounts of N when subjected to the application of 20 kg/ha at budding + 20 kg/ha at full flowering. This shows these are the best application times, promoting increased nutrient use by the grapevines.

In combination with application times, providing adequate nutrient doses can also increase nutrient use and yields. Lorensini et al. (2015) found that, even in sandy soil with low organic matter content, the highest grape yields in Cabernet Sauvignon grapevines were obtained when subjected to the application of 20 kg N/ha. However, the authors highlight that part of the N absorbed by the grapevines was most likely from the mineralization of soil organic matter, which is not physically protected (Brunetto et al., 2016).

The nutrient source (e.g., N) can also determine the amount absorbed by the grapevines and the impact on plant parameters. An example of this is the results obtained by Lorensini et al. (2012). The authors evaluated the release of mineral N in sandy textured soil of a vineyard subjected to the application of urea, polymer-coated urea, and organic compost for 141 days. The release of N was faster in soils subjected to urea and polymer-coated urea. On the other hand, more gradual release of N was observed in soil subjected to organic compost application, which may desirable as it provides greater synchronicity with N uptake by the grapevines.

Maintaining cover crops in vineyards can also be a strategy to increase nutrient use efficiency by the grapevines. This is not only because cover crops provide increased nutrient (e.g., N, P, K, Ca, and Mg) cycling but also because of biological N fixation, especially when legumes can promote the increase of N content in the soil. An example of this is the study by Ferreira et al. (2014), who observed that increased N release from black oat (*Avena strigosa*) and vetch (*Vicia villosa*) residues occurred up to 33 days after plant deposition. This corresponded approximately to grapevine flowering, which is one of the phenological stages of increased nutrient demand. However, N mineralized by cover crop residues in vineyards cannot always be absorbed by the grapevines. An example of this is the results obtained by Brunetto et al. (2014). The authors evaluated the decomposition of ^{15}N-labeled rye (*Secale cereale*) residues deposited on the vineyard soil and found that most of the N absorbed by the grapevines was derived from other N sources than from the decomposing residue itself. Also, N derived from rye residues was accumulated preferentially in annual organs. Reduced nitrogen use by the grapevines is likely associated with residue N losses by volatilization, denitrification, runoff, and leaching (Brunetto et al., 2014). Still, cover crops should be used to protect the soil, thus preventing water erosion.

Selecting grapevine rootstocks/cultivars may also be an alternative strategy to increase nutrient use. In a study conducted in nutrient solution with different rootstocks in Sauvignon Blanc and Chardonnay grapevines, Tomasi et al. (2015) evaluated kinetic uptake parameters and high- and low-affinity nitrate transporters. The authors found that nitrate uptake is different in each rootstock and influenced by graft characteristics and that the performance of the transporters depends not only on the rootstock but also on nitrate concentration in the solution. Thus, more efficient rootstocks/cultivars in absorbing less available N can be grown in soils with a lower organic matter content or with a history of less nitrogen fertilization.

9 Contamination of vineyard soils

9.1 Contamination of vineyard soils with P

Phosphorus applied via fertilizers in vineyard soils tends to be adsorbed to the reactive particles of the colloidal fraction, especially Fe and Al oxides. However, continuous P fertilization, without technical criteria and above the export by the fruits (e.g., clusters), may result in soil P accumulation (Schmitt et al., 2014). This increases the likelihood of P loss to water (Gatiboni et al., 2015), where it can cause the proliferation of algae, through the process known as eutrophication.

P losses can occur in vineyard soils both by surface runoff (Ramos et al., 2015; Napoli et al., 2017) and leaching (Esteller et al., 2009). These losses can be either of inorganic forms (Napoli et al., 2017) or organic forms (Ramos et al., 2015). This will depend on the type of fertilization carried out in the vineyards (Ramos et al., 2015). P losses by surface runoff can reach 11 kg P/ha/yr, which corresponds to approximately 26% of the P applied (Martínez-Casasnovas and Ramos, 2006) and equivalent to P exported by the clusters (Tecchio et al., 2011). These P losses are more pronounced in soils with low clay content, due to the smaller amount of adsorptive sites compared with clayey soils.

In general, studies show the increase of P in soils grown with grapevines using P fractionation (Brunetto et al., 2013; Schmitt et al., 2014). Thus, it is possible to evaluate P uptake dynamics in soil, as some studies have shown increases of P in all fractions but mainly in labile fractions (Brunetto et al., 2013; Schmitt et al., 2014), which potentiates the P losses. However, the technique of fractionation is complex, making it difficult to establish a laboratory routine. Consequently, its applicability is restricted to more advanced scientific studies.

Due to problems of excess P, several environmental agencies have established values of P considered safe in the soil. These items consider several factors such as amount of soil P, application form, soil texture, soil slope, and proximity to water resources. For instance, in soils of Southern Brazil, the critical limit is related to the clay content and the amount of P extracted by Mehlich-1, which results in clay content (%) + 40 (Gatiboni et al., 2015). This information is typically included in laboratory reports of soil analysis. However, this methodology can be better tested in soils, including those of vineyards, with different physical characteristics, declivities, and so on.

9.2 Contamination of vineyard soils with nitrate

N fertilization along the grapevine cycle increases the mineral N content in soils (especially nitrate) (Rogeri et al., 2015), even in vineyard soils. Part of nitrate may be absorbed by the grapevines, but because it forms outer sphere complex with functional groups of reactive soil particles, it can be very mobile in the soil profile. Moreover, N forms may also be lost by surface runoff (Napoli et al., 2017), causing decreased soil fertility and promoting eutrophication, along with P, of the watercourses.

N is usually supplied to grapevines in growth and maintenance fertilization. For example, N applications in maintenance fertilization can be done to add up to 100 kg N/ha (Comissão de Química e Fertilidade do Solo-RS/SC, 2016). When a single dose of N is applied, it can increase the tendency of losses of N forms, both by leaching and surface runoff. N fertilizations may increase nitrate losses, especially in sandy soils (Tahir and Marschner, 2017). N losses may reach half of the N dose, especially at high doses.

9.3 Contamination of vineyard soils with Cu and Zn

Frequent applications of Cu-based fungicides such as Bordeaux [$Ca(OH)_2 + CuSO_4$] and copper oxychloride [$CuCl_2 \cdot 3Cu(OH)_2$] are used to control foliar fungal diseases in vineyards, especially mildew (*Plasmopara viticola*) (Brunetto et al., 2017). In a grapevine cycle, the amount of fungicide applications may reach 30 kg Cu/ha (Casali et al., 2008), which often leads to an increase in soil Cu contents considered toxic to plants. For this reason, producers have started using Zn-based fungicides instead of copper fungicides. Bordeaux mixture also contributes to the increase of other heavy metals such as zinc in the soil, although to a lesser extent (Mirlean et al., 2007). The occurrence of high contents of both Cu and Zn in vineyard soils has been reported, especially in the last decade (Brunetto et al., 2014, 2017; Tiecher et al., 2017, 2018).

The maximum contents of Cu and Zn in soil are related to sorption capacity, which depends on several components. In general, sorption capacity is greater with increasing contents of clay minerals, oxides and hydroxides, carbonates, and soil organic matter (Bradl, 2004). Cation exchange capacity (CEC) is also an important factor that regulates Cu and Zn sorption and bioavailability in soil (Brun et al., 2001; Arias et al., 2005). Furthermore, soils with high pH values can withstand higher Cu and Zn contents in comparison with acidic soils, where the bioavailability of these elements is higher (Chaignon et al., 2003).

As a result of the interaction between these physical and chemical factors, Cu and Zn are found in different fractions in the soil, as they are adsorbed at different degrees of energy (Brunetto et al., 2008). However, increased Cu and Zn contents caused by frequent fungicide application can interfere in fraction distribution in soil. In general, Cu and Zn sorption in the soil occurs primarily at the most avid binding sites. Then, the remaining ions are redistributed into fractions that are retained with less energy and greater availability and mobility (Brunetto et al., 2014). One should consider that Cu and Zn accumulation in soils depends on the competitive interaction with the adsorption sites and with other existing chemical species in the environment. The simultaneous addition of these elements to the soil typically favors the increase of Cu and Zn contents in more labile fractions (Tiecher et al., 2016a,b).

Successive fungicide applications and the consequent increase in Cu and Zn bioavailability in vineyard soils increase the risk of environmental contamination, including surface and groundwater (Brunetto et al., 2014, 2017). This was observed by Mirlean et al. (2007), who verified the leaching of Cu and Zn in a sandy soil and attributed this fact to the simultaneous occurrence of heavy rainfall, coarse soil particle size, acidic soil, and shallower water table depth. In addition to water contamination, increasing Cu and Zn contents in forms with higher bioavailability can cause toxicity to grapevines and other existing plants within the vineyards, damaging the entire production system. This toxicity is maximized in sandy, acidic soil with low soil organic matter, conditions found in various sites where grapevines are grown (Tiecher et al., 2016a,b, 2017, 2018).

10 Future research and considerations

In viticulture, selecting suitable rootstocks and varieties is a decisive step for the success of the activity, which has high profitability per area. Understanding the role of specific nutrient transporters is a key to better explain why plants exhibit symptoms of nutrient deficiency or excess and why distinct cultivars perform differently depending on nutritional conditions. This may also contribute to selecting cultivars for different types of soils or even identify the actual demand of the grapevines for a specific nutrient. Furthermore, it is important to have recommendation systems in the wine regions of the world, which provide clear information on correcting soil acidity and fertilization, thus allowing rapid plant growth and adequate grape yields, in addition to of high-quality grapes. However, further research should be done to find the most suitable criteria for estimating soil acidity parameters, adequate nutrient contents in soil, and plant nutritional status (soil and tissue analysis and expected yield have been typically used). Defining the reference values of these criteria is equally important. For example, critical nutrient levels in soil and in tissue enable increased yields and high-quality grapes (wine, other by-products, etc.) for grapevines in general and for specific groups (e.g., wine grapes, table grapes), as well as for rootstocks and cultivars, associating the climatic peculiarities of the growing region. This will allow better definition of the actual need for fertilizer application and doses. This information can be better obtained with the advance in the studies of bivariate methods, such as diagnosis and recommendation integrated system (DRIS) and compositional nutrient diagnosis (CND) in grapevine. Lastly, the technical criteria on nutrient doses and correction, growth, and maintenance fertilization should be followed to obtain plants with adequate nutritional status and to avoid contamination of soil, subsurface waters, or water bodies adjacent to the vineyards. Thus, grape production with desired yields and quality will be possible, in addition to reducing the negative impact on the environment.

References

Aitchison, J., 1986. The Statistical Analysis of Compositional Data. Chapman Hall.

Ambrosini, V.G., Rosa, D.J., Basso, A., Borghezan, M., Pescador, R., Miotto, A., de Melo, G.W.B., Soares, C.R.F.S., Comin, J.J., Brunetto, G., 2016. Liming as an ameliorator of copper toxicity in black oat (*Avena strigosa* Schreb.). J. Plant Nutr. 40 (3), 404–416.

Araújo, F., Williams, L.E., Matthews, M.A.A., 1995. Comparative study of young Thompson Seedless grapevines (*Vitis vinifera* L.) under drip and furrow irrigation. II. Growth, water use efficiency and nitrogen partitioning. Sci. Hortic. 60, 251–265.

Arias, M., Pérez-Novo, C., Osorio, F., et al., 2005. Adsorption and desorption of copper and zinc in the surface layer of acid soils. J. Colloid Interface Sci. 288, 21–29.

Arroyo-Garcia, R., Ruiz-Garcia, L., Bolling, L., et al., 2006. Multiple origins of cultivated grapevine (*Vitis vinifera* L. ssp. sativa) based on chloroplast DNA polymorphisms. Mol. Ecol. 15, 3707–3714.

Baldi, E., Miotto, A., Ceretta, C.A., et al., 2018. Soil-applied phosphorous is an effective tool to mitigate the toxicity of copper excess on grapevine grown in rhizobox. Sci. Hortic. 227, 102–111.

Balzergue, C., Puech-Pagès, V., Bécard, G., et al., 2011. The regulation of arbuscular mycorrhizal symbiosis by phosphate in pea involves early and systemic signalling events. J. Exp. Bot. 62, 1049–1060.

Beaufils, E.R., 1973. Diagnosis and recommendation integrated system (DRIS). In: Soil Science Bulletin 1. University of Natal.

Bortolon, L., Gianello, C., Schlindwein, J.A., 2009. Avaliação da disponibilidade de fósforo no solo para o milho pelos métodos Mehlich-1 e Mehlich-3. Sci. Agrár. 10, 305–312.

Bowers, J., Boursiquot, J.M., This, P., et al., 1999. Historical genetics: the parentage of Chardonnay, Gamay and other wine grapes of Northeastern France. Science 285, 1562–1565.

Bradl, H.B., 2004. Adsorption of heavy metal ions on soils and soils constituents. Colloid Interface Sci. 277, 1–18.

Brun, L.A., Maillet, J., Hinsinger, P., et al., 2001. Evaluation of copper availability to plants in copper-contaminated vineyard soils. Environ. Pollut. 111, 293–302.

Brunetto, G., Kaminski, J., Melo, G.W.B., et al., 2006. Destino do nitrogênio em videiras Chardonnay e Riesling Renano quando aplicado no inchamento das gemas. Rev. Brasil. Frutic. 28, 497–500.

Brunetto, G., Ceretta, C.A., Kaminski, J., et al., 2007. Aplicação de nitrogênio em videiras na Campanha Gaúcha: produtividade e características químicas do mosto da uva. Ciênc. Rural 37, 389–393.

Brunetto, G., Bordignon, C., Mattias, J.L., et al., 2008. Produção, composição da uva e teores de nitrogênio na folha e no pecíolo em videiras submetidas à adubação nitrogenada. Ciênc. Rural 38, 2622–2625.

Brunetto, G., Melo, G.W.B., Kaminski, J., 2011. Critérios de predição da adubação e da calagem em frutíferas. Bol. Inf. Soc. Brasil. Ciênc. Solo 36, 24–29.

Brunetto, G., Lorensini, F., Ceretta, C.A., et al., 2013. Soil phosphorus fractions in a sandy typic hapludaft as affected by phosphorus fertilization and grapevine cultivation period. Commun. Soil Sci. Plant Anal. 44, 1937–1950.

Brunetto, G., Miotto, A., Ceretta, C.A., Schmitt, D.E., Heinzen, J., Moraes, M.P., Canton, L., Tiecher, T.L., Comin, J.J., Girotto, E., 2014. Mobility of copper and zinc fractions in fungicide-amended vineyard sandy soils. Arch. Agron. Soil Sci. 60, 609–624.

Brunetto, G., Ceretta, C.A., De Melo, G.W.B., et al., 2016. Contribution of nitrogen from urea applied at different rates and times on grapevine nutrition. Sci. Hortic. 207, 1–6.

Brunetto, G., Ferreira, P.A.A., Melo, G.W., et al., 2017. Heavy metals in vineyards and orchard soils. Rev. Brasil. Frutic. 39, 263–274.

Cambrollé, J., García, J.L., Figueroa, M.E., et al., 2015. Evaluating wild grapevine tolerance to copper toxicity. Chemosphere 120, 171–178.

Carmo, C.A.F.S., Araújo, W.S., Bernardi, A.C.C., et al., 2000. Métodos de Análises de Tecidos Vegetais Utilizados na Embrapa Solos. Circular Técnica, 6Embrapa Solos, Rio de Janeiro.

Casali, C.A., Moterle, D.F., Rheinheimer, D.S., et al., 2008. Copper forms and desorption in soils under grapevine in the Serra Gaúcha of Rio Grande do Sul. Rev. Brasil. Ciênc. Solo 32, 1479–1487.

Chaignon, V., Sanchez-Neira, I., Herrmann, P., et al., 2003. Copper bioavailability and extractability as related to chemical properties of contaminated soils from a vine-growing area. Environ. Pollut. 123, 229–238.

Cochetel, N., Escudié, F., Cookson, S.J., et al., 2017. Root transcriptomic responses of grafted grapevines to heterogeneous nitrogen availability depend on rootstock genotype. J. Exp. Bot. 68, 4339–4355.

Comissão de Química e Fertilidade do Solo-RS/SC, 2016. Manual de adubação e calagem para os Estados do Rio Grande do Sul e Santa Catarina. Núcleo Regional Sul—Sociedade Brasileira de Ciência do Solo, Santa Maria.

Conn, S.J., Gilliham, M., Athman, A., et al., 2011. Cell-specific vacuolar calcium storage mediated by CAX1 regulates apoplastic calcium concentration, gas exchange, and plant productivity in Arabidopsis. Plant Cell 23, 240–257.

Cuéllar, T., Azeem, F., Andrianteranagna, M., et al., 2013. Potassium transport in developing fleshy fruits: the grapevine inward K(+) channel VvK1.2 is activated by CIPK-CBL complexes and induced in ripening berry flesh cells. Plant J. 73, 1006–1018.

Dalbó, M.A., Bettoni, J.C., Gardin, J.P.P., et al., 2015. Productivity and quality of grapes of cv. Isabel (*Vitis labrusca* L.) under potassium fertilization. Rev. Brasil. Frutic. 37, 789–796.

Davies, C., Shin, R., Liu, W., et al., 2006. Transporters expressed during grape berry (*Vitis vinifera* L.) development are associated with an increase in berry size and berry potassium accumulation. J. Exp. Bot. 57, 3209–3216.

de Souza, W.J.O., Rozane, D.E., de Souza, H.A., et al., 2018. Machine traffic and soil penetration resistance in guava tree orchards. Rev. Caatinga 31, 980–986.

Duchene, E., Schneider, C., Gaudillere, J.P., 2011. Effects of nitrogen nutrition timing on fruit set of grapevine, cv. Grenache. Vitis-Geilwelerhof 40 (1), 45–46.

Entav, A., Inra, A., Ensam, et al., 1995. Catalogue des variétés et clones de vigne cultivés en France.

Esteller, M.V., Martínez-Valdés, H., Garrido, S., et al., 2009. Nitrate and phosphate leaching in a Phaeozem soil treated with biosolids, composted biosolids and inorganic fertilizers. Waste Manag. 29, 1936–1944.

Ferreira, P.A.A., Girotto, E., Trentin, G., et al., 2014. Biomass decomposition and nutrient release from black oat and hairy vetch residues deposited in a vineyard. Rev. Brasil. Ciênc. Solo 38, 1621–1632.

Fuentes, S., Sullivan, W., Tilbrook, J., et al., 2010. A novel analysis of grapevine berry tissue demonstrates a variety-dependent correlation between tissue vitality and berry shrivel. Aust. J. Grape Wine Res. 16, 327–336.

Galet, P., 1990. Cépages et vignobles de France. In: Tome II: L'ampélographie Française. second ed. Dehan, Montpellier.

Gatiboni, L.C., Smyth, T.J., Schmitt, D.E., et al., 2015. Soil phosphorus thresholds in evaluating risk of environmental transfer to surface waters in Santa Catarina, Brazil. Rev. Brasil. Ciênc. Solo 39 (4), 1225–1234.

Giehl, R.F., von Wirén, N., 2014. Root nutrient foraging. Plant Physiol. 166, 509–517.

Hanana, M., Cagnac, O., Yamaguchi, T., et al., 2007. A grape berry (*Vitis vinifera* L.) cation/proton antiporter is associated with berry ripening. Plant Cell Physiol. 48, 804–811.

Hannan, J.M., 2011. Potassium-magnesium antagonism in high magnesium vineyard soils. (Master's thesis)Iowa State University, Ames, IA.

Hocking, B., Tyerman, S.D., Burton, R.A., et al., 2016. Fruit calcium: transport and physiology. Front. Plant Sci. 7, 569.

Jaillon, O., Aury, J.M., Noel, B., et al., 2007. The grapevine genome sequence suggests ancestral hexaploidization in major angiosperm phyla. Nature 449, 463–467.

Kasimatis, A.N., Bearden, B.E., Bowers, K., 1979. Wine Grape Varieties in the North Coast Counties of Califórnia. 4069ANR Publications, Berkeley, CA.

Khiari, L., Parent, L.E., Trembaly, N., 2001a. The phosphorus compositional nutrient diagnosis range for potato. Agron. J. 93, 815–819.

Khiari, L., Parent, L.E., Trembaly, N., 2001b. Critical compositional nutrient indexes for sweet corn at early growth stage. Agron. J. 93, 809–814.

Klein, V.A., 2008. Física do solo. Universidade de Passo Fundo, Passo Fundo.

Lagatu, H., Maume, L., 1934a. Le diagsontic foliaire de la pomme de terre. In: Annual Ecole Nationale Superieure Agronomique de Montpellier 22. pp. 50–158.

Lagatu, H., Maume, L., 1934b. Recherches sur lê diagnostic foliaire. In: Annual Ecole Nationale Superieure Agronomique de Montpellier 22. pp. 257–306.

Lecourt, J., Lauvergeat, V., Ollat, N., et al., 2015. Shoot and root ionome responses to nitrate supply in grafted grapevines are rootstock genotype dependent. Aust. J. Grape Wine Res. 21, 311–318.

Leng, X., Jia, H., Sun, X., et al., 2015. Comparative transcriptome analysis of grapevine in response to copper stress. Sci. Rep. 5, 17749.

Lorensini, F., Ceretta, C.A., Girotto, E., et al., 2012. Lixiviação e volatilização de nitrogênio em um Argissolo cultivado com videira submetida à adubação nitrogenada. Ciênc. Rural 42, 1173–1179.

Lorensini, F., Ceretta, C.A., Lourenzi, C.R., et al., 2015. Nitrogen fertilization of Cabernet Sauvignon grapevines: yield, total nitrogen content in the leaves and must composition. Acta Sci. 37, 321–329.

Malavolta, E., Vitti, G.C., de Oliveira, S.A., 1997. Avaliação do estado nutricional das plantas: princípios e aplicações, second ed Potafospp, Piracicaba, pp. 115–230.

Marschner, H., 2012. Mineral Nutrition of Higher Plants. Academic Press, New York.

Martinez, H.E.P., Carvalho, J.G., Souza, R.B., 1999. Diagnose foliar. In: Comissão de Fertilidade do Solo do Estado de Minas Gerais. Recomendações para o uso de corretivos e fertilizantes em Minas Gerais: 5ª aproximação. Universidade Federal de Viçosa, Viçosa.

Martínez-Casasnovas, J.A., Ramos, M.C., 2006. The costs of soil erosion in vineyard fields in the Penedès-Anoia Region (NE Spain). Catena 68, 194–199.

Martins, V., Bassil, E., Hanana, M., et al., 2014. Copper homeostasis in grapevine: functional characterization of the Vitis vinifera copper transporter 1. Planta 240, 91–101.

Mehlich, A., 1953. Determination of P, Ca, Mg, K, Na and NH₄ by North Carolina Soil Testing Laboratories. University of North Carolina, Raleigh.

Melo, G.W., 2003. Correção de deficiência de boro em videira—Circular Técnica (INFOTECA-E). Embrapa Uva e Vinho, Bento Gonçalves.

Mirlean, N., Roisenberg, A., Chies, J.O., 2007. Metal contamination of vineyard soils in wet subtropics (southern Brazil). Environ. Pollut. 149, 10–17.

Mpelasoka, B.S., Schachtmam, D.P., Treeby, M.T., et al., 2003. A review of potassium nutrition in grapevines with special emphasis on berry accumulation. Aust. J. Grape Wine Res. 9, 154–158.

Napoli, M., Marta, A.D., Zanchi, C.A., et al., 2017. Assessment of soil and nutriente losses by runoff under different soil management practices in a Italian hilly vineyard. Soil Tillage Res. 168, 71–80.

O'Brien, J.A., Vega, A., Bouguyon, E., et al., 2016. Nitrate transport, sensing, and responses in plants. Mol. Plant 9, 837–856.

Oliveira, F.F., dos Santos, R.E.S., de Araujo, R.C., 2018. Dinâmica, agentes causadores e fatores condicionantes de processos erosivos: aspectos teóricos. Rev. Brasil. Iniciaç. Cient. 5 (3), 60–83.

Parent, L.E., 2011. Diagnosis of the nutrient compositional space of fruit crops. Rev. Brasil. Frutic. 33, 321–334.

Parent, L.E., Dafir, M.A., 1992. Theorical concept of compositional nutrient diagnosis. J. Am. Soc. Hortic. Sci. 117, 239–242.

Parent, L.E., Natale, W., Ziadi, N., 2009. Compositional nutrient diagnosis of corn using the Mahalanobis distance as nutrient imbalance index. Can. J. Soil Sci. 89, 383–390.

Parent, S.E., Parent, L.E., Egozcue, J., et al., 2013. The plant ionome revisited by the nutrient balance concept. Front. Plant Sci. 4, 1–10.

Pauletti, V., Motta, A.C.V., 2017. Manual de adubação e calagem para o estado do Paraná. In: Sociedade Brasileira de Ciência do Solo. Núcleo Estadual Paraná, Curitiba.

Pérez-Castro, R., Kasai, K., Gainza-Cortés, F., et al., 2012. VvBOR1, the grapevine ortholog of AtBOR1, encodes an efflux boron transporter that is differentially expressed throughout reproductive development of Vitis vinifera L. Plant Cell Physiol. 53, 485–494.

Pii, Y., Alessandrini, M., Guardini, K., et al., 2013. Induction of high-affinity NO₃– uptake in grapevine roots is an active process correlated to the expression of specific members of the NRT2 and plasma membrane H+-ATPase gene families. Funct. Plant Biol. 41, 353–365.

Pittman, J.K., Hirschi, K.D., 2016. CAX-ing a wide net: cation/H+ transporters in metal remediation and abiotic stress signalling. Plant Biol. 18, 741–749.

Porro, D., Stefanini, M., Palladini, L.A., et al., 2016. Tecnologias para o Desenvolvimento da Vitivinicultura de Santa Catarina, first ed. Palma & Association, Trento.

Raij, B.V., Cantarela, H., Quaggio, J.A., et al., 1997. Recomendações de adubação e calagem para o estado de São Paulo, second ed. Instituto Agronômico de Campinas e Fundação IAC, Boletim técnico, p. 100.

Ramos, M.C., Benito, C., Martínez-Casasnovas, J.A., 2015. Simulating soil conservation measures to control soil and nutrient losses in a small, vineyard dominated, basin. Agric. Ecossyst. Environ. 213, 194–208.

Robinson, J., 1996. Guide to Wine Grapes. Oxford University Press, Oxford, New York.

Rogeri, D.A., Ernani, P.R., Lourenço, K.S., et al., 2015. Mineralização e nitrificação do nitrogênio proveniente da cama de aves aplicada ao solo. Rev. Brasil. Eng. Agríc. Ambient. 19 (6), 534–540.

Rogiers, S.Y., Coetzee, Z.A., Walker, R.R., et al., 2017. Potassium in the grape (Vitis vinifera L.) berry: transport and function. Front. Plant Sci. 8, 1629.

Rozane, D.E., Natale, W., Prado, R.M., et al., 2009. Tamanho da amostra foliar para avaliação do estado nutricional de goiabeiras com e sem irrigação. Rev. Brasil. Eng. Agríc. Ambient. 13, 233–239.

Rustioni, L., Grossi, D., Brancadoro, L., et al., 2018. Iron, magnesium, nitrogen and potassium deficiency symptom discrimination by reflectance spectroscopy in grapevine leaves. Sci. Hortic. 241, 152–159.

Schmitt, D.E., Gatibon, L.C., Girotto, E., et al., 2014. Phosphorus fractions in the vineyard soil of the Serra Gaúcha of Rio Grande do Sul, Brazil. Rev. Brasil. Eng. Agríc. Ambient. 18, 134–140.

Schreiner, R.P., Scagel, C.F., 2006. Nutrient uptake and distribution in a mature 'Pinot Noir' vineyard. HortScience 41, 336–345.

Siqueira-Silva, A.I., da Silva, L.C., Azevedo, A.A., et al., 2012. Iron plaque formation and morphoanatomy of roots from species of resting subjected to excess iron. Ecotoxicol. Environ. Saf. 78, 265–275.

Tagliavini, M., Rombolà, A.D., 2001. Iron deficiency and chlorosis in orchard and vineyard ecosystems. Eur. J. Agron. 15, 71–92.

Tahir, S., Marschner, P., 2017. Clay addition to sandy soil reduces nutrient leaching—effect of clay concentration and ped size. Commun. Soil Sci. Plant Anal. 48, 1813–1821.

Tecchio, M.A., Teixeira, L.A.J., Terra, M.M., et al., 2011. Extração de nutrientes pela videira "Níagara Rosada" enxertada em diferentes portaenxertos. Rev. Brasil. Fruticult. 33, 736–742.

Tecchio, M.A., Terra, M.M., Maia, J.D.G., 2012. Nutrição, calagem e adubação da videira Niágara. In: O cultivo da Videira Niagara no Brasil, Cap 8. Embrapa Uva e Vinho, Bento Gonçalves.

Tedesco, M.J., Gianello, C., Bissani, C.A., et al., 1995. Análise de solo, plantas e outros materiais, second ed. Universidade Federal do Rio Grande do Sul, Porto Alegre.

This, P., Lacombe, T., Thomas, M.R., 2006. Historical origins and genetic diversity of wine grapes. Trends Genet. 22, 511–519.

Tiecher, T.L., Ceretta, C.A., Tiecher, T., et al., 2016a. Effects of zinc addition to a copper-contaminated vineyard soil on sorption of Zn by soil and plant physiological responses. Ecotoxicol. Environ. Saf. 129, 109–119.

Tiecher, T.L., Tiecher, T., Ceretta, C.A., et al., 2016b. Physiological and nutritional status of black oat (Avena strigosa Schreb.) grown in soil with interaction of high doses of copper and zinc. Plant Physiol. Biochem. 106, 253–263.

Tiecher, T.L., Tiecher, T., Ceretta, C.A., et al., 2017. Tolerance and translocation of heavy metals in young grapevine (Vitis vinifera) grown in sandy acidic soil with interaction of high doses of copper and zinc. Sci. Hortic. 222, 203–212.

Tiecher, T.L., Soriani, H.H., Tiecher, T., et al., 2018. The interaction of high copper and zinc doses in acid soil changes the physiological state and development of the root system in young grapevines (*Vitis vinifera*). Ecotoxicol. Environ. Saf. 148, 985–994.

Tomasi, N., Monte, R., Varanini, Z., et al., 2015. Induction of nitrate uptake in Sauvignon Blanc and Chardonnay grapevines depends on the scion and is affected by the rootstock. Aust. J. Grape Wine Res. 21, 331–338.

Tsay, Y.F., Chiu, C.C., Tsai, C.B., et al., 2007. Nitrate transporters and peptide transporters. FEBS Lett. 581, 2290–2300.

Valat, L., Deglène-Benbrahim, L., Kendel, M., et al., 2018. Transcriptional induction of two phosphate transporter 1 genes and enhanced root branching in grape plants inoculated with *Funneliformis mosseae*. Mycorrhiza 28, 179–185.

Vannozzi, A., Donnini, S., Vigani, G., et al., 2017. Transcriptional characterization of a widely-used grapevine rootstock genotype under different iron-limited conditions. Front. Plant Sci. 7, 1994.

Velasco, R., Zharkikh, A., Troggio, M., et al., 2007. A high quality draft consensus sequence of the genome of a heterozygous grapevine variety. PLoS One 2 (12), e1326.

Further reading

Rozane, D.E., Brunetto, G., Melo, G.W.B., et al., 2016. Avaliação do estado nutricional de videiras pela Diagnose da Composição Nutricional-CND. In: Melo, G.W.B., Zalamena, J., Brunetto, G. et al., (Eds.), Calagem, adubação e contaminação em solos cultivados com videiras. 100th ed. Embrapa-Uva e Vinho, Bento Gonçalves, pp. 45–60.

CHAPTER

48

Diagnosis and management of nutrient constraints in guava

William Natale[a],*, Danilo Eduardo Rozane[b], Márcio Cleber de Medeiros Corrêa[a], Léon Etienne Parent[c], José Aridiano Lima de Deus[d]

[a]Federal University of Ceará, Fortaleza, Brazil
[b]São Paulo State University, UNESP, Registro, Brazil
[c]Department of Soil and Agri-Food Engineering, Université Laval, Québec, QC, Canada
[d]Institute of Technical Assistance and Rural Extension of Paraná (EMATER-PR), Curitiba, Brazil
*Corresponding author. E-mail: natale@ufc.br

1 Introduction

Yield and quality of fruit crops are determined by several growth factors. Where growers maximize those factors, the plant can express its genetic potential, reaching economic profitability and environmental protection, avoiding problems associated with factor excess or misbalance. In tropical regions, infertile acid soils prevail due to advanced weathering. Soil capacity to meet plant nutrient demand is supported by liming and fertilization. According to Roberts (2009), up to 60% of the global food production results from the use of fertilizers.

Fertilization, irrigation, and labor costs represent a large proportion of production costs in guava orchards. However, determining the nutritional requirements of a crop like guava remains a challenge. The right dosage of essential elements depends on cultivar, soil and climatic conditions, irrigation, production capacity, crop cycle, etc. With genetic improvement resulting in highly productive guava cultivars, it was necessary to improve management practices, adopting drastic pruning and irrigation and optimizing fertilization to achieve three harvests every 2 years.

This chapter presents lime and fertilizer management from planting to orchard maturity for guava production at high-yield level.

A.K. Srivastava, Chengxiao Hu (eds.)
Fruit Crops: Diagnosis and Management of Nutrient Constraints
https://doi.org/10.1016/B978-0-12-818732-6.00048-4

2 Guava production

2.1 Economic importance

Guava crop is one of the major fruit crops from tropical regions of the terrestrial globe, being known by many as the apple of tropics (Maity et al., 2006; Small, 2012) although its cultivation extends to subtropical regions. Besides, it is one of the cheapest, popular, and good source of ascorbic acid (Osman and Abd El-Rahman, 2009) and other important phytochemicals for human health (Rojas-Garbanzo, 2018) in countries with expressive cultivated area.

According to data from the Food and Agriculture Organization of the United Nations (FAO, 2018) in 2016, the main guava producers in the world were India, China, Thailand, Mexico, and Indonesia. However, it is important to highlight that these data consider the group (mangoes, mangosteens, and guavas), because guava is considered as a minor tropical fruits.

In particular, it is reported that India, Pakistan, China, Brazil, and Indonesia were the main producing countries between the years 2015 and 2017. India is ranked as the major guava producing country, accounting for an estimated 56% of total global output in 2017 (Altendorf, 2018).

Fruticulture differs from other agribusiness activities in the country in three aspects. The first is social: fruit cultivation requires an enormous amount of labor from orchard planting to harvest. The second is economic: there is a strong demand for high-quality fruits, but the production must be viable to maintain fruit producers in rural areas. The third is environmental: cultivation of fruit plants allows managing soils once considered inadequate for conventional agriculture, thus contributing to improve soil quality from its initial infertile state.

Although there is no consensus on its origin in tropical America, its center of origin covers Mexico and Central and South America (possibly from Mexico to Peru) where it is found cultivated and growing wild (Menzel, 1985; Small, 2012). It has high nutritive value and is consumed fresh or processed as candies, juices, jellies, etc. The literature provides extensive information about nutritional facts and qualitative properties such as taste, color, aroma, form, size, appearance, resistance to pests and diseases, and postharvest storage (Amorim et al., 2015a,b; Brunini et al., 2003; Quintal et al., 2017).

3 Soils as growing media for guava production

Fruticulture is an important activity on ferralsols (oxisols) and acrisols (ultisols) (IUSS Working Group WRB, 2015; Soil Survey Staff, 2010) that are deep and permeable, an ideal soil condition for the deep rooting system of perennial plants. However, ferralsols have a strong acid reaction and are infertile, requiring both liming and fertilization to sustain fruit production. However, perennial crops respond to liming and fertilization differently from annual crops. The reasons are many and varied, as follows (adapted from Gros, 1974):

(a) Roots of perennial plants such as guava explore a large volume of soil that increases with plant age, but little is known about reserves of plant-available nutrients in deeper soil layers.

(b) The perennial plant is a substantial reservoir of nutrients. Hence, the tree does not immediately suffer from nutrient deficiencies detected by soil analysis and responds slowly to correcting measures.

(c) Regular pruning makes it difficult to set apart the effects of liming and fertilization from that of pruning that restricts vegetative growth. Pruning is essential because "hunger for light" is as harmful as "hunger for nutrients."

(d) Liming and fertilization are important not only for vegetative development and fruiting but also to sustain future harvest by forming new fruit branches for future harvesting and building reserves of nutrients in belowground and aboveground organs for the next fruiting stages.

Some fruit crops, especially those native to tropical regions such as guava, were considered as rustic plants, as is still the case in some countries, like Pakistan, which are observed: limited of superior varieties, low implementation of good cultivation practices, less developed technology, and resistance to follow recommendations made by technics (Khushk et al., 2009; Hassan et al., 2012). For this reason, they were grown regardless of soil and climate conditions. However, it is not possible to imagine that a soil can be mined indefinitely by a crop without replacing exported nutrients. Due to the specific growth patterns of perennial fruit plants, it remains difficult to implement long-term experimentation. In the short run, soil and plant tissue analyses allow detecting growth-limiting nutrients. Rational liming and fertilization programs can then be elaborated to optimize the efficient use of nutrients, increase fruit yield and quality, and minimize economic costs and environmental risks.

3.1 Soil analysis

There is large variation in soil capacity to supply nutrients. Soils are highly complex and interactive systems. Most soils, especially those located in tropical regions, cannot supply adequate amounts of nutrients to meet the demand of fruit crops. Soil chemical analysis can assess soil fertility status to assist quantifying fertilizer and lime requirements by cropping systems. Soil analyses are easy to perform, standardized, rapid and inexpensive, executable at any moment, and reproducible. Although there is a consensus that soil chemical analysis is the most practical way to evaluate soil fertility, its value as decision tool depends on representative sampling.

In perennial crops, such as fruit crops, there is little consensus on soil sampling procedures. While some bulletins recommend sampling the whole area, research revealed closer relationship between soil nutrients in the inter row and leaf and fruit nutrient contents (Natale et al., 2007; Prado and Natale, 2008). This is due to the large and efficient root system of fruit crops. If soil samples are collected either in the fertilized area within canopy projection or in inter rows, then sample from which location should lead to recommendations? Another very controversial issue is sampling depth, considering the large soil volume explored by the root system. Finally, should the soil be sampled the same way before planting and in established orchards?

Before planting guava orchards, soil sampling should be performed randomly by collecting 15–20 subsamples in representative areas that are uniform in terms of color, slope, topographic position, soil type and management, etc. Because the deep and extensive root system of guava explores the same volume of soil for many years, soils are sampled separately in the top (0–20 cm) and subsurface (20–40 cm) layers to identify growth factor limitations. After compositing and mixing soil subsamples, an approximate amount of 300 g is sent to a registered laboratory well before planting for in-time interpretation of the results, lime and fertilizer recommendation, and input purchase, application, and incorporation into soil before planting.

In established guava orchards, soil samples are collected in strips within fertilized areas that are uniform in plant age, cultivar, yield potential, soil type, fertilization regime, crop management, etc. (Fig. 48.1). The most adequate period for soil sampling is close to the end of the harvest season, allowing enough time between the last and the next fertilization operation and allowing liming as required. Soil in inter rows should be analyzed every 2–3 years in the 0–20-cm and 20–40-cm layers.

Based on a wide database, the interpretation of soil chemical analyses against expected yield has been facilitated by the development of a software to establish liming and fertilization programs, for example, the *Fert-Goiaba* software (Silva et al., 2009) developed by Brazilian researchers that can be accessed for free at http://www.registro.unesp.br/#!/sites/cnd/.

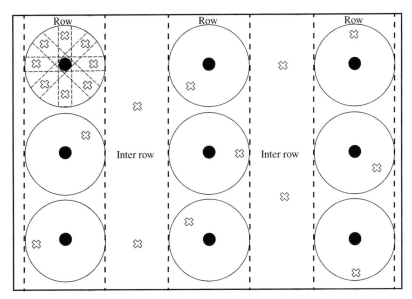

FIG. 48.1 Examples of collection points for soil sampling in strips within fertilized areas (rows) and inter row.

3.2 Leaf diagnosis

Because soil chemical (acidity) or physical (compaction) impediments may limit rooting volume of perennial crops beyond soil sampling depth, the only practical way to assess the overall benefits of liming and fertilization is to "ask the plant" through leaf analysis. Nutritional status monitoring of fruit crops using foliar analysis allows fine adjustment of the mineral nutrition across several nutrients to reach high fruit yield with fertilizers and liming materials (Anjaneyulu et al., 2008). In this matter, soil and plant diagnoses are complementary. The cost of soil and plant analyses is low compared with expected yield increase after rebalancing fertilization. Saving of 30 kg/ha/yr of fertilizer (from over 1400 kg applied) or yield increase of about 100 kg/ha/yr of guava (from over 60,000 kg produced) is sufficient (e.g., in Brazil) to cover the cost of leaf analysis.

The choice of the diagnostic tissue is based on the assumption, confirmed by research, that a key tissue is a representative of carbon assimilation. Because leaves can elaborate substances for growth and fruiting, their nutrient content must reflect the general nutritional status of plants. The diagnostic leaf is a well-defined organ sensitive to change in nutrient concentration and photosynthetic capacity. Leaf diagnosis compares leaf sample analytical results with nutrient standards defined from a "normal" plant or group of plants where all nutrients are in quantities and proportions thought to be adequate to achieve high yields. The interpretation of leaf analytical results is based on causal relationships between leaf nutrient concentrations, on the one hand, and fertilization or liming, on the other, that impact crop productivity.

In the specific case of guava, recently mature leaves (third pair from the tip of the branch), including petiole (Fig. 48.2), is collected at full bloom around the tree, approximately 1.5 m high from soil surface (Natale et al., 2002). Full bloom stage allows enough time for possible correction. Sampling is conducted by grouping areas similar in terms of cultivar, age, yield, management, soil, and fertilization regime. Four pairs of undamaged diagnostic leaves are collected from 20 guava trees in irrigated areas and 40 pairs in rainfed areas (Rozane et al., 2009b). Leaves are collected at least 30 days after the last spraying to minimize contamination, placed in paper bags, and sent to the laboratory within a maximum of 2 days. If longer time is necessary, leaves are placed in a refrigerator at ≈5°C (Souza et al., 2010).

Nutrient budgets compare nutrient removal through harvest with nutrient inputs. It is frequently recommended to fertilize using a "replacement" concept, which is a serious mistake because nutrients necessary to form guava shoots after pruning and the natural renewal of the root system and soil fertility level are not taken into account. In addition, the order of importance of nutrients for fruits is rarely the same as for the whole plant (Amorim et al., 2015a,b). Another aspect to consider is that nutrient-use efficiency of use of the nutrients is generally low in the tropics, despite adequate management techniques, corresponding to 50%, 10%, and 40% for N, P, and K, respectively (Baligar and Bennett, 1986a,b) (Table 48.1). The nutrients exported per ton of fruit and leaf nutrient concentrations is considered as adequate for cultivar "Paluma" (Natale et al., 2002).

FIG. 48.2   Example the diagnostic leaf for sampling (third pair from the tip of the branch). *From Natale, W., Rozane, D.E., 2018. Análise de solo, folhas e adubação de frutíferas. UNESP, Registro.*

TABLE 48.1 Quantity of macro- and micronutrients exported per ton of fresh fruits and contents of leaf macro- and micronutrients considered as adequate for the guava cultivar "Paluma," Brazil.

Nutrients	N	P	K	Ca	Mg	S	B	Cu	Fe	Mn	Zn
	g/t										
Export[a]	1179	121	1554	94	107	107	0.67	1.34	1.88	1.88	1.88
	g/kg						*mg/kg*				
Adequate content	20–23	1.4–1.8	14–17	7–11	3.4–4.0	2.5–3.5	20–25	20–40	60–90	40–80	25–35

[a] *Considering mean yield of 52,475 kg/ha. Dry matter represented, on average, 13.8% of the fresh matter.*
Data from Natale, W., Coutinho, E.L.M., Pereira, F.M., Boaretto, A.E., 2002. Nutrients foliar content for high productivity cultivars of guava in Brazil. Acta Hortic. 594, 383–386. https://doi.org/10.17660/ActaHortic.2002.594.48.

4 Lime requirement, liming, and liming materials

4.1 Liming before planting

Soil acidity is a useful concept relating to pH, aluminum toxicity, base saturation of cation exchange capacity, cationic ratios, lime requirement, and nutrient availability that impact crop yield in tropical regions. Because the roots of fruit trees explore the same volume of soil during most of their lifetime, a uniform and deep incorporation of limestone to at least 30 cm could provide a suitable root environment for rapid plant establishment and efficient use of water and nutrients. Orchard planting offers a unique opportunity to incorporate limestone deeply into soil to facilitate rooting, promote plant development, and foster early fruit production. Before planting seedlings, it is recommended to apply evenly and incorporate 90 days before planting limestone of coarse particle size in the 0–20-cm or 0–30-cm layers to increase soil base saturation (*V*) to 65%. Where lime requirements exceed 4 t/ha, it is recommended to apply half rate before plowing, followed by cross harrowing, and then apply the other half, repeating both operations.

Before planting new orchards, the best liming strategy is to select a limestone with long residual effect in terms of grain size distribution and mineral composition. The Brazilian legislation, for example, allows commercializing limestone with low total relative neutralizing power (TRNP) of at least 45% that contains coarse limestone particles (100% particles passing through of a sieve with 2-mm openings) and have long-lasting effects. If a liming material with 45% TRNP is incorporated into acid soil, a proportion 45% reacts within 90 days leaving the residual proportion of 55% for acid neutralization in future years. Limestone dosage requires not only the TRNP value but also mineral composition enabling to reach the proper cationic base saturation level. The liming material should be incorporated as deep as possible due to its low solubility and slow mobility across the soil profile. Application and incorporation into the 0–30-cm layer is generally sufficient, allowing for rapid establishment and early production of guava trees (Natale et al., 2012). Note that limestone dosage for incorporation down to 30 cm must be enhanced by 50% compared with dosage recommended for incorporation down to 20 cm.

Natale et al. (2007) measured the effect of liming on soil fertility and guava nutrition and yield in Brazil. The soil was a ferralsols (oxisols) with clay texture and initial base saturation of 26% in the 0–20-cm layer. Liming material was applied manually, half before incorporation with a moldboard plow down to 0–30 cm and the other half surface incorporated with a disk harrow. Lime doses were 0, 1.85, 3.71, 5.56, and 7.41 t/ha. Soil chemical analyses were monitored for 78 months following lime application. Nutritional status and yield of guava trees were evaluated across five agricultural harvests. Liming increased pH, Ca, Mg, sum of bases (SB), and base saturation (*V*), besides decreased potential acidity (H + Al) down to 60 cm. Highest cumulative fruit production was obtained where *V* was close to 50% in the row and to 65% in the inter row in the 0–20-cm layer. During the first years after planting, there were generally close correlations between Ca and Mg concentrations in leaf and in inter row and on-the-row soils. Yield increased linearly or nonlinearly with limestone dose during the experimental period (2002–06). The Ca and Mg supply by liming increased Ca and Mg concentrations in the soil and the leaf. Cumulative fruit production showed a quadratic relationship with leaf Ca and Mg. Leaf Ca/Mg ratio close to 4:1 led to largest fruit production.

4.2 Liming established orchards

Limestone incorporation is not recommended in established orchards, due to potential phytosanitary problems resulting from root injury. Considering low limestone solubility; root injury and reduced rooting; risk of infecting the whole orchard; pest invasion, especially nematodes; and loss in soil physical quality, neutralizing soil acidity in the root zone is challenging.

In orchards already established, ammonium-based fertilizers applied within canopy projection contribute to soil acidity, requiring regular soil analyses (0–20-cm and 20–40-cm layers). Sampling is usually performed in the 0–20-cm layer. It is also important to assess calcium and aluminum concentrations in deeper layers. Limestone applied to correct acidity in the arable layer (0–20 cm) moves slowly to the subsurface layer (20–40 cm) depending on soil type, rainfall and irrigation, dosage, elapsed time, and subsequent fertilization.

An experiment carried out in Brazil evaluated doses and types of surface-applied limestone on soil fertility, plant nutrition, and guava yield in an established orchard (Corrêa et al., 2018). There were two types of limestone (common, 80% TRNP, and calcined, 131% TRNP) surface applied at 0, 0.5, 1.0, 1.5, and 2.0 times the recommended dosage to elevate V (%) to the value recommended for guava. Soil acidity was reduced proportionally to lime dosage in the 0–10-cm and 10–20-cm layers. The 20–40-cm and 40–60-cm layers were not affected by liming 24 months following lime application. Common limestone and calcined limestone reduced soil acidity in the 10–20-cm layer, respectively, 24 months and 6–12 months after liming. Leaf and fruit mineral composition and guava yield were impacted 14 and 20 months, respectively, following liming. Hence, surficial liming established orchards could correct soil acidity in the 0–20-cm layer within 24 months. Calcined limestone produced similar effects in orange orchards 12–24 months following surficial liming (Silva et al., 2007). It is recommended to sample soil every year in planting rows below canopy projection and apply limestone in strips, that is, in within fertilized areas (rows). Agricultural implements used for limestone application have devices that adapt to strip application, but the rate must be adjusted to strip area because recommended dosage is per hectare.

Maintenance is performed by applying limestone of fine particle size to raise V to 50% in the row and to 65% in the inter row where V (%) is below 40% and 55%, respectively (Natale et al., 2007). In high-yielding orchards, annual limestone applications in the planting row (up to 1 t/ha) every 2–3 years in the inter row are based on soil analysis. Where soil Mg concentration is less than 9 mmol$_c$/dm^3, dolomitic limestone requires MgO concentration above 12%.

Lime requirement is calculated by using the base saturation method as follows (van Raij et al., 1997):

$$LR = \frac{(V_2 - V_1) \times CEC}{TRNP \times 10}$$

where LR is lime requirement (t/ha), V_2 is ideal base saturation for the crop (%), V_1 is base saturation of the soil (%), CEC is cation exchange capacity (mmol$_c$/dm^3) computed as the sum of exchangeable cations (K, Ca, and Mg) and acidity (H$^+$ + Al^{3+}), and TRNP is total relative neutralizing power, considering carbonate concentration and particle size of the liming material (%). Total acidity (H$^+$ + Al^{3+}) is quantified using the SMP buffer pH method (Shoemaker et al., 1961) converted into mmol$_c$ (H$^+$ + Al^{3+})/dm^3 as follows (Quaggio et al., 1985):

$$\left(H^+ + Al^{3+}\right) = 10^{7.76 + 1.053 pH_{SMP}}$$

Gypsum application is an interesting alternative when the objective is to increase the Ca/Al ratio, neutralize toxic Al, and increase the concentrations of calcium and sulfur in subsurface layers, given the greater mobility of gypsum compared with limestone. The presence of Ca and elimination of toxic Al improve root system development in a larger soil volume. Gypsum alters soil pH by no more than 0.3 units at normal to high dosage because its reaction in soil does not release hydroxyl or carbonate ions (Meurer, 2012). The SO$_4^{2-}$ anions move downward with accompanying cations such as Ca^{2+}, Mg^{2+}, and K$^+$. Gypsum is indicated for crops in general and fruit crops in particular where soil analysis of the 20–40-cm layer shows Ca concentrations less than 4 mmol$_c$/dm^3, Al^{3+} concentrations more than 5 mmol$_c$/dm^3, or aluminum saturation (%) exceeding 40%. Gypsum requirement is estimated as follows (van Raij et al., 1997):

$$GR = 6 \times clay$$

where GR is gypsum requirement (kg/ha) and $clay$ is clay content (g/kg). An alternative product is single superphosphate that contains large amount of gypsum, providing both calcium and sulfur to the crop and neutralizing toxic aluminum.

5 Fertilization

5.1 Fertilization at planting

Phosphate fertilization depends on soil test P using the resin method (Table 48.2). Due to low P mobility in soils, phosphate fertilizers should be dropped in pits or furrows 1 month before planting. Pits are $60 \times 60 \times 60$ cm^3 in size for standard tree spacing of 7×5 m. Pits are filled with 20–30 L of composted cattle manure or 7–10 L of composted poultry manure besides phosphate fertilizer and mixed with excavated soil before filling (Table 48.2). Furrows about 50 cm

TABLE 48.2 Phosphorus dose at planting of guava seedlings based on soil test P.

Fertility classes	P₂O₅ gram per pit or furrow
Very low	270
Low	230
Medium	180
High	140

Obs.: Preferentially use magnesium thermal phosphate + B and Zn or single superphosphate + B and Zn.

TABLE 48.3 Recommendation of fertilization with nitrogen, phosphorus, and potassium for young guava trees cv. "Paluma," according to soil analysis and plant age.

Age Years	Nitrogen N (g/plant)	Phosphorus Very low P₂O₅ (g/plant)	Low	Medium	High	Potassium Very low K₂O (g/plant)	Low	Medium	High
0–1	100	0	0	0	0	200	150	100	50
1–2	200	150	100	50	30	250	200	150	100
2–3	400	200	150	100	50	300	250	200	150

deep are filled with the same mixture (phosphorus, manure, and surface soil). Liming material must not be mixed to avoid phosphate immobilization. If soil analysis shows boron concentration less than 0.21-mg B/dm³ using hot water extraction and colorimetric determination, and zinc concentration less than 0.80-mg Zn/dm³ using DTPA extraction and quantification by atomic absorption spectrophotometry, 2.0 g of boron and 5.0 g of zinc are applied per pit or in furrow, and 3.0-kg B/ha and 6.0-kg Zn/ha are applied broadcast.

5.2 Fertilization of young, nonbearing plants

Between guava planting and 3 years of age, fertilizer recommendations for cv. "Paluma" depend on soil test P (resin method) or K (exchange resin) (Table 48.3).

Fertilizers are applied annually under canopy projection around the plant (Fig. 48.1), within a radius of 0.3 m, increasing up to 0.6 m with age. Fertilizers should be time-split into at least six applications at 30-day intervals during the rainy period. Where possible, especially on sandy soils, 25 L of cured cattle manure/plant/yr or 8 L of poultry manure/plant/yr are applied. Where boron concentrations is less than 0.21-mg B/dm³ and zinc concentrations less than 0.8–mg Zn/dm³, it is recommended to apply 2-g B/plant and 5-g Zn/plant along with manure application.

5.3 Fertilization of established orchards

A modified fertilization program is recommended from the third year after planting when guava trees reach full production. Fertilization must meet crop nutrient requirements to sustain orchard productivity and offset nutrient removal by fruits while maintaining nutrient balance and fruit quality. Nutrient requirements are assessed from soil and leaf analyses, considering plant age, crop management, and expected yield (Table 48.4). Fertilizer dosage is adjusted in intensively exploited orchards under irrigation managed with drastic pruning and three harvests every 2 years.

Fertilizers should be applied every season below canopy projection and around the plant within a radius of 0.6 m from stem. There are at least four applications of equal doses, at pruning, at full bloom, about 30 days later when the fruits reach 1.5 cm in diameter, and about 60 days after flowering when the fruits reach about 2.5–3.0 cm in diameter. Whenever possible, especially in sandy soils, add 25 L of cured cattle manure/plant/yr or 8 L of poultry manure/plant/yr. Considering frequent shortage of boron and zinc and their removal by fruits, foliar application of

TABLE 48.4 Fertilizer recommendation for nitrogen, phosphorus, and potassium in established orchards of cv. "Paluma" according to soil (and leaf) analysis and yield classes.

Production classes	Nitrogen	Phosphorus Fertility classes				Potassium Fertility classes			
		Very low	Low	Medium	High	Very low	Low	Medium	High
t/ha	N (g/plant)	P$_2$O$_5$ (g/plant)				K$_2$O (g/plant)			
<-50	800	200	150	100	50	800	600	400	200
50–60	1000	250	200	150	100	1000	800	600	400
60–70	1200	[a]	250	200	150	[a]	1000	800	600
>70	1400	[a]	300	250	200	[a]	1200	1000	800

[a] It is hardly possible to reach such production class with such low level of phosphorus or potassium.
Obs. 1: Where N leaf concentration exceeds 23-g N/kg, reduce N fertilization regardless of last application.
Obs. 2: Where K leaf concentration exceeds 17-g K/kg, reduce K fertilization regardless of last application.

B and Zn is recommended in fruit-bearing orchards. Each 1 L of liquid mixture should contain 0.6 g of boric acid and 5.0 g of zinc sulfate.

Boron and zinc are monitored by leaf analysis to avoid overreaching the narrow nutrient ranges between deficiency and excess. Plants should be sprayed once at early vegetative development and at flowering. It is advisable to take advantage of phytosanitary treatments performed during these periods, combining pesticides and micronutrients in a single operation.

5.4 Alternative fertilizer sources: By-products of the guava processing agroindustry

The increasing activity of fruit processing generates large amounts of wastes as by-products. Lack of scientific information based on long-term field experiments prevented disposing by-products as fertilizers. By-products can be recycled to sustain the guava production systems viable both environmentally and economically. By-products are composed primarily of guava seeds, which are a "clean" waste. This by-product represents on average 8% of fruit fresh weight and is high in nitrogen and other nutrients (Souza et al., 2014a,b, 2016). A long-term field experiment was conducted in Brazil, applying different doses of guava by-product to measure their effect on soil, plants, and fruit production in a commercial guava orchard (Souza et al., 2014a,b, 2016). The by-product improved soil fertility; leaf N, Mg, and Mn concentrations; and crop production. By-product dosage between 18 and 27 t/ha applied annually was sufficient to maintain leaf N contents at adequate levels.

5.4.1 Contribution pruning to nutrient cycling

Being a perennial plant guava has some peculiarities that make it react differently to liming and fertilization compared with annual crops (Natale et al., 2012). The soil exploration volume is large, and the nutrient vertical distribution is highly variable. There is also a large nutrient reserve in roots, branches, trunk, and leaves. Mobile nutrients can be translocated, preventing or slowing down the occurrence of nutritional deficiencies (Epstein and Bloom, 2005). The biogeochemical cycle that transfers nutrients from soil to plant and from plant to soil is accelerated by drastic pruning of the trees, slowing down plant reaction to lime and fertilizers, and their ensuing effects on plant nutrient status (Hernandes et al., 2012).

In guava orchards, correctly performed pruning requires deep understanding of plant physiology (Simão, 1998). Reducing the number of leaves by pruning weakens the photosynthetic capacity of the plant. However, the more drastic and well conducted is the pruning, the greater is the plant vigor because pruning stimulates sap circulation and plant vegetative development. The vigor of the buds depends on their position and their number of branches. When the plant reaches its maturity, fruiting pruning will be carried out as it favors the production of fruiting branches. The fruiting is a consequence of the accumulation of carbohydrates, which tends to be greater in new branches with smaller diameter, compared with old and larger branches (Piza Júnior, 2002).

Thus, to conduct pruning operations correctly (Fig. 48.3), knowledge about plant vigor, phenological phase, sap flow, canopy relationship, and root system, among others, is mandatory, as outlined by Rozane et al. (2009a):

- Eliminate parts attacked by pests and diseases to keep plants healthy;
- Increase insolation inside the canopy to reach high photosynthetic activity;

FIG. 48.3 Example the pruning technique in guava.

- Sustain high plant productivity with adequate architecture of the canopy by balancing vegetative and fruiting branches;
- Stabilize the production, avoiding alternate bearing and improving fruit distribution and quality (size, color, and weight).

It's important to highlight that the correct time for guava crop pruning differs to different regions of the world due to the geographic location of the area directly influence the plant phenology and flowering cycles (Singh et al., 2001).

6 Evolution of guava nutrient diagnostic methods

Plant mineral nutrition has evolved a lot since the 19th century, where the essentiality of nutrients was demonstrated and the foundations of agricultural chemistry were established based on Sprengel's law of minimum followed by Liebscher's law of optimum stating that production factors perform best where other factors are close to their optimum (De Wit, 1992). The beginning of the 20th century, the idea of using leaf mineral content as criterion for assessing plant nutritional status emerged. The chemical analysis of diagnostic tissues is now an essential tool to determine the nutrient status of perennial crops after reaching nutritional stability.

In India, Hundal et al. (2007) checked that monitoring nutrient status of guava was equally effective by Diagnosis and Recommendation Integrated System (DRIS) and sufficiency range for diagnosing deficiencies of nitrogen, phosphorus, potassium, calcium, sulfur, manganese, zinc, and copper. However, researchers acknowledged that maximum yield depends not only on nutrient concentrations but also on balances between nutrients defined as ratios. Due to myriads of nutrient interactions (Parent, 2011), optimum nutrient concentration ranges taken in isolation are far from sufficient to achieve high yield. Nutrients must be combined in a system approach (Nowaki et al., 2017).

Compositional data are intrinsically multivariate, influencing each other in a system close to the unit of measurement (e.g., kilogram of leaf dry matter). Multivariate analysis has been applied to guava crop databases elaborated over the last 25 years to diagnose nutrients interactively and globally. To facilitate the execution of computations and the interpretation of the results, a team of researchers developed the program "CND-Goiaba" (Rozane et al., 2013) that evaluates the nutritional status of guava based on the *compositional nutrient diagnosis* (CND) method (Parent and Dafir, 1992; Parent, 2011). Based on leaf analysis, this tool allows fruit growers to assess nutrient adequacy, deficiency, or excess accounting for their interactions (Rozane et al., 2012). The program is available for free at https://web.registro.unesp.br/sites/cnd_goiaba/. Fig. 48.4 presents how leaf mineral composition is diagnosed, interpreted, and transcribed using the program "CND-Goiaba." Needless to say, the efficiency of computational tool depends on

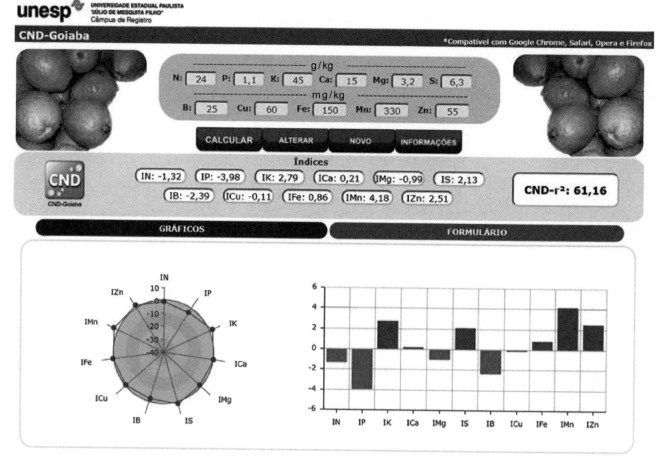

FIG. 48.4 Diagnosis of leaf nutrient composition transcribed and calculated by the program "CND-Goiaba."

the robustness, size, and quality of the database to integrate a progressively larger number of interacting factors, and this requires constant research efforts.

Guava is a very important crop for producing countries. It is a perennial crop that explores a large volume of soil. Because the orchard is established for several decades, the soil must be properly characterized chemically, morphologically, and physically before planting the orchard, and corrective measures must be implemented immediately. Trees must be pruned regularly to maintain its vigor.

In this chapter, we describe and quantify liming, fertilization, and pruning operations from tree planting to orchard maintenance. Nutrient diagnosis is required to adjust the fertilization regime after accounting for soil and plant reserves and for the contribution of pruned branches.

References

Altendorf, S., 2018. Minor tropical fruits: mainstreaming a niche market. In: Food and Agriculture Organization of the United Nations (FAO). Food outlook—Biannual report on global food markets. FAO Trade and Markets Division, pp. 69–76. Available at: http://www.fao.org/3/CA0239EN/ca0239en.pdf.

Amorim, D.A., Rozane, D.E., Souza, H.A., Modesto, V.V., Natale, W., 2015a. Adubação nitrogenada e potássica em goiabeiras 'Paluma': I. Efeito na produtividade e na qualidade dos frutos para industrialização. Rev. Bras. Frutic. 37, 201–209. https://doi.org/10.1590/0100-2945-051/14.

Amorim, D.A., Souza, H.A., Rozane, D.E., Modesto, V.V., Natale, W., 2015b. Adubação nitrogenada e potássica em goiabeiras 'Paluma': II. Efeito no estado nutricional das plantas. Rev. Bras. Frutic. 37, 210–219. https://doi.org/10.1590/0100-2945-052/14.

Anjaneyulu, K., Raghupathi, H.B., Chandraprakash, M.K., 2008. Compositional nutrient diagnosis norms (CND) for guava (*Psidium guajava* L.). J. Hortic. Sci. 3, 132–135.

Baligar, V.C., Bennett, O.L., 1986a. Outlook on fertilizer use efficiency in the tropics. Fertil. Res. 10, 83–86. https://doi.org/10.1007/BF01073907.

Baligar, V.C., Bennett, O.L., 1986b. NPK-fertilizer efficiency—a situation analysis for the tropics. Fertil. Res. 10, 147–164. https://doi.org/10.1007/BF01074369.

Brunini, M.A., Oliveira, A.L., Varanda, D.B., 2003. Avaliação da qualidade de polpa de goiaba 'Paluma' armazenada a -20 °C. Rev. Bras. Frutic. 25, 394–396. https://doi.org/10.1590/S0100-29452003000300008.

Corrêa, M.C.M., Natale, W., Prado, R.M., Banzatto, D.A., Queiroz, R.F., Silva, M.A.C., 2018. Surface application of lime on a guava orchard in production. Rev. Bras. Ciênc. Solo 42, e0170203. https://doi.org/10.1590/18069657rbcs20170203.

De Wit, C.T., 1992. Resource use efficiency in agriculture. Agric. Syst. 40, 125–151. https://doi.org/10.1016/0308-521X(92)90018-J.

Epstein, E., Bloom, A.J., 2005. Mineral Nutrition of Plants: Principles and Perspectives, second ed. Sinauer Associates, Sunderland, MA.

FAO (Food and Agriculture Organization of the United Nations, Statistics Division —FAOSTAT), Available at: http://www.fao.org/faostat/en/#data/QC, 2018.

Gros, A., 1974. Engrais: Guide Pratique de la Fertilisation, sixth ed. Librairie de l'Académie d'Agriculture, Paris.

Hassan, I., Khurshid, W., Iqbal, K., 2012. Factors responsible for decline in guava (Psidium guajava) yield. J. Agric. Res. 50, 129–134.

Hernandes, A., Parent, S.É., Natale, W., Parent, L.É., 2012. Balancing guava nutrition with liming and fertilization. Rev. Bras. Frutic. 34, 1224–1234. https://doi.org/10.1590/S0100-29452012000400032.

Hundal, H.S., Singh, D., Singh, K., 2007. Monitoring nutrient status of guava fruit trees in Punjab, Northwest India through the diagnostic and recommendation integrated system approach. Commun. Soil Sci. Plant Anal. 38, 2117–2130. https://doi.org/10.1080/00103620701548894.

IUSS Working Group WRB, 2015. World Reference Base for Soil Resources 2014. update 2015 International Soil Classification System for Naming Soils and Creating Legends for Soil Maps. World Soil Resources Reports No. 106, FAO, Rome.

Khushk, A.M., Memon, A., Lashari, M.I., 2009. Factors affecting guava production in Pakistan. J. Agric. Res. 47, 201–210.

Maity, P.K., Das, B.C., Kundu, S., 2006. Effect of different sources of nutrients on yield and quality of guava cv. L-49. J. Crop Weed 2, 17–19.

Menzel, C.M., 1985. Guava: an exotic with potential in Queensland. Qld. Agric. J. 111, 93–98.

Meurer, E.J., 2012. Fundamentos de química do solo, fifth ed. Genesis, Porto Alegre.

Natale, W., Coutinho, E.L.M., Pereira, F.M., Boaretto, A.E., 2002. Nutrients foliar content for high productivity cultivars of guava in Brazil. Acta Hortic. 594, 383–386. https://doi.org/10.17660/ActaHortic.2002.594.48.

Natale, W., Prado, R.M., Rozane, D.E., Romualdo, L.M., 2007. Efeitos da calagem na fertilidade do solo e na nutrição e produtividade da goiabeira. Rev. Bras. Ciênc. Solo 31, 1475–1485. https://doi.org/10.1590/S0100-06832007000600024.

Natale, W., Rozane, D.E., Parent, S.-É., Parent, L.E., 2012. Soil acidity and liming in tropical fruit orchards. In: Issaka, R.N. (Ed.), Soil Fertility. IntechOpen, https://doi.org/10.5772/53345.

Nowaki, R.H.D., Parent, S.-É., Cecílio Filho, A.B., Rozane, D.E., Meneses, N.B., Silva, J.A.S., Natale, W., Parent, L.E., 2017. Phosphorus over-fertilization and nutrient misbalance of irrigated tomato crops in Brazil. Front. Plant Sci. 8, 825. https://doi.org/10.3389/fpls.2017.00825.

Osman, S.M., Abd El-Rahman, A.E.M., 2009. Effect of slow release nitrogen fertilization on growth and fruiting of guava under mid Sinai condition. Aust. J. Basic Appl. Sci. 3, 4366–4375.

Parent, L.E., 2011. Diagnosis of the nutrient compositional space of fruit crops. Rev. Bras. Frutic. 33, 321–334. https://doi.org/10.1590/S0100-29452011000100041.

Parent, L.E., Dafir, M., 1992. A theoretical concept of compositional nutrient diagnosis. J. Am. Soc. Hortic. Sci. 117, 239–242.

Piza Júnior, C.T., 2002. A poda da goiabeira "de mesa", second ed. CATI, Campinas. 44 p. (Boletim Técnico, 222).

Prado, R.M., Natale, W., 2008. Effect of liming on the mineral nutrition and yield of growing guava trees in a typic hapludox soil. Commun. Soil Sci. Plant Anal. 39, 2191–2204. https://doi.org/10.1080/00103620802137613.

Quaggio, J.A., van Raij, B., Malavolta, E., 1985. Alternative use of the SMP-buffer solution to determine lime requirement of soils. Commun. Soil Sci. Plant Anal. 16, 245–260.

Quintal, S.S.R., Viana, A.P., Campos, B.M., Vivas, M., Amaral Júnior, A.T., 2017. Selection via mixed models in segregating guava families based on yield and quality traits. Rev. Bras. Frutic. 39, e-866. https://doi.org/10.1590/0100-29452017866.

Roberts, T.L., 2009. The role of fertilizer in growing the world's food. Better Crops Plant Food 93, 12–15.

Rojas-Garbanzo, C., 2018. Psidium fruits: endemic fruits of Latin America with a wide variety of phytochemicals. Clin. Oncol. 3, 1479.

Rozane, D.E., Brugnara, V., Souza, H.A., Amorim, D.A., 2009a. Condução arquitetura e poda da goiabeira para 'mesa' e/ou 'indústria'. In: Natale, W., Rozane, D.E., Souza, H.A., Amorim, D.A. (Eds.), Cultura da goiaba do plantio à comercialização. In: vol. 2. FUNEP, Jaboticabal, pp. 407–428 (chapter. 17).

Rozane, D.E., Natale, W., Prado, R.M., Barbosa, J.C., 2009b. Tamanho da amostra foliar para avaliação do estado nutricional de goiabeiras com e sem irrigação. Rev. Bras. Eng. Agríc. Ambient. 13, 233–239.

Rozane, D.E., Natale, W., Parent, L.E., Santos, E.M.H., 2012. The CND-Goiaba 1.0 software for nutritional diagnosis of guava (Psidium guajava), Paluma cultivar, in Brazil. Acta Hortic. 959, 161–166. https://doi.org/10.17660/ActaHortic.2012.959.19.

Rozane, D.E., Natale, W., Parent, L.E., Parent, S.-É., Santos, E.M.H., 2013. CND-Goiaba. BR5120130003792. Apr. 18.

Shoemaker, H.E., McLean, E.O., Pratt, P.F., 1961. Buffer methods for determining lime requirement of soils with appreciable amounts of extractable aluminium. Soil Sci. Soc. Am. Proc. 25, 274–277.

Silva, M.A.C., Natale, W., Prado, R.M., Corrêa, M.C.M., Stuchi, E.S., Andriolli, I., 2007. Aplicação superficial de calcário em pomar de laranjeira Pêra em produção. Rev. Bras. Frutic. 29, 606–612.

Silva, S.H.M.G., Natale, W., Santos, E.M.H., Bendini, H.N., 2009. FERT-Goiaba: software para recomendação de calagem e adubação para goiabeira cultivar Paluma, irrigada e manejada com poda drástica. In: Natale, W., Rozane, D.E., Souza, H.A., Amorim, D.A. (Eds.), Cultura da goiaba do plantio à comercialização. FCAV, Capes, CNPq, FAPESP, Fundunesp, SBF, Jaboticabal, pp. 281–283.

Simão, S., 1998. Tratado de fruticultura. FEALQ, Piracicaba. 760 p.

Singh, G., Singh, A.K., Rajan, S., 2001. Influence of pruning date on fruit yield of guava (Psidium guajava L.) under subtropics. J. Appl. Hortic. 3, 37–40.

Small, E., 2012. Top 100 Exotic Food Plants. CRC Press, Boca Raton, FL.

Soil Survey Staff, 2010. Keys to Soil Taxonomy, 11th ed. Natural Resources Conservation Service, United States Department of Agriculture, Washington, DC.

Souza, H.A., Rozane, D.E., Hernandes, A., Romualdo, L.M., Natale, W., 2010. Variação no teor de macronutrientes de folhas de goiabeira, em função do tipo e tempo de armazenamento. Rev. Biotemas 23, 25–30.

Souza, H.A., Rozane, D.E., Amorim, D.A., Modesto, V.C., Natale, W., 2014a. Uso fertilizante do subproduto da agroindústria processadora de goiabas. I—Atributos químicos do solo. Rev. Bras. Frutic. 36, 713–724.

Souza, H.A., Rozane, D.E., Amorim, D.A., Modesto, V.C., Natale, W., 2014b. Uso fertilizante do subproduto da agroindústria processadora de goia-bas. II—Estado nutricional e produção de goiabas. Rev. Bras. Frutic. 36, 725–730.

Souza, H.A., Parent, S.-É., Rozane, D.E., Amorim, D.A., Modesto, V.C., Natale, W., Parent, L.E., 2016. Guava waste to sustain guava (*Psidium guajava*) agroecosystem: nutrient "balance" concepts. Front. Plant Sci. 7, 1252. https://doi.org/10.3389/fpls.2016.01252.

van Raij, B., Cantarella, H., Quaggio, J.A., Furlani, A.M.C., 1997. Recomendações de adubação e calagem para o Estado de São Paulo, second ed. Fundação IAC, Campinas.

Further reading

Natale, W., Rozane, D.E., 2018. Análise de solo, folhas e adubação de frutíferas. UNESP, Registro.

Diagnosis and management of nutrient constraints in citrus

Tripti Vashisth*, Davie Kadyampakeni

University of Florida, Institute of Food and Agricultural Sciences, Citrus Research and Education Center,
Lake Alfred, FL, United States
*Corresponding author. E-mail: tvashisth@ufl.edu

1 Introduction

1.1 Citrus origin and its production

Citrus is one of the most widely cultivated fruit crops in the world. The interest in citrus continues to increase due to its numerous health benefits, refreshing flavor, aroma, and taste (Baldwin et al., 2014). Citrus fruits are rich source of vitamin C and number of other health benefiting compounds such as flavonoids and carotenoids (Lv et al., 2015). The genus *Citrus* and related genera (*Fortunella*, *Poncirus*, *Eremocitrus*, and *Microcitrus*) are angiosperm, belonging to Rutaceae family and subfamily Aurantioideae. Citrus is believed to have originated from Southeast Asia, especially in foothills of Himalaya. A newly published genomics study in "Nature" by Wu et al. (2018) suggests that citrus is originated in Southeast Asia and flourished under positive effect of monsoon climate. However, as the monsoon effect weakened in late Miocene period, citrus moved to different parts of Asia and Australia (Fig. 49.1). With progression to different geographical locations and climate, citrus diversified to distinct species. Currently, according to Food and Agriculture Organization (FAO), more than 80 countries produce citrus and the main production region ranging from zero latitude

A.K. Srivastava, Chengxiao Hu (eds.)
Fruit Crops: Diagnosis and Management of Nutrient Constraints
https://doi.org/10.1016/B978-0-12-818732-6.00049-6

FIG. 49.1 Proposed origin of citrus and ancient dispersal routes. *Arrows* suggest plausible migration directions of the ancestral citrus species from the center of origin—the triangle formed by northeastern India, northern Myanmar, and northwestern Yunnan. *Adapted from Wu, G.A., Terol, J., Ibanez, V., López-García, A., Pérez-Román, E., Borredá, C., Domingo, C., Tadeo, F.R., Carbonell-Caballero, J., Alonso, R., Curk, F., 2018. Genomics of the origin and evolution of citrus. Nature 554 (7692), 311.*

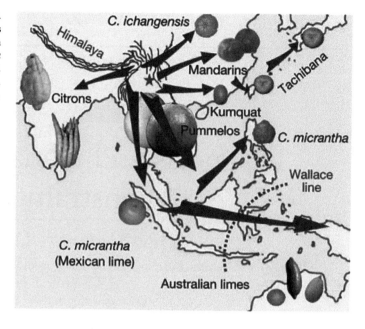

(equator) to 40° north-south latitude. The world citrus production has been consistently increasing. According to FAO 2016–17, China is the largest producer of total citrus, whereas Brazil is the largest producer of oranges; interestingly, 10 years ago, Brazil was the largest producer of citrus and China was the second largest (China's citrus production has doubled in 10 years). Interestingly, in Brazil, orange is the only type of citrus grown; for more than 20 years, Brazil has predominated the orange production worldwide, whereas rest of the major citrus-producing countries grows a variety of citrus.

2 Types and main cultivars of citrus

The genus *Citrus* encompasses a wide range of species, ranging from ornamental shrubs to fruit trees. Some of the common types of citrus fruit are the following:

Oranges: Often interchangeably referred to as "sweet orange," it is one of the most common and widely grown types of citrus. Oranges are believed to have originated and cultivated in China and were brought to western hemisphere by Columbus in 15th century. Orange fruit are spherical to oblong, with low number of seeds, and good sugar content. Oranges are categorized in four groups: round oranges, navel oranges, blood oranges, and acidless oranges. Round oranges are widely used for processing juice; "Hamlin" and "Valencia" are the most popular round oranges. Navel are the second most popular orange and are primarily consumed fresh; cara cara and Washington navels are the most popular navel cultivars.

Mandarins: Mandarins and tangerines are often used interchangeably to refer to easy peeling, well-segmented, orange-colored fruit that are primarily consumed fresh. Mandarin fruit are well distinct from fruit of sweet oranges, and the trees are more cold hardy than sweet oranges. Satsuma mandarin is native to Japan and is one of the most cold hardy citrus species. Some of the common mandarin cultivars are Ponkan, Nagpur, Temple, Dancy, Murcott, and Minneola. Recently, a newly released mandarin cultivar, LB8-9 (Sugar Belle), is gaining popularity due to its tolerance to a bacterial disease called "Huanglongbing" (HLB; discussed at the end of this chapter). In addition, a tangerine hybrid, *Citrus clementine*, has become popular due to its easy peel, small size, low seed, and great flavor characteristics.

Grapefruit: Grapefruit are a distinct commercial species, a hybrid of "pummelo" and sweet orange. Grapefruit are very popular in many cultures and are a high value crop. Grapefruit are known for their distinct flavor, with low acid, high sugars and flavonoids (imparting bitter-astringent flavor), and large fruit size. Duncan, Marsh, white-flesh, and red-flesh grapefruit are the most common grapefruit varieties.

Acid fruits: This group primarily comprises lemons and limes. They mostly are grown in warm climate, where freeze event is unlikely. Lemons and limes are vigorous, thorny trees, with multiple fruit-set in a year. Lemons and limes are known for their sourness/high acidity. Some of the common acid fruit cultivars are Meyer lemon, Key lime, and Tahiti lime.

Microcitrus: This group is native to Australia. They grow in warm, arid climate and are extremely thorny shrub. They are distinct from any other kind of citrus due to their loosely grouped small, round/oblong juice vesicles. Australian finger lime cultivar is one of the well know microcitrus; recently, they have been gaining popularity for two reasons: culinary use and tolerance to HLB.

3 Soil type of citrus-producing regions

The main citrus-producing areas in the world are found in soil orders such as oxisols, ultisols, entisols and alfisols, inceptisols, and mollisols, with high organic matter predominate in temperate areas like Argentina. For example, the soils found in citrus-producing areas of Brazil, predominantly oxisols, are heterogeneous, presenting characteristics that give them different potentials and limitations for sustainable use and management. These soils are highly weathered and inherently infertile, holding low organic matter and levels of potassium (K), calcium (Ca), and magnesium (Mg), low cation exchange capacity (CEC; pH-dependent charge), high phosphorus (P) fixing capacity, and low availability of micronutrients (Mattos et al., 2019). Furthermore, they are acidic with high levels of toxic aluminum (Al) and present low water holding capacity when texture is sandy and susceptible to compaction when clayey (Table 49.1).

In Florida with a humid subtropical region, citrus is grown in extremely well-drained sandy soils low in mineral nutrients and organic matter (entisols); in fine-textured lowland soils with a sandy surface, often developed on calcareous marine deposits (alfisols); or in sandy, acidic, coarse-textured, and poorly drained soils (spodosols) (Alva and Tucker, 1999) (Table 49.1). Entisols are sandy mineral soils low in organic matter, natural fertility, and water-holding capacity (Weil and Brady, 2016). They have weak or no diagnostic subsurface layers and are well to excessively well drained (Obreza and Collins, 2008). Spodosols are sandy mineral soils low in organic matter and natural fertility in the surface layer that contain an acidic subsurface restrictive layer composed of Al and iron (Fe) "cemented" together with organic matter (Obreza and Collins, 2008). Alfisols are sandy mineral soils low in organic matter in the surface layer but higher in relative natural fertility compared with spodosols (Obreza and Collins, 2008). Alfisols contain a subsurface layer of loamy material (a mixture of mostly clay and sand with little silt) that has a relatively high water-holding capacity (Weil and Brady, 2016).

Citrus is cultivated in more than half of the Mediterranean-climate regions. Other similar regions with citrus are found in California and the Western Cape Province of South Africa. In the Mediterranean basin, the main

TABLE 49.1 Typical root zone soil physical and chemical properties for common soil orders found in citrus-producing regions of some parts of the world.

| Soil orders | Soil texture | | | Organic matter | Water-holding capacity | | pH | Cation exchange capacity |
	Sand g/kg	Silt	Clay	g/dm^3	cm/m	cm in the root zone		mmol$_c$/dm^3
BRAZIL[a]								
Oxisol	160–270	660–790	40–64	5–11	–	–	4.1–4.2	24–54
Ultisol	120–270	680–810	34–44	8–10	–	–	5.2–5.7	40–66
Entisol	120–135	840–860	13–18	6–10	–	–	4.3–4.9	32–49
FLORIDA[b]								
Entisols	970–985	50–125	75–125	5–10	2.5–6.6	1.5–5.3	3.6–7.3	20–40
Alfisols	850–965	20–60	15–90	5–3	2.5–10.7	1.3–4.6	4.5–8.4	20–180
Spodosols	960–985	10–35	05–10	10–30	2.5–6.6	1.3–3.8	3.6–7.3	20–60
SPAIN[c]								
Alfisol	502	230	268	26	–	–	6.4	17.8
Inceptisol	345	272	383	26	–	–	7.8	30.3
Vertisol	93	370	537	31	–	–	7.7	46.5

[a] *Adapted from Corá et al. (2005).*
[b] *Adapted from Obreza and Collins (2008). Characteristics were measured in the top 90 cm of soil for central Ridge Entisols and top 45 cm of soil for Flatwoods alfisols and spodosols.*
[c] *Adapted from Duiker et al. (2001).*

citrus-producing countries are Spain, Egypt, Italy, Turkey, and Morocco. Although differences exist regarding soils, in California (the United States), the soils are generally more fertile than in the other Mediterranean-climate regions such as South Africa and the Mediterranean basin where they are alkaline.

The most common soil pattern in Mediterranean areas, expressed in terms of temperature and soil moisture regimes, is xeric, whereby most of the rainfall is observed during winter, which is followed by an important dry period during summer (Mattos et al., 2019). The mean annual soil temperature ranges between 15°C and 22°C (Soil Survey Staff, 2014). CEC and base saturation are generally high, except of some small areas where lithology is dominated by acid minerals (Table 49.1). Since the CEC is almost saturated, most of these soils are well provided by nutrients as P and K, usually well supplied from the mineral weathering with high illitic clay content and adequate levels of Ca and Mg. However, iron (Fe), copper (Cu), manganese (Mn), zinc (Zn), and boron (B) are frequently deficient in calcareous soils. Additionally, in irrigated land, salinization may become an important problem, due to mainly bad irrigation quality water and/or water-saving systems, in combination with the Mediterranean-climate characteristics. In some parts of California (the United States), the moderate soil weathering and clay illuviation to deeper B horizons are the main processes to build-up the alfisol type of soils. In these soils, the hematite-induced reddening of the clays due to summer dehydration of free Fe-oxyhydroxides. Other type of soils such as xerochrept (calcisols) is found in semiarid region where the main soil forming processes is the carbonate dissolution and reprecipitation, which build-up calcic horizons. Other minority type of soils where citrus is cultivated in Spain are the vertisols, mostly in lowlands, where deep layers of swelling/cracking clays have sedimented.

4 Role of mineral nutrition on yield and fruit quality

Like other plant species, 17 essential nutrients are required for growth, development, and normal functioning of citrus trees. The tree uptakes, carbon, hydrogen, and oxygen from the surrounding but the rest of the 14 essential nutrients should be supplied in form of fertilizer. The 14 essential mineral nutrients that are critical for successful citrus production are the following: nitrogen (N), phosphorus (P), potassium (K), calcium (Ca), magnesium (Mg), sulfur (S), iron (Fe), zinc (Zn), manganese (Mn), boron (B), copper (Cu), molybdenum (Mo), chlorine (Cl), and nickel (Ni).

Mineral nutrients are divided into macronutrients, which are elements that plants require in large amounts (N, P, K, Ca, Mg, and S), and micronutrients, which are needed only in small amounts (Fe, Zn, Mn, B, Cu, Mo, Ni, and Cl). Nitrogen, phosphorus, and potassium are referred to as primary macronutrients, while Ca, Mg, and S are considered secondary macronutrients. Citrus, being evergreen, perennial tree with long fruit-growing period, has a high nutrient requirement throughout the year. It is essential that the citrus fertilizer should have all the essential macro- and micronutrients. Recently, complete and balanced fertilizer has been showing promising effect on managing HLB-affected trees (further discussed in Section 7); these observations underline the critical role of mineral nutrition in citrus production.

4.1 Nitrogen

Similar to other fruit crops, nitrogen is one of the most well-studied nutrients in citrus production, largely due to its high requirement by the tree. Nitrogen is required in many processes including vegetative and reproductive growth. In citrus, new leaves and developing fruit accumulates high amount of nitrogen (Feigenbaum et al., 1987); therefore, year round fertilization of nitrogen is beneficial for the tree. The citrus tree requires nitrogen in optimum concentration for good yield and fruit quality. Deficiency and excessive amount of nitrogen can be detrimental for fruit quality and non-beneficial for yield. For example, under Florida conditions, 200 kg/ha/yr is recommended for optimal yield with good fruit quality (Fig. 49.2; Alva et al., 2006); higher rate of nitrogen does not improve yield however and can reduce the fruit quality (Alva et al., 2006; He et al., 2003).

4.2 Potassium

Citrus tree requires potassium in same or higher magnitude as nitrogen. For example, in Florida conditions, nitrogen/potassium ratio is 1:1 to 1:1.25 (Obreza and Morgan, 2008) for optimal yield and fruit quality. Potassium is required for many physiological and metabolic processes; nevertheless, K^+ role in osmosis, stomatal opening and closing are well recognized. A tree deficient in potassium is relatively more drought susceptible (Gimeno et al., 2014) than a tree receiving optimal potassium, and therefore, K^+ deficient tree will have decreased photosynthesis (Vu and Yelenosky, 1991). Potassium also improves fruit size and weight, total soluble solid (TSS; Brix), and peel thickness (Ashraf et al., 2010).

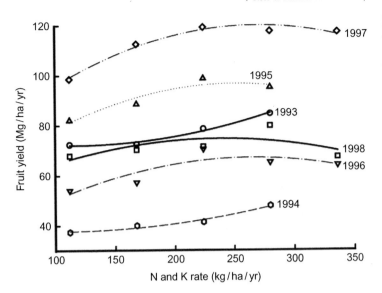

FIG. 49.2 Relationship between fruit yields and N and K rates during 1993–98 of >20-year-old "Hamlin" orange trees on "Cleopatra mandarin" rootstock planted in a Tavares fine sand. *Reproduced from Alva, A.K., Paramasivam, S., Obreza, T.A., Schumann, A.W., 2006. Nitrogen best management practice for citrus trees: I. Fruit yield, quality, and leaf nutritional status. Sci. Hortic. 107 (3), 233–244, with permission from Elsevier.*

4.3 Phosphorus

Phosphorus is a macronutrient; however, its application is generally at low rates in most of citrus production, since phosphorus does not leach out of soil that is above soil pH 6 and the fruit removes very low amount of phosphorus. However, good phosphorus fertilization is recommended when a virgin land is used to set up an orchard of citrus. The effect of phosphorus fertilizer on the citrus production is dependent on the inherent phosphorus concentration of soil. When the soil phosphorus is very low to low, addition of phosphorus fertilization improves yield linearly; however, no significant improvement in yield occurs, when soil phosphorus is high to moderate (Quaggio et al., 1998; Mattos et al., 2006).

4.4 Magnesium

In citrus, magnesium deficiency can reduce carbon dioxide assimilation, alter carbohydrate metabolism, and result in accumulation of starch by affecting invertase activity (Lavon et al., 1995; Yang et al., 2012). Magnesium fertilization is recommended for citrus production at a rate of 15%–30% of nitrogen. Often magnesium is applied as dolomite to increase soil pH, when the soil pH is low.

4.5 Calcium

Calcium has a very important role in citrus tree growth and fruit quality. Calcium application on fruit can improve the tensile strength of peel and can reduce various splitting disorders and improve postharvest shelf life of fruit (Zaragoza et al., 1996; Agusti et al., 2002). Calcium like magnesium is often applied to increase the soil pH when the soil is acidic. In addition to correcting soil pH for optimal tree growth, calcium availability through liming improves tree growth and thereby improves bearing habits of the citrus tree (Anderson, 1987).

4.6 Micronutrients

Manganese, zinc, iron, and boron are the main micronutrients that have been extensively studied in citrus production. Application of these micronutrients improves the yield, fruit quality, and tree growth habits (Khan et al., 2015). In contrast to macronutrients, mild deficiencies of micronutrients (manganese and zinc) for a short period do not affect the yield of orange and grapefruit fruit trees (Swietlik and LaDuke, 1991). Nonetheless, in case of deficiency of these nutrients, foliar application is recommended and has been proven effective as compared with granular fertilization (Pestana et al., 2005; Obreza and Morgan, 2008). Chelated form of micronutrients as compared with sulfate form improves tree growth, fruit yield, and quality in citrus (Sourour, 2000).

Sulfur, copper, and molybdenum are essential for citrus production; however, their specific roles in yield and fruit quality are not well studied. The effect of all the nutrients on fruit and juice quality under Florida conditions is summarized (Table 49.2).

5 Diagnosis of deficiency

When any essential element is in short supply, tree function is restricted. A severe shortage of an element typically produces a characteristic deficiency symptom exhibited by the leaves, which usually persists until the deficiency is corrected. Sometimes, twigs and fruits may also exhibit characteristic symptoms. Occasionally, two or three elements can be deficient in varying degrees, resulting in confusing visual symptoms. Conversely, excessive amounts of some elements may be present in the soil and may prevent the tree from functioning properly resulting in visible stress due to toxicity. Visual symptoms and leaf and soil analysis are all useful to evaluate nutritional status. Fig. 49.3 shows the common symptoms that can be observed in citrus leaf when the tree is deficient in nutrients.

TABLE 49.2 Specific internal and external fruit quality effects resulting from macronutrient, micronutrient, and irrigation applications to Florida citrus groves.

Measurement	Macronutrient element					Micronutrient element					Irrigation
	N	P	K	Ca	Mg	Mn	Zn	Cu	Fe	B	
JUICE QUALITY											
Juice content	+	o	−	o	o	o	o	o	o	o	+
Soluble solids (SS)	+	o	−	o	+	o	o	o	+	o	−
Acid (A)	+	−	+	o	o	o	o	o	o	o	−
SS/A ratio	−	+	−	o	+	o	o	o	o	o	+
Juice color (red)	+	o	−	?	?	?	?	?	?	?	o
Juice color (yellow)	+	o	−	?	?	?	?	?	?	?	+
Solids/box	+	o	−	o	+	o	o	o	+	o	−
Solids/acre	+	+	+	o	+	o	o	o	o	o	+
EXTERNAL FRUIT QUALITY											
Size	−	o	+	o	+	o	o	o	o	o	+
Weight	−	o	+	o	+	o	o	o	o	o	+
Green fruit	+	+	+	o	o	o	o	o		o	+
Peel thickness	−[a]	−	+	o	−	o	o	o	o	o	−
PEEL BLEMISHES											
Wind scar	−	+	o	?	?	?	?	?	?	?	+
Russet	−	−	o	?	o	o	o	o	o	o	o
Creasing	+	o	−	?	?	?	?	?	?	?	o
Plugging	−	o	−	?	?	?	?	?	?	?	−
Scab	+	o	o	?	?	?	?	?	?	?	+
Storage decay											
Stem-end rot	−	o	−	?	?	?	?	?	?	?	−
Green mold	−	o	o	?	?	?	?	?	?	?	+
Sour rot	o	o	o	?	?	?	?	?	?	?	o

[a] Except in young trees where peel may be thicker.
Increase (+), decrease (−), no change (o), no information (?)
From Obreza, T.A., Morgan, K.T., 2008. Nutrition of Florida Citrus Trees. UF/IFAS SL. Sep; 253.

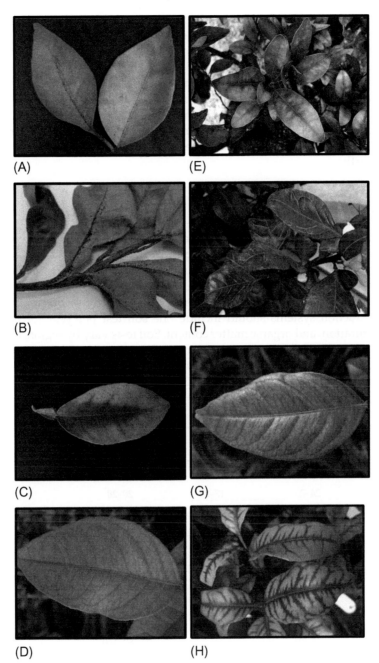

FIG. 49.3 Picture of citrus leaves showing deficiency of the essential mineral nutrients. (A) *Nitrogen deficiency symptom*: Leaf yellowing (dark gray in the print version) of old leaves. (B) *Copper deficiency symptom*: 'S' curved branching, internodal stem gumming, twig dieback. (C) *Magnesium deficiency symptom*: Inverted 'V' pattern at base of leaf. (D) *Iron deficiency symptom*: Green veins on a light green leaf (light gray in the print version); symptoms appear first on new foliage. (E) *Potassium deficiency symptom*: Yellowing (light gray in the print version) of the tips and margins which becomes broader. (F) *Boron deficiency symptom*: Corky veins. (G) *Manganese deficiency symptom*: Dark green bands (dark gray in the print version) along midrib and main veins surrounded by light green (light gray in the print version) interveinal areas. (H) *Zinc deficiency symptom*: Leaf is yellow (light gray in the print version) with green veins (dark gray in the print version). *Courtesy: Tripti Vashisth.*

The fertilizer requirement of citrus depends whether the purpose is to grow the crop (prebearing stage) or feed the crop (bearing stage). Based on these objectives, two types of fertilization, namely, corrective and preventive, are usually adopted. According to Gallasch (1992), an optimum fertilizer program is the one in which the cost of each unit of fertilizer applied is at least covered by an extra return through fruit yield obtained in both, the short- and long-term life of a citrus orchard.

A successful nutrient management program in citrus can be separated into four major components. The four components are monitoring, program development, application, and evaluation. Monitoring can be qualitative (visual observations of orchard performance in terms of growth and yield) or quantitative (laboratory-based analysis of soil or leaf samples). In the program development, the factors like type of fertilizer sources, the rate, timing, and frequency are considered. The application phase concentrates on methods of fertilizer application, for example, basin application, foliar spray, or fertigation (discussed in next section). Following fertilizer application, the evaluation step determines the crop response through improvement in tree growth, fruit yield, and quality. Nutrient management can become a complex task, if all the factors affecting the efficiency of fertilizer use are considered. Therefore, relative sensitivity of citrus to various nutritional factors is of utmost importance. The sensitivity of citrus trees to shortage or excess of individual nutrients differs greatly.

5.1 Leaf and soil analysis and optimum levels

The optimum levels for nutrients in leaf tissue are indicated in Table 49.3 from South America (Brazil), Asia (India and China), Africa (South Africa), and Europe (Spain). Typically, these levels are described for 4–6-month-old leaves. Usually, the high range to about slightly above the optimum range, beyond which any excess, results in toxicity that lowers citrus fruit yield and retards canopy development. Leaf nutrient concentration below the optimum range result in yield losses and need to be corrected through adjustment of the fertilizer or nutritional program.

Soil tests for citrus production systems should include one soil test per year for nutrient content, pH, cation exchange capacity, base saturation, and organic matter content. Soil tests vary by region, country, and standard equipment used. In Florida, for example, inorganic nitrogen forms such as nitrate and ammonium are determined using the 2M KCl extraction method, while as other macronutrients such as P, K, Ca, Mg, and S and micronutrients such as Fe, B, Zn, Mn, and Cu can be determined using the Mehlich 1 or 3 extraction methods. Further tests for P can also use Bray 1

TABLE 49.3   Optimum leaf nutrient concentrations in citrus trees.

Nutrient	Brazil[a,b]	China[c,d]	India[c,e]	South Africa[a,f]	Spain[c,g]	The United States[c,h]
	g/kg					
N	25–30[i]	24–26	17–28	20–28	28–30	25–27
P	1.2–1.6	1.3–1.5	0.7–1.7	1.1–1.6	1.3–1.6	1.2–1.6
K	12–16	–	10–26	7–15	7.1–10	12–17
Ca	35–50	34–48	13–33	35–55	30–50	30–49
Mg	3.5–5.0	2.0–2.9	2.8–9.2	3.0–5.5	2.5–4.5	3.0–4.9
S	2.0–3.0	2.3–2.5	–	–	2.0–3.0	–
	mg/kg					
B	75–150	17–19	–	50–150	31–100	36–100
Cu	10–20	3.7–10	2–19	5–16	6–14	5–16
Fe	50–150	40–46	70–249	80–300	61–100	60–120
Mn	35–70	14–23	42–112	30–150	26–60	25–100
Zn	50–75	23–30	12–39	20–70	26–70	25–100
Mo	0.5–2.0	–	–	–	0.1–3.0	0.1–2.0

[a] *Based on 4–6-month-old spring flush leaves from fruiting terminals with fruit 2–4 cm in diameter.*
[b] *Adapted from Quaggio et al. (2010).*
[c] *Based on 4–6-month-old spring flush leaves.*
[d] *Adapted from Srivastava et al. (1999) and Srivastava and Singh (2004).*
[e] *Adapted from Menino (2012).*
[f] *Du Plessis and Koen (1992).*
[g] *Quiñones et al. (2012).*
[h] *Obreza and Morgan (2008), Obreza et al. (2008).*
[i] *For lemons and acid limes, optimum range = 20–24 g/kg.*
–, not available.

or 2 extraction methods. Except for N, K, S, and the micronutrients, there are some guidelines for P, Ca, and Mg as to how much fertilizer to add or no further fertilization is warranted based on the soil test.

5.2 Role of soil pH on nutrient uptake and availability

Soil pH is the characteristic that determines biological and chemical reactions in soil and ultimate availability and solubility of nutrients. Soil acidity, also called low pH (pH < 7), is usually a result of intense industrial or agricultural activity where primary pollutants include SO_2, NH_3, and various NO_x gases such as nitric oxide (NO), nitrogen dioxide (NO_2), and nitrous oxide (N_2O) (Havlin et al., 2005). The global sources of NO_x gases include fossil fuel combustion, biomass combustion, lightning, soil microbial activity, and chemical oxidation. Soil organic matter can release CO_2 during decomposition, which reacts with water to form H^+ and HCO_3^-, thereby lowering pH. Leaching of NO_3^- and SO_4^{2-} result in a decrease in pH as basic cations such as Ca, Mg, and K also leach. As plant roots absorb cations, electrical neutrality is maintained through uptake of an anion or extrusion of H. When anions are absorbed, uptake of actions or extrusion of OH^- or HCO_3^- occurs to maintain electrical neutrality. When cation exceeds anion uptake, excess H^+ is released into the rhizosphere, while OH^-/HCO_3^- is released when anion exceeds cation uptake. To address the low pH problem, apply lime or gypsum to supply some Ca to raise the pH (Obreza and Morgan, 2008).

6 Methods of fertilizer application

As critical as it is to choose the right fertilizer and rate for growing healthy and productive citrus trees, choosing the right method of fertilization is absolutely critical and indispensable for efficient grove management. The method of fertilizer application depends on the type of fertilizer and available resources. In citrus production, often it is advisable not to rely on sole method of fertilization. Citrus, being an evergreen perennial tree with long fruit growth period, has high nutritional requirement all year round; therefore, it is critical to meet those nutritional for optimal production and fruit quality. The method of fertilizer application depends on the type of fertilizer. Two types of fertilizer application are common in citrus production.

6.1 Granular fertilizer

It is often referred to as "dry fertilizer" as opposed to "liquid fertilizer" (discussed in next section). This kind of fertilizer is often in pellet form and most commonly used fertilizer in citrus production. Typically, a dry fertilizer is a physical mix of different nutrients at specific rates and a filler material. There are multiple methods of applying granular fertilizer as shown in Fig. 49.4. Hand placement and planting hole are common for new/young planting. In these methods, fertilizer is carefully placed over or around the root zone to ensure the fertilizer is available the plant; as the newly planted plants have small root system, they benefit more from such careful placement of fertilizer. In broadcast method, the fertilizer is uniformly spread all over the row of trees; it is a common practice for mature tree groves. As the broadcast application lacks any kind of precision, it is often inexpensive method but can result in fertilizer spread at undesirable spots. Recently, in Florida citrus production, the variable rate technology (VRT) spreader has become very popular. Variable rate spreader adjusts the amount of fertilizer applied based on the tree canopy, therefore, can potentially reduce excessive fertilizer application and cost (Schumann et al., 2006).

Granular fertilizer can be divided into two main categories in citrus production:

6.1.1 Conventional fertilizer

It is traditional, dry fertilizer, which is readily available to the plants. Commonly used in citrus production to provide N, P, K, Mg, and Ca, split in multiple applications per year. Relatively inexpensive but have high potential for leaching.

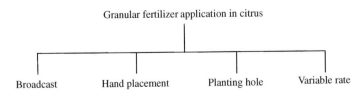

FIG. 49.4 Popular methods of applying granular fertilizer in citrus production.

FIG. 49.5 Popular methods of applying liquid fertilizer in citrus production.

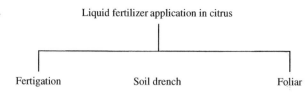

6.1.2 Controlled release fertilizer

In citrus production, the use of controlled release fertilizer (CRF) has gained popularity in last two decades, mainly due to improved nutrient availability and reduced leaching to the environment. The external coating of CRF with a polymer or resin makes the nutrient release controlled; the nutrient release depends on temperature and humidity combination; therefore, the nutrient leaching from the soil reduces, and the fertilizer efficiency improves. In commercial citrus production, the nitrogen uptake is greater when applied as CRF in comparison with conventional urea fertilizer, thereby improving plant growth (Dou and Alva, 1998). Use of CRF is highly desirable where the soil type is poor and constant availability of nutrient is required. However, CRF are relatively more expensive; therefore, their use has been limited mostly to young planting in citrus production.

6.2 Liquid fertilizer

The methods of liquid fertilizer application comprises of fertigation, soil drenching, and foliar application (Fig. 49.5). Use of liquid fertilizer offers an advantage of easy and immediate uptake as compared with granular fertilizer. In both fertigation and soil drench, the fertilizer is applied to the roots, and the plant uptakes the nutrient with the water stream. Fertigation is the preferred method of soil-applied liquid fertilizer. However, often the growers, who do not have the infrastructure for fertigation, apply soil drench. Soil drench and fertigation only differ in method of application; nevertheless, the nutrient solution applied are very similar in both cases.

6.2.1 Fertigation

Fertigation is the practice of applying fertilizer with irrigation water (Obreza and Morgan, 2008) in amounts, form, and time when needed (Quiñones et al., 2012). Advantages of fertigation include fertilizer placement in the wetted area (where plant roots grow) and frequent application of fertilizer in small doses, therefore, increases fertilizer use efficiency and reduces leaching, resulting in better tree growth, greater yield, and fruit quality with less fertilizer compared with conventional practices. Recent modification in planting and orchard configuration such as advanced citrus production systems (ACPS) and citrus undercover protective screen[a] (CUPS; Schumann et al., 2017) use advanced fertigation practices to cope with the devastating nature of Huanglongbing (discussed latter in this chapter). The goal of ACPS is a sustainable, profitable citrus grove designed and managed in a way that produces higher, earlier yields to reach economic payback sooner and improve disease and pest management efficiency (Schumann et al., 2009a,b; Morgan et al., 2009). In addition to high tree densities (220–1400 trees per hectare), the ACPS is based on the open hydroponic system (OHS) already being used for intensive production of citrus in climates different from Florida, including Spain and South Africa (Schumann et al., 2009a; Morgan and Kadyampakeni, 2012).

6.2.2 Foliar fertilizer

Foliar fertilizer is applied to the foliage using a sprayer. In citrus production, foliar fertilizer is commonly used for applying micronutrients. Micronutrients are generally required in a small amount; therefore, foliar method of application has better tree coverage than granular fertilizer. Moreover, depending on the soil chemistry, granular fertilizer may precipitate in the soil, becoming unavailable to the plant for uptake; hence, foliar fertilization has potential to circumvent such problems. Use of foliar nutrients has been well documented to improve tree growth and fruit quality

---

[a] CUPS is an innovation, which represents further advancement beyond ACPS where trees are grown in enclosures. Citrus can be grown under protective screen structures for fresh fruit production to completely exclude the Asian citrus psyllid (ACP, *Diaphorina citri*) and therefore Huanglongbing (HLB) disease, or citrus greening. The benefits of eliminating HLB are immediate and include rapid, normal tree growth, higher yields of premium quality fruit, negligible fruit drop, and uncomplicated fertilizer and irrigation requirements (Schumann et al., 2017; Ferrarezi et al., 2017a,b). Because CUPS is a relatively new citrus production system with new challenges, current guidelines are preliminary and undergoing constant refinement through research.

as well as yields (Khan et al., 2015; Razzaq et al., 2013; Ullah et al., 2012). Razzaq et al. (2013) demonstrated that the use of high rates of zinc sulfate not only improved tree growth parameters but also helped in reducing the fruit drop as the foliar fertilization provided the nutrients directly to the fruit (site of abscission). Currently, with advent of Huanglongbing, the use of foliar fertilization has become very popular in citrus production (see Section 7). Foliar fertilization is also beneficial in situations when the foliage is displaying nutrient deficiency symptoms, as the foliar fertilizer applies the nutrient at the site of deficiency symptom (foliage/leaves), and therefore, it suffices the need of nutrient immediately. However, foliar fertilization should not be the sole method of fertilizer application; it provides optimum results when used to complement a root-applied fertilizer program.

7 Huanglongbing

Huanglongbing (HLB; aka citrus greening) is one of the most devastating diseases of citrus and threatens the citrus production wherever it is present. HLB was first reported almost a century ago in China; now it is widespread in countries such as the United States, Brazil, Mexico, India, China, and South Africa. Interestingly, the name "Huanglongbing" originates from Chinese word that literally means "yellow dragon disease" as the tree suffering from HLB trees often have yellow (chlorotic) foliage sectored in one area of canopy (Fig. 49.6). HLB is a bacterial disease, caused by *Candidatus* Liberibacter asiaticus (CLas) and primarily spread by an insect vector, Asian citrus psyllid. CLas is phloem-limited bacteria; once a tree is infected by CLas, plugging in phloem sieve pores can be observed, resulting in an accumulation of starch in symptomatic leaves and the aerial stem. Visible symptoms of HLB include blotchy mottle and/or chlorotic patterns of leaves resembling those induced by nutrient deficiencies and small upright leaves. In HLB-affected tree, the disruption of vascular function, loss of roots, and altered mineral nutrition leads to arrested plant and fruit growth and decline in the production. Fruit produced by HLB-affected trees are often smaller, lopsided, poorly colored, with aborted seeds and drop prematurely leading to decline in yield and economical losses for the growers.

7.1 Huanglongbing and nutrient deficiencies

HLB is often confused with nutrient deficiencies and vice-versa. HLB-affected leaves often display chlorotic patters and islands, vein corking, and yellowing. Symptoms like vein corking can be confused with boron deficiency and

FIG. 49.6 Picture of a HLB-affected sweet orange tree, showing sectored chlorotic branch.

FIG. 49.7 "Pen test" to distinguish nutrient deficiency from a HLB-caused blotchy mottle in leaf.

chlorotic islands with zinc. The most distinguishing visible foliar symptom of HLB is random chlorotic blotchy mottles on leaf blade, which are formed as a result disruption of chloroplast by accumulated starch in the leaf. Therefore, the most common and easy way to distinguish between nutrient deficiency and HLB symptoms is a "pen test" (Fig. 49.7). According to this test, take the leaf in question and draw a circle with pen at the same position on the both sides of the leaf blade. If the same pattern of chlorosis is observed in the area encompassed in the circle, then most likely it is a nutrient deficiency; if the chlorosis is random, then it can be HLB. It is important to consider that HLB-affected trees often display nutrient deficiencies as well; therefore, it is necessary to distinguish deficiencies from HLB symptoms and rectify the deficiencies with right fertilizer formulation.

7.2 Water uptake in HLB-affected trees

With HLB, irrigation scheduling is becoming more important and critical. Growers cannot afford water stress or water excess. Citrus trees affected by HLB are known to lose substantial foliage and up to 80% of the root mass depending on disease severity, thus potentially negatively influencing water and nutrient uptake. Premature fruit drop is increased if slight water stress is experienced by citrus trees and canopy size is reduced as is the number of fruit and fruit size. In addition, growth of shoots, leaves, and roots is reduced. Benefits of proper irrigation scheduling include reduced loss of nutrients through leaching due to excess water applications and reduced pollution of groundwater or surface waters. A study was conducted in a Florida commercial citrus grove from 2011 to 2015 with the objective of determining irrigation requirements of HLB-affected citrus trees compared with healthy trees (Hamido et al., 2017b). Results from the study indicated that healthy trees consumed approximately 22%–25% more water than HLB-affected trees possibly as a result of greater root density and leaf area in the former (Hamido et al., 2017a). Reduced water uptake by HLB-affected trees resulted in significantly greater soil water content. The relationship between leaf area and water uptake indicated that diseased trees with lower canopy density and corresponding lower leaf area index take up less water and consequently less nutrients from the soil. The elevated soil water content may partially explain higher rates of root infection with *Phytophthora* spp. observed in some HLB-affected trees. In a field study, researchers also showed that daily irrigation was more effective in improving water use of HLB-affected trees compared with infrequent irrigation (Kadyampakeni et al., 2014a,b; Hamido et al., 2017a; Kadyampakeni and Morgan, 2017).

8 Balanced and constant nutrition (BCN)

Traditionally, citrus growers try to achieve optimum nutrition through direct soil management. Currently with the introduction of HLB in Florida, many growers and production managers consider foliar fertilization to complement soil-applied fertilization program to ensure nutrient availability to foliage and to improve fruit yield, and fruit quality. Hence, foliar nutrition programs are becoming very common and extensively used in Florida to supplement soil-applied fertilizer program to deliver all of the essential nutrient elements to citrus trees. Research has demonstrated that HLB symptoms can be reduced by foliar applications of micronutrients, especially Mg, Mn, and Zn (Morgan et al., 2016; Uthman et al., 2019). These responses have promoted development and use of enhanced foliar nutritional programs in Florida. Efficacy of these programs has been a topic of considerable discussion and debate. Fertilization programs have varied considerably among growers and have consisted of various rates and application schedules of essential macro- and micronutrients.

A 5-year-long study of foliar applications of Mn, Zn, and B on 5–7-year-old Valencia trees on Swingle rootstock was recently concluded in a commercial grove with the goal of determining the effect of improved leaf nutrient status on canopy density and yield volume (Morgan et al., 2016). This approach has provided the citrus industry with new information regarding fertilization practices to support continued production of existing citrus groves affected by HLB. However, foliar nutrient applications are not likely to lead to past production levels in the short term. Despite some essential nutrients being low in the leaves, the nontreated control trees continued to increase in canopy volume and yield during the course of the study. The first analysis conducted was to determine whether the foliar application of potassium nitrate (KNO_3) affected foliar concentrations of N and K and growth and productivity of the trees. The lack of an increase in foliar N after application indicated dilution as N moves out of mature leaves to new growth. Unlike leaf N, foliar K concentration of trees receiving KNO_3, which was below the optimum range prior to foliar application, was within the optimum range after KNO_3 application. The application of KNO_3 increased canopy volume compared with the controls. However, yields for KNO_3-treated trees were not significantly greater than yields for the controls. One interesting result of this study was that the amount of Mn and Zn taken up into the leaf was not affected by KNO_3 as some have speculated.

The tree nutrients, Mn, Zn, and B, were applied to separate trees at three rates plus nonsprayed controls. The three rates were 0.5, 1.0, and 2.0 times the current UF/IFAS foliar recommendations. The nutrient sprays were applied three times per year following flushes in March, May, and September. Thus, the three rates (0.5, 1.0, and 2.0 times UF/IFAS) resulted in a total of 1.5, 3.0, and 6.0 times UF/IFAS recommendations on an annual basis. For example, the UF/IFAS recommendation for Mn and Zn is 5-pound metallic per acre per year; thus, trees receiving three times UF/IFAS recommendation would receive 15-pound metallic per acre per year. The highest rates of Mn and Zn application had the greatest increase in those foliar nutrients. Increase in leaf Mn and Zn concentration right after application disappeared such that no difference was found compared with controls prior to the next foliar application.

Canopy volume increased with increased application of Mn and Zn but not B. Yield increased with 1.5 and 3.0 annual rates of Mn and Zn but was lower for the 6.0 rate compared with the 3.0 rate. These results indicate increased growth of trees proportional to Mn and Zn within the range tested but reduced yield at the highest rate. Reduced yield at high rates of nutrients are common as excess nutrient results in increased growth at the expense of yield and could partially explain variability in tree response to nutrient applications by citrus growers. However, the current maximum optimum range should not be exceeded unless larger and/or more dense tree canopies are desired at the risk of lower yields.

Overall, currently, there is no fixed recommendation for HLB-affected trees; however, a number of researches are underway to develop these recommendations. HLB-affected trees have smaller and weaker root systems than healthy trees. Therefore, frequent and small doses of fertilizer are beneficial as this maintains a constant supply of nutrients and reduces potential nutrient leaching. Controlled release fertilizer and fertigation can be strategic alternatives to multiple applications of conventional dry granular fertilizer. Similarly for micronutrient, a slightly higher than standard rate of nutrient is beneficial. For commercial production, it highly recommended that leaf and soil nutrient analysis be performed and taken into consideration before making any changes to a fertilizer program.

8.1 Soil pH/irrigation water pH

From field and greenhouse studies, it appears there is strong interaction between pH and HLB. In some citrus-producing regions such as Florida, California, and Israel (and other Mediterranean regions), high soil pH (>7.2) might result due to the use of alkaline, saline, and sodic water. Irrigation water high in bicarbonates increases soil pH in the wetted area under the microsprinkler or dripper, which causes adverse effects on feeder root functioning, expression of yellow shoots and die-back, premature fruit and leaf drop, and reduced fruit size and yield in in HLB-affected citrus (Graham and Morgan, 2017). Strategies to lower the pH include application of elemental sulfur, periodic injection of dilute sulfuric or phosphoric acid, and ammonium polysulfide (Havlin et al., 2005; Graham and Morgan, 2015, 2017).

Vashisth et al. (2019) have shown that HLB-affected plants perform better when irrigated with low pH (moderately acidic) water. The HLB-affected plants tended to perform better when soil pH was close to 6.0. When pH was above 6.5–7.0, the HLB plants began to decline. Healthy plants performed well at a wider pH range (6.0–7.0).

To conclude, citrus is a high nutrient-demanding crop and is generally grown in soils that are not considered rich in soil nutrients. Therefore, it is critical to have a good nutrition management program to obtain optimum yield and fruit quality. HLB is a devastating citrus disease and often display symptoms similar to nutrient deficiency. Therefore, it is critical to differentiate HLB symptoms from nutrient deficiency. Complete and balanced nutrition program is indispensable for HLB-affected and healthy citrus trees.

9 Future research

Field of citrus nutrition is vast especially because citrus is pretty much grown on every continent in wide variety of soil. However, considering current situation, few future directions are as follows:

- Developing fertilization recommendation for HLB-affected trees.
- A number of new rootstocks have been developed for HLB-tolerance. It is critical to determine the optimum pH for their growth.
- Role of mineral nutrients in fruit quality and postharvest life.

References

Agustí, M., Martinez-Fuentes, A., Mesejo, C., 2002. Citrus fruit quality. Physiological basis and techniques of improvement. Agrociencia 6 (2), 1–16.

Alva, A.K., Tucker, D.P.H., 1999. Soils and citrus nutrition. In: Timmer, L.W., Duncan, L.W. (Eds.), Citrus Health Management. APS Press, St. Paul, pp. 59–71.

Alva, A.K., Paramasivam, S., Obreza, T.A., Schumann, A.W., 2006. Nitrogen best management practice for citrus trees: I. Fruit yield, quality, and leaf nutritional status. Sci. Hortic. 107 (3), 233–244.

Anderson, C.A., 1987. Calcium: fruit yields, tree size, and mineral nutrition relationships in 'Valencia' orange trees as affected by liming. J. Plant Nutr. 10 (9–16), 1907–1916.

Ashraf, M.Y., Gul, A., Ashraf, M., Hussain, F., Ebert, G., 2010. Improvement in yield and quality of Kinnow (Citrus deliciosa x Citrus nobilis) by potassium fertilization. J. Plant Nutr. 33 (11), 1625–1637.

Baldwin, E.A., Bai, J., Plotto, A., Ritenour, M., 2014. Citrus fruit quality assessment: producer and consumer perspectives. Stewart Postharvest Rev. 10 (2), 1–7.

Corá, J.E., Silva, G.O., Martins Filho, M.V., De Negri, J.D., 2005. Manejo do solo sob citros. In: Mattos Jr., D., Pio, R.M., Pompeu Jr., J. (Eds.), Citros. InstitutoAgronômico e Fundag, Campinas, pp. 347–368.

Dou, H., Alva, A.K., 1998. Nitrogen uptake and growth of two citrus rootstock seedlings in a sandy soil receiving different controlled-release fertilizer sources. Biol. Fertil. Soils 26 (3), 169–172.

Du Plessis, S.F., Koen, T.J., 1992. Leaf analysis norms for lemons [Citrus limon (L.) Burm.]. In: Proceedings of Seventh International Citrus Congress 2, pp. 551–552.

Duiker, S.W., Flanagan, D.C., Lal, R., 2001. Erodibility and infiltration characteristics of five major soils of southwest Spain. Catena 45, 103–121.

Feigenbaum, S., Bielorai, H., Erner, Y., Dasberg, S., 1987. The fate of 15 N labeled nitrogen applied to mature citrus trees. Plant Soil 97 (2), 179–187.

Ferrarezi, R.S., Wright, B.R., Boman, B.J., Schumann, A.W., Gmitter, F.G., Grosser, J.W., 2017a. Protected fresh grapefruit cultivation systems: antipsyllid screen effects on environmental variables inside enclosures. HortTechnology 27 (5), 675–681.

Ferrarezi, R.S., Wright, A.L., Boman, B.J., Schumann, A.W., Gmitter, F.G., Grosser, J.W., 2017b. Protected fresh grapefruit cultivation systems: antipsyllid screen effects on plant growth and leaf transpiration, vapor pressure deficit, and nutrition. HortTechnology 27 (5), 666–674.

Gallasch, P.T., 1992. The use of leaf analysis in Southern Australia to develop more efficient fertilizer programs. Proceedings of the International Society of Citriculture, March 1992, 8–13.

Gimeno, V., Díaz-López, L., Simón-Grao, S., Martínez, V., Martínez-Nicolás, J.J., García-Sánchez, F., 2014. Foliar potassium nitrate application improves the tolerance of Citrus macrophylla L. seedlings to drought conditions. Plant Physiol. Biochem. 83, 308–315.

Graham, J., Morgan, K., 2015. Managing excessive bicarbonates with acidification. Citrus Ind. 96 (5), 8–11.

Graham, J., Morgan, K., 2017. Why bicarbonates matter for HLB management. Citrus Ind. 98 (4), 16–21.

Hamido, S.A., Morgan, K.T., Ebel, R.C., Kadyampakeni, D.M., 2017a. Improved irrigation management of sweet orange with Huanglongbing. HortScience 52 (6), 916–921.

Hamido, S.A., Morgan, K.T., Kadyampakeni, D.M., 2017b. The effect of Huanglongbing on young citrus tree water use. HortTechnology 27 (5), 659–665.

Havlin, J.L., Beaton, J.D., Tisdale, S.L., Nelson, W.L., 2005. Soil Fertility and Fertilizers: An Introduction to Nutrient Management (No. 631.422/H388). Pearson Prentice Hall, New Jersey.

He, Z.L., Calvert, D.V., Alva, A.K., Banks, D.J., Li, Y.C., 2003. Thresholds of leaf nitrogen for optimum fruit production and quality in grapefruit. Soil Sci. Soc. Am. J. 67 (2), 583–588.

Kadyampakeni, D.M., Morgan, K.T., 2017. Irrigation scheduling and soil moisture dynamics influence water uptake by Huanglongbing affected trees. Sci. Hortic. 224, 272–279.

Kadyampakeni, D.M., Morgan, K.T., Schumann, A.W., Nkedi-Kizza, P., Obreza, T.A., 2014a. Water use in drip- and microsprinkler-irrigated citrus trees. Soil Sci. Soc. Am. J. 78, 1351–1361.

Kadyampakeni, D.M., Morgan, K.T., Schumann, A.W., Nkedi-Kizza, P., 2014b. Effect of irrigation pattern and timing on root density of young citrus trees infected with Huanglongbing disease. HortTechnology 24, 209–221.

Khan, A.S., Nasir, M., Malik, A.U., Basra, S.M., Jaskani, M.J., 2015. Combined application of boron and zinc influence the leaf mineral status, growth, productivity and fruit quality of 'Kinnow' Mandarin (*Citrus nobilis Lour × Citrus deliciosa Tenora*). J. Plant Nutr. 38 (6), 821–838.

Lavon, R., Goldschmidt, E.E., Salomon, R., Frank, A., 1995. Effect of potassium, magnesium, and calcium deficiencies on carbohydrate pools and metabolism in citrus leaves. J. Am. Soc. Hortic. Sci. 120 (1), 54–58.

Lv, X., Zhao, S., Ning, Z., Zeng, H., Shu, Y., Tao, O., Xiao, C., Lu, C., Liu, Y., 2015. Citrus fruits as a treasure trove of active natural metabolites that potentially provide benefits for human health. Chem. Cent. J. 9 (1), 68.

Mattos Jr., D., Quaggio, J.A., Cantarella, H., Alva, A.K., Graetz, D.A., 2006. Response of young citrus trees on selected rootstocks to nitrogen, phosphorus, and potassium fertilization. J. Plant Nutr. 29 (8), 1371–1385.

Mattos Jr., D., Kadyampakeni, D.M., Quiñones, A., Boaretto, R.M., Morgan, K., Quaggio, J.A., 2019. Soil and nutrition interactions. In: Talon, M., Caruso, M., Gmitter, F. (Eds.), The Genus Citrus, first ed. (in press).

Menino, R., 2012. Leaf analysis in citrus: interpretation tools. In: Kumar, A.K. (Ed.), Advances in Citrus Nutrition. Springer, Dordrecht, pp. 59–79.

Morgan, K.T., Kadyampakeni, D., 2012. Open field hydroponics: concept and application. In: Srivastava, A.K. (Ed.), Advances in Citrus Nutrition. Springer, Dordrecht, pp. 271–280.

Morgan, K.T., Schumann, A.W., Castle, W.S., Stover, E.W., Kadyampakeni, D., Spyke, P., Roka, F.M., Muraro, R., Morris, R.A., 2009. Citrus production systems to survive greening: horticultural practices. Proc. Fla. State Hortic. Soc. 122, 114–121.

Morgan, K.T., Rouse, R.E., Ebel, R.C., 2016. Foliar applications of essential nutrients on growth and yield of 'Valencia' sweet orange infected with Huanglongbing. HortScience 51 (12), 1482–1493.

Obreza, T.A., Collins, M.E., 2008. Common Soils Used for Citrus Production in Florida. SL 193, Florida Coop. Ext. Serv., Gainesville.

Obreza, T.A., Morgan, K.T., 2008. Nutrition of Florida Citrus Trees. UF/IFAS. SL. Sep; 253.

Obreza, T.A., Boman, B.J., Zekri, M., Futch, S., 2008. In:Obreza, T.A., Morgan, K.T. (Eds.), Nutrition of Florida Citrus Trees. Soil and Water Science Department, Institute of Food and Agricultural Sciences, University of Florida, Gainesville, pp. 43–47. SL253.

Pestana, M., de Varennes, A., Abadía, J., Faria, E.A., 2005. Differential tolerance to iron deficiency of citrus rootstocks grown in nutrient solution. Sci. Hortic. 104 (1), 25–36.

Quaggio, J.A., Cantarella, H., Van Raij, B., 1998. Phosphorus and potassium soil test and nitrogen leaf analysis as a base for citrus fertilization. Nutr. Cycl. Agroecosys. 52 (1), 67–74.

Quaggio, J.A., Mattos Jr., D., Boaretto, R.M., 2010. Citros. In: Prochnow, L.I., Casarin, V., Stipp, S.R. (Eds.), Boas práticas para uso eficiente de fertilizantes. International Plant Nutrition, Piracicaba, pp. 371–409.

Quiñones, A., Martínez-Alcántara, B., Primo-Millo, E., Legaz, F., 2012. Fertigation: concept and application in citrus. In: Advances in Citrus Nutrition. Springer, Dordrecht, pp. 281–301.

Razzaq, K., Khan, A.S., Malik, A.U., Shahid, M., Ullah, S., 2013. Foliar application of zinc influences the leaf mineral status, vegetative and reproductive growth, yield and fruit quality of 'Kinnow' mandarin. J. Plant Nutr. 36 (10), 1479–1495.

Schumann, A.W., Miller, W.M., Zaman, Q.U., Hostler, K.H., Buchanon, S., Cugati, S., 2006. Variable rate granular fertilization of citrus groves: spreader performance with single-tree prescription zones. Appl. Eng. Agric. 22 (1), 19–24.

Schumann, A.W., Syvertsen, J.P., Morgan, K.T., 2009a. Implementing advanced citrus production systems in Florida—early results. Proc. Fla. State Hortic. Soc. 122, 108–113.

Schumann, A., Morgan, K., Castle, B., Syvertsen, J., 2009b. Advanced citrus production systems in Florida. Citrus Ind. 91 (8), 1–3.

Schumann, A.W., Singerman, A., Wright, A.L., Ferrarezi, R.S., 2017. Citrus Under Protective Screen (CUPS) Production Systems. EDIS Publication, UF/IFAS Extension, Horticultural Sciences Department, #HS1304.

Soil Survey Staff, 2014. Keys to Soil Taxonomy, 12th ed. USDA-Natural Resources Conservation Service, Washington.

Sourour, M.M., 2000. Effect of foliar application of some micronutrient forms on growth, yield, fruit quality and leaf mineral composition of Valencia orange trees grown in North-Sinai. Alex. J. Agric. Res. 45 (1), 269–285.

Srivastava, A.K., Singh, S., 2004. Leaf and soil nutrient guides in citrus—a review. Agric. Rev. 25 (4), 235–251.

Srivastava, A.K., Kohli, R.R., Dass, H.C., Huchche, A.D., Ram, L., Singh, S., 1999. Evaluation of the nutritional status of Nagpur mandarin (*Citrus reticulata* Blanco) by foliar sampling. Trop. Agric. 76 (2), 93–98.

Swietlik, D., LaDuke, J.V., 1991. Productivity, growth, and leaf mineral composition of orange and grapefruit trees foliar-sprayed with zinc and manganese. J. Plant Nutr. 14 (2), 129–142.

Ullah, S., Khan, A.S., Malik, A.U., Afzal, I., Shahid, M., Razzaq, K., 2012. Foliar application of boron influences the leaf mineral status, vegetative and reproductive growth, yield and fruit quality of 'Kinnow' mandarin (*Citrus reticulata* Blanco.). J. Plant Nutr. 35 (13), 2067–2079.

Uthman, Q., Kadyampakeni, D., Nkedi-Kizza, P., 2019. Response of Huanglongbing (HLB) affected citrus trees to zinc fertilization. Proc. Fla. State Hortic. Soc. 132, (in press).

Vashisth, T., Kadyampakeni, D., Ghimire, L., 2019. Irrigation water pH can make the difference! Citrus Ind. 100 26–29.

Vu, J.C., Yelenosky, G., 1991. Photosynthetic responses of citrus trees to soil flooding. Physiol. Plant. 81 (1), 7–14.

Weil, R.R., Brady, N.C., 2016. The Nature and Properties of Soils. Pearson, Columbus, p. 1104.

Wu, G.A., Terol, J., Ibanez, V., López-García, A., Pérez-Román, E., Borredá, C., Domingo, C., Tadeo, F.R., Carbonell-Caballero, J., Alonso, R., Curk, F., 2018. Genomics of the origin and evolution of citrus. Nature 554 (7692), 311.

Yang, G.H., Yang, L.T., Jiang, H.X., Li, Y., Wang, P., Chen, L.S., 2012. Physiological impacts of magnesium-deficiency in citrus seedlings: photosynthesis, antioxidant system and carbohydrates. Trees 26 (4), 1237–1250.

Zaragoza, S., Almela, V., Tadeo, F.R., Primo-Millo, E., Agusti, M., 1996. Effectiveness of calcium nitrate and GA3 on the control of peel-pitting of 'Fortune' mandarin. J. Hortic. Sci. 71 (2), 321–326.

50

Diagnosis and management of nutrient constraints in pineapple

Victor Martins Maia[a],*, Rodinei Facco Pegoraro[b], Ignácio Aspiazú[a], Fernanda Soares Oliveira[a], Danúbia Aparecida Costa Nobre[a]

[a]State University of Montes Claros, Montes Claros, Brazil
[b]Federal University of Minas Gerais, Belo Horizonte, Brazil
*Corresponding author. E-mail: victormartinsmaia@gmail.com

1 Introduction

Pineapple is among the five most important tropical fruits in the world. About 90 countries have areas of pineapple cultivation; however, the 10 largest producers account for 70% of total world production. Among the main pineapple producers are Costa Rica, Brazil, Philippines, Thailand, and Indonesia (Food and Agriculture Organization of the United Nations, 2017). The main importers are the North American countries, especially the United States, as the world's largest importer of fresh and canned pineapple and juice, followed by countries in Europe and Japan. Countries as Brazil, India, and China, despite high production, have their own domestic market as main destination (Food and Agriculture Organization of the United Nations, 2017). To ensure the high demand of a competitive fruit market, it is worth noting that the nutritional status of the plant has a high influence on its growth and development. A balanced pineapple nutrition can be considered one of the main determinants for greater yield, quality, and weight of the fruits (Amorim et al., 2011).

The economic importance of pineapple is related to not only its wide commercialization but also the employment activities that involve the production of this crop, which requires the intensive use of labor. In addition, pineapple represents a good model for plant evolution and genetic research because it presents crassulacean acid metabolism (CAM), a carbon fixation mechanism that allows high water use efficiency and drought tolerance (Xu and Liu,

A.K. Srivastava, Chengxiao Hu (eds.)
Fruit Crops: Diagnosis and Management of Nutrient Constraints
https://doi.org/10.1016/B978-0-12-818732-6.00050-2

2015). Therefore, the success of pineapple as a cultivated plant is the result of its wide adaptability in tropical and subtropical areas, high rusticity, asexual propagation efficiency, and great consumer acceptance, which justifies its dispersion throughout the world (Crestani et al., 2010).

2 Nutritive value

Pineapple is a very appreciated fruit, consumed in natural, canned, frozen, in syrup, crystallized, and in the form of raisins and pickles; used in the confection of sweets, ice creams, yoghurts, candies, and cakes; and also consumed in the form of juice, soda, syrup, liqueur, wine, vinegar, and brandy. It also serves as raw material for the extraction of alcohol and animal feed, with the use of residues from industrialization (Crestani et al., 2010; Debnath et al., 2012). The fruit of the pineapple presents variation in its chemical composition, depending on cultivated variety, stage of maturation, climate, time of production, and processing of fruits, among other factors (Sanches and Matos, 2013; Taussig and Batkin, 1988). However, the nutritional value is related to the content of soluble sugars (13–15 °Brix); of minerals, such as potassium, calcium, magnesium, phosphorus copper, and iron, and vitamins, especially A, B_1, and C (Matsuura and Rolim, 2002; Sanches and Matos, 2013).

The pineapple also contains several active ingredients of pharmacological value, with immense health benefits, such as bromelain, a proteolytic enzyme that is effective for vascular, inflammatory, and digestive problems, and has antioxidant and anticancer effects (Hossain et al., 2015; Lee et al., 2019; Taussig and Batkin, 1988). This enzyme is also used in the brewing of meat, clarification of beers, cheese making, textile industry (leather and wool treatment), and the preparation of children's and dietetic foods (Sanches and Matos, 2013). The remains of pineapple processing (peel and pulp), still very nutritious, are used for various purposes, as in animal feed, with positive results in the production of milk (Gutiérrez et al., 2003; López-Herrera et al., 2014, 2009). After the fruit and seedlings are harvested, pineapple remains are rich in fiber, minerals, and an appreciable amount of starch in the stem and can be used either by direct grazing or by cutting the plants for direct supply for the animals or for silage (Sanches and Matos, 2013). There are some caveats and necessary precautions regarding this use due to the large amount of nutrients extracted from the area (Pegoraro et al., 2014a).

3 Geographical distribution

Early reports and studies of phylogeny indicate that pineapple (*Ananas comosus* var. *comosus*) had its origin, domestication, and initial distribution in South America (Collins, 1951; Loison-Cabot, 1992). The expansion of pineapple around the world was followed by the opening of sea lanes by the Spaniards and Portuguese during the 16th century, and the navigators were responsible for this diffusion, with the fruits being loaded on board during the voyages and the abandonment of the crowns in the different ports of landing in Africa and Asia, serving as natural multiplication material (Medina, 1978). At the end of the 19th century, it arrived in Hawaii, from where it was distributed to the Greater Antilles, Mexico, the Philippines, Taiwan, and Kenya (Collins, 1960).

Because it is a plant of tropical origin, it has a good development in hot and humid places. These conditions are generally recorded in the range between the parallels of 30° north and south, where temperatures remain above 10°C (considered the lower basal temperature) and below 40°C. The development and production of pineapple must be done in low-latitude regions, and the most suitable areas for cultivation are located between the 25° north and 25° south parallels, between the Tropics of Cancer and Capricorn (Sanches and Matos, 2013). Currently, pineapple is extensively produced in all tropical countries, but the expansion of a cultivar in certain areas depends on its acclimatization and local market interest.

4 Major cultivars

When choosing a pineapple variety, consideration should be given to its adaptation to the planting site, market requirements, and availability and quality of the seedling. Thus, pineapple breeding programs aim to obtain more productive cultivars, adapted to different climatic conditions and resistant to pests and diseases. In addition, breeding programs usually look for fast-growing genotypes: with leaves with none or a few spines; early blight located at the base of the plant; cylindrical fruit; yellow and little fibrous peel; flat eyes; yellow flesh, firm but not fibrous; moderate acidity; medium to small crown; high total soluble solid content; and high ascorbic acid (vitamin C) content (Cabral, 2000; Cunha, 2007; Crestani et al., 2010).

Finding a variety that has all the favorable characteristics has not been an easy task, so the ideal is that the variety is adapted to the place of cultivation and that it has the characteristics required by the market for which it is intended. Another important factor is that there is no predominance of a single cultivar in production, because, thus, pineapple becomes vulnerable to the occurrence of biotic factors such as the appearance of new pests and diseases and negative effects such as the loss of diversity (genetic erosion) and the disappearance of local varieties (Sanches and Matos, 2013). In view of this, variety diversification is important for the sustainability of the crop. The greatest diversity of pineapple cultivars, found in the Amazon region, remained unknown until the end of the 20th century. With efforts for varietal diversification, they focused on hybrid breeding to develop a cultivar that outperformed "Smooth Cayenne." However, even the best hybrids failed in the final evaluations (Leal and D'eeckenbrugge, 2018).

As a result, the predominantly cultivated variety in the world continues to be "Smooth Cayenne," however, the commercial production of pineapple is also based on other varieties such as MD2 (Gold), Singapore Spanish, Queen, Red Spanish, Perolera, and Pérola, described later, according to Cabral and Junghans (2003); Medina and García (2005); and Moretti-Almeida (2018).

Smooth Cayenne: the most planted variety in the world, both in terms of area and latitude range, has many favorable characteristics. It is a robust, semierect plant whose leaves have few spines at the apical edge of the border. The fruit is attractive and slightly cylindrical and weighs 1.5–2.5 kg, presenting a yellow-orange peel when ripe, yellow flesh, rich in sugars (13–19 °Brix) and of a higher acidity than the other varieties; the crown is relatively small, and the plant produces few slips. It is susceptible to the wilt associated with cochineal (*Dysmicoccus brevipes*) and fusariosis (*Fusarium subglutinans*). These characteristics make it suitable for industrialization and export as fresh fruit.

MD2 (Gold): this variety is a result of the cross between "58–1184" and "59–443" by the Pineapple Research Institute of Hawaii (PRI) (Williams and Fleisch, 1993; Sanewski et al., 2018). Due to the quality and great acceptance in the international market of fresh fruits, this variety has become the standard of the market. The fruits are large, cylindrical, and of yellow flesh and, when ripe, reach values of brix ranging from 15% to 17%, with lower acidity than the fruits of the Smooth Cayenne variety. The leaves present only a spiny tip. The MD2 variety accounts for at least 80% of the pineapple commercialized in the international market (Sanewski et al., 2018).

Singapore Spanish: the second variety of importance for industrialization. The plant presents medium size, with dark-green leaves whose length varies from 35 to 70 cm. The spines are variable, with clones completely without spines and others with few spines at the edges of the leaves. The fruit is small, weighing from 1.0 to 1.5 kg, cylindrical, and with low sugar content (10–12 °Brix) and low acidity. The plant is vigorous, with a regular production of slips and ground suckers. The occurrence of multiple crowns is frequent. It shows some resistance to pests and diseases.

Queen: small plant, 60–80 cm high, vigorous, with silvery leaves, small, and with occurrence of dense spines. It produces a large number of ground suckers, but the number of slips is variable, and they are usually poorly developed. The fruit is small (0.5–1.0 kg) with yellow peel. The pulp is yellow and sweet (14–16 °Brix) and has low acidity, excellent flavor, and long shelf life. This cultivar exhibits some characteristics similar to the Pérola variety.

Red Spanish: the plants are medium sized, vigorous, with dark-green leaves, with small and short spines, and being prickly or partially prickly. The fruit is of medium size (1.2–2.0 kg) in the shape of a barrel, white or pale yellow, juicy, sweet-tasting pulp (total soluble solids around 12 °Brix) and low acidity, and with a pleasant aroma. Generally, it produces few slips and suckers.

Perolera: adapted to altitudes up to 1500 m, behaves as resistant to fusariosis. The plant has a height (distance from soil level to fruit base) of 51 cm, long peduncle, a length of 29.2 cm, leaf of dark-green color, and slick edge, evidencing a little pronounced silver strip, producing 1 or 2 suckers and 8–10 slips, and presents a long peduncle that can favor the tipping of fruits. Its fruit is of cylindrical shape, weighing 1.5–3.0 kg, of yellow peel and pulp, with total soluble solid content around 13 °Brix, titratable acidity around 10 mEq/100 mL, and high content of ascorbic acid.

Pérola: the plant has medium size and erect growth; it is vigorous, with leaves about 65 cm in length and spines at the edges. The peduncle of the fruit is long (around 30 cm). It produces many slips (5–15) attached to the peduncle, close to the base of the fruit, which presents a conical shape, yellowish peel (when ripe), white pulp, juicy, and with total soluble solids of 14–16 °Brix, pleasant to the Brazilian taste. The fruit weighs 1.0–1.5 kg, has a large crown, and has not been much used for export in nature and industrialization in the form of slices. Presents tolerance to wilt associated with cochineal and susceptibility to fusariosis.

There are numerous other varieties cultivated on a reduced scale for local and regional markets in the world, especially in Latin American countries. Variants of the established cultivars are being selected and improved genetically, allowing to obtain new characteristics of interest, fruits of good quality, and resistance to pests and diseases.

5 Commercial belts

About 43.3% of all pineapples in the world come from Asia, while 36.4% come from the Americas and 19.9% from Africa. The fruit is produced in many countries all around the world, and its trade generates about 2 billion dollars per year (International Trade Center. Trade Map, 2018). Costa Rica is the main pineapple producer, with over 3 million tons. Exports increased from approximately 41 thousand tons in 1986 to over 2 million tons in 2016 (Food and Agriculture Organization of the United Nations, 2017), and now, the country is also the main exporter of this fruit, sending about 48% of the exported volume to the United States and the other 52% to Europe, which generated almost one billion dollars in revenue in 2018 (International Trade Center. Trade Map, 2018).

Philippines is the second main pineapple producer, with 2.67 million tons in 2017 (Food and Agriculture Organization of the United Nations, 2017). The country exports about 567 thousand tons of the fruit, generating over 228 million dollars in revenue, setting it as the second main exporter. About 31% of these fruits go to Japan, 20% to China, and 17% to South Korea (International Trade Center. Trade Map, 2018). Brazil, despite being the center of origin of this species and showing increasing production of this fruit since the middle of the 1970s (2.64 million tons in 2017), is also the main consumer market for pineapples, and this causes the export volume to be very low, about three thousand tons in 2017 (Food and Agriculture Organization of the United Nations, 2017).

6 Major soil types with taxonomical distribution

Pineapple is cultivated on several soil types around the world, all of them with peculiar characteristics and attributes, and it is very important to know them very well to maximize their use potential. Soils for this crop should be sand-clayey, well drained, preferably flat, and with good depth and pH around 5.5. They should not be too heavy or subjected to waterlogging. Clayey soils can be used, provided that they have good aeration and drainage conditions. Soil preparation consists of plowing and harrowing operations, which should be done to facilitate the proper development of the fragile root system of the plants (Cunha et al., 1995).

In Costa Rica, the main producer, pineapple is cultivated on a variety of soils, but the ones used for pineapple cultivation are mostly ultisols (acrisols). Some other regions show soils derived from volcanic rocks. These volcanic soils, mostly cambisols (inceptisols), usually show high fertility (Bertsch et al., 2000), but some of them present minerals such as allophane and imogolite, which have great phosphate adsorption capacity. According to Parfitt (2009), in this kind of soils, large amounts of labile P are required to neutralize the high absorption capacity of allophane and other Al compounds to ensure adequate P supply for plant growth.

Pineapple is grown mainly in southern regions of Thailand, especially in the provinces of Prachuap Khiri Khan, Rayon, Chumphon, and others (Anupunt et al., 2000), where there are many types of soils. The peats show usually little to no plasticity and cohesion and very high compressibility and are commonly black and fibrous. Parts of these soils have developed under salty water conditions. Therefore, they may be acid or very acid, showing high sulfur contents (Land Development Department, 2014). Other soil class frequently used for pineapple cultivation in Thailand is Gray Podzolic. These are usually sandy loam or loamy sand in the surface and sandy loam to loam in the subsurface, with good drainage (Land Development Department, 2014). Ultisols (acrisols), which are mostly acidic and show high levels of Al, especially, in the subsoils are also frequent in Thai pineapple areas and, when fine-loamy, present a very good water retention capacity. They also show low-activity clays, low CEC, and consequently low nutrient availability for the plants (West et al., 1997). This soil class is also the most frequent under pineapple plantations in another top producer, the Philippines. Other types of soil in this country are cambisols and luvisols (Food and Agriculture Organization of the United Nations, 2017).

In some states of Brazil, the third main world producer, the crop is grown on coastal tablelands, as in the state of Paraíba, in which the soils are mostly sandy and acidic and have low natural fertility (Souza, 2000). Although apparently uniform, this ecosystem presents a great diversity of soil classes, being the most important the yellow ferralsols (oxisols), the yellow podzolics, and the gray podzolics (Cintra and Libardi, 1998). In the state of Minas Gerais, most of the soils used for planting pineapples are oxisols, which are highly intemperized and, due to a low cation exchange capacity (CEC), usually have a very low availability of nutrients for the plants, especially phosphorus. Most of these soils are acid, with pH values ranging from 4 to 5.5, and have low fertility. However, they are frequently located in flat areas and are usually friable and well drained, which is important when it comes to soil preparation and crop management.

7 Soil property-fruit quality relationship

Soil is considered one of the main factors of production in pineapple, being responsible for the dynamics in the supply of water and nutrients to the plants. Pineapple is cultivated worldwide in soils with a wide range of physical and chemical characteristics, with high organic matter content, such as in Malaysian soils; on volcanic soils in Hawaii, Costa Rica, and the Philippines; on sandy soils of Queensland (Australia) and in South Africa (Uriza-Ávila et al., 2018); and in soils with low fertility, the presence of exchangeable aluminum, and acidic pH such as Brazilian Cerrado soils and coastal tablelands. The physical and chemical properties of soils interfere with the production and quality of pineapple fruits. These properties include soil texture, organic matter content, nutrients, pH, Al content, and CEC.

The cultivation of pineapple should preferably be done in deep and well-drained soils, since it is a species that has low tolerance to flooded environments or with low aeration in the root system (Guinto and Inciong, 2012). The taste and quality of fruits grown on sandy soils (<15% clay) and medium soils (15%–35% clay) are considered superior (Hossain, 2016). This can be explained by the fact that these soils present a balanced dynamics in the supply of nutrients and water during the cycle of cultivation of pineapple especially after fertilization. They also enable deepening of the root system of pineapple (Manica, 1999) and better aeration and drainage conditions (Uriza-Ávila et al., 2018). The quality of the clays present in the mineral fraction of the soils also interferes in the production and quality of fruits. According to Vásquez-Jiménez and Bartholomew (2018), pineapple cultivation in weathered soils with kaolinite mineralogy, where 1:1 minerals predominate, lead to the formation of microaggregation (granular structure) of clay particles and greater porosity and infiltration of the water in the profile. These factors imply higher yield and fruit quality after fertilization compared with temperate soils with 2:1 mineralogy. Still according to these authors, the ultisols, alfisols, and oxisols present the best physical characteristics for the pineapple cultivation.

The production of larger pineapple fruits correlates positively with root growth of the plants and their distribution in the soil profile. However, the morphology of the pineapple roots and their distribution in the soil profile are dependent on their physical characteristics, since the plant presents a superficial and fragile root system, with the majority of the roots distributed in the first 25–35 cm, being able to develop up to 60 cm depth (Manica, 1999), in aerated and permeable soils. However, plantings in clayey soils present a greater propensity to compaction and flooding, limiting plant development and fruit growth. However, pineapples grown on different types of soil may have different postharvest quality. For example, in Malaysia, the pineapple grown on mineral soil is sweeter than those grown on organic peat. However, "Josapine" pineapple grown on mineral soil is more susceptible to bacterial heart rot disease (Hassan and Othman, 2011).

Soil organic matter (SOM) content interferes with soil water and nutrient dynamics for cultivated plants, especially in tropical soils, where cation exchange capacity is dependent on its content in the soil. The application of organic matter coupled with mineral fertilization in pineapple fields has benefited nutrient absorption and chlorophyll content in the leaves (Leonardo et al., 2013), weight and quality of the fruits, and soil nutrient content (Darnaudery et al., 2018; Weber et al., 2010), contributing to the improvement of the physical, chemical, and biological characteristics of the soil (Jin et al., 2015; Liu et al., 2013; Primo et al., 2017; Sampaio et al., 2008; Singh et al., 2010). However, in tropical environments under conventional cultivation systems, with soil tillage and only mineral fertilization, it is difficult to maintain or increase the SOM content (Amaral et al., 2015).

In Mexico, in pineapple-producing regions, located in Playa Vicente and Isla, Uriza-Ávila et al. (2018) describe that the majority of soils with intense agricultural exploitation have lower levels of organic matter, citing that 30% of pineapple soils are poor (0.6%–1.2%) or extremely poor (<0.6%) in organic matter and 50% present moderately poor (1.2%–1.8%) to medium (1.81%–2.4%) contents. The authors point out that these soils are located in areas with slopes higher than 5%, lack mechanical or cultural techniques to control the erosion caused by rainwater, and have favorable climatic conditions for rapid decomposition of organic matter and soil degradation. In regions with sandy soils and sloping topography (land with more than 5% slope), this process of degradation of SOM may be considered internal. The intense mineralization process of SOM in tropical soils dominated by pineapple crops in the world, and the low production capacity of plant residues in the first 6 months of cultivation has contributed to the reduction of SOM in conventional crops with constant soil rotation and without addition of organic compounds (organic mulching, organic fertilizer, etc.). This phenomenon can reduce crop sustainability and fruit quality. However, cultural management practices, such as the maintenance of crop residues from the harvest, the use of organic mulching, the addition of organic compounds as fertilizers, and the crop rotation, can reverse this process of soil degradation and contribute to the sustainable pineapple production.

The use of organic fertilizers, such as biosolids (nitrogen source) in the production and nutrition of the cultivars Pérola, Vitória, Smooth Cayenne, and IAC Fantástico, was equal to mineral fertilization in relation to the production and postharvest quality of pineapple fruits, with positive effects in the soil, with an increase in organic matter content, in the availability of phosphorus, calcium, iron, and zinc, and, consequently, increased cation exchange capacity and soil saturation (Mota, 2016). The addition of organic compounds (composted pineapple residue) to the soil increased the contents of chlorophyll, soluble sugars, and soluble protein, as well as root vigor, fruit transverse and longitudinal diameters, weight, and yield of next-cropped pineapple (Liu et al., 2013). The soluble sugar content after composted pineapple residue addition was increased by 39.6% and that for soluble protein by 29.5% in positive response of several factors related to soil organic compounds, such as decreasing the bulk density, increasing the fertility, the abundance of microorganisms (increased the abundance of bacteria, fungi, and actinomycetes), and activity of enzymes (urease, catalase, acid phosphatase, and invertase) of the soil where the next-cropped pineapple grown (Liu et al., 2013).

The application of 250-g poultry manure, azospirillum, and phosphobacteria at 650 mg each along with N, P_2O_5, and K_2O at 8:4:8/g plant recorded higher values in terms of growth of plants, juice percentage, and quality parameters of fruits. Soil fertility parameters such as available P and K of the experimental plots increased after 1 year. Organic carbon of soil also increased significantly, when compared with the control plots (Devadas and Kuriakose, 2005).

Research on organic (chicken manure) and mineral (urea) using a control with NPK presents the following results: the lowest doses of N applied (2.62 and 4.50 g/plant) resulted in the contents of total soluble and nonreducing sugars different for the control. Without it, the use of 152 g/plant chicken manure resulted in ascorbic acid accumulation. Lower doses of urea resulted in the highest yellow flavonoids, and when combined with chicken manure, it resulted in higher antioxidant activity and ascorbic acid content. Principal components analysis explained 64.7% of variability covering most of the variables analyzed, except for total antioxidant activity by the DPPH method. All together, the use of chicken manure combined with urea was effective in improving the quality of "Vitória" pineapple at doses of up to 4.5 g/plant of N (Dantas et al., 2015).

However, it should be emphasized that the isolated use of organic compounds without the supplementation of mineral fertilizers can reduce the production and quality of pineapple fruits. Darnaudery et al. (2018) described that with organic fertilization (organic: *Mucuna pruriens* green manure incorporated into the soil and foliar applications of sugarcane vinasse from a local distillery), rich in K (14.44 g/L), pineapple growth was slower, 199 days after planting versus 149 days for integrated *M. pruriens* green manure (240.03 kg/ha N, 18.62 kg/ha P, and 136.11 kg/ha K) incorporated into the soil or conventional fertilizations (NPK fertilizer at recommended doses: 265.5 kg/ha N, 10.53 kg/ha P, and 445.71 kg/ha K), and fruit yield was lower, 47.25 t/ha versus 52.51 and 61.24 t/ha, respectively. Interestingly, organic fertilization significantly reduced leathery pocket disease and produced the best quality fruit with the highest total soluble solid contents (TSS) and the lowest titratable acidity (TTA). Fruit quality was also significantly improved with integrated fertilization, with fruit weight similar to that of conventional fertilization (Darnaudery et al., 2018).

A study published by Guinto and Inciong (2012) described the existence of positive correlation between Mg content and organic matter of the soils of the Philippines with the production of pineapple fruits. Exchangeable Mg and organic C are closely positively related to yield. Magnesium is a component of chlorophyll, the green pigment in leaves that uses sunlight energy to convert carbon dioxide to carbohydrates. Organic matter acts as a source and sink of nutrients in soils, and it appears that it is also a sensitive indicator for crop yield. Thus, any changes in these variables are likely to be good predictors of pineapple productivity. In summary, the presence of organic matter in soils cultivated with pineapple is essential for maintaining or increasing the production and postharvest quality of pineapple fruits. Table 50.1 shows the main positive effects of SOM on soil and pineapple production.

Pineapple presents adequate growth in soils with pH between 4.5 and 5.5. Therefore, it is one of the few agricultural crops well adapted to soil conditions with relatively high acidity (Reinhardt et al., 2000; Vásquez-Jiménez and Bartholomew, 2018). Under soil conditions with pH above or below the mentioned range, there are negative changes in the availability and absorption of nutrients by pineapple plants (Reinhardt et al., 2000). When the pH rises above 5.5, there are deficiencies in micronutrient utilization by the plant. These factors may reduce the accumulation of nutrients in fruits, implying changes in their quality.

Agricultural crops in soils with pH between 4.5 and 5.5 provide the presence of Al^{3+}, potentially phytotoxic to most cultivated plants. However, pineapple plants present high tolerance to Al^{3+} present in the soil solution. This phenomenon is attributed to several adaptation mechanisms of the species, which can be divided into two main groups: The first one is related to mechanisms of exclusion, with exudation of organic ligands (mucilage, organic compounds of low molecular weight, etc.) by the roots, which are capable of complexing Al by the efflux of the Al accumulated in the roots and by the alteration in the pH of the rhizosphere (Langer et al., 2009). The second group of tolerance mechanisms is related to internal detoxification, by the fixation of Al in the cell wall, by the complexation in the symplast via organic ligands, and by the accumulation of Al in the vacuole (Ryan et al., 1994).

Despite the adaptation of the cultivation of pineapples in acid soils and of the presence of Al^{3+}, there are distinct responses of pineapple varieties to the presence of this element. According to Le Van and Masuda (2004),

TABLE 50.1 Positive effects of soil organic matter on pineapple cultivation.

Soil	Plant
Positive	
Increases soil aggregation	Greater growth and deepening of the root system in the soil profile
Lowers soil compaction	Greater absorption of water and nutrients
Reduction of the propensity of soil loss by erosive processes	Higher chlorophyll content, shoot and fruit growth
Increases soil CEC	Decreases cultivation cycle
Increases soil fertility	Increased efficiency of water and nutrients use by plants
Promotes the transport of nutrients to the plant roots	More and heavier fruits
Neutralization of Al^{3+}	Higher sugar and protein content in fruits
Aids in the complexation and precipitation of toxic metals	Higher antioxidant activity and ascorbic acid content
Increase in soil water availability	Higher content of nutrients and vitamins in fruits
Increases microbiological activity (bacteria, fungi, and actinomycetes)	Higher content of antioxidant substances in fruits
Increases enzymatic activity of soil	Better (balanced) sugar/acidity ratio in fruits (TSS/TTA)

considering the cultivars Cayenne, Queen, Soft Touch, Honey Bright, Bogor, Red Spanish, and Cream Pine, the Cayenne was the most Al-tolerant and Soft Touch the most Al-sensitive cultivar after the application of a highly saturated Al concentration in a nutrient solution (300-µM $AlCl_3$ or 90.5 µM of inorganic monomeric Al^{3+} activity). In addition to organic acids, variations in the proteins in root apices are regarded as the mechanism involved in Al resistance.

Other studies confirmed the difference in response between pineapple varieties and the presence of Al^{3+}. Mota et al. (2016) reported that "IAC Fantástico" was less affected by Al concentration than "Vitória." Lin (2010) studying the cultivars Cayenne and Tainung No. 17 observed that the absorptions of macronutrients and micronutrients were not affected in Al-resistant Smooth Cayenne but the absorptions of Ca, Mg, and K were inhibited when $AlCl_3$ was 200 µM and Fe, Mn, and Cu absorptions were inhibited significantly when $AlCl_3$ was 300 µM in the Al-sensitive Tainung No. 17. Lin and Chen (2011) studying cultivars Cayenne, Tainung No. 6, Tainung No. 13, and Tainung No. 17 found greater tolerance of cv. Cayenne to Al^{3+} and cited the maintenance of Ca and Mg contents in the roots and leaves of this cultivar as an adaptation mechanism to aid in cell protection (Ca) and synthesis of phosphoglycerate kinase (Mg) was one of the important proteins of plants in an unfavorable environment.

Al^{3+} phytotoxicity initially affects the root system of pineapple plants (Fig. 50.1), impairing root growth and the absorption of water and nutrients, causing the reduction of shoot growth and size and quality of the fruits. In this sense, the moderate application of soil correctives (gypsum or limestone) to reduce the solubility of Al^{3+} in soils with a high exchangeable content of this element can be considered as an alternative to increase fruit production and quality. Silva et al. (2006) observed that the application of gypsum or limestone did not increase the weight of fruits of the cultivar D10 in acid soil (pH of 1:1 soil-water paste, 4.5) of the Wahiawa series (very fine clayey, kaolinitic, isohyperthermic, and Rhodic Haplustox) from central Oahu, Hawaii, but increased calcium levels in the soil and in D-leaf and fruit tissues. Mite et al. (2010) found positive effect in the growth of fruits of the cultivar MD2 after applying 1.5 t/ha of several pH amendments (dolomitic, calcitic, and magnesian limestone, plus gypsum) in soil of volcanic origin (pH [H_2O] 4.4 and Al^{3+} 1.5 cmol$_c$/kg) in Ecuador. Again, the effect of amendment application on soil Al^{3+} explains the response. Once Al^{3+} has been precipitated or complexed, there is no need for higher application rates. Actually, fruit yield was reduced with higher amendment rates (>1.5 t/ha), due to the presence of *Phytophthora* sp., a known risk of overapplying lime. For this reason, it is difficult to use general lime recommendation for all the sites based only on Al^{3+} content of the soil as is common practice in ultisols and oxisols (Mite et al., 2010).

However, most studies on acid soils with or without the neutralization of Al^{3+} ionic activity in the soil solution did not identify changes in size and most postharvest characteristics of the fruits (Silva et al., 2006).

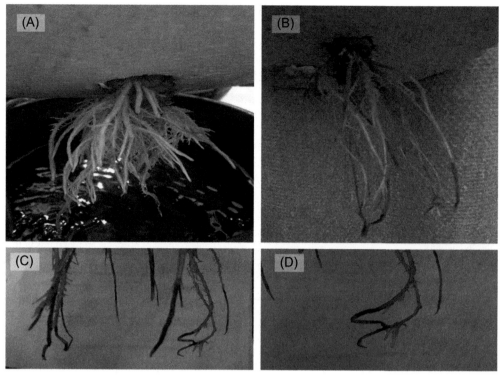

FIG. 50.1　Normal (A) and damaged (B–D) pineapple roots by the addition of 400 μmol/L of $AlCl_3$ in hydroponics.

8 Diagnosis of nutrient constraints

Pineapple is considered to be nutrient demanding, and mineral or organic supplementation with macro- and micronutrients is required during its growing cycle to obtain high productivity and fruit quality. The main essential nutrients considered limiting for cultivation are N, P, K, Ca, Mg, S, Fe, Mn, Zn, Cu, B, and Mo.

Nitrogen is one of the most demanded macronutrients by pineapple and the one that is more related to fruit weight and productivity. To obtain 72 t/ha of "Vitória" pineapple fruits, the uptake of 452 kg/ha of N was observed at the end of the growing cycle (Pegoraro et al., 2014a). The "Pérola" pineapple, in turn, uptook 764 kg/ha of N to produce 66 t/ha of fruits (author's data). Uptake values ranging up to 450 kg/ha are observed for cultivar Smooth Cayenne. These differences between varieties are due to variations in fruit growth and yield (Hiroce et al., 1977; Py et al., 1987).

Nitrogen deficiency in commercial pineapple crops (Fig. 50.2) may occur due to several biotic and abiotic factors related to the management, climate, soil, and nutritional requirements of the varieties. Among these factors, we highlight planting crops in soils with low fertility, poor in organic matter, sandy, and under the wrong nutritional management.

Pineapple cultivation in sandy soils is common around the world. These soils are well drained; however, they are poor in available nutrients and organic matter. These conditions imply the need to adopt cultural management with organic or green fertilization, with the aim of increasing the availability of nitrogen for the plants. The adequate nutrition with nitrogen provides fruits of larger size and higher productivity, besides an adequate ratio between acidity and soluble solids. However, in a deficiency condition, the plants present generalized chlorosis (yellowing; light gray in the print version). This symptom occurs initially in old leaves, since nitrogen is considered a mobile element in the plant (Hawkesford et al., 2012; Marschner, 2012; Taiz et al., 2017).

The omission of nitrogen in nutrient solution with the Imperial variety results, initially (26 days after the macronutrient concentration reduction to 10% of the complete solution), in leaves with yellowish-green coloration (light gray in the print version), with a higher intensity of this symptom in the older ones. Six months after planting, the leaves present progressive yellowing (dark gray in the print version) in a generalized way in the plant, which produces small fruits (reduced the fruit mass with crown at 58%), with chlorosis even in the leaves of the crown. In addition, fruits produced from nitrogen deficient plants had higher pulp firmness, higher titratable acidity (TA) and vitamin C, reduced SS/TA ratio, lower pH, and whitish pulp (Ramos and Rocha Pinho, 2014; Ramos et al., 2009, 2010).

FIG. 50.2   Symptoms of nitrogen deficiency in pineapple cv. Vitória. (A) Plants without nitrogen deficiency. (B–D) Nitrogen-deficient plants.

8.1 Nitrogen

Nitrogen deficiency causes a decrease in the synthesis of amino acids and, consequently, of proteins, resulting in reduced growth and accumulation of nonnitrogen metabolites, promoting greater availability of photoassimilates to be used in the synthesis of compounds of secondary metabolism, ascorbic acid, among other organic acids (Marschner, 2012; Taiz et al., 2017). However, it should be considered that the visual diagnosis of deficiency symptoms serves only to guide the occurrence of possible nutrition-related problems, because there are several nutrients responsible for the formation of color in the leaves of plants. Iron and magnesium are also responsible for the synthesis of chlorophyll in leaves. Therefore, the successful use of leaf color as an index is dependent on eliminating, minimizing, or recognizing all factors other than N that also can influence leaf color (Vásquez-Jiménez and Bartholomew, 2018).

8.2 Phosphorus

Phosphorus is not among the most absorbed macronutrients by pineapple. However, it is the third most used macronutrient in the crops, because its dynamics in the soil is considered to be impaired, due to the low natural availability and the presence of specific adsorption with oxidic clay minerals in tropical weathered soils. This specific adsorption implies in the unavailability (nonlabile P) for the plants of considerable fraction of the phosphorus coming from the fertilization in the complex of exchange of the clay minerals. Phosphorus fixation can occur within 35 days after application (Vásquez-Jiménez and Bartholomew, 2018), especially after the application of soluble forms. However, there is a consensus about the lack of response of plants to phosphate fertilization, indicating that the amount of P available in the soil, even in tropical soils, is sufficient, in most cases, to meet the demand for pineapple.

Phosphorus plays a number of roles in plants, among which it is notable for being a component of sugar phosphates, nucleic acids, nucleotides, coenzymes, phospholipids, and phytic acid, and the central role in reactions involving ATP (Taiz et al., 2017). Its deficiency in pineapples decreases the size of plants and fruits, besides altering postharvest quality. The most common visual symptoms of phosphorus deficiency are the yellowing of the leaves, followed by the appearance of red-purplish coloration, initially on old leaves and progressing to new leaves and fruits, as the severity of the deficiency increases.

The cultivation of the Imperial variety in nutrient solution with omission of phosphorus results in the appearance of the red-purplish coloration in the central part of the limbo of young and medium leaves with well-defined green edges. The fruits of these plants have the reddish peel in contrast to the yellow-orange coloration (Ramos et al., 2009). Considering pineapple "Jupi" under P deficiency, the fruit mass with crown was reduced by 26.8% and the fruit length up to 33.7%. However, that condition did not influence the organoleptic characteristics of the fruits (Ramos and Rocha Pinho, 2014).

8.3 Potassium

Potassium is the most absorbed macronutrient by pineapple. To obtain 72 t/ha of "Vitória" pineapple fruits, the uptake of 898 kg/ha of potassium was observed at the end of the growing cycle (Pegoraro et al., 2014a). The "Pérola" pineapple, in turn, uptakes 796 kg/ha of potassium to produce 66 t/ha (author's data). The adequate potassium nutrition in commercial crops has favored weight and organoleptic quality of fruits, especially by increasing soluble solids. However, most pineapple crops are grown in weathered soils with low availability of nutrients, which favors the occurrence of symptoms of nutritional deficiency, requiring the application of high doses of this nutrient via mineral fertilization.

"Smooth Cayenne" pineapple achieved the maximum fruit yield (66.6 t/ha) when fertilized with 700 kg/ha of K_2O (source: K_2SO_4). Under the same conditions, the fruit yield achieved 51.6 t/ha with no potassium fertilization. This trial was carried out on an ultisol (Teixeira et al., 2011a). The elevation of the potassium doses to 410.4 kg/ha increased infructescence (fruit) mass with crown, yield, fruit length and diameter, and soluble solid content (Rios et al., 2018a,b). The developmental changes in fruit potassium were significantly correlated with fruit acidity and fruit soluble solids in both high and low acid clones of "Smooth Cayenne," possibly due to the promotion of sugar translocation to the fruit (Saradhuldhat and Paull, 2007). In addition, there are some reports showing that potassium fertilization does increase pineapple titratable acidity (Py et al., 1987).

Potassium is required by plants as a cofactor of more than 40 enzymes, being the most abundant cation in the cytosol and responsible for establishing cellular turgor and maintaining cellular electroneutrality (Hawkesford et al., 2012; Taiz et al., 2017). When K is deficient, growth is retarded, and net transport of K^+ from mature leaves and stems is enhanced. Under severe deficiency, these organs become chlorotic and necrotic, depending on the light intensity to which the leaves are exposed (Hawkesford et al., 2012).

Visual symptoms of potassium deficiency in pineapple leaves are characterized by green to dark-green (light to dark gray in the print version) foliage, more pronounced with nitrogenized fertilization (Leonel and dos Reis, 2012). The leaves show small yellow dots (light gray dots in the print version) that grow, multiply, and may concentrate on the limb margins (Leonel and dos Reis, 2012) (Fig. 50.3), also characterized by presenting the apex of the older leaves browned and necrotic (Ramos et al., 2009).

Potassium deficiency reduced by 23% the fruit mass with crown, caused bleaching of fruit pulp, and reduced fruit acceptance (Ramos and Rocha Pinho, 2014). Potassium deficiency can also lead to the appearance of dark spots on the fruit pulp, corresponding to the symptoms of internal browning. In addition, this condition increases fruit firmness but reduces the percentage of juice, soluble solids (SS), vitamin C, pH, and sensory acceptance (Ramos et al., 2009, 2010). In

FIG. 50.3 Symptoms of potassium deficiency in pineapple cv. Vitória. (A) Anatomy of plants with potassium deficiency. (B) Anatomy of leaves with potassium deficiency.

summary, the reduction in postharvest quality of pineapple fruits caused by potassium deficiency is related to imbalance in the metabolic activity of plants, as a consequence of the inadequate process of osmotic regulation and cellular electrochemical imbalances, besides the alteration of the enzyme activity and the loading and unloading of sucrose in the phloem.

8.4 Calcium

Pineapple has a low demand for calcium, and the availability in soils for optimum growth should be higher than 100 mg/kg (neutral ammonium acetate extraction) (Vásquez-Jiménez and Bartholomew, 2018) or higher than 0.50 cmol$_c$/kg. The pineapple accumulates, at the end of the growing cycle, in the plant shoots, an average content of 80–129 kg/ha of calcium (Amaral et al., 2014; Pegoraro et al., 2014a). However, due to the differences between the varieties, there may be greater accumulation, reaching values of 398 kg/ha of calcium, becoming the third most absorbed nutrient, only behind potassium and nitrogen.

The occurrence of nutritional deficiencies and the need for fertilization with calcium sources is not common in commercial crops, especially due to adequate growth capacity in acid soils. When necessary, the application of calcium is usually associated with the use of soil acidity correctives (limestone, gypsum, calcium silicates, steel slag, etc.) or via fertilization with mineral sources of calcium. In weathered and sandy soils, extremely acid (pH <4.5), with high levels of Al^{3+} and lower natural availability of calcium, there is a need for soil correction for adequate fruit production. Thus, the application of moderate doses of limestone (1.5 t/ha) increases the production of roots and fruits of MD2 pineapple cultivated in an andisol with 1.5 cmol$_c$/kg of Al^{3+} and pH (H$_2$O) equal to 4.4 (Mite et al., 2010).

Calcium is the constituent of the middle lamella of the cell walls; it is also required as a cofactor by some enzymes involved in the hydrolysis of ATP and phospholipids, besides acting as a secondary messenger in metabolic regulation (Taiz et al., 2017). However, calcium mobility in symplasm and phloem is low (Hawkesford et al., 2012). Its supply should be done especially during the initial stages of crop establishment, because cell division and differentiation are dependent on an adequate supply. In addition, calcium is also important after forcing because it also is a period of rapid cell division and growth although calcium uptake is reduced after that practice.

Calcium may also improve cell structure and reduce fruit translucence (Vásquez-Jiménez and Bartholomew, 2018), and its application to the soil is an effective method to control or reduce internal browning (blackheard) development in "Mauritius" pineapple (Herath et al., 2003). According to these authors, the application of calcium as basal and top dressings is more effective than applying only a basal dressing in controlling internal browning and for the maintenance of fruit quality under cold storage. The application of calcium fertilizer as basal dressing (150 kg/ha) followed by a top dressing (100 kg/ha) 6 months after planting is more effective to control internal browning in cold stored "Mauritius" pineapple. Similar results have been described by Hewajulige et al. (2006), who studied the application of CaO and CaCl$_2$ as spray in combination with basal dressing. However, Pusittigul et al. (2014) reported that calcium contents in pineapples harvested from various growing areas in Thailand showed inconsistent correlation with internal browning development. Pre- or postharvest application of calcium could raise calcium content in the fruit, but its effect on reducing internal browning was not reliable. It is suggested that internal browning is a result of multifactorial aspects and calcium content is one of the factors influencing this disorder.

Calcium deficiency has less influence on the postharvest characteristics of fruits compared with nitrogen and potassium. The application of calcium in the "Imperial" variety results in an increase in the soluble solid content (SS) in the fruit pulp. This result can be attributed to higher potassium uptake, due to the lower competition between potassium and calcium by absorption sites (Ramos et al., 2010, 2011). On the other hand, there is no interference of calcium deficiency in the quality and organoleptic evaluation in the cultivar of "Jupi" pineapple fruits (Ramos and Rocha Pinho, 2014). However, it should be noted that calcium deficiency in pineapples implies a severe reduction in plant growth and fruit production.

8.5 Magnesium

Pineapple can be considered a demanding culture in magnesium. Scientific results report the absorption of similar contents of magnesium and calcium for pineapple cultivars, indicating that the availability in the soil and the supply through fertilization should establish a 1:1 ratio of Ca/Mg. Results from studies with the Vitória variety confirm this statement, since this genotype uptakes 126 kg/ha of magnesium and 129 kg/ha of calcium (Pegoraro et al., 2014a). Hanafi et al. (2009) obtained a similar use efficiency ratio of magnesium and calcium for cultivars N-36 and Josapine, corresponding, respectively, to 1.18 and 0.43 g of dry matter/mg of magnesium and 1.31 and 0.45 g of dry matter/mg of calcium absorbed, respectively. However, there may be differences in behavior between cultivars with those that require a higher Ca/Mg ratio (i.e., 2:1).

Magnesium is considered a mobile element in plants, and its main functions are related to the constitution of many enzymes involved in the transfer of phosphates and in the constitution of the chlorophyll molecule. Magnesium ions also play a specific role in the activation of enzymes involved in respiration, photosynthesis (Rubisco), and the synthesis of DNA and RNA (Marschner, 2012; Taiz et al., 2017). A characteristic symptom of magnesium deficiency is chlorosis between leaf veins, occurring, first, in older leaves, because of the mobility of this cation (Fig. 50.4). This pattern of chlorosis occurs because chlorophyll in the vascular bundles remains unchanged for longer periods than that in the cells between the bundles. If the deficiency is long, the leaves may become yellow or white (light gray in the print version). An additional symptom of magnesium deficiency may be premature senescence and leaf abscission (Marschner, 2012; Taiz et al., 2017).

Studies carried out with the omission of magnesium have described the lack of alteration in the postharvest quality of pineapple fruits because it does not alter the juice, total soluble solids, vitamin C content, acidity, and sensory properties of "Imperial" (Ramos et al., 2010) and "Jupi" (Ramos and Rocha Pinho, 2014) pineapples. However, magnesium deficiency in pineapples increased the absorption of calcium and potassium (Ramos et al., 2011), indicating the existence of an antagonistic effect on the absorption of these cations in pineapple plants. The uptake of Mg^{2+} can be strongly depressed by other cations, such as K^+, NH_4^+, Ca^{2+}, and Mn^{2+} or even low pH (H^+) (Hawkesford et al., 2012). Notably, in tropical soils, a higher proportion of calcium in relation to magnesium has impaired magnesium absorption, being necessary the use of mineral fertilization with magnesium ($MgSO_4$) for soils, the use of higher doses of potassic fertilizers, and the use of soil correctives and correction of deficiencies caused by nutritional imbalance in the soil. The use of magnesium silicates can also help correcting these deficiencies.

Magnesium deficiency in plants may lead to reduced chlorophyll formation, reducing the photosynthetic activity of plants. This phenomenon in the pineapple is characterized in the visual diagnosis by the presence of old leaves with bright yellow limb, in leaves exposed to sunlight, and maintenance of the dark-green color in the leaves or in the part shaded by the younger leaves, located in the upper part of the plant (Vásquez-Jiménez and Bartholomew, 2018). This symptom is characteristic for magnesium deficiency in pineapple and, under severe conditions of deficiency, can progress to generalized yellowing in the leaves after flowering, reducing CO_2 assimilation capacity, stem diameter and length, root volume, acidity, sugar content, and the aroma of the fruits (Py et al., 1987).

Sulfur deficiency in pineapple crops is considered to be uncommon, mainly due to the lower nutritional requirement and its use as a secondary source in nitrogen, phosphate, and potassium fertilizers. However, in sandy and in poor organic matter soils, it is possible to observe the nutritional deficiency of this nutrient. Sulfur is found in certain amino acids such as cystine, cysteine, and methionine and is a constituent of several coenzymes and vitamins, such as coenzyme A, S-adenosylmethionine, biotin, vitamin B1, and pantothenic acid, which are essential for metabolism (Marschner, 2012; Taiz et al., 2017). For a production of 72 t/ha of pineapple, the sulfur content accumulated in pineapple shoots can correspond to 134 kg/ha (Pegoraro et al., 2014a). In the Pérola variety, the plant accumulates 49 kg/ha of sulfur to produce 66 t/ha. These values indicate that this element is little absorbed by the plant, which may explain the small occurrence of its deficiency in experimental and field conditions.

FIG. 50.4 Deficiency symptoms of magnesium in pineapple. (A) Mg deficiency in the plant. (B) Deficiency of Mg in leaves.

Sulfur is considered an immobile element in the plant, and the symptoms of deficiency are also associated with generalized chlorosis. However, chlorosis occurs initially on new leaves. Many of the symptoms of sulfur deficiency are similar to those of nitrogen deficiency, including chlorosis, reduced growth, and accumulation of anthocyanins, as both nutrients are constituents of proteins (Marschner, 2012; Taiz et al., 2017).

However, there were no symptoms of sulfur deficiency in leaves and fruits of "Imperial" pineapple with the omission of this nutrient (Ramos et al., 2009). On the other hand, the omission of sulfur in "Jupi" pineapple increases the total soluble solid content (Ramos and Rocha Pinho, 2014). This result is due to the probable reduction in the synthesis of proteins, provoking the accumulation of soluble carbohydrates.

Considering the absorption of all macronutrients by the pineapple, there is an order of accumulation that varies very little among the most diverse cultivated varieties. A sequence of accumulated macronutrients is normally $K > N > Ca > Mg > S > P$. Micronutrients deficiency of iron, zinc, boron, copper, manganese, molybdenum, and chlorine in pineapple crops is associated with variations in soil and climatic conditions and the adoption of misguided cultural management practices such as the lack of or excessive fertilization with macronutrients, cultivation of plants outside the ideal range of soil pH (4.5–5.5), excessive liming, removal of plant residues from the growing area, and the absence of crop rotation. Therefore, the use of soil or foliar fertilization with micronutrients implies positive results in the production and quality of the fruits of pineapple.

The soil and foliar micronutrient application increased the concentrations of carbohydrates and N-aminosoluble and reduced the leaf pH, especially during flowering and fruit development (Amorim et al., 2013). The pineapple has an increase in dry-matter production after the application of micronutrients via soil or foliar route. This increase in shoot dry matter can be up to 234% when compared with plants that did not receive fertilization with micronutrients (Feitosa et al., 2011). Therefore, the importance of micronutrient fertilization on fruit quality is evident, but little is known about the effects of micronutrients on the characteristics of pineapple (Amorim et al., 2013).

Considering the quantities absorbed by the pineapple plant and therefore demanded by the pineapple, the following decreasing order follows: $Fe > Mn > Zn > B > Cu$, with some variations depending on the cultivars. Iron, which is the most absorbed micronutrient by pineapple, is essential for the synthesis of complexes constituted by chlorophyll and protein in the chloroplast, being a constituent of enzymes involved in the transfer of electrons, such as cytochromes. In this process, iron is reversibly oxidized from Fe^{2+} to Fe^{3+}, aiding in electron transfer (Marschner, 2012; Taiz et al., 2017).

In plants, iron is considered an immobile micronutrient, and the initial symptoms of deficiency in fresh leaves include internerval chlorosis and yellow leaves with green spots (Newett and Rigden, 2015; Py et al., 1987). Under conditions of extreme or prolonged deficiency, the veins may also become chlorotic, causing the entire leaf to become white (Marschner, 2012; Taiz et al., 2017). Fruits on plants with severe iron deficiency will be small, hard, and reddish in color, with cracks between the fruitlets, and the crowns will be light yellow or creamy white in color (Vásquez-Jiménez and Bartholomew, 2018).

Pineapple plants growing in acid soils tolerate high levels of both soluble manganese and aluminum, where other plants either will not grow or show symptoms of toxicity (Vásquez-Jiménez and Bartholomew, 2018). However, acidic soils with high natural manganese contents affect the absorption of other cationic micronutrients, such as iron. In this condition, it is important to establish an ideal relationship between iron and manganese in soils and plants. According to Vásquez-Jiménez (2010), iron deficiency can occur when the Fe/Mn ratio is less than 0.4 (in D-leaf) for "Smooth Cayenne" while a 0.2 ratio (BG) is recommended for "MD-2." Proportions lower than these can cause iron deficiency in plants. The earliest reports of manganese interference in iron absorption date back to the beginning of the last century (Gile, 1916).

Iron sulfate sprays, often applied fortnightly on plants grown in soils high in soluble manganese, are used to manage iron deficiency. To be effective, the iron in iron sulfate sprays must be in reduced form, and dry storage conditions are required to prevent oxidation (Vásquez-Jiménez and Bartholomew, 2018). Another alternative to reduce the excessive content of manganese in the soil would be the application of soil correctives to raise the pH and increase calcium supply.

Cultivation in alkaline soils or the use of irrigation water that presents carbonates and raises soil pH promotes iron deficiency in pineapple. As a solution to this problem, it is recommended to apply a foliar application of this micronutrient, which can be up to 6 kg/ha (Py et al., 1987), or even the use of substances that lower soil pH, such as elemental sulfur, or the pH of water, such as nitric acid or other acid available at the planting site. To determine the amount of sulfur, it is necessary to take into account the pH of the soil, the pH units to be lowered, and the soil texture. If the option is the pH correction of an alkaline water, the amount of acid applied in the irrigation water is a function of the pH, of the presence of carbonates in the water, and of the pH to be reached. This amount can be calculated as a function of the stoichiometry of the acid-base reaction that will occur.

The weathered and acidic tropical soils present high natural manganese contents. However, degradation conditions of organic matter, the lack of moisture in the soil, cultivations in sandy soils, and elevation of pH of the soil (above 6.5) or calcium content, associated with the absence of fertilization with manganese, may imply deficiency of manganese in pineapple plants, although this phenomenon is not common. Manganese is considered an immobile element in plants, and the main symptoms of deficiency are characterized by chlorosis between the veins associated with the development of small necrotic spots. This chlorosis can occur in young or older leaves, depending on plant species and growth speed. These symptoms are due to the main functions in the plant, since manganese acts in the photosynthetic process as an essential cofactor in the process of water oxidation and O_2 generation (Marschner, 2012; Taiz et al., 2017). Although it is an essential nutrient for pineapple, manganese is neglected in scientific studies on this element and its relationship with pineapple. The papers and reports are usually focused on the interference of this nutrient in the absorption of iron.

The zinc ion (Zn^{2+}) is required for the activity of several enzymes and in the biosynthesis of plant chlorophyll. It also acts as a catalyst in oxidation and reduction processes and has great importance in sugar metabolism in pineapple (Kumari and Deb, 2018). This nutrient is considered to have low mobility in the plant, and its deficiency symptoms occur initially on new leaves. Its growth occurs in rosette, forming circular grouping that radiates in the soil or next to it. The leaves may also be small and twisted, with wrinkled appearance margins (Taiz et al., 2017). When the deficiency is severe, the center cluster of leaves is mildly to sharply curved. When the deficiency develops in older plants, the surface of older leaves develops yellowish-brown pinhead-sized dashes. The center leaves may have rips or serrations on their margins (Vásquez-Jiménez and Bartholomew, 2018).

The nutritional deficiency of zinc in pineapple crops is associated with, in sandy soils and poor organic matter soils and in regions of acid or semiarid climate, the presence of soil pH higher than 6.0, because this condition reduces the availability of zinc in the soil and its absorption by the pineapple. The mismanagement of fertilization with excessive doses of phosphorus also causes a reduction in the availability of zinc to the plants, due to the formation of zinc phosphate ($Zn-HPO_4$), a complex considered of low solubility. The higher absorption of phosphorus also reduces the translocation of zinc from the root to the shoot, among others.

Zinc deficiency reduces the size and alters the organoleptic properties of pineapple fruits. In this context, the correction of deficiencies with the foliar application of zinc sulfate (0.5%) in association with borax (0.5%) after the flowering of pineapple cv. Mauritius showed maximum TSS/acid ratio (21.46), total sugars (8.66%), and reducing sugars (1.72%) and lowest acidity (0.67%) along with higher TSS (14.4 °Brix). However, Maeda et al. (2011) did not observe changes in pineapple fruits with zinc fertilization. After the application of zinc and boron at 7 and 9 months of planting the cultivar Smooth Cayenne, these authors did not verify effects on soluble solids, titratable acidity, average fruit diameter, fruit length without crown, and fruit maturation index. Only boron, zinc, and potassium contents in the leaf were influenced by the treatments. Such variations in scientific results may be associated to different demands of zinc by pineapple cultivars.

Copper is a redox-active transition element with roles in photosynthesis, respiration, C and N metabolism, and protection against oxidative stress (Broadley et al., 2012). Copper deficiencies in pineapple crops are usually observed in weathered soils with sandy texture, limestone soils, and soils with high organic matter content, since copper presents specific adsorption with organic compounds humidified from organic matter, preventing its absorption by plants. The initial symptom of copper deficiency in many plant species is the production of dark-green leaves, which may contain necrotic spots. These spots appear first in the apices of young leaves and then extend toward the base of the leaf along the edges. The sheets may also become twisted or malformed (Marschner, 2012; Taiz et al., 2017). Foliar applications of copper in the form of inorganic salts, oxides, or chelates can be used to rapidly correct its deficiency in soil grown plants (Broadley et al., 2012). However, excessive application of copper may lead to phytotoxicity in plants.

Boron, like iron, is one of the most important micronutrients for the production of fruits in pineapple, playing important roles in sugar transport, cell wall synthesis, lignification, cell wall structure, carbohydrate metabolism, RNA metabolism, respiration, indole acetic acid (IAA) metabolism, phenol metabolism, and membranes (Broadley et al., 2012). Boron is considered a mobile micronutrient in the phloem of pineapple plants. According to Siebeneichler et al. (2005), pineapple plants show the ability to synthesize mannitol and sorbitol, which would allow the formation of mobile complexes in the phloem, like mannitol-B-mannitol or sorbitol-B-sorbitol. However, depending on the severity of the deficiency, visual symptoms can be observed in old and new leaves. Thus, 12 months after planting "Imperial" pineapple, Ramos et al. (2009) observed that the deficiency of boron caused deformation on the leaf of the slips, which had a serrated margin and a deformation in the leaf of the fruit crown, with the formation of corky excrescence and cracks between fruit.

Boron fertilization in the cultivar Pérola increases leaf boron content but does not interfere with fruit weight and size (Siebeneichler et al., 2008). This is probably due to the availability of this element in the soil in quantity to meet the boron demand of this cultivar. The application of boron and zinc (110 g/ha of B and 250 g/ha of Zn), via two foliar sprays, at 7 and 9 months after planting of cv. Smooth Cayenne, had no effect on average diameter of the fruit, length of the fruit without crown, contents of TSS and TTA, and ratio TSS/TTA (Maeda et al., 2011). These results suggest that plant factors, such as the nutritional requirement of the cultivar and edaphic conditions related to soil texture, organic matter content, mineralogy, and soil pH, interfere in the dynamics of boron for plants and in the need for fertilization via soil or leaves.

Molybdenum is considered a micronutrient little demanded by the pineapple culture. Normally, the levels present in the soil are considered sufficient for adequate molybdic nutrition. However, it is a component of several enzymes, including nitrate reductase, nitrogenase, xanthine dehydrogenase, aldehyde oxidase, and sulfite oxidase (Marschner, 2012; Taiz et al., 2017). The first indicative of molybdenum deficiency is widespread chlorosis between the veins and necrosis of older leaves. Deficiency of molybdenum may cause high nitrate levels in fruits, which leads to detinning in canned pineapple (Hassan and Othman, 2011). Research in Australia failed to demonstrate any change in fruit juice nitrate levels as a result of spraying plants with molybdenum (Scott, 2000).

9 Management of nutrient constraints

Fertilization in pineapple crops is considered a responsible cultural practice for increasing productivity and fruit quality. However, the balanced management of plant nutrition is dependent on the adoption of soil fertility diagnosis criteria and nutritional demand of the cultivars and the adoption of application practices that follow the nutritional demand of the crops in the different stages of growth and phenology of the plant (Pegoraro et al., 2014b). The pineapple shows sigmoidal growth (Fig. 50.5), with greater vegetative increase after the first 3–6 months of planting, intensifying its growth until the period of floral induction. In phases V2, V3, and V4, there is a greater increase of the vegetative growth, indicating a greater nutritional demand, which may guide the fertilization to the adequate growth of the culture, so that, by the moment of forcing, the plant has the correct size and weight to produce good commercial fruits (Maia et al., 2016; Pegoraro et al., 2014b).

The production of plant biomass presents a positive linear correlation with the weight and size of pineapple fruits. To obtain a fruit weight greater than or equal to 1.2 kg/plant, flower induction in the "Vitória" pineapple was suggested for those plants having a minimum weight for *D*-leaf fresh matter of 70 g or with a minimum stem diameter of 8.5 cm (Vilela et al., 2015). In fact, plants with higher biomass production produce proportionally heavier fruits (Guarçoni and Aires Ventura, 2011; Razzaque and Hanafi, 2001; Rodrigues et al., 2010; Santos et al., 2018; Vilela et al., 2015).

Mineral and organic fertilization plus the combination of these and the use of fertirrigation are management techniques that assist in the balanced supply of nutrients during the pineapple cultivation cycle. The adoption of top dressings and the integration between organic and mineral fertilization, for allowing a continuous supply of nutrients during the growing season, provide good results from the point of view of fruit quality and productivity. Some producing regions have adopted foliar fertilization applied by spray boom to supply all the needs of the crop, which implies in the formulation of a spray solution with all the necessary nutrients to the pineapple, in a balanced way and in quantity of diluted solute that does not cause necrosis of the leaves. Some growers use nutrient spraying

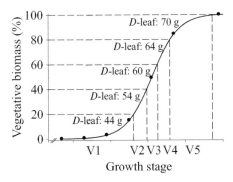

FIG. 50.5 Indication of phenological stages by means of relative production of vegetative biomass (leaves, stems, and roots) for the vegetative cycle of the "Vitória" pineapple and characterized by average *D*-leaf fresh weight. Five phenological stages are proposed based on vegetative biomass production: ≤20% biomass production (V1), 21%–40% (V2), 41%–60% (V3), 61%–80% (V4), and >80% (V5). *From Pegoraro, R.F., Souza, B.A.M.D., Maia, V.M., Amaral, U.D., Pereira, M.C.T., 2014. Growth and production of irrigated Vitória pineapple grown in semi-arid conditions. Rev. Bras. Frutic. 36 (3), 693–703. doi:https://doi.org/10.1590/0100-2945-265/13 (web archive link).*

solutions with up to 9% diluted solutes, but it is a very risky practice. Values of up to 5% are well tolerated, especially in regions with high rainfall or irrigated irrigation. These cultivation techniques require the pineapple grower to have specific knowledge regarding the management of fertilization.

Conventional fertilization in pineapple crops via mineral fertilization can be adopted with the use of formulated (mixed) or single fertilizers. The fertilization with formulated fertilizers has the preference of use by the rural producers, because it allows in a single application the supply of several nutrients. However, in the planting phase, in many crops, there is a need to use only a simple source of fertilizer, as in the use of phosphate fertilizer. It can be noticed that the choice of fertilization sources should mainly follow the nutritional demand of the crops and the soil.

The use of organic fertilization provides several positive effects for soil and pineapple, as previously mentioned. We highlight the gradual supply of nitrogen to the plants and the contribution of organic matter to the soil. However, its isolated use does not provide sufficient and balanced amounts of macronutrients for the plants, being necessary the complementation of the fertilization via mineral fertilizers (Mota et al., 2018). In addition, it depends on the availability of organic fertilizer in the growing region. Therefore, it is considered that the optimal nutrition management in pineapple should advocate the use of organic fertilization in conjunction with mineral fertilization.

Due to the anatomical conformation of the pineapple leaves (Fig. 50.6), which converge to the center of the plant, and the presence of adventitious roots in the stem, much of the mineral fertilization can be provided via fertirrigation or via foliar spraying, as already mentioned. In this management condition, top dressing during the vegetative period has provided good results in increasing productivity and fruit quality.

The nutritional requirement varies according to the pineapple cultivars and their productive potential (Table 50.2) and usually follows this absorption order: $K > N > Ca > Mg > S > P$. Due to these variations in nutritional demand among cultivars (Table 50.3), different fertilization responses are observed in the production of pineapple fruits. These variations are further accentuated by the population of plants used in each field. This can be minimized when the fertilization recommendation changes to grams per plant in substitution of quantities per area (hectare). Table 50.3 can be used as reference for fertilization of several pineapple cultivars and the different populations used.

As discussed in this chapter, nitrogen is the nutrient that has the greatest effect on fruit size and pineapple productivity. This effect can be visualized in Fig. 50.7, which shows the result of the application of increasing doses of nitrogen up to 20 g/plant, in the form of urea, on the size of the fruits of the "Vitória" cultivar (Fig. 50.2). These results confirm the high response potential of nitrogen fertilization for this fruit (Cardoso et al., 2013).

10 Use of sewage sludge and treated waste water on pineapple nutrition

Increases in population, urbanization, and industrialization in the world have resulted in increased levels of liquid and solid waste generated in sewage treatment processes. These residues, when inadequately handled and treated, are a source of contamination to the environment and to living beings, especially when directed to dumps and water sources. In this context, studies have been intensified in the last years aiming at a correct disposal of these wastes, so that they are not only an environmental problem but also an economically feasible alternative for agriculture, since

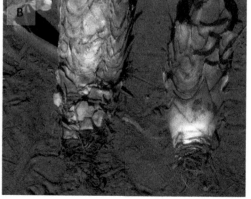

FIG. 50.6   Anatomical conformation of the pineapple leaves converging to the center of the plant (A) and presence of adventitious roots in the stem of the pineapple (B).

TABLE 50.2 Macronutrient accumulation (kg/ha) in shoots of pineapple cultivars.

Source	N	P	K	Ca	Mg	S
"Vitória"[a]	451.71	107.26	898.32	129.17	126.41	134.27
"Vitória"[b]	453.00	36.00	1703.00	76.00	124.00	112.00
"Pérola"[c]	763.65	14.04	795.72	397.84	49.80	48.90
"Gandul"[d]	504.39	91.31	947.28	70.81	34.79	–
"N-36"[d]	530.48	98.14	993.87	39.75	43.48	–
"Josapine"[d]	337.30	50.31	754.10	65.22	69.57	–
"Pérola"[e]	238.00	13.50	1234.00	253.00	157.00	17.00
"Smooth Cayenne"[f]	252.00	13.00	441.00	161.00	33.00	35.00
Py et al. (1987)	450.00	75.00	650.00	205.00	120.00	–

[a] Pegoraro et al. (2014a).
[b] Souza et al. (2019)—"Vitória" with the lowest nitrogen dose and plant population (51,282 plants/ha)—article in press.
[c] Oliveira (2013).
[d] Hanafi et al. (2009), estimated for a population of 62,117 plants/ha.
[e] Paula et al. (1985).
[f] Paula et al. (1985), estimated for a population of 50,000 plants/ha.

TABLE 50.3 Yield and average doses of macronutrients (kg/ha) observed for pineapple cultivars.

Source	Yield (t/ha)	Population (plants/ha)	N	P	K
"Vitória"[a]	72.00	51,280	769	154	769
"Smooth"[b]	72.00	30,300	498	80	394
"Smooth"[c]	54.44	36,000	391	161	450
Imperial[d]	26.36	41,666	285	80	410
MD2[c]	56.14	36,000	460	161	480
Pérola[e]	62.00	–	397	130	583
Pérola[f]	66.41	47,619	714	87	860

[a] Pegoraro et al. (2014a): Foliar sprayings were also carried out with 0.07% de B, 0.1% Zn, and 0.1% Cu, in the forms of boric acid, zinc sulfate, and copper sulfate, respectively (six applications during the vegetative cycle).
[b] Spironello et al. (2004).
[c] García et al. (2017).
[d] Rios et al. (2018b): Borax (1.9 kg/ha), zinc sulfate (8 kg/ha), copper sulfate (8 kg/ha), and iron sulfate (16 kg/ha).
[e] Agbangba et al. (2011).
[f] Oliveira (2013).

they are a source of nutrients, being an alternative to reduce the high costs of inputs required for agricultural production. Considering the importance of pineapple farming to Brazil and the world and its high nutritional demand, it is proposed that the use of these residues can help increase productivity, reduce production costs, and improve the environmental sustainability of waste production systems, as well as pineapple residues.

This practice becomes interesting because most of the nutritional demand of the pineapple occurs until the time of floral induction, preventing the direct contact of the effluent and the sewage sludge with the fruit, which reduces the risk of microbiological contamination. As previously written, pineapple presents high nutritional demand compared with other crops (Pegoraro et al., 2014a; Teixeira et al., 2011b). Therefore, wastewater and sludge may partially or totally replace fertilizers. Sewage sludge is defined as waste from the treatment of both domestic and industrial wastewater, with the aim of reducing its pollutant load and, consequently, its impact on the environment. Sludge, after treatment involving sanitation, stabilization, and drying stages, is called biosolid and can be used in agriculture (Melo et al., 2001).

FIG. 50.7 Nitrogen doses applied in the cultivation of "Vitória" pineapple.

This residue has large amounts of nutrients, especially nitrogen, and is very rich in organic matter. Therefore, the use of sewage sludge in the cultivation of pineapple can meet the demand in a significant way in relation to fertilization with conventional fertilization. However, sewage sludge, in addition to nutrients, has mineral contaminants such as lead, nickel, cadmium, chromium, copper, and zinc (Zuba Junio et al., 2019) and biological contaminants such as pathogens (bacteria, protozoa, and viruses), as well as a range of synthetic organic compounds (polycyclic aromatic hydrocarbons, dioxins, furans, pesticides, and synthetic and natural hormones), and it is necessary to manage this residue in a judicious manner (Nascimento et al., 2015). Due to the economic and social importance associated with the high demand of pineapple inputs, especially nitrogen, the use of sewage sludge as a source of nutrients can totally replace nitrogenous fertilization in the form of urea, without affecting production and quality of fruits and without microbiological contamination of them (Mota et al., 2018). In addition, it contributes to the reduction of environmental pollution caused by the inadequate disposal of this waste.

Considering treated wastewater, the presence of nutrients in the sanitary sewer can be a problem that is not always easily solved, since it is necessary to comply with the requirements of the environmental laws of each country. On the other hand, nutrients can mean a substantial advantage for water reuse, since they are necessary inputs for plant cultivation, maintaining fertility and soil productivity levels by partially or totally substituting fertilization with mineral fertilizers. Treated wastewater (TWW) is seen as a renewable resource, and its agricultural use is common in several countries, such as Singapore, India, Barbados, Philippines, Spain, Australia, Japan, Jordan, the United States, and Belgium (U.S. Environmental Protection Agency (USEPA), 2012). In agriculture, the use of these waters can improve the physical and chemical properties of the soil, by providing essential nutrients, such as nitrogen, phosphorus, and potassium for plant growth and fruit quality (Bourazanis et al., 2016), and by increasing organic matter content, main conditioner of tropical soils and responsible for the increase in water retention capacity (Hespanhol, 2003).

The application of this water as a fertirrigation of pineapple becomes quite feasible, due to the high nutritional demand of this crop especially in relation to nitrogen and potassium, which are normally present in high concentrations in this type of effluent. The use of this technology can reduce the need to apply nitrogen, potassium, and clean water in 15%, 16%, and 25%, respectively (unpublished data). However, it is important that an adequate dose of residual water is defined for the cultivation of pineapple, to avoid imbalance in the supply of nutrients and problems such as excessive vegetative growth, with reduction of qualitative and quantitative yield of the crop (Pedrero et al., 2010). As with sewage sludge, wastewater can be a source of pathogenic organisms and chemicals such as bacteria, viruses, drugs, and hormones, as well as toxic mineral elements such as sodium, chlorine, and heavy metals (Gatta et al., 2015), although the latter are uncommon. As for sodium, as the plant presents CAM carbon fixation, this nutrient may probably be essential for the conversion of pyruvate to phosphoenolpyruvate.

11 Future line of research

Despite considerable advances in pineapple nutrition, there are still gaps and opportunities for research lines in underexploited areas specific to this crop. Among these, the use of microorganisms can be highlighted, especially nitrogen-fixing plants or diazotrophs associated with specific growth-promoting bacteria and selected for pineapple. These studies are only increases in pineapple demand and may occur in production cycles that can be reduced in terms of improving cash flow and the ability to do so in addition to reducing the environmental impact of pineapple farming. The seeds can be incorporated into the use of species for the green cover, being incorporated or even cultivated simultaneously with the pineapple.

Another point to consider, especially to enable pineapple cultivation in regions with low water availability, is studies related to the use of treated domestic sewage water, from secondary, tertiary, or quaternary treatment, evaluating agronomic viability and the long-term environmental impacts and the possible contamination of the fruits with microorganisms, drugs, and heavy metals. Add to this the studies with sewage sludge or biosolids that also come from sewage treatment plants.

In addition, although pineapple crops are predominantly grown in acid soils, the study of cultivation of this species in alkaline soils or irrigation with water containing dilute calcium carbonate results in a rise in soil pH, and possible mitigation measures of this problem will allow the extension of cultivated areas to previously unexploited sites. Among the mitigation measures that can be studied, jointly or separately, are the use of elemental sulfur applied directly to the soil and the definition of doses according to soil type and pH, correction of pH of water with weak acids, and the foliar fertilization with sulfated sources as well as the uptake of nutrients applied with this technology.

Finally, in spite of some references in the literature, more studies of pineapple cultivation in saline soils with the presence of chlorine and sodium are necessary, identifying possible salinity-tolerant genotypes, as well as the function of these nutrients in the physiology and metabolism of pineapple, especially sodium, allowing the characterization of this element as toxic, beneficial, or essential for pineapple.

References

Agbangba, E.C., Olodo, G.P., Dagbenonbakin, G.D., Kindomihou, V., Akpo, L.E., Sokpon, N., 2011. Preliminary DRIS model parameterization to access pineapple variety Perola nutrient status in Benin (West Africa). Afr. J. Agric. Res. 6 (27), 5841–5847. https://doi.org/10.5897/AJAR11.889.

Amaral, U.D., Maia, V.M., Pegoraro, R.F., Kondo, M.K., Aspiazú, I., 2014. Water depths and macronutrients accumulation in 'pérola' pineapple irrigated by drip. Rev. Bras. Frutic. 36 (3), 755–760. https://doi.org/10.1590/0100-2945-187/13.

Amaral, U., Martins Maia, V., Facco Pegoraro, R., Koiti Kondo, M., Carvalho Brant Maia, L., 2015. Matéria seca, conteúdo de carbono e nitrogênioem cultivo de abacaxizeiro 'Pérola' irrigado. Interciencia 40 (9), 639–643.

Amorim, A.V., Lacerda, C.F., Moura, C.F.H., Gomes Filho, F., 2011. Fruit size and quality of pineapples cv. Vitória in response to micronutrient doses and way of application and to soil covers. Rev. Bras. Frutic. 505–510 (special issue), https://doi.org/10.1590/S0100-29452011000500068.

Amorim, A.V., Lacerda, C.F.D., Marques, E.C., Ferreira, F.J., Júnior, S., da Costa, R.J., Gomes-Filho, E., 2013. Micronutrients affecting leaf biochemical responses during pineapple development. Theor. Exp. Plant Physiol. 25 (1), 70–78. https://doi.org/10.1590/S2197-00252013000100009.

Anupunt, P., Chairidchai, P., Kongswat, A., Isawilanon, S., Subhadrabhandu, S., Vasunun, S., Siripat, S., 2000. The pineapple industry in Thailand. Acta Hortic. 529, 99–110. https://doi.org/10.17660/ActaHortic.2000.529.11.

Bertsch, F., Alvarado, A., Henriquez, C., Mata, R., 2000. Properties, geographic distribution, and management of major soil orders of Costa Rica. In: Quantifying Sustainable Development. Academic Press, pp. 265–294. https://doi.org/10.1016/B978-012318860-1/50016-2.

Bourazanis, G., Roussos, P.A., Argyrokastritis, I., Kosmas, C., Kerkides, P., 2016. Evaluation of the use of treated municipal waste water on the yield, oil quality, free fatty acids' profile and nutrient levels in olive trees cv Koroneiki, in Greece. Agric. Water Manag. 163, 1–8. https://doi.org/10.1016/j.agwat.2015.08.023.

Broadley, M., Brown, P., Cakmak, I., Rengel, Z., Zhao, F., 2012. Function of nutrients: micronutrients. In: Marschner's Mineral Nutrition of Higher Plants. Academic Press, pp. 191–248.

Cabral, J.R.S., 2000. Variedades. In: Reinhardt, D.H., Souza, F.S., Cabral, J.R.S. (Eds.), Abacaxi. Produção: aspectos técnicos. Embrapa Mandioca e Fruticultura (Cruz das Almas, BA). Embrapa, Brasília, pp. 15–18.

Cabral, J.R.S., Junghans, D.T., 2003. Variedades de abacaxi. In: Embrapa Mandioca e Fruticultura-Circular Técnica. INFOTECA-E.

Cardoso, M.M., Pegoraro, R.F., Maia, V.M., Kondo, M.K., Fernandes, L.A., 2013. Crescimento do abacaxizeiro 'Vitória' irrigado sob diferentes densidades populacionais, fontes e doses de nitrogênio. Rev. Bras. Frutic. 35 (3), 769–781.

Cintra, F.L.D., Libardi, P.L., 1998. Caracterização física de uma classe de solo do ecossistema do tabuleiro costeiro. Sci. Agric. 55 (3), 367–378.

Collins, J.L., 1951. Notes on the origin, history, and genetic nature of the Cayenne pineapple. Pac. Sci. 5 (1), 3–17.

Collins, J.L., 1960. The Pineapple: Botany, Cultivation, and Utilization. Hill, London. 294 p.

Crestani, M., Barbieri, R.L., Hawerroth, F.J., Carvalho, F.I.F., Oliveira, A.C., 2010. Das Américas para o Mundo—origem, domesticação e dispersão do abacaxizeiro. Cienc. Rural 40 (6), 1473–1483.

Cunha, G.A.P., 2007. Equipe técnica de abacaxi comemora 30 anos de atividades e realizações. Embrapa Mandioca e Fruticultura Tropical, Cruz das Almas. 18 p. (Documentos, 170), https://www.infoteca.cnptia.embrapa.br/infoteca/bitstream/doc/654393/1/documentos170.pdf. (Accessed 9 May 2019).

Cunha, G.A., Matos, A.P., Sanches, N.F., Reinhardt, D.H., Souza, L.F.S., Cabral, J.R.S., Almeida, O.A.A., 1995. A cultura do abacaxi: práticas de cultivo. Embrapa Mandioca e Fruticultura-Circular Técnica. INFOTECA-E.

Dantas, A.L., de Melo Silva, S., Dantas, R.L., Pereira, W.E., Lima, R.P., Mendonça, R.M.N., Santos, D., 2015. Influence of combined sources of nitrogen fertilization on quality of cv. Vitria pineapple. Afr. J. Agric. Res. 10 (40), 3814–3824.

Darnaudery, M., Fournier, P., Lechaudel, M., 2018. Low-input pineapple crops with high quality fruit: promising impacts of locally integrated and organic fertilisation compared to chemical fertilisers. Exp. Agric. 54 (2), 286–302. https://doi.org/10.1017/S0014479716000284.

Debnath, P., Dey, P., Chanda, A., Bhakta, T.A., 2012. Survey on pineapple and its medicinal value. Sch. Acad. J. Pharm. 1 (1), 24–29.

Devadas, V.S., Kuriakose, K.P., 2005. Evaluation of different organic manures on yield and quality of pineapple var. Mauritius. Acta Hortic. 666, 185–189. https://doi.org/10.17660/ActaHortic.2005.666.17.

Feitosa, H.D.O., Amorim, A.V., Lacerda, C.F.D., Silva, F.B.D., 2011. Growth and micronutrients extraction by 'vitoria' pineapple. Rev. Bras. Frutic. 33 (SPE1), 706–712. https://doi.org/10.1590/S0100-29452011000500099.

Food and Agriculture Organization of the United Nations, 2017. FAOSTAT Database. http://www.fao.org/faostat/en/#data/QC/visualize. (Accessed 8 May 2019).

García, S.S., López, D.J.P., Cruz, J.Z., García, C.F.O., Espinoza, L.C.L., Estrada, M.C., Peña, A.G., Ceballos, A.I.O., Sánchez, S.C., 2017. Integrated system for recommending fertilization rates in pineapple. (*Ananas comosus* (L.) Merr.) crop. Acta Agron 66 (4), 566. https://doi.org/10.15446/acag.v66n4.62257.

Gatta, G., Libutti, A., Gagliardi, A., Beneduce, L., Brusetti, L., Borruso, L., Disciglio, G., Tarantino, E., 2015. Treated agro-industrial wastewater irrigation of tomato crop: effects on qualitative/quantitative characteristics of production and microbiological properties of the soil. Agric. Water Manag. 149, 33–43. https://doi.org/10.1016/j.agwat.2014.10.016.

Gile, P.L., 1916. Chlorosis of pineapples induced by manganese and carbonate of lime. Science 44 (1146), 855–857.

Guarçoni, M.A., Aires Ventura, J., 2011. Adubação NPK e o desenvolvimento, produtividade e qualidade dos frutos do abacaxi 'gold' (MD-2). Rev. Bras. Ciênc. Solo 35 (4), 1367–1376. https://doi.org/10.1590/S0100-06832011000400031.

Guinto, D.F., Inciong, M.M., 2012. Soil quality, management practices and sustainability of pineapple farms in Cavite, Philippines: Part 1. Soil quality. J. S. Pac. Agric. 16 (1), 30–41.

Gutiérrez, F., Rojas Bourrillón, A., Dormond, H., Poore, M., Wing Ching Jones, R., 2003. Características nutricionales y fermentativas de mezclas ensiladas de desechos de piña y avícolas. Agron. Costarric. 27 (1), 79–89.

Hanafi, M.M., Selamat, M.M., Husni, M.H.A., Adzemi, M.A., 2009. Dry matter and nutrient partitioning of selected pineapple cultivars grown on mineral and tropical peat soils. Commun. Soil Sci. Plant Anal. 40 (21–22), 3263–3280.

Hassan, A., Othman, Z., 2011. Pineapple (*Ananas comosus* L. Merr.). In: Yahia, E.M. (Ed.), Postharvest Biology and Technology of Tropical and Subtropical Fruits: Mangosteen to White Sapote. Elsevier, pp. 194–217.

Hawkesford, M., Horst, W., Kichey, T., Lambers, H., Schjoerring, J., Møller, I.S., White, P., 2012. Functions of macronutrients. In: Marschner's Mineral Nutrition of Higher Plants. Academic Press, pp. 135–189.

Herath, H.M.I., Bandara, D.C., Abeysinghe Banda, D.M.G., 2003. Effect of pre-harvest calcium fertilizer application on the control of internal browning development during the cold storage of pineapple 'Mauritius' (*Ananas comosus* (L.) Merr.). J. Hortic. Sci. Biotechnol. 78 (6), 762–767. https://doi.org/10.1080/14620316.2003.11511696.

Hespanhol, I., 2003. Potencial de reuso de água no Brasil: agricultura, indústria, município e recarga de aquíferos. In: Mancuso, P.C.S., Santos, H.F. (Eds.), Reuso de água. Manole, Barueri, pp. 37–95.

Hewajulige, I.G., Wilson Wijeratnam, S., Wijesundera, R.L., 2006. Pre-harvest application of calcium to control black heart disorder in Mauritius pineapples during low-temperature storage. J. Sci. Food Agric. 86 (3), 420–424. https://doi.org/10.1002/jsfa.2361.

Hiroce, R., Bataglia, O.C., Furlani, P.R., Furlani, A.M.C., Giacomelli, E.J., Gallo, J.R., 1977. Composição química inorgânica do abacaxizeiro (*Ananas comosus* 'Cayenne') da região de Bebedouro, SP. Ciênc. Cult. 29, 323–326.

Hossain, M.F., 2016. World pineapple production: an overview. Afr. J. Food Agric. Nutr. Dev. 16 (4), 11443–11456. https://doi.org/10.18697/ajfand.76.15620.

Hossain, M.F., Akhtar, S., Anwar, M., 2015. Nutritional value and medicinal benefits of pineapple. Int. J. Nutr. Food Sci. 4 (1), 84–88. https://doi.org/10.11648/j.ijnfs.20150401.22.

International Trade Center. Trade Map, 2018. ITC. Available at:https://www.trademap.org/Index.aspx. (Accessed 8 May 2019).

Jin, V.L., Potter, K.N., Johnson, M.V.V., Harmel, R.D., Arnold, J.G., 2015. Surface-applied biosolids enhance soil organic carbon and nitrogen stocks but have contrasting effects on soil physical quality. Appl. Environ. Soil Sci. 2015 (3), 1–10. https://doi.org/10.1155/2015/715916.

Kumari, U., Deb, P., 2018. Effect of foliar application of zinc and boron on quality of pineapple cv. Mauritius. J. Pharmacogn. Phytochem. 7 (6), 1166–1168.

Land Development Department, 2014. Classification and Kind of Soil. Available at:http://www.ldd.go.th/ldd_en/en-US/classification-and-kinds-of-soil/. (Accessed 8 May 2019).

Langer, H., Cea, M., Curaqueo, H., Borie, F., 2009. Influence of aluminum on the growth and organic acid exudation in alfalfa cultivars grown in nutrient solution. J. Plant Nutr. 32, 618–628. https://doi.org/10.1080/01904160802715430.

Le Van, H., Masuda, T., 2004. Physiological and biochemical studies on aluminum tolerance in pineapple. Soil Res. 42 (6), 699–707.

Leal, F., D'eeckenbrugge, G.C., 2018. History, distribution and world production. In: Sanewski, G.M., Bartholomew, D.P., Paull, R.E. (Eds.), The Pineapple: Botany, Production and Uses. CAB International, Wallingford, pp. 1–10.

Lee, J.H., Lee, J.T., Park, H.R., Kim, J.B., 2019. The potential use of bromelain as a natural oral medicine having anticarcinogenic activities. Food Sci. Nutr. 7, 1656–1667. https://doi.org/10.1002/fsn3.999.

Leonardo, F.D.A., Pereira, W.E., Silva, S.D.M., Costa, J.D., 2013. Teor de clorofila e índice spad no abacaxizeiro cv. Vitória em função da adubação nitrogenada. Rev. Bras. Frutic. 35 (2), 377–383.

Leonel, S., dos Reis, L.L., 2012. Potassium fertilization on fruits orchards: a study case from Brazil. In: Soil Fertility. IntechOpenhttps://doi.org/10.5772/53210.

Lin, Y.H., 2010. Effects of aluminum on root growth and absorption of nutrients by two pineapple cultivars [*Ananas comosus* (L.) Merr.]. Afr. J. Biotechnol. 9 (26), 4034–4041.

Lin, Y.H., Chen, J.H., 2011. Effects of aluminum on nutrient uptake in different parts of four pineapple cultivars. Afr. J. Agric. Res. 6 (6), 1438–1446. https://doi.org/10.5897/AJAR10.820.

Liu, C.H., Liu, Y., Fan, C., Kuang, S.Z., 2013. The effects of composted pineapple residue return on soil properties and the growth and yield of pineapple. J. Soil Sci. Plant Nutr. 13 (2), 433–444. https://doi.org/10.4067/S0718-95162013005000034.

Loison-Cabot, C., 1992. Origin, phylogeny and evolution of pineapple species. Fruits 47 (1), 25–32.

López-Herrera, M.L., Wing Ching-Jones, R., Rojas-Bourrillón, A., 2009. Características fermentativas y nutricionales del ensilaje de rastrojo de piña (*Ananas comosus*). Agron. Costarric. 33 (1), 1–15.

López-Herrera, M., Wing Ching-Jones, R., Rojas-Baurrillón, A., 2014. Meta-análisis de los subproductos de piña (*Ananas comosus*) para la alimentación animal. Agron. Mesoam. 25 (2), 383–392.

Maeda, A.S., Buzetti, S., Boliani, A.C., Benett, C.G.S., Teixeira Filho, M.C.M., Andreotti, M., 2011. Foliar fertilization on pineapple quality and yield. Pesqui. Agropecu. Trop. 41 (2), 248–253. https://doi.org/10.5216/pat.v41i2.8810.

Maia, V.M., Oliveira, F.S., Pegoraro, R.F., Souza, B.A.M., Ferreira, L.B., Aspiazú, I., 2016. Vegetative growth stages of irrigated 'Pérola' pineapple. Acta Hortic. 1111, 275–280. https://doi.org/10.17660/ActaHortic.2016.1111.39.

Manica, I., 1999. Fruticultura Tropical 5: Abacaxi. Cinco Continentes.

Marschner, P., 2012. Marschner's Mineral Nutrition of Higher Plants, third ed. Academic Press 651 p.

Matsuura, F.C.A.U., Rolim, R.B., 2002. Avaliação da adição de suco de acerola em suco de abacaxi visando à produção de um "blend" com alto teor de vitamina C. Rev. Bras. Frutic. 24 (1), 138–141.

Medina, J.C., 1978. A cultura do abacaxi. In: Medina, J.C. et al., (Ed.), Frutas Tropicais 2. Canton, São Paulo, pp. 06–68.

Medina, J.D., García, H.S., 2005. Pineapple Post-Harvest Operations. Instituto Tecnologico de Veracruz.http://www.fao.org/fileadmin/user_upload/inpho/docs/Post_Harvest_Compendium_-_Pineapple.pdf. (Accessed 9 May 2019).

Melo, W.J., Marques, M.O., Melo, V.P., 2001. O uso do lodo de esgoto e as propriedades do solo. In: Tsutyia, M.T., Comparini, J.B., Alem Sobrinho, P., Hespanhol, I., Carvalho, P.C.T., Melfi, A.J., Melo, M.O. (Eds.), Lodo de esgotos na Agricultura. SABESP, São Paulo, pp. 289–363.

Mite, F., Espinosa, J., Medina, L., 2010. Liming effect on pineapple yield and soil properties in volcanic soils. Better Crop. Plant Food 94 (1), 7–9.

Moretti-Almeida, G., 2018. Pineapple taxonomy. In: Bogsan, C.S., Todorov, S.D. (Eds.), Tropical Fruits: From Cultivation to Consumption and Health Benefits, Pineapple. Food Science and Technology: Nova Science Publishers, New York, pp. 1–14.

Mota, M.F., 2016. Atributos químicos do solo e produção de abacaxizeiro adubado com lodo de esgoto. (Dissertação de Mestrado)Programa de Pós-Graduação em Produção Vegetal no Semiárido, Universidade Estadual de Montes Claros, Janaúba. 84 p.

Mota, M.F., Pegoraro, R.F., Batista, P.S., Pinto, V.D.O., Maia, V.M., Silva, D.F.D., 2016. Macronutrients accumulation and growth of pineapple cultivars submitted to aluminum stress. Rev. Bras. Eng. Agric. Ambient. 20 (11), 978–983. https://doi.org/10.1590/1807-1929/agriambi.v20n11p978-983.

Mota, M.F., Pegoraro, R.F., Santos, S.R.D., Maia, V.M., Sampaio, R.A., Kondo, M.K., 2018. Contamination of soil and pineapple fruits under fertilization with sewage sludge. Rev.Bras. Eng. Agric. Ambient. 22 (5), 320–325. https://doi.org/10.1590/1807-1929/agriambi.v22n5p320-325.

Nascimento, A.L., Junio, Z., Geraldo, R., Sampaio, R.A., Fernandes, L.A., Carneiro, J.P., Barbosa, C.F., 2015. Metais pesados no solo e mamoneira adubada com biossólido e silicato de cálcio e magnésio. Rev.Bras. Eng. Agric. Ambient.. 19(5)https://doi.org/10.1590/1807-1929/agriambi.v19n5p505-511.

Newett, S., Rigden, P., 2015. The Pineapple Problem Solver Field Guide. State of Queensland, Queensland.

Oliveira, F.S., 2013. Marcha de absorção de nutrientes, crescimento e produção do abacaxizeiro 'pérola'. (Dissertação de Mestrado)Programa de Pós-Graduação em Produção Vegetal no Semiárido, Universidade Estadual de Montes Claros, Janaúba. 81 p.

Parfitt, R., 2009. Allophane and imogolite: role in soil biogeochemical processes. Clay Miner. 44, 135–155. https://doi.org/10.1180/claymin.2009.044.1.135.

Paula, M.B., Carvalho, J.G., Nogueira, F.D., Silva, C.R.R., 1985. Exigências nutricionais do abacaxizeiro. Inf. Agropecu. 11 (130), 27–32.

Pedrero, F., Kalavrouziotis, I., Alarcón, J.J., Koukoulakis, P., Asano, T., 2010. Use of treated municipal wastewater in irrigated agriculture—review of some practices in Spain and Greece. Agric. Water Manag. 97 (9), 1233–1241. https://doi.org/10.1016/j.agwat.2010.03.003.

Pegoraro, R.F., Souza, B.A.M.D., Maia, V.M., Silva, D.F.D., Medeiros, A.C., Sampaio, R.A., 2014a. Macronutrient uptake, accumulation and export by the irrigated 'vitória' pineapple plant. Rev. Bras. Ciênc. Solo 38 (3), 896–904. https://doi.org/10.1590/S0100-06832014000300021.

Pegoraro, R.F., Souza, B.A.M.D., Maia, V.M., Amaral, U.D., Pereira, M.C.T., 2014b. Growth and production of irrigated Vitória pineapple grown in semi-arid conditions. Rev. Bras. Frutic. 36 (3), 693–703. https://doi.org/10.1590/0100-2945-265/13.

Primo, D.C., Menezes, R.S., Silva, W.T., Oliveira, F.F., Júnior, J.C., Sampaio, E.V., 2017. Characterisation of soil organic matter in a semi-arid fluvic Entisol fertilized with cattle manure and/or gliricidia by spectroscopic methods. Soil Res. 55 (4), 354–362. https://doi.org/10.1071/SR16106.

Pusittigul, I., Siriphanich, J., Juntee, C., 2014. Role of calcium on internal browning of pineapples. Acta Hortic. 1024, 329–338. https://doi.org/10.17660/ActaHortic.2014.1024.45.

Py, C., Lacoeuilhe, J.J., Teison, C., 1987. The Pineapple, Cultivation and Uses. G.P. Maisonneuve et Larose, Paris. 568 p.

Ramos, M.J.M., Rocha Pinho, L.G., 2014. Physical and quality characteristics of Jupi pineapple fruits on macronutrient and boron deficiency. Nat. Resour. 5 (08), 359. https://doi.org/10.4236/nr.2014.58034.

Ramos, M.J.M., Monnerat, P.H., Carvalho, A.D., Pinto, J.D.A., Silva, J.D., 2009. Sintomas visuais de deficiência de macronutrientes e de boro em abacaxizeiro "imperial" Rev. Bras. Frutic. 31 (1), 252–256.

Ramos, M.J.M., Monnerat, P.H., Pinho, L.D.R., Carvalho, A.D., 2010. Qualidade sensorial dos frutos do abacaxizeiro 'imperial' cultivado em deficiência de macronutrientes e de boro. Rev. Bras. Frutic. 32 (3), 692–699.

Ramos, M.J.M., Monnerat, P.H., Pinho, L.D.R., Silva, J.D., 2011. Deficiência de macronutrientes e de boro em abacaxizeiro 'Imperial': composição mineral. Rev. Bras. Frutic. 33 (1), 261–271.

Razzaque, A.H.M., Hanafi, M.M., 2001. Effect of potassium on growth, yield and quality of pineapple in tropical peat. Fruits 56 (1), 45–49. https://doi.org/10.1051/fruits:2001111.

Abacaxi produção: aspectos técnicos. Reinhardt, D.H., Souza, L.D.S., Cabral, J.R.S. (Eds.), 2000. Embrapa Informação Tecnológica. Embrapa Mandioca e Fruticultura, Cruz das Almas. 77 p.

Rios, C., Santos, E., Nunes Mendonça, R.M., de Almeida Cardoso, E., da Costa, J.P., de Melo Silva, S., 2018a. Quality of 'Imperial' pineapple infructescence in function of nitrogen and potassium fertilization. Braz. J. Agric. Sci./Rev. Bras. Ciênc. Agrár. 13(1). https://doi.org/10.5039/agraria.v13i1a5499.

Rios, É.S.C., Mendonça, R.M.N., Fernandes, L.F., de Figueredo, L.F., de Almeida Cardoso, E., de Souza, A.P., 2018b. Growth of leaf D and productivity of 'Imperial' pineapple as a function of nitrogen and potassium fertilization. Braz. J. Agric. Sci./Rev. Bras. Ciênc. Agrár. 13 (2), 1–9.

Rodrigues, A.A., Mendonça, R.M.N., Silva, A.D., Silva, S.D.M., Pereira, W.E., 2010. Desenvolvimento vegetativo de abacaxizeiros 'Pérola' e 'Smooth Cayenne' no estado da paraíba. Rev. Bras. Frutic. 32 (1), 126–134.

Ryan, P.R., Kinraide, T.B., Kochian, L.V., 1994. Al^{3+}-Ca^{2+} interactions in aluminum rhizotoxicity. I. Inhibition of root. Planta 192, 98–103.

Sampaio, D.B., Araújo, A.S.F., Santos, V.B., 2008. Avaliação de indicadores biológicos de qualidade do solo sob sistemas de cultivo convencional e orgânico de frutas. Cienc. Agrotecnol. 32 (2), 353–359.

Sanches, N., Matos, A.P., 2013. Abacaxi: o produtor pergunta, a Embrapa responde. In: Área de Informação da Sede-Col Criar Plantar ABC 500P/ 500R Saber. INFOTECA-E.

Sanewski, G.M., D'eeckenbrugge, G.C., Junghans, D.T., 2018. Varieties and breeding. In: Sanewski, G.M., Bartholomew, D.P., Paull, R.E. (Eds.), The Pineapple: Botany, Production and Uses. CAB International, Wallingford, pp. 43–84.

Santos, M.P.D., Maia, V.M., Oliveira, F.S., Pegoraro, R.F., Santos, S.R.D., Aspiazú, I., 2018. Estimation of total leaf area and D leaf area of pineapple from biometric characteristics. Rev. Bras. Frutic. 40(6), https://doi.org/10.1590/0100-29452018556.

Saradhuldhat, P., Paull, R.E., 2007. Pineapple organic acid metabolism and accumulation during fruit development. Sci. Hortic. 112 (3), 297–303. https://doi.org/10.1016/j.scienta.2006.12.031.

Scott, C., 2000. The effect of molybdenum applications on the juice nitrate concentration of pineapples. In: Pineapple Field Day Notes. Queensland Fruit and Vegetable Growers, Beerwah, Queensland, pp. 23–27.

Siebeneichler, S.C., Monnerat, P.H., Carvalho, A.D., Silva, J.D., Martins, A.O., 2005. Mobilidade do boro em plantas de abacaxi. Rev. Bras. Frutic 27 (2), 292–294. https://doi.org/10.1590/S0100-29452005000200026.

Siebeneichler, S.C., Monnerat, P.H., Carvalho, A.D., Silva, J.D., 2008. Boro em abacaxizeiro Pérola no norte fluminense-teores, distribuição e características do fruto. Rev. Bras. Frutic. 30 (3), 787–793.

Silva, J.A., Hamasaki, R., Paull, R., Ogoshi, R., Bartholomew, D.P., Fukuda, S., Hue, N.V., Uehara, G., Tsuji, G.Y., 2006. Lime, gypsum, and basaltic dust effects on the calcium nutrition and fruit quality of pineapple. Acta Hortic. 702, 123–131. https://doi.org/10.17660/ActaHortic.2006.702.15.

Singh, V., Singh, B., Singh, Y., Thind, H.S.E., Gupta, R.K., 2010. Need based nitrogen management using the chlorophyll meter and leaf colour chart in rice and wheat in South Asia: a review. Nutr. Cycl. Agroecosyst. 88, 361–380. https://doi.org/10.1007/s10705-010-9363-7.

Souza, L.F.S., 2000. Adubação. In: Reinhardt, D.H., Souza, L.F.S., Cabral, J.R.S. (Eds.), Abacaxi. Produção: Aspectos técnicos. Embrapa, Brasília, pp. 30–34.

Souza, R.P.D., Pegoraro, R.F., Reis, S.T., Maia, V.M., Sampaio, R.A., 2019. Partition and macronutrients accumulation in pineapple under nitrogen doses and plant density. Comun. Sci. 10(2). https://doi.org/10.14295/CS.v10i2.2604.

Spironello, A., Quaggio, J.A., Teixeira, L.A.J., Furlani, P.R., Sigrist, J.M.M., 2004. Pineapple yield and fruit quality effected by NPK fertilization in a tropical soil. Rev. Bras. Frutic. 26 (1), 155–159. https://doi.org/10.1590/S0100-29452004000100041.

Taiz, L., Zeiger, E., Moller, I.M., Murphy, A., 2017. Physiology and Plant Development. vol. 6 Artmed, Porto Alegre. 858 p.

Taussig, S.J., Batkin, S., 1988. Bromelain, the enzyme complex of pineapple (Ananas comosus) and its clinical application. An up-date. J. Ethnopharmacol. 22 (2), 191–203.

Teixeira, L.A.J., Quaggio, J.A., Cantarella, H., Mellis, E.V., 2011a. Potassium fertilization for pineapple: effects on plant growth and fruit yield. Rev. Bras. Frutic. 33 (2), 618–626. https://doi.org/10.1590/S0100-29452011000200035.

Teixeira, L.A.J., Quaggio, J.A., Cantarella, H., Mellis, E.V., 2011b. Potassium fertilization for pineapple: effects on soil chemical properties and plant nutrition. Rev. Bras. Frutic. 33 (2), 627–636. https://doi.org/10.1590/S0100-29452011000200036.

U.S. Environmental Protection Agency (USEPA), 2012. Global Experiences in Water Reuse. Guidelines for Water Reuse. EPA/600/R-12/618. Environmental Protection Agency, Washington, D.C. http://nepis.epa.gov/Adobe/PDF/P100FS7K.pdf. (Accessed 25 April 2019).

Uriza-Ávila, D.E., Torres-Ávila, A., Aguilar-Ávila, J., Santoyo-Cortés, V.H., Zetina-Lezama, R., Rebolledo-Martínez, A., 2018. La piña mexicana frente al reto de la innovación. Avances y retos en la gestión de la innovación. Colección Trópico Húmedo. UACh, Chapingo.

Vásquez-Jiménez, J., 2010. Evaluación de la Necesidad de Hierro del Cultivo de Piña Ananas comosus (L) Merr., var MD-2 en Tres Órdenes de Suelo del Norte y Caribe Norte de Costa Rica. MSc thesis. Costa Rica University, San José, Costa Rica 151 pp.

Vásquez-Jiménez, J., Bartholomew, D.B., 2018. Plant nutrition. In: Sanewski, G.M., Bartholomew, D.P., Paull, R.E. (Eds.), The Pineapple: Botany, Production and Uses. CAB International, Wallingford, pp. 175–202.

Vilela, G.B., Pegoraro, R.F., Maia, V.M., 2015. Predição de produção do abacaxizeiro 'Vitória' por meio de características fitotécnicas e nutricionais. Ciênc. Agron. 46 (4), 724–732. https://doi.org/10.5935/1806-6690.20150059.

Weber, O.B., Lima, R.N., Crisostomo, L.A., Freitas, J.A.D., Carvalho, A.C.P.P., Maia, A.H.N., 2010. Effect of diazotrophic bacterium inoculation and organic fertilization on yield of Champaka pineapple intercropped with irrigated sapota. Plant Soil 327 (1–2), 355–364.

West, L.T., Beinroth, F.H., Sumner, M.E., Kang, B.T., 1997. Ultisols: characteristics and impacts on society. Adv. Agron. 63, 179–236.

Williams, D.D.F., Fleisch, H., 1993. Historical review of pineapple breeding in Hawaii. Acta Hortic. 334, 67–76. https://doi.org/10.17660/ActaHortic.1993.334.7.

Xu, Q., Liu, Z.J., 2015. A taste of pineapple evolution through genome sequencing. Nat. Genet. 47 (12), 1374–1376.

Zuba Junio, G.R., Sampaio, R.A., Fernandes, L.A., Pegoraro, R.F., Maia, V.M., Cardoso, P.H.S., Sousa, I.P., Vieira, I.T.R., 2019. Content of heavy metals in soil and in pineapple fertilized with sewage sludge. J. Agric. Sci. 11(9), https://doi.org/10.5539/jas.v11n9p281.

Index

Note: Page numbers followed by *f* indicate figures and *t* indicate tables.